W9-BEF-565

Handbook of CONCRETE ENGINEERING

Handbook of CONCRETE ENGINEERING

edited by Mark Fintel

VNR
VAN NOSTRAND REINHOLD COMPANY
NEW YORK CINCINNATI TORONTO LONDON MELBOURNE

Van Nostrand Reinhold Company Regional Offices:
New York Cincinnati

Van Nostrand Reinhold Company International Offices:
London Toronto Melbourne

Library of Congress Catalog Card Number: 74-4045
ISBN: 0-442-22393-5

Manufactured in the United States of America

Published by Van Nostrand Reinhold Company
135 West 50th Street, New York, N.Y. 10020

Published simultaneously in Canada by Van Nostrand Reinhold Ltd.

15 14 13 12 11 10 9

Library of Congress Cataloging in Publication Data

Fintel, Mark.
Handbook of concrete engineering.

Includes bibliographical references.
1. Concrete construction–Handbooks, manuals, etc.
I. Title.
TA682.F56 624'.1834 74-4045
ISBN 0-442-22393-5

Preface

This handbook contains up-to-date information on planning, design, analysis, and construction of engineered concrete structures. Its intention is to provide engineers, architects, contractors, and students of civil engineering and architecture with authoritative practical design information.

The tremendous progress and changes in all the areas of concrete engineering in the last two decades seemed to indicate a need for a new *Handbook of Concrete Engineering*. In addition, the many inquiries on subjects of concrete engineering received daily by the editor in the course of his professional activity, heightened his enthusiasm in accepting the proposal from Van Nostrand Reinhold to assemble this handbook.

Much of the information contained in this book has evolved during the last 15 to 20 years. The subjects of a number of chapters are so recent that the material has never before been published in book form.

The following traditional engineering subjects are covered in chapters on proportioning of members (ultimate strength design); deflections; flat plates and flat slabs; foundations; properties of materials for reinforced concrete (geared particularly to design engineers); joints in buildings; thin shells; prestressed concrete; chimneys; silos and bunkers; concrete masonry; sanitary structures; pipes, marine structures; paving; construction methods and equipment; and structural analysis by force, displacement and finite element methods.

A number of chapters are of particular interest because they represent areas of recent technological advances: fire resistance, multistory structures; tubular structures; ductility of reinforced concrete members; earthquake resistance; large panel structures; and finally, computer applications and computer software.

Because of space limitations, a careful selection of topics has been made. The 27 chapters have been written by 28 experts from industry and universities, recognized as outstanding authorities in their respective fields. Their wide experience has resulted in concise chapters geared toward practical application in planning, design and construction of engineered concrete structures. Each chapter contains the general philosophy, the basic concepts, and applications elaborated by design examples. While ready-to-use formulas and design approaches are presented in the chapters, theoretical development of formulas has been omitted, and references for additional source material have been appended to most chapters.

The editor gratefully acknowledges the efforts of the authors in preparing high quality manuscripts, and their cooperation and patience with the editor and the publishers in all the stages of producing the handbook. Thanks to their unstinting cooperation, this book has been produced in a relatively short number of years.

MARK FINTEL

September, 1974

Contents

Handbook of CONCRETE ENGINEERING

1

Proportioning of Sections—Ultimate Strength Design

NOEL J. EVERARD*

1.1 INTRODUCTION

This chapter is intended to serve as a guide to the structural engineer in the proportioning of reinforced concrete structural elements. Its purpose is not to present basic theoretical concepts and comprehensive derivations of design equations, but to present the appropriate equations for specific situations and to show the procedures to be followed in using those equations. In order to simplify the design process, design aids are included. Because of space limitations, design aids generally include only the most used material strengths, except in some cases in which the author considered it advisable to include other material strengths because other design aids presently available do not cover those conditions. Where existing design aids apply, the sources are listed so the reader can obtain all of the available material necessary to simplify the design of reinforced concrete structural elements using ultimate strength design methods.

The material in this chapter is presented in the following order: (1) General discussion, (2) appropriate equations, (3) general design examples, (4) design aids, and (5) design examples using design aids.

The reason for presenting equations and general design examples is that design aids do not always cover all situations, and because the designer should understand the methods of solving problems before using design aids, to avoid using design aids in cases where they do not apply.

It is assumed that the reader has available a copy of ACI Standard 318-71, *Building Code Requirements for Reinforced Concrete*, and the *Commentary on Building Code Requirements for Reinforced Concrete* (ACI 318-71), both of which provide background material on which this chapter is based. Those documents may be obtained from the American Concrete Institute, P. O. Box 4754, Redford Station, Detroit, Michigan, 48219. Design aids quoted may be obtained from the American Concrete Institute (hereinafter referenced as ACI), the Portland Cement Association (PCA), 5420 Old Orchard Road, Skokie, Illinois, 60076, and the Concrete Reinforcing Steel Institute (CRSI), 228 North LaSalle St., Chicago, Illinois, 60601.

The equations contained in this chapter and the related design aids are based on the "equivalent rectangular stress block" described in the *ACI Code* and its *Commentary*, and in the textbooks listed in the references for this chapter.

In order to make use of the equations provided in this chapter, the reader must have a fundamental understanding of basic principles related to reinforced concrete. The novice should therefore use the material contained in this chapter only after studying the background material presented in one of the standard textbooks listed in the reference list.

*Consulting Engineer, and Professor and Chairman, Department of Civil Engineering, University of Texas at Arlington, Arlington, Texas.

Certain design tables apply to several sections of this chapter. In order to make those tables easy to locate, they appear at the end of this chapter. Table 1-12 provides steel areas for given numbers of specific bar sizes, and Table 1-13 provides minimum web widths related thereto.

The nomenclature and variable designations conform wherever possible to those given in the *ACI Code*. When new variables are used, they conform with International Notation for Reinforced Concrete, mutually agreed upon by the American Concrete Institute and the European Concrete Committee.

1.1.1 Design Methods

The design procedure used in this chapter is the strength design method of the *1971 ACI Code*. While the *Code* permits the use of an alternate design method, that process is not used in this chapter.

During the early stages of the development of procedures for reinforced concrete design, the ultimate strength method (now called the strength design method) was used. When the first reinforced concrete design code was published in 1909, two procedures were proposed. These were the ultimate strength design method and the elastic design method. The elastic design method was adopted by the Cement Manufacturer's Association by a very narrow margin, and that method was codified in 1909. That adoption hindered the development of reliable reinforced concrete design procedures for nearly 40 years.

In the early 1930's, tests at the University of Illinois and Lehigh University proved that elastic design methods were incorrect for compression members. Thus, after 1940, column design methods were based on ultimate strength methods which were stated in terms of elastic design.

In 1937 Charles Whitney developed equations for beam and column design based on the ultimate strength method. Those equations appeared in the Appendix to the *1956 ACI Code* as an alternate design method. The work of Eivind Hognestad and others at the Portland Cement Association in the late 1950's formed the basis of the *1963 ACI Code*. Except for flexure without axial load, all other working stress design methods were based on the ultimate strength design method. Working stress design and ultimate strength design received equal coverage in the *1963 ACI Code*. The only true elastic design method remaining at that time related to flexure without axial load.

The *1971 ACI Code* emphasized ultimate strength design and relegated working stress design a small section in the *Code*. For the most part, working stress design (or service load design, as it is now called) is actually ultimate strength design with increased safety factors. This is logical, since service load design does not take into account the reliability of computations of various load types. Nor, does it consider the variability of construction techniques, understrength of materials and other factors which, while individually within tolerances, collectively may add to considerably less strength than that anticipated. It is also logical to emphasize ultimate strength methods because those methods have been based on test data. It has never been possible to obtain even reasonable correlation of elastic design methods with test data. Safety factors for that method in flexure have ranged from 2.0 to 4.0 for flexural members, and from less than 1.0 to more than 10.0 for columns. Flexural members in shear have had safety factors often less than 1.0. Hence, elastic design methods have produced flexural members that have usually had excessive safety factors and on occasion, has produced unsafe members. Because flexural members designed by the elastic method have always had adequate safety factors, the *1971 ACI Code* permits such members to be designed in the traditional way. Because the elastic method has been found to be unsatisfactory except in flexural design, modified ultimate strength methods replace elastic design in the form of the alternate design method.

Undoubtedly, revisions to the *1971 ACI Code* will be made in the future. However, the revisions should never be drastic, and undoubtedly will be slight, updating *Code* provisions to conform with new knowledge and new information obtained by experimentation and comprehensive studies made using electronic computers.

1.2 NOTATION

a = depth of equivalent rectangular stress block
a = shear span, distance between concentrated load and face of support
A_b = area of an individual bar, sq in.
A_c = area of core of spirally reinforced column measured to the outside diameter of the spiral, sq in.
L = live loads, or their related internal moments and forces
M_c = moment to be used for design of compression member
M_{cr} = cracking moment
M_m = modified bending moment
M_{max} = maximum bending moment due to externally applied design loads
M_u = applied design load moment at a section, in.-lb
M_1 = value of smaller end moment on compression member calculated from a conventional elastic frame analysis, positive if member is bent in single curvature, negative if bent in double curvature
M_2 = value of larger end moment on compression member calculated from a conventional elastic frame analysis, always positive
n = modular ratio = E_s/E_c
N_u = design tensile force on bracket or corbel acting simultaneously with V_u
N_u = design axial load normal to the cross section occurring simultaneously with V_u to be taken as positive for compression, negative for tension, and to include the effects of tension due to shrinkage and creep
P_b = axial load capacity at simultaneous assumed ultimate strain of concrete and yielding of tension steel (balanced conditions)
P_c = critical load
P_u = axial design load in compression member
r = radius of gyration of the cross section of a compression member
s = tie spacing, in.
s = shear or torsion reinforcement spacing in a direction parallel to the longitudinal reinforcement, in.
s = spacing of stirrups, in.
s_1 = spacing of vertical reinforcement in a wall
s_2 = shear of torsion reinforcement spacing in a direction perpendicular to the longitudinal reinforcement —or spacing of horizontal reinforcement in a wall
T = cumulative effect of temperature, creep, shrinkage, and differential settlement
T_u = design torsional moment
U = required strength to resist design loads or their related internal moments and forces
v_c = nominal permissible shear stress carried by concrete
v_{ci} = shear stress at diagonal cracking due to all design

loads, when such cracking is the result of combined shear and moment

v_{cw} = shear stress at diagonal cracking due to all design loads, when such cracking is the result of excessive principal tensile stresses in the web

v_{tc} = nominal permissible torsion stress carried by concrete

v_{tu} = nominal total design torsion stress

v_u = nominal total design shear stress

V_d = shear force at section due to dead load

V_1 = shear force at section occurring simultaneously with M_{max}

V_u = total applied design shear force at section

w = design load per unit length of beam or per unit area of slab

w = weight of concrete, lb per cu ft

w_d = design dead load per unit area

w_l = design live load per unit area

W = wind load, or its related internal moment and forces

x = shorter overall dimension of a rectangular part of a cross section

x_1 = shorter center-to-center dimension of a closed rectangular stirrup

y = longer overall dimension of a rectangular part of a cross section

y_t = distance from the centroidal axis of gross section, neglecting the reinforcement, to the extreme fiber in tension

y_1 = longer center-to-center dimension of a closed rectangular stirrup

α = angle between inclined web bars and longitudinal axis of member

α_t = a coefficient as a function of y_1/x_1

β_b = ratio of area of bars cut off to total area of bars at the section

β_d = the ratio of maximum design dead load moment to maximum design total load moment, always positive

β_1 = ratio of depth of rectangular stress block to the distance from the compression face to the neutral axis

δ = moment magnification factor for columns

ρ = A_s/bd

ρ' = A_s'/bd

ρ_b = reinforcement ratio producing balanced conditions

ρ_{min} = minimum reinforcement ratio

ρ_s = ratio of volume of spiral reinforcement to total volume of core (out-to-out of spirals) of a spirally reinforced concrete or composite column

ρ_v = $(A_s + A_h)/bd$

ρ_w = A_s/b_wd

ϕ = capacity reduction factor

ω = $\rho f_y/f_c'$

ω' = $\rho' f_y/f_c'$

1.3 LOAD FACTORS AND UNDERSTRENGTH FACTORS

In ultimate strength design, shears, moments, and axial loads obtained using an elastic method of analysis (using service level loads) are modified to reflect conditions that would cause failure due to the development of ultimate stresses and strains. This method insures adequate safety and eliminates excessive strength beyond the selected load factors.

Adequate strength is provided by two modifications. First, the service load shears, moments and axial loads are increased by multiplying by load factors. The load factors have been established on the basis of probability of occurrence of given loads and the accuracy of computation of loads. In this chapter we are concerned primarily with dead loads (D), live loads (L), and lateral loads such as wind loads (W). Since earthquake loads, soil pressure loads, blast loads, and similar loads relate to specialized conditions of analysis and design, they are not covered in this chapter.

Then, the possibility of understrength due to minor variations in dimensions of sections and strength of materials and other factors which lead to strength less than that theoretically assumed are taken into account.

The load factors are accounted for by using three equations given in the *1971 ACI Code*. Here U is the ultimate strength to be used in design for shear, bending moment, or axial load. In order to provide continuity with the *ACI Code*, the equation numbers used in Chapter 9 of the *Code* will be used below.

The required strength (U) must be at least

$$U = 1.4D + 1.7L \tag{9-1}$$

when dead load (D) and live load (L) govern the design.

However, when wind loads are present, it is also necessary to investigate two separate conditions, as follows:

$$U = .75(1.4D + 1.7L + 1.7W) \tag{9-2}$$

and

$$U = 0.9D + 1.3W \tag{9-3}$$

Equation (9-2) conforms with long time practice of permitting an increase of one-third for stresses when wind load is combined with dead load and live load. It can be readily seen that the ultimate capacity in (9-2) corresponds to that required in (9-1) reduced by one-third, since $1/(1 + 0.33) = 0.75$.

Equation (9-3) is intended to cover conditions where tension forces develop due to overturning moments. This equation governs only in very tall structures where wind loads are high. Dead loads are discounted by 10% to account for situations in which dead loads may be overestimated.

The magnitudes of the load factors are based on the probability of correctly predicting the magnitude of the load. Since dead loads are more likely to be in the order of magnitude calculated using known weights of materials and dimensions previously set, a load factor of 1.4 is presumed to be realistic.

Since live loads are based on less reliable information than dead loads, the factor 1.7 is used. The wind load factor is based on similar considerations.

Regardless of the ratio of dead load to live load, the average "safety factor" against failure is very close to 1.65. This conforms closely with designs using other materials under the provisions of modern building codes.

While the load factors are used to increase service load level shears, moments, and axial forces to the ultimate level, the theoretical capacities of the cross-sections are reduced by using the understrength factor, ϕ. The ϕ factor varies in accordance with the severity of catastrophic effects of failure of a structural element, as well as the degree of accuracy one can expect in predicting the theoretical strength of the element.

Since flexural members and members subjected primarily to tension forces, when tested to failure, show excellent agreement with test data, the understrength factor ϕ chosen is 0.9.

Members subjected to shear and/or torsion have a lesser conformance with test data at failure and their mode of failure is abrupt, without warning, therefore a ϕ factor of 0.85 is considered to be advisable.

Spirally reinforced compression members exhibit a great deal of toughness and ductility when tested to failure. Compression members with noncontinuous ties show considerably less ductility and toughness. Hence, the general understrength factors (ϕ) have been selected as $\phi = 0.70$, for tied compression members, and $\phi = 0.75$, for spiral reinforced compression members.

As axial loads become small in a compression member, it acts more as a beam than a compression member. Thus, a transition from "column action" to "beam action" is permitted by the ACI Code. For simplicity, the transition from compression member action to beam action is permitted when the ultimate axial load p_u is $0.1 f'_c A_g$.

This value represents a statistical average for the most common conditions, obtained using numerous solutions by electronic computer and comparisons with test data. This condition applies when the cross section is symmetrically reinforced, and the ratio of the dimension (γh) between end face reinforcement and the overall depth (h) is not less than ($0.7h$), and the steel yield strength does not exceed 60 ksi.

1.4 SERVICE LOAD DESIGN*

In the *1971 ACI Code*, Section 8.10 permits an alternate design method to be used in lieu of the general method of ultimate strength design. For members subjected to flexure only, the design method conforms with the traditional straight line theory, which will not be covered in this chapter. For those interested in that method, the references provide sufficient material for review.

For serviceability, shear, torsion, and anchorage, the design methods provided in this chapter apply to "Service Load Design" (Working Stress Design) under Section 8.10 of the *1971 ACI Code*, except that constant load factors apply. When dead load and live load only are considered, 40% of the ultimate strength can be used. Thus, a load factor of 2.5 is applied to the combination of dead load plus live load. When wind load is involved, 75% of the combination of $D + L + W$ may be used with 40% of the ultimate strength. Thus, the constant load factor for $D + L + W$ is 1.875. After the constant load factors are multiplied by the service loads, ultimate strength values are used with the understrength factor ϕ taken as 1.0.

The alternate design method of Section 8.10 of the 1971 ACI Code may be used as explained above with all problems provided in this chapter except those for flexure without axial load, for which straight line theory may be used. Those who are interested in that method should review the references.

1.5 BALANCED DESIGN

In the past, using Service Load Design, balanced design has always referred to conditions for which the concrete developed maximum permissible stresses (usually $0.45 f'_c$) at the same time as the steel stresses reached some permissible value, such as 20 ksi.

Under present day concepts, it is realized that strains are of more importance than the related stresses. Hence, in considering ultimate loads, ultimate strains are recognized as very important factors to be considered.

Comprehensive tests of reinforced concrete members indicate that a strain of 0.003 in./in. is the average value for concrete at ultimate strength in compression. For reinforcing steel, the yield strain is equal to the yield strength divided by the elastic modulus for the steel, which averages 29×10^6 psi.

At ultimate strength, balanced design occurs for beam type members when the concrete strain of 0.003 in./in. is reached at the same time the tension steel just reaches its yield strain. For compression members, balanced design is not specifically defined in the *ACI Code*. However, ACI Committee 340, CRSI and PCA, who have developed computer programs and design aids for compression members have mutually agreed to define balanced conditions as those for which the outermost compression fiber of the concrete reaches a strain of 0.003 in./in. and the reinforcing bar farthest from the neutral axis (on the tension side) yields. It is important to note here that certain cross-sections having given material strengths require tension axial loads in order to produce balanced conditions.

1.6 BRITTLE AND DUCTILE FAILURES

If any member develops its ultimate strength while the tension side steel is within the elastic stress range or just at yield stress, the member would fail in a brittle manner without warning of impending failure. On the other hand, if the tension steel strains are greater than the yield strain at ultimate strength, ductile failure will occur. That is, the member fails gradually over a reasonably long period of time, showing definite distress and impending collapse. Under such circumstances, failure will not be catastrophic because of sufficient warning of impending failure. In many cases, failure can be prevented by taking corrective measures.

For beam type members, the *ACI Code* requires ductility in the interest of safety. This is insured by limiting the quantity of reinforcement to 75% of that which would correspond to balanced conditions. Appropriate equations for balanced steel ratios for various types of members are provided in this chapter, without proof. The derivations are given in the *Commentary for the 1971 ACI Code*.

Figure 1-1 illustrates the stresses, forces, and strains related to a beam reinforced for tension only, where the concrete develops all of the compression force due to the ultimate moment, M_u. When the steel strain, ϵ_s, is equal to or greater than the yield strain, and the concrete strain is 0.003 in./in., the ultimate strength is

$$M_u = \phi b d^2 f'_c \omega [1 - 0.59\omega] \tag{1-1}$$

where

$$\omega = \frac{\rho f_y}{f'_c} = \frac{A_s}{bd}\frac{f_y}{f'_c} \tag{1-2}$$

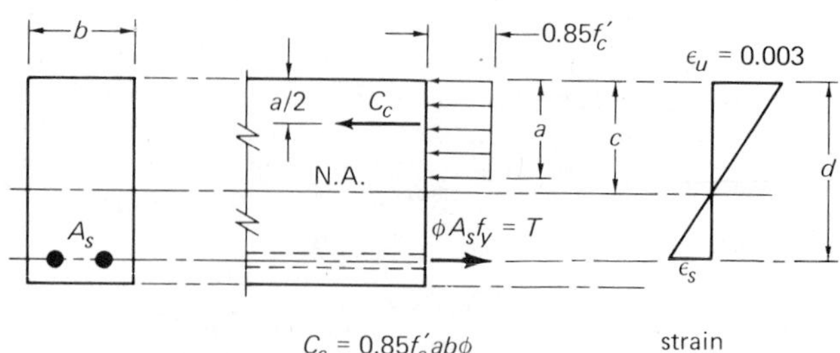

Fig. 1-1 Beam reinforced for tension only.

*Service Load Design, also customarily called Working Stress Design is presented in the *1971 ACI Code* as the Alternate Design Method in Section 8.10.

An alternative expression is

$$M_u = \phi A_s f_y \left(d - \frac{a}{2}\right) \tag{1-3}$$

where

$$a = \frac{A_s f_y}{0.85 f'_c b} \tag{1-4}$$

The steel ratio

$$\rho = \frac{A_s}{bd} \tag{1-5}$$

may not exceed 75% of the balanced ratio

$$\rho_b = \frac{0.85 \beta_1 f'_c}{f_y} \left[\frac{87{,}000}{87{,}000 + f_y}\right] \tag{1-6}$$

where $\beta_1 = 0.85$ for $f'_c \leqslant 4000$ psi, and $0.85 - 0.05(f'_c - 4000)/1000$ for $f'_c \geqslant 4000$ psi. This variable defines the depth of the equivalent rectangular stress block, $a = \beta_1 C$. The factor β_1 relates the equivalent rectangular stress block to the true stress-strain diagram which is actually parabolic, rising to maximum stress f'_c and then descending to approximately $0.85 f'_c$. In order to simplify design calculations, the rectangular stress block has been developed to closely approximate the total compression force and its point of application. As concrete strength (f'_c) increases beyond 4000 psi, the initial portion of the stress-strain diagram becomes increasingly linear and the descending portion of the stress-strain diagram diminishes. Limited test* data indicates that the rectangular stress block should be replaced by a parabolic stress block when f'_c is equal to or greater than 8000 psi. The parabolic stress block originates at the neutral axis.

It is apparent from eq. (1-1) that many combinations of the steel ratio ρ and cross section dimensions will provide a given ultimate moment M_u. The minimum dimensions occur when the steel ratio ρ is maximum. This condition is not always the most economical, and may often be associated with relatively large deflections.

The *ACI Code* requires that deflections be computed if the overall depth (h) of a one-way system is less than the appropriate value provided in Table 1-1. The depths given in the table are based on normal weight concrete (145 pcf) and grade 60 steel ($f_y = 60{,}000$ psi). For other materials, the footnotes provide equations for modifying the minimum depth. The depths in the table are usually on the conservative side. When many repetitive members are involved it may be more economical to use lesser depths that satisfy deflection requirements. To simplify the design process, Table 1-2 provides the ultimate moment computed for flexural members using eq. 1-1. The table is appropriate for sections one foot wide. For other widths, multiply the actual moment by $12/b$ and enter the table as though the width is 12 in. Since eq. (1-1) is not linear in d, direct interpolation is not strictly applicable. A simple plot may be used for interpolation, however. Note that the understrength factor, $\phi = 0.9$, has been included in the tables.**

TABLE 1-1 Minimum Thickness of Beams or One-Way Slabs unless Deflections are Computed.[a,b,c,d]

Member	Minimum thickness, t							
	Simply supported		One end continuous		Both ends continuous		Cantilever	
Solid one-way slabs	L/23	L/25	L/27	L/30	L/32	L/35	L/11.5	L/12.5
	L/18	L/20	L/22	L/24	L/25.5	L/28	L/9	L/10
Beams or ribbed one-way slabs	L/18	L/20	L/21	L/23	L/23.5	L/26	L/9	L/10
	L/14.5	L/16	L/17	L/18.5	L/19	L/21	L/7.5	L/8

[a] Key to table:

	110 to 120	120 to 150
Concrete wt. (pcf) →	110 to 120	120 to 150
f_y = 40,000 psi →	L/23	L/25
f_y = 60,000 psi →	L/18	L/20

[b] The span length L and t must be in the same units.

[c] Non-prestressed reinforced concrete

[d] Valid only for members not supporting or attached to partitions or other construction likely to be damaged by large deflections.

EXAMPLE 1-1: For $f'_c = 4000$ psi and $f_y = 60{,}000$ psi, find the balanced steel ratio, ρ_b, and the maximum steel ratio permitted.

SOLUTION: With $\beta_1 = 0.85$ and using eq. (1-6)

$$\rho_b = \frac{(0.85)(0.85)(4000)}{60{,}000}\left[\frac{87{,}000}{87{,}000 + 60{,}000}\right] = 0.0285$$

$$\rho_{max} = 0.75\rho_b = (0.75)(0.0285) = 0.0214$$

(Note in Table 1-2 that the largest steel ratio provided is 0.021. Use of the table automatically guarantees that ρ_{max} will not be exceeded.)

EXAMPLE 1-2: For $f'_c = 4000$ psi and $f_y = 60{,}000$ psi, determine the theoretically required effective depth, d, and the steel area required, if the steel ratio is approximately $0.5\rho_b$, the beam width is 10 in. and M_u is 79.5 ft-kips.

SOLUTION: $\rho = (0.5)(0.0285) = 0.01425$ (Say 0.014). Using eq. (1-1), $\phi = 0.9$, $\omega = \rho f y / f'_c = (0.014)(60{,}000/4000) = 0.21$

$$M_u = (0.9)(10)d^2(4000)(0.21)[1 - (0.59)(0.21)]$$

$$= (79.5)(12)(1000) = 6623d^2, \text{ so } d = 12 \text{ in.}$$

$$A_s = \rho bd = (0.014)(10)(12) = 1.68 \text{ sq in.}$$

EXAMPLE 1-3: Use eqs. (1-3) and (1-4) to investigate the moment capacity of the beam designed in ex. (1-2).

SOLUTION: $A_s = 1.68$ sq in., $f_y = 60{,}000$ psi, $f'_c = 4000$ psi. Using eq. (1-4), the depth of the stress block is

$$a = \frac{A_s f_y}{0.85 f'_c b} = \frac{(1.68)(60{,}000)}{(0.85)(4000)(10)} = 2.965 \text{ in.}$$

and using eq. (1-3)

$$M_u = \phi A_s f_y \left(d - \frac{a}{2}\right) = (0.9)(1.68)(60{,}000)\left(12.0 - \frac{2.965}{2}\right)$$

$$= 954 \text{ in.-kips} = 79.5 \text{ ft-kips}$$

EXAMPLE 1-4: Use Table 1-2 to design the beam described in Example 1-1.

SOLUTION: $0.5\rho_b = 0.14$, $b = 10$ in., so enter the table with $M_u = (12/10)(79.5) = 95.4$ ft-kips. For $\rho = 0.014$, $M_u = 95.4$ ft-kips, find $d = 12.0$ in.

1.8 BEAMS REINFORCED FOR TENSION AND COMPRESSION

Often a beam size is limited by space or by aesthetic considerations, and the cross section with tension reinforcement only is insufficient to provide the required strength. In such cases, compression reinforcement with additional

*Test data obtained by Howard Nedderman at the University of Texas at Arlington, Arlington, Texas, 1973.

**It is important to determine whether or not the understrength factor (ϕ) has been included before making use of design aids. Many design aids have been prepared using $\phi = 1.0$.

TABLE 1-2 Ultimate Resisting Moments in Sections One-Foot Wide (Slabs).

f'_c = 4. KSI FY = 60. KSI

EFFECTIVE DEPTH d INCHES

ρ	3.0	3.5	4.0	4.5	5.0	5.5	6.0	6.5	7.0	7.5
.0015	.8	1.0	1.3	1.7	2.0	2.5	2.9	3.4	4.0	4.5
.0020	1.0	1.3	1.7	2.2	2.7	3.3	3.9	4.5	5.2	6.0
.0025	1.2	1.7	2.2	2.7	3.4	4.0	4.8	5.6	6.5	7.5
.0030	1.5	2.0	2.6	3.2	4.0	4.8	5.7	6.7	7.8	8.9
.0035	1.7	2.3	3.0	3.8	4.6	5.6	6.6	7.8	9.0	10.4
.0040	1.9	2.6	3.4	4.3	5.3	6.4	7.6	8.9	10.3	11.8
.0045	2.1	2.9	3.8	4.8	5.9	7.1	8.4	9.9	11.5	13.2
.0050	2.4	3.2	4.2	5.3	6.5	7.9	9.3	11.0	12.7	14.6
.0055	2.6	3.5	4.6	5.8	7.1	8.6	10.2	12.0	13.9	15.9
.0060	2.8	3.8	5.0	6.3	7.7	9.3	11.1	13.0	15.1	17.3
.0065	3.0	4.1	5.3	6.7	8.3	10.1	12.0	14.0	16.3	18.7
.0070	3.2	4.4	5.7	7.2	8.9	10.8	12.8	15.0	17.4	20.0
.0075	3.5	4.7	6.1	7.7	9.5	11.5	13.7	16.0	18.6	21.3
.0080	3.7	5.0	6.5	8.2	10.1	12.2	14.5	17.0	19.7	22.6
.0085	3.9	5.2	6.8	8.6	10.7	12.9	15.3	18.0	20.8	23.9
.0090	4.1	5.5	7.2	9.1	11.2	13.6	16.2	18.9	22.0	25.2
.0095	4.3	5.8	7.6	9.6	11.8	14.3	17.0	19.9	23.1	26.5
.0100	4.5	6.1	7.9	10.0	12.4	14.9	17.8	20.8	24.2	27.7
.0105	4.7	6.4	8.3	10.5	12.9	15.6	18.6	21.8	25.3	29.0
.0115	5.1	6.9	9.0	11.3	14.0	16.9	20.1	23.6	27.4	31.4
.0125	5.5	7.4	9.7	12.2	15.1	18.2	21.7	25.4	29.5	33.8
.0135	5.8	7.9	10.3	13.0	16.1	19.5	23.2	27.2	31.5	36.2
.0145	6.2	8.4	11.0	13.9	17.1	20.7	24.6	28.9	33.5	38.4
.0155	6.5	8.9	11.6	14.7	18.1	21.9	26.0	30.6	35.4	40.7
.0165	6.9	9.4	12.2	15.5	19.1	23.1	27.4	32.2	37.3	42.9
.0175	7.2	9.8	12.8	16.2	20.0	24.2	28.8	33.8	39.2	45.0
.0185	7.6	10.3	13.4	17.0	20.9	25.3	30.1	35.3	41.0	47.0
.0195	7.9	10.7	14.0	17.7	21.8	26.4	31.4	36.9	42.7	49.1
.0205	8.2	11.2	14.5	18.4	22.7	27.5	32.7	38.3	44.5	51.0
.0214	8.5	11.5	15.0	19.0	23.5	28.4	33.7	39.6	45.9	52.7

f'_c = 4. KSI FY = 60. KSI

EFFECTIVE DEPTH d INCHES

ρ	8.0	9.0	10.0	11.0	12.0	13.0	14.0	15.0	16.0	17.0
.0015	5.2	6.5	8.0	9.7	11.6	13.6	15.7	18.0	20.5	23.1
.0020	6.8	8.6	10.7	12.9	15.3	18.0	20.8	23.9	27.2	30.7
.0025	8.5	10.7	13.3	16.0	19.1	22.4	25.9	29.8	33.8	38.2
.0030	10.1	12.8	15.8	19.1	22.8	26.7	31.0	35.5	40.4	45.6
.0035	11.8	14.9	18.4	22.2	26.4	31.0	35.9	41.3	46.9	53.0
.0040	13.4	16.9	20.9	25.3	30.1	35.3	40.9	46.9	53.4	60.3
.0045	15.0	18.9	23.4	28.3	33.6	39.5	45.8	52.5	59.8	67.5
.0050	16.6	21.0	25.9	31.3	37.2	43.7	50.6	58.1	66.1	74.6
.0055	18.1	22.9	28.3	34.2	40.7	47.8	55.4	63.6	72.4	81.7
.0060	19.7	24.9	30.7	37.2	44.2	51.9	60.2	69.1	78.6	88.7
.0065	21.2	26.8	33.1	40.1	47.7	56.0	64.9	74.5	84.7	95.7
.0070	22.7	28.8	35.5	43.0	51.1	60.0	69.5	79.8	90.8	102.5
.0075	24.2	30.7	37.9	45.8	54.5	64.0	74.2	85.1	96.8	109.3
.0080	25.7	32.6	40.2	48.6	57.9	67.9	78.7	90.4	102.8	116.1
.0085	27.2	34.4	42.5	51.4	61.2	71.8	83.2	95.6	108.7	122.7
.0090	28.7	36.3	44.8	54.2	64.5	75.6	87.7	100.7	114.6	129.3
.0095	30.1	38.1	47.0	56.9	67.7	79.5	92.1	105.8	120.3	135.8
.0100	31.6	39.9	49.3	59.6	70.9	83.2	96.5	110.8	126.1	142.3
.0105	33.0	41.7	51.5	62.3	74.1	87.0	100.9	115.8	131.7	148.7
.0115	35.7	45.2	55.8	67.5	80.4	94.3	109.4	125.6	142.8	161.3
.0125	38.5	48.7	60.1	72.7	86.5	101.5	117.7	135.1	153.7	173.5
.0135	41.1	52.0	64.2	77.7	92.5	108.5	125.9	144.5	164.4	185.6
.0145	43.7	55.3	68.3	82.6	98.3	115.4	133.8	153.6	174.8	197.3
.0155	46.3	58.5	72.3	87.4	104.0	122.1	141.6	162.5	184.9	208.8
.0165	48.7	61.7	76.1	92.1	109.6	128.6	149.2	171.3	194.8	219.9
.0175	51.2	64.7	79.9	96.7	115.1	135.0	156.6	179.7	204.5	230.9
.0185	53.5	67.7	83.6	101.1	120.4	141.2	163.8	188.0	213.9	241.5
.0195	55.8	70.6	87.2	105.5	125.5	147.3	170.8	196.1	223.1	251.8
.0205	58.0	73.4	90.7	109.7	130.5	153.2	177.7	203.9	232.0	261.9
.0214	60.0	75.9	93.7	113.3	134.8	158.2	183.5	210.7	239.7	270.6

TABLE 1-2 *(continued)*

f'_c = 4. KSI FY = 60. KSI

EFFECTIVE DEPTH d INCHES

ρ	18.0	19.0	20.0	21.0	22.0	23.0	24.0	25.0	26.0	27.0
.0015	25.9	28.9	32.0	35.3	38.7	42.3	46.1	50.0	54.1	58.3
.0020	34.4	38.3	42.5	46.8	51.4	56.2	61.2	66.4	71.8	77.4
.0025	42.8	47.7	52.9	58.3	63.9	69.9	76.1	82.6	89.3	96.3
.0030	51.1	57.0	63.1	69.6	76.4	83.5	90.9	98.6	106.7	115.0
.0035	59.4	66.2	73.3	80.8	88.7	96.9	105.5	114.5	123.9	133.6
.0040	67.6	75.3	83.4	91.9	100.9	110.3	120.1	130.3	140.9	151.9
.0045	75.6	84.3	93.4	102.9	113.0	123.5	134.4	145.9	157.8	170.1
.0050	83.7	93.2	103.3	113.9	124.9	136.6	148.7	161.3	174.5	188.2
.0055	91.6	102.0	113.1	124.7	136.8	149.5	162.8	176.6	191.0	206.0
.0060	99.5	110.8	122.8	135.3	148.5	162.3	176.8	191.8	207.4	223.7
.0065	107.2	119.5	132.4	145.9	160.2	175.0	190.6	206.8	223.7	241.2
.0070	114.9	128.1	141.9	156.4	171.7	187.6	204.3	221.7	239.7	258.5
.0075	122.6	136.6	151.3	166.8	183.1	200.1	217.8	236.4	255.7	275.7
.0080	130.1	145.0	160.6	177.1	194.3	212.4	231.3	250.9	271.4	292.7
.0085	137.6	153.3	169.8	187.2	205.5	224.6	244.5	265.3	287.0	309.5
.0090	145.0	161.5	179.0	197.3	216.5	236.7	257.7	279.6	302.4	326.1
.0095	152.3	169.7	188.0	207.3	227.5	248.6	270.7	293.7	317.7	342.6
.0100	159.5	177.7	196.9	217.1	238.3	260.4	283.6	307.7	332.8	358.9
.0105	166.7	185.7	205.8	226.9	249.0	272.1	296.3	321.5	347.7	375.0
.0115	180.8	201.4	223.2	246.0	270.0	295.1	321.3	348.7	377.1	406.7
.0125	194.6	216.8	240.2	264.8	290.6	317.6	345.8	375.3	405.9	437.7
.0135	208.0	231.8	256.8	283.1	310.7	339.6	369.8	401.2	434.0	468.0
.0145	221.2	246.4	273.1	301.0	330.4	361.1	393.2	426.6	461.4	497.6
.0155	234.0	260.8	288.9	318.5	349.6	382.1	416.0	451.4	488.2	526.5
.0165	246.6	274.7	304.4	335.6	368.3	402.6	438.3	475.6	514.4	554.7
.0175	258.8	288.4	319.5	352.3	386.6	422.5	460.1	499.2	539.9	582.3
.0185	270.7	301.6	334.2	368.5	404.4	442.0	481.3	522.2	564.8	609.1
.0195	282.3	314.6	348.6	384.3	421.7	461.0	501.9	544.6	589.0	635.2
.0205	293.6	327.2	362.5	399.7	438.6	479.4	522.0	566.4	612.6	660.6
.0214	303.3	338.0	374.5	412.9	453.1	495.2	539.2	585.1	632.8	682.4

f'_c = 4. KSI FY = 60. KSI

EFFECTIVE DEPTH d INCHES

ρ	29.0	31.0	33.0	35.0	37.0	39.0	41.0	43.0	45.0	47.0
.0015	67.3	76.9	87.1	98.0	109.5	121.6	134.4	147.8	161.9	176.6
.0020	89.3	102.0	115.6	130.0	145.3	161.4	178.4	196.2	214.9	234.4
.0025	111.1	126.9	143.8	161.8	180.8	200.8	222.0	244.1	267.4	291.7
.0030	132.7	151.6	171.8	193.2	215.9	239.9	265.1	291.6	319.4	348.4
.0035	154.1	176.1	199.5	224.4	250.8	278.6	307.9	338.7	370.9	404.6
.0040	175.3	200.3	226.9	255.3	285.3	317.0	350.3	385.3	422.0	460.3
.0045	196.3	224.3	254.1	285.9	319.5	354.9	392.3	431.5	472.5	515.5
.0050	217.1	248.0	281.1	316.2	353.3	392.5	433.8	477.2	522.6	570.1
.0055	237.7	271.6	307.7	346.2	386.9	429.8	475.0	522.5	572.2	624.2
.0060	258.1	294.9	334.1	375.9	420.1	466.7	515.8	567.3	621.3	677.8
.0065	278.3	318.0	360.3	405.3	452.9	503.2	556.1	611.7	669.9	730.8
.0070	298.3	340.8	386.2	434.4	485.5	539.4	596.1	655.7	718.1	783.3
.0075	318.0	363.4	411.8	463.2	517.7	575.2	635.7	699.2	765.7	835.3
.0080	337.6	385.8	437.2	491.8	549.6	610.6	674.8	742.3	812.9	886.8
.0085	357.0	408.0	462.3	520.0	581.1	645.7	713.6	784.9	859.6	937.7
.0090	376.2	429.9	487.1	548.0	612.4	680.4	751.9	827.1	905.8	988.1
.0095	395.2	451.6	511.7	575.6	643.3	714.7	789.9	868.8	951.5	1038.0
.0100	414.0	473.1	536.1	603.0	673.9	748.7	827.5	910.1	996.8	1087.3
.0105	432.6	494.3	560.1	630.1	704.1	782.3	864.6	951.0	1041.5	1136.2
.0115	469.2	536.1	607.5	683.3	763.7	848.5	937.7	1031.4	1129.6	1232.2
.0125	504.9	577.0	653.8	735.4	821.9	913.1	1009.2	1110.1	1215.7	1326.2
.0135	539.9	616.9	699.1	786.4	878.8	976.4	1079.1	1186.9	1299.9	1418.0
.0145	574.0	656.0	743.3	836.1	934.4	1038.2	1147.4	1262.0	1382.2	1507.7
.0155	607.4	694.1	786.5	884.7	988.7	1098.5	1214.0	1335.4	1462.5	1595.4
.0165	640.0	731.3	828.7	932.1	1041.7	1157.4	1279.1	1406.9	1540.9	1680.9
.0175	671.7	767.5	869.8	978.4	1093.4	1214.8	1342.6	1476.7	1617.3	1764.2
.0185	702.7	802.9	909.8	1023.5	1143.8	1270.7	1404.4	1544.8	1691.8	1845.5
.0195	732.8	837.3	948.9	1067.4	1192.8	1325.3	1464.7	1611.0	1764.4	1924.7
.0205	762.1	870.9	986.9	1110.1	1240.6	1378.3	1523.3	1675.5	1835.0	2001.8
.0214	787.3	899.6	1019.4	1146.7	1281.5	1423.8	1573.6	1730.8	1895.6	2067.8

tension reinforcement can supply the excess bending moment.

When compression reinforcement is used, the total tension steel area should not exceed 75% of the balanced ratio for beams reinforced for tension and compression. The expression for this case, as derived in the *ACI Code Commentary*, is

$$\rho_b = \bar{\rho}_b + \rho' f'_{sb}/f_y \qquad (1\text{-}7)$$

where ρ_b is given by eq. (1-6) for tension reinforcement only, $\rho' = A'_s/bd$, the compression steel ratio, and *at balanced conditions only*, the stress in the compression steel is

$$f'_{sb} = 87{,}000 \left[1 - \frac{d'}{d}\left(\frac{87{,}000 + f_y}{87{,}000}\right)\right] \leqslant f_y \qquad (1\text{-}8)$$

The appropriate variables are illustrated in Fig. 1.2.

Table 1-3 provides values of ρ_{BAL} for beams reinforced for tension and compression. The solid line separates values for which the compression steel has a stress (f'_s) less than or equal to f_y at balanced conditions. The tabular values relate to $f'_c = 4000$ psi and $f_y = 60{,}000$ psi. Similar tables can be prepared for other material strengths.

The most general solution for beams reinforced with tension and compression steel is obtained by superposition of M_1 and M_2, as shown in Fig. 1-2. M_1 is the moment capacity of a beam reinforced with tension steel only. M_2 is the moment couple developed by the compression steel and additional tension steel. Following this approach, the beam capacity can be obtained using Table 1-2 or the related equations, with tension steel area, A_{s1}. Then, the additional moment capacity required can be obtained using an appropriate quantity of compression steel (A'_s) and the related necessary additional quantity of tension steel, A_{s2}. The sum, $A_{s1} + A_{s2}$ should not exceed that corresponding to ρ_{max} in Table 1-2. Since precise values calculated for A'_s, A_{s1} and A_{s2} may not always be provided due to available bar sizes and their areas, the maximum steel ratio check must be made after the bars have been selected and actual areas of steel provided are known.

To select the proper quantities of steel, the *actual stress* in the compression steel must be known. Therefore, the location of the neutral axis must be known. The most simple approach is to consider that the neutral axis location will not change after the compression steel and the additional tension steel have been added and to design accordingly. The stress in the tension steel will always be f_y, since ductile conditions are required by the ACI Code.

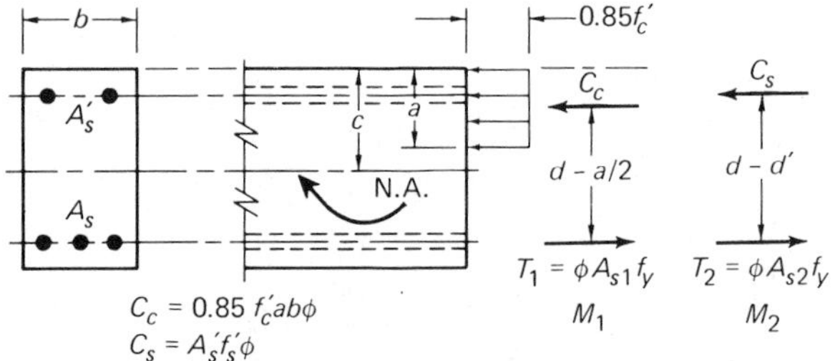

Fig. 1-2 Beams reinforced with tension and compression steel.

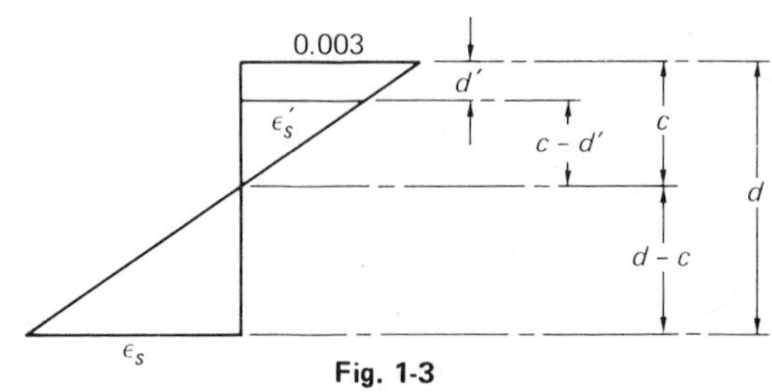

Fig. 1-3

Figure 1-3 shows the strain diagram from which the stress in the compression steel may be computed. Using similar triangles, note that $f'_s = \epsilon'_s E_s \leqslant fy$ with $E_s = 29 \times 10^6$ psi, it is found that

$$f'_s = 87{,}000 \left[1 - \frac{d'}{c}\right] \leqslant fy \qquad (1\text{-}9)$$

The force in the compression steel is $A'_s f'_s$ and that in the additional tension steel is $A_{s2} f_y$. Since those forces are equal and opposite, it follows that*

$$A'_s = \frac{A_{s2} f_y}{f'_s} \qquad (1\text{-}10)$$

If the total moment is M_u and the capacity of the section with tension reinforcement only is M_1, then the additional moment capacity required is

$$M_2 = M_u - M_1 \qquad (1\text{-}11)$$

and the tension steel area, A_{s2}, is obtained as

$$A_{s2} = \frac{M_2}{\phi f_y (d - d')} \qquad (1\text{-}12)$$

Thus, A_{s2} is obtained directly using eqs. (1-11) and (1-12). The actual value of f'_s is then obtained from eq. (1-9), and A'_s from eq. (1-10). In order to find the neutral axis location (c) for use in Eq. (1-9), eq. (1-4) may be used and divided by β_1. That is

$$c = \frac{a}{\beta_1} = \frac{A_{s1} f_y}{0.85 f'_c b \beta_1} \qquad (1\text{-}13)$$

The foregoing discussion applies to the *design* of beams reinforced for tension and compression. When one wishes to *investigate* such a beam, a different approach must be used, since the neutral axis location is not known, and the total tension steel area is known, rather than A_{s1} and A_{s2}.

*More precisely, $A'_s = A_{s2} fy/(f'_s - 0.85 f'_c)$, accounting for concrete displaced by steel.

TABLE 1-3 Balanced Steel Ratio—Beams with Tension and Compression Steel.

$f'_c = 4$ ksi $f_y = 60$ ksi

d'/d \ ρ'/ρ	.000	.100	.200	.300	.400	.500	.600	.700	.800	.900	1.000	
.05	.029	.031	.034	.037	.040	.043	.046	.048	.051	.054	.057	
.10	.029	.031	.034	.037	.040	.043	.046	.048	.051	.054	.057	$f'_s = f_y$
.15	.029	.031	.034	.037	.040	.043	.046	.048	.051	.054	.057	
.20	.029	.031	.034	.037	.039	.042	.045	.048	.050	.053	.056	
.25	.029	.031	.033	.036	.038	.040	.043	.045	.048	.050	.052	$f'_s < f_y$
.30	.029	.031	.033	.035	.037	.039	.041	.043	.045	.047	.049	

It can be shown that the compression steel yields if

$$\rho = \rho' \geqslant 0.85\beta_1 \frac{f_c'd}{f_yd}\left[\frac{87{,}000}{87{,}000 - f_y}\right] \quad (1\text{-}14)$$

When eq. (1-14) is satisfied, $A_{s2} = A_s'$, and this condition applies to both design and investigation of beams reinforced for tension and compression.

When eq. (1-14) is not satisfied $f_s' < f_y$, so the neutral axis is located by solving the quadratic equation

$$C^2 + \left[\frac{87{,}000A_s' - A_sf_y}{0.85\beta_1\, bf_c'}\right]C - \frac{87{,}000A_s'd'}{0.85\beta_1\, bf_c'} = 0 \quad (1\text{-}15)$$

Once the distance (c) of the neutral axis from the compression face is known, f_s' can be obtained, and a solution developed.

EXAMPLE 1-5: Design a rectangular beam using $f_c' = 4000$ psi and $f_y = 60{,}000$ psi for an ultimate moment of 600 ft-kips. The dimensions are limited to $b = 12$ in., $d = 24$ in. and $d' = 2.5$ in. The effective depth (d) is based on preliminary calculations which indicate that two layers of tension reinforcement will be required.

SOLUTION: Using the maximum permissible quantity of tension steel without compression steel

$$\rho_{max} = 0.75\rho_b = 0.0214 \quad \beta_1 = 0.85$$

$$A_{s1} = \rho_{max}\, bd = (0.0214)(12)(24) = 6.16 \text{ sq in.}$$

$$C_c = T = \phi A_sf_y = (0.9)(6.16)\left(\frac{60{,}000}{1000}\right) = 332.6 \text{ kips}$$

$$C_c = 0.85\phi f_c'ab = 332.6 \quad \text{(See Fig. 1-1)}$$

$$a = \frac{332.6}{(0.85)(4)(12)(0.9)} = 9.056 \text{ in.}, \; c = \frac{a}{\beta_1} = 10.65 \text{ in.}$$

$$M_1 = \phi A_{s1}f_y\left(d - \frac{a}{2}\right) = T\left(d - \frac{a}{2}\right)$$

$$M_1 = 332.6\left(24.0 - \tfrac{9.056}{2}\right)\left(\tfrac{1}{12}\right) = 539.7 \text{ ft-kips}$$

$$M_2 = 600.0 - 539.7 = 60.3 \text{ ft-kips}$$

$$A_{s2} = \frac{M_2}{(d - d')\phi f_y} = \frac{(60.3)(12)}{(21.5)(0.9)(60)} = 0.62 \text{ sq in.}$$

From eq. (1-9), $d'/c = 2.5/10.65 = 0.2347$, so $f_s' = 87{,}000(1 - 0.2347) = 66{,}580$ psi, $> f_y$, so use $f_s' = 60{,}000$ psi $= f_y$
Since

$$f_s' = f_y,\; A_s' = A_{s2} = 0.62 \text{ sq in.}$$

$$A_s = A_{s1} + A_{s2} = 6.16 + 0.623 = 6.78 \text{ sq in.}$$

Table 1-11 shows that three No. 11 bars and three No. 8 bars provide $A_s = 7.05$ sq in., and Table 1-12 indicates that the three No. 11 bars will fit in a section 12 in. wide, as will three No. 8 bars. This satisfies requirements for the tension steel. The larger bars, No. 11, will be placed in the lower layer, and the smaller bars in the upper layer.

WARNING: Care must be exercised in the field to insure that the bars are placed in the proper layers. The *ACI Code* requires that bars in the upper layer be placed directly above those in the lower layer. This requires special consideration in the selection of bars in order to insure symmetry about a vertical axis. Lack of such symmetry will cause torsion to develop. Further, use of several layers of reinforcement sometimes causes violation of the ductility requirements of the *ACI Code*. When the effective depth is measured to the centroid of the reinforcing steel, it is possible in shallow beams that the upper layer of tension reinforcement will have stresses (at ultimate load) that are less than the yield stress. The beam is then less ductile than required by the code.

For the compression steel (A_s'), two No. 5 bars will provide an area of 0.62 sq. in., which is exactly what is needed. The combination of A_s and A_s' selected, however, may possibly not satisfy the Code requirements for ρ_{max}. This should be checked.

$$\rho = \frac{A_s}{bd} = \frac{7.05}{(12)(24)} = 0.0244$$

$$\rho^1 = \frac{A_s'}{bd} = \frac{0.62}{(12)(24)} = 0.0021$$

Equation (1-8) is used to find f_{sb}' at balanced conditions, so for $d'/d = 2.5/24 = 0.104$

$$f_{sb}' = 87{,}000\left[1 - 0.104\frac{(87{,}000 + 60{,}000)}{60{,}000}\right] \leqslant f_y$$

$$= 64{,}830 \text{ psi} > f_y \quad \text{use } f_{sb}' = f_y$$

thus

$$\rho_b = \bar{\rho}_b + \frac{\rho' f_s'}{f_y} = \rho_b + \rho'$$

or

$$\rho_b = 0.0285 + (0.0021)(1.0) = 0.0306$$

and

$$\rho_{max} = (0.75)(0.0306) = 0.023$$

Note that the steel area selected provides $\rho = 0.0244$, which is greater than the permissible value, 0.023.

The solution can be modified to satisfy the *ACI Code* in a simple manner, by increasing the area of compression steel. Using two No. 7 bars in the compression zone, ρ' becomes 0.0041, and ρ_b becomes 0.0326 and ρ_{max} is 0.0245.

The resulting capacity of the cross section is slightly greater than that required, but all provisions of the *Code* will then be satisfied.

This problem was selected to illustrate the fact that difficulties will usually arise when A_{s1} is based on the maximum permissible tension steel ratio for the cross section without compression reinforcement when it is known that compression reinforcement will be required.

It is advisable to base A_{s1} on a maximum steel ratio of approximately $0.65\rho_b$ when it is anticipated that compression reinforcement will be required. This will take care of the usual increases in steel area provided when bars are selected.

ALTERNATE SOLUTION: When A_{s1} is based on $0.75\bar{\rho}_b$, the maximum steel ratio will not be exceeded if A_s' is selected using the equation:

$$A_s' = \tfrac{4}{3}A_{s2}\, f_y/f_{sb}'$$

Here,

$$A_s' = (\tfrac{4}{3})(0.62)(60/60) = 0.827 \text{ sq in.}$$

Two No. 6 bars will suffice, $A_s = 0.88$

Total steel area = 7.05 + 0.88 = 7.93 sq. in.

EXAMPLE 1-6: Solve Example 1-5 starting with $0.65\,\rho_b$ in determining M_1.

$$\rho = (0.65)(0.0285) = 0.0185$$

$$A_{s1} = \rho bd = (0.0185)(12)(24) = 5.33 \text{ sq in.}$$

$$C_c = T = \phi A_sf_y = (0.9)(5.33)(60) = 287.8 \text{ kips}$$

$$C_c = 0.85\phi f_c'ab = 287.8$$

$$a = \frac{287.8}{(0.85)(4)(12)(0.9)} = 7.838 \text{ in.}$$

$$c = \frac{a}{\beta_1} = \frac{7.838}{0.85} = 9.22 \text{ in.}$$

$$M_1 = \phi A_{s1}f_y\left(d - \frac{a}{2}\right) = T\left(d - \frac{a}{2}\right)$$

$$= 287.8\left(24.0 - \frac{9.22}{2}\right)\left(\frac{1}{12}\right) = 465.0 \text{ ft-kips}$$

$$M_2 = 600.0 - 465.0 = 135.0 \text{ ft-kips}$$

$$A_{s2} = \frac{M_2}{(d - d')\phi f_y} = \frac{(135.0)(12)}{(21.5)(0.9)(60)} = 1.40 \text{ sq in.}$$

Using eq. (1-9), $f_s' = f_y$, so $A_s' = A_{s2}$.

Thus, $A_s = A_{s1} + A_{s2} = 5.33 + 1.40 = 6.73$ sq in. and $A'_s = 1.40$ sq in. Using Tables 1-11 and 1-12, select three No. 10 and three No. 9, $A_s = 6.81$ sq in. Use two rows with the No. 10 bars in the lower layer. For A'_s, select five No. 5, $A'_s = 1.55$ sq in.

$$\rho = \frac{A_s}{bd} = \frac{6.81}{288} = 0.0236$$

$$\rho' = \frac{A'_s}{bd} = \frac{1.55}{288} = 0.0053$$

$$\rho_b = \rho_b + \frac{\rho' f'_s}{f_y} \text{ and } \frac{f'_s}{f_y} = 1.0,$$

$$\rho_b = 0.0285 + 0.0053 = 0.0338$$

$$\rho_{max} = (0.75)(0.0338) = 0.0254 > 0.0236$$

Total steel area, $A_s + A'_s = 6.81 + 1.55 = 8.36$ sq in. From Example 1-5, $A_s + A'_s = 7.05 + 0.88 = 7.93$ sq in. (Alternate Solution)

The procedure followed in Example 1-5 requires slightly less steel, but will usually require an upward adjustment of A'_s in order to meet the ductility requirements of the *ACI Code*, when bar sizes are selected and actual steel areas are obtained.

EXAMPLE 1-7: A beam has $b = 12$ in., $d = 12$ in., $d' = 2.4$ in., $f'_c = 5{,}000$ psi, $f_y = 60{,}000$ psi, $A'_s = 0.877$ sq in. and $A_s = 4.28$ sq in. Determine the ultimate strength capacity of the cross section.

SOLUTION: Since this is a review problem, it is first necessary to determine whether or not the compression steel will yield. Equation (1-14) applies.

$$\rho - \rho' \geqslant 0.85 \frac{f'_c d'}{f_y d} \left[\frac{87{,}000}{87{,}000 - f_y}\right] \quad \text{for } f'_s = f_y$$

$$(\rho - \rho') \geqslant \frac{(0.85)(5)(2.4)}{(60)(12)} \left[\frac{87{,}000}{27{,}000}\right] \geqslant \frac{709{,}920}{19{,}440{,}000} \geqslant 0.0456$$

$$\rho = \frac{A_s}{bd} = \frac{4.28}{144} = 0.0297$$

$$\rho' = \frac{A'_s}{bd} = \frac{0.877}{144} = 0.0061$$

$$\rho - \rho' = 0.0297 - 0.0061 = 0.0236 < 0.0456$$

Therefore, $f'_s < f_y$ so eq. (1-15) must be used.

$$0.85 \beta_1 b f'_c = (0.85)(0.85)(12)(5000) = 43{,}350 \text{ psi}$$

so

$$C^2 + C\left[\frac{(87{,}000)(0.877) - (4.28)(60{,}000)}{43{,}350}\right] - \left[\frac{(87{,}000)(0.877)(2.4)}{43{,}350}\right] = 0$$

or

$$C^2 - (4.16)C - 4.224 = 0$$

Using the quadratic formula, find $C = 5.0$ in. Next, find f'_s, using eq. (1-9).

$$f'_s = 87{,}000 \left(\frac{1 - d'}{c}\right) \leqslant f_y \qquad (1\text{-}9)$$

Since $d'/c = 2.4/5.0 = 0.48$ and $1 - d'/c = 0.52$, $f'_s = 45{,}240$ psi.

Figure 1-4 shows the free body diagram for this problem, with $a = \beta_1 c = (0.85)(5.0) = 4.25$ in., and $a/2 = 2.58$ in.

Summing moments about the tension steel (A_s)

$$M_u = (\phi)\,(0.85 f'_c ab)\,(d - a/2) + (\phi)\,(A'_s f'_s)\,(d - a/2) \text{ (in.-kips)}$$

So

$$M_u = (0.9)(0.85)(5)(4.25)(12)(9.875) + (0.9)(0.877)(45.24)(9.6)$$

$$M_u = 1{,}926 + 343 + 2269 \text{ in.-kips}$$

$$M_u = 189.0 \text{ ft-kips}$$

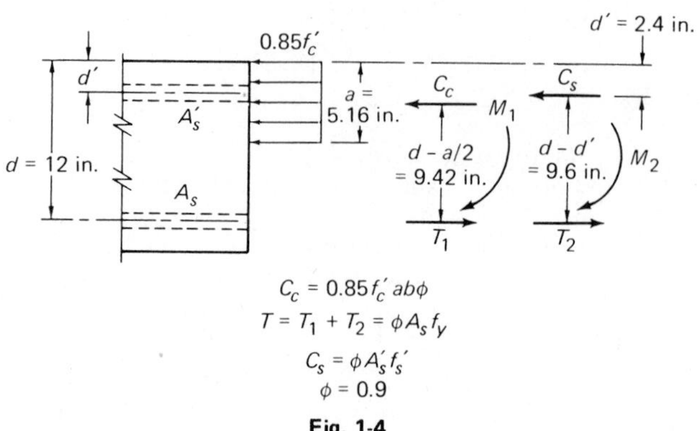

Fig. 1-4

1.9 INCREASE IN MOMENT CAPACITY DUE TO COMPRESSION STEEL

Frequently, compression reinforcement is furnished automatically due to continuing of some positive moment reinforcement into the support. In many cases, the increased capacity of a cross section is more than that required, and the area of tension reinforcement may be reduced. In such cases, Table 1-4 is useful. The table provides the ratio of ultimate moment M_u with compression reinforcement to M_u without compression reinforcement. Adjustment of the neutral axis location for addition of compression steel and the stress in the compression steel are accounted for automatically. The maximum permissible steel ratio is also accounted for in the table.

EXAMPLE 1-8: A beam is 12 in. wide and has an effective depth, d, 14 in. For $f'_c = 4000$ psi, $f_y = 60{,}000$ psi, find its capacity if $A_s = 3.36$ sq in., $A'_s = 0.672$ sq in., and $d'/d = 0.10$.

SOLUTION:

$$\rho = \frac{3.36}{168} = 0.02$$

$$\rho' = \frac{0.672}{168} = 0.004$$

$$\rho'/\rho = \frac{0.004}{0.02} = 0.2$$

From Table 1-2, $\phi M_u = 181.7$ ft-kips for $b = 12$ in., $d = 14$ in., $\rho = 0.02$. From Table 1-4, for $\rho'/\rho = 0.2$, $\rho = 0.02$, coefficient $= 1.062$. So, with compression steel

$$\phi M_u = (1.062)(181.7) = 193 \text{ ft-kips}$$

1.10 T-BEAMS

Members having a flange in compression are called T-beams. These may be part of a floor or roof system in which a number of T-beams exist side by side as shown in Fig. 1-5, or they may be isolated members.

Spandrel beams, having a flange on one side only, fall into the same category as T-beams.

Because of the limited "spread" of forces laterally in the

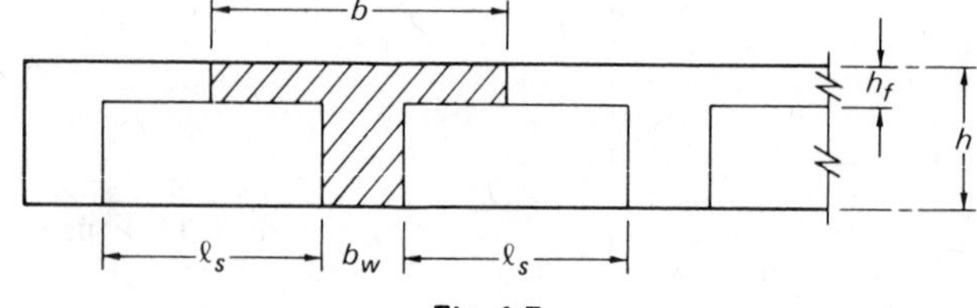

Fig. 1-5

TABLE 1-4a Ratio of M_u with Compression Steel to M_u without Compression Steel. (Note: Where Values are .000, ρ Exceeds 0.75 ρ_b.)

d'/d = .05 f'_c = 4. KSI FY = 60. KSI

ρ \ ρ'/ρ	.1	.2	.3	.4	.5	.6	.7	.8	.9	1.0
.0033	1.001	1.001	1.001	1.001	1.001	1.001	1.001	1.001	1.001	1.001
.0040	1.001	1.002	1.003	1.003	1.003	1.003	1.003	1.003	1.003	1.004
.0045	1.002	1.003	1.004	1.005	1.005	1.005	1.006	1.006	1.006	1.006
.0050	1.003	1.005	1.006	1.007	1.007	1.008	1.008	1.008	1.009	1.009
.0055	1.004	1.006	1.008	1.009	1.010	1.011	1.011	1.011	1.012	1.012
.0060	1.005	1.008	1.010	1.012	1.013	1.014	1.014	1.015	1.015	1.015
.0065	1.006	1.010	1.012	1.014	1.016	1.017	1.018	1.018	1.019	1.019
.0070	1.007	1.011	1.015	1.017	1.019	1.020	1.021	1.022	1.022	1.023
.0075	1.008	1.013	1.017	1.020	1.022	1.024	1.025	1.026	1.026	1.027
.0080	1.009	1.015	1.020	1.023	1.026	1.028	1.029	1.030	1.030	1.031
.0085	1.010	1.017	1.023	1.027	1.029	1.031	1.033	1.034	1.034	1.035
.0090	1.011	1.019	1.025	1.030	1.033	1.035	1.037	1.038	1.039	1.039
.0095	1.012	1.021	1.028	1.033	1.037	1.039	1.041	1.042	1.043	1.044
.0100	1.013	1.023	1.031	1.037	1.041	1.043	1.045	1.047	1.048	1.048
.0105	1.014	1.025	1.034	1.040	1.045	1.048	1.050	1.051	1.052	1.053
.0110	1.015	1.027	1.037	1.044	1.049	1.052	1.054	1.056	1.057	1.058
.0115	1.016	1.029	1.040	1.048	1.053	1.056	1.059	1.060	1.062	1.062
.0120	1.017	1.030	1.042	1.051	1.057	1.061	1.063	1.065	1.066	1.067
.0125	1.017	1.032	1.045	1.055	1.061	1.065	1.068	1.070	1.071	1.072
.0130	1.018	1.034	1.047	1.059	1.066	1.070	1.073	1.075	1.076	1.077
.0135	1.019	1.036	1.050	1.062	1.070	1.075	1.078	1.080	1.081	1.082
.0140	1.020	1.038	1.053	1.065	1.074	1.079	1.083	1.085	1.087	1.087
.0145	1.021	1.040	1.056	1.069	1.079	1.084	1.088	1.090	1.092	1.093
.0150	1.022	1.042	1.058	1.072	1.083	1.089	1.093	1.095	1.097	1.098
.0155	1.024	1.044	1.061	1.076	1.087	1.094	1.098	1.101	1.102	1.103
.0160	1.025	1.046	1.064	1.079	1.092	1.099	1.103	1.106	1.108	1.109
.0165	1.026	1.048	1.067	1.083	1.096	1.104	1.109	1.111	1.113	1.114
.0170	1.027	1.050	1.070	1.086	1.100	1.110	1.114	1.117	1.119	1.120
.0175	1.028	1.052	1.072	1.090	1.104	1.115	1.120	1.123	1.124	1.126
.0180	1.029	1.054	1.075	1.094	1.109	1.120	1.125	1.128	1.130	1.131
.0185	1.030	1.056	1.078	1.097	1.113	1.125	1.131	1.134	1.136	1.137
.0190	1.031	1.058	1.081	1.101	1.117	1.130	1.136	1.140	1.142	1.143
.0195	1.032	1.060	1.084	1.105	1.122	1.135	1.142	1.146	1.148	1.149
.0200	1.033	1.062	1.087	1.109	1.126	1.140	1.148	1.151	1.154	1.155
.0205	1.034	1.064	1.091	1.113	1.131	1.145	1.154	1.157	1.160	1.161
.0210	1.036	1.067	1.094	1.117	1.136	1.150	1.160	1.164	1.166	1.167
.0215	1.037	1.069	1.097	1.121	1.140	1.156	1.166	1.170	1.172	1.174
.0220	1.038	1.071	1.100	1.125	1.145	1.161	1.172	1.176	1.178	1.180
.0225	1.039	1.073	1.103	1.129	1.150	1.166	1.178	1.182	1.185	1.186
.0230	1.040	1.076	1.106	1.133	1.154	1.172	1.184	1.189	1.191	1.193
.0235	1.042	1.078	1.110	1.137	1.159	1.177	1.190	1.195	1.198	1.199
.0240	.000	1.080	1.113	1.141	1.164	1.183	1.196	1.202	1.204	1.206
.0245	.000	1.083	1.116	1.145	1.169	1.188	1.202	1.208	1.211	1.213
.0250	.000	1.085	1.120	1.150	1.174	1.194	1.209	1.215	1.218	1.219
.0255	.000	1.088	1.123	1.154	1.179	1.200	1.215	1.222	1.225	1.226
.0260	.000	1.090	1.127	1.158	1.185	1.205	1.221	1.229	1.231	1.233
.0265	.000	1.092	1.130	1.163	1.190	1.211	1.227	1.236	1.238	1.240
.0270	.000	.000	1.134	1.167	1.195	1.217	1.234	1.243	1.246	1.247
.0275	.000	.000	1.137	1.172	1.200	1.223	1.240	1.250	1.253	1.255
.0280	.000	.000	1.141	1.176	1.206	1.229	1.247	1.257	1.260	1.262

flange of a T-beam or spandrel beam, the ACI Code limits the usable flange width.

For symmetrical T-beams as shown in Fig. 1-5, b may not exceed one-fourth the span length ($l_n/4$), $(b - b_w)/2$ may not exceed $8h_f$, and $(b - b_w)/2$ may not exceed one-half the distance to the next beam ($l_s/2$). The lesser of these values determines the effective flange width, b.

For isolated T-beams, the flange thickness (h_f) may not be less than one-half the web width (or $h_f \geqslant b_w/2$), and b may not exceed four times the web width (or $b \leqslant 4b_w$).

For spandrel beams (flange on one side only) the overhanging flange $(b - b_w)$ may not exceed $l_n/12$, nor $6h_f$, nor one-half the distance to the next beam ($l_s/2$).

In all cases, the most severe limitation applies.

The *ACI Code* states that a T-beam exists when the neutral axis is in the web, considering a generalized stress-

TABLE 1-4b Ratio of M_u with Compression Steel to M_u without Compression Steel. (Note: Where Values are .000, ρ Exceeds 0.75 ρ_b.)

d'/d = .10 f_c' = 4. KSI FY = 60. KSI

ρ \ ρ'/ρ	.1	.2	.3	.4	.5	.6	.7	.8	.9	1.0
.0033	1.003	1.005	1.006	1.007	1.007	1.008	1.008	1.009	1.009	1.010
.0040	1.001	1.002	1.002	1.003	1.003	1.003	1.004	1.004	1.004	1.004
.0045	1.001	1.001	1.001	1.001	1.001	1.001	1.002	1.002	1.002	1.002
.0050	1.001	1.001	1.001	1.001	1.001	1.001	1.001	1.001	1.001	1.001
.0055	1.000	1.000	1.000	1.000	1.000	1.000	1.000	1.000	1.000	1.000
.0060	1.001	1.001	1.001	1.001	1.001	1.001	1.001	1.001	1.001	1.001
.0065	1.001	1.001	1.001	1.001	1.001	1.001	1.001	1.001	1.001	1.001
.0070	1.001	1.002	1.002	1.002	1.003	1.003	1.003	1.003	1.003	1.003
.0075	1.002	1.003	1.003	1.004	1.004	1.004	1.004	1.005	1.005	1.005
.0080	1.002	1.004	1.005	1.005	1.006	1.006	1.006	1.007	1.007	1.007
.0085	1.003	1.005	1.006	1.007	1.008	1.008	1.009	1.009	1.009	1.009
.0090	1.004	1.006	1.008	1.009	1.010	1.011	1.011	1.011	1.012	1.012
.0095	1.005	1.008	1.010	1.011	1.012	1.013	1.014	1.014	1.014	1.015
.0100	1.006	1.009	1.012	1.014	1.015	1.016	1.017	1.017	1.018	1.018
.0105	1.006	1.011	1.014	1.016	1.018	1.019	1.020	1.020	1.021	1.021
.0110	1.007	1.012	1.016	1.019	1.021	1.022	1.023	1.024	1.024	1.024
.0115	1.008	1.014	1.018	1.021	1.024	1.025	1.026	1.027	1.028	1.028
.0120	1.009	1.016	1.021	1.024	1.027	1.028	1.030	1.031	1.031	1.032
.0125	1.010	1.018	1.023	1.027	1.030	1.032	1.033	1.034	1.035	1.036
.0130	1.011	1.020	1.026	1.030	1.033	1.035	1.037	1.038	1.039	1.040
.0135	1.013	1.022	1.028	1.033	1.037	1.039	1.041	1.042	1.043	1.044
.0140	1.014	1.024	1.031	1.036	1.040	1.043	1.045	1.046	1.047	1.048
.0145	1.015	1.026	1.034	1.040	1.044	1.047	1.049	1.050	1.052	1.052
.0150	1.016	1.028	1.037	1.043	1.048	1.051	1.053	1.055	1.056	1.057
.0155	1.017	1.030	1.040	1.047	1.051	1.055	1.057	1.059	1.060	1.061
.0160	1.019	1.032	1.043	1.050	1.055	1.059	1.062	1.064	1.065	1.066
.0165	1.020	1.035	1.046	1.054	1.059	1.063	1.066	1.068	1.070	1.071
.0170	1.021	1.037	1.049	1.057	1.063	1.068	1.071	1.073	1.074	1.075
.0175	1.022	1.039	1.052	1.061	1.068	1.072	1.075	1.078	1.079	1.080
.0180	1.023	1.042	1.055	1.065	1.072	1.077	1.080	1.082	1.084	1.085
.0185	1.024	1.044	1.058	1.069	1.076	1.081	1.085	1.087	1.089	1.090
.0190	1.025	1.047	1.062	1.073	1.081	1.086	1.090	1.092	1.094	1.096
.0195	1.026	1.049	1.065	1.077	1.085	1.091	1.095	1.098	1.099	1.101
.0200	1.028	1.051	1.069	1.081	1.090	1.096	1.100	1.103	1.105	1.106
.0205	1.029	1.053	1.072	1.085	1.094	1.101	1.105	1.108	1.110	1.112
.0210	1.030	1.055	1.076	1.089	1.099	1.106	1.110	1.113	1.116	1.117
.0215	1.031	1.057	1.079	1.094	1.104	1.111	1.116	1.119	1.121	1.123
.0220	1.032	1.059	1.082	1.098	1.109	1.116	1.121	1.124	1.127	1.128
.0225	1.033	1.062	1.086	1.102	1.113	1.121	1.126	1.130	1.132	1.134
.0230	1.034	1.064	1.089	1.107	1.118	1.127	1.132	1.136	1.138	1.140
.0235	1.036	1.066	1.092	1.111	1.124	1.132	1.138	1.141	1.144	1.146
.0240	.000	1.068	1.095	1.116	1.129	1.137	1.143	1.147	1.150	1.152
.0245	.000	1.071	1.098	1.121	1.134	1.143	1.149	1.153	1.156	1.158
.0250	.000	1.073	1.102	1.125	1.139	1.149	1.155	1.159	1.162	1.164
.0255	.000	1.075	1.105	1.130	1.145	1.154	1.161	1.165	1.168	1.170
.0260	.000	1.078	1.108	1.134	1.150	1.160	1.167	1.172	1.175	1.177
.0265	.000	.000	1.112	1.138	1.156	1.166	1.173	1.178	1.181	1.183
.0270	.000	.000	1.115	1.142	1.161	1.172	1.179	1.184	1.188	1.190
.0275	.000	.000	1.119	1.147	1.167	1.178	1.186	1.191	1.194	1.196
.0280	.000	.000	1.122	1.151	1.173	1.184	1.192	1.197	1.201	1.203

strain diagram. If the neutral axis falls within the flange, a rectangular section having a width equal to b, exists.

However, when using the equivalent rectangular stress block, rectangular beam theory provides a correct solution as long as the stress block lies in the flange. Equation (1-16) provides a simple test to determine whether or not to use rectangular beam theory or T-beam theory. If h_f is equal to or greater than a, rectangular beam theory governs, where

$$a = \frac{1.18 \rho f_y d}{f_c'} \tag{1-16}$$

If h_f is less than a, T-beam theory applies. Note that $\rho = A_s/bd$, regardless of whether T-beam action or rectangular beam action exists.

When the area of steel is not known, the steel ratio (ρ) is unknown, and eq. (1-16) can not be applied. In such a case,

TABLE 1-4c Ratio of M_u with Compression Steel to M_u without Compression Steel. (Note: Where Values are .000, ρ Exceeds 0.75 ρ_b.)

d'/d = .15 f'_c = 4. KSI FY = 60. KSI

ρ \ ρ'/ρ	.1	.2	.3	.4	.5	.6	.7	.8	.9	1.0
.0033	1.013	1.021	1.026	1.031	1.034	1.037	1.040	1.042	1.044	1.045
.0040	1.008	1.013	1.017	1.020	1.022	1.024	1.026	1.027	1.028	1.029
.0045	1.006	1.010	1.012	1.014	1.016	1.017	1.018	1.019	1.020	1.021
.0050	1.004	1.007	1.008	1.010	1.011	1.012	1.013	1.013	1.014	1.014
.0055	1.003	1.004	1.006	1.007	1.007	1.008	1.008	1.009	1.009	1.009
.0060	1.002	1.003	1.004	1.004	1.005	1.005	1.005	1.005	1.006	1.006
.0065	1.001	1.002	1.002	1.002	1.003	1.003	1.003	1.003	1.003	1.003
.0070	1.001	1.001	1.001	1.001	1.001	1.001	1.001	1.001	1.002	1.002
.0075	1.001	1.001	1.001	1.001	1.001	1.001	1.001	1.001	1.001	1.001
.0080	1.000	1.000	1.000	1.000	1.000	1.000	1.000	1.000	1.000	1.000
.0085	1.000	1.000	1.000	1.000	1.000	1.000	1.000	1.000	1.000	1.000
.0090	1.001	1.001	1.001	1.001	1.001	1.001	1.001	1.001	1.001	1.001
.0095	1.001	1.001	1.001	1.001	1.001	1.002	1.002	1.002	1.001	1.001
.0100	1.001	1.002	1.002	1.002	1.003	1.003	1.003	1.003	1.003	1.003
.0105	1.002	1.003	1.003	1.004	1.004	1.004	1.004	1.004	1.004	1.004
.0110	1.002	1.003	1.004	1.005	1.005	1.006	1.006	1.006	1.006	1.006
.0115	1.003	1.005	1.006	1.006	1.007	1.007	1.008	1.008	1.008	1.008
.0120	1.004	1.006	1.007	1.008	1.009	1.009	1.010	1.010	1.010	1.010
.0125	1.004	1.007	1.009	1.010	1.011	1.012	1.012	1.012	1.013	1.013
.0130	1.005	1.008	1.010	1.012	1.013	1.014	1.014	1.015	1.015	1.015
.0135	1.006	1.010	1.012	1.014	1.015	1.016	1.017	1.018	1.018	1.018
.0140	1.007	1.011	1.014	1.016	1.018	1.019	1.020	1.021	1.021	1.021
.0145	1.007	1.013	1.016	1.019	1.021	1.022	1.023	1.024	1.024	1.024
.0150	1.008	1.014	1.018	1.021	1.023	1.025	1.026	1.027	1.027	1.028
.0155	1.009	1.016	1.021	1.024	1.026	1.028	1.029	1.030	1.031	1.031
.0160	1.010	1.018	1.023	1.027	1.029	1.031	1.033	1.034	1.034	1.035
.0165	1.011	1.019	1.025	1.029	1.032	1.034	1.036	1.037	1.038	1.039
.0170	1.012	1.021	1.028	1.032	1.036	1.038	1.040	1.041	1.042	1.042
.0175	1.014	1.023	1.030	1.035	1.039	1.041	1.043	1.045	1.046	1.046
.0180	1.015	1.025	1.033	1.038	1.042	1.045	1.047	1.049	1.050	1.050
.0185	1.016	1.027	1.035	1.041	1.046	1.049	1.051	1.053	1.054	1.055
.0190	1.017	1.029	1.038	1.045	1.049	1.053	1.055	1.057	1.058	1.059
.0195	1.018	1.031	1.041	1.048	1.053	1.057	1.059	1.061	1.062	1.063
.0200	1.019	1.033	1.044	1.051	1.057	1.061	1.063	1.065	1.067	1.068
.0205	1.021	1.036	1.047	1.055	1.061	1.065	1.068	1.070	1.071	1.072
.0210	1.022	1.038	1.050	1.058	1.065	1.069	1.072	1.074	1.076	1.077
.0215	1.023	1.040	1.053	1.062	1.069	1.073	1.077	1.079	1.081	1.082
.0220	1.024	1.043	1.056	1.066	1.073	1.078	1.081	1.084	1.086	1.087
.0225	1.026	1.045	1.059	1.070	1.077	1.082	1.086	1.089	1.090	1.092
.0230	1.027	1.047	1.063	1.074	1.081	1.087	1.091	1.093	1.095	1.097
.0235	.000	1.050	1.066	1.077	1.086	1.092	1.096	1.099	1.101	1.102
.0240	.000	1.053	1.069	1.081	1.090	1.096	1.101	1.104	1.106	1.107
.0245	.000	1.055	1.073	1.086	1.095	1.101	1.106	1.109	1.111	1.113
.0250	.000	1.058	1.076	1.090	1.099	1.106	1.111	1.114	1.117	1.118
.0255	.000	1.060	1.080	1.094	1.104	1.111	1.116	1.120	1.122	1.124
.0260	.000	.000	1.083	1.098	1.109	1.116	1.121	1.125	1.128	1.129
.0265	.000	.000	1.087	1.103	1.114	1.121	1.127	1.131	1.133	1.135
.0270	.000	.000	1.091	1.107	1.119	1.127	1.132	1.136	1.139	1.141
.0275	.000	.000	1.095	1.112	1.124	1.132	1.138	1.142	1.145	1.147
.0280	.000	.000	1.099	1.116	1.129	1.138	1.144	1.148	1.151	1.153

the total moment developed by the entire flange may be computed, considering the stress block to cover the flange only. If the resulting moment exceeds M_u, a is less than h_f, and rectangular beam theory applies. If the moment developed by the entire flange is less than M_u, the stress block covers part of the web, T-beam theory applies, and eq. (1-17) may be used.

$$M_{uh} = (0.85\ \phi f'_c h_f b)\,(d - h_f/2) \tag{1-17}$$

Figure 1-6 illustrates the general conditions for T-beams. The moments and forces on the cross section are divided into two groups. One group includes the forces and moment developed by the web, acting as a rectangular beam having a

Table 1-4d Ratio of M_u with Compression Steel to M_u without Compression Steel. (Note: Where Values are .000, ρ Exceeds 0.75 ρ_b.)

d'/d = .20 f'_c = 4. KSI FY = 60. KSI

ρ \ ρ'/ρ	.1	.2	.3	.4	.5	.6	.7	.8	.9	1.0
.0033	1.029	1.047	1.060	1.070	1.078	1.084	1.090	1.095	1.099	1.103
.0040	1.021	1.034	1.044	1.051	1.057	1.061	1.066	1.069	1.072	1.075
.0045	1.017	1.027	1.034	1.040	1.045	1.049	1.052	1.055	1.057	1.059
.0050	1.013	1.021	1.027	1.032	1.035	1.038	1.041	1.043	1.045	1.046
.0055	1.010	1.016	1.021	1.025	1.027	1.030	1.032	1.033	1.035	1.036
.0060	1.008	1.013	1.016	1.019	1.021	1.023	1.025	1.026	1.027	1.028
.0065	1.006	1.010	1.012	1.014	1.016	1.017	1.019	1.020	1.020	1.021
.0070	1.004	1.007	1.009	1.011	1.012	1.013	1.014	1.015	1.015	1.016
.0075	1.003	1.005	1.007	1.008	1.009	1.009	1.010	1.011	1.011	1.011
.0080	1.002	1.004	1.005	1.005	1.006	1.006	1.007	1.007	1.008	1.008
.0085	1.002	1.002	1.003	1.004	1.004	1.004	1.005	1.005	1.005	1.005
.0090	1.001	1.002	1.002	1.002	1.002	1.003	1.003	1.003	1.003	1.003
.0095	1.001	1.001	1.001	1.001	1.001	1.001	1.001	1.001	1.001	1.002
.0100	1.001	1.001	1.001	1.001	1.001	1.001	1.001	1.001	1.001	1.001
.0105	1.000	1.000	1.000	1.000	1.000	1.000	1.000	1.000	1.000	1.000
.0110	1.000	1.000	1.000	1.000	1.000	1.000	1.000	1.000	1.000	1.000
.0115	1.001	1.001	1.000	1.000	1.000	1.000	1.000	1.000	1.000	1.000
.0120	1.001	1.001	1.001	1.001	1.001	1.001	1.001	1.001	1.001	1.001
.0125	1.001	1.001	1.001	1.002	1.002	1.002	1.002	1.002	1.002	1.001
.0130	1.001	1.002	1.002	1.002	1.003	1.003	1.003	1.003	1.003	1.003
.0135	1.002	1.003	1.003	1.003	1.004	1.004	1.004	1.004	1.004	1.004
.0140	1.002	1.003	1.004	1.005	1.005	1.005	1.005	1.005	1.006	1.006
.0145	1.003	1.004	1.005	1.006	1.007	1.007	1.007	1.007	1.007	1.007
.0150	1.003	1.005	1.007	1.008	1.008	1.009	1.009	1.009	1.009	1.009
.0155	1.004	1.006	1.008	1.009	1.010	1.011	1.011	1.011	1.012	1.012
.0160	1.005	1.008	1.010	1.011	1.012	1.013	1.013	1.014	1.014	1.014
.0165	1.005	1.009	1.011	1.013	1.014	1.015	1.016	1.016	1.016	1.017
.0170	1.006	1.010	1.013	1.015	1.016	1.017	1.018	1.019	1.019	1.019
.0175	1.007	1.012	1.015	1.017	1.019	1.020	1.021	1.021	1.022	1.022
.0180	1.008	1.013	1.017	1.019	1.021	1.023	1.024	1.024	1.025	1.025
.0185	1.009	1.015	1.019	1.022	1.024	1.026	1.027	1.027	1.028	1.028
.0190	1.010	1.016	1.021	1.024	1.027	1.028	1.030	1.031	1.031	1.032
.0195	1.011	1.018	1.023	1.027	1.030	1.032	1.033	1.034	1.035	1.035
.0200	1.012	1.020	1.025	1.030	1.033	1.035	1.036	1.037	1.038	1.039
.0205	1.013	1.021	1.028	1.032	1.036	1.038	1.040	1.041	1.042	1.042
.0210	1.014	1.023	1.030	1.035	1.039	1.041	1.043	1.045	1.046	1.046
.0215	1.015	1.025	1.033	1.038	1.042	1.045	1.047	1.048	1.049	1.050
.0220	1.016	1.027	1.035	1.041	1.045	1.048	1.051	1.052	1.053	1.054
.0225	1.017	1.029	1.038	1.044	1.049	1.052	1.055	1.056	1.057	1.058
.0230	1.018	1.031	1.041	1.048	1.052	1.056	1.059	1.060	1.062	1.063
.0235	.000	1.033	1.044	1.051	1.056	1.060	1.063	1.065	1.066	1.067
.0240	.000	1.036	1.046	1.054	1.060	1.064	1.067	1.069	1.070	1.072
.0245	.000	1.038	1.049	1.058	1.064	1.068	1.071	1.073	1.075	1.076
.0250	.000	.000	1.052	1.061	1.068	1.072	1.075	1.078	1.080	1.081
.0255	.000	.000	1.055	1.065	1.072	1.076	1.080	1.082	1.084	1.086
.0260	.000	.000	1.059	1.069	1.076	1.081	1.085	1.087	1.089	1.090
.0265	.000	.000	1.062	1.072	1.080	1.085	1.089	1.092	1.094	1.095
.0270	.000	.000	1.065	1.076	1.084	1.090	1.094	1.097	1.099	1.101
.0275	.000	.000	.000	1.080	1.089	1.095	1.099	1.102	1.104	1.106
.0280	.000	.000	.000	1.084	1.093	1.099	1.104	1.107	1.109	1.111

width, b_w. The other group includes the forces and moment developed by the overhanging flanges. The area of tensile reinforcement that develops the compressive force in the web is A_{sw}. The area of tensile reinforcement that develops the compressive forces in the overhanging flanges is A_{sf}. The total tension reinforcement, A_s, is the sum of A_{sw} and A_{sf}. Figure 1-7 illustrates the two conditions.

From Fig. 1-7(a),

$$C_w = 0.85 d f'_c a b_w = T_w = \phi A_{sw} f_y \tag{1-18}$$

$$M_w = C_w (d - a/2) = T_w (d - a/2) \tag{1-19}$$

and from Fig. 1-7(b),

$$C_f = 0.85 d f'_c h_f (b - b_w) = T_f = A_{sf} f_y \tag{1-20}$$

TABLE 1-4e Ratio of M_u with Compression Steel to M_u without Compression Steel. (Note: Where Values are .000, ρ Exceeds 0.75 ρ_b.)

d'/d = .25 f'_c = 4. KSI FY = 60. KSI

ρ \ ρ'/ρ	.1	.2	.3	.4	.5	.6	.7	.8	.9	1.0
.0033	1.052	1.082	1.104	1.121	1.135	1.147	1.157	1.166	1.174	1.181
.0040	1.039	1.063	1.080	1.093	1.104	1.113	1.121	1.127	1.133	1.138
.0045	1.032	1.052	1.066	1.077	1.086	1.093	1.100	1.105	1.110	1.114
.0050	1.027	1.043	1.055	1.064	1.071	1.077	1.083	1.087	1.091	1.094
.0055	1.022	1.036	1.045	1.053	1.059	1.064	1.068	1.072	1.075	1.078
.0060	1.018	1.029	1.037	1.044	1.049	1.053	1.057	1.060	1.062	1.065
.0065	1.015	1.024	1.031	1.036	1.040	1.044	1.047	1.049	1.051	1.053
.0070	1.012	1.020	1.025	1.030	1.033	1.036	1.038	1.040	1.042	1.043
.0075	1.010	1.016	1.020	1.024	1.027	1.029	1.031	1.033	1.034	1.035
.0080	1.008	1.013	1.016	1.019	1.022	1.023	1.025	1.026	1.027	1.028
.0085	1.006	1.010	1.013	1.015	1.017	1.019	1.020	1.021	1.022	1.022
.0090	1.005	1.008	1.010	1.012	1.013	1.014	1.015	1.016	1.017	1.017
.0095	1.004	1.006	1.008	1.009	1.010	1.011	1.012	1.012	1.013	1.013
.0100	1.003	1.005	1.006	1.007	1.007	1.008	1.009	1.009	1.010	1.010
.0105	1.002	1.003	1.004	1.005	1.005	1.006	1.006	1.006	1.007	1.007
.0110	1.002	1.002	1.003	1.003	1.004	1.004	1.004	1.004	1.005	1.005
.0115	1.001	1.001	1.002	1.002	1.002	1.002	1.003	1.003	1.003	1.003
.0120	1.001	1.001	1.001	1.001	1.001	1.001	1.001	1.001	1.001	1.002
.0125	1.001	1.001	1.001	1.001	1.001	1.001	1.001	1.001	1.001	1.001
.0130	1.000	1.000	1.000	1.000	1.000	1.000	1.000	1.000	1.000	1.000
.0135	1.000	1.000	1.000	1.000	1.000	1.000	1.000	1.000	1.000	1.000
.0140	1.000	1.000	1.000	1.000	1.000	1.000	1.000	1.000	1.000	1.000
.0145	1.001	1.001	1.001	1.000	1.000	1.000	1.000	1.000	1.000	1.000
.0150	1.001	1.001	1.001	1.001	1.001	1.001	1.001	1.001	1.001	1.001
.0155	1.001	1.001	1.002	1.002	1.002	1.002	1.002	1.002	1.002	1.002
.0160	1.001	1.002	1.002	1.002	1.003	1.003	1.003	1.003	1.003	1.003
.0165	1.002	1.003	1.003	1.003	1.004	1.004	1.004	1.004	1.004	1.004
.0170	1.002	1.003	1.004	1.005	1.005	1.005	1.005	1.005	1.005	1.005
.0175	1.003	1.004	1.005	1.006	1.006	1.007	1.007	1.007	1.007	1.007
.0180	1.003	1.005	1.006	1.007	1.008	1.008	1.009	1.009	1.009	1.009
.0185	1.004	1.006	1.008	1.009	1.010	1.010	1.011	1.011	1.011	1.011
.0190	1.005	1.007	1.009	1.011	1.011	1.012	1.013	1.013	1.013	1.013
.0195	1.005	1.009	1.011	1.012	1.013	1.014	1.015	1.015	1.015	1.016
.0200	1.006	1.010	1.012	1.014	1.016	1.016	1.017	1.018	1.018	1.018
.0205	1.007	1.011	1.014	1.016	1.018	1.019	1.020	1.020	1.021	1.021
.0210	1.008	1.013	1.016	1.018	1.020	1.021	1.022	1.023	1.023	1.024
.0215	1.008	1.014	1.018	1.021	1.023	1.024	1.025	1.026	1.026	1.027
.0220	1.009	1.015	1.020	1.023	1.025	1.027	1.028	1.029	1.029	1.030
.0225	1.010	1.017	1.022	1.025	1.028	1.030	1.031	1.032	1.033	1.033
.0230	.000	1.019	1.024	1.028	1.031	1.033	1.034	1.035	1.036	1.036
.0235	.000	1.020	1.026	1.031	1.034	1.036	1.037	1.038	1.039	1.040
.0240	.000	1.022	1.029	1.033	1.037	1.039	1.041	1.042	1.043	1.043
.0245	.000	.000	1.031	1.036	1.040	1.042	1.044	1.045	1.046	1.047
.0250	.000	.000	1.034	1.039	1.043	1.046	1.048	1.049	1.050	1.051
.0255	.000	.000	1.036	1.042	1.046	1.049	1.051	1.053	1.054	1.055
.0260	.000	.000	1.039	1.045	1.050	1.053	1.055	1.057	1.058	1.059
.0265	.000	.000	.000	1.048	1.053	1.057	1.059	1.061	1.062	1.063
.0270	.000	.000	.000	1.051	1.057	1.060	1.063	1.065	1.067	1.068
.0275	.000	.000	.000	1.055	1.060	1.064	1.067	1.069	1.071	1.072
.0280	.000	.000	.000	1.058	1.064	1.068	1.071	1.074	1.075	1.077

$$M_f = C_f(d - h_f/2) = T_f(d - h_f/2) \quad (1\text{-}21)$$

The traditional procedure is to use eq. (1-16) or (1-17) to determine whether or not T-beam theory applies. If T-beam theory applies, then the overhanging flange moment capacity is obtained using eqs. (1-20) and (1-21). The remaining moment capacity is used in conjunction with eqs. (1-18) and (1-19), or other procedures previously described for rectangular sections. In this case, the rectangular section has a width equal to the web width, b_w.

The total steel area, $A_s = A_{sf} + A_{sw}$ may not exceed the maximum quantity, $\rho_{max} b_w d$, where ρ_{max} is 75% of the balanced value for T-beams, derived in the Code Commen-

TABLE 1-4f Ratio of M_u with Compression Steel to M_u without Compression Steel. (Note: Where Values are .000, ρ Exceeds 0.75 ρ_b.)

d'/d = •30 f_c' = 4• KSI FY = 60• KSI

ρ \ ρ'/ρ	•1	•2	•3	•4	•5	•6	•7	•8	•9	1•0
•0033	1•080	1•126	1•159	1•184	1•206	1•224	1•239	1•253	1•265	1•275
•0040	1•063	1•099	1•126	1•146	1•163	1•177	1•190	1•200	1•210	1•218
•0045	1•053	1•084	1•107	1•124	1•138	1•150	1•161	1•170	1•177	1•184
•0050	1•045	1•071	1•091	1•106	1•118	1•128	1•137	1•144	1•151	1•157
•0055	1•038	1•061	1•077	1•090	1•101	1•110	1•117	1•123	1•129	1•134
•0060	1•032	1•052	1•066	1•077	1•086	1•094	1•100	1•105	1•110	1•114
•0065	1•027	1•044	1•057	1•066	1•074	1•080	1•086	1•090	1•094	1•098
•0070	1•023	1•038	1•048	1•057	1•063	1•068	1•073	1•077	1•080	1•083
•0075	1•020	1•032	1•041	1•048	1•054	1•058	1•062	1•066	1•069	1•071
•0080	1•017	1•027	1•035	1•041	1•046	1•050	1•053	1•056	1•058	1•060
•0085	1•014	1•023	1•030	1•035	1•039	1•042	1•045	1•047	1•049	1•051
•0090	1•012	1•019	1•025	1•029	1•032	1•035	1•038	1•040	1•041	1•043
•0095	1•010	1•016	1•021	1•024	1•027	1•029	1•031	1•033	1•034	1•036
•0100	1•008	1•013	1•017	1•020	1•022	1•024	1•026	1•027	1•029	1•030
•0105	1•007	1•011	1•014	1•016	1•018	1•020	1•021	1•022	1•023	1•024
•0110	1•005	1•009	1•011	1•013	1•015	1•016	1•017	1•018	1•019	1•019
•0115	1•004	1•007	1•009	1•010	1•012	1•013	1•014	1•014	1•015	1•015
•0120	1•003	1•005	1•007	1•008	1•009	1•010	1•010	1•011	1•012	1•012
•0125	1•003	1•004	1•005	1•006	1•007	1•007	1•008	1•008	1•009	1•009
•0130	1•002	1•003	1•004	1•004	1•005	1•005	1•006	1•006	1•006	1•007
•0135	1•001	1•002	1•003	1•003	1•003	1•004	1•004	1•004	1•004	1•004
•0140	1•001	1•001	1•002	1•002	1•002	1•002	1•002	1•003	1•003	1•003
•0145	1•001	1•001	1•001	1•001	1•001	1•001	1•001	1•001	1•001	1•002
•0150	1•001	1•001	1•001	1•001	1•001	1•001	1•001	1•001	1•001	1•001
•0155	1•000	1•000	1•000	1•000	1•000	1•000	1•000	1•000	1•000	1•000
•0160	1•000	1•000	1•000	1•000	1•000	1•000	1•000	1•000	1•000	1•000
•0165	1•000	1•000	1•000	1•000	1•000	1•000	1•000	1•000	1•000	•999
•0170	1•000	1•000	1•000	1•000	1•000	1•000	1•000	1•000	1•000	1•000
•0175	1•001	1•001	1•001	1•001	1•000	1•000	1•000	1•000	1•000	1•000
•0180	1•001	1•001	1•001	1•001	1•001	1•001	1•001	1•001	1•001	1•001
•0185	1•001	1•001	1•002	1•002	1•002	1•002	1•002	1•002	1•002	1•002
•0190	1•001	1•002	1•002	1•003	1•003	1•003	1•003	1•003	1•003	1•003
•0195	1•002	1•003	1•003	1•004	1•004	1•004	1•004	1•004	1•004	1•004
•0200	1•002	1•003	1•004	1•005	1•005	1•005	1•005	1•005	1•005	1•005
•0205	1•003	1•004	1•005	1•006	1•006	1•007	1•007	1•007	1•007	1•007
•0210	1•003	1•005	1•006	1•007	1•008	1•008	1•008	1•009	1•009	1•009
•0215	1•004	1•006	1•008	1•009	1•009	1•010	1•010	1•010	1•011	1•011
•0220	1•005	1•007	1•009	1•010	1•011	1•012	1•012	1•012	1•013	1•013
•0225	1•005	1•008	1•011	1•012	1•013	1•014	1•014	1•015	1•015	1•015
•0230	•000	1•010	1•012	1•014	1•015	1•016	1•017	1•017	1•017	1•017
•0235	•000	1•011	1•014	1•016	1•017	1•018	1•019	1•019	1•020	1•020
•0240	•000	•000	1•015	1•018	1•019	1•021	1•021	1•022	1•022	1•023
•0245	•000	•000	1•017	1•020	1•022	1•023	1•024	1•025	1•025	1•026
•0250	•000	•000	1•019	1•022	1•024	1•026	1•027	1•028	1•028	1•028
•0255	•000	•000	•000	1•024	1•027	1•028	1•030	1•030	1•031	1•032
•0260	•000	•000	•000	1•027	1•029	1•031	1•033	1•034	1•034	1•035
•0265	•000	•000	•000	1•029	1•032	1•034	1•036	1•037	1•038	1•038
•0270	•000	•000	•000	1•032	1•035	1•037	1•039	1•040	1•041	1•042
•0275	•000	•000	•000	•000	1•038	1•040	1•042	1•044	1•044	1•045
•0280	•000	•000	•000	•000	1•041	1•044	1•046	1•047	1•048	1•049

tary as

$$\rho_b = \left(\frac{b_w}{b}\right)(\bar{\rho}_b + \rho_f) \qquad (1\text{-}22)$$

where $\bar{\rho}_b$ is the balanced quantity for a rectangular beam of width (b_w) and effective depth (d), and $\rho_f = A_{sf}/b_w d$. Similarly, for the web, $\rho_w = A_{sw}/b_w d$. Note that those ratios ρ for T-beams are now based on the web width. However, the commentary derivation uses $\rho_b bd$ for the balanced steel area for the total cross section. Thus, in comparing $\rho_{max} = 0.75\rho_b$ to the actual ratio of total tension steel, the latter is calculated using $\rho = A_s/(bd)$.

Note that the rectangular stress block depth (a) is a variable in both eqs. (1-18) and (1-19). When the steel area is unknown and to be found, a quadratic equation results in order to find a. Equation (1-19) applies, and

$$a^2 - (2d)a + \frac{2M_w}{0.85 f_c' \phi b_w} = 0 \qquad (1\text{-}23)$$

TABLE 1-4e Ratio of M_u with Compression Steel to M_u without Compression Steel. (Note: Where Values are .000, ρ Exceeds 0.75 ρ_b.)

d'/d = .25 f'_c = 4. KSI FY = 60. KSI

ρ \ ρ'/ρ	.1	.2	.3	.4	.5	.6	.7	.8	.9	1.0
.0033	1.052	1.082	1.104	1.121	1.135	1.147	1.157	1.166	1.174	1.181
.0040	1.039	1.063	1.080	1.093	1.104	1.113	1.121	1.127	1.133	1.138
.0045	1.032	1.052	1.066	1.077	1.086	1.093	1.100	1.105	1.110	1.114
.0050	1.027	1.043	1.055	1.064	1.071	1.077	1.083	1.087	1.091	1.094
.0055	1.022	1.036	1.045	1.053	1.059	1.064	1.068	1.072	1.075	1.078
.0060	1.018	1.029	1.037	1.044	1.049	1.053	1.057	1.060	1.062	1.065
.0065	1.015	1.024	1.031	1.036	1.040	1.044	1.047	1.049	1.051	1.053
.0070	1.012	1.020	1.025	1.030	1.033	1.036	1.038	1.040	1.042	1.043
.0075	1.010	1.016	1.020	1.024	1.027	1.029	1.031	1.033	1.034	1.035
.0080	1.008	1.013	1.016	1.019	1.022	1.023	1.025	1.026	1.027	1.028
.0085	1.006	1.010	1.013	1.015	1.017	1.019	1.020	1.021	1.022	1.022
.0090	1.005	1.008	1.010	1.012	1.013	1.014	1.015	1.016	1.017	1.017
.0095	1.004	1.006	1.008	1.009	1.010	1.011	1.012	1.012	1.013	1.013
.0100	1.003	1.005	1.006	1.007	1.007	1.008	1.009	1.009	1.010	1.010
.0105	1.002	1.003	1.004	1.005	1.005	1.006	1.006	1.006	1.007	1.007
.0110	1.002	1.002	1.003	1.003	1.004	1.004	1.004	1.004	1.005	1.005
.0115	1.001	1.001	1.002	1.002	1.002	1.002	1.003	1.003	1.003	1.003
.0120	1.001	1.001	1.001	1.001	1.001	1.001	1.001	1.001	1.001	1.002
.0125	1.001	1.001	1.001	1.001	1.001	1.001	1.001	1.001	1.001	1.001
.0130	1.000	1.000	1.000	1.000	1.000	1.000	1.000	1.000	1.000	1.000
.0135	1.000	1.000	1.000	1.000	1.000	1.000	1.000	1.000	1.000	1.000
.0140	1.000	1.000	1.000	1.000	1.000	1.000	1.000	1.000	1.000	1.000
.0145	1.001	1.001	1.001	1.000	1.000	1.000	1.000	1.000	1.000	1.000
.0150	1.001	1.001	1.001	1.001	1.001	1.001	1.001	1.001	1.001	1.001
.0155	1.001	1.001	1.002	1.002	1.002	1.002	1.002	1.002	1.002	1.002
.0160	1.001	1.002	1.002	1.002	1.003	1.003	1.003	1.003	1.003	1.003
.0165	1.002	1.003	1.003	1.003	1.004	1.004	1.004	1.004	1.004	1.004
.0170	1.002	1.003	1.004	1.005	1.005	1.005	1.005	1.005	1.005	1.005
.0175	1.003	1.004	1.005	1.006	1.006	1.007	1.007	1.007	1.007	1.007
.0180	1.003	1.005	1.006	1.007	1.008	1.008	1.009	1.009	1.009	1.009
.0185	1.004	1.006	1.008	1.009	1.010	1.010	1.011	1.011	1.011	1.011
.0190	1.005	1.007	1.009	1.011	1.011	1.012	1.013	1.013	1.013	1.013
.0195	1.005	1.009	1.011	1.012	1.013	1.014	1.015	1.015	1.015	1.016
.0200	1.006	1.010	1.012	1.014	1.016	1.016	1.017	1.018	1.018	1.018
.0205	1.007	1.011	1.014	1.016	1.018	1.019	1.020	1.020	1.021	1.021
.0210	1.008	1.013	1.016	1.018	1.020	1.021	1.022	1.023	1.023	1.024
.0215	1.008	1.014	1.018	1.021	1.023	1.024	1.025	1.026	1.026	1.027
.0220	1.009	1.015	1.020	1.023	1.025	1.027	1.028	1.029	1.029	1.030
.0225	1.010	1.017	1.022	1.025	1.028	1.030	1.031	1.032	1.033	1.033
.0230	.000	1.019	1.024	1.028	1.031	1.033	1.034	1.035	1.036	1.036
.0235	.000	1.020	1.026	1.031	1.034	1.036	1.037	1.038	1.039	1.040
.0240	.000	1.022	1.029	1.033	1.037	1.039	1.041	1.042	1.043	1.043
.0245	.000	.000	1.031	1.036	1.040	1.042	1.044	1.045	1.046	1.047
.0250	.000	.000	1.034	1.039	1.043	1.046	1.048	1.049	1.050	1.051
.0255	.000	.000	1.036	1.042	1.046	1.049	1.051	1.053	1.054	1.055
.0260	.000	.000	1.039	1.045	1.050	1.053	1.055	1.057	1.058	1.059
.0265	.000	.000	.000	1.048	1.053	1.057	1.059	1.061	1.062	1.063
.0270	.000	.000	.000	1.051	1.057	1.060	1.063	1.065	1.067	1.068
.0275	.000	.000	.000	1.055	1.060	1.064	1.067	1.069	1.071	1.072
.0280	.000	.000	.000	1.058	1.064	1.068	1.071	1.074	1.075	1.077

$$M_f = C_f (d - h_f/2) = T_f (d - h_f/2) \quad (1\text{-}21)$$

The traditional procedure is to use eq. (1-16) or (1-17) to determine whether or not T-beam theory applies. If T-beam theory applies, then the overhanging flange moment capacity is obtained using eqs. (1-20) and (1-21). The remaining moment capacity is used in conjunction with eqs. (1-18) and (1-19), or other procedures previously described for rectangular sections. In this case, the rectangular section has a width equal to the web width, b_w.

The total steel area, $A_s = A_{sf} + A_{sw}$ may not exceed the maximum quantity, $\rho_{max} b_w d$, where ρ_{max} is 75% of the balanced value for T-beams, derived in the Code Commen-

TABLE 1-4f Ratio of M_u with Compression Steel to M_u without Compression Steel. (Note: Where Values are .000, ρ Exceeds 0.75 ρ_b.)

d'/d = .30 f'_c = 4. KSI FY = 60. KSI

ρ \ ρ'/ρ	.1	.2	.3	.4	.5	.6	.7	.8	.9	1.0
.0033	1.080	1.126	1.159	1.184	1.206	1.224	1.239	1.253	1.265	1.275
.0040	1.063	1.099	1.126	1.146	1.163	1.177	1.190	1.200	1.210	1.218
.0045	1.053	1.084	1.107	1.124	1.138	1.150	1.161	1.170	1.177	1.184
.0050	1.045	1.071	1.091	1.106	1.118	1.128	1.137	1.144	1.151	1.157
.0055	1.038	1.061	1.077	1.090	1.101	1.110	1.117	1.123	1.129	1.134
.0060	1.032	1.052	1.066	1.077	1.086	1.094	1.100	1.105	1.110	1.114
.0065	1.027	1.044	1.057	1.066	1.074	1.080	1.086	1.090	1.094	1.098
.0070	1.023	1.038	1.048	1.057	1.063	1.068	1.073	1.077	1.080	1.083
.0075	1.020	1.032	1.041	1.048	1.054	1.058	1.062	1.066	1.069	1.071
.0080	1.017	1.027	1.035	1.041	1.046	1.050	1.053	1.056	1.058	1.060
.0085	1.014	1.023	1.030	1.035	1.039	1.042	1.045	1.047	1.049	1.051
.0090	1.012	1.019	1.025	1.029	1.032	1.035	1.038	1.040	1.041	1.043
.0095	1.010	1.016	1.021	1.024	1.027	1.029	1.031	1.033	1.034	1.036
.0100	1.008	1.013	1.017	1.020	1.022	1.024	1.026	1.027	1.029	1.030
.0105	1.007	1.011	1.014	1.016	1.018	1.020	1.021	1.022	1.023	1.024
.0110	1.005	1.009	1.011	1.013	1.015	1.016	1.017	1.018	1.019	1.019
.0115	1.004	1.007	1.009	1.010	1.012	1.013	1.014	1.014	1.015	1.015
.0120	1.003	1.005	1.007	1.008	1.009	1.010	1.010	1.011	1.012	1.012
.0125	1.003	1.004	1.005	1.006	1.007	1.007	1.008	1.008	1.009	1.009
.0130	1.002	1.003	1.004	1.004	1.005	1.005	1.006	1.006	1.006	1.007
.0135	1.001	1.002	1.003	1.003	1.003	1.004	1.004	1.004	1.004	1.004
.0140	1.001	1.001	1.002	1.002	1.002	1.002	1.002	1.003	1.003	1.003
.0145	1.001	1.001	1.001	1.001	1.001	1.001	1.001	1.001	1.001	1.002
.0150	1.001	1.001	1.001	1.001	1.001	1.001	1.001	1.001	1.001	1.001
.0155	1.000	1.000	1.000	1.000	1.000	1.000	1.000	1.000	1.000	1.000
.0160	1.000	1.000	1.000	1.000	1.000	1.000	1.000	1.000	1.000	1.000
.0165	1.000	1.000	1.000	1.000	1.000	1.000	1.000	1.000	1.000	.999
.0170	1.000	1.000	1.000	1.000	1.000	1.000	1.000	1.000	1.000	1.000
.0175	1.001	1.001	1.001	1.001	1.000	1.000	1.000	1.000	1.000	1.000
.0180	1.001	1.001	1.001	1.001	1.001	1.001	1.001	1.001	1.001	1.001
.0185	1.001	1.001	1.002	1.002	1.002	1.002	1.002	1.002	1.002	1.002
.0190	1.001	1.002	1.002	1.003	1.003	1.003	1.003	1.003	1.003	1.003
.0195	1.002	1.003	1.003	1.004	1.004	1.004	1.004	1.004	1.004	1.004
.0200	1.002	1.003	1.004	1.005	1.005	1.005	1.005	1.005	1.005	1.005
.0205	1.003	1.004	1.005	1.006	1.006	1.007	1.007	1.007	1.007	1.007
.0210	1.003	1.005	1.006	1.007	1.008	1.008	1.008	1.009	1.009	1.009
.0215	1.004	1.006	1.008	1.009	1.009	1.010	1.010	1.010	1.011	1.011
.0220	1.005	1.007	1.009	1.010	1.011	1.012	1.012	1.012	1.013	1.013
.0225	1.005	1.008	1.011	1.012	1.013	1.014	1.014	1.015	1.015	1.015
.0230	.000	1.010	1.012	1.014	1.015	1.016	1.017	1.017	1.017	1.017
.0235	.000	1.011	1.014	1.016	1.017	1.018	1.019	1.019	1.020	1.020
.0240	.000	.000	1.015	1.018	1.019	1.021	1.021	1.022	1.022	1.023
.0245	.000	.000	1.017	1.020	1.022	1.023	1.024	1.025	1.025	1.026
.0250	.000	.000	1.019	1.022	1.024	1.026	1.027	1.028	1.028	1.028
.0255	.000	.000	.000	1.024	1.027	1.028	1.030	1.030	1.031	1.032
.0260	.000	.000	.000	1.027	1.029	1.031	1.033	1.034	1.034	1.035
.0265	.000	.000	.000	1.029	1.032	1.034	1.036	1.037	1.038	1.038
.0270	.000	.000	.000	1.032	1.035	1.037	1.039	1.040	1.041	1.042
.0275	.000	.000	.000	.000	1.038	1.040	1.042	1.044	1.044	1.045
.0280	.000	.000	.000	.000	1.041	1.044	1.046	1.047	1.048	1.049

tary as

$$\rho_b = \left(\frac{b_w}{b}\right)(\bar{\rho}_b + \rho_f) \qquad (1\text{-}22)$$

where $\bar{\rho}_b$ is the balanced quantity for a rectangular beam of width (b_w) and effective depth (d), and $\rho_f = A_{sf}/b_w d$. Similarly, for the web, $\rho_w = A_{sw}/b_w d$. Note that those ratios ρ for T-beams are now based on the web width. However, the commentary derivation uses $\rho_b bd$ for the balanced steel area for the total cross section. Thus, in comparing $\rho_{max} = 0.75\rho_b$ to the actual ratio of total tension steel, the latter is calculated using $\rho = A_s/(bd)$.

Note that the rectangular stress block depth (a) is a variable in both eqs. (1-18) and (1-19). When the steel area is unknown and to be found, a quadratic equation results in order to find a. Equation (1-19) applies, and

$$a^2 - (2d)a + \frac{2M_w}{0.85 f'_c \phi b_w} = 0 \qquad (1\text{-}23)$$

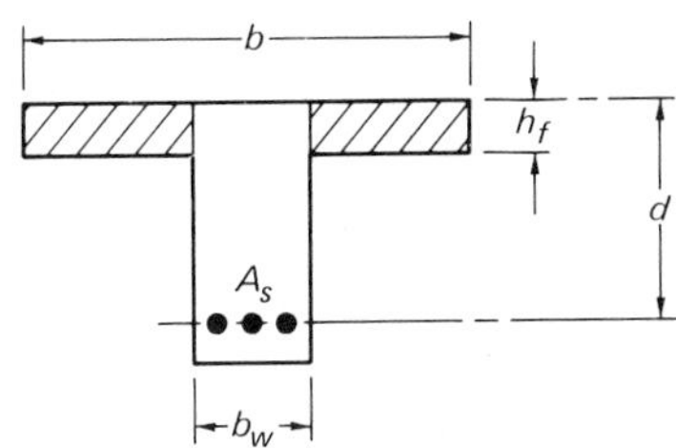

Fig. 1-6

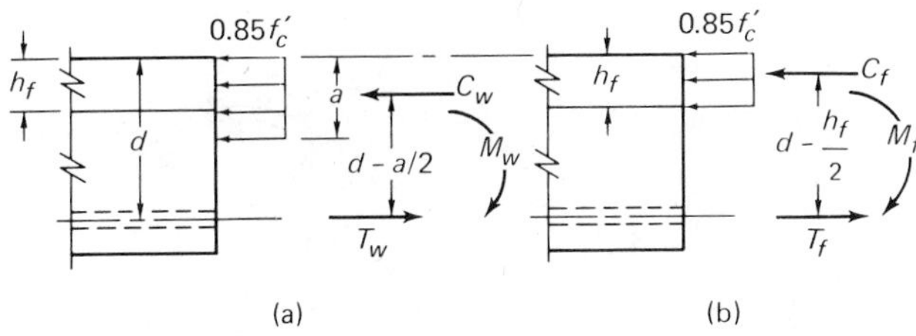

Fig. 1-7

EXAMPLE 1-9: A T-shaped section has the following characteristics: $b = 30$ in., $b_w = 10$ in., $d = 19$ in., $h_f = 2.5$ in., $f_c' = 4000$ psi, $f_y = 60{,}000$ psi. Determine the area of tension steel required, and check to establish whether or not the steel ratio satisfies the *1971 ACI Code*. $M_u = 400$ ft-kips.

SOLUTION: A_s is unknown, so use eq. (1-17) to determine whether or not T-beam action applies.

$$M_{uh} = (0.85)(0.9)(4)(2.5)(30)(19 - 1.25) = 4{,}074 \text{ in.-kips}$$
$$= 339.5 \text{ ft-kips}$$

Since this is less than $M_u = 400$ ft-kips, T-beam theory applies. The force capacity of the overhanging flanges is found using eq. (1-20) and the flange moment (M_f) is found using eq. (1-21).

$$T_f = C_f = (0.85)(0.9)(4)(2.5)(30 - 10) = 153 \text{ kips}$$
$$M_f = (153)(19 - 1.25) = 2{,}716 \text{ in.-kips} = 226 \text{ ft-kips}$$
$$A_{sf} = \frac{T_f}{\phi f_y} = \frac{153}{(0.9 \times 60)} = 2.83 \text{ sq in.}$$
$$\rho_f = \frac{A_{sf}}{b_w d} = \frac{2.83}{190} = 0.0148$$

The web moment is

$$M_w = M_u - M_f = 400 - 226 = 174 \text{ ft-kips}$$

The depth of the stress block is found using eq. (1-23)

$$a^2 - (2)(19)a + \frac{(2)(174)(12)}{(0.85)(4)(0.9)(10)} = 0$$

or

$$a^2 - (38)a + 136.5 = 0$$

from which, by the quadratic formula

$$a = 4.0 \text{ in.}$$

Using eq. (1-18)

$$C_w = 0.85\phi f_c' a b_w = (0.85)(0.9)(4)(4)(10) = 122.4 \text{ kips}$$
$$T_w = C_w,$$

so

$$A_{sw} = \frac{T_w}{(\phi f_y)} = \frac{122.4}{54} = 2.27 \text{ sq in.}$$
$$\text{Total } A_s = A_{sf} + A_{sw} = 2.83 + 2.27 = 5.10 \text{ sq in.}$$
$$\rho = \frac{A_s}{bd} = \frac{5.10}{570} = 0.00895$$

Using eq. (1-22)

$$\bar{\rho}_b = \left(\frac{b_w}{b}\right)(\bar{\rho}_b + \rho_f) = \left(\frac{10}{30}\right)(0.0285 + 0.0148) = 0.0144$$
$$\rho_{max} = 0.75\rho_b = 0.0108$$

Since $\rho = 0.00895$ is less than 0.0108, the *1971 ACI Code* is satisfied.

When reinforcement is selected, care must be exercised in order to avoid selection of excess reinforcement, which might lead to conditions that violate the *1971 ACI Code*.

In this example, two No. 11 bars and two No. 9 bars provide $A_s = 5.12$ sq in., which satisfies the code. Use two layers of steel with the No. 11 bars in the lower layer. The effective depth, d, is measured to the centroid of the reinforcing steel.

EXAMPLE 1-10: A T-beam was designed in strict accord with the 1963 ACI Code with the following values and results: $M_u = 1130$ ft-kips, $h_f = 4$ in., $f_c' = 4000$ psi, $f_y = 60{,}000$ psi, $b = 70$ in., $d = 16$ in., $A_{sf} = 9.1$ sq in., $b_w = 30$ in., $A_s = 18.6$ sq in. Determine whether or not this section satisfies the *1971 ACI Code* for maximum steel ratio.

SOLUTION: Since A_s is known, eq. (1-16) may be used to determine whether or not T-beam theory applies. Thus, if the stress block is in the flange, with $\rho = A_s/bd = 18.6/1120 = 0.0166$, $a = 1.18\rho f_y d/f_c' = (1.18)(0.0166)(60)(19)/4 = 5.58$ in., greater than $h_f = 4.0$ in., so the stress block extends down into the web. Thus, T-beam theory applies and

$$\rho_f = \frac{A_{sf}}{b_w d} = \frac{9.1}{480} = 0.0189$$
$$\rho_b = \left(\frac{b_w}{b}\right)(\bar{\rho}_b + \rho_f) = \left(\frac{30}{70}\right)(0.0285 + 0.0189) = 0.0203$$
$$\rho_{max} = 0.75\rho_b = 0.0152$$

Since actual $\rho = A_s/bd = 0.0166$ exceeds $\rho_{max} = 0.0152$, the design does not satisfy the *1971 ACI Code*.

Example 1-10 was presented to illustrate the concern stated by Professor Ferguson in his textbook, *Reinforced Concrete Fundamentals*, John Wiley Publishing Co., Second Edition, page 66. The provision of the *1963 ACI Code* permitted nonductile T-beam design in some cases. This is the reason for modification of the definition of "balanced steel ratio" for T-beams and beams reinforced for compression in the *1971 ACI Code*.

When the stress block extends below the flange, a T-beam may be investigated using Tables 1-2 and 1-4. An equivalent compression steel area, A_{sf}, may be used in place of A_s', where A_{sf} is the steel area required to balance the compression force in the overhanging flange elements, shown shaded in Fig. 1-8.

The ratio ρ_f is:

$$\rho_f = \frac{A_{sf}}{b_w d} = \frac{(b - b_w)(h_f)(0.85f_c')}{b_w d f_y} \tag{1-24}$$

Table 1-2 is used with b_w, d and ρ to obtain the section capacity without the overhanging flanges. Then, Table 1-4 is used with $\rho'/\rho = \rho_f/\rho$, and ρ, to obtain the multiplier which indicates the total capacity ratio, M_u with overhanging flanges to M_u without overhanging flanges. The ratio d'/d is replaced by $h_f/2d$.

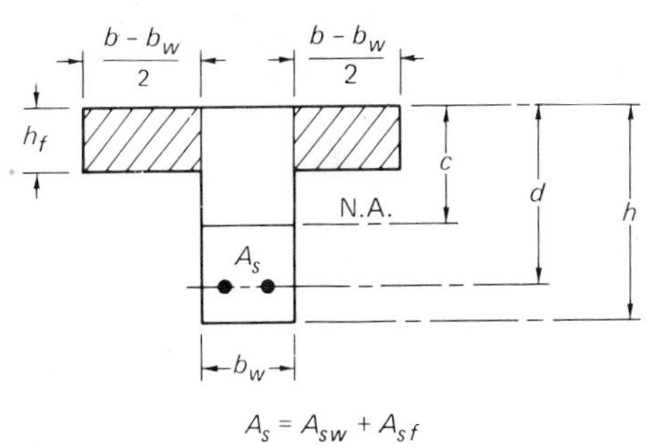

Fig. 1-8

1.11 COMPRESSION MEMBERS

The intent of the *1971 ACI Code* is to encourage more precise methods of structural analysis than have been used in the past. This is particularly true for compression members subjected to the effects of axial load and bending.

The *Code* refers to "second-order" methods of solution for obtaining axial forces and bending moments in frame members. Such methods of analysis always include the stiffnesses of the members and the rotation and translation of joints. The structure is analyzed as a whole, and not by isolating portions of the structure for separate analysis. The effects of sustained loads must be included. For structural frames that do not exceed about 10 stories, existing time-sharing computer programs are adequate to qualify for second-order analysis if the cracked-section moment of inertia is used for beams and the effects of reinforcement are included for columns. For the latter, the *Code* equation

$$EI = \frac{E_c I_g/5 + E_s I_s}{1 + \beta_d} \tag{1-25}$$

is satisfactory. The creep factor, β_d, is defined as M_u dead load/M_u total load. The elastic modulus for the concrete, E_c, is taken as

$$E_c = 33\,W^{1.5}\sqrt{f_c'} \tag{1-26}$$

where W is the unit weight of the concrete, pounds per cubic foot. I_g is the gross section moment of inertia for the column, $bh^3/12$ for rectangular sections and $\pi h^4/64$ for circular sections.

E_s is the elastic modulus for the steel, 29,000 ksi. I_s is the moment of inertia of the steel in the column.

Equation (1-24) can be defined in general as

$$EI = \frac{E_c I_g}{1 + \beta_d}\,[0.2 + \alpha_i\, np_t \gamma^2] \tag{1-27}$$

where α_i is a coefficient dependent on the steel pattern and the shape of the cross section, $n = E_s/E_c$, $\rho_t = A_s/A_g$ and γ is illustrated in Fig. 1-9. Table 1-5 provides values of α_i.

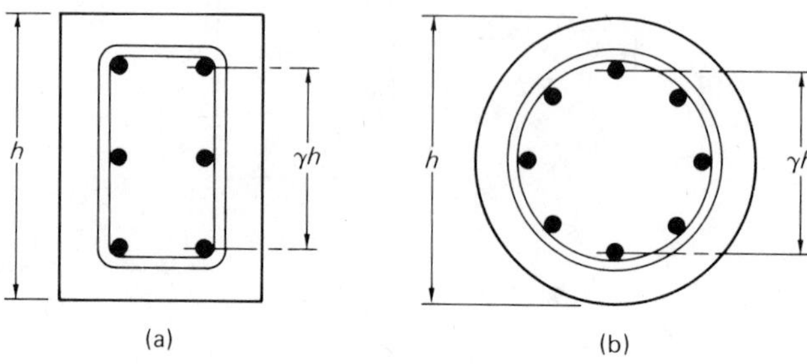

Fig. 1-9

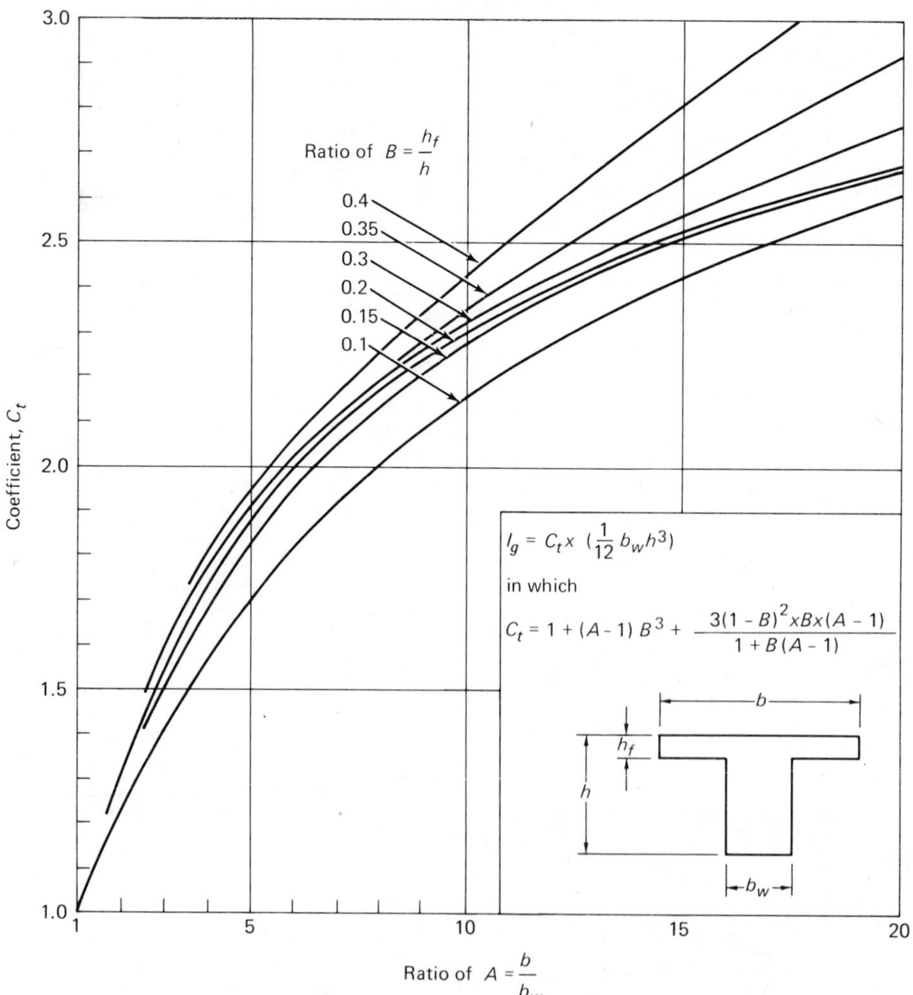

Fig. 1-10

TABLE 1-5 Values of α_i

Shape	*Steel Pattern*	α_i
Square	Equal distribution on four faces	2.0
Rectangular	Equal distribution on end faces	3.0
Rectangular	Equal distribution on side faces	1.0
Square	Circular, equal spacing	1.5
Circle	Circular, equal spacing	1.273

The value of EI obtained using eq. (1-25) satisfies the use of "moment-curvature relationships" as recommended by ACI-ASCE Committee 441 (Columns). Later, when approximate methods of analysis are discussed, eq. (1-25) will be used again with β_d equal to zero.

The "cracked-section" moment of inertia may be obtained in a reasonably simple manner for the beams, using Fig. 1-10 and Table 1-6. Figure 1-10 provides a multiplier, (β_t), for the gross moment of inertia for T-beams, when the gross moment of inertia of the web is known. Thus, $I_g = (b_w h^3/12)\beta_t$, for a T-beam having known ratios b/b_w and h_f/h.

Tables 1-6a–d provide multipliers, μ_i, for obtaining the cracked-section moment of inertia for a beam, knowing the dimensions, steel areas and concrete strength. These tables will also be used later in connection with approximate methods of design when a second-order analysis is not used.

When EI for either beams or columns exceeds $E_c I_g$, the gross section properties are used instead of the respective values EI which include the effects of steel for the columns, and the effects of steel and cracking for beams. For continuous beams, the code permits (but does not demand) the use of an average of positive and negative moment cross section EI for the beams. Engineering judgement must be used to determine whether or not the average moment of inertia should be used. Ordinarily, the average value of EI will be greater than the value at the column when T-beam action is included in the positive moment region at midspan.

For members having variable moments of inertia, modified stiffness factors can be obtained from the *Handbook of Frame Constants*, Portland Cement Association, Skokie, Illinois. For unusual cases, the procedure described by Maugh* can be used.

The *ACI Code* permits the use of approximate methods of analysis, providing that column moments are magnified to approximate the effects of joint displacement. In this case, moment coefficients for beams are not applicable. For gravity loads, a moment distribution process is used to obtain beam and column moments.** For wind or other lateral forces, either the portal method or the cantilever method is used. Since the effects of joint translation are not included, column moments must be magnified. This, then, requires that additional moments be resisted by the beams. Hence, for preliminary design, the beams should be proportioned for moments somewhat larger than the calculated values. An increase in beam moments of 20% is often realistic. However, since the *ACI Code* permits redistribution of moments, a 10% overdesign in the positive and negative moment regions will normally suffice.

When a second-order analysis has been used, compression member design is relatively simple. While the approximate formulas for column design contained in the *1963 ACI Code* are no longer considered as being valid, extensive design aids*** are available for use.

Example 1-11 is typical when a second-order analysis method has been used.

EXAMPLE 1-11: A second-order analysis has been used, and the results for a 20 in. diameter column show that: P_{DL} = 120 kips, P_{LL} = 180 kips, M_{DL} = 100 ft-kips, M_{LL} = 150 ft-kips. The analysis

*L. Maugh, *Statically Indeterminate Structures*, John Wiley and Sons Publishing Co., New York: 1964.

**See *Continuity in Concrete Building Frames*, Portland Cement Association, 5420 Old Orchard Road, Skokie, Illinois.

***See SP-17-A, "Ultimate Strength Design Handbook, Columns, American Concrete Institute, Detroit, Michigan. Also, see Portland Cement Association Design Aids and Computer Programs.

TABLE 1-6a Ratio of Cracked Moment of Inertia to Gross Moment of Inertia (μ_i) for Rectangular Sections, d'/d = 0.050.

$r = (n-1)\,\rho'/n\rho$

$n\rho$	.0	.2	.3	.4	.5	.6	.7	.8	.9
.020	.160	.160	.161	.161	.161	.162	.162	.162	.162
.040	.287	.290	.292	.293	.295	.296	.297	.299	.300
.060	.398	.405	.408	.411	.414	.418	.420	.423	.426
.080	.496	.509	.515	.520	.526	.531	.536	.541	.545
.100	.586	.605	.614	.622	.630	.638	.645	.653	.659
.120	.668	.694	.707	.718	.730	.740	.751	.761	.770
.140	.743	.779	.795	.810	.825	.839	.853	.866	.878
.160	.814	.858	.879	.899	.917	.935	.952	.968	.983
.180	.879	.935	.960	.984	1.007	1.029	1.049	1.068	1.087
.200	.941	1.008	1.038	1.067	1.094	1.120	1.144	1.167	1.189
.220	.999	1.078	1.114	1.147	1.179	1.209	1.238	1.264	1.290
.240	1.054	1.145	1.187	1.226	1.263	1.297	1.330	1.360	1.389
.260	1.105	1.211	1.258	1.303	1.344	1.384	1.421	1.455	1.488
.280	1.155	1.274	1.328	1.378	1.425	1.469	1.511	1.549	1.586
.300	1.201	1.335	1.395	1.452	1.504	1.553	1.599	1.643	1.683
.320	1.246	1.395	1.462	1.524	1.582	1.637	1.688	1.735	1.780
.340	1.289	1.453	1.527	1.596	1.660	1.719	1.775	1.827	1.876
.360	1.329	1.510	1.591	1.666	1.736	1.801	1.862	1.918	1.971
.380	1.368	1.566	1.654	1.736	1.811	1.882	1.948	2.009	2.066
.400	1.406	1.621	1.716	1.804	1.886	1.962	2.033	2.099	2.161
.420	1.441	1.674	1.777	1.872	1.960	2.042	2.118	2.189	2.255
.440	1.476	1.726	1.837	1.939	2.034	2.121	2.203	2.278	2.349
.460	1.509	1.778	1.897	2.006	2.107	2.200	2.287	2.367	2.442
.480	1.541	1.829	1.955	2.072	2.179	2.279	2.371	2.456	2.536
.500	1.572	1.879	2.013	2.137	2.251	2.357	2.454	2.545	2.629

TABLE 1-6b Ratio of Cracked Moment of Inertia to Gross Moment of Inertia (μ_i) for Rectangular Sections, d'/d = 0.100.

$r = (n - 1)\rho'/n\rho$

$n\rho$	.0	.2	.3	.4	.5	.6	.7	.8	.9
.020	.139	.139	.139	.139	.139	.139	.140	.140	.140
.040	.250	.251	.252	.253	.253	.254	.255	.255	.256
.060	.346	.350	.352	.353	.355	.357	.358	.360	.361
.080	.432	.439	.442	.446	.449	.452	.455	.458	.460
.100	.509	.521	.526	.532	.537	.541	.546	.550	.554
.120	.581	.597	.605	.613	.620	.626	.633	.639	.645
.140	.646	.669	.680	.690	.699	.708	.717	.725	.733
.160	.708	.737	.751	.763	.776	.787	.798	.809	.819
.180	.765	.802	.818	.834	.850	.864	.878	.890	.903
.200	.818	.863	.884	.903	.921	.939	.955	.971	.985
.220	.869	.922	.947	.970	.991	1.012	1.031	1.049	1.066
.240	.916	.979	1.008	1.034	1.060	1.083	1.106	1.127	1.147
.260	.961	1.034	1.067	1.097	1.126	1.154	1.179	1.203	1.226
.280	1.004	1.087	1.124	1.159	1.192	1.223	1.251	1.279	1.304
.300	1.045	1.138	1.180	1.220	1.256	1.291	1.323	1.353	1.382
.320	1.084	1.188	1.235	1.279	1.320	1.358	1.394	1.427	1.459
.340	1.121	1.237	1.289	1.337	1.382	1.425	1.464	1.501	1.535
.360	1.156	1.284	1.342	1.395	1.444	1.490	1.533	1.574	1.611
.380	1.190	1.331	1.393	1.451	1.505	1.555	1.602	1.646	1.687
.400	1.223	1.376	1.444	1.507	1.566	1.620	1.671	1.718	1.762
.420	1.254	1.420	1.494	1.562	1.625	1.684	1.738	1.789	1.837
.440	1.284	1.464	1.543	1.616	1.684	1.747	1.806	1.860	1.911
.460	1.313	1.506	1.591	1.670	1.743	1.810	1.873	1.931	1.985
.480	1.340	1.548	1.639	1.723	1.801	1.873	1.940	2.002	2.059
.500	1.367	1.589	1.686	1.776	1.859	1.935	2.006	2.072	2.133

TABLE 1-6c Ratio of Cracked Moment of Inertia to Gross Moment of Inertia (μ_i) for Rectangular Sections, d'/d = 0.150.

$r = (n - 1)\rho'/n\rho$

$n\rho$	.0	.2	.3	.4	.5	.6	.7	.8	.9
.020	.121	.121	.121	.122	.122	.122	.122	.122	.122
.040	.219	.219	.219	.220	.220	.220	.220	.221	.221
.060	.303	.305	.305	.306	.307	.308	.309	.309	.310
.080	.378	.382	.384	.385	.387	.389	.390	.392	.393
.100	.446	.452	.456	.458	.461	.464	.467	.469	.472
.120	.508	.518	.523	.527	.531	.535	.539	.543	.547
.140	.566	.580	.586	.592	.598	.604	.609	.614	.619
.160	.619	.638	.646	.654	.662	.669	.676	.683	.689
.180	.669	.693	.704	.714	.724	.733	.742	.750	.758
.200	.716	.745	.759	.771	.783	.795	.805	.815	.825
.220	.760	.796	.812	.827	.841	.855	.867	.879	.891
.240	.802	.844	.863	.881	.897	.913	.928	.942	.956
.260	.841	.890	.912	.933	.952	.971	.988	1.004	1.020
.280	.879	.935	.960	.984	1.006	1.027	1.047	1.065	1.083
.300	.914	.978	1.007	1.034	1.059	1.083	1.105	1.126	1.145
.320	.948	1.020	1.053	1.083	1.111	1.137	1.162	1.185	1.207
.340	.981	1.061	1.097	1.131	1.162	1.191	1.219	1.244	1.268
.360	1.012	1.101	1.141	1.178	1.212	1.245	1.275	1.303	1.329
.380	1.041	1.140	1.184	1.224	1.262	1.297	1.330	1.361	1.389
.400	1.070	1.178	1.225	1.270	1.311	1.349	1.385	1.418	1.449
.420	1.097	1.215	1.267	1.315	1.359	1.401	1.439	1.475	1.509
.440	1.123	1.251	1.307	1.359	1.407	1.452	1.493	1.532	1.568
.460	1.149	1.286	1.347	1.403	1.454	1.502	1.547	1.588	1.627
.480	1.173	1.321	1.386	1.446	1.501	1.553	1.600	1.645	1.686
.500	1.196	1.355	1.425	1.489	1.548	1.603	1.653	1.700	1.744

was based on an assumed column steel ratio $\rho_t = 0.04$ and $\gamma = 0.8$, with $f'_c = 4$ ksi and $f_y = 50$ ksi. Design the column.

SOLUTION: Since a second-order analysis was used, moment magnification does not apply.

$$P_u = (1.4)(120) + (1.7)(180) = 474 \text{ kips}$$

$$M_u = (1.4)(100) + (1.7)(150) + 395 \text{ ft-kips}$$

The interaction diagrams, Figs. 1-11, 1-12 and 1-13 (see end of chapter) are plotted in terms of $\kappa = P_u/f'_c bh$ and $\kappa e/h = M_u \times 12/f'_c bh^2$ for rectangular sections and in terms of $\kappa = P_u/f'_c h^2$ and $\kappa e/h = M_u \times 12/f'_c h^3$ for circular sections.

In order to illustrate the use of various types of interaction diagrams that are available, diagrams for circular columns and rectangular columns with steel on four faces, and diagrams for rectangular columns with steel on two faces are plotted in terms of the variable $\rho_t m$, where $m = f_y/(0.85 f'_c)$.

TABLE 1-6d Ratio of Cracked Moment of Inertia to Gross Moment of Inertia (μ_i) for Rectangular Sections, $d'/d = 0.200$.

$r = (n - 1)\ \rho'/n\rho$

$n\rho$	.0	.2	.3	.4	.5	.6	.7	.8	.9
.020	.107	.107	.107	.107	.107	.107	.107	.107	.107
.040	.192	.192	.193	.193	.193	.193	.193	.193	.193
.060	.266	.267	.267	.268	.268	.268	.269	.269	.269
.080	.333	.334	.335	.336	.337	.337	.338	.339	.340
.100	.392	.396	.397	.399	.400	.402	.403	.404	.405
.120	.447	.453	.455	.458	.460	.462	.464	.466	.468
.140	.498	.506	.510	.513	.516	.520	.523	.526	.528
.160	.545	.556	.561	.566	.570	.575	.579	.583	.587
.180	.589	.603	.610	.616	.622	.628	.633	.638	.643
.200	.630	.648	.657	.665	.672	.679	.686	.692	.698
.220	.669	.691	.702	.711	.720	.729	.737	.745	.752
.240	.706	.733	.745	.756	.767	.777	.787	.796	.805
.260	.741	.772	.787	.800	.813	.825	.836	.847	.857
.280	.773	.810	.827	.843	.857	.871	.884	.896	.908
.300	.805	.847	.866	.884	.901	.917	.931	.945	.958
.320	.835	.883	.905	.925	.944	.961	.978	.993	1.008
.340	.863	.918	.942	.965	.986	1.005	1.024	1.041	1.057
.360	.890	.951	.978	1.003	1.027	1.049	1.069	1.088	1.106
.380	.917	.984	1.014	1.042	1.067	1.092	1.114	1.135	1.155
.400	.942	1.016	1.049	1.079	1.107	1.134	1.158	1.181	1.203
.420	.966	1.047	1.083	1.116	1.147	1.175	1.202	1.227	1.250
.440	.989	1.077	1.116	1.152	1.186	1.217	1.246	1.273	1.298
.460	1.011	1.107	1.149	1.188	1.224	1.258	1.289	1.318	1.345
.480	1.032	1.136	1.181	1.223	1.262	1.298	1.332	1.363	1.392
.500	1.053	1.164	1.213	1.258	1.300	1.338	1.374	1.407	1.438

Hence, for the circular cross-section under consideration

$$\kappa = \frac{P_u}{f_c'h^2} = \frac{474}{(4)(400)} = 0.296, \quad m = \frac{f_y}{(0.85 f_c')} = 17.65$$

$$\frac{\kappa e}{h} = \frac{M_u \times 12}{f_c' h^3} = \frac{395 \times 12}{(4)(8000)} = 0.148$$

From Fig. 1-13(c), $\gamma = 0.8$, find $\rho_t m = 0.5$, and $m = 17.65$, so $\rho_t = 0.0283$. Thus $A_{st} = \rho_t A_g = (0.0283)(\pi/4)(20)^2 = 8.89$ sq in. Using radial splices,* fifteen No. 7 bars suffice and satisfy the code requirements for cover and clearance, using No. 4 spiral bars.

The design and selection of spiral reinforcement is a detailing problem primarily, with the spiral reinforcement providing strength equal to that of the concrete shell outside of the spiral. Tables and design examples are provided in the *USD Handbook** and will not be repeated herein.

If the method of analysis does not qualify as a second-order method, the moments in the columns may require magnification before the interaction diagrams are used to develop a design. The effective length (kl_u) is an equivalent length in comparison with the actual length (l) in computing the buckling load, P_c, where:

$$P_c = \frac{\pi^2 EI}{(kl_u)^2} \tag{1-28}$$

For the column, EI is that given by either eq. (1-24) or eq. (1-26). The *Code* permits the use of an alternate equation

$$EI = E_c I_g / 5 \tag{1-29}$$

but this equation is valid only for preliminary designs since the effects of reinforcement must be included.

In order to obtain the effective length (kl_u) it is first necessary to compute the rigidities of the columns and beams at both ends of the column. The factor ψ is defined as:

$$\psi = \frac{\Sigma EI/l \text{ columns}}{\Sigma EI/l \text{ beams}} \tag{1-30}$$

*See Table, Reinforcement 5.1.1, *Ultimate Strength Design Handbook*, Vol. 2, SP-17-A, American Concrete Institute, Detroit, Michigan.

For the columns, eq. (1-26) is used to obtain EI. The cracked section EI for the beams is obtained using Table 1-6. For example, for rectangular beams without T-beam action, $EI = \mu_i E_c I_g$, where μ_i is a multiplier obtained from Table 1-6. In all cases where column or beam EI exceeds EI_g, EI_g is used.

The factor ψ corresponds to the factor r' used in connection with the *1963 ACI Code*. The factor ψ is computed at the top and bottom of a column for single column action, and either Fig. 1-14 or 1-15 is used to obtain the effective length, kl_u. Figure 1-14 applies when the columns are braced against sidesway. Such bracing may be assumed if bracing elements or shear walls provide stiffness, EI, at least six times that of all columns in the story. Otherwise, the columns are considered to be free to translate, and Fig. 1-15 applies. The factor k is obtained by alignment of a straightedge on ψ_A for the base of the column and ψ_B for the top of the column. Where the straightedge intersects the nomograph line for k, that quantity is read. If a column has absolute fixity, ψ is taken as 1.0 rather than the theoretical value zero. This is done because it is practically impossible to construct a perfectly fixed connection.

ACI-ASCE Committee 441 (Columns) advises that it is not realistic to assume that (kl) will be less than $1.2l$ for columns in sway frames. Hence, it is advisable to make preliminary designs assuming $k \geqslant 1.2$. Further, since the steel ratio (ρ_t) for columns must range between 0.01 and 0.08, it is advisable to make the assumption that ρ_t is at least 0.02 for preliminary calculations.

Once the length factor, k, is known for individual columns in a story, the moment magnification factor can be determined. However, if kl/r is less than 22 for unbraced frames or less than $34 - 12M_1/M_2$ for braced frames, moment magnification is not required. In the previous expression, M_1/M_2 is the ratio of the smaller and larger end moments on the column. The ratio is positive if the member is bent in single curvature and negative if bent in double curvature. Here, the radius of gyration is based on the gross concrete cross section, so $r = \sqrt{I_g/A_g}$. For rectangular sections r may be approximated as $0.3h$. For circular sections, r is precisely $0.25h$.

A moment coefficient C_m is used to compute the magnification factor, δ. For members braced against sidesway, without transverse loads between the supports

$$C_m = 0.6 + 0.4 M_1/M_2 \geqslant 0.4 \tag{1-31}$$

For all other cases, $C_m = 1.0$.

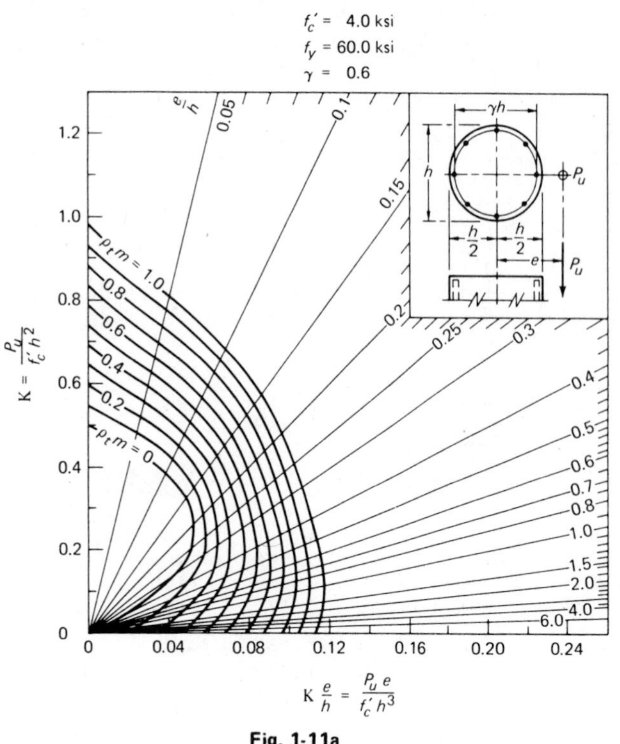

Fig. 1-11a

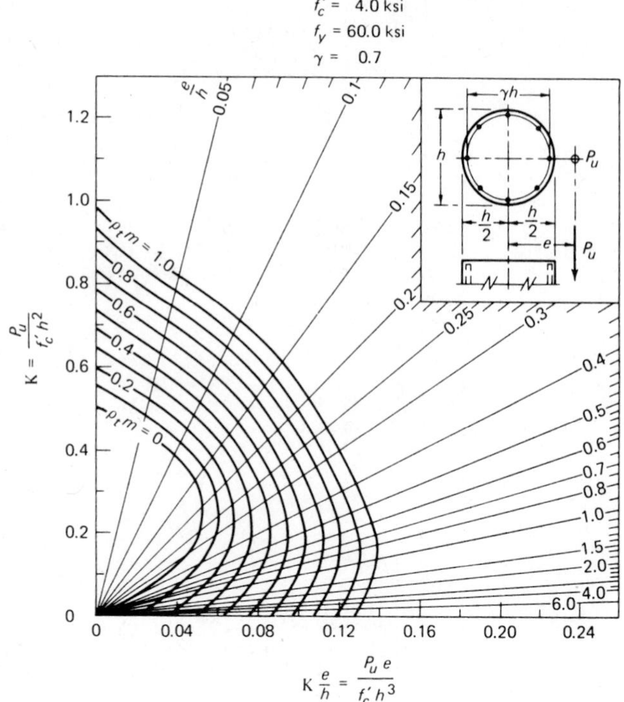

Fig. 1-11b

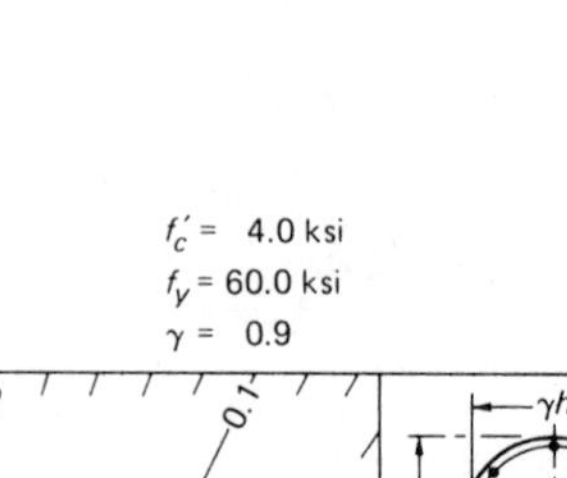

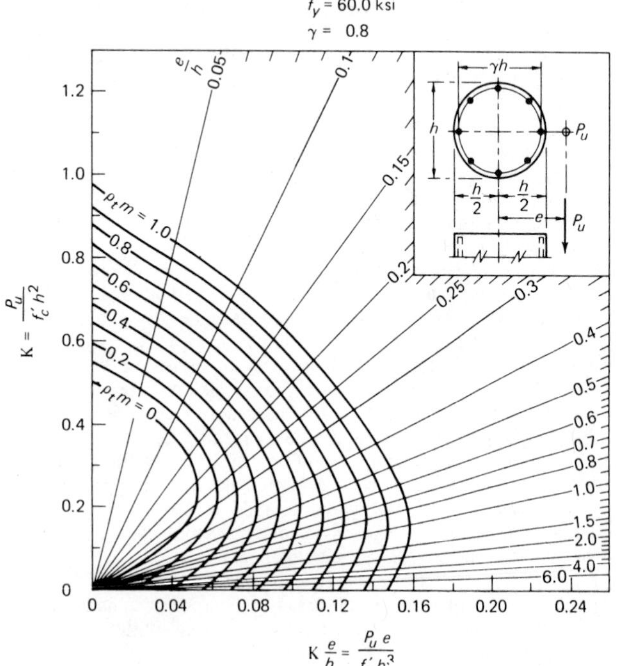

Fig. 1-11c

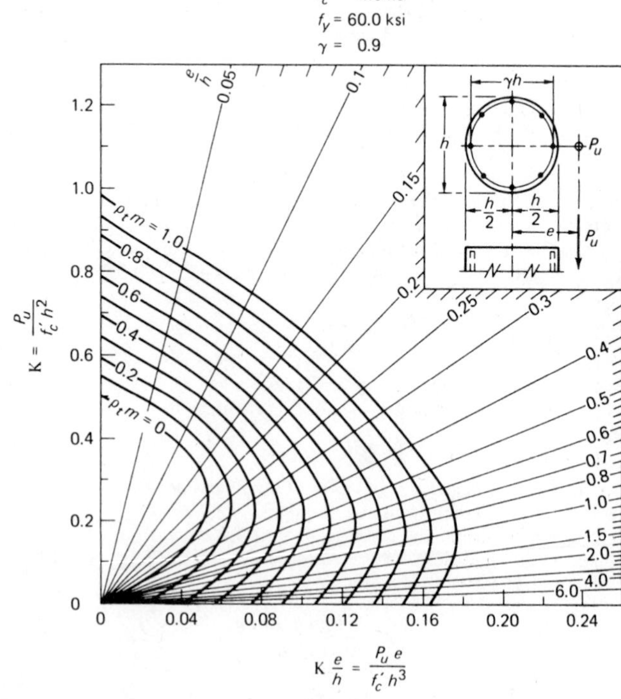

Fig. 1-11d

Spirally Reinforced Columns, $\phi = 0.75$

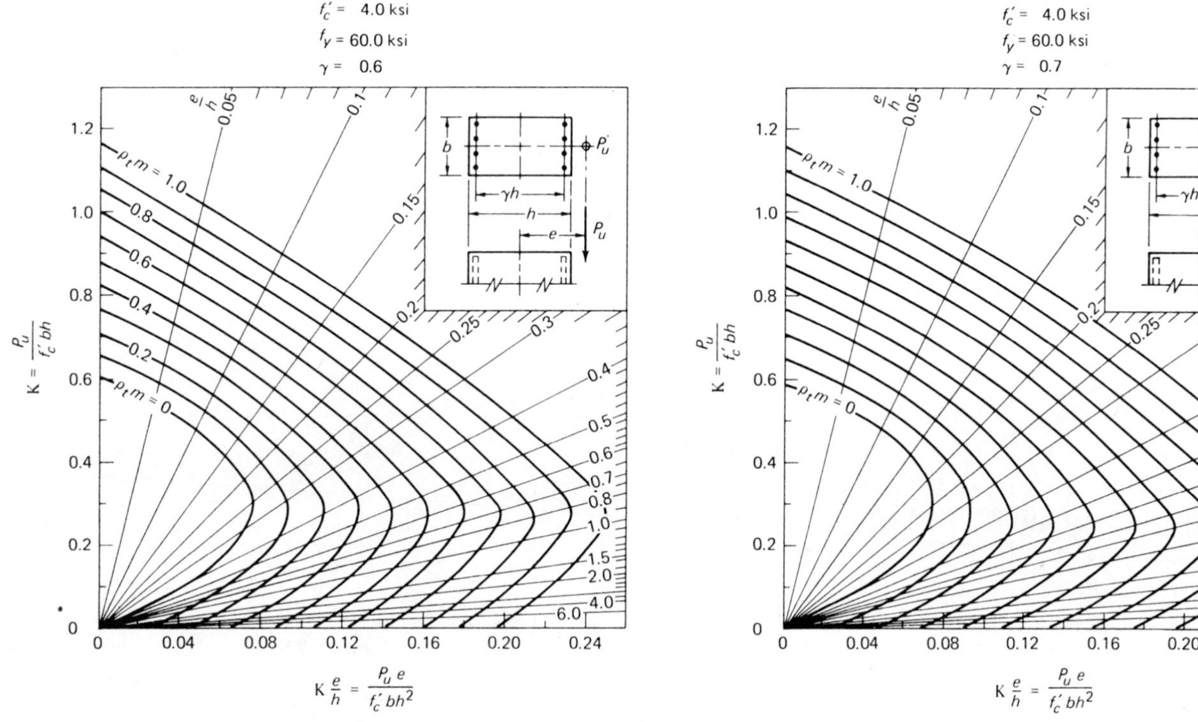

Fig. 1-12a

Fig. 1-12b

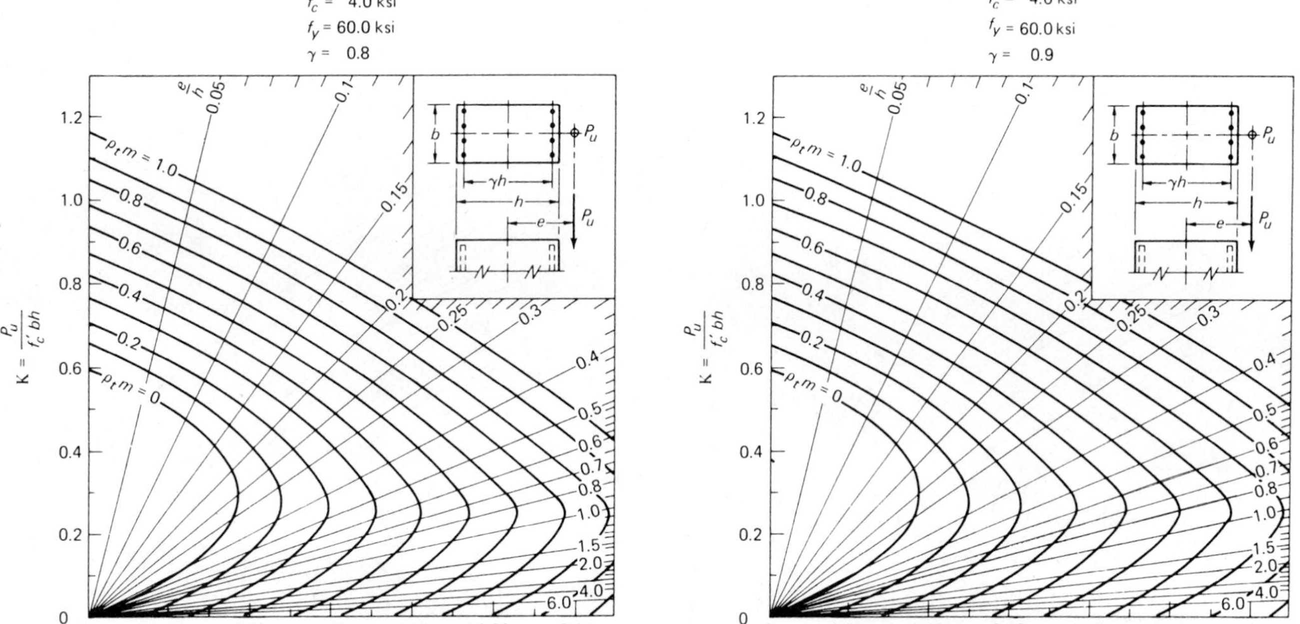

Fig. 1-12c

Fig. 1-12d

Rectangular Tied Columns, ϕ = 0.7

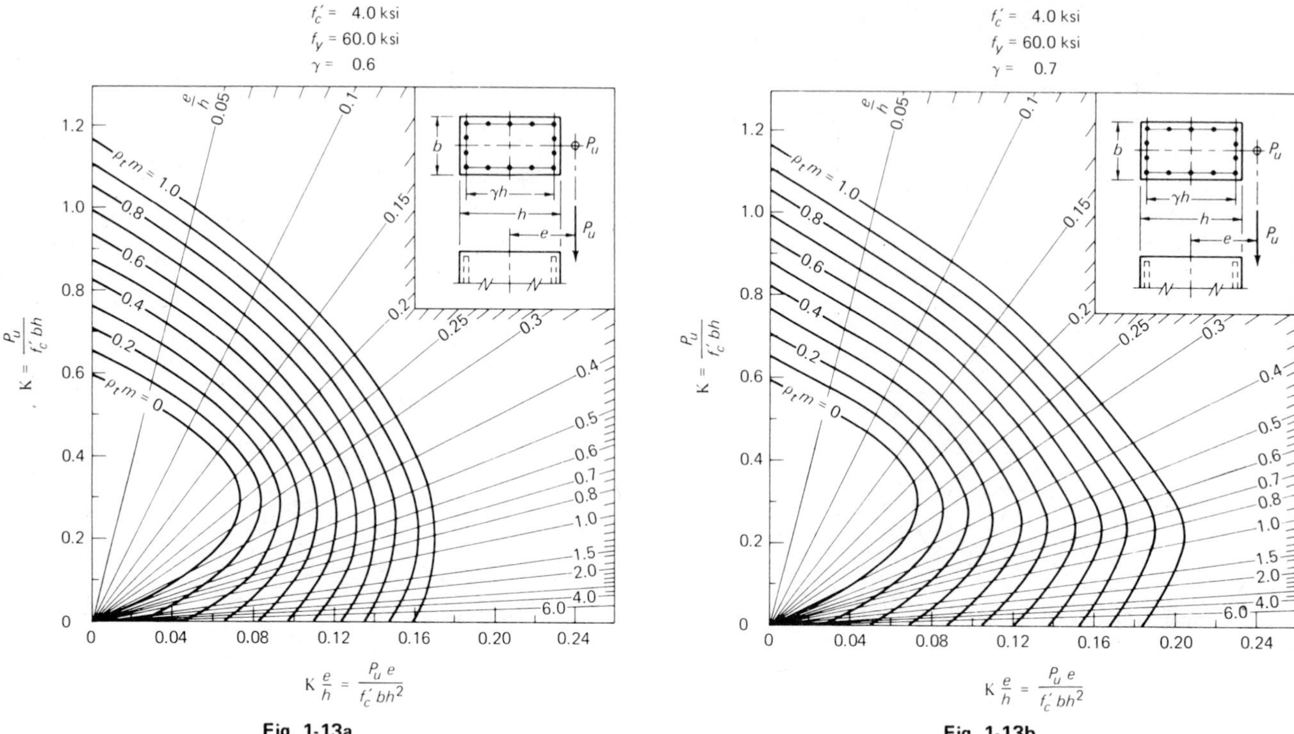

Fig. 1-13a

Fig. 1-13b

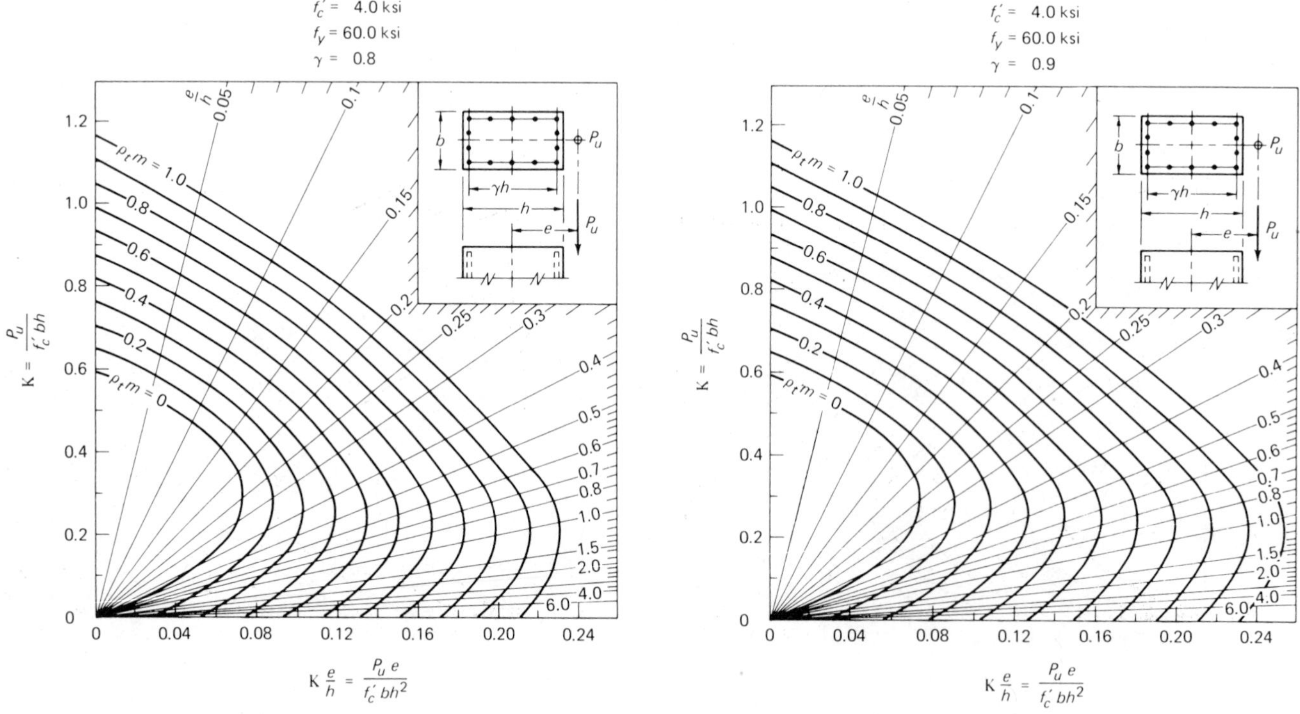

Fig. 1-13c

Fig. 1-13d

Rectangular Tied Columns, $\phi = 0.7$

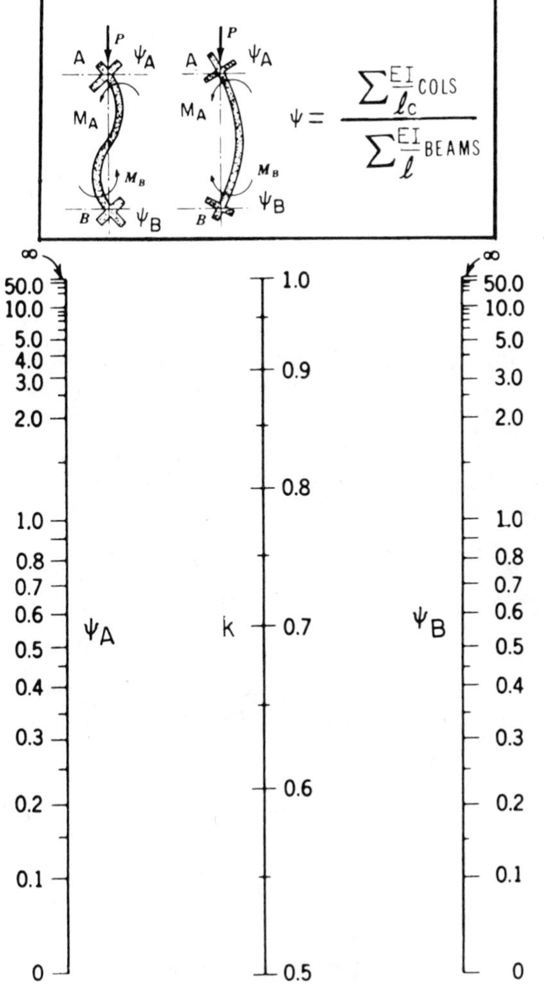

Fig. 1-14 Effective length factors for braced members.

Also, the Euler buckling load, P_c, enters into the computation, where

$$P_c = \frac{\pi^2 EI}{(kl_u)^2} \tag{1-32}$$

Then the magnification factor is

$$\delta = \frac{C_m}{1 - \dfrac{P_u}{\phi P_c}} \geqslant 1.0 \tag{1-33}$$

Finally, the design moment M_c is

$$M_c = \delta M_2 \tag{1-34}$$

The magnification factor (δ) is computed for each individual column in any story, considering the column to be braced against sidesway, and also for the entire story, assuming that the story is free to sway unless shear walls are used. In the latter case, δ is computed by substituting $\Sigma P_u/\Sigma \phi P_c$ for all columns for $P_u/\phi P_c$ in eq. (1-32).

It is therefore possible that single column action could govern for an individual column, while story action could govern for other columns in a story.

When bending occurs about two principal axes, the moment magnifiers are computed separately for each axis. The individual moments are then magnified separately. This case will be treated in detail later.

Once the moment magnifier (δ) has been determined, the larger end moment (M_2) is magnified so that $M_c = \delta M_2$. Then, the factors κ and $\kappa e/h$ are computed, and the interaction diagrams used to obtain the steel ratio, ρ_t. If assumed dimensions and an assumed steel ratio (ρ_t) differ from those determined, the design must be checked.

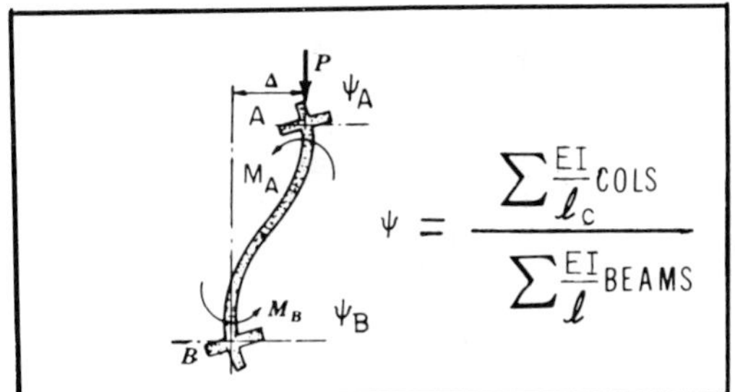

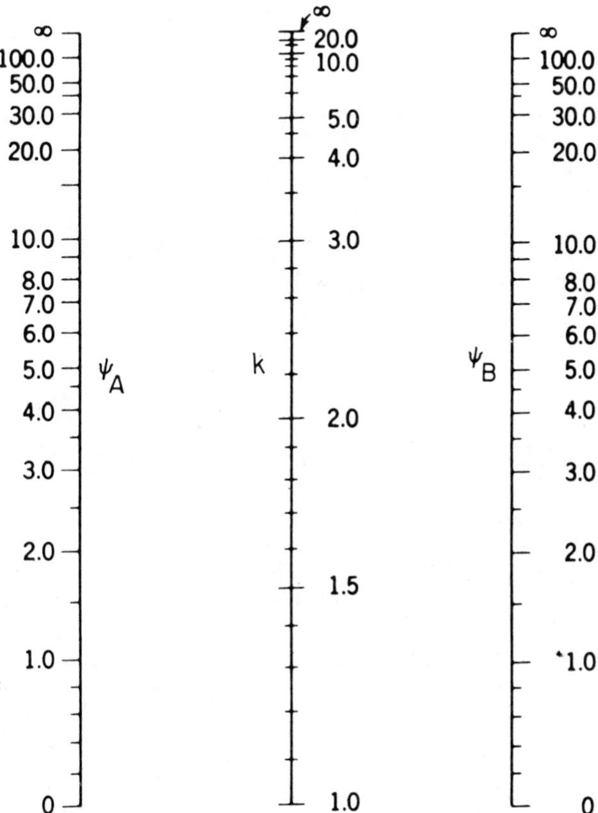

Fig. 1-15 Effective length factors for unbraced members.

EXAMPLE 1-12: Fig. 1-16 shows a typical story for a frame that

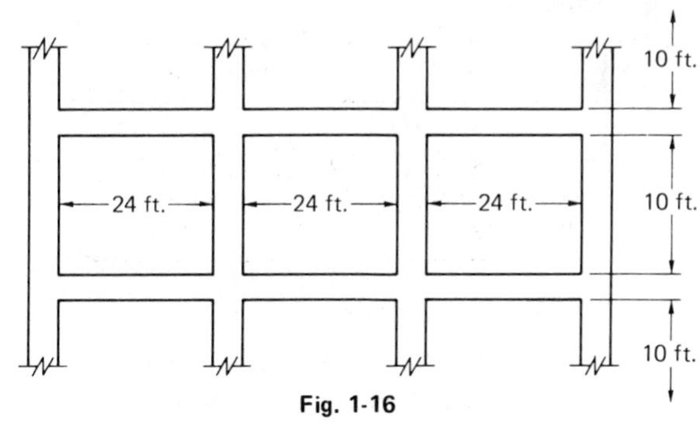

Fig. 1-16

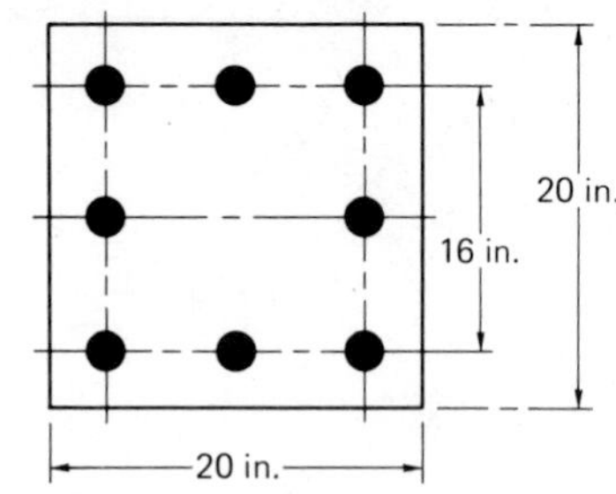

Fig. 1-17

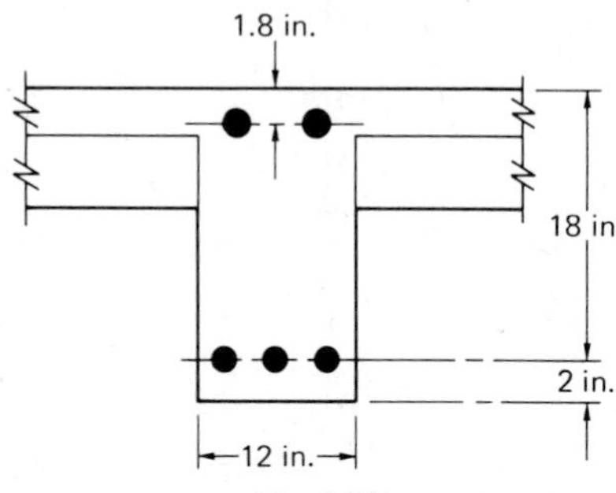

Fig. 1-18

has been analyzed using first order (approximate) methods. The columns were assumed to be 20 in. × 20 in., steel equally distributed on four faces, $f_c' = 4000$ psi and $f_y = 60{,}000$ psi. The cross-section of the column is shown in Fig. 1-17, and that of the beam at the column in Fig. 1-18. Design the columns, if the steel ratios in the beams are $\rho = 0.02$ and $\rho' = 0.007$. Axial loads and moments at service loads are shown in Table 1-7. The column steel ratio has been assumed to be 0.02.

The *1971 ACI Code* specifies that, in considering frame action, the ΣP_u and $\Sigma \phi P_c$ must be computed for all columns in the story. For example, if there were ten lines of frames in the story, there would be forty columns to consider. If there are several different sizes of columns and several different bracing systems, there will be a number of different values of P_u. Similarly, if P_u varies in the different frames, all values of P_u must be included.

For the purpose of illustration, it is assumed in this problem that the frame shown and the loads given are typical throughout the entire story. Thus, considering two interior columns and two exterior columns provides a solution identical to that obtained by considering forty exterior columns and forty interior columns. If the ϕ factor varies, this must be considered in obtaining $\Sigma \phi P_c$.

TABLE 1-7

Exterior Column	*Interior Column*
P_{DL} = 120 kips	P_{DL} = 190 kips
P_{LL} = 180 kips	P_{LL} = 220 kips
M_{DL} = 140 ft-kips	M_{DL} = 90 ft-kips
M_{LL} = 230 ft-kips	M_{LL} = 150 ft-kips

SOLUTION: For exterior columns

$$P_u = (1.4)(120) + (1.7)(180) = 474 \text{ kips}$$
$$M_u = (1.4)(140) + (1.7)(230) = 587 \text{ ft-kips}$$

For interior columns

$$P_u = (1.4)(190) + (1.7)(220) = 640 \text{ kips}$$
$$M_u = (1.4)(90) + (1.7)(150) = 381 \text{ ft-kips}$$

The beams were assumed to be identical at all supports. Interior and exterior columns were assumed to be identical. Conditions at the tops and bottoms of all columns were assumed to be identical.

For the beams, the gross moment of inertia fo the concrete cross section is

$$(12)\,\frac{(20)^3}{12} = 8{,}000 \text{ in.}^4$$

Using Table 1-6, the cracked-section moment of inertia for the beams may be determined. Since $d' = 1.8$ in. and $d = 18$ in., then $d'/d = 0.10$. For $f_c' = 4000$ psi, and $w = 145$ pcf, $E_c = 3.6 \times 10$ ksi, so $n = 29 \times 10^3/(3.6 \times 10^3) = 8.0$. Thus, $n\rho = (8)(0.02) = 0.16$ and $(n-1)\rho' = (7)(0.007) = 0.049$. Hence, $r = (n-1)\rho'/(n\rho) = 0.049/0.016 = 0.3$.

From Table 1-6b, $\mu_i = 0.75$, so

$$I_{cr} = (0.75)(8000) = 6{,}000 \text{ in.}^4 \text{ for beams}$$

Then

$$E_c I_{cr} = (3.6 \times 10^3)(6000) = 21.6 \times 10^6 \text{ k-in.}^2$$

For the columns, EI is obtained using eq. (1-26), with $\beta_d = 0$, or

$$EI = \frac{E_c I_g}{1 + \beta_d}\,\{0.2 + \alpha_i n \rho_t \gamma^2\}$$

where $\alpha_i = 2.0$ for square columns with the steel equally distributed along all four faces. Thus

$$I_g = (20)\,\frac{(20)^3}{12} = 13{,}333 \text{ in.}^4$$

so

$$E_c I_g = (3.6 \times 10^3) \times (13.333) \times 10^3 = 48 \times 10^6 \text{ kips-in.}^2$$

Since $\gamma = 16/20 = 0.8$, then

$$EI = (48 \times 10^6)[0.2 + (2)(8)(0.02)(0.8)^2] = 48 \times 10^6\,[0.2 + 0.205] = 19.44 \times 10^6 \text{ kips-in.}^2$$

For exterior columns there are two columns and one beam, so

$$\psi = \frac{\Sigma EI/l \text{ cols}}{\Sigma EI/l \text{ beams}} = \frac{(2)(19.44 \times 10^6)/120}{21.6 \times 10^6/288}$$

or

$$\psi = \frac{0.324 \times 10^6}{0.75 \times 10^5} = 4.32 \text{ at both ends of the column}$$

For interior columns there are two columns and two beams, so

$$\psi = \frac{0.324 \times 10^6}{1.5 \times 10^5} = 2.16 \text{ at both ends of the column}$$

Using the nomograph for columns braced against sway, Fig. 1-14, with (a) Exterior columns, $k = 0.92$, and (b) Interior columns, $k = 0.86$, the radius of gyration of the gross cross section is $0.3h = (0.3)(20) = 6.0$ in. Thus, $kl_u/r = (0.92)(120)/6 = 18.4$ for exterior columns, and $kl_u/r = (0.86)(120)/6 = 17.2$ for interior columns.

Since kl_u/r is less than $34 - 12M_1/M_2 = 22$ in both cases, the moments do not have to be magnified unless the action of the entire story requires moment magnification. Here, $M_1 = M_2$.

The critical load, P_c, must be computed for all columns and summed for story action. For exterior columns, unbraced frame, $k = 2.1$. $P_c = \pi^2 EI/(kl_u)^2$ where EI is obtained using eq. (1-26) considering β_d. Now, $\beta_d = M_{ud}/M_{ut}$ or

$$\beta_d = \frac{(140)(1.4)}{(140)(1.4) + (230)(1.7)} = 0.334$$

So

$$EI = \frac{19.44 \times 10^6}{1.334} = 14.59 \times 10^6 \text{ k-in.}^2$$

Then

$$P_c = \frac{(9.87)(14.59)(10^6)}{(2.1 \times 120)^2} = \frac{143.8 \times 10^6}{63.5 \times 10^3} = 2.265 \text{ kips}$$

$$\frac{P_u}{f_c' A_g} > 0.1, \text{ so } \phi = 0.7$$

For interior columns, unbraced frame, $k = 1.6$

$$\frac{P_u}{f_c' A_g} > 0.1, \text{ so } \phi = 0.7$$

$$P_c = \frac{(9.87)(14.59)(10^6)}{(1.6 \times 120)^2} = \frac{143.8 \times 10^6}{36.864 \times 10^3} = 3{,}900 \text{ kips}$$

It is now necessary to consider the translational action of the entire story. There are two exterior columns and two interior columns. Hence, for the entire story

$$\Sigma P_u = (2)(474) + (2)(640) = 2228 \text{ kips}$$

$$\Sigma \phi P_c = 2(0.7)(2265) + (2)(0.7)(3900) = 8631$$

For the entire story

$$\delta = \frac{1.0}{1 - \dfrac{2228}{8631}} = \frac{1.0}{1 - 0.258} = 1.35$$

The design procedure hereafter is the same for interior and exterior columns. Hence, the design of an interior column will serve to illustrate the procedure.

For an interior column, $P_u = 640$ kips, $M_u = 381$ ft-kips, and $\delta = 1.35$. Thus

$$M_c = \delta M_u = (1.35)(381) = 514 \text{ ft-kips}$$

$$\kappa = P_u / f_c' bh = \frac{640}{(4)(400)} = 0.40, \quad m = f_y/(0.85 f_c') = 17.65$$

$$\kappa e/h = \frac{12 M_c}{f_c' bh^2} = \frac{(12)(514)}{(4)(8000)} = 0.193$$

From the interaction chart for $\gamma = 0.8$, find $\rho_t m = 0.89$, so $\rho_t = 0.05$. Then $A_{st} = (0.05)(20)^2 = 20$ sq in.

Using radial splices*, sixteen No. 10 bars, A_{st} provided = 20.27 sq in. The design should be reviewed since ρ_t was assumed as 0.02.

1.12 BIAXIAL BENDING AND AXIAL LOAD

When bending about two axes occurs, the design process is similar to that for bending about one axis.

For circular columns the resultant moment is computed as $M_{u\theta} = \sqrt{(M_{ux})^2 + (M_{uy})^2}$ and the usual interaction charts are used since a circular column has the same capacity about all diameters. For other shapes, three-dimensional interaction must be considered. Figure 1-19 illustrates a typical three-dimensional interaction chart.

For a constant P_u, the plane $OABC$ represents the contour of M_u for bending about any axis. If the interaction diagrams for bending about the x-axis (FCD) and about the y-axis (FAE) are available, it is usually satisfactory to use linear interpolation between the two diagrams to design for bending about any intermediate axis.

For a given axial load P_u, a particular cross section will

*See Table, Reinforcement 3.1.1, *Ultimate Strength Design Handbook*, Vol. 2, SP-17-A, American Concrete Institute, Box 4754, Redford Station, Detroit, Michigan 48219.

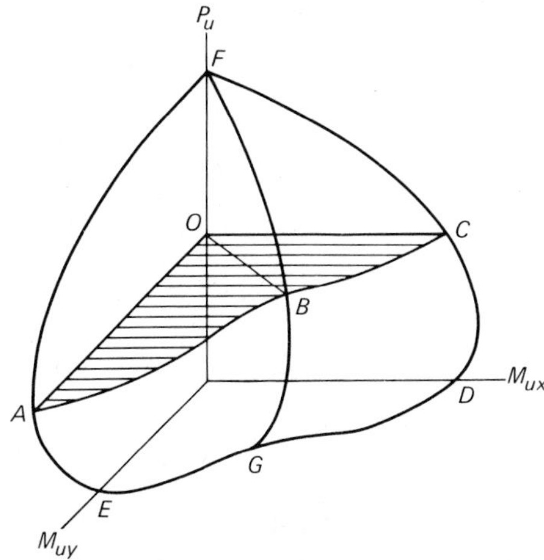

Fig. 1-19

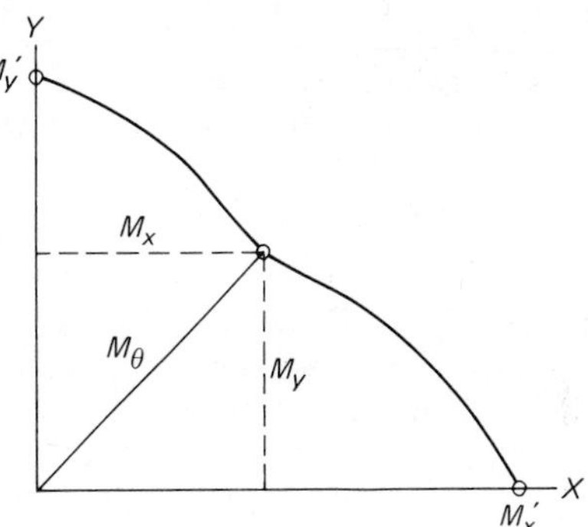

Fig. 1-20

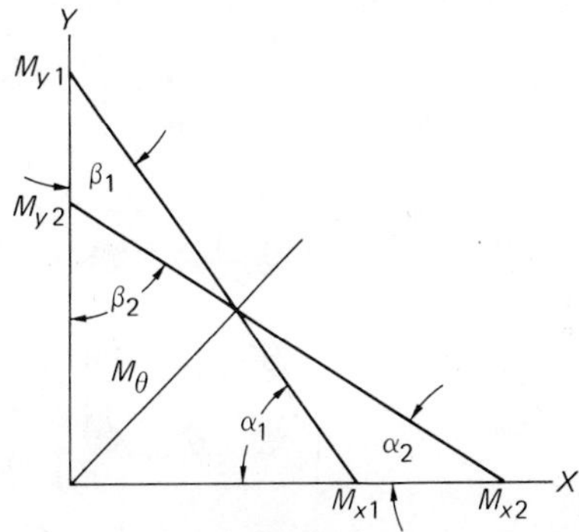

Fig. 1-21

provide a moment contour as shown in Fig. 1-20. In order to provide the required moment M_θ for biaxial bending, appropriate values of moment capacity M_x' and M_y' must be furnished. Many combinations of M_x' and M_y' will satisfy the situation. Assuming a linear moment variation in M from M_x' to M_y', two possible combinations that provide capacity M_θ are illustrated in Fig. 1-21.

The relationship between moments may be expressed in dimensionless form, as shown in Fig. 1-22, where

$$\frac{M_x}{M_x'} + \frac{M_y}{M_y'} = 1.0 \tag{1-35}$$

This expression can also be transformed to

$$M_x' = M_x + M_y(M_x'/M_y') = M_x + M_y \cot \alpha \tag{1-36}$$

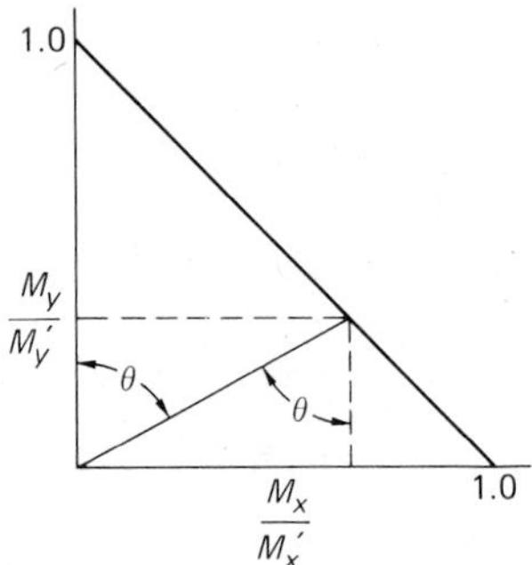

Fig. 1-22

and

$$M_y' = M_y + M_x(M_y'/M_x') = M_y + M_x \tan\alpha \quad (1\text{-}37)$$

Thus, many cross sectional capacities will satisfy the criteria. For example, if the cross section is square and symmetrically reinforced with steel along four faces, $M_x' = M_y'$, so, since $\tan\alpha = M_y'/M_x'$, $\alpha = 45$ degrees. Hence, $M_x' = M_x + M_y$ and $M_y' = M_x + M_y$.

In the case of a rectangular cross section having width (b) and depth (h), an appropriate estimate of M_y'/M_x' must be made. It is safe but uneconomical to design the cross section so that the capacity about the weak axis is $M_x + M_y$, since the capacity thus provided about the strong axis is several times $M_x + M_y$.

A reasonable procedure is to assume that $(M_x'/M_y') \approx (h/b)$. While this is not extremely accurate, it is more economical than designing for maximum capacity about the weak axis. The final design must satisfy eq. 1-34.

It should be noted that the moments are magnified for long column action separately about the x- and y-axes.

EXAMPLE 1-13: A square tied column is to be designed for $f_c' = 4000$ psi and $f_y = 60{,}000$ psi. Preliminary calculations indicate that kl/r is less than 22 so moment magnification is not required. The axial loads and moments for an exterior column are shown in Table 1-8. The column and beam cross sections are shown in Fig. 1-23.

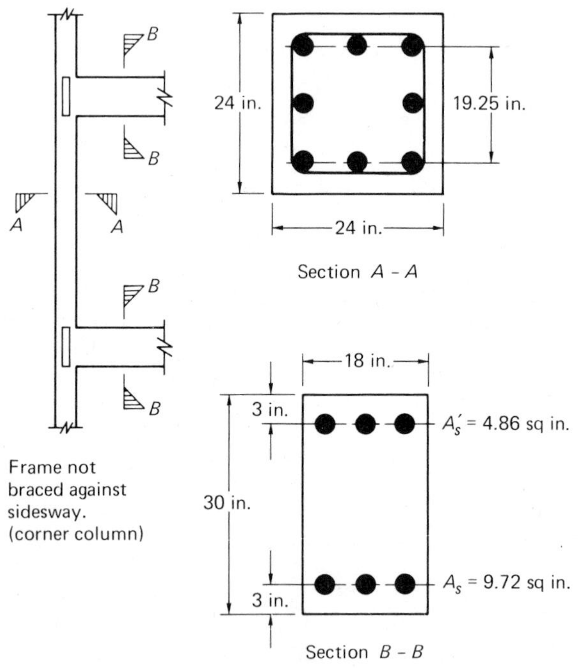

Fig. 1-23

TABLE 1-8 Moments and Forces at Service Loads

	Dead Load	*Live Load*
P	100 kips	140 kips
M_x	110 ft-kips	150 ft-kips
M_y	90 ft-kips	130 ft-kips
(Moments are identical at both ends of the columns)		

SOLUTION: $P_u = (1.4)(100) + (1.7)(140) = 378$ kips

$$M_{ux} = (1.4)(110) + (1.7)(150) = 409 \text{ ft-kips}$$

$$M_{uy} = (1.4)(90) + (1.7)(130) = 347 \text{ ft-kips}$$

$$\frac{P_u}{f_c'A_g} = \frac{378}{(4)(576)} = 0.164 > 0.1, \text{ so } \phi = 0.7 \text{ (tied column)}$$

Because of biaxial bending, the design moment about the x- or y-axis is $M_{ux} + M_{uy} = 409 + 347 = 756$ ft-kips. Hence,

$$\kappa e/h = \frac{M_u}{f_c'bh^2}$$

or

$$\kappa e/h = \frac{(756)(12000)}{(4000)(24)(576)} = 0.164$$

and

$$\kappa = \frac{P_u}{f_c'bh} = \frac{378{,}000}{(4000)(576)} = 0.164$$

$$\gamma = 19.25/24 = 0.8$$

Using the interaction diagram, Fig. 1-13c, $\rho_t m = 0.6$, and $m = f_y/0.85f_c' = 60/(0.85 \times 4) = 17.65$. Hence

$$\rho_t = \rho_t m/m = 0.6/17.65 = 0.034$$

so

$$A_{st} = (0.034)(24)(24) = 19.6 \text{ sq in.}$$

The steel is selected using Table Reinforcement 3.3.2, page 186, *Ultimate Strength Design Handbook Vol. II*, American Concrete Institute Publication SP-17A. Sixteen No. 10 bars are selected.

1.13 LATERAL REINFORCEMENT FOR COLUMNS

The foregoing problems concerning compression members dealt with longitudinal reinforcement only. The *ACI Code* requires the use of lateral reinforcement to contain the longitudinal reinforcement. Two types of lateral reinforcement are used: (1) individual ties, and (2) continuous, closely spaced lateral reinforcement.

Lateral ties normally consist of individual bars bent to conform with the shape of the column, spaced along the longitudinal bars to provide confinement. Fig. 1-24 illustrates a typical tie pattern for a rectangular column.

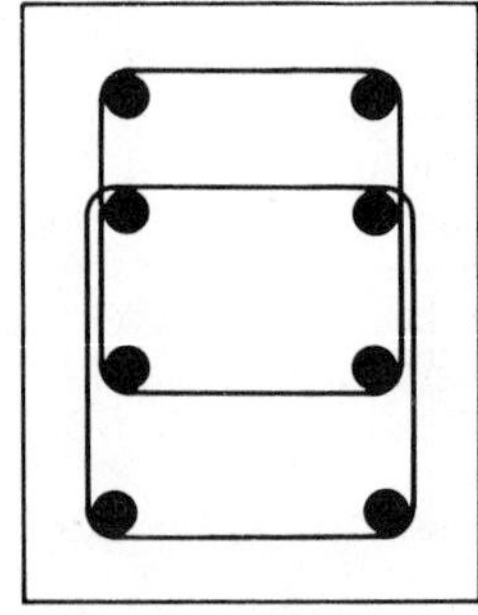

Fig. 1-24

Lateral ties for columns consist of #3 bars (or larger) for use with longitudinal bars #10 or smaller. For larger longitudinal bars, ties are #4 (or larger). This also applies to bundled longitudinal bars. The spacing of ties may not exceed 16 longitudinal bar diameters, 48 tie bar diameters, nor the least dimension of the column.

Continuous, closely spaced lateral reinforcement is normally used in the form of circular spirals. However, such reinforcement bent into rectangular form is used, particularly in earthquake resistant design. The *ACI Code* contains provisions concerning circular spirals, but does not include continuous bars bent in other shapes.

Spiral reinforcement consists of evenly spaced continuous hoops that provide confinement for the main reinforcement. While spirals are sometimes used for shear and torsional reinforcement and for confining splices, they are primarily used as lateral reinforcement in columns.

For cast-in-place construction, spirals must have a minimum diameter of 3/8 in. The clear spacing between spirals shall not exceed 3 in. or be less than 1 in. The clear spacing shall also be at least 1 1/3 times the maximum aggregate size.

The column core is that inner portion measured outside-to outside- of the spirals. The ratio of spiral reinforcement (ρ_s) is the ratio of the volume of spiral bar per turn to the volume of the core per turn. This ratio is computed as

$$\rho_s = \frac{4a_s(D_c - d_s)}{SD_c^2} \qquad (1\text{-}38)$$

The ratio ρ_s shall not be less than the value given by

$$\rho_s = 0.45(A_g/A_c - 1)f'_c/f_y \qquad (1\text{-}39)$$

The yield strength of the spiral reinforcement (f_y) may not be taken as greater than 60,000 psi.

EXAMPLE 1-14: A rectangular column 16 in. × 20 in. has eight No. 9 bars. Using No. 3 ties, design the tie system.

SOLUTION: Given (a) Least dimension of the column = 16 in., (b) 16 Longitudinal bar diameters = 16 × 1.128 = 18 in., (c) 48 tie diameters = 48 × 0.375 = 18 in.

Maximum spacing = 16 in. For eight bars use the tie pattern shown in Fig. 1-25. For other patterns see the detailing manual of the American Concrete Institute.

EXAMPLE 1-15: A round column, 20 in. diameter, 1 1/2 in. cover outside spiral, f'_c = 4000 spi, f_y = 60,000 psi, No. 3 spiral bars. Design the spiral. Maximum aggregate size is 1 in.

SOLUTION: The core diameter D_c is 20 − 2(1.5) = 17 in., d_s is 0.375 in., and a_s = 0.11 sq in. Using eq. (1-39) with

$$A_g = 0.7854(20)^2 = 314 \text{ sq in.}$$
$$A_c = 0.7854(17)^2 = 227 \text{ sq in.}$$

then

$$\rho_s = 0.45(314/227 - 1)(4/60)$$

or

$$\rho_s = 0.0115 \text{ minimum}$$

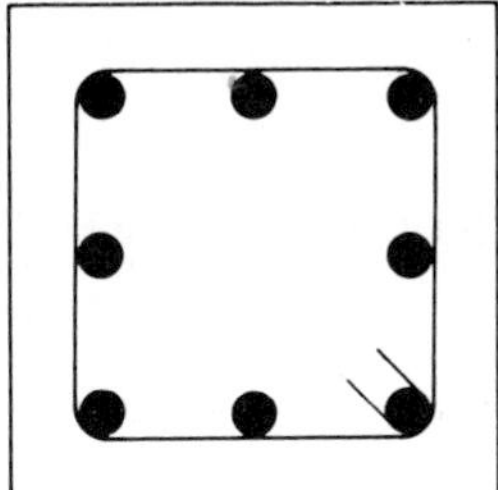

Fig. 1-25 Tie pattern for eight vertical bars.

If eq. (1-38) is rearranged to solve for the pitch (s), then

$$s = \frac{4a_s(D_c - d_s)}{\rho_s D_c^2}$$

and by substitution of appropriate values

$$s = \frac{(4)(0.11)(20 - 0.375)}{(0.0115)(17)^2} = 2.6 \text{ in.}$$

The spacing may not be less than 1 1/3 times the maximum aggregate size or 1.33 × 1.0 = 1.33 in., nor less than 1 in. nor more than 3 in. A clear spacing of 2.5 in. is practical, and satisfies all criteria.

Article 7.12 of the *1971 ACI Code* should be reviewed to determine splice and anchorage requirements for the spiral.

1.14 WALLS DESIGNED AS COLUMNS

Article 10.16 of the *1971 ACI Code* permits walls to be designed as columns, with certain limitations and exceptions as follows:

1. The minimum ratio of vertical reinforcement to gross concrete area shall be (a) 0.0012 for deformed bars not larger than #5 and with a specified yield strength of 60,000 psi or greater; or (b) 0.0015 for other deformed bars; or (c) 0.0012 for welded wire fabric not larger than 5/8 in. in diameter.
2. Vertical reinforcement shall be spaced not farther apart than three times the wall thickness nor 18 in.
3. Vertical reinforcement need not be provided with lateral ties if such reinforcement is 0.01 times the gross concrete area or less, or where such reinforcement is not required as compression reinforcement.
4. The minimum ratio of horizontal reinforcement to gross concrete area shall be (a) 0.0020 for deformed bars not larger than #5 and with a specified yield strength of 60,000 psi or greater; or (b) 0.0025 for other deformed bars; or (c) 0.0020 for welded wire fabric not larger than 5/8 in. in diameter.
5. Horizontal reinforcement shall be spaced not farther apart than one and one-half times the wall thickness nor 18 in.

The column design equations that were provided in the *1963 ACI Code* are known to be inaccurate for the usual steel ratios required for columns, since the steel on the compression side is frequently stressed below the yield strength. Since those equations assume that the compression steel has yielded, the solution is incorrect when moderately large steel ratios are used.

While it is always desirable to have interaction charts based on a more correct solution which accounts for statics and strain compatibility, the *1963 Code* equations are adequate for wall design since the steel ratio will be low. When the axial load is large in wall design, compression controls and the appropriate equation is

$$P_u = \phi\left[\frac{A'_s f_y}{\dfrac{e}{d - d'} + 0.5} + \frac{bhf'_c}{(3he/d^2) + 1.18}\right] \qquad (1\text{-}40)$$

If the bending moment is large, tension may control the design and the appropriate equation is

$$P_u = \phi(0.85f'_c bd\,\{\rho'm' - \rho m + (1 - e'/d) + \sqrt{(1 - e'/d)^2 + 2[(e'/d)(\rho m - \rho'm') + \rho'm'(1 - d'/d)]}\,\}) \qquad (1\text{-}41)$$

where $m = f_y/(0.85f'_c)$ and $m' = m - 1$. The eccentricity of the axial load, e, is measured from the centroid of the section, and the eccentricity, e', is measured from the centroid of the tension steel.

In almost all cases, eq. (1-40) will govern. However, the lesser P_u obtained using both equations governs the design. The equations provide satisfactory solutions for symmetrical steel along two faces, unsymmetrical steel or steel along one face only.

The interaction diagrams for tied columns with symmetrical steel along two faces provide precise results for that case, within interpolation accuracy.

EXAMPLE 1-16: Design a wall as a column using $f_c' = 4000$ psi and $f_y = 60{,}000$ psi. The wall is 12 ft high, 8 in. thick and supports the following forces and moments at service loads:

	P	*M*
Dead Load	11.0 kip/ft	4.0 ft-kip/ft
Live Load	15.0 k/ft	5.0 ft-k/ft

Concrete cover from the centerlines of reinforcing bars is 1 in.

SOLUTION: Since the moment is small, moment magnification is not important.

$$P_u = (1.4)(11) + (1.7)(15) = 40.9 \text{ kip/ft}$$

$$M_u = (1.4)(4) + (1.7)(5) = 14.1 \text{ ft-kip/ft}$$

$$e = 12M_u/P_u = (12)(14.1)/40.9 = 4.14 \text{ in.}$$

$$\gamma = 6/8 = 0.75, \quad \kappa = \frac{40.9}{(4)(12)(8)} = 0.1065$$

$$\kappa e/h = \frac{(14.1)(12)}{(4)(12)(64)} = 0.055$$

From Fig. 1-12c, for $\gamma = 0.8$, $\rho_t m = 0.04$, and from Fig. 1-12b, for $\gamma = 0.7$, $\rho_t m = 0.05$. Thus, for $\gamma = 0.75$, $\rho_t m = 0.045$. Since $m = f_y/0.85 f_c' = 60/3.4 = 17.65$, $\rho_t = \rho_t m/m = 0.05/17.65 = 0.0028$. For vertical bars not larger than No. 5, the minimum steel ratio is 0.0012, so the steel ratio $\rho_t = 0.0028$ will be used. Thus, $A_s = \rho_t bh = (0.0028)(12)(8) = 0.269$ sq in./ft. Since the steel is used on two faces, each face has $0.269/2 = 0.1345$ sq in.

A No. 3 bar has an area equal to 0.11 sq in., so the required spacing is $(12)(0.11)/0.1345 = 9.8$ in. For practical purposes, space bars at $9\frac{3}{4}$ in. center to center along each face vertically.

The steel ratio for horizontal steel is 0.002 for bars not larger than No. 5. Hence, $A_s = (0.002)(12)(8) = 0.192$ sq in./ft. Using No. 3 bars along two faces, the area per face is 0.095 sq in., and the spacing is $(12)(0.11)/0.095 = 13.89$ in.

The maximum spacing of horizontal reinforcement is the lesser of 18 in. or 1.5 times the wall thickness, or $(1.5)(8) = 12$ in. Thus, the No. 3 horizontal bars are spaced at 12 in. center to center.

1.15 SHEAR AND TORSION

Recognizing the need for shear reinforcement to insure ductile failure due to flexure, ACI Committee 318 requires minimum reinforcement for shear in most cases. Exempt are: (a) Slabs and footings; (b) concrete joist floor construction (thin slabs); (c) beams in which the total depth does not exceed 10 in., or 2.5 times the flange thickness, or ½ the web width, whichever is greater; and (d) such cases where v_u is less than $v_c/2$.

For the cases listed above, however, the torsional stress may not exceed $1.5\sqrt{f_c'}$. In other cases, the minimum shear reinforcement area must be at least

$$A_v = 50 b_w s/f_y \tag{1-42}$$

Since stirrup sizes are normally limited to bar sizes No. 3, No. 4 and No. 5, eq. (1-42) is best expressed in the form

$$s = \frac{A_v f_y}{50 b_w} \tag{1-43}$$

Equations (1-42) and (1-43) apply when the torsional shear stress does not exceed $1.5\sqrt{f_c'}$.

When the torsional shear stress v_{tu} exceeds $1.5\sqrt{f_c'}$ the shear reinforcement *must* consist of closed stirrups, and the spacing may not exceed

$$s = \frac{(A_v + 2A_t)f_y}{50 b_w} \tag{1-44}$$

A_t is defined later by eq. (1-58). The spacing, s, may not exceed 24 in.

However, eq. (1-44) permits greater spacing of stirrups than eq. (1-43). In the presence of significant torsion, eq. (1-43) is more demanding than eq. (1-44), so the former should be applied, but the stirrups should be closed when the torsional stress exceeds $1.5\sqrt{f_c'}$.

The nominal shear stress is computed as

$$v_u = \frac{V_u}{\phi bd} \tag{1-45}$$

where $\phi = 0.85$ for shear. Note that the understrength factor (ϕ) is applied when computing the shear stress. Hence, the ϕ factor is omitted when computing the permissible shear stress. The design shear force (V_u) is computed at a distance (d) from the face of the support, and the design shear stress is then computed using eq. (1-45).

The permissible shear stress (v_c) is computed by the simplified method as

$$v_c = 2\sqrt{f_c'} \tag{1-46}$$

which does not take into account the effects of the longitudinal reinforcement and the shear span $M_u/V_u d$. When those factors are considered, the permissible shear stress is

$$v_c = 1.9\sqrt{f_c'} + 2500 \rho_w V_u d/M_u \tag{1-47}$$

where $V_u d/M_u$ is never taken as greater than 1.0, and is always positive. The larger value of v_c obtained from eqs. (1-46) and (1-47) is used as the permissible shear stress.

For conditions other than short cantilevers and corbels, shear reinforcement required at a distance (d) from the face of the support is also used from that section to the face of the support, when the reaction is compressive. When the end reaction is tensile, the critical section for shear is taken at the face of the support and stirrups are spaced accordingly. This is implied by Section 11.2.2 of the *1971 ACI Code*.

All expressions involving $\sqrt{f_c'}$ relate to normal weight concrete. When lightweight aggregate concrete is used, the term $\sqrt{f_c'}$ may be replaced by $f_{ct}/6.7$, where f_{ct} is the splitting tensile strength of the concrete. Alternatively, $\sqrt{f_c'}$ may be replaced by $0.75\sqrt{f_c'}$ for lightweight aggregate concrete or $0.85\sqrt{f_c'}$ for sand-lightweight aggregate concrete.

Axial compression reduces diagonal tension stresses. Hence, in the presence of axial compression, eq. (1-47) may be modified so that

$$v_c = 1.9\sqrt{f_c'} + 2500 \rho_w V_u d/M_m \tag{1-48}$$

where

$$M_m = M_u - N_u(4h - d)/8 \tag{1-49}$$

or alternatively, v_c may be computed as

$$v_c = 2(1 + 0.0005 N_u/A_g)\sqrt{f_c'} \tag{1-50}$$

However, considering eqs. (1-49) and (1-50), v_c may not exceed

$$v_c = 3.5\sqrt{f_c'}\sqrt{1 + 0.002 N_u/A_g} \tag{1-51}$$

When the axial force N_u is a tension force, N_u/A_g is negative. Further, when N_u/A_g is used, the quantity is expressed in psi, since N_u is in pounds.

If torsional stresses exceed $1.5\sqrt{f_c'}$ the shear stress in the concrete is limited to

$$v_c = \frac{2\sqrt{f_c'}}{\sqrt{1 + (v_{tu}/1.2v_u)^2}} \tag{1-52}$$

In eq. (1-52), the modifications previously stated for lightweight and sand-lightweight aggregate concrete apply.

If the actual shear stress (v_u) computed using eq. (1-45) exceeds the permissible shear stress (v_c), shear reinforcement must be used. The added effectiveness of bent-up bars and inclined stirrups may be utilized according to the *ACI Code*. However, such refinements are beyond the scope of this chapter. Vertical shear reinforcement only is considered herein.

When the actual shear stress (v_u) exceeds the permissible shear stress (v_c), shear reinforcement must be provided. If $v_u' = v_u - v_c$, then the spacing of vertical stirrups is

$$s = \frac{A_v f_y}{v_u' b_w} \tag{1-53}$$

In all cases involving shear and torsion web reinforcement, f_y may not be considered as greater than 60,000 psi, regardless of the steel grade used.

In addition to the maximum spacing of shear reinforcement required by eq. (1-38), the maximum spacing may not exceed $d/2$ when v' does not exceed $4\sqrt{f_c'}$. When v' exceeds that quantity, the spacing may not exceed $d/4$.

The effects of torsion may be neglected when the torsional shear stress (v_{tu}) does not exceed $1.5\sqrt{f_c'}$.

The torsional shear stress is computed as

$$v_{tu} = \frac{3T_u}{\phi\Sigma(x^2 y)} \tag{1-54}$$

where x is the smaller dimension and y is the larger dimension of component rectangles of a flanged member. Flanges or other overhanging members are limited so that the effective width, y, does not exceed $3x$. (Overhang $\leqslant$ three times flange thickness).

Torsional stresses are computed beginning at a critical section which exists at a distance d from the face of the support. While the *Code* does not so state, when tension axial forces are present, it is advisable to compute the torsional stresses at the face of the support and provide torsional reinforcement accordingly.

The torsional stress taken by the concrete cross section (v_{tc}) shall not exceed

$$v_{tc} = \frac{2.4\sqrt{f_c'}}{\sqrt{1 + (1.2v_u/v_{tu})^2}} \tag{1-55}$$

If tension stresses are present with torsion, the permissible torsional stress in the concrete (v_{tc}) obtained using eq. (1-55) and the permissible shear stress in the concrete obtained using eq. (1-52) shall each be multiplied by the quantity

$$(1 + 0.002N_u/A_g) \tag{1-56}$$

where N_u is negative for tension, and is expressed in pounds.

The total torsional stress (v_{tut}) may not exceed

$$v_{tut} = \frac{12\sqrt{f_c'}}{\sqrt{1 + (1.2v_u/v_{tu})^2}} \tag{1-57}$$

When torsional web reinforcement is required, closed stirrups must be used. The stirrup area required for shear and torsion may be combined, providing that the stirrups are closed. Torsion reinforcement consists of closed stirrups and longitudinal reinforcement in addition to that required for flexure.

When torsion reinforcement is required, it is supplied in addition to that required to resist shear, flexure and axial load. The required area of closed stirrups is

$$A_t = \frac{(v_{tu} - v_{tc})s\,\Sigma x^2 y}{3\alpha_t x_1 y_1 f_y} \tag{1-58}$$

where

$$\alpha_t = 0.66 + 0.33y_1/x_1 \leqslant 1.5 \tag{1-59}$$

The spacing (s) may not exceed

$$(x_1 + y_1)/4 \tag{1-60}$$

nor 12 in., whichever is the smaller.

It is important to note that A_t represents *one leg* of a closed stirrup, in comparison to the total number of legs referenced by A_v for shear reinforcement.

In addition to the stirrups, longitudinal steel must be provided for torsion in the amount.

$$A_l = 2A_t(x_1 + y_1)/s \tag{1-61}$$

or

$$A_l = \frac{400xs}{f_y}\,\frac{v_{tu}}{v_{tu} + v_u} - 2A_t\frac{x_1 + y_1}{s} \tag{1-62}$$

whichever is the greater.

However, $50b_w s/f_y$ may be substituted for $2A_t$ in eqs. (1-61) and (1-62).

Longitudinal bars must be at least No. 3 in size, distributed around the perimeter of the stirrups, spacing shall not exceed 12 in., and at least one bar must be placed in each corner of the stirrups.

EXAMPLE 1-17: A beam has a clear span of 28 ft, supports a service live load of 2.4 kips/ft and a service dead load of 2.0 kips/ft, including its own weight. The beam has dimensions b = 12 in., d = 24 in. The beam is simply supported. The supports are 12 in. wide. Design the beam for shear if f_c' = 4000 psi and f_y = 60,000 psi. The overall depth, h, is 26.5 in.

SOLUTION: $w_u = (1.4)(2.0) + (1.7)(2.4) = 6.88$ k/ft.

The reactions at the centers of supports are

$$R = (6.88)(28 + 1)/2 = 99.76 \text{ kips}$$

At the critical section, the shear force is

$$V_u = 99.76 - 6.88(0.5 + 2.0) = 82.55 \text{ kips}$$

The actual shear stress is

$$v_u = \frac{82{,}550}{(0.85)(12)(24)} = 337 \text{ psi}$$

The permissible shear stress taken by the concrete is

$$v_c = 2\sqrt{4000} = 126.5 \text{ psi} < 337 \text{ psi}$$

so stirrups are required. Since $v_u - v_c = 337 - 126.5 = 210.5$ psi which is less than $4\sqrt{f_c'}$ or 235 psi, the maximum spacing of the stirrups at the critical section is $d/2$ or 12 in.

It is convenient to draw a shear stress diagram as shown in Fig. 1.26, and to label the diagram with key values. The distance over which stirrups are theoretically required from the critical section is 12.0 − 12(126.5/337) = 7.5 ft, but stirrups are required by the Code to the point at which $v_u = v_c/2$ or 9.75 ft from the critical section.

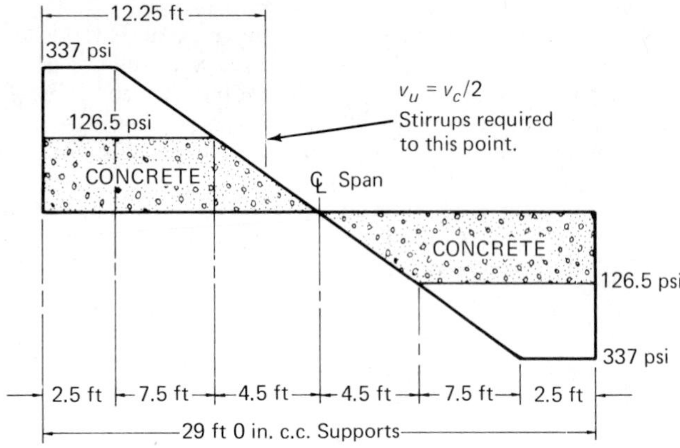

Fig. 1-26

If No. 4 stirrups with two vertical legs are used, $A_v = 0.4$ sq in. At the critical section, the spacing is

$$s = \frac{Afy}{(v_u - v_d)b} = \frac{(0.4)(60{,}000)}{(210.5)(12)} = 9.5 \text{ in.}$$

and the maximum spacing is

$$\frac{A_v f_y}{50b} = \frac{(0.4)(60{,}000)}{(50)(12)} = 40 \text{ in.}$$

If No. 3 stirrups are used, $A_v = 0.22$ and the spacing at the critical section is $(0.22)(9.5)/0.4 = 5.2$ in.

The maximum spacing is

$$s = \frac{A_v f_y}{50b} = \frac{(0.22)(60{,}000)}{(50)(12)} = 22 \text{ in.}$$

Using No. 3 stirrups, the 22 in. maximum spacing is less than $d/2 = 12$ in., so the maximum spacing is 12 in.

In order to simplify stirrup spacing, a table can be developed. Noting that $W_u = 6.88$ kip/ft, the shear stress decreases by

$$\frac{6880}{(0.85)(12)(24)} = 28.1 \text{ psi/ft}$$

In Table 1-9, the distance is measured from the critical section toward the center of the span. At any location, the theoretical spacing of No. 3 stirrups is

$$s = \frac{(0.22)(60{,}000)}{(v_u - v_c)(12)} = \frac{1100}{v_u - v_c}$$

The table is used as a guide to select practical spacing, beginning at the face of the support. An example of spacing is

seven spaces @5″ = 35″ = 2′ - 11″
two spaces @6″ = 12″ = 1′ - 0″
three spaces @8″ = 24″ = 2′ - 0″
5′ - 11″

TABLE 1-9

Distance, ft	$v_u - v_c$ *psi*	*Theoretical s, in.*
0	210.5	5.2
1	182.4	6.0
2	154.3	7.1
3	126.2	8.7
4	98.1	11.2
5	70.0	15.7
6	41.9	26.3
7	13.8	79.7
7.5	0	∞

with the first stirrup 1 in. from the face of the support, a total of 6.0 ft has been taken care of. The remaining 1.5 ft over which stirrups are theoretically required, plus stirrups over an additional distance 2.25 ft is cared for by spacing stirrups at 12 in. on centers. Thus, the stirrups extend over a distance of 11.5 ft from the face of the support.

EXAMPLE 1-18: Determine the stirrup spacing at the critical section for Example 1-17 if a compressive axial load, N, is applied at the ends of the span. For dead load, N is 100 kips and for live load, N is 120 kips.

SOLUTION: $N_u = (1.4)(100) + (1.7)(120) = 344$ kips.

The permissible shear stress, v_c, is $2(1 + 0.0005N_u/A_g)\sqrt{f_c'}$, since $N_u/A_g = 344{,}000/288 = 1194$ psi, and $0.0005N_u/A_g = 0.597$, then

$$v_c = 2(1.597)\sqrt{f_c'} = 202 \text{ psi}$$

The shear stress v_c may not exceed

$$v_c = 3.5\sqrt{f_c'}\sqrt{1 + 0.002N_u/A_g}$$

Since $0.002N_u/A_g = (0.002)(1194) = 2.388$ and $\sqrt{3.388} = 1.55$, then v_c may not exceed

$$v_c = (3.5\sqrt{4000})(1.55) = 343 \text{ psi}$$

which is greater than 202 psi. Then, $v_u - v_c = 337 - 202 = 135$ psi and the spacing at the critical section is

$$s = \frac{(0.22)(60{,}000)}{(135)(12)} = 8.1 \text{ in.}$$

The spacing throughout the span is then adjusted to reflect the permissible increase in shearing stress due to a compressive axial load.

If the axial load had been tension, the permissible shearing stress v_c would decrease, and the stirrup spacing would decrease.

EXAMPLE 1-19: The beam of Example 1-17 is subjected to torsion. At service loads, $T = 9.0$ ft-kips for dead load and 8.0 ft-kips for live load. Design the shear and torsion reinforcement.

SOLUTION: At ultimate load

$$T_u = (1.4)(9) + (1.7)(8) = 26.2 \text{ ft-kips}$$

The torsional stress is

$$v_{tu} = \frac{3T_u}{\phi\Sigma x^2 y}$$

Now,

$$\Sigma x^2 y = (12)^2(26.5) = 3816 \text{ in.}^3$$

and

$$v_{tu} = \frac{(3)(26.2)(12)(1000)}{(0.85)(3816)} = 291 \text{ psi}$$

If v_{tu} does not exceed $1.5\sqrt{f_c'}$ it may be disregarded, but $1.5\sqrt{f_c'} = 95$ psi. Since $291 > 95$, torsion must be considered.

The nominal torsion stress taken by the concrete may not exceed

$$v_{tc} = \frac{2.4\sqrt{f_c'}}{\sqrt{1 + (1.2v_u/v_{tu})^2}}$$

From ex. 1-17, $v_u = 337$ psi, so

$$v_u/v_{tu} = 337/291 = 1.158$$

and

$$\sqrt{1 + [1.2(1.158)]^2} = \sqrt{2.93} = 1.71$$

Thus

$$v_{tc} = (2.4)(63.25)/1.71 = 89 \text{ psi}$$

The total torsional stress may not exceed that given by eq. (1-57), which computes to be 445 psi, and is greater than the actual v_{tu},

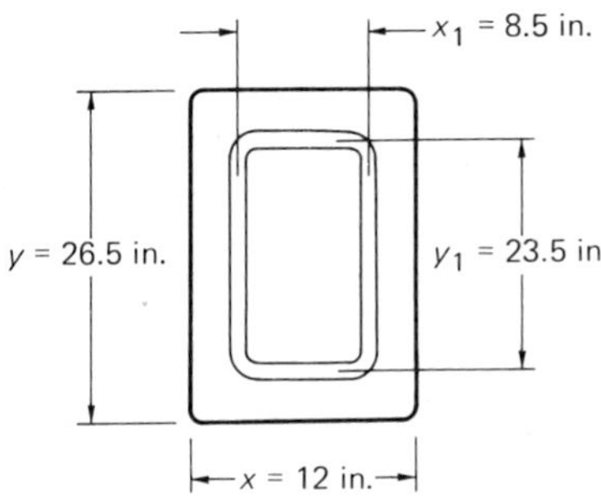

Fig. 1-27

291 psi. Thus, it is not necessary to increase the dimensions of the cross section.

Since the actual torsional shear stress is greater than that permitted on the concrete, torsion reinforcement is required.

The required area of one leg of a closed vertical stirrup is

$$A_t = \frac{(v_{tu} - v_{tc})s\Sigma x^2 y}{3\alpha_t x_1 y_1 f_y}$$

where $\alpha_t = 0.66 + 0.33\,(y_1/x_1)$ but not more than 1.5.

The center to center distances x_1 and y_1 are shown in Fig. 1-27. Thus, $\alpha_t = 0.66 + 0.33(23.5/8.5) = 1.57 > 1.5$, use 1.5. But in the presence of torsional stress exceeding $1.5\sqrt{f_c'}$, the permissible shear stress is decreased to

$$v_c = \frac{2\sqrt{f_c'}}{\sqrt{1 + (v_{tu}/1.2v_u)^2}}$$

Now,

$$v_{tu}/1.2v_u = 291/(1.2 \times 337) = 0.72$$

so,

$$v_c = \frac{126.5}{\sqrt{1 + (0.72)^2}} = \frac{126.5}{1.23} = 103 \text{ psi}$$

Without a torsional stress exceeding $1.5\sqrt{f_c'}$, v_c was 126.5 psi. Hence, the stirrup spacings found for shear only in Example 1-17 are no longer valid, since $v_u - v_c = 337 - 103 = 234$ psi. The theoretical spacing of No. 3 stirrups at the face of the support decreases to $(5.2)(210.5)/234 = 4.68$ in., and for No. 4 stirrups decreases to $(9.5)(210.5)/234 = 8.55$ in. Considering the 4.68 in. spacing for No. 3 bars, the area of one leg of a torsion stirrup is

$$A_t = \frac{(291 - 89)(4.68)(3816)}{(3)(1.5)(8.5)(23.5)(60{,}000)} = 0.067 \text{ sq in.}$$

For shear, $A_v/s = 0.22/4.68 = 0.047$ for 2 legs. For torsion, $A_t/s = 0.067/4.68 = 0.0143$ for 1 leg. $(A_v + 2A_t)/s = 0.047 + 2(0.0143) = 0.0756$ for 2 legs. Using No. 4 closed stirrups, $(A_v + 2A_t) = 0.4$ sq in. Hence, $s = 0.4/0.0756 = 5.29$ in. maximum. From the face of the support to the critical section, a spacing of 5.0 in. may be used, since the maximum spacing of torsion stirrups is $(x_1 + y_1)/4 = (8.5 + 23.5)/4 = 8.0$, which is less than the alternate maximum, 12 in.

Additional longitudinal steel is required for torsion. Equations (1-61) and (1-62) apply. Thus, since $A_t = 0.067$ sq in., and $s = 5.29$ in. theoretically,

$$A_l = (2)(0.067)(8.5 + 23.5)/5.29 = 0.808 \text{ sq in.}$$

or

$$A_l = \frac{(400)(12)(5.29)}{60{,}000}\left[\frac{291}{291 + 337} - (2)(0.067)\frac{(8.5 + 23.5)}{4.5}\right]$$
$$= 0.27 \text{ sq in.}$$

If this value governed, it could be reduced by using $50bs/f_y$ for $2A_t$ in the equation. The larger value governs, so $A_l = 0.808$ sq in. This may be added to the longitudinal steel areas required for flexure, and the steel selected accordingly. Longitudinal bars for torsion must be at least No. 3 bars spaced around the perimeter of the cross section at spaces not exceeding 12 in. Hence, two No. 4 bars are placed at the upper corners and at midheight of the beam, providing an area $(4)(0.2) = 0.8$ sq in. The remaining 0.08 sq in. can be added to the tension steel area. However, since the two upper layers of torsion reinforcement each contain 0.4 sq in., it would be desirable to maintain symmetry of torsion steel and add 0.4 sq in. to the tension steel area required. This additional area does not increase the tension steel ratio for flexure, since a portion is used only for torsion. Further, the top layer of torsion steel cannot be considered as compression reinforcement.

EXAMPLE 1-20: A T-beam has dimensions shown in Fig. 1-28. Compute the value of $\Sigma x^2 y$ for determining torsional stresses.

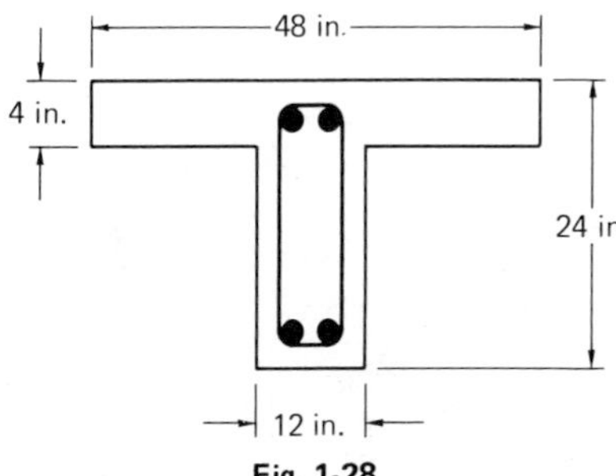

Fig. 1-28

SOLUTION: Since only the web is reinforced with torsion reinforcement, in separating the beam into rectangles, the web is considered to extend through the flange, as shown in Fig. 1-29. Further, the overhanging flange width is limited to three times the flange thickness. Thus, $\Sigma x^2 y = (12)^2(24) + (2)(4)^2(12) = 3456 + 384 = 3840$ in.^{3}

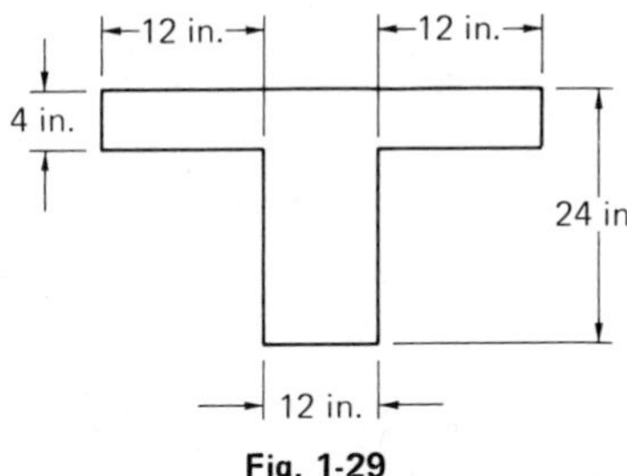

Fig. 1-29

EXAMPLE 1-21: A special T-beam is illustrated in Fig. 1-30. Compute $\Sigma x^2 y$ for torsional stress computations.

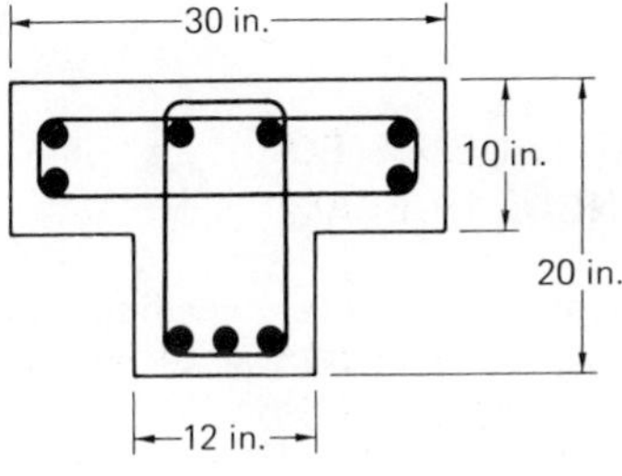

Fig. 1-30

SOLUTION: Since web reinforcement is provided in both, the web and the flange, two arrangements of component rectangles can be considered, as shown in Fig. 1-31.

For (a)

$$\Sigma x^2 y = (12)^2(20) + (2)(9)^2(10) = 4500 \text{ in.}^3$$

and for (b)

$$\Sigma x^2 y = (10)^2(30) + (12)^2(10) = 4440 \text{ in.}^3$$

The larger value may be used.

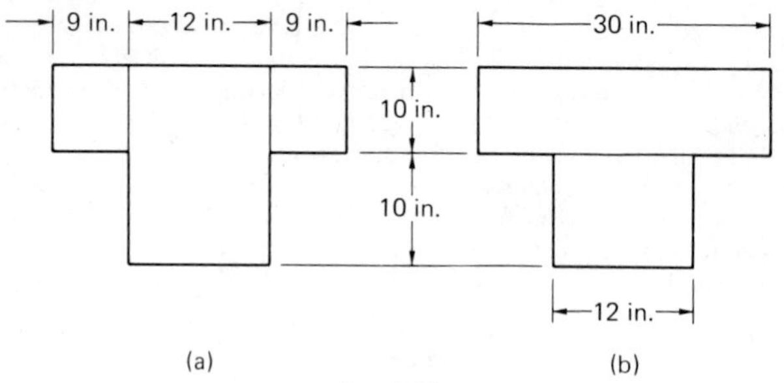

9 in. 12 in. 9 in. 10 in. 10 in. 30 in. 12 in.

(a) (b)

Fig. 1-31

EXAMPLE 1-22: Figure 1-32 shows a box section. Compute $\Sigma x^2 y$ for torsion.

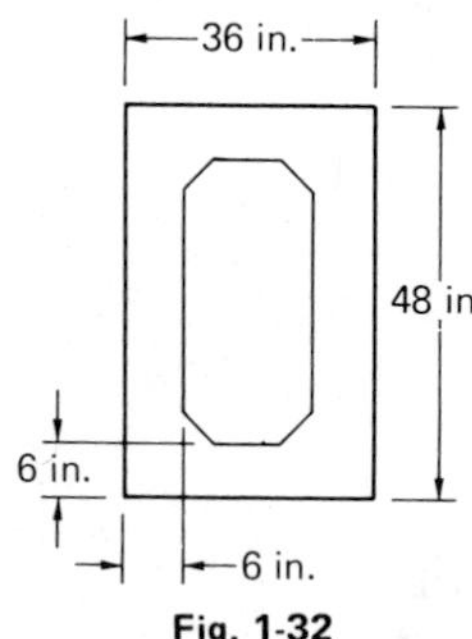

Fig. 1-32

SOLUTION: The larger dimension, 48 in., is y and the smaller dimension, 36 in., is x. If the wall thickness h is at least $x/4 = 9$ in., the section may be taken as a solid rectangle 36 in. × 48 in. However, $h = 6$ in. $< x/4$, so further considerations are necessary. Since $h = 6$ in. and is greater than $x/10 = 3.6$ in., $\Sigma x^2 y$ is that for the solid section multiplied by $4h/x$. Thus

$$\Sigma x^2 y = (36)^2 (48)(4)(6)/36 = 41{,}472 \text{ in.}^3$$

If the wall thickness is less than $x/10$ (or 3.6 in. here), torsional buckling must be considered. Such a consideration is beyond the scope of this chapter.

In general, it is good practice to make the wall thickness at least 1/10 of the height or width of the segment. In this case, the top and bottom should not be less than 3.6 in. thick, and the vertical walls not less than 4.8 in. thick.

In all cases, fillets must be used at the interior corners of hollow sections. The *Code* does not specify fillet sizes, but it is not unreasonable to have the side dimensions of the fillet at least equal to one-half of the wall thickness.

1.16 BIAXIAL BENDING FOR RECTANGULAR BEAMS

Occasionally a problem involves bending about two axes. In such a case a general solution must be used, and this involves successive trials. For other than rectangular beams, the solution is quite complex. However, illustration of the procedure for rectangular beams will provide insight to the designer for solving more complex cases.

Figure 1-33 shows two possibilities for a rectangular beam. The concrete area subjected to the compression block may be triangular or trapezoidal.

For Case (a), there are three equations of statics and one of geometry. Only the statics equations are independent. The relationships are

$$A_c = 0.5 b_1 b_2 \tag{1-63}$$

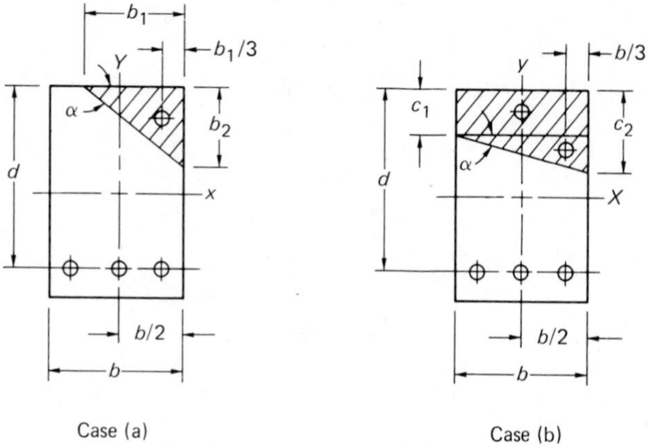

Fig. 1-33

$$0.85 f_c' A_c - A_s f_y = P_u/\phi = 0 \tag{1-64}$$

$$M_{uy}/\phi = A_s f_y (b/2 - b_1/3) \tag{1-65}$$

$$M_{ux}/\phi = A_s f_y (d - b_2/3) \tag{1-66}$$

A cross section must be assumed, as well as A_s. Using eq. (1-64), solve for A_c. Then, using eq. (1-65), find b_1 and using eq. (1-66), find b_2. The solution is checked using eq. (1-63). If the final A_c is less than the first value, the cross section is adequate, and may be excessive. This will be due to slightly greater steel area than necessary to resist the moments. Adjustments can be made to obtain a more economical section if desired or possible.

If b_1 exceeds b, the solution is not valid and Case (b) governs.

For Case (b), there are also four equations available:

$$A_c = (c_1 + c_2)(b/2) \tag{1-67}$$

$$0.85 f_c' A_c - A_s f_y = P_u/\phi = 0 \tag{1-68}$$

$$M_{uy}/\phi = 0.85 f_c' b (c_2 - c_1)(0.5)(b/2 - b/3) \tag{1-69}$$

$$M_{ux}/\phi = 0.85 f_c' c_1 b (d - c_1/2) + 0.85 f_c' (c_2 - c_1)(b/2)[d - c_1 - (c_2 - c_1)/3] \tag{1-70}$$

The solution is developed in the same manner as that outlined for Case (a).

For both cases, it is necessary to investigate strain compatibility to find the stresses in the steel. The stress in all bars must be f_y. Otherwise, the member is not ductile. In order to investigate strain compatibility it is necessary to know the angle of tilt of the neutral axis. This is determined as follows:

For Case (a)

$$\tan \alpha = b_2/b_1 \tag{1-71}$$

and for Case (b)

$$\tan \alpha = (c_2 - c_1)/b \tag{1-72}$$

The angle α is used with Fig. (1-34) where, for Case (a),

$$a = b_2 \cos \alpha \tag{1-73}$$

and for Case (b)

$$a = c_2 \cos \alpha \tag{1-74}$$

For both cases,

$$c = a/\beta_1 \tag{1-75}$$

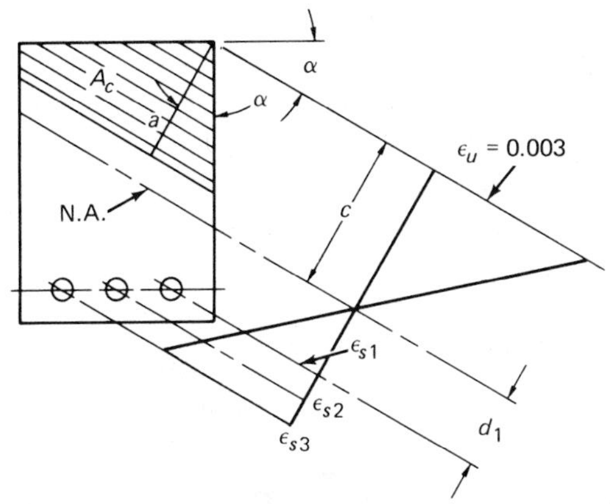

Fig. 1-34

The critical strain is ϵ_{s1}, and f_{s1} is obtained as

$$f_{s1} = 87{,}000 d_1/c \leqslant f_y \tag{1-76}$$

If an axial load is present, it is assumed acting at the centroid of the concrete. M_{uy}/ϕ does not change, but M_{ux}/ϕ includes the term $P_u/\phi(d - h/2)$, where h is the overall height of the beam. Also, for the forces

$$P_u/\phi = 0.85 f_c' A_c - A_s f_y \tag{1-77}$$

Thus, A_c increases when P_u is compressive and decreases when P_u is tensile or negative.

The example problem will illustrate the procedure.

EXAMPLE 1-23: A beam has properties $b = 10$ in., $d = 20$ in., $f_c' = 4000$ psi, $f_y = 60{,}000$ psi, and is reinforced with two No. 8 bars. $M_{ux} = 128$ ft-kip and $M_{uy} = 15$ ft-kip. Investigate the adequacy of the beam.

SOLUTION: Using the equations previously given, assuming that Case (a) governs, then

$$A_s = 1.58 \text{ sq in. for two No. 8 bars}$$

$$(0.85)(4.0)A_c - (1.58)(60) = 0 \qquad \text{eq. (1-64)}$$

$$A_c = 95/3.4 = 28 \text{ sq in.}$$

$$(15)(12)/0.9 = (1.58)(60)(5 - b_1/3) \qquad \text{eq. (1-65)}$$

$$200 = 95(5) - 95b_1/3$$

$$b_1 = 3(495 - 200)/95 = 9.3 \text{ in.}$$

Since $b_1 < b$, Case (a) governs, and

$$(128)(12)/0.9 = (1.58)(60)(20 - b_2/3)$$

$$1710 = 95(20) - 95b_2/3 \qquad \text{eq. (1-66)}$$

$$b_2 = 3(1900 - 1710)/95 = 6.0 \text{ in.}$$

Recheck:

$$A_c = (0.5)(9.3)(6.0) = 27.9 \text{ sq in.} \qquad \text{eq. (1-63)}$$

This is very close to the value $A_c = 28$ in., obtained by summing forces to zero.

The steel stresses must now be checked to insure ductility. Figure 1-35 illustrates the conditions.

$$\tan\alpha = 6.0/9.3 = 0.645, \alpha = 32.8 \text{ deg}$$

$$\cos\alpha = 0.8406$$

$$a = (6)(0.8406) = 5.04 \text{ in.}$$

$$c = 5.04/0.85 = 5.92 \text{ in.}$$

$$f_{s1} = 87000(11.98/5.92) = 176{,}000 > 60{,}000$$

Therefore, $f_{s1} = f_y$ and the beam is satisfactory.

It is necessary to determine whether or not the steel ratio exceeds $0.75\rho_b$. For $f_c' = 4000$ psi and $f_y = 60{,}000$ psi, $0.75\rho_b = 0.0213$. Here, $\rho = 1.58/200 = 0.0079$, so the section is satisfactory.

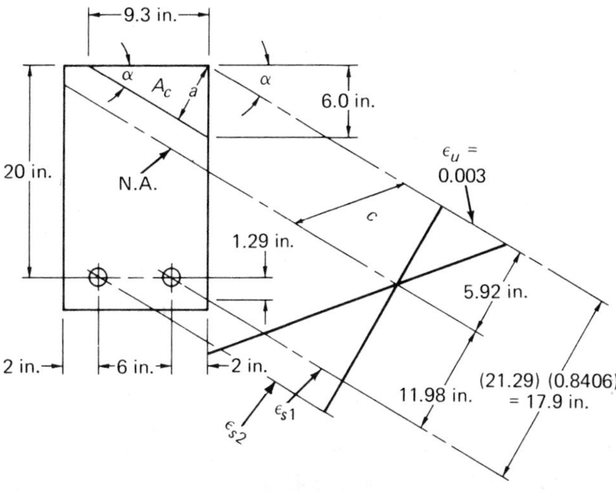

Fig. 1-35

Another necessary check is to determine whether or not ρ is less than the minimum ratio, $\rho = 200/f_y$, or $200/60{,}000 = 0.0033$. The steel ratio furnished is satisfactory.

EXAMPLE 1-24: All values are the same as in the previous example, except $M_{ux} = 232$ ft-kip and $M_{uy} = 9.7$ ft-kip, and $A_s = 3.0$ sq in. for three No. 9 bars. Investigate the cross section.

SOLUTION: Using equations previously stated,

$$(0.85)(4)A_c - (3)(60) = 0$$

$$A_c = \frac{180}{3.4} = 52.94 \text{ sq in.} \qquad \text{eq. (1-64)}$$

$$(12)(9.7)/0.9 = (3)(60)(5 - b_1/3)$$

$$129.3 = (180)(5) - 60b_1 \qquad \text{eq. (1-65)}$$

$$b_1 = (900 - 129.3)/60 = 12.85 \text{ in.}$$

This is greater than $b = 10$ in., so Case (b) applies. The concrete area is correct, since eqs. (1-64) and (1-68) are identical.

$$(12)(9.7)/0.9 = (0.85)(4)(10)(c_2 - c_1)(0.5)(5 - 3.33) \qquad \text{eq. (1-69)}$$

$$129.3 = 28.39(c_2 - c_1)$$

$$(c_2 - c_1) = 129.3/28.39 = 4.55 \text{ in.}$$

$$(232)(12)/0.9 = (0.85)(4)(c_1)(10)(20 - c_1/2) + (0.85)(4)(4.55)(5)(20 - c_1 - 4.55/3) \qquad \text{eq. (1-70)}$$

This reduces to the quadratic equation

$$c_1^2 - (35.45)c_1 + 97.84 = 0$$

and the solution is $c_1 = 3.0$ or 32.43 in. The second value exceeds d, and is extraneous.

Since $c_1 = 3$ and $(c_2 - c_1) = 4.55$, then $c_2 = 7.55$ in. Checking A_c by eq. (1-67)

$$A_c = (3.0 + 7.55)(10/2) = 52.75 \text{ sq in.}$$

This is slightly less than the concrete area corresponding to $A_s f_y$, indicating that A_s is slightly large. An A_s of 2.99 sq in. satisfies $A_c = 52.75$ sq in. However, in selecting practical bar sizes, 3.0 sq in. is selected.

It is now necessary to investigate the steel stresses. Figure 1-36 illustrates the conditions.

$$\tan\alpha = (7.55 - 3)/10 = 0.455 \quad \alpha = 24.47 \text{ deg}$$

$$\cos\alpha = 0.9102$$

$$a = (7.55)(0.9102) = 6.87 \text{ in.}$$

$$c = a/\beta_1 = 6.87/0.85 = 8.08 \text{ in.}$$

$$f_{s1} = (10.95/8.08)(87000) = 117{,}900 \text{ psi} > 60{,}000 \text{ psi}, f_{s1} = f_y$$

The beam is ductile.

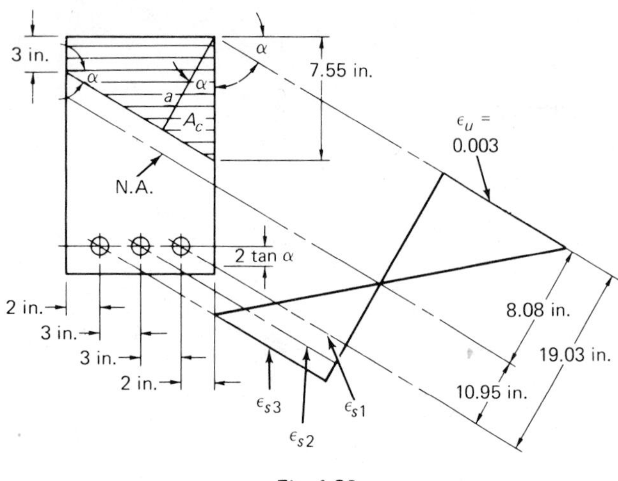

Fig. 1-36

1.17 DEVELOPMENT LENGTH AND ANCHORAGE OF REINFORCEMENT

The concept of development length and anchorage of reinforcement replaces the former practice of satisfying permissible bond stresses. There is little difference in the two concepts, because the new process was devised by utilizing the old process, and rearranging the variables. The present method, however, is more direct since the final product is stated in terms of the additional length of a bar of a given diameter that must be provided beyond a given critical section.

Critical sections occur generally at the face of a support, at a point of maximum stress and at points where bars may theoretically be terminated. In addition, points of contraflexure in continuous beams provide critical sections with regard to development length. The force in any reinforcing bar must be transmitted to the concrete by bond stress before the bar is terminated. Different criteria govern for bars in tension and bars in compression.

Splices are used to transfer forces from one bar to another by lapping or by direct compression. In general, for splices other than those in compression, bond stress plays an important part in transmitting the force from one bar to another via the concrete. Hence, splice lengths are closely related to development lengths.

The *1971 ACI Code* requires that the force in any bar must be developed on each side of any cross section by an embedment length or anchorage or a combination thereof.

When a bar is no longer required to resist flexure, it may be bent into the compression zone and made continuous with the compression reinforcement. Or it may be terminated at a distance, d, or 12 bar diameters (whichever is the greater) beyond the point at which it is no longer needed.

However, at the point of proposed termination, if the remaining reinforcement is in tension, one of the following conditions must be satisfied:

(a) The shear at the cutoff point may not exceed two-thirds of the total shear capacity at the cross section, including the shear strength of web reinforcement.

(b) Stirrup area in excess of that required for shear and torsion is provided along each terminated bar from the point of cutoff a distance $^3/_4\ d$. The additional stirrups shall be designed to provide a value $A_v b_w / f_y \geqslant 60$ psi. The resulting spacing shall not exceed $d/8\beta_b$, where β_b is the ratio of areas of discontinued reinforcement to that of the total reinforcement before cutoff.

(c) For No. 11 and smaller bars, the continuing bars provide twice the area of steel required at the point of cutoff, and the shear may not exceed three-fourths of the total shear capacity of the cross section, including that of web reinforcement.

The most simple method of satisfying the foregoing criteria is to insure that the cross sections along the extended length beyond theoretical cutoff points have shear capacity 1.5 times that required.

Although reinforcement may be theoretically discontinued at certain points in a flexural member, at least one-third of the positive moment reinforcement and at least one-fourth in continuous members shall extend along the same face into the support. For beams, that reinforcement shall extend into the support at least 6 in.

Flexural reinforcement in primary supporting members shall extend into the support and must be anchored so as to develop its yield strength at the face of the support.

Special conditions exist at simple supports and at points of contraflexure. Bar sizes are limited so that the development length satisfies the condition

$$l_d \leqslant M_t / V_u + l_a \qquad (1\text{-}78)$$

where M_t is the total capacity of the cross section assuming all reinforcement to be stressed to f_y, and V_u is the maximum shear force applied.

At simple supports, l_a is the sum of the embedment length beyond the center of the support and the equivalent length of hooks or other mechanical anchorage devices.

At points of contraflexure, l_a is limited to the effective depth (d) of the member or 12 d_b, whichever is greater.

If the ends of the reinforcement are confined by a compressive reaction, M_t/V_u in eq. (1-78) may be increased by 30%. The expression is then

$$l_d \leqslant 1.3 M_t / V_u + l_a \qquad (1\text{-}79)$$

Tension reinforcement in a continuous, restrained, or cantilever member or in any member of a rigid frame, must be anchored in or through the supporting member by embedment length, hooks or mechanical anchorage.

Negative moment reinforcement must have an embedment length into the span for a distance, d, or 12 bar diameters, whichever is greater. At least one-third of the total reinforcement provided for negative moment at the support must extend beyond the point of contraflexure an embedment length not less than d, 12 bar diameters or $^1/_{16}$ the clear span, whichever is greater.

In brackets, deep beams, stepped, sloped or tapered footings and other special members where reinforcement stress is not directly proportional to moment or the reinforcement is not parallel to the compression face, special effort must be made to insure adequate end anchorage.

The development length l_d, in inches, of deformed bars and deformed wire in tension shall be computed as the product of the basic development length of (a) and the applicable modification factor or factors of (b), (c), and (d), but l_d shall be not less than 12 in.

(a) The basic development length shall be as follows:

For #11 or smaller bars	$0.04 A_b f_y / \sqrt{f'_c}$*	(1-80)
but not less than	$0.0004 d_b f_y$**	(1-81)
For #14 bars	$0.085 f_y / \sqrt{f'_c}$***	(1-82)
For #18 bars	$0.11 f_y / \sqrt{f'_c}$***	(1-83)

*The constant carries the unit of 1/in.
**The constant carries the unit of in.^{3}/lb.
***The constant carries the unit of in.

For deformed wire $\quad 0.03 d_b f_y / \sqrt{f_c'} \quad$ (1-84)

(b) The basic development length shall be multiplied by the applicable factor or factors for:

Top reinforcement* $\quad 1.4 \quad$ (1-85)

Bars with f_y greater than 60,000 psi $\quad 2 - 60{,}000/f_y \quad$ (1-86)

Table 1-10 provides development lengths in inches for all bar sizes, concrete strengths of 3, 4, 5 and 6 ksi, and steel strengths 40, 50 and 60 ksi. All combinations of basic development length are provided for normal weight concrete, all-lightweight concrete and sand-lightweight concrete.

(c) When lightweight aggregate concrete is used, the basic development lengths in (a) shall be multiplied by 1.33 for "all-lightweight" concrete and 1.18 for "sand-lightweight" concrete with linear interpolation when partial sand replacement is used, or the basic development length may be multiplied by $6.7\sqrt{f_c'}/f_{ct}$, but not less than 1.0, when f_{ct} is specified and the concrete is proportioned according to Section 4.2 of the *Code*. The factors of (b) and (d) shall also be applied.

(d) The basic development length, modified by the appropriate requirements of (b) and (c), may be multiplied by the applicable factor or factors for the following:

Reinforcement being developed in the length under consideration and spaced laterally at least 6 in. on center and at least 3 in. from the side face of the member $\quad$ 0.8

Reinforcement in a flexural member in excess of that required $\quad (A, \text{required})/(A, \text{provided})$

Bars enclosed within a spiral which is not less than 1/4 in. diameter and not more than 4 in. pitch $\quad$ 0.75

*Top reinforcement is horizontal reinforcement so placed that more than 12 in. of concrete is cast in the member below the bar.

TABLE 1-10a Development Length for Tension Bars–Other than Top Bars, Normal Weight Concrete.

f_c'	3.0	4.0	5.0	6.0	3.0	4.0	5.0	6.0	3.0	4.0	5.0	6.0
FY	40.0	40.0	40.0	40.0	50.0	50.0	50.0	50.0	60.0	60.0	60.0	60.0
BAR NO.												
3	12.0	12.0	12.0	12.0	12.0	12.0	12.0	12.0	12.0	12.0	12.0	12.0
4	12.0	12.0	12.0	12.0	12.0	12.0	12.0	12.0	12.0	12.0	12.0	12.0
5	12.0	12.0	12.0	12.0	12.5	12.5	12.5	12.5	15.0	15.0	15.0	15.0
6	12.9	12.0	12.0	12.0	16.1	15.0	15.0	15.0	19.3	18.0	18.0	18.0
7	17.5	15.2	14.0	14.0	21.9	19.0	17.5	17.5	26.3	22.8	21.0	21.0
8	23.1	20.0	17.9	16.3	28.8	25.0	22.3	20.4	34.6	30.0	26.8	24.5
9	29.2	25.3	22.6	20.7	36.5	31.6	28.3	25.8	43.8	37.9	33.9	31.0
10	37.1	32.1	28.7	26.2	46.4	40.2	35.9	32.8	55.6	48.2	43.1	39.3
11	45.6	39.5	35.3	32.2	57.0	49.3	44.1	40.3	68.4	59.2	52.9	48.3
14	62.1	53.8	48.1	43.9	77.6	67.2	60.1	54.9	93.1	80.6	72.1	65.8
18	80.3	69.6	62.2	56.8	100.4	87.0	77.8	71.0	120.5	104.4	93.3	85.2

TABLE 1-10b Development Length for Tension Bars–Other than Top Bars, All Lightweight Concrete.

f_c'	3.0	4.0	5.0	6.0	3.0	4.0	5.0	6.0	3.0	4.0	5.0	6.0
FY	40.0	40.0	40.0	40.0	50.0	50.0	50.0	50.0	60.0	60.0	60.0	60.0
BAR NO.												
3	12.0	12.0	12.0	12.0	12.0	12.0	12.0	12.0	12.0	12.0	12.0	12.0
4	12.0	12.0	12.0	12.0	13.3	13.3	13.3	13.3	16.0	16.0	16.0	16.0
5	13.3	13.3	13.3	13.3	16.6	16.6	16.6	16.6	19.9	19.9	19.9	19.9
6	17.1	16.0	16.0	16.0	21.4	19.9	19.9	19.9	25.6	23.9	23.9	23.9
7	23.3	20.2	18.6	18.6	29.1	25.2	23.3	23.3	35.0	30.3	27.9	27.9
8	30.7	26.6	23.8	21.7	38.4	33.2	29.7	27.1	46.0	39.9	35.7	32.6
9	38.9	33.6	30.1	27.5	48.6	42.1	37.6	34.3	58.3	50.5	45.1	41.2
10	49.3	42.7	38.2	34.9	61.7	53.4	47.8	43.6	74.0	64.1	57.3	52.3
11	60.6	52.5	46.9	42.9	75.8	65.6	58.7	53.6	90.9	78.7	70.4	64.3
14	82.6	71.5	64.0	58.4	103.2	89.4	79.9	73.0	123.8	107.2	95.9	87.6
18	106.8	92.5	82.8	75.5	133.6	115.7	103.4	94.4	160.3	138.8	124.1	113.3

TABLE 1-10c Development Length for Tension Bars–Other than Top Bars, Sand Lightweight Concrete.

f_c'	3.0	4.0	5.0	6.0	3.0	4.0	5.0	6.0	3.0	4.0	5.0	6.0
FY	40.0	40.0	40.0	40.0	50.0	50.0	50.0	50.0	60.0	60.0	60.0	60.0
BAR NO.												
3	12.0	12.0	12.0	12.0	12.0	12.0	12.0	12.0	12.0	12.0	12.0	12.0
4	12.0	12.0	12.0	12.0	12.0	12.0	12.0	12.0	14.2	14.2	14.2	14.2
5	12.0	12.0	12.0	12.0	14.7	14.7	14.7	14.7	17.7	17.7	17.7	17.7
6	15.2	14.2	14.2	14.2	19.0	17.7	17.7	17.7	22.8	21.2	21.2	21.2
7	20.7	17.9	16.5	16.5	25.9	22.4	20.6	20.6	31.0	26.9	24.8	24.8
8	27.2	23.6	21.1	19.3	34.0	29.5	26.4	24.1	40.8	35.4	31.6	28.9
9	34.5	29.9	26.7	24.4	43.1	37.3	33.4	30.5	51.7	44.8	40.1	36.6
10	43.8	37.9	33.9	31.0	54.7	47.4	42.4	38.7	65.7	56.9	50.9	46.4
11	53.8	46.6	41.7	38.0	67.2	58.2	52.1	47.5	80.7	69.9	62.5	57.0
14	73.2	63.4	56.7	51.8	91.6	79.3	70.9	64.7	109.9	95.2	85.1	77.7
18	94.8	82.1	73.4	67.0	118.5	102.6	91.8	83.8	142.2	123.1	110.1	100.5

TABLE 1-10d Development Length for Tension Bars—Top Bars, Normal Weight Concrete.

f'_c	3.0	4.0	5.0	6.0	3.0	4.0	5.0	6.0	3.0	4.0	5.0	6.0
FY	40.0	40.0	40.0	40.0	50.0	50.0	50.0	50.0	60.0	60.0	60.0	60.0
BAR NO.												
3	12.0	12.0	12.0	12.0	12.0	12.0	12.0	12.0	12.6	12.6	12.6	12.6
4	12.0	12.0	12.0	12.0	14.0	14.0	14.0	14.0	16.8	16.8	16.8	16.8
5	14.0	14.0	14.0	14.0	17.5	17.5	17.5	17.5	21.0	21.0	21.0	21.0
6	18.0	16.8	16.8	16.8	22.5	21.0	21.0	21.0	27.0	25.2	25.2	25.2
7	24.5	21.3	19.6	19.6	30.7	26.6	24.5	24.5	36.8	31.9	29.4	29.4
8	32.3	28.0	25.0	22.8	40.4	35.0	31.3	28.6	48.5	42.0	37.5	34.3
9	40.9	35.4	31.7	28.9	51.1	44.3	39.6	36.1	61.3	53.1	47.5	43.4
10	51.9	45.0	40.2	36.7	64.9	56.2	50.3	45.9	77.9	67.5	60.3	55.1
11	63.8	55.3	49.4	45.1	79.7	69.1	61.8	56.4	95.7	82.9	74.1	67.7
14	86.9	75.3	67.3	61.5	108.6	94.1	84.1	76.8	130.4	112.9	101.0	92.2
18	112.5	97.4	87.1	79.5	140.6	121.7	108.9	99.4	168.7	146.1	130.7	119.3

TABLE 1-10e Development Length for Tension Bars—Top Bars, All Lightweight Concrete.

f'_c	3.0	4.0	5.0	6.0	3.0	4.0	5.0	6.0	3.0	4.0	5.0	6.0
FY	40.0	40.0	40.0	40.0	50.0	50.0	50.0	50.0	60.0	60.0	60.0	60.0
BAR NO.												
3	12.0	12.0	12.0	12.0	14.0	14.0	14.0	14.0	16.8	16.8	16.8	16.8
4	14.9	14.9	14.9	14.9	18.6	18.6	18.6	18.6	22.3	22.3	22.3	22.3
5	18.6	18.6	18.6	18.6	23.3	23.3	23.3	23.3	27.9	27.9	27.9	27.9
6	23.9	22.3	22.3	22.3	29.9	27.9	27.9	27.9	35.9	33.5	33.5	33.5
7	32.6	28.3	26.1	26.1	40.8	35.3	32.6	32.6	49.0	42.4	39.1	39.1
8	43.0	37.2	33.3	30.4	53.7	46.5	41.6	38.0	64.5	55.8	49.9	45.6
9	54.4	47.1	42.1	38.5	68.0	58.9	52.7	48.1	81.6	70.7	63.2	57.7
10	69.1	59.8	53.5	48.8	86.3	74.8	66.9	61.1	103.6	89.7	80.3	73.3
11	84.9	73.5	65.7	60.0	106.1	91.9	82.2	75.0	127.3	110.2	98.6	90.0
14	115.6	100.1	89.5	81.7	144.5	125.1	111.9	102.2	173.4	150.1	134.3	122.6
18	149.6	129.5	115.9	105.8	187.0	161.9	144.8	132.2	224.4	194.3	173.8	158.7

TABLE 1-10f Development Length for Tension Bars—Top Bars, Sand Lightweight Concrete.

f'_c	3.0	4.0	5.0	6.0	3.0	4.0	5.0	6.0	3.0	4.0	5.0	6.0
FY	40.0	40.0	40.0	40.0	50.0	50.0	50.0	50.0	60.0	60.0	60.0	60.0
BAR NO.												
3	12.0	12.0	12.0	12.0	12.4	12.4	12.4	12.4	14.9	14.9	14.9	14.9
4	13.2	13.2	13.2	13.2	16.5	16.5	16.5	16.5	19.8	19.8	19.8	19.8
5	16.5	16.5	16.5	16.5	20.6	20.6	20.6	20.6	24.8	24.8	24.8	24.8
6	21.2	19.8	19.8	19.8	26.5	24.8	24.8	24.8	31.9	29.7	29.7	29.7
7	29.0	25.1	23.1	23.1	36.2	31.3	28.9	28.9	43.4	37.6	34.7	34.7
8	38.1	33.0	29.5	27.0	47.7	41.3	36.9	33.7	57.2	49.5	44.3	40.4
9	48.3	41.8	37.4	34.1	60.3	52.2	46.7	42.7	72.4	62.7	56.1	51.2
10	61.3	53.1	47.5	43.3	76.6	66.3	59.3	54.2	91.9	79.6	71.2	65.0
11	75.3	65.2	58.3	53.2	94.1	81.5	72.9	66.5	112.9	97.8	87.5	79.8
14	102.5	88.8	79.4	72.5	128.2	111.0	99.3	90.6	153.8	133.2	119.2	108.8
18	132.7	114.9	102.8	93.8	165.9	143.7	128.5	117.3	199.1	172.4	154.2	140.8

TABLE 1-10g Development Length for Compression Bars.

f'_c	3.0	4.0	5.0	6.0	3.0	4.0	5.0	6.0	3.0	4.0	5.0	6.0
FY	40.0	40.0	40.0	40.0	50.0	50.0	50.0	50.0	60.0	60.0	60.0	60.0
BAR NO.												
3	8.0	8.0	8.0	8.0	8.0	8.0	8.0	8.0	8.2	8.0	8.0	8.0
4	8.0	8.0	8.0	8.0	9.1	8.0	8.0	8.0	11.0	9.5	9.0	9.0
5	9.1	8.0	8.0	8.0	11.4	9.9	9.4	9.4	13.7	11.9	11.2	11.2
6	11.0	9.5	9.0	9.0	13.7	11.9	11.2	11.2	16.4	14.2	13.5	13.5
7	12.8	11.1	10.5	10.5	16.0	13.8	13.1	13.1	19.2	16.6	15.7	15.7
8	14.6	12.6	12.0	12.0	18.3	15.8	15.0	15.0	21.9	19.0	18.0	18.0
9	16.5	14.3	13.5	13.5	20.6	17.8	16.9	16.9	24.7	21.4	20.3	20.3
10	18.5	16.1	15.2	15.2	23.2	20.1	19.1	19.1	27.8	24.1	22.9	22.9
11	20.6	17.8	16.9	16.9	25.7	22.3	21.1	21.1	30.9	26.8	25.4	25.4
14	24.7	21.4	20.3	20.3	30.9	26.8	25.4	25.4	37.1	32.1	30.5	30.5
18	33.0	28.5	27.1	27.1	41.2	35.7	33.9	33.9	49.4	42.8	40.6	40.6

The development length l_d for bars in compression shall be computed as $0.02 f_y d_b / \sqrt{f_c'}$ but shall not be less than $0.0003 f_y d_b$ or 8 in. Where excess bar area is provided, the l_d length may be reduced by the ratio of required area to area provided. The development length may be reduced 25% when the reinforcement is enclosed by spirals not less than 1/4 in. in diameter and not more than 4 in. pitch.

The development length of each bar of bundled bars shall be that for the individual bar, increased by 20% for a three-bar bundle, and 33% for a four-bar bundle.

Standard hooks shall be considered to develop a tensile stress in bar reinforcement $f_h = \xi\sqrt{f_c'}$ where ξ is not greater than the values in Table 1-11. The value of ξ may be increased 30% where enclosure is provided perpendicular to the plane of the hook.

TABLE 1-11 ξ **Values**

	$f_y = 60$ ksi		$f_y = 40$ ksi
Bar size	*Top bars*	*Other bars*	*Bars all*
#3 to #5	540	540	360
#6	450	540	360
#7 to #9	360	540	360
#10	360	480	360
#11	360	420	360
#14	330	330	330
#18	220	220	220

An equivalent embedment length l_c shall be computed using eqs. (1-80) through (1-86) by substituting f_h for f_y and l_c for l_d.

Hooks shall not be considered effective in adding to the compressive resistance of reinforcement.

Development length l_d may consist of a combination of the equivalent embedment length of a hook or mechanical anchorage plus additional embedment length of the reinforcement.

The yield strength of smooth longitudinal wires of welded wire fabric shall be considered developed by embedding at least two cross wires with the closer one at least 2 in. from the point of critical section. An embedment of one cross wire at least 2 in. from the point of critical section may be considered to develop half the yield strength.

The development length of welded deformed wire fabric may be computed as a deformed wire using eqs. (1-80) through (1-86) by substituting $(f_y - 20{,}000n)$ for f_y, where n is the number of cross wires within the development length which are at least 2 in. from the critical section. The minimum development length shall be $250 A_w / S_w$ where A_w is the area of one tension wire and S_w is the spacing of wires. Factors (b), (c) and (d) always apply when eqs. (1-80) through (1-86) are used, even with modifications stated above.

Any mechanical device capable of developing the strength of the reinforcement without damage to the concrete may be used as anchorage. Test results showing the adequacy of such devices shall be presented to the Building Official.

Web reinforcement shall be carried out as close to the compression and tension surfaces of the member as cover requirements and the proximity of other steel will permit, and in any case the ends of single leg, simple U-, or multiple U-stirrup shall be anchored by one of the following means:

(a) A standard hook plus an effective embedment of $0.5 l_d$. The effective embedment of a stirrup leg shall be taken as the distance between the middepth of the member $d/2$ and the start of the hook (point of tangency).

(b) Embedment above or below the middepth, $d/2$, of the beam on the compression side for a full development length l_d but not less than 24 bar diameters.

(c) Bending around the longitudinal reinforcement through at least 180°. Hooking or bending stirrups around the longitudinal reinforcement shall be considered effective anchorage only when the stirrups make an angle of at least 45° with deformed longitudinal bars.

For each leg of welded plain wire fabric forming simple U-stirrups, either:

(a) Two longitudinal wires running at a 2 in. spacing along the beam at the top of the U.

(b) One longitudinal wire not more than $d/4$ from the compression face and a second wire closer to the compression face and spaced at least 2 in. from the first. The second wire may be beyond a bend or on a bend which has an inside diameter of at least eight wire diameters.

Between the anchored ends, each bend in the continuous portion of a transverse simple U- or multiple U-stirrup shall enclose a longitudinal bar.

Pairs of U-stirrups or ties so placed as to form a closed unit shall be considered properly spliced when the laps are $1.7 l_d$. In members at least 18 in. deep, such splices having $A_b f_y$ not more than 9000 lb per leg may be considered adequate if the legs extend the full available depth of the member.

EXAMPLE 1-25: For normal weight concrete, compute the basic tension development length for a No. 9 bar for $f_c' = 4000$ psi and $f_y = 60{,}000$ psi. The bar is not a top bar.

SOLUTION: Equations (1-80) and (1-81) apply, with bar area, $A_b = 1.0$ sq in., diameter, $d_b = 1.128$ in., and $\sqrt{f_c'} = 63.25$

$$l_d = (0.04)(1.0)(60{,}000)/63.25 = 37.94 \text{ in.} \qquad (1\text{-}80)$$

but not less than

$$l_d = (0.0004)(1.128)(60{,}000) = 27.07 \text{ in.} \qquad (1\text{-}81)$$

and not less than 12 in. Equation (1-80) governs and $l_d = 37.94$ in. This may be verified in Table 1-10, where $l_d = 37.9$ in.

In general, the development length should be rounded off to the next higher 1/2 inch. Thus, use 40 in.

EXAMPLE 1-26: Compute the development length for the No. 9 bar of Example 1-25 if lightweight aggregate concrete is used.

SOLUTION: The development length previously found is multiplied by 1.33 for all-lightweight aggregate concrete. Thus, $l_d = (1.33)(37.94) = 50.46$ in. Table 1-10 shows $l_d = 50.5$ in.

EXAMPLE 1-27: Figure 1-37 shows a cantilever beam extending from a column. The bars are No. 9, $f_c' = 4000$ psi and $f_y = 60{,}000$ psi. Lightweight aggregate concrete is used.

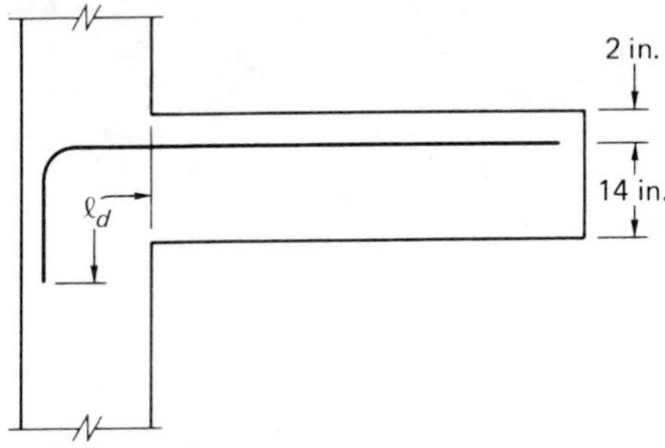

Fig. 1-37

SOLUTION: The bars are classified as top bars since there is more than 12 in. of concrete cast below them. (See footnote, page 43, *1971 ACI Code.*)

Thus, since f_y does not exceed 60,000 psi, the additional multiplier is 1.4. In the previous example for other than top bars, $l_d = 50.46$ in. Thus, for top bars, $l_d = (1.4)(50.46) = 70.64$ in.

Table 1-10 shows that $l_d = 70.7$ in. The difference occurs because the electronic computer used to prepare the tables uses eight significant figures to compute the development lengths, and then rounds off the answer when it is printed.

EXAMPLE 1-28: Figure 1-38 shows a portion of a continuous beam with top bars in tension, $f_c' = 4000$ psi, $f_y = 60{,}000$ psi and bars are No. 14. The concrete is normal weight. Compute the anchorage length into the next beam and the required extension of at least one-third of the bars beyond the point of contraflexure (P.C.). The diameter of a No. 14 bar is 1.693 in.

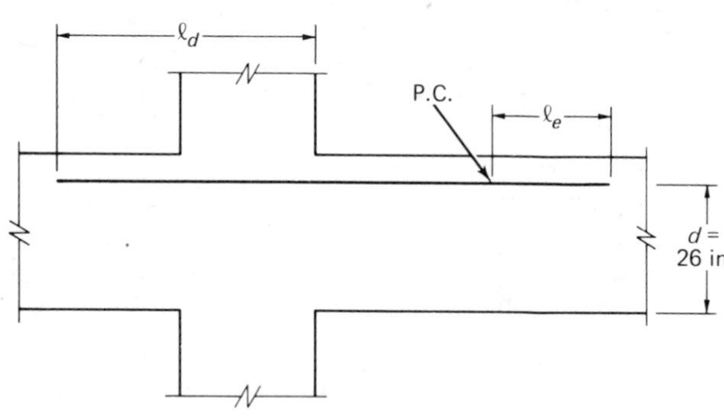

Fig. 1-38

SOLUTION: The anchorage into the adjacent beam (l_a) must equal a development length. For top bars, No. 14 in size, eq. (1-82) applies, and the multiplier for top bars is 1.4. Thus,

$$l_a = l_d = (1.4)(0.085)(60{,}000)/63.25 = 112.89 \text{ in.}$$

Table 1-10 shows that $l_d = 112.9$ in. (Use 113 in. = 9 ft − 5 in.)

For the extension length (l_e) for at least 1/3 of the bars beyond the point of contraflexure, the greater of three conditions applies:

(1) l_e = effective depth = 26 in.
(2) $l_e = 12d_b = (12)(1.6393) = 19.67$ in.
(3) $l_e = l_d/16 = 112.89/16 = 7$ in.

Thus, the effective depth governs, and $l_e = 26$ in.

EXAMPLE 1-29: Figure 1-39 shows conditions at a simple support where two No. 8 bars have been extended from the maximum moment region. For normal weight concrete, $f_c' = 4000$ psi, $f_y = 60{,}000$ psi and a compressive reaction, determine whether or not the bar size is satisfactory. $V_u = 60$ kips, $d = 20$ in. and $b = 12$ in.

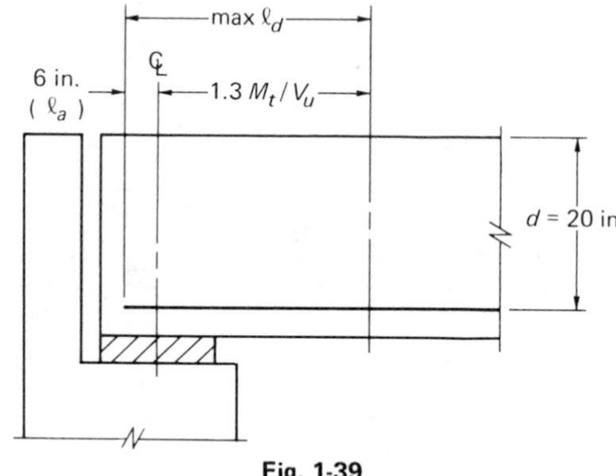

Fig. 1-39

SOLUTION: The furnished anchorage length l_a is 6 in. The moment capacity of three No. 8 bars must be computed for $d = 20$ in. and $A_s = 1.58$ sq in.

$$a = A_s f_y/(0.85 f_c' b) = \frac{(1.58)(60)}{(0.85)(4)(12)} = 2.32 \text{ in.}$$

Since the reaction is compressive and therefore confines the

TABLE 1-12 Area of Steel for Given Number of Bars of Various Sizes.

NO. OF BARS	BAR SIZE DESIGNATION									
	4	5	6	7	8	9	10	11	14-S	18-S
1	.20	.31	.44	.60	.79	1.00	1.27	1.56	2.25	4.00
2	.40	.62	.88	1.20	1.57	2.00	2.54	3.12	4.50	8.00
3	.60	.93	1.32	1.80	2.36	3.00	3.81	4.68	6.75	12.00
4	.80	1.24	1.76	2.40	3.14	4.00	5.08	6.24	9.00	16.00
5	1.00	1.55	2.20	3.00	3.93	5.00	6.35	7.80	11.25	20.00
6	1.20	1.86	2.64	3.60	4.71	6.00	7.62	9.36	13.50	24.00
7	1.40	2.17	3.08	4.20	5.50	7.00	8.89	10.92	15.75	28.00
8	1.60	2.48	3.52	4.80	6.28	8.00	10.16	12.48	18.00	32.00
9	1.80	2.79	3.96	5.40	7.07	9.00	11.43	14.04	20.25	36.00
10	2.00	3.10	4.40	6.00	7.85	10.00	12.70	15.60	22.50	40.00
11	2.20	3.41	4.84	6.60	8.64	11.00	13.97	17.16	24.75	44.00
12	2.40	3.72	5.28	7.20	9.42	12.00	15.24	18.72	27.00	48.00
13	2.60	4.03	5.72	7.80	10.21	13.00	16.51	20.28	29.25	52.00
14	2.80	4.34	6.16	8.40	11.00	14.00	17.78	21.84	31.50	56.00
15	3.00	4.65	6.60	9.00	11.78	15.00	19.05	23.40	33.75	60.00
16	3.20	4.96	7.04	9.60	12.57	16.00	20.32	24.96	36.00	64.00
17	3.40	5.27	7.48	10.20	13.35	17.00	21.59	26.52	38.25	68.00
18	3.60	5.58	7.92	10.80	14.14	18.00	22.86	28.08	40.50	72.00
19	3.80	5.89	8.36	11.40	14.92	19.00	24.13	29.64	42.75	76.00
20	4.00	6.20	8.80	12.00	15.71	20.00	25.40	31.20	45.00	80.00
21	4.20	6.51	9.24	12.60	16.49	21.00	26.67	32.76	47.25	84.00
22	4.40	6.82	9.68	13.20	17.28	22.00	27.94	34.32	49.50	88.00
23	4.60	7.13	10.12	13.80	18.06	23.00	29.21	35.88	51.75	92.00
24	4.80	7.44	10.56	14.40	18.85	24.00	30.48	37.44	54.00	96.00
25	5.00	7.75	11.00	15.00	19.64	25.00	31.75	39.00	56.25	100.00
26	5.20	8.06	11.44	15.60	20.42	26.00	33.02	40.56	58.50	104.00
27	5.40	8.37	11.88	16.20	21.21	27.00	34.29	42.12	60.75	108.00
28	5.60	8.68	12.32	16.80	21.99	28.00	35.56	43.68	63.00	112.00
29	5.80	8.99	12.76	17.40	22.78	29.00	36.83	45.24	65.25	116.00
30	6.00	9.30	13.20	18.00	23.56	30.00	38.10	46.80	67.50	120.00

reinforcement

$$l_d \leqslant 1.3M_t/V_u + l_a$$

where

$$M_t = A_s f_y(d - a/2) = (1.58)(60)(20 - 1.16) = 1786 \text{ in.-kips}$$
$$M_t/V_u = 1786/60 = 29.4 \text{ in.}$$

Thus

$$l_d \leqslant (1.3)(29.4) + 6 = 44.2 \text{ in.}$$

For a No. 8 bar, normal weight concrete, $f_c' = 4$ ksi and $f_y = 60$ ksi, other than top bars, $l_d = 30.0$ in. in Table 1-10.

Since 30 in. < 44.2, the bar size is satisfactory and not too large. If the conditions were not satisfied, the permissible maximum l_d

TABLE 1-13 Minimum Web Widths for Various Combinations of Bars.

Width of beam in inches based on use of #3 stirrups. Width required for joist webs: 2 in. less than table values. Large bar(s) assumed to be placed on outside face of beam. Aggregate size assumed ≤3/4".

Beam: $1\frac{1}{2}$"; $\frac{3}{8}$"

Joist: $\frac{3}{4}$"; $\frac{1}{8}$"

D_1 — D_2

1" minimum spacing

D for bars of one size

$\frac{D_1 + D_2}{2}$ or $\frac{D_1 + 1}{2}$ for bars of two sizes

Columns headed [0][5] contain data for bars of one size in groups of one to ten. Columns headed [1][2][3][4][5] contain data for bars of two sizes with from one to five of each size. Data are in accordance with ACI 318-63.

Bar	No.	Bar	0	5	Bar	1	2	3	4	5	Bar	1	2	3	4	5	Bar	1	2	3	4	5
#3 or #4	1	#3 or #4	---	12.0																		
	2		6.0	13.5																		
	3		7.5	15.0																		
	4		9.0	16.5																		
	5		10.5	18.0																		
#5	1	#5	---	12.5	#4	6.0	7.5	9.0	10.5	12.0												
	2		6.0	14.5		7.5	9.0	10.5	12.0	13.5												
	3		8.0	16.0		9.5	11.0	12.5	14.0	15.5												
	4		9.5	17.5		11.0	12.5	14.0	15.5	17.0												
	5		11.0	19.0		12.5	14.0	15.5	17.0	18.5												
#6	1	#6	---	13.5	#5	6.5	8.0	9.5	11.0	13.0	#4	6.0	7.5	9.0	10.5	12.0						
	2		6.5	15.0		8.0	9.5	11.5	13.0	14.5		8.0	9.5	11.0	12.5	14.0						
	3		8.0	17.0		10.0	11.5	13.0	14.5	16.5		9.5	11.0	12.5	14.0	15.5						
	4		10.0	18.5		11.5	13.0	15.0	16.5	18.0		11.5	13.0	14.5	16.0	17.5						
	5		11.5	20.5		13.5	15.0	16.5	18.0	20.0		13.0	14.5	16.0	17.5	19.0						
#7	1	#7	---	14.0	#6	6.5	8.5	10.0	12.0	13.5	#5	6.5	8.0	9.5	11.5	13.0	#4	6.5	8.0	9.5	11.0	12.5
	2		6.5	16.0		8.5	10.0	12.0	13.5	15.5		8.5	10.0	11.5	13.0	15.0		8.0	9.5	11.0	12.5	14.0
	3		8.5	18.0		10.5	12.0	14.0	15.5	17.5		10.0	12.0	13.5	15.0	16.5		10.0	11.5	13.0	14.5	16.0
	4		10.5	20.0		12.0	14.0	15.5	17.5	19.0		12.0	13.5	15.5	17.0	18.5		12.0	13.5	15.0	16.5	18.0
	5		12.5	21.5		14.0	16.0	17.5	19.5	21.0		14.0	15.5	17.0	19.0	20.5		14.0	15.5	17.0	18.5	20.0
#8	1	#8	---	15.0	#7	7.0	8.5	10.5	12.5	14.5	#6	6.5	8.5	10.0	12.0	13.5	#5	6.5	8.0	10.0	11.5	13.0
	2		7.0	17.0		9.0	10.5	12.5	14.5	16.5		8.5	10.5	12.0	14.0	15.5		8.5	10.0	12.0	13.5	15.0
	3		9.0	19.0		11.0	12.5	14.5	16.5	18.5		10.5	12.5	14.0	16.0	17.5		10.5	12.0	14.0	15.5	17.0
	4		11.0	21.0		13.0	14.5	16.5	18.5	20.5		12.5	14.5	16.0	18.0	19.5		12.5	14.0	16.0	17.5	19.0
	5		13.0	23.0		15.0	16.5	18.5	20.5	22.5		14.5	16.5	18.0	20.0	21.5		14.5	16.0	18.0	19.5	21.0
#9	1	#9	---	16.5	#8	7.0	9.0	11.0	13.0	15.0	#7	7.0	9.0	11.0	12.5	14.5	#6	7.0	8.5	10.5	12.0	14.0
	2		7.5	18.5		9.5	11.5	13.5	15.5	17.5		9.5	11.0	13.0	15.0	17.0		9.0	11.0	12.5	14.5	16.0
	3		9.5	21.0		11.5	13.5	15.5	17.5	19.5		11.5	13.5	15.5	17.0	19.0		11.5	13.0	15.0	16.5	18.5
	4		12.0	23.0		14.0	16.0	18.0	20.0	22.0		14.0	15.5	17.5	19.5	21.5		13.5	15.5	17.0	19.0	20.5
	5		14.0	25.5		16.0	18.0	20.0	22.0	24.0		16.0	18.0	20.0	21.5	23.5		16.0	17.5	19.5	21.0	23.0
#10	1	#10	---	18.0	#9	7.5	10.0	12.0	14.5	16.5	#8	7.5	9.5	11.5	13.5	15.5	#7	7.0	9.0	11.0	13.0	14.5
	2		8.0	20.5		10.0	12.5	14.5	17.0	19.0		10.0	12.0	14.0	16.0	18.0		9.5	11.5	13.5	15.5	17.0
	3		10.5	23.0		12.5	15.0	17.0	19.5	21.5		12.5	14.5	16.5	18.5	20.5		12.0	14.0	16.0	18.0	19.5
	4		13.0	25.5		15.0	17.5	19.5	22.0	24.0		15.0	17.0	19.0	21.0	23.0		15.0	16.5	18.5	20.5	22.5
	5		15.5	28.0		17.5	20.0	22.0	24.5	26.5		17.5	19.5	21.5	23.5	25.5		17.5	19.0	21.0	23.0	25.0
#11	1	#11	---	19.5	#10	8.0	10.5	13.0	15.5	18.0	#9	7.5	10.0	12.0	14.5	16.5	#8	7.5	9.5	11.5	13.5	15.5
	2		8.0	22.5		11.0	13.5	16.0	18.5	21.0		10.5	12.5	15.0	17.5	19.5		10.0	12.0	14.0	16.0	18.0
	3		11.0	25.0		13.5	16.0	18.5	21.0	23.5		13.5	15.5	18.0	20.0	22.5		13.0	15.0	17.0	19.0	21.0
	4		14.0	28.0		16.5	19.0	21.5	24.0	26.5		16.0	18.5	20.5	23.0	25.0		16.0	18.0	20.0	22.0	24.0
	5		16.5	31.0		19.0	22.0	24.5	27.0	29.5		19.0	21.0	23.5	25.5	28.0		18.5	20.5	22.5	24.5	26.5
#14	1	#14	---	22.5	#11	8.5	11.5	14.0	17.0	20.0	#10	8.0	11.0	13.5	16.0	18.5	#9	8.0	10.0	12.5	14.5	17.0
	2		9.0	26.0		12.0	14.5	17.5	20.5	23.0		11.5	14.0	16.5	19.0	22.0		11.5	13.5	16.0	18.0	20.5
	3		12.5	29.5		15.5	18.0	21.0	23.5	26.5		15.0	17.5	20.0	22.5	25.0		14.5	17.0	19.0	21.5	23.5
	4		16.0	33.0		18.5	21.5	24.5	27.0	30.0		18.5	21.0	23.5	26.0	28.5		18.0	20.5	22.5	25.0	27.0
	5		19.0	36.0		22.0	25.0	27.5	30.5	33.5		22.0	24.5	27.0	29.5	32.0		21.5	23.5	26.0	28.5	30.5
#18	1	#18	---	29.5	#14	10.0	13.5	17.0	20.5	24.0	#11	9.5	12.5	15.5	18.5	21.0	#10	9.0	12.0	14.5	17.0	19.5
	2		11.5	34.0		14.5	18.0	21.5	25.0	28.5		14.0	17.0	20.0	23.0	25.5		13.5	16.5	19.0	21.5	24.0
	3		16.0	38.5		19.0	22.5	26.0	29.5	33.0		18.5	21.5	24.5	27.5	30.0		18.0	21.0	23.5	26.0	28.5
	4		20.5	43.0		23.5	27.0	30.5	34.0	37.5		23.0	26.0	29.0	32.0	34.5		22.5	25.5	28.0	30.5	33.5
	5		25.0	47.5		28.0	31.5	35.0	38.5	42.0		27.5	30.5	33.5	36.5	39.0		27.0	30.0	32.5	35.0	38.0

could be increased by increasing the end anchorage length l_a. This could be done by adding a standard hook at the end.

EXAMPLE 1-30: It is desired to discontinue two No. 8 bars of six bars in a tension zone. The shear (V_u) at a distance d = 20 in. beyond the theoretical cutoff point is 14 kips, b = 12 in., d = 20 in., f'_c = 4000 psi and f_y = 60,000 psi. The moment at the cross section is 290 ft-kips. Design the details required to cut off the two bars. The beam does not contain stirrups at this point.

SOLUTION: From the previous problem, the moment capacity of two bars is 1786 in.-kips or 148.8 ft-kips. Thus, for four continuing bars, the moment capacity is 297.6 ft-kips. Since this is not twice the moment at the section, the bars may not be cut off without additional considerations.

The permissible shear force without stirrups is

$$V_u = (0.85)(63.25)(12)(20)/1000 = 15.17 \text{ kips}$$

$$\text{and } 2/3\, V_u = (2/3)(15.17) = 10.11 \text{ kips.}$$

Since the actual shear force is 14 kips and is greater than 10.11 kips, stirrups must be provided in order to discontinue the bars. Stirrups must satisfy the conditions

$$A_v f_y / b_w \geqslant 60 \text{ psi}$$

and

$$s \leqslant d/(8\beta_b)$$

where β_b is the ratio of the area of bars cut off to the total area, or $^2/_6 = {^1/_3}$. Hence,

$$s = 20/(8 \times 1/3) = 7.5 \text{ in.}$$

and $A_v = 60 b_w / f_y = (60)(12)/60{,}000 = 0.0012$ sq in. minimum.

Use No. 3 stirrups (A_v = 0.22 sq in.) at 7.5 in. on centers. The total distance over which stirrups are required is $3/4d$ or (3/4)(20) = 15 in.

Use three No. 3 stirrups at 7.5 in. on centers.

1.18 SPLICES

Details of reinforcement that must be established during the design stage include splices of reinforcement in beams, slabs, columns, etc. The basic types of splices are lap splices and end bearing splices.

Figure 1-40 illustrates two types of lap splices. In (a), one bar is bent to lap with the other so that the centerlines of the continuing bars coincide. This type of splice is frequently used in columns.

In (b), the bars are lap spliced out of line. This type of splice is used frequently in beams. However, when column sizes change at a floor level, this type of splice is often used. The bars may or may not be in contact. There may be several inches of concrete in between the bars. However, the clear distance may not exceed $^1/_5$ the lap length nor 6 in., whichever is smaller.

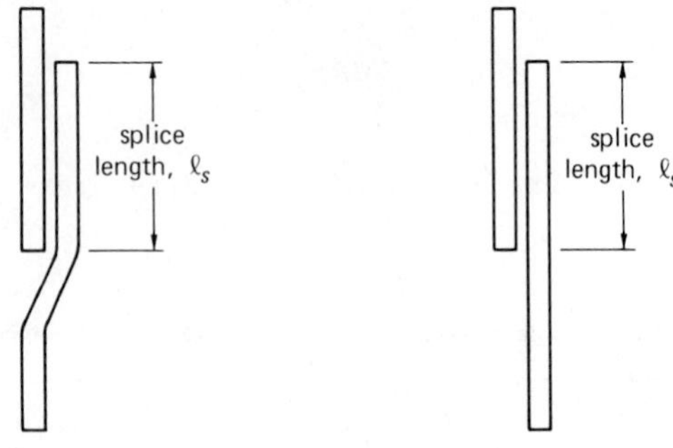

Fig. 1-40

No. 14 and No. 18 bars may not be lap spliced, except to footing dowels. Then, the dowel bars extending out of the footing must be smaller in size than the No. 14 or No. 18 bar. The splice length is the development length of the larger bar.

End bearing splices are used in columns when the bars are always in compression. Care must be exercised to insure proper bearing between bars. The ends must be flat within $1^1/_2$ degrees and shall have full bearing within 3° after assembly.

Other types of splices permitted include welded splices and mechanical splices. For welded splices, the bars must be composed of a weldable steel. The weld must develop 125% of the yield strength of the bars in tension. Mechanical splice devices are patented, and information may be obtained from the manufacturer. (For example, G-Loc Splices).

Tension splices are classified as Class A, B, C and D. The splice length is expressed in terms of multiples of development length previously discussed, and are as follows

Class A, $l_s = 1.0 l_d$
Class B, $l_s = 1.3 l_d$
Class C, $l_s = 1.7 l_d$
Class D, $l_s = 2.0 l_d$

When Class D splices are required the splice must be enclosed in a spiral, and the ends of the spliced bars must be hooked 180° when larger than No. 4.

Splices should be avoided in regions of maximum moment. If not more than one-half of the bars in regions of high tension stress are lapped over a given lap length, the splices are Class B. If more than one-half of the bars are lapped in regions of high tension stress, the splices are

TABLE 1-14 Properties of Reinforcing Bars.

	Nominal dimensions— round sections											
	Bar designation No.											
	2	3	4	5	6	7	8	9	10	11	14	18
Unit Weight per ft., lb.	0.167	0.376	0.668	1.043	1.502	2.044	2.670	3.400	4.303	5.313	7.65	13.60
Diameter in.	0.250	0.375	0.500	0.625	0.750	0.875	1.000	1.128	1.270	1.410	1.693	2.257
Cross-sectional area, sq. in.	0.05	0.11	0.20	0.31	0.44	0.60	0.79	1.00	1.27	1.56	2.25	4.00
Perimeter, in.	0.786	1.178	1.571	1.963	2.356	2.749	3.142	3.544	3.990	4.430	5.32	7.09

Class C. In the above statements, regions of high tension stress are those where f_s exceeds $0.5f_y$.

If the tension stress is always less than $0.5f_y$, a region of low stress is assumed to exist. The following requirements govern splice design:

(a) If not more than 3/4 of the bars are lap spliced within a given lap length, the splices are Class A.
(b) If more than 3/4 of the bars are lap spliced within a given lap length, the splices are Class B.

Compression splices must have lap lengths at least l_d, but not less in inches than

$$l_d = 0.0005 f_y d_b \quad (1\text{-}87)$$

for $f_y \leqslant 60{,}000$ psi, and not less than

$$(0.0009 f_y - 24) d_b \quad (1\text{-}88)$$

for f_y greater than 60,000 psi. In no case may a splice be less than 12 in. in length.

In tied compression members where ties throughout the lap length have an area at least $0.0015hs$, the lap length computed by eqs. (1-87) and (1-88) may be multiplied by 0.83, but the resulting lap length may not be less than 12 in.

When lap splices are used within the spiral of a spirally reinforced compression member and the bars are in compression, the lap length computed using eqs. (1-87) and (1-88) may be multiplied by 0.75, but the resulting lap length may not be less than 12 in.

The specifications for splices are simple and easy to apply. Hence, it is not necessary to provide design examples related thereto.

2

Deflections

RUSSELL S. FLING*

2.1 BASIC THEORY

2.1.1 General

The vast majority of concrete members are statically indeterminate whereas most structures framed in structural steel and wood are designed as statically determinate frames. Therefore, a working knowledge of any one of several computational methods for determining the deflection of indeterminate structures is essential. These methods are thoroughly discussed elsewhere, hence they need not be repeated here.

In addition to continuity, a major difference in the deflection behavior of concrete members from structural steel members is the time dependent strain of concrete due to sustained loads and ambient temperature and humidity. Another complicating factor is the variation in flexural rigidity along the length of the member. (In the following discussion, flexural rigidity is defined as the product of EI and is referred to as "rigidity"). Each of these subjects is covered in a discussion below.

2.1.2 Variation in Rigidity

Since concrete is so easily molded, many concrete members are built with a variable cross section. The amount of reinforcement is frequently varied along the length of the member. In addition, a fully cracked concrete cross section could have a rigidity as little as 10% of the uncracked rigidity in extreme cases. The cracked rigidity will typically fall between 30% and 80% of the uncracked rigidity of the same cross section and depends on the configuration of the concrete, the amount of reinforcing steel, and the bending moment. The amount of reinforcing steel will affect the cracked rigidity in an almost linear fashion but has a much smaller effect on the uncracked rigidity.

As the external moment on a concrete member increases, the member remains at its uncracked rigidity until the combined shrinkage and tensile bending stresses reach the tensile strength of the concrete. Subsequently the rigidity does not drop immediately to the cracked section rigidity but diminishes gradually and approaches the fully cracked rigidity only under very high moments. Obviously, an exact method for computing deflections must take into account the degree of cracking and the variation in rigidity along the length of the beam.

2.1.3 Time-Dependent Effects

Under most practical conditions of service, concrete will shrink. Since steel does not shrink, this leads to a differential strain across the cross section which causes warping similar to deflection from external loads. The warping will usually cause a deflection in the same direction as from load

*Principal, R. S. Fling and Partners, Consulting Engineers, Columbus, Ohio.

effects because the positive reinforcement exerts a restraining effect on the shrinkage. Occasionally, warping could be contrary to the gravity load deflection as in the case of a T-beam cantilever with heavy steel in the bottom of the stem. The greater the distance from the centroid of the steel to the neutral axis of the concrete cross section, the greater will be the shrinkage warping.

Shrinkage is affected by ambient humidity and temperature, aggregates, amount of mixing water, temperature of the concrete mix, and size of the flexural member. As the ambient relative humidity rises, the amount of shrinkage decreases, and becomes almost nil at a relative humidity of 100%. A high ambient temperature is frequently associated with a low relative humidity and would tend to drive moisture out of the concrete, and hence, increase the shrinkage. Certain aggregates cause concrete shrinkage to increase markedly. Fortunately, most aggregates do not exhibit this characteristic. The shrinkage of a given concrete mix is almost directly proportional to the amount of mixing water. Higher slump, retempering, or anything else that increases the mixing water will usually increase the shrinkage. As the ratio of the volume to surface area of the member increases, drying of the concrete becomes more inhibited and shrinkage is reduced. Tests have shown that the volume/surface ratio is a good index to this effect and the shape of the member has little effect on shrinkage.[2-1]

Creep is affected by many of the same factors that affect shrinkage. In addition, the amount of creep is affected by the age of the concrete when loaded, the duration of loading, the modulus of elasticity of the concrete, and the presence of reinforcement in the compressive zone.

Creep increases if the concrete is loaded at an early age. The creep of concrete loaded at any age continues indefinitely but at a decreasing rate. After about five years, the increase in creep strain is usually negligible. The total amount of creep strain is affected by the modulus of elasticity when the load is first applied because the ratio of ultimate creep strain to initial strain is nearly constant if all other factors are constant. This statement holds true for concrete members which are already stressed from loads applied earlier. Thus, the ratio of ultimate creep strain to initial strain might be 4 or 5 for loads applied to a member when it is seven days old, and might only be 2 or 3 for loads applied several months later.

Many cases of excessive deflection of concrete members have been traced to a severe loading of the concrete at too early an age. A common cause for early overloading is lack of adequate reshoring in multistoried buildings. This is especially critical for thin flat plates designed for a small superimposed load as is the case for residential buildings. Early loading of a concrete member increases its ultimate deflection because the modulus of elasticity is lower than at later ages. The modulus of rupture is also lower and thus the member will more likely be cracked with a consequent reduction in the rigidity.

A detailed evaluation of the various material properties is presented in papers by ACI Committee 209[2-2] and by Branson and Christiason[2-3] in the *ACI Publication SP-27*, "Designing for Effects of Creep, Shrinkage and Temperature in Concrete Structures."

2.1.4 Allowable Deflections

Except for arches, folded plates, slender columns, and certain other structures, the deflection of a concrete member has no effect on the strength of the member, per se. However, there are many limitations on the deflection of a concrete member imposed by the use of the structure of which the member is a part. The deflection might be limited by the sensitivity of persons using the structure; by the fragility of partitions and other non-structural elements supported by or attached to the structure; by the tolerances built into adjacent nonstructural elements; or by the operation of the structure. Table 9.5(b) of the ACI 318 *Code* gives numerical deflection limits.[2-4] A detailed discussion of these limitations is found in an ACI Committee 435 report on *Allowable Deflections*.[2-5] The following discussion excerpts the more important features from this report.

1. The most serious deflection problem associated with concrete floors is the damage to fragile masonry partitions caused by the time dependent deflection occurring after the partitions are constructed. Partitions vary greatly in their resiliency. Walls constructed of gypsum board are relatively insensitive to deflection and can be made even more adaptable to changes in curvature of supporting beam members. On the other hand, exposed concrete block walls are highly susceptible to damage from deflection of members supporting them because of their high modulus of elasticity, their high bending stiffness, and low tensile strength. Furthermore, the available tensile strength is frequently consumed by drying shrinkage of the concrete block themselves. If such partitions have door or window openings which weaken the wall, their susceptibility to damage from deflection of supporting members is further increased. The incremental* deflection of members supporting very fragile walls should be limited to a span/deflection ratio of at least 600 and preferably higher or the walls should be detailed to accept floor deflections without distress.

2. Many times, fragile nonstructural elements of a structure are built below or connected to a concrete member with a connection that allows a specific amount of movement before the nonstructural element would be damaged. The tolerance built into such nonstructural elements should be more than the anticipated absolute incremental deflection of the concrete member. In this case, the span/deflection ratio has no significance. Experience indicates that members spanning more than about 30 or 40 ft frequently have critical absolute deflections.

3. Operation of the structure may impose limitations on incremental deflections of structural members occurring after the structure is put into use. For example, concrete members supporting bowling alleys or printing presses will be required to remain sufficiently stable so as to avoid undue interference with the bowling or operation of the printing presses. As another example, level concrete roofs should not pond objectionable quantities of water by deflecting excessively.

4. A structure of any material must be sufficiently rigid and vibration free so as to impart a feeling of security and safety to persons using the structure. The allowable live load deflection depends on the human activities carried out in the structure, the sensitivity of the persons using it and the amount of structural damping. For example, the allowable span/deflection ratio in a prestigious public building should be as high as 1,000 or 2,000. In a low cost office building where deflections are damped by many partitions, a span/deflection ratio of 360 or 400 would be adequate. In a roof or a warehouse floor the span/deflection ratio could be as low as 180. These ratios are based on the live load deflection and not the total load. In concrete structures, live load deflection is rarely a problem.

*The incremental deflection is that portion of the total deflection occurring during a limited time period. It will include time-dependent deflections for that period and elastic deflections for loads applied or removed during the period. As used here, the incremental deflection time period starts with construction of the fragile wall and continues for the life of the wall.

2.2 DESIGN PROCEDURES

2.2.1 Reinforced Concrete Linear Members—Abbreviated Procedure

For a quick estimate of probable deflection, the procedure outlined below will suffice. It should be accurate within -50% to +100%. If the computed deflection is excessive, the engineer may want to recompute the deflection using the extended procedure in the next section, before revising the member sizes.

Step 1 Assemble the necessary data.

a. Dead load.
b. Live load.
c. Actual moments produced by the loading condition for which deflection is to be computed. Frequently, these moments will be available from the design for strength but care must be taken to insure that they are not overstated such as would be the case if an approximate frame analysis is made using the moment coefficients given in Section 8.4 of the ACI 318 *Code*.[2-4]
d. Physical dimensions of the concrete. Size, quantity and location of flexural reinforcement.

Step 2 Compute the percentage of the principal tensile reinforcement at the point of maximum midspan moment (or at the support for cantilevers).

Step 3 Compute the effective rigidity at the point of maximum midspan moment (or at the support for cantilevers). The effective rigidity is the gross uncracked rigidity $(E_c I_g)$*, 60% of the gross uncracked rigidity or the cracked rigidity $(E_c I_{cr})$, depending on the percentage of tensile reinforcement as given in Table 2-1.

The gross uncracked rigidity can be computed from fundamental principles of strength of materials. For T-beams, the curves in Fig. 2-1 will greatly simplify the computation.

The cracked flexural rigidity[2-6] is

$$E_c I_{cr} = E_s A_s (1 - k) j d^2 \tag{2-1}$$

where k = ratio of height of compression block by elastic theory to effective depth, d, and j = ratio of internal lever arm by elastic theory to effective depth, d.

The equation can be solved from fundamental principles for the elastic design of reinforced concrete or by using tables for k and j. The curves in Fig. 2-2 simplify the computation for rectangular, singly-reinforced members. The curves are also valid for T-beams if the neutral axis falls within the flange and are accurate within a few percent if the neutral axis is just below the flange. Only in cases where the flange is considerably thinner than $k \cdot d$ for a rectangular cross section will it be necessary to compute k and j from fundamental principles.

*The percentage of principal reinforcement indicating the line of demarcation between an uncracked and a partially cracked cross section can be computed by equating the moment at first cracking to the moment capacity of the reinforcing steel. Thus:

$$f_r b_w h^2 \gamma_1 / 6 = M_{cr} = \rho b_w \left[\left(\frac{d}{h}\right) h\right]^2 \cdot \left(\frac{0.9 f_y}{\gamma_2}\right)\left(1 - \frac{a}{2d}\right)$$

and

$$\rho = \frac{\gamma_1 \gamma_2}{5.4(1 - a/2d)(d/h)^2} \times \frac{f_r}{f_y} \tag{2-2}$$

in which: f_r = modulus of rupture of concrete
γ_1 = ratio of section modulus for an uncracked tee beam to section modulus for the stem of the uncracked T-beam
γ_2 = average load factor
$(1 - a/2d)$ = internal lever arm

TABLE 2-1* Limits of Principal Tensile Steel Percentage for the Determination of Effective Rigidity

	Effective Rigidity		
Type of Member	$E_c I_g$	$0.6 E_c I_g$	$E_c I_{cr}$
Solid Slabs and Rectangular Beams	$\rho < 0.4\%$**	$0.4\% < \rho < 0.8\%$	$\rho > 0.8\%$
T-beams with Tension in the Stem	$\rho_w < 0.6\%$	$0.6\% < \rho_w < 1.2\%$	$\rho_w > 1.2\%$
T-beams with Tension in the Flange	$\rho_w < 1.5\%$	$1.5\% < \rho_w < 3.0\%$	$\rho_w > 3.0\%$

**Table 2-1 is based on the following assumptions:*

1. *In T-beams the area of the flange is equal to or greater than the area of the stem. For T-beams with smaller flanges, values may be interpolated between the values for rectangular beams and values for T-beams given here.*
2. *The member has a prismatic concrete cross section.*
3. $f_r = 10\sqrt{f'_c} = 550$ *psi where* $f'_c = 3000$ *psi.*
4. $\gamma_1 = 1.00$ *for rectangular beams, 1.5 for T-beams with tension in the stem and 3.75 for T-beams with tension in the flange.*
5. $d/h = 0.83$.
6. $f_y = 60$ *ksi. The values of* ρ *are inversely proportional to other values of* f_y.
7. *The average load factor,* $\gamma_2 = 0.5(1.4 + 1.7) = 1.55$, *assuming live load equals dead load. The values of* ρ *are directly proportional to other load factors.*
8. $(1 - a/2d) = 0.95$.

***For example, using eq. (2.2):* $\rho = \dfrac{1.00 \times 1.55}{5.4(0.95)(0.83)^2} \times \dfrac{550}{60{,}000} = 0.0040 = 0.4\%$.

Step 4 Compute the deflection which occurs immediately on application of load by the usual methods or formulas for elastic deflections. The deflection formulas for most common loading and support conditions are shown in Table 2-2. The computation can be further simplified by an appropriate fraction of the deflection for a simple beam under a uniformly distributed load. These fractions are also given in Table 2-2.

The additional long-time deflection can be estimated by multiplying the immediate deflection caused by the sustained load considered by the factor $[2 - 1.2(A'_s/A_s) \geq 0.6]$ to allow for creep and shrinkage of the concrete.

Step 5 Assess the acceptability of the computed deflection. If the deflection appears to be critical, compute incremental deflections for the critical time period as determined by the projected loading history or specify the loading history that is necessary to obtain satisfactory results. In some cases it may be necessary to revise the width or depth of the member or to add compressive reinforcement to obtain a satisfactory deflection response.

EXAMPLE 2-1: The slab in Fig. 2-3 carries a superimposed load of 200 psf including an allowance for masonry partitions. Assume normal weight concrete. Since this is a thinner slab than permitted by Table 9.5(a) of the ACI 318 *Code*,[2-4] the deflections should be computed.

Step 1 From a frame analysis, the maximum midspan moment is 90 k-in./ft.

Step 2 $\rho = 0.31 \text{ in.}^2/(8 \text{ in.} \times 7 \text{ in.}) = 0.55\%$, where $d = 7$ in.

Step 3 From Table 2-1 the effective rigidity is 60% of the gross uncracked rigidity

$E_c = 57{,}000\sqrt{3000 \times 1.15} = 3.35 \times 10^6$ psi (assuming the average concrete strength is 15% higher than the minimum)

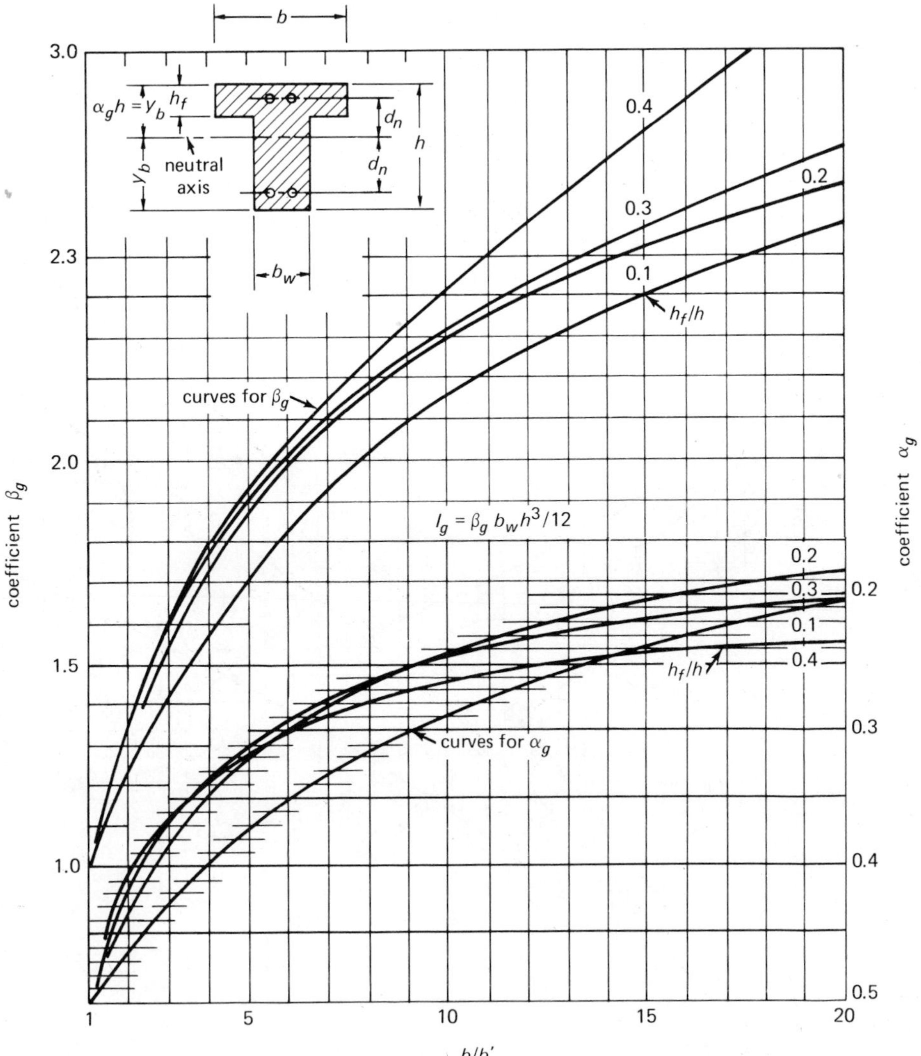

Fig. 2-1 Curves for flexural rigidity and location of neutral axis of uncracked reinforced concrete T-beams.

$I_g = 12 \times (8)^3/12 = 512$ in.^{4}/ft

$E_c I_g = 0.60 \times 3.35 \times 10^6 \times 512 = 1.03 \times 10^9$ lb-in.^{2}/ft

Step 4 Referring to Table 2-2, the slab in this example falls somewhere between cases 1 and 4 since loading on one span only results in a negative moment smaller than the fixed ended moment. Therefore, it can be assumed a fraction of 0.80 will be sufficiently accurate.

$$a_i = 0.8 \times \frac{5 \times 90{,}000 \times (18 \times 12)^2}{48 \times 1.03 \times 10^9} = 0.34, \text{ say } 0.3 \text{ in.*}$$

of which the dead load $a_i = 0.3(100 \text{ psf}/300 \text{ psf}) = 0.1$ in. and the superimposed load $a_i = 0.3(200 \text{ psf}/300 \text{ psf}) = 0.2$ in.

If one half the superimposed load is dead load from partitions, the additional long-time deflection, a_{lt}, is

$$a_{lt} = (0.1 \text{ in.} + 0.2 \text{ in.}/2) \times 2 = 0.4 \text{ in.}$$

the total deflection from all sources $a_t = 0.7$ in.

Step 5 The live load deflection of 0.2 in. is less than the 0.6 in. (18 × 12/360) permitted by Table 9.5(b) in the ACI 318 *Code*.[2-4] The deflection which occurs after the partitions are built is 0.4 in. + 0.2 in./2 = 0.5 in. which is slightly more than the (18 × 12/480) = 0.45 in. permitted by Table 9.5(b). To rectify this situation, the structural engineer can

1) Compute deflections by the extended procedure which might, in some cases, result in a lower computed deflection;

2) Make the slab thicker or add compressive reinforcement;

3) Delay construction of the partitions until a portion of the long term dead load deflection has taken place;

4) Construct partitions in such a way that they will not be damaged by slab deflections; or

5) Make changes or adopt procedures appropriate for the actual construction situation.

EXAMPLE 2-2: The frame for a telephone equipment building shown in Fig. 2-4 and Fig. 2-5 is designed to carry a superimposed (live) load of 175 psf of which the owner expects an average permanent load of 75 psf. What will be the expected live load and total load deflections and the critical incremental deflection of the beam? Will the deflection response be satisfactory?

SOLUTION:

Step 1 Loads and moments are taken from the design for strength.

Dead load of structure		1.54 kip/ft
Permanent live load	1.20	
Transient live load	1.60	
Total superimposed load		2.80
Total load		4.34 kip/ft

*Deflection values to the nearest 0.1 in. represent the highest level of accuracy suggested for the abbreviated procedure.

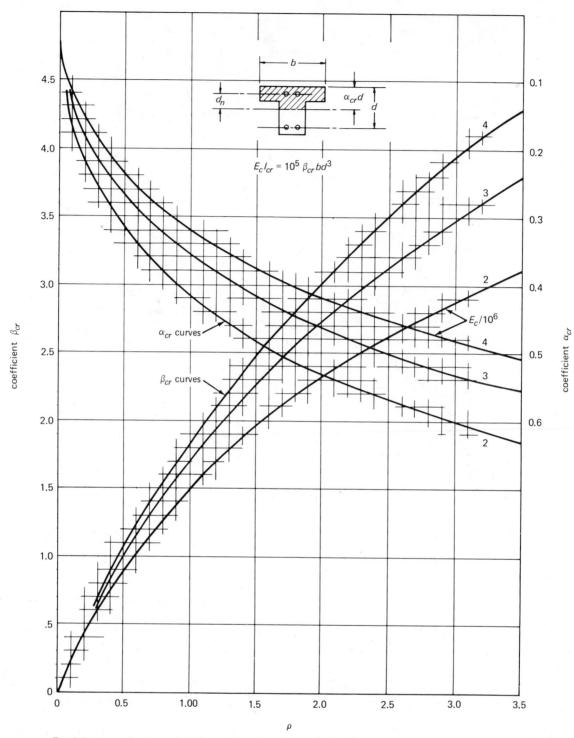

Fig. 2-2 Curves for flexural rigidity and location of neutral axis of cracked reinforced concrete T-beams.

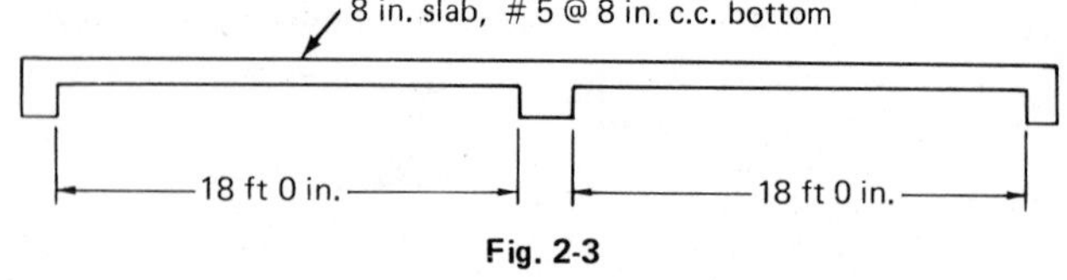

Fig. 2-3

Total load negative moment at the face of each support =1420 kip-in. (In this example, the negative moment was taken from a frame analysis using the total load.)

To prepare an accurate estimate of deflection, amplified or "safe" moments used in design for strength should be avoided. Therefore the actual net positive moment is computed as follows:

$$0.125 \times 4.34 \text{ k/ft} \times (27 \text{ ft})^2 \times 12 \text{ in./ft} = 4750 \text{ kip-in.}$$

$$\text{less average negative moment} = \underline{1420}$$

$$\text{Net positive moment} = 3330 \text{ kip-in.}$$

TABLE 2-2 Beam Diagrams and Deflection Formulas for Static Loading Conditions

*Ratio**	
1.00	1. Simple Beam—uniformly distributed load 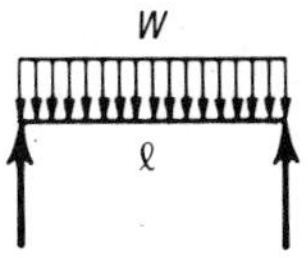 W, ℓ $a = \frac{5}{384} \times \frac{Wl^3}{EI} = \frac{5}{48} \times \frac{M_a l^2}{EI}$
0.80	2. Simple Beam—concentrated load at center 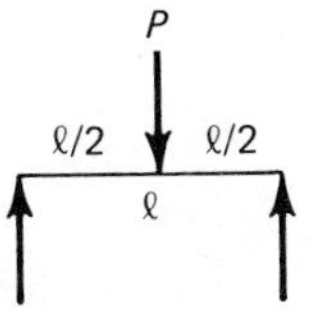 P, ℓ/2, ℓ/2, ℓ $a = \frac{Pl^3}{48EI} = \frac{1}{12} \times \frac{M_a l^2}{EI}$
1.02	3. Simple Beam—two equal concentrated loads at third points 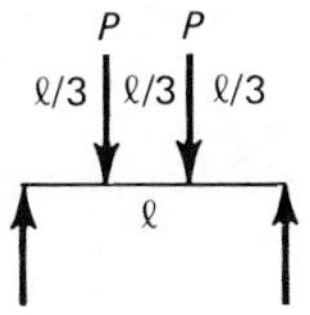 P, P, ℓ/3, ℓ/3, ℓ/3, ℓ $a = \frac{23Pl^3}{648EI} = \frac{23}{216} \times \frac{M_a l^2}{EI}$
0.74	4. Beam fixed at one end, supported at other—uniformly distributed load 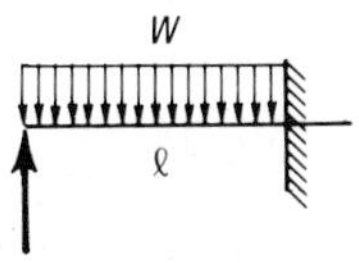 W, ℓ $a = \frac{Wl^3}{185EI} = \frac{128}{1665} \times \frac{{}^{+}M_a l^2}{EI}$
0.57	5. Beam fixed at one end, supported at other—concentrated load at center 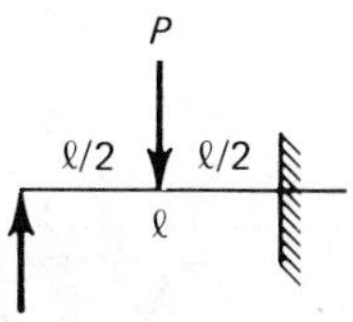 P, ℓ/2, ℓ/2, ℓ $a = \frac{Pl^3}{48\sqrt{5}\,EI} = 0.00932\,\frac{Pl^3}{EI}$ $a = \frac{2}{5\sqrt{5}} \times \frac{M_a l^2}{EI} = 0.0596\,\frac{M_a l^2}{EI}$
0.60	6. Beam fixed at both ends—uniformly distributed loads 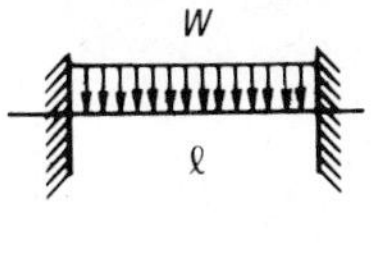 W, ℓ $a = \frac{Wl^3}{384EI} = \frac{1}{16} \times \frac{{}^{+}M_a l^2}{EI}$
0.40	7. Beam fixed at both ends—concentrated load at center 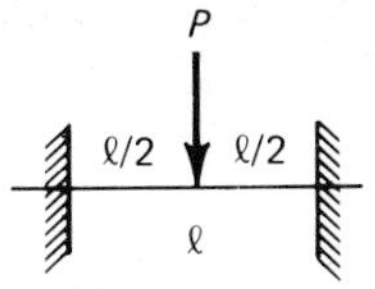 P, ℓ/2, ℓ/2, ℓ $a = \frac{Pl^3}{192EI} = \frac{1}{24} \times \frac{M_a l^2}{EI}$

TABLE 2.2 *(Continued)*

*Ratio**	
0.67	8. Beam fixed at both ends—two equal concentrated loads at third points $a = \frac{5Pl^3}{648EI} = \frac{5}{72} \times \frac{{}^{+}M_a l^2}{EI}$
2.4	9. Cantilever beam—uniformly distributed load $a = \frac{Wl^3}{8EI} = \frac{1}{4} \times \frac{{}^{-}M_a l^2}{EI}$
3.2	10. Cantilever beam—concentrated load at free end $a = \frac{Pl^3}{3EI} = \frac{1}{3} \times \frac{{}^{-}M_a l^2}{EI}$
0.6	11. Simple beam—moment at one end $a = \frac{1}{16} \times \frac{M_a l^2}{EI}$

**Ratio of deflection for the subject case to deflection for a simple beam with uniformly distributed load causing an equivalent maximum moment.*

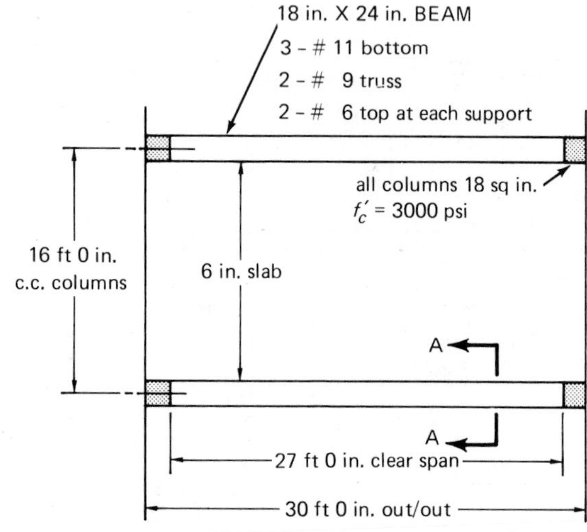

Fig. 2-4

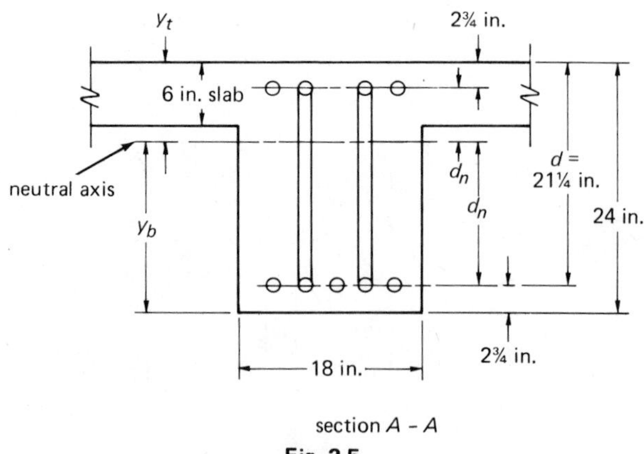

Fig. 2-5

Step 2 $\rho_w = 6.68 \text{ in.}^2/18 \text{ in.} \times 21.25 \text{ in.} = 1.75\%$.

Step 3 Since $\rho_w = 1.75\% > 1.2\%$, use the cracked rigidity.

$A_s = 6.68$ sq in., $b = (30 - 1.5)/4 = 7.125$ ft $= 85.5$ in.

$\rho = 6.68/85.5 \times 21.25 = 0.37\%$

from Fig. 2-2

$$\beta_{cr} = 0.80, \text{ using } E_c \cong 3.5 \times 10^6 \text{ psi}$$

then

$$E_c I_{cr} = 10^5 \times 0.80 \times 85.5(21.25)^3 = 66 \times 10^9 \text{ lb-in.}^2$$

Step 4 In this case, the beam is about midway between a simple span and a beam fixed at both ends. Therefore, a fraction of 0.80 can be interpolated between case 1 and case 6 in Table 2-2.

$$a_i = 0.8 \frac{5 \times 3{,}330 \times 10^3 \times (27 \times 12)^2}{48 \times 66 \times 10^9} = 0.44 \text{ in.} \quad \text{Say } 0.4 \text{ in.}$$

of which dead and permanent superimposed load

$$a_i = \frac{2.74}{4.34} \times 0.4 = 0.3$$

and live load

$$a_i = \frac{1.60}{4.34} \times 0.4 = 0.1$$

The additional long-term deflection is

$$a_{lt} = 0.3 \text{ in.} \times 2 = 0.6$$

The total deflection, $a_t = 0.4 + 0.6 = 1.0$ in.

Step 5 The live load deflection is considerably less than the 0.9 in. (27 × 12/360) permitted by the ACI 318 *Code*.[2-4] The incremental deflection occurring after partitions have been built is 0.7 in. (0.6 + 0.1) which is more than the 0.675 in. (27 × 12/480) permitted by the ACI 318 *Code* even though the depth is greater than minimum allowable. (Table 9.5(a) of the ACI 318 *Code* requires a depth of only 28.5 × 12/21 = 16.5 in.) The dilemma this situation poses is explored further in Example 2-3.

2.2.2 Reinforced Concrete Linear Members—Extended Procedure

If greater accuracy than is possible by the procedure outlined in section 2.2.1 is desired, the following extended procedure can be used at the expense of more computation time. Even so, accuracy greater than about −20% to +30% should not be expected.[2-7] The extended procedure can also be used to study the effect of various parameters on the final deflection.

Step 1 Assemble the necessary data.

a. Dead load of the concrete structure.

b. The permanent superimposed load which could be either additional dead load or a portion of the live load.

c. Transient superimposed load which is the remaining live load and sometimes other loads such as wind.

d. Actual moments as required in Step 1 of section 2.2.1.

e. Physical dimensions of the concrete. Size, quantity and location of flexural reinforcement.

f. The sequence and timing of the placing of the superimposed loading. This information may be design requirements of the ultimate user of a structure, or assumed from the engineer's general knowledge of construction procedures. In some cases the sequence and timing of superimposed loading will be modified by the engineer after he assesses the results of deflection computations.

g. Properties of the concrete: modulus of elasticity, modulus of rupture, specific creep and shrinkage.

Step 2 Compute the uncracked moment of inertia and section modulus at each critical section, generally at midspan and at each end for continuous members. Use fundamental principles or the curves of Fig. 2-1. The effect of reinforcing steel can be conservatively estimated by using 80% of the moment of inertia created by transferring the transformed area of steel to the computed neutral axis of the plain concrete section. High percentages of steel have a significant effect on the moment of inertia especially in T-beams.

Step 3 Determine whether the member will be cracked, and if so, in what regions. This can be done by comparing the computed flexural stresses in the plain concrete to the modulus of rupture of the concrete. Although most codes and textbooks recommend a modulus of rupture for concrete that is on the low side, it is generally more realistic to assume an average or even a high value for the modulus of rupture when computing deflections. This is because the presence of a single crack barely affects the deflection response of a concrete member. Furthermore, an overstatement of deflection will many times be as troublesome as an understatement. A value one-third higher than that specified in Section 9.5.2.2 of the ACI 318 *Code*[2-4] is recommended unless more specific data for the concrete in question is available. For empirical calculations the average cylinder strength should be used instead of the specified minimum. Similarly, the average modulus of rupture should be used rather than the minimum value obtained from a set of test specimens.

Step 4 Compute the cracked rigidity of each section of the member that was determined in the previous step to be cracked. Use eq. (2-1) or Fig. 2-2 as described in Step 3 of Section 2.2.1. As with an uncracked moment of inertia, the effect of compressive reinforcement on the rigidity can be estimated with sufficient accuracy by including 80% of the transfer moment of inertia of the transformed steel area times the modulus of elasticity of concrete. The tensile steel area should not be so included.

Step 5 Compute the effective rigidity in the positive and negative moment regions of the beam that will be cracked. The cracked rigidity can be used with sufficient accuracy if the cracking moment is less than about 50% of the actual moment. For other cases, the following equation can be used:

$$E_c I_e = E_c I_{cr} + \left(\frac{M_{cr}}{M_a}\right)^3 (E_c I_g - E_c I_{cr}) \tag{2-3}$$

This equation is a restatement of eq. 26 from the ACI 435 report.[2-6] It is shown graphically in Fig. 2-6.

Step 6 Compute the average rigidity for the full length of the member. It will simply be the effective rigidity at midspan for simple span members and the effective rigidity at the support for cantilevers. The average rigidity of continuous members can also be taken as the effective rigidity at midspan if the ratio of end region to midspan rigidity is no larger than 2.0 and the moment pattern is between that for a simple span and fixed-ended span. For other continuous members, the average rigidity should include the contribution of the effective rigidities at midspan and each support in proportion to the magnitude of the moments at these

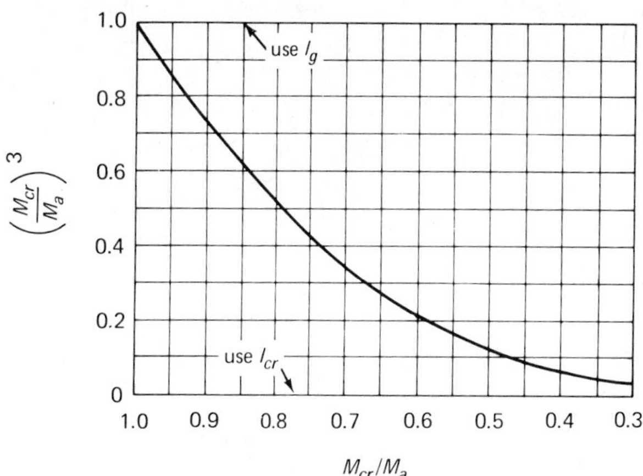

Fig. 2-6

points. This can be done by using

$$(EI)_a = (E_c I_e)_c \left[1 - \left(\frac{M_{e1} + M_{e2}}{2M_0}\right)^4\right] + \frac{(E_c I_e)_{e1} + (E_c I_e)_{e2}}{2} \left(\frac{M_{e1} + M_{e2}}{2M_0}\right)^4 \quad (2\text{-}4)$$

(Despite its length, eq. 2.4 is relatively simple to use as shown in Example 2-3.) While this procedure differs from that recommended in the ACI 318 Code,[2-2] analytical studies indicate it is more accurate over a wider range of conditions than is the simple average of the moment of inertia for the positive and negative moment sections.

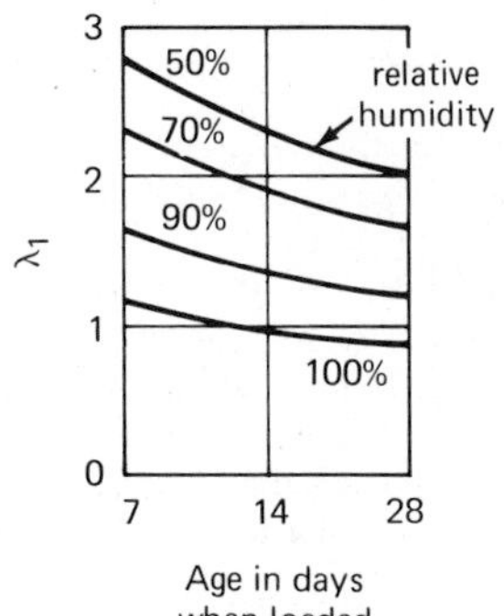

Fig. 2-7

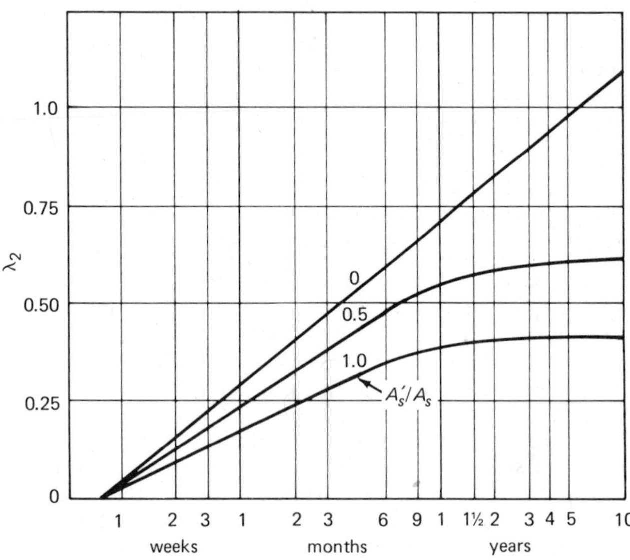

Fig. 2-8

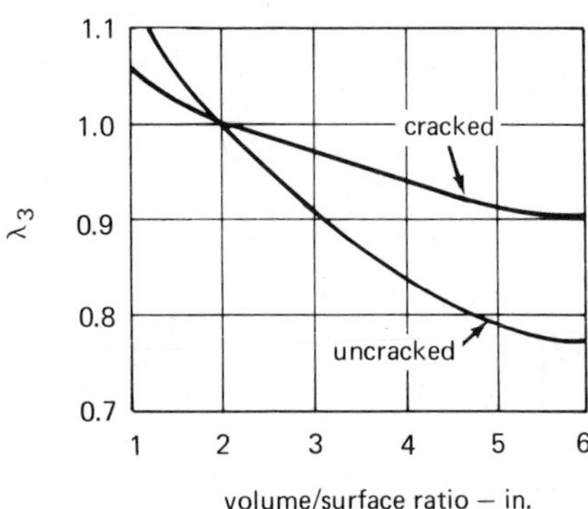

Fig. 2-9

Step 7 Compute the deflection which occurs immediately on application of load by the usual methods or formulas for elastic deflections. Compute the additional long-time deflection by multiplying the immediate deflection by factors.

$$a_{lt} = a_i \lambda_1 \lambda_2 \lambda_3 \quad (2\text{-}5)$$

where

λ_1 = a coefficient taking into account the age when loaded and the ambient relative humidity. (See Fig. 2-7 which has been prepared from data in Ref. 2-2.)

λ_2 = a coefficient taking into account the duration of loading and the amount of compressive reinforcement. (See Fig. 2-8 which has been drawn from Table 2 in Ref. 2-8.)

λ_3 = a coefficient taking into account the volume to surface ratio of the concrete member. (See Fig. 2-9 which has been prepared from data in Ref. 2-1.)

If further accuracy is desired in the computation of long term deflections, or if the shrinkage and creep values of the concrete to be used in the structure are different from normal or average values, the sustained modulus method can be used and shrinkage warping computed separately. The engineer is referred to the report by ACI Committees 435[2-6] and 209[2-2] for details of this approach.

Step 8 Assess the acceptability of the computed deflection as in Step 5 of section 2.2.1.

EXAMPLE 2-3: Using the same beam as in ex. 2-2, compute the deflections and assess their acceptability.

SOLUTION:

Step 1 Use the same data as in ex. 2-2.

Step 2 The uncracked moment of inertia is computed using Fig. 2-1.

$$\text{beam T} = (30 - 1.5)/4 = 7.125 \text{ ft} = 85.5 \text{ in.}$$
$$b/b_w = 85.5/18 = 4.75$$
$$h_f/h = 6/24 = 0.25, \text{ from Fig. 2-1, } \beta_g = 1.85, \alpha_g = 0.32$$

Therefore the neutral axis location is $\alpha_g h = 0.32 \times 24 = 7.7$ in.

$$\text{Concrete } I_g = \beta_g b_w h^3/12 = 1.85 \times 18 \times (24)^3/12 = 38{,}400$$

$$\text{Steel } I_g = 0.8(n - 1)A_s(d_n)^2 = 0.8(9 - 1)6.68(21.25 - 7.7)^2 = 7{,}850$$

Midspan I_g 46,250 in.^{4}

End Region I_g = 38,400 in.^{4} (ignoring slight effect of steel at ends)

$$y_t = \alpha_g h = 7.7 \text{ in.,} \quad S_t = \frac{38{,}400}{7.7} = 5{,}000 \text{ in.}^3 \text{ at ends}$$

$$y_b = h - y_t = 16.3 \text{ in.,} \quad S_b = \frac{46{,}250}{16.3} = 28{,}400 \text{ in.}^3 \text{ at midspan}$$

Since

$$E_c = 57{,}000\sqrt{3000 \times 1.15} = 3.35 \times 10^6$$
$$E_c I_g = 46{,}250 \times 3.35 \times 10^6 = 155 \times 10^9 \text{ at midspan}$$
$$= 38{,}400 \times 3.35 \times 10^6 = 129 \times 10^9 \text{ at ends}$$

Step 3 Compute concrete stresses to determine if the beam is cracked

$$\text{at midspan} \quad f_t = \frac{M_a}{S_b} = \frac{3330 \text{ kip-in.}}{28{,}400 \text{ in.}^3} = 1175 \text{ psi}$$

$$\text{at the ends} \quad f_t = \frac{M_a}{S_t} = \frac{1420 \text{ kip-in.}}{5000 \text{ in.}^3} = 284 \text{ psi}$$

The modulus of rupture should be about $10\sqrt{f_c'} = 10\sqrt{3000 \times 1.15} =$ 590 psi, therefore the beam will be cracked at midspan but not at the ends. Even under the average permanent load, the beam will be partially cracked since the tensile stress at midspan is higher than the modulus of rupture. (By ratio, the tensile stress at midspan is $(4.34 - 1.60)/4.34 \times 1175 = 744$ psi under permanent loads.)

Step 4 From Example 2-2, the cracked rigidity $E_c I_{cr} = 66 \times 10^9$ lb-in.^{2}

Step 5 Using Fig. 2-6, the effective rigidity at midspan is computed as $M_{cr}/M_a = f_r/f_t = 590/1175 = 0.50$. Then, $(M_{cr}/M_a)^3 = .125$, therefore the effective rigidity is

$$E_c I_{cr} = 66 \times 10^9$$
$$(155 \times 10^9 - 66 \times 10^9) \times .125 = 11 \times 10^9$$
$$E_c I_e = 77 \times 10^9 \text{ lb-in.}^2 \text{ at midspan}$$

Step 6 Even though it is not necessary, compute an average rigidity. To average the rigidities at each section of the beam, use eq. (2-4).

$$\left(\frac{M_{e1} + M_{e2}}{2M_0}\right)^4 = \left(\frac{1420}{4750}\right)^4 = 0.008$$

$$(E_c I_e)_c: \quad 77 \times 10^9 (1 - 0.008) = 76.4 \times 10^9$$
$$(E_c I_e)_{e1} + (E_c I_e)_{e2}: \quad 129 \times 10^9 \times 0.008 = 1.0$$
$$(EI)_a = 77.4 \times 10^9$$
$$\text{Say } 77 \times 10^9 \text{ lb-in.}^2$$

Step 7 The deflection is then computed using conventional procedures and the average rigidity.

Simple span deflection

$$\frac{5wl^4}{384(EI)_a} = \frac{5 \times 4.34 \times 10^3 \times (27)^4 \times 1728 \text{ in.}^3/\text{ft}^3}{384 \times 77 \times 10^9} = 0.67 \text{ in.}$$

less deflection due to end moments

$$-\frac{M_e l^2}{8(EI)_a} = -\frac{1420 \times 10^3 \times 27^2 \times 144 \text{ in.}^2/\text{ft}^2}{8 \times 77 \times 10^9} = -0.24$$

Immediate deflection $\quad a_i \quad = 0.43$ in.

of which structure dead load deflection

$$= \frac{1.54}{4.34} \times 0.43 = 0.15 \text{ in.}$$

other dead load deflection

$$= \frac{1.20}{4.34} \times 0.43 = 0.12 \text{ in.}$$

and transient live load deflection

$$= \frac{1.60}{4.34} \times 0.43 = 0.16 \text{ in.}$$

Assuming that the beam will be loaded for the first time at the age of 28 days, from Fig. 2-7, $\lambda_1 = 2.0$ for indoor conditions. From Fig. 2-8, $\lambda_2 = 1.0$ since no compressive reinforcement is provided. The volume/surface ratio = $(67.5 \times 6 + 18 \times 14)/(85.5 + 18)2 = 4$. From Fig. 2-9, $\lambda_3 = 0.9$ for a partially cracked beam. The ratio of additional long-time deflection to immediate deflection = $\lambda_1 \lambda_2 \lambda_3 = 2.0 \times 1.0 \times 0.9 = 1.8$. Thus the total long-time deflection would be

structure deal load	$0.15 \times (1 + 1.8)$	= 0.42
other dead load	$0.12 \times (1 + 1.8)$	= 0.34
transient live load	0.16×1	= 0.16
Total long-time deflection	a_{lt}	= 0.92 in.

If camber is built into the beam, it should be approximately ¾ in. to maintain a level surface.

Step 8 If the permanent superimposed load consists of masonry partitions which can be damaged by excessive deflection, the incremental deflection is computed as follows:

CASE I: If the partitions are constructed before shores are removed from under the concrete, the incremental deflection is the full total deflection of 0.92 inches. The walls might be damaged by this much deflection since the span/deflection ratio is 350. $(27 \times 12/0.92)$.

CASE II: If the walls are constructed immediately after the shores are removed, the incremental deflection is the total long-time deflection less the short-time dead load deflections. Thus the incremental deflection would be $(0.92 - 0.15 - 0.12 =)0.65$ in. The span/deflection ratio of 500 is slightly better than the ACI *Code* minimum of 1/480. The walls might still suffer some damage if they are especially fragile.

CASE III: If the walls are constructed three months after the shores are removed, the incremental deflection is the total long-time deflection less the immediate dead load deflections and less the portion of the long-time dead load deflections occurring before the partitions are constructed.

From Fig. 2-8, factor "λ_2" is 0.45 at three months, thus

total long-time deflection	=	0.92 in.
structure dead load deflection = $0.15(1 + 2 \times .45 \times .9)$	= −0.27	
other dead load short-time deflection	= −0.12	
Subtotal		−0.39
Net incremental deflection =		0.53 in.

The span/deflection ratio of 610 indicates the walls will probably not suffer significant damage in this case.

CASE IV: If still further assurance is desired that deflection will not damage the walls, add top steel for the full length of the span equal to 50% of the bottom steel and assume that construction of the walls will be delayed as in Case III.

From Fig. 2-8, factor "λ_2" is 0.4 at three months and 0.6 at 10 years, thus

long-time deflection:		
structure dead load	$0.15(1 + 2 \times .6 \times .9) = 0.31$	
other dead load	$0.12(1 + 2 \times .6 \times .9) = 0.25$	
transient live load	$0.16 \times 1.0 = 0.16$	
Total =		0.72 in.

Deflection prior to completion of partitions is

structure dead load	$0.15(1 + 2 \times .4 \times .9) = 0.26$	
other dead load	$0.12 \times 1.0 = 0.12$	
Subtotal		−0.38
Net incremental deflection =		0.34 in.

The span/deflection ratio of 950 is sufficiently conservative to satisfy most criterion for deflection limits.

CASE V: Now suppose the contractor strips the formwork at seven days and does not reshore. He starts work on the masonry partitions immediately and has them completed by the time the concrete is 14 days old. What will be the effect of deflection on the partitions?

Assume f_c' at seven days = 70% f_c' at 28 days. Therefore

$$E_{c7} = 57{,}000\sqrt{.70 \times 1.15 \times 3000} = 2.82 \times 10^6$$

and

$$f_{r7} = 10\sqrt{.70 \times 1.15 \times 3000} = 490 \text{ psi}$$
$$E_c I_g = 46{,}250 \times 2.82 \times 10^6 = 131 \times 10^9 \text{ at midspan}$$
$$E_c I_g = 38{,}400 \times 2.82 \times 10^6 = 108 \times 10^9 \text{ at the ends}$$

From Fig. 2-2, $\beta_{cr} = 0.75$ for $E_c = 2.82 \times 10^6$ and $\rho = 0.37\%$

then

$$E_c I_{cr} = 10^5 \times 0.75 \times 85.5(21.25)^3 = 61.5 \times 10^9 \text{ lb-in.}^2$$
$$M_{cr}/M_a = f_r/f_t = 490/1175 = 0.417$$

and

$$(M_{cr}/M_a)^3 = 0.073$$

therefore the effective rigidity is

$$E_c I_{cr} = 61.5 \times 10^9$$
$$(131 - 61.5)10^9 \times 0.073 = \underline{5.1} \times 10^9$$
$$E_c I_e = 66.6 \times 10^9 \text{ lb-in.}^2 \text{ at midspan}$$

Use the midspan rigidity as the average rigidity and round off to $(EI)_a = 67 \times 10^9$ lb-in.^{2}.

Compute deflections by multiplying the deflections computed in Step 7 of ex. 2-2, by the inverse ratio of flexural rigidities.

$$(EI)_{a\text{-}28\text{ day}}/(EI)_{a\text{-}7\text{ day}} = 77/67 = 1.15$$

The total immediate deflection, $a_i = 0.43 \times 1.15$ = 0.49 in.
of which, structure dead load deflection 0.15×1.15 = 0.17 in.
other dead load deflection 0.12×1.15 = 0.14 in.
and transient live load deflection 0.16×1.15 = 0.18 in.

From Fig. 2-7, $\lambda_1 = 2.8$ for loading at 7 days
and $\lambda_1 = 2.3$ for loading at 14 days
From Fig. 2-8, $\lambda_2 = 0.05$ for one week duration of loading. Thus, the ultimate long-time deflection would be:

structure dead load, $0.17(1 + 2.8 \times 1.0 \times 0.9)$ = 0.61
other dead load, $0.14(1 + 2.3 \times 1.0 \times 0.9)$ = 0.43
transient live load, 0.18×1.0 = 0.18
Total long-time deflection = 1.22 in.

Deflection prior to completion of partitions is

structure dead load, $0.17(1 + 2.8 \times 0.05 \times 0.9)$ = 0.19
other dead load, 0.14×1.0 = .14
Subtotal = −0.33
Net incremental deflection = 0.89 in.

Note that this is a higher incremental deflection than computed previously except for Case I. The span/deflection ratio of 360 indicates that a fragile masonry partition will probably sustain visible damage from deflection of the beam.

REFERENCES

2-1 Hansen, T. C., and Mattock, A. H., "The Influence of Size and Shape of Member on the Shrinkage and Creep of Concrete," *ACI Journal, Proceedings* **63**, (2), 267–290, Feb. 1966.

2-2 ACI Committee 209, Subcommittee 2, "Prediction of Creep, Shrinkage, and Temperature Effects in Concrete Structures," ACI Publication, SP-27, 1971.

2-3 Branson, D. E., and Christiason, M. L., "Time-Dependent Concrete Properties Related to Design—Strength and Elastic Properties, Creep and Shrinkage," ACI Publication, SP-27, 1971.

2-4 ACI Committee 318, *Building Code Requirements for Reinforced Concrete* (ACI 318-71), American Concrete Institute, 1971.

2-5 ACI Committee 435, Subcommittee 1, "Allowable Deflections," *ACI Journal, Proceedings* **65**, (6), 433–444, June 1968.

2-6 ACI Committee 435, "Deflections of Reinforced Concrete Flexural Members," *ACI Journal, Proceedings* **63**, (6), 637–674, June 1966.

2-7 ACI Committee 435, Subcommittee 2, "Variability of Deflections of Simply Supported Reinforced Concrete Beams," *ACI Journal, Proceedings* **69**, (1), 29–35, January 1972.

2-8 Yu, Wei-Wen, and Winter, George, "Instantaneous and Long-Time Deflections of Reinforced Concrete Beams Under Working Loads," *ACI Journal, Proceedings* **57**, (1), 29–50, July 1960.

NOTATION

(excepting terms defined in *ACI 318 Code*[2-4])

a_i = deflection of a member relative to a line joining the ends of the span, or of the free end of a cantilever relative to its support, which occurs immediately upon application of the load, in inches

a_{lt} = deflection of a member relative to a line joining the ends of the span, or of the free end of a cantilever relative to its support, which occurs after application of the load due to time-dependent strains, in inches

a_t = the total deflection from all sources

d_n = distance from neutral axis of a reinforced concrete beam to centroid of reinforcement (See Figs. 2-1 and 2-2)

$(E_c I_e)_c$ = the effective flexural rigidity of flexural members at midspan

$(E_c I_e)_{e1}, (E_c I_e)_{e2}$ = the effective flexural rigidity of flexural members at ends No. 1 and No. 2

$(EI)_a$ = average effective flexural rigidity of flexural members

f_t = stress in concrete at extreme tensile fiber in an uncracked concrete section

h_f = thickness of flange of a T-beam (see Fig. 5-1)

j = ratio of internal lever arm by elastic theory to effective depth, d

k = ratio of height of compression block by elastic theory to effective depth, d

M_{e1}, M_{e2} = maximum moment in continuous concrete member at ends No. 1 and No. 2, respectively

S_t, S_b = section modulus of an uncracked concrete section with top or bottom fiber in tension, respectively, $S_t = I_g/y_t$. $S_b = I_g/y_b$

y_t, y_b = distance from the centroidal axis of reinforced concrete section, neglecting the reinforcement, to the extreme top or bottom fiber, respectively (See Fig. 2-1)

α_{cr} = ratio of the distance from extreme compression fiber to neutral axis, to the effective depth, d, of a cracked beam (See Fig. 2-2)

α_g = ratio of the distance from the neutral axis of an uncracked T-beam to the extreme fiber in the T, to the overall thickness of the beam (See Fig. 2-1)

β_{cr} = a coefficient of bd^3 giving the cracked flexural rigidity of a reinforced concrete beam

β_g = ratio of moment of inertia for an uncracked T-beam to moment of inertia for the stem of the uncracked T-beam

γ_1 = ratio of section modulus for an uncracked T-beam to section modulus for the stem of the uncracked T-beam

γ_2 = average load factor

$\lambda_1, \lambda_2, \lambda_3$ = coefficients (See Sect. 2.2.2)

3

One- and Two-way Slabs

SIDNEY H. SIMMONDS, Ph.D.*

3.1 INTRODUCTION

Reinforced concrete slabs are one of the most widely used structural elements. In many structures the slab, in addition to providing a versatile and economical method of supporting vertical loads, also forms an integral portion of the structural frame to resist lateral forces. In spite of their widespread use, due to the complexity of plate behavior, most slabs over the past sixty years have been designed using simplified methods of analysis, and many slabs which have had excellent performance records have been proportioned from sets of design coefficients. However, the presence of unusual support conditions can substantially alter the behavior and stress patterns in the slab system, and occasionally the use of simplified methods of design for support conditions outside the range originally intended has resulted in slabs with unsatisfactory performance in the form of excessive deflection and cracking. These two serviceability factors provide the criteria on which most slab designs are ultimately judged and their importance cannot be overly emphasized.

Slabs are classified by the way they are supported. Those slabs which are supported such that they can deflect essentially in only one direction are known as one-way slabs. Those slabs which are supported by isolated supports (columns) arranged in more or less regular rows and which permit the slab to deflect in two orthogonal directions are classified as two-way slabs.**

To improve performance and economy, two-way slabs may be stiffened by the addition of beams between columns, the thickening of the slab in the vicinity of the column (drop panels), or by the flaring of the column cross section directly under the slab (column capital). Slab systems on supports which are random in size and location are known as irregular slabs and may exhibit both one-way and two-way behavior. Such slabs require special attention.

The most common building slab is the two-way slab supported on columns forming reasonably rectangular panels and supporting mainly distributed loads. To compensate for the lack of rigorous analysis most building codes for reinforced concrete contain design procedures for this type of slab system. In the past there has been no single procedure on which all designers have agreed so that several methods of design were given for essentially the same slab system. Recently, a better understanding of the behavior of two-way slab systems has led to the development of design procedures which are more consistent with the current philosophy of reinforced concrete design, namely a prescribed factor of safety by the use of load factors and ultimate strength design. These new procedures have been

*Prof. of Civil Engineering, The University of Alberta, Edmonton, Alberta, Canada.

**Not to be confused with past usage of "two-way" to designate a slab system with beams in both directions (see section 3-2).

adopted by several building codes such as, for example, ACI 318-71[3-1]*. The requirements of the new procedures and examples of their use are presented in subsequent sections.

3.2 HISTORICAL DEVELOPMENT OF SLAB DESIGN

Before examining the new slab design requirements, it is advantageous to review the origin of the previous code requirements for slab design. Immediately one is confronted with the decision to examine the basis for "two-way" slabs or the basis for "flat" slabs, since each was treated as a separate structural component even to the extent of placing the requirements in different code chapters. The reason for these very different requirements is that each system has a unique genesis.

Although patents were issued for reinforced concrete slabs as early as 1854 and 1867, the analyses were based on the reinforcing acting as either the tie to an arch or as a catenary with the concrete used primarily as a filler. Such slabs were restricted to small spans on a gridwork of closely spaced beams and were not economical. Lacking a means of analyzing such highly redundant structures, it was not until the end of the nineteenth century that designers showed an interest in reinforced concrete slabs.

Not surprisingly, early designers turned to classical plate theory for a means of analysis, the governing differential equation for plate bending having been presented by Lagrange as early as 1807. However, it was not until 1899 that Levy, using an infinite series of hyperbolic functions, solved this equation for certain single rectangular panels supporting a distributed load. In 1909, similar problems were solved by Ritz using energy methods. Since these solutions assumed simply supported or clamped edges only, the panel boundaries were nondeflecting and the slab bending moments based on these solutions are valid only when stiff beams are present on all four sides of each panel. This type of solution formed the basis for the design of "two-way" slab systems wherein the slab moments were obtained from a set of coefficients based on a modified elastic analysis and the beams were proportioned for the remainder of the static moment. Consistent with the original assumptions of nondeflecting panel boundaries the required beams were generally quite large.

In 1903, Turner received a patent for a "mushroom" slab, that is, a slab supported directly by columns with no beams spanning between columns. Two years later he built the first such slab in Minneapolis but, since there was no rigorous mathematical basis for the design, a performance load test was required. While much has been written regarding the validity of these load tests, suffice it to state here that his "flat" slab and the similar slabs of many of his competitors were deemed satisfactory. By 1913 over 1000 such slabs had been built. In 1914, Nichols[3-2] established a simple criteria for the minimum moment that must exist across critical sections to satisfy basic equilibrium which seemed contrary to the widely accepted understanding of the behavior of "flat" slabs. This difference between Nichols' theoretical requirements and the supposedly measured moments in the load tests are partially reconciled by Westergaard and Slater in their now famous paper[3-3]. Based on their work the provisions for flat slabs were introduced into the 1921 edition of the *ACI Building Code*, and except for the introduction of the frame method in the 1941 edition, there were no major changes until 1971.

*References in this chapter to a building code or to a specified value will be to "Building Code Requirements for Reinforced Concrete," ACI 318-71 unless noted otherwise.

Although the two design procedures existed side by side for half a century, it was known that neither procedure was entirely satisfactory. To begin with the very names "two-way" and "flat" were meaningless (see Fig. 3-1), since both systems carry load by two-way action, both have flat top surfaces and except for the special case of flat plates, both have projections below the soffit. In the 1971 edition of the code, and for the remainder of this chapter, these terms have been replaced by the simpler, more descriptive terms of "slabs with beams" and "slabs without beams," respectively. Previously there was no provision for designing slab systems which fell between the two defined systems, for example, a slab supported on shallow beams of specified depth. However, the most unsatisfactory aspect of the two systems was the great difference in the factors of safety based on ultimate load carrying capacity. To obtain reliable data of failure and to serve as the basis of evaluating several analytical investigations which were in progress, a series of test slabs was undertaken at the University of Illinois[3-4,3-5,3-6]. These test slabs, which were chosen to be typical of their respective prototypes, yielded the following results:

Type of Slab (*former designation*)	*Design Live Load*	*F. of S.* $= \frac{w_f - 1.5w_d}{w_l}$
Two-way (with beams)	50 psf	6.4
Flat plate	40 psf	3.3
Flat slab	100 psf	2.1

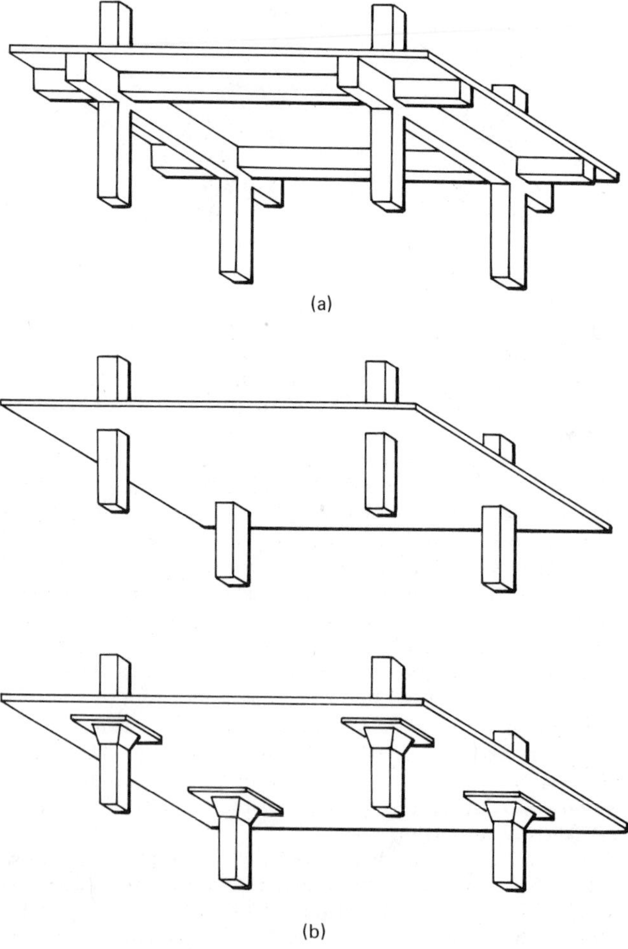

Fig. 3-1 Types of slab systems. (a) Slabs with beams (formerly two-way); and (b) slabs without beams (formerly flat slab and flat plate).

Obviously such discrepancies in the factor of safety at failure for similar structural components are not consistent with the philosophy of ultimate strength design which is used for all other structural elements.

It has been observed repeatedly that reinforced concrete slabs are capable of substantial redistribution of the stress resultants, and that, in general, a slab can support any load without collapse if capacity corresponding to any statically admissable moment configuration is provided and failure due to shear is prevented. It has also been observed that this redistribution is accompanied by cracking and deflection of the slab which may or may not be tolerable. Hence, it is essential that some restriction be placed on the proportioning of the slab bending moments to ensure that serviceability requirements are satisfied. It is on such observations that the design procedure outlined in Section 3.6 is based.

3.3 METHODS OF SLAB ANALYSIS

The extent of analysis required is very much dependent on the type of loading and the geometry of the supports. An exact analysis for reinforced concrete slabs with arbitrary geometry and support conditions due to the highly indeterminate character of the problem, requires the use of numerical techniques such as finite difference[3-7] and finite element[3-8, 3-9] procedures. These methods permit complex analyses but are not, as yet, suitable for routine design use. Fortunately most slabs encountered in practice are supported so that simplified procedures may be used. Where the supports are such that the slab deflects in essentially single curvature (one-way slabs) the analysis may be approximated by that for a continuous beam. Such slabs are discussed in section 3.5. Building slabs consisting of reasonably rectangular panels supported on columns in more or less straight rows (beams, drop panels, and column capitals may or may not be provided) may be proportioned for distributed loading using the code procedures discussed in section 3.6. It is, of course, understood that the presentation of approximate design procedures does not preclude the use of a more exact analysis if one is available and the resulting design meets all specified load factors and serviceability requirements including limits on the slab deflections.

Since the initial step in any analysis of an indeterminate structure is to assign relative stiffnesses, the selection of slab and beam dimensions is discussed prior to a detailed examination of the simplified methods of slab analysis.

3.4 STIFFNESS REQUIREMENT FOR SLAB ELEMENTS

A well designed slab is judged on the basis of its performance under service conditions. The selection of slab thickness and the decision to use auxiliary stiffening elements such as column capitals and beams are therefore of paramount importance in the design of slabs since, to a great extent, these items determine the degree of cracking and the magnitude of the slab deflections. As a general rule, economy of construction dictates that the slab should be as thin as practical and have as few auxiliary stiffening elements as possible. On the other hand the slab must provide sufficient shear capacity and flexural rigidity to keep deflections within acceptable limits. In many instances it is both practical and economical to obtain the required shear capacity and also decrease the slab deflections by providing either column capitals or beams between the columns. We shall first examine the choice of slab thickness and then the use of auxiliary stiffening elements.

With the development of such numerical techniques as finite element and finite difference methods, it is possible to compute slab deflections taking into account the size and shape of the panel; the condition of support; the nature of restraint at the panel edges; the modulus of elasticity; and variations in the effective moment of inertia of the concrete sections. When such factors are considered the designer may use any slab section providing the computed deflections are less than acceptable maximum values. A set of maximum allowable computed deflections for various types of application are listed in the code.

For most design situations the computing of slab deflections is not practical and a minimum thickness which will satisfy serviceability requirements is used. The following expression, which attempts to consider the effects of the major variables on the slab deflection, is specified for minimum slab thickness for all slabs reinforced for moment in two directions:

$$h = \frac{l_n(800 + 0.005f_y)}{36000 + 5000\,\beta[\alpha_m - 0.5(1 - \beta_s)(1 + 1/\beta)]}$$

By specifying the minimum thickness in terms of the clear span, l_n, advantage is taken of the beneficial effect of large column sections or column capitals in reducing slab deflections. The term in the numerator in parenthesis provides the greater slab thickness required to compensate for the decrease in stiffness caused by the greater degree of cracking associated with the higher allowable steel stresses. In ACI 318-63 this increase in thickness was 10% for each 10,000 psi increase in f_y above 40,000 psi, an increase which is retained for one-way slabs. However, in ACI 318-71 for two-way slabs this increase is reduced to 5%, since under service load conditions this type of slab is less likely to be cracked and thus is less sensitive to changes in steel stresses.

The influence of the shape and position of the panel on the slab deflections is considered by the terms β and β_s, respectively. β is defined as the ratio of long to short clear spans and β_s as the ratio of length of continuous edges to total perimeter of a slab panel. The effect of the presence of stiffening beams is given by α_m which is defined as the average value of the ratios of beam to slab flexural stiffnesses on all sides of the panel. The effects of these variables on the required minimum thickness are presented graphically in Fig. 3-2 for corner and interior panels.

For very flexible beams, or no beams, the latter term in the denominator becomes negative and the thickness by the above equation is greater than that required. For this reason it is specified that the thickness need not be greater than

$$h = \frac{l_n(800 + 0.005f_y)}{36,000}$$

which for $f_y = 40,000$ psi, reduces to $l_n/36$.

On the other hand, for very stiff beams, the required thickness can become less than desirable, so a limiting minimum thickness of

$$h = \frac{l_n(800 + 0.005f_y)}{36,000 + 5000(1 + \beta_s)}$$

is specified. For square panels, this corresponds to beams having an average flexural stiffness ratio of 2.0.

In addition, for panels having discontinuous edges, either an edge beam having a minimum stiffness of $\alpha = 0.8$ shall be provided or the minimum thicknesses shall be increased by 10%. However, if a drop panel having dimensions at least one-third of the clear spans and a projection below the

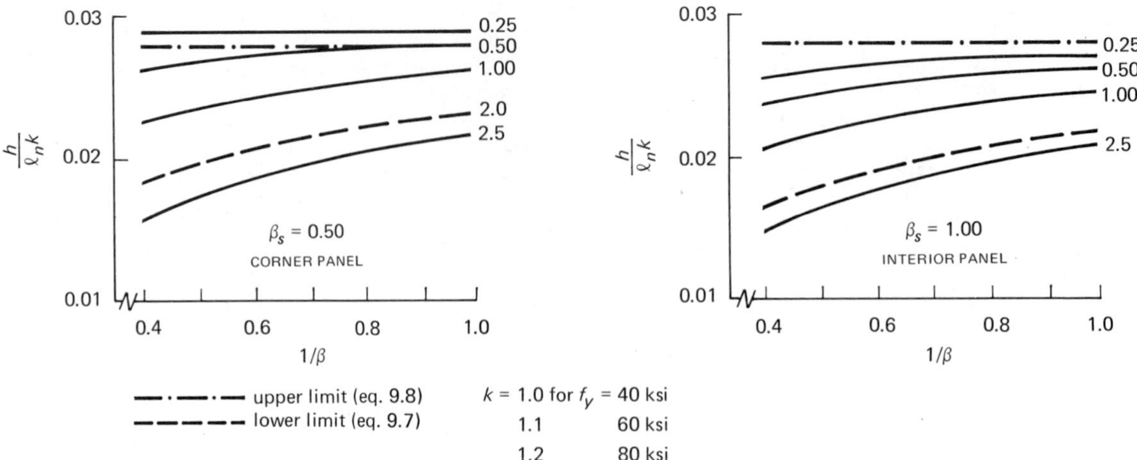

Fig. 3-2 Minimum thickness to clear span ratios.

slab of at least $h/4$ is provided, the required minimum thickness may be reduced by 10%.

Notwithstanding the computed thicknesses from the above expressions, the thickness shall not be less than the following values:

For slabs without beams or drop panels	5 in.
For slabs without beams but with drop panels	4 in.
For slabs with beams on all four edges with a value of α_m at least equal to 2.0	3½ in.

When the slab thickness chosen is such that the shear capacity of the slab for the given column size is not sufficient, it is generally good practice to provide a column capital. It should be noted that only that portion of the column capital which lies inside the largest right circular cone or pyramid, with a 90° vertex, which can be included within the outlines of the supporting element, can be considered as structurally effective.

Under certain conditions it is desirable to stiffen the slab by providing beams spanning between the columns. Examples of such conditions are where the vibrational properties of the slab are important or where heavy partitions or equipment are placed near column lines. The most common example is the use of beams along the outside edges to support the exterior walls directly and to provide the additional flexural rigidity required to reduce the slab thickness.

The flexural rigidity of such beams is considered in the design procedure outlined in section 3.6 by the dimensionless ratio α, which is defined as the ratio of flexural stiffness of beam section to the flexural stiffness of a width of slab bounded laterally by the center line of the adjacent panel, if any, on each side of the beam. In this definition, for monolithic or fully composite construction, beam section as shown in Fig. 3-3 is taken to mean not only the beam stem but also that portion of the slab on each side of the beam extending a distance equal to the projection of the beam above or below the slab, whichever is greater, but not greater than four times the slab thickness. The value of α may be obtained directly from the slab geometry using Fig. 3-4.

The effects of the torsional stiffness of the transverse beams is given either by K_t, when computing the equivalent

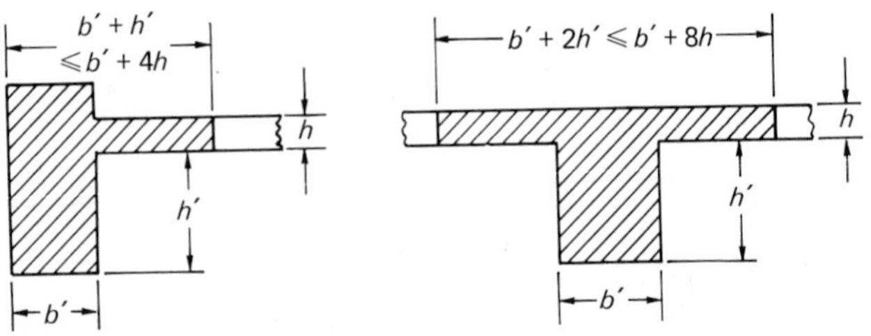

Fig. 3-3 Beam sections used in the definition of α.

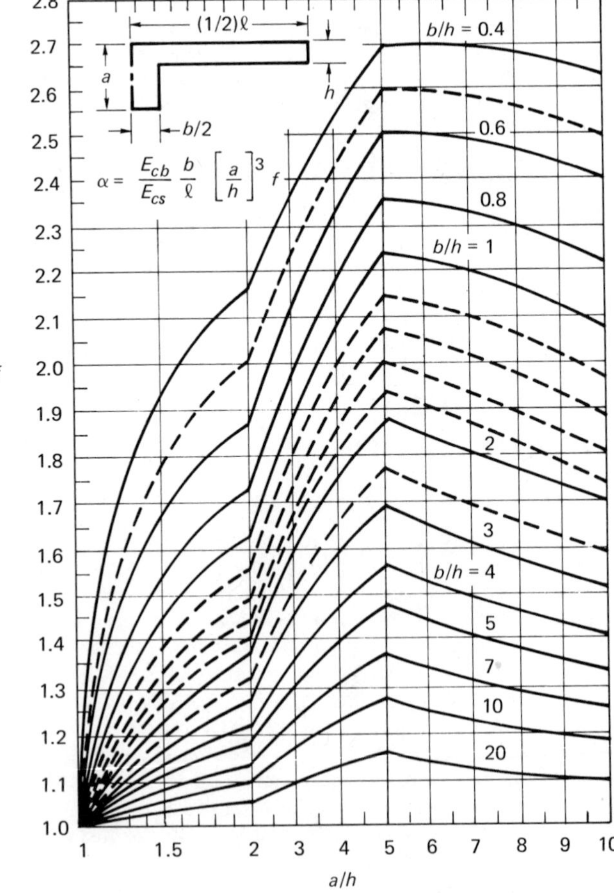

Fig. 3-4 Beam to slab ratio, α.

column stiffness, or by β_t when distributing design moments at the exterior edge, where K_t is defined as the torsional stiffness of the transverse beam section, moment per unit rotation, and β_t as the ratio of torsional stiffness of the edge beam section to the flexural stiffness of a width of slab equal to the span length of the beam, center to center of supports. In both these definitions the beam section includes the largest of: (a) when no beam stem is provided, that portion of the slab having a width equal to the dimension of the smaller column or column capital into which the slab frames; (b) for monolithic or fully composite construction that portion of the slab specified in (a) plus that part of the transverse beam above and below the slab; or (c) the beam section defined in the preceding paragraph for flexural stiffness.

3.5 ONE-WAY SLABS

One-way slabs, as the name implies, are those slabs which resist loading by deflecting in single curvature and hence have bending moments primarily in just one direction. While this type of behavior is obviously associated with isolated slabs which are supported only along opposite edges of a panel, for example, a stairway slab supported only at the top and bottom treads, it also occurs in slab panels which may be supported on all sides and may even be continuous in both directions but due to the geometry of the supports the curvatures in one direction are very much greater than in the other. A common example of this latter case is in buildings with either structural steel or reinforced concrete framing which make use of a grid of beams and girders between columns. When the ratio of the long to short dimension of the slab panel is large, say greater than 2, the slab may be designed as a one-way slab. An example of this type of slab system is given later in this section.

One-way behavior will result whenever a slab has substantially greater flexural stiffness in one direction than in the other. As an alternative to providing a grid of girders and beams the weight of the slab may be reduced and the required stiffness provided for the longer spans by ribbing the slab in one or both directions. Such slabs are easily formed with readily available pan forms. An example of such a slab is shown in Fig. 3-5.

In addition, localized one-way behavior can occur in a slab system, which would otherwise be considered exhibiting two-way behavior, due to the presence of isolated stiff beams or columns with cross sections elongated in one direction only so that they act as isolated wall sections. The designer is cautioned to be alert to such irregularities in the framing and to make the necessary modifications to the design to ensure satisfactory performance in these regions. A suggested procedure for isolated beams is given in section 3.9.

One-way slabs which have no stiffening elements such as beams spanning in the primary direction offer no peculiar difficulties to the designer. An analysis is made for moments and shears in the primary direction by considering a strip of slab, generally of unit width, as a continuous beam. The design also follows that for continuous beams and only minimum temperature and shrinkage reinforcement is placed in the transverse direction. Where isolated supporting elements such as beams or portions of a wall exist parallel to the primary direction, significant transverse moments will occur locally for which additional transverse reinforcing is required. These points are illustrated in Example 3-1.

Fig. 3-5 One-way joist floor.

EXAMPLE 3-1: Design a typical floor slab system for a retail sales store as shown on the Design Sheet. On the basis of display of merchandise and movement of customers the architect has recommended column spacing of 24 ft in the transverse direction and 30 ft in the longitudinal direction. For floors to be used for general merchandise the basic live loading is generally specified as 100 psf. In addition to the dead weight of the structural slab, consideration must also be given to permanent loads such as a topping to act as a wearing surface or to reduce cleaning costs and mechanical services such as fire sprinkling systems, ventilation ducts, etc. With these loads and dimensions the designer may suggest the grid of beams and girders shown to give a clear span of the slab of approximately 11 ft which is in the range of economical spans for solid one-way slabs.

The analysis consists of considering a strip of unit width in the primary direction of the slab. Final selection of the slab thickness is seen to depend on deflection considerations. The bending moments may be determined by either a moment distribution procedure in which case loading patterns must conform to those specified in section 8.5 of the *ACI Code* or if the requirements of code section 8.4.2 are satisfied by the use of the frame coefficients. In the example the latter method is used which gives rise to the moment envelope shown. Reinforcing is chosen in an identical manner as for a continuous beam (see Chapter 1 of this handbook). For this reason calculations for bend-up points and embedment lengths are not shown.

The supporting beams form an integral part of the slab system. All beams and girders would be poured integrally with the slab and would be designed as T-beams. Exterior beams and girders will have to resist in torsion the exterior negative moments from the slab. A detailed calculation for these torsion stresses is shown for a typical exterior beam. For interior beams and girders the torsional shear stress computed from eq. 11-16, of the *ACI Code* are such that torsional effects may be neglected. It is interesting to note that although the interior beams carry more vertical load, the width of beam stem need be only 12 in. for the same depth to permit a reasonable design for shear and flexural reinforcing whereas for the exterior beam it was economical to increase the stem width to 14 in. to assist in reducing torsional shear stresses and permit a greater minimum spacing of torsional stirrups.

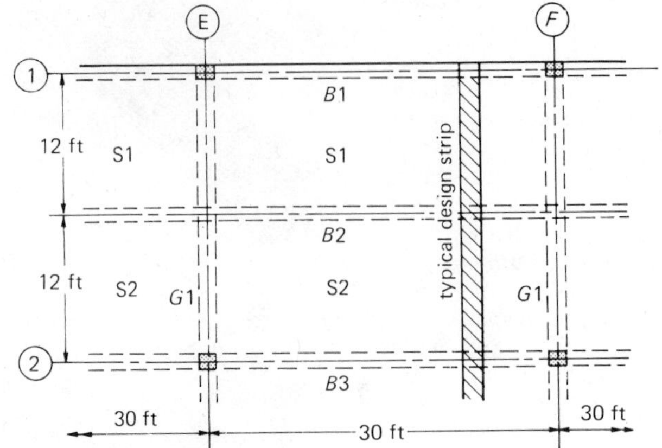

Materials

Concrete–normal weight	$f'_c = 3000$ psi
Reinforcing–intermediate	$f_y = 40{,}000$ psi

Loading

Live load (specified)	$w_l = 100 \times 1.7 = 170$ psf
Dead load–slab (5″)	63 psf
Topping (2″)	25
Ceiling and mech.	20
	$w_d = 108 \times 1.4 = 151$ psf
	$w = 321$ psf

Design of slab. Panel aspect ratio $= l_2/l_1 = 30/12 = 2.5 > 2.0$, design as one-way slab. Assume beam stems $b_w = 12''$, then $l_n = 12 - 1 = 11$ ft $= 132$ in. Use method of coefficients (*Code*–section 8.4).*

Choose slab thickness. Design conditions for strip 1 ft. wide.

$$\begin{cases} M_{max} = \frac{1}{10} \times 321 \times 11^2 = 3880 \text{ ft lbs} \\ V_{max} = 1.15 \times 321 \times \frac{11}{2} = 2030 \text{ lbs} \end{cases}$$

Assume: $q = 0.18$, $K_u = 434$, then

$$d^2 = \frac{M_u}{K_u b} = \frac{3880 \times 12}{434 \times 12} = 8.95 \text{ in.}^2 \; d \cong 3.0 \text{ in.}$$

since $h_{min} \equiv d$ (computed) + cover (specified) + 1/2 bar diameter (estimated), we then get $h_{min} = 3 + 3/4 + 1/4 = 4$ in.

Trying $h = 4$ in., we revise our conditions.

revised $w = 303$ psf, $M_{max} = 3660$ ft lbs, $V_{max} = 1920$ lbs

(a) Check for shear (section 11.2.1)

$$v_u = \frac{V_u}{\phi bd} = \frac{1920}{0.85 \times 12 \times 3} = 63 \text{ psi} < 110 \text{ psi}$$

*Section and equation numbers in parentheses refer to *ACI Code* 318-71.

(b) Check for deflection (section 9.5.2.1)

From table 9.5(a) of *Code* and exterior span

$$h_{min} = (0.4 + 0.1 f_y) \frac{l_n}{24} = (0.4 + 0.1 \times 40) \frac{132}{24} = 4.4 \text{ in.}$$

since $h_{min} > h_{selected}$, we must compute deflections. Consider strip 12 in. wide for typical interior panel, then

$$M_{cr} = 7.5\sqrt{3000} \times 12 \times (4)^3/12 \times 2 = 13{,}200 \text{ in. lbs}$$

(section 9.5.2.2)

$$M_a = \frac{wl_n^2}{12} = \frac{100 \times (11)^2}{12} \times 12 = 12{,}100 \text{ in. lbs}$$

Since $M_{cr}/M_a > 1$, use $I_{eff} = I_g = 12 \times (4)^3/12 = 64$ in.^{4}

$$\text{midspan defl.} = \frac{wl_n^4}{384\,EI} = \frac{100 \times (11 \times 12)^4}{384 \times 3.14 \times 10^6 \times 64} = 0.393 \text{ in.}$$

NOTE: w in above calculations is live load without load factor. Thus computed deflection is for live loads only based on elastic behavior. (section 9.5.2.2)

$$\text{max. all. defl.} = \frac{l}{360} = \frac{11 \times 12}{360} = 0.357 \text{ in.} \quad \text{(section 9.5.4.3)}$$

Since deflection in exterior panel will be even greater, we must increase slab thickness.

Now repeat, trying $h = 5$ in.

$$d = 5 - \tfrac{3}{4} - \tfrac{1}{4} = 4 \text{ in.}$$

From above $h = 5$ in. will be adequate for shear and deflection.

$$A_s = \frac{M_u \times 12}{\phi f_y \gamma_u d} = 0.0925\,M_u$$

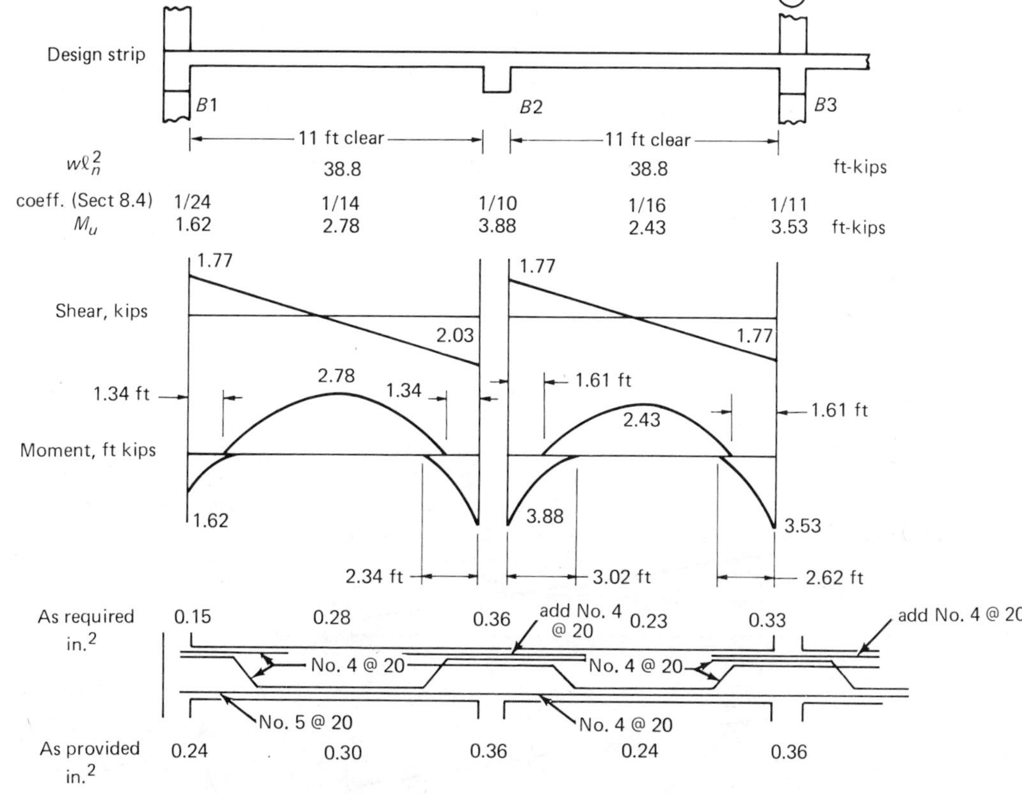

NOTE: Bend-up points computed from design moment diagram; cut-off points on basis of required anchorage lengths as specified in Chapter 12 of the *Code*.

Additional reinforcing

(1) Transverse reinforcing over girder. (section 8.7.5)

Girder flange width. (section 8.7.4)

$$\frac{l}{4} = \frac{24}{4} \times 12 = 72 \text{ in.}$$

$$2 \times (8 \times 5) + 12 = 92 \text{ in.}$$

$$\frac{30 \times 12}{2} = 180 \text{ in.}$$

Assume web width 12 in.

For calculating moment, overhang = (72 − 12)/2 = 30 in. and so

$$M = 0.321 \times \left(\frac{30}{12}\right)^2 \times \frac{1}{2} = 1.01 \text{ ft-kips/ft}$$

$$d_{\text{eff}} = 5 - \tfrac{3}{4} - \tfrac{1}{2} - \tfrac{1}{4} = 3.5 \text{ in.}$$

30 in. max

ℓ_d

$$A_s = \frac{1.01 \times 12}{0.9 \times 40 \times 0.87 \times 3.5} = 0.111 \text{ in.}^2/\text{ ft}$$

max. spacing = 18″ (section 8.7.5)

(2) Minimum reinforcing. (section 7.13)

$$A_s = 0.0020 \times 12 \times 5 = 0.12 \text{ in.}^2/\text{ft}$$

max. spacing = 18″

Therefore for transverse reinforcing use #4 bars at 18 in. spacing.

Design of Beam B_1:

Loading

Exterior wall = 660 lb/ft (see note 1)

Slab $w_d = 5.5 \times 151 = 830$

$w_l = 5.5 \times 170 = 935$

Assumed wt of beam = 490 (see note 2)

$$(M_t^s)_d = \frac{w_d l_n^2}{24} = \frac{151 \times 11^2}{24} = 760 \text{ ft-lb/ft}$$

$$(M_t^s)_l = \frac{w_l l_n^2}{24} = \frac{170 \times 11^2}{24} = 860 \text{ ft-lb/ft}$$

NOTE: 1. Assume wall weight acts 4 in. from exterior beam face.
2. Beam dimensions determined on basis of moment capacity at support. Assume $b = 14$, $h = 24$.

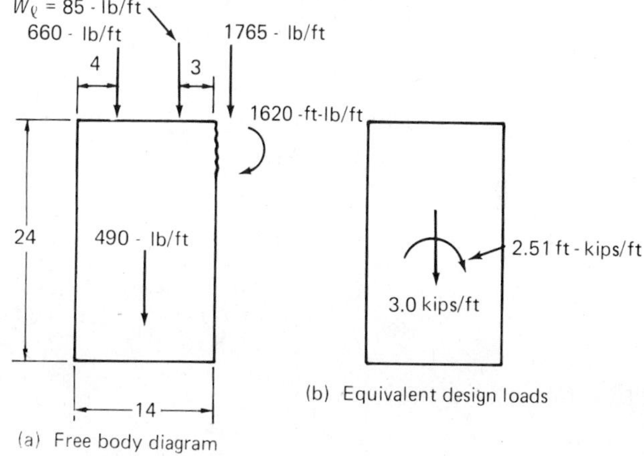

(a) Free body diagram

(b) Equivalent design loads

$$w = 660 + 85 + 1765 + 490 = 3.0 \text{ kips/ft}$$

$$M_t = 1620 + 1765 \times \frac{7}{12} + 85 \times \frac{4}{12} - 660 \times \frac{3}{12} = 2.51 \text{ ft-kips/ft}$$

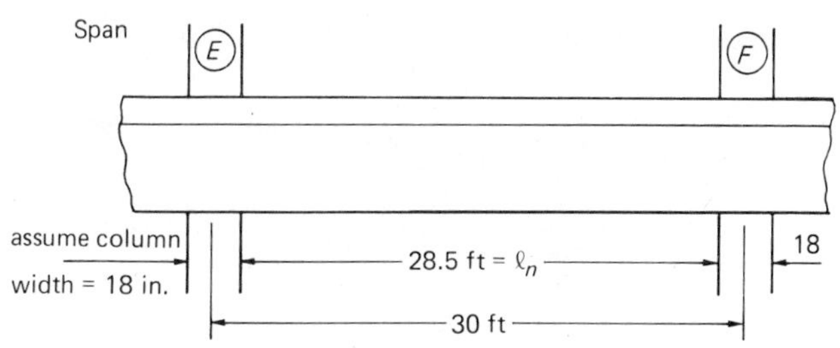

Design envelopes:

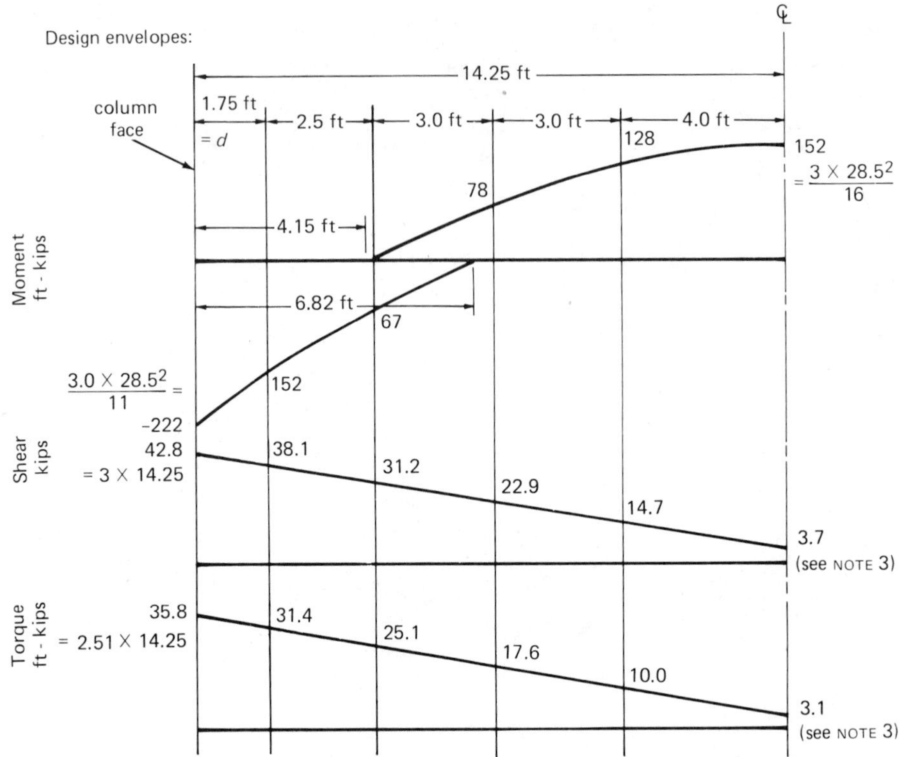

Design sections

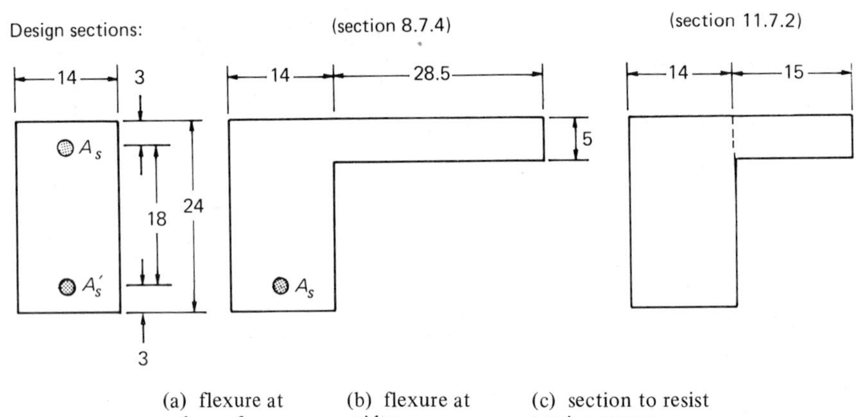

(a) flexure at column face $M_u = -222^{1K}$

(b) flexure at midspan $M_u = +152^{1K}$

(c) section to resist torsion stresses.

Flexural and torsional shear stresses:

design shear

$$V_u = 38.1 \text{ kips} \qquad \text{(section 11.2.2)}$$

design shear

$$T_u = 31.4 \text{ ft-kips} \qquad \text{(section 11.7.4)}$$

For torsion design section $\Sigma x^2 y = 14^2 \times 24 + 5^2 \times 15 = 5075 \text{ in.}^3$

$$v_u = \frac{V_u}{\phi bd} = \frac{38.1}{0.85 \times 14 \times 21} = 153 \text{ psi} \qquad \text{(eq. 11-3)}$$

$$v_{tu} = \frac{3T_u}{\phi \Sigma x^2 y} = \frac{3 \times 31.4 \times 12}{0.85 \times 5075} = 262 \text{ psi} \qquad \text{(eq. 11-16)}$$

$$v_{tu\,\min} = 1.5\sqrt{f'c} = 1.5\sqrt{3000} = 82 \text{ psi} < 262 \qquad \text{(section 11.7.1)}$$

$$v_{tu\,\max} = \frac{12\sqrt{f'c}}{\sqrt{1 + (1.2v_u/v_{tu})^2}} = \frac{12\sqrt{3000}}{\sqrt{1 + (1.2 \times 153/262)^2}}$$

$$= 538 > 262 \qquad \text{(eq. 11-18)}$$

Therefore we cannot neglect torsion, but can use stirrups to resist flexural and torsional shear stresses.

Allowable concrete values:

$$v_c = \frac{2\sqrt{f'c}}{\sqrt{1 + (v_{tu}/1.2v_u)^2}} = \frac{2 \times \sqrt{3000}}{\sqrt{1 + (262/1.2 \times 153)^2}} = 63 \text{ psi}$$

(eq. 11-9)

$$v_{tc} = \frac{2.4\sqrt{f'c}}{\sqrt{1 + (1.2v_u/v_{tu})^2}} = \frac{2.4\sqrt{3000}}{\sqrt{1 + (1.2 \times 153/262)^2}} = 108 \text{ psi}$$

(eq. 11-17)

$V_c = \phi v_c bd = 0.85 \times 63 \times 14 \times 21 = 15.7$ kips

$$T_c = \phi v_{tc} \frac{\Sigma x^2 y}{3} = 0.85 \times 108 \times \frac{5075}{3 \times 12} = 12.9 \text{ ft-kips}$$

Web reinforcement: Use 1/2 in. diameter closed stirrups and 1½ in. clear cover.

$x_1 = 14 - 2(1.5 + 0.25) = 10.5, \quad y_1 = 24 - 0.2(1.5 \times 0.25) = 20.5$

This gives maximum stirrup spacing as

(a) $s_{max} = d/2 = 21/2 = 10.5$ (section 11.1.4)

(b) $s_{max} = (x_1 + y_1)/4 = (10.5 + 20.5)/4 = 7.75$ (section 11.8.3)

(c) $s_{max} = 12$ (section 11.8.3)

$s_{max} = 7.75$ in. based on torsion

Summary of Reinforcement

	Distance From Column Face–ft						
	0	*d = 1.75*	*4.25*	*7.25*	*10.25*	*14.25*	*Units*
Web Reinforcement							
v_u (eq. 11-3)	Not	153	125	92	58	15	psi
$v_u - v_c$	Critical	96	68	35	1	–	psi
v_{tu} (eq. 11-16)		262	209	147	83	26	psi
$v_{tu} - v_{tc}$		154	101	39	–	–	psi
$A_v/2s$ (eq. 11-13)		0.0168	0.0119	0.0061	0.0001	–	in.²/in.
A_t/s (eq. 11-19)		0.0232	0.0152	0.0059	–	–	in.²/in.
$A_v/2s + A_t/s$		0.0400	0.0271	0.0120	0.0001	–	in.²/in.
s (½ in. stirrups)		5.0	7.4	16.6	200	∞	in.
Spacing required		5.0	7.4	7.75*	–	–	in.
Longitudinal Reinforcing							
$-M_u$	-222	-152	-67	0	0	0	ft-kips
$+M_u$	0	0	2	78	128	152	ft-kips
$-A_s = A_{s\,top}$	3.95	2.71	1.19	0	0	0	in.²
$+A_s = A_{s\,bot.}$	0	0	0.03	1.26	2.06	2.46	in.²
A_l (eq. 11-20)	1.44	1.44	0.94	0.37	–	–	in.²
A_l (eq. 11-21)	1.30	1.30	1.77	2.13	2.01	2.20	in.²
$A_{l\,top} = A_{l\,bot.}$**	0.41	0.41	.58	0.76	0.70	0.79	in.²
$(A_s + A_l)_{top}$	4.36	3.12	1.77	0.76	0.70	0.79	in.²
$(A_s + A_l)_{bot.}$	0.41	0.41	0.61	2.02	2.76	3.25	in.²
Bars top	2-#9,3-#8	–	–	–	–	2-#8†	–
Bars bot.	2-#7‡	–	–	–	–	4-#7,2-#6	–

*Max. spacing governs.
†See (section 11.8.5).
‡See (section 12.2.1).
**See calculations below.

At $d = 1.75$ ft. from support face.

$$A_l = 2\frac{A_t}{s}(x_1 + y_1) = 2(0.0232)(10.5 + 20.5) = 1.44 \text{ in.}^2 \qquad \text{(eq. 11-20)}$$

$$A_l = \left[\frac{400 \times 14}{40{,}000}\left(\frac{262}{262 + 153}\right) - 2(0.0232)\right](10.5 + 20.5) = 1.30 \text{ in.}^2 \qquad \text{(eq. 11-21)}$$

Bars required at midheight. (section 11.8.5)

Assume that two #5 bars are placed at midheight to act as torsional longitudinal reinforcement. Remainder of required torsional longitudinal reinforcement is proportioned equally at top and bottom.

$$A_{l\,top} = A_{l\,bot.} = \frac{1.44 - 0.62}{2} = 0.41 \text{ in.}^2$$

3.6 DESIGN PROCEDURE FOR TWO-WAY SLABS

The 1971 edition of the *ACI Code* permits for the first time a single procedure for the design of all two-way slab systems consisting of reasonably rectangular panels supported on isolated columns with or without beams, drop panels and column capitals, and supporting predominately distributed loading. This procedure is summarized in five steps.

1. Choose slab thickness and dimensions of auxiliary elements, if any.
2. Compute the slab design moments at the critical positive and negative sections and the support design moments.
3. Proportion slab design moments between column and middle strips.
4. Design for shear and torsion.
5. Choose reinforcing.

Step 1 Slab Dimensions: The choice of the slab thickness and the use of auxiliary elements such as column capitals and beams will depend on such factors as panel location and geometry, and stiffness and shear requirements. These problems are discussed in detail in section 3.4.

Step 2 Design Moments at the Critical Sections: This step may be performed by either of two methods. For the more regular slab systems meeting the limitations listed for this method, the simpler Direct Design Method may be used, otherwise the more general Equivalent Frame Method is required. The basic features of these methods are as follows:

Direct Design Method—The limitations for use of the direct design method are given in section 13.3.1 of the *Code* as

1. There shall be a minimum of three continuous spans in each direction.
2. The panels shall be rectangular with the ratio of longer to shorter spans within a panel not greater than 2.0.
3. The successive span lengths in each direction shall not differ by more than one-third of the longer span.
4. Columns may be offset a maximum of 10% of the span in direction of the offset, from either axis between center lines of successive columns.
5. The live load shall not exceed three times the dead load.
6. If a panel is supported by beams on all sides, the relative stiffness of the beams in the two perpendicular directions $\alpha_1 l_2^2/\alpha_2 l_1^2$ shall not be less than 0.2 nor greater than 5.0.

Experienced designers will recognize some of these limitations as being similar to those for the former empirical method for flat slabs, but it should be noted that the limitations for the direct design method are not as restrictive and that this method is also applicable for slabs with beams.

Some explanation of limitation 6 may be desirable. During the compilation of the results of the slab program at Illinois[3-10] it became apparent that it would be impractical to incorporate all possible variables in a simplified design procedure. Two dimensionless ratios, namely $\alpha_1 l_2^2/\alpha_2 l_1^2$ and $\alpha_1 l_2/l_1$, were defined for use in determining slab moments. The first ratio is equal to unity when the beam flexural stiffness in each direction is proportional to the span length, a condition that was assumed throughout the slab program and is approximately satisfied by most slabs in practice. Limits are placed on this ratio to prevent the use of this design procedure when very stiff beams are provided in one direction compared to the stiffness of the beams in the other direction since, for this condition, the slab tends to behave in one-way action for which this method is not applicable. The second ratio, $\alpha_1 l_2/l_1$, is the parameter used in defining effective beam flexural stiffness.

The total static design moment is computed for each span for a strip bounded laterally by the center line of the panel on each side of the supports and is defined as the absolute sum of the positive and average negative moments. By equilibrium this moment is

$$M_0 = \frac{w l_2 l_n^2}{8}$$

The critical section for negative moment is the face of the column, capital or wall. The clear span, l_n, is also measured from this position but not less than 0.65 times the corresponding center to center span. Since this method is based on ultimate strength design, the design load, w, is the sum of the dead and live loads with the appropriate load factors included.

This total moment, M_0, is distributed to the negative and positive critical sections on the basis of the stiffnesses of supporting columns as shown in Fig. 3-6. For interior panels, by assuming that the rotation of interior columns are small, the proportioning is similar to that for uniformly loaded continuous beams, namely 0.65 M_0 for negative moments and 0.35 M_0 for positive moments. For exterior panels studies have shown that the moments are sensitive to the degree of restraint provided at the discontinuous edge[3-11]. For example, when the exterior column has essentially no flexural stiffness, the exterior negative design moment is approximately zero. However, even if the exterior column has infinite flexural stiffness, the moments will still not be distributed as for an interior panel unless the rotation of the discontinuous edge is prevented by providing an edge beam which is extremely stiff in torsion. For this reason, the exterior panel moments are expressed in terms of the equivalent column stiffness ratio, α_{ec}, which is a function of both the exterior column stiffness and the torsional stiffness of the edge slab and beam strips. As this equivalent column is an integral part of the equivalent frame method, details of obtaining the equivalent stiffness are deferred to the discussion of this method.

Since the above distribution is based on small rotations of interior columns, these columns above and below the slab must be designed to resist, in direct proportion to their stiffnesses, a minimum moment equal to

$$M = \frac{0.08\,[(w_d + 0.5\,w_l)\,l_2\,l_n^2 - w_d^1\,l_2^1\,(l_n^1)^2\,]}{1 + \dfrac{1}{\alpha_{ec}}}$$

where w_d^1, l_2^1 and l_n^1 refer to the shorter span.

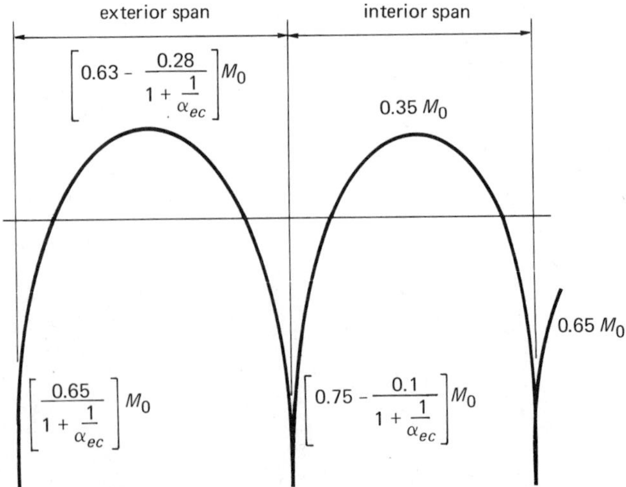

Fig. 3-6 Proportioning M_0 into negative and positive moments.

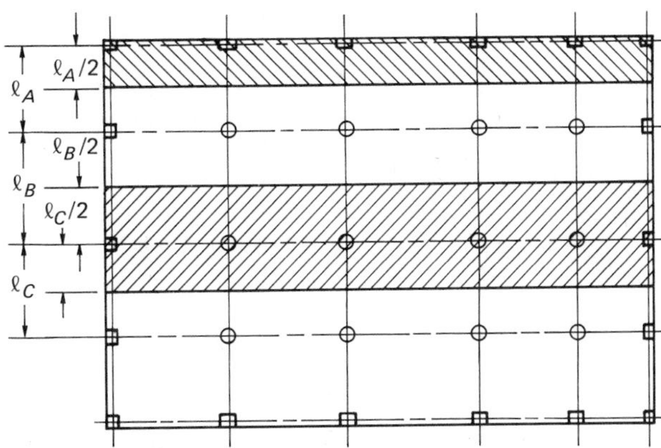

Fig. 3-7 Design strips for equivalent frame method.

In addition, when the ratio of live load to dead load exceeds 0.5 the effects of pattern loading must be considered by either (a) providing a sum of flexural stiffnesses of the columns above and below the slab, α_c, greater than the minimum stiffness α_{min} given in Table 3-1; or by (b) multiplying the positive bending moments in the panels supported by those columns by the factor

$$\delta_s = 1 + \frac{2 - \beta_a}{4 + \beta_a}\left(1 - \frac{\alpha_c}{\alpha_{min}}\right)$$

It is seen that the ratio δ_s will always be greater than 1.0 when its use is required.

TABLE 3-1 α_{min}*

β_α	*Aspect ratio* l_2/l_1	*Relative beam stiffness,* α				
		0	0.5	1.0	2.0	4.0
2.0	0.5–2.0	0	0	0	0	0
1.0	0.5	0.6	0	0	0	0
	0.8	0.7	0	0	0	0
	1.0	0.7	0.1	0	0	0
	1.25	0.8	0.4	0	0	0
	2.0	1.2	0.5	0.2	0	0
0.5	0.5	1.3	0.3	0	0	0
	0.8	1.5	0.5	0.2	0	0
	1.0	1.6	0.6	0.2	0	0
	1.25	1.9	1.0	0.5	0	0
	2.0	4.9	1.6	0.8	0.3	0
0.33	0.5	1.8	0.5	0.1	0	0
	0.8	2.0	0.9	0.3	0	0
	1.0	2.3	0.9	0.4	0	0
	1.25	2.8	1.5	0.8	0.2	0
	2.0	13.0	2.6	1.2	0.5	0.3

**From Ref. 3-1.*

Equivalent Frame Method—When the limitations listed for the direct design method are not satisfied the design moments in Step 2 may be determined using the equivalent frame method. In this method the structure is considered as a series of frames taken in both the longitudinal and

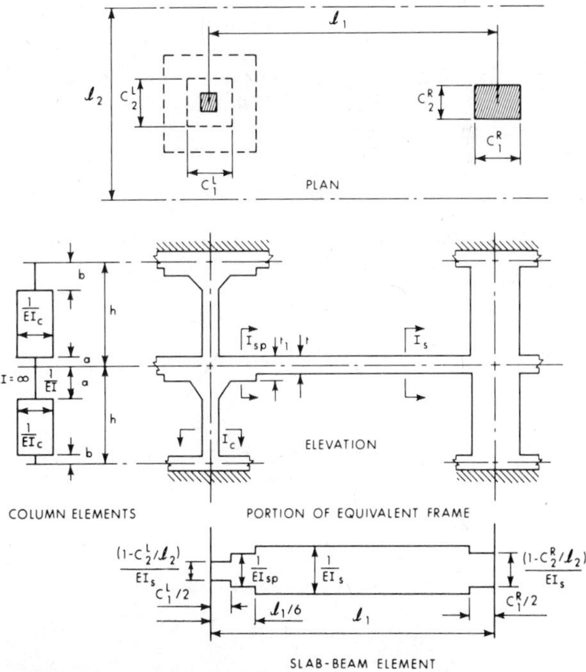

Fig. 3-8 Column and slab-beam elements for the equivalent frame method.[3-15]

transverse directions. Each frame consists of a row of equivalent columns or supports and slab-beam strips bounded laterally by the center line of the panel on each side of the center line of the columns or supports (Fig. 3-7). For vertical loading, each floor together with its column above and below may be analyzed separately and the far ends of the columns may be assumed fixed. However, the stiffnesses assigned to the various members of these frames to obtain the equivalent frame are substantially different from previous frame procedures to permit a more accurate representation of the actual three-dimensional slab-beam system by the equivalent two-dimensional elastic frame.[3-12] The factors required for the cross moment distribution, namely the fixed-end moment coefficient, the stiffness factor and the carry-over factor, must be evaluated for each element of the equivalent frame.

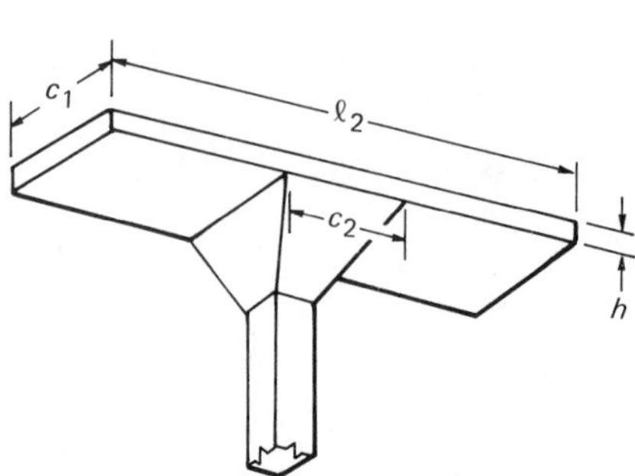

Fig. 3-9 Equivalent column.[3-15]

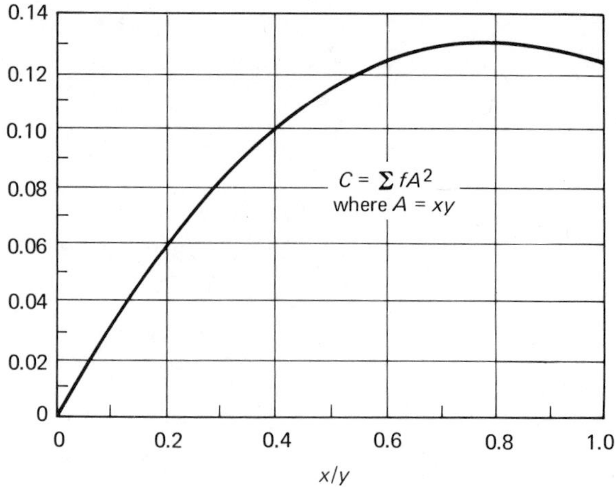

Fig. 3-10 Torsion factor curve.

A typical slab-beam element for a portion of an equivalent frame is shown in Fig. 3-8. The moment of inertia for this element between the faces of the columns, brackets or columns is based on the uncracked cross section of the concrete including beam or drop panel, if any. The moment of inertia between the center of the column and the face of the column, bracket or capital is considered finite and is dependent on the transverse dimensions of the panel and support.

To approximate actual slab behavior under unequal loading on adjacent panels, or unequal adjacent panel lengths, the equivalent column is considered to consist of the actual column plus an attached torsional member running transverse to the direction in which the moments are being determined as shown in Fig. 3-9. The transverse slab-beam may rotate even though the column is infinitely stiff, thus permitting consideration of the moment redistribution between adjacent panels. This equivalent column element is also used for the exterior columns in the direct design method.

The stiffness of the equivalent column, K_{ec}, is determined from the expression

$$\frac{1}{K_{ec}} = \frac{1}{\Sigma K_c} + \frac{1}{K_t}$$

where

$$K_c = \frac{k_c E_{cc} I_c}{l_c} = \text{stiffness of attached column}$$

$$K_t = \frac{\Sigma 9 E_{cs} C}{l_2 (1 - c_2/l_2)^3}$$

= torsional stiffness of attached transverse slab-beam elements

In computing the stiffness of the column, K_c, the moment of inertia is assumed infinite from the top to the bottom of the slab-beam or capital as indicated in Fig. 3-8. The attached torsional members for monolithic construction are considered to include, in addition to the beam, that portion of the slab on each side of the beam extending a distance equal to the beam projection either above or below the slab, whichever is greater, but not more than four times the slab thickness. The torsional constant, C, can be computed for this transverse member by dividing the cross section into separate rectangular parts and carrying out the summation $C = \Sigma[1 - 0.63(x/y)]\ (x^3 y/3)$ or by the use of Fig. 3-10.

After evaluating the necessary factors from the slab-beam and column elements, the equivalent frame is analyzed by elastic analysis. Where live load does not exceed three-quarters of the dead load or the nature of live load is such that all panels must be loaded simultaneously, maximum moments are obtained assuming full live load over entire system. For other conditions maximum moments may be obtained by considering pattern loading using three-quarters of full design live load on appropriate panels.

Since the analysis of the frame is based on center to center lengths the negative moments may be reduced to those occurring at an assumed critical section. This critical section is taken at the face of rectilinear supports except that for interior supports of large dimension the negative moments can become unrealistically low. For this reason the critical section cannot be considered further than $0.175 l_1$ from the support center line. At exterior supports provided with brackets or capitals the critical section for negative moment in the direction perpendicular to the edge shall be taken at a distance not greater than one-half the projection of the bracket or capital beyond the face of the column. It should be noted that this is different than the location of this critical section by the direct design method. Circular or regular polygonal supports are treated as square supports having the same area. Where the slab system is within the limitations of geometry required for the direct design method and analysis has been performed by the equivalent frame method, the absolute sum of the positive and average negative design moments in any panel need not exceed the total static moment required by the former method.

For interior columns or supports the design moment is equal to the maximum difference between the negative moments for the slab-beam elements on either side, and for exterior columns or supports to the maximum exterior negative moments of the exterior slab-beam element. These moments are assigned to the support elements above and below the slab in the ratio of their stiffnesses.

Step 3 Proportion Moment to Column and Middle Strips: Once the design moments at the negative and positive critical sections of the slab have been determined by either the direct design method or the equivalent frame method, they must be distributed across the width of the panel. In common with previous design methods it is advantageous to use the concept of dividing the slab into a series of transverse and longitudinal strips parallel to the column lines, as shown in Fig. 3-11. That portion of the design moment assigned to a particular strip may be assumed to

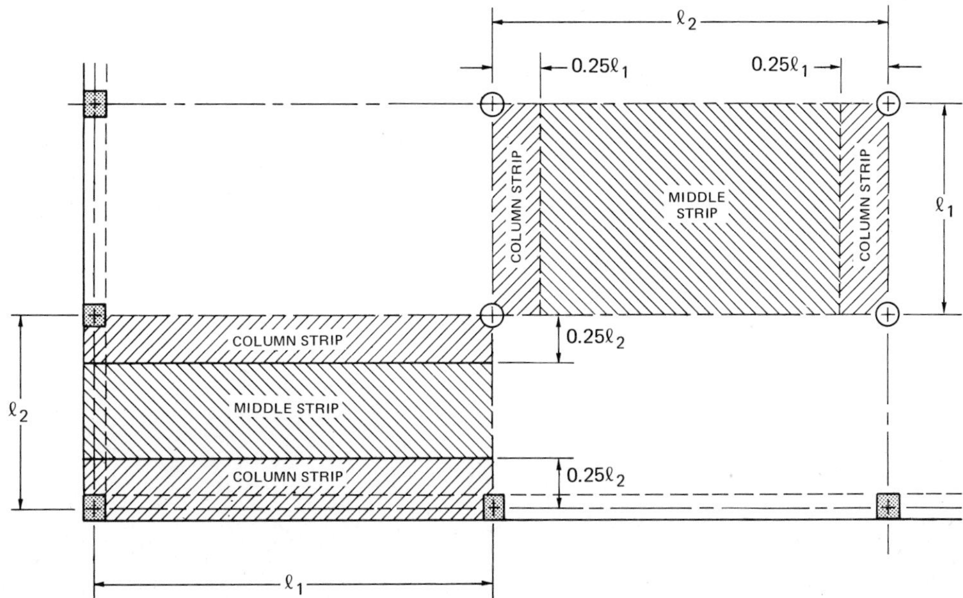

Fig. 3-11 Definition of column and middle strips.

be distributed uniformly across the strip. Previously the column strip in either direction was considered to be that portion of the slab bounded by the quarter panel on each side of the column line and the middle strip consisted of the remaining central half panel bounded by two column strips. When the panel is rectangular the negative moments in the short direction are observed to be more concentrated in the vicinity of the column than is implied by this definition. For this reason the above definition of strips in the long direction is retained but the width of the column strip in the short direction is considered not to exceed half of the short span. This results in a wider middle strip in the short direction.

The portion of the design moment assigned to the column strip is dependent on the effective column strip stiffness and the panel aspect ratio, as shown in Fig. 3-12. Across the exterior support the proportion assigned to the

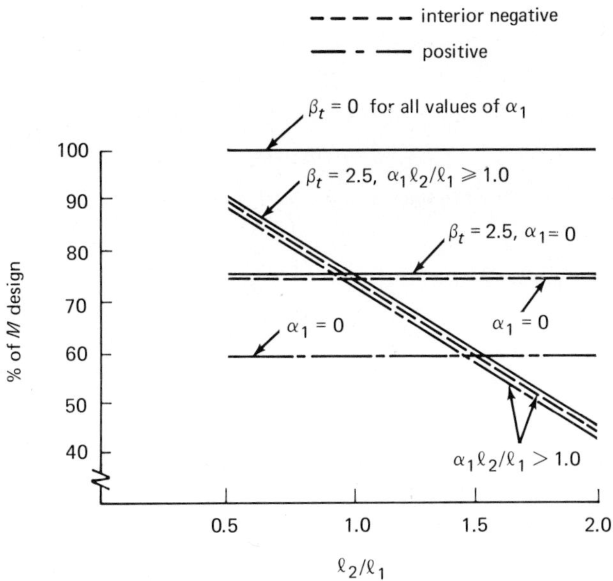

Fig. 3-12 Percentage of M_{design} assigned to column strip.

column strip is reduced as the torsional stiffness of the edge members is increased, reflecting the more uniform distribution when a torsionally stiff edge member is provided.

When beams are part of the column strip, the beam is proportioned to resist 85% of the column strip moment if $\alpha_1 l_2/l_1$ is equal to or greater than 1.0. For values of this ratio between 1.0 and 0.0, the proportion resisted by the beam is obtained by linear interpolation between 85 and 0%. The beam must also be designed to carry any loading that is applied directly to the beam.

Middle strips are designed for the remainder of the design moment not assigned to the adjacent column strips. The code also permits modifying any moment value up to 10% providing the total moment in any panel remains unaltered.

For the special case of when the exterior support consists of a wall or column extending for a distance equal to or greater than three-quarters of the transverse span, l_2, the exterior negative moment is considered to be uniformly distributed across l_2. In addition, the middle strip adjacent to and parallel with an edge supported by a wall is proportional to resist twice the moment assigned to the half middle strip corresponding to the first row of interior supports.

Step 4 Design for Shear and Torsion: The total shear occurring in each panel is distributed between the slab and beam, if any, in the following manner. When beams with an effective stiffness of $\alpha_1 l_2/l_1$ equal to or greater than 1.0 are provided, they shall be proportioned to resist the shear caused by loads in the tributary areas bounded by 45° lines drawn from the corners of the panels and the center line of the panels parallel to the long sides. Beams with values of $\alpha_1 l_2/l_1$ less than 1.0 are considered to resist the shear obtained from linear interpolation, assuming that when $\alpha_1 = 0$ the beams carry no load. In addition all beams must be designed to resist the shear caused by loads applied directly to the beam. The slab is designed to resist the shear not assigned to the beams.

Edge beams or, in their absence, the strip of slab parallel to the edge must also be designed to resist the torsion caused by the assumed distribution of exterior negative design moments.

Slabs without beams may fail in shear (diagonal tension) along one of two failure surfaces. The first mode of failure is to consider the slab as a wide beam with the surface of potential diagonal cracking extending across the entire width of slab. For this condition the allowable shear stress in the concrete, v_u, is the same as for a beam namely $2\sqrt{f_c'}$. A second mode of failure is possible in the region of concentrated loads or columns where the potential diagonal cracking may occur along the surface of a truncated cone or pyramid around the concentrated load or column. This mode of failure is generally referred to as punching shear and the failure surface is assumed to be perpendicular to the plane of the slab and to have a periphery, b_o, that is a minimum but approaches no closer than $d/2$ to the periphery of the concentrated load or reaction area. Due to the confining effect on the concrete caused by the two-way action the allowable shear stress at this section ($v_u = V_u/\phi b_o d$) is limited to $4\sqrt{f_c'}$. This value may be increased by 75% when appropriate shear reinforcing is provided. The provisions for shear reinforcing for slabs is discussed in section 3.9.

For slabs without beams a significant portion of the design shear stress comes from the moment transferred between the slab and column. This moment may be caused by lateral loads applied to the structure or by unequal gravity loads on adjacent spans. For design purposes a portion of this moment is considered to be transferred by shear and a portion by flexure. The critical section for shear is assumed to be one-half of the effective depth ($d/2$) from each face of the column or column capital as shown in Fig. 3-13. The shear stresses resulting from portion transferred by shear are added directly to the shear stresses obtained from resisting the vertical load.

The portion of the unbalanced moment transferred by shear is given by

$$\alpha_v = 1 - \frac{1}{1 + \frac{2}{3}\sqrt{\frac{c_1 + d}{c_2 + d}}}$$

or may be obtained more simply by the use of Fig. 3-14. The shear stress at any distance c from the centroidal line of the shear section is given by $\alpha_v Mc/\phi J$ where M is the total moment transferred from slab to column, ϕ is the capacity reduction factor for shear given as 0.85 and J is the sum of the second centroidal moments of the shear areas.

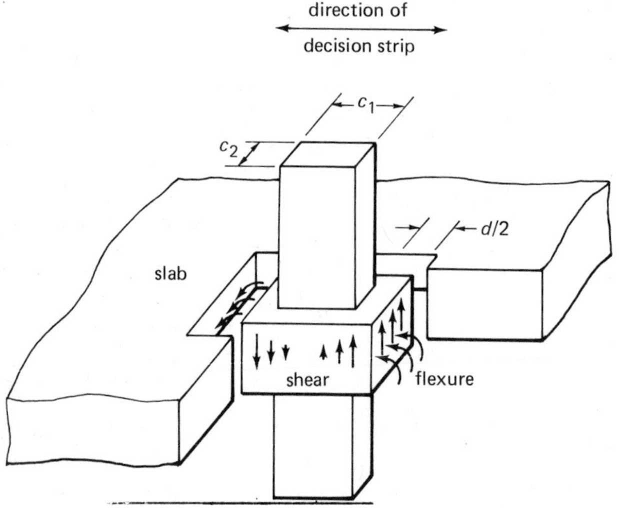

Fig. 3-13 Moment transfer between slab and column.

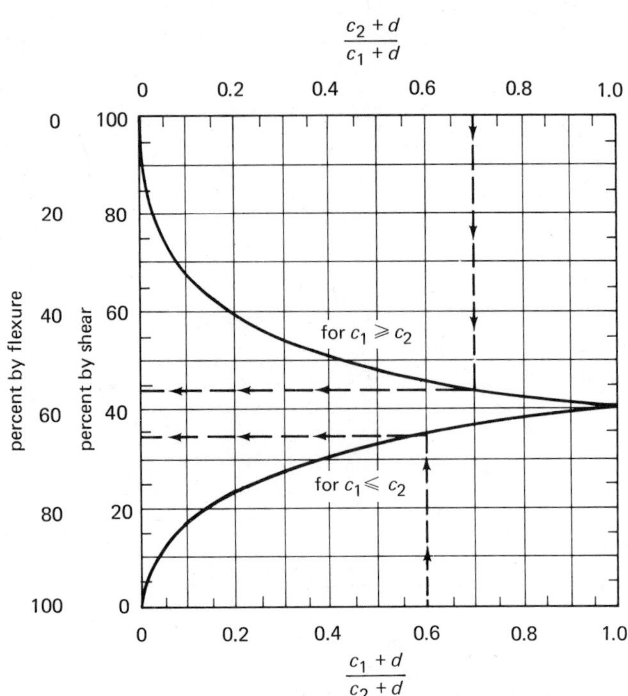

Fig. 3-14 Percentage of moment transferred by shear and flexure.

For an interior column with dimensions c_1 and c_2 in the longitudinal and transverse directions, respectively, the critical shear section will have plan dimensions of $c_1 + d$ and $c_2 + d$ and depth d. The design parameters for this case are

$$J = 2\left\{\frac{d(c_1 + d)^3}{12} + \frac{(c_1 + d)\,d^3}{12} + d(c_2 + d)\left(\frac{c_1 + d}{2}\right)^2\right\}$$

$$c = \frac{(c_1 + d)}{2}$$

For most slabs the second term in the expression for J will be small compared to the other two terms. Neglecting this term the ratio of J/c reduces to $J/c = [d(c_1 + d)/3] \times (c_1 + 3c_2 + 4d)$.

For exterior columns with exterior face flush with exterior of building the corresponding ratios are:

strip parallel to the exterior edge

$$J/c = [d(c_1 + d)/6]\ (c_1 + 6c_2 + 4d)$$

strips perpendicular to the exterior edge

$$J/c = [d(2c_1 + d)/12]\ (2c_1 + 4c_2 + 5d)$$

corner columns

$$J/c = [d(c_1 + d)\,(c_1 + c_2 + 2d)]/6$$

An example of computing shear stresses due to moment transfer between slab and column is given in ex. 3-2.

Step 5 Reinforcing: Flexural reinforcing for the slab and beam is proportioned in the usual manner with attention given to required minimum areas and lengths of embedment. However, for the slab to behave in a manner consistent with the assumptions used in this simplified design procedure, there are several additional restrictions which influence the selection of the reinforcing and these are discussed briefly.

For solid two-way slabs the spacing of bars at critical sections shall not exceed two times the slab thickness. Where drop panels are used their presence does not affect the distribution of moments. However, to be considered effective in reducing the amount of negative reinforcement over the column of a slab without beams, the drop must extend in each direction a distance of at least one-sixth of the span length in that direction and have a projection below the slab of at least one-quarter of the slab thickness beyond the drop. For determining reinforcement, the thickness of the drop panel below the slab shall not be considered to be greater than one-fourth of the distance from the edge of the drop panel to the edge of the column capital.

Minimum lengths of reinforcing for each strip are given in Fig. 3-15, which corresponds to Fig. 13-5.6 in ACI 318-71. In addition it is required that both positive and negative reinforcing perpendicular to a discontinuous edge extend to the edge of the slab and have an embedment, straight or hooked, of at least 6 in. in spandrel beams, walls, or columns.

For slabs supported on beams having a value of α greater than 1.0 additional reinforcing proportioned to resist a moment equal to the maximum positive moment per foot of width in the slab is required at exterior corners in both top and bottom of the slab to resist the twisting moments not otherwise accounted for. The direction of this moment is parallel to the diagonal from the corner in the top of the slab and perpendicular to the diagonal in the bottom of the slab, hence the required reinforcing may be placed in a single layer in the top of the moment or in two layers parallel to the sides of the slab. This reinforcement should extend in each direction from the corner equal to one-fifth of the longer span.

Considerable economy in placing slab reinforcing can be achieved if no bent bars are used and the negative reinforcing is placed as a preformed mat. To achieve maximum economy it has been the practice of some designers to place no negative steel in the middle strips and concentrate all of the panel negative steel in the column strips, a placing sequence which is in violation with the requirements of past and present ACI codes. To investigate the effects of this practice a floor of an apartment building was load tested by Cardenas and Kaar[3-13] in which one portion of the slab was reinforced according to ACI requirements and an identical portion had all of the panel negative steel concentrated in the column strips. Under the standard test load $(0.3w_d + 1.7w_l)$ there was no significant difference in the deflections of the two portions and all crack widths were small with no major effects on stiffness or serviceability.

The writer is familiar with several buildings in which the slab's negative steel mat was extended into the middle strips such that although the required negative steel area in the middle strip was provided, it was concentrated in the portion adjacent to the column strips rather than uniformly distributed over the middle strip. These slabs have performed well. Thus it would appear that the negative steel can be concentrated in the column strips without problems in serviceability. However, the reader is cautioned that the above slabs were all proportioned using ACI 318-63 in which the column strip was defined as a quarter of the panel width on each side of the column line. For rectangular panels the definition of column strip in the short direction by ACI 318-71 will be less than this width. For this reason, when the panel aspect ratio is appreciably greater or less than 1.0 it would be prudent in the short direction to extend the negative steel mat into adjacent portions of the middle strip. This will not increase steel costs but will ensure a more satisfactory distribution of cracks.

3.7 TWO-WAY SLABS WITHOUT BEAMS

Slabs which carry load by two-way action but without the use of beams are one of the most efficient structural systems. Such slabs are economical since they can be constructed with minimum field labor resulting from the use of the simplest formwork and reinforcing steel arrangements. In addition, these slabs provide the most flexibility in the layout of columns and partitions, require the least story height for a specified clear headroom, and in many buildings the ceiling finish may be applied directly to the slab soffit eliminating costly ceiling support systems. Slabs without beams have been used extensively for multistory hotels, hospitals, dormitory, and apartment buildings. In most cases the load capacity of such slabs is governed by the shear capacity available. For industrial loading or for longer spans the use of column capitals and/or drop panels may be required. For longer spans the dead load may be decreased and the stiffness increased by using a waffle flat plate rather than a solid slab.

The design of two-way slabs without beams is demonstrated by exs. 3-2 through 3-4.

EXAMPLE 3-2: Design a typical floor for an apartment building with the floor framing indicated. To achieve minimum story height, a slab with constant thickness and no auxiliary stiffening elements such as capitals or drop panels is chosen. The exterior wall consists of precast insulated panels. To support these panels the slab is extended 3 in. beyond the outside face of the exterior columns.

The first step is to determine the slab thickness. The critical panel in this case is the corner panel. By satisfying the minimum thickness requirements of section 3.4, no deflection calculations are required. Due to the longer span in the N–S direction the reinforcing in this direction was placed in the outer layer. The average effective depth was used in shear calculations.

Although a final calculation of shear stresses at the columns must be deferred until the bending moments to be transferred are established, a preliminary check on the shear stresses should be made especially if no beams or capitals are to be used. These preliminary calculations are made more uncertain if column sizes must be estimated. In general, for economy, the slab thickness should be controlled by deflection requirements rather than by shear requirements.

The second step is to determine the design moments in the slab. For this example the direct design method is used since all of the requirements for the use of this method are satisfied.

To proportion the panel moment in the exterior panels the equivalent stiffness of the exterior column is required. In the example the coefficient for determining the column and slab stiffnesses was taken as "4" as suggested in the commentary[3-14] when using the direct design method. The value of α_{ec} for the exterior strips is almost identical with the value for corresponding interior strips because of the ratio of the column sizes assumed but such close agreement will not always occur. However, the commentary suggests that where the dimensions of the corner column are the same as for the adjacent exterior column, the same value of α_{ec} may be used, buf the exterior negative slab moment in the exterior strip will be somewhat underestimated.

Design moments for the exterior columns are obtained from the exterior negative design moments in the slab strips. For interior columns the moments are obtained by considering partial live loading on the longer of the adjacent spans.

The next step is to distribute the slab design moments at the critical sections between the column and middle strips. Since there are no beams in this example the ratio $\alpha_1 l_2/l_1$ is everywhere zero. The moments assigned to each strip are shown in the diagram in Step 3 of this example. Also shown on this diagram are the areas of reinforcing required for each strip.

Step 4 consists of checking the slab for adequacy in shear. There are two distinct modes of shear failure which must be examined. The first is to consider the slab strips as wide beams and compute the flexural shear stresses across these strips. The computations and allowable shear stresses are identical with those for beams and since this mode of failure will rarely govern for rectangular slabs without beams, the computations are not shown. The second mode of failure is in the vicinity of the support columns and the design calculations are shown in detail for a typical exterior column. Similar calculations are required for each column type.

In the slab column connection some of the slab moment is transferred to the column by flexure on the column face perpendicular to the slab strip being considered while the remainder is transferred by shear as discussed in section 3-6, Step 4. Some assumption is also required as to the distribution of shear stresses caused by the weight of the exterior wall. In the example given, due to the location of the wall with respect to the column, this distribution was assumed to have a triangular distribution across the critical section *A-B*. Another approach which is also reasonable would be to reduce the moment to be transferred by subtracting the moment due to the weight of wall about the column centerline.

In this example the shear stresses at the exterior column were as high as permitted by *ACI 318-71*, and the shear stresses resulting from the unbalanced moment were approximately half of the final shear stresses. While it is common practice to cantilever the slab

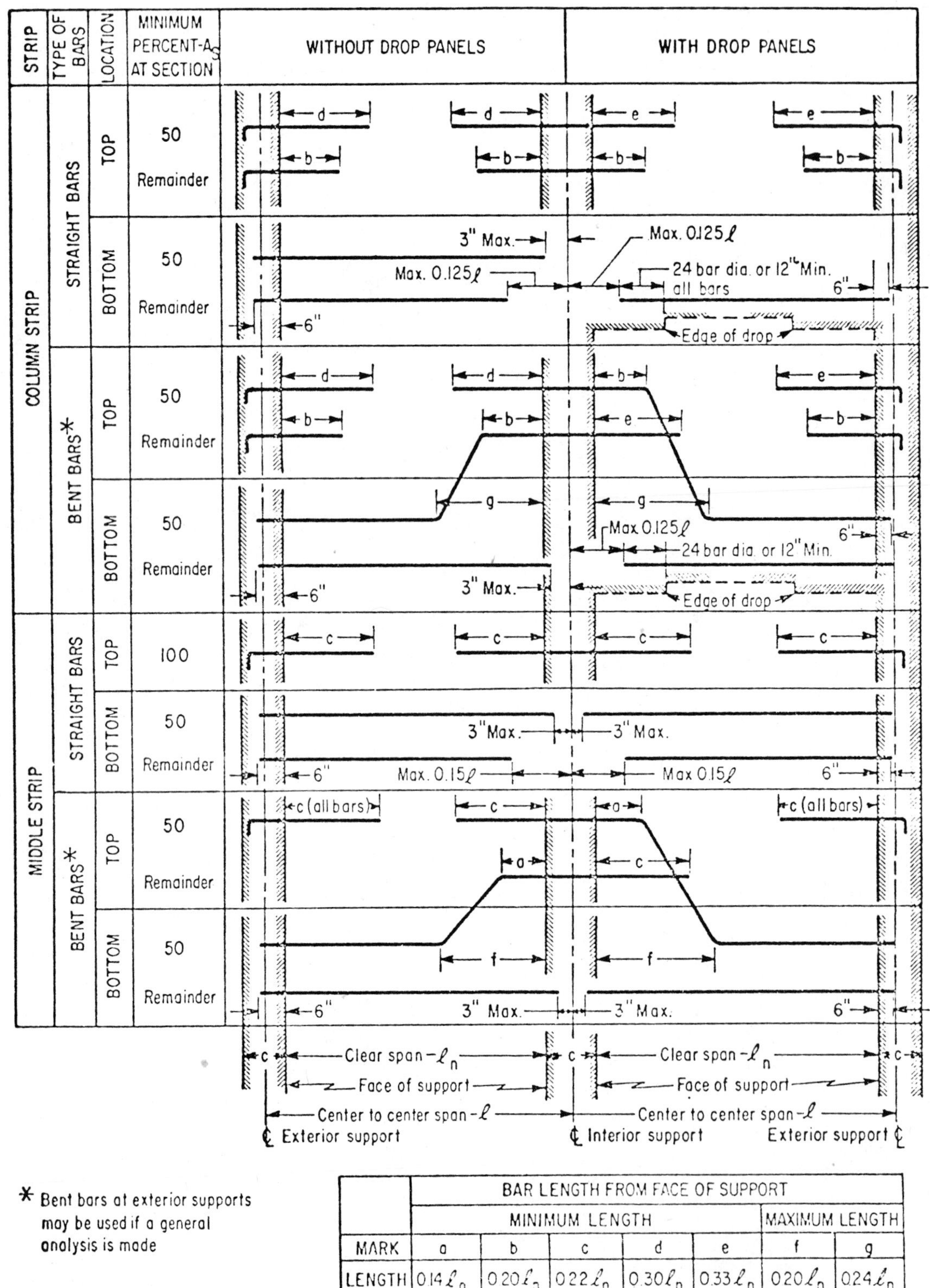

* Bent bars at exterior supports may be used if a general analysis is made

	BAR LENGTH FROM FACE OF SUPPORT						
	MINIMUM LENGTH					MAXIMUM LENGTH	
MARK	a	b	c	d	e	f	g
LENGTH	$0.14\ell_n$	$0.20\ell_n$	$0.22\ell_n$	$0.30\ell_n$	$0.33\ell_n$	$0.20\ell_n$	$0.24\ell_n$

Fig. 3-15 Minimum length of slab reinforcement, slabs without beams.[3-1]

beyond the exterior column face to facilitate placing precast wall units, in this example the projection was necessary to assist in resisting shear. Had the edge of the slab been flush with the outside column face, also common practice, the shear stresses would have substantially exceeded allowable values. This would have meant either increasing the thickness of the slab, providing an edge beam or column capital or increasing the column size. In general, to satisfy the shear requirements at exterior columns for slabs without beams it is usually necessary for economy to provide auxiliary stiffening elements such as capitals or to extend the slab beyond the outside face of the columns.

The final step is to choose the reinforcing for each strip. For the areas of steel required and the restriction on spacing, the diameter of bar for most slabs will be small. This coupled with the fact that shear stresses are small in areas not immediately adjacent to columns means that bond stresses in slabs without beams are not critical. For such slabs the bend-up points and imbedment lengths in terms of the clear span are given in Fig. 3-15. Generally the reinforcing required for each strip is spaced uniformly across the strip. The major exception will be for the exterior negative steel in the column strips where the reinforcing required to resist that portion of the moment transferred to the column by flexure must be placed within a slab width equal to one-half of the slab or drop panel thickness on each side of the column or capital, otherwise additional reinforcing is required.

The equivalent frame method could have been used in Step 2 to determine the design moments and the steps following would have proceeded in the same manner as for the direct design method. Not surprisingly, since the slab geometry is very regular in this example, the design moments obtained by the two methods will agree closely.

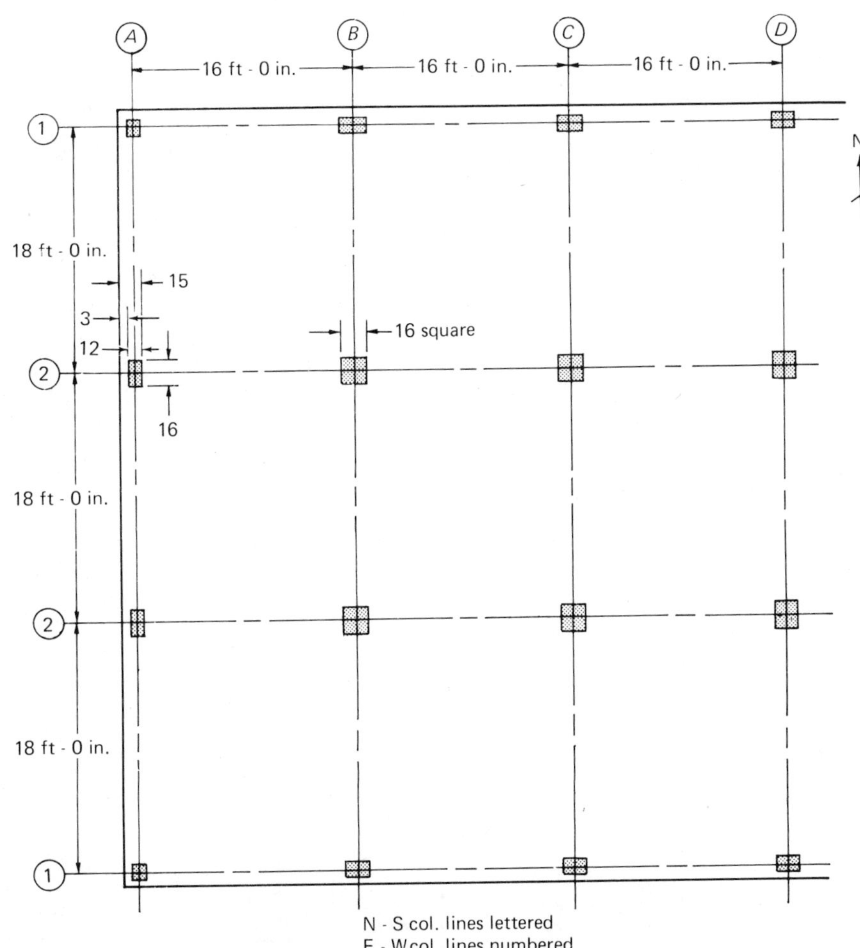

Design data:
Live loading specified = 40 lb/ft^2
Partition loading = 15 lb/ft^2
Exterior wall, 200 lb/ft
NOTE: slab extended 3 in. beyond external column face to support insulated precast panels for exterior wall
No column capitals or edge beams
Lateral forces resisted by shear walls
Story height, center to center = 9 ft

Material:
Concrete

$$f'_c = 4000 \text{ psi};\ f_y = 40{,}000 \text{ psi}$$

$$\gamma = 145 \text{ lb/ft}^3$$

Step 1: Assume constant thickness for all panels. Therefore governing condition is corner panel.

Corner panel—clear spans

$$\text{N-S}\quad l_n = 18 \times 12 - (6 + 8) = 216 - 14 = 202 \text{ in.}$$

$$\text{E-W}\quad l_n = 16 \times 12 - (6 + 8) = 192 - 14 = 178 \text{ in.}$$

$$\therefore \beta = \frac{202}{178} = 1.13;\ \beta_s = 0.5;\ \alpha_m = 0$$

Minimum slab thickness

(eq. 9-6, section 9.5.3.1)

$$h = \frac{202(800 + 0.005 \times 40{,}000)}{36{,}000 + 5000 \times 1.13\,[0 - 0.5(1 - 0.5)(1 + 1/1.13)]}$$

$$= \frac{202}{36 - 5.05(0.47)} = 6.07 \text{ in.}$$

but not less than

(eq. 9-7, or 5 in.)

$$h_{min} = \frac{202(800 + 0.005 \times 40{,}000)}{36{,}000 + 5000 \times 1.13(1 + 0.5)} = \frac{202}{44.5} \text{ in.} < 5 \text{ in.}$$

but need not be greater than

$$h = \frac{202(800 + 0.005 \times 40{,}000)}{36{,}000} = 5.62 \text{ in.}$$

by (section 9.5.3.3) use $h = 1.1 \times 5.62 = 6.2$, say 6½ in.

Effective depths

N–S $d = 6.5 - 0.25 - 0.75 = 5.5$ in.

E–W $d =$ 5.0 in.

for shear

$d_{ave} = 5.25$ in.

Design loads

weight of slab $= \frac{6.5}{12} \times 150 = 81.5 \text{ lb/ft}^2$

partitions (not moveable) $\underline{15.0}$

96.5 lb/ft^2

$w_d = 1.4 \times 96.5 = 135 \text{ lb/ft}^2$

$w_l = 1.7 \times 40 = \underline{68}$

$w = 203 \text{ lb/ft}^2$

Step 2: (By Direct Design Method)

Transverse strips, N–S

(a) Column lines Ⓑ, Ⓒ, etc.—interior

SYMM.

	①			②		
l_1		18.0 ft = 216 in.			18.0 ft = 216 in.	
l_n		202 in. = 16.8 ft			200 in. = 16.7 ft	
$M_o = \frac{0.203}{8} \times 16 \times l_n^2$		115			113	ft-kips
K_c (above) $= 4I_c/l_c$	$\frac{4 \times 2300}{108} = 85$			$\frac{4 \times 5680}{108} = 210$		
K_c (below) $= 4I_c/l_c$	85			210		
ΣK_c	170			420		
C (eq. 13-7)	724			1090		
$x = (1 - c_2/l_2)^3$	0.745			0.745		
$K_t = \Sigma(9/l_2)(C/x)$	91			137		
K_{ec} (eq. 13-5)	59			103		
$K_s = 4I_s/l_1$	$\frac{4 \times 4400}{216} = 82$		82	82		
$\alpha_{ec} = K_{ec}/\Sigma K_s$	$\frac{59}{82} = 0.72$			$\frac{103}{164} = 0.63$		
$\%M_o$ (section 13.3.3)	27	51	71	65	35	
design M slab	−31	+59	−81	−73	+40	ft-kips
column	31			5*		ft-kips

(b) Column lines Ⓐ—exterior

	①			②		
$M_o = \frac{0.203}{8} \times 8.8\, l_n^2$		63			62	ft-kips
K_c (above)	$\frac{4 \times 1728}{108} = 64$			$\frac{4 \times 4100}{108} = 152$		
K_c (below)	64			152		
ΣK_c	128			304		
K_t (½ for column line Ⓑ)	45			68		
K_{ec}	33			55		
K_s	$\frac{4 \times 2420}{216} = 45$		45	45		
$\alpha_{ec} = K_{ec}/\Sigma K_s$	$\frac{33}{45} = 0.73$			$\frac{55}{90} = 0.61$		
$\%M_o$ (section 13.3.3)	27	51	91	65	35	
design M slab	−17	+32	−45	−40	+22	
column	17					
$M_{wall} = \frac{0.2 \times l_n^2}{8} (\%M_o)$	−2	+3.5	−5	−4.5	+2.5	ft-kips

*Interior column design moments obtained from eq. (13-3).

$$M = \frac{0.08\,[(0.135 + 0.5 \times 0.68)\, 16 \times 16.8^2 - (0.135)\, 16 \times 16.7^2]}{1 + \frac{1}{0.63}} = 5 \text{ ft-kips}$$

NOTE: $\beta_a = \frac{96.5}{40} = 2.41 > 2.0$ ∴ Provisions of section 13.3.6.1 do not apply.

Longitudinal strips, E–W

(a) Column lines ② and ③–interior

	Ⓐ			Ⓑ		Ⓒ	
l_1		16.0 ft = 192 in.			16.0 ft = 192 in.		
l_n		178 in. = 14.8 ft			176 in. = 14.7 ft		
$M_o = \frac{0.203}{8} \times 18 \times l_n^2$		80			79		ft-kips
K_c (above) = $4I_c/l_c$	$\frac{4 \times 2300}{108} = 85$			$\frac{4 \times 5680}{108} = 210$		210	
K_c (below) = $4I_c/l_c$	= 85			= 210		210	
ΣK_c	170			420		420	in.^{3}
C (eq. 13-7)	724			1090		1090	in.^{4}
$x = (1 - c_2/l_2)^3$	0.795			0.795		0.795	
$K_t = \Sigma 9/l_2 C/x$	76			114		114	in.^{3}
K_{ec} (eq. 13-5)	53			89		89	in.^{3}
$K_s = 4I_s/l_1$	$\frac{4 \times 4940}{192} = 103$		103	103		103	in.^{3}
$\alpha_{ec} = K_{ec}/\Sigma K_s$	$\frac{53}{103} = 0.51$			$\frac{89}{206} = 0.43$		0.43	
$\%M_o$ (section 13.3.3)	22	53.5	71.5	65	35	65	
design M slab	−17.5	+43	−57		+28	−51.5	ft-kips
column	17.5		3.5	(eq. 13-3)			ft-kips

(b) Column lines ① and ④–exterior

	Ⓐ			Ⓑ		Ⓒ	
$M_o = \frac{0.203}{8} \times 9.8 \times l_n^2$		54			54		ft-kips
K_c (above)	$\frac{4 \times 1728}{108} = 64$			$\frac{4 \times 4100}{108} = 152$			
K_c (below)	64			152			
K_c	128			304			
K_t (½ for column line ②)	38			57			
K_{ec} (eq. 13-5)	29			48			
K_s	$\frac{4 \times 2690}{192} = 56$			56			
$\alpha_{ec} = K_{ec}/\Sigma K_s$	0.52			$\frac{48}{112} = 0.43$		0.43	
$\%M_o$ (section 13.3.3)	22	53	71.5	65	35	65	%
design M slab	−12	+29	−38	−35	+19	−35	ft-kips
column	12		2	(eq. 13-3)			ft-kips
$M_{wall} = \frac{0.2 \times l_n^2}{8} (\%M_o)$	−1.0	+3.0	−4	−3.5	+2.0		ft-kips

Step 3: Moments due to weight of wall are assigned to exterior column strips. Assign all negative moment at discontinuous edge to column strips. (section 13.3.4.2)

$$\text{N–S direction} \quad \beta_t = \frac{c}{2I_s} = \frac{724}{2 \times 4400} = 0.082$$

$$\text{E–N direction} \quad \beta_t = \frac{724}{2 \times 4940} = 0.075$$

N–S:
Moments ft-kips
Area of steel ( ) in.^{2}
*designates min. steel governs.

Note–steel in N–S direction placed in first layer from bottom and top.
d = 5.5 in.

Strip	Ⓐ (4.75′)	(8′)	Ⓑ (8′)	(8′)	Ⓒ (8′)
① 4.75′	−17 −2 −19 (1.20)	0 0 0 (1.25)*	−31 (1.92)	0 0 0 (1.25)*	−31 (1.92)
10′	+19.0 + 3.5 +22.5 (1.35)	+13 +12 +25 (1.68)	+35 (1.79)	+12 +12 +24 (1.60)	+35 (1.79)
② 8′	−34 − 5 −39 (2.44)	−11 −10 −21 (1.44)	−61 (3.77)	−10 −10 −20 (1.36)	−61 (3.77)
5′ (SYMM.)	+13.0 + 2.5 +15.5 (1.04)	+9 +8 +17 (1.25)*	+24 (1.60)	+8 +8 +16 (1.25)	+24 (1.60)

E–W direction
Steel in E–W direction placed in second layer from bottom and top.
d = 5.0 in.

Strip	Ⓐ (4.75′)	(8′)	Ⓑ (8′)	(8′)	Ⓒ (8′)
① 4.75′	−12 − 1 −13 (0.90)	+17 + 3 +20 (1.40)	−29 − 4 −33 (2.35)	+11 + 2 +13 (0.90)	−26 − 4 −30 (2.14)
10′	0 0 0 (1.56)*	+9 +12 +21 (1.60)	−7 −9 −16 (1.56)*	+6 +8 +14 (1.56)*	−6 −9 −15 (1.56)*
② 8′	−18 (1.44)	+25 (1.76)	−43 (3.14)	+16 (1.25)*	−39 (2.64)
10′ (SYMM.)	0 0 0 (1.56)*	+9 +9 +18 (1.56)*	−7 −7 −14 (1.56)*	+6 +6 +12 (1.56)*	−6 −6 −12 (1.56)*

Step 4:

(a) Edge column–typical column Ⓒ on line ①

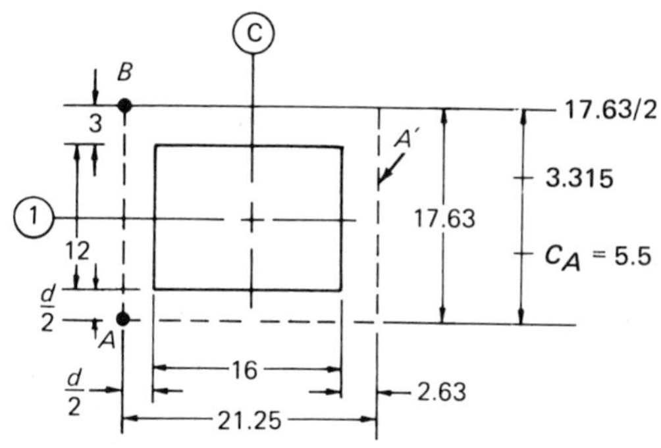

M about column line ① = 31 ft-kips

M about column line Ⓒ = 2 ft-kips

$$V_u = 0.203\left((9.75 \times 16) - \frac{21.25 \times 17.63}{144}\right) = 31.2 \text{ kips}$$

Assume wt. of wall carried only on side faces and linearly distributed shear stresses such that $v_A = 0$ and $v_B = \max$.

$$V_{wall} = (1.4 \times 0.2) \times \frac{16}{2} = 2.24^{K}$$

$$A' = [2(17.63) + 21.25]\ 5.25 = 297 \text{ in.}^2$$

$$A_{AB} = 17.63 \times 5.25 = 92.5 \text{ in.}^2$$

$$C_A = \frac{2 \times (17.63)^2/2}{2(17.63) + 21.25} = 5.5 \text{ in.}$$

$$J'_{①} = 2\left\{\frac{5.25}{12}(17.63)^3 + \frac{17.63}{12}(5.25)^3 + (5.25)(17.63)(3.315)^2\right\} + (5.25)(21.25)(5.5)^2 = 10630 \text{ in.}^4$$

$$J'_{②} = \frac{5.25}{12}(21.25)^3 + \frac{21.25}{12}(5.25)^3 + 2\left\{(5.25)(17.63)\left(\frac{21.25}{2}\right)^2\right\} = 25350 \text{ in.}^4$$

$$\overset{①}{\alpha_v} = 1 - \frac{1}{1 + \frac{2}{3}\sqrt{\frac{17.63}{21.25}}} = 1 - 0.62 = 0.38 \quad \text{(section 11.13.2)}$$

$$\overset{Ⓒ}{\alpha_v} = 1 - \frac{1}{1 + \frac{2}{3}\sqrt{\frac{21.25}{17.63}}} = 1 - 0.58 = 0.42$$

TABLE 3-2 Moment Distribution Factors for Slab-Beam Elements (Flat Plate)[3-15]

c_1/l_1 \ c_2/l_2		0.00	0.05	0.10	0.15	0.20	0.25	0.30	0.35	0.40	0.45	0.50
0.00	M	0.083	0.083	0.083	0.083	0.083	0.083	0.083	0.083	0.083	0.083	0.083
	k	4.000	4.000	4.000	4.000	4.000	4.000	4.000	4.000	4.000	4.000	4.000
	C	0.500	0.500	0.500	0.500	0.500	0.500	0.500	0.500	0.500	0.500	0.500
0.05	M	0.083	0.084	0.084	0.084	0.085	0.085	0.085	0.086	0.086	0.086	0.086
	k	4.000	4.047	4.093	4.138	4.181	4.222	4.261	4.299	4.334	4.368	4.398
	C	0.500	0.503	0.507	0.510	0.513	0.516	0.518	0.521	0.523	0.526	0.528
0.10	M	0.083	0.084	0.085	0.085	0.086	0.087	0.087	0.088	0.088	0.089	0.089
	k	4.000	4.091	4.182	4.272	4.362	4.449	4.535	4.618	4.698	4.774	4.846
	C	0.500	0.506	0.513	0.519	0.524	0.530	0.535	0.540	0.545	0.550	0.554
0.15	M	0.083	0.084	0.085	0.086	0.087	0.088	0.089	0.090	0.090	0.091	0.092
	k	4.000	4.132	4.267	4.403	4.541	4.680	4.818	4.955	5.090	5.222	5.349
	C	0.500	0.509	0.517	0.526	0.534	0.543	0.550	0.558	0.565	0.572	0.579
0.20	M	0.083	0.085	0.086	0.087	0.088	0.089	0.090	0.091	0.092	0.093	0.094
	k	4.000	4.170	4.346	4.529	4.717	4.910	5.108	5.308	5.509	5.710	5.908
	C	0.500	0.511	0.522	0.532	0.543	0.554	0.564	0.574	0.584	0.593	0.602
0.25	M	0.083	0.085	0.086	0.087	0.089	0.090	0.091	0.093	0.094	0.095	0.096
	k	4.000	4.204	4.420	4.648	4.887	5.138	5.401	5.672	5.952	6.238	6.527
	C	0.500	0.512	0.525	0.538	0.550	0.563	0.576	0.588	0.600	0.612	0.623
0.30	M	0.083	0.085	0.086	0.088	0.089	0.091	0.092	0.094	0.095	0.096	0.098
	k	4.000	4.235	4.488	4.760	5.050	5.361	5.692	6.044	6.414	6.802	7.205
	C	0.500	0.514	0.527	0.542	0.556	0.571	0.585	0.600	0.614	0.628	0.642
0.35	M	0.083	0.085	0.087	0.088	0.090	0.091	0.093	0.095	0.096	0.098	0.099
	k	4.000	4.264	4.551	4.864	5.204	5.575	5.979	6.416	6.888	7.395	7.935
	C	0.500	0.514	0.529	0.545	0.560	0.576	0.593	0.609	0.626	0.642	0.658
0.40	M	0.083	0.085	0.087	0.088	0.090	0.092	0.094	0.095	0.097	0.099	0.100
	k	4.000	4.289	4.607	4.959	5.348	5.778	6.255	6.782	7.365	8.007	8.710
	C	0.500	0.515	0.530	0.546	0.563	0.580	0.598	0.617	0.635	0.654	0.672
0.45	M	0.083	0.085	0.087	0.088	0.090	0.092	0.094	0.096	0.098	0.100	0.101
	k	4.000	4.311	4.658	5.046	5.480	5.967	6.517	7.136	7.336	8.625	9.514
	C	0.500	0.515	0.530	0.547	0.564	0.583	0.602	0.621	0.642	0.662	0.683
0.50	M	0.083	0.085	0.087	0.088	0.090	0.092	0.094	0.096	0.098	0.100	0.102
	k	4.000	4.331	4.703	5.123	5.599	6.141	6.760	7.470	8.289	9.234	10.329
	C	0.500	0.515	0.530	0.547	0.564	0.583	0.603	0.624	0.645	0.667	0.690
$X = (1 - c_2/l_2)^3$		1.000	0.856	0.729	0.613	0.512	0.421	0.343	0.274	0.216	0.166	0.125

FEM (uniform load w) $= Mwl_2l_1^2$

K (stiffness) $= kEl_2t^3/12l_1$

Carry-over factor $= C$

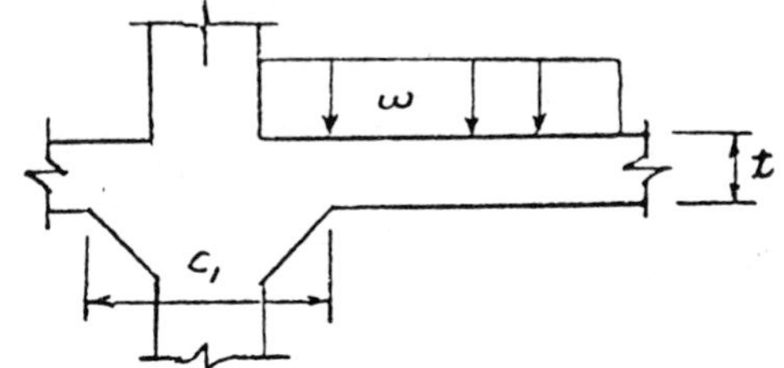

Shear stress at point A

$$v_A = \frac{31.2}{(0.85 \times 297)} + \frac{(0.38)(31 \times 12)(5.5)}{(0.85)(10630)} + \frac{(0.42)(2 \times 12)(10.63)}{(0.85)(25350)}$$

$$= 0.124 + 0.086 + 0.005 = 0.215 \text{ k/in.}^2 = 215 \text{ psi}$$

$$v_B = \frac{31.2}{(0.85)(297)} - \frac{(0.38)(31 \times 12)(17.63 - 5.5)}{(0.85)(10630)}$$

$$+ \frac{(0.42)(2 \times 12)(10.63)}{(0.85)(25350)} + \frac{(2 \times 2.24)}{(0.85)(92.5)}$$

$$= 0.124 - 0.190 + 0.005 + 0.057 = -0.004 \text{ kip/in.}^2 = -4 \text{ psi}$$

$$v_{all} = 4\sqrt{4000} = 253 \text{ psi} \quad \therefore \text{satisfactory}$$

Similar calculations can be made to show adequacy in shear for corner and interior columns.

EXAMPLE 3-3: To illustrate the features of the equivalent frame method and to obtain some comparison between the two design methods consider a slab system that is less regular, that is, a slab with unequal span lengths and with a variety of auxiliary stiffening elements. The design moments for a typical strip of such a slab are determined in this example. To perform the moment distribution it is necessary to evaluate the fixed end moment, carry-over factor and distribution factors taking into account the variations in the cross sections of the equivalent frame members, a tedious task indeed by elastic theory. It was found[3-15] that by introducing the assumption that the ratios c_1/l_1 and c_2/l_2 at the far end were equal to the respective ratios at the near end, the number of variables could be reduced to permit a reasonable tabulation of the required quantities without significant difference from code definitions. Such tables are reproduced in Tables 3-2 to 3-6. The detailed calculations for the moment distribution factors are presented in ex. 3-3 with the quantities obtained from these tables enclosed in square brackets. To obtain the final design moments, the moments computed at the column center line are reduced to the critical sections.

If it is assumed that the slab system is continuous in the direction perpendicular to the strip considered above, the direct design method may also be used. As noted in ex. 3-2, the equivalent column stiffness may be computed for exterior columns when using this method, by assuming stiffness coefficients equal to 4. Since the equivalent column stiffnesses using the stiffness definitions given in code were already computed for the equivalent frame method, the design moments were computed for both values of equivalent column stiffness. It is seen that although there is a substantial difference in the corresponding values of α_{ec}, there is not a great difference in the final design moments.

A comparison of the design moments obtained by the direct design method with those obtained by the equivalent frame method indicates differences in values at corresponding locations, although the total moment in corresponding panels is the same. These differences in moments are due primarily to neglecting the rotation of the interior columns in the direct design method. If the span lengths had been more equal, these differences would have been less.

The design moments obtained by either method are distributed to column and middle strips in the same manner as for ex. 3-2. For this reason, this and the remaining steps in the design are not presented.

TABLE 3-3 Moment Distribution Factors for Slab-Beam Elements[3-15]
$t_1 = 1.5t$

c_1/l_1 \ c_2/l_2		0.00	0.05	0.10	0.15	0.20	0.25	0.30
0.00	M	0.088	0.088	0.088	0.088	0.088	0.088	0.088
	k	4.795	4.795	4.795	4.795	4.795	4.795	4.797
	C	0.542	0.542	0.542	0.542	0.542	0.542	0.542
0.05	M	0.088	0.088	0.089	0.089	0.089	0.089	0.090
	k	4.795	4.846	4.896	4.944	4.990	5.035	5.077
	C	0.542	0.545	0.548	0.551	0.553	0.556	0.558
0.10	M	0.088	0.088	0.089	0.090	0.090	0.091	0.091
	k	4.795	4.894	4.992	5.039	5.184	5.278	5.368
	C	0.542	0.548	0.553	0.559	0.564	0.569	0.573
0.15	M	0.088	0.089	0.090	0.090	0.091	0.092	0.092
	k	4.795	4.938	5.082	5.228	5.374	5.520	5.665
	C	0.542	0.550	0.558	0.565	0.573	0.580	0.587
0.20	M	0.088	0.000	0.090	0.091	0.092	0.093	0.094
	k	4.795	4.978	5.167	5.361	5.558	5.760	5.962
	C	0.542	0.552	0.562	0.571	0.581	0.590	0.590
0.25	M	0.088	0.089	0.090	0.091	0.092	0.094	0.095
	k	4.795	5.015	5.245	5.485	5.735	5.994	0.261
	C	0.542	0.553	0.565	0.576	0.587	0.598	0.600
0.30	M	0.088	0.089	0.090	0.092	0.093	0.094	0.095
	k	4.795	5.048	5.317	5.601	5.902	0.219	0.550
	C	0.542	0.554	0.567	0.580	0.593	0.605	0.618

FEM (uniform load w) $= Mwl_2l_1^2$

K (stiffness) $= kEl_2t^3/12l_1$

Carry-over factor $= C$

TABLE 3-4 Moment Distribution Factors for Slab-Beam Elements[3-15]
($t_1 = 1.5t$)

c_1/l_1 \ c_2/l_2		0.00	0.05	0.10	0.15	0.20	0.25	0.30
0.00	M	0.093	0.093	0.093	0.093	0.093	0.093	0.093
	k	5.837	5.837	5.837	5.837	5.837	5.837	5.837
	C	0.589	0.589	0.589	0.589	0.589	0.589	0.589
0.05	M	0.093	0.093	0.093	0.093	0.094	0.094	0.094
	k	5.837	5.890	5.942	5.993	6.041	6.087	6.131
	C	0.589	0.591	0.594	0.596	0.598	0.600	0.602
0.10	M	0.093	0.093	0.094	0.094	0.094	0.095	0.095
	k	5.837	5.940	6.042	6.142	6.240	6.335	6.427
	C	0.589	0.593	0.598	0.602	0.607	0.611	0.615
0.15	M	0.093	0.093	0.094	0.095	0.095	0.096	0.096
	k	5.837	5.986	6.135	6.284	6.432	6.579	6.723
	C	0.589	0.595	0.602	0.608	0.614	0.620	0.626
0.20	M	0.093	0.093	0.094	0.095	0.006	0.096	0.097
	k	5.837	6.027	6.221	6.418	6.616	6.816	7.015
	C	0.589	0.597	0.605	0.613	0.621	0.628	0.635
0.25	M	0.093	0.094	0.094	0.095	0.096	0.097	0.098
	k	5.837	6.065	6.300	6.543	6.790	7.043	7.298
	C	0.589	0.598	0.608	0.617	0.626	0.635	0.644
0.30	M	0.093	0.094	0.095	0.096	0.097	0.098	0.090
	k	5.837	0.099	6.372	6.657	6.953	7.258	7.571
	C	0.589	0.599	0.610	0.620	0.631	0.641	0.651

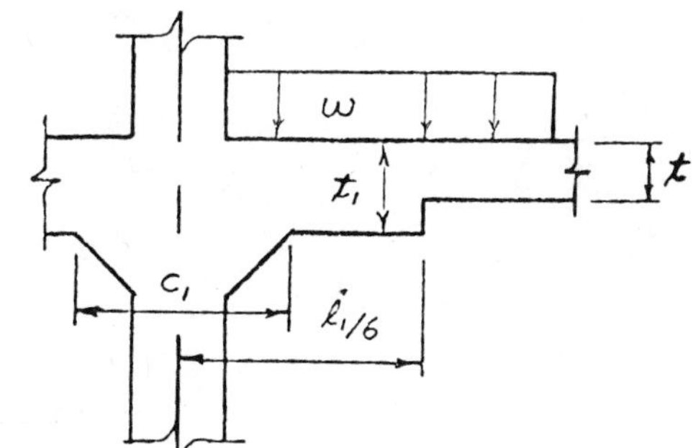

FEM (uniform load w) $= Mwl_2l_1^2$

K (stiffness) $= kEl_2t^3/12l_1$

Carry-over factor $= C$

TABLE 3-5 Column Stiffness Coefficients, k_c*[3-15]

a/h \ b/h	0.00	0.02	0.04	0.06	0.08	0.10	0.12	0.14	0.16	0.18	0.20	0.22	0.24
0.00	4.000	4.082	4.167	4.255	4.348	4.444	4.545	4.651	4.762	4.878	5.000	4.128	5.263
0.02	4.337	4.433	4.533	4.638	4.747	4.862	4.983	5.110	5.244	5.384	5.533	5.690	5.856
0.04	4.709	4.882	4.940	5.063	5.193	5.330	5.475	5.627	5.787	5.958	6.138	6.329	6.533
0.06	5.122	5.252	5.393	5.539	5.693	5.855	6.027	6.209	6.403	6.608	6.827	7.060	7.310
0.08	5.581	5.735	5.898	6.070	6.252	6.445	6.650	6.868	7.100	7.348	7.613	7.897	8.203
0.10	6.091	6.271	6.462	6.665	6.880	7.109	7.353	7.614	7.893	8.192	8.513	8.859	9.233
0.12	6.659	6.870	7.094	7.333	7.587	7.859	8.150	8.461	8.796	9.157	9.546	9.967	10.430
0.14	7.292	7.540	7.803	8.084	8.385	8.708	9.054	9.426	9.829	10.260	10.740	11.250	11.810
0.16	8.001	8.291	8.600	8.931	9.287	9.670	10.080	10.530	11.010	11.540	12.110	12.740	13.420
0.18	8.796	9.134	9.498	9.888	10.310	10.760	11.260	11.790	12.370	13.010	13.700	14.470	15.310
0.20	9.687	10.080	10.510	10.970	11.470	12.010	12.600	13.240	13.940	14.710	15.560	16.490	17.530
0.22	10.690	11.160	11.660	12.200	12.800	13.440	14.140	14.910	15.760	16.690	17.210	18.870	20.150
0.24	11.820	12.370	12.960	13.610	14.310	15.080	15.920	16.840	17.870	19.000	20.260	21.650	23.260

$$*K_c = \frac{k_c E_c I_c}{h}$$

Note:

a = length of rigid column section at near end
b = length at rigid column section at far end

TABLE 3-6 Values of Torsion Constant, C^{*}[3-15]

y \ x	4	5	6	7	8	9	10	12	14	16
12	202	369	592	868	1,188	1,538	1,900	2,557	—	—
14	245	452	736	1,096	1,529	2,024	2,566	3,709	4,738	—
16	388	534	880	1,325	1,871	2,510	3,233	4,861	6,567	8,083
18	330	619	1,024	1,554	2,212	2,996	3,900	6,013	8,397	10,813
20	373	702	1,167	1,782	2,553	3,482	4,567	7,165	10,226	13,544
22	416	785	1,312	2,011	2,895	3,968	5,233	8,317	12,055	16,275
24	548	869	1,456	2,240	3,236	4,454	5,900	9,469	13,885	19,005
27	522	994	1,672	2,583	3,748	5,183	6,900	11,197	16,628	23,101
30	586	1,119	1,888	2,926	4,260	5,912	7,900	12,925	19,373	27,197
33	650	1,243	2,104	3,269	4,772	6,641	8,900	14,653	22,117	31,293
36	714	1,369	2,320	3,612	5,284	7,370	9,900	16,381	24,860	35,389
42	842	1,619	2,752	4,298	6,308	8,828	11,900	19,837	30,349	43,581
48	970	1,869	3,184	4,984	7,332	10,286	13,900	23,293	35,836	51,773
54	1,098	2,119	3,616	5,670	8,356	11,744	15,900	26,749	41,325	59,965
60	1,226	2,369	4,048	6,356	9,380	13,202	17,900	30,205	46,813	68,157

$^{*}C = (1 - 0.63x/y)\,\dfrac{x^3y}{3}$

x is smaller dimension of rectangular cross section.

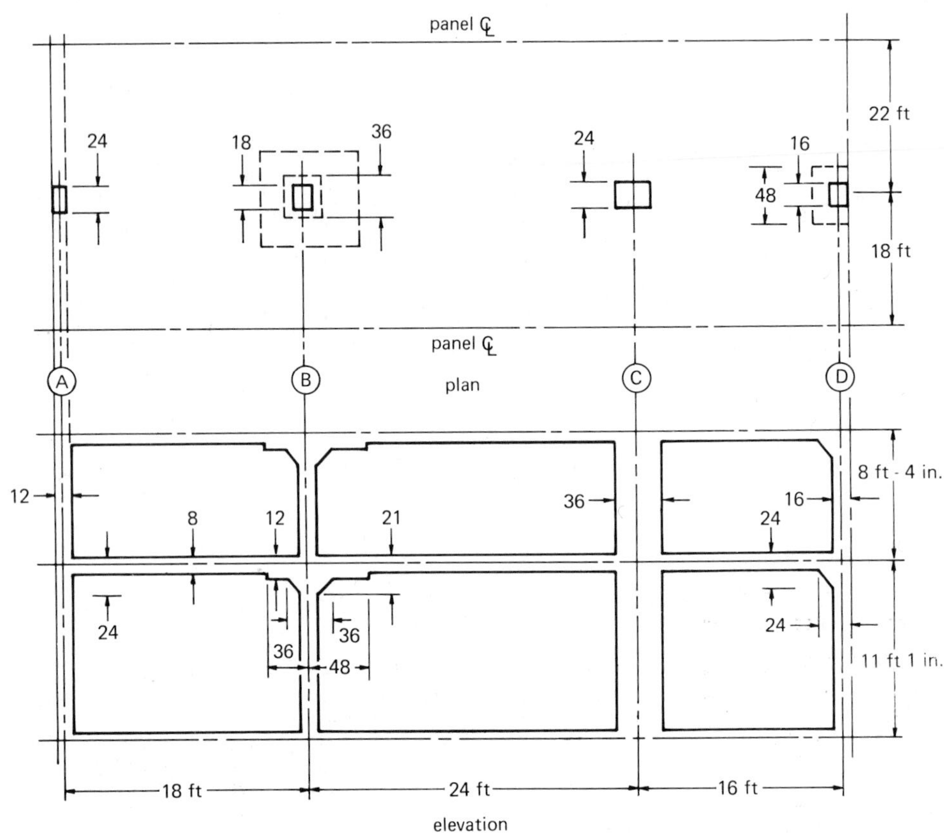

Taking the above sketch as a typical transverse design strip, we have for the loads.

Loading:

$$w_{\alpha} = 1.4 \times 100 = 140 \text{ lb/ft}^2$$

$$w_l = 1.7 \times 75 = 128 \text{ lb/ft}^2$$

$$w = \qquad = 268 \text{ lb/ft}^2$$

Step 2 Equivalent Frame Method

	A	A–B left	A–B mid	A–B right	B	B–C left	B–C mid	B–C right	C	C–D left	C–D mid	C–D right	D	
GEOMETRY														
Column above b/h	20/100 = 0.20				17/100 = 0.17				4/100 = 0.04				20/100 = 0.20	
Column above a/h	4/100 = 0.04				4/100 = 0.04				4/100 = 0.04				4/100 = 0.04	
Slab c_1/l_1 (left)	——				36/216 = 0.17				36/288 = 0.13				32/192 = 0.17	
Slab c_2/l_2	24/240 = 0.10				36/240 = 0.15				24/240 = 0.10				48/240 = 0.20	
Slab c_1/l_1 (right)	12/216 = 0.06				36/288 = 0.13				36/192 = 0.19				——	
Column below a/h	20/133 = 0.15				17/133 = 0.13				4/133 = 0.03				20/133 = 0.15	
Column below b/h	4/133 = 0.03				4/133 = 0.03				4/133 = 0.03				4/133 = 0.03	
STIFFNESS														
$K_c^{above} = [k_c]\ I_c/h$	[6.138] 3456/100 = 212				[5.872] 8748/100 = 514				[4.940] 93312/100 = 4609				[6.138] 5461/100 = 335	
$K_c^{below} = [k_c]\ I_c/h$	[8.045] 3456/133 = 209				[7.327] 8748/133 = 482				[4.676] 93312/133 = 3281				[8.045] 5461/133 = 330	
ΣK_c	421				996				7890				665	
$\Sigma[c]$	[9469] + [1871] = 11340				[5284]				[5284]				[4601]	
$[x]$	[0.729]				[0.613]				[0.729]				[0.512]	
$K_T = \Sigma(9/l_2)c/x$	1167				646				544				674	
$K_{ec} = 1/(1/\Sigma K_c + 1/K_T)$	310				392				509				335	
$K_s^{left} = [k_s]\ I_s/l_1$	——				[6.338] 10240/216 = 300				[4.233] 10240/288 = 151				[4.611] 10240/192 = 246	
$K_s^{right} = [k_s]\ I_s/l_1$	[4.174] 10240/216 = 198				[6.227] 10240/288 = 221				[4.330] 10240/192 = 231				——	
ΣK	508				913				891				581	
FACTORS														
Moment Coefficient $[M]$		[0.084]		[0.095]		[0.094]		[0.085]		[0.086]		[0.087]		
Carry-Over Factor		[0.509] ⟶		⟵ [0.610]		[0.606] ⟶		⟵ [0.515]		[0.521] ⟶		⟵ [0.538]		
Slab Distribution Factor	— 0.39				0.33 0.24				0.17 0.23				0.42 —	
Column Distribution Factor	0.61				0.43				0.57				0.58	
DISTRIBUTION														
$FEM = [M]\ Wl_2 l_1^2$		−146		+165		−290		+262		−118		+119		
bal	+89	+ 57		+ 41	+54	+ 30		− 25	−82	− 37		− 53	−69	
co		+ 25		+ 29		− 13		+ 18		− 27		− 19		
bal	−15	− 10		− 5	− 7	− 4		+ 2	+ 5	+ 2		+ 8	+11	
co		− 3		− 5		+ 1		− 2		+ 4		+ 1		
bal	+ 2	+ 1		+ 1	+ 2	+ 1			− 1	− 1			− 1	
Moments at Column	+76	− 76		+226	+49	−275		+253	−78	−167		+ 59	−59	ft-kips
Design Moments		− 57	+67	−149		−183	+121	−166		− 90	+54	− 14		ft-kips

Step 2: (by direct design method)

Column line	Ⓐ		Ⓑ		Ⓒ		Ⓓ	
l_1 (center to center span)		18		24		16		ft
l_n (clear span)		16		21		12.5		ft
l_2/l_1 (panel aspect ratio)		1.11		0.83		1.25		
$M_0 = \dfrac{w l_2 l_n^2}{8}$		172		296		105		ft-kips
$K_c = \Sigma \dfrac{4I_c}{h}$ $(k_c = 4)$	242		613		6500		383	in.^{4}/in.
K_t (see calc. EFM)	1167		646		544		674	in.^{4}/in.
K_{ec} (eq. 13-5)	200		315		502		245	in.^{4}/in.
$K_s = \Sigma \dfrac{4I_s}{l_1}$ $(k_s = 4)$		190		142		213		in.^{4}/in.
α_{ec}	1.05		0.95		1.42		1.15	
% of M_0 (section 13.3.33)	33	49	70 | 65	35	65 | 70	48	35	%
M_{Des} slab	−57	+84	−120 −192	+104	−192 −73	+50	−37	ft-kips
M_{Des} col (eq. 13-3)	57		27.5		42.5		37	ft-kips
Alternate solution using α_{ec} from stiffnesses accounting for variations in cross section.								
α_{ec} (from EFM)	1.57		0.75		1.33		1.36	
M_{Des} slab	−68	+79	−118 −192	+104	−192 −73	+49	−40	ft-kips
M_{Des} col	68		24		41		40	ft-kips

EXAMPLE 3-4: With two way slabs, as the spans become longer, the increased thickness required to limit deflections can result in the condition where the dead weight of the slab becomes the major portion of the design load. In such cases it is frequently economical to rib the slab in both directions to provide greater stiffness with less weight. Such slabs known as "waffle" slabs may be designed using the procedures outlined in Section 3.6.

The ribs in waffle slabs are obtained using generally square pans to form the void or recess between the ribs. These pans can be supported on standard scaffolding to provide a simple and economical forming system. Standard pans are widely available for rent to form recesses on a 24 in. or 36 in. module. For a 24 in. module the standard pan has a plan dimension at the base of 19 × 19 in. and is available in heights of 6, 8, 10, and 12 in. Pans for the 36 in. module measure 30 × 30 in. at the base and is available in heights of 8, 10, 12, 14, 16, and 20 in. To facilitate stripping all standard pans have sloping sides forming a bevel of 1½ in./ft from the vertical. In many localities other pan sizes are available and frequently custom pans are fabricated from fibre glass, plywood, or sheet metal to meet requirements. In other cases the recesses between ribs are formed by lightweight products which are left in place after pouring. Except for the additional weight there is no difference in the design procedure if the strength of this filler material is neglected.

The design follows the steps outlined in section 3-6. The minimum thickness of the slab between ribs is specified as 1/12 of the clear span between ribs. In ex. 3-4 this thickness was chosen to provide sufficient cover for two layers of reinforcing steel. In some instances this thickness may be governed by shear if concentrated loads on small areas is possible. The choice of overall thickness depends primarily on shear so it is desirable to check the adequacy of the sections in shear near the beginning of the design. To increase the shear capacity in the vicinity of the columns it is usual to fill the recesses with structural concrete to form a solid slab. In some cases a column capital will also be desirable.

It would appear from section 9.5.3 of ACI 318-71 that it is not necessary to compute deflections for waffle slabs if the overall depth provided is greater than the computed minimum thicknesses. In the design example calculations are included for the immediate deflection caused by live load only for comparison with specified maximum deflection. The method of calculation is the simplified procedure for finding the maximum deflection for a typical interior panel for a solid slab without beams but with an appropriate modification to the moment of inertia to account for the recesses. The coefficient α is dependent on the long to short span ratio b/a and is given as follows:

b/a	1.0	1.1	1.2	1.3	1.4	1.5
α	0.063	0.052	0.045	0.040	0.036	0.033

It is admitted that such a computation is only approximate for a waffle slab but it does indicate that this slab is sufficiently stiff for use in a parking structure. An upper limit to the deflection may have been obtained by basing the average value for the moment of inertia on the cracked section instead of the uncracked section as assumed in the sample calculation.

The design for moment follows in a similar manner to ex. 3-2 except that the moment away from the column must be resisted by individual ribs. It is assumed that each rib across a column or middle strip participates equally in resisting the total moment assigned to that strip.

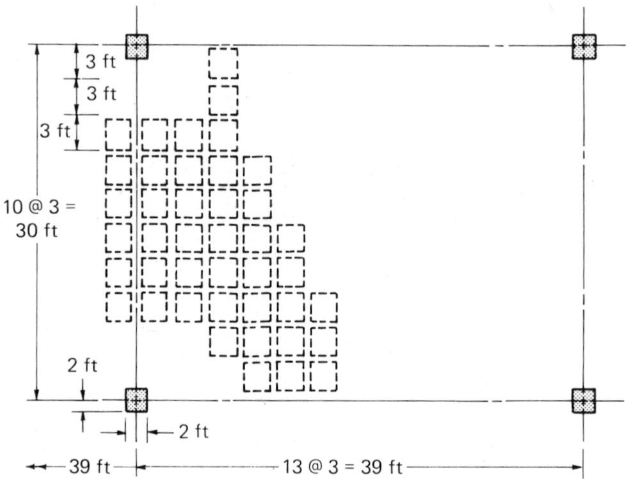

In this example, given the floor plan above, first collect the necessary data.

Loading: Passenger cars

distributed loading = 50 lb/ft^2

concentrated load = 2500 lbs on area 2.5 ft^2

Material:

concrete–(normal weight) $f'_c = 4000$ psi

reinforcing steel $f_y = 60{,}000$ psi

Due to spans and to obtain minimum height of structure consider use of waffle slab. Start with a trial section of standard 30 in. pans with pan depth of 16 in., resulting ribs to be 36 in. center to center.

To accommodate reinforcement in two directions and to maintain minimum cover of ¾ in., use 3 in. top slab.

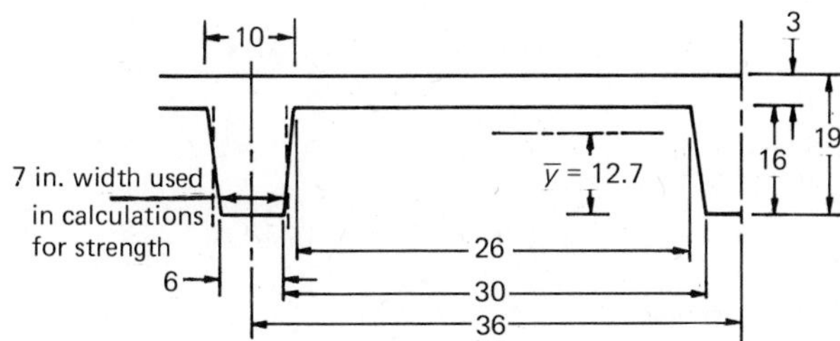

Assume initially that the slab is solid in the vicinity of the column by filling the 16 cells adjacent to the column.

Dead weight of slab:

Waffle slab

$$w_w = \frac{1}{9}\left(\frac{36 \times 36}{144} \times \frac{19}{12} \times 150 - \frac{28 \times 28}{144} \times \frac{16}{12} \times 150\right)$$

$$= \tfrac{1}{9}(2140 - 1090) = 116 \text{ lb/ft}^2$$

Solid slab

$$w_s = \tfrac{19}{12} \times 150 = 238 \text{ lb/ft}^2$$

Design loads:

w_d	$1.4 \times 116 = 163$ lb/ft²	$1.4 \times 238 = 333$ lb/ft²
w_l	$1.7 \times 50 = 50$ lb/ft²	$1.7 \times 50 = 85$ lb/ft²
w	248 lb/ft²	418 lb/ft²

Check shear: Concentrated load of 2500 lbs will not govern since it is distributed over area greater than area of single cell.

(a) Punching shear at column

$$\text{Area enclosed by } b_0 = \frac{41 \times 41}{144} = 11.7 \text{ ft}^2$$

Area of solid slab = $12 \times 12 = 144$ ft²

Area of waffle slab = $(30 \times 39) - 144 = 1026$ ft²

$b_0 = 4 \times 41 = 164$ in.

$V_u = 1026 \times 0.248 + (144 - 11.7) \times 0.418 = 254 + 55 = 309$ kips

$$v_c = \frac{V_u}{\phi b_0 d} = \frac{309{,}000}{0.85 \times 164 \times 17} = 130 \text{ psi}$$

$$(v_c)_{\text{allowable}} = 4\sqrt{f'_c} = 4\sqrt{4000} = 252 \text{ psi}$$

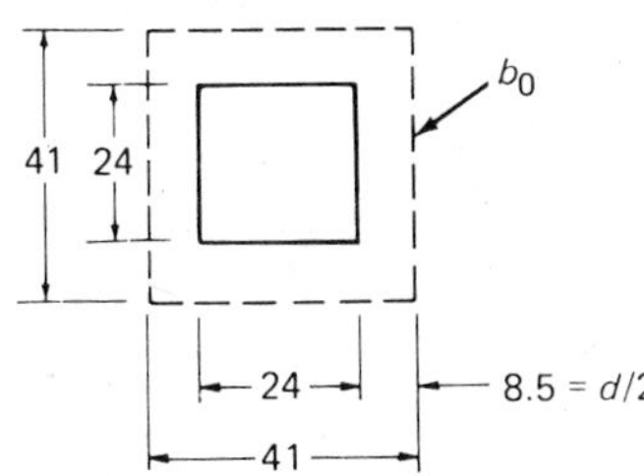

Since 130 < 252, satisfactory for punching shear at column.

(b) Shear at periphery of solid portion

$V_u = 1026 \times 0.248 = 254^{K}$

N_0 of ribs to resist shear = $4 \times 5 = 20$

V_u per rib = $254/20 = 12.7^{K}$

$$v_c = \frac{V_u}{\phi b_w d} = \frac{12{,}700}{0.85 \times 7 \times 17} = 126 \text{ psi}$$

$$(v_c)_{\text{allowable}} = 1.1(2\sqrt{f'_c}) = 1.1 \times 2\sqrt{4000} = 138 \text{ psi}$$

Since 126 < 138, satisfactory in shear.

Deflection: (see comments in text, ex. 3-4, section 3.7)

$$h = \frac{l_n(800 + 0.005 f_y)}{36{,}000} = \frac{37 \times 12 \times (800 + 0.005 \times 60{,}000)}{36{,}000} = 13.6 \text{ in.}$$

From Table 9-5(b) ACI318-71, for floors not supporting or attached to non-structural elements likely to be damaged by large deflections, the immediate deflection due to the live load (no load factor) is $l/360$.

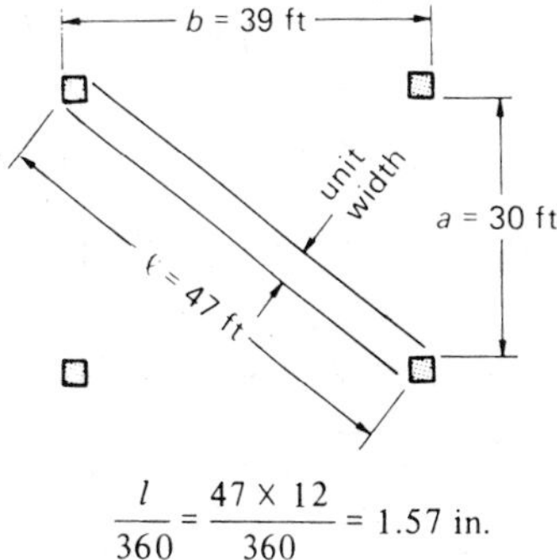

$$\frac{l}{360} = \frac{47 \times 12}{360} = 1.57 \text{ in.}$$

For interior panels

$$\delta \,\underline{\Omega}\, \alpha \frac{wb^4}{Eh^3} \,\underline{\Omega}\, \alpha \frac{wb^4}{12EI}$$

$w = 50$ lb/ft² (live load); $E = 57{,}000\sqrt{f'_c} = 3.6 \times 10^6$ psi

$b = 39$ ft; $I_{\text{Ave}} = 2060$ in.⁴/ft (uncracked section for portion with cells)

$$\alpha = 0.40, \text{ when } \frac{b}{a} = \frac{39}{30} = 1.3 \text{ (See Table, section 3.7)}$$

$$\delta = \frac{0.040 \times 50 \times (39)^4 \times 1728}{12 \times 3.6 \times 10^6 \times 2060} = 0.11 \text{ in.}$$

NOTE: Although above calculation is approximate for waffle slabs, it indicates that deflections under immediate live load are much below the specified maximums. Indication of dead load deflection can also be obtained by using appropriate w.

Design for moment: Use direct design method. For moment calculations use constant load obtained by weighting loads in proportion to their areas.

$$w = \frac{144}{1170} \times 418 + \frac{1026}{1170} \times 248 = 52 + 217 = 269 \text{ lb/ft}^2$$

(a) Long span direction

$l_n = 39 - 2 = 37$ ft $\quad l_2 = 30$ ft

$$M_0 = \frac{0.269 \times 30 \times 37^2}{8} = 1380 \text{ ft-kips}$$

x = distance from face of support to face of solid section

$= 6 \times 12 + 3.5 - 12 = 63.5$ in. $= 5.3$ ft

M_0ES – 0.65 M_0 $= -900^{1K}$ $\qquad$ **+ 0.35 M_0 $= +484^{1K}$**

$w = .269 \times 30 = 8.03$ kip/ft.

5.3 ft

$$M_x = -900 - \frac{8.03\,(5.3)^2}{2} + (149)\,(5.3)' = -208^{1K}$$

$$V = \frac{37}{2} \times 8.03 = 149^{K}$$

Consider a column strip of width = $l_2/2 = 30/2 = 15$ ft.

$$M_{neg} \text{ (at face of support)} = -0.75 \times 900 = -675 \text{ ft-kips}$$

Width of solid slab in column strip = $12 \times 12 + 2(3.5) = 151$ in.

$$M_{neg}/\text{ft width solid slab} = \frac{675 \times 12}{151} = 53.6 \text{ ft-kips/ft}$$

for $d = 17$ in.

$$A_s = 0.73 \text{ in.}^2/\text{ft}$$

$$M_{neg} \text{ (at face of solid slab)} = -0.75 \times 208 = -156 \text{ ft-kips}$$

Must be resisted by five joist webs

$$M_{neg}/\text{joist} = \frac{156}{5} = 31.2 \text{ ft-kips}$$

for $d = 17$ in.

$$A_s = 0.42 \text{ in.}^2/\text{joist}$$

$$M_{pos} \text{ (midspan)} = +0.6 \times 484 = 290 \text{ in.-kip}$$

Must be resisted by five T-beams

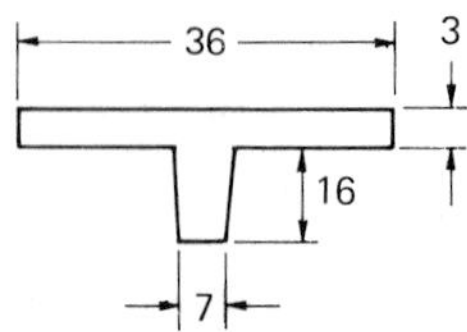

$$M/\text{T-beam} = \frac{290}{5} = 58 \text{ ft-kips}$$

for $b = 36$ in., $d = 17$ in.

$$F = \frac{bd^2}{12{,}000} = 0.867$$

$$K_u = \frac{M_u}{F} = \frac{58}{0.867} = 67$$

$$c/d = 0.027, \quad \frac{h}{d} = \frac{3}{10} = 0.187$$

Since $(h/d) > (c/d)$, neutral axis is in flange, design as rectangular beams.

for $f_c' = 4000$ psi,

$$a_u = 4.46, \quad A_s = \frac{58}{4.46 \times 17} = 0.77 \text{ in.}^2$$

$$p_{min} = \frac{200}{f_y} = \frac{200}{60{,}000} = 0.0033$$

$$(A_s)_{min}\ 19 \times 7 \times 0.0033 = 0.45 \text{ in.}^2, \quad \therefore \text{use } A_s = 0.77 \text{ in.}^2$$

column strip sections:

No. 6 @ 7 in.
$A_s = 0.75$ in.^{2}

neg M at support

2 - No. 6
$A_s = 0.44$

at face of solid section

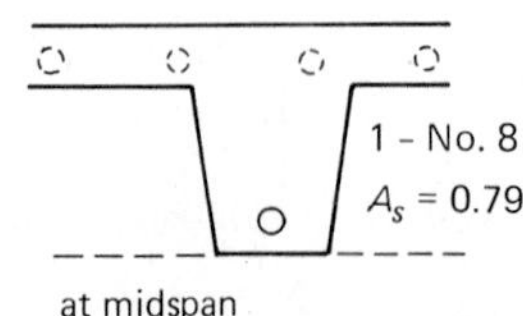

at midspan

Consider middle strip of width = $l_2/2 = 15$ ft

$$M_{neg} = -0.25 \times 900 = -225 \text{ ft-kips}$$

moment resisted by five joist webs

$$M_{neg}/\text{joist} = \frac{225}{5} = 45 \text{ ft-kips}$$

$$d = 17 \text{ in.}, \quad A_s = 0.6 \text{ in.}^2$$

$$M_{pos} = 0.4 \times 484 = 193 \text{ ft-kips}$$

$$\text{Moment/T-beam} = \frac{193}{5} = 38.6 \text{ ft-kips}$$

from column strip

$$A_s = \frac{38.6}{58} \times 0.77 = 0.51 > A_{min}.$$

Middle strip sections:

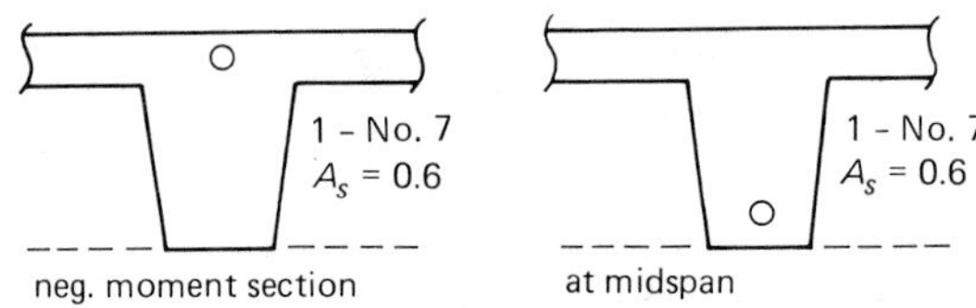

NOTES:
1. Reinforcing must be anchored by providing sufficient embedment at ends.
2. Moment design in short direction is similar. Reinforcing will be approx. same since d in short direction will be less, say $d = 15$ in.
3. Minimum reinforcement of $0.002 \times 3 \times 12 = 0.072$ in.^{2}/ft to be placed in slab between joists. This can be either bars or welded wire mesh.

3.8 TWO-WAY SLAB WITH BEAMS

There are many cases in which it is advantageous to stiffen the slab by providing beams between columns. Such beams will decrease the slab deflections and so permit longer spans with thinner slab sections. In addition, the presence of beams greatly reduces the problem of shear and moment transfer between the columns and slab, and the use of beams to support unusual loading conditions, such as concentrated or line loads, is frequently an economical solution. The most common use of beams is along exterior edges where the beam is used to carry the exterior wall load directly and, by stiffening the discontinuous edge, to permit a smaller slab thickness.

Where the panel shape and beam stiffnesses are such as to develop two-way behavior, the slab system with beams can be designed in a manner very similar to such slabs without beams. Since the new provisions permit consideration of any beam cross section it is necessary to begin with the selection of the beam dimensions. The ratio of depth to width of the beam stem is arbitrary and the depth may be selected on basis of span to depth ratios. It must be remembered that a stiffer beam will attract more load and the beam size initially selected must be capable of resisting the moments and shears which are assigned to it. It may be desirable for economy to modify the initial selection of beam dimensions to reduce steel requirements and/or concrete stresses.

EXAMPLE 3-5: The design procedure for slabs with beams is

illustrated by the floor system given. On the basis of clear height and forming considerations a stem width of 11½ in. and a depth below the slab soffit of 12 in. were arbitrarily selected.

The next step is to determine a slab thickness. The general equation for minimum slab thickness given in section 3.4 requires a value for the average beam to slab flexural stiffness ratio, α_m, but the calculation of this ratio depends on the slab thickness. For this reason the initial slab thickness is estimated by assuming the second term in the denominator is zero. Using the slab thickness and the beam sections previously chosen the value of α_m is easily obtained using Fig. 3.4. While the slab thickness could be made to vary from panel to panel it is usual practice to use the same thickness in all panels unless spans differ significantly.

The determination of the slab moments at the critical design sections is demonstrated by considering the design strip defined by a half panel width on each side of column line B. Since all of the requirements for use of the direct design method are satisfied in the example layout this method is used. The method proceeds as in ex. 3-2 except that the stiffnesses of the beams are included. It is noted that only the load that comes onto the slab is considered, that is loads that come onto the beam directly such as dead weight of the stem are considered later. If desired, the equivalent frame method could also have been used for this step.

These design moments are distributed transversely to the slab and to the beam in proportion to the equivalent beam stiffness. The percentage of the design moment assigned to the column strip which includes the beam as well as a portion of the slab is given in Fig. 3-12, or *ACI Code*, section 13.3.4. The remainder of the moment is assigned to the two half middle strips of the slab. The portion of the column strip moment assigned to the beam is also dependent on the equivalent beam stiffness as given in section 13.3.4.4. Added to the beam moment are the moments from load considered to be carried directly by the beam such as walls or stem weight.

There are two critical locations for shear. In the slab the critical section is d from the face of the beams with the slab periphery considered as a wide beam, hence the allowable shear stress in the concrete is $2\sqrt{f_c'}$. This shear will rarely govern. The other critical section is in the beams which must be designed to resist the shear caused by the loads assumed to act on them. These loads are obtained by assuming loaded areas bounded by 45° lines drawn from the panel corners and the center line of the panels parallel to the long sides as sketched in the calculations. Again shear forces resulting from loading applied to the beam must be added to that obtained from the slab loading.

Since the design strip considered is interior there will be no appreciable torsion in the beam along column line B. However there will be torsion in the edge beam along column line 1 caused by the exterior negative moments in the design strip considered. The negative moment assigned to the half middle strips and to the slab in the column strip (Step 3) can be assumed to be distributed uniformly across these strips. By assuming zero torsional moment at midspan the design torsional moment is obtained by assuming the exterior slab moments across the column face. The design of the beams for the moments, shear and torsion follows the same procedure outlined in Example 3-1.

Slab reinforcing is provided in each strip to resist the moments computed for that strip and is generally distributed uniformly across the strip. In addition, since α is greater than 1.0, special reinforcing as described under Step 5 of section 3.6 is required at the exterior corners.

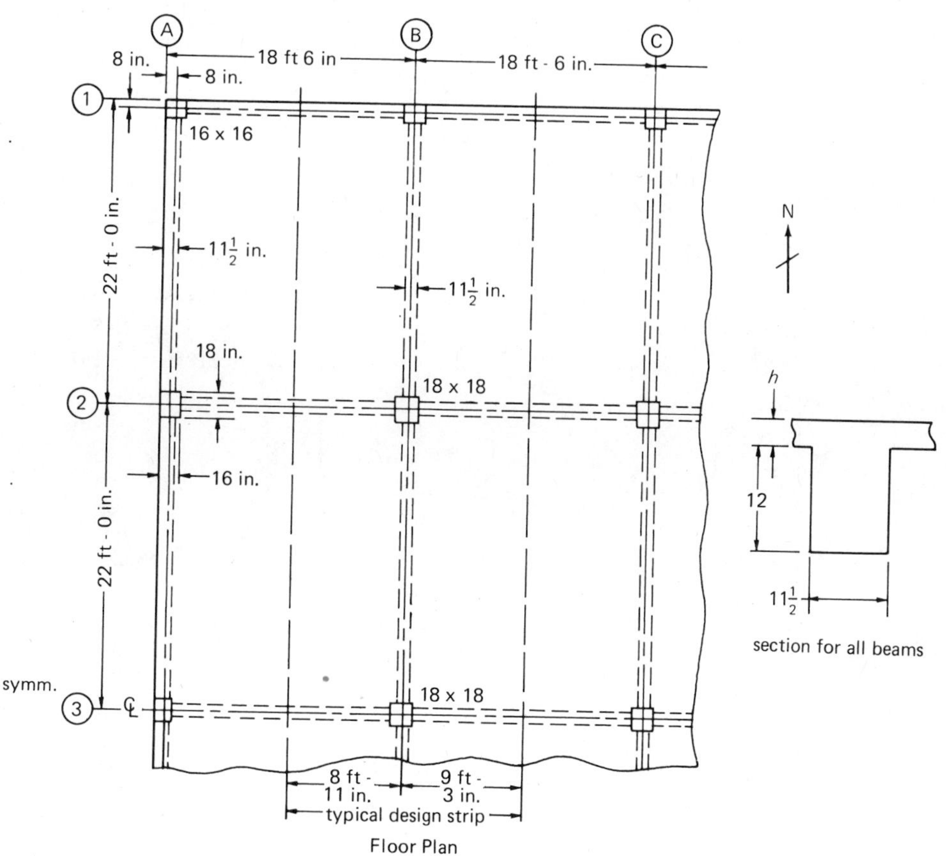

Floor Plan

FLOOR PLAN

Design data:

nominal live load	150 lb/ft^2
mechanical services	10 lb/ft^2
exterior wall	500 lb/ft
story height, center to center	14 ft

Material:

$f_c' = 4000$ psi $\gamma = 150$ lb/ft^3

$f_y = 60{,}000$ psi

beam depth–12 in. below slab soffit

Step 1: Initial selection based on requirements for interior panel.

$$\text{N-S } l_n = (22 \times 12) - (9 + 9) = 246 \text{ in.} \qquad \beta_s = 1$$

$$\text{E-W } l_n = (18.5 \times 12) - (9 + 9) = 204 \text{ in.,} \quad \beta = \frac{246}{204} = 1.21$$

$$h = \frac{246(800 + 0.005 \times 60{,}000)}{36{,}000 + 5000 \times 1.21\,(1 + 1)} = 5.64 \text{ in.} \qquad \text{(eq. 9-7)}$$

$$h_{max} = \frac{246(800 + 0.005 \times 60{,}000)}{36{,}000} = 7.5 \text{ in.} \qquad \text{(eq. 9-8)}$$

Try $h = 6.0$ in. (Check later for adequacy of corner panel.)

Beam sections:

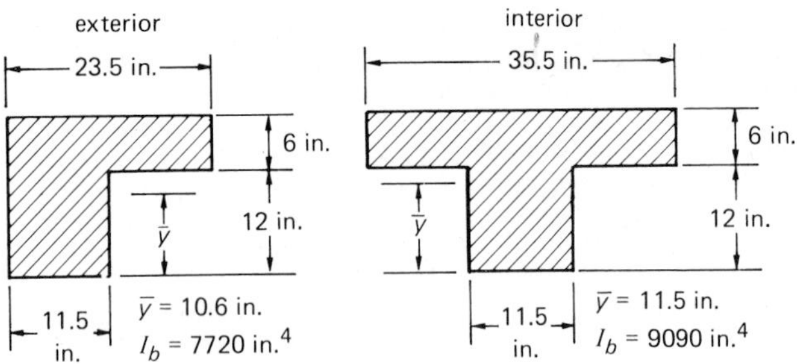

$\alpha = I_b/I_s$, for different slab strips

$$\text{Column line ①} \quad \alpha = \frac{7720 \times 12}{(11.33 \times 12) \times 6^3} = 3.15,$$

$$\text{Column line Ⓑ} \quad \alpha = \frac{9090 \times 12}{(18.16 \times 12) \times 6^3} = 2.32$$

$$\text{Column line Ⓐ} \quad \alpha = \frac{7720 \times 12}{(9.58 \times 12) \times 6^3} = 3.72,$$

$$\text{Column line ②} \quad \alpha = \frac{9090 \times 12}{(21.63 \times 12) \times 6^3} = 1.94$$

Check corner panel:

$$\text{N-S } l_n = (22 \times 12) - (16 + 9) = 239 \text{ in.,} \qquad \beta_s = 0.5$$

$$\text{E-W } l_n = (18.5 \times 12) - (16 + 9) = 197 \text{ in.,} \qquad \beta = \frac{239}{197} = 1.22$$

$$\alpha_m = 2.78$$

$$h = \frac{239(800 + 0.005 \times 60{,}000)}{36{,}000 + 5000 \times 1.22\left[2.78 - 0.5(0.5)\left(1 + \frac{1}{1.22}\right)\right]} = 5.3 \text{ in.} \qquad \text{(eq. 9-6)}$$

.. use $h = 6$ in. all panels

Loading:

$$\text{slab} \qquad w_d = 1.4\left(\frac{6}{12} \times 150 + 10\right) = 119 \text{ lb/ft}^2$$

$$w_l = 1.7(150) = 255$$

$$w = 374 \text{ lb/ft}^2$$

$$\text{beam stem} \qquad w = 1.4\left(\frac{11.5 \times 12}{144} \times 150\right) = 202 \text{ lb/ft}$$

Step 2: Interior panel

$$\frac{\alpha_1 l_2^2}{\alpha_2 l_1^2} = \frac{2.32(18.16)^2}{1.94(22)^2} = 0.815, \; >0.2 \text{ and } <5.0. \qquad \text{(section 13.3.1.6)}$$

Can use direct design method.

N-S slab strip about column line B

	①				②				③ SYMM.	
l_1	256 in. = 21.33 ft					264 in. = 22.0 ft				
l_n	239 in. = 19.9 ft					246 in. = 20.5 ft				
l_2 (8′ − 11″ + 9′ − 3″)	218 in. = 18.167 ft					218 in. = 18.167 ft				
$M_o = \frac{(18.167 \times 0.374)}{8} l_n^2$		338					356			ft-kips
K_c (column above) $= \frac{4I_c}{l_c}$	147				208			208		in.^{3}
K_c (column below) $= \frac{4I_c}{l_c}$	147				208			208		in.^{3}
ΣK_c	294				416			416		in.^{3}
c ⊠⊠ or ⊠⊠⊠	6042				6634			6634		in.^{4}
$x = (1 - c_2/l_2)^3$	0.772				0.772			0.772		
K_t	645				710			710		in.^{2}
K_{ec}	202				262			262		in.^{3}
$K_s = 4I_s/l_1$	61		61		59			59		in.^{3}
$K_b = 4I_b/l_1$	142		142		138			138		in.^{3}
$\alpha_{ec} = K_{ec}/\Sigma(K_s + K_b)$	0.995			0.655					0.665	
$\%M_o$ (section 13.3.3.3)	32	49	70		65	35	65			%
design M due to slab	−110	+165	−236		−231	+125	−231			ft-kips
column moments	110 + 3 = 113 ↑ beam stem			(eq. 13-2)	32.3*		31.3		(eq. 13-3)	ft-kips

$$*M_{col} = \frac{0.08[0.119 + 0.5 \times 0.255)\, 18.167(20.5)^2 - 0.119 \times 18.167(19.9)^2\,]}{1 + \dfrac{1}{0.655}} = 32.3 \text{ ft-kips}$$

Step 3:

N–S slab strip about column line Ⓑ

	①				②				③ SYMM.	
l_2/l_1			0.85				0.825			
$\alpha l_2/l_1$			1.98				1.92			
$\beta_t = \dfrac{6042 \times 12}{2(21.33 \times 12)6^3} =$	0.655									
$\%M_{des}$ assigned to column strip (section 13.3.4)		94.7	79.5	79.5		80.2	80.2	80.2		%
M_{des} (step 2)		−110	+165	−236		−231	+125	−231		ft-kips
$M_{col.\ strip}$		−104	+131	−188		−185	+100	−185		ft-kips
slab moments: column strip		−16	+20		−28		+15	−28		ft-kips
slab moments: middle strip		−6	+34		−48		+25	−46		ft-kips
beam moments: from slab		−88	+111		−160		+85	−157		ft-kips
beam moments: beam loads (stem)		− 3	+ 5		− 7		+ 4	− 7		ft-kips
beam moments: total beam M		−91	+116		−167		+89	−164		ft-kips

Step 4:

(a) Slab panel shear $d/2$ from beam lines; check interior panel.

$$V_u = 20.5 \times 17.0 \times 0.374 = 130 \text{ kip},\quad b_0 = 2(20.5 + 17.0)12 = 900 \text{ in.},\quad d_v = 4.75$$

$$v_u = \frac{130{,}000}{0.85 \times 900 \times 4.75} = 35.8 \text{ psi},\quad v_{all} = 2\sqrt{f'_c} = 126 \text{ psi.}$$

(b) Shear on beams

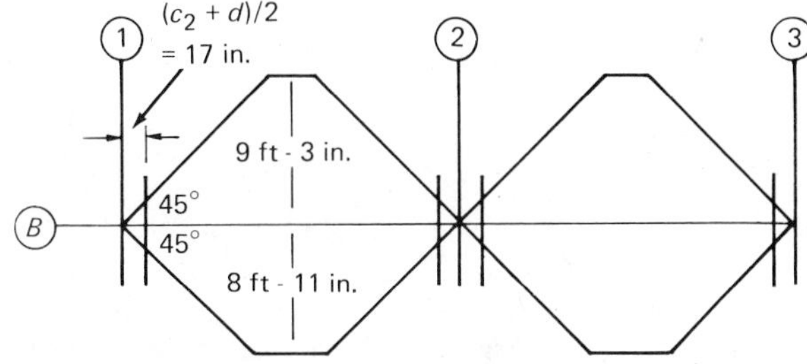

Tributary areas

critical shear at
$d/2$ from column face

	①	②	②	③	
V at critical sections	43	43	45	45	kips
V due to M end span	−4	+4	–	–	
V due to beam leads (stem)	+2	+2	2	2	
design shear, V_u	41	49	47	47	kips

3.9 SPECIAL SLAB PROBLEMS

Most slab systems consist of reasonably rectangular panels with no irregularities to cause serious discontinuities in the moment patterns and may be designed using the procedures outlined in the previous sections. When the designer is faced with conditions which violate the basic assumptions of these design procedures, modifications are required. Some of the more commonly encountered irregularities are discussed briefly in this section.

3.9.1 Staggered Column Spacing

In certain floor arrangements architectural considerations require the location of columns such that the resulting panels are not rectangular. The code design procedures are restricted to those cases where the maximum column offset from either axis between center lines of successive columns does not exceed 10% of the span in the direction of the offset. Solutions for slabs with completely random placed columns can only be obtained using numerical techniques and are beyond the scope of this discussion but attempts have been made to modify the basic procedures when the column rows are staggered in a more or less regular manner. Such a solution has been proposed by Van Buren 3-16, and consists of dividing the slab into design strips in a manner similar to the direct design method except that for the strip containing the staggered columns the width of the column strip is increased in the positive moment region as shown in Fig. 3-16. It is demonstrated that the moment coefficients can be based on the clear span of the strips and correspond with respective coefficients obtained from the direct design method. For example, with equal column spacing in both directions and a value of $c/l = 0.1$, the moment coefficients from the critical sections are given in Table 3-7. It is seen that, when K = 0.5 (i.e., maximum stagger), the positive moments at midspan are uniformly distributed across the panel. The length of reinforcement with the staggered columns is the same as for typical square columns except for the middle strips perpendicular to the direction of stagger and designated as M. The negative moment in these strips will occur over the shorter diagonal lines with the positive moments midway between these lines. When the degree of stagger is small the top bars across the shorter diagonal line may be arranged symmetrically with a length equal to half the span plus the stagger. For greater stagger it may be preferred to center each top bar across the shorter diagonal.

TABLE 3-7 Moments in Strips

$c/l = 0.1$

MK.	*Negative Moment*	*Width*	*Positive Moment*	*Width*
PN3	$-.0494\, wl^3$	$l/2$	$+ wl^3(.0212 + .0284K)$	$l/2 + Kl$
N3	$-.0164\, wl^3$	$l/2$	$+ wl^3[.0142(1 - 2K)]$	$l/2 - Kl$
A	$-.0494\, wl^3$	$l/2$	$+ .0212\, wl^3$	$l/2$
M	$-.0164\, wl^3$	$l/2$	$+ .0142\, wl^3$	$l/2$

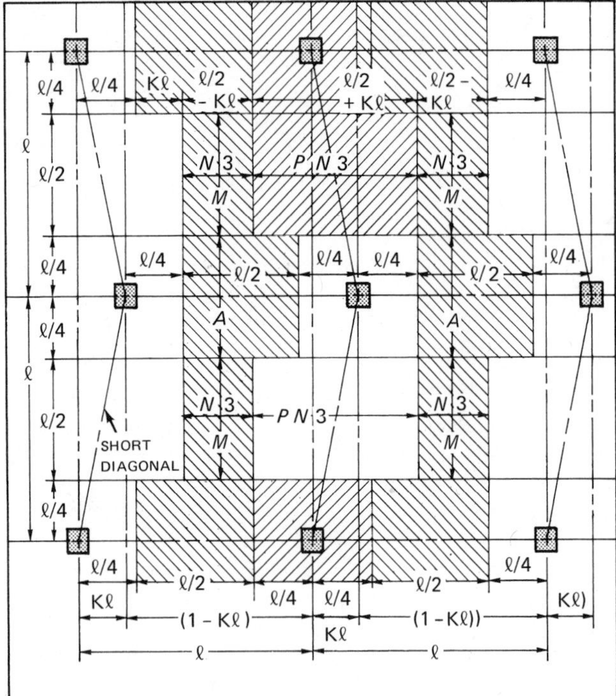

Fig. 3-16 Typical interior flat plate panel modified or partial stagger; $c/l = 0.1$.

3.9.2 Concentrated Loads

The common occurrence of concentrated loading is from wheel loads in parking structures or in warehouses where forklift trucks are used. Since in these cases the loads can move about there is the additional problem of where to place the load to obtain the critical design moments and shears. Generally such loads are considered to act over a finite area using the recommendations for distributing wheel loads as given in highway specifications for bridge decks. This practice is satisfactory for determining the adequacy of the slab against failure by punching shear.

To determine the most unfavorable location for the load for bending moments in the slab, use is made of influence surfaces. The most complete study of the moments caused by concentrated loads in the interior panels of continuous slabs with and without beams was made by Woodring and Seiss 3-17. In keeping with the concept of dividing the slab into middle and column design strips, they obtained influence surfaces for the positive and negative design moments in each strip for concentrated loads and for area loads extending to one-third of the span length. They concluded the effect on the design moments of spreading a concentrated load over a larger area is small unless the load is in close proximity to the design moment locations. However the position of the concentrated load with respect to the moment location was the most important variable. They found that the critical moments in the middle strip could be reversed in sign from the moments usually associated with these positions. For example, when a concentrated load is placed on a column line midway between two columns, the moment in the middle strip directly under this load normally referred to as the negative middle strip moment is positive. Moreover the magnitude of this positive moment is more than twice as great as the maximum negative moment produced at this section by the same concentrated load located anywhere in the panel even if Poisson's ratio is taken as zero. This explains why certain warehouse slabs reinforced only for a uniformly distributed load have shown excessive cracking due to moment reversals caused by moving loads for which they are not reinforced. In an attempt to simplify the calculations for concentrated loads Woodring proposed an equivalent load factor, C, which when multiplied by the concentrated load gave an equivalent uniform panel load. Plots giving this load factor were prepared for each design moment location for different concentrated load positions and beam stiffnesses but the number of plots required is extensive.

In summary the common building code requirement of considering a load of 2500 lbs distributed over a 2.5 sq ft area and located to produce maximum effect will generally not govern design of floor slabs of usual dimensions. Where the magnitude of the concentrated load is large compared to the dead load, consideration should be given when detailing reinforcing to changing locations of points of inflection and possible reversal of moment signs in the middle strips.

3.9.3 Openings in the Slab

Generally openings in the slab do not constitute any difficulty in design. *ACI 318-71* permits openings of any size in the area within the middle half of the span in each direction; or in the area common to two column strips openings or not more than one eighth of the width of strip in either span; or in the area common to one column strip and one middle strip openings, such that not more than one fourth of the reinforcing in either strip is interrupted without special analysis, providing the equivalent amount of reinforcing interrupted is added on all sides of the opening, and shear requirements are satisfied. Where larger openings occur it is generally necessary to stiffen the slab with the use of beams around the openings. It is important that provision be made to carry the total panel moment in both directions without excessive deflections.

When openings are located within a distance less than ten times the slab thickness, or when, for slabs without beams, openings are located in the column strips, the effective periphery, b_o, for shear is to be reduced by that portion which is enclosed by radial projections from the centroid of the column or loaded area to the outer edges of the openings as shown in Fig. 3-17. Where shearheads are pro-

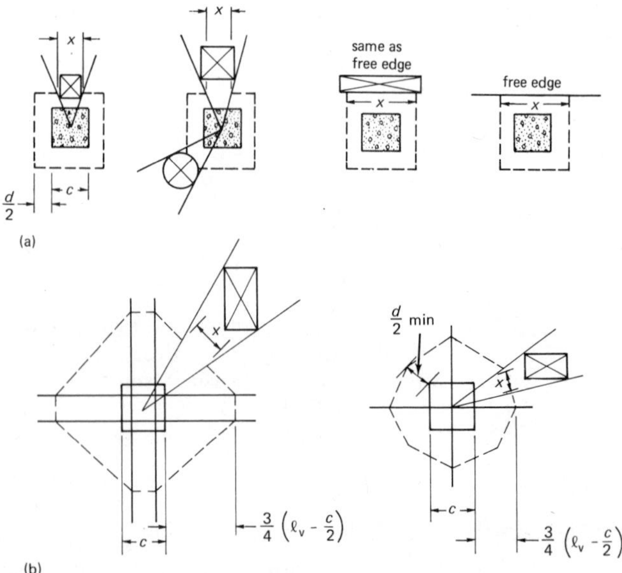

Fig. 3-17 Effect of openings in slabs on effective shear periphery. (a) Slabs without shear reinforcements, effective periphery = $b_0 - x$; (b) slabs with shear reinforcement, effective periphery = $b_0 - x/2$.

vided only one half of this reduction is required (see section 3.9.5).

3.9.4 Isolated Beams

The design procedures outlined in ACI 318-71 are applicable to slab systems with beams only when the ratio $\alpha_1 l_2^2/\alpha_2 l_1^2$ is not less than 0.2 nor greater than 5.0. Clearly then, where a beam is placed only along one edge of an interior panel the code procedures will not apply and the design must be modified in the region of this discontinuity. A suggested procedure for this modification follows. 3-18 Consider the slab system, illustrated in Fig. 3-18, in which a single line of beams is placed in the y-direction and all panels are similar in shape, although the panel aspect ratio can vary from 0.5 to 2.0. Strips are considered in each direction as in the direct design method. The total static moment for the strip in each panel can be expressed as $Cwl_2l_1^2$ (the value l_n can be substituted for l_1 without invalidating the procedure providing values of c/l are not excessive).

First consider the strip containing the beam. As the flexural stiffness of the beam is increased the portion of the load carried by this strip increases with a corresponding decrease in the parallel adjacent strips. This increase in load and hence panel static moment M_o depends on the panel shape as seen in Fig. 3-19 where values of the coefficient C are given. It is recommended that the corresponding decrease in M_o for the adjacent strips be neglected and the moments in these strips be designed using code values.

The total panel moment must now be proportioned between negative and positive sections and between column and middle strips. The percentages of M_o assigned to each slab and beam section are given in Table 3-8.

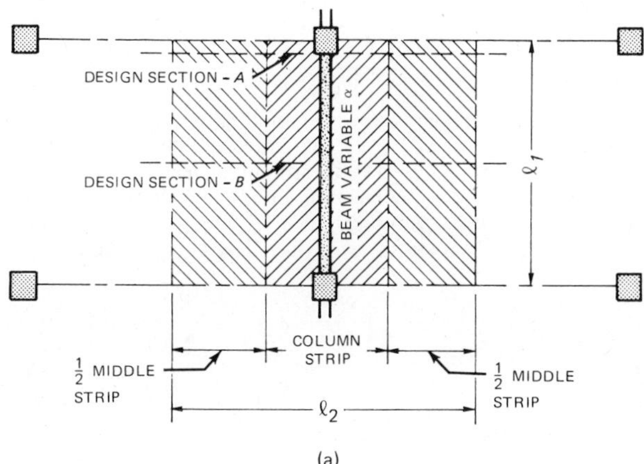

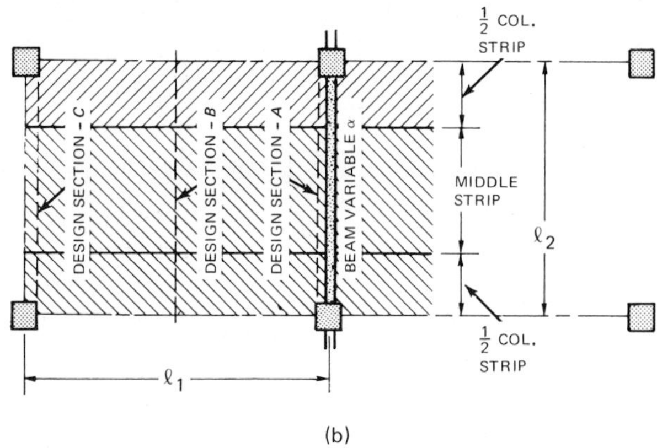

Fig. 3-18 Design sections and strip definitions for isolated beam. (a) Design sections, table 1; (b) design sections, table 2.

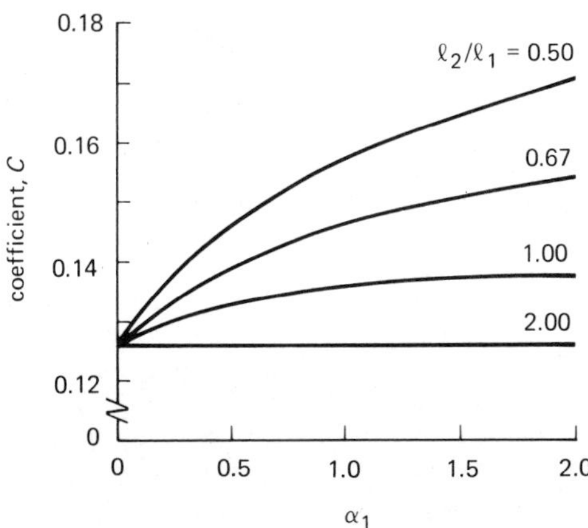

Fig. 3-19 Coefficient, C, for panel static moment parallel to beam.[3-18]

TABLE 3-8 Distribution of Moments in Strip Parallel to Beam as Percentages of M_o.*[3-18]

α_1	0.0		0.5		1.0		2.0	
Section**	A	B	A	B	A	B	A	B
Beam Moment								
l_2/l_1 = 0.5	–	–	27	12	37	17	45	23
0.67	–	–	30	12	39	17	47	23
1.0	–	–	35	14	45	19	52	24
2.0	–	–	46	20	54	25	60	28
Slab Moment, Column Strip								
0.5	42	17	22	11	16	9	10	6
0.67	45	18	22	11	15	9	9	6
1.0	51	20	21	12	14	8	8	6
2.0	50	20	14	8	8	5	5	3
Slab Moment, Middle Strip								
0.5	25	16	16	12	12	9	9	7
0.67	21	16	14	11	11	9	8	7
1.0	16	13	9	9	7	7	5	5
2.0	16	14	6	6	4	4	2	2

*$M_o = Cwl_2l_1^2$; values of C given in Fig. 3-19.
**For location of sections see Fig. 3-18.

TABLE 3-9 Distribution of Moments in Strip Perpendicular to Beam as Percentages of M_o.*[3-18]**

α_1	0.0			0.5			1.0			2.0		
Section**	A	B	C	A	B	C	A	B	C	A	B	C
Slab Moment, Column Strip												
l_2/l_1 = 2.0	50	20	50	41	17	49	36	15	48	31	14	47
1.5	50	20	50	42	17	50	38	16	49	34	15	48
1.0	51	20	51	46	19	50	43	19	50	41	19	50
0.5	42	17	42	37	17	42	36	17	42	35	17	42
Slab Moment, Middle Strip												
2.0	16	14	16	44	13	10	59	12	7	76	11	4
1.5	16	14	16	36	14	13	44	14	11	53	14	10
1.0	16	13	16	26	14	14	30	14	14	33	14	14
0.5	25	16	25	30	17	25	32	17	25	33	17	25

***$M_o = 0.125\ wl_2l_1^2$.
**For location of sections see Fig. 3-18.

Consider now the design strips perpendicular to the beam. In all cases the total panel static moment from elastic analysis is $0.125\ wl_2/l_1^2$. The distribution of this moment to the critical design sections is given in Table 3-9.

By examining the distribution of moments in Tables 3-8 and 3-9, it is seen that, if an isolated beam line is placed in an otherwise regular slab system without beams, it will attract moment in the direction in which the beam spans so that the total static moment of a strip bounded by the center line of the panel adjacent on each side will exceed the static moment caused by the load acting on this area. In addition, the moments in the vicinity of the beam will be altered as the beam stiffness increases to approach those of a one-way slab spanning perpendicular to the beam.

3.9.5 Slab Shear Reinforcement (shearheads)

When the shear strength (punching shear) of a two-way slab without beams is not sufficient we have seen that the designer may increase the slab thickness, use column capitals, and/or drop panels, or increase the strength of the concrete to improve the shear capacity of the slab. An alternative permitted by ACI 318-71 is to use shear reinforcement consisting of reinforcing bars, when the actual shear stress does not exceed the allowable concrete shear stress by more than 50%; or structural steel shearheads, when the actual shear stress does not exceed the allowable by more than 75%. The procedure for designing such reinforcement is illustrated in ex. 3-6.

EXAMPLE 3-6: Check the adequacy in shear of an interior panel of a slab without beams (flat plate), and if it is not sufficient provide suitable shear reinforcement. You are given the following data:

$f'_c = 3000$ psi (normal weight concrete)

f_y (reinforcing bars) = 60,000 psi

F_y (structural steel) = 36,000 psi

$d = 8$ in., $c_1 = c_2 = 12$ in., $V_u = 170$ kips

NOTE: It is assumed that the capacity of the slab in shear as a wide beam has been found satisfactory and that two-way (punching) shear governs.

SOLUTION:

(a) Determine adequacy of slab in shear without shear reinforcement

$b_o = 4(12 + 8) = 80$ in., $\varphi = 0.85$ (specified in ACI 318-71)

$$v_u = \frac{V_u}{\varphi b_o d} = \frac{170{,}000}{.85 \times 80 \times 8} = 312 \text{ psi}$$

$$(v_c)_{max} = 4\sqrt{f'_c} = 4\sqrt{3000} = 219 \text{ psi}$$

Since 219 < 312, capacity of slab in shear is not adequate without shear reinforcement.

(b) Shear reinforcement consisting of reinforcing bars

Maximum allowable v_u for use of reinforcing bars = $1.5 \times 4 \times \sqrt{f'_c} = 328$ psi. Since 328 > 312, use reinforcing bars for shear reinforcement.

Shear carried by concrete (code section 11.11.1)

$$= v_c = 2\sqrt{f'_c} = 109 \text{ psi}$$

Shear carried by reinforcing bars = $(v_u - v_c) = 312 - 109 = 203$ psi

Try #3 stirrups

$$A_v = 0.22 \text{ in.}^2 \quad b_w = (c_2 + d) = 12 + 8 = 20 \text{ in.}$$

Stirrup spacing

$$s = \frac{A_v f_y}{(v_u - v_c)\, b_w} = \frac{0.22 \times 60{,}000}{203 \times 20} = 3.25 \text{ in.}$$

$$s_{max} = d/2 = 4 \text{ in.}$$

Now calculate length over which stirrups required (code section 11.11.1) Stirrups required until concrete shear stress at section b_o indicated by dashed line is equal to or less than $v_c = 2\sqrt{f'_c} = 109$ psi.

$$b_o = 4(c + a\sqrt{2})$$

$$\text{From } v_c = \frac{V_u}{\varphi b_o d} = \frac{V_u}{4\varphi(c + a\sqrt{2})d}$$

$$a = \frac{V_u}{4\varphi\sqrt{2}\,V_c d} - \frac{c}{\sqrt{2}}$$

$$= \frac{170\ 000}{4 \times .85 \times \sqrt{2} \times 109 \times 8} - \frac{12}{\sqrt{2}}$$

$$= 32 \text{ in.}$$

length required = $a + d$

$$a + d = 32 + 8 = 40 \text{ in.}$$

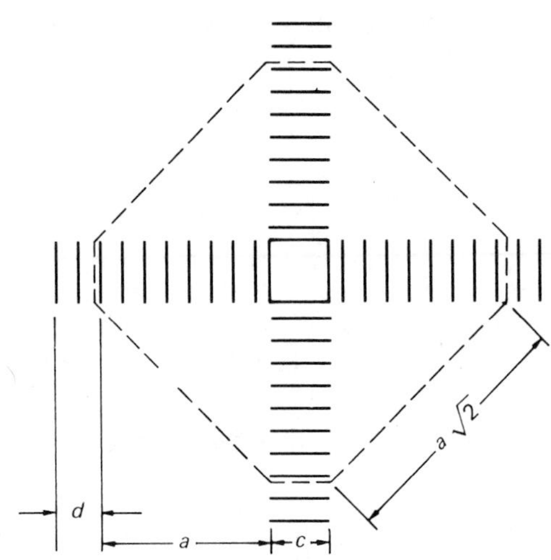

(c) Shear reinforcing consisting of structural steel section

Maximum allowable v_u for use of structural steel = $1.75 \times 4 \times \sqrt{f'_c} = 384$ psi. Since 384 > 312, use structural steel sections. Length of shearhead arm, l_v, (Code section 11.11.2) such that at critical section, b_o, located at 3/4 of distance $[l_v - (c_1/2)]$ from column face, the concrete shear stress does not exceed $4\sqrt{f'_c}$ see sketch.

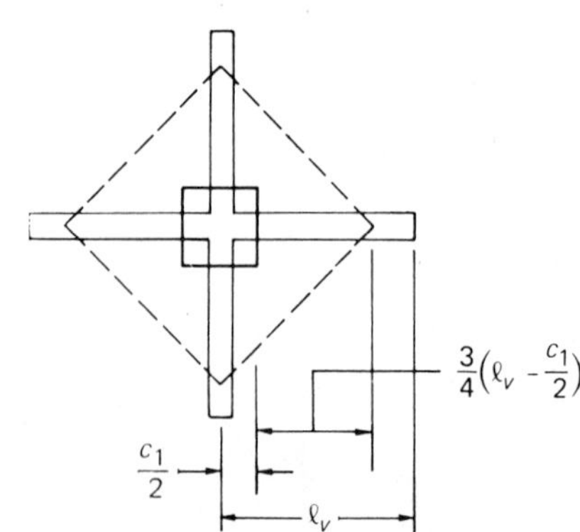

$$b_{o,\,min} = \frac{V_u}{4\varphi d\sqrt{f'_c}} = \frac{170\ 000}{4 \times .85 \times 8 \times \sqrt{3000}} = 114 \text{ in.}$$

$$b_o = 4\sqrt{2}\left[\frac{c_1}{2} + \frac{3}{4}(l_v - c_1/2)\right]$$

$$\therefore l_v = \frac{1}{3}\left[\frac{b_o}{\sqrt{2}} - \frac{c_1}{2}\right] = \frac{1}{3}\left[\frac{114}{\sqrt{2}} - \frac{12}{2}\right]$$

$$= \tfrac{1}{3}(80.5 - 6) = 24.5 \text{ in.}$$

To ensure that the shearhead reinforcing does not fail in flexure before the shear strength of the slab is reached each arm must be capable of resisting the full plastic moment M_p given in the code as

$$M_p = \frac{V_u}{48} [h_v + \alpha_v (l_v - c_1/2)]$$

where h_v = depth of steel section; and α_v = ratio of stiffness of single shearhead arm to composite cracked slab section of width $(c_2 + d)$, and α_v must be not less than 0.15.

Assume h_v = 5 in., α_v = 0.2, then

$$M_p = \frac{170}{0.9 \times 8} [5 + 0.2(24.5 - 12/2)] = 205 \text{ in.-kips}$$

Try S5 I 10.0 Z_x = 5.6 in.^{3}

$$M_p = Z_x F_y = 5.6 \times 36{,}000 = 202 \text{ in.-kips}$$

Although 202 < 205, it is sufficiently close if α_v is satisfactory.

$$EI(\text{S5 I 10.0}) = 12.1 \times 29{,}000 = 350{,}000 \text{ kip-in.}^2$$

$$EI \text{ (cracked section) } 1{,}680{,}000 \text{ kip-in.}^2$$

(assumed five #5 bars)

$$\alpha_v = \frac{350{,}000}{1{,}680{,}000} = .208$$

$$\text{revised } M_p = \frac{170}{0.9 \times 8} (5 + 208(24.5 - 12/2)) = 209 \text{ in.-kips}$$

About 4% underdesign, use anyway. Maximum depth of steel section permitted is (70 × web thickness)

$$\text{maximum depth} = 70 \times .210 = 14.7 \text{ in.} > 5 \text{ in.}$$

Allowable distance from compressive face of concrete to compression flange of section = 0.3d = 2.4 in.

Use S5 I 10; length 24.5 in. + 5 in.; say, 30 in.

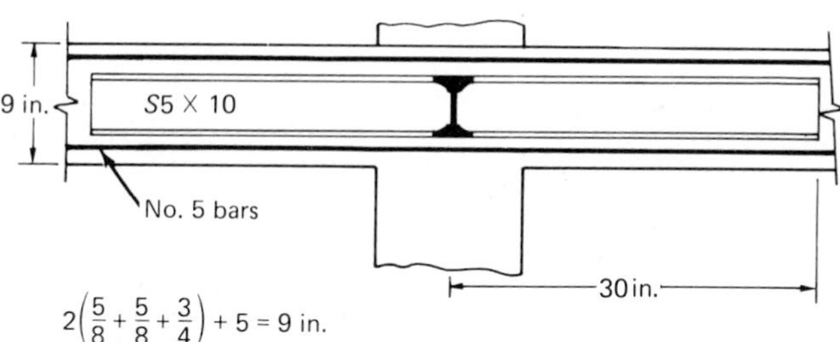

$$2\left(\frac{5}{8} + \frac{5}{8} + \frac{3}{4}\right) + 5 = 9 \text{ in.}$$

∴ no bars need be cut to fit I section.

3.10 ULTIMATE STRENGTH DESIGN PROCEDURES

The design methods discussed previously in this chapter are based on elastic analysis and so do not indicate the ultimate load capacity of the slab. For most slabs it is not possible to determine the collapse load exactly but procedures which provide upper and lower bounds to this load have been developed. These procedures, which conceptually are much simpler than elastic analysis, deal with moments alone and so do not guarantee safety against failure by diagonal tension (punching shear), cracking, or excessive deflection. Independent checks must be made to ensure satisfying shear and serviceability requirements. However, in spite of this shortcoming, these procedures provide a powerful tool for designing slabs of irregular geometry and loading which justifies their introduction here.

The two basic ultimate strength design procedures are the yield-line method which provides an upper bound solution and is attributed to Johansen 3-19 and the strip method which provides a lower bound solution and is attributed to Hillerborg. 3-20 The two methods are presented separately.

3.10.1 Yield-Line Method

The basic assumption of yield-line theory is that a reinforced concrete slab will, under increasing overload, form yield hinges but will not collapse until a mechanism is formed. This assumption is justified since most slabs are underreinforced permitting the reinforcement to yield with the corresponding large angle change and only slight increase in moment capacity. Because the slab is highly indeterminate this initial yielding causes a redistribution of the resisting moments and the yielding progresses to form lines of yielding. The load can be increased until these lines of yielding subdivide the slab into segments which form a collapse mechanism. Since the angle changes across the lines of yielding are much greater than the elastic curvatures within the segments, the segments are assumed to be plane which means that the yield lines may be considered straight lines.

Frequently the exact configuration of the yield lines is not known and a series of possible patterns must be considered by trial and error. Generally yield lines will occur along lines of fixed support, lines of symmetry, and, for slabs supported directly on columns, along an axes passing through the column. Typical yield-line patterns are shown in Fig. 3-20.

Once the yield line pattern is assumed two approaches are possible. The first is an energy method in which the external work done by the loads during a virtual displacement of the collapse mechanism is equated to the internal work. The second is to consider the equilibrium of the various parts of the slab into which the slab is divided by the yield lines.

Slabs are generally reinforced in two perpendicular directions, say x and y, with corresponding ultimate moment capacities m_x and m_y. The bending (m) and twisting (m_t) moments on a yield line making an angle θ with the x-axis as shown in Fig. 3-21 are

$$m = m_y \cos^2 \theta + m_x \sin^2 \theta$$

$$m_t = (m_y - m_x) \sin \theta \cos \theta$$

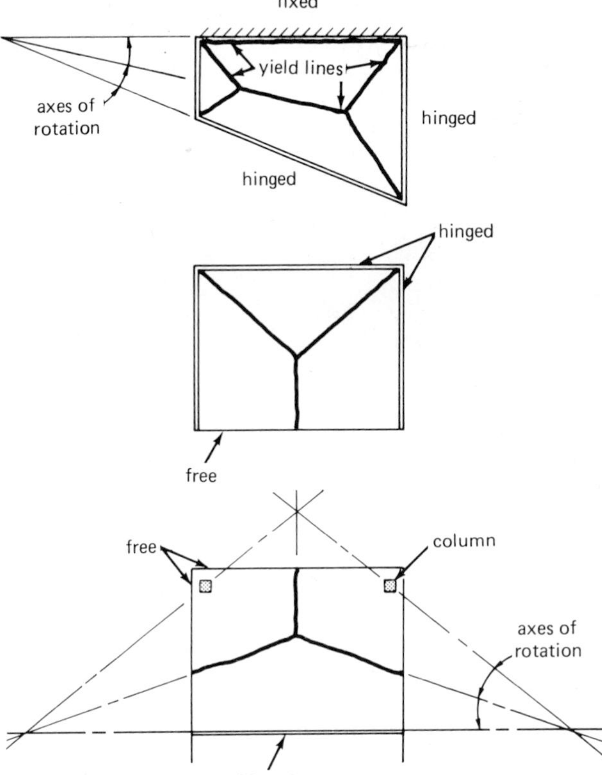

Fig. 3-20 Yield-line patterns.

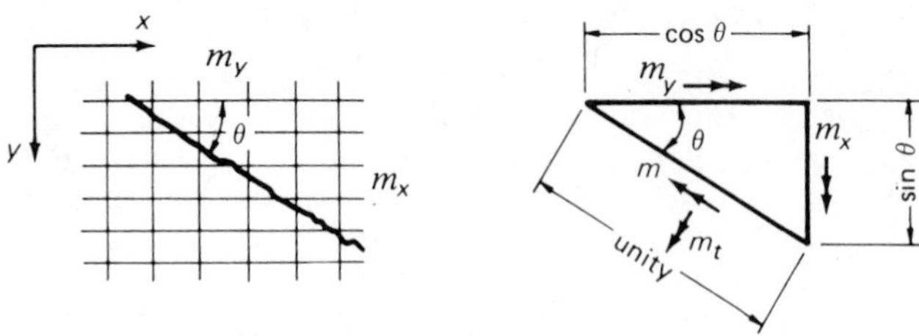

Fig. 3-21 Moment transformation.

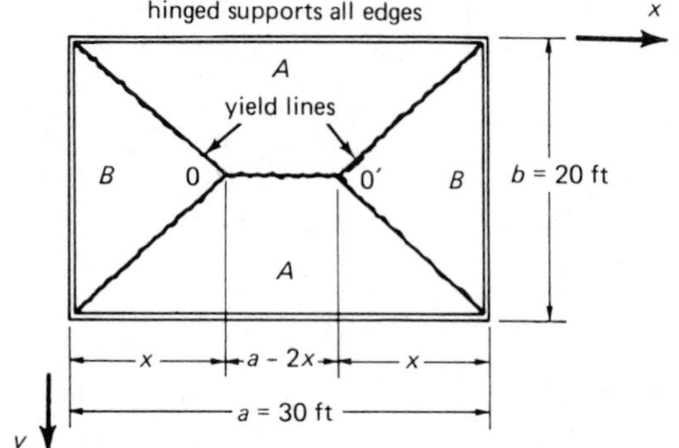

Fig. 3-22 Rectangular slab used in yield-line example.

For the special case of an isotropic slab, that is, where $m_x = m_y$, $m = m_x = m_y$ and $m_t = 0$, for all values of θ.

The basic principles of yield-line theory are illustrated in the example of a rectangular, simply supported panel supporting a uniformly distributed load as shown in Fig. 3-22. The problem is to determine the collapse load when the positive moment capacity in both directions is given as 8000 ft-lbs/ft. From experience, the general yield-line pattern is known to be similar to that shown in Fig. 3-22. The position of the intersection of the yield lines, "O," is not known precisely, but by symmetry, can be defined in terms of the parameter, x. The slab is divided into four segments by the yield lines which are labeled either A or B. We will first consider the equilibrium approach in which the equilibrium of each segment is considered. The trial yield pattern is found by assuming values of x until the load required for equilibrium is the same for each segment.

Trial 1: Assume $x = 10$ ft.

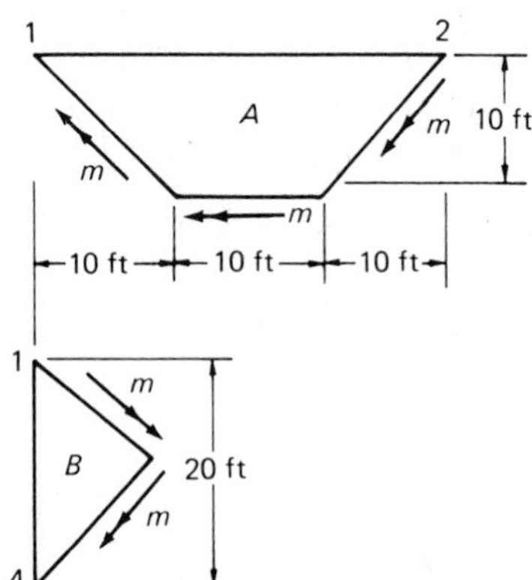

Equilibrium about line 1–2:

$$30 \times 8000 - 10 \times 10\, w_A \times \tfrac{10}{2} - 2(\tfrac{1}{2} \times 10 \times 10\, w_A \times \tfrac{10}{3}) = 0$$

$$w_A = 288 \text{ psf}$$

Equilibrium about line 1–4:

$$20 \times 8000 - \tfrac{1}{2} \times 20 \times 10\, w_B \times \tfrac{10}{3} = 0$$

$$w_B = 480 \text{ psf}$$

It is seen that the load, w_A, on segment A, required to satisfy equilibrium of this segment is not equal to the corresponding load on segment B. Since w_B is greater, we must increase the size of segment B.

Trial 2: Assume $x = 12$ ft

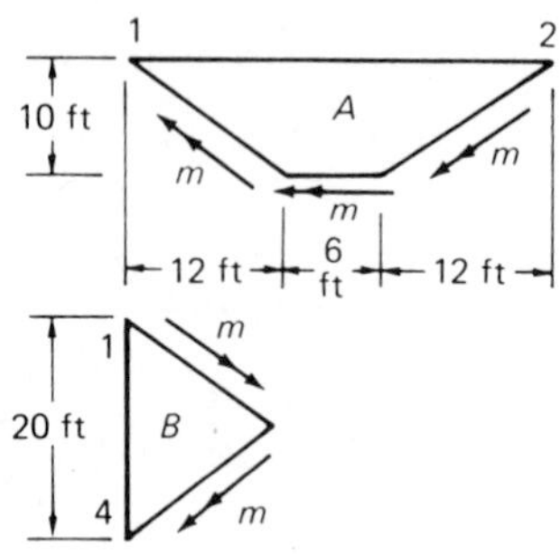

$$30 \times 8000 - 6 \times 10\, w_A \times \tfrac{10}{2} + 2(\tfrac{1}{2} \times 12 \times 10 \times w_A \times \tfrac{10}{3}) = 0$$

$$w_A = 343 \text{ psf}$$

$$20 \times 8000 - \tfrac{1}{2} \times 20 \times 12\, w_B \times \tfrac{12}{3} = 0$$

$$w_B = 324 \text{ psf}$$

The agreement between w_A and w_B in the second trial is seen to be much closer, but now w_B is less than w_A. A third trial can be made with x just less than 12 ft, if closer agreement is desired.

For this simple case an exact value of x can be obtained by using the energy approach. Here the segments are considered to rotate and the work done in this rotation is computed. For our example consider the line O–O′ to deflect a distance δ. The rotation of segment A is then $\theta_A = 2\delta/b$.

The energy input to the slab for segment A is

$$E_A = w\,(a - 2x)\,\frac{b}{2}\,\frac{\delta}{2} + 2\left(w\,\frac{x}{2}\,\frac{b}{2}\,\frac{\delta}{3}\right)$$

and for segment B

$$E_B = w\,\frac{xb}{2}\,\frac{\delta}{3}$$

The work done by m acting on segment A is

$$W_A = ma\,2\,\frac{\delta}{b}$$

and for segment B is

$$W_B = m\,\frac{b\delta}{x}$$

From the principle of virtual work, the energy lost by the load in the rotation of each segment is equal to the work done by m in resisting this rotation. Therefore $E_A = W_A$ and $E_B = W_B$. Substituting $a = 30$ ft; $b = 20$ ft; $m = 8000$ ft-lb/ft; we obtain $x = 11.9$ ft and $w = 340$ lb/ft^2.

The results are seen to agree with the equilibrium method. It must also be noted that the yield-line method, like all ultimate strength procedures gives the ultimate load, w, and appropriate load factors must be applied to obtain actual superposed loading.

If in our example we had considered the edges as fixed then yield lines would have had to form around the edges to permit the segments to rotate. The procedure would be the same in both approaches with the work done by the positive moments. Thus if a slab of the same dimensions but with fixed edges had been reinforced such that in each direction the negative moment capacity was 5000 ft-lb/ft, and the positive moment capacity was 3000 ft-lb/ft, we would have obtained the same yield pattern and ultimate load as we did for the 8000 ft-lb/ft positive moment capacity and no negative moment capacity. It should also be mentioned that the yield lines for the simply supported panel assumed in our example neglected the corner effects which would reduce the ultimate load by approximately 10%. A further discussion is beyond the scope of this handbook and the reader is referred to Wood 3-21 for further details.

3.10.2 Strip Method

The strip method is a design method in which the designer provides a bending moment field in equilibrium with the applied loading at all points in the slab. This procedure is greatly simplified if the designer assumes that the twisting moment is everywhere zero.

The equilibrium equation which must be satisfied at all points is 3-22

$$\frac{\partial^2 m_x}{\partial_x^2} + \frac{\partial^2 m_y}{\partial_y^2} - 2\frac{\partial^2 m_{xy}}{\partial_x \partial_y} = -w$$

where w is the intensity of distributed load. By placing $m_{xy} = 0$ this equation can be reduced to

$$\frac{\partial^2 m_x}{\partial_x^2} = -\alpha w$$

and

$$\frac{\partial^2 m_y}{\partial_y^2} = -(1-\alpha)w$$

where α is the proportion of load taken in the x-direction and $(1 - \alpha)$ in the y-direction. While α may be assigned any value, it is usual to consider α as either 0 or 1. The procedure is demonstrated by considering the example of the simply supported rectangular slab carrying a uniformly distributed load.

The slab is subdivided into regions in which the load is assumed to be carried in different directions as shown in Fig. 3-23. Note that although the pattern appears similar the yield pattern for this slab the two are quite independent of each other. The location of the yield lines was unique and could be computed from conditions of equilibrium whereas the load-dispersion lines can be assigned quite arbitrarily, but the choice will affect the amount of reinforcing and hence the final economy of the design.

Loads in region (1) are considered to be carried solely in the x-direction (i.e., $\alpha = 1$) whereas loads in region (2) are considered to be carried entirely in the y-direction. Since there are no twisting moments strips acting like beams can be considered to span in both directions. Typical strips with their corresponding loading and bending moment diagrams are shown in Fig. 3-23.

From the bending moment diagram for a strips such as a–a it is seen that reinforcing is required over the whole span in an almost constant amount. However in strips such as c–c the bars can be terminated near the ends in the same manner as for a simple beam. The strips shown were assumed to have unit width. In practice to consider such narrow strips would be tedious and a greater width would be used. This would imply a stepwise load-dispersion diagram which would violate the lower bound concept. However if the widths of the strips are kept small in comparison to the slab dimensions the error will be negligible.

Obviously if a different load-dispersion diagram or other values of α were assumed different bending moment diagrams for the strips would be obtained. The designer therefore has considerable freedom in the choice of reinforcing he may use and normally would choose the load-dispersion

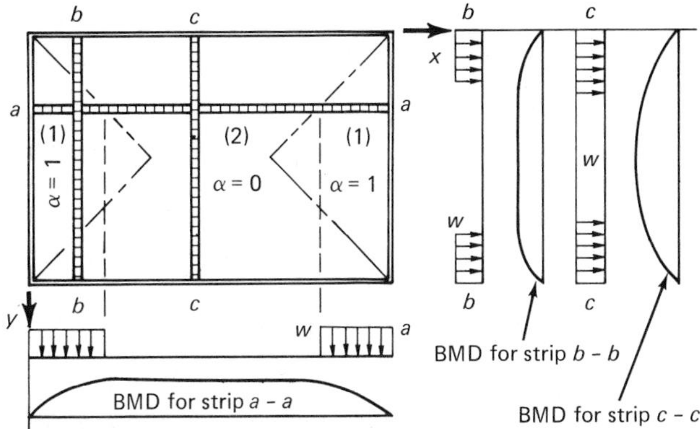

Fig. 3-23 Load dispersion diagram for strip method, showing typical strips.

diagram which would give the minimum amount of reinforcing.

The concept of the strip method is most useful and has been extended to consider slabs with openings, slabs on point supports and slabs with irregular panel configurations. 3-23, 3-24

REFERENCES

3-1 ACI Committee 318, "Building Code Requirement for Reinforced Concrete," (ACI 318-71), American Concrete Institute, 1971.

3-2 Nichols, J. B., "Statistical Limitations Upon the Steel Requirement in Reinforced Concrete Flat Slab Floors," *Transactions ASCE,* **77**, 1670–1736, 1914.

3-3 Westergaard, H. M. and Slater, W. A., "Moments and Stresses in Slabs," *Proceedings ACI,* **17**, 415, 1921.

3-4 Gamble, W. L.; Sozen, M. A.; and Siess, C. P., "Test of a Two-Way Reinforced Floor Slab," *Proceedings ASCE,* **95** (ST6), 1073–1096, June 1969.

3-5 Hatcher, D. S.; Sozen, M. A.; and Siess, C. P., "Test of a Reinforced Concrete Flat Plate," *Proceedings ASCE,* **91** (ST5), 205–231, Oct. 1965.

3-6 Hatcher, D. S.; Sozen, M. A.; and Siess, C. P., "Test of a Reinforced Concrete Flat Slab," *Proceedings ASCE,* **95** (ST6), 1051–1072, June 1969.

3-7 Ang, A. H. S., and Prescott, W. S., "Equations for Plate-Beam Systems in Transverse Bending," *Proceedings ASCE,* **87** (EM6), December 1961.

3-8 Clough, R. W., "Stress Analysis," New York: John Wiley and Sons, 1965, pp 85–93.

3-9 Gallagher, R. H., "Analysis of Plate and Shell Structures," *Application of Finite Element Methods in Civil Engineering,* Publication of ASCE, 1969, pp 155–205.

3-10 Gamble, W. L.; Sozen, M. A.; and Siess, C. P., "Measured and Theoretical Bending Moments in Reinforced Concrete Floor Slabs," Civil Engineering Studies, *Structural Research Series No. 246,* University of Illinois, June 1962.

3-11 Simmonds, S. H., and Siess, C. P., "Effects of Column Stiffness on the Moments in Two-Way Floor Slabs," Civil Engineering Studies, *Structural Research Series No. 253,* July 1962.

3-12 Corley, W. G., and Jirsa, J. O., "Equivalent Frame Analysis for Slab Design," *Proceedings ACI,* **67** (11), 875–884, Nov. 1970.

3-13 Cardenas, A. E., and Kaar, P. H., "Field Test of a Flat Plate Structure," *Proceedings ACI,* **68** (1), 50–59, January 1971.

3-14 "Commentary on Building Code Requirements for Reinforced Concrete," (ACI 318-71), American Concrete Institute, 1971.

3-15 Simmonds, S. H., and Misic, J., "Design Factors for the Equivalent Frame Method," *Proceedings ACI,* **68** (11), 825–831, Nov. 1971.

3-16 Van Buren, M. P., "Staggered Columns in Flat Plates," *Proceedings ASCE,* **97** (ST6), 1791–1797, June 1971.

3-17 Woodring, R. E., and Siess, C. P., "An Analytical Study of the Moments in Continuous Slabs Subjected to Concentrated Loads," Civil Engineering Studies, *Structural Research Series No. 264,* University of Illinois, May 1963.

3-18 Simmonds, S. H., "Effects of Supports on Slab Behavior," presented at ASCE National Structural Engineering Meeting, April 24–28, 1972, Cleveland, Ohio, meeting preprint 1697.

3-19 Johanson, K. W., *Pladeformler,* Polyteknish Forening, Copenhagen, 2nd ed. 1949.
See also *Yield-Line Theory,* English Translation, Cement and Concrete Association, London, 1962.

3-20 Hillerborg, A., "Jämviksteori för armerade betongplatter," *Betong,* **41** (4), 171–182, 1956.
See also Building Research Station Translation LC1082.

3-21 Wood, R. H., *Plastic and Elastic Design of Slabs and Plates,* Thames and Hudson, London, 1961.

3-22 Timoshenko, S., and Woinowsky-Krieger, S., *Theory of Plates and Shells,* McGraw-Hill, New York, 1959.

3-23 Hillerborg, A., *Strimlemetoden,* Svenska Riksbyggen, Stockholm, 1959.
See also CSIRO, Division of Building Research, Translation No. 2, by F. A. Blakey, Melbourne, 1964.

3-24 Wood, R. H., and Armer, G. S. T., "The Strip Method for Designing Slabs," *Building Research Station Publication CP 39/70,* London, 1970.

NOTATION

b_o = periphery of critical section for shear
b_w = width of beam web, in.
c = distance from centroid of support to critical section
c_1 = size of rectangular or equivalent rectangular column, capital or bracket measured in the direction in which moments are being determined
c_2 = size of rectangular or equivalent rectangular column, capital or bracket measured transverse to the direction in which moments are being determined
C = cross sectional property to define torsional stiffness
C = equivalent load factor (section 3.9.2)
C = coefficient for determining panel moment (section 3.9.4)
d = distance from extreme compression fibre to centroid of tension reinforcement, in.
E_{cb} = modulus of elasticity for beam concrete
E_{cc} = modulus of elasticity for column concrete
E_{cs} = modulus of elasticity for slab concrete
f_c' = specified compressive strength of concrete, psi
f_y = specified yield strength of nonprestressed reinforcement
F_y = specified yield strength of structural steel (section 3.9.5)
h = slab thickness, in.
h_v = total depth of shearhead cross section (section 3.9.5)
I_b = moment of inertia about centroidal axis of gross beam section as defined in Fig. 3-3
I_c = moment of inertia of gross cross section of columns
I_s = moment of inertia about centroidal axis of gross section of slab
J = sum of the second centroidal moments of the shear area
K = measure of column stagger (section 3.9.1)
K_b = flexural stiffness of beam; moment per unit rotation
K_c = flexural stiffness of column; moment per unit rotation
K_{ec} = flexural stiffness of an equivalent column; moment per unit rotation
K_s = flexural stiffness of slab; moment per unit rotation
K_t = torsional stiffness of torsion member; moment per unit rotation
l_n = length of clear span in long direction of two-way construction measured face to face of columns in slabs without beams and face to face of beams or other supports in other cases when determining minimum slab thickness
l_n = length of clear span, in the direction moments are being determined, measured face to face of supports
l_v = length of shearhead arm from centroid of reaction (section 3.9.5)
l_1 = length of span in the direction moments are being determined measured center to center of supports
l_2 = length of span transverse to l_1 measured center to center of supports
M = unbalanced moment transferred from slab to column
M_o = total static design moment in a panel
T_u = torsional design moment
v_c = nominal permissible shear stress carried by concrete
v_{tu} = nominal total design torsion stress
v_u = nominal total design shear stress
V_u = total applied design shear force at critical section
w = design load per unit area of slab
w_d = design dead load per unit area of slab
w_l = design live load per unit area of slab
x = shorter overall dimension of rectangular part of a cross section
y = longer overall dimension of rectangular part of a cross section
α = portion of load taken in x-direction (section 3.10.2)
α = ratio of flexural stiffness of beam section (Fig. 3.3) to the flexural stiffness of a width of slab bounded laterally by the center line of the adjacent panel, if any, on each side of the beam
α_c = ratio of flexural stiffness of the columns above and below the slab to the combined flexural stiffness of the slabs and beams at a joint taken in the direction moments are being determined
α_{ec} = ratio of flexural stiffness of the equivalent column to the combined flexural stiffness of the slabs and beams at a joint taken in the direction moments are being determined
α_m = average value of α for all beams on the edges of a panel
α_v = ratio of stiffness of shearhead arm to surrounding composite slab section (section 3.9.5)
α_1 = α in the direction of l_1

α_2 = α in the direction of l_2
β = ratio of clear spans in long to short direction
β_a = ratio of dead load per unit area to live load per unit area (in each case without load factors)
β_s = ratio of length of continuous edges to total perimeter of a slab panel
β_t = ratio of torsional stiffness of edge beam section to the flexural stiffness of a width of slab equal to the span length of the beam, center to center of supports
δ_s = factor by which the positive moments in a panel are increased to account for pattern loading
ϕ = capacity reduction factor
= 0.85 for shear
= 0.90 for flexure

4

Joints in Buildings

MARK FINTEL*

4.1 INTRODUCTION

Concrete is subject to changes in length, plane, and volume caused by changes in its temperature moisture content, reaction with atmospheric carbon dioxide, or by the imposition or maintenance of loads. The effects may be permanent contractions, due, for example, to initial drying shrinkage, carbonation, and irreversible creep. Other effects are transient and depend on environmental fluctuations in humidity and temperature, or the application of loads, and may result in either expansions or contractions.

The results of these changes are movements, both permanent and transient, of the extremities of concrete elements. If contraction movements are restrained, then cracking may occur within the unit. The restraint of expansion movement may result in distortion and crushing within the units, or crushing of its ends in the transmission of unanticipated forces to abutting units. In most concrete structures these effects are objectionable from a structural or an appearance viewpoint; one of the means of handling them is to provide joints at which movement can be accommodated without loss of integrity of the structure.

The occurrence of cracks in concrete construction due to volume changes has long been a problem. It is generally accepted that some crack formation in slabs and exterior walls is unavoidable, except in very low structures. However, much can be done to reduce or control cracking. Steel reinforcement can be used to distribute cracks. Whether or not these distributed cracks can be seen readily depends upon the magnitude of the volume change, the percent of reinforcement and the size and shape of the structure.

The designer usually wants to encourage as many narrow, closely spaced cracks as possible in reinforced flexural members. Wide cracks are objectionable for aesthetic reasons, and because they permit the entrance of water or aggressive solutions which might corrode the reinforcement. Since the sum of all the crack widths is more or less constant and determinate, the width of a crack is inversely proportional to the number of cracks which can be encouraged to form in a certain length.

The problem of cracking can not be ignored. It is handled either by hiding the cracks in preformed grooves, or by using reinforcement to insure a large number of hairline cracks.

The use of joints, particularly in large buildings, is, therefore, inevitable and rarely do we find a concrete structure built without the inclusion of either construction joints, control joints, expansion joints, shrinkage strips, isolation joints, or a combination of these. Although joints are placed in concrete so that cracks do not occur elsewhere, it is seemingly almost impossible to prevent occasional cracks

*Director, Engineering Services Department Portland Cement Association, Skokie, Illinois.

between joints. The provision of joints to take up the movement occurring in concrete structures is a subject which does not always receive the consideration its importance should demand. Often the decision of whether or not to provide the joint becomes a matter of opinion, while in most cases it can be the result of a logical consideration of effects. Because of their great influence on correct detailing of the job, on the progress schedule, and on the appearance of the finished building, the location of joints must not be left to chance.

4.2 MOVEMENTS IN STRUCTURES

To understand the action of the various types of joints and the demands put on them by the structure, it is necessary to have some knowledge of the movements to which concrete structures are subjected during their life.

The movements in hardened concrete which can cause cracking can originate from:

(a) the properties inherent in concrete as a material which are independent of the type of structure; these properties include initial shrinkage.

(b) movements depending on the type of structure and consisting of effects of all imposed loads such as self-weight and lateral loads of wind and earthquakes. Such movements may be deflections, elastic strains, and strains due to creep caused by permanent loads or by the applied prestressing forces.

(c) movements depending on the location of the structure caused by changes in temperature and humidity. The severity of the environment indicates the magnitude of the movement to be expected which is influenced by the effects of relative exposure to sun and prevailing winds. Also, in this group should be included (although somewhat indefinite) the effects of the shape of the structure and its relation to other buildings in the vicinity.

Only some of the foregoing qualitative considerations of movements can be considered quantitatively by designers. If the individual movements are known, the final effects can be considered by adding them together. However, the accuracy of such assessment may be very poor. Some of the basic parameters which are needed to compute the individual movements are still poorly defined; for example, the modulus of elasticity of concrete, under conditions of stress close to the state of rupture. The magnitude of movements may frequently be rendered more uncertain by the presence of restraints such as friction and the interaction of structural elements which are sometimes not measurable. This is more true of complex structures such as large buildings, than of more simple structures such as bridges. Fortunately, however, limited quantitative accuracy can be accepted since the overall knowledge of the movement in complex structures is of more interest to the designer than its actual magnitude. Knowledge of overall movements allows provision of details to accommodate the movements and relieve a buildup of stresses which may otherwise cause distress. Only rarely is it possible to predict accurately the amount of movement in a complex structure, and a precise prediction of the magnitude of a number of superimposed movements is imprudent. However, an exact analysis is sometimes possible, and in a simple structure may be of practical value. But in a complex structure, calculations should be relied upon only to a limited extent, because while theoretically the magnitude of the movements can be obtained, those movements are sure to differ from the reality due to fluctuations of environmental conditions, the various effects of restraint of the structure that are not considered in the analysis, and the variation of the shrinkage and creep properties of the concrete. It is of value, however, to consider qualitatively the effects of the various movements and to assess their magnitude for comparison of different construction processes.

Only movements due to shrinkage and creep are specific to concrete as a material; all other movements, be they due to loads or in response to temperature variations, are characteristic to structural steel as well.

4.3 TENSILE STRESSES LEADING TO CRACKING AND THEIR ORIGIN

Cracking of a concrete section occurs when the tensile stress acting on the section is larger than its tensile strength, or when tensile strain exceeds the tensile strain capacity. Since the tensile strength of concrete may be only about 1/10th of its compression strength, its tendency is to crack under relatively low tensile stresses. It should be appreciated that strains due to shrinkage, creep, and temperature variations do not result in stressing of the material in the structural sense unless restraint against free movement is provided. In floor construction, for example, external means of restraint against volume changes may be caused by columns and walls; while internal restraint is caused by the reinforcement, particularly against shrinkage and creep strains; and in case of slabs on the ground, by subgrade restraint.

The actual mechanism by which cracks form may be quite complicated, but the basic causes involved are straightforward. If a structure were freely supported in space, if all its parts had the same rate of volume change due to shrinkage and temperature, and if, furthermore, all parts of the structure were exposed to the same atmospheric conditions, no differential volume changes and therefore no cracks would result. In an actual structure, however, restraint may be applied to beams or slabs in lower stories of buildings by rigid columns and by the foundation, since volume changes in the foundation are minimized by the insulating effect of the surrounding soil. In higher stories, as the restraint is diminished, cracking is reduced.

Observations show that in buildings with basements, or large heavy foundations under walls, the most numerous cracks are in the first story walls, due to the restraining action of the portion of the building below ground where shrinkage is the least. Buildings with freestanding columns in the first story will crack less because the freestanding columns will accommodate the movements due to differential shrinkage between the basement and the stories above.

The use of different materials in conjunction with each other also may lead to cracking, since different coefficients of expansion and/or shrinkage result in movements of one material relative to the other.

Buildup of tensile stresses resulting from restrained shortening may be due to shrinkage and temperature drop. It has long been accepted that within reasonable limits contraction stresses are relieved by the beneficial effects of creep. This is true in many cases, but in others it occurs by chance, and not by good judgment.

4.3.1 Shrinkage

Shrinkage during the curing and drying of concrete is unavoidable unless the concrete is submerged in water. The amount of shrinkage varies with the mix and with atmospheric conditions. A typical coefficient of shrinkage for average thickness structural slabs may be up to about 600×10^{-6} in./in. Thus, for a 100 ft long slab, the contraction could be about 3/4 in. The rate at which shrinkage occurs depends on the rate of loss of moisture. This rate is important since a slow loss will allow time for the concrete to gain strength, and also to undergo a certain amount of plastic flow. More rapid drying will greatly reduce these beneficial effects. Thus, for any particular slab, a more rapid moisture loss (i.e., drying at higher air temperature and lower atmospheric humidity) will result in an increase in the width and number of shrinkage cracks.

Cracks may also result from differential moisture and temperature gradients. Tensile stresses are thus induced in the outer skin of the concrete mass when the surface is shrinking and cooling more rapidly than the interior. This effect is not too critical for thin structures such as slabs and walls, but it can be a major consideration in massive concrete work or thick pavements. Since the rate of advancing shrinkage diminishes as a square from the depth of the drying surface, the thick element may never dry out and reach its ultimate shrinkage because during the changes of seasons it will be absorbing moisture while the thin element will dry out and crack before the change in season arrives. However, the thick element drying from one face only may crack at the outer face due to internal restraint, but the cracks most likely will be shallow. A building with exterior columns and 6 in. thick walls with openings will have more "leaky" cracks than a building with 12 in. thick bearing walls with similar openings.

4.3.2 Temperature Changes

The stresses introduced in concrete by temperature variations can lead to serious cracking. A reduction in temperature is more serious than an increase because the stresses induced are tensile and because they combine with shrinkage stresses. Also, the maximum temperature of the concrete during the first day controls the amount of subsequent expansion and contraction of the concrete. The higher the temperature during the first day, the greater the contraction during cold weather. Consequently, from this standpoint alone, it is preferable to place concrete in cool weather, Under these conditions, the difference between the temperature of the concrete after curing and drying, and the lowest temperature to which it will be subjected, will be reduced.

The amount of thermal expansion and contraction of concrete varies with factors such as type and amount of aggregate, richness of mix, water-cement ratio, temperature range, concrete age, and degree of saturation of concrete. Of these, aggregate type has the greatest influence. A typical value is on the order of 5 to 6×10^{-6} in./in. per °F, i.e., very nearly the same as for steel. If an unrestrained slab or wall 100 ft long is exposed to a temperature variation from summer to winter of 100°F, the total thermal movement might be about 0.6–0.7 in. Movements occur at the exposed surface of the concrete, which cools off quicker, before they occur in the interior of the section, leading frequently to additional warping or curling effects. Observations of buildings in service indicate the total movement is somewhat less than half of that which might be anticipated by combining the contraction due to temperature drop with the shrinkage. This is due to restraining effects of the reinforcing steel and restraining effects of columns, walls and the foundation. Also, at low temperature, moisture content is less than at high temperature. Thus, the large temperature shortening is offset by a lesser shrinkage shortening.

4.4 REINFORCING STEEL

The introduction of reinforcement produces internal restraint and, thereby, reduced movement of the concrete but not necessarily an elimination of cracking. The attempt of the concrete to reduce in size places the steel in compression and concrete in tension. A highly reinforced section approaches the condition of full restraint and can cause cracking in the concrete. Average reinforced sections (under 2% steel) usually have an apparent shrinkage potential of 0.02 to 0.04%, compared to 0.03 to 0.08% for plain concrete.

The presence of properly distributed reinforcement causes numerous small cracks to occur, rather than a few wide cracks. The reinforcement, whether as a means of controlling cracks, or provided for structural reasons, can partially resist some of the movements. Prestressing which can prevent cracks from occurring can play an important part in the control of movement. It must be noted, however, that reinforcement cannot suppress entirely length changes of concrete from whatever causes.

The ACI 318-71 *Building Code* recommends that the following minimum ratios of shrinkage and temperature reinforcement be provided perpendicular to the main reinforcement in structural floors and roof slabs:

1. where grade 40 and 50 deformed bars are used — 0.0020
2. where grade 60 deformed bars or welded wire fabric, deformed or plain, are used — 0.0018

4.5 TYPES OF JOINTS

It is convenient to categorize into two groups the many types of joints in reinforced and plain concrete structures which have been devised to accommodate construction needs and the various movements.

4.5.1 Construction joints which are installed to break up the structure into smaller units in accordance with the production capacity of the construction site. True construction joints are not designed to provide for any movements, but are merely separations between consecutive concreting operations.

4.5.2 Movement (functional) joints which are installed to accommodate volume changes:

1. expansion joints
2. control (contraction) joints
3. shrinkage strips

Movement joints serve to prevent restraints which would otherwise occur as a result of differences in deformation of the adjacent parts. In some cases, such joints are interposed between the structure and its foundations, since the superstructure deforms due to external loads, internal forces, and to temperature variations while the foundation normally remains immovably secured in the ground.

Since properly functioning joints are usually expensive to build, it is desirable to install one joint to serve a dual purpose whenever possible. By the very nature of its construction, an expansion joint acts also as a contraction joint; and obviously expansion joints can also function as construction joints. A further saving in the number of joints required may be effected by arranging to have a contraction joint coincide with a construction joint, thus eliminating one joint.

4.6 JOINT SPACING

Spacing of functional (contraction and expansion) joints depends upon a great number of factors: shrinkage properties of the concrete, type of exposure to temperature and humidity, resistance to movement (restraint), thickness of members, amount of reinforcement, structural function of the member, external loads, soil conditions, structural configurations, and other conditions. Many of these factors are elusive variables, sometimes difficult to establish. As a consequence, both experience and opinion on joint spacing vary greatly.

In reinforced concrete elements, joint spacing and reinforcement are interrelated variables and the choice of one should be related to the other. As yet, however, a reliable relationship between the two quantities does not appear to have been established. Sufficient steel must be included to control cracking between the joints. If the joint spacing is increased, the reinforcement must be increased correspondingly to control cracking over the longer distance.

The shape of a building has a definite effect on joint locations. Any change in direction in such buildings shaped as T, L and Y may require a close examination of the necessity of joints at the junctions which usually create stress concentrations. Also, at the junction between tall and low buildings, the differential settlements require a stress relief mechanism, either in the form of an expansion joint or shrinkage strip (Fig. 4-1).

Joints may also be needed at any location at which stress concentrations may occur, such as large openings in walls or slabs, changes in thickness of walls, slabs, etc. In each of these cases a sufficient amount of reinforcement (in lieu of joint) across the potential crack may prevent cracking.

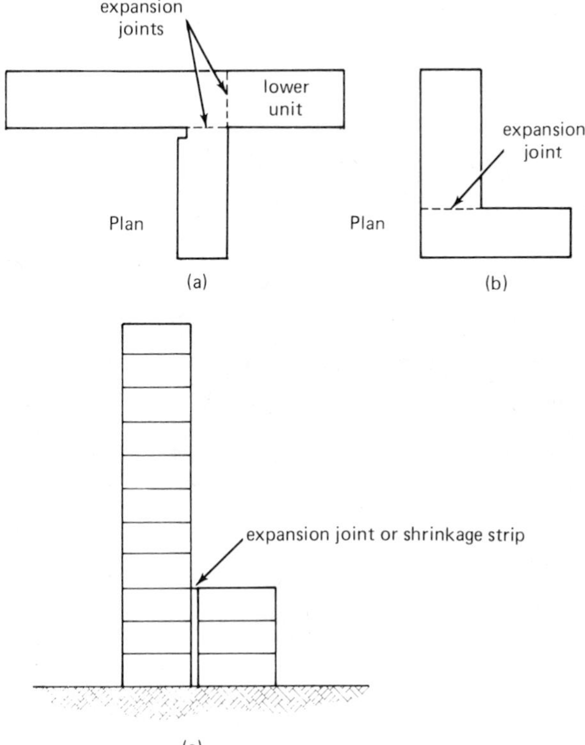

Fig. 4-1 Joints related to the shape of buildings. (a) and (b) Joints are placed at major changes of building run; in (c) joint is placed at junction of tall and low buildings.

The type of construction and the climate are also important in determining the spacing of movement joints. For example, a building which has uninsulated walls or is unheated, must have joints at more frequent intervals than a heated building. Thermal movements are reduced when an insulation layer is placed on the exterior face of the structure.

There is a considerable divergence of opinion on spacing of movement joints (expansion and contraction) with recommendations for expansion joints varying from 100 to 200 ft. while for contraction joints, they vary from a few feet up to 80 ft.

The distance between construction joints depends on the production capacity of the construction site, being limited either by the formwork or casting capacity. When the distance between construction joints becomes too large, intermediate control joints are introduced.

4.7 NONSTRUCTURAL ELEMENTS

The need for movement joints must be investigated in the building as a whole and not in the structure alone. For the owner and user of the building it is not enough that the structural frame performs to the selected criteria when the doors do not close, the windows leak, the partitions crack and the external cladding buckles away from the structure. The completed building, and not the structural frame alone, is the responsibility of the professionals involved in the design process.

To prevent development of distress in the nonstructural elements, it is often necessary to limit the extent of their movement independently of overall expansions, contractions, and deflections occurring in the concrete frame.

Brittle partitions and finishes are particularly sensitive to deflections of beams supporting them. Since the deflections due to creep and shrinkage are dependent on time, it is advisable to give consideration to controlling the time of installation of fragile partitions and finishes connected to concrete beams. For instance, a certain concrete beam deflects 0.25 in. immediately after removal of the shores. Eventually it may deflect an additional 0.50 in. due to creep and shrinkage. However, half of the above added deflection will take place during about two months after removal of the shores. Thus, only 0.25 in. of the deflection will be left to affect the nonstructural fragile elements, if they are installed two months after removal of the shores.

If brittle partitions are used in flexible frames, or if the frames have exposed columns which move up and down in response to temperature variations, the partitions must either be separated from the frame or they must have the same flexibility as the frame. Such flexibility would allow the partitions to follow the distortions of the frame without being distressed and cracked. Unless intentionally separated from the frame, the partitions will distort with the building and contribute to the rigidity of the structure in resisting any movements of the frame.

To leave the partitions unaffected from the distortion of the structural frame, details around the edges of partitions should be provided to allow vertical as well as horizontal slippage. One of the simplest ways to achieve this is to provide a channel enclosure for partition walls, where the partitions meet the columns and ceiling. The partition detail shown in Fig. 4-2 originally suggested by the dry wall partition manufacturers, have been extensively used in the design of many projects. This detail allows a partition to float and provides the necessary restraint against lateral loads acting on the partition.

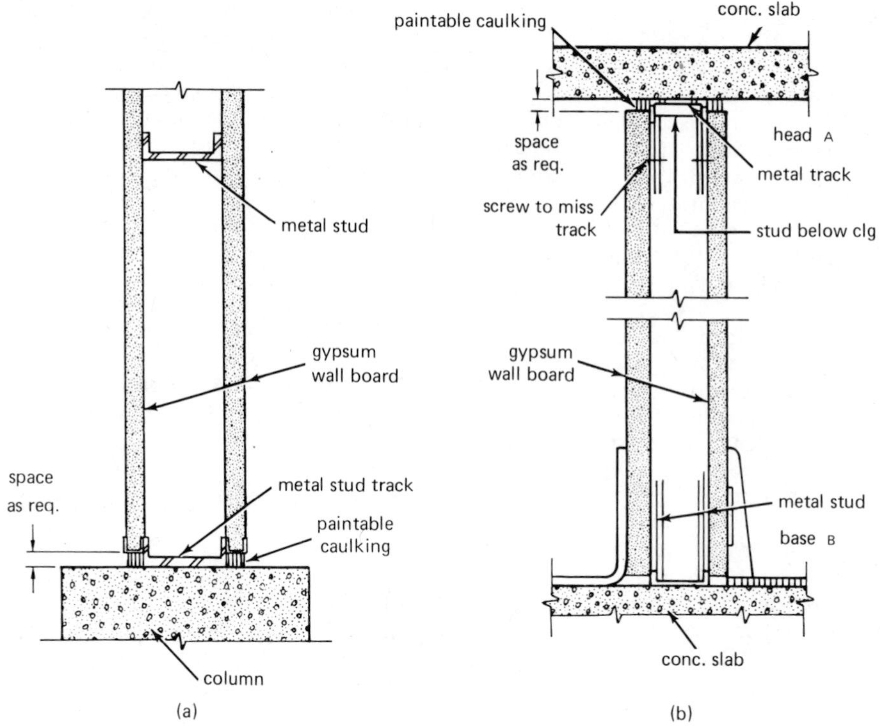

Fig. 4-2 Details of floating partitions. (a) Joint at exterior column, plan view; and (b) joint at slab and ceiling, section view.

4.8 JOINTS IN CLADDING

A large number of high-rise reinforced concrete buildings are being built with the exterior columns and end shearwalls clad with clay masonry or natural stone, such as travertine or granite. Although in the traditional buildings of the early 20th century such cladding was detailed by the architects themselves, most of these buildings were made of structural steel, and therefore, did not have to consider shrinkage and creep of the frame after construction. Also the ability of the heavy cladding to carry substantial loads reduced considerably stresses and movements.

External cladding supported by a multistory structural frame is a typical example of the need to examine vertical as well as horizontal movements. Exterior clay masonry cladding as well as thin stone cladding has been known to suffer from movement distress and cause buckling failures. Such failures in the mid-fifties were initially attributed to the shrinkage and creep shortening of the lightweight concrete columns and walls of the structural frame. Further study, however, showed that the expansion of the brickwork was a factor additive to the shortening of the concrete. Expansive movement in clay brickwork is caused mainly by gain in moisture content over a period of years after the bricks leave the kiln. Whereas clay brick expands due to a gain in moisture after leaving the kiln, concrete and concrete block lose moisture and consequently shrink in volume when exposed to the atmosphere. Clay bricks start to expand as soon as they are removed from the kiln and their rate of expansion decreases with time. Clay bricks which go directly from the kiln into walls will, therefore, have a much larger expansion than similar bricks which have aged a few months. Thus, the amount of expansion to be provided for, i.e., width and spacing of expansion joints, depends on the time that has elapsed since the bricks were removed from the kiln.

Brickwork may expand as much as ½ to 3 in. in 100 ft. Therefore, vertical expansion joints are also needed to accommodate horizontal expansion.

No single recommendation on the positioning and spacing of expansion joints can be applicable to all structures. Each building design should be analyzed to determine the potential movements and provisions should be made to relieve excessive stress which might be expected to result from such movement.

Total unrestrained temperature expansion of clay masonry walls may be estimated from the formula:

$$w = [0.0002 + 0.000004\,(T_{max} - T_{min})]\,L$$

where

L = length of wall in in.;
T_{max} = maximum mean wall temperature in °F;
T_{min} = minimum mean wall temperature in °F; and
w = total expansion of wall in in.

However, this will be reduced by indeterminate compensating factors such as restraint, shrinkage and plastic flow of mortar, and variations in workmanship. The recommended spacing of expansion joints in masonry walls seems to range between 20 and 75 ft and is affected by the severity of the environment and by the tensile and shear strength of the walls. The means provided for resisting differential movement also affect the joint spacing.

When cladding attached to a multistory frame (column or walls) expands and/or the frame shortens due to shrinkage and creep, the vertical load is transferred from the columns or walls to the cladding until buckling of the cladding occurs. This problem can be eliminated by providing horizontal expansion joints at every floor level or every alternate floor level at shelf angles as shown in Fig. 4-3. This joint should consist of a compressible material immediately below the angle. The joint should be sealed with a mortar colored elastic sealant. The shelf angles should be secured against any rotation and against deflections over 1/16 in. A

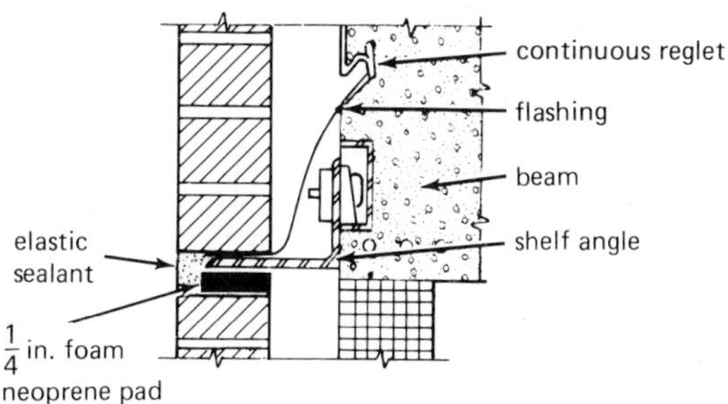

Fig. 4-3 Horizontal expansion joint at shelf angle.

small space ($^1/_2$ in.) should be left between length of angles to allow for thermal movement. A joint width of $^1/_4$ in. per story should be adequate for most cases to accommodate the elastic shortening of the column caused by progress of construction after placing the cladding, creep and shrinkage effects of the column, and the expansion of the cladding due to extreme summer temperature and possible moisture absorption. The individual distribution of each of these effects are as follows, assuming a 12 ft (144 in.) story height:

elastic strain	$= 200 \times 10^{-6} \times 144$	$= .0288$ in.
creep	$= 300 \times 10^{-6} \times 144$	$= .0432$ in.
shrinkage	$= 450 \times 10^{-6} \times 144$	$= .0650$ in.
temperature	$= 70^\circ\mathrm{F} \times 6 \times 10^{-6} \times 144$	$= .0605$ in.
	total	$= .1974 = 3/16$ in.

Vertical joints between individual stone or precast cladding units should be provided to accommodate thermal expansion of the cladding and the distortions of the frame due to lateral forces.

4.9 CONSTRUCTION JOINTS

Construction joints are stopping places in the process of placing concrete, and are required because it is impractical to place concrete in a continuous operation, except for very small structures or special types of structures built with slip forms. Construction joints should not be confused with expansion joints, which, if considered necessary, will allow for free movement of parts of a building and should be designed for complete separation.

The main problem in the formation of a good construction joint is that of obtaining a well-bonded watertight joint between the hardened and the fresh concrete. For a sound joint, the reinforcing should be cleaned and the aggregate of the hardened concrete should be exposed by brushing, waterblasting or sandblasting before placing the new concrete.

If there should be any doubt as to the adequacy of the bond between the old and new concrete, the reinforcement crossing the construction joints should be supplemented by dowels.

The simplest type of construction joint is a butt type formed by the usual bulkhead board, as in Fig. 4-4a. This joint is suitable for thin slabs.

Slabs can use a type of joint which resembles tongue and groove lumber construction. The keyway may be formed by fastening metal, wood, or premolded key material to a wood bulkhead. Concrete above the joint should be hand tooled or sawed to match a control joint in appearance. The second placing of concrete later enters the groove to form the tongue and thus allow for shear forces to be transmitted through the joint, as in Fig. 4-4b. In plain slabs on ground this ensures that future slabs will remain level with previously cast concrete.

Joints should be made straight, exactly horizontal or vertical, and should be placed at suitable locations. In walls, horizontal construction joints can be made straight by nailing a 1 in. wood strip to the inside face of the form (Fig. 4-4c). Concrete is then placed to a level about $^1/_2$ in. above the bottom of the strip. After the concrete has settled and just before it becomes hard, any laitance which has formed on the top surface is removed. The strip is then removed and irregularities in the joint are leveled off.

The forms are usually removed at construction joints and then re-erected for the next lift of concrete, as illustrated in Fig. 4-4c. A variation of this procedure is to use a rustication strip instead of the 1 in. wood strip and to form a groove in the concrete for architectural effect. Rustication strips may be V-shaped or rectangular with a slight bevel. If V-shaped, the joint should be made at the point of the V. If a rectangular strip is used, the joint should be made at the top edge of the inner face of the strip.

Continuous or intermittent keyways in either vertical or horizontal construction joints of reinforced slabs or walls are of questionable value. While they seemingly contribute little added resistance to the joints, they may contribute to spalling, and they interfere with getting the best quality of concrete and maximum strength at the joint. When proper concreting procedures are followed, the bond between the old and new concrete plus the doweling effect of the reinforcement can be made good enough to provide shear resistance equivalent to that of concrete placed monolithically.

Joints should be perpendicular to the main reinforcement. All reinforcement should be continued across construction joints. In providing construction joints it is essential to minimize the leakage of grout from under stop-end boards. If grout does escape, forming a thin wedge, it should be removed before subsequent concreting commences to avoid weakening the structure. For water tightness a continuous waterstop of plastic, copper, or rubber is essential. In wall construction and other reinforced concrete work, it may not be convenient to sandblast or to use water jets for cleaning joint surfaces. Good results have been obtained by constructing the form to the level of the joint, overfilling the forms an inch or two, and then removing the excess concrete just before setting occurs. The concrete then can be finished with stiff brushes.

Hardened concrete should be moistened thoroughly before new concrete is placed on it. Where the concrete has dried out it may be necessary to saturate it for a day or more. No pools of water should be left standing on the wetted surface when the new concrete is placed.

Where concrete is to be placed on hardened concrete or on rock, a layer of mortar on the hard surface is needed to provide a cushion against which the new concrete can be placed. The fresh mortar prevents stone pockets and assists in securing a tight joint. The mortar should have a slump of less than 6 in. and should be made of the same materials as the concrete, but without the coarse aggregate. It should be placed to a thickness of $^1/_2$ to 1 in. and should be worked well into the irregularities of the hard surface.

Depending on the structural design, construction joints may be required to later function as expansion or contraction joints, or they may be required to be monolithic; that is, the second placement must be soundly bonded to the first so as to prevent movement and to be essentially as strong as the section without a joint.

Construction joints at which the concrete of the second placement is intentionally separated from that of the preceding placement by a bond breaking membrane, but with-

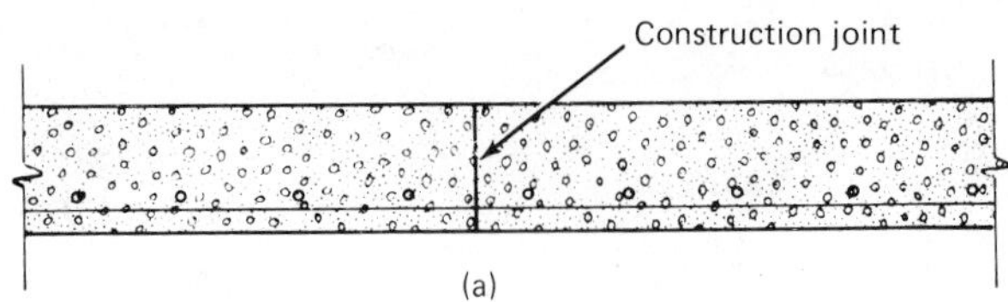

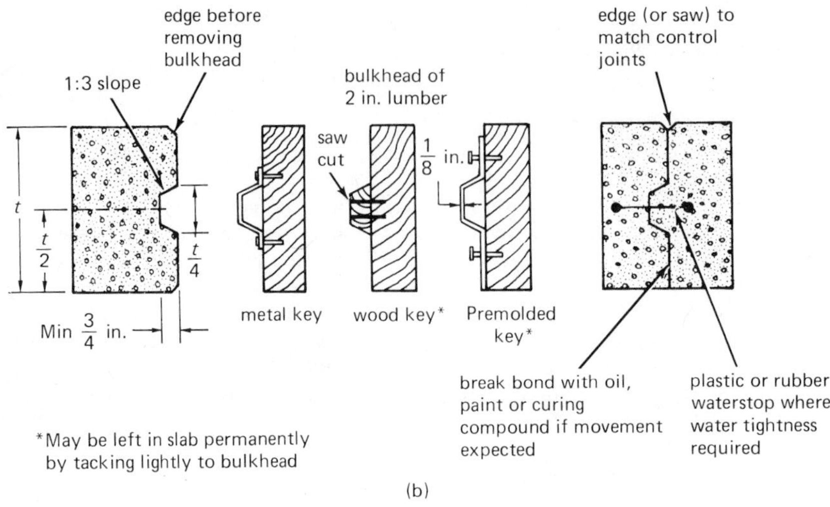

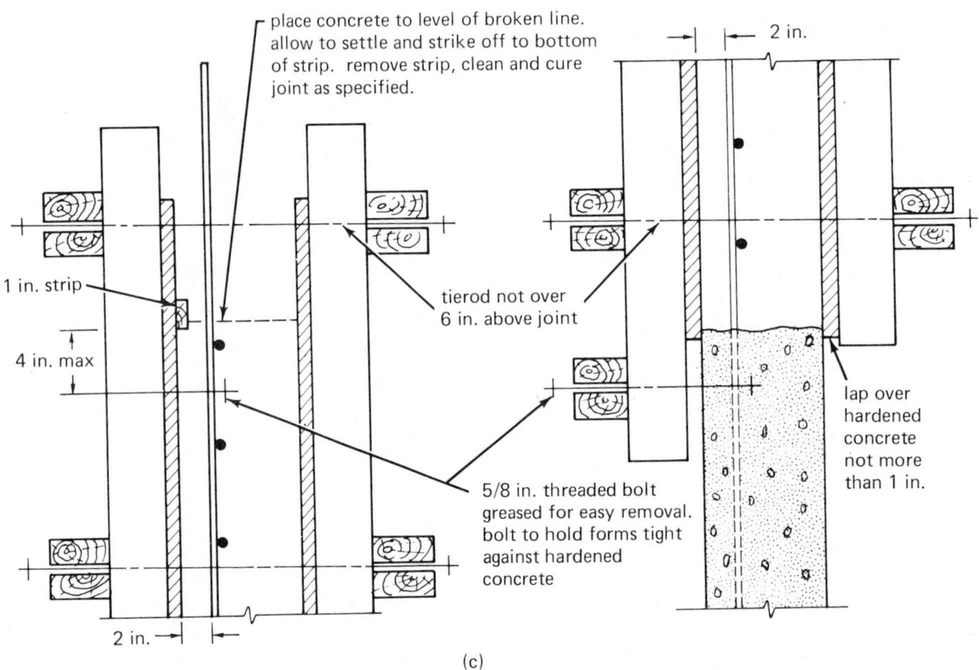

Fig. 4-4 Construction joints. (a) Butt joint in structural slabs; (b) tongue and groove joint in slabs; and (c) straight horizontal joint.

out space to accommodate expansion of the abutting units, function as contraction joints. Similarly, construction joints in which a filler is placed, or a gap is otherwise formed by bulkheading or the positioning of precast units, function as expansion joints. Construction joints may run horizontally or vertically, depending on the placing sequence prescribed by the design of the structure.

4.9.1 Location of Construction Joints

Location of construction joints is usually predetermined by agreement between the architect-engineer and the contractor, so as to limit the work that can be done at one time to a convenient size, with least impairment of the strength of the finished structure, though it may also be necessitated by unforeseen interruptions in concreting operations. In walls a horizontal length of placement in excess of 40 ft is not normally recommended.

Generally it is impractical to place concrete in lifts higher than one story. This fact should be recognized when locating horizontal construction joints. For buildings in which the concrete walls are to be exposed, joints may be located at bends of ornamentation, ledges, rustications, or other architectural details. It is convenient to locate horizontal joints at the floor line or in line with window sills. In the design of hydraulic structures, construction joints usually are spaced at shorter intervals than in nonhydraulic structures to reduce shrinkage and temperature stresses.

As construction joints are provided to accommodate progress of construction, their spacing is determined by the type of work, site conditions, and the production capacity of the construction site. It is also important that the spacing of construction joints be planned such that the joints, while in accordance with production capacity, occur only where they may be properly constructed and do not occur where they may create stress concentrations.

Construction joints should be located by the designer so as to provide logical separation between segments of the structure. As a rule, construction joints are allowed only where shown on the drawings. If the placing of concrete is involuntarily stopped for a time longer than the initial setting time of the concrete used, the old surface is to be considered as a construction joint, and treated as such before casting is resumed.

It is best to avoid locating vertical construction joints at or near the corners of the building, since their presence may make it difficult to tie the corners together securely; it is, however, desirable to have a control joint within 10 to 15 ft of a corner if possible.

The appearance of a structure can be influenced by the location of the construction joints and the aim should be to install them in a position which renders them as inconspicuous as possible; the alternative is to make them clearly visible as a feature of the structure. The joints should fit into the architectural design and their location should facilitate the construction of forms and placing of concrete. However, from the point of view of strength of the structure, it is desirable to position construction joints at points of minimum shear. For slabs and beams it is, therefore, usual to have construction joints at midspan or in the middle third of the span. These rules are based on the assumption that a construction joint may result in less than 100% of shear capacity in the interface. If it would be practicable to have such joints at the supports for slabs and beams, it would improve appearance and result in a considerable saving on the cost of the formwork.

Joints in bearing walls and columns should be located on the underside of floor slabs, beams or girders, and at the tops of footings or floor slabs. Columns should be filled to a level preferably a few inches below the junction of a beam or haunch before making a construction joint. To avoid cracking due to settlement, concrete in columns and walls should be allowed to stand for at least two hours, and preferably overnight, before concrete is placed in slabs, beams, or girders framing into them. Haunches, drop panels, and column capitals are considered as part of the floor, or roof, and should be placed integrally with them.

4.10 EXPANSION JOINTS

Expansion joints are used to allow for expansion and contraction of concrete during the curing period and during service; to permit dimensional changes in concrete due to load; to separate, or isolate, areas or members that could be affected by any such dimensional changes; and to allow relative movements or displacements due to expansion, contraction, differential foundation movement, or applied loads. Obviously, expansion joints can also function as construction joints.

Expansion joints are frequently used to isolate walls from floors or roofs; columns from floors or cladding; pavement slabs and decks from bridge abutments or piers; and in other locations where restraint or transmission of secondary forces is not desired. Many designers consider it good practice to place expansion joints where walls change direction as in L-, T-, Y-, and U-shaped structures, and where different cross sections develop (Fig. 4-1). Expansion joints in structures are sometimes called isolation joints because they are intended to isolate structural units that behave in different ways.

Expansion joints are made by providing a space for the full cross section between abutting cast in place structural units by the use of filler strips of the required thickness or by leaving a gap when precast units are positioned. Expansion joints usually start about the foundation level, and continue throughout the height of the structure.

Expansion joints are installed mainly to control the effects of temperature increase, and to a lesser degree, to allow independent structural action of adjacent units.

The term "expansion joint" is a misnomer so far as cast-in-place concrete is concerned, since the initial shrinkage a structural element underwent from the time it was cast will usually be greater than any possible subsequent expansion due to temperature rise, increase in humidity, or other factors. The term expansion joint is more appropriate for structures assembled from materials such as brick, stone or structural steel with which material the term possibly originated. However, in sanitary structures designed to contain liquids, the contraction may be halted when the structures are placed in use, and even reversed during hot humid weather.

The use of expansion joints in buildings is a controversial issue. There is a great divergence of opinion concerning the importance of expansion joints in concrete construction. Some experts recommend joint spacings as low as 30 ft while others consider expansion joints entirely unnecessary. Joint spacings of roughly 100 to 200 ft for concrete structures seem to be typical ranges recommended by various authorities. Divergent viewpoints are reflected both in private practice and in building codes. The existence of such opposing opinions which, obviously, cannot both be equally valid as a consideration in a single structure, is nonetheless understandable, since it is based on divergence of previous experience.

Those who advocate the complete omission of expansion joints in concrete construction state that the initial shrinkage is greater than the expansion caused by a 100°F increase in temperature; therefore, any temperature increase will tend to close up shrinkage cracks and there will be practically no compressive stress in the concrete due to thermal expansion.

In a 1940 report of a joint committee (AIA, ASCE, ACI, AREA, PCA) it was suggested that, in localities with large temperature ranges, expansion joints should be provided every 200 ft. In milder climates 300 ft was suggested.

A very conservative (in respect to expansion joints) ACI report of August 1971 on Sanitary Engineering Structures recommended that in general, expansion joints should be provided in any noncircular structure having a dimension of 120 ft or more in any principal direction. It is desirable that expansion joint spacing be not more than 50 to 60 ft for members exposed to the atmosphere, or 80 to 100 ft for members completely underground. Recommended expansion joint widths are shown in their Table 2.8.2 as follows:

RECOMMENDED EXPANSION JOINT WIDTHS (IN.) FOR SANITARY STRUCTURES

Temperature Range	*Spacing Between Expansion Joints*			
	40 ft	*60 ft*	*80 ft*	*100 ft*
Underground, 40° F	½	¾	⅞	1
Partly protected–above ground, 80° F	¾	⅞	1	*
Unprotected–exposed roof slabs, etc. 120° F	⅞	1	*	*

**Not recommended.*

The 1972 "Minimum Property Standards–Manual of Acceptable Practices" by the FHA recommends that spacing of expansion joints for buildings not exceed the following values:

Type of Building	*Outside Temperature Variations*	*Maximum Joint Spacing (ft)*
Heated	Up to 70° F	600
	above 70° F	400-500
Unheated	up to 70° F	300
	above 70° F	200

In the 1940's a distinct trend started towards the elimination of expansion joints in long buildings. In the 1950's and 60's this trend continued. Even in localities with large temperature ranges, buildings up to 400 and 500 ft have been constructed without expansion joints, and seemingly the performance has been satisfactory. The following are examples of such buildings:

Military Personnel Records Center, St. Louis, Missouri, is a six story, flat slab building built in 1956, 728 X 282 ft in plan with 22 X 22-ft bays. Control joints were spaced every second bay (44-ft centers) in each direction. The concrete was placed between control joints in a checkerboard fashion with a 48 hour interval between adjacent sections. The slab reinforcement is continuous through the joints with a #4 dowel 3 ft long at 12 in. center added at the top of the slab. A technical paper with background information on this building and expansion joints in general appeared in the ACI Journal.[1]

The *General Accounting Office Building*, built in 1951,[2] Washington, D.C. is an eight story, flat slab building 638 X 389 ft in plan, with 25 X 25-ft bays. Control joints were used on 50-ft centers. No evidence of distress resulting from the omission of expansion joints was found in this building in 1956.[1]

The *Los Angeles Union Terminal* has a seven story, flat slab warehouse 550 X 100 ft in plan, with 20 X 20-ft bays, as well as a four story building 440 X 100 ft in plan. According to an inspection report of 1958, both buildings were in excellent condition after 40-50 years, showing no distress due to lack of expansion joints.

If expansion joints are to be provided, they must be introduced in the preliminary planning stage. It is generally difficult to form breaks in structures when the design has reached an advanced stage without provision for such breaks. Experience and intuitive understanding of how the completed structure behaves may help in the positioning of such joints. Calculations should support the selection of a joint width, and be related to the anticipated movements. The elastic range of the filler material employed in a joint must also be considered.

Expansion joints should not be provided unless they are clearly necessary since they can be an embarrassment to the structural and architectural designer, as they are often incompletely detailed and frequently badly constructed. There is no doubt that the proper course to adopt, where practicable, is to control a movement without permanent joints; for example, with shrinkage strips, described later in this chapter. When, however, a designer considers that danger of unacceptable cracking exists, the structure must be divided into controllable units by providing expansion joints at suitable places.

Expansion joints are designed for relative movement of adjacent sections and should be located so that

(1) they act as stress relief planes;
(2) the concrete between the joints is not subjected to substantial volume change stresses;
(3) other elements supported by the concrete, such as partitions, exterior cladding, window frames, and others in the building, are not subjected to movement distress; and
(4) the shape, size and type of joint will function correctly for all conditions of movement.

Factors that should be considered in the design and detailing of expansion joints are: shrinkage, creep, thermal movements, foundation settlements, and elastic deformations of adjacent structural units.

Thermal expansion of concrete roofs caused by solar radiation is a common cause of distress to buildings. Such distress can be minimized by applying thermal insulation on top of the roof to reduce the temperature differentials. Otherwise, either expansion joints at required intervals should be arranged, or the roof should be able to slide on top of the supporting walls (see Fig. 4-13) with suitable separation of the plaster finish at the junction between roof and walls.

Other than in long buildings, expansion joints may be necessary at the junction of tall and short buildings (Fig. 4-1) to avoid distress due to differential settlements. If, for example, a 40-story tower has a 2-story base, the base will have fully settled elastically when the second story is completed, and the construction progress on the tower will create differential settlements at the junction as the tower will continue to settle. An expansion joint at the junction will allow each of the two different adjacent building units to settle individually without distressing their connection. Incidentally, a shrinkage strip may fulfill the same function. Also, where a new building unit is attached to an existing building, an expansion joint may be desirable.

4.10.1 Expansion Joint Details

To be effective, an expansion joint should separate the two adjacent units into completely independent structures. Joints should extend through foundation walls, but column footings need not be cut at a joint unless the columns are short and rigid. No reinforcement should pass through these

[1] Cohn, Earl B., and Wahl, W. A., "Military Personnel Records Center Built Without Expansion Joints," *ACI Journal*, **54**, June 1958.
[2] Reynolds, W. E., "Ways to Cut Building Costs Shown by General Accounting Office Building, Wash. D.C.," *Civil Engineering*, 22, 50-54, June 1952.

joints; it should terminate 2 in. from the face of the joint. Dowels with bond breaker may be used to maintain plane.

Joint location is generally dictated by structural considerations, with architectural treatment of walls at the joints developed accordingly.

The simplest expansion joint is one on a column line with double columns (see Fig. 4-5a). This may be costly and complicated as it may involve the construction of two independent sets of columns and beams at the joints, separated by a 1 in. space. Such costs should, however, be considered against the risk and cost of extensive trouble which might occur year after year if the joints were omitted.

Expansion joints without a double column may be used by introducing them in the third or quarter point in the slab in a manner indicated in Fig. 4-5b. The introduction of such a hinge near the point of inflection should, of course, be taken into account in the structural analysis. Water tightness of such joints may be difficult to achieve.

Expansion joints in buildings are usually $^3/_4$ in. to 1 in. wide. Judging from past experience, it appears safe to consider that the maximum movement at joints located 200 ft apart will not exceed 1 in. under the most unfavorable conditions.

Expansion joints are filled with a compressible elastic, nonextruding material which is intended to accommodate the movements occurring, and at the same time provide an adequate seal against water and foreign matter. Generally, no single material has been found yet which will completely satisfy both conditions mentioned. Essentially three types of jointing materials are used: joint fillers (such as strips of asphalt-impregnated fibreboard), sealers, and waterstops. Sealers (sealing compounds) and waterstops (rubber, plastic, or metal) are used where a joint has to be sealed against the passage or pressure of water.

Expansion joints must be carefully detailed and carried through the finishes of the floor and ceiling as shown in Figs. 4-6a and 4-6b.

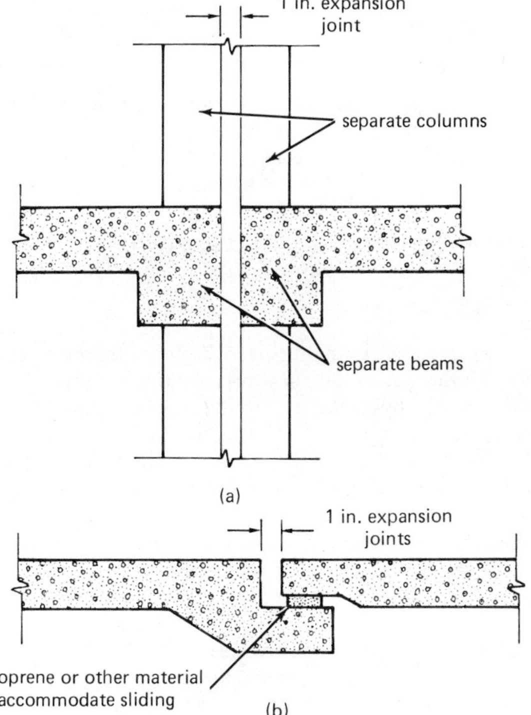

Fig. 4-5 Expansion joints. (a) Joint with separate columns and beams; and (b) joints in the slab at one-third or one-quarter of the span.

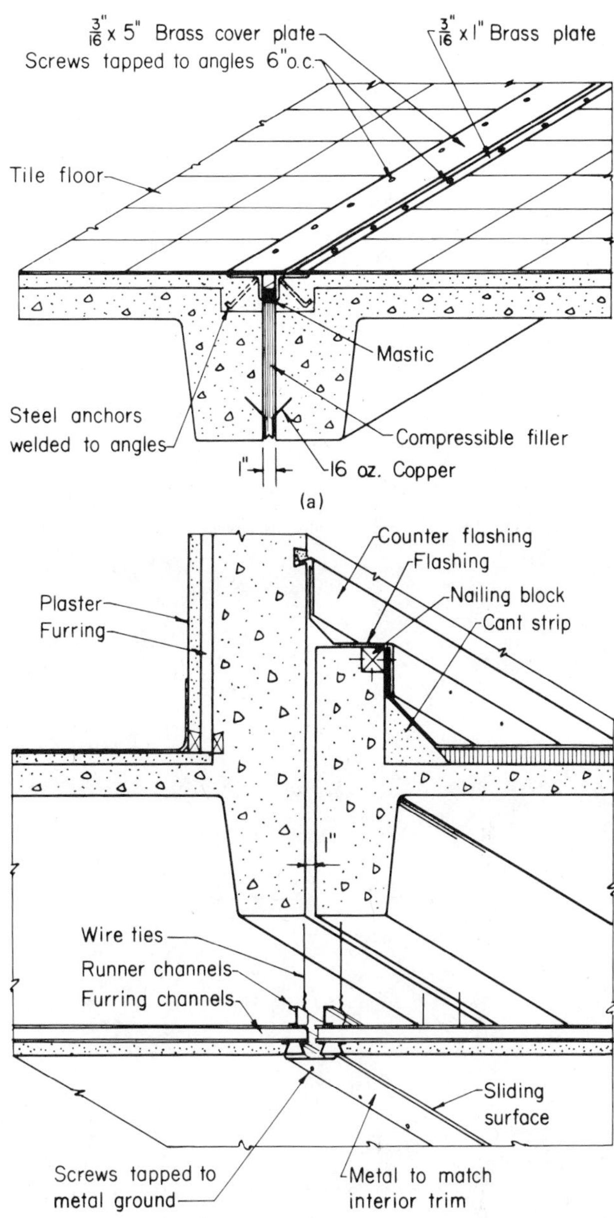

Fig. 4-6 Details of expansion joints through finishes. (a) Covering of expansion joint in floor; and (b) expansion joint in ceiling and on the roof.

4.10.2 Seismic Joints

Seismic joints are wide expansion joints provided to separate portions of buildings dissimilar in mass and in stiffness. The joints are provided to allow the adjacent buildings or building units which have various vibration characteristics to oscillate during an earthquake without hammering at each other. The alternative to a seismic joint is to tie the buildings or units together so that they can vibrate as a single system.

The width of a seismic joint should be equal to the sum of the total deflections at the level involved from the base of the two buildings, but not less than the arbitrary rule of 1 in. for the first 20 ft of height above the ground, plus $^1/_2$ in. for each 10 ft additional height. The determination of these deflections will be the summation of the story drift in addition to the building's flexural deflection (column lengthening and shortening) to the level involved.

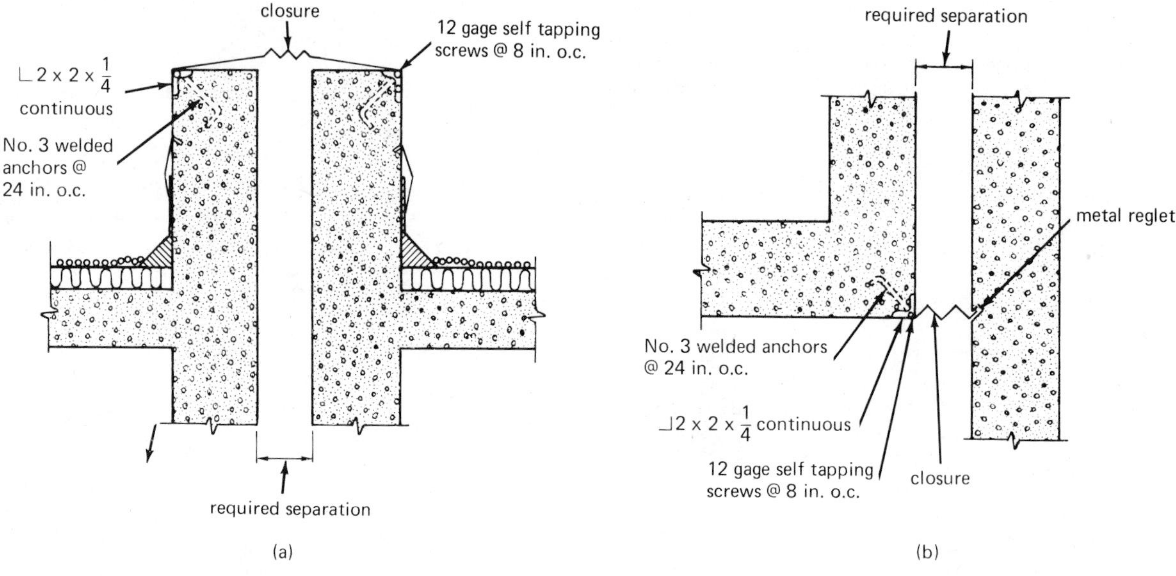

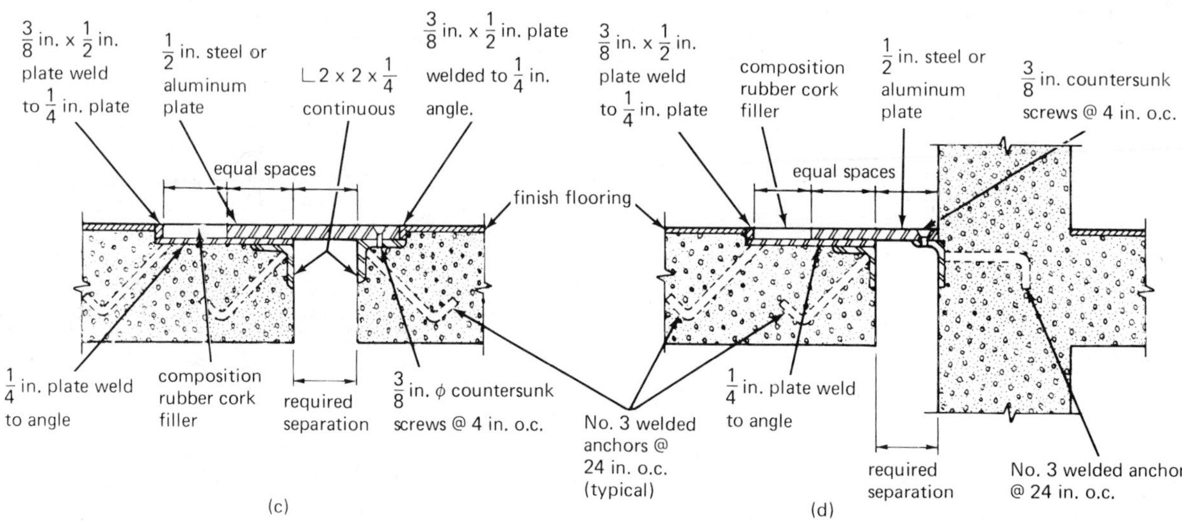

Fig. 4-7 Seismic separation joint details. (a) Roof parapet separation; (b) plan at exterior vertical closure; (c) typical floor plate closure at building separation; and (d) typical building separation floor plate closure at walls.

Shearwall buildings, being much stiffer, need a seismic joint only, say, half as wide, since the earthquake oscillations of shearwall buildings will be much smaller than those of framed buildings.

Figure 4-7 shows seismic separation joints at various locations in buildings. As shown in the figure, the seismic joint coverages must allow movement, be waterproof, and architecturally acceptable. The details of the joint coverages should allow at least a doubling of the joint opening when subjected to an earthquake.

4.11 CONTROL JOINTS

The tensile stresses caused by shrinkage and temperature drop in the concrete which is not free to contract can be relieved or reduced to tolerable limits by control joints.

A control joint is a purposely made plane of weakness of a section so that cracking and contraction will occur along these preselected straight lines which are inconspicuous and less objectionable. Control joints are inexpensive and can be formed easily by tacking wooden, rubber, plastic, or metal strips to the inside of the forms as shown in Figs. 4-8a and 4-8b. After removal of the form and the tacked on strips, a narrow vertical groove is left in the concrete on the inside and/or outside surfaces.

Control joints can also serve as partial expansion joints, limited to the extent of contraction that occurred in the joint. The terms "contraction joints" and "dummy contraction joints" are often used interchangeably for control joints, causing some confusion.

Numerous ways have been devised for forming control joints. In all cases it is recommended that the combined depth of inside and outside grooves be not less than about $1/5$ to $1/4$ of the thickness of the slab or wall to make the joint effective. In slabs and pavements, the necessary weakening of the section can be made by sawing a groove in the

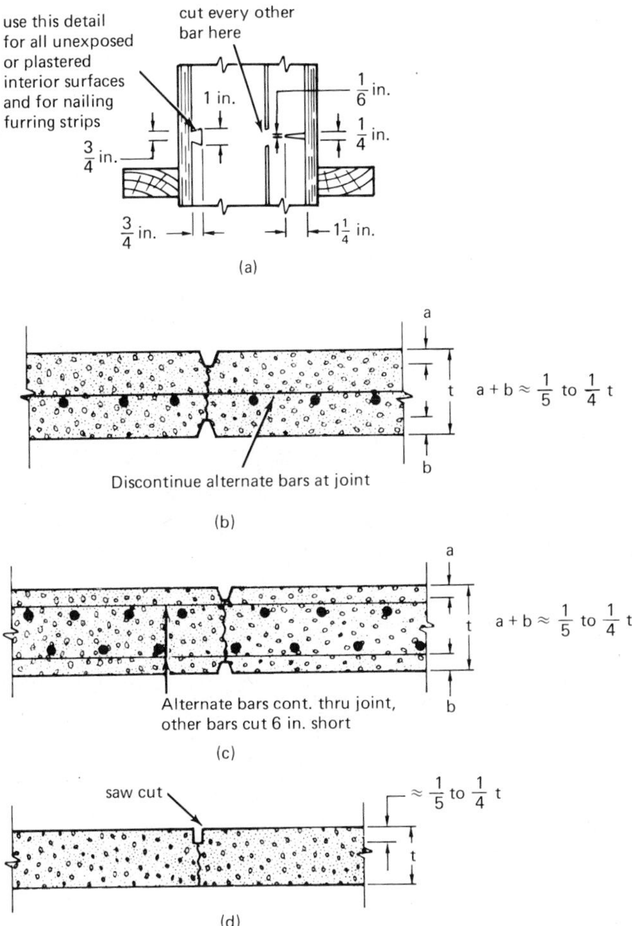

Fig. 4-8 Control joints. (a) Control joints should reduce the effective wall thickness by at least one-fourth; (b) control joints in walls 8 in. thick, or less; (c) control joints in walls over 8 in. thick; and (d) saw cut joint in a slab or pavement.

concrete soon after it has hardened (Fig. 4-8d). The groove may also be formed with a hand tool before hardening.

Quite often wooden or metal strips forming control joints are left permanently in the concrete as shown in Fig. 4-8a.

Control joints are appropriate only where the net result of the contraction and any subsequent expansion during service is such that the units abutting are always shorter than at the time the concrete was placed. They are frequently used to divide large, relatively thin units; for example, pavements, floors, retaining, and other walls into smaller panels. They are called control joints because they are intended to control crack location.

Control joints may form a complete break, dividing the original concrete unit into separate units. Where the joint does not open wide, some shear transfer may be maintained by aggregate interlock. Where greater shear transfer is required without restricting freedom to open and close, coated dowels may be used. If restriction of the joint opening is required for structural stability, appropriate tie bars or continuation of all or part of the reinforcing steel across the joint may be provided. Reinforcement across control joints is usually partially or completely cut.

Where small amounts of reinforcing steel or wire mesh are provided in an element, it is desirable to interrupt them at joints at least 3 in. away from the joint on either side.

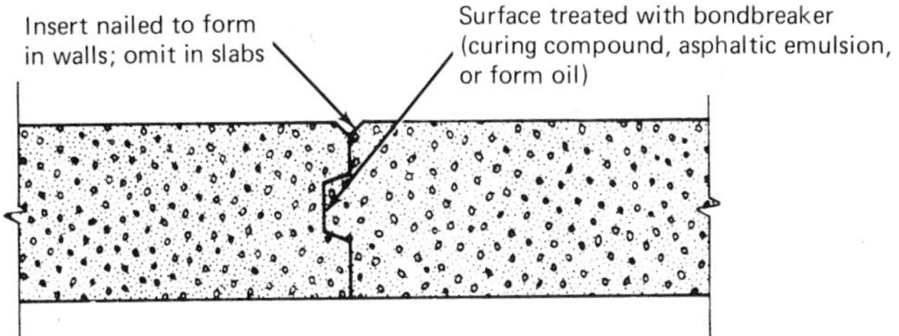

Fig. 4-9 Construction joint used as contraction joint in slab or wall.

If a construction joint serves as a control joint, an insert in the form provides a better finish of the casting operation which at the same time forms a straight control joint. Obviously, the surface of the joint should be treated with a bond breaker to assure free movement (Fig. 4-9).

4.11.1 Spacing of Control Joints

No exact rules for the location of control joints can be made. Each job must be evaluated individually to determine where and how far apart joints should be placed, taking into account the specific structural design. Joint spacings commonly used show considerable variation ranging from a few feet up to 80 ft. The spacing of joints depends on the exposure and severity of climatic conditions, and on the internal restraint (reinforcing) and external restraint (columns, walls, friction with the ground) provided by other parts of the structure. Obviously, the shrinkage of the concrete and temperature range to be expected have a major influence on the spacing. Joints should be made where abrupt changes in thickness or height occurs. To some degree there is a choice between joints at short intervals with normal, or no, reinforcement and larger distances between joints with increased reinforcement. Many of the recommendations favor a 15 to 25-ft spacing between control joints in walls and slabs on grade. First joints should occur 10 to 15 ft from a corner. In walls and slabs, the optimum ratio of panel dimensions enclosed by joints is 1 to 1, with a maximum aspect ratio of $1\frac{1}{2}$ to 1.

In walls having frequent openings, spacing control joints 20 ft apart is considered maximum. The spacing in walls without windows should not be more than 25 ft and a joint within 5 to 15 ft of each corner is desirable.

Observations show that building fronts having about 60% or more of openings crack least, but solid blank walls and walls with relatively small openings (say 25% of gross area) present problems. Cracks develop at about 10 to 15 ft on centers; the greater the spacing of the cracks, the wider the cracks.

Walls with frequent openings should have smaller joint spacings than solid walls. Cracks usually form at windows and doors where the concrete sections are weakest. Additional reinforcing should, therefore, be placed at the corners of these openings to intercept potential cracks.

Many of the references and specifications available seem to favor a 20 to 25-ft spacing between control joints in walls. If a wall is less than 10 ft high, however, and restrained at the foundation by anchorage to more massive concrete or a rock foundation, it may even crack near the center of a 20-ft section. Under such conditions, joints at 10 ft intervals may be desirable. For walls of 10 to 20 ft in height, joint spacing should be approximately equal to the wall height.

Long slabs that form external balconies are prone to develop cracks originating from their edges. The cracks are caused by shrinkage and thermal differentials between the slab on the inside and the concrete near the free edge of the balcony. Control joints at sufficiently close intervals (less than 10 ft) will usually attract all the shrinkage and temperature cracking. If such joints are extended only partly across the balcony slab, the concrete near the end of the joint should be locally reinforced to prevent the prolongation of the joint as a crack.

In cases of continuous forming (slip forming or extrusion) it may be advisable to make larger units between joints (if impractical to form units as one would wish) as a justifiable risk, and to rely on the ability of the unit to resist the forces due to frustrated movements caused by volumetric changes. In such cases a certain amount of shrinkage and temperature reinforcement is required to improve the resistance. It is at times more prudent to trust an unjointed concrete to resist its own movement than to provide an incomplete joint which the structure will probably complete by itself in due course and in a most unsightly manner.

4.12 SHRINKAGE STRIPS

Shrinkage strips are temporary joints which are left open for a certain time during the construction so as to allow a significant part of the shrinkage to take place during construction without inducing stresses. Such joints have been used to a considerable extent both in massive structures and in thin walls and slabs for the purpose of reducing shrinkage stresses and minimizing shrinkage cracks. Such strips divide the structure into parts which shrink independently until they are connected by casting the strip.

In recent years shrinkage strips have replaced expansion joints to a great extent in long multistory buildings. In enclosed multistory buildings the slabs are subjected only to shrinkage while the roof slab is subjected to temperature length changes in addition to shrinkage. Therefore, the roof slab, if not insulated sufficiently, may need an expansion or sliding joint (see Figs. 4-12 and 4-13) in addition to the shrinkage strips. Experience has shown that in long multistory buildings shrinkage strips spaced about 100 to 150 ft apart perform successfully. The distance between horizontal shrinkage strips should be less in slabs with stiff supports (large columns or reinforced concrete walls parallel to the direction of shrinkage) or if the expected shrinkage coefficient is unusually high.

The shrinkage strip is usually 2 to 3 ft wide across the entire building and is cast two to four weeks later than the adjacent units to reestablish continuity. During the period the strip remains open, a significant part of the shrinkage takes place (about 40% of the total shrinkage may occur in the first month) while in the meantime the concrete will gain tensile strength to resist the remaining shrinkage with little or no cracking.

The flexural reinforcement crossing the shrinkage strip is generally made to act continuously from section to section after the joint is closed. However, to run the flexural reinforcement in one piece continuously from one unit through the strip into the other unit would impede unrestrained shrinkage of the concrete units on either side of the strip, thus defeating the objective of the shrinkage strip. Therefore, it is necessary to lap the reinforcing bars within the strip as shown in Fig. 4-10a. An alternate method of detailing the reinforcement to avoid excessive stresses and possible slippage of the bars on either side of the strip while the strip is still open is special expansion bends within the strip, as shown in Fig. 4-10b, so as to allow for a change in the length of the strip. The plane of the bends must be parallel to the face of the slab, and the bends placed alternately left hand and right hand, so as to avoid accumulation of stresses when they tend to straighten out due to the loads superimposed upon the completed structure.

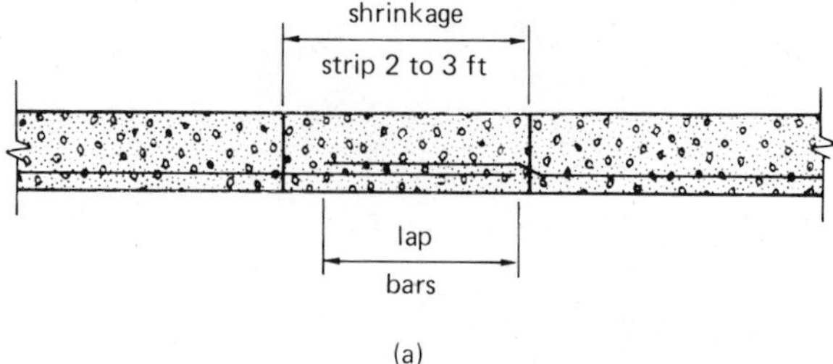

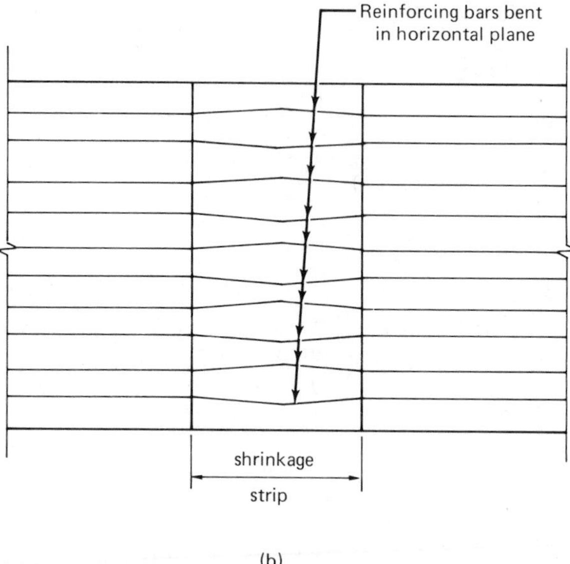

Fig. 4-10 Reinforcement details in shrinkage strips. (a) Section—lapped reinforcement; and (b) plan—bent reinforcement.

Shrinkage strips can also be used in walls. In such cases the walls are constructed in lengths not exceeding 25 ft with internal gaps of 2 ft left between each length. In order to allow unrestrained contraction to take place in each length of wall, the gap should be left open for three to four weeks after the casting of the adjacent length of wall. As in slabs the reinforcement should be lapped within the gap. Another method of reducing shrinkage stresses in walls is to construct them on an alternate panel system (similar to checkerboard casting of slabs on ground), the length of each panel not exceeding about 25 ft. Both examples, or variations of them, are in common use and they form an acceptable practical approach to limiting the amount of shrinkage cracking.

4.13 SPECIAL PURPOSE JOINTS

4.13.1 Elastic Control Joints

To improve the performance of control joints, the bond between reinforcement passing through the joint and the concrete can be eliminated for an appropriate length across the joint as shown in Fig. 4-11a. This prevents yielding of the reinforcement in the vicinity of the joint (which might otherwise occur) and as a result the joint always remains elastic. The breaking of the bond can be achieved either by a paper wrapping, asphalt coating, plastic tubing, or other means. The length of bar over which bond should be eliminated is determined with due regard to maximum joint

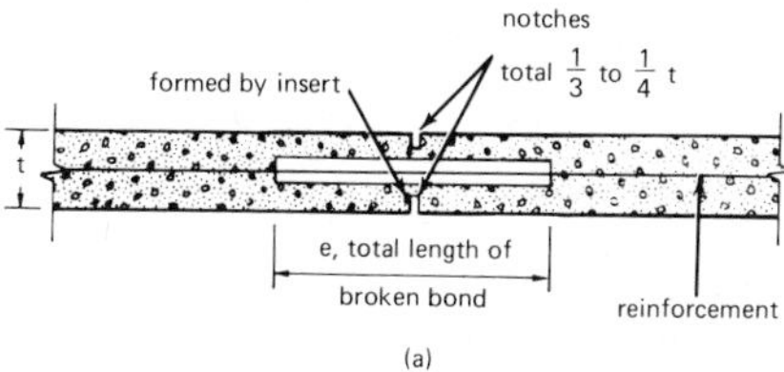

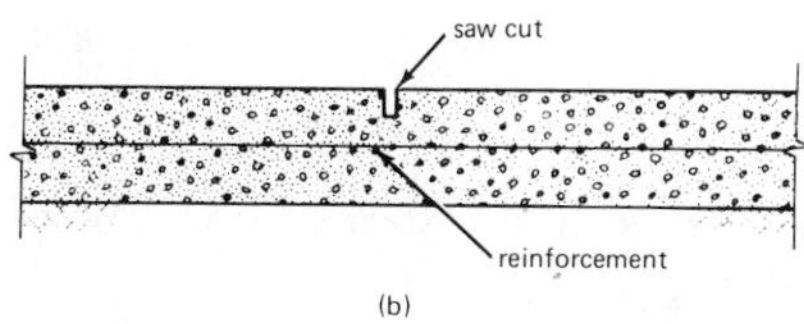

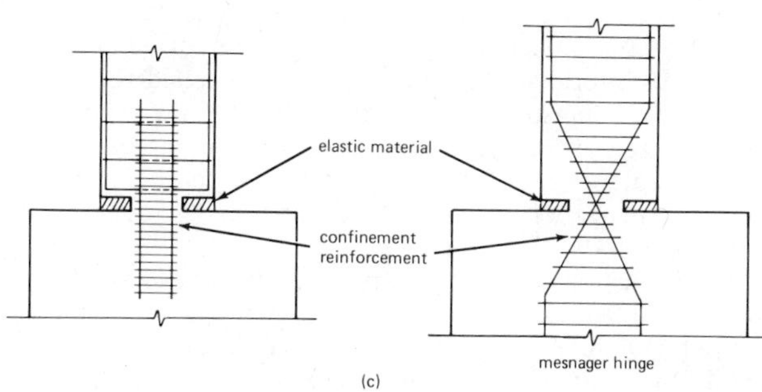

Fig. 4-11 Hinged joints. (a) Elastic control joint; (b) hinged joint in pavement; and (c) hinged joints in structures.

TABLE 4-1 Stresses in Reinforcement of Elastic Joints

Temperature Drop $t\,(^\circ F)$	*Length of Broken Bond to Joint Spacing* (c/a)	*Reinforcing Ratio* ρ	*Steel Stress at Joint* f_s *in ksi*
25	0.50	0.01	9
		0.02	8
	0.25	0.01	15
		0.02	13
50	0.50	0.01	17
		0.02	16
	0.25	0.01	30
		0.02	26
100	0.50	0.01	34
		0.02	32
	0.25	0.01	60
		0.02	52

opening, maximum concrete temperature drop, and the yield strength of the bars. At the maximum joint opening, the bars should be stressed below their elastic limit. The maximum joint opening varies with joint spacing, other conditions being similar. The concrete temperature drop is the difference between the concrete temperature at time of set and the lowest temperature to which the concrete will be subjected.

Elastic control joints were first used in an experimental highway project in Indiana more than 30 years ago. In recent years these joints have been used experimentally in pavements in Sweden and Germany.

Table 4-1 illustrates the approximate relationship between concrete temperature drop, t, length of broken bond to joint spacing ratio, c/a, reinforcement ratio, ρ, and steel stress at a joint, f_s.

4.13.2 Hinge Joints

Hinge joints permit rotation within a section, however, separation of the abutting units is limited by tie bars or the continuation of reinforcing steel across the joint. This term has wide usage in, but is not restricted to, pavements where longitudinal joints (Fig. 4-11b) function in this manner to overcome warping effects by forcing the same deflections on both sides of the joint due to wheel loads or settlement of the subgrade. The reinforcement usually yields in the vicinity of the joint if shrinkage and temperature drop forces are present.

In structures, hinge joints are often referred to as articulated joints and are arranged as shown in Fig. 4-11c. Such joints are arranged to preclude a moment transfer between two elements, mostly between the foundation and column above. It is usually achieved when the section is substantially reduced by blocking out notches which are filled with an elastic, easily compressible material. The reinforcement through the joint can consist of either parallel bars or the bars can cross each other in an X-shape (Mesnager hinge).

In both cases a sufficient amount of confinement reinforcement (usually spirals) is necessary to increase the strain and strength capacity of the concrete in the hinge. Obviously, the hinge section must be sufficient to carry all the vertical loads and the shear forces.

4.13.3 Sliding Joints

Sliding joints may be required where one unit of a structure must move perpendicularly to another unit. The joints are usually made with a layer of a bond breaking material such as a bituminous compound, paper, neoprene, felt, or other materials that facilitate sliding. Figure 4-12 shows a sliding joint utilizing a neoprene pad. Occasionally steel plates are

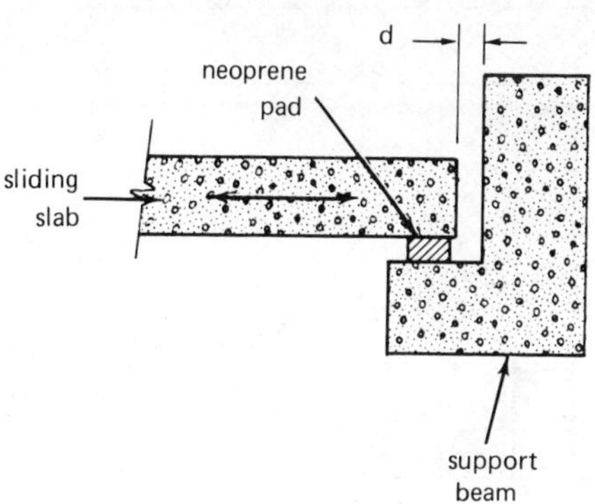

Fig. 4-12 Sliding joint on a beam support.

embedded in both sliding surfaces. In such cases a lubricant is used between the plates to prevent the plates from "freezing" together. Sliding joints permit creep, shrinkage and temperature drop shortening of elements without buildup of stresses. A distance, d, should be left (Fig. 4-12) for the joint to accommodate elongation of the element due to temperature rise thus acting like an expansion joint. With neoprene as a connecting material, this joint can also facilitate rotation due to deflection of the connected element without significantly shifting the support reaction towards the edge. The required thickness of the neoprene depends on the amount of sliding to be provided for the joint and the durometer of the neoprene. Particular attention should be paid to the sealing of such joints where watertightness is required.

4.13.4 Slip Joints Between Masonry Walls and Concrete

Concrete slabs and foundations have considerably different moisture and thermal movements than do the masonry walls with which they must work. Slabs and foundations also are usually under different states of stress than are the walls due to different temperature and humidity environments. Therefore, it is important to break bond and provide a slip joint between these elements by positive means. With horizontal layers of material facilitating sliding between slabs and walls, each element will be able to move somewhat independently while still providing the necessary support to the other. Among materials that should be used as slip joints in structures are two sheets of building paper bearing pads of polytetrafluoroethylene and two layers of galvanized steel coated with grease.

Horizontal cracks have been observed in numerous loadbearing masonry walls supporting reinforced concrete slabs. Such cracking is generally attributed to thermal expansion, curling or horizontal shrinkage of the slab. If the movement of the slab pulls the top of the wall sideways, the wall usually cracks several courses below the underside of the slab. A typical method of breaking bond between the walls and slabs to prevent horizontal cracks is shown in Fig. 4-13.

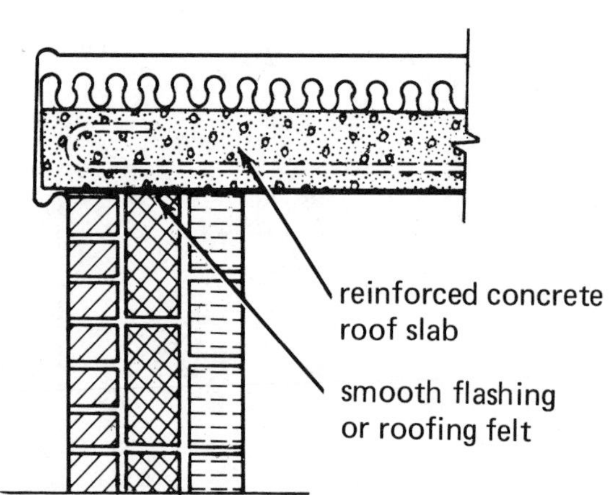

Fig. 4-13 Slip joint between masonry walls and concrete roof slab.

4.14 JOINTS FOR SLABS ON GRADE IN BUILDINGS

4.14.1 Types of Joints

There are three types of joints that will provide crack control in slabs on grade. They are defined by the character and direction of the movement that will occur in the joint:

1. Control joints which allow differential movement only in the plane of the floor.
2. Isolation joints which allow differential movement in all directions.
3. Construction joints which allow no movement in the completed floor. As this is difficult to achieve, construction joints should be made to act as control joints.

Figure 4-14 shows a plan of a slab on ground in which all the above joints are indicated.

4.14.2 Control Joints

Control joints allow horizontal movement of the adjoining slabs, but do not allow differential vertical movement. This means that vertical loads on one side of the joint are transferred to the adjacent slab across the joint usually through aggregate interlock. Control joints are installed to allow for contraction caused by drying shrinkage, and temperature drop. If no control joints were used, random cracking in the floor would occur if the shrinkage and temperature drop tensile stresses within the concrete exceed the concrete tensile strength.

Joint spacing to control drying and thermal contraction should be from 15 to 25 ft in unreinforced or lightly reinforced concrete floors. Under certain favorable conditions joints up to 30 ft, and more, have performed well without cracking. It is usual practice to place control joints on the column lines and between column lines as shown in Fig. 4-14. Variation in spacing is the result of differences in

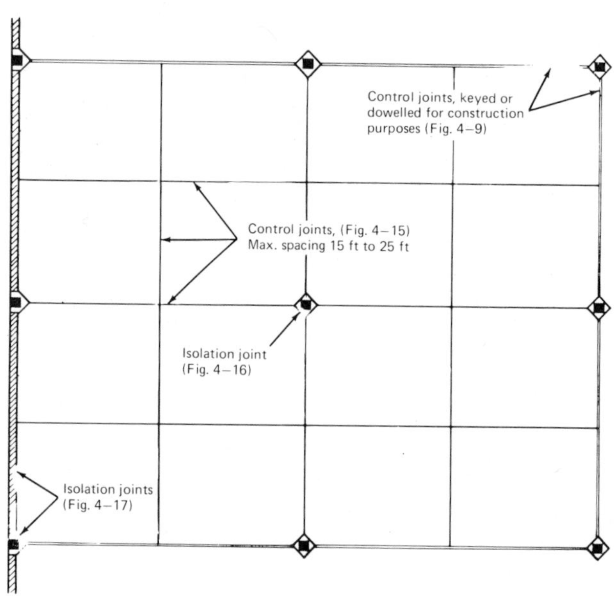

Fig. 4-14 Plan for a typical floor on ground. The double line around the perimeter indicates a key or dowelled joint.

local conditions such as the concrete itself, climate, construction practices, and the type of soil. A thin slab in a dry environment, placed over a dry soil, will experience much more shrinkage, and therefore, will require a much closer spacing of joints (even less than 10 ft) than a thick slab in an environment of high humidity, placed over wet soil. Also, the joint spacing should be chosen so that the panels are approximately square. Panels with excessive length to width ratios (more than $1\frac{1}{2}$ to 1) are likely to crack.

Occasionally, joint spacing at much shorter distances than suggested above may be necessary to satisfy the particular geometry of a specific job. Joints may be necessary at abrupt changes in slab thickness. Reentrant corners are particularly sensitive to cracking and require either joints or reinforced bars placed diagonally. The importance of proper joint locations cannot be left to chance, and therefore, to insure adequate follow through on the job, joints should be planned in advance and detailed on the plans.

Control joints are made by purposely creating a vertical plane of weakness in the slab. The cracking then occurs at this weakened plan rather than at random locations in a slab. All control joints should be continuous, not staggered or offset. One of the most economical ways to make a control joint is by sawing a slot in the top of the finished slab, as shown in Fig. 4-15. When a jagged crack forms then under the joint, it allows vertical load to be transferred to the adjoining slab without differential vertical movement. Because the joint is made after the concrete is placed and finished, control joints do not interfere with concreting operations. The saw cut should be made as early as possible, prior to the buildup of shrinkage stresses, and preferably during rising temperature. Therefore, it is generally done in the early afternoon for morning construction, or the following morning for concrete that was placed in the afternoon. The width of the cut should not exceed $^{3}/_{16}$ in. and the depth should be $^{1}/_{5}$ to $^{1}/_{4}$ of the slab depth, but not less than the maximum size of the aggregate. Immediately after sawing, the joint should be flushed with water or air under pressure to remove the sawing residue. It is good practice to seal the joints to avoid infiltration of foreign material which may cause spalling at the joints. If hard-wheeled traffic is expected, the joint should be sealed with lead.

Another method of forming a control joint is creating a joint completely through the slab. Such joints are made by following a checkerboard pattern. In checkerboard bay placing, concrete is placed in alternate squares as in the squares of one color on a checkerboard. In this way partial shrinkage of half the squares occurs prior to placement of the alternate ones. The resulting joints between the previously placed concrete and the freshly placed concrete could be bonded as in a construction joint and later sawed to form the control joint. However, it is usually simpler to use the nonbonded keyed control joint with or without dowels as shown in Fig. 4-4b. Breaking the bond between the new and previously placed slabs by spraying or painting with a curing compound, asphaltic emulsion, or form oil allows for the freedom of horizontal movement necessary for crack control.

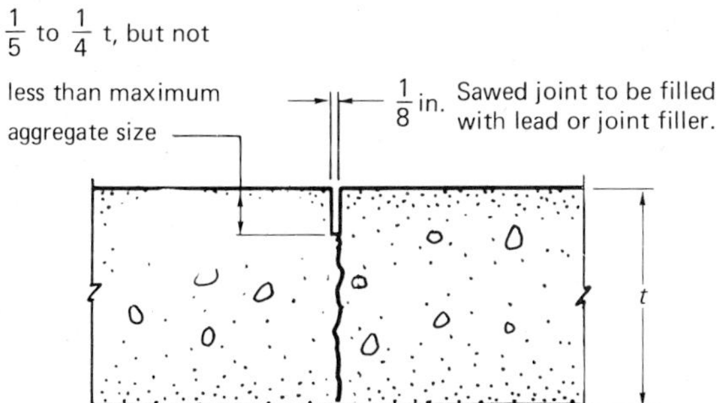

Fig. 4-15 Control joint in slabs on ground.

The key provides the same type of vertical load transfer achieved by the aggregate interlock in the sawed control joint. Experience has shown that the tongue and groove need not be as deep as given in Fig. 4-4b.

If wire mesh or reinforcing steel is used in the slab, it should be reduced or eliminated at a control joint, regardless of the joint construction method. Wire mesh is best reduced by cutting out alternate wires where the joint will be. If reinforcing bars are used, half the number of bars are eliminated across the joint. Reinforcing steel is sometimes used to increase the distance between joints. However, the greater the joint spacing the more the joint will open. Where joints open wide, jolting by vehicles may result and increased joint maintenance will be required. The easiest joint to maintain is one that is narrow and straight.

Welded wire fabric is often placed in the slab to control cracking. However, unless the steel is prestressed, the cracks will not be reduced because there is no stress in the steel to counter the tension in the slab until after the cracks occur. The steel fabric may prevent shrinkage cracks opening too wide, but the amount of steel normally used has little, if any, effect on cracking.

4.14.3 Isolation Joints

Isolation joints separate or isolate concrete slabs from columns, footings or walls. In addition to horizontal movements of the slab caused by shrinkage, they also permit vertical movement that occurs due to differences in unit soil pressure under floors, walls, columns and machinery footings. They should also be used at other points of restraint such as drain pipes, sumps or stairways. Because the isolation joint is used to allow freedom of movement, there should be no connection across the joint by reinforcement, keyways, or bond.

Interior and exterior columns should be boxed out as shown in Fig. 4-16. Circular fiberboard forms can be used in place of the square wood forms. The rectangular form of the boxout is placed so that its corners point at the control joints along the column lines making it easier to saw up to the column. The isolation joint eliminates cracks radiating from the corners.

An isolation joint between a floor and the wall may be made by attaching an asphalt-impregnated sheet not more than $^{1}/_{4}$ in. thick to the wall prior to placing the concrete as shown in Fig. 4-17. The joint shown in Figure 4-17b is not recommended unless the unit load on the ground under the footing is greater than under the slab and special care is taken to consolidate the fill for some distance from the wall.

4.14.4 Construction Joints

Construction joints are temporary stopping places in concreting. They are necessary in large projects because there is a limit to how much concrete can be placed and finished in one work shift.

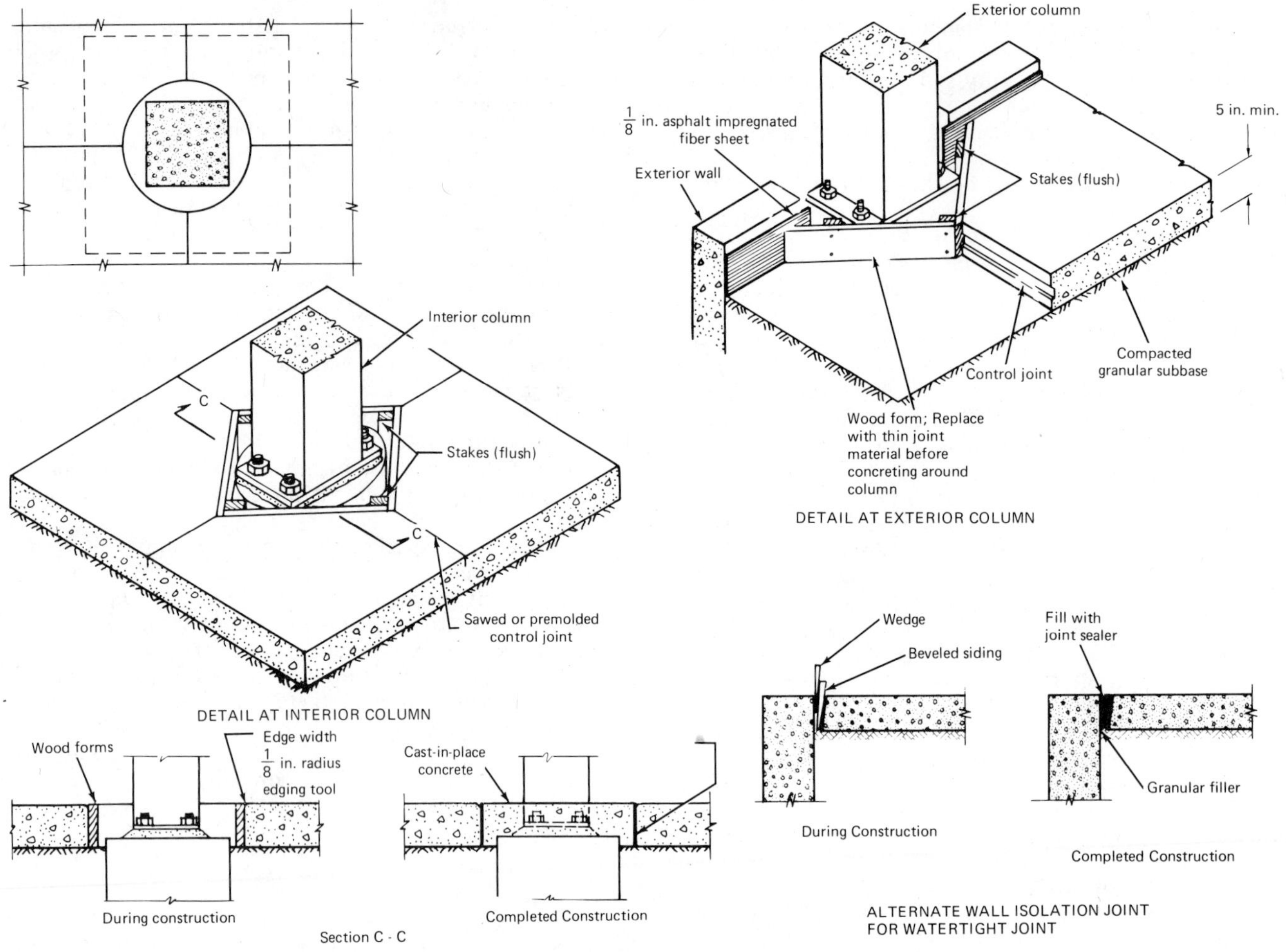

Fig. 4-16 Isolation joints at interior and exterior columns.

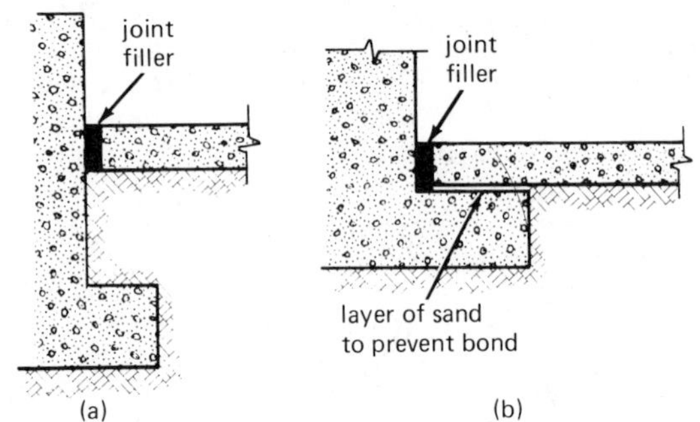

Fig. 4-17 Isolation joints at exterior wall. (a) Correct, and (b) incorrect.

A true construction joint should constitute neither a plane of weakness nor an interruption in the homogeneity of the concrete; therefore, every effort must be made to insure bonding of the newly placed concrete to that previously placed. To assure a bond, the old concrete should be thoroughly cleaned and dampened before the new concrete is placed. To avoid the possibility of breaking the bond after the floor is in use, tie bars are frequently used to assist in vertical load transfer across the joint. Tie bars have the added advantage of also being able to resist shearing stresses along the joint caused by differential shrinkage of the old and new concrete. However, since it is difficult to achieve monolithic joints in floors on grade, construction joints should be spaced and made to act as control joints.

5

Footings

FRITZ KRAMRISCH, D.Eng.*

5.1 GENERAL

The purpose of a footing is to transfer safely to the ground the dead load of the superstructure (weight), and all other external forces acting upon it. The latter ones consist not only of general live load (as in dwellings, warehouses, industrial buildings, etc.) or fill (as in silos, bunkers, tank supporting structures, and similar), but also of the ground effects of various lateral forces such as wind, blast, or earthquake. In case any of these forces are of dynamic character, an allowance for impact should be included in the estimate of their magnitude unless a more exact evaluation is required. Footings also include foundation elements that are designed to resist uplift and overturning, and comprise in general all structural elements that will provide a stable base for the superstructure and safe transfer of all applied forces down to the ground.

The ways and means by which the transfer of these forces into the subsoil can be obtained are manyfold. They depend primarily on the type and magnitude of the loading, on the stiffness and structural behaviour of the superstructure, on the bearing capacity of the soil in general and on that of a certain stratum in particular, and on the depth below ground surface where this soil stratum is encountered, and on the depth of the groundwater level below grade. The type of foundation is also influenced, though to a lesser degree, by the geographical location and climatic condition of the site; by the time of the year when the construction is to be executed; by the speed of construction that is desired; by the availability of material, equipment and manpower; by the prevailing economical conditions; and by many other circumstances. The importance of these apparently secondary considerations can grow with the magnitude of the job and may, in certain cases, influence the successful outcome of an entire project. It is, therefore, not enough to approach the selection of the type of foundation with a cool, calm engineering mind; very often, experienced judgment that can envision advantages and foresee difficulties deserves serious consideration.

The type and magnitude of the loading will usually be furnished by the engineer designing the superstructure. It is up to the foundation engineer to collect all information regarding the purpose of the superstructure, the materials that will be used in its construction, its sensitivity to settlements in general and to differential settlements in particular, and all other pertinent information that may influence the successful selection and execution of the foundation design.

The evaluation of the maximum bearing capacity of the soil is to be made on the basis of soil mechanical considerations. The engineer designing the foundation should select the soil stratum that is most suitable for the support of the

*Chief Civil Engineer, Albert Kahn Associates, Inc., Architects and Engineers, Detroit, Michigan.

superstructure; he must assume the appropriate safety factor to arrive at the allowable bearing pressure; and finally, decide on the most economical type of foundation to be used. For this reason it is essential for a foundation engineer to possess a good knowledge of the problems that are involved in the design and behaviour of the superstructure, a certain familiarity with the basic principles of soil mechanics, and a good understanding of the interaction between both.

A detailed treatment of above topics does not fall within the scope of this handbook; however, a short discussion of the basic considerations affecting the evaluation and distribution of the bearing pressures under footing bases is given below.

5.2 EVALUATION OF BEARING PRESSURES AT FOOTING BASES

5.2.1 General Principles

The distribution of the bearing pressures under a concentrically loaded, infinitely stiff footing, with frictionless base, resting on an ideal, cohesionless or cohesive subsoil,[1,2] is generally known, and shown in Fig. 5-1. Under ordinary conditions few soils will exhibit such a behavior; no footing could be considered to be infinitely stiff. The distribution of the bearing pressure under somewhat flexible footings and ordinary soil conditions will be similar to those shown in Fig. 5-2; or it may assume any intermediate distribution. The assumption of a uniform bearing pressure over the entire base area of a concentrically loaded footing, as shown in Fig. 5-3, seems to be justified, therefore, for reasons of simplicity, and is common design practice. This assumption not only represents an average condition, but is usually on the safe side because most of the common soil types will produce bearing pressure distributions similar to that shown in Fig. 5-2a. The foundation designer, however, shall keep in mind that the assumption of a uniform bearing pressure distribution was primarily made for reasons of simplicity and may, in special cases, require adjustment.

Any footing that is held in static equilibrium solely by bearing pressures acting against its base has to satisfy the following basic requirements regardless of whether it is an isolated or a combined footing:

1. The resultant of all bearing pressures, acting against the footing base (reaction), must be of equal intensity and opposite direction as the resultant of all loads and/or vertical effects due to moments and lateral forces, acting on the footing element (action).

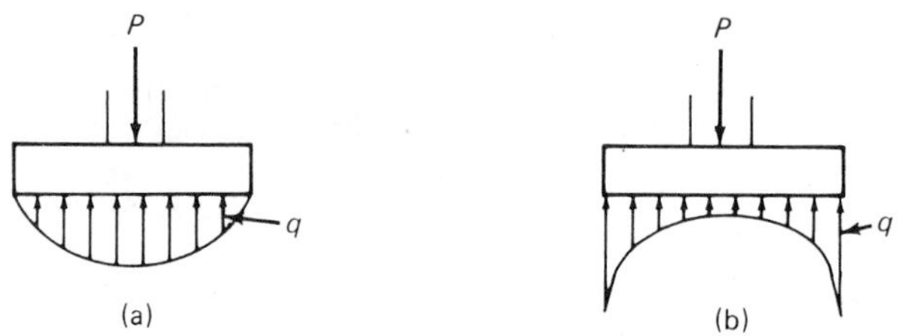

Fig. 5-1 Bearing pressure distribution for a stiff footing with frictionless base on ideal soil. (a) On cohesionless soil (sand); and (b) on cohesive soil (clay).

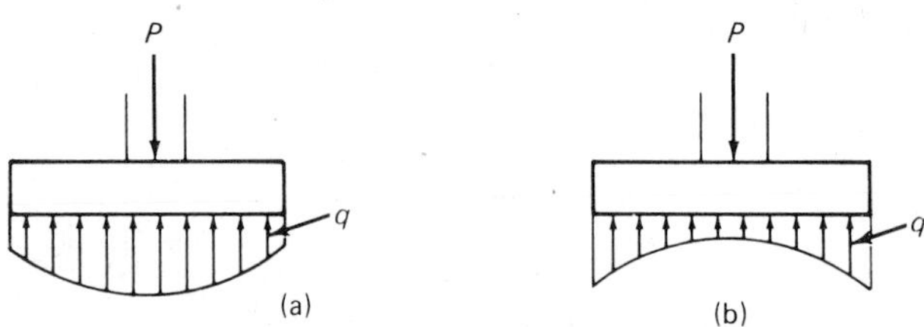

Fig. 5-2 Bearing pressure distribution for a flexible footing on ordinary soil. (a) On granular soil; and (b) on clayey soil.

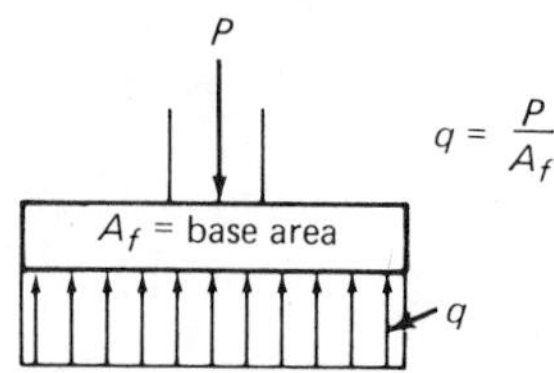

Fig. 5-3 Simplified bearing pressure distribution (commonly used).

2. The location where the resultant vector of the reaction intersects the footing base must coincide with the location where the resultant vector of the action is applied. Action and reaction are as defined under (1) above.

3. The maximum intensity of the bearing pressures under the most severe combination of service loads must be smaller than, or equal to, the maximum bearing pressure allowed for this kind of loading and type of soil, as determined by principles of soil mechanics.

4. The resultant vector of the least favorable combination of vertical loads, horizontal shears, and bending moments that may occur under service load conditions, including wind or earthquake, must intersect the footing base within a maximum eccentricity that will provide safety against overturning.

The method most commonly used for the design of footings and related elements for ordinary building construction, is the one where static equilibrium is obtained by bearing pressures against the footing base only. This method is also the standard method that has been included in the "Building Code Requirements for Reinforced Concrete" ACI 318-71.

For zero eccentricities, the bearing pressures will be uniformly distributed over the entire base area of the footing as shown in Fig. 5-3 and will have the intensity of $q = P/A_F$.

If the footing shall restrain the column base, i.e., if a bending moment has to be resisted by the subsoil alongside with a concentric load, or if the column load is applied outside of the centroid of the base area of the footing, the bearing pressure distribution will vary depending on the magnitude of the eccentricity and its relationship to the kern distance c_k. The kern distance can generally be evaluated as shown in Fig. 5-4.

When the eccentricity is equal to, or smaller than, the kern distance c_k, the extreme (maximum or minimum) bearing pressures $q_{\substack{max\\min}}$ can be found by superposing the flexural bearing pressures over the axial bearing pressures, see Fig. 5-5a.

When the eccentricity becomes greater than the kern distance superposition cannot be applied anymore, because it would result in tensile stresses between soil and footing near the lifted edge of the base. Equilibrium can, however, be attained by resisting the load resultant by a bearing pressure resultant of equal magnitude and location. In this case the extreme bearing pressures at the edge of the base can be evaluated as shown in Fig. 5-5b. The maximum edge pressure q_{max} must, under all conditions, be smaller or equal than the maximum allowable soil pressure, q_a.

This condition applies until the excentricity, e, of the load, P, reaches the edge of the footing base. Any greater eccentricity will result in overturning. Such a condition, however, can only occur on rock or on very hard, stiff soils. For most practical cases, edge-yielding can make a footing

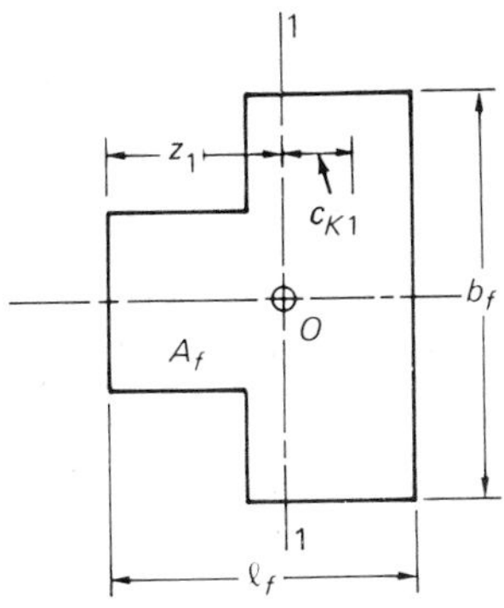

Fig. 5-4 Kern distance. (a) $c_{k1} = I_{F1}/A_{F1}z_1$; (b) I_{F1} = moment of inertia of footing base about neutral axis *I-I*; (c) z_1 = distance of extreme fiber at opposite side of desired kern distance; (d) for strips, $c_k = I_f/6$.

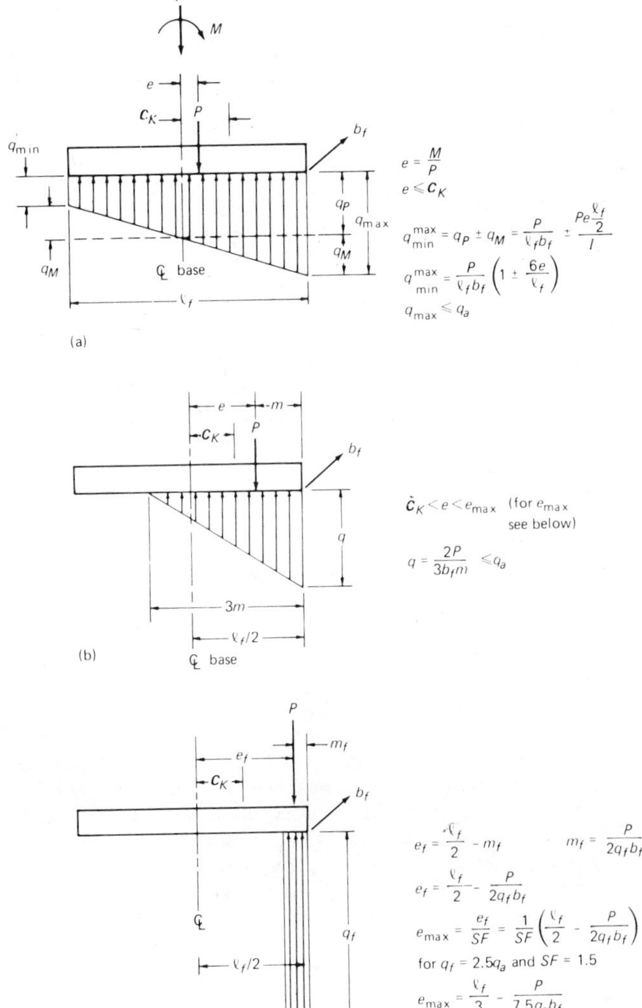

Fig. 5-5 Bearing pressure distribution under eccentric loading.

unusable and produce a condition that is equivalent to overturning. Edge-yielding will occur when the extreme bearing pressure at the pressed edge will cause failure in the bearing capacity of the subsoil. The eccentricity causing this condition will, therefore, limit the maximum useful eccentricity. Unless actual test results are available, the failure condition in the bearing capacity of the soil, q_f, can be assumed with about 2.5 times the allowable bearing capacity, q_a; the minimum safety factor against overturning is usually specified as 1.50; although somewhat greater safety factors are sometimes desirable. Introducing these requirements, we arrive in Fig. 5-5c at a maximum eccentricity, $e_{\max}$, that can safely be utilized. How far the design engineer will take advantage of this condition will depend on his judgment of the soil and on the sensitivity of the superstructure to tolerate lateral tilting that may occur if a loading, causing such an eccentricity, is applied for a longer period.[5-3]

Moments occurring alongside with concentric loads, may be uniaxial or biaxial. If they occur in oblique directions it is most practical to have their influence divided into two perpendicular components, each of them parallel to the main axes of the footing mat, and superpose the resulting bearing pressures. Such conditions occur not only with isolated spread footings, but also with strip footings of limited length, as in the case of shear walls and similar.

Combined footings, (i.e., footings supporting more than one column load, such as exterior double-column footings), strip footings supporting spaced column loads, rafts, or mats can be designed as described above, as long as the entire foundation can be considered as infinitely stiff. In this case the resultant of all bearing pressures must be equal to the resultant of all loads, and its location must coincide with the eccentricity of the resultant. This approach is statically correct, but not necessarily close to the actual condition.

In certain cases it may be advisable to consider the footing as a beam on an elastic foundation and utilize the elasticity of the footing mat as well as that of the soil in the evaluation of the bearing pressures. The bearing pressures obtained by this method no longer follow a straight line distribution across the contact area. They show maximum accumulations immediately below, and in the vicinity of, concentrated loads and greatly reduced intensities between them, as shown in Fig. 5-6. Such a pressure distribution can reduce the maximum design moments of a foundation considerably and is therefore in many cases quite economical. This method is, in addition, intriguing in its setup and appealing especially to the mathematically inclined engineer.[5-4]

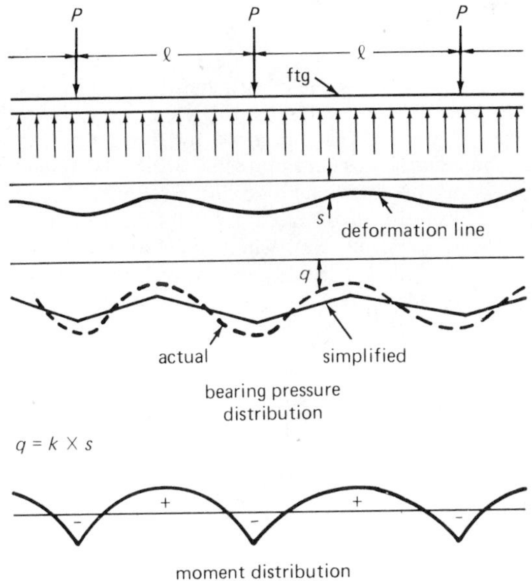

Fig. 5-6 General conditions for a beam on an elastic foundation.

In the analysis of a beam on elastic foundation, the soil is introduced as an elastic medium and the intensity of its reaction (bearing pressure) is assumed to be proportional to the deformation (settlement) of the footing, or soil, under the load.

Soils behave like elastic materials only to a limited degree and only small portions of the settlements can be recovered in case of unloading the superstructure. Settlements of the subsoil occur very seldom immediately or shortly after the load application. Most soils are rather insensitive to sudden load changes and react more to average loadings applied over an extended period. Variable configurations of the line of deformations during this period will cause a variable distribution of corresponding bearing pressures and internal stresses of the footing or foundation mat. In addition, the factor relating the magnitude of the bearing pressure to that of the deformation is not a constant but varies with the magnitude of the settlements.

Some of these characteristics have been overcome by refinements that have been introduced into this method, making it much more complex. If, however, foundations under combined footings are not too stiff and they follow the deformations of the subsoil, the beam on elastic foundation-method will furnish bearing pressure distributions that are closer related to the actual condition and more economical than the straight line distribution of the bearing pressures.

Due to the basic deficiencies of a foundation design executed with the help of elasticity relationships, as described in this chapter, it is advisable that results obtained by this method are not taken as exact solutions but more as a guide to arrive at a reasonable distribution of the bearing pressure; such foundations shall be designed with ample reserve capacity to sustain deviations from the theoretical findings. Under such considerations, the method can provide an excellent tool and a valuable aid in the design of sometimes rather difficult foundation problems.[5-5]

Figure 5-6 shows the basic relationships between loading, deformation and bearing pressure as well as the resulting moment distribution. Formulas giving the bearing pressures and bending moments for a strip designed as a beam on elastic foundation are given later in Fig. 5-22.

So far, bearing pressures against the footing base have been considered as the only means to resist external forces. Static equilibrium of a footing element subjected to lateral forces and/or moments in addition to vertical loads, can also be attained with the help of passive earth pressure, as shown in Fig. 5-7. In this case lateral (passive) pressure of the soil is developed while resisting either lateral movements or tilting tendencies of the footing element. This method is sometimes used in the design of footings supporting tall, light superstructures such as transmission towers, light poles, etc. One of the drawbacks of this method is that foundations are usually surrounded by backfill of questionable lateral resistance; but even where the footing is cast tightly against the original subsoil, passive earth pressure will only develop (unless the soil is very stiff or hard) after a certain movement or tilting of the superstructure has taken place.

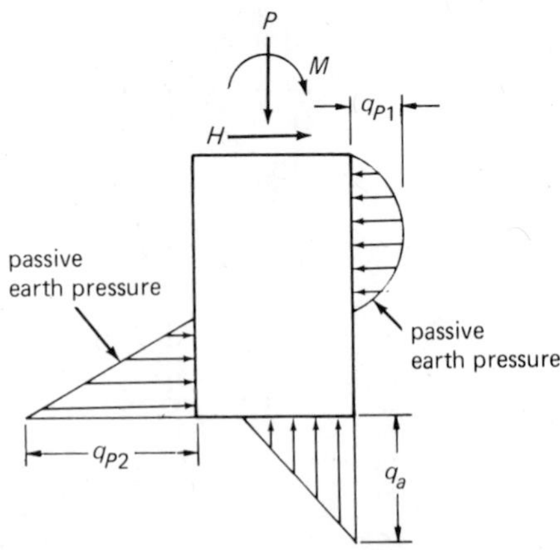

Fig. 5-7 General distribution of base and side pressures of a footing resisting overturning moments, where q_{P_1} and q_{P_2} are passive earth pressures due to loading and q_{P_1} and q_{P_2} are to be $\leqslant$ the maximum allowable passive earth pressure that can be utilized at each depth.

5.2.2 Strip Footings

-1. Concentrically loaded strip footings—The loadings of strip footings considered in the following section consist primarily of continuous line loads such as walls, or closely spaced, concentrated loads arranged in one direction. In many cases, the element causing the load, e.g., the wall itself, may provide sufficient stiffness to permit the assumption of a continuous line load, even if the intensity of the loading slightly varies (Fig. 5-8). Concentrated loads at close spacings sometimes require a stiffening member to transform the effects of the concentrated loads into a continuous loading. For small variations in the load intensity, the stiffening member may be the footing itself; or a foundation wall resting on top of the footings; or a beam formed by composite action of both. In either case the stiffening member has to be designed as an inverted continuous beam, upwardly loaded by a line load made up of the bearing pressure, q, and intermittently supported where the loads are applied, as shown in Fig. 5-9.

-2. Eccentrically loaded strip footings (or strip footings with concentric load and moment at base)—For stability considerations, a free standing wall is equivalent to one that is fully restrained at its base. Any shear and bending moment caused by lateral forces such as wind, fill, earth-pressure, simulated earthquake loadings, etc., as well as all moments and shears caused by eccentrically applied vertical loads acting on projecting piers or brackets, have to be transferred into the subsoil, see Fig. 5-10.

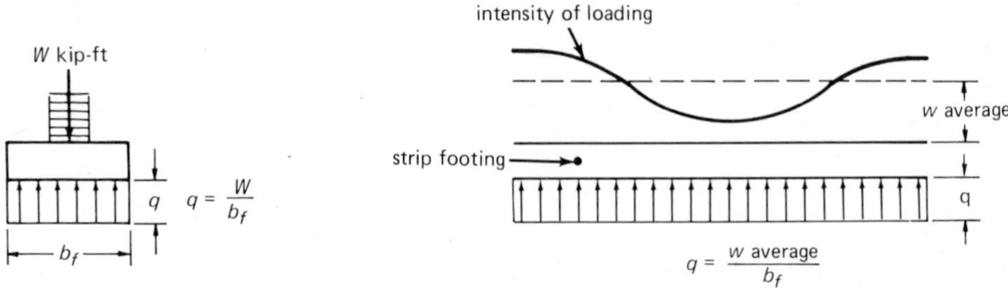

Fig. 5-8 Equalizing strip loadings of variable intensity.

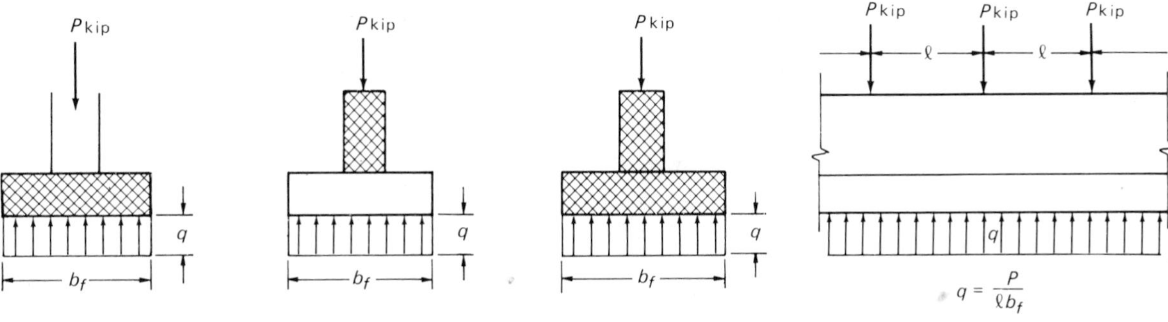

Fig. 5-9 Transforming concentrated loads into equal bearing pressures.

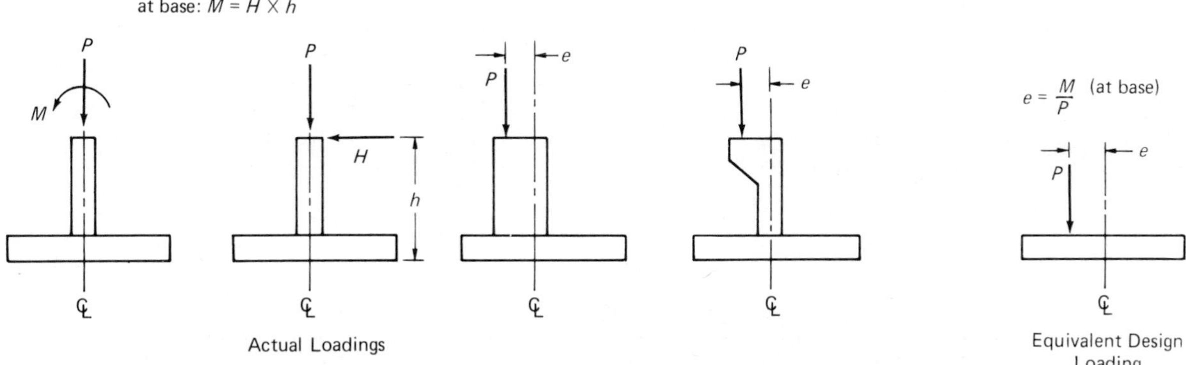

Fig. 5-10 Analogy between moments and eccentric loading conditions.

In cases where walls are not free standing but form a part of a frame work, such as in monolithic floor-wall constructions, the wall may transfer a certain amount of moment into the ground. How successful such a design assumption may be, deserves a brief discussion.

It can be proven by statics that only small, lopsided settlements are needed to reduce the restraining effect at the base of a frame work to such a degree that it may be rendered useless. Such a rotation would produce a condition at the base that is in effect similar to that of a hinge and would increase the moments in the upper frame by the amount lost at the base. Cautious designers assume and design the base of such frameworks as hinges and make the structure this way independent of the deformation of the subsoil.

Bending moments or eccentric loadings caused by wind or earthquake simulating forces may very well be taken by restraint at the base of the footing mat; such forces are of short duration and stable soils usually respond rather elastically to loads of short duration. It is however recommended that bending moments and eccentric forces caused by dead loads, fill, or average service live load conditions should not be transferred into the soil unless the soil is very stiff or well consolidated.

For intermediate conditions, partial restraint in an amount selected by judgement, may be utilized.

5.2.3 Isolated Spread Footings (square and oblong)

-1. Concentrically loaded isolated spread footings—Isolated spread footings are used to resist column loads that are concentric with the centroid of the footing base. Theoretically such footings may have any shape. Square and oblong bases are most common, but round and polygonal shapes are being used under certain conditions. Footings may be solid or without center portion, like a ring. We speak of a concentric loading if the resultant of all acting forces coincides with the centroid of the footing mat or bearing area, regardless of its shape. In this case the bearing pressure distribution to be used for the design is usually assumed to be of uniform intensity over the entire area, see Fig. 5-3.

-2. Eccentrically loaded isolated spread footings or isolated footings with concentric load and moment at base—In case of eccentric loadings caused by a moment at the column base, or by an eccentrically applied load, or by an unsymmetrical footing base, the bearing pressures will deviate from the uniform distribution and gradually vary across the footing base. The distribution of the bearing pressures will follow the rules as described above and shown in Fig. 5-5.[5-3] Fig. 5-11 gives the base area, moment of inertia; and kern distance for various shapes of footing bases[5-6]. For irregular shapes, it can be found as indicated in Fig. 5-4.

Round and polygonal footings, whether solid or ring-shaped, can only have a uniaxial eccentricity under the resultant of such a loading. Square and rectangular footings may have eccentricities in the direction of either one or both main axes, if the eccentricity falls in an oblique direction.

Figures 5-12 to 5-16 give the intensity of the extreme bearing pressure, and the location of the zero pressure, for various shapes of footings under eccentric loading.[5-6, 5-7]

Moments affecting the bearing pressures shall be determined about the elevation of the footing base. For evaluating the bearing pressures by means of Figs. 5-12 to 5-16, moments shall be converted first into eccentric loadings having the magnitude of the resultant loading, P, acting at the eccentricity $e = M_B/P$. In the evaluation of the resultant P it is important to investigate all conditions of loading that may occur simultaneously with the moment under consideration, i.e., the maximum as well as the minimum. In

Shape of Base Area	Base Area A_F	Moment of Inertia I_x	Kern Distance c_k
CIRCLE	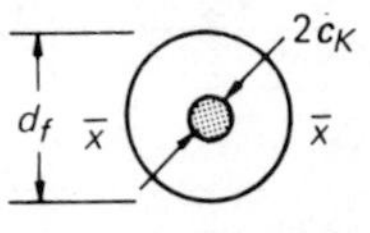 $0.785d_f^2$	$0.049d_f^4$	$0.125d_f$
OCTAGON	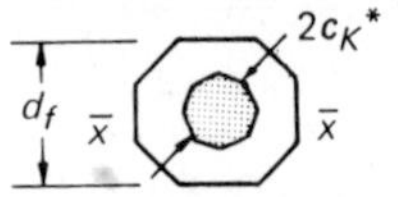 $0.828d_f^2$	$0.055d_f^4$	$0.122d_f$
HEXAGON	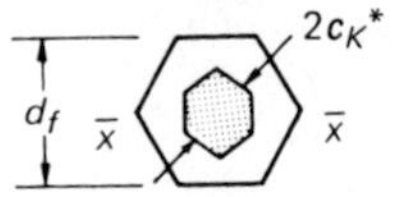 $0.866d_f^2$	$0.060d_f^4$	$0.120d_f$
SQUARE	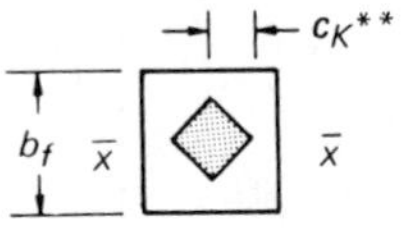 b_f^2	$0.083b_f^4$	$0.167d_f$
RECTANGLE	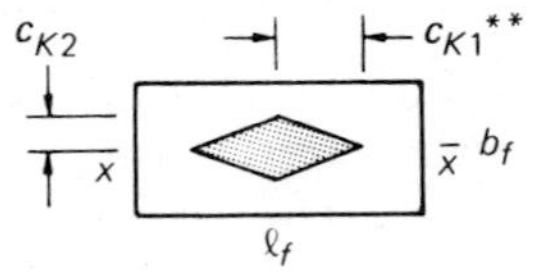 $l_f b_f$	$0.083 l_f b_f^3$	$c_{k1} = 0.167 l_f$ $c_{k2} = 0.167 b_f$
RING	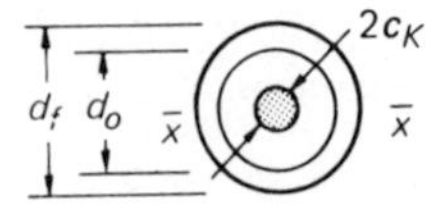 $0.785(d_f^2 - d_0^2)$	$0.049(d_f^4 - d_0^4)$	$\frac{d_f}{8}\left(1 + \frac{d_0^2}{d_f^2}\right)$

Fig. 5-11 Area properties of various cross sections. $^*c_k$ = radius of circle inscribed in polygonal kern area. $^{**}c_k$ = kern distance on main axis. Shaded area in diagrams indicates kern.

Values of C_1 and C_2 for Various Values of e/l_f

e/l_f	C_1	C_2
0.000	–	1.000
0.025	–	1.150
0.050	–	1.300
0.075	–	1.450
0.100	–	1.600
0.125	–	1.750
0.150	–	1.900
0.167	1.000	2.000
0.175	0.975	2.051
0.200	0.900	2.222
0.225	0.825	2.424
0.250	0.750	2.667
0.275	0.675	2.962
0.300	0.600	3.333
0.325	0.525	3.809
0.333	0.500	4.000
0.350	0.450	4.444
0.375	0.375	5.333
0.400	0.300	6.667
0.425	0.225	8.889
0.450	0.150	13.333
0.475	0.075	26.667
0.500	0.000	∞

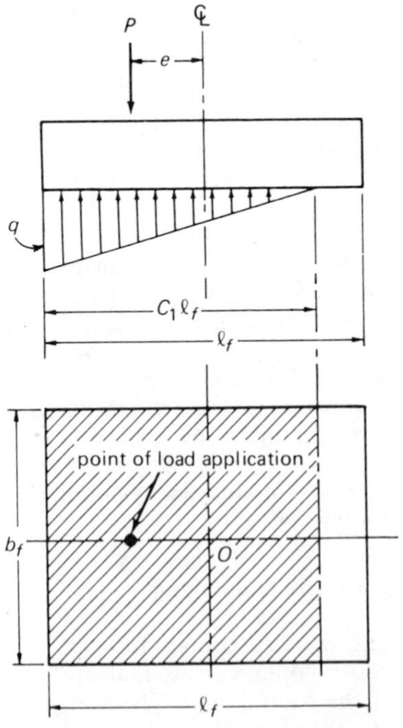

Fig. 5-12 Bearing pressure under square and oblong bases. Notation used includes: $A_F = l_f b_f$; $q_P = P/A_F$; $e = M/P$; $q = C_2 q_p$; and q_P = bearing pressure under concentric loading.

Values of C_1 and C_2 for Various Values of $e/b_f\sqrt{2}$

$e/b_f\sqrt{2}$	C_1	C_2
0.000	–	1.000
0.025	–	1.30
0.050	–	1.60
0.075	–	1.90
0.083	1.000	2.00
0.100	0.915	2.21
0.110	0.870	2.34
0.120	0.832	2.48
0.130	0.796	2.63
0.140	0.766	2.80
0.150	0.736	2.97
0.160	0.707	3.16
0.170	0.680	3.37
0.180	0.654	3.61
0.190	0.628	3.85
0.200	0.604	4.14
0.210	0.582	4.44
0.220	0.561	4.77
0.230	0.540	5.14
0.240	0.521	5.54
0.250	0.500	6.00
0.300	0.400	9.38
0.350	0.300	16.67
0.400	0.200	37.50
0.450	0.100	150.00
0.500	0.000	∞

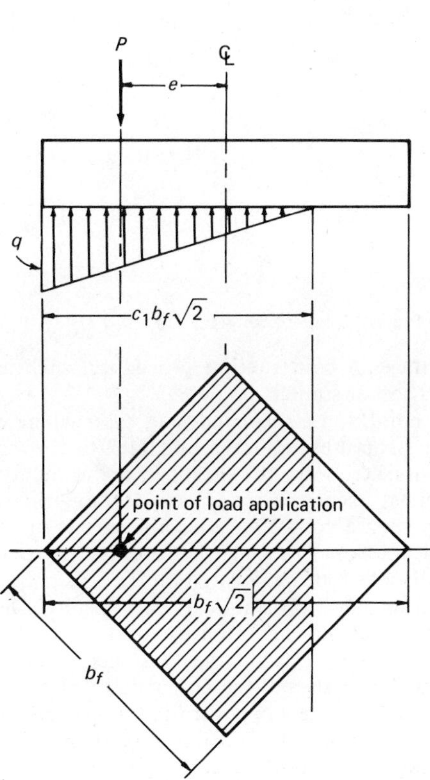

Fig. 5-13 Bearing pressure under square bases (about diagonal axis). Notation same as in Fig. 5-12 except for $A_F = b_f^2$.

Values of C_1 and C_2 for Various Values of e/d_f

e/d_f	C_1	C_2
0.000	–	1.00
0.025	–	1.20
0.050	–	1.40
0.075	–	1.60
0.100	–	1.80
0.125	1.000	2.00
0.150	0.910	2.23
0.175	0.830	2.48
0.200	0.755	2.76
0.225	0.685	3.11
0.250	0.615	3.55
0.275	0.550	4.15
0.294	0.500	4.69
0.300	0.485	4.96
0.325	0.420	6.00
0.350	0.360	7.48
0.375	0.295	9.93
0.400	0.235	13.87
0.425	0.175	21.08
0.450	0.120	38.25
0.475	0.060	96.10
0.500	0.000	∞

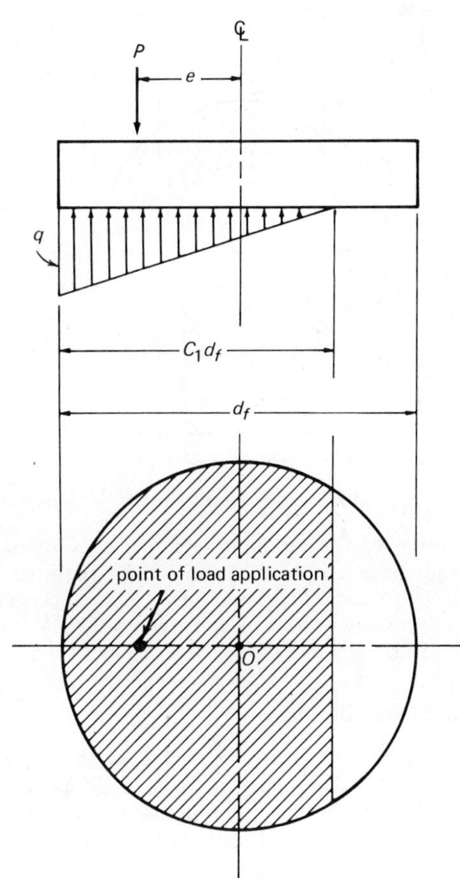

Fig. 5-14 Bearing pressure under circular and polygonal bases. Notation same as in Fig. 5-12 except for $A_F = d_f^2\pi/4$.

NOTE: For hexagonal bases use $d_f = 1.077\ d_{fo}$.
For octagonal bases use $d_f = 1.041\ d_{fo}$, where d_{fo} = diameter of inscribed circle.

Values of C_1 and C_2 for Various Values of e/d_f and d_o/d_f

	C_1 – Values d_o/d_f						
e/d_f	0.0	0.5	0.6	0.7	0.8	0.9	1.0
0.125	2.00	–	–	–	–	–	–
0.150	1.82	–	–	–	–	–	–
0.175	1.66	1.89	1.98	–	–	–	–
0.200	1.51	1.75	1.84	1.93	–	–	–
0.225	1.37	1.61	1.71	1.81	1.90	–	–
0.250	1.23	1.46	1.56	1.66	1.78	1.89	2.00
0.275	1.10	1.29	1.39	1.50	1.62	1.74	1.87
0.300	0.97	1.12	1.21	1.32	1.45	1.58	1.71
0.325	0.84	0.94	1.02	1.13	1.25	1.40	1.54
0.350	0.72	0.75	0.82	0.93	1.05	1.20	1.35
0.375	0.59	0.60	0.64	0.72	0.85	0.99	1.15
0.400	0.47	0.47	0.48	0.52	0.61	0.77	0.94
0.425	0.35	0.35	0.35	0.36	0.42	0.55	0.72
0.450	0.24	0.24	0.24	0.24	0.24	0.32	0.49
0.475	0.12	0.12	0.12	0.12	0.12	0.12	0.25
0.500	0.00	0.00	0.00	0.00	0.00	0.00	0.00

	C_2 – Values d_o/d_f						
e/d_f	0.0	0.5	0.6	0.7	0.8	0.9	1.0
0.000	1.00	1.00	1.00	1.00	1.00	1.00	1.00
0.025	1.20	1.16	1.15	1.13	1.12	1.11	1.10
0.050	1.40	1.32	1.29	1.27	1.24	1.22	1.20
0.075	1.60	1.48	1.44	1.40	1.37	1.33	1.30
0.100	1.80	1.64	1.59	1.54	1.49	1.44	1.40
0.125	2.00	1.80	1.73	1.67	1.61	1.55	1.50
0.150	2.23	1.96	1.88	1.81	1.73	1.66	1.60
0.175	2.48	2.12	2.04	1.94	1.85	1.77	1.70
0.200	2.76	2.29	2.20	2.07	1.98	1.88	1.80
0.225	3.11	2.51	2.39	2.23	2.10	1.99	1.90
0.250	3.55	2.80	2.61	2.42	2.26	2.10	2.00
0.275	4.15	3.14	2.89	2.67	2.42	2.26	2.17
0.300	4.96	3.58	3.24	2.92	2.64	2.42	2.26
0.325	6.00	4.34	3.80	3.30	2.92	2.64	2.42
0.350	7.48	5.40	4.65	3.86	3.33	2.95	2.64
0.375	9.93	7.26	5.97	4.81	3.93	3.33	2.89
0.400	13.87	10.05	8.80	6.53	4.93	3.96	3.27
0.425	21.08	15.55	13.32	10.43	7.16	4.50	3.77
0.450	38.25	30.80	25.80	19.85	14.60	7.13	4.71
0.475	96.10	72.20	62.20	50.20	34.60	19.80	6.72
0.500	∞	∞	∞	∞	∞	∞	∞

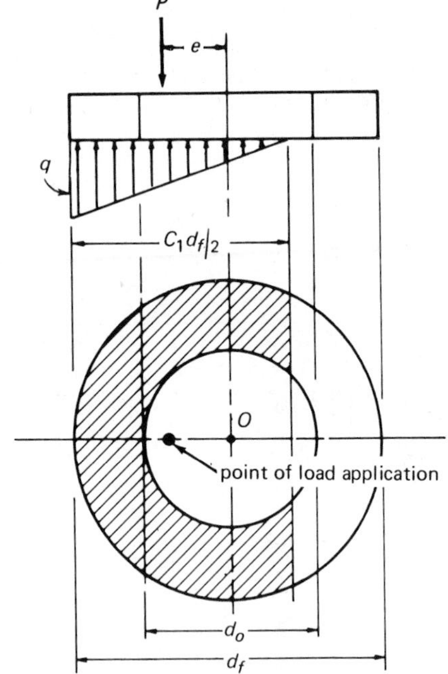

Fig. 5-15 Bearing pressure under ring-shaped bases. Notation same as in Fig. 5-12 except for $A_F = (d_f^2 - d_o^2)\,\pi/4$.

the calculation of the minimum loading it is advisable to reduce dead loads arbitrarily, similarly as required in the *ACI Building Code 318-71*, section 9.3, to ascertain that the bearing pressure and/or eccentricity resulting from this loading condition is a maximum. See also section 5.3.1.

5.2.4 Combined Footings

Independent, isolated footings for each column load, usually provide the most economical method to design a foundation. Support of two or more column loads on one combined footing should be attempted only when required by one of the following conditions:

a. Proximity of the building line or any other space limitation adjacent to a building column.
b. Overlapping of adjacent isolated column footings.
c. Insufficient bearing capacity of subsoil requiring large bearing areas.
d. Sensitivity of the superstructure to differential settlements.
e. Advantage in construction procedure, such as trench excavation or similar.

The first condition also includes all cases where elements of existing structures, underground facilities, excavations, embankments, etc., limit the extent of the foundation to be constructed; it is a common case for exterior building columns. If such a condition does not permit the design of a symmetrical column footing, the footing can be combined with the adjacent footing, located on the axis perpendicular to the limiting line, to balance its eccentricity. Depending on the difference in the intensity of the two combined column loads, as well as on the existence of any simultaneous moments or shears, the eccentricity of the resultant load and its magnitude will govern the shape and size of the required base area for the footing. Any shape of base area can be used for the footing as long as it will satisfy the basic requirements discussed in section 5.2.1. Rectangular, trapezoidal, and T-shaped footings are most commonly used. Figures 5-17 through 19 provide the necessary formulae to determine the required areas and geometrical

(continued on p. 121)

Values of C_3 for Various Values of e_l/l_f; e_b/b_f

e_l/l_f	e_b/b_f 0.50	0.40	0.30	0.20	0.18	0.16	0.14	0.12	0.10	0.08	0.06	0.04	0.02	0.00
0.50	∞	∞	∞	∞	∞	∞	∞	∞	∞	∞	∞	∞	∞	∞
0.40	∞	37.5	18.8	12.5	11.6	10.9	10.2	9.6	9.0	8.5	8.0	7.5	7.1	6.7
0.30	∞	18.8	9.4	6.2	5.8	5.4	5.1	4.8	4.5	4.2	4.0	3.8	3.5	3.3
0.20	∞	12.5	6.2	4.1	3.9	3.6	3.4	3.2	3.0	2.8	2.7	2.5	2.4	2.2
0.18	∞	11.6	5.8	3.9	3.6	3.4	3.2	3.0	2.8	2.6	2.5	2.3	2.2	2.1
0.16	∞	10.9	5.4	3.6	3.4	3.2	3.0	2.8	2.6	2.5	2.3	2.2	2.1	2.0
0.14	∞	10.2	5.1	3.4	3.2	3.0	2.8	2.6	2.5	2.3	2.2	2.1	2.0	1.8
0.12	∞	9.6	4.8	3.2	3.0	2.8	2.6	2.5	2.3	2.2	2.1	2.0	1.8	1.7
0.10	∞	9.0	4.5	3.0	2.8	2.6	2.5	2.3	2.2	2.1	2.0	1.8	1.7	1.6
0.08	∞	8.5	4.2	2.8	2.6	2.5	2.3	2.2	2.1	2.0	1.8	1.7	1.6	1.5
0.06	∞	8.0	4.0	2.7	2.5	2.3	2.2	2.1	2.0	1.8	1.7	1.6	1.5	1.4
0.04	∞	7.5	3.8	2.5	2.3	2.2	2.1	2.0	1.8	1.7	1.6	1.5	1.4	1.2
0.02	∞	7.1	3.5	2.4	2.2	2.1	2.0	1.8	1.7	1.6	1.5	1.4	1.2	1.1
0.00	∞	6.7	3.3	2.2	2.1	2.0	1.8	1.7	1.6	1.5	1.4	1.2	1.1	1.0

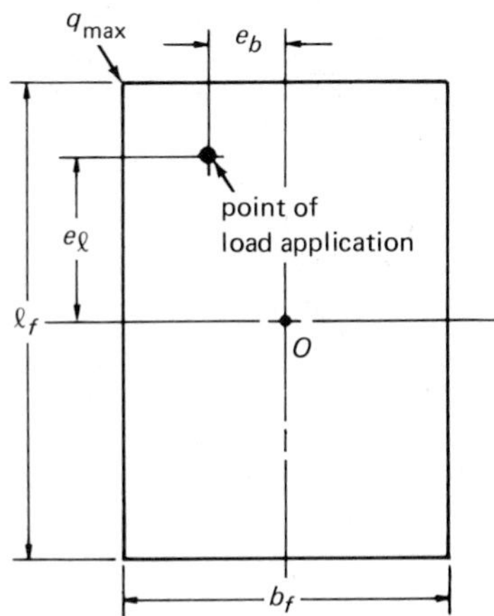

Fig. 5-16 Maximum corner pressure for biaxial eccentricities. Notation same as Fig. 5-12 except for $q_{max} = C_3 q_p$.

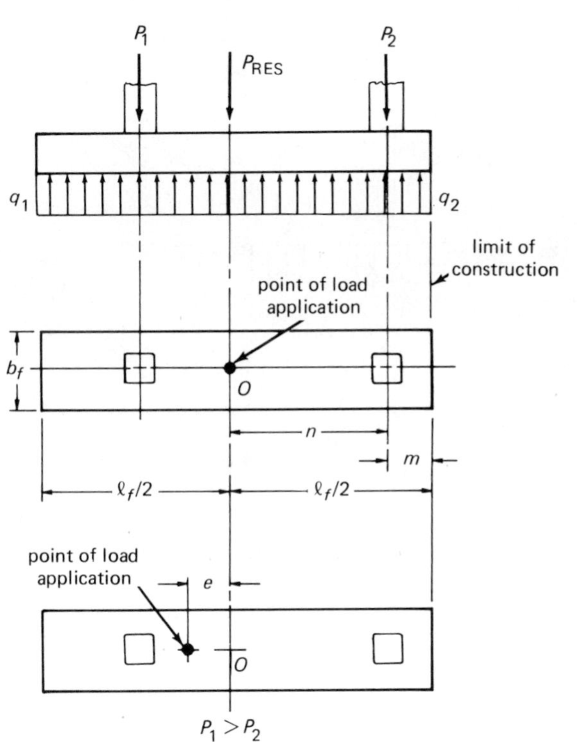

1) For uniform bearing pressure

$$l_f = 2(m+n), \quad P_{RES} = P_1 + P_2$$

$$b_f = \frac{P_{RES}}{q\,l_f}, \qquad q_1 = q_2 = q$$

2) For eccentric loading e on either side of O; q_1 = bearing pressure at loaded edge

for $e \leqslant \dfrac{l_f}{6}$

$$\frac{q_1}{q_2} = \frac{P_{RES}}{b_f\,l_f}\left(1 \pm \frac{6e}{l_f}\right)$$

for $e > \dfrac{l_f}{6} < e_{max}$

$$q_1 = \frac{2P_{RES}}{3\left(\dfrac{l_f}{2} - e\right) b_f}$$

$q = 0$ at a distance $3\left(\dfrac{l_f}{2} - e\right)$ from loaded edge (for e_{max} see Fig. 5-5c).

Fig. 5-17 Strap footing.

For uniform bearing pressure (q) (general relationship)

$$\frac{b_f}{b_{f1}} = \frac{3(n+m) - l_f}{2l_f - 3(n+m)}$$

$$(b_f + b_{f1}) = \frac{2P_{RES}}{ql_f}$$

$$C_1 = \frac{l_f(2b_f + b_{f1})}{3(b_f + b_{f1})}; \quad C_2 = \frac{l_f(b_f + 2b_{f1})}{3(b_f + b_{f1})}$$

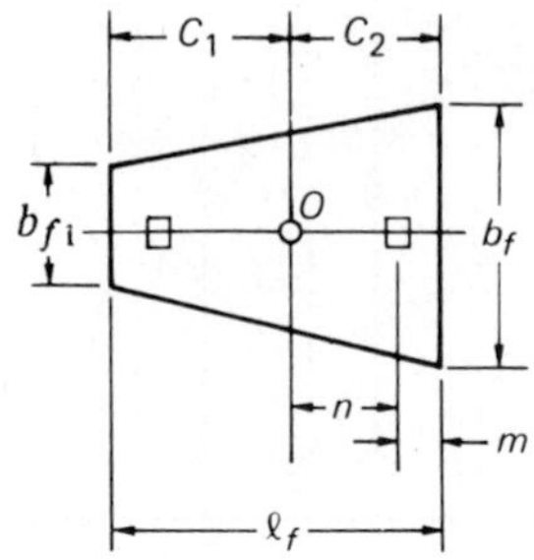

for uniform bearing pressure of known magnitude (q).*

for $P_1 > P_2$

$$b_f = \frac{6P_{RES}}{l_f^2 q}\left(n_1 + m - \frac{l_f}{3}\right)$$

$$b_{f1} = \frac{2P_{RES}}{ql_f} - b_f$$

for $P_1 < P_2$

$$b_f = \frac{6P_{RES}}{l_f^2 q}\left[\frac{2}{3}l_f - n_2 - m\right]$$

$$b_{f1} = \frac{2P_{RES}}{ql_f} - b_f$$

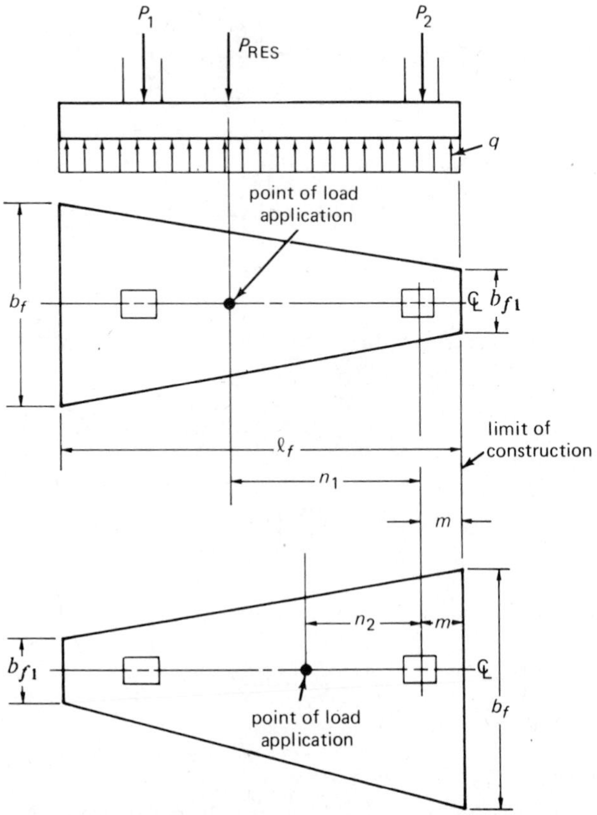

Fig. 5-18 Trapezoidal footing. *For non-uniform bearing pressure under eccentric loadings see: A. Zweig, "Eccentrically loaded trapezoidal or round footings," *Proceedings ASCE*, ST-1, 161-168, Feb. 1966.

$$P_1 < P_2$$

for uniform bearing pressure $q_1 = q_2 = q$

$$b_1 = \frac{P_{RES}}{q}\left[\frac{2(n+m) - l_2}{l_1(l_1 + l_2)}\right]$$

$$b_2 = \frac{P_{RES}}{l_2 q} - \frac{l_1 b_1}{l_2}$$

area $A_F = l_1 b_1 + l_2 b_2$

$$n + m = \frac{l_1^2 b_1 + 2l_1 b_1 l_2 + l_2^2 b_2}{2(l_1 b_1 + l_2 b_2)}$$

$$c_{K1} = \frac{I}{(n+m)A_F}, \quad c_{K2} = \frac{I}{[l - (n+m)]A_F}$$

for eccentric load application:

$$e \leqslant c_k$$

$$q_1 = \frac{P_{RES}}{A_F} \mp \frac{P_{RES}\, e[l - (n+m)]}{I}$$

$$q_2 = \frac{P_{RES}}{A_F} \pm \frac{P_{RES}\, e(n+m)}{I}$$

The upper sign applies if e is towards wide side. The lower sign applies if e is towards narrow side.

$$e > c_K$$

if e is towards wide side

$$q_2 = \frac{2P_{RES}}{3b[(n+m) - e]}$$

if e is towards narrow side

$$q_1 = \frac{2P_{RES}}{3b_1[l - (n+m+e)]}$$

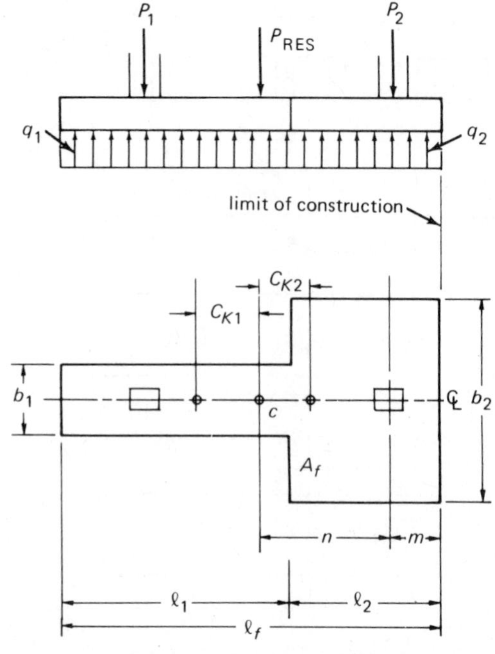

Fig. 5-19 T-shaped footings.

relations to design such footings for uniform bearing pressure; or to evaluate the extreme bearing pressures for a given footing shape and size in case of eccentric load applications. It is recommended to maintain bearing pressures over the entire base area of a combined footing, i.e., to design it in such a way that the eccentricity does not exceed the kern distance. Under extreme conditions, however, combined footings may also be designed for partial bearing with exclusion of tensile stresses at the footing base, as indicated.

If the nearest column, which can be utilized for the design of a combined footing, is too far away to permit a combined footing to be built economically, counterweights, or deadmen, can be provided to balance the eccentric loading of a footing as shown in Fig. 5-20. In such a case it is advantageous to design the footing so as to establish concentric loading for each element; otherwise, edge pressure conditions have to be investigated carefully. In the evaluation of the available weight on a deadman, it is recommended to rely primarily on the actual developed weight, and disregard load spreading or internal friction as far as possible. The safety factor to be used in the design of such foundations shall be at least the same as used for overturning.

Referring to condition (b), if overlapping of isolated single footings occurs, combination of the two footings into one will permit greater lateral spreading at maintaining a symmetrical base area, see Fig. 5-21.

In (c) and (d) above, it can easily be seen that a combination of two or more column loads by means of strips, rafts, or mats will not only spread the load over a bigger area, but will also give the foundation a monolithic quality which will

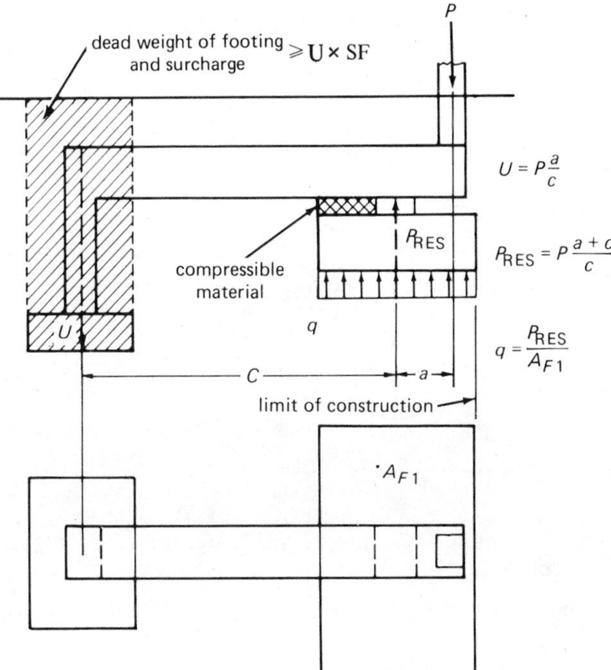

Fig. 5-20 Footing with deadman.

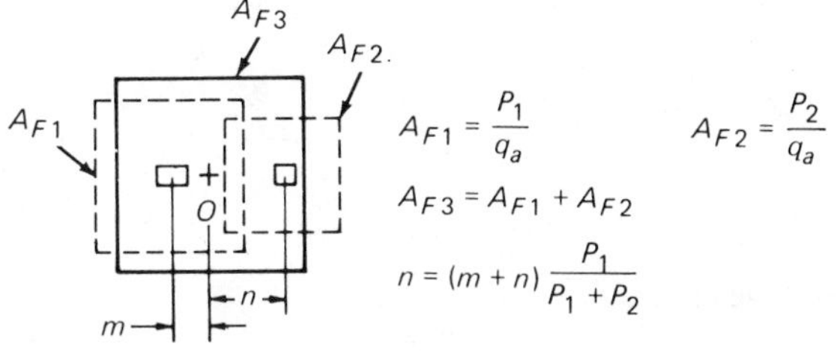

Fig. 5-21 Combined footing areas.

help to bridge over soft spots in the subsoil. This will reduce the risk of differential settlements. Such a design can be structurally desirable and may also be economical on soft or irregular subsoils. If the foundation is stiff enough (either by means of its own thickness or by means of well placed, monolithic basement walls), then the foundation may be considered as one unit. The shape of its base and the resulting bearing pressures can be evaluated as a combined footing. The stiffness of a strip, raft, or mat foundation alone is often not great enough to produce sufficient rigidity; in this case, the footing is best treated as a beam on elastic foundation. Figure 5-22 gives the basic equations required to design and evaluate the bearing pressures and moments using the simpified method.[5-3, 5-5]

As far as the last condition goes, the advantage that may be gained is primarily economic.

5.3 BASIC DESIGN PROCEDURE–REINFORCED CONCRETE FOOTINGS WITH EQUALLY DISTRIBUTED BEARING PRESSURES

5.3.1 Footing Size

It is important to keep the determination of the footing size completely separate from the design of the footing strength. The determination of the footing size, or of its width, as in the case of a strip footing or foundation raft, depends on the following criteria:

a. The evaluation of the service loads. (The term service is used here in order to differentiate the actual quantities from the factored loads to be used in the strength design.)
b. The evaluation of the bearing pressures at the footing base under the service loads, and the maximum allowable soil pressure.

In the evaluation of the service loads, the entire dead load (weight) of the superstructure, including the weight of the footing with surcharge and all floor live loads, should be used in their actual intensity. Floor live loads of multistory buildings, having an intensity of 100 lb/ft^2 or less may, according to the requirements of many local building codes, be reduced on the assumption that the full live load will hardly ever occur simultaneously on all floors. Reductions for live loads exceeding 100 lb/ft^2 are seldom permitted by building codes because such loads usually apply to warehouses or storage facilities.

Floors of industrial buildings are often designed for heavy loads in anticipation of machines that may have to be moved across the floor or will have to be set up at certain unforeseen locations. It is up to the judgment of the designing engineer and his personal knowledge of the manufacturing processes, to suggest a justified live load reduction to be used for the design of columns and footings and have it approved by the respective authority.

Crane loads, hoist loads, equipment loads; and the like have to be included in full, even if they are only of short duration.

Impact caused by the occasional passing of a crane need not be considered in the design load of a footing; however, impact caused by continuously operating hammers or recipprocating machines shall be considered in the loading. Footings resting on loose, granular subsoil may require special provisions to prevent the transfer of impact or vibrations to the subsoil.

Loadings caused directly or indirectly by fill, lateral earth pressure, or water pressure have to be considered in full.

Lateral loads due to wind and their related effects, such as vertical forces and moments caused by them, have to be determined and considered in the design.

In certain geographical areas, earthquake-simulating lateral

(*continued on p. 123*)

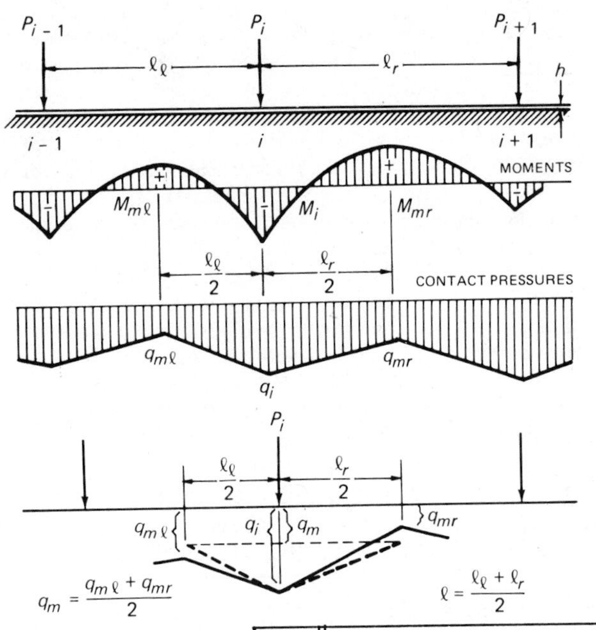

$$l = \frac{l_l + l_r}{2}, \qquad \frac{1.75}{\lambda} < l < \frac{3.50}{\lambda}$$

minimum total length: $\frac{5.25}{\lambda}$ or 3 bays

$$\lambda = 2.9 \sqrt[4]{\frac{K_{si} F}{f'_c h^3}}$$

Form Factors, F

Sand	$b(ft)$	5	10	15	LARGE
	F	0.36	0.30	0.275	0.25
Clay	l/b	1	2	3	STRIP
	F	1	0.83	0.77	0.67

	Sand			*Clay*		
N	LOOSE 4-10	MEDIUM 10-30	DENSE 30-50	STIFF 8-15	VERY STIFF 15-30	HARD >30
K_{si}	20-60	60-300	300-1000	50-100	100-200	>200

at column (i): $$M_i = -\frac{P_i}{4\lambda}(0.24\lambda l + 0.16) \leqslant -\frac{P_i l}{12}$$

for about equal spans

$$q_1 = \frac{5P_i}{l} + \frac{48 M_i}{l^2}$$

at midspan left: $$q_{ml} = 2P_i \frac{l_r}{l_l l} - q_i \frac{l}{l_l}$$

for equal spans $q_m = \frac{2P_i}{l} - q_i$

at midspan right: $$q_{mr} = 2P_i \frac{l_l}{l_r l} - q_i \frac{l}{l_r}$$

at midspan left: $$M_{ml} = M_{ol} + \frac{M_l + M_i}{2}$$

$$M_{ol} = \frac{l_l^2}{48}(q_l + 4q_{ml} + q_i)$$

at midspan right: $$M_{mr} = M_{or} + \frac{M_r + M_i}{2}$$

$$M_{or} = \frac{l_r^2}{48}(q_r + 4q_{mr} + q_i)$$

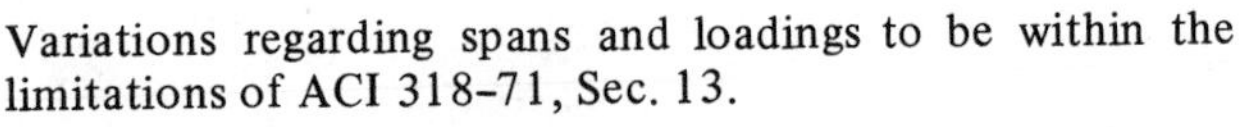
Variations regarding spans and loadings to be within the limitations of ACI 318-71, Sec. 13.

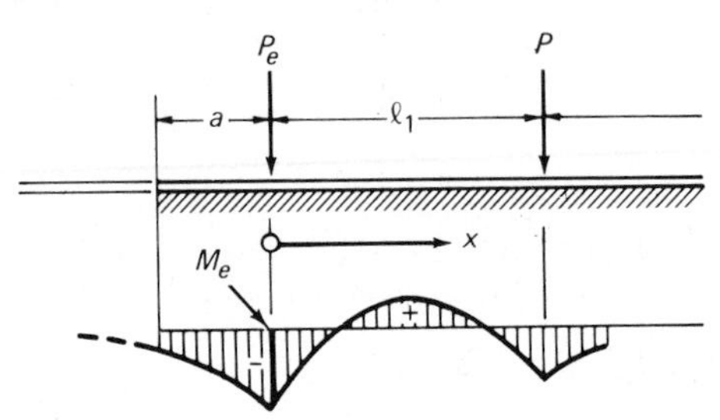

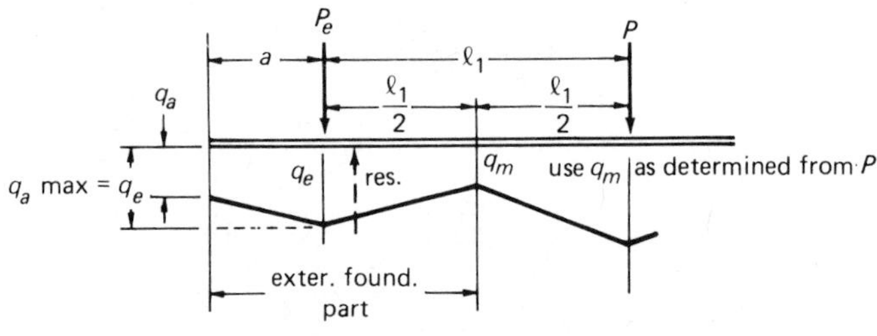

<u>End Condition:</u> (The index e in P_e, M_e, and q_e refers to the end column.)

at end column: 1) $M_e = -\frac{P_e}{4\lambda}(0.13\lambda l_1 + 1.06\lambda a - 0.50)$

$$q_e = \frac{4P_e + 6M_e/a - q_m l_1}{a + l_1} \qquad q_a = -\frac{3M_e}{a^2} - \frac{q_e}{2}$$

2) $$M_e = -\left(\frac{4P_e - q_m l_1}{4a + l_1}\right)\frac{a^2}{2}$$

$$q_e = q_a = \frac{4P_e - q_m l_1}{4a + l_1}$$

The smaller M_e governs.

Fig. 5-22 Simplified design of combined footings.[5-5]

forces have to be applied at all mass centers of the structure and have to be transmitted through the foundation into the subsoil. These forces can act in any direction, but do not have to be considered simultaneously with the wind forces. Whichever force will have the greater effect on the element under consideration will govern.

Building codes usually permit an increase in the allowable stresses by 33 percent where wind or seismic forces are included. The load combination resulting in the largest required base area shall be used.

In general, footings shall be designed to be at least as strong as the design loads of the columns which they support; all possible combinations of forces that can act simultaneously on the footing under consideration have to be considered. For the design of footings for equal bearing pressures under average service load conditions, see Section 5.4.2.

The "Building Code Requirements for Reinforced Concrete," ACI 318-71, stipulates the following combinations of load effects as generally applicable:

1. $(D + L)$ The resulting maximum bearing pressure must be smaller than or equal to the maximum allowable soil pressure, $q \leqslant q_a$.

2. $(D + L + W)$, or $(D + L + E)$ The resulting maximum caused by either loading (W or E), must be smaller than $1.33q_a$. This increase of 33% above the allowable bearing pressure is usually accepted but ought to be checked with the local building code requirements.

In case of uplift or overturning the most critical combinations shall be investigated. Live loads shall only be considered with uplift forces where they contribute to the overturning; wherever dead load counteracts the overturning it shall be introduced with only 0.9 of its actual value to be on the safe side. The required safety factor against overturning shall not be less than 1.50, but shall be checked with the local building code requirements.

The evaluation of the intensity and distribution of the bearing pressures shall be done along the lines discussed in Section 5.2.

The maximum allowable soil pressure shall be determined by principles of soil mechanics. Unless the engineer designing the foundation has determined the maximum allowable soil pressure himself, it is important for him to understand the basis on which it was determined and the safety factor which was used in its evaluation. It is also important for him to find out whether the weight of the overlying soil surcharge was included in the evaluation of the allowable bearing pressure, and what the minimum depth below ground surface is at which this bearing pressure can be developed. He should also know the influence that raising or lowering of a footing may have on the magnitude of the allowable bearing pressure, and the elevation and possible fluctuations of the groundwater level. All these factors can be of important influence on the design of the foundation.

5.3.2 Footing Strength

-1. General principles–The strength design, also called Ultimate Strength Design (USD), of practically all types of footings can proceed along the lines described below.

The size of the footing or foundation, and the resulting bearing pressures, are determined from the service loads (actual loads). In order to perform the strength design of a footing, the various types of loads have to be multiplied by the respective load factors; and the bearing pressures have to be reevaluated for these factored loads. It has to be kept in mind, however, that these newly evaluated bearing pressures are of purely mathematical nature and have no soil-mechanical significance. They are calculated reactions to an imaginary factored loading condition resisted by the strength capacity of the foundation element.

The strength design requires that the minimum capacity of every structural element be sufficient to resist the factored loadings in their most severe combination. For this purpose it is advisable to assemble all different types of loadings and their related effects, such as shears and bending moments, independently for dead load, live load, wind, and earthquake (if so required), so that the sum of each type of loading can be multiplied by the respective load factor. ACI 318-71 requires in Section 9.3 that the following load factors be used:

1.4, for dead loads, fills and liquid loads
1.7, for live loads, wind loads and earth pressures
1.87, for earthquake loads

After the most severe strength combinations have been determined for axial loads, bending moments, and shears according to Section 9.3 of the *ACI Building Code*, the footing has to be designed strong enough to resist these factored load combinations and their resulting bearing pressures.

It has to be kept in mind, however, that factored loadings do not always furnish the same eccentricities as service loads. This is true where lateral loads, or other loads causing eccentricities or overturning, are of a different origin from the loads or load combinations causing the concentric loading. Consequently, the various loadings will have to be multiplied by different load factors, and will thus result in bearing pressure distributions that are, in principle, different from those obtained for the unfactored service load conditions. Since this is not the intent of the design, it is advisable to determine the resulting bearing pressure distribution from the service loads and then multiply it by an appropriate load factor for the strength design. In such a case, either the maximum load factor of 1.7 may be used, or an approximate load factor between 1.7 and 1.4 may be selected.

A footing, depending on its type, can be compared with a heavily loaded bracket, or with a column capital, or with a beam or slab, all in inverted position.

Because of the heavy (inverted) loading due to the bearing pressure, the thickness (depth) of a reinforced concrete footing is, with the exception of long-span rafts and mats, usually governed by shear. Due to the different working conditions under which footings often have to be constructed, it is common practice to design them without shear reinforcement; however, the *ACI Building Code* permits the use of shear reinforcement, if so required, except for mats or slabs less than 10 in. thick. Since the column, pier, or pedestal is usually much narrower than the size of the footing or the width of the footing strip, oneway shear action (also called beam shear), *and* two-way shear action (also called slab or perimeter shear), have to be investigated. With the exception of oblong footings or long rafts, two-way shear action will usually govern the design. It is therefore advisable to investigate it first.

-2. Investigation for two-way shear action (slab or perimeter shear)–The investigation for two-way shear action is the same whether shear reinforcement is provided or not; only the permissible value of the nominal shear stress varies depending on whether shear reinforcement is used or not. The nominal shear stress is to be determined along a line concentric with the loaded area, usually provided by the pier or pedestal, and is located at a distance $d/2$ from the face of the loaded area. The line forms the perimeter of the base of a truncated cone or pyramid, through which bearing

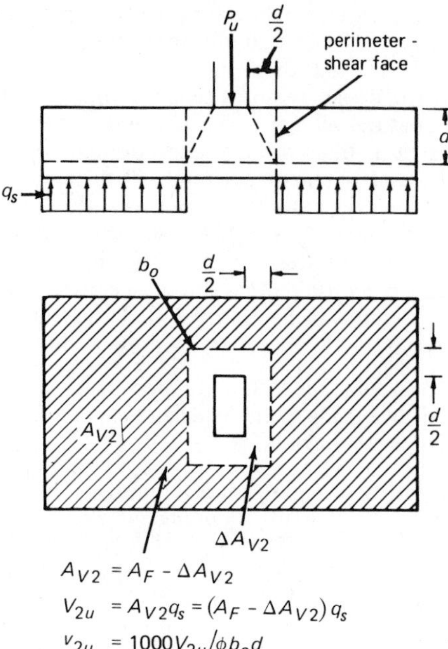

Fig. 5-23 Two-way (perimeter) shear action.

stresses are considered to be transferred straight into the supporting subsoil. Considering the case of a concentrically loaded, isolated spread footing with uniform bearing pressure, the force V_{2u}, as shown in Fig. 5-23, is pressing the footing portion located outside of ΔA_{v2} upward. This force is resisted by the two-way shear of average intensity, v_{2u}, acting along the perimeter area $b_0 d$.

The nominal shear stress for two-way shear action is therefore

$$v_{2u} = \frac{V_{2u}}{\phi b_0 d} 1000$$

where ϕ is the capacity reduction factor for shear ($\phi = 0.85$) from Section 9.2.1.3 of ACI 318-71.

The maximum permissible two-way shear stress under ultimate loading is $v_{2u} = 4\sqrt{f'_c}$ and is applicable to all concrete footings without shear reinforcement. Here, as well as in the following sections and examples, all concrete is assumed to be normal weight concrete. Where sand-lightweight, or all-lightweight concrete is used (which is rather unusual in connection with footings), the permissible shear values have to be reduced in accordance with the requirements of Chapter 11 of ACI 318-71. If shear reinforcement is provided in accordance with Section 11.11.1 of ACI 318-71, the maximum permissible shear stress is

$$v_{2u} = 6\sqrt{f'_c}$$

If shearhead reinforcement is provided in accordance with Section 11.11.2 of ACI 318-71

$$v_{2u} = 7\sqrt{f'_c}$$

However, for practical reasons, footings ought to be designed whenever possible without the use of shear reinforcement.

In the following, reference is frequently made to tables contained in other chapters of this handbook, or in the Strength Design Handbook, ACI SP-17. Numbers of tables contained in other handbooks have been omitted herein and the tables are referred to by their general name only, with ACI SP-17 added. Tables that are similar to those referred to in the ACI SP-17, can also be found in other design aids and textbooks, and can be used in a similar manner.

Tables to determine the footing depth as required by perimeter shear have been provided in ACI SP-17 to simplify the calculation. These tables are prepared for square footings, concentrically loaded by square, round, or polygonal pedestals or piers. If the loading pressure, q_c, at the base of the column, pier, or pedestal, and the bearing pressure at the footing base, q_s, due to the factored load are known; then the effective depth of the reinforced concrete footing can be determined from the ratio d/t which can be taken from the tables.

Similar tables are also provided in the ACI SP-17 for rectangular footings loaded by square or rectangular columns, piers, or pedestals. These tables can also be used for round and polygonal columns, piers, and pedestals, if their loading areas are transformed into squares of equal cross sectional area.

-3. Investigation for one-way shear action—The nominal shear stress due to one-way shear action shall be calculated as in an ordinary reinforced concrete beam, along a plane perpendicular to the footing, located at a distance d from the face of the column, pier, or pedestal. This plane shall extend across the entire footing and the shear stresses shall be assumed to be uniformly distributed over this plane, as shown in Fig. 5-24. The governing shear force, V_{1u}, for one-way shear action consists, therefore, of the sum of all bearing pressures, q_s, acting outside of the critical section.

The average nominal shear stress acting along the critical section is then

$$v_{1u} = \frac{1000 V_{1u}}{\phi (12 b_f) d}$$

where ϕ is the capacity reduction factor for shear ($\phi = 0.85$) from Section 9.2.1.3 of ACI 318-71. If no shear reinforcement is provided, which is the usual case, and the concrete is of normal weight taking all of the shear stresses, then the maximum permissible stress is

$$v_{1u} \leq v_c = 2\sqrt{f'_c}$$

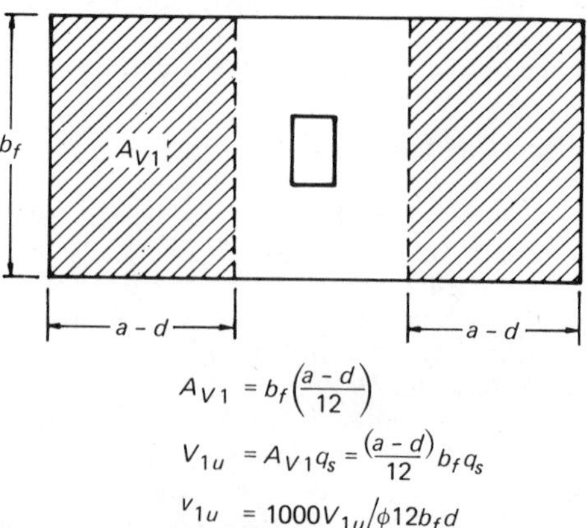

Fig. 5-24 One-way (beam) shear action.

In the exceptional cases, where shear reinforcement is provided in footings, it needs to be designed only for the excessive shear stress $(v_{1u} - v_c)$ in accordance with Section 11.6 of the *ACI Code*. In no case shall $(v_{1u} - v_c)$ exceed $8\sqrt{f'_c}$.

-4. Anchorage development of column dowels–After the minimum footing thickness that will satisfy both shear requirements has been determined, it should be checked to see whether it provides sufficient depth for the development of the column dowels. Anchorage development in compression is the most common condition for column bars; however where bending moments or uplift forces have to be transmitted to the footing, the bars must also satisfy the anchorage requirements for tension bars. Right-angle, or 90°, hooks at the bottom end of column dowels (common practice), are of no help in the development of anchorage for compression bars, see Fig. 5-25. Chapter 1 of this handbook contains Tables 1-10a to g with minimum required anchorage lengths for bar development in compression and in tension, with and without hooks. Dowels of smaller diameter than the column bars can be used to reduce the required anchorage length, and, therefore, thickness of the footing. Dowels of bigger diameter than the column bars may be used also if desirable; the *Code*, however, does not permit the diameter of the dowel to exceed that of the column bar by more than 0.15 in. ACI 318-71 provides in Section 15.6 that under certain conditions not all longitudinal column reinforcement needs to be extended into the footings, but only enough to cover the excess beyond the permissible bearing stress of the supporting or the supported member, whichever is smaller. In this connection, column steel, which has to be counted on at or above the contact area, has to be extended or developed in the column above and in the footing below. This also applies to areas where high edge pressures are caused by eccentric loadings or moments. See also ACI SP-17A.

-5. Investigation for flexure–The flexural strength of a footing can be determined in a similar manner to that of an ordinary beam or cantilever member. The condition is shown in Fig. 5-26.

The reference lines about which the bending moments are to be determined shall extend all the way across the footing and shall be located as follows:

a. At the face of the supported element, for footings supporting columns, piers or pedestals.

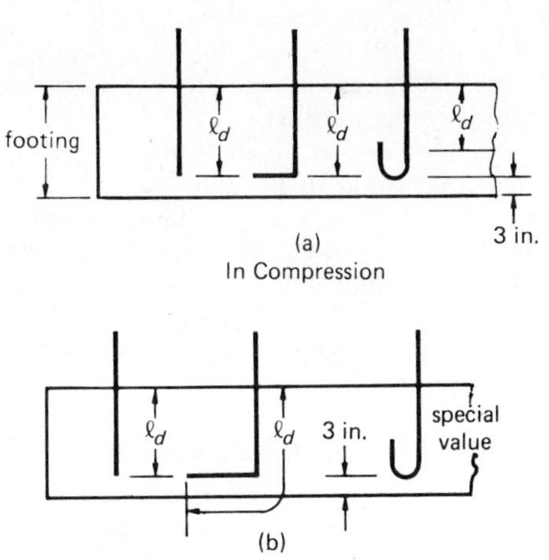

Fig. 5-25 Development (anchorage) of column bars or dowels. (a) In compression; and (b) in tension.

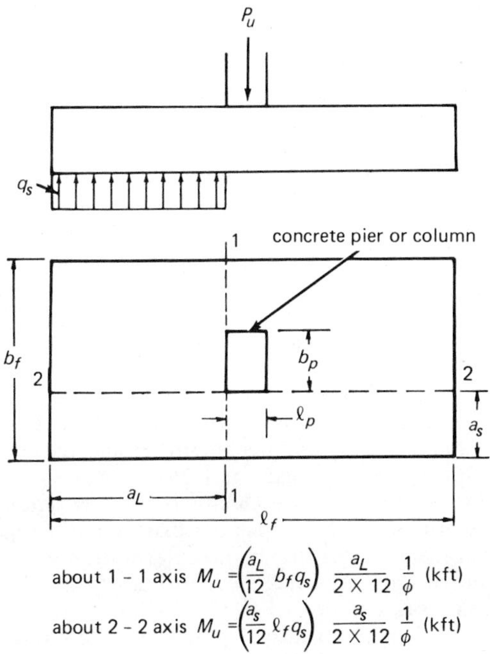

Fig. 5-26 Flexure.

b. Halfway between the center and the edge of the supported masonry, for footings supporting masonry construction.

c. Halfway between the face of the column and the edge of the base plate, for footings supporting steel base plates and steel columns.

The flexural moment, M_u, to be used in the calculation is therefore

$$M_u = \left(\frac{a}{12} b_f q_s\right)\left(\frac{a}{2 \times 12}\right)\frac{1}{\phi} \quad \text{in kip-ft}$$

For rectangular footings or for square footings supporting rectangular columns, pedestals, or piers, the larger moment, i.e., the moment in the direction of the longer projection, may influence the selection of the footing depth; to determine the reinforcement, however, the moments have to be calculated for each direction. The capacity reduction factor, ϕ, in the above equation is 0.9 for flexure according to Section 9.2.1.1 of ACI 318-71. If Tables 1-2 in Chapter 1 of this handbook, or if tables or graphs of ACI SP-17 are used, the ϕ-factor can be disregarded because it has generally been included in the handbook.

The flexural strength shall also be investigated at all changes in the cross section of the footing element. Such checks are important in the design of stepped or sloping footing mats.

The Flexural Tables 1-2 provided in Chapter 1 of this handbook, for the design of ordinary beam and slab problems, can be used just as well for the flexural design of a footing. In this case, the F-factor can be determined from the selected size of the footing as

$$F = (12b_f)d^2/12000, \quad \begin{matrix} b_f \text{ in ft} \\ d \text{ in in.} \end{matrix}$$

or can be read from the F-table. With the help of $K_u = M_u/F$, the corresponding a_u or ρ can be taken from the tables and the cross sectional area of the reinforcement be determined from

$$A_s = M_u/a_u d \quad \text{or} \quad A_s = \rho 12 b_f d$$

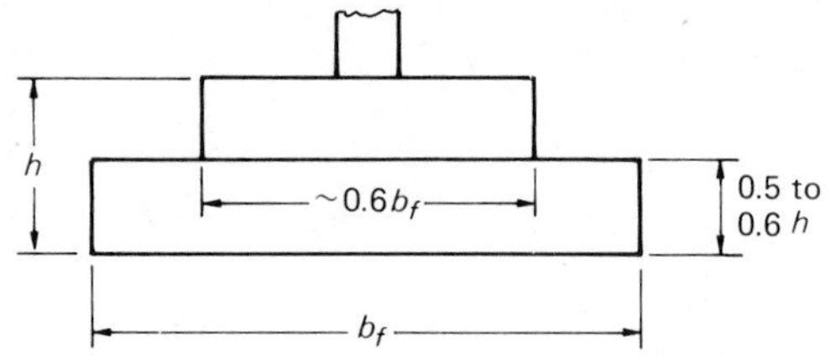

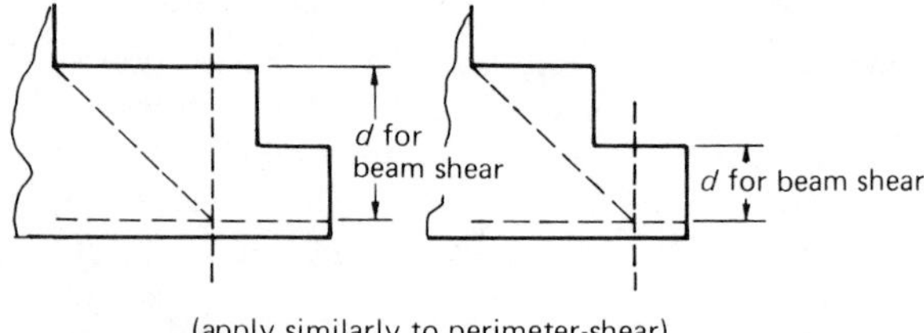

Fig. 5-27 Stepped footing.

If the required percentage exceeds the maximum permissible percentage $\rho_{max} = 0.75\rho_b$ or if the size and spacing of the reinforcing bars appears to be impractical, the footing thickness has to be increased to satisfy the requirements.

If the Flexural Graphs of USD SP-17 are used, the moments, M_u, must be determined either for a 1 ft wide strip if the slab graphs are used; or for a 10 in. wide strip if the beam graphs are used; and the obtained reinforcement must be multiplied by b_f for slabs or $1.2b_f$ for beams to find the total amount required in each direction. Attention must be drawn to the selection of the bar diameter to be used for the reinforcement, because it must, in addition to providing the required cross sectional area, also satisfy the anchorage requirements (see 6. below).

USD SP-17 contains also a special set of Footing Tables which are rather practical to use. These tables cover a wide range of bearing pressures due to factored loadings from $q_s = 3.33$ kip/ft^2 to $q_s = 26.67$ kip/ft^2 and can be used for both structural plain and reinforced concrete footings. Since the base area (size) of the footing must be known before the strength design is attempted, its greatest projection beyond the critical line is also known. The effective footing depth, d, can be selected so that the permissible perimeter shear and beam shear values are satisfied for the given projection. After these checks have been made, the required amount of reinforcement can be read from the table.

-6. *Development of footing reinforcement*—Most of the footings or portions thereof consist of short, heavily loaded cantilever sections. In the design of such elements it is important to see that the reinforcing bars are sufficiently anchored on either side of the critical section, as described above, to develop the full tension required. Since the projection of the footing beyond the critical section is a given length, and to satisfy the anchorage requirements, the diameters of the reinforcement selected should be small enough to provide sufficient embedment length on either side of the critical section. Most footing tables prepared for Strength Design, including those in ACI SP-17, contain the maximum diameters of bars that may be used to satisfy whatever projection the footing has.

5.4 SPECIAL CONDITIONS

5.4.1 Stepped Footings

In the design of isolated spread footings, the calculated thickness is required at the various critical locations near the center of the footing, but not at the edge of it. It has been common practice to give footings a tapered or stepped cross section in order to save the concrete in areas where it is not structurally needed. This method is structurally sound but getting out of practice for the following reasons:

a. The extra cost of the formwork is often greater than the saving in the amount of concrete used; and
b. the monolithic action between the upper (cap) and lower (mat) portion of the footing, which is structurally essential, is in practice difficult to obtain if cap and mat are not cast simultaneously, see Fig. 5-27. Unless special provisions are made, this method of construction will develop a "cold" joint and a cleavage plane will separate the two pieces.

Steps, however, are still in use in cases where the mat is getting excessively thick (usually more than 3 ft). If footings are not cast monolithically, key ways or shear-friction reinforcement have to be provided to transfer the horizontal shear and obtain monolithic action.

The size and thickness of the caps have to be designed in such a way that at each step (change in cross section) all shear stresses and flexural requirements are satisfied.

5.4.2 Footings Designed for Equal Bearing Pressures

Since settlements are practically independent of short time fluctuations in the loading, foundations for apartment houses, office buildings, institutional buildings, and the like are often designed for equal bearing pressures under average service load conditions, with the intent to obtain equal settlements over the entire building area. Full dead load plus one-half of the live load are often considered to represent an "average service load"; however, other ratios may be substituted depending on the judgement of the designing engineer. Such an approach will, at its best, reduce the amount of differential settlements to some degree because mutual influence, dishing, and, especially, variations in footing sizes will influence the settlement of each footing in a different way, regardless of the equal bearing pressure.

If such a design is desired, proceed as follows:

a. Determine the live load to dead load ratio for each column footing.
b. Determine the average (reduced service load for all column footings usually assumed with full dead load plus one-half live load).
c. Select the column with the greatest ratio of live load to dead load (from item a) and design its footing for the maximum allowable soil pressure under full load.
d. Determine the bearing pressure q_{av} for the same footing under the average service load (as determined under item b).
e. Design the size of all other footings for the average service load (as determined under b) and the average bearing pressure (as determined under d).
f. Ascertain for all footings, that the bearing pressure under the maximum load does not exceed the maximum allowable bearing pressure.
g. Make strength design of all footings at least for the factored maximum load and the bearing pressure caused by it or preferably for the q_s determined from the maximum allowable bearing pressure factored according to the applicable dead load to live load ratio.

The final result will be a somewhat overdesigned foundation, having equal bearing pressures under average service load conditions.

5.4.3 Footings of Structural Plain Concrete

Structural plain concrete footings on soil are permitted by the code; such footings are designed without reinforcement. The critical sections at which shear and flexure are to be determined are the same as for reinforced concrete footings, so are also the permissible shear stresses assigned to the unreinforced concrete. For all ordinary cases, flexure will govern the design.

The flexural stresses are calculated as for a homogeneous monolithic section. In this case the maximum tensile stress becomes

$$f_t = M_u z_t / I_c$$

The maximum tensile stress, f_t, in a structural plain concrete footing must not exceed $5\phi\sqrt{f'_c}$. Since a capacity reduction factor of $\phi = 0.65$ is prescribed by the *ACI Building Code* in Section 9.2.1.5, the maximum permissible tensile stress for structural plain concrete assumes the value of

$$f_t = 3.25\sqrt{f'_c}.$$

Footing Tables in ACI SP-17 contain also values to select the minimum thickness of footings made of structural plain concrete for given projections and bearing pressures under the factored loading.

The following design examples were prepared with and without the help of tables and graphs. In the first part of the design examples, the procedure was presented in easy-to-follow steps. A considerable amount of explanatory text was also added to assist in the understanding of the basic requirements. It cannot be stressed enough that a full understanding of these basic approaches makes them applicable to footings of any type or character. This may make the procedure appear rather lengthy, it is, however, evident that a great part of the extra steps will become unnecessary in the solution of an actual design problem; however, they were included here for illustrative purposes.

EXAMPLE 5-1: Without using footing tables and graphs, design a *concentrically loaded, square, spread footing* for the following conditions:

column load

$$P_D = 350.0 \text{ kips}, \quad P_L = 275.0 \text{ kips}$$

pier

$$b_p = 18 \text{ in.}, \quad l_p = 24 \text{ in.}$$

$$q_a = 4.50 \text{ k/ft}^2, \quad h_s = 5 \text{ ft}$$

$$w_L = 100 \text{ lb/ft}^2$$

$$f'_c = 3000 \text{ psi, normal weight concrete}, \quad f_y = 40{,}000 \text{ psi}$$

Design footing by strength design method (ACI).

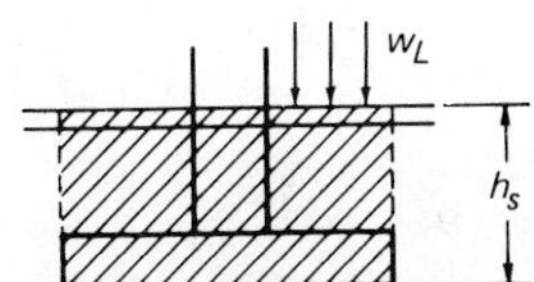

Fig. 5-28 Evaluation of surcharge.

Step 1: Determine footing size from service loads. The approximate weight of the surcharge is $\Delta q = \gamma_s h_s + w_L$ where γ_s is the assumed average unit weight for all material above the footing base. Since it consists of concrete and soil it can be estimated close enough between 150 and 100 lbs/ft³. In this example it is assumed to be 130 lbs/ft³, see Fig. 5-28.

$$\Delta q = \gamma_s h_s + w_L = 0.13 \times 5 + 0.1 = 0.75 \text{ kip/ft}^2$$

Determine effective soil pressure (allowable soil pressure that can be utilized)

$$q_e = q_a - \Delta q = 4.50 - 0.75 = 3.75 \text{ kip/ft}^2$$

to find the minimum required base area of footing

$$A_{F\min} = \frac{P_D + P_L}{q_e} = \frac{350 + 275}{3.75} = 167 \text{ ft}^2$$

selected footing size

$$\text{use 13-ft square}, \quad A_F = 169 \text{ ft}^2$$

Step 2: Determine bearing pressure to be used for strength design (ACI).

$$q_s = \frac{P_{Du} + P_{Lu}}{A_F} = \frac{350 \times 1.4 + 275 \times 1.7}{169} = 5.70 \text{ kip/ft}^2$$

Step 3: Determine thickness of footing mat. In the case of a square spread footing, concentrically loaded by a square column, the mat is always governed by two-way (slab or perimeter) shear action.[8] It is, therefore, advisable to investigate such a footing first for this condition. Only if the footing is oblong, or if the column or pier has a rectangular cross section, does the footing thickness have to be checked also for one-way (beam) shear action (see also Step 3 of Ex. 5-2), and may be governed by it. The tentatively selected mat thickness must also be checked for dowel embedment length and flexural requirements.

a. When investigating for two-way shear action, the most practical design approach is to assume a footing thickness and check it for its required strength; if incorrectly assumed, it can easily be adjusted to the correct thickness. The location of the base of the truncated shear cone or pyramid is concentric with that of the pier and located at a distance $d/2$ outside of it, as shown in Fig. 5-29.

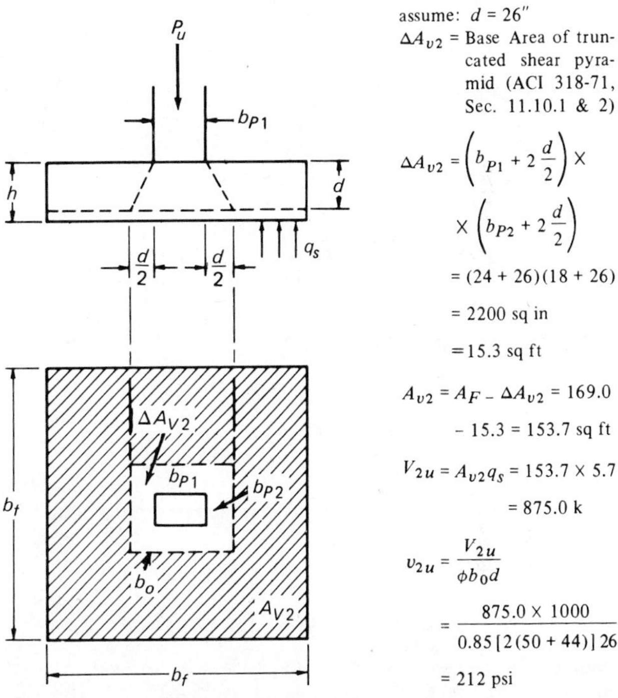

assume: $d = 26''$

ΔA_{v2} = Base Area of truncated shear pyramid (ACI 318-71, Sec. 11.10.1 & 2)

$$\Delta A_{v2} = \left(b_{P1} + 2\frac{d}{2}\right) \times \left(b_{P2} + 2\frac{d}{2}\right)$$

$$= (24 + 26)(18 + 26)$$

$$= 2200 \text{ sq in}$$

$$= 15.3 \text{ sq ft}$$

$$A_{v2} = A_F - \Delta A_{v2} = 169.0 - 15.3 = 153.7 \text{ sq ft}$$

$$V_{2u} = A_{v2} q_s = 153.7 \times 5.7 = 875.0 \text{ k}$$

$$v_{2u} = \frac{V_{2u}}{\phi b_o d} = \frac{875.0 \times 1000}{0.85\,[2(50 + 44)]\,26} = 212 \text{ psi}$$

Fig. 5-29 Two-way shear action.

The permissible two-way shear stress (ACI 318-71, Section 11.10.3) is

$$v_{2u} = 4\sqrt{f'_c} = 220 \text{ psi}$$

NOTE: It is advisable to design a footing thick enough to satisfy the permissible shear stresses for unreinforced concrete. When, under extreme conditions, the footing cannot be made thick enough, the shear capacity of the footing has to be strengthened by reinforcement the same way as in a column-slab intersection. If shear reinforcement (usually bent up bars) is used, the permissible average shear stress may be increased by 50%; if a steel shearhead reinforcement is provided, the permissible average shear stress may be increased by 75%. In each case the concrete section can only be stressed to the permissible stress value of $4\sqrt{f'_c}$ and the remainder has to be carried by the reinforcement.

b. Investigation for one-way shear action begins by investigating the footing mat in the direction in which the distance between footing edge and critical section is largest. The critical section for

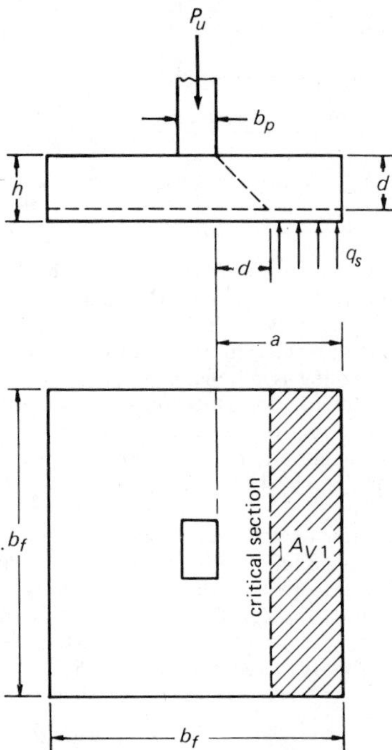

Fig. 5-30 One-way shear investigation.

one-way shear runs parallel to each pier face, and at the distance d away from it, across the entire footing mat, see Fig. 5-30.

$$a = \frac{b_f - b_p}{2} = \frac{13.0 - 1.5}{2} = 5.75 \text{ ft} = 69 \text{ in}$$

$$A_{v1} = \frac{(a-d)}{12} b_f = \frac{69 - 26}{12} 13.0 = 46.6 \text{ ft}^2$$

$$V_{1u} = A_{v1} q_s = 46.6 \times 5.7 = 268 \text{ kips}$$

$$v_{1u} = \frac{V_{1u}}{\phi b_f d} = \frac{268.0 \times 1000}{0.85(13 \times 12)26} = 78 \text{ psi}$$

The permissible shear stress for one-way action is

$$v_{1u} = 2\sqrt{f_c'} = 110 \text{ psi}$$

NOTE: In essence, the note provided at the end of Step 3a applies here also. (ACI 318-71, Section 11.10.3)

Step 4: Investigation for flexure. If the projection of the footing beyond the critical section varies, the flexural investigation has to be made for each direction. In each case the critical section extends across the entire footing.

For concrete piers or columns, the critical section is located at the face of the pier (ACI 318-71, section 15.4.2). (See also Fig. 5-26, $l_f = b_f$.)

a. Flexural computation in the direction of the longer footing projection

$$a_L = \frac{l_f - l_p}{2} = \frac{13.0 - 1.5}{2} = 5.75 \text{ ft} = 69 \text{ in.}$$

$$M_u = b_f a_L q_s \frac{a_L}{2} = 13.0 \times 5.75 \times 5.7 \frac{5.75}{2} = 1220 \text{ kip-ft}$$

$$F = \frac{b_f d^2}{12{,}000} = \frac{(13 \times 12)\, 26^2}{12{,}000} = 8.75, \quad K_u = M_u/F = \frac{1220}{8.75} = 139$$

$$\rho = 0.004. \; A_s = \rho b_f d = 0.004\,(13 \times 12)\,26 = 16.2 \text{ in.}^2$$

Minimum reinforcement required (ACI 318-71, section 7.13)

$$\rho_{min} = 0.0018$$

$$A_{s\,min} = \rho_{min} b_f h = 0.0018\,(13 \times 12)\,31 = 8.8 \text{ in.}^2$$

For evaluation of h see below.

$$A_s = 16.2 \text{ in.}^2 \text{ governs.}$$

b. Flexural computation in the direction of the shorter footing projection

$$a_s = \frac{b_f - b_p}{2} = \frac{13.0 - 2.0}{2} = 5.5 \text{ ft} = 66 \text{ in.}$$

$$M_u = l_f a_s q_s \frac{a_s}{2} = 13.0 \times 5.5 \times 5.7 \frac{5.5}{2} = 1120 \text{ kip-ft}$$

$$F = 8.75 \text{ as above}$$

$$K_u = \frac{M_u}{F} = \frac{1120}{8.75} = 128, \quad \rho = 0.0037$$

Minimum reinforcement required (ACI 318-71, section 7.13)

$$\rho_{min} = 0.0018$$

$$A_s = \rho b_f d = 0.0037\,(13 \times 12)\,26 = 15.0 \text{ in.}^2 \text{ governs.}$$

NOTE: It would be theoretically correct to use a different d-value for the flexural computation in each direction because the reinforcement is placed in two layers. The requirement of placing a certain layer below the other one is, economically, only seldom worth the effort, and practically difficult to control, unless the character of the two layers is drastically different; e.g., in an oblong footing where the longer bars are usually specified to be placed in the first layer from the bottom. For square footings it is advisable to stay on the safe side, and design the reinforcement in both directions for the shorter effective depth d. Care has to be exercised in the selection of the bar size that can be used for the reinforcement, because bar development (anchorage) is always critical in members which are highly stressed by shear. Since every bar has to be fully anchored at either side of the critical section, the shorter length, which is (a − 3 in.) will govern the design, see Fig. 5-31.

It can be seen from Table 1-10a in this handbook showing the bar development lengths for the various bar sizes, that for the example under consideration, any deformed bar up to size No. 11 can be used for this purpose. Bar sizes greater than No. 11 are commonly not used for footings, although there is no *Code* restriction in this respect. 17-#9 ($A_s = 17$ in.^{2} > 16.2 in.^{2}) will be provided in each direction.

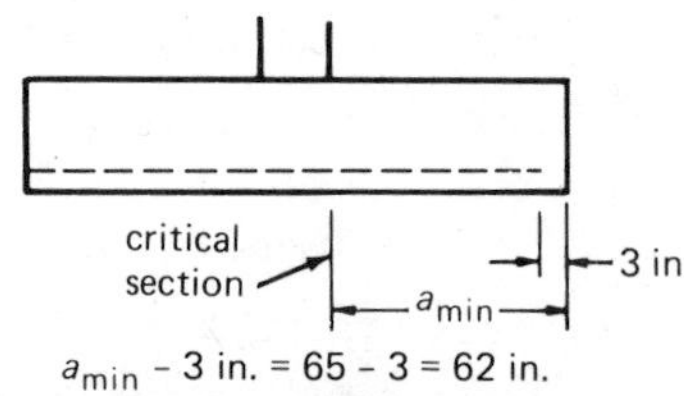

a_{min} − 3 in. = 65 − 3 = 62 in.

Fig. 5-31 Bar development (anchorage).

The total thickness of the footing can be found, under consideration of the above, as shown in Fig. 5-32.

$$h = d + 1.5 d_b + c_c = 26 + 1.5 \times 1.13 + 3 = 30.7 \text{ in.}$$

$$\cong 31.0 \text{ in.}$$

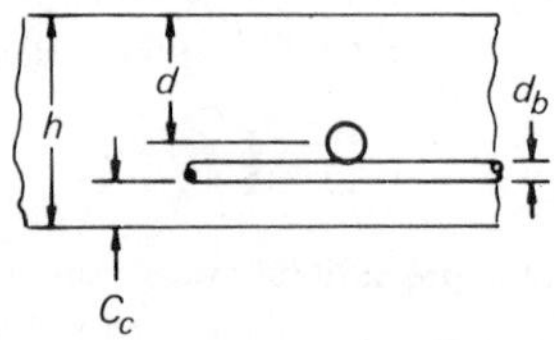

Fig. 5-32 Effective depth of spread footings.

EXAMPLE 5-2: Below, ex. 5-1 is solved with the help of footing tables and graphs contained in the "Strength Design Handbook SP-17." *Step 1* and *Step 2* have to be performed as before.

Step 3:

a. Use Footing Graphs of USD SP-17.

$$\frac{P_u}{A_c} = \frac{P_{Du} + P_{Lu}}{l_p \times b_p} = \frac{350.0 \times 1.4 + 275.0 \times 1.7}{24 \times 18} = 2.21 \text{ kip/in.}^2$$

enter graph with $P_u/A_p = 2.21$ and read at intersection with $q_s = 5.7$, a ratio of $d/b_p = 1.25$.

The graph was prepared for square piers, but it also can be used for slightly rectangular piers if the rectangle is transformed into a square of equal area. The size of the equivalent square is then $b_p = \sqrt{24 \times 18} = 20.7$, $d = 1.25 \times 20.7 = 26$ in. (Since the intersection with the q_s-curves cannot be read with accuracy it is recommended to verify the selected value by calculation.)

b. From Footing Tables of USD SP-17, select the table which is closest to the required q_s, or interpolate if necessary. Footing Table for $q_s = 6.67$ will be on the safe side. By entering this table with the selected depth $d = 26$ in. (interpolate between 24 and 28), we find the appropriate a_b-value equal to 78.25 in. This value represents the maximum projection that can be used for one-way shear action. Since the maximum projection, a, in the example is 69 in. (which is smaller than 78.25), the selected depth of 26 in. is satisfactory.

Step 4: The footing must also be deep enough to develop the column dowels, or the column dowels must be selected so that their necessary development length is satisfied by the depth of the footing. In case the dowels have to be anchored for compression only, the l_d must be smaller than or equal to 26 in., which is satisfied by #11 bars. In case of tension, the anchorage can be increased through hooks or bends.

Step 5: The same footing table can be used to determine the reinforcement. Enter the table with the value of $d = 26$ and $a = 69$ in. Double interpolation furnishes a value of 1.22 in.^{2}/ft for the reinforcement in each direction. The total required reinforcement is therefore

$$A_s = A_s/\text{ft (from Table)} \times b_f = 1.22 \times 13 = 15.9 \text{ in.}^2,$$

against 16.2 in.^{2} found by calculation in Step 4a) of Ex. 5-1. The column at the right hand side of the Table indicates the maximum bar size that can be used corresponding to each projection a.

EXAMPLE 5-3: Without the help of footing tables and graphs, design an oblong, concentrically loaded, spread footing for the following conditions: all data identical with those of ex. 5-1 except that the width of the footing is restricted to 8 ft, and the long side of the pier is, for architectural reasons, perpendicular to the long-side of the footing.

Step 1: Determine footing size from service loads. Evaluation of the minimum base area remains unchanged

$$A_F = 167 \text{ ft}^2$$

Selected footing size

$$b \times l_f = 8 \times 21 \text{ ft}, \quad A_F = 168 \text{ ft}^2$$

Step 2: Determine bearing pressure to be used in strength design. Unchanged from example 8.5.1

$$q_s = 5.7 \text{ kip/ft}^2$$

Step 3: Determine the thickness of the footing. In the case of an oblong, concentrically loaded, spread footing, the footing thickness is often governed by one-way (beam) shear action, depending on the length to width ratio of the footing. It is therefore necessary to investigate the footing for this condition first, and check the tentatively selected thickness afterwards for two-way (slab) shear action (see also Step 3 of Example 5-1). The tentatively selected footing thickness must also be checked for dowel anchorage and flexural requirements.

a. When investigating for one-way shear action, consider that the critical section for one-way shear extends across the entire footing parallel to each pier face at the distance d away from it. Similarly to Step 3 of Example 5-1, we assume the footing depth, in this case to be 36 in., see Fig. 5-33.

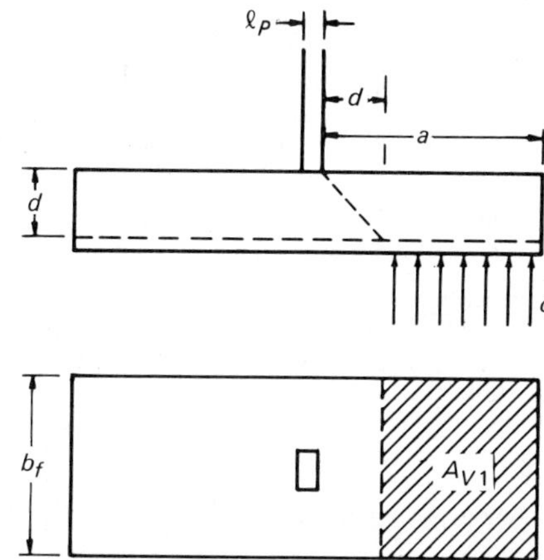

Fig. 5-33 One-way shear in oblong footings.

$$a = \frac{l_f - l_p}{2} = \frac{21.0 - 1.0}{2} = 10.0 \text{ ft} = 120 \text{ in.}$$

$$A_{V1} = \frac{(a-d)}{12} b_f = \frac{120 - 36}{12} 8.0 = 56.0 \text{ ft}^2$$

$$V_{1u} = A_{V1} q_s = 56.0 \times 5.7 = 312.0 \text{ kip}$$

$$v_{1u} = \frac{V_{1u}}{\phi b_f d} = \frac{312 \times 1000}{0.85(8 \times 12) \times 36} = 108 \text{ psi}$$

The permissible shear stress for one-way action is

$$2\sqrt{f'_c} = 2\sqrt{3000} = 110 \text{ psi}$$

b. Investigation for two-way shear.

$$\Delta A_{V2} = (l_p + 2d/2)(b_p + 2d/2) = (18 + 36)(24 + 36)$$
$$= 3240 \text{ in.}^2 = 22.4 \text{ ft}^2$$

$$V_{2u} = A_{V2} q_s = (l_f \times b - \Delta A_{V2}) q_s$$
$$= (8.0 \times 21.0 - 22.4)\, 5.7 = 830.0 \text{ kip}$$

$$v_{2u} = V_{2u}/\phi b_0 d = 830 \times 1000/0.85[2(54 + 60)]\, 36 = 120 \text{ psi}$$

which is well below the permissible value of $4\sqrt{f'_c} = 220$ psi. The correctness of this result is questionable, if we look at the plan of the footing in Fig. 5-34 which is illustrating the condition. An even distribution of the two-way shear along the perimeter is hardly probable if we consider the narrow width of the influence area parallel to the short sides of the pier.

A more reasonable result is obtained if we divide the influence area in four portions, and investigate the greatest shear stress caused by each part separately.

$$A'_{V2} = \frac{l_c - (l_p + d)}{2} = \frac{21.0 \times 12 - (18 + 36)}{2 \times 12} \times 8 -$$
$$- 2 \times \frac{18^2}{2 \times 144} = 63.75 \text{ ft}^2$$

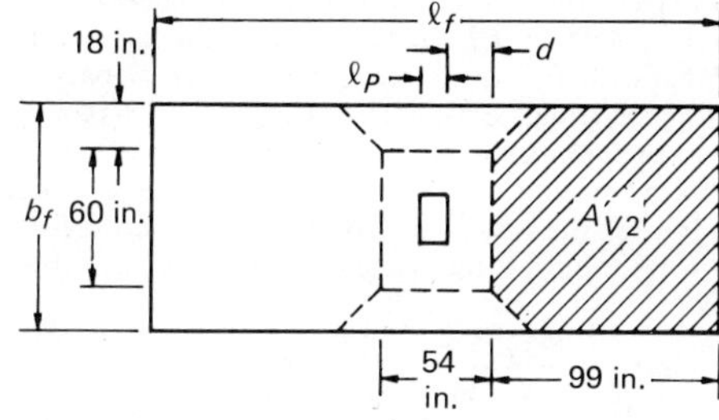

Fig. 5-34 Two-way shear in oblong footings.

$V'_{2u} = A'_{V2} q_s = 63.75 \times 5.7 = 365$ kip

$$v'_{2u} = \frac{V'_{2u}}{\phi(b_p + d)d} = \frac{365 \times 1000}{0.85(24 + 36)\,36} = 200 \text{ psi}$$

which is considerably greater and more realistic than the first value of v_{2u} found above, but still within the permissible limit of 220 psi.

Step 4: The depth of 36 in. will permit the use of any bar size for dowels, regarding their development in compression.

Step 5: Investigation for flexure. The oblong shape of the footing requires the design to be made independently for both directions. It is good practice, and reasonable, to assume that the reinforcement in the long direction will be placed at the bottom in order to utilize a greater depth. It is safer, however, and recommended, to design both layers for the shorter effective depth to stay independent of the field inspection.

a. In the long direction of the mat

$$a_L = \frac{21.0 - 1.0}{2} = 10.0$$

$$M_u = a_L\, b_f q_s \frac{a_L}{2} = 10.0 \times 8.0 \times 5.7 \times \frac{10}{2} = 2280 \text{ kip-ft}$$

$$F = b_f d^2/12{,}000 = (8 \times 12) \times 36^2/12{,}000 = 10.4,$$

$$K_u = 2280/10.4 = 218$$

$a_u = 2.85$, from Flexure Table 1-2 in Chapter 1 of this handbook

$$A_s = \frac{M_u}{a_u d} = \frac{2280}{2.85 \times 36} = 22.2 \text{ in.}^2$$

The minimum available anchorage length in this direction is $a_L - 3 = 10 \times 12 - 3 = 117$ in. It can be seen from Tables 1-10 in Chapter 1 of this handbook that this length permits the use of any bar size up to #11. Select

$$A_s = 15 - \#11, \quad A_s = 23.4 \text{ in.}^2$$

b. In the short direction of the mat

$$a_s = \frac{8.0 - 1.5}{2} = 3.25 \text{ ft}$$

$$M_u = a_s l_f q_s \frac{a_s}{2} = 3.25 \times 21 \times 5.7 \times \frac{3.25}{2} = 630 \text{ kip-ft}$$

$$F = \frac{l_f d^2}{12{,}000} = \frac{(21 \times 12) \times 36^2}{12{,}000} = 27.2, \quad K_u = \frac{540}{27.2} = 20$$

$$a_u = 2.96$$

only minimum reinforcement will be required.

$$A_s = \frac{1.33 M_u}{a_u d} = \frac{1.33 \times 540}{2.96 \times 36} = 6.8 \text{ in.}^2$$

or $A_{s\min} = \rho_{\min} l_f h = 0.0018\,(21 \times 12)41 = 18.7$ in.^{2}

For evaluation of h see below. The smaller value can be used.

The minimum available anchorage length in this direction is $a_s - 3 = (3 \times 12) - 3 = 33$ in. According to Tables 1-10 in Chapter 1 of this handbook, the maximum permissible bar size which can be used in this respect is #9. Because of the large footing length a greater number of bars is desirable and we select #7 bars.

In an oblong footing, section 15.4.4 of ACI 318-71 requires that a portion of the reinforcement in the short direction A_{s1}, be distributed over a width b_f and the balance of the reinforcement be spread evenly over the rest of the footing length, eq. (15-1).

If A_{sT} = total required reinforcement in the short direction, then

$$A_{s1} = \frac{A_{sT} \times 2}{(S + 1)} = \frac{6.8 \times 2}{(2.65 + 1)} = 3.7 \text{ in.}^2$$

$$S = l_f/b_f = 21/8 = 2.65$$

To avoid unequal spacings of reinforcement, the total reinforcement A_{sT} may be increased to A_{s2} in order to be spread evenly over the entire length of the footing.

$$A_{s2} = \frac{2A_{sT}S}{(S + 1)} = \frac{2 \times 6.8 \times 2.65}{(2.65 + 1)} = 9.8 \text{ in.}^2 \text{ still less than } 18.7 \text{ in.}^2$$

Reinforcement used is 31 #7 bars, $A_s = 18.6$ in.^{2} and the total thickness of the footing is

$$h = d + 3 + d_{bL} + \frac{1}{2} d_{bs} = 36 + 3 + 1.41 + \frac{0.88}{2} = 40.85 \cong 41 \text{ in.}$$

where d_{bL} and d_{bs} are the bar diameters in the long and short direction.

EXAMPLE 5-4: Stepped footing. If the thickness of the footing investigated in ex. 5-3 is considered to be uneconomical or otherwise excessive, it may be designed as a stepped footing and required to be cast monolithically, see Fig. 5-27.

It is common practice to make the cap (the upper portion) about $0.6\, b_f$ long, however this length should be checked as described below. The cap thickness is usually assumed with a fraction of the total footing thickness depending on the number of steps to be used. In this case the shear investigations of ex. 5-3, for one-way and for two-way shear, are applicable only as long as the critical sections fall within the extent of the cap; otherwise, the depth occurring at the critical section has to be used. The shear-cone or pyramid must not intersect the step, or the size of the cap has to be adjusted as needed. The entire shear investigation has to be repeated along the perimeter of the cap.

In the flexural investigation, care has to be exercised so that only that portion of the cross section that is in compression is utilized for determining the F-value, $bd^2/12{,}000$, and consequently the amount of reinforcement. In addition to the maximum moment occurring at the critical section, the amount at the edge of each step has to be investigated, and the amount of reinforcement at these locations checked for the reduced available depth.

EXAMPLE 5-5: Structural plain concrete footing

column dead load:	40 kips
column live load:	60 kips
total column load: (service)	100 kips

allowable soil pressure

$$P_a = 4.0 \text{ kip/ft}^2$$

$$f'_c = 3000 \text{ psi}, \quad \text{pier size } 12 \times 12 \text{ in.}$$

footing size

$$A_F = \frac{100}{4} = 25 \text{ ft}^2, \quad \text{or} \quad 5 \times 5 \text{ ft}$$

Start with the flexural investigation because it governs the design of a square, structural plain concrete footing supporting a square column or pier. See Fig. 5-35.

$$P_{uD} = 40.0 \times 1.4 = 56.0 \text{ kips}$$

$$P_{uL} = 60.0 \times 1.7 = 102.0 \text{ kips}$$

$$P_u = 158.0 \text{ kips}$$

$$q_s = \frac{P_u}{A_F} = \frac{158}{25} = 6.35 \text{ kips-ft}^2$$

$$M_u = b_f a q_s \frac{a}{2} = 5 \times \frac{24}{12} \times 6.35 \times \frac{24}{2 \times 12} = 63.5 \text{ kip-ft}$$

$$S_F = \frac{12 b_f \times h^2}{6} = 10h^2, \quad \text{for } b_f = 5.0$$

The permissible flexural strength is, according to sections 15.7.2 and 9.2.1.5 of ACI 318-71

$$f_t = 5.0 \times \phi \times \sqrt{f'_c} = 5 \times 0.65 \times \sqrt{3000} = 179 \text{ psi}$$

$$S_F(\text{required}) = \frac{M_u \times 12{,}000}{f_t} = \frac{63.5 \times 12{,}000}{179} = 4230$$

$$4230 = 10h^2, \quad h^2 = 423, \quad h = 20.5 \text{ in., say } 21 \text{ in.}$$

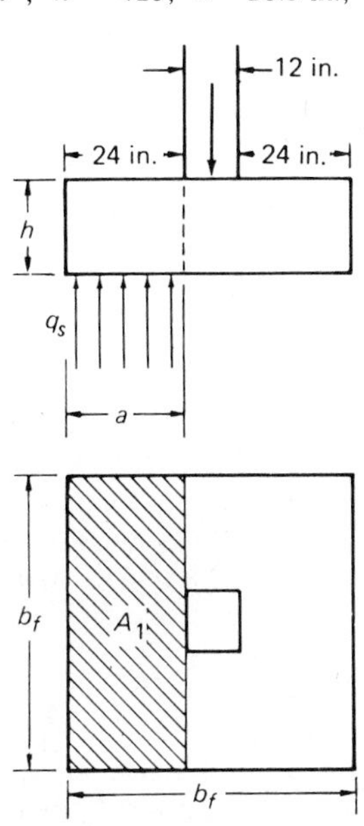

Fig. 5-35 Flexural design of structural plain concrete footings.

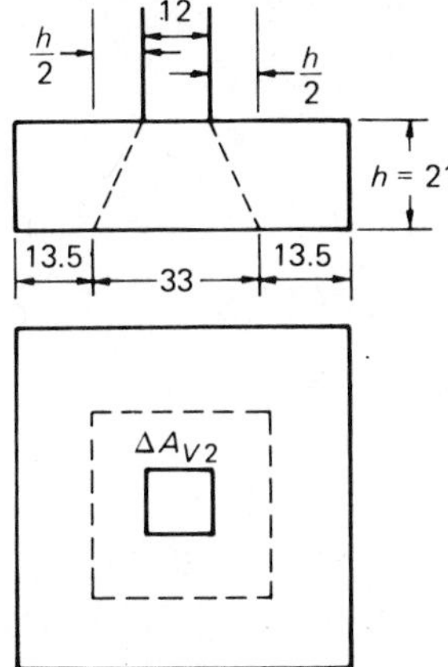

Fig. 5-36 Two-way shear in structural plain concrete footings.

Check footing thickness for two-way shear (for illustrative purposes only). See Fig. 5-36.

$$\Delta A_{V2} = \left(\frac{33}{12}\right)^2 = 7.55 \text{ ft}^2$$

$$\Delta V_{2u} = \Delta A_{V2} \times q_s = 7.55 \times 6.35 = 48 \text{ kips}$$

$$P_u - \Delta P_u = 158 - 48 = 110 \text{ kip}$$

$$v_{2u} = \frac{(P_u - \Delta P_u)\,1000}{b_o d} = \frac{110 \times 1000}{4 \times 33 \times 21} = 40 \text{ psi}$$

permissible

$$v_{2u} = 4\phi\sqrt{f_c'} = 4 \times 0.85 \times \sqrt{3000} = 187 \text{ psi}$$

5.5 REINFORCED CONCRETE FOOTINGS WITH CONCENTRATED REACTIONS (PILE CAPS)

5.5.1 General Principles

Where soil conditions do not favor the design or construction of shallow foundations (spread footings), but a firm soil stratum can be found at greater depth, piles can be used to transfer the loads from the superstructure down to the soil stratum, where the required resistance is available. The piles may develop this resistance by end bearing (bearing piles) on the firm stratum; or by skin friction (friction piles) developed by driving the piles into the firm stratum. Foundation piers or caissons can also be used for similar purposes but do not form a part of this discussion.

Similar to the action of a spread footing, a footing on piles (commonly called pile cap) has to distribute the column load to the piles in each group, which in turn will transmit it to the subsoil. The main difference between the two types of footings lies in the application of the base reactions which, in the case of a footing on piles, consists of a number of concentrated loads. If we divide the sum of all pile reactions in a group, just for reasons of comparison, by the base area of the pile cap, we obtain an equivalent bearing pressure caused by the bearing capacities of the individual piles. Such an average bearing pressure would be quite high because of the large bearing capacities of the individual piles. These large pile capacities were brought about by great progress made in the theoretical understanding of the soil resistance; by improvement in the quality of the materials used; and by the higher power and reliability of modern driving procedures and equipment.

The allowable bearing capacity that can be expected from a pile is usually based on the information gained from exploratory soil borings, and evaluated with the help of soil mechanical principles; it should be confirmed, however, by performance tests made on the site to ascertain the actual conditions. Depending on the availability of rock, hardpan, or other firm soil stratum and on their distance below grade, the engineer will decide whether bearing piles can be used economically. Otherwise, he has to resort to friction piles of some sort to utilize the available soil condition.

Lack of a firm soil stratum at reasonable depth can sometimes be treated also with the help of floating (boatlike) foundations which do not form a part of this discussion. The structural design of a pile cap is, in principle, not affected by the type of pile to be used, because it is primarily dependent on the magnitude of the pile reaction; however, a few explanations are necessary for a better understanding in the evaluation of the basic design approach.

5.5.2 Number of Piles Required

In the case of a spread footing, the size of the footing is determined from the total load on the footing and the allowable bearing pressure; hence, the size of the footing is rather made to order. In the case of a pile load, however, the number of piles is determined from the total load and the allowable load bearing capacity of each individual pile. Since the addition of a pile will raise the capacity of the whole group by a considerable amount, some of the pile groups may have, in order to be on the safe side, a capacity that exceeds that of the column load by a substantial amount. Furthermore, it is common practice to use, for reasons of stability, a minimum of three piles in a free standing pile group; a minimum of two piles if a foundation beam or similar provides lateral support; and a single pile only if lateral support can be provided in two directions. These minimum requirements have to be satisfied even if the capacity provided by the pile group far exceeds the amount of the load to be supported. It is good practice to design the pile caps in any case for the full allowable capacity of the group. This is done whether required by the column load or not, and in spite of the waste that may be connected with it, to permit full utilization of the pile capacity under any circumstances.

In the case of bearing piles, every pile in a group may be considered to act as an independent pier down to the bearing stratum and to share equally in the carrying of the load. In the case of friction piles, the number of piles in a group affects their carrying capacity, especially that of the interior piles. Although this deficiency is usually averaged over the entire pile group, as far as the capacity of the group is concerned, the variations in the capacity of each individual pile requires, sometimes, consideration in the design of the pile cap.

In either case, whether we are dealing with bearing piles or friction piles, there is always a chance that some piles in the group may develop a smaller (or greater) resistance than others; a pile cap ought to be stiff enough to equalize this condition. It is therefore advisable not to keep the effective depth of a pile cap down to the minimum required, but to increase it somewhat wherever possible.

The design of a pile cap follows in general the same rules and regulations as that of a spread footing, except that the base reactions (pile reactions) are applied as concentrated loads in the center of each pile. Attention is drawn to section 15.5.5 of ACI 318-71 which states that "in computing the external shear on any section through a footing supported on piles, the entire reaction from any pile whose center is located $d_p/2$ (d_p is the pile diameter at the upper end) or more outside the section shall be assumed as producing shear on the section. The reaction from any pile whose center is located $d_p/2$ or more inside the section shall be assumed as producing no shear on the section. For intermediate positions of the pile center, the portion of the pile reaction to be assumed as producing shear on the section shall be based on straight line interpolation between full value at $d_p/2$ outside the section and zero value at $d_p/2$ inside the section."

For evaluation of pile reactions under various loading conditions see the following section.

The considerable intensity of the concentrated pile reactions requires that more than usual attention be given to the design for shear in the concrete cap and the development (anchorage) of the reinforcement in the section. Due to the importance of a crack free entity of a pile cap in the distribution of the column load to the supporting pile group, the use of plain concrete is not permitted for pile caps.

5.5.3 Evaluation of Pile Reactions

-1. Concentric loading conditions—After the allowable pile reaction, R_{pa} (often incorrectly called "allowable pile capacity"), has been determined or evaluated by principles of soil mechanics,* the minimum number of piles for each column load can be determined as follows:

The effective pile reaction, R_{pe} (kips), consists of the allowable pile reaction, R_{pa} (tons), less the weight of the pile cap per pile, W_p. Any eventual surcharge shall be added to the weight of the pile cap.

$$R_{pe} \text{ (in kips)} = 2R_{pa} - W_p$$

The number of piles, n_p, required to support the unfactored total column load P is then $n_p = P/R_{pe}$, where n_p is to be rounded up to the next whole number.

*Verification of the validity of this "allowable pile reaction" is usually established by one or more pile loading tests performed at the site under actual driving conditions and at the beginning of construction. A safe assumption of the allowable pile reaction, however, has to be made, at a much earlier date to enable the engineer to design the foundation ahead of the actual construction.

Unless special conditions require a spreading of the piles, they are assembled in tight patterns to arrive at the most economical design for the pile caps. An often recommended spacing, c_p, is about three times the butt diameter of the pile, usually not less than 2½ ft. The most common spacing for piles of an average pile reaction ranging from 30 to 70 tons is 3 ft.

-2. Eccentric loading condition or concentric loading with moment at base—To transform eccentric loading conditions into concentric loadings with moment at base proceed as follows:

a. . find pile reaction R_p for concentric loading condition
b. find pile reaction R_{pM} for moment at base
c. superpose 1 and 2

$$R_p + R_{pM} \leqslant 2R_{pa}$$

Where wind or earthquake are included, the R_{pa} can be increased by 33% if so allowed by the local building code. The extreme pile reaction due to a moment M is

$$R_{pM} = \frac{M}{I_{pG}/z_{pG}}$$

To calculate the moment of inertia of a pile group I_{pG}, first find the centroid of the pile group and moment of inertia of all units in the group about the centroidal axis.

$$I_{pG} = \sum_{1}^{n} y^2$$

where y is the distance of each pile in the group from the centroidal axis.

Where a pile group consists of m equal, parallel rows of piles, the moment of inertia of the entire group is

$$I_{pG} = mI_p/\text{Row} = m\,\frac{n_{pr}(n_{pr}^2 - 1)}{12}\,c_p^2$$

and the section modulus for the extreme piles in the group is

$$S_{pG} = m\,\frac{n_{pr}(n_{pr} + 1)}{6}\,c_p$$

However, if the parallel rows are not of the same configuration, sum up the moments of inertia for the various rows and find the section modulus of the extreme pile by dividing the moment of inertia of the entire group by the distance of the extreme pile from the centroid, as

$$S_{pG} = I_{pG}/z_{pG}$$

EXAMPLE 5-6: As discussed in section 5.3.2 for ordinary spread footings, the number of piles or their arrangement in the pile group depends only on the unfactored loading conditions, as shown in Fig. 5-37, and the strength design of the pile cap has to be done by converting all loads and reactions to the factored conditions.

column load:

$$\begin{aligned} D &= 400 \text{ kip} \\ L &= 520 \\ \hline \text{total} &= 920 \text{ kip} \end{aligned} \qquad R_{pa} = 50 \text{ tons}$$

$$R_{pe} = 2R_{pa} - W_p = 2 \times 50 - 6.5 = 93.5 \text{ kip}$$

$$n_p = \frac{920}{93.5} = 9.8 \cong 10 \text{ piles}$$

$$W_p = A_p(\Delta q) = 3^2(50 + 75 + 150 + 450)$$

$$= 6525 \text{ lb} \cong 6.5 \text{ kip}$$

where $\Delta q = [w_L + \text{slab} + \text{fill} + \text{cap}]$. See Fig. 5-37.

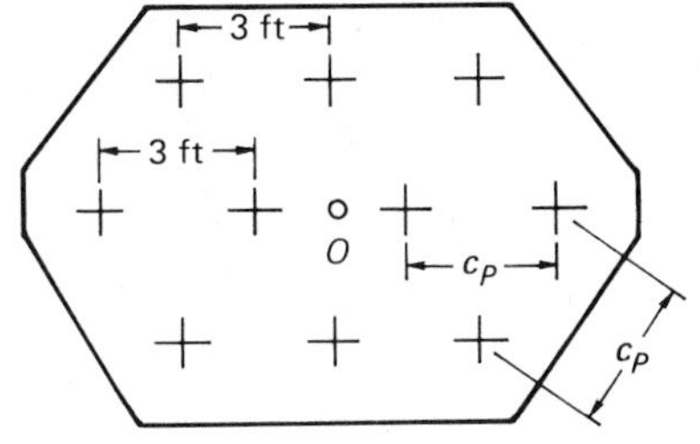

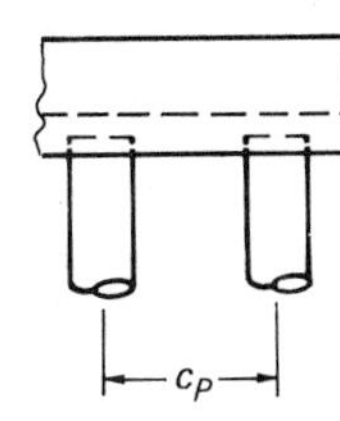

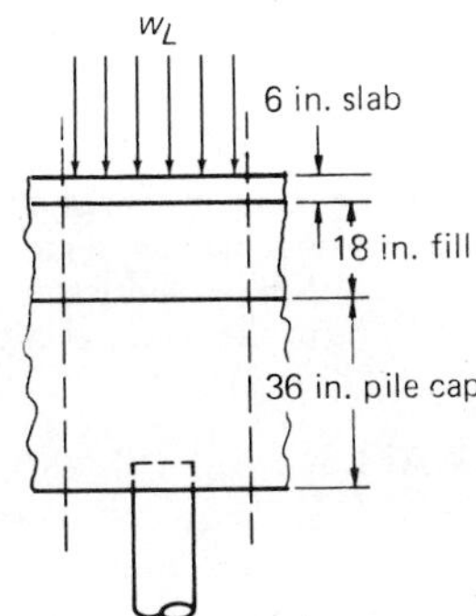

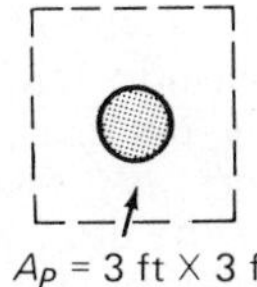

Fig. 5-37 Pile cap. Conventional pile arrangement in ten-pile cap.

EXAMPLE 5-7: Investigate ex. 5-6 for an additional wind moment of 450 kip-ft in the long direction of the pile group.

The moment of inertia of the entire pile group can be considered as the sum of the moments of inertia of each row of piles or

$$I_{pG} = \sum \frac{n_{pr}(n_{pr}^2 - 1)}{12} c_p^2 = \left[2 \times \frac{3(3^2 - 1)}{12} + 1 \times \frac{4(4^2 - 1)}{12}\right] 3^2 = 81 \text{ ft}^3$$

The section modulus of the extreme pile in longitudinal direction is then

$$S_{pG} = I_{pG}/1.5\, c_p = 81/1.5 \times 3 = 18 \text{ ft}^2$$

and the reaction on this pile due to the wind moment is

$$R_{pM} = \frac{M}{S_{pG}} = \frac{450}{18} = 25 \text{ kip}$$

Summing up, we obtain a total maximum pile reaction under wind of

$$R_p + R_{pM} = 98.5 + 25.0 = 123.5 \text{ kip}$$

Since the maximum allowable pile reaction under wind is

$$R_{pa(w)} = 1.33 \times R_{pa} = 1.33 \times 100 = 133 > 123.5 \text{ kip}$$

no increase in the number of piles is required due to wind.

EXAMPLE 5-8: Strength design of pile cap. The column load and allowable pile capacity is the same as in ex. 5-6.

$$f_c' = 3000 \text{ psi}, \quad \text{and} \quad f_y = 60{,}000 \text{ psi}$$

Pier size is 22 X 22 in., the butt diameter of the piles is 14 in. Determine the thickness and reinforcement of the pile cap.

The strength design of the pile cap is based on the R_{pu} which is determined from the factored loading, similar to the q_s for the spread footings, and has also here no other significance.

from dead load

$$\frac{400 \times 1.4}{10} = 56.0 \text{ kip}$$

from live load

$$\frac{520 \times 1.7}{10} = 88.4$$

The factored pile reaction is then 56.0 + 88.4 = 144.4 kips; it is, however, recommended to design the pile cap for the maximum factored pile reaction based on the average load factor.

average load factor

$$\frac{400 \times 1.4 + 520 \times 1.7}{920} \cong 1.6$$

maximum factored pile reaction due to column load is

$$93.5 \times 1.6 \cong 150 \text{ kip}$$

Fig. 5-38 shows the layout for a ten-pile cap and the various approaches that need to be followed in the evaluation of its strength design.

Step 1: For two-way shear section (a), as indicated in the lower left quadrant of Fig. 5-38, let us assume that the necessary depth has been evaluated with 30 in. and is checked herewith: the critical

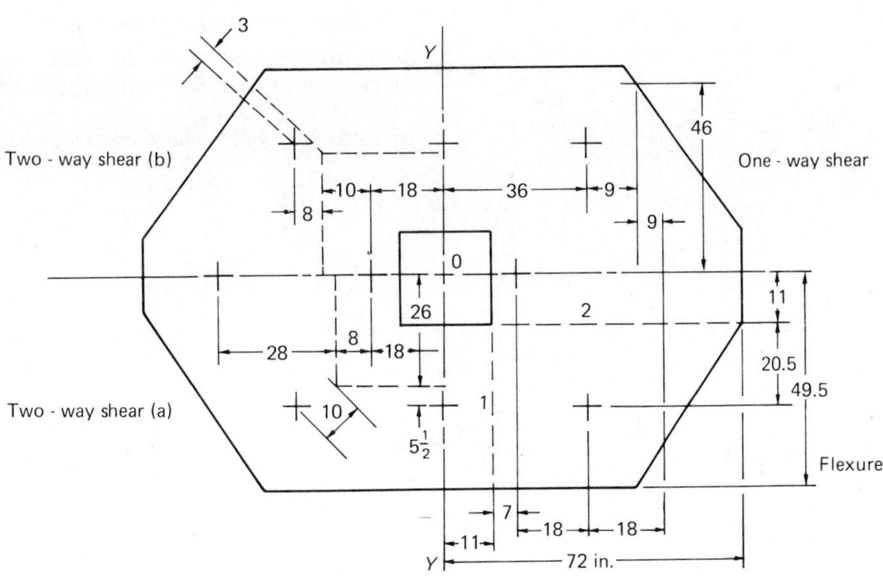

Fig. 5-38 Stress evaluation in pile caps.

base size of the truncated pyramid is 22 + 2 × 30/2 = 52 in. The critical shear force V_{2u} is then

$$V_{2u} = 6 \times 150.0 + 2 \times 16 = 932 \text{ kip}$$

where the contribution of the outer piles located on the $y - y$ axis is

$$150 \times \frac{1.5}{14} = 16 \text{ kip}, \quad d_p = 14 \text{ in.}, \quad \frac{d_p}{2} - 5.5 = 1.5 \text{ in.}$$

$$v_{2u} = \frac{V_{2u}}{b_0 d\phi} = \frac{932 \times 1000}{(4 \times 52) \times 30 \times 0.85} = 170 \text{ psi}$$

which is smaller than $4\sqrt{f_c'} = 220$ psi.

Two-way shear action (b), is indicated in the upper left quadrant of Fig. 5-38. It can be realized, by inspection of Fig. 5-38, that the actual shear distribution is unequal and will be much greater in the long direction. If we take an approach similar to ex. 5-3(b) we require a much greater cap thickness, as evaluated in the approach (a) described above. In this respect we divide the shear action again into two portions separated by a 45° line placed at the corner of the truncated pyramid base. Let us assume again that the necessary thickness of 34 in. nas been evaluated before and is checked below.

The critical base size of the truncated pyramid is here 22 + 2 × 34/2 = 56 in. and the shear force for the most stressed quadrant becomes

$$V'_{2u} = 1 \times 150 + 2 \times 107 = 364.0 \text{ kip}$$

where the contribution of the outer piles is

$$150 \times 10/14 = 107.0 \text{ kip}, \quad d_p/2 + 3 = 10 \text{ in.}$$

$$v_{2u} = \frac{V'_{2u}}{b'_0 d\phi} = \frac{364 \times 1000}{56 \times 34 \times 0.85} = 225 > 220 \text{ psi}$$

but acceptable. The greater depth of 34 in. is, therefore, selected.

Step 2: One-way shear action, as indicated in the upper right quadrant of Fig. 5-38. The critical line is 22/2 + 34 = 45 in. away from the $y - y$ axis. The critical shear force is then V_{1u} = 150.0 kip,

$$v_{1u} = \frac{V_{1u}}{b_f d\phi} = \frac{150 \times 1000}{92 \times 34 \times 0.85} = 56 \text{ psi}$$

which is smaller than $2\sqrt{f_c'} = 110$ psi.

Step 3: The critical sections for flexure, as indicated in the lower right quadrant of Fig. 5-38, are at the face of the pier; the moments and reinforcements are determined for these sections.

critical section 1:

$$M_u = 150.0\left(\frac{7 + 2 \times 25 + 43}{12}\right) = 1250 \text{ kip-ft}$$

$$F = \frac{99 \times 34^2}{12{,}000} = 9.6, \quad K_u = 1250/9.6 = 130, \quad a_u = 4.37$$

$$A_s = \frac{M_u}{a_u d} = \frac{1250}{4.37 \times 34} = 8.4 \text{ in.}^2, \quad \rho_{min} = \frac{200}{60{,}000} = 0.0033$$

$$A_{s\,min} = 0.0033 \times 99 \times 34 = 11.3 \text{ in.}^2, \text{ or } 1.33 \times 8.4 = 11.2 \text{ in.}^2$$

which governs

critical section 2:

$$M_u = 3 \times 150 \times 20.5/12 = 770 \text{ kip-ft}$$

$$F = \frac{144 \times 34^2}{12{,}000} = 13.7, \quad K_u = \frac{770}{13.7} = 56, \quad a_u = 4.45$$

$$A_s = \frac{M_u}{a_u d} = \frac{770}{4.45 \times 34} = 5.1 \text{ in.}^2$$

$$A_{s\,min} = \rho_{min}\, bd = 0.0033 \times 144 \times 34 = 16.1 \text{ in.}^2, \text{ or}$$

$$1.33 \times 5.1 = 6.8 \text{ in.}^2 \text{ which governs}$$

The selection and distribution of the bars is done as described in Example 5-1, Step 4b. The maximum bar size that may be used has to be selected in such a way that the development (anchor) length of the bar is smaller or equal than the shortest available embedment length of the bar at either side of the critical section.

5.6 RETAINING WALLS

5.6.1 General

A retaining wall is a structure designed for the purpose of providing one-sided lateral confinement of soil or fill.

All retaining walls, with the exception of true cantilever walls anchored to rock, are in principle gravity walls, i.e., their action depends primarily on their developed weight. In common practice, however, only those retaining walls are called gravity walls where the dead weight required to make the resultant vector intersect the base within safe allowable limits is made up solely by the dead weight of the concrete. Such walls are usually designed unreinforced, Fig. 5-39a. The commonly called cantilever walls are in principle gravity walls where reinforcement is used to reduce and modify the cross section of the concrete in such a way that portions of the soil or fill are utilized for developing the necessary rightening moment, Fig. 5-39b.

In every retaining wall design, regardless of the type used, three resultant forces, namely, the lateral confinement pressure, Q, the total developed weight, P, and the soil reaction or bearing resistance, R, have to be brought into equilibrium Fig. 5-40a; in addition all internal stresses in the structure and all external soil reactions have to be within the permissible limits.

Retaining walls of the commonly called cantilever type can be subdivided into two main groups:

1. Continuous walls of constant cross section, where every foot of wall length is providing its own equilibrium, Fig. 5-39b.
2. Sectional walls, where crosswalls introduced at certain

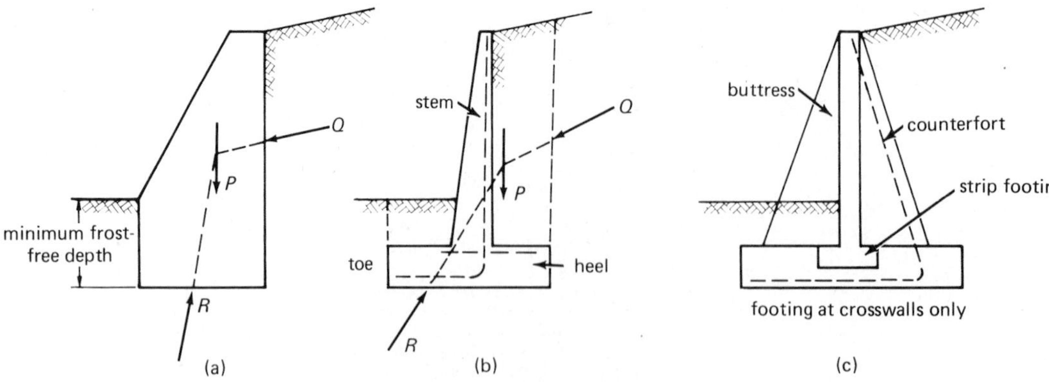

Fig. 5-39 Types of retaining walls.

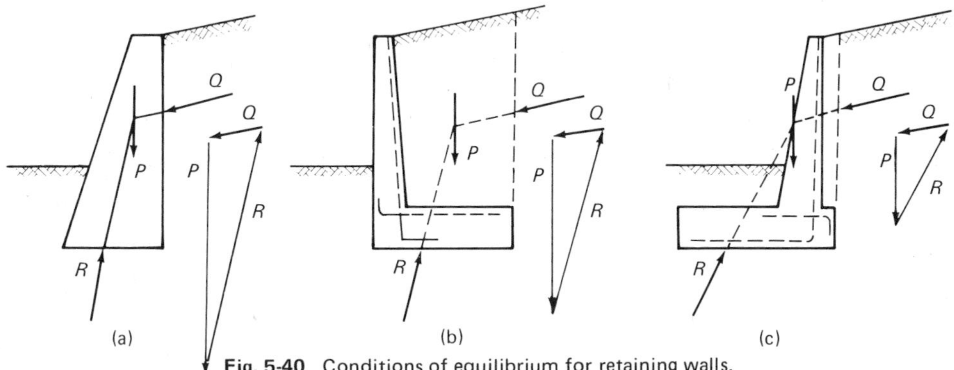

Fig. 5-40 Conditions of equilibrium for retaining walls.

spacings, provide all stability requirements and the walls between them act only as intermediate elements, Fig. 5-39c.

Where the crosswalls are visible in front they are called buttresses, where they are behind the wall and inside the soil, they are called counterforts. Some retaining walls are designed to have both.

The portion of a continuous cantilever wall or crosswall which is pressed downward into the soil is called the toe, and the portion which is lifted upward is called the heel. The vertical portion is called the stem, Fig. 5-39b.

The footing of a continuous retaining wall or crosswall must be large enough—

1. to resist the resultant vector due to confinement pressures, dead weight of the concrete and developed weight of the soil by means of safe bearing pressures; and
2. to keep the wall safely from overturning.

5.6.2 Confinement Pressure

The magnitude and distribution of the lateral pressures exerted by the confined soil or backfill depends on the kind of material, its moisture content, existence and depth of the groundwater, slope of backfill, and eventual surcharge due to live loads, storage, building loads, etc., applied close enough to be of influence. These pressures and their distribution have to be evaluated by principles of soil mechanics and do not form a part of this discussion. Most textbooks on Soil Mechanics contain detailed information on this subject.[5-1, 5-2]

The resultant of these pressures is located at the centroid of the pressure wedge. The angle of inclination between the resultant and a line perpendicular to the back of the wall indicates the wall friction and is usually expressed as a fraction of the angle of internal friction, ϕ, of the fill material; it is often assumed with $\phi/2$. It is, however, important to keep in mind that this angle is also influenced by the slope, material and compaction of the backfill, by the surface texture of the concrete (at the back of the wall), by the existence of ground water behind the wall, or merely, by moisture in the soil which can act like a lubricant and reduce the friction angle to a minimal value. It is common practice with many designers to disregard the wall friction and to apply the resultant perpendicular to the back of the wall, in order to be on the safe side.

Spaced weep holes or continuous backdrains are often provided to alleviate the heavy pressure condition that can be caused by groundwater accumulating behind the back of the wall. Since weep holes and other drainage provisions may be clogged, it is recommended to investigate a retaining wall for a condition of full, or at least increased, water pressure. Such loading, however, represents an emergency condition and it is up to the engineer's judgement to reduce the safety factor for such a design as he sees fit.

5.6.3 Bearing Pressure

The pressure distribution under the footing of a retaining wall follows the same rules and is determined by the same methods as the pressure distribution under an eccentrically loaded footing. It is of course desirable, and to be attempted wherever possible, to keep the intersection of the resultant of all active forces (confinement pressures and developed weight) within the kern of the footing base. (In the case of a strip footing under a continuous retaining wall, the kern distance is $1/6$ of the footing size in the direction of the loading.) However, in many cases this is not economically feasible and greater eccentricities have to be accepted with a pressure distribution extending only over a part of the footing. The maximum edge pressure is then determined as discussed in sections 5.2.2 and 5.2.3 and shown in Fig. 5-5.

5.6.4 Overturning

Overturning can also be treated similarly as for regular column or wall footings. Where the base of the retaining wall is resting on rock or very hard soil, overturning may be calculated about the pressed edge and the safety factor can be expressed by the ratio $SF = M_Q/M_R$, where M_Q is the overturning moment caused by the confinement pressures acting about the pressed edge, and M_R is the resisting moment consisting of the dead weight of the retaining structure plus the developed weight of the fill material and any other frictional or passive resistances in the soil that may be mobilized during the overturning. The safety factor may also be expressed according to the "Suggested Design Procedures for Combined Footings and Mats"[3] as the ratio of the distance of the pressed edge from the base centroid to the eccentricity, e. In this ratio, $e = M_B/P$, where M_B is the sum of all moments about the base centroid and P is the sum of all forces acting perpendicular to the base. The safety factor against overturning should customarily be not less than 1.5.

Where the retaining wall rests on soil, the investigation regarding overturning may proceed along similar lines, except that the critical line about which the overturning and resisting moments are to be calculated is not at the pressed edge but somewhat inside, at the distance e_f from the centroid which is the center of gravity of the pressure block evaluated for an ultimate soil bearing condition. The evaluation of this condition is similar to the one described in section 5.2.1 and shown in Fig. 5-5c. Where the ultimate soil

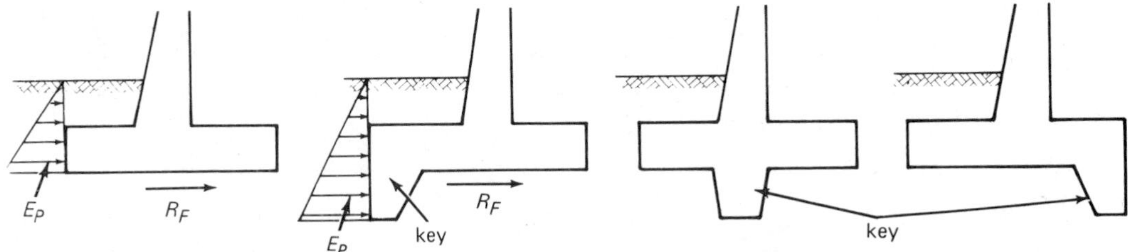

Fig. 5-41 Lateral resistance at base for retaining walls.

pressure is not known, it may be assumed to be 2.5 times the allowable soil pressure.[5-3]

5.6.5 Sliding Resistance

Lateral resistance against the horizontal component of the confinement pressure, commonly called resistance against sliding, has to be supplied by static friction at the footing base and by passive earth pressure against the embedded front portion of the retaining wall. Where this resistance is insufficient, the passive pressure can be increased by extending a key or lug into the soil below the footing base. Since the preference for the location of the lug can be argued about, it is probably located best where it is most practical with regard to construction and placement of reinforcement, Fig. 5-41.

5.6.6 Type Selection

The quality of the available subsoil and its allowable bearing pressure, as well as the height of the wall itself and the available construction space in front of the slope, can influence the selection of an economical type of wall to a great degree.

Where the construction space is ample, the allowable soil pressure is reasonable, and the height of the structure is not excessive, gravity walls of structural plain concrete can be used with advantage. Their weight is usually large enough to develop all the necessary base friction; however, the provision of keys or lugs projecting into the firm subsoil are rather common.

Where the available bearing pressure is large and construction space is available, the designer can utilize a long heel and develop much back fill weight to deflect the horizontal pressure resultant sharply down to the ground Fig. 5-40b. However, where the available bearing pressure is small, too much of it would have to be used to carry the developed weight of the backfill. In this case, the designer will have to be satisfied with a smaller total dead weight which will decrease the inclination of the resultant. Consequently he will have to increase the toe to keep the resultant sufficiently within the base to satisfy the allowable bearing pressure, and to keep the wall from overturning Fig. 5-40c. For ordinary conditions it may be considered good practice to keep the sizes of heel and toe about the same. The base of the toe has to be placed at frost free level.

Butresses and/or counterforts are used presently only for retaining walls of greater height (more than 20 to 25 ft, unless buttresses are architecturally desired) because of the great expense usually involved in the formwork. This type of design may see a kind of revival in combination with the use of precast wall panels.

For the enclosure of underground storage or extra large basements areas, freestanding retaining walls are sometimes used as basement walls.

In certain cases retaining walls are designed to act as such only temporarily until a full tie-in with the rest of the structure is achieved. In other cases such walls are designed only for the final loading condition, but not for the loading during construction, and will therefore require temporary shoring to maintain their stability. Some retaining walls may during construction change from a fully independent free standing structure to a top and bottom supported basement wall and in such a case every condition of loading ought to be considered for their stability as well as for their strength.

The use of precast sections may also here see a considerable field of application in the future. Such sections may, in the form of sheetings or in the form of entire wall panels, be driven or inserted into the ground before the main excavation has taken place. In such a case the same unit may serve successively as protective sheeting, retaining wall and final basement enclosure.[5-9]

Free standing retaining walls may also be constructed entirely of precast units. Such designs may simulate either the action of a sectional cantilever wall as shown in Fig. 5-39c, or that of a gravity wall in which case they are called cribbings.

5.6.7 Cribbings[5-10]

Cribbings consist in general of two types of units, namely, face and anchor units, also called stretchers and headers.

The structural design of cribbings is usually based on an empirical evaluation. The open faced units provide excellent drainage; in other cases drainage has to be provided to prevent groundwater from backing up and exerting pressure and unsightly leakage.

The satisfactory behavior of such precast cribbings depends to a considerable degree on the quality of the compacted backfill, which shall be installed in close coordination with the placement of the sections.

The face units are straight precast members of various cross sections, often with protruding lugs at their ends to connect them to the adjacent face and/or anchor members. Face members can be designed to open, closed, and flush-type manner, depending on the architectural requirements. They are usually set with a batter of about 1:6 at the front and can also be placed so as to form a curved face, up to 20°, without special units.

The anchor units usually come in two types of design, fishtail units or continuous back units. A fishtail unit is a T-shaped, precast section, placed perpendicular to the face of the wall with the purpose of tying two adjacent face units together and anchor them back to the fill material. Continuous back units are used together with cross wall units to form box like openings that are filled with compacted soil. There exist also combined units, where face and anchor units are cast together, simplifying erection where transportation permits.

5.6.8 Pile Supported Retaining Walls

Retaining walls may also be supported on piles which is often the case along waterfronts, or where the subsoil does not have a sufficient bearing capacity or lateral stability.

In such designs the vertical component of the pressure resultant is to be taken by the piles similarly as for a strip footing and the horizontal component is to be taken either by batter piles, tie backs, or similar devices.

Where the subsoil is also capable of providing uplift anchorage for the piles, the necessary developed weight may be reduced accordingly.

5.6.9 Design Procedure

The static design and the stability investigation of a retaining wall shall be based on service load conditions; the size of the footing and its location with regard to the wall itself is, therefore, entirely governed by the actual fill and/or liquid pressures under consideration of the allowable soil bearing pressures and allowable values for soil friction and passive resistance. The structural design of the stem and footing sections and their reinforcement, however, shall be based on the strength design method. For this purpose, the actual loads and active confinement pressures have to be multiplied by the appropriate load factors, and the resulting bearing pressures and other resistances have to be evaluated from these factored loading conditions. The procedure is not always easy to be executed because of variations in eccentricities that may be caused by it (see ex. 5-9, section 7). It is important to keep in mind, just as in the case of the ordinary footing, that these reactions are only caused by assumed factored loading conditions, but have no relationship to ultimate soil bearing values or similar.

After the ultimate loadings and corresponding reactions have been determined, every part of the retaining wall (stem, toe, or heel) has to be designed independently as a fully restrained cantilever section protruding from a mass center and carrying all applied loads.

The design to be performed for the service load conditions consists of a trial-and-error approach. The result is not too sensitive to slightly incorrect assumptions and can usually be adjusted in a second trial, if so desired. Due to the multitude of possible combinations it is rather difficult to arrive at reliable recommendations in this respect, however, if the conditions are not too much out of the ordinary, the footing size perpendicular to the wall directions can be assumed to be about $^2/_5$ to $^2/_3$ of the total wall height, the footing thickness about $^1/_8$ to $^1/_{12}$ of the footing size, and the base of the stem about $^1/_{10}$ to $^1/_{12}$ of its height. The following example has not been cut down to the minimum size, i.e., it could be adjusted and recalculated for further savings; such a procedure would be mostly repetitive and more confusing than helping. The top of the stem should not be made less than 8 in. to be able to place into it two layers of reinforcement, if so required. It is customary to place a second layer of reinforcement inside the exposed face for shrinkage and cracking control. It is practical to coordinate the reinforcements of stem and toe in such a way that the reinforcing bars of the stem can be bent right into the toe. For greater wall heights it is economical to run only a part of the vertical wall reinforcement to the top and stop the remainder at lower elevations.

EXAMPLE 5-9: Design a continuous retaining wall (cantilever wall) for the following conditions:

a. The difference between upper and lower level is 12 ft.
b. The upper level shall sustain a surcharge of 200 lb/ft^2 (medium heavy parking).
c. Soil of satisfactory bearing capacity is encountered at a depth of 4 ft below lower level (must be equal or greater than frost-free depth).
b. Provide weep holes to relieve water pressure; maximum expected ground water level 3 ft above lower level.
e. Use concrete with minimum f'_c = 3000 psi, and reinforcing steel having a minimum yield point of 60,000 psi.
f. Soil mechanical considerations provided the following values: The bearing soil is a medium dense, silty sand with an allowable bearing pressure of 3000 lb/ft^2. The weight of the moderately dry soil can be assumed with 100 lbs/ft^3 and its angle of internal friction ϕ = 32°.

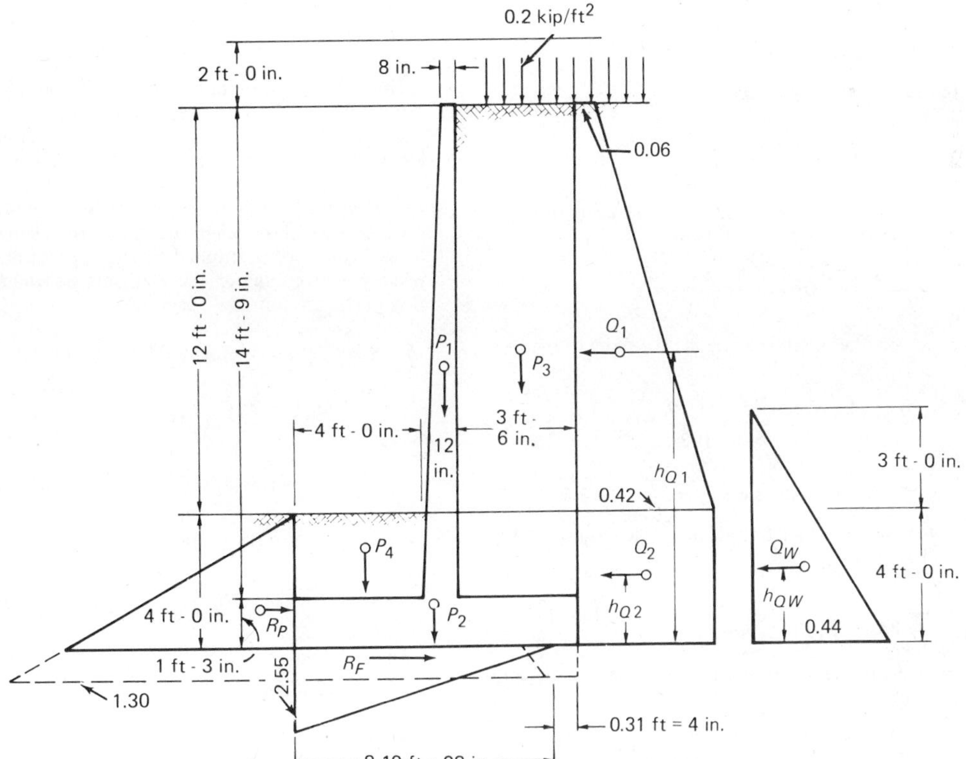

Fig. 5-42 Loadings and reactions for retaining wall, Example 5-8.

-1. Assumption of size and evaluation of loadings—Based on the requirements and other informations given above, the designer makes an assumption for the shape, sizes, and other relationships, and arrives at an arbitrary cross section as given in Fig. 5-42. Using these assumptions he arrives at the following values:

$$P_1 = \frac{(8 + 12)}{12 \times 2} \times 14.75 \times 0.15 \qquad = 1.85 \text{ kip}$$

$$P_2 = 8.50 \times 15/12 \times 0.15 \qquad = 1.60 \text{ kip}$$

$$P_3 = 3.50\,(14.75 \times 0.1 + 0.20) \qquad = 5.90 \text{ kip}$$

$$P_4 = 2.75 \times 4.00 \times 0.1 \qquad = 1.10 \text{ kip}$$

total dead load $P = P_1 + P_2 + P_3 + P_4 = 10.45$ kip/foot of wall

eccentricity of total dead load P from centroid of base area, o;

$$e_P = \frac{P_1 \times 0.05 - P_3 \times 2.50 + P_4 \times 2.25}{P} = -1.15 \text{ ft}$$

the active earth pressure

$$p_a = \gamma h \tan^2 (45^\circ - \phi/2)$$

Introducing above values the pressure increment can be evaluated with $p_a = 0.03$ kip/ft^2 per foot of depth. Transforming the given surcharge into an equivalent height of additional soil we obtain, for the active earth pressure at the upper level, $p_{a1} = 2 \times 0.03 = 0.06$ kip/ft^2, and at the lower level $p_{a2} = 14 \times 0.03 = 0.42$ kip/ft^2.

Below this elevation the active earth pressure remains constant.

$$Q_1 = \frac{0.06 + 0.42}{2} \times 12 = 2.88 \text{ kips}$$

$$Q_2 = 0.42 \times 4.0 \qquad = 1.68 \text{ kips}$$

total active earth pressure is

$$Q = Q_1 + Q_2 = 4.56 \text{ kip/lin-ft of wall.}$$

The elevations at which they are applied are for Q_1

$$h_{Q1} = 4.0 + \frac{12}{3}\left(\frac{2 \times 0.06 + 0.42}{0.06 + 0.42}\right) = 4.0 + 4.5 = 8.5 \text{ ft}$$

for Q_2

$$h_{Q2} = 2.0 \text{ ft}$$

and for the resultant active earth pressure, Q

$$h_Q = \frac{2.88 \times 8.5 + 1.68 \times 2.0}{4.56} = 6.1 \text{ ft}$$

-2. Calculation of bearing pressures—The resultant eccentricity at the base of the retaining wall can be found from

$$e = \frac{M_Q \pm M_P}{P} = \frac{Q \times h_Q \pm P \times e_P}{P}$$

The (±) sign depends on the location of e_P with regard to the base centroid, o.

$$e = \frac{4.56 \times 6.1 - 10.45 \times 1.15}{10.45} = \frac{15.9}{10.45} = 1.52 \text{ ft}$$

which is slightly outside the kern distance $c_k = \frac{8.5}{6} = 1.43$ ft.

The maximum toe pressure can be calculated according to Fig. 5-5 as

$$q = \frac{2P}{3mb} = \frac{2 \times 10.45}{3(4.25 - 1.52)\,1} = 2.55 \text{ kip/ft}^2 < 3.0 \text{ kip/ft}^2$$

-3. Overturning—Overturning according to Fig. 5-5c can be computed as follows:

$$e_f = \frac{l_f}{2} - \frac{P}{2q_f b_f} = \frac{8.5}{2} - \frac{10.45}{2(2.5 \times 3)\,1} = 3.55 \text{ ft}$$

$$SF = \frac{e_f}{e} = \frac{3.55}{1.52} = 2.30 > 1.5$$

Overturning under full water pressure is computed using: the total water pressure above base elevation

$$Q_W = (7.0 \times 0.0624)\,\frac{7.0}{2} = 1.55 \text{ kip}$$

and the additional moment due to the waterpressure about the base

$$M_W = 1.55 \times \frac{7.0}{3} = 3.6 \text{ kip-ft}$$

The eccentricity under this extreme condition would be

$$e_W = \frac{M_Q \pm M_P + M_W}{P_{\text{RED}}} = \frac{28.0 - 12.1 + 3.6}{9.57} = 2.0 \text{ ft}$$

The P in this equation was reduced to account for the buoyancy of the submerged portions.

$$SF = \frac{e_f}{e_W} = \frac{3.55}{2.0} = 1.77 > 1.5$$

-4. Sliding—A friction coefficient of 0.45 was assumed from soil mechanical considerations.

The total horizontal force is $Q = 4.56$ kip, then the frictional resistance that can be developed at the base is

$$R_F = 0.45P = 0.45 \times 10.45 = 4.75 \text{ kip}$$

The passive earth pressure that can be developed at the front of the retaining wall is

$$p_{p1} = \gamma \tan^2 (45^\circ + \phi/2)h = 0.1 \tan^2\left(45^\circ + \frac{32^\circ}{2}\right) 4.0 = 1.3 \text{ kip/ft}^2$$

$$R_p = p_{p1}\, h/2 = 1.3 \times \frac{4.0}{2} = 2.6 \text{ kip}$$

The total resistance is therefore

$$R = R_F + R_P = 4.75 + 2.60 = 7.35 \text{ kip}$$

and the safety factor

$$SF = \frac{7.35}{4.56} = 1.6 > 1.5$$

This safety factor appears to be satisfactory. However, under full water pressure the horizontal force will be larger and the frictional resistance smaller due to the lubricating effect of the water. Provision of a key at the base is therefore recommended.

-5. Design of concrete thicknesses and reinforcement—In order to proceed with the strength design of the elements the actual service loads have to be multiplied by the appropriate load factors, and the respective strengths of the elements be evaluated under consideration of the appropriate ϕ-factors.

-6. Stem design—Moment about base of stem

$$M_u = 1.7\left[Q_1 (h_{Q1} - 1.25) + p_{a2}\,\frac{(2.75)^2}{2}\right]$$

$$= 1.7\left[2.88(8.50 - 1.25) + 0.42 \times \frac{(2.75)^2}{2}\right]$$

$$= 38.2 \text{ kip-ft}$$

for 2 in. concrete protection,

$$d_{\text{eff}} = 12 - 2.5 = 9.5 \text{ in.}$$

$$F = \frac{bd^2}{12{,}000} = \frac{12 \times 9.5^2}{12{,}000} = 0.09, \quad K_u = \frac{M_u}{F} = \frac{38.2}{0.09} = 425$$

a_u from Table 1-2 of Chapter 1, $a_u = 4.03$

$$A_s = M_u/a_u d = 38.2/4.03 \times 9.5 = 1.01 \text{ in.}^2 \ \#7 @ 7$$

Check for full water pressure:

$$M_u = 38.2 + 1.4\left[\frac{(7-1.25)^2}{2}\,0.0624 \times \frac{(7-1.25)}{3}\right]$$

$$= 38.2 + 1.4\,[1.04 \times 1.9] = 40.95 \text{ kip-ft}$$

$$K_u = \frac{40.95}{0.09} = 455, \quad a_u = 4.0, \quad A_s = \frac{40.95}{4.0 \times 9.5} = 1.07$$

$$A_s \text{ provided} = \#7 @ 7 = 1.03 \text{ in.}^2; \ \frac{1.07}{1.03} = 1.04, \text{ O.K.}$$

-7. *Toe design*—It is advisable to check the shear condition first. In connection with bearing pressures, it is sometimes difficult to apply the corresponding load factors properly if the bearing pressures were caused by loadings with different load factors. Such a procedure may sometimes cause relocation of the resulting eccentricities and lead to pressure distributions that are in principle different from those obtained under service load conditions. Since this is not considered to be the intent of the design it is recommended to select in such cases either the highest load factor, 1.7, in order to be on the safe side, or to select by judgment an approximate load factor between 1.4 and 1.7, to apply to the bearing pressures obtained from the investigation of the service load conditions.

In this example a load factor of 1.7 was used.

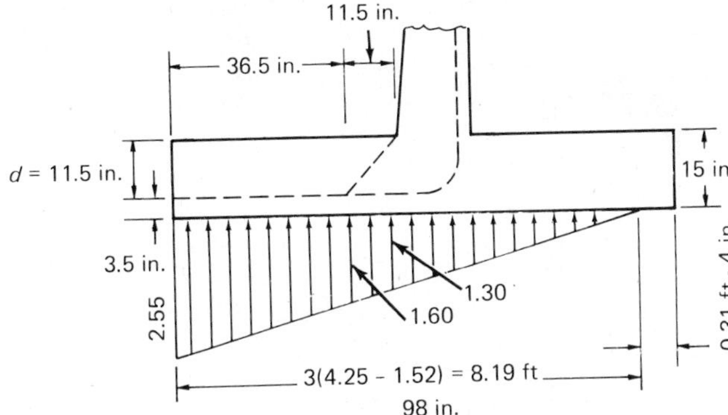

Fig. 5-43 Toe Design

$$d_{eff} = 15 - 3.5 = 11.5 \text{ in.}$$

$$V_{1u} = 1.7\left(\frac{2.55 + 1.60}{2}\right)\frac{36.5}{12} = 10.8 \text{ kips}$$

$$v_{1u} = \frac{1000\,V_{1u}}{\phi Bd} = \frac{1000 \times 10.8}{0.85 \times 12 \times 11.5} = 92 \text{ psi}$$

$$v_{c\,all} = 2\sqrt{f_c'} = 110 \text{ psi}$$

$$92 < 110$$

For flexure, the bearing pressure outside the critical line (face of stem) is

$$1.7\left(\frac{2.55 + 1.30}{2}\right)4 = 13.0 \text{ kip}$$

The centroid of the trapezoidal pressure distribution is 2.25 ft away from the face of the stem. Hence

$$M_u = 13.0 \times 2.25 = 29.5 \text{ kip-ft}, \quad F = \frac{12 \times 11.5^2}{12{,}000} = 0.132$$

$$K_u = \frac{29.5}{0.132} = 220, \quad a_u = 4.27, \quad A_s = \frac{29.5}{4.27 \times 11.5} = 0.62 \text{ in.}^2$$

Since it is practical to bend the stem reinforcement right into the toe, the provided stem reinforcement, $A_s = 1.03$ in.^{2}, is compared with the required one and found to be ample.

-8. *Heel design*—

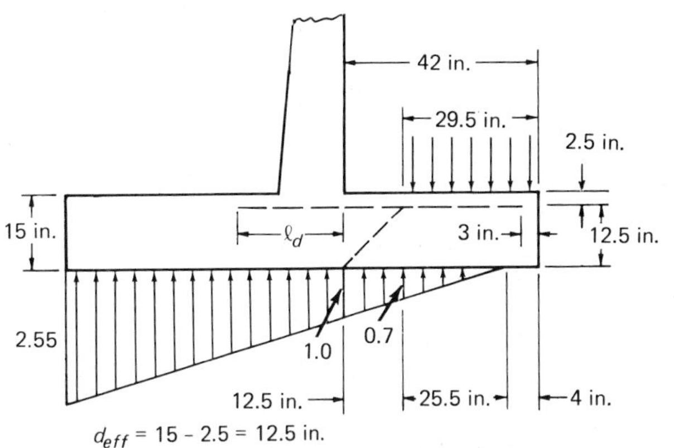

Fig. 5-44 Heel Design.

The shear force is

$$V_{1u} = 1.7\left(\frac{P_3 \times 29.5}{42} - \frac{0.7}{2} \times \frac{25.5}{12}\right) = 6.0 \text{ kip}$$

$$v_{1u} = \frac{1000\,V_{1u}}{\phi bd} = \frac{1000 \times 6.0}{0.85 \times 12 \times 12.5} = 47 \text{ psi} < 110 \text{ psi}$$

The flexural moment to be resisted is

$$M_u = 1.4 \times P_3 \times 1.75 - 1.7\left[\frac{1.0}{2}\left(\frac{38}{12}\right)^2 \frac{1}{3}\right] = 14.5 - 2.8$$

$$= 11.7 \text{ kip-ft}$$

$$F = \frac{12 \times 12.5^2}{12{,}000} = 0.155, \quad K_u = \frac{11.7}{0.155} = 76 \text{ less than min.}$$

$$A_s = \frac{1.33\,M_u}{a_u d}, \text{ or } \rho_{min}\,bd$$

$$A_s = \frac{1.33 \times 11.7}{4.42 \times 12.5} = 0.28 \text{ in.}^2, \text{ or } 0.0033 \times 12 \times 12.5 = 0.49 \text{ in.}^2$$

$$A_s = 0.28 \text{ in.}^2 \text{ governs}$$

A bar size has to be selected for which the development length is smaller than the available embedment 42 − 3 = 39 in. Use #5 bars @ 12 in. o.c. See Tables 1-10 of Chapter 1. The bars have to be extended for the full development length beyond the face of the stem.

NOTATION

A_F = base area of footing, ft^2

A_P = cross sectional area of pier, in.^{2}

A_{ip} = average influence area per pile, ft^2

A_{V1} = influence area of bearing pressure for one-way shear, ft^2

A_{V2} = influence area of bearing pressure for two-way shear, ft^2

ΔA_{V2} = base area of truncated cone or pyramid for two-way shear, ft^2

a = footing projection in general, ft

a_L = projection of footing beyond critical face in long direction, in.

a_s = projection of footing beyond critical face in short direction, in.

a_u = factor used in determining $A_s = M_u/a_u d$

b = width of combined footing, ft

b_f = side of square footing, ft

b_f = short side of oblong footing, ft

b_o = base perimeter of truncated cone or pyramid for two-way shear, in.
b_p = dimension of pier parallel to footing side b_f, in.
c_c = concrete protection, in.
c_k = kern distance, ft
c_P = pile spacing, ft
D = dead loads or their related internal moments and forces
d = effective depth of section, in.
d_b = bar diameter, in.
d_f = diameter of round or polygonal footing, ft
d_o = diameter of opening in circular or polygonal footings, ft
d_P = pile diameter, in.
E = earthquake loads or their related moments and forces
e = eccentricity of load resultant from footing centroid, ft
e_f = eccentricity, as under e, causing failure pressure q_f at edge, ft
e_{max} = maximum permissible eccentricity to prevent overturning, ft
F = $bd^2/12{,}000$
γ_s = average unit weight of footing + surcharge, kip/ft^3
H = horizontal force, kip
h = lever arm of horizontal force, ft
h_f = thickness of footing, in.
h_Q = lever arm of force Q, ft
h_s = depth of footing base below floor, ft
I_c = moment of inertia of concrete cross section, in.^{4}
I_F = moment of inertia of footing base area, ft^4
I_{PG} = moment of inertia of pile group, ft^2
K_u = factor used in determining $F = M_u/K_u$
K_{si} = coefficient of vertical subgrade reaction for a 1 ft^2 area, tons/ft^2/ft
L = live loads and their related internal moments and forces
l = long dimension of combined footing, ft
l = distance between column centers, ft
l_d = development length (anchorage) of bar, in.
l_f = long side of rectangular footing, ft
l_P = dimension of pier parallel to footing side l_f, in.
M, M_u = flexural moment, kip-ft
M_B = moment about base, kip-ft
m = distance of eccentric load from pressed edge of footing, ft
m = general footing dimension, ft
m = number of pile rows of equal configuration
N = blow count of standard penetration test
n = general footing dimension, ft
n_P = number of piles in group
n_{Pr} = number of piles in row
o = centroid of footing or pile group
P, P_u = total column load at base, kip
P_D, P_{Du} = column dead load, kip
P_L, P_{Lu} = column live load, kip
P_{RES} = resultant load, kip
p_a = active earth pressure, kip/ft^2
p_p = passive earth pressure, kip/ft^2
Q = lateral force against retaining wall, kip
q = bearing pressure, kip/ft^2
Δq = weight of footing + surcharge, kip/ft^2
q_a = allowable bearing pressure, kip/ft^2
q_c = bearing pressure at base of column or pier, kip/in.^{2}
q_e = effective bearing pressure $(q - \Delta q)$, kip/ft^2
q_f = bearing pressure at failure, kip/ft^2
q_m = bearing pressure due to moment, kip/ft^2
q_P = bearing pressure under concentric loading, kip/ft^2
q_s = bearing pressure at footing base due to factored loading, kip/ft^2
R = lateral resistance, kip
R_F = frictional resistance at base, kip
R_M, R_{Mu} = pile reaction due to moment, kip
R_P, R_{Pu} = pile reaction due to concentric loading, kip
R_{P_a} = allowable pile reaction, tons
R_{P_e} = effective pile reaction $(2R_{P_a} - W_P)$, kip
ρ = percentage of reinforcement
S = ratio of long side to short side of footing
s = soil deformation
S_{PG} = section modulus of pile group, ft
S_{PR} = section modulus of pile row, ft
SF = safety factor
U = uplift force, kip
V_{1u} = shear force for one-way shear, kip
V_{2u} = shear force for two-way shear, kip
v_{1u} = average shear stress for one-way shear, psi
v_{2u} = average shear stress for two-way shear, psi
W = wind loads or their related moments and forces
W_P = weight of pile footing + surcharge, per pile, kip
w = line load, kip/ft
w_L = live load on floor, kip/ft^2
z = distance of extreme fiber from centroid, ft
z_{PG} = distance of extreme pile from centroid of pile group, ft
z_t = distance of extreme tensile fiber from centroid, in.
z_x = distance of reference line from centroid, ft

REFERENCES

5-1 Terzaghi, K., "Theoretical Soil Mechanics," John Wiley and Sons, Inc., New York, 1942.
5-2 Terzaghi, K., and Peck, R. B., "Soil Mechanics in Engineering Practice," 2nd ed, John Wiley and Sons, Inc., New York, 1967.
5-3 ACI Committee 436, "Suggested Design Procedures for Combined Footings and Mats," Title 63-49 *ACI Proceedings*, **63** (10), 1041–1057, Oct. 1966.
5-4 Hetenyi, M., "Beams on Elastic Foundations," University of Michigan Press, Ann Arbor, Michigan, 1946.
5-5 Kramrisch, F., and Rogers, P., "Simplified Design of Combined Footings," *Proceedings ASCE*, Paper No. 2959, **87** (SM5), Oct. 1961.
5-6 Molitor, D. A., "A practical Treatise on Chimney Design," Peters Company, Detroit, 1938.
5-7 "Beton Kalender," Wilhelm Ernst and Sohn, Berlin, 1939.
5-8 Furlong, R., "Design Aids for Square Footings," *ACI Proceedings*, **62**, 363–371, March 1965.
5-9 Kramrish, F., "Structural Walls of Precast Concrete Sheet Piling," *Building Construction*, **5** (11), 64–67, Nov. 1964.
5-10 "Concrete Crib Retaining Walls," Portland Cement Association, Concrete Information No. ST-46.
5-11 Gaylord, E. H., and Gaylord, C. N., "Structural Engineering Handbook," McGraw-Hill, New York, 1968.
5-12 Urquhart, L. C., "Civil Engineering Handbook," 4th ed., McGraw-Hill, New York, 1959.
5-13 Waddell, J. J., "Concrete Construction Handbook," McGraw-Hill, New York, 1968.
5-14 Winter, G., et al., "Design of Concrete Structures," 7th ed., McGraw-Hill, New York, 1964.
5-15 Everard, N. J., and Tanner, J. L., "Theory and Problems of Reinforced Concrete Design," Schaum Publishing Co., New York, 1966.

6

Properties of Materials for Reinforced Concrete

SIDNEY FREEDMAN*

6.1 INTRODUCTION

The properties of hardened concrete have considerable significance to designers of concrete structures or products. The physical properties of concrete depend upon a number of factors including mix proportions, aggregates, type of cement, curing conditions, and age. They are also affected by environmental conditions such as temperature and relative humidity. The durability of concrete is related to these same factors with particular emphasis on the cement content and amount of entrained air.

Tests to determine strengths are undoubtedly the most common tests made to evaluate the properties of the hardened concrete. There are three reasons for this: (1) the strength of concrete in compression or tension has in most cases a direct influence on the load carrying capacity of both plain and reinforced structures; (2) of all the properties of hardened concrete, strength can usually be determined most easily; and (3) the results of strength tests can be used as a qualitative indication of other important qualities of hardened concrete.

*Director, Architectural Precast Division, Prestressed Concrete Institute, Chicago, Illinois; formerly, Manager, Concrete Technology Section, Portland Cement Association, Skokie, Illinois.

6.2 COMPRESSIVE STRENGTH

6.2.1 Quality of Cement Paste and Bond

The compressive strength of concrete made with aggregate of adequate strength is governed, in general, by the strength of either the cement paste or of the bond between the paste and the aggregate particles. At early ages the bond strength is lower than the paste strength; at later ages the reverse may be the case.

The water-cement ratio and the degree to which hydration has progressed to a large extent determine the quality of the portland cement paste. The water-cement ratio rule is as follows: for a given cement and acceptable aggregates the strength that may be developed by a workable, properly placed mixture of cement, aggregate, and water (under the same mixing, curing, and testing conditions) is influenced by the (a) ratio of mixing water to cement, (b) ratio of cement to aggregate, (c) grading, surface texture, shape, strength, and stiffness of aggregate particles, and (d) maximum size of the aggregate. Mix factors partially or totally independent of water-cement ratio which affect the strength are (1) type and brand of cement, (2) amount and type of admixture or pozzolan, and (3) mineralogic makeup of the aggregate.

Figure 6-1 presents a series of band curves which were

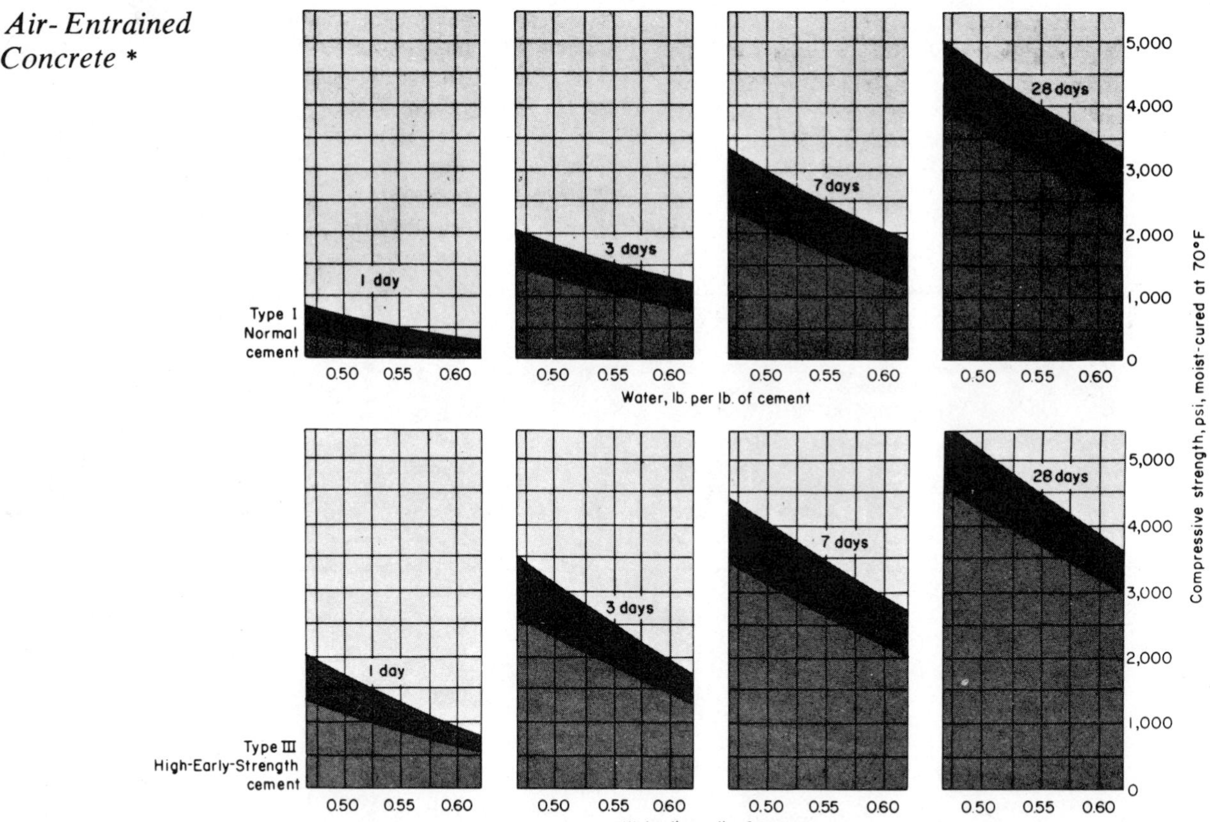

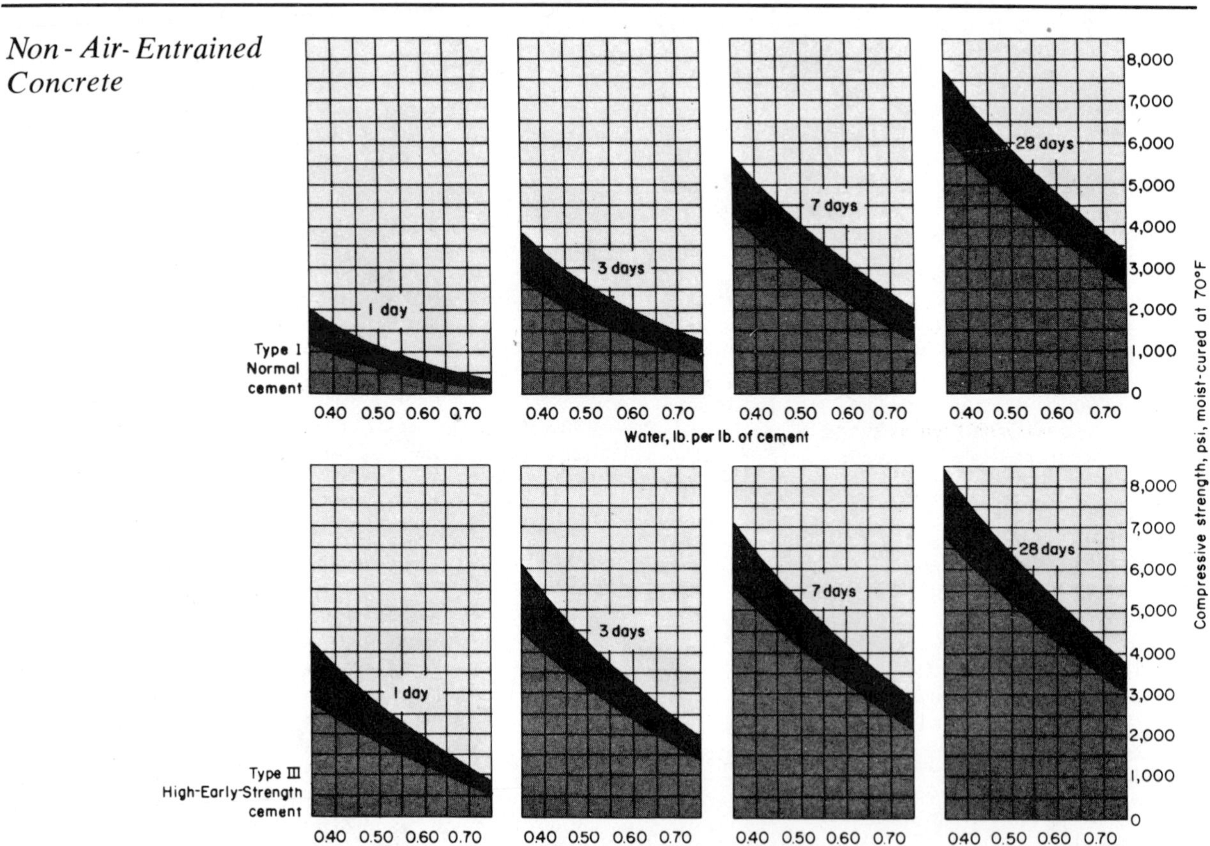

Fig. 6-1 Relationship between water-cement ratio and compressive strength for portland cements at different ages. These relationships are approximate and should be used only as a guide in lieu of data on job materials. (From: *Design and Control of Concrete Mixtures* 11th ed., Portland Cement Association, Skokie, Ill., 1968.)

TABLE 6-1 Approximate Relative Strength of Concrete As Affected by Type of Cement*

Type of Portland Cement	*Compressive Strength–percent of strength of Type I Portland Cement concrete*			
ASTM	1 day	7 days	28 days	3 months
I	100	100	100	100
II	75	85	90	100
III	190	120	110	100
IV	55	55	75	100
V	65	75	85	100

**From: "Design and Control of Concrete Mixtures," Portland Cement Association, Skokie, Ill., 1968.*

developed from a large number of compressive strength tests on 6 X 12 in. cylinders made by many laboratories using a variety of materials. Note that strengths increase as the water-cement ratios decrease, and that strengths increase with age. These saturated specimens show 20 to 30% lower strength than companion specimens tested dry.

Table 6-1 shows the relative strengths of concretes made with different cements; Type I portland cement concrete is used as the basis for comparison. These values are characteristic for concretes that are moist-cured until tested.

6.2.2 Curing Temperatures

Figure 6-2 shows the age-compressive strength relationship for concrete that has been mixed, placed and cured at temperatures between 40 and 73°F. At temperatures below 73°F, strengths are lower at early ages but higher at later periods.

Higher early strengths may be achieved through use of Type III or High-Early-Strength cement. Principal advantages occur prior to 7 days. At 40°F curing temperatures, the early advantages of this type of mixture are more pronounced and persist longer than at higher temperatures.

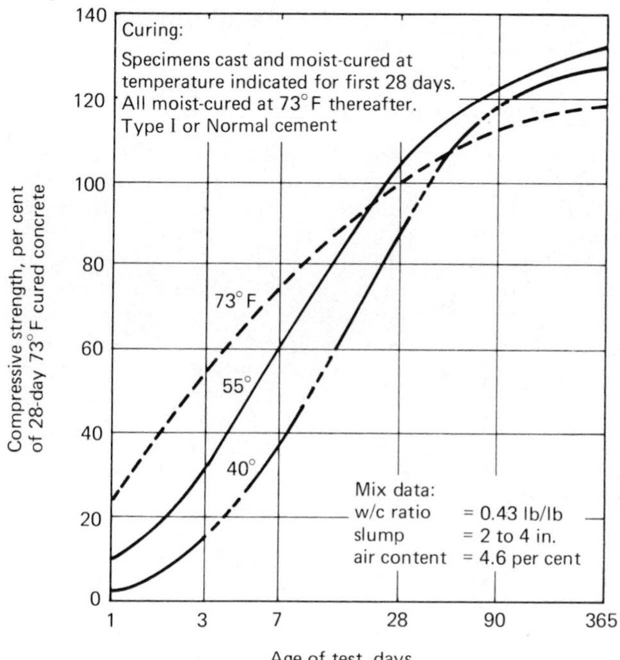

Fig. 6-2 Effect of low temperatures on concrete compressive strength at various ages. (From: *Design and Control of Concrete Mixtures*, PCA.)

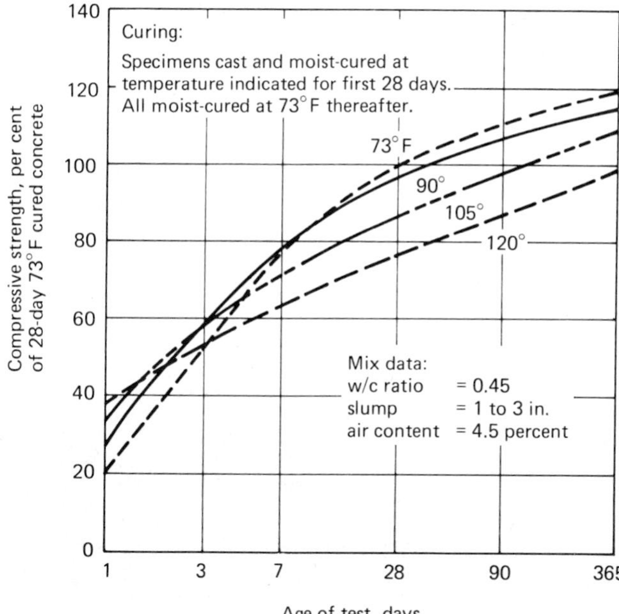

Fig. 6-3 Effect of high temperatures on concrete compressive strength at various ages. (From: *Design and Control of Concrete Mixtures*, PCA.)

Figure 6-3 shows the effect of high concrete curing temperatures on compressive strength. These tests, using identical concretes of the same water-cement ratio, show that while higher concrete temperatures increase early strength, at later ages the reverse is true. If the water content had been increased to maintain the same slump (without changing the cement content), the reduction in strength would have been even greater.

Strength gain practically stops when moisture required for curing is no longer available. Concrete that is placed at low temperatures (but above freezing) may develop higher strengths than concrete placed at high temperatures, but curing must be continued for a longer period. It is not safe to expose concrete to freezing temperatures at early periods. If freezing is permitted within 24 hours, much lower strength will result.

6.2.3 Air-Entrained Concrete

Strength of air-entrained concrete depends principally upon the voids-cement ratio. For this ratio, "voids" is defined as the total volume of water plus air (entrained and entrapped). Hence, when air content is maintained constant, strength varies inversely with the water-cement ratio. As air content is increased, a given strength generally may be maintained by holding to a constant voids-cement ratio. Some increase in cement content may be necessary in richer mixes in order to accomplish this.

At a given water-cement ratio, the compressive strength of air-entrained concrete is reduced by approximately 5% for every 1% of entrained air. Flexural strength is reduced considerably less by air entrainment. Air-entrained as well as non-air-entrained concrete can usually be proportioned to provide any desired strength. Air-entrained concretes have lower water-cement ratios than non-air-entrained concretes, therefore, reductions of strength that generally accompany entrainment of air are minimized. However, some reductions in strength may be tolerable in view of other benefits. These reductions become significant only in mixes containing more than about 565 lb of cement per cubic yard. In lean or harsh

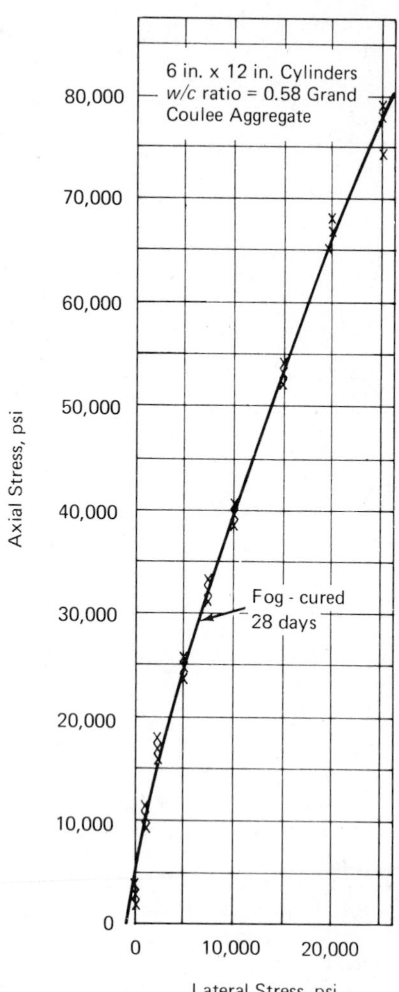

Fig. 6-4 Relations of axial stress to lateral stress at failure in triaxial compression tests of concrete. (From: Balmer, G. A., "Shearing Strength of Concrete under High Triaxial Stress—Computation of Mohr's Envelope as a Curve," *Structural Research Laboratory Report No. SP-23*, U. S. Department of the Interior, Bureau of Reclamation, Oct. 28, 1949.)

mixes, strengths are generally increased by entrainment of air in proper amounts.

Attainment of high strength with air-entrained concretes may be difficult at times. Both air-entrained and non-air-entrained concretes have increased water demands when slumps are maintained constant as concrete temperatures rise. Even though a reduction in mixing water is associated with air entrainment, mixtures with high cement contents require more mixing water; hence, the increase in strength expected from the additional cement is offset somewhat by the additional water added. Also, with some aggregates it is not possible to secure extremely high strengths with air-entrained concrete.

6.2.4 Confining Pressures

The axial compressive strength of concrete is appreciably increased by the existence of lateral confining pressures such as those exerted by the spiral reinforcement in a column or by that portion of column concrete surrounded by floor concrete, Fig. 6-4. (Also see Fig. 6-17 and 6-18.) The increase in axial strength over the unconfined compressive strength ranges from as much as four to five times the value of the lateral stress for small confining pressures (0 to 1500 psi) to about two and one-half or three times the lateral stress for large confining pressures.

6.2.5 Effect of High Temperature Exposure

The influence of heat exposure per se on the structural properties of concrete is critically dependent on the moisture content of the concrete at the time of heating and is quite different for concrete in which the moisture is free to evaporate as compared to concrete sealed against moisture loss. Deterioration of structural properties also can be expected from any test variable that produces large temperature differentials in the concrete on heating (rapid heating or cooling or use of very large specimens). Partial loss of chemically combined water (dehydration of cement paste) occurs above 250°F and primarily affects the flexural strength. For practical situations in which the free moisture can evaporate as heating commences, it can be expected that little or no change or a slight beneficial change will take place in the compressive strength on heating to 500°F provided the conditions for thermal shock are not present, Fig. 6-5. When free moisture is retained in the concrete during heating, deterioration in structural properties can be expected in increasing degrees of severity at all levels of heating up to 500°F.

Carbonate aggregate concrete or sanded lightweight concrete retain more than 75% of their original strengths at temperatures up to 1200°F when heated without load and tested hot. The corresponding temperature for the siliceous aggregate concrete is about 800°F, Fig. 7-7.* Strengths of specimens stressed in compression during heating generally are 5 to 25% higher than those of companion specimens which are not stressed during heating and residual strengths are somewhat lower than the strengths of companion specimens tested at high temperatures. Strengths of specimens stressed in compression during heating are not significantly affected by the applied stress level ranging from 25 to 55% of the original strength. Original strength of the concrete also has little effect on the percentage of strength retained after heating concretes to high temperatures and then cooling the concrete to room temperature. (See Chapter 7, Fire Resistance, for practical implications.)

6.2.6 Effect of Low Temperature Exposure

The strength of concrete at a given low temperature will depend on the cement content and the water-cement ratio of the mix, the aggregate, the age, and especially the moisture condition.

The results of strength determinations at low temperatures are shown in Fig. 6-6a for various concrete mixtures in a moist condition. Although there is some scatter of points, the trends are similar for the different mixes. Tests on the 5.5-bag mix indicate very little difference in strengths between 75°F and 35°F and it is felt that the same would be true for the other concretes. In general, strengths increase as the temperature is lowered from 75°F, reach maximum values in the range from -75 to -150°F, then decrease as the temperature approaches -250°F.

The increases in compressive strength are related to the evaporable water contents and Fig. 6-6b shows that the compressive strength of oven-dry concrete increases by only about 20% as the temperature is lowered from 75 to -150°F, whereas the increase in comparable moist concrete is 240%. The increase for concrete equilibrated in an atmo-

*Chapter 7, Fire Resistance.

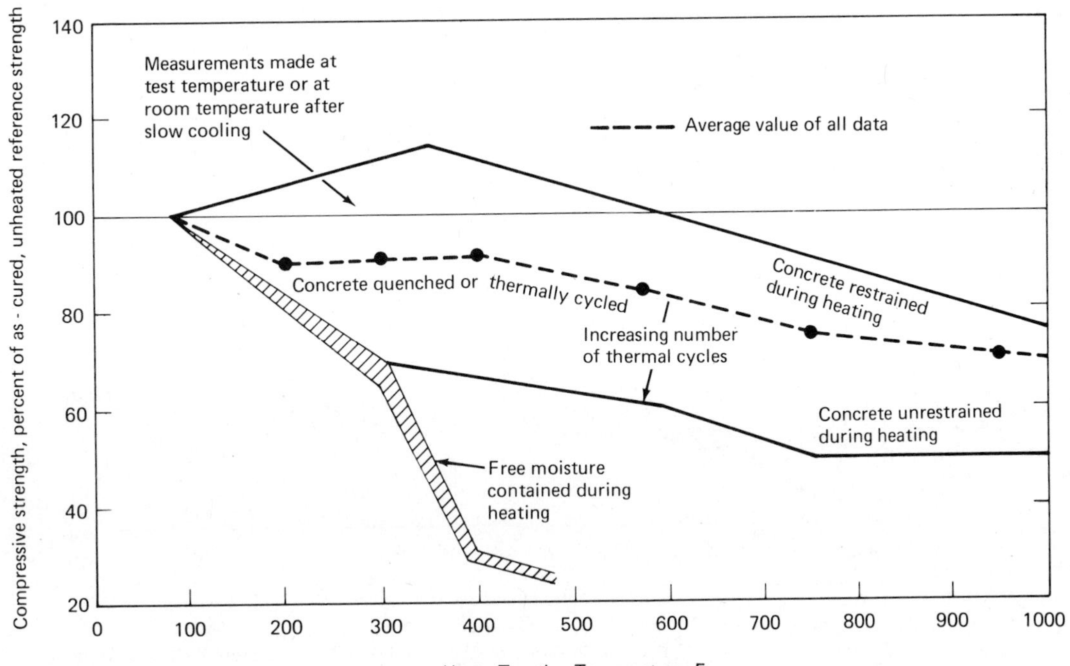

Fig. 6-5 Extremes of the influence of heat exposure on the compressive strength of concrete as determined by various investigators. (From: Lankard, et al., "Effects of Moisture Content on the Structural Properties of Portland Cement Concrete Exposed to Temperatures up to 500° F," ACI SP-25 *Temperature and Concrete*, 1971.)

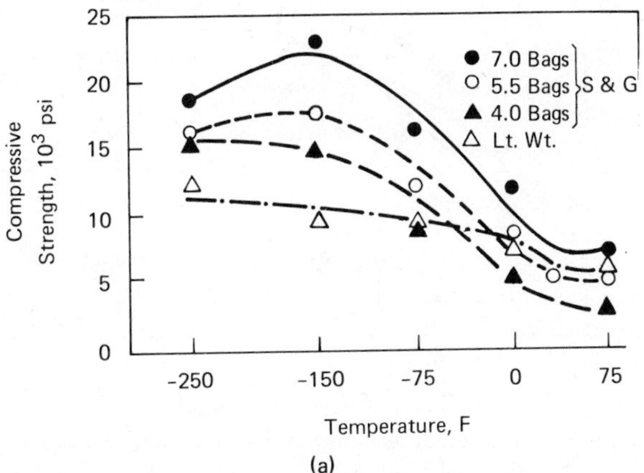

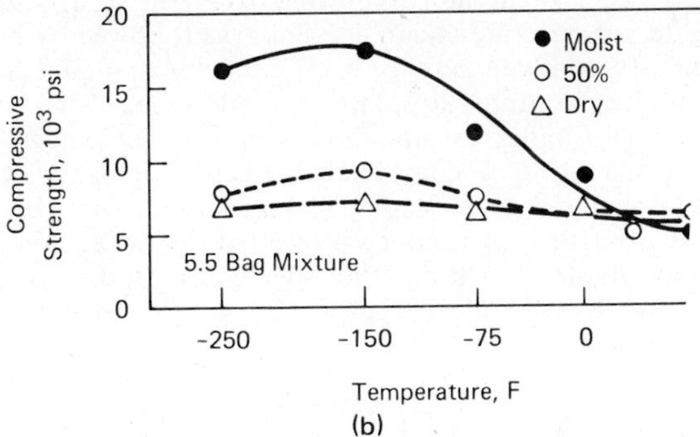

Fig. 6-6 (a) Effect of low temperatures on strength of moist concrete; and (b) effect of moisture condition on strengths at low temperatures. (From: Monfore, G. E., and Lentz, A. E., "Physical Properties of Concrete at Very Low Temperatures," *PCA Research Bulletin 145*, 1962.)

sphere of 50% relative humidity is only slightly greater than that for oven-dry concrete. These increases in strength of moist concrete with decrease in temperature below freezing must be largely due to the formation of ice which increases progressively as the temperature is lowered.

6.2.7 Impact Loading

Loads which are applied at rates considerably higher than 10,000 psi per second are referred to as impact loads. The dynamic compressive strength of a given type of concrete is higher than the static strength, becoming relatively greater as the duration of impact decreases Fig. 6-7. There are significant increases in compressive strengths, elastic moduli and ability to absorb strain energy under impact loading.

6.3 TENSILE STRENGTH

Tensile strength is one of the fundamental properties of concrete. Although reinforced concrete structures are not normally designed to resist direct tension, a knowledge of the tensile strength of their members helps in understanding the behavior of these structures and designers use this property to resist loads (flexural tension), shear, shrinkage, and temperature stresses. Significant principal tension stresses may be associated with multiaxial states of stress in walls, shells, or deep beams. In determining the tensile strength of concrete three types of tests have been used: (1) direct, (2) flexural, and (3) splitting tension.

6.3.1 Direct Tension

A direct application of a pure tension force free from eccentricity is difficult and is further complicated by secondary stresses induced by the grips or embedded studs. No standard test involving the application of direct tension

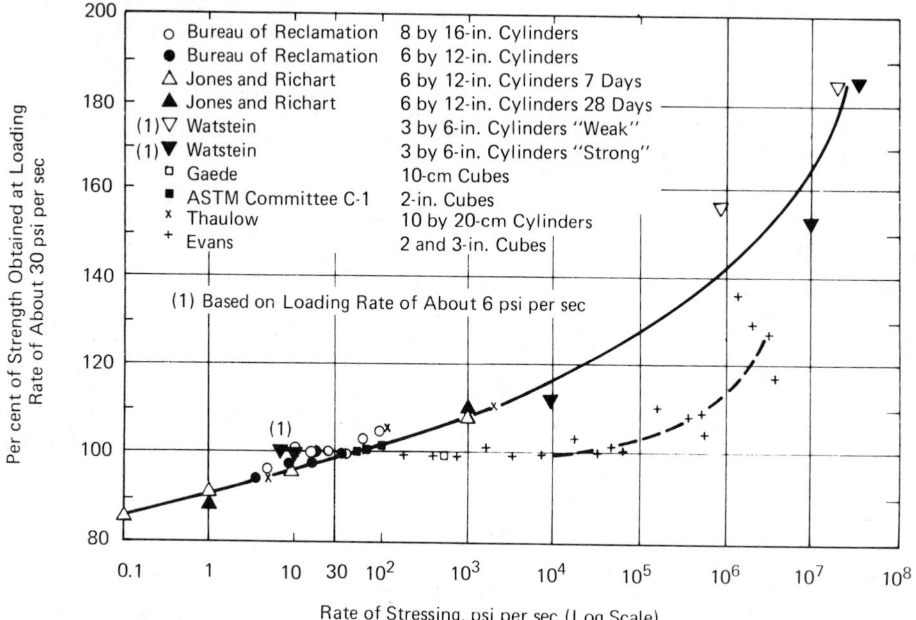

Fig. 6-7 Effect of rate of stressing on the compressive strength of concrete. (From: McHenry, D., and Shideler, J. J., "Review of Data on Effect of Speed in Mechanical Testing of Concrete," *PCA Development Bulletin D9*, 1956.)

exists and it is usual to determine the tensile strength of concrete in flexure or by the application of splitting tension. However, the available data indicates that direct tension is about 12 to 7% of the compressive strength when strengths vary from 2500 to 7000 psi; as compressive strength increases the ratio of direct tension to compressive strength decreases.

6.3.2 Flexural Tension

The flexural test does not measure the true tensile strength of concrete but determines what is known as the modulus of rupture. The tensile strength as determined by the modulus of rupture of a given concrete depends to a considerable extent on the distribution of load, size of the specimen, moisture condition, and the rate of loading.

The modulus of rupture may be found by cantilever, centerpoint, or third point loading. The third point method is a better estimate of flexural strength and is normally used for design.

The modulus of rupture of a 6 in. deep beam is about 20% greater than that of a 10 in. beam. This phenomenon is thought to be caused mainly by difference in strain gradient; the smaller the depth, the steeper the gradient and the higher the modulus of rupture.

Modulus of rupture tests are sensitive to shrinkage, temperature, and minor surface defects, such as large pieces of aggregate near the surface. The outer fibers of concretes undergoing drying are extremely sensitive to the transient moisture content, and under these conditions may not furnish data that is satisfactorily reproducible. Flexural and direct tension specimens register a higher strength when completely dry than when surface wet, however, partially dry flexural specimens may show a reduction of 40% over wet specimens.

Flexural strengths of lightweight and sand and gravel concretes do not differ greatly at early ages but after 28 days lightweight shows less strength gain with continuous moist curing than does sand and gravel. Figure 6-8 shows the increase in modulus of rupture for moist cured sand

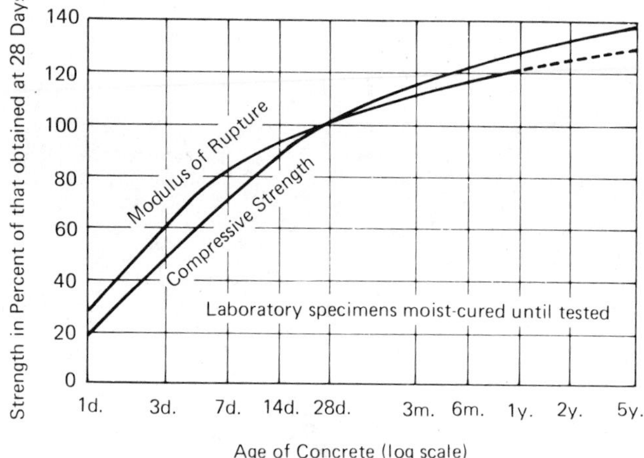

Fig. 6-8 Age-strength relationships for moist-cured concrete.

and gravel concrete. Until about 28 days the modulus gains strength faster than the compressive strength. Lightweight aggregate concrete moist-cured seven days followed by drying and 50% relative humidity has a much lower modulus of rupture than continuously moist cured beams. Sand and gravel show similar results; however, the percentage of strength reduction is considerably greater for lightweight concrete than for sand and gravel concrete. The loss of flexural strength is attributed to internal stresses caused by moisture gradients within the specimens. Shrinkage or rapid cooling will reduce tensile strength of an element because of the greater initial contraction of the surface. This leaves tensile stress at the surface, compressive stress in the interior.

Increasing the rate of application of stress results in an increase in the modulus of rupture. A straightline relationship exists between the modulus of rupture and the log of the rate of increase of stress.

The flexure test yields a considerably higher value of tensile strength than either the direct or splitting tension test because the formulas to calculate modulus of rupture assume a linear tensile stress distribution which is known to be incorrect.

Flexural strength is substantially affected not only by the strength of the mortar but since failure is in tension also by the tensile strength of the coarse aggregate particles and by the bond between the coarse aggregate particles and the mortar. Certain aggregates may produce high compressive strength but be incapable of producing high flexural strength. Similarly, certain aggregates are adequately strong in all respects, but may have surface characteristics which are not favorable to a good bond between the mortar and the coarse aggregate, and therefore, have poor flexural strength producing ability. The relative level of tensile resistance of lightweight concrete is a definite characteristic of each particular aggregate when the concretes are compared on an equal compressive strength basis. The relationship between water-cement ratio and strength, which is helpful for compressive strength, is less definitive in its application to flexural strength.

Concrete exposed to temperatures well above normal room temperatures generally show a deterioration in flexural strength, Fig. 6-9. The rate of strength deterioration is highest at temperatures above 400°F. Cycling of the concrete produces large reductions in tensile strength with cracks becoming noticeable on the concrete surfaces after a number of cycles. These cracks have their origin in the differential movements of paste and aggregate during the heating and cooling periods.

Flexural strength is not a constant proportional part of the compressive strength but the proportionality ratio decreases from about 22 to 10% as compressive strength increases. Angular coarse aggregate improves the flexural strength much more than the compressive strength so that the tensile to compressive strength ratio is greater for concretes made with crushed coarse aggregate. Other factors influencing the ratio are age, mix proportions, properties of fine aggregate, and curing.

With given materials and procedures a relationship can be established between the compressive and flexural strength. There being no general relationship between compressive and flexural strength which is valid for all conditions. Where flexural strength is of importance, the compressive strength corresponding to the required flexural strength might well be determined in the laboratory and compression tests used thereafter for field control and evaluation purposes. However, it is generally accepted that the modulus

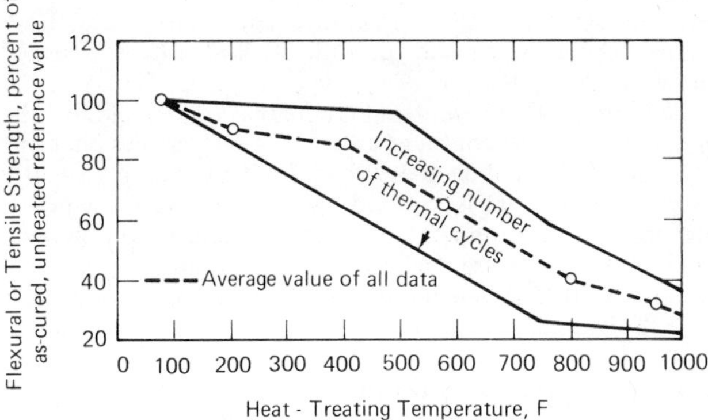

Fig. 6-9 Extremes of the influence of heat exposure on the flexural or tensile strength of concrete as determined by various investigators. (From: Lankard, *et al.*, "Effects of Moisture Content on the Structural Properties of Portland Cement Concrete Exposed to Temperatures up to 500°F," ACI SP-25 *Temperature and Concrete*, 1971.)

of rupture is approximately a function of the square root of compressive strength

$$f_r = K\sqrt{f_c'}$$

where

f_r = modulus of rupture, psi, third point loading
f_c' = compressive strength, psi
K = a constant usually between 8 and 10

6.3.3 Splitting Tension

The splitting tensile strength of concrete cylinders (ASTM C496) is a convenient relative measure of tensile strength. The test is performed by application of diametrically opposite compressive loads to a concrete cylinder laid on its side in the testing machine. Fracture, or "splitting", occurs along a diametral plane which is subjected largely to a uniform tensile stress. The theory on which the splitting tensile strength is calculated assumes that the concrete is elastic and that a state of plain stress exists. Neither assumption is true and the calculated splitting tension is slightly higher than the true axial tensile strength.

The cylinder splitting method is a more reliable measure of tensile strength (lower coefficient of variation) than the modulus of rupture beam test, particularly for concretes that are subjected to drying conditions. Perhaps the most important advantages of the split cylinder are the approximate uniformity of tensile stress over the diametral area of the cylinder and the simplicity of the test which affords the opportunity to test economically a large number of specimens. The large number of samples may offset the variation of results that must always be expected in tensile investigations. The split cylinder test results will satisfactorily reflect such variables as compressive strength, age of concrete and type of curing.

The tensile splitting strength of continuously moist cured lightweight and normal weight concretes of equal compressive strength are equal. In all cases the tensile strength of steam cured lightweight concrete is considerably less than that of normal weight concrete. The tensile strength of concretes which undergo drying is more relevant to behavior of concrete in structures. During drying of the concrete, moisture loss progresses at a slow rate into the interior of concrete members, resulting in the probable development of tensile stresses at the exterior faces and balancing compressive stresses in the still moist interior zones. Thus the tensile resistance of drying concrete will be reduced from that indicated by continuously moist-cured concrete. This effect is more pronounced for concrete made with lightweight aggregate than for concrete made with natural aggregates. The splitting tensile strengths of normal weight concretes actually increase when subjected to air drying during the curing period, Fig. 6-10.

The splitting tensile strength of air-dried all-lightweight concrete varies from approximately 70 to 100% that of the normal weight concrete, when comparisons are made at equal compressive strength. The reduction is more pronounced for the higher strength concrete than for lower and is also greater for all-lightweight concrete than for sand-lightweight. Replacement of lightweight fines by sand generally increases the splitting tensile strength of lightweight concrete subjected to drying. In some cases this increase is nonlinear with respect to the sand content, so that with some aggregates partial sand replacement is as beneficial as complete replacement. Also air-dried all-lightweight concretes with high splitting strengths show little improvement as the sand content is increased. The results of splitting tensile strength tests on air-dried lightweight concrete may

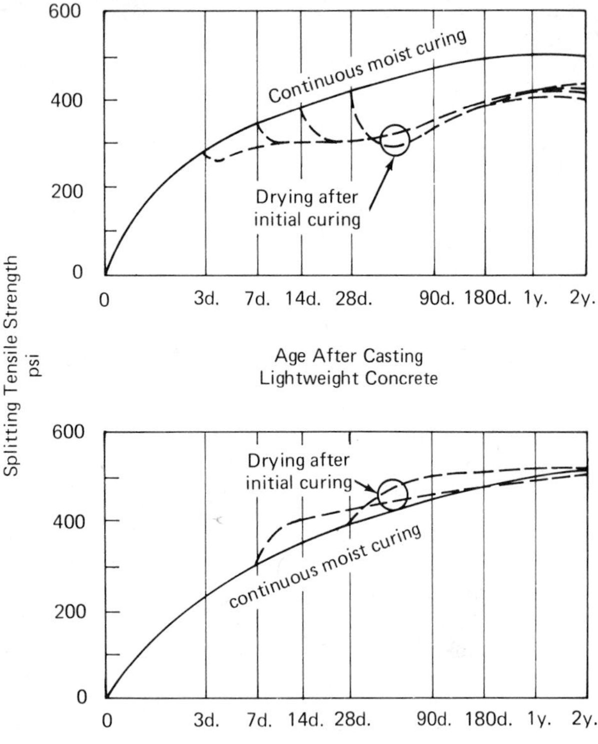

Fig. 6-10 Effect of moist curing and drying time on splitting tensile strength (Series I). (From: Hanson, J. A., "Effect of Curing and Drying Environments on Splitting Tensile Strength of Concrete," *ACI Journal Proceedings*, **65** (7), July, 1968.)

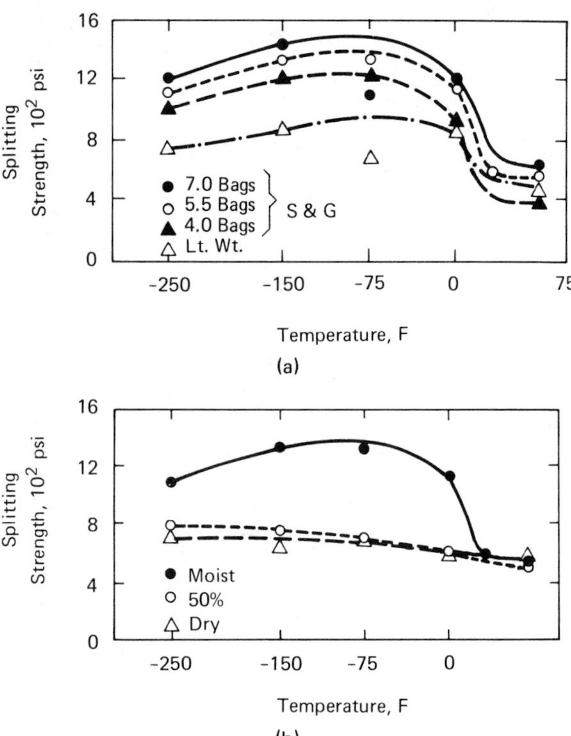

Fig. 6-11 (a) Effect of low temperatures on strength of moist concrete; and (b) effect of moisture condition on strengths at low temperatures. (From: Monfore and Lentz, "Physical Properties of Concrete at Very Low Temperatures." *PCA Research Bulletin* 145, 1962.)

be used as a reliable measure of the unit shear capacity of lightweight concrete beams and slabs.

Tensile splitting tests are not required in ACI 318-71 if shear and torsion stresses, cracking moment, modulus of rupture, and bar development lengths are based on the reasonable assumption that, for a given compressive strength, the tensile strength of lightweight aggregate concrete (with or without sand replacement) is a fixed proportion of that for normal weight concrete. The percentage of normal weight concrete shear stress permitted is 75 if all-lightweight aggregate is used, or 85 if natural sand is combined with lightweight coarse aggregate to produce sand-lightweight concrete. Linear interpolation is used for partial sand replacement of fine aggregate. Alternatively these values may be upgraded if splitting tensile tests demonstrate that the tensile strength is higher than the assumed conservative percentages. In any case, the test for splitting tensile strength is used only for laboratory determination of its relationship to the compressive strength. It is not intended for control of, or acceptance of, strength properties of the concrete in the field. If use of the splitting tensile strength of lightweight aggregate concrete yields calculated permissible shear values greater, or bar development lengths less, than allowed for normal weight concrete, the values for normal weight concrete should be used.

Most structural lightweight aggregate producers have sufficient test data available to recommend splitting tensile strength value to the engineer for concrete in which the aggregate consists entirely of lightweight aggregate and for concrete containing lightweight aggregate and natural sand.

There is a general indication that the splitting test of a drilled core provides a better measure of the flexural strength of thick concrete pavements than can be obtained by testing beams sawed from the pavements. Since the cost of procuring and testing sawed beams as compared to drilled cores is in the order of 50 to 1, a much larger number of cores can be tested to increase the reliability of the splitting strength.

The results of splitting tension determinations at low temperatures are shown in Fig. 6-11a for various concrete mixtures in a moist condition. Although there is some scatter of points, the trends are similar for the different mixes. Tests on the 5.5-bag mix indicate very little difference in strengths between 75 and 35°F and it is felt that the same would be true for the other concretes. In general, strengths increase as the temperature is lowered from 75°F, reach maximum values in the range from -75 to -150°F, then decrease as the temperature approaches -250°F. A rather abrupt increase in the splitting strengths is observed between 75 and 0°F.

The increases in splitting strengths are related to the evaporable water contents and Fig. 6-11b shows that the splitting strength of oven-dry concrete increases only slightly as the temperature is lowered from 75 to -150°F whereas the increase in comparable moist concrete is considerably greater. The increase for concrete equilibrated in an atmosphere of 50% relative humidity is only slightly greater than that for oven dry concrete.

6.3.4 Relationships Between Tests

The ratio between the direct tensile strength and the splitting strength changes with variations in the composition of the concrete and depends to a considerable degree on the type and properties of the aggregates. The relationship for lightweight concrete is shown in Fig. 6-12.

The ratio of direct tensile strength to modulus of rupture

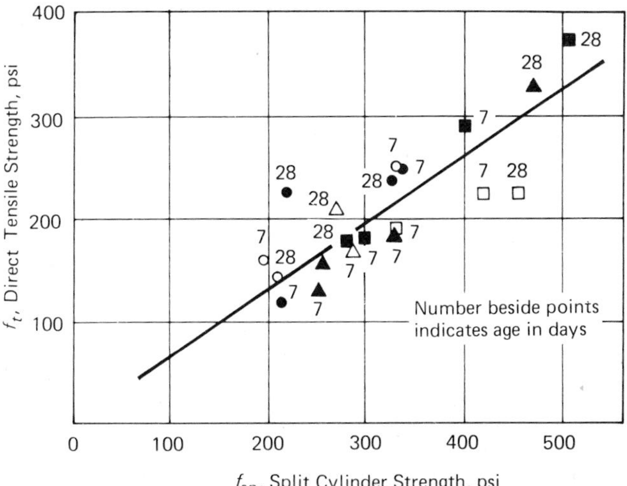

Fig. 6-12 Relationship between direct tensile and splitting strengths of lightweight concrete. (From: Ledbetter, W. B., and Thompson, J. N., "A Technique for Evaluation of Tensile and Volume Change Characteristics of Structural Lightweight Concrete," *ASTM Proceedings*, **65**, 712–726, 1965.)

increases with increasing specimen size and compressive strength. When the size of the modulus of rupture beam becomes very large the strain gradient diminishes and the modulus of rupture approaches the tensile strength.

For a given coarse aggregate and method of curing, a linear relation exists between the splitting tensile strength and the flexural strength of concrete and this relationship is not affected by the cement content of the concrete or the age at test Fig. 6-13. However, there is an appreciable reduction in the flexural-splitting strength ratio with increase in tensile splitting strength of the concrete. For a given tensile splitting strength, the flexural strength of concrete increases as the maximum size of aggregate decreases. Also for a given tensile splitting strength, the corresponding flexural strengths are higher for rounded aggregates than for angular aggregates, which in turn are higher than lightweight aggregate concrete.

The ratio of the splitting tensile strength to the compressive strength is curvilinear, Fig. 6-14. The ratio decreases as the compressive strength increases (Fig. 6-15); the ratio is affected by both the cement content, and the age at test.

6.4 MODULUS OF ELASTICITY

The stress-strain relation and modulus of elasticity of concrete are important design properties. Static determinations of the modulus of elasticity provide one of the values needed for design purposes such as determining deformation and stress distribution between concrete and steel in reinforced or prestressed concrete members and the buckling effect in long columns. The static modulus of elasticity also is useful for calculating the stresses resulting from shrinkage, settlement, or other distortions such as floor slab deflections and it affects camber control or prestress loss.

6.4.1 Stress-Strain Relation

Although concrete is not a truly elastic material, the theory of elasticity can be applied to it within limits of stress (up to 35 to 50% of ultimate stress) and time (creep effects). However, marked departure from linearity is noted at near-ultimate stresses. Concrete has no range of proportionality, no pronounced elastic limit, and no marked yield point, though it does creep under constant stress.

Unreinforced concrete has a certain amount of ductility. This ductility, however, decreases with increasing concrete

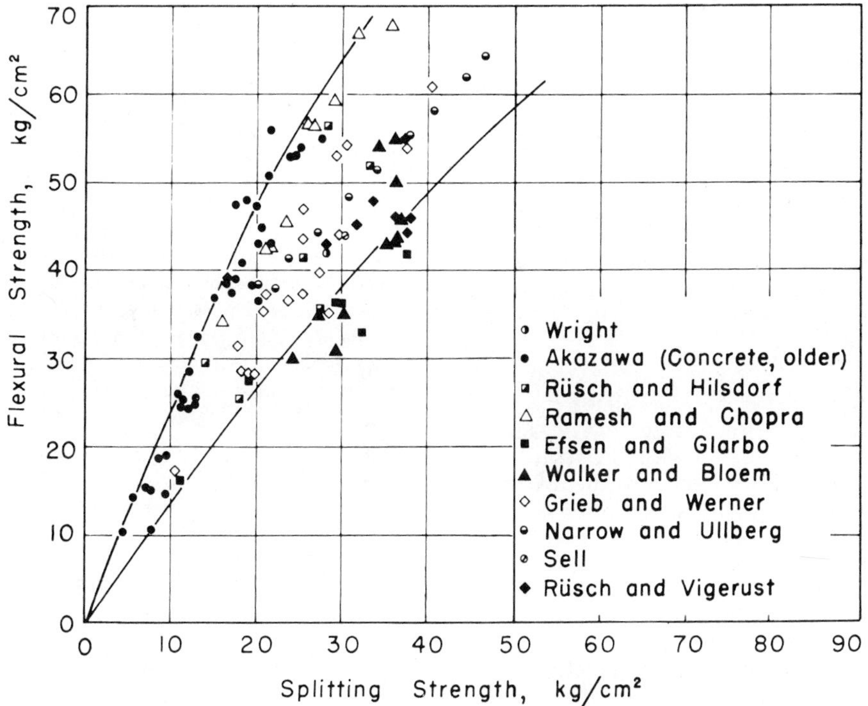

Fig. 6-13 Experimental results obtained by various investigators for the relationship between splitting strength and flexural strength of concrete (after Bonzel). Note: to convert to psi, multiply by 14.22. (From: Popovics, S., "Relations between Various Strengths of Concrete," Highway Research Record 210, 1967.)

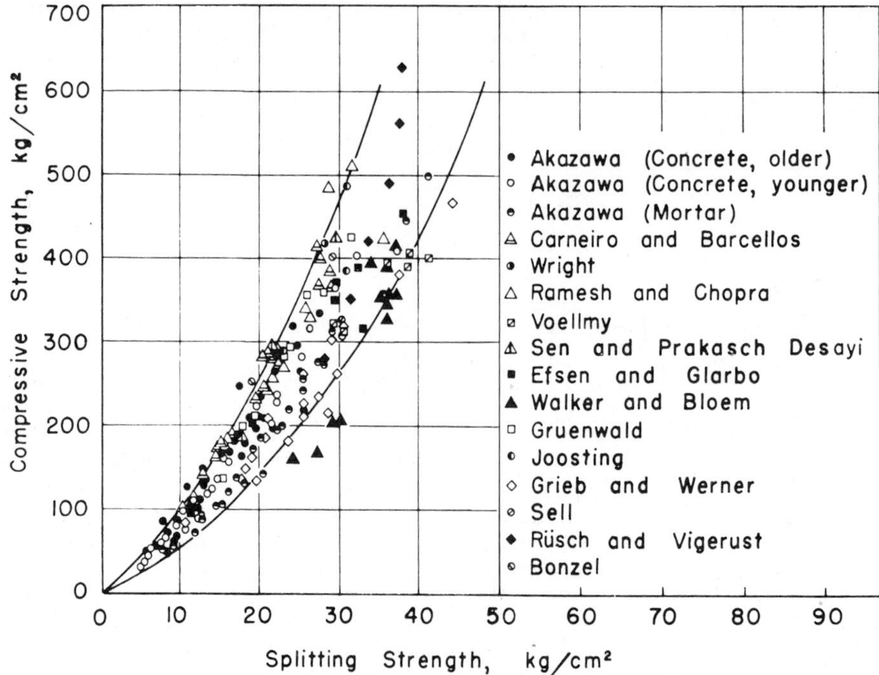

Fig. 6-14 Experimental results obtained by various investigators for the relationship between splitting strength and compressive strength of concrete (after Bonzel). Note: to convert to psi, multiply by 14.22. (From: Popovics, S., "Relations between Various Strengths of Concrete," Highway Research Record 210, 1967.)

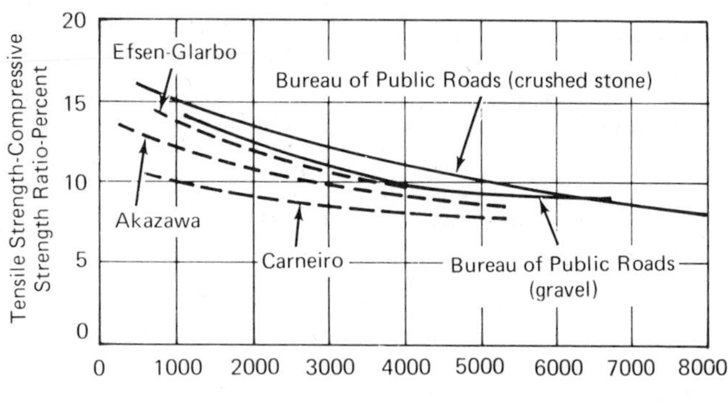

Fig. 6-15 Comparison of relation between ratios of splitting tensile to compressive strengths and compressive strengths. (From: Grieb, W. E., and Werner, G., "Comparison of Splitting Tensile Strength of Concrete with Flexural and Compressive Strengths," *ASTM Proceedings*, **62**, 972-990, 1962.)

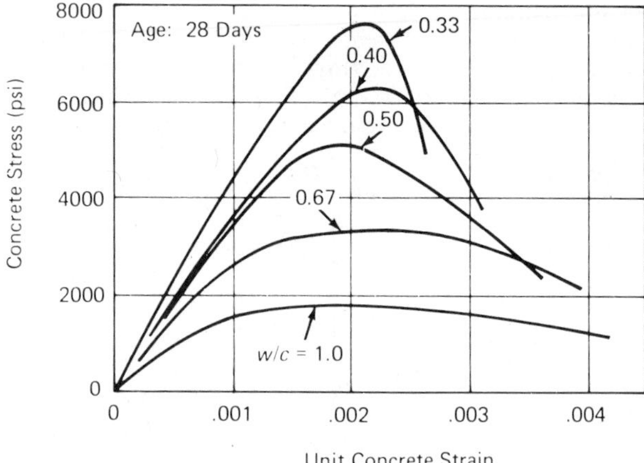

Fig. 6-16 Stress-strain curves for concrete. (From: Hognestad, E., Hanson, N. W., and McHenry, D., "Concrete Stress Distribution in Ultimate Strength Design," *Journal of ACI* 27 (4), 445-479, Dec. 1955; Proc. 52; PCA Research and Development Laboratory, Development Department Bulletin D6.)

strength. The stress-strain relation becomes almost a straight line as the concrete strength increases, (Fig. 6-16). A feature of these curves is that each shows a descending branch after the maximum stress has been reached. Also the maximum strain at failure is lower at higher concrete strengths. The maximum ultimate strain for design, at the extreme compression fiber may be below 0.003 for higher strength concretes.

At a load representing the same proportion of the ultimate strength, the higher strength concrete has a higher strain. However, when comparing high-strength concretes to conventional-strength concretes, the ratio of strains is smaller than the ratios of strengths, so that the modulus of elasticity is greater for higher strength concrete.

Another important feature of these curves is that their peaks occur—that is, the maximum stresses are reached—at strains of about 0.0020 to 0.0025, for all tested mix proportions and cylinder strengths. In design it is desirable that the concrete not crush or substantially enter the descending branch of the curve before the steel has reached its yield point. In steel, an elastic strain of 0.001 corresponds to a stress of 30 ksi. Therefore, reinforcement with yield strengths between 60 and 75 ksi will reach these yield points before the concrete starts weakening.

Stress-strain curves for unconfined and triaxial strength tests are shown in Fig. 6-17. The results indicate the effect of pressure at various moisture contents on the relation of stress to strain in concrete. Considering each mixture individually, the initial portion of the stress-strain curves for each confining pressure and moisture content approximately coincide. Since the modulus is usually computed from the initial portion of the curve, Young's modulus of

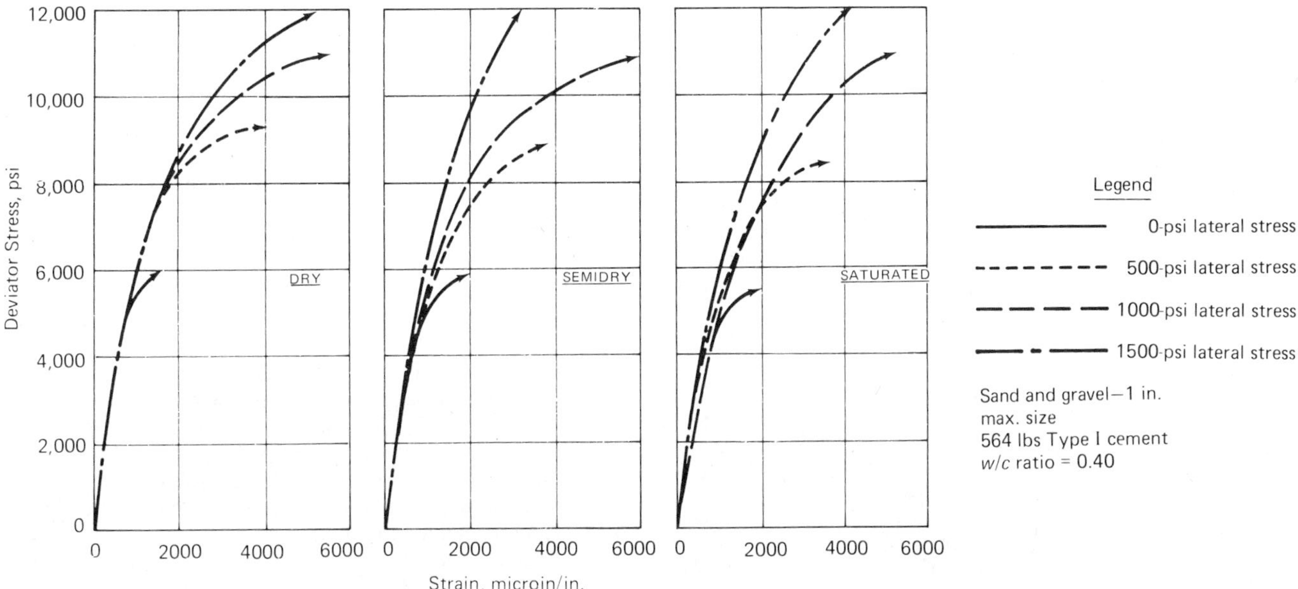

Fig. 6-17 Stress vs strain at various moisture conditions. (From: Saucier, K. L., "Correlation of Hardened Concrete Test Methods and Results," U.S. Army Engineer Waterways Experiment Station Technical Report C69-2, March 1969.)

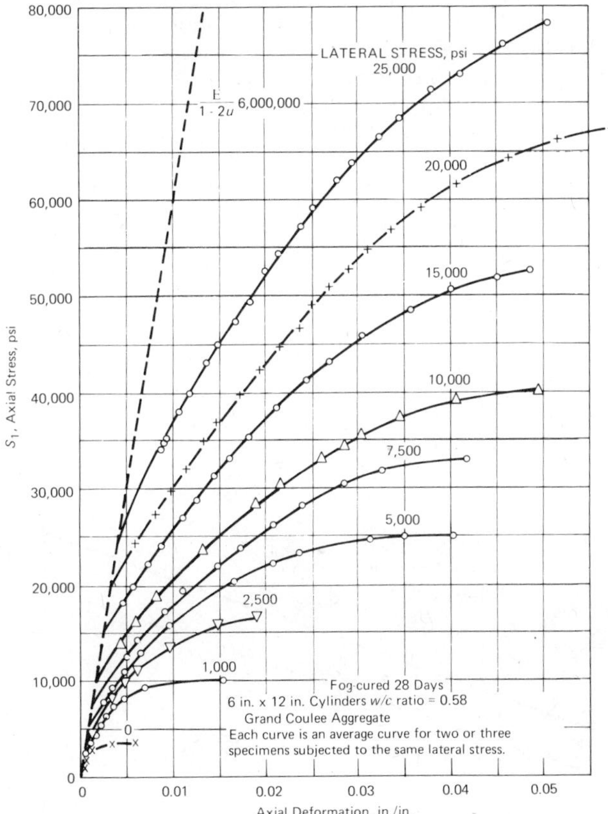

Fig. 6-18 Stress-strain curves for concrete under high triaxial stresses illustrating large strains. (From: Balmer, "Shearing Strength of Concrete under High Triaxial Stress—Computation of Mohr's Envelope as a Curve," *Structural Research Laboratory Report No. SP-23,* U. S. Department of the Interior, Bureau of Reclamation, Oct. 28, 1949.)

elasticity would probably not be appreciably affected by confining pressure up to 1500 psi or by moisture condition at time of test if pore pressure effects were not present. Fig. 6-18 shows that considerably higher strains can be attained as the lateral stress increases.

6.4.2 Modulus of Elasticity of Normal Weight Concrete

The modulus of elasticity of concrete, E_c is defined as the ratio of normal stress to corresponding strain for tensile or compressive stresses. It is also known as elastic modulus, Young's modulus or Young's modulus of elasticity. It depends on the modulus of the cement paste, the modulus of the aggregate, and the relative amounts of paste and aggregate. The modulus also varies according to its definition, whether initial tangent, tangent, chord, or secant modulus; rate of loading; number of cycles of load application; the time the load is sustained; shrinkage and creep; and to some extent all the other possible variable parameters of the concrete specimen. The secant modulus is commonly used and because it decreases with an increase in stress, the stress at which the modulus is determined should be stated.

The modulus of the paste increases as the degree of hydration increases. Changes in modulus for a given concrete occur because of changes in the modulus of elasticity of the paste and increased bond with aggregate as curing continues. As the modulus of the paste increases, the concrete strength also increases, and for any given concrete mixture and curing condition there is a general empirical relationship between strength and modulus of elasticity. The formula, $E_c = w^{1.5}\ 33\sqrt{f_c'}$ defines this relationship for normal weight concrete; where w = unit weight and f_c' = compressive strength, (Fig. 6-19).

Concrete may comply with this formula only within ±15 to 20%. Modulus of elasticity for a given concrete exhibits a much higher coefficient of variation than the compressive strength. The greater variation results in part from the greater inaccuracies of the test procedures used to measure the small strains. Depending on how critically values for E_c will affect the design (as in buckling of long columns), the engineer should decide if the values determined by formula are sufficiently accurate, or if he should determine secant modulus values from tests made on the specified concrete in accordance with ASTM C469, "Method of Test for Static Young's Modulus of Elasticity and Poisson's Ratio in Compression of Cylindrical Concrete Specimens." In general, normal weight concrete mixes would be expected to yield similar tensile and compressive secant moduli of

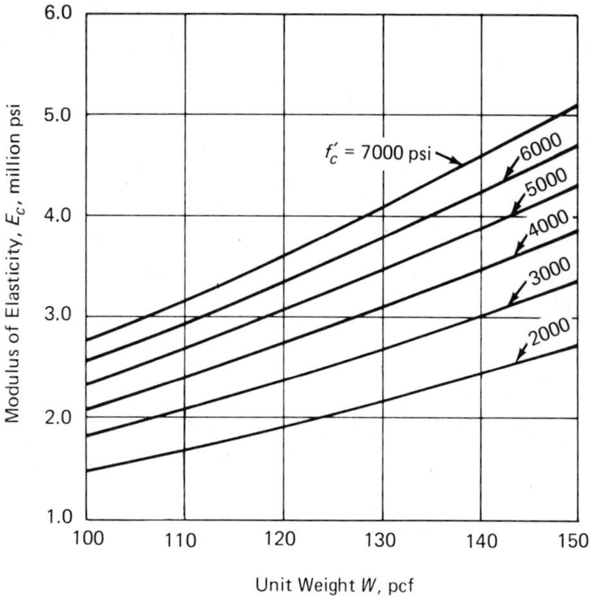

Fig. 6-19 Modulus of elasticity as a function of strength and air-dry unit weight of normal weight concrete.

elasticity values up to 40-50 percent of the ultimate strength.

In the high-strength region, the modulus may not increase as rapidly as strength and may approach some limiting value, Fig. 6-20 (not shown in this data). The modulus of elasticity of mass concrete representative of various dams ranges from 3.5 to 5.5 X 10^6 psi at 28 days and from 4.3 to 6.8 X 10^6 psi at 1 year.

6.4.3 Modulus of Elasticity of Lightweight Concrete

The modulus of elasticity of concretes made with lightweight aggregates is not influenced materially by the volumetric concentration of aggregate. This is because the modulus of the aggregate is generally about the same as that of the paste. Normal weight sand often is used with lightweight aggregates to increase the modulus of elasticity. Depending on the type of lightweight aggregate and sand content, the modulus of elasticity of lightweight concrete is generally 20 to 50% lower than that for normal weight concrete of equal strength, with the greater difference occurring in the low-strength range. A modulus of 3.5 X 10^6 psi has been obtained for sand-lightweight concrete of 7000 psi, 28-day strength, and 118 pcf wet concrete weight.

An approximate relationship that can be used to estimate the modulus of elasticity of structural lightweight concrete, E_c, in pounds per square inch, is

$$E_c = Cw^{1.5}\sqrt{f_c'}\ \text{psi}$$

in which w is the air-dry unit weight of the concrete in pounds per cubic foot, f_c' is the compressive strength in pounds per square inch as determined from 6 X 12 in. cylinders, and C is a factor dependent upon the value of f_c'. Values of C corresponding to different values of f_c' are given in Fig. 6-21. This empirical formula is reasonably reliable for structural lightweight concretes with compressive strengths of 2,500 to 7,000 psi.

Lake Point Tower
● 7500 psi
○ 6000 psi
design strength

Waterways Experiment Sta.
X 3/8 in. Aggregate
Δ 1 in. Aggregate
U. S. Air Force Weapons Lab. RTD-TDR-63-3114, Feb, 1964

– – – 145 pcf
—— 150 pcf
ACI equation $E_c = 33W^{3/2}\sqrt{f_c'}$

predicted

E_c X10^6 psi: 4.0, 4.2, 4.4, 4.6, 4.8, 5.0, 5.2, 5.4, 5.6, 5.8, 6.0, 6.2, 6.4, 6.6, 6.8, 7.0, 7.2, 7.4, 7.6

f_c', psi: 5000, 6000, 7000, 8000, 9000, 10000, 11000, 12000

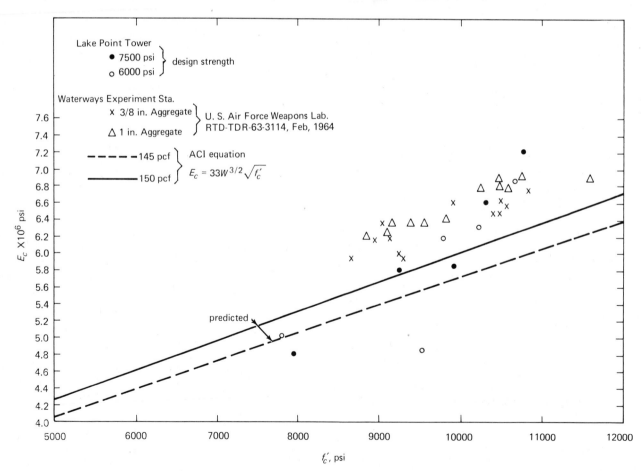

Fig. 6-20 E_c vs f_c' for high strength concrete.

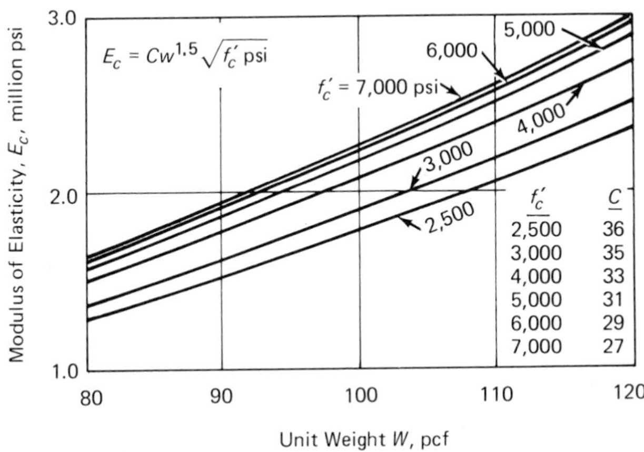

Fig. 6-21 Modulus of elasticity as a function of strength and air-dry unit weight of structural lightweight concrete. (From: Pfeifer, D. W., and Meinheit, D. F., "Structural Lightweight Aggregate Concrete—Modulus of Elasticity," Portland Cement Association, unpublished.)

Static modulus and Poisson's ratio are of greater significance to the design engineer than the corresponding dynamic values (ASTM C215) because these elastic constants are stress-dependent. Static tests involve the application of stresses of the same order as those in practice whereas the stresses induced in the dynamic tests are very small. The sustained modulus is approximately one-half that of the instantaneous modulus when load is applied at early ages and is a slightly higher percentage of the instantaneous modulus when the loading age is 90 days or greater. Structural lightweight concrete has a smaller difference than normal weight. Saturated concrete has a higher static modulus than dry or moist concrete since pore water introduces a triaxial state of stress while the reverse is true for the dynamic modulus, so that as concrete dries, the difference between methods is reduced. Also the higher the compressive strength or modulus (around 6 X 10^6 psi), the closer the results.

For both normal weight and structural lightweight concretes, the modulus of elasticity increases faster than compressive strength at early ages, as shown by the fact that all curves are located above the line of equality in Fig. 6-22. The modulus of steam-cured concrete may vary from that of concrete moist-cured at normal temperatures.

6.4.4 Effect of High Temperature Exposure

Exposure to elevated temperatures causes a reduction in the elastic modulus of concrete, Fig. 6-23. At temperatures up to about 600°F the elastic modulus appears to be the structural property of concrete most sensitive to the effects of heat exposure. For a given concrete, after dehydration, the modulus tends to be independent of age or curing conditions. However, the lower the water-cement ratio, the

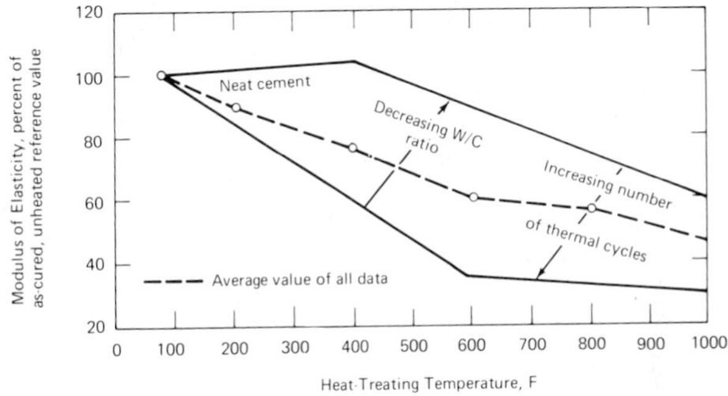

Fig. 6-23 Extremes of the influence of heat exposure on the modulus of elasticity of concrete as determined by various investigators. (From: Lankard, et al., "Effects of Moisture Content on the Structural Properties of Portland Cement Concrete Exposed to Temperatures up to 500°F," ACI SP-25 *Temperature and Concrete*, 1971.)

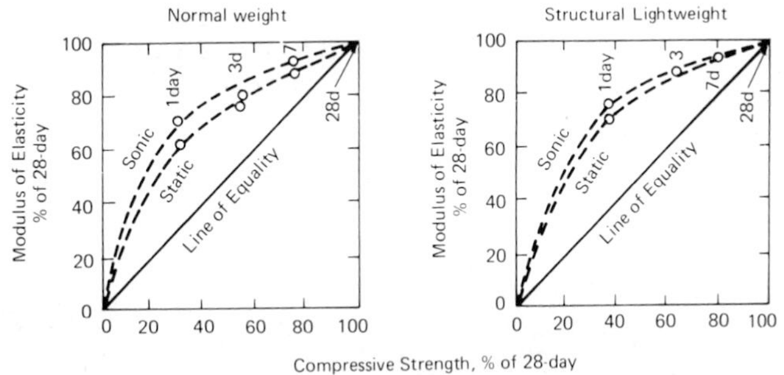

Concrete	Compressive Strength, 28 days	Modulus of Elasticity, 28 days	
		Sonic	Static
Normal Weight (145 pcf) Type I cement–517 lbs/c.y. sand and gravel–1 in. max. size w/c ratio–0.45 Slump–2 to 3 in.	6610 psi	6.09 X 10^6 psi	4.52 X 10^6 psi
Lightweight (99 pcf) Type I cement–611 lb/c.y. Expanded shale fine and coarse Air content–5.8% Slump–3 to 4 in.	5020	2.57	2.36

Fig. 6-22 Modulus of elasticity at early ages. (From: PCA, unpublished data.)

TABLE 6-2 Young's Modulus of Elasticity at Various Temperatures*

Series No.	Aggregate	Cement Content, bags per cu yd	Resonance Modulus, 10^4 psi					
			75 F	35 F	0 F	-75 F	-150 F	-250 F
2	Sand and Gravel	7.0	6.6	6.6	7.0	8.0	9.0	9.0
5 (Moist)	*	5.5	5.5	5.8	6.6	7.6	8.4	8.5
8 (50%)	*	5.5	5.0	5.1	5.1	5.1	5.3	5.4
10 (Dry)	*	5.5	4.3	4.3	4.3	4.3	4.3	4.3
12	*	4.0	5.2	5.2	6.2	6.7	8.2	8.3
15	Expanded Shale	6.0	3.0	3.0	3.3	3.8	4.2	4.3

*From: *Monfore, G. E. and Lentz, A. E., "Physical Properties of Concrete at Very Low Temperatures,"* Portland Cement Association Research Bulletin *145, Portland Cement Association, Skokie, 1962.*

higher the modulus of elasticity after dehydration. The effects of aggregate type are not clearly defined but appear to be small providing the aggregate does not melt or decompose. If the concrete is subjected to cyclic temperatures, the decrease in modulus is even more severe, probably as a result of thermal stresses.

6.4.5 Effect of Low Temperature Exposure

Data on modulus of elasticity by the resonance frequency method (ASTM C215) on concretes cooled from room temperature to -250°F are listed in Table 6-2. The modulus of sand and gravel concretes in a moist condition increases about 50% as the temperature is decreased to -250°F. The 5.5-bag sand and gravel mix, in equilibrium with an atmosphere of 50% relative humidity, increases only 8% for this same temperature range; and an oven dried specimen exhibits a constant modulus as the temperature decreases.

6.4.6 Repetitive Loading

The stress-strain relationship for concrete subjected to repeated compressive loads is cycle dependent while the relationship for reinforcing steel is cycle independent. During the first cycle the stress-strain curve for concrete is the same as that obtained in static tests, but after a few cycles the curve becomes almost linear. As the number of repetitions of load increases, the lines representing the stress-strain curve become concave upward. There also is a general trend towards a decrease in secant modulus (Fig. 6-24) and an increase in non-elastic strain with an increase in the number of loading cycles, presumably due to the development of microcracks in the concrete. Changes in modulus are also accompanied by a reduction in natural frequency.

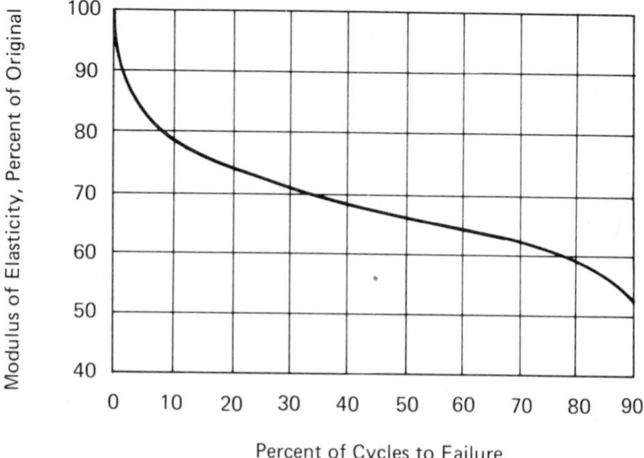

Fig. 6-24 Reduction in secant modulus of elasticity as the result of repeated loads. (From: Linger, D. A., and Gillespie, H. A., "A Study of the Mechanism of Concrete Fatigue and Fracture," *Highway Research News, No. 22,* 40-51. Highway Research Board, Feb. 1966.)

6.5 POISSON'S RATIO

Poisson's ratio is the ratio of transverse (lateral) strain to the corresponding axial (longitudinal) strain resulting from uniformly distributed axial stress. Values of Poisson's ratio, μ, like modulus of elasticity, vary with the Poisson's ratio of the aggregate, the cement paste, and the relative proportions of the two. It also varies with moisture condition and age of the concrete. Values generally range between 0.11 and 0.27. For elastic strains under normal working stresses, Poisson's ratio is taken as 0.20. It is approximately the same for structural lightweight and normal weight concretes. With an increase in strength, age, or aggregate content, Poisson's ratio tends to decrease.

At room temperature, lower values of Poisson's ratio are obtained for higher strength concrete. Results at higher temperatures are erratic and no general trend for the effect of temperature on Poisson's ratio is indicated. Poisson's ratios, calculated from longitudinal and torsional resonant frequency tests remain essentially constant as the temperature of the concrete is lowered from 75 to -250°F.

Dynamic determinations of Poisson's ratio are consistently higher than the corresponding static values, especially at early ages. For saturated concretes, the dynamic values are 0.02 higher for lightweight and 0.03 higher for normal weight concretes, with the difference becoming negligible for dry concretes.

6.6 SHEAR AND TORSION

The resistance of concrete to pure shearing stress has never been directly determined. In addition, the case of pure shear acting on a plane is seldom, if ever, encountered in actual structures. Whenever a state of pure shearing stress is produced in a specimen, it follows from the laws of mechanics that principal tensile stresses, equal in magnitude to the shearing stresses, must also exist on another plane. Since the strength of concrete in tension is less than its strength in shear, failure inevitably occurs as a result of tensile stresses before the strength in shear is reached. Thus, a pure shear test is of no value for determining shearing strength.

A reliable indication of the strength of concrete in pure shear can be obtained only from tests under combined stresses. The most widely used and accepted method of combined-stress testing consists of confined compressive triaxial tests. Mohr's theory has been successfully applied for analysis of triaxial results; however, some modification is required to explain the curvature of the failure envelope. The shearing strength determination obtained from the Mohr's concept appears valid, and, therefore, makes for a convenient and encompassing method of examining all strength parameters. It also more closely approximates the true shear strength of concrete than does that determined from direct tests.

In Mohr's theory, a combination of principal stresses at failure are transformed into a system of concurrent normal and shear stresses that define an envelope of failure conditions. A graphical solution for these stress combinations at failure may be obtained by plotting Mohr's circle for each of the test results on a common axis and then estimating the location of the common tangent of these circles.

The strength of concrete in pure shear—that is, when no normal stresses are present on the plane of failure—is the stress measured to the intersection of the rupture line or limiting curve with the vertical axis, Fig. 6-25. On the basis of the available data, the value of the strength of concrete in pure shear is approximately 20% of the compressive strength.

Shear failures can be abrupt in nature with little or no advance warning of their occurrence. Failure may occur with the formation of a critical diagonal crack or if redistribution of internal forces is accomplished, failure may occur by shear-compression destruction of the compression zone at a higher load. Shear is resisted by the uncracked concrete or by concrete located above the inclined cracks, by aggregate interlock along the inclined cracks, by dowel action of the reinforcement crossing the inclined cracks, and by contribution of any shear reinforcement. Uncracked concrete is very strong in direct shear; however, there is always the possibility that a crack will form in an unwanted location. The approach is to assume that a crack will form in an unfavorable location, and then to provide reinforcement that will prevent this crack from causing undesirable consequences. This discussion only considers the contribution of the concrete and not that of the reinforcement.

Shear stresses along a crack may be resisted by friction. The shear transfer strength is a function of the roughness of the fracture surface and is independent of the concrete strength. However, in some very high strength concretes, fractures cross the aggregates leaving a relatively smooth fracture surface, and consequently a relatively low shear transfer at failure. When the crack is rough and irregular, the apparent coefficient of friction, μ is 1.4 for concrete cast monolithically, 1.0 for concrete placed against hardened concrete, and 0.7 for concrete placed against as-rolled structural steel. To develop friction, however, a normal force must be present. This normal force may be obtained by placing reinforcing steel perpendicular to the assumed crack. As shear slip occurs along the crack, the irregularities of the crack will cause the opposing faces to separate, stressing the reinforcing steel in tension. A balancing compressive stress will then exist in the concrete, and friction will be developed along the confined crack.

The magnitude of shear at initial diagonal tension cracking is a function mainly of concrete strength, percentage of tension reinforcement, beam dimensions, and ratio of moment to shear. Diagonal tension is a combined stress problem in which horizontal tensile stresses due to bending as well as shearing stresses must be considered.

The resistance of concrete to diagonal tension stress is related to the tensile strength of concrete which in turn may be approximated as a function of the square root of the compressive strength. This latter function is used for design since only compressive strength is usually specified and controlled in construction of concrete structures.

The provision of ACI 318-71 for nominal shear stress and nominal torsion stress carried by the concrete apply to normal weight concrete. When structural lightweight aggregate concretes are used, provisions for shear and torsion must be modified.

For normal weight concrete, the splitting tensile strength, f_{ct}, is approximately equal to $6.7\sqrt{f'_c}$. Therefore, when f_{ct} is specified and determined (ASTM C496) for a particular lightweight aggregate concrete, the value of $f_{ct}/6.7$ may be substituted for all values of $\sqrt{f'_c}$ affecting shear and torsion. If use of the splitting tensile strength of lightweight aggregate concrete yields calculated permissible shear values greater than allowed for normal weight concrete, the values for normal weight concrete must be used. Once this value is determined for a particular lightweight aggregate, it is representative of structural concrete made with that aggregate.

The modulus of elasticity in shear, G, which is the ratio of unit shearing stress to the corresponding unit shearing strain may be calculated when the modulus of elasticity and Poisson's ratio are known: $G = E/2(1 - \nu)$. Typical values at

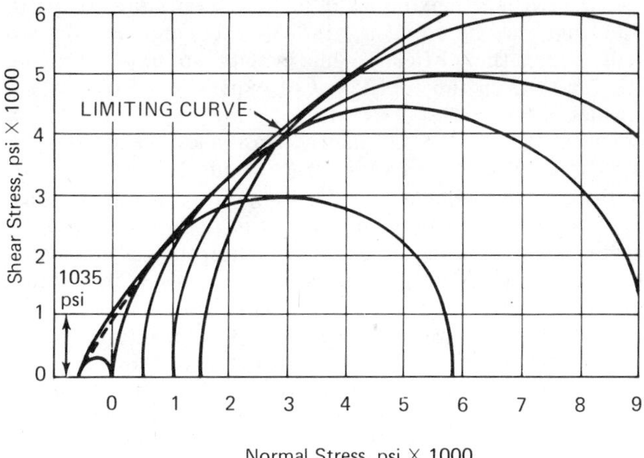

Fig. 6-25 Typical Mohr rupture diagram for normal weight concrete.

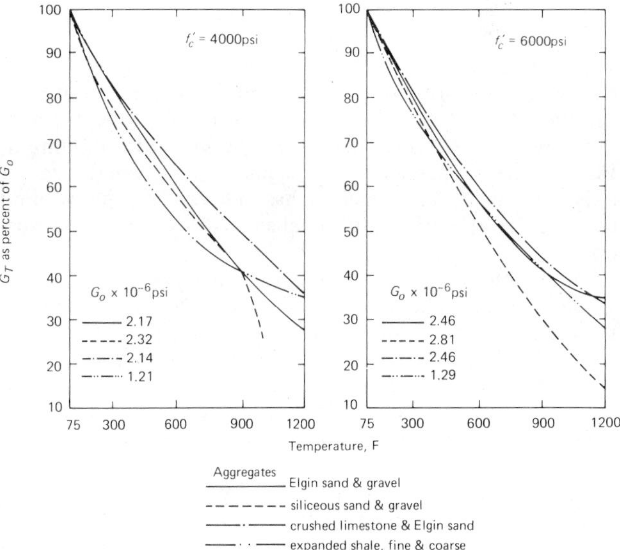

Fig. 6-26 Effect of temperature on the modulus of shear (G) of concrete. (From: Cruz, C. R., "Elastic Properties of Concrete at High Temperatures," *PCA Research Bulletin 191*, 1966.)

ordinary temperatures, 75°F, (G_o) are shown in Fig. 6-26 for various aggregate types. The modulus at elevated temperatures, G_T decreases markedly as temperature increases from 75 to 1200°F. In the case of concrete members, G is difficult to measure and it is not a constant because the stress-strain curve of concrete is not linear.

Behavior of concrete in pure torsion is analagous to behavior in pure shear. Members subjected to pure torsional loads develop diagonal tensile stresses due to the torsional shear stresses. The member fails by bending about an axis parallel to the wider face of the cross section and inclined at 45° to the longitudinal axis of the member. Failure is reached when the tensile stresses induced by a 45° bending component of torque on the wider face reaches a reduced modulus of rupture. This tensile strength is related to some function of $\sqrt{f_c'}$.

The relationship between torsional strength and tensile strength of concrete is dependent on the size and shape of the torsion specimen and on the method used to calculate torsion strength. The torsion strength determined on specimens of circular section and calculated on the basis of rigid-plastic stress distribution (plastic theory) is 75% of the torsion strength calculated from a linear stress distribution (elastic theory). Therefore, the magnitude of the pure tensile strength of a concrete is expected to be less than its comparable torsion strength which was calculated from a linear stress distribution.

6.7 VOLUME CHANGES

Concrete changes in volume slightly during and after the hardening period. Understanding the nature of these changes is necessary for the designer. Frequently, stresses causing cracking can be prevented or minimized by controlling the variables that affect volume changes.

6.7.1 Factors Affecting Volume Changes

Normal volume changes of concrete are caused by variations in temperature and moisture (drying shrinkage), and by sustained stress (creep). Reliable information is available on the most important factors that affect the magnitude of volume changes. When several unfavorable factors act at the same time, the net result is the product, rather than the sum, of the individual effects.

The magnitude of volume changes in concrete is directly related to the properties of the constituent materials. By careful selection of materials and the proportions in which they are used, volume changes can be kept to a minimum. In areas where properties of economically available materials are such that concrete has inherently high volume change, the effects of additional adverse factors can be very critical.

If concrete were free to deform, uniform volume changes would be of little consequence, but since concrete is usually restrained by foundations, subgrades, steel reinforcement or connecting members, significant stresses may develop. This is particularly true when tension is developed; thus restrained contractions causing tensile stresses in concrete are usually more important than restrained expansions which cause compressive stresses.

For convenience, the magnitude of volume changes is generally stated in linear rather than volumetric units. The volume changes that ordinarily occur are small, ranging in terms of change in length from a few up to about 1,000 millionths. Changes in length are often expressed in "millionths" (1 millionth is simply 0.000001). For example, a change in length of 600 millionths may also be expressed as 0.000600. This can also be expressed as 0.06% or 0.72 in. per 100 ft.

6.7.2 Thermal Properties

Thermal properties of concrete are significant in connection with keeping differential volume or length changes at a minimum, extracting excess heat from the concrete, and dealing with similar operations involving heat transfer. The basic properties involved are coefficient of thermal expansion, or contraction, conductivity, diffusivity, and specific heat. The last three are largely interrelated.

The main factor affecting the thermal properties of a concrete is the mineralogic composition of the aggregate. Since the selection of the aggregate to be used is based on other considerations little or no control can be exercised over the thermal properties of the concrete and tests for thermal properties are conducted only for providing constants to be used in behavior studies. Specification requirements for cement, pozzolan, percent sand, and water content are modifying factors but with negligible effect. Entrained air is an insulator and reduces thermal conductivity but other considerations which govern the use of the entrained air outweigh the significance of its effect on thermal properties.

-1. Coefficient of expansion or contraction—The coefficient of thermal expansion or contraction, α, represents the change of concrete volume or, as usually measured on test specimens, the change in length with change in temperature.

An increase in temperature may cause concrete to expand or contract, the latter depending upon a change in moisture content. As with most materials, concrete tends to expand with increasing temperature; however, this normal thermal expansion may be overshadowed by shrinkage due to moisture loss.

Thermal expansion and contraction of concrete vary with factors such as type and amount of aggregate, richness of mix, water-cement ratio, temperature range, concrete age, and relative humidity (degree of saturation of concrete). Of these, aggregate type has the greatest influence.

Values of the coefficient of thermal expansion of cement paste may vary from 5.5 to 12.0×10^{-6} in./in. per °F, depending upon the differences in cement fineness, composition, water-cement ratio, age, and moisture content. The coefficient of the paste is generally higher than the coefficient of aggregate. Pastes which are oven dry or saturated have similar coefficients of expansion. Pastes with intermediate moisture contents have higher coefficients of expansion with a maximum at relative humidities of 50 to 70% that may be twice as large. Since concrete is only partially paste, these effects, while evident, are moderated and the variation in the coefficient of expansion of concrete is smaller.

The coefficient of thermal expansion of the most commonly used aggregates varies from about 2.0 to 7.0×10^{-6} in./in. per °F. Aggregates with a high quartz content, such as quartzite and sandstone, have the highest coefficients; aggregates containing little or no quartz, such as limestone, have the lowest coefficients; and aggregates with medium quartz content, such as igneous rocks, (granite, basalt, etc.) have intermediate values, see Table 6-3.

The coefficient of expansion of concrete varies approximately in proportion to the thermal coefficients and quantity of aggregate in the mixture. Since the coefficient of the paste is generally higher than that of the aggregate, the coefficient of the concrete will be somewhat higher than that of the aggregate, Table 6-3. The coefficients of expan-

TABLE 6-3 Average Coefficients of Linear Thermal Expansion of Rocks (Aggregates) and Concrete (Within Normal Temperature Ranges)

Type of Rock (Aggregate)	*Average Coefficient of Thermal Expansion* $\alpha \times 10^{-6}$ in./in per deg. F	
	Aggregate	*Concrete**
Quartzite, Cherts	6.1-7.0	6.6-7.1
Sandstones	5.6-6.7	5.6-6.5
Quartz Sands and Gravels	5.5-7.1	6.0-8.7
Granites and Gneisses	3.2-5.3	3.8-5.3
Syenites, Diorites, Andesite Gabbros, Diabase, Basalt	3.0-4.5	4.4-5.3
Limestones	2.0-3.6	3.4-5.1
Marbles	2.2-3.9	2.3
Dolomites	3.9-5.5	–
Expanded Shale, Clay and Slate	–	3.6-4.5
Expanded Slag	–	3.9-6.2
Blast-Furnace Slag	–	5.1-5.9
Pumice	–	5.2-6.0
Perlite	–	4.2-6.5
Vermiculite	–	4.6-7.9
Barite	–	10.0
Limonite, Magnetite	–	4.6-6.0
None (Neat Cement)		10.3
Cellular Concrete		5.0-7.0
1:1 (Cement: Sand) **	–	7.5
1:3 **		6.2
1:6 **		5.6

**Coefficients for concretes made with aggregates from different sources vary from these values, especially those for gravels, granites, and limestones. Fine aggregates generally the same material as coarse aggregate.*
***Tests made on 2-yr old samples.*

sion of concrete can be estimated from the coefficients of the paste and aggregates if due consideration is given to their relative solid volumes in the concrete.

Ranges for normal weight concretes are 5 to 7×10^{-6} for those made with siliceous aggregates and 3.5 to 5×10^{-6} for those made with limestone or calcareous aggregate. The values in each case depend on the mineralogy of specific aggregates. Approximate values for structural lightweight concretes are 3.6 to 6×10^{-6} depending on the aggregate type and amount of natural sand used. When a precise value is not required, a coefficient of 5.5×10^{-6} is frequently used. If greater accuracy is needed, tests should be made on the specific concrete mix.

The thermal coefficient for steel is $(6.1 + 0.002\theta)\ 10^{-6}$ in./in./°F, where θ is the temperature in °F; which is comparable to that for concrete. The thermal coefficient for reinforced concrete can be assumed as six millionths per °F, the average for concrete and steel.

The effect of temperature on the thermal coefficient is difficult to separate from shrinkage effects. Only small changes take place in the coefficient as temperature is increased to 500°F or until the temperature at which nearly all the water is removed from the concrete and carbon dioxide is driven off from aggregates containing calcium carbonate. Above this temperature, the transition to a higher expansion coefficient occurs, Table 6-4. Concrete with lower coefficients of expansion have a great advantage in resisting stresses due to severe temperature changes or cycling over a short period of time.

Coefficients of thermal contraction for the range of temperature from 75 to −250°F varies from 3.3×10^{-6} for lightweight aggregate concrete to 4.5×10^{-6} for 7.0-bag sand and gravel concrete, Table 6.5.

-2. Conductivity—Conductivity, k, represents the uniform flow of heat through a unit thickness over a unit area of concrete subjected to a unit temperature difference between the two faces, Btu/ft hr sq ft °F.

Designers usually compute the heat transfer coefficient of an assembly on the basis of values for the thermal conductivity of the *dry* component materials. In service, some moisture is usually present. Moisture contents of materials in service are variable and unpredictable. It is difficult to measure heat transfer in moist materials and the effect on heat transfer of a given amount of moisture is influenced by its distribution within the materials, which in turn is controlled by the properties of the materials, their arrangement in the construction, and variations of temperature imposed upon them by climatic exposure.

Generally, conductivity is essentially a function of the cement paste and aggregate, and, within these narrow limits, is a function of the unit weight of the concrete.

Table 6-6 correlates conductivity with unit weight for oven-dry, normally dry, and saturated concretes. Conductivity of water at 75°F is quite high, about 56. Concretes, which vary in oven-dry unit weight from 20 to 150 pcf will absorb from 43 to 7 pcf of water and produce saturated unit weights varying from 63 to 157 pcf. However, high unit weight of concrete does not necessarily indicate high conductivity. For instance, concrete made with barite aggregate has a k value of 0.79 with a unit weight of 182 pcf; while concrete made with hematite aggregate and with a unit weight of 177 pcf has a k of 1.52.

Mix proportions of cement, aggregate, and free water influence the conductivity of concrete. If the water content of concrete mixture is kept constant the effect of richness of the mix will depend on the type of aggregate: in the case of lightweight aggregates, where k factor of aggregate is generally less than that of cement paste, the richer the mix the higher the conductivity; while in the case of conventional aggregate, where k factor of aggregate is greater than that

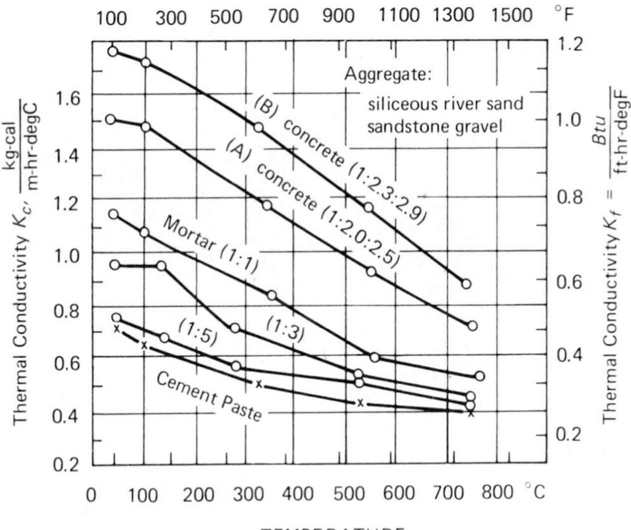

Fig. 6-27 Effect of elevated temperatures on thermal conductivity (after Harada). (From: Zoldners, N. G., "Thermal Properties of Concrete Under Sustained Elevated Temperatures," ACI SP-25, *Temperature and Concrete*, 1971.)

TABLE 6-4 Linear Coefficients of Thermal Expansion

Concrete Containing Elgin Sand and Gravel					
Water-Cement Ratio, by weight	Cement Content, sacks per cu yd	Linear Coefficient of Thermal Expansion at 28 days, per deg F		Linear Coefficient of Thermal Expansion at 90 days, per deg F	
		Below 500 F	Above 800 F	Below 500 F	Above 800 F
Moist Cured–average of three specimens per test					
0.4	7.82	4.2×10^{-6}	11.3×10^{-6}	3.6×10^{-6}	6.2×10^{-6}
0.6	5.52	7.1×10^{-6}	11.4×10^{-6}	4.7×10^{-6}	12.5×10^{-6}
0.8	4.43	6.1×10^{-6}	11.7×10^{-6}	9.3×10^{-6}	18.2×10^{-6}
Air Dried–one specimen per test					
0.4	7.82	4.3×10^{-6}	10.5×10^{-6}	6.8×10^{-6}	11.5×10^{-6}
0.6	5.52	4.3×10^{-6}	11.7×10^{-6}	4.9×10^{-6}	11.2×10^{-6}
0.8	4.43	5.3×10^{-6}	11.5×10^{-6}	6.5×10^{-6}	12.0×10^{-6}
Concrete Containing Expanded Shale Aggregate–one specimen per test					
Water-Cement Ratio	Cement Content, sacks per cu yd	Linear Coefficient of Thermal Expansion at 14 days, per deg F		Linear Coefficient of Thermal Expansion at 21 days, per deg F	
		Below 500 F	Above 800 F	Below 500 F	Above 800 F
Moist Cured					
0.68	6.36	3.4×10^{-6}	4.2		
Air Dried					
0.68	6.36	2.6×10^{-6}	5.4	2.8×10^{-6}	4.9

Note: Air-dried specimens were moist cured 3 days, then dried at 50 percent relative humidity.
From: Phillio, R., "Some Physical Properties of Concrete at High Temperatures," PCA Research Bulletin 97, *Portland Cement Association, Skokie, 1958.*

TABLE 6-5 Thermal Contraction at Various Temperatures

Series No.	*Aggregate*	*Cement Content, bags per cu yd*	*Contraction, Millionths, in./in.*					
			75 F	*35 F*	*0 F*	*-75 F*	*-150 F*	*-250 F*
3	Sand and Gravel	7.0	0	210	410	680	1110	1470
6 (Moist)	*	5.5	0	250	400	740	1090	1420
8 (50%)	*	5.5	0	210	420	770	1020	1280
10 (Dry)	*	5.5	0	170	430	680	890	1240
13	*	4.0	0	170	250	350	720	1280
16	Expanded Shale	6.0	0	180	390	220	510	1060

From: Monfore, G. E., and Lentz, A. E., "Physical Properties of Concrete at Very Low Temperatures," PCA Research Bulletin 145, *Portland Cement Association, Skokie, 1962.*

of the paste, the richer the mix the lower the conductivity. Concrete mixes with high water-cement ratios will produce porous concrete with a low conductivity, especially when dry, however, strengths will be low.

Since moisture content decreases with increasing temperature and exposure time, the conductivity of concrete decreases with increasing temperature. At temperatures above that at which complete dehydration occurs, a gradual disintegration of cement paste occurs, resulting in a further decrease in thermal conductivity, see Fig. 6-27 and Table 6-7. Conductivity values obtained on cooled concrete test specimens, after being exposed to elevated temperatures, are higher by approximately 10 and 20%, respectively, than values obtained during heating.

Thermal conductivities of various concretes at temperatures ranging from −250 to 75°F are listed in Table 6-8. The conductivity of lightweight aggregate concrete is nearly independent of temperature. An increase in moisture content causes an increase in conductivity of all the concretes and at all temperatures.

-3. Specific heat—Specific heat is the amount of heat required to raise the temperature of a unit mass of concrete

TABLE 6-6 Correlation of Free Water, Unit Weight, and Conductivity of Concrete*

Concrete							
Oven Dry		*Normally Dry***			*Saturated*		
Unit Wt, ρ_o	Conductivity, k_o	Free Water, $(\Delta\rho)_n$	Unit Wt, ρ_n	Conductivity, k_n	Free Water, $(\Delta\rho)_s$	Unit Wt, ρ_s	Conductivity, k_s
20	0.60	3.0	23.0	0.77	43.0	63.0	4.2
30	0.81	2.8	32.8	1.02	34.0	64.0	4.3
40	1.05	2.6	42.6	1.28	27.0	67.0	4.5
50	1.30	2.4	52.4	1.58	21.7	71.7	4.9
60	1.60	2.2	62.2	1.91	17.3	77.3	5.5
70	1.94	2.0	72.0	2.30	15.0	85.0	6.5
80	2.34	1.8	81.8	2.80	13.2	93.2	7.7
90	2.82	1.7	91.7	3.40	12.0	102.0	9.5
100	3.47	1.6	101.6	4.18	11.0	111.0	12.0
110	4.30	1.5	111.5	5.40	10.1	120.1	15.0
120	5.62	1.4	121.4	7.10	9.3	129.3	19.0
130	7.30	1.3	131.3	9.30	8.6	138.6	24.0
140	9.40	1.2	141.2	12.05	7.8	147.8	29.0
150	12.00	1.1	151.1	15.40	7.0	157.0	34.0
160	15.00						

**From: Brewer, H. W., "General Relation of Heat Flow Factors to the Unit Weight of Concrete,"* PCA Development Bulletin 114, *Portland Cement Association, Skokie, 1967.*
***In equilibrium with normal ambient weather conditions, 50% R.H.*

TABLE 6-7 Thermal Conductivity of Concretes Made with Type I Portland Cement and Various Aggregates, Btu/ft hr deg F[a]*

Expanded Slag (d = 98)[b]		*Gravel (d = 144)*		*Limestone (d = 143)*		*Sandstone (d = 137)*	
Mean Temp, deg F	*k*	*Mean Temp, deg F*	*k*	*Mean Temp, deg F*	*k*	*Mean Temp, deg F*	*k*
...	...	212	0.883	...	...	202	1.317
359	0.293	327	0.883	327	0.563	336	1.313
749	0.287	756	0.746	740	0.670	783	0.887

[a]W/C *ratio by weight, 0.65 ± 0.03. Aggregate-cement ratio by volume, 7.3 ± 0.07.*
[b]*d = density, lb/ft.*^{3}
**From: Zoldners, N. G., "Effect of High Temperatures on Concrete Incorporating Different Aggregates,"* Proceedings, *American Society for Testing and Materials, Vol. 60, 1960.*

by one degree, (heat capacity of the concrete) and is identified by symbol C.

The specific heat of concrete is equal to the summation of the weighted specific heats of the constituents. The common range of values for normal weight concrete is 0.20 to 0.28 Btu/lb–°F. The specific heat increases with a decrease in unit weight of concrete, but is affected very little by the mineralogical character of the aggregates or by variations in the aggregate or cement content of the mix. In general, the specific heat varies directly with variation in moisture content of the concrete because of the high specific heat of water. It also varies approximately linearly with an increase in temperature, over the range 68 to 212°F, the increase in specific heat may be as much as 35%.

-4. Diffusivity—Diffusivity is an index of the facility with which concrete will undergo temperature change, and it is identified by the diffusion constant a or h^2. It is expressed in m^2/hr or ft^2/hr and may be determined from the formula:

$$a = \frac{k}{C\rho}$$

where k = thermal conductivity; C = specific heat; and ρ = density of concrete in kg/m^3 or lb/ft^3.

The value of diffusivity is largely affected by the aggregate type used in the concrete. Table 6-9 shows diffusivities for concretes made of a number of rock types. The range of typical values of diffusivity varies between 0.02 and 0.08 ft^2/hr.

The higher the value of diffusivity, the more readily heat will move through the concrete. If the rock type is not known, an average value of diffusivity can be taken as 0.04

TABLE 6-8 Thermal Conductivity of Concrete at Various Temperatures*

Moisture Condition at Test	*Cement Content, bags/cu yd*	*Density lb/cu ft*	*Thermal Conductivity at Indicated Temperature, BTU/(hr) (sq ft) (F/in.)*				
			−250 F	*−150 F*	*−75 F*	*0 F*	*75 F*
Crushed Marble Concrete							
Moist	5.5	152	22	22	17	16	15
50% RH		148	21	18	16	16	15
Oven dry		143	13	14	13	13	12
Crushed Sandstone Concrete							
Moist	5.5	133	35	30	26	23	20
50% RH		124	17	16	16	14	15
Oven dry		120	11	11	11	10	10
Crushed Limestone Concrete							
Moist	5.5	141	24	20	18	16	15
50% RH		130	18	14	13	12	11
Oven dry		126	14	13	11	11	10
Elgin Sand and Gravel (3/4-inch max size)							
Moist	7.0	149	32	30	26	23	23
Moist	5.5	147	32	31	26	23	23
50% RH	5.5	143	25	26	21	20	19
Oven dry	5.5	141	18	18	17	15	16
Moist	4.0	143	34	31	26	24	21
Expanded Shale (3/4-inch max size)							
Moist	6.0	99	6.5	6.4	6.6	6.6	5.9
50% RH		96	5.1	5.6	5.6	5.3	5.5
Oven dry		89	3.4	3.6	3.7	3.9	4.3

**From: Lentz, A. E. and Monfore, G. E., "Thermal Conductivity of Concrete at Very Low Temperatures,"* PCA Research Bulletin 182, *1965 and "Thermal Conductivities of Portland Cement Paste, Aggregate and Concrete Down to Very Low Temperatures,"* PCA Research Bulletin 207, *Portland Cement Association, Skokie, 1966.*

TABLE 6-9 Diffusivity and Rock Type

Coarse Aggregate	*Diffusivity of Concrete, ft² per hour*
Quartzite	0.0579
Limestone	0.0508
Dolomite	0.0500
Granite	0.0429
Rhyolite	0.0350
Basalt	0.0321

ft^2 per hour although as can be seen from the table the value of diffusivity varies from this average value. In general, the diffusivity increases with an increase in aggregate content or decrease in water-cement ratio; and it decreases with an increase in temperature of the concrete.

6.7.3 Adiabatic Temperature Rise

Heat of hydration is the heat generated when cement and water react. The amount of heat generated is dependent upon the chemical composition of the cement (heat of hydration of the cement used), water-cement ratio and the cement content; the rate of heat generation is affected by fineness of cement and temperature of curing as well as cement chemical composition.

Adiabatic conditions are defined as occurring without loss or gain of heat; i.e., as isothermal. The extent to which hydration of the cement paste heats the concrete depends on the size of the structural element and its environment. Assumptions of adiabatic temperature rise become more significant as the size of the structural elements or dimensions of the placement become larger. Heat dissipation depends on the type of form, amount of exposed surface, and ambient temperatures at the various surfaces of concrete. The low thermal conductivity of concrete results in a slow rate of heat exchange between concrete and its surroundings. Therefore, at early ages, heat is generated in concrete at a higher rate than it can be transmitted to exposed surfaces. The result is a heat buildup approximating adiabatic conditions. Departure from adiabatic conditions is greatest near the cooling surfaces.

In certain structures such as those with considerable mass, the rate and amount of heat generated are important. If this heat is not rapidly dissipated, a significant rise in concrete temperature may occur accompanied by thermal expansion. Subsequent cooling of the hardened concrete to

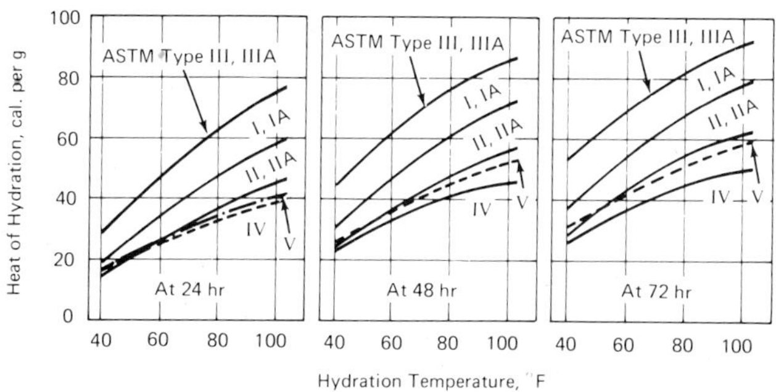

Fig. 6-28 Effect of temperature on the heat of hydration at early ages. Average heat of hydration of the different types of portland cement-conduction calorimeter. (From: Verbeck, G. J., and Foster, C. W., "Long-Time Study of Cement Performance in Concrete with Special Reference to Heats of Hydration," *PCA Research Bulletin 32*, 1949.)

ambient temperature may create undesirable stresses. On the other hand, a rise in concrete temperature caused by heat of hydration is often beneficial in cold weather since it helps maintain favorable curing temperatures.

Typical heat of hydration curves for various portland cements are shown in Fig. 6-28. Most of the heat is generated within the first seven days, during which time the concrete also gains early strength. Both the rate and total adiabatic temperature rise in concrete differ among the various types of cement. In addition, low water-cement ratios reduce the rate of heat generation. An estimate of the temperature rise can be made using the equation

$$T = \frac{CH}{S}$$

where

T = Temperature rise of the concrete due to heat generation of cement under adiabatic conditions, °F.
C = Proportion of cement in the concrete, by weight
H = Heat generation due to hydration of cement, Btu/lb. (To convert calories per gram to Btu's per pound, multiply by 1.8; i.e., cal/g × 1.8 = Btu/lb)
S = Specific heat of the concrete, Btu/lb.-°F.

The temperature-rise equation provides the basis for computing maximum temperature rise and/or temperature rise at various time intervals corresponding to the rate of cement hydration time intervals. The maximum concrete temperature is determined by adding to the computed temperature the initial placement temperature of the concrete mix. Several methods of step-by-step integration have been devised for more detailed analysis of the temperature rise. For final temperature control studies, the actual heat generation should be obtained from laboratory tests. The laboratory study is made using the actual cement, concrete mix proportions, and specimen size that simulates adiabatic conditions for the element under consideration.

When a portion of the cement is replaced by a pozzolan, the temperature rise curves are greatly modified, particularly in the early ages. While the effects of pozzolans differ greatly, depending on the composition and fineness of the pozzolan and cement used in combination, a rule of thumb that has worked fairly well on preliminary computations has been to assume that pozzolan gives off about 50% as much heat as the cement that is replaces.

In general, the effects of water-reducing retarders in concrete are felt only during the first few hours after mixing and can be neglected in preliminary computations.

Placing concrete at lower temperatures (about 50°F) will lower the early rate of heat generation but total heat generation will be equal in the long run. However, the lower initial temperature of the concrete results in a lower maximum temperature of concrete. The chief means for limiting temperature rise is controlling the type and amount of cementitious materials, although temperature rise may be controlled by embedded pipe cooling systems, placement in shallow lifts or flooding the surface of the concrete for about a week to encourage maximum heat loss from the surface by evaporative cooling.

Factors, in addition to the above, subject to at least some degree of control by the designer include the overall temperature drop from the maximum concrete temperature to the final stable temperature, the rate of temperature drop, and the age of the concrete when the concrete is subjected to the temperature change.

6.7.4 Thermal Stresses and Cracking

Thermal stresses develop in two ways: (1) from temperature variations due to the heat of hydration, and (2) from temperature variations due to periodic cycles of ambient temperature.

The temperature in concrete and hence the thermal stresses depend upon the rate at which this heat is dissipated. The properties of concrete relevant to the determination of temperature distribution are: (1) thermal conductivity-for the determination of steady-state conditions, and (2) thermal diffusivity-for the determination of transient conditions. In general, thermal stresses are created due to restraint or in unrestrained member whenever the thermal gradient varies from linear within the concrete mass.

Heat can escape from a body inversely as the square of its least dimension. Consider a number of walls, made of average concrete and exposed to cool air on both faces. For a wall 6 in. thick, 95% of the heat in the concrete will be lost to the air in 1-½ hr. For a 5-ft thick wall, this same amount of heat would be lost in a week. For a 50-ft thick wall, which might represent the thickness of an arch dam, it would take two years to dissipate 95% of the heat stored, and for a 500-ft thick dam, it would take 200 years to dissipate this amount of heat. Thus in most building structural members most of the heat generated by the hydrating cement is dissipated almost as fast as it is generated and there is little temperature differential from the inside to the outside of the member.

Temperature drop may cause tensile stresses and eventual cracking if a member is restrained from shortening or if the cooling affects only part of a member (partially exposed walls or columns). Tensile stresses build up on the face as it is cooled. If the member is not restrained, uniform cooling causes only shortening but no stress. When cooling of an unrestrained member results in a linear gradient through the section, the result is shortening and curvature of the member but no stresses are created.

Extreme differences between internal and outside temperatures may result in surface cracking. For example, temperature stresses occur as the temperature of the concrete rises due to heat of hydration and then drops essentially to the temperature of its surroundings. As the outer surface cools and tends to shrink, compressive stresses are set up in the center and tensile stresses in the cooler outer surfaces. When these tensile stresses become greater than the tensile strength of the concrete, cracking occurs. Concrete cracks at a strain of about 10^{-4} to 2×10^{-4}. The rate of cooling should be regulated to control the temperature drop to prevent extreme temperature differences. The rate may range from less than 1°F per day for periods of less than 25 days, with ½ to ¾°F per day perferable for massive members, to 5°F in any 1 hr. or 50°F in any 24-hr. period for thin members.

The use of insulation on exposed surfaces of massive structural elements tends to decrease the temperature fluctuations within the mass; it also serves, however, to retain more heat so that a higher maximum temperature is usually attained.

The maximum permissible gradient for a nominally reinforced structure is generally given as less than 100°F. However, as percentage of reinforcement increases, maximum permissible temperature difference across the element may be increased. This should be evaluated in terms of permissible deflections and restraint conditions.

Exposed structural members are affected by daily and annual cycles of temperature, as well as sustained elevated temperatures. Figure 6-29 gives a general idea of what happens in a concrete member subjected to an exterior sine-form temperature oscillation with a frequency n. At a distance x from the face, the oscillation arrives with a time lag $\Delta\tau$. The surface amplitude, θ_{oa}, attenuates at the distance x to the amplitude θ_{xa}. The effects of time lag and attenuation can be combined into a single expression for the amplitude (range of temperature variation) at a distance x from the surface at, τ

$$\theta_{xa} = \theta_{oa}\, e^{-x\sqrt{\pi n/h^2}} \sin\left(2\pi n\tau - x\sqrt{\frac{\pi n}{h^2}}\right)$$

where

n = the frequency of temperature change
h^2 = thermal diffusivity

Since the interior reacts so much more slowly than the surface to cycles of temperature, it is as though the surface was restrained by the interior concrete.

In a radiation shield, attenuation of radiation results in a rise in temperature of the shielding concrete as the absorbed energy of radiation is converted to heat. The thermal distribution curve is assumed to be nonlinear with the maximum temperature occurring some distance in from the surface exposed to the radiation source, due to the dissipation of energy of the absorbed neutrons. In general, reinforced concrete should not be exposed to an incident energy flux greater than 2×10^{11} MeV/cm^2*, in order to ensure acceptable thermal stresses. This flux corresponds to approximately 100 Btu/ft^2/hr and would cause about 90°F temperature rise in the shield.

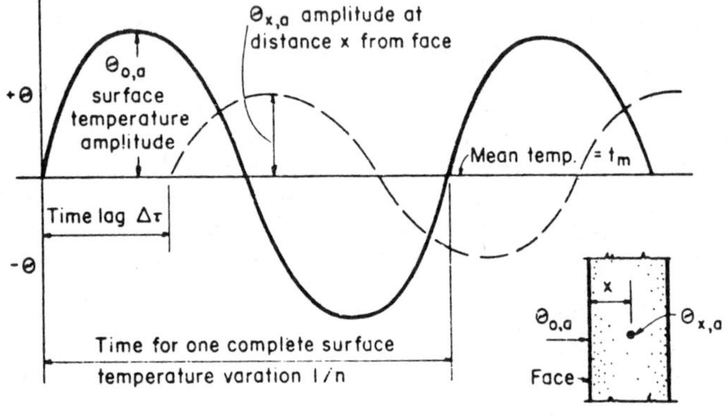

Fig. 6-29 Temperature amplitude at surface and at distance, x. (From: Fintel, M., and Khan, F. R., "Effects of Column Exposure in Tall Structures," PCA, EB018.01D, 1968.)

*1 MeV (million-electron-volt) is the amount of energy which would be acquired by an electron in falling through a potential of 1,000,000 v. It is equal to 1.6×10^{-3} watt-sec.

Tensile stresses across a section due to nonlinear temperature gradients can be calculated by the following equation and compared with the tensile strength of the concrete to determine if thermal cracking will occur

$$\sigma = \frac{P + P_\theta}{A} + \frac{M + M_\theta}{I} y - E\alpha\Delta\theta(y)$$

where

$$P_\theta = E\alpha \int_A \Delta\theta(y)dy$$

$$M_\theta = E\alpha \int_A \Delta\theta(y)y\, dy$$

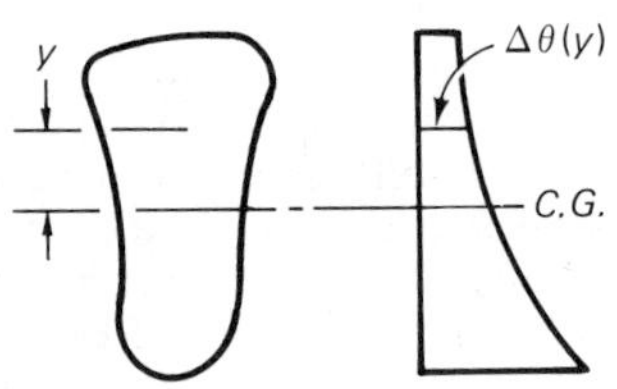

where**

E = modulus of elasticity
α = thermal coefficient of expansion
$\Delta\theta(y)$ = distribution of rise in temperature
P, M = external axial load and moment
y = distance from center of gravity
A = total area of cross section
I = moment of inertia of cross section

Where the temperature variation, $\Delta\theta(y)$, cannot be expressed analytically, the indicated integrations can be performed numerically by the use of Simpson's Rule.

Generally, creep modifies the thermal stresses considerably but little quantitative information is available.

In the above discussion, the temperature was assumed to vary only vertically, that is as a function of y. In reality, it can vary along the width of a beam as well as the depth. The equations could be extended to include the effects of two-dimensional distribution of temperature but such a refinement is not considered warranted.

6.7.5 Drying Shrinkage and Creep

Concrete expands with a gain in moisture and contracts with a loss in moisture. If kept continuously in water, concrete slowly expands for several years, but the total amount of expansion is normally so small that it is unimportant (usually less than 150 millionths). Concrete that is not continuously wet is subject to water loss with resulting contraction or shrinkage. Since concrete exposed to the atmosphere loses some of its original water, it normally exists in a somewhat contracted state compared to its original dimensions.

The consideration of shrinkage in design may be highly critical for some structures and unimportant for others.

**Equation reference is from REFERENCE: Gustaferro, A. H., Abrams, M. S., and Salse, E. A. B., *Fire Resistance of Prestressed Concrete Beams, Study C: Structural Behavior During Fire Tests* (RD009.01B), Portland Cement Association, 1971, and C. L. Townsend "Control of Cracking in Mass Concrete Structures" U.S. Dept. of the Interior Bureau of Reclamation Engineering Monograph No. 34, 1965.

Shrinkage can affect performance and appearance—extent of cracking, prestress loss, effective tensile strength, and warping. The amount of shrinkage that is tolerable depends on jointing and design details of the structure.

Unit length change due to drying shrinkage of small plain concrete (no reinforcement) specimens ranges from about 400 to 800 millionths when exposed to air at 50% humidity. The extremes, however, may range from less than 200 millionths for low slump lean mixes with good quality aggregates to over 1000 millionths for rich mortars or some concretes containing poor quality aggregates and an excessive amount of water. Concrete with a unit shrinkage of 600 millionths shortens about 0.72 in. per 100 ft while drying from a saturated condition to a state of moisture equilibrium in air at 50% relative humidity. This equals approximately the thermal contraction caused by a decrease in temperature of 100°F.

When concrete is loaded, the deformation caused by the load may be divided into two parts: a deformation which occurs immediately, and a time-dependent deformation which begins immediately but continues for years. This latter deformation is called creep.

Creep of concrete is the dimensional change or increase in strain with time due to a sustained stress. Magnitude of creep in tension or compression is the same. Creep of concrete may be either beneficial or detrimental, depending on the prevalent structural conditions. Concentrations of stress, either compressive or tensile may be reduced by stress transfer through creep, or creep may lead to excessive long-time deflection, prestress loss, or growing camber. The effects of creep along with those of drying shrinkage should be considered, (shrinkage may act with or against creep), and, if necessary, compensated for in structural designs. For example, in structures subjected to sustained thermal gradients, creep and shrinkage will reduce the initial thermal stresses but give rise to undesirable reversals of stress on cooling.

Creep consists of two components:

a. Basic (or true) creep occuring under conditions of hygric equilibrium, which means that no moisture movement occurs to or from the ambient medium. In the laboratory basic creep can be reproduced by sealing the specimen in copper foil.
b. Drying creep resulting from exchange of moisture between the stress member and its environment. Drying creep has its effect primarily during the initial period under load.

For structural engineering practice it is convenient to consider specific creep, ϵ'_c, which is defined as the ultimate creep strain per unit of sustained stress. The ultimate magnitude of creep of plain concrete per unit stress (psi) can range from 0.2 to 2.0 millionths in terms of length, but is ordinarily about 1 millionth or less.

In newly placed concrete, the change in volume or length due to creep is largely unrecoverable. However, creep that occurs in old or dry concrete is largely recoverable and creep recovery appears to be essentially independent of temperature and stress.

-1. Important parameters—Shrinkage and creep of concrete are related phenomena and are controlled by similar parameters. The needed parameters are those which are known or predictable by the designing engineer. These parameters affecting shrinkage and creep include: (1) water-cement ratio of the cement paste; (2) physical characteristics of the aggregate; (3) cement paste content and characteristics; (4) age of concrete when exposed to drying or when an external load is applied; (5) size and shape of the structural member; (6) amount of steel reinforcement; (7) environmental exposure conditions such as relative humidity, temperature, and carbon dioxide content of the air; and (8) curing conditions.

Most research data considers the basic creep and shrinkage characteristics of concrete as measured on small unreinforced prisms or cylinders in a controlled laboratory environment. However, the shrinkage and creep of full-scale structural elements is considerably reduced because of size effect, sequence of loading and amount of reinforcement. In addition, most creep research is based on application of loads in one increment. Such creep information, therefore, is applicable to flexural elements of reinforced concrete and to elements of prestressed concrete. In the construction of a high-rise building, columns are loaded in as many increments as there are stories above the level under consideration and research has shown that incremental loading over a long period of time makes a considerable difference in the magnitude of creep. Therefore, only after the creep and shrinkage strains have been determined separately and modified for the conditions of the designed structure can their combined effect on the structure be considered.

In comparing behavior under load of concretes made with different cements, the ratio of the applied stress to the strength at the time of loading should be considered. With respect to the cement, the magnitude and rate of creep stain is influenced by the strength attained by the cement paste at the time of loading. This is controlled to some extent by the chemical composition and fineness of the cement. For concretes loaded to a given stress, the creep will be less for a cement with the highest rate of strength gain. But for concretes loaded to a particular stress-strength ratio, the cement will be of less importance. Cements of different composition and fineness also have variable effects on drying shrinkage. Such differences have been moderated considerably in recent years in most cements by providing the optimum amount of gypsum in the cement. At optimum gypsum content, the type of cement *per se* does not significantly influence creep or drying shrinkage. The optimum amount of gypsum is unique for each cement and increases as the anticipated hydration temperature increases.

The role of the aggregate in concrete is to dilute the paste matrix and to restrain greatly the shrinkage and creep of the paste, thereby reducing the overall shrinkage and creep of the concrete. The effectiveness of aggregate in reducing shrinkage and creep increases as the volumetric fraction of coarse aggregate in the concrete increases. The mineralogical and physical properties of coarse aggregate are important in providing restraint to shrinkage and creep. Because of the great variation in aggregate within any mineralogical or petrological type, it is not possible to make a general statement about the magnitude of shrinkage or creep of concrete made with aggregates of different types. Hard aggregates with high density and high modulus of elasticity coupled with moderate porosity or absorption produce concrete with the lowest drying shrinkage and creep. Concrete for radiation shielding containing hydrous aggregates (limonite, bauxite, or serpentine) will shrink more than concrete containing aggregates such as granite or magnetite.

As excessive amount of clay or other fine filler material in an aggregate tends to increase shrinkage. Although an increase in water content results, the increase in drying shrinkage appears greater than can be accounted for on this basis alone. Clay particles or coatings on aggregates cause significant increases in creep, principally because of loss of bond, loss of restraint and moisture movement.

Other aggregates variables such as grading, maximum size, and particles shape have their main influence on shrinkage or creep through the effect they have on the paste content required for adequate workability. For example, shrinkage can be minimized by using aggregate of the largest practicable size, since concrete mixes with large-size aggregates have a greater total aggregate content than mixes made with small-size aggregates.

Structural lightweight concretes made and cured at normal temperatures have drying shrinkage ranging from about slightly less to 30% more and creep ranging from about the same to 50% more than that of some normal-weight concretes. High-strength lightweight concrete has about the same shrinkage and creep as comparable normal-weight concrete. Partial or full replacement of lightweight fines by a good grade of natural sand usually reduces shrinkage and creep for concretes made with most lightweight aggregates.

Little information is available on the effect of admixtures on drying shrinkage and creep, possibly because of the multiplicity of admixtures available and their frequent modification. At present, it is not possible to predict which combinations of admixtures and cements will influence shrinkage or creep. When drying shrinkage and creep are important, the admixtures should be evaluated using job materials, mixing and curing. However, the use of accelerators such as calicum chloride and triethanolamine results in substantial increases in drying shrinkage and creep of concrete. Despite reductions in water content, some chemical admixtures of the water-reducing type also increase drying shrinkage and creep substantially (age of loading is important) particularly those which contain an accelerator to counteract the retarding effect of the admixture. The materials commonly used to entrain air in concrete have little effect on shrinkage or creep.

Concrete containing pozzolanic admixtures which require more water may have increased shrinkage, but most low carbon fly ash will not appreciably affect shrinkage. However, pozzolans may increase creep because of the increase in paste content resulting from the increased volume of hydration products brought about by the reaction of cement hydration products and the pozzolans.

The influence of proportions is best understood by considering inter-related factors such as water-cement ratio, cement content, aggregate content, and total water content. Both the creep and the shrinkage of concrete increase with an increase in water content. Shrinkage is approximately proportional to the percentage of water by volume in the concrete, while creep varies with the water-cement ratio. Thus, shrinkage can be minimized by keeping the water content of the paste as low as possible and the total aggregate content of the concrete as high as possible. Use of low slumps and placing methods that minimize water requirements are thus major factors in controlling shrinkage. Any practice that increases the water requirements of the cement paste such as the use of high slumps, excessively high fresh concrete temperatures, or smaller coarse aggregate, increases shrinkage.

For a given stress, creep increases with an increase in water-cement ratio for a given paste content. However, for a constant water-cement ratio, creep increases with increasing paste content. Generally speaking for a particular stress, lean mixes creep more than rich mixes.

However, specimens loaded to the same stress-strength ratio and having the same paste content will have the same creep regardless of the water-cement ratio. Under this loading condition, a lean mix will exhibit less creep than a rich mix, although in a lean mix, a larger portion of the ultimate creep occurs at an early age.

The method of curing of concrete has a marked effect on the amount of shrinkage and creep. Atmospheric steam curing may reduce the shrinkage from 10 to 40% and creep about 25 to 40%. The effects of length of moist curing period on drying shrinkage are not as significant as their effects upon creep.

Shrinkage of concrete is caused by evaporation of moisture from the surface. The rate and amount of evaporation and consequently of shrinkage depend greatly upon relative humidity of the environment and ambient temperature. Since evaporation occurs only from the surface of a structural element, the volume-to-surface ratio (size and shape) of a member has a pronounced effect on the amount of its shrinkage.

The rate at which the internal relative humidity of concrete decreases as the member is subjected to a drying atmosphere is of considerable importance. Moisture migrates very slowly through concrete so that cement continues to hydrate beneath the outer surface for a long period of time after exposure to drying; hydration ceases when relative humidity is less than about 80%. Also, the relative humidity in concretes having low water-cement ratios decreases more slowly than concretes having high water-cement ratios and moisture diffuses considerably slower in structural lightweight concrete than in normal weight concrete.

The variation of internal relative humidity with time at various radial distances from the center of 6 × 12-in. solid cylinders is shown in Fig. 6-30. The internal humidity at any location dropped from the initial 100% relative humidity at a decreasing rate toward an asymptotic value of 50% relative humidity, the humidity of the environment. The initial rate of decrease of internal humidity is fastest near the surface, and later the rate of decrease is approximately the same for all locations. In a dry atmosphere, moderate-size members (24-in. diameter) may undergo 50–70% of their ultimate shrinkage within two to four months while identical members kept in water may exhibit growth instead of shrinkage. In moderate-size members the inside relative humidity has been measured at 80% after four years of storage in a laboratory in 50% relative humidity.

Concrete, such as an interior column, subjected to a continuously dry atmosphere will exhibit greater drying shrinkage than concrete exposed to high humidities. Since

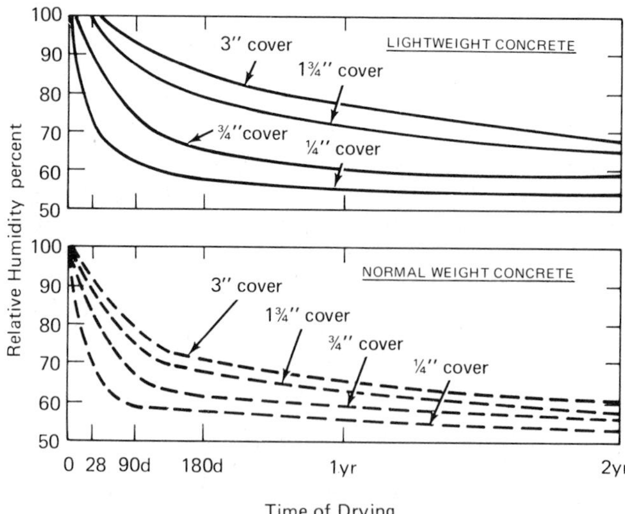

Fig. 6-30 Typical relative humidity distribution of 6 × 12 in. cylinders—moist-cured seven days, then dried at 73°F and 50% relative humidity. (From: Hanson, J. A., "Effects of Curing and Drying Environments on Splitting Tensile Strength of Concrete," *PCA Development Department Bulletin D141*, 1968.)

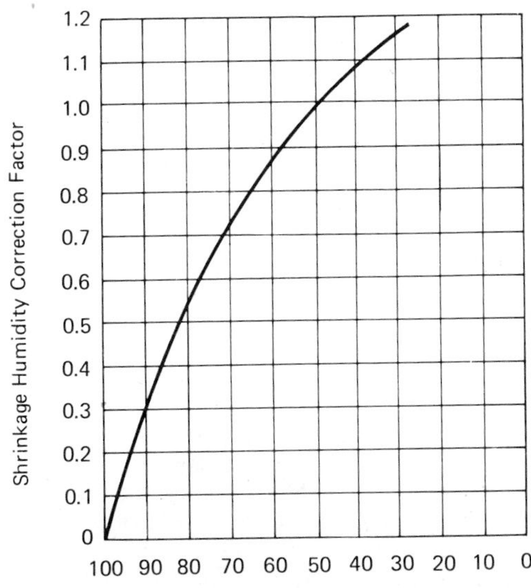

Fig. 6-31 Shrinkage humidity correction factor. (From: Freyermuth, C. L., "Design of Continuous Highway Bridges with Precast, Prestressed Concrete Girders," PCA EB014.01E, 1969.)

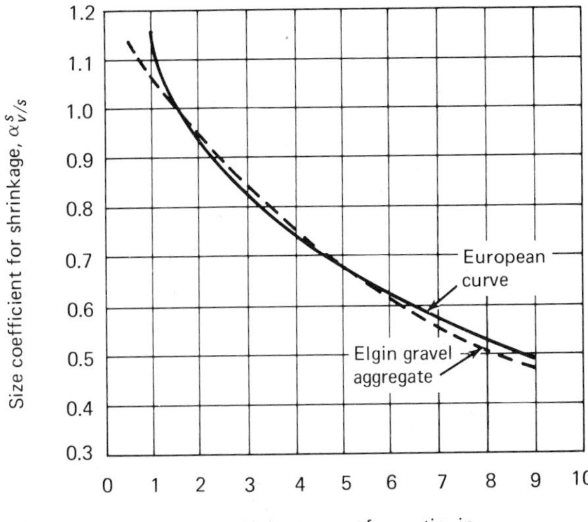

Fig. 6-32 Shrinkage vs volume to surface ratio. (From: Fintel, M., and Khan, F. R., "Effects of Column Creep and Shrinkage in Tall Structures—Prediction of Inelastic Column Shortenings," *ACI Journal Proceedings*, 66 (12), Dec. 1968.)

the rate and amount of shrinkage greatly depend upon the relative humidity of the environment, the shrinkage specimen should be stored under conditions similar to those for the actual structure, If this is not possible, the shrinkage results of a specimen not stored under field humidity conditions must then be modified to account for the humidity conditions of the structure by multiplying by a humidity correction factor as presented in Fig. 6-31.

Alternating the ambient relative humidity within two limits results in higher shrinkage and creep than that obtained at a constant humidity within the given limits. Laboratory tests, therefore, may underestimate creep and shrinkage under conditions of practical exposure.

The rate and ultimate amount of shrinkage are smaller for large masses of concrete than for smaller masses, although shrinkage continues longer for the large mass. Much of the shrinkage data available in the literature is obtained on 11-in. long prisms of a 3 X 3-in. section (volume-to-surface ratio, $V/S = 0.75$ in.) or on 6-in. diameter cylinders ($V/S = 1.5$ in.). Obviously, such data cannot be applied to usual size members without considering the effect of size.

The relationship between the magnitude of shrinkage and the volume to-surface ratio has been plotted in Fig. 6-32 based on laboratory data. Also plotted is a curve based on European investigations. The coefficient $\alpha^s_{v/s}$ shown in the figure is used to convert shrinkage data obtained on 6-in cylinders ($V/S = 1.5$ in.) to any other size columns. A similar curve can also be plotted to convert laboratory data obtained from 3 X 3-in. prisms ($V/S = 0.75$ in.).

Thus, the amount of shrinkage, ϵ_s, of a nonreinforced column is

$$\epsilon_s = \epsilon_{s,\ \text{test}}\, \alpha^s_{v/s}$$

where $\epsilon_{s,\ \text{test}}$ is the shrinkage obtained from 6-in. cylinder specimens made of the concrete mix to be used in the structure and stored under job-site conditions and $\alpha^s_{v/s}$ is the coefficient from Fig. 6-32 for the volume-to-surface ratio of the column being designed.

For many outdoor applications, concrete reaches its maximum moisture content during the season of low temperature. Thus, the volume changes due to moisture and temperature variations frequently tend to offset each other.

For constant relative humidity, changes in temperature have a negligible effect on shrinkage of concrete. Humidity of the air, however, tends to vary inversely with temperature. Hence shrinkage strains tend to be less for lower temperatures because of both the higher relative humidity and the lower rate of evaporation. In general, high temperatures tend to accelerate the rate of moisture removal from concrete. High temperature accompanied by drying may lead to initial shrinkage followed by expansion once the effect of thermal expansivity is greater than the shrinkage associated with moisture loss. There is a good correlation between internal relative humidity and drying shrinkage.

Creep is less sensitive to member size than shrinkage since only the drying creep component of the total creep is affected by size and shape of members, whereas basic creep in independent of size and shape. It appears that drying creep has its effect only during the initial three months. Beyond 100 days, the rate of creep is equal to the basic creep. Creep of sealed specimens (basic creep) is about 20% less than similar unsealed specimens, although values up to about 65% have been obtained after three years of loading.

In Fig. 6-33 the relationship between creep and volume-to-surface ratio has been plotted. Also plotted is the curve based on European experience. The curves are almost identical. It is seen from the curves that in members with a volume-to-surface ratio of 10, drying creep is negligible and only basic creep occurs. Only smaller members indicate any significant drying creep. In structural members of substantial dimension the rate of moisture diffusion is sufficiently slow so that their behavior approximates that of sealed laboratory specimens.

Creep increases as the ambient temperature increases. Sealed or water-stored specimens generally exhibit less creep than unsealed specimens, and creep decreases with increasing degree of hydration and increases with increasing moisture content of the specimen at loading. Creep at 120°F is approximately two to three times as great as creep at room temperatures. For temperatures of 120°F to about 212°F, some controversy exists about whether or not there is a further increase of total creep with increasing temperature. Some investigators have found a definite maximum of total

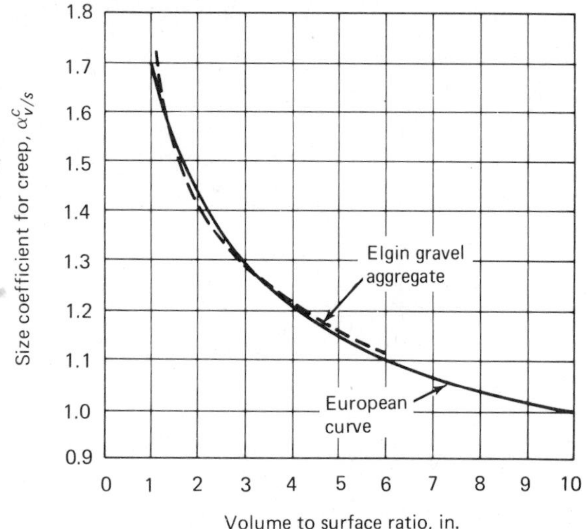

Fig. 6-33 Creep vs volume to surface ratio. (From: Fintel and Khan, "Effects of Column Creep and Shrinkage in Tall Structures—Prediction of Inelastic Column Shortenings," *ACI Journal Proceedings*, **66** (12), Dec. 1968.)

creep in the range of 120°F to 180°F, but most have not and have concluded that creep increases with temperature up to around 212°F, the creep at 212°F being on the order of four to six times as great as the creep at room temperature (at the end of a 60- to 100-day loading period). On the other hand, in an apparent contradiction, most investigators have also found a definite maximum for the creep rate between 120 and 180°F, if the creep rate is computed for some period between 1 and 107 days under load. This seems to indicate that as the temperature increases, a larger portion of the (larger) total creep deformation occurs during the first few hours under load.

Limited data is available for creep at temperatures exceeding 212°F. Tests on unsealed specimens show no appreciable change in creep within the temperature range of 212 to 280°F; the creep rate for a 1 to 100 day loading period appears to decline. Beyond 280°F, both creep rate and creep magnitude increase with temperature (unsealed specimens). For example, creep of concrete subjected to constant sustained compressive stress of 1800 psi for five hours while at high temperatures showed the following:

*Creep of Concrete at High Temperatures Expressed as a Multiple of Creep at 75° F**
(5-hr Test)

Temperature, °F	75	300	600	900	1200
Ratio	1.0	3.3	6.4	14.9	32.6

*From: Cruz, C. R. "Appartus for Measuring Creep of Concrete at High Temperatures," PCA Research Bulletin 225, 1968.

These changes in creep at high temperatures may be related to the inversions in strength and elasticity properties with increasing temperature.

Creep of concrete is a linear function of the stress-strength ratio for stress up to about 50% of the ultimate strength of the concrete. (Normal working stresses fall below 50%). Beyond that level creep becomes a nonlinear function of stress, increasing as second or even third order power of the sustained stress. In the extremely high stress range (85 to 90% of ultimate strength), the creep rate increases rapidly and leads to failure at stresses well below the instantaneous ultimate strength. Under continuous load, creep continues for many years, but the rate decreases with time as the concrete increases in strength while under load. Drying shrinkage is not appreciably affected by the magnitude of the sustained stress.

During the initial period of loading the rate of creep is significant. The rate diminishes as time progresses until it eventually approaches zero. Figures 6-34(a) shows a typical creep versus time curve drawn on a standard scale. The same curve plotted on semilogarithmic graph paper is shown in Fig. 6-34(b), with time on the logarithmic abscissa.

For a given mix of concrete the amount of creep depends not only upon the total stress but also to a great extent upon the loading history. A concrete element with its load applied at an early age exhibits a much larger specific creep than a specimen loaded later. This means that creep decreases with increase in strength-to-stress ratio at the time of loading. Also because of the gradual increase of the modulus of elasticity with age, the elastic shortening per unit stress of older concrete is smaller than that of concrete loaded at an earlier age. Therefore, concrete specimens of equal strength but of different age will have different creep characteristics. Figure 6-35 shows elastic and creep strains of columns loaded at various ages. The elastic response to

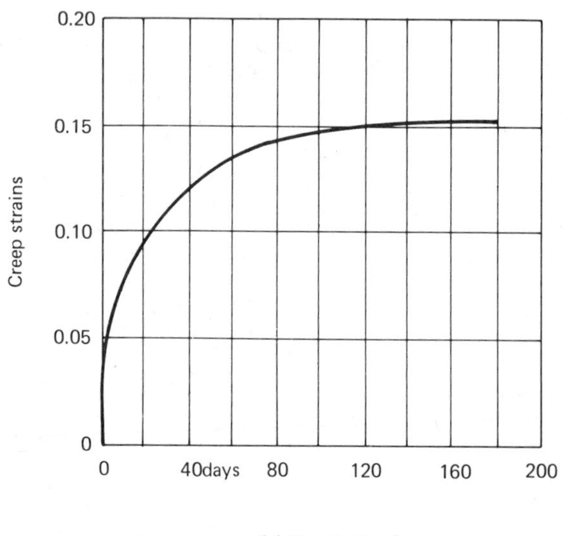

(a) Standard scale

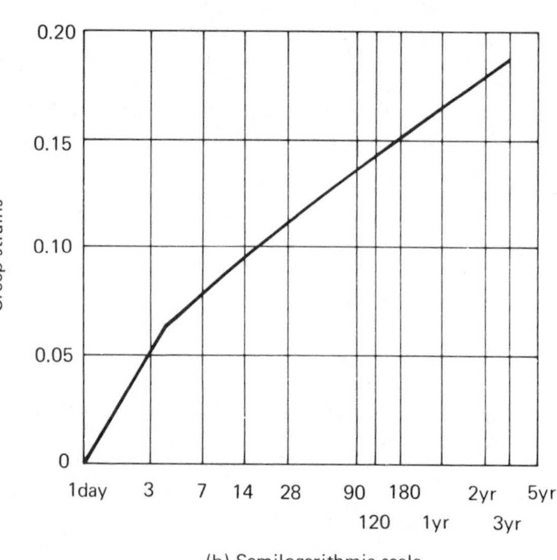

(b) Semilogarithmic scale

Fig. 6-34 Creep strains vs time under load. (From: Fintel and Khan, "Effects of Column Creep and Shrinkage in Tall Structures—Prediction of Inelastic Column Shortenings," *ACI Journal Proceedings*, **66** (12), Dec. 1968.)

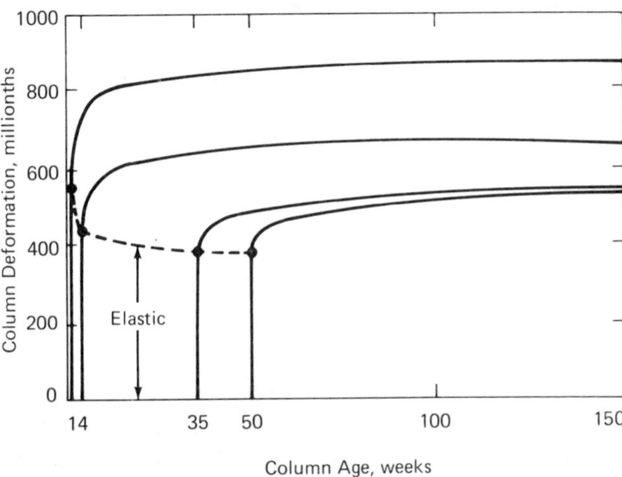

Fig. 6-35 Elastic and time-dependent creep deformation of instantaneously loaded reinforced columns. (From: Pfeifer, D. W., and Hognestad, E., "Incremental Loading of Reinforced Lightweight Concrete Columns," Final Report of the Eighth Congress, International Association of Bridge and Structural Engineers, New York, Sept. 1968, 1055–1063.)

load is in accord with elastic theory, and the significant increase in modulus of elasticity of concrete as a function of curing time is quite evident in the measured elastic response of the concrete. The measured time-dependent creep characteristics also reflect the influence of age of concrete at loading on the time-dependent behavior of concrete.

Loading history is particularly significant for columns of multistory buildings which are loaded in as many increments as there are stories above the level under consideration. Since creep decreases with age of the concrete at load application, each subsequent incremental loading contributes a smaller specific creep to the final average specific creep of the column, Fig. 6-36. The computed elastic shortening, taking into account the increased modulus of elasticity of the concrete, is also shown in the figure. It can be seen that the influence of creep is small when sealed reinforced columns are incrementally loaded during a long-time period.

The postulated and confirmed principle of superposition of creep states that: Strains produced in concrete at any time by a stress increment are independent of the effects of any stress applied either earlier or later. The stress increment may be either positive or negative, but stresses which

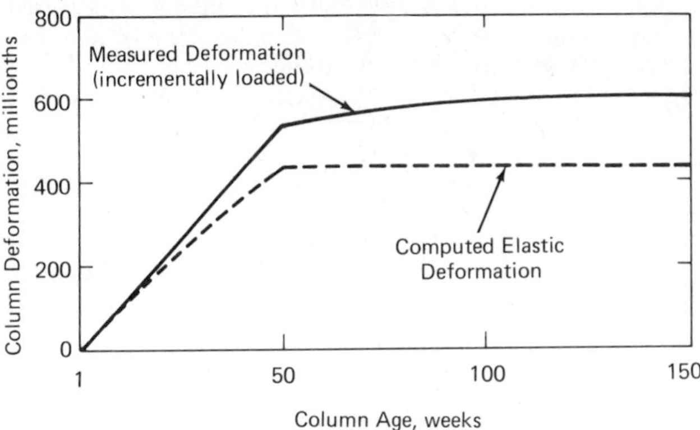

Fig. 6-36 Elastic and time-dependent creep deformation of incrementally loaded reinforced columns. (From: Pfeifer and Hognestad, "Incremental Loading of Reinforced Lightweight Concrete Columns," Final Report of the Eighth Congress, International Association of Bridge and Structural Engineers, New York, Sept. 1968, 1055–1063.)

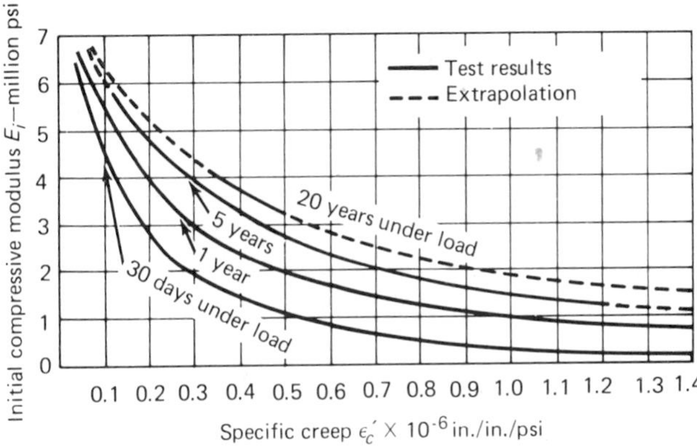

Fig. 6-37 Prediction of basic creep from elastic modulus. (From: Fintel and Khan, "Effects of Column Creep and Shrinkage in Tall Structures—Prediction of Inelastic Column Shortenings," *ACI Journal Proceedings*, **66** (12), Dec. 1968.)

approach the ultimate strength are excluded. Thus, each load increment causes a creep strain corresponding to the strength-to-stress ratio at time of its application, as if it were the only loading to which the column is subjected.

The specific creep values corresponding to the ages at which incremental loadings are applied in an intended multistory structure can be obtained by extrapolation from a number of laboratory samples prepared in advance from the actual mix to be used in the structure. It is obvious that sufficient time for such tests must be allowed prior to the start of construction, since reliability of the prediction improves with length of time over which creep is actually measured.

An alternate method to predict basic creep (without testing) from elastic modulus of elasticity has been proposed. Results of limited tests on normal weight concrete indicate that creep can be predicted easily from the initial modulus at time of load application. Curves in Fig. 6-37 give the creep magnitude as related to the initial modulus of elasticity for different load durations. For design purposes the 20-year creep can be regarded as the ultimate creep. The dashed lines in Fig. 6-37 are extrapolations while the solid lines are test results. Thus from the specified 28-day strength, the basic specific creep for loading at 28 days can be determined and then modified for construction time, member size and percentage of reinforcement as discussed later.

The creep-time relationship plotted on semilog graph paper in Fig. 6-34 (b) consists of three straight line segments. Such a curve is mathematically represented by an equation which is a sum of three exponential components. Only one of the three components is effective beyond about 10 days, the two others having an effect only during the very early days.

For the solution of creep due to incremental loading the primary interest is in the final creep value. The exponential expression for creep, represented graphically in Fig. 6-34 (b), has a particular advantage for the structural engineer. It allows interpolation and extrapolation with as few as two points, since beyond about 10 days it is represented as a straight line when time is plotted on a logarithmic scale.

The curve in Fig. 6-38 gives the relationship between creep and age at loading. The coefficient α_{age} relates the creep for any age at loading to the creep of a specimen loaded at the age of 28 days. The 28-day creep is used as a basis for comparison (with $\alpha_{age,28} = 1.0$). Thus, a loaded 6-in. diameter standard cylinder wrapped in foil can supply the basic information which is subsequently modified to consider any age at loading. The wrapped specimen isolates the

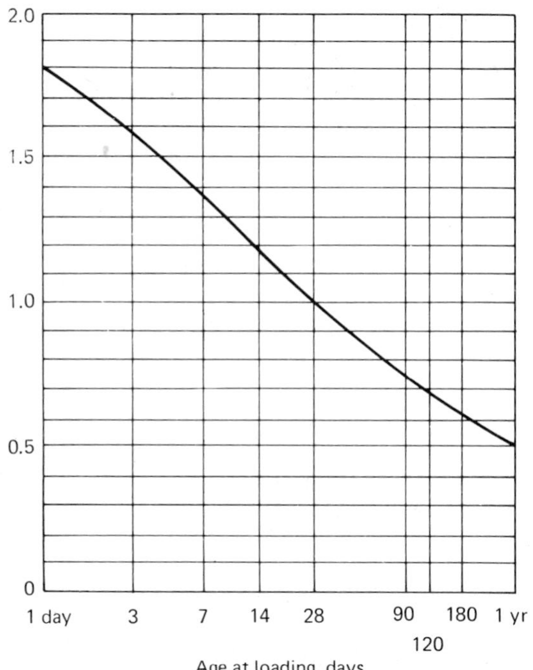

Fig. 6-38 Creep vs age at loading. (From: Fintel and Khan, "Effects of Column Creep and Shrinkage in Tall Structures—Prediction of Inelastic Column Shortenings," *ACI Journal Proceedings,* **66** (12), Dec. 1968.)

basic (true) creep excluding drying creep and shrinkage, thus simulating a very large column section. It is advantageous to use 6-in. standard cylinders for uniformity with other testing such as compression, split cylinder (tensile) and shrinkage, although any size specimens (foil wrapped) would produce the same specific creep for the same concrete mix and the same loading. Individual values for specific creep can be obtained from Fig. 6-37 or from the creep of a test specimen loaded at 28 days and then modified for the

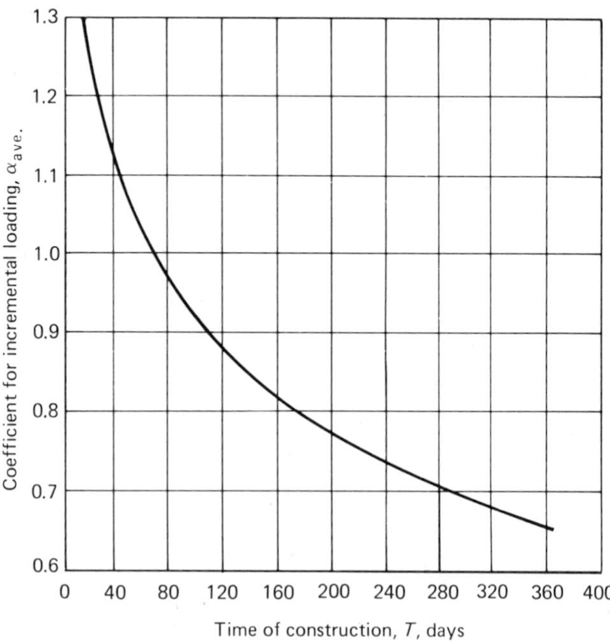

Fig. 6-39 Creep vs construction time. (From: Fintel and Khan, "Effects of Column Creep and Shrinkage in Tall Structures—Prediction of Inelastic Column Shortenings," *ACI Journal Proceedings,* **66** (12), Dec. 1968.)

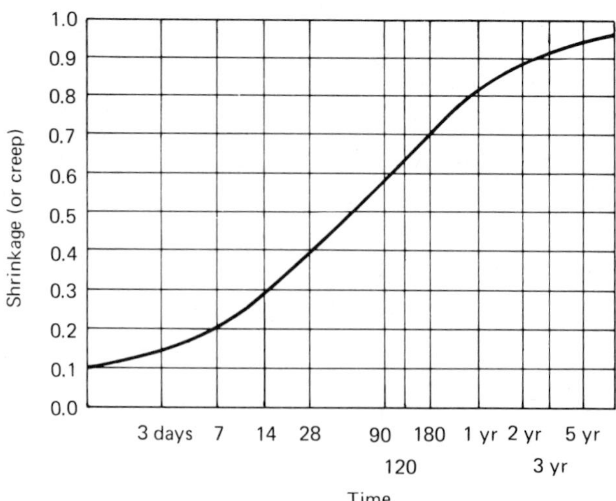

Fig. 6-40 Shrinkage or creep vs time. (From: Fintel and Khan, "Effects of Column Creep and Shrinkage in Tall Structures—Prediction of Inelastic Column Shortenings," *ACI Journal Proceedings,* **66** (12), Dec. 1968.)

various ages at loading using the coefficient α_{age} from Fig. 6-38.

The coefficient, α_{ave}, plotted in Fig. 6-39 is used to convert the 28-day creep into the average specific creep for a column loaded with equal load increments at equal time intervals. The curve in this figure shows the relationship between creep and the total time of construction, T, during which N equal load increments have been applied. If the entire load were applied to the column at the age of seven days, it would have twice the creep ($\alpha_{ave} = 1.4$) than in incremental loading over a period of 262 days. It is evident from Fig. 6-39 that columns loaded in many increments during a long period of time have much smaller creep than columns loaded during a short period of time with only a few increments.

Both creep and shrinkage have a similarity regarding the rate of progress with respect to time. Figure 6-40 shows an average curve for the ratio of creep or shrinkage at any time to the final value at time t_∞.

It can be seen from Fig. 6-40 that at 28 days about 40% of the inelastic strains have taken place. After three and six months, 60 and 70% respectively, of all the creep and shrinkage have taken place.

This curve can be used to extrapolate the ultimate creep and shrinkage values from laboratory testing of a certain duration of time. For example, if a specimen was measured to have a shrinkage strain of ϵ in./in. at 90 days, we can estimate the ultimate shrinkage to be

$$\epsilon_s = \frac{\epsilon}{0.60}$$

Conversely, the curve can also be used to estimate the creep or shrinkage at any time from the given ultimate value. For a further discussion of the structural implications of drying shrinkage and creep see Chapter 10, "Multistory Structures."

6.8 BOND OF CONCRETE TO STEEL

Bond can be thought of as the shearing stress or force between a bar and the surrounding concrete. The force in the bar is transmitted to the concrete by bond, or vice versa. Bond is made up of three components: (a) chemical adhesion, (b) friction, and (c) mechanical interaction between concrete and steel. Bond of plain bars depends primarily on the first two elements, although there is some mechanical interlocking due to the roughness of the bar surface. Deformed bars, however, depend primarily on mechanical interlocking for superior bond properties. This does not mean friction and chemical adhesion are negligible in case of deformed bars, but that they are secondary.

With the use of deformed bars, bond failures could result from concrete crushing at the bearing face of the deformations; by shearing of the concrete around the outer extremities of the bar; by longitudinal splitting of the concrete cover in the vicinity of the bar; or by a combination of these three failure modes.

The bond of deformed bars is developed mainly by the bearing pressure of the bar ribs against the concrete. Bars having ribs with steep face angles (larger than about 40° with the bar axis) slip mainly by compressing the concrete in front of the bar rib. The concrete is crushed and a concrete wedge forms in front of the bar rib. Bars with flat ribs (less than 30°), however, slip with the ribs sliding relative to the concrete.

With little confinement, large deformed bars may fail in bond by splitting of the concrete along the plane of the bar. This type of failure depends primarily on the load on the concrete and not much on the bar stress and the bar diameter or the bar perimeter. As the confinement around a bar improves, usually by the use of external concrete or transverse reinforcement, the ultimate load per unit length depends increasingly on the bar diameter. Small bars, top cast bars or bars which are confined to the extent that bond failure generally occurs by shear failure of the concrete keys instead of splitting, will carry a maximum unit load proportional to the bar perimeter (hence to the bar diameter).

Based on the above reasoning, the maximum bond force per unit length depends primarily on the bar diameter, the amount of enclosure provided by transverse reinforcement (hoops, ties, or stirrups) placed around the bar or confinement provided by external concrete (concrete cover), as well as the concrete tensile strength, ($\sqrt{f_c'}$).

The bond of welded wire fabric embedded in concrete is dependent on the ability of the welded transverse wire to provide anchorage. The important variables are the size ratio between the transverse and longitudinal wire of the fabric and the quality of the weld connecting them. For deformed wire, which has been developed recently, crushing of the concrete against the deformations or the shearing of the concrete core at the outer periphery of the wire may be more critical than splitting. The bearing area also is important for controlling slip at a given load.

The development of bond in prestressed concrete is also attributed to adhesion, friction and mechanical resistance. Friction between concrete and tensioned tendons (or grout and tendons) is the principal factor responsible for the transfer of the prestress force from steel to the concrete, and the other two factors are of less importance in bond development.

When the external prestressing force in the tendon is released, the tendon diameter tends to increase, (so-called Hoyer effect), thus producing high radial pressure against the concrete, which in turn produces high frictional resistance in the transfer zone. Prestress transfer bond is present from the ends of a prestressed member to the beginning of a region in which the steel tension is constant. The length over which this transfer is made is termed the prestress transfer length, and is a function of the perimeter configuration, area and surface condition of the steel, the stress in the steel, and the method used to transfer the steel force to the concrete. Tendons with a slightly rusted surface can have an appreciably shorter transfer length than clean tendons. However, the danger of careless rusting allowing localized pitting should be considered. Gentle release of the tendon will permit a shorter transfer length than abruptly cutting the tendons. There is relative movement of steel and concrete, and accordingly adhesion cannot contribute to prestress transfer. Mechanical resistance probably contributes little to prestress transfer in the case of individual smooth wires, but it might be argued that strand will offer some mechanical resistance because of the helical grooves in the stranded configuration, but this type of deformation may not be depended upon for any appreciable contribution to the development of bond. A significant innovation in prestressing strand is the development of deformed strand, which is desirable for reducing the transfer length in some structural elements.

An increase in wire tension due to flexure reduces the diameter, relieves the radial pressure, and reduces the bond near the ends of a beam. Bond failure in prestress concrete also may occur due to too close spacing of the tendons.

Bond to concrete may be prevented for some pretensioned reinforcement in the end regions. Bond may be prevented by various means; one method is the use of plastic tubing which is often referred to as "blanketing."

Metal reinforcement should be free from loose, thick rust, mill scale, mud, oil, grease, paint, and loose dried mortar at the time concrete is placed, as these materials adversely affect or reduce bond.

A normal amount of rust increases bond. Normal rough handling of bars generally removes most of the loose rust and mill scale. However, in some instances it may be necessary to rub with a coarsely woven sack or to use a wire brush. Metal reinforcement, except prestressing steel, with rust, mill scale, or a combination of both can be considered as satisfactory, provided the minimum dimensions, including height of deformations, and weight of a hand wire brushed test specimen are not less than applicable ASTM specification requirements. Prestressing steel should be clean and free of excessive rust, oil, dirt, scale and pitting. A light oxide coating which can be removed by a soft dry cloth is permissible.

Slip caused by relative movement (in addition to that caused by crushing) also occurs when the frictional properties of the rib face are reduced by grease. The extent to which slip properties are affected by grease depends on the face angle; ribs with flatter face angles are more affected by poor frictional properties.

Bond is increased by a tight adherent cement paste or mortar coating on the bars. In some cases, bond of hot dip galvanized reinforcement may be reduced due to the attack of fresh concrete on the zinc and evolution of hydrogen with the resultant formation of gas pockets next to the bars. Suspensions of bentonite and water are commonly used for construction convenience to support side walls of foundation trenches. Reinforcement in cages immersed in the suspension and then surrounded by tremie placed concrete may suffer reduced bond strength.

In freshly placed concrete, bleeding or water gain results in formation of a water (or water and entrapped air) space beneath the surfaces of solids, including the undersides of reinforcing bars. If reinforcing bars are rigidly positioned or restrained from settling with the concrete, an additional break of bond occurs at the underside of the reinforcement due to settlement shrinkage. This loss of bond may be expected to be about proportional to the depth of fresh concrete beneath the bars; the bond capacity of top bars is less than that of bottom bars. Settlement of concrete in the form results in better concrete consolidation on top of the lugs of a vertical bar than beneath the lugs. The early slippage and ultimate bond strength are thus more favorable when the bar is pulled against the direction in which the concrete settled.

During the earliest stages of hardening, after the concrete loses its plasticity, bond may be impaired if projecting reinforcement is subjected to impact or rough handling. Exposed portions of bars that are only partly embedded should not be struck or carelessly handled, and workmen should not be permitted to climb on bar extensions until the concrete is at least seven days old. Forms to which embedded parts are fastened or through which they protrude should not be stripped until the concrete has hardened sufficiently to avoid injury to bond.

Bond between concrete and deformed steel reinforcement is principally a function of the compressive strength of concrete or of its splitting tensile strength. Bond strength tests of some lightweight concretes yield bond strength values ranging from equal to 20% less than those of normal-weight concrete of equal compressive strength. ACI 318-71, *Building Code Requirements for Reinforced Concrete*, provides that the bond capacity of all-lightweight structural concrete shall be taken as 75% of that of normal weight concrete, and may be increased to 85% by the replacement of lightweight fines with natural sand. Results from a limited number of tests indicate that the bond strength of pretensioned strand in lightweight aggregate concrete is not different from that in normal weight concrete.

The age of the concrete at the time of test affects the bond strength result. Concretes which reach the same compressive strength at different ages will indicate slightly different bond values with the same bar. Tests at early ages should give bond strength as a higher proportion of the concrete compressive strength. Concrete strength (up to 5500 psi) at transfer of prestress has little influence on the transfer length of clean seven wire strands up to and including 1/2 in. diameter. The average increase in transfer length over a period of one year following prestress transfer is quite low for all strand sizes, with the increase in transfer length with time independent of the concrete strength at the time of transfer.

The concrete mix and slump, rate of casting and amount of vibration employed are each important factors affecting bond. Mix ingredients and proportions that result in low bleeding of concrete are beneficial. Air entrainment improves bond of concretes which tend to bleed badly. For nominal amounts of entrained air (4 or 5% upper limit) no major effect on bond results, as compared with non-air-entrained concrete. With higher percentages of air, the bond strength of horizontal bars drops off rapidly as concrete strength decreases and air bubbles may have the tendency to collect beneath the bars in exactly the same manner as does excess water. Low slump concrete gives higher bond values than high slump concrete even when the mixes are adjusted to give the same compressive strength. Proper vibration of concrete markedly increases bond. However, overvibration can increase water gain and reduce bond. Shrinkage of concrete may destroy the bond between the concrete and reinforcing steel and cause a high loss in bond strength at low values of slip. However, the ultimate bond strength is little affected when deformed steel is used, since with higher values of slip, the concrete is brought to bear against the protrusions of the reinforcing steel.

6.9 FATIGUE OF CONCRETE

Like other construction materials, concrete is subject to the effects of fatigue. Concrete, when subjected to repeated loads, may exhibit excessive cracking and may eventually

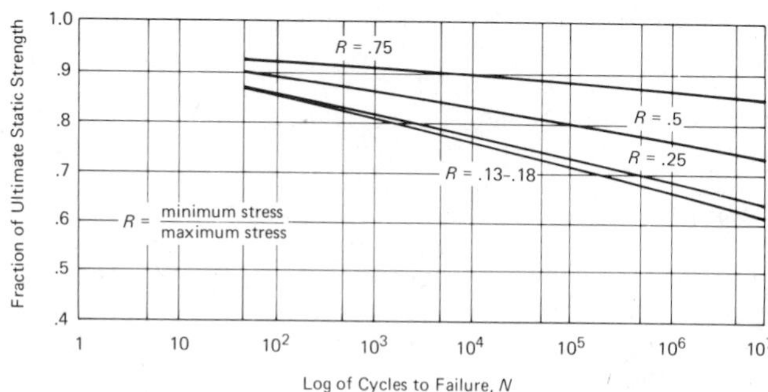

Fig. 6-41 *S-N* curves for plain concrete beams showing various ratios of minimum to maximum stress in the loading cycles. (From: Murdock, J. W., and Kesler, C. E., "Effect of Range of Stress on Fatigue Strength of Plain Concrete Beams," *ACI Journal*, **30** (2), Aug. 1958. Also, *Proceedings* **55**, pp. 221-231.)

fail after a sufficient number of load repetitions, even if the maximum stress is less than the static strength of a similar specimen. The fatigue strength of concrete is defined as a fraction of static ultimate strength that it can support repeatedly for a given number of cycles.

Fatigue is a process of progressive, permanent internal structural (microcracking) change in a material subjected to fluctuating stresses or strains. The internal changes may be damaging and result in progressive growth of cracks and complete fracture if stresses and fluctuations are sufficiently large. Fatigue fracture of concrete is characterized by a considerably larger microcracking and strains as compared to fracture of concrete under static loading.*

Although the design of flexural members of concrete is based almost entirely on data from static tests, many concrete structures both plain and reinforced, are subjected to fatigue loadings. Included in this category are pavements, bridges, docks, crane girders and other structures supporting oscillating machinery. However, the only known cases of fatigue failures have been in pavements and railway ties.

Fatigue data is usually shown by stress-fatigue life curves, known as *S-N* curves, Fig. 6-41. The repeated stress, *S*, expressed as a ratio or percentage of the static ultimate strength is the ordinate, while the abscissa gives the life or number of cycles until failure on a logarithmic scale. The following may be observed from this figure: (1) the fatigue strength of concrete decreases with an increasing number of cycles, (2) a decrease of the range between maximum and minimum load results in increased fatigue strength for a given number of cycles, and (3) plain concrete has no endurance limit up to 10 million cycles. (Endurance limit is the maximum fraction of ultimate static strength at which concrete exhibits essentially elastic behavior, thereby permitting unlimited stress repetitions without loss in fatigue resistance.)

Fatigue strength of paste, mortar and concrete are about the same when expressed as a fraction of their static ultimate strength. Many variables such as cement content, water-cement ratio, curing, entrained air, and aggregate that affect the static strength influence fatigue strength in a similar proportionate manner. Fatigue strength of concrete for a life of ten million cycles—for compression, tension, or flexure—is approximately 55% of static ultimate strength. This result is valid when the loads vary from near zero to some predetermined maximum and not when the minimum load is a significant percentage of the maximum load.

*Much of this discussion on fatigue is based on "Fatigue and Fracture of Concrete" Stanton Walker Lecture by C. E. Kesler, Nov. 18, 1970.

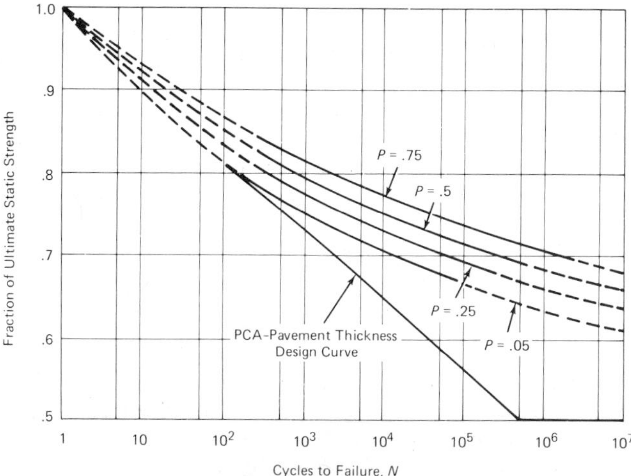

Fig. 6-42 *S-N* curve for constant probabilities of failure adjusted for use with Miner hypothesis. (Adapted from: Hilsdorf, H. K., and Kesler, C. E., "Fatigue Strength of Concrete Under Varying Flexural Stresses," *ACI Journal Proceedings,* **63,** 1059-1076, 1966.)

The results of fatigue tests usually exhibit substantially larger scatter than static tests. This inherent statistical nature of fatigue test results can best be accounted for by applying probabilistic procedures: for a given maximum load, minimum load, and number of cycles, various probabilities of failure can be calculated from the test results. By repeating this for several number of cycles, a relationship between probability of failure and number of cycles until failure at a given level of maximum load can be obtained. From such relationships, *S-N* curves for various probabilities of failure can be plotted, Fig. 6-42.

The usual fatigue curve is that shown for a probability of failure of 0.5. Design should probably be based on a lower probability of failure.

Design for fatigue is facilitated by use of a modified Goodman diagram as shown in Fig. 6-43. This diagram is based on the observation that the fatigue strength of plain concrete is essentially the same whether the mode of loading is tension, compression, or flexure, as long as no reversal of stress is involved. The diagram also incorporates the influence of range of loading. For a zero minimum stress level, the maximum stress level the concrete can support for ten million cycles without failure is taken conservatively as 50% of the static strength. As the minimum stress level is increased, the stress range that the concrete can support decreases.

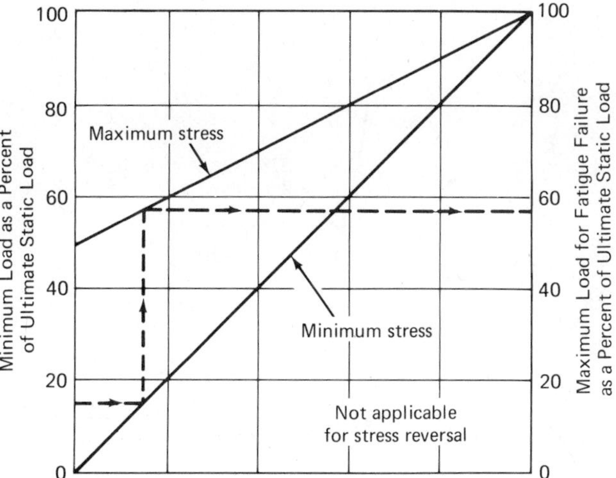

Fig. 6-43 Modified Goodman diagram showing fatigue strength of plain concrete in tension, compression, or flexure under 10 million cycles of repeated loading.

From Fig. 6-43, one can determine the maximum stress in tension, compression, or flexure that concrete can withstand for ten million repetitions and for a given minimum stress. For example, consider a structural element to be designed for 10 million repetitions. If the minimum stress is 15% of the static ultimate strength, then the maximum load that will cause fatigue failure is about 57% of static ultimate load. This value refers to 50% probability of failure.

Varying stress rates during fatigue tests change the fatigue response from that found when stress rates are constant. The fatigue strength and life of concrete subjected to repeated loads of varying magnitude is influenced by the sequence in which these loads are applied. Consider a test in which the maximum stress level is changed only once. Then the fatigue life of a specimen is larger if the higher stress level has been applied first compared to the fatigue life of a specimen in which the lower stress level was applied first. A relatively low number of cycles of high loads can in fact increase the fatigue strength of concrete under a lower load beyond the fatigue strength of concrete which has not been previously loaded. When the load is repeatedly varied, the fatigue strength decreases with increases in the ratio of the number of cycles at high stress level to those at low stress level. Data is not available showing the effect of randomly varying loads on fatigue behavior of concrete.

Accumulation of the effects of repeated loads of varying stress can be made on the basis of the Miner hypothesis with sufficient conservatism to incorporate a very low probability of failure. The Miner hypothesis is that fatigue resistance not consumed by repetitions of one load is available for repetitions of other loads. Theoretically, the total fatigue used should not exceed 100%. Design curves, Fig. 6-42, incorporating the probability of failure have been developed so that when the Miner hypothesis is applied to them reasonable results are obtained. For example, the curve most applicable to pavement design has a constant probability of 0.05, which means that for 100 fatigue tests not more than five would show fatigue strengths below the curve. However, because the research to develop this curve was too limited to justify general use of the 0.05 curve for design, the PCA curve has been adopted for pavement thickness design.

Repetitions of loads with stress ratios below the fatigue strength (understress) increase concrete's ability to carry loads with stress ratios above the fatigue strength; i.e., decrease in pavement deflection with increase in load repetitions. This understress may be a sustained load or rest period (up to five minutes) between repeated load cycles. No additional benefits are derived when rest periods extend beyond five minutes. Also, if the maximum stress at sustained load is above 80% of the static strength, then sustained loading may have detrimental effects on fatigue life.

The frequency of loading, between 70 and 900 cycles per minute, has no significant effect on fatigue strength. However, frequency as low as 10 cycles per minute may result in slightly lower fatigue lives because of increased significance of creep effects.

Stress gradient influences the fatigue strength of concrete in a manner similar to the influence of a gradient on static strength. Fatigue strength of eccentrically loaded specimens is higher than concentrically loaded otherwise similar specimens. For the purpose of design of flexural members limited by concrete fatigue in compression, it may be safe to assume that fatigue strength of concrete with stress gradient is the same as that of uniformly stressed specimens.

While bond fatigue has generally not been a problem in reinforced concrete structures, it has not been thoroughly explored in research, and the limitations it could present are not known. Bond fatigue is the progressive deterioration of bond and the slip of tensile reinforcement under some form of repetitive loading. Such slip could possibly have the same effect as reinforcement with a reduced elastic modulus and could lead to premature concrete crushing and collapse. Based on the local conditions near the steel-concrete interface, it has been hypothesized that bond fatigue is related directly to the dissipation of energy in slippage.

Special fatigue tests are essentail to evaluate the strength of structural elements, such as railroad crossties, where bond is critical (available transfer length of pretensioned tendons is short) and repeated loading is the main design consideration.

6.10 PERMEABILITY

Movement of water or air through concrete can be produced by various combinations of air or water pressure differentials, humidity differentials, and solutions of different concentrations (osmotic effects) or by temperature differentials. Various tests have been devised to determine permeability. Although these procedures may reveal the relative characteristics of the concretes involved, the quantitative value obtained may depend considerably upon details of the experimental conditions. This means that the conditions anticipated in service should be used for the test conditions.

6.10.1 Water Permeability

The water permeability of high quality concrete depends mostly on the respective permeabilities of the cement paste and aggregate. Since the aggregate particles are surrounded by hardened cement paste, the permeability of concrete to water under hydrostatic pressure is principally a function of the permeability of the cement paste component of the concrete, providing the concrete is intact—not previously damaged by frost or rapid drying and not containing excessive under-aggregate fissures or honeycomb. All of the permeating water must pass through the paste component of the concrete (the continuous phase) regardless of the relative porosity or permeability of the aggregate. If the paste is of low permeability, the concrete will show similar characteristics.

The permeability to water of hardened cement paste is dependent primarily on the capillary porosity of the paste. The capillary porosity is a function of the original water-cement ratio and the length of the curing period (extent of hydration).

The water permeability of a well-cured paste is reduced approximately a thousandfold by reduction in water-cement ratios from 0.8 to 0.4 by weight. The change in permeability of a given paste with length of curing time also is enormous.

Introduction of aggregate particles into cement paste should tend to reduce the permeability by reducing the number of channels per unit cross section and by lengthening the path of flow per unit linear distance in the general direction of flow. However, during the plastic period the paste settles more than the aggregate and thus fissures under the aggregate particles develop. In saturated concrete, these void spaces are paths of low resistance to hydraulic flow and thus increase the permeability of concrete. In general, with paste of a given composition, permeability of concrete is greater the larger the maximum size of the aggregate. In addition, the permeability of hardened concrete can never be as low as that of hardened cement paste because of the imperfect bond between aggregate and paste.

Significant quantities of air voids in the concrete should increase the permeability of saturated concrete roughly in proportion to their quantity, provided other factors remain constant. However, other factors seldom remain constant—it is commonly observed that air entrainment in most concretes will increase workability and reduce segregation and bleeding and permit reductions in the water-cement ratio—with the result that the concrete may actually be more impermeable despite the presence of the air voids. Therefore, all concrete which must be watertight should be air entrained.

Figure 6-44 shows the effect of duration of moist curing on relative permeabilities of different concrete mixtures and something of the effects of cement content. Effects of air entrainment do not seem to be large with cement contents held constant.

Test results obtained by subjecting non-air-entrained mortar discs to 20 psi water pressure are shown in Fig. 6-45. Mortar discs moist-cured for 7 days had no leakage when made with a water-cement ratio of 0.50. Leakage occurred in mortars made with higher water-cement ratios. In discs with a water-cement ratio of 0.80 the mortar still had leakage after curing for a month. For each water-cement ratio, leakage became less as the curing period was lengthened. The permeability of steam-cured concrete is generally higher than that of wet-cured concrete. Supplemental moist curing may be required to achieve an acceptably low permeability.

When made with normal weight aggregate, concrete that is intended to be watertight should have a maximum water-cement ratio or water-cement plus pozzolan ratio of 0.48 for exposure to fresh water and 0.44 for exposure to sea water. In addition it should have a specified 28-day com-

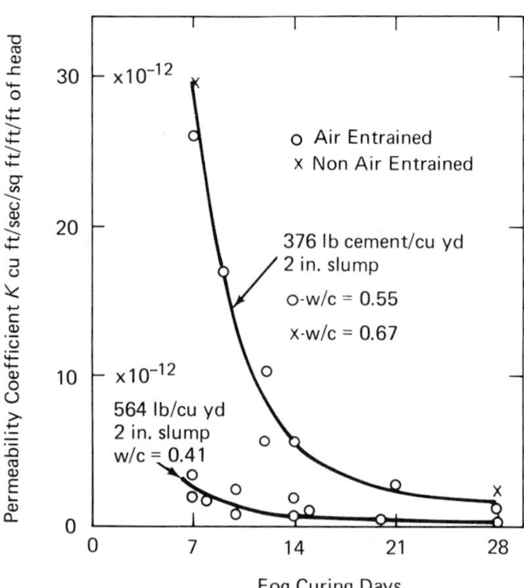

Fig. 6-44 Permeability vs curing time. (From: Tyler, I. L., and Erlin, B., "A Proposed Simple Test Method for Determining the Permeability of Concrete," *PCA Research Bulletin 133,* 1961.)

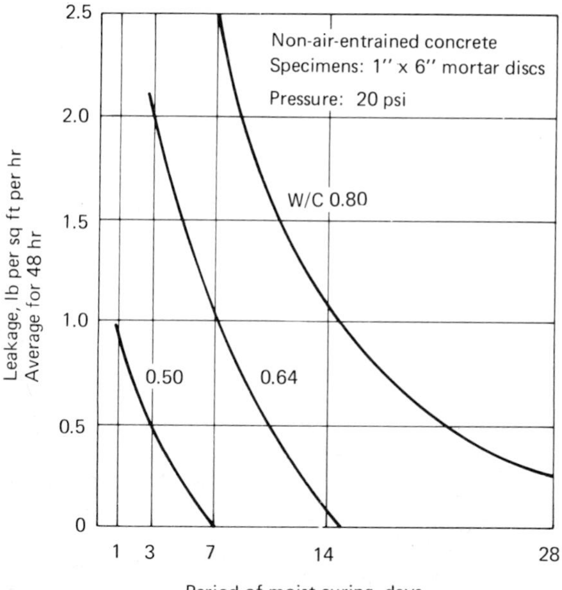

Fig. 6-45 Effect of water-cement ratio and curing on watertightness. Note that leakage is reduced as the water-cement ratio is decreased and the curing period increased. (From: *Design and Control of Concrete Mixtures* 11th ed., Portland Cement Association, Skokie, Ill., 1968.)

pressive strength of 3500 psi where concrete is not exposed to severe and frequent freezing and thawing of 4000 psi where concrete is exposed to severe and frequent freezing and thawing, except where special structural or other considerations require concrete of greater strength. With lightweight aggregate, the specified compressive strength, f_c', should be at least 3750 psi for exposure to fresh water and 4000 psi for exposure to seawater.

Concrete to be watertight needs adequate cement paste therefore it should have a minimum cement content as follows:

Coarse aggregate No.	lb/yd^3
467 (1½ in. maximum)	517
57 (1 in. maximum) or 67 (¾ in. maximum)	564

Admixtures known as permeability-reducing agents are usually either water-repellents or pozzolanic materials. They may reduce the permeability of concretes that have low cement contents and/or a deficiency of fines in the aggregate. However, their use in well-proportioned mixes with the above cement contents may increase the mixing water required, resulting in increased rather than reduced permeability. However, in lean mass concrete, age becomes a more important factor in permeability than cement content because of slow pozzolanic reaction.

Dampproofing admixtures, usually water-repellent materials, are sometimes used to reduce the capillary flow of moisture through concrete that is in contact with water or damp earth. Many so-called dampproofers are not effective, especially when used in concretes that are in contact with water under pressure.

Watertight concrete structures are not difficult to construct. Where there have been faulty structures, careful examination invariably has shown that leakage is confined to relatively small areas and the rest of the structure is sound.

Principal causes of leaks are improper or careless construction methods rather than improper design or poor materials. Of the various imperfections caused by defective construction methods, segregation of ingredients stands out as the prime reason for non-uniformity and local porousness of concrete. Segregation can be minimized by proper handling, placing and consolidation procedures. To be watertight concrete must also be free from honeycomb and cracks (or the crack width minimized). To aid in preventing honeycomb, concrete should have a minimum slump of 1 in.; with a 3 in. maximum for footings, caissons, substructure walls; and a 4 in. maximum for slabs, beams, reinforced walls, and columns.

In some cases, inadequate foundations have permitted unequal settlement with subsequent cracking and leakage. Careful surveys of ground conditions will indicate proper procedures for preparing subgrades and foundation.

It is believed that the principal mode of transport of water through quality concrete is not ordinary laminar flow but a mode that might be called surface diffusion. An estimate of the amount of water passing through a concrete section can be obtained from Darcy's Law for viscous flow (although viscosity cannot be treated as a constant but as a function of the mean size of the pores in the paste):

$$Q = K\,i\,A$$

where

Q = total volume of fluid, percolating in unit time, (cm^3).
K = permeability coefficient, (Darcy or cm/sec).
i = hydraulic gradient or difference of energy heads in the fluid at any two points along a flow distance. It represents the energy lost through viscous friction as the fluid flows around the particles and through irregular void passages. Velocity head is neglected in this gradient. (i) can be regarded as the driving force, especially since phenomena like capillarity are not considered.
A = cross sectional area of the element subject to percolation, (cm^2).

or the equation

$$Q = KA\frac{H}{L}T$$

where

$\frac{H}{L}$ = hydraulic gradient
H = hydraulic head, (cm)
L = thickness of concrete, (cm)
T = time, (sec)

As can be expected for an empirical formula, the units present a confused picture and care must be taken in determining Q. Within the ranges of concretes normally used in practice, the coefficient of permeability, K, ranges from 10^{-12} to 10^{-10} cm/sec–(the unit rate of discharge at unit hydraulic gradient with the dimensions of a velocity at a temperature of 70°F). Permeability values for mass concrete (dams) is shown in Table 6-10. Additional moist curing takes place as water penetrates and this could conservatively reduce fluid outflow volume about 1/4 or 1/5 in a years time.

6.10.2 Water Vapor Permeability

Water vapor in air is a gas which exerts its own vapor pressure and which can diffuse through concrete, independently

TABLE 6-10 Permeability of Mass Concrete*

Dam Structure	*Permeability Kq***
Hoover	0.62×10^{-4}
Grand Coulee	–
Angostura	–
Kortes	–
Hungry Horse	1.85×10^{-4}
Canyon Ferry	1.93×10^{-4}
Monticello	8.20×10^{-4}
Anchor	45.2×10^{-4}
Glen Canyon	1.81×10^{-4}
Flaming Gorge	11.09×10^{-4}
Yellow Tail	1.97×10^{-4}†

**From: "Mass Concrete for Dams and Other Massive Structures" ACI Committee 207, ACI Journal, April 1970.*
***18 × 18-in. specimen, standard correction to age of 60 days. Kq is in cu ft/sq ft/yr/ft (head); it is a relative measure of the flow of water through concrete.*
†Preliminary mix investigations.

of the air with which it is mixed, until the partial pressures of the water vapor are equalized. However, when the air is moved suddenly or is heated or cooled, the water vapor present is similarly affected so that it is usually necessary to consider it as a part of an air-vapor mixture. Differences in total pressure of the air may result in a transfer of vapor with air, augmenting, and at times over-riding, the effects of the flow produced by vapor-pressure gradients alone. This can be particularly important in the transfer of vapor through cracks and pinholes or through air-permeable building constructions. It will seldom be important in constructions without air spaces and having parged or plastered surfaces. It may, however, be an important means of vapor transfer through constructions lacking in air tightness.

Water vapor transmission through materials under the influence of a water vapor pressure gradient is called vapor diffusion. The design equation commonly used in calculating water-vapor transmission through materials is based on a form of Fick's Law, and is as follows:

$$W = \bar{\mu} A \theta \frac{\Delta p}{l}$$

where

W = total weight of vapor transmitted, grains.
$\bar{\mu}$ = average permeability, grains per (hour)(square foot)(inch of mercury vapor pressure difference per inch of thickness).
A = area of cross section of the flow path, ft^2.
θ = time during which the transmission occurred, hours
Δp = difference of vapor pressure between ends of the flow path, inches of mercury. (at 73.4F: 100–0 percent RH, $\Delta p = 0.82948$; for 100–20 percent RH, $\Delta p = 0.8(0.82948) = 0.66358$; for 100–50 percent RH, $\Delta p = 0.5(0.82948) = 0.41474$)
l = length of flow path (or thickness of specimen), in.

The actual transmission of vapor through a material is extremely complex, so that the coefficient, $\bar{\mu}$, is not a simple one but is actually a function of relative humidity and temperature, and may vary along the flow path through the concrete.

Whenever it is convenient to deal with a concrete of a thickness other than the unit thickness to which $\bar{\mu}$ refers, use may be made of the permeance coefficient M, where $M = \bar{\mu}/l$. The designation perm for the unit of permeance is now widely used and is a convenient substitute for the unit, 1 grain per (hour)(square foot)(inch of mercury vapor pressure difference). The corresponding unit of permeability is perm-inch, since it is the permeance of unit thickness.

The resistance of water vapor flow through concrete is the reciprocal of its permeance and is equal to the thickness divided by its permeability. The overall resistance of a section to water vapor flow is the sum of the resistances of the various materials in series comprising the section. The overall permeance of a section is equal to the reciprocal of the overall resistance.

The simplest method of finding the permeance of a specimen is to seal it over the top of a cup containing desiccant or water, placing it in a controlled atmosphere, and weighing it periodically. The steady rate of weight gain or loss is normally the water vapor transfer. When the cup contains a desiccant the procedure is called the dry-cup method and when the cup contains water, the wet-cup method. Usually the surrounding atmosphere is held at 50% relative humidity, thus providing, in either method, substantially the same difference of vapor pressure, but the results obtained by the two methods for the same specimens are likely to be much different, the wet-cup method producing the higher values. Because concrete is sorptive, the rate of weight loss is high during the early period of the wet-cup test and decreases asymptotically as equilibrium conditions are obtained. The interval needed to establish diffusion equilibrium varies not only with the type of concrete but also with vapor pressure and specimen thickness. In the dry-cup test, virtually all transmission takes place by diffusion, and capillary flow is absent.

Typical permeability values for concrete are as follows:

*Concrete**	*Thickness in.*	*Permeance, M*	*Resistance 1/M*
Normal Weight Aggregate–cast as a plastic, workable mix:	Varies		
w/c = 0.40 by wt.	2 to 6	0.5	2.0
w/c = 0.50 "		0.8	1.25
w/c = 0.60 "		1.1	0.91
w/c = 0.70 "		1.3	0.77

**Assumes conventional curing and six months age. Concretes of lightweight aggregates having similar compressive strength may be regarded as having similar water vapor properties.*

Forces other than water vapor pressure, such as hydraulic pressure, absorption, adsorption, hygroscopicity, and capillarity, affect water vapor flow and cause moisture migration. These various forces are interrelated but the actual relationship is very complex, and has not been expressed in a simple way for design use. Evidence to date indicates that in porous materials partially saturated with water there is also likely to be a migration of moisture to the cold side under the influence of the temperature gradient. This can occur by a process of evaporation, vapor flow, and condensation within the material.

The vapor flow calculations can be considered reasonably accurate when the primary mechanism for moisture migration is vapor diffusion and when coefficients are available that define the rate of vapor flow for the material under the conditions of use. When, however, the material is capable of holding substantial quantities of adsorbed water, as in concrete, the diffusion approach may be inadequate or even inappropriate, depending on the situation. Moisture is apparently transferred through partially dry concrete in the absorbed or condensed state by surface diffusion and does not move as vapor through concrete. Also, where good practice has been followed in the design and curing of concrete, it is believed that moisture does not move through the concrete by capillarity as it is usually understood. For example, the transfer process is considered to be the same where either liquid water or water vapor are present directly beneath a slab-on-ground. However, other conditions being similar, there is believed to be a much slower transfer to the absorbed state when vapor rather than liquid water is in contact with the slab. This results in a slower rate of moisture transfer through the slab where only vapor is present.

Due to the hygroscopic nature of concrete it is believed incorrect to consider the rate of moisture transfer proportional to the vapor pressure differential between a moist and a dry side of a slab. For water moving through concrete in a condensed state, the driving force is not computed from the difference between the two pressures but from the ratio between the two relative humidities.

It is recommended that the water vapor permeability of concrete not be expressed in terms of permeance. Computation of permeance is based on the assumption that moisture travels through partially dry concrete as vapor, whereas, under the conditions of concrete tests, it is transferred in the adsorbed state by surface diffusion. Motive force is proportional to the logarithm of the ratio between the vapor pressure on the moist side and that on the dry side; it is not equal to the difference between the two vapor pressures as assumed in the ASTM definition.

In a number of tests, vapor flow through concrete was measured as water vapor transmission (*WVT*), *WVT* being defined as the rate of migration of water through concrete, ($WVT = Wl/A\theta$) using a wet cup. The units of *WVT* are usually grains per in.^{2} per day or grains per ft^2 per hr.

Water vapor transmission of concrete decreases with increased age of the concrete and approaches an asymptotic value, Fig. 6-46. Also, shown in these figures is that *WVT* decreases with an increase in the strength (decrease in water-cement ratio) of the concrete. Admixtures have no appreciable effect on moisture migration when compared with plain concretes of the same *w/c*. This conclusion applies to concrete at all ages.

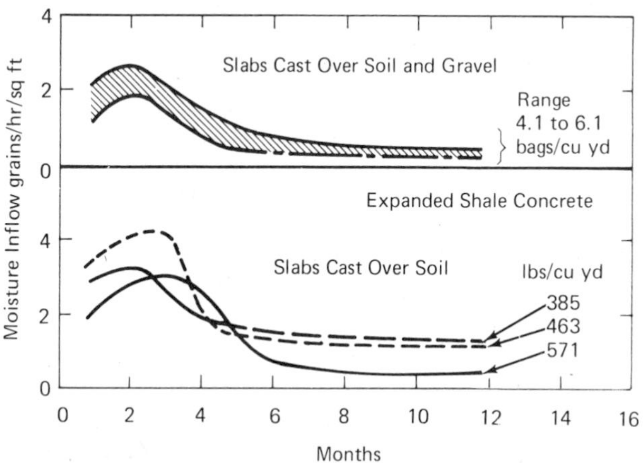

Fig. 6-47 Effect of cement content and gravel capillary layer for lightweight concrete. (From: Brewer, H. W., "Moisture Migration—Concrete Slab-on-Ground Construction," *PCA Development Bulletin 89*, 1965.)

Moisture migration through good quality concrete is less than 0.3 grains/hr/ft^2; but the flow through low quality concrete often exceeds 2.0 grains. For slabs-on-ground, a gravel capillary break between the soil and the slab reduces *WVT*. No significant correlation exists between types of soil beds and moisture migration rates through slabs-on-ground. However, an increase in the slab thickness up to about 6 in. will decrease *WVT*.

Structural lightweight concretes show considerably less *WVT* than most sand-gravel concretes, possibly because of greater sorptive properties of the aggregate, Fig. 6-47. Slabs cast over soil show reduced inflow with increasing cement content. This relationship is not well defined for companion slabs cast over soil with a gravel capillary break, as all of the curves fall into a relatively narrow band. Comparison of the curves of Fig. 6-47 reveals the advantage of the gravel capillary break, particularly for slabs with lower cement content.

The outflow or rate of moisture loss from the surface of the same concretes is much higher at early ages—approximately four times the above values at 28 days, and ten

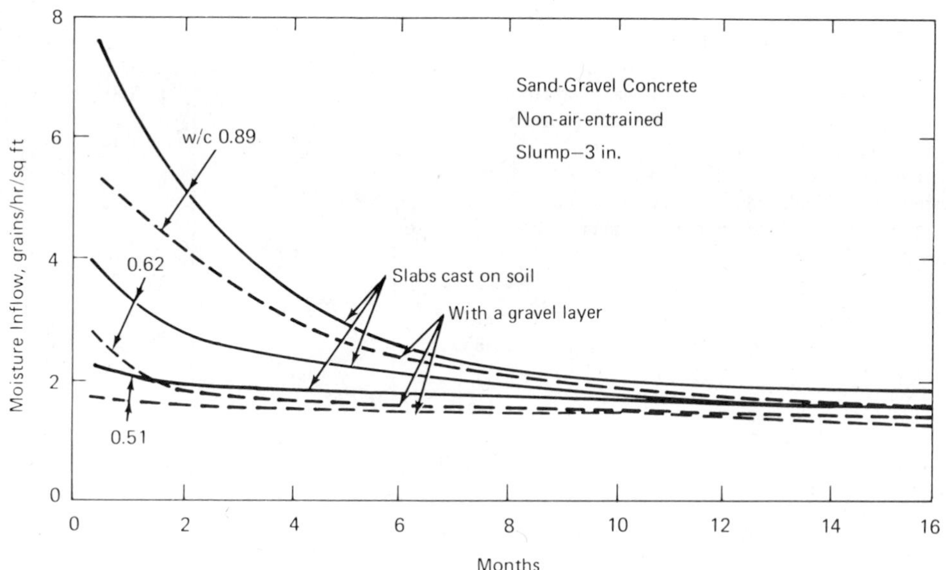

Fig. 6-46 Effect of water-cement ratio and gravel capillary layer. (From: Brewer, H. W., "Moisture Migration—Concrete Slab-on-Ground Construction," *PCA Development Bulletin 89*, 1965.)

times at seven days. This outflow at early ages is the flow that contributes to the problems encountered upon installation of floor coverings. For water-vapor exposure the loss from the surface is due almost entirely to loss of original mixing water, with very little moisture passing through the slab. After extended exposure periods, exceeding one year under the constant temperature and humidity conditions, inflow and outflow will become equal; at that time stabilized moisture migration through concrete is attained.

Concretes made with poor quality aggregate show higher *WVT* rates than concretes made with high quality aggregates, other factors being equal. *WVT* decreases with increases in maximum aggregate size.

Present knowledge and understanding regarding the transmission of water vapor through concrete is still rather incomplete. Although the values listed for concrete derive from standard test conditions they may not adequately describe behavior under other conditions or test methods. For example, vapor permeability of concrete decreases as the mean relative humidity increases. The reason for this is that an increase in relative humidity decreases the air-filled pore space available for diffusion.

6.10.3 Air Permeability

Air leakage is the uncontrolled movement of air through walls and roofs both into a structure (infiltration) and out of it (exfiltration) and the interchange of air within spaces in the building envelope. Air movement occurs as a result of air pressure differences produced by wind, chimney effects (when the temperature in a building differs from that outside, pressure differences occur between inside and outside as a result of the difference in density of the air), and the operation of mechanical ventilation systems.

The rate of flow of air through concrete depends on the thickness of the concrete moisture content, and on the pressure applied. There is no known relation between the air and water permeabilities of all concretes. The rate of flow of air through concrete having a low rate of flow may attain equilibrium within an hour but, if the rate is high, it may continue to increase for many hours. The rate appears to be inversely proportional to the thickness of the slab and directly proportional to the pressure.

Concrete Thickness, in.*	4	7	8	9
Rate of flow, in.^{3}/psi-hr-ft^2	0.58	0.37	0.30	0.24

**Water-cement ratio = 0.62.*

Based on tests with pressures up to about 30 in. water-gauge and for thickness from 4 in. to 9 in. the rate of seepage in cubic inches per square foot per hour is equal to 12½ times the ratio of the pressure (psi) to the thickness (in.).

A dense concrete having a minimum cement content of 500 lb per cu yd, a very good aggregate grading, a fly ash or pozzolan (where necessary to add fines), air entrainment, and a low water-cement ratio can produce concrete impermeable to air under high pressures. Slump does not appear to affect permeability.

Prolonged curing reduces the air permeability but drying at any age significantly increases permeability, see Specimen A, Table 6-11. The numerous pores in concrete are normally lined with water—relatively immobile water—adsorbed on the pore walls. Presumably the pore area available for the relatively free flow of fluid is greatly reduced by this immobile lining, particularly in the small pores where it may constitute most or all of the cross-sectional area of the pore. However, when concrete is air-dried, the coarser capillary spaces are completely emptied of water, and the finer capillary spaces and gel pores partially emptied due to decreased water adsorption on the

TABLE 6-11 Air Permeability of Concrete Specimens at Various Moisture Contents

Specimen Test Condition†	*Pressure Difference Across Specimen During Test (mm Hg)*	*Permeability Coefficient (K)*††	
		Air and Water Vapour	*Dried Air*
Specimen A			
Oven-dried beforehand, for 1 week at 105 C.	709	0.23	0.18
57% saturated beforehand, exposed to water vapour during test.	700	0.04	0.04
Vacuum end exposed for 4 hours beforehand, and during test, to water vapour at 1 psia.	701	0.06	0.06
Vacuum end exposed to water vapour, overnight at atmospheric pressure beforehand, and at 1 psia during test.	707	0.11	0.11
Vacuum end exposed for 7 hours beforehand, and during test, to water vapour at 1 psia.	707	0.11	–
Specimen B			
Oven-dried beforehand, for 1 week at 105 C, 48 hours at 70 C.	712	0.21	0.16
Vacuum end exposed beforehand for 2 hours to water vapour at 1 psia.	712	0.14	0.12
Specimen C			
Oven-dried beforehand, for 72 hours at 105 C, 48 hours at 70 C.	711	0.15	0.10
Vacuum end exposed beforehand for 65 hours to water vapour at atmospheric pressure.	711	0.09	0.07

(From: Loughborough, M. T., "Permeability of Concrete to Air," Ontario Hydro Research Quarterly, *First Quarter, 1966)*

†Concrete with 4000 psi compressive strength at 28 days made with ¾ in. max size aggregate, 500 lb. cement per cu. yd and a water-cement ratio of 0.50.

††Cu in./min/sq ft area/in. of concrete/psi pressure differential.

pore walls. This results in a drastically increased transmission area available for the movement of air or other gas. Concretes that have been dried are very permeable to air, thousands of times more permeable than they are to water.

Little information is available on permeability of concrete to various gases but even if a given concrete is relatively permeable to air it may be substantially impermeable to some other gases.

Air leakage of a typical concrete structure occurs at joints and through cracks in the concrete. Specially detailed joints are required in low leakage structures to minimize leakage, allow movement of the concrete to accommodate thermal expansion and to permit maintenance. Cracks in concrete will have a leakage rate according to their number, spacing, width and penetration into the concrete. Leakage of air through cracks which do not completely penetrate the concrete can be computed by assuming that the leakage is only that which results from the thickness of the uncracked concrete.

When cracks penetrate completely through the concrete, the leakage rate through the cracks can be computed as follows:

$$q_T = 3.9 \times 10^4 \frac{b^3 LP}{x}$$

where

q_T = volumetric leak rate (cfm)
b = crack width (in.)
L = crack length (in.)
x = crack depth (in.)
P = pressure differential (in. H_2O)

Mass flow of gases through joints or cracks is characterized by a flow proportional to the pressure differential or the square root of the pressure differential. Flow through a porous membrane such as concrete is more nearly represented by diffusion equations and is dependent on the partial pressure of the gas or vapor being considered on one side of the membrane compared to the other side. In such a case, it is possible for the flow of two different gases (air or water vapor) through a membrane to be in opposite directions if the partial pressure gradients are of opposite sign. Various combinations of diffusion and mass flow are possible.

6.11 DURABILITY

A durable concrete is one which will withstand the effects of service conditions to which it will be subjected, such as weathering, chemical action, and wear. Numerous laboratory tests have been devised for measuring durability of concrete, but it is extremely difficult to obtain a direct correlation between laboratory tests and field service.

6.11.1 Weathering Resistance

Disintegration of concrete by weathering is caused mainly by the disruptive action of freezing and thawing and by expansion and contraction, under restraint, resulting from temperature variations and alternate wetting and drying.

The most destructive factor of weather is freezing and thawing while the concrete is wet or moist. Deterioration may be caused by expansion of the water in the paste or in the aggregate particles, or by a combination of both. The resistance of concrete to the action of freezing and thawing is dependent upon the following factors: entrained air, cement content, total water content, and moisture condition of the aggregate. Entrained air improves the resistance of concrete to damage from frost action and should be specified for all concrete subject to cycles of freezing and thawing.

The effect of cement content and total water content on durability of normal weight or lightweight concrete is approximately the same—increasing the cement content and decreasing the total water content improves durability.

Tests have indicated that the freeze-thaw resistance of structural lightweight concrete of compressive strengths less than 5,000 psi may be increased by partial or full replacement of fine aggregates with normal weight sand.

Moisture condition of lightweight aggregates at the time of mixing has a significant effect on freeze-thaw resistance of concrete. Non-air-entrained concrete made with air-dried aggregates usually is more resistant to freezing and thawing than concrete made with soaked aggregates. Vacuum-treated lightweight aggregates, when used in concrete subject to a freezing environment, must be allowed an extended air-drying period prior to freezing weather. The influence of moisture condition of aggregate is not as pronounced for air-entrained concretes.

To evaluate freeze-thaw resistance, laboratory freeze-thaw tests of concrete should be used, supplemented by field performance records.

6.11.2 Resistance to Deicers

Deicing chemicals* used for snow and ice removal can cause surface scaling. The damage is primarily a physical action and is not caused by chemical reactions or crystal pressures. Deicer scaling of inadequately air-entrained or non-air-entrained concrete is caused, as in normal freezing, by high hydraulic pressures. The presence of a deicer solution in concrete during freezing causes a buildup of hydraulic pressures in excess of the normal hydraulic pressures produced when water in concrete freezes. These pressures become critical and scaling will result when entrained air voids are not present to act as relief valves. The extent of scaling depends upon the amount of salts used and the frequency of application. Unfortunately, deicers can be applied indirectly in various ways, such as by drippings from the undersides of vehicles. Scaling is much more severe in poorly drained areas because the deicer solution is retained on the concrete surface during freezing and thawing.

Concrete requires a minimum curing period of seven days at or above 70°F to ensure adequate strength and a durable scale-resistant surface. When temperatures fall to near 40°F, a 14-day curing period may be necessary when using Type I cement. However, the required curing period can be reduced to seven days by the use of Type III (High-Early-Strength) cement or Type I or II cement with an accelerator. In addition, air-entrained concrete's resistance to deicers is greatly increased by a period of air drying after the curing period. Concrete placed in the spring or summer have drying periods in the normal course of aging. Concretes placed in the fall season, however, often do not dry out enough before the use of deicing agents become necessary. This is especially true of fall paving cured by membrane-forming compounds. These membranes remain intact until worn off by traffic and thus adequate drying may not occur before the onset of winter. Curing methods that allow drying at the completion of the curing period are preferable for fall paving on all projects where deicers will be used.

*Ammonium nitrate and ammonium sulfate have been sold as deicers. These materials in the presence of water react chemically with all forms of concrete and cause objectionable disintegration—even at room temperatures. Their use should be strictly prohibited.

The required time for sufficient drying to take place cannot be pinpointed due to variations in climate and weather conditions. A good rule of practice, however, is: Age for safe use of deicers–curing period plus 30 days.

If surface scaling (an indication of an inadequate air void system) should develop during the first frost season, a surface treatment may be used to protect the concrete against further damage. These treatments are usually made with linseed oil or neutral petroleum oil.

6.11.3 Chemical Resistance

Quality concrete must be assumed in any discussion of the effect of various substances on concrete and protective treatments. In general, achievement of adequate strength and sufficiently low permeability to withstand many exposures indefinitely requires proper proportioning, placing and curing.

-1. Design considerations–Whenever concrete is to be coated for corrosion protection, the forms should be coated with materials that will remain on the forms when they are stripped. Hence, forms coated with form oils or waxes should not be used against surfaces to be coated. Curing membranes may be weakly bonded to the concrete and may in turn develop little or no bond to coatings applied over them. If form oils, waxes, or curing membranes are present, they should be removed by acid washing, sandblasting, scarifying, or other such processes.

Where spillage of corrosive substances is likely to occur, the floor should slope to drains approximately 1/8 to 1/4 in. per linear foot to facilitate washing down of the floor. The slope required depends on the distance between drains and the corrosive substance involved.

Many solutions that have no chemical effect on concrete, such as brines and salts, may crystallize upon drying. It is especially important that concrete subject to alternate wetting and drying of such solutions be impervious. When free water in concrete is saturated with salts, the salts crystallize in the concrete near the surface during the process of drying and this crystallization may exert sufficient pressure to cause scaling. Structures exposed to brine solutions and having a free surface of evaporation should therefore be provided with a protective treatment on the side exposed to the solution.

In addition, movement of salts into the concrete may result in corrosion of reinforcing steel. It is important that sufficient concrete coverage be provided for reinforcement where the surface is to be exposed to corrosive substances. Metal chairs for support of reinforcement should not extend to the concrete surface. Deep recesses in the concrete should be provided for form ties, and they should be carefully filled and pointed with mortar.

Concrete in contact with soil or water containing moderate sulfate concentrations should be made with cement having less than 8% tricalcium aluminate (C_3A). A moderate sulfate condition exists when the water soluble sulfate (as SO_4) in soil is from 0.10 to 0.20%, or the sulfate (as SO_4) in groundwater is from 150 to 1,000 parts per million. Cement with tricalcium aluminate limited to 5% should be used when the water-soluble sulfate (as SO_4) in soil exceeds 0.20%, or sulfate (as SO_4) in groundwater exceeds 1,000 parts per million. If this type is not available, a cement with a C_3A content between 5 and 8 may be used with a 10% reduction in water-cement ratio.

Sulfate resistance of concrete is also improved by use of entrained air. Concrete made with a low water-cement ratio, adequate cement content (660 lb/cu yd), entrained air, and cement having a low tricalcium aluminate content will be most resistant to attack from sulfate containing soil or waters, or seawater.

Deterioration of concrete in sea water exposures may occur through freezing and thawing action, corrosion of steel reinforcement and by chemical attack. Other causes for deterioration include wave action and erosion, crystallization of salts brought within permeable concrete by capillary action and subsequent evaporation, particularly in and above tidal levels. Average sea water contains about 3% sodium chloride (NaCl) and 0.5% magnesium sulfate ($MgSO_4$). Solutions containing as little as 0.5% $MgSO_4$ will attack concrete at warm temperatures. The presence of chlorides (as in sea water and some brines) minimizes or inhibits the expansion of concrete in sulfate solutions. Several plausible reasons have been offered to account for this, but they must be deemed conjectural.

Low C_3A cement (C_3A potential less than 8%) is indicated for concrete in seawater or exposed to sea spray. Aggregates should be resistant to alkali-aggregate reaction. If this requirement cannot be met then the cement should contain not more than 0.60% alkali (as Na_2O). If the concrete will be exposed to frost action it should contain proper air entrainment. The use of sea water for mixing water should be prohibited unless the concrete is continually and completely submerged in seawater. Steel in reinforced concrete completely and permanently immersed in sea water will not corrode at any appreciable rate, as oxygen and carbon dioxide are virtually excluded. Steel reinforcement should be protected by at least 3 in. of concrete cover (AASHO requires 4 in. cover for precast piles in sea water) except at corners where 4 in. of cover should be allowed.

Portland-pozzolan and portland blast-furnace slag cements are not always resistant to sulfate solutions, and much depends on the C_3A content of the portland cement constituent. If the portland cement constituent of cements of either of these types contains sufficient C_3A as to make it, when used alone, vulnerable to chemical attack by sulfates or sea water, the replacement of some of such cement by a good pozzolan or good blast-furnace slag may be expected to improve the resistance of the composite cement. If, on the other hand, the portland cement constituent contains less than say 8% C_3A then partial replacement by pozzolan or blast-furnace slag would not be expected to result in an increased resistance.

Acids attack concrete by dissolving both hydrated and unhydrated cement compounds as well as calcareous aggregate. In certain acid waters it may be impossible to apply an adequate protective treatment to the concrete, and the use of a "sacrificial" calcareous aggregate should be considered. Replacement of siliceous aggregate by limestone or dolomite having the equivalent of a calcium oxide concentration of at least 50% will aid in neutralizing the acid. The acid will attack the entire exposed surface more uniformly, reducing the rate of attack on the paste and preventing loss of aggregate particles at the surface. The use of calcareous aggregate will also retard expansion resulting from sulfate attack caused by some acid solutions.

The rate of attack on concrete may be directly related to the activity of the aggressive chemical. Solutions of high concentration are generally more corrosive than those of low concentration–but with some, the reverse is true. The rate of attack may sometimes be affected by the solubility of the reaction products of the particular concrete in the corrosive solution. Lowering of the hydrogen ion concentration, pH, generally causes more rapid attack in the concrete. Also, high temperatures usually accelerate any possible attack and thus better protection is required than for normal temperatures.

-2. Surface preparation—Proper preparation of the concrete surface and good workmanship are essential for the successful application of any protective treatment. Concrete should normally be well cured (28 days to six months, depending on service conditions and coatings used) and dry before the protective coatings are applied. Moisture in the concrete may cause excessive internal vapor pressure that can result in the treatment's blistering and peeling.

Precautions should be taken to eliminate objectionable voids in the surface that may cause pinholes in the coating. Good vibration and placing techniques will reduce the number of these surface imperfections. The surface should be smoothed immediately after removal of forms by applying grout or mortar, or by grinding the surface and then working a grout into it.

It is important to have a firm base free of grease, oil, efflorescence, laitance, dirt, and loose particles. Removal of chemical contaminants must be accomplished before any other surface cleaning, such as acid-etching or sandblasting, takes place.

Concrete cast against forms is sometimes so smooth as to make adhesion of protective coatings very difficult to obtain. Such surfaces should be acid-etched, sandblasted lightly, or ground with silicon carbide stones to provide a slightly roughened surface.

-3. Choosing the treatment—Protective treatments for concrete are available for almost any degree of protection required. The coatings vary so widely in composition and performance that no one material will serve best for all conditions.

Every coating is formulated to render a certain performance under specified conditions. Its quality is not determined solely by the merits of any one raw material since minor variations in formulation can make very substantial changes in performance. Coating performance also depends upon the surface preparation, method and quality of coating application, conditions during application, and film thickness. Any general discussion of chemical resistance and other properties of coatings must assume optimum formulation and proper use. The producers of the various coatings can provide valuable information on the merits of their products for a particular use and on the proper and safe procedure for application. Many coatings contain solvents that are fire, explosion, or toxic hazards.

The more common protective treatments are indicated in the following tables which are taken from the publication "Effect of Various Substances on Concrete and Protective Treatments, Where Required" (IS001.03T), Portland Cement Association, 1968. For most substances, several treatments are suggested. They will provide sufficient protection in most cases. *The information in the tables is only a guide for determining when to consider various coatings for chemical resistance.*

The numbers and letters in the table refer to the following materials:

1- Magnesium fluosilicate or zinc fluosilicate
2- Sodium silicate (commonly called water glass)
3- Drying oils
4- Coumarone-indene
5- Styrene-butadiene
6- Chlorinated rubber
7- Chlorosulfonated polyethylene (Hypalon)
8- Vinyls
9- Bituminous paints, mastics, and enamels
10- Polyester
11- Urethane
12- Epoxy
13- Neoprene
14- Polysulfide
15- Coal tar-epoxy
16- Chemical-resistant masonry units and mortars:
  a. asphaltic and bituminous mortars
  b. epoxy resin mortars
  c. furan resin mortars
  d. hydraulic cement mortars
  e. phenolic resin mortars
  f. polyester resin mortars
  g. silicate mortars
  h. sulfur mortars
17- Sheet rubber
18- Resin Sheets
19- Lead sheet
20- Glass

ACIDS

Material	*Effect on concrete*	*Protective treatments*
Acetic		
<10%	Slow disintegration	1, 2, 9, 10, 12, 14, 16 (b, c, e, f, g, h)
30%	Slow disintegration	9, 10, 14, 16, (c, e, f, g)
100% (glacial)	Slow disintegration	9, 16, (e, g)
Acid waters (pH of 6.5 or less)	Slow disintegration.* Natural acid waters may erode surface mortar but then action usually stops	1, 2, 3, 6, 8, 9, 10, 11, 12, 13, 16, (b, c, e, f, g, h), 17
Arsenious	None	
Boric	Negligible effect	2, 6, 7, 8, 9, 10, 12, 13, 15, 16 (b, c, e, f, g, h), 17, 19
Butyric	Slow disintegration	3, 4, 8, 9, 10, 12, 16 (b, c, e, f)
Carbolic	Slow disintegration	1, 2, 16 (c, e, g), 17
Carbonic (soda water)	0.9 to 3 ppm of carbon dioxide dissolved in natural waters disintegrates concrete slowly	2, 3, 4, 8, 9, 10, 12, 13, 15, 16 (b, c, e, f, h), 17
Chromic:		
5%	None*	2, 6, 7, 8, 9, 10, 16 (f, g, h), 19
50%	None*	16 (g), 19
Formic:		
10%	Slow disintegration	2, 5, 6, 7, 12, 13, 16 (b, c, e, g), 17
90%	Slow disintegration	2, 7, 13, 16 (c, e, g), 17

**In porous or cracked concrete, it attacks steel. Steel corrosion may cause concrete to spall.*

ACIDS *(Cont.)*

Material	*Effect on concrete*	*Protective treatments*
Humic	Slow disintegration possible, depending on humus material	1, 2, 3, 9, 12, 15, 16 (b, c, e)
Hydrochloric:		
10%	Rapid disintegration, including steel	2, 5, 6, 7, 8, 9, 10, 12, 14, 16 (b, c, e, f, g, h), 17, 19, 20
37%	Rapid disintegration, including steel	5, 6, 8, 9, 10, 16 (c, e, f, g, h)
Hydrofluoric:		
10%	Rapid disintegration, including steel	5, 6, 7, 8, 9, 12, 16 (carbon and graphite brick; b, c, e, h), 17
75%	Rapid disintegration, including steel	16 (carbon and graphite brick; e, h), 17
Hypochlorous, 10%	Slow disintegration	5, 8, 9, 10, 16 (f, g)
Lactic, 5%	Slow disintegration	3, 4, 5, 7, 8, 9, 10 11, 12, 13, 15, 16 (b, c, e, f, g, h), 17
Nitric:		
2%	Rapid disintegration	6, 8, 9, 10, 13, 16 (f, g, h), 20
40%	Rapid disintegration	8, 16 (g)
Oleic, 100%	None	
Oxalic	No disintegration, It protects concrete against acetic acid, carbon dioxide, and salt water. POISONOUS, it must not be used on concrete in contact with food or drinking water.	
Perchloric, 10%	Disintegration	8, 10, 16 (e, f, g, h)
Phosphoric:		
10%	Slow disintegration	1, 2, 3, 5, 6, 7, 8, 9, 10, 11, 12, 13, 14, 15, 16 (b, c, e, f, g, h), 17, 19
85%	Slow disintegration	1, 2, 3, 5, 7, 8, 9, 10, 13, 14, 15, 16 (c, e, f, g, h), 17, 19
Stearic	Rapid disintegration	5, 6, 8, 9, 10, 11, 12, 13, 15, 16 (b, c, e, f, g, h), 17
Sulfuric:		
10%	Rapid disintegration	5, 6, 7, 8, 9, 10, 12, 13, 14, 15, 16 (b, c, e, f, g, h), 17, 19, 20
110% (oleum)	Disintegration	16 (g), 19
Sulfurous	Rapid disintegration	6, 7, 9, 11, 12, 13, 16 (b, c, e, h), 19, 20
Tannic	Slow disintegration	1, 2, 3, 6, 7, 8, 9, 11, 12, 13, 16 (b, c, e, g), 17
Tartaric, solution	None. See wine under "Miscellaneous."	

SALTS AND ALKALIES (SOLUTIONS)*

Material	*Effect on concrete*	*Protective treatments*
Bicarbonate:		
Ammonium Sodium	None	
Bisulfate:		
Ammonium** Sodium	Disintegration	5, 6, 7, 8, 9, 10, 11, 12, 13, 14, 15, 16 (b, c, e, f, h), 17
Bisulfite:		
Sodium	Disintegration	5, 6, 7, 8, 9, 10, 12 13, 16 (b, c, e, f, h), 17
Calcium sulfite solution)	Rapid disintegration	7, 8, 9, 10, 12, 13, 16 (b, c, e, f, h), 17
Bromide, sodium	Slow disintegration	1, 2, 5, 6, 7, 8, 9, 10, 11, 12, 13, 14, 16 (b, c, e, f, h), 17
Carbonate:		
Ammonium Potassium Sodium	None	
Chlorate, sodium	Slow disintegration	1, 4, 6, 7, 8, 9, 10, 16 (f, g, h), 17, 19
Chloride:		
Calcium† Potassium Sodium† Strontium	None, unless concrete is alternately wet and dry with the solution**	1, 3, 4, 5, 6, 7, 8, 9, 10, 11, 12, 13, 15, 16 (b, c, e, f, g, h), 17
Ammonium Copper Ferric (iron) Ferrous Magnesium Mercuric Mercurous Zinc	Slow disintegration**	1, 3, 4, 5, 6, 7, 8, 9, 10, 11, 12, 13, 15, 16 (b, c, e, f, g, h), 17
Aluminum	Rapid disintegration**	1, 3, 4, 5, 6, 7, 8, 9, 10, 11, 12, 13, 15, 16 (b, c, e, f, h), 17
Chromate, sodium	None	
Cyanide:		
Ammonium Potassium Sodium	Slow disintegration	7, 8, 9, 12, 13, 16 (b, c), 17
Dichromate:		
Sodium	Slow disintegration with dilute solutions	1, 2, 6, 7, 8, 9, 10, 11, 12, 13, 15, 16 (b, c, e, f, h), 17
Potassium	Disintegration	1, 2, 6, 7, 8, 9, 10, 11, 12, 13, 15, 16 (b, c, e, f, h), 17
Ferrocyanide, sodium	None	
Fluoride:		
Ammonium Sodium	Slow disintegration	3, 4, 8, 9, 13, 16 (a, c, e, h), 17
Fluosilicate, magnesium	None	

**Dry materials generally have no effect.*

***In porous or cracked concrete, it attacks steel. Steel corrosion may cause concrete to spall.*

†Frequently used as a de-icer for concrete pavements. If the concrete contains insufficient entrained air or has not been air-dried for at least 30 days after completion of curing, repeated application may cause surface scaling. See de-icers under "Miscellaneous."

SALTS AND ALKALIES (SOLUTIONS) *(Cont.)*

Material	*Effect on concrete*	*Protective treatments*
Hexametaphosphate, sodium	Slow disintegration	5, 6, 7, 8, 9, 12, 13, 15, 16 (b, c, e), 17
Hydroxide: Ammonium Barium Calcium Potassium, 15%‡ Sodium, 10%‡	None	
Potassium, 25% Sodium, 20%	Disintegration. Use of calcareous aggregate lessens attack.	5, 7, 8, 12, 13, 14, 15, 16 (carbon and graphite brick; b, c) 17
Nitrate: Calcium Ferric Zinc	None	
Lead Magnesium Potassium Sodium	Slow disintegration	2, 5, 6, 7, 8, 9, 10, 11, 12, 13, 16 (b, c, e, f, g, h), 17, 20
Ammonium	Disintegration**	2, 5, 6, 8, 9, 10, 11, 12, 13, 16 (b, c, e, f, g, h), 17, 20
Nitrite, sodium	Slow disintegration	1, 2, 5, 6, 7, 8, 9, 12, 13, 16 (b, c), 17
Orthophosphate, sodium (dibasic and tribasic)	None	
Oxalate, ammonium	None	
Perborate, sodium	Slow disintegration	1, 4, 7, 8, 9, 10, 13, 16 (d, f, g, h), 17
Perchlorate, sodium	Slow disintegration	6, 7, 8, 10, 16 (f, g, h), 17
Permanganate, potassium	None	
Persulfate, potassium	Disintegration of concrete with inadequate sulfate resistance	1, 2, 5, 7, 8, 9, 10, 12, 13, 16 (b, c, e, f, h), 17
Phosphate, sodium (monobasic)	Slow disintegration	5, 6, 7, 8, 9, 12, 15, 16 (b, c), 17
Pyrophosphate, sodium	None	
Stannate, sodium	None	
Sulfate: Ammonium	Disintegration**	5, 6, 7, 8, 9, 10, 11, 12, 13, 14, 15, 16 (b, c, e, f, g, h), 17
Aluminum Calcium Cobalt Copper Ferric Ferrous (iron vitriol) Magnesium (epsom salt) Manganese Nickel Potassium	Disintegration of concrete with inadequate sulfate resistance. Concrete products cured in high-pressure steam are highly resistant to sulfates.	1, 3, 4, 5, 6, 7, 8, 9, 10, 11, 12, 13, 15, 16 (b, c, e, f, g, h), 17
Potassium aluminum (alum) Sodium Zinc	Disintegration of concrete with inadequate sulfate resistance. Concrete products cured in high-pressure steam are highly resistant to sulfates.	1, 3, 4, 5, 6, 7, 8, 9, 10, 11, 12, 13, 15, 16 (b, c, e, f, g, h), 17
Sulfide: Copper Ferric Potassium	None unless sulfates are present	7, 8, 9, 10, 12, 13, 15, 16 (b, c, e, f, h), 17
Sodium	Slow disintegration	6, 7, 8, 9, 11, 12, 13, 15, 16 (b, c), 17
Ammonium	Disintegration	7, 8, 9, 12, 13, 15, 16 (a, b, c, e), 17
Sulfite: Sodium	None unless sulfates are present	1, 2, 5, 6, 7, 8, 9, 11, 12, 13, 15, 16 (b, c, e), 17
Ammonium	Disintegration	8, 9, 12, 15, 16 (b, c, e, h), 17
Superphosphate, ammonium	Disintegration**	8, 9, 12, 13, 15, 16 (b, c, e), 17, 19
Tetraborate, sodium (borax)	Slow disintegration	5, 6, 7, 8, 9, 10, 11, 12, 13, 15, 16 (b, c, e, f, g, h), 17
Thiosulfate Sodium	Slow disintegration of concrete with inadequate sulfate resistance	1, 2, 5, 6, 7, 8, 9, 10, 12, 13, 15, 16 (b, c, e), 17
Ammonium	Disintegration	8, 9, 12, 13, 15, 16 (b, c, e), 17

‡*If concrete is made with reactive aggregates, disruptive expansions may occur.*

PETROLEUM OILS

Material	*Effect on concrete*	*Protective treatments*
Heavy oil below 35° Baume* Paraffin	None	
Gasoline Kerosene Light oil above 35° Baume* Ligroin Lubricating oil* Machine oil* Mineral spirits	None. Impervious concrete is required to prevent loss from penetration, and surface treatments are generally used.	1, 2, 3, 8, 10, 11, 12, 14, 16 (b, c, e, f), 17, 19
Gasoline, high octane	None. Surface treatments are generally used to prevent contamination with alkalies in concrete.	11, 14, 17

**May contain some vegetable or fatty oils and the concrete should be protected from such oils.*

COAL TAR DISTILLATES

Material	Effect on concrete	Protective treatments
Alizarin Anthracene Carbazole Chrysen Pitch	None	
Benzol (benzene) Cumol (cumene) Phenanthrene Toluol (toluene) Xylol (xylene)	None. Impervious concrete is required to prevent loss from penetration, and surface treat treatments are generally used.	1, 2, 10, 11, 12, 16 (b, c, e, f, g), 19
Creosote Cresol Dinitrophenol Phenol, 5 to 25%	Slow disintegration	1, 2, 16 (c, e, g), 17, 19

SOLVENTS AND ALCOHOLS

Material	Effect on concrete	Protective treatments
Carbon tetrachloride	None*	1, 2, 10, 12, 16 (b, c, e, g)
Ethyl alcohol	None* (see de-icers under "Miscellaneous")	1, 2, 5, 7, 10, 12, 13, 14, 16, (b, c, e, f, g, h), 17, 19
Ethyl ether	None*	11, 12, 16 (c, e), 19
Methyl alcohol	None*	1, 2, 5, 7, 10, 12, 13, 14, 16 (b, c, e, f, g, h), 17, 19
Methyl ethyl ketone	None*	16 (c, e), 17, 19
Methyl isoamyl ketone	None*	16 (c, e), 17
Methyl isobutyl	None*	16 (c, e), 17
Perchloroethylene	None*	12, 16 (b, c, e)
t-Butyl alcohol	None*	1, 2, 5, 7, 10, 12, 13, 14, 16 (b, c, e, f, g, h), 17, 19
Trichloroethylene	None*	1, 2, 12, 16 (b, c, e, g)
Acetone	None.* However, acetone may contain acetic acid as impurity (see under "Acids").	1, 2, 10, 16 (c, e, g), 17, 19
Carbon disulfide	Slow disintegration possible	1, 2, 11, 16 (c, e, g)
Glycerin (glycerol)	Slow disintegration possible	1, 2, 3, 4, 7, 11, 12, 13, 16 (b, c, e, f, g), 17
Ethylene glycol**	Slow disintegration	1, 2, 7, 10, 12, 13, 14, 16 (b, c, e, f, g, h), 17

*Impervious concrete is required to prevent loss from penetration, and surface treatments are generally used.
**Frequently used as de-icer for airplanes. Heavy spillage on concrete containing insufficient entrained air may cause surface scaling.*

VEGETABLE OILS

Material	Effect on concrete	Protective treatments
Rosin and rosin oil	None	
Turpentine	Mild attack and considerable penetration	1, 2, 11, 12, 14, 16 (b, c, e)
Almond China wood* Linseed* Olive Peanut Poppy seed Soybean* Tung* Walnut	Slow disintegration	1, 2, 8, 10, 11, 12, 14, 16 (b, c, e, f), 17. For expensive cooking oils, use 20.
Margarine	Slow disintegration—faster with melted margarine	1, 2, 8, 10, 11, 12, 13, 16 (b, c, e, f)
Castor Cocoa bean Cocoa butter Coconut Cottonseed Mustard Rapeseed	Disintegration, especially if exposed to air	1, 2, 8, 10, 11, 12, 14, 16 (b, c, e, f), 17

Applied in thin coats, the material quickly oxidizes and has no effect. The effect indicated above is for constant exposure to the material in liquid form.

FATS AND FATTY ACIDS (ANIMAL)

Material	Effect on concrete	Protective treatments
Fish liquor	Disintegration	3, 8, 10, 12, 13, 16 (b, c, e, f), 17
Fish oil	Slow disintegration with most fish oils	1, 2, 3, 8, 10, 12, 13 16 (b, c, e, f), 17
Whale oil	Slow disintegration	1, 2, 3, 8, 10, 12, 13, 16 (b, c, e, f), 17
Neatsfoot oil Tallow and tallow oil	Slow disintegration	1, 2, 3, 8, 10, 12, 13, 16 (b, c, e, f), 17
Beef fat Horse fat Lamb fat	Slow disintegration with solid fat—faster with melted	1, 2, 3, 8, 10, 12, 13, 16 (b, c, e, f), 17
Lard and lard oil	Slow disintegration—faster with oil	1, 2, 3, 8, 10, 12, 13, 16 (b, c, e, f) 17
Slaughterhouse wastes	Disintegration due to organic acids	8, 12, 13, 16 (b, c, e)

MISCELLANEOUS

Material	*Effect on concrete*	*Protective treatments*
Alum	See sulfate, potassium aluminum, under "Salts and Alkalies"	
Ammonia:		
Liquid	None, unless it contains harmful ammonium salts (see under "Salts and Alkalies")	
Vapors	Possible slow disintegration of moist concrete and steel attacked in porous or cracked moist concrete	8, 9, 10, 12, 13, 16 (a, b, c, f), 17
Ashes:		
Cold	Harmful if wet, when sulfides and sulfates leach out (see sulfate, sodium, under "Salts and Alkalies")	1, 2, 3, 8, 9, 12, 13, 16 (b, c, e)
Hot	Thermal expansion	16 (calcium, aluminate cement, fireclay, and refractory-silicate-clay mortars)
Automobile and diesel exhaust gases	Possible disintegration of moist concrete by action of carbonic, nitric, or sulfurous acid (see under "Acids")	1, 5, 8, 12, 16 (b, c, e)
Baking soda	None	
Beer	No progressive disintegration, but in beer storage and fermenting tanks a special coating is used to guard against beer contamination. Beer may contain, as fermentation products, acetic, carbonic, lactic, or tannic acids (see under "Acids").	8, 10, 11, 12, 16 (b, c, f), 17, coatings made and applied by Borsari Tank Corp. of America, 605 Third Ave., New York, N. Y. 10016
Bleaching solution	See the specific chemical, such as hypochlorous acid, sodium hypochlorite, sulfurous acid, etc.	
Borax (salt)	See Tetraborate, sodium, under "Salts and Alkalies"	
Brine	See chloride, sodium, or other salts under "Salts and Alkalies"	
Bromine	Disintegration if bromine gaseous—or if a liquid containing hydrobromic acid and moisture	10, 13, 16 (f, g)
Buttermilk	Slow disintegration due to lactic acid	2, 3, 4, 7, 8, 9, 10, 11, 12, 13, 16 (b, c, e, f), 17
Butyl stearate	Slow disintegration	8, 9, 10, 16 (b, c, e)
Carbon dioxide	Gas may cause permanent shrinkage. See carbonic acid under "Acids."	1, 2, 3, 6, 8, 9, 10, 11, 12, 13, 15, 16 (b, c, e, f, h), 17
Caustic soda	See hydroxide, sodium, under "Salts and Alkalies"	
Chile saltpeter	See nitrate, sodium, under "Salts and Alkalies"	
Chlorine gas	Slow disintegration of moist concrete	2, 8, 9, 10, 16 (f, g), 17
Chrome plating	Slow disintegration	7, 8, 9, 10, 16 (f, g), 20
Cider	Slow disintegration. See acid under "Acids."	1, 2, 9, 10, 12, 14, 16 (b, c, e, f, g), 17
Cinders cold and hot	See ashes above	
Coal	None, unless coal is high in pyrites (sulfide of iron) and moisture. Sulfides leaching from damp coal may oxidize to sulfurous or sulfuric acid, or ferrous sulfate (see under "Acids" and "Salts and Alkalies"). Rate is greatly retarded by deposit of an insoluble film.	1, 2, 3, 6, 7, 8, 9, 12, 13, 16 (b, c, e, h), 17
Coke	Sulfides leaching from damp coke may oxidize to sulfurous or sulfuric acid (see under "Acids").	1, 2, 3, 6, 7, 8, 9, 12, 13, 16 (b, c, e, h)
Copper plating solutions	None	
Corn syrup (glucose)	Slow disintegration	1, 2, 3, 7, 8, 9, 12, 13, 16 (b, c, e), 17
De-icers	Chlorides (calcium and sodium), urea, and ethyl alcohol cause scaling of non-air-entrained concrete.	50% solution of boiled linseed oil in kerosene, soybean oil, modified castor oil, cottonseed oil, sand-filled epoxy, or coal-tar epoxy
Distiller's slop	Slow disintegration due to lactic acid	1, 8, 9, 10, 12, 13, 15, 16 (b, c, e, f, h), 17
Fermenting fruits, grains, vegetables, or extracts	Slow disintegration. Industrial fermentation processes produce lactic acid (see under "Acids").	1, 2, 3, 8, 9, 12, 16 (b, c, e), 17
Flue gases	Hot gases (400–1100° F.) cause thermal stresses. Cooled, condensed sulfurous, hydrochloric acids disintegrate concrete slowly.	9 (high melting), 16 (g, fireclay mortar)
Formaldehyde, 37% (formalin)	Slow disintegration due to formic acid formed in solution	2, 5, 6, 8, 10, 11, 12, 13, 14, 16 (b, c, e, f, g, h), 17, 20
Fruit juices	Little if any effect for most fruit juices as tartaric and citric acids do not appreciably affect concrete. Sugar and hydrofluoric and other acids cause disintegration.	1, 2, 3, 6, 7, 8, 9, 11, 12, 16 (b, c, e), 17
Gas water	Ammonium salts seldom present in sufficient quantity to disintegrate concrete	9, 12, 16 (b, c)
Glyceryl tristearate	None	

MISCELLANEOUS *(Cont.)*

Material	*Effect on concrete*	*Protective treatments*
Honey	None	
Hydrogen sulfide	Slow disintegration in moist oxidizing environments where hydrogen sulfide converts to sulfurous acid	1, 2, 5, 6, 7, 8, 9, 10, 11, 12, 13, 16, (b, c, e, f, g, h), 17, 19
Iodine	Slow disintegration	1, 2, 6, 12, 13, 16 (b, c, e, g), 17
Lead refining solution	Slow disintegration	1, 2, 6, 8, 9, 12, 16, (carbon and graphite brick; b, c, e, h), 17 20
Lignite oils	Slow disintegration if fatty oils present	1, 2, 6, 8, 10, 12, 16 (b, c, e, f)
Lye	See hydroxide, sodium and potassium, under "Salts and Alkalies"	
Manure	Slow disintegration	1, 2, 8, 9, 12, 13, 16 (b, c, e)
Mash, fermenting	Slow disintegration due to acetic and lactic acids and sugar	1, 8, 9, 10, 12, 13, 16 (b, c)
Milk	None, unless milk is sour. Then lactic acid disintegrates concrete slowly.	3, 4, 8, 9, 10, 11, 12, 13, 16 (b, c, f), 17
Mine water, waste	Sulfides, sulfates, or acids present disintegrate concrete and attack steel in porous or cracked concrete	1, 2, 5, 8, 9, 10, 12, 13, 15, 16 (b, c, e, f, h), 17
Molasses	Slow disintegration at temperatures ⩾120°F	1, 2, 7, 8, 9, 12, 13, 16 (b, c, e), 17
Nickel plating solutions	Slow disintegration due to nickel ammonium sulfate	2, 5, 6, 7, 8, 9, 10, 13, 16 (c, e, f), 17
Niter	See nitrate, potassium, under "Salts and Alkalies"	
Ores	Sulfides leaching from damp ores may oxidize to sulfuric acid or ferrous sulfate (see under "Acids" and "Salts and Alkalies")	2, 9, 10, 12, 13, 15, 16 (b, c, e, f, g), 17
Pickling brine	Steel attacked in porous or cracked concrete. See salts, boric acid, or sugar.	1, 7, 8, 9, 12, 13, 16 (b, c, e, h), 17
Sal ammoniac	See chloride, ammonium, under "Salts and Alkalies"	
Sal soda	See carbonate, sodium, under "Salts and Alkalies"	
Saltpeter	See nitrate, potassium, under "Salts and Alkalies"	
Sauerkraut	Slow disintegration possible due to lactic acid. Flavor impaired by concrete.	1, 2, 8, 9, 10, 12, 13, 16 (b, c, e, f), 17
Sea water	Disintegration of concrete with inadequate sulfate resistance and steel attacked in porous or cracked concrete	1, 2, 5, 6, 7, 8, 9, 10, 11, 12, 13, 14, 15, 16 (b, c, e, f), 17
Sewage and sludge	Usually not harmful. See hydrogen sulfide above.	
Silage	Slow disintegration due to acetic, butyric, and lactic acids, and sometimes fermenting agents of hydrochloric or sulfuric acids	3, 4, 8, 9, 10, 12, 16 (b, c, e, f)
Sodium hypochlorite	Slow disintegration	7, 8, 9, 10, 13, 16 (d, f), 17
Sugar (sucrose)	None with dry sugar on thoroughly cured concrete. Sugar solutions may disintegrate concrete slowly.	1, 2, 3, 7, 8, 9, 10, 12, 13, 15, 16 (b, c, e, f), 17
Sulfite liquor	Disintegration	1, 2, 3, 5, 6, 8, 9, 10, 12, 13, 16 (b, c, e, f, h), 17, 19
Sulfur dioxide	None if dry. With moisture, sulfur dioxide forms sulfurous acid.	2, 5, 6, 8, 9, 10, 12, 13, 16 (b, c, e, f, g, h), 17, 19
Tanning bark	Slow disintegration possible if damp. See tanning liquor below.	1, 2, 3, 6, 8, 9, 11, 12, 13, 16 (b, c, e), 17
Tanning liquor	None with most liquors, including chromium. If liquor is acid, it disintegrates concrete.	1, 2, 3, 5, 6, 8, 9, 11, 12, 13, 16 (b, c, e), 17
Tobacco	Slow disintegration if organic acids present	1, 8, 9, 10, 12, 13, 16 (b, c, e, f), 17
Trisodium phosphate	None	
Urea	None (see de-icers)	
Urine	None, but steel attacked in porous or cracked concrete	7, 8, 12, 13, 16 (b, c, e)
Vinegar	Slow disintegration due to acetic acid	9, 12, 16 (b, c, e, h), 17
Washing soda	None	
Water, soft (<75 ppm of carbonate hardness)	Leaching of hydrated lime by flowing water in porous or cracked concrete	2, 3, 4, 8, 9, 10, 12, 13, 16 (b, c, e, f, h), 17
Whey	Slow disintegration due to lactic acid	3, 4, 5, 7, 8, 9, 10, 12, 13, 16 (b, c, e, f, h), 17
Wine	None–but taste of first batch may be affected unless concrete has been given tartaric acid treatment	For fine wines, 2 or 3 applications of tartaric acid solution (1 lb. tartaric acid in 3 pints water), 2, 8, 12, 16 (b), 20
Wood pulp	None	
Zinc refining solutions	Disintegration if hydrochloric or sulfuric acids present	8, 9, 10, 12, 13, 16 (b, c, e, f, h), 17
Zinc slag	Zinc sulfates (see under "Salts and Alkalies") may be formed by oxidation	8, 9, 10, 12, 13, 16 (b, c, e, f, h), 17

6.11.4 Abrasion and Skid Resistance

Concrete surfaces must be capable of withstanding many abrasive forces with a minimum of wear. Use of quality concrete and proper attention to design, construction, and maintenance of the surfaces can do much to minimize or eliminate wear of the concrete.

-1. Classification of wear—Wear of concrete by abrasion can be classified as follows:

a. Wear on concrete floors due to foot traffic, light trucking with small-wheeled carts, and the skidding or sliding of objects on the surface. This type of wear is essentially a rubbing action and is usually caused by the introduction of foreign particles, such as sand, metal scraps, or similar materials.
b. Wear on concrete pavements due to heavy trucking and automobiles, with or without chains or metal studs. This type of wear is caused by a rubbing plus an impact-cutting action.
c. Wear on underwater construction, due to the action of abrasive materials carried by flowing or turbulent waters. The abrasive materials are generally sand particles which often may be softer than the concrete, but the force exerted by rapidly moving water carrying the sand is such that a cutting action (erosion) is produced. Erosion varies with the type and amount of abrasive material, its velocity, the angle of contact and changes in the direction of flow or the presence of eddies.
d. Wear of concrete in hydraulic structures due to cavitation. Cavitation is very destructive and one to which even the best quality concrete offers very little resistance. The magnitude of cavitation damage varies with the mean velocity of flow raised, perhaps, to the sixth or seventh power. However, other factors, such as boundary roughness and the growth and form of boundary layers, together with stream turbulence, also have some influence on the intensity of cavitation damage.

-2. Concrete requirements—Quality concrete is a prerequisite to abrasion and erosion resistance and to the retention of pavement skid resistance. The compressive strength of concrete is the most important single factor related to wear. Wear resistance increases as compressive strength increases up to a certain point (usually about 6000 psi), Fig. 6-48, depending upon aggregate, mix composition and type of test, but beyond this point increases in strength have very little effect on wear resistance. Concrete with large aggregate erodes less than mortar of equal strength. Soft aggregates are not desirable because after the film of paste has been worn away the forces of erosion are resisted to a large extent by the aggregate. For best resistance to cavitation, the maximum size of aggregate near the surface should not exceed 3/4 in. because cavitation tends to remove large particles.

Comparisons of results of aggregate abrasion tests with those of abrasion resistance of concrete do not generally show a direct correlation. The wear resistance of concrete can be determined more accurately by abrasion tests of the concrete itself, (ASTM C779).

Because of the porous structure of structural lightweight aggregates, the resistance of each thin wall or shell to load and/or impact may be low compared to the point load and impact resistance offered by concrete made with solid particles of similar composition. Therefore, the abrasion resistance of all-lightweight aggregate concretes may not be sufficient for steel-wheeled or exceptionally heavy industrial traffic. As the severity of wear becomes less, the abrasion resistance should be as satisfactory as that of normal weight concrete. The use of natural sand in lightweight concrete improves resistance to abrasion.

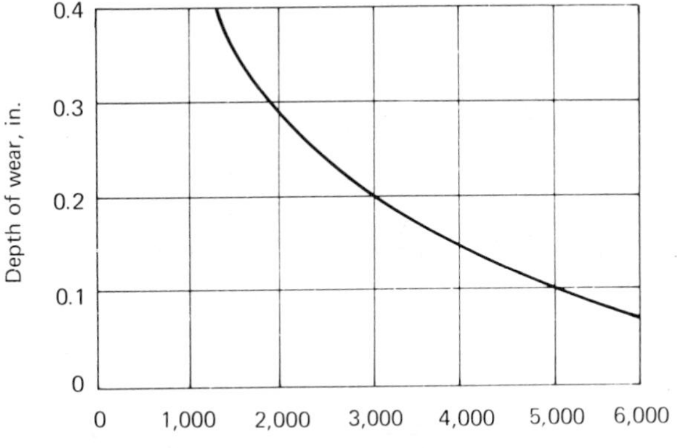

Fig. 6-48 Effect of compressive strength on the abrasion resistance of concrete. High-strength concrete is highly resistant to abrasion. (From: Sawyer, J. L., "Wear Tests on Concrete Using the German Standard Method of Test and Machine," ASTM Presentation at 60th Annual Meeting, June 1957.)

Operations and procedures such as proper finishing and adequate curing, removal of water by the vacuum process or absorptive form lining, low slump and low water-cement ratio, all of which tend to increase the strength of the concrete at the surface or throughout the mass, also increase the abrasion resistance of the concrete.

Wear resistance of concrete increases as the cement content is increased and when concrete surfaces are exposed to highly abrasive action or carry high traffic volumes, the cement content should be greater than 564 lbs per cu yd. Also, significant reduction in wear occurs as the length of the moist curing period is increased, Fig. 6-49. Wear resistance increases as the water cement or voids (air plus water) to cement ratio is decreased and the maximum water-cement ratio should not exceed 0.50. Air-entrained concrete is as resistant to wear as non-air-entrained concrete providing they are of equal compressive strength.

When conditions are such as to cause rapid loss of moisture from the freshly placed concrete surface, either a uniform fog spray of water, just enough to restore the surface sheen or a monomolecular film to retard evaporation may be applied during finishing operations. The application of excessive amounts of water to the surface during finishing operations must be avoided. Excess water from any source reduces the possibility of obtaining wear resistant surfaces.

Final finishing should be delayed as long as possible. The surface of a floor should be struck off slightly higher than the finished dimension and when bleeding has stopped, the surface should be struck off to proper elevation. This will eliminate a possible weak top layer of mortar formed with excessive bleed water. Pressure applied to the surface by several hard trowelings after the concrete begins to set increases the surface density and abrasion resistance. Curing should be started as soon as possible without marring the surface and continued for at least seven days. Curing compounds, if used, should be applied after finishing operations are completed and between the end of the bleeding and the beginning of the drying of the concrete.

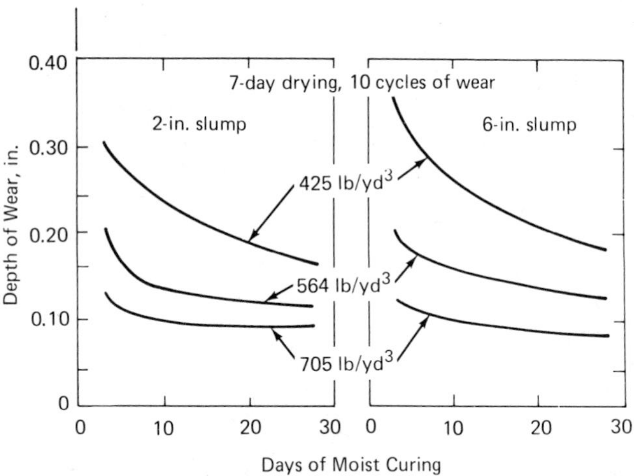

Fig. 6-49 Effect of cement content and cure on concrete wear. (From: Sawyer, J. L., "Wear Tests on Concrete Using the German Standard Method of Test and Machine," ASTM Presentation at 60th Annual Meeting, June 1957.)

-3. Wear on floors—Floors that are to be subjected to severe abrasion should preferably be constructed with a heavy-duty concrete floor topping. Other satisfactory methods of densifying the surface are the use of $^3/_8$ in. traprock cast on the surface, or the use of dry shakes consisting of iron filings, corundum, silicon carbide, or heat-treated aluminum oxides. The use of surface hardeners are more effective in increasing abrasion resistance on poorly cured or porous concrete surfaces than on high density properly cured concrete.

Concrete that is subject to abrasion while in a moist condition is more liable to show wear than the same concrete maintained in a dry condition.

-4. Cavitation—Quality concrete is not affected by a steady, tangential, high-velocity flow of water. On horizontal or sloping surfaces subjected to high-velocity flow (40 feet per second in open channels or as low as 25 feet per second in closed conduit) or on vertical surfaces past which water flows, an obstruction, or abrupt change in surface alignment causes a zone of severe subatmospheric pressure to be formed against the surface immediately downstream from the obstruction or abrupt change. When the local absolute pressure approaches the ambient vapor pressure of the water, this zone is promptly filled with turbulent water interspersed with large, single voids, which later break up, or clouds of small fast-moving bubblelike cavities of water vapor. These cavities flow downstream with the water, and upon entering an area of higher pressure collapse with great impact. Water from the boundaries of the cavities rushes toward their centers at high speed when the collapse takes place generating extremely high pressures (as high as 100,000 psi). Repeated collapse of the cavities near the surface of the concrete will pit it. Pitting resulting from cavitation can be easily distinguished by its honeycombed appearance as contrasted to the smoother surfaces resulting from damage by abrasion or erosion.

Although the advent of cavitation depends primarily on pressure changes (and consequently also on velocity changes), it is especially likely to occur in the presence of small quantities of undissolved air in the water. These bubbles of air behave as nuclei at which the change of phase from liquid to vapor can more readily occur. Dust particles have a similar effect. On the other hand, free air in large quantities (up to 2% by volume), while promoting cavitation, may cushion the collapse of the cavities and hence reduce the cavitation damage. Deliberate air entrainment in water may therefore be advantageous.

TABLE 6-12 Offset and Grinding Tolerances for High-Velocity Flow*

Velocity Range, fps	*Distance of Treatment Downstream from Gate Frame, ft*	*Grinding Bevel, Ratio of Height to Length*
40 to 90	15	1 to 20
90 to 120	30	1 to 50
Over 120	50	1 to 100

**From: Concrete Manual, U.S. Dept. of Interior, Bureau of Reclamation, 7th edition, 1963.*

Cavitation can be reduced by constructing smooth and well aligned surfaces free from irregularities such as depressions, projections, joints, and misalignments, and by the absence of abrupt changes in slope or curvature that tend to pull the flow away from the surface. For example, high velocity flow passing through a gate opening causes the boundary layer of the flow to be disrupted; and a certain length of continuous surface contact, depending upon the velocity, is required for it to be reestablished. Surface irregularities must be completely eliminated for specific distances downstream from the ends of gate frames, by grinding to specified levels according to flow velocity, Table 6-12.

In extreme cases where low pressures cannot be avoided critical areas are sometimes protected by facing with metal or resilient coatings which have good adhesion to concrete, and an ability to be easily repaired.

-5. Skid resistance—Skid resistance of concrete pavements, for simplicity, can be defined as the force, relative to the weight of the vehicle, which is available to permit stopping within a reasonable distance and to permit cornering without sideways displacement. It is influenced by both the tire and pavement surface. Regarding the tire, skid resistance is a function of its rubber characteristics and tread design. Tread design is not too significant on dry pavements because even smooth tires produce good skid resistance. However, when surfaces are wet, depth of tire tread gains in importance. The contribution of a concrete pavement to skid resistance is a function mainly of surface texture, concrete mix design, and aggregates needed to minimize wear. The durability of the surface texture is a function of the wear resistant qualities of the concrete and the character and volume of the traffic.

The surface of a pavement should include both fine and coarse texture. The fine texture (grittiness) is formed by the sand in the cement-mortar layer and imparts the adhesion component in the tire-pavement interaction. The coarse texture is formed by the ridges of mortar left by the method of finishing and has the dual role of imparting the hysteresis* component and providing drainage under the tire.

Finishing of a concrete pavement has generally been accomplished using a burlap drag, a broom, or a belt. Texture depths obtained with a burlap drag vary considerably with the time of finishing. One way of obtaining a texture when finishing with a burlap drag would be to make three passes, each at a different time interval after casting. A

*The hysteresis component is a function of the energy losses within the rubber as it is deformed by the textured pavement surface.

pavement with a deep texture will retain skid resistance longer than a pavement with a shallow texture; other factors being equal, however, the finishing method selected should be compatible with the environment, speed and density of traffic, topography and layout of the pavement, and economics of vehicle operations.

The skid resistance of a concrete pavement is controlled mainly by the fine aggregate in the mortar or wearing layer that is textured during finishing rather than the coarse aggregate that seldom functions as a portion of the surface. For example, a coarse aggregate with high polishing characteristics can be permitted in a mix if the mortar surface has adequate skid resistance. (Polish resistant coarse aggregates include siliceous gravels, granites, diabase, quartzite, sandstones and expanded shale). To provide good skid-resistance the proportion of fine aggregate in the concrete mix should be near the upper limit of the range that permits proper placing, finishing, and texturing. The siliceous particle content of the fine aggregate, as determined by an insoluble residue test, should be not less than 25%. The siliceous particle content is very important, and where economically feasible, a higher percentage should be required. Where suitable materials conforming to these requirements are not economically available, alternate methods of achieving a skid-resistant surface should be investigated. These might include blending of fine aggregates or applying wear-resistant particles to the surface. ASTM C33 cautions that certain manufactured sands produce slippery pavement surfaces and should be investigated for acceptance before use. Consideration should be given to replacing limestone fines with expanded shale fines.

To improve skid resistance at critical areas such as toll plazas, near busy intersections, or in areas where frequent braking, traction, or cornering occurs, ½ to 1 lb per sq yd of abrasives such as aluminum oxide or silicon carbide of 12 grit or smaller can be spread over the surface and floated into the plastic concrete.

Methods of increasing the skid resistance of old or worn surfaces include acid etching, mechanical abrading, sawing, and resurfacing. The choice of method depends on the economics and the desired result.

6.12 ACOUSTICAL PROPERTIES

Good acoustical design utilizes both absorptive and reflective surfaces, sound barriers and vibration isolators. Some surfaces must reflect sound so that the loudness will be adequate in all areas where listeners are located. Other surfaces absorb sound to avoid echoes, sound distortion and long reverberation times. Sound is isolated from rooms where it is not wanted by selected wall and floor-ceiling constructions. Vibrations generated by mechanical equipment must be isolated from the structural frame of the building.

6.12.1 Sound Absorption

Sound absorption control involves reduction of sound emanating from a source within a room. The extent of control depends on the efficiency of the room surfaces in absorbing rather than reflecting sound waves. The amount of sound absorbed by concrete depends upon its porosity. Sound waves reaching a porous surface cause air to flow into and from the surface. The friction of the air flowing through the pores converts a portion of the sound energy to heat.

The sound absorption coefficient is an indication of the sound absorbing efficiency of a surface. A surface which could theoretically absorb 100% of impinging sound would have a sound absorption coefficient of 1. Similarly, a dense, non-porous concrete surface (cast-in-place) typically absorbs 1 to 2% of incident sound and has a coefficient of 0.015. (Most materials are tested for sound absorption at frequencies from 125 to 4,000 cps in octave steps.) Another designator, termed the noise reduction coefficient (NRC) is calculated by taking a mathematical average of the sound absorption coefficients obtained at frequencies of 250, 500, 1,000 and 2,000 Hertz (cps). It is expressed to the nearest integral multiple of 0.05. Cast-in-place concrete would have an NRC of 0.02, while precast units with lightweight aggregates may have an NRC of as high as 0.60. Where additional sound absorption is desired, cast-in-place concrete can be coated with spray applied acoustical materials, acoustical tile can be applied with adhesive, or an acoustical ceiling can be hung below.

Many surfaces must be acoustically reflective, not absorptive, i.e., a band shell. Hard reflective surfaces on walls and ceiling near the stage in an auditorium serve as megaphones to direct sounds to the audience. Concrete is an excellent material for this use.

6.12.2 Sound Transmission Loss

Airborne sound reaching a wall, floor or ceiling produces vibrations in the wall and is reradiated with reduced intensity on the other side. The ratio of sound pressure level on the source side to the sound pressure level on the reradiating side is known as the sound transmission loss (STL). The ratio is expressed in decibels, db.

Sound transmission loss measurements are made at 16 frequencies at one-third octave intervals covering the range from 125 to 4000 Hz (ASTM E90). Although test data are reported to a fraction of a decibel, a difference of three or less db is not especially significant because the human ear cannot detect a change in sounds of less than three decibels. Single figure ratings are obtained by comparing plots of STL vs frequency data to standard contours, (STC contour), ASTM E413. The matching system permits performance to fall below the contour to a degree which is limited in amount at any single test frequency and for all 16 frequencies. The highest contour meeting the requirements is selected. The sound transmission class, or STC rating, is the STL of the STC contour at 500 Hz. The STC provides a convenient means for comparing various systems of construction, but for specific applications the data at specific frequencies may be of greater importance.

The STL of a wall or a floor-ceiling assembly is a function of the weight, stiffness and vibration damping characteristics. Weight is concrete's greatest asset when it is used as a sound insulator. Stiffness increases the sound transmission loss at low frequencies. But sometimes the stiffness controlled region is below the lowest test frequency. Resonant and coincident frequencies frequently may be present in the critical range. At these frequencies the sound transmission loss is reduced–there are dips in the sound transmission loss curve. If a wall or floor is highly damped the dips are smaller. Small changes in weight or stiffness can shift the critical frequencies. Such shifts can either eliminate or introduce overlaps. Significant changes then result in sound transmission at those frequencies. This in turn can produce large changes in STC's.

The weight per square foot of a simple wall or floor construction can be used as a guide for determining sound transmission loss. Sound transmission loss should increase 6 db when either the weight of the panel or the frequency is doubled. This is known as the Mass Law, Weight Law, or Limp Wall Law. Unfortunately the frequency range over

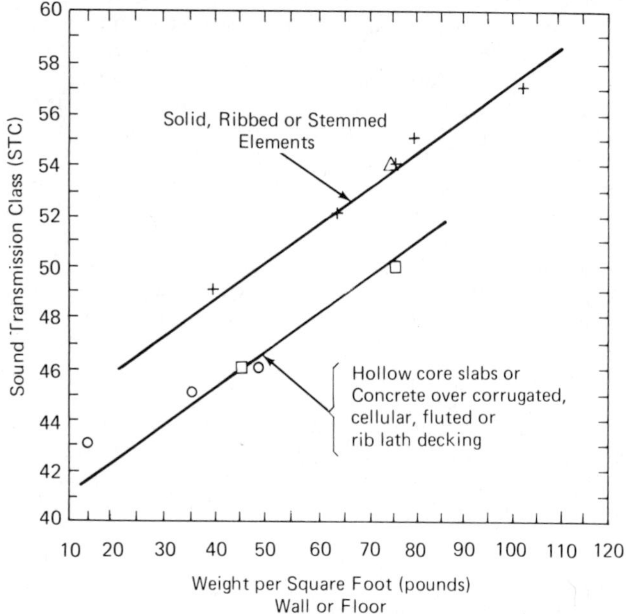

Fig. 6-50 Practical mass law curves.

TABLE 6-13 Typical Improvements with Floor and Ceiling Treatments*

Type of Treatment	*Change in Ratings*	
	Airborne (STC)	*Impact (INR or IIC)*
2 in. concrete topping, 24 psf	3	0
Standard 44 oz. carpet and 40 oz. pad	0	48
Other carpets and pads	0	44 to 56
Vinyl tile	0	3
½ in. wood block adhered to concrete	0	20
½ in. wood block and resilient fiber underlayment adhered to concrete	4	26
Floating concrete floor on fiberboard	7	15
Wood floor, sleepers on concrete	5	15
Wood floor on fiberboard	10	20
Acoustical ceiling resiliently mounted	5	27
Acoustical ceiling added to floor with carpet	5	10
Plaster or gypsum board ceiling resiliently mounted	10	8
Plaster or gypsum board ceiling with insulation in space above ceiling	13	13
Plaster direct to concrete	0	0

**From:* PCI Design Handbook—Precast and Prestressed Concrete, *Prestressed Concrete Institute, Chicago, 1971.*

which the mass law controls the sound transmission loss of most concrete walls and floors is not as great as the range which is important for sound control, (100 to 4000 Hz.)

When the STC's of numerous constructions are plotted against weight per square foot, a Practical Mass Law curve is obtained, Fig. 6-50. Concrete floor slabs cast on steel supporting members such as fluted or cellular steel decks or metal lath over bar joists perform acoustically very much like plain structural concrete floor slabs. However, dips result in STC's about three points lower than those of slabs not cast on steel supports of equal total weight.

When a single concrete wall is used as a sound barrier it is sometimes desirable to provide a resilient connection between the wall and the building frame to reduce the sound transmission to adjacent walls, floors and ceilings. A five to seven db improvement can result. Usually a one-half inch fiber board will provide the desired isolation. When a resiliently-mounted plaster or drywall surface finish is added to a concrete wall or ceiling, the STL is further improved. The effect of various floor and ceiling treatments can be predicted from results of previous tests. Table 6-13 indicates typical improvements. The improvements are usually additive, but in some cases the total effect may be slightly less than the sum. It should also be noted that carpets and acoustical ceilings can further reduce the sound level by absorbing sound in source or receiving room.

The performance of a building section with an otherwise adequate STL can be seriously reduced by a relatively small hole or any other path which allows sound to by pass the wall. Common flanking paths are openings around doors, pipes, electric boxes and air ducts. Suspended ceilings in rooms where walls do not extend from the ceiling to roof or floor above allow sound to travel to adjacent rooms.

6.12.3 Impact Noise

Footsteps, dragged chairs, dropped objects, slammed doors, and plumbing generate impact noises. Even when airborne sounds are adequately controlled there can be severe impact noise problems.

The test method used to evaluate systems for impact sound insulation is described in "Field and Laboratory Measurements of Air-Borne and Impact Sound Transmission," International Standards Organization Recommendation R140-1960. A standard "tapping machine" is used. The floor ceiling structure is tested by measuring the impact sound pressure levels (ISPL'S) in decibels in successive $^1/_3$-octave frequency bands of the noise generated in the room below by the standard tapping machine. A plot of the ISPL vs frequency data then is compared to standard contours similar to the STC used for airborne sound transmission. The single figure Impact Noise Rating (INR) is determined from FHA publication No. 750 "Impact Noise Control in Multifamily Dwellings," January 1963. In this method a higher positive INR means better impact noise isolation, e.g., a +5 rating is better than 0, which in turn is better than −14. The newer rating system, Impact Insulation Class (IIC), is described in the publication, "U.S. Department of Housing and Urban Development, FT/TS-24, Airborne, Impact, and Structure Borne Noise Control in Multi-family Dwellings," January, 1968. The IIC criteria were selected to avoid negative ratings. In most cases, the IIC is 51 points higher than the INR rating, but other differences in the systems result in a spread of about ±2 points about this difference. The IIC of a floor-ceiling assembly is usually of the same magnitude as the STC.

IIC's, like NRC's and STC's, should not be used as a complete substitute for performance data at specific frequencies. Sources of impact noise and acceptable noise spectra differ. In many instances, ISPL's at critical frequencies identify floor systems which give superior performance more accurately than IIC's.

The validity of both the INR and IIC ratings is still questioned by some architectural acousticians because the ISPL's produced by the ISO tapping machine differ from the ISPL's generated by footfalls. When tests are carried out using walking people as the sound source, bare concrete floors and concrete with asphalt or vinyl floor coverings are found to give somewhat more favorable performances than indicated by the INR or IIC ratings. They are still inferior, however, to floors with resilient coverings or other resilient components.

In general, thickness or weight per sq ft of concrete does not greatly affect the transmission of impact sounds as shown in the following table:

Thickness, in.	*Unit Weight of Concrete, pcf*	*INR*
5	79	-28
	114	-27
	144	-27
10	79	-23
	114	-21
	144	-20

The most important floor construction or surface treatment items in the insulation of impact sounds are the resiliency, damping, and method of attachment to the frame. Structural concrete floors in combination with resilient materials effectively control impact sounds. One simple solution consists of good carpeting on resilient padding. The overall efficiency varies according to the characteristics of the carpeting and padding such as resilience, thickness and weight. So called resilient flooring materials such as linoleum, rubber, asphalt vinyl, vinyl asbestos, etc. are not entirely satisfactory directly on concrete nor are parquet or strip wood floors when applied directly. Impact sounds also may be controlled by providing a discontinuity in the structure such as would be obtained by adding a resiliently-mounted plaster or drywall suspended ceiling or a floating floor consisting of a second layer of concrete cast over resilient pads, insulation boards or mastic. The thickness of floating slabs is usually controlled by structural requirements, however, eight pounds per sq ft is adequate in most instances.

The improvements to be made to a bare concrete floor having an INR of -27 or IIC of 24 by adding various materials or constructions are shown in Table 6-13.

Solid connections between concrete floor slab, frame and wall increase impact sound transmission at low frequencies. Impact sound would be reduced by soft joints between structural members or interior walls if structural stability is not impaired.

6.12.4 Vibration Isolation

The isolation of vibrations produced by equipment with unbalanced operating or starting forces can frequently be accomplished by mounting the equipment on a heavy concrete block placed on resilient supports. A slab of this type is called an inertia block. Inertia blocks can provide a desirable low center of gravity and compensate for thrusts such as those generated by large fans. For equipment with less unbalanced weight, a "housekeeping" slab is sometimes used below the resilient mounts to provide a rigid support for the mounts and to keep them above the floor where they remain cleaner and easier to inspect. This slab may also be mounted on pads of precompressed glass fiber.

The natural frequency of the total load on resilient mounts must be well below the frequency generated by the equipment. The required weight of an inertia block depends on the total weight of the machine and the unbalanced force. For a long-stroke compressor, five to seven times its weight might be needed. For high pressure fans, one to five times the fan weight is frequently sufficient.

Simplified theory shows that for 90% vibration isolation a single resiliently-supported mass should have a natural frequency about one-third the driving frequency. For 99% isolation the natural frequency should be one-tenth the driving frequency. Fortunately the natural frequency of a resiliently mounted mass is a function of the static deflection of the resilient supports; this makes determination of the natural frequency simple. A floor supporting resiliently mounted equipment must be stiff and heavy. If the static deflection of a floor supporting resilient mounts approaches the static deflection of the mounts, the floor then becomes a part of the vibrating system and little vibration isolation is achieved. Therefore, floors supporting heavy rotating or reciprocating equipment often must be stiffer than floors designed to meet ordinary structural design criteria with floor deflection not more than about $^1/_6$ or $^1/_8$ of mount deflections. For example: a 20-ft span, with the usual $^1/_{360}$ of span deflection, will deflect 0.67 in. at maximum allowable load. This is the static deflection required of isolation mounts for 90% isolation of an 800 CPM machine. To obtain 90% isolation on such a floor the static deflection of the resilient mounts has to be increased by six to eight times. However, floors are seldom loaded to the design limits and loading at center of span can often be avoided. With a moderate floor load good isolation is usually provided if the static deflection of the mounts is increased by an amount equal to $^1/_{360}$ of the floor span.

6.13 ELECTRICALLY CONDUCTIVE FLOORS

Concrete floors in hospital operating rooms and other places, where sparking would be a hazard, may be made electrically conductive with an integral acetylene carbon-black admixture, or with specially prepared metallic aggregates or powders which are worked into the fresh concrete surface. For terrazzo work, 2% acetylene carbon-black by weight of cement has been used satisfactorily in a marble-chip topping over a sand bedding course which also contains 3% acetylene carbon-black by weight of cement. Detailed installation recommendations for conductive terrazzo are available from the National Terrazzo and Mosaic Association. Construction techniques are somewhat specialized for conductive floors.

The electrical resistance of the floors may be determined by reference to the U.S. Navy Bureau of Yards and Docks Specification 48Y, "Static Disseminating and Spark-Resistant Floors" or to *Code for the Use of Flammable Anesthetics* (Recommended Safe Practice, for Hospital Operating Rooms), Publication No. 56, National Fire Protection Association, Boston Mass. Electrically conductive floors often are also required to be spark resistant under abrasion or impact; a typical test for this property is given in the Bureau of Yards and Docks specification.

Concrete made for conductive flooring gives standard test resistance of about 0.03 to 0.3 megohms; regular concretes (perhaps approximately equilibrated at 50% humidity) give resistances of 8 to 80 megohms, as determined at one fixed voltage. High alkali portland cements show lower resistance than low alkali cements, and at 0.9% alkalies the high alkali cements gave 8 to 11 megohms. At 15% relative humidity the regular concretes gave values about 5 to 8 times as high as at 50% humidity.

Cleaning procedures for conductive floors should not adversely affect the conductivity characteristics. It is preferable not to use any type of coating or wax on the floor unless it has been pretested to maintain conductivity. The floor should be pre-rinsed with plain water before each washing and only a non-ionic conductive floor cleaner should be used for washing. After each washing, the floor should be rinsed with plain water. Phenolic-based germicides should not be used as they may form insoluble salts with the calcium in the floor, resulting in the build-up of an insulating layer.

6.14 REINFORCEMENT

6.14.1 Reinforcing Bars*

U.S. standard specifications for reinforcing bars are established by the American Society for Testing and Materials (ASTM). These standards govern strength grades, rib patterns, sizes, and markings of bars. All bars are furnished "deformed"—that is, with lugs, ribs or protrusions rolled into the bar, which increase bond performance with concrete.

-1. Specifications and sizes—Reinforcing bars are produced from three kinds of steel—new billet, axle, or rail—in four grades of yield strengths. The yield strength of a bar is its useful strength. The three ASTM bar specifications are:

Specifications for Deformed Billet-Steel Bars for Concrete Reinforcement (ASTM A615),
Specifications for Rail-Steel Deformed Bars for Concrete Reinforcement (ASTM A616), and
Specifications for Axle-Steel Deformed Bars for Concrete Reinforcement (ASTM A617).

*Properties of welded wire fabric are not covered, for information refer to *Manual of Standard Practice—Welded Wire Fabric*, Wire Reinforcement Institute, 1972.

Table 6-14 summarizes the main properties of these bars. If steel from a foreign source is to be used, its quality should be checked carefully, since it may fall below ASTM requirements for uniformity; it may also tend to be more brittle, thereby introducing serious problems in bending, when welding, or during handling at temperatures below 40°F.

There are 11 standard deformed reinforcing bar sizes. These are designated #3 through #11, #14 and #18. The size of reinforcing steel bars is designated by a number which refers to its nominal diameter (equivalent to the diameter of a plain bar having the same weight per foot as the deformed bar). For bar sizes #3 through #8 this designation represents the nominal diameter of the bar in eighths of an inch, Table 6-15.

Manufacturers mark the bar size, producing mill identification and type of steel on the reinforcing bars. Only high strength steel bars, Grades 60 and 75, are also identified by a grade marking, Fig. 6-51.

-2. Stress-strain curves—Ultimate strength design of reinforced concrete sections in accordance with the 1971 and previous *ACI Codes* is based on the assumption of an ideal elasto-plastic (flat-top) stress-strain relationship for the reinforcement. However, the actual stress-strain relationship for high-strength reinforcement may not be ideally elasto-

TABLE 6-14 Physical Requirements for ASTM Deformed Reinforcing Bars (Modified by ACI 318)

Type of Steel and ASTM Specification	*Size Nos. Inclusive*	*Grade*	*Yield*[a] *min. psi*	*Tensile Strength min. psi*	*Elongation in 8 in. min. %*	*Cold-Bend Test*[b,c] *90°*
Billet Steel A615[d]	3-11 14 18	40	40,000	70,000	Note *e*	3, 4, 5 $d = 3t$ 6, 7, 8 $d = 4t$ 9, 10, 11 $d = 5t$
	3-11 14 18	60	60,000	90,000	Note *f*	3, 4, 5 $d = 4t$ 6, 7, 8 $d = 5t$ 9, 10, 11 $d = 6t$
	11, 14, 18	75	75,000	100,000	5	11 $d = 8t$
Rail Steel A616[d]	3-11	50	50,000	80,000	1,000,000[g] / tens. str.	Same as A615, Grade 60
	3-11	60	60,000	90,000	1,000,000[h] / tens. str.	Same as A615, Grade 60
Axle Steel A617[d]	3-11	40	40,000	70,000	Note *i*	Same as A615, Grade 40
	3-11	60	60,000	90,000	Note *j*	Same as A615, Grade 60

[a] *Yield point for A615 Grade 40. Yield strength for all others. Yield strength corresponds to that determined by tests on full-size bars and for reinforcing bars with a specified yield strength, f_y, exceeding 60,000 psi, f_y should be the stress corresponding to a strain of 0.35 percent.*
[b] *d = diameter of pin around which specimen is bent; t = nominal diameter of specimen.*
[c] *If #14 or #18 bars are to be bent, they should also be capable of being bent 90° at a minimum temperature of 60 F, around a 10-bar-diameter pin without cracking transverse to the axis of the bar.*
[d] *Weldability not a part of the specification; may be subject to agreement with supplier.*
[e] *Bar #3, 11%; 4-6, 12%; 7, 11%; 8, 10%; 9, 9%; 10, 8%; 11-18, 7%.*
[f] *Bar #3-6, 9%; 7-8, 8%; 9-18, 7%.*
[g] *Deduct 1% for bar #3. For #7, 8, 9, 10, 11 deduct respectively, 1, 2, 3, 4 and 5%, except that minimum elongation shall not be less than 5%.*
[h] *Limited to 5% on bar #3-8 and to 4½% on #9-11.*
[i] *Bar #3, 11%; 4-6, 12%; 7, 11%; 8, 10%; 9, 9%; 10, 8%; 11, 7%.*
[j] *Bar #3-7, 8%; 8-11, 7%.*

AMERICAN STANDARD BAR MARKS

Lower Strength Bars Show Only 3 Marks (no grade mark):
- 1st - Producing Mill (usually an initial)
- 2nd - Bar Size Number (# 3 through # 18)
- 3rd - Type (N for New Billet; A for Axle; I for Rail)

High Strength Bars Must Also Show Grade Marks:
- 60 or One (1) Line for 60,000psi Strength
- 75 or Two (2) Lines for 75,000psi Strength

(Grade mark lines are smaller and between the two main ribs which are on opposite sides of all American Bars)

CONTINUOUS LINE SYSTEM-GRADE MARKS

Main Ribs
Initial of Producing Mill
Bar Size # 11
Type Steel (New Billet)
H
II
N
Two Lines
One Line
Grade Marks
GRADE 40
GRADE 50
GRADE 60
GRADE 75

NUMBER SYSTEM-GRADE MARKS

Main Ribs
Initial of Producing Mill
Bar Size # 11
Type Steel (New Billet)
H
II
N
60
75
Grade Mark
GRADE 40
GRADE 50
GRADE 60
GRADE 75

Fig. 6-51 Bar markings. (From: Concrete Reinforcing Steel Institute.)

TABLE 6-15 ASTM Standard Reinforcing Bars

Bar Size Designation	*Weight Pounds, per ft*	*Nominal Dimensions–Round Sections*		
		Diameter, in.	*Cross-Sectional Area, sq. in.*	*Perimeter, in.*
#3	0.376	0.375	0.11	1.178
#4	0.668	0.500	0.20	1.571
#5	1.043	0.625	0.31	1.963
#6	1.502	0.750	0.44	2.356
#7	2.044	0.875	0.60	2.749
#8	2.670	1.000	0.79	3.142
#9	3.400	1.128	1.00	3.544
#10	4.303	1.270	1.27	3.990
#11	5.313	1.410	1.56	4.430
#14	7.65	1.693	2.25	5.32
#18	13.60	2.257	4.00	7.09

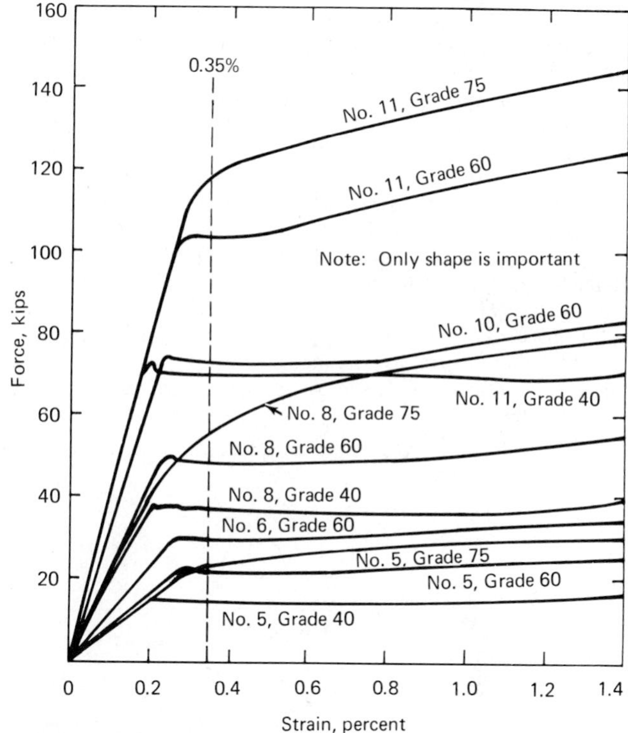

Fig. 6-52 Representative force-strain curves for test bars. (From: Hanson, J. M., et al., "Fatigue Strength of High-Yield Reinforcing Bars," *NCHRP Report Project No. 4-7*, PCA, Feb. 1970.)

plastic. Some reinforcing bars, particularly those meeting the ASTM specifications for Grade 75 steel, and some European and British steels do not always have a well-defined yield point and a flat-plateau. Moreover, even those steels which do have a well-defined yield point, including most Grade 40 and some Grade 60 steels, exhibit strain-hardening; some of these have a very short flat-plateau, and others a considerably long one.

Representative force-strain curves for five sizes in each grade of steel are shown in Fig. 6-52. While the curves for the Grade 60 bars are typically shown as having a flat yield plateau, some of the curves for these bars exhibit characteristics of the curves for the Grade 75 bars. The modulus of elasticity, E_s, for all reinforcing bars is practically the same and is taken as $E_s = 29 \times 10^6$ psi.

-3. Bending—Some of the specified minimum bend radii permitted by the American Concrete Institute's *Building Code Requirements for Reinforced Concrete* (ACI 318-71) are more severe than those required by applicable ASTM specifications. In such cases, occasional breakage can be expected unless the steel, by chance, is more ductile than required by the specifications. In any case, breakage may be expected when bending is attempted at low temperatures. Heating of bars to facilitate bending or straightening should be allowed only with the permission of designer but may not be detrimental provided the temperature of the steel is not permitted to exceed 600 to 800°F., and provided the cooling rate is slow and uniform (normal still-air cooling).

Rapid quenching of heated bars, such as might occur if the bars were dropped in snow or cold water, is detrimental to the steel.

-4. Fatigue—Fatigue of steel reinforcing bars has not been a significant factor in their application as reinforcement in concrete structures. Fatigue strength of deformed bars has been investigated by testing samples of the bar, either by themselves in axial tension, or embedded in concrete beams in tension or flexural bending. The fatigue strength of a bar is a hypothetical value of stress for failure at exactly N cycles as determined from an S-N diagram, see section 6.9. The hypothetical value of S referred to will be the stress range, S_r, at which 50% of the specimens of a given sample would be expected to survive N cycles. Disagreement exists as to whether a bar embedded in concrete has a lower or higher fatigue strength than one tested in air. However, there are indications that any difference should be small if the bar has transverse lugs meeting the requirements of ASTM Designation A615-68 and good bond is obtained between the steel and concrete.

Reinforcing bars have a practical fatigue limit. The transition from the finite-life region to the long-life region of the stress range-fatigue life relationship occurs at about 1 million cycles, see Fig. 6-53. Stress range is the predominant variable affecting the fatigue strength of the test bars in the finite-life region, while stress range accounts for only a small part of the total variation in fatigue life of bars included in the long-life region. Therefore, that fatigue limit, for practical purposes, could be considered as a stress range at which an infinite number of loadings could be sustained.

Design provisions for the fatigue life of reinforcing bars should be based on stress range and minimum stress level. They should not be based on grade of bar or bar size. Fatigue life is relatively insensitive to yield or tensile strength of bar and slightly decreases with increasing bar size.

The stress range in straight deformed hot-rolled reinforcing bars, under repeated application of service loads, should not be permitted to exceed the value given by the following equation unless a higher value is substantiated by tests.

$$S_r = [40 - 19(\log N - 5)]\left(1.13 - 0.67\frac{S_{min}}{f_y}\right)$$

where

S_r = stress range, in ksi
N = design fatigue life
S_{min} = minimum stress level
f_y = specified yield strength of bar

The value of S_r at N equal to 1 million cycles may be used for all greater values of N. At a minimum stress level of 0.2 of the yield strength of Grade 60 bars, the stress range at one million cycles computed from the equation is 21 ksi. This value of fatigue strength should be considered as the fatigue limit for these conditions. These design recommendations are conservative for most North American made bars.

A limited amount of information available regarding the effect of bends—from tests on bars made by one manufacturer—indicates that the fatigue strength of deformed bars with bends in a maximum stress region may be reduced by roughly one-half. While the practical answer to this problem is simply to avoid bends in high-stress regions, further research may be needed to assess the fatigue strength of commonly used bend details.

Except for stress range, most of the variables which designers can readily control—bar size, type of beam, mini-

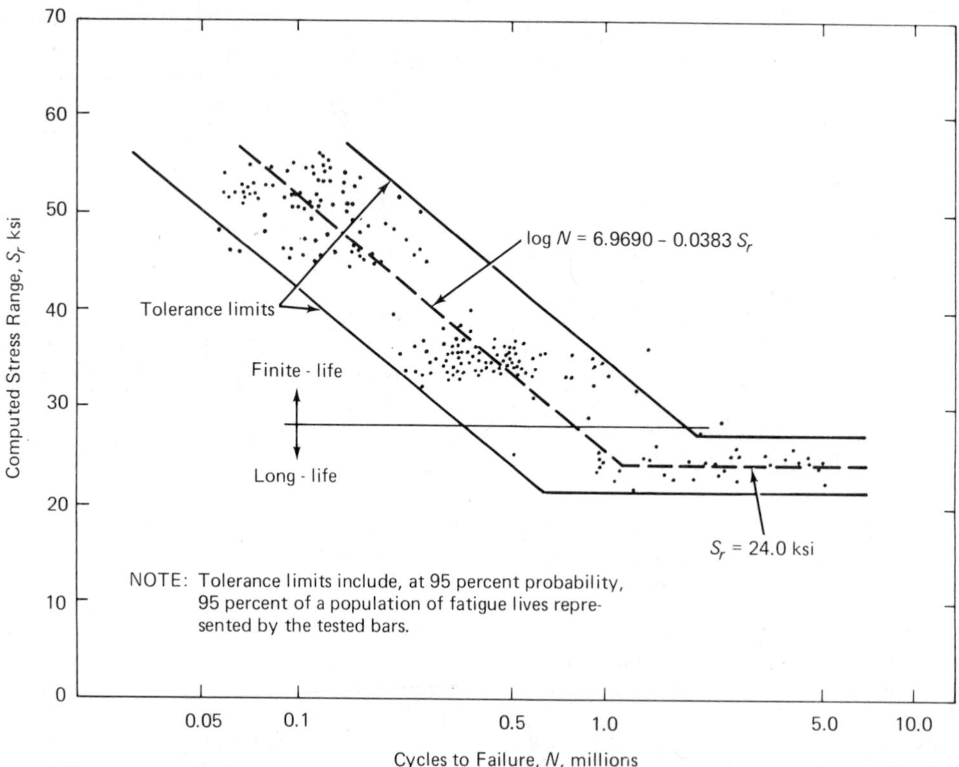

Fig. 6-53 Experimental test data on fatigue of deformed bars from a single U.S. manufacturer. Lower limit is also lower limit to previously published test data on North American-made bars. (From: Hanson, et al., "Fatigue Strength of High-Yield Reinforcing Bars," *NCHRP Report Project No. 4-7*, PCA, Feb. 1970.)

mum stress, bar orientation and grade of bar—do not have a large effect on fatigue strength. Other variables related to reinforcement manufacture and fabrication—deformation geometry, bending, and tack welding—are much more significant.

6.14.2 Prestressing Tendons

High-tensile steel is almost the universal material for producing prestress and supplying the compressive force to prestressed concrete. High-strength tendons are required to allow for stress losses due to inelastic deformation of the concrete and other causes. The tendon should behave elastically (stress-strain curve is straight line) up to at least the initial tensioning force, so that its behavior as a tendon is predictable and so that the measurement of elongation can be used to check the force in the tendon. A tendon should also have a large amount of plastic deformation before failure to provide a steel which cannot fail abruptly without prior indication of distress.

High-tensile steel for prestressing (pre-tensioned and post-tensioned) takes one of three forms: wires, strands, or bars. The strength and bond of wire decreases with increasing diameter, leading to practical limits of about $^9/_{32}$ in. diameter in prestressing applications. For post-tensioning, wires are grouped in parallel into tendons utilizing as many as 186, $^1/_4$ in. wires. Strands are fabricated in the factory by twisting wires together. Strands also can be grouped into multi-strand tendons. This decreases the number of units to be handled in tensioning operations and may reduce anchorage costs. By grouping strands it is also possible to obtain large steel cross sections. Strands have almost completely replaced wires in pretensioning operations. Heat-treated alloy steel bars are used for post-tensioning in many applications where the advantages of large diameter offset the reduced strength capability.

The stress-strain characteristics of wire in the as-drawn condition are nonelastic (Fig. 6-54) and the wire has a

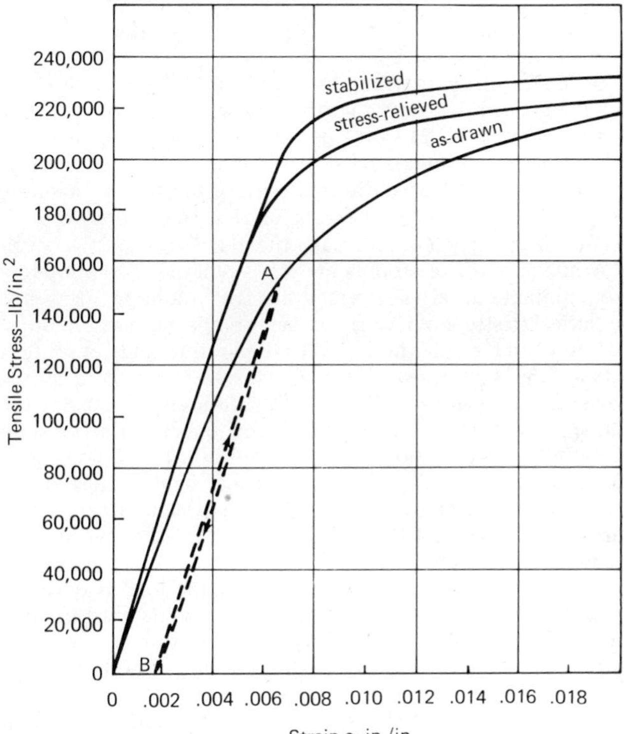

Fig. 6-54 Effect of postdrawing treatments on wire.

permanent curvature. To overcome these disadvantages the wire is mechanically straightened and then subjected to low-temperature heat-treatment (about 600°F). This process, referred to as stress relieving, causes the wire to behave elastically up to a high proportion of the maximum stress. The as-drawn wire can also be subjected to combined low-temperature heat-treatment and high tension. This process, known as stabilization, improves the maximum strength and the elongation at the maximum stress and typically reduces losses due to relaxation.

There is no yield point for prestressing steels as exhibited in the curve for structural carbon steel; at least, there is no marked yield point before the ultimate strength is reached. Instead, between the proportional limit and the ultimate strength, the elongation of the steel takes place in a regularly increasing manner, without any point where strain increases suddenly without a corresponding increase in stress. However, the idea of yield point strength is so well established in thinking of carbon steel members that an arbitrary means of fixing a yield strength, for comparing high strength steels of the sort used for prestressing and for use in design equations, has been established. This is generally called the 0.2% offset yield strength. Specifications for prestressing wire and strand require that the yield strength be measured by the 1.0% extension under load method.

-1. Specifications—Wire and strands for tendons in prestressed concrete conform to "Specifications for Uncoated Seven-Wire Stress-Relieved Strand for Prestressed Concrete" (ASTM A416) or "Specifications for Uncoated Stress-Relieved Wire for Prestressed Concrete" (ASTM A421). Strands or wire not specifically itemized in ASTM A416 or A421 may be used provided they conform to the minimum requirements of these specifications and have no properties which make them less satisfactory than those listed in ASTM A416 or A421. Wires used in making strands should be cold-drawn and either stress-relieved, in the case of uncoated strands, or hot-dip galvanized, in the case of galvanized strands.

Heat-treated cold-worked alloy bars are made from two alloy steels which are commerically classified as AISI 9260 and AISI 5160 and ASTM specifications A29 and A322 are often applied to the steel. High strength alloy steel bars for post-tensioning tendons are proof-stressed (cold stretched) during manufacture to 85% of the minimum guaranteed tensile strength to eliminate any bars having surface imperfections or metallurgical defects. After proof-stressing, bars are subjected to a stress-relieving heat treatment to produce the prescribed physical properties. After processing, the physical properties of the bars when tested on full sections, should conform to the following minimum properties:

Yield strength (0.2% offset)	
Smooth bars	$0.87 f_{pu}$
Deformed bars	$0.85 f_{pu}$
Elongation at rupture in 20 diameters	4%
Reduction of area at rupture	20%

-2. Wire—The stress-strain characteristics in typical tensile tests of uncoated stress relieved wires are shown in Fig. 6-55. Galvanized wire of 0.196 in. diameter with a minimum ultimate strength of 220,000 psi has also been used in prestressed concrete, although other sizes also can be manufactured. The modulus of elasticity of uncoated wire is approximately 29×10^6 psi. While this value is sufficiently accurate for some design calculations, it is recommended that the steel fabricator's load-elongation curve for the particular wire be used in computing the elongation required in jacking wires to a specified tension.

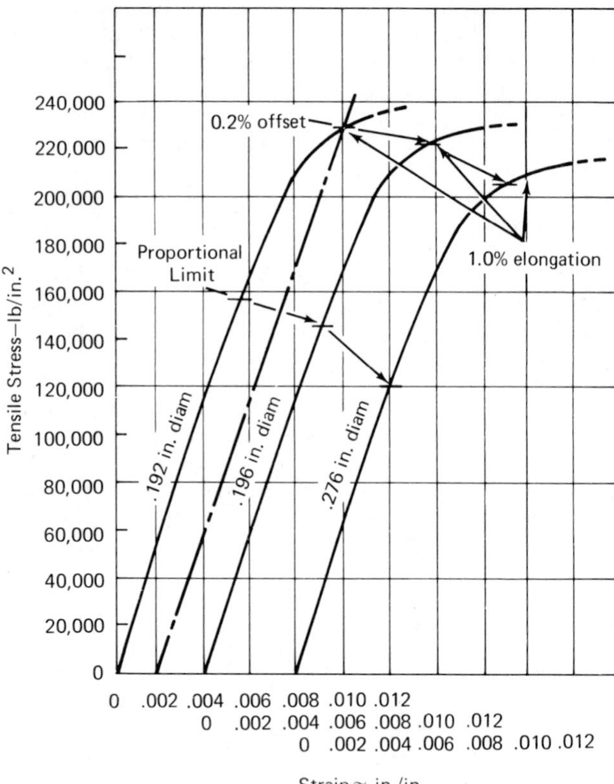

Fig. 6-55 Typical tensile tests in elastic region on uncoated stress-relieved wire for prestressed concrete.

Table 6-16 shows a number of properties of prestressing wire used in the U.S. Wires are manufactured according to the U.S. Steel Wire Gage, No. 2 of which has a diameter of 0.2625 in. and No. 6 has a diameter of 0.1920 in. Neither of these is the exact equivalent of the millimeter counterparts. Hence, when the European types of anchorages are adopted, 0.276-in. and 0.196-in. wires are often specified. For post-tensioning systems developed in the United States, 1/4-in. wires have been most commonly incorporated.

-3. Strand—When strand is loaded, a considerable constructional stretch (a combination of direct tension, shear, torsion and bending stresses) occurs due to the tendency of the helical wires to straighten and the resulting compaction of the strand. Since a strand with a long lay will have an appreciably higher E_s than a strand with a short lay, specifications require that the "strand have a uniform lay"; i.e., the lay shall be constant for the full length of the strand. ("Lay" is the distance along the strand in which an outer wire makes one complete turn around the center wire.) For a given stress, a greater elongation is obtained than with a solid wire, and this gives a lower modulus of elasticity. For approximate calculations, a modulus of elasticity of 27×10^6 psi is often used for 250K grade and 28×10^6 psi for 270K grade.

Typical load-strain curves for 270 ksi seven-wire stress-relieved and stabilized strands are shown in Fig. 6-56, which are also typical for strands of all sizes. The properties of the various strands available in the U.S. are shown in Table 6-16.

For post-tensioning, multi strand tendons (with 1/2 in. or 0.6 in. strands) considerably larger than those listed in Table 6-16, 1 11/16 in. in diameter and above, are often employed. Beyond this size the problems of strand stiffness, heavy jacking equipment, etc., more than offset any saving due to handling fewer strands. In those few cases where larger strands must be used to get a large force into a small space, the strand fabricator should be consulted for advice on the best size to choose.

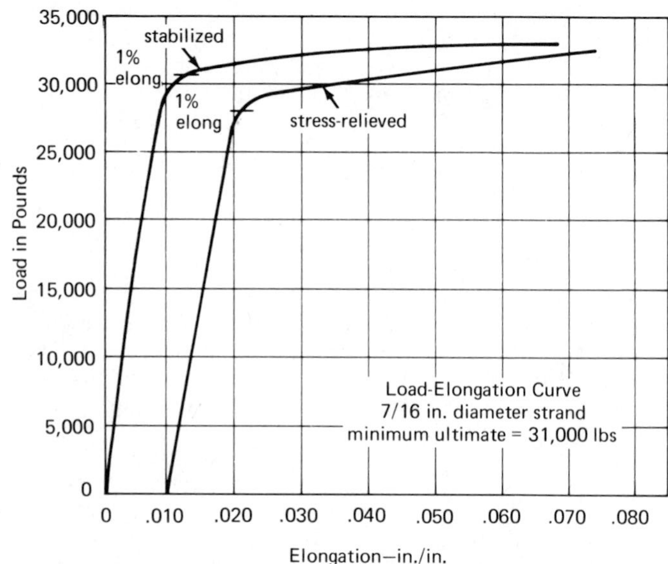

Fig. 6-56 Typical load-elongation curve for 7/16-in. diameter stress-relieved and stabilized strand.

A galvanized prestressed concrete strand is composed of seven or more hot-dip galvanized wires. The temperature of the zinc bath and the speed of the wire passing through it are regulated so that the wire is stress-relieved as it is galvanized. Galvanizing can result in hydrogen embrittlement of prestressing steel and therefore its use in structures where fatigue is a consideration is not recommended. The smallest seven-wire galvanized strand listed in Table 6-16 is 0.600 in. diameter. Larger strands are made by adding extra layers of wire. A 1-in. diameter strand is a 0.600-in. diameter seven-wire strand with one more layer composed of 12 wires for a total of 19 wires. Different sizes are made by varying either the number or size of the wires or both to produce the most economical strand. However, very little galvanized strand is being used today.

When the ultimate strengths given in Table 6-16 are divided by the areas, it is found that the stresses at ultimate load are only 202,000 to 214,000 psi instead of the much higher stresses given for uncoated strand. This is partly because it is standard practice to list the gross area of the strand including the zinc, even though the zinc carries practically no load. Galvanizing may also increase the relaxation of the steel.

While seven-wire strands are stress-relieved after stranding to eliminate internal stresses due to stranding, it is not practical to stress-relieve the larger strands because the outer wires would reach the critical temperature and revert to a crystalline structure before the inside wires were stress-relieved. To achieve uniform elongation characteristics the large strands are prestretched. In this operation a full shop length (usually 3,600 ft) is stretched in a tensioning rig to approximately 70% of its ultimate and held at that load a short time. During this period of high stress the fibers which are stressed beyond their yield point because of the applied load added to internal stresses will continue to elongate until they reach a point of stability. After the load is released, the strand has good elastic properties up to its prestretching load. The modulus of elasticity of these strands, when prestretched, ranges between 24,000,000 and 26,000,000 psi.

-4. Bars—A typical stress-strain curve for bars is shown in Fig. 6-57 from which it can be noticed that a constant modulus of elasticity exists only for a limited range (up to

TABLE 6-16 Properties of Prestressing Tendons*

Prestressing Wire

Diameter	0.105	0.120	0.135	0.148	0.162	0.177	0.192	0.196	0.250	0.276
Area, sq in.	0.0087	0.0114	0.0143	0.0173	0.0206	0.0246	0.0289	0.0302	0.0491	0.0598
Weight, plf	0.030	0.039	0.049	0.059	0.070	0.083	0.098	0.10	0.17	0.20
Ult. strength, f_{pu}, ksi	279	273	268	263	259	255	250	250	240	235

Seven Wire Strand, f_{pu} = 270 ksi

Nominal Diameter, in.	3/8	7/16	1/2	9/16	0.600
Area, sq in.	0.085	0.115	0.153	0.192	0.215
Weight, plf	0.29	0.40	0.53	0.65	0.74

Seven Wire Strand, f_{pu} = 250 ksi

Nominal Diameter, in.	1/4	5/16	3/8	7/16	1/2	0.600
Area, sq in.	0.036	0.058	0.080	0.108	0.144	0.215
Weight, plf	0.12	0.20	0.27	0.37	0.49	0.74

Three and Four Wire Strand, f_{pu} = 250 ksi

Nominal Diameter, in.	1/4	5/16	3/8	7/16
No. of wire	3	3	3	4
Area, sq in.	0.036	0.058	0.075	0.106
Weight, plf	0.13	0.20	0.26	0.36

Galvanized Prestressing Strands

Diameter, in.	0.600	0.835	1.000	1 1/8	1 1/4	1 3/8	1 1/2	1 9/16	1 5/8	1 11/16
Area, sq in.	0.215	0.409	0.577	0.751	0.931	1.12	1.36	1.48	1.60	1.73
Weight, plf	0.737	1.412	2.00	2.61	3.22	3.89	4.70	5.11	5.52	5.98
Ult. strength, f_{pu}, ksi	46	86	122	156	192	232	276	300	324	352

Smooth Prestressing Bars, f_{pu} = 145 ksi

Nominal Diameter, in.	3/4	7/8	1.000	1 1/8	1 1/4	1 3/8
Area, sq in.	0.442	0.601	0.785	0.994	1.227	1.485
Weight, plf	1.50	2.04	2.67	3.38	4.17	5.05

Smooth Prestressing Bars, f_{pu} = 160 ksi

Nominal Diameter, in.	3/4	7/8	1.000	1 1/8	1 1/4	1 3/8
Area, sq in.	0.442	0.601	0.785	0.994	1.227	1.485
Weight, plf	1.50	2.04	2.67	3.38	4.17	5.05

Deformed Prestressing Bars

Nominal Diameter, in.	5/8	1.000	1 1/4	1 1/2
Area, sq in.	0.307	0.852	1.295	1.63
Weight, plf	1.06	2.96	4.55	5.74
Ult. strength, f_{pu}, ksi	215	150	150	150

**From: Adapted from* PCI Design Handbook—Precast and Prestressed Concrete, *Prestressed Concrete Institute, Chicago, 1971.*

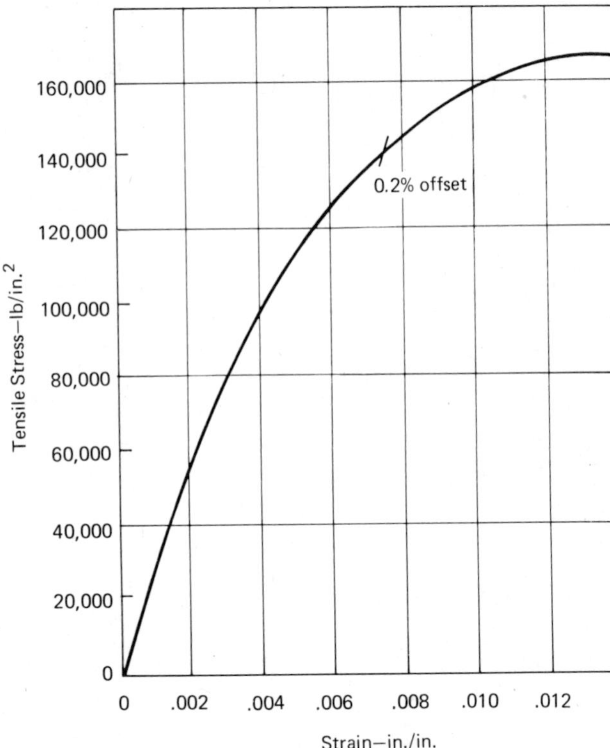

Fig. 6-57 Typical tensile test in elastic region for high-strength alloy bars.

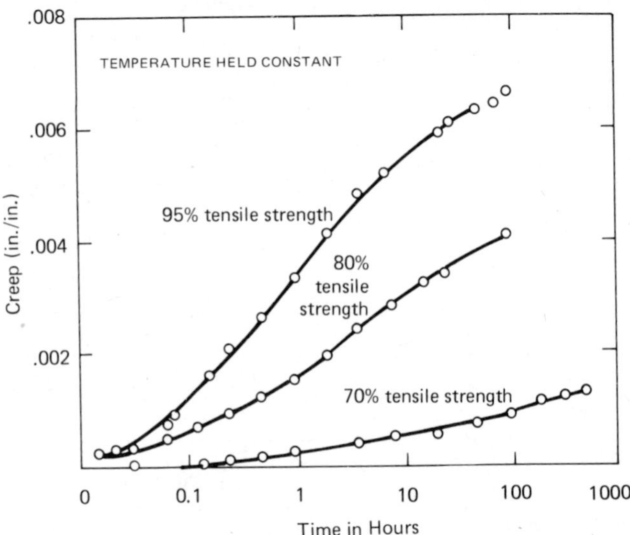

Fig. 6-58 Creep of stress-relieved wire loaded to different stress levels. (From: Podolny, W. Jr., and Melville, T., "Understanding the Relaxation in Prestressing," *PCI Journal*, **14** (4), 43–54, Aug. 1969.)

about 80,000 psi stress) with a value between 25 and 28 $\times$ 10^6 psi. Some specifications determine the yield strength on the basis of 0.7% extension. Statistically the values so obtained are almost identical to those determined by the 0.2% offset method. Common sizes and properties of high strength bars are listed in Table 6-16. Although only about 60% as strong per unit cross section, bars compete successfully with wire and strand for a range of applications.

-5. Creep and relaxation—In the design of prestressed concrete structures, a very important consideration is the loss of initial prestress force that may be expected. Losses are due to various factors which are part of the stressing operation as well as to the immediate and long-term strains in the structural materials.

The creep behavior of steel wires under long-term tensile stress is measured by one of two methods: constant stress (creep) or constant strain (relaxation). The creep phenomenon in steel varies with chemical composition, as well as with mechanical and thermal treatment applied during the manufacturing process.

Stress-relieved wire will show no appreciable creep when stressed up to 50% of its tensile strength. Beyond 50%, creep begins to become noticeable, Fig. 6-58, and gradually increases as the stress approaches tensile strength. The rule governing the rate of creep of stress-relieved high tensile wire is more complicated than a straight line plot on a semi-logarithmic scale. At low stresses, the curves are almost straight, becoming parabolic within the time span investigated. At high stresses the curves are hyperbolic. The S-shaped curve indicates that the rate of creep reaches an inflection point and then tends to level off. Tests over longer periods of time indicate that the S-shaped curve (hyperbolic) also occurs at lower stress levels but over much longer periods of time.

Because of the logarithmic time nature of creep phenomenon in wire and a doubt as to its validity when applied to conditions prevailing in prestressed concrete, a more valid approach to the creep behavior of prestressing steels for the designer is the measurement of stress relaxation when the wire is held at constant strain. As might be expected, relaxation, like creep, follows a substantially straight logarithmic law at normal stress and temperature.

For a given steel, the rate of relaxation is a function of initial stress, temperature and duration of load application. At normal initial stress levels and temperatures, relaxation is predictable and is of relatively minor significance in terms of other losses imposed upon the structure or member. At elevated temperatures, the losses due to relaxation of the prestressing steel is of greater significance. The engineer must apply relaxation data to the actual conditions applying to the structure he is designing with rational reasoning and judgment. For all practical purposes, in prestressed concrete, 1000 hours seems to be a practical measure of an end point. If one considers that the shrinkage, creep and elastic shortening of the concrete itself will reduce rather quickly the initial tension in the steel, and the approximate logarithmic nature of the steel relaxation, nothing very important can be expected to happen after 1000 hours.

In examining the 1000-hour value of stress relaxation for stress-relieved wires, Fig. 6-59, it is evident that relaxation is very small up to values of about 55% of ultimate strength. Beyond this point the rate of relaxation gradually increases, but reaches a peak at about 80% of tensile strength and then becomes less beyond that point. A stabilized wire or strand has considerably less relaxation than the stress-relieved tendon as shown in Fig. 6-60.

Tendons composed of stress-relieved wires have relaxation losses of about the same magnitude as stress-relieved strand. Relaxation of strand is greater than in the straight constituent wire due to the combined stress relaxation in the helical wires. In tendons at elevated temperatures, Fig. 6-61, or subjected to large lateral loads, relaxation loss is greater. Increasing temperatures have the same effect as increasing the initial value of stress.

In certain types of structures, where there may be considerable prestress losses due to causes other than relaxation, such as slip at anchorage, creep, shrinkage, elastic shortening of the concrete, and frictional loss due to intended or unintended curvature in the tendons, it may become economically feasible to re-stress tendons after some period of time. Relaxation losses may be considerably less after re-tensioning.

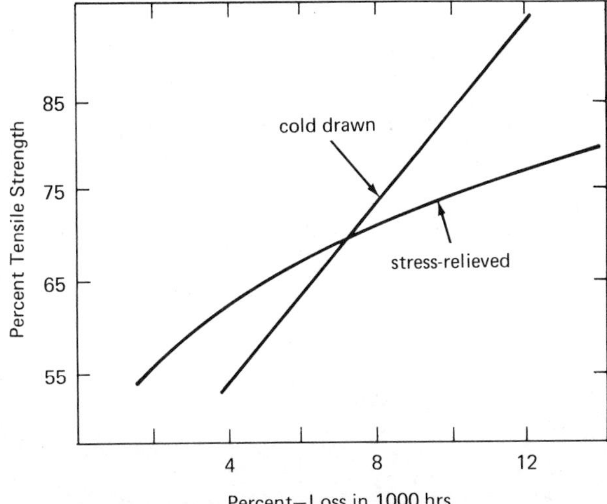

Fig. 6-59 Relaxation of high-strength, 0.196-in. diameter wires at various stress levels. (From: Podolny, W., "Understanding the Steel in Prestressing," *PCI Journal*, **12** (5), 54–56, Oct. 1967.)

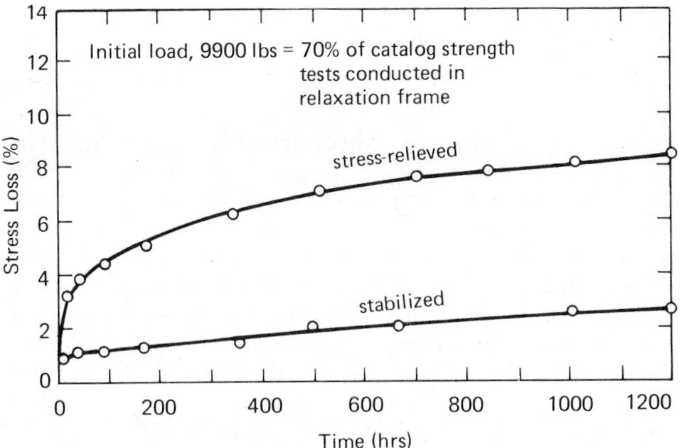

Fig. 6-60 Relaxation test in stress-relieved and stabilized, 0.276-in. diameter, prestressed-concrete wire.

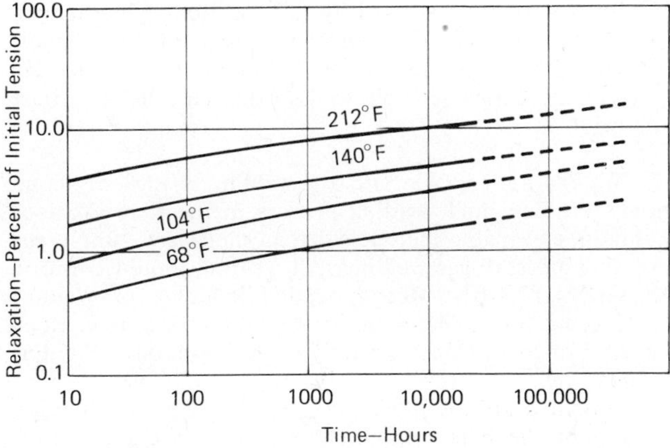

Fig. 6-61 Stress loss due to relaxation of stabilized strand at various temperatures. Initial tension = 70% of guaranteed ultimate strength.

-6. *Fatigue*—If the precompression in a prestressed concrete member is sufficient to ensure an uncracked section throughout the service life of the member, the fatigue characteristics of the prestressing steel and anchorages are not likely to be critical design factors. Consequently, fatigue considerations have not been a major factor in either the specification of steel for prestressed concrete or the development of anchorage systems. Cracks presumably cause severe stress concentrations in the concrete and in the strands contiguous to the cracks rendering them vulnerable to fatigue. It, therefore, appears necessary and desirable to limit the nominal tensile stress in the precompressed tensile zone to $6\sqrt{f'_c}$. Creep and shrinkage affect fatigue in that the prestress may be reduced, and thus the cracking load of a structure is also reduced—the stress range in the prestressing steel may be increased. Depressed strands (although stress in strands is increased by bending at hold-down points) do not cause a reduction in fatigue life.

Tests have indicated that beams partially prestressed with pre-tensioned strand show no detrimental effects from repeated loads. However, more fatigue tests to evaluate bond fatigue are warranted, especially on partially prestressed beams and beams with strand anchorages away from zones of normal concrete compressive stress.

The characteristics of the anchorage in a post-tensioning system and not the prestressing system, control the fatigue characteristics of the unbonded tendon. For unbonded construction, stress changes in the prestressing steel are transmitted directly to the anchorage. Although most anchorages can develop the static strength of the prestressing steel, they are unlikely to develop its fatigue strength. Further, bending at an anchorage can cause higher local stresses than those calculated from the tensile pull in the prestressing steel. Bending is likely where the prestressing steel is connected to the member at a few locations only throughout its length or where there is angularity of the prestressing steel at the anchorage. Fatigue characteristics based on tests of single wire or strand anchorages are likely to overestimate the strength of multi-wire or multi-strand anchorages.

The fatigue characteristics of wire vary greatly with the manufacturing process, the tensile strength of the wire and the presence or type of ribs, (prevalent on European wires).

The fatigue life of the high strength ½-in., 270 ksi, seven-wire prestressing strand compares favorably with that of the conventional 7/16 in., 250 ksi strand, as shown in Fig. 6-62. At stresses in the working load range, the fatigue life of a specimen was in the vicinity of approximately 2,000,000 cycles of loading. There are indications of a decrease in fatigue strength with increasing wire size in the strand. Some prestressed concrete flexural members have shown that the

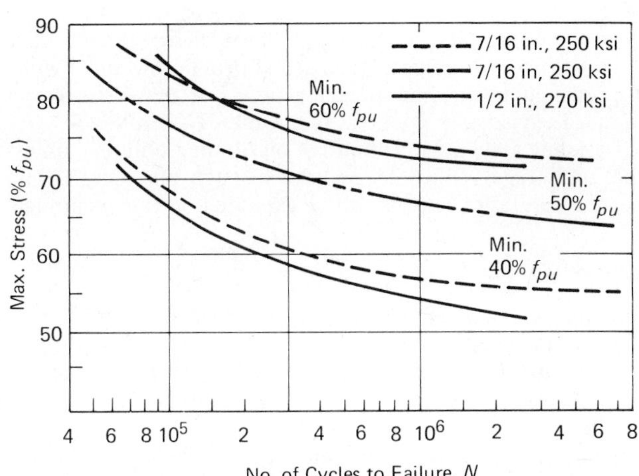

Fig. 6-62 Summary of fatigue tests on seven-wire, prestressing strand of U.S. manufacture. (From: Tide R. H. R., and Van Horn D. A., "A Statistical Study of the Static and Fatigue Properties of High Strength Prestressing Strand," Lehigh Univ., Fritz Eng. Lab. Report 309.2, June 1966.)

prestressing strand failed in fatigue at a significantly lower number of cycles when cracks in concrete cross the strand.

The fatigue life increases slightly as the temperature of the test specimen is reduced from 70°F to 0°F. This temperature effect is somewhat in agreement with the metallurgical concept which indicates an increase in fatigue life with lowered temperature, when the temperatures considered fall between 500°F and -250°F. The range of temperatures considered has not been great enough to produce a conclusive result.

Unless test data is available to justify higher values, the stress range in prestressed reinforcement should not exceed the following:

strand and bars	$0.10 f_s'$
wires	$0.12 f_s'$

The following specification is proposed for testing the capability of an unbonded tendon assembly (prestressing steel and anchorages) in resisting cyclic loading:

A dynamic test shall be performed on a representative specimen and the tendon shall withstand, without failure, 500,000 cycles of stressing varying from 60% to 66% of its minimum specified ultimate strength, and when the structure will be subject to earthquake loading another specimen shall withstand without failure 50 cycles of loading, corresponding to the following percentages of the minimum specified ultimate strength–

$$60 \pm \frac{2000}{L + 100}$$

where L is the length in feet of the tendon to be used in the structure. The period of each cycle involves the change from the lower stress level to the upper stress level and return to the lower.

Systems utilizing multiple strands, wires or bars may be tested utilizing a test tendon of smaller capacity than the full-size tendon. The test tendon shall duplicate the behavior of the full-size tendon and generally shall not have less than 10% of the capacity of the full-size tendon. It is not necessary to require tests on samples taken from the material being used for the specific project, tests from lots made to the same specifications should be sufficient. With the recent advent of very large post-tensioning tendons it was recognized that the availability of testing machines capable of performing a true fatigue test of full capacity tendons was very uncertain. The main feature to be tested is the possible effect of the anchorage in reducing the fatigue strength of the prestressing steel itself. The design of the different anchorage components is a separate mechanical problem, which can be approached in an analytical manner. For these reasons, a reduced capacity fatigue test is considered acceptable.

Dynamic tests are not required on bonded tendons, unless the anchorage is located or used in such manner that repeated load applications can be expected on the anchorage.

6.14.3 Splices

Splicing of reinforcing bars is only for the purpose of providing continuity. Splices are needed because of (1) limitations set up by the physical length of the bar, (2) transitions from a larger to a smaller bar, and (3) construction requirements, i.e., construction joints.*

The choice of methods available for splicing reinforcing bars has been widened in recent years. The traditional methods are lap splice by bond and welded lap or butt splices. Newer methods include couplers (for tension or compression) and end-bearing (for compression only). Semiautomatic butt-welded splices may be performed with commercially available equipment.

The entire responsibility for design, specifications and performance of splices ultimately rests upon the Structural Engineer. Only the man familiar with the structural analysis, probable construction conditions, and final conditions of service can properly evaluate the variables to select the most efficient and economical splice method. The drawings, notes, and specifications should clearly show or describe all splice locations, types permitted or required, and, for lap splices, lengths of lap.

There is no such thing as a perfect splice or a perfect splicing system. Therefore, the designing engineer must carefully consider these points.

a. Decide on a splice system to meet his design requirements.
b. Check availability for that particular project.
c. Cost of material, supplies and tools.
d. Cost of labor, insurance, taxes, etc.
e. Equipment needed–welding machines, temporary electrical supply, etc.
f. Time to make splices–for example one floor.
g. Inspection under job conditions.
h. Cost of inspection.
i. Union regulations.

-1. Lap splices–Lapped splices may be effected with bars either spaced or in contact. For bar to bar splices, contact splices are preferred for the practical reasons that, wired together, they are more easily secured against displacement during concreting. Caution must be exercised against designing lap spliced bars spaced too widely apart, permitting a zigzag crack across an unreinforced section. Spacing of bars in spaced lap splices shall not exceed one-fifth the lap length nor 6 in. Splices are usually staggered to prevent concentration in one plane, especially in seismic areas. Unless staggered, splices capable of carrying some tension are required to meet increasingly severe criteria. Lapped splices are generally used for bar sizes #11 and smaller, except that in columns #11 bars are frequently butt spliced. If lapped splices are employed in critical tension, increased lap lengths or spirals must be provided at this location. When lapped splices are provided, the bars that extend up from below come inside of the bars above. Code requirements, for lap splice length should be followed so that all spliced bars are fully utilized.

-2. Welded splices–Electric arc welding is the only commonly used manual welding process in the field. Various techniques available include shielded metal arc, submerged arc, and pressure gas welding. All welding should conform to AWS D12.1-61 "Recommended Practice for Welding Reinforcing Steel, Metal Inserts and Connections in Reinforced Concrete Construction" of the American Welding Society.

For welded splices it is necessary to know the chemical analysis of the bars being welded and use the appropriate welding procedures. The important consideration is that the specified welding procedure, including method, material, amount of preheat, if any, etc., be compatible with the chemical analysis. Bars with a carbon content exceeding about 0.35% cannot *easily* be welded in the field. Arc welding should be prohibited on bars with carbon equivalent greater than 0.55%. Bars are furnished to ASTM physical specifications only, and the chemical analysis can vary widely. This variation requires careful control of procedures at the job

*Most of the discussion on splices is based on "Reinforcing Bar Splices" Concrete Reinforcing Steel Institute, 2nd ed., 1971.

site. It may be desirable to restrict steel chemistry (usually carbon and manganese) to a given range to suit a specified welding procedure. Also, the ends of the bars to be butt welded must be prepared with square and/or bevel cuts. Where arc welded splices are used, the following are required:

a. Mill test analysis of the steel.
b. Adjustment of welding techniques to suit analysis.
c. Correct strength, grade, and size low-hydrogen electrodes kept oven-dry.
d. Qualification tests to certify all welders before beginning a project and periodically during long projects.
e. Continuous supervision of all welding operations.
f. Magnetic particle, radiography, or other nondestructive inspection of welds.
g. Occasional quality control tests of actual welds removed from structure when nondestructive test results are unsatisfactory.

Connection of crossing bars or stirrups by small arc welds, known as tack welds is not recommended. Unless these welds are made in conformance with all requirements of AWS D12.1-61, they tend to cause a metallurgical notch effect and may affect the ductility, fatigue and strength of the bars. Tie wire will do the job without harm to the bars. Bars that have been joined by butt-welding also may have reduced fatigue strength. Good practice dictates that no welding should be done within 8 in. of a cold bend to prevent loss in bar strength. In fact, it is preferable to weld only straight bars.

A recent development is the availability of large diameter reinforcing bars conforming to ASTM A615, Grade 60, which can be welded satisfactorily by the shielded metal-arc, the gas metal-arc or the flux-cored-arc processes using appropriate procedures. Suggested practices for welding butt joints are contained below.

Suggested Welding Practice

Electrode	*Bar Size*	*Suggested Minimum Preheat or Interpass Temp, °F Butt Joints*
Low-hydrogen type electrodes of AWS A5.5 Class E90XX-D1, G or M for shielded metal-arc welding or an electrode for gas metal-arc or flux-cored-arc welding that provides filler metal similar to that of the above mentioned electrodes for shielded metal-arc welding.	Numbers 10 & 11	None*
	Numbers 14 & 18	60

NOTE 1 No welding should be done when ambient temperature is below 0°F.
NOTE 2 Low-hydrogen electrodes for shielded metal-arc welding, as well as gases for gas metal-arc or flux-cored-arc welding, must be properly dry.
**When the steel temperature is below 32°F, preheat the steel to at least 60°F and maintain this minimum temperature during welding.*

Thermite welding is a process in which the ends of the bars are fusion-welded. The bars are aligned with a gap between them. Refractory molds are assembled on the bars and sealed in place. Exothermic powders are filled into a separate cavity in the molds. The powders are ignited and burn with enough heat to form superheated molten steel. The steel flows through the gap between the bars and some flows into a second cavity beyond the bars, preheating them. Subsequent flow completes the fusion. Finishing involves breaking off the gates.

Thermite welding is not as sensitive to chemistry or end preparation of the bar. It has been used with success in making butt-welded joints in the large size bars, #14 and #18. This process has been successful in joining hard-to-weld steels because it welds the entire cross-section at the same time and automatically provides preheat and slow cooling.

-3. Couplers—These are mechanical devices joining two reinforcing bars to resist both tensile and compressive forces. The code requirement that such couplers develop 125% of the specified yield point usually controls design of these devices. Couplers are generally equivalent to butt welds although their resistance to fatigue, stress reversal, dynamic load, longtime creep, and other special conditions may vary. For specific performance information on the couplers considered, test data should be secured.

The metal filled sleeve bar splice is a mechanical butt splice in which filler metal (metallic grout) interlocks the grooves inside the splice sleeve with the deformations on the rebar. Their performance is not dependent on the bar chemistry. End preparation is limited to checking that bar ends will fit into the sleeve.

Shear cut, flame cut or saw cut ends can be used as this same metallic grout fills the space between the ends of the bars.

The filler metal flows into the splice sleeve and around the ends of the rebars. As this filler metal cools and solidifies, shrinkage bubbles and shrinkage fissures develop. A shrinkage bubble may be visible at the tap hole (in the center of the sleeve) where the molten metal is introduced. Shrinkage fissures and pin holes may be visible at the top of a vertical column splice. Extensive testing has proved that these casting flaws do not adversely affect the physical performance of the splice. Design of the splice compensates for these voids and flaws.

Properly made splices will have filler metal visible at both ends of the sleeve and at the tap hole in the center of the sleeve.

In one system the bar ends are machined to remove the deformation ribs. The coupler is then tightened and the bars are gripped by a wedging action of the serrated sleeve sections inside the coupler. An uneven split sleeve, serrated inside and specially threaded outside, is attached to the ends of the bar. If designed to a particular condition and tightened properly it should give good results.

The use of end-bearing splices to transfer compression from bar-to-bar requires that the ends of the bars be cut within 1.5° of square to the long axis. In field assembly such splices when erected must fit within 3°. Commercial devices are used to ensure concentric bearing.

In one system, adjustable sleeve clamps are independent of temporary positioning devices. They will hold a free standing bar 25 to 30 ft in length. Inspection holes are at the center to check on the proper meeting of the saw cut bar ends. The sleeve and wedge have grout holes. The concrete in and around the assembly plus the clamping action gives a nominal, but indeterminate, tension value to the splice. When the two ends of the bars are brought into contact it takes two to three minutes to complete a splice. Tension must be taken by other bars or by added separate loose splice bars across the joint.

The use of a butt splice with a pipe sleeve can produce very sloppy connections. They are generally held in place by tack welding. Uncontrolled tack welding is dangerous and not recommended. They should only be used with complete inspection to assure conformance with AWS recommendations.

Prestressing tendon manufacturing and/or shipping conditions are often such that some tendons must be spliced. All splices have certain disadvantages, but they cannot be eliminated.

Couplings of tendons should be used only at locations specifically indicated and/or approved by the engineer. Couplings should not be used at points of sharp tendon curvature. All couplings should develop at least 95% of the minimum specified ultimate strength of the prestressing steel without exceeding anticipated set. The coupling of tendons should not reduce the elongation at rupture below the requirements of the tendon itself. Couplings and/or coupling components should be enclosed in housings long enough to permit the necessary movements, and fittings should be provided to allow complete grouting of all the coupling components in bonded tendons or should be completely protected with a coating material prior to final encasement in concrete in unbonded tendons.

Splices are not permitted in the wires of parallel-wire cables. There are two reasons for this. Since these wires are manufactured in lengths to 7,000 ft, it is a simple matter to make a tendon without splices. When wires are spliced, they are spliced by welding, which damages the fibers and therefore lowers the strength of the wire. In a parallel-wire tendon each individual wire must carry its full share of the tension at all times. It cannot distribute part of its load to adjacent wires at a point where it is weak. As a result if a welded wire is used in a parallel-wire cable, it usually fails during the tensioning operation.

Seven-wire strands are produced in endless lengths and are most often shipped on reels containing 22,000 ft. In use the full length required is cut from the reel in one length and no splice is needed. When the last length on the reel is not long enough to reach the full length of the casting bed, some operators find it desirable to salvage this length by splicing it to another short length. Splicing two short lengths of strand together in this manner is permissible if the coupling used will develop the full strength of the strand, will not cause failure of the strand under the fatigue loadings which will be applied to the structure being fabricated, and does not weaken the member by replacing too much of the concrete in the cross section. Strand tendon couplers are available and used occassionally for multistrand tendons in post-tensioning. The engineer must determine whether or not a particular splice has the necessary qualifications. Information on this subject is also available from some of the fabricators of seven-wire strands.

Splices are used in the individual wires of seven-wire strands during shop fabrication. ASTM Specification A416 says "During fabrication of the strand, butt-welded joints may be made in the individual wires, provided there is not more than one such joint in any 150-ft section of the completed strand." Although the strength of the welded wire is reduced in the vicinity of the weld, the total strength of the strand is not seriously reduced. The welded wire is treated in the vicinity of the weld so that part of its load is transferred to the six other wires.

High-strength bars are spliced at the job site with couplers when necessary. Bars are fabricated in maximum lengths of up to 100 ft and are sometimes furnished in shorter lengths with splices because of shipping difficulties. Sleeve couplers are available to splice the bars to any desired length. The couplers are larger in diameter than the bar and are tapped to take a threaded bar in each end. They develop the full strength of the bar. The disadvantage in the use of couplers is the large hole which must be cored in the concrete at the location of the coupler. If concrete stresses are high, the net section should be checked with the area of the hole deducted.

6.14.4 Corrosion of Steel and Nonferrous Metals

Corrosion refers to the destruction of a metal or alloy by chemical change, electrochemical reaction, or physical dissolution due to its environment. Many years of experience have shown that reinforcing steel is protected from corrosion by concrete. The degree of protection afforded by the concrete depends, of course, upon the thickness of the concrete—the cover over the steel; and the quality of the concrete—its water-cement ratio, and hence, its permeability.

The exact mechanism by which concrete prevents corrosion of encased steel is not completely understood. When steel is encased in concrete, a protective iron oxide film forms at the steel-concrete interface as a result of the high alkalinity (pH) of concrete or mortar. The pH of concrete tends to be buffered at a value of about 12.5 corresponding to a saturated lime solution. Traces of sodium and potassium oxides may increase the pH to 13.2. As long as the alkalinity is maintained, this film is effective in preventing corrosion due to anions such as chlorides, sulfides, bromides and iodides as long as bubbles of free oxygen do not occur at the steel surface.

Corrosion of metals in concrete is an electrochemical process and requires that moisture be present. The moisture conditions of concrete required to support active galvanic corrosion of susceptible metals is not known with sufficient accuracy. Moisture in concrete may be available from two sources: (1) when fresh concrete is placed, it contains mixing water which is only partly used up as the cement hydrates. The remaining water migrates to exposed surfaces and evaporates. The length of time it takes for concrete to dry out varies. It depends on such factors as relative humidity of the surrounding air, size and shape of the concrete mass, and concrete mix; (2) if hardened concrete is not dense and impervious, moisture may penetrate into the interior from the environment.

Galvanic corrosion is influenced by differences in: (1) the composition of the solution at the two electrodes; (2) the nature of the metals of the electrodes; and (3) environmental conditions, such as the presence of oxygen, chloride concentration, alkalinity, moisture, temperature, or large air- or liquid-filled spaces next to the metals. These conditions can be brought about by differences in external exposure, permeability, thickness, or uniformity of the concrete.

Localized galvanic corrosion may occur when one metal is placed in two different electrolytes, or in an electrolyte of varying concentration. For example, metal piping should not rest directly on a sand, earth, or insulating concrete base when concrete is to be placed over it. The piping in this situation makes contact with the base material and this contact can set up a flow of corrosion-causing current.

Corrosion reactions are strongly promoted by the presence of some of the halogen ions, particularly chlorides, which make moist concrete a strong electrical conductor electrolyte. Concentrations exceeding a certain value can, in the presence of oxygen and moisture, destroy the passivating oxide film on the surface of the steel and as a result the steel rusts. The amount of calcium chloride which may be safely used depends upon the composition of the cement, particularly upon its tricalcium aluminate content. The following threshold values have been suggested as a general guide.

*Chloride Ion Concentration, ppm**	*pH of Concrete*
71	11.6
710	12.4
8900	13.2

Concentrations below these values probably do not cause corrosion of steel in the particular alkaline environment. It

*Table reference from D. A. Hausmann, "Electrochemical Behavior of Steel in Concrete," *Journal, ACI,* February 1964.

has been suggested that a chloride ion content greater than 400 or 500 ppm might be considered dangerous in prestressed concrete or in concrete with aluminum embedment and ACI Committee 222, Corrosion of Metals in Concrete, recommends that levels well below these values be maintained, if practicable. At concentrations above these values, the steel may corrode if oxygen and moisture are available.

Chlorides in concrete may originate from concrete admixtures, sea water, beach sand, or salt-containing aggregate. It is therefore recommended that any material for use in making concrete that is suspected of containing chloride be analyzed and rejected if chlorides present will exceed threshold value.

The best protection against diffusion of chloride ions from the concrete exterior to the reinforcing steel is proper cover of good quality concrete. For concrete exposed in marine environments or to deicer salts, corrosion after long-time exposure is always a possibility. Even the best concrete is not completely impermeable to chloride solutions and given sufficient time, chloride ions will eventually reach the steel.

Porous concrete may permit the passage of carbon dioxide from the air into the concrete. This reacts with calcium hydroxide to form calcium carbonate which lowers the pH. If this carbonation reaches the steel the passivating oxide film can be impaired or destroyed. Normally, with good-quality concrete, the effect of carbonation does not penetrate more than 1/2 in. even when exposed to the weather for a great many years. High water-cement ratios, low cement contents, thin cover over the reinforcing steel, and poor consolidation of the concrete should be avoided to preserve alkalinity. This also reduces the possibility of penetration to the steel of chlorides and acid-producing gases such as sulfides which are in the atmosphere in some industrial areas.

Where stress is associated with corrosion, the effect may be considerably more serious. Apart from the relatively simple case of stress-assisted corrosion, there could also be the phenomena of stress-corrosion and corrosion-fatigue. The distinctions between these phenomena are discussed below.

-1. Stress-assisted corrosion—Steel wire that is not susceptible to stress corrosion may, under faulty conditions such as cracks in the concrete or a poorly made joint, undergo severe localized corrosion. The mechanism is relatively simple in that the penetration of the corrosive medium is assisted by the tendency of any initial corrosion pit to be enlarged or pulled open by the tensile stress. Provided the stressed steel is protected by well-made concrete there is no danger of this type of corrosion. Where prestressing tendons are not bonded to concrete, such corrosion is quite possible if not guarded against.

-2. Stress-corrosion—The phenomenon of stress-corrosion is more complex. Not only do materials differ in their susceptibility, but also a limited number of corrosive media (i.e., chlorides, nitrates, and sulfides) are known to give rise to the phenomenon. Given the appropriate conditions of susceptible material (high strength steels used in prestressing), stress, and corrosive medium, stress-corrosion is characterized by quite rapid development of deep cracks in the material. Quenched and oil-tempered, high-carbon steel wire is more susceptible to stress-corrosion cracking than cold-drawn wire. The mechanism of the attack is thought to be related to electrical potential differences between grains, or areas of the metal, which will assist intergranular attack. Under some circumstances, molecular hydrogen may be formed in the grain boundaries to exert pressures that weaken the material and encourage cracking under stress. For stress-corrosion to be progressive, it is necessary for the corrosive medium to be able to penetrate to the steel face. In general terms, if the corrosion product forms a hard but brittle film that will crack easily under stress, or if the corrosion products are removed in solution (e.g., acid attack), stress-corrosion may occur. If the conditions are such that the corrosion product forms a relatively soft blanket on the steel surface with consequent minimization of cracking or flaking of the film under stress, the possibility of stress-corrosion is minimized.

-3. Corrosion-fatigue—When a material is subjected to stress cycles in a corrosive environment, the effect of corrosion and fatigue acting together is synergistic and much more serious than that of the two effects acting independently. Not only may the corrosion produce pitting and give rise to stress concentration, but the alternating stresses tend to disperse the corrosion products and so enable the corrosive medium to be replenished at the metal surface. Once a fatigue crack has commenced, corrosion penetration combines to produce a rapid extension of the fissure. Under such circumstances there will be no true limiting fatigue-stress endurance limit, and low amplitude stress cycles may eventually cause failure.

-4. Corrosion prevention—Various methods have been advanced for reducing the corrosion occurring on reinforcing steel. One method is the use of galvanized steel reinforcement. However, zinc is susceptible to attack by fresh concrete. This results in the evolution of hydrogen and the formation of calcium zincate which occupies a greater volume than the original metal and may exert expansive pressures around the embedded element. The hydrogen may cause a stress corrosion failure of prestressing steel.

The film of zinc covering reinforcement is so thin that the expansive pressures generally do not cause any damage to the surrounding concrete. Whether cracking of the concrete occurs subsequently or not depends on many factors such as strength of concrete, amount of concrete cover, size of galvanized member, exposure conditions, and chemical composition of cement. To what extent the steel will be corroded cannot be stated with certainty, but galvanizing does furnish sacrificial protection to the steel.

The use of calcium chloride admixtures should be avoided in concrete containing galvanized steel exposed to corrosive or wet environments as their use may lead to severe cracking and spalling of the surrounding concrete.

If it is desirable to avoid any reaction between zinc and alkaline fresh concrete, the metal should be protected by an organic coating or passivated by a chromate treatment.

The effect of addition of chromates to concrete on its performance is not yet fully known. It is therefore recommended, whenever a chromate treatment is required, to apply it to the galvanized surfaces, rather than add chromates to the concrete mixture. Chromate coating of galvanized surfaces can be readily done in most galvanizing plants. The chromate treatment is recommended for all galvanized elements which will be embedded in or will come in contact with concrete and mortar. Treated surface should be protected from water washing away coating before concrete is placed.

Nickel-plated steel will not corrode when embedded in chloride-free concrete and will provide protection to steel as long as no breaks or pinholes are present in the coating. The coating should be 3 to 5 mils thick to resist rough handling. Minor breaks in the coating may not be very detrimental in the case of embedment in chloride-free concrete; however, corrosion of the underlying steel would be strongly accelerated in the presence of chlorides.

Cadmium coatings will satisfactorily protect steel embedded in concrete, even in the presence of moisture and normal chloride concentrations. Minor imperfections or breaks in the coating will generally not promote corrosion of the underlying steel.

Various nonmetallic coatings have been applied to reinforcing steel to reduce corrosion. An asphalt-epoxy is said to show promise. Aside from the obvious problem of developing bond with some materials, the loss of the protection of the alkaline environment at the steel would seem to be a major shortcoming of all nonmetallic coatings except a dense cement mortar coating. (Similar to the approach of applying a cement slurry to stressed wires prior to grouting or in the case of pretensioned members prior to casting of the concrete.)

Where the penetration of chlorides into the interior of a reinforced or prestressed member is expected to be damaging, the use of a paint coat to seal the concrete surface against penetration may be feasible but it adds a maintenance cost. The same effect may be gained by using a thicker cover over the steel of up to 2 or 2-$^1/_2$ in.

Metals of dissimilar composition should not be embedded near or in direct contact with each other in moist or saturated concrete unless experience has shown that no detrimental galvanic action will occur. When it is not possible to separate the metals, impervious protective organic coatings such as bituminous coatings, phenolic varnish, chlorinated rubber, or coal tar-epoxies, should be used on the metal surfaces to prevent galvanic action. Chlorides, sulfides, nitrates or other harmful chemicals must not be introduced by decomposition of the organic materials. In posttensioning, it is important that no galvanic action occur between the duct material and tendon steel or the tendon anchorages. (Copper and aluminum should be avoided.)

The presence of stray currents flowing through the earth may lead to serious galvanic corrosion of long metallic elements encased in concrete in the ground. Corrosion by stray currents may be prevented by several methods such as conducting the stray currents away from the embedded metals back to their source or by interrupting the continuity of the embedded metallic elements by means of closely spaced insulating joints.

Cathodic protection, a process for reducing or eliminating corrosion on a metallic structure in contact with a corrosive electrolyte by introducing an electrical potential greater in strength and opposite in direction to the electrical potential that would result from electrolytic action, has been used successfully in a few instances to protect the steel in reinforced concrete structures where cover over the steel has been damaged to the extent that its protective value was lost. The application of such protection to prestressing cables does not appear to be justified economically or technologically. For reinforcing steel, a current requirement of 100 microamperes per square foot has been suggested but long-time polarization effects will reduce this requirement. (Polarization effects result from the chemical and physical changes at the surfaces of the anode or cathode, or both, and reduce the current flowing in the system. Reducing the current requirement reduces the long-time cost of the cathodic protection.)

Nonferrous metals embedded in concrete may corrode in two ways: (1) by direct oxidation in strong alkaline solutions normally occurring in fresh concrete and mortar, or (2) by galvanic currents which occur when two dissimilar metals are in contact in the presence of an electrolyte; when an alloy or metal is not perfectly homogeneous; or when different parts of a metal have been subjected to different heat treatments or mechanical stresses.

Copper and copper alloys are practically immune to action from fresh concrete. No destructive action will occur on the embedded area even when the concrete is kept saturated with moisture. The presence of soluble chlorides, however, may lead to corrosion, and it is therefore advisable to avoid embedding copper in concrete containing chlorides, especially if the concrete and metal are to be exposed to moisture. Galvanic corrosion should be expected when copper and steel reinforcement are connected or in close proximity. Copper is strongly cathodic and will accelerate the corrosion of the steel if chlorides are present. When copper is used in conjunction with steel, it should be electrically insulated from the steel by means of an impervious organic coating or by use of short lengths of polyethylene tubing slit and slipped over the copper.

Copper coatings for the protection of a basic metal such as steel can protect only when they form a complete envelope. Breaks in the coating may create local galvanic cells and cause the exposed basic metal to corrode much more rapidly than it would if it were not coated.

Aluminum suffers attack when embedded in concrete. Initially, when aluminum is placed in fresh concrete, a reaction occurs resulting in the formation of aluminum oxide and the evolution of hydrogen. The greater volume occupied by these oxidation products causes expansive pressures around the embedded metal and may lead to damage to the surrounding concrete.

Corrosion will also occur if aluminum is galvanically connected to steel, with both metals embedded in concrete. If aluminum is to be embedded in reinforced concrete, it should be electrically insulated by a permanent coating. Bituminous paint, alkali-resistant lacquer such as methacrylate or zinc chromate paint can be used.

If electrical insulation is not permanently maintained, the presence of chlorides will greatly accelerate corrosion of aluminum and lead to serious damage. Due to the difficulty of assuring the permanency of the coating at present, aluminum should not be embedded in or come in contact with concrete containing chlorides. Also, aluminum should not be used in concrete in or near sea water.

Where unpainted aluminum is not embedded but is in contact with concrete under conditions of condensation or dampness, corrosion may occur. This can be prevented by coating as above or by the use of a moisture-proof membrane such as plastic film, bitumen-impregnated paper or felt.

Lead is attacked by fresh concrete and is converted to lead oxide or to a mixture of lead oxides. This corrosion tends to stop as the concrete cures and dries, but will continue in the presence of moisture and may cause total destruction of an embedded lead pipe in a few years. If the lead is coupled to reinforcing steel in the concrete, galvanic cell action may be accelerated, and depending on the particular circumstances, either the lead or the steel will be attacked.

When lead is partially embedded in concrete and the remainder exposed to the air, a condition known as differential aeration occurs. The embedded lead has a different electrical potential than that exposed to the atmosphere and in the presence of water will form the anodic (positive) element of an electric cell. The portion in the air forms the cathodic (negative) element of the couple. The current flow will cause corrosion and gradual disintegration of the embedded lead.

Where it is necessary to embed lead in concrete, protection of the embedded portion with organic coatings is suggested. Where a lead strip connects two pieces of concrete such as for an expansion joint, the air space between the two sections should be as small as possible so as to reduce differential aeration.

Generally, no damage will be observed in concrete because of the softness of the lead which will absorb the expansive pressures caused by the formation of corrosion products.

6.14.5 Corrosion Protection of Bonded Tendons

The two main objectives when grouting the ducts of post-tensioned concrete members are: (1) to prevent corrosion of the prestressing steel, and (2) to provide efficient bond between the prestressing steel and the concrete member so as to control the spacing of cracks at heavy overload and thus increase the ultimate strength of a structural member. Both of these objectives require complete filling of the void space within the duct. Filling will be dependent on the production of a grout mix having the desired properties together with efficient equipment for its injection, and careful workmanship and supervision on the site.

Temporary coatings which are applied to the tendon steel to protect it from corrosion during transit, storage and initial installation should allow the grout to bond uniformly to all parts of the tendon steel. Oil base coatings are not recommended for this application. The temporary corrosion prevention coating should be easily removed in the field with the use of nonchlorinated petroleum solvents for the installation of field attached anchorages.

The essential properties of grout are sufficient consistency so it can be readily injected to completely fill the voids within the duct, low water content to ensure high strength and low shrinkage characteristics, low bleeding characteristics to prevent segregation and formation of water pockets which may later become air voids in contact with the steel and increase corrosion hazards, and expansion of grout during the first few hours after mixing.

For details and procedures for grouting refer to "Recommended Practice for Grouting of Post-Tensioned Prestressed Concrete" PCI Committee on Post-Tensioning, *Journal of the Prestressed Concrete Institute*, 17, (6), Nov/Dec. 1972.

6.14.6 Corrosion Protection of Unbonded Tendons

Unbonded tendons must be permanently protected against corrosion by a properly applied coating of galvanizing, or cadmium plating, epoxy, grease, wax, plastic, bituminous, or other suitable material. The coating should (1) remain free from cracks and not become brittle or fluid over the entire anticipated range of temperatures. In the absence of specific requirements, this is usually taken as 0 to 160°F; (2) be chemically and physically stable for the life of the structure; (3) nonreactive with the surrounding materials such as concrete, tendons, wrapping, or ducts; (4) noncorrosive or corrosion inhibiting; and (5) impervious to moisture.

Tendons are usually coated by the manufacturer with a wax designed primarily to give protection during transportation and storage on site and also to assist in lubricating the tendon during tensioning. During manufacture, the tendon is also coated with a grease primarily to provide additional lubrication for the tendon, but also to provide corrosion protection for the tendon up to the time it is stressed. Rust inhibitors are usually added to the grease together with additional compounds to ensure its uniform consistency in extremes of temperature. Asbestos fibers are often added to the grease to hold it together during application. Plastic or paper sheathing should be wound with wires or tapes at frequent intervals. Plastic tubing is extruded on the strand or plastic tubes of the split type are overlapped and taped along the seams, so as to seal them against any leakage of mortars, which might bind the tendons to the tubes. When papers are spirally wrapped around the tendons, care should be exercised in wrapping so as to avoid jamming of the papers when the tendons are tensioned. Wrapping must be continuous over the entire zone to be unbonded, and should prevent intrusion of cement paste or the loss of coating materials during casting operations.

If the tendon is to be stored for any length of time, some means must be taken to protect against corrosion. Vapor phase corrosion inhibitors (fine crystals of amine nitrite) are being incorporated into wrapping papers and this is effective in preventing localized pitting. These crystals may also be placed in the duct.

After stressing, the ducts should be filled with a petroleum-based jelly containing a corrosion inhibitor to protect the tendons. The jelly is pumped through a central hole in the concrete anchor block after holes through the anchorages have been sealed. The anchorages and projecting strand ends usually are enclosed by fabric-reinforced plastic sheaths, which are secured around the female cones. The use of wax and grease may reduce the efficiency of wedge anchorages. Anchorage zones should be encased in drypack concrete, grout or epoxy mortar and the encasement should be free from any chlorides. Shrinkage cracks will permit moisture penetration, therefore, details of mix and application are important. When encasement cannot be used, anchorage and end fittings should be completely coated with a corrosion-resistant paint or grease equivalent to that applied to the tendons. A suitable enclosure will then be necessary to prevent entrance of moisture or deterioration or removal of the coating.

APPENDIX

A6.1 SPECIFYING CONCRETE

"Specifications for Ready-Mixed Concrete," ASTM C94 provides three alternate bases for specifying concrete.

(1) Under Alternate 1, the so-called prescription basis for specifying quality is followed. In this case, the purchaser accepts responsibility for the design of the concrete mixture and specifies cement content, aggregate sizes, maximum water-cement ratio, slump, and (in the case of air-entrained concrete) air content. In the case of governmental agencies with a staff of engineers and testing facilities available, this procedure may go so far as to specify batch weights of cement, of each of the sizes of the aggregates to be used, and of water. The philosophy followed under this procedure is that a properly designed prescription produced in specified equipment according to carefully laid-out procedures will give the desired results.

(2) Under Alternate 2, the ready mixed concrete producer is responsible for the design of the concrete mixture. The purchaser merely specifies the minimum compressive strength desired along with usual aggregate sizes, slump and air content. This procedure comes as close as any to a performance-type specification. But even here, the producer must furnish evidence in advance of delivery that the specified strength will be attained. The philosophy followed here is that, if the strength is adequate and requirements for aggregate size, slump and air content are met, the desired properties of concrete will be obtained. Unfortunately, specified strengths can be obtained by means which may be prejudicial to other properties of concrete such as durability, shrinkage, resistance to abrasion, and long-time strength. Also, it is unfortunate that this procedure has been abused by producers who were overly optimistic in their estimates of what strengths can be obtained over a period of time from materials from given sources.

(3) Under Alternate 3, the ready mixed concrete producer is responsible for the design of the concrete mixture but the purchaser specifies both a minimum cement content and a minimum compressive strength. It is suggested that the minimum cement content be at about the same level as would ordinarily be required for the specified strength while at the same time the amount of cement should be sufficient to assure durability, watertightness, surface texture and density for the expected service conditions. The philosophy followed under this procedure is to obtain the properties of concrete desired by the use of the right amount of cement while at the same time using strength tests to check on uniformity and quality.

A6.2 CONSTRUCTION PRACTICES*

Ready-mixed concrete should be batched, mixed and transported in accordance with "Specifications for Ready-Mixed Concrete" (ASTM C94). Plant equipment and facilities should conform to the "Check List for Certification of Ready Mixed Concrete Production Facilities" of the National Ready Mixed Concrete Association. Reference should also be made to ACI 614 "Recommended Practice for Measuring, Mixing, Transporting and Placing Concrete" *ACI Journal* July 1972.

Cost of forming is a major item and must be kept to a minimum to achieve construction economy. This is accomplished by considering various types of forms for their flexibility, re-usability, and ease of handling. Because proper design, construction, and removal of forms is an involved subject, refer to the work of ACI Committee 347 in "Recommended Practice for Concrete Formwork (ACI 347-68)" and *Formwork for Concrete*, ACI Special Publication No. 4.

Consolidation is very important in achieving the potential strength of concrete. Figure A6-1 shows the substantial effects that incomplete consolidation has on several properties of concrete.

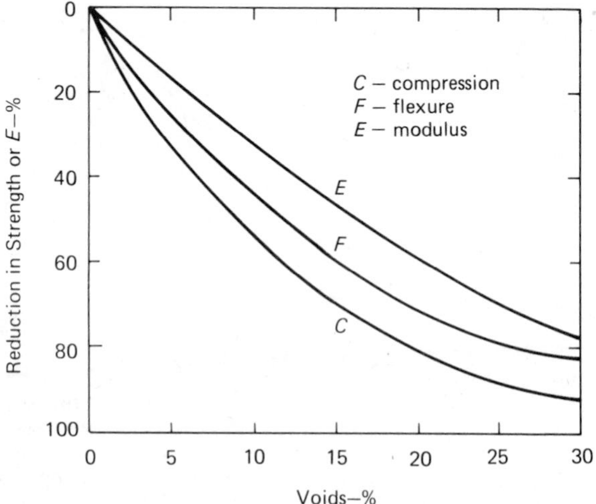

Fig. A6-1 Comparative effects of incomplete consolidation on concrete properties. (From: Kaplan, M. F., "Effects of Incomplete Consolidation on Compressive and Flexural Strength, Ultrasonic Pulse Velocity, and Dynamic Modulus of Elasticity of Concrete," *ACI Proceedings*, **56**, 853–867, March 1960.)

Jarring of concrete during hardening may occur from blasting, pile driving, railway train movements or vehicular traffic, for example, a truck passing over a bridge while other sections of the deck are being paved.

The damage potential of intermittent or continuous jarring depends on the amplitude and frequency of the vibration and the age of the concrete. The particle velocity of earthborne vibrations, which relates amplitude and frequency, is a useful indicator of damage potential; damage potential is greater with increased velocity. Particle velocity decreases exponentially with depth below the surface of the earth. In addition, the particle velocity varies with the amount of energy transmitted to the soil, the physical properties of the soil, and the distance between the fresh concrete and the source of vibration. The safe level of particle velocity for continuous vibration is usually less than that of intermittent vibration; this is due to the possible magnification at resonance which depends primarily on the inherent damping characteristics of the structure. However, no definite threshold value has been established for a damaging particle velocity.

Jarring of concrete or reinforcement generally will not be detrimental to compressive strength if such jarring does not continue beyond the time of initial set of the concrete. In fact jarring may actually be beneficial to strength due to improved concrete consolidation and reduced water-cement ratio from bleeding which is stimulated by the vibration. It may also increase the concrete-to-steel bond through the removal of entrapped air and water from underneath the reinforcing bars. However, violent impacts or vibration may cause fresh slip-formed concrete to slump. Also, concrete is still very weak after initial set and jarring can produce cracks that subsequently will not close and heal. Such cracking may be more of a problem with certain types of construction, for example thin lightly reinforced slabs on ground would be more prone to damage than a heavy foundation.

*The reader should refer to one of the many excellent tests on concrete technology such as "Design and Control of Concrete Mixtures," available from the Portland Cement Association.

A6.3 CURING

For most structural uses, the curing period for cast-in-place concrete is usually three days to two weeks, depending on such conditions as air temperature, cement type, mix proportions, etc. More extended curing periods are desirable for bridge decks and other slabs exposed to weather and chemical attack.

Forms should be removed in such manner as to insure the complete safety of the structure. In determining the time for removal of forms, consideration should be given to the construction loads and to the possibilities of deflections. The construction loads are frequently at least as great as the design live loads. At early ages, a structure may be strong enough to support the applied load but may deflect sufficiently to cause permanent damage. Where the structure as a whole is adequately supported on shores, the removable floor forms, beam and girder sides, column forms, and similar vertical forms may be removed after 24 hr, provided the concrete is sufficiently strong not to be injured. Forms also should be designed and constructed with some thought as to their removal with a minimum of danger to the concrete.

For the development of sufficient strength to carry imposed loads the judgment of the design engineer may be guided by strength tests of job-cured cylinders or beams. Moisture retention measures may be discontinued when the average strength reaches 70% of the specified strength. Still more valid indications of the rate of strength gain can be obtained from strength-time curves which may have been developed for the set of materials used on the job. Pavements should not be opened to traffic until the concrete has attained a flexural strength of 550 psi when tested by the third-point loading method or a compressive strength of 3,500 psi.

Without careful simultaneous reshoring, it is hazardous in freezing weather to remove shores even temporarily before suitable tests show conclusively that the specified strength has been attained. Ordinarily, for temporary removal of support from an entire panel during reshoring, attainment of 55 to 65% of the design strength is sufficient. The strength may be determined from field cured test specimens or estimated for properly cured concrete by determining the length of the curing period and comparing it to the age of laboratory-cured specimens which reached the desired strength. The length of time the field concrete has been cured should be determined by the cumulative number of days or fractions thereof, not necessarily consecutive, during which the temperature of the air in contact with the concrete is above 50°F and the concrete has been damp or thoroughly sealed from evaporation and loss of moisture.

Reshores should be left in place as long as necessary to safeguard each member and, consequently, the entire structure. The number of tiers reshored below the tier being placed and the length of time reshores remain in place are dependent on the development of sufficient strength to carry dead loads and any construction loads with adequate factors of safety.

To ensure that concrete achieves the desired durability during cold weather, it should be maintained at the temperature shown in line 1 of Table A6-1(b), for the period of time shown in Table A6-1(a). The concrete should not be subjected to freezing in a saturated condition before reaching the design strength. The minimum curing period for adequate scale resistance to chemical deicers generally corresponds to the time required to develop the design strength of the concrete. An air drying period of 30 days which enhances resistance to scaling, should then elapse before application of deicing salts.

At the end of the necessary heating (protection) period, concrete should be cooled slowly. Rapid cooling can cause cracking (especially in mass concrete configurations such as bridge piers, abutments, dams, and large structural members). By simply shutting off the source of heat, concrete will cool gradually to the outside air temper-

TABLE A6-1 Recommended Duration of Protection for Concrete Placed in Cold Weather*

Air-entrained concrete recommended.

(a)

Degree of Exposure to Freeze-Thaw	*Normal Concrete***	*High-Early-Strength Concrete****
No Exposure	2 days	1 day
Any Exposure	3 days	2 days

(b)

Line	*Curing Condition*	*Thin Sections*	*Moderate Sections*	*Massive Sections*
1	Min. Temp. fresh concrete, °F	55	50	45
2	(a) Max. allowable gradual drop in temp. throughout first 24 hours after end of protection, °F	50	40	30
	(b) Max. allowable temp. drop in any 1 hr, °F	5	4	3

**Adapted from* Recommended Practice for Cold Weather Concreting *(ACI 306-66)*
***Made with Type I or II portland cement.*
****Made with Type III or High-Early-Strength cement, or an accelerator, or an extra 100 lb of cement.*

ature, but the depth of the different sections of concrete determines the cooling period. Concrete should be permitted to cool at a rate not greater than that shown in line 2, Table A6-1(b).

A6.4 QUALITY CONTROL

A comprehensive quality control program is required at both the concrete plant and the site to insure that concrete quality and uniformity are acceptable and meet job requirements. Inspection of concreting operations from stockpiling of aggregate through completion of curing is important. Also, routine sampling of all mix materials may be necessary to control uniformity of the concrete. Usually, normal field control can be accomplished adequately by testing fresh concrete for slump, temperature, air content, and unit weight; the hardened concrete is tested for strength.

The architect or engineer should select the testing laboratory and the owner should pay for testing by separate contract with the laboratory. This quality assurance to the owner in no way relieves the contractor of his responsibility for quality materials and workmanship required to meet the specifications.

A6.4.1 Sampling

Samples must be taken on a strictly random basis if they are to measure properly the acceptability of the concrete. Samples should be obtained and handled in accordance with "Method of Sampling Fresh Concrete (ASTM C172)." The choice of times of sampling or the batches of concrete to be sampled must be made on the basis of chance alone within the period of placement in order to be representative. If batches to be sampled are selected on the basis of appearance, convenience, or other possibly biased criteria, statistical concepts lose their validity. Obviously, not more than one strength test (average of two cylinders made from a sample) should be taken from a single batch, and water may not be added after the sample is taken.

A predetermined sampling plan (chance approach) should be set up before the start of production or deliveries by establishing the intervals at which samples will be taken. The intervals may be set either in terms of time elapsed or yardage placed.

The number of samples will be based generally on the degree of control required. The degree of control that can be realized practically and the level of quality aimed at are directly linked to the economy of the situation. One can control to closer and closer tolerances, until a point is reached at which further control costs outweigh the benefits to be realized. Job conditions will determine the most practical number of samples required. The minimum number of samples is usually set forth in specifications. Requirements are usually a minimum of one sample for each 100 or 150 cu yd for each class of concrete, and at least one sample to be taken daily from each class of concrete. In addition, "*Building Code Requirements for Reinforced Concrete, ACI 318-71*" requires in the case of flatwork that samples be taken for each 5000 sq ft of area placed. For pavements, two beams should be made for each 2000 sq yd with not less than two beams per day. These values are a minimum and adequate for general construction. However, concrete used in vital structural sections such as columns, beams or slabs should be sampled more frequently. Additional specimens may be required when temperature or moisture content changes suddenly or when materials or sources of materials are changed.

When the frequency of testing will provide less than five strength tests for a given class of concrete, tests should be made from at least five randomly selected batches or from each batch if fewer than five are used. When the total quantity of a given class of concrete is less than 50 cu yd, the strength tests may be waived if adequate evidence of satisfactory strength is provided, such as strength test results from the same type of concrete supplied on the same day by the same supplier and under comparable conditions in other work.

The slump test should be made at the start of operations each day, whenever the appearance of the concrete indicates a change in consistency, and whenever making strength and air content tests. Temperature of concrete should be recorded when strength tests are made, at frequent intervals in hot or cold weather conditions, and at start of operations each day. The frequency of air content tests depends on the severity of exposure conditions. It may vary from one test for each truckload to one test for each five truck loads of concrete.

On work such as pavements and large bridge decks where very rapid rates of concrete placement occur, i.e., > 50 cubic yards per hour, it is impractical to continue a frequency of one test per truckload for very long. Under such conditions and once satisfactory control has been established, the frequency of testing can be reduced for as long as each test result falls within specification requirements. When a test result falls outside the specified limits, the testing frequency should revert to one test per load of concrete until satisfactory control is reestablished. A sudden loss in slump, the development of finishing difficulties, the appearance of bleeding, a change in temperature or aggregate grading, or a loss in yield calls for a check on the air content. Unit weight tests are made at the same time as the slump test.

A6.4.2 Consistency Tests

The slump test is the common method for controlling concrete consistency. It is also a good indicator of variations in water content from batch to batch. A change in slump of structural lightweight concrete may be indicative of a change in air content, moisture content of the aggregates, or a change in gradation or density. The slump test for consistency of concrete should be made in accordance with the Method of Test for Slump of Portland Cement Concrete (ASTM C143). The slump test should be made first before any other test when concrete arrives at the job since it frequently determines at once whether a batch of concrete will be acceptable or should be rejected. A tolerance of up to 1 in. above the maximum specified slump may be allowed for individual batches provided the average for all batches or the most recent 10 batches, whichever is fewer, does not exceed the maximum slump. Concrete of lower than usual slump may be used provided it can be properly placed and consolidated.

Another method of measuring consistency is the ball penetration test, Method of Test for Ball Penetration in Fresh Portland Cement (ASTM C360). When calibrated for a particular set of materials, the results can be related directly to slump. However, the relationship must be checked at least once a day thereafter if the ball penetration test is used for control. This test has the advantages of being relatively simple and of not requiring a molded specimen. The test can be made on fresh concrete in any open container provided the minimum lateral dimension of the sample is about 18 in. and the depth at least 8 in. This test is not accurate in high-slump concrete (5 in. or more), concrete containing large coarse aggregate (2 in. or more), or thin slabs.

A6.4.3 Temperature Measurement

Because of the important influence of concrete temperature on the properties of fresh and hardened concrete, many specifications place limits on the temperature of fresh concrete. Temperature measurements are made primarily during hot or cold weather to make certain that the concrete arrives ready for placing at a temperature suitable for prevailing weather conditions. Although there is no standard method for measuring temperature of fresh concrete, certain simple precautions should be observed. Armored thermometers are available. The thermometer should be accurate to ±2°F and should remain in a representative sample of concrete until the reading becomes stable.

A6.4.4 Tests for Air Content

Available methods for determining air entrainment in freshly mixed concrete measure only air volume and not the air void characteristics. However, it has been shown that the volume of entrained air in a concrete mixture is generally indicative of the adequacy of the air void system when using air-entraining materials meeting ASTM specifications.

A number of methods for measuring air content of fresh concrete are in use. Standards cover the pressure method (ASTM C231), the volumetric method (ASTM C173), and the gravimetric method (ASTM C138). Variations of the first two methods are also in use. Air content is best measured by the pressure method for normal weight concrete and the volumetric method for structural lightweight concrete. Variations from the specified value of air content should not exceed ±1% for normal weight concrete or ±1.5 percentage points for structural lightweight to avoid adverse effects on compressive strength, workability, or durability.

Rapid and frequent indications of the air content can be obtained with a Chace pocket air meter. The air content as measured by this method is an approximation, usually within one-half a percentage point of that measured by the pressure method. Therefore, the pocket air meter is used as a control to indicate changes in air content but it should be supplemented by tests from another method, particularly when the air content approaches the specified limits. It may also be especially useful in checking air contents in small areas near the surface that may have suffered reductions in air content because of faulty finishing procedure.

A6.4.5 Unit Weight

The unit weight is a quick and useful measurement for controlling quality. A change in unit weight generally indicates a change in either air content or aggregate weight. For lightweight concrete a maximum allowable unit weight is usually specified. The unit weight of freshly mixed lightweight concrete is correlated with the 28-day (design) air-dry unit weight and used as a basis for placement and control and acceptance during construction. When unit weight measurements (ASTM C138 or C567) indicate a variation of more than a few lb per cu ft for normal weight concrete or ±2% for structural lightweight concrete the air content should be checked first to establish if the correct amount of air has been entrained. If air contents are correct, then a check should be made on the aggregates to make certain that the unit weight, gradation, and moisture content have not changed. Results of these checks generally will reveal the cause of the variations in unit weight of concrete and indicate what mix adjustments are in order.

A6.4.6 Tests for Strength

Strength specimens must be fabricated, cured, handled, and tested in strict conformance to standard procedures. Specimens should be made and cured in accordance with Method of Making and Curing Concrete Compressive and Flexure Test Specimens in the Field (ASTM C31). Compression specimens should be cast in reusable steel molds or in disposable tin molds rather than paraffined cardboard molds (ASTM C470). Most cardboard molds appear to produce specimens with strengths 2 or 3% less than those molded in steel molds, and occasionally strength reductions of 10 to 15% may occur.

The strength of a specimen is greatly affected by disturbances, changes in temperature and exposure to drying, particularly within the first 24 hours after casting. Thus, cylinders should be cast in locations where subsequent movement is unnecessary and where protection is possible. Cylinders should not be moved after they are 15 to 20 minutes old. Because of the danger of producing cracks and weakened planes, concrete with slumps less than about 1 in. should not be moved even in the first 15 minutes. Cylinders and test beams should be protected from rough handling at all ages.

After cylinders are molded they must be stored carefully at 60 to 80°F during the first 24 hours and moisture loss prevented by covering them with an oiled metal or glass plate and a double layer of wet burlap. Storage at lower temperatures but above freezing would tend to increase 28-day strengths while storage at higher temperatures would produce lower 28-day strengths. In the wintertime plywood curing boxes equipped with light bulbs or small thermostatically controlled electric heaters have been used with great success. Maximum-minimum thermometers should be used. Coverings of damp burlap are highly recommended in curing boxes. Curing boxes with a small cake of ice are a practical means of obtaining the desired temperature control in hot weather. When curing boxes are not used in the summertime, coverings of damp burlap will provide some cooling due to the evaporation of water in the shade. In the summertime, coverings of damp burlap with sheet polyethylene over the outside must be avoided since the plastic sheeting prevents evaporation and the burlap provides insulation for the retention of heat due to hydration during the first 24 hours.

Within 24 hours the specimens should be taken carefully to the laboratory and placed in standard moist curing. The failure to bring the specimens into the laboratory for standard curing within 20 ± 4 hours is probably the most frequent and perhaps the most serious of all potential violations of standard testing procedures. If tin can or cardboard molds are used, it is preferable to ship the cylinders in the molds for protection. Cylinders removed from the molds should be packed in wet sawdust, wet sand, or wet burlap to prevent drying. These materials will also act to cushion any jolts. Rough handling and lack of protection from too rapid drying will tend to produce erratic and low-strength results.

Cylinder ends should be capped or ground in accordance with the requirements of "Method of Capping Cylindrical Concrete Specimens (ASTM C617)."

The age of test specimens should be 28 days (unless using accelerated curing) but 7-day specimens may be used, provided that the relationship between the 7- and 28-day strength of the concrete is established by test for the materials and proportions used. Usually if three specimens are made, two are tested at 28 days for acceptance and one is tested at 7 days for information.

Testing of specimens should be done in accordance with Method of Test for Compressive Strength of Cylindrical Concrete Specimens (ASTM C39) and Method of Test for Flexural Strength of Concrete–Using Simple Beam with Third-Point Loading (ASTM C78). Modulus of rupture tests by center-point (ASTM C293) or cantilever loading, or compressive strength tests, may be used for job control of pavements if relationships to third-point tests are determined before construction starts. Typical relationships are shown in Fig. A6-2.

The moisture content of the specimen has a considerable effect on the result; a saturated cylinder will show 20 to 30% lower strength than that of a companion specimen tested dry. This is important when comparing cores taken from concrete in service with molded specimens that are tested as they are taken from the moist room.

The flexural strength tests are very sensitive to deviations in specified standard moist curing conditions. Since the high stresses in a flexural specimen are at the surface, these specimens are highly sensitive to small amounts of drying which induce tensile stresses in the extreme fibers of the specimen thereby greatly reducing the measured tensile stresses at failure. For this reason, when a moist specimen is removed from curing for testing, it must be maintained in a moist condition until testing is completed. As little as 30 minutes of drying may reduce measured flexural strength as much as 10% and a period of drying prior to testing of three or four days reduces flexural strengths 40%.

Tests on concrete specimens when carefully conducted only show the potential strength of the concrete in the structure. When test specimens are to be cured entirely under field conditions in accordance with ASTM C31–as a check on the adequacy of curing and protection of the concrete in the structure–they should be molded at the same time and from the same samples as the laboratory-cured acceptance test specimens. Field-cured cylinders must be cured under conditions identical to those of the structure. If the structure is protected from the elements, the cylinder should be protected similarly. That is, cylinders related to members not directly exposed to the weather should be cured adjacent to those members and provided with the same degree of protection and type of curing. Obviously, the field clyinders should not be treated more favorably than the elements they represent.

Procedures for protecting and curing the concrete should be improved when the strength of field cured cylinders at the test age designated for measuring f'_c is less than 85% of that of the companion standard laboratory moist-cured cylinders. This percentage has been set merely as a rational basis for judging the adequacy of field curing. The comparison is made between the actual measured strengths of companion job-cured and laboratory-cured cylinders, not between job-cured cylinders and the specified value of f'_c. However, results for the job-cured cylinders are considered satisfactory if they exceed the specified f'_c by more than 500 psi even though they fail to reach 85% of the strength of companion laboratory-cured specimens.

Field cured specimens also are useful in determining when sufficient strength has developed to permit safe removal of forms or when a structure may be put into service.

Instead of using field-cured cylinders to approximate conditions of strength development in the structure, cylinders can be cast integrally with the slabs in special plastic inserts that permit them to be pushed out and tested for strength. These cores can be used to form service voids in the concrete and their installation, of course, must be planned prior to concrete placement. Push-out cylinders cast in slabs provide a fairly reliable measure of core strengths at all ages irrespective of the way the slab is cured. The cylinders must be removed carefully from the slabs in order to obtain satisfactory test results.

Present day speed of construction requires knowledge of concrete quality earlier than the usual 7, 14, or 28 days, otherwise erection will have proceeded far beyond the placement in question. Since strength develops with the hydration of cement, concrete specimens can be exposed to elevated temperatures and moisture conditions adequate to develop a significant portion of their ultimate strength within 24 to 48 hours.

ASTM in 1971 adopted Method of Making and Accelerating Curing of Concrete Compression Strength Test Specimens (ASTM C684) which includes three procedures. The procedure to use is chosen by the engineer on the basis of his experience, and local conditions. The three procedures are: (1) Warm Water method, (2) Boiling Water method and (3) Autogenous Curing method. All have some particular merits and all have certain disadvantages. Methods (1) and (2) have the merit of avoiding the need for a delay because, in Boiling method, 24 hours of initial moist curing is used before the 3½ hour boiling period; and in the Warm Water method, the water external to the specimen is kept at 95°F throughout the 24-hour curing period. However, both these methods require curing tanks and the compression testing machine to be on site or close by. Both provide results by next day, after about 28½ hours for one and 24 hours for the other.

While it may be desirable to have strength results in about 24 hours, this is not essential on most jobs and a compensating advantage of the autogenous method is that testing at 72 hours or 96 hours is possible. Appropriate correction factors can be developed which allow cylinders made on a Thursday or Friday to simply sit in their containers until Monday, thus avoiding all overtime weekend work. Valid relationships exist for non-standard hydration conditions and it is simply a matter of calibration. Starting temperatures over the normal range 50°F to 90°F have little effect. Though the initial container cost is high, the containers are designed for many re-uses. At 100 re-uses the extra cost is only 20 to 30 cents per test.

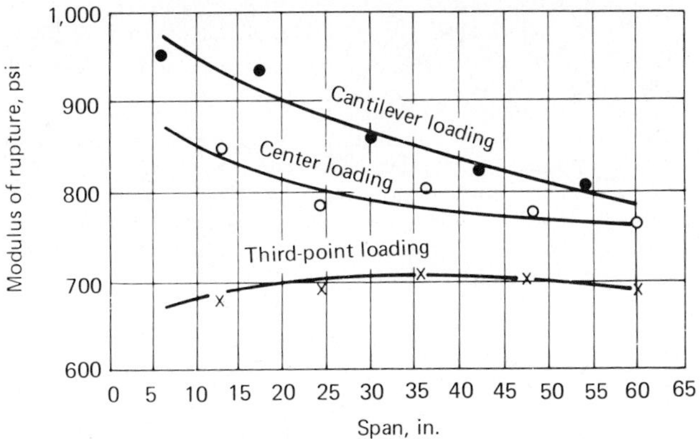

Fig. A6-2 Relationships between modulus of rupture test methods. (From: *Design and Control of Concrete Mixtures* 11th ed., Portland Cement Association, Skokie, Ill., 1968.)

A6.4.8 Investigation for Structural Adequacy

Procedures used in the design of concrete structures allow for the normal variability of strength tests as well as for the favorable treatment of laboratory test specimens in comparison with the concrete in place by specifying strength levels greatly in excess of stresses to which the concrete will actually be subjected.

The concrete strength is considered to be satisfactory as long as averages of any three consecutive tests remain above the specified f'_c and no individual test falls below the specified f'_c by more than 500 psi or the specified modulus of rupture by more than 75 psi. Strength tests failing to meet these criteria will occur occasionally (probably about once in 100 tests) even though strength level and uniformity are satisfactory. Allowance should be made for such statistically normal deviations in deciding whether or not the strength level being produced is adequate.

Occasionally an isolated test result deviates so far from the average values as to be highly improbable. A test cylinder that is one of a set of three or more companion cylinders may be discarded if the compressive strength deviates more than 1500 psi from the average of the remaining cylinders. However, an attempt should be made to explain and correct the cause of the deviation. If the suspect cylinder is one of a set of two companion cylinders, it may be discarded only if the cylinder shows evidence of questionable practices in sampling or in molding, curing, capping, or testing.

The significance of low strength measurements should be investigated in the following sequence:

Step 1 Verifying testing accuracy–Sampling and testing of concrete in the field are often not in strict accord with the standard procedures prescribed for acceptance testing. Stern measures should be taken whenever deviations from standard procedures are discovered. Inaccuracies may lead to expensive delays or supplementary investigation. The testing laboratory should be held responsible for deficiencies in the testing procedures and services it is asked to provide. Laboratories should not be employed unless they meet the requirements of ASTM Recommended Practice E329 and unless they have been inspected within three years by the Cement and Concrete Reference Laboratory of the National Bureau of Standards and have corrected deficiencies disclosed in the inspection.

If the testing of the concrete is found to have deviated from the standard methods, it may be possible to terminate the investigation at this point. If the testing has been satisfactory, or if this phase of the findings is inconclusive, it may be necessary to continue with Step 2 or Step 3, or possibly both simultaneously.

Step 2 Compare structural requirements with measured strength–Lower strength may, of course, be tolerated under many circumstances but this is a matter of judgment. The structural engineer must determine if the load carrying capacity has been significantly reduced. If time and conditions permit, an effort can be made to improve the strength of the concrete in place by supplemental wet curing. Effectiveness of such a treatment must, of course, be verified by further strength evaluation using procedures to be discussed.

Step 3 Non-destructive tests–Several devices are available for obtaining estimates of concrete strength in place. Generally speaking, they do not provide values that can be translated directly into cylinder or core strengths. However, in the hands of a skilled operator, they can yield useful information on the concrete in place by comparing readings on portions of the structure represented by the low strength tests with other portions that are considered acceptable. Such readings can indicate quite convincingly whether or not the questioned concrete differs appreciably from concrete judged acceptable.

a. Rebound hammer

One such test involves the use of the Swiss (Schmidt) hammer–an impact testing device. The hammer is not intended to replace the standard cylinder nor is it intended to give an accurate measure of the compressive strength of concrete. Its limitations and proper use should be understood by all concerned prior to its acceptance as a testing tool. It is valuable for field use to determine relative levels of strength, uniformity of concrete in place, and where and when cores should be drilled. But it must be properly calibrated and it must be used competently.

The Swiss hammer, essentially a concrete hardness tester, measures the rebound of a spring loaded plunger after it has struck a smooth concrete surface. A reading then is taken from a scale on the side of the test hammer while holding the hammer firmly against the concrete. A button usually is provided to lock the pointer on the scale after impact it if is not convenient to take a reading while holding the hammer against the concrete. High rebound readings are associated with high concrete strength.

At least 10 individual readings are necessary to obtain a good representative mean of a given test. The significance of the mean is not improved much by taking more than 20 individual readings. The two or three highest and lowest readings should be rejected when evaluating the mean. The rebound number is considered reliable when the 10 readings do not deviate more than ±2-½ to 3-½ on the scale. Whenever possible, the readings representing one test (between 10 or 20) should be confined to an area of about 1 sq ft. A rough grid should be marked out on a 2-in. spacing and the intersections used for test points. The test position should be 2 in. or more from the edge of the concrete or low readings will result. A repeated test on the same spot or near it will result in lower readings due presumably to local partial crushing of the concrete.

The rebound of the hammer is affected by the hardness of the surface; smoothness of the surface; moisture content of the near surface; age of the specimen; aggregate type and concentration; size, shape, and rigidity of the specimen, and the strength of the concrete.

The calibration curve provided with the instrument to indicate the compressive strength from the reading obtained from the test hammer scale is not always precise. To be most useful, the test hammer should be calibrated for the particular materials and mix designs on the job or the hammer can be used only to indicate the difference between questionable and acceptable portions of the structure.

One method of calibration is to obtain rebound readings on standard test cylinders or cores immediately prior to their being tested in compression. Fifteen readings should be taken on the middle two-thirds of the cylinder surface along three vertical lines, 120° apart, while the cylinder is anchored in the testing machine by a load of at least 300 psi. The average reading (from nine numbers) after discounting the three highest and three lowest readings then is assigned to the cylinder compressive strength and a curve developed. The test hammer can be used horizontally, vertically upward or downward or at any intermediate angle; however, readings in these various directions require separate calibration or correction charts. Rebound readings with the test hammer in a horizontal position generally are higher than those obtained in the vertical position. Even after a calibration curve is developed, individual points will deviate from the relationship by ±500 to 1000 psi. The method of curing and age at test should be kept constant when calibrating the Swiss hammer.

The area to be tested should be smooth and uniform and preferably formed rather than finished. Areas exhibiting honeycombing, scaling, rough texture or high porosity should be avoided. Rock, sand, or air pockets will result in very low readings and the presence of air holes, large rock particles or reinforcement near the surface will cause erratic results. Concrete placed against wooden forms usually produces rebound numbers 10 to 20% lower than concrete cast against metal forms or finished with a metal trowel. This is especially evident with concrete of high compressive strength. However, the scatter of individual results on formed surfaces is lower. When necessary, a smooth surface may be produced by uniformly rubbing an area about 2 in. in diameter with a carborundum stone. This is a cumbersome procedure and can lead to large errors if not carried out uniformly.

Usually, old, dry concrete (age greater than 90 days) in which the surface is harder (carbonation effects) than the interior registers higher rebound numbers than normal. Conversely, new, moist concrete surfaces (age less than seven days) give a low rebound value. It is recommended that the test hammer be used only on concrete from 7 to 90 days old. Within this general range of ages, a wet concrete surface will give about a 20% lower mean reading than a dry surface of the same concrete. The differences between wet and dry specimens are smaller at the earlier ages. It should be remembered that a standard moist cured cylinder will show lower strength than that of a companion cylinder tested dry. However, the rebound number changes faster than strength with a change in moisture condition. Wetting concrete areas to be tested and covering them with damp burlap and polyethylene for several days can be used to minimize the effect when relatively young and relatively old concrete must be compared.

Different aggregates, in general, will give different calibration curves. A lightweight aggregate concrete will give a lower rebound reading than normal weight concrete. Gravel coarse aggregate concrete often produces a higher reading than crushed stone concrete (particularly soft limestones) of the same compressive strength. In addition, the coefficient of variation of individual readings increases with the size and amount of coarse aggregate.

Low rebound numbers will result if small, long, or thin specimens (less than 4 in. thick) are not rigidly supported to prevent movement under impact.

The Swiss hammer is designed to measure compressive strength above 1000 psi and, therefore, is not applicable to concrete at very early ages. At early ages the increase in concrete surface hardness lags behind the gain in strength but after three months, outstrips it. Therefore, use of the hammer at early ages causes severe pocking and scarring of the concrete. The coefficient of variation of individual readings constituting a mean decreases with increase in the strength of the concrete. An understanding of all of these factors will aid in interpreting the results of a Swiss hammer test.

b. Penetration probe

The Windsor probe is a recent development which also appears to be basically a hardness tester. The technique of using the Windsor probe consists of forcing a hardened alloy probe into the concrete. The input energy is developed by the expanded gases of a precision loaded industrial type powder cartridge.

The test probe is used as follows: (1) three probes, spaced in a 7-in. triangular pattern by means of a templet, are driven into the concrete. The driver is loaded separately for each probe; (2) the 9-in. equilateral triangular, lower gauge plate is placed over the three probes and can be secured by compression springs for overhead work; (3) an upper gauge plate is placed over the three probes. Both lower and upper plates provide a mechanical average of the concrete surface and the projection of the probe respectively; and (4) the depth gauge is inserted through the appropriate hole in the upper gauge plate and the exposed height of the probes is recorded in inches. Finally, refer to a graph for compressive strength. However, reliance should be placed upon comparisons between accepted and questioned portions of the structure rather than upon numerical strength levels.

The variables affecting the test results have not been fully investigated. However, tests during World War II on projectiles shot into concrete showed that moisture content of concrete and aggregate type had significant effects on penetration. In addition, penetration decreased with an increase in volume, fineness modulus, maximum size, and crushing strength of the aggregate (Moh's hardness of aggregate). These same variables may be of significance with the Windsor probe.

Preliminary test results indicate that the probe should be calibrated for the particular materials used on the job and that:

(1) the depth of penetration does indicate the development of strength;
(2) the shape of the probe is important with a step (lightweight concrete) and nonstep probe (normal weight concrete) utilized;
(3) maximum size of aggregate does make a difference;
(4) the probe will have a greater penetration into mortar than into concrete;
(5) the probe will indicate a difference in wet and dry strength of concrete; and
(6) the probe can measure compressive strength in the range of 1500 to 7000 psi when the minimum thickness of the sample is around 6 in.

Since the Windsor probe is a hardness tester similar to the Swiss hammer it probably will be affected by the same variables and the advantages of one testing system over the other should be considered. The probe measures hardness to a much greater depth than the Swiss hammer and test results may be influenced less by surface texture and carbonation effects. The initial cost of the Windsor probe test apparatus is more than the cost of the Swiss hammer. In addition, the probes and charges are a continual expense and the Windsor system requires cleaning of the gun. The test may not be exactly nondestructive since it may cause spalling. A spall may result when the probe is used closer than 5 or 6 in. from the edge of the concrete or if the probe hits a flat or hard aggregate particle immediately beneath the surface and deflects. If the probes are withdrawn properly, little or no damage results. Surface damage usually consists of not over $1^1/_8$ in. disturbance on the surface and a $^5/_{16}$ in. hole in the center for the depth of the probe. The dimples left in the concrete by the test hammer may also be objectionable.

c. Pulse velocity

The condition of concrete in service may be obtained by measuring the velocity of pulses of sonic or ultrasonic vibrations transmitted through it. The pulse velocity method (ASTM C597) was developed to test large masses of concrete. This method also may be used to assess concrete uniformity but should not be considered as a means of measuring strength unless a velocity-strength calibration curve is developed from cylinders made with the concrete mix used on the job. The pulse velocity method is a specialized technique which requires experience in using the apparatus and in interpreting the results.

The pulse velocity method consists of measuring the time of travel of a pulse or train of waves through a measured path length in the concrete. The measurements may be made on any size or shape object. The receiving and transmitting transducers need not be positioned exactly opposite each other. Knowing the distance between transducers, transmission times can be measured across corners of structures or along one face. When using the amplitude of the received pulse as a criterion, the attenuation characteristics of the concrete can be determined. The higher frequency components of the pulse are more rapidly attenuated (dampened) than the lower frequencies and the pulses's shape changes as it passes through the concrete, becoming more rounded as the distance increases. Therefore, choice of the transducers for a particular application is important. Care should be taken in coupling the transducers onto rough or uneven surfaces or it may be difficult to obtain reproducible, accurate measurements of pulse velocity and attenuation. Often it is recommended that a note be made of the signal strength for each observation for use in interpreting the pulse velocity data.

There are a number of factors affecting the pulse velocity. Some of these are: (1) mix proportions, (2) type of aggregate, (3) moisture content, (4) compaction, (5) reinforcement, and (6) age of concrete. There does not appear to be complete uniformity of opinion as to the degree of importance of these factors in the estimation of strength.

Concretes of different mix proportions may have the same compressive strength for different pulse velocities. If the water-cement ratio is held constant and the proportions of cement paste and aggregate varied, the strength will remain essentially constant whereas the velocity will not. The more aggregate added, the more closely will the pulse velocity of the concrete approach that of the aggregate. An increase in aggregate-cement ratio results in an increase in pulse velocity. The fine aggregate does not affect the velocity as much as does the coarse aggregate. For a given change in compressive strength, the change in velocity is higher for concretes with a high aggregate-cement ratio. However, this increase is nominal when high strength concrete is used.

Coarse aggregate has a velocity greater than that of the concrete in which it is contained. Pulse velocities through neat cement paste are appreciably lower than those through concrete or coarse aggregate. More cement increases the strength of concrete but tends to decrease the pulse velocity. For the same maximum size and type of aggregate, the pulse velocity decreases as the water-cement ratio increases.

For the same maximum size aggregate, water-cement ratio, and aggregate-cement ratio, the pulse velocity may vary considerably with aggregate type. This is due to the wide ranges in aggregate elasticity and density. The type of coarse aggregate also may have a larger effect on the damping coefficient than on the strength.

Pulse attenuation (damping) of concrete increases when the concrete dries out but is less sensitive to changes in moisture content than is pulse velocity. Pulse velocity may vary as much as 10 percent depending on the amount of moisture in the concrete. Voids due to poor consolidation have much less effect on pulse velocity than on compressive strength due to transmission through the coarse aggregate. The effect of steel on pulse velocity depends upon the proximity of the steel to the transmission path, the size and concentration of reinforcing, the distance between transducers and the relative velocities in concrete and steel. Only a small amount of energy is coupled into reinforcement, but the pulse should not be propagated near the axis of thick reinforcing bars.

Pulse velocity increases with age at a decreasing rate similar to a strength age curve. At strengths less than 4000 psi, the relationship between velocity and strength appears to be less dependent on age and water-cement ratio than it does at higher strengths. Low pulse velocity at an early age forecasts low strength at later ages. It is generally agreed that very high velocities (more than 12,000 fps) are indicative of very good concrete and that very low velocities (less than 10,000 fps) are indications of poor concrete. It is further agreed that periodic, systematic changes in velocity are indicative of similar changes in the quality of the concrete. When test cylinders are available for development of a calibration curve, the strength may be estimated to within ±10 to 25%. The pulse velocity test has particular application as an accurate and economical method of quality control in prestressing and precast operations because of the uniformity of materials.

In some cases, use of all of the non-destructive tests should be considered. A combination of several tests will improve the estimate of the concrete strength for which full information on composition and age is not available and will assist in locating areas of low, medium and high strength prior to coring. The investigator of concrete quality in field structures must have accurate knowledge of the factors affecting the test results to adequately interpret the variations in results. However, it is a dangerous procedure to try and develop a general relationship which will hold for all concretes on the basis of any of these nondestructive tests.

Step 4. Core and beam tests–Compressive strength tests of cores drilled from the structure provide a measure of concrete strength in place, but not one that can be equated to or translated into an equivalent strength for molded cylinders. The testing of cores requires great care in the operation itself and in the interpretation of the results.

The procedures for obtaining, preparing, and testing cylindrical cores drilled from concrete are covered in ASTM C42 "Method of Obtaining and Testing Drilled Cores and Sawed Beams of Concrete." Controversy frequently develops over interpreting the results of core tests. Concrete in the area represented by the core tests should be considered structurally adequate if the average of three cores is equal to at least 85% of f_c' and if no single core is less than 75% of f_c'. To expect core tests to be equal to f_c' is not realistic, since differences in the size of specimens, conditions of obtaining samples, and procedures for curing do not permit equal values to be obtained.

The fact that core strengths may not equal the strength specified for molded cylinders should not be cause for concern. As mentioned earlier specified cylinder strengths allow a large margin for the unknowns of placement and curing conditions in the field as well as for normal variability. For cores actually taken from the structure, the unknowns have already exerted their effect, and the margin of measured strength above expected working stresses can logically be reduced. The structural engineer should examine cases where core strength values are below 85% to determine if, in his judgment, there is cause for concern. In many cases, values considerably below 85% might be obtained from acceptable concrete due only to normal variation.

Three cores should be taken for each case of a cylinder test more than 500 psi below f_c'. The location of the core in the concrete member is important. The location of cores should be determined by the architect/engineer to least impair the strength of the structure. When possible, the cores should be cut so that the test load is applied in the same direction as the service load. Horizontally drilled cores may be up to 15% weaker than vertically drilled cores. Slabs generally show an increase in strength from top to bottom, possibly due to the migration of water toward the surface while the concrete is plastic. In columns, there is a sizeable decrease (15 to 20%) in strength near the top 12 to 18 in. and an increase (5 to 7%) near the bottom. The strength in between is nearly constant. These differences are decreased when using a relatively low slump air-entrained concrete mix.

It has been found that sawing significantly reduces the flexural strength of beams below that obtained on the same size of molded specimens at all strength levels, Fig. A6-3. For similar curing conditions the modulus of rupture of the sawed specimens averages about 25–30% less than molded beams. Intermittent drying, such as might be experienced by specimens from a structure, is more damaging to sawed specimens than to molded specimens. The effect of sawing is not related to type of aggregate or size of beam but is related to the location of the sawed surface with respect to the tension side, the

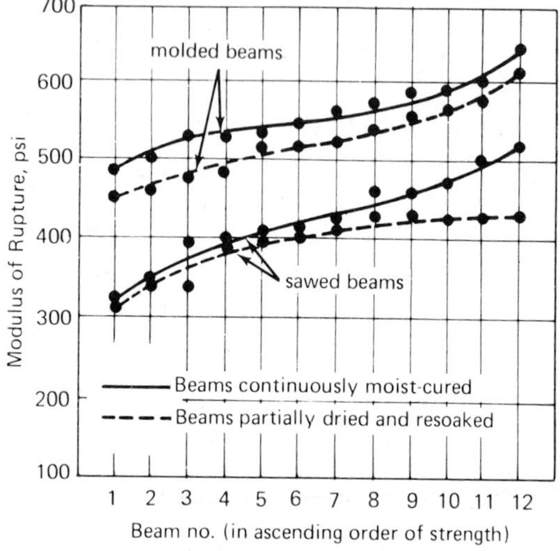

Fig. A6-3 Relationship between flexural strengths of sawed and molded beams of same concrete. (From: Walker, S., and Bloem, D. L., "Studies of Flexural Strength of Concrete," *ASTM Proceedings*, 57, 1957.)

strength reduction being greatest when a sawed surface is in tension. Because of the importance of curing and moisture distribution, it is doubtful that sawed beam specimens will give reliable indications of the potential load-carrying capacity of a pavement structure.

The use of the splitting tensile test (ASTM C496) has been proposed for determining the tensile strength of drilled cores. Certain aspects of this test appear to recommend it over the flexural test for determining tensile strength of concrete where cut specimens are involved. Because the zone of maximum stress in the splitting tensile test does not occur at the surface of the specimen, it is unlikely that there would be any effect on strength as a result of the cutting of aggregate particles during drilling. However, because the bearing edges of such a specimen would be along a cut surface, the effect of such an edge, or the need for special leveling treatment of the specimen in this regard, have to be determined.

Cores should be drilled with a diamond bit to avoid irregular cross-section and damage from drilling. If possible, drill completely through the member to avoid having to break out the core. If the core must be broken out, use wooden wedges to minimize the likelihood of damage and allow two extra inches of length at the broken end to permit sawing off ends to plane surfaces before capping.

The inclusion of reinforcing steel in the cores reduces the strength. The reductions tend to be larger for cylinders containing two bars rather than one. The cores, therefore, should be trimmed to eliminate the reinforcement whenever possible.

The length (L) of the core, when capped, should be twice its diameter (D). Seldom will it be possible to obtain this ratio of 2. ASTM C42 gives correction factors for converting the strength of a test core to that of a cylinder with an L/D ratio of 2. These correction factors apply to lightweight as well as to normal weight concrete in the strength range of 2000 to 6000 psi. They are not necessarily applicable to concrete dry at the time of loading, (greater corrections are necessary). Length to diameter ratios between 2.0 and 1.5 are to be preferred and cores with an L/D ratio of less than 1.0 should be avoided.

The smallest core used should have a diameter of 2½ in., while ASTM C42 states that the diameter of the core specimen should be at least three times the maximum nominal size of the coarse aggregate and must be at least twice the maximum coarse aggregate size. Some tests have shown that 2-, 4-, and 6-in. diameter cores, all having an L/D ratio of 2, give essentially equal strength. While other tests showed that 6 × 12 in. cores have about 20% higher strength than 4 × 8 in. cores. The variability of compressive strength results of drilled cores increases with decreasing core diameter. For most structural concretes, 3½ to 4 in. diameter cores seem to be the most logical compromise between cost, reliability, and potential damage to the structural element.

There is a wide difference of opinion over the relative merits of soaking the cores in water for 40 hr to obtain a uniform moisture content and testing them wet or approximating the moisture content existing in the structure. The moisture condition chosen will depend upon the purpose of the test, but it should be recognized that dry samples of concrete test 20 to 30% stronger than similar saturated samples. Usually, if the concrete in the structure will be dry under service conditions, the cores should be air dried (temperature 60 to 80° F, relative humidity less than 60%) for seven days before test and should be tested dry. If the concrete in the structure will be more than superficially wet under service conditions, the cores should be immersed in water for at least 40 hr and tested wet.

Curing is a critical factor in strength development of all concretes but it affects thinner elements to a greater extent than massive sections. Poorly cured slabs may be 10 to 30% weaker than well-cured slabs, although both will be less than standard cylinders. Cores from porous lightweight aggregate concretes are less adversely affected by poor curing possibly due to the absorbed moisture in the aggregate providing additional internal curing.

In addition to indicating compressive strength, an examination of a core may provide information on the water, cement, and air contents of the concrete. This data should help answer the question of why the strength was low initially.

If the results of properly made core tests are so low as to leave structural integrity in doubt, further action may be required as described in Steps 5 and 6.

Step 5. Load tests—As a last resort, a structural strength investigation by analytical methods or by means of load tests, or by a combination of these methods, may be required to check the capacity of structural members in which adequacy of strength is seriously in doubt. Generally, such tests are suited only to flexural members—slabs, beams, and the like—but they may in some cases be applied to other types.

In some cases the analytical procedure is preferable to load testing. In other cases, analytical evaluation may be the only practicable procedure. Certain members, such as columns and walls, may be difficult to load and the interpretation of the load test results equally as difficult unless severe damage or actual collapse occurs. In any event, the selection of the portion of the structure to be tested, the test procedure and the interpretation of the results should be done under the direction of a qualified engineer experienced in structural investigations.

Load testing procedures and criteria for their interpretation are to be found in ACI 318-71.

Step 6. Corrective measures—In those rare cases where a structural element fails the load test or where structural analysis of untestable members indicates an inadequacy, appropriate corrective measures must be taken. The alternatives, depending upon individual circumstances, are:

a. Reduce the load rating to a level consistent with the concrete strength actually obtained.
b. Augment the construction to bring its load-carrying capacity up to original expectations. This might involve adding new structural members or increasing the size of existing members.
c. Replace the unacceptable elements.

Step 7. Assignment of responsibility—Usually, violation of strength specifications is not of such a nature that repair or replacement of the concrete is required. However, with or without actual condemnation of the structural element, considerable expense may accrue. Assignment of financial responsibility for such expense can become very complicated because of overlapping responsibilities among the principals—the specifier, contractor, producer, and testing laboratory. The National Ready-Mixed Concrete Association has developed the following guidelines for determining responsibility:

"A sensible method of adjudicating controversies over the consequences of low strength tests is described in the ASTM Specification for Ready-Mixed Concrete, Designation C94. In cases where direct negotiation does not provide a solution, C94 states that 'a decision shall be made by a panel of three qualified engineers, one of whom shall be designated by the purchaser, one by the manufacturer (of the concrete), and the third chosen by these two members of the panel. The question of responsibility for the cost of such arbitration shall be determined by the panel. Its decision shall be binding, except as modified by a court decision.' This system is intended for controversies between the concrete supplier and his customer, but the principle can logically be extended to include other parties to a dispute. If agreement to use such a procedure cannot be reached, it is recommended that controversies of this kind be submitted to the American Arbitration Association.

The following suggestions are offered as guidelines in the allocation of responsibility:

1. If the molding, curing, or testing of strength cylinders was demonstrably faulty and the concrete later found to be acceptable, whoever was responsible for the faulty procedures should absorb the costs of any supplemental testing required. The panel or arbitration referee should decide whether or not any related costs are also to be assigned to those responsible for the improper testing.
2. If the testing of the concrete as delivered is found to have been properly performed, the costs of additional testing required to verify the acceptability of the structure should be borne by the producer. Related costs, such as structural modifications if needed, should be assigned by the panel or arbitration referee to the producer or contractor, depending upon the extent to which low strength is attributable to improper manufacture of the concrete or to deficiencies in its handling, placing and curing.

Obviously, clear-cut establishment of responsibility is not always possible, and compromise may be the appropriate solution. The engineer must realize his responsibility to prepare lucid, noncontradictory specifications and to apply sound judgment in their application and interpretation. This factor should be taken into account in the assignment of responsibility for costs resulting from questioned strength."

7

Fire Resistance

ARMAND H. GUSTAFERRO*

7.1 INTRODUCTION

7.1.1 Definitions

The following definitions apply to certain terms used in this chapter:

Fire resistance—the property of a material or assembly to withstand fire or to give protection from it. As applied to elements of buildings, it is characterized by the ability to confine a fire or to continue to perform a given structural function or both. (Defined in ASTM E 176)

Fire resistive—having fire resistance. (ASTM E 176)

Fire endurance—a measure of elasped time during which a material or assembly continues to exhibit fire resistance under specified conditions of test and performance. As applied to elements of buildings, it is measured by the methods and to the criteria defined in the "Methods of Fire Tests of Building Construction and Materials" (ASTM Designation: E 119). (ASTM E 176)

Fire resistance rating (or "fire rating")—a legal term defined in building codes. Fire ratings are assigned by building codes for various types of construction and are usually given in half hour increments. Fire ratings are usually based on fire endurances.

Fire rate—the amount of insurance premium per unit of valuation, usually expressed in cents per hundred dollars valuation. Fire rates are based on type of structure, occupancy, location, and to some extent, on the fire rating of component parts.

7.1.2 Fire Rating Requirements in Building Codes

Table 7-1 summarizes some of the requirements for fire resistance ratings that appear in model building codes. For simplicity, only certain values are shown, and qualifying statements have been omitted. As a result, Table 7-1 should not be used to determine specific rating requirements. Instead, the table is included merely to indicate the range and trends in building code requirements.

The fire resistance ratings shown in Table 7-1 are given in hours, and are based on the criteria of ASTM E 119.

7.1.3 Standard Fire Tests of Building Construction and Materials (ASTM E 119)

The fire resistive properties of building components are measured and specified according to this common standard. Performance is defined as the period of resistance to standard fire exposure elapsing before the first critical end point is reached.

The standard fire exposure is defined in terms of a time-temperature relationship of the fire shown in Fig. 7-1. The

*Consulting Engineer, The Consulting Engineers Group, Inc., Glenview, Illinois.

TABLE 7-1 Typical Fire Resistive Rating Requirements in Model Building Codes

Structural Element	Type of Construction			Model Code*
	Highest Fire Resistive	2nd Highest Fire Resistive	Highest Non-Combustible	
Columns–supporting more than one floor	4 hrs	3 hrs	1 hr	National
	4	3	2	Basic
	4	3	2	Southern
	3	2	1	Uniform
Girders, Beams, and Trusses	4	3	1	National
	4	3	2	Basic
	4	3	2	Southern
	3	2	1	Uniform
Floors	3	2	1	National
	3	2	1.5	Basic
	2.5	1.5	N.C.**	Southern
	2	2	1	Uniform
Roofs	2	1.5	1	National
	2	1.5	¾	Basic
	1.5	1	N.C.**	Southern
	2	1	1	Uniform

National = National Building Code *promulgated by the American Insurance Association.*
Basic = Basic Building Code *promulgated by the Building Officials Conference of America, Inc.*
Southern = Southern Standard Building Code *promulgated by the Southern Building Code Congress.*
Uniform = Uniform Building Code *promulgated by the International Conference of Building Officials.*
***Noncombustible*

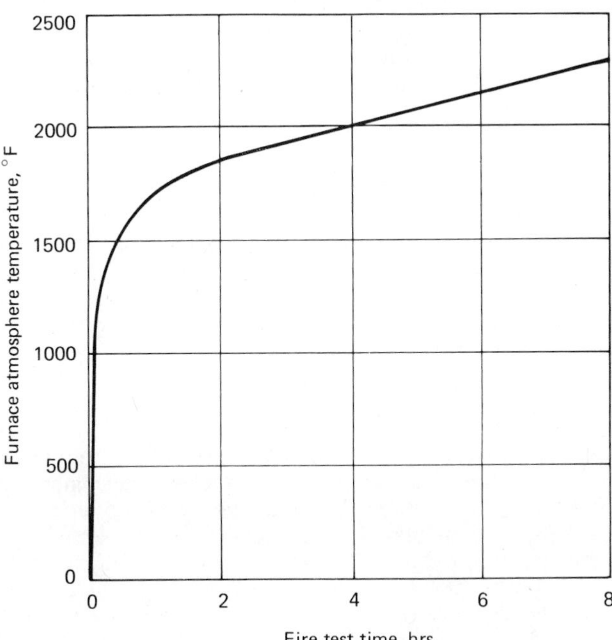

Fig. 7-1 Furnace atmosphere temperature specified in ASTM E119.[7-2]

(Permission to reprint granted by the American Society for Testing and Materials, Copyright ASTM.)

fire represents combustion of about 10 lb of wood (with a heat potential of 8,000 BTU per lb) per sq ft of exposure area per hr of test. Actually, the fuel consumed during a fire test is dependent on the furnace design and on the heat capacity of the test assembly. For ,example, the amount of fuel consumed during a fire test of an exposed concrete floor specimen is likely to be 10 to 20% greater than that used for test of a floor with an insulated ceiling, and considerably greater than that for a combustible assembly.

The standard, ASTM E 119, specifies minimum sizes of specimens to be exposed in fire tests. For floors and roofs, at least 180 sq ft must be exposed to fire from beneath, and neither dimension can be less than 12 ft. For tests of walls, either loadbearing or nonloadbearing, the minimum specified area is 100 sq ft with neither dimension less than 9 ft. The minimum length for columns is specified to be 9 ft, while for beams it is 12 ft.

During fire tests of floors, roofs, beams, loadbearing walls, and columns, the maximum permissible superimposed load is applied. The standard permits alternate tests of large steel beams and columns in which a superimposed load is not required, but the test end point criteria are modified for such tests.

Floor and roof specimens are exposed to fire from beneath, beams from the bottom and sides, walls from one side, and columns from all sides.

End Point Criteria:

a. Loadbearing specimens must sustain the applied loading–collapse is an obvious end point.
b. Holes, cracks, or fissures through which flames or gases hot enough to ignite cotton waste must not form.
c. The temperature of the unexposed surface of floors, roofs, or walls must not rise an average of 250°F or a maximum of 325°F at any one point.
d. Walls must sustain a hose stream test (simulating in a specified manner a fire fighter's hose stream) and twice the superimposed load after the fire test.
e. In alternate tests of large steel beams (not loaded during test) the end point occurs when the steel temperature reaches an average of 1,000°F or a maximum of 1,200°F at any one point.

In 1970, new end point criteria were tentatively added to ASTM E119 for floors, roofs, and beams fire tested in a "restrained" condition. Restrained in this case means that

TABLE 7-2 Construction Classification Restrained and Unrestrained[7-2]

I. *Wall bearing:*	
Single span and simply supported end spans of multiple bays:[a]	
(1) Open-web steel joists or steel beams, supporting concrete slab, precast units, or metal decking	unrestrained
(2) Concrete slabs, precast units, or metal decking	unrestrained
Interior spans of multiple bays:	
(1) Open-web steel joists, steel beams or metal decking, supporting continuous concrete slab	restrained
(2) Open-web steel joists or steel beams, supporting precast units or metal decking	unrestrained
(3) Cast-in-place concrete slab systems	restrained
(4) Precast concrete where the potential thermal expansion is resisted by adjacent construction[b]	restrained
II. *Steel framing:*	
(1) Steel beams welded, riveted or bolted to the framing members	restrained
(2) All types of cast-in-place floor and roof systems (such as beam-and-slabs, flat slabs, pan joists and waffle slabs) where the floor or roof system is secured to the framing members	restrained
(3) All types of prefabricated floor or roof systems where the structural members are secured to the framing members and the potential thermal expansion of the floor or roof system is resisted by the framing system or the adjoining floor or roof construction[b]	restrained
III. *Concrete framing:*	
(1) Beams securely fastened to the framing members	restrained
(2) All types of cast-in-place floor or roof systems (such as beam-and-slabs, flat slabs, pan joists and waffle slabs) where the floor system is cast with the framing members	restrained
(3) Interior and exterior spans of precast systems with cast-in-place joints resulting in restraint equivalent to that which would exist in condition III(1)	restrained
(4) All types of prefabricated floor or roof systems where the structural members are secured to such systems flooring and the potential thermal expansion of the floor or roof systems is resisted by the framing system or the adjoining floor or roof construction.[b]	restrained
IV. *Wood construction:*	
All types	unrestrained

[a]Floor and roof systems can be considered restrained when they are tied into walls with or without tie beams, which walls are designed and detailed to resist thermal thrust from the floor or roof system.
[b]For example, resistance to potential thermal expansion is considered to be achieved when:
(1) Continuous structural concrete topping is used,
(2) The space between the ends of precast units or between the ends of units and the vertical face of supports is filled with concrete or mortar, or
(3) The space between the ends of precast units and the vertical faces of supports, or between the ends of solid or hollow core slab units does not exceed 0.25 per cent of the length for normal weight concrete members or 0.1 per cent of the length for structural lightweight concrete members.
[Permission to reprint granted by the American Society for Testing and Materials, Copyright ASTM.]

thermal expansion of the specimen is restricted during the fire test. Two classifications can be derived from fire tests of restrained specimens, unrestrained and restrained. Only unrestrained assembly classifications can be obtained from tests of unrestrained specimens. The tentative revision of ASTM E119 includes a guide for classifying constructions as restrained or unrestrained, Table 7-2. It can be noted that cast-in-place and most precast concrete constructions are considered to be restrained.

7.2 PROPERTIES OF STEEL AND CONCRETE AT HIGH TEMPERATURES

The physical properties of steel and concrete are affected by temperature. At temperatures encountered in fires, strength and modulus of elasticity are reduced.

7.2.1 Steel

Figure 7-2 shows stress-strain diagrams for ASTM A-36 and A-421 steels at various temperatures.

Figure 7-3 shows the range of yield strengths at high temperatures for structural grade steels,[7-27] and Fig. 7-4 shows the relationship between temperature and ultimate strength of ASTM A-421 steel and high strength alloy steel bars.[7-9, 7-29] The information in these graphs can be used in calculations of the capacities of reinforced or prestressed concrete members at high temperatures. Note that for structural steels half the room temperature strength is retained at about 1,100°F, while for cold-drawn prestressing steel the comparable temperature is 800°F.

Figure 7-5 shows the relationship between the temperature and modulus of elasticity of structural steels.[7-27]

Figure 7-6 shows the thermal expansion of steel at elevated temperature.[7-37]

7.2.2 Concrete

The properties of concrete at high temperatures are affected to some extent by the type of aggregate. Most structural concretes can be classified into three aggregate types: carbonate, siliceous, and lightweight. Carbonate aggregates include limestone and dolomite, and are grouped together because they undergo chemical changes at temperatures in the range of 1,300°F to 1,800°F. Siliceous aggregates, which include granite, quartzite, sandstones, schists, and other material consisting largely of silica, do not undergo chemical changes at temperatures commonly encountered in fire. Although an abrupt volume change accompanies the inversion of quartz from the α- to β-form at about 1,060°F, siliceous aggregate concretes do not exhibit the same abrupt change in volume or other physical properties. Both the carbonate and siliceous aggregate concretes generally have unit weights in the 140 to 150 pcf range. Structural lightweight concretes have strengths in excess of about 2,500 psi and unit weights within the range of about 90 to 115 pcf. Lightweight aggregates are either manufactured by expanding shale, slate, clay, slag, or fly ash, or occur naturally. The expanded shales, clays, and slates are heated to about 1900°F to 2000°F during manufacture. At these temperatures the aggregates become molten. As a result, such lightweight aggregates near the surface of concrete subjected to standard fire tests begin to soften after about four hours exposure. In most practical situations, this softening is not significant.

Figure 7-7 shows data on the strength of concrete at high temperatures.[7-28] The data show strengths of concrete specimens stressed to 40% of their compressive strength during the heating period. After the test temperature was

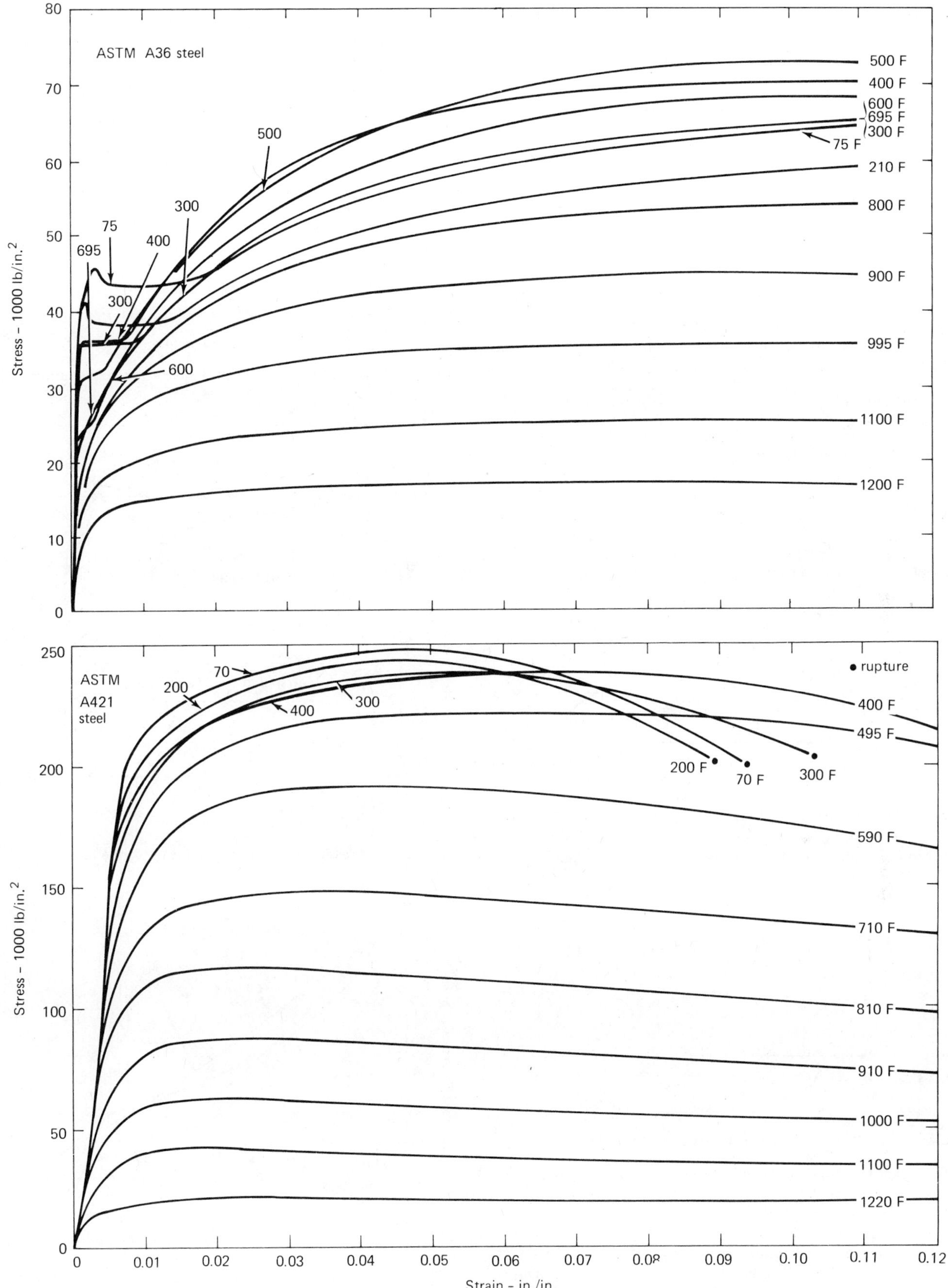

Fig. 7-2 Stress-strain curves for ASTM A36 and A421 steels at various elevated temperatures.[7-4] (Permission to reprint granted by the American Society for Testing and Materials, Copyright ASTM.)

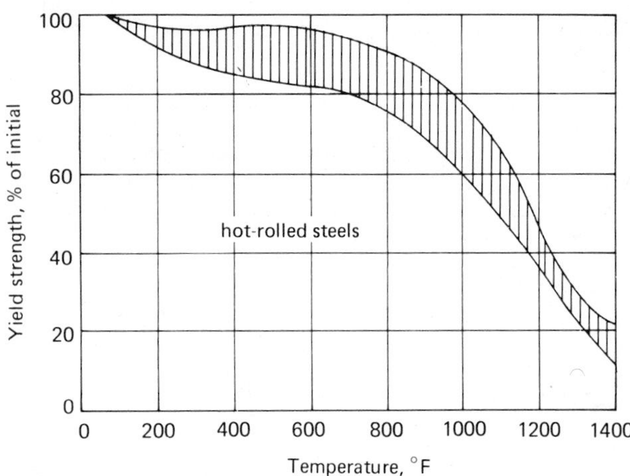

Fig. 7-3 Range of yield strengths of structural grade steels at high temperatures. Data may be used for reinforcing bars.[7-27]

[Figures from (USS) Steel Design Manual (Copyright 1968).]

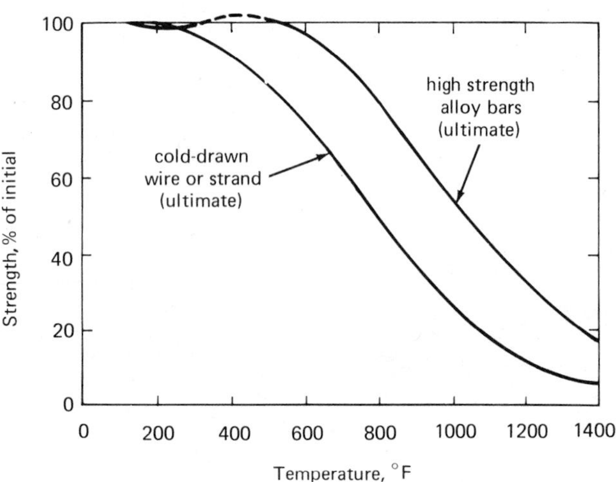

Fig. 7-4 Ultimate strength of prestressing steels at high temperatures.[7-9, 7-27]

[Figures from (USS) Steel Design Manual (Copyright 1968).]

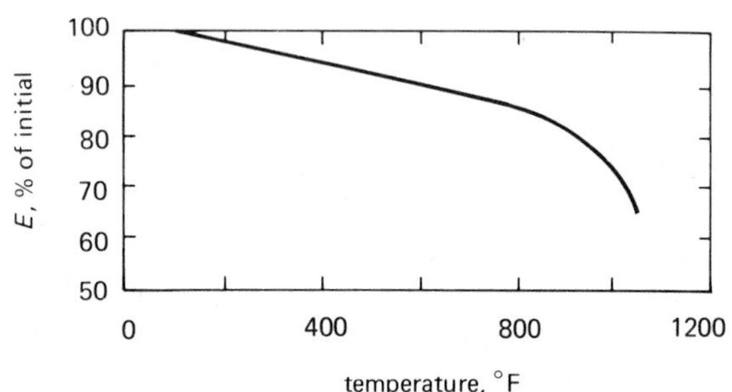

Fig. 7-5 Modulus of elasticity of steel at high temperatures.[7-27]

[Figures from (USS) Steel Design Manual (Copyright 1968).]

reached, the load on the specimen was increased until failure occurred. Note that the strengths of carbonate and lightweight aggregate concretes are only slightly reduced at temperatures up to 1,200°F. The strength of siliceous aggregate concrete is reduced to about 55% at 1,200°F.

Figure 7-8 shows the modulus of elasticity of concretes at high temperatures,[7-13] and Fig. 7-9 shows data on thermal expansion of concretes. A comparison of Fig. 7-6 and

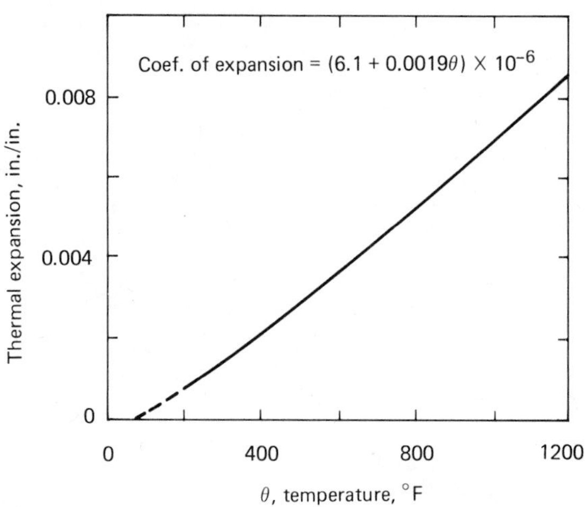

Fig. 7-6 Thermal expansion of steel at elevated temperatures.[7-37]

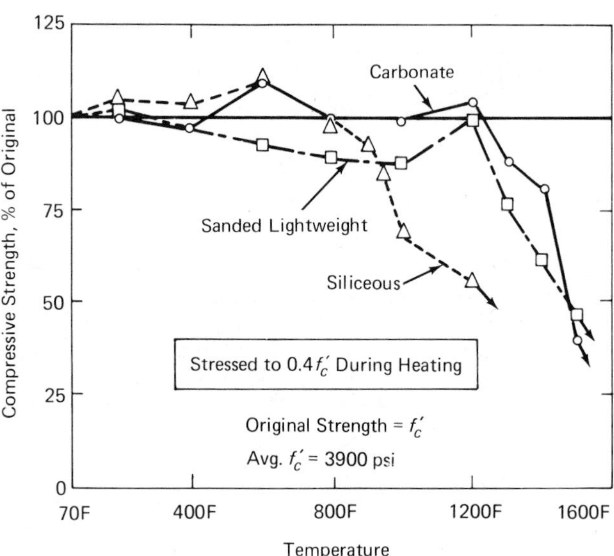

Fig. 7-7 Compressive strength of concrete at high temperatures.[7-28]

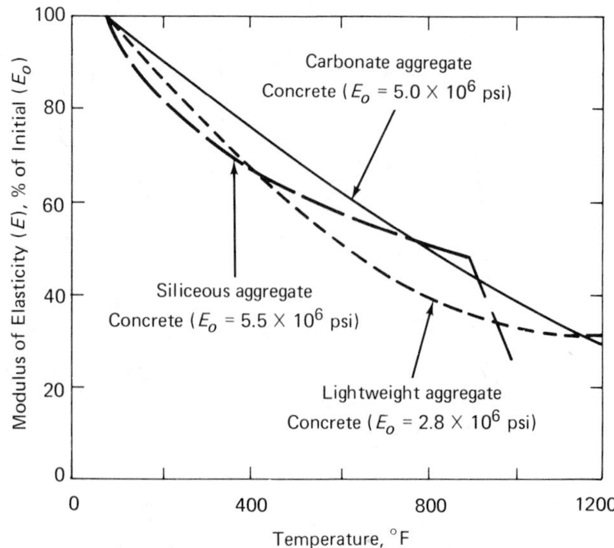

Fig. 7-8 Modulus of elasticity of concrete at high temperatures.[7-13]

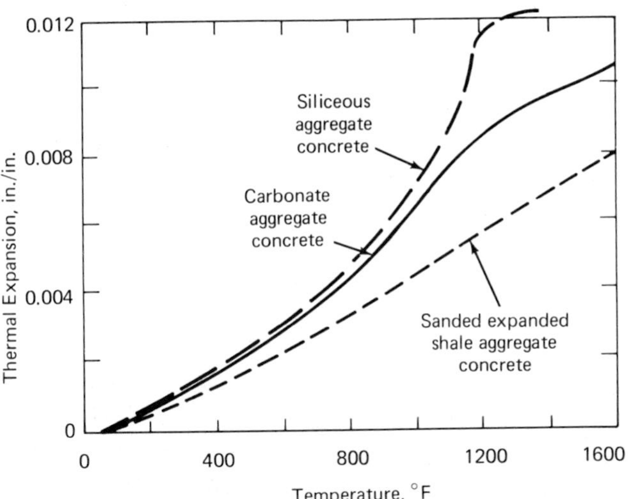

Fig. 7-9 Thermal expansion of concrete at high temperatures.

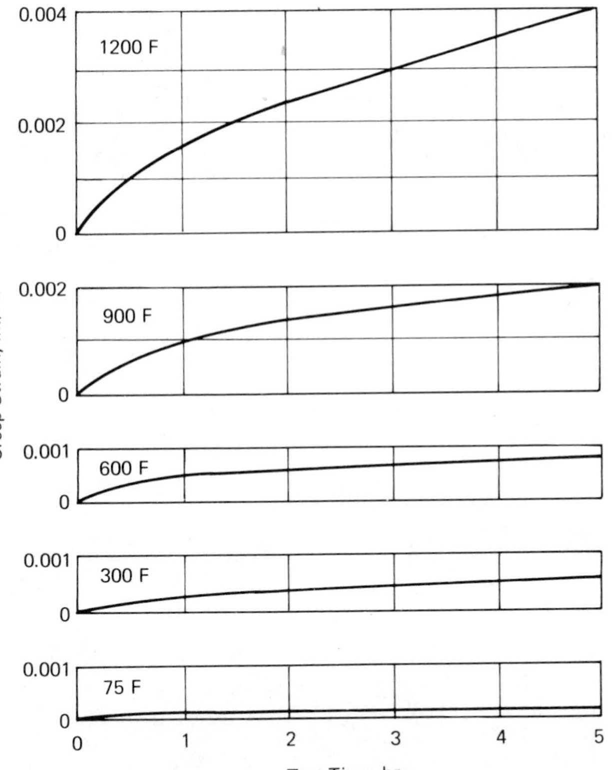

Fig. 7-10 Creep of a carbonate aggregate concrete at various temperatures (applied stress = 1800 psi, f'_c = 4000 psi.)[7-23]

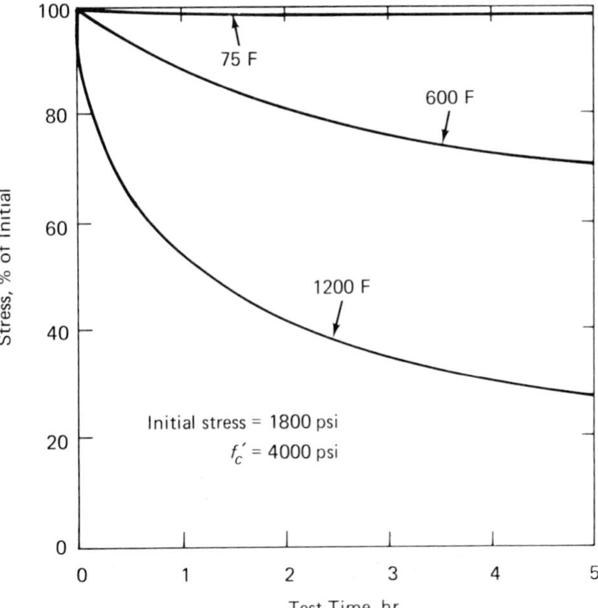

Fig. 7-11 Stress relaxation of a carbonate aggregate concrete at various temperatures.[7-23]

Fig. 7-9 indicates that the thermal expansion of concrete and steel are about the same throughout the range encountered in fires.

At high temperatures, creep and relaxation of concrete increase significantly. Figures 7-10 and 7-11 show data on creep and relaxation of carbonate aggregate concretes.[7-23]

7.3 TEMPERATURES WITHIN CONCRETE MEMBERS DURING FIRES

Figure 7-12 shows data on temperatures within solid concrete slabs exposed to a standard fire (ASTM E 119) on one side.[7-19] The curves in Fig. 7-12 are applicable to slabs of any thickness provided that the slab thickness is at least about 1 in. thicker than the curve in question. If a steel bar is centered 1 in. above the underside of a carbonate aggregate concrete slab exposed to an ASTM E 119 fire from beneath, its temperature will reach 1,100°F at about 2 hr 20 min. Thus, if the critical temperature of a bar is 1,100°F, the fire endurance of the slab would be 2 hr 20 min.

Graphs of temperatures within concrete beams are not as simple as those for slabs because beams are heated from more than one face. In standard fire tests, beams are fired from beneath, so the sides as well as the bottom are generally exposed to fire.

Figures 7-13 and 7-14 show isothermal plots of temperatures within concrete beams after various periods of fire exposure. These are used to estimate temperatures within beams of any size during standard fire exposure by interpolation, if necessary.

7.4 FIRE ENDURANCE DETERMINED BY 250° RISE OF UNEXPOSED SURFACE

The standard fire test method (ASTM E 119) limits the average temperature rise of the unexposed surface of floors, roofs, and walls to 250°F. Some building codes modify this requirement. For example, the *National Building Code of Canada* waives the temperature rise criteria for roofs. The *Wisconsin State Building Code* reduces the fire endurance requirement determined by heat transmission (i.e., 250°F rise) to half that required for structural integrity for many occupancies. For concrete assemblies, this temperature rise depends mainly on the thickness and unit weight of the concrete, but it is also influenced by aggregate type, concrete moisture condition, air content, maximum aggregate size, and aggregate moisture content at time of mixing. Figure 7-15 shows typical unexposed surface temperatures recorded during standard fire tests.[7-19] The data in Fig. 7-15 can be used to determine the fire endurance of concrete slabs as affected by the temperature rise of the unexposed

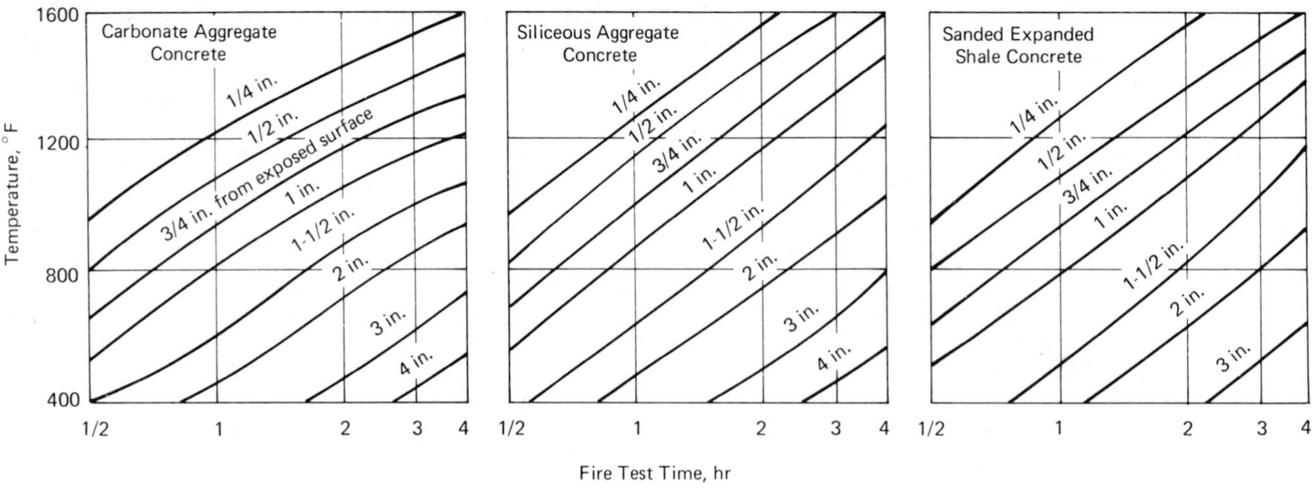

Fig. 7-12 Temperatures within concrete slabs during fire tests.[7-19]

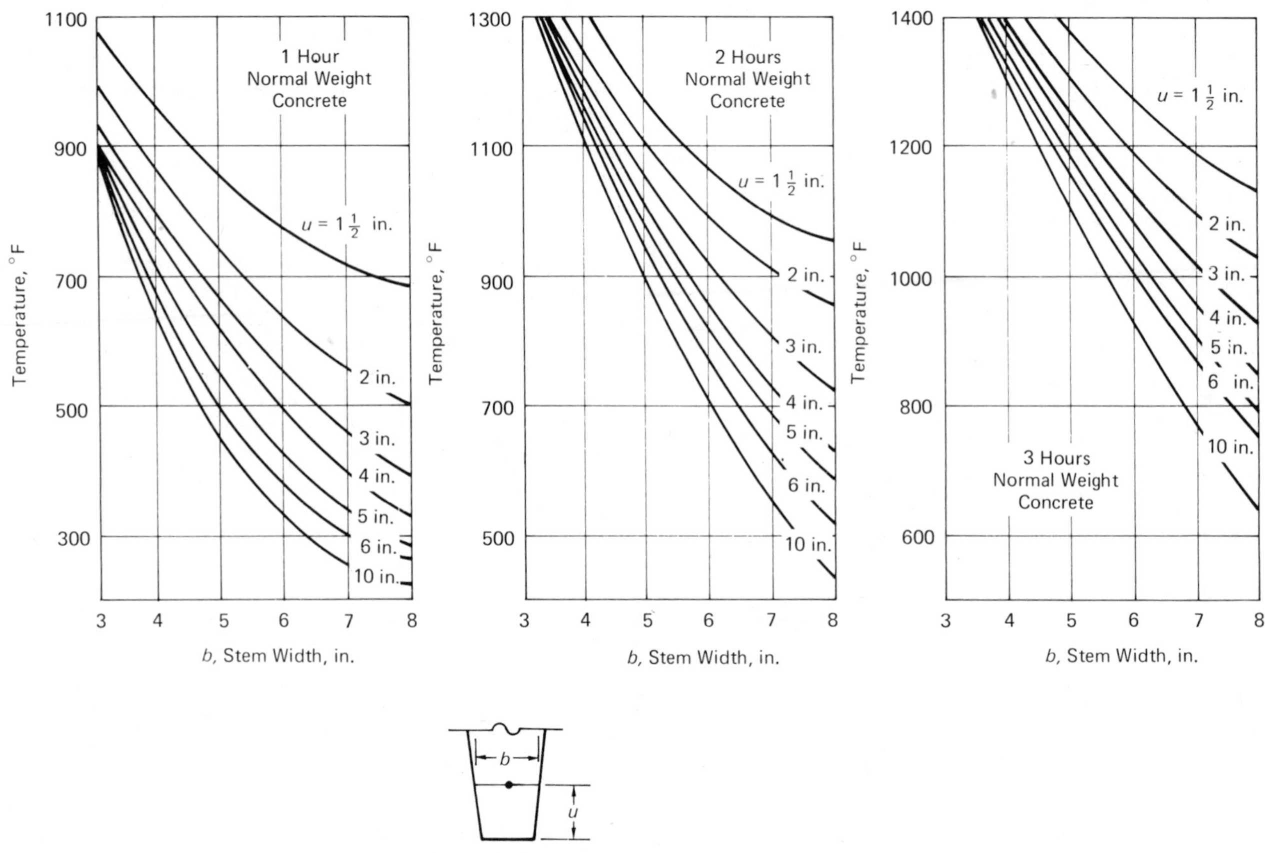

Fig. 7-13 Temperatures within beams during fire tests—normal weight concrete.

surface. In addition, the data can be used for solving special problems. For example, in a storage vault for computer tapes it may be desirable to limit the unexposed surface temperature to, say 200°F for a 2-hr fire duration. If the ambient temperature is 70°F, the rise to 200°F is 130°F. To determine the floor slab thickness to meet these requirements, enter the charts in Fig. 7-15 with a time of 2 hr and a rise of 130°F. Slab thicknesses of about 6.75 in. for carbonate aggregate concrete, 7 in. for siliceous, or 5.5 in. for sanded lightweight concrete would satisfy the conditions in the example.

7.4.1 One-Course Slabs

Figure 7-16 shows the relationship between slab thickness and fire endurance for a wide range of structural concretes.[7-19] The values shown represent air-entrained concretes fire tested when the concrete was at the standard moisture condition (75% R.H. at middepth), made with air-dried aggregates having a nominal maximum size of 0.75 in. It can be seen that, in general, fire endurance increases with a decrease in unit weight. Lightweight concretes generally weigh about 95 to 105 pcf while sanded lightweight con-

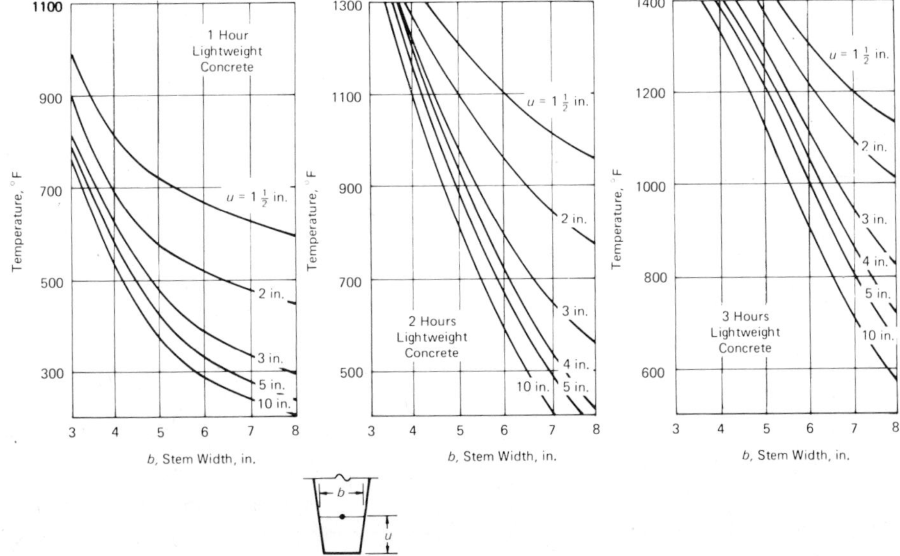

Fig. 7-14 Temperatures within beams during fire tests—lightweight concrete.

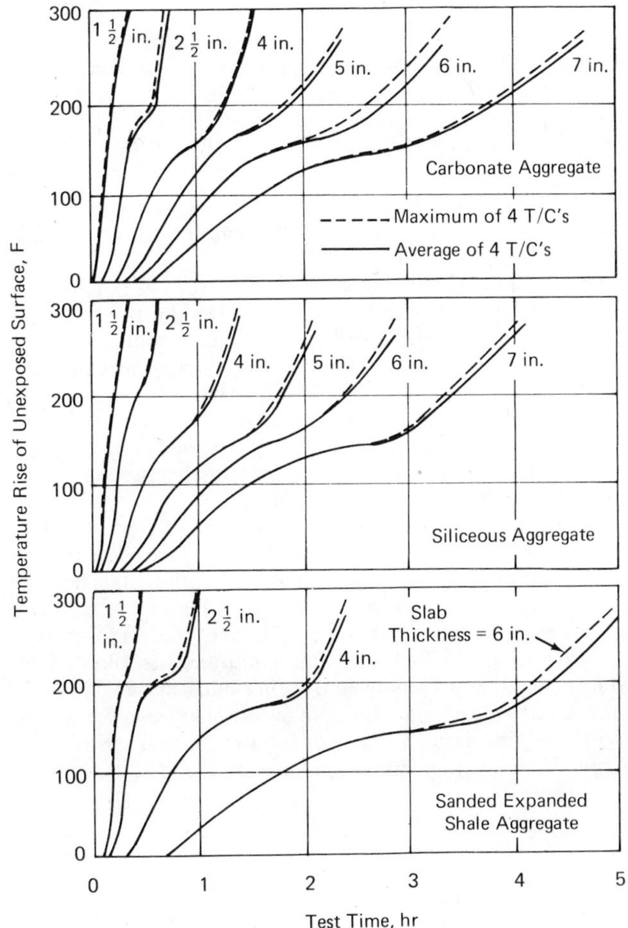

Fig. 7-15 Typical temperature rise of unexposed surface during fire tests of concrete slabs.[7-19]

cretes generally weigh about 105 to 120 pcf. Also, aggregate type influences fire endurance. Note that for 2 hr, a siliceous aggregate concrete slab must be about 5 in. thick. For carbonate aggregate concrete, the comparable thickness is about 4.7 in., while for structural lightweight concretes

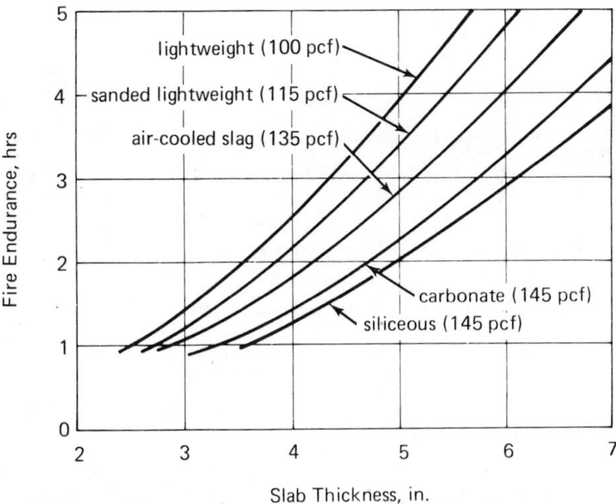

Fig. 7-16 Effect of slab thickness and aggregate type on fire endurance of concrete slabs, based on 250°F rise of unexposed surface.[7-19, 7-35]

the slab thickness is between 3.5 and 3.9 in., depending on the aggregate type and concrete unit weight.

Fire endurance increases with increasing moisture content of the concrete. An appendix to ASTM E 119-71 gives a method for calculating the effect of moisture on fire endurance.

The fire endurance of a concrete slab increases with an increase in air content, particularly for air contents above 10%. However, this effect is not significant for air contents less than about 7%.

The fire endurance of concrete slabs increases as the amount of mortar in the concrete increases. Thus, concretes with small maximum aggregate sizes have longer fire endurances than those with larger sizes.

Factors such as water-cement ratio, cement content, and slump have almost no influence on fire endurance within the normal ranges for structural concretes.

7.4.2 Two-Course Floors and Roofs

Floors and roofs frequently consist of concrete base slabs with insulating undercoatings or overlays of other types of

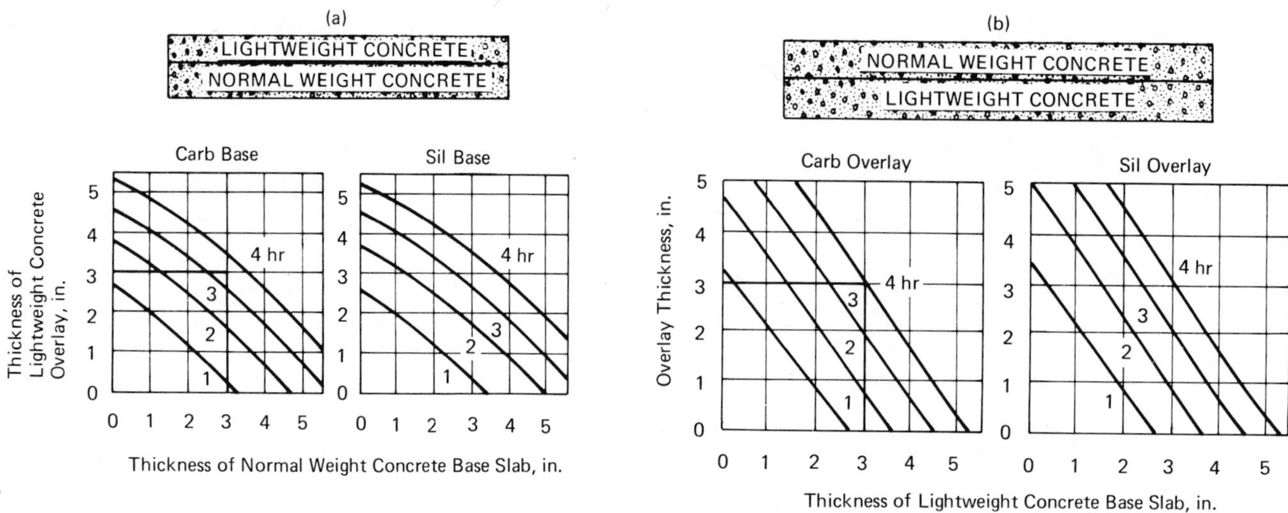

Fig. 7-17 Fire endurance of two-course floors—combinations of normal weight concrete and lightweight concrete. (a) Lightweight concrete overlays on normal weight concrete base slabs; and (b) normal weight concrete overlays on lightweight concrete base slabs.[7-24]

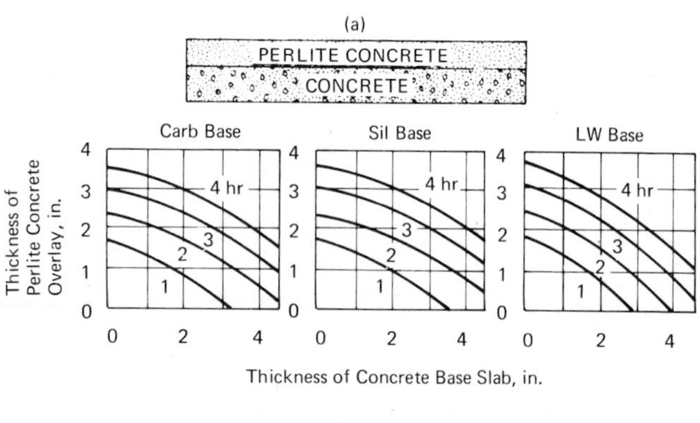

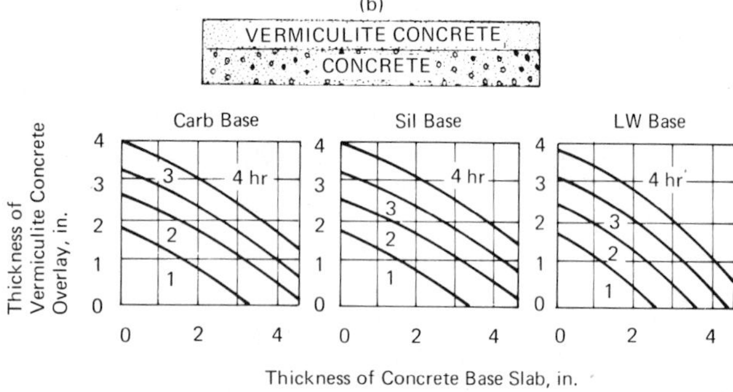

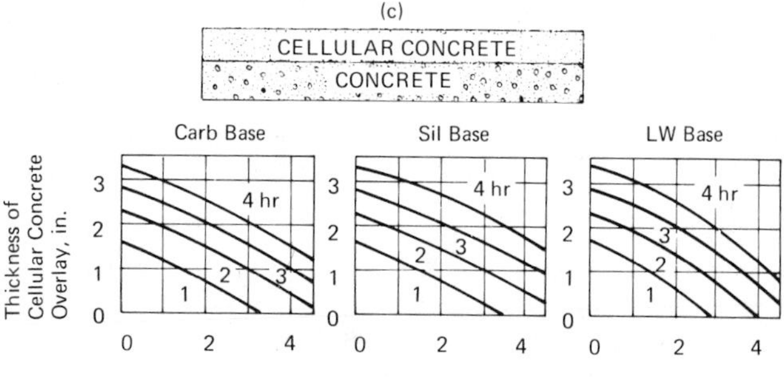

Fig. 7-18 Fire endurance of two-course roofs—concrete slabs with insulating concrete (30 pcf) overlays. (a) Perlite concrete overlays on concrete base slabs; (b) vermiculite concrete overlays on concrete base slabs; and (c) cellular concrete overlays on concrete base slabs.[7-24]

concrete or insulating materials. Also, roofs generally have built-up roofing.

If the fire endurances of the individual courses are known, the fire endurance of the composite assembly can be estimated by the formula:

$$R = (R_1^{0.59} + R_2^{0.59} + \cdots R_n^{0.59})^{1.7}$$

where R = the fire endurance of the composite assembly in minutes, and R_1, R_2, R_n = the fire endurances of the individual courses in minutes.[7-38]

This formula has shortcomings in that it does not account for the location of the course relative to the fired surface. The course nearest the fire has the greatest influence on the fire endurance. Thus, if a two-course floor consists of equal thicknesses of normal weight and lightweight concretes, the fire endurance would be longer if the lightweight concrete were exposed to fire, than if the courses were reversed. Nevertheless, the formula generally estimates the fire endurance within about 10% of a test value.[7-24]

Figure 7-17 thru 7-20 show fire endurances of various two-course floors.[7-24] From Fig. 7-17(a) it can be seen that a floor consisting of a 3-in. base slab of carbonate aggregate concrete and a 3-in. overlay of lightweight concrete will have a fire endurance of about 3.5 hours. With the courses reversed, Fig. 7-17(b), the fire endurance is about 4 hr. Standard three-ply built-up roofing adds about 10 to 20 minutes to most concrete roof assemblies, but with some rigid board insulations, the addition can be an hour or more. Asphalt floor tile does not affect the fire endurance significantly.

7.4.3 Walls

The structural fire endurance of concrete walls is seldom a governing factor, because it is generally much longer than the fire endurance determined by temperature rise of the unexposed surface.

Walls made of cast-in-place or precast structural concrete have about the same fire endurance (250°F rise of unexposed surface) as floors or roofs made of the same materials. Thus the data in Figs. 7-16 through 7-19 are applicable for walls as well as floors and roofs. The fire endurance of masonry walls depends mainly on the equivalent thickness and aggregate type. Table 7-3 shows typical equivalent thicknesses for hourly fire endurances of masonry walls made of various

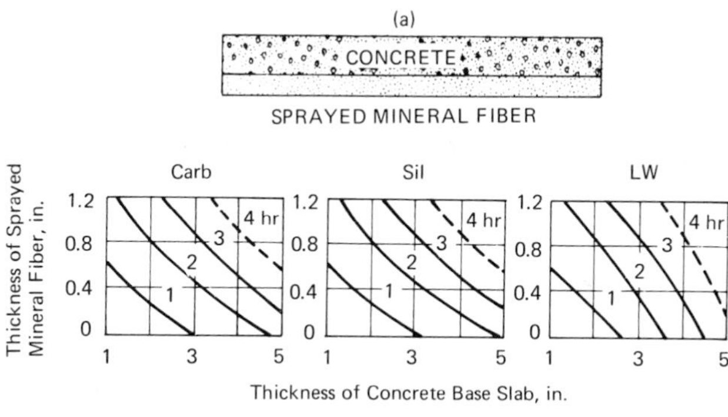

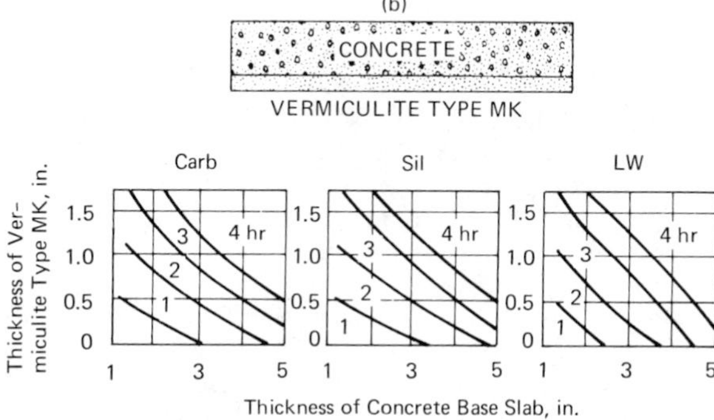

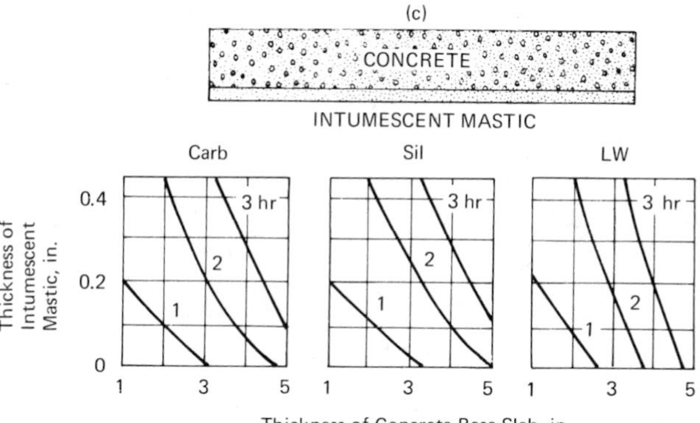

Fig. 7-19 Fire endurance of two-course floors—slabs with sprayed insulation. (a) Concrete floors undercoated with sprayed mineral fiber; (b) concrete floors undercoated with sprayed vermiculite type MK; and (c) concrete floors undercoated with sprayed intumescent mastic.[7-24]

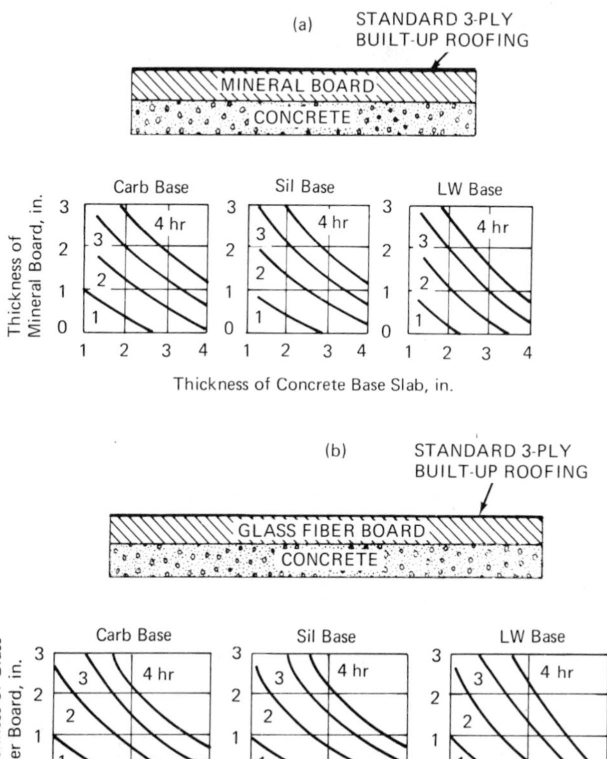

Fig. 7-20 Fire endurance of two-course roofs—rigid board insulation with built-up roofing. (a) Mineral board insulation on concrete roofs; and (b) glass fiber board insulation on concrete roofs.[7-24]

TABLE 7-3 Minimum Equivalent Thickness of Concrete Masonry Walls for Various Fire Endurances

Coarse Aggregate Type	*Minimum Equivalent Thickness, in., for Fire Endurance of*			
	1 Hr	*2 Hr*	*3 Hr*	*4 Hr*
Expanded Slag or Pumice	2.1	3.2	4.0	4.7
Expanded Clay or Shale	2.6	3.8	4.8	5.7
Limestone or Air-Cooled Slag	2.7	4.0	5.0	5.9
Calcareous Gravel	2.8	4.2	5.3	6.2
Siliceous Gravel	3.0	4.5	5.7	6.7

(a) From Reference[7-32]

Coarse Aggregate Type	*Minimum Equivalent Thickness, in., for Fire Endurance of*			
	1 Hr	*2 Hr*	*3 Hr*	*4 Hr*
Siliceous	2.9	4.4	5.5	6.6
Non-Siliceous Normal Weight	2.7	4.1	5.2	6.2
Lightweight, partially sanded	2.5	3.8	4.8	5.7
Lightweight, unsanded	2.4	3.6	4.4	5.2

(b) Adapted from Reference[7-33]

aggregates. The values shown are those included as parts of two different codes, the *National Building Code* (AIA),[7-32] and the *National Building Code of Canada*.[7-33] Equivalent thickness is determined by dividing the solid volume of a masonry unit by its face area. The volume can be determined by a number of methods including the immersion method, the sand or shot method, or by accurate measurements. Results of the three methods do not agree precisely, partly because the immersion method is influenced by the unit's surface texture and absorption, and the measurement method depends on accuracy of determining fillet radii and core taper. The shot method is reported to be accurate and reproducible to within about 1%.

EXAMPLE 7-1: What is the fire endurance of a wall made of 8 × 8 × 16-in. block if the actual dimensions of the block are 7.65 in. high by 15.60 in. long, and the volume is 471 cu in.? The aggregate is expanded shale.

SOLUTION:

$$\text{equivalent thickness} = \frac{471}{15.60(7.65)} = 3.95 \text{ in.}$$

From Table 17-3, the fire endurance can be estimated to be slightly greater than two hours.

7.5 FIRE ENDURANCE OF SIMPLY SUPPORTED (UNRESTRAINED) SLABS AND BEAMS

7.5.1 Structural Behavior

Figure 7-21 shows a simply supported reinforced concrete slab. The rocker and roller supports indicate that the ends of the slab are free to rotate and expansion can occur without resistance. The reinforcement consists of straight bars located near the bottom of the slab. If the underside of the slab is exposed to fire, the bottom of the slab will expand more than the top, resulting in a deflection of the slab. The strength of the concrete and steel near the bottom of the slab will decrease as the temperature increases. When the strength of the steel reduces to that of the stress in the steel, flexural collapse will occur.[7-16]

Figure 7-22 illustrates the behavior of a simply supported slab exposed to fire from beneath. If the reinforcement is straight and uniform throughout the length, the ultimate moment capacity will be constant throughout the length.

$$M_u = A_s f_y \left(d - \frac{a}{2}\right)$$

where

- A_s is the area of the reinforcing steel;
- f_y is the yield strength of the reinforcing steel;
- d is the distance from the centroid of the reinforcing steel to the extreme compressive fiber;
- a is the depth of the equivalent rectangular compressive stress block at ultimate load, and is equal to $A_s f_y / 0.85 f_c' b$ where f_c' is the cylinder compressive strength of the concrete and b is the width of the slab.

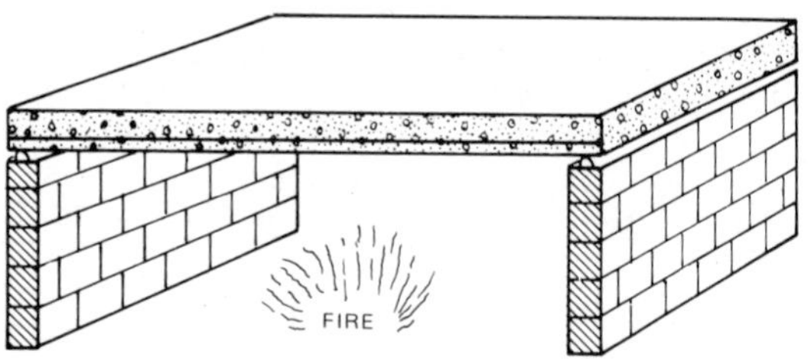

Fig. 7-21 Simply supported reinforced concrete slab subjected to fire from below.

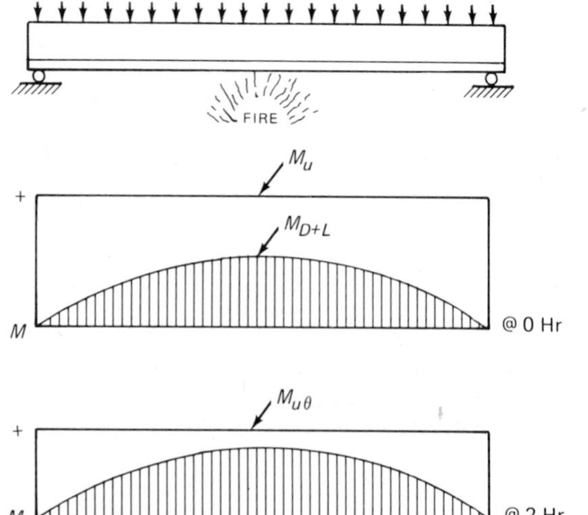

Fig. 7-22 Moment diagram for simply supported beam or slab before and during fire exposure.

If the slab is uniformly loaded, the moment diagram will be parabolic with a maximum value at midspan

$$M_{D+L} = \frac{w_{D+L}\, l^2}{8}$$

where

w_{D+L} = dead plus live load per unit of length, and
l = span length.

It is generally assumed that during a fire the dead and live loads remain constant. However, the material strengths are reduced so that the retained moment capacity is

$$M_{u\theta} = A_s f_{y\theta} \left(d - \frac{a_\theta}{2}\right)$$

in which θ signifies the effects of elevated temperatures. Note that A_s and d are not affected, but $f_{y\theta}$ is reduced. Similarly a_θ is reduced, but the concrete strength at the top of the slab, f_c', is generally not reduced significantly. If, however, the compressive zone of the concrete is heated, an appropriate reduction should be assumed.

Flexural failure can be assumed to occur when $M_{u\theta}$ is reduced to M_{D+L}. From this expression, it can be noted that the fire endurance depends on the load intensity and the strength-temperature characteristics of steel. In turn, the duration of the fire until the critical steel temperature is reached depends upon the protection afforded to the reinforcement. Usually the protection consists of the concrete cover, i.e., the thickness of concrete between the fire-exposed surface and the reinforcement. In some cases additional protective layers of insulation or membrane ceilings might be present.

For prestressed concrete the ultimate capacity formulas must be modified by substituting f_{ps} for f_y, and A_{ps} for A_s where f_{ps} is the stress in the prestressing steel at ultimate, and A_{ps} is the area of the prestressing steel. In lieu of an analysis based on strain compatability the value of f_{ps} can be assumed to be

$$f_{ps} = f_{pu} \left(1 - \frac{0.5\, A_{ps} f_{pu}}{bd\, f_c'}\right)$$

where f_{pu} is the ultimate tensile strength of the prestressing steel.

7.5.2 Estimating Fire Endurance

Figure 7-23 shows the fire endurance of simply supported concrete slabs as affected by type of reinforcement (hot-rolled reinforcing bars and cold-drawn wire or strand), type of concrete (carbonate, siliceous, and lightweight aggregate), moment intensity, and cover thickness.[7-16] If the reinforcement is distributed over the tensile zone of the cross section, the cover is the average of the covers to the individual bars or strands in the tensile zone. The curves are applicable to hollow-core slabs as well as solid slabs.

The graphs in Fig. 7-23 can be used to estimate the fire endurance of simply supported prestressed concrete beams by using "effective cover" rather than "cover." Effective cover accounts for beam width by assuming that the cover of corner bars or tendons are reduced by one half for use in calculating the average cover.

EXAMPLE 7-2: Given a simply supported one-way slab reinforced with #4 Grade 60 bars on 6-in. centers. Slab is made of carbonate aggregate concrete with $f_c' = 4000$ psi. Bars are centered 1 in. from bottom of 6-in. slab. Span is 15 ft and live load is 100 psf. Determine the fire endurance.

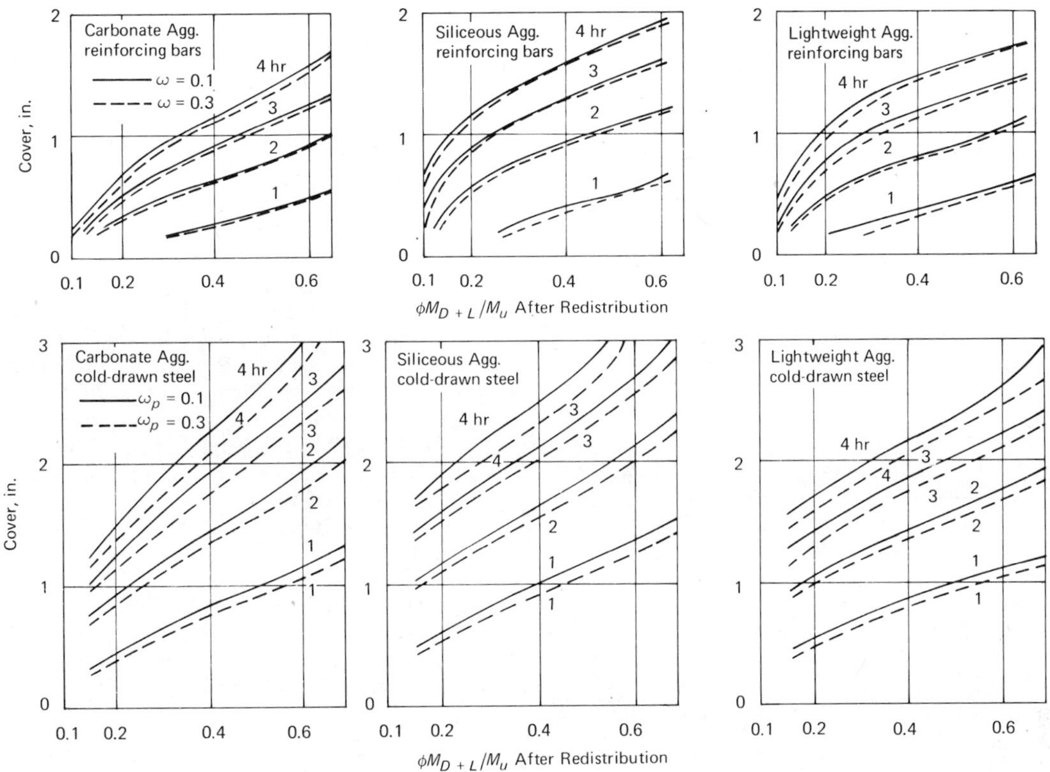

Fig. 7-23 Fire endurance of concrete slabs as influenced by aggregate type, reinforcing steel type, moment intensity, and cover.

SOLUTION:

$$w_d = \frac{6}{12}(150) = 75 \text{ psf}$$

$$M_{D+L} = (0.075 + 0.100)(15)^2/8 = 4.92 \text{ ft-kips}$$

$$M_u/\phi = A_s f_y \left(d - \frac{a}{2}\right)$$

$$a = 0.4(60)/0.85(4)(12) = 0.59 \text{ in.}$$

$$M_u/\phi = 0.4(60)(5.00 - 0.30)/12 = 9.40 \text{ ft-kips}$$

$$\frac{M_{D+L}}{M_u/\phi} = \frac{\phi M_{D+L}}{M_u} = \frac{4.92}{9.40} = 0.523$$

$$\omega = A_s f_y/bd f_c' = \frac{0.4}{12(5)} \frac{(60)}{4} = 0.10$$

From upper left graph of Fig. 7-23, with $\phi M_{D+L}/M_u = 0.523$, cover = 0.75 in., and $\omega = 0.10$, the fire endurance is about 1 hr 50 min.

EXAMPLE 7-3: Determine the fire endurance for the prestressed concrete beam shown. Lightweight aggregate concrete with $\phi M_{D+L}/M_u = 0.50$ and $\bar{\omega}_p = A_{ps} f_{pu}/bd f_c' = 0.306$

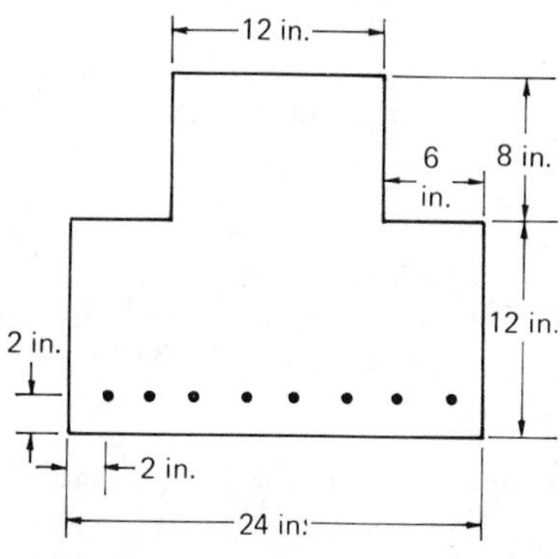

SOLUTION:

$$\text{Effective cover} = \frac{6(1.75) + 2(0.875)}{8} = 1.53 \text{ in.}$$

From lower right graph of Fig. 7-23, fire endurance is about 2 hr.

7.6 FIRE ENDURANCE OF CONTINUOUS BEAMS AND SLABS

7.6.1 Structural Behavior

Structures that are continuous or otherwise statically indeterminate, undergo changes in stresses when subjected to fire.[7-14, 7-15, 7-17, 7-26] Such changes in stress result from temperature gradients within structural members, or changes in strength of structural materials at high temperatures, or both.

Figure 7-24 shows a continuous beam whose underside is exposed to fire. The bottom of the beam becomes hotter than the top and tends to expand more than the top. This differential heating causes the ends of the beam to tend to lift from their supports thus increasing the reaction at the interior support. This action results in a redistribution of

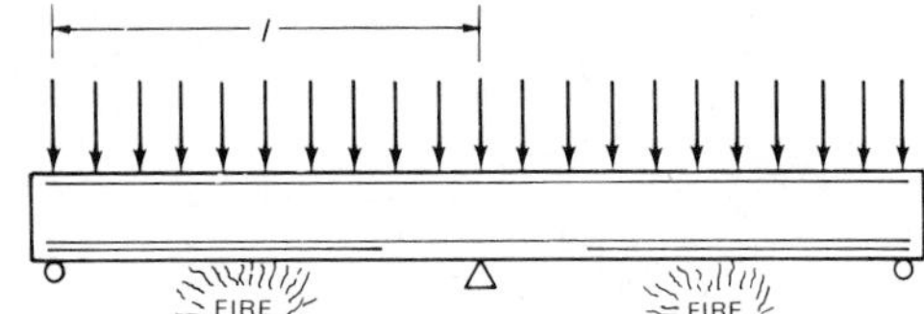

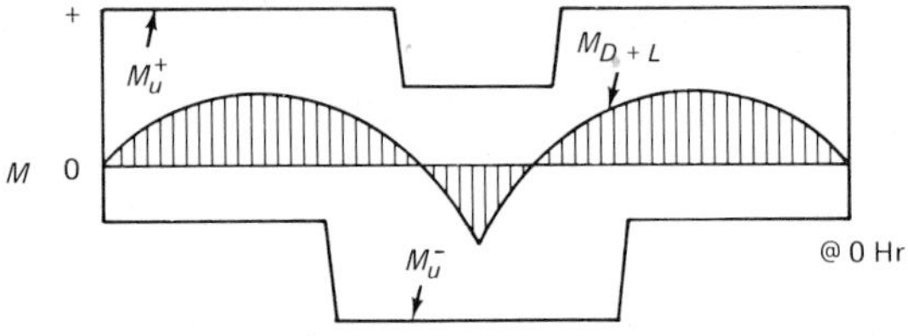

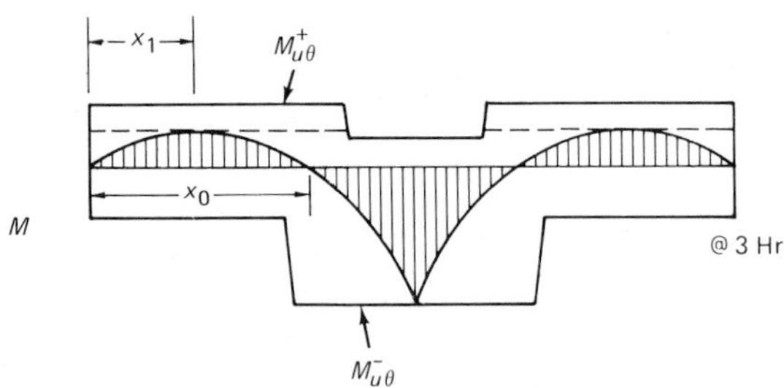

Fig. 7-24 Moment diagrams for continuous two-span beam before and during fire exposure.

moments, i.e., the negative moment at the interior support increases while the positive moments decrease.

During the course of a fire, the negative moment reinforcement (Fig. 7-24) remains cooler than the positive moment reinforcement because it is further from the fire. Thus the increase in negative moment can be accommodated. Generally the redistribution that occurs is sufficient to cause yielding of the negative moment reinforcement. The resulting decrease in positive moment means that the positive moment reinforcement can be heated to higher temperature before failure will occur. Thus, it is apparent that the fire endurance of a continuous reinforced concrete beam is generally significantly longer than that of a simply supported beam having the same cover and loaded to the same moment intensity.

7.6.2 Detailing Precautions

It should be noted that the amount of redistribution that occurs is sufficient to cause yielding of the negative moment reinforcement. Since by increasing the amount of negative moment reinforcement, a greater negative moment will be attained, care must be exercised in designing the member to assure that a secondary type of failure will not occur. To avoid a compressive failure in the negative moment region, the amount of negative moment reinforcement should be small enough so that ω, i.e., $A_s f_y / bd f_c'$, is less than about 0.30 even after reductions due to temperature in f_c', b, and d are taken into account. Furthermore, the negative moment reinforcing bars must be long enough to accommodate the complete redistributed moment and change in the location of inflection points. It is recommended that at least 20% of the maximum negative moment reinforcement in the span extend throughout the span.

7.6.3 Estimating Fire Endurance

The charts in Fig. 7-23 can be used to estimate the fire endurance of continuous beams and slabs. To use the charts, first estimate the negative moment at the supports taking into account the temperatures of the negative moment reinforcement and of the concrete in compressive zone near the supports. Then estimate the maximum positive moment after redistribution. Entering the appropriate chart with the ratio of that positive moment to the initial positive moment capacity, the fire endurance for the appropriate cover (positive moment region) can be estimated. If the resulting fire endurance is considerably different than that originally assumed in estimating the steel and concrete temperatures, a more accurate estimate can be made by trial-and-error. Usually such refinement is unnecessary.

It is also possible to design the reinforcement in a continuous beam or slab for a particular fire endurance period. From the lowermost diagram of Fig. 7-24, the beam can be expected to collapse when the positive moment capacity, $M_{u\theta}^+$, is reduced to the value indicated by the dashed horizontal line, i.e., when the applied moment at a point x, from the outer support, $M_{x1} = M_{u\theta}$.

For a uniform applied load, w,

$$M_{x_1} = \frac{wlx_1}{2} - \frac{wx_1^2}{2} + \frac{M_{u\theta}^- x_1}{l}$$

$$x_1 = \frac{M_{u\theta}^-}{wl} + \frac{l}{2}$$

and

$$M_{u\theta}^- = -\frac{wl^2}{2} + wl^2 \sqrt{\frac{2M_{u\theta}^-}{wl^2}}$$

also

$$x_0 = 2x_1$$

For a symmetrical interior bay,

$$x_1 = l/2$$

$$M_{x_1}^- = wl^2/8 + M_{u\theta}^-$$

or

$$M_{u\theta}^- = M_{u\theta}^+ - wl^2/8$$

EXAMPLE 7-4: Given a two-span reinforced concrete slab 6 in. thick reinforced for positive moment with #4 Grade 40 bars on 6-in. centers with 0.75 in. cover. Each span is 18 ft and superimposed load is 40 psf. Concrete is made with siliceous aggregate, and $f_c' = 4000$ psi. Determine amount of negative moment reinforcement required for a 2-hr fire endurance.

SOLUTION:

a. Determine positive moment capacity at 2 hr. From Fig. 7-12, $\theta = 1280°$F. From Fig. 7-3, $f_{y\theta} = 0.24$, $f_y = 9.6$ ksi, $A_s = 0.20\ (12)/6 = 0.40$ in.^{2}/ft

$$a_\theta = 0.40\ (9.6)/0.85\ (4)(12) = 0.09 \text{ in.}$$

$$M_{u\theta}^+ = 0.40\ (9.6)(5.00 - 0.05) = 19.02 \text{ in.-kip} = 1586 \text{ ft-lb}$$

b. Determine required negative moment capacity at interior support.

$$M_{u\theta}^- = -\frac{wl^2}{2} \pm wl^2 \sqrt{\frac{2M_{u\theta}^+}{wl^2}}$$

for a 1-ft strip, $w = 40 + 75 = 115$ lb/ft

$$M_{u\theta}^{-} = -\frac{115\,(18)^2}{2} \pm 115\,(18)^2\sqrt{\frac{2\,(1586)}{115\,(18)^2}} = -7760 \text{ ft-lb}$$

$$= 93.1 \text{ in.-kips}$$

c. Determine required amount of negative moment reinforcement. From Fig. 7-12, it can be estimated that the negative moment reinforcement will be about 200°F at 2 hr, so $f_y \approx 40$ ksi. Neglect the compressive zone concrete with a temperature above 1300°F (see Fig. 7-7); $d = 4.25$ in., and assume that the average f_c' in the compressive zone is 3000 psi, and that $a_\theta = 0.8$ in.

$$A_s = \frac{93.1}{40\,(4.25 - 0.40)} = 0.60 \text{ in.}^2\text{/ft}$$

use #4 Grade 40 on 4 in. centers. Check a_θ

$$a_\theta = \frac{0.60\,(40)}{0.85\,(3)(12)} = 0.78 \text{ in. ok}$$

Check

$$\omega = 0.6\,(40)/12\,(4.25)(3) = 0.157 < 0.30 \text{ ok}$$

d. Determine lengths of top bars

$$x_0 = 2x_1 = 2\left(\frac{M_{u\theta}^{-}}{wl} + \frac{l}{2}\right)$$

$$= 2\left(\frac{-7760}{115\,(18)} + \frac{18}{2}\right) = 10.5 \text{ ft}$$

Theoretically, bars could be cut off 18.0 – 10.5 = 7.5 ft (plus development length) on either side of the intermediate support, but it is suggested that 20% of the bars be continued to the support, 40% of the bars be cut off at 7.5 ft (plus development length), and the others 3.75 ft (plus development length).

7.7 FIRE ENDURANCE OF FLOORS AND ROOFS IN WHICH RESTRAINT TO THERMAL EXPANSION OCCURS

7.7.1 Structural Behavior

If a fire occurs beneath a small interior portion of a large reinforced concrete slab, the heated portion will tend to expand and push against the surrounding part of the slab. In turn, the unheated part of the slab exerts compressive forces on the heated portion. The compressive force, or thrust, acts near the bottom of the slab when the fire first occurs, but as the fire progresses the line of action of the thrust rises as the heated concrete deteriorates.[7-11] If the surrounding slab is thick and heavily reinforced, the thrust forces that occur can be quite large, but considerably less than those calculated by use of elastic properties of concrete and steel together with appropriate coefficients of expansion. At high temperatures, creep and stress relaxation play an important role. Nevertheless, the thrust is generally great enough to increase the fire endurance significantly. In most fire tests of restrained assemblies, the fire endurance is determined by temperature rise of the unexposed surface rather than by structural considerations, even though the steel temperatures are often in excess of 1500°F.[7-11, 7-14, 7-25, 7-30]

The effects of restraint to thermal expansion can be characterized as shown in Fig. 7-25. The thermal thrust acts in a manner similar to an external prestressing force, which, in effect, increases the positive moment capacity.

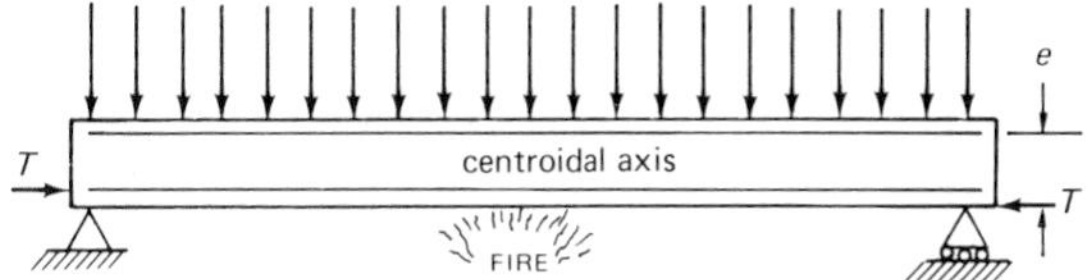

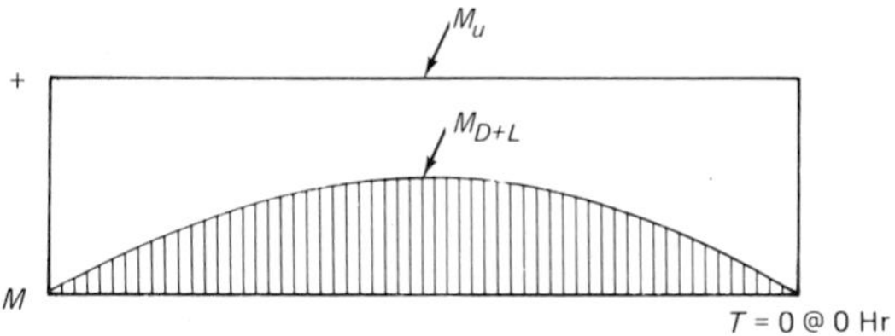

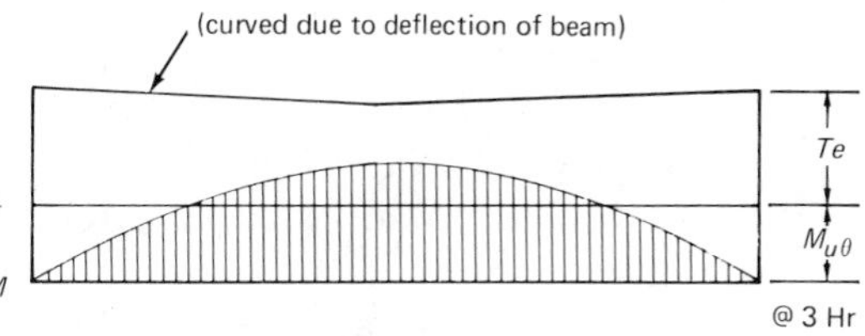

Fig. 7-25 Moment diagrams for axially restrained beam during fire exposure. Note that at three hours, M_u is less than M_{D+L} and effects of axial restraint permit beam to continue to support load.[7-14]

7.7.2 Calculation Procedure

The increase in bending moment capacity is similar to the effect of "fictitious reinforcement" located along the line of action of the thrust.[7-39] It can be assumed that the fictitious reinforcement has a yield strength (force) equal to the thrust. By this approach, it is possible to determine the magnitude and location of the required thrust to provide a given fire endurance. The procedure for estimating thrust requirements is: 1) determine temperature distribution at the required fire test duration; 2) determine the retained moment capacity for that temperature distribution; 3) if the applied moment, M_{D+L}, is greater than the retained moment capacity $M_{u\theta}$, estimate the midspan deflection at the given fire test time (if $M_{u\theta}$ is greater than M_{D+L}, no thrust is needed); 4) estimate the line of action of the thrust; and 5) calculate the magnitude of the required thrust, T; 6) calculate the thrust parameter, T/AE, where A is the gross cross sectional area of the section resisting the thrust and E is the concrete modulus of elasticity prior to fire exposure;[7-36] 7) calculate $Z = A/S$ in which S is the heated perimeter defined as that portion of the perimeter of the cross section resisting the thrust exposed to fire; 8) enter Fig. 7-26 with the appropriate thrust parameter and Z value and determine the strain parameter, L_a/L;[7-9] calculate L_a by multiplying the strain parameter by the heated length of the member; and 10) determine if the surrounding or supporting structure can support the thrust T with a displacement no greater than L_a.

The above explanation is greatly simplified because in reality restraint is quite complex, and can be likened to the behavior of a flexural member subjected to an axial force. Interaction diagrams similar to those for columns can be constructed for a given cross section at a particular stage of a fire, e.g., 2-hr of a standard fire exposure.

The guidelines in ASTM E119-71 given for determining conditions of restraint are useful for preliminary design purposes. Basically, interior bays of multi-bay floors or roofs can be considered to be restrained and the magnitude and location of the thrust are generally of academic interest only.

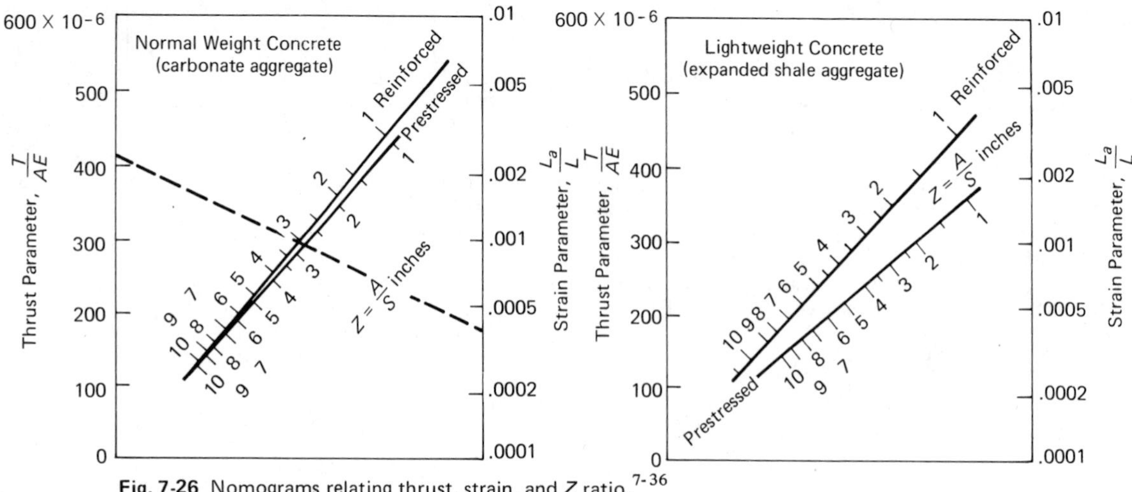

Fig. 7-26 Nomograms relating thrust, strain, and Z ratio.[7-36]
(Permission to reprint granted by the American Society for Testing and Materials, Copyright ASTM.)

7.8 FIRE ENDURANCE OF CONCRETE COLUMNS

In a standard fire test, ASTM E119-71, a column is concentrically loaded and exposed to fire on all sides. The fire endurance is the duration of fire exposure prior to structural collapse.

Historically, reinforced concrete columns have had excellent performance records in fires. Because of structural requirements, building columns are seldom smaller than 12-in. in diameter or 12-in. square. Many building codes assign ratings of four hours for 12-in. columns made of carbonate or structural lightweight aggregate concrete, and ratings of two and three hours for similar columns made of siliceous aggregate concrete. Also, for siliceous aggregate concrete, the minimum dimension for a four-hour rating is generally 16 in. in many building codes. Fire tests conducted in Europe indicate that smaller columns can be used for shorter fire endurances.

Table 7-4 shows the requirements for fire ratings of reinforced concrete columns assigned by Supplement No. 2 of the *National Building Code of Canada*.[7-33]

Even though the standard fire test for columns calls for fire exposure on all sides of a column, a more severe situation may exist when a column is exposed on one, two, or three sides. To evaluate the structural consequences of such a fire exposure, an estimate of the temperature distribution must be made. The isotherms in rectangular columns exposed to fire on three sides will be similar to those for beams of the same dimensions. For columns exposed to fire on one side only, the isotherms can be assumed to be similar to those in a slab. The temperature-strength relationships of steel and concrete can then be applied, and the capacity estimated.

During a standard fire test of a column, a constant concentric load is applied. In many situations the load on a column exposed to fire will increase due to thermal expansion of the column and the restraint of the members framing into the column. This situation is likely to occur in lower story columns of multistory buildings. In general, such columns are large and can withstand fires of long duration while sustaining considerable overloads.

NOTATION—CHAPTER 7

A = gross cross sectional area of a section
A_{ps} = area of prestressing steel
A_s = area of reinforcing steel
a = depth of equivalent rectangular stress block at ultimate load
b = width of the compressive zone of a flexural member
d = distance between the centroid of the reinforcing steel and the extreme compressive fiber

TABLE 7-4 Minimum Size and Cover Requirements for Columns as Given in the *National Building Code of Canada*, 1965.[7-33]

Item	Concrete Aggregate Type	Minimum Value in Inches for Fire Resistance Rating of						
		0.50 hr	0.75 hr	1 hr	1.5 hr	2 hr	3 hr	4 hr
Minimum Dimension of Column	Carbonate, Siliceous, or Lightweight	6	6	6	8	10	12	16
Minimum Cover to Reinforcement	Siliceous	1.5	1.5	1.5	1.5	2 1.5*	2.5 1.5*	2.5 2*
	Carbonate or Lightweight	1.5	1.5	1.5	1.5	1.5	1.5	1.5

**Reinforcing steel enclosed in wire mesh or steel ties.*

E = modulus of elasticity
E_0 = modulus of elasticity of concrete prior to heating
e = distance between the line of thrust and the centroidal axis of a member
f_c' = compressive strength of concrete
f_{ps} = stress in prestressing steel at ultimate load
f_{pu} = ultimate tensile strength of prestressing steel
f_y = yield strength of reinforcing steel
L = heated length
l = span length
L_a = thermal expansion of member
M_{D+L} = applied dead plus live load moment
M_u = ultimate moment capacity of a flexural member
M_u^+ or M_u^- = positive or negative ultimate moment capacity of a flexural member
R = fire endurance of composite assembly as determined by criteria for temperature rise of unexposed surface
R_1, R_2, or R_n = fire endurance of individual courses of a composite assembly as determined by criteria for temperature rise of unexposed surface
S = "heated perimeter," i.e., the heated portion of the perimeter of the cross section resisting the thrust
T = thermal thrust
w = applied load
w_d = applied dead load per unit of length
w_{D+L} = applied dead plus live load per unit of length
x_0 and x_1 = distances measured along a span from the exterior support (see Fig. 7-24) to points of zero moment and maximum positive moment, respectively
Z = A/S
θ = temperature (subscript θ signifies the effects of elevated temperatures)
ϕ = capacity reduction factor = 0.9 for flexure
ω = $A_s f_y / bdf_c'$
$\overline{\omega}_p$ = $A_{ps} f_{pu} / bdf_c'$

BIBLIOGRAPHY

7-1. "Fire Protection Handbook," 13th Edition, National Fire Protection Association, Boston, Mass.

7-2. ASTM Designation: E119-71, "Standard Methods of Fire Tests of Building Construction and Materials," Part 14, *ASTM Book of Standards*, American Society for Testing and Materials.

7-3. Kordina, K., and Bornemann, P., "Brandversuehe an Stahlbeton-platten," *Deutscher Ausschuss fur Stahlbeton*, Heft 181, 1966, Wilhelm Ernst and Sohn, Berlin.

7-4. Harmathy, T. Z., and Stanzack, W. W., "Elevated-Temperature Tensile and Creep Properties of Some Structural and Prestressing Steels," *Fire Test Performance*, ASTM STP 464, 186-208, American Society for Testing and Materials, 1970.

7-5. Philleo, Robert, "Some Physical Properties of Concrete at High Temperatures," *Proceedings, American Concrete Institute*, **54**, 857, April 1958.

7-6. Thompson, John P., "Fire Resistance of Reinforced Concrete Floors," PCA Publication T140 (1963), 32 pp.

7-7. Commissie Voor Uitvoering Van Research (Neatherlands), "Fire Tests of Prestressed Concrete Beams," *CUR Report 13* (In Dutch), January 1958, 54 pp.

7-8. Ashton, L. A., and Bate, S. C. C., "The Fire Resistance of Prestressed Concrete Beams," *Journal of the American Concrete Institute,* **32**, 1417-1440, May 1961.

7-9. Abrams, M. S., and Cruz, C. R., "The Behavior at High Temperature of Steel Strand for Prestressed Concrete," *Journal of the PCA Research and Development Laboratories*, **3** (3), 8-19 Sept. 1961,; *PCA Research Department Bulletin 134.*

7-10. Carlson, C. C. "Fire Resistance of Prestressed Concrete Beams, Study A–Influence of Thickness of Concrete Covering Over Prestressing Steel Strand," *PCA Research Department Bulletin 147*, July 1962.

7-11. Selvaggio, S. L., and Carlson, C. C., "Effect of Restraint on Fire Resistance of Prestressed Concrete," *Fire Test Methods, ASTM STP No. 344*, American Society for Testing and Materials, 1962.

7-12. Selvaggio, S. L., and Carlson, C. C., "Fire Resistance of Prestressed Concrete Beams. Study B–Influence of Aggregate and Load Intensity," *Journal of the PCA Research and Development Laboratories*, **6** (1), 41-64 Jan. 1964, and **6** *No.* (2), 10-25 May 1964; *PCA Research Department Bulletin 171.*

7-13. Cruz, C. R., "Elastic Properties of Concrete at High Temperatures," *Journal, PCA Research and Development Laboratories*, **8** (1), 37-45, Jan. 1966; *PCA Research Department Bulletin 191.*

7-14. Carlson, C. C. et al., "A Review of Studies of the Effects of Restraint on the Fire Resistance of Prestressed Concrete," *Proceedings, Symposium on Fire Resistance of Prestressed Concrete*, Braunschweig, Germany, 1965, International Federation for Prestressing (F.I.P.). Bauverlag GmbH, Wiesbaden, Germany. *PCA Research Department Bulletin 206.*

7-15. Ehm, H., and vonPostel, R., "Tests of Continuous Reinforced Beams and Slabs Under Fire," *Proceedings, Symposium on Fire Resistance of Prestressed Concrete*, Translation available at "S.L.A. Translation Center," John Crerar Library, Chicago, Ill.

7-16. Gustaferro, A. H., and Selvaggio, S. L., "Fire Endurance of Simply Supported Prestressed Concrete Slabs," *Journal, Prestressed Concrete Institute*, **12** (1) 37-52, Feb. 1967. *PCA Research Department Bulletin 212.*

7-17. Report No. B1-59-22, "Fire Test of a Simple, Statically Indeterminant Beam," TNO Institute for Structural Materials and Building Structures, Delft, Holland. Tranlsation available at "S.L.A. Translation Center," John Crerar Library, Chicago, Ill.

7-18. Selvaggio, S. L., and Carlson, C. C., "Restraint in Fire Tests of Concrete Floors and Roofs," *ASTM STP 422*, American Society for Testing and Materials. *PCA Research Department Bulletin 220.*

7-19. Abrams, M. S., and Gustaferro, A. H., "Fire Endurance of Concrete Slabs as Influenced by Thickness, Aggregate Type, and Moisture," *Journal*, PCA Research and Development Laboratories, V. 10, No. 2, May 1968, pp. 9-24. *PCA Research Department Bulletin 223.*

7-20. Menzel, Carl A., "Tests of the Fire Resistance and Thermal Properties of Solid Concrete Slabs and Their Significance," *Proceedings, ASTM*, **43**, 1099-1153, 1943.

7-21. Malhotra, H. L., "The Effect of Temperature on the Compressive Strength of Concrete," *Magazine of Concrete Research* (London), **8** (22), 85–94, 1956.

7-22. Harmathy, T. Z., and Berndt, J. E., "Hydrated Portland Cement and Lightweight Concrete at Elevated Temperatures," *ACI Journal Proceedings* **63** (1), 93–112, Jan. 1966.

7-23. Cruz, C. R., "Apparatus for Measuring Creep of Concrete at High Temperatures," *PCA Research Department Bulletin 225.*

7-24. Abrams, M. S. and Gustaferro, A. H., "Fire Endurance of Two-Course Floors and Roofs," *Journal of the American Concrete Institute*, **66** (2), 92–102, February 1969.

7-25. Gustaferro, A. H., and Carlson, C. C., "An Interpretation of Results of Fire Tests of Prestressed Concrete Building Components," *Journal of the Prestressed Concrete Institute*, **7** (5), 14–43, October 1962.

7-26. Gustaferro, A. H., "Temperature Criteria at Failure," *Fire Test Performance*, ASTM STP 464, 68–84, American Society for Testing and Materials, 1970.

7-27. Brockenbrough, R. L., and Johnston, B. G., "Steel Design Manual," U.S. Steel Corp., Pittsburgh, Pa., 1968, 246 pp.

7-28. Abrams, M. S., "Compressive Strength of Concrete at Temperatures to 1600 F," *Symposium on Effect of Temperature on Concrete*, American Concrete Institute Publication ST-25, Detroit, Mich., 1971.

7-29. Gustaferro, A. H. et al., "Fire Resistance of Prestressed Concrete Beams. Study C: Structural Behavior During Fire Tests," *PCA Research and Development Bulletin (RD009.01B)*, Portland Cement Association, 1971.

7-30. Allen, L. W., "Fire Endurance of Selected Non-Load-bearing Concrete Masonry Walls," *Fire Study No. 25 of the Division of Building Research*, National Research Council of Canada, Ottawa, Canada.

7-31. Abrams, M. S. and Gustaferro, A. H., "Fire Endurance of Prestressed Concrete Units Coated with Spray-Applied Insulation," *Journal of the Prestressed Concrete Institute*, Jan.-Feb., 1972.

7-32. "Fire Resistance Ratings," American Insurance Association, New York, N.Y.

7-33. "Fire Performance Ratings, 1972," Supplement No. 2 to the *National Building Code of Canada.*

7-34. "Fire Resistance Index," Underwriters' Laboratories, Inc., Northbrook, Ill., January, 1972.

7-35. Gustaferro, A. H., et al., "Fire Resistance of Lightweight Insulating Concretes," ACI Special Publication 29, *Lightweight Concrete*, American Concrete Institute, Detroit, Mich.

7-36. Issen, L. A., et al., "Fire Tests of Concrete Members: An Improved Method for Estimating Restraint Forces," *Fire Test Performance*, ASTM, STP 464, 153–185, American Society for Testing and Materials, 1970.

7-37. "Manual of Steel Construction," 7th Ed., American Institute of Steel Construction, New York, N.Y.

7-38. *Report BMS 92*, "Fire Resistance Classifications of Building Constructions," National Bureau of Standards, Washington, D.C., 1940, 70 pp.

7-39. Salse, E. A. B. and Gustaferro, A. H., "Structural Capacity of Concrete Beams During Fires as Affected by Restraint and Continuity," Proceedings 5th CIB Congress, Paris, France, 1971, Available from Centre Scientifique et Technique du Batiment, Paris, when published.

8

Ductility of Structural Elements

J. G. MacGREGOR, Ph.D.*

8.1 INTRODUCTION

A ductile material is one that can undergo large strains while resisting loads. When applied to reinforced concrete members and structures, the term ductility implies the ability to sustain significant inelastic deformations prior to collapse. A brittle material or structure is one that fails suddenly upon attaining its maximum load. Force-deformation relationships are presented in Fig. 8-1 for brittle and ductile behavior.

8.1.1 Importance of Ductility

When a ductile statically indeterminate structure is subjected to an unexpected settlement or overload it will tend to deform inelastically and in doing so will redistribute some of the effects of the overload to other parts of the structure. This concept is utilized in different ways in various applications:

1. The strength design procedures in the *ACI Code* assume that all critical sections in the structure will reach their ultimate moment at the ultimate or design load for the structure. For this to occur, all splices and joints must be able to withstand bar forces and deformations corresponding to yielding of the reinforcement.
2. If a structure is ductile it can be expected to adapt to unexpected overloads, impact, load reversals, and structural movements due to foundation settlements and volume changes. Many of these things are ignored in our simplified design loadings and analyses which count on the existence of some ductility in the final structure.
3. If a structure is designed to fail in a ductile manner, the users or occupants will have warning of the impending failure, thus reducing the probability of loss of life in the collapse. This is recognized in most building codes. Thus, the *ACI Code* limits the steel percentage in beams to ensure a minimum amount of ductility. Similarly, the values of the *ACI Code* ϕ-factors of 0.70, 0.75, and 0.90 for tied columns, spiral columns, and beams respectively, are due in part to the increasing ductility of these types of members.
4. Since the response of a structure to a seismic or blast loading is a function of its stiffness, ductile inelastic action tends to reduce the inertia forces on the building. As a result, lower equivalent seismic design forces are permitted for ductile structures than for elastic structures in current design codes.
5. Plastic design methods based on the limit state of the formation of plastic mechanisms require ductile plastic

*Professor of Civil Engineering, University of Alberta, Edmonton, Alberta, Canada.

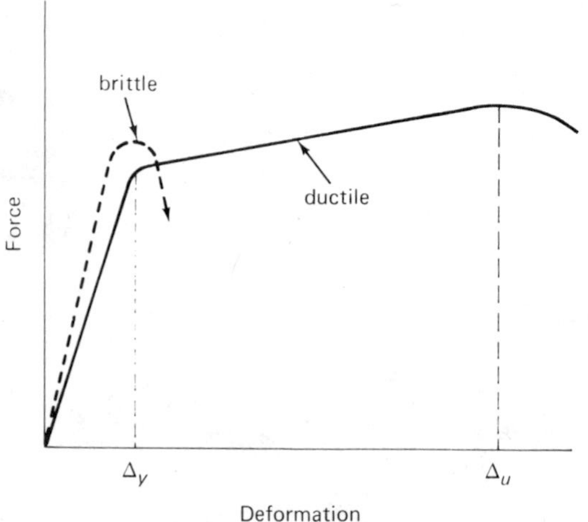

Fig. 8-1 Ductile and brittle behavior.

hinges. Although the subject of plastic or limit design is outside the scope of this chapter, Reference 8.2 presents a highly practical method of using limit design concepts.

8.1.2 Methods of Defining Ductility

A typical force-deformation relationship is plotted in Fig. 8-1 for a ductile member. The force may be load, moment or stress while the deformation could be elongation, curvature, deflection or twist. In this figure Δ_y is the yield deflection corresponding to yielding of the reinforcement in a cross section, or to a major deviation from the linear load deflection curve for a member or structure. Δ_u is the ultimate deflection beyond which the load-deflection curve has a negative slope. The ultimate deflection may be determined by local failure of the compression zone at some point in the member, by instability, or by any other set of conditions that leads to a failure of the section, member, or structure in question.

Referring to Fig. 8-1, the most common measures of ductility are:

1. The absolute value of the failure deformation, Δ_u or the inelastic deformation $\Delta_u - \Delta_y$. Frequently the overall rotation of a hinge is required in limit design analyses.
2. The "ductility ratio," $\mu = \Delta_u/\Delta_y$, or some form of this ratio $(\phi_u/\phi_y, \theta_u/\theta_y)$. Ductility ratio is relatively easy to define and is widely used at present.
3. The energy absorbed by the element or structure as given by the area under the force deformation diagram. This definition is of great importance, at least conceptually, in the field of design for earthquake resistance.

To some extent, each of these measures are empirical since the significance of ductility in a particular application may not correspond to the deformations assumed in its calculation. Thus, the failure deformation Δ_u will be larger in a specimen loaded monotonically to failure than in a specimen subjected to several cycles of high load prior to loading to failure. Generally, the ductility ratio, μ, is clearly defined only for the case of monotonic loading to failure. The rate of loading affects the ductility and is not normally considered in definitions.

It is also important to understand that ductility can refer to either the total structure or only part of it. The amount of ductility will differ in each case. Thus one may define the ductility of a material, the ductility of a cross section of a beam or column, and the ductility of an entire structure. At the material level, for example, plain concrete is not very ductile, having a ductility ratio of 1 to 2. A lightly reinforced cross section made of this concrete may be very ductile with μ = 10 to 20 while a beam with this cross section will often be less ductile, particularly if bond, or shear failures occur, or the beam is subjected to load reversals. When such beams are combined in a structure the ductility of the assembly may be large but will depend on the location of the plastic hinges and the behavior of the joints in the structure.

8.1.3 Scope of Chapter 8

As may be seen from sections 8.1.1 and 8.1.2, ductility is a loosely defined but important concept. Because of this, the design of reinforced concrete structures to ensure ductility relies on engineering judgement and good practice more than mathematically defined rules. This chapter differs from the balance of the Handbook in that the majority of the chapter is a review of the behavior of reinforced concrete to provide a background for ductility concepts. Sections 8.2 to 8.6 briefly discuss important parameters affecting the ductility of concrete and steel as materials, the ductility of the cross-sections of members and of the members themselves. Section 8.7 presents some general rules for incorporating ductility into structures. The reader is referred to Chapter 1 of this handbook for details of proportioning of members, Chapter 12 for a discussion of limit analysis of reinforced concrete frames. Blume, Newmark, and Corning have presented a very complete and practical discussion of ductility in Ref. 8-3.

8.2 PROPERTIES AND DUCTILITY OF CONCRETE AND REINFORCEMENT

8.2.1 Plain Concrete

a. Uniaxial loading

Plain concrete is a relatively brittle material having stress-strain curves similar to the ones shown in Fig. 8-2. In tension, concrete is brittle with very little inelastic action. In compression the stress-strain curve is approximately linear to about 40% of the strength, f_c', reaches a maximum, and has a descending branch. If the specimen is unloaded and reloaded in compression, stress-strain curves similar to the one shown by dashed lines in Fig. 8-2 are obtained.

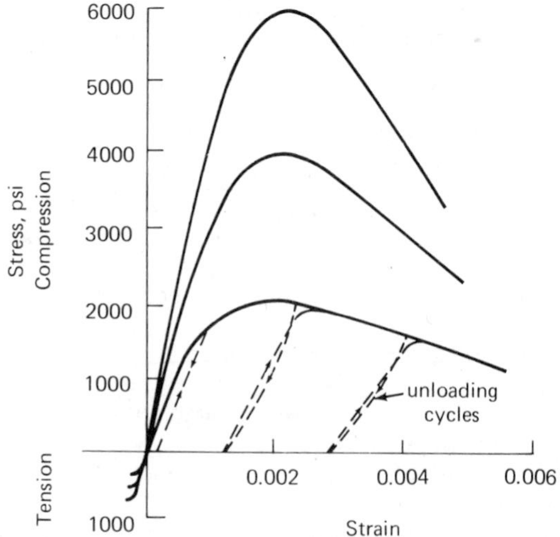

Fig. 8-2 Typical stress-strain curves for normal weight concrete.

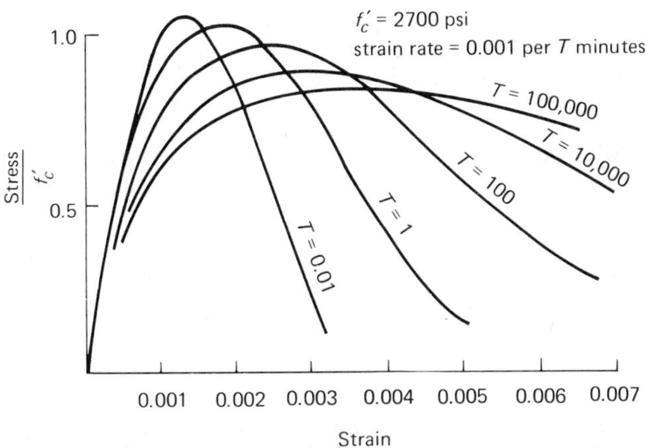

Fig. 8-3 Effect of rate of loading on stress-strain curves for concrete.

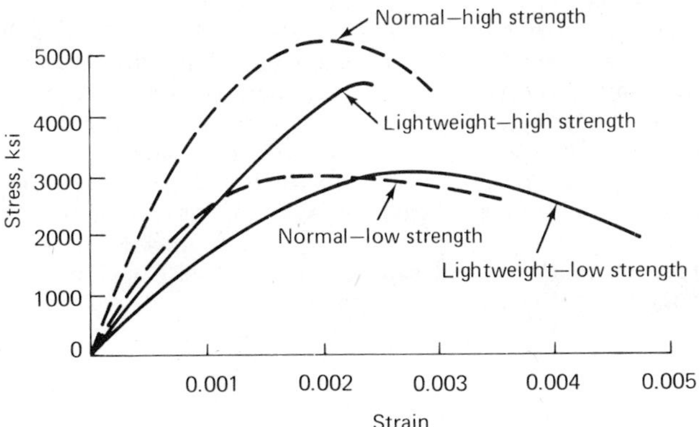

Fig. 8-4 Stress-strain curves for lightweight concrete.[8-4]

Under sustained high axial compression loads the strength of concrete may be reduced to about 80% of the short term strength. This effect is offset to some extent by the strength gain of the concrete with time and is alleviated by nonaxial loadings and compression reinforcement. As a result it has relatively little effect on ductility and will not be considered further in this chapter.

As the rate of loading increases, the compressive strength of concrete increases and the strain at the maximum stress decreases as shown in Fig. 8-3. For seismic loadings (assumed to correspond to the loading of a cylinder to failure in one second) the compressive strength increases to about 110% of that in a standard ASTM cylinder test.

Structural lightweight concrete displays two types of stress-strain behavior as shown in Fig. 8-4. Low strength lightweight concrete ($f'_c \leqslant 4000$ psi) fails in a more ductile manner than high strength lightweight concrete ($f'_c > 4000$ psi) which frequently fails suddenly at its maximum stress.[8-4] As indicated by the area under the stress-strain curves in Fig. 8-4, the energy absorbing capacity of normal concrete is generally greater than of lightweight concrete having the same strength. Reinforced concrete made from low strength lightweight concrete will generally be more ductile than similar members made from normal concrete of the same strength. The reverse is generally true for the more brittle high strength lightweight concretes.[8-4]

b. Triaxial loading

When plain concrete is subjected to a lateral confining pressure its compressive strength is increased by about four times the value of the lateral confining pressure. Equally important, the strains at which the enhanced strength is reached are several times those achieved by unconfined concrete indicating that the confinement also increases the ductility of the confined concrete.

In reinforced concrete a triaxial pressure can be approximated by means of closely spaced spiral reinforcement or by hoops at close spacing. As the concrete expands laterally the hoop or spiral reinforcement is stressed in tension and the concrete in turn, is stressed in compression as shown in Fig. 8-5. Tests suggest that rectangular hoops are about half as efficient as spiral or circular hoops because the sides of the ties deflect outward leading to a reduction in the confining pressure.

In regions where large deformations are anticipated, the strength and strain capacity of concrete can be increased by providing closely spaced spirals or hoops referred to as Special Transverse Reinforcement[8-5] or Confinement Reinforcement.[8-1] Confined Concrete is defined as concrete which is confined by closely-spaced special transverse reinforcement which is provided to restrain the concrete in directions perpendicular to the applied stresses.[8-5]

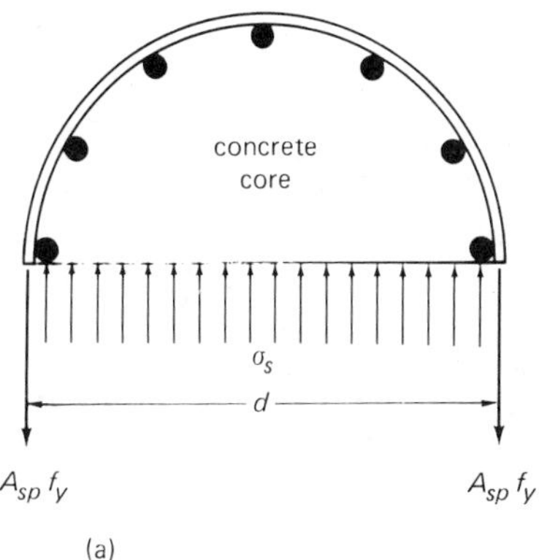

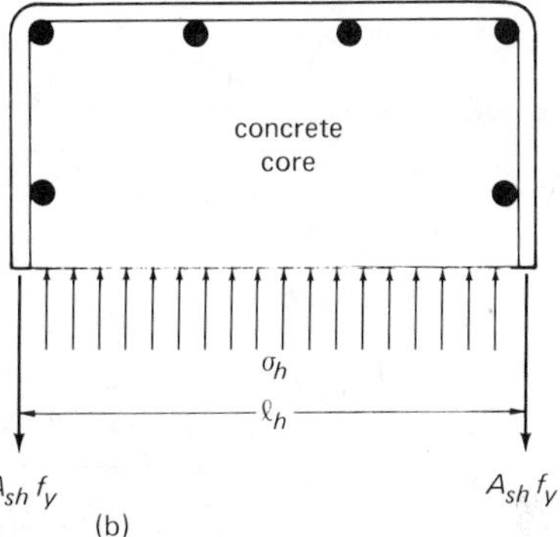

Fig. 8-5 Confining pressure due to spirals and hoops. (a) Spiral; and (b) hoop or stirrup tie.

8.2.2 Reinforcement

Reinforcing steel has much more ductility than concrete as shown by the stress-strain curves in Fig. 8-6. It is this ductility that makes it possible for reinforced concrete sections to display ductility. The ductility and the length of the yield plateau both decrease as the grade of steel increases.

Each of the steel stress-strain curves plotted in Fig. 8-6 exhibit strain-hardening. Under large seismic load reversals it is not uncommon for reinforcement to undergo strain-hardening in beams. As a result, the reinforcement stresses may exceed the yield strength of the bars in hinging regions.

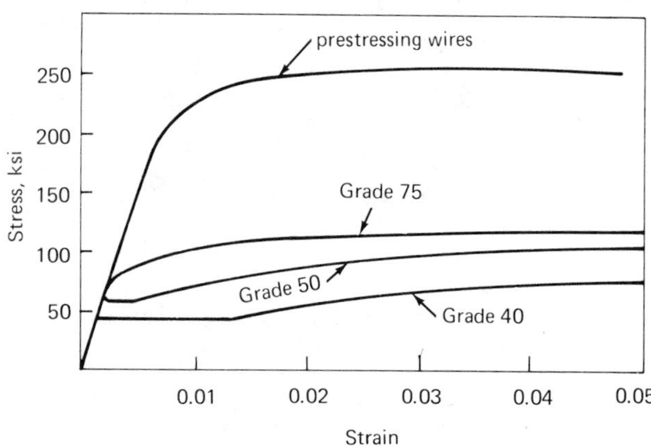

Fig. 8-6 Stress-strain curves for reinforcement.

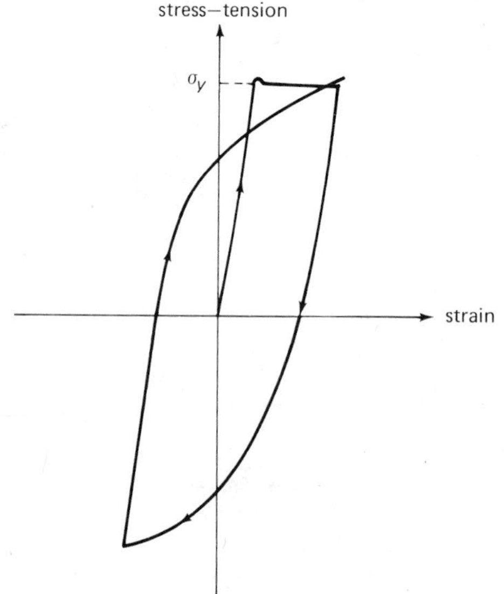

Fig. 8-7 Stress-strain curve for reinforcing bar showing Bauschinger effect.[8-7]

Extreme care must be taken in providing anchorage for such bars because of their high strains and stresses.

On cyclic tensile and compressive loading cycles reinforcing bars exhibit a pronounced Bauschinger effect as shown in Fig. 8-7.[8-7] This tends to reduce the energy absorbed by the reinforcement below that corresponding to an elastic-plastic steel. In addition, the steel has a low modulus of elasticity during some stages of the loading cycle and compression bars may buckle unless well tied.

The yield strength of the reinforcement is affected by the rate of loading. The rate of straining associated with earthquake loadings is about equal to that in a rapid mill test and corresponds to a yield strength about 7% greater than that achieved in a static loading.

8.3 THE DUCTILITY OF REINFORCED CONCRETE BEAMS

The ductility of beams may be defined in terms of the behavior of individual cross sections or the behavior of entire beams. Although the latter definition is more important in limit design and earthquake design, the behavior of cross sections is much better defined and is easier to compute. For these reasons the cross sectional ductility is widely used as a measure of beam ductility.

Sections 8.3.1 to 8.3.3 deal with the ductility of cross sections and serve as an introduction to sections 8.3.4 to 8.3.6 which consider entire beams.

8.3.1 Calculation of Ductility Ratio of Beam Cross Sections, $\mu = \phi_u/\phi_y$

The easiest and most common method of defining the ductility of a cross section is in terms of a ratio of the computed ultimate curvature to the computed yield curvature for a cross section. The ductility ratio computed in this way depends only on the cross section considered and does not include any effects of the distribution of moments, curvatures, and shears along the member. As a result, additional assumptions and calculations are required to determine the ductility of the entire beam. Nevertheless, this form of ductility ratio is widely used in design and other calculations and, unless otherwise stated, the term ductility ratio, μ, will be taken to mean $\mu = \phi_u/\phi_y$ in this chapter.

For research purposes it is possible to determine ϕ_u and ϕ_y from moment-curvature diagrams similar to the solid lines in Fig. 8-8. These curves were derived in Reference 8.6 using an iterative calculation procedure based on accurately defined stress-strain properties for concrete and steel.

For design purposes the curvatures are generally computed using standard beam theory.[8-3] Using the symbols in Fig. 8-9(a) the yield curvature, ϕ_y, of a beam without compression reinforcement can be computed with adequate accuracy from the straight line theory

$$\phi_y = \frac{\epsilon_y}{d(1 - k)} \tag{8-1}$$

where

ϵ_y is the yield strain of the tensile reinforcement
$= f_y/E_s$ for steels with a well defined yield point
d is the effective depth
kd is the depth of the compression zone computed using straight line theory
$k = -\rho n + \sqrt{2\rho n + \rho^2 n^2}$
$\rho = A_s/bd$
$n = E_s/E_c$

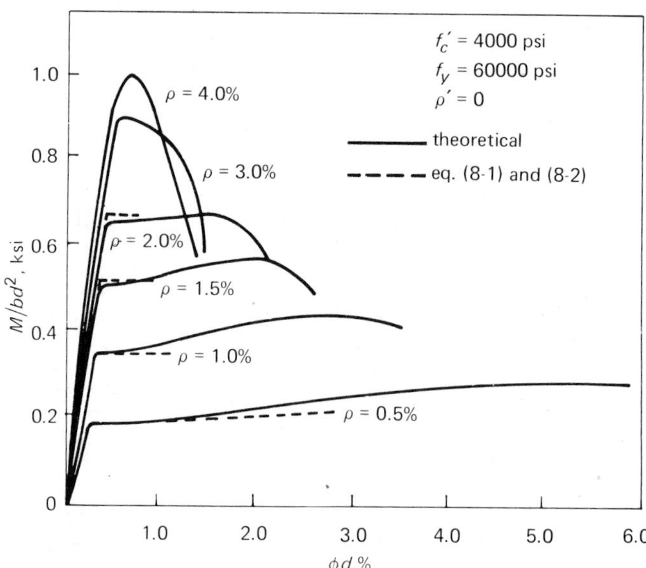

Fig. 8-8 Moment-curvature diagrams for beam cross section.[8-6]

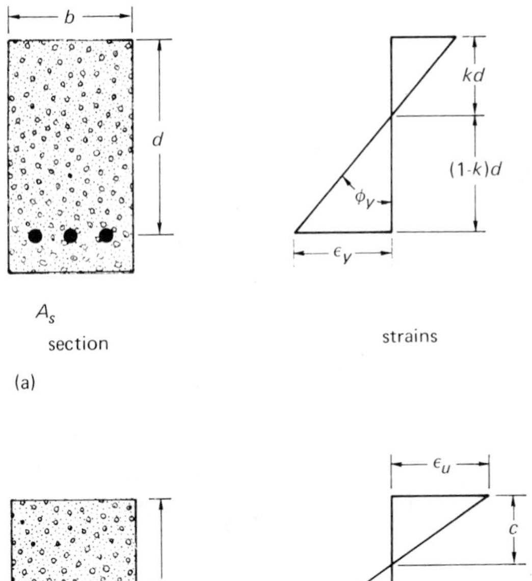

Fig. 8-9 Calculation of curvatures. (a) Yield curvature, ϕ_y; and (b) ultimate curvature, ϕ_u.

TABLE 8-1 Steel Percentages for Various Ductility Ratios[8-2]

f'_c psi	Ductility Ratio μ	$f_y = 40000$ psi ρ or $(\rho - \rho')$	$f_y = 60000$ psi ρ or $(\rho - \rho)'$
3000	4	0.017	0.0088
	5	0.014	0.0073
	6	0.012	0.0063
4000	4	0.022	0.012
	5	0.018	0.0095
	6	0.016	0.0082
5000	4	0.026	0.013
	5	0.021	0.011
	6	0.019	0.0095

Similarly, from Fig. 8-9(b) the ultimate curvature, ϕ_u, can be computed as

$$\phi_u = \frac{\epsilon_u}{c} \tag{8-2}$$

where

ϵ_u = concrete compression strain at crushing of concrete
c = depth of compression zone at ultimate

$$= \frac{\rho f_y}{f'_c}\,\frac{d}{0.85\beta_1} = \omega\,\frac{d}{0.85\beta_1}$$

β_1 = coefficient given in Table 8-2

Combining eqs. (8-1) and (8-2) gives

$$\mu = \frac{\epsilon_u\,(0.85\,\beta_1 f'_c)\,E_s\,(1 + \rho n - \sqrt{2\,\rho n + \rho^2 n^2})}{\rho f_y^2} \tag{8-3}$$

The value of μ calculated using this equation is directly affected by the value of ϵ_u assumed in the calculation. The ultimate strain, ϵ_u, is a function of a number of variables including the concrete strength, the rate of loading, the moment to shear ratio and the strengthening effect of binders or stirrups in the beam.[8-7] The *1971 ACI Code*[8-1] assumes $\epsilon_u = 0.003$. Statistical studies have shown that there is considerable scatter in the actual values of μ for beams.[8-8] If ϵ_u is taken equal to 0.003 in eq. (8-3), only about 5% of actual beams can be expected to have μ values less than the calculated values. Thus, for design use it is recommended that $\epsilon_u = 0.003$ be used unless special binding reinforcement is provided to increase ϵ_u.[8-1,8-2,8-9] Accordingly, substituting $\epsilon_u = 0.003$ and $E_s = 29000$ ksi into eq. (8-3) gives eq. (8-4) which can be used to compute design values of the ductility ratio[8-2]

$$\mu = \frac{74\beta_1 f'_c\,(1 + \rho n - \sqrt{2\rho n + \rho^2 n^2})}{\rho f_y^2} \tag{8-4}$$

where f'_c and f_y are in ksi

Values of μ calculated using this equation are presented in Table 8-1 and Fig. 8-10.[8-2] The *ACI Code* limits the maximum allowable steel percentage to 0.75 ρ_b except in seismic regions where the upper limit is 0.5 ρ_b. These two limits are also shown in Fig. 8-10. Moment curvature diagrams based on eqs. (8-1) and (8-2) are plotted with dashed lines in Fig. 8-8 for comparison with theoretically derived diagrams based on the actual stress strain curves of steel and concrete. As can be seen, ϕ_y is accurately predicted but ϕ_u is underestimated.

Although it is known that the presence of confinement reinforcement increases the ultimate compression strain of concrete, as yet no method of calculating this effect has been widely accepted for design use.[8-9] However, the concept that hoops, spirals, or other binders lead to more ductility is very important in the design of ductile structures.

The significance of a particular value of the ductility ratio, μ, is difficult to define since this term is only an arbitrary measure of the inelastic rotation capacity of a cross section. The ductility actually required in a particular situation will vary depending on the situation involved and the value of ϵ_u used in computing μ. Blume, Newmark, and Corning[8-3] suggest that a minimum ductility ratio of 4 to 6 is a reasonable criteria for ductile frames in earthquake resisting areas. Cohn and Ghosh[8-6] and Furlong[8-2] accept $\mu = 5$ and $\mu = 1 + l_n/d$, respectively as minimum values for hinging sections in plastically designed beams. The *ACI Code* indirectly requires μ at least 2 in all beams and allows 10% redistribution of moments in beams with $\mu = 3$. For design use, it is recommended that the ductility ratio from eq. 8-4 or Fig. 8-10 should be at least 4 in structures in seismic regions and at least 3 in other structures requiring limited ductility.

8.3.2 Variables Affecting the Ductility of Beam Cross Sections

a. Percentage of longitudinal reinforcement, ρ

As shown in Fig. 8-8 and 8-10 and Table 8-1 the ductility of a beam cross section increases as the steel percentage, ρ (or $\rho - \rho'$), decreases. If too much reinforcement is provided the concrete will crush before the steel yields, leading to a brittle failure corresponding to $\mu = 1.0$. This case is illustrated by the moment curvature diagrams for $\rho = 3\%$ and

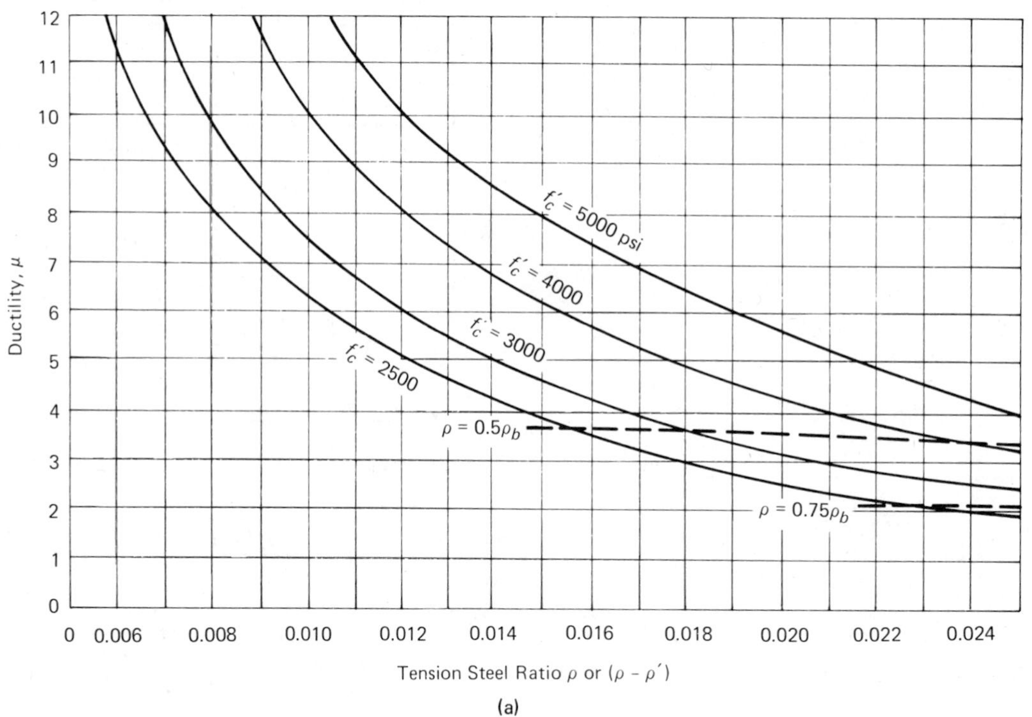

Fig. 8-10 (a) Relationship between ρ and ductility, f_y = 40,000 psi.[8-2]

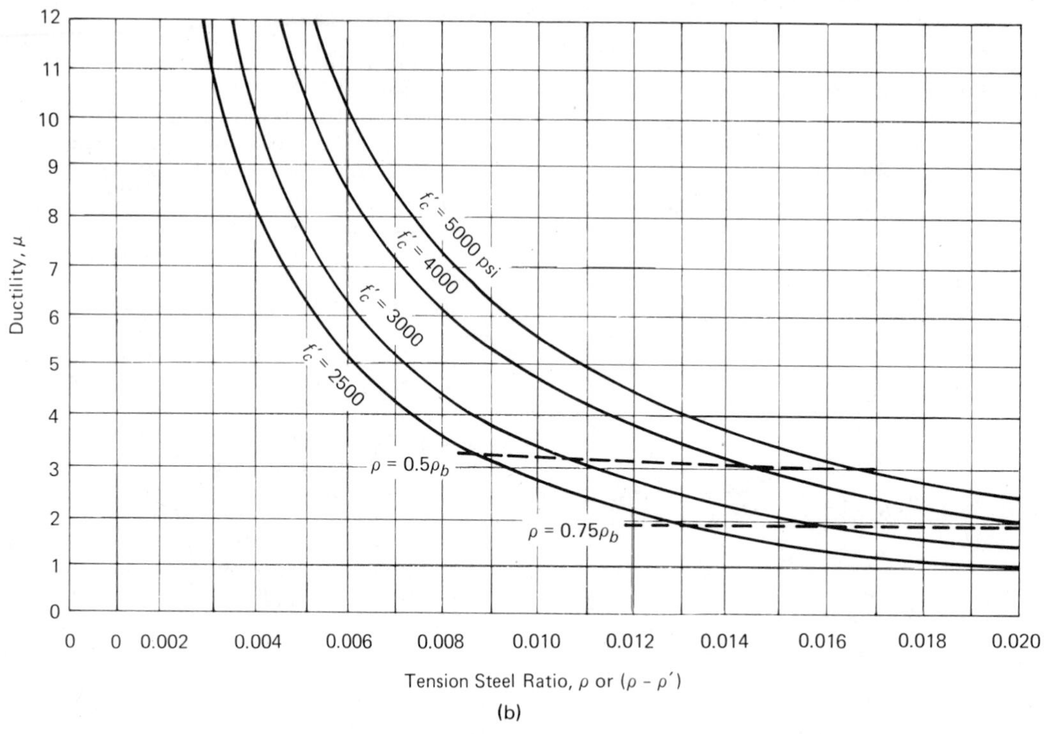

Fig. 8-10 (b) Relationship between ρ and ductility, f_y = 60,000 psi.[8-2]

4% in Fig. 8-8. For this reason, as stated earlier, the *ACI Code* limits the maximum amount of tension reinforcement to three-quarters of the amount, ρ_b, which corresponds to simultaneous yielding of the reinforcement and crushing of the concrete. In seismic regions ρ is limited to 0.5 ρ_b to ensure still more ductility. The dashed lines in Fig. 8-10 show that ρ = 0.75 ρ_b and 0.5 ρ_b correspond to nominal ductility ratios of about 2 and 3, respectively.

For a rectangular beam without compression reinforcement

$$\rho_b = \bar{\rho}_b = \left(\frac{0.85\ \beta_1 f_c'}{f_y}\right)\frac{87000}{87000 + f_y} \tag{8-5}$$

where β_1 = coefficient given in Table 8-2 and *ACI Code*.

Values of $\bar{\rho}_b$, 0.75 $\bar{\rho}_b$, and 0.5 $\bar{\rho}_b$ are given in Table 8-2.

TABLE 8-2 Balanced Ratio of Reinforcement, ρ_b, for Rectangular Sections with Tension Reinforcement Only*

f_y \ f_c'		2500 psi $\beta_1 = 0.85$	3000 psi $\beta_1 = 0.85$	4000 psi $\beta_1 = 0.85$	5000 psi $\beta_1 = 0.80$	6000 psi $\beta_1 = 0.75$
Grade 40	$\bar{\rho}_b$	0.0309	0.0371	0.0495	0.0582	0.0655
40,000 psi	$0.75\ \bar{\rho}_b$	0.0232	0.0278	0.0371	0.0437	0.0492
	$0.50\ \bar{\rho}_b$	0.0155	0.0186	0.0247	0.0291	0.0328
Grade 50	$\bar{\rho}_b$	0.0229	0.0275	0.0367	0.0432	0.0486
50,000 psi	$0.75\ \bar{\rho}_b$	0.0172	0.0206	0.0275	0.0324	0.0365
	$0.50\ \bar{\rho}_b$	0.0115	0.0138	0.0184	0.0216	0.0243
Grade 60	$\bar{\rho}_b$	0.0178	0.0214	0.0285	0.0335	0.0377
60,000 psi	$0.75\ \bar{\rho}_b$	0.0134	0.0161	0.0214	0.0252	0.0283
	$0.50\ \bar{\rho}_b$	0.0089	0.0107	0.0143	0.0168	0.0189
Grade 75	$\bar{\rho}_b$	0.0129	0.0155	0.0207	0.0243	0.0274
75,000 psi	$0.75\ \bar{\rho}_b$	0.0097	0.0116	0.0155	0.0182	0.0205
	$0.50\ \bar{\rho}_b$	0.0065	0.0078	0.0104	0.0122	0.0137

*$\bar{\rho}_b = (0.85\ \beta_1 f_c'/f_y)\ (87{,}000/87{,}000 + f_y)$
$\beta_1 = 0.85$ for f_c' up to 4,000 psi, above 4,000 psi reduce β_1 by 0.05 per 1000 psi (ACI Code Sec. 10.2.7)
Max. $\rho = 0.75\ \rho_b$ (ACI Code Sec. 10.3.2)
Max. $\rho = 0.50\ \rho_b$ (ACI Code Sec. 8.6)

b. Percentage of compression reinforcement, ρ

Figure 8-11 presents a series of moment-curvature diagrams for beams with various amounts of compression reinforcement. The addition of compression reinforcement to a beam has relatively little effect on its yield strength or yield curvature. It does greatly increase the ultimate curvature. This is because the term c in eq. (8-2) becomes

$$c = \left(\rho - \frac{\rho' f_s'}{f_y}\right) \frac{f_y d}{0.85\ \beta_1 f_c'} \tag{8-6}$$

where

$d = A_s'/bd$ = ratio of compression reinforcement

and

f_s' = the stress in the compression reinforcement

If $f_s' = f_y$, eq. (8-6) becomes

$$c = \frac{(\rho - \rho')\ f_y d}{0.85\ \beta_1 f_c'} \tag{8-7}$$

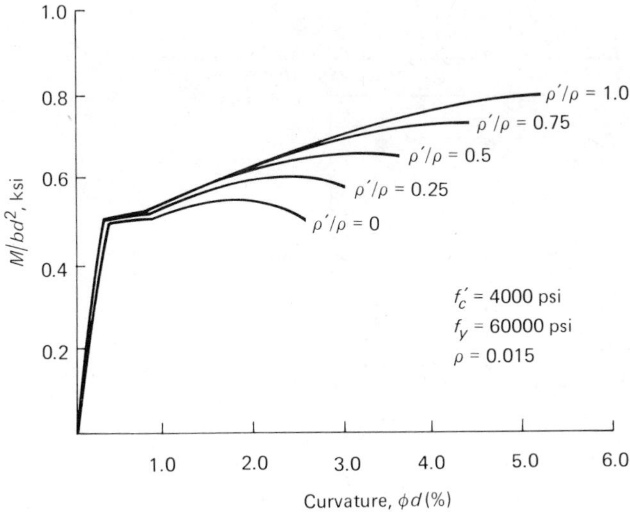

Fig. 8-11 Effect of compression reinforcement on ductility.[8-6]

This suggests that $(\rho - \rho')$ is the important parameter in defining the ductility ratio. For design estimates of μ it is sufficiently accurate to enter Table 8-1 with $\rho - \rho'$ (which equals ρ when $\rho' = 0$) to get ductility factors. Theoretical calculations have shown that for steel ratios $\rho \leqslant 0.03$ and $\rho' = \rho$, the design value of the ductility of beams without axial loads will always exceed $400/f_y$ where f_y is in ksi.

The *ACI Code* requires that the steel percentage, ρ, in beams in ductile frames for seismic regions be less than 0.50 of the balanced steel ratio. For a beam with compression reinforcement the *1971 ACI Code* defines the balanced steel ratio ρ_b as

$$\rho_b = \bar{\rho}_b + \rho' \frac{f_s'}{f_y} \tag{8-8}$$

where

$\bar{\rho}_b$ = value given by eq. (8-5) and Table 8-2.
f_s' = f_y if the value of $(\rho - \rho')\ d/d'$ for the given cross section is greater than the value given in Table 8-3. If $(\rho - \rho')\ d/d'$ is less than the value given in the table, $f_s' < f_y$ and can be found using strain compatibility.
d' = distance from extreme compression fiber to centroid of compression reinforcement.

TABLE 8-3 Determination of Stress in Compression Reinforcement

Minimum values of $(\rho - \rho')\ d/d'$

f_y, psi	f_c', psi			
	3000	4000	5000	6000
40000	0.0371	0.0495	0.0582	0.0655
50000	0.0275	0.0367	0.0432	0.0486
60000	0.0214	0.0285	0.0335	0.0377

If the value of $(\rho - \rho')\ d/d'$ for a beam with compression reinforcement equals or exceeds the values given, $f_s' = f_y$. If it is less, use strain compatibility to compute f_s'.

c. Material strengths

As shown by Table 8-1 or eq. (8-4) the ductility increases with an increase in concrete strength and a decrease in the yield strength of the reinforcement because both trends corresponds to a reduction in the depth of the neutral axis at failure and hence an increase in ϕ_u.

d. Cross-sectional shape

The presence of an enlarged compression flange in a T-beam reduces the depth of the compression zone at ultimate and thus increases the ductility. Generally, the neutral axis will fall in the flange in such a beam and Tables 8-1 and 8-2 will give the ductility and ρ_b directly. For a T-beam these tables are entered with $\rho = A_s/bd$ where b = flange width. In the negative moment region of a conventional T-beam the flange is stressed in tension and hence has no effect on ductility.

e. Lateral reinforcement

Lateral reinforcement tends to improve ductility by preventing premature shear failures, restraining the compression reinforcement against buckling and by constraining or binding the compression zone, thus increasing its strength and deformation ability.

The solid lines in Fig. 8-12 present moment-curvature diagrams for the constant moment regions of two beams which were similar except that one had no stirrups and the other one had closed stirrups at $d/4$ on centers.[8-10] The closed stirrups had no effect on the strength but greatly increased the ductility of the beam. Because of the significant increase in ductility resulting from the use of closed stirrups, seismic design codes require such stirrups or ties in all regions where ductility may be required.

Although continuous spirals are the most efficient binding reinforcement, it is generally necessary to use closed rectangular ties or stirrups in beams or columns. As lateral expansion develops in the bound concrete, however, the straight side of a tie will deflect outwards and its binding effect will decrease. Ties or stirrups are most efficient as binders when the distance between their corners is relatively small or when supplementary cross-ties are used to reduce the unsupported length of the side of the tie (Fig. 8-13). Note the difference between supplementary crossties which hook around the main tie and supplementary ties which do not.

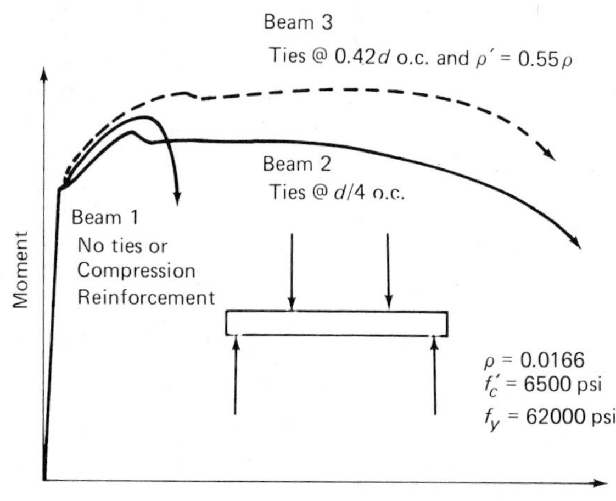

Fig. 8-12 Effect of lateral and compression reinforcement on ductility.[8-10]

Stirrup-ties are only effective as binders if their ends remain anchored after the compression concrete starts to spall. For this reason the *ACI* and *SEAOC Codes* both require anchorage by a 135° bend around a longitudinal bar and a 10-diameter extension. For nonseismic regions a 6-diameter but not less than 2.5 in. extension should be sufficient.

Analytical studies have shown that pound for pound, compression reinforcement is more efficient than lateral reinforcement in improving the ductility of a section.[8-6] This is borne out by the measured moment curvature diagram plotted by the dashed line in Fig. 8-12. This beam had less binding reinforcement than Beam 2 in this figure but had compression reinforcement. Not only was it more ductile but it was also stronger.

Closed stirrups are also required to prevent the buckling of the compression reinforcement. In non-seismic regions the spacing of such stirrups should conform to *ACI Section 7.1.5*. In regions of a member where the reinforcement may be yielded alternately in tension and compression the tie spacing should be reduced to $d/4$ or eight bar diameters to prevent buckling of the compression bars and the requirements of either the *ACI Code*[8-1] or the *SEAOC Code*[8-5] should be satisfied.

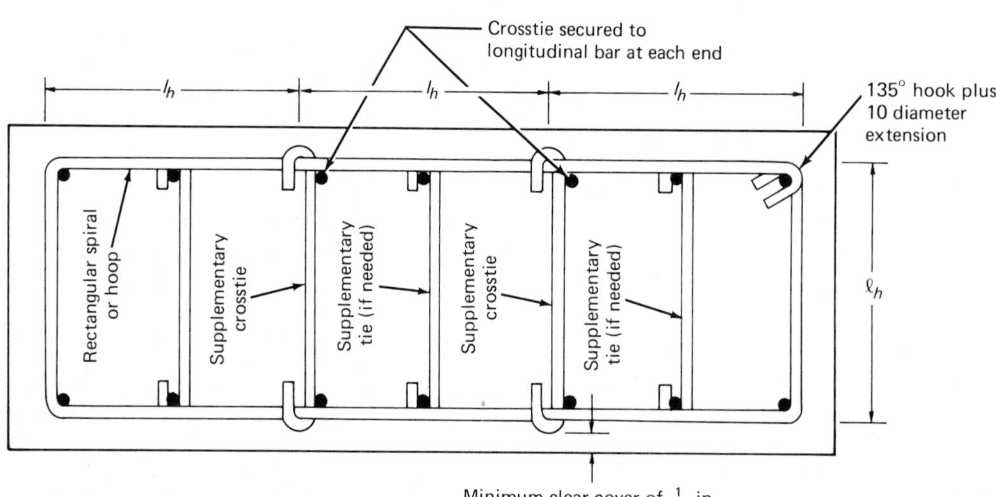

Fig. 8-13 Sectional view of column hoop showing supplementary ties and supplementary crossties.

8.3.3 Example of Computation of Ductility of Beam Cross Sections

EXAMPLE: Calculate the ductility, μ, and balanced steel ratio, ρ_b, at mid-span and adjacent to the columns for the beam shown in Fig. 8.14a. Assume f'_c = 3000 psi and f_y = 60,000 psi

SOLUTION

a. Negative moment section

The negative moment section is shown in Fig. 8-14b.

$$b = 13 \text{ in.}, \; d = 22 \text{ in.}, \; d' = 2 \text{ in.}$$

$$A_s = 6 \# 7 = 3.60 \text{ in.}^2, \; \rho = A_s/bd = \frac{3.60}{13 \times 22} = 0.0126$$

$$A'_s = 2 \# 7 = 1.20 \text{ in.}^2, \; \rho' = A'_s/bd = 0.0042$$

(*i*) Compute ductility

$$\rho - \rho' = 0.0084$$

From Table 8-1 the ductility ratio is: $\mu = 4.2$.

This value should be adequate for structures in seismic regions and would certainly be adequate in structures in non-seismic regions. If desired, the ductility could be improved by

1. Add more compression reinforcement
2. Increase f'_c and/or reduce f_y
3. Place closely spaced stirrups at the end of the beam

(*ii*) Compute balanced steel ratio

Check whether compression reinforcement is yielding. $(\rho - \rho') d/d' = 0.0084 \times \dfrac{22 \text{ in.}}{2 \text{ in.}} = 0.0925$. Since this exceeds the value for for f'_c = 3000 psi and f_y = 60,000 psi in Table 8-3 [Min $(\rho - \rho')\, d/d'$ = 0.0214] the compression reinforcement will yield, therefore by eq. (8-8)

$$\rho_b = \bar{\rho}_b + \rho' \frac{f_y}{f_y}$$

from Table 8-2, for f'_c = 3000 psi and f_y = 60,000 psi

$$\bar{\rho}_b = 0.0214$$

therefore

$$\rho_b = 0.0214 + 0.0042 \times \frac{60}{60} = 0.0256$$

The negative moment section has $\rho = 0.0126$ which is 0.49 ρ_b.

b. Positive moment section

The positive moment section is shown in Fig. 8-14(c) and $b = 78$ in. since the compression zone is entirely in the flange.

$$d = 22 \text{ in.}$$

$$A_s = 4 \# 7 = 2.40 \text{ in}^2, \; \rho = \frac{2.40}{78 \times 22} = 0.0014$$

from Table 8-1, μ exceeds 12

from Table 8-2, $\rho_b = 0.0214$.

8.3.4 Definition of the Ductility Ratio for a Beam

The ductility ratio, $\mu = \phi_u/\phi_y$ applies to only a single cross section. When speaking of a beam or column, the ductility is usually presented in terms of the total angle change, θ_u, at a hinge, the ratio of angle changes $\mu_r = \theta_u/\theta_y$ or the ratio of deflections $\mu_d = \Delta_u/\Delta_y$. The angle changes, θ, or the deflections, Δ, are a function of the first, or second integral of the curvatures along the member. In a beam with a plastic hinge, a major part of the total rotation occurs at the hinge so that the ductility based on angle changes is similar to that based on curvatures provided that bond or shear failures do not occur.

When a beam is loaded into the inelastic range its end rotations or deflections can be computed from the distribution of curvatures along the beam by means of moment-area or a similar technique. Figures 8-15(a), (b), and (c), respectively, show a beam containing flexural and shear cracks, its moment diagram and the distribution of curvatures along this member. For calculation purposes the actual distribution of curvatures is generally simplified to the

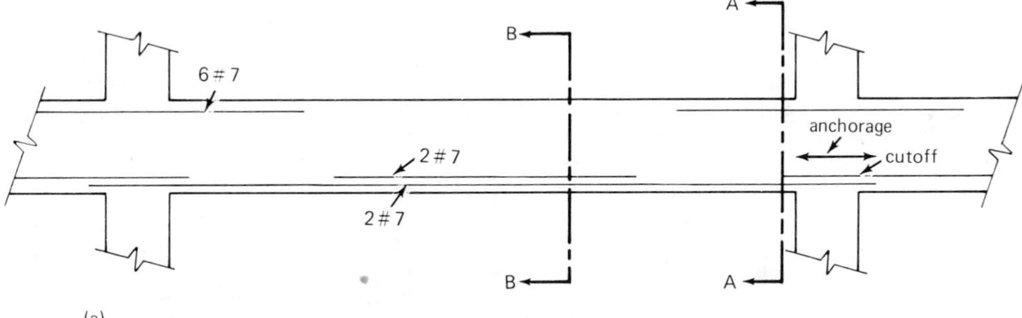

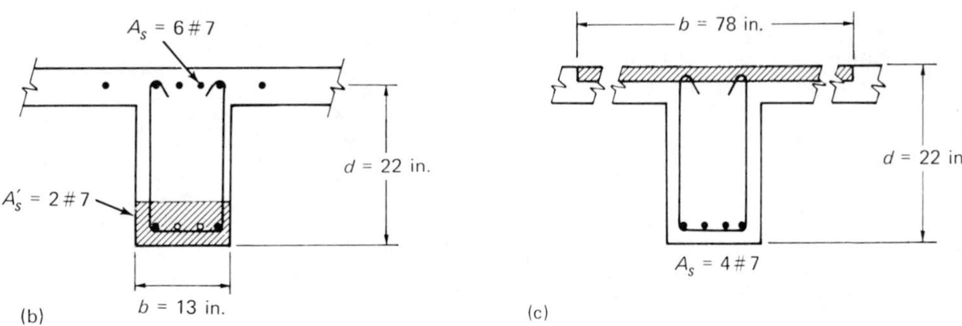

Fig. 8-14 Example, section 8.3.3. (a) Longitudinal reinforcement; (b) negative moment, section A-A; and (c) positive moment, section B-B.

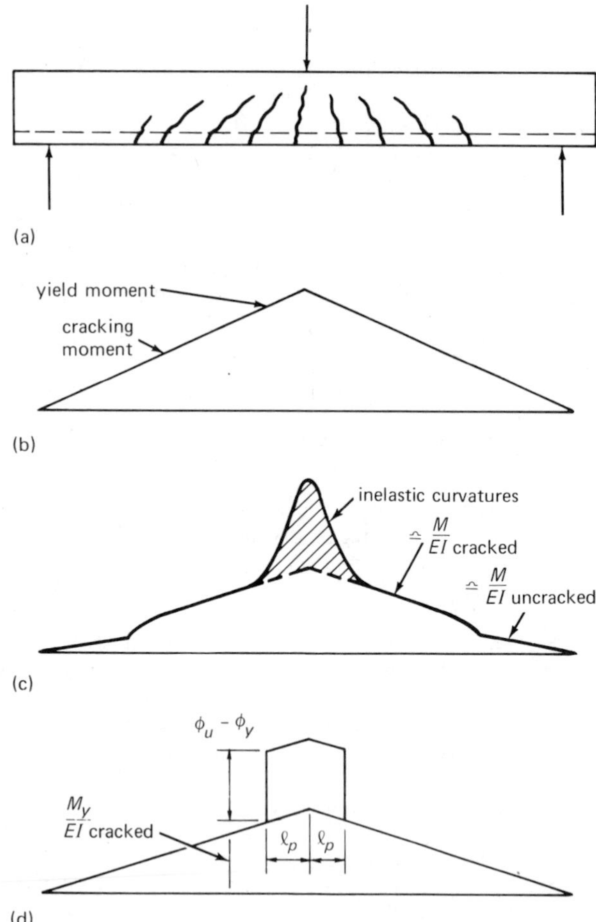

Fig. 8-15 Curvature diagram for a simple beam. (a) Beam; (b) moment diagram; (c) curvatures; and (d) curvatures assumed in calculation.

shape shown in Fig. 8-15(d). This diagram consists of elastic curvatures plus a region of uniform inelastic curvature for a distance of l_p on either side of the section of maximum moment. The value of l_p depends on the shape of the moment diagram and the distribution of flexural and shear cracks. In Section 8.3.6(b) it is recommended that:

$$\text{If} \quad v_u < v_c, \quad l_p = d/4$$

$$v_u \geqslant v_c, \quad l_p = 2d/3 \tag{8-9}$$

assuming adequate stirrups are provided for the actual shear at ultimate.

8.3.5 Behavior of Beams Loaded to Failure

The behavior of three beams loaded to failure will be discussed here to illustrate various aspects of behavior. The first represents a member developing inelastic action in a positive moment region, the second and third represent members forming hinges at the face of a column.

Figure 8-16 presents a measured load deflection diagram for a simply supported beam with loaded at two points with a constant moment region resembling a positive moment region in a beam.[8-11] In such a loading, crushing develops at the top of the beam over a considerable length of the constant moment region and failure occurs soon after the onset of crushing. A load deflection diagram computed from eqq. (8-1) and (8-2) and section 8.3.3 is plotted on this graph with dashed lines. The agreement is quite good.

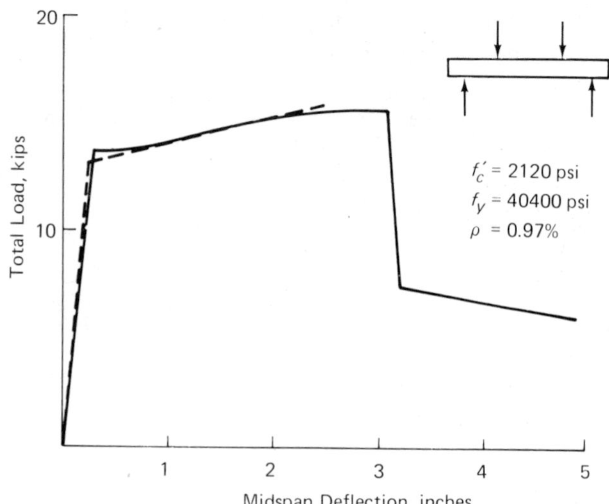

Fig. 8-16 Load-deflection diagram for a beam with a constant moment region.[8-11]

On the other hand, when failure occurs adjacent to a column in a negative moment region the moment gradient and the presence of the column adjacent to the section of maximum moment restrain the failure of the compression zone, allowing considerable increases in load and deflection to occur after the onset of crushing.

Figure 8-17 shows a measured load deflection diagram for a simply supported beam with a stub column at midspan.[8-12] This specimen simulates a beam-column connection and the parts of the adjacent beams between their points of inflection and the column. The beam did not have compression reinforcement. As can be seen from Fig. 8-17 crushing of the concrete adjacent to the column had no appreciable effect on the load deflection diagram. In addition, the results indicate that the removal and reapplication (but not reversal) of load had little apparent effect on either the load capacity or ductility. The ultimate deflection was 17 times the yield deflection. In this case, the computed load-deflection diagram based on eqq. (8-1) and (8-2) did not agree with the measured diagram because the computed diagram terminated at the onset of crushing of the compression zone. Means of correcting this error are proposed by a number of authors.[8-9]

The measured load deflection diagram is typical for such specimens. Although the reinforcing steel had a well defined yield plateau, and the corresponding moment curvature diagram also showed a plastic portion, there is no plastic region in the load deflection diagram. This is because the plastic curvatures were developed at only one section at a time and thus generated only a small amount of plastic deformation before strain-hardening started. For significant inelastic action to develop at this "hinge" it is necessary for the hinge to extend into the span. This in turn requires strain hardening to develop adjacent to the column. The spread of the hinge is also affected by the state of inclined cracking.

It may be concluded that a substantial portion of the ductility of a hinge in a negative moment region depends on the development of strain-hardening of the reinforcement. The ultimate load was 130% of the yield load for the beam shown in Fig. 8-17. This implies that the reinforcing bars adjacent to the column were stressed approximately 30% over the yield point and must be anchored for such a stress. Furthermore, joints and adjacent members must be designed for the moments or shears resulting from bar stresses higher than the yield point.

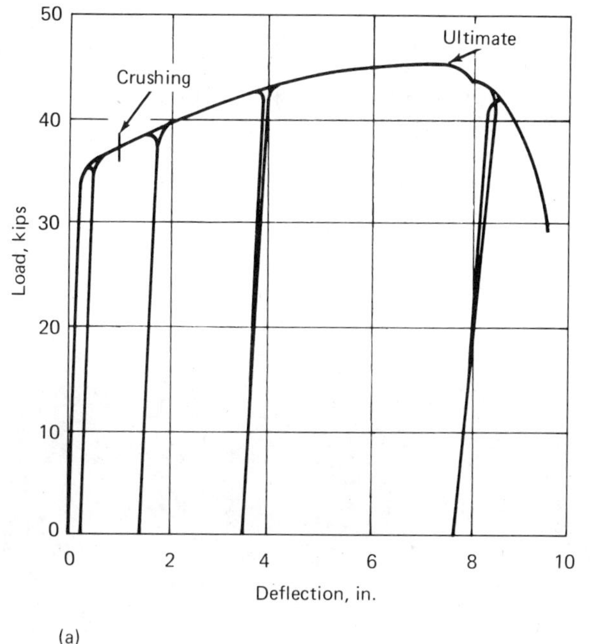

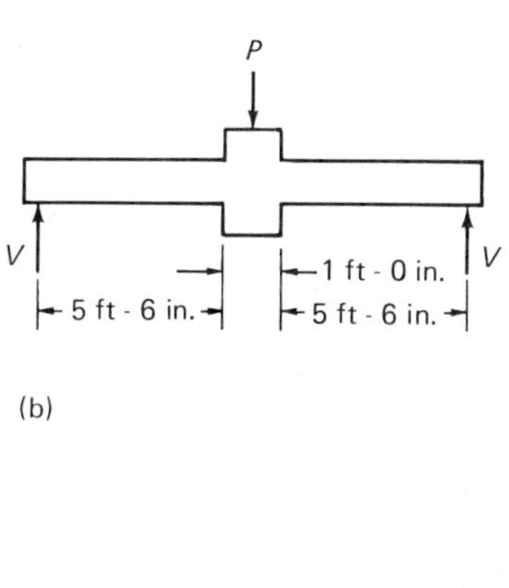

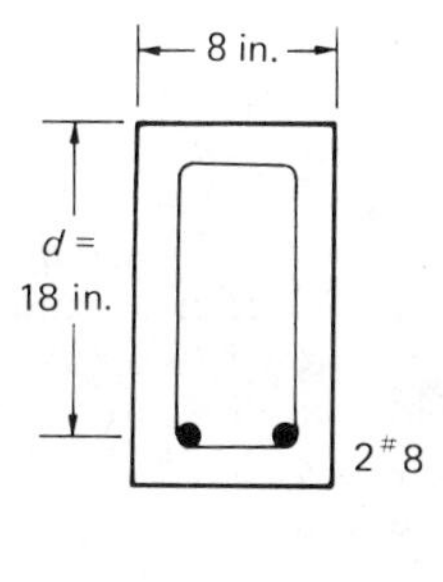

Fig. 8-17 Load-deflection diagram for beam-column connection without compression reinforcement.[8-12] (a) Load-deflection diagram; (b) loading arrangement; and (c) section.

Figure 8-18 compares the load deflection diagram for two beams similar to the beam shown in Fig. 8-17 except that they had the same reinforcement for the top and bottom. Beam J-6 was loaded downwards to failure. Beam J-7 was loaded downwards and upwards several times. The load-deflection curve for J-6 has been plotted in both quadrants for comparison. The envelope load-deflection curve for J-7 agrees with that for beam J-6 which was loaded in one direction only, except that the ultimate deflection of the beam subjected to load reversals was only two thirds of that for the other beam. The portions of the load deflection diagram in the lower right and upper left quadrants show the Bauschinger effect in the reinforcement and the effect of closing and opening the crossed flexural and inclined cracks. Again, it should be noted that strain hardening of the reinforcement has obviously occurred and the failure loads and the corresponding bar stresses are about 30% over the values at yield.

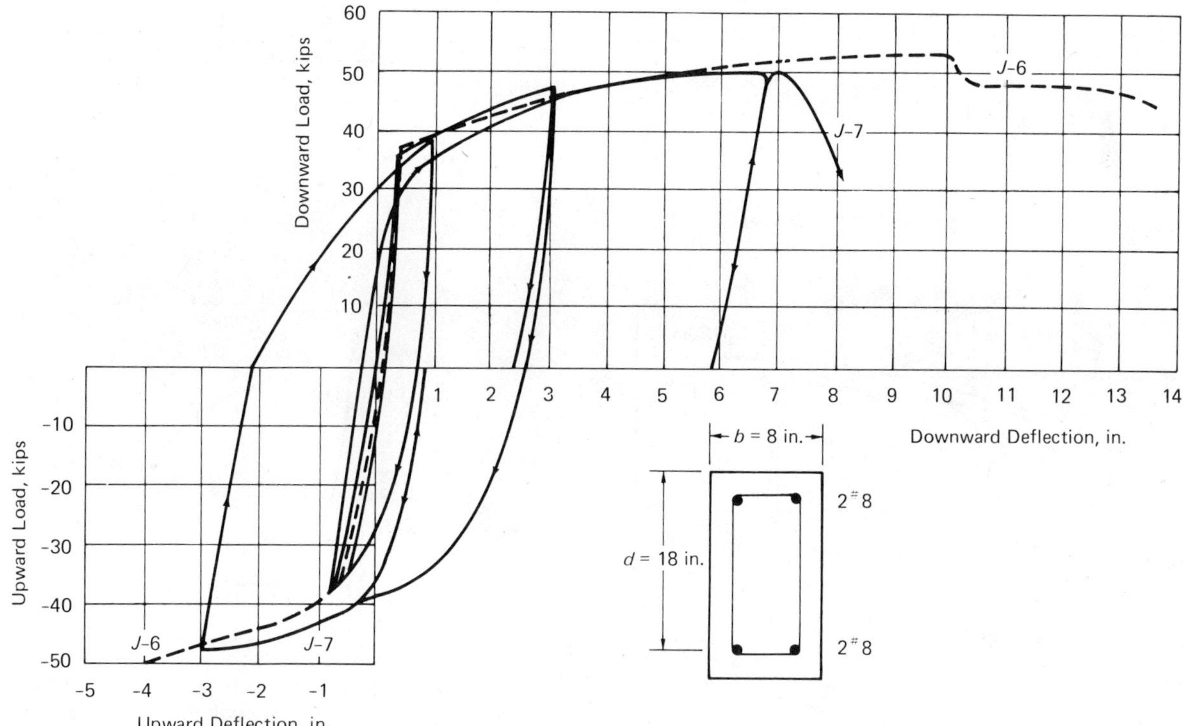

Fig. 8-18 Load-deflection curve for beams J-6 and J-7.[8-13]

8.3.6 Variables Affecting the Ductility of Beams

a. Moment curvature diagrams of beam cross-sections

The moment curvature diagram of the beam cross sections and the variables affecting it have a major effect on the ductility of a given beam. Thus, as indicated in section 8.3.2:

1. The ductility increases as the steel percentage is decreased.
2. The ductility increases as the percentage of compression reinforcement, ρ', is increased. This is a very practical way of increasing ductility as shown by Fig. 8-11 and the measured load deflection diagrams for beam-column connection specimens given in Fig. 8.19. In Fig. 8-19, it should be noted that the final failure of beam J-13 resembled a shearing failure except that the longitudinal reinforcement was stressed well beyond the yield point. This again illustrates the need to base the design shears and bond stresses on loads corresponding to some degree of strain hardening.
3. Increases in concrete strength and/or decreases in reinforcement yield strength lead to increases in ductility.
4. Constraint of the compression zone by closed stirrups is an effective way of improving the ductility as shown by Beams 1 and 2 in Fig. 8-12.

b. Shear

Shear has two effects on the ductility of a member. First, a shear failure generally occurs at a smaller deflection than a flexural failure and hence absorbs much less energy than a flexural failure. For this reason shear failures cannot be tolerated in a ductile structure. It is essential therefore, to provide sufficient web reinforcement to prevent shear failures from occurring at loads up to 120 to 130% of the yield load. In nonseismic regions the provisions of Chapter 11 of the *ACI Code* are satisfactory in this regard provided amplified loads are considered. In seismic regions, load reversals may cause crossing inclined cracks which will tend to reduce the shear carried by the concrete. Accordingly, Section A.5.9 of the *ACI Code* requires additional web reinforcement within $4d$ from the ends of beams. The amount required was derived assuming the concrete carried no shear in this region.[8-3]

Second, the total amount of rotation which can be developed at a plastic hinge is a function of the shearing stresses in the beam.[8-9, 8-14] In a region of pure moment the rotations developed are governed by a concrete compression strain of $\epsilon_u = 0.003$ and the length of the region of pure moment. If the shear is large enough for inclined cracks to occur the inelastic rotations may be fairly large (if stirrups prevent a shear failure) because compression strains greater than 0.003 can be developed and, equally important, because the steel can yield at several sections as shown in Fig. 8-20(a) leading to a spread of the length of the hinge. If, on the other hand, the shears are so small that inclined cracks do not occur, the yield stress may be only reached at the crack at the point of maximum moment as shown in Fig. 8-20(b) and the inelastic rotations will be very much smaller. The rotations at these two types of hinges are

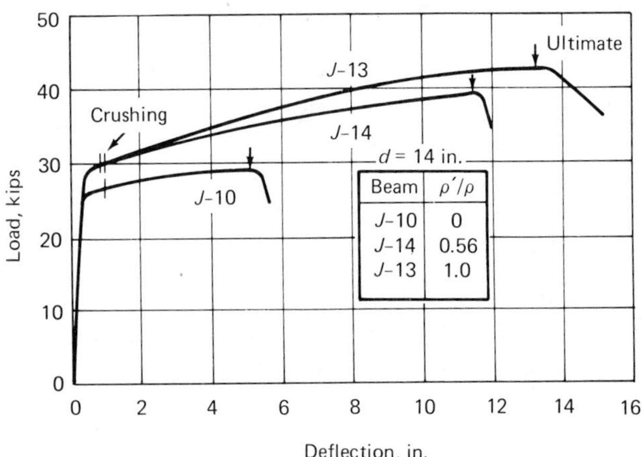

Fig. 8-19 Effect of compression reinforcement.[8-12]

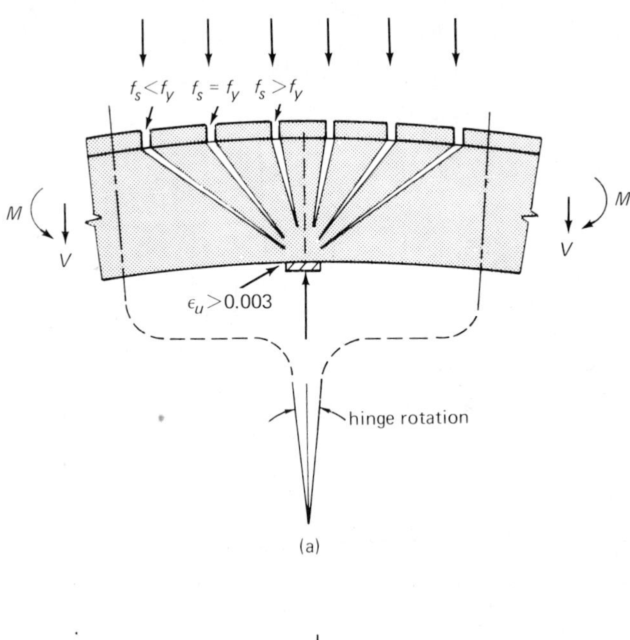

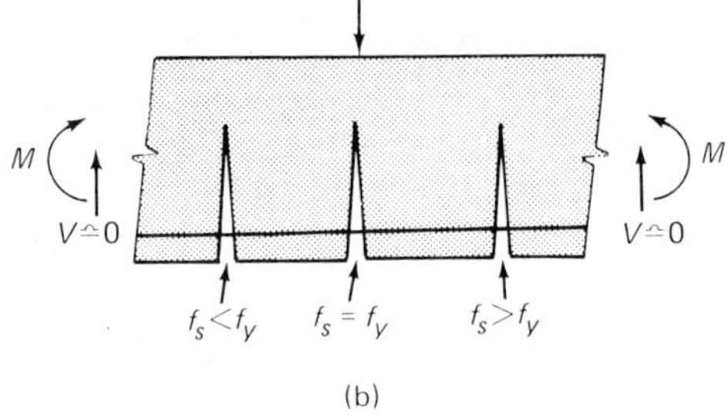

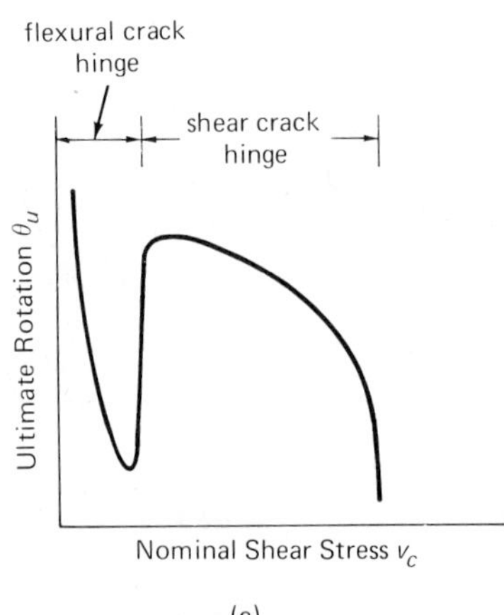

Fig. 8-20 Effect of shear on hinge rotations.[8-14] (a) Effect of shear cracks on hinge rotations, "shear-crack hinge"; (b) hinge in a region of small shears, "flexural-crack hinge;" and (c) curve of nominal shear vs ultimate rotation.

plotted in Fig. 8-20(c).[8-14] The values of effective hinge length in eq. (8-9) are based on this concept.

c. Bond and anchorage

Bond failures and anchorage failures are generally sudden and brittle and special attention must be given to details to prevent them from occurring in structures which must behave in a ductile manner. This is especially true in the case of structures in seismic regions.

The *ACI Code* provisions for bond and anchorage were derived on the assumption that the reinforcement should yield and strain-harden slightly prior to bond failure and will therefore allow a certain amount of ductility to develop. Wherever possible, however, splices should be avoided at points of maximum moment or stress. The addition of stirrups to the zones near splices or anchorages will generally improve the bond strength.

8.4 THE DUCTILITY OF COLUMNS

Generally speaking, tied columns do not have much ductility, especially in tall buildings where the eccentricity ratio, e/t, may be small. Spiral columns have much more ductility and can absorb major amounts of energy. This is illustrated by the appearance of adjacent tied and spiral columns in the Olive View Hospital following the 1971 San Fernando earthquake as shown in Fig. 8-21. Although both columns had undergone the same deformations, only the spiral column was still able to carry load.

The effect of axial loads on moment-curvature diagrams for tied columns is shown in Fig. 8-22a for a column having the interaction diagram shown in Fig. 8-22b.[8-6] The balanced failure load is $P/P_o = 0.3$ which corresponds to 0.34 $f'_c A_g$. For axial loads less than about 40% of the balanced failure load, or less than about 0.12 $f'_c A_g$, which is essentially the same load, this column has ductility ratios of at least 4 to 6. For axial loads greater than these values the ACI and SEAOC, respectively, require special design provisions to offset the drop in ductility. For axial loads greater than P_b the column has no ductility as shown in Fig. 8-22a.

Measured moment-curvature diagrams are presented in Fig. 8-23 for tied and spiral columns loaded with a large eccentricity ratio, $e/h = 1.25$.[8-15] The concrete in the tied column was unconfined while the core concrete in the

(a)

(b)

Fig. 8-21 Behavior of tied and spiral columns in an earthquake. (a) Tied column; and (b) spiral column.

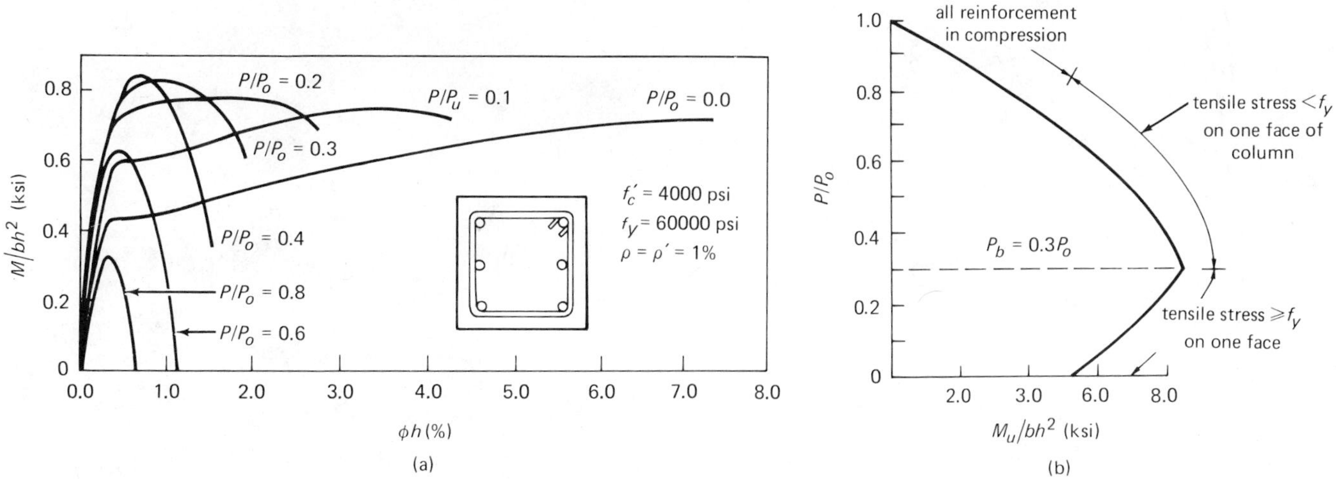

Fig. 8-22 Ductility of a tied column.[8-6] (a) Effects of axial loads on moment-curvature diagrams; and (b) interaction diagram.

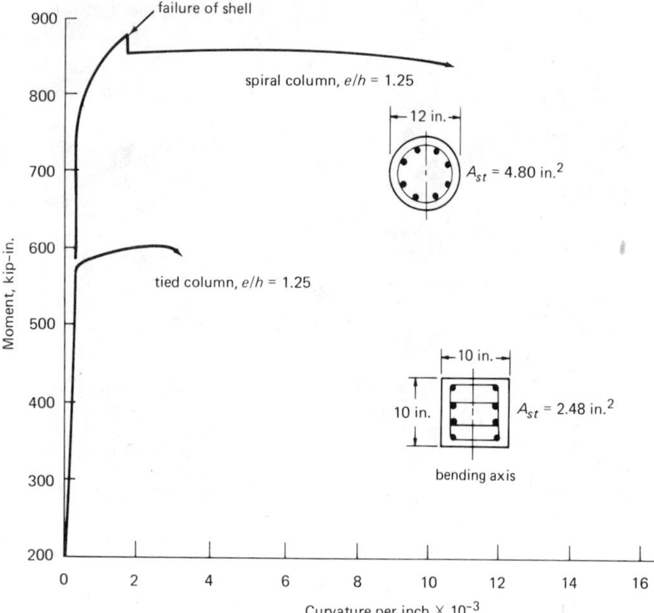

Fig. 8-23 Measured moment-curvature diagrams for spiral and tied columns failing in tension.[8-15]

spiral column was confined by a spiral designed according to the *ACI Code*. Because of the large eccentricity the tied column failed at 0.22 P_b in a ductile manner. The moment curvature diagram for the spiral column shows considerably more ductility. The sudden drop in moment resistance in this diagram corresponds to the failure of the shell concrete. When loading was discontinued the core concrete was still intact.

Measured load-deflection curves are presented in Fig. 8-24 for tied and spiral columns having an eccentricity ratio, e/h, of 0.25. This corresponded to P_u/P_o about 0.53 for both columns. As would be expected from Fig. 8-22(a), the tied column was not ductile since P_u exceeded P_b. Although the spiral column never regained its initial capacity after the spiral was lost, it was still carrying 60% of its maximum capacity when the test was terminated at a deflection of 3 in.

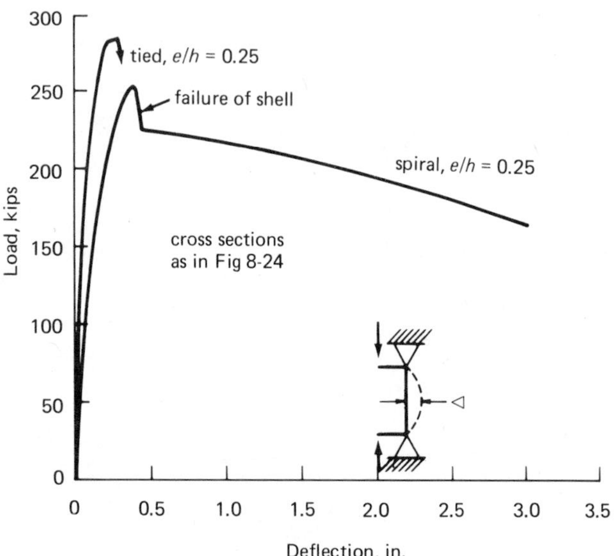

Fig. 8-24 Measured load-deflection diagrams for spiral and tied columns failing in compression.[8-15]

The ductility of a tied column can be improved by providing closely spaced stirrup ties or hoops and supplementary cross ties (Fig. 8-13) in critical regions to restrain the concrete in the transverse directions.[8-3] The action of such ties has been discussed in section 8.3.2(e). Generally speaking, extra confining ties are only required in the vicinity of the regions of high moment at the ends of the column. The *ACI Code* requirements for confinement ties in seismic areas are discussed in Chapter 12 of this handbook.

In seismic areas columns may develop significant shear forces and frequently fail in shear in severe earthquakes (Fig. 8-25).[8-16] To ensure ductility of the column, shear reinforcement must be provided for the applied shears at the formation of plastic hinges in the frame due to lateral displacements and gravity loads. The calculation of column shears is illustrated in Fig. 8-26. Under seismic loads the two column end moments M_c in Fig. 8-26a should both be clockwise or both anticlockwise. Their magnitude depends on whether plastic hinges develop in the columns or beams. If the columns hinge, M_c is the plastic moment capacity of the column. The maximum possible column shear corresponds to hinges at the top and bottom of the column. When plastic hinges develop in the beams the column moments, M_c, can be evaluated from the simplified free body diagrams in Fig. 8-26b. Where two beam moments are shown in this drawing they both act in the same direction.

The *SEAOC Code* recognizes the possibility of strain-

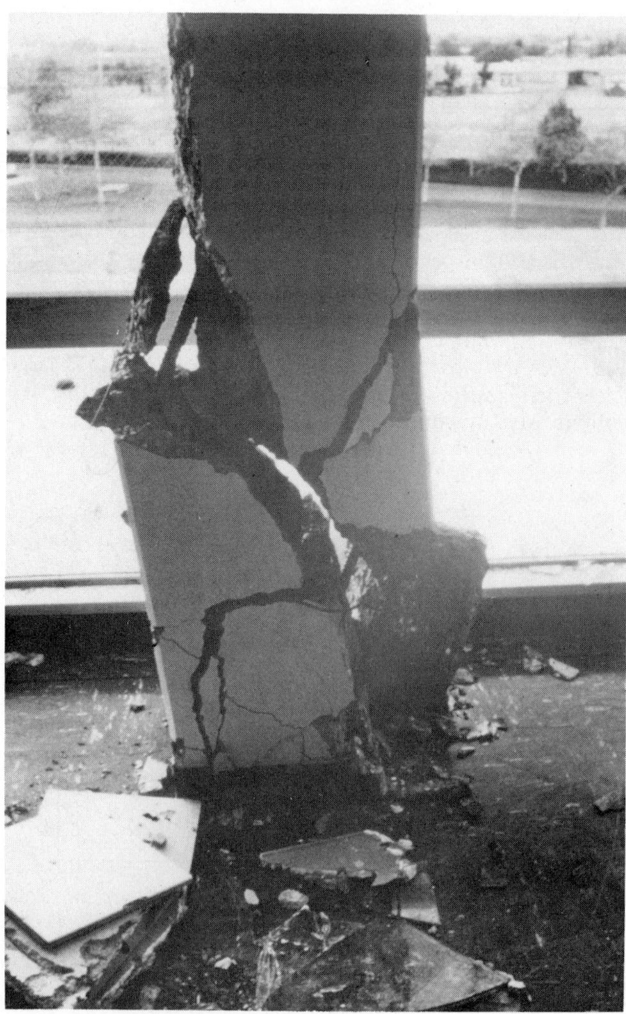

Fig. 8-25 Shear failure in a column.

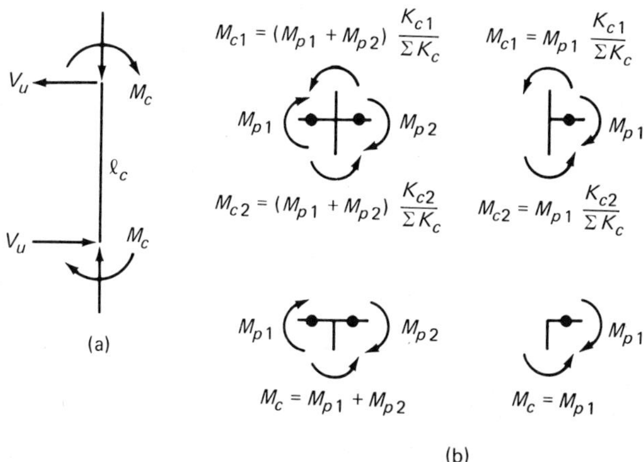

Fig. 8-26 Shear in column under seismic loading. (a) Maximum total column shear under seismic loading; and (b) column moments in joints where plastic hinges develop in beams.

hardening in the reinforcement at hinges near the joint and requires that the various values of the M_B and M_c in Fig. 8-26 be evaluated using 125% of the specified yield strength and the understrength ϕ-factor equal to 1.0

In conventional reinforced concrete construction, the column reinforcement is generally spliced just above the floor level. This is an extremely poor location for such a splice since it is a point of maximum column moment. Figure 8-27 presents load deflection curves for three specimens simulating column splices.[8-17] The splices corresponded to an ACI Class C splice which would be required in this instance because all the bars in tension are spliced at this point. The splice enclosed by normal column ties reached a ductility ratio of about 5, while that enclosed by closely spaced ties developed a much higher ductility. The deflection of the specimen without a splice was larger because in the spliced specimens the major inelastic rotations were essentially restricted to the space between the face of the column and the splice plus some slip in the splice.

Thus, column splices at sections of maximum column moment are undesirable. If possible such splices would be located at mid-height of the story. Lapped column splices should be enclosed by confining reinforcement to improve their ductility.

Finally, tension stresses frequently occur in the bars in a column as shown by the interaction diagram in Fig. 8-22. As a result, column splices and column-footing dowels often have to be designed as tensile splices.

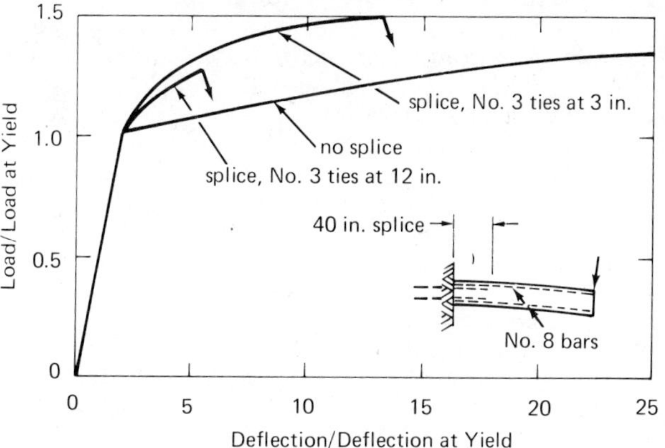

Fig. 8-27 Effect of reinforcement splices on the ductility of a connection.[8-17]

8.5 THE DUCTILITY OF BEAM-COLUMN JOINTS

In a beam-column frame it is essential that the joints withstand the forces and deformations imposed by inelastic action in the beams or columns. Frequently joints are points of weakness due to the lack of adequate anchorage for bars entering the joint from the columns and beams.

The forces acting on the exterior beam column joint are shown in Fig. 8-28a and b for two stages in a seismic loading cycle. These loadings could also occur due to gravity loads or settlement of the exterior columns. By cutting the horizontal plane X-X shown in Fig. 8-28b through the joint, the shear in the joint is seen to be:

$$V = A_s f_y - V_{col} \tag{8-10}$$

This shear force causes diagonal cracks and diagonal compression forces in the joint, see Fig. 8-28c. With every reversal of loading this shear changes sign causing cracks in both directions. In addition, major reversing bond stresses exist in the joint both in the beam reinforcement and in the column reinforcement. These cause splitting stresses in the concrete around the bars.

Unless the core of a beam column joint is confined by stirrup ties or by beams on all sides of the joint, the joint will be destroyed by the combined effects of the column load, the diagonal compression forces, and the splitting forces caused by anchoring the bars. Such a failure is not

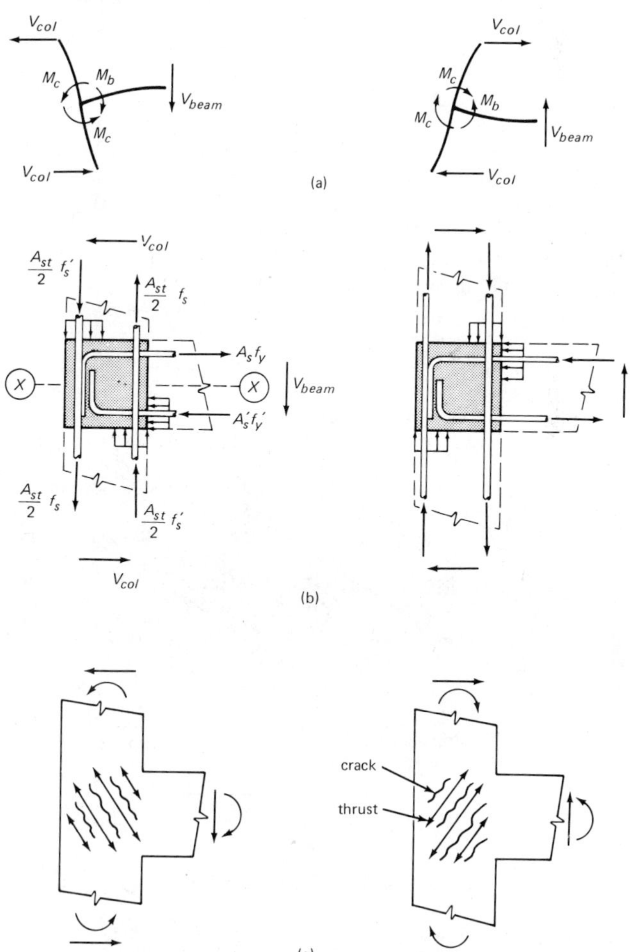

Fig. 8-28 Forces in an exterior beam-column joint. (a) Moments and shears on joint; (b) forces on joint; and (c) cracks and compression thrusts in joint.

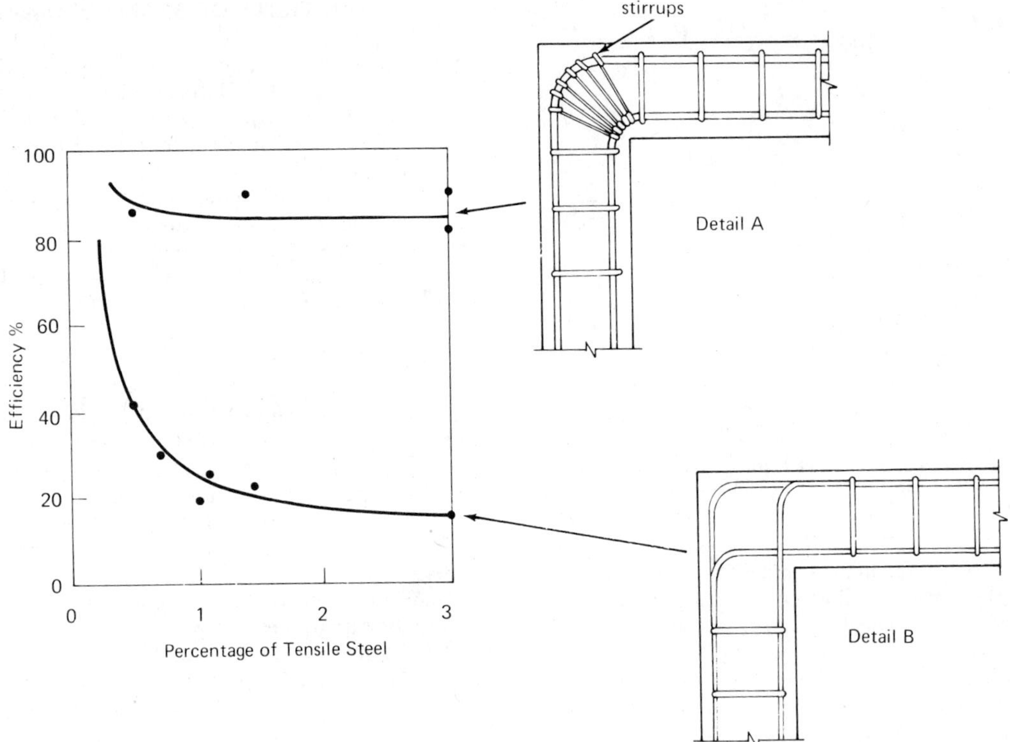

Fig. 8-29 Relative strengths of beam-column joints.[8-19]

ductile and may occur at loads lower than the hinging capacity of the beams or columns.

The *ACI Code* calls for stirrup ties in the joints of special ductile frames for seismic areas. The ties are designed for the shear given by eq. (8-10) and are checked to ensure they are large enough to confine the concrete in the joint. In addition the designer must ensure that the beam reinforcement is adequately anchored. Beam column joints designed in this manner using the beam bar anchorage detail shown in Fig. 8-28b developed ductility ratios of 5 when subjected to reversed cycles provided hinges did not develop in the columns.[8-18] The design and detailing of such joints are discussed in this handbook in Chapter 12, "Earthquake Resistant Structures."

A critical situation may also exist in the exterior roof beam to column joint. Figure 8-29 shows the relative strengths in tests[8-19] of two common types of joints when they were subjected to moments that tended to open the corner. The strengths are expressed as a fraction of the moment capacity of the beam entering the joint. The reinforcement was equivalent to Grade 60. In no case were the joints able to fully develop the beam strength in the tests shown. It is recommended that Detail A in Fig. 8-29 or a similar detail with a fillet and fillet reinforcement be used when such a joint occurs in a structure.[8-19]

Flat slab to column joints are very seldom sufficiently ductile to allow their use in ductile moment resisting frames. The ductility of such a joint can be improved by providing shear reinforcement of the form shown in Fig. 8-30. This detail has not been adequately tested to provide design rules, for combined shear and moment, however.

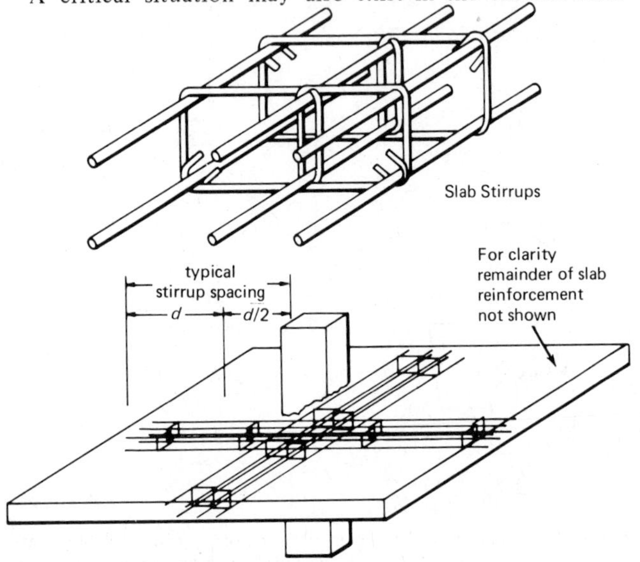

Fig. 8-30 Shear reinforcement in slabs.

8.6 DUCTILITY OF SHEAR WALLS

A reinforced concrete shear wall in a multistory reinforced concrete building is essentially a tall cantilever beam subject to axial and transverse forces. The ductility of such an element can be estimated using the techniques described earlier. Generally, however, the vertical reinforcement is spread over the full depth of the wall, either uniformly, or with a portion concentrated at the two edges. In the latter case the concentrated reinforcement is often placed in a boundary member which resembles a tied column with confining reinforcement.

Measured moment curvature diagrams for two walls are presented in Fig. 8-31.[8-20] The wall with reinforcement concentrated at the ends yielded at a higher load and had the same ultimate strength but more ductility than the other wall even though it had less vertical reinforcement. In addition, its moment-curvature diagram more nearly

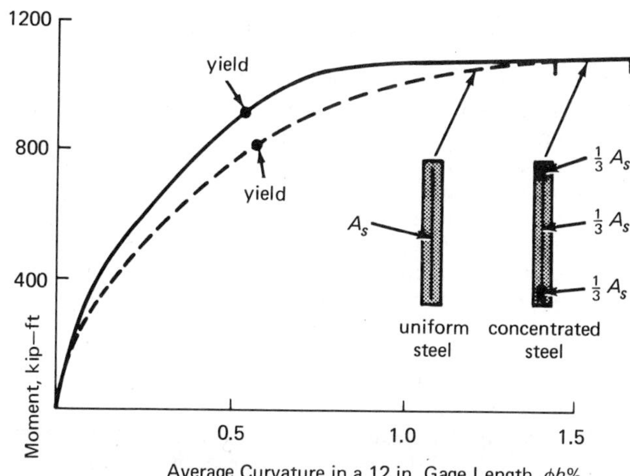

Fig. 8-31 Measured moment-curvature diagrams for shear walls.[8-20]

approaches an elastic-plastic diagram and hence this wall has more energy absorbing potential than the wall with a uniform distribution of reinforcement. The considerable load increase following initial yield is partly due to the gradual yielding of bars farther and farther from the face of the wall and partly due to strain hardening of the reinforcement near the tension faces of the walls. Calculations show that the ductility ratio increases as the portion of the total reinforcement which is placed in the flange is increased and decreases, as expected, as the axial load approaches the balanced load, P_b.

For ductility to be realized in a shear wall there must not be a shear or anchorage failure prior to a bending failure.[8-21] Thus, the shearing strength of the wall should be enough to resist shears corresponding to the failure moment based on strain hardening in the reinforcement. Similarly, the wall anchorage and foundations must be designed for the actual anticipated failure moment.

Very large ductility ratios are required in connecting beams coupling two shear walls together.[8-22] Tests of such beams have suggested that the total shear force in a connecting beam in a wall subjected to load reversals should be taken by shear reinforcement. Good ductility was obtained in tests of connecting beams having diagonal main reinforcement.

8.7 DESIGN FOR DUCTILITY

8.7.1 Philosophy Of Design

The design of structures which have sufficient ductility either to undergo moments due to settlement or seismic actions or to redistribute moments due to local overloads requires a different design approach. It is relatively easy to choose cross sections which will have adequate strength. It is much more difficult to achieve adequate strength and at the same time have the desired ductility. To ensure ductility the design engineer must also consider bar cutoffs, splices, and even more important, the joint details. Fortunately, a considerable amount of ductility can be ensured by a few relatively simple and inexpensive design details.

A few basic guidelines in design for ductility apply to most structures:

1. The structural layout should be as regular and simple as possible, avoiding such things as offsets of beams to columns or offsets of columns from floor to floor. In general a highly indeterminate but regular rectangular frame will absorb large amounts of energy. Changes in stiffness should be gradual from floor to floor. Extra strength should generally be provided in areas of high stiffness.
2. Because beams are generally more ductile than columns, the sum of the moment capacities of the columns for the design axial loads at a beam-column connection should exceed the sum of the moment strengths of the beams along each principal plane.
3. The amount of tensile reinforcement in beams should be limited and compression reinforcement is desirable in areas where ductility is required. If compression reinforcement is used it must be enclosed by ties to prevent it from buckling.
4. Closed stirrup ties or spirals should be used to confine the concrete at sections of maximum moment to increase the ductility of the members. This is especially important at the upper and lower ends of columns and within beam-column joints which do not have beams on all sides. The use of spiral columns or tied columns with a confined core should be considered if the axial load exceeds 0.4 times the balanced load.
5. Shear reinforcement must be adequate to ensure that the strength in shear exceeds the strength in flexure. The effect of strain-hardening of the reinforcement should be considered in calculating the design shears if large amounts of inelastic action are anticipated.
6. Splices and bar anchorages must be adequate to prevent bond failures. Wherever possible splices should be located at points of less than maximum stress. The reversal of stress in beams and columns should be considered. Frequently, column or wall splices must resist tensile stresses equal to or greater than the yield strength, especially in upper stories. Closed stirrups or ties are highly desirable in the region of splices.
7. Joints must be analyzed and designed to resist the forces which they will receive if the beams or columns develop yield hinges.
8. Special attention should be given to details to avoid planes of weakness due to excessive splicing, bending or terminating of reinforcement. All possible failure planes should be intercepted by reinforcement.

Two broad classes of structures can be identified. The structures in these classes require different levels of ductility.

1. Ductile structures designed for severe seismic action
   This class of structures requires a high degree of ductility. Special design provisions for such structures are presented in Appendix A of the *ACI Code*[8-1] and in the *SEAOC Code*[8-5] and are discussed in Chapter 12 of this handbook and in Ref. 8-3.
2. Structures requiring a limited amount of ductility
   Almost all other statically indeterminate reinforced concrete structures would fall into this class especially if any of the conditions listed in section 8.1.1 exist. Structures designed for moment redistribution or limit design also fall in this class. Current design procedures in the *ACI Code* assume that a number of sections in a structure will reach their ultimate capacities at the ultimate or design load. Although relatively little ductility as such is required for this to occur, the bond, shear and anchorage strengths and the joint details must be adequate.

8.7.2 Design Details for Limited Ductility Structures

a. Girders

1. To ensure adequate ductility the ratio of longitudinal tension reinforcement, ρ, in beams and girders should be

limited to:

$$\rho - \rho' \leqslant 0.5\,\bar{\rho}_b$$

but

$$\rho \leqslant 0.025$$

where $\bar{\rho}_b$ = the balanced reinforcement ratio for rectangular beam of width b (or b_w) without compression reinforcement as given in Table 8-2 and eq. (8-5)

Placement and splicing problems may require even lower values of ρ or $\rho - \rho'$ than those specified here.

2. The minimum limit on the longitudinal reinforcement ratio $\rho = A_s/bd$ for rectangular beams or $\rho_w = A_s/b_w d$, for beam webs should be $200/f_y$.

3. If moments have been computed by the elastic theory (not by moment coefficients) the calculated negative moments at the supports of continuous beams for each assumed loading arrangement can be increased or decreased by up to:

$$20\left(1 - \frac{\rho - \rho'}{\rho_b}\right)\text{percent}$$

When negative moments are changed, the revised moments must be used to calculate the other moments within the span.

4. The normal detailing rules in Chapter 12 of the *ACI Code* are adequate but should be carefully followed. If moment redistribution has been considered in the design or if spans are not uniform it will be necessary to calculate cut-off points for the reinforcement. When a beam or girder is part of the lateral load carrying system or may be subject to differential settlements, etc., at least one-quarter of the positive moment reinforcement should be extended into the support and anchored to develop its yield stress in tension at the face of the support. It is not sufficient to use more bars at lower stresses to satisfy this requirement. The recommended arrangements of longitudinal reinforcement are summarized in Fig. 8-32.

5. Stirrup design follows normal practice. In all cases the minimum web reinforcement required by the *ACI Code* should be provided. Where compression reinforcement has been assumed in the design it must be restrained from buckling by ties or stirrups. The customary tie spacings in Section 7.12.3 of the *ACI Code* are satisfactory provided the structure is not subjected to repeated loading cycles.

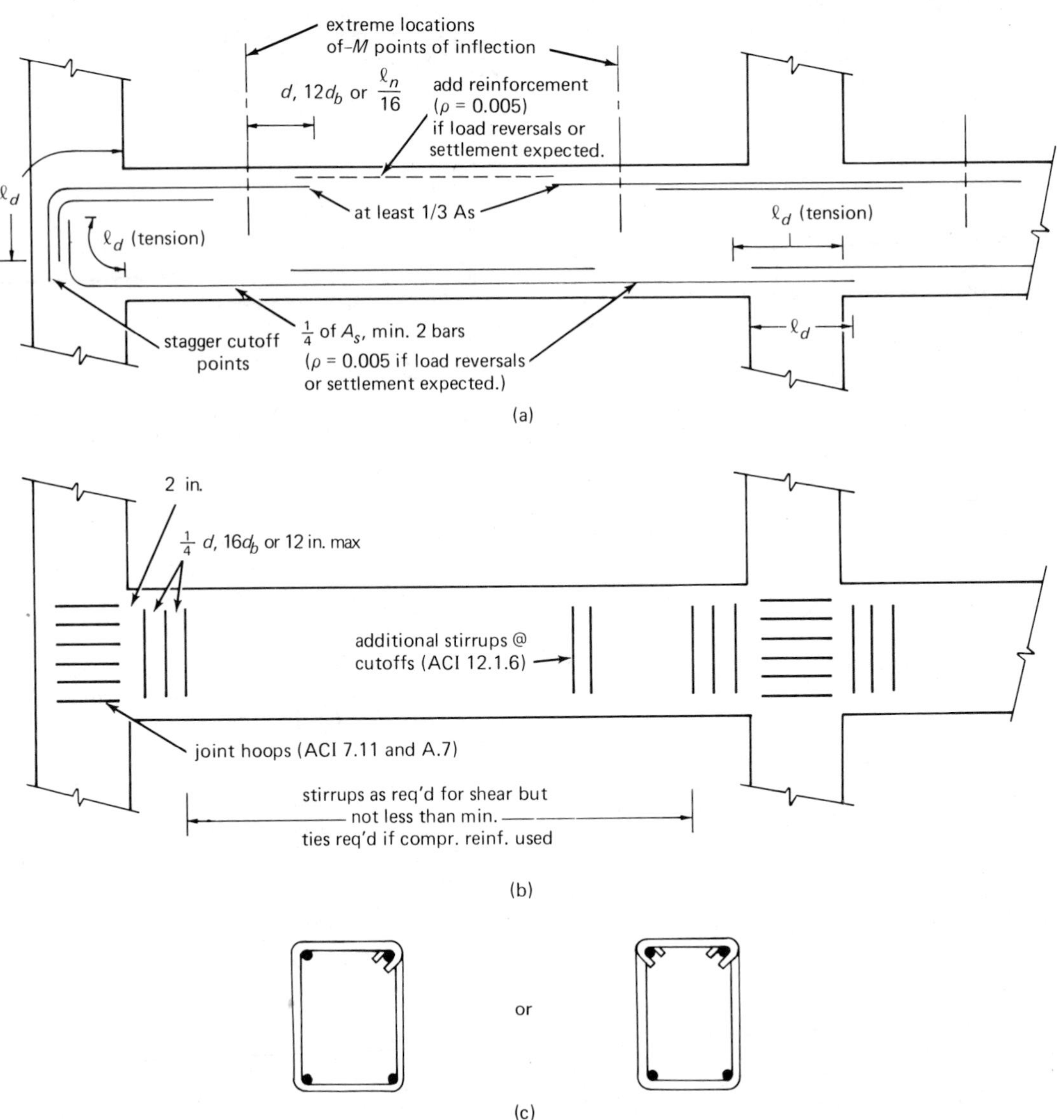

Fig. 8-32 Recommended reinforcement details for limited ductility sections. (a) Minimum longitudinal reinforcement provisions; (b) minimum shear reinforcement provisions; and (c) ties for compressive reinforcement in positive moment region, or if load reversals expected.

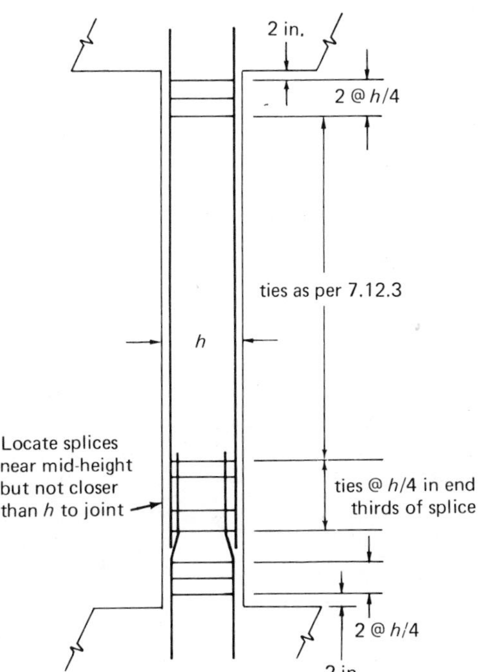

Fig. 8-33 Recommended column details for limited ductility structures.

b. Columns

The design of columns for limited ductility should consider the following items:

1. Spiral columns should be used wherever possible if $P \geqslant 0.12\ f'_c A_g$ or $P \geqslant 0.4\ P_b$.
2. If tied columns are used, a reduced tie spacing should be provided at the upper and lower ends of the column as shown in Fig. 8-33.
3. Lapped or other splices should not be located closer than h to the end of a column. Such splices should be enclosed in closely spaced ties.
4. The possibility of tensile stresses in the reinforcement often equal to the yield point must be investigated when designing splices and anchorages for columns. Figure 8-22b illustrates the regions of the interaction diagram for a column in which tensions may exist in column bars.

c. Beam-Column joints

Frequently the beam column joints are the weakest links in a structure. To avoid frame failures due to inadequate joints the joint details must be considered by the structural engineer during the frame design. Sometimes it is necessary to provide scale drawings (not line drawings) of the bar placement in critical joints. The following items must be considered in the design:

1. Anchorage of beam reinforcement in the joint.
2. Shearing stresses and the resulting diagonal tensions.
3. Hoops around the joint will improve the anchorage of beam reinforcement and will carry shear forces.

REFERENCES

8-1 Committee 318, *Building Code Requirements for Reinforced Concrete* (ACI 318-71), American Concrete Institute, Detroit, 1971.

8-2 Furlong, R. W., "Design of Concrete Frames by Assigned Limit Moments," *Proc. American Concrete Institute*, **67**, 341-353, April 1970.

8-3 Blume, J. A.; Newmark, N. M.; and Corning, L. H., "Design of Multi-story Reinforced Concrete Buildings for Earthquake Motions," Portland Cement Assn., 1961.

8-4 Bresler, B., "Lightweight Aggregate Reinforced Concrete Columns," Lightweight Concrete, *ACI Special Publication No. 29*, Detroit, 1971.

8-5 Seismology Committee, "Recommended Lateral Force Requirements and Commentary," Structural Engineers Association of California, 1967, 1-90; Revisions and Addendum 1968, 91-100; 1971 Revisions, 106-112.

8-6 Cohn, M. Z. and Ghosh, S. K., "The Flexural Ductility of Reinforced Concrete Sections," S. M. Report No. 100, Solid Mech. Div., Univ. of Waterloo and *Publications* International Association of Bridge and Structural Engineers, Vol. 32-2, 1972.

8-7 Singh, A.; Gerstle, K. H.; and Tulin, L. G., "The Behavior of Reinforcing Steel Under Reversed Loading," Materials Research and Standards, ASTM, January 1965, 12-17.

8-8 Allen, D. E., "Probabilistic Study of Reinforced Concrete in Bending," *Proc. American Concrete Institute*, **67**, 989-992, December 1970.

8-9 Corley, W. G., "Rotational Capacity of Reinforced Concrete Beams," *Proc. American Soc. of Civil Engineers*, J. Struct. Div., **92** (ST5), 121-146, October 1966.

8-10 Bertero, V. V., and Felippa, C., Discussion of "Ductility of Concrete," *Flexural Mechanics of Reinforced Concrete*, ACI SP12, Detroit, 1965, 227-234.

8-11 Gaston, J. R., Siess, C. P. and Newmark, N. M., "An Investigation of the Load Deformation Characteristics of Reinforced Concrete Beams up to the Point of Failure," Civil Eng. Studies, *Structural Research Series No. 40*, Univ. of Illinois, Urbana, December 1952.

8-12 Burns, N. H. and Siess, C. P., "Plastic Hinging in Reinforced Concrete," *Proc. ASCE*, J. Struct. Div., **92** (ST5), 45-64, October 1966.

8-13 Burns, N. H., and Siess, C. P., "Repeated and Reversed Loading in Reinforced Concrete," *Proc. ASCE*, J. Struct. Div., **92** (ST5), 65-78, October 1966.

8-14 Bachmann, H., "Influence of Shear and Bond on Rotation Capacity of Reinforced Concrete Beams," Publications, International Assn. for Bridge and Structural Engineering, No. 30-11, 1970, 11-28.

8-15 Hognestad, E., "A Study of Combined Bending and Axial Load in Reinforced Concrete Members," *Univ. of Illinois Bulletin, No. 399*, Univ. of Illinois Engrg. Experiment Stn., November 1951, 128 pp.

8-16 Lew, H. S.; Leyendecker, E. V.; and Dikkers, R. D., "Engineering Aspects of the 1971 San Fernando Earthquake," National Bureau of Standards (U.S.), Bldg. Sci Ser No. 40, December 1971, pp. 419.

8-17 Colaco, J. P. and Siess, C. P., "Behavior of Splices in Beam-Column Connections," *Proc. ASCE*, J. Struct. Div. **93** (ST5), 175-194, October 1967.

8-18 Hanson, N. W., "Seismic Resistance of Concrete

Frames With Grade 60 Reinforcement," *Proc. A.S.C.E.*, (ST6), 1685–1700, June 1971.

8–19 Balint, P. S., and Taylor, H. P. J., "Reinforcement Detailing of Frame Corner Joints with Particular Reference to Opening Corners," Tech. Report 42.462, Cement and Concrete Assn., London, February 1972, 16 pp.

8–20 Cardenas, A. E., and Magura, D. D., "Strength of High-Rise Shear Walls–Rectangular Cross Sections," *Response of Multistory Concrete Structures to Lateral Forces*, American Concrete Institute Special Publication SP-36, Detroit, 1973.

8–21 Allen, C. M.; Jaeger, L. G.; and Fenton, V. C., "Ductility in Reinforced Concrete Shear Walls," *Response of Multistory Concrete Structures to Lateral Forces*, American Concrete Institute Special Publication SP-36, Detroit, 1973.

8–22 Paulay, T., "Simulated Seismic Loading of Spandrel Beams," *Proc. ASCE*, J. Struct. Div., 97 (ST9), 2407–2419, September 1971.

9

Prestressed Concrete

JAMES R. LIBBY*

9.1 INTRODUCTION

9.1.1 General

Prestressing can be defined as the application of a predetermined force or moment to a structural member in such a manner that the combined internal stresses in the member, resulting from this force or moment and from any anticipated condition of external loading, will be confined within specific limits. Prestressing concrete, the subject of this chapter, is the result of applying this principle to concrete structural members, with a view toward eliminating or materially reducing the tensile stresses in the concrete.

The prestressing principle is believed to have been well understood since about 1910, although patent applications relating to types of construction which involved the principle of prestressing date back to 1888.[9-1]

The early attempts at prestressing were abortive, however, due to the poor quality of the materials that were available in the early days and also to a lack of understanding of the action of creep in concrete. Eugene Freyssinet, the eminent French engineer, is generally regarded as the first to discover the nature of creep in concrete and to realize the necessity of using high-quality concrete and high-tensile steel to ensure that adequate prestress is retained. Freyssinet applied prestressing in structural applications during the early 1930's. The history and evolution of prestressing are controversial and not well documented, and for this reason, they are not discussed further in this chapter. The reader who is interested may find additional historical details in the references.[9-2, 9-3]

Many experiments have been conducted to demonstrate that prestressed concrete has properties that differ from those of reinforced concrete. Diving boards and "fishing poles" have been made of prestressed concrete to demonstrate the ability of this material to withstand large deflections without cracking. The more significant fact, however, is that prestressed concrete has proved to be economical in buildings, bridges, and other structures that would not be practical or economical in reinforced concrete under conditions of span and loading.

Prestressed concrete was first used in the U.S. (except in tanks) in the late 1940's. At that time, most U.S. engineers were completely unfamiliar with this mode of construction. Design principles of prestressed concrete were not taught in the universities, and the occasional structure that was constructed with this new material received wide publicity.

The amount of construction utilizing prestressed concrete has become tremendous and certainly will increase in the future. The contemporary structural engineer must be well informed on all facets of prestressing concrete.

Practical considerations prevent all facets of prestressed concrete being covered in this chapter. The data presented

*President, Libby-Perkins Engineers, San Diego, California.

are from a comprehensive reference book written by the author of this chapter.[9-4] The reader who wishes more detailed information should refer to a reference book of this type.

9.1.2 General Design Principles

Prestressing, in its simplest form, can be illustrated by considering a simple prismatic, flexural member (rectangular in cross section) prestressed by a concentric force, as shown in Fig. 9-1. The distribution of the stresses at midspan are as indicated in Fig. 9-2. It is readily seen that if the flexural tensile stresses in the bottom fiber, due to the dead and live loads, are to be eliminated, the uniform compressive stress due to prestressing must be equal in magnitude to the sum of these tensile stresses.

There is a time-dependent reduction in the prestressing force, due to the creep and shrinkage of the concrete and the relaxation of the prestressing steel. If no tensile stresses are to be permitted in the concrete, it is necessary to provide an *initial* prestressing force that is larger than would be required to compensate for the flexural stresses resulting from the external loads alone. These losses, which are discussed in greater detail in section 9.7.2, generally result in a reduction of the initial prestressing force by 10 to 30%. Therefore, if the stress distributions shown in Fig. 9-2 are desired after the loss of stress has taken place (under the effects of the *final* prestressing force), the distrubition of stresses under the initial prestressing force would have to be as shown in Fig. 9-3.

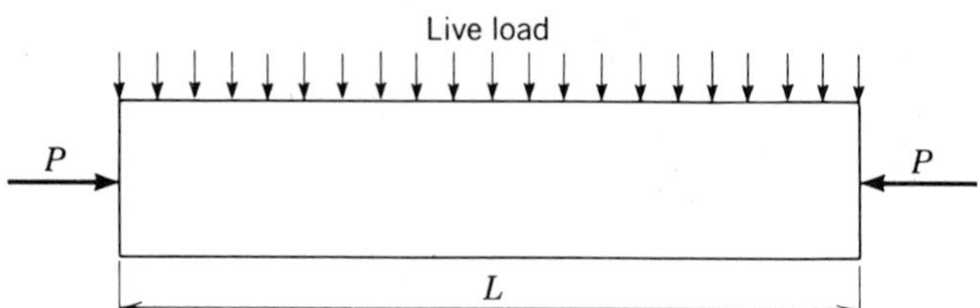

Fig. 9-1 Simple rectangular beam prestressed concentrically.

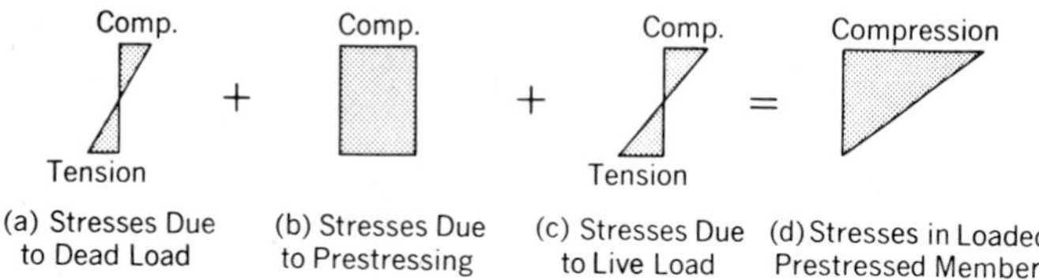

Fig. 9-2 Distribution of stresses at midspan of a simple beam concentrically prestressed. (a) Stresses due to dead load; (b) stresses due to prestressing; (c) stresses due to live load; (d) stresses in loaded prestressed member.

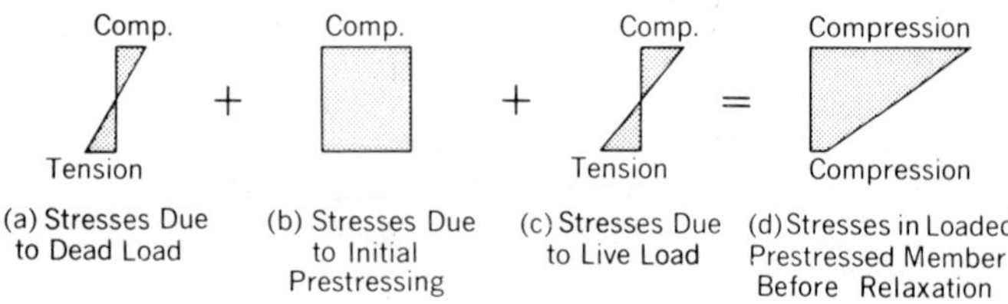

Fig. 9-3 Distribution of stresses at midspan of a simple beam under initial concentric prestressing force. (a) Stresses due to dead load; (b) stresses due to initial prestressing; (c) stresses due to live load; (d) stresses in loaded prestressed member before relaxation.

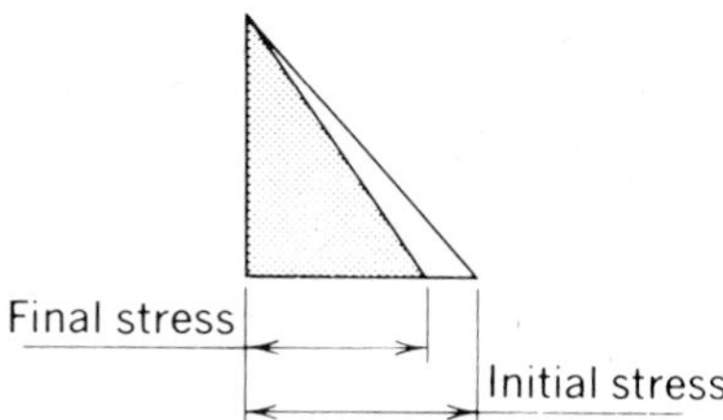

Fig. 9-4 Distribution of stresses due to prestressing forces applied at lower third point of rectangular cross section.

Prestressing with the concentric force just illustrated has the disadvantage that the top fiber is required to withstand the compressive stress due to prestressing in addition to the compressive stresses resulting from the design loads. Furthermore, since sufficient prestressing must be provided to compress the top fibers, as well as the bottom fibers, if sufficient prestressing is to be supplied to eliminate all of the flexural tensile stresses, the average stress due to the prestressing force (P/A) must be equal to the maximum flexural tensile stress resulting from the design loads.

If this same rectangular member were prestressed by a force applied at a point one-third of the depth of the beam from the bottom of the beam, the distribution of the stresses due to prestressing would be as shown in Fig. 9-4. In this case, as in the previous example, the final stress in the bottom fiber due to prestressing should be equal in magnitude to the sum of the tensile stresses resulting from the design loads. By inspection of the two stress diagrams for prestressing (Figs. 9-2b and 9-4), it is evident that the average stress in the beam, prestressed with the force at the third point, is only one-half of that required for the beam with concentric prestress. Therefore, the total prestressing force required to develop the desired prestressing of the second example will be only one-half of the amount required in the first example. In addition, the top fiber is not required to carry any compressive stress due to prestressing when the force is applied at the third point.

The economy that results from applying the prestressing force eccentrically is obvious. Further economy can be achieved when small tensile stresses are permissible in the top fibers—these tensile stresses may be due to prestressing alone or to the combined effects of prestressing and any external loads that may be acting at the time of prestressing. This is because the required bottom-fiber prestress can be attained with a smaller prestressing force, which is applied at a greater eccentricity under such conditions.

9.1.3 Prestressing with Jacks

The prestressing force in the above examples could be the result of placing jacks horizontally at the ends of the member, if there were abutments at each end of the beam which were sufficiently strong to resist the prestressing force that would be developed by such jacks. Prestressing with jacks, which may or may not remain in the structure, depending upon the circumstances, has been used on dams, dry docks, pavements, and other special structures. This method has been used to a limited degree, because extremely careful control of the design (including the study of the behavior under overloads), construction planning, and execution of the construction is required if the results obtained are to be satisfactory. Furthermore, the loss of prestress that results from this method is much larger than when other methods are used, unless frequent adjustments of the jacks are made, because the concrete is subjected to constant strain in this method rather than to nearly constant stress as in the case in other methods. For these reasons, and because the types of structures to which this method of prestressing can be applied are very limited and beyond the scope of usual generalities, subsequent consideration of this method is not given in this chapter.[9-5]

9.1.4 Pre-Tensioning

Another method of creating the necessary prestressing force is referred to as pre-tensioning. Pre-tensioning is accomplished by stressing steel wires or strands, called tendons, to a predetermined amount, and then, while the stress is maintained in the tendons, placing concrete around the tendons. After the concrete has hardened, the tendons are released and the concrete, which has become bonded to the tendons, is prestressed as a result of the tendons attempting to regain the length they had before they were stressed. In pre-tensioning, the tendons are usually stressed by the use of hydraulic jacks. The stress is maintained during the placing and curing of the concrete by anchoring the ends of the tendons to abutments that may be as much as 500 ft or more apart. The abutments and appurtenances used in this procedure are referred to as a pre-tensioning bed or bench. In some instances, rather than using pre-tensioning benches, as mentioned above, the steel molds or forms that are used to form the concrete members are designed in such a manner that the tendons can be safely anchored to the mold after they have been stressed. The results obtained with each of these methods is identical, and the factors involved in determining which method should be used are of concern to the fabricator of prestressed concrete, but do not usually affect the designer.

The tendons used in pre-tensioned construction must be relatively small in diameter, because the bond stress between the concrete and the tendon is relied upon to transfer the stress from the tendon to the concrete. It should be recognized that the ratio of bond area to cross sectional area for a circular wire or bar is

$$\frac{\text{Bond Area}}{\text{Cross Section Area}} = \frac{4L}{d} \tag{9-1}$$

in which d is the diameter and L is the length. For a unit length, it will be seen from eq. (9-1) that the ratio, which is also the ratio of the bond area available to the force the tendon can withstand, decreases as the diameter increases. Therefore, a number of the small tendons are normally required to develop the required prestressing force.

Pre-tensioning is a major method used in the manufacture of prestressed concrete in this country. The basic principles and some of the methods that are currently used domestically were imported from Europe, but much has been done here to develop and adapt the procedures to the North American market. One of the more recent developments in this country has been the use of pre-tensioned tendons that do not pass straight through the concrete member, but which are deflected or draped into a trajectory that approximates a curve. This procedure was first used on light roof slabs, but it is now commonly used on large structural members including precast bridge girders. In other countries, this method is used to a much smaller extent.

Although many of the devices used in pre-tensioned construction are patented, the basic principle is in the public domain. This partially accounts for the very rapid rate of increase in the use of this method in this country.

9.1.5 Post-Tensioning

When a member is fabricated in such a manner that the tendons are stressed and each end is anchored to the concrete section after the concrete has been cast and has attained sufficient strength to safely withstand the prestressing force, the member is said to be post-tensioned. In this country, when using post-tensioning, a common method used in preventing the tendon from bonding to the concrete during placing and curing of the concrete is to encase the tendon in a mortar-tight, metal tube (or flexible metal hose) before placing it in the forms. The metal hose or tube is referred to as the sheath or duct and remains in the structure. After the tendon has been stressed, the void between the tendon and the sheath is filled with grout. In this manner, the tendon becomes bonded to the concrete section and corrosion of the steel is prevented.

Rather than use a metal tube or hose, a rubber hose (which may be inflated with air or water or may be stiffened during the placing of the concrete by the insertion of a metal rod) has been used to form a hole through the concrete section. After the concrete is sufficiently set, the rubber tube is removed by pulling from one end. The tendon is then inserted in the duct that was formed by the tube. The construction is completed by stressing the tendon, anchoring each end, and grouting it in place. Alternatively, a rigid metal tube which remains in place can be used to form a duct for the tendon.

Another method of preventing post-tensioning tendons from becoming bonded to the concrete before stressing is to coat the tendons with grease or a bituminous material, after which, the tendon is wrapped in waterproof paper or plastic. Tendons of this type are not pressure grouted after stressing. This type of post-tensioning is usually referred to as unbonded construction.

Post-tensioning offers a means of prestressing on the job site. This procedure may be necessary or desirable in some instances. Very large building or bridge girders that cannot be transported from a precasting plant to the job site (due to their weight, size, or the distance between plant and job site) can be made by post-tensioning on the job site. Post-tensioning is used in precast as well as in cast-in-place construction. In addition, fabricators of pre-tensioned concrete will frequently post-tension the members for small projects on which the number of units to be produced does not warrant the expenditures required to set-up pre-tensioning facilities. There are other advantages inherent in post-tensioned construction—these will be discussed later.

In post-tensioning, it is necessary to use some type of device to attach or anchor the ends of the tendons to the concrete section. These devices are referred to as end anchorages. The end anchorages, together with the special jacking and grouting equipment used in accomplishing the post-tensioning by one of the several available methods, are referred to as post-tensioning systems. Many of the systems used in the U.S. were invented and developed in Europe or were modeled after such a system. The various systems are or were patented; this deterred somewhat the early use of the method. Post-tensioning systems undergo almost constant revision and for this reason details of the various systems available are not given here. The interested reader should contact the firms who market post-tensioning materials. A list of these firms can be obtained from the Prestressed Concrete Institute.*

9.1.6 Pre-Tensioning vs Post-Tensioning

It is generally considered impractical to use post-tensioning on very short members, because the elongation of a short tendon (during the stressing) is small and would require

*Prestressed Concrete Institute, 20 N. Wacker Dr., Chicago, Ill. 60606.

very precise measurement by the workmen. In addition, many of the post-tensioning systems do not function well with very short tendons. A number of short members can be made in series on a pre-tensioning bench without difficulty and without the necessity of precise measurement of the elongation of the tendons during stressing. This is because a relatively long tendon length results from making a number of short members in series.

It has been pointed out that very large members may be more economical when cast-in-place and post-tensioned or when precast and post-tensioned near the job site, rather than attempting to transport and handle large pre-tensioned structural elements.

Post-tensioning allows the tendons to be placed, with little difficulty, through the structural elements on smooth curves of any desired trajectory. Pre-tensioned tendons can be employed on other than straight trajectories, but not without expensive plant facilities and somewhat complicated construction procedures.

The cost of post-tensioned tendons, measured in either cost per pound of prestressing steel or in cost per pound of effective prestressing force, is generally significantly greater than the cost of pre-tensioned tendons. This is due to the larger amount of labor required in placing, stressing, and grouting post-tensioned tendons and to the cost of the special anchorage devices and stressing equipment. A post-tensioned member may require less total prestressing force than an equally strong pre-tensioned member; however, and for this reason, care must be exercised when comparing the relative cost of these modes of prestressing.

The basic shape of an efficient pre-tensioned flexural member may be different from the most economical shape that can be found for a post-tensioned design. This is particularly true of moderate and long-span members and complicates any generalizations about which method is best under such conditions.

Post-tensioning is generally regarded as a method of making prestressed concrete at the job site, yet post-tensioned beams are often made in precasting plants and transported to the job site. Pre-tensioning is often thought of as a method of manufacturing that is limited to permanent precasting plants. Yet on very large projects where pre-tensioned elements are to be utilized, it is not uncommon for the general contractor to set up a temporary pre-tensioning plant at or near the job site. Each method of making prestressed concrete has particular theoretical and practical advantages and disadvantages, which will be more apparent after the principles are well understood. The final determination of the mode of prestressing that should be used on any particular project can only be made after careful consideration of the structural requirements and the economic factors that prevail for the particular project.

9.1.7 Linear vs Circular Prestressing

The subject of prestressed concrete is frequently divided into linear prestressing, which includes the prestressing of elongated structures or elements such as beams, bridges, slabs, piles, etc.; and circular prestressing, which includes pipe, tanks, pressure vessels, and domes. There are no generally recognized criteria for the design and construction of circularly prestressed structures. The theory of such construction is relatively simple and is adequately covered in the literature (see refs. 9-6 through 9-13). This chapter has been confined to the structural design and analysis of linear prestressed structures and the methods of prestressing used in this type of construction.

9.1.8 Application of Prestressed Concrete

Prestressed concrete, when properly designed and fabricated, can be virtually crack-free under normal service loads as well as under moderate overload. This is believed to be an advantage in a structure that is exposed to an especially corrosive atmosphere. Prestressed concrete efficiently utilizes high-strength concretes and steels and is economical even with long spans. Reinforced concrete flexural members cannot be designed to be crack-free, cannot efficiently utilize high-strength materials, and are not economical on long spans.

A number of other statements can be made in favor of prestressed concrete, but there are bona fide objections to the use of this material under specific conditions. Among the more significant points to be kept in mind about this material are (1) in many structural applications, prestressed concrete is more economical in first cost than other types of construction; and (2) in many cases, if the reduced maintenance costs that are inherent with concrete construction are taken into account, prestressed concrete offers the most economical solution for the structure. This fact has been well confirmed by the very rapid increase in the use of prestressed concrete that has taken place in the United States since 1953. It is well known that the advantage of real economy outweighs all intangible advantages that may be claimed, except for very special conditions. It should also be kept in mind that prestressing is not a panacea for the construction industry, but has definite limitations. Precautions that must be observed in designing and constructing prestressed concrete structures differ from those required for reinforced concrete structures. Some of these precautions are discussed subsequently in section 9.8.

9.1.9 Materials for Prestressed Concrete

The materials used in prestressed concrete construction include concrete, common reinforcing steel, and high tensile-strength steel. Each of these is discussed in detail in chapter 6.

It should be emphasized that concretes with higher compressive strengths are more commonly used in prestressed concrete. This is because the shrinkage and creep of higher strength concretes is normally less than those of concretes having low or moderate compressive strengths. In addition, efficient structural use can be made of high strength concrete when it is prestressed and this is not the case for ordinary reinforced concrete.

High tensile strength steel is employed for the principal reinforcing in prestressed concrete for two basic reasons. The first is that the strain which occurs during stressing of a high-strength steel is greater than that which can be obtained in a lower-strength steel (without exceeding the yield point). Hence, the losses of strain due to concrete shrinkage and creep are a smaller percentage of the total strain when high-strength steel is used. This minimizes the loss of prestress. Secondly, a large prestressing force can be developed in a confined area when the steel strength is high because the steel area required to develop a specific force is smaller.

9.1.10 Design Criteria

The design of prestressed concrete members that are to be used in building construction are normally made to conform to the requirements of *Building Code Requirements for Reinforced Concrete* (ACI 318-71) which hereinafter is referred to as ACI 318-71.[9-14] The basic prestressed concrete criteria appears in Chapter 18 of ACI 318-71 whereas other criteria are covered in other chapters (i.e., deflections in Chapter 9, shear in Chapter 11, etc.). Hence, the designer should be familiar with all of the provisions of ACI 318-71.

The design of prestressed concrete members that are to be used in bridge construction are most frequently made to conform to the requirements of *Standard Specifications for Highway Bridges* adopted by the American Association of State Highway Officials[9-15] which hereinafter is referred to as the AASHO Specifications. Prestressed concrete is covered in Section 6 of the design portion of the AASHO Specifications.

9.2 BASIC PRINCIPLES OF FLEXURAL DESIGN

9.2.1 Introduction

The basic principles and mathematical relationships used in the design and analysis of prestressed-concrete flexural members are not unique to this type of construction. Virtually all of the fundamental relationships are based upon the normal assumptions of elastic design, which forms the basis of study of the strength of materials. The form in which the relationships appear in a discussion of prestressed concrete may be somewhat modified to facilitate their application. Little difficulty should be experienced in understanding these modified relationships.

Two major forms of design problems are encountered by the engineer engaged in the design of prestressed concrete flexural members. These are frequently referred to as the *review* of a member and as the *design* of a member.

The review of a member actually consists of the determination of the concrete flexural stresses, under all conditions of service loading and prestressing, in order to confirm the compliance of these stresses with the applicable design criteria. In addition, the review of a member *must* include a study of the ultimate moment that the section can be expected to develop (this is done to ensure adequate safety against a flexural failure). An investigation of the shear stresses must be made and the adequacy of the web reinforcing that is specified for the member must be confirmed. It should be apparent that in order to review a member as described here, the dimensions of the concrete section, the properties of the materials, the amount and eccentricity of the prestressing steel, the amount of nonprestressed reinforcing, as well as the amount of web reinforcing, must be known.

The design of a member consists of selecting and proportioning a concrete section in which the stresses in the concrete do not exceed the permissible values under any condition of service loading or prestressing. Design also includes the determination of the amount and eccentricity of the prestressing force that is required for the specific section. The design of a member must include a study of the moment that the section will develop at design load, and the determination of the amount of nonprestressed reinforcing that may be required. Additionally, a study of the shear stresses must be made and the amount of web reinforcing that may be required for adequate shear strength at design load must be determined. It must be emphasized that the design of a flexural member is normally a trial and error procedure. The designer must assume a concrete section and compute the prestressing force and eccentricity that are required to confine the concrete stresses within the allowable limits under all service loading conditions. In the design of a member, several adjustments of the trial section are normally required before a satisfactory solution is found.

In either the review or design of a member, the engineer should make a study of the deflection of the member under service loads. (See section 9.7.3)

This section is devoted to the consideration of the fundamental principles pertaining to the determination of the concrete stresses due to prestressing, the determination of the prestressing force and eccentricity required for a specific distribution of stresses due to prestressing, the consideration of the pressure line in simple flexural members that are loaded in the elastic range, and other topics related to flexural analysis and design. The problems presented in this chapter are confined to the "review" type.

The elastic analysis and design of prestressed flexural members can be done rapidly and accurately only after the fundamental theorems and axioms have been thoroughly mastered. Many of the operations discussed in this chapter can be done more rapidly by the use of expedients which become apparent with design experience. These classical methods should be well understood, however, before attempting the use of expedients. The design and analysis of continuous prestressed members (see section 9.6) also require complete familiarity with the principles presented in this section.

9.2.2 Mathematical Relationships for Prestressing Stresses

The stresses due to prestressing alone are generally combined stresses due to a direct load eccentrically applied. Therefore, these stresses are computed using the following well-known relationship for combined stresses

$$f = \frac{P}{A} \pm \frac{My}{I} \tag{9-2}$$

in which f is the fiber stress at the distance y from the centroidal axis. P is the axial force, A is the area of the cross section, M is the moment acting on the section, and I is the moment of inertia of the cross section.

Since the moment due to the prestressing is equal to the prestressing force multiplied by the eccentricity of this force (i.e., $M = Pe$), and since the square of the radius of gyration is equal to the moment of inertia divided by the area of the cross section ($r^2 = I/A$), the above relationship can be rewritten

$$f = \frac{P}{A}\left(1 \pm \frac{ey}{r^2}\right) \tag{9-3}$$

Using y_t and y_b to denote the distances from the centroidal axis to the top and bottom fibers, respectively, and by assuming the eccentricity to be positive when it is on the same side of the center of gravity as the fiber under consideration, the top and bottom fiber stresses for a prestressing force applied eccentrically below the center of gravity are expressed by

$$f_t = \frac{P}{A}\left(1 - \frac{ey_t}{r^2}\right) \tag{9-4}$$

$$f_b = \frac{P}{A}\left(1 + \frac{ey_b}{r^2}\right) \tag{9-5}$$

where f_t and f_b are the stresses in the top and bottom fibers due to the prestressing alone, respectively. A positive stress is compressive in the above relationships.

These relationships are the same for the stresses resulting from the initial and the final prestressing forces. In computing these stresses, one would of course use the initial prestressing force when computing the initial stresses and the final force when computing the final stresses. Frequently, the designer assumes a ratio between the final and the initial prestressing forces for design purposes, since the relaxation of the prestressing force cannot be accurately estimated until the design is nearly complete (see section 9.7.2). Therefore, if the designer bases his computation on the final prestressing force and has assumed that the total relaxation will be 15% of the initial force, for example, the stresses resulting from the initial prestressing force can be determined by dividing the final stresses by 0.85.

The experienced designer generally prefers to design with the final prestressing force assumed to be from 75% to 90% of the initial force. A comprehensive study of the losses of stress cannot be made until the basic design is finalized. If, when this study is made, it is found the loss will be greater than assumed, the initial prestressing force can be increased so that the final force will be satisfactory. The advantage of this procedure will be apparent after experience is gained in the analysis of prestressed concrete members.

EXAMPLE 9-1: Compute the stresses due to prestressing alone in a beam with a rectangular cross section 10 in. wide and 12 in. high that is prestressed by a final force of 120 k at an eccentricity of 2.5 in. State whether the stresses are compressive or tensile. Compute the stresses due to the initial prestressing force, if the ratio between the final force and the initial force is 0.85.

SOLUTION:

$$A = 120 \text{ sq. in.} \quad I = \frac{10 \times 12^3}{12} = 1440 \text{ in.}^4 \quad r^2 = \frac{1440}{120} = 12 \text{ sq. in.}$$

Final stresses

$$f_t = \frac{120{,}000}{120}\left(1 - \frac{2.5 \times 6}{12}\right) = -250 \text{ psi (tension)}$$

$$f_b = \frac{120{,}000}{120}\left(1 + \frac{2.5 \times 6}{12}\right) = +2250 \text{ psi (compression)}$$

Initial stresses

$$f_t = \frac{-250}{0.85} = -294 \text{ psi (tension)}$$

$$f_b = \frac{+2250}{0.85} = +2650 \text{ psi (compression)}$$

EXAMPLE 9-2: Compute the prestressing force and eccentricity that would be necessary in the beam of Example 9-1 in order to obtain a bottom-fiber compression of 2400 psi and a top-fiber tension of 350 psi, by equating the relationships for stresses due to prestressing in the top and bottom fibers.

SOLUTION:

$$\frac{P}{A}\left(1 - \frac{ey_t}{r^2}\right) = -350 = \frac{P}{120}\left(1 - \frac{e \times 6}{12}\right)$$

$$\frac{P}{A}\left(1 + \frac{ey_b}{r^2}\right) = +2400 = \frac{P}{120}\left(1 + \frac{e \times 6}{12}\right)$$

$$2400 - \frac{2400\, e \times 6}{12} = -350 - \frac{350\, e \times 6}{12}$$

$$1025\, e = 2750$$

$$e = 2.68 \text{ in.}$$

$$P = \frac{120 \times 2400}{1 + \dfrac{2.68 \times 6}{12}} = 123{,}000 \text{ lb}$$

The familiar principle of superposition is used to determine the combined effect of the prestressing and the other loads that may be acting simultaneously on a prestressed beam. Although it is possible to write a single equation that will accurately define the stress at any particular point in a beam, it is normally less confusing if the effect of each load (or prestressing) is computed separately and the net effect is determined by algebraically adding the effects of the several loads.

EXAMPLE 9-3: Compute the net initial and final concrete stresses in the extreme top and bottom fibers at the center line of a beam that is 10 in. wide, 12 in. deep and on a span of 25 ft. The beam is to support an intermittent, uniformly distributed live load of 0.45 k/ft and is to be prestressed with a final force of 120 k positioned with an eccentricity of 2.50 in. The ratio between the final and initial prestressing forces is assumed to be 0.85.

SOLUTION:

$$A = 120 \text{ in.}^2 \quad I = \frac{10 \times 12^3}{12} = 1440 \text{ in.}^4$$

$$S = \frac{1440}{6} = 240 \text{ in.}^3 \quad r^2 = \frac{I}{A} = \frac{1440}{120} = 12 \text{ in.}^2$$

Stresses due to final prestress

$$f_t = \frac{120{,}000}{120}\left(1 - \frac{2.5 \times 6}{12.0}\right) = -250 \text{ psi (tension)}$$

$$f_b = \frac{120{,}000}{120}\left(1 + \frac{2.5 \times 6}{12.0}\right) = +2250 \text{ psi (compression)}$$

Stresses due to initial prestress

$$f_t = \frac{-250 \text{ psi}}{0.85} = -294 \text{ psi (tension)}$$

$$f_b = \frac{+2250}{0.85} = +2650 \text{ psi (compression)}$$

Stresses due to the dead load of the beam alone

$$w_{DL} = \frac{120}{144} \times 0.150 = 0.125 \text{ k/ft,}$$

$$M_{DL} = 0.125 \times \frac{(25)^2}{8} = 9.78 \text{ k-ft}$$

$$f_t = \frac{9.78 \times 12{,}000}{240} = +488 \text{ psi (compression)}$$

$$f_b = \frac{9.78 \times 12{,}000}{240} = -488 \text{ psi (tension)}$$

Stresses due to live load alone

$$w_{LL} = 0.45 \text{ k/ft,}$$

$$M_{LL} = 0.45 \times \frac{(25)^2}{8} = 35.2 \text{ k-ft}$$

$$f_t = \frac{+35.2 \times 12{,}000}{240} = +1760 \text{ psi (compression)}$$

$$f_b = \frac{-35.2 \times 12{,}000}{240} = -1760 \text{ psi (tension)}$$

Combined stresses:

	Top Fiber	*Bottom Fiber*
Initial prestress	- 294 psi	+2650 psi
Beam dead load	+ 488 psi	- 488 psi
Initial prestress plus dead load	+ 194 psi	+2162 psi
Live load	+1760 psi	-1760 psi
Initial prestress plus total load	+1954 psi	+ 402 psi

	Top Fiber	*Bottom Fiber*
Final prestress	− 250 psi	+2250 psi
Dead load of beam	+ 488 psi	− 488 psi
Final prestress plus dead load	+ 238 psi	+1762 psi
Live load	+1760 psi	+1760 psi
Final prestress plus total load	+1998 psi	+ 2 psi

9.2.3 Pressure Line in a Beam with a Straight Tendon

At any section of a beam, the combined effect of the prestressing force and the externally applied load will result in a distribution of concrete stresses that can be resolved into a single force. The locus of the points of application of this force in any beam or structure is called the pressure line.

This can be illustrated by considering a rectangular beam prestressed by an eccentric, straight tendon, as is shown in Fig. 9-5. Such a beam would have a distribution of stresses due to prestressing alone at every cross section, as is shown in Fig. 9-6(a). It is readily seen that the force resulting from the distribution of internal prestressing stresses (C) is equal in magnitude to the prestressing force. In addition, it is applied at the same point as the prestressing force at every section, since the prestressing force and the eccentricity are both constant throughout the length of the beam.

If a uniform load of such magnitude, that results in the bottom-fiber prestress being nullified at midspan, is applied to the beam, the resulting stress distrubiton would be as indicated in Fig. 9-6(b) and the pressure line at this point would then be applied at a point $d/6$ above the centroidal axis of the beam. At the quarter point of this beam, under the same loading conditions, the stresses due to the external load are only 75% as much as those at midspan. The stress distribution resulting from the combination of prestressing and the flexural stresses due to the external load would be as shown in Fig. 9-6(c). At this point the pressure line is located at a distance of $d/12$ above the centroidal axis. At the support, since there are no flexural stresses resulting from the external load, the pressure line remains at the level of the steel. Plotting the location of the pressure line for this loading reveals that it is a parabola with its vertex at the center of the beam, as shown in Fig. 9-7.

In a similar manner, it can be shown that a larger uniform load would result in the pressure line being moved up even higher, and for a uniform load applied upward rather than

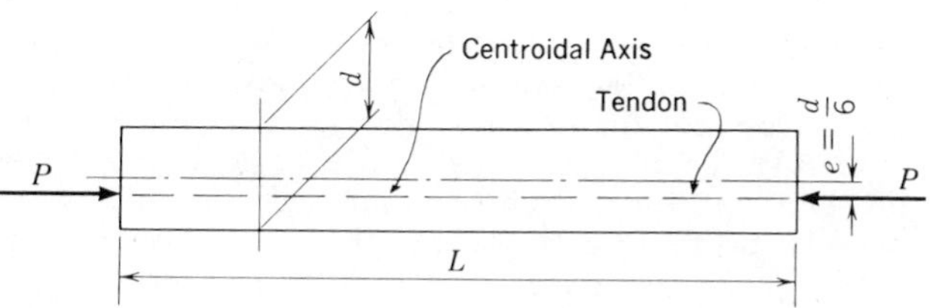

Fig. 9-5 Simple rectangular beam prestressed by an eccentric straight tendon.

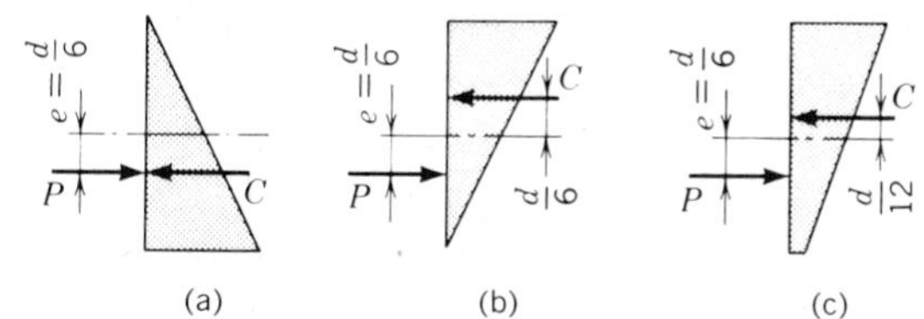

Fig. 9-6 Stress distributions and pressure-line locations for a simple rectangular beam prestressed with a straight eccentric tendon (a) due to prestressing alone; (b) at midspan under full design load; and (c) at quarter point under full design load.

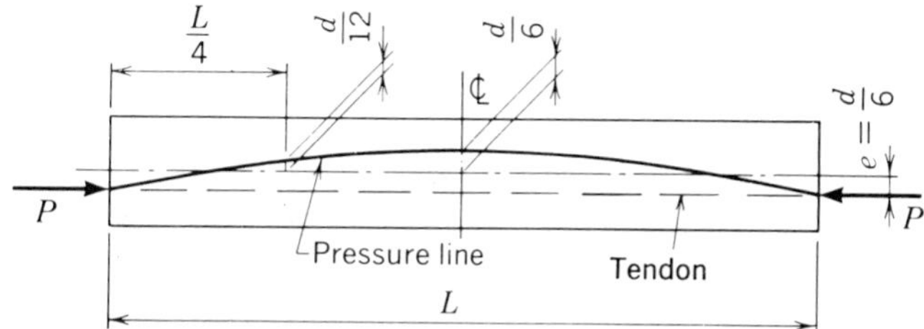

Fig. 9-7 Location of pressure line in a simple beam of rectangular cross section, prestressed by a force at $e = d/6$ and under uniform load resulting in zero bottom-fiber stress at the midspan.

downward, the result would be a downward movement of the pressure line. Therefore, it is apparent that the location of the pressure line in simple prestressed beams is dependent upon the magnitude and direction of the moments applied at any cross section and the magnitude and distribution of stress due to prestressing: *A change in the external moments in the elastic range of a prestressed beam results in a shift of the pressure line rather than an increase in the resultant force in the beam as is the case in beams composed of other materials.*

Due to the change in the strain in the concrete at the level of the steel (assuming the flexural bond strength between the steel and concrete is adequate, as it is in pre-tensioned and bonded post-tensioned beams), there is an increase in the stress in the prestressing steel as a result of applying an external load. This is rarely of importance and the effect is normally disregarded.

EXAMPLE 9-4: Compute and draw to scale the location of the pressure line for a rectangular beam 10 in. wide and 12 in. deep that is prestressed with a force of 120 k at a constant eccentricity of 2.5 in. and that is supporting a 15 k concentrated force at midspan of a span of 10 ft. Use an exaggerated vertical scale in a sketch, and dimension the location of the pressure line at midspan, quarter point, and end of the beam. Neglect the deadweight of the beam.

$$A = 120 \text{ sq in.} \qquad \frac{r^2}{y} = \frac{I}{Ay} = 2 \text{ in.} \qquad S = \frac{I}{y}\ 240 \text{ in.}^3$$

SOLUTION:
Stresses due to prestressing

$$f_t = \frac{120{,}000}{120}\left(1 - \frac{2.5 \text{ in.}}{2.0 \text{ in.}}\right) = -250 \text{ psi}$$

$$f_b = \frac{120{,}000}{120}\left(1 + \frac{2.5 \text{ in.}}{2.0 \text{ in.}}\right) = +2250 \text{ psi}$$

At the end of the beam, moment = zero. Therefore, the pressure line is at $e = 2.50$ in.

At the midspan

$$M = \frac{PL}{4} = \frac{15 \text{ k} \times 10 \text{ ft}}{4} = 37.5 \text{ k-ft}$$

$$f = \pm \frac{37.5 \times 12{,}000}{240} = \pm 1880 \text{ psi}$$

Stress distribution at midspan

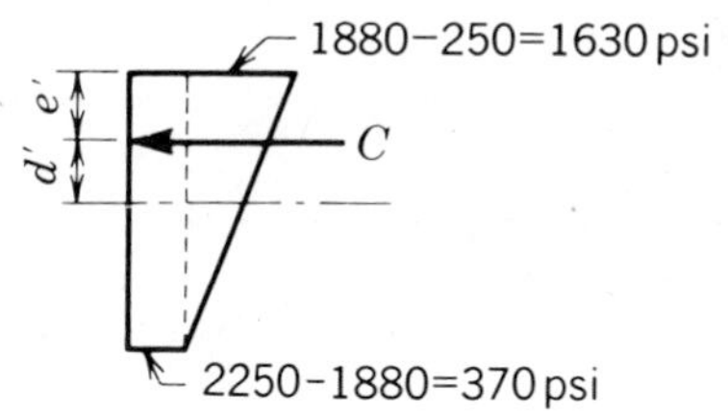

$$d'' = \frac{6 \times 370 \times 120 + (1260/2) \times 120 \times 4}{370 \times 120 + (1260/2) \times 120}$$

$$d'' = 4.73 \text{ in.}$$

$$e' = 6.00 \text{ in.} - 4.73 \text{ in.} = 1.27 \text{ in.}$$

At the quarter point, the moment due to the external load is only one-half that at the midspan. Therefore, the flexural stresses due to the applied load are only one-half of those at the center line, or ± 940 psi.

Stress distribution at quarter point

$$d' = \frac{6 \times 690 \times 120 + (620/2) \times 120 \times 8}{690 \times 120 + (620/2) \times 120} = 6.62 \text{ in.}$$

$$e' = 6.62 \text{ in.} - 6.00 \text{ in.} = 0.62 \text{ in. below the centroid}$$

The results are shown plotted in Fig. 9.8.

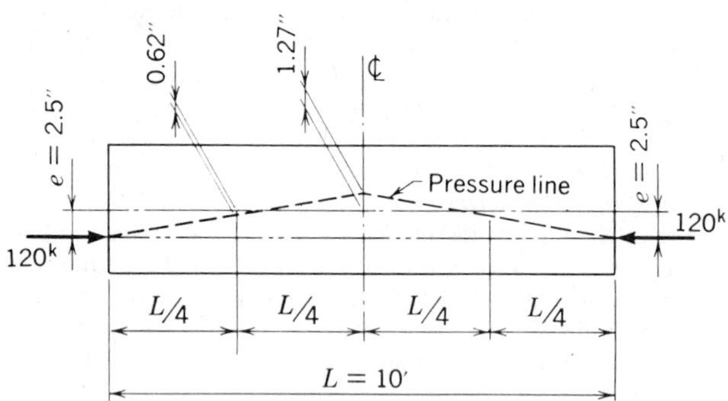

Fig. 9-8 Location of pressure line for ex. 9-4.

9.2.4 Variation in Pressure-Line Location

If tensile stresses are not permitted in the bottom fibers of a simple prestressed concrete beam, when it is subjected to service loads, the distribution of stresses will be as shown in Fig. 9-9. Also shown in Fig. 9-9 is the cross section of the beam. The force C is the resultant of the stresses in the concrete (pressure line) and it obviously must be equal in magnitude and opposite in direction to the prestressing force P, since the horizontal forces acting on the cross section must be in equilibrium. In addition, from the relationship for combined stresses developed in section 9.2.2, we can write the relationship for the stress in the bottom fibers as follows:

$$\frac{C}{A}\left(1 - \frac{e'y_b}{r^2}\right) = 0 \qquad (9\text{-}6)$$

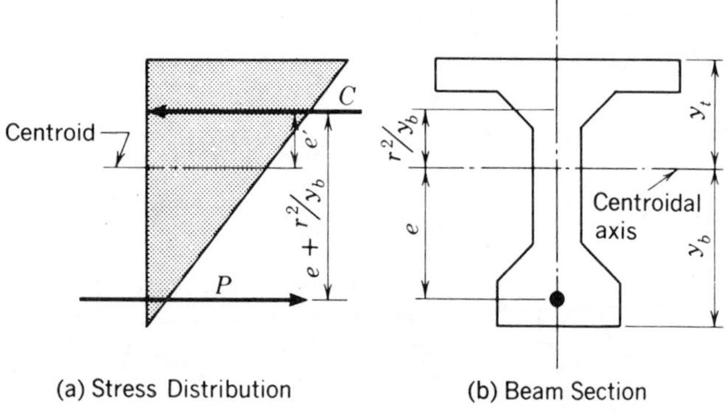

Fig. 9-9 Relationship between prestressing force, pressure line, and section properties of a beam having zero stress in bottom fiber under design load.

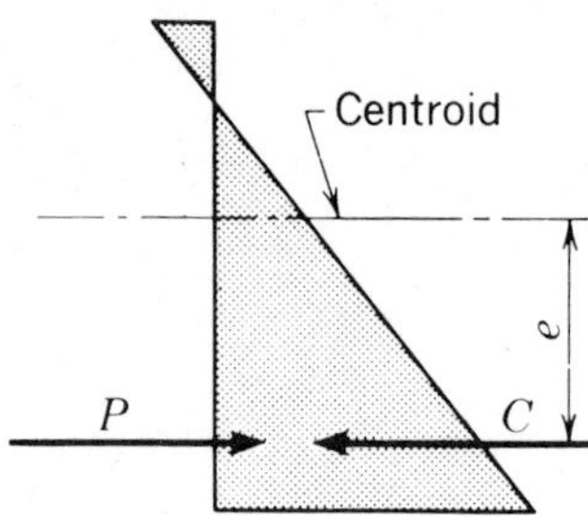

Fig. 9-10 Distribution of stress and location of C when external moment = 0 (prestress alone).

from which we obtain

$$e' = r^2/y_b \qquad (9\text{-}7)$$

The eccentricity e' of the resultant C should not be confused with the eccentricity of the prestressing force.

Another requirement of equilibrium is that the internal and external moments are equal in magnitude and opposite in direction at every section. It follows, then, that the total external moment that the beam is resisting at this section is numerically equal to

$$M_T = M_{DL} + M_{LL} = C(e + r^2/y_b) = P(e + r^2/y_b) \qquad (9\text{-}8)$$

in which e is the normal eccentricity of the prestressing force.

The above example further illustrates that prestressed beams, functioning in the elastic range, resist the moment due to externally applied loads by the movement of the resultant of the stresses in the concrete, rather than by an increase in the prestressing stress, as was brought out in section 9.2.3. From eq. (9-8), it is apparent that if $M_T = 0$, the product of C multiplied by the quantity $(e + r^2/y_b)$ must also be equal to zero and the concrete stresses would be distributed as shown in Fig. 9-10. If the external moment (M_T) were some value less than that which nullifies the precompression of the bottom fibers, the force C would be applied above the location of the prestressing steel at a distance (d) equal to

$$d = \frac{M_T}{P} \qquad (9\text{-}9)$$

This condition is illustrated in Fig. 9-11.

The relationship given by eq. (9-8) is extremely useful in the preliminary design of beams as well as in checking the final design. Since the value of $(e + r^2/y_b)$ is normally of the order of 65% of the depth of the beam section (it varies between the approximate limits of 33 to 80% for different cross sections) for a given superimposed moment, the designer can assume a dead weight for the beam and estimate the prestressing force required for different depths of construction. The use of this relationship is illustrated in ex. 9-5.

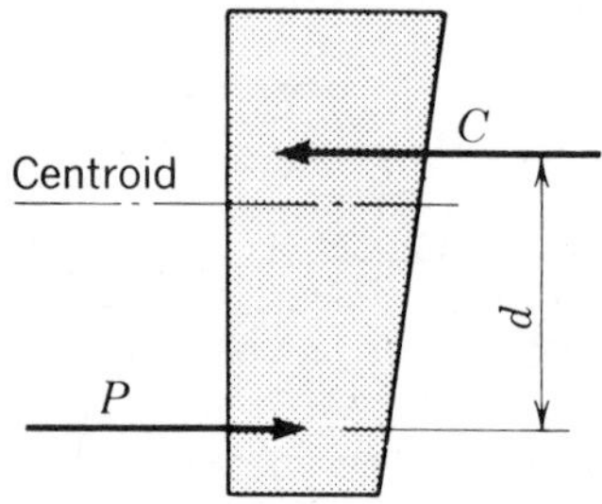

Fig. 9-11 Distribution of stress and location of resultant C when external moment is of nominal magnitude.

EXAMPLE 9-5: Compute the maximum concentrated load that can be applied at midspan of a beam that is 10 in. wide, 12 in. deep, prestressed with 120 k at an eccentricity of 2.5 in., and is to be used on a span of 10.0 ft center to center of bearings without tensile stresses resulting in the bottom fibers.

SOLUTION: (Using the basic relationships for flexural design)

$$A = 120 \text{ in.}^2 \qquad \frac{r^2}{y} = \frac{I}{Ay} = 2.0 \text{ in.} \qquad S = \frac{I}{y} = 240 \text{ in.}^3$$

$$P = 120 \text{ k} \qquad e = 2.5 \text{ in.} \qquad f_b = \frac{120{,}000}{120}\left(1 + \frac{2.5}{2.0}\right) = 2250 \text{ psi}$$

$$\text{Maximum allowable moment} = \frac{2250 \times 240}{12{,}000} = 45.0 \text{ k-ft}$$

$$\text{Moment due to dead load} = \frac{wl^2}{8} = \frac{120}{144} \times 0.15 \times \frac{(10)^2}{8} = 1.56 \text{ k-ft}$$

Moment due to concentrated load

$$= \frac{PL}{4} = 45.0 \text{ k-ft} - 1.6 \text{ k-ft} = 43.4 \text{ k-ft}$$

$$P = \frac{43.4 \times 4}{10} = 17.4 \text{ k}$$

Using eq. (9-8) the computation of the maximum permissible moment becomes

$$M_T = 120 \text{ k} \, \frac{2.5 \text{ in} + 2.0 \text{ in.}}{12} = 45.0 \text{ k-ft}$$

9.2.5 Pressure-Line Location in a Beam with a Curved Tendon

It has been shown in section 9.2.3 that the pressure line for prestressing alone in a prismatic beam is coincident with the prestressing force when the beam is prestressed with a straight tendon. This can also be demonstrated for a beam prestressed with a curved tendon as shown in Fig. 9-12.

It is readily seen that in stressing the tendon, the natural tendency for the tendon to straighten out is resisted by the concrete. If a short segment of the tendon is studied as a free body, as shown in Fig. 9-13, forces must be present normal to the tendon (neglecting friction) in order to prevent this straightening. If friction is neglected, the force acting throughout the tendon is uniform, and since the tendon is flexible, it cannot support any bending moments.

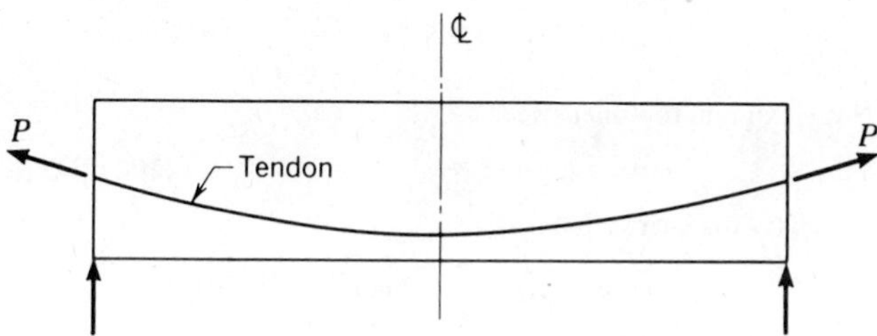

Fig. 9-12 Simple beam with curved tendon.

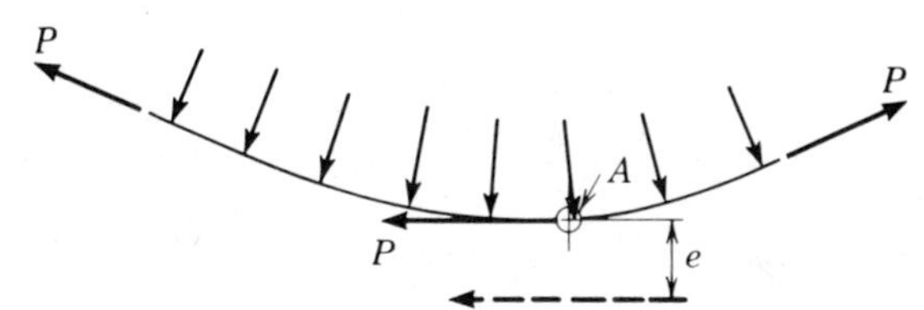

Fig. 9-13 Free-body diagram of portion of curved prestressing tendon.

Therefore, at every point such as point A, the force in the tendon is equal to P and is located on the trajectory of the tendon. If the force were not coincident with the tendon at A, but were located at some distance from A (as shown by the dashed vector), the tendon would have to withstand the moment Pe caused by this eccentric force.

From this analysis then, it can be concluded that the pressure line for prestressing alone in a simple beam, prestressed with a curved tendon, is coincident with the trajectory of the tendon, since the forces in the concrete must be equal and opposite to those in the steel in order to maintain equilibrium. Furthermore, it can be shown that the pressure line moves when an external load is applied to a beam with a curved tendon, just as it does in a beam with a straight tendon.

EXAMPLE 9-6: Compute and plot to scale the location of the pressure line for a 10 × 12 in. rectangular beam, if prestressed with a force of 120 k, which is on a parabolic curve and which has an eccentricity of 2.5 in. and zero, at midspan, and end, respectively, if the beam is spanning 10 ft and subjected to a uniformly distributed load of 3.5 k/ft. Neglect dead load of the beam.

SOLUTION: At midspan

$$M = 3.5 \text{ k} \times \frac{(10)^2}{8} = 43.8 \text{ k-ft}$$

$$\text{Movement} = \frac{43.8 \text{ k-ft} \times 12 \text{ in./ft.}}{120 \text{ k}} = 4.38 \text{ in.}$$

$$\text{Location} = 4.38 \text{ in.} - 2.50 \text{ in.} = 1.88 \text{ in. above c.g.s}$$

At quarter point

$$M = 0.75 \times 43.8 \text{ k-ft} = 32.8 \text{ k-ft}$$

$$\text{Movement} = \frac{32.8 \times 12 \text{ in./ft}}{120 \text{ k}} = 3.28 \text{ in.}$$

$$\text{Location} = 3.28 \text{ in.} - 0.75 \times 2.50 \text{ in.} = 1.40 \text{ in. above c.g.s.}$$

At end

$$M = 0 \qquad e = 0$$

(See Fig. 9-14.)

EXAMPLE 9-7: Calculate the maximum, uniformly distributed load that can be applied to the beam of ex 9-6, if the span is 10 ft and the bottom-fiber stress is zero at midspan.

SOLUTION: Under the loaded condition, the pressure line will be at r^2/y_b above the centroid (the upper limit of the kern zone).

$$r^2/y_b = 2.0 \text{ in.} \qquad e = 2.5 \text{ in.}$$

$$M_T = 120 \text{ k} \, \frac{4.5 \text{ in.}}{12} = 45.0 \text{ k-ft}$$

$$W_{max} = \frac{45 \times 8}{(10)^2} = 3.60 \text{ k/ft}$$

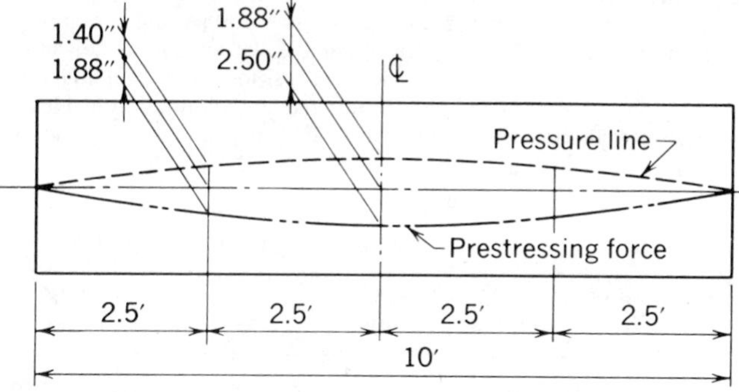

Fig. 9-14 Location of pressure line for ex. 9-6.

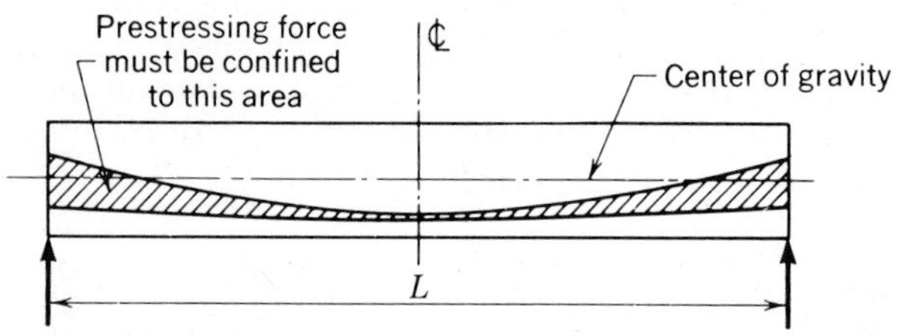

Fig. 9-15 Schematic diagram showing area in which prestressing force must be confined in order to satisfy initial and final stress requirements.

9.2.6 Limiting Eccentricities

It should be apparent that a greater eccentricity of the prestressing force can frequently be allowed at the midspan of a beam than at the ends, without exceeding the permissible stresses, due to the dead weight of the beam itself which acts at the time of stressing. The permissible stresses that must be satisfied in a normal, simple beam include a maximum compressive stress in the bottom fiber and a maximum tensile stress in the top fiber under the combined action of the initial prestressing force (before relaxation) and the dead weight of the beam. In addition, a maximum top-fiber compressive stress and a maximum bottom-fiber tensile stress under the combined effects of the total external load and the final prestressing force (after relaxation) are normally specified. For most beams, a number of combinations of prestressing force and eccentricity can be found that will satisfy these conditions of stress. In the interest of economy, however, the minimum force that satisfies the above conditions of stress at the most highly stressed section is usually selected.

For the force that is selected, one can compute maximum and minimum eccentricities that can be used at various locations along the length of the beam without exceeding the permissible stresses enumerated above. Plotting these eccentricities in a schematic elevation of the beam, which has an exaggerated vertical scale, reveals the limiting dimensions in which the center of gravity of the prestressing force must remain in order to satisfy the conditions of allowable stress. Such a schematic diagram is shown in Fig. 9-15, where the area in which the selected prestressing force must be confined is cross-hatched. It is generally not necessary to make such a diagram in designing beams subjected to normal loading, since by placing the tendons on parabolic (or near parabolic) curves, the stress conditions can generally be satisfied without difficulty. However, when nonprismatic beams, continuous beams, or beams that have acute and unusual stress conditions are encountered, such diagrams facilitate the design.

EXAMPLE 9-8: Compute the limits of the eccentricity of the prestressing force of 550 k at midspan, quarter point, and end, for a simple beam, if the allowable tensile stress is 200 psi and zero in the top and bottom fibers, respectively, and if the maximum allowable compressive stress is 2000 psi and 2200 psi in the bottom and top fibers, respectively. The maximum and minimum external-load stresses (total load and beam dead load alone, respectively), are as follows:

Location	*Max. top (psi)*	*Min. top (psi)*	*Max. bottom (psi)*	*Min. bottom (psi)*
Quarter point	+1350	+453	−1530	−328
Midspan	+1800	+605	−2038	−438

The area of the beam is 445 in.^{2} and r^2/y_b and r^2/y_t are equal to 8.99 and 6.50 in., respectively.

SOLUTION:

$$\frac{P}{A} = \frac{550{,}000}{445} = 1235 \text{ psi}$$

At midspan
Maximum allowable stress due to prestress in bottom fiber

$$f_b = 2438 = 1235\left(1 + \frac{e}{8.99}\right), \qquad e = 8.70 \text{ in.}$$

for

$$e = 8.70 \text{ in.}, \qquad f_t = 1235\left(1 - \frac{8.70}{6.50}\right) = -420 \text{ psi}$$

Net top-fiber stress = −420 + 605 = +185 psi > −200 psi O.K.

Minimum allowable stress due to prestress in bottom fiber

$$f_b = 2038 = 1235\left(1 + \frac{e}{8.99}\right), \qquad e = 5.83 \text{ in.}$$

for

$$e = 5.83 \text{ in.}, \qquad f_t = 1235\left(1 - \frac{5.83}{6.50}\right) = +123 \text{ psi}$$

Net maximum top-fiber stress

$$= +123 + 1800 = +1923 < +2200 \text{ psi O.K.}$$

Summary for center line

$$e_{max} = 8.70 \text{ in.}, \qquad e_{min} = 5.83 \text{ in.}$$

At quarter point
Maximum allowable stress due to prestress in bottom fiber

$$f_b = 2328 = 1235\left(1 + \frac{e}{8.99}\right), \qquad e = 8.00 \text{ in.}$$

for

$$e = 8.00 \text{ in.}, \qquad f_t = 1235\left(1 - \frac{8.00}{6.50}\right) = -284 \text{ psi}$$

Net top-fiber stress = −284 + 453 = +169 psi > −200 psi O.K.

Minimum allowable stress due to prestress in bottom fiber

$$f_b = 1530 = 1235\left(1 + \frac{e}{8.99}\right), \qquad e = 2.16 \text{ in.}$$

for

$$e = 2.16 \text{ in.}, \qquad f_t = 1235\left(1 - \frac{2.16}{6.50}\right) = +825 \text{ psi}$$

Net maximum top-fiber stress

$$= +825 + 1350 = +2175 < +2200 \text{ psi O.K.}$$

Summary for quarter point

$$e_{max} = 8.00 \text{ in.}, \qquad e_{min} = 2.16 \text{ in.}$$

At the support
Maximum allowable stress due to prestress in bottom fiber

$$f_b = 2000 = 1235\left(1 + \frac{e}{8.99}\right), \qquad e = 5.57 \text{ in.}$$

for

$$e = 5.57 \text{ in.}, \qquad f_t = 1235\left(1 - \frac{5.57}{6.50}\right) = +173 \text{ psi} > -200 \text{ psi O.K.}$$

Minimum allowable stress due to prestress in bottom fiber

$$f_b = 0 = 1235\left(1 + \frac{e}{8.99}\right), \qquad e = -8.99 \text{ in.}$$

for

$$e = -8.99 \text{ in.,}$$

$$f_t = 1235\left(1 + \frac{8.99}{6.50}\right) = +2940 \text{ psi} > +2200 \text{ psi N.G.}$$

for

$$f_t = +2200 \text{ psi,} \quad e = \left(\frac{2200}{1235} - 1\right) 6.50 = -5.07 \text{ in.}$$

for

$$e = -5.07 \text{ in.,} \quad f_b = 1235\left(1 - \frac{5.07}{8.99}\right) = +537 \text{ psi}$$

Summary at center line of support: e_{max} = 5.57 in., e_{min} = −5.07 in.

The limits in which the center of gravity of the tendons must fall are shown in Fig. 9-16.

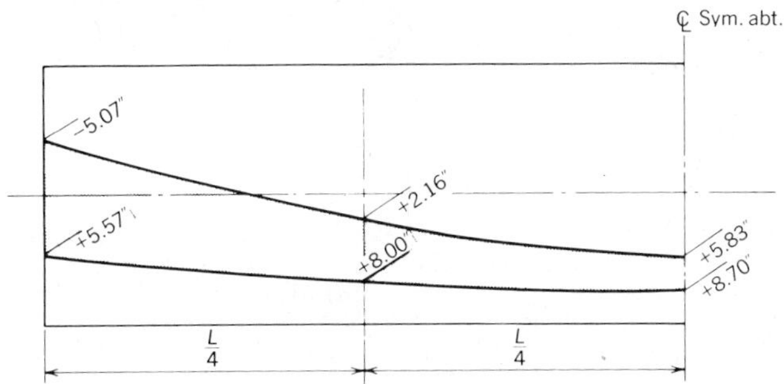

Fig. 9-16 Plot of the limits of the prestressing force for ex. 9-8.

9.2.7 Cross Section Efficiency

In a rectangular beam the distribution of the unit flexural stresses in the concrete under prestress alone and under total load at the midspan may be as is illustrated in Fig. 9-17. The distribution of the forces in this beam will have the identical shape as the distribution of the unit stresses, and the conversion of the unit stresses to forces can be made by multiplying the unit stresses by the width of the cross section. As has been explained, the total moment to which this member is subjected can be computed by determining the distance between the points of application of the resultant forces in the concrete, under the conditions of prestressing alone and when under full load, and by multiplying this distance by the prestressing force.

Analysis of a beam with an I-shaped cross section, such as illustrated in Fig. 9-18, will reveal that the distribution of unit stresses varies linearly, as in the case of the rectangular cross section; however, due to the variable width of the cross section, the distribution of forces is variable as illustrated. It is apparent that the resultants of the force diagrams for the I-shaped member will be nearer the extreme fibers of the cross section. For this reason, the resultant

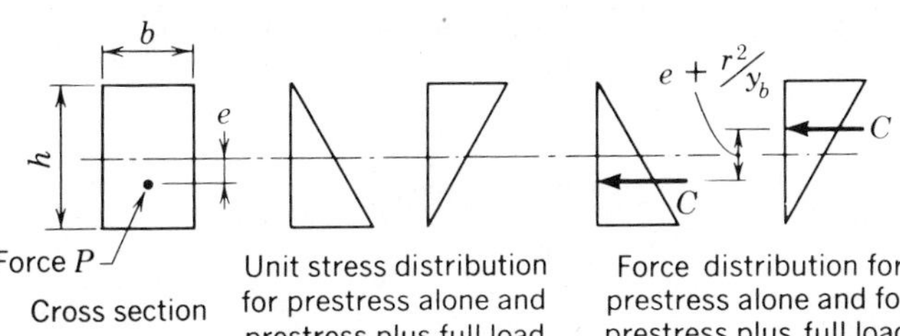

Fig. 9-17 Distribution of unit stresses and forces in a rectangular beam under prestress alone and under prestress plus full load.

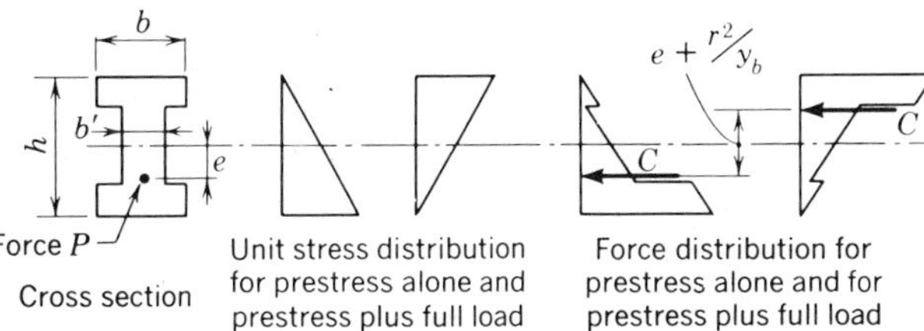

Fig. 9-18 Distribution of unit stresses and forces in a beam with I-shaped cross section, under prestress alone and under prestress with full load.

force in the I-shaped concrete section moves through a greater distance when external load is applied, to nullify the bottom-fiber prestress, than is the case for a rectangular cross section of equal depth. From this consideration, it is obvious that the I-shape will be more efficient and is capable of withstanding a greater load than a rectangular section of equal depth, providing each section is prestressed with a force of equal magnitude and tensile stresses are not allowed in the section.

This consideration is the primary reason for using I, T, and hollow shapes in prestressed flexural members in which major tensile stresses must be avoided and in which construction depth is of importance and must be minimized. Solid slabs and rectangular beams are economical under some conditions of span, loading, and design criteria, but the more complicated shapes generally result in the minimum quantities of prestressing steel and concrete required to carry a particular load condition, and as a result, they are frequently more economical.

The effect of allowing tensile stresses in the top or bottom fiber under any specific condition of beam cross section, magnitude and eccentricity of the prestressing force is to permit the pressure line to move through wider limits. Hence, the section is able to withstand a larger moment due to applied dead and live loads. A review of section 9.2.4 should make this apparent.

EXAMPLE 9-9: Determine the maximum total moments that can be imposed upon the I-shaped and rectangular cross sections in Fig. 9-19 if each is prestressed with a straight tendon having an effective force of 200 k and if tensile stresses are not allowed under any condition of loading. (Neglect the effect of initial prestressing force.)

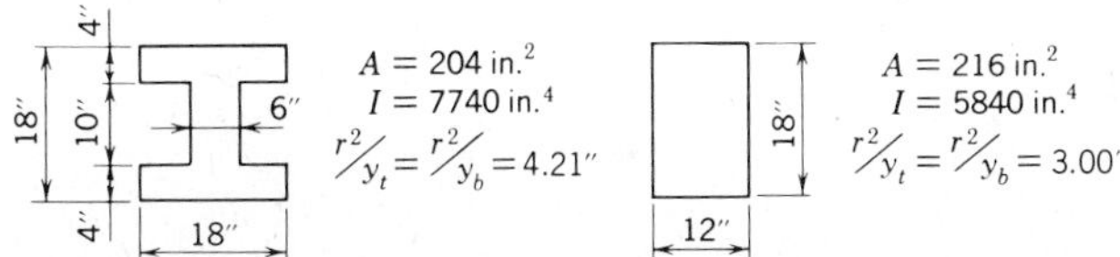

Fig. 9-19 Cross sections compared in ex. 9-9.

SOLUTION: Since tensile stresses are not allowed under any conditions, the maximum eccentricity of the prestressing force is r^2/y_t, see eq. (9-8). Therefore,
for the I-shape

$$M_T = 200 \text{ k}\left(\frac{4.21 + 4.21}{12}\right) = 140 \text{ k-ft}$$

for the rectangular shape

$$M_T = 200 \text{ k}\left(\frac{3.00 + 3.00}{12}\right) = 100 \text{ k-ft}$$

9.2.8 Effective Cross Section

In the past, the most commonly used procedure in prestressed-concrete design has been to base the flexural computations in the elastic range upon the section properties of the gross concrete section. The gross section is defined as the concrete section from which the area of steel, or ducts in the case of post-tensioning, has not been deducted and to which the transformed area of the steel has not been added. This procedure has been considered to render sufficiently accurate results in the usual applications of prestressed concrete. The change in the stresses that would result by basing the computations on the net or transformed section properties is not normally significant, in view of the fact that concrete is not a completely elastic material. Furthermore, the modulus of elasticity of concrete is not generally known precisely and a value must be assumed in computing transformed section properties. It is important, however, that the designer of prestressed concrete be aware of the nature of the section that is theoretically involved in the various types of construction, since it can be important under special conditions.

In the case of pre-tensioning, when the prestress is applied the deformation of the concrete is a function of the net section, since the concrete is compressed by the steel, which does not assist the concrete in resisting the prestressing force. The net section is defined as the section that results when the area occupied by the tendons (or ducts in the case of post-tensioning) is deducted from the gross section. Since the pre-tensioned tendons are bonded to the concrete, when there is a change of strain in the concrete at the level of the steel, there must be a corresponding and equal change of strain in the steel. Therefore, when external loads, other than the dead load of the beam, which is acting at the time of prestressing, are applied, the deformation of the member is a function of the transformed section. The transformed section can be defined as the section that results when the area of steel is transformed into an elastically equivalent area of concrete, by multiplying the steel area by the modular ratio and adding this transformed area to the net section at the proper location. Because in normal pre-tensioned practice, the tendons are straight and spaced out in order to achieve adequate bond, the effect of the transformed section is small and little is normally gained by taking these effects into account. The effect of the transformed section will normally be greater in large members with bundled pre-tensioned tendons. However, little is to be gained under normal conditions by including these refinements in the computations.

In the case of post-tensioning, the deformation of a member is a function of the net section under all conditions of prestressing and external load, until such time as grout is injected into the ducts and allowed to harden and bond the tendons to the concrete section. After bond is established, the deformation of the member is a function of the transformed section. As in the case of pre-tensioning, under normal conditions little is gained by including these effects in the computations.

The use of the net section for the computation of stresses that occur before the bonding of the tendons is required by ACI 318-71, whereas the use of the transformed section is optional for stresses that occur after bonding.

The net and transformed sections should be used in computing stresses in long-span, post-tensioned girders that have large concentrations of ducts in relatively small bottom flanges. In such a case, the ducts can have a significant influence on the compressive stresses, in the bottom flange, which result from the prestressing, since the area occupied by the ducts may be a large portion of the total bottom flange area. Additionally, the area of the prestressing steel is generally large and has a significant effect upon the stresses due to superimposed loads under such conditions.

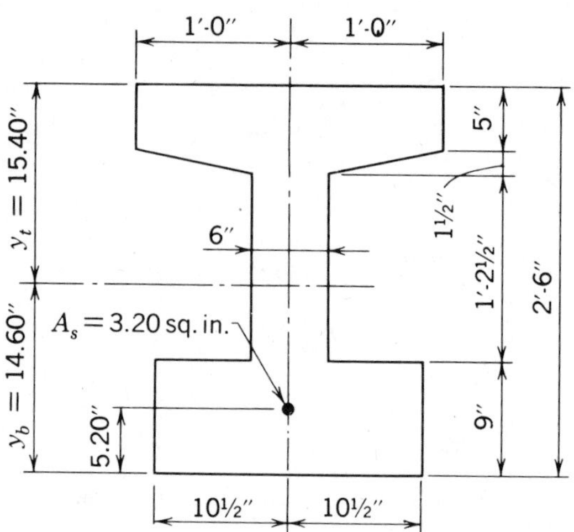

Fig. 9-20 Cross section used to demonstrate the effect of the transformed and net beam cross sections as compared to gross cross section in ex. 9-10.

EXAMPLE 9-10: For the pre-tensioned girder illustrated in Fig. 9-20, compute the stresses in the concrete due to prestressing, based upon the gross- and net-section properties. In addition, compute the concrete stresses, based upon the gross and transformed sections at the center of a 40 ft span, when the externally applied load is 3.13 k/ft. The section properties of the gross section are:

A = 419 in.^{2} | r^2/y_t = 6.94 in.
A_s = 3.20 in.^{2} | r^2/y_b = 7.30 in.
y_t = 15.40 in. | w = 0.44 k/ft.
y_b = 14.60 in. | n = 6
P = 440 k | S_t = 2900 in.^{3}
I = 44,670 in.^{4} | S_b = 3060 in.^{3}
e = 14.60 in. − 5.20 in. = 9.40 in.

SOLUTION:

Stresses due to prestressing based upon gross section

$$f_t = \frac{440{,}000}{419}\left(1 - \frac{9.40}{6.94}\right) = -373 \text{ psi}$$

$$f_b = \frac{440{,}000}{419}\left(1 + \frac{9.40}{7.30}\right) = +2400 \text{ psi}$$

Compute the section properties for the net section (moment about the top).

A	$\bar{y}$	$A\bar{y}$	$\bar{y}'$	$\bar{y}'^2$	$A_y'^2$	I_0	$I_0 + A_y'^2$
419.0 ×	15.4 =	6450	0.10 in.	0.01	4	44,670	44,674
−3.2 ×	24.8 =	−79	9.50 in.	90.0	−288	0	−288
415.8		6371					44,386

y_t = 15.3 in. r^2/y_t = 6.97 in. S_t = 2900 in.^{3}

y_b = 14.7 in. r^2/y_b = 7.25 in. S_b = 3020 in.^{3}

e = 14.7 − 5.20 = 9.50 in.

Stresses due to prestressing based upon the net section

$$f_t = \frac{440{,}000}{416}\left(1 - \frac{9.50}{6.97}\right) = -380 \text{ psi}$$

$$f_b = \frac{440{,}000}{416}\left(1 + \frac{9.50}{7.25}\right) = +2440 \text{ psi}$$

Stresses due to prestressing, dead and superimposed loads, based upon the gross section, are computed as follows:

$$w_d + w_{SL} = 0.44 + 3.13 = 3.57 \text{ k/ft}$$

$$M_D + M_{SL} = 0.44 \times \frac{(40)^2}{8} + 3.13 \times \frac{(40)^2}{8} = 88 + 627 = 715 \text{ k-ft}$$

$$f_t = \frac{715 \times 12{,}000}{2900} - 373 = +2577 \text{ psi}$$

$$f_b = -\frac{715 \times 12{,}000}{3060} + 2400 = -400 \text{ psi}$$

Compute the transformed section properties (moments about the top).

	A	$\bar{y}$	$A\bar{y}$	$\bar{y}'$	$\bar{y}'^2$	$A\bar{y}'^2$	I_0	$I_0 + A\bar{y}'^2$
	416 ×	15.3 =	6371	0.40	0.16	66	44,386	44,452
6 × 3.2 =	19 ×	24.8 =	471	9.10	83	1575	0	1,575
	435		6842					46,027

$y_t = 15.7$ in., $S_t = 2930$ in.^{3}

$y_b = 14.3$ in., $S_b = 3220$ in.^{3}

Stresses in the top and bottom fibers, respectively, due to prestressing, dead and superimposed loads, based upon the net and transformed sections are

$$f_t = -380 + \frac{88 \times 12{,}000}{2900} + \frac{627 \times 12{,}000}{2930} = +2554 \text{ psi}$$

$$f_b = 2440 - \frac{88 \times 12{,}000}{3020} - \frac{627 \times 12{,}000}{3220} = -250 \text{ psi}$$

EXAMPLE 9-11: Compute the stresses due to prestressing in the top and bottom fibers for the girder of Fig. 9-21 based upon an effective prestressing force of 2380 k located 5.3 in. above the soffit based upon: (1) the gross section properties and (2) the net section properties if the area of the ducts is 39.0 sq. in. Also, determine the allowable superimposed live load on the girder based upon: (3) the gross section properties and (4) the transformed section properties if $nA_s = 83.5$ in.^{2}. The design span is 200 ft. What is the ratio between the allowable superimposed live loads of (3) and (4)?

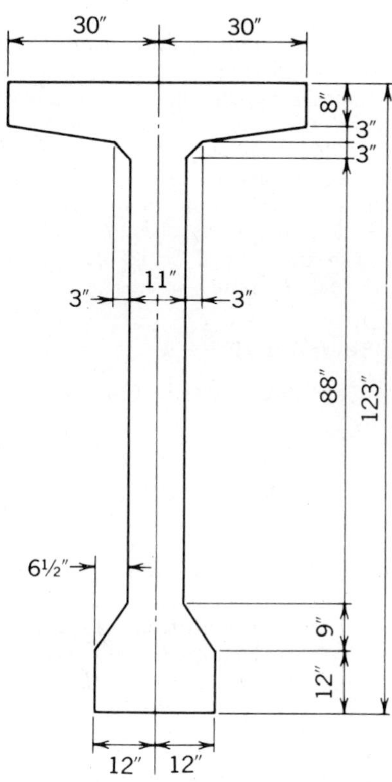

Fig. 9-21 Girder cross section for ex. 9-11.

Gross section properties

$A = 2051$ in.^{2} $I = 3{,}735{,}950$ in.^{4}
$y_t = 53.7$ in. $S_t = 69{,}700$ in.^{3} $r^2/y_t = 34.0$ in.
$y_b = 69.3$ in. $S_b = 54{,}000$ in.^{3} $r^2/y_b = 26.3$ in.

Net section properties

$A = 2012$ in.^{2} $I = 3{,}572{,}900$ in.^{4}
$y_t = 52.5$ in. $S_t = 68{,}000$ in.^{3} $r^2/y_t = 33.8$ in.
$y_b = 70.5$ in. $S_b = 50{,}750$ in.^{3} $r^2/y_b = 25.2$ in.

Transformed section properties

$A = 2096$ in.^{2} $I = 3{,}915{,}500$ in.^{4}
$y_t = 55.0$ in. $S_t = 71{,}300$ in.^{3}
$y_s = 68.0$ in. $S_b = 57{,}600$ in.^{3}

SOLUTION:

1.

$$f_t = \frac{2380}{2051}\left(1 - \frac{64.0}{34.0}\right) = -1021 \text{ psi}$$

$$f_b = \frac{2380}{2051}\left(1 + \frac{64.0}{26.3}\right) = +3980 \text{ psi}$$

2.

$$f = \frac{2380}{2012}\left(1 - \frac{65.2}{33.8}\right) = -1100 \text{ psi}$$

$$f_b = \frac{2380}{2012}\left(1 + \frac{65.2}{25.2}\right) = +4250 \text{ psi}$$

3.

$$w_d = 2.14^{k/1} \quad M_d = 2.14 \times 5000 = 10{,}700'^k$$

$$\frac{l^2}{8} = 5000 \text{ ft}^2$$

$$f_{dt} = \frac{10{,}700 \times 12{,}000}{69{,}700} = +1{,}840 \text{ psi}$$

$$f_{db} = \frac{10{,}700 \times 12{,}000}{54{,}000} = -2{,}380 \text{ psi}$$

Final top fiber stress = + 1840 − 1020 psi = +820 psi

Final bottom fiber stress = + 3980 − 2380 = 1600 psi

$$w_a = \frac{1600 \times 54{,}000}{12{,}000 \times 5000} = 1.44^{k/1}$$

4.

$$f_{dt} = \frac{10{,}700 \times 12{,}000}{68{,}000} = +1890 \text{ psi}$$

$$f_{db} = \frac{10{,}700 \times 12{,}000}{50{,}750} = -2530 \text{ psi}$$

Final top fiber stress = +1890 − 1100 = +790 psi

Final bottom fiber stress = +4250 − 2530 = +1720 psi

$$w_a = \frac{1720 \times 57{,}600}{12{,}000 \times 5000} = 1.65^{k/1}$$

Ratio = 1.15 (more for 4).

9.2.9 Variation in Steel Stress

Since the prestressing steel is never located at the extreme fiber of a prestressed beam, but is at some distance from the surface of the concrete, the maximum change in concrete stress that can normally be expected to occur at the level of the center of gravity of the steel is approximately 70 to 80% of the bottom-fiber stress that results from superimposed loads. With concrete that has a cylinder strength of 5000 psi, the stress change in the concrete at the level of the steel could be expected to be of the order of 1500 psi. The modular ratio between the prestressing steel and the concrete can be assumed to be 6 for loads of short duration. As a result, the application of the short-duration, superimposed load would cause an increase in steel stress of approximately 9000 psi, providing the steel and the concrete were adequately bonded. If the steel is not bonded to the concrete, but is anchored at the ends of the member only, the increase in steel stress resulting from the application of the superimposed load would be less than 9000 psi, since the steel can slip in the ducts. The steel increase in stress in unbonded tendons tends to be proportional to the *average* change in the concrete stress at the level of the steel.

It should be noted that the increase in stress of 9000 psi, which results from the application of the superimposed load, is only about 7% of the final stress normally employed in wire or strand tendons and about 11% of the final stress normally employed in bar tendons. The reduction in the stress in the prestressing steel due to the relaxation of the steel, shrinkage of the concrete, and creep of the concrete is of the order of 10 to 30% under average conditions (see section 9.7.2). Hence, the stress that exists in the tendon under the superimposed load after all of the losses of prestress have taken place is not as high as the initial stress in the steel.

The small variation in steel stress that occurs in a normal prestressed member subjected to frequent applications of the design load is responsible for the high resistance to fatigue failure that is associated with this material.

EXAMPLE 9-12: Compute the increase in the stress in the steel at the midspan of the beam in ex. 9-10.

SOLUTION: The concrete stress at the level of the steel due to the external load of 3.13 k/ft is

$$y_{c.g.s.} = 14.30 \text{ in.} - 5.20 \text{ in.} = 9.10 \text{ in.}$$

$$f_c = \frac{627 \times 12{,}000}{46{,}027} \times 9.1 = 1490 \text{ psi}$$

The increase in steel stress due to the superimposed load is

$$\Delta f_s = nf_c = 6 \times 1490 = 8940 \text{ psi}$$

9.3 SHEAR

9.3.1 Introduction

Shear related cracking of two types are recognized to exist in prestressed concrete flexural members. These are illustrated in Fig. 9-22. These will be referred to as types I and II cracking. Type I cracking is associated with flexural cracking. Some authorities believe that for this type of crack to adversely affect the capacity of a member, it must be cracked in such a manner that the horizontal projection of the crack is equal in length to the depth of the member, and for this reason, it is believed a flexural crack that occurs at a distance equal to the depth of the member away (in the direction of lesser moment) from the section being investigated may lead to a critical crack.[9-16] In addition, principal tensile stresses along a potential crack may be aggravated by flexural cracks that may occur in the vicinity of the potential type I crack, and since principal tensile stresses are normally maximum at the center of gravity of a beam (can be taken to be approximately equal to one-half of the depth of the beam), a flexural crack at one-half the beam depth from the section under consideration can be considered to cause a type I crack. Shear cracking in reinforced- and prestressed-concrete beams are generally type I cracks. These *flexural-shear* cracks begin as flexural cracks extending approximately vertically into the beam. When a critical combination of flexural and shear stresses develop near the top of a flexural crack, the inclined crack forms. Type II cracking is associated with principal tensile stresses, in areas where there are no flexural cracks, and originates in the web of the member near the centroid where the principal tensile stresses are the greatest; it subsequently extends towards the flanges.[9-16] Type II cracking is fairly unusual. It may appear near the supports of highly prestressed beams that have thin webs. It may also appear near inflection points and bar cut-off points of continuous, reinforced-concrete members under axial tension.[9-17]

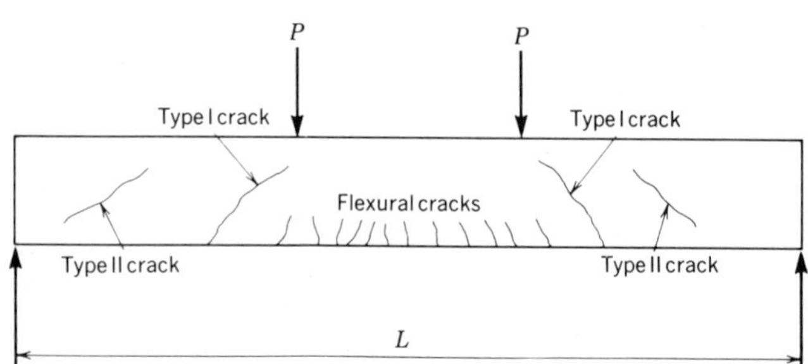

Fig. 9-22 Illustration of flexural cracks and type I and type II shear cracks.

A complete analysis of a prestressed concrete flexural member for shear stresses requires a determination of the portions of the member which may be subject to each type of cracking after which the amount of reinforcing steel required to render the member safe for the design loads must be determined. The shear analysis is required for design loads only. An analysis under service loads is not generally made.

Chapter 11 of *The Building Code Requirements for Reinforced Concrete* (ACI 318-71) provides an approximate method for determining the shear stress that the concrete section can carry without making a detailed analysis. The approximate method can only be used when the effective prestressing force equals or exceeds 40% of the tensile strength of the flexural reinforcement. Provisions for a detailed analysis are also given. Hence, the designer must decide if an approximate analysis is sufficient for each case or if a detailed analysis should be made.

9.3.2 Shear Stress Analysis

The nominal shear stress at the section under consideration is computed by

$$v_u = \frac{V_u}{\phi b_w d} \tag{9-10}$$

in which

b_w = web width, or diameter of circular section, in.
d = distance from extreme compression fiber to centroid of tension reinforcement, in. (but not less than 0.80 of the overall thickness of member)
V_u = total applied design shear force at section
ϕ = capacity reduction factor. See Section 9.2 of ACI 318-71.

The maximum shear stress can be taken as that which is computed at a distance of $h/2$ from the support when the reaction introduces compression into the region of the member (normally the case).

The nominal shear stress carried by the concrete can be computed using the approximate method when the effective prestressing force is at least equal to 40% of the tensile strength of the flexural reinforcement. The relationship for the nominal shear stress v_c, is

$$v_c = 0.6\sqrt{f_c'} + 700\,\frac{V_u d}{M_u} \tag{9-11}$$

in which

d = distance from extreme compression fiber to centroid of tension reinforcement, in.
f_c' = specified compressive strength of concrete, psi
M_u = applied design load moment at a section, in.-lb
V_u = total applied design shear force at section, lb.

When using eq. (9-11), v_c need not be taken as less than $2\sqrt{f_c'}$ and cannot be taken as more than $5\sqrt{f_c'}$. In addition, the value of $V_u d/M_u$ is not permitted a value greater than 1.

When a more detailed analysis is made, the nominal shear stress carried by the concrete, v_c, is taken as being the lesser of v_{ci} and v_{cw} which are the shear stresses at diagonal cracking due to all design loads when such cracking is due to flexural-shear and principal tensile stress respectively. The value of v_{ci} is taken as $1.7\sqrt{f_c'}$ or

$$v_{ci} = 0.6\sqrt{f_c'} + \frac{V_d + \left(\dfrac{V_l M_{cr}}{M_{\max}}\right)}{b_w d} \tag{9-12}$$

whichever is greater. In eq. (9-12) the terms are as previously defined and as follows:

M_{cr} = cracking moment.
$M_{\max}$ = maximum bending moment due to externally applied design loads
V_d = shear force at section due to dead load
V_l = shear force at section occurring simultaneously with $M_{\max}$.

The cracking moment is defined as follows:

$$M_{cr} = (I/y_t)\,(6\sqrt{f_c'} + f_{pe} - f_d) \tag{9-13}$$

in which

f_d = stress due to dead load, at the extreme fiber of a section at which tensile stresses are caused by applied load, psi
f_{pe} = compressive stress in concrete due to prestress only after all losses, at the extreme fiber of a section at which tensile stresses are caused by applied loads, psi
I = moment of inertia of section resisting externally applied design loads.
y_t = distance from the centroidal axis of gross section, neglecting the reinforcement, to the extreme fiber in tension

The value of v_{cw} may be taken to be the shear stress corresponding to a multiple of dead load plus live load which results in a computed principal tensile stress of $4\sqrt{f_c'}$ at the centroidal axis of the member, or at the intersection of the flange and web when the centroidal axis is in the flange. Alternatively, v_{cw} can be taken as:

$$v_{cw} = 3.5\sqrt{f_c'} + 0.3 f_{pc} + \frac{V_p}{b_w d} \tag{9-14}$$

in which

f_{pc} = compressive stress in the concrete, after all prestress losses have occurred, at the centroid of the cross section resisting the applied loads or at the junction of the web and flange when the centroid lies in the flange, psi. (In a composite member, f_{pc} will be the resultant compressive stress at the centroid of the composite section, or at the junction of the web and flange when the centroid lies within the flange, due to both prestress and to the bending moments resisted by the precast member acting alone)
V_p = vertical component of the effective prestress force at the section considered

and the remaining terms are as previously defined.

For non-composite uniformly loaded members, eq. (9-12) can be simplified to

$$v_{ci} = 0.6\sqrt{f_c'} + \frac{V_u M_{cr}}{M_u b_w d} \tag{9-15}$$

It should be recognized that for simply supported members which are loaded with uniformly distributed loads, the value of V_u/M_u or $V_l/M_{\max}$ is equal to $(l - 2x)/x\,(l - x)$. This fact simplifies the computation of v_c with eq. (9-11) as well as for v_{ci} with eqq. (9-12) or (9-15).

In eqs. (9-12) and (9-14), d is the distance from the extreme fiber to the centroid of the prestressing tendons or eight-tenths of the overall thickness of the member whichever is greater. The value of $M_{\max}$ (and V_l) is to be computed from the load distribution which results in the maximum moment at the section under consideration.

In pretensioned members consideration must be given to the transfer length when determining the value of f_{pe} to be used in eq. (9-14). The transfer length can be taken as 50 diameters for strand tendons and 100 diameters for wire tendons. The prestressing force can be assumed to vary linearly from zero at the end of the tendon to a maximum at a distance from the end equal to the transfer length. If the value of $h/2$ (section for maximum shear) is less than the transfer length, this effect must be considered. The term h is defined as the overall thickness of member.

When the value of v_u as computed by eq. (9-10), exceeds the value of v_c, the minimum area of shear reinforcing perpendicular to the longitudinal axis of the member is computed as

$$A_v = \frac{(v_u - v_c) b_w s}{f_y} \tag{9-16}$$

in which s is the spacing and f_y is the specified yield strength of the shear reinforcement respectively. Inclined stirrups or bent bars are not permitted in prestressed members. (ACI 318-71 section 11.1.4.)

Minimum shear reinforcement is required for prestressed members with the same exceptions that apply to reinforced concrete member (see ACI 318-71, section 11.1.1). The minimum area of shear reinforcement can be computed from

$$A_v = 50\,\frac{b_w s}{f_y} \tag{9-17}$$

or from

$$A_v = \frac{A_{ps}}{80}\,\frac{f_{pu}}{f_y}\,\frac{s}{d}\,\sqrt{\frac{d}{b_w}} \tag{9-18}$$

in which A_{ps} equals the area of prestressed reinforcement in the tension zone.

EXAMPLE 9-13: For the member shown in cross section in Fig. 9-23, compute the shear reinforcing required using the approximate method with the 3 in. topping not being composite. Assume f'_c = 4000 psi, the tendon is straight and located 7.50 in. from the beam soffit. The beam is on a simple span of 30 feet and supports a live load of 3.00 kips/foot. Use load factor of $1.4D + 1.7L$. Use f_y = 40,000 psi, A_{ps} = 1.38 in.^{2} and f_{pu} = 270,000 psi

For noncomposite section

$$\text{Area} = 400 \text{ sq in.} \quad I = 51{,}897 \text{ in.}^4 \quad y_b = 22.48 \text{ in.}$$

For composite section

$$\text{Area} = 508 \text{ sq in.} \quad I = 71{,}162 \text{ in.}^4 \quad y_b = 25.67 \text{ in.}$$

Loads

Beam section	417 plf
Topping	113 plf
Superimposed deadload	580 plf
Live load	3000 plf

SOLUTION:
Design shear force at X feet from the support is

$$V_{ux} = [1.40\,(1110 \text{ plf}) + 1.7\,(3000 \text{ plf})]\left[\frac{30'}{2} - X\right]$$

$$= 6654\,[15 - X]$$

$$v_{ux} = \frac{6654\,[15 - X]}{0.85 \times 8 \times 28.5} = 34.33\,[15 - X]$$

Because load is uniformly distributed

$$\frac{V_u}{M_u} = \frac{30 - 2X}{X(30 - X)}$$

and from eq. (9-11)

$$v_c = 0.6\sqrt{4000} + \frac{700 \times 28.5}{12}\left[\frac{30 - 2X}{X(30 - X)}\right]$$

$$= 37.9 + 1662.5\left[\frac{30 - 2X}{X(30 - X)}\right]$$

$$v_{c\,max} = 5\sqrt{f'_c} = 316 \text{ psi}$$

$$v_{c\,min} = 2\sqrt{f'_c} = 126 \text{ psi}$$

The computations can be summarized as follows:

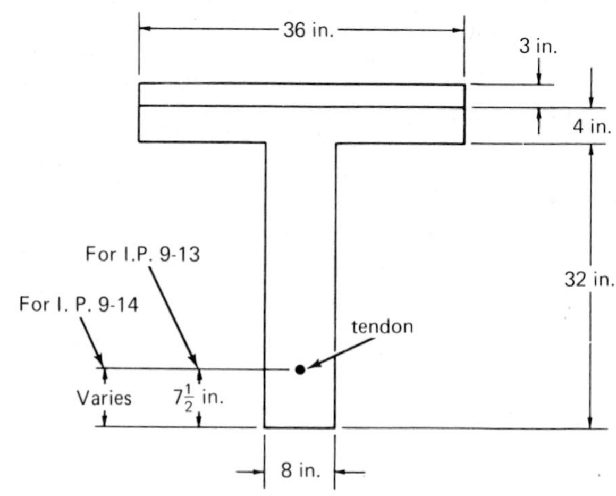

Fig. 9-23 Cross section of member in ex. 9-13.

or

$$A_v = \frac{1.38}{80} \cdot \frac{270000}{40000} \cdot \frac{12}{28.5}\sqrt{\frac{28.5}{8}} = 0.09 \text{ sq in. per ft}$$

EXAMPLE 9-14: Compute the shear reinforcing for the member of Prob. 9-13 using the detailed analysis assuming the section to be composite and the tendon is on a parabolic trajectory from 7.75" above the soffit at midspan to 22.50 in. at the support. The effective prestress is 225,600 pounds. The remaining details are as in Prob. 9-13.

SOLUTION:
At midspan:

Bottom fiber dead load stress

$$= \frac{530 \times 30^2 \times 12 \times 22.48}{8 \times 51{,}897} = -310 \text{ psi}$$

Bottom fiber super. dead load stress

$$= \frac{580 \times 30^2 \times 12 \times 25.67}{8 \times 71{,}162} = -282 \text{ psi}$$

$$-592 \text{ psi}$$

$$\text{Bottom fiber effective prestress} = \frac{225600}{400}\left(1 + \frac{14.73}{5.77}\right) = +2003 \text{ psi}$$

$$\text{Top fiber effective prestress} = \frac{225600}{400}\left(1 - \frac{14.73}{9.60}\right) = -301 \text{ psi}$$

x ft	v_u psi	$\frac{30 - 2X}{X(30 - X)}$	v_c (by eq. 9-11) psi	Effective v_c psi	$v_u - v_c$ psi	A_v (by eq. 9-16) sq in./ft	Provide A_u sq in./ft
0	515	∞	∞	316	199	0.48	0.48
3	412	0.296	530	316	96	0.23	0.24
6	310	0.125	246	246	64	0.15	0.24
9	206	0.063	143	143	63	0.15	0.24
12	103	0.028	84	126	—	0	0.09
15	0	0.000	38	126	—	0	0.09

For Grade 40 stirrups

$$A_v = \frac{(v_u - v_c)\,8 \times 12}{40{,}000 \text{ psi}} = 0.0024\,(v_u - v_c) \text{ sq in. per ft}$$

$$\text{Minimum } A_v = \frac{50 \times 8 \times 12}{40{,}000} = 0.12 \text{ sq in. per ft}$$

$$\text{Top fiber dead load stress} = \frac{310 \times 13.52}{22.48} = +186 \text{ psi}$$

$$-115 \text{ psi}$$

$$f_{pc} = (1693 + 115)\frac{10.33}{36.00} - 115 = 404 \text{ psi}$$

$$V_p = 0 \quad V_d = 0$$

At support:

$$\text{Effective prestress (uniform)} = \frac{225{,}600}{400} = +564 \text{ psi}$$

$$f_{pc} = 564 \text{ psi}$$

$$V_p = P \sin \alpha = \frac{225{,}600 \times 2(22.50 - 7.75)}{12 \times 15} = 36{,}973 \text{ lbs}$$

$$V_d = 1110 \times 15 = 16{,}650 \text{ lbs}$$

The values of f_d, f_{pe}, M_{ci}, and f_{pc} are assumed to vary parabolically between midspan and the support. V_d and V_p vary linearly.

$$M_{cr} = \frac{71.162}{25.67 \times 12000}(6\sqrt{4000} + f_{pe} - f_d) = 0.231(379 + f_{pe} - f_d)$$

$$V_1/M_{max} = \frac{l - 2x}{x(1-x)}$$

$$v_{cw} = 3.5\sqrt{4000} + 0.3 f_{pc} + \frac{V_p}{8 \times 31.20} = 221.3 + 0.3 f_{pc} + 0.004\ V_p$$

$$0.80h = 31.20 \text{ in.} \simeq 39.25 - 7.75 = 31.25$$

$$v_c \text{ min} = 1.7\sqrt{4000} = 107.5 \text{ psi}$$

$$\text{Development length} = 50 \times 0.50 = 25 \text{ in.}$$

$$\frac{h/2}{\text{development length}} = \frac{19.5}{25} = 0.78$$

At support

$$f_{pc} = 0$$

$$v_{cw} = 221.3 + 0.004 \times 36{,}973 = 369 \text{ psi}$$

At $x = 1.5$ ft

$$v_{cw} = 221.3 + 0.3 \times 0.78 \times 534 + 33278 \times 0.004 = 479 \text{ psi}$$

$$v_{ux} = \frac{[1.4 \times 1110 + 1.7 \times 3000]\ [15 - x]}{0.85 \times 8 \times 31.2} = [15 - x]\ [31.36]$$

$$A_v = \frac{(v_u - v_c)\ 8 \times 12}{40{,}000} = 0.0024\ (v_u - v_c)$$

$$\text{Minimum } A_v = \frac{50 \times 8 \times 12}{40{,}000} = 0.12 \text{ sq in./foot}$$

or

$$A_v = \frac{1.38}{80} \cdot \frac{270{,}000}{40{,}000} \cdot \frac{12}{d} \cdot \sqrt{\frac{d}{8}}$$

x ft	Bottom Fiber, D. L. Stress (f_d) psi	Bottom Fiber, Eff. Prestress (f_{pe}) psi	M_{cr} ft-kips	V_d lbs	V_l/M_{max} ft^{-1}	$\frac{V_l M_{cr}}{M_{max}}$ ft-lbs	v_{ci} psi	V_p lbs	f_{pc} psi	v_{cw} psi	v_c psi	v_u psi	d in.	A_v min (by eq. 9-18)	$v_u - v_c$ psi	A_v (by eq. 9-16) lb^2/ft	Provide A_u sq in./ft
0 support	0	+564	218	16,650	∞	∞	∞	36,973	564	369	369	470	16.50	0.1216	101	0.24	0.24
1.5	-112	+838	255	14,985	0.632	161,160	743	33,278	534	470	470	423	19.30	0.1124	—	0	0.24
3.0	-213	+1082	288	13,320	0.296	102,297	433	29,578	506	491	433	376	21.81	0.1058	—	0	0.11
4.5	-302	+1298	318	11,655	0.182	58,194	318	25,881	482	469	318	329	24.02	0.1008	11	0.026	0.11
6.0	-379	-1485	343	9990	0.125	42,875	250	22,184	462	449	250	282	25.94	0.0970	32	0.077	0.11
7.5	-444	+1644	365	8325	0.889	32,449	201	18,487	444	429	201	235	27.56	0.0941	34	0.082	0.11
9.0	-497	+1774	383	6660	0.0635	24,320	162	14,789	430	410	162	188	28.89	0.0919	27	0.065	0.11
10.5	-539	+1874	396	4995	0.0440	17,424	128	11,092	418	391	128	141	29.92	0.0903	14	0.034	0.11
12.0	-568	+1946	406	3330	0.0278	11,287	97	7,395	410	374	108	94	30.66	0.0892	—	0	0.09
13.5	-586	+1990	412	1665	0.0135	5,562	67	3,697	406	358	108	47	31.10	0.0886	—	0	0.09
15.0 midspan	-592	+2003	414	0	0	0	38	0	404	343	108	0	31.25	0.0884	—	0	0.09

9.4 BOND OF PRESTRESSING TENDONS

9.4.1 General

Two types of bond stress must be considered in the case of prestressed concrete. The first of these is referred to as "transfer bond stress" and has the function of transferring the force in a pre-tensioned tendon to the concrete. Transfer bond stresses come into existence when the prestressing force in the tendons is transferred from the prestressing beds to the concrete section. The second type of bond is termed "flexural bond stress" and comes into existence in pre-tensioned and bonded, post-tensioned members when the members are subjected to external loads. Flexural bond stress does not exist in unbonded, post-tensioned construction, which accounts for the term "unbonded."

9.4.2 Transfer Bond Stress

When a prestressing tendon is stressed, the elongation of the tendon is accompanied by a reduction in the diameter due to Poisson's effect. When the tendon is released, the diameter increases to its original diameter at the ends of the prestressed member where it is not restrained. This phenomenon is generally regarded as the primary factor that influences the bonding of pre-tensioned wires to the concrete. The stress in the wire is zero at the extreme end and is at a maximum value at some distance from the end of the member. Therefore, in the length of the tendon from the extreme end to the point where it attains maximum stress, called the transmission length, there is a gradual decrease in the diameter of the tendon, giving the tendon a slight wedge shape over this length. This wedge shape is often referred to as the Hoyer Effect after the German engineer E. Hoyer, who was one of the early engineers to develop this theory. Hoyer, and others more recently, derived elastic theory to compute the transmission length as a function of Poisson's ratio for steel and concrete, the moduli of elasticity of steel and concrete, the diameter of the tendon, the coefficient of friction between the tendon and the concrete, and the initial and effective stresses in the steel. Laboratory studies of the transmission lengths have indicated a relatively close agreement between the theoretical and actual values. There can be wide variation, however, due to the different properties of concrete and steel and due to surface conditions of the tendons, which affect the coefficient of friction.

There is reason to believe that the configuration of a seven-wire strand (i.e., six small wires twisted about a slightly larger center wire) results in very good bond characteristics. It is believed the Hoyer Effect is partially responsible for this, but the relatively large surface area and twisted configuration is believed to result in a significant mechanical bond.

Although these theoretical relationships are of academic interest, they have little practical application, due to the inability of designers and fabricators of prestressed concrete to control the several factors that influence the transmission length. Fortunately, there has been sufficient research into the magnitude of transmission lengths under both laboratory and production conditions for the following significant conclusions to be drawn.

1. The bond of clean three and seven-wire prestressing strands and concrete is adequate for the majority of pre-tensioned concrete elements.
2. Members that are of such a nature that high moments may occur near the ends of the members, such as short cantilevers, may require special consideration.
3. Clean smooth wires of small diameter are also adequate for use in pre-tensioning, but the transmission length for tendons of this type should be expected to be approximately double that for seven-wire strands (expressed as a multiple of the diameter).
4. Under good conditions, the initial transmission length for clean seven-wire strands can be assumed to be 50 to 75 times the diameter of the strand.
5. The transmission length of tendons can be expected to increase from 5 to 20% within one year after release as a result of relaxation.
6. The transmission length of tendons released by flame cutting or with an abrasive wheel can be expected to be from 20 to 30% greater than tendons that are released gradually.
7. Hard non-flaky surface rust and surface indentations effectively reduce the transmission lengths required for strand and some forms of wire tendons.
8. Concrete compressive strengths between 1500 and 5000 psi at the time of release result in transmission lengths of the same order, except for strand tendons larger than ½ in., in which case strengths, less than 3000 psi result in larger transmission lengths.
9. It would seem prudent to use 3000 psi as a minimum release strength in pre-tensioned tendons, except for very unusual cases. Higher strengths may be required for tendons larger than ½ in.
10. Because of relaxation, a small length of tendon (± 3 in.) at the end of a member can be expected to become completely unstressed.
11. The degree of compaction of the concrete at the ends of pre-tensioned members is extremely important if good bond and short transmission lengths are to be obtained. Honeycombing must be avoided at the ends of the beams.
12. There is little if any reason to believe that the use of end blocks improves the transfer bond of pre-tensioned tendons, other than to facilitate the placing and compacting of the concrete at the ends. Hence, the use of end blocks is considered unnecessary in pre-tensioned beams, if sufficient care is given to this consideration.
13. Tensile stresses and strains develop in the ends of pre-tensioned members along the transmission length as a result of the wedge effect of the tendons. Little if any beneficial results can be gained in attempting to reduce these stresses and strains by providing mild reinforcing steel around the ends of these tendons, since the concrete must undergo large deformations and would probably crack before such reinforcing steel could become effective.
14. Lubricants and dirt on the surface of tendons has a detrimental effect on the bond characteristics of the tendons.

A curve showing the typical variation of stress along the length of a pre-tensioned tendon near the end of a beam is given in Fig. 9-24. It will be seen that this curve is approximately hyperbolic. The stress is zero at the extreme end and for a distance of 4 in., as is assumed to be the case in most applications. This should be considered in the design.

The transfer bond provisions of ACI 318-71 are given in Section 11.5.3 in which it is stated that the prestressing force in pre-tensioned members may be assumed to vary linearly from zero at the end of the member to a maximum value at a distance assumed to be 50 strand diameters or 100 wire diameters. This provision is used in designing for shear. There are of course no transfer bond stresses in post-tensioned members, because the end anchorage devices accomplish the transfer of stress and (normally) remain permanently in the construction.

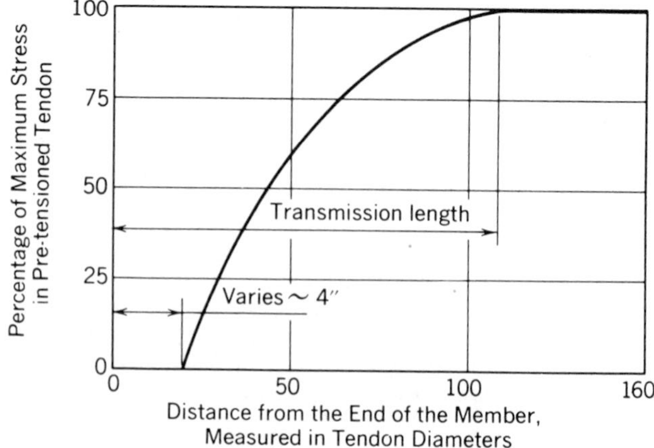

Fig. 9-24 Variation in stress in pre-tensioned tendon near end of beam after relaxation.

9.4.3 Flexural Bond Stresses

Bond stresses also occur between the tendons and the concrete in both pre-tensioned and bonded, post-tensioned members, as a result of changes in the external load. Although it is known that flexural-bond stresses are relatively low in prestressed members for loads less than the cracking load, there is an abrupt and significant increase in these bond stresses after the cracking load is exceeded. Because of the indeterminancy which results from the plasticity of the concrete for loads exceeding the cracking load, accurate computation of the flexural-bond stresses cannot be made under such conditions. Again, tests must be relied upon as a guide for design.

The effect of flexural bond is most evident when two identical post-tensioned members, one grouted and one unbonded, are tested to destruction and the results are compared. The load-deflection curve for such tests, when plotted together, would appear as in Fig. 9-25. From these curves, it will be seen that the grouted beam does not deflect as much under a specific load as the unbonded beam. The explanation for this is that the tendon in the bonded beam must undergo changes in strain equal to the strain changes in the concrete to which it is bonded, whereas the unbonded tendon can slip in the duct and the strain changes are averaged. Hence, the grouted beam deforms and deflects as a function of a transformed section. This difference generally results in the cracking load of the grouted beam being from 10 to 15% higher than that of the unbonded beam, while the ultimate load may be as much as 50% higher. The presence of flexural bond results in many very fine cracks in a bonded member in which the cracking load is exceeded, while, in an identical unbonded member subjected to the same load, only a few wide cracks occur. This is significant because removal of the load from the bonded member will result in the fine cracks closing completely, while in the unbonded member, the large cracks do not completely close.*

The flexural bond stress (development length) provisions for prestressing strand of ACI 318-71 are only given for three-wire and seven-wire pre-tensioned strands. This provision (Section 12.11) provides the development length shall be not less than

$$\left(f_{ps} - \frac{2}{3} f_{se}\right) d_b \qquad (9\text{-}19)$$

where d_b is the nominal diameter in inches, f_{ps} and f_{se} are expressed in kips per square inch, and the expression in the parenthesis is used as a constant without units.

The tendon must extend beyond the critical section by a distance which equals or exceeds the development length. For tendons which are not bonded to the end of the member, the development length specified by eq. (9-19) must be doubled.

9.5 ULTIMATE MOMENT CAPACITY

9.5.1 Action Under Overloads

It has been shown that a variation in the external load acting on a prestressed beam results in a change in the location of the pressure line for beams in the elastic range. This is a fundamental principle of prestressed construction. In a normal prestressed beam, this shift in the location of the pressure line continues at a relatively uniform rate, as the external load is increased, to the point where cracks develop in the tension fiber. After the cracking load has been exceeded, the rate of movement in the pressure line decreases as additional load is applied, and a significant increase in the stress in the prestressing tendon and the resultant concrete force begins to take place. This change in the action of the internal moment continues until all movement of the pressure line ceases. The moment caused by loads that are applied thereafter is offset entirely by a corresponding and proportional change in the internal forces, just as in reinforced-concrete construction. The range of loading that is characterized by these different actions is illustrated in the load deflection curve of Fig. 9-26. This fact, that the load in the elastic range and the plastic range is carried by actions that are fundamentally different, is very significant and renders ultimate-moment computations essential for all designs in order to ensure that adequate safety factors exist. This is true even though the stresses in the elastic range may conform to a recognized elastic design criterion.

It should be noted that the load deflection curve in Fig. 9-26 is close to a straight line up to the cracking load and that the curve becomes progressively more curved as the load is increased above the cracking load.** The curvature of the load-deflection curve for loads over the cracking load is due to the change in the basic internal resisting moment

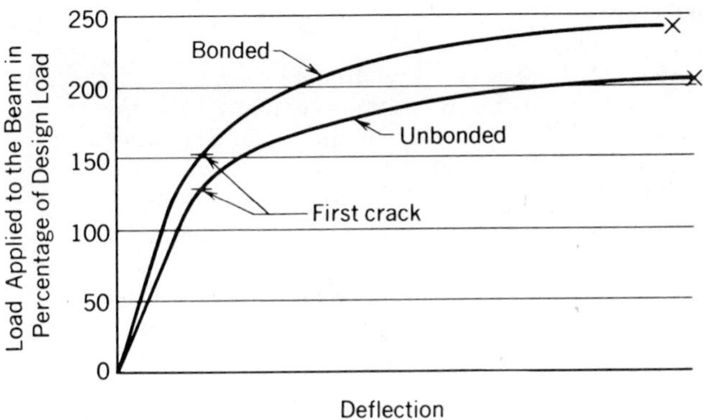

Fig. 9-25 Comparison of load-deflection curves for bonded and unbonded post-tensioned construction.

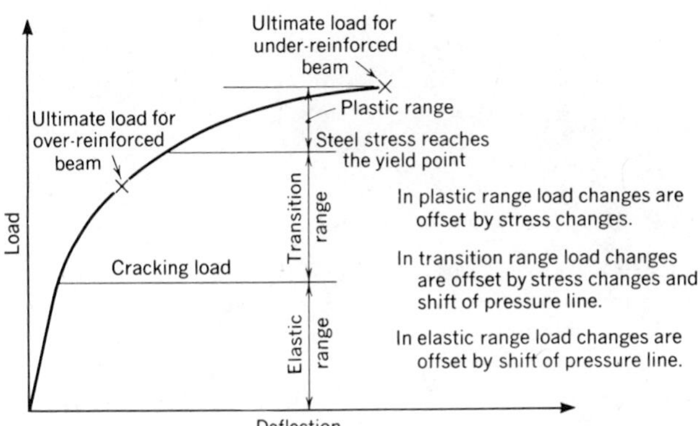

Fig. 9-26 Load-deflection curve for the prestressed beam.

*The provision of non-prestressed reinforcement, if in sufficient quantity, will result in a non-bonded beam having deflection and cracking characteristics similar to a bonded beam.

**The presence of non-prestressed reinforcing in the tensile flange will tend to make the cracking load more difficult to detect from a load-deflection curve as well, as from observations of a beam during loading.

action that counteracts the applied loads, as described above, as well as to plastic strains that begin to take place in the steel and the concrete when stressed to high levels.

In some structures it may be essential that the flexural members remain crack free even under significant overloads. This may be due to the structures being exposed to exceptionally corrosive atmospheres during their useful life. In designing prestressed members to be used in special structures of this type, it may be necessary to compute the load that causes cracking of the tensile flange, in order to ensure that adequate safety against cracking is provided by the design.

Many tests have demonstrated that the load-deflection curves of prestressed beams are virtually linear up to and slightly in excess of the load that causes the first cracks in the tensile flange. For this reason, normal elastic-design relationships can be used in computing the cracking load by simply determining the load that results in a net tensile stress in the tensile flange (prestress minus the effects of the applied loads) that is equal to the tensile strength of the concrete. It is customary to assume that the tensile strength of the concrete is equal to the modulus of rupture of the concrete when computing the cracking load.

It should be recognized that the performance of bonded prestressed members is actually a function of the transformed section rather than the gross concrete section (see section 9.2.8). If it is desirable to make a precise estimate of the cracking load, such as is required in some research work, this effect should be considered.

EXAMPLE 9-15: Compute the total uniformly distributed load required to cause cracking in a beam that is 10 in. wide, 12 in. deep, and is supported on a simple span of 25 ft, if the final prestressing force is 120,000 lb applied at an eccentricity of 2.50 in. Assume the modulus of rupture equals $7.2\sqrt{f_c'}$ and $f_c' = 5000$ psi.

SOLUTION:

$$A = 120 \text{ in.}^2, \quad I = 1440 \text{ in.}^4, \quad r^2/y = 2.00 \text{ in.}$$

$$S = \frac{1440}{6} = 240 \text{ in.}^3$$

$$\text{Modulus of rupture} = 7.2\sqrt{5000} = 509 \text{ psi}$$

$$f_b = \frac{120{,}000}{120}\left(1 + \frac{2.50}{2.00}\right) = 2250 \text{ psi}$$

Therefore, the moment that causes cracking must result in a bottom-fiber stress equal to: 509 psi + 2250 psi = 2759 psi.

$$M_t = \frac{w_t L^2}{8} = f_b S = \frac{2759 \times 240}{12{,}000} = 55.18 \text{ k-ft}$$

$$w_t = \frac{55.18 \times 8}{25.0^2} = 0.706 \text{ k/ft}$$

9.5.2 Ultimate Moment of Bonded Members

When prestressed flexural members that are stronger in shear and bond than in bending are loaded to failure, they fail in one of the following modes:

1. *Failure at cracking load.* In very lightly prestressed members, the cracking moment may be greater than the moment the member can withstand in the cracked condition and, hence, the cracking moment is the ultimate moment. This condition is rare and is most likely to occur in members that are prestressed concentrically with small amounts of steel. Determination of the possibility of the occurrence of this type of failure is accomplished by comparing the estimated moment that would cause cracking to the estimated ultimate moment. When the estimated cracking load is larger than the computed ultimate load, this type of failure would take place if the member were subjected to the required load. Because this type of failure is a brittle failure, it occurs without warning—designs that would yield this mode of failure should be avoided.

2. *Failure due to rupture of steel.* In lightly reinforced members subjected to ultimate load, the ultimate strength of the steel may be attained before the concrete has reached a highly plastic state. This type of failure is occasionally encountered in the design of structures with very large compression flanges in comparison to the amount of prestressing steel, such as a composite bridge stringer. Computation of the ultimate moment of a member subject to this type of failure can be done with a high precision.

3. *Failure due to strain.* The usual underreinforced, prestressed structures that are encountered in practice are of such proportions that, if loaded to ultimate, the steel would be stressed well into the plastic range and the member would evidence large deflection. Failure of the member will occur when the concrete attains the maximum *strain* that it is capable of withstanding. It is important to understand that research into the ultimate bending strength of reinforced and prestressed concrete has led most investigators to the conclusion that concrete, of the quality normally encountered in prestressed work, fails when the limiting strain of 0.003 is attained in the concrete. Since the ultimate bending capacity is limited by strain rather than stress in the concrete, it is a function of the elastic moduli of the concrete and steel. The magnitude of the ultimate moment for members of this category can also be predicted, as a rule, within the normal tolerances expected in structural design. The ultimate moment of underreinforced sections cannot be predicted with the same precision as the lightly reinforced members described above, since the ultimate moments of underreinforced members are a function of the elastic properties of the steel and the effective stresses in the prestressing steel, whereas the ultimate moment capacities of lightly reinforced members are not.

4. *Failure due to crushing of the concrete.* Flexural members that have relatively large amounts of prestressing steel or relatively small compressive flanges are referred to as being overreinforced. Overreinforced members, when loaded to destruction, do not attain the large deflections associated with underreinforced members—the steel stresses do not exceed the yield point and failure is the result of the concrete being crushed. Computation of the ultimate moments of overreinforced members is done by a trial and error procedure, involving assumed strain patterns, as well as by empirical relationships.[9-4]

ACI 318-71 provides that ultimate moment computations be based upon stress and strain compatibility using the assumptions of Section 10.2 of the *Code*. The basic assumptions are as follows:

1. Plane sections are assumed to remain plane.
2. The stress in reinforcing steel is assumed to be equal to the product of the steel strain and the steel elastic modulus for stresses below the yield stress and shall be considered to be independent of strain and equal to the yield stress for strains larger than the strain at the yield point.
3. The stress-strain curve for the prestressing steel is assumed to be a smooth curve without a definite yield point.
4. The limiting strain of the concrete is taken to be 0.003 regardless of concrete strength.
5. The steel and concrete are completely bonded.
6. The distribution of concrete stresses at ultimate strength of the section is taken to be either a rec-

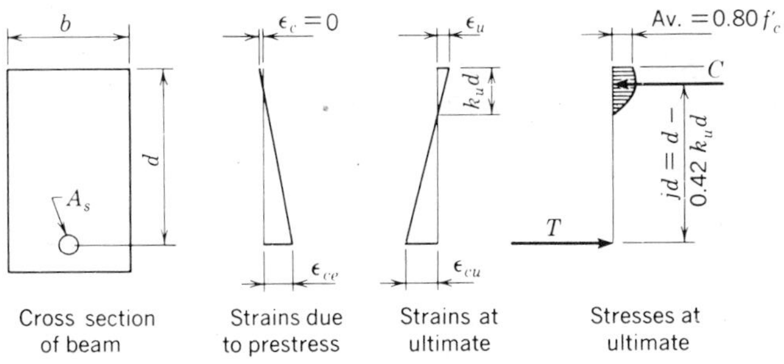

Fig. 9-27 Strain and stress distributions assumed in ultimate moment computations.

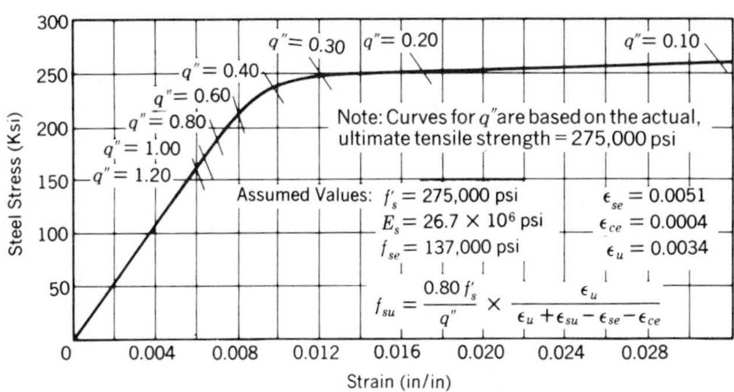

Fig. 9-28 Stress-strain diagram with f_{su} vs ϵ_{su} for various values of ω_p superimposed.

tangle, trapezoid or parabola within certain restrictions (see ACI 318-71, Sections 10.2.6 and 10.2.7).

The derivation of the basic flexural strength relationships using these assumptions is available in the literature (see Ref. 9-4, p. 111). Two important relationships result from the derivation. The first of these establishes the ratio of the depth of the compression block to the effective depth (k_u) in terms of strain. This relationship is

$$k_u = \frac{\epsilon_u}{\epsilon_u + \epsilon_{su} - \epsilon_{se} - \epsilon_{ce}} \tag{9-20}$$

in which (see Fig. 9-27)

ϵ_c = concrete strain due to prestressing (assumed = 0)
ϵ_u = concrete strain at ultimate (assumed = 0.003)
ϵ_{ce} = concrete strain at the level of the steel due to prestressing
ϵ_{cu} = concrete strain at the level of the steel at ultimate
ϵ_{se} = steel strain due to the effective prestress
ϵ_{su} = steel strain at ultimate

Equation (9-20) is useful in the analysis of overreinforced members. The other basic relationship is

$$f_{p\,i} = \frac{0.80\, f_{pu}}{\rho_p\, f_{pu}/f'_c} \times \frac{\epsilon_u}{\epsilon_u + \epsilon_{su} - \epsilon_{se} - \epsilon_{ce}} \tag{9-21}$$

in which

A_{ps} = area of prestressed reinforcement in tension zone
b = width of compression face of member
d = distance from extreme compression fiber to centroid of prestressing steel, or to combined centroid when nonprestressing tension reinforcement is included, in.
f'_c = specified compressive strength of concrete, psi
f_{pu} = ultimate strength of prestressing steel, psi
$\rho_p = A_{ps}/bd$

Equation (9-21) is used in a trial and error procedure, the results of which can be plotted on a stress-strain diagram for the particular prestressing steel that is to be used. The intersection of the curves obtained with the trial and error procedure with the stress-strain curve reveals values of stress and strain which are compatible at ultimate moment. This is illustrated in Fig. 9-28 from which it will be seen that for reinforcing indexes (ω_p), which are above 0.30, the steel is stressed above the yield point and hence can be considered to be underreinforced. From Fig. 9-28 it should be apparent that the reinforcing index (ω_p) (see definition below), is a useful parameter in determining the action of a member at ultimate moment.

Further derivations results in the general equation for underreinforced sections which is

$$M_u = (1 - 0.60\,\omega_p)\, A_{ps}\, f_{ps}\, d \tag{9-22}$$

in which

f_{ps} = calculated stress in prestressing steel at design load, psi

$\omega_p = \rho_p f_{ps}/f'_c$ = reinforcing index

Equation (9-22) is the general equation for underreinforced sections.

9.5.3 *Code* Provisions for Bonded Members

ACI 318-71 provides (Section 18.3.3) that the ultimate moment capacity of a section shall be at least 1.20 times the moment which causes cracking based upon the modulus of rupture specified in Section 9.5.2.2. (i.e., $f_r = 7.5\sqrt{f'_c}$ for normal concrete). This provision is to guard against failure at the cracking load which is a sudden type of failure that occurs with little or no warning. Furthermore, Section 18.2.2 provides that when the reinforcement index exceeds 0.30, the section shall be assumed to be overreinforced and the flexural strength based upon the capacity of the compression portion of the internal couple. For reinforcement indices of 0.30 and less, the *Code* provides the ultimate moment capacity be based upon stress and strain compatibility using the assumptions of Section 10.2. The basic relationship for an underreinforced, rectangular, prestressed flexural member is

$$M_u = \phi\, A_{ps} f_{ps} \left(d - \frac{a}{2}\right) \tag{9-23}$$

in which

a = depth of equivalent rectangular stress block, defined by Section 10.2.7
A_{ps} = area of prestressed reinforcement in tension zone
d = distance from extreme compression fiber to centroid of prestressing steel, or to combined centroid when nonprestressing tension reinforcement is included, in.
f_{ps} = calculated stress in prestressing steel at design load, psi
ϕ = capacity reduction factor. See Section 9.2 of ACI 318-71.

The depth of the equivalent rectangular stress block can normally be taken as

$$a = \frac{f_{ps} A_{ps}}{0.85\, f'_c b} \tag{9-24}$$

The capacity reduction factor is to be taken as 0.90 for flexure.

The *Code* further provides that for bonded members having an effective prestressing stress that is not less than one-half of the ultimate tensile strength, unless a strain compatibility analysis is made the stress in the prestressing steel at design (ultimate) load shall be taken as

$$f_{ps} = f_{pu}\left(1 - 0.5\rho_p \frac{f_{pu}}{f_c'}\right) \tag{9-25}$$

in which

f_{pu} = ultimate strength of prestressing steel, psi
$\rho_p = A_{ps}/bd$ = ratio of prestressed reinforcement

The method for including the effects of nonprestressed reinforcing and of flanged sections is given in the example problems.

EXAMPLE 9-16: Compute the ultimate moment capacity for the section shown in Fig. 9-29. The concrete strengths are 5000 psi and 4000 psi for the stringer and slab respectively. The prestressing steel is bonded.

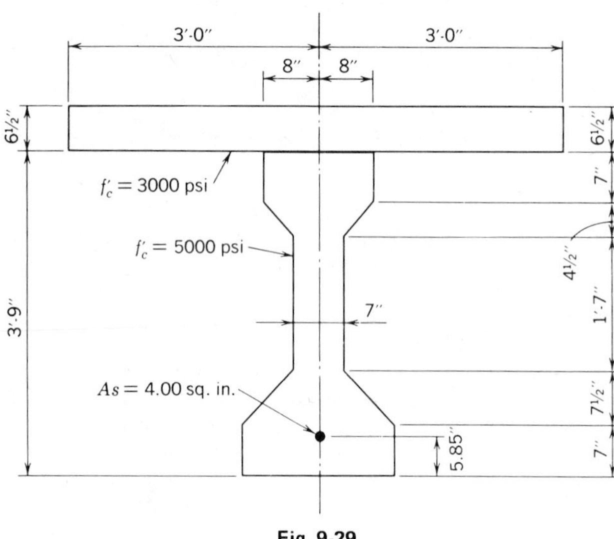

Fig. 9-29

SOLUTION:

$$\rho_p = \frac{4.00}{72 \times 45.65} = 0.00122$$

$$f_{ps} = 275{,}000\left[1 - 0.5 \times 0.00122\,\frac{275{,}000}{4{,}000}\right] = 263{,}500 \text{ psi}$$

$$\omega_p = 0.00122 \times \frac{263{,}500}{4{,}000} = 0.0804 < 0.30$$

The section can be analysed as a rectangular section if the flange thickness is greater than $1.18\,\omega_p\, d/\beta_1$ (see *Code* Section 10.2.7 and Ref. 9-18).

$$\frac{1.18 \times 0.0804 \times 45.65}{0.85} = 5.09 \text{ in.} < 6.50 \text{ in.}$$

$$a = \frac{263{,}500 \times 4.00}{0.85 \times 72 \times 4000} = 4.31 \text{ in.}$$

$$M_u = \frac{0.90 \times 4.00 \times 263{,}500}{12}\left(45.65 - \frac{4.31}{2}\right) = 3438 \text{ k-ft}$$

EXAMPLE 9-17: Compute the ultimate moment capacity assuming the slab is not composite and the steel has the characters illustrated in Fig. 9-28.

SOLUTION: By inspection the section will be overreinforced. The average concrete stress at ultimate can be taken as: $0.85\, f_c' = 4250$ psi.

Assume

$$\epsilon_{se} = 0.0050, \quad \epsilon_{ce} = 0.0004 \quad \epsilon_u = 0.0030$$

$$\epsilon_{su} = \epsilon_{se} + \epsilon_{ce} + \epsilon_u\left(\frac{d - k_u d}{k_u d}\right)$$

Try

$$k_u d = 15 \text{ in.} \quad \beta_1 = 0.80 \text{ for } f_c' = 5000 \text{ psi}$$

$$\beta_1 k_u d = 0.80 \times 15 = 12.00 \text{ in.} = \text{depth of equivalent stress block}$$

Compression force

$$C = \left[7 \times 12 + 9 \times 7 + \frac{9 \times 4.5}{2}\right] \times 4.250 \text{ ksi} = 711\text{k}$$

$$\epsilon_{su} = 0.0054 + 0.0030\left(\frac{39.15 - 15.00}{15.00}\right) = 0.0102$$

from Fig. 9-28

$$f_{ps} = 245{,}000 \text{ psi}; \quad f_{ps}A_{ps} = 980\text{ k} > 711\text{ k}$$

Try

$$k_u d = 22 \text{ in.} \quad \beta_1 k_u d = 17.60 \text{ in.}$$

$$C = 711 + 5.60 \times 7 \times 4.25 = 878\text{ k}$$

$$\epsilon_{su} = 0.0077 \quad f_{ps} = 220{,}000 \text{ psi}$$

$$f_{ps}A_{ps} = 880\text{ k} \cong 878\text{ k} = C$$

Compute location of resultant compressive force.

7 × 17.6	=	123.20 × 8.80	=	1084
9 × 7	=	63.00 × 3.5	=	220
9 × 4.5 / 2	=	20.25 × 8.5	=	172
		206.45		1278

Distance from the top fiber

$$\frac{1278}{206.45} = 6.19$$

$$M_u = \frac{880}{12}(39.15 - 6.19) = 2417 \text{ k-ft}$$

EXAMPLE 9-18: Compute the ultimate moment capacity for the beam with bonded reinforcing shown in Fig. 9-30. Assume

$$f_c' = 4{,}000 \text{ psi}, \quad f_{pu} = 270 \text{ ksi}, \quad f_y = 60 \text{ ksi}$$

SOLUTION:

$$\rho_p = \frac{3.50}{60 \times 27.25} = 0.00214$$

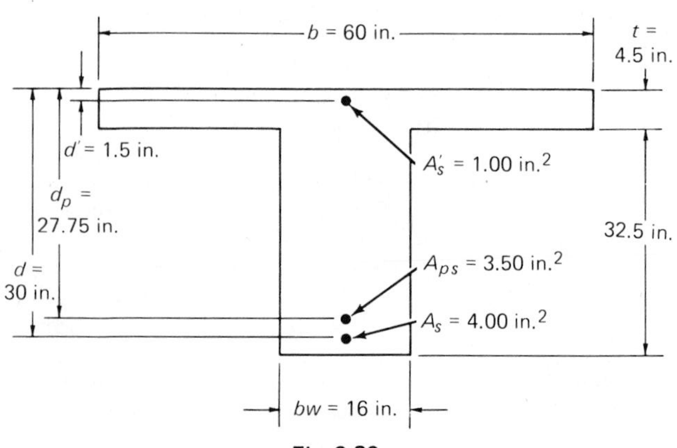

Fig. 9-30

$$f_{ps} = 270\left(1 - \frac{0.5 \times 0.00214 \times 270}{4}\right) = 250.5 \text{ ksi}$$

$$= \frac{A_{ps}f_{ps}d_p + A_sf_yd}{A_{ps}f_{ps} + A_sf_y} =$$

Average depth to tensile reinforcement

$$= \frac{3.50 \times 250.5 \times 27.25 + 4 \times 60 \times 30}{3.50 \times 250.5 + 4 \times 60} = \frac{31091}{1116.75} = 27.84 \text{ in.}$$

$$\omega_p = \frac{3.50 \times 250.5}{60 \times 27.84 \times 4} = 0.131 < 0.30 \therefore \text{ underreinforced}$$

$$\omega = \frac{4 \times 60}{60 \times 27.84 \times 4} = 0.036$$

$$\omega' = \frac{1 \times 60}{60 \times 27.84 \times 4} = 0.009$$

$\omega_p + \omega - \omega' = 0.158 < 0.30 \therefore$ underreinforced

Force to develop the total flange $= 0.85 f'_c bt + A'_sf_y = 978$ k

Total tensile force = 1116.75 k > 978 k – Analyze as a T-beam.

Force to develop flange overhangs and compressive reinforcing

$$F_{sp} + A'_sf_y = 0.85 \times 4\ (60 - 16)\ 4.5 + 1 \times 60 = 733.2$$

Force to be developed by the web (F_{sw}).

$$(A_{ps}f_{ps} + A_sf_y) - (F_{sf} + A''_sf_y) = 1116.75 - 733.2 = 383.55 \text{ k}$$

ω_w, ω_{pw}, ω'_w which are the reinforcement indices for flanged sections must be computed to determine if the web is overreinforced. They are computed as for ω, ω_p, and ω' except that b is the web width, and the steel area is that required to develop the compressive strength of the web only.

$$\omega_w + \omega_{pw} - \omega'_w = \frac{383.55}{16 \times 27.84 \times 4}$$

$$= 0.215 < 0.30 \therefore \text{ underreinforced}$$

$$M_u = \phi\left[F_{sw}d\left(1 - \frac{0.59\,F_{sw}}{b_wdf'_c}\right) + 0.85\,f'_c\,(b - b_w)t\left(d - \frac{t}{2}\right) + A'_sf_y\,(d - d')\right]$$

$$= \frac{0.90}{12}\left\{383.55 \times 27.84\left[1 - \left(\frac{0.59 \times 383.55}{16 \times 27.84 \times 4}\right)\right] + (0.85)(4)(44)(4.5)(25.59) + (1)(60)(26.34)\right\} = 2110 \text{ k-ft}$$

9.5.4 *Code* Provisions for Unbonded Members

Unbonded members are treated (by *the Code*) in the same manner as bonded members with two exceptions. The value of the stress in the prestressing steel at design load is specified to be lower for unbonded members. The specified value is

$$f_{ps} = f_{se} + 10{,}000 + \frac{f'_c}{100\rho_p} \tag{9-26}$$

but not more than

$$f_{py} \text{ or } f_{se} + 60{,}000.$$

In the above

f_{py} = specified yield strength of prestressing steel, psi

f_{se} = effective stress in prestressing steel, after losses, psi

A minimum amount of bonded reinforcement is required in members prestressed with unbonded tendons. The amount required for beams and one-way slabs is

$$A_s = \frac{N_c}{0.5f_y} \tag{9-27}$$

or

$$A_s = 0.004A \tag{9-28}$$

whichever is larger. The term N_c is defined as the tensile force in the concrete under a load of dead load plus 1.2 times the live load. The term A in eq. (9-28) is defined as the area of that part of the cross section between the flexural tension face and the center of gravity of the gross section. In eq. (9-27) the value of f_y shall not exceed 60,000 psi.

The *Code* further provides that in two-way slabs the minimum amount of bonded reinforcement provided by eqq. (9-27) and (9-28) shall be provided except the amount may be decreased (by an unspecified amount) if there is no tension in the precompressed tensile zone at service loads.

9.6 CONTINUITY

9.6.1 Introduction

The theoretical reduction in moments, stresses and hence cost of materials, achieved through the use of continuity in prestressed concrete, is comparable to that which can theoretically be made with other structural materials. The actual economy of materials and cost of construction resulting from the use of continuity is greatly influenced by the design criteria used, the magnitude of the spans involved, the type of structure under consideration, the type of loading to be carried by the structure in question, and the methods of prestressing available.

The economy associated with the majority of prestressed structural elements would be non-existent if the elements were not precast. Although precast elements can be field-connected in order to form fully continuous prestressed members, this procedure has proved economical only under special conditions.

One of the important uses of cast-in-place, continuous prestressed concrete in the United States has been in the construction of roof and floor slabs. The slabs may be flat plates which are continuous in both directions or may be one-way slabs supported by beams or walls. Another important use of cast-in-place continuous prestressed concrete members has been in the construction of highway and railroad bridges. Continuity has been used extensively in cast-in-place long span bridge construction in the Western United States.

Precast, prestressed elements, which are rendered continuous by cast-in-place concrete and normal reinforcing steel, have been widely used. In this type of construction, the precast, prestressed elements act as simple beams resisting a portion of the dead load, but the live load, as well as the the dead load which is applied after the hardening of the cast-in-place concrete, is carried by the continuous beam. This type of construction has proved to be economical in the American market, and it is expected that the use of this method will continue. The method results in a structure with the fundamental advantages of precast, prestressed concrete, as well as those resulting from the use of continuity. The basic structural analysis of this type of construction does not involve principles unfamiliar to structural engineers.

Continuous prestressed spans frequently have depth-to-span ratios of the order of 1 to 30 for prismatic members and as little as 1 to 80 at the section of minimum depth in members which have variable depths. The greater rigidity of continuous prestressed members also results in less vibration from moving or alternating loads. A significant advantage gained through the use of continuity in prestressed concrete construction, as is the case with other materials, is the reduction of deflections. Over-all structural stability and resistance to longitudinal and lateral loads is normally improved through the introduction of continuity, in any structural system.

9.6.2 Disadvantages of Continuity

The construction of cast-in-place, continuous, prestressed-concrete box girder bridges of conventional design do not present any unusual or serious construction problems. The more sophisticated methods of constructing continuous prestressed bridges, such as cantilevered construction, require a great deal of technical skill on the part of the contractor. Additionally, these methods frequently require more construction labor than is required with conventional construction methods. These factors have deterred the use of almost all of the more sophisticated methods of constructing continuous prestressed concrete members in the United States. Except for very special cases, it is expected this situation will continue as the industry seeks methods of reducing the amount of labor, as well as the degree of skill required in construction.

Many engineers are under the impression that continuous prestressed structures are difficult to design and analyze, because of the secondary moments that result from the deformation of the structure during prestressing. As will be seen in the following discussion, the analysis of continuous prestressed structures is not particularly complex and involves only the familiar principles used in the analysis of ordinary statically indeterminate structures.

9.6.3 Elastic Analysis

The moments due to dead and live loads in an indeterminate prestressed-concrete structure are calculated using the same classical methods employed in analyzing indeterminate structures composed of other materials. The one significant difference in a prestressed structure is that secondary moments may or may not result from the prestressing. The secondary moments are the result of the deformation of the structure and are also calculated by the usual methods of indeterminate analysis. In most areas of structural design, the term secondary moments denotes undesirable moments that are to be avoided if possible. In prestressed concrete design, the secondary moments are not always undesirable and can be quite helpful. It is essential that the designer be aware that such moments do exist and that they must be included in the design of indeterminate prestressed structures.

In the design and analysis of continuous prestressed beams, the following assumptions are generally made:

1. The concrete acts as an elastic material within the range of stresses permitted in the design.
2. Plane sections remain plane.
3. The effects of each cause of moments can be calculated independently and superimposed to attain the result of the combined effect of the several causes (the principle of superposition).
4. The effect of friction on the prestressing force is small and can be neglected.
5. The eccentricity of the prestressing force is small in comparison to the span and, hence, the horizontal component of the prestressing force can be considered uniform throughout the length of the member.
6. Axial deformation of the member is assumed to take place without restraint.

Research into the performance of continuous prestressed-concrete beams has revealed that these assumptions do not introduce significant errors in normal applications. If the cracking load of a beam is exceeded, and in cases where the effect of friction during stressing is significant, special attention should be given to the effects of these conditions. The axial deformation that results from prestressing can have a significant effect on the moments and stresses when such deformations are restrained, as in rigid frames; hence, special investigation into the effects of this phenomenon may be required. Some of these effects will be considered subsequently, but for the general discussion which follows, the above assumptions will be assumed to be valid.

The magnitude and nature of secondary moments can be illustrated by considering a two-span, continuous, prismatic beam that is not restrained by its supports, but which must remain in contact with them, as is illustrated in Fig. 9-31. This beam is prestressed with a straight tendon that has a force of P and eccentricity of e. This would tend to make the beam deflect away from the center support by the amount

$$\delta = \frac{Pe(2l)^2}{8EI} = \frac{Pel^2}{2EI} \tag{9-29}$$

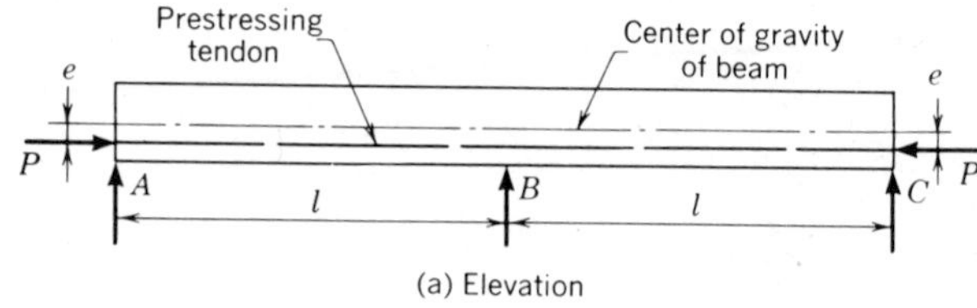

(a) Elevation

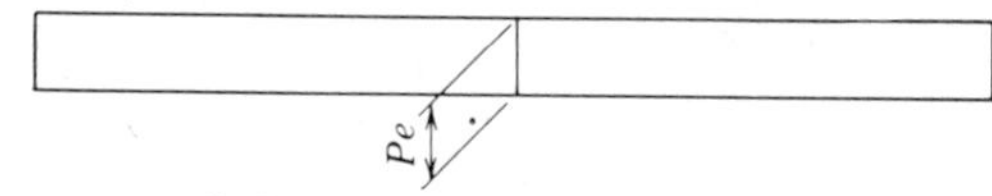

(b) Primary Moment Due to Prestressing

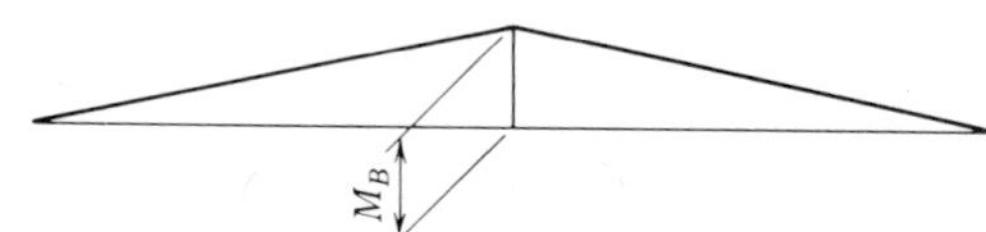

(c) Secondary Moment Due to R_b (downward)

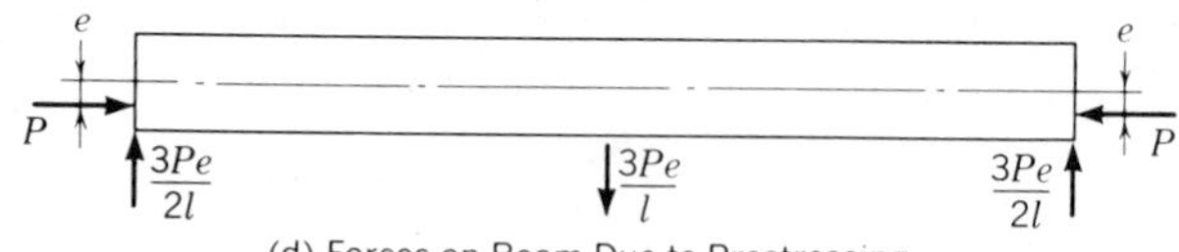

(d) Forces on Beam Due to Prestressing

$-\frac{Pe}{2}$

(e) Moment Due to Prestressing

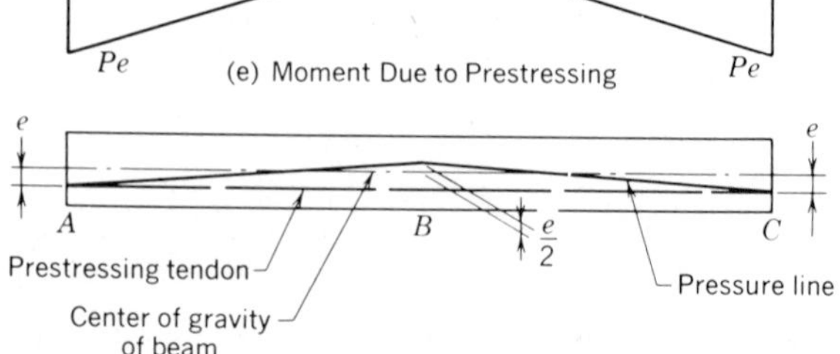

(f) Location of Pressure Line

Fig. 9-31

Because the beam must remain in contact with the center support, a downward reaction must exist at the location of the center support to cause an equal but opposite deflection. The deflection at the center of a beam which has a span of $2l$, due to a concentrated load (R_b) applied at the center, is equal to

$$\delta = \frac{R_b l^3}{6EI} \qquad (9\text{-}30)$$

Since the deflections must be equal in magnitude, by equating eqq. (9-29) and (9-30), the value of R_b is found to be

$$R_b = \frac{3PE}{l} \qquad (9\text{-}31)$$

Therefore, by applying the rules of statics, it can be shown that the forces that are acting upon the beam must be as shown in Fig. 9-31(d) in order to maintain equilibrium, and the moment diagram due to prestressing alone is as shown in Fig. 9-31(e).[9-19]

By dividing the moment due to prestressing at each section by the prestressing force P, the eccentricity of the pressure line, e, can be found and plotted as in Fig. 9-31(f). At the ends, as would be expected, the pressure line is seen to be coincident with the location of the prestressing force, while at the center support, the pressure line is at $-e/2$ above the center of gravity of the section. The effect of the secondary reaction, R_b, at the center support has been to move the pressure line from an eccentricity of e below the center of gravity of the section to an eccentricity of $-e/2$ above the center of gravity. Since there are no additional loads between the end support and the center support, the pressure line is a straight line, as shown in Fig. 9-31(f). It will be seen from this example that the secondary moment is secondary in nature, but not in magnitude.

As the above example illustrates, one of the effects of the prestressing may be the creation of secondary reactions, and these reactions cause a linear moment diagram, as shown in Fig. 9-31(c), which, when combined with the primary, prestressing-moment diagram, Fig. 9-31(b), results in the actual moment due to prestressing, Fig. 9-31(e). Because the secondary reactions can only cause a moment which varies linearly, the effect of the secondary moment is to displace the pressure line linearly from the center of gravity of the steel, in direct proportion to the distance from the support. It is also apparent from this example that, if the reactions which result from the prestressing are known, the location of the pressure line can be determined and the stresses due to prestressing at any point can be determined thereby.

The stresses due to prestressing in continuous beams are calculated from the basic relationship

$$f = \frac{P}{A}\left(1 \pm \frac{ey}{r^2}\right) \qquad (9\text{-}32)$$

in which e is the eccentricity of the pressure line and not the eccentricity of the tendon (although it may be both, as will be seen). An important axiom that is illustrated by the above is that the pressure line due to prestressing is not necessarily coincident with the center of gravity of the steel in indeterminate prestressed structures.

If the position of the tendon in the above example is revised so that it is coincident with the location of the pressure line, which was computed above, the resulting tendon location and moment diagram due to prestressing alone would be as shown in Figs. 9-32(a) and (b), respectively. Removing the reaction at B, in order to render the structure statically determinate, and using the principle of elastic

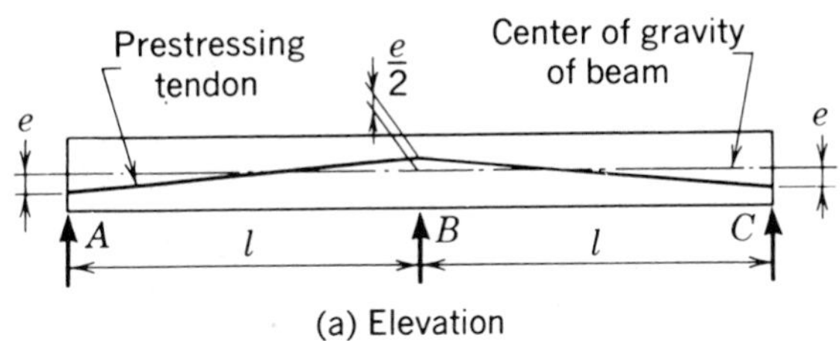

(a) Elevation

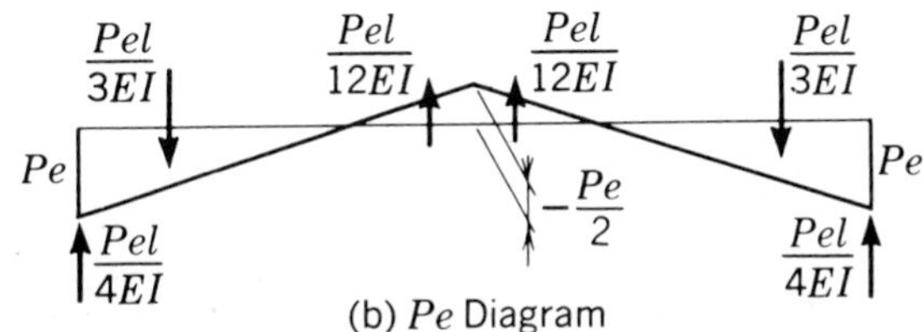

(b) Pe Diagram

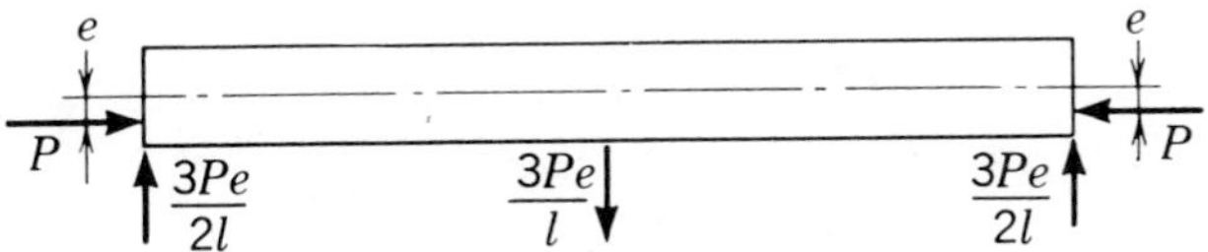

(c) Forces on Beam Due to Prestressing

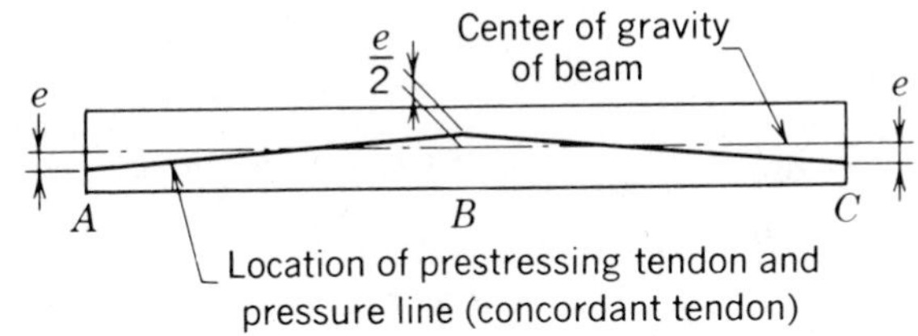

(d) Location of Pressure Line

Fig. 9-32

weights, the reactions and forces acting on the beam are as shown in Fig. 9-32(c). The deflection of the beam at B due to the prestressing is equal to the moment at B resulting from the elastic weights or

$$\delta = \frac{Pel}{4EI} \times l - \frac{Pel}{3EI} \times \frac{7l}{9} + \frac{Pel}{12EI} \times \frac{l}{9}$$

$$= \frac{27Pel^2}{108EI} - \frac{28Pel^2}{108EI} + \frac{Pel^2}{108EI} = 0$$

Since the deflection due to prestressing at B is equal to zero, no secondary reaction is required at B to keep the center support in contact with the beam, and there are no secondary moments. In this example, the pressure line and the center of gravity of the steel are coincident and the tendon is said to be "concordant." In the first example, the pressure line and the center of gravity of the prestressing were not coincident and the tendon is said to be "nonconcordant."

If the tendon is placed in the trajectory shown in Fig. 9-33(a), it can be shown that the moment diagram due to prestressing, and hence the location of the pressure line, is identical to that in the above two examples. This tendon is also nonconcordant.

This series of three examples illustrates several principles that are extremely useful in the elastic design and analysis of statically indeterminate prestressed structures. In the three examples, the only variable is the eccentricity of the prestressing tendon at the center support. The force in the tendon, P, as well as the eccentricity of the tendon, e, at the

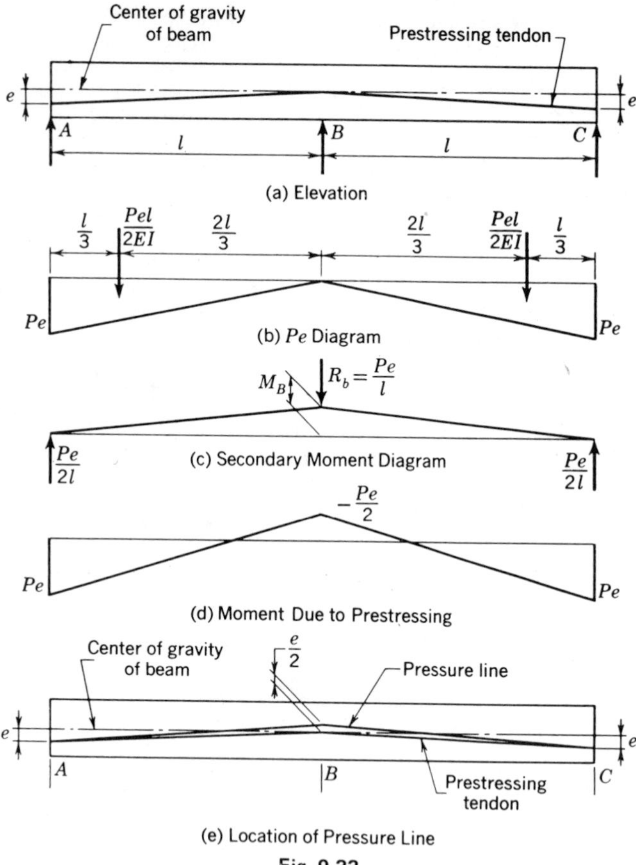

Fig. 9-33

end supports was held constant. The inspection of the three solutions reveals that the moments on the concrete section which resulted from each trajectory of the tendon are identical, although the secondary reactions and secondary moments are not equal. It is apparent, from inspection of the moment diagrams used in the three examples, that if the eccentricity at the end of the member were changed to another value, e', the moment due to prestressing and, therefore, the location of the pressure line, would be changed.

The above examples also illustrate the very important principle of linear transformation, which can be defined as follows: The trajectory of the prestressing force in any continuous prestressed beam is said to be linearly transformed when the location of the trajectory at the interior supports is altered without altering the position of the trajectory at the end supports and without changing the basic shape (straight, curved, or series of chords) of the trajectory between any supports. Linear transformation of any tendon can be made without altering the location of the pressure line.

The only difference between the three tendon trajectories in the above three examples is that they are displaced from each other, at every section, by an amount which is in direct (linear) proportion to the distance of the section from the end of the member. The eccentricity of the tendon at the end supports and the shape of the tendon between the supports were not changed. This principle of linear transformation is equally applicable to tendons that are curved. From from the above examples, it is apparent that the principle applies to concordant tendons as well as to nonconcordant tendons.

The principle of linear transformation is particularly useful in designing continuous beams when it may be desirable to adjust the location of the tendon in order to provide more protective cover over the prestressing tendons, without altering the location of thr pressure line.

Another significant principle, apparent from these examples, is that the location of the pressure line in beams stressed by tendons alone is not a function of the elastic properties of the concrete. The elastic modulus of the concrete did not appear in the values of the secondary reactions. Therefore, the only effect of changes in the elastic properties of the concrete (i.e., creep) is a reduction in the magnitude of the prestressing force, just as it is in statically determinate structures. This has been proven by tests.[9-20]

On the other hand, if the location of the pressure line were to be altered by the application of an additional reaction to a beam, such as by moving the beam up or down at one or more supports with jacks, the reactions and moments induced thereby are a function of the elastic properties of the concrete. Therefore, if such methods are used, the effect of creep must be included in the design of the structures by employing an analysis similar to the type used in studying the effects of differential settlement.

The introduction of curvature to the tendon does not affect the basic methods or principles of analysis in any way; the calculation of the secondary reactions or moments can be made using the principle of elastic weights, the theorem of three moments, or other classical methods, if desired. The familiar moment distribution method is considered among the easier methods for analyzing prismatic beams that have simple tendon trajectories.[9-21] For beams with variable moments of inertia, the theorem of three moments, which may be simplified by using a semigraphical method of computing the static moments of the M/I diagram, is rapid and easily understood.

The method used in the analysis of a continuous prestressed-concrete member does not affect the results obtained, and the selection of the methods to be used in actual design should be determined by the designer on the basis of the ease of application of the various methods for the particular conditions at hand.

In using the moment distribution method, the end eccentricities, curvature, and abrupt changes in slope of the prestressing tendons are converted into specific, equivalent end moments, uniformly applied loads, and concentrated loads, respectively, and the fixed end moments that result from the equivalent loads are distributed in the usual manner. The conversion of end eccentricity and curvature of the tendon into equivalent loading is illustrated in example problems which follow.

EXAMPLE 9-19: Compute the moments due to prestressing and draw the pressure line for the prismatic beam shown in Fig. 9-34. Use moment distribution and the theorem of three moments.

SOLUTION:

$$\text{Equivalent uniform load} = w = \frac{8Pe}{l^2}$$

$$e = 0.80 + \frac{0.50 + 1.20}{2} = 1.65 \text{ ft}$$

$$w = \frac{8 \times 500 \times 1.65}{(100)^2} = 0.66 \text{ k/ft}$$

Fixed end moments.

$$M_{AB}^F = -\frac{0.66 \times (100)^2}{12} = -550 \text{ k-ft} = M_{BC}^F$$

$$M_{BA}^F = +\frac{0.66 \times (100)^2}{12} = +550 \text{ k-ft} = M_{CB}^F$$

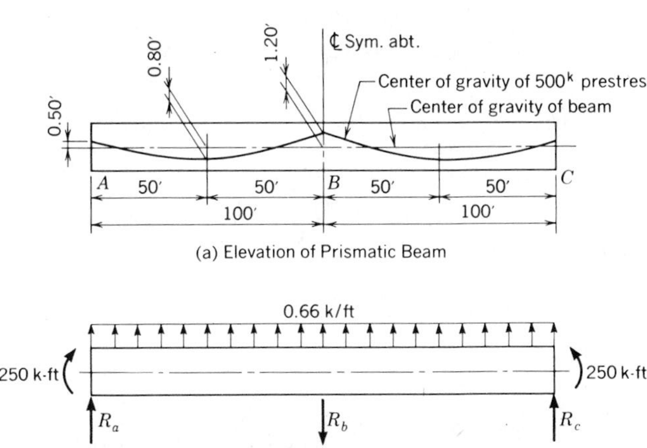

(a) Elevation of Prismatic Beam

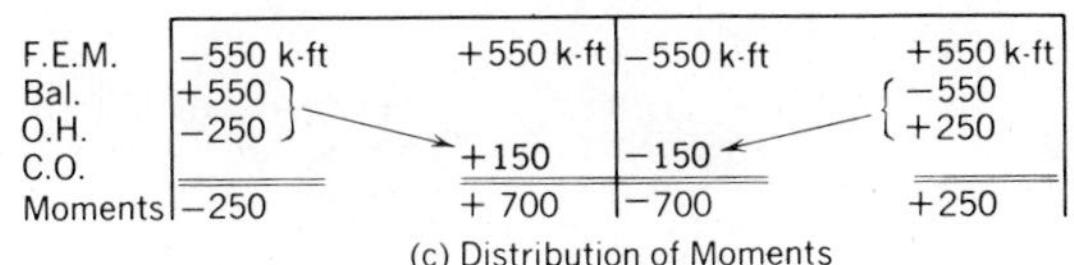

(b) Equivalent Loading Due to Prestress

F.E.M.	−550 k-ft	+550 k-ft	−550 k-ft	+550 k-ft
Bal.	+550			−550
O.H.	−250			+250
C.O.		+150	−150	
Moments	−250	+700	−700	+250

(c) Distribution of Moments

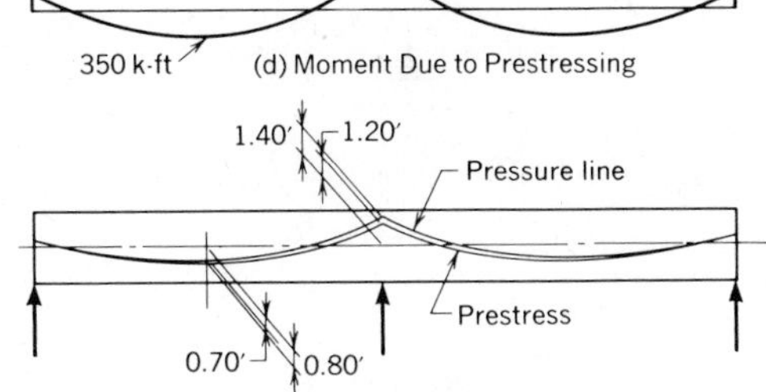

(d) Moment Due to Prestressing

(e) Location of Pressure Line

Fig. 9-34 Analysis of two-span continuous beam with moment distribution of equivalent loading.

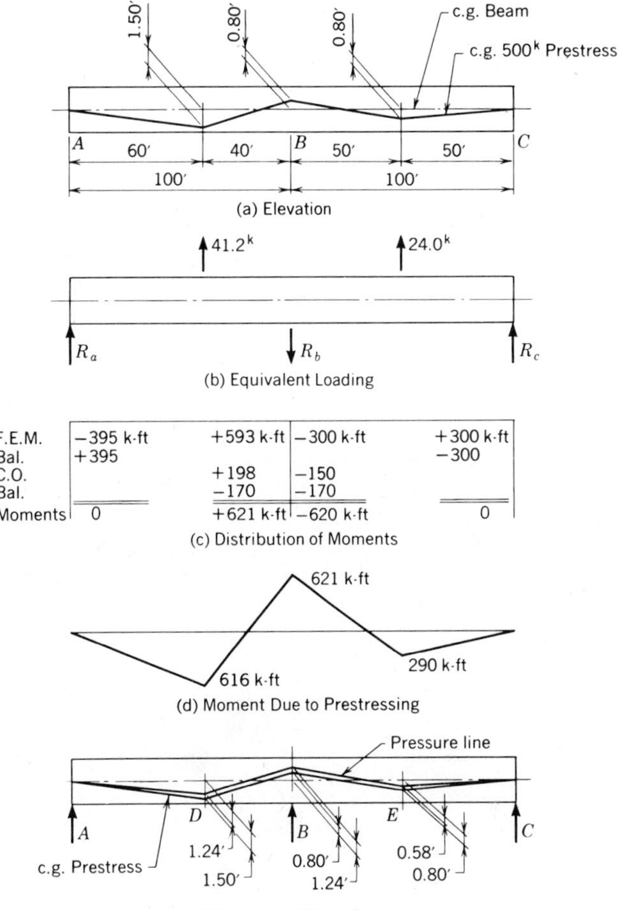

Fig. 9-35 Analysis of two-span continuous beam with moment distribution of equivalent loading.

Overhang moments.

$$M_A = -0.5 \text{ ft} \times 500 = -250 \text{ k-ft}$$

$$M_C = +0.5 \text{ ft} \times 500 = +250 \text{ k-ft}$$

The distribution of moments is performed in Fig. 9-34 and the moment diagram due to prestressing is plotted in Fig. 9-34(d). The computed moment at B is 700 k-ft and the eccentricity of the pressure line is computed as follows:

$$e = \frac{700}{500} = 1.40 \text{ ft}$$

The pressure line is then 1.40 ft − 1.20 ft = 0.20 ft above the tendon at B, since the pressure line is linearly transformed from the tendon trajectory. At the center of the span, the pressure line is 0.10 ft higher than the tendon trajectory since (50 × 0.20) ÷ 100 = 0.10 ft. This is shown in Fig. 9-34(e).

Due to the symmetry of the beam and the loading, the three-moment equation for this example is reduced to

$$M_A + 4M_B + M_C = -\frac{wL^2}{2}$$

$$250 + 4M_B + 250 = -\frac{0.66 \times (100)^2}{2}$$

$$M_B = -700 \text{ k-ft}$$

EXAMPLE 9-20: Compute the moments due to prestressing for the prismatic beam and condition of loading illustrated in Fig. 9-35. Use moment distribution and the theorem of three moments.

SOLUTION:

$$\text{Concentrated load in span } AB = 500\left(\frac{1.50}{60} + \frac{2.30}{40}\right) = 41.2 \text{ k}$$

$$\text{Concentrated load in span } BC = 500\left(\frac{0.80 + 1.60}{50}\right) = 24.0 \text{ k}$$

Fixed end moments.

$$M_{AB}^F = \frac{41.2 \times 60 \times (40)^2}{(100)^2} = -395 \text{ k-ft}$$

$$M_{BA}^F = \frac{41.2 \times (60)^2 \times 40}{(100)^2} = +593 \text{ k-ft}$$

$$M_{BC}^F = \frac{24 \times 50 \times (50)^2}{(100)^2} = -300 \text{ k-ft}$$

$$M_{CB}^F = +300 \text{ k-ft}$$

The moments are distributed in Fig. 9-35, and the moment at B is found to be 621 k-ft. The eccentricity of the pressure line at B is then

$$e = \frac{621}{500} = 1.24 \text{ ft}$$

Therefore, the pressure line is 1.24 ft − 0.80 ft = 0.44 ft above the trajectory of the steel at B. The distance from the pressure line to the tendon trajectory at points D and E are found from the principle of linear transformation as

At point D $\qquad \dfrac{0.44 \text{ ft} \times 60}{100} = 0.264 \text{ ft}$

At point E $\qquad \dfrac{0.44 \text{ ft} \times 50}{100} = 0.220 \text{ ft}$

Using the theorem of three moments, the computation of the moment at B for the above example is as follows:

$$M_A L_1 + 2M_B (L_1 + L_2) + M_C L_2 = -\frac{41.2 \times 60}{100}[(100)^2 - (60)^2]$$

$$- \frac{24 \times 50}{100}[(100)^2 - (50)^2]$$

$$400\ M_B = -158.000 - 90,000$$

$$M_B = -620 \text{ k-ft}$$

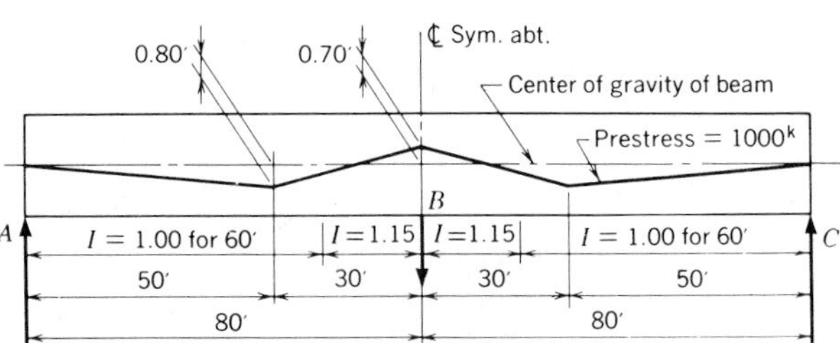

(a) Elevation of Beam with Variable Moment of Inertia

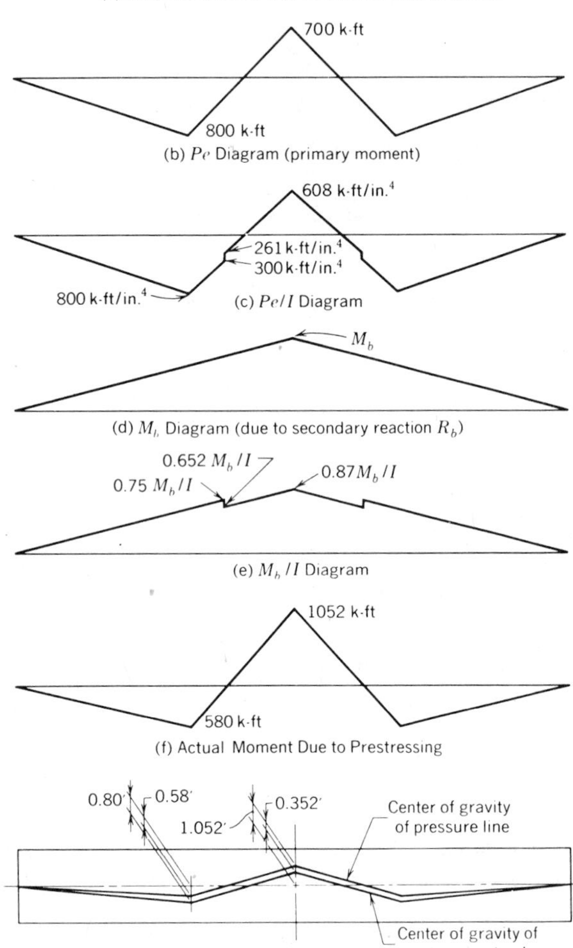

(b) Pe Diagram (primary moment)

(c) Pe/I Diagram

(d) M_b Diagram (due to secondary reaction R_b)

(e) M_b/I Diagram

(f) Actual Moment Due to Prestressing

(g) Elevation of Beam Showing Location of Pressure Line

Fig. 9-36 Analysis of two-span continuous beam with a variable moment of inertia, using the principle of elastic weights.

EXAMPLE 9-21: Compute the moments due to prestressing for the beam illustrated in Fig. 9-36. Note that the relative moment of inertia is 1.00 for the outermost of 60 ft of each span and 1.15 for the center 40 ft of the beam. The center of gravity of the section is a straight line (the variable moment of inertia is the result of an abrupt change in web thickness or abrupt, symmetrical change in top and bottom flange thicknesses, or both).

SOLUTION: In Fig. 9-36(b), (c), (d), and (e) are plotted the Pe diagram, the Pe/I diagram, the assumed M_b diagram, and the M_b/I diagram, respectively. The magnitude of M_b can be determined rapidly by employing the principle of elastic weights. The former is used here by computing and equating the moments (deflections) at the center of the span AC of the beam loaded with the M_b/I diagrams for Pe and M_b.

The upward deflection due to Pe/I diagram (δpe) is

$$\begin{aligned}
-800 \times 50.2 &= -20{,}000 \times 46.70 = -\ 933{,}000\\
-300 \times 10\ \ &= -\ \ 3{,}000 \times 25.00 = -\ \ 75{,}000\\
-500 \times 10.2 &= -\ \ 2{,}500 \times 26.70 = -\ \ 66{,}700\\
-261 \times \ \ 6.2 &= -\ \ \ \ 783 \times 18.00 = -\ \ 14{,}100\\
+608 \times 14.2 &= +\ \ \underline{4{,}256} \times \ \ 4.67 = +\ \ \underline{19{,}800}\\
& \quad -22{,}027 \qquad\qquad\quad -1{,}069{,}000
\end{aligned}$$

$$\delta_{pe} = +22.027 \times 80 \text{ ft} - 1{,}069{,}000 = 691{,}000 \text{ k-ft}^3/\text{in.}^4$$

The downward deflection due to M_b/I diagram (δ_{Mb}) is

$$\begin{aligned}
+0.75\ M_b \times 60.2 &= \ \ 22.5M_b \times 40.00 = +\ 900.0M_b\\
+0.652M_b \times 20\ \ &= \ \ 13.0M_b \times 10.00 = +\ 130.0M_b\\
+0.218M_b \times 20.2 &= \ \ \ \underline{2.2M_b} \times \ \ 6.67 = +\ \ \underline{14.6M_b}\\
& \quad +37.7M_b \qquad\qquad\quad +1044.6M_b
\end{aligned}$$

$$\delta_{Mb} = -37.7M_b \times 80 + 1044.6M_b = -1965M_b$$

Equating the two deflections

$$\delta_{pe} = \delta_{Mb}$$

$$M_b = \frac{691{,}000}{-1965} = -352 \text{ k-ft}$$

The combined or actual moment diagram due to prestressing and the location of the pressure line are as shown in Fig. 9-36(f) and (g), respectively.

It should be noted that the total moment due to prestressing and not the secondary moment was computed in exx. 9-19, and 9-20. When equivalent loads are used for the analysis of moment due to prestressing, the total moment due to prestressing is computed. In ex. 9-21 the secondary moment and not the total moment due to prestressing was computed. The secondary moment due to prestressing is computed when the effects of prestressing are analyzed, using the basic principles of indeterminate structural analysis with the primary moment due to prestressing being considered as the initial loading condition.

9.6.4 Limitations of Elastic Action

As has been stated previously, prestressed-concrete continuous structures perform substantially as elastic structures under loads which do not result in stresses that exceed the normal working stresses permitted by recognized prestressed-concrete design criteria. Adequate experimental data are available to substantiate this.

Under normal conditions, when the design load is exceeded, the concrete stresses remain reasonably elastic up to the load which causes visible cracking in the structure. The first cracks, which are not visible to the unaided eye, do not materially affect the performance of the structure. As a matter of fact, when the load at which the first crack observed with the unaided eye during a test of a beam is plotted on a load-deflection curve, it is often below the point at which pronounced deviation of the tangent (or plasticity) to the elastic deflection curve takes place.

In most continuous structures, the cracking load is not reached simultaneously in all highly stressed sections, since the magnitude of the moments vary at different sections along the member and, when members are designed for moving live loads, the largest moments that may occur at each section under the assumed design loads do not occur under the same condition of loading. Furthermore, once the cracking load has been significantly exceeded at a particular section, there is a reduction in the effective moment of inertia, and hence, stiffness of the member in the cracked area. At loads which result in one or more areas of a beam being stressed substantially above cracking, the effective modulus of elasticity of the concrete in the cracked areas may be considerably lower than in areas that are still stressed in the elastic range—this further contributes to the reduction in the stiffness of the member in the cracked areas.

Because of the localized changes in stiffness of a continuous member subjected to significant overload, the distribution of moments are no longer proportional to the distribution of moments in the elastic range. This is explained by the fact that, after cracking has reached a significant degree at one or more areas in a beam, the moment that results from the application of additional loads is carried in greater proportion by the portion of the member that remains uncracked. The areas first to attain a highly cracked and highly stressed condition yield more upon the application of additional load than areas that remain uncracked, and hence, the cracked areas resist less of the additional loads than would be indicated by purely elastic analysis. The phenomenon is called "redistribution of the moments."

Redistribution of the moments is the phenomenon that results in continuous beams which are designed on a purely elastic basis, frequently, but not always, having very high factors of safety.

The redistribution of moments takes place to various degrees. Redistribution is said to be complete when the various critical sections of a beam all attain a high degree of plasticity and attain the ultimate moments that would be indicated by the ultimate moment analysis developed in section 9.5.2. Some beams, if loaded to destruction, would fail before redistribution is complete. It is not completely understood why this occurs.

The provisions of ACI 318-71 relative to continuity in prestressed concrete members are contained in Section 18.12. They permit a nominal redistribution of the moments at design load (ultimate) for members with "sufficient bonded steel to assure control of cracking." No guide lines are given for determining the quantity of bonded reinforcing that is sufficient for this purpose. Redistribution of moments is only permitted when the net effective steel index $(\omega + \omega_p - \omega')$ in which

ω = reinforcing index for nonprestressed tensile reinforcing,

ω' = reinforcing index for nonprestressed compression reinforcing,

and ω_p = reinforcing index for prestressed tensile reinforcing,

is equal to or less than 0.20. The negative moments are permitted an increase or decrease in the amount of

$$20\left[1 - \frac{(\omega + \omega_p - \omega')}{0.30}\right] \tag{9-33}$$

in percent. Moment decreases or increases which occur at other locations in the member must be considered when the negative moments are varied from those indicated by an elastic analysis. The effect of moments due to prestressing (secondary moments) is to be neglected in an ultimate moment analysis.

9.7 ADDITIONAL DESIGN CONSIDERATIONS

9.7.1 Introduction

Included in this section is a discussion of several factors pertaining to the elastic design of simple prestressed flexural members, which may not at first be apparent to the student of prestressed concrete. The topics treated in this section are often important in problems encountered in practice, and it is well for the designer and student to be familiar with these principles.

The engineer who is often engaged in the design of prestressed structures will become familiar with these relationships as a result of his design experience. On the other hand, the engineer who is only occasionally involved in the design or review of prestressed members will find that this section contains valuable, concise reference material presented in a manner intended to facilitate its use.

9.7.2 Losses of Prestress

The final stress that is required in the prestressing steel at *all* of the critical sections in a prestressed member should be specified by the designer. If the system of prestressing to be used is specified, the complete stressing schedule and initial stresses required should be computed and indicated on the drawings or in the specifications. In order to do this, it is necessary to either compute or assume a value for the loss of stress, in the prestressing tendons, which results from the several contributing phenomena.

The losses of prestress which require consideration include the following.

1. Frictional loss due to intended and unintended curvature in the tendons
2. Slip at anchorage
3. Elastic shortening of concrete
4. Creep of concrete
5. Shrinkage of concrete
6. Relaxation of steel stress

The losses of stress due to frictional loss during stressing and slip at anchorage have traditionally been computed in American practice. Since the "Tentative Recommendations for Prestressed Concrete" by ACI-ASCE Joint Committee 323 appeared in 1958, it has been common practice for American designers to use assumed values (35,000 psi for pre-tensioned members and 25,000 psi for post-tensioned members) for the other losses. Another approach that has been used is to assume the loss of prestress to be a fraction of the initial prestressing stress—the assumed losses being from 10 to 30%.

A strict interpretation of ACI 318 since the 1963 edition would prohibit one from using an assumed loss of prestress and require a computation of the loss based upon the specific details of the design being prepared. The current requirements of AASHO include an approximate method of determining the loss of prestress which includes the effects of average humidity during service and the average stress in the concrete at the level of the steel.

The computation of the losses of prestress due to the shrinkage of the concrete, elastic deformation of the concrete and relaxation of the prestressing steel is straight forward, provided that the parameters governing these phenomena are known. The loss due to creep is not made accurately as easily, because the creep of concrete is a function of both time and the level of stress in the concrete. Since the stress is constantly changing, as a result of the losses in prestress that are occurring, the most accurate method of computing stress loss is to employ a numerical integration that takes into account the several variables. This procedure consists of employing a unit creep curve, the assumed shrinkage curve, and the relaxation curve for the materials under consideration and computing the changes in stress over a large number of increments of relatively short duration. The curves are treated as step functions to facilitate the computations.[9-4]

Sophisticated methods of estimating the loss of prestress utilizing modern electronic calculating devices are available.[9-4, 9-22] Space limitations prevent a detailed description of these methods in this section. These methods include expressing creep, shrinkage and relaxation as time functions and intergrating their effects together with the effects of the applied loads. This type of analysis should be used on major structures and is generally included in the deflection study (section 9.7.3).

When the losses of prestress are to be computed, one must be informed as to the several parameters which affect the losses. The various phenomena which affect the losses are discussed below.

1. Elastic shortening of the concrete When the prestress is applied to the concrete, an elastic shortening of the concrete takes place. This results in an equal and simultaneous shortening of the prestressing steel. The loss in stress in the steel from this shortening is equal to the product of the stress in the concrete at the level of the steel and the ratio of the moduli of elasticity of the steel and concrete. In a simple beam, the critical section for flexural stress, and hence the section at which the losses of prestress should be considered, is normally at the midspan of the beam. When pre-tensioning is used, the concrete stress that should be used in computing the reduction in prestress (due to elastic shortening) is equal to the net, initial concrete stress that results from the algebraic sum of the stress due to initial prestressing and the stress due to the dead load of the beam at the level of the steel. In post-tensioning, the first tendon that is stressed is shortened by the subsequent stressing of all other tendons, and the last tendon is not shortened by any subsequent stressing. Therefore, in post-tensioning, an average value of stress change can be computed and applied equally to all tendons.

2. Creep of concrete The loss in steel stress resulting from the creep of the concrete should also be computed on the basis of stresses that occur at the critical section for flexure rather than average values, since the greatest factor of safety against cracking is generally required at the section of maximum moment. In bonded construction, since creep is time dependent and does not take place to any significant degree until after bond has been established between the steel and the concrete, the strains along the tendons are not averaged in their effect upon the loss of steel stress. In the case of post-tensioned construction, in which the tendons are not bonded to the concrete after stressing, the creep strains become averaged, since the tendon can slip in the member, and for this reason, the average concrete stress at the center of gravity of the steel should be used in the computation of this stress loss.

The magnitude of the creep of concrete, as well as the rate at which it occurs, can be estimated using the data from Chapter 6. When possible, the concrete stress at the level of the steel, taking into account the stress-history of the member, should be used in computing the loss of stress due to this phenomenon.

TABLE 9-1

Type of Tendon	*Type of Duct*	*Design Values* μ	*Design Values* K
Uncoated wire or large diameter strands	Bright flexible metal sheath	0.30	0.0020
	Galvanized flexible metal sheath	0.25	0.0015
	Galvanized rigid metal sheath	0.25	0.0002
	Mastic coated, paper or plastic wrapped	0.05	0.0015
Uncoated seven-wire strand	Bright flexible metal sheath	0.30	0.0020
	Galvanized flexible metal sheath	0.25	0.0015
	Galvanized rigid metal sheath	0.25	0.0002
	Mastic coated, paper or plastic wrapped	0.08	0.0014
Bright metal bars	Bright flexible metal sheath	0.20	0.0003
	Galvanized flexible metal sheath	0.15	0.0002
	Galvanized rigid metal sheath	0.15	0.0002
	Mastic coated, paper or plastic wrapped	0.05	0.0002

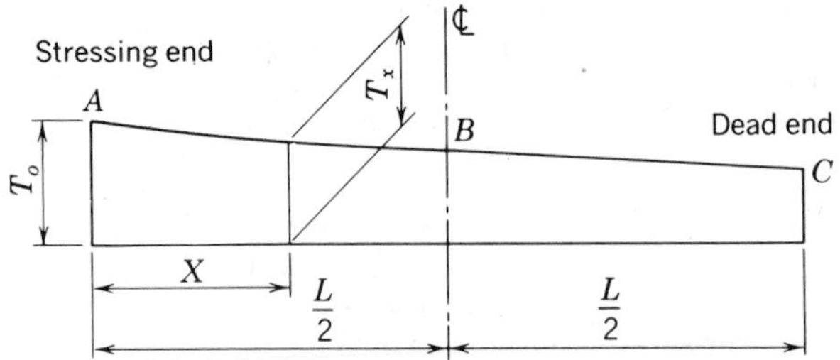

Fig. 9-37 Variation in stress due to friction in a tendon that is post-tensioned from one end.

3. Shrinkage of the concrete The rate at which concrete shrinks as well as the magnitude of the ultimate shrinkage can be estimated using the data of Chapter 6. The entire shrinkage strain is effective in reducing the steel stress in pre-tensioned construction, while only the amount of shrinkage that occurs after the stressing is of significance in this respect in the case of post-tensioning.

4. Relaxation of the prestressing steel Relaxation of the prestressing steel should be estimated on the basis of the data presented in Chapter 6.

9.7.3 Effect of Friction During Stressing

The variation in stress in post-tensioned tendons due to friction between the tendon and the duct during stressing is considered to be the function of two effects. These are the primary curvature or *draping* of the tendon, which is intentional, and the *secondary curvature* or wobble which is the unavoidable minor horizontal and vertical deviations of the tendon from the theoretical position. Coefficients for each of these effects, which are commonly specified by post-tensioning design criteria in this country, are listed in Table 9-1.

The stress at any point in a tendon can be determined by substituting the proper coefficients in the relationship.*

*This relationship can be written for unit stresses as follows:

$$f_o = f_x e^{(\mu\alpha + KX)}$$

$$T_0 = T_x e^{\mu\alpha + KX} \qquad (9\text{-}34)$$

in which: T_0 = the force at the jacking end, T_x = the force at point X feet from the jacking end, e = base of the Naperian logarithm, K = secondary curvature coefficient, μ = primary curvature coefficient, α = total angle change between the tangents to the tendon at the end and at point X (sum of the horizontal and vertical angles) in radians, and X = distance in feet from the jacking end to the point under consideration.

For low values of $\mu\alpha + KX$, eq. (9-34) is approximately equal to:

$$T_0 = T_x(1 + \mu\alpha + KX) \qquad (9\text{-}35)$$

Equation (9-35) is permitted for use with values of $\mu\alpha + KX$ as high as 0.3 by ACI 318-71.

The variation in stress in a post-tensioned tendon stressed from one end, according to the above relationship, will be as is illustrated in Fig. 9-37. The maximum value of stress in the tendon is at the jacking end and the minimum value is at the dead end. If the tendon were stressed from each end simultaneously, the curve would by symmetrical about the center line and would have the shape of AB on each side.

The effects of friction during post-tensioning should be investigated by the designer at the time the trajectory of the post-tensioning tendons is selected. The computations will reveal the magnitude of the friction loss that can be expected and will allow the designer to determine if special precautions should be required in the specifications in order to reduce the friction loss. Special procedures to reduce the effects of friction include using galvanized sheath in lieu of bright sheath, using rigid sheath in lieu of flexible, reducing the curvature of the tendons, or using a water-soluble oil on the tendons. The oil is removed by flushing the ducts with water before the tendon is grouted. The use of water-soluble oil to reduce the friction is generally employed only as a last resort when serious friction is encountered on the job site. It is not recommended that this procedure be relied upon during design.

EXAMPLE 9-22: Compute the force which would be expected to result at midspan of a tendon that is 100 ft long and which is on a parabolic curve having an ordinate of 3 ft at the center line. Assume the sheath is to be a bright flexible metal hose and that the tendons are to be composed of parallel wires.

SOLUTION: Since the tangent of the angle between the tangents to the tendon can be assumed to be numerically equal to the value of the angle expressed in radians, the value of α is found by

$$\alpha = \tan\alpha = \frac{4e}{L} = \frac{4 \times 3 \text{ ft}}{100 \text{ ft}} = 0.12$$

Using the coefficients from Table 9-1

$$\mu\alpha = 0.30 \times 0.12 = 0.036$$

$$KX = 0.0002 \times 50 \text{ ft} = \frac{0.100}{0.136}$$

$$T_0 = T_x e^{0.136} = 1.15\,T_x$$

$$T_x = 0.873\,T_0$$

The loss of stress due to friction is 13%.

With galvanized sheath rather than bright sheath, the loss of prestress would be computed as follows:

$$\mu\alpha = 0.25 \times 0.12 = 0.030$$

$$KX = 0.0015 \times 50 = \frac{0.075}{0.105}$$

$$T_0 = 1.111\,T_x$$

$$T_x = 0.900\,T_0$$

9.7.4 Elastic Deformation of Post-tensioning Anchorages

As explained above, the variation in stress along a post-tensioned tendon at the time of stressing is assumed to follow a curve such as ABC in Fig. 9-38. The curve $EDBC$ in this figure indicates the assumed variation in stress after the tendon has been stressed and anchored. It will be noted that a reduction in the stress at the end of the tendon resulted from the anchoring procedure. The reduction in stress occurs when the prestressing force is transferred from the jack to the anchorage device, at which time, a portion of the elongation of the tendon obtained during stressing is lost due to the deformation of the anchorage device. Although some positive-type anchorage devices, such as the buttonhead systems, when properly applied, have no appreciable deformation of anchorage, the wedge or cone-type anchorages often deform significantly as the load is applied to them. Other types of anchorage which have components stressed in the plastic range may require several hours to reach the limiting value of anchorage deformation. The manufacturers of the anchorage devices that are to be allowed on any project should be consulted in order to determine the limits of the anchorage deformation that may be encountered in the field. If the computations indicate that the deformation may result in a significant reduction in stress of the tendons, the deformation should be measured in the field as a means of ensuring that the desired results are obtained.

In order to simplify the computation of the effect of the anchorage deformation on the stress in the tendon, the curves AB, AD, BC and DE in Fig. 9-38 are assumed to be straight lines. The slopes of lines AB and DE are assumed to be of equal magnitude, but of opposite sign. Tests have shown that this assumption is approximately correct.

The state of stress indicated in Fig. 9-38 will be referred to as condition I. This condition is characterized by the fact that the stress at the center of the tendon is not affected by the deformation of the anchorage, which means that the length X is less than one-half of the length of the tendon. This condition generally occurs in long tendons or in mem-

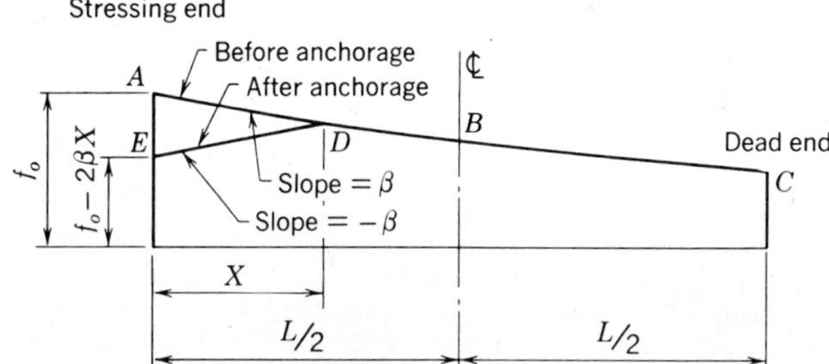

Fig. 9-38 Variation in stress in a post-tensioned tendon before and after anchorage. Condition I.

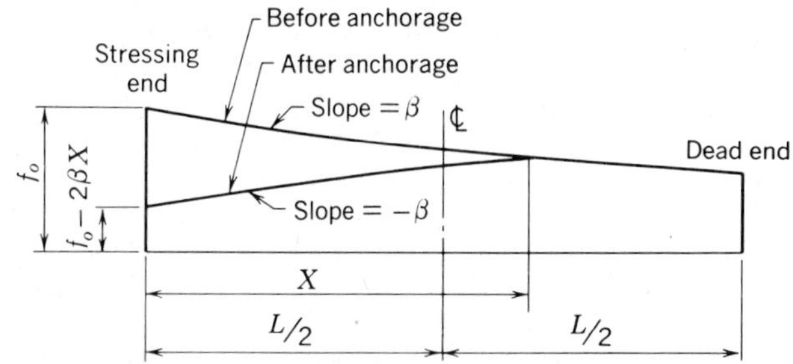

Fig. 9-39 Variation in stress in a post-tensioned tendon before and after anchorage. Condition II.

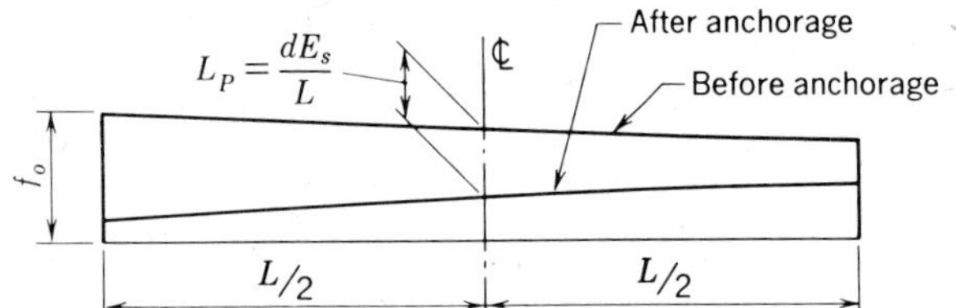

Fig. 9-40 Variation in stress in a post-tensioned tendon before and after anchorage. Condition III.

bers that have high friction, such as those with high primary curvatures. In some instances, it is desirable to stress tendons having stress distributions of this type from one end only, but, since this condition generally exists only in tendons of considerable length, they are more frequently stressed from both ends simultaneously.

The assumed variation in stress referred to as condition II is shown in Fig. 9-39. In this case, the stress at the center line is affected by the deformation of the anchorage, since the distance X is greater than one-half the length of the tendon. There is generally no advantage to be gained in stressing a tendon having this type of stress distribution from each end simultaneously.

The most severe effects from the deformation of the anchorage are found in short cables and those with very low friction. In this case, the stress at the dead end is reduced by the deformation of the end anchorage. This condition is characterized by the computed length of distance X being greater than the length of the tendon. The assumed distribution of stress for condition III is illustrated in Fig. 9-40.

Determination of the effect of anchorage deformation on the stress at midspan, which is generally the most critical section from the designer's viewpoint, will be developed. A similar procedure is used in structures having the critical sections at other locations. The procedure is as follows:

1. Determine the ratio between the stress at the end and midspan resulting from the friction, as explained above. This can be expressed as

$$f_0 = f_{\text{℄}} e^{\phi}$$

where

$$\phi = \mu\alpha + \frac{KL}{2}$$

2. Assume the deformation of anchorage does not affect the stress at midspan of the beam (condition I) and compute the slope of the curve between the center line and stressing end as follows:

$$\text{Slope} = \beta = \frac{2f_{\text{℄}}(e^{\phi} - 1)}{L} \tag{9-36}$$

3. Compute the length of the tendon on which the stress is reduced by the anchorage deformation

$$X = \sqrt{\frac{dE_s}{\beta}} \tag{9-37}$$

where X = length of tendon which is affected in in. β = slope in psi/in., E_s = modulus of elasticity of the steel in psi, d = deformation of the anchorage in in.

4. If X is less than $L/2$, condition I exists, the stress at the center line is not affected, and the stress at each end is readily computed if desired. If X is greater than $L/2$, but less than L, condition II exists. The stress loss at midspan resulting from the anchorage deformation is equal to

$$L_p = 2\beta(X - L/2) \tag{9-38}$$

If this loss is too great to be tolerated, higher initial stresses should be investigated in a trial and error procedure until a satisfactory solution is found. If X is found to be greater than L, condition III exists and the loss at the center of the tendon is equal to

$$L_p = \frac{dE_s}{L} \tag{9-39}$$

and the stress after anchorage at the center of the tendon can be computed directly.

It must be emphasized again that the anchorage deformation for some post-tension systems is very small and can be reasonably neglected. Anchorage deformations as high as 1 in. have been observed with wedge-type anchors under very unusual conditions. It is important to recognize that this phenomenon exists and that it must be taken into consideration during the design. This is particularly true when short tendons are to be used.

EXAMPLE 9-23: Determine the effect of the elastic deformation of the anchorage cone on the stress at midspan of a tendon 100 ft long with a maximum ordinate of 3 ft, if the desired stress at the midspan is 165,000 psi, the deformation of the anchorage is 0.50 in., the modulus of elasticity of the steel is 28,000,000 psi, and a bright metal sheath is used. Also determine the effect if the sheath is galvanized rather than bright.

SOLUTION: The effect of friction for these conditions was determined in ex. 9-22. Using the calculated value for e^{ϕ}, the computation of β and X becomes

$$\beta = \frac{2 \times 165{,}000 \times 0.15}{1200 \text{ in.}} = 41.2 \text{ psi/in.}$$

$$X = \sqrt{\frac{0.50 \times 28 \times 10^6}{41.2}} = 584 \text{ in.} < 600 \text{ in.} = \frac{L}{2}$$

Therefore, condition 1 exists and the deformation of the anchorage of 0.50 in. does not reduce the stress in the tendon at the center line.

Using galvanized sheath

$$\beta = 41.2 \times \frac{0.11}{0.15} = 30.2 \text{ psi/in.}$$

$$X = \sqrt{\frac{0.50 \times 28 \times 10^6}{30.2}} = 680 \text{ in.} > 600 \text{ in.} = \frac{L}{2}$$

The reduction in stress at midspan due to the deformation of the anchorage is

$$L_p = 2\beta\left(X - \frac{L}{2}\right) = 2 \times 30.2 \times 80 = 4830 \text{ psi}$$

Therefore, the stress in the tendon at the center after the tendon is anchored is

$$f = 165{,}000 - 4800 = 160{,}200 \text{ psi}$$

If the stress at midspan is increased to 169,000 psi at the time of stressing, the values of β and X and the loss of the prestressing stress in the tendon at midspan become

$$\beta = \frac{2 \times 169{,}000 \times 0.11}{1200} = 31.0 \text{ psi}$$

$$X = \sqrt{\frac{0.50 \times 28 \times 10^6}{31.0}} = 672 \text{ in.}$$

$$L_p = 2 \times 31.0 \times 72 = 4500 \text{ psi}$$

Therefore, if the stress at midspan is increased to 169,500 psi at the time of stressing, the deformation of the anchorage of 0.50 in. will reduce the stress to approximately 165,000 psi, which is the desired value.

It should be noted that the stress at the end of the tendon which is required to obtain 165,000 psi in the tendon at the center is equal to 165,000 × 1.15 = 190,000 psi when a bright sheath is used. When galvanized sheath is used, the stress at the end of the tendon which is required to obtain the desired stress at midspan after anchoring is equal to 169,500 × 1.11 = 188,000 psi. It will be seen that the required tendon stress at the end of the tendon is not materially reduced by the use of galvanized sheath in this case.

9.7.5 Deflection and Camber

The computations of short term deflections in prestressed-concrete flexural members are made with the assumption that the concrete section acts as an elastic and homogeneous material. This assumption is only approximately correct, since the elastic modulus for concrete is not a constant value for all stress levels, and in addition, the elastic modulus varies with the age of the concrete and is influenced by other factors. As a result, deflection computations for prestressed concrete are approximate.

The deflections for dead and live loads are calculated in the same manner as they are for steel or reinforced concrete members. Normally the moment of inertia of the gross section is used in the computations. In members that have a large amount of reinforcing steel, the moment of inertia of the transformed section can be used. The deflection resulting from the prestressing can be readily calculated for prismatic members with known prestressing force and eccentricity by use of the area-moment principle. The results of such a calculation for a prismatic member with straight tendons (see Fig. 9-41) is

$$\delta = -\frac{PeL^2}{8\,EI} \tag{9-42}$$

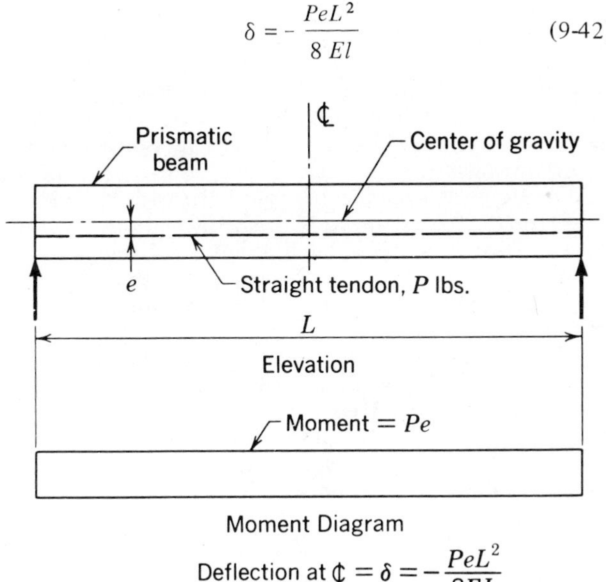

Fig. 9-41 Layout and prestressing moment diagram for a beam and a straight tendon.

where P is the prestressing force in pound, e is the eccentricity in inches, L is the span in inches, E is the modulus of elasticity of the concrete in psi, and I is the moment of inertia of the gross section in inches to the fourth power. The negative sign indicates that the deflection of the beam due to prestressing alone is upward.

In Fig. 9-42, the moment diagram and corresponding deflection due to prestressing is shown for a member prestressed with a tendon that is on a parabolic curve having zero eccentricity at the ends. Finally, in Fig. 9-43 the moment diagram and corresponding deflection is indicated for a prismatic member prestressed with a tendon that has a trajectory composed of three straight lines that are symmetrical about midspan and has zero eccentricity at each end.

It is assumed that the deflections due to the various loads can be superimposed in order to determine the net deflection or camber of the member. In this manner, the deflection of a beam under the effects of its own dead load and due to

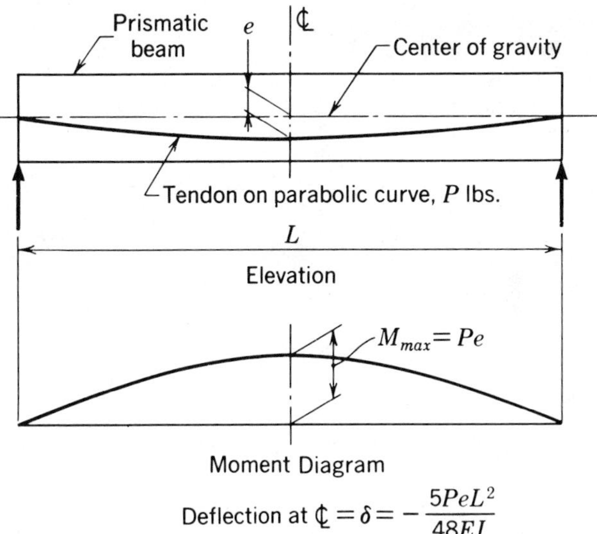

Fig. 9-42 Layout and prestressing-moment diagram for a beam with a parabolic tendon having no eccentricity at the supports.

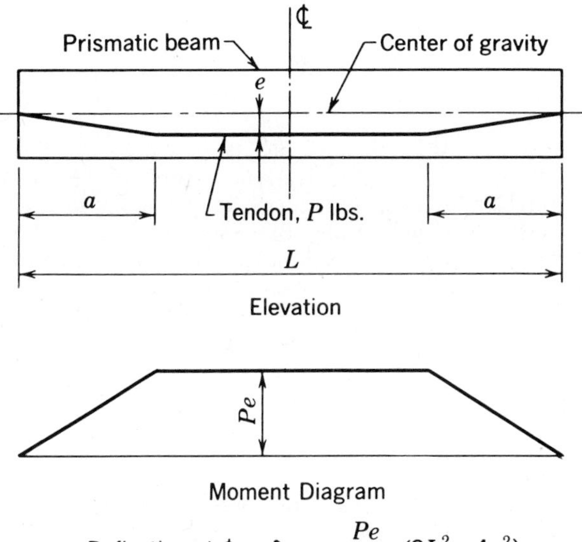

Fig. 9-43 Layout and prestressing-moment diagram for a beam with a deflected tendon having a constant eccentricity in the center position and zero eccentricity at the ends.

prestressing is computed as the algebraic sum of the deflections due to dead load of the beam and due to prestressing. In a similar manner, if a beam were prestressed with a tendon that was on a parabolic curve with an eccentricity at the ends, the deflection of the beam due to prestressing could be determined by computing the algebraic sum of the deflections indicated for a member prestressed with a straight tendon and for a member with a parabolic tendon as indicated in Figs. 9-41 and 9-42, respectively. In applying the principle of superposition as described, it is necessary to divide the moment due to prestressing into two portions that can be substituted into the appropriate relationships for the terms *Pe*. For unusual prestressing-moment diagrams, or if the designer questions the results obtained through the use of the superposition principle, the deflection can be easily calculated from the basic, area-moment principle. When members with variable moments of inertia are used, it is necessary to compute the deflections by use of basic principles.

The deflections at the ends of members that have overhanging ends are frequently large and often result in an undesired appearance. Because of this, many engineers and contractors avoid the use of cantilevers when ever possible. In the case of overhanging beams, the deflection at the end of the overhang is the algebraic sum of the deflection the cantilevered end would have if it were a fixed cantilever and the product of the length of the cantilever and the rotation at the support. It is frequently found that the effect of the rotation at the support is much greater than that of the cantilever deflection by itself. It is strongly recommended that deflection computations always be made for beams that have an overhanging end.

It is well known that in reinforced concrete the tendency is for the deflection of a member to increase with time as a result of creep of the compression flange. In prestressed concrete, the variation in deflection is a function of time as well as of the average distribution of stress in the member under the normal condition of loading. For example, if the effects of the prestressing and the external loads at the average section of a member were such that the distribution of stress was a uniform compression, the effect of creep would be to shorten the member without change in the deflection. If under the same conditions, the stress in the bottom flange were greater than the average compression, the tendency would be for the member to increase in camber, whereas, if the top-fiber compressive stress under the normal loading were higher than the average compression, the tendency would be for the deflection to increase as a result of the creep.

It is interesting to note that for the deflection due to prestressing alone, the effects of concrete shrinkage and steel relaxation are to reduce the deflection, since these two effects tend to reduce the prestressing force. The effect of creep is to alter the deflection for cases where the resultant force in the concrete is significantly eccentric, because the rotational changes due to creep are greater than is the creep shortening effect on reducing the prestress.

9.7.6 Composite Beams

Flexural members that are formed of precast and cast-in-place elements are frequently employed in bridge and building construction. Beams of this type are referred to as composite beams. An illustration of a typical, composite, bridge beam is given in Fig. 9-44.

Composite beams are used to permit the precasting of the elements that are difficult to form and that have the bulk of the reinforcing. Falsework is not normally needed with

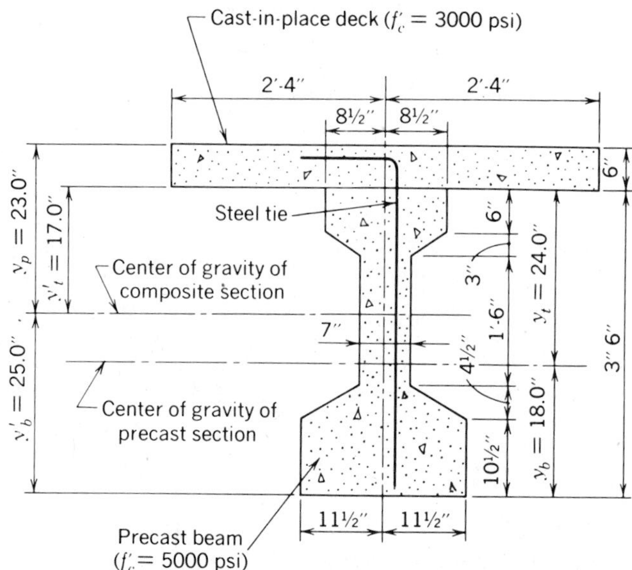

Fig. 9-44 Typical cross section of a composite beam.

composite construction, since the precast elements are usually designed to carry the dead load of the precast beams, the diaphragms as well as the composite slab, without the composite top flange. Dead load and live load, which are applied after the cast-in-place deck has hardened, are carried by the composite beam.

The use of large, composite top flanges results in high, ultimate resisting moments and greater flexural strength in the elastic range, but does not improve the shear strength of a prestressed beam to a significant degree. For this reason, there is little if any advantage to be gained by using composite construction for short-span members in which shear stresses are more significant than flexural stresses.

In designing composite beams, it is necessary to compute the section properties of the precast and composite sections. The flexural stresses due to the various causes at the critical fibers are then computed using the appropriate section properties. In addition, the designer must provide adequate means of transferring the shear stress from the precast to the cast-in-place elements. Nominal shear stresses can be transferred by bond alone, if the concrete surface receiving the composite topping is clean and rough. Higher shear stresses can be transferred if, in addition to the above, steel dowels are extended from the precast section into the cast-in-place section. Shear keys are not required to transfer shear in composite members.

The current ACI 318 permits the full transfer of horizontal shear forces to be assumed without calculations if: (1) the contact surfaces are clean and intentionally roughened; (2) the minimum amount of reinforcing connecting the components conform to eqq. (9-17), or (9-18), with a spacing not exceeding four times the least dimension of the supported element nor 24 in. (b' used in eqq. (9-17), or (9-18) should be the width of the cross section being investigated for horizontal shear); (3) web members are designed to resist the entire vertical shear; and (4) all stirrups are fully anchored into all intersecting components. If this is not done, the horizontal shear stresses must be fully investigated.

The horizontal shear stress may be computed from

$$v_{dh} = \frac{V_u}{\phi b_v d} \tag{9-41}$$

in which V_u is the total design (ultimate) shear force at the section, ϕ is the strength reduction factor and equals 0.85, b_v is the width of the cross section being investigated for horizontal shear, and d is the depth of the entire composite section. In lieu of using eq. (9-41), the designer may compute the actual force that must be transferred within any incremental length and make the necessary provisions for transferring the force by horizontal shear to the supporting element.

Permissible design (ultimate) shear stresses are:

1. When ties are not provided, but the contact surfaces are clean and intentionally roughened — 80 psi
2. When vertical bars or extended stirrups, proportioned to be equal to or exceed the requirements of eqq. (9-17), or (9-18), are provided at spacings that do not exceed four times the least dimension of the supported element nor 24 in., the contact surfaces are clean but not intentionally roughened — 80 psi
3. When the conditions of (2) are met, but the contact surfaces are intentionally roughened — 350 psi
4. If an area of reinforcing equal to

$$A_{vf} = \frac{V_u}{\phi f_y \mu} \tag{9-42}$$

is provided ($\phi = 0.85$, $\mu = 1.0$ and $f_y = 60{,}000$ psi max.) — $0.2f_c'$ (800 psi max.)

Of course, all ties must be fully anchored into the components.

It is recommended that the reader review the complete requirements of ACI 318-71 regarding composite concrete flexural member, since only the more important points have been discussed here.

Differential-shrinkage stresses in composite construction can result in tensile stresses being developed in the cast-in-place concrete and a reduction in the precompression of the tensile flange of the precast element. The differential shrinkage has no effect on the ultimate flexural strength of the composite beam, but it does slightly reduce the load required to crack the tensile flange of the precast element. This effect should be considered in structures in which the cracking load is significant. The effect is normally ignored.

When computing the properties of the transformed section, the difference in the elastic properties of the cast-in-place concrete and the concrete in the precast element should be taken into account by adjusting the width of the composite flange in proportion to the modular ratio of the two concretes. This is exemplified in ex. 9-24.

EXAMPLE 9-24: Compute the flexural stresses in the precast and cast-in-place concrete for the composite section shown in Fig. 9-44, if the moment due to the dead load of the beam, slab, and diaphragms is 673 k-ft and the moment due to the future wearing surface, live load, and impact is 830 k-ft. The section properties for the precast and composite sections are as follows:

Precast section

$y_t = 24.0$ in. $S_t = 4450$ in.^{3} $y_b = 18.0$ in. $S_b = 5950$ in.^{3}

Composite section properties are based upon the transformed cast-in-place flange width of 0.6 × 56 in. = 33.6 in. It is assumed that the ratio of the elastic modulus of the slab and girder concrete is 0.60.

$y_p = 23.0$ in. $S_p = 9400$ in.^{3} $I = 216{,}000$ in.^{4}
$y_t = 17.0$ in. $S_t = 12{,}700$ in.^{3}
$y_b = 25.0$ in. $S_b = 8650$ in.^{3}

SOLUTION:

		Top Fiber Stress in Cast-in-Place Slab	Stresses in Precast Section	
			Top	Bottom
Dead load	$\frac{673 \times 12{,}000}{4450/5950}$		+1810 psi	−1360 psi
$L + I$	$\frac{830 \times 12{,}000}{9400/12{,}700/8650}$	+1060 psi	+ 785 psi	−1150 psi
Totals		+1060 psi	+2595 psi	−2510 psi

9.7.7 Buckling Due to Prestressing

The danger of buckling of columns or other long, slim compression members is known to all structural engineers. The question of the possibility of a prestressed member buckling as a result of the prestressing force is frequently raised. Obviously, when prestressing is done by the application of external load such as jacking against abutments and where no tendons are used through the member, a possibility exists that the member may buckle. In such a case, it is essential that buckling be investigated in the conventional manner. In addition, if tendons are used to prestress the member and the tendons are placed externally in such a manner that they are in contact with the member at the ends only, there would be a possibility of buckling.

When the tendons are placed internally in a way that they are in contact with the member at points between the ends of the member, the tendency to buckle is reduced a significant degree. When the tendons are in intimate contact with the member throughout the length of the member, as is the normal case, in post-tensioning and in pre-tensioning, there is no possibility of buckling. This fact has been demonstrated experimentally and mathematically and can be understood by considering the difference between the action of prestressing and column action.

Column action is characterized by an increase in eccentricity of the load as the load is increased over a critical value. This is illustrated in Fig. 9-45, in which it is seen that

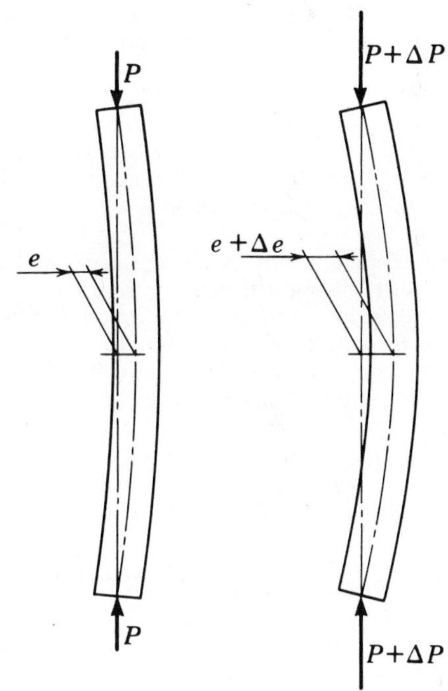

Fig. 9-45 Illustration of column action.

the column load has an eccentricity of e at load P, and if the load is increased to ΔP, the member deflects an additional amount, Δe. This action continues until the critical value of $P + \Delta P$ is reached, and the column buckles. If there were no eccentricity of the load, the column would fail by crushing, as is indeed the case for short columns.

Prestressing action results in a specific distribution of stresses in a member. The eccentricity of the prestressing force remains constant, even if the member is deflected laterally, providing, as was mentioned above, the tendons and concrete are in intimate contact with each other. If the concrete section were cast slightly curved or crooked, as is often the case, the effect of the prestressing alone would be to straighten the concrete member (opposite to column action), since the taut tendon would attempt to assume a straight path.

Prestressed columns and piles, which are pre-tensioned or post-tensioned with the tendons in ducts through the members in the normal manner, can of course buckle, due to externally applied loads, and these members must be designed with care.

The top flanges of flexural members that do not have adequate lateral support can also fail as a result of buckling. For this reason, the designer should give attention to the conditions of support and loading when selecting the dimensions of the concrete section.

9.8 DESIGN DETAILS

9.8.1 Introduction

Prestressed concrete differs from reinforced concrete in several respects. The designer should be aware that these differences exist as well as how to cope with them. Some of the properties and details of prestressed concrete construction that have caused problems are discussed in this section. Detailed discussions of these and other potential problems are to be found elsewhere.[9-4]

9.8.2 Restraint of Volume Changes

If a structural member can deform in accordance with natural laws without restraint, such deformations do not result in stresses. This applies equally to strain changes resulting from temperature variations, elastic deformation, shrinkage, and creep. If fully restrained, the forces developed by strain changes due to these effects can be enormous and are generally only limited by the ultimate capacity of the weakest portion of the structural member or the restraining elements.

Prestressed concrete members are considerably more critical with respect to restraint than is reinforced concrete, due to the fact the sections are generally crack-free and creep deformations tend to shorten the length of the members, as does shrinkage and the elastic shortening due to prestressing. The creep deformation tends to change the deflection of normal reinforced concrete flexural members, but does not tend to shorten the member. For this reason, the designer must give particular attention to the problems of restraint when designing prestressed-concrete structures.

In Fig. 9-46, plan views showing the structural framing of two buildings are shown. In one plan the shear walls, which are provided to give stability against lateral loads, are at the corners of the building with a maximum distance of 150 ft between the walls. Assuming the prestressed roof members are precast, much of the shortening due to shrinkage and creep may take place before the members are erected and all of the elastic shortening will take place before erection.

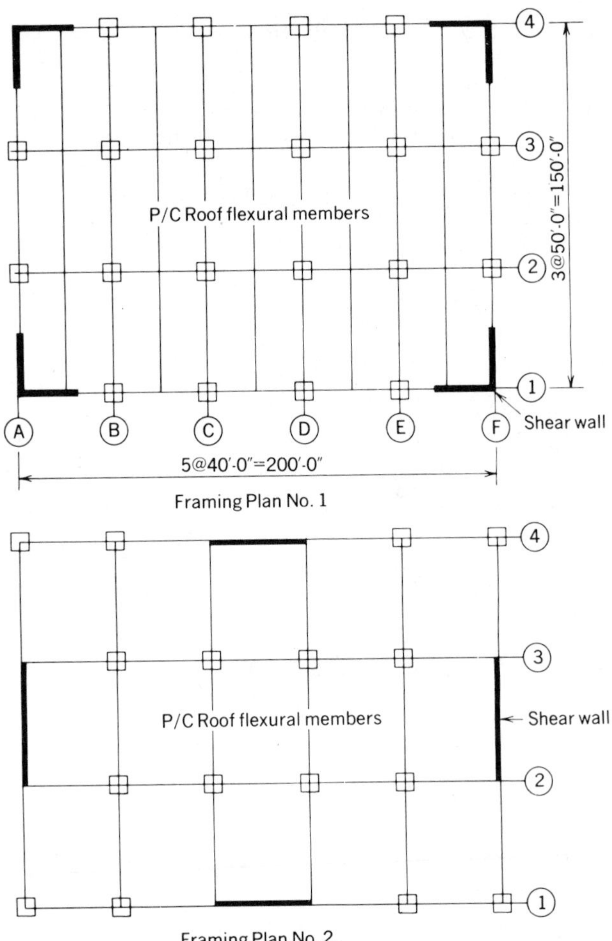

Fig. 9-46 Framing plans with different shear wall layouts.

For the purposes of this illustration, assume the deferred deformation for the members after they are erected is 600×10^{-6} in./in. Deferred deformation is defined as the deformation that would take place, due to volume changes, if the concrete were unrestrained. If the elastic modulus of the concrete were 4×10^6 psi, the deferred deformation would result in a unit stress of 2400 psi if the concrete were fully restrained. It is obvious that unit stresses of this order would develop tremendous forces which could easily be expected to exceed the strength of some portion of the structure between the shear walls. The unrestrained deformation would be $600 \times 10^{-6} \times 150 \times 12 = 1.08$ in. and, if the stress in the prestressed member due to the restraint is to be avoided, a movement of this amount must take place somewhere between the shear walls.

In the other plan of Fig. 9-46, the shear walls are located near the center of each wall rather than near the corners. With this layout, the deformation of the prestressed concrete for all practical purposes is not restrained by the walls and the forces due to creep and shrinkage are avoided.

The basic principal one must keep in mind is that one must avoid having shear walls on line with each other that are some distance apart and which are rigidly connected to each other by prestressed concrete. The author has observed buildings in which pilasters have been pulled away as much as one-half inch from the shear walls to which they were attached as a result of volume changes with framing similar to that shown in Framing Plan No. 1 of Fig. 9-46.

Failure of double-T roof slabs which have had each end embedded in cast-in-place beams that were unyielding, due to the presence of shear walls, have been observed. The failures were either in the form of embedded ends pulled from the cast-in-place beam, in which case a large piece of the beam spalled out with the T leg, or the T legs had cracked completely through. The T legs should have been provided with a reliable expansion bearing detail at one end or the shear walls should have been detailed in such a way as to remove the restraint.

Cantilevered roof spans, particularly those with long slender interior spans, can be subject to large vertical movements due to solar heat. The vertical movements of the cantilever can cause severe damage to walls to which they may be attached or cracking may result in the beam itself. The deflection characteristics of cantilevered spans should be considered carefully before they are used.

Severe movements, some of which can cause serious cracking, can be developed in columns which form a portion of a rigid frame. These movements can result from shortening due to initial prestressing, creep and shrinkage, or from unanticipated movements induced by solar heat. The effect of such movements should be evaluated by the designer and avoided by the provision of hinges and slip joints or adequate reinforcing where required.

9.8.3 Camber-Deflection

The camber of prestressed members can result in construction difficulties in instances when it is significantly greater or less than that which was anticipated. In composite cast-in-place bridge construction, it is customary to use a detail as shown in Fig. 9-47. This detail anticipates the camber of the girder not being exactly as computed and gives a means of compensating for the deviations from the computed camber as well as for the variations in camber between adjacent girders. Occasionally, a detail as shown in Fig. 9-48 is used, but this detail is not considered as good, since it frequently requires field adjustment of the finished grade.

Variation in camber between adjacent elements in building construction can present difficulties in constructing structures of precast elements, unless provision is made for the variation. Roofing cannot be applied directly to precast concrete surfaces that have abrupt edges or joints between elements. If it is, there is danger that the roofing will become damaged and will leak. Therefore, provision should be made to eliminate sharp edges or joints which could cause roof damage.

Fig. 9-47 Recommended bridge deck detail.

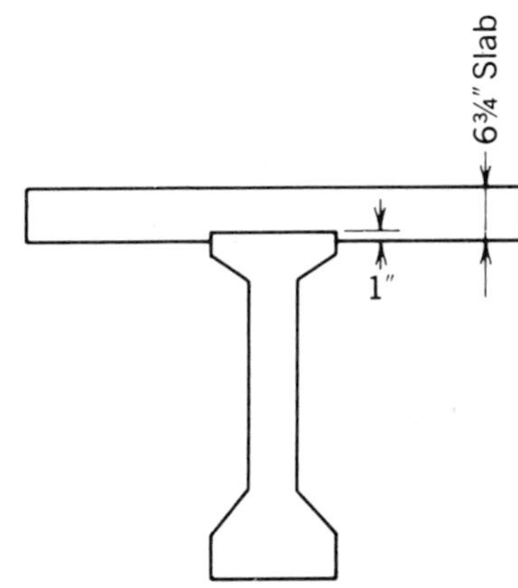

Fig. 9-48 Bridge deck detail not considered as good as that of Fig. 9-47.

Floor construction often consists of precast elements over which a topping is placed in order to achieve a smooth level wearing surface. Electrical conduits and other items are often embedded in the topping and the prudent designer should take this, as well as the estimated camber and variation in camber, into account when specifying the minimum thickness of the topping.

9.8.4 Dimensional Tolerances

The designer should give consideration to the dimensional tolerances that might be expected in the construction of prestressed concrete. Special allowances may be required to provide for the fact that concrete members cannot be made to exact dimension, as is the case of members made of other materials.

Cast-in-place prestressed concrete can be expected to be built within the same dimensional tolerances as one would expect for reinforced concrete construction. In the case of precast concrete, the designer should specify the maximum dimensional tolerances he is willing to accept. It should be recognized that exceptionally small permissible tolerances would be expected to increase the cost of precast members.

The tolerances considered to be standard by the Prestressed Concrete Manufacturers Association of California are as follows:[9-23]

Tolerances

Cross Sectional Dimensions:		
Less than 24 in.		± 1/4 in.
24 to 36 in.		± 3/8 in.
Over 36 in.		± 1/2 in.
Length		
Less than 25 ft		± 1/2 in.
25 to 50 ft		± 3/4 in.
Over 50 ft		± 1 in.
Deviation in squareness of ends:		
	Vertical	*Horizontal*
Less than 12 inches	1/32 in. per in.	1/64 per in.
Over 12 inches	3/16 in. + 1/64 per in. Max. ± 3/4 in.	1/16 + 1/64 per in. Max. ± 1/2 in.
Deviation from straight line (Sweep) 3/16 in. per 10 ft × Total Length		
Deviation from mean camber (as installed) ± 1/8 in. per 10 ft × Total Length		
Prestressing Force:		
Deviation in location from specified C.G.		± 3%
Individual tendon force or elongation		± 5%
Total Prestress, force or elongation		± 5%
Concrete cover over reinforcing		± 1/4 in.

9.8.5 Connections

The connections between precast members should be designed in such a manner that they are capable of withstanding the ultimate vertical *and horizontal* loads for which the structure is proportioned, without failure, excessive deformation or rotation. It is normally preferable that the strength of the connections exceed those of the members connected. The details of the connection should be such that they are readily adjusted to accommodate construction tolerances. The tolerances to be considered are not only variations in length, width and elevation, but also possible deviation from the anticipated planes of bearing. Connections must also be detailed to provide for the necessary erection clearances, reinforcing bar bends, and clearances that may be required for special requirements such as post-tensioning tendons after erection.

In the interest of economy, connections should be as simple as possible and should be such that the precast members can be set and disconnected from the erection equipment quickly with the erected members being stable. The connections should be of such a nature that they are easily inspected after they are completed.

The designer should pay attention to the details of the connections in order to be sure the structure will act as has been assumed in the design of the individual elements. If the beams that connect to the opposite sides of a column are designed as simple beams, the connection details should not result in the members being continuous or partially continuous as a result of a cast-in-place reinforced slab or topping, or due to other mechanical fasteners. On the other hand, if the members have been designed as continuous under certain conditions of loading, the connections should be carefully detailed to achieve the required continuity.

Flexural members undergo rotations and deflections due to the application of transverse loads. Rotations and deflection due to other effects such as temperature variation, creep, shrinkage, etc. may be encountered. Connection details, particularly for simple spans, should be made in such a manner that the necessary rotations can take place without restraint and the risk of spalling from the member being supported unintentionally at an unreinforced edge.

Connections that incorporate welded reinforcing steel should be used with caution, since reinforcing steel frequently contains relatively high amounts of carbon, which necessitates special welding procedures. It is recommended that all welding be done in conformance with the recommendations of the American Welding Society.

The requirements for fireproofing of connections must be kept in mind. Elastomeric bearing pads, which are combustible, and steel bearings will not be permitted by most building codes to be exposed and unprotected in fire-rated construction. If connections are fireproofed by rigid materials, the fireproofing must not restrain the connection and cause it to act in an unintended manner.

Booklets showing standard connection details are available from prestressed-concrete manufacturer's associations.

REFERENCES

9-1 "Proceedings of the Conference Held at the Institution," The Institution of Civil Engineers, 5–10, Feb. 1949.

9-2 Abeles, P. W., *Principles and Practice of Prestressed Concrete,* 18–20, Crosby Lockwood & Son, Ltd., London, 1949.

9-3 Dobell, C., "Patents and Code Relating to Prestressed Concrete," *Journal of the American Concrete Institute,* **46**, 713–724, May 1950.

9-4 Libby, James R., *Modern Prestressed Concrete Design Principals and Construction Methods,* Van Nostrand Reinhold Company, New York, 1971.

9-5 Guyon, Y., *Prestressed Concrete,* pp. 4, 49, John Wiley & Sons, Inc., New York, 1953.

9-6 Dobell, C., "Prestressed Concrete Tanks," Proc. First U.S. Conference on Prestressed Concrete, 9, 1951.

9-7 Hendrickson, J. G., "Prestressed Concrete Pipe," Proc. First U.S. Conference on Prestressed Concrete, 21, 1951.

9-8 Kennison, H. F., "Prestressed Concrete Pipe—Discussion," Proc. U.S. Conference on Prestressed Concrete, 25, 1951.

9-9 "Proceedings of the Conference Held at the Institution," The Institution of Civil Engineers, 52–56, Feb. 1959.

9-10 Timoshenko, S., *Theory of Plates and Shells,* McGraw-Hill Book Company, Inc., New York, 1940.

9-11 "Circular Concrete Tanks Without Prestressing," Bulletin, Portland Cement Association.

9-12 "Design of Circular Domes," Bulletin, Portland Cement Association.

9-13 Crom, J. M., "Design of Prestressed Tanks," *Proc. A.S.C.E.*, 37, Oct. 1950.

9-14 *Building Code Requirements for Reinforced Concrete,* ACI Committee 318-71, American Concrete Institute, 1971.

9-15 Standard Specifications for Highway Bridges, 11th Ed., The American Association of State Highway Officials, 1973.

9-16 Portland Cement Association, Notes from the Building Code Seminar, Chapter 63S-28, 8–31, 1963.

9-17 MacGregor, J. G.; and Hanson, J. M., "Proposed Changes in Shear Provisions for Reinforced and Prestressed Concrete Beams," *Journal of the American Concrete Institute,* **66** (4), 276–288, April 1969.

9-18 *Building Code Requirements for Reinforced Concrete,* ACI-318-63, American Concrete Institute, 1963, p. 69.

9-19 Muller, Jean, "Continuous Prestressed Concrete Structural Design," *Proc. Western Conf. on Prestressed Concrete,* 109–132, Nov. 1952.

9-20 Saced-Un-Din, K., "The Effect of Creep Upon Redundant Reactions in Continuous Prestressed Concrete Beams," *Mag. of Concrete Research,* **10**, 109, Nov. 1958.

9-21 Parme, A. L.; and Paris, G. H., "Designing for Continuity in Prestressed Concrete Structures," *Journal of the American Concrete Institute,* **23**, 45–64, Sept. 1951.

9-22 Subcommittee 5, ACI Committee 435, "Deflection of Prestressed Concrete Members," *Journal of the American Concrete Institute,* **60** (12), Dec. 1963.

9-23 "Prestressed Concrete Inspectors' Manual," Prepared by the Prestressed Concrete Manufacturers Association of California.

10

Multistory Structures

MARK FINTEL*

10.1 DEFINITION

The definition of a tall building depends first on such considerations as to where the building is located. A 20-story building may be considered tall in Evanston, Illinois, but is not considered tall in New York City. It also depends on who is asking the question. For a transportation engineer it is tall when elevator and escalator problems are to be solved; for an architect, when tallness becomes a consideration in planning, esthetics, and environment; for a fire prevention specialist when the building is taller than the fire fighting equipment or when the building has to be divided into horizontal fire zones, etc.

For the structural engineer a tall building can be defined as one whose structural system must be modified to make it sufficiently economical to resist lateral forces due to wind or earthquakes within the prescribed criteria for strength, drift, and comfort of the occupants.

10.2 HISTORY OF DEVELOPMENT OF HIGH-RISE BUILDINGS

Throughout history, man has always tried to express his greatness by building monuments as high as he could practically achieve, with the Egyptian pyramids as the best example. The largest of these monumental tombs of the Pharaohs is the Pyramid of Cheops which was constructed by thousands of slaves who piled huge masonry blocks one on top of the other to a peak of 481 ft—equivalent to a 50-story building.

Through the centuries there were two basic materials used in construction: wood and masonry. Neither of these two materials had qualities that were suited to reaching more than a few stories. Wood lacked the strength for buildings more than 50 to 60 ft., and any tall building of wood presented a fire hazard.

Masonry, on the other hand, had the strength and was fire resistant, but had, however, the drawback of weight. The mass of masonry is too great, thus putting too much weight on the lower supports of a tall structure. In the pyramids there is little usable space, since most of the area of the lower levels is used to support the parts above.

Although masonry construction reached its peak in 1891 when the 17-story, 210 ft high Monadnock Building was constructed in Chicago by the architects Holabird and Root, the walls in the ground floor were up to 7 ft thick. The walls of the load bearing masonry structure needed the thickness to provide the strength required to support the building above and to resist the overturning due to wind, since the walls had little help for wind resistance from other parts of the building.

*Director, Engineering Services Department, Portland Cement Association, Skokie, Ill.

A wide utilization of multistory buildings for practical purposes had to wait until many of the technical problems which blocked the commercial utilization and the construction of multistory buildings were solved. The most important of the technical developments were efficient and safe vertical elevators and construction materials with a higher strength.

The socio-economic problems that came with the industrialization of the nineteenth century and the insatiable need for space in the cities gave the big impetus towards high-rise construction. By the latter part of the nineteenth century, land in the rapidly developing cities was already so scarce that the only way was to go up.

The first modern multistory building using a steel skeleton was constructed in 1883 in Chicago. It was the ten-story Home Insurance Building utilizing steel I-beams and masonry walls. The big step forward in this building was that the weight of the outer shell was not carried on the masonry but was supported on the steel I-beams.

By the turn of the century the downtown business section of New York had established itself as the nation's financial center, and there was a great demand for office space around Wall Street. The first skyscraper at the turn of the century was the 20-story steel skeleton Battery Place Building, followed by the 21-story Trinity Building, and further uptown, the 21-story triangular Flatiron Building was constructed. Other cities, too, had their skyscraper beginnings at that time; in Cincinnati, the Ingalls Building rose to a height of 16 stories.

In 1913 the 60-story, 792 ft high Woolworth Building in lower Manhattan was completed. A product of the soaring imagination of architect Cass Gilbert and financing by Woolworth, this gothic cathedral style building with a high-rise tower remained the queen of New York's skyline for two decades. After 60 years of service and some modernization (self-service elevators and air conditioning) the Woolworth Building is still as vigorous and useful as on the day it was completed. Concrete was used extensively for its foundations and in the floor slabs.

World War I slowed down the progress in high-rise construction and only after the War's conclusion did a new and larger period of high-rise buildings start which continued through the 1920s and into the 30s. This period produced many excellent structures such as:

the 67-story 60 Wall Tower in New York,
the 71-story Bank of Manhattan Building in New York,
the 77-story, 1046 ft Chrysler Building in New York,
the 33-story Tribune Tower in Chicago,
the 32-story Saving Fund Society Building in Philadelphia.

In this period the tall building became a symbol of American development in the eyes of the world. Also at home, every new skyscraper became a source of amazement and pride.

The crowning glory of this 'golden age' of the skyscraper was, obviously, the Empire State Building in New York, which was announced on August 30, 1929. The 102 stories rise to a height of 1250 ft, and with its modern appendage, the TV antenna, now has a height of 1472 ft. The building used 60,000 tons of steel and 62,000 cubic yards of concrete. Concrete was used for the floor slabs, staircases, and other places where strength, stiffness and fire protection were needed.

Also, the Rockefeller Center, a new concept of a city within a city, started in this period—still during the Depression. The Center contains the towering RCA Building, with its 70 stories, a number of other high-rise office and commercial buildings, and the world's largest movie house.

The Depression put an end to the great era of skyscrapers, and only in the late 1940s in the wake of World War II did a new era of modern high-rise building set in again.

Breaking with tradition, the Secretariat Building of the United Nations completely departed from the lofty towers of the early period, creating a slab type tower with walls of glass.

As reinforced concrete penetrated the construction field at the turn of the century, its use in high-rise concrete structures also became more widespread. Although prior to World War I high-rise structures were mostly in the domain of structural steel, the foundations and, at times, floor slabs were concrete. After World War I, reinforced concrete multistory structures appeared sporadically, mostly for loft buildings using flat slabs with column capitals and in a very few instances for apartment buildings up to 12–14 stories in height. The structural systems of reinforced concrete buildings of those times were developed as an imitation of steel structures and consisted of columns, girders, beams, and slabs.

A quiet revolution was taking place in the field of concrete buildings. A group of reinforced concrete apartments, 12 to 14 stories high, constructed in the early '40s in Brooklyn, known as the Clinton Houses, used a beamless type of floor system: the flat plate. The Clinton Houses were a radical departure and set the pattern for future development of the flat plate construction. At the same time shear walls were introduced as an economical efficient bracing system for multistory buildings. Both of these elements—the flat plate and shear walls—became the major structural systems in all residential buildings of any height.

Also, during the post World War II period, and especially in the '50s, concrete structures developed a personality of their own, reflecting the properties inherent in the material, such as its strength, moldability, fire-safety, handsome appearance, and weather resistance. Contributing to this trend was the widespread use of flat-plate slabs, which utilize two-dimensional load transfer instead of the traditional one-way action. Also, many reinforced concrete buildings of this period abolished the conventional boxlike form and developed new concepts. An example of a structure with a strongly pronounced "concrete personality" is the Marina City complex in Chicago, Ill. (Fig. 10-1) which, with its sculpture-like appearance, is a milestone in the development of concrete high-rise structures.

As a result of new technology of materials, new construction methods, and bigger cranes, by the early fifties, reinforced concrete buildings reached 20–22 stories; in 1958 they went to 38 stories, and in 1962 the Americana Hotel in New York City rose to a 50-story height—while in the same year Marina City in Chicago rose to 60 stories.

In 1968 the 70-story Lake Point Tower apartment building in Chicago was completed as the tallest building utilizing the economical flat plate-shear wall combination.

The fifties and sixties witnessed an unprecedented surge in concrete high-rise buildings which soared rapidly from record to record.

The development of framed tube type buildings applied first in the 43-story Chestnut De Witt apartment building in Chicago in 1965 and the 38-story CBS building in New York in 1965 was followed by the tube-in-tube concept used in the 38-story Brunswick Building in Chicago, and many more buildings since, including the 52-story One Shell Plaza in Houston, Texas, completed in 1966.

In the field of steel structures, the late sixties brought the diagonally braced 100-story John Hancock Building in Chicago; the twin 110-story World Trade Center in New York City; and the crowning glory of the steel high-rise structures. the 120-story Sears Tower in Chicago, which

TABLE 10-1 Partial List of Concrete Framed Buildings 400 Feet and more in Height (Unofficial)

	Height Feet	*Stories*	*Year Built*	*Framing for Lateral Resistance*	*Type*	*Name and Location*
1.	850	77	u. c.	Multicell-tube	Apts-hotel commercial	Water Tower Inn Building, Chicago, Ill.
2.	760	62	u. c.	Shear wall-frame interaction	office	MLC Center, Sydney, Australia
3.	714	52	1970	Tube-in-tube	office	One Shell Plaza, Houston, Texas
4.	660	63	1962	Coupled shear walls	office	Edificio Puegeot, Buenos Aires, Argentina
5.	660	30	1972	Shear wall-frame interaction	office	Carlton Centre, Johannesburg, South Africa
6.	645	70	1965	Reinforced concrete core & flat plate	apts.	Lake Point Tower, Chicago, Ill.
7.	640	55	1964	Shear wall-frame interaction	apts.	1000 Lake Shore Plaza, Chicago, Ill.
8.	624	47	1964	Shear wall-frame interaction	office	Place Victoria, Montreal, Can.
9.	612	50	1973	Shear wall-frame interaction	office	A.M.P. Society II, Sydney
10.	610	49	1973	Tubular ex. bearing wall	office	Tour Fiat, Paris, France
11.	605	50	u. c.	Shear wall-frame interaction	office	Quantas, Sydney, Australia
12.	600	45	u. c.	Shear wall-frame interaction	office-hotel	Collins Tower, Melbourne, Australia
13.	600	45	u. c.	Shear wall-frame interaction	office	ANZ Tower, Melbourne
14.	590	50	1971	Composite (concrete tube steel interior)	office	One Shell Square, New Orleans, Louisiana
15.	588	60	1962	Shear wall-frame interaction	apts-park'g	Marina City, Chicago, Ill.
16.	582	50	1972	Shear wall-frame interaction	office	Mid-Continental Plaza, Chicago
17.	580	51	u. c.	Shear wall-frame interaction	office	Nauru House, Melbourne, Australia
18.	562	45	1968	Shear wall-frame interaction	office	Australia Square Tower, Sydney
19.	560	58	1972	Shear wall-frame interaction	apts-commercial	Newberry House, Chicago
20.	546	42	1972	Shear wall-frame interaction	office-school	Central Manhattan, New York
21.	542	42	1973	Slipformed conc. core, steel frame exterior	office	CB 21, Paris, France
22.	540	51	1971	Coupled shear walls	residential-commercial	Manufacturer's Life Center Toronto, Canada
23.	533	55	1973	Shear wall-frame interaction	apts.	Frontier Towers, Chicago, Ill.
24.	525	42	u. c.	Shear wall-frame interaction	office	Center Square Complex West Tower, Philadelphia, Pa.
25.	520	57	1972	Shear wall-frame interaction	apts.	111 E. Chestnut, Chicago, Ill.
26.	518	40	1968	Shear wall-frame interaction	office	964 3rd Ave., New York City
27.	515	42	1972	Shear wall-frame interaction	office	Royal Centre Office Bldg. Vancouver, B.C. Canada
28.	507	34	1946	Frame	office	Branco de Estado, Sao Paulo, Brazil
29.	500	50	1962	Coupled shear walls	hotel	Americana Hotel, New York City
30.	496	45	1972	Shear wall-frame interaction	hotel	Landmark Hotel, Vancouver, B.C.
31.	491	38	1964	Tube-in-tube	office	CBS, New York City
32.	483	47	1966	Shear wall-frame interaction	apts.	Excelsior House, New York City
33.	478	38	1964	Tube-in-tube	office	Brunswick Bldg, Chicago
34.	471	42	1969	Shear wall-frame interaction	apts.	Broadway & 64th St. New York City
35.	469	32	1968	Shear wall-frame interaction	office	FDR PO & Tower, New York City
36.	467	40	1970	Shear wall-frame interaction	office	1250 Broadway, New York City
37.	465	38	u. c.	Shear walls at service core	office	Project 200 Office Bldg. Vancouver, B.C., Canada

TABLE 10-1 (Cont'd)

	Height Feet	*Stories*	*Year Built*	*Framing for Lateral Resistance*	*Type*	*Name and Location*
38.	461	43	1970	Shear wall-frame interaction	apts-hotel	36 Central Pk. So., New York City
39.	459	40	1971	Suspended Structure	office	Standard Bank, Johannesburg, S.A.
40.	456	40	u. c.	Shear wall-frame interaction	office	Miami, Florida
41.	450	51	1968	Shear wall-frame interaction	apts.	Park Regis, Sydney, Australia
42.	450	38	1967	Shear wall core, steel external columns	office	State Government Office Block Sydney, Australia
43.	450	34	1968	Shear wall-frame interaction	office	Main Place, Dallas, Texas
44.	443	45	1956	Frame	office	Italia Building, Sao Paulo, Brazil
45.	440	46	u. c.	Shear wall-frame interaction	apts.	Harlem River, New York City
46.	440	40	1972	Exter. frame & core	office	LaSalle Plaza, Chicago, Ill.
47.	437	35		Shear wall-frame interaction	office	St. Martin's Tower, Sydney, Aust.
48.	430	42	1971	Shear wall-frame interaction	apts.	Jerome Ave., New York City
49.	430	38	1971	Shear wall-frame interaction	apts.	Madison & 89th St. New York City
50.	430	37	u. c.	Shear walls at service core	office	2 Illinois Center, Chicago
51.	430	38			office	North Point, Sydney, Australia
52.	430	30		Framed tube	office	Xerox Building, Rochester, N.Y.
53.	429		1973	Shear wall-frame interaction	office	Canoco Building, Houston, Texas
54.	428	30	1970	Shear wall-frame interaction	office	National Life Bldg., Nashville, Tenn.
55.	426	42	1971	Shear wall-frame interaction	apts.	East End Ave., New York City
56.	425	45	u. c.	Shear wall-frame interaction	gar'g-apts	Colony North, New Jersey
57.	422	31	1963	Shear wall-frame interaction	office	L.T.V. Tower, Dallas, Texas
58.	420	41	1965	Shear wall-frame interaction	office-apt.	860 UN Plaza, New York City
59.	420	36	1970	Shear wall-frame interaction	office	111 Wacker Drive, Chicago
60.	420	31	1969		office	Laclede Gas Bldg., St. Louis, Mo.
61.	418	42	1967	Shear wall-frame interaction	apts.	2626 Lakeview Av., Chicago, Ill.
62.	418	31	1958		office	110 William St., New York City
63.	417	24	1963		office	Nonalco (Arrowhead) Mexico City
64.	414	32	1955	Frame	office	Pirelli Building, Milan, Italy
65.	411	42	u. c.	Shear wall-frame interaction	apts.	3rd & 92nd, New York City
66.	410	42	1957		apt-office	Atlas Building, Buenos Aires
67.	410	40	1962	Shear wall core	apts	Carlyle Apts, Chicago
68.	410	39			apts.	Demetrio Eliades, Buenos Aires
69.	410	36		Shear wall-frame interaction	hotel	Lanray, Sydney, Australia
70.	410			Shear wall	educational	Institute of Technology, Sydney
71.	409	43			housing	Barbican Redevelopment, London
72.	409	38	1970	Shear wall-frame interaction	apts.	1st Ave. & 55th St., New York City
73.	408	35	1971	composite (conc. tube-int. steel framing)	office	Gateway III, Union Sta., Chicago
74.	407	33	1965		office	Intern'l Trade Cent. New Orleans
75.	405	40	1964	Shear wall-frame interaction	apts.	Outer Drive East, Chicago
76.	405	33	1972	Shear wall-frame interaction	office-theatre	Broadway & 44th. New York City
77.	404	27	1973	R/c core-peripheral ductile space frame	office	Bentall Centre, Tower Ill Vancouver, B.C. Canada
78.	402	40	1962	slab and shear walls	apts	1300 N. LakeShore Dr. Chicago
79.	402	39	1955	shear wall	apts.	FOCSA Bldg., Vedada, Cuba
80.	400	44	1965	shear wall-flat plate	apts	1130 S. Michigan, Chicago

TABLE 10-1 (Cont'd)

	Height Feet	Stories	Year Built	Framing for Lateral Resistance	Type	Name and Location
81.	400	44	1966	Shear wall-flat plate	apts	East Point Apt., Chicago
82.	400	43	1965	Framed tube	apts.	DeWitt Chestnut Apt., Chicago
83.	400	43	1966	Shear wall	apts-office	Brooks Tower, Denver, Colorado
84.	400	40	1969	Shear wall	apts.	1020 Rush St., Chicago
85.	400	40	1966	Shear wall-frame interaction	apts.	Edgewater Twin Towers, Chicago
86.	400	33	1970	Shear wall-frame interaction	office	1010 Commons, New Orleans
87.	400	30	1969	Shear wall-frame interaction	office	Time-Life Bldg., Chicago
88.	400	30	1972	Shear wall-frame interaction	hotel	Carlton Centre Hotel, Johannesburg

Fig. 10-1 The 60-story Marina City Twin Towers built in Chicago in 1962.

utilizes bundled tubes as the structural system, soaring to a world record of 1450 ft.

Table 10-1 gives a partial list of concrete buildings taller then 400 feet all over the world.

The tallest reinforced concrete building now under construction is the 77-story Water Tower Inn Building in Chicago, which will be 850 ft tall when completed. It will have a 63-story residential and hotel tower on top of a 14-story base of commercial space. The wind resistance of the tower is supplied by the exterior framed tube stiffened by a shear wall connecting the two longitudinal faces.

The shear wall or diagonally braced structure (similar to the John Hancock) seems to have good technical and economical potential to reach heights in excess of 100 stories.

10.3 CONTRIBUTING FACTORS TO CONCRETE HIGH-RISE BUILDINGS

The tremendous blossoming of reinforced concrete multistory construction in the past several decades has been attributable in large part to three major factors which are only briefly summarized below, since a complete description of these factors appears in other chapters throughout the Handbook: (1) development of high-strength materials; (2) development of new design concepts; (3) development of new structural systems; and (4) improved construction methods.

10.3.1 High-Strength Materials

Only 15 years ago, concretes of 2,000 to 3,000 psi cylinder strength at 28 days were commonplace. Today, in multistory construction for lower story columns and shear walls, a strength of 5,000 psi is seen on most jobs, and occasionally up to 7,000 psi. A compressive strength of 9,000 psi at 90 days has already been used in a 50-story building in Chicago. In some precasting plants, strengths up to 10,000 psi are achieved, primarily to speed the casting cycle to achieve lower production costs.

The development and acceptance of structural lightweight concrete has helped overcome some design difficulties of multistory structures. Dead load savings of up to 20% are important to critical column size, in earthquake design, and where poor soil conditions limit the weight of the structure. Also, the use of lightweight concrete often proved more economical than normal weight concrete. The moderately higher price for lightweight concrete usually is more than offset by a saving in the slab and column reinforcement, and by savings in the cost of foundations.

Development of reinforcing bars is continuing in several ways. Steels with yield strengths of 60 and 75 ksi were developed and put on the market. Their use in multistory structures started with the introduction of ultimate strength design for the first time in the *ACI Building Code* of 1956, and is gaining steadily. About two-thirds of all reinforcing steel used today has a yield strength of 60 ksi. High steel stresses utilized in structures designed by this method result in considerable savings in the amounts of reinforcing steel per square foot of floor area.

The use of prefabricated welded wire fabric and welded reinforcing mats has helped save time and labor on slabs, shells, walls, and other plate-type construction.

The use of bundled bars, and the introduction of large size bars—#14S and #18S with diameters of 1.70 and 2.26 in. respectively—help to satisfy spacing requirements, particularly in heavily reinforced members. A further help in the same direction is the use of column splices by welding and couplings. Several kinds of couplings are on the market. Some have only compression transfer capacity, while others are good also for the transfer of tensile forces.

10.3.2 New Design Concepts

Cast-in-place reinforced concrete structures, by virtue of their monolithic character, are designed with consideration of continuous frame action. This results in a more economical design than in structural systems in which most of the members are designed as simply supported.

Two recent developments are especially worthy of mention: ultimate strength design, and limit design. In USD, which has become the major design method of the *1971 ACI Code* under the name of "strength design," the ultimate capacity of a section is computed to correspond to the service load increased by a selected factor of safety. Limit design considers the behavior of a structural assembly, as it approaches its ultimate load capacity.

Although many developments of limit design methods have been advanced by researchers in recent years, the possibility does not appear that in the near future a formalized limit design method will be developed for high-rise reinforced concrete structures. The moment redistribution of up to 20% allowed in the *ACI Code* (to be made after an elastic analysis is carried out) is a partial limit design.

In both methods, (strength design and limit design) the inelastic behavior of materials is considered rather than considering only elastic behavior according to the classical working stress design theory, the elastic analysis of structures. These new methods help to close the gap that exists between strength computations and actual test results.

The above design developments permit a better utilization of materials, through methods that more closely reflect the actual behavior of the materials and of the structure in which they are used. They allow a more uniform factor of safety throughout the structure than is possible with working stress design and elastic analysis.

10.3.3 New Structural Systems

Concrete started off as an imitation of structural steel systems, and the early buildings consisted of columns, beams, and slabs. It can be compared to the early stages in the automotive industry when the motor cars were patterned after the horse drawn buggies.

The development and popularization of the flat plate is an important contributor to the wide spread use of multistory reinforced concrete apartment buildings. Its major technical advantages are a smooth, beamless ceiling which requires only a skim coat of plaster to provide a perfect finish; and reduced story height, which permits an additional story every 10 stories as compared with structural systems using beam and slab construction. The simplified formwork required for the flat plate results in an efficient job operation. No other structural system can offer all these technical and economic advantages to residential high-rise buildings. In recent years, casting a floor every second day in tall structures has not been uncommon in metropolitan areas.

The introduction of shear walls as a bracing element in the late 1940s in housing projects in New York had an important impact on concrete high-rise buildings. Fulfilling simultaneously the functions of a load carrying element, a wall within or between apartments, and an efficient bracing element, it became the most important element for all multistory buildings. Today there is rarely a building of more than 15–18 stories without shear walls as its major bracing element.

Utilization of shear panels—one or several story shear walls—scattered throughout the plan and shifted in location to offer architectural flexibility while supplying sufficient rigidity is another way to reduce cost and adapt to the functional requirements.

The staggered wall-beam system developed in the late 1960s is another innovation suitable for residential buildings. Although only a number of high-rise buildings have been built with this system, its advantages of large unobstructed areas in the typical floor, and column-free areas under the building for parking, may eventually lead to a broader application.

Finally, the introduction of the framed tube, a peripheral system of closely spaced exterior columns, interconnected with spandrel beams, creates a highly rigid structure for tall buildings. A further step is the tube-in-tube resulting from adding a core element to the framed tube.

In the explosive development of the high-rise structural systems, which is now only in its infancy, it appears safe to predict that many other new and revolutionary concepts will be developed by architects and engineers in the years ahead.

10.3.4 Improved Construction Methods

Improved construction techniques, a few of which are mentioned below, and improved equipment probably are the strongest contributors to the development of high-rise structures. Improvement of construction methods has provided better technical solutions to many problems and, in most cases, has resulted in reduced construction costs.

Lift-slab construction appeared on the scene in the early fifties. With this method, slabs are cast at the ground level, one on top of the other. Only side forms are necessary. Specially designed collars for column connections are cast in the slab. After the slabs attain the desired strength, they are lifted hydraulically upon the columns (Fig. 10-2) and fastened to the columns after reaching their final position. Buildings up to 15 floors in height have been constructed by this method. Stability of such structures usually is achieved by shear walls, or by reinforced concrete cores. Although the idea of lift slabs seems to eliminate expensive operations of shoring and forming, lift slabs have not attained the widespread utilization warranted by their apparent advantages.

In some cases lift slabs have been designed as post-tensioned, prestressed structures.

Slip-forming is a method of constructing walls in which 3- to 4-ft high forms are raised continuously, while concrete

Fig. 10-2 Lift slab method of construction.

Fig. 10-4 Slipforming of exterior and interior walls of a 25-story building in Milwaukee, Wisconsin (Bay Shore Apartments).

is placed between them. Progress of 10 to 12 in. per hour is regarded as average today. This method, originally developed for silos and bins, has been used recently on many high-rise structures in which the cores were slip-formed (Fig. 10-3). In the example shown in Fig. 10-4 a 25-story apartment building in Milwaukee, Wisconsin was built using slip-forming for the exterior and interior walls, at a speed of a story per day. The floor slabs followed three stories behind at the same speed, consisting of precast beams and slabs cast-in-place over nonrecoverable plywood forms. Even the sculptured balconies were made by the slip-form sliding past a stationary sculptured plastic form (Fig. 10-5).

Two advantages result from slip-forming: saving of time and cost. However, to satisfy an economical slip-forming operation, a certain minimum wall height is required, and this appears to be only 35 feet.

Fig. 10-3 Slipforming of a core for a 20-story building.

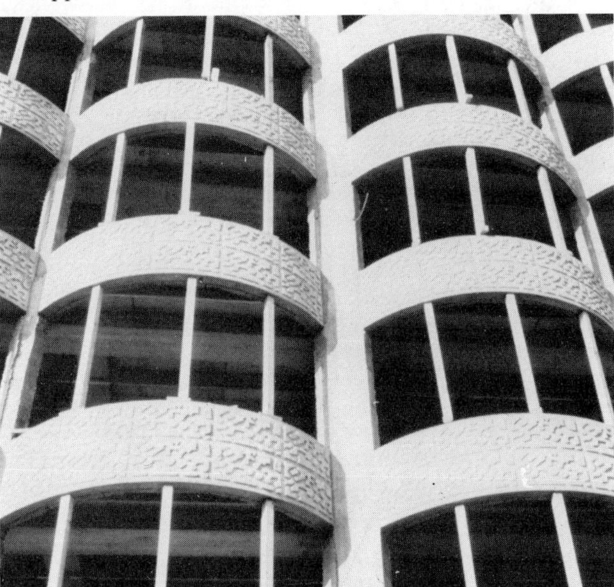

Fig. 10-5 Balconies in the Milwaukee building.

Fig. 10-6 Construction utilizing slipforming and lift slabs.

A combination of slip-forming with lift slabs has a good potential for lowering the cost of multistory buildings. Figure 10-6 shows an apartment building in San Francisco with three slip-formed cores and lift slab floors.

Plastic lined forms are another innovation that provides smooth concrete surfaces, and more reuses of formwork. Also, fiber glass reinforced plastic forms provide an excellent concrete surface, and make possible the molding of the most complicated shapes. Such forms were still in perfect condition after 60 reuses each on the Marina City project (Fig. 10-7). An additional advantage is that no finish was required for column surfaces. Fiber glass forms are particularly advantageous where complicated shapes are to be executed, or where the same shape is to be repeated many times.

The use of prefabricated members, which has grown rapidly in the bridge, housing and small structures field, has entered the field of multistory buildings. The tendency has been to bring production work into a plant where weather is immaterial, and working conditions are stable. However, precasting on or near the site has also found favor because of the greater quality control possible, and the savings possible through repetition in production and avoiding the expensive formwork. Precast construction combines the advantages of reinforced concrete, such as weather, fire, and corrosion resistance, with the advantages of prefabrication, shorter construction time, and ease in obtaining occuracy in dimensions through an industrialized process.

Fig. 10-7 Reinforced fiber glass forms (Marina City).

In recent years the developments progressed to the point of precasting entire rooms or apartments, including almost all finishes, as in the example of the Palacio Del Rio in San Antonio, Texas (Fig. 10-8) thus leaving for the field only the fastening together of such units and connecting utilities. The majority of prefabricated multistory structures, however, consist of large panel components (slabs and walls) or systems in which prefabrication has included primarily the

Fig. 10-8 Palacio Del Rio, San Antonio, Texas.

Fig. 10-9 Large panel structure.

structural frame and exterior envelope, the finishes (some also prefabricated) being applied in the field. Figures 10-9 and 10-10 show structures consisting entirely of large precast elements except for cast-in-place cores and shear walls.

Between the two types of structural systems, fully prefabricated skeleton and the cast-in-place structure, there is a wide variety of composite structures, in which prefabricated elements (thin slabs or shells of columns) are used as formwork, sometimes containing the main reinforcement. The cast-in-place part ranging from about 20 to 80%, ties the structure together into a monolithic unit.

Sucessful examples of such buildings are the many buildings in Hawaii (like the 27-story Ilikai Hotel, Fig. 10-11) and the 27-story Gulf-Life Building in Jacksonville, Florida (Fig. 10-12).

Another important factor contributing to the growing use of concrete for multistory structures is improvement in construction equipment, such as hoists, cranes, self-rising cranes, and concrete mixing, conveying, and pumping equipment. This subject is discussed in detail in the chapter "Construction Methods and Equipment."

10.4 TYPES OF OCCUPANCY

The function of a building has a considerable effect on the selection of the structural type. As a result of the particular function of the building, such considerations as the size of the open space required, mechanical services (in particular, the forced air distribution system), and the need for future flexibility all enter the picture when the most adaptable structural framing system is chosen.

With regard to type of occupancy, the following categories can be distinguished as specifically affecting the structural systems of tall structures:

1. residential buildings, including apartment buildings, hotels and dormitories
2. office buildings and commercial space
3. mixed occupancy–commercial and residential
4. industrial buildings and garages.

10.4.1 Residential Buildings

Residential buildings (apartment buildings, hotels, and dormitories) are characterized by the presence of partitions which are designed during the planning stage, are constructed with the progress of the structure, and usually remain in the same location for the life of the building. This is obviously the present method of construction. It is possible that in the future apartment buildings will be

Fig. 10-10 Large panel structure.

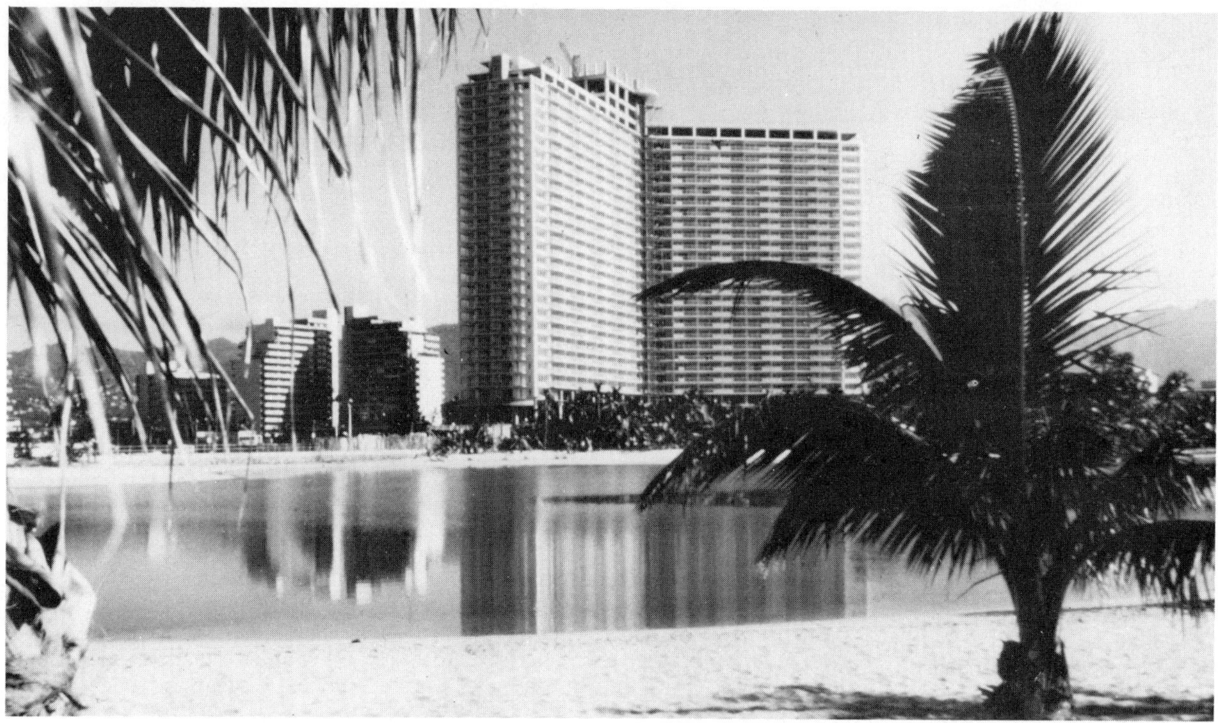

Fig. 10-11 The 27-story Ilikai Hotel in Honolulu, Hawaii.

Fig. 10-12 Gulf-Life Building in Jacksonville, Florida.

designed with movable partitions which may then change our structural philosophy.

-1. Structural systems for residential buildings—The presence of permanent partitions allows the column layout to correspond to the architectural plan. The present architectural layouts prevalent in the United States can be handled with spans up to 20–22 ft. This also includes the category of luxury apartments. It is worth noting that in Western Europe the average spans in residential occupancy are in the range of 12–15 ft.

Among mechanical systems affecting the structure, distribution of air for heating and cooling has the strongest influence on the selection of the structural system. In residential buildings, due to permanency of partition location, air is handled in vertical ducts from which a room or several rooms are fed using short horizontal stubs through the walls from the vertical ducts. Such air supply arrangement does not require ducts under the ceiling except sometimes in vestibule areas, thus making is possible to use the structural slab as the ceiling. As a result, the flat plate has become the most adaptable slab system for residential construction. The smooth ceiling, which needs only a skim coat of plaster, the possibility to have the columns in scattered locations to suit the architectural layout, and the absence of beams, thus giving absolute freedom to the partitions layout, has made the flat plate the slab system that allowed economical residential buildings up to 70 stories high.

Experience has shown that in tall buildings the best economy is achieved with the thinnest slab, as the slab weight also affects the load on columns and on foundations.

The typical modern apartment slab type building with a double loaded corridor is 50–60 feet wide, uses three transversal spans, and requires a 7- to 7½-in. flat plate. Customarily, hotels also have the same transversal three-span layout while in the longitudinal direction the columns are located about 24 ft apart to accommodate two rooms. The 24 ft span may require a minimum slab thickness of 8 in.

When one-way joist systems are used for residential occupancy, (usually with somewhat larger spans) a hung ceiling provides a smooth finish. Since the space between the webs is generally not needed for mechanical purposes in residential construction, the joist system is economical only if the cost of the hung ceiling is offset by economies of the joist slab against other slab systems.

Column size in residential buildings is usually not a critical design consideration (except for lobbies and public areas) since they can be hidden in closets, kept in corners, blended into the architectural layout and in the extreme case, utilized as walls. Experience shows that in tall residential buildings columns with the lowest percentage of reinforcing result in the best overall structural economy. If larger columns are acceptable to the architect, they not only are economical but also add substantially to the rigidity of the structure.

The lateral resistance of residential buildings is provided mostly by the frames in buildings up to 15–18 stories. Since flat plate slab systems are predominant, spandrel beams around the periphery carry a substantial part of the lateral forces. In recent years spandrel beams have been eliminated in the majority of apartment buildings to accommodate the use of the more economical flying forms. In such buildings, shear walls have been introduced to provide the required rigidity. Residential buildings above 20 stories, except for shear wall buildings, always have a shear wall-frame interactive system for resistance to lateral loads. The tallest building in this category is the 70-story Lake Point Tower (see Fig. 10-13) which has a shear wall-frame system in which an extremely large core resists the bulk of the wind forces and thus the resistance provided by the frame amounts to an insignificant portion.

Fig. 10-13 The 70-story Lake Point Tower in Chicago.

A major objective of the design engineer is to devise a structural system which is designed for vertical loads only while the resistance to lateral forces is provided by the 33% increase in allowable stresses. This can be achieved if the shear walls carry a sufficient amount of vertical loads, so that the overturning moments due to lateral loads do not result in substantial net tensile forces. Tensile resultants in shear walls cause serious difficulties at the foundation level, since they have to be either anchored into the ground, or the weight of adjacent columns has to be mobilized to overcome the tensile forces by designing very heavy girders to transfer the loads. The ideal case is created when all the gravity load of the building is used to counteract the overturning moment—which is possible usually only in shear wall buildings. The practical optimum case is to have sufficient shear walls to limit the resultant tension under gravity plus wind to an insignificant value. Such approach should result in no premium for height in structures. Most of the well engineered concrete buildings above 40 stories have been in this category.

10.4.2 Office Buildings

Office buildings are characterized by the absence of partitions during design and construction, since office space is designed rather than offices, with subsequent partitioning to accommodate the needs of a particular tenant. The partition layout usually changes when tenants change. This procedure requires flexibility in the supply of heating, cooling, power, and telephone connections to the entire area which can be used as office space. As a result, many services are usually carried vertically within the service core, and a distribution network of air ducts, telephone, and power supply lines is carried horizontally under the structural slab to distribute the services over the entire office space. Since a hung ceiling is required to cover up the ducts and lines, the structural slab does not require a smooth ceiling and the slab system can be selected for structural efficiency and economy. The flat slab with drop panels is the prevalent system for spans up to 23–25 ft, due to its low overall thickness and structural efficiency. For two-way spans in the range of 30–35 ft, the waffle slab provides good efficiency, either as a flat plate type with a solid portion around the columns or as a two-way slab with solid beams on the column lines in both directions of the building; such slabs have been used for spans up to 60 ft. The joists or waffle slabs are either 2 or 3 ft on centers, with 5 ft on centers introduced recently. The waffle slab provides a very attractive ceiling where a hung ceiling is not required.

In recent years many of the prestigious office buildings have been built without interior columns—the structure consisting of an interior core and an exterior peripheral tubular system of closely spaced columns interconnected with spandrel beams. The column-free span between core and periphery ranges between 35 and 40 ft. The Brunswick Building shown in Fig. 10-14 is a typical example of this building type. The slab system consists of one-way joists connecting the core with the periphery, while the corners

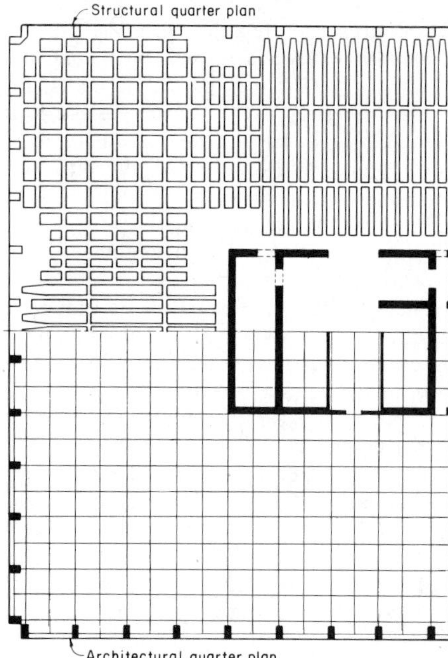

Fig. 10-14 Half of the typical floor of the Brunswick Office Building in Chicago (structural and architectural).

are two-way joists (waffle slabs). It is desirable to have a correlation between the module of the column spacing and the spacing of the joists.

-1. Structural systems for office buildings—With the increasing height of buildings in the fifties and sixties, new concepts evolved to economically provide resistance to lateral forces due to wind and in many cases also due to earthquake. This evolution is shown in Fig. 10-15.

As can be seen in the figure, in buildings up to 15 stories, frame action usually suffices to provide lateral resistance, except for very slender or unusual buildings. The 33% increase in allowable stresses provided by codes is in most cases sufficient to accommodate the wind forces and moments.

When we approach the 20-story height, the rigidity of frame buildings is mostly insufficient and sway due to wind may begin to control the design. Introduction of shear walls which interact with the frame (see section 10.5.7) materially increases the total rigidity of the building beyond the sum of the two individual components. With proper spandrel beam exterior frames such shear wall frame interactive systems can be used for office buildings of up to 50 stories if a large enough core is utilized.

The framed tube, first introduced in the sixties, consists of a closely spaced grid of exterior columns, connected with beams. It is an efficient system to provide lateral resistance without premium for height in buildings without interior columns starting at 30-35 stories. The efficiency of this system is derived from the great number of rigid joints acting along the periphery, creating a large tube. The framed tube represents a logical evolution of the conventional frame structure, possessing the necessary lateral stiffness with excellent torsional qualities while retaining the planning flexibility which isolated interior columns allow.

To visualize the action of a framed tube, it is simple to start with a solid peripheral wall (a solid tube) which obviously will act as a cantilever with a moment deflection. When the wall is penetrated with small round openings (windows) it will still behave as a cantilever. When the openings become larger and rectangular, instead of round, part of the lateral forces are resisted by shear distortion of the columns and beams, and only the rest by moment (cantilever) deflection of the tube.

The ratio of moment deflection to shear deflection depends on the stiffness relationship between beams and columns. The two sides parallel to the wind act as frames while the windward and leeward sides have a considerable drop off in participation. The rate of shear lag (drop-off from the corner column) depends on the beam-to-column stiffness ratio and the number of floors. A typical distribution of column axial forces in such a structure is shown in Fig. 10-16. The effect of shear lag will readily be noted. With increasing beam stiffness and increasing number of stories, a higher participation of the windward and leeward sides may bring the framed tube closer to a rigid tube. In addition, the framed tube has an unusually high torsional resistance owing to the location of the stiffness around the periphery.

When the sway or wind stresses begin controlling the design (it may be around 40 stories), the framed tube is supplemented by a core, to create the tube-in-tube system which is essentially a shear wall-frame interactive system with all its advantages. The tallest tube-in-tube is the 714 ft

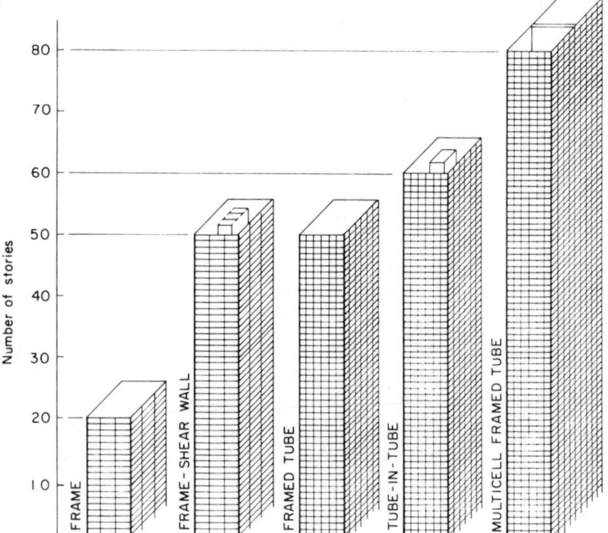

Fig. 10-15 Structural systems for various height office buildings.

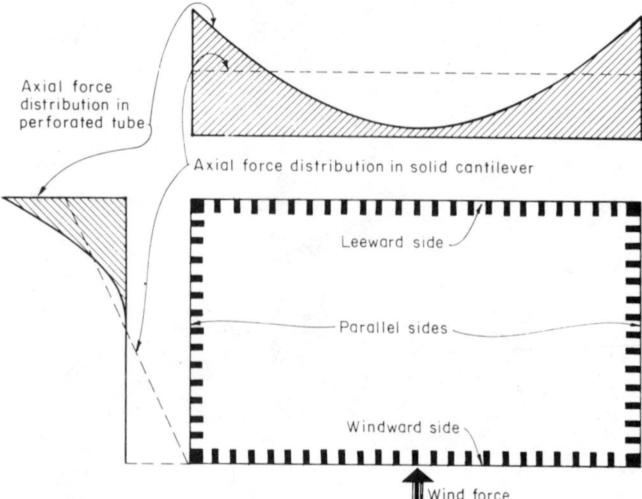

Fig. 10-16 Shear lag in the windward and leeward sides of a framed tube.

Fig. 10-17 The 714-ft tall, 52-story Shell Oil Plaza Building in Houston, Texas.

tall, 52-story Shell Oil Building in Houston, Texas (see Fig. 10-17).

When we reach 70–80 stories, the tube-in-tube may no longer be a sufficiently rigid system, and the exterior tube must be made rigid to act in 100% flexure as a cantilever. Either a set of diagonal members within the peripheral tube creates an exterior cantilevered truss system, or interior connecting shear walls act as webs to tie the opposite faces of the tube into a single unit.

Although the use of diagonals in the exterior tube has not yet been applied in a reinforced concrete building, their use has been investigated in a research project at the Illinois Institute of Technology in a 115-story building 1450 ft high. The thorough study[10-1] investigated all the architectural, structural, mechanical, construction and cost aspects of the building with the conclusion that the building is technically feasible at an estimated cost lower than that of the 100-story John Hancock Building, its steel counterpart. The alternate method of creating a rigid tube is a multicell arrangement as shown in Fig. 10-18 through the use of cross shear walls in one or both directions. The 77-story Water Tower Inn in Chicago utilizes a two cell arrangement as shown in Fig. 10-18. Although the dividing shear wall is penetrated by door openings it still provides close to 100% efficiency of the exterior tube.

10.4.3 Mixed Occupancy–Commercial, Residential

Although multistory parking garages under offices and apartment buildings have been used for many years, urban planning and financing considerations recently led to multistory buildings combining such occupancies as: offices and hotels, stores and apartments, theater and apartments, in addition to parking floors under all of these buildings with mixed occupancies. The usual architectural treatment is to

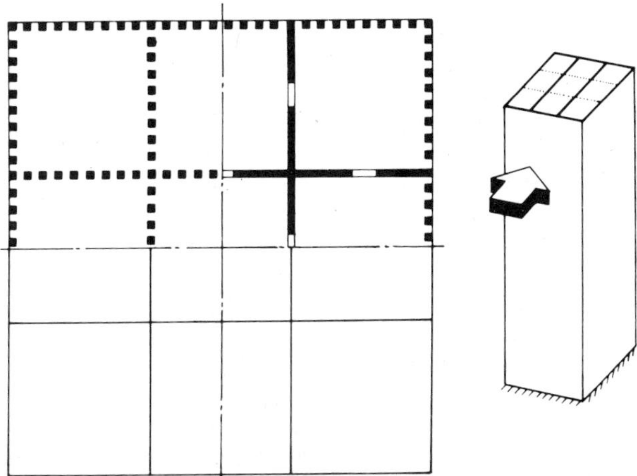

Fig. 10-18 Multicell framed tubes (plan).

place the residential occupancy (apartment or hotel) in the upper part of the building while using the lower part for office floors, department store floors, and parking underground.

Since commercial space requires larger spans than residential occupancy, the planner is faced with the dilemma of either choosing a span suitable for only one of the two occupancies, or to seek the difficult and costly technical solution of a transitional floor between the two occupancies. There are numerous examples of story high transfer girders of reinforced or prestressed concrete picking up columns of multistory buildings up to 50 stories high. In most cases the entire frame above participates together with the transfer girder as a vierendeel girder carrying the load, regardless of whether the designer has considered this or not. The cost of transfer girders is very high and should be weighed against the long term financial advantages offered by the improved function of a structure planned for anticipated occupancy with proper spans.

10.4.4 Industrial Buildings and Garages

Industrial and warehouse multistory buildings are mostly characterized by heavy floor loads, sometimes up to 1000 pounds per square foot of floor area for which the flat slab with column capitals is very well suited. Also, two-way solid slabs supported on beams are efficient for this type of use. Only moderate spans are economically feasible for heavy floor loading.

Multistory parking garages have been using flat slabs with drop panels with or without column capitals extensively. In the 1950s and 1960s, spans of 29–30 feet were widely used since they provide efficient right-angle parking with a 22 ft roadway. Also, one-way joists spanning 50–60 ft supported by shorter spanning beams and beams with one-way solid slabs have been used for parking garages. With the popularization of prestressing in recent years, many garages have been constructed using long-span prestressed, precast elements. Two-way post-tensioning of parking decks and use of shrinkage compensating cements helps to reduce the problem of leakage on the cars on the floor below.

Lateral resistance of industrial buildings and garages is usually provided by frame action, since the number of stories is generally in the lower range. If cores are available in such buildings, they contribute substantially to the lateral resistance.

10.5 RESISTANCE TO WIND LOADS*

In the 1950s and 1960s our buildings underwent a complete transformation in appearance and in the makeup of their components; as a result, their resistance to wind and earthquake forces was completely changed.

Traditionally, the primary concern of the structural engineer designing a building has been the provision of a structurally safe and adequate system to support vertical loads. This is understandable, since the vertical load-resisting capability of a building is the reason for its existence. Any calculations undertaken to check the adequacy of the design with regard to lateral loads were often cursory in nature and more as an afterthought than as an essential and integral part of the total design effort. This attitude did not appear to affect the resulting designs significantly, as long as the buildings involved were not too tall, were not in seismic zones, or were constructed with adequate built-in safety margins in the form of substantial nonstructural masonry walls and partitions.

With the increasing use of light curtain walls, dry-wall partitions, and high-strength concrete and steel reinforcement in tall buildings, the effects of wind loads have become more significant.

Figure 10-19 shows a schematic plot of the lateral resistance (as represented by the lateral load versus story deflection) in a conventional building. It can readily be seen that although the frame may have been designed to provide the lateral resistance, in reality the bulk of the lateral resistance was supplied by the heavy brick walls, stairs, and partitions. It is evident that under these circumstances only a load many times the design wind load (such as may occur in an earthquake) could exhaust the strength of the so-called nonstructural elements, in which case the frame takes over the resistance of lateral loads and eventually yields after reaching the elastic limit at quite large story deflections.

Figure 10-20 shows a schematic plot of the lateral resistance in a modern structure in which the exterior enclosure is a light curtain wall, the partitions are movable, and the stairs are separated from the structure. As a result, the lateral resistance of the building is only slightly larger than that of the frame alone, and no reserve capacity against lateral loads is available.

If, in the past, we would have neglected to consider wind loads in our traditional buildings, they would still probably have behaved well even in an extremely intense wind, due to the enormous overstrength of the heavy cladding. In today's average structure, the sensitivity to wind forces has materially increased, and neglecting wind forces may cause distress. Despite the fact that we have stripped our buildings to the bare skeleton, we may still get away with neglect of wind forces in cases of wide structures, as a result of our still generous factors of safety. However, in our multistory structures of tomorrow, the consideration of lateral resistance will be the prime consideration, since the gradual reduction of member stiffnesses will cause the secondary effects of today to become primary effects of our future structures.

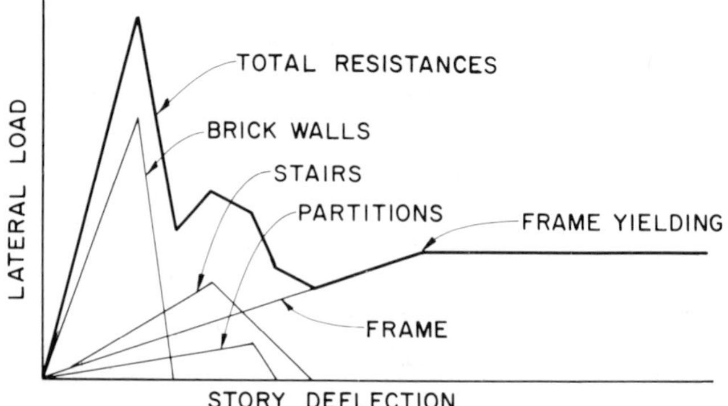

Fig. 10-19 Schematic plot of load versus story-deflection in a conventional building.

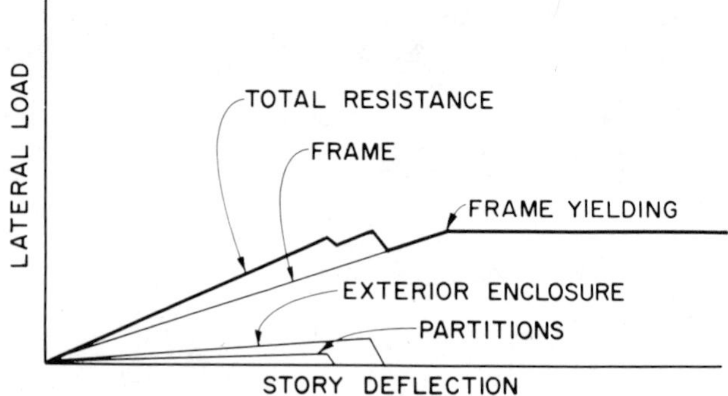

Fig. 10-20 Schematic plot of load versus story-deflection in a modern building.

10.5.1 Wind Forces

Wind is the general word for air naturally in motion which, by virtue of the mass and velocity, possesses kinetic energy. If an obstacle is placed in the path of the wind so that the moving air is stopped or deflected from its path, then all or part of the kinetic energy of the moving air is transformed into the potential energy of pressure. The intensity of pressure at any point on an obstacle depends on the shape of the obstacle, the angle of incidence of the wind, the velocity and density of the air, and the lateral stiffness of the engaged structure.

Under the action of a natural wind, a tall building will be continually buffeted by gusts and other aerodynamic forces. The structure will deflect toward a mean position and will oscillate continuously.

If the wind energy that is absorbed by the structure is larger than the energy dissipated by the structural damping, then the amplitude of oscillation will continue to increase and will finally lead to destruction; it will become aerodynamically unstable.

As discussed previously, the structural forms used today have greater flexibility combined with less mass and damping than those used for traditional structures of the past. These factors have increased the importance of wind as a design consideration. For estimations of the overall stability of a structure and of the local pressure distribution on the building, a knowledge of the maximum steady or time-averaged wind loads is usually sufficient.

The determination of wind design forces on a structure is basically a dynamic problem. However, for reasons of tradition and for simplicity, it has been usual practice to use a quasi-static approach and to treat the wind as a statically applied pressure neglecting its dynamic nature.

The designer's source of information on steady wind loads is usually a building code, but the data contained in such codes are of necessity presented in a generalized form and may not be adequate for a special case, especially if the structure departs from the conventional building forms.

Some of the considerations which enter into the choice of a design wind pressure are:

*Design for earthquake loads is handled in Chapter 12, "Earthquake Resistant Structures."

a. the anticipated lifetime of the structure and its relation to the return period of maximum wind velocity,
b. the duration of gusts,
c. the magnitude of gusts,
d. variation of wind speed with height,
e. angle of incidence of the wind,
f. influence of the ground effect,
g. influence of the architectural features,
h. influence of internal pressures, and
i. lateral resistance of structure.

A comprehensive study of existing information on wind forces was made by an ASCE task committee on wind forces, and the final report[10-2] is a compact source of information which can be of use to the design engineer. Recent studies[10-3] have shown that an equivalent static design wind loading p, can be obtained from the expression

$$p = \frac{1}{2} C_s C_a C_g \, \rho \, V_h^2 \left(\frac{H}{h}\right)^{2/\alpha}$$

where

C_s = coefficient dependent on the shape of the structure
C_a = coefficient dependent upon nearby topographic features
C_g = gust coefficient which is dependent upon the magnitude of gust velocities and the size of the structure
ρ = air density
V_h = basic design wind velocity at height h
H = height above the ground at which p is evaluated, or a characteristic height of the structure
h = height at which base velocity was determined
α = an exponent for velocity increase with height determined by the surface roughness in the vicinity of the site

The quasi-static approach to wind load design has generally proved sufficient for most structures. However, it is unrealistic in several respects, and a more detailed analysis, including wind tunnel studies, may be appropriate for special structures. The above equation for wind may not be satisfactory for ultra high-rise buildings, especially with respect to the comfort of the occupants (in very flexible structures) and the permissible horizontal movement, or drift, which might result in cracking of partitions and glass. These important factors are related to the frequency and amplitude of the vibrations, which depend on the natural frequencies of the building and gust fluctuations of the wind, rather than a steady wind pressure.

10.5.2 Serviceability Criteria

With respect to wind design, the following serviceability conditions have to be considered to ensure the satisfactory performance of the structure under service conditions:

a. lateral deflection of the structure, particularly as this affects the stability and cracking of nonstructural elements and structural members, and
b. motion of structure as it affects comfort of occupants.

-1. Lateral deflection or drift is the magnitude of displacement at the top of a building relative to its base. The ratio between the total lateral deflection to the building height, or the story deflection to the story height, is referred to as the "deflection index." The imposition of a maximum allowable lateral sway (drift) is based on the need to limit the possible adverse effects of lateral sway on the stability of individual columns as well as the structure as a whole, the integrity of nonstructural partitions, glazing, and mechanical elements in the building. No systematic study has yet been published to determine the precise relationship between drift and the above factors. Cracking associated with lateral deflections of nonstructural elements such as partitions, windows, etc., may cause serious maintenance problems (loss of acoustical properties, leakage, etc.). Therefore, a drift limitation should be selected to minimize such cracking.

In the absence of code limitations in the past, buildings were designed for wind loads with arbitrary values of drift, ranging from about 1/300 to 1/600, depending on the judgment of the engineer. Deflections based on drift limitation of about 1/300 used several decades ago were computed assuming the wind forces to be resisted by the structural frame alone. In reality, as mentioned previously, the heavy masonry partitions and exterior cladding common to buildings of that period considerably increased the lateral stiffness of such structures. In contrast, in most buildings that have been constructed in recent years, the frame alone resists the lateral forces. The dry-wall interior partitions and light curtain wall exterior contribute little to the lateral force resistance of modern buildings.

To date (1973) only the *Uniform Building Code* and the *National Building Code of Canada 1970*, among North American model building codes, specify a maximum value of the deflection index of 1/500, corresponding to the design wind loading. Also, ACI Committee 435[10-4] recommends a drift limit of 1/500.

The performance of modern reinforced concrete buildings designed in recent years to meet this criterion appears to have been satisfactory with respect to the stability of the individual columns and the structure as a whole, the integrity of nonstructural elements, and the comfort of the occupants of such buildings.

Most of the modern tall reinforced concrete buildings containing shear walls have computed deflections ranging between $H/800$ to $H/1200$ due to the inherent rigidity of the shear wall-frame interaction.

It is realized, of course, that the method of calculating the drift as well as the degree to which the assumptions used in such calculations corresponded to the actual structure, may have varied widely from one building to another. Since computing drift is a highly complex procedure, many simplifying assumptions are usually made, which cast doubt on the reliability of such computations. Even between two highly sophisticated computer programs, (both incorporating axial shortening of columns), the magnitude of computed deflections may vary 30 to 40% as a result of one assuming center-to-center distances of its line network, and the other considering finite dimension of its joists. Neglecting the axial deformation in the columns of a frame leads to a stiffer structure. This results in computed values for the lateral deflection which are less than those which would be obtained if axial deformation were not suppressed. This effect, although relatively small (about 10–15%) for most multistory buildings, can be significant for tall slender buildings. The increasing availability of computer programs for lateral load analysis should help eliminate the differences in the manner of computing deflections and should lead to a more precise definition of an allowable drift.

To establish a realistic relationship between drift and the pertinent design parameters based on existing buildings, an elaborate field study would have to be undertaken to evaluate their behavior considering the actual wind forces, the method of calculation, and the assumptions made in computing lateral deflections.

For cases where excessive drift is expected, floating partitions with a capability to accept relative movement between skeleton and partition may be required. Floating partitions do not contribute to the lateral rigidity of buildings.

-2. *Perception*—The sway motion of a tall building under turbulent wind, if perceptible, may produce psychological effects which render the building undesirable from the user's viewpoint. The reduction of such perceptible motion to acceptable levels may thus become an important criterion in the design of any tall building. It is apparent that the sensation of motion which can be disturbing to an occupant of a building can result either from the visual perception of relative displacement with respect to some reference object, or, if visual effects are excluded, from the acceleration of the floor on which the observer stands. A number of tests have confirmed the effect of acceleration in producing a sensation of motion. Also, if the acceleration is very small, but changes frequently from negative to positive, the rate of change of acceleration (which is commonly known as jerk) can equally produce the sensation of motion.

Determination of minimum tolerable values of acceleration for the typical or normal person needs further studies. It is obvious that the acceptability of a design with respect to perception of sway motion can only be assessed through a dynamic analysis of the building under a set of a probable range of wind exposures. No perceptible motion has been reported in concrete buildings to date.

10.5.3 Analysis for Lateral Forces

The degree of sophistication to which a structural analysis is carried out obviously depends on the importance of a project. A wide range of approaches have been used for buildings of varying heights and importance, from simple approximate methods which can be carried out manually, or with the aid of a desk calculator, to more refined techniques involving computer solutions, as well as model studies.

While the use of approximate methods for the lateral load analysis of frames may, by current standards of engineering practice, be adequate for regular frames with a few stories or for the preliminary analysis of tall frames, the increasing availability of computer programs argues against the use of such approximate methods when more reliable solutions are obtainable.

Only a few years ago most of the multistory buildings were analyzed by approximate methods such as the portal or the cantilever method. The recent advent of computers and the abundance of publicly available sophisticated programs has de-emphasized the approximate methods which at present are useful for preliminary estimating only.

Simplified methods, when used for the preliminary analysis of tall structures, should generally be followed by a computer analysis which includes the effects of axial deformations as well as bending.

For major high-rise structures, a computer analysis may be the most economical means of arriving at a reliable design. A logical procedure for lateral load analysis is:

a. carry out a simple preliminary analysis by longhand calculations with approximate methods and determine the required member sizes for the combination of gravity and lateral loads;
b. if necessary, use a computer program to achieve a more accurate analysis.

In view of the increasing availability of computer programs for frame analysis, longhand methods which involve simplifying assumptions and lengthly arithmetical work must be considered as obsolete.

When investigating the use of a program, there are many features which can be included and may be important in a frame analysis. For example:

1. plane frame or space frame,
2. flexible data input for irregular structures, but simplified input for regular structures;
3. axial deformation of vertical members;
4. shear deformation of members;
5. finite size of joints between members;
6. floors fully rigid in-plane;
7. nonlinear material behavior;
8. second order geometry effects (i.e., P-Δ effect and decrease in rotational stiffness of members with axial load).

It is important to know the significance of any feature which makes a significant increase in data input, storage requirement, or execution time. For example, if column axial deformation is negligible, it is worthwhile to neglect it in the analysis by suppressing the vertical degrees of freedom of the structure.

At present, the scope of the analysis is normally governed by the features available in the program to be used. As more programs become available and further features can be included in the analysis, it will become increasingly important for the designer to be able to choose the program or program features which give an efficient solution to his problem.

The following assumptions are common to the analysis and design of all high-rise structural systems:

a. The floor slabs distribute load to the lateral load resisting units, mainly through forces in their own plane. The actual in-plane deformation of the floors will seldom have a significant effect on this distribution and the assumption that the floors are fully rigid in-plane is widely used.

b. Out-of-plane bending of floor slabs can be important. A very small resistance to bending in a beam connecting two vertical units can have a significant effect on the behavior of the entire system. It is most important to ensure that a connecting member possesses the necessary flexural strength. Slab-column joints in flat-plate structures should be given careful attention.

c. The effect of torsion should be considered if the layout is unsymmetrical or if the stiff vertical units are located close to the center of the structure.

Any combination of frames, walls or tubes can be idealized as a space frame. If a space frame computer program which considers all six degrees of freedom at each joint is used in the analysis, any torsional effects will automatically be accounted for. Space frame analysis of multistory buildings can be costly and is presently used only in unusual cases. The discussion in this chapter will be confined to analysis techniques which assume full in-plane rigidity of the floor slabs.

The effect of relative deflections due to temperature and creep can be important, as discussed further in this chapter. P-Δ moments may be significant in tall structures. A method of including these in analyses is described briefly below in the section on Frames.

With respect to lateral resistance, there are three basic types of structural systems: frames, shear walls, and tubes. The basis for the classification is the mode of deformation of the unit when subjected to lateral loading.

Frames deform in a predominantly shear mode where relative story deflections depend on the shear applied at the story level. Shear mode deformation is illustrated in Fig. 10-21(a).

Walls deform in an essentially bending mode as illustrated in Fig. 10-21(b).

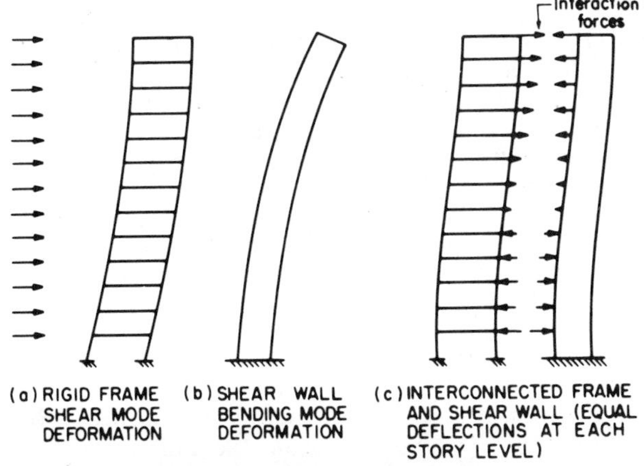

Fig. 10-21 Deformation modes.

Tubes, if unperforated, behave in the same way as walls. However, openings which are normally present in units of this type produce a behavior intermediate between that of a frame and that of a wall.

Any of the three types of units described above, singly or combined, form a *structural system* (for lateral load). Generally, as the height of a building increases, a point is reached beyond which the lateral sway under wind loading, and hence consideration of stiffness and not strength, will govern the design of the structural system. The point, in terms of height at which this condition is reached depends on the type of structural system. Ideally, that system should be chosen which does not require an increase in the sizes of the members beyond that required to support the design vertical loads, i.e., a "premium-free" building.

Table 10-2 from Ref. 10-5 is presented to guide in choosing a suitable structural system for a particular building. The ranges of applicability shown may vary somewhat depending upon the use of the building, the story heights, and the design live and wind loads.

The present methods of lateral load analysis for average buildings disregard the dynamic character of the wind-structure interaction. A major reason for this is that all of the presently used building codes specify loads which are to be considered as static loads upon a structure, and which the structure is assumed to resist within its elastic range. Only unusual structures may require a dynamic investigation (analytical and/or experimental) involving the periods of vibration.

The methods of analysis applicable for the individual types of structures will be discussed in the corresponding sections, together with the other aspects of these structural types.

TABLE 10-2 Guide to Selection of Structural Systems

	*Number of stories**	
Structural System	*Office Buildings*	*Apartment Buildings, hotels, etc.*
Frame	up to 15	up to 20
Shear Wall (egg crate)		up to 150
Staggered Wall Beam		up to 40
Shear Walls acting with Frames	up to 40	up to 70
Single Framed-Tube	up to 40	up to 60
Tube-in-tube	up to 80	up to 100

**The values given here are based on present day practice as well as trends indicated by current thinking.*

10.5.4 Frame Structures

The term "frame" denotes a structure which derives its resistance to lateral loading from the rigidity of the connections between columns and beams or slabs.

-1. Components of drift—In a frame-type structure the drift may be thought of as consisting of two parts: one due to bending in the columns and beams, and the other due to axial deformation of the columns. As the height-to-width ratio of the structure increases, the effect of column axial deformation assumes greater significance.

The effect of secondary moments caused by the axial forces and the deflections (P-Δ) will also tend to increase the lateral deflection.

-2. Hand calculation methods—Simplified methods of analysis for building frames are often used for purposes of preliminary designs. Such methods should be able to furnish rough numerical data with a minimum of effort.

Classical methods of analysis, such as the slope-deflection method or the moment distribution method, have been superseded by more refined methods using matrix formulation which are suitably programmed for computers.

-3. Portal method—The basic assumption of the portal method is that points of contraflexure are located at mid-length of all columns and beams. In addition, an assumption concerning the distribution of shears in the columns of a story is made. These assumptions reduce a highly statically indeterminate problem to a statically determinate one. The method neglects the effect of axial deformation in the columns.

The assumptions associated with the portal method result in errors in the vicinity of the base and top of the frame and at setbacks or locations where significant changes in member stiffness occur. In these locations large errors in the calculated member moments may be expected. This can be particularly serious in bottom stories, where the combination of large axial forces and moments in the columns can lead to instability problems. This type of error can be partially corrected by performing a more exact analysis for the localized discontinuity regions or by using tabulated information on the location of inflection points in the bottom and top stories.[10-6]

The errors resulting from disregarding column axial deformations increase with the increase in the number of bays and the number stories in a frame, and are reflected most markedly in the moments in the exterior columns and girders of the upper stories of tall frames.

Drift calculated on basis of moments obtained by the portal method is subject to errors as the assumed locations of the points of contraflexure lead to predictions of drift larger than based on exact analysis. Additional errors arise from the neglect of axial column deformation.

5.4.4 Computer Programs for Frames

A number of computer programs have been developed in recent years. STRUDL, for example, is capable of carrying out elastic analysis of either plane frames or three-dimensional reticulated structures under static and dynamic loading conditions.

The more important frame program features which have relevance to lateral load analysis are as follows:

a. Axial deformation. Axial deformation of columns may be important in tall slender frames or in frames with stiff connecting beams. While no definite rules can be given, the effect of column axial deformation will generally be important if the frame height-to-width ratio exceeds about three to four. Axial deformation of beams is always negligible.

b. Shear deformation is usually neglected.
c. Finite size of joints between members. Manual methods and the early computer programs based the analysis on the centerline network mathematical model. This, however, results in erroneous moments and drifts, since the real spans are smaller than the center-to-center distances. A number of possibilities exist (in particular in computer programming) to reduce these errors.

In most cases it is realistic to assign a greater stiffness to the area within the joint between beams and columns than that assigned to the connecting members.

For symmetrical members with equal end rotations, an equivalent EI value can be used.

$$(EI)_e = \frac{K(1+C)L}{6}$$

where

$(EI)_e$ = equivalent EI
K = end rotational stiffness
C = carry-over factor
L = center-to-center span of member

where K and C are evaluated for a member with infinite flexural stiffness within the joints.[10-7]

Where a computer program has the capability of considering members with variable moments of inertia, the joint rigidity can be simulated by increasing the moments of inertia of the column and beam sections within the joint areas; they should be increased to about 10 or 20 times the normal column or beam moment of inertia to realistically simulate the joint rigidity. Excessive increase of stiffness in the joint area may lead to significant numerical error in the solution and should be avoided.

An efficient approach is to treat the ends of the members as being fully rigid, and then calculate the stiffness properties of the combined elements before the frame analysis is undertaken. Some frame programs incorporate a finite joint facility of this type.

d. Foundation movement. Ability to include elastic spring supports is a useful program feature (especially for shear wall foundations).
e. Second order geometry effects (P-Δ effect). When a frame sways laterally by an amount, Δ, the columns are subjected to an eccentric moment equal to $P\Delta$, where P is the total vertical load at the level at which Δ is measured. This P-Δ effect can be significant, particularly in tall unbraced frames.

-5. Effects of masonry infill walls and partitions—In many frame-type structures, walls and partitions are made of precast or masonry units. Although such elements are often considered to be non-structural and perhaps—because of very light reinforcement—may not contribute significantly to the ultimate strength of the structure, they generally contribute substantially to the lateral stiffness of the structure under working load conditions. The behavior of multistory infilled frames, i.e., frames with the space between the members filled with a masonry or cast-in-place concrete wall, under lateral loading is essentially that of a vertical cantilever beam. The effect of walls and partitions can be particularly well observed on the response of structures subjected to earthquake motions. Walls filling the space between frame members not only tend to increase the stiffness, but may altogether alter the mode of response of the frame, changing it to a shear wall and as a result, changing the entire structure and the resulting distribution of forces among the different frame components.

Studies[10-8, 10-9] have indicated the possibility of taking full advantage of infill partitions by constructing them to ensure their participation in resisting frame distortions with particular reference to the satisfaction of the drift criterion.

The stiffening effect of the infill panel on the frame can be represented fairly well by a diagonal strut having the same thickness as the panel and an effective width which depends on a number of factors. Studies of frames in which the infilling was not bonded to the frame indicate that the effective width of the diagonal strut is influenced by the relative stiffness of the column and the infill; the height-to-length ratio of the infill panel; the stress-strain relationship of the infill material; and the diagonal load on the infill. The effective width of the strut increases with increasing column stiffness and panel height-to-length ratio and decreases with increasing value of the load and modulus of elasticity of the infill material. Reference 10-9 gives curves for the approximate equivalent strut width in terms of the geometrical and physical properties of the frame and infill and the applied diagonal load. While it may be conservative to disregard the stiffening effect of infilling panels on the behavior of frames subjected to the wind loading, neglecting their effect on the response of a structure subjected to earthquakes can be dangerous. Walls filling the space between frames not only increase the stiffness (and damping) and hence the frequency of vibration of the building, but also alter its mode of deformation under lateral load. A major effect of the change in behavior from a predominantly shearing type to a cantilever mode of deformation is a significant increase in the axial forces to which the exterior columns are subjected.

-6. Effective slab width—Structures consisting of flat slabs or plates and columns are utilized as frames to resist lateral loads by considering the slabs as equivalent beams. The effective width of the slab determines the stiffness of the one-dimensional beam elements to be used in a plane frame analysis.

Analytical and limited experimental studies dealing with the problem of effective slab width have considered the column-to-panel width ratio as the principle variable, and have indicated effective slab width ratios ranging from 0.50 to values greater than unity, the effective width increasing with increasing values of the column-to-panel width ratio. However, it appears that other parameters not yet considered in studies, such as the stiffness of the slab, the aspect ratio of the slab panel, the dimension of the column in the plane of the frame relative to the slab span, as well as the distribution of the slab reinforcement, may also have significant influence on the effective slab width.

The problem of determining the stiffness properties of an "equivalent beam" to properly model the action of a slab for use in a plane frame analysis needs additional study. A distinction may have to be drawn between the purpose of equivalence to be used, depending upon the object of the analysis. Thus, an analysis undertaken to provide values of slab design moments may require different equivalence criteria from an analysis intended to serve as a basis for column design or from an analysis aimed at studying overall frame deformation

It is presently customary among designers to consider only half the panel width as equivalent slab width for *stiffness* when carrying out an analysis for lateral loads. However, only a narrow strip of the slab (one and a half slab thicknesses on every side of the column) is considered effective for the transfer of the unbalanced moments between columns and slabs resulting from the analysis.

The strength of joints between column and flat plate floors designed for vertical loads only can be relied on to a limited extent only for resistance to lateral loads. This is one of the main reasons for adding shear walls in flat plate structures above 15 stories. To enhance the joint stiffness, drop panels or beams can be added. In some cases, the total floor thickness has been increased to satisfy lateral resistance requirements.

The slab-column joints of lift-slab structures are not normally assumed to participate in lateral load resistance.

-7. *Analysis*—If all the plane frames parallel to the direction of loading are the same, the total lateral load in that direction can be distributed equally to each frame. The frames can be analyzed by the methods discussed previously in section 10.5.3. Alternatively, a distribution of shear to each column based on relative column and beam stiffness[10-10] or column stiffness[10-11] can be used. Parallel frames (with no torsion) can be idealized in a manner similar to the shear wall-frame method illustrated further in Fig. 10-32.

-8. *Range of applicability-number of stories*—Although frame-type structural steel buildings of up to 60 stories have been built, economic considerations tend to limit the height of such structures in reinforced concrete to about 10 to 15 stories, particularly for flat plate type structures without spandrel beams. This limitation has become apparent with the increasing use of lightweight elements or glass for walls and lightweight partitions. For buildings of greater height, shear walls, or shear walls acting in conjunction with adjoining frames, generally provide a more efficient solution.

10.5.5 Shear Wall Structures

The term shear wall is actually a misnomer as far as high-rise buildings are concerned, since a slender shear wall when subjected to lateral forces has predominantly moment deflections and only very insignificant shear distortions.

Figure 10-22 shows various ways of utilizing shear walls in multistory buildings. The scheme 10-22(a) has shear walls without openings across the building with an access through a gallery running alongside the building. Each wall accepts a share of the lateral load proportional to its stiffness. The calculation of lateral stiffness is simple and stresses in such shear walls without openings involve simple bending theory only.

Schemes (b) and (c) of Fig. 10-22 have interior corridors, and, therefore, the shear walls are interrupted by an opening on each floor, and they are interconnected either by slabs or beams. Although the major shear walls are usually in the transversal direction of the building, separating the individual apartments, stability in the longitudinal direction is normally provided by elevator shafts or some longitudinal shear walls. Such structural systems as shown in Fig. 10-22 are known as egg-crate or crosswall buildings; they are extremely rigid in the direction of the shear walls.

10.5.6 Coupled Shear Wall Structures

Walls with openings present a much more complex problem to the analyst. Openings normally occur in vertical rows throughout the height of the wall and the connection between the wall sections is provided by either connecting beams which form part of the wall, or floor slabs, or a combination of both.

The terms "coupled shear walls," "pierced shear walls" and "shear wall with openings" are commonly used to describe such units as shown in Fig. 10-23.

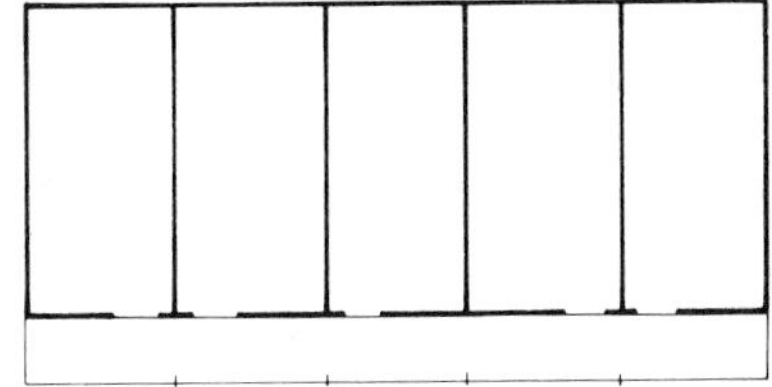

(a) No openings in walls: access from outside

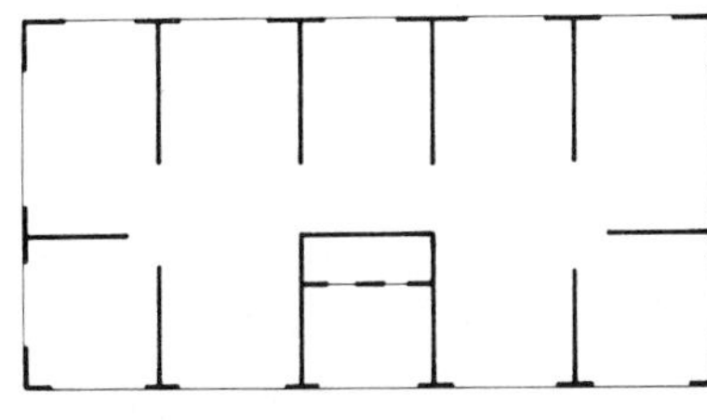

(b) Coupled walls

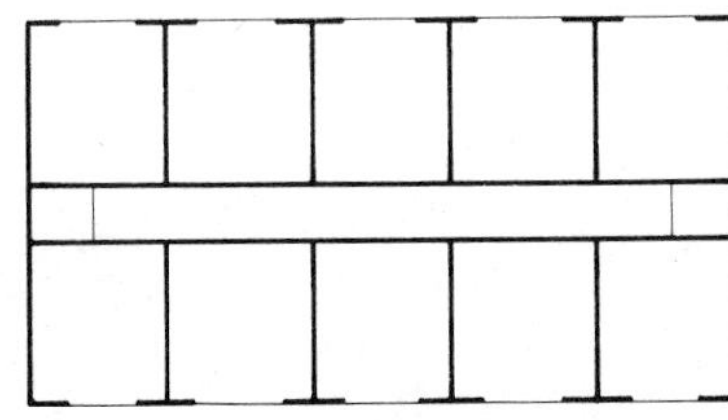

(c) Longitudinal corridor walls

Fig. 10-22 Schematic layouts of shear wall structures.

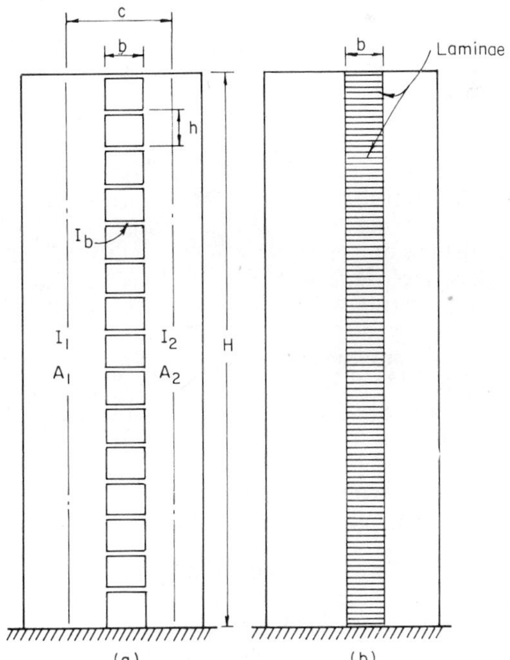

Fig. 10-23 Coupled shear walls.

If the openings are very small, their effect on the overall state of stress in a shear wall is minor. Larger openings have a more pronounced effect and, if large enough, result in a system in which typical frame action predominates. The degree of coupling between two walls separated by a row of openings has been conveniently expressed in terms of a geometrical parameter α (having a unit 1/length) which gives a measure of the relative stiffness of the connecting beams with respect to that of the walls. The parameter α appears in the basic differential equation of the so-called continuum approach.[10-12–10-15] A study by Marshall[10-16] indicated that when the dimensionless parameter, αH, (H being the total height of the walls) exceeds 13, the walls may be analyzed as a single homogeneous cantilever. When $\alpha H < 0.8$, the walls may be treated as two separate cantilevers. For intermediate values of αH (i.e., $0.8 < \alpha H < 13$), the stiffness of the connecting beams should be considered.

The effectiveness of the coupling of shear walls can be clearly seen from the Fig. 10-28 (ex. 10-1) where the cantilever moment of the shear wall acting alone is compared with the moment reduced due to frame action caused by coupling.

Prior to the early 1960s no analytical techniques for coupled shear walls were available. During the 1950s and into the 1960s, coupled shear walls were designed as individual cantilevers, each carrying the amount of wind in proportion to its stiffness. Although the shear walls were greatly overdesigned, this approach resulted in underdesigned connecting links (beams or slabs). Since the 1960s a considerable amount of research information has become available, and a number of practical approaches have been suggested.[10-12–10-17]

-1. Continuum approach—In its most basic form the theory assumes that elastic structural properties of the coupled wall system remain constant throughout, that both walls are founded in a common stiff footing, and that the points of contraflexure of all beams are at midspan.

In this method, the individual connecting beams of Fig. 10-23(a) are replaced by a continuous connection of laminae as in Fig. 10-23(b). Under horizontal loading, the walls deflect and induce shear forces in the lamina. A second order differential equation is set up and solved to give shears, moments and deformations throughout the wall. Several papers use this approach with differing choice of variables, all yielding essentially the same results.

Tests on plastic models have generally confirmed the accuracy of the continuum approach for walls conforming to the basic assumptions of the method. In practice some of these assumptions do not hold. Slight inaccuracies of the method may stem from local wall deformations and reduction in stiffness of coupling beams due to their cracking.

Solutions are available for uniformly distributed loads, triangularly distributed loads and for concentrated point loads at the top of the building. Perhaps the most readily available and convenient are those developed by Coull and Choudhury.[10-14, 10-15]

In some cases, elastic analysis will result in moments which cannot be developed by beam sections restricted in size by architectural reasons. The possibility of designing the beams for ultimate moments in proportion to the elastic moment distribution has been suggested[10-17] and a theory has also been developed for the elasto-plastic analysis of coupled shear walls[10-18] which would result in substantial simplifications and economies.

Analytical methods employing computerized frame analysis are not only more accurate but are considerably more flexible and can take into account many more variables than

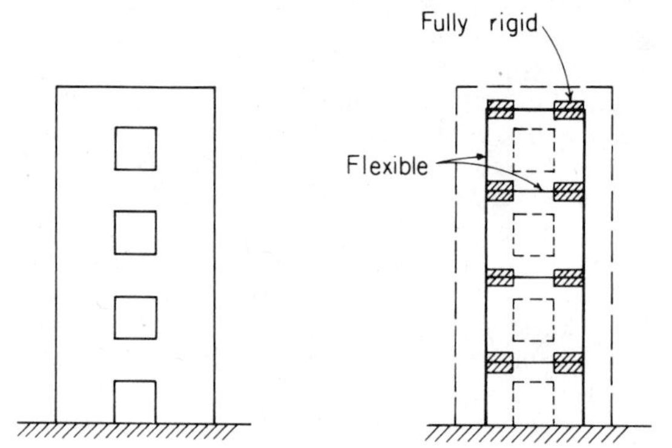

Fig. 10-24 Substitute frame.

the continuum approach which was originally developed for manual operation.

For analysis of multistory shear walls, a good computer program considers a frame with finite joints in which the wall is analyzed as a frame except that the finite width of the columns in comparison with the beam is recognized. The analogy, in which the beams are assumed to be infinitely stiff from the centerline of the column (wall) to the edge of the actual opening, is illustrated in Fig. 10-24(b). The calculation can take into account changes in wall thickness, story height and concrete strength at various locations within the height of the building.

Width of slab to be considered in frame action between wall and slab has received by far insufficient research using reinforced concrete specimens (there have been a number of research projects using elastic models) despite this important aspect of shear wall design; in fact, available test experience is contradictory. Values less than the full width, equal to the full width, and greater than the full width have all been shown to be valid under different circumstances on elastic models and analytical studies. Clearly, there are many factors affecting the behavior of more complex wall systems which cannot yet be handled easily by theoretical means. The consequences of choosing an effective width should be fully understood by the designer (see author's closure to discussion of Ref. 10-17).

-2. Proportioning of shear walls—Although the analysis of shear wall systems has advanced considerably in the last decade, the proportioning (design) of shear walls has not kept pace with the analytical developments. A number of experimental studies[10-20, 10-21] carried out mainly on short shear walls have ascertained the shear strength and the results have been incorporated into Sections 11.16 and A.8 of the ACI 318-71. As for the flexural strength, it becomes obvious that the traditional method of treating the wall as a homogeneous section may be valid only for low eccentricities resulting in uncracked sections, while the slender walls of tall buildings which, by their geometry, are slender cantilevers, should be proportioned for moment similarly to any flexural element, preferably using the Ultimate Strength Design Method. Such method of flexural treatment will lead to concentrations of flexural reinforcement at the extreme ends of the shear walls with minimum reinforcement both vertically and horizontally (for shear and shrinkage control) in the remainder of the section.

Tests on long shear walls[10-22] and the previously discussed short wall tests indicate that only very low and long shear walls with height-to-depth ratios (H/D) of about less than one, will fail in shear. Contrary to the common opinion and the misleading name, the strength of taller shear walls, and particularly those of multistory buildings, will invariably be controlled by flexure; they behave like slender cantilevers. As a matter of fact, it is quite difficult to provide sufficient flexural reinforcement to force a shear failure on a slender shear wall; only extremely heavy reinforced columns at the ends of thin shear walls (dumbbells) can force a shear failure under heavy lateral loads.

The tensile reinforcement, obviously, also acts in compression at moment reversals and should preferably be enclosed with ties (as in a column) to improve the strain capacity of the concrete in compression and at the same time improve the ductility of the shear wall.

All portions of a shear wall should be designed to resist the combined effects of axial load, bending, and shear, determined from the analysis of the structural system.

Flexural reinforcement should be provided in accordance with the requirements of the *ACI Building Code*. Walls with proportions in which a linear strain distribution does not apply should be designed as short cantilevers.

Minimum amount of reinforcement in the vertical direction should be as required by flexural calculations and in the horizontal direction as specified in the provisions for shear strength and shrinkage control. In addition to providing the necessary amounts of reinforcement, it is essential that reinforcement details in every shear wall receive careful attention to insure optimum performance.

If there are tensile forces resulting from the most severe combination of vertical loads and overturning moments due to lateral loads, they must be anchored into the foundation medium, unless they can be overcome by gravity loads mobilized from neighboring elements.

-3. Methods of analysis—The proportion of the total lateral load which each wall resists depends on its stiffness relative to that of all walls or coupled wall systems of the building. The lateral stiffness of each wall or coupled wall can be based on the deflection at the top when subjected to a uniformly distributed unit lateral loading. Calculation of stiffnesses can be made quickly through the use of tables.[10-14, 10-15] The method of distributing the load is described in Reference 10-23 for low structures. When the walls do not remain constant over the height of the structure, the use of a plane frame program with walls connected by link bars (as illustrated in Fig. 10-32) is advantageous for irregular configura-

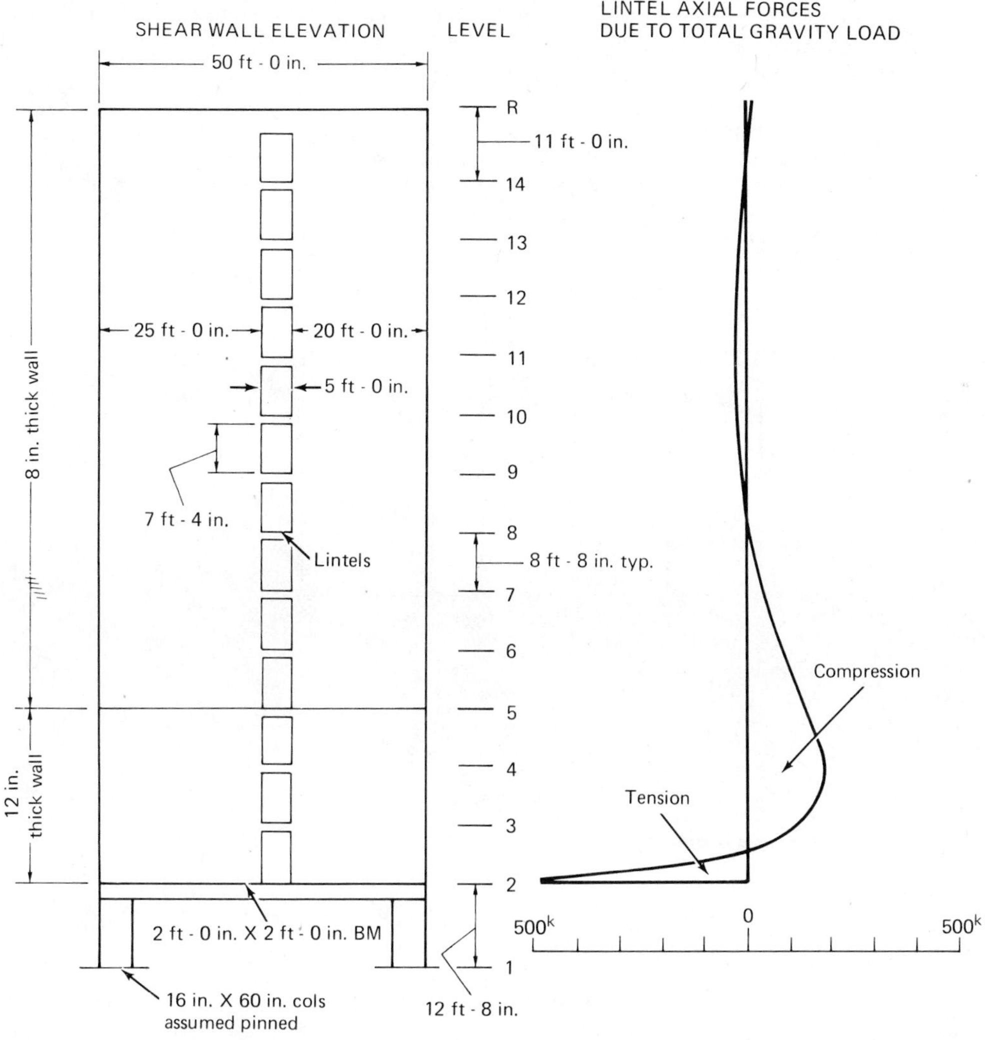

Fig. 10-25 Coupled shear walls supported on exterior columns.

tions in multistory structures. Once the loading on each wall is determined, wall stresses can be calculated and thicknesses modified, if necessary.

-4. *Application*–Shear wall buildings are used in apartment, hotel and other residential buildings where walls are customarily spaced between 15 and 24 ft apart with floor slab thicknesses proportioned according to span. Spans up to 32 ft have been used with prestressed concrete slabs.

The shear wall structure is used only in buildings where permanent partitions and the lack of flexibility for future modifications can be tolerated. Its major advantages lie in the speed of construction, low reinforcing steel content and acoustical privacy.

In current North American practice, shear wall buildings are mainly cast-in-place but trends to systems building are leading to an increase in the number of buildings being constructed using large panel, precast components for floors and/or walls.

Shear wall structures are well suited for construction in earthquake areas and they have performed well during recent disasters.[10-24] While costs vary from city to city, shear wall buildings usually become economical as soon as lateral forces affect the design and proportioning of flat plate or beam and column structures. Buildings of up to 70 stories have been built using shear walls. Feasibility studies for projects up to 200 stories utilizing shear walls have been made and found workable.

-5. *Coupled shear walls supported on exterior columns only*–Parking areas under residential buildings require different spans than the apartments above. For this reason the shear walls must be stopped and supported on exterior columns, thus leaving the entire parking area column-free. The lower portion of the solid shear walls, in buildings with outside corridors, acts as a deep beam spanning between the exterior columns.

The majority of shear wall buildings, however, have coupled shear wall systems to accommodate corridors in the middle. A yet unpublished study carried out at the Portland Cement Association showed the feasibility of supporting a coupled shear wall on exterior columns as shown in Fig. 10-25. The study, carried out using the IBM 1130 version of the STRESS program, showed that the second floor beam (supporting the shear walls) acts like the tension member for the coupled shear wall above. The lintels over the doors for the next five stories act as compression struts; above that level the forces in the lintels are minimal, as can be seen in Fig. 10-25. The study also indicated that the shear wall must be supported during construction only till about the fifth floor is cast; after that the structure is self supporting.

EXAMPLE 10-1: The 20-story apartment building designed with coupled shear walls is shown in its typical plan and elevation in Fig. 10-26. The typical wall section shows the pair of coupled shear walls each 22 ft long and having a thickness which varies from 8 in. for the top six stories to 10 in. for the intermediate seven stories, and 12 in. for the bottom seven stories.

The coupled shear wall linked by (1) 8 in. slab; (2) the same slab plus 8 × 9 in. stem.

The objective is to determine if there is a benefit in coupling the walls, and, to what degree the addition of a beam to the slab improves the coupling.

An analysis for a uniform wind load of 20 psf was carried out using a frame computer program[10-19] which can accommodate the finite width of the shear walls. The results of the lateral load analysis are shown in Figs. 10-27 and 10-28.

Two cases are shown in Figs. 10-27 and 10-28. The first case has slabs for the connecting elements while the second case has the same slabs but with an additional beam in the plane of the coupled walls, so that the coupling elements–having a T-section–have a stiffness twice that of the slab in the first case. The choice between the slab

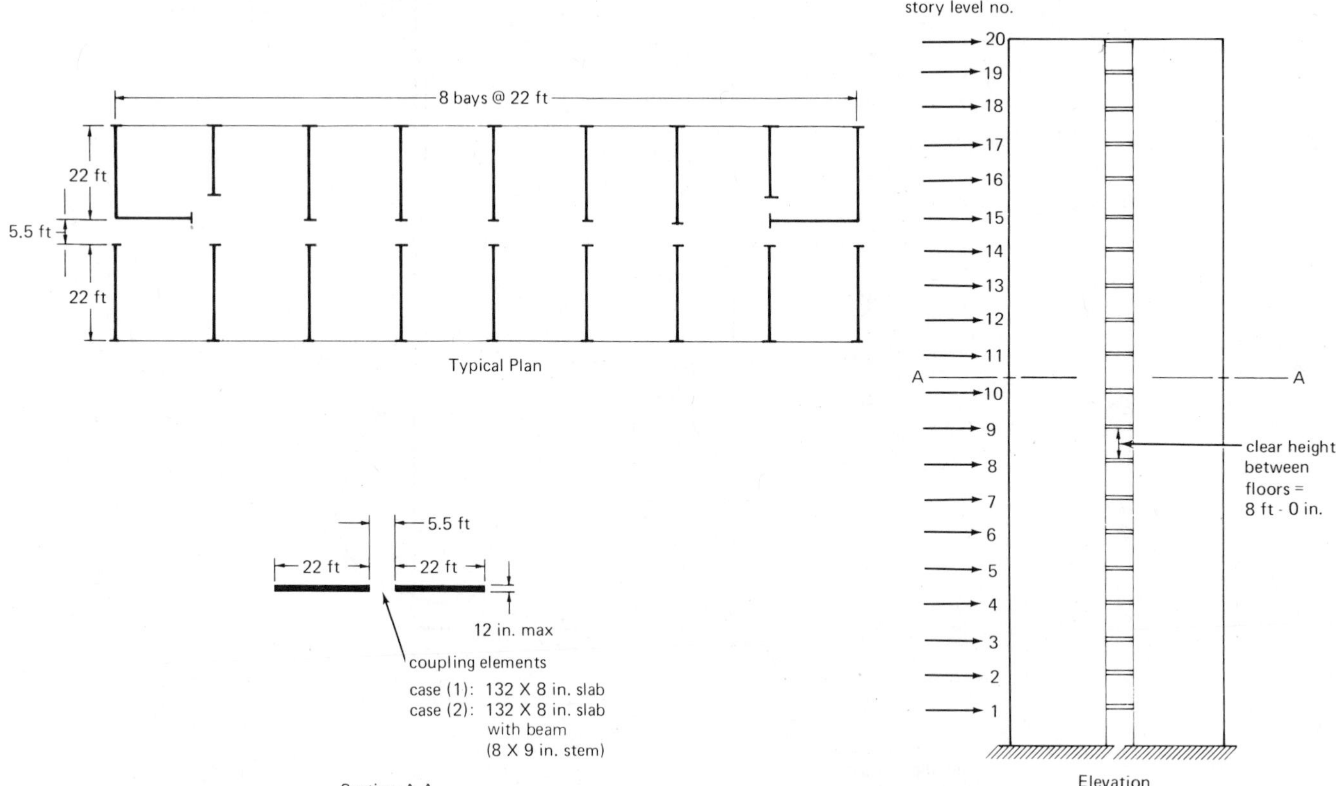

Fig. 10-26 Typical plan and elevation of a 20-story, coupled shear wall building.

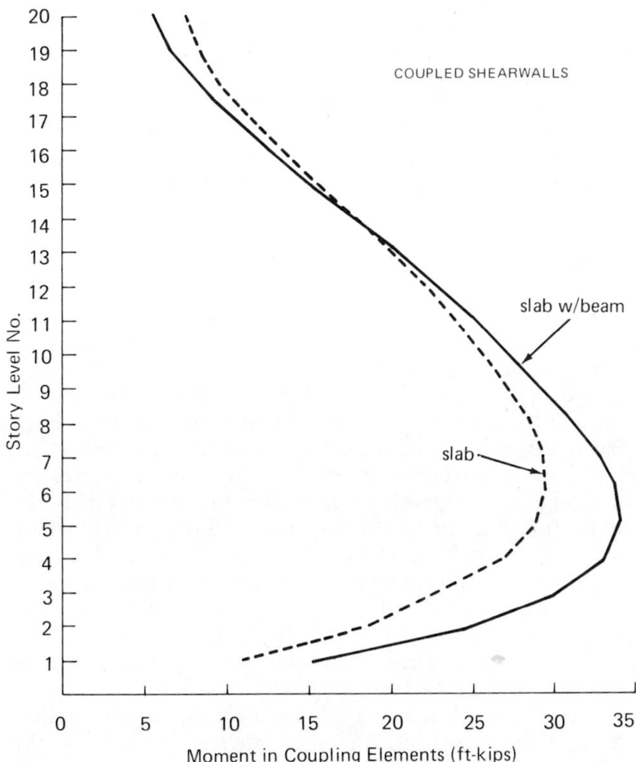

Fig. 10-27 Moments in the coupling elements.

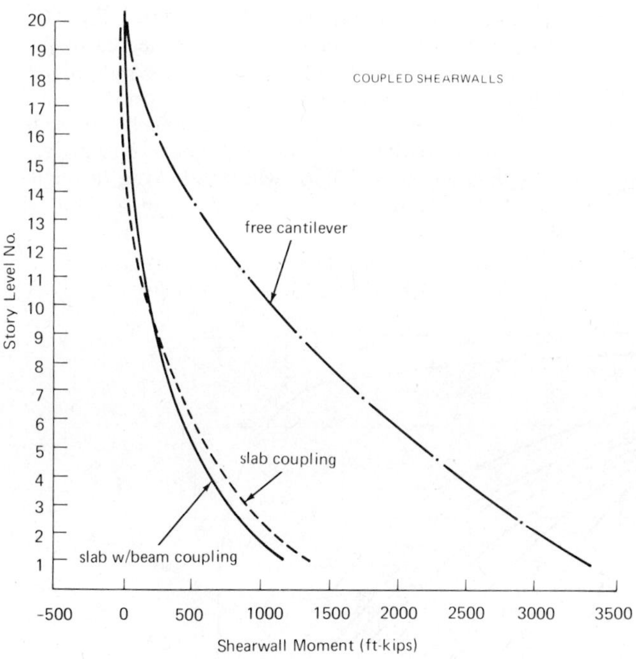

Fig. 10-28 Shear wall moments for the coupled system.

as coupling or beam coupling should be made on the basis of the capability of the linkage to transfer into the shear wall the coupling moments and shears, in addition to the moments and shears due to gravity.

Figure 10-27 shows, as expected, that increasing the stiffness of the coupling elements increases the moments which these elements take. In this case, doubling the stiffness of the coupling elements increased the maximum moment in the coupling element by about 15%. Increasing the stiffness of the coupling elements in turn decreases the moments in the shear walls. This is shown in Fig. 10-28. Also shown in this figure are the overturning moments in a shear wall when acting as a free cantilever, i.e., without coupling. Note the very significant reduction (to about one-third) in the shear wall moments resulting from the coupling of the shear walls. The effect of the doubling of the coupling stiffness on the shear wall moments can be seen as only secondary within the total reduction of the free cantilever moments.

10.5.7 Shear Wall-Frame Buildings

Since the late 1940s, when the first shear walls were introduced, their use in high-rise buildings to resist lateral loads has been extensive, in particular to supplement frames which, if unaided, often could not be efficiently designed to satisfy lateral load requirements.

The great majority of multistory buildings today are, in fact, shear wall-frame structures, since elevator shafts, stairwells and central core units of tall buildings are mostly treated as shear walls, in addiiton to isolated concrete walls (if some are used). Frame structures depend primarily on the rigidity of member connections for their resistance to lateral forces, and they tend to be uneconomical beyond 10 to 15 stories. To improve the rigidity and economy, shear walls are introduced in buildings exceeding 10 to 15 stories in height.

The term shear wall-frame structure is used here to denote any combination of frames and shear walls. Included in this category would be the typical frame structure with shear walls appropriately located about the plan, as shown in Fig. 10-29 (b, c, d) and the so-called hull-core or framed-tube-with-core structure shown in Fig. 10-29(a). The shear wall can have any plan shape and may be linear, angular, rectangular, or circular in plan.

The common assumption to neglect the frame and assume that all the lateral load is taken by the shear walls may not always be conservative, since due to the interactive forces, the frame is usually subjected to forces higher than the exterior applied wind forces in the upper stories. Therefore, distributing the applied wind to the different resisting elements in proportion to their relative stiffness in the case of shear wall-frame interaction can lead to grossly erroneous results.

Consideration of shear wall-frame interaction leads to a

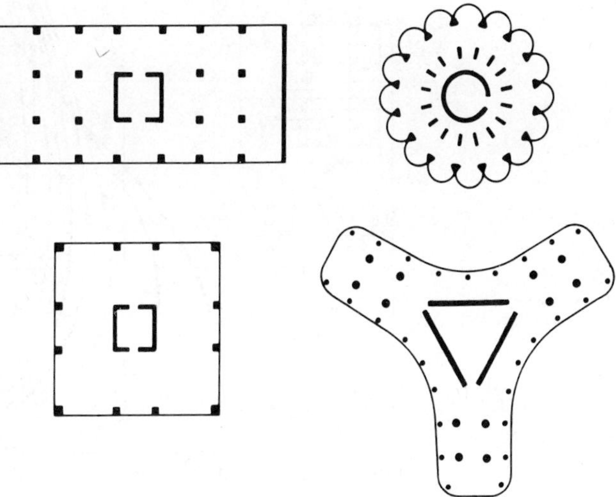

Fig. 10-29 Typical layouts of high-rise buildings with shear wall-frame interaction.

more economical design. Since the shear wall moments are reduced, less reinforcing is needed due to the frame taking over some of the lateral load moments. In most cases the frame can accept the additional moments due to lateral loads within the 33% increase in allowable stresses, except for the top stories of the frame which often require additional reinforcing. Shear walls are efficiently utilized if they are distributed about the plan so that they carry their proportional share of the vertical load, rather than having them function mainly as lateral load resisting elements. This condition may, however, conflict with the desirability of locating the principal lateral load resisting elements along or near the periphery of a building.

The main function of a shear wall for the type of structure being considered here is to increase the rigidity for lateral load resistance. Shear walls also resist vertical load, and the difference between a column and a shear wall may not always be obvious. The distinguishing features are the much higher moment of inertia of the shear wall than a column and the width of a shear wall which is not negligible in comparison with the span of adjacent beams. The moment of inertia of a shear wall would normally be at least 50 times greater than that of a column, and a shear wall would be at least 5 feet wide.

The introduction of deep vertical elements (shear walls) represents a structurally efficient solution to the problem of stiffening a frame system. The frame deflects predominantly in a shear mode shown in Fig. 10-21(a) while the shear wall deflects predominantly in a bending mode as shown in Fig. 10-21(b).

In a building, the in-plane rigidity of the floor slabs forces the deflection of the walls and the frames to be identical at each story level. To force the walls and the frame into the same deflected shape, internal forces are generated which equalize the deflected shapes of each. Thus, the frame in the upper stories pulls back the wall, while in the lower stories the wall pushes back the frame. These internal interactive forces, shown in Fig. 10-21(c), greatly reduce the deflection of the overall combined system, creating a considerably higher overall stiffness than would be the sum of the individual components, each resisting a portion of the exterior loads. In that distinctive feature of increasing the stiffness through a set of internal forces lies the great advantage of shear wall-frame interactive systems. In addition, the shear wall in the combined system reduces the shear deflections of the columns.

-1. Methods of analysis—Methods of analysis for manual or desk calculator application of shear wall-frame interaction were first developed in the early 1960s.[10-11, 10-25] As distinct from many other methods, the Khan-Sbarounis[10-11] analytical treatment is based on a forced-convergence procedure which causes both systems to have the same deflected shape by the use of an iterative process. The authors present a family of curves for the distribution of the wind forces between the walls and the frames for a wide range of stiffness ratios between a shear wall and a one-bay equivalent frame.

To enter the charts, an equivalent shear wall-frame system must first be established by lumping together all the frames (columns and beams) into a one-bay equivalent 10-story frame and all shear walls into an equivalent shear wall. Then, the stiffness ratio of the shear wall to column is determined and a corresponding chart entered either for a uniformly or triangularly distributed lateral load. From the example chart (Fig. 10-30) the percentages of the base shear resisted by the frame at every level can be read and then translated into actual forces in the equivalent frame. Subsequently, these forces of the equivalent frame are distributed to the component elements from which the equivalent frame was composed. The left side of the chart in Fig. 10-30 gives the percentage of the lateral load at any particular level resisted by the shear wall. Thus, at a glance, the effectiveness of the shear wall within the system can be evaluated.

With the forces determined, the moments and the deflections of the frame and the shear wall are computed; the deflections of the two systems must obviously be the same if the procedure was carried out correctly.

An alternate method can be used with the information given in the same paper by reversing the order of the procedure. A family of deflected shapes for the interactive system is given for the various stiffness ratios between the shear wall and column and for the different loadings (see Fig. 10-31 for an example chart*). From the known deflected shape of the frame and the shear wall, the moments and forces can be computed which would cause these deflections.

Although the procedure[10-11] may seem tedious by present day standards, it offers the advantage of giving a clear physical understanding of the behavior when following through the computational techniques. If organized in a tabular

*Both example charts are for a column-to-beam ratio S_c/S_b = 5 for the equivalent one-bay frame and for a uniform wind loading.

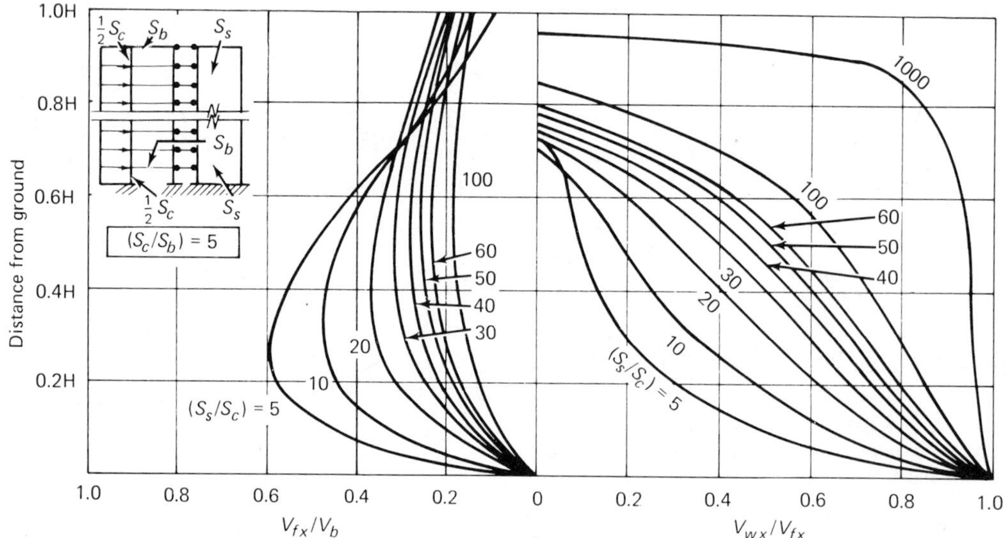

Fig. 10-30 Shear wall-frame lateral load distribution curves.

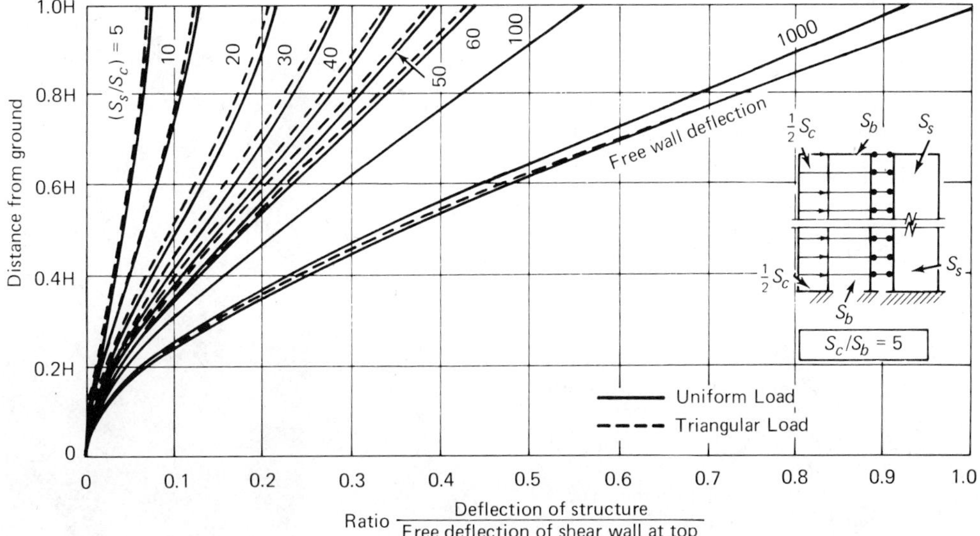

Fig. 10-31 Deflected shape of shear wall-frame interactive system.

form, it does not present insurmountable difficulties in an office routine. In the 1960s this method was the most popular with designers for manual handling of the shear wall-frame interaction problem.

Another method based on a number of simplifying assumptions was published by MacLeod in 1970[10-26] with the purpose of offering a less time-consuming procedure for office use. Its results are slightly less accurate, however, they are more than sufficient for handling the average building of average height.

In buildings with extermely large shear walls the frames take only a small proportion of the lateral load; a simplified analysis may be adequate, or the shear wall assumed resisting all the lateral load. When further analysis is required, a computer program should be used. The analysis of frame-shear wall structures under lateral load using one of a number of computer programs now available is a fairly straightforward procedure.

Discontinuity of a shear wall (both at the top and bottom) is usually accompanied by a shifting of large shear forces between the wall and the slabs.

In frame-shear wall structures, neglecting the axial deformation in the columns will tend to throw a greater proportion of the lateral load to the frames, and underestimate the load carried by the shear wall.

Where torsion may be neglected, structures can be idealized as shown in Fig. 10-32 and plane-frame programs used such as STRESS, STRUDL, PCA programs, or the many other programs publicly available. Because torsion due to lateral loads can produce significant stresses, particularly in corner columns, a more or less symmetrical arrangement of the principal lateral load-resisting elements with respect to the entire plan should be aimed for. In this connection, the torsional resistance of a structure would be much improved by locating the lateral load resisting elements where they do the most good, i.e., along the plan periphery of the building.

-2. Shear wall-frame interaction through large girders—A significant increase in lateral stiffness of tall buildings can be achieved by tieing a centrally located shear wall with the peripheral columns through deep (usually story high) flexural members at the top and possibly at other intermediate levels.

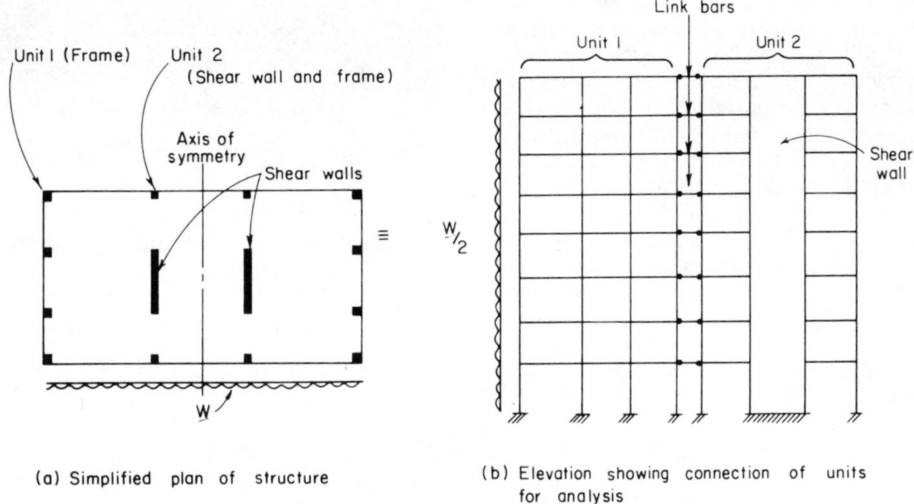

Fig. 10-32 Idealization for plane frame analysis.

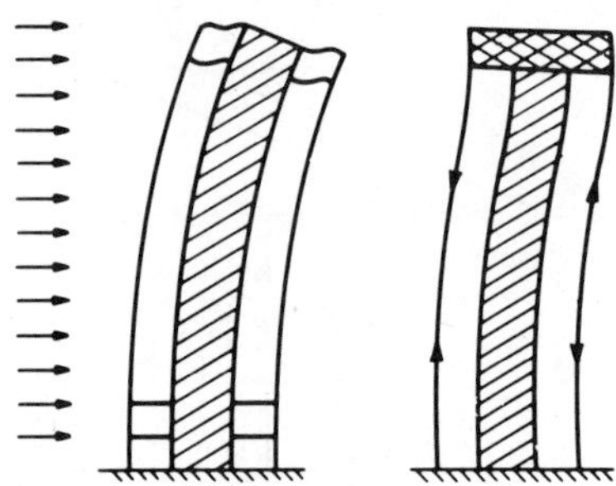

Fig. 10-33 Core connected with heavy girders at the top to exterior columns.

Such linkage, as in Fig. 10-33, mobilizes the longitudinal (axial) stiffness of the peripheral columns in resisting the lateral loads.

The first reinforced concrete building utilizing this concept was the 51-story Place Victoria, Victoria Towers in Montreal (Fig. 10-34) constructed in 1964 in which the X-shaped core is linked at four levels (mechanical floors) by story-high girders to the heavy corner columns.

The U.S. Steel Building in Pittsburgh, completed in 1970, is another example with a heavy girder at the roof level connecting the core with the exterior columns.

The main objective is to cause the structure to act more as a vertical cantilever beam and so resist a larger proportion of the lateral loads by axial forces rather than by bending in the columns. Obviously, the axial load resistance represents the more efficient use of the structural material.

-3. Application—The plan in Fig. 10-29(a) represents the typical modern hull-core type office building without interior columns and using the periphery either as a framed tube or with conventional column spacing. Nonetheless, many office buildings are still constructed with the traditional circa 25-foot spans in both directions. The concrete core is almost always present in office buildings as the backbone of the structural system. No interior shear walls are used in office buildings, as they would interfere with the rentable space.

In apartment buildings, Fig. 10-29(b, c, d), the amount of interior shear walls depends on the architectural layout. Since shear walls fulfill the function of partitions or walls separating apartments, there is usually no conflict with the architectural layout of the typical floor and the number of walls can be generous, thus providing sufficient rigidity.

Shear walls constructed with conventional forming techniques are more expensive than other types of walls, and therefore, there is usually the tendency to use the least amount of them. In addition, the walls may be an architectural obstruction, since even in apartment buildings where shear walls are part of the floor plan in the typical floor, they may create layout difficulties in the lower stories of the building in parking garages and lobbies. Also, in the case of future typical floor modifications, the presence of immovable shear walls may create difficulties. Experienced designers have a feel of how many shear walls are necessary for a particular building height and wind load intensity, based on their previous design experience and also observation of the performance of buildings. The final choice of the amount and location of shear walls usually represents a compromise between the architectural and structural requirements which meet the cost objective.

-4. Amount and size of shear walls—The least amount of shear walls sufficient structurally for a particular building is determined with the following considerations in mind:

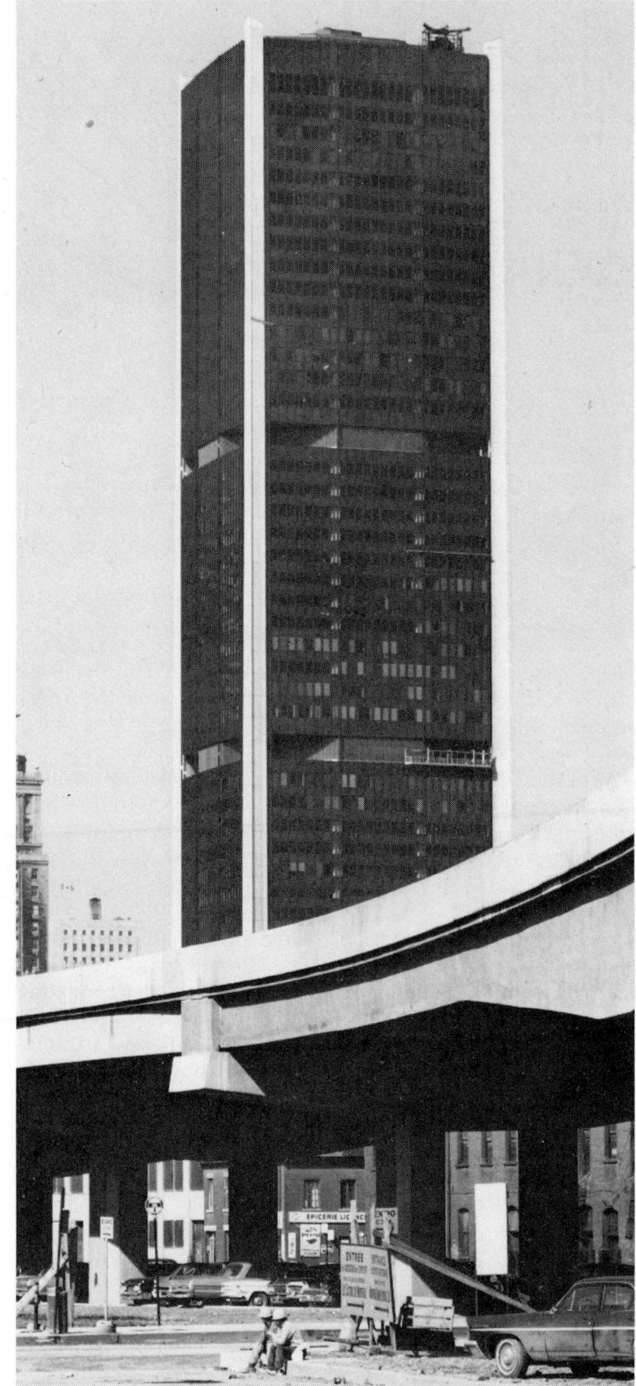

Fig. 10-34 The 51-story high Place Victoria, Montreal, Canada.

(a) to provide adequate rigidity to meet the imposed deflection criteria; and

(b) no tensile forces in the shear walls should result from the combination of gravity (dead) load and overturning wind moments. If there are resulting tensile forces, they must be resisted either by mobilizing load from adjacent columns using transfer girders (usually in the basement), or by anchoring the shear walls into the foundation medium. The economic consequences of the tensile forces should be investigated, if they are unavoidable, against increasing the size or number of shear walls.

Structures or elements of structures in which the ratio of wind stress to dead load stress is high are very sensitive to wind.

The most economical shear wall-frame structure (no premium for height) is achieved when there is a proper balance between the gravity load (dead load) and the overturning moment on each of the shear walls; ideally, if the overturning stresses can be accommodated within the 33% increase in the allowable stresses. Such a condition is achieved in shear wall structures where the shear walls carry the entire gravity load. At the other extreme, many buildings have only a central core as a shear wall which carries sometimes only a small portion of the dead load while resisting the majority of the overturning moments. If such buildings have a properly balanced shear wall-frame interaction, they may have sufficient rigidity to resist lateral forces. However, cases have been observed where a slip-formed core had around it a one-bay frame in which the flat plate had neither sufficient connection to the exterior columns nor a moment connection to the core; the result was an intolerable flexibility of the total building in response to wind.

EXAMPLE 10-2: Shear wall-frame interaction–The objective of this example is to determine, through a step-by-step optimization, the minimum amount of shear walls required for a given apartment building. The typical half-floor plan, section and column sizes (36-story building) are shown in Fig. 10-35. The basic structural system denoted as structure 'A' consists of a centrally located corewall extending throughout the entire height of the building and ten 3-bay open frames in the transverse direction. The assumed uniform wind load is 20 psf on the 60 × 220 ft building.

The structure was analyzed for wind using the computer program described in Ref. 10-19. Although structure 'A' has more than sufficient stiffness (the computed drift is 1/850) the net tension in the extreme windward "fiber" at the base of the corewall is 700 psi.

To reduce this tension in the corewall a 20-foot wide, 24-story high shear wall was introduced in an exterior bay of one of the open frames as shown in Fig. 10-36(a). This new structural layout was denoted structure 'B'. The analysis of this case (see Table 10-3) shows that although the tension at the base of the corewall is reduced significantly–to 395 psi from 700 psi in 'A'–substantial tensile stresses occur at the base of the added shear wall (515 psi). This indicates that the added shear wall, due to its stiffness, attracted a high moment relative to its dead load–resulting in significant net tensile stresses.

In an effort to further reduce the tension at the base of the corewall a third structure, structure 'C', with a pair of 20-foot wide, 18-story high shear walls along an exterior column line, Fig. 10-36(b), was next analyzed. As might be expected, this further stiffened the structure, bringing the drift (deflection index) down from 1/873 for structure 'B' to 1/928. In addition, the stress at the base of the corewall was reduced to the point where a net compressive stress of 30 psi occurs in the extreme windward fiber. However, the tensile stresses at the base of the additional shear walls have increased to 615 psi–from 515 psi in structure 'B'.

When the shear walls in structure 'C' are assumed to be located along an interior column line, such as line 2 in Fig. 10-36(b), the net tensile stress at the base is reduced from 615 psi to 192 psi (as shown for the case 'C-1' in Table 10-3). This is due to the added dead load on the shear wall. In structure 'C' and 'C-1' only the slab strips were considered as linking the additional pair of shear walls.

Structure 'D', a fourth structure considered, is essentially the same as structure 'C-1' except that beams were introduced to link the additional shear walls along column-line 2 such that the stiffness of the coupling elements connecting the pair of shear walls is three times that of the slab strips in 'C'. Table 10-3 indicates that the increase in stiffness of the coupling between the pair of shear walls not only increased the compressive stress at the base of the corewall, but also

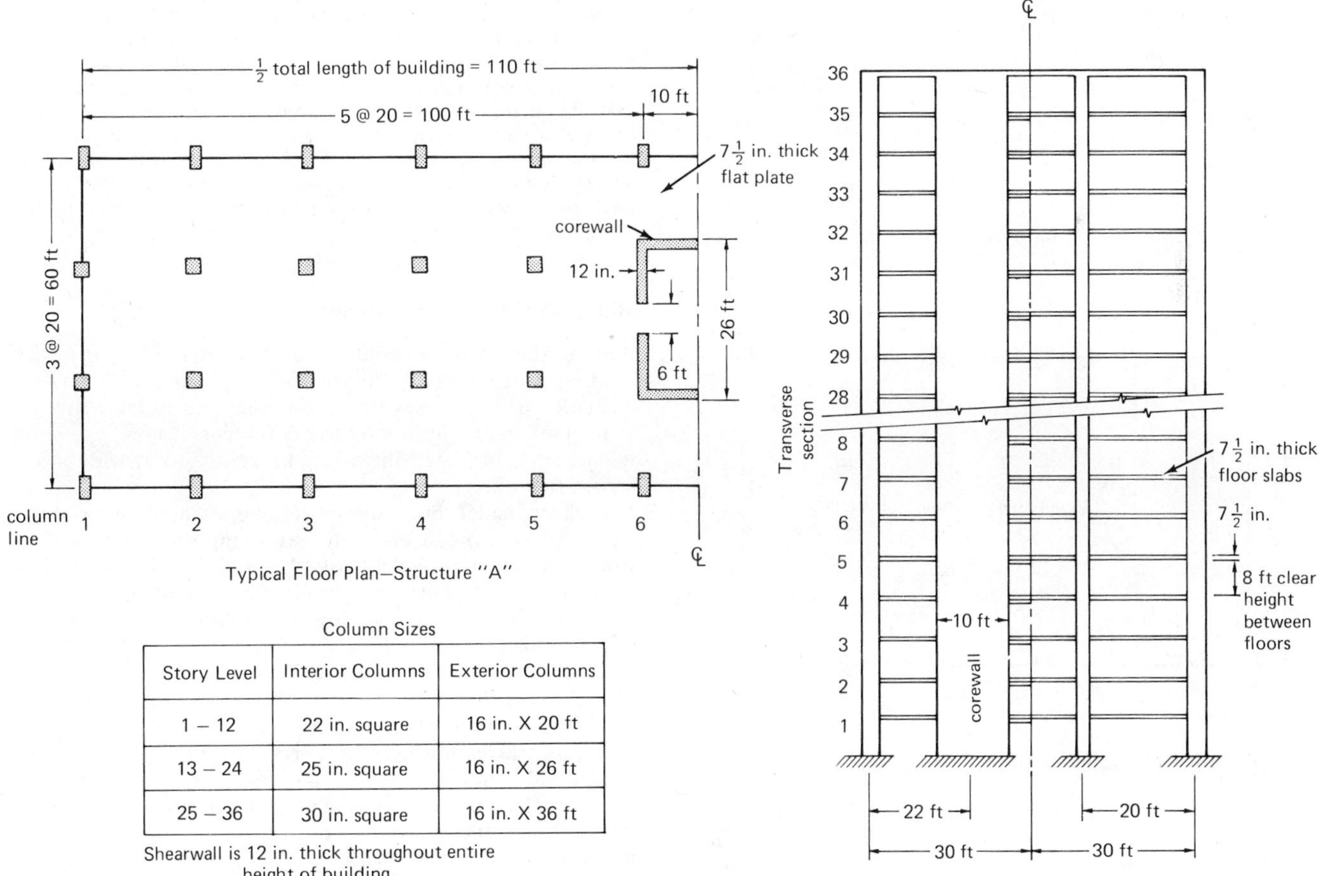

Column Sizes

Story Level	Interior Columns	Exterior Columns
1 – 12	22 in. square	16 in. X 20 ft
13 – 24	25 in. square	16 in. X 26 ft
25 – 36	30 in. square	16 in. X 36 ft

Shearwall is 12 in. thick throughout entire height of building.

Fig. 10-35 Structure 'A'–plan, section, and column sizes.

TABLE 10-3 Summary of Results of Lateral Load Analysis of 36-Story Apartment Building.

Structure designation	Description	Drift	Stress in extreme fiber-windward side: Corewall	Stress in extreme fiber-windward side: Additional shear wall
A	Basic Structure 1 frame-shear wall 4 open frames	1/820	+700 psi (tension)	–
B	Basic structure 'A' with one open frame replaced by frame shear wall (w/single 20′ wide, 24-story high shear wall)	1/873	+395 psi	+515 psi
C	'A' w/one open frame replaced by frame-shear wall (w/2-20′ wide 18 story high shear walls on column line 1)	1/928	−30 psi compression	+615 psi
C-1	Same shear walls as in 'C' moved to interior column line 2	1/928	−30 psi compression	+192 psi
D	'C-1' with beams linking 18-story shear walls 3 times as stiff as in 'C'	1/992	−130 psi	+180 psi

Considering effect of dead load only.

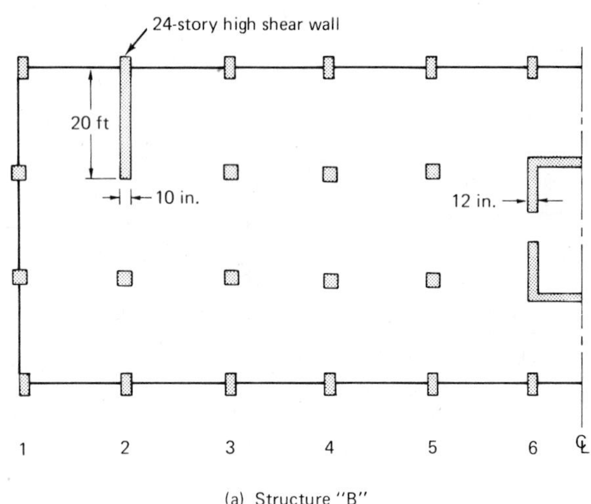

(a) Structure "B"

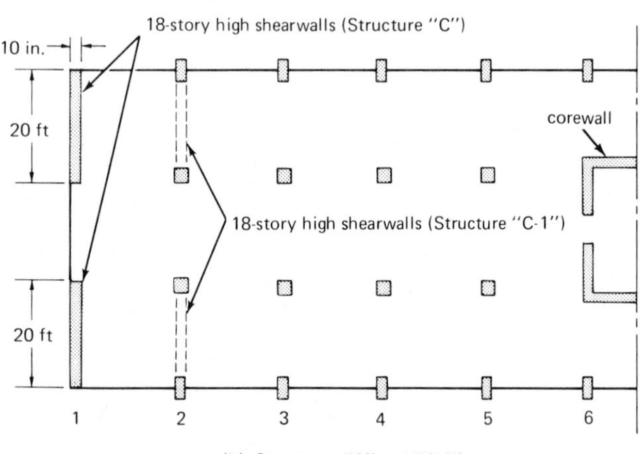

(b) Structures "C" and "C-1"

Fig. 10-36 Typical floor plans for structures 'B' and 'C'.

slightly decreased the tensile stress under the additional (coupled) shear walls. The decrease of tensile stress in the coupled shear walls is only slight since the reduced shear wall moments are accompanied by axial forces (tension and compression) resulting from coupling.

It is obvious that the tensile force resulting from the net tensile stresses has to be either anchored into the foundation material, or it must be shifted with the help of shear beams to the neighboring columns to be overcome by their gravity loads. If the tensile load cannot be accommodated, the shear wall will rotate at the base, resulting in a different distribution of wind shears and moments throughout the structure.

Figure 10-37 shows the variation of the story shears in the corewall along the height of the building for the four cases considered, plotted in terms of the percentage of the total applied story shear. Note the relatively abrupt changes in the magnitude of the story shears at locations where changes in stiffness occur. Also shown in the figure is a tabulation of the overturning moments at the first floor resisted by the corewall in terms of percentages of the total overturning moment. It can be seen from the figure that in the upper stories the corewall has negative shears due to frame-shear wall interaction–which means that the internal interactive forces cause a considerable reduction of the overturning moments in the corewall. As can also be seen, the corewall in structure 'A' carries 82% of the shear at the first floor, while carrying only 33% of the total overturning moment at this level; the remainder of the overturning moment is resisted by the overturning of the frame.

10.5.8 Shear Panel Buildings

Regular shear walls extending throughout the height of the building force a discipline on the architect and it may be difficult, in some cases, to accommodate a suitable mix of apartment sizes required by the developer. A new system of shear panels has been introduced recently to provide more layout flexibility.

A shear panel building is defined as containing shear panels of reinforced concrete extending on one or several stories within the height of the building and scattered throughout the plan. The panels are used mainly as walls separating apartments. In lower buildings, say up to 15 stories, the same function of limiting the drift could be accomplished by clay or concrete masonry panels, if they fit tightly within the frame. It should be noted that in a shear panel building resistance to lateral loads is provided primarily through their shear resistance. The panels can accept only limited overturning moments, since they have no vertical continuity throughout the height of the building. In this respect, shear panels are shear walls in the true meaning of the word, while continuous shear walls in tall buildings resist the lateral forces predominantly by moment resistance.

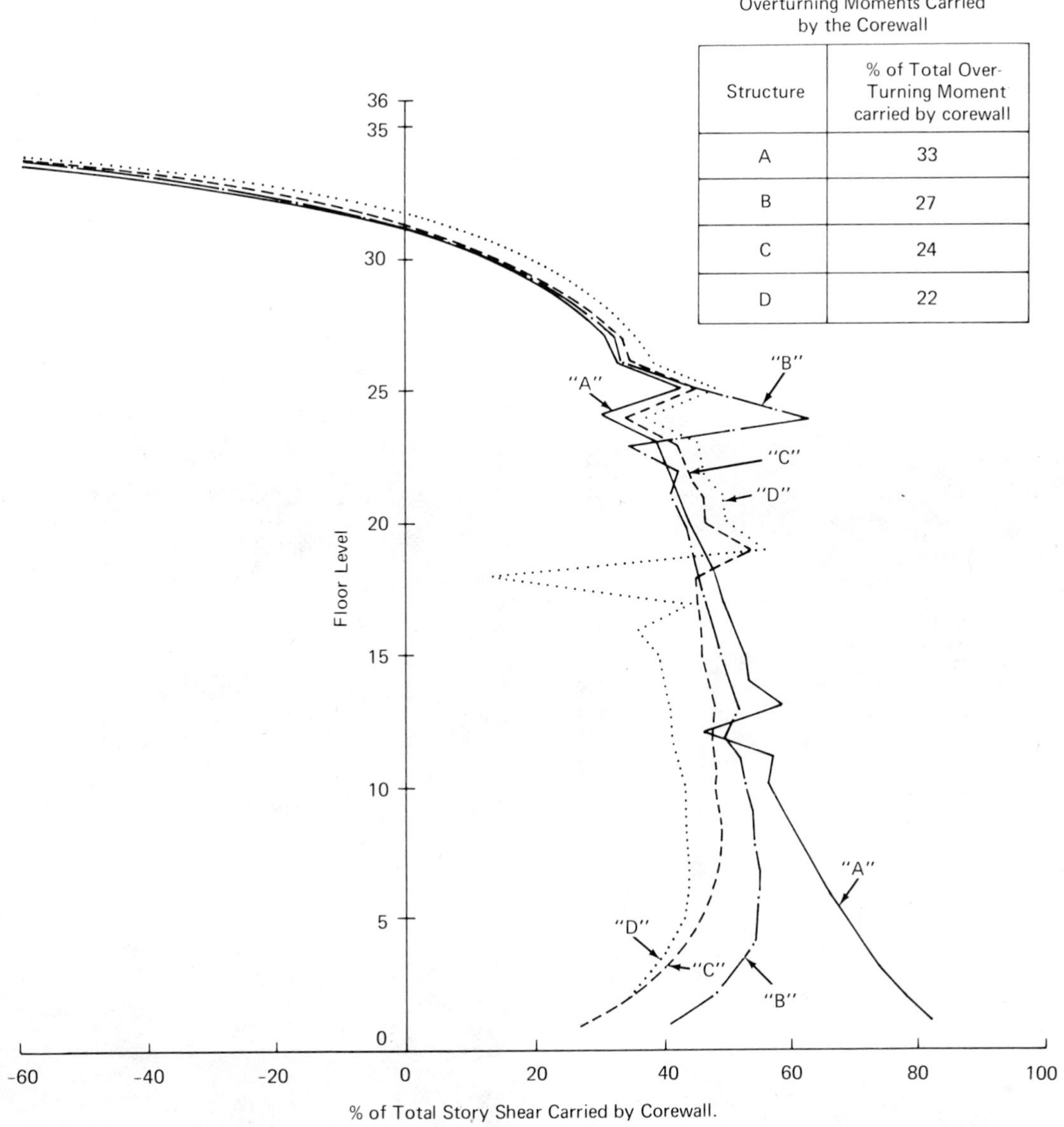

Overturning Moments Carried by the Corewall

Structure	% of Total Over-Turning Moment carried by corewall
A	33
B	27
C	24
D	22

Fig. 10-37 Comparison of shears and moments in the corewall due to wind.

It is desirable to locate shear panels between columns so that their weight can be carried directly by the columns and not by the slabs in bending. If located between columns, they act as deep beams supporting their own weight and the slabs.

At tops and bottoms of shear panels, the slab acting as a membrane transfers shear from panel to panel. Thus, the slabs are subjected in their planes to large forces at every discontinuity of a shear panel. These forces may require reinforcement of the slab as a horizontal plate. If the panel layout is not symmetrical in plan, the torsion of the building should be considered.

Since no shear wall-frame interaction can take place to any significant degree, the effectiveness of shear panels may be comparatively low.

Shear panels have been used on several flat plate apartment buildings in the 20-story range. Fulfilling the double function of a wall and a lateral resisting element for the building, they may be economically attractive, particularly if prefabricated in a precasting yard or on site and erected as the casting of the slabs progresses. Further studies are required on this potentially advantageous system before it can be widely utilized.

10.5.9 Lateral Resistance Contributed by the Shape of the Structure

It is well known that the shape of structures (other than rectangular blocks) have a substantial effect on the lateral resistance. Unfortunately, to the structural designer this is still elusive quantitatively at present, since in the majority of cases (even with computerized analysis) our analytical models are plane frames usually lumped together for computer time economy. Figure 10-38 shows a number of building shapes which, by the nature of their shape, are assumed to increase the lateral resistance. Only a three-dimensional computer analysis can offer a realistic appraisal of the influence of shape. However, for most buildings, such sophistication is still beyond the economical reach. Only very few structures have been analyzed to date using a three-dimensional computerized procedure.

A structure in which the shape was the prime source of lateral resistance is the new City Hall of Toronto (Fig. 10-39). The two towers were analyzed as vertical cylindrical shells stiffened by the slab system.

The example of the curved 19-story Caromay building in Caracas, Venezuela (Fig. 10-40) is only one of many cases

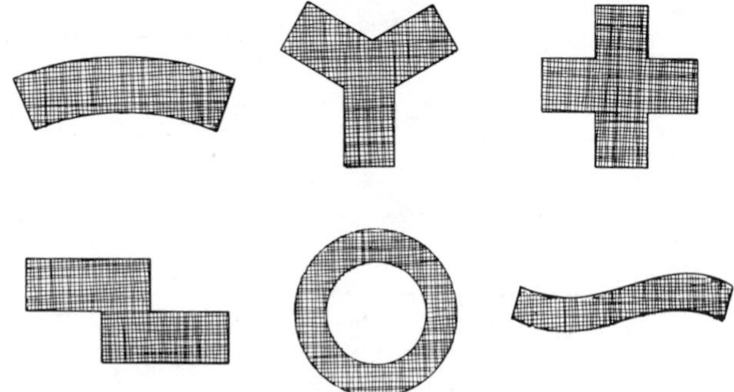

Fig. 10-38 Shapes of buildings affecting lateral resistance.

Fig. 10-39 Toronto City Hall.

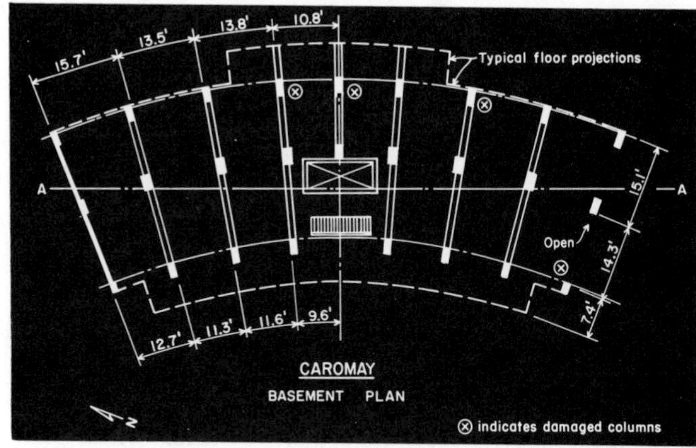

Fig. 10-41 Caromay framing plan.

of structures which, during an earthquake, exhibited the influence of the shape on the lateral resistance. The structural system of the Caromay consists of radial frames as shown in Fig. 10-41, spanned between the frames by a light joist floor with hollow clay tile fillers. Although the building was designed with radial frames, it behaved as a vertical cylindrical shell as a result of its shape. The motion of the building was perpendicular to axis *A-A* (Fig. 10-41). As a result, the columns located furthest from the neutral axis of the building were subjected to high axial forces due to racking and failed in classical compression at their mid-heights. No evidence of frame moments could be seen at the tops and bottoms of the columns. These observations led to the conclusion that the lateral resistance of this structure was due to shell action and no frame action (for which the building was designed and reinforced) took place.

10.6 DESIGN FOR VERTICAL LOADS

Although the lateral resistance is what constitutes the difference between a tall building and a low building, the design for vertical load is nevertheless the basic aspect of engineering, even for tall structures, in its effect on the economy of the project. Proper column layout which leads to economical spans while not impeding the functional requirements of the building is the key to an economical slab system to support the gravity loads. The slab system constitutes between 60 and 85% of the structural cost, depending upon the height and function of the building.

10.6.1 Analysis for Vertical Loads

Although computers have considerably advanced the sophistication of our designs for lateral loads, the soundest approach to design slab systems for gravity loads is the traditional method of considering each floor slab with its columns above and below assumed fixed at their remote ends. This procedure, which has been in use for a number of decades, proved itself both technically valid and easy to handle in a design office. Most of the available procedures for the design of the various slab systems are based on these assumptions.

Although many of the comprehensive computer programs can be used for gravity load analysis, there is an inherent susceptibility of receiving misleading results when such an analysis is carried out, since these programs are based on the input of the entire structural system with its loading, and then an analysis is performed for the entire structure as one unit. The dead load in a real building, however, is built up gradually and, for example, in a 40-story building, a dead load of the 10th floor cannot be resisted by a 40-story frame, since at the time the 10th floor load is applied, there is only a 10-story frame available to resist this load, and not a 40-story frame. Therefore, the procedure of simultaneous analysis of an entire structure is correct only for live loads and loads applied after the structure is completed. For dead loads a simultaneous analysis is correct only if all vertical elements of the structure have identical stress levels. Simultaneous loading of the structure can lead to large errors in cases where neighboring columns are designed to different stress levels which result in differential elastic shortenings. For example, if we consider a flat plate slab or a continuous beam in the 50th story, supported by a highly stressed column while the neighboring columns are lightly stressed, the differential elastic shortening over the height of the structure will result in the slab not having any negative moments over the highly stressed column; on the contrary, the analysis will show substantial positive moments over the column. In reality, when the 50th floor slab is cast, all the elastic shortening of its supports due to gravity from 49 slabs has already taken place and no elastic settlement stress will affect the slab when it is cast. However, the slab will be subjected to elastic differential support shortening for all the dead loads which will be applied above the 50th floor.

Similarly, if a transfer girder is designed with the entire structure above considered as a vierendeel girder in a simultaneous analysis, the load on the transfer girder will be underestimated since the lower story loads are carried mainly by the girder (no frame above is yet available at the time they are cast) and not by the entire vierendeel. (See section 10.6.5.)

Computer programs need to be developed which will treat construction time as an additional variable, which means the vertical load will be considered as applied gradually story-by-story while at the same time the framework is being progressively built up. Until the time when such programs are available, the traditional method of designing one story at a time produces much more reliable results.

10.6.2 Selection of Slab Systems

The choice of a slab system for vertical loads depends upon length of span, loading intensity, and the function of a building. As discussed in a previous section of this chapter, the need for a smooth ceiling in residential occupancies (apartments, hotels, dormitories) resulted in the flat plate being

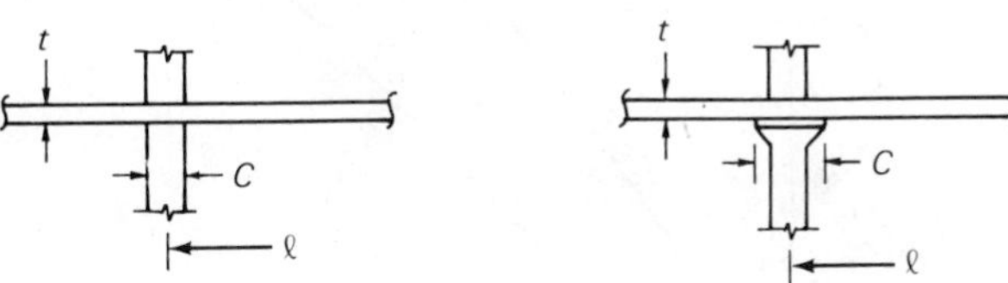

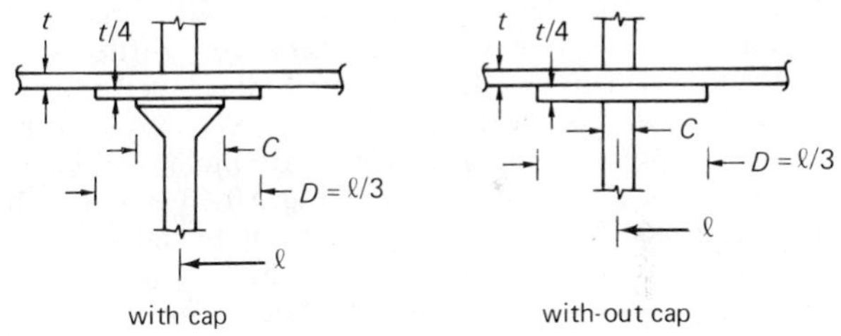

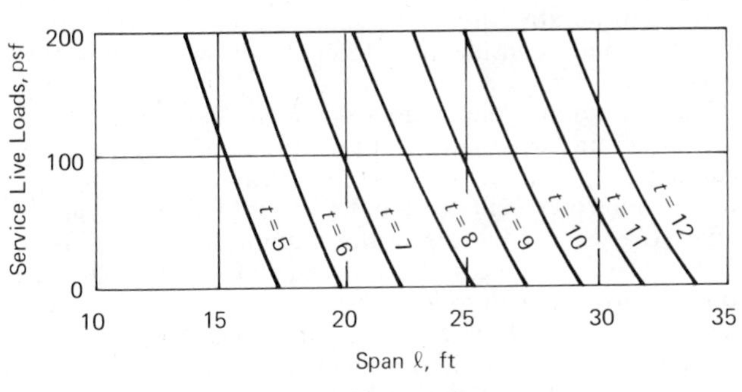

(a) Flat Plates – Square Bays

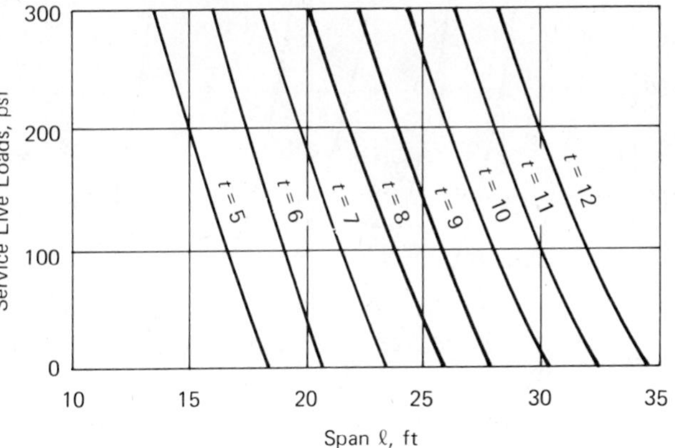

(b) Flat Slabs – Square Bays

Fig. 10-42 Preliminary selection charts for flat plates and flat slabs.

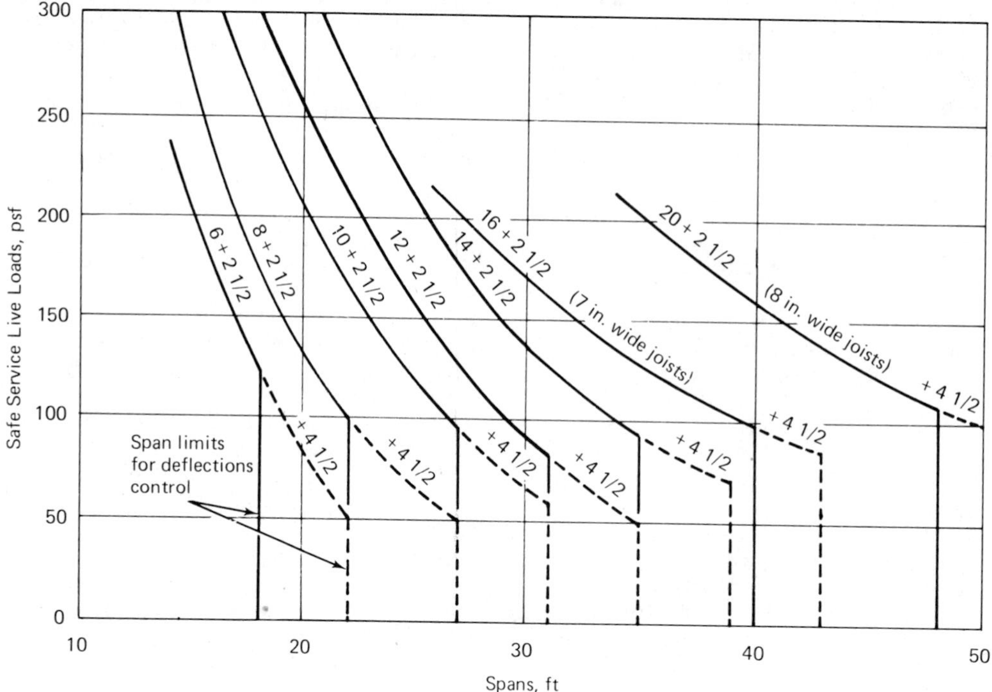

Fig. 10-43 Preliminary selection chart for one-way joist slabs.

the most prevalent and economical slab type. Also, post-tensioned flat plates and flat slabs received wide usage in recent years, in particular for longer spans in apartments, commercial uses, and parking garages.

In office buildings, where a hung ceiling is used to cover mechanical and electrical ducts, a ribbed ceiling (either one-way joists or waffle slabs) may be economical.

Figures 10-42, 10-43, and 10-44, show preliminary selection charts for flat plates, flat slabs with and without column capitals, one-way joists, and for waffle slabs on 3-ft centers. Figure 10-45 gives a preliminary selection chart for thicknesses of post-tensioned flat plates and flat slabs. While the previous selection charts are plots of thicknesses as functions of span versus load, Fig. 10-45 gives the thickness as a band versus span for the normal range of loadings. These selection charts are based on the longtime deflection limit of $L/360$ stipulated by the *Code*. The charts can be used for preliminary sizing only.

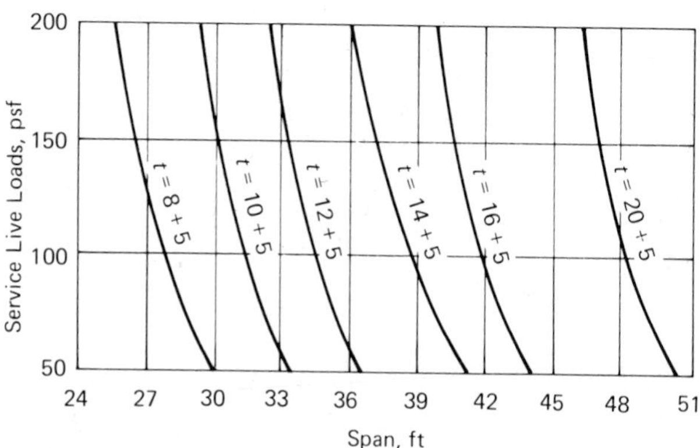

*Thickness, t, is based on longtime deflection limit $\ell/360$ from combined weight of slab and 30 psf superimposed load; ribs 6 in. wide spaced at 36 in.; normal-weight concrete, strength 4,000 psi; grade 60 steel.

Fig. 10-44 Preliminary selection chart for 3-ft waffle slabs.

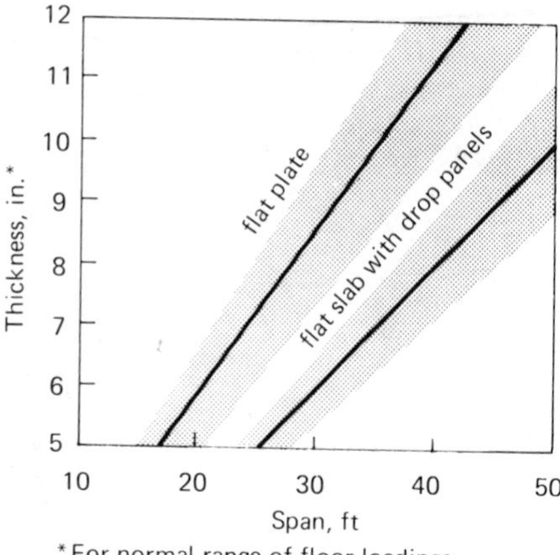

*For normal range of floor loadings

Fig. 10-45 Preliminary selection chart for post-tensioned flat plates and flat slabs.

10.6.3 Columns

Column loads are customarily determined on the basis of tributary areas, despite the fact that such effects as differential creep, differential elastic shortening, high differences in adjacent negative moments, and creep of the slab system, may sometimes considerably upset the distribution by tributary areas.

The amount of column reinforcement has a very dominant effect on the economy of high-rise reinforced concrete buildings. Therefore, the lowest possible percentage of vertical reinforcement should be attempted. This can be achieved either by increasing the column size (if there is no conflict with architectural requirements), by increasing the concrete strength (up to 9000 psi concrete has already been used) or by both. High strength reinforcement with a yield strength of 60 ksi is almost exclusively used for columns. Premiums sometimes have to be paid for the use of 75 ksi steel if the material is not available in stock.

Quite often in the upper stories of buildings the columns

TABLE 10-4 Column Live Load Reduction Provisions

UBC, ANSI[1,2], NBC[1,2], BOCA[3]

Reduction Factors		
Storage Load	*Assembly Load*	*Parking Load*
Min. { 0.08 times supported area (%); 23.1 $(1 + D/L)$ (%); A max value depending on load type.		
Max 60%	0%	40%

CAN

No. of Story	Reduced Loads		
	Storage Blgds.	*Industrial Garages*	*Other*
1	1.00	0.90	0.85
2	0.95	0.85	0.80
3	0.90	0.80	0.75
4	0.85	0.75	0.70
5	0.80	0.70	0.65
6	0.80	0.70	0.60
7	0.80	0.70	0.55
>8	0.80	0.70	0.50

SBC, CHG[3]

No. of Story	Reduction Loads		
	Storage Bldg.	*Industrial Garages*	*Other*
1	1.00	1.00	1.00
2	0.90	0.90	0.90
3	0.80	0.80	0.80
4	0.70	0.70	0.70
5	0.70	0.70	0.60
>6	0.70	0.70	0.50

NYC

No. of Story	Reduced Loads		
	Storage Loads	*Assembly & Parking*	*Other Loads*
1	1.00	1.00	1.00
2	0.85	0.85	0.85
3	0.80	0.80	0.80
4	0.80	0.80	0.75
5	0.80	0.80	0.70
6	0.80	0.80	0.65
7	0.80	0.80	0.60
8	0.80	0.80	0.55
>9	0.80	0.80	0.50

Abbreviations:
UBC Uniform Building Code
ANSI American National Standards Institute
NBC National Building Code (US)
SBC Southern Building Code
CHG Chicago Building Code
CAN National Building Code of Canada
BOCA Building Officials, Code Administrators
NYC New York City

NOTES: 1. Parking and roof loads are not reduced.
2. Reduction based on the area of the floor supported is calculated one story at a time rather than for all stories.
3. Parking loads are not specifically mentioned.

are much larger than required to carry the load. The oversized columns result either from architectural requirements or from economical considerations to have as few changes in the column section as possible. In such cases the *Code* stipulates that the percentage of reinforcement be less than the required minimum of 1% by considering only that part of the column section which is needed to carry the load.

Most all codes allow reduction of column live loads for multistory buildings. However, storage loads, parking loads and assembly loads are classified in the various codes for special treatment. Assembly loads are considered to be nonreducible live loads. An example of such load would be a public meeting hall load for which no reduction is allowed according to most building codes.

Storage and parking loads are reducible live loads and special load reduction procedures are specified in the various codes. A tabulation of the live load reduction provisions of the major codes is summarized in Table 10-4.

10.6.4 Selection of Normal Weight vs Lightweight Concrete

Structural lightweight concrete is widely available and the present knowledge of its physical properties and behavior has made it an important factor in high-rise construction. It is used in many and varied construction applications; however, for multistory construction its use may be particularly attractive. The use of normal weight versus lightweight concrete in multistory construction involves the study of several variables. In most areas of the country the lightweight concrete used in buildings has a weight of 110-115 lbs per cubic yard, since the lightweight fines are replaced by natural sand.

The reduced dead load resulting from using lightweight concrete produces the following structural advantages: (a) a reduction of sizes of flexural members, columns, and foundations; (b) equivalent fire ratings are obtained with thinner lightweight concrete sections; and (c) when equal member thicknesses are used, dead load creep deflections may be smaller for lightweight concrete due to lower sustained stresses.

Lightweight concrete is more costly per cubic yard, with price differentials ranging between \$3.00 to \$7.00 per cubic yard, depending on location and market conditions.

Figure 10-46 shows the summary of a series of charts carried out in a study of cost comparison[10-27] between normal weight and lightweight concrete for multistory flat plate residential buildings in which average material and labor prices in the Chicago area were used.

A price differential of \$5.00 for lightweight over normal

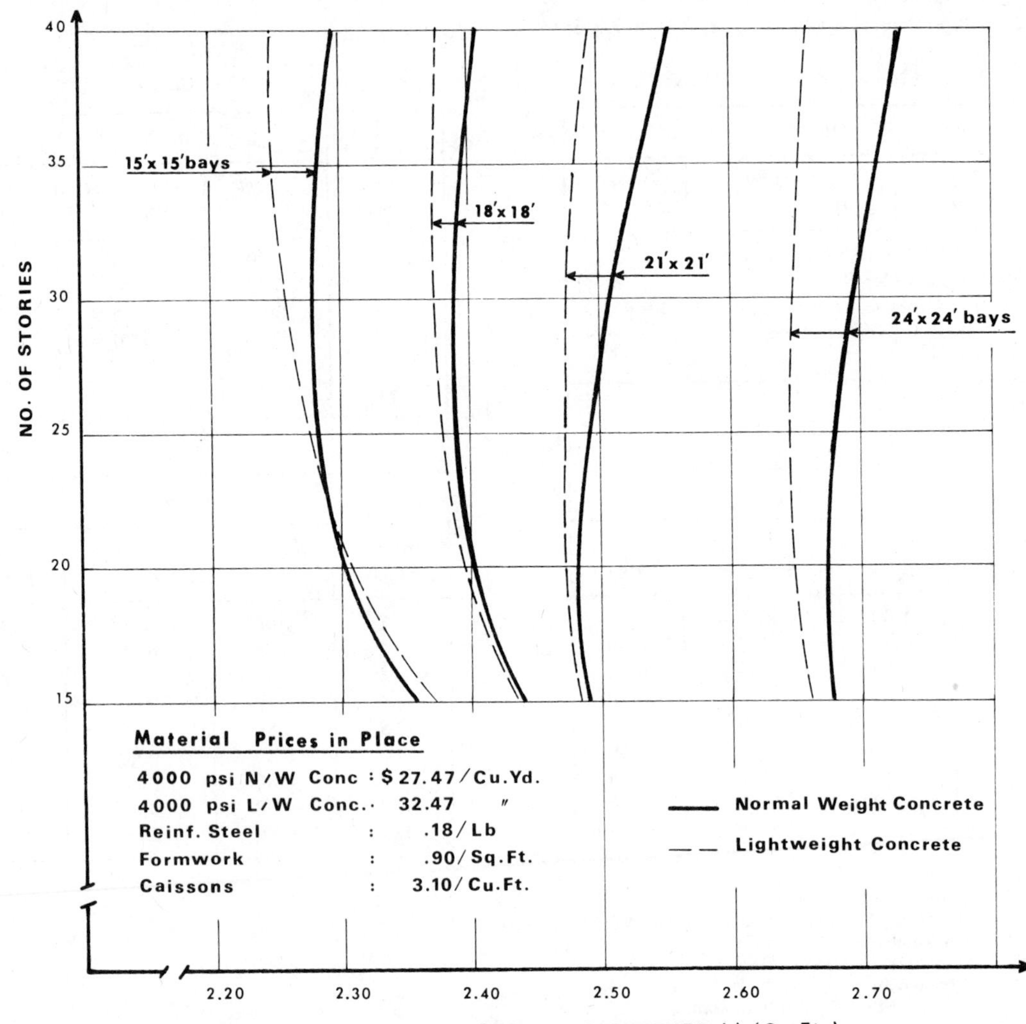

Fig. 10-46 Cost comparison between normal weight and lightweight structures for flat plate apartment buildings.

weight concrete was assumed for the study. Contractor's overhead and profit were not included in the prices. The different structural elements were analyzed and designed (static loadings only), following local codes and standard engineering practices. The main variables were number of stories and bay sizes in a typical building containing 3 X 6 square bays. The buildings were assumed to be supported by caissons, about 100 feet long, typical for the Chicago area. Comparative costs are given in dollars per square foot of construction. The details of the study reveal that the slab reinforcing shows a difference in favor of lightweight concrete in all bay sizes, while the slab concrete shows an additional cost for lightweight from 8 to 16 cents per square foot.

The concrete used in the columns for this study was normal weight concrete in all cases, as is presently customary in the Chicago area. It should be noted, however, that recent research has shown that lightweight concrete can be safely used in columns of multistory buildings. The column curves show that column cost per square foot decreases as the bay size increases and that the column costs show a difference in favor of lightweight concrete slabs in all bay sizes as compared to normal weight concrete slabs.

The final cost curves shown in Fig. 10-46 were compiled by adding the cost of caissons, columns, form work, slab reinforcing, and slab concrete at prices shown in the figure. They show an obvious total cost increase when the bay sizes increase for both normal weight and lightweight concrete.

When soil conditions or other structural requirements do not demand the use of lightweight concrete, the local price differential and availability in the area are the major considerations in the choice of normal weight versus lightweight concrete for a given structure.

10.6.5 Transfer Girders

In recent years many buildings have been constructed with mixed occupancies, such as apartments or hotels in the upper stories while the lower stories contained commercial space, theaters, schools, or other nonresidential space. Additionally, in metropolitan locations most of the new buildings contain parking garages in the lower stories.

Often closely spaced peripheral columns of multistory buildings cannot continue within the ground story where wide open spaces are required in lobbies and entrances. In other cases buildings are constructed over highways or railroads, and the normally laid out columns of a multistory building must be transferred to a system of columns compatible with the highway or railroad.

Functional considerations require different spans for the different parts of the building. On the other hand, cost per square foot rises with increases of span. If there are many floors of shorter spans to be constructed over an area with larger spans, it is advisable to investigate the technical and economical feasibility of a transfer floor separating the two different structural systems.

There are numerous examples of buildings with transfer girders in reinforced concrete, prestressed concrete and structural steel. In one example an entire 60-story residential building has been placed over a commercial building with only a few of the upper columns continuing directly through the lower part, while most of the 60-story columns were supported by a story high grid which, in turn, rested on columns laid out to the requirements of the commercial space.

In general, transfer of substantial loads is a costly operation and should be done only after careful consideration. The least costly transfer can be accomplished by using inclined columns within the transfer floor, as shown in Fig. 10-47. In this particular case, each of the quadrupeds 'collected' four columns of the 10-story office building into one column below the ground floor. It is obvious that such transfer of forces is accompanied by tensile and compressive horizontal components in the slabs at both ends of the inclined columns. These forces, which can be very substantial, have to be accommodated within the slabs. Post-tensioning offers a very effective means, both technically and economically, to accommodate such tensile forces.

The usual way to transfer column loads is through transfer girders, sometimes one or several stories high, to give them sufficient rigidity not only to reduce the elastic deflections, but also to minimize the creep deflections which would cause distortions within the entire building above. Shear and creep considerations are important design aspects to keep in mind when designing transfer girders.

The transfer girders are usually designed to carry the entire dead and live load of the columns above. This condition of full load can actually occur only if the structure above consists of simply supported members. In reinforced concrete, where all members are usually continuous, the live load is carried not only by the transfer girder but by the entire structure above acting together with the transfer girder as a vierendeel structure. However, in supporting the dead load, participation of the structure above is only partial in sharing the load with the transfer girder, since the dead load is applied gradually, as the structure is being built up. Thus, for example, the load of the lower 10 stories of a 60-story frame resting upon a transfer girder, is resisted by a structure consisting *only* of the transfer girder and 10 stories above (not 60 stories). Therefore, to use a frame program to simultaneously handle all the vertical loads for vierendeel action of the transfer girder with the entire structure will lead to an underestimated load on the transfer girder. Tables and curves to estimate the load carried by transfer girders carrying multistory columns with consideration of the above mentioned loading history were prepared in Ref. 10-28.

Figure 10-48 shows an elevation of a 32-story building in which some of the ground floor columns have been eliminated and the entire elevation designed as a vierendeel girder. It is obvious that in such a case spandrel beams in a number of lower stories are subjected to extremely high shears to transfer the load gradually to the column below; consequently their shear reinforcing must be carefully designed, or rather, overdesigned.

Figure 10-49 shows a "natural solution" for the problem of shear transfer by the spandrel beams of the lower stories,

Fig. 10-47 Inclined columns as a means of changeover from regular to large spans (Australia Square Office Building in Sydney, Australia).

Fig. 10-48 Some of the ground-story columns have been eliminated and the entire elevation participates as a vierendeel girder (Lincoln Park Building).

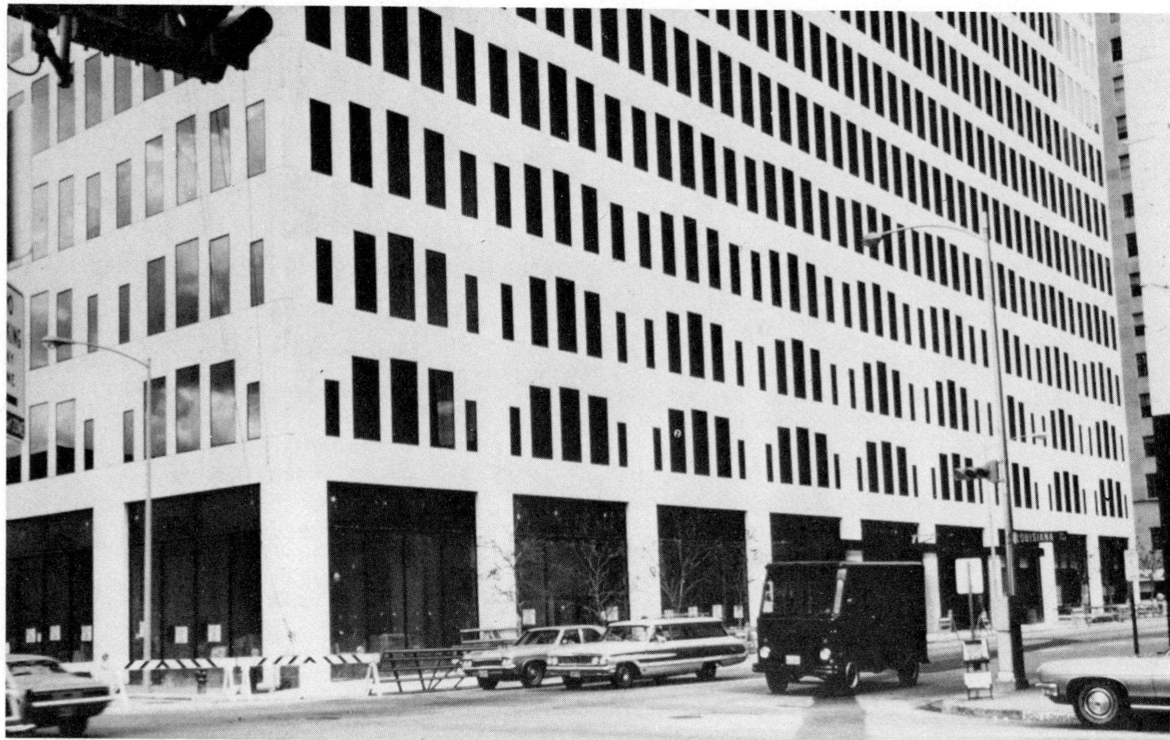

Fig. 10-49 The natural look of the transfer to a limited number of columns in the ground story.

resulting in a tree-like appearance–the support columns branch out into the closely spaced columns of the structure above. The thickness changes of the spandrel beams followed the results of the analysis, while the shear stresses were kept about constant.

Due to its size, the construction of a transfer girder may pose field problems. For example, the transfer girder of the Brunswick Building in Chicago (see Fig. 10-50), which is about 7 ft thick by 26 ft high, had to be cast in three horizontal layers to avoid excessive generation of heat of hydration. The decision of such unusual horizontal layering of a girder was made only after a series of tests verified the validity of this procedure. The actually measured temperature of the concrete in the massive girder never exceeded 155°F.

10.7 STAGGERED WALL-BEAM STRUCTURES

This new structural system developed for both structural steel and reinforced concrete, was introduced in the 1960s.[10-29, 10-30] The basic concept is illustrated in Fig. 10-51. The system employs story-high, pierced walls extending across the entire width of the building to lines of columns placed along the exterior faces. By staggering the locations of these wall beams on alternate floors, large clear areas are created on each floor, yet the floor slabs span only half the spacing between adjacent wall beams from the top of one to the bottom of the next. Within certain structural limitations, the wall beams can be pierced for corridor and door openings at locations required by architectural planning.

The staggered wall-beam system is best suited for rectangular plans, but it may be adapted almost equally to other plan shapes, e.g., Y-shape, cruciform, annular, serpentine, or broken rectangle. The system is suitable for most types of multistory construction having permanent interior partitions such as apartments, hotels, student residences, etc. Although the system was initially developed for cast-in-place construction, it can very well be built with precast elements both for the walls and the slabs, or either. Obviously very careful planning must take care of the fact that the main reinforcement for the beams is located in the slabs.

A great advantage of the wall-beam building is the ease with which a large open area can be created underneath for parking, commercial use or even to allow a highway to pass under the building. If such clear areas are desired, a nontypical slab will be created for the slab above the open space.

It is possible to maintain the basic module of, say, 12 ft as the span of the structural slabs and exterior column spacing and still achieve different clear spaces from those in Fig. 10-51. Figure 10-52 shows some of the many possible variations in wall-beam layout to accommodate the architectural planning of various apartment units within a building. Although the wall-beam layouts in Figs. 10-52(a) and 10-52(b) are shown as regular both horizontally and vertically, the pattern can be changed along the length of the building to accommodate various combinations of apartments on any one floor. Examples of such variations are shown in Figs. 10-52(c) and 10-52(d). Thus, considerable latitude in arrangement of units is possible within the basic structural requirements of the system, namely, wall beams on alternate floors at each column line (to keep floor slab spans on the basic module) and staggered with respect to those in the other–though not necessarily adjoining–column lines (to provide lateral rigidity).

10.7.1 Architectural Design

It can be seen from Figs. 10-53(a) and 10-53(b) that the wall-beam system will always have two typical floors, the layout on any one floor being repeated two floors above and two floors below. The existence of two separate typical floors makes it easy to create attractive building ex-

Fig. 10-50 The Brunswick Building in Chicago.

teriors of either very simple or highly complex design. This is true particularly with cantilevered balconies as shown in Fig. 10-54.

The wall-beam is well suited for two-story apartment units to separate living and bedroom areas.

10.7.2 Mechanical Design

Since the apartment units on consecutive floors are staggered, more vertical plumbing lines are required as kitchens and bathrooms are above one another on alternate floors and not consecutive floors. A detailed study of the plumbing, ventilating, heating and air conditioning requirements associated with an apartment building using the staggered wall-beam system shows that these can be satisfied at a cost only slightly higher than that required for a flat plate or a shear wall building.

10.7.3 Structural Analysis and Design

In the longitudinal direction, the floors act as continuous one-way slabs either resting on or hung from alternate wall beams. The slabs, in combination with the wall beams, form a concrete I-beam. The alternating or staggered system of beams results in each floor slab being alternately subjected to tension and compression forces in the transverse direction as it forms first the bottom and then the top flange of wall beams. As a result, shearing forces are created in the horizontal plane of the floor slabs which tend to reduce the external bending moment applied to the individual wall beam. These beneficial shearing forces are not considered in the analysis and design.

Virtually all wall beams will contain openings. The presence of an opening reduces the shear capacity of the beam at that section. Since the lower slab is relatively flex-

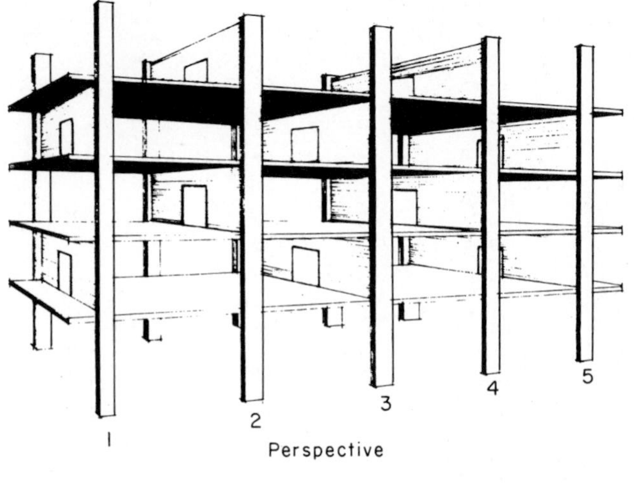

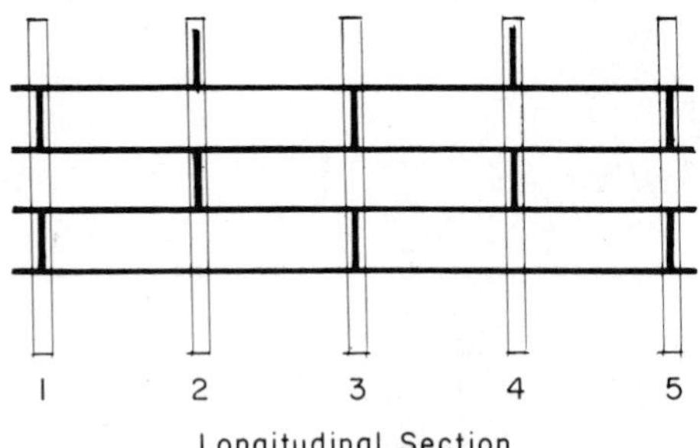

Fig. 10-51 Regularly staggered wall-beam building.

ible, it can be assumed that the beam over the opening carries the entire vertical shear. The action of gravity loading on the wall beams creates a state of combined bending and axial compression in this beam.

In buildings say, less than 12 stories, due to the relatively flexible columns, the wall-beams may be analyzed as simply-supported beams.

Where the column stiffness is of a similar order of magnitude as that of the wall beam, it may be desirable to consider the restraining effect of the columns, particularly since this can result in a reduced area of reinforcement required at the center of the beam span. Expressions for the rotational stiffness of a wall beam with openings in its web and worked out examples illustrating the analysis and design of a typical wall beam have been worked out in Ref. 10-31.

10.7.4 Simply Supported Wall Beams with a Single Opening in Web

The analysis for the moments and forces in the sections of a simply-supported wall beam with a web opening may be carried out by considering Fig. 10-55. The wall beam has an I-section consisting of a story-high wall acting as a web and portions of floor slabs forming the top and bottom flanges.

By assuming that the lower framing flange (LFF) acts mainly as a tension member, developing negligible moment and shear, the following relation holds at any vertical section through the opening:

$$M^\circ = M' + M''$$

where

M° = total external moment at any section through the opening

M' = Tf = primary moment, produced by the tension, T, in the LFF and the compressive force, C, in the UFF (upper framing flange); f being the effective lever arm of the C–T couple

M'' = secondary bending moment in the UFF

It is convenient to carry out the analysis by determining the influence line for the uniform tension, T, in the lower framing flange. From the known value of T, the moment M''_x at any section in the upper framing flange may be obtained from the preceding relationship, i.e.,

$$M''_x = M^\circ_x - M' = M^\circ_x - Tf$$

where M°_x is the simple beam external bending moment at the section considered.

The influence line for the tension, T, in the lower framing flange reinforcing steel is obtained by determining the deflected shape of the beam axis due to a unit shortening of the lower flange. The deflected shape then represents the influence line for T (Muller-Breslau's principle).

Curves from Ref. 10-31 for calculating the tension T in the lower framing flange for a wall-beam with a single rectangular opening under a uniform load over the entire span

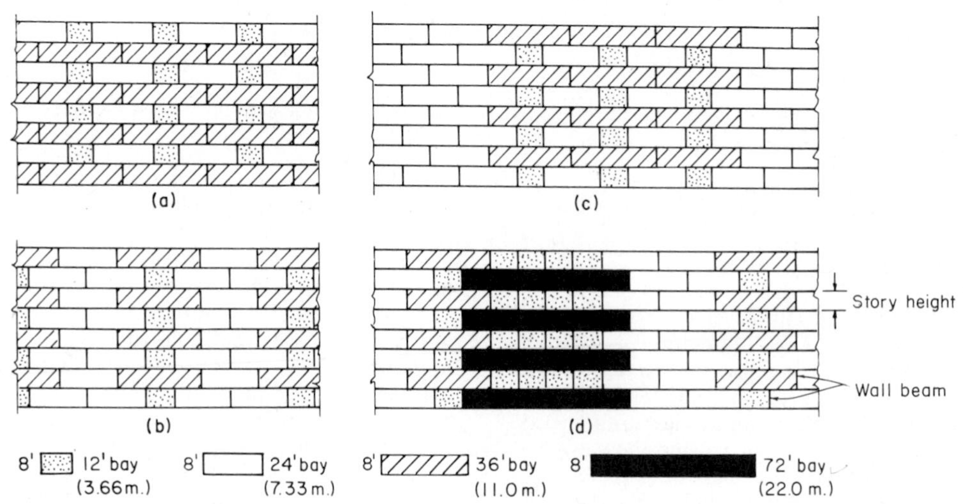

Fig. 10-52 Longitudinal section showing possible wall-beam layouts based on a 12-ft basic module.

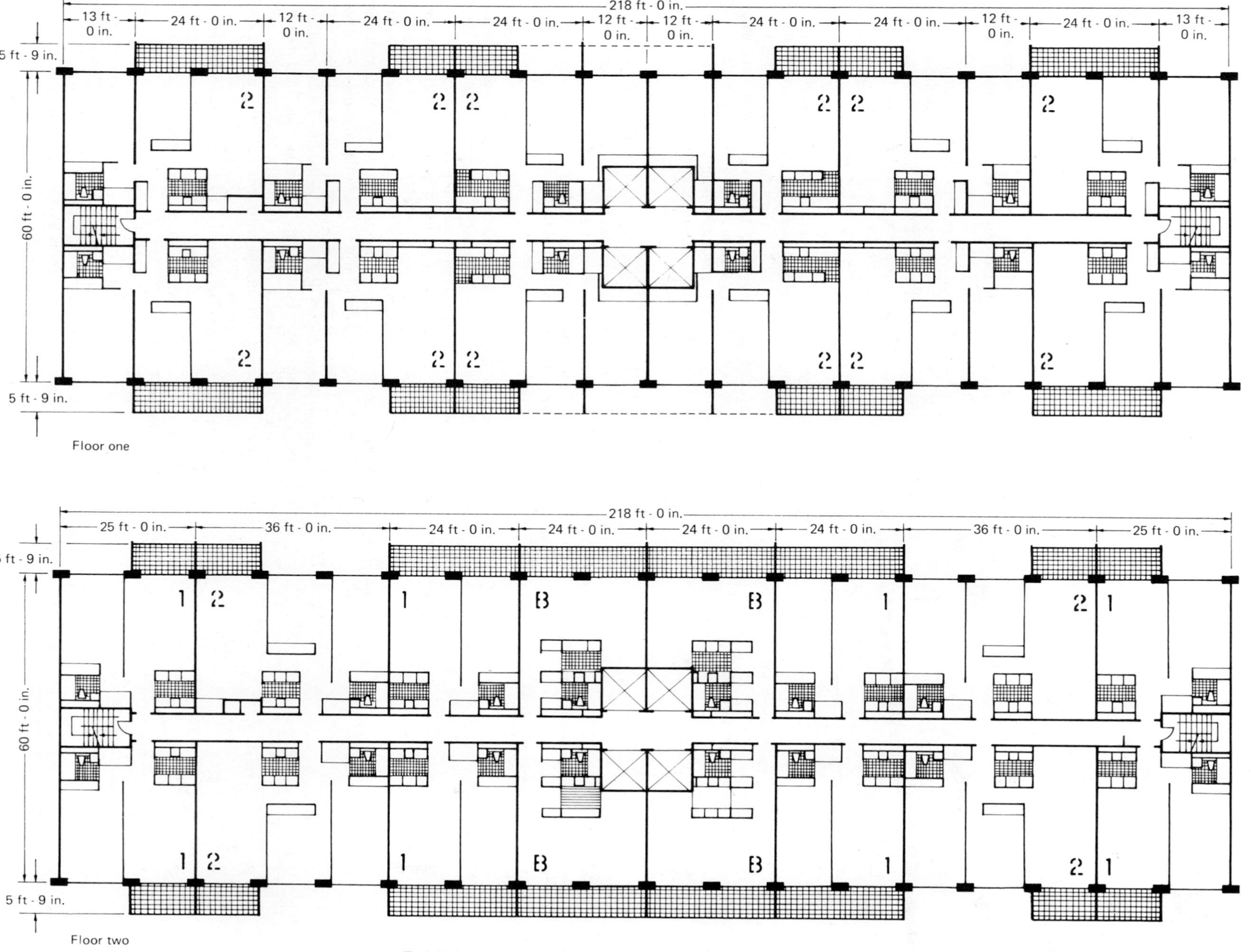

Fig. 10-53 Two typical floor layouts for staggered wall-beams.

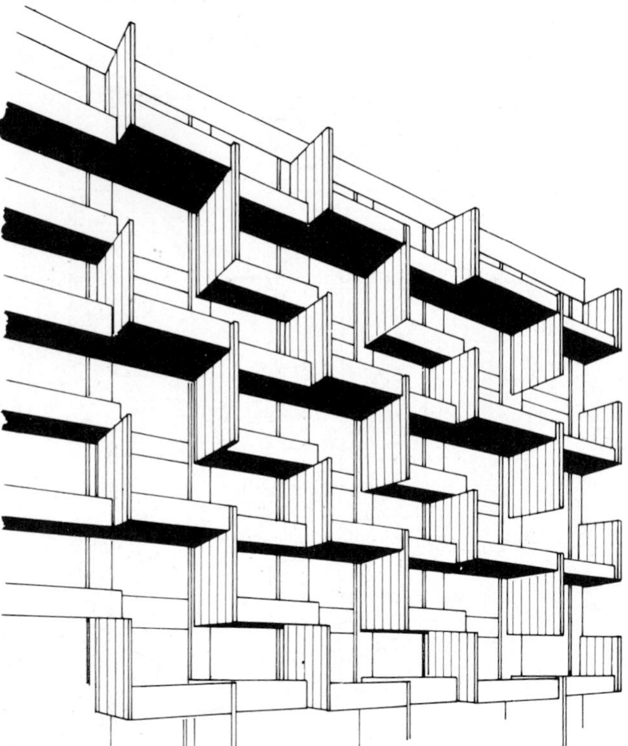

Fig. 10-54 A possible facade treatment for two-story units with balconies.

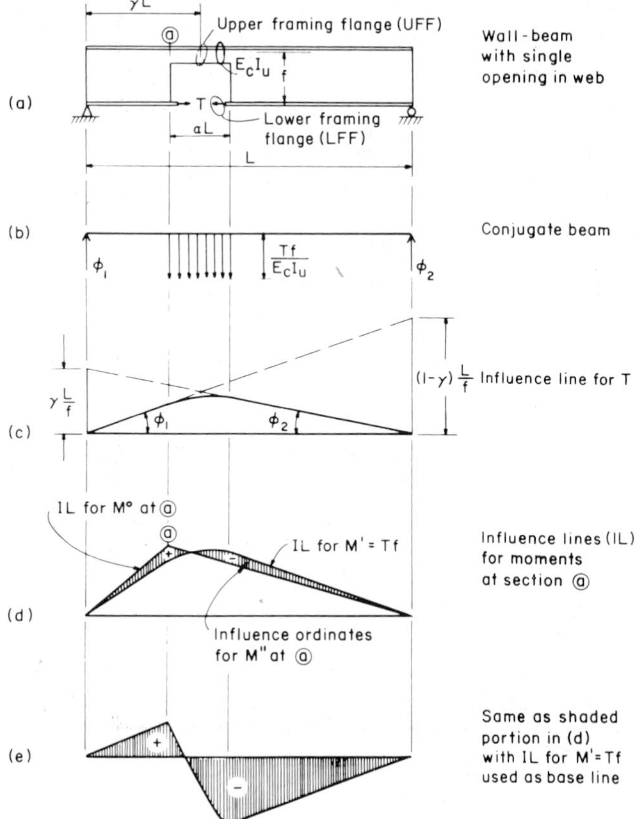

Fig. 10-55 Influence line for secondary moments at left end of upper framing flange.

are given in Fig. 10-56(a). In a wall-beam building, corridor walls will usually run longitudinally along each side of the corridor opening. The dead weight of these walls will represent concentrated loads on the wall-beam. The value of the tension, T, corresponding to concentrated loads on the span may be obtained using the influence lines for T shown in Fig. 10-56(b).

Reference 10-31 also has curves for calculating moments in the upper framing flanges due to uniform loads over the span, concentrated loads, for partial loads, and charts for stiffness of wall beams to be used in frame analysis.

10.7.5 Shear Stresses

The shear stresses in the beams above the web openings may be the critical design condition. Consequently, openings in the web should be so located as to avoid regions of high shear which means preferably in the center of the span. Also wind action causes added shears in the UFF in particular in the lower part of the building. Shear strength considerations may require thickening of the UFF at times.

10.7.6 Deflections

The wall beam is an extremely rigid element and under usual loading conditions will present no deflection problem. The presence of openings in the beam causes a slight increase in the deflection due to vertical loads over what would be expected in a solid-web beam. In view of this, and to estimate possible camber requirements, a check of the deflections of the wall beam may be desirable.

10.7.7 Lateral Load Design

To understand the behavior under lateral loads, it is essential to consider the combined action of adjacent transverse frames. Assuming that the floor slabs act as infinitely stiff horizontal diaphragms, all points on any one floor slab will have equal horizontal deflections.

Considering each transverse frame separately, it would appear at first glance that each bent would undergo the stiff beam-flexible column behavior illustrated in Fig. 10-57(a). However, if the adjacent bents are also considered, the horizontal deflections at each floor level would not be equal, so that this behavior is not possible. The deflected shape must, therefore, be of a form shown in Fig. 10-57(b) which results in equal deflections at each floor with the columns in a single curvature. The horizontal shears at each floor are resisted by the wall-beam system acting in the manner illustrated in Fig. 10-57(b) inducing axial loads in the columns. The floor slabs, acting as rigid diaphragms, transfer the horizontal shears from the wall beams on one floor to the wall beams on adjacent column lines on the floors above and below.

The building may be analyzed by lumping together identical frames and considering each of these composite frames separately under lateral loads proportional to their respective tributary areas.

Because of the restraint provided by the adjacent wall beams, lateral (shear) deformation of the columns in each of the above composite frames is prevented. The frames thus act essentially as vertical cantilevers. Because the columns function principally as axially loaded members under lateral loads directed parallel to the wall beams, they can be oriented more effectively by having their long sides parallel to the longitudinal axis of the building. This alignment also will provide greater freedom in locating the interior partitions behind the columns.

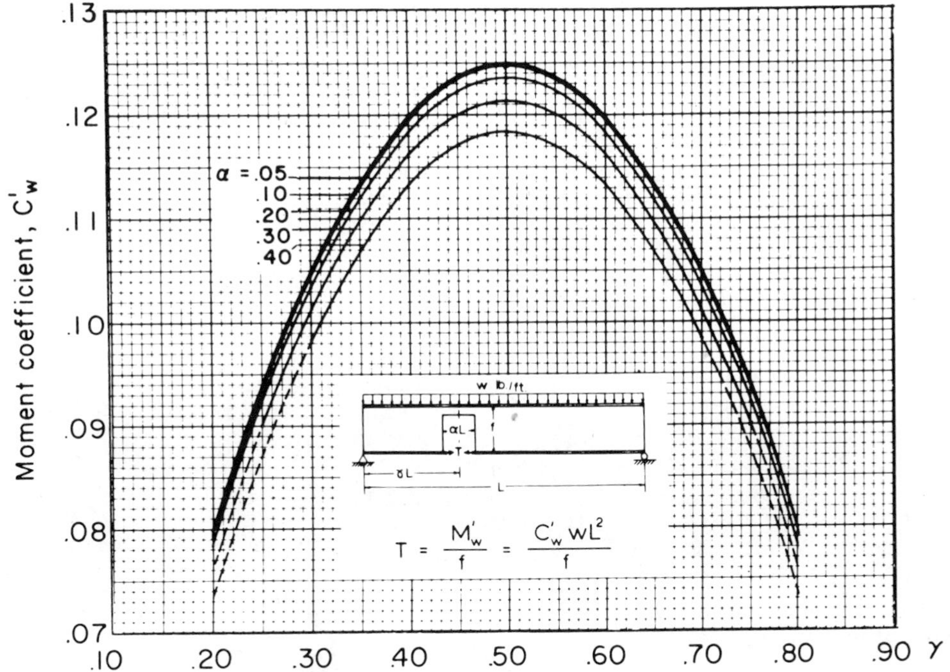

(a) Coefficient for the tension, T, in the lower framing flange due to a uniformly distributed load, w, over entire span

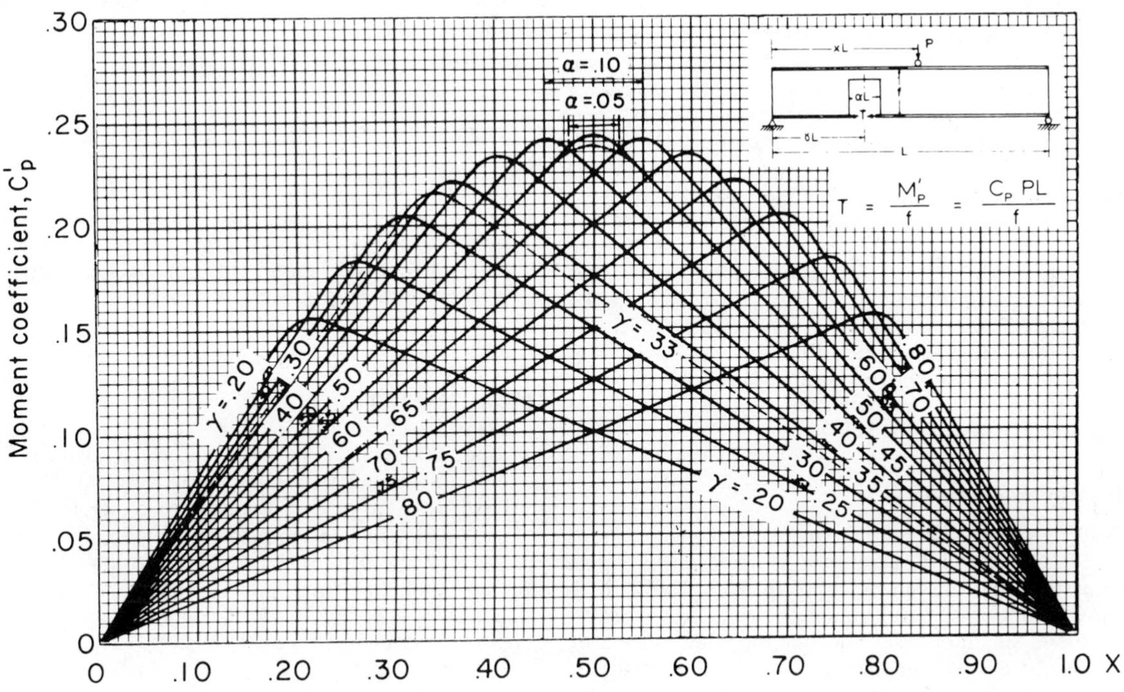

(b) Influence lines for the tension, T, in lower framing flange (Connecting curves over openings for $\alpha \neq .05$ omitted

Fig. 10-56 Coefficients for calculating the tension in lower framing flange.

Lateral loads applied in a longitudinal direction of the building can be handled in a number of ways. With the floor slabs, the outside columns constitute a frame which would be sufficient to carry the horizontal loading in a building of moderate height. For taller buildings, this frame could be stiffened by the introduction of shallow spandrel beams at the column lines. If these frames were insufficient to carry the horizontal loading, then the elevator shaft or

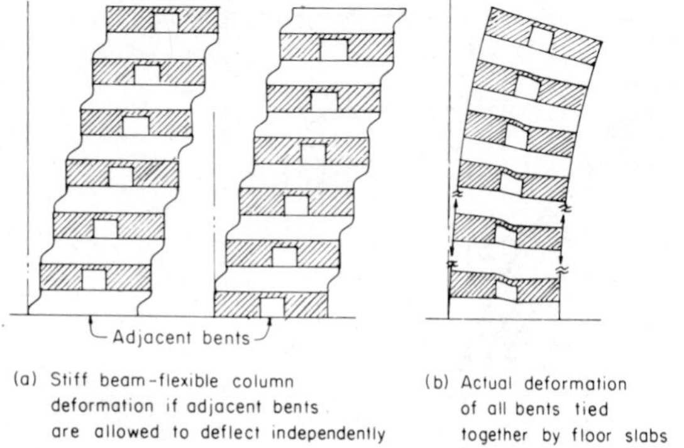

(a) Stiff beam-flexible column deformation if adjacent bents are allowed to deflect independently

(b) Actual deformation of all bents tied together by floor slabs

Fig. 10-57 Deformation of staggered wall-beam frames under transverse lateral loads.

stairwells could be made load-bearing to create interior shearwalls. In this case, the analysis should take into account the interaction of the shearwall and the frame systems.[10-11, 10-25, 10-26]

10.7.8 Slab and Wall Minimum Thickness and Construction Sequence

The short slab spans may require a minimum slab thickness of only 4 in. to satisfy strength and deflection considerations. However, for ease of reinforcement placement and to reduce transmission of airborne noise between floors to acceptable levels it is recommended that slabs be a minimum of 5 in. thick. Impact noise is almost solely a function of the type of floor covering.

For ease of construction and to satisfy the minimum thickness requirement, a 6-in. web is recommended. This thickness permits one layer of reinforcing and adequate sound insulation. Under certain loading conditions thicker webs may be necessary to satisfy shear requirements.

The construction sequence will usually involve casting floor slabs and webs separately. The slabs contain the main tensile steel for the wall beams above them. Because the slab forming the lower flange of a wall beam is suspended from the beam web, the interface between web and flange is subjected to some tension in addition to the horizontal shearing forces. In view of this, it is advisable to check the capacity of this interface. Generally, the connection provided by a properly roughened construction joint, tied together by the vertical web reinforcement, is adequate as verified in the tests described below.

10.7.9 Reinforcing Details

Figure 10-58 shows the reinforcement arrangement for a 60 ft wall-beam with two openings in an apartment building as presented in the design example in Ref. 10-31. It is important to note the bars around the openings as in the flanges above the openings. The side opening had to be made lower to satisfy the critical shear condition in the flange above it. The flexural reinforcement of the wall beam is placed within the slab, so as not to interfere with the openings. Since the slab containing the flexural reinforcement of the wall is cast first, a cold joint is unavoidable. The joint must be carefully roughened and additional dowels provided through the interface.

10.7.10 Testing

The tests to failure which were carried out at the PCA laboratories[10-32] on a half-scale model (Fig. 10-59) showed the ultimate load capacity to be in excess of the flexural capacity predicted by the *Code* formulas. This apparent increase should be attributed to strain hardening of the reinforcing steel. Also the shear behavior above the doorway was by far better than predicted by theory. For the beam over the smaller doorway the maximum applied shear was nearly double the capacity implied by the *ACI Code* formulas.

A secondary but significant result of the tests was to indicate the desirability of increasing the web reinforcement in the solid web sections up to 100 percent above that required by the *Code* in order to reduce the width of diagonal tension cracks if and when they occur. The additional web reinforcement is not required for structural reasons but rather by esthetic considerations of the crack width, if the wall has no other finishes than plaster or paint, since these are at eye level.

10.7.11 Economics

An economic study of apartment buildings ranging from 16 to 36 stories to compare the wall beam system with the traditional flat plate and shearwall buildings showed that considering the reinforced concrete skeleton and walls between apartments, the wall beam system uses by far the least amount of concrete and is, therefore, the lightest as far as foundations are considered. The amount of reinforcement is about the same as in the flat plate system. The study also showed that since the wall beam system resists wind by cantilever action, the premium for height is smaller than in flat plate buildings. Figure 10-60 shows that each of the three systems is most economical for one quantity only; amount of formwork, volume of concrete, or the amount of reinforcing steel.

10.8 PIPE COLUMNS–FLAT PLATE STRUCTURES

In search for improved economy for low rise apartments in reinforced concrete a new system has recently been introduced.

The structural system introduced by Wiesinger and Holland, consulting engineers, Chicago, consists of a 5 in. thick, reinforced concrete flat plate supported on pipe columns. The columns are capped by I-beam shear heads, located below the slab in the direction of the partition as shown on Fig. 10-61. The columns are 3 in. diameter pipes and the shear heads are 6 I standard (12.5 lb/ft) beams about 2 ft long. Connection between columns above the floor is provided by means of a pipe sleeve welded on top of the shear head and embedded within the concrete floor thickness and serving as a receptacle of the column above. This pipe sleeve is the only mechanical shear connector between the floor and the shear head. The shear strength which is usually the governing design factor for thin flat plates supported on small columns is entirely eliminated as a problem by the size of the shear heads.

In a recent modification of the system, a shearhead (consisting of two channel-sections) has been located within the slab thickness, instead of under the slab. Thus, all the protrusions have been eliminated, resulting in a simpler operation.

The size of the pipe columns and the protruding shear heads (if any) are such that they can be incorporated within the thickness of the standard 4-in. partitions. Obviously, the orientation of the shear head must follow the direction of the partition containing the column. The 5-in. slab is theoretically sufficient for spans up to 15 ft. A number of

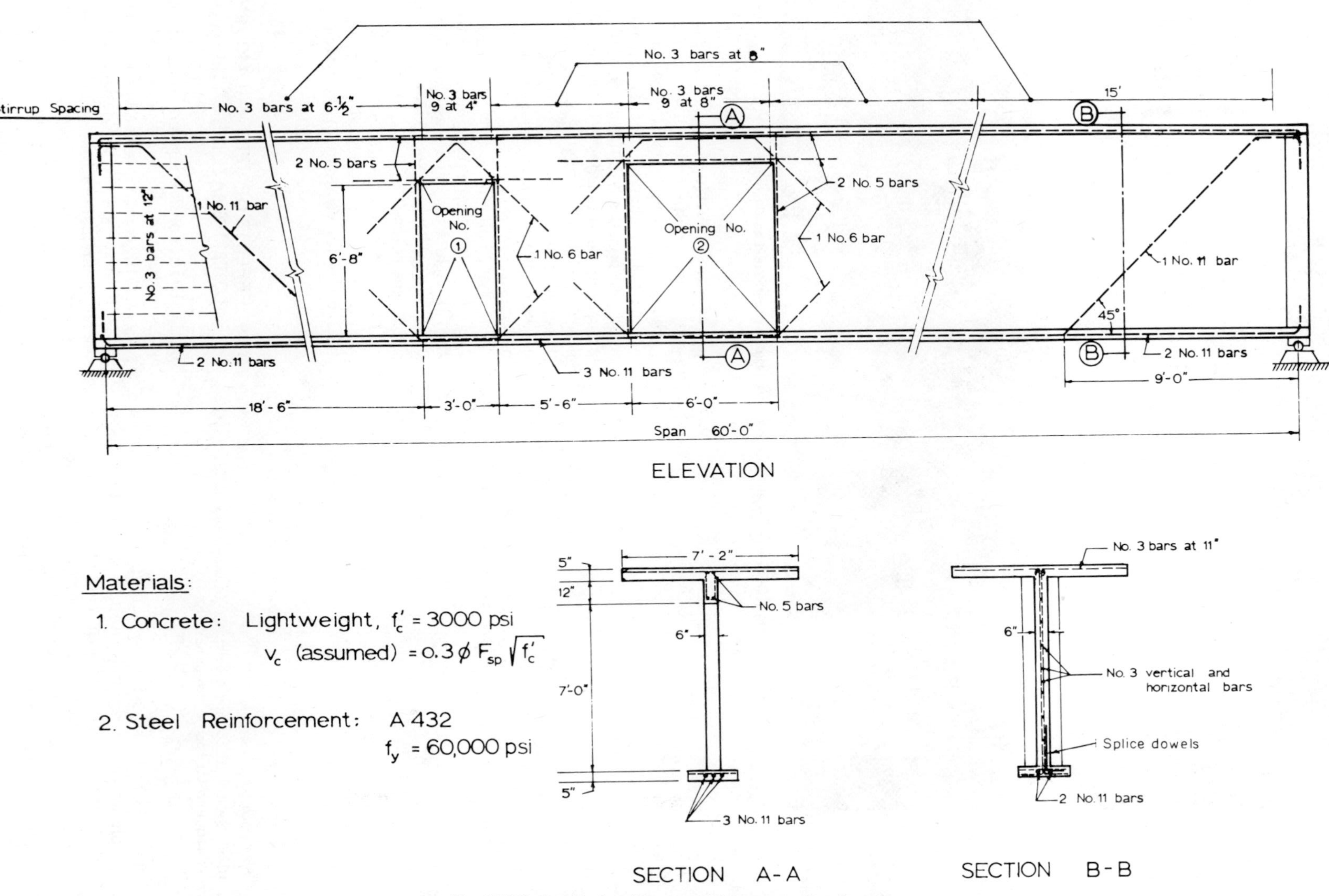

Fig. 10-58 Detail of wall-beam with two openings in web.

Fig. 10-59 Half-scale model of wall-beam during test to destruction.

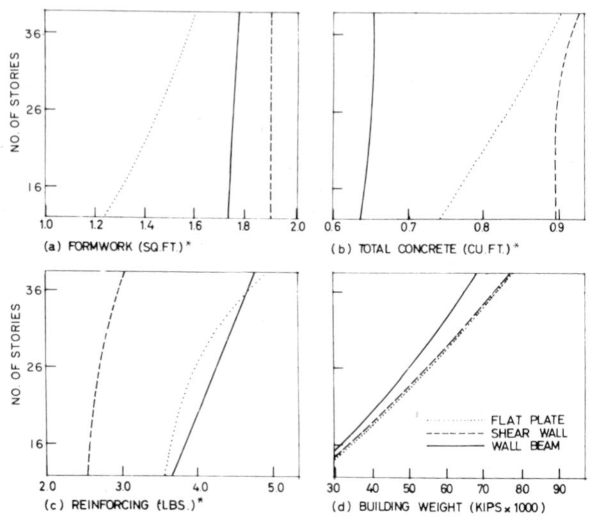

Fig. 10-60 Results of comparative study on quantities of structural material required for each of three building types.

buildings in which the system was utilized were designed with a column layout resulting in spans of about 13–14 ft without interference with the apartment layout.

The slabs were designed in accordance with the ACI 318-71 requirements for flat plates. A simplification of the usual reinforcement arrangement was achieved by concentrating the entire negative reinforcement over the column heads, thus eliminating the middle strip negative reinforcement. To verify that no excessive cracking would occur in the areas where the negative reinforcement was removed, in violation of the accepted practice and current codes, a field test was carried out[10-33] which proved that neither deflection nor cracking was affected by the changed location of the negative reinforcement. The fact that strength of flat plates is not affected by steel distribution has been verified some years ago in a series of tests in Australia.

The pipe column–flat plate system described above does not have any inherent stability. Therefore, in some of the buildings constructed, masonry walls in both directions proceeded simultaneously with the slabs, providing the necessary lateral strength, while in others a system of diagonal cables between the columns provided the stability during construction, supplemented later by the peripheral masonry.

The 3 in. columns are sufficient to carry about four to five floor slabs, depending upon the tributary areas of the slab. In eight story buildings, double columns were used in some locations where the load was beyond the capacity of a single pipe column.

The forming and shoring of the slabs was accomplished with large area prefabricated flying forms contributing considerably to the economy of the system.

10.9 EFFECTS OF VERTICAL MOVEMENTS DUE TO TEMPERATURE, CREEP, AND SHRINKAGE

In the 1950s considering the effects of temperature, creep, and shrinkage on the behavior of buildings was an academic subject. Although the researcher was familiar with some aspects of these subjects, the designer of buildings rarely considered them quantitatively in his designs.

Traditionally, the effect of temperature, creep, and shrinkage was considered in horizontal structures, such as long span bridges. These effects were usually neglected in multistory concrete buildings since, in the past, such structures seldom exceeded 20 stories. Some tall reinforced concrete buildings, built in the 1960s without consideration of creep, shrinkage, and temperature effects in the columns and shear walls have developed partition distress, as well as structural overstress, and in some cases, cracking in horizontal elements. It has become necessary for the structural engineer to consider the various differential movements and

Fig. 10-61 Pipe columns with I-beam shear heads.

to develop acceptable structural, as well as architectural, details for the satisfactory performance of the buildings.

The differential elastic and inelastic length changes are cumulative over the height of a structure. They start with zero at ground level and reach a maximum at the roof. Therefore, they become more critical with increasing height of a structure. For example, a realistic strain differential between an adjacent column and wall of 150×10^{-6} in./in. would produce the insignificant amount of about 0.016 in. per story of 9-ft height; which, however, for an 80-story building, would mean a maximum cumulative differential shortening of 1.3 in. at the roof level. Such distortions can obviously not be neglected, neither structurally nor in architectural details.

This sensitivity of ultra high-rise buildings to the effects of volume changes has, in the late 1960s, brought about the the development of a number of procedures to consider these effects in the design of buildings. Also, architectural and mechanical details to accommodate the distortions of the building have been developed.

10.9.1 Vertical Movements due to Temperature Changes of Exposed Columns

In the 1960s a large number of multistory apartments and office buildings were built with exterior columns partially, and in some cases fully, exposed to the weather. This was done initially for architectural expression of the structural frame and later for its significant economy, since about 15 to 20% of the elevation is replaced by structural elements. Also, removing protruding columns from the inside helped in accepting the trend of exposing columns.

Exposed columns, when subjected to seasonal temperature variations, change their length relative to the interior columns which remain unchanged in a controlled environment. For low buildings this causes insignificant structural problems which can be ignored. However, in taller buildings with partially or fully exposed columns this creates distortions of the slabs in the exterior bays, and temperature stresses become significant and must be considered due to the cumulative nature of the problem.

In the mid sixties, several 30-story buildings in the eastern United States developed serious partition cracking which focused attention on the problem of vertical movements due to column exposure.

The thermal movement of the exposed columns causes a racking of floor slabs and consequently a distortion of partitions. The column shortening within a story is only secondary in its effects on partitions. As can be seen in Fig. 10-62, the major effect on partitions is the rotation of the partition due to the cumulative change in length of the exposed column. The partition pivots around Point B when the exposed column shortens. This rotation causes separation between partition and exterior columns. The separation, δ, is the widest at Points A and C, decreasing gradually to zero.

The width of separation depends not only on the amount of movement of the exposed column, but also on the ability of the partition to distort elastically and inelastically. As could be observed from a number of buildings with partially exposed columns, there was generally insignificant distress in most of the common partition assemblies in the range below 20 stories.

It should be mentioned that all tall buildings with columns protruding beyond the glass line, clad or unclad, will experience temperature movements. Even some multistory apartment buildings with exterior columns completely within the building cladding developed partition distress in upper floors attributable to temperature movements.

Between the years 1965 and 1968 a comprehensive and relatively simple engineering procedure was established[10-34, 10-35, 10-36] for the solution of the structural and architectural considerations of column exposure. Also, at the same time, measurements of thermal movements were taken on a number of high-rise structures with exposed columns to verify the validity of the developed analytical procedure.

By the late sixties, the manufacturers of dry wall partitions improved the resiliency and details so that partitions could take a sufficient amount of frame distortions (temperature, creep, wind) without visible signs of distress and without loss of their acoustical properties.

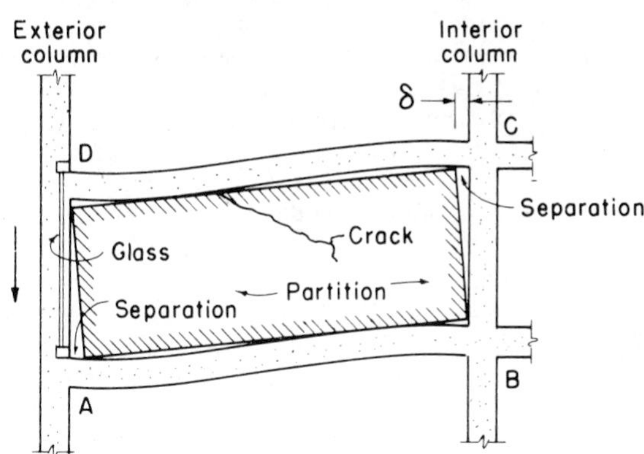

Fig. 10-62 Partition rotation and distress.

Two basic planning and design concepts have evolved for handling structures with exposed columns:

a. to accept large thermal movements and to design the structure for the resulting stresses providing all the needed architectural details to accept distortions without distress, and
b. to limit thermal movement to an acceptable magnitude as related to the span by limiting the column exposure through changing the glass line or by applying insulation.

The implications and applications of these two concepts will be discussed in subsections 10.1 and 10.2 of this chapter.

The engineering solution of the column exposure problem consists of:

a. determination of a realistic design temperature;
b. determination of isotherms, gradients and average temperatures for exposed columns;
c. determination of length changes of exposed columns and bowing stresses as a result of thermal gradients;
d. a frame analysis and design of the structure for moments and load transfer due to differential length changes of columns; and
e. details for partitions, cladding, windows, and other structural elements subjected to distortions as a result of thermal movements.

A design example including determination of isotherms, gradients and a complete analysis for moments, movements, load transfer and stress is given in Ref. 10-35.

-1. Design temperature—The highest and lowest effective temperatures of all partially exposed columns or walls (having one face inside the building at a constant temperature and the other face outside subjected to ambient temperature variations) for the full seasonal cycle are to be established. Weather conditions rarely remain steady over a long period of time. Since a steady state of heat conduction through any section is only possible if the boundary temperatures remain constant for a sufficient time, an equivalent steady state has to be derived.

There are two factors that influence the equivalent steady state of temperature: (a) the time lag of penetration of exterior temperature fluctuation into the concrete, and (b) the attenuation of their intensity (damping) as the distance from the face increases.

The time lag and attenuation of amplitude depend upon the frequency of temperature change and upon the thermal properties of the material (concrete), such as conductivity and specific heat. Materials with a higher thermal conductivity respond more quickly to temperature change.

The attenuation of exterior temperature amplitudes at various distances from the face is such that rapid temperature changes penetrate only skin deep. Studies show that amplitude of the 24-hour cycle does not stay long enough to penetrate sufficiently into the concrete. For example, 6 in. away from the face the daily amplitude from the mean attenuates to only 28%; the 7-day amplitude to 62%, and the 90-day amplitude to 88%. Thus, only the amplitudes of slowly changing temperatures penetrate considerably into the depth of a member. Beyond a certain distance from the surface the temperature of the concrete is affected by fluctuations of long cycles only.

It follows that to use a peak temperature lasting for only a few hours as the exterior design temperature for reinforced concrete members is too severe since the thermal inertia prevents the concrete from responding to quick changes of the surrounding temperatures.

Based on studies[10-34] it was recommended to use the minimum mean daily temperature with a frequency of recurrence once in 40 years as the equivalent steady state exterior winter temperature for design purposes of reinforced concrete members of usual size range subjected to exterior temperature variations. Such climatic temperature data is readily available from the guide books for heating, ventilating and air conditioning. Also, some local weather bureaus or building codes may be a good source to obtain the mean temperature.

A study of the lowest temperatures with a 40-year recurrence shows that the lowest daily mean temperatures for the United States range from -40°F. for parts of Montana, Wyoming and North Dakota to +40°F. for tips of California and Florida. The highest mean daily temperatures, however, show very little fluctuations over the entire United States; they range from 95°F to 105°F.

-2. Temperature isotherms, gradients, and average temperatures—After an equivalent steady state temperature has been selected, isotherms and gradients through the partially exposed columns have to be determined. For sections other than the infinitely long wall, the temperature gradient under a steady state of conduction is not linear. The problem is to determine realistic isotherms and an average temperature of partially exposed columns as well as the lowest and highest surface temperatures.

One of the better known methods of solving two- and three-dimensional steady-state heat flow problems is the relaxation method. The time involved in the analysis by the relaxation method may not always justify its use. A useful known graphical method of flow of water through soil was adapted for construction of isotherms and gradients. The application is based on the concept of flow net construction known from soil mechanics. The heat flow is represented by two sets of intersecting lines. One set of lines represents the heat flow lines, the other set of lines, perpendicular to the first, is the set of isothermal lines. An isothermal line is a contour of equal temperatures. The two sets of lines must intersect at right angles to form curvilinear squares with equal average sides in each direction. The difference in the final gradient between a very accurately drawn flow net and a less accurate net is insignificant. Usually, a few successive trial sketches are sufficient to construct a reasonably accurate flow net. The basic requirements are: (a) that the two sets of curves intersect at right angles and (b) that each field be as nearly curvilinear square as possible. The construction of isotherms is shown in Fig. 10-63.

The average temperature is given theoretically in a three-dimensional representation by a horizontal plane which divides the total volume created by the isotherms into equal parts. In most cases of prismatic columns, this may be simplified into a two-dimensional problem. A mean gradient through the column is drawn and the average temperature computed by a numerical summation as shown in Fig. 10-64.

$$t_{\text{average}} = \frac{\Sigma(\Delta d \times t)}{d} = \frac{\Sigma t}{n}$$

where Δd are vertical strips into which the column width, d, is subdivided. The graphically measured mean temperature of each strip is denoted t. If the column is subdivided into n equal strips, each d/n wide, the equation is simplified as indicated above.

To account for the thermal properties of the applied finishes and surface conductances, an equivalent column section instead of the actual section is used. The finishes

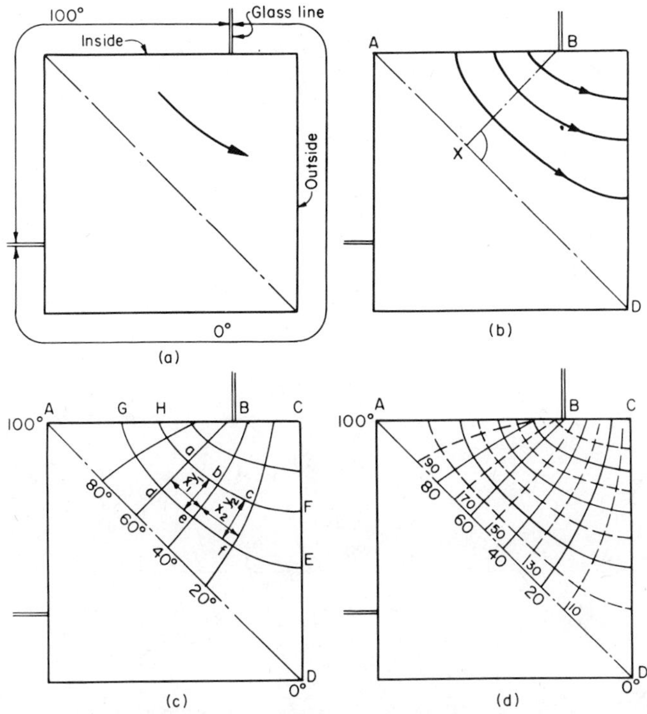

Fig. 10-63 Graphical construction of the heat flow net.

and surface conductances are converted into equivalent thicknesses of concrete (normal weight or lightweight) having the same thermal conductance. These equivalent thicknesses are added to the actual column section, to provide the equivalent column section used for construction of the isotherms. The equivalent thicknesses, t_e, of concrete are:

$$t_e = \frac{\text{conductivity of concrete, } k}{\text{conductance of finish or surface resistance, } C}$$

The following are some of the commonly assumed thermal properties of the concrete, finishes, and surface conductances:

Conductance of inside surface (still air)	$C = 1.46$
Conductance of outside surface (winter, 15 mph wind)	$C = 6.00$
Conductance of gypsum plaster $^1/_2$ in. thick	$C = 3.12$
Conductance of gypsum board $^3/_8$ in. thick	$C = 3.12$
Conductance vertical spaces $^3/_4$ to 4 in. (winter)	$C = 1.03$
Conductivity lightweight concrete (110 psf)	$k = 5.0$
Conductivity normal weight concrete	$k = 13.0$

The units for thermal conductance, C, are
B hr^{-1} sq ft^{-1} F^{-1}

The units for thermal conductivity, k, are
B in. hr^{-1} sq ft^{-1} F^{-1}

The equivalent column section is used only to construct the isotherms. The temperatures at the face of the plaster

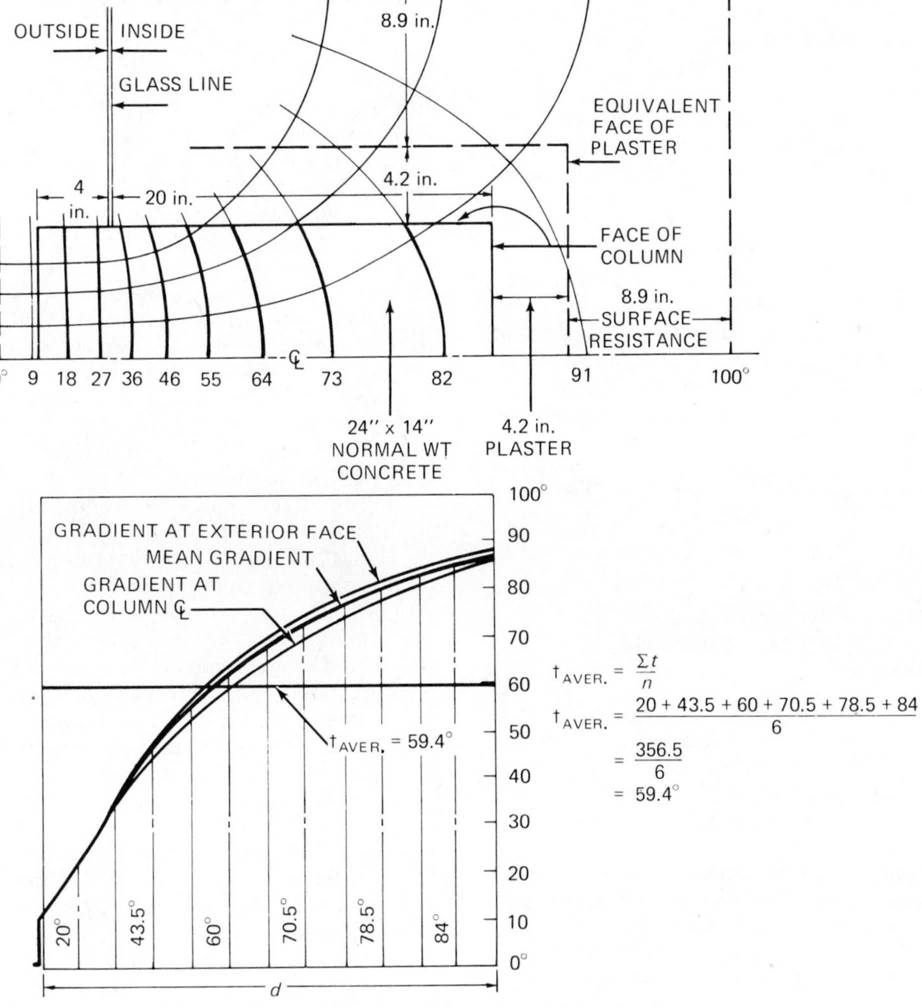

Fig. 10-64 Isotherms, gradients, and average temperature.

are those along the line of equivalent concrete thickness of the plaster as shown in Fig. 10-64. Similarly, the outside concrete surface temperature is at the line of the concrete thickness. The inside finish in Fig. 10-64 consists of $^1/_2$ inch plaster. The temperatures used in the formula for the computations of the average temperature are at the actual column faces and not of the equivalent column. Comparison of lightweight and normal weight columns indicates that the surface temperature drop for lightweight columns is much smaller than for normal weight columns. Consequently, the average temperature of lightweight columns is higher. Due to lower thermal conductivity, lightweight columns have better insulation capacity.

For fully exposed columns no isotherms need be drawn, since all faces will have the same temperature. It is, however, necessary to choose a design temperature based on the section dimension and the time lag studies.[10-34] It is obvious that larger sections will have a mean temperature of a period longer than the mean daily temperature.

-3. *Condensation control*—The interior surface of an exterior, partially-exposed column is much colder than the temperature inside the room. This is due to surface conductances which produce a temperature drop. However, this may not be objectionable, since adjacent windows are usually much colder, and within any space, condensation occurs on the surface with the lowest temperature. Insulation, air space, or plaster are used to raise the temperature of the inside surface to avoid condensation. This may not always be the best solution as far as the structural problems resulting from thermal movements are concerned. Columns with insulation on the inside will have a lower average temperature and larger thermal movements and stresses will result. Studies show that no plaster or other finishes are required on columns in rooms with single glass windows; plaster on columns may be required in rooms with double-glazed windows; air spaces or rigid insulation may be necessary for windowless rooms.

-4. *Length changes and bowing*—When a freestanding infinite wall is subjected to a temperature drop, a straight line temperature gradient will result. A strip of wall, if not restrained, will bow as a result of different temperatures at opposing faces and will also shorten by Δl as shown in Fig. 10-65. The face with lower temperature shortens as compared with the warm face. If the wall is restrained, the bending moment caused by this restraint will be:

$$M = \frac{\Delta t \alpha EI}{d}$$

with

Δt = temperature differential at opposing faces
α = coefficient of thermal expansion
d = wall thickness

When the wall element is forced back by an applied moment M of the same magnitude as restraint moment into its original shape, stresses will develop and obviously there will be no resulting strains. If the wall is freestanding with no bending moments applied, strains will develop; however there will be no stresses.

In an exposed column when the faces are subjected to different temperatures, a curvilinear gradient will result. If a restraint is applied as shown in Fig. 10-66(a) tensile forces will build up since the column is prevented from shortening. When the longitudinal restraint is removed and replaced by rollers, Fig. 10-66(b), the column shortens and com-

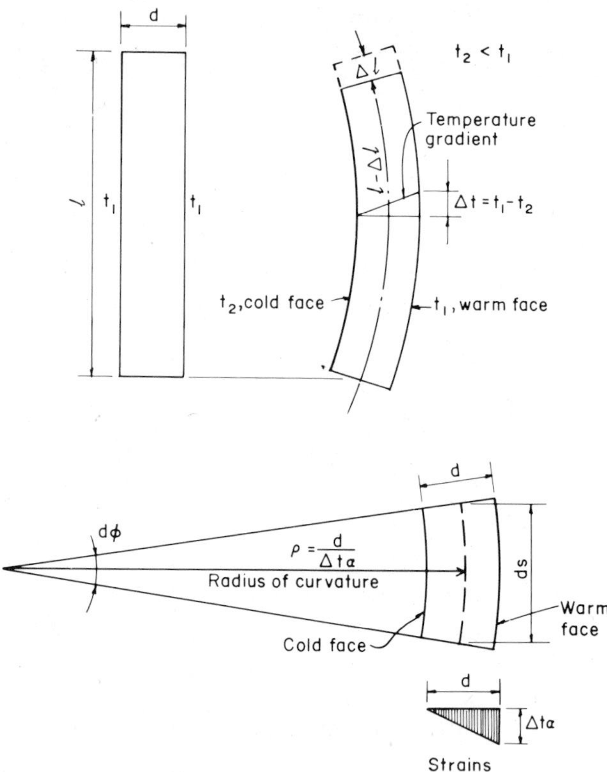

Fig. 10-65 Bowing of a wall strip, radius of curvature, and strains.

pressive stresses build up on the opposite side until the compressive force equals the tension force; equilibrium of forces is established and the shortening stops. The tensile and compressive forces P form a couple over the distance e', and establish the bowing moment $M_B = Pe'$ which is resisted by the rollers. In the building the entire frame prevents the bowing of the columns and as a result the columns are subjected to tensile and compressive stresses on the cold and warm faces respectively, as shown on Fig. 10-66(b). The bowing moment depends upon the depth of the column and is independent of column length. Throughout the height of the structure bowing affects the frame only at the roof where there is no column above and where the exposed column changes its depth considerably.

If the rollers are removed, Fig. 10-66(c), the column is free to bow until a new equilibrium of forces and moments is established on the section.

The two effects caused by a temperature gradient through a column, the bowing and axial length change (shortening or elongation) should be considered separately and the resulting stresses superimposed.

-5. *Analysis and design for length changes of exposed columns*—The exposed columns respond to exterior temperature variations by changing their length to conform to the average temperature. This results in distortions of the exterior bay. The deflecting slabs, in turn, develop resistance shears which act on the exterior and interior columns, decreasing the length changes. This rebound of exterior and interior columns presents the actual resistance of the structure to the thermal distortions, and depends upon slab or beam stiffness and upon the axial stiffness of the columns.

In multistory frames with partially or fully exposed exterior columns, the effect of temperature movement of the exterior column is significant only in the elements of the exterior bay: the exterior column, the exterior bay slab or beam, and the first interior column.

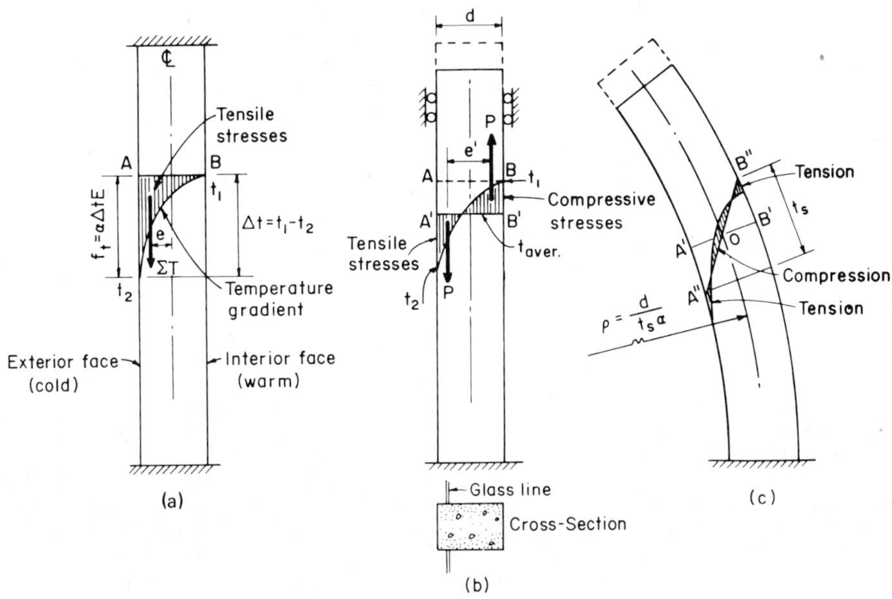

Fig. 10-66 Effects of curvilinear gradient on a column.

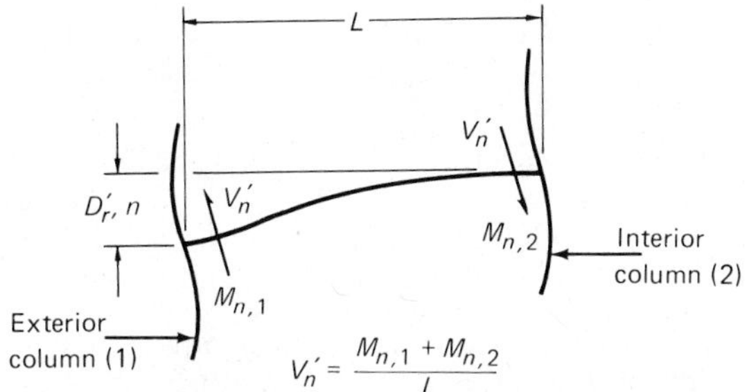

Fig. 10-67 Distributed moments and shears in beam.

An iterative analysis has been developed[10-35] to be performed in the following steps (see Fig. 10-67):

a. Compute the free vertical movement D_{fn} of the exterior column. These are assumed as the initial relative movements between the exterior (1) and interior (2) columns at each floor.
b. Compute fixed-end moments at horizontal floor members at each floor. Zero rotation of joints is assumed.
c. Distribute moments using moment distribution or slope-deflection method.
d. Compute resistance shears of the slabs applied to the columns, V_n'.
e. Using shear forces as vertical loads at exterior and interior columns, compute the rebound of columns.
f. Compute total story rebound, consisting of the sum of exterior and interior column rebounds.
g. Apply convergence correction and compute corrected relative movements for next cycle. The analysis is repeated from step b. using the new corrected movements until convergence to within several percent is reached.

The presented iterative method can be used to obtain a fairly accurate solution within two or three cycles for buildings having relatively flexible girders. Even for very stiff girders, the convergence is quite rapid.

-6. *Simplified design method*—A simplified design method has been developed[10-35] using the following three assumptions:

a. Sizes of columns and horizontal members do not change drastically from floor to floor.
b. Story heights are fairly constant.
c. Points of contraflexure in all columns are at mid-height of each story.

In most buildings these assumptions are fulfilled fairly well. Using the above assumptions, the general iterative method is greatly simplified by introducing the equivalent shear stiffness, V_{eff}, which is the shear required to cause a unit relative movement of the two beam ends. Figure 10-68 shows curves for the equivalent shear stiffness for a wide range of practical stiffness ratios. The moments and shears at columns are then computed directly without the use of moment distribution. As a further simplification, the exterior and interior columns may be replaced by a single equivalent column as shown in Fig. 10-69 which has the same rebound as the total story rebound of the two columns. The simplified design method has been used in conjunction with the one-column, 10-story reference frame to develop curves for direct prediction of residual relative movements between the exterior and interior columns.

Since the exterior and interior columns are interconnected by the floor structure, the computed free movement of the

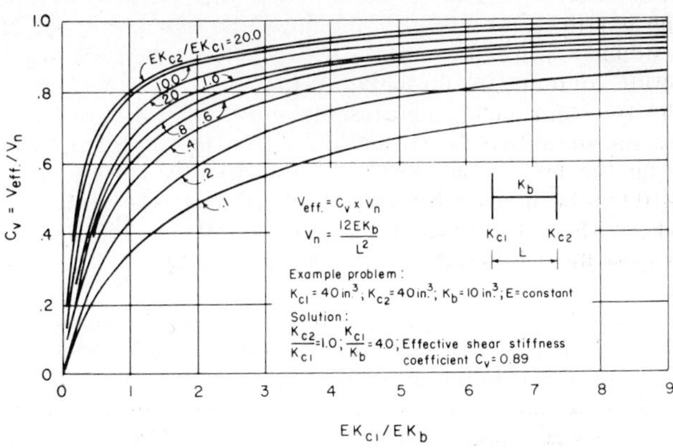

Fig. 10-68 Equivalent shear stiffness as a portion of nominal shear stiffness.

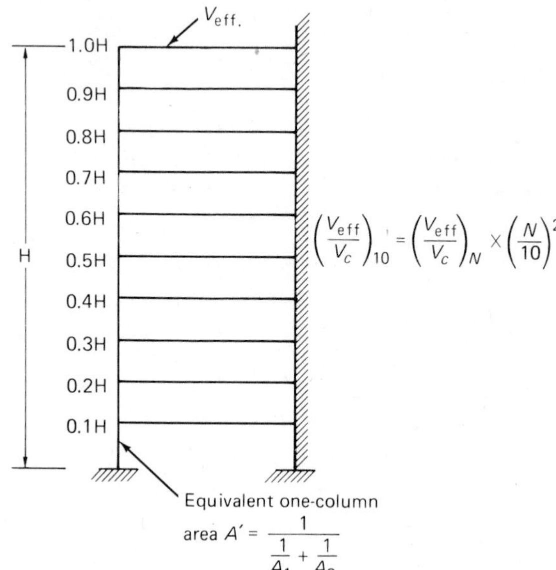

Fig. 10-69 One-column, 10-story reference frame for influence curves.

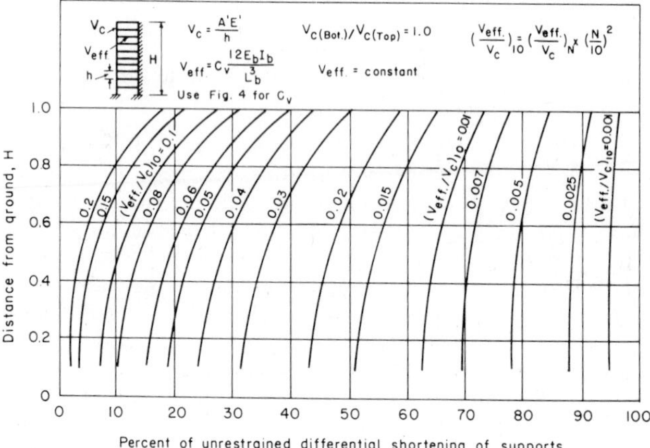

Fig. 10-70 Residual movements between exterior and interior columns as percent of free movement (10-story equivalent structure)—constant column stiffness.

exterior column will be reduced to a residual value, due to the stiffness of the floors. The resistance of the structure to length changes of the exterior columns depends not only upon the floor stiffness, but also upon the axial stiffness of the exterior and interior columns. Curves for prediction of residual relative movements between the exterior and interior columns are presented in Fig. 10-70 as a percent of the free movement. The residual movements as computed by the simplified design method differ only insignificantly from the more exact results of the iterative method. The differences increase towards the bottom of the structure, where they are less significant since the moments and stresses due to thermal movements are small.

-7. *Load transfer between exterior and interior columns*—Vertical shears generated in the deflected slabs (see Fig. 10-67) cause transfer of loads from the shortened exposed column to the interior column. During the summer, when the exposed column elongates, the load transfer is reversed. The accumulation of shears starts at the roof level proceeding downward, the biggest accumulation of load occurring at the lowest level. The amount of load transfer can be extremely high for very rigid connecting beams. Flexible slab systems follow the column length changes more easily and, therefore, transfer only insignificant amounts of load from one column to the other; at the same time their restraining effect is limited on the thermal movement of the exposed column.

-8. *Stresses*—Stresses in structural elements due to temperature length changes and bowing should be combined with effects of gravity loads and lateral loads due to wind or earthquakes. However, it would be unrealistic to design a structure for the full combined effect of extreme temperature, highest wind velocity, snow load, and gravity loads. To establish design criteria, consideration should be given to the following probabilities:

a. The highest stresses due to temperature and wind do not usually occur in the same member within the structure. While temperature effects are highest in the upper parts of the frame, the wind stresses are usually highest in the lower parts.
b. Although extreme temperature conditions may exist for a longer period, coincidence of full temperature and highest wind effects is extremely improbable during the lifetime of a structure.
c. Some temperature effects, due to their occurrence over a longer period of time, are partially reduced by creep.

To assess the coincidence of high wind and extreme temperatures, local weather condition records should be studied. These conditions may be substantially different from one geographic location to the other. Usually, severe temperatures are associated with milder winds; the stronger winds are associated with milder temperatures.

A study of the coincidence of high winds and low temperatures of the 40-year Chicago weather records[10-35] led to the conclusion that for a similar climate, in addition to the usual combination of vertical loads, the following combinations of wind and temperature (whichever is larger) should provide adequate safety:

a. full temperature effects plus 20% of wind effects,
b. full wind effects plus 25% of temperature effects.

-9. *Partition behavior*—The differential movement of exterior columns in high-rise buildings affects primarily partitions perpendicular to the exposed columns. Observations and studies of a number of existing structures with partially exposed columns[10-36] indicate only insignificant distress in buildings with less than 20 stories. The movements of the exposed column cause racking of floor slabs which may result in distress of partitions in upper floors (Fig. 10-62). Column shortening within a story is only secondary in its effects on partitions. However, the major effect on partitions is the rotation of the partitions due to the cumulative change in length of the exposed column. The partition pivots at the interior column when the exposed column shortens in winter, and separates from both columns. In winter the separation is widest at the lower corner of the exterior column and at the upper corner of the interior column; in summer, the separation is opposite to that in winter. The width of separation and possible magnitude of distress depends on the amount of movement of the exposed column, and on the ability of the partition to distort. Dry-wall partitions, composed of a series of individual vertical panels attached flexibly to studs can absorb at each joint a certain amount of displacement which together with some distortion of each panel may provide sufficient adaptability to frame distortions. However, with large movements, hairline cracks may develop in the paint along the vertical joints of the panels. Rigid partitions constructed without consideration of vertical floor movements may suffer distress in the form of diagonal cracking.

There is a significant difference in behavior of partitions in apartment and office buildings. In apartment buildings, partitions are usually stacked over the entire height of the building. They are tightly fitted into the frame. Unless intentionally separated from the frame, the partitions will distort with the building and contribute to the rigidity of the structure in resisting any movements of the frame. In apartment buildings, solid partitions because of their rigidity, (even if very small for some types) will reduce the residual movement of the exposed column. This additional resistance to thermal movements is usually not considered in the analysis.

In office buildings, on the other hand, partitions are mostly of the movable type and do not reach the underside of the slab. Consequently, they provide only insignificant resistance to distortions of the structure.

Some of the other possible causes of partition distress are:

a. volumetric changes of the partition material due to moisture changes;
b. wind, shrinkage, and creep distortions of the frame;
c. horizontal roof slab movements due to temperature changes of insufficiently insulated roof slabs;
d. deflection of slabs due to gravity loads.

Partition detailing should, therefore, have adequate provisions to accommodate the particular conditions not related to temperature movement of columns. Effects of thermal and other distortion of the frame on partitions should be anticipated by the designers and due consideration given in partition detailing.

Details of partitions and partition layout are of major influence on their behavior. For example, door openings carried to the ceiling may eliminate common location of cracks. The distortion of the structural frame induces shear and bending stresses in the partitions beyond their strength capacity, and if not relieved by proper details, can cause unsightly as well as acoustically unacceptable cracking. In order to avoid such cracking of partitions, details around the edges of partitions should be provided to allow vertical as well as horizontal slippage. One of the simplest ways to achieve this is to provide a channel enclosure for partitions, where they meet columns and ceiling.

Manufacturers of dry-wall partitions have developed a number of details to let the building distort without straining the partitions. Details of the floating-type partitions are usually provided in upper floors only. Such details are described in the chapter "Joints in Buildings," and sketches given there in Fig. 4-2.

Although partition cracks seem to be the primary problem in some buildings, because they are visible, the structure itself can be considerably overstressed. Therefore, it is not sufficient to design and detail only the partitions for the temperature movements; it is also necessary to design and detail the slabs and columns for temperature deformations and stresses. It may also be desirable in some cases of very tall structures, or where the columns are fully exposed, to provide insulation at the outer faces of the columns if they are clad with stone or precast concrete panels.

-10. Planning and design concepts—Buildings with exposed columns can be designed either (a) to accommodate large expected relative temperature movements, or (b) with controlled temperature movement.

-10.1 Buildings designed for large movements—Large movements result usually when architectural considerations dictate a largely recessed glass line in tall buildings or if the column exposure is complete even in moderately tall buildings.

Structurally there are two alternative solutions to solve problems in buildings designed with large exposure.

a. To provide strength for accommodation of relatively large movements of exterior columns, or
b. to provide details to relieve the stresses in floors which have excessive rotations.

To provide strength, heavy restraining girders connecting the exterior and interior columns may be provided at the roof, or at intermediate floors (mechanical floors). The restraining girders are generally very rigid and a full story in depth. These girders obviously also improve the resistance to wind loads, as discussed previously in the section on "Resistance to Wind Loads" in this chapter.

The alternative is to provide details to relieve the stresses in the floor structure. Hinging of floors at the interior columns or shear walls for severe winter shortenings may assure proper functioning of the structural system.

The first building known to be designed for significant structural movements was the 38-story Brunswick Building in Chicago (Fig. 10-50). All typical exterior columns were exposed about 70% of their area, resulting in an anticipated maximum computed movement of 1.25 in. at the top floor. To relieve high bending stresses, hinges between the slab and the shear wall were incorporated in the upper 12 floors, as shown in Fig. 10-71. The hinge details should include dowels to maintain lateral restraint between slabs and walls. By using elastomeric material around a portion of dowels, the slabs are free to rotate, but lateral restraint is provided. Observations indicate that the structure is performing as planned; no architectural or structural problems have been encountered.

-10.2 Buildings designed for controlled movements—Although it is possible to accommodate structurally a considerable amount of movement of exposed columns, serviceability and economy may set an upper limit on the amount of thermal movement. Serviceability criteria are logically expressed as a ratio of the movement to the slab, or beam, length (angular distortion). Any movement limitation should relate to the type of structure and building material used. It is clear that apartment buildings will impose different movement limitations from warehouses, industrial buildings and similar types of structures.

Structures with brickwork, masonry partitions and plaster walls will require more stringent limitations than buildings with metal cladding. It should be noted that slab deflections due to column exposure are additive to deflections due to gravity loads. For buildings without masonry or plastered partitions, a thermal movement of $L/200$ (of exterior column relative to interior column) may be tolerable. For a thermal movement limitation of $L/200$, the center of the span will be subjected to a movement of $L/400$ only, which should be added to vertical load deflections and then compared to specified deflection limitations of the Code.

In some cases it may be desirable to develop more stringent criteria for maximum temperature movements. Reference 10-36 suggests that in office buildings a reasonable limit for temperature movements may be taken as 0.75 in. up or down from the horizontal position. Assuming clear span in office buildings of about 36 ft, a relative movement of 0.75 in. corresponding to $L/600$ between the exterior column and the shear core will generally not cause excessive stresses requiring special structural details. These stresses

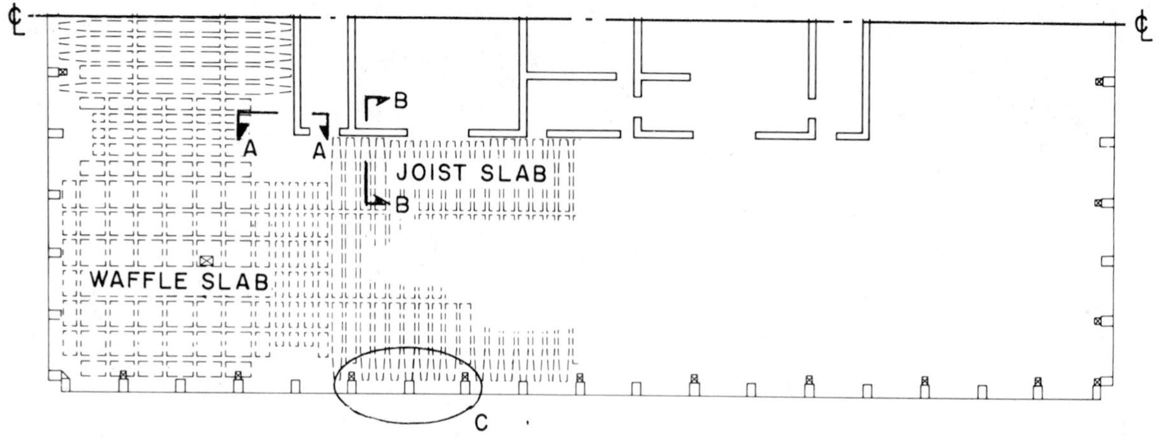

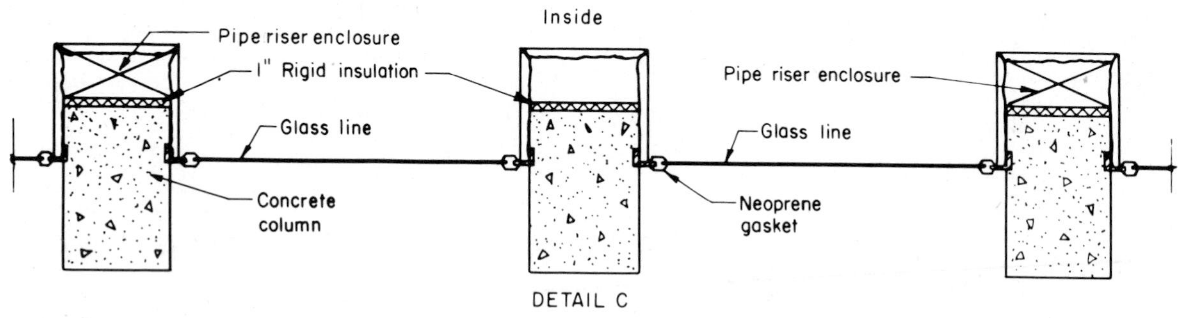

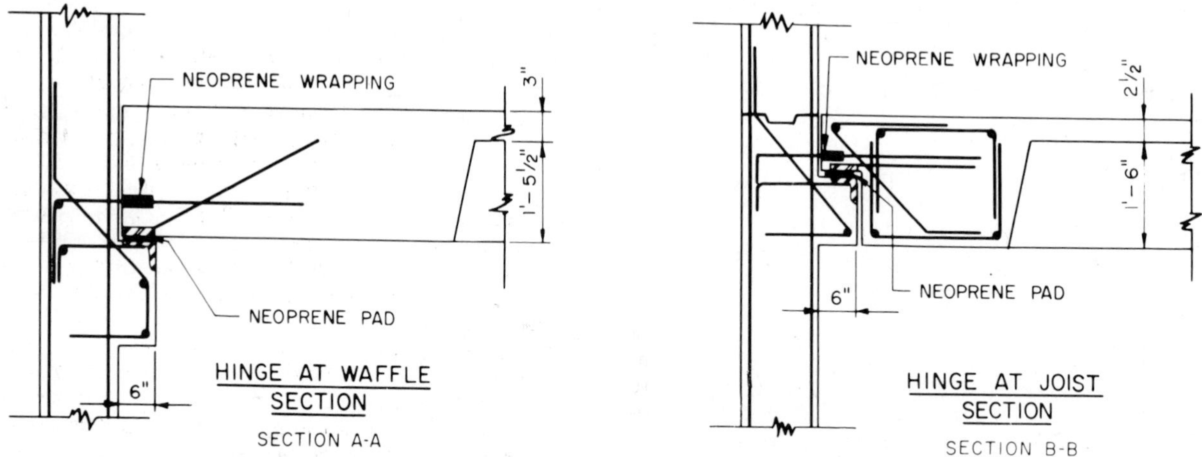

Fig. 10-71 Details of hinges and the glassline in the Brunswick Office Building.

normally will be less than the allowable overstress due to temperature as proposed earlier. Furthermore, a movement of $L/600$ can be reasonably taken care of by simple partition details without serious economic implications.

For the 52-story Shell Oil Building in Houston, the 0.75 in. limitation was accomplished by setting the glass line to achieve the desired average temperature of the column. Nominal modification of typical partition details were needed to avoid stresses in the partitions for such movements.

In apartment buildings where the bay sizes are considerably smaller, a relative movement between the exterior column and the interior column of 0.75 in. corresponding to $L/300$ for an 18-ft span, may require structural and partition modifications. To minimize the need for special structural and partition details in apartment buildings, it may be advisable to limit computed temperature movements to about 0.5 to 0.625 in. by placing the glass line as far out as possible. The actual movements in apartment buildings are usually smaller than computed, due to the unaccounted contribution of partitions to the resistance of the frame.

10.9.2 Vertical Movements Due to Creep and Shrinkage

With the increasing height of buildings, the importance of time-dependent shortening of columns and shear walls becomes more critical due to the cumulative nature of such shortening. It is known that columns with varying percentage of reinforcement and varying volume-to-surface ratio will have different creep and shrinkage strains. An increase in percentage of reinforcement and in volume-to-surface ratio reduces strains due to creep and shrinkage under similar stresses.

In a multistory building, adjacent columns may have different percentage of reinforcement due to different tributary areas or different wind loads. As a result, the differential elastic and inelastic shortening will produce moments in the connecting beams or slabs and will cause load transfer to the element that shortens less. As the number of stories increases, the cumulative differential shortening also increases, and the related effects become more severe. A common example is the case of a large, heavily reinforced column attracting additional loads from the adjacent shear wall which has higher creep and shrinkage due to lower percentage of reinforcement and lower volume-to-surface ratio. Significant differential shortening may also occur due to a time gap between a slip-formed core and the slabs. In this case the columns are subjected to the full amount of creep and shrinkage while the core may have had the bulk of its inelastic shortening occurring prior to casting of the adjacent columns.

It is customary, at present, to neglect the effect on the frame of elastic and inelastic shortening of columns and walls. For low and intermediate height structures this may be acceptable, however, neglecting the differential shortening in ultra high-rise buildings may lead to distress in the structure and in nonstructural elements of the building.

In a number of tall buildings in the United States built in the early sixties, structural cracking and partition distress were observed as a result of differential creep between shear walls and highly reinforced columns in close proximity to each other. Another example of the reality of differential creep and shrinkage of vertical elements is a 50-story building in Australia in which the measured differential at the roof level between the concrete core and peripheral columns was 1.1 in. after about four and one-half years. Fortunately, no problems were experienced—the long spans of about 35 ft between the core and peripheral columns caused only small slab rotations. The elevator rails had to be adjusted twice over the years to accommodate the changed height of the elevator shafts.

Buildings up to 30 stories with flexible slab systems, such as flat plate slabs of average spans, or long-span joist systems, are usually not adversely affected structurally by differential shortening of supports. In these cases the knowledge of the total shortening is needed to make allowance in the architectural details to avoid future distress of partitions, windows, cladding, and other nonstructural elements. Differential shortening can be minimized by proportioning adjacent columns or walls to have similar stresses of the transformed section and similar percentage of reinforcement. Volume-to-surface ratio has a lesser effect on differential shortening.

Although a large amount of research information is available on shrinkage and creep strains, it is not directly applicable to columns of high-rise buildings. The available shrinkage data must be modified since it is obtained on small standard prisms or cylinders stored in a controlled laboratory environment. The available creep research is based on application of loads in one increment. Such creep information, therefore, is applicable to flexural elements of reinforced concrete and to elements of prestressed concrete. In the construction of a high-rise building, however, columns are loaded in as many increments as there are stories above the level under consideration. If a 50-story building is constructed in 50 weeks, then the first story columns receive 2% of their design load every week during the construction period. Incremental loading over a long period of time makes a considerable difference in the magnitude of creep.

An engineering procedure was established in the late 1960s[10-37, 10-38] for the solution of the structural considerations involved in the effects of differential column shortening due to elastic shortening, creep, and shrinkage in tall buildings consisting of

a. determining the amount of creep and shrinkage occurring in columns and walls with consideration of the loading history, size of the member, percentage of reinforcement, and environment;
b. establishing the amount of elastic shortening in columns and walls if necessary for analysis; and
c. analysis and design for the structural effects of differential elastic and inelastic shortenings of vertical load-carrying members in the structure.

The first part of the developed procedure[10-37] handles the predictions of the inelastic (creep and shrinkage) shortening as a function of the incremental loading sequence, the volume-to-surface ratio, and the effect of the percentage of reinforcement. The second part[10-38] handles the analysis of multistory structures for the structural effects of differential elastic and inelastic shortening between adjacent columns or walls; from the residual differential movements, the corresponding moments in the frame are computed, as well as the load transfer from the support member that shortens more to the support member that shortens less.

A rigorous frame analysis for the elastic and inelastic strains in the supports may be needed only in ultra high-rise buildings or in rigid slab systems connecting vertical elements with high differential shortening. Also, structures where elements which shorten differentially are closely spaced may need special investigation.

The computation of elastic strains does not present difficulties as it is carried out by established procedures. The prediction of the creep and shrinkage strains is more complex and requires a quantitative consideration of the effects on creep and shrinkage of section size, loading sequence and amount of reinforcement. Example 10-3 shows how to compute the creep and shrinkage strains for a reinforced concrete column protected from the weather.

-1. Computation of creep and shrinkage shortening—The nature of creep and shrinkage and their dependence upon the constituent materials is explained in the chapter on "Properties of Materials for Reinforced Concrete" of this handbook. Since creep depends upon sequence and intensity of loading while shrinkage proceeds independently of construction time, creep and shrinkage strains should be computed separately, and modified for the condition of the designed structure and then their combined effect on the structure considered.

For structural engineering practice, it is convenient to consider specific creep ϵ_c' which is defined as the ultimate creep strain per unit of sustained stress loaded at any specified age, say, 28 days.

Determination of specific creep can be done either in the laboratory on 6-in. cylinders or can also be roughly predicted from the initial modulus of elasticity[10-39] from charts.

-2. *Effect of construction time*—For a given mix of concrete, the amount of creep depends not only upon the total stress but also to a great extent upon the loading history. It is well established that a concrete specimen with its load applied at an early age exhibits a much larger specific creep than a specimen loaded at a later age. Also due to the gradual increase of the modulus of elasticity with age, the elastic shortening per unit stress of older concrete is smaller than of concrete loaded at an earlier age. Figure 10-72 from Ref. 10-40 shows elastic and creep strains of specimens loaded at various ages.

Loading history is particularly significant for columns of multistory buildings which are loaded in as many increments as there are stories above the level under consideration. Since creep decreases with age of the concrete at load application, each subsequent incremental loading contributes a smaller specific creep to the final average specific creep of the column.

Each load increment causes a creep strain as if it were the only loading to which the column is subjected.[10-41] In Fig. 10-73 the relationship of creep to the age of loading is shown using the 28-day specific creep as unity (as a basis of comparison). It can be seen that a cylinder loaded at the age of one year has only one-half the creep of one loaded at 28 days of age. If we load a column with a number of equal

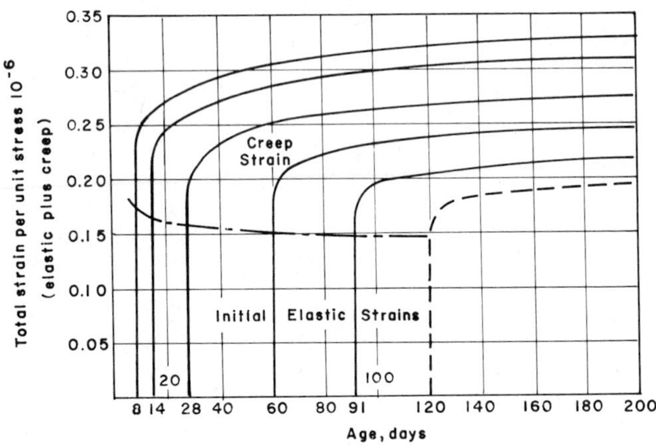

Fig. 10-72 Elastic and creep strains versus time from Ref. 10-40.

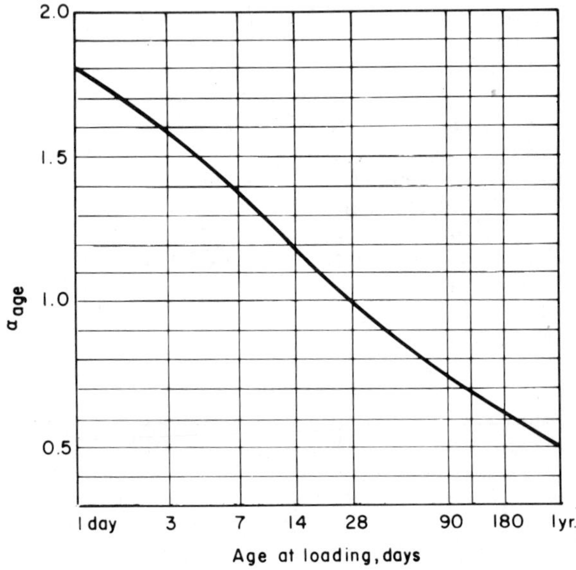

Fig. 10-73 Creep versus age at loading.

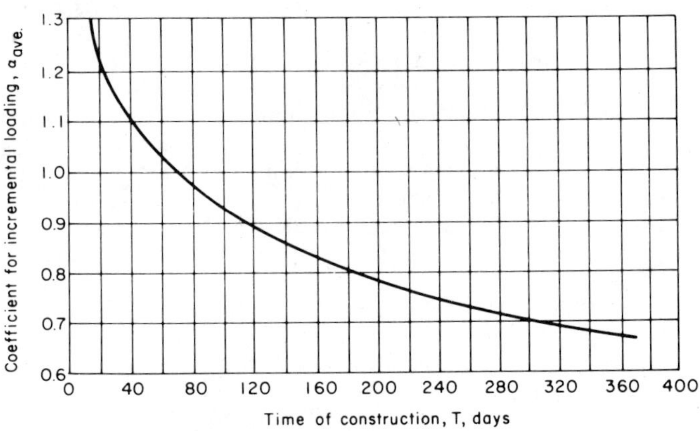

Fig. 10-74 Creep versus construction time.

incremental loads over a period of time, T, we could then determine the average creep for the total load by a summation of the incremental load with the corresponding creep values. Figure 10-74 gives the coefficient, α_{ave} to be used to convert the 28-day creep into the average creep for a specific length of construction time. We can see from the curve that if the entire load is applied at the age of seven days, it would have twice the creep than applied in incremental loading over a period of 262 days.

-3. *Effect of member size*—Both creep and shrinkage depend on the member size, however, not to the same degree. Creep is less sensitive to member size than shrinkage which is caused by evaporation of moisture from the surface. The rate and amount of evaporation and consequently of shrinkage depend greatly upon the relative humidity of the environment, the size of the member and the mix proportions of the concrete. In moderate-size columns (30-in. diameter) the inside relative humidity has been measured at 80% after four years of storage in a laboratory in 50% relative humidity. Since evaporation occurs only from the surface of members, the volume-to-surface ratio of a member has a pronounced effect on the amount of shrinkage.

In Fig. 10-75 the relationship between creep and the volume-to-surface ratio is plotted, based on Elgin gravel aggregate concrete. A plot of shrinkage versus the volume-

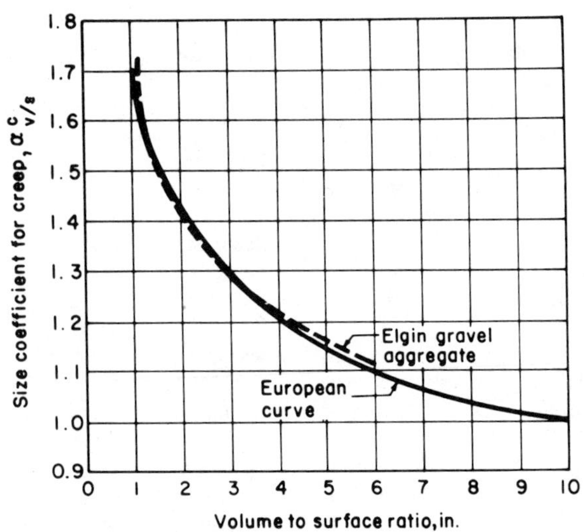

Fig. 10-75 Creep versus volume-to-surface ratio.

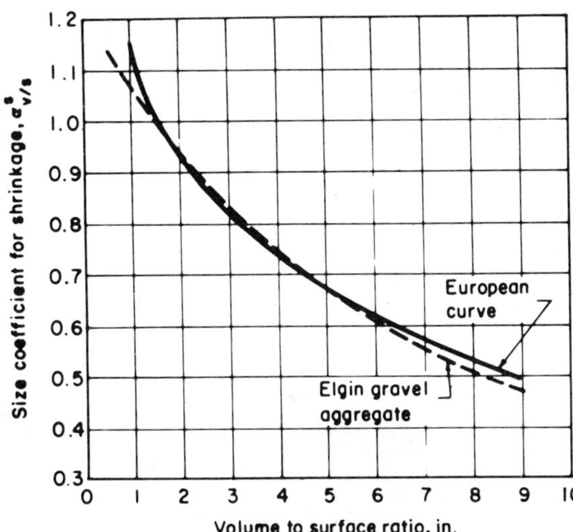

Fig. 10-76 Shrinkage versus volume-to-surface ratio.

to-surface ratio is given in Fig. 10-76. The European experience on the surface-to-volume effects on creep and shrinkage is also plotted for comparison, since the U.S. information is based on only one research investigation. The coefficient $\alpha^s_{v/s}$ is used to convert shrinkage data obtained on 6-in. cylinders (v/s = 1.5 in.) to any other size columns.

-4. *Progress of creep and shrinkage with time*—Both creep and shrinkage have a similarity regarding the rate of progress with respect to time. Figure 10-77 shows an average curve for the ratio of creep or shrinkage at any time to the final value at time, t_∞. It can be seen from the curve that at 28 days about 40% of the inelastic strains have taken place. After three and six months, 60 and 70%, respectively, of all the creep and shrinkage have taken place. The curve can be used to extrapolate the ultimate creep and shrinkage values from laboratory testing of a certain duration of time. Conversely, the curve can also be used to estimate the creep or shrinkage at any time from the given ultimate value.

-5. *Effect of reinforcement on creep and shrinkage*—The longitudinal reinforcement has by far the most predominant restraining effect on creep and shrinkage. Although the magnitude of creep and shrinkage of plain concrete specimens may vary considerably depending on the concrete properties and climatic conditions, the final inelastic strains in reinforced concrete columns and walls have much less variation due to the restraining effect of the reinforcement.

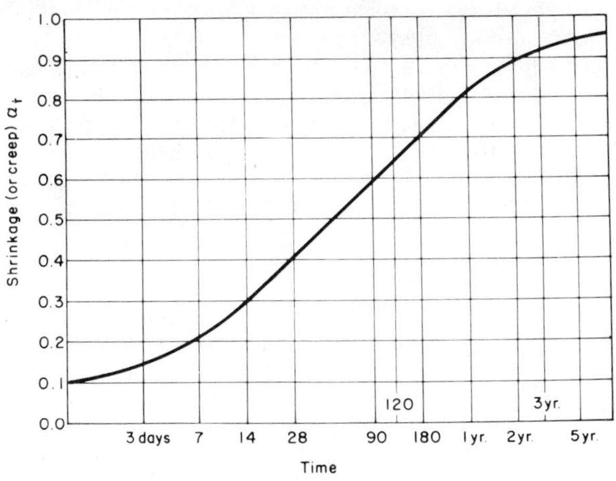

Fig. 10-77 Shrinkage or creep versus time.

Tests have shown that when reinforced concrete columns are subjected to sustained loads there is a tendency for additional stress to be gradually transferred to the steel with a simultaneous decrease in the concrete stress. Long-term tests by Troxell, et al.[10-42] showed that in columns with low percentage of reinforcement the stress in the steel increased until yielding; while in highly reinforced columns after the entire load had been transferred to the steel, further shrinkage actually caused some tensile stresses and cracking of the concrete. It should be noted, however, that despite the redistribution of load between concrete and steel, the ultimate load capacity of the column remains unchanged.

The change in stress in the concrete, Δf_c, and in the steel, Δf_s, due to creep and shrinkage can be calculated with the following formulas developed by Dishinger[10-43] which have been verified by tests in the United States.[10-44]

$$\Delta f_c = \left(f_c + \frac{\epsilon_s}{\epsilon_c'}\right) \quad 1 - e^{-\frac{pn}{1+pn}\epsilon_c' E_c}$$

$$\Delta f_s = \frac{\Delta f_c}{p} = \frac{f_c + \epsilon_s/\epsilon_c'}{p} \quad 1 - e^{-\frac{pn}{1+pn}\epsilon_c' E_c}$$

$$= \frac{(f_c\epsilon_c' + \epsilon_s)}{p\epsilon_c'} \quad 1 - e^{-\frac{pn}{1+pn}\epsilon_c' E_c}$$

in which

f_c = initial elastic stress in the concrete

ϵ_s = total shrinkage strain of plain concrete adjusted for v/s ratio

ϵ_c' = ultimate specific creep of plain concrete (in./in./psi)

E_c = modulus of elasticity of concrete

p = reinforcement ratio of the section

n = modular ratio E_s/E_c

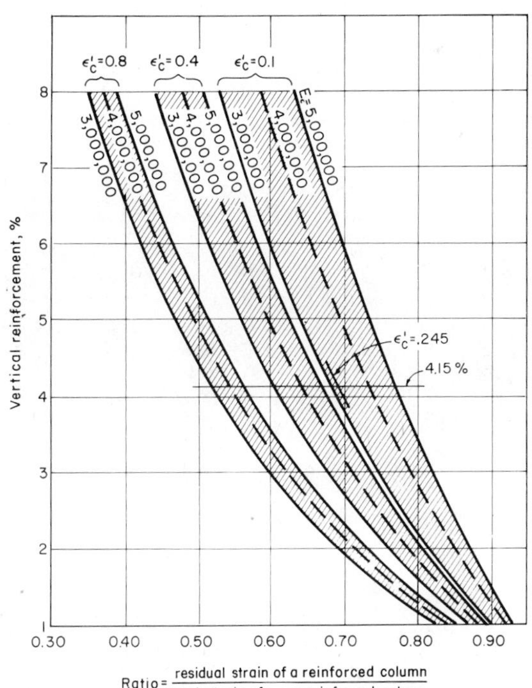

Fig. 10-78 Restraining effect of the reinforcement.

The ratio of residual creep and shrinkage strains of a reinforced column to the total creep and shrinkage strain of the identical column without reinforcement is presented in Fig. 10-78 for various percentages of reinforcement, varying specific creep and modulus of elasticity of concrete. It is evident from the curves that the residual creep and shrinkage decreases with increased percentage of reinforcement.

The function $1 - e^{-\frac{pn}{1+pn}\epsilon_c' E_c}$ has been plotted in Ref. 10-37 for convenience of numerical handling of the Dishinger equations given above for computing the changes in the concrete and steel stresses.

EXAMPLE 10-3: Assume an inside column 36 stories below the roof. Floor-to-floor height is 9.0 ft. The 20 X 49-in. column is reinforced with 26 #11 bars (4.15%) of A431; $f_y = 75{,}000$ psi.

Concrete:

$$f_c' = 5000 \text{ psi at 28 days, normal weight}$$

$$E_c = 33w^{3/2}\sqrt{f_c'} = 4.05 \times 10^6$$

$$n = \frac{E_s}{E_c} = 7.2$$

Transformed column area:

$$A_t = A_g + (n-1)A_s$$

$$= 20 \times 49 + (7.2 - 1) \times 40.6 = 1232 \text{ in.}^2$$

The planned construction progress is one floor in eight calendar days—total time for 36 floors (load increments) is

$$T = 36 \times 8 = 288 \text{ days}$$

The dead load of the typical floor is 37 kips.

Twenty-year specific basic creep for loading at age of 28 days was estimated from previous experience to be $\epsilon_c' = 0.33 \times 10^{-6}$ in./in./psi for $E_c = 4.05 \times 10^6$ psi.

Shrinkage determined on the same mix of a previous job was 630×10^{-6} in./in. during the first 90 days. The 6-in. cylinders were moist cured for seven days and then stored in the laboratory in 50% relative humidity and 70°F.

Required: Compute the ultimate residual creep and shrinkage strains of the reinforced concrete column and the additional stress in reinforcing steel.

The following steps will be carried out:

a. compute for the plain concrete column the total ultimate creep strains, considering effects of incremental loading of column size; and shrinkage considering volume-to-surface ratio;
b. compute for the reinforced concrete column the residual creep and shrinkage strains;
c. compute the additional stress in the vertical reinforcing steel due to creep and shrinkage.

SOLUTION:

a. Creep strains for nonreinforced column—

Conversion of specific creep for loading at 28 days to consider incremental loading over a period $T = 288$ days using $\alpha_{ave} = 0.70$ from Fig. 10-74.

$$\epsilon_{c,ave}' = \epsilon_{c,28}' \times \alpha_{ave}$$

$$= 0.33 \times 10^{-6} \times 0.70 = 0.231 \times 10^{-6} \text{ in./in./psi}$$

Modification of specific creep for size effect using Fig. 10-75.

Volume-to-surface ratio

$$V/S = \frac{20 \times 49}{2(20+49)} = 7.1 \text{ in.} \rightarrow \alpha_{V/S}^c = 1.06$$

$$\epsilon_c' = 0.231 \times 10^{-6} \times 1.06 = 0.245 \times 10^{-6} \text{ in./in./psi}$$

Sustained stress on the concrete

$$f_c = \frac{P}{A_T} = \frac{36 \times 37{,}000}{1232} = 1080 \text{ psi}$$

Total creep strain

$$\epsilon_c = \epsilon_c' \times f_c = 0.245 \times 10^{-6} \times 1080 = 265 \times 10^{-6} \text{ in./in.}$$

Shrinkage strains for nonreinforced concrete column—

Conversion of 90-day measured shrinkage to ultimate shrinkage (coefficient from Fig. 10-77 representing the ratio of shrinkage at 90 days to ultimate shrinkage).

$$\epsilon_{s_\infty} = \frac{630 \times 10^{-6}}{0.60} = 1035 \times 10^{-6} \text{ in./in.}$$

Conversion of shrinkage measured on the 6-in. cylinder to account for size of real column using Fig. 10-76.

$$V/S = \frac{20 \times 49}{2(20+49)} = 7.1 \text{ in.} \rightarrow \alpha_{V/S}^s = 0.57$$

$$\epsilon_s = 1035 \times 10^{-6} \times 0.57 = 590 \times 10^{-6}$$

Total creep and shrinkage strains for nonreinforced column

$$\epsilon = \epsilon_c + \epsilon_s = (265 + 590)\,10^{-6} \text{ in./in.}$$

b. Residual creep and shrinkage for columns from Fig. 10-78 for a reinforcing ratio of 4.15% for $E_c \approx 4 \times 10^6$, interpolating for $\epsilon_c' = 0.245$

$$\text{Ratio} = \frac{\text{residual strain of a reinforced column}}{\text{total strain of a nonreinforced column}} = 0.69$$

c. Additional stress in the reinforcing steel corresponds to the residual strain of the reinforced column, therefore Δf_s = residual strain $\times E_s = 590 \times 10^{-6} \times 29 \times 10^6 = 17{,}100$ psi. The same stress could be computed by solving the Dishinger formula for Δf_s. From the residual creep and shrinkage strain the shortening of the column over the entire height of the structure can be computed.

-6. *Analysis for elastic and inelastic differential movement*—After the elastic and inelastic differential shortenings of the columns and walls are determined, a frame analysis can be carried out.

It should be pointed out that each slab is affected only by the differential shortenings of its support which will occur after the slab has become a part of the frame. All elastic and inelastic shortenings to which the supports were subjected prior to the casting of a slab are of no consequence to this particular slab. Thus, a fourth dimension enters the analysis; namely, construction time and sequence. Solutions to the problem of construction time in conjunction with creep and shrinkage do not seem imminent at present and they also do not seem warranted for the majority of structures. There is no simple and easy procedure ready for application, even for complex structures where it may be warranted.

A relatively simple analysis for differential shortening is considered sufficient in view of the existing uncertainties in the assumptions of magnitude of creep and shrinkage, and their still unexplored dependence upon changes in climate.

It is reasonable to suggest at present the simple frame analysis for differential shortening similar to that suggested in sections 10.9.1.5. and 10.9.1.6. of this chapter for the effects of temperature changes in exposed columns by using the charts in Figs. 10-68, 10-69, and 10-70.

-7. *Effects of slow differential settlements of slabs*—The elastic and inelastic differential shortening of supports of a slab in a multistory building do not occur instantaneously.

The elastic shortening occurs during the period it takes to construct the structure above the slab under consideration. The creep and shrinkage shortening continues for years at a progressively decreasing rate. During this time creep of the concrete of the slab will cause relaxation of the moments caused by differential shortening; in other words, the elastic moments in the slab (due to slab settlement) will continue 'creeping out'.

Experimental and analytical studies by Ghali, et al[10-45] have shown that for settlements applied over a period of more than 30 days the amount of relaxation is about 50%. Which means that for practical design of buildings, it seems reasonable to assume the maximum value of the differential settlement moments at 50% of the elastic moments that would occur without creep relaxation of the slab. This reduced moment should then be used with appropriate load factors in combination with other loads. The 50% reduction accounts only for creep relaxation during the period of settlement. Beyond this time a further creeping-out of the settlement moments takes place.

-8. Load transfer between adjacent differentially shortening elements—The girders and slabs tend to equalize the column stresses; hence, unless the columns are specifically designed for equal axial stresses, columns adjacent to each other over short spans usually carry substantially different (greater or smaller) loads than can be accounted for on the basis of tributary areas at each floor.

In a frame with a differential settlement of supports, the support which settles less will receive additional load from the support which settles more. The transferred load is

$$V = \frac{M_1 + M_2}{L}$$

where M_1 and M_2 are the settlement moments (reduced due to creep relaxation) at the two ends of the horizontal element and L is the span. The load transferred over the entire height of the structure is a summation of load transferred on all floors.

The load transfer is cumulative starting from the top of the structure progressing down to its base.

-9. Stresses—The stresses due to differential shortening should be treated as equivalent dead load stresses with appropriate load factors before combining them with other loading conditions. When choosing a load factor, it should be considered that the design shortening-moments in slabs or beams occur only for a short while during the life of the structure and they continue to creep out.

-10. Field observations—A field investigation on columns and shear walls of the second floor of a 38-story building in Chicago was carried out over a seven-year period, to obtain actual creep and shrinkage strains of columns and walls and to compare them with the analytically predicted values according to the procedure described in this section. The measured strains compare reasonably well with the analytically computed values. The measured creep and shrinkage strains ranged between 335 and 464 $\times 10^{-6}$ in./in. for the columns and walls. The percentages of reinforcement were about 4.5 and 2.50 for columns and walls, respectively.

REFERENCES

10-1 Hodgkison, R. L. "An Ultra High-Rise Concrete Office Building," Master of Science thesis in School of Architecture, Illinois Institute of Technology, Chicago, Ill. June, 1968.

10-2 "Wind Forces on Structures," *Transactions, ASCE* **126**, Part 2, 1124–1198, 1961.

10-3 Davenport, A. G. "Wind Loads on Structures," *Technical Paper No. 88.* National Research Council of Canada, Ottawa, March 1960, 81 pp.

10-4 ACI Committee 435 "Allowable Deflections," *ACI Journal, Proceedings* **65** (6), 433–444, June 1968.

10-5 ACI Committee 442, "Response of Buildings to Lateral Forces," Mark Fintel, Chairman, *ACI Journal Proceedings* **68** (2), 81–106, Feb. 1971.

10-6 "Frame Constants for Lateral Loads on Multistory Concrete Buildings" *Advanced Engineering Bulletin No. 5*, Portland Cement Association, 1962.

10-7 "Handbook of Frame Constants," T-32-2, Portland Cement Association, 1947, 32 pp.

10-8 Fiorato, A. E.; Sozen, M. A.; and Gamble, W. L., "Behavior of Five-Story Reinforced Concrete Frames with Filler Walls," Interim Report to the Department of Defense, Office of the Secretary of the Army, Office of Civil Defense, January 1968.

10-9 Smith, C. S., and Carter, C., "A Method of Analysis for Infilled Frames," *Proceedings Inst. of Civ. Eng.* (London), **44**, Sept. 1969.

10-10 "Continuity in Concrete Building Frames," Fourth Edition, Portland Cement Association, 1959.

10-11 Khan, F. R., and Sbarounis, J. A., "Interaction of Shear Walls and Frames," *Proceedings, ASCE* **90** (St 3), 285–335, June 1964.

10-12 Beck, H., "Contribution to the Analysis of Coupled Shear Walls," *ACI Journal Proceedings*, **59** (8), 1055–1070, August 1962.

10-13 Rosman, R., "Approximate Analysis of Shear Walls Subject to Lateral Loads," *ACI Journal, Proceedings*, **61** (6), 717–732, June 1964.

10-14 Coull, A., and Choudhury, J. R., "Stresses and Deflections in Coupled Shear Walls," *ACI Journal, Proceedings* **64** (2), 65–72, Feb. 1967.

10-15 Coull, A., and Choudhury, J. R., "Analysis of Coupled Shear Walls," *ACI Journal, Proceedings*, **64** (9), 587–593, Sept. 1967.

10-16 Marshall, M. G., "The Analysis of Shear Wall Structures," M. Sc. Thesis, University of Waterloo, Waterloo, Ontario, Sept. 1968.

10-17 Barnard, P. R. and Schwaighofer, Jr., "Interaction of Shear Walls Connected Solely through Slabs," *Tall Buildings*, Pergamon Press Limited, London, 1967, pp. 157–173.

10-18 Pauley, T., "The Coupling of Shear Walls," Ph.D. Thesis, University of Canterbury, Christ Church, New Zealand, 1969.

10-19 Derecho, A. T., "Computer Program for the Analysis of Plane Multistory Frame-Shear Wall Structures Under Lateral and Gravity Loads," Portland Cement Association, 1971.

10-20 Cardenas, A. E.; Hanson, J. M,; Corley, W. G., and Hognestad, E., "Design Provisions for Shear Walls," *ACI Journal* (to be published)

10-21 Tsuboi, Y.; Suenaga, Y.; and Shigenobu, T., "Fundamental Study on Reinforced Concrete Shear Wall

Structures–Experimental and Theoretical Study of Strength and Rigidity of Two-Directional Structural Walls Subjected to Combined Stresses M. N. Q.," *Transactions of the Architectural Institute of Japan*, No. 131, Jan. 1967, *PCA Foriegn Literature Study No. 536*, November, 1967.

10-22 Cardenas, A. E., Magura, D. D., "Strength of High-Rise Shear Walls–Rectangular Cross Sections," *Response of Structures to Lateral Forces*, Publication SP-31, American Concrete Institute, Detroit, Mich., 1973.

10-23 Benjamin, J. R., *Statically Indeterminate Structures*, McGraw-Hill Book Company, New York, 1959, 347 pp.

10-24 Fintel, M., "The Behavior of Reinforced Concrete Structures in the Caracas Earthquake of July 29, 1967," XS6731, Portland Cement Association, 1968, 52 pp.

10-25 "Design of Combined Frames and Shear Walls," *Advanced Engineering Bulletin No. 14*, Portland Cement Association, 1965.

10-26 MacLeod, I. A., "Shear Wall-Frame Interaction–A Design Aid," Portland Cement Association, 1970.

10-27 Moreno, J, consulting engineer, Material Service Corporation, Chicago, private communication, July 1972.

10-28 Colaco, J. P., Lombajian, Z. H., "Analysis of Transfer Girder System," *ACI Journal, Proceedings*, **68** (10), October 1971.

10-29 "New Steel Framing System Promises Major Savings in High-Rise Apartments," *Architectural Record*, **139**, 191–196, June 1966.

10-30 Fintel, M., "Staggered Transverse Wall Beams for Multistory Concrete Buildings," *ACI Journal, Proceedings*, **65** (5), 366–378, May 1968.

10-31 Fintel, M.; Barnard, P. R.; and Derecho, A. T., *Staggered Transverse Wall Beams for Multistory Concrete Buildings, A Detailed Study*, Portland Cement Association, Skokie, Ill. 1968.

10-32 Carpenter, J. E., and Hanson, N. W., "Tests of Reinforced Concrete Wall Beams with Large Web Openings," *ACI Journal, Proceedings* **66** (9), 756–766, September 1969.

10-33 Cardenas, A. E., Kaar, P. H. "Field Test of a Flat Plate Structure," *ACI Journal, Proceedings*, **68** (1), January 1971.

10-34 Fintel, Mark, and Khan, F. R., "Effects of Column Exposure in Tall Structures, Temperature Variations and Their Effects," *ACI Journal Proceedings*, **62** (12), December 1965.

10-35 Khan, Fazlur R., and Fintel, Mark, "Effects of Column Exposure in Tall Structures, Analysis for Length Changes of Exposed Columns," *ACI Journal, Proceedings*, **63** (8), August 1966.

10-36 Khan, Fazlur R., and Fintel, Mark, "Effects of Column Exposure in Tall Structures, Design Considerations and Field Observations of Buildings," *ACI Journal, Proceedings*, **65** (2), February 1968.

10-37 Fintel, Mark, and Khan, F. R., "Effects of Column Creep and Shrinkage in Tall Structures–Prediction of Inelastic Column Shortening," *ACI Journal Proceedings*, **66** (12), December 1968.

10-38 Fintel, M., and Khan, F. R., "Effects of Column Creep and Shrinkage in Tall Structures - Analysis for Differential Shortening of Columns and Field Observations of Structures," *Designing for Effects of Creep, Shrinkage and Temperature in Concrete Structures*, Publication SP-27 American Concrete Institute, Detroit, Mich. 1971.

10-39 Hickey, K. B., *Creep of Concrete Predicted from Elastic Modulus Tests*, Report No. C-1242, U.S. Dept. of the Interior, Bureau of Reclamation, Denver, Colorado, January 1968.

10-40 Ross, A. D., "Creep of Concrete Under Variable Stress," *ACI Journal* **54** (9), 739–758, Mar. 1958.

10-41 McHenry, D., "A New Aspect of Creep in Concrete and its Application to Design," *Proceedings ASTM* **43**, p. 1069, 1943.

10-42 Troxell, G. E.; Raphael, J. M.; and Davis, R. E., "Long-Time Creep and Shrinkage Tests of Plain and Reinforced Concrete," *Proceedings ASTM*, **58**, 1101–1120, 1958.

10-43 Dishinger, F., *Der Baningenieur (Berlin)* **18** (39/40), 595–621, Oct. 1937.

10-44 Pfeifer, D. W., "Reinforced Lightweight Concrete Columns," *Journal of the Structural Division, ASCE*, **95** (ST 1), 57–82, Jan. 1969; and Pfeifer, D. W. "Full-size Lightweight Concrete Columns," *Journal of the Structural Division, ASCE*, **97**, (ST 2), 495–508, Feb. 1971.

11

Tubular Structures for Tall Buildings

FAZLUR R. KHAN, Ph.D. *

11.1 INTRODUCTION

It is only in the last 20 years that reinforced concrete has found increasing use in the construction of tall buildings. In its initial development in the early 1900s reinforced concrete buildings were limited to only a few stories in height. The structural type used was the traditional beam-column frame system which made the construction of taller buildings relatively expensive and, therefore, economically unfeasible. In the early 1950s, the introduction of shear wall type of construction opened up the possibility of using concrete in apartment and office buildings as high as 30 stories. Taller buildings still remained economically unattractive, and technically inadequate, because the shear walls which were mostly used in the core of the building were relatively small in dimension compared to the height of such buildings, leading to insufficient stiffness to resist lateral loads. It was obvious that the overall dimensions of the interior cores were too small to economically provide the stability and stiffness for buildings over 30 or 40 stories.

The natural tendency then was to find new systems that would utilize the perimeter configurations of such buildings rather than to rely on the core configurations alone. The development of the framed tube system was, therefore, a logical outcome of this challenge. The framed tube system in its simplest form consists of closely spaced exterior columns tied at each floor level with relatively deep spandrel beams, thereby creating the effect of a hollow concrete tube perforated by openings for the windows. Since the system simulated a hollow tube using perimeter closely spaced frame elements, it is referred to as "framed tube" (Fig. 11-1). This system was apparently first applied on the design of the 43-story DeWitt-Chestnut apartment building in Chicago in 1963 (Fig. 11-2). Since then the system has received wide acceptance among designers all over the world, and many variations are being used in a number of buildings under construction.

From the point of view of construction economy, the framed tube compares favorably with the normal shear wall type of construction for medium-rise buildings, but provides a distinct economic advantage for taller buildings. Moreover, the closely spaced column system has the additional advantage of also being the window wall system, thus replacing the vertical mullions for the support of the glass windows. In some recent buildings the elimination of the traditional curtain wall with its metallic mullions was in itself the justification for choosing this structural system.

The center-to-center spacing of the exterior columns in the framed tube structural system is generally from 4 ft to a maximum of about 10 ft. Depending on the overall proportion and height of the building, the maximum center-to-

*Dr. Fazlur R. Khan, Partner, Skidmore, Owings and Merrill, Chicago, Ill.

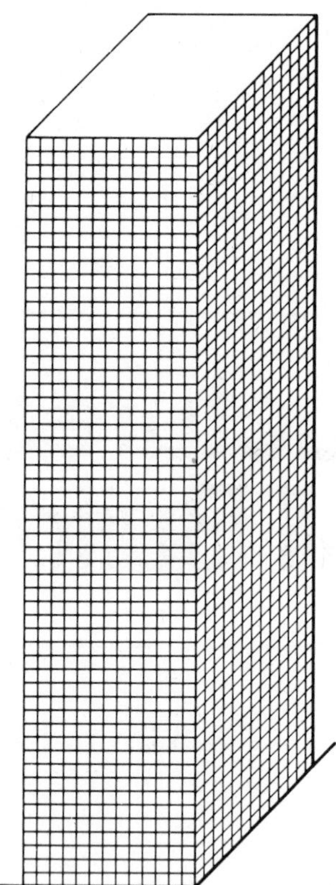

Fig. 11-1 Framed tube.

Fig. 11-2 The 43-story DeWitt-Chestnut apartment building in Chicago.

center spacing of the peripheral columns can probably be increased to 15 ft. The spandrel beams interconnecting the closely spaced columns generally vary from 2 ft in depth to about 4 ft in depth with widths from 10 in. up to 3 ft. In designing the framed tube structural system it is necessary to keep the proper balance of stiffness between the spandrels and the columns so that both of these elements are efficiently utilized to provide stiffness of the structure against lateral sway, and to assure the overall strength of the tube system to resist lateral forces. In most recent structures stiffness for limiting lateral sway controlled the proportions more often than strength requirements. It is obviously advisable to optimize the subsystem represented by two columns and two spandrels as shown in Fig. 11-3 to arrive at the optimum spacing of the columns and the proportion of the spandrels and the columns in terms of overall architectural program. In the DeWitt-Chestnut apartment building the columns were spaced on 5 ft 6 in. centers and the spandrels were 2 ft deep. The spacing of the columns in this case was also related to the module for interior planning of the apartment floors.

The framed tube structural system has expanded in its application and in its variation over the last five years. One of these variations is its application with an interior shear wall, commonly referred to as 'tube-in-tube' system, (discussed in section 11.8 below) as used in the 52-story One Shell Plaza Building in Houston, reaching a height of 714 ft. Another variation of the framed tube system has been used as the exterior envelope in conjunction with the traditional steel framing for the interior of the building. This system, known as the SOM Composite System, has been used in three major tall buildings in the country, namely: the 24-story CDC Building in Houston, the 35-story Union Station Building in Chicago, and the 50-story One Shell Square Building in New Orleans.

In view of the wide and varied application of the framed tube system, there is an obvious need for developing a preliminary analysis and design method for such a system. Of course, with availability of large computer programs, the final solution may be easily refined; however, the use of computers for preliminary design of even a moderate size framed tube may prove too expensive. Therefore, a method of preliminary analysis and design (using influence curves) for a wide range of proportions of the framed tube elements and geometry is presented.

This chapter is primarily intended for a clearer and better understanding of the behavior of the framed tube system.

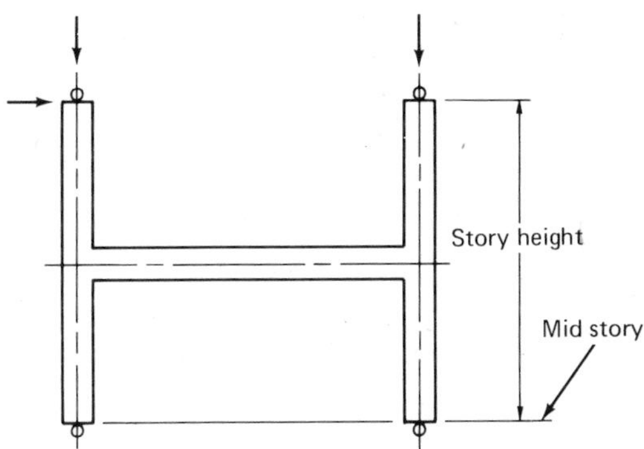

Fig. 11-3 Column-spandrel subsystem.

11.2 BEHAVIOR OF THE FRAMED TUBE SYSTEM

The framed tube system combines the behavior of a true cantilever, such as a shear wall, with that of a beam-column-frame. The overturning under lateral load is resisted by the tube form causing compression and tension in the columns, while the shear from the lateral load is resisted by bending in columns and beams primarily in the two sides of the building parallel to the direction of the lateral load (Fig. 11-4). Therefore, for all practical purposes the bending moments in these columns can be determined by judicious choice of the point of contraflexure in each story. While it is true that in the lower few stories, as well as in the upper few stories, the point of contraflexure does not remain in the middle of story height, the intermediate stories which constitute the major portion of the building generally have the point of contraflexure at mid-height of each story. It is, therefore, possible to compute the bending moments in these columns with reasonable accuracy for any known lateral shear at each story. One can, of course, make a simple iterative, Maney-Goldberg type, slope-deflection solution, or a modified moment distribution solution to determine more accurate moments in these columns. In fact, such an iterative solution will also give a good approximation of that portion of the total deflection which is caused by the frame action only. To this the additional overturning deflection caused by tension or compression in the column must be added to compute the total lateral deflection.

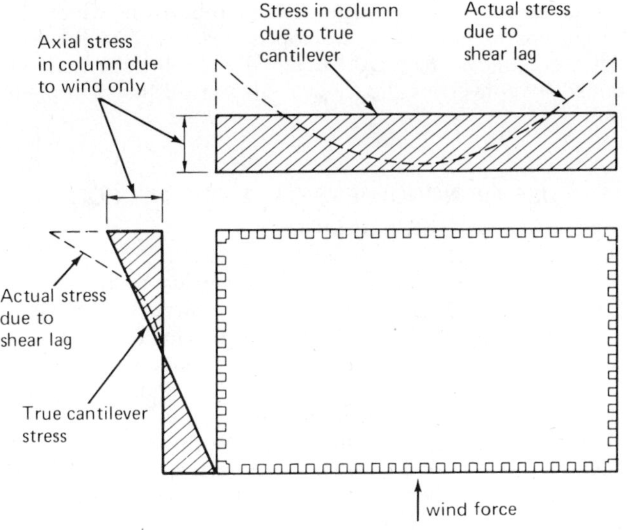

Fig. 11-4 Axial stress distribution in the columns due to wind—true cantilever vs actual stress due to shear lag.

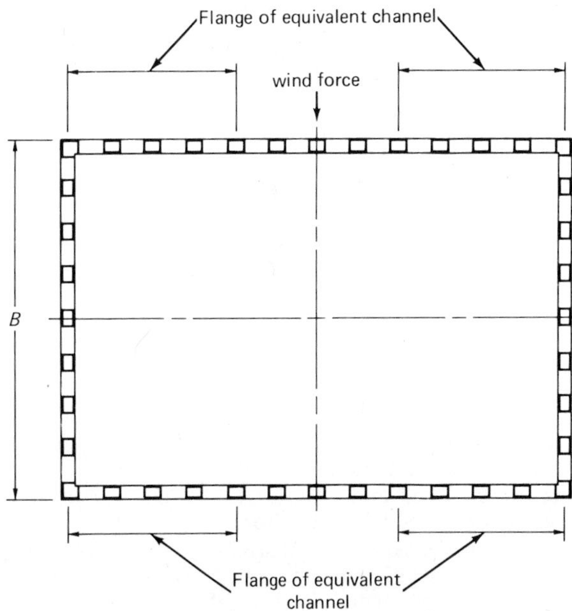

Fig. 11-5 Framed tube plan.

The cantilever tube type behavior becomes significant when the overturning of the entire building due to lateral load is considered. For analyzing the overturning of the entire framed tube, the exterior column system can be considered as part of a rigidly-diaphragmed hollow tube. However, in recognition of the fact that the webs of the hollow tube, that is, the two sides parallel to the direction of the lateral force, are not truly solid webs but are, in fact, grid frames, one must consider then the effect of loss of efficiency due to the flexibility of this web-frame causing what is known as shear lag, as shown in Fig. 11-4. For a very preliminary estimate of the overall resistance, as well as the deflection of the building, the effective configuration of the tube could be reduced to two equivalent channels resisting the total overturning moments (Fig. 11-5). Experience has indicated that for preliminary designs channel flanges normally should not be more than half the depth of the web (walls parallel to lateral load), or more than about 10% of the height of the building. These approximate rules have generally given conservative values of shear and moment as compared to the actual forces in the exterior columns obtained by the exact analysis performed subsequently by a generalized computer program such as STRESS, STRUDL, or others.

11.3 INITIAL PRELIMINARY DESIGN APPROACH

The overturning moment resisted by the two equivalent channels will produce axial forces in the closely spaced columns of these channels, as well as shear forces in the connecting spandrels. The preliminary estimate of the axial forces in the columns as well as the shears in the connecting spandrel can be based on the classical beam theory and can be expressed as follows:

$$P_w = \frac{M \times C \times A_c}{I_e}$$

and

$$V_s = \frac{V_w \times Q \times h}{I_e}$$

where

P_w = axial force due to wind
M = overturning moment
I_e = effective moment of inertia of the tube
V_s = spandrel shear
V_w = total wind shear
Q = sum of first moment of column areas about the neutral axis
C = distance of any column from the neutral axis
h = story height
A_c = cross sectional area of column

From the structural point of view, the framed tube differs from the solid tube by many large openings. Its behavior is, therefore, of a hybrid nature showing characteristics of the pure frame as well as of the pure tube. Even though one may expect the framed tube behavior to be more like a tube resisting the lateral forces through axial forces in the columns, significant moments develop in the columns in the two walls parallel to the wind direction.

In a framed tube-type structure the preliminary design method will generally indicate a reasonably uniform shear force in the spandrel beams along the two exterior walls parallel to the wind force. The preliminary moments in the spandrel beams are consequently derived from these shears. The preliminary design of the closely spaced columns should be based on the known dead and live loads added to the axial forces due to overturning; the moments caused by story shears should also be considered.

The study of a number of actual framed tube structures indicates that the lateral deflection, or sway, of these structures due to lateral load is primarily contributed by the frame action of the two walls parallel to the wind forces. For example, in the 43-story DeWitt-Chestnut Building, out of a total of 7 in. of lateral sway under the Chicago wind loading, approximately 5 in. (about 70%) was contributed by the frame action and only 2 in. was contributed by the overturning moment. This particular case shows that the efficiency of a framed tube diminishes as the building gets taller. As the height reaches a certain point, the lateral sway, and not the strength, of the building will control the design of the structural system. Therefore, the structural efficiency producing a premium-free building will gradually diminish with increasing number of stories.

11.4 PRELIMINARY DESIGN USING INFLUENCE CURVES

To achieve a more accurate design than the equivalent channel method proposed earlier for framed tube structures of any proportion and height within practical range, the author developed influence curves which can be directly used for a relatively accurate preliminary design. These curves have been developed on the basis of a number of computer runs on a 10-story equivalent framed tube with variable nondimensional parameters representing ratios of shear stiffness, S_b, of the spandrel beam to the axial stiffness, S_c, of the columns, and a linearly varying ratio of bending stiffnesses of columns to spandrels. A check of a number of framed tube buildings showed that this ratio ranged from 0.95 at the roof to 0.4 at the ground level. For this chapter a ratio of 0.75 at the roof to 0.5 at ground level is assumed. A separate computer run, with a constant ratio of bending stiffness of columns to spandrels of 1.0, did not change the qualitative results.

The significant structural properties affecting the tube action are

1. Bending stiffness:

$$K_c \text{ for column} = \frac{I_c}{H}$$

$$K_b \text{ for spandrel beam} = \frac{I_b}{L}$$

2. Shear stiffness of the spandrel beams (defined as the force required to displace one end of the spandrel a unit distance at right angles to the axis of the beam):

$$S_b = \frac{12\,EI_b}{L^3}$$

3. Axial stiffness of the column (defined as the axial force required to shorten the column a unit distance along the axis of the column):

$$S_c = \frac{A_c E}{H}$$

where

I_c = moment of inertia of the column
I_b = moment of inertia of the spandrel beam
A_c = cross-sectional area of the column
H = height of column
L = effective span of the spandrel beam
E = modulus of elasticity

The controlling parameters of framed tubes are.

$$\text{stiffness ratio} = \frac{K_e}{K_b}$$

$$\text{stiffness factor, } S_f = \frac{S_b}{S_c}$$

$$\text{aspect ratio, R} = \frac{\text{flange frame}}{\text{web frame}}$$

To plot influence curves, framed tubes have been analyzed for uniform lateral load and for the following ranges of variables: aspect ratio values of 0.5, 0.666, 1.0, 1.5 and 2.0; stiffness factor values of 0.1, 1.0, 10.0, and a linearly-varying stiffness ratio of 0.75 at roof to 0.5 at ground level for all values of aspect ratio and stiffness factor combinations.

Analysis of the equivalent framed tube from which the design curves have been developed was based on the configuration shown in Fig. 11-6, and direct solutions were obtained by converting the framed tubes into equivalent plane frames.

11.5 USE OF NONDIMENSIONAL CURVES FOR PRELIMINARY DESIGN

A total of nine nondimensional preliminary design curves, Figs. 11-7 to 11-15 are presented. These curves are primarily for computing column axial force coefficients for flange and web frame columns and shear force coefficients for the web frame beams. All these coefficients relate to unit values for the corresponding forces of the ideal tube. The main purposes of developing the curves was to provide the design engineer a tool to determine the tubular characteristics of any given framed tube and quickly compute the total deflection and bending moments and shears in the beams caused by the tubular nature of the entire structure. To make all the curves applicable to a wide range of realistic

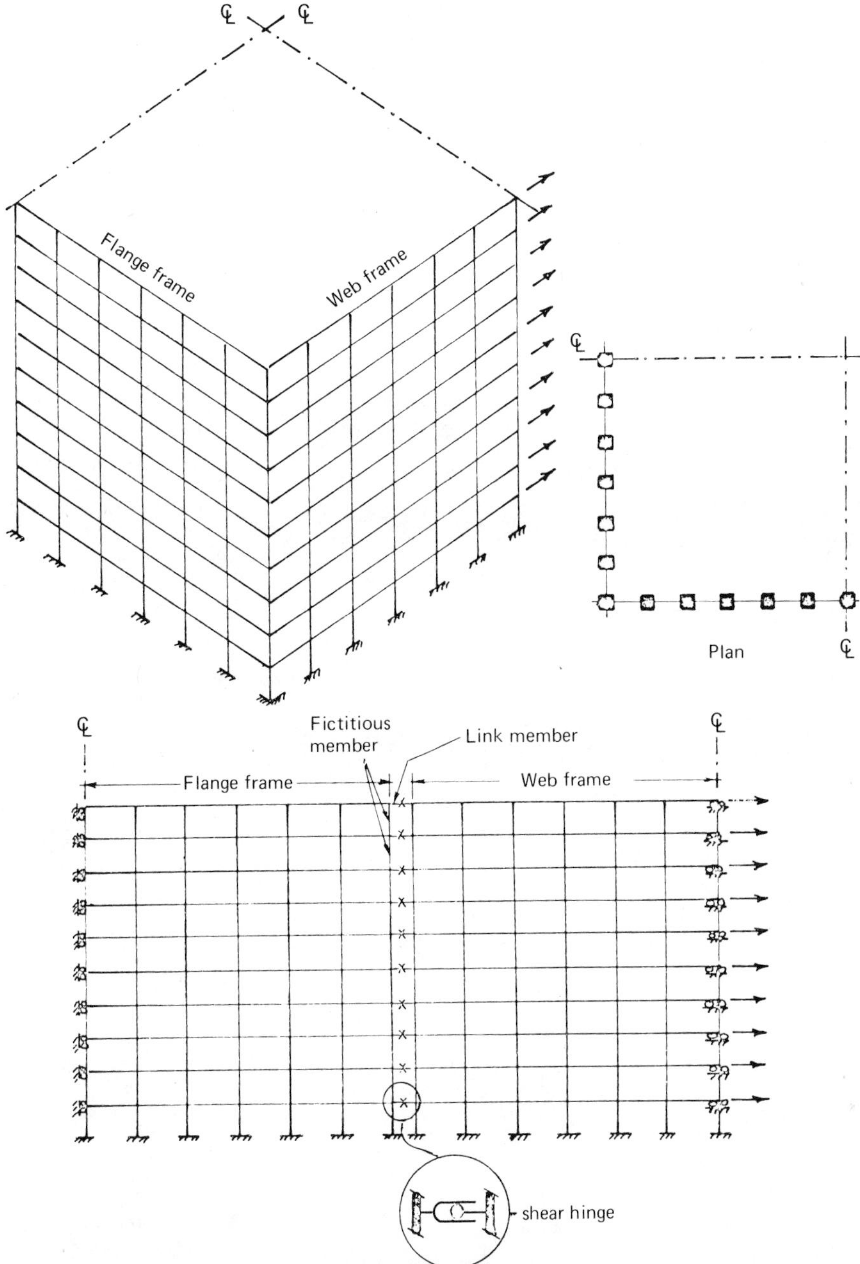

Fig. 11-6 Substituting an equivalent plane-frame for the framed tube.

proportions of actual foreseeable buildings, these curves have been plotted against nondimensional parameters representing the basic properties of the column and beam elements and aspect ratios.

To make the curves usable for any number of stories as well as any number of columns around the perimeter of a building, reduction model techniques similar to those used in previous developments[11-1-11-5] have been used as described in the following sections.

11.6 REDUCTION MODELING APPROACH

To achieve uniformity of reference to the design curves, computer solutions were obtained on a number of specific 10-story frame tubes, each representing stiffness factor, aspect ratios, and stiffness ratio indicated on the curves. Although the solutions were obtained on 10-story hypothetical framed tubes, it can be shown that the tubular behavior of any framed tube of any number of stories can be simulated by converting it into an equivalent framed tube of the same height, but having a fewer or greater number of stories. For any given plan proportion, commonly referred to as the aspect ratio, R, of a building, the variables that directly affect the tubular stress distribution are the shear stiffness, S_b, of the spandrel beams, and the axial stiffness, S_c, of the columns. For reasonably uniform spacing of perimeter columns, S_b and S_c in the graphs represent the sum total of all columns and beams around the perimeter. If, however, stiffness factor, S_f, varies on each face, an average value can be taken for a reasonable approximation.

It can be shown that any framed tube of N stories can be reduced to a 10-story equivalent tube of the same height by

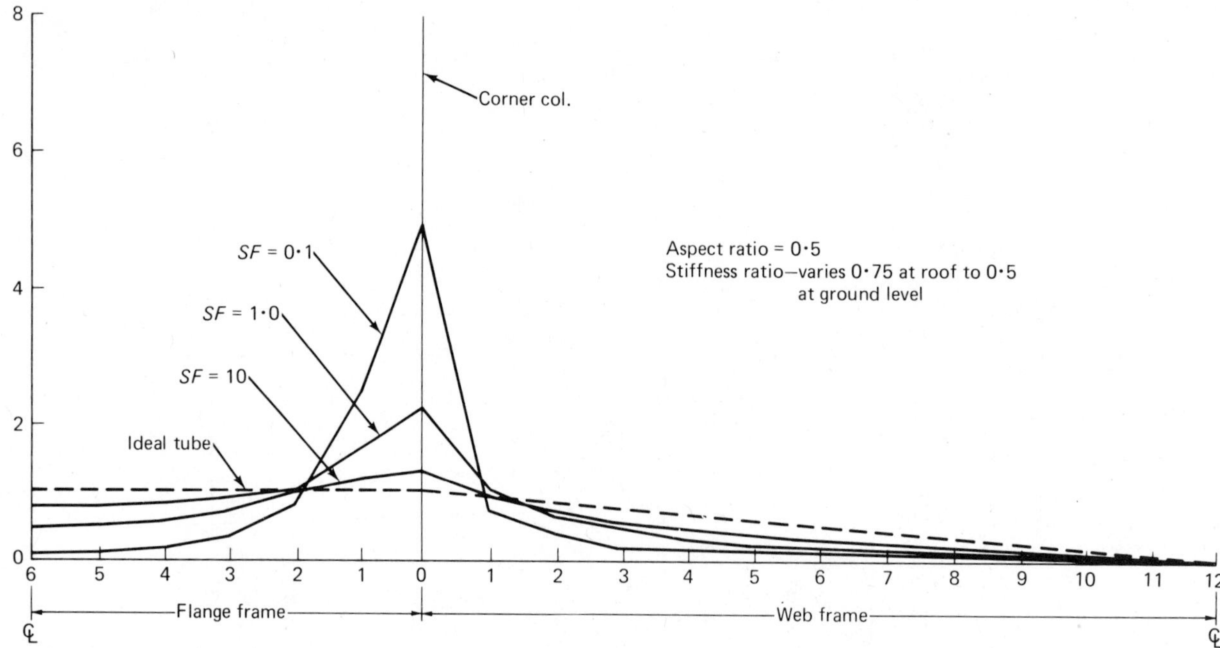

Fig. 11-7 Column axial force coefficients—level 1, aspect ratio = 0.5.

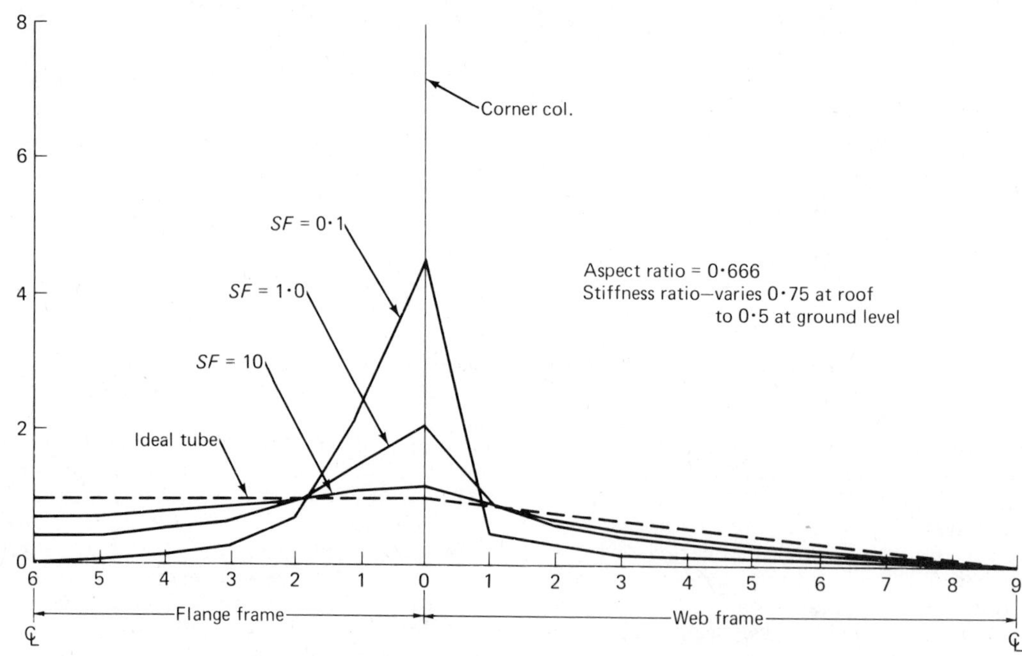

Fig. 11-8 Column axial force coefficients—level 1, aspect ratio = 0.666.

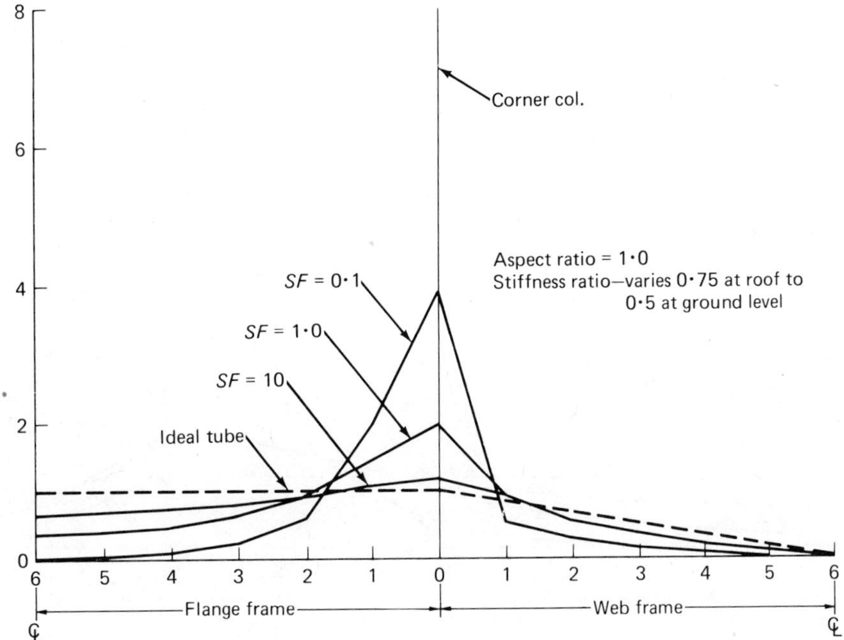

Fig. 11-9 Column axial force coefficients—level 1, aspect ratio = 1.0.

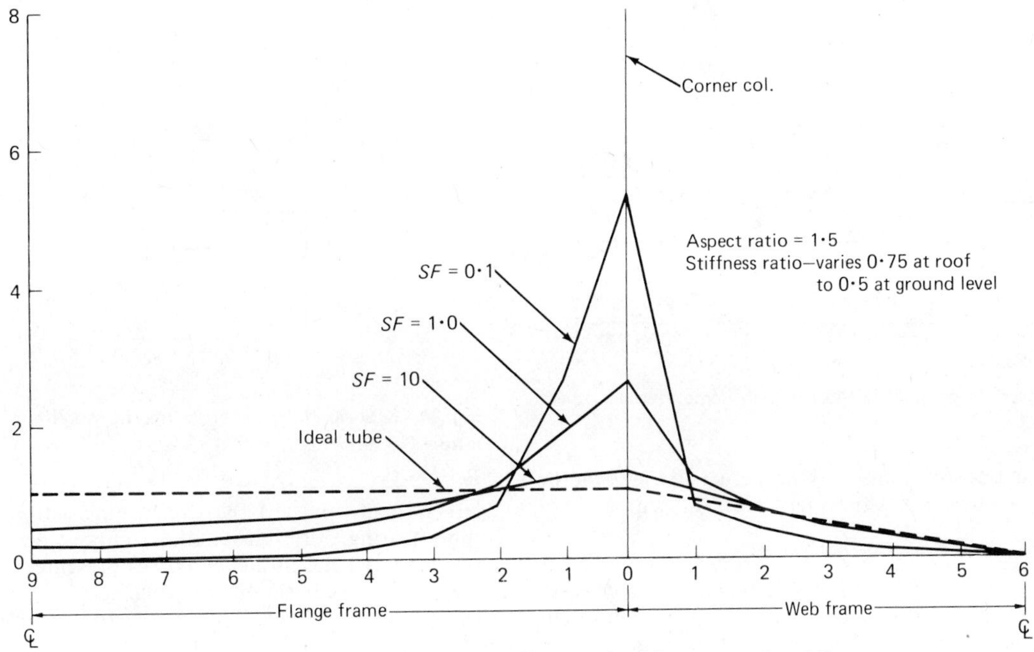

Fig. 11-10 Column axial force coefficients—level 1, aspect ratio = 1.5.

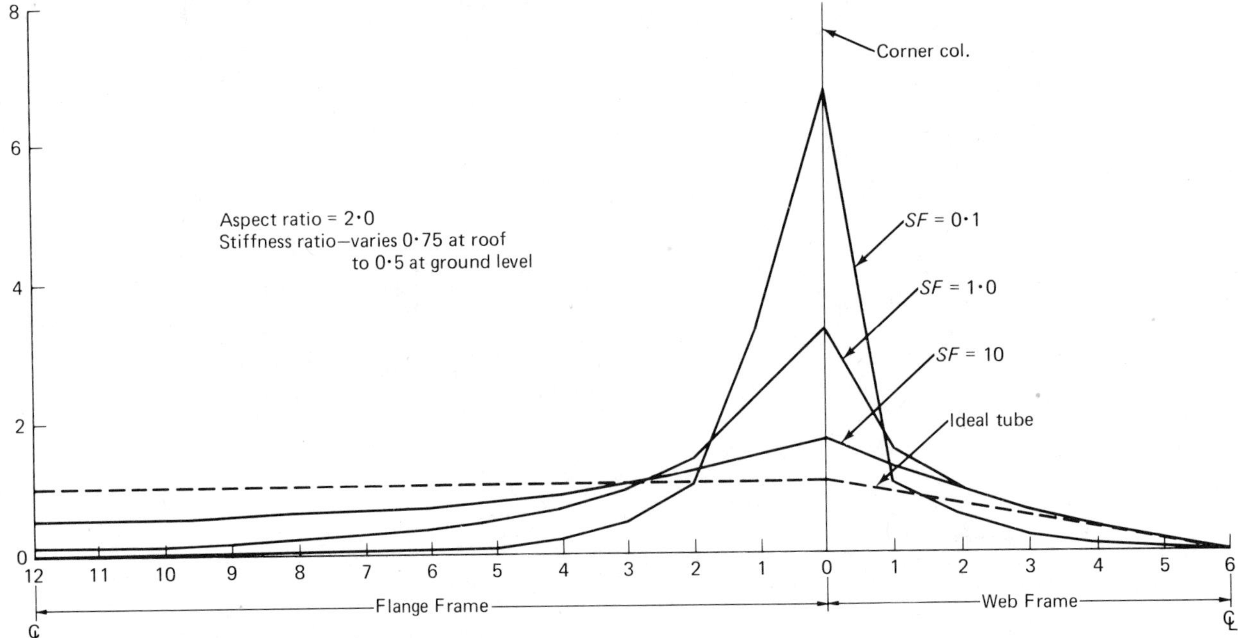

Fig. 11-11 Column axial force coefficients—level 1, aspect ratio = 2.0.

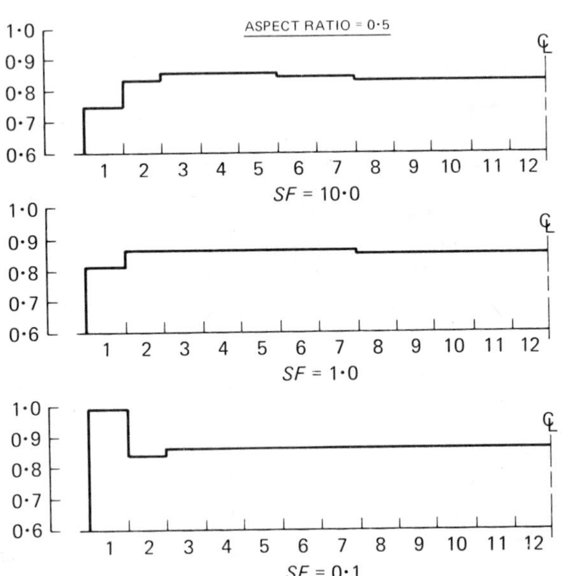

Fig. 11-12 Shear force coefficients in web-frame beams, aspect ratio = 0.5.

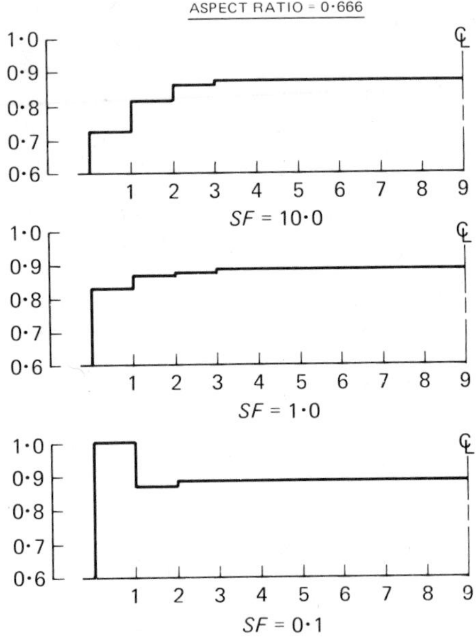

Fig. 11-13 Shear force coefficients in web-frame beams, aspect ratio = 0.666.

considering a transformation of the actual stiffness factor to a 10-story equivalent stiffness factor of S'_{f10} where

$$S'_{f10} = S_f \times (N/10)^2$$

Using this simple reduction model technique, any given actual framed tube can first be converted to an equivalent 10-story framed tube and then analyzed by using the influence curves.

11.7 TOTAL BEHAVIOR OF A FRAMED TUBE

As was pointed out earlier in this chapter, the framed tube system always has two components of its behavior: (a) the frame action of the two sides parallel to the direction of the lateral load, and (b) the overturning action of the entire tube causing only tension and compression in the exterior columns. All the influence curves presented in this chapter are for the evaluation of the tube action only, although the use of these curves will allow one to compute the approximate moments and shears in the spandrels which will define one boundary of the column moments. The other boundary of the column moments at any typical story should be obtained by assuming the point of contraflexure at mid-height of each story. For preliminary design the higher of the two values should be used.

To compute the total deflection, the additional deflection due to frame action must be calculated separately and added to the deflection caused by tube action.

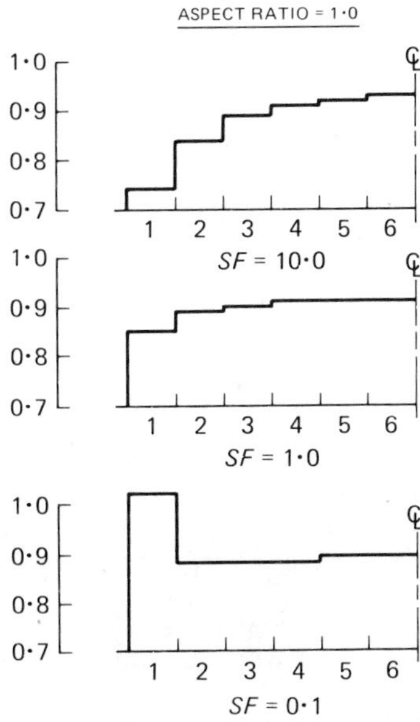

Fig. 11-14 Shear force coefficients in web-frame beams, aspect ratio = 1.0.

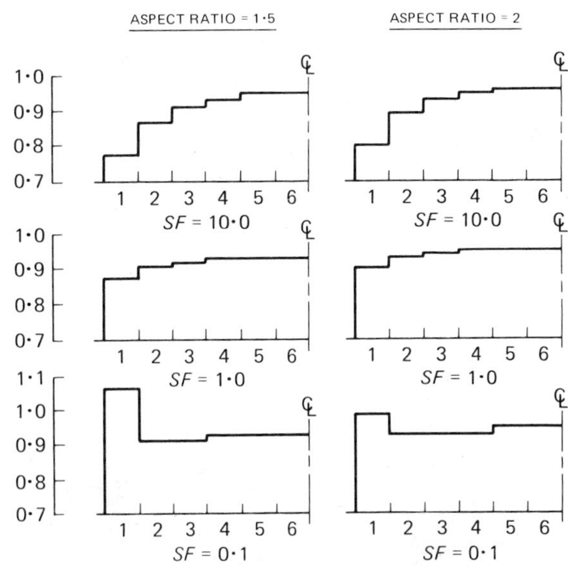

Fig. 11-15 Shear force coefficients in web-frame beams, aspect ratio = 1.5.

EXAMPLE 11-1: To illustrate the use of influence curves for preliminary designs, a hypothetical framed tube 50 stories high and 130 × 180 ft in plan with columns at 10 ft on center and a 13-ft floor to floor height is considered.

Before using the influence curves, the basic parameters of the structure, like aspect ratio, average stiffness factor, average stiffness ratio, and frame deflection, must be computed. Also, the deflection and forces in columns and beams for an ideal tube of the same Aspect Ratio must be known. Basic section properties and parameters are presented in Tables 11-1 and 11-2.

Frame deflection can be computed by any standard method and is approximately 3.79 in. From Table 11-2, by conjugate-beam method, cantilever deflection is computed as 7.5 in. Any other similar method could also give reliable results.

The actual forces in columns, shears in beams, and deflection of the system, can now be computed using the influence curves. For this example, Fig. 11-10, SF = 10 curve is used for computing the column axial forces in first story columns, and Fig. 11-15 (c), SF = 10 curve is used for computing the beam shears in the first story beams of a 10-story model. The above results are summarized in Table 11-3 and compared with the results obtained from a more exact computer solution.

The deflection of the total system is the sum of the frame deflection plus the ideal tube deflection times the magnification factor. This magnification factor is defined as the ratio of the sum of the ideal tube column forces to the actual column forces of the flange columns. For this example, the sum of the ideal tube column forces is 1280 × 10 = 12800 kips and the sum of the actual column forces is 10825 kips. Hence, the magnification factor is 1.17.

$$\Delta \text{ Total } = 3.79 + 7.5 \times 1.17 = 12.49 \text{ in.}$$

$$\Delta \text{ Actual} = 11.51 \text{ in.}$$

Comparing the values of the above example problem with an exact computer analysis shows that the use of influence lines gives fairly reliable results with an accuracy of 90 to 95%. The shear forces in the spandrel beams of the web frames (parallel to the wind direction) differ by 5% only from the exact analysis at the center of

TABLE 11-1

	Column Properties			*Spandrel Beam Properties*					
Floor	*Area* $in.^2$	*M. of I.* $in.^4$	*Axial Stiffness* $S_cK/in.$	*Area* $in.^2$	*M. of I.* $in.^4$	*Shear Stiffness* $S_bK/in.$	*Stiffness Factor* $S_f = S_b/S_c$	$S_{f10} = S_f(N/10)^2$	*Stiffness Ratio* = K_c/K_b
1-5	972	462000	24400	1455	452000	13000	0.53	13.2	0.795
6-10	810	447000	20400	1200	372000	10700	0.525	13.1	0.925
11-15	810	447000	20400	1200	372000	10700	0.525	13.1	0.925
16-20	810	447000	20400	1200	372000	10700	0.525	13.1	0.925
21-25	717	400000	18000	1058	330000	9520	0.525	13.1	0.935
26-30	615	363000	15400	940	282000	8140	0.52	13.0	1.0
31-35	615	363000	15400	940	282000	8140	0.52	13.0	1.0
36-40	510	307000	12800	754	234000	6750	0.52	13.0	1.01
41-45	510	307000	12800	754	234000	6750	0.52	13.0	1.01
46-50	510	307000	12800	754	234000	6750	0.52	13.0	1.01

Average Stiffness Factor for ten-story model = 13.05

Average Stiffness Ratio = 0.95

$$\text{Aspect Ratio} = \frac{180}{130} = 1.4$$

For this example use Stiffness Factor = 10.0
and Aspect Ratio = 1.5

TABLE 11-2

Floor	*1*	*5*	*10*	*15*	*20*	*25*	*30*	*35*	*40*	*45*
Overturning Moment, ft-kip	975000	789750	624000	477750	351000	243750	156000	87750	39000	9750
Tubular Moment of Inertia = $\Sigma\ AR^2$, ft^4	333444	277870	277870	277870	245966	210975	210975	174955	174955	174955

$$\text{Column axial force at Level 1} = \frac{M \times C}{I} \times A_c = 1280 \text{ kip}$$

$$\text{Spandrel beam shear at center line spandrel} = \frac{VQ}{I} \times H = 2850 \text{ kip}$$

TABLE 11-3

Flange Column Axial Forces			*Shear Force in Web Frame Beams*		
Col. Location	*Using Influence Curves kips*	*Actual, kips*	*Beam Location*	*Using Influence Curves, kips*	*Actual, kips*
0	1660	1583	1	2340	1954
1	1530	1426	2	2620	2257
2	1280	1237	3	2750	2450
3	1150	1100	4	2810	2565
4	1020	1001	5	?850	2640
5	900	929	6	2850	2690
6	830	877	7	2850	2700
7	830	841			
8	830	818			
9	795	806			

the frame; the difference, increasing progressively towards the corners. However, since in a typical framed tube structure the size of the spandrel beam is generally constant at any level, the inaccuracy of the shears in the corner spandrel beam during the preliminary design is of little consequence.

11.8 TUBE-IN-TUBE STRUCTURAL SYSTEM

Tall office buildings have generally a reasonably large service core and it is advantageous to use a shear wall enclosing the entire service core. Because of the demand for column-free office space, it is then a natural solution to eliminate all interior supports and provide closely spaced exterior columns, thus creating a framed tube. In most such recent office buildings this clear span generally varies from 35 to 40 ft. The resulting structural system consists of an inner tube created by the shear walls, and the outer tube consisting of the closely spaced column system, creating what may be called a tube-in-tube system.

The tube-in-tube system has the advantage of both the framed tube structure, as well as the shear wall type structure. In fact, the shear wall inner tube greatly enhances the structural characteristics of the exterior framed tube by considerably reducing the shear deflection of the columns of the framed tube. The tube-in-tube system is a refined and unique version of the shear wall-frame interaction type structures as described in section 10.5.7 of this handbook.

11.8.1 Approximate Analysis

Before making an approximate analysis for preliminary design of this type of structure, it is first necessary to define the factors contributing to its total lateral deflection. The tube-in-tube structure subjected to lateral loads will primarily act as a shear wall-frame interactive system, interaction being between the interior shear wall tube and the walls of the framed tube parallel to the direction of wind. An additional deflection will occur due to column shortening of the exterior tube caused by overturning moments.

For approximate analysis of the shear wall-frame interaction type behavior of this system, one of the most direct approaches would be to use the influence charts presented in Ref. 11-2. Results from these charts should be sufficient for a preliminary design, as the more sophisticated, exact analysis of actual structures have shown that these influence charts normally give results within about 5% accuracy. The column shortening of the framed tube can be taken into account with reasonable accuracy as suggested in Ref. 11-2. The method suggests compensation for the column shortening by adjusting the actual moment of inertia of the inner shear wall tube to an equivalent moment of inertia slightly larger than its actual value. This procedure may be summarized as follows:

$$I_e = I \times (1 + K')$$

where

I_e = equivalent increased moment of inertia of inner tube
I = actual moment of inertia of inner tube
K' = λ_1/λ_2
λ_1 = pure frame stiffness of framed tube
λ_2 = pure rigid tube stiffness of framed tube

After the shear wall-frame interaction charts are used to determine the lateral shear distribution between the shear wall and the exterior frame tube, the two separate systems, viz., the shear wall and the framed tube, can be designed

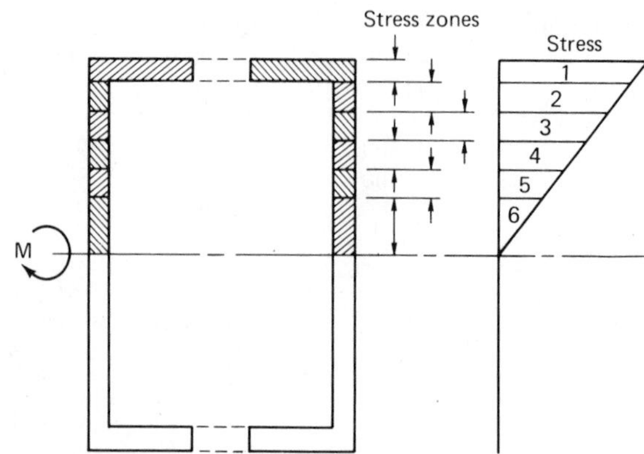

Fig. 11-16 Dividing a shear wall into stress zones.

separately. The preliminary design of the exterior framed tube has already been discussed in detail earlier. The design of the shear wall may require distribution of the shear force in the shear wall to the different elements which make up the shear wall tube. In designing the shear wall, the overturning moment in the shear wall may cause large axial stresses which can be handled in the design by dividing the shear wall into small segments and considering each of these segments as a column unit as shown in Fig. 11-16. This rational design method will allow distribution of vertical reinforcement throughout the shear walls according to the expected stress intensity due to overturning moments. Another method is to treat the shear wall as a flexural reinforced concrete element and to concentrate the tensile reinforced the furthest distance away from the neutral axis as possible (see section 10.5.6.2).

11.8.2. Exact Method of Analysis

It is not practical at the present time to attempt a truly exact analysis in view of its complexity due to the large number of members and joints in the exterior framed tube, and due to an extremely large moment of inertia of the shear wall tube which is interconnected to the tube by a rather flexible floor structure. However, in recognition of the nature and philosophy of structural design, a somewhat idealized mathematical model may be used for the tube-in-tube structure. One possibility of analysis is to apply an iterative method using a forced-convergence technique[11-2, 11-4] to the idealized structure consisting of the framed tube interconnected with the inner shear wall tube.

While the presently available generalized type computer programs such as STRESS and STRUDL are easily adaptable to framed tube type structures as discussed earlier, their adaptation to tube-in-tube type structures becomes somewhat undesirable because of the problems involved in simulation of this system into a space-frame. It is possibly easier to simulate the tube-in-tube structure as a plane-frame composed of an equivalent frame representing the framed tube and an equivalent shear wall representing the inner tube; both connected by the floor structure represented by a link beam as shown in Fig. 11-17. The plane-frame modeling will generally give acceptable results except for the following two possible problems:

a. If the shear wall moment of inertia is extremely high compared to the columns of the framed tube, there may be a possibility of ill-conditioned matrix and round-off errors.
b. If the structural floor slab is sufficiently rigid, then the rotation of the shear walls will cause vertical displacements at the junction of the shear wall and the link beam which must be compensated. This can be done by simulation of the link beam as shown in Fig. 11-18.

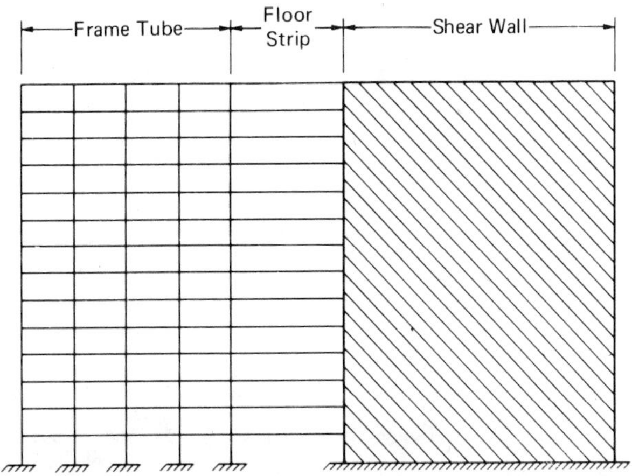

Fig. 11-17 Plane-frame model of tube-in-tube system.

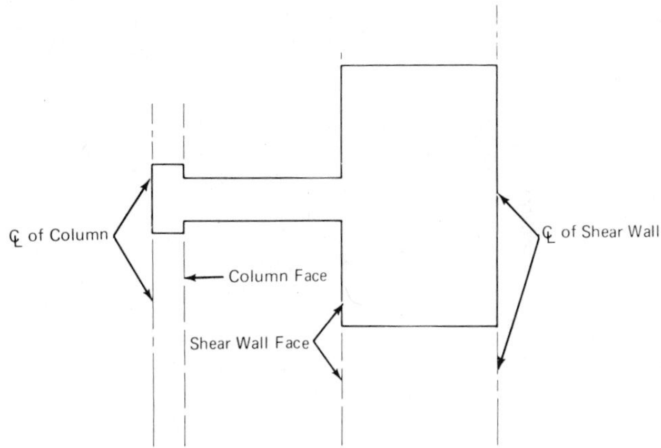

Fig. 11-18 Simulation of the link beam.

REFERENCES

11-1 Khan, F. R., "Current Trends in Concrete High Rise Buildings," *Proceedings*, Symposium on Tall Buildings, University of Southhampton, England, April, 1966.

11-2 Khan, F. R., and Sbarounis, J. A., "Interaction of Shear Walls and Frames," *Proceedings, ASCE*, **90** (St 3), 285–335, June 1964.

11-3 Khan, F. R., "On Some Special Problems of Analysis and Design of Shear Wall Structures," *Proceedings*, Symposium on Tall Buildings, University of Southhampton, England, April, 1966.

11-4 Khan, Fazlur R., and Fintel, Mark, "Effects of Column Exposure in Tall Structures, Analysis for Length Changes of Exposed Columns," *ACI Journal, Proceedings*, **63** (8), August 1966.

11-5 Khan, F. R. and Navinchandra, R. Amin, "Analysis and Design of Framed Tube Structures for Tall Concrete Buildings," *ACI Journal*, SP36, 1973.

12

Earthquake-Resistant Structures

ARNALDO T. DERECHO, Ph.D.* **and MARK FINTEL****

12.1 INTRODUCTION

Although the incidence of earthquakes of destructive intensity has been confined to a relatively few areas of the world, the catastrophic consequences attending the few that have struck near centers of population have focused attention on the need to provide adequate safety against this most awesome of nature's quirks.

The satisfactory performance of a large number of reinforced concrete structures subjected to severe earthquakes in different areas of the world has demonstrated that it is possible to design such structures to successfully withstand earthquakes of major intensity.

Early attempts to provide for earthquake resistance in buildings were based on rather crude assumptions about structural behavior and were handicapped by the lack of the proper analytical tools as well as reliable earthquake records. Observations of the behavior of reinforced concrete structures subjected to actual earthquakes, analytical studies, and laboratory experiments by a number of investigators over the last two decades or so have all contributed toward putting the subject of earthquake-resistant design on a firm rational basis.

*Principal Structural Engineer, Design Research Section, Portland Cement Association, Skokie, Illinois.
**Director, Engineering Services Department, Portland Cement Association, Skokie, Illinois.

The present state of knowledge regarding earthquake-resistant design and construction has benefited from efforts aimed at understanding the various aspects of the problem, the major ones being

1. The earthquake phenomenon—seismological studies, observations, and records have not only yielded insight into the immediate cause of earthquakes but have also provided data on earthquake distribution and ground acceleration.
2. Behavior of structures and structural components under earthquake loading—Analytical studies of the dynamic response of idealized models of structures to earthquake excitation have provided much valuable information which has helped explain the observed behavior of actual structures subjected to earthquakes. Such dynamic response studies, using actual, modified, or artificially generated accelerograms as input data, represent one of the most useful tools for the design of earthquake-resistant structures. The sophistication with which these analyses have been carried out has been made possible only recently by the availability of large electronic digital computers.

Experimental studies aimed at understanding the behavior of structural members subjected to earthquake-type loadings have helped provide the basis for design specifications on earthquake-resistant construction.

On-site observations of structures subjected to earthquakes have provided valuable lessons on the effects of deficiencies in design and construction, and a basis for assessing the adequacy of earthquake design specifications.

Much still remains to be done, especially in the area of seismic regionalization. Various aspects of the problem still await further clarification, such as damping, foundation structure interaction, and three-dimensional nonlinear behavior, and a large amount of information still requires to be accumulated in order to be able to treat earthquake analysis and design on a sound probabilistic basis. However, the experience of the last two decades has indicated that proper analysis and design, guided by a knowledge of the basic requirements to be achieved and a particular attention to design details, can provide safe and economical structures to resist earthquakes.

The material which follows will be concerned principally with earthquake engineering as it applies to the structural design of reinforced concrete buildings. It should be noted, however, that many of the observations made in relation to buildings also apply, in some degree, to other civil engineering structures.

12.2 EARTHQUAKES

12.2.1 Causes of Earthquakes

The exact chain of events which leads to earthquakes has been the subject of much theorizing and speculation. However, seismologists are agreed on two distinct mechanisms which directly result in earthquakes. Thus, *tectonic* earthquakes are those disturbances (waves of distortion) resulting from rupture or a sudden movement along an existing fault in the earth's crust,* while *volcanic* earthquakes are those associated with volcanic eruptions or subterranean movements of magma. The major portion of observed activity has been tectonic in origin.

A widely held theory of the tectonic mechanism of earthquakes, considered well-substantiated for earthquakes accompanied by visible faulting, is the "elastic rebound theory" first proposed by H. F. Reid in 1906.[12-1] According to the theory, earthquakes are caused by the relatively sudden release of accumulated strains in the crustal rocks through fracture along areas of weakness. The release of energy produces waves of distortion which propagate in all directions, with accompanying reflections from the earth's surface as well as reflections and refractions as they traverse the earth's interior. The exact process by which forces are generated within the earth's interior which ultimately result in crustal upheavals (mountain building and earthquakes) is not yet well understood.

The point of fracture or the point where the initial slip or rupture occurs, which triggers a large scale slippage (in some cases extending for hundreds of miles) is called the earthquake *focus* or *hypocenter*. The point on the surface vertically above the focus is called the *epicenter* (see Fig. 12-1).

Earthquake foci may occur anywhere from near the surface to a depth of 700 kilometers or more. The term normal or shallow-focus earthquakes has been used to denote shocks with focal depths of about 60 km. (0–44 miles) or less, while those with greater focal depths fall under the broad classification of deep-focus earthquakes. Under the latter term are included two subclasses: intermediate, with foci between 70 to 300 km. (44–188 miles) and deep, for focal depths greater than 300 km. Most ordinary earthquakes have shallow foci, i.e., they originate in the crustal layer, above the Mohorovicic discontinuity.

*A term usually applied to the upper layer of the earth, from the surface to the so-called Mohorovicic discontinuity (the "Moho"), which separates the crust from the mantle. The thickness of the earth's crust varies from about 5 to 60 kilometers, being thicker in the continental areas and thinner under the sea. Another seismic discontinuity which has been detected at levels above the Mohorovic discontinuity is the Conrad discontinuity.

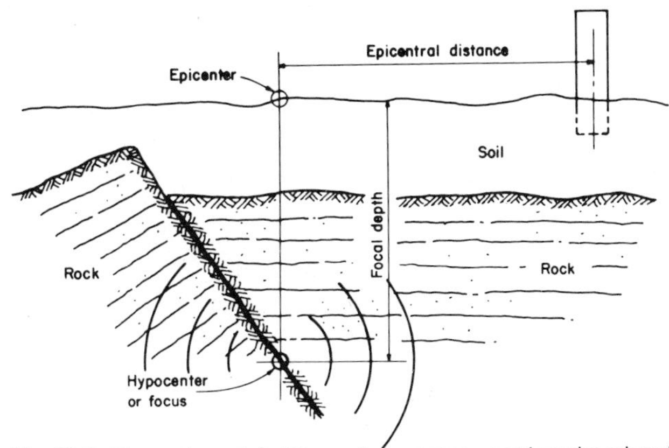

Fig. 12-1 Illustrating definition of common earthquake-related terms.

Shallow-focus earthquakes tend to produce only local effects. The ground motion in the localized area near the epicenter, however, can be expected to be more intense than that which would result if the focus were located deeper and the effects felt over a much wider area.

12.2.2 Geographic Distribution of Earthquakes

Almost all major earthquakes and approximately 95% of all earthquakes have occurred in two long and relatively narrow zones. The principal zone, which accounts for about 80% of all earthquakes, is the *circum-Pacific seismic zone*, which borders the Pacific ocean, running up from the west coast of South, Central, and North America, through the Aleutians, and down the coast of Asia through Japan, the Philippines, to New Zealand. Paralleling the circum-Pacific belt of major tectonic activity, but generally a hundred or so miles to one side of it, is the chain of active volcanoes making up what has often been referred to as the Pacific 'circle of fire'. The closeness of these two zones has once led to the mistaken belief that earthquakes are due principally to volcanic activity.

The second major seismic zone, called the *Alpide zone*, runs in a generally east-west direction across Europe, from the Azores through Spain, Italy, Greece, Turkey, northern India and turning southeastward through Burma and Sumatra to join the circum-Pacific belt in New Guinea. The Alpide belt accounts for about 15% of all earthquakes. A third zone, accounting for most of the remaining five or so percent, is a fairly narrow zone along mid-oceanic rifts—along the center of the mid-oceanic ridges—particularly those in the Atlantic and Indian oceans.

A plot of earthquake epicenters all over the world as calculated by the U.S. Coast and Geodetic Survey for the year 1966 is shown in Fig. 12.2. Plots for other years show a similar distribution of earthquakes as that indicated in the figure. The map can be considered complete for earthquakes with Richter magnitude above 4 (see Section 12.2.5).

12.2.3 Nature of Earthquake Motions—Seismic Waves

The effects of earthquakes are usually classified as primary, if due directly to the causative process, such as faulting or

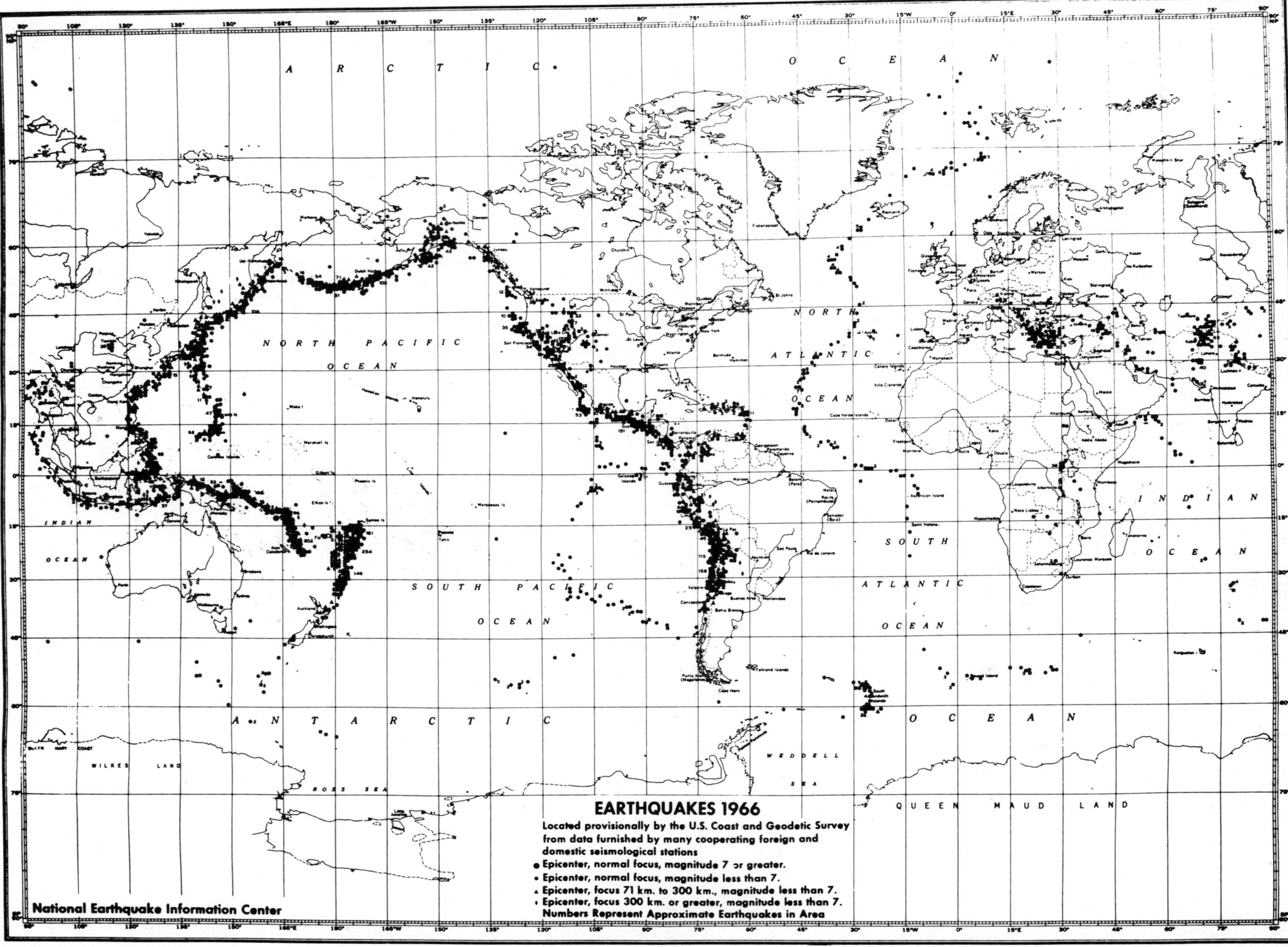

Fig. 12-2 Map showing global seismicity for the year 1966. The epicenters of the earthquakes were calculated by the U.S. Coast and Geodetic Survey from the observed travel times of seismic *P*-waves to the seismographic stations. The map can be considered almost complete for earthquakes with a magnitude above 4.[12-2]

(a)

(b)

Fig. 12-3 (a) Apartment buildings at Kawagishi-cho, Niigata, Japan, after the Niigata earthquake of June 16, 1964.[12-3] (b) Close-up of building shown in (a), Niigata, Japan.[12-4]

volcanic action; and secondary, when due to the ground motion resulting from the passage of seismic waves. Of the secondary effects, two types may be distinguished: one where inertial effects in structures are negligible, such as those associated with landslides or soil consolidation or liquefaction triggered by earthquake motions and the other in which dynamic (inertial) effects are predominant. Except in unusual conditions,* most of the damage associated with earthquakes has been due to the latter type of effects and engineering efforts aimed at designing earthquake-resistant structures are concerned mainly with such effects.

The ground motion at a particular site is influenced by three factors: (a) source parameters, such as the earthquake magnitude (energy released), depth of focus and geological conditions at and near the focus; (b) transmission path parameters, i.e., epicentral diatance and properties and geologic character of the intervening ground; and (c) local site parameters or the geologic configuration and properties of the ground at the site.

The complex and erratic motion of the ground characterizing an earthquake is the result of the passage of waves of distortion through the affected area. Such motions recorded by seismographs are juxtapositions of a number of different wave types,** among the more significant of which are longitudinal (compressional-dilatational) or P-waves, transverse (shear) or S-waves, Rayleigh or R-waves, and Love or Q-waves.

Both P- and S-waves are body waves, traversing the interior of the earth and making up the high frequency components of the strong ground motion near the epicenter. P-waves travel at velocities ranging from 5 to 13 km/sec compared to a range of from 3 to 8 km/sec for S-waves, the velocity in each case increasing with depth.*** In a P-wave, the particles vibrate in the direction of propagation, while in an S-wave the particles oscillate in a plane perpendicular to the direction of propagation.

The R- and Q-waves are surface waves which propagate along the surface of the earth at velocities only slightly less than that of S-waves. Rayleigh waves are generated whenever waves propagating through a continuous medium (such as the body waves, P and S) reach a free surface. The energy input to structures on or near the ground surface comes predominantly from Rayleigh waves. These tend to attenuate rapidly with distance from the surface. Love waves result from the presence of a distinct discontinuity at some depth from the free surface of the medium through which body waves propagate. Whenever several layers having different material properties occur, other types of waves are generated, e.g., the so-called guided waves in layered media. In the R-wave the vibration is in the direction of propagation while in the Q-wave the particle displacement is transverse to the direction of travel, with no vertical or longitudinal component. Surface waves form a significant portion of the strong ground motion in epicentral regions of shallow-focus earthquakes and also make up the dominant long-period phases of shallow-focus teleseisms, i.e., earthquakes recorded at distances over 1000 km (621 miles) from the epicenter.

Generally, the amplitude of recorded ground motion diminishes with increasing distance from the epicenter. Under certain geologic conditions, particularly in areas where a deep layer of saturated alluvium overlays bedrock, a substantial amplification of wave motion transmitted from the bedrock may occur. Also, the short period (or high frequency), high energy components of seismic waves tend to be damped out much faster than the longer period waves, so that at some distance from the epicenter the ground motion consists predominantly of long-period oscillations.

The ground motion at a particular point is completely determined by three translational and three rotational components along three perpendicular directions. The rotational components of earthquake motions are, however, generally negligible. Most seismographic stations are equipped with instruments designed to record the horizontal components along two perpendicular directions and the vertical component of the ground motion.

12.2.4 Strong-Motion Earthquake Records

Of particular interest to the structural engineer as far as earthquake motions are concerned are instrumental records of the ground motion close to the epicenter. These usually consist of acceleration traces of the components of motion along two perpendicular horizontal directions and one in the vertical direction. Such records are obtained by using strong-motion accelerographs (SMACs).

Strong-motion seismographs or accelerographs are basically simple, damped oscillators, the recorded response (i.e., the relative displacement between the movable mass and a fixed base) being proportional to the input acceleration. Unlike the more sensitive standard seismographs used by seismologists to record motions resulting from distant earthquakes, SMACs are less-sensitive instruments designed to record motions in the epicentral regions of earthquakes. In such locations, the high magnification of a conventional seismograph usually results in a jumbled trace, if not a broken instrument. Also, while standard seismographic stations usually have their instruments set up on massive concrete blocks founded on solid rock, SMACs, by the very nature of their purpose, are set up at widely different locations, from open fields to basements and upper floors of buildings. Another difference is that while conventional seismographic records are continuous (the recording paper being set at a comparatively low speed), with particular emphasis placed on the time at which signals are received, SMACs set up for engineering purposes do not start recording until triggered by an initial motion, with the time scale used mainly for the proper reduction of the data. The first strong-motion seismograph of modern type was introduced in 1931 and the first strong-motion records were obtained during the March 1933 Long Beach (California) earthquake.

Although ground motions recorded at a given location can hardly be expected to be repeated in the future, strong-motion records, if available over a long period, can provide valuable insight into the general character of the ground motion and the effect of geologic conditions at the particular site. Where a number of accelerograms are available for a limited region, a set obtained by judicious sampling from the available data can serve as the basis of dynamic response studies of proposed structures, when such studies are deemed desirable. In an attempt to remedy the present lack of strong-motion records for structural response studies, several methods have recently been proposed for generating

*Such as occurred in the city of Niigate, Japan, during the earthquake of June 16, 1964, when a number of buildings were rendered useless—though some were practically undamaged—by excessive tilting due to the liquefaction of the sand on which many of the buildings were founded (see Fig. 12-3).

**The dominance of one type of wave in the combination often producing a distinct characteristic or "phase" in the recorded signal.

***The velocity increase occurs from the surface to the boundary of the mantle and the core, at which depth an abrupt decrease in wave velocity occurs. From this point down to the inner core, there is again a slight increase in the P-wave velocity. There have been no shear wave rays known to have traversed the earth's outer core, which is a basis for the belief that the material in this region is liquid in character.

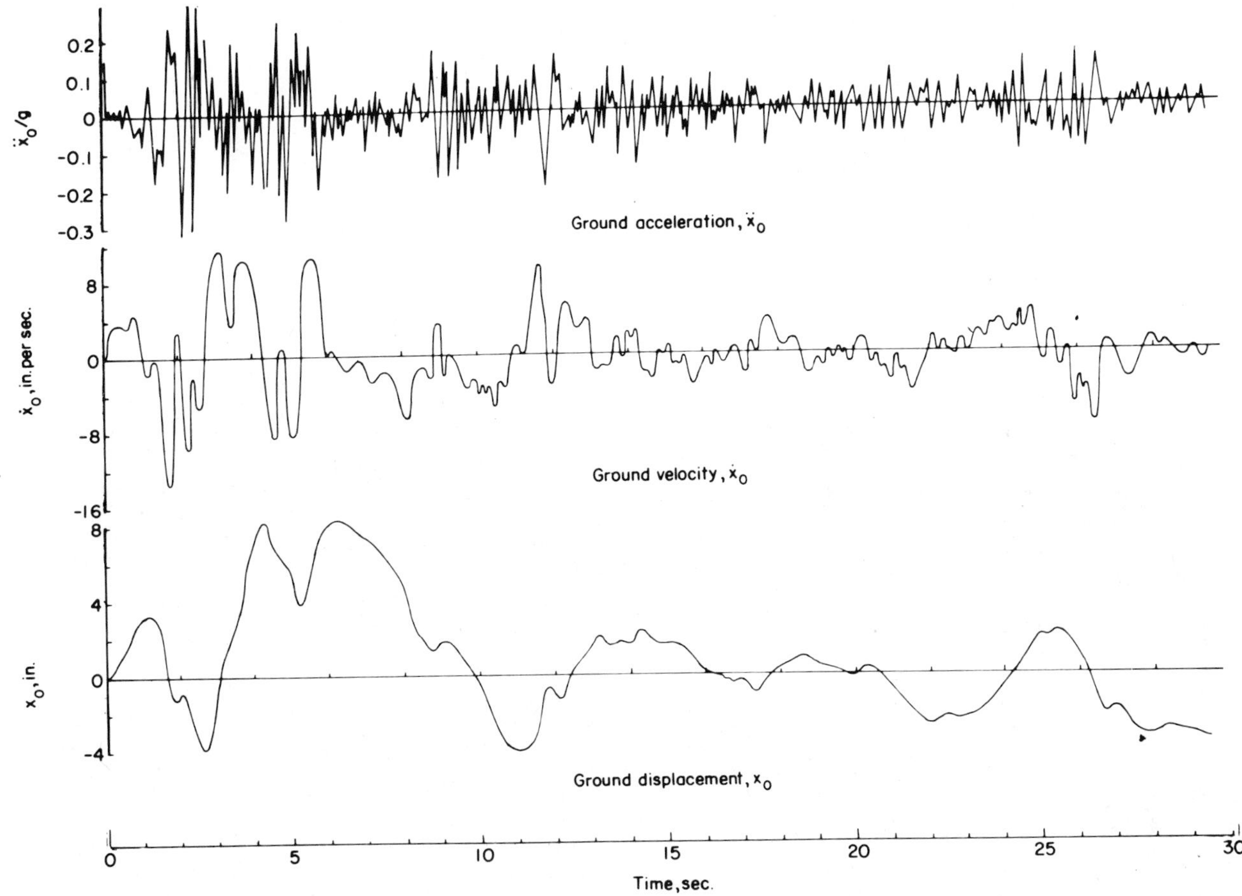

Fig. 12-4 May 18, 1940 El Centro (Calif.) Earthquake accelerogram, N-S component—and corresponding derived velocity and displacement plots.[12-7]

accelerograms artificially, using nonstationary stochastic models, i.e., filtered 'shot noise'.[12-5, 12-6] The methods consist essentially in assuming certain properties of actual earthquake accelerograms as basic characteristics and in using these parameters as constraints in the generation of random numbers representing ground acceleration amplitudes.

A set of acceleration traces which has gained widespread popularity through repeated use in dynamic response studies is that of the Imperial Valley (California) earthquake of May 18, 1940. This was recorded at the El Centro instrument site which rests on some 5000 feet of alluvium and located some 4 miles from the causative fault break. The 1940 El Centro record represents the strongest ground motion thus far recorded. Although several other earthquakes and magnitudes and maximum recorded ground accelerations greater than the 1940 El Centro (see next section and Table 12-2), none so far* has exhibited the combination of high-frequency, large-amplitude pulses lasting over as long a duration as the El Centro record.

A plot of the variation of the north–south component of horizontal ground acceleration during the first 30 seconds of the 1940 El Centro earthquake is shown in Fig. 12-4. Also shown are plots of the ground velocity and displacement as obtained from the acceleration trace by successive integration. The maximum value of the recorded ground acceleration is about 0.33 g, (1.0 g being equal to the acceleration due to gravity), the maximum calculated ground velocity about 13.7 in./sec and the corresponding recorded vertical acceleration was about 0.28 g.

The reason for the use of acceleration traces instead of velocity or displacement traces in strong-motion records is the greater accuracy with which the latter two can be obtained from the former by a process of integration rather than vice versa by differentiation. This will be obvious from an examination of Fig. 12-4.

Since actual strong-motion accelerograph records do not show a zero-acceleration line, such a base line has to be assumed when reducing the data for digitization and integration purposes. Several methods, satisfying different criteria, have been used to adjust the accelerogram base line. The method used in such a base line adjustment can affect the calculated maximum displacement significantly although the maximum velocity and response spectra (see section 12.3.3.5) derived from the resulting accelerogram are relatively insensitive to the method of base line adjustment. Significant errors in the calculated net ground displacement can also be expected as a result of the uncertainty concerning the initial part of the shock, which is not recorded due to accelerograph triggering and start-up delays.

12.2.5 Earthquake Magnitude and Intensity

Magnitude and intensity scales have been developed in attempts to describe earthquakes in quantitative terms.

*Except perhaps the record taken at the abutment of Pacoima Dam during the San Fernando earthquake of February 9, 1971. There is, however, some doubt as to the cause of the extremely high accelerations recorded at this particular site, where the rock on which the instrument rested exhibited fractures resulting from the earthquake.

TABLE 12-1 The Modified Mercalli Intensity Scale and corresponding approximate ground accelerations.[12-8]

	Ground Acceleration a (cm/sec)	a/g
I - Not felt except by a very few under especially favorable circumstances.		
II - Felt only by a few persons at rest, especially on upper floors of buildings. Delicately suspended objects may swing.	2, 3	
III - Felt quite noticeably indoors, especially on upper floors of buildings, but many people do not recognize it as an earthquake. Standing motor cars may rock slightly. Vibration like passing truck. Duration estimated.	4, 5, 6	.005g
IV - During the day felt indoors by many, outdoors by few. At night some awakened. Dishes, windows, doors disturbed; walls make creaking sound. Sensation like heavy truck striking building. Standing motor cars rocked noticeably.	7, 8, 9, 10	.01g
V - Felt by nearly everyone; many awakened. Some dishes, windows, etc., broken; a few instances of cracked plaster; unstable objects overturned. Disturbances of trees, poles, and other tall objects sometimes noticed. Pendulum clocks may stop.	20, 30	
VI - Felt by all; many frightened and run outdoors. Some heavy furniture moved; a few instances of fallen plaster or damaged chimneys. Damage slight.	40, 50, 60, 70	.05g
VII - Everybody runs outdoors. Damage negligible in buildings of good design and construction; slight to moderate in well-built ordinary structures; considerable in poorly built or badly designed structures; some chimneys broken. Noticed by persons driving motor cars.	80, 90, 100	0.1g
VIII - Damage slight in specially designed structures; considerable in ordinary substantial buildings, with partial collapse; great in poorly built structures. Panel walls thrown out of frame structures. Fall of chimneys, factory stacks, columns, monuments, walls. Heavy furniture overturned. Sand and mud ejected in small amounts. Changes in well water. Disturbs persons driving motor cars.	200, 300	
IX - Damage considerable in specially designed structures; well-designed frame structures thrown out of plumb; great in substantial buildings, with partial collapse. Buildings shited off foundations. Ground cracked conspicuously. Underground pipes broken.	400, 500, 600	0.5g
X - Some well-built, wooden structures destroyed; most masonry and frame structures destroyed with foundations; ground badly cracked. Rails bent. Landslides considerable from river banks and steep slopes. Shifted sand and mud. Water splashed over banks.	700, 800, 900, 1000	1g
XI - Few, if any, (masonry) structures remain standing. Bridges destroyed. Broad fissures in ground. Underground pipelines completely out of service. Earth slumps and land slips in soft ground. Rails bent greatly.	2000, 3000	
XII - Damage total. Waves seen on ground surfaces. Lines of sight and level distorted. Objects thrown upward into the air.	4000, 5000, 6000	5g

The magnitude of an earthquake is a measure of its size in terms of the energy released and radiated in the form of seismic waves. It is determined on the basis of instrumental (i.e., microseismic) data, particularly, the amplitude of ground motion at a specified distance from the epicenter.

Of several magnitude scales which have been proposed, the best known is the Gutenberg-Richter or Richter scale. On this scale, magnitude is defined in terms of the amplitude of the trace in a standard type seismograph located 100 kilometers from the epicenter. Estimates of the radiated energy based on the recorded amplitudes and periods of seismic waves during a number of earthquakes have led to correlations between the arbitrarily defined magnitude and the energy released during an earthquake. The accepted relationship has changed over the years with the accumulation of more data, the current (1970) one being given by

$$\log E = 11.4 + 1.5M$$

where E is the energy released in ergs and M is the magnitude on the Richter scale. The above equation indicates an energy increase of about 32 times for every unit increase in magnitude.

The largest earthquakes recorded* have had a Richter magnitude of 8.9. This is believed to be the largest possible magnitude on the Richter scale, the limit being set by the maximum amount of strain energy which can be stored by crustal rocks without rupture.

Although earthquakes of magnitude less than about 5 are believed not capable of producing damaging effects, the magnitude of an earthquake, per se, does not provide an indication of its destructiveness at any given location. The destructiveness of an earthquake, although dependent partly on its magnitude, is a function of other equally important factors such as focal depth, epicentral distance, local geology, and structural characteristics and resistance to damaging oscillations.

*One on the Columbia-Ecuador border in 1906 and another in Japan in 1933.

The word "intensity" when used to describe earthquakes is generally intended to denote the potential destructiveness of an earthquake at a particular location. If it were possible to predict intensity as such with respect to specific structural types, the structural engineer would have a good basis for assessing his design earthquake loads. Since the destructiveness of earthquake motions is determined not only by the character of the ground motion at a site but also by the type of structure and geology of the site, it is readily seen that intensity as a measure of potential destructiveness is not easy to define, much less predict, quantitatively.

If one considers only the ground motion at a given site, the principal parameters characterizing it (particularly as it affects the dynamic response of structures) are the amplitude of the motion, its frequency characteristics, and the duration of the large-amplitude pulses. Early attempts at quantifying earthquake intensity have sought to relate the maximum ground acceleration with the observed effects on structures and other natural features at specific locations.

The most commonly quoted intensity scale, at least on the American continent, is the Modified Mercalli (MM) scale, which is given in Table 12-1. The scale, which has 12 grades of intensity, was adopted, with modifications to suit conditions in America, by H. O. Wood and F. Neumann in 1931 from a proposal originally devised by G. Mercalli for use in Italy. Also shown in the table are approximate values of the ground acceleration, a, corresponding to the different intensity ranges, as obtained from the relation $I = 3 \log a + 1.5$ proposed by Gutenburg and Richter.[12-8]

The estimated magnitudes and reported MM intensity ratings of some recent earthquakes are shown in Table 12-2.

The Modified Mercalli (MM) intensity of an earthquake is usually assessed by distributing questionnaires to or interviewing persons in affected areas, in addition to observations of earthquake effects (i.e., macroseismic data) by experienced personnel. It is readily apparent that the determination of the intensity at a particular site involves considerable subjective judgment and is influenced greatly

TABLE 12.2 Statistics Of Some Recent Earthquakes

Name of Earthquake	*Date of Occurrence*	*Richter Magnitude*	*Maximum MM Intensity*	*Max. Recorded Hor. Grnd. Acceleration*
San Francisco, California	April 18, 1906	8.3	XI	
Imperial Valley, (El Centro) California	May 18, 1940	6.7–7.1	X	0.33 g
Olympia, Washington	April 13, 1949	7.1	VIII	0.31 g
Kern County (Taft), California	July 21, 1952	7.7	XI	0.18 g
Chile (Concepcion & Southern Chile)	May 22, 1960	7.5–8.5		
Agadir, Morocco	February 29, 1960	5.6		
Alameda Park, Mexico	May 11, 1962			0.098 g
Skopje, Yugoslavia	July 26, 1963	5.4–6.0	VIII	
Prince William Sound, Alaska	March 27, 1964	8.4		
Niigata, Japan	June 16, 1964	6.2–7.5	VII	0.19 g
Parkfield, California	June 27, 1966	5.5	VIII	0.50 g
Lima, Peru	October 7, 1966			0.40 g
Caracas, Venezuela	July 29, 1967	5.7–6.5		
Koyna, India	December 11, 1967	6.25–7.5	VIII	0.63 g
San Fernando, California	February 9, 1971	6.6	VIII	1.20 g (Pacoima Dam)

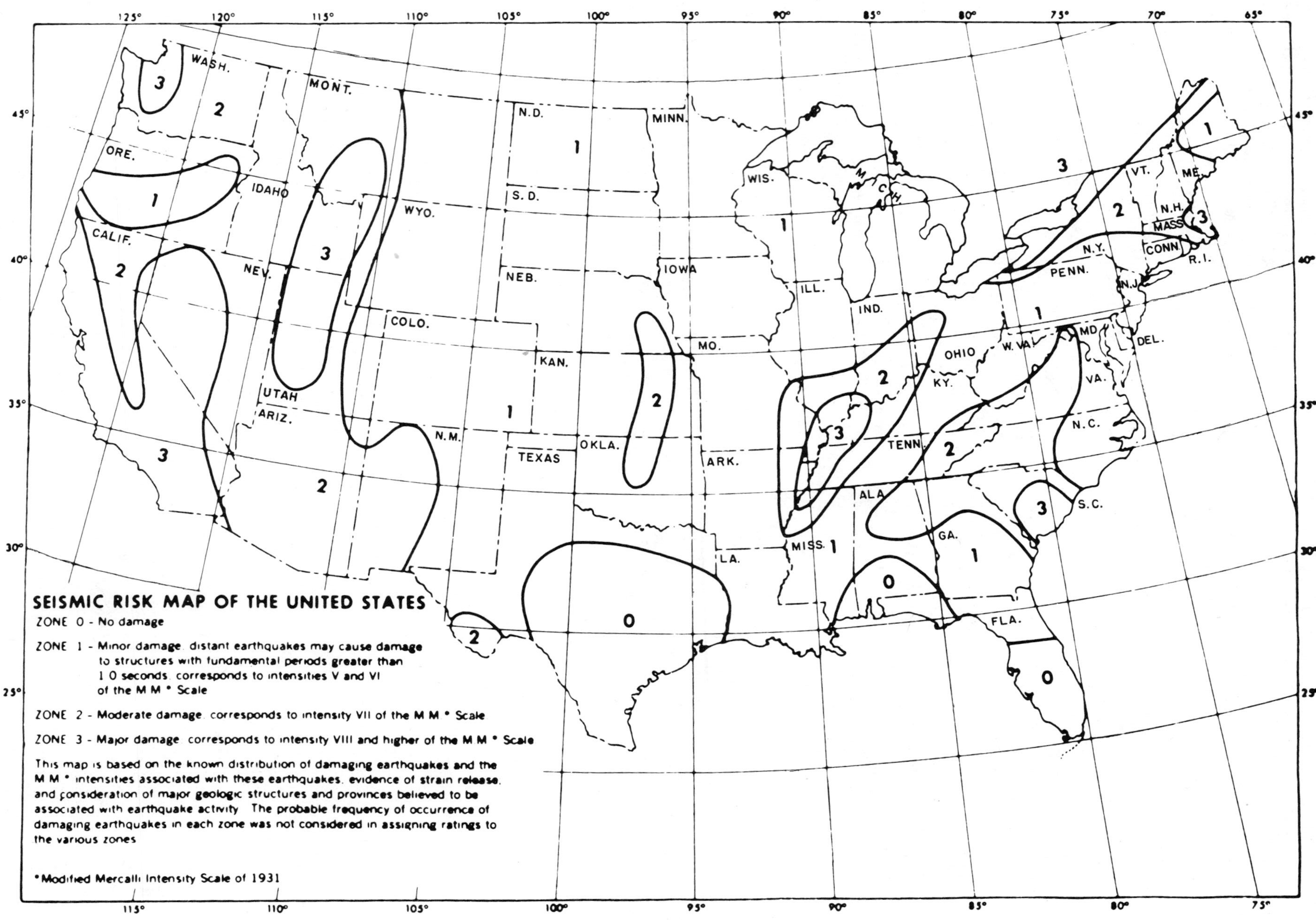
SEISMIC RISK MAP OF THE UNITED STATES
ZONE 0 - No damage
ZONE 1 - Minor damage, distant earthquakes may cause damage to structures with fundamental periods greater than 1.0 seconds, corresponds to intensities V and VI of the M.M.* Scale
ZONE 2 - Moderate damage, corresponds to intensity VII of the M.M.* Scale
ZONE 3 - Major damage, corresponds to intensity VIII and higher of the M.M.* Scale
This map is based on the known distribution of damaging earthquakes and the M.M.* intensities associated with these earthquakes, evidence of strain release, and consideration of major geologic structures and provinces believed to be associated with earthquake activity. The probable frequency of occurrence of damaging earthquakes in each zone was not considered in assigning ratings to the various zones.
*Modified Mercalli Intensity Scale of 1931
WASH.
ORE.
CALIF.
IDAHO
NEV.
MONT.
WYO.
UTAH
ARIZ.
COLO.
N.M.
N.D.
S.D.
NEB.
KAN.
OKLA.
TEXAS
MINN.
IOWA
MO.
ARK.
LA.
WIS.
ILL.
MICH.
IND.
OHIO
KY.
TENN.
MISS.
ALA.
GA.
FLA.
S.C.
N.C.
VA.
W. VA.
MD.
DEL.
N.J.
PENN.
N.Y.
CONN.
R.I.
MASS.
N.H.
VT.
ME.
125°
120°
115°
110°
105°
100°
95°
90°
85°
80°
75°
70°
65°
45°
40°
35°
30°
25°

Fig. 12-5 Seismic risk map of the United States—*UBC 1970.*[12-7]

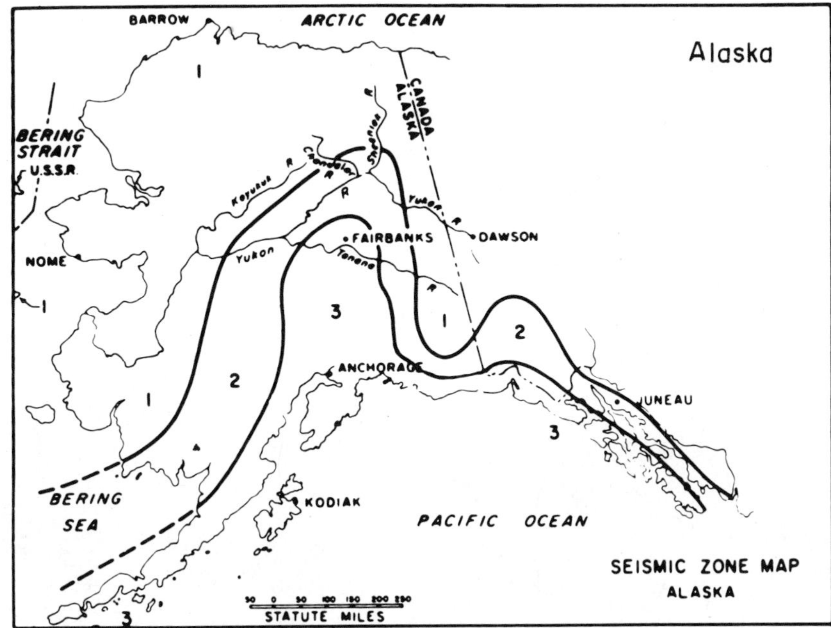

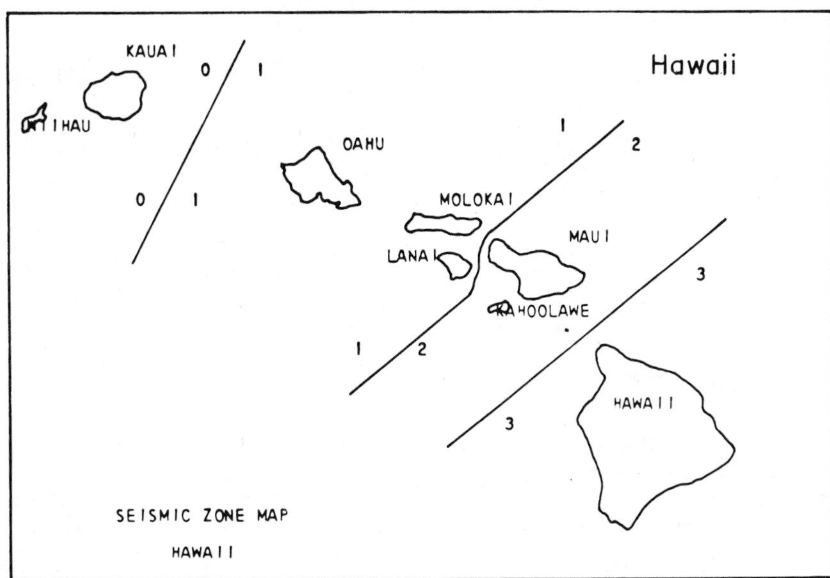

Fig. 12-5 *Continued*

by the type and quality of structures as well as the geology of the area. Thus, comparisons of intensity ratings made by different persons in different countries or during periods of time when different conditions prevail can be misleading. However, intensity ratings, when assessed uniformly over a sufficiently large area affected by a single earthquake, can yield valuable information on the geology of the site. Irregularities in the shape of lines connecting plotted points of equal intensity, i.e., isoseismals, may indicate unusual geologic conditions.

From the point of view of the structural engineer, reported Modified Mercalli intensities can be considered only as a very crude quantitative measure of the destructiveness of an earthquake since they do not provide, in terms of relevant structural parameters, specific information on damage corresponding to structures of interest.

The recognition of the shortcomings of the MM intensity scale, particularly in relation to structural engineering applications, has prompted a number of investigators to propose other measures of earthquake intensity.

12.2.6 Seismic Regionalization

The problem of designing economical earthquake-resistant structures rests heavily on the determination of reliable quantitative estimates of expected earthquake intensities corresponding to particular regions. The task of obtaining such estimates of local seismic risk is necessarily a long-term project, with the reliability of the estimates improving as more data on earthquakes and local tectonic features are accumulated.

The subdivision of an area into regions each associated with a known or assigned seismic probability or risk has served as a useful basis for the implementation of code provisions on earthquake-resistant design. The present seismic zoning maps used in United States and Canadian model building codes show each area divided into zones of approximately equal seismic probability or risk. Associated with each zone is a factor which enters into the expression for determining the total design base shear.

The Seismic Risk Map shown in Fig. 12-5 was prepared

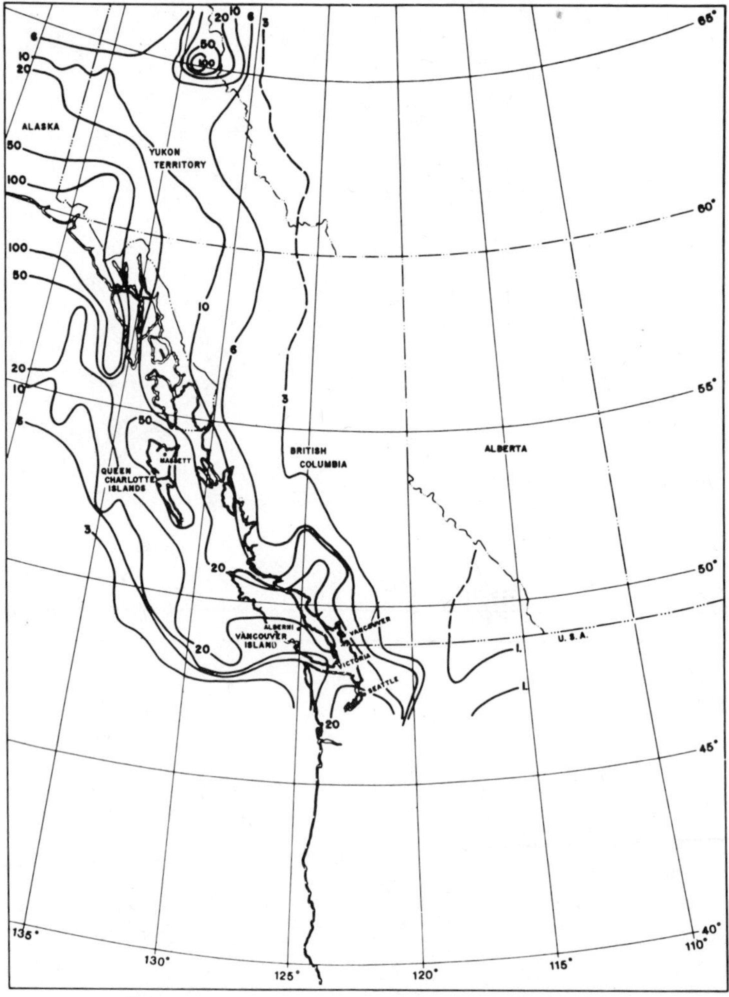

Fig. 12-6 Accelerations, in percent of g, with a 100-year return period for Canada.

by S. T. Algermissen[12-9] of the U.S. Coast and Geodetic Survey* and has been adopted by the International Conference of Building Officials for inclusion in the 1970 edition of the *Uniform Building Code*.[12-10] The map replaces the Seismic Probability Map which first appeared in the 1952 edition of the *Uniform Building Code* and in subsequent editions without change. As a part of the *Code*, the map serves as a guide in determining the minimum equivalent earthquake forces to be used in designing structures located in each of the four zones indicated. As noted in the map, the probable frequency of occurrence of damaging earthquakes in each zone was not considered in assigning ratings to the various zones. The lack of sufficient strong-motion and other data has been given as a reason for not attempting to present seismic risk in terms of measurable quantity less subjective than MM intensity.

One way of presenting the result of seismic risk studies in quantitative terms is a seismic probability map showing contours of equal maximum ground acceleration corresponding to specific return periods. A series of such maps has been prepared by W. G. Milne and A. G. Davenport[12-11] for Canada based on a statistical analysis of earthquake data and assumed relationships between maximum ground acceleration and magnitude and epicentral distance or reported earthquake intensities. Figure 12-6 shows contours of estimated maximum accelerations corresponding to a 100-year return period for Western Canada. Although, as noted earlier, the maximum ground acceleration alone does not provide a reliable index of the destructiveness of an earthquake, such a map represents a significant step towards the characterization of earthquake risk in terms of a quantitative parameter, i.e., maximum ground acceleration corresponding to a specific return period.

*Now the National Oceanic and Atmospheric Administration.

12.3 RESPONSE OF STRUCTURES TO EARTHQUAKE MOTIONS

12.3.1 General

The loads or forces, which a structure subjected to earthquake motions is called upon to resist, result directly from the distortions induced by the motion of the ground on which it rests. The response (i.e., the magnitude and distribution of the resulting forces and displacements) of a structure to such a base motion is influenced by the properties of both the structure and the surrounding foundation as well as the character of the exciting motion.

A simplified picture of the behavior of a building during an earthquake may be obtained by considering Fig. 12-7. As the ground on which the building rests is displaced, the base of the building moves with it. However, the inertia of the building mass resists this motion and causes the building

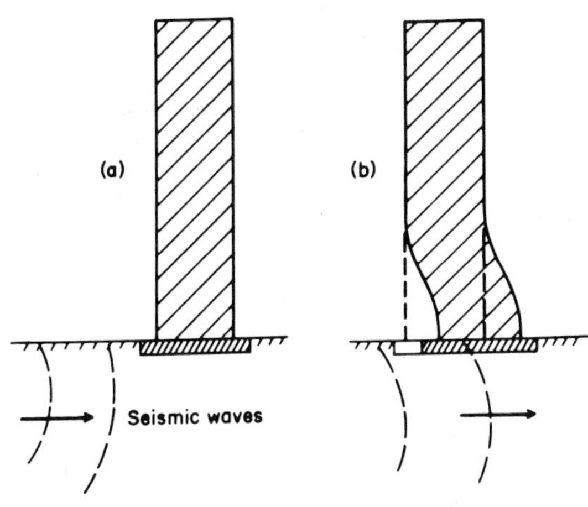

Fig. 12-7

to suffer a distortion (greatly exaggerated in the figure). This distortion wave travels along the height of the structure in much the same manner as a stress wave in a bar with a free end. The continued shaking of the base causes the building to undergo a complex series of oscillations.

It is important to draw a distinction between forces due to wind and those produced by earthquakes. Occasionally, even engineers tend to think of both these loadings as belonging to the same category, just because codes specify loadings for both cases in terms of equivalent static forces. Although both wind and earthquake loads are dynamic in character, a basic difference exists in the manner in which these loads are induced in a structure. Whereas wind loads are external loads applied, and hence proportional, to the exposed surface of a structure, earthquake loads are essentially inertial forces. The latter result from the distortion produced by both the earthquake motion and the inertial resistance of the structure. Their magnitude is then a function of the mass of the structure rather than its exposed surface. Also, in contrast to structural response to essentially static gravity loading or even to wind loads, which can often be validly treated as static loads, the dynamic character of the response to earthquake excitation can seldom be ignored. Thus, whereas in designing for static loads one would feel greater assurance about the safety of a structure made up of members of heavy section, in the case of earthquake loading, the stiffer and heavier structure does not necessarily represent the safer design.

Uncertainties in the determination of the proper earthquake design loads to be used for a proposed structure arise from a number of factors, among the more important of which are:

a. The difficulty of predicting the character of the critical earthquake motions (i.e., amplitude, frequency characteristics, and duration) to which a proposed structure may be subjected during its lifetime; and
b. the difficulty of ascertaining the values of the structural parameters affecting the dynamic response (e.g., stiffness and damping), as well as the dynamic properties of the soil or supporting medium.

Insofar as the earthquake motion is concerned, the quantity most commonly used in analysis is the timewise variation of the ground acceleration in the immediate vicinity of a structure. For earthquakes of tectonic origin, the character of the variation of the surface acceleration is primarily dependent on the magnitude and focal depth of the earthquake, the epicentral distance of the site considered and the properties of the intervening as well as the surrounding ground. The most significant parameters describing the ground acceleration which influence structural response are the maximum acceleration, frequency characteristics and duration of the large-amplitude pulses. Except that the earthquake motions will generally be random in character, there is no way to ascertain the exact nature of any future earthquake at a particular site. However, continuing studies of the seismic history and geology of a region should yield valuable estimates of the expected range of the significant ground acceleration parameters. The Davenport-Milne isoacceleration maps[12-11] (see Fig. 12-6) corresponding to different return periods represent a good start toward the quantification of seismic zoning studies. With information from such maps as prepared by Davenport-Milne as a basis and some knowledge of the geologic conditions at a proposed site, a set of accelerograms[12-5, 12-6] can be artificially generated as representative of probable future earthquakes to serve as input excitation in dynamic analyses. The range of variation of each of the significant parameters to be used in such an input set of accelerograms will obviously depend on the degree of uncertainty attached to each parameter. The range will doubtless also be influenced by cost considerations.

At any particular point, the ground acceleration may be described by horizontal components along two perpendicular directions and a vertical component. In addition, rocking and twisting (rotational) components may be involved. As mentioned earlier, the rotational components of earthquake ground motion are usually negligible. Rocking and torsional effects in structures due to the horizontal components of ground motion do occur, however, as a result of ground compliance and the noncoincidence of the centers of mass and of rigidity.

Because buildings and most structures are more sensitive to horizontal or lateral distortions, it has been the practice in most instances to consider only the structural response to the horizontal components of ground motion. The effects of the vertical component of ground motion* has generally not been considered significant enough to merit special attention. Here, the principal motivation has been simplification and the reduction of the analytical work involved, which can be considerable, even when considering only the horizontal components. For most structures, experience seems to have justified this viewpoint. In most instances, a further simplification of the actual three-dimensional response of structures is made by assuming the design horizontal acceleration components to act nonconcurrently in the direction of each principal plan-axis of a building. It is tacitly assumed that a building designed by this approach will have adequate resistance against the resultant acceleration acting in any direction.

Insofar as the structure is concerned, the principal properties which affect its dynamic response are its mass, stiffness, and damping characteristics. Under certain conditions, the effect of the foundation or supporting medium may have to be considered.

When yielding is expected to occur under moderate to strong earthquake excitation, the yield level, or alternatively, the ratio of yield-to-design moments becomes a significant factor. Under this condition, the capacity of a structure to deform well beyond the initial yield displacement without significant loss of strength, i.e., ductility, assumes utmost importance.

The determination of the mass and initial stiffness

*In several instances, the vertical component of ground acceleration has been of the same order of magnitude as the horizontal components.

properties of a particular structure generally does not present any difficulty. However, the effective stiffness and damping of a structure, and hence its characteristic period of vibration, can change during vibration as a result of cracks occurring in members, even before the onset of yielding. Yielding further increases the period of vibration of a structure.

The evaluation of the nature, magnitude, and distribution of a structure's damping has, up to the present, been a rather vague process and is a problem that is only beginning to receive the systematic study that it deserves. Convenience in mathematical modelling and solution of the dynamic response problem has required the assumption of a viscous-type damping in place of the actual mechanism. Vibration tests of actual structures have yielded estimates of the equivalent viscous damping ranging from 5% to 20% of critical.*

Most of the dynamic analyses of multistory structures which have been carried out have assumed the free-field ground motion to be applied, unaltered, to the building foundation, which is assumed to be rigid. While this approximation may be valid for most structures founded on soil which is relatively stiff with respect to the structure, cases arise where soil-structure interaction effects produce significantly different results from those based on a rigid foundation assumption. The effect of soil-structure interaction on the earthquake response can vary depending upon the properties of the soil and the structure as well as the character of the input motion. The effect is largest on the fundamental mode and period of vibration and diminishes rapidly in the higher modes.

12.3.2 Dynamic Analysis of Single-degree-of-freedom (SDF) Systems

An often convenient way of studying the dynamic response of structures is by considering the total response in terms of component modal responses. The response in the elastic range, and during sufficiently short intervals in the inelastic range, may be thought of as the superposition of the responses of the modes of vibration characterizing the system. In general, a system will have as many modes of vibration as it has significant degrees of freedom. In each such mode of vibration, the masses all vibrate in a phase, i.e., they maintain the same positions relative to each other. Associated with each mode is a characteristic frequency or period of vibration. Each mode of vibration may thus be considered as a single-degree-of-freedom system having a particular period of frequency of vibration. Figure 12-8 shows typical shapes for the first three modes of a multistory building where the significant degrees of freedom are related to the lateral displacement of the floors. The first or fundamental mode corresponds to the longest or fundamental period (or lowest frequency) of vibration.

The response of most multistory buildings is made up predominantly of the first few modes, with the higher modes contributing only a small portion of the total, except at the top of relatively flexible buildings. Studies of the elastic response of multistory frame buildings indicate that for most buildings the fundamental mode contributes about 80%, while the second and third modes about 15%, of the total response. Because of the dominant influence of a single (i.e., the fundamental) mode of vibration on the response of multistory structures and the comparative ease of analyzing the response of single-degree-of-freedom (SDF) systems, much can be learned of basic response characteristics of actual structures from a consideration of the dynamic response of simple systems. This is the main justification for studying the response of SDF systems here. Of course, as mentioned earlier, each of the higher modes of vibration can be considered as an SDF system.

*The value of damping which would just cause an initial displacement to decay to zero without any oscillation.

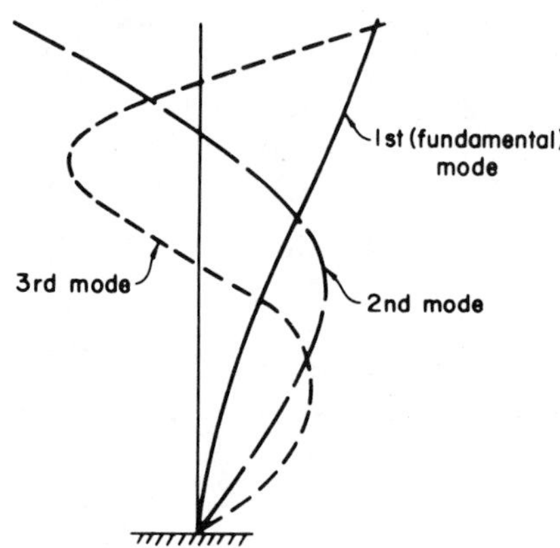

Fig. 12-8 Typical shapes of the first three natural modes of vibration of a multistory structure.

12.3.3 Linear SDF Systems

The force-deformation curve for a linearly elastic system is shown in Fig. 12-9(a). Also shown, in Fig. 12-9(b), is the corresponding curve for an inelastic bilinear system. The solid-line curve is often used as a convenient approximation of the actual force-deformation curve shown as a dashed line in the same figure. The unloading branch of the idealized inelastic curve is usually assumed to have the same slope as the initial loading branch.

-1. Formulation of equation of motion—A schematic representation of a SDF system with a fixed base is shown in Fig. 12-10a. The system consists of a rigid mass M, a weightless resisting element with spring constant k, and a dashpot with damping coefficient c. The dashpot, which represents the energy-absorbing mechanism in the system, is assumed here to be of the viscous-type (in which the damping force is proportional to the velocity of the mass). The mass is assumed to be constrained to move in only one direction, i.e., along the direction of the time-varying applied force $P(t)$. The system may be thought of as representing the single-story frame shown in Fig. 12-10b with the mass of the frame assumed concentrated in the infinitely stiff girder and where the only motion considered significant is the lateral displacement of the mass.

Although the fixed-base structure shown in Fig. 12-10 does not correspond to the earthquake situation where the base of the structure undergoes motion, it is considered here in order to point out the similarity between the externally applied force in a fixed-base system and the equivalent effective force corresponding to the moving base system.

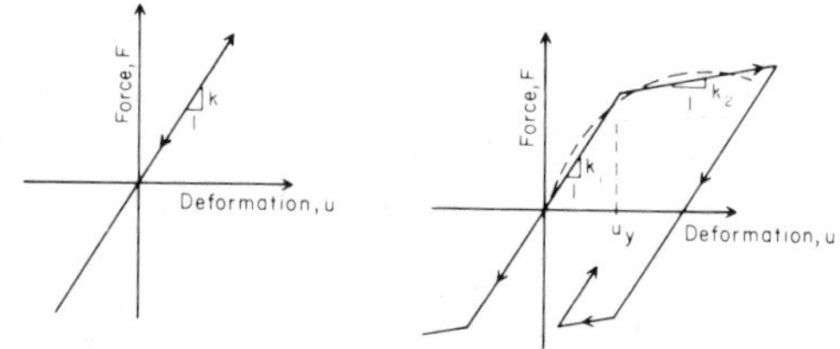

Fig. 12-9 (a) Linear and (b) bilinear restoring-element force-deformation relationships.

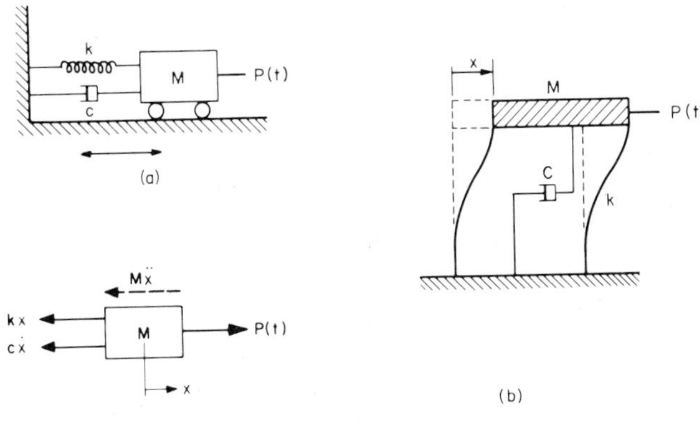

PROPERTIES OF UNDAMPED SYSTEM

Natural circular frequency, ω (radians / sec.) $= \sqrt{\frac{k}{M}}$

Frequency, f (cycles / sec.) $= \frac{\omega}{2\pi}$

Natural period of vibration, T (seconds) $= \frac{1}{f} = \frac{2\pi}{\omega} = 2\pi\sqrt{\frac{M}{k}}$

Fig. 12-10 Fixed-base, single-degree-of-freedom (SDF) system subjected to a time-varying force.

The differential equation of motion obtained by considering the equilibrium of the free body shown in Fig. 12-10c is given by

$$M\ddot{x} + c\dot{x} + kx = P(t) \qquad (12\text{-}1)$$

where the dot ($\dot{}$) indicates differentiation with respect to time.

Consider next the moving-base SDF system shown in Fig. 12-11 with the displacement of the base described by x_o. Note that there is no applied force acting on the mass, just as in a structure subjected to earthquake motion. Equilibrium of the free body in Fig. 12-11 now gives

$$M\ddot{x} + c(\dot{x} - \dot{x}_o) + k(x - x_o) = 0 \qquad (12\text{-}2)$$

Expressing eq. (12-2) in terms of the relative displacement of the mass with respect to the base, $u = x - x_o$, yields

$$M\ddot{u} + c\dot{u} + ku = -M\ddot{x}_o(t) \qquad (12\text{-}3)$$

A comparison of eqq. (12-1) and (12-4) indicates that the relative displacement, u, of the mass in a system subjected to a time-varying base acceleration, $\ddot{x}_o(t)$, is equal to the absolute displacement, x, of the mass in a fixed-base system in which the mass is subjected to a time-varying force

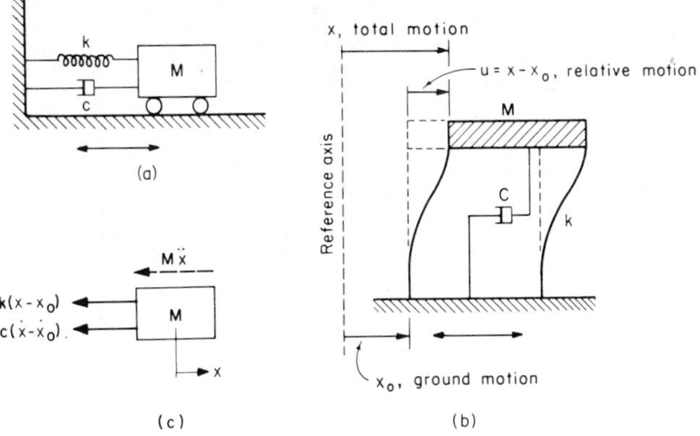

Fig. 12-11 Movable-base SDF system subjected to a time-varying base acceleration.

equal to $-M\ddot{x}_o(t)$. The quantity $-M\ddot{x}_o(t)$ may be called the effective load resulting from the base motion.

-2. Closed form solution of equation of motion—For the purpose of solving eq. (12-3) the following notation will be introduced for convenience:

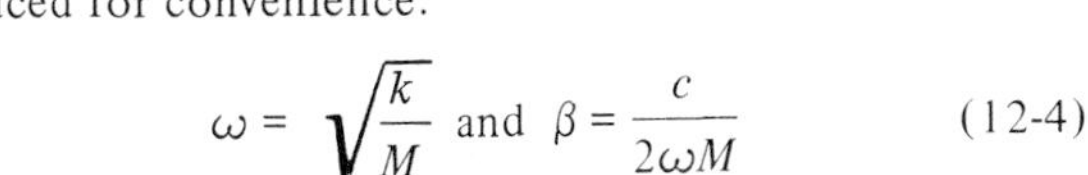

$$\omega = \sqrt{\frac{k}{M}} \quad \text{and} \quad \beta = \frac{c}{2\omega M} \qquad (12\text{-}4)$$

where ω represents the circular frequency of undamped vibration (in radians/sec) and β the damping ratio relative to the critical value of the damping coefficient. ω is related to the natural frequency, f, in cycles per second, and period of vibration, T, by the relations

$$f = \frac{\omega}{2\pi} \quad \text{and} \quad T = \frac{1}{f} = \frac{2\pi}{\omega} \qquad (12\text{-}5)$$

With the above notation, eq. (12-3) may be rewritten as

$$\ddot{u} + 2\beta\omega\dot{u} + \omega^2 u = -\ddot{x}_o(t) \qquad (12\text{-}6)$$

a. Homogeneous Case—The solution for the homogeneous case of eq. (12-6) when the right hand side of the equation representing the applied force is equal to zero, i.e., the free vibration case, is given by

$$u(t)_h = e^{-\beta\omega t}(A\cos\omega_d t + B\sin\omega_d t) \qquad (12\text{-}7)$$

where

$$\omega_d = \omega\sqrt{1 - \beta^2}$$

may be referred to as the damped circular frequency of vibration. The solution given by eq. (12-7) corresponds to the case when $\beta < 1$. When $\beta = 1$, the system is said to be critically damped (so that $c_{cr} = 2\omega M$); when $\beta > 1$, it is overdamped. In either case, the motion is nonoscillatory. Since for most civil engineering structures β ranges from about 0.02 to 0.20, the preceding two cases need not be considered here.

b. *Particular solution for a linearly varying base acceleration*—The input motions used in earthquake response analyses usually consist of digitized data based on actual strong-motion earthquake accelerograph records or artificially generated accelerograms. The data consist of time-acceleration coordinates defining the accelerogram, with the variation of the acceleration assumed to be linear between adjacent points of definition. A general case of an earthquake input acceleration segment may thus be described by the linearly varying acceleration pulse of duration. Δt shown in Fig. 12-12 and described by the relation

$$\ddot{x}_o(t) = \ddot{x}_{o1} + \frac{t}{\Delta t}(\ddot{x}_{o2} - \ddot{x}_{o1}) \qquad (12\text{-}8)$$

where t is reckoned from the beginning of the pulse. Solution of eq. (12-6) with the expression of $\ddot{x}_o(t)$ as defined by eq. (12-8) substituted on the right hand side, yields

$$u(t)_p = -\frac{1}{\omega^2}\left[\ddot{x}_{o1} + \left(\frac{t}{\Delta t} - \frac{2\beta}{\omega\Delta t}\right)\Delta\ddot{x}_o\right] \qquad (12\text{-}9)$$

where

$$\Delta\ddot{x}_o = \ddot{x}_{o2} - \ddot{x}_{o1}$$

The complete solution of the differential equation of motion for a SDF system with damping ratio β, when subjected to a piecewise linear base acceleration, is given by the sum of the right-hand sides of eqq. (12-7) and (12-9), i.e.,

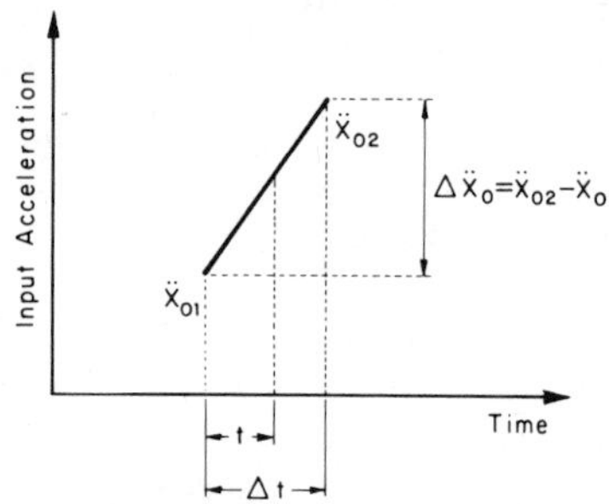

Fig. 12-12 A linearly-varying acceleration pulse representing a typical segment of an earthquake acceleration input.

$$u(t) = u(t)_h + u(t)_p$$
$$= e^{-\beta\omega t}(A\cos\omega_d t + B\sin\omega_d t$$
$$-\frac{1}{\omega^2}\left[\ddot{x}_{01} + \left(t - \frac{2\beta}{\omega}\right)\frac{\Delta\ddot{x}_o}{\Delta t}\right] \quad (12\text{-}10)$$

Note that if the relative displacement at the end of the interval is desired, as is usually the case, $t = \Delta t$.

Using the initial conditions $u = u_1$ and $\dot{u} = \dot{u}_1$ at $t = 0$, the constants of integration, A and B, are determined, as follows:

$$A = u_1 + \frac{1}{\omega^2}\left(\ddot{x}_{01} - \frac{2\beta}{\omega\Delta t}\Delta\ddot{x}_o\right) \quad \text{(a)}$$

and (12-11)

$$B = \frac{1}{\omega_d}\left(\dot{u}_1 + \beta\omega A + \frac{\Delta\ddot{x}_o}{\omega^2\Delta t}\right) \quad \text{(b)}$$

where the A appearing in eq. (12-11b) is given by eq. (12-11a).

Equation (12-10)–with A and B substituted from eq. (12-11)–gives the relative displacement of the mass at any time during the interval Δt. Successive differentiations of eq. (12-10) yield the corresponding expressions for the relative velocity, $\dot{u}(t)$, and relative acceleration, $\ddot{u}(t)$. The dynamic response analysis of a linearly elastic SDF system subjected to earthquake motion may be undertaken by repeated application of eq. (12-10) and the corresponding expressions for $\dot{u}(t)$ and $\ddot{u}(t)$ for each segment of the base acceleration record, the response values obtained for the end of one time interval being used as the initial values of the succeeding interval. The calculations are most conveniently carried out with the aid of a digital computer. For completeness, the expressions for the relative velocity and acceleration are given below.

$$\dot{u}(t) = -\ [(A\omega_d + B\beta\omega)\sin\omega_d t$$
$$+ (A\beta\omega - B\omega_d)\cos\omega_d t]\ e^{-\beta\omega t} - \frac{\Delta\ddot{x}_o}{\omega^2\Delta t} \quad (12\text{-}12a)$$

$$\ddot{u}(t) = (\beta^2\omega_d^2)(A\cos\omega_d t + B\sin\omega_d t)\,e^{-\beta\omega t}$$
$$+ 2\beta\omega\omega_d(A\sin\omega_d t - B\cos\omega_d t)\,e^{-\beta\omega t} \quad (12\text{-}12b)$$

where A and B are given by eqq. (12-11).

-3. Numerical integration of D.E. of notion assuming a linearly varying response acceleration–Equation (12-10) above represents an exact solution of the equation of motion for a linearly elastic damped SDF system; exact in the sense of involving no assumption concerning the variation of the response parameters, i.e., the displacement, velocity or acceleration of the mass.

A commonly used method of arriving at a solution to the equation of motion eq. (12-6) is a numerical procedure based on the assumption that over a sufficiently small interval of time, the response acceleration may be assumed to vary linearly without significant error. A limited number of studies have indicated that the method gives fairly good results when the interval is taken at most equal to about one-tenth of the period of vibration of the system. The accuracy of the results improves with decreasing values of the interval. The procedure has the advantage of involving a smaller number of operations in evaluating the response parameters compared to the method of the preceding section and may be readily extended to the case of inelastic multi-degree-of-freedom systems.

As in the method of the proceding section, the determination of the response history consists in a step-by-step advance in the time domain, with the response values at the end of one interval being used as starting values for the next interval.

To apply the method, eq. (12-6) is expressed in incremental form, as follows:

$$\Delta\ddot{u} + 2\beta\omega(\Delta\dot{u}) + \omega^2(\Delta u) = -\Delta\ddot{x}_o \quad (12\text{-}13)$$

Equation (12-13) is essentially a statement of equilibrium among incremental forces, i.e., if the system remains in dynamic equilibrium during the interval Δt, the changes in the damping and restoring (spring) forces equal the change in the inertia force during the interval. In eq. (12-1),

$$\Delta\ddot{u} = \ddot{u}_2 - \ddot{u}_1$$
$$\Delta\dot{u} = \dot{u}_2 - \dot{u}_1$$
$$\Delta u = u_2 - u_1$$
$$\Delta\ddot{x}_o = \ddot{x}_{o2} - \ddot{x}_{o1}$$

where the subscripts 1 and 2 correspond to the beginning and end, respectively, of the interval $\Delta t = t_2 - t_1$.

The use of incremental values of the parameters instead of the actual parameters is a more convenient approach in that it makes the stiffness independent of the previous loading history.

To determine the response parameters at the end of an interval Δt satisfying eq. (12-13), the following relations, based on the assumption of a linearly varying response acceleration during the interval Δt, are used:

$$\dot{u}_2 = \dot{u}_1 + \frac{\Delta t}{2}(\ddot{u}_1 + \ddot{u}_2) \quad \text{(a)}$$

(12-14)

$$u_2 = u_1 + \dot{u}_1\Delta t + \frac{(\Delta t)^2}{6}(2\ddot{u}_1 + \ddot{u}_2) \quad \text{(b)}$$

Equation (12-14a) follows readily from the conditions of the above-mentioned assumption. Equation (12-14b) results from the successive integration of the relation expressing the basic assumption, i.e.,

$$\ddot{u}(t) = \ddot{u}_1 + \frac{\ddot{u}_2 - \ddot{u}_1}{\Delta t}(t - t_1).$$

Substituting eq. (12-14) into eq. (12-13), one obtains, upon solving for $\ddot{u}_2$.

$$\ddot{u}_2 = \frac{1}{C}[D\ddot{u}_1 - \omega^2(\Delta t)\dot{u}_1 - \Delta\ddot{x}_o]$$

where

$$C = 1 + \beta\omega\Delta t + \frac{1}{6}\omega^2(\Delta t)^2 \quad (12\text{-}15)$$

and

$$D = 1 - \beta\omega\Delta t - \frac{1}{3}\omega^2(\Delta t)^2$$

Equation (12-15) gives the relative acceleration of the mass at the end of the time interval Δt when the relative acceleration and relative velocity at the beginning of the interval as well as the increment in the base (input) acceleration are known. Once the acceleration at the end of the interval, $\ddot{u}_2$, is known, the corresponding relative velocity and displacement at the end of the interval may be obtained using eqq. (12-14).

The initial relative acceleration, $\ddot{u}_1$, may be calculated from eq. (12-6)—written for time $t = t_1$—when the initial displacement, u_1, and velocity, $\dot{u}_1$, are known, i.e.,

$$\ddot{u}_1 = -(\ddot{x}_{o1} + 2\beta\omega\dot{u}_1 + \omega^2 u_1)$$

-4. *Improving the stability of the linear acceleration method of numerical integration*—The accuracy of the results obtained by numerical integration, such as the linear acceleration method discussed in the preceding sections, is heavily dependent on the length of the time interval used relative to the shortest period of the system. The stability of different methods of numerical integration, i.e., the tendency of the results obtained by a particular technique to converge to the correct solution, can vary widely but is generally improved by using a reasonably short time step. Investigators have used time increments of one-tenth to one-twentieth of the shortest period of vibration of the structure considered. It is readily appreciated that where a very short time interval is used, the computing time required to carry out an analysis can be substantial, particularly for non-linear systems.

A study by Wilson[12-12] has shown that the stability of the linear acceleration method can be considerably improved by a slight modification of the method previously discussed. Since the instability in the linear acceleration method is initiated by oscillations of the calculated displacements about the true solution, the tendency for the oscillation to develop is reduced by taking the value of the displacement corresponding to the midpoint of the time interval as a better approximation to the correct solution.

Thus, in order to obtain an improved value of the response at the end of a time interval $\Delta t = t_2 - t_1$, one considers an interval $2\Delta t$, twice as long (see figure below).

t_1 t_2 t_x

$\vdash \Delta t \dashv \vdash \Delta t \dashv$

Using the procedure discussed previously, the response values at time t_x are determined on the basis of the initial values at time t_1. The improved value for time t_2 is then obtained as the average of the values corresponding to t_1 and t_x, e.g.,

$$u_2 = \frac{1}{2}(u_1 + u_x)$$

The same procedure is repeated for the succeeding time step, using t_2 now as the initial point.

The use of such a simple modification has been reported to eliminate all stability problems associated with the linear acceleration method. The technique, however, has a tendency to damp out the higher mode responses in the analysis of multi-degree-of-freedom (MDF) systems. This partial truncation of the higher mode responses may be considered tolerable in most earthquake analyses since the higher frequencies in the input acceleration, which tend to excite these modes the most, are generally not accurately recorded.

-5. *Response spectra, pseudo-velocity, and pseudo-acceleration*—A response spectrum is a plot showing the variation of the maximum value of a particular response parameter (e.g., displacement, velocity, acceleration, stress, etc.) with the frequency (or period) of a linear SDF system subjected to a specific forcing function. In earthquake response studies, the forcing function is generally identified with the earthquake accelerogram record. Although the term response spectrum was originally used, and is still commonly used only with reference to linear SDF systems, its use has been extended in later work to linear as well as nonlinear MDF systems. A point on a response spectrum curve is obtained by analyzing a particular SDF system—as defined by its frequency and damping ratio—using a specific base acceleration record. The maximum value of the response parameter of interest, which generally occurs within the duration of the exciting motion, is then noted. For a fixed value of the damping ratio, β, the spectral value of the parameter and the associated frequency (or period) of the system, together define a point on the response spectrum curve. Other points are obtained by using different values of the frequency. A response spectrum thus shows the maximum response of different SDF systems, as defined by their frequencies and damping ratios. In the following discussion, the term response spectrum will be assumed to refer to linear SDF systems, unless otherwise noted.

An examination of eqq. (12-11) and (12-12) shows that the response of a SDF system to base excitation is dependent on the properties of the system as given by its natural circular frequency, ω, and its damping ratio, β, as well as the character of the exciting motion $\ddot{x}(t)$. It is thus seen that for the same value of β and the same range of frequencies, the response spectra for different earthquake motions may be expected to differ from each other. As such, they provide a means of characterizing earthquake motions.

Furthermore, because the dynamic behavior of more complex systems, such as multistory buildings, is strongly influenced by the fundamental mode response, spectral plots provide the designer with a convenient means of assessing the response of a structure of known period to a specified ground motion.

Figure 12-13 shows acceleration spectra for different values of the damping ratio, β, corresponding to the 1940 El Centro, California, earthquake (N-S component).

The sharp peaks in certain ranges of frequency shown in Fig. 12-13, especially for the case of $\beta = 0$, reflects the resonant behavior of the system in these frequency ranges. In this respect, the lightly damped SDF system acts as a narrow band filter which amplifies the input motion in ranges where the dominant frequency approaches that of the system. A response spectrum plot may thus be used to infer the character of the input motion, with particular regard to frequency content. It is significant to note, however, that even a moderate amount of damping tends not only to reduce the response but also smooth out the jagged character of the spectral plot. Since most structures possess some amount of damping, the sharp peaks assume lesser importance in practice. Also, where response spectra are used for design purposes, it is customary to allow for earthquake motions of varying frequency characteristics by using

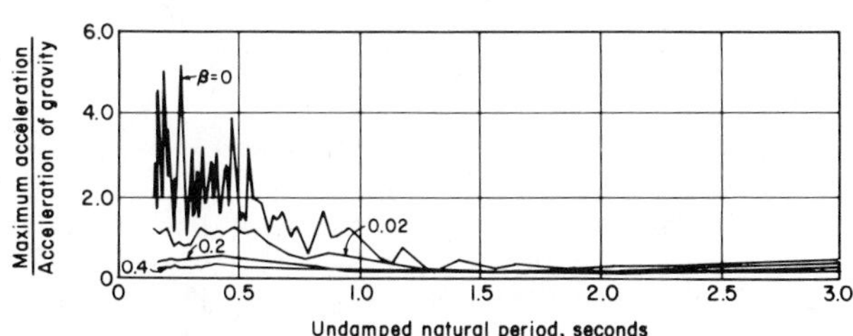

Fig. 12-13 Acceleration response spectra for 1940 El Centro, Calif., earthquake (N-S component).[12-13]

smoothed or averaged spectra based on a number of earthquake records—all reduced or normalized to a reference intensity level.

The response parameters most often considered in earthquake response studies and in terms of which response spectra are plotted are the maximum relative displacement, maximum relative velocity and the absolute acceleration of the mass. The maximum relative displacement assumes importance in design since the strains in a structure are proportional to the relative displacement. As mentioned earlier and as will be shown subsequently, the maximum relative velocity provides a measure of the elastic energy stored in the system while the absolute acceleration can be shown to be directly related to the seismic or lateral force coefficient appearing in most design codes. Also, where experimental measurements are to be taken as a check on calculations, the absolute acceleration serves as a good basis for comparison, since it is the structural response most easily measured.

Spectral pseudo-velocity and pseudo-acceleration—In the study of the earthquake response of simple systems, it has been found most useful and convenient to present response spectra in terms of velocity and acceleration quantities that are not quite equal to the actual quantities they are meant to represent. These have, however, been shown to provide fairly good measures of the quantities they represent and bear such a simple relationship to each other and to the spectral relative displacement as to allow convenient representation.

Thus, if the maximum (spectral) value of the relative displacement is denoted by S_d, the spectral pseudo-velocity is defined as

$$S_v = \omega S_d \tag{12-16}$$

and the spectral pseudo-acceleration as

$$S_a = \omega^2 S_d \tag{12-17}$$

For zero damping, the difference between the maximum relative velocity and the spectral pseudo-velocity, S_v, is generally negligible. For exciting motions of very short duration containing relatively short period peaks, however, the difference between these two quantities can be significant over some frequency ranges. As damping increases, a deviation of the order of 20% may be expected. Generally, the spectral velocity, S_v, is nearly equal to the maximum relative velocity in the higher frequency range, with the difference tending to be greater for the low frequency (long period) systems.

The spectral velocity can be shown to be directly related to the maximum energy per unit mass in the system. Thus, denoting the strain energy in the spring by U, we have

$$\frac{U_{max}}{M} = \frac{\frac{1}{2}k(x - x_o)^2_{max}}{M} = \frac{1}{2}\omega^2 S_d^2 = \frac{1}{2}S_v^2$$

The spectral acceleration is actually the maximum absolute acceleration of the mass in an undamped SDF system. Thus, for such a system, using Newton's Second Law of Motion

$$F = k(x - x_o)_{max} = kS_d = M\ddot{x}$$

$$\ddot{x} = \frac{k}{M}S_d = \omega^2 S_d = S_a$$

The spectral pseudo-acceleration, however, does not differ greatly from the maximum absolute acceleration for systems with moderate damping, for the entire frequency range. The relation between the spectral acceleration, S_a, and the maximum force acting on the spring, which in codes is usually expressed as a seismic or lateral force coefficient, C, multiplied by the weight of the structure, W, i.e., $F_{max} = CW$, may be shown as follows:

$$F_{max} = k(x - x_o)_{max} = kS_d = CW$$

$$C = \frac{k}{W/g}\cdot\frac{S_d}{g} = \frac{\omega^2 S_d}{g} = \frac{S_a}{g}$$

Figure 12-14 shows averaged velocity spectra for ground

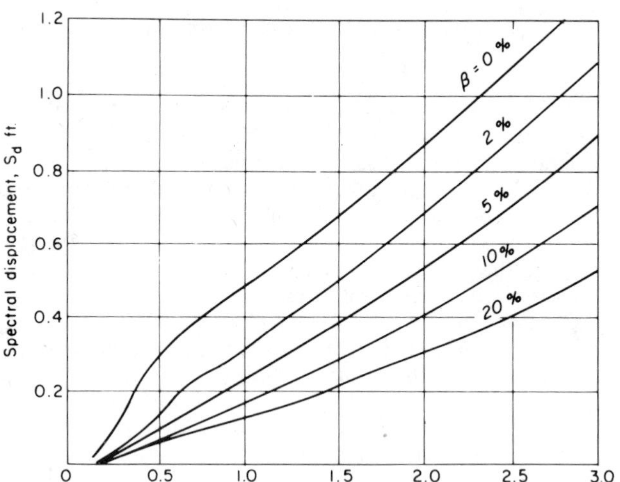

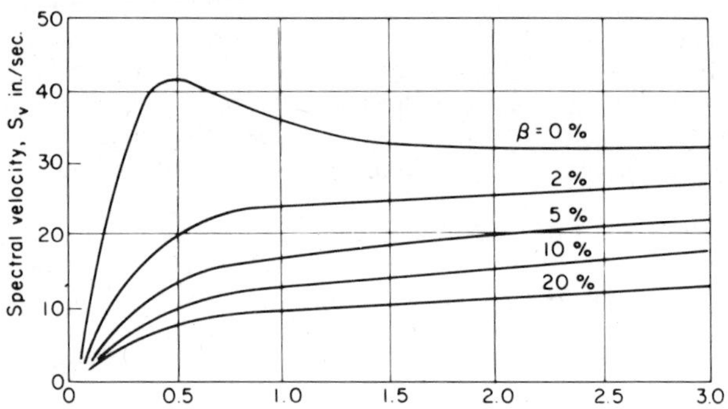

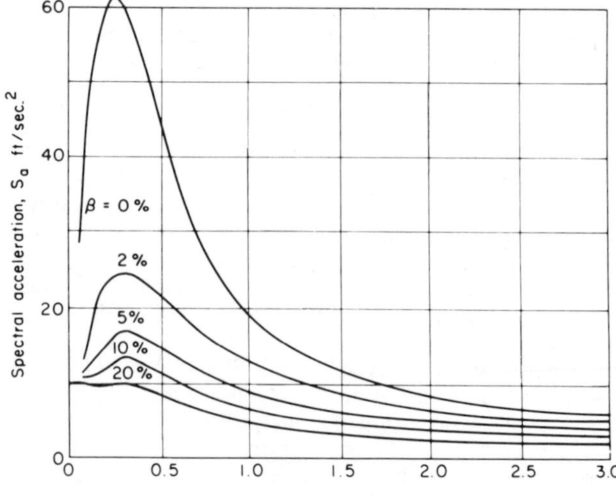

Fig. 12-14 Averaged response spectra.[12-38]

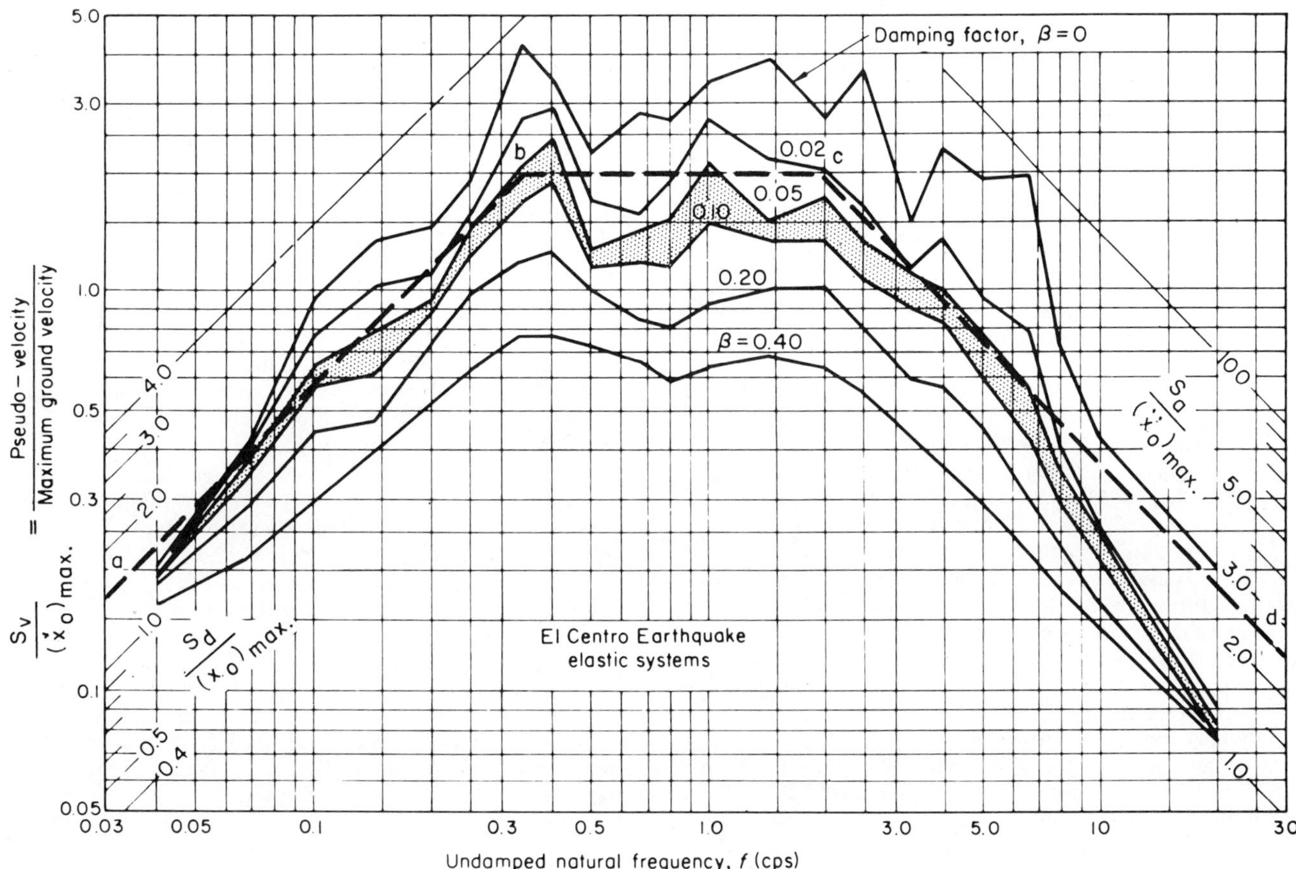

Fig. 12-15 Deformation spectra for linearly elastic SDF systems subjected to 1940 El Centro earthquake (N-S component).[12-14]

motions, all adjusted to have the same spectral intensity* as the 1940 El Centro, California, earthquake. It will be noted that, with the relationships given by eqq. (12-16) and (12-17) all three response parameters can be obtained when one of them is known. Thus, both spectral displacement and spectral acceleration can be obtained from the spectral pseudo-velocity plot; the first by dividing and the second by multiplying the velocity spectrum ordinate by the corresponding circular frequency ω.

Because of the linear relationship between the logarithms of both the spectral displacement and acceleration with the logarithms of the spectral velocity and circular frequency ω, as indicated by eqq. (12-16) and (12-17), it is possible to have a single plot showing the variation of all three response quantities with the frequency (or period). Figure 12-15 shows such a plot, with log scales in all axes, the spectral displacement and acceleration being read along diagonal scales.

In Fig. 12-15, the scales have been nondimensionalized to indicate the ratios of the spectral values of the response parameters to the corresponding maximum ground motion parameter. The plot shown is typical of earthquake response spectra and provides analytical support to some aspects of the dynamic response of simple systems which would appear intuitively obvious. Thus, for low-frequency (long-period) systems—corresponding to a heavy mass supported by a light spring—the mass remains practically motionless when the base is subjected to an earthquake-type motion. For this case, the relative displacement of the mass or the deformation in the spring is essentially equal to the base displacement. This is indicated by the spectral lines which approach the diagonal line $S_d/(x_o)_{\max} = 1.0$ on the left side of the plot. On the right side of the plot, i.e., for high-frequency systems, the spectral lines are shown converging towards the diagonal line $S_a/(\ddot{x}_o)_{\max} = 1.0$, indicating that for systems exemplified by a light mass supported by a very stiff spring, the mass simply moves with the base, with the spring hardly deforming. Hence the mass acquires essentially the same absolute acceleration as the base.

In the intermediate frequency range, some modification of the response parameters characterizing the motion of the mass relative to that of the base, occurs. Generally, the amplification factor for displacement is less than that for velocity, which in turn is less than that for acceleration. Figure 12-15 indicates that for linear systems with damping ratios of 5 to 10% subjected to the 1940 El Centro motion, the maximum amplification factors for displacement, velocity and acceleration are about 1.5, 2.0 and 2.5, respectively.

A typical response spectrum curve in Fig. 12-15 may be represented by three line segments marking the approximate envelope of response for systems having a particular damping ratio and subjected to a specific ground motion. The three line segments, shown as the dashed broken line a-b-c-d in the figure for systems with 5–10% damping and subjected to the 1940 El Centro (N–S) motion, effectively divides the frequency axis into three regions, i.e., a low-frequency region from a to b, a medium-frequency region from b to c and a high-frequency region from c to d. The width of each frequency region will generally vary with the nature of the ground motion.

*First defined by Housner[12-38] as the area under the velocity spectrum curve between ordinates representing periods of 0.1 to 2.5 seconds. The spectrum intensity provides a measure of the maximum stress induced in linearly elastic structures by the particular ground motion.

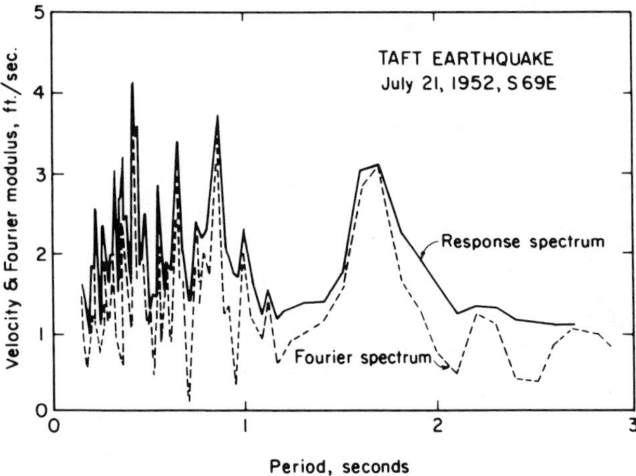

Fig. 12-16 Comparison of Fourier spectrum and undamped velocity spectrum for the 1952 Taft earthquake.[12-15]

-6. *Fourier spectra*—A means of representing earthquake ground motions which is related to response spectra is the Fourier spectrum or Fourier transform. The Fourier spectrum is essentially a frequency decomposition of the complex ground motion and shows directly the significant frequency characteristics of the motion considered.

A typical Fourier spectrum plot gives the variation of the Fourier modulus (a measure of the wave amplitude) with the frequency of the component waves. Figure 12-16 shows the Fourier amplitude spectrum of the ground acceleration for the Taft, California, earthquake of July 21, 1952 (S69E component).

It will be noted that the Fourier spectrum for acceleration, which is obtained by integrating the acceleration record with respect to time, has units of velocity. Hudson[12-15] has shown that when the maximum response of a system occurs at the end of an earthquake, the undamped relative velocity response spectrum has a form identical to the Fourier amplitude spectrum of the ground acceleration. When the maximum response of a SDF system does not occur at the end of the earthquake record, the plots for the above two quantities are only roughly similar. Figure 12-16 also shows the relative velocity response spectrum for the 1952 Taft earthquake for comparison. Note that the abscissa in a Fourier spectrum represents the frequency of the various wave components of the acceleration trace while the abscissa in the response spectrum plot represents the frequency of the SDF system.

The Fourier spectrum provides a useful means of visualizing the predominant frequencies of a particular earthquake record.

12.3.4 Bilinear Hysteretic SDF Systems

When the force-deformation characteristics of the restoring element (spring) in a SDF system is other than linear (i.e., with k constant throughout the response), the governing differential equation of motion of the system, corresponding to eq. (12-1), becomes nonlinear. However, during sufficiently small intervals of time the system may be assumed to be linear, with the value of the spring stiffness depending upon the magnitude and direction of the deformation during the interval considered. A step-by-step solution using such an approach allows the practical evaluation of the response of nonlinear systems for arbitrary force-deformation system characteristics and base motion.

-1. *Numerical solution for the nonlinear SDF system assuming a linearly varying response acceleration*—If, in addition to assuming a structure with nonlinear material properties to behave linearly during a small time interval so that the analysis can be carried out by considering a succession of generally different linear structures, the assumption is further made that the response acceleration varies linearly during each interval, eqq. (12-14) and (12-15) developed for the linearly elastic case may be applied directly provided the appropriate value of $\omega_i = \sqrt{k_i/M}$ is used in eq. (12-15). Note that since the incremental form of the differential equation, eq. (12-13), applies only to the changes in the forces during each interval considered, the term in the equilibrium equation corresponding to the spring force need involve the effective spring constant k_i associated with the current state of deformation of the spring element only rather than the total spring force.

-2. *Some results of analyses of bilinear, inelastic (hysteretic) SDF systems*—A system having a force-deformation curve similar to that shown in Fig. 12-9(b) differs from a linearly elastic system in two respects: (1) the hysteretic action which occurs after yielding allows it to dissipate energy (as measured by the area under the F-u curve). This effect tends to decrease the maximum deformation of the system; and (2) it represents a softer system, with a lower effective frequency of vibration than a linear system having the same initial stiffness or small-amplitude frequency. The reduction in effective frequency which occurs after yielding may either decrease or increase the maximum deformation depending upon the frequency and the yield level of the system, i.e., the force required to produce first yield.

The determination of the response of nonlinear systems involves a significantly greater amount of time than that required for linear systems. Because of this and the availability of extensive data on linear systems, efforts have been directed toward formulating approximate design rules relating the behavior of nonlinear systems to that of linear systems.

A special case of the bilinear hysteretic system is one in which the slope of the second, post-yield, branch of the force-deformation curve, k_2, is equal to zero. This is commonly referred to as an elasto-plastic system.

Figure 12-17 from Ref. 12-16 shows the variation of the maximum spring deformation, u_{max}, with the undamped small-amplitude frequency of elasto-plastic systems subjected to the 1940 El Centro (N-S) earthquake, for different values of the yield level. A constant damping ratio, $\beta = 0.02$ has been used throughout. The maximum spring deformation is given as the ratio $u_{max}/(x_o)_{max}$, while the yield level is expressed as the ratio $F_y/(F_e)_{max}$, of the yield force (or deformation) in the elasto-plastic system to the maximum force (or deformation) in the associated linear system. Note that the curve marked $F_y/(F_e)_{max} = 1.0$ corresponds to a linear system.

The following points will be noted in Fig. 12-17. For extremely low-frequency systems, the maximum deformation of linear and elasto-plastic systems are practically equal to each other and to the maximum ground displacement, $(x_o)_{max}$. Thus, in this frequency range, the softness of the elasto-plastic system overshadows any effect of the yield level on the response. As one moves to the right in the spectral plot, there is a region in which the effect of the yield level is most pronounced, a lower yield level generally resulting in lower values of the maximum spring deformation, u_{max}. Next to the right is a region in which the maximum deformation of both linear and elasto-plastic systems are

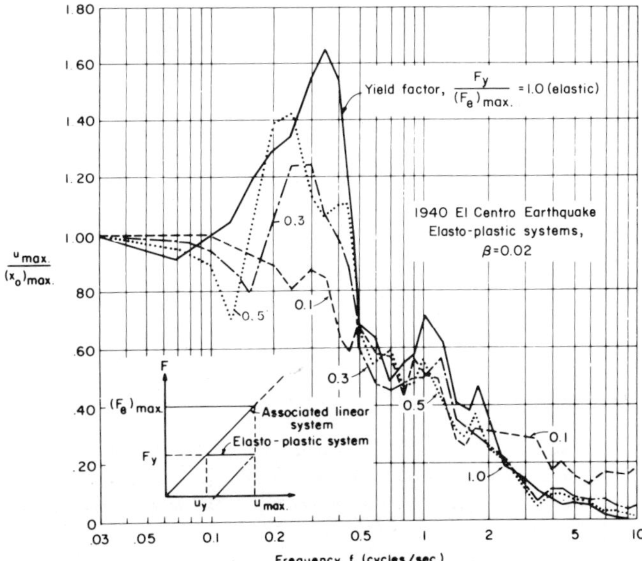

Fig. 12-17 Displacement spectra for elasto-plastic SDF systems having different yield levels—1940 El Centro (N-S) earthquake.[12-16]

approximately the same, the maximum deformation decreasing steadily relative to the maximum ground displacement with increasing frequency.

It will be noted, however, that while the effect of a decreasing yield level is to decrease the maximum deformation in the moderately-low and medium-frequency regions, the same decrease in yield level produces an increase, though slight, in the maximum deformation of high-frequency systems. Thus, for very stiff systems, a lowering of the yield level below that required for elastic response may produce significant inelastic components of deformation. This observation points to a qualitative distinction between the effect of inelastic action and that of viscous damping on a linear system since the latter tends to decrease the maximum deformation of a linear system over the entire frequency range.

The bilinear hysteretic system in which the slope of the second, post-yield, branch of the *F-u* curve is nonzero and positive represents an intermediate case between the elasto-plastic and the linear cases just discussed. As might be expected, the spectral curves for bilinear systems with low values of the ratio k_2/k_1 do not differ appreciably from those ofr elasto-plastic systems. As the value of the ratio k_2/k_1 increases, i.e., as the bilinear *F-u* curve approaches the linear case, the response of the bilinear system tends to approach that of the associated linear system. The response of bilinear systems appears to be most sensitive to changes in the ratio k_2/k_1 in the high-frequency region, where a slight increase in the value of the ratio above zero can produce a significant decrease in the maximum deformation.

As the yield level in a bilinear hysteretic system approaches zero, the maximum deformation of the system tends to the value corresponding to a linear system having a stiffness equal to k_2.

12.3.5 Multi-Degree of Freedom (MDF) Systems

-1. Some terms related to the dynamic analysis of MDF systems—Space does not allow a detailed discussion of the dynamic analysis of structures with multiple degrees of freedom (MDF). For this, the reader can refer to any of a number of textbooks on structural dynamics.[12-17,18,19] Reference 12-20 also gives a clear account of the dynamic analysis of frame structures.

In order to provide the reader with some background on the basic ideas involved in the dynamic analysis and response of MDF systems, a few of the terms commonly encountered in the literature on the subject are discussed briefly below.

The number of degrees of freedom of a system is defined as the least number of independent displacement coordinates necessary to completely determine the configuration of the system at any instant. For frame structures, the significant degrees of freedom are generally associated with the displacement components of the joints defined by the intersection of member axes. Thus, for a plane frame, the degrees of freedom are usually identified with the rotation and the vertical and horizontal displacement components of the joints. Often, due to the relatively high axial stiffness of the floor system, only one horizontal degree of freedom is assumed for all joints of a particular floor.

The use of lumped-mass models to represent the actual distributed mass structure is a convenient device for reducing the infinite number of degrees of freedom of the structure to a manageable few. This makes possible the formulation of the force equilibrium of the system in terms of a set of ordinary differential equations instead of the partial differential equations which would be required for the continuous system. With proper care in the lumping procedure, model response values corresponding closely to that of the real structure should be obtainable. In multistory building response studies, for instance, it is generally sufficient to assume the masses as concentrated at the floor levels and to formulate the problem in terms of the displacements of these masses.

The natural or principal modes of vibration of a linear structure are free, undamped, periodic vibrations, a unique linear combination of which make up the actual motion of the structure at any instant. Each mode of vibration is characterized by the masses or points defining the geometry of the structure all vibrating in phase, i.e., with the displacements of the masses or defining points, as measured from the initial position, always having the same relationship with respect to each other. Thus, all points of a structure vibrating in one of its natural modes pass the equilibrium position at the same time and reach their extreme positions at the same instant. A structure can be made to vibrate freely in one of its natural modes alone if carefully started in that particular mode.

A structure will have as many natural modes of vibration as it has degrees of freedom—each mode being associated with a characteristic frequency or period of vibration. The mode having the lowest frequency (or longest period) is called the first or fundamental mode. Generally, each mode of vibration and the associated frequency is distinct from all the others, although cases often arise in which two or more modal frequencies have values very close to each other.

Figure 12-8 shows typical shapes for the first three modes of vibration of a multistory building in terms of the lateral displacements of the floors. Note that the curves intersect the vertical axis at as many points or nodes (counting the one at the base) as the number of the mode. The displacements defining a particular mode shape are only relative and do not indicate a specific magnitude. As an example, the characteristic vectors, $\{\phi_i\}$, for a three-degrees-of-freedom system representing a three-story structure may be given by

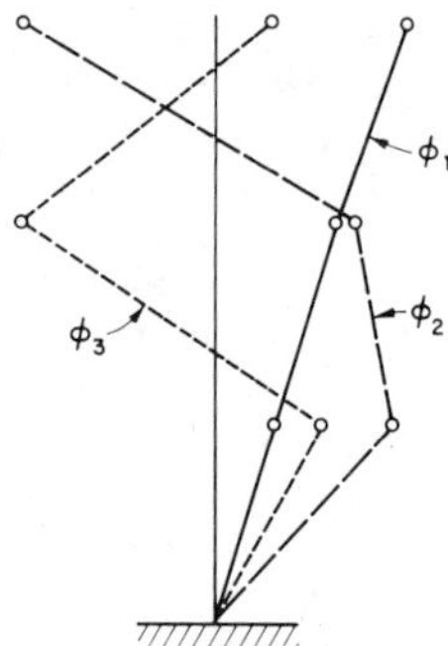

Fig. 12-18 Natural modes of vibration of a three-degree-of-freedom system.

$$\{\phi_1\} = \begin{Bmatrix} 1.0 \\ 0.62 \\ 0.30 \end{Bmatrix} \quad \{\phi_2\} = \begin{Bmatrix} -1.0 \\ 0.70 \\ 0.92 \end{Bmatrix} \quad \{\phi_3\} = \begin{Bmatrix} -0.28 \\ 1.0 \\ -0.52 \end{Bmatrix}$$

and are shown plotted in Fig. 12-18. The maximum relative displacement in each case has been taken as unity. What is important to note is that the displacements characterizing a normal mode are always in the same proportion, regardless of the cause of the vibration.

A damped MDF system possesses natural modes of vibration only if the character of the damping satisfies a certain condition.* Among systems satisfying this condition are those in which the damping matrix is proportional to either the stiffness or the mass matrix. For such systems, the same (normal mode) coordinates which uncouple the equations of motion of the undamped system also uncouple the equations corresponding to the damped system.

-2. Two connotations of dynamic analysis as used in earthquake response studies—The term dynamic analysis as applied to MDF systems, and particularly in earthquake engineering, has been associated with two slightly different analytical approaches. In the first one, sometimes referred to as response spectrum superposition or mode superposition using response spectrum data, approximate values of the maximum response parameters are determined by summing up the maximum responses corresponding to the different modes of vibration of a system. Since each component mode of vibration behaves as an independent SDF system with a characteristic frequency, the maximum modal response values may be obtained from appropriate response spectra for SDF systems (see section 12.3.3.5).

Several methods of combining the modal contributions have been proposed to allow for the fact that the individual modal maxima generally do not occur simultaneously. A commonly used method takes the square root of the sum of the squares of the modal maxima, treating these in the manner of random quantities. This method tends to yield low values when the frequencies corresponding to two or more significant modes of vibration have very close values. Taking the sum of the absolute values of the modal contributions provides an upper bound on the response. Often, only the contribution of the first few modes are considered, since these usually make up the major part of the response.

The response spectrum superposition method has been employed by many for design purposes using averaged or smoothed response spectra based on a number of earthquake records, all reduced to a standard or reference intensity. The use of average spectra provides, in one simple step, for the probable variation in response to different earthquake input. It is pointed out that response spectrum superposition is applicable only to a linearly elastic system and, if damped, only when the damping satisfies a certain condition so that independent uncoupled modes exist.

The second method associated with the term dynamic analysis involves the determination of the time history of response of the mathematical model of a structure to a particular earthquake. The analysis may employ modal superposition or a direct numerical integration[12-20] of the equations of motion. In both cases, the calculations are carried out for the total response at the end of a short interval of time, the analysis proceeding in a step-by-step fashion using the conditions prevailing at the end of one time interval as the initial conditions for the succeeding interval.

In modal analysis, the displacement, $u_i(t)$, of the ith mass of an N-degrees-of-freedom system is expressed as a linear combination of the characteristic modal displacements and a time-varying function, $q_j(t)$, i.e.,

$$u_i(t) = \sum_{j=1}^{N} \phi_{ij}\, q_j(t)$$

where j is the index denoting mode number. In matrix notation, the vector of displacements

$$\{u(t)\} = [\Phi]\ \{q(t)\}$$

where $[\Phi]$ is the $(N \times N)$ modal matrix, each column, ϕ_j, of which represents the characteristic displacements corresponding to a natural mode of vibration. The time-varying function $q_j(t)$ corresponding to the jth mode of vibration essentially gives the response of a SDF system having the same frequency as the jth mode when subjected to an effective load (see Eq. 12-3) given, in matrix notation, by

$$\frac{\{\phi_j\}^T [m] \{1\}}{\{\phi_j\}^T [m] \{\phi_j\}}\ \ddot{x}_0(t)$$

In the above expression, $[m]$ is the $(N \times N)$ mass matrix and $\ddot{x}_0(t)$ is the actual input base acceleration. The superscript T in $\{\phi_j\}^T$ indicates a transpose of the vector (or matrix).

As with the response spectrum superposition method, modal analysis is applicable only to linearly elastic systems in which damping, if present, satisfies certain conditions. Also, where external forces are applied to a structure, these must all have the same time variation in order for the uncoupling which characterizes modal analysis to be possible.

For a viscous-damped system, the decomposition of the response of a structure into its component modal responses is possible only when the damping matrix satisfies the relation**

$$(M^{-1}C)(M^{-1}K) = (M^{-1}K)(M^{-1}C) \tag{12-18}$$

where M, C and K represent the mass, damping and stiffness matrices of the system, respectively, and the (-1) superscript on M indicates the inverse of the matrix M. Equation

*See eq. (12-18).

**A method of uncoupling the equations of motion corresponding to a viscous damped system not satisfying eq. (12-18) was presented by K. A. Foss in Ref. 12-21. The method is based on obtaining orthogonality relations between the homogeneous solutions of the governing differential equations of motion. From these are developed a set of coordinates in terms of which completely uncoupled equations of motion can be written.

(12-18) essentially states that the product of M^{-1} with K and C commute. Among the forms of the damping matrix which satisfy the above condition are those in which it is proportional to either the stiffness on the mass matrix of the system, or is a linear combination of these, i.e.,

$$C = \alpha_1 K + \alpha_2 M$$

where α_1 and α_2 are scalar proportionally constants.

For a linearly elastic system with a viscous damping which allows the uncoupling of the equations of motion, modal analysis permits a significant saving in computing time, since the determination of the modes of vibration and the associated frequencies is carried out only once. The rest of the calculations involve the determination of the responses of single degree-of-freedom systems.

In the direct approach,[12-20] the set of simultaneous differential equations in incremental form is transformed into a set of algebraic equations in terms of one of the response parameters, using the assumption of a linearly varying response acceleration. The resulting set of algebraic equations, which can be cast in tri-diagonal or banded form is then solved using a recursion relation. The direct integration method, which does not require the uncoupling of the equations of motion, is necessary in nonlinear dynamic analysis.

The time history response analysis of a typical multistory structure, which will generally require the use of a large computer, can be time-consuming and expensive, particularly when inelastic response is considered. Its use may be justifiable only for a few important projects. It does, however, provide more reliable values than response spectrum superposition and is the only means of determining the ductility requirements of members corresponding to a particular design and earthquake input.

For design purposes, the use of a single earthquake record as input may not provide sufficient assurance of the behavior of a proposed structure under future earthquakes. For this purpose, the use of a set of earthquake records, possibly including artificially generated accelerograms would be advisable.

-3. Shear and flexure beam lumped-mass models for multistory buildings—A shear beam model is one in which the shear force acting on any mass depends only on its displacement relative to adjacent masses or supports. This model, which is an example of a close-coupled system, is shown schematically in Fig. 12-19a, and is typified by the idealized flexible column-rigid girder frame shown in Fig. 12-19b in which the effect of column axial deformation is neglected. The assumption of rigid girders, which implies no rotation at the joints and the neglect of axial deformation in the columns restricts the coupling between stories and results in a relatively simple model. Because of its simplicity, the shear beam model has often been used to approximate the behavior of low, open-frame structures.

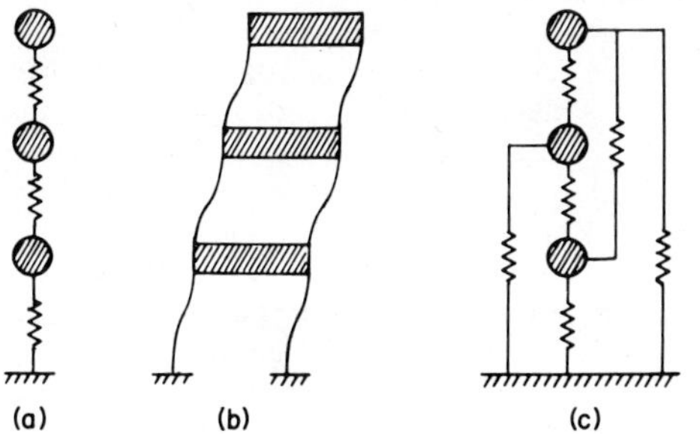

Fig. 12-19 (a) and (b) Shear beam model and (c) flexure beam model.

The effect of assuming no joint rotation can result in errors of the order of 100% in both the mode shapes and frequencies, particularly of low frequency structures, while the neglect of axial deformation in the columns can lead to errors of about 10% in the calculated frequencies. Both of the above assumptions result in a stiffer structure and hence tend to yield frequencies of vibration which are greater than the actual.

A flexure beam model corresponds to a typical cantilever beam with masses concentrated at discrete points along its length and represents an example of a far-coupled system. As indicated in the schematic diagram shown in Fig. 12-19c, the force acting on any mass is dependent on its displacement relative to all the other masses (and supports) of the system. An analytical model of a multistory building which considers joint rotations as well as axial deformation in the columns would correspond to a flexure beam model

12.3.6 Nonlinear Systems

For analyses considering nonlinear response, the procedure outlined for SDF systems may be used. This approach considers the response of a nonlinear system as being essentially linear over small intervals of time. The value of the stiffness during each short interval is taken as the slope of the local tangent to the force-displacement curve. Thus, as yielding occurs in members and the stiffness of a structure changes, the response of the nonlinear system is obtained as the response of successively different linear systems. Since each change in member stiffness, which may occur when a member yields or unloads from a post-yield condition, theoretically changes the stiffness of the structure, it is readily seen that the nonlinear dynamic analysis of even a plane multistory, multibay structure can be relatively time-consuming.

Computer programs developed to undertake the nonlinear dynamic analysis of multistory frames can be made to generate a variety of useful information such as maximum deformations and corresponding forces at all significant locations, as well as ductility requirements and time history records or plots of deformations at particular points in the structure. In spite of the limitations of the method, associated mainly with uncertainties in the values of stiffness and damping to be used in the model, nonlinear dynamic analysis represents a powerful tool in the study of the earthquake response of structures. Its use has provided insight into the basic mechanism of earthquake resistance of framed structures as well as a measure of the magnitude of the structural response to particular earthquakes.

The relatively large amount of numerical work involved and considerable computer storage capacity required in a typical inelastic dynamic analysis have made it necessary, in most instances, to either assume fairly simple force-deformation characteristics for the members of multistory, multibay frames or to consider only single-bay frames a few stories high. Allowance is usually made only for yield hinges forming at the ends of members; the hinging being governed by the magnitude of the bending moment only. It is pointed out that the neglect of axial load-moment interaction effects on the assumed force-deformation curves of members would be reasonable in strong column-weak beam designs where little or no yielding is expected in the columns. However, where the axial loads on the columns are relatively high, the effect of axial load on the yield moment could have appreciable effects on the response.

12.4 SOME RESULTS OF ANALYTICAL STUDIES ON THE BEHAVIOR OF MULTISTORY STRUCTURES SUBJECTED TO EARTHQUAKE EXCITATION

Some of the more significant results of analytical studies of the earthquake response of multistory frame and frame-shear wall structures are presented briefly below. In view of the limited scope of these studies, particularly with respect to the range of the parameters considered, the observations drawn from their results can serve only as indications of what might be expected under closely similar conditions. The trends exhibited by the results presented below, however, should provide useful guides in the preliminary planning of multistory structures.

The principal basis of comparison used in the following discussion is the ductility requirement associated with each parameter examined. The provision of adequate ductility, in addition to strength and stiffness, gains particular importance in view of the current philosophy of earthquake-resistant design as embodied in the earthquake design provision of most codes.* This design approach recognizes the impracticability of proportioning structures to resist moderate-to-strong earthquakes within the elastic range of stresses and instead takes advantage of the energy-dissipative action which occurs when some members of a structure are allowed to yield under strong earthquake excitation. Minimum ductility requirements are then provided to ensure the integrity of a structure some members of which may undergo deformations well beyond that corresponding to yield.

Much of the data presented below have been taken from a study by Clough and Benuska.[12-20] This fairly extensive study involved the computer evaluation of the response of regular multistory open frames as well as frame-shear wall combinations to the 1940 El Centro (N-S component) earthquake, considering both linear and nonlinear phases of the response. Among the parameters studied were the period of vibration, height of building, relative strength of girders and columns, stiffness and weight distribution, and the intensity and duration of the earthquake. It should be pointed out that the observed variations of the response apply strictly only within the range of values of the parameters considered. The trends exhibited by the results presented below can serve as useful guides in preliminary design; however, their extrapolation to parameter values far beyond those covered by the study should be carried out with caution. When a unique structural configuration is involved, careful consideration should be given to the effects of various details while bearing in mind the general trends exhibited by these results.

The effects of other parameters not covered by the Clough-Benuska study are discussed briefly in section 12.4.3.

12.4.1 Open Frame Structures

The configuration and relative member stiffnesses of the basic 20-story frame structure considered in the study[12-20] are shown in Fig. 12-20. The typical tapering of the member stiffnesses from the base to the top will be noted. The

*For instance, the *SEAOC Recommendations*[12-22], the *Uniform Building Code*,[12-10] the *National Building Code of Canada 1970*[12-23] and Appendix A of the *ACI Building Code, ACI 318-71*.[12-24]

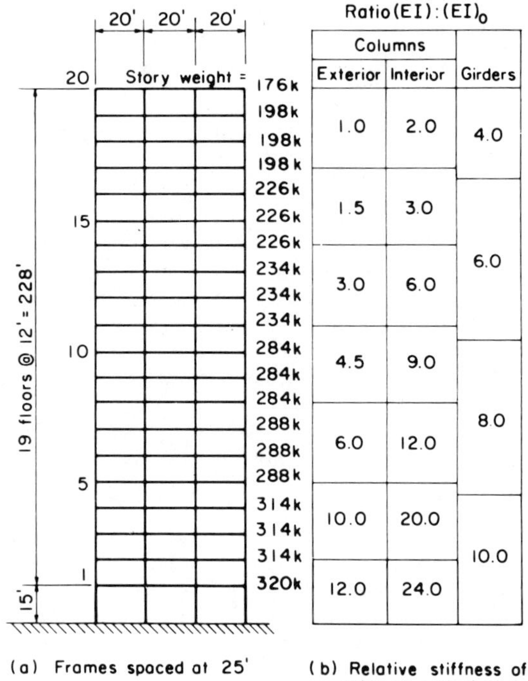

Ratio (EI):(EI)$_0$

Columns		Girders
Exterior	Interior	
1.0	2.0	4.0
1.5	3.0	6.0
3.0	6.0	6.0
4.5	9.0	8.0
6.0	12.0	8.0
10.0	20.0	10.0
12.0	24.0	10.0

(b) Relative stiffness of columns and girders

Fundamental period, 2.2 sec.

$(EI)_0 = 133{,}500$ k-ft.^{2}

Fig. 12-20 Properties of standard open-frame structure considered in study reported in Ref. 12-20.

frames are assumed spaced at 25-ft intervals, all the lowest columns being assumed fixed at the base. The frames were designed for vertical loads plus the lateral forces prescribed by the *Uniform Building Code*,[12-10] 1964 Edition, using simple approximate procedures. Then, using the resulting relative stiffness properties indicated in the figure, a computer analysis for these same statically applied forces was carried out. The computed member forces were designated as design forces. The yield moments were then taken as twice the design values for the girders and six times the corresponding design values for the columns. The relatively high value of the column yield-to-design moment ratio results from the fact that for the average column, a significant portion of the load capacity of the section is provided to take care of the axial load while the yield moment is calculated on the basis of purely flexural action. It is pointed out that the effect of axial load on the yield capacity of the columns was not considered in the study.

In the nonlinear dynamic response analysis, the moment-rotation characteristics of the members was assumed to be of the bilinear type with the second, post-yield branch having a slope equal to 5% that of the first (elastic) branch. The term ductility factor as used in the following discussion is defined as the ratio of the maximum total (elastic + plastic) rotation at the end of a yielded member to the yield rotation angle. The yield rotation angle is defined as that corresponding to a moment acting at the end of a simply-supported member having the same section but with a span equal to one-half that of the actual member. The use of a half-span is based on the antisymmetrical mode of deformation of members in a frame due to lateral displacement. It is pointed out that because the moment at a member end depends not only on the deformation at that end but also on the deformation at the other end, the actual yield rotation for a particular member end will vary with the state of deformation of the member ends. Because of this, some investigators[12-25] prefer using curvature instead of rotation as a basis for defining the moment-deformation characteristics and ductility factors for members.

Where damping was considered, an equivalent damping ratio, β_1, equal to 10% of the critical for the first mode of vibration was assumed, with smaller damping in the higher modes (proportional to the period of vibration).

Except where the duration of the earthquake was the parameter considered, the analytical results shown in Figs. 12-21 to 12-34 were obtained by subjecting the structures to the first four seconds of the 1940 El Centro earthquake (N-S component) applied directly to the base. As can be seen in Fig. 12-4, the first four seconds of the 1940 El Centro record contains the most intense portion of the motion. The limited duration considered was dictated by the need to minimize computer time and costs since the nonlinear dynamic analysis requires considerable processing time. It is important to note that other earthquake records with different frequency characteristics may produce results which vary significantly from those obtained using the 1940 El Centro (N-S component) record. A brief discussion of the effect of the frequency characteristics of earthquake motions is given in section 12.4.3.5.

-1. Comparison of linear (elastic) and nonlinear response— As with SDF systems (see Fig. 12-17), the maximum lateral displacements for both linearly elastic and nonlinear frames having the same period* and damping are approximately equal over a broad range of fundamental period values. Figure 12-21a shows the maximum lateral displacements for elastic and nonlinear frames having a fundamental period, T_1 = 2.2 seconds and a damping ratio β_1 = .10. This similarity in the maximum displacement envelopes for both cases should not, however, be interpreted as implying that similar maximum deformations will be developed in corresponding members of the two frames. Figure 12-21b shows the girder ductility requirement for the nonlinear case varying from 2 at mid-height to 5 at the top, compared to a maximum-to-yield moment ratio of about 2 for the elastic case. Clearly, deducing member ductility requirements on the basis of elastic response values can be grossly unconservative. An analysis assuming completely elastic response will generally overestimate the inelastic deformations in the columns and underestimate these in the girders of a typical strong column-weak beam code-designed frame.

Also shown in Fig. 12-21a is the computed lateral deflection of the elastic frame when subjected to the static lateral forces specified by the *Uniform Building Code***. It will be noted that, for this particular frame, the deflection under the code-specified lateral forces is only about one-fifth of the maximum dynamic displacement produced by the N–S component of the 1940 El Centro earthquake.

It is important to draw a distinction between the ductility factors associated with the lateral displacement of a frame and the member ductility factors. Since the former is achieved through inelastic deformations at the critically stressed portions of a relatively few members (mostly beams), the corresponding member ductility factors will generally be greater than that associated with the lateral displacement. The dashed curve in Fig. 12-21a shows the approximate deflected shape corresponding to first yield in the statically loaded, *UBC*-designed frame. The curve was obtained by multiplying the displacement values from a static analysis under *UBC*-forces (the -Δ- curve shown in the same figure) by the ratio of yield to design stress of the weakest element—in this case, 2.0. If one were to use the average ratio of maximum displacements for either elastic or nonlinear dynamic response to the above 'static first-yield' displacements as an indication of ductility requirements, one would arrive at a value of about 2.5 for this case. A look at Fig. 12-21b, as well as similar figures which follow, clearly indicates that the maximum displacement

*The (fundamental) period of the nonlinear frame, as used in this discussion, refers to the period corresponding to the initial stiffness, before yielding occurs.

**The 1964 edition of the *Code* was used for this study.[12-20] Both 1964 and 1970 editions of the *Code* provide for the same total design base shear and differ only in the manner in which this base shear is to be distributed along the height of the building. In the 1964 *Code*, the total base shear is distributed over the height of the building in a triangular manner, i.e., with a maximum intensity at the top and zero at the base. The 1970, as well as 1967, editions incorporated one of the principal recommendations resulting from the Clough-Benuska study, and provides for the application of a maximum of 15% of the specified base shear as a concentrated load at the top, with the balance to be distributed over the height of the building in a triangular fashion. The provision for the application of a portion of the base shear to the top of slender structures (with height-to-depth ratios greater than 3) is an attempt to approximate the observed whiplash effect.

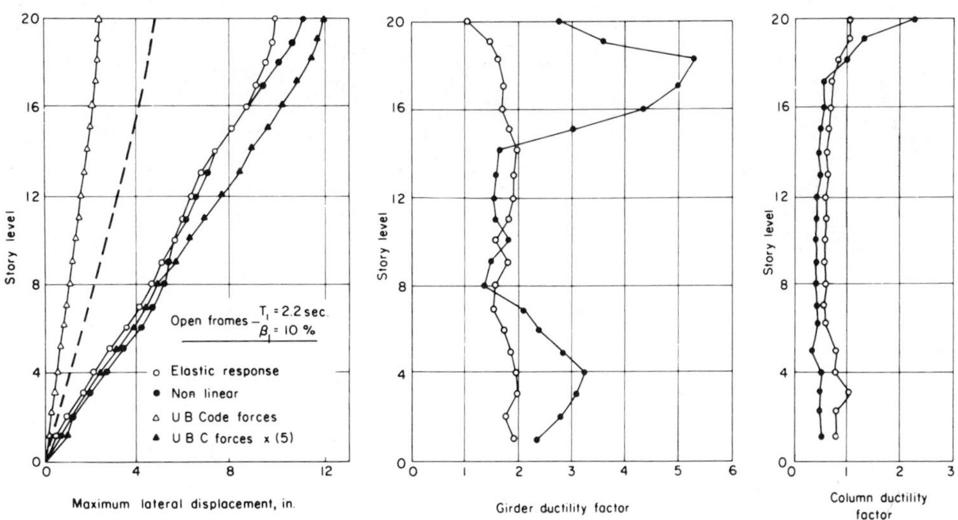

Fig. 12-21 Comparison of elastic and nonlinear response.[12-20]

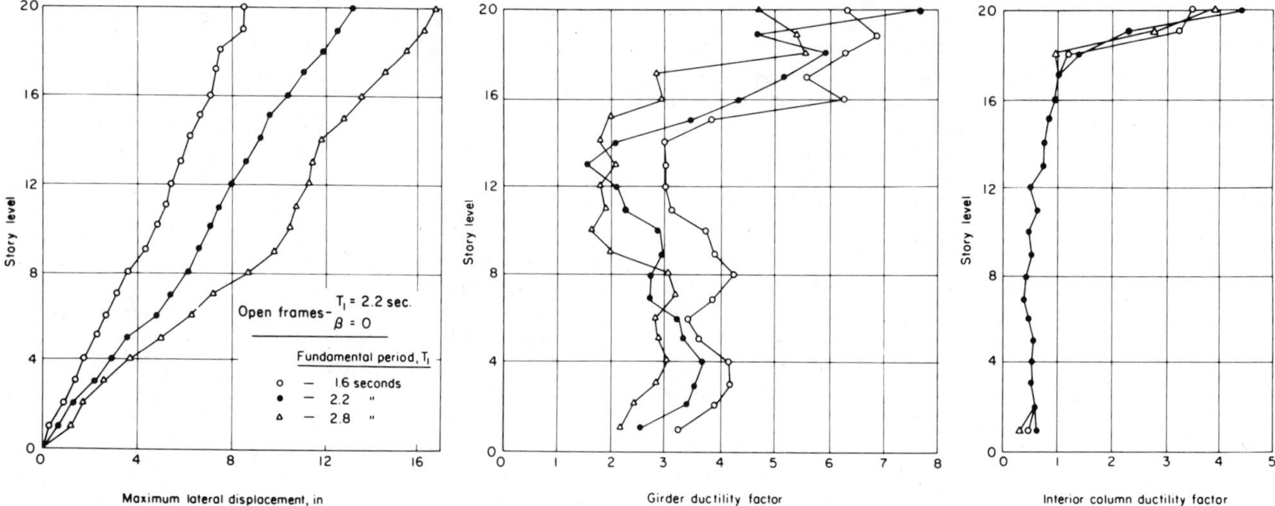

Fig. 12-22 Effect of period of vibration.[12-20]

does not give a good measure of the actual member ductility requirements.

In Fig. 12-21c, as well as in the following figures, a ductility factor less than unity indicates the ratio of the maximum moment to the yield moment in a member.

It is of interest to note that a superposition of the contributions of the various modes of the elastic structure, using the response spectrum for the 1940 El Centro earthquake (N-S component), provides an upper bound on the maximum lateral displacement for both the elastic and nonlinear cases.

-2. *Effect of period of vibration*—In order to examine the effect of this variable, two other 20-story frames having fundamental periods of 1.6 and 2.8 sec were considered in addition to the standard 2.2-sec frame. All three frames were the same except that the basic stiffness parameter, EI_o, was varied to obtain the other periods of 1.6 and 2.8 seconds. The results, shown in Fig. 12-22, indicate that the maximum lateral displacement, in the period range considered, increases almost in proportion to the period of the buildings (though not with the flexibility, which varies as the square of the period). The general tendency of the maximum displacement to increase with the period of a structure (or with a decrease in its fundamental frequency of vibration) in the period range considered is also apparent in the response spectra for both elastic and elasto-plastic SDF systems shown in Fig. 12-17.

There is a slight decrease in girder ductility requirements for the more flexible (long-period) structures. However, a study by Goel and Berg[12-25] of the response of 10-story, single-bay frames to three different earthquake records showed that this particular trend can be reversed in the case of earthquake records characterized by dominant velocity spectrum peaks in the 2-3 second range (see section 12.4.3.5). Otherwise, the effect of period of vibration in the range considered in the Clough-Benuska study does not appear to be significant as far as ductility requirements are concerned.

For stiff, short-period structures, the variation in ductility requirements over the height of the frame differs from that shown in Fig. 12-22b. A probabilistic study by Ruiz and Penzien[12-26] of the response of eight-story, shear-beam models subjected to a number of artificially generated accelerograms showed that in structures with a fundamental period of about 0.5 second, the ductility requirements tend to decrease toward the top of the structure. This contrasts with the variation typical of the more flexible frames shown

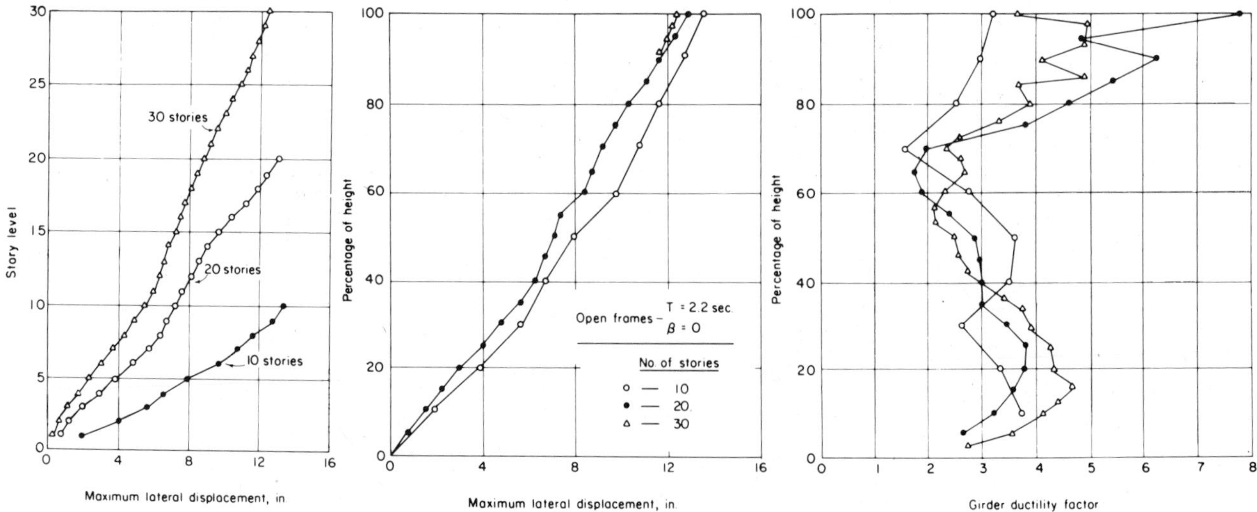

Fig. 12-23 Effect of building height.[12-20]

in Fig. 12-22b in which the influence of the higher modes of vibration causes a significant increase in the ductility requirements in the top stories. The same study by Ruiz and Penzien showed that the ductility requirements at the base of a stiff structure can be significantly greater than that for a flexible structure subjected to the same excitation.

An interesting result of the Ruiz-Penzien study is that the whiplash effect in flexible structures, i.e., with fundamental period of approximately 2.0 seconds, due to the influence of the higher modes of vibration still occurs even though the model frame was designed with 10% of the *UBC*-specified base shear applied at the top of the frame.

-3. Effect of number of stories—Figure 12-23 shows the results obtained from analyzing three frames, i.e., a 10-story and a 30-story frame in addition to the standard 20-story frame—all having fundamental periods of 2.2 seconds. It is worth noting that all three frames have about the same maximum deflection at the top, indicating a decreasing relative story displacement with increasing height of frame. When the maximum deflection and ductility requirements are plotted as a function of relative story height, however, the curves corresponding to the three frames are very similar. The more pronounced whiplash effect on the member ductility requirements of the taller structures is evident in the figure.

In actual practice, the fundamental period of vibration generally increases with the height of a building. On the basis of the results presented in the preceding section, one can expect the maximum lateral displacement to increase with the height of a building. The same study by Goel and Berg[12-26] mentioned earlier, which considered the response of three buildings of different height—where the 10- and 25-story models were identical to the top 10 and 25 stories, respectively, of the 40-story model—showed this to be the case. This study also showed a slight but consistent decrease in the column maximum-to-yield moment ratio with an increase in the number of stories.

-4. Effect of strength of girders—Since most of the inelastic deformation in a frame designed according to present earthquake code provisions may be expected to occur in the girders, it is important to know the effect of girder strength (as given by the yield strength, in the elasto-plastic case considered here) on the earthquake behavior of

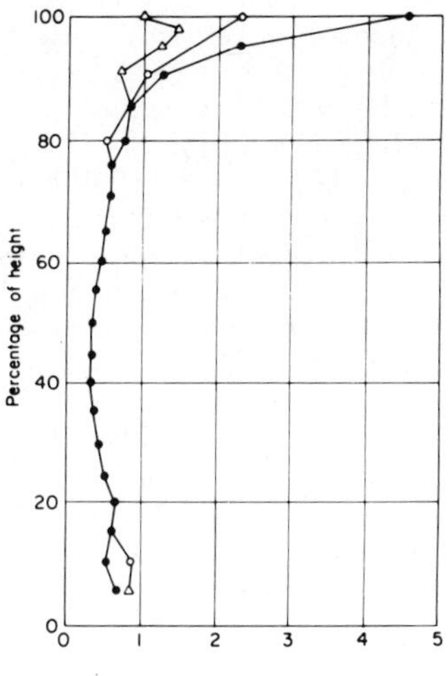

Fig. 12-23 *Continued*

multistory buildings. To obtain an indication of this effect, three frames were considered: the standard one with a girder yield-to-design moment ratio of 2.0 and two other frames, identical to the first except that the yield moments were 1.5 and 4.0 times the design moments.

Figure 12-24 shows the results of the analysis. Figure 12-24a indicates that the maximum lateral deflection of the three frames are practically the same except in the upper stories where an increase in deflection accompanies an increase in girder strength. Figure 12-24c shows that this increase in deflection in the upper stories is a result of the increased inelastic deformation in the columns of these stories, indicating that the maximum deflection is primarily controlled by the columns rather than the girders.

As might be expected, the girder ductility requirements decrease with increasing girder strengths. This is shown in Fig. 12-24b. More significant, however, is the fact that the increase in girder strength forces more of the inelastic deformation to occur in the columns, as indicated in Fig. 12-24c. In general, decreasing the yield strength of a member type, i.e., column or girder, with respect to another tends to attract inelastic deformation toward the weaker member, resulting in reduced yielding in the stronger member type.

-5. Effect of column strength—This variable was studied by considering two other frames, having column yield-to-design moment ratios of 2.0 and 10.0, in addition to the standard frame which had a moment ratio of 6.0. Figure 12-25 indicates that increasing the column strength beyond that corresponding to a ratio of 6.0 does not materially affect the response. This follows from the fact that the columns in the standard building remain essentially elastic during the response. Thus, increasing the column strength beyond that used in the standard frame does not produce any significant changes in the distribution of inelastic deformation between columns and girders.

Figure 12-25c, however, shows that a reduction in column strength can have a significant effect on the distribution of ductility requirements. The results indicate that if the columns do not have a sufficient margin of strength above the design strength, most of the inelastic deformation will tend to occur in the columns. Because of the danger of instability associated with excessive yielding in the columns, such a condition should be avoided.

Appendix A of ACI 318-71[12-24] provides for yielding to occur in the beams rather than in the columns by requiring that "at any beam-column connection, the sum of the moment strengths of the columns at the design axial load shall be greater than the sum of the moment strengths of the beams along each principal plane at that connection unless the sum of the moment strengths of the confined cores of the columns is sufficient to resist the applied design load" (Section A.6.2). Similar provisions are found in the *SEAOC Recommendations*[12-22], *UBC-1970*[12-10] and *NBC-1970.*[12-23] The use of the relatively high yield-to-design moment ratio of 6 for the columns in the Clough-Benuska study results in a behavior essentially similar to that intended by the above provision. Compliance with the strong column-weak beam provision of the *Codes* would have forced all of the yielding above the base in the elasto-plastic frames considered to occur in the girders. In frames designed to such a provision, the sum of the yield moments of the girders framing into a joint would determine the maximum moment which can be developed in the joint, so that the columns would remain elastic throughout the response, except perhaps at the bases of the lowest story columns. This observation has led some investigators[12-25] to use analytical models in which the columns are con-

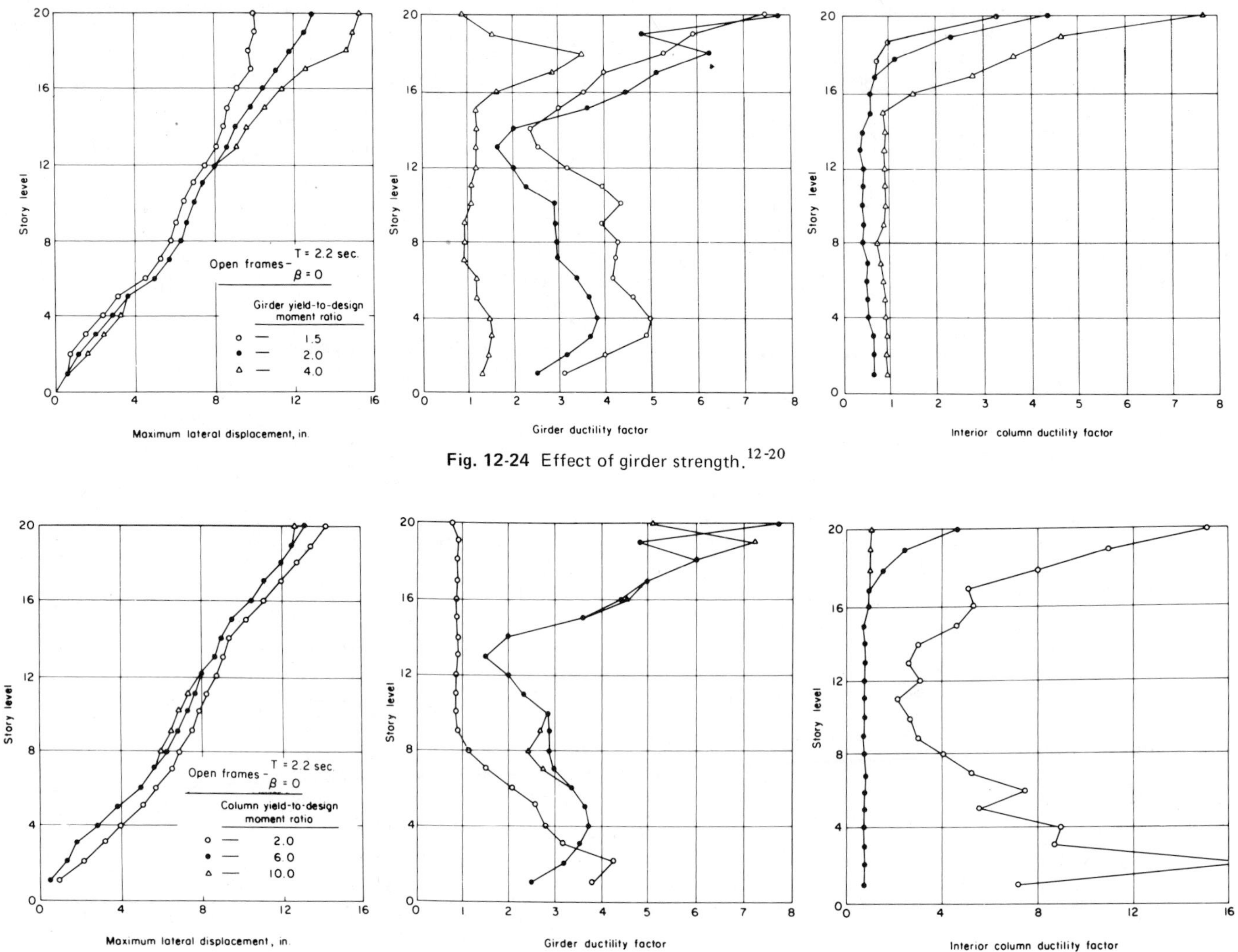

Fig. 12-24 Effect of girder strength.[12-20]

Fig. 12-25 Effect of column strength.[12-20]

sidered not to yield regardless of the intensity of the ground motion, all of the inelastic deformation being confined to the girders.

-6. *Effect of stiffness taper*—In order to assess the effect of the decrease in stiffness with height common in multistory buildings on their dynamic response, a building having uniform stiffness and mass properties over its full height was considered in addition to the standard building which had a stiffness taper. The uniform building had relative member stiffnesses equal to that of the central stories of the standard building, the reference stiffness being adjusted to give the same 2.2 sec fundamental period as the standard building.

Figure 12-26 shows that the whiplash effect, with its associated increase in deflection and ductility requirements in the upper stories, is not as pronounced in the uniform stiffness building as in frames with tapering stiffness. A similar beneficial effect due to the use of uniform column stiffnesses throughout the height of a building instead of the typical tapered column stiffness has been reported by Muto.[12-27]

-7. *Effect of intensity of earthquake*—The results shown in Fig. 12-27 were obtained by subjecting the standard open frame building to the first four seconds of the 1940 El Centro (N-S) earthquake, the intensity—here, used to denote the amplitude of the ground acceleration—being varied by multiplying the actual ground acceleration by 0.7 in one case and by 1.3 in another. As might be expected, there is a general increase in response with increasing intensity. It will be noted that, except in the top stories, the columns remain elastic even under an earthquake with intensity equal to 1.3 times that of the 1940 El Centro.

It is pointed out that although the 1940 El Centro earthquake represented the strongest ground motion then (1965) recorded, it does not necessarily represent the strongest which might be expected at any given site. Actually, no record has yet (1971) been obtained within 100 miles of the epicenter of an earthquake of Richter magnitude greater than 7.7.[12-28] Some investigators[12-25] have used the 1940 El Centro motion, magnified by a factor of 1.5, in their studies as a means of allowing for stronger earthquakes. Records[12-29] taken at one of the abutments of the Pacoima Dam in California during the February 9, 1971 San Fernando earthquake (Richter magnitude, 6.6) indicate an intensity of ground motion much greater than the 1940 El Centro.

As mentioned earlier, the maximum amplitude of the ground acceleration represents only one of the three major

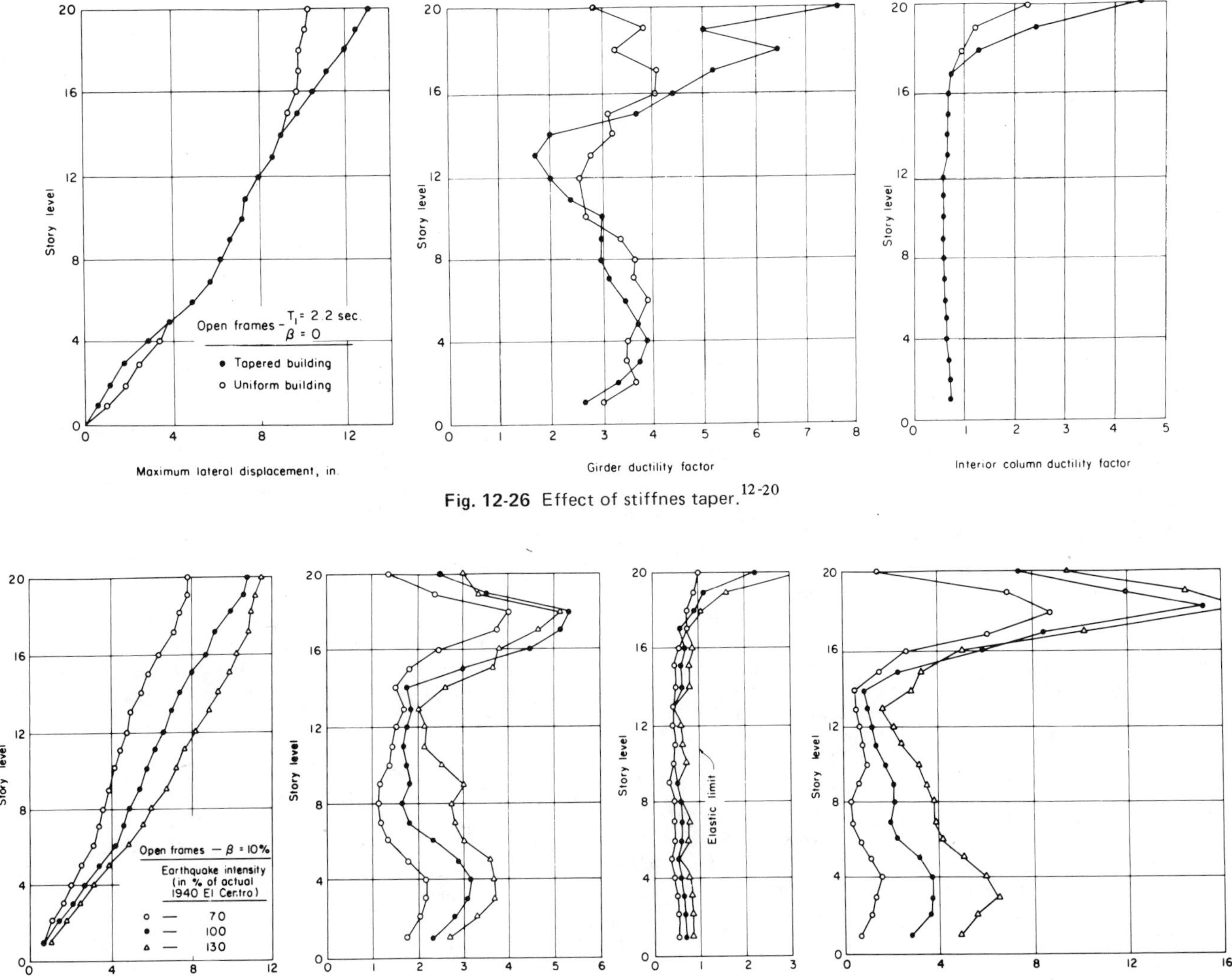

Fig. 12-26 Effect of stiffnes taper.[12-20]

Fig. 12-27 Effect of earthquake intensity.[12-20]

properties of earthquake ground motion which affect structural response. The effect of the duration of the large amplitude pulses will be discussed in the next section. Furthermore, the frequency characteristics of the ground motion, particularly as these relate to the fundamental frequency of vibration of a building, can vary depending upon a number of factors, with corresponding differences in structure response. This is discussed in section 12.4.3.5.

-8. Effect of duration of earthquake—To study this effect, the standard building was subjected to a synthetic 12-sec base motion consisting of the first 6 sec of the El Centro earthquake followed by the same 6-sec motion and also to the first 8 sec of the El Centro record.

Figure 12-28 shows the results, which also includes those corresponding to the first 4 sec of the El Centro for comparison. Figures 12-28a and b indicate that the effect of the second 4 sec of the El Centro motion, which have relatively smaller amplitude pulses, is not significant as far as maximum displacement and ductility requirements are concerned. However, the 12-sec synthetic earthquake of large amplitude pulses produced a significant increase in both maximum lateral displacements and girder ductility requirements. The effect of duration is also apparent on the total plastic rotation or accumulated ductility requirements of the girders, as shown in Fig. 12-28c. The total plastic rotation provides an indication of the number of times yielding occurs and/or the extent of such yielding and is significant in view of the deterioration in strength which may accompany repeated loading into the plastic range, or the instability which may result from extensive yielding. The damage potential of an earthquake thus increases with the duration of the large amplitude motions. This observation is also borne out by the results of a statistical study by Husid[12-30] of the earthquake response of simple (SDF) bilinear systems subjected to gravity loads which indicated that the intensity of earthquakes of short duration required to produce collapse would have to be greater than that of longer lasting earthquakes.

For the structure and base motions considered, the column ductility requirements are not significantly affected by the duration of the ground motion. This follows from the fact that most of the inelastic deformation occurs in the girders, the columns remaining essentially elastic except at the top stories.

-9. Effect of damping—To assess the effect of the energy-dissipating mechanism which is present in structures even before the onset of yielding, a viscous type damping was assumed. A damping ratio, β_1, equal to 10% of critical was

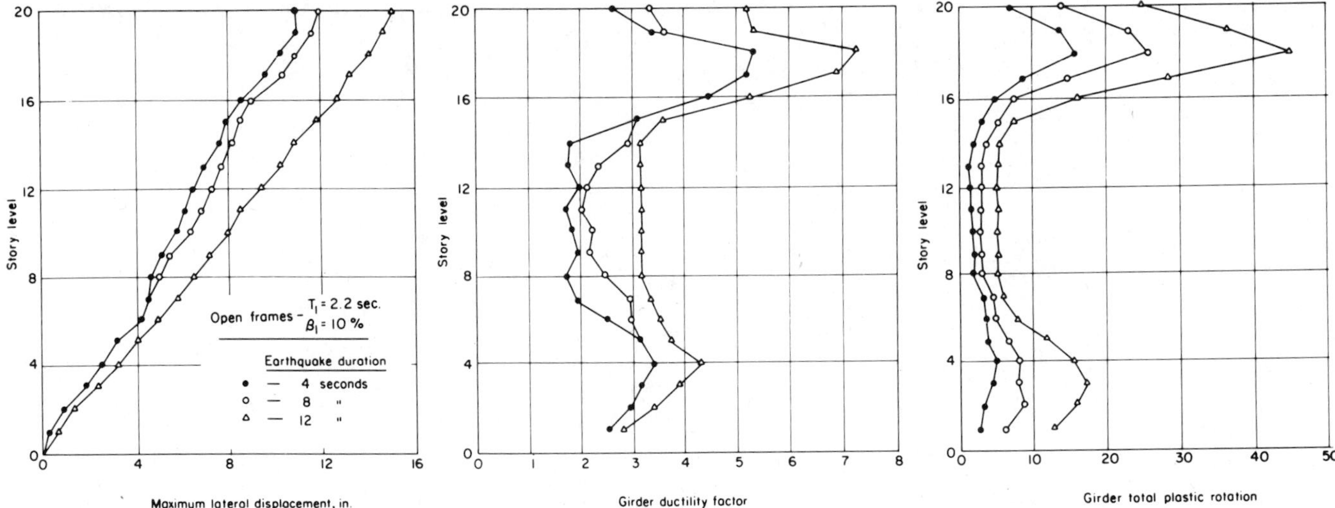

Fig. 12-28 Effect of earthquake duration.[12-20]

assumed for the first mode of vibration, with smaller damping in the higher modes. Although experimental studies[12-31,12-32] indicate that the percentage of equivalent viscous damping increases in the higher modes, the above assumption is not considered unreasonable and leads to conservative results.

Figure 12-29 shows the results for the standard 2.2-sec period frame. The effect of damping on the response of sample buildings with periods of 1.6 and 2.8 sec (not shown) was about the same as that for the standard building shown in the figure, i.e., a reduction of approximately 20% in the maximum displacements, corresponding to the assumed damping ratio. The percentage reduction in the girder ductility requirements due to damping, shown in Fig. 12-29b, is about the same as that in the displacement, with slightly greater percentages at mid-height and at the top. The effect of damping on the column ductility requirement for the buildings considered was relatively small; here again because the columns remain essentially elastic throughout the response.

Results of the Ruiz-Penzien study[12-26] of eight-story, shear-beam models mentioned earlier indicated greater reductions in ductility requirements due to damping in stiff (fundamental period, $T_1 \approx 0.5$ sec) structures than in flexible ($T_1 \approx 2.0$ sec) structures. The study also showed that increasing the damping reducing yielding and hence the amount of energy dissipated by inelastic action. The result was that the total energy dissipated remained practically independent of the amount of viscous damping in the structure.

12.4.2 Frame-Shear Wall Structures

The buildings considered in this part of the Clough-Benuska study[12-20] consist of a parallel arrangement of essentially identical frames and shear walls such that in the direction of motion, and neglecting torsional effects, four frames may be assumed to act together with a shear wall. In Fig. 12-30, the four frames have been lumped into the single frame shown coupled with the shear wall. Also shown in Fig. 12-30 are the relative stiffnesses of the members of the standard structure in terms of a reference $(EI)_0$. The value of $(EI)_0$ has been adjusted to give the standard structure a fundamental period of 2.2 sec.

As in the open frame buildings discussed earlier, the de-

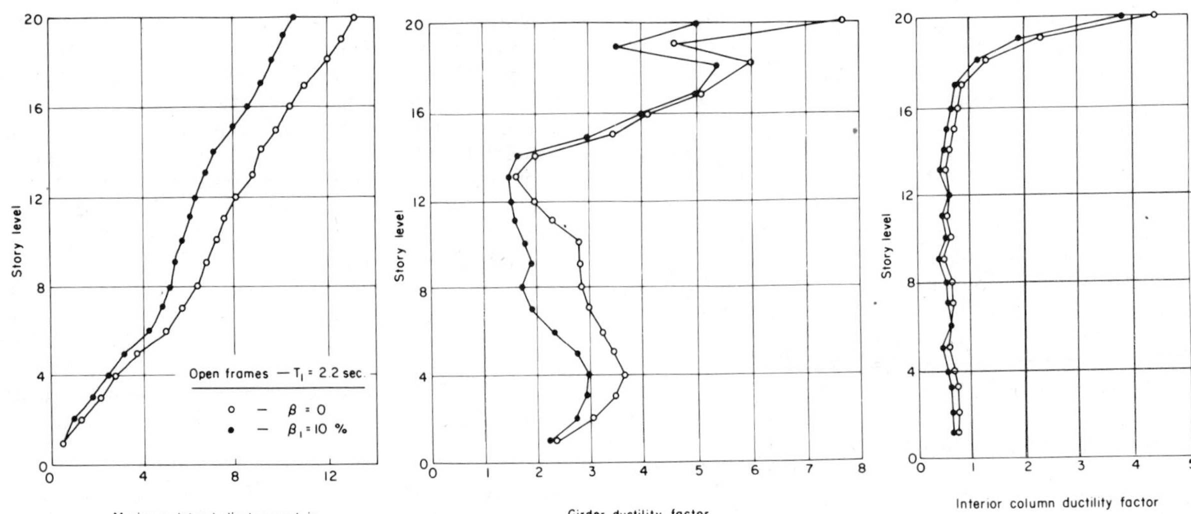

Fig. 12-29 Effect of damping.[12-20]

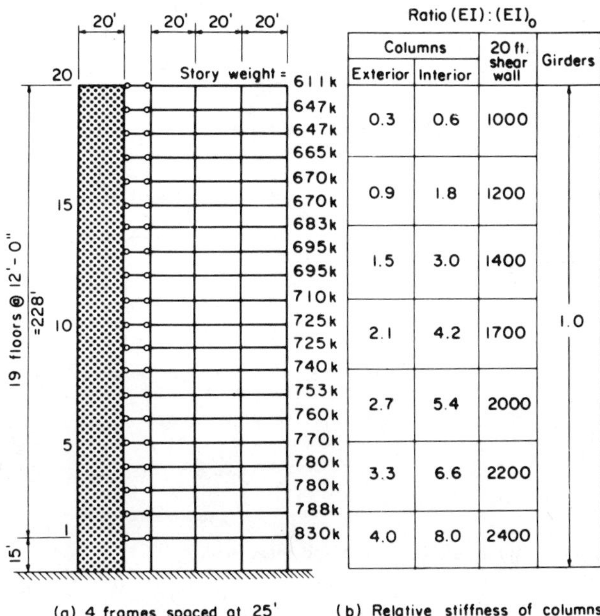

Ratio $(EI):(EI)_0$

Columns Exterior	Columns Interior	20 ft. shear wall	Girders
0.3	0.6	1000	1.0
0.9	1.8	1200	
1.5	3.0	1400	
2.1	4.2	1700	
2.7	5.4	2000	
3.3	6.6	2200	
4.0	8.0	2400	

(a) 4 frames spaced at 25' per shear wall

(b) Relative stiffness of columns and girders

Fundamental period, 2.2 sec.

$(EI)_0 = 390{,}000$ k-ft^2

Fig. 12-30 Properties of standard frame-shear wall structure considered in study reported in Ref. 12-20.

sign moments in this case were determined by a computer analysis for the static vertical and code seismic forces.

-1. Effect of design assumption on distribution of lateral loads between frame and shear wall—Designs corresponding to three different methods of distributing lateral loads found in practice, were considered. A first design, which will be referred to as the gravity-load frame building, was based on the assumption that the entire lateral load is carried by the shear wall. The frame for this building is thus designed only for vertical loads, with the girder moments being uniform and the column moments increasing from top to bottom. A second design was based on the *Uniform Building Code* provision requiring the frame to be designed for (at least) 25% of the total lateral force. This leads to both girder and column moments which increase from top to bottom. This second design will be referred to as the 25% lateral-load frame building. A third design was based on the true interaction behavior of the frame-shear wall system. In this case, the girder and column moments are largest at mid-height and decrease both upward and downward. This will be called the interaction-frame building.

It is pointed out that because of the different deformation patterns exhibited by frames and shear walls when acting separately under lateral loads, interacting forces which vary in magnitude and direction along the height of the structure are developed when both elements are tied together by floor slabs. The floor slabs are usually assumed to be stiff enough in their planes to produce equal lateral displacements at the floor levels, in the absence of torsion. This is illustrated in Fig. 12-31. It will be noted that the shear wall behaves essentially as a vertical cantilever beam while the frame exhibits the deformation typical of a shear beam under transverse loads. The interaction between the two elements is such that the frame tends to reduce the lateral deflection of the shear wall at the top while the wall helps support the frame near the base.

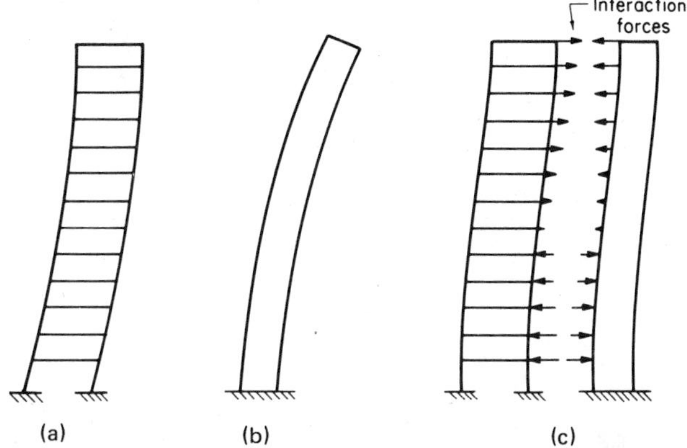

Fig. 12-31 (a) Rigid-frame shear mode deformation; (b) shear wall bending mode deformation; and (c) interconnected frame and shear-wall (equal deflection at each story level).

Because of the different bases of design used in proportioning the members of each structure, a different set of member yield moments corresponds to each building. In each case the ratio of yield-to-design moments was set equal to 2 for the girders and 6 for the columns and shear walls. In all cases, the reference stiffnesses were adjusted to yield a fundamental period of 2.2 sec.

Figure 12-32 shows the maximum lateral displacements corresponding to the three cases considered. All three types of buildings have essentially the same maximum displacement at the top. Also shown in Fig. 12-32a is a curve corresponding to a 25% lateral-load frame building with the shearwall hinged at the base. The flexure beam mode is apparent in both the gravity-load frame building and the 25% lateral-load frame building with the shear wall fixed at the base—indicating the predominant influence of the shearwall relative to the frame. The curve for the interaction frame building, on the other hand, indicates an almost linear variation with height, representing an intermediate case between the flexural mode of the shear wall and the shear-beam mode of the frame.

The girder and column ductility requirements corresponding to the four buildings considered are shown in Fig. 12-32b and 12-32c. The very favorable distribution of strength in the interaction frame building, resulting in significantly lower ductility requirements for girders over the entire height and for columns at the top is evident. The relatively low design strength of the frame in the gravity-load frame building is reflected in the high girder ductility requirements. It is worth noting that designing for frame-shear wall interaction tends to eliminate yielding of the columns at the top stories.

Figure 12-32d shows that the maximum overturning moment in the shear wall for the four buildings considered are roughly of the same order of magnitude. It is significant to note that, for the yield-to-design moment ratio assumed, none of the shear walls was stressed beyond the elastic range.

An idea of the relative strengths of the different frames corresponding to the three design assumptions used is given in Fig. 12-32e, which shows the maximum column axial forces in the various frames. Since the magnitude of the column axial forces developed is a direct reflection of the strength of the girders, it is readily seen that the interaction frame is almost twice as strong as the 25% lateral-load frame and about three times as strong as the gravity-load frame.

The above comparisons clearly demonstrate the desirabil-

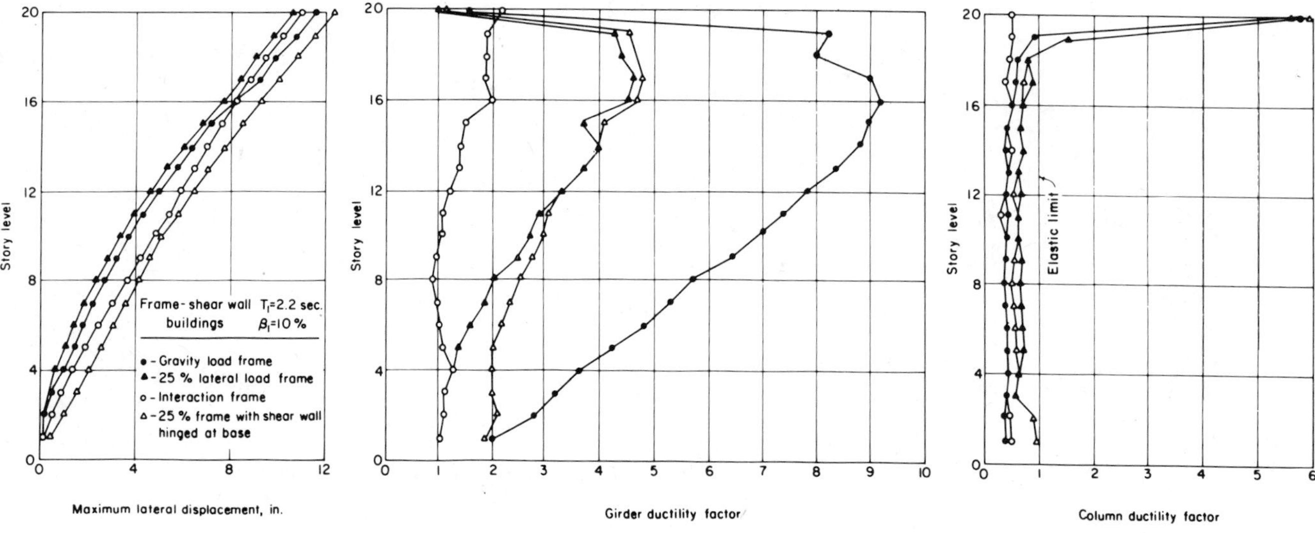

Fig. 12-32 Effect of design assumption on distribution of lateral forces between frame and shear wall.[12-20]

ity of considering the true interaction behavior in designing frame-shear wall structures.

-2. Effect of period of vibration and frame-to-shear wall stiffness ratio—In order to examine the effect of the period of vibration two other structures with fundamental periods of 1.6 sec and 2.6 sec were considered in addition to the standard 25% lateral-load frame building, having a fundamental period of 2.2 sec. The 1.6 and 2.6-sec period buildings were obtained by varying the width of the shearwall in the standard building. Thus, the 2.6-sec building had a 10-ft wide shearwall with a stiffness ratio relative to the shear wall in the standard building of 0.2, while the 1.6-sec building had a 38-ft wide shear wall and a stiffness ratio of 5.0.

The maximum lateral displacements for the three buildings considered are shown in Fig. 12-33a. The increase in displacement with increasing flexibility (longer period) observed earlier in the case of open frames is also evident here. Figures 12-33b and 12-33c show the girder and column ductility requirements, respectively. The latter does not show any significant difference in behavior among the three buildings. A slight decrease in girder ductility requirements occurs in the stiffer (shorter-period) structures. This trend is contrary to that noted in open frame structures. It should be noted, however, that the periods of the three structures considered differ not because of a change in stiffness of the frames, as was the case in the open frame study discussed earlier, but was accomplished by changing the width and hence the stiffness of the shear walls. In all three structures, the frame portions were identical. The above observed difference in girder ductility requirements could thus be interpreted as reflecting more the effect of the shear wall-to-frame stiffness ratio than of the period of vibration.

A plot of the shear wall overturning moments, shown in Fig. 12-33d, indicates a decreasing ductility requirement (M_{max}/M_y for ductility ratios less than unity) for the stiffer structures. More important, however, is the relatively large ductility requirement indicated for the 2.6-sec structure compared to the elastic behavior of the other two structures considered. This points to the potential danger of rupture in such stiffening elements in structures with low shear wall-to-frame stiffness ratios.

-3. Effect of number of stories—To study this effect, a 10-story building consisting of the top ten stories of the standard 25% lateral-load frame building was considered in

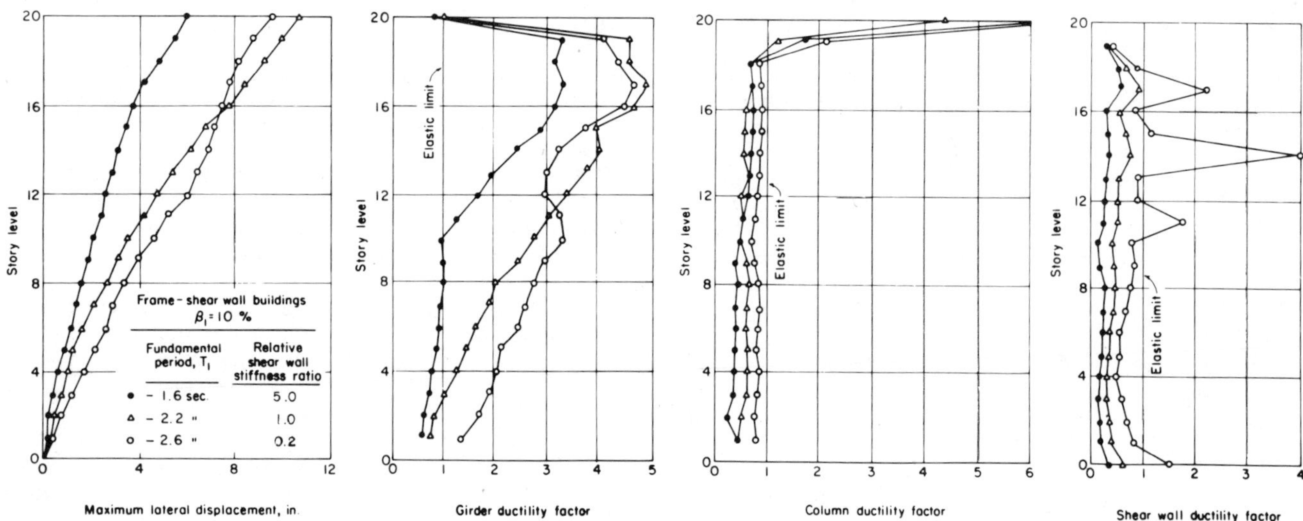

Fig. 12-33 Effect of period of vibration.[12-20]

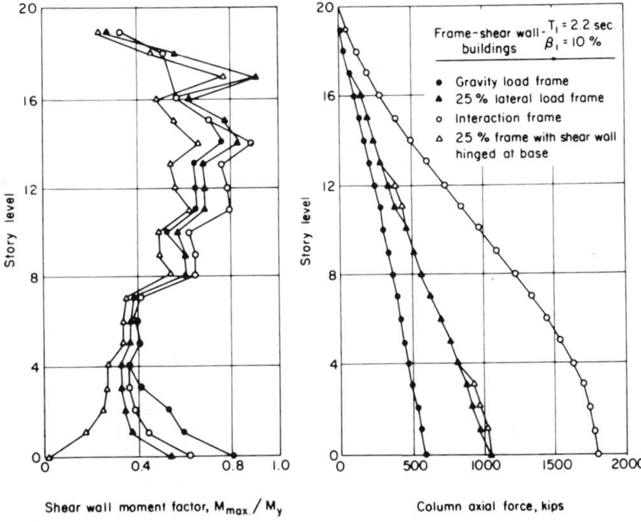

Fig. 12-32 Continued

addition to the standard 20-story building. The reference stiffness was adjusted in the 10-story building to give the same 2.2-sec fundamental period as the 20-story structure. This results in a structure that is very flexible for the 10-story building.

The results are shown in Fig. 12-34. The maximum lateral displacements, shown plotted against relative story height, are similar for both structures. The nearly linear variation of displacement with height in the 10-story building, a behavior also noted in the 20-story interaction-frame building (see Fig. 12-32), indicates the reduced influence of the shear wall on the building response.

The reduced influence of the shear wall in the 10-story building evident in Fig. 12-34a is also reflected in the corresponding girder ductility requirements, shown in Fig. 12-34b, which are significantly less than those for the 20-story structure with the same period. The column ductility requirements, as well as the overturning moments in the shear wall, shown in Figs. 12-34c and d, are essentially similar for both cases. As was noted in the case of open frames, the columns remain elastic except at the top few stories.

The above comparison indicates that, as in open frames, the number of stories by itself is a relatively unimportant factor in the response of frame-shear wall structures.

12.4.3 Effects of Other Variables on Dynamic Response

The study reported in Ref. 12-20, the principal results of which were presented in the preceding sections, represents the most extensive single study of the effects of the basic parameters considered in current design practice on the nonlinear response of building frames, using a fairly sophisticated analytical model. Among the variables not considered in the study which may have significant influence on the earthquake response of structures are: (a) effect of soil conditions, (b) effect of gravity loads, (c) three-dimensional action; torsion, (d) effect of irregular building configuration; setbacks, (e) frequency characteristics of ground motion, and (f) shape of load-deformation curve of the structure. A brief discussion of these factors will be given below.

-1. Effect of soil condiitons—The effect of ground conditions at the site on the earthquake response of structures involves two aspects: (1) the amplifying (or attenuating) effect of the local geology on the intensity as well as its filtering effect on the frequency characteristics of the transmitted seismic waves, and (2) the effect of soil properties in the immediate vicinity of a structure on the structural response. The latter aspect has been referred to as soil-structure interaction.

The effect of local geology on the amplification or diminution of transmitted seismic waves will vary with the properties of the soil such as stiffness, strength and layering characteristics, as well as the intensity and frequency characteristics of the input waves. Observations[12-33, 12-34] have generally indicated a tendency of the ground displacement and acceleration to be amplified as seismic waves pass from bedrock to softer surface layers. This observed amplification has been particularly pronounced in cases where deep alluvial deposits overlay bedrock. Measurements[12-35] of earthquake motion made in the basements of buildings situated at different depths below the ground surface and at different locations in the Tokyo area indicated an increase in intensity with decrease in depth below the surface. On the other hand, when weak soils are overstressed by seismic waves, so that energy absorption takes place, a diminution of intensity may be expected. Thus, while amplification may occur for input motions of low

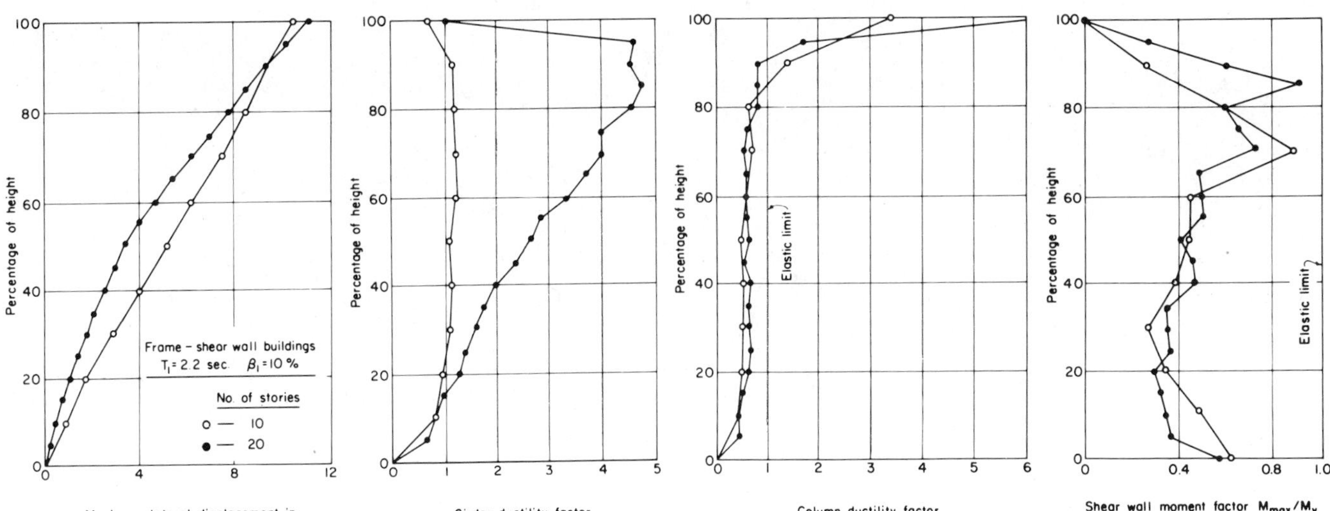

Fig. 12-34 Effect of building height.[12-20]

intensity in relatively soft soils, there may be a negligible amplification or even a diminution of the incoming motion for intense input motions. The amplitude of the resulting motion is also influenced by the frequency of the component waves of the incoming motion.

Studies by Seed and Idriss[12-36,12-37] point to the tendency of deep deposits of relatively soft material to transmit wave motions having predominantly long-period characteristics. Such surface motion would be expected to produce their maximum effect on long-period structures such as multistory buildings (with fundamental periods in the range of 1.5 to 3 sec) and smallest effect on low, stiff, short-period structure. On the other hand, shallow deposits of stiff soils tend to transmit waves having predominantly short period characteristics which would have their largest effect on short-period structures, with smaller effects on tall, multistory buildings.

Closely related to local geology is the probable epicentral distance. Records of earthquake motions at different sites indicate that short-period components tend to be filtered out faster as seismic waves spread out from the energy source, with the result that longer-period waves tend to predominate at greater distances from the epicenter. The effect of epicentral distance on the period or frequency characteristics of seismic waves is illustrated in Fig. 12-35, from Ref. 12-38. This figure shows smoothed undamped velocity response spectrum curves corresponding to the July 21, 1952 Kern County (Tehachapi) earthquake (Richter magnitude = 7.7) and the March 22, 1957 San Francisco earthquake (magnitude = 5.5). Curves A and B were obtained from records of the Tehachapi earthquake, curve A being based on a record taken at Taft, Calif., about 25 miles from the epicenter and curve B on a record obtained at the Hollywood Storage Co. building, some 70 miles from the epicenter. Curve A shows the relatively large response in the short-period range (indicating the dominant character of the component waves in this period range) for points close to the epicenter. The flattening of the spectrum in the short-period range for points at greater distances from the epicenter, as shown by curve B, clearly demonstrates the attenuation of the high-frequency components of ground motion with increasing distance from the epicenter. Curve C in Fig. 12-35 shows the velocity spectrum based on a record obtained about 10 miles from the epicenter of the relatively small 1957 San Francisco earthquake.

The above discussion suggests that for shocks originating from a particular region, the ground motion at a given site may be expected to exhibit dominant periods within a relatively narrow range. For instance, in a large part of Mexico City which is built on soft volcanic sediments to a depth of about 1000 ft in what was once Lake Texcoco in the Valley of Mexico, recorded motions originating from the Acapulco region some 260 km away have indicated dominant periods in the 1.8- to 3.0-sec range (see Fig. 12-38(d)). Most of the damage during the July 28, 1957 Mexico earthquake (magnitude = 7.5), for instance, occurred in tall, long-period buildings.[12-39] This and similar observations[12-40] indicate the desirability of avoiding resonance effects, whenever possible, by designing buildings in which none of the significant modes of vibration (usually the first few) have periods close to that of the underlying soil deposit.

The above observations should be borne in mind when assessing the probable effects of local geology on the expected character of the ground motion at a particular site. Whenever possible, use should be made of local measurements and observations of the behavior of existing structures under previous earthquakes to gain an indication of the effect of local geology on the expected earthquake motion.

The effect of the foundation medium in modifying the response of a structure, in relation to its behavior when founded on an essentially rigid base, has been the subject of a number of analytical studies.[12-41–12-43] These studies indicate that, for buildings founded on firm or moderately firm ground, the effect of soil-structure interaction is relatively small. This is particularly true for multistory buildings which are relatively flexible compared to the supporting medium. For such cases, the assumption of a structure fixed to a rigid base to which the input motion is applied is justifiable.

For stiff structures, particularly those resting on relatively soft ground, the effects of soil-structure interaction can be significant. No definite trends are, however, shown by the analytical studies mentioned. Thus, soil-structure interaction can produce an increase in one case and a decrease in another, depending upon the properties of the structure, the supporting soil, as well as the earthquake excitation. It is known that the coupling (or the lack of it) between a structure and its supporting soil generally results in a system which has a longer fundamental period than the same structure fixed to a rigid base. Also, the damping in the soil-structure system, which consists of internal damping within both the structure and the soil immediately adjacent to it and radiation damping resulting from the dissipation of energy through dispersion to the surrounding ground, will generally be greater than that for the structure alone. On the basis of the above observation, it has been suggested by Whitman[12-44] that from the design standpoint, the effect of soil-structure interaction on the stresses in a structure is always beneficial. This statement can best be understood by considering Fig. 12-36, which shows a sketch of response

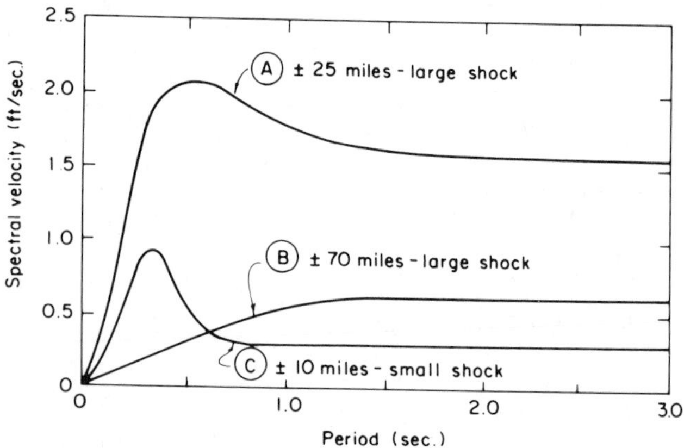

Fig. 12-35 Showing effect of epicentral distance on frequency characteristics of earthquake motion.[12-38]

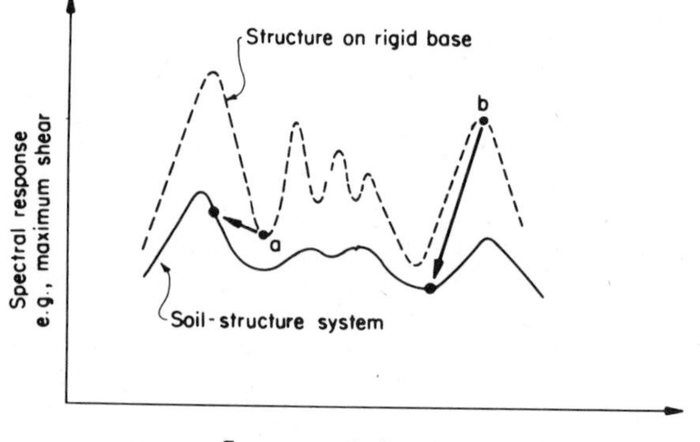

Fig. 12-36 Typical effect of soil-structure interaction.[12-44]

spectra corresponding to a single earthquake for a structure founded on a rigid base and also for a soil-structure system. Note that because of the greater damping in the soil-structure system, the corresponding curve everywhere lies below that associated with a rigid-based structure. If the structure has a period corresponding to a low point in the spectrum, such as point 'a' in Fig. 12-36, the result of considering soil-structure interaction may be a slight increase in the response. On the other hand, if the point on the spectrum corresponding to the period of the system lies near a peak, such as point 'b' in the figure, soil-structure interaction will tend to decrease the response. Since, in designing for earthquakes, it is necessary to allow for slightly different earthquake motions, the effect of which may be approximated by smoothed or averaged spectra, the shift in period indicated in Fig. 12-36 will have little effect on the maximum response. With such smoothed spectra, consideration of soil-structure interaction will always tend to produce a decrease in response.

It should be pointed out that because of the predominant influence of the rocking component of motion of the structure, soil-structure interaction may cause the absolute motion of the top of a structure to increase slightly. This will have to be considered in relation to connecting or nearby structures.

-2. Effect of gravity loads—A study by Anderson and Bertero[12-45] to determine the effect of gravity loads, among others, on the response of 10- and 20-story, single-bay, unbraced frames, showed that the inclusion of gravity loads, as well as the effects of the associated P-Δ moments* and flexure-axial load interaction in the columns, causes yielding to occur much earlier in the response so that the stiffness of the frame is reduced earlier and more energy is dissipated through inelastic deformation. For the base excitation considered in the study, an average of about 80% of the input energy was dissipated through inelastic action and damping, the former making up more than half of the dissipated energy. The increased inelastic deformation is reflected in a significant increase in the girder ductility requirements, compared to a case when gravity loads are absent. A substantial increase in the column ductility requirements, particularly in the lower stories, can also occur as a result of the action of gravity loads and the reduction in effective yield moment due to flexure-axial load interaction.** The presence of gravity loads does, however, have some beneficial results in that it tends to reduce the maximum lateral displacements as well as the whiplash effect in the top stories. A reduction in the maximum lateral displacement due to the P-Δ moments was also observed by Hanson and Fan[12-46] in a study of the earthquake response of braced multistory single-bay frames.

Figure 12-37, from Ref. 12-45, shows the results for a 20-story, single-bay frame designed according to the *Uniform Building Code* (1968 Edition) and subjected to the 8-sec portion of the 1940 El Centro (N-S component) record between 0.72 and 8.72 sec. An equivalent viscous damping equal to 5% of the critical for the fundamental mode, with lower values for the higher modes, was assumed. Figure 12-37a shows the maximum lateral displacements for identical frames with and without gravity loads. The reduction in the whiplash effect is apparent from Fig. 12-37b, which shows the variation of the story deflection index, Δ/h, i.e., the ratio of the relative story displacement to the story height, along the height of the frame. The ductility factor shown plotted in Fig. 12-37c is expressed in terms of curvature rather than rotation as in the Clough-Benuska study.[12-20]

It is pointed out that for frames in which the columns are relatively weaker than the girders the maximum lateral displacements may be increased by the presence of gravity loads. The girder ductility requirements, however, always tend to increase due to gravity loads.

The frames originally considered in the study reported in Ref. 12-45 were designed to satisfy strength requirements only, hence the large deflections under the code-specified shears shown in Fig. 12-37a. For these frames, the effect of the P-Δ moments varied from 4% to 8% of the overturning forces. When the frames were stiffened to satisfy the drift criterion (i.e., $\Delta/h \approx 0.0025$ under working loads), the P-Δ effect became negligible. The same conclusion was arrived at by Goel[12-47] in a study of the response of 10-

*The secondary moment produced by a vertical column load, P, acting through a relative lateral displacement, Δ, between the column ends.

**The flexure-axial load interaction curve assumed in this study[12-45] however, corresponds to steel members where the presence of an axial load always reduces the flexural capacity of a member. It is pointed out that the shape of the flexure-interaction curve for a typical reinforced concrete member differs from that for steel members in that for values of the axial load below the balance point, an increase in the axial load is accompanied by an increase in the ultimate moment capacity of a 'short-column' section.

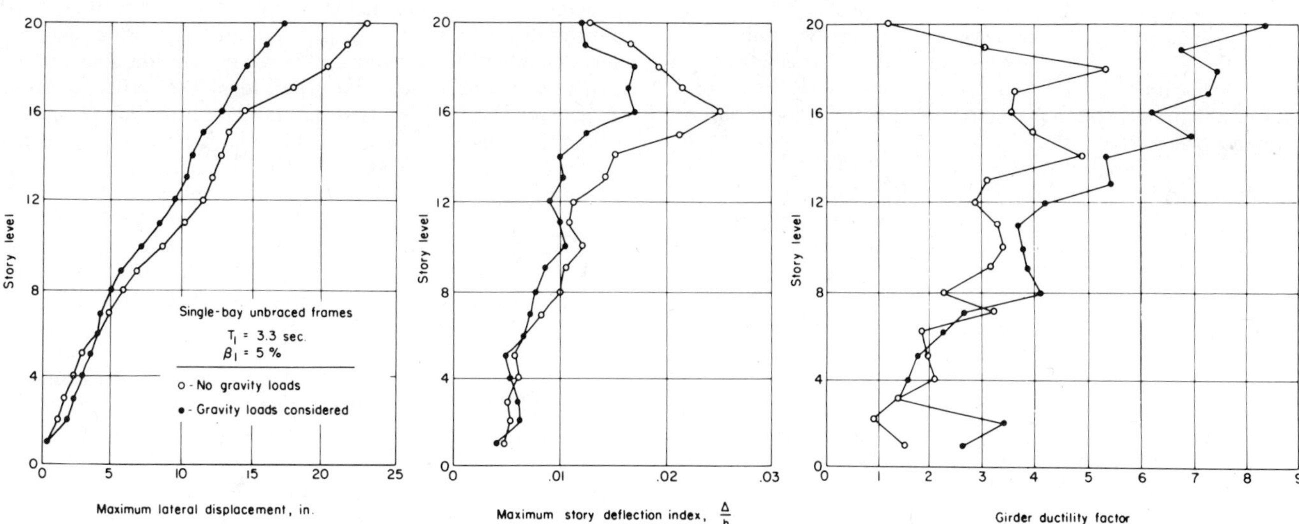

Fig. 12-37 Effect of gravity loads.[12-45]

and 25-story, single-bay symmetrical frames to the 1940 El Centro (N-S component and the 1952 Taft, Calif., (S 21°W component) earthquakes, with the acceleration ordinates magnified by 1.5 and 3.0, respectively. This study showed that although the maximum lateral displacements were increased by about 10% by the P-Δ effect when the response was elastic, the effect became a negligible 1% when the girders were assumed to respond inelastically, the columns remaining elastic throughout the response.

In a study of the effect of gravity loads on the collapse of simple (SDF) bilinear systems, Husid[12-30] noted that gravity loads significantly increased the permanent set which a structure suffered when subjected to strong ground motion.

-3. Effect of three-dimensional action; torsion—The analysis of a structure by considering separate sets of planar frames lying in mutually perpendicular planes and subjected to horizontal components of base motion parallel to their respective planes assumes no interaction between the forces acting on members common to both sets of frames and neglects torsional effects. In addition, a two-dimensional analysis can only allow for the stiffness contribution of elements lying normal to the plane considered in an approximate manner. Such an approach is consistent with present design practice and should yield reasonable results for most cases, particularly if a predominantly linear response can be assumed and no appreciable torsional effects are present. The *SEAOC-UBCode* requires that a structure be designed to withstand the specified lateral forces considered as acting nonconcurrently in the direction of each of the main axes of the structure.

When considering inelastic response, however, the interaction of forces resulting from components of motion parallel to each of the principal planes in a structure may cause early yielding in some members and modify the response of the structure significantly. The determination of the interaction effects in such a case will require a three-dimensional analysis of the response of the entire structure. A study by Nigam and Housner[12-48] of the earthquake response of a simple, single story, three-dimensional frame model showed that the interaction of moments along two mutually perpendicular directions resulted in early yielding in the columns with a consequent reduction in the input energy and the response velocity. The interaction also tended to produce greater drift of the line of zero oscillation, this being a measure of the permanent set with respect to lateral displacement.

The study of dynamic torsional effects in buildings, particularly in multistory structures where this effect is more pronounced, has been possible only with the recent development of programs for the dynamic analysis of three-dimensional frame structures.[12-49-54] Torsion occurs when the center of mass does not coincide with the center of twist in a story level. This can be the result of a lack of symmetry in the building plan or from the random disposition of live loads in an otherwise symmetrical structure. Torsion can also be induced in symmetrical structures by the rotational component of ground motion.

SEAOC-UBC requires the consideration of a minimum torsional moment equal to the story shear acting with an eccentricity of five percent of the maximum building dimension at that level in buildings which depend on the diaphragm action of the floor slabs to distribute the horizontal shears to the various vertical resisting elements. A study by Newmark[12-55] of the elastic torsion in symmetrical buildings, based on an approximate analysis of torsional ground displacements corresponding to different building plan dimensions and shear wave velocities,* indicate that buildings with short periods of torsional vibration should be designed for larger values of "accidental" eccentricity than buildings with longer periods. Newmark further suggests that an accidental design eccentricity of 10% of the longer plan dimension would appear reasonable for frame buildings with periods shorter than about 0.6 sec or shear wall buildings with periods shorter than about 1.0 sec.

The magnitude of the torsional shear induced in the exterior elements of a structure will obviously depend on the disposition of the various resisting elements in plan. Clearly the most efficient arrangement insofar as torsion is concerned is the fully symmetrical building with as much of the lateral load-resisting elements located along the periphery. This, of course, assumes that the floor slabs provide an effective means of transmitting the horizontal forces to the major resisting elements and make these act as a unit. A study by Koh, et. al.,[12-56] of the elastic dynamic response of four actual buildings in Japan ranging from 15 to 40 stories indicated that torsional effects arising from an eccentricity of from 5% to 10% of the longer plan dimension of a building, may produce increases in the shears acting on the exterior vertical elements of up to 20% over that which would occur if torsion were not present.

In buildings with L-, Y-, U-, H-, or T-shaped plans which are built integrally as a unit, large forces may be developed at the junction of the 'arms' of such buildings as a result of vibrational components directed normal to the axes of the arms. In addition, there will be horizontal torsional effects on each arm arising from the differential lateral displacement of the two ends of each arm.

It is pointed out that yielding in a corner column or end shear wall in a building due to torsional stresses tends to destroy the (stiffness) symmetry in an originally symmetrical building or increase the eccentricity in an unsymmetrical building, as the center of resistance moves away from the yielded member. The increase in eccentricity causes yielding to develop further. This tendency toward magnification of torsional effects by yielding in corner or end elements suggests that such elements should be designed more conservatively than other members where torsional vibrations can be significant.

-4. Effects of setbacks in buildings—Setbacks or appreciable changes in the overall plan size or dimensions in elevation in buildings represent major discontinuities in geometry, mass, and stiffness and, as such, give rise to force concentrations at and in the immediate vicinity of the discontinuity. The increase in the maximum dynamic shears at the base of a setback structure** can be considerable, particularly when the fundamental period of the setback is close to the period of the first lateral vibration mode of the supporting base structure. This resonance effect occurs because the motion transmitted to the base of the setback structure is an essentially forced harmonic vibration which is influenced to a large extent by the dynamic properties of the base structure. In many instances, the magnitude of the acceleration in multistory structures at levels above the ground is significantly greater than that at the base.

A study by Jhaveri[12-57] of the elastic response of shear beam models of 15-story buildings, with symmetrically disposed setbacks having different heights and masses relative to the base structure, showed that the ratio of the maximum shear at the base of the setback to the maximum

*Velocities of propagation in the supporting medium.

**i.e., the upper portion of a structure where a change—generally a decrease—in the lateral dimension(s) occurs.

shear at the same height in a uniform building (without a setback) tends to increase with decreasing height and mass of the setback.

In addition to the increase in shear at the base of the setback structure, the same study showed that an increase in the maximum shear at the base of the building over that which would occur in a uniform building of the same total height can result from the presence of a setback.

It is clearly desirable to avoid major setbacks in multistory buildings in seismic areas, particularly unsymmetrical setbacks where torsional forces can further aggravate an unfavorable situation.

-5. *Effect of frequency characteristics of earthquake motion*—The Clough-Benuska study[12-20] considered the effect of two of the three major characteristics of the earthquake input motion, namely, intensity (i.e., acceleration amplitude) and duration of large amplitude pulses. The frequency characteristics of the motion, which generally vary with each earthquake, may have a significant effect on the response of a structure, depending upon the significant periods of vibration of the latter.

The frequency characteristics of a particular earthquake motion is most conveniently studied by means of a Fourier spectrum, which is essentially a frequency decomposition of the complex ground motion and shows directly the significant frequency components. As mentioned earlier, when the maximum response of a system occurs at the end of an earthquake, the undamped relative velocity response spectrum has a form identical to the Fourier amplitude spectrum of the ground acceleration. When the maximum response of a SDF system does not occur at the end of the earthquake, the plots of the two quantities are only roughly similar. The close relationship between the Fourier amplitude spectrum of the ground acceleration and the undamped velocity spectrum, together with the more common application of the latter in response studies, has led to the use of the velocity spectrum to give an indication of the frequency characteristics of earthquake accelerograms.

Figures 12-38a, b and c, from Ref. 12-25, shows the acceleration traces of three earthquakes having distinct frequency characteristics. The amplitudes of the 1952 Taft (S 21°W) and the 1962 Alameda Park, Mexico (N 10°46′W) accelerograms shown have been multiplied by factors of 3.0 and 2.4, respectively, to give them about the same spectral intensity* as the 1940 El Centro (N-S) accelerogram magnified by a factor of 1.5. The latter magnified value of the 1940 El Centro record, incidentally, is usually taken as representing a major earthquake. The corresponding undamped velocity response spectra are shown in Fig. 12-38d.

Figure 12-38d shows that the 1952 Taft (S 21° W) accelerogram has its dominant frequencies in the short period range (0.5–1.0 sec) while the 1962 Alameda Park record has its peaks in the longer period range (2–3 sec). The 1940 El Centro record, on the other hand, has a relatively uniform distribution of peaks over the period range of interest (i.e., 0.5 to 3.0 sec) compared to the other two records. This partly explains the description of the 1940 El Centro as being the most severe earthquake then (up to 1970) recorded. It is readily seen that a strong ground motion having the frequency characteristics exhibited by the 1940 El Centro is more likely to excite a greater number of the

*Spectral intensity is defined as the area under the undamped velocity response spectrum between ordinates including period values of practical interest. In this study[12-25], the period range used was from 0.5 to 3.0 sec.

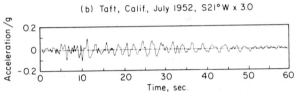

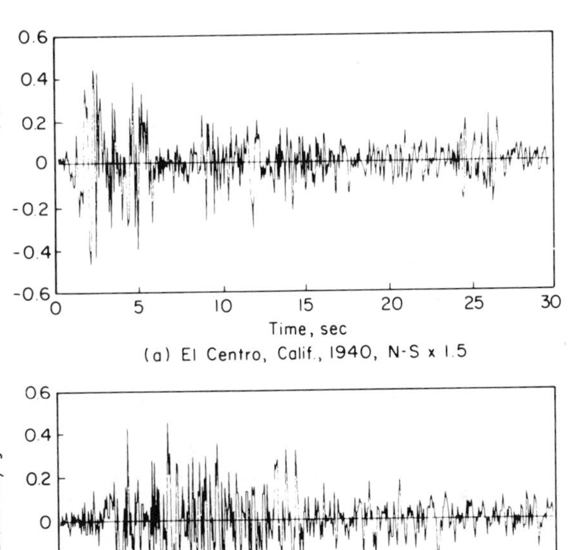

(a) El Centro, Calif., 1940, N-S x 1.5

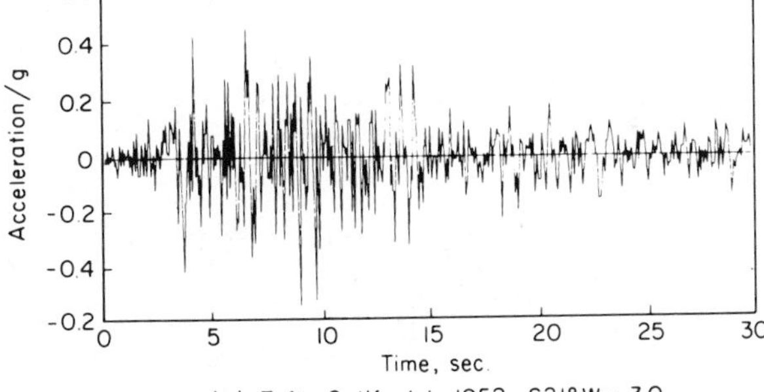

(b) Taft, Calif., July 1952, S21°W x 3.0

Acceleration/g
0.2
0
-0.2
0 10 20 30 40 50 60
Time, sec

(c) Alameda Park, Mexico, May 11, 1962, N10°46′W x 2.4

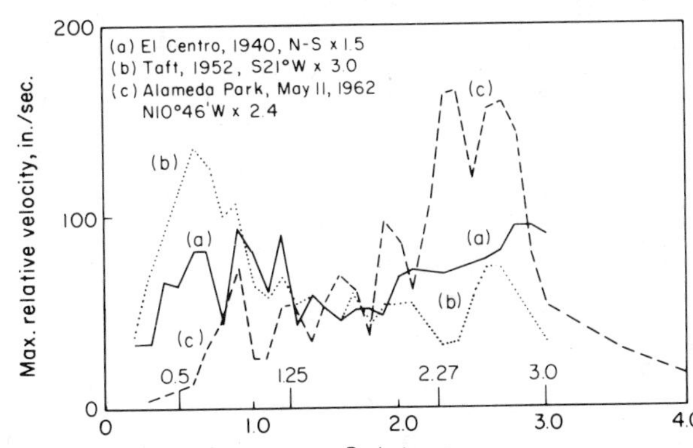

(d) Undamped relative velocity spectra

Fig. 12-38 Frequency characteristics of three sample earthquake accelerograms.

vibration modes which contribute significantly to the response of a structure than an earthquake in which the dominant frequencies are restricted to a narrow range. Note that the Alameda Park record has a 60-sec duration compared to the 30-sec duration of the other two records shown in Fig. 12-38a-c.

As mentioned in section 12.4.1.2, the trend of decreasing column and girder ductility requirements with increasing period of vibration observed for the 1940 El Centro earthquake is reversed in the case of the Alameda Park earthquake. This is shown in Fig. 12-39, from a study by Goel and Berg[12-25] of the response of 10-story, single-bay, symmetrical shear frames to the three earthquake motions shown in Fig. 12-38. The columns in this study were assumed to behave elastically throughout the response, with

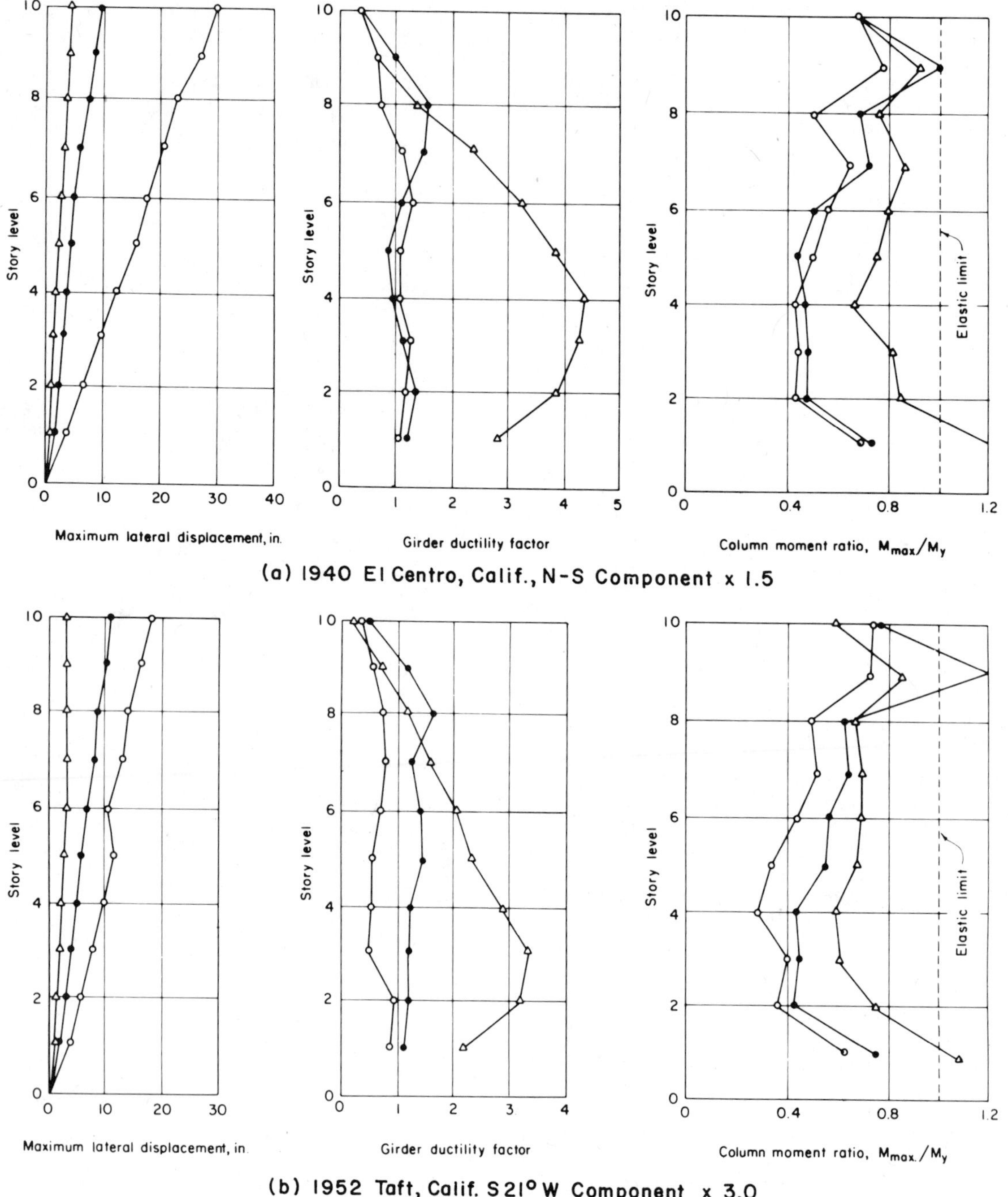

Fig. 12-39 Effect of frequency characteristics of earthquake motion.[12-25]

inelastic action allowed only in the girders. The whiplash effect, which is apparent in the column maximum-to-yield moment ratio curves for the El Centro and Taft earthquakes, is practically absent in the Alameda Park response. Since the whiplash effect reflects the contribution of the higher modes of vibration and none of the structures considered have higher modes with periods in the 2–3 sec range where the Alameda Park velocity spectrum has its dominant peaks, the results shown in Fig. 12-39 would be expected. The relatively lesser ductility requirements associated with the Alameda Park earthquake, even for the 2.27-sec structure, indicates that the higher modes of vibration also play a significant role in determining the magnitude of member ductility requirements.

It is significant to note that in spite of the smaller amplitude of the acceleration pulses in the Alameda Park earthquake, the maximum lateral displacement of the 2.27-sec structure is greater for this earthquake than for either El Centro or Taft earthquake. This clearly indicates the major influence of the fundamental mode on the maximum lateral displacement—since for this structure the period of the fundamental mode falls within the range of the dominant frequency components of the Alameda Park earthquake. The same trend of increasing maximum displacements with increasing period will be noted for all three earthquakes.

For design purposes, the variability in frequency characteristics of earthquake motions may be provided for by using either average response spectra or, where a time-

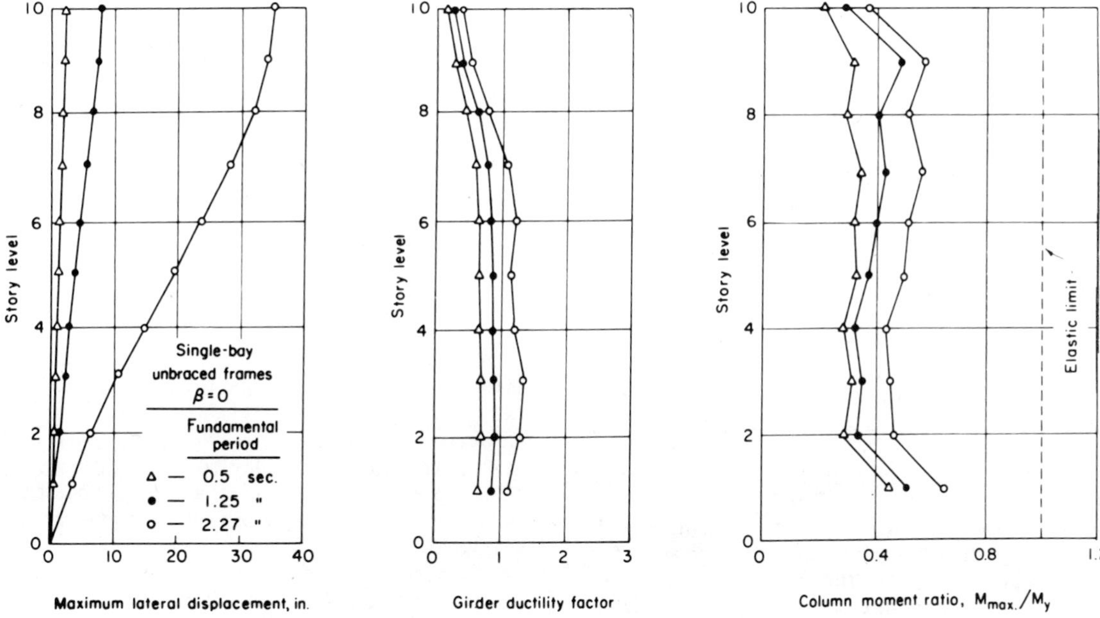

(c) 1962 Alameda Park, Mexico, N10°-46'W Component x 2.4

Fig. 12-39 *(Cont.)*

history response is desirable, a set of input accelerograms with frequency characteristics covering the range of expected variation.

12.5 SOME RESULTS OF INVESTIGATIONS ON THE BEHAVIOR OF REINFORCED CONCRETE MEMBERS AS RELATED TO EARTHQUAKE LOADING

Summarized below are some of the more significant results of analytical and experimental studies of the behavior of reinforced concrete members under load, particularly as these relate to earthquake-resistant design requirements.

12.5.1 Effect of Rate of Loading

The relatively greater rate of loading associated with earthquakes, as compared to static loading, results in a slight increase in the strength of reinforced concrete members, due primarily to the increase in the yield strength of the reinforcement. The calculation of both strength and deformation capacity (ductility) of reinforced concrete members for earthquake loads on the basis of material properties obtained by static tests is thus reasonable and conservative.

12.5.2 Providing Against Diagonal Tension and Other Non-ductile Types of Failures

In buildings subjected to earthquake motions, both beams and columns are loaded in transverse shear in addition to bending, axial load (generally significant only in columns) and possibly torsion. Because of the nonductile and relatively abrupt nature of a failure due to shear (i.e., diagonal tension) or loss of anchorage of reinforcement, members should be designed and detailed so that their strength and ductility is controlled by flexure rather than these non-ductile types of failure mechanisms. Code design requirements are framed with the intent of providing members with an inclined cracking load which is greater than their flexural capacity. Provision is in turn made to ensure that flexural failure be of the ductile type* by imposing limits on the maximum amount of longitudinal tensile reinforcement which may be used so that failure is preceded by yielding of such reinforcement. Codes also specifiy a minimum amount of flexural reinforcement. This minimum reinforcement, given in terms of $\rho = 200/f_y^{**}$ in ACI 318-71, in intended to provide a tensile capacity at least equal to the tensile force carried by the concrete in the tension zone of an unreinforced beam and so prevent a sudden brittle type flexural failure in lightly loaded members when tension cracking occurs.

Reinforced concrete being a composite material, the successful behavior of structures made up of such a material depends heavily on the effective interaction of the principal elements comprising it. Effective interaction between the compression-resisting concrete and the tension-carrying steel requires that adequate anchorage be provided in order to develop the strength of the reinforcement. The relatively abrupt character associated with bond or anchorage failures requires careful attention in design to avoid the occurrence of this mode of failure before the flexural capacity of a member is reached.

Tests have indicated that the inclined cracking load of a beam without shear reinforcement is influenced principally by the tensile strength of the concrete (usually expressed in terms of the compressive strength f_c'), the amount of longitudinal reinforcement and the ratio of the applied shear to moment acting at the critical section.[12-58, 12-59] The presence of a compressive load increases the apparent tensile strength of the concrete and hence increases the inclined cracking load. Once inclined tensile cracking starts, the load on a beam is transferred across a crack by a combination of shear on the compression zone, aggregate interlock along the crack and dowel action of the longitudinal reinforcement. At this stage, the integrity of a beam and its

*As is well known, the compression zone failure associated with an over-reinforced beam is a relatively sudden, nonductile type of failure.

**Where $\rho = A_s/bd$, is the ratio of the cross-sectional area of the reinforcement to the area of the member section.

capacity to carry increasing amounts of load has to be maintained through the use of shear reinforcement.

12.5.3 Confinement Reinforcement

Apart from the transverse reinforcement required by shear (i.e., diagonal tension), tests have indicated the need to provide sufficient transverse confinement reinforcement at critical regions of a structure where inelastic action might occur. This is particularly true for structures which may be subjected to moderate-to-strong earthquakes, in which the loading at critical regions can take the form of repeated, reversed cycles of deformation well into the inelastic range. Locations where plastic hinges can occur, such as the ends of frame members, require confinement reinforcement to insure adequate deformation capacity in such members. The connections or joints between such frame members also require confinement in order that the plastic moment capacity of the members can be developed.

The use of sufficient confinement reinforcement in such regions ensures their ductile performance by increasing the strength and strain capacity of the confined concrete in compression and by reducing the disruptive effect of large amplitudes of reversed loading on the concrete core. In beam-column connections, such confinement assures the maintenance of the vertical load carrying capacity of the joint as well as adequate anchorage of any longitudinal beam steel embedded within the core. The increase in compressive strength of the confined concrete also helps to compensate for the strength loss corresponding to the spalled outer shell.

Appendix A of ACI 318-71, for instance, provides for separate and independent calculations to determine transverse reinforcement requirements for shear and for confinement. However, only the larger of the two amounts required need be supplied, i.e., the transverse reinforcement indicated for confinement may be assumed to serve also as shear reinforcement and vice versa.

Shear reinforcement lying in planes transverse to the longitudinal axis of a member would be the most efficient since they function equally well for shear in either direction and can also serve as hoops to confine the concrete and as ties to restrain the longitudinal reinforcement. This may take the form of spirals (circular or rectangular) or individual stirrups. Tests of reinforced concrete beams under cyclic loading have clearly demonstrated the considerable energy-absorbing capacity of members which are well provided with such transverse reinforcement.

12.5.4 Effects of Different Variables on the Flexural Ductility of Reinforced Concrete Members

In view of the design philosophy presently used in relation to multistory buildings (see section 12.6.2), which relies heavily on the inelastic action of members (primarily beams) to absorb the vibrational energy resulting from strong ground motion and hence reduce the forces to which a structure would otherwise be subjected to, the available ductility in structural members and joints assumes the utmost importance. The emphasis on ductility distinguishes the design for resistance to strong earthquake motion from the conventional design for essentially static loads.

It should be recognized that the ductility of a member or structure is a function not only of the material used but also of such factors as temperature, type and rate of loading, as well as the configuration of the system. Thus, members tend to exhibit brittle behavior at low temperatures and rapid rates of loading or when they are subjected to a triaxial state of tensile stresses. The particular range of temperature and rate of loading when the structural behavior changes from essentially ductile to brittle will depend on the material used. In earthquake-resistant reinforced concrete design, the major attention insofar as ensuring ductile performance is concerned has been focused on the proper choice of materials (strength of concrete and grade of reinforcement) and the adequate design and detailing of structural members and their connections, these being the most conveniently controlled factors affecting ductility. In order to provide a basis for the intelligent design of reinforced concrete members for earthquake resistance, a discussion of the effects of the different parameters on the sectional ductility of members is given below.

The major variables affecting sectional ductility may be classified under three groups: (a) material variables, specifically, the stress-strain curve of concrete, particularly as this is affected by lateral confinement, and grade of reinforcement; (b) geometric variables, such as the amount of tension and compression reinforcement and the shape of the section; and (c) loading variables, mainly, the presence or absence of axial load.

The most commonly used measure of ductility, μ, of structural members is that relating to their flexural action and is defined as the ratio of the ultimate curvature, ϕ_u, attainable without significant loss of strength, to the curvature corresponding to first yield in the tension reinforcement, ϕ_y, i.e.,

$$\text{ductility}, \mu = \frac{\phi_u}{\phi_y}$$

Figure 12-40, based on the assumption of a linear distribution of strain across the section of a typical member and for the usual case where the deformation capacity of the member is governed by the useful limit of strain for concrete, ϵ_{cu}, shows that

$$\phi_u = \frac{\epsilon_{cu}}{k_u d} \qquad (12\text{-}19)$$

where $k_u d$ is the distance from the extreme compressive fiber to the neutral axis or the depth of the compressive block when $k_u \leqslant 1.0$. Equation (12-19) indicates that factors which tend to increase ϵ_{cu} or decrease $k_u d$ contribute toward increasing the ductility of a section. A major variable affecting the value of ϵ_{cu} is the degree of lateral confinement of the critical region. Lateral confinement, whether from active forces such as transverse compressive loads, or passive restraints from other framing members or lateral reinforcement, tends to increase ϵ_{cu}. Tests have also indicated that ϵ_{cu} increases as the depth to the neutral axis decreases or as the strain gradient across the critical section increases[12-60, 12-61] and as the moment gradient along the span of a member increases*[12-62, 12-63] or as the shear span

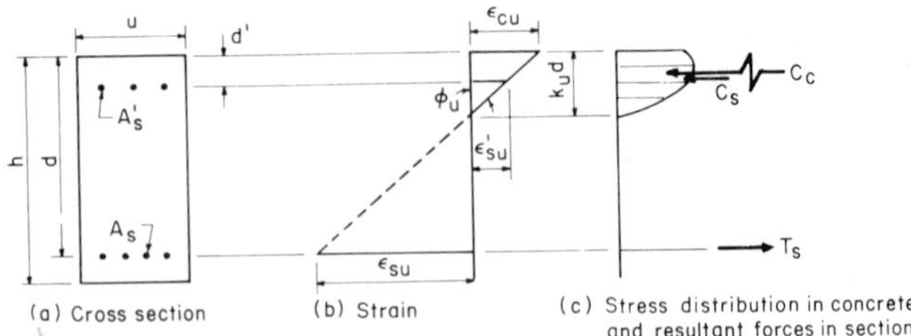

Fig. 12-40 Distribution of strain and stress in a typical section at ultimate capacity.

*For a given maximum moment, the moment gradient increases as the distance from the point of zero moment to the critical section decreases.

decreases. The value of ϵ_{cu} can range from about 0.0025 for unconfined concrete to about 0.01 for confined concrete subjected to predominantly axial (concentric) load. Under eccentric loading, values of ϵ_{cu} for confined concrete of 0.04 and more have been observed.[12-64, 12-65]

The presence of compressive reinforcement and the use of concrete with a high compressive strength, as well as the use of relatively wide flange sections tend to reduce the required depth of the compressive block, $k_u d$, for a given combination of bending moment and axial load and hence tend to increase the ultimate curvature, ϕ_u. On the other hand, compressive axial loads and large amounts of tensile reinforcement, especially tensile reinforcement with a high yield stress, tend to increase the required $k_u d$ and thus decrease the ultimate curvature, ϕ_u. These factors will be discussed in more detail below.

-1. Confinement of concrete; maximum usable compressive strain in concrete—Generally, the lateral confinement of concrete increases its compressive strength and deformation capacity in the longitudinal direction, whether such longitudinal stress represents a purely axial load or the compressive component of a bending couple.

In actual members where the confinement commonly takes the form of lateral ties or spiral reinforcement covered by a thin shell of concrete, the confining effect of the reinforcement is not mobilized until the concrete undergoes sufficient lateral deformation. At this stage, the outer shell of concrete usually has reached its useful load limit and starts to spall. Because of this, the net increase in strength of the section due to the confined core may not amount to much as a result of the loss in capacity of the spalled outer shell. In many cases, the total strength of the confined core may be slightly less than that of the original section. The increase in ductility due to confining reinforcement, however, is significant.

The confining action of rectangular hoops mainly involves reactive forces at the corners with only minor restraint provided along the straight unsupported sides. Because of this, such hoops are generally not as effective as circular spiral reinforcement in confining the concrete core of members subjected to compressive loads. Tests[12-64] have shown that circular spiral reinforcement can be twice as effective as rectangular hoops as confinement reinforcement. Square spirals, because of their continuity, are slightly better than separate rectangular hoops. The use of rectangular hoops of the proper amount and spacing, however, still produces a significant increase in the ductility of the confined core. Appendix A of ACI 318-71 specifies a minimum size of rectangular hoop which is designed to provide the same average compressive stress in the rectangular core as would exist in the core of an equivalent circular spiral compression member having the same core area. In addition, most codes specify the use of supplementary crossties, engaging the peripheral hoop with a standard semicircular hook and secured to a longitudinal bar, if necessary to maintain a certain minimum unsupported length of the peripheral hoop.

For a particular type of transverse reinforcement, the degree of confinement is influenced by the amount (usually expressed in terms of a volumetric ratio relative to the confined core) as well as the spacing and grade of reinforcement. Because the confining effect of hoops results from an arching action of the concrete between the peripheral reinforcement, the spacing of the transverse reinforcement has a very significant effect on the efficiency of the confinement. Appendix A of ACI 318-71 specifies a minimum hoop spacing of 4 in. Studies[12-64, 12-65] have shown that the confining effect of transverse reinforcement becomes negligible for spacing greater than the least dimension of the confined core. The presence of longitudinal compressive reinforcement also improves the confining effect of transverse reinforcement.

An idea of the effect of confinement on the strength and ductility of concrete can best be obtained from the stress-strain relationship of short concrete specimens. Typical curves corresponding to unconfined and confined concrete specimens are shown in Fig. 12-41. The curve for laterally confined concrete is marked by a slightly greater maximum stress and a significantly greater maximum strain relative to the unconfined specimen. The effect of confinement becomes apparent only after the longitudinal strain in the concrete approaches 80–90% of the strain corresponding to the maximum stress, and so the initial portion of the curves for both confined and unconfined concrete is essentially the same. The slope of the descending branch of the curve for confined concrete tends to decrease (i.e., flatten out) with increasing amounts of transverse reinforcement. The curves shown in Fig. 12-41 are for concentrically loaded specimens. Studies[12-60] indicate that for eccentrically loaded specimens, a slight increase in both the maximum stress and the associated strain, relative to the curve for concentrically loaded specimens, may be expected due to the presence of a strain gradient across the specimen section.

The character of the stress-strain curve for the concrete in compression is important in designing for ductility because in most instances, the available sectional ductility of a member is controlled by the concrete. However, other factors also influence the ductility of a section; factors which may augment or diminish the effects of confinement on the ductility of concrete. Note the distinction between the ductility of concrete as affected by confinement and the ductility of a section (i.e., sectional ductility) in a member as influenced by the ductility of the concrete as well as other factors.

Figure 12-42 shows ultimate load interaction curves for a section subjected to combined bending and axial load, assuming both confined and unconfined conditions. Such curves give the combinations of bending moment, M, and axial load, P, which a given section can carry, corresponding to a particular value of the maximum compressive strain in the concrete. A point on an interaction curve is obtained by calculating the forces M and P associated with an assumed linear strain distribution across the section considered, account being taken of the appropriate stress-strain rela-

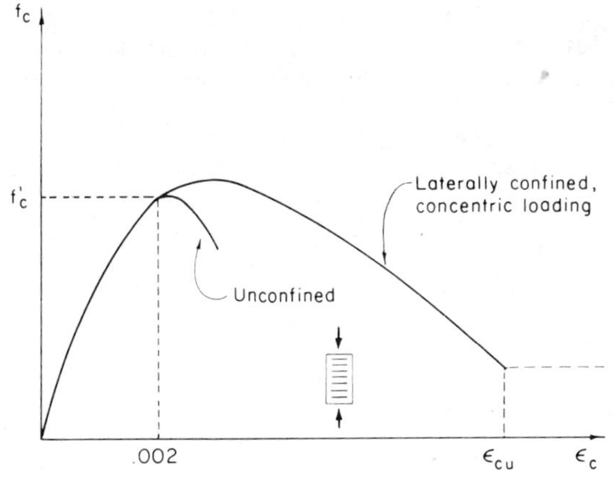

Fig. 12-41 Typical stress-strain curve for concrete.

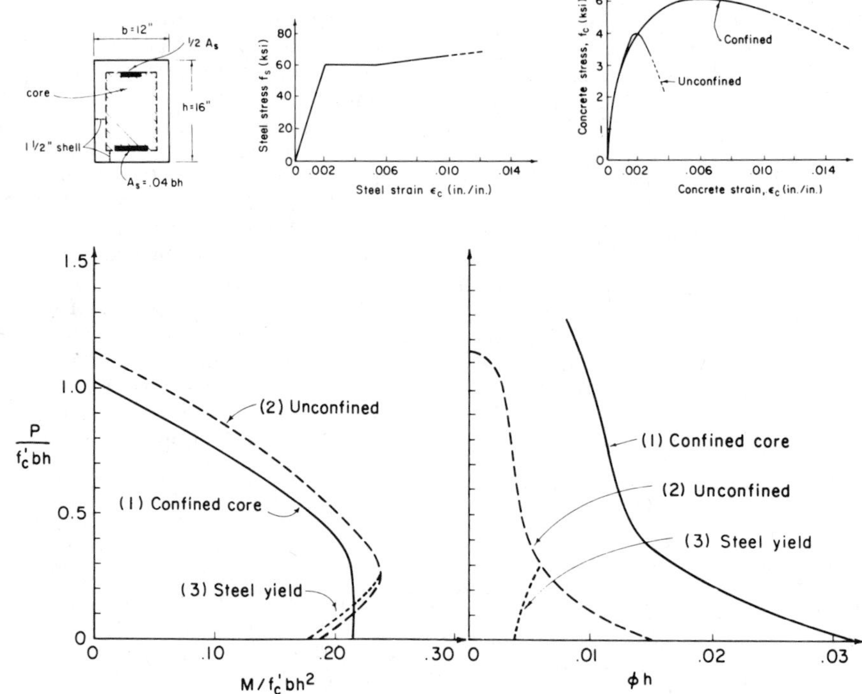

Fig. 12-42 Interaction (*P-M*) and *P-φ* curves for a rectangular section assuming both confined and unconfined cores.

tionship for concrete and steel. For an ultimate load curve, the concrete strain, ϵ_c, is assumed to be at the useful limit, ϵ_{cu}, while the strain in the tensile reinforcement, ϵ_s, varies. The balanced point in the interaction curve shown in Fig. 12-42a corresponds to the condition in which the tensile reinforcement is stressed to its yield point at the same time that the extreme concrete fiber reaches its useful limit of compressive strain. Points on the interaction curve above the balance point represent conditions in which the strain in the tensile reinforcement is less than its yield strain, ϵ_y, while for those points on the curve below the balance point, $\epsilon_s > \epsilon_y$.

The stress-strain curves assumed for the confined concrete core as well as the unconfined concrete are shown in Fig. 12-42. The maximum compressive strain in the concrete has been taken as 0.003 for the unconfined* case and 0.01 for the confined core. The slightly lesser ultimate capacity indicated by curve (1) in Fig. 12-42a for confined concrete relative to curve (2) for unconfined concrete is due to the reduced depth of the section after the outer shell has failed.

Figure 12-42b shows the variation of the curvature, ϕ (in units of $1/h$) with the axial compressive load, P. The relatively greater ultimate curvature (being a measure of the ductility of the section) associated with values of P less than that corresponding to the balanced point, for both unconfined and confined conditions should be noted. Figure 12-42b also clearly shows the significant increase in the ultimate curvature due to confinement.

The dashed curves (3) in Fig. 12-42 corresponds to first yield in the tension reinforcement, i.e., $\epsilon_s = \epsilon_y$, $\epsilon_c \leqslant \epsilon_{cu}$, and represents the limit of elastic deformation. The difference in horizontal ordinates to curves (2) and (3) in Fig. 12-42b thus is a measure of the amount of inelastic bending deformation in the unconfined section.

In discussing the effects of the other design parameters on the flexural ductility of a section, use will be made of the results of an extensive analytical study by Cohn and Ghosh[12-66] of the flexural ductility of reinforced concrete sections. This study involved, among others, the determination of the effects of the different material, geometric and loading variables on the sectional ductility ratio ϕ_u/ϕ_y. The basic procedure used in the study consisted essentially of considering the equilibrium of forces across a typical section, such as is shown in Fig. 12-40, corresponding to different strain distributions or locations of the neutral axis, using the appropriate stress-strain relationships for concrete and steel. Typical material stress-strain relationships used in the study are shown in Fig. 12-43, the concrete being assumed as unconfined.

In contrsat to the common practice of assuming a constant value of ϵ_{cu} in such analyses, however, the Cohn-Ghosh study[12-66] defines the ultimate concrete strain, ϵ_{cu}, as the strain corresponding to the maximum load or moment carrying capacity of a section. The ultimate curvature, ϕ_u, is that associated with the strain ϵ_{cu} at the maximum load (P_u or M_u). This definition allows different values of ϵ_{cu}, depending upon the section properties and the loading and appears to be a more satisfactory approach since the ultimate strain is directly related to the load carrying capacity of the section.

-2. Effect of maximum concrete stress, f_c'—Figure 12-44, from Ref. (12-66), shows that, for a particular section and reinforcement ratio, ρ, an increase in the concrete strength, f_c', increases the ductility of a section. As pointed out earlier, this is due primarily to the reduction in the required depth of the concrete compression zone, for a given moment acting on the section.

-3. Yield point of tensile reinforcement, f_y—A comparison of the curves corresponding to the different values of f_y in Fig. 12-44 shows that an increase in the yield point of the tensile reinforcement decreases the ductility of a section. This can be explained in terms of the greater depth of the compressive block required to balance the tensile force

*Concrete with only nominal transverse reinforcement.

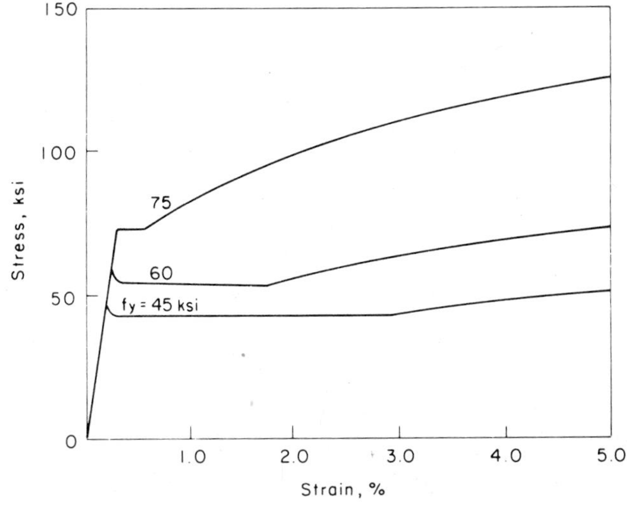

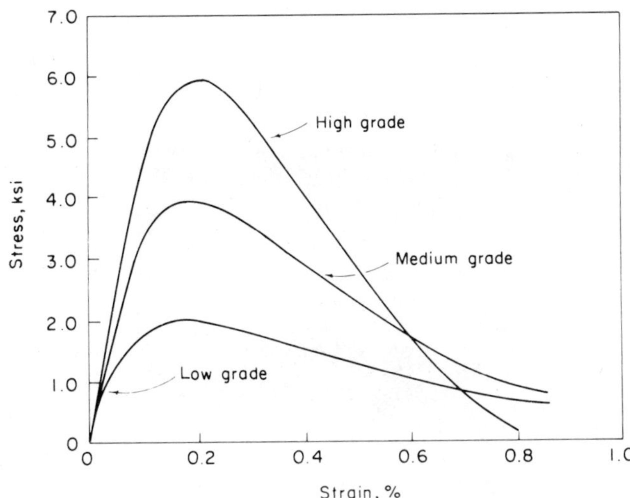

Fig. 12-43 Material stress-strain relationships used in study reported in Ref. 12-66.

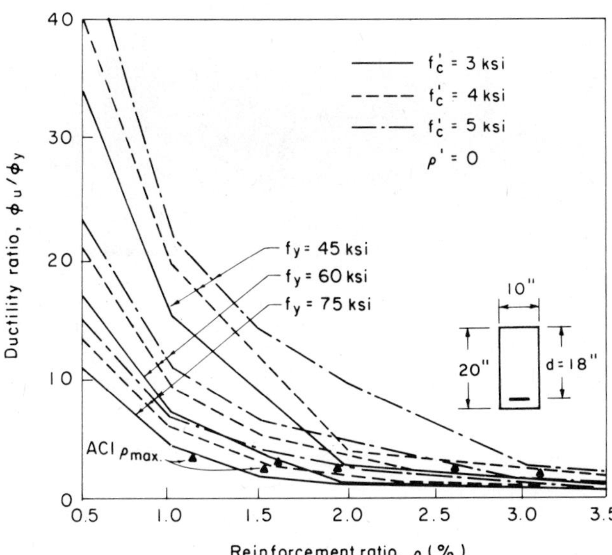

Fig. 12-44 Effect of concrete strength f'_c, steel yield strength, f_y, and reinforcement ration, ρ.[12-86]

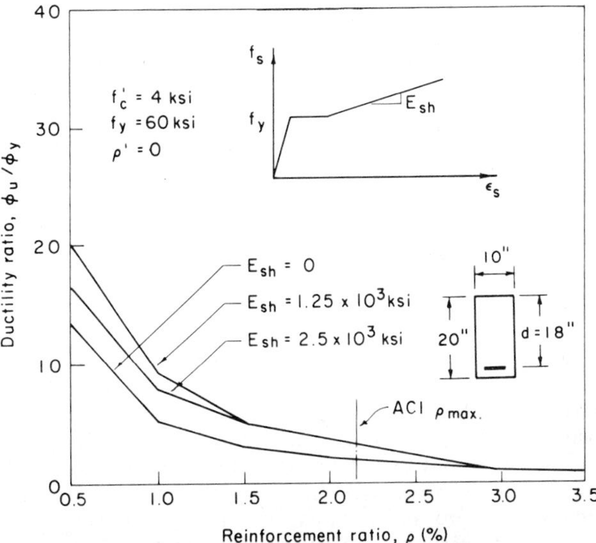

Fig. 12-45 Effect of strain-hardening in tension reinforcement.[12-66]

associated with the yield strain in the tension reinforcement (which does not vary significantly with increasing f_y), relative to a similar section having tensile reinforcement with lower f_y.

-4. Effect of strain-hardening in tension reinforcement—Figure 12-45, also from Ref. (12-66), shows the results of varying the strain-hardening modulus, E_{sh} (i.e., the slope of the σ-ϵ curve in the strain-hardening range) on the ductility ratio. It is seen that strain-hardening improves the ductility of lightly reinforced sections but has a negligible effect on heavily reinforced sections. Note that the ductility ratio increases as E_{sh} is increased from zero (elasto-plastic case) to 1.25×10^3 ksi but decreases with a further increase in E_{sh} to 2.5×10^3, indicating an optimal value of E_{sh} which maximizes the sectional ductility.

-5. Amount of longitudinal tensile reinforcement, A_s—An increase in the amount of tensile reinforcement increases the strength but decreases the ductility of a section. The latter effect is shown in Fig. 12-44, which indicates decreasing ductility with increasing values of the reinforcement ratio $\rho = A_s/bd$. The very limited ductility available in sections with relatively high steel percentages (i.e., $\rho \geqslant 0.04$) is readily apparent from the figure. This explains why most codes specify an upper limit on the amount of tension reinforcement which may be used in a flexural member. The arrowheads attached to the different curves shown in the figure indicate the corresponding maximum allowable steel ratio by ACI 318-71.*

-6. Amount of compression reinforcement, A'_s—The addition of compression reinforcement to a section subjected to bending increases its ductility. This is clearly indicated by the increase in the ductility ratio, ϕ_u/ϕ_y, with curves associated with increasing values of ρ'/ρ shown in Fig. 12-46, $\rho' = A'_s/bd$, being the compressive reinforcement ratio. This effect is due to the lesser depth of the concrete compression zone required to resist a given bending moment. As was

*Section 10.3.2 of the *Code* requires that the reinforcement ratio, for members under flexure or combined flexure and axial load shall not exceed 0.75 of that ratio which will produce balanced conditions in the same section when subjected to flexure only. The value 0.75 is reduced to 0.50 for flexural members in ductile frames in Section A.5 of Appendix A to the *Code*.

noted earlier, the ultimate curvature, ϕ_u, (see Fig. 12-40) increases with decreasing depth to the neutral axis, $k_u d$, for a particular value of the maximum compressive strain, ϵ_{cu}.

The need for compressive reinforcement is particularly critical at sections near the supports of continuous beams where negative moments subject the stem of the usual T-section beams to compression. As will be shown in the next section, the presence of flanges in the compression zone of a member subjected to flexure substantially increases its ductility. Appendix A of ACI 318-71 requires the use of compression steel at the supports such that a positive moment capacity equal to at least 50% of the negative moment capacity at the same section is provided.

In addition to the increase in sectional ductility associated with the increased compressive capacity of the concrete compression zone due to the presence of compression steel discussed above, the presence of compression reinforcement improves the confining effect of lateral reinforcement. A comparison of curves (2) and (4) of Fig. 12-46a, from a study by Bertero and Felippa,[12-67] shows the increase in both the strength and ductility of a reinforced concrete beam due to the presence of compression reinforcement. A major benefit arising from the use of compression steel, besides that mentioned in the preceding paragraph, is the spread of the inelastic deformation over a greater length that would be possible in a beam without such reinforcement. The photograph of beam (2) in Fig. 12-46b shows that in a beam without compression steel, the inelastic action is confined to a relatively short region compared to beam (4), which is essentially the same as beam (2) except for the addition of compressive reinforcement. This study[12-67] showed that, with the proper amount of transverse and compression reinforcement, the rotation capacity of a reinforced concrete member can be increased to the point where it is actually controlled by the ductility of the reinforcement rather than the concrete, as in beam (4) in Fig. 12-47b.

-7. *Effect of flanges in compression zone.*—Figure 12-48 shows the increase in sectional ductility which occurs when flanges are added to the compression side of an otherwise rectangular section. As in the case of the use of compression reinforcement or high strength concrete, this increase in ductility results from the reduced depth of the concrete compression zone required to resist a particular ultimate moment.

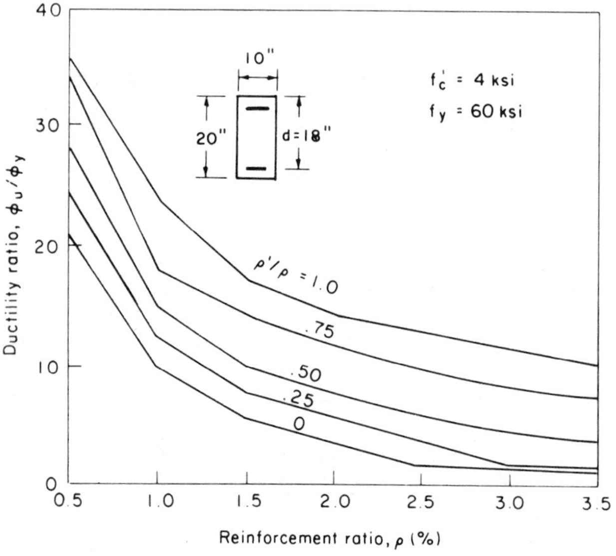

Fig. 12-46 Effect of compressive reinforcement.[12-66]

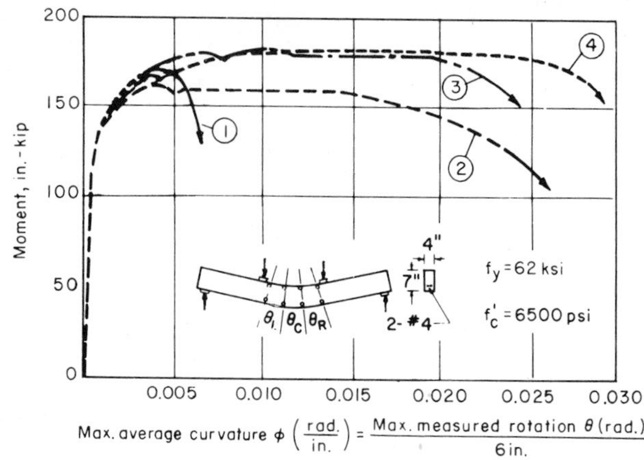

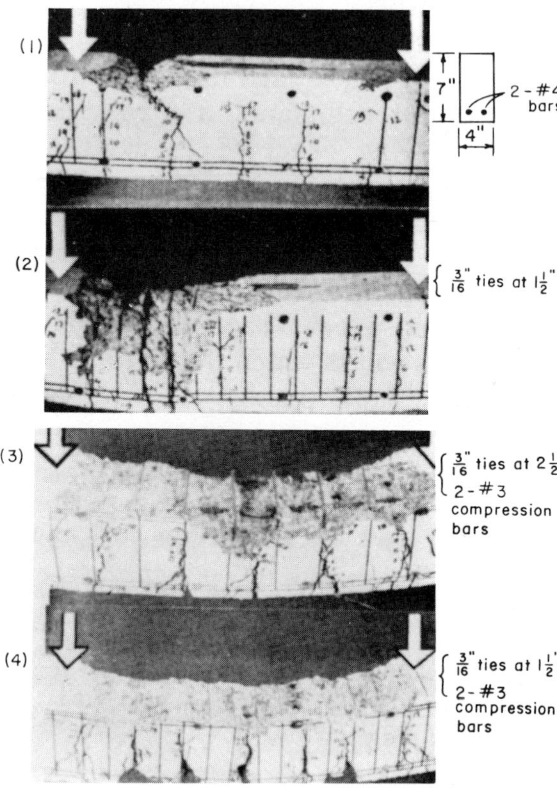

Fig. 12-47 Effect of longitudinal compression reinforcement on extent of hinging region and ductility.[12-67]

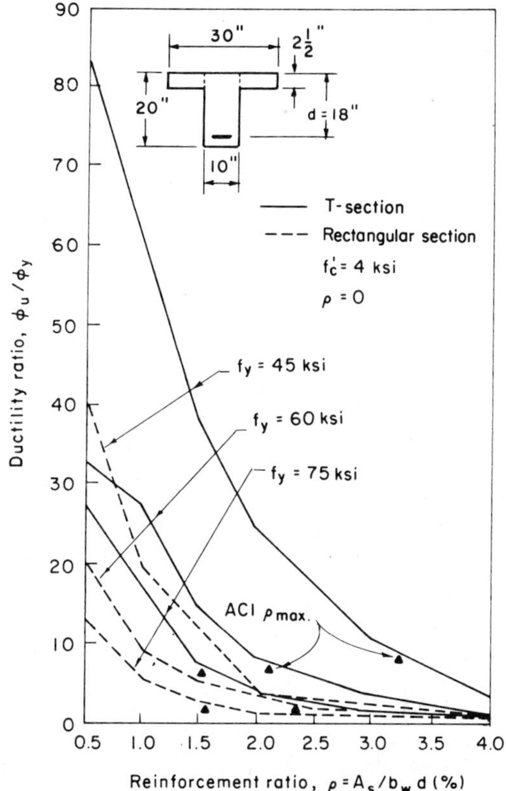

Fig. 12-48 Effect of additional compressive flange in T-section.[12-66]

-8. *Effect of axial compressive load*—Both curves (1) and (2) in Fig. 12-42b clearly indicate the significant decrease in ultimate curvature accompanying an increase in the axial load, P, the ultimate curvature being relatively small for loads larger than that corresponding to the balanced point. The same trend is shown in Fig. 12-49, from Ref. 12-66.

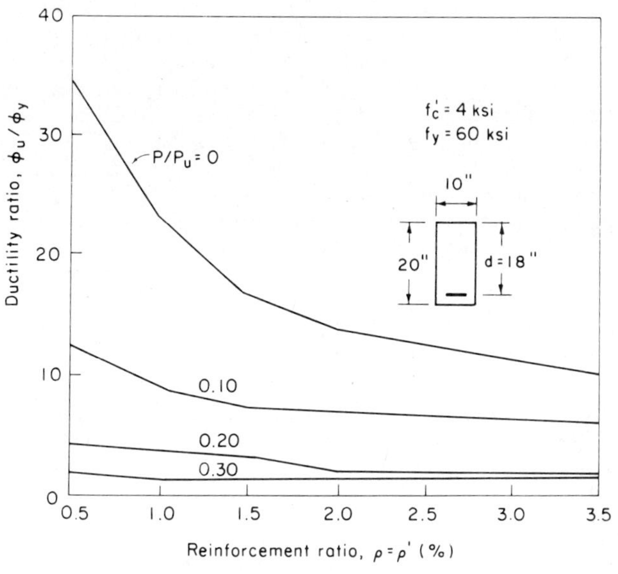

Fig. 12-49 Effect of compressive axial load.[12-66]

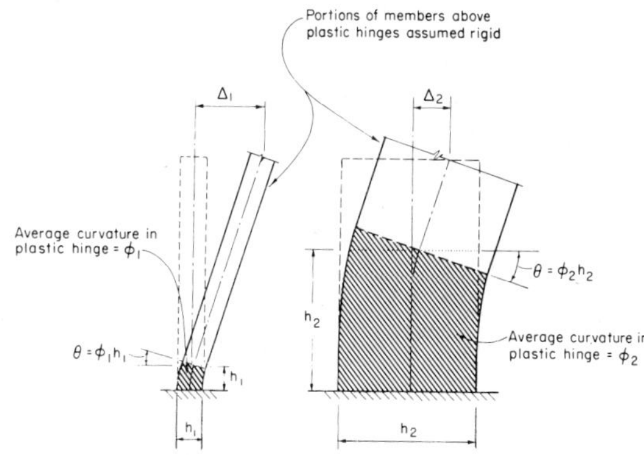

Fig. 12-50 Lateral-deformation capacity in a given length of shallow vs deep members.

-9. *Effect of member depth on lateral deformation capacity*—The use of the curvature ratio ϕ_u/ϕ_y as a measure of the flexural ductility or deformation capacity of members provides a meaningful basis of comparison for members of approximately the same depth. However, when used in connection with members having significantly different depths, the ratio can be misleading. Because of the relatively smaller curvature associated with first yielding in a deep member, such as a shearwall, a ductility ratio of say, 4, for such a member would represent a relatively lower lateral deformation capacity over a given length than the same ratio would indicate when used in connection with a regular-sized column. In Fig. 12-42, for instance, the curvature ϕ_u corresponding to a particular intensity of load, P/bhf'_c, has to be less when the depth, h, is increased in order to obtain the same product $\phi_u h$. Because of the lesser curvature attainable in deep members, a greater length would be required in such members to develop a given total rotation or lateral displacement than in a shallower member. Thus, if the ultimate curvature, ϕ_u, is assumed to be the average curvature over a hinge length equal to the depth, h, of a member, the total hinge rotation in each case would be equal to $\phi_u h$. The lateral displacement at a particular height due to such a total base rotation (see Fig. 12-50) will then be greater for a shallow member than for a deep member.

This distinction between the sectional ductility ratio, ϕ_u/ϕ_y, and the lateral deformation capacity of a member is important in structures consisting of connected elements having substantially different depths, such as frame-shearwall structures. The design of such structures should be concerned with ensuring that the deeper elements or shearwalls possess sufficient deformation capacity to enable them to resist their share of the lateral loads throughout the entire range of expected structure deformation.

12.5.5 Beam-Column Connections

Beam-column connections in frames represent geometric and stiffness discontinuities and hence regions of force concentration. Under the repeatedly reversed loading characteristic of earthquake response, and especially when the framing beams are stressed beyond yield, these connections are subjected to severe stresses which, if not properly provided for, can result in undermining the load capacity of the columns and beams forming the joint. The region at and near joints have been observed to be common points of distress in frame buildings subjected to earthquakes.

Tests[12-68, 12-69] have shown that properly designed beam-column connections can develop the ductility which would be required under a major earthquake without appreciable loss of strength. The main objectives in the design of a beam-column connection for a ductile frame are:

a. the provision of adequate lateral confinement of the concrete to allow transmission of the column load through the joint should large deformations cause the concrete cover to spall off;

b. the use of shear reinforcement to resist any internal shear force in excess of that carried by the concrete; and

c. the provision of adequate anchorage for the flexural reinforcement to allow these to develop their yield strength.

Tests[12-68, 12-70] have indicated that when the beam framing into a connection undergoes inelastic flexural deformations, the concrete within the joint is subjected to large bursting pressures which often produce yielding in the confining transverse reinforcement. By preventing the disruption of the concrete core, particularly under reversed cycles of loading, the transverse reinforcement not only maintains the vertical load carrying capacity of the joint, but allows the development of the ultimate moment capacity of the beam through bond, shear, and a diagonal compression-strut action across the joint. Because the confining action of the transverse reinforcement, as distinguished from its shear-resisting function, plays a dominant role in maintaining the integrity of the joint concrete, the circular spiral would appear to be the most efficient form of transverse reinforcement. However, construction difficulties arising from the congestion of reinforcement in the joint area may preclude the use of circular spirals. The use of rectangular spirals would not only provide a slightly better performance than single rectangular hoops but would also eliminate the congestion which sometimes occurs with the use of single hoops at the corner bars where hooks are located.

From the point of view of confinement, the isolated beam-column connection, i.e., with only one beam framing into the column, which, incidentally rarely occurs in practice, represents the weakest type of joint. This type of joint, as well as the corner column joint with only two beams framing into the column at right angles with each other, require particular attention with respect to confinement to ensure that they perform satisfactorily. An interesting detail which has been found to significantly improve the performance of an isolated beam-column connection[12-71] is the anchorage of the beam reinforcement into a short extension of the beam beyond the opposite face of the column.

The beneficial confining effect of beams framing into the other sides of a column on the performance of a beam-column connection has been amply demonstrated by tests.[12-68, 12-71] Appendix A of ACI 318-71 recognizes this effect by allowing a reduction of the required transverse reinforcement by 50% when beams satisfying certain requirements on relative width and depth frame into all four sides of a column.

12.5.6 Infilled Frames

The behavior of multistory infilled frames, i.e., frames with the space between the members filled with a masonry or cast-in-place concrete wall, under lateral load is essentially that of a vertical cantilever beam, at least until the failure of the infill occurs. While it may be conservative to disregard the stiffening effect of infilling panels on the behavior of frames subjected to wind loading, neglecting their effect on the response of a structure subjected to earthquakes can be dangerous. Walls filling the spaces between frame members not only increase the stiffness (and damping) and hence the frequency of vibration of a building, but also change the mode of deformation under lateral load.

The use of relatively brittle materials, such as unreinforced clay or masonry blocks for infill panels should be avoided since the sudden and often explosive failure of these elements, when subjected to large diagonal compressive forces accompanying the lateral displacement of the frame constitutes a potential hazard to occupants and passersby. In addition, the abrupt failure of such brittle infilling can lead to a progressive collapse if the confining frame elements are unable to resist the sudden imposition of additional forces resulting from the failure of the infill panels.

Unless concrete or masonry walls or partitions are purposely isolated from the structural frame, the resulting infilled frame should be designed as a shearwall with proper regard to its ductile performance and its interaction with other elements in the structure. Isolation of a wall from the structural frame may be achieved by providing a sufficient gap between the wall and the surrounding frame or by having the wall offset from the line of columns. In the latter case, gaps should be provided between the top of the wall and the floor slab above. Channel sections may be used to provide lateral support to the wall as well as conceal the gaps. In all cases, the width of the gap should be sufficient to allow for the maximum expected lateral displacement of the structure.

12.5.7 Shear Walls

The behavior of shear walls, with particular reference to their typical mode of failure is, as in the case of beams, influenced by their proportions as well as their support conditions. Thus, low shear walls, characterized by relatively small height-to-depth ratios, may be expected to fail in shear, just like deep beams.* Walls occurring in multistory buildings, on the other hand, generally behave as vertical cantilever beams with their strength controlled by flexure rather than by shear. Such elements may therefore be designed in the same manner as regular flexural elements. A portion of shear walls which interact with frames, however, may behave as low shear walls, depending upon the proportions of the wall and the location of the point of contraflexure along the height of the wall. The latter is dependent primarily on the relative stiffnesses of the frame and shear wall elements in a structure.

When acting as a vertical cantilever beam, the behavior of a shear wall which is properly reinforced for shear (i.e., diagonal tension) will be governed by the yielding of the tension reinforcement located near the vertical edge of the wall and to some degree by the vertical reinforcement distributed along the central portion of the wall.

Since the ductility of a flexural member such as a tall shear wall can be significantly affected by the maximum usable strain in the compression zone concrete, confinement of the concrete at the ends of the shear wall section should improve the performance of such shear walls. Such confinement can take the form of enlarged boundary elements with adequate confinement reinforcement or may result from the presence of other walls running at right angles to the shear wall at its ends. In both cases, the effect of the additional compression flanges (see section 12.5.4.7) also contributes to the increase in ductility.

Appendix A of ACI 318-71 requires that shear walls subjected to axial loads greater than $0.4P_b$ corresponding to balanced conditions be provided with vertical boundary elements capable of resisting all the vertical forces resulting

*Note that the term shear wall derives from the usual function of such elements in resisting a major portion of the horizontal shear in buildings and is not meant to be indicative of the mode of failure.

from gravity and lateral loads on the shear wall. In addition such boundary elements are to be provided with transverse confinement reinforcement throughout their entire height. The requirement that the boundary elements in such special ductile shear walls with $P_e \geqslant 0.4P_b$ be designed to resist the entire gravity load on the shear wall is an attempt to provide an adequate support structure in the event of failure of the panel between the boundary elements. The requirement that the vertical (longitudinal) forces resulting from horizontal (seismic) loads also be resisted entirely by the boundary elements is similar to the design approach used in proportioning thin-web steel I-beams or trusses in which the flanges resist all the flexural stresses while the web (the wall panel in the case of the shear wall) carries all the shear. The object of a multistory shear wall design, which treats the shear wall as a vertical cantilever beam, should be to provide enough shear capacity so that a shear failure does not precede a flexural failure.

12.6 CURRENT PRACTICE IN DESIGN FOR EARTHQUAKE MOTIONS

12.6.1 General

The design of economical earthquake-resistant structures should aim at providing the appropriate dynamic and structural characteristics so that acceptable response levels result under the design earthquake(s). The structural properties over which the designer exercises some degree of control and which he can modify to achieve the desired results are parameters such as magnitude and distribution of stiffness and mass, relative strengths of members and their ductility, among others. Perhaps the most convenient measure of response, particularly insofar as it affects the design of a structure, is the maximum lateral displacement of the structure as well as the maximum deformation of its component elements. The range of deformation, relative to the force-deformation characteristic of a particular structure or element, which may be considered acceptable will vary with the type of structure and/or its function.

In some structures, such as slender free-standing towers or smoke stacks or the more recent suspension-type buildings* which depend for their stability on the stiffness of the single element making up the structure, or in nuclear containment structures, where a more-than-usual conservatism in design is desirable, yielding of the principal elements in the structure would not be tolerated. In such cases, the design would be based on an essentially elastic response to moderate-to-strong earthquakes, with the critical stresses limited to the range below yield.

In most buildings, particularly those consisting of frame members and other multiply-redundant structures, however, economy is achieved by allowing yielding to take place in some members under moderate-to-strong earthquake motion.

The performance criteria implicit in most earthquake code provisions require that a structure be able to[12-22]

a. resist earthquakes of minor intensity without damage. A structure would be expected to resist such frequent but minor shocks within its elastic range of stresses;
b. resist moderate earthquakes with minor structural and some nonstructural damage. With proper design and construction, it is expected that structural damage due to the majority of earthquakes will be limited to repairable damage; and
c. resist major catastrophic earthquakes without collapse.

The above performance criteria allow only for the effects of a typical ground shaking. The effects of slides, subsidence or active faulting in the immediate vicinity of the structure, which may accompany an earthquake, are not considered. Of course, proper design should aim at minimizing the effects of these other factors, when investigation reveals a reasonable chance of such actions occurring.

While no clear quantitative definition of the above earthquake intensity ranges has been given, their use implies the consideration not only of the actual intensity level but also of their associated probability of occurrence with reference to the expected life of a structure. The quantitative definition of such earthquake intensity ranges would have to consider all the significant ground motion characteristics affecting structural response, i.e., the magnitude of acceleration pulses, frequency characteristics, and duration of the significant portion of the ground motion. The recurrence interval associated with each intensity range would then have to be established for each particular site. The present lack of adequate data on earthquakes renders such an approach beyond immediate realization, although a number of attempts[12-9, 12-11] in this direction have been made. Such a quantitative definition of earthquake intensity ranges, however, would find its greatest use in establishing seismic zones. The principal concern in earthquake-resistant design is the provision of adequate strength and ductility for the most intense earthquake which may reasonably be expected at a site during the life of a structure.

12.6.2 Design Philosophy Behind Earthquake Code Provisions for Buildings**

The availability of dynamic analysis programs[12-20, 12-49–54] designed for use with large electronic computers has made possible the analytical determination of earthquake-induced forces in reasonably realistic models of most structures. However, except perhaps for the relatively simple analysis by modal superposition using response spectrum data, such dynamic analyses, which can range from an elastic time-history analysis for a single earthquake record to nonlinear analyses using a representative ensemble of accelerograms, can be costly and may be economically justifiable as a design tool only for a few large and important structures. For most practical applications, reliance will usually have to be placed on the simplified prescriptions found in most codes.[12-72, 12-73] Although necessarily approximate in character—a feature imposed by the need for simplicity and ease of application—the provisions of such codes and the philosophy behind them gain in reliability as design guides with continued application and modification to reflect the lessons derived from observations of actual structural behavior under earthquakes. Code provisions must, however, be viewed in the proper perspective, i.e., as minimum requirements covering a broad class of structures of more or less conventional configuration. Unusual structures must still be dealt with special care and may call for procedures beyond that normally required by codes.

The basic form and values of modern code provisions on earthquake-resistant design have evolved from rather simplified concepts of the dynamic behavior of structures and have been influenced greatly by observations of the per-

*Usually consisting of a centrally located corewall which is the principal load-carrying element and from which the floor slabs are suspended by means of hangers along the periphery of the building.

**The remarks which follow refer mainly to the *SEAOC Recommended Lateral Force Requirements*,[12-22] and the provisions on earthquake-resistant design of the *Uniform Building Code*[12-10], which are essentially based on the former.

formance of structures subjected to actual earthquakes. It has been noted, for instance, that many structures built in the 1930s and designed on the basis of more or less arbitrarily chosen lateral forces have successfully withstood severe earthquakes. The satisfactory performance of such structures has been attributed to one or more of the following:[12-74, 12-75] (1) yielding in critical sections of members. Yielding not only increased the period of vibration of such structures but allowed them to absorb greater amounts of the input energy from an earthquake; (2) the greater actual strength of such structures resulting from so-called nonstructural partitions–which were generally ignored in analysis–and the significant energy-dissipation capacity that cracking in such members represented; and (3) the reduced response of the structure due to yielding of the foundation.

The code-specified design lateral forces have the same general distribution as the typical envelope of maximum horizontal shears indicated by an elastic dynamic analysis. However, the code forces, which are assumed to be resisted by a structure within its elastic (working) range of stresses, are substantially smaller than those which would be developed in a structure subjected to an earthquake of intensity equal to that of the 1940 El Centro, if the structure were to respond elastically to such ground excitation. Thus, buildings designed under the present codes would be expected to undergo fairly large deformations (about five times the lateral displacements resulting from the code-specified, statically applied shears–see Fig. 12-21) when subjected to an earthquake with the intensity of the 1940 El Centro. These large deformations will be accompanied by yielding in many members of the structure, and in fact, such is the intent of the codes. The acceptance of the fact that it is economically unwarranted to design buildings to resist major earthquakes elastically and the recognition of the capacity of structures possessing adequate strength and ductility to withstand major earthquakes by responding inelastically to these, lies behind the relatively low forces specified by the codes coupled with the special requirements for ductility particularly at and near member connections. Analytical studies[12-20] have shown that building frames designed to resist the code-specified lateral forces elastically do in fact respond inelastically to an earthquake of El Centro (1940) intensity, with maximum lateral displacements comparable to that of a purely elastic dynamic response. The ductility ratio associated with lateral displacement, i.e., the ratio of the maximum to the yield displacement, for the inelastic response, ranged from 4 to 6 (see Fig. 12-21).

The capacity of a structure to deform in a ductile manner, that is, to deform beyond the yield limit without significant loss of strength, allows such a structure to absorb a major portion of the energy from an earthquake without serious damage. Laboratory tests[12-68, 69 and 12-76, 77] have demonstrated that properly designed reinforced concrete members do possess the necessary ductility to allow a structure designed by the present codes to respond inelastically to earthquakes of major intensity without significant loss of strength.

This, briefly, then, is the rationale behind the present code provisions on earthquake-resistant design of buildings: A set of forces is specified, the magnitude of these forces depending upon the weight, the gross dimensions and the type of structure, as well as the seismicity of the area in which it is to be built. These forces are assumed to be statically applied to the structure, which must be designed to resist them within the elastic (working) range of stresses. Although the specified distribution of forces is similar to that of the maximum shears obtainable from a dynamic analysis of the same structure, their magnitudes are four to six times less than the dynamic shears which would be developed if the structure were to respond elastically to an earthquake of intensity such as that recorded in 1940 at El Centro, California. A structure designed to resist the code-specified static forces elastically, does in fact yield at a number of places when subjected to an earthquake of moderate intensity, and thus responds inelastically to such excitation. The maximum deformations during such an inelastic dynamic response can range from four to six times that corresponding to the code-specified loading. The codes specify additional requirements for the design of members and their connections in order to ensure sufficient deformation capacity in the inelastic range.

Because of the relatively large inelastic deformations which a building designed by the present codes may undergo during a strong earthquake, proper provision must be made to ensure that the structure maintains its integrity or does not become unstable under the vertical loads, while undergoing large lateral displacements.

The present philosophy of the codes is to have the yielding confined to the beams while the columns remain elastic throughout the response, i.e., the so-called strong column-weak beam design. This is achieved by requiring that the sum of the moment strengths of the columns meeting at a joint, under the design axial load, be greater than the sum of the moment strengths of the framing beams in the same plane. This provision is intended to preclude instability problems due to large inelastic deflection of the columns. A comparative study by Anderson and Bertero[12-45] of the dynamic response of frames designed according to this philosophy with those of frames designed using allowable stress methods with yielding allowed in the columns and also frames designed to achieve minimum weight (and hence, minimum stiffness) showed that the present code philosophy in this regard results in better structural performance under earthquake loading than the other design methods.

It is pointed out that the design provisions contained in the main body of the *ACI Building Code*[12-24] as well as the regular provisions in most other codes do in fact provide some ductility which should be sufficient for structures subjected only to minor earthquakes that may occur frequently, such as those associated with Zone I areas. For such cases, however, the design forces to be used in proportioning structural members should be the actual expected dynamic forces, since no special provision is made for inelastic deformations. For structures which may be subjected to earthquakes of moderate intensity (Zone II) some additional confinement, anchorage and shear reinforcement details may be required. For structures which may be subjected to strong intensity earthquakes, and designed to reduced forces such as provided in the *SEAOC Recommendations*[12-22] and the UBC Zone III requirements, appreciable inelastic deformations can be expected so that substantial ductility is required. The design provisions contained in the *SEAOC Recommendations* and the UBC Zone III requirements, as well as Appendix A of ACI 318-71, are intended to provide this additional ductility.

The discussion which follows will be concerned mainly with the design of structures which have to satisfy special ductility requirements in order to ensure satisfactory performance under moderate-to-strong earthquake intensities.

12.6.3 Isolation Concepts for Earthquake Resistance

Apart from problems assocaited with the current approach to earthquake-resistant design, there is the important question of whether we have chosen the best means of adapting our structures to the earthquake phenomenon. The concern over the soundness of our traditional approach to earthquake-resistant design and the results of observations of structures subjected to recent earthquakes have brought about the examination of earthquake-adaptive systems and the revival of shock-isolation concepts as applied to earthquake-resistant structures. This concept represents a radical departure from current aseismic design practice. Its successful implementation on a broad scale promises significant simplifications in the design of tall reinforced concrete structures.

In contrast to the present philosophy of designing an entire structure to withstand the distortions resulting from earthquake motions, an adaptive system is designed to isolate the upper portions of a structure from destructive vibrations by confining the severe distortions to a specially designed portion at its base. A number of isolation devices or mechanisms have been proposed including a soft story with hinging columns,[12-78] a combination ball bearing and rod system,[12-79] steel balls on ellipsoidal cavities,[12-80] etc.[12-81]

A principal consideration in an isolation system for a multistory structure is the provision of a resisting element which possesses adequate stiffness and exhibits essentially linear elastic behavior under the maximum wind loading but yields when subjected to earthquake forces slightly greater than that corresponding to the maximum wind loading. By allowing the isolating mechanism at the base of a structure to yield at a predetermined lateral load, the structure above it is effectively isolated or shielded from forces which would otherwise cause inelastic deformations. The isolating mechanism thus sets an upper limit to the forces which can be transmitted to the structure from the foundation.

The structure supported on an isolating mechanism then need only be designed for vertical and wind loads, with special attention for earthquake resistance focused only on the isolating mechanism at its base. This clearly offers economic and technical advantages when compared to the present method of analyzing a complex structure under earthquake motion and providing ductile members and connections throughout the entire structure.

The isolating effect which a yielding base can provide to a superstructure during earthquakes has been borne out by observations of buildings in earthquake-damaged areas. In a number of very illuminating cases, the upper portions of multistory buildings in which a lower story apparently suffered extensive inelastic deformation early in the shaking remained virtually unscathed.

An effective isolation system not only allows the structure above to remain elastic during a strong earthquake but spares the nonstructural elements from extensive distress. Since the nonstructural components (e.g., partitions, glazing, mechanical equipment, etc.) in a typical multistory structure account for about 80% of the building's cost, a positive means of preventing distress in such elements during strong earthquakes would represent significant savings in the repair and replacement costs which would otherwise be required.

12.6.4 Preliminary Design—Some Guides to Planning and Detailing

A reasonably good basis for a preliminary design of an earthquake-resistant building would be a structure proportioned to satisfy the requirements of gravity and wind loading. The planning and layout of the structure, however, must be undertaken with proper consideration of the dynamic character of earthquake response. Thus, modifications in both configuration and proportions to anticipate earthquake requirements may be incorporated immediately into the design for gravity and wind. This is done before an earthquake analysis and design is undertaken for compliance with the applicable code. For most structures, the design process ends with compliance with the earthquake code provisions. When the importance of a structure or its unusual features warrant it, a dynamic analysis is undertaken at this stage. The object of the dynamic analysis would be to check the adequacy of the design and/or indicate additional requirements which have to be satisfied. As with all converging, iterative processes, the design procedure becomes more efficient with a preliminary design which constitutes a 'good, first guess'. Setting up a good preliminary design can be facilitated by keeping in mind the basic requirements to be achieved. For this purpose, the results of the analytical studies presented in the preceding sections will serve as useful guides. Also essential to the finished design is a particular attention to detail which can often mean the difference between a severely damaged structure and one with but minor repairable damage.

To help in the preparation of a preliminary design for earthquake resistance, the points given below have been collected for ready reference. The list reflects the results of the studies discussed in the preceding sections as well as some of the more important lessons gained from observations of the behavior of structures subjected to actual earthquakes. They should be considered where deemed applicable, in addition to the minimum requirements implied in earthquake code provisions. The list is by no means exhaustive nor generally applicable to all types of buildings.

-1. Basic planning and design considerations

(1) A building which is simple in both plan and elevation, with a minimum of setbacks or changes in section, is generally preferable to an irregularly shaped structure. This is because the effects of force concentrations which occur at major discontinuities in either geometry or stiffness, which is well known even for static loading cases, tend to be aggravated under dynamic conditions. The required ductility at such regions of discontinuity is usually substantially greater than at other portions of a structure.

Although it may not be practicable to plan a fully symmetrical building, any effort to reduce the eccentricity of the effective inertial force due to the noncoincidence of the centers of mass and of rigidity will pay off in reduced torsional stresses, which can be critical in corner columns and end walls.

The provision for horizontal torsional effects, as well as noneccentric lateral loads is best served by having the major lateral stiffening elements—whether shearwalls or grids of closely spaced frame elements—located near or along the plan periphery of a building. The so-called hull-core or framed-tube-with-core structure laid out on a square plan, as shown schematically in Fig. 12-51a, is well suited for high-rise buildings. For buildings of moderate height, a frame shear wall arrangement such as is shown in Fig. 12-51b may be considered. For relatively low buildings, a structure consisting exclusively of conventional frame elements may be used. The introduction of shearwalls will generally be dictated by the need to limit the drift or lateral deflection under wind as well as earthquake loading to safe and tolerable values.

(2) The useful range of deformation of the principal resisting elements in a structure should be of about the same order of magnitude, when practicable. This makes for a more efficient design wherein all the resisting elements participate in carrying the induced forces in proportion to

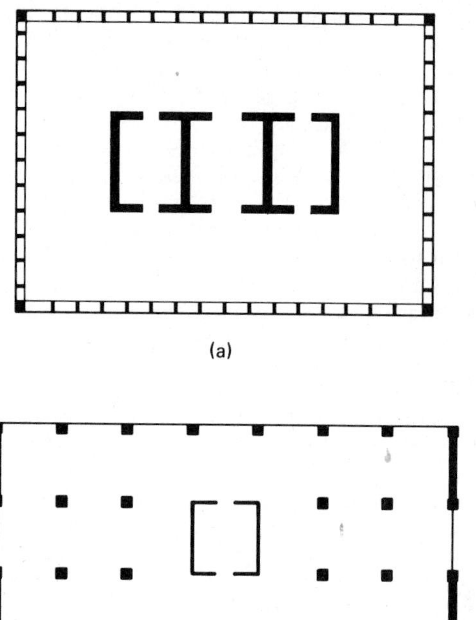

Fig. 12-51 (a) Framed tube with core and (b) frame-shear wall structure.

their relative stiffnesses over the entire range of deformation. This is illustrated in Fig. 12-52a, for a structure with two resisting elements ① and ②. In Fig. 12-52b, the resisting elements ① and ② not only possess different initial stiffnesses (as given by the slopes of the curves) but, more important, exhibit different ductilities or deformation capacities. In such a case, the following points have to be considered:

a. If the structure cannot survive with only element ① supporting it, care must be taken to ensure that the maximum probable deformation or lateral displacement under dynamic conditions does not exceed the deformation capacity of element ②, Δ_2.
b. If the expected maximum deformation could exceed Δ_2, then the resisting element ① must be so designed that it can support the additional load which may come upon it when element ② fails.

It should be borne in mind that the failure or partial failure of a major element in a structure will alter its dynamic characteristics and hence its response to a particular

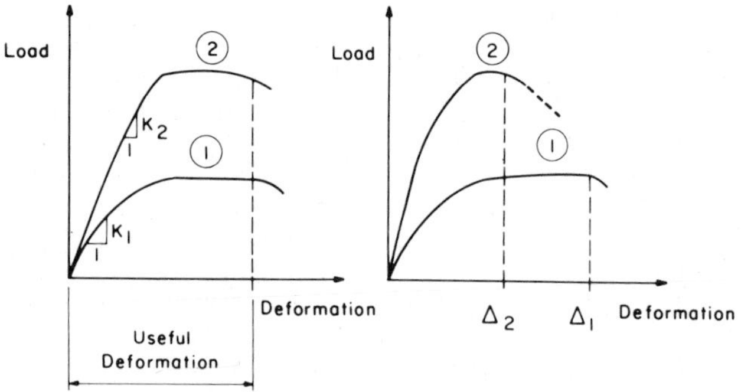

Fig. 12-52 Ductility of elements in structure.

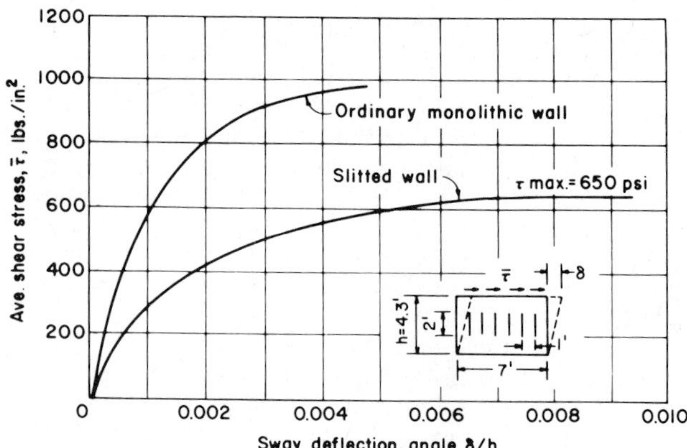

Fig. 12-53 Experimental load-deflection curves for solid and slitted walls.[12-83]

ground motion. A partial failure of an element which does not lead to general instability will usually result in a significant increase in the period of vibration (due mainly to a decrease in stiffness) as well as damping.

The case illustrated in Fig. 12-52b is typical of a combination frame-shear wall structure. In this connection, the use of slitted walls[12-82,12-83] instead of the solid-panel shear wall may be considered. The "slitted wall" consists essentially of columns placed side by side with small gaps in between. The closeness of the columns and hence the considerable rotational restraint provided by the horizontal connecting members obviously increases the lateral stiffness of the columns above that which they would possess if connected by fairly flexible beams. Figure 12-53 shows the load-deflection curve for an experimental slitted wall panel reported in Ref. 12-83, as compared to a solid-wall panel. Note the significantly greater deformation capacity, but reduced stiffness and strength, of the slitted-wall panel.

(3) The need to adequately tie together all the structural elements making up a building or a portion of it which is intended to act as a unit cannot be overemphasized. This applies to superstructure as well as foundation elements, particularly in buildings founded on relatively soft soil. Here, attention should be focused on the design of the segments of elements at and near the joints since these are generally the regions which are most critically stressed.

Adequate connections should be provided across construction joints* which may be required between portions of a building or between the main element of a structure and an appendage, e.g., stairway enclosure, carport, etc.

(4) The different portions of a building should either be tied together adequately or separated from each other by a sufficient distance to prevent their hammering against each other.

Expansion or similar joints used to separate parts of a building which differ considerably in height, plan size, shape, or orientation should be sufficient to allow for their swaying independently or out of phase of each other without impact. Any required passageway, corridor or bridge linking structurally separated parts of a building should be so detailed as to allow their free, unhindered movement during an earthquake (Fig. 12-54). Multidirectional rollers

*A distinction should be made between a construction joint and an expansion joint. The former marks the termination of one placement of concrete and the beginning of the next placement, the stoppage usually being required by the impracticality of doing a large concreting job in one continuous operation. An expansion joint, on the other hand, is a plane of separation specifically provided to allow free movement between parts of a structure.

(a)

(b)

Fig. 12-54 (a) The Olive View Medical Center in Los Angeles after the February 9, 1971 San Fernando earthquake; and (b) close-up of one of the three toppled stair towers, Olive View Medical Center, L.A.

or pads of low-friction material (e.g., Teflon) under one end of such connecting elements may be used.

In order to avoid hammering between adjoining buildings or separate portions of a building when vibrating out of phase of each other, a gap (perhaps filled with readily crushable material) equal to from four to six times the sum of the calculated lateral deflections of the two structures under the design (code) seismic forces or the sum of the maximum deflections of the two structures as indicated by a dynamic analysis, would be desirable.

(5) The use of very stiff walls to fill the spaces between relatively more flexible frame elements should be considered carefully during the preliminary design stage.

If the infilling material is intended to act in combination with the enclosing frame, then it should be designed and constructed to ensure this composite action. Proper reinforcement and connection to the enclosing frame are essential. The analysis should likewise consider the increased stiffness of the infilled frames.

If the infill is made of fairly brittle material, such as glass or lightly reinforced brick masonry, and cannot be expected to contribute significantly to the lateral resistance of the frame, then it should be effectively isolated from the surrounding frame by gaps or readily crushable or yielding material (e.g., deep rubber gaskets for glass panels) to allow sufficient relative movement between the frame and such elements. Because the nonstructural elements in a building usually represent a major percentage of the total cost, huge

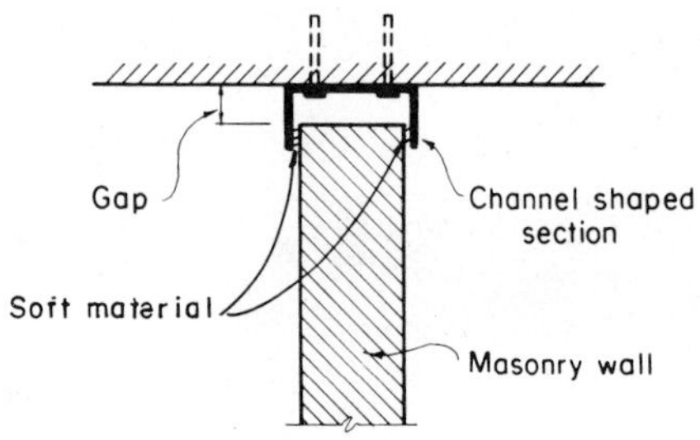

Fig. 12-55 Detail of gap between masonry wall and adjoining structure.

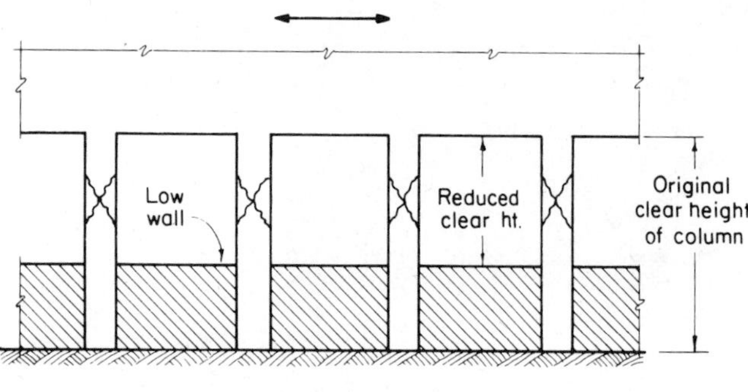

Fig. 12-56 Effect of introducing low walls between columns originally designed for longer clearance height.

losses can result from the failure of such elements. In addition, the failure of many of these brittle elements, when subjected to compressive forces, can sometimes be explosive in nature and hence potentially injurious to occupants and passersby.

The isolation of a masonry wall may be accomplished by providing a gap between the wall and the surrounding structural members. A channel attached to the latter may be used to provide lateral support to the wall while allowing free movement within the concealed gap (see Fig. 12-55). Alternatively, masonry walls may be run along the line slightly offset from the plane of the columns, with a gap provided between the top of the wall and the floor above.

The *SEAOC Recommendations*,[12-22] for example, call for connections and joints between exterior precast non-load-bearing wall panels and similar elements and the supporting frame sufficient to allow for relative movement between stories of not less than two times the drift caused by wind or seismic forces, or 1/4 in., whichever is greater.

The effect of introducing low walls between columns, as shown in Fig. 12-56, should be noted. The reduction in height of the columns increases their stiffness with respect to bending in the plane of the wall. This will cause the columns to be subjected to greater horizontal shears than they would be expected to develop if the walls were absent. This is in addition to the effect which a decrease in the period of vibration of the structure—due to the increase in stiffness of the columns—will bring. The reduction in height also reduces the lateral deformation capacity of the columns in the plane of the wall. The use of such walls without allowing for their effects on the columns has been known to cause severe distress in those portions of the columns above the wall.

6. In order to avoid excessive internal forces in the structure due to differential vertical displacements resulting from foundation yielding or liquefaction, buildings, particularly multistory buildings, should be founded on a firm stratum.

12.6.5 Principal Earthquake Design Provisions of *SEAOC, UBC, NBCC* and Appendix A of ACI 318-71

-1. Principal design steps—The principal steps involved in the design of typical single-story or multistory reinforced concrete structures, with particular reference to the application of the earthquake-resistant design provisions of nationally accepted North American building codes, will be discussed below. The provisions of *SEAOC*[12-22] *-UBC*[12-10] will be used as a basis for the discussion on the design forces to be used in analysis. Except for relatively minor differences, the earthquake design provisions of *NBCC*[12-23] are essentially the same as those found in *SEAOC-UBC*. Although ACI Appendix A[12-24] does not specify the magnitude of the earthquake forces to be used in analysis, its provisions are intended for structures which are designed to reduced forces such as those specified in *SEAOC* or the *UBC* requirements for Zones II and III. The provisions of ACI Appendix A will be used as a basis for the discussion on member design.

Insofar as compliance with earthquake code provisions is concerned, the design process may be divided into the following principal steps:

a. Determination of design 'earthquake' forces—
   (i) calculation of base shear corresponding to computed or estimated fundamental period of vibration of the structure. (A preliminary design of the structure is assumed here.)
   (ii) distribution of the base shear over the height of the building.
b. Analysis of the structure under the (static) lateral forces calculated in step (a), as well as under gravity and wind loads, to obtain member design forces. The lateral load analysis can be carried out most conveniently by using a computer program for frame analysis.
c. Designing members and joints for the most unfavorable combination of gravity and lateral loads. The emphasis here is on the design and detailing of members and their connections to ensure their ductile behavior.

Changes in section dimensions of some members may be indicated during the design phase under step (c) above. However, unless the required changes in dimensions are appreciable and of such an extent as to materially affect the overall distribution of forces in the structure, a reanalysis of the structure using the new member dimensions need not be undertaken. Uncertainties in the actual magnitude and distribution of the seismic forces as well as the effects of yielding in redistributing forces in the structure would make such a refinement unwarranted. It is, however, most important to design the members and their connections to ensure their ductile behavior and thus allow the structure to sustain the severe distortions which may occur during a major earthquake without collapse. The code provisions intended to ensure adequate ductility in structural elements represent the major difference between the design requirements for conventional, non-earthquake-resistant structures and those located in regions of high earthquake risk.

The major details involved in the above steps are discussed below.

-1.1 Design base shear—This represents the total horizontal seismic force (service load level) which may be assumed acting parallel to the axis of the structure considered. The force corresponding to the other horizontal axis of the structure is assumed to act nonconcurrently, and may thus be considered separately. The total lateral force or base shear, V, is given by

$$V = ZKCW \tag{12-20}$$

where

Z = a numerical coefficient whose value depends on the seismic zone in which the structure is located, as determined from the map shown in Fig. 12-5 (for the United States). The *Uniform Building Code* assigns values of 0.25 to Z for areas in Zone I, 0.5 for those in Zone II, and 1.0 for locations in Zone III.

It is pointed out that the SEAOC expression for the base shear does not include a seismic zoning factor, Z, (this being taken equal to 1.0) since the *SEAOC Recommendations* are meant to cover only California. This effectively puts the entire area of California under the Zone III classification of the *UBC*.

K = a factor depending on the type of structural system used. This factor is intended to account for differences in the available ductility or energy absorption capacity of various structural systems. Thus, systems which have been observed to perform well under actual earthquakes have been assigned lower values of K.

The values of K corresponding to different types of structures are listed in Table 12-3, and range from

TABLE 12-3 Horizontal Force Factor "K" for Buildings or Other Structures (1).[12-10]

TYPE OR ARRANGEMENT OF RESISTING ELEMENTS	VALUE[2] OF K
All building framing systems except as hereinafter classified	1.00
Buildings with a box system as specified in Section 2314 (b)	1.33
Buildings with a dual bracing system consisting of a ductile moment resisting space frame and shear walls using the following design criteria: (1) The frames and shear walls shall resist the total lateral force in accordance with their relative rigidities considering the interaction of the shear walls and frames (2) The shear walls acting independently of the ductile moment resisting portions of the space frame shall resist the total required lateral forces (3) The ductile moment resisting space frame shall have the capacity to resist not less than 25 per cent of the required lateral force	0.80
Buildings with a ductile moment resisting space frame designed in accordance with the following criteria: The ductile moment resisting space frame shall have the capacity to resist the total required lateral force	0.67
Elevated tanks plus full contents, on four or more cross-braced legs and not supported by a building[3]	3.00[4]
Structures other than buildings and other than those set forth in Table No. 23-I	2.00

[1]Where wind load as specified in Section 2307 would produce higher stresses, this load shall be used in lieu of the loads resulting from earthquake forces.

[2]See maps on pages 122 and 123 for seismic probability zones and definition of "Z" as specified in Subsection (c).

[3]The minimum value of "KC" shall be 0.12 and the maximum value of "KC" need not exceed 0.25.

[4]The tower shall be designed for an accidental torsion of five per cent as specified in Section 2314 (g). Elevated tanks which are supported by buildings or do not conform to type or arrangement of supporting elements as described above shall be designed in accordance with Section 2314 (d) 2 using "C_p" = .2.

0.67 for a ductile moment resisting space frame designed to resist the entire lateral load to 3.0 for elevated tanks.

Note that frame-shear wall structures which are designed for a K-factor of 0.80 have to satisfy three independent loading requirements, each of which calls for a separate analysis. Thus, in addition to the forces to which each element of such a system is subjected when considering frame-shear wall interaction, the capacity of each component element has to be checked separately: the shear wall for the entire lateral load and the frame for at least 25% of the same load, when these elements are considered as acting independently of each other.

As will be discussed in connection with the coefficient C, the factor K also accounts for the differences in the fundamental period between various types of structures having the same outside dimensions, for which the code formula gives the same value of the period.

C = a coefficient related to the flexibility of the structure.

$$C = \begin{cases} 0.10 & \text{for all one- and two-story buildings} \\ \dfrac{0.05}{\sqrt{T}} & \text{but not greater than 0.10, for all other buildings} \end{cases} \tag{12-21a}$$

The fundamental period of vibration, T, to be used in eq. (12-21a) above may be established by analysis or experiment, or, in the absence of such data, may be determined from the formula

$$T\,(\text{sec}) = \frac{.05h_n}{\sqrt[3]{D}} \tag{12-21b}$$

where

h_n = height (ft) of the structure considered above the base

D = dimension (ft) of building in the direction parallel to the applied forces

For "buildings in which the lateral force resisting system consists of a moment resisting space frame which resists 100% of the required lateral forces and which frame is not enclosed by or adjoined by more rigid elements which would tend to prevent the frame from resisting the lateral forces," T is to be taken equal to 0.10 N, where N is the number of stories.

Values of the coefficient C corresponding to a range of values of the parameter $h_n/\sqrt{D}$ or the period, T, may be obtained directly from Fig. 12-57.

The above expression for C is essentially an adaptation, with slight modifications, of similar expressions appearing in earlier codes and has been used principally because of its simplicity. That it is meant to incorporate the effects on the dynamic structural response of such factors as damping and inelastic action as well as the effects of the higher modes of vibration, which are not reflected in the fundamental period of vibration, has long been recognized as a basic deficiency.

W = the total dead load, except that in storage and warehouse occupancies, W shall include the total dead load plus 25% of the floor live load.

The *National Building Code of Canada 1970* includes other factors in the calculation of the base shear. In this

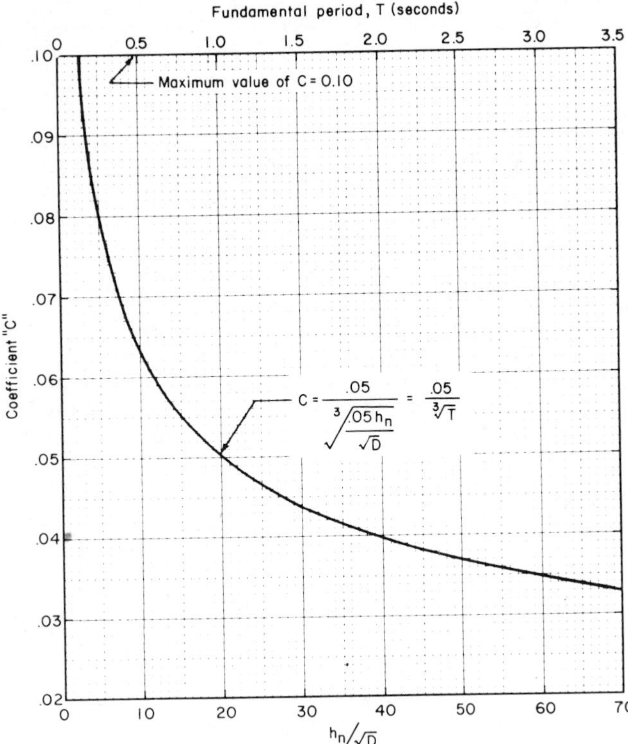

Fig. 12-57 Variation of seismic coefficient, C, with fundamental period, T, and height-to-depth ration, $h_n/\sqrt{D}$.

code, the base shear is given by the expression

$$V = \frac{1}{4} R\,K\,C\,I\,F\,W \tag{12-22}$$

The quantity ($^1/_4$ R) in eq. (12-22) corresponds to Z of the *UBC* expression in eq. (12-20), the seismic regionalization factor R having values of 0, 1, 2, and 4, while K, C, and W are identical to the same quantities in eq. (12-20).

The importance factor

$$I = \begin{cases} 1.3 & \text{for all buildings designed for postdisaster services and schools} \\ 1.0 & \text{for all other buildings} \end{cases}$$

and the foundation factor

$$F = \begin{cases} 1.5 & \text{for structures founded on soils having low dynamic shear modulus such as highly compressible soils} \\ 1.0 & \text{for all other soils} \end{cases}$$

-1.2 Distribution of base shear—The total lateral force or base shear, V, is to be divided into two parts: a concentrated load, F_t, not exceeding 0.15 V, applied at the top of the structure (to simulate the whiplash effect due to the higher modes of vibration) for buildings with a height-to-depth ratio greater than 3, and the balance, $(V - F_t)$, to be distributed over the entire height of the building, generally as concentrated loads at the floor levels. The latter is to be distributed in a triangular manner (a distribution corresponding essentially to the fundamental mode of vibration response) increasing from zero at the base to a maximum at the top.

The top load, F_t, is given by

$$F_t = \begin{cases} 0.004\, V \left(\dfrac{h_n}{D_s}\right)^2 \text{ but not} > 0.15\, V, \quad \text{for } \left(\dfrac{h_n}{D_s}\right) > 3 \\ 0 \quad \text{when } \left(\dfrac{h_n}{D_s}\right) \leqslant 3 \end{cases} \tag{12-23}$$

where

D_s = the plan dimension (ft) of the lateral force resisting system in a direction parallel to the applied forces. When the lateral force resisting system is composed of several elements of different plan dimensions, an equivalent dimension should be determined which will represent the action of the building as a whole.

The magnitude of the distributed forces, F_x, making up the balance of the total lateral force, i.e., $(V - F_t)$, is given by

$$F_x = \begin{cases} \dfrac{(V - F_t)\, w_x h_x}{\sum\limits_{i=1}^{n} w_i h_i}, & \text{in general} \\ \dfrac{(V - F_t)\, w_x}{\sum\limits_{i=1}^{n} w_i}, & \text{for two story buildings (optional)} \end{cases} \tag{12.24}$$

where

w_x, w_i = that portion of the total weight, W, which is located at or assigned to level x or i, respectively

h_x, h_i = the height (ft) above the base to level x or i, respectively

Level n = the uppermost level in the main portion of the structure

At each level x, the force F_x is to be applied over the area of the building in accordance with the mass distribution at that level.

At the top floor, the total horizontal load will consist of the sum of F_t and the distributed load $(F_x)_{x=h_n}$. Figure 12-58 shows a typical distribution of forces for a multistory structure having a uniformly distributed dead weight.

-1.3 Setbacks—When the plan dimension of a setback structure in each direction is at least 75% of the correspond-

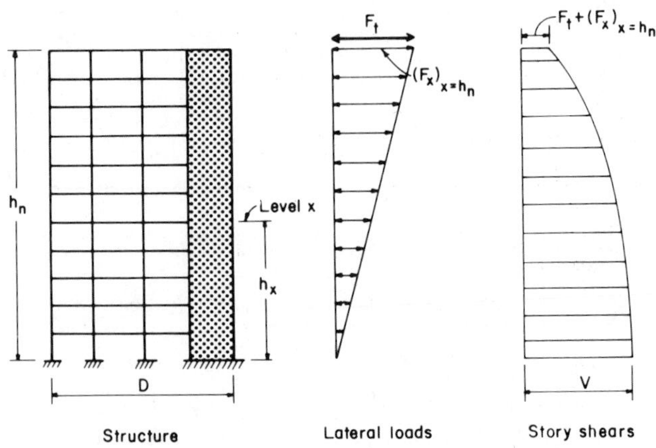

Fig. 12-58 Typical distribution of code-specified static lateral forces and story shears in a building with uniform mass-distribution.

ing plan dimension of the supporting structure, the codes allow the determination of the design seismic forces by considering the building as uniform.

Setbacks which have dimensions less than 75% of the corresponding dimensions of the base structure are to be designed as separate structures, using for the base shear the larger value obtained by considering the tower or setback either as a separate building for its own height or as part of the overall structure.

Appendix C of the *SEAOC Recommendations* gives more detailed recommendations for the treatment of buildings with setbacks.

-1.4 Limitations in height of buildings using certain structural systems–*SEAOC-68,* as well as *UBC-70,* requires that buildings exceeding 160 ft in height have ductile moment resisting space frames capable of resisting at least 25% of the specified lateral forces on the entire structure. This requirement is intended to provide such buildings with an extra margin of safety in the event that the deformations induced by an earthquake should exceed the capacity of the less ductile elements, such as shear walls. The requirement implies that buildings without such 25% lateral-load frames, such as box-type or shear wall buildings consisting of assemblages of interconnected walls, may have a height exceeding 160 ft. No height limit is imposed on frame buildings or on frame-shear wall buildings which have a moment resisting space frame satisfying the above 25% lateral-load-resisting capacity requirement.

NBCC-70 uses a greater height limit of 200 ft for structures not having a 25% lateral-load frame. In addition, it allows buildings exceeding 200 ft in height located in Canadian seizmic Zone I to have concrete shear walls instead of the required moment resisting space frames, provided such shear walls are designed to ensure their ductile behavior.

On the basis of the requirement for frame-shear wall buildings designed for a K-factor of 0.80 [see eq. (12-20) and Table 12-3], the above 25% lateral-load capacity requirement may be understood as specifying a minimum design lateral load on the moment resisting space frame, when acting independently of any shear wall or other resisting element, equal to 25% of that corresponding to the entire structure.

-1.5 Drift–Appendix A of the *SEAOC Recommendations* suggests a maximum drift or lateral deflection under the code-specified lateral forces twice that normally used for wind. In terms of the deflection index, i.e., the ratio of the drift at the top of a building to its total height, or alternatively, the ratio of the relative story displacement to the story height, the values which have been used as a basis for wind design range from 1/400 to 1/600. The SEAOC Commentary also states that drift need not be considered for buildings with 13 stories or less, or when the height-to-width ratio is 2.5 or less.

In tall buildings, the drift limitation under wind loading, rather than stress considerations, often governs the design. To satisfy the drift limitation under wind loading, additional stiffness may have to be provided–by increasing frame member sizes (generally, the beams) or introducing shear walls–beyond that required to support the gravity loads. The limitation of drift has been used as a convenient criterion to keep within safe and tolerable levels the effects of lateral sway under wind loading on (a) the stability of the individual columns as well as the structure as a whole, (b) the integrity of nonstructural partitions and glazing, and (c) the comfort of the occupants of such buildings. Under strong earthquake motion, the major consideration insofar as drift is concerned is the stability of the structure under the action of gravity loads when undergoing significant lateral displacements.

-2. *Analysis for member forces*–With the lateral loads at the floors levels determined, the next step would be an analysis of the structure for the corresponding member forces. Before undertaking an analysis of the structure under the design seismic loads, however, it is advisable to determine the corresponding design wind loads and compare these with the former. In many instances, it becomes readily apparent which of the two loading cases is the more critical, so that the lateral load analysis of the structure need be carried out only for the critical loading. Where the critical loading condition is not immediately apparent because of the different distribution of loads which are of about the same order of magnitude, analyses of the structure considering both loading cases may be desirable.

For relatively low buildings, an approximate analysis, such as by the portal method, may be carried out. When using the portal method, however, allowance must be made for the fact that in regions of discontinuity, such as the base and top of a frame, the point of contraflexure (ordinarily assumed at mid-height) in the columns generally does not occur at mid-height. The results of the portal method for these regions may be improved by adjustments in the assumed location of the point of contraflexure in the columns. For this purpose, the data found in Ref. 12-84 will be found useful.

In structures consisting of frames and shear walls or of linked shear walls the variation in stiffness along the height of which differ significantly, the distribution of the applied lateral load in proportion to the relative stiffnesses of the resisting elements at each story level can lead to erroneous results. For these cases, the interaction between connected elements having different deformation patterns under lateral load must be considered.

Figure 12-59 shows the results of a static elastic analysis of a typical 20-story frame-shear wall structure under lateral load. The curves indicate the distribution of the horizontal story shears between frame and shear wall along the height of the structure. It is worth noting that the total shear carried by the frame at the top stories can exceed the applied story shear at these levels. It is clear that distributing the applied shear at each level in proportion to the relative stiffnesses of the resisting elements can lead to errone-

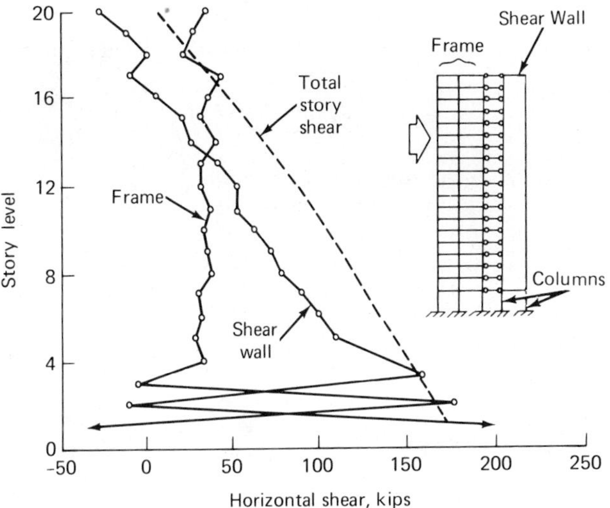

Fig. 12-59 Distribution of horizontal story shears in a typical frame-shearwall structure.[12-85]

ous results in this case. On the other hand, assuming the shear wall to carry all the lateral load can result in underdesigned frame members.

The same figure shows the sizable increase in the horizontal shears which occurs in the lower stories of a system in which the shear wall is supported at the base by columns. The shears shown for the lowest story of the shear wall are those corresponding to the two supporting columns. The character of the shear distribution curves shown for the lower stories is typical of the force concentrations which occur in regions of major geometric or stiffness discontinuities in any structural system. While yielding of the columns will tend to relieve the high stresses accompanying such large shears, the fact that these forces can exceed the total base shear for such configurations is worth noting. In structures subjected to earthquakes, such an arrangement can lead to early yielding in the columns supporting the shear wall, with corresponding large ductility requirements.

For tall structures, an analysis which takes account not only of the axial deformation in columns but also the finite width of members would be desirable. For this purpose, the use of a computer program for frame analysis would be most convenient.

-2.1 Overturning moment—The lateral loads determined from a distribution of the design base shear are intended to represent, to some measure, the envelope of maximum forces at the different story levels. Earlier versions of the *SEAOC-UBC* allowed a reduction in the overturning moments calculated on the basis of these forces, because these maximum forces generally do not occur simultaneously. Such a reduction would reflect the effect of the higher modes of vibration on the response of the structure since in the higher modes, the lateral displacements, and hence the associated forces, change direction at each node.

The latest (1971) revisions of the *SEAOC Recommendations* and the *Uniform Building Code,* however, no longer allow such a reduction of the overturning moment calculated on the basis of the distributed base shear. This essentially assumes that the response of a structure is described entirely by the fundamental mode response.

The *National Building Code of Canada 1970* (*NBCC*) allows a reduction in the overturning moment as calculated from the distributed base shear. The reduction factor for the base overturning moment, J, decreases in value as the period, T, of a structure increases, and is given by

$$J = 0.5 + 0.25/\sqrt[3]{T^2} \leqslant 1.0.$$

The overturning moment at the base is then calculated as

$$M = J\left(F_t h_n + \sum_{i=1}^{n} F_i h_i\right).$$

At intermediate story levels, a variable reduction factor, J_x, which varies from 1.0 at the top of a building to J at the base is specified, i.e.,

$$J_x = J + (1 - J)\left(\frac{h_x}{h_n}\right)^3,$$

so that the overturning moment at any level h_x above the base is obtained as

$$M_x = J_x\left[F_t(h_n - h_x) + \sum_{i=x}^{n} F_i(h_i - h_x)\right].$$

The preceding discussion has considered the determination of design forces on the elements forming the structure under the action of loads directed parallel to one of the principal plan axes of the structure. The analysis of the structure under loads acting in the perpendicular direction can be carried out similarly. For this purpose, the structure may be assumed to consist of two sets of mutually perpendicular plane frames, each set of frames being analyzed separately for the corresponding lateral loads. The codes require that a structure be designed for lateral loads, acting nonconcurrently, in the direction of each of the main axes of the structure. Members of a structure common to both directions, i.e., columns, may be proportioned for the lateral loads acting along one axis (the more critical direction) and then checked for adequacy with respect to the lateral loads acting in the perpendicular direction. Note, however, that columns, and particularly corner columns, will generally be subjected to biaxial moments due to the gravity loads.

-2.2 Minimum horizontal torsional moments—*SEAOC-UBC* requires that where the vertical lateral load resisting elements are connected together by a floor system which is fairly rigid in its own plane, such as a monolithically cast reinforced concrete slab, such elements should be designed for the direct shears plus torsional shears corresponding to a minimum eccentricity of the story lateral loads not less than 0.05 of the maximum building dimension at that level. Torsional shears which are directed opposite to the direct shears are to be neglected.

NBCC, on the other hand, requires the use of a design eccentricity at each story level equal to "1.5 times the computed eccentricity, increased by an accidental eccentricity equal to 0.05 times the plan dimension in the direction of the computed eccentricity considered."

The additional shears due to horizontal torsional moments may be calculated approximately by assuming the vertical elements at each story to be fixed at their ends to parallel rigid plates. The torsional shear force acting on each element may then be taken as proportional to its lateral stiffness and its distance from the center of rigidity of the story considered.

The location of the center of rigidity or center of rotation, c_r, under the above assumption, is determined by a method similar to that used to obtain the center of rotation of a rivet group, except that one uses the lateral stiffness* of the vertical elements in place of the cross-sectional area of the rivets. This recognizes the flexural action of the vertical elements as the principal mode of resistance to horizontal torsion. Thus, referring to Fig. 12-60, if k_x and k_y are the lateral stiffnesses of a particular element along the x- and y-axis, respectively, then the coordinates of the center of rotation, x_r and y_r, with respect to an arbitrary origin, such as the point O in the figure, is given by

$$x_r = \frac{\sum k_y x}{\sum k_y} \quad \text{and} \quad y_r = \frac{\sum k_x y}{\sum k_x},$$

the summation being taken over all the vertical elements in the story.

The shear due to torsion along each coordinate axis resisted by a particular element will then be proportional to the lateral stiffness of the element, relative to the total rotational stiffness of the story, and its distance from the center of rigidity. The total rotational stiffness of the story, J_r, about the center of rigidity, is given by

$$J_r = \sum (k_x \bar{y}^2 + k_y \bar{x}^2)$$

where $\bar{x}$ and $\bar{y}$ are the coordinates of the centroid of a particular element (in plan) referred to the center of rigidity. Note the similarity between J_r and the polar moment of inertia of a rivet group about its centroid. The shears along

*i.e., the force required to displace one end of a vertical element a unit horizontal distance relative to the other end.

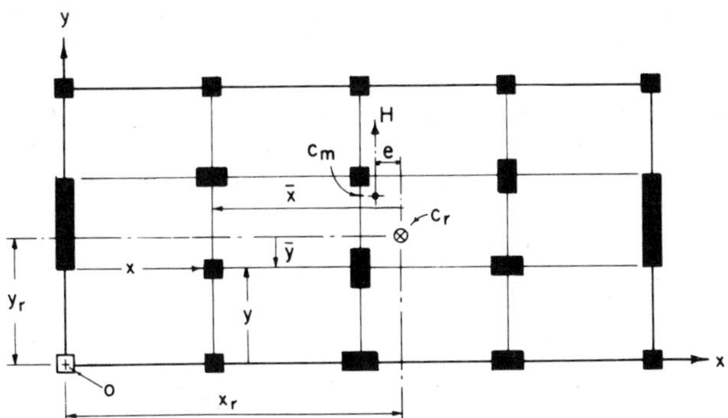

Fig. 12-60 Calculating shears due to horizontal torsional moments.

each axis due to a horizontal torsional moment $M_r = He$, for a particular element with stiffness k_x and k_y is then obtained from

$$V_x = \frac{M_r \bar{y}}{J_r} k_x, \text{ and } V_y = \frac{M_r \bar{x}}{J_r} k_y.$$

-2.3 Lateral forces on parts or portions of buildings—Although the failure of ornamentations, marble veneers and parapets or similar appendages in buildings seldom affect the structural integrity of the building, they represent a serious menace to the safety of the occupants of such buildings as well as passersby. Numerous cases of such failures have clearly indicated the need to ensure that portions of buildings, such as parapets, be designed to resist the expected earthquake-induced forces and, more important, that they be adequately tied to the rest of the structure.

SEAOC-UBC, as well as *NBCC*, specify design forces for parts or portions of buildings and other structures in terms of a coefficient, C_p, and the weight of the part considered. thus,

$$F_p = C_p W_p.$$

In the above equation, F_p is the design earthquake force and W_p is the weight of the portion of the structure considered. Values of the coefficient C_p as given in the *SEAOC Recommendations* are shown in Table 12-4. Note that the design force on cantilevered parapets and walls are to be applied normal to the plane of such members while for other parts or appendages the specified force is to be applied in any direction. In all cases, the intent is to design the part and its anchorage for the most unfavorable loading condition.

TABLE 12-4 Horizontal Force Factor "C_p" for Parts or Portions of Buildings or Other Structures.[12-10]

PART OR PORTION OF BUILDINGS	DIRECTION OF FORCE	VALUE OF C_p
Exterior bearing and nonbearing walls, interior bearing walls and partitions, interior nonbearing walls and partitions over 10 feet in height, masonry or concrete fences over 6 feet in height[1]	Normal to flat surface	0.20
Cantilever parapet and other cantilever walls, except retaining walls	Normal to flat surface	1.00
Exterior and interior ornamentations and appendages	Any direction	1.00
When connected to or a part of a building: towers, tanks, towers and tanks plus contents, chimneys, smokestacks, and penthouses	Any direction	0.20[2]
When resting on the ground, tank plus effective mass of its contents	Any direction	0.10
Floors and roofs acting as diaphragms[3]	Any direction	0.10
Connections for exterior panels or for elements complying with Section 2314 (k) 5	Any direction	2.00
Connections for prefabricated structural elements other than walls, with force applied at center of gravity of assembly[4]	Any horizontal direction	0.30

[1]See also Section 2312 (b) for minimum load on deflection criteria for interior partitions.

[2]When "h_n/D" of any building is equal to or greater than five to one increase value by 50 per cent.

[3]Floors and roofs acting as diaphragms shall be designed for a minimum value of "C_p" of 10 per cent applied to loads tributary from that story unless a greater value of "C_p" is required by the basic seismic formula $V = ZKCW$.

[4]The "W_p" shall be equal to the total load plus 25 per cent of the floor live load in storage and warehouse occupancies.

-3. *Load factors and loading combinations to be used as bases for design*—Codes generally require that the ultimate strength of a structure and its component elements be equal to or greater than the forces corresponding to any of a number of loading combinations which may reasonably be expected during the life of the structure. The usual method for allowing for the differing variabilities of specific loading conditions is to apply different load factors to the working loads (or their induced forces) before combining these to obtain a particular design ultimate load. Different load factors are usually employed for different combinations of the separate types of loads.

Thus, ACI 318-71 requires that a structure be designed for an ultimate strength* not less than that resulting from any of the following combinations of loads:

$$U = \begin{Bmatrix} 1.4D + 1.7L \\ 0.75\ [1.4D + 1.7L \pm (1.7W \text{ or } 1.87E)] \\ 0.9D \pm (1.3W \text{ or } 1.43E) \\ 0.75\ [1.4\ (D + T) + 1.7L] \end{Bmatrix} \tag{12-25}$$

where

U = required strength to resist the design* loads or their related internal moments and forces
D = dead loads or their related internal forces
L = live loads or their related internal forces
W = wind load or its related internal forces
E = load effects of earthquake or their related internal forces
T = load effects of contraction and expansion caused by temperature changes, creep, shrinkage, or differential settlement.

The *SEAOC Recommendations*, as well as the *Uniform Building Code*, calls for designs to satisfy the following ultimate load combinations:

$$U = \begin{Bmatrix} 1.40\ (D + L \pm E) \\ 0.90D \pm 1.40E \end{Bmatrix} \tag{12-26}$$

with the added proviso that a value of $2.80E$ shall be used instead of $1.40E$ in the second of the above equations for

*It will be noted that the word 'ultimate' is no longer used in ACI 318-71, but instead, the word 'strength' is used when referring to the ultimate strength design method. The adjective 'design' is used, as in 'design moments,' to indicate the ultimate or (load) factored moments. Working stress design is referred to as the 'alternate design method.'

calculating shear and diagonal tension in buildings without a 100% moment resisting space frame. Note that the first of eqq. (12-26) above applies a load factor of 1.4 to the dead loads, when acting in combination with live loads and earthquake loads. This contrasts with the 1.05 factor used in the second of eqq. (12-25) of ACI 318-71, which, when multiplied out, gives: $U = 1.05D + 1.28L \pm (1.28W \text{ or } 1.4E)$.

Supplement No. 4* to the *National Building Code of Canada 1970* requires the use of the most unfavorable of the following loading combinations for the ultimate strength design of structural members:

$$U = \left\{\begin{array}{l} 1.5D + 1.8\ (L \text{ or } W \text{ or } E \text{ or } T) \\ 1.15D + 1.35\ [L \pm (W \text{ or } E \text{ or } T)] \\ 0.9D \pm 1.35\ (W \text{ or } E) \\ D + 1.20\ [L + T \pm (W \text{ or } E)] \end{array}\right\} \qquad (12\text{-}27)$$

The ± sign before W and E in the above expressions allows for the possibility of a critical combination of loads arising from the reversal in direction of these loads due to a change in the direction of the steady wind component and the vibratory character of earthquake response.

In the process of proportioning structural members, capacity reduction factors, ϕ, designed to account for variations in material strength, the nature of the expected failure mode as well as the importance of a member to the safety of the structure as a whole, are used either in calculating the effects of the design forces or in determining the permissible stresses. Values of ϕ for the principal types of members or modes of resistance are identical in all three codes considered here, except that *NBCC* makes no distinction between spiral and tied columns and uses a common factor of 0.75 for both cases. ACI 318-71 allows a linear increase in the ϕ-factor from 0.70 or 0.75, corresponding to axial compression, to 0.90 as the axial load decreases from P_b at balanced condition to zero (i.e., pure flexure), for members subjected to combined bending and axial load.

The proposed 1973 edition of *UBC* requires the use of a ϕ-factor of 1.0 and the use of $1.25f_y$ instead of f_y when calculating the moment capacity of flexural members for the purpose of designing the shear reinforcement in the members.

-4. Code provisions designed to ensure ductility in reinforced concrete members—The principal code provisions designed to ensure the necessary ductility in reinforced concrete structures which may be subjected to strong earthquake motions as contained in Appendix A of ACI 318-71 will be discussed below.

The inclusion of provisions for earthquake-resistant design of reinforced concrete structures in the 1971 revision of the *ACI Building Code* is an entirely new addition to the code. The provisions of Appendix A supplement the main body of the *Code* and cover the design of ductile moment-resisting space frames and special shear walls of cast-in-space reinforced concrete.

Although Appendix A does not specify the magnitude of the earthquake forces to be used in design, its provisions are intended for structures which are designed to so-called equivalent static forces which are substantially less than the dynamic inertial forces which may be expected to be developed if these structures were to respond elastically to an earthquake of moderate-to-strong intensity. Such structures, observations and analyses have shown, must depend for survival on their capacity to absorb a significant portion of the vibrational energy resulting from an earthquake through inelastic deformations at critically stressed regions. Thus, the principal aim of the provisions of Appendix A is to ensure adequate ductility in structures, that is, to provide structures which are capable of undergoing relatively large inelastic deformations in local regions without significant loss of strength.

For structures located in regions of relatively minor earthquake intensity and designed for the estimated earthquake-induced forces, very little inelastic deformation may be expected. In such cases, the ductility provided by designing to the provisions contained in the main body of the code will generally be sufficient.

As in the main body of the Code, a major object of the design provisions in Appendix A is to have the strength of a structure determined by a ductile type of flexural failure mechanism. The main difference lies in the relatively greater range of deformation—with yielding actually expected—and hence greater ductility required in designs for resistance to major earthquakes. Again, it is pointed out that the need for greater ductility results from the design philosophy which allows the use of reduced forces in proportioning members and provides for the inelastic deformations which are expected under severe earthquakes by special ductility requirements. This philosophy is based on the recognition that it is generally uneconomical to proportion structures, and buildings in particular, to resist major earthquakes within their elastic range of stresses. The strongest justification of this philosophy has been the satisfactory performance of many structures so designed which have been subjected to actual earthquakes.

The provisions for ductility given in Appendix A are aimed at preventing the brittle or abrupt types of failures associated with inadequately reinforced and overreinforced members failing in flexure, as well as shear (i.e., diagonal tension) and reinforcement anchorage failures. In addition, because the critical regions in frame structures generally occur at and near joints, Appendix A requires the use of adequate confinement reinforcement to insure the proper performance and integrity of these regions when subjected to large amplitudes of reversed loading.

A provision unique to earthquake-resistant design, insofar as it applies to frames, is the requirement that the sum of the moment strengths of the columns in a beam-column connection be greater than that of the beams framing into the joint. This is intended to ensure that yielding in such frames occurs in the beams rather than in the columns and hence preclude any instability effects which may result from plastic hinges forming in the columns. This requirement often results in column sizes which are larger than would otherwise be required, particularly in the upper floors of multistory buildings with appreciable beam spans.

The major provisions of Appendix A of ACI 318-71 will be discussed below in relation to the type of element to which they apply.

-4.1 Limitations on material strengths—Appendix A requires a minimum specified concrete strength, f'_c, of 3000 psi and a maximum specified yield strength of reinforcement, f_y, of 60,000 psi. These limits are imposed as reasonable bounds on the variation of the material properties, particularly with respect to their unfavorable effects on the sectional ductility of members in which they are used. As was pointed out in section 12.5.4, a decrease in the concrete strength and an increase in the yield strength of the tensile reinforcement tend to decrease the ultimate curvature and hence the sectional ductility of a member subjected to flexure. Also, an increase in the yield strength of

*"Canadian Structural Design Manual—1970," NRC No. 11530, issued by the Associate Committee on the National Building Code, National Research Council of Canada, Ottawa.

the steel reinforcement is generally accompanied by a decrease in the ductility—as measured by the maximum deformation—of the material itself. A decrease in material ductility, whether in the compression zone concrete or in the tensile reinforcement, will be reflected directly in a decrease in the sectional ductility of a member.*

-4.2 Flexural members (beams)—The significant provisions relating to flexural members have to do with

a. Limitation on flexural reinforcement ratio:

$$\rho_{min} = 200/f_y;\ \rho_{max} = 0.50\ \rho_{balanced}$$

b. Minimum positive moment capacity at column connections:

$$M_y^+\ (\text{end}) \geqslant .50\ M_y^-$$

c. Anchorage of longitudinal steel.

d. Web reinforcement for both shear and confinement—to preclude shear failures prior to the development of plastic hinges at ends.

a. Flexural reinforcement limitations.—Because the ductility of a flexural member decreases with increasing values of the reinforcement ratio, ρ, the *Code* limits the maximum value of this ratio of 0.50 of the ratio corresponding to the balanced condition. This limitation is intended to ensure adequate ductility even when the actual yield strength of the reinforcement exceeds the specified strength by a substantial margin. From a practical standpoint, the lower steel ratios should be used whenever possible.

It is pointed out that in the 'strong column-weak beam' frame intended by the *Code,* the relationship of moment capacity between columns and beams may be upset if the beams turn out to have greater moment capacity than intended by the designer. Thus, the substitution of 60-ksi steel of the same area for 40-ksi steel in beams which may be called for in the plans can be detrimental. Also, the shear strength of beams and columns, which is generally based on the condition of plastic hinges forming at the ends of the beams, may become inadequate if the moment capacity of the beams becomes greater than intended as a result of the use of a higher strength longitudinal steel.

The selection of the number, size and arrangement of flexural reinforcement must be made with full consideration of the detailing and construction aspects. This is particularly important in relation to beam-column connections, where construction difficulties can arise as a result of reinforcement congestion. The preparation of large-scale detail drawings of the connections, showing all beam, column and joint reinforcement will help eliminate unanticipated problems in the field. The inclusion of such large-scale details in the plans will pay dividends in terms of lower bid prices and a smooth-running construction job.

b. Positive moment capacity at beam ends.—In order to allow for the possibility of the positive moment at one end of a beam due to earthquake-induced lateral displacement exceeding the negative moment capacity due to gravity loads, the *Code* requires a minimum positive moment capacity at beam ends equal to fifty percent of the negative moment capacity.

c. Anchorage of flexural reinforcement.—Except in very large columns, it is usually not possible to develop the yield strength of a reinforcing bar from a framing beam within the width of a column. Where beam reinforcement can extend through a column, its capacity is developed within the compression zone of the beam on the far side of the connection (see Fig. 12-86). Where no flexural member occurs on the opposite side of a column, such as in exterior columns, the flexural reinforcement in a framing beam has to be developed within the confined regions of the column. For such cases, the *Code* requires every bar to have a standard 90° hook plus whatever extension is necessary to develop the bar, the development length being measured from the near face of the column. The *Code* allows a 33% reduction in the required development length (as specified in Chapter 12 of the *Code*) when bars are anchored within confined regions.

In order to allow for load combinations not accounted for in design, the *Code* requires at least one-third of the negative moment reinforcement near the support to be extended beyond the extreme position of the point of contraflexure by the required anchorage length. In addition, one-quarter of the larger amount of the tension reinforcement required at either end of the beam is to be made continuous throughout the top of the beam. (see Fig. 12-61).

Splices in regions of tension or reversing stresses should be avoided whenever possible. Where necessary, the *Code* allows splices (equal to 24 bar diameters or 12 in.) provided these are confined by stirrup ties, a minimum of two stirrup ties being required at each splice.

d. Web reinforcement.—Because the ductile behavior of an earthquake-resistant frame designed to the present codes is premised on the ability of the beams to develop plastic hinges with adequate rotational capacity, it is essential to ensure that the members do not fail in shear before the flexural capacity of the beams has been developed.

The stirrups or stirrup ties, which must be placed perpendicular to the longitudinal reinforcement, are designed for the maximum shear resulting from the simultaneous

*Although plain unconfined concrete of low strength tends to exhibit slightly greater ductility than higher strength concrete, the net effect of an increase in the concrete strength is generally an increase in the sectional ductility.

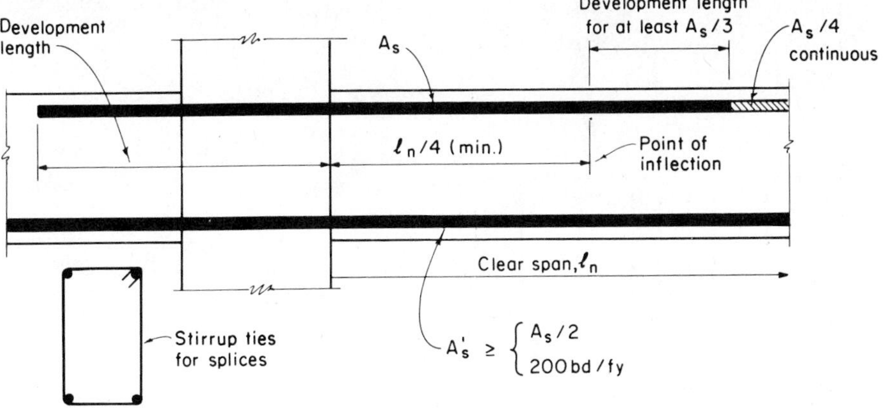

Fig. 12-61 Anchorage of beam reinforcement in beam on opposite side of column.

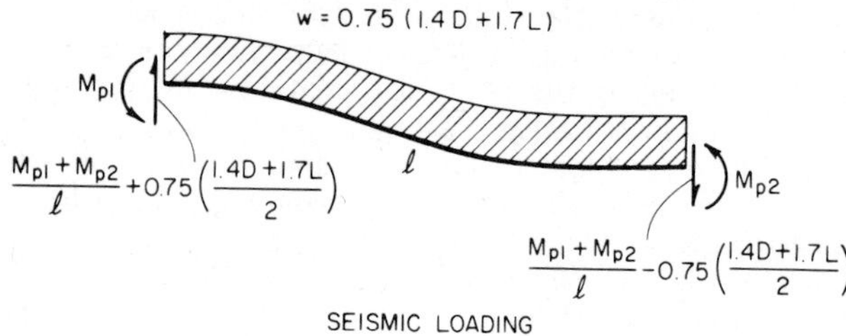

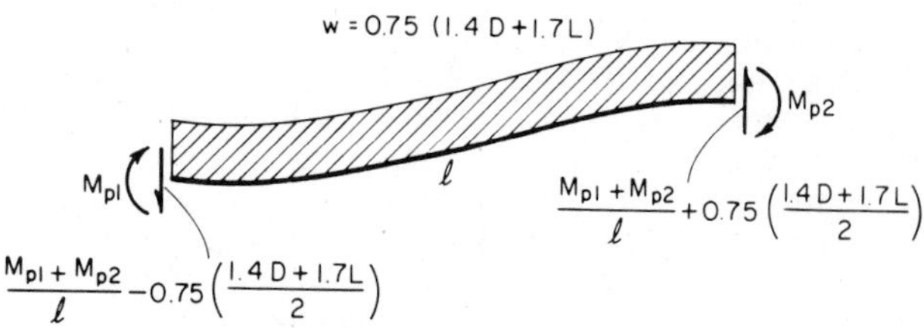

Fig. 12-62 Loadings for design of transverse reinforcement in beams.

action of the gravity loads and the shear corresponding to plastic hinges at both ends, due to the lateral displacement of the frame in either direction. The two conditions to be considered are shown in Fig. 12-62 for the case of a uniformly distributed gravity load.

Because of the direct dependence of the web reinforcement design on the specified yield strength of the flexural reinforcement, any unintended overstrength in the latter could result in a nonductile shear failure preceding the development of the full flexural capacity of a member.

To allow for load combinations unaccounted for in design, a minimum amount of web reinforcement is required throughout the length of all flexural members. Within regions of potential hinging, a closer spacing of stirrups is required (see Fig. 12-63).

Stirrup ties, which are stirrups with closed tops, are required in regions where longitudinal bars are required to act in compression. These not only restrain the longitudinal compression bars from buckling but also help confine the compression zone concrete.

e. Splices.—Splices in regions of tension or reversing stresses, and particularly at regions of potential hinging, should be avoided whenever possible. When necessary at such locations, the *Code* allows splices (equal to at least 24 bar diameters or 12 in.) provided these are confined by stirrup ties, a minimum of two such stirrup ties being required at each splice.

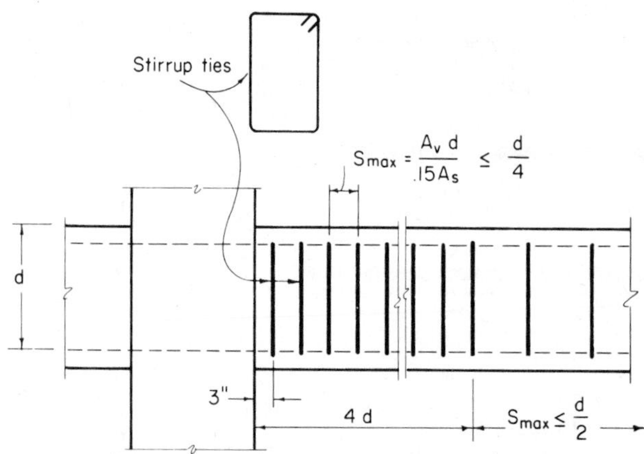

Fig. 12-63 Beam web reinforcement limitations, min. bar size—#3.

-4.3 Frame columns—The main provisions in Appendix A of ACI 318-71 governing the design of frame columns relate to

a. Limitation on reinforcement ratio:

$$\rho_{min} = 1\%, \ \rho_{max} = 6\%$$

b. Strong column-weak beam frame:

$$\sum M_y^{columns} > \sum M_y^{beams}$$

c. Special transverse (confinement) reinforcement when $P_e > 0.4 P_b$

d. Transverse reinforcement for shear, to develop plasting hinging in frame (generally in beams only)

a. Reinforcement ratio limitation.—Appendix A specifies a reduced upper limit for the reinforcement ratio in columns from the 8% of Chapter 10 to 6%. However, the designer will find that construction considerations will tend to place the practical upper limit on the reinforcement ratio, ρ, near 4%. Convenience in detailing and placing reinforcement in beam-column connections make it desirable to keep the column reinforcement ratio low.

b. Relative column-to-beam strength requirements.—In order to ensure the stability of a frame and maintain its vertical load-carrying capacity while undergoing large lateral displacements, the *Code* requires that inelastic deformation be generally restricted to beams. This is done by requiring that the sum of the flexural strengths of the columns meeting at a joint, under the design axial load, be greater than the sum of the moment strengths of the framing beams in the same plane (see Fig. 12-64).

A beam-column connection need not comply with the above provision when the strengths of the confined cores of the columns is sufficient to resist the applied design loads or when the capacity of the remainder of the structure is sufficient to resist the entire shear at that level, i.e., without reliance on the resisting capacity of the nonconforming connection. In the latter case, consideration has to be given to the changes in the distribution of forces in the structure which may result from the action, e.g. premature failure, of the nonconforming connection.

c. Special transverse (confinement) reinforcement when $P_e > 0.4 P_b$.—Columns subjected to axial loads less than 0.4 of that corresponding to balanced condition function essentially as flexural members and may be designed in accordance with the provisions governing beams, as discussed in the preceding section.

When the axial load on a column exceeds 0.4 P_b special transverse reinforcement has to be provided near the ends to ensure adequate strength after loss of the shell, should hinging take place in the columns (see Fig. 12-65). A minimum volumetric ratio of spiral reinforcement, ρ_s, equal to $0.12 f_c'/f_y$ is specified. This value will usually govern in the

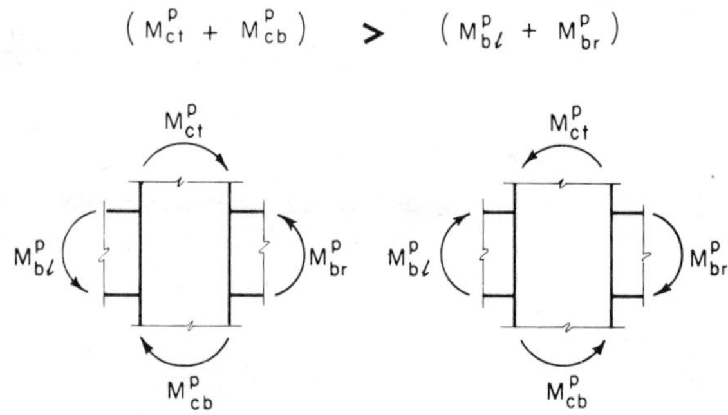

Fig. 12-64 "Strong column—weak beam" frame.

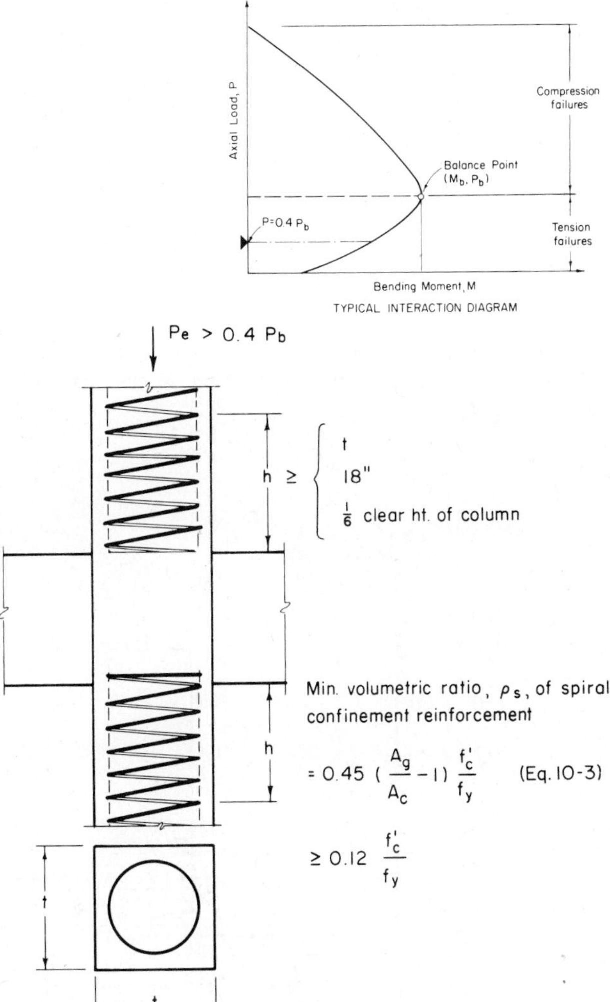

Fig. 12-65a Confinement reinforcement at column ends–spirals.

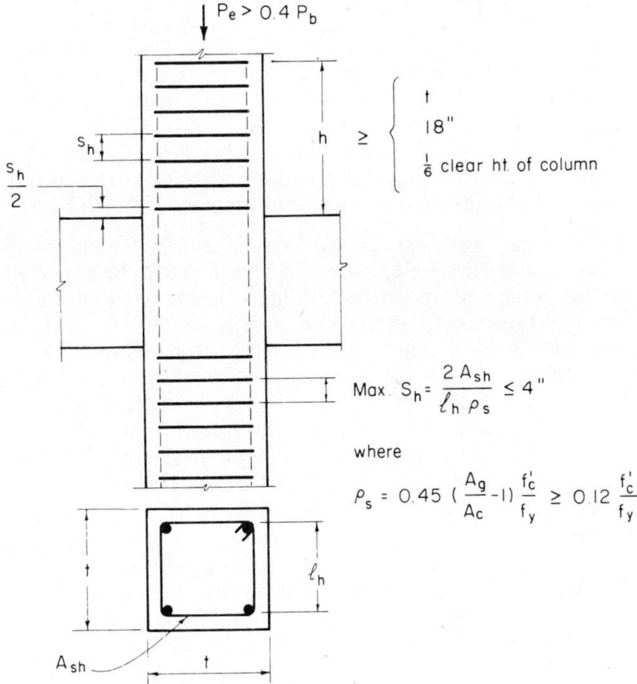

Fig. 12-65b Confinement reinforcement at column ends—rectangular hoops.

case of large columns, for which the alternative expression, as given by eq. (10-3) of the *Code*, i.e.,

$$\rho_s = 0.45\left(\frac{A_g}{A_c} - 1\right)\frac{f'_c}{f_y}$$

provides insufficient confinement. Circular spirals provide the most efficient form of confinement reinforcement. The extension of such spirals into the beam-column joint, however, may cause some construction difficulties.

Rectangular hoops, when used in place of spirals, are assumed to be only 50% as efficient as spirals with respect to confinement of the concrete core. Their efficiency is increased, however, by the use of supplementary crossties, which are to be bent around the peripheral hoop and secured to longitudinal bars (see Fig. 12-73). The *Code* specifies a maximum center-to-center spacing of 4 in. for hoops, the smallest permissible bar size being No. 3. The sequence of bar placement must be considered when using supplementary crossties since these will generally have to be in place on the hoops before the longitudinal bars are threaded through.

d. Transverse reinforcement–shear.—Transverse shear reinforcement in columns is designed to resist the maximum shear corresponding to the formation of plastic hinges in the frame. This will usually be governed by the plastic hinge capacity of the beams, but where column hinging is anticipated, the column hinging capacity should be used. If column hinging governs, the moment capacity to be used in calculating the design shear should be the highest value which can occur under the possible range of design axial loads. The axial load corresponding to the maximum moment capacity should then be used in computing the permissible shear stress in concrete, v_c.

The transverse reinforcement provided to satisfy the confinement requirement may be considered effective also in satisfying the shear requirement. Similarly, supplementary crossties may be used to satisfy the need for additional shear reinforcement or for lateral support of longitudinal bars. However, if only additional shear reinforcement is required, and no supplementary crossties are provided which may be used to satisfy the requirement, supplementary ties, bent around the longitudinal bars, may be conveniently used. Note the difference—mainly in the manner of placement—between supplementary *crossties* and supplementary *ties*.

e. Columns supporting discontinuous walls.—Because columns supporting discontinuous shear walls or stiff partitions are subjected to large shear (see Fig. 12-59) and compressive forces and can be expected to suffer significant inelastic deformations during strong earthquakes, the *Code* requires confinement reinforcement throughout the entire height of such columns (Fig. 12-66).

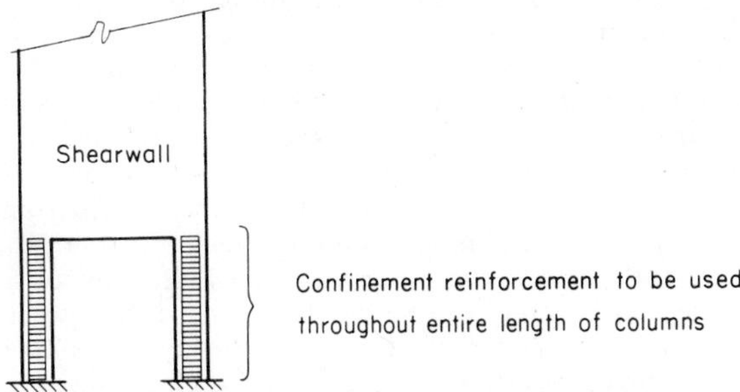

Fig. 12-66 Columns supporting discontinued shearwall.

f. Splices.—Appendix A provides for a minimum splice length in vertical reinforcement of 30 bar diameters or 16 in. Where welding or mechanical devices are used to establish continuity, not more than one-fourth of the bars are to be spliced at any level, the minimum distance between levels of splices of adjacent bars being 12 in.

-4.4 Beam-column connections—Generally, the beam-column connections in reinforced concrete buildings are not designed by the structural engineer, the detailing of bars within the joint usually being relegated to a draftsman or detailer. In buildings which may be subjected to major earthquakes and hence to significant member deformations, however, the design of beam-column connections requires as much attention as the design of the members themselves, since the integrity of the structure may well depend on the proper functioning of such connections. As mentioned earlier, beam-column joints represent regions of geometric and stiffness discontinuities in a frame and as such tend to be subjected to relatively high force concentrations. A substantial portion of the damage in frame structures subjected to strong earthquakes has been observed to occur at and near such connections.

Because of the congestion of reinforcement which may arise as a result of too many bars converging within the limited space of the joint, the proportioning of the frame columns and beams must be undertaken with due regard to the design of the beam-column connection. Little difficulty is usually encountered if the designer keeps the amount of longitudinal reinforcement used in the frame members to the low steel percentages. Also, the preparation of large scale detail drawings showing the bar arrangement within the joint will help much toward avoiding unexpected difficulties in the field.

The provisions in Appendix A relating to beam-column connections have to do mainly with:

a. Transverse reinforcement for
   (i) confinement
   (ii) shear
b. 50% allowable reduction in required transverse reinforcement for joints with four framing beams

a. Transverse reinforcement for confinement and shear.—The transverse reinforcement in a beam-column connection is intended to provide adequate confinement of the concrete to ensure its ductile behavior and allow it to maintain its vertical load-carrying capacity even after spalling of the outer shell, as well as provide sufficient shear strength to resist the forces transmitted by the framing members. The confinement also improves the bond between the steel and concrete within the connection. Thus, two separate requirements have to be met by the transverse reinforcement within a connection: one relating to confinement, as specified for the end portions of columns subjected to axial loads $P_e > 0.4 P_b$, and the other relating to shear within the connection.

As indicated in Fig. 12-67, the design of the transverse reinforcement for shear is based on the most critical combination of horizontal shears transmitted by the framing beams and columns, as well as the column axial load. Unless hinging occurs in the columns, the horizontal shears from the beams are determined from the condition of yielding of the negative reinforcement in at least one of the framing beams. Because the axial compressive load on the column significantly affects the permissible shear stress in concrete, v_c, the lowest value of the axial load consistent with the location of the joint in the frame and the column moments, where hinging occurs in the columns, should be used.

In a manner similar to the design of web reinforcement in a flexural member, the shear reinforcement in a beam-column joint is calculated on the basis of the truss analogy, using a nominal shear stress as a measure of diagonal tension. For most frame member proportions, the mode of shear resistance within a reinforced joint will most likely consist of a combination of a truss-type action, characterized by a diagonal compressive strut across the joint, and direct shear resistance.

When the calculated nominal shear stress in the connection exceeds the permissible stress in the concrete, the designer may either increase the column size or increase the depth of the beams. The former will increase the shear capacity of the joint section while the latter will tend to reduce the required amount of flexural reinforcement in the beams, with accompanying decrease in the shear transmitted to the joint.

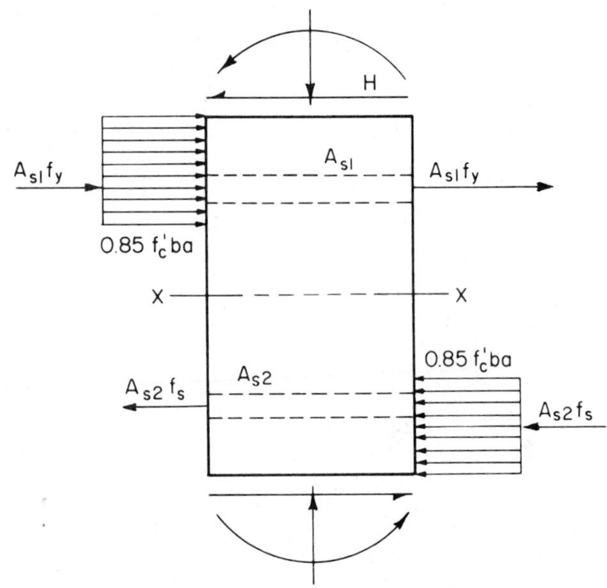

Fig. 12-67 Beam-column connection.

b. Reduction in transverse reinforce for joints with four framing beams.—The *Code* allows a 50% reduction in the amount of transverse reinforcement within a connection when beams satisfying certain relative size requirements frame on all four sides of a column. This provision recognizes the confinement and restraint provided by such beams.

-4.5 Shear walls—Shear walls are normally much stiffer than regular frame elements and therefore can be expected to be subjected to correspondingly greater lateral forces during response to earthquake motions. Because of their relatively greater depth, their lateral-deformation capacity is limited, so that, for a given amount of lateral displacement, shear walls will tend to exhibit greater apparent distress than frames. However, as indicated in Figs. 12-22 and 12-33, over a broad period range, a shear wall structure which is substantially stiffer, and hence having a shorter period, than a frame structure will suffer a lesser lateral displacement than the frame under the same ground motion. Shear walls with height-to-depth ratios greater than about 3 behave essentially as vertical cantilever beams and should therefore be designed as flexural members, with their strength governed by flexure rather than shear.

Because of the relatively lower lateral-deformation capacity of shear walls and the fact that such elements possess a lower degree of structural redundancy than frames of comparable stiffness, codes have tended to penalize shear walls by specifying larger K-factors for structures using shear walls.

The satisfactory performance of many shear wall and frame-shear wall structures[12-86] subjected to actual earthquakes attests to the reliability of structures which rely on properly designed and well founded shear walls for their lateral support.

The principal provisions in Appendix A relating to shear walls deal with the following:

a. Shear reinforcement to be designed to *un*reduced forces.
b. When $P_e < 0.4\,P_b$—minimum longitudinal reinforcement required at ends when maximum tensile stress $< 0.15\,f_r$.
c. When $P_e > 0.4\,P_b$—boundary elements to resist all vertical loads—confinement of longitudinal reinforcement in boundary elements throughout entire height.

a. Shear reinforcement in shear walls designed to unreduced forces.—Appendix A provides that no reduction in the design horizontal forces, as permitted in the case of ductile frames, be allowed in the design of the shear reinforcement for shear walls. The reduction mentioned here refers to the use of a K-factor less than unity in eq. (12-20) for the design base shear. This increase in the design force to be used in shear or diagonal tension calculations is in the same spirit as the required doubling of the load factor for earthquake loads in the second of eqs. (12-26) specified by *SEAOC-UBC* for buildings without a 100% moment resisting space frame, i.e., for buildings with shear walls. In Appendix A of ACI 318-71, however, the increase is less and applies only to the shear design of shear walls.

b. Shear walls with axial load $P_e < 0.4\,P_b$.—For shear walls with low axial load, i.e., with $P_e < 0.4\,P_b$, the reinforcement requirements are no different from those for ordinary walls as provided in the main body of the *Code*, except that a minimum steel area is required to be concentrated near the ends of the wall when the minimum tensile stress exceeds 15% of the modulus of rupture of the concrete, f_r, (see Fig. 12-68). This minimum steel requirement is similar to that for ordinary flexural members and is intended to prevent a sudden flexural failure after tension cracking occurs.

c. Shear walls with axial load $P_e > 0.4\,P_b$.—When the axial load on a shear wall exceeds $0.4\,P_b$, the *Code* requires the provision of vertical boundary elements capable of resisting all the vertical forces resulting from the design loads. As indicated in Fig. 12-68, confinement of the longitudinal reinforcement over the entire height of the boundary elements is also required.

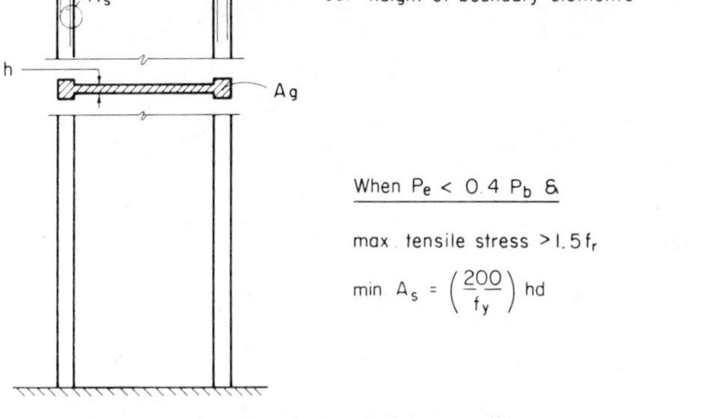

Min. distributed reinforcement each way = .0025 A_g

When $P_e > 0.4\,P_b$

(a) boundary elements to carry all vertical loads

(b) confinement reinforcement to be used throughout height of boundary elements

When $P_e < 0.4\,P_b$ &

max. tensile stress $> 1.5\,f_r$

min $A_s = \left(\frac{200}{f_y}\right) hd$

Fig. 12-68 Special shear walls.

12.7 DESIGN EXAMPLES—SAMPLE DESIGNS FOR ELEMENTS OF A 12-STORY FRAME-SHEAR WALL BUILDING

EXAMPLE 12-1: Determination of design forces—The application of the earthquake-resistant design provisions of *UBC-70* with respect to design loads and of ACI Appendix A in connection with the design of typical members of a 12-story frame-shear wall building located in seismic Zone III will be illustrated below.

The typical plan and elevations of the structure considered are shown in Figs. 12-69a, b and c. The columns and shear walls have a constant cross section throughout the height of the building,* the bases of the lowest story segments being assumed fixed. The beams and slabs also have the same dimensions at all floor levels. It is pointed out that the structure considered is a hypothetical one and has been chosen mainly for illustrative purposes. The other pertinent design data are as follows:

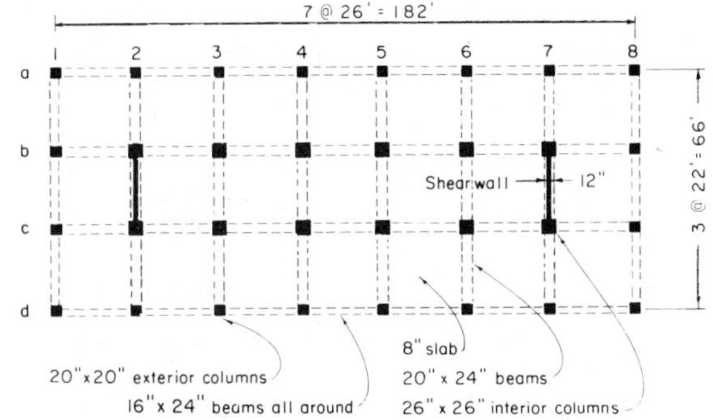

Fig. 12-69a Typical plan of building considered.

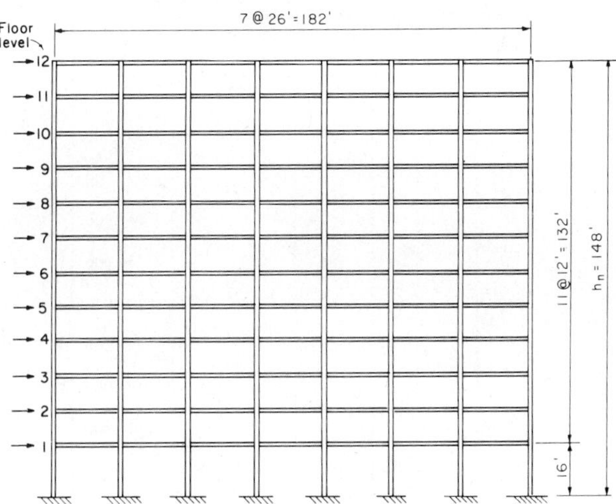

Fig. 12-69b Longitudinal section.

*The beneficial effect of such a uniform stiffness along the height on the dynamic response of a structure is discussed in section 12.5.1.6. Otherwise, the uniformity in member dimensions used in the above example has been adopted mainly for simplicity in the calculations and to avoid complicating the illustration needlessly.

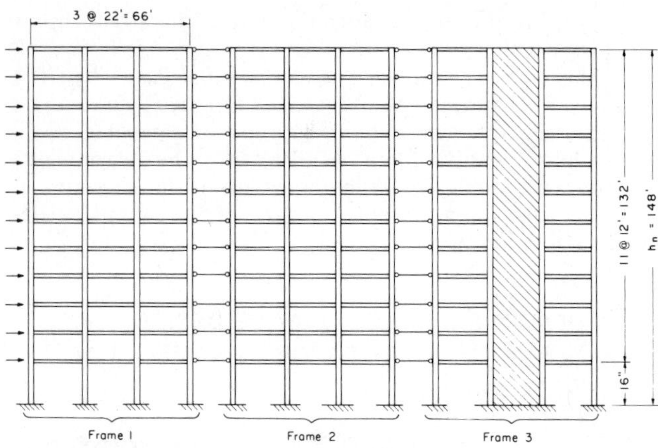

Fig. 12-69c Analytical model of building for lateral load analysis.

Service loads–vertical:

Live load $\begin{cases} \text{basic} = 50 \text{ psf} \\ \text{additional average load to allow for heavier} \\ \text{basic load on corridors} = 25 \text{ psf} \\ \text{Thus, total average live load} = 75 \text{ psf} \end{cases}$

Superimposed Dead load $\begin{cases} \text{average for partitions} = 20 \text{ psf} \\ \text{ceiling and mechanical} = 10 \text{ psf} \\ \text{Thus, total superimposed dead load} = 30 \text{ psf} \end{cases}$

The roof load will be assumed to be the same as that for a typical floor.

Material properties:

Concrete: $f_c' = 4000$ psi; $w = 145$ pcf

Reinforcement: $f_y = 60$ ksi

SOLUTION: On the basis of the above data and the dimensions shown in Fig. 12-69, the weight of a typical floor* and the roof have been estimated and tabulated in Tables 12-5 and 12-6. The calculation of the base shear, V, using eq. (12-20) for the transverse and longitudinal directions are shown at the bottom of Tables 12-5 and 12-6, respectively. Note that a value of $K = 1.0$ has been used for the transverse direction where one has a frame-shear wall structure** while a value of $K = 0.67$ is used for the longitudinal direction since one has a moment resisting frame along this axis. In the longitudinal direction, where the structure consists entirely of frames, the fundamental period, T, is taken equal to $0.10\,N$, where N is the number of stories.

It is interesting to note that an 'exact' calculation*** of the undamped natural periods of vibration of the structure in the transverse direction, using the story weights listed in Table 12-5 and member stiffnesses based on the gross concrete section, gave a value for the fundamental period of 1.21 sec. The mode shapes as well as the corresponding periods of the first five modes of vibration of the structure in the transverse direction are shown in Fig. 12-70. Note that the mode shapes indicate only the relative displacements of the story masses (assumed concentrated at the floor levels). The maximum displacement for each mode has been taken equal to unity.

The lateral seismic design forces resulting from the distribution of the base shear in accordance with eq. (12-24) are listed in Tables 12-5 and 12-6. Since $h_n/D_s < 3$ for both transverse and longitudinal directions, no concentrated force, F_t, is required at the top of the structure. Thus, for example, the seismic lateral force F_x at the 10th floor in the transverse direction is obtained from

$$F_x = \frac{Vw_x h_x}{\Sigma w_i h_i} = \frac{1400(2200)\,124}{2,144,300} = 177 \text{ kips}$$

Also shown in the tables are the story shears corresponding to the distributed seismic forces.

For comparison, the wind forces and story shears calculated on the assumption that the structure is located in an area with a specified wind pressure of 25 lbs/sq ft, are shown for each direction in Tables 12-5 and 12-6.

*The weight of a typical floor includes all elements located between the two imaginary parallel planes passing through the midheight of the columns above and below the floor considered.

**Note that a value of $K = 0.80$ is permissible for this type of structure provided that the three criteria specified in Table 12-3 are satisfied. In the example discussed above, only the first criteria stated in Table 12-3, i.e., forces corresponding to frame-shear wall interaction, is considered, hence the use the larger K-value of 1.0.

***Using the computer program described in Ref. 12-87.

TABLE 12.5 Design Lateral Forces in Transverse (Short) Direction (corresponding to entire structure)

		Seismic Forces				Wind Forces		
Floor Level (from base)	*Height h_x feet*	*Story Weight, w_x kips*	*$w_x h_x$ ft-kips*	*Lateral Force, F_x kips*	*Story Shear, ΣF_x kips*	*Wind Pressure (average) psf*	*Lateral Force, H_x kips*	*Story Shear, ΣH_x kips*
12 (Roof)	148	2100	311,000	203		40	43.8	
					203			43.8
11	136	2200	299,000	195		40	87.5	
					398			131.3
10	124		273,000	177		40	87.5	
					575			218.8
9	112		246,000	160		40	87.5	
					735			306.3
8	100		220,000	143		35	76.7	
					878			383.0
7	88		193,000	126		30	65.8	
					1004			448.8
6	76		167,000	109		30	65.8	
					1113			514.6
5	64		141,000	92		30	65.8	
					1205			580.4
4	52		114,000	75		26.6	58.3	
					1280			638.7
3	40		88,000	57		25	54.8	
					1337			693.5
2	28		61,500	50		21.6	47.3	
					1377			740.8
1	16		30,800	23		20	51.0	
					1400			791.8
Σ		26,300	2,144,300	1400			791.8	

Base shear, $V = ZKCW$, *where* $C = \dfrac{0.05}{\sqrt[3]{T}}$ *and* $T = \dfrac{0.05\,h_n}{\sqrt{D}}$ *in transverse direction,* $h_n = 148'$ *and* $D = 66' \Rightarrow T = 0.91,\ C = 0.052$

Thus, $V = (1)\,(1.0)\,(0.052)\,W = 0.052\,W = 0.052\,(26,300) = 1730$, *say* <u>1400 kips</u>

TABLE 12.6 Design Lateral Forces in Longitudinal Direction (corresponding to entire structure)

		Seismic Forces				Wind Forces		
Floor Level (from base)	*Height h_x feet*	*Story Weight, w_x kips*	*$w_x h_x$ ft-kips*	*Lateral Force, F_x kips*	*Story Shear, ΣF_x kips*	*Wind Pressure (average) psf*	*Lateral Force, H_x kips*	*Story Shear, ΣH_x kips*
12 (Roof)	148	2100	Same as for transverse direction	120	120	Same as for transverse direction	15.9	15.9
11	136	2200		116	236		31.6	47.5
10	124			106	342		31.6	79.1
9	112			94	436		31.6	110.7
8	100			85	521		27.8	138.5
7	88			75	596		23.8	162.3
6	76			65	661		23.8	186.1
5	64			55	716		23.8	209.9
4	52			44	760		21.1	231.0
3	40			34	794		19.9	250.9
2	28			24	818		17.1	268.0
1	16			12	830		18.5	286.5
Σ		26,300	2,144,300	830			286.5	

Base shear (in longitudinal direction), $V = ZKCW$ with $T = 0.10\,N = 1.2$ sec.
$C = 0.047$
Thus, $V = (1)(0.67)(0.047)\,W = 0.0315\,W = 0.0315\,(26{,}300) = 828$, say 830 kips

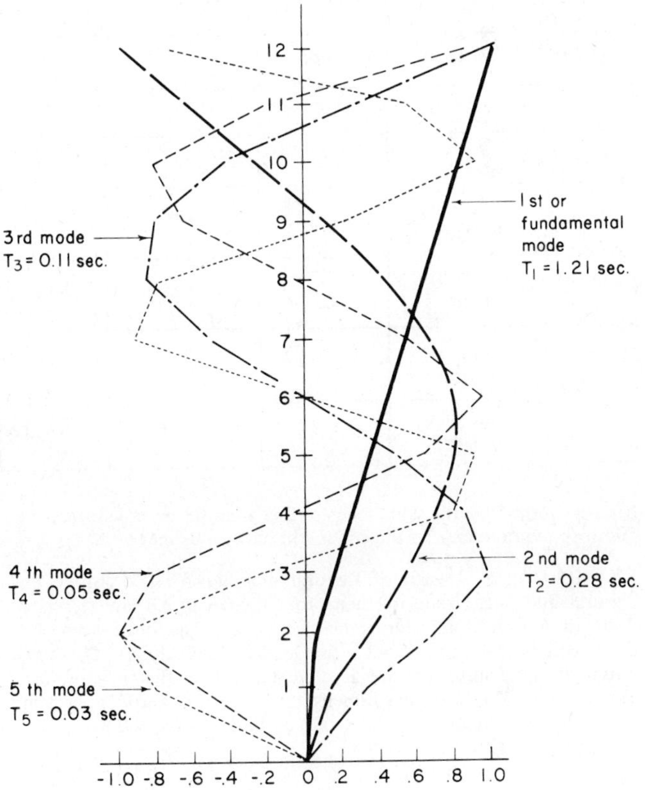

Fig. 12-70 Undamped natural modes and periods of vibration of structure in transverse direct ion.

Analyses of the structure for both directions under the respective seismic and wind loads, and assuming no torsional effects, was carried out using the computer program described in Ref. 12-88. For the purpose of analyzing the structure in the transverse direction, the model shown in Fig. 12-69c was used. This model consists of three different frames linked by hinged rigid bars at the floor levels to impose equal lateral deformations at these levels.* Frame T-1 represents the four identical interior frames along lines 3, 4, 5, and 6 which have been lumped together in this single frame, while Frame T-2 represents the two exterior frames along lines 1 and 8, with 16-in. wide beams. The third frame, T-3, represents the two identical frame-shear wall systems along lines 2 and 7.

In the longitudinal direction, two linked frames, each similar to that shown in Fig. 12-69b, were used to represent the two identical exterior frames L-1 along lines 'a' and 'd' and the two identical interior frames L-2 along lines 'b' and 'c'.

The lateral displacements due to both seismic and wind forces have been plotted in Fig. 12-71. Although the seismic forces used to obtain the curves of Fig. 12-71 are approximate, the results shown still serve to draw the distinction between wind and seismic forces, i.e., the fact that the former are external forces the magnitudes of which are proportional to exposed surface while the latter are inertial forces depending, among others, on the mass and stiffness properties of the structure. Thus, while the ratio of the total wind force in the transverse direction to that acting in the longitudinal direction (see Tables 12-5 and 12-6) is about 2.8, the corresponding ratio for seismic forces is only 1.7. As a result of this and the lesser stiffness of the structure in the longitudinal direction, the displacement due to seismic forces in the longitudinal direction is significantly greater than that in the transverse direction, although the displacements due to wind in both directions are about the same. The typical deflected shapes associated with predominantly cantilever or flexure type structures (as in the transverse direction) and shear (open-frame) type buildings (as in the longitudinal direction) are evident in the figure. The average deflection indices, i.e., the ratio of the lateral displacement at the top to the total height of the structure, are 1/3000 for wind and 1/1100 for seismic load.

An idea of the distribution of lateral loads among the different frames making up the structure in the transverse direction may be obtained from an examination of Table 12-7 which lists the portion of the total story shear at each level resisted by each type of frame. Note that at the base, most of the horizontal shear in the transverse direction is carried by the pair of shear walls.

To illustrate the design of a flexural member, the results of the analysis of the structure under seismic loads in the transverse direc-

*A device used to model the effect of floor slabs which may be assumed as rigid in their own planes.

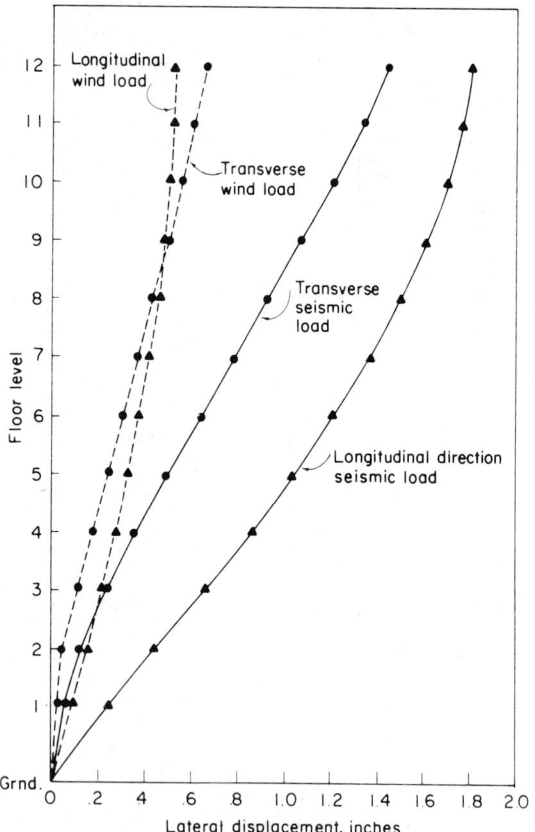

Fig. 12-71 Lateral displacements under seismic and wind loads.

TABLE 12-7 Distribution of Horizontal Seismic Story Shears Among the 3 Transverse Frames Shown in Fig. 12-69c

	Frame T-1 (4 interior frames)		*Frame T-2 (2 exterior frames)*		*Frame T-3 (2 interior frames with shearwalls)*		
Story Level	*Story Shear*	*% of Total*	*Story Shear*	*% of Total*	*Story Shear*	*% of Total*	*Total Story Shear (kips)*
12	316	156	140	69	-253	-125	203
11	227	57	96	24	75	19	398
10	263	46	114	20	198	34	575
9	273	37	117	16	345	47	735
8	286	33	123	14	469	53	878
7	293	29	125	13	586	58	1004
6	293	26	125	11	695	63	1113
5	284	23	121	10	800	67	1205
4	264	20	112	9	904	71	1280
3	225	17	96	7	1016	76	1337
2	191	14	82	6	1104	80	1377
1	45	3	20	2	1335	95	1400

tion have been combined with those obtained from a gravity load analysis of a single story bent* for two typical beams in the sixth floor of an interior frame, using eq. (12-25). The results are listed in Table 12-8. Similar values for typical exterior and interior columns on the second floor of the same interior frame are shown in Table 12-9. Corresponding design forces for the shear wall section at the first floor of frame T-3 in Fig. 12-69c are also shown in Table 12-10.

It is pointed out that even in the case where the design of a mem-

*Using the computer program described in Ref. 12-89.

TABLE 12-8 Summary of Ultimate Moments for Typical Beams on 6th Floor of Interior Transverse Frames Along Lines 3 through 6

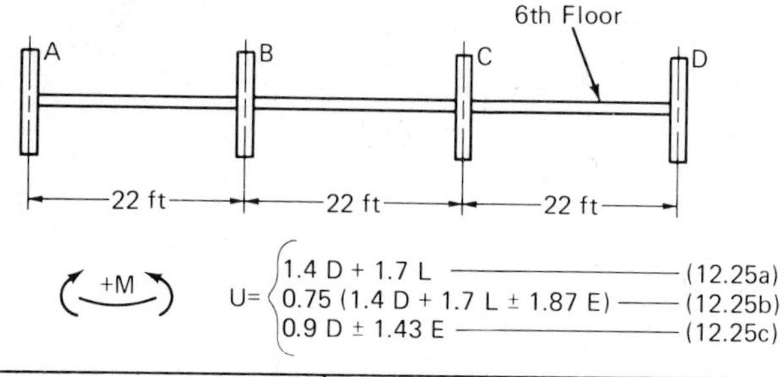

Beam AB		*Ultimate Moment in ft-kips*		
		A	*Near Midspan of AB**	*B*
Eq. (12-25a)		- 78	(+58)	-134
Eq. (12-25b)	sidesway to right	+111	+43	(-274)
	sidesway to left	(-229)	+45	+ 74
Eq. (12-25c)	sidesway to right	(+133)	+28	-259
	sidesway to left	-213	+28	(+ 87)

**10.4 ft from centerline of left support.*

Beam BC		*B*	*Midspan*	*C*
Eq. (12-25a)		-118	(+46)	-118
Eq. (12-25b)	sidesway to right	+123	+35	(-301)
	sidesway to left	(-301)	+35	+123
Eq. (12-25c)	sidesway to right	(+144)	+22	-288
	sidesway to left	-288	+22	(+144)

ber is governed by the wind loads, rather than the seismic loads, the ductility requirements of the codes still have to be met.

EXAMPLE 12-2: Design of flexural member AB–Determine the flexural and shear reinforcement for the beam AB on the sixth floor of a typical interior transverse frame. The beam carries a dead load of 2.7 kips/ft and a live load of 0.80 kips/ft. The maximum ultimate moments are as indicated in the figure below (see Table 12-8). The beam has dimensions b = 20 in. and d = 21.5 in. The slab is 8 in. thick. Use f_c' = 4,000 psi and f_y = 60,000 psi.

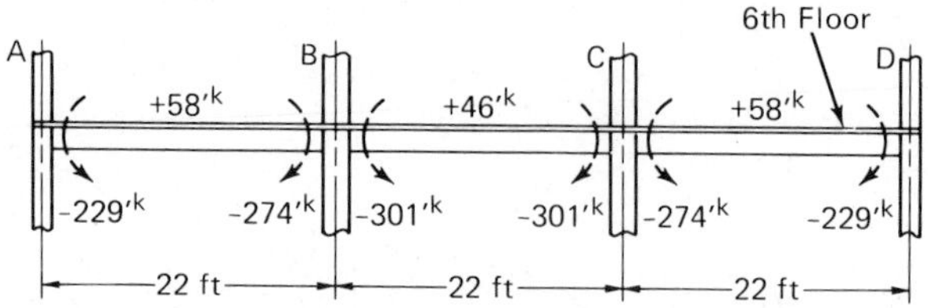

SOLUTION: (the boxed-in sections on the right-hand margin correspond to ACI 318-71)

TABLE 12-9 Summary of Ultimate Moments and Axial Loads for Typical Columns on 2nd Floor of Interior Transverse Frames Along Lines 3 through 6

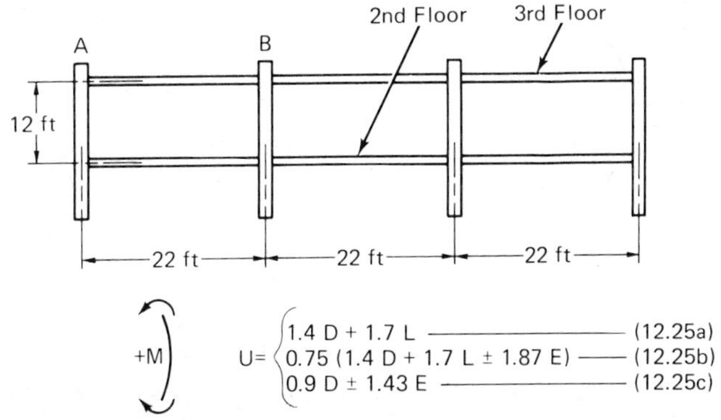

		Exterior Column A			Interior Column B		
		Axial Load, kips	*Moment, ft-k.**		*Axial Load, kips*	*Moment, ft-k.**	
			Top	*Bottom*		*Top*	*Bottom*
Eq. (12-25a)		-852	- 46	+ 46	-1623	+ 43	- 43
Eq. (12-25b)	sidesway to right	-489	+ 31	- 31	-1198	+186	-186
	sidesway to left	-791	-100	+100	-1282	-122	+122
Eq. (12-25c)	sidesway to right	-300	+ 48	- 48	- 812	+163	-163
	sidesway to left	-610	- 86	+ 86	- 898	-151	+151

**Including moments corresponding to horizontal torsional shears due to minimum story shear eccentricity.*

TABLE 12-10 Summary of Ultimate loads on Shear Wall at First Floor Level of Transverse Frame along Line 2 (or 7)

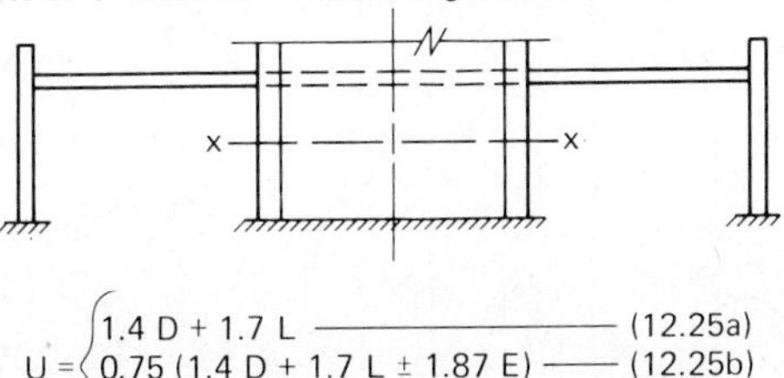

U = 1.4 D + 1.7 L (12.25a)
0.75 (1.4 D + 1.7 L ± 1.87 E) (12.25b)
0.9 D ± 1.43 E (12.25c)

	Design Forces Acting on Entire Shear Wall			
Loading Condition	*Axial Load (kips)*	*Bending (Overturning) Moment (ft-kips)*	*Horizontal Shear (kips)*	*Axial Load** on Boundary Element (kips)*
Eq. (12-25a)	4340	nominal	nominal	2170
Eq. (12-25b)	3260	46,800	930	3760
Eq. (12-25c)	2340	47,850	950	3350

***On the assumption that the vertical boundary elements resist all the vertical stresses due to the design dead, live and horizontal (seismic) loads, as provided in Section A 8.5.1.*

a. Determine limitations on longitudinal steel percentage. A.5.1

$$\rho_{max} = 0.5\,\rho_b$$

for a member with compression reinforcement (required by Sec. A.5.3),

$$\rho_b = \bar{\rho}_b + \rho' \frac{f_s'}{f_y} \quad \text{(see ACI 318-71 \textit{Commentary}, p. 35)}$$

$$= \frac{0.85\,\beta_1 f_c'}{f_y}\left(\frac{87{,}000}{87{,}000 + f_y}\right) + \rho' \frac{f_s'}{f_y} \qquad \text{(E2-1)}$$

Assuming $f_s' = f_y$, and $\rho' = \frac{1}{2}\rho_{max} = 0.25\,\rho_b$, and substituting into eq. (E2-1) above

$$\rho_b = \frac{1}{0.75}\left[\frac{0.85\,(0.85)\,(40)}{60}\left(\frac{87}{87 + 60}\right)\right] = 0.038$$

$$\rho_{max} = 0.5\,(0.038) = 0.019$$

$$\rho_{min} = \frac{200}{f_y} = \frac{200}{60{,}000} = 0.0033$$

b. Determine required flexural reinforcement.
   (1) Negative moment reinforcement at support B. Since the negative flexural reinforcement for both beams AB and BC at joint B will be provided by the same continuous bars, the larger negative moment at joint B will be used. The effect of any compressive steel will be disregarded in in the calculations below.

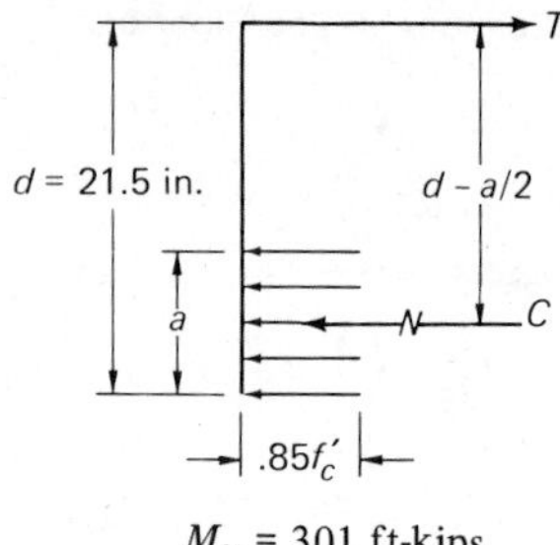

$$M_u = 301 \text{ ft-kips}$$

assuming

$$a = 3.5 \text{ in.}$$

$$A_s(\text{req'd}) = \frac{M_u}{\phi f_y(d - a/2)}$$

$$= \frac{12(301)}{0.90(60)(19.75)} = 3.4 \text{ in.}^2$$

Use three #10 bars, $A_s = 3.79 \text{ in.}^2$

$$\rho = \frac{A_s}{bd} = \frac{3.79}{20(21.5)} = 0.0088 > \rho_{min} \text{ and } < \rho_{max}$$

check value of 'a':

$$a = \frac{A_s f_y}{0.85 f'_c b_w} = \frac{3.79(60)}{0.85(4)(20)} = 3.35 \text{ in.}$$

with $a_{(actual)} \approx 3.4$ in., negative moment capacity of beam

$$M_u = \phi A_s f_y(d - a/2)$$

$$= 0.90(3.79)(60)(21.5 - 1.7)$$

$$= 4050 \text{ in.-kips or } 338 \text{ ft-kips}$$

(2) Negative moment reinforcement at support A.

$$M_u = 229 \text{ ft-kips}$$

assuming $a = 2.5$ in.

$$A_s(\text{req'd}) = \frac{12(229)}{0.90(60)(20.25)} = 2.51 \text{ in.}^2$$

Area of two #10 bars = 2.53 in.^{2} Thus, the two #10 bars can be made continuous along the top.

$$\rho = \frac{2.53}{20(21.5)} = 0.0059 > \rho_{min}$$

check value of 'a':

$$a = \frac{2.53(60)}{0.85(4)(20)} = 2.23 \text{ in.}$$

with $a_{(actual)} \approx 2.3$ in., negative moment capacity at A

$$M_u = 0.90(2.53)(60)(21.5 - 1.2)/12 = 232 \text{ ft-kips}$$

(3) Positive moment reinforcement at support.
A positive moment capacity at the supports equal to at least 50% of the negative moment capacity is required, i.e.,

$$\min M_u^+(\text{at support B}) = \frac{338}{2} = 169 \text{ ft-kips}$$

$$\min M_u^+(\text{at support A}) = \frac{232}{2} = 116 \text{ ft-kips}$$

Note that both of the above required capacities are greater than the design positive moment near the midspans of both beams AB and BC.

(4) Positive moment reinforcement at midspan–to be made continuous to supports. Use three #7 bars ($A_s = 1.80 \text{ in.}^2$)

$$\rho = \frac{A_s}{bd} = \frac{1.80}{20(21.5)} = 0.0042 > \rho_{min}$$

$$b(\text{T-beam}) = \tfrac{1}{4} \text{ span} = 12(20.1)/4 = 60 \text{ in.} \qquad \boxed{8.7.2}$$

$$a = \frac{1.80(60)}{0.85(4)(60)} = 0.53 \text{ in.}$$

$$M_u = 0.90(1.8)(60)(21.5 - 0.27)/12$$

$$= 172 \text{ ft-kips} > 169, \text{ OK}$$

c. Calculate required length of anchorage for flexural reinforcement in exterior column.

Anchorage length (in confined region of beam-column joint) shall include a standard 90° hook, and $\geqslant$ 2/3 development length, l_d, as required by Sec. 12.5(a), i.e., [A.5.5] [A.5.6]

$$l_d = 0.04 A_b f_y/\sqrt{f'_c} \text{ but not } < 0.0004\, d_b f_y \qquad \boxed{12.5(a)}$$

For #10 bars

$$l_d = 0.04(1.27)(60{,}000)/\sqrt{4000} \text{ but not}$$

$$< 0.0004(1.27)(60{,}000)$$

$$= 48 \text{ in. (governs)} > 31 \text{ in.}$$

Required anchorage length = $\frac{2}{3}(48)$ = 32 in.

Anchorage provided (measured from face of column) (see figure below),

embedment	11 in.
bend	8
extension (12 dia.)	14
	Total = 33 in.

Similar calculations for the #7 bottom bars yield a required anchorage length of 16 in.

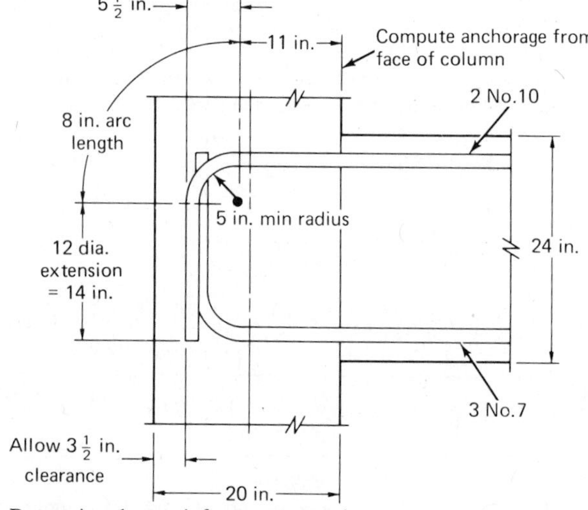

d. Determine shear reinforcement requirements:
(1) Design for shears corresponding to yield moment capacity at beam ends plus design gravity loads (see Fig. 12.62) [A.5.7]
The table below shows the values of the design shears corresponding to the two loading cases to be considered. In the table,

$$w_u = 0.75(w_D + w_L) = 0.75[1.4(2.7) + 1.7(0.8)]$$

$$= 3.86 \text{ kips/ft}$$

At end B of beam, nominal shear stress

$$v_u = \frac{V_u}{\phi b_w d} = \frac{64{,}100}{0.85(20)(21.5)} = 175 \text{ psi} \quad \boxed{11.2.1}$$

$$\text{Using } v_c(\text{allowable}) = 2\sqrt{f_c'} = 2\sqrt{4000} = 126 \text{ psi} \quad \boxed{11.4.1}$$

at the ends of a member, e.g., due to concentrated loads near midspan, the spacing of stirrups will have to be determined in accordance with the requirements applying to locations near the supports. In the above example,

TABLE 12.11

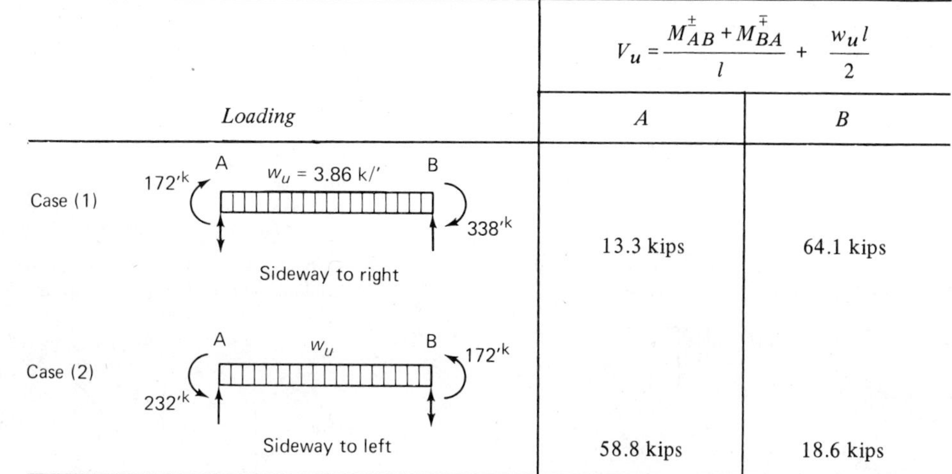

Loading	$V_u = \dfrac{M_{AB}^{\pm} + M_{BA}^{\mp}}{l} + \dfrac{w_u l}{2}$ A	B
Case (1): w_u = 3.86 k/′; 172′k at A, 338′k at B; Sideway to right	13.3 kips	64.1 kips
Case (2): w_u; 232′k at A, 172′k at B; Sideway to left	58.8 kips	18.6 kips

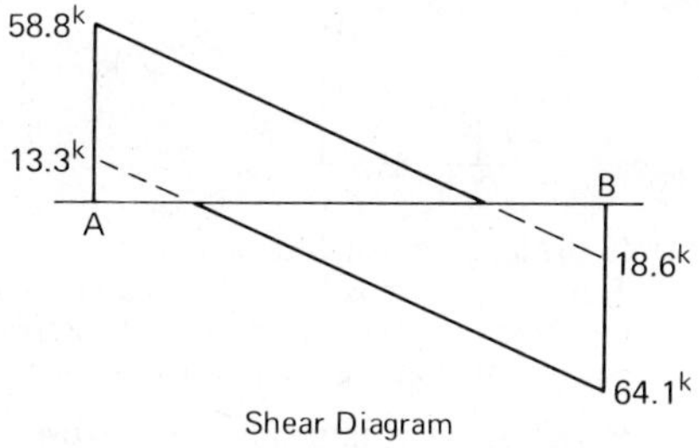

Shear Diagram

Required spacing of #3 stirrups [A_v(2 legs) = 0.22 in.^{2}]

$$s = \frac{A_v f_y}{(v_u - v_c)\, b_w} = \frac{0.22(60{,}000)}{(175-126)(20)} = 135 \text{ in.} \quad \boxed{11.6.1}$$

(2) Check minimum allowable stirrup spacing

$$s_{max} = \frac{d}{2} = 10.7 \text{ in.} \quad \boxed{A.5.8}$$

Within a distance of $4d$ from the ends of the beam

$$s_{max} = \frac{A_v d}{0.15 A_s} \text{ but not} > \frac{d}{4} \quad \boxed{A.5.9}$$

$$= \frac{(0.22)(21.5)}{(0.15)(3.79)} \text{ but not} > \frac{21.5}{4}$$

$$= 8.3 \text{ in.} > 5.4 \text{ in. (governs)}$$

Similar calculations for the left end of the beam at A also indicate that the maximum allowable stirrup spacing of $d/4$ = 5.4 in. governs.

USE #3 stirrups spaced as shown below

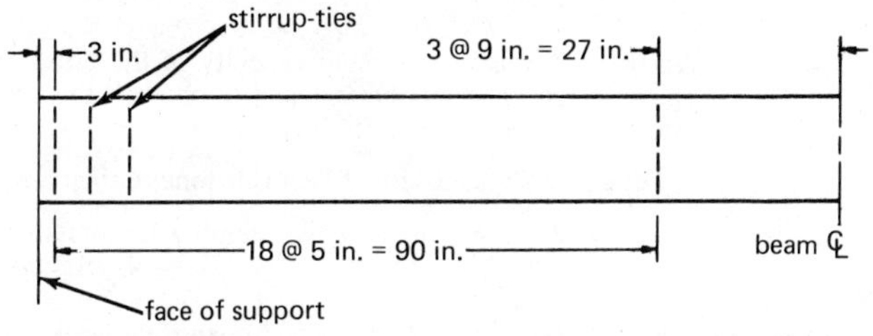

Where the loading is such that inelastic deformation may occur at locations other than $\boxed{A.5.11}$

the maximum positive moment near midspan (see Table 12-8) is well below the positive moment capacity of the section.

Note that the longitudinal bars have not been designed to act as compression reinforcement. $\boxed{A.5.10}$

e. Negative reinforcement cutoff points.

For the purpose of determining the cutoff points for negative reinforcement, a moment diagram corresponding to plastic end moments and 0.90 of dead load will be used.

In the case considered, the cutoff point for the third #10 bar at the top, near support B, will be determined.

With the negative moment capacity of a section with 2 #10 bars ≈ $\frac{2}{3}$(338) = 225 ft-kips, the distance from the right support where the applied moment under the loading considered equals 225 ft-kips is readily obtained by summing moments about section a-a in the figure below.

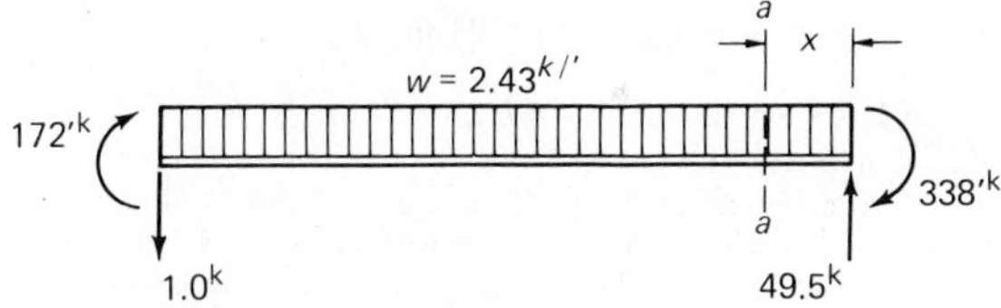

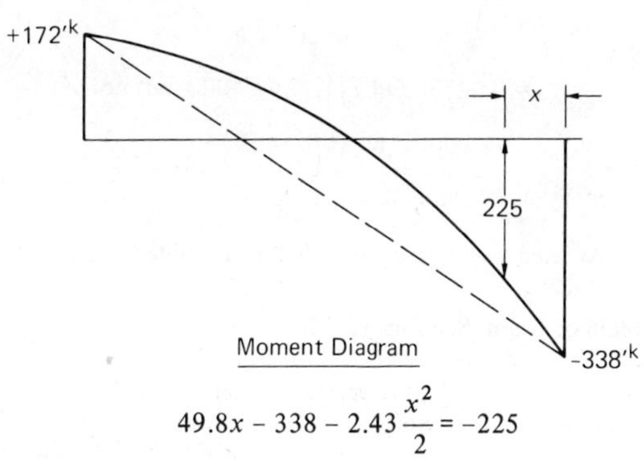

Moment Diagram

$$49.8x - 338 - 2.43\frac{x^2}{2} = -225$$

Solution of the above equation gives x = 2.4 ft, hence one #10 bar at the top may be cut off at

$$x + d = 2.4 + \frac{21.5}{12} = 4.2 \text{ ft from the face of the right support B} \quad \boxed{12.1.4}$$

Fig. 12-72a Flexural-member reinforcement.

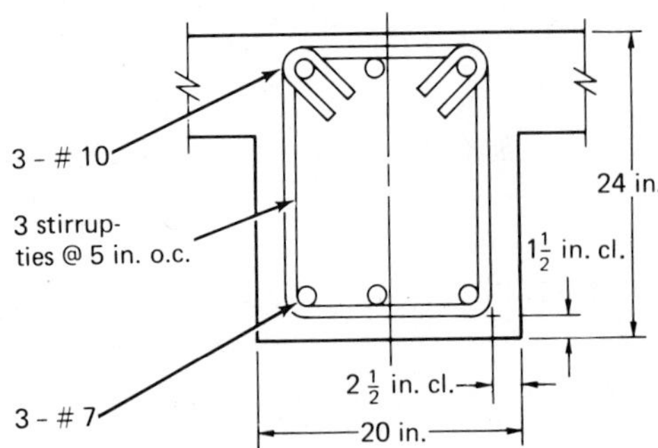

Fig. 12-72b Typical beam section.

f. Flexural reinforcement splices.
   (1) Bottom bars, #7
   The bottom bars along most of the length of the beam may be subjected to maximum stress. USE Class C splice [7.6.3.1.1]

$$l_d = 0.04 A_b f_y / \sqrt{f'_c} \text{ but not } < 0.0004 d_b f_y \quad \boxed{12.5(a)}$$

$$= 0.04(0.60\ (60{,}000)/\sqrt{4000} \text{ but not}$$

$$< 0.0004(0.875)\ (60{,}000)$$

$$= 23 \text{ in. (governs)} > 21 \text{ in.}$$

   Class C splice length = $1.7\, l_d$ = 39 in.
   (2) Top bars, #10
   Since the midspan portion of the beam is always subject to a positive bending moment (see Table 12-8), splices in the top bars should be located at or near midspan.

$$l_d = 0.02 f_y d_b / \sqrt{f'_c} \text{ but not} \quad \boxed{7.7.1 \; 12.6}$$

$$< 0.0005 f_y d_b \text{ or } 12 \text{ in.}$$

$$= 0.02(60{,}000)\ (1.27)/\sqrt{4000} \text{ but not}$$

$$< 0.0005\,(60{,}000)\ (1.27)$$

$$= 24 \text{ in.} < 38 \text{ in. (governs)}$$

   At least two stirrup ties shall be provided at all splices. [A.5.12]

g. Detail of beam. See Fig. 12-72.

EXAMPLE 12-3: Design of frame column A—Design the transverse reinforcement for the exterior tied column on the second floor of a typical interior frame, i.e., Frame T-1 in Fig. 12-59c. The column dimension has been established as 20 in. square and, on the basis of the different combinations of axial load and moment corresponding to the three loading conditions listed in Table 12-9, eight #7 bars arranged in a symmetrical pattern have been found adequate. Assume the same beam section framing into the column as considered in section 12.7.2. Use f'_c = 4000 psi and f_y = 60,000 psi.

SOLUTION:

a. Check limitations on amount of vertical reinforcement.

$$0.01 < \rho < 0.06 \quad \boxed{A.6.1}$$

$$\rho = \frac{A_s}{h^2} = \frac{8(0.60)}{(20)^2} = 0.012 \text{ OK}$$

Check if sum of column moment strengths exceeds beam moment capacity. [A.6.2]

(1) In transverse direction—
From ex. 12-1, $M_{u(\max)}$ of beam = 232 ft-kips

From ACI SP-17a* (pp. 98–99), moment capacity of column section when P_u = 852 kips (corresponding to f'_c = 4 ksi, f_s = 60 ksi, g = 0.76, $K = P_u/f'_c bt$ = 0.532 and $\rho_t m = f_y/0.85 f'_c$ = 0.212) = 2090 in-kips or 174 ft-kips.

With the same size column above and below the beam, total moment capacity of columns = 2(174)
= 348 ft-kips > 232, OK

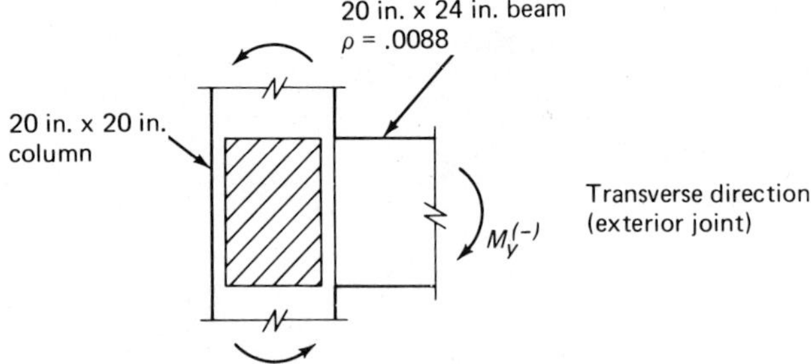

(2) In the longitudinal direction—
The moment capacity of the columns in the longitudinal direction in this case may be conservatively assumed as equal to that in the transverse direction, i.e., 348 ft-kips.

If we assume a ratio for the negative moment reinforcement of 0.0075 in the beams along the periphery of the building (b = 16 in., d = 21.5 in.),

$$A_s = \rho\, bd = 0.0075(16)\ (21.5) = 2.58 \text{ in.}^2$$

$$a = \frac{A_s f_y}{0.85 f'_c b_w} = \frac{2.58(60)}{0.85(4)\ 16} = 2.84 \text{ in.}$$

Negative moment capacity of beam,

$$M_u = \phi A_s f_y (d - a/2)$$

$$= 0.90(2.58)\ (60)\ (21.5 - 1.42)$$

$$= 2800 \text{ in.-kips, or } 232 \text{ ft-kips}$$

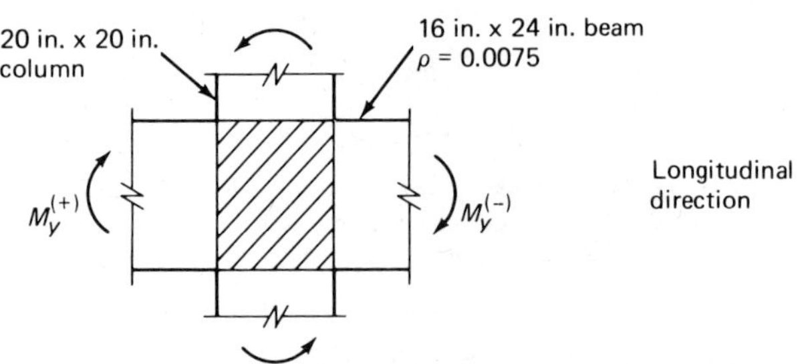

Assume the positive moment capacity of the beam at the opposite side of the column equal to one-half the negative moment capacity or 117 ft-kips.

Total moment capacity of beams in longitudinal direction

= 232 + 116 = 348 ft-kips = combined moment capacity of columns

*ACI Committee 340, *Ultimate Strength Design Handbook, Vol. 2, Columns*, SP-17A, American Concrete Institute, Detroit, 1970, 226 pp.

The reinforcement ratio of 0.0075 would then represent the upper limit on the amount of negative moment reinforcement in the exterior peripheral beams—under the assumption of equal moment capacity of the column section in both transverse and longitudinal directions—if the strong column-weak beam provision of the code is to be satisfied.

Note that the above condition need not be satisfied provided the sum of the moment strengths of the confined cores of the columns is equal to or greater than the design moment.

c. Determine special transverse reinforcement requirements for confinement. A.6.4

(1) Check floor level in column stack below which special confinement reinforcement is required (assuming the same column section throughout entire height).

For a 20 in. square column with $\rho_t = 0.012$, $P_b \approx 450$ kips

$$0.4\,P_b = 0.4\,(450) = 180 \text{ kips}$$

$$P_u \text{ for } 11^{th} \text{ floor column} = 107 \text{ kips}$$

$$P_u \text{ for } 10^{th} \text{ floor column} = 180 \text{ kips} = 0.4\,P_b$$

Therefore, special transverse reinforcement will be required on all exterior columns of Frame T-1 below the 11th floor.

(2) Compute minimum volumetric ratio, ρ_s, for spiral reinforcement.

$$\rho_s = 0.45\left[\frac{A_g}{A_c} - 1\right]\left(\frac{f_c'}{f_y}\right) \text{ but not} < 0.12\left(\frac{f_c'}{f_y}\right) \quad \text{A.6.4.2}$$

$$= 0.45\left[\frac{(20)^2}{(17)^2} - 1\right]\left(\frac{4}{60}\right) \text{ but not} < 0.12\left(\frac{4}{60}\right)$$

$$= 0.0114 \text{ (governs)} > 0.008$$

(3) Compute required size and spacing of rectangular hoop reinforcement, as required for confinement.

$$A_{sh} = l_h\,\rho_s s_h/2 \quad \text{A.6.4.3}$$

Try #4 hoops with one supplementary crosstie in each direction as shown in Fig. 12-73.

$$l_h = 9.5 \text{ in.}$$

Required spacing,

$$s_h = \frac{2A_{sh}}{l_h\,\rho_s} = \frac{2(0.20)}{9.5(0.0114)} = 3.7 \text{ in.}$$

Therefore, #4 rectangular hoops with one supplementary crosstie in each direction, spaced at 3.5 in. (< 4 in. max.) satisfy the requirements for confinement reinforcement. Such transverse reinforcement shall extend from the face of the connection to a distance equal to the largest of the following:

h = 20 in. (governs), 18 in., or A.6.4.1

1/6 clear height of column = (10 × 12)/6 = 20 in.

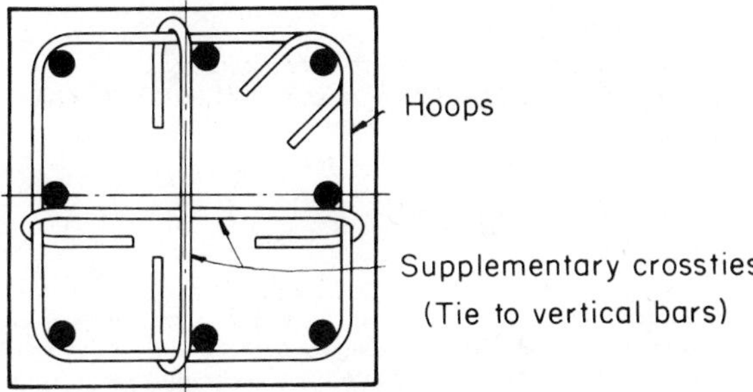

Fig. 12-73 Column hoops with supplementary crossties.

d. Determine transverse reinforcement required for shear:

For the case considered, in which the moment capacity of the column exceeds that of the framing beam, the maximum shear that may be developed in the column is determined by the moment capacities of the beams framing into the column ends.*

(1) Calculate design shear for column. For this purpose, a ϕ-factor of 1.0 will be used in computing the moment capacities of the beams.** See Fig. 12-74.

$$V_u = \frac{(M_{y2} + M_{y3})/\phi}{2l_n} = \frac{232 + 232}{0.90(2)\,(12)} = 21.5 \text{ kips}$$

Nominal shear stress

$$v_u = \frac{V_u}{\phi b_w d} = \frac{21{,}500}{0.85\,(20)\,(17)} = 75 \text{ psi}$$

Using

$$v_c\,(\text{allowable}) = 2\sqrt{f_c'} = 126 \text{ psi} > v_u$$

Since only nominal shear reinforcement is required, i.e., #3 spaced at $d/2$ = 8.5 in., the confinement requirement governs. Therefore, USE #4 hoops with supplementary crossties spaced at 3.5 in. o.c. for a distance of 20 in. from the face of the connection. A.6.6 Beyond this distance, the minimum shear reinforcement of #3 hoops spaced at 8.5 in. o.c. will be used. 11.1.4(b)

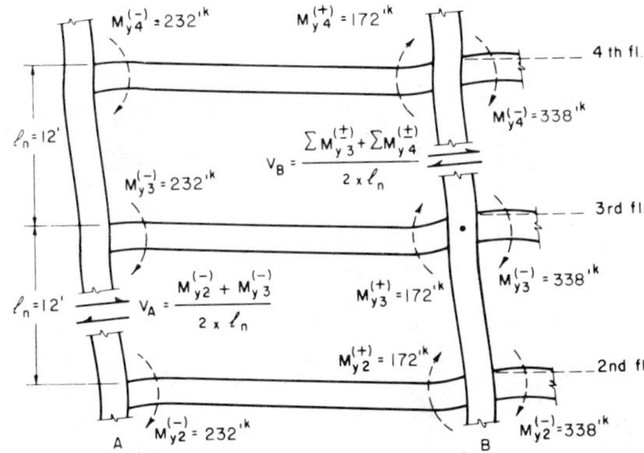

Fig. 12-74 Yield moments in interior transverse frame for design of transverse reinforcement in columns and beam-column connections.

e. Minimum length of lap splice for column bars, #7 bars:

$$l_d = 0.02 f_y d_b/\sqrt{f_c'} \text{ but not} < \begin{cases} 0.0005 f_y d_b, \\ 30 \text{ bar diameter,} \\ \text{or } 16 \text{ in.} \end{cases} \quad \begin{matrix} 12.6 \\ 7.7.1.1 \\ \text{A.6.7} \end{matrix}$$

*Where the sum of the moment capacities of the columns at a joint is less than that of the framing beam(s), so that hinging may occur in the columns rather than in the beam(s), the design shear for the columns would have to be based on the moment capacities of the columns. In this case, it is conservative to use as the maximum column moment the value corresponding to the balanced condition, i.e., M_b, but not to exceed the moment capacity of the beams. This will allow for some variation in the column axial load.

**The use of a ϕ-factor of 1.0 is not required by ACI Appendix A. The use of ϕ = 1.0 for the particular purpose noted above is in conformance with a recent revision of the *UBC* (amendments introduced at 1971 Annual Business Meeting of the ICBO).

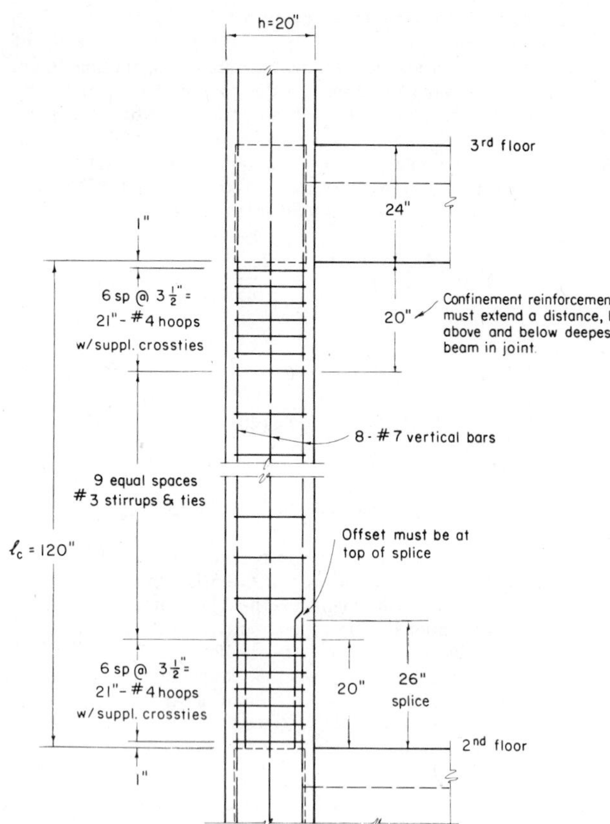

Fig. 12-75 Column reinforcing details.

$$= \frac{0.02(60{,}000)(0.875)}{\sqrt{4000}} \text{ but not} < \begin{cases} 0.0005(60{,}000)(0.875), \\ 30(0.875), \text{ or} \\ 16 \text{ in.} \end{cases}$$

$= 17$ in. < 26 in. (governs)

f. Detail of column.

See Fig. 12-75

EXAMPLE 12-4: Design of exterior beam-column connection–Design the transverse reinforcement for the exterior beam-column connection between the beam considered in Example 12-2 and the column of Example 12-3. Assume the joint to be located on the sixth floor where the design axial load on the column varies from $N_{u(max)} = 525$ kips to $N_{u(min)} = 186$ kips. Use $f_c' = 4000$ psi and $f_y = 60{,}000$ psi.

SOLUTION: Note that two different loading combinations arise with respect to the design of the transverse reinforcement for shear in the beam-column joint–

(1) For sidesway to the left, as shown in Fig. 12-76a, the horizontal shear force transmitted by the beam is equal to the yield strength of the negative moment reinforcement, $A_s f_y$, and the axial force from the column above due to the earthquake motion is compressive.

(2) For sidesway to the right (Fig. 12-76b), the shear force transmitted by the beam is equal to the yield strength of the positive moment reinforcement, $A_s' f_y$, and the axial load from the column due to earthquake motion is tensile.

It can be easily verified that the same two loading combinations occur whether the beam frames into the exterior column from the left or the right side. In case (1), a large shear force from the beam occurs simultaneously with an increased compressive load from the column, which increases the shear capacity of the connection. In case (2), a relatively lower horizontal shear from the beam is combined with a reduced compressive load from the column, which reduces the shear capacity of the connection.

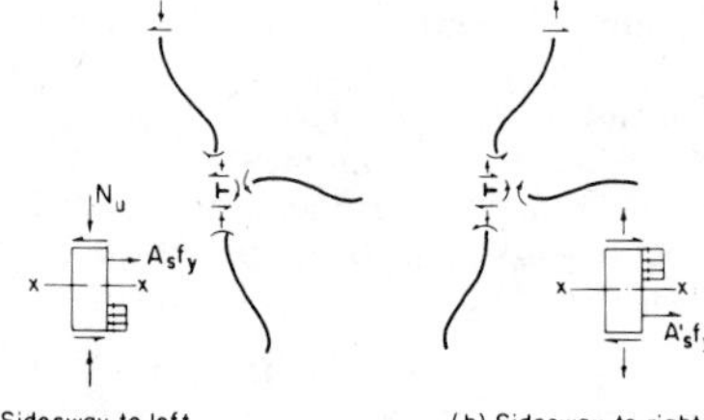

Fig. 12-76 Loading combinations for exterior beam-column connection.

The effect of a reversal in the sign of the axial load due to earthquake motion is worth noting particularly in exterior frame columns, where the axial forces due to lateral loads can be significant.

a. Calculate shear across section x-x of the connection (see Fig. 12-77) transmitted by the beam negative moment reinforcement at yield [case (1) above]. A.7.1

$$A_s f_y = 2.53(60) = 152 \text{ kips}$$

Shear from column above connection (with $M_{y(beam)} = 232$ ft-kips)

$$V_{col} = 232/12 = 19 \text{ kips}$$

Net shear in connection

$$V_u = 152 - 19 = 133 \text{ kips}$$

Nominal shear stress across section x-x, using $d = 20$ in.*

$$v_u = \frac{V_u}{\phi b_w d} = \frac{133{,}000}{0.85(20)(20)} = 391 \text{ psi}$$

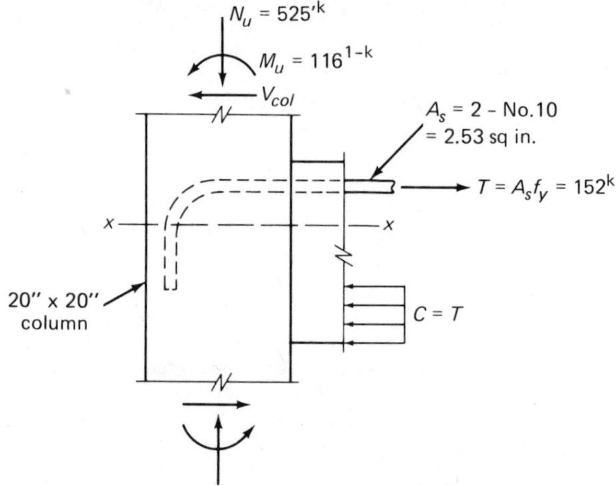

Fig. 12-77 Shear forces in exterior beam-column connection, Case (a).

b. Determine allowable shear stress in concrete, v_c.
Using eqq. (11-4) and (11-5) of ACI 318-71, $d = 20$ in. and with

$$M_u = 232/2 = 116 \text{ ft-kips},$$

$$N_u = 525 \text{ kips},$$

*Because of the presence of the beam across section x-x, it is considered permissible and conservative to use the full depth of the column rather than the effective depth for the purpose of determining v_u (see Fig. A-7, page 92 of the Commentary to ACI 318-71).

$$M_m = M_u - N_u \left(\frac{4h - d}{8}\right) \qquad \boxed{11.4.3}$$

$$= 116(12) - 525 \left[\frac{4(20) - 20}{8}\right] = -2550$$

Since M_m is negative, eq. (11-5) does not apply, and v_c may be determined using eq. (11-7), i.e.,

$$v_c = 3.5\sqrt{f'_c}\ \sqrt{1 + 0.002 N_u/A_g}$$

$$= 3.5\sqrt{4000}\ \sqrt{1 + 0.002(525{,}000)/(20)^2} = 442 \text{ psi} > v_u$$

Similar calculations for loading case (2), i.e., using $A'_s f_y$ and N_u = 186 kips, yield v_u = 262 psi and v_c = 308 psi, > v_u.

The above results indicate that only nominal shear reinforcement is required for the connection.

c. Determine transverse reinforcement required for confinement. A.6.4.3

In order to avoid congestion, no supplementary crossties will used.

Minimum allowable volumetric ratio for spirals

$$\rho_s = 0.45 \left(\frac{A_g}{A_c} - 1\right) \frac{f'_c}{f_y} \text{ but not} < 0.12 \frac{f'_c}{f_y} \qquad \boxed{A.6.4.2}$$

$$= 0.0114 \text{ (governs)} > 0.008$$

Using eq. (A-4) with l_h = 17 in. and assuming a hoop spacing of 4 in., required cross-sectional area of hoop (one leg)

$$A_{sh} = \frac{l_h \rho_s s_h}{2} = \frac{17(0.0114)(4)}{2} = 0.39 \text{ in.}^2 \qquad \boxed{A.6.4.3}$$

USE #6 hoops (A_{sh} = 0.44 in.^{2}) spaced at 4 in. o.c.

d. A check on the adequacy of the transverse reinforcement determined above, considering the forces acting in the longitudinal direction, assuming ρ = 0.0075 for the negative reinforcement of the beams along the exterior of the building, indicates that the net shear in the connection

$$V_u = 126 \text{ kips} < 133 \text{ kips (in transverse direction)}$$

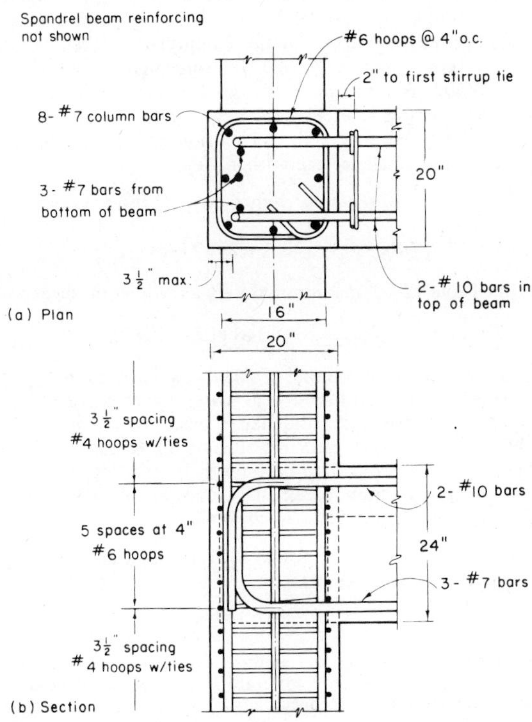

Fig. 12-78 Detail of exterior beam-column connection.

Hence, the design for forces in the transverse direction of the building governs, since the column section in this case is square.

e. Detail of joint. See Fig. 12-78.

EXAMPLE 12-5: Design of interior beam-column connection–Design the transverse reinforcement for the interior beam-column connection at the sixth floor of the interior transverse frame T-1. The column is 26 in. square and is reinforced with eight #11 bars. The design axial load on the column ranges from $N_{u(max)}$ = 1050 kips to $N_{u(min)}$ = 482 kips. The beams have dimensions b = 20 in. and d = 21.5 in. and are reinforced as determined in Example 12-2. Use f'_c = 4000 psi and f_y = 60,000 psi.

SOLUTION: Figure 12-79 below shows that for this case, the critical shear condition within the connections results from the combination of the yield strength of the negative flexural reinforcement in the interior beam BC and the yield strength of the positive moment reinforcement in the exterior beam AB, acting together with a reduced axial load on the column. This condition arises in the interior beam-column joint B located on the left of the frame centerline when swaying to the right (as shown in (b) below) and in the right beam-column joint C when the frame sways to the left.

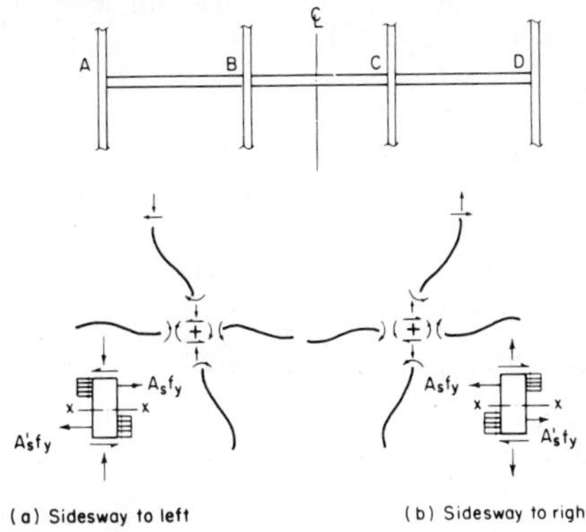

Fig. 12-79 An interior beam-column joint located left of building centerline.

a. Calculate shear stress across section x-x of connection. A.7.1

Shear developed by negative moment reinforcement in interior beam at yield

$$A_s f_y = 3.79(60) = 227 \text{ kips}$$

Shear developed by compressive block on other side of the column = yield strength of positive moment reinforcement in exterior beam,

$$A'_s f_y = 1.80(60) = 108 \text{ kips}$$

Shear from column above connection (see Fig. 12-80),

$$V_{col} = \frac{\Sigma M_{y6}^{(\pm)} + \Sigma M_{y7}^{(\pm)}}{2 l_n} = \frac{2(172 + 338)}{2(12)}$$

$$= 43 \text{ kips}$$

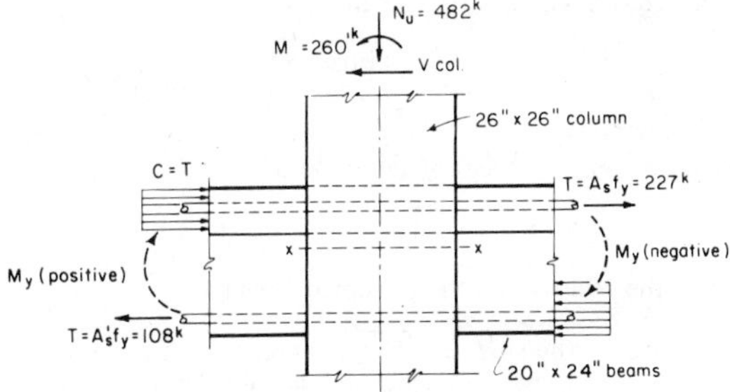

Fig. 12-80 Shear forces in an interior beam-column connection of transverse frame.

Net shear in connection

$$V_u = (227 + 108) - 43 = 292 \text{ kips}$$

Nominal shear stress, using $d = 26$ in.,

$$v_u = \frac{V_u}{\phi b_w d} = \frac{292{,}000}{0.85(26)(26)} = 510 \text{ psi} \qquad \boxed{11.2.1}$$

b. Determine allowable shear stress in concrete, v_c.

Using eqq. (11-4) and (11-5), ACI 318-71, with $d = 26$ in., and

$M_u = (338 + 172)/2 = 260$ ft-kips, and

$N_u = 482$ kips,

$$M_m = M_u - N_u \left(\frac{4h - d}{8}\right) \qquad \boxed{11.4.3}$$

$$= 260(12) - 482 \left[\frac{4(26) - 26}{8}\right] = -1580 \text{ in.-kips}$$

Since M_m is negative, eq. (11-4) does not apply, and v_c may determined using eq. (11-7)

$$v_c = 3.5\sqrt{f_c'}\,\sqrt{1 + 0.002N_u/A_g}$$

$$= 3.5\sqrt{4000}\,\sqrt{1 + 0.002(482{,}000)/(26)^2}$$

$$= 346 \text{ psi} < v_u$$

NOTE: Maximum allowable contribution of shear reinforcement to shear capacity of section,

$$(v_u - v_c)_{max} = 8\sqrt{f_c'} = 8\sqrt{4000} = 506 \text{ psi} \qquad \boxed{11.6.4}$$

Maximum possible shear stress which can be resisted by shear reinforced section,*

$$8\sqrt{f_c'} + v_c = 506 + 346 = 852 \text{ psi} > v_u = 510, \text{ OK}$$

c. Determine required shear reinforcement, assuming the maximum allowable spacing of 4 in.

$$A_v = \frac{(v_u - v_c)\,b_w s}{f_y} = \frac{(510 - 346)(26)(4)}{60{,}000} = 0.284 \text{ in.}^2$$

$$\boxed{11.6.1}$$

d. Compare the above requirement for shear with the confinement requirement as specified by Sec. A.6.4.3, assuming no supplementary crossties.

Minimum allowable volumetric ratio for spirals

$$\rho_s = 0.45\left(\frac{A_g}{A_c} - 1\right)\frac{f_c'}{f_y} \text{ but not} < 0.12\frac{f_c'}{f_y} \qquad \boxed{A.6.4.2}$$

$$= 0.45\left[\frac{(26)^2}{(23)^2} - 1\right]\frac{4}{60} \text{ but not} < 0.12\left(\frac{4}{60}\right)$$

$$= 0.0084 \text{ (governs)} > 0.008$$

Using eq. (A-4), with $l_h = 23$ in.,

$$A_{sh} = \frac{l_h \rho_s s_h}{2} = \frac{(23)(0.0084)(4)}{2} = 0.39 \text{ in.}^2 \qquad \boxed{A.6.4.3}$$

Required area of 2 legs of square hoop

$$3(0.39) = 0.78 \text{ in.}^2 > A_v = 0.284 \text{ in.}^2 \text{ (required for shear)}$$

Thus, the confinement requirement governs.

*In certain cases, the sum $(8\sqrt{f_c'} + v_c)$ may be less than v_u, indicating that the required shear capacity cannot be developed by the section in accordance with *Code* provisions. In such cases, compliance with the *Code* provisions would require an increase in the section dimensions.

e. Determine required hoop size and spacing:

(1) Check compliance of relative member dimensions with provisions of Sec. A.7.2 for possible reduction in transverse reinforcement requirements.

The beams in both transverse and longitudinal directions have the same section dimensions: 20 × 24 in.

Is beam $b_{min} > 1/2$ column width? $\boxed{A.7.2}$

20 in. > 13 in., OK

Is min. beam depth > 3/4 of max. beam depth?

Yes, since all beams have same size.

The connection complies, so that the transverse reinforcement may be reduced to one-half that required by Sec. A.7.1, as determined above.

(2) Reduce the required area of transverse reinforcement determined in Step (d) above by 50%.

USE #4 hoops spaced at 4 in. o.c.–$2A_{sh} = 0.40$ in.^{2}

EXAMPLE 12-6: Design of shear wall–Design the shear wall section at the first floor of the transverse frame T-3 (See Fig. 12.69c). The preliminary design, as shown in Fig. 12.69a, is based on a 12 in. thick wall with 26 in. square, vertical boundary elements, each of the latter being reinforced with eight #11 bars. Use $f_c' = 4000$ psi and $f_y = 60$ ksi.

SOLUTION: Preliminary calculations indicated that the vertical boundary elements at the lower floor levels have to be increased in section to 32 × 44 in. reinforced with 22–#11 bars. A 14 in. thick wall will be assumed with the above boundary elements. This section will now be investigated and the other reinforcing requirements determined. The design forces on the shear wall at the first floor level are listed in Table 12-10.

Note that because the axis of the shear wall coincides with the centerline of the frame, lateral loads do not induce any vertical (axial) loads on the shear wall.

a. Determine P_b for the shear wall section, i.e., the axial load corresponding to balanced conditions in the particular section considered (see Fig. 12-81).

A good estimate of P_b may be obtained by assuming all the longitudinal (vertical) reinforcement in the boundary elements to be concentrated at the center of the boundary element section, as shown in Fig. 12-81. The balanced condition is then characterized by an extreme compression fiber strain of $\epsilon_{cu} = 0.003$ and a tensile reinforcement yield strain of $\epsilon_y = 0.002$. For the present case, the corresponding strain in the compression steel is 0.0026, i.e., greater than ϵ_y so that the longitudinal forces in the tension and compression reinforcement just balance each other. Hence,

$$P_b = \phi C_c \approx 0.70(9550) = 6700 \text{ kips}^{**}$$

$$0.4P_b = 0.4(6700) = 2680 \text{ kips}$$

From Table 12-10, maximum design axial load on shear wall,

$$P_{e(max)} = 4340 \text{ kips} > 0.4P_b$$

Thus, the vertical boundary elements have to be designed to resist all the vertical forces due to gravity as well as horizontal (earthquake) loads. $\boxed{A.8.5}$

b. Check vertical load carrying capacity of boundary elements. These will be assumed to be essentially concentrically loaded, with only minimum eccentricity considered in the direction perpendicular to the plane of the wall.

Maximum axial compressive load on boundary element (see Table 12-10) $\boxed{A.8.5.1}$

$$= 3760 \text{ kips}$$

**A more accurate (computer) analysis of the section using a parabolic stress-strain curve for concrete and considering the distribution of longitudinal reinforcement as well as the effect of the longitudinal reinforcement in the shear wall web, gave P_b = 7300 kips.

From Chart No. 64, page 100 of ACI SP-17a, one obtains, for a 32 × 44-in. section with 22-#11 bars (A_s = 34.3 in.^{2}, = 0.0266) and minimum eccentricity = $0.10h$,

$$P_u = 3830 \text{ kips} > 3760, \text{OK}$$

Also,

$$M_u = 12{,}250 \text{ in.-kips}$$

32"

44"

Alternate location

14"

22-#11 bars

c. Determine transverse reinforcement required for confinement in vertical boundary elements.

(1) Minimum volumetric ratio, ρ_s, for spiral reinforcement:

$$\rho_s = 0.45\left(\frac{A_g}{A_c} - 1\right)\frac{f'_c}{f_y},$$

$$\text{but not} < 0.12\,\frac{f'_c}{f_y}$$ A.6.4.2

$$\rho_s = 0.45\left[\frac{(32)\,(44)}{(29)\,(41)} - 1\right]\frac{4}{60} \text{ but not} < 0.12\left(\frac{4}{60}\right)$$

$$= 0.0055 < 0.008 \text{ (governs)}$$

(2) Compute required size and spacing of rectangular hoops, as required for confinement. A.8.5.2

Try #4 hoops with supplementary cross-ties as shown in the figure above 7.12.3 A.6.4.3

$$l_h = 11 \text{ in.}$$

Required spacing,

$$s_h = \frac{2A_{sh}}{l_h\,\rho_s}$$ A.6.4.3

$$s_h = \frac{2(0.20)}{11(0.008)} = 5.0 \text{ in.}$$

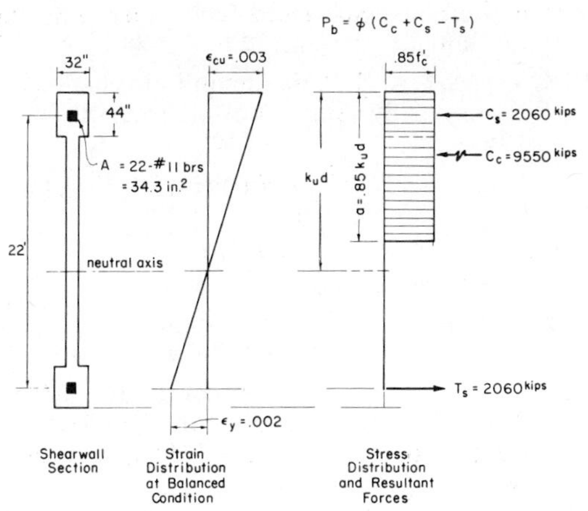

Fig. 12-81 Axial load capacity of shear wall section at balanced condition, P_b.

USE #4 rectangular hoops with supplementary crossties, as illustrated, spaced at 4.0 in. o.c. A.6.4.3

d. Check adequacy of transverse reinforcement as determined above with respect to shear in the longitudinal direction.

In the longitudinal direction, 20 × 20-in. beams—reinforced with three #10 top bars (M_u = 338 in.-kips) and three #7 bottom bars (M_u = 172 in.-kips)—frame into the shear wall vertical boundary elements.

$$V_u = \frac{2(338 + 172)}{2(12)} = 42.3 \text{ kips}$$ A.6.6

Nominal shear stress

$$v_u = \frac{42{,}300}{0.85(44)\,(29)} = 39.3 \text{ psi} < v_c = 2\sqrt{f'_c} = 126 \text{ psi}$$

Note that the sum of the moment capacities of the vertical boundary element sections [= 2(12,250) = 24,500 in.-kips] is greater than the sum of the moment capacities of the framing beams (= 338 + 172 = 510 in.-kips). A.6.2

e. Determine horizontal reinforcement in web of shear wall required by shear.

(1) Max. horizontal shear on shear wall (see Table 12.10),

$$V_u = 950 \text{ kips}$$

Nominal shear stress on 14 in. thick web,

$$v_u = \frac{V_u}{\phi h(0.8l_w)} = \frac{950{,}000}{0.85(14)\,(0.8 \times 308)} = 322 \text{ psi}$$

11.16.1

(2) Permissible shear stress in concrete, v_c, using eqq. (11-32) and (11-33) of ACI 318-71: 11.16.2

Eq. (11-32):

$$v_c = 3.3\sqrt{f'_c} + \frac{N_u}{4l_w h}$$

$$= 3.3\sqrt{4000} + \frac{2{,}340{,}000}{4(308)\,(14)} = 345 \text{ psi}$$

Eq. (11-33):

$$v_c = 0.6\sqrt{f'_c} + \frac{l_w\left(1.25\sqrt{f'_c} + 0.2\,\dfrac{N_u}{l_w h}\right)}{\dfrac{M_u}{V_u} - \dfrac{l_w}{2}}$$

$$= 0.6\sqrt{4000} + \frac{30\left[1.25\sqrt{4000} + 0.2\left(\dfrac{2{,}340{,}000}{308 \times 14}\right)\right]}{\dfrac{47{,}850(12)}{950} - \dfrac{308}{2}}$$

$$= 166 \text{ psi (governs)}$$

Required spacing of #4 horizontal bars in two rows

$$s = \frac{A_v f_y}{(v_u - v_c)\,b_w} = \frac{2(0.20)\,(60{,}000)}{(322 - 166)\,(14)} = 11 \text{ in.}$$ 11.16.4.1 11.6.1

f. Calculate minimum area of distributed horizontal and vertical reinforcement in shear wall web. A.8.2 11.16.4.1 11.16.4.2

(1) Min. vertical shear reinforcement ratio, using eq. (11-34), ACI 318-71:

$$\rho_n = 0.0025 + 0.5\left(2.5 - \frac{h_w}{l_w}\right)(\rho_h - 0.0025)$$

$$= 0.0025, \text{ since } \rho_h = \rho_{h(\min)} = 0.0025$$

Consider using #4 bars bothways placed in two rows: [14.2(g)]

$$\text{Required spacing} = 2\left[\frac{0.20(12)}{0.0025(12)(14)}\right] \text{but not}$$

$$> \{l_w/3, 3h, \text{ or } 18 \text{ in.} \quad [11.16.4.2]$$

$$= 11.4 \text{ in. (governs)} < 18 \text{ in.}$$

USE #4 bars bothways in two rows, spaced at 11 in. o.c.

g. Minimum length of lap splice for vertical reinforcement in boundary elements, #11 bars:
Assuming no more than one-half of the bars are lap spliced within the required lap length, splices shall conform to the requirements for Class B tension splices, that is, with a minimum length of lap = $1.3 l_d$. [7.6.3.1.1 7.6.1]

$$l_d = 0.04 A_b f_y / \sqrt{f'_c} \text{ but not}$$

$$> \{30 \text{ bar diameters, or } 16 \text{ in.}$$

$$= 0.04(1.56)(60{,}000)/\sqrt{4000} \text{ but not} \quad [12.6 \; A.6.7]$$

$$> \{30(1.41), \text{ or } 16 \text{ in.}$$

$$= 59 \text{ in. (governs)} > 42 \text{ in.}$$

Minimum requires splice length = $1.3 l_d$
= 1.3(59) = 77 in.

h. Detail of shear wall. See Fig. 12-82.

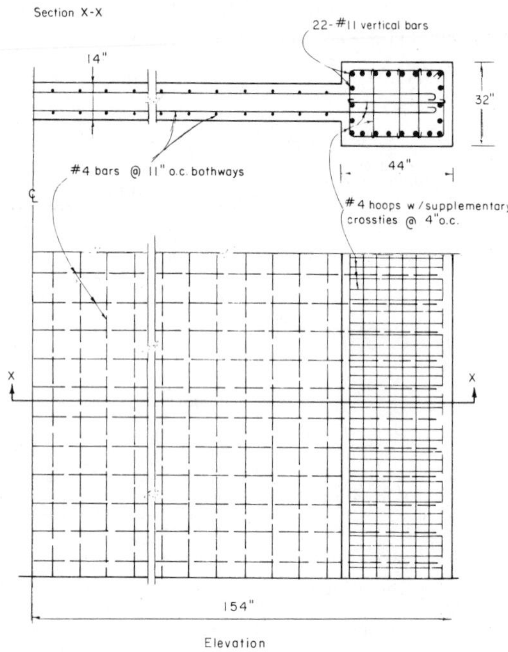

Fig. 12-82 Detail of shear wall reinforcement.

REFERENCES

The following abbreviations will be used to denote commonly recurring reference sources:

BSSA *Bulletin of Seismological Society of America*
JEMD *Journal of Engineering Mechanics Division, American Society of Civil Engineers (ASCE)*
JSTR *Journal of the Structural Division, ASCE*
WCEE World Conference on Earthquake Engineering

12-1 Richter, C. F., *Elementary Seismology*, W. H. Freeman & Co., San Fran., 1958.

12-2 *BSSA*, Vol. 57, No. 3, June 1967.

12-3 Japan National Committee on Earthquake Engineering, "Niigata Earthquake of 1964," Vol. III, 3rd WCEE, New Zealand, 1965.

12-4 Kawasumi, H. (editor), "General Report on the Niigata Earthquake of 1964," Tokyo Electrical Engineering College Press, March 1968.

12-5 Jennings, P. C.; Housner, G. W.; and Tsai, N. C., "Simulated Earthquake Motions for Design Purposes," Vol. 1, Proc. 4th WCEE, Chile, 1969.

12-6 Ruiz, P., and Penzien, J., "Probabilistic Study of the Behavior of Structures During Earthquakes," Report No. EERC 69-3, March 1969, Earthquake Eng. Res. Center, Univ. of California, Berkeley.

12-7 Blume, J.; Newmark, N. M.; and Corning, L., "Design of Multistory Reinforced Concrete Buildings for Earthquake Motions," Portland Cement Association, 1970.

12-8 Gutenberg, B., and Richter, C. F., "Earthquake Magnitude, Intensity, Energy and Acceleration," *BSSA*, 32 (3), July, 1942.

12-9 Algermissen, S. T., "Seismic Risk Studies in the United States," Vol. I, Proc. 4th WCEE, Santiago, Chile, 1969.

12-10 International Conference of Building Officials, 5260 South Workman Mill Road, Whittier, California 90601, *Uniform Building Code*. The latest edition of the *Code* is the 1973 Edition.

12-11 Milne, W. G. and Davenport, A. G., "Earthquake Probability," Vol. I, 4th WCEE, Santiago, Chile, 1969.

12-12 Wilson, E. L., "A Method of Analysis for the Evaluation of Foundation-Structure Interaction," Vol. III, Proc. 4th WCEE, Chile, 1969.

12-13 Housner, G. W.; Martel, R. R.; and Alford, J. L., "Spectrum Analysis of Strong-Motion Earthquakes," *BSSA*, **43** (2), April 1953.

12-14 Newmark, N. M., "Current Trends in the Seismic Analysis and Design of High-Rise Structures," Chap. 16 of *Earthquake Engineering*, R. Weigel, Ed., Prentice-Hall, 1970.

12-15 Hudson, D. E., "Some Problems in the Application of Response Spectrum Techniques to Strong-Motion Earthquake Analysis," *BSSA*, 52 (2), April 1962.

12-16 Veletsos, A. S., "Maximum Deformations of Certain Nonlinear Systems," Vol. II, Proc. 4th WCEE, Chile, 1969.

12-17 Biggs, J. M., *Structural Dynamics*. McGraw-Hill Book Co., New York, 1964.

12-18 Rogers, G. L., *Dynamics of Framed Structures*, John Wiley & Sons, Inc., New York, 1959.

12-19 Hurty, W. C., and Rubinstein, M. F., *Dynamics of Structures*, Prentice-Hall, New York, 1965.

12-20 Clough, R. W., and Benuska, K. L., "FHA Study of Seismic Design Criteria for High-Rise Buildings," Report HUD TS-3, Federal Housing Administration, Wash., D.C., August 1966.

12-21 Foss, K. A., "Coordinates Which Uncouple the Equations of Motion of Damped Linear Dynamic Systems," *Journal of Applied Mech., ASME*, **25**, Sept. 1958.

12-22 Seismology Committee, Structural Engineers Association of California, "Recommended Lateral Force Requirements and Commentary," 171 Second Street, San Francisco, Calif., 94105.

12-23 Associate Committee on the National Building Code, *National Building Code of Canada 1970*, National Research Council, Ottawa, Canada.

12-24 ACI Committee 318, *Building Code Requirements for Reinforced Concrete* (ACI 318-71), American Concrete Institute, P.O. Box 4754, Redford Station, Detroit, Michigan 48219.

12-25 Goel, S. C., and Berg, G. V., "Inelastic Earthquake Response of Tall Steel Frames," *JSTR*, Paper No. 6061, August 1968.

12-26 Ruiz, P., and Penzien, J., "Stochastic Seismic Response of Structures," *JEMD*, Paper No. 8050, April 1971.

12-27 Muto, K., "Recent Trends in High-Rise Building Design in Japan," Vol. I, Proc. 3rd WCEE, New Zealand, 1965.

12-28 Cloud, W. K., and Perez,. V., "Strong Motion–Records and Acceleration," Vol. I, Proc. 4th WCEE, Chile, 1969.

12-29 Hudson, D. E. (Editor), "Strong-Motion Data on the San Fernando Earthquake of Feb. 9, 1971," Earthquake Eng. Res. Lab., Calif. Inst. of Tech. & Seismological Field Survey, Nat'l. Oceanic and Atmospheric Adm., U.S. Dept. of Commerce, Sept. 1971.

12-30 Husid, R., "The Effect of Gravity on the Collapse of Yielding Structures with Earthquake Excitation," Vol. II, Proc. 4th WCEE, Chile, 1969.

12-31 Katayama, T., "A Review of Theoretical and Experimental Investigations of Damping in Structures," UNICIV Report I-4, Univ. of New South Wales, Kensington, Australia, 1965.

12-32 Jennings, P. C.; Matthiesen, R. B.; and Hoerner, J. B., "Forced Vibration of a 22-Story Steel Frame Building," Calif. Inst. of Tech. Earthquake Eng. Res. Lab. Report EERL 71-01, February 1971.

12-33 Gutenberg, B., "Effects of Ground on Earthquake Motion," *BSSA*, **47** (3), July 1957.

12-34 Alcock, E. D., "The Influence of Geologic Environment on Seismic Response," *BSSA*, **59** (1), February 1969.

12-35 Ohsaki, Y., and Hagiwara, T., "On Effects of Soils and Foundations Upon Earthquake Inputs to Buildings," Bldg. Res. Inst. (Japan) *Research Paper No. 41*, June 1970.

12-36 Seed, H. B., and Idriss, I. M., "Influence of Soil Conditions on Ground Motions During Earthquakes," *Jour. of the Soil Mech. and Foundation Div., ASCE*, Paper No. 6347, January 1969.

12-37 Seed, H. B., and Idriss, I. M., "Influence of Soil Conditions on Building Damage Potential During Earthquakes," *JSTR*, Paper No. 7909, February 1971.

12-38 Housner, G. W., "Behavior of Structures During Earthquakes," *JEMD*, Paper No. 2220, October 1959.

12-39 Duke, C. M., and Leeds, D. J., "Soil Conditions and Damage in the Mexico Earthquake of July 28, 1957," *BSSA*, **49** (2), April 1959.

12-40 Minami, J. K., and Sakurai, J., "Some Effects of Substructure and Adjacent Soil Interaction on the Seismic Response of Buildings," Vol. III, Proc. 4th WCEE, Chile, 1969.

12-41 Merritt, R. G., and Housner, G. W., "Effect of Foundation Compliance on Earthquake Stresses in Multistory Buildings," *BSSA*, **44** (4), October 1954.

12-42 Parmelee, R. A.; Perelman, D. S.; and Lee, S. L., "Seismic Response of Multiple-Story Structures on Flexible Foundations," *BSSA*, **59** (3), June 1969.

12-43 Khanna, J., "Elastic Soil-Structure Interaction," Vol. III, Proc. 4th WCEE, Chile, 1969.

12-44 Whitman, R. V., "Soil-Structure Interaction," *Seismic Design for Nuclear Power Plants*, Ed. by R. J. Hansen, MIT Press, Cambridge, Mass., 1970.

12-45 Anderson, J. C., and Bertero, V. V., "Seismic Behavior of Multistory Frames Designed by Different Philosophies," Earthquake Eng. Res. Center Report no. EERC 69-11, Univ. of California, Berkeley, Oct. 1969.

12-46 Hanson, R. D. and Fan, W. R. S., "The Effect of Minimum Cross Bracing on the Inelastic Response of Multi-story Buildings," Vol. II, Proc. 4th WCEE, Chile, 1969.

12-47 Goel, S. C., "P-Δ and Axial Column Deformation in Aseismic Frames," *JSTR*, Paper No. 6738, August 1969.

12-48 Nigam, N. C., and Housner, G. W., "Elastic and Inelastic Response of Frame Structures During Earthquakes," Vol. II, Proc. 4th WCEE, Chile, 1969.

12-49 Weaver, W., Jr.; Nelson, M. F.; and Manning, T. A., "Dynamics of Tier Buildings," *JEMD*, Paper No. 6293, December 1968.

12-50 Wen, R. K., and Farhoomand, F., "Dynamic Analysis of Inelastic Space Frames," *JEMD*, Paper No. 7621, October 1970.

12-51 Utku, S., "ELAS–Program for the Analysis of Structures," Dept. of Civil Engineering, Duke University, 1968.

12-52 Wilson, E. L., "SAP–A General Structural Analysis Program," Structural Eng. Lab., Univ. of Calif., Berkeley, Sept. 1970.

12-53 NASTRAN (NASA Structural Analysis Computer System), Computer Software Management and Information Center (COSMIC), Barrow Hall, Univ. of Georgia, Athens, Georgia 30601.

12-54 Wilson, E. L., and Dovey, H. H., "TABS–Static and Earthquake Analysis of Three-Dimensional Frame and Shearwall Buildings," Earthquake Engineering Research Center Report No. EERC 72-1, College of California, Berkeley, May 1972.

12-55 Newmark, N. M., "Torsion in Symmetrical Buildings," Vol. II, Proc. 4th WCEE, Santiago, Chile, 1969.

12-56 Koh, T.; Takase, H.; and Tsugawa, T., "Torsional Problems in Design of High-Rise Buildings," Vol. II, Proc. 4th WCEE, Chile, 1969.

12-57 Blume, J. A., and Jhaveri, D., "Time-History Response of Buildings with Unusual Configurations," Vol. II, Proc. 4th WCEE, Chile, 1969.

12-58 ACI-ASCE Committee 326, "Shear and Diagonal Tension," *ACI Journal, Proc.*, **59** Jan., Feb., Mar., 1962.

12-59 Bresler, B., and MacGregor, J. G., "Review of Concrete Beams Failing in Shear," *JSTR*, Feb. 1967.

12-60 Sturman, G. M.; Shah, S. P.; and Winter, G., "Effects of Flexural Strain Gradients on Microcracking and Stress-Strain Behavior of Concrete," Title No. 62-50, *ACI Journal*, July 1965.

12-61 Clark, L. E.; Gerstle, K. H.; and Tulin, L. G., "Effect of Strain Gradient on the Stress-Strain Curve of Mortar and Concrete," Title No. 64-50, *ACI Journal*, Sept. 1967.

12-62 Mattock, A. H., "Rotational Capacity of Hinging Regions in Reinforced Concrete Beams," *Proc. Int'l. Symposium on Flexural Mechanics of Reinforced Concrete*, ASCE, 1965, pp. 143-181. Also *PCA Development Dept. Bulletin 101*.

12-63 Corley, W. G., "Rotational Capacity of Reinforced Concrete Beams," *JSTR*, *Proc.* **92** (ST5), 121-146, Paper No. 4939, October 1966. Also *PCA Development Dept. Bulletin 108*.

12-64 Iyengar, K. T. S. R.; Desayi, P.; and Reddy, K. N., "Stress-Strain Characteristics of Concrete Confined in Steel Binders," *Mag. of Conc. Res.*, **22** (72), Sept. 1970.

12-65 Sargin, M.; Ghosh, S. K.; and Handa, V. K., "Effects of Lateral Reinforcement Upon the Strength and Deformation Properties of Concrete," *Mag. of Conc. Res.*, **23** (75-76), June-Sept. 1971.

12-66 Cohn, M. Z., and Ghosh, S. K., "The Flexural Ductility of Reinforced Concrete Sections," *IABSE Publications*, **32** (2), 1972.

12-67 Bertero, V. V., and Felippa, C., discussion of "Ductility of Concrete," by Roy, H. E. H. and Sozen, M. A., Proceedings of the International Symposium on Flexural Mechanics of Reinforced Concrete, ASCE-ACI, Miami, Nov. 1964, pp. 227-234.

12-68 Hanson, N. W., and Conner, H. W., "Seismic Resistance of Reinforced Concrete Beam-Column Joints," *JSTR*, Paper No. 5537, October 1967.

12-69 Bertero, V., and Bresler, B., "Seismic Behavior of Reinforced Concrete Framed Structures," Vol. I, Proc. 4th WCEE, Chile, 1969.

12-70 Megget, L. M., and Park, R., "Reinforced Concrete Exterior Beam-Column Joints Under Seismic Loading," *New Zealand Engineering*, 15 Nov. 1971.

12-71 Patton, R. N.; Park, R.; and Paulay, T., "Preliminary Report on Beam-Column Joint Tests," Univ. of Canterbury, Christchurch, New Zealand, Sept. 1971.

12-72 Clough, R. W., "Dynamic Effects of Earthquakes," *Transactions, ASCE*, **126**, Part II, Paper No. 3252, 1961.

12-73 Binder, R. W., and Wheeler, W. T., "Building Code Provisions for Aseismic Design," Vol. III, Proc. 2nd WCEE, Japan, 1960.

12-74 Blume, J. A., "Structural Dynamics in Earthquake-Resistant Design," Paper No. 3054, Part I, *ASCE Transactions*, **125**, 1960.

12-75 Berg, G. V., "Response of Multistory Structures to Earthquake," *JEMD*, Paper No. 2790, April 1961.

12-76 Agrawal, G. L.; Tulin, L. G.; and Gerstle, K. H., "Response of Doubly Reinforced Concrete Beams to Cyclic Loading," *ACI Journal, Proc.*, **62** (7), 823-835, July 1965.

12-77 Burns, N. H., and Siess, C. P., "Repeated and Reversed Loading in Reinforced Concrete," *JSTR, Proc.*, **92** (ST5), 65-78, Paper No. 4932, October 1966.

12-78 Fintel, M., and Khan, F. R., "Shock-Absorbing Soft-Story Concept for Multistory Earthquake Structures," *ACI Journal, Proc.*, **66** (5), 381-390, May 1969.

12-79 Caspe, M. S., "Earthquake Isolation of Multistory Concrete Structures," *ACI Journal, Proc.*, **67** (11), Nov. 1970.

12-80 Anonymous, "Ball Bearing Seismic Resistance," (on Matsushita system), *Engineering News Record*, 73, March 16, 1967.

12-81 Newmark, N. M. and Rosenblueth, E., *Fundamentals of Earthquake Engineering*, Prentice-Hall, New York, 1971.

12-82 Muto, Kiyoshi, "Earthquake Resistant Design of 36-Storied Kasumigaseki Building," Special Report, 4th WCEE, Chile, 1969.

12-83 Muto, Kiyoshi, "Earthquake Proof Design Gives Rise to First Japanese Skyscrapers," *Civil Engineering*, 49-52, ASCE, March 1971.

12-84 "Frame Constants for Lateral Loads on Multistory Concrete Buildings," *Advanced Engineering Bulletin No. 5*, Portland Cement Association, 1962.

12-85 Derecho, A. T., "Frames and Frame-Shearwall Structures," paper presented at Tall Building Symposium, ACI 1971 Fall Convention, Buffalo, N.Y., Nov. 1971.

12-86 Fintel, M. et al., "Preliminary Report—The Behavior of Reinforced Concrete Structures in the Caracas, Venezuela Earthquake of July 29, 1967," *Portland Cement Association XS6731*, 1967.

12-87 Derecho, A. T., "DYMFR—Program for the Dynamic Analysis of Plane Rectangular Frames," Portland Cement Association, 1970 (unpublished).

12-88 Derecho, A. T., "IBM 1130 Computer Program for the Analysis of Plane Multistory Frame-Shearwall Structures under Lateral and Gravity Loads," Portland Cement Association, 1971.

12-89 Lee, B. H., "IBM 1130 Computer Program for the Analysis and Design of Flat Plates, Flat Slabs, Waffle Slabs and Continuous Frames," *Portland Cement Association SR113.01D*, 1972.

13

Large Panel Structures

IAIN A. MacLEOD, Ph.D.*

13.1 INTRODUCTION

This chapter deals with the design of high-rise buildings whose main supporting structure consists of story-height, precast concrete panels. Large panel structures is one of the most important areas in which industrialized building techniques have proved advantageous and has been the most popular form of construction for high-rise apartment blocks in Europe for many years. The maximum number of stories in a large panel building, so far, is of the order of 26. Figure 13-1 is a photograph of a 20-story large panel block. Figure 13-2 shows a wall panel being placed.

The basic large panel is the wall unit, but the floor slabs are normally large precast concrete panels also. Large panels are used in low-rise construction where the problems are similar, though less critical from the structural point of view.

There is not a lot of published information on design of large-panel buildings. Reference 13-2 which is a good general text together with Refs. 13-3 and 13-4 are the main published sources.

In this chapter we consider the main problems involved and discuss what appears to be common practice with this form of construction. One should always bear in mind however that the methods which are common in practice are not necessarily the best.

13.1.1 Reasons for Using Large Panels

The low construction time for large panel buildings is probably the most important advantage of this method of construction. By precasting the main parts of the structure, the amount of site labor is greatly reduced in comparison with traditional methods. Most of the work is done in a factory where it can be carried out more efficiently. Since the amount of work which is dependent on favorable weather is minimized, critical path-type planning methods can be used with much greater success. Another major advantage of large panel construction is that the quality and finish of precast concrete can be controlled to a fine degree.

Probably the main unfavorable factor is that the connection of the precast panels can be troublesome. However, in the light of past experience, large panel buildings can nowadays be constructed whose behavior will be satisfactory in all respects.

It is inappropriate here to make any general statement about the cost of large panel structures. The cheapest method of construction at a given location is dependent upon local cost and availability of materials, and above all, on labor sources. Efficient large panel construction needs specially trained personnel both in manufacture and erection. Thus,

*Professor and Head Dept. of Civil Engineering, Paisley College of Technology, England.

Fig. 13-1 Large panel buildings, Glasgow, U.K.

if it is to be adopted in an area, future continuity of this type of construction would be most advantageous.

13.1.2 Layout

All large panel structures are by definition shear wall structures. (A shear wall structure is defined as one which derives its lateral stability exclusively from high 'shear' walls.)

Fig. 13-2 Erection of external panel.

Figure 13-3 shows a typical layout. This is a 'crosswall' type of building which has one way span floors. Two way span floors, though less common, are used. There is wide scope for variation in the layout. (See refs. 13-29, 13-30, 13-31.)

Large panel buildings consisting as they do of systems of interconnected shear walls are laterally stiff and can normally be reinforced without difficulty to withstand specified loading conditions. As the height of such buildings increases, however, the layout of the walls becomes more important from a structural standpoint; for taller buildings it becomes essential that an engineer is involved in the layout planning. For example, suitable arrangement of the layout can draw load away from stiff, yet weak, units. The layout is also important with respect to susceptibility to progressive collapse (section 13.2).

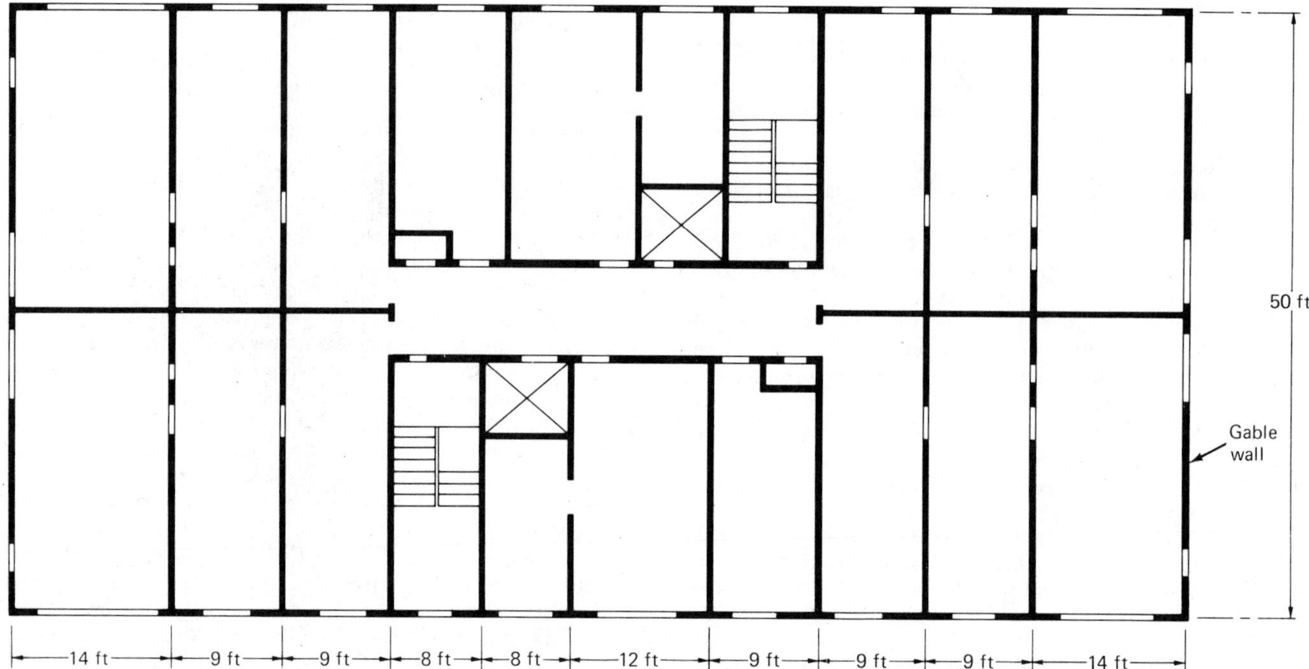

Fig. 13-3 Plan of large panel structure showing structural units.

13.2 LOADING

In this section information about the various types of loading which must be considered in design are discussed. Within the term 'loading' are included all sources of normal and abnormal stress.

Vertical loading–Under normal circumstances this is due to dead and superimposed loading. Most building codes allow a reduction in live load for the design of supporting members at lower floors in multi-story buildings.

Lateral load–is mainly due to wind or the horizontal component of an earthquake motion. These are assumed to act nonconcurrently and are important factors in the design of large panel buildings. When designing for earthquake since it is essential that the structure is ductile, particular attention must be paid to connections. The bulk of existing large panel buildings are not in such areas although satisfactory performance of such buildings during severe earthquakes has been reported. Cast-in-situ shear wall structures are known to behave well in earthquakes[13-20] and therefore, provided the connection detailing is good, large panel structures should behave similarly. In fact the energy absorption capacity of properly reinforced connections may be an advantage under such conditions. The analysis of large panel buildings under lateral load is described in section 13.3.

Progressive collapse–We class this under loading since it must be initiated by a loading. In the case of the Ronan Point Collapse[13-1] the triggering action was a gas explosion. Lack of sufficient steel to connect the precast panels together allowed the floor and wall units above and below the explosion area to collapse progressively. The C.E.B. Recommendations[13-4] do state that it is important to guard against progressive collapse in large-panel structures. However, bylaws or codes of practice relevant to Ronan Point at the time of construction made no mention of this possibility. A recently issued *British Code of Practice for Large-Panel Structures*[13-5] specifies that either the structure be designed to withstand explosive pressures of 5 lbs/sq in. or it must be ensured that the removal of a structural element (or elements) will not cause progressive collapse of the structure. Designers normally find it better to take the latter course.

Large-panel structures can be designed to withstand serious loss of structural components without progressive collapse. The designer should consider the consequences of the removal of a support at any story level and ensure that the vertical load tributary to that support has an acceptable alternative path. Any structure properly designed to resist earthquakes will normally be resistant to progressive collapse.

Differential settlement of foundations–Since large panel structures are basically stiff, a small amount of foundation movement can cause appreciable stress and cracking throughout the structure. The real source of cracking is normally most difficult to pinpoint and we have no record of structural cracking specifically due to differential foundation settlement. The relationship between the stiffness of the foundations, stiffness of the structure and stiffness of the soil, has an important effect on the distribution of vertical load as well as on the differential settlement. For large panel buildings the structure would normally have a controlling influence and this should be taken into account when designing the foundations.

Misalignment–Incorrectly placed units can cause significant stresses due to misalignment. The effect this has on the design of the wall is discussed in section 13.5.

Fire–The Ronan Point Report[13-1] suggested that fire could cause serious movement at horizontal joints in the external walls due to 'arching' of the slab. C.E.B.[13-4] suggests that thermal gradients in walls whose ends are restrained from rotation could significantly reduce their buckling strength.

One of the most important factors which governs the fire resistance of a panel is the amount of in-plane restraint provided by the adjacent structure.[13-6] Interior panels therefore have a much greater fire resistance than panels adjacent to the exterior of the building. The provision of tie reinforcement within the joints (sections 13.7 and 13.8) will have some effect in restraining in-plane movement of the slabs and walls and hence will enhance their fire resistance. Fire is another source of local failure which could result in progressive collapse. In general however, the fire resistance of large panel structures should be most favorable.

13.3 ANALYSIS TECHNIQUES FOR LATERAL LOAD

When making an overall analysis of a large panel structure it is necessary to make assumptions about the effect of connections. For ultimate strength requirements, to neglect the effect of a connection will normally be a conservative assumption. When considering serviceability, however, a weak connection which nevertheless will take load can cause undesirable cracking. Therefore, it is best to design the connections so that wherever possible the structure is effectively monolithic.

For those parts of the structure which are assumed to be monolithic bending theory is used to estimate stress and deformation.

To simplify the analysis, the building is divided into a series of wall units which are analyzed separately and then connected together. A wall unit may be a single wall, an elevator or stairwell shaft, or a series of walls in a plane connected by beams. Such units can be readily identified in Fig. 13-3. The connecting beams are formed from lintel beams or floor slabs or a combination of both. Careful attention must be given to their design.

The techniques used for the analysis of single-wall units are basically similar to those used for conventional frames, the important additional factor being that the finite width of the walls must be taken into account. These techniques are now briefly reviewed.

13.3.1 Analysis of Wall Units

Walls which do not have openings can be idealized as simple vertical cantilevers to which simple bending theory is applicable.

For walls with openings, the following techniques are used:*.

a. Shear connection method–For walls with a single row of openings (or two rows symmetrically placed) and constant properties with height, the row of openings can be effectively treated as a continuous medium in pure shear (Fig. 13-4b). A second order differential equation is formed and solved to give deflections, moments and shears. Probably the easiest way to use this method is to use the charts given in reference.[13-8] Another set of charts which include movement at the base of the walls is given in reference.[13-9] Rosman[13-10 to 13-12] has done much to develop this method. References 13-13 and 13-14 discuss some sources of error. This method can be used for walls with several rows of openings and other variations although the arithmetic becomes cumbersome.

*(See Ref. 13-7 for a more detailed comparison.)

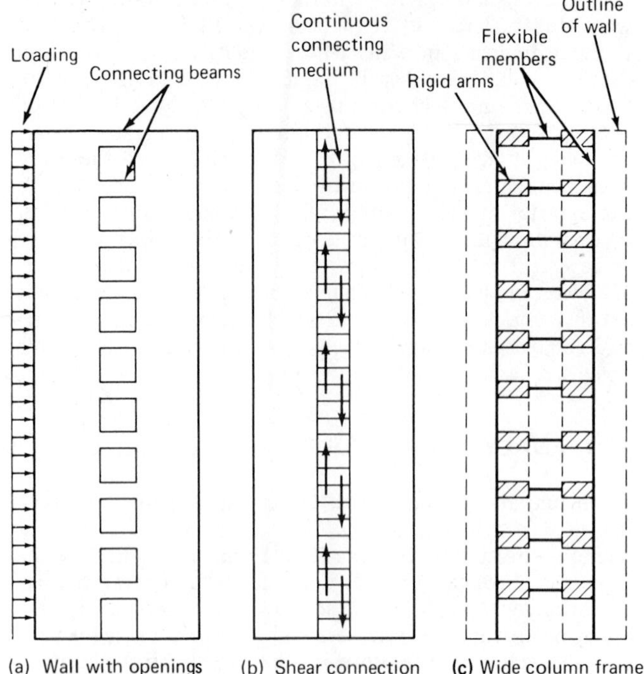

Fig. 13-4 Analysis techniques for shear walls with openings.

b. Wide-column frame analysis—This is the most versatile approach and requires the use of a plane frame computer program. Figure 13-4c demonstrates the idealization. It is preferable to treat the rigid portions of the frame as being fully rigid. (They can be given a high but finite stiffness but this is susceptible to numerical error in the solution.) This requires a special facility within the program which, although not common at present (1971), must become increasingly available as the demand for realistic analysis of shear walls increases. The frame program to be used should allow axial deformation of columns,* spring supports to model foundation effects, and, a feature which can be useful is a rotational spring connection between the rigid and flexible parts of the frame (to model the action of connecting beams which may not be fully fixed at their ends). A wide range of wall shapes can be efficiently analyzed using this analogy—in particular, variation of properties with height presents little difficulty.. Also with some further modification to a standard plane frame program, the full three-dimensional behavior of a shear wall building can be modeled.[13-15]

c. Finite element analysis—For specially important problems where a frame analysis would be inadequate, the use of a finite element model can be practicable (see section 13.9). Quadrilateral or triangular elements are most useful. There will be a significant increase in both the amount of data required and the solution time in comparison with a frame analysis.

13.3.2 Distribution of Loads to Units

The floors of a tall building are normally assumed to be rigid within their own planes so that under lateral load, (provided there is no torsion) each shear wall takes up the same deflected shape. Consider the hypothetical building shown in plan in Fig. 13-5a. First, if none of the three shear walls has any openings, the distribution with height of the lateral load will have the same form as that of the applied load, e.g., for a uniformly distributed applied load W lbs/ft, the loading will be distributed in the proportion:

$$\frac{I_1}{\Sigma I}W : \frac{I_2}{\Sigma I}W : \frac{I_3}{\Sigma I}W$$

to each wall. I_i is the moment of inertia of wall i and $\Sigma I = I_1 + I_2 + I_3$ (shear deformation is neglected here).

Now suppose the centre wall has a single row of openings and the uniformly distributed load is applied again. The distribution of load to each wall is no longer uniform and will take the form illustrated in Fig. 13-5d (computer analysis). This is because the walls behave differently but are constrained by the floor slabs to take up the same deflection at each story level. Therefore, for best accuracy an analysis should account for this fact.[13-16 to 13-19] However, it is normally found to be sufficiently accurate to connect the walls analytically at only a few stories and, in fact, preferably (as far as the arithmetic is concerned) only at one location, namely the top. In other words the stiffness, K_w, of each shear wall, is defined as the load to cause unit deflection at the top; the lateral load is distributed in proportion to these stiffnesses.

Another way to approach the same solution is to calculate an equivalent EI for the walls with openings

$$(EI)_e = \frac{H^3 K_w}{8} \tag{13-1}$$

where H is the height of the wall and $(EI)_e$ is the equivalent EI. The load is then distributed in proportion to these EI values.

EXAMPLE 13-1: The idealized 20-story structure of Fig. 13-5a is composed of two plain walls (Fig. 13-5c) and one wall with openings (Fig. 13-5b). Denoting the wall with openings as Wall A and the two plain walls together as Wall B we want to calculate the distribution of lateral load to these walls and hence calculate the stresses. Consider a uniformly distributed load of 10 lb/ft height on the structure and assume that this does not induce any torsion. Units used are feet and lbs, $E = 5 \times 10^8$ lb/ft² and shear deformation is neglected.

SOLUTION: Simplified Method.

Step 1: Analyze Walls A and B separately. Wall B analysis is straightforward but Wall A is more complex. The charts of Ref. 13-8 could be used but since a frame program was readily available on analysis of the frame idealization of Fig. 13-5e was carried out. This is called Run 1.

Step 2: Calculate wall stiffnesses.

$$K_{WA} = \frac{10^4}{3.082} \quad \text{(from Run 1)} \quad = 3245$$

$$K_{WB} = \frac{8\,EI}{H^4} = \frac{8 \times 5 \times 10^8 \times 0.5 \times 28^3}{180^4 \times 12} = 3486$$

$$\Sigma K = 6731$$

Step 3: Distribute load to walls.

$$\text{to Wall A} = \frac{K_{WA}}{\Sigma K}W = \frac{3245}{6731} \times 10 = 4.82 \text{ lb/ft}$$

$$\text{to Wall B} = \frac{K_{WB}}{\Sigma K}W = \frac{3486}{6731} \times 10 = 5.18 \text{ lb/ft}$$

This distribution is shown in Fig. 17-6d.

Step 4: Calculate the top deflection of structure.

$$\Delta = \frac{\text{Load on Wall A}}{K_{WA}} = \frac{4.82}{3245} = 1.482 \times 10^{-3} \text{ ft}$$

*Axial deformation of connecting beams may be neglected. This should be done if possible so as to reduce the order of solution and improve the conditioning of the equations to be solved.

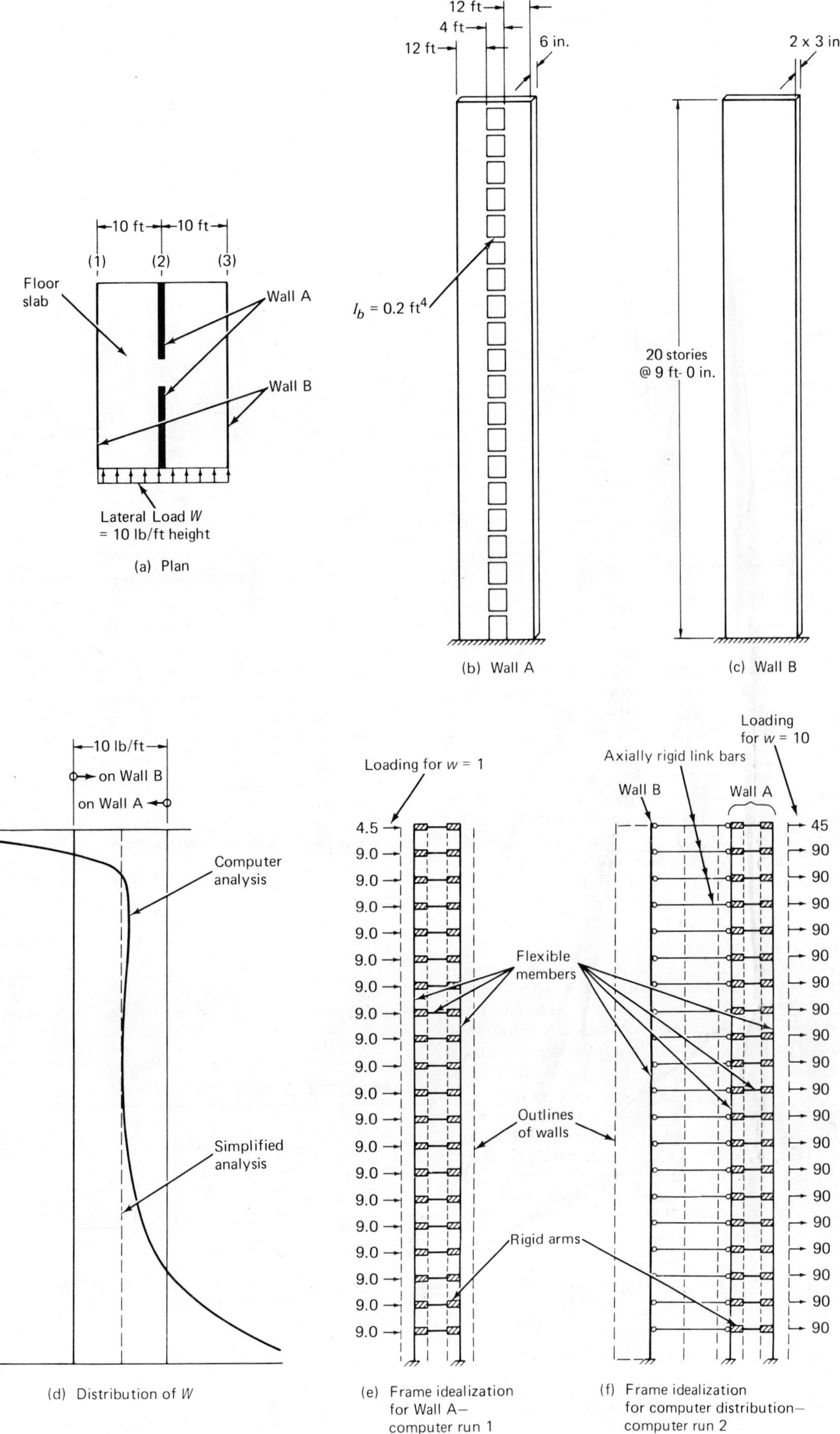

Fig. 13-5 Interconnected shear wall problem.

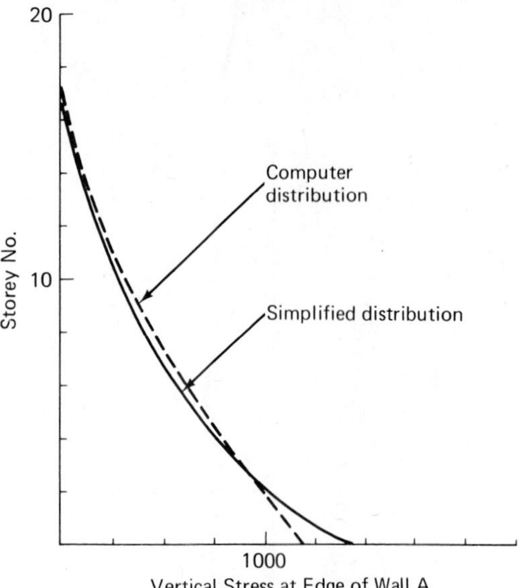

Fig. 13-6 Vertical edge stress in Wall A.

Step 5: Calculate stresses.

e.g., Max. bending stress in Wall B

$$= \frac{5.18 \times H^2}{2} \times \frac{6}{0.5 \times 28^2} = 1284 \text{ lb/ft}^2$$

-1. Computer distribution—Fig. 13-5f shows how a plane frame program can be used to carry out a no-torsion analysis of a shear wall structure taking account of the interaction at each story level. The distribution of load to each wall is illustrated in Fig. 13-5d.

-2. Comparison of accuracy—The distribution of load by the simplified analysis is significantly different from the computer distribution. This is not important from the point of view of design. The important factors are as follows:

a. Vertical stress in the walls—Fig. 13-6 shows the vertical edge stress in Wall A. Accuracy for Wall B is slightly better than this (not shown).
b. Maximum beam shear—Table 13-1 shows that the accuracy is adequate.
c. Deflection—Table 13-1 shows that the accuracy for estimation of top deflection is very good. The closeness of the results here is to some extent fortuitous but estimates of deflection tend to be always more accurate than the moments and forces which are derived from them.

The properties of the structure used for this example are favourable to accuracy for the simplified method. The results shown in Fig. 13-6 and Table 13-1 for the simplified method would be suitable for design. However, they probably represent an upper limit to the accuracy for the method when walls with and without openings are connected. Techniques for determining when the simplified method is inadequate have not yet been established.

TABLE 13-1 Accuracy of Simplified Analysis

	(a) *Simplified Analysis*	*(b)* *Computer Run 2*	*% difference* $\frac{(a)-(b)}{(b)} \times 100$
Maximum Beam Shear (Wall A)	313.0	269.2	+16%
Top Deflection	1.479×10^{-3}	1.482×10^{-3}	−0.2%

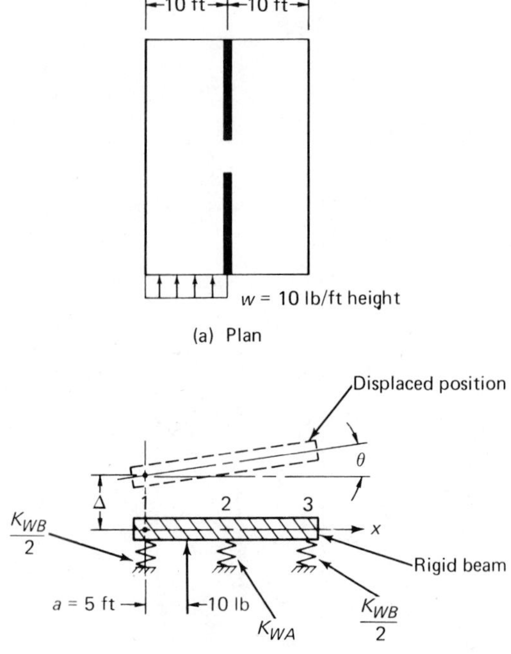

Fig. 13-7 Example problem with torsion.

EXAMPLE 13-2: Torsion—An analysis with torsion can be made in a similar fashion to the distribution previously described. For example, assume the load acts off-center on the structure of Fig. 13-5a, i.e., as in Fig. 13-7a.

SOLUTION: Consider the top of the structure as a rigid beam supported by springs and loaded as shown in Fig. 13-7b. Movement of this beam can be defined by a deflection at 1(Δ) and a rotation (Θ). Equilibrium gives the simultaneous equations

$$\Sigma K_i\,(\Delta + x_i\Theta) = W \tag{13-2}$$

$$\Sigma K_i x_i\,(\Delta + x_i\Theta) = Wa \tag{13-3}$$

These can be solved for Δ and Θ. The load on each wall is then

$$W_i = WK_i\,(\Delta + x_i\Theta) \tag{13-4}$$

The calculations for the system of Fig. 13-7b are set out in Table 13-2.

TABLE 13-2 Analysis with Torsion

Wall	K_i	X_i	X_i^2	K_iX_i	$K_iX_i^2$
1	1743	0	0	0	0
2	3245	10	100	32450	324500
3	1743	20	400	34860	697200
Σ =	6731			67310	102170

The equilibrium equations, eqq. (13-2) and (13-3), are then

$$6731\,\Delta + 67310\,\Theta = 10$$

$$67310\,\Delta + 102170\,\Theta = 10 \times 5$$

Solving this gives

$$\Delta = 61.0 \times 10^{-5}$$

$$\Theta = 8.76 \times 10^{-5}$$

Hence the loading on the separate walls, using eq. (13-4), will be

$$W_1 = 10 \times 1734 \times 10^{-5}\ (61.0 + 0) = 1.06\ \text{lb/ft}$$

$$W_2 = 10 \times 3245 \times 10^{-5}\ (61.0 + 10 \times 8.76) = 4.84\ \text{lb/ft}$$

$$W_3 = 10 \times 1734 \times 10^{-5}\ (61.0 + 20 \times 8.76) = 4.10\ \text{lb/ft}$$

These are the approximate loads on the walls assumed to be uniformly distributed.

This procedure gives satisfactory answers when the free deflection of the walls are close to that of a cantilever. When openings induce frame behavior in some of the walls, comparison of results between this simplified analysis and one where all the floors are considered may show significant differences.

13.3.3 Flange Effect of Walls Perpendicular to Plane of Loading

Lateral stiffening units are composed of:

(1) walls in the plane of the loading, and
(2) walls at right angles to the plane of the loading.

The latter walls form flanges to the former, provided joints between the two are capable of transmitting the necessary shear. This should be checked in the design. No systematic study on what should be taken as the effective flange width under such circumstances is known. Therefore, a conservative estimate should be made, i.e., the effective flange width on each side should not be greater than one-half of the distance to the next wall nor the distance to the edge of the nearest opening (Fig. 13-8). See Ref. 13-4, clause 52.25.

13.3.4 Effective Width of Floor Slabs

The width of floor effective in transfering vertical shear between adjacent wall sections has been studied[13-21, 13-22] but no general recommendations can yet be made. The floor slabs, if properly reinforced, will certainly have a significant effect on the behavior of the wall. To overestimate the effectiveness of the floors will underestimate the stresses in the walls (and vice versa). Similarly an overestimate will be conservative as far as the load tributary to that wall is concerned but unconservative as far as the other walls are concerned.

13.4 STRUCTURAL AND FUNCTIONAL DESIGN—GENERAL

In the sections which follow, the main problems associated with the design of large panel structures are discussed. In practice the design of the wall and floor panels cannot be considered separately from the design of the vertical and horizontal connections. They are discussed separately here in order to emphasize the importance of the connection design. The word 'connection' is used here to describe the region where panels are connected and the 'joint' is specifically the area between the connected parts, see Fig. 13-9.

The discussion covers mainly structural problems which cannot however be considered in isolation from functional factors such as thermal, weather, and sound insulation.

13.5 DESIGN OF WALL PANELS

13.5.1 General

In order to limit differential vertical movements it is important that all the load bearing walls be constructed from the same type of concrete within each story height. For the same reason, the vertical stress due to dead load should ideally be constant in all the walls. This will not normally be possible but at least precast and cast-in-situ load bearing walls should not be present together at the same story level.

13.5.2 Internal Walls

The thickness of internal walls is normally governed by sound-insulation requirements—a solid thickness of 6 in. normal weight concrete being a common minimum. The thickness for structural requirements depends on several

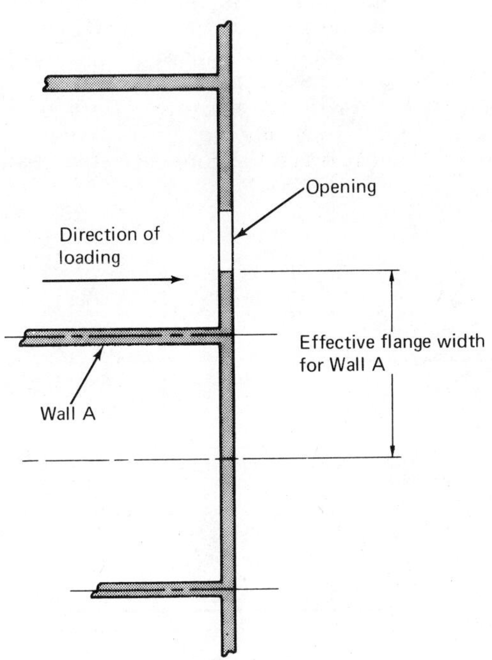

Fig. 13-8 Effective wall flange width.

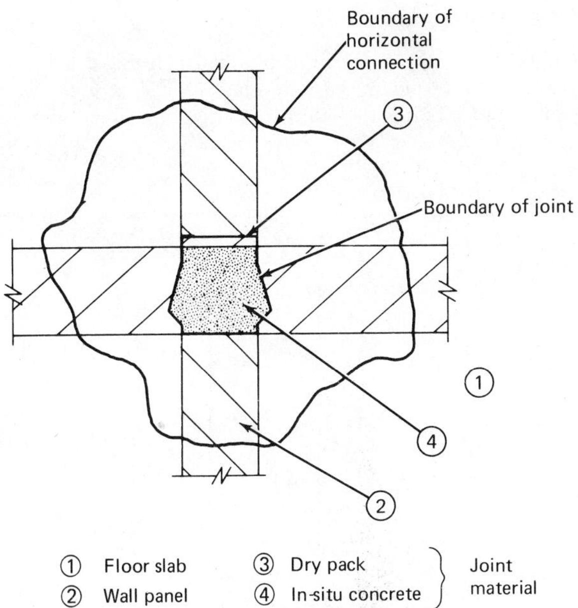

Fig. 13-9 Definition of 'joint' and 'connection'.

factors as discussed below but 6 in. is normally adequate for this purpose. This is an important consideration in the choice of a wall as against a column support system for apartment blocks.

13.5.3 External Walls

Figure 13-10 shows a section through a typical external wall panel. Where thermal insulation is required, a sandwich panel appears to be the standard method of construction. Such panels tend to be the most expensive part of the structural system and cheaper alternatives may be possible in mild climates. The structural part of the sandwich panel is kept to the inside so as to minimize the effect of diurnal temperature changes. Typical dimensions are given in Fig. 13-10.

The weather seal at the connections between external, precast panels must be given careful attention. The basic principle normally adopted is to provide a drainage path behind an outer insert. The insert and the space behind it protect the main weather seal from the pressure of driving rain. This problem is discussed in detail in Ref. 13-2.

13.5.4 Structural Design

The walls should be designed to withstand the combined effect of vertical and lateral loads with due consideration given to eccentricity caused by lack of alignment, end fixity, etc., and to the possibility of buckling. General recommendations to take account of the various types of eccentricity, distribution of point loads, and reduction in stress to limit buckling are given in Reference 13-4. See also reference 13-2.

13.5.5 Reinforcement

Unreinforced wall panels have been common in the past but some steel is useful for handling stresses and to limit the effect of shrinkage. Unreinforced panels should certainly not be used in regions susceptible to earthquakes. Reference 13-5 specifies a minimum of 0.1% of reinforcement both horizontally and vertically in all wall panels. Reinforcement around openings is also necessary, especially to cover handling stresses.

13.5.6 Uplift and Tension

In taller blocks it is not uncommon for design calculations to indicate a tendency to develop tensile stresses under severe wind loading at the bottom edge of a shear wall unit. The analytical techniques used do not normally recognize the true three-dimensional interaction of the walls so that predicted tensions may not, in fact, be realized in practice. However, if the calculations say that there may be tension, something must be done about it. It is difficult to provide vertical tensile reinforcement between the horizontal joint and the panel above. Such reinforcement will be provided within the vertical joints (see section 13-8). Rearrangement of the wall system to redistribute dead load and/or lateral load stresses could be an alternative solution if considered early enough.

13.6 DESIGN OF THE FLOOR SLABS

Unlike wall design, the design of the floors for large panel buildings involves consideration of several alternative possibilities. The designer must weigh up these interacting factors and decide on a system which will satisfy design criteria of strength, serviceability, sound insulation, and economy.

The possible alternatives for floor construction include the following:

a. Precast or cast-in-situ systems do exist which have cast-in-situ floors with precast wall panels. The main advantage of casting the floors in-situ is that they can easily be made continuous over the wall supports—hence, they can be made thinner or less reinforcement can be used. However the slab thickness is often governed by sound insulation requirements and may exceed the optimum thickness based on purely structural considerations.

   Some of the points made in the discussion which follows are pertinent to cast-in-situ construction but this chapter is concerned specifically with precast work.

b. One-way or two-way span—A two-way span is obviously more efficient structurally than a one-way system but can only be used (a) if the layout permits it, and (b) if room-size floor units are used.

   With less than room-size floor units, the unsupported joint between adjacent units should include a shear key. Tests by Lewicki[13-2] show that this is essential for spreading load and preventing undesirable cracking in the line of the connection. The actual shape of this key does not appear to be of great importance. The main thing is that some shear transfer is provided. Figure 13-11 shows some typical keys.

c. Continuous or simply supported—When a connection is designed to be simply supported, cracking at the top of the joint between the floor concrete and the in-situ

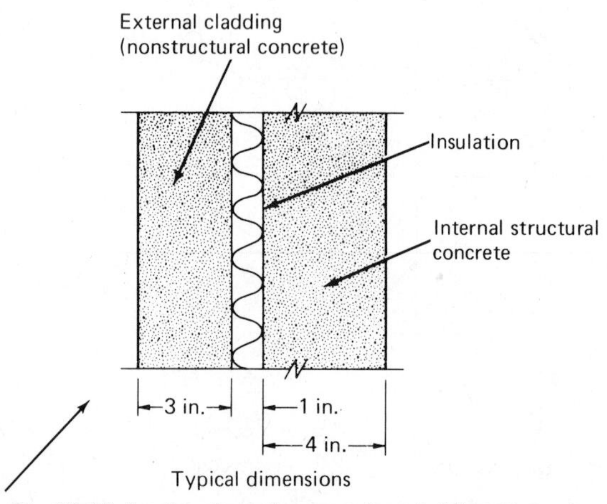

Fig. 13-10 Section through external sandwich wall panel.

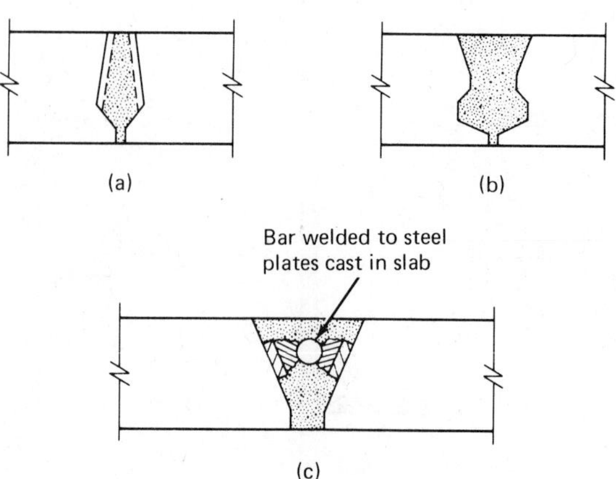

Fig. 13-11 Joints between floor units—parallel to span.

joint concrete may prove troublesome. Some continuity is desirable to prevent this.

The provision of moment continuity at the wall-floor junction will obviously enhance its resistance to progressive collapse. It may however induce undesirable eccentricities in the wall panels.

The type of floor-wall connection detail which is becoming commonly used (Fig. 13-13) will result in moment continuity at the slab support whether considered in the design or not. With this type of detail, standard practice appears to be to design the slab as simply supported for dead load and as continuous for live load.

d. Normal weight or lightweight concrete–Lightweight concrete might be used in preference to normal weight for two reasons:
   (1) To reduce vertical stress. This is seldom necessary in large panel structures since the walls are not normally fully stressed. For high blocks where wall stresses are significant there could be some advantage in using lightweight concrete, but sometimes *extra* dead load is an advantage in limiting any tendency for tension to be developed in the walls.
   (2) To allow larger precast units to be erected for a given crane capacity. This is discussed in section 13.10.

   Any advantage the lightweight concrete might have must of course be set against its normal disadvantages of increased cost and decreased rigidity.
e. Hollow or solid–In order to reduce weight one-way span slabs are normally cored. The hollow cores can be used as service ducts.
f. Reinforced or prestressed–The need for prestressing is of course dependent on the span. It is advantageous to keep the floor slab thickness at each story level the same. Having decided what this is to be, the longer spans can be prestressed to limit deflection. A problem sometimes arises with precast, prestressed floor units in trying to match initial deflection when the top and/or the bottom is to represent the finished surface. In fact a screed on top and a plaster or false ceiling below may be essential with this type of floor construction.
g. Strength–Floor thickness based on sound insulation and deflection criteria is normally adequate to ensure that the slab will be underreinforced (i.e., flexural yield of the steel will precede concrete failure).

   If the building is to be constructed so as to withstand internal explosive forces, then top reinforcement in the slab will be necessary.

13.7 DESIGN OF HORIZONTAL CONNECTIONS

Floors and walls are connected by a horizontal joint as shown diagrammatically in Fig. 13-12. This connection must transmit the following force actions:

a. vertical load (including vertical resultants of lateral load)
b. horizontal shear due to lateral load
c. shear in the plane of the floors
d. bending stress in the plane of the floors
e. tie forces
f. transverse bending due to floor loadings
g. shear due to floor loadings

If all these stress actions act concurrently (as they can do), then the state of stress within the joint will be most complex. Thus, it is not rational to treat the stresses resulting from the different loadings separately. However, very little published information is available on wall-floor connections and there is no design procedure available which will account for the interaction of even two of these force actions. Horizontal connections are therefore designed mainly on the empirical basis.

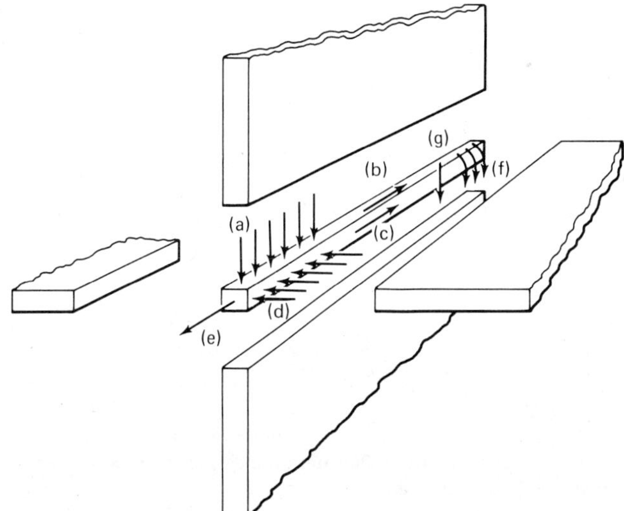

Fig. 13-12 Exploded view of horizontal connection showing force actions.

13.7.1 Vertical Load (a)

This is probably the most important of the loadings mentioned above. The presence of the connection may cause the wall panel to fail at a lower load than it would otherwise take. Laboratory testing of the connection under vertical load is normally recommended. If experimental results are not available, Ref. 13-4, (Clause 47.21) recommends (somewhat tentatively) values for a reduction factor on the strength of the concrete in the wall panels to estimate the strength of the joints. It is implied that the separate strengths of the wall and joint concretes should be similar.

The bearing of the panel on the joint should be as uniform as possible. To achieve this (1) in the direction of the thickness of the panel the joint should not be stepped or consist of different materials, (see discussion of floor support later in this section) and (2) the erection technique is important (section 13-10).

Two important factors affecting the vertical strength of the connection are the ratio of strength of the in-situ concrete to that of concrete in the wall, and the method of supporting the floor slabs. These two factors are related in that the floor slab support detail will affect the confinement, and hence, the strength of the in-situ material.

Three types of connection are shown in Fig. 13-13. Of these, the platform detail of Fig. 13-13a is the only one for which design rules have been formulated.

Defining

f_c = cylinder strength of wall concrete;
f_m = cylinder strength of mortar;
f_w = strength of wall;
t = thickness of wall panel;
a = penetration distance of floor slab, i.e., bearing distance of floor slab on wall;

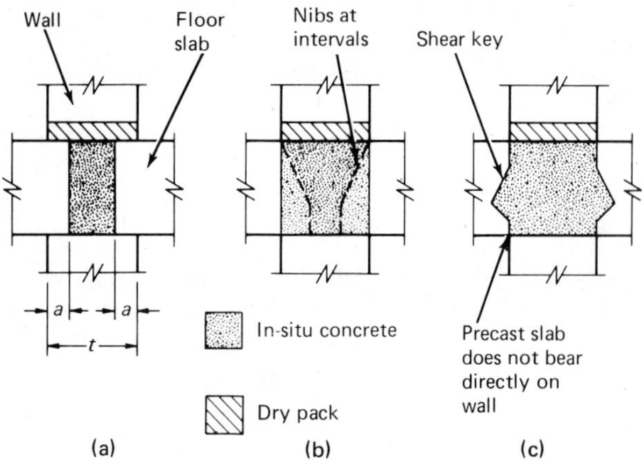

Fig. 13-13 Vertical section through horizontal connections (reinforcement not shown). (a) Platform support; (b) nibs; and (c) shear transfer via in-situ concrete.

Fig. 13-14 Photograph of floor-leveling device.

an emperical formula for the compressive strength of the connection is

$$\frac{f_w}{f_c} = \left[0.6\left(\frac{a}{0.5t}\right)^{3/2} + \frac{0.4f_m}{f_c}\right] \qquad (13\text{-}5)$$

This is only applicable to connections where the floor support is continuous. This formula is proposed in Ref. 13-3 and also quoted in Ref. 13-2. It appears to give a conservative estimate of strength in the range $0.2 < a/t < 0.5$. Presumably at $a/t = 0$ the strength of the joint would not be as low as $0.4f_m$ but we do not have information about the strength for $a/t < 0.2$.

13.7.2 Transmission of Shear Due to Lateral Load (b)

Horizontal shear due to lateral load must be transmitted along the crack (preformed by the method of construction) between the bottom of the wall and the joint. Since reinforcement does not normally extend up from the horizontal joint into the wall panel, this shear must be transmitted by friction along the bottom of the wall panels and by dowel action of the steel in the vertical joints. Reference 13-4, Clause 54.51, allows the use of a higher coefficient of friction for joints formed entirely in concrete (0.35) than for joints with a layer of mortar (0.2).

The mechanism of shear transfer in the wall is not as simple as this, of course. The tie steel in the horizontal joints will act as shear reinforcement in a similar fashion to stirrups in a conventional reinforced concrete beam. Care must be taken if there is a likelihood of vertical tension in the wall but normally shear failure of shear walls is most unlikely.

The transmission of vertical shear at the junction with a vertical joint is discussed in section 13.8.

13.7.3 Force Actions in the Plane of the Floors due to Lateral Load (c), (d), and (e)

Lateral load is distributed between the vertical units by in-plane actions in the floor slabs. These actions are seldom if ever calculated on a mathematical basis and the in-plane reinforcement is specified on an empirical basis. (See under 'Reinforcement of Horizontal Connections' in this section).

13.7.4 Transverse Bending and Shear due to Floor Loadings (f) and (g)

The provision of moment continuity between floor slabs and walls is discussed in section 13.6.

Figure 13-13 shows three common details for shear transfer between floor slabs and walls.

The detail of Fig. 13-13a where the floor slab rests continuously on top of the wall has been common in the past. It has the advantage of providing an even bearing stress along the length of the connection and is easy to cast and erect. However it does not leave much space for in-situ joint concrete and reinforcement. Hence, despite its apparent solidity it will not allow a sufficiently strong connection to be formed to prevent failures of the Ronan Point type.

The intermittent nib support, Fig. 13-13b, does provide space for more in-situ joint material and can allow adequate connection strength. It does not provide uniform bearing stress on the wall and the casting molds are more complicated. A positive advantage is that it provides keys which will resist shear forces in the plane of the slabs.

A detail of the type shown in Fig. 13-13c is becoming increasingly popular. The floor panel initially rests either on special removable levelling devices (see Fig. 13-14) or is supported by the projecting bars which bear on the top of the wall panels. The in-situ concrete is then poured and when set will allow vertical shear to be transmitted via the shear key on to the wall. The floor slab does not rest directly on the wall. Thus a uniform bearing stress will be achieved together with adequate space for joint concrete and reinforcement.

13.7.5 Reinforcement of Horizontal Connections

It is a basic principle that reinforcing bars should extend horizontally from the floor slabs and vertically up from the wall panels below into the in-situ joint material. A positive mechanical connection should be made between these bars possibly by welding, sleeving, or threading but most commonly by the type of detail shown in Fig. 13-15. Note the use of the terms 'projecting bars' and 'tie bars' to define bars which are cast into the precast units or lie longitudinally within the joint material respectively. The design of the reinforcement thus introduced into the joint must be based on three competing factors, namely:

a. the need for plenty of steel,
b. the cost of the steel, and
c. the difficulty involved in fixing the steel and placing in-situ concrete around it.

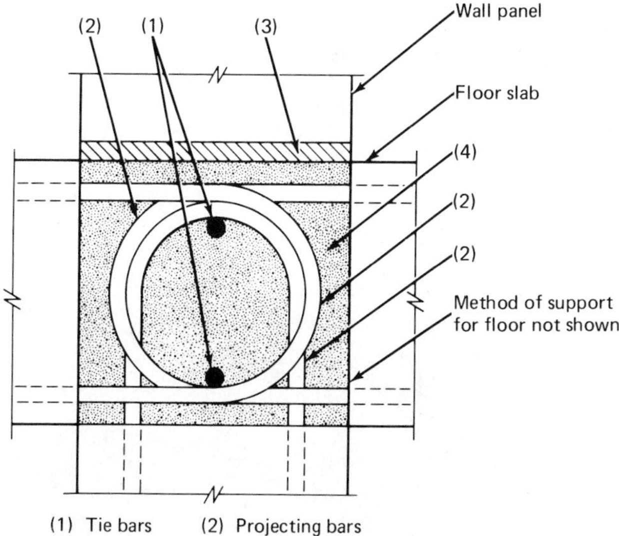

Fig. 13-15 Reinforcement of a horizontal connection. (1) Tie bars; (2) projecting bars; (3) drypack; and (4) in-situ concrete.

The most important function of the reinforcement within the joint is to provide a connection which will prevent a progressive collapse of the type which occurred in the Ronan Point disaster.[13-1] Under such circumstances ductility is important and mild steel reinforcement is preferable.

13.7.6 Design of Projecting Bars

The bars which project from the top of the wall panels into the joint are normally used as lifting hooks. Therefore at least two are needed per panel. The bars from the floor slabs, if suitably anchored into the joint, may be required to

a. take the live load support continuity moment and
b. act as ties in the plane of the floor. With the connection detail of Fig. 13-13c they may also be required to support the slab during erection as discussed previously.

13.7.7 Design of Tie Bars

The tie steel performs three functions.

a. It acts as a tie to the vertical connections (section 13-8).
b. It provides reinforcement in the plane of the floors and the walls, improving their ability to act as monolithic units rather than as separate panels.
c. At ultimate load conditions if detailed as in Fig. 13-5 it can form a mechanical connection between the precast panels.

Equation (13-9) gives an estimate of steel requirement under (a).

References 13-4 and 13-5 give recommendations for tie steel under (b) above. In this respect the bars in the external horizontal connections are the most important. These should be effectively continuous around the periphery of the building at each story level. Internal ties are also required. Reference 13-5 recommends that in the direction of the span of the floor slabs the internal ties should be encased in the precast floor units. Some difficulty is involved in making these effectively continuous (Fig. 13-15). The ties within the internal horizontal joints are also important. The detailing of these tie bars may in fact be more important than the actual choice of bar size. Tie bars should be properly lapped and at their ends should preferably be hooked around a vertical tie bar in a vertical joint.

The tie steel specified under (b) above is not specified in addition to that used for other purposes.

13.8 DESIGN OF VERTICAL CONNECTIONS

13.8.1 Structural Action

The force actions associated with a vertical connection between wall panels are illustrated in Fig. 13-16.

The main action is vertical shear, V, due to lateral load together with a horizontal tie restraint. This tie action may be concentrated at the floor levels or may be distributed throughout the height of the connection as shown in Fig. 13-16.

CEB[13-4] states that the V may be assumed to be transmitted along the height of the vertical joint or at the wall-

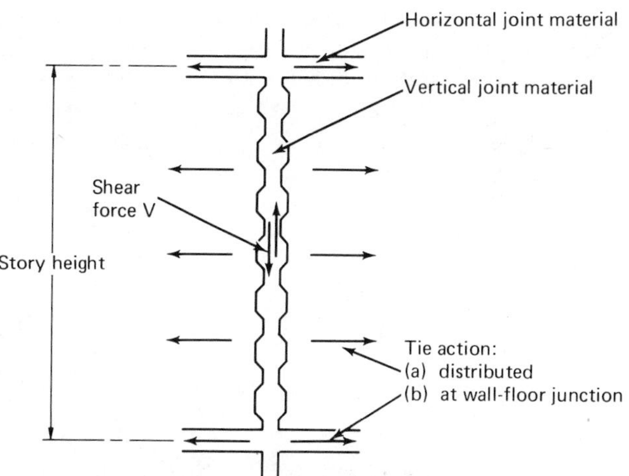

Fig. 13-16 Force actions for transmission of shear at vertical connections.

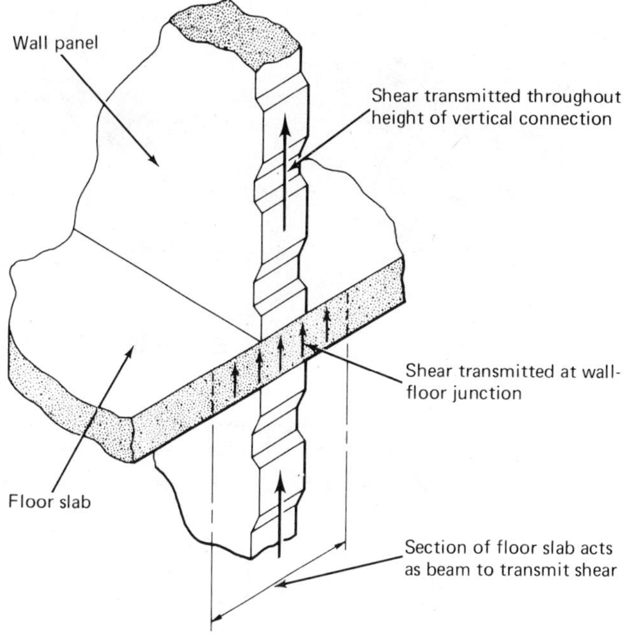

Fig. 13-17 Transmission of vertical shear.

floor junction* (Fig. 13-17). The effect of the wall-floor junction may in some cases be significant but, in view of the lack of published information, we will assume that V is transmitted only by the vertical joints between the panels and neglect the effect of the wall-floor junction.

The strut hypothesis normally used to explain shear transfer in beams is also used to describe shear transfer in such a joint.[13-4, 13-25] Inclined struts are assumed to form within the joint as shown in Fig. 13-18. The inclined strut reaction may be resolved into horizontal and vertical components. The horizontal thrust T must be resisted by tie steel distributed as discussed above. The vertical component must be resisted by shear keys formed in the edges of the wall panels. Thus, three important factors are (1) the angle of inclination of the struts, (2) the tie steel, and (3) the shear keys.

It is normal practice to calculate the vertical shear, V, from the shear stress using bending theory, assuming the wall to be monolithic across the line of the vertical connection. This shear is assumed constant over the height of each story.

-1. Angle of inclination of the struts—Consider the two rigid planes AB and CD (Fig. 13-19) connected by n struts with angle of inclination, θ, and with vertical spacing, d. If a total shear of V is applied as shown, then the shear per strut, Q, is given

$$Q = \frac{V}{n} \tag{13-6}$$

This causes an

$$\text{axial load in the strut} = Q/\sin\theta \tag{13-7}$$

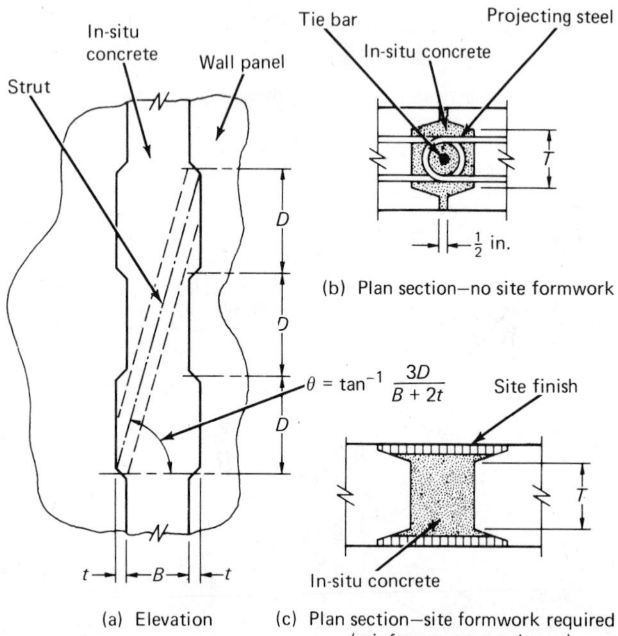

Fig. 13-18 Castellated joints. (a) Elevation; (b) plan section—no site formwork; and (c) plan section—site formwork required (reinforcement not shown).

*Ref. 13-4 is an English translation of the original French and uses the term 'elastic locking element' to describe the part of the floor-wall junction which takes vertical shear.

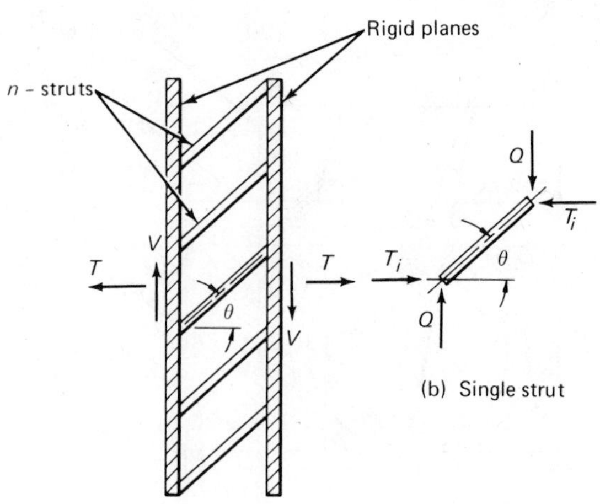

Fig. 13-19 Transmission of shear by strut action.

with the resulting horizontal component

$$H_i = Q/\tan\theta \tag{13-8}$$

Therefore the total horizontal component is

$$H = nQ/\tan\theta = V/\tan\theta \tag{13-9}$$

This is the horizontal tensile force which must be resisted by tie reinforcement. The smaller the assumed θ, the larger will be the predicted tie and strut forces. CEB[13-4] recommends $\theta = 45°$. Recent test results[13-27] indicate that $\theta = \tan^{-1}(3D/(B + 2t))$ may be acceptable for castellated joints (see Fig. 13-18).

-2. Tie steel—If V is the total vertical shear over a story height as calculated using bending theory, then eq. (13-9) gives the total tie force over the story height. Steel to resist this can be placed in the horizontal joint and should be continuous across the line of the vertical joint. In addition, loops within the vertical joint will also resist the tie forces—see later in this section under "Reinforcement of Vertical Connections."

-3. Shear keys—The most common type of shear key is the castellated type shown in Fig. 13-18. A considerable amount of testing work has been carried out on such joints.[13-27, 13-28, 13-29, 13-35] It appears that within normal limits the dimensions of the keys are not critical to the strength of the connection. The width of the joint B should be kept as small as possible consistent with adequate anchorage for loop reinforcement and satisfactory placing of concrete. A depth of castellation, D, of the order 3 to 12 in. is normal. The distance, t, should not be less than 0.5 in., and the sloping edges to the key allow better filling of the joint without significantly affecting its strength.

The thickness of the shear key, T, depends on the method of forming the joint. The detail of Fig. 13-18b has been common in the past. This requires a minimum of site formwork but has not proved altogether satisfactory. A detail of the type shown in Fig. 13-18c is becoming popular. This allows almost the full thickness of wall to act as a shear key. Extra site finishing is required but this is not felt to be a disadvantage.[13-23]

The strength of such joint may be expressed as[13-27]

$$Q = \beta D T f_{ct} \tag{13-10}$$

where f_{ct} is the tensile splitting strength of the concrete and β is an empirical constant. Experimental results are not sufficiently comprehensive as yet to make a firm recom-

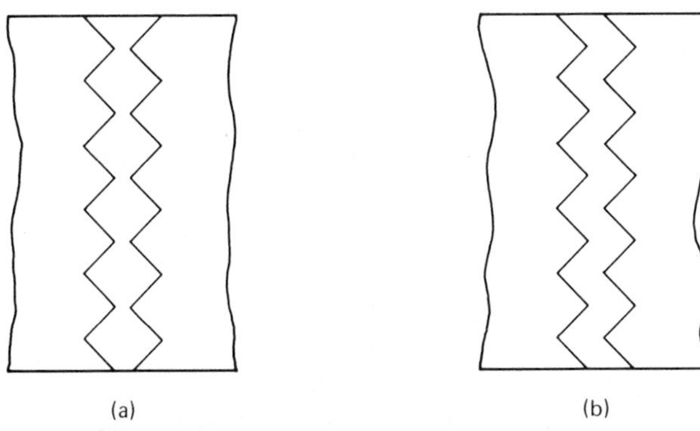

Fig. 13-20 Alternative shapes for shear keys.

mendation for the choice of β. However, for cracking shear, (Q_c), $\beta = 0.5$, and for ultimate shear, (Q_u) $\beta = 1.0$, should give conservative estimates.

Formerly the shear strength of keys was assessed on the basis of

$$Q_{allowable} = DTf_s \tag{13-11}$$

where f_s is the allowable shear stress in the concrete. This procedure did not recognize the basic behavior of the shear transmission but was conservative. The use of eq. (13-10) with the above values of β will normally be equivalent to using a slightly higher allowable shear stress.

The castellated shape may not be the most efficient. Studies are at present being undertaken to assess the value of using keys of the type shown in Fig. 13-20.

13.8.2 Reinforcement of Vertical Connections

-1. Loop bars (Fig. 13-18b)—These will contribute to the tie resistance and help to limit long term cracking as discussed below. Some designers allow a contribution to the shear strength of the joint from loop bars acting as dowels. Tests[12-27] indicate that the presence of the loops may not significantly affect the cracking strength but may increase the ultimate strength and make the behavior more ductile.

-2. Vertical tie bars (Fig. 13-18c)—This bar in combination with the loops will form a mechanical connection between adjacent units which will help to prevent progressive collapse. The vertical tie may also provide tensile reinforcement in the walls. It does not appear to affect the cracking or ultimate strength of the connection.

-3. Long term movements—Trouble has been experienced in some large panel buildings with long term cracking particularly at the vertical joints. The mortar infill in a vertical joint is normally placed with a wet consistency so as to ensure that the joint is properly filled. This will tend to cause shrinkage. Other sources may be creep, differential settlement, wind stresses, or temperature movements.

The provision of loop reinforcement within the vertical joint will help to reduce such effects. Practical experience tends to support this conclusion.[13-23] Also there is some evidence to suggest that cracking is more prominent in vertical joints which do not have castellations than in those which do (they certainly have a much lower cracking strength).[13-27]

13.9 DESIGN OF TRANSFER BEAMS

A common feature of all types of tall buildings is to have a system of support at the first story level which is different from that of the upper floors. This is done for architectural considerations, often to facilitate parking. In the case of large panel structures, the shear walls are often discontinued at second floor level and a transfer beam distributes the loading to a column structure. Figure 13-21 shows a wall support of this type.

13.9.1 Structural Action of Transfer Beams

It is not realistic to consider the transfer beams simply as bending members carrying the weight of the wall. For the system of Fig. 13-21 there must be a flow of stress from the uniformly distributed value to more concentrated values above the supports. This involves an 'arching' action, and hence, tensile stresses in the transfer beam. The tensile action tends to predominate over any bending action which may develop.

Therefore, as far as design is concerned, it may be necessary to reinforce the wall above the columns to resist the concentrated loads there. It will also be necessary to reinforce the transfer beam to take the tension.

As a very rough guide to this tie force the formula[13-36]

$$T = \frac{wL}{4} \tag{13-12}$$

will normally give a conservative estimate. T is the tie force, L is the distance between the column centres, and w is the uniformly distributed vertical load. The reinforcement to take this load should extend right across the beam and should be suitably anchored at the ends. Reference 13-36 gives more general formulae for this purpose together with formulae for estimating beam bending moment and wall stress concentrations.

Where the transfer beam is continuous over several supports, differential settlement may have an important effect and a finite element analysis may be necessary.[13-37] Such an analysis may also be desirable to assess the effect of lateral loading in the region of the transfer beam.

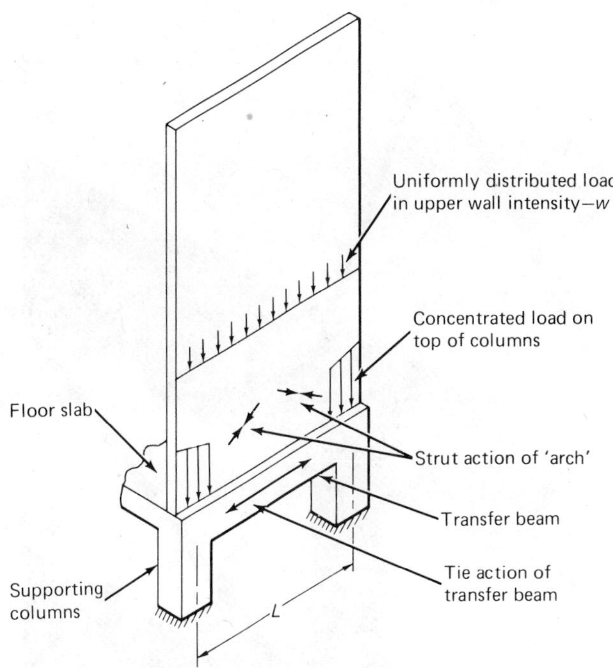

Fig. 13-21 Transfer beam at base of wall.

13.10 CONSTRUCTION PROBLEMS

As with all tall buildings, factors affecting the organization of the construction of large panel buildings are important for the economic success of the structure. In this section we briefly discuss some of these factors.

13.10.1 Location of Casting Plant

Some contractors have movable casting plants which can be set up on the job site. This minimizes transport costs but requires a fairly large work area at the job site. Transport costs do not appear to be a critical factor and the use of a fixed casting plant is the most common solution. Haulage distances of 100 miles are not uncommon.

13.10.2 Size of Panels

The amount of site connection should be minimized in order to keep site work down and to maintain even wall and floor surfaces. Hence, the larger the precast unit the better. The capacity of the erection crane used on the site together with the physical size of the units for transportation are factors which limit the size of the units. Also, if a large unit is damaged when being moved it will be more expensive to replace than a smaller one. Fifteen-foot long wall panels are not uncommon.

13.10.3 Erection Details

The stability of the structure under erection is most important. The wall panels are normally temporarily propped by stays which are fixed to the slab lifting-hooks and to special brackets on the walls (Fig. 13-22). These stays should be carefully designed to withstand wind loading since the consequences of a heavy panel coming adrift can be serious.

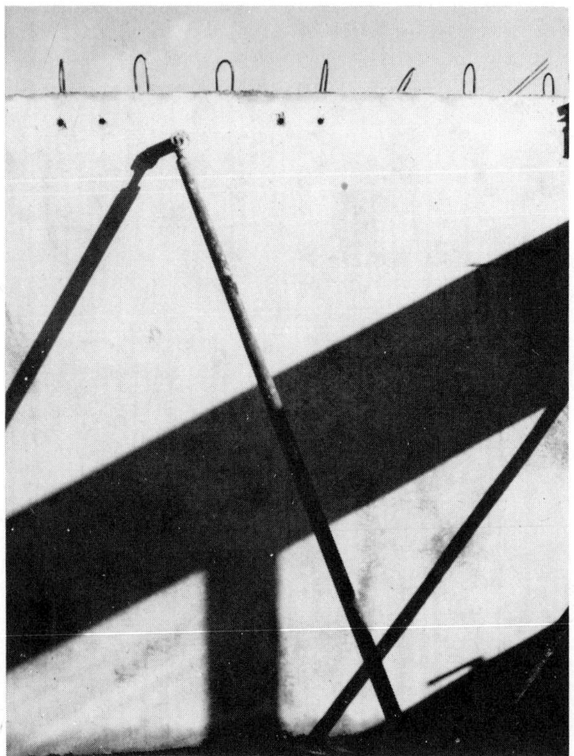

Fig. 13-22 Photograph showing wall stay.

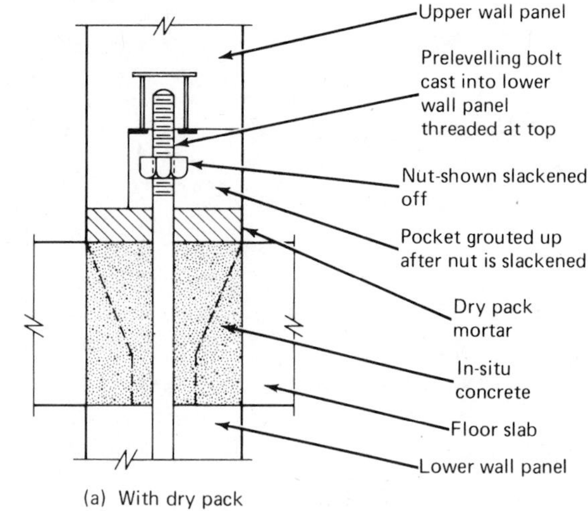

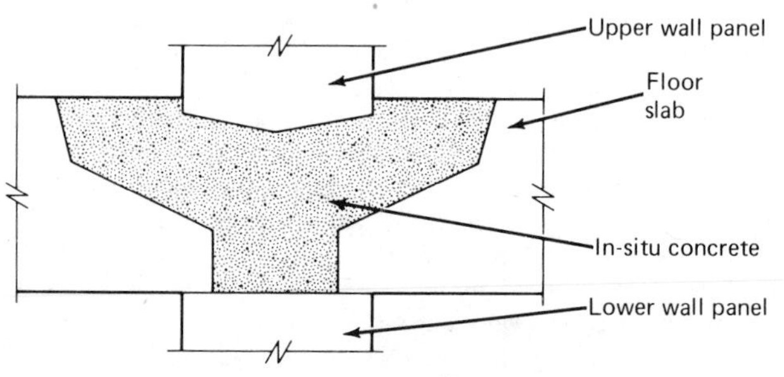

Fig. 13-23 Methods of supporting wall panels. (a) With dry pack; and (b) without dry pack, leveling equipment not shown.

13.10.4 Leveling of the Units

The most common approach is to level the walls carefully and then sit the floors on them directly. With the type of floor support detail shown in Fig. 13-13c, it becomes possible to level the floors using special temporary support brackets (Fig. 13-14), and then to sit the walls on the floors without the necessity for careful leveling. However, the leveling and alignment of the external wall units is most important since the joints between the exposed faces are normally featured and must line up. This is not only governed by aesthetic considerations. The joints usually form a drainage path which must not be blocked up.

The most popular method of leveling walls is to sit them on a preleveling bolt. Figure 13-23a shows a typical detail.

The construction sequence here would be:

(1) The lower panel is in position.
(2) The floor slabs are placed.
(3) The in-situ concrete is placed.
(4) The nuts on the preleveling bolts (two per panel) are leveled.
(5) The upper panel is placed to sit on the nuts (Fig. 13-2).
(6) The dry-pack mortar is placed.
(7) When the dry pack has set the nuts are slackened off.

Slackening off the nuts ensures even bearing pressure at the base of the panel. Wedges are often used instead of preleveling bolts especially for internal walls. The wedges must also be removed when the dry pack has set. In some systems the wall support is as shown in Fig. 13-23b. The bottom of the wall panel is shaped for better compaction of the in-situ concrete and dry-pack mortar* is not used.

13.10.5 Method of Casting

-1. Vertical—A common method of casting wall panels is in vertical battery molds. A series of vertical, double-faced molds are laid side by side so that the back of one mold forms the casting face for the next unit. This is especially useful when the casting is done on-site where space is normally restricted. Disadvantages are that it is difficult to seal the molds from leakage of wet concrete at projecting bars and the placing of the concrete is more difficult to control. Also, there may be a marked variation in the concrete strength within the height of the panel if cast vertically.

-2. Horizontal—This is the best casting method in principle, but requires a larger working area.

Special devices for sealing the molds at projecting bars may be desirable.[12-23] Consideration of other factors associated with casting procedures is beyond the scope of this chapter.

13.10.6 Handling of Units

The handling stresses in large panels is an important factor in the design of the reinforcement and there is scope for ingenuity in devising methods which keep such stresses to a minimum. For wall panels it is obviously advantageous to minimize out-of-plane bending by keeping the units vertical at all times.

Floor panels must be suspended horizontally for erection and the position of the lifting hooks is a most important factor. Ideally extra reinforcement to cover handling stresses should not be required.

13.10.7 Crainage for Erection

Optimization of the erection procedure is dependent on the choice of crane system used. Factors to be considered are:

(1) Number of cranes: normally only one is used but for large buildings two cranes might prove advantageous.
(2) The weight of the precast units in combination with the required reach.
(3) The crane base can be fixed or it can move on rails.
(4) It may be desirable to provide lateral support for the crane against the building.

13.11 ACKNOWLEDGMENT

I am most grateful to the following individuals for advice on the material contained in this chapter: Dr. P. Bhatt of Glasgow University who read the manuscript, Mr. R. Girardau formerly of Reema (Scotland) Ltd., Dr. D. Green, Mr. H. M. Nelson of Glasgow University, Mr. I. Munro of Reema (Scotland) Ltd. and Mr. D. Orme, formerly of Concrete (Scotland) Ltd.

REFERENCES

13-1 Griffiths, H.; Pugsley, A.; and Saunders, O., "Report of the Inquiry into the Collapse of Flats at Canning Town," H.M.S.O., London, U.K., 1968.

*Dry-pack mortar has a high cement content and a dry consistency (water/cement ratio of the order of 0.35).

13-2 Lewicki, B., *Building with Large Prefabricates*, Elsevier Publishing Co., 1966.

13-3 Sebestyen, G., *Large Panel Buildings*, Publishing House of the Hungarian Academy of Sciences, Budapest, 1965.

13-4 "International Recommendations for the Design and Construction of Large-Panel Structures," Comité Europeen du Beton (CEB), 1967. English Translation No. 137, Cement and Concrete Association, 52 Grosvener Gardens, Longon, S.W.1, U.K.

13-5 "Large Panel Structures and Structural Connections in Precast Concrete," Addendum No. 1 (1970) to CP116, The Council for Codes of Practice, British Standards Institution, 2 Park St., London, W1A 2BS, U.K.

13-6 Selvaggio, S. L., and Carlson, C. C., "Restraint in Fire Tests of Concrete Floors and Roofs," *Research Department Bulletin No. 220*, Portland Cement Association, Skokie, Illinois.

13-7 MacLeod, I. A., "Lateral Stiffness of Shear Walls with Openings," *Tall Buildings*, (Coull and Stafford Smith, Eds.), 223–244, Pergamon Press, Elmsford, N.Y., 1967.

13-8 (a) Coull, A., and Choudhury, J. R., "Analysis of Coupled Shear Walls," *Journal of the American Concrete Institute Proceedings*, **64**, 587–593, Sept. 1967.
(b) Coull, A., and Chouldhury, J. R., "Stresses and Deflections in Coupled Shear Walls," *ACI Journal, Proceedings*, **64**, 65–72, Feb. 1967.

13-9 Magnus, D., "Pierced Shear Walls," *Concrete and Construction Engineering*, **60**, 89–98, 127–136, 177–185, 1965.

13-10 Rosman, Riko, "Approximate Analysis of Shear Walls Subject to Lateral Loads," *ACI Journal, Proceedings*, **61**, 717–732, June, 1964.

13-11 Rosman, R., "Tables for the Internal Forces of Pierced Shear Walls Subject to Lateral Load," *Bauingenieur-Praxis*, Heft 66. (In German and English) W. Ernst and Sohn, Berlin, 1966.

13-12 Rosman, R., *Statik und Dynamik der Scheibensysteme des Hochbaues*, Springer Verlag, 1968 (in German).

13-13 Michael, D., "The Effect of Local Wall Deformations on the Elastic Interaction of Cross Walls Coupled by Beams," *Tall Buildings*, (Coull and Stafford Smith Eds.), 253–271, Elmsford, New York, Pergamon Press, 1967.

13-14 MacLeod, I. A., "Connected Shear Walls of Unequal Width," *Jour. ACI*, **67** (5), 408–412, May 1970.

13-15 MacLeod, I. A., "Analysis of Shear Wall Buildings by the Frame Method," *Proc. Instn. Civ. Engrs.* **55**, 593–603, September, 1973.

13-16 Winokour, A., and Gluck, J., "Lateral Loads in Asymmetric Multistory Structures, *Proc. ASCE*, **94** (ST3), 645–656, March, 1968.

13-17 Webster, J., "The Static and Dynamic Analysis of Orthogonal Structures Composed of Shear Walls and Frames," *Tall Buildings*, (Coull and Stafford Smith, Eds.), 377–395, Pergamon Press, Elmsford, New York, 1967.

13-18 Coull, A., and Irwin, A. W., "Analysis of Load Distribution in Multistory Shear Wall Structures," *The Structural Engineer*, **8** (8), 301–306, August, 1970.

13-19 Clough, R. W.; King, I. P.; and Wilson, E. L., "Structural Analysis of Multistory Buildings," *Jour. Struct. Div. ASCE*, **90** (ST3), 19-34, 1964.

13-20 "Report on the Behaviour of Reinforced Concrete Structures in the Caracas, Venezuala Earthquake of July 29, 1967," Portland Cement Association, Skokie, Illinois.

13-21 Barnard, P., and Schwaighofer, J., "Interaction of Shear Walls Connected Solely Through Slabs," *Tall Buildings*, (Coull and Stafford Smith, Eds.), 157-173, Pergamon Press, Elmsford, New York, 1967.

13-22 Quadeer, A., and Stafford Smith, B., "The Bending Stiffness of Slabs Connecting Shear Walls," *Jour. ACI*, 464-473, June, 1967.

13-23 Peacock, J. D., "Large Precast Wall Panels. A Manufacturers Review of Jointing Problems–with some Solutions," Proceedings; International Symposium on Load Bearing Walls held in Warsaw, June, 1969.

13-24 "Design Philosphy and its Application to Precast Concrete Structures," Proceedings of a Symposium held at Church House, London, SW1, U.K., May, 1967, International Association for Bridge and Structural Engineering. Published by the Cement and Concrete Association, 52 Grosvener Gardens, London, SW1.

13-25 Despeyroux, J., Comments on paper "Shear Wall Construction in System Building," *Tall Buildings*, 129 and 130.

13-26 "Industrialized Building and the Structural Engineer," Proceedings of a Symposium organized by the Institution of Structural Engineers, 11 Upper Belgrave St., London, SW1, U.K.

13-27 Bhatt, P., and Nelson, H. M., "Strength and Deformation of Castellated Vertical Joints in Shear Walls," Proceedings, Conference on Joints in Structures, University of Sheffield, U.K., July, 1970.

13-28 Hansen, K., and Olesen S. Ø., "Failure Load and Failure Mechanism of Keyed Shear Joints," *Report No. 69/22*, Damarks Ingeniørakademi, Civil Engineering Department, Structural Laboratory, 10 Øster Volgade, Copenhagen, Denmark.

13-29 von Haslasz, R., and Tantwo, G., "Grosstafelbauten–Konstruction und Berechnung," *Bauingenieur Praxis*, Heft 55, W. Ernst, Berlin, 1966. (General treatment on design of Large Panel Buildings. In German).

13-30 Berndt, K., "Die Montagebauarten des Wohnungsbaues in Beton," Bauverlag GMBH, Weisbaden, 1969. (Details of German Large Panel Systems. In German).

13-31 Diamant, R. M. E., *Industrialized Buildings*, 3 Vols., Iliffe Books, London, 1964.

13-32 Proceedings, International Conference on Load Bearing Walls, Warsaw, 1969.

13-33 Shapiro, G. A., and Sokolov, M. E., "The Strength and Deformation of Horizontal Joints in Large Panel Buildings," *Library Communication No. 1217*, August, 1964, Building Research Station, Watford, U.K., 1964. (Translated from Russian).

13-34 Pommeret, M., "Contreventement des Batiments par Grands Panneaux," (Wind Bracing of Large Panel Buildings), *Annales de L'Institut Technique du Batiments et des Travaux Publics*, **20** (234), 795-796 June 1967, and **21** (246), 940-941, June 1968.

13-35 Pommeret, M., "Les Joints Verticaux Organises Entre Grands Panneaux Coplanaires," (Vertical Joints Between Coplanar Large Panels), *Annales de L'Institute Technique de Batiments et des Travaux Publics*, **22** (258) 997-999, June 1969.

13-36 Green, D. R., MacLeod, I. A., and Girardan, R. S., "Force Actions in Shearwall Supports Systems." Response of Multistory Concrete Structures to Lateral Load. American Concrete Institute Publication SP-36, 1971.

13-37 Green, D., "The Interaction of Solid Shear Walls and their Supporting Structure," *Building Science* **7**, 239-248, 1972.

14

Thin Shell Structures

W. C. SCHNOBRICH, Ph.D.*

14.1 INTRODUCTION

A precise classification of a structure as being a thin shell or a nonshell is not possible as many structures could be classified either way. Within this chapter a thin shell is considered to be a surface structure, one of whose dimensions, its thickness, is much smaller than the other dimensions. This structure is constructed as a curved, faceted, or folded surface so that geometry activates axial forces to become a significant load-carrying stress-system. Thin shells can be classified according to several different systems, two of which are their shape and their method of formation.

14.1.1 Shell Geometry

Shells may have a curvature of the surface in one or two directions. The singly-curved surfaces are developable. Cylinders and cones are examples. The doubly-curved surfaces are nondevelopable. They are of positive curvature if for any point on the surface the origin of both principal radii of curvature of that point are on the same side of surface, that is curvatures are in the same direction. If the radii are on opposite sides of the surface the shell has a negative curvature. Hyperbolic paraboloids and conoids are examples of this latter type of shell. Positive curvature shells react to pressures normal to the surface with direct forces of the same sign in any two orthogonal directions, while negative curvature shells have stresses of opposite sign. The sign of the curvature is therefore indicative of the structural behavior.

Negative curvature shells have two sets of asymptotic lines; singly-curved and zero curvature shells have one such set of lines; while the positive curvature shells have no real asymptotic lines. If an edge of the shell is one of the asymptotic lines of the surface there is a tendency for any disturbance or force applied to the edge to propagate deep into the shell.

Shells may also be classified according to the manner of generating them. Rotational shells are formed by revolving a plane curve about an axis in its plane. Translational shells are generated by moving one curve along another. Classification on the basis of the method of generation however tells little about the structural behavior and is therefore not too meaningful. All translational shells do not behave in a similar fashion; elliptical and hyperbolic paraboloids, although formed in the same manner, have markedly different load-carrying characteristics.

14.1.2 Shell Usage

Concrete thin shells are used for many structures. Roof, foundation, and containment structures have all been built

*Professor of Civil Engineering, University of Illinois at Urbana.

using thin shell systems. Their use as roof structures is motivated primarily by span and esthetic reasons. Employment as foundation and containment structures is done primarily for economic reasons.

The cost factors for roof shells vary considerably with shape, support conditions, etc. An extensive discussion of cost factors for various shells is presented in paper by Gensert, Kirsis, and Peller.[14-15] The high cost of form work or scaffolding can severely restrict the size of spans considered economic. Form reuse can considerably lower the cost for the shell.

As the span length increases buckling may become the controlling factor. Use of boxlike or hollow cross sections then becomes necessary to further extend the possible span sizes.[14-12]

14.1.3 Shell Behavior

The ideal behavior of the shell is to carry its load by only inplane or membrane forces and have these forces nearly constant over the shell. For arches, this requires the structure be funicular with the load. For shells, this is not completely necessary. They will carry load predominately by membrane forces if the support conditions are appropriate. Sharp variations in either loading or shell stiffness will result in bending moments developing either to carry the load, or restore compatibility. How widespread this region of bending is depends upon the shell geometry.

Positive curvature, dome-like shells transmit loads to the supports primarily by compressive arching forces, provided some support exists along each edge. Disturbances applied to the edges tend to damp out quickly. Negative curvature shells utilize inplane shear as a prime mechanism; singly curved shells behave as curved beams when longitudinal edges are not supported, or as arches if those edges are supported. These latter shells tend to propagate edge disturbances in the form of moments much further into the shell than do the positive curvature shells.

14.2 SHELLS OF REVOLUTION

The most frequent use of the shell of revolution is for a containment or reservoir structure, but it has been em-

TABLE 14-1 Geometry of Shells of Revolution

Doubly Curved Shells of Revolution

Increment of Arc

$$ds^2 = a_1^2 d\phi^2 + a_2^2 d\theta^2$$

where

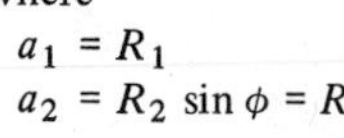

$a_1 = R_1$

$a_2 = R_2 \sin\phi = R$

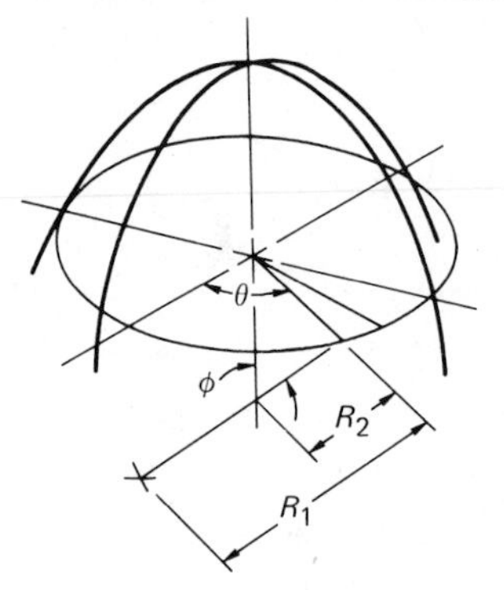

Shell Shape		R_1	R_2
sphere	$\gamma = 0$	R_o	R_o
paraboloid	$\gamma = 1$	$\dfrac{R_o}{(1 + \gamma \sin^2\phi)^{3/2}}$	$\dfrac{R_o}{(1 + \gamma \sin^2\phi)^{1/2}}$
ellipsoid	$\gamma > -1$	R_o = radius of curvature at $\phi = 0$	
hyperboloid	$\gamma < -1$	γ = shell shape parameter	

Developable Shells of Revolution

Increment of Arc

$$ds^2 = a_1^2 dx^2 + a_2^2 d\theta^2$$

where

$a_1 = 1$

$a_2 = R_2$

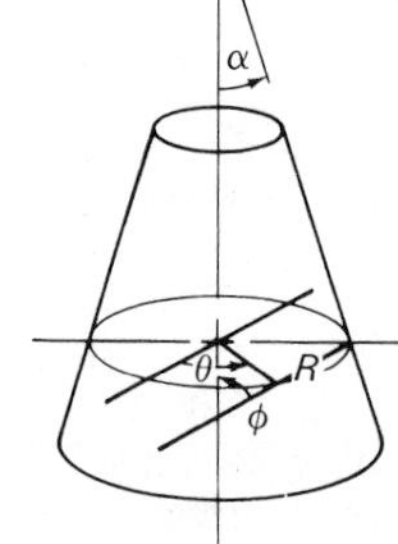

	R_1	R_2
cylinder	∞	R
cone	∞	$\dfrac{R}{\sin(90^\circ - \alpha)}$

ployed for roof structures when a circular plan is being used.

14.2.1 Surface Equation

A surface of revolution is obtained by rotating a plane curve about an axis lying in that plane. The curve is called the meridian; its plane the meridian plane. Planes normal to the axis of revolution intersect the surface on parallel circles. For these shells of revolution the lines of principal curvature are the meridians and the parallels; therefore, positions of points on these lines are convenient coordinates to use.

The radii of principal curvature are those of the meridian denoted by R_1 and the distance along the normal from the axis of revolution to the surface denoted by R_2, Fig. 14-1.

Table 14-1 presents expressions for the radii of curvature for several common shell types.

14.2.2 Method of Analysis

The method of analysis that can be used depends upon the geometry of the shell. If the shell is shallow, opening angle $\phi < 30^\circ$, or if the rise, H, is less than one-eighth the base diameter, then special shell equations must be used.[14-4,14-20] If the shell is steep it can be analyzed in the following manner:

1. Compute the membrane forces and displacements for the shell.
2. The values of these quantities will in general violate the junction or support conditions.
3. Compute the influences of applying unit values of edge reaction and of edge moment.
4. Write the compatibility equations for the junction or support and solve for the necessary corrective forces.
5. Superimpose these solutions upon the original membrane solution.

In view of the complexity of the computations involved even in the simplified approach described above, a number of computer programs have been written to solve this problem.[14-5,14-22]

14.2.3 Membrane Theory

The stress resultants or internal forces acting in the shell are shown in Fig. 14-2. Since there are only three resultants, the membrane equilibrium equations are determinate. For the axisymmetric case, axial equilibrium gives

$$N_\phi = -\frac{P}{2\pi r_o \sin \phi_o} \tag{14-1}$$

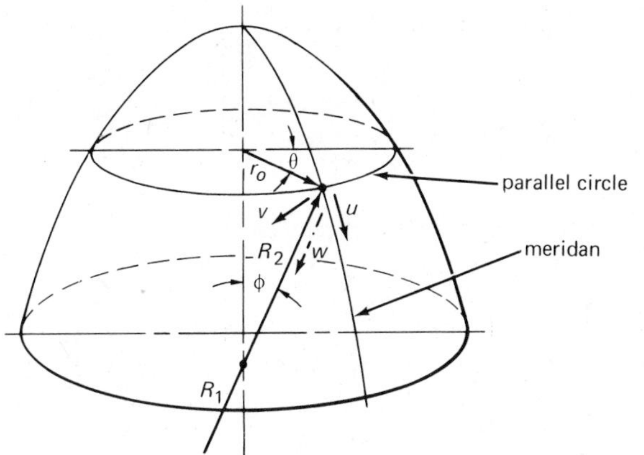

Fig. 14-1 Shell of revolution.

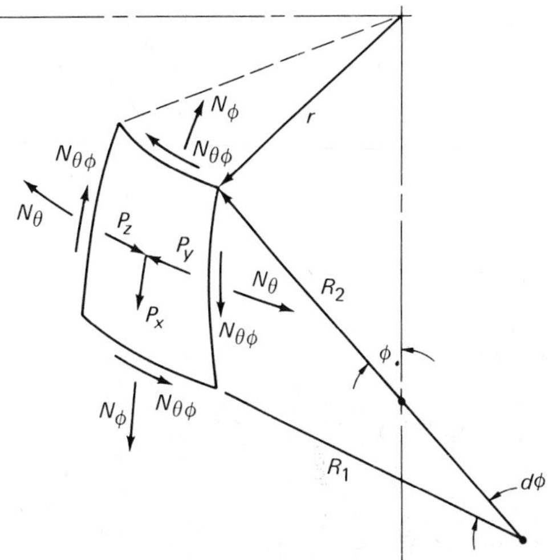

Fig. 14-2 Membrane forces.

where P is the integrated axial force resultant of all loads applied to the shell above the parallel section ϕ_o. Radial equilibrium of the element shown in Fig. 14-2 gives

$$\frac{N_\phi}{R_1} + \frac{N_\theta}{R_2} = -p_z \tag{14-2}$$

From these equations the membrane solutions for a variety of shell types and loadings can be found, Table 14-2. Further tables can be found in Ref. 14-1.

With the membrane forces established, the displacements can be calculated as shown in Ref. 14-14 from

$$u = \left[\int \frac{f(N)}{\sin \phi} d\phi + C\right] \sin \phi \tag{14-3a}$$

$$w = u \cot \phi - \frac{R_2}{Et}(N_\theta - \nu N_\phi) \tag{14-3b}$$

$$\Delta_\phi = \frac{u}{R_1} + \frac{dw}{R_1 d\phi} \tag{14-3c}$$

where $f(N) = (1/Et)\,[N_\phi(R_1 + \nu R_2) - N_\theta(R_2 + \nu R_1)]$, ν is Poisson's ratio, and C is a constant rigid body motion to adjust to support conditions.

14.2.4 Edge Corrections

If the forces and/or displacements, as computed from the above equations, do not agree with the support conditions of the structure, a correction must be applied. This correction involves solving the governing equations of bending theory. Several approximate solutions[14-3,14-19] have been reported in the literature. If the opening angle exceeds thirty degrees[14-20] the Geckeler or equivalent cylinder solution is adequate. Table 14-3 gives the important internal forces and displacements when unit edge forces are applied. Reference 14-1 contains a more extensive set of tables drawn from many sources. The functions F_1 to F_4 are given in Table 14-4.

14.2.5 Stability Considerations

Buckling has long been an area of active experimental and analytical research.[14-17,14-18] Since the buckling of many shells is very sensitive to initial imperfections and the type of support conditions, sizable differences can exist between computed and actual buckling values. For reinforced con-

TABLE 14-2 Membrane Stress Resultants—Spherical shell

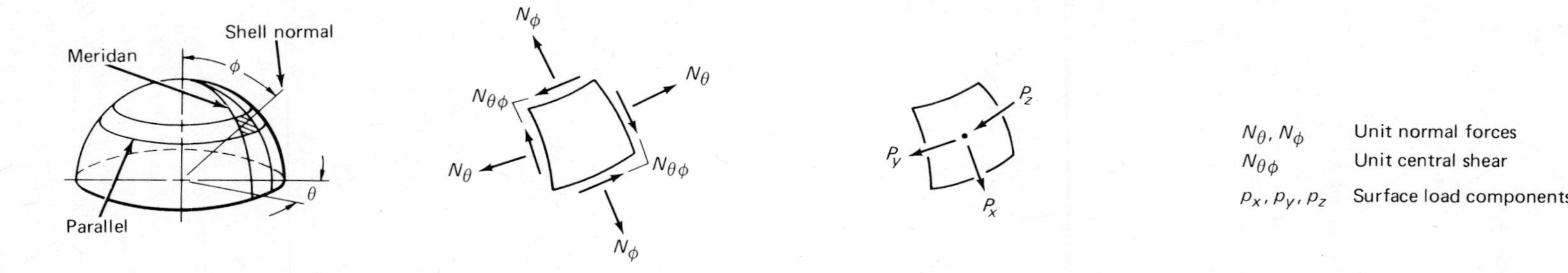

System	*Loading*	N_φ	N_θ	$N_{\theta\varphi}$
	$p_x = p_E \sin\varphi$ $p_z = p_E \cos\varphi$	$-p_E r \dfrac{\cos\varphi_o - \cos\varphi}{\sin^2\varphi}$	$p_E r \left(\dfrac{\cos\varphi_o - \cos\varphi}{\sin^2\varphi} - \cos\varphi\right)$	0
		For $\varphi_o = 0$ (no vertex opening)		
		$-p_E r \dfrac{1}{1 + \cos\varphi}$	$p_E r \left(\dfrac{1}{1 + \cos\varphi} - \cos\varphi\right)$	0
	$p_x = p_s \sin\varphi \cos\varphi$ $p_z = p_s \cos^2\varphi$	$-p_s \dfrac{r}{2}\left(1 - \dfrac{\sin^2\varphi_o}{\sin^2\varphi}\right)$	$p_s \dfrac{r}{2}\left(1 - \dfrac{\sin^2\varphi_o}{\sin^2\varphi} - 2\cos^2\varphi\right)$	0
		For $\varphi_o = 0$ (no vertex opening)		
		$-p_s \dfrac{r}{2}$	$-p_s \dfrac{r}{2} \cos 2\varphi$	0
	$p_z = p$	$-p \dfrac{r}{2}\left(1 - \dfrac{\sin^2\varphi_o}{\sin^2\varphi}\right)$	$-p \dfrac{r}{2}\left(1 + \dfrac{\sin^2\varphi_o}{\sin^2\varphi}\right)$	0
		For $\varphi_o = 0$ (no vertex opening)		
		$-p \dfrac{r}{2}$	$-p \dfrac{r}{2}$	0
	Edge Load p_L Vertex Load P_L	$-p_L \dfrac{\sin\varphi_o}{\sin^2\varphi}$	$p_L \dfrac{\sin\varphi_o}{\sin^2\varphi}$	0
		For $\varphi_o = 0$ (no vertex opening)		
		$-P_L \dfrac{1}{2\pi r \sin^2\varphi}$	$P_L \dfrac{1}{2\pi r \sin^2\varphi}$	0

System	Loading	N_φ	N_Θ	$N_{\Theta\varphi}$
	$p_z = p_w \sin\varphi \cos\theta$	$-p_w \frac{r}{3} \frac{\cos\varphi \cos\theta}{\sin^3\varphi} \times$ $\times [3(\cos\varphi_o - \cos\varphi) - (\cos^3\varphi_o - \cos^3\varphi)]$	$p_w \frac{r}{3} \frac{\cos\theta}{\sin^3\varphi} \times$ $\times [\cos\varphi(3\cos\varphi_o - \cos^3\varphi_o) - 3\sin^2\varphi$ $- 2\cos^4\varphi]$	$-p_w \frac{r}{3} \frac{\sin\theta}{\sin^3\varphi} [3(\cos\varphi_o - \cos\varphi) -$ $- \cos^3\varphi_o + \cos^3\varphi]$
		For $\varphi_o = 0$ (no vertex opening)		
		$-p_w \frac{r}{3} \frac{\cos\varphi \cos\theta}{\sin^3\varphi} \times$ $\times (2 - 3\cos\varphi + \cos^3\varphi)$	$p_w \frac{r}{3} \frac{\cos\theta}{\sin^3\varphi} (2\cos\varphi -$ $- 3\sin^2\varphi - 2\cos^4\varphi)$	$-p_w \frac{r}{3} \frac{\sin\theta}{\sin^3\varphi} (2 - 3\cos\varphi +$ $+ \cos^3\varphi)$

Membrane Stress Resultants—Other shells of revolution with curved meridian

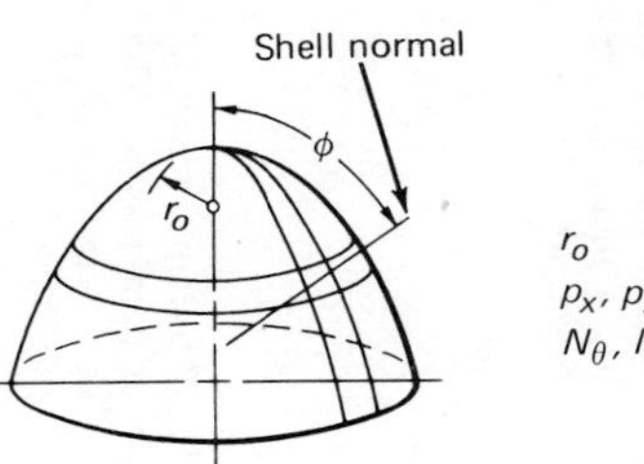

r_o Radius of curvature at vertex
p_x, p_z Surface load components
N_θ, N_ϕ Unit normal forces

	System	Loading	N_φ	N_Θ	$N_{\Theta\varphi}$
Parabola		$p_x = p_E \sin\varphi$ $p_z = p_E \cos\varphi$	$-p_E \frac{r_o}{3} \frac{1 - \cos^3\varphi}{\sin^2\varphi \cos^2\varphi}$	$-p_E \frac{r_o}{3} \frac{2 - 3\cos^2\varphi + \cos^3\varphi}{\sin^2\varphi}$	0
		$p_x = p_s \sin\varphi \cos\varphi$ $p_z = p_s \cos^2\varphi$	$-p_s \frac{r_o}{2} \frac{1}{\cos\varphi}$	$-p_s \frac{r_o}{2} \cos\varphi$	0
	h	$p_s = \gamma\left(h + \frac{r_o}{2}\tan^2\varphi\right)$	$-\gamma \frac{r_o}{2}\left(h + \frac{r_o}{4}\tan^2\varphi\right)\frac{1}{\cos\varphi}$	$-\gamma \frac{r_o}{2} [h(2\tan^2\varphi + 1) +$ $+ r_o \tan^2\varphi\left(\tan^2\varphi + \frac{3}{4}\right)]\cos\varphi$	0
		$p_z = p$	$-p \frac{r_o}{2} \frac{1}{\cos\varphi}$	$-p \frac{r_o}{2} \frac{1 + \sin^2\varphi}{\cos\varphi}$	0

TABLE 14-2 (cont.)

System	Loading	N_φ	N_θ	$N_{\theta\varphi}$
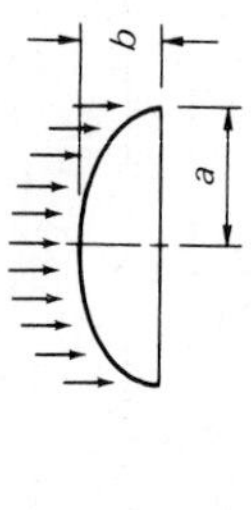 Ellipse: a, b; $\frac{\sqrt{a^2-b^2}}{a}=\epsilon$	$p_x = p_E \sin\varphi$; $p_z = p_E\cos\varphi$	$-\frac{p_E}{2}\frac{\sqrt{a^2\tan^2\varphi+b^2}}{a^2\sin\varphi\tan\varphi}\left[a^2 - \frac{a^2b^2\sqrt{1+\tan^2\varphi}}{b^2+a^2\tan^2\varphi} + \frac{b^2}{\epsilon}\ln\frac{(1+\epsilon)\sqrt{b^2+a^2\tan^2\varphi}}{b(\epsilon+\sqrt{1+\tan^2\varphi})}\right]$	$p_E\left[\frac{(b^2+a^2\tan^2\varphi)^{3/2}}{2\tan^2\varphi\sqrt{1+\tan^2\varphi}}\times\left(\frac{1}{\epsilon a^2}\ln\frac{(1+\epsilon)\sqrt{b^2+a^2\tan^2\varphi}}{b(\epsilon+\sqrt{1+\tan^2\varphi})}+\frac{1}{b^2}-\frac{\sqrt{1+\tan^2\varphi}}{b^2+a^2\tan^2\varphi}\right)-\frac{a^2}{\sqrt{b^2+a^2\tan^2\varphi}}\right]$	0
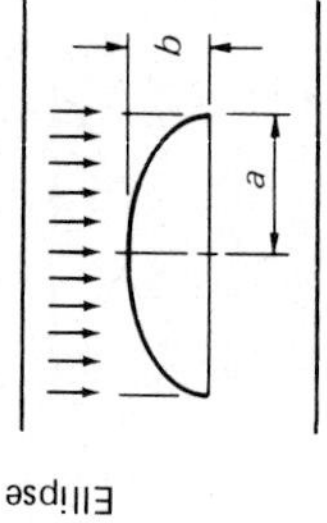 Ellipse: a, b	$p_x = p_S\sin\varphi\cos\varphi$; $p_z = p_S\cos^2\varphi$	$-\frac{p_S}{2}\frac{a^2\sqrt{1+\tan^2\varphi}}{\sqrt{b^2+a^2\tan^2\varphi}}$	$-\frac{p_S a^2}{2\,b^2}\frac{b^2-a^2\tan^2\varphi}{\sqrt{b^2+a^2\tan^2\varphi}\sqrt{1+\tan^2\varphi}}$	0

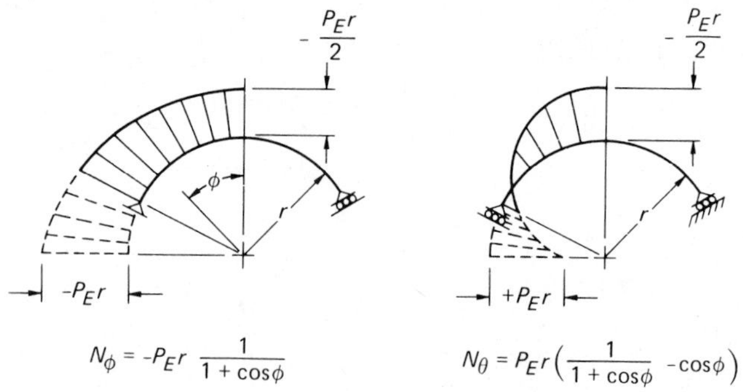

$-\frac{P_E r}{2}$ $-P_E r$ $N_\phi = -P_E r\,\frac{1}{1+\cos\phi}$ (a)

$-\frac{P_E r}{2}$ $+P_E r$ $N_\theta = P_E r\left(\frac{1}{1+\cos\phi}-\cos\phi\right)$ (b)

Fig. 14-3 Membrane stresses due to dead weight. (a) Meridianal stress; and (b) hoop stress.

crete, shrinkage and tensile cracking, as well as creep, can further reduce the buckling load. Therefore generous factors of safety are recommended.

The buckling pressure for a spherical shell can be expressed as

$$P_{cr} = CE\left(\frac{t}{R}\right)^2 \tag{14-4}$$

where E is the tangent modulus for the material and C is a coefficient to account for boundary conditions, imperfections, etc. The classical value of C for an ideal clamped shell is $2/\sqrt{3(1-\nu^2)}$. This value is reduced to a recommended value of 0.35 for metal shells. For concrete shells a further reduction of no less than one-half is recommended. The tangent modulus value can be that suggested by Griggs.[14-17]

$$E = \frac{2\sigma_o}{\epsilon_o}\sqrt{\left(1-\frac{\sigma}{\sigma_o}\right)} \tag{14-5}$$

where the subscript o indicates the maximum quantity.

EXAMPLE 14-1: Analyze a spherical shell of constant thickness.

SOLUTION: For a spherical shell loaded by its own weight membrane theory (Table 14-2) predicts stresses of

$$N_\phi = -\frac{p_E r}{1+\cos\phi}$$

$$N_\theta = p_E r\left(\frac{1}{1+\cos\phi} - \cos\phi\right) \tag{14-6}$$

where p_E is the weight of the shell in force per unit area of shell surface. If the opening angle of the shell is less than 52° then the entire shell would be in compression provided the support conditions are as required by membrane theory. Figure 14-3 shows the variation of these membrane forces along the meridian of the shell.

If the spherical shell is loaded by a load uniformly distributed over a plane perpendicular to the axis of the shell the membrane stresses are

$$N_\phi = -\frac{p_s r}{2}$$

$$N_\theta = -\frac{p_s r}{2}\cos 2\phi \tag{14-7}$$

If the opening angle is less than 45°, then, again, the entire shell is in compression. The variation of the stresses along the meridian is shown in Fig. 14-4. The stress trajectories of such a system are as shown in Fig. 14-5.

The displacements of the spherical shell under dead weight, found

TABLE 14-3 Influence of Edge Forces on Forces and Displacements of Shell of Revolution

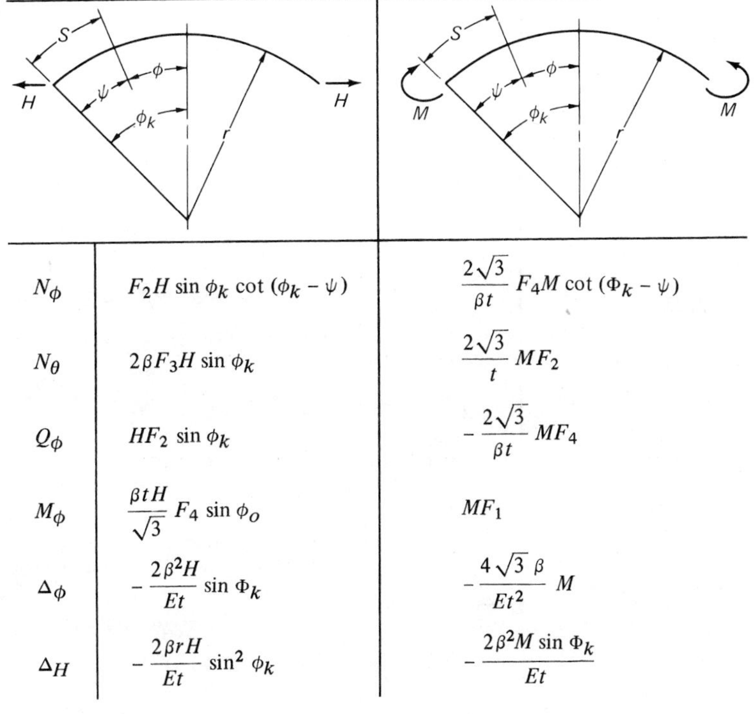

N_ϕ	$F_2 H \sin\phi_k \cot(\phi_k - \psi)$	$\dfrac{2\sqrt{3}}{\beta t} F_4 M \cot(\Phi_k - \psi)$
N_θ	$2\beta F_3 H \sin\phi_k$	$\dfrac{2\sqrt{3}}{t} M F_2$
Q_ϕ	$H F_2 \sin\phi_k$	$-\dfrac{2\sqrt{3}}{\beta t} M F_4$
M_ϕ	$\dfrac{\beta t H}{\sqrt{3}} F_4 \sin\phi_o$	$M F_1$
Δ_ϕ	$-\dfrac{2\beta^2 H}{Et}\sin\Phi_k$	$-\dfrac{4\sqrt{3}\,\beta}{Et^2} M$
Δ_H	$-\dfrac{2\beta r H}{Et}\sin^2\phi_k$	$-\dfrac{2\beta^2 M \sin\Phi_k}{Et}$

where:

$$\psi = \phi_k - \phi \qquad \beta^4 = 3(1-\nu^2)\left(\frac{r}{t}\right)^2$$

and the functions F_1 to F_4 are given in Table 14-4.

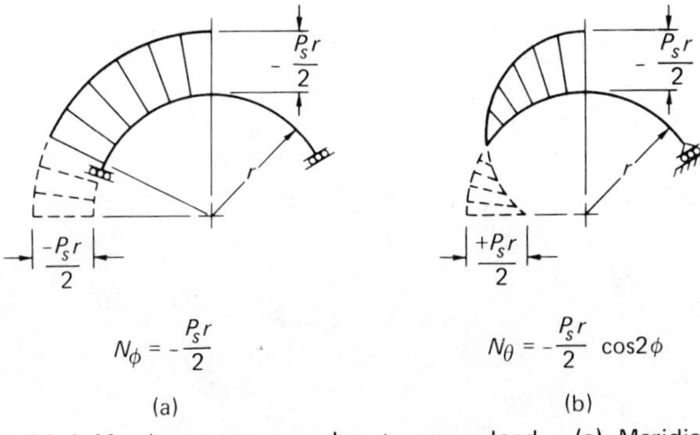

Fig. 14-4 Membrane stresses due to snow load. (a) Meridianal stress; and (b) hoop stress.

from eqs. (14-3a b c) are

$$u = \frac{p_E r^2(1+\nu)}{Et}\left[\ln\frac{1+\cos\phi}{1+\cos\phi_c} + \frac{1}{1+\cos\phi_c} - \frac{1}{1+\cos\phi}\right]\sin\phi$$

$$w = \frac{p_E r^2(1+\nu)}{Et}\left[\ln\frac{1+\cos\phi}{1+\cos\phi_c} - \frac{\cos\phi_c}{1+\cos\phi_c}\right]\cos\phi + \frac{p_E r^2}{Et}\cos\phi \tag{14-8}$$

so that the displacements of the edge become

$$\Delta_H = -\frac{p_E r^2}{Et}\left(\frac{1+\nu}{1+\cos\phi_c} - \cos\phi_c\right)\sin\phi_c$$

$$\Delta_\phi = -\frac{p_E r}{Et}(2+\nu)\sin\phi_c \tag{14-9}$$

A similar set of displacements can be found for the projected load.

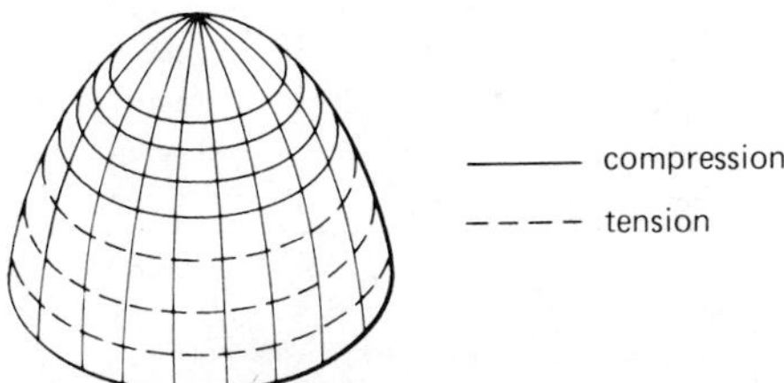

Fig. 14-5 Stress trajectories.

EXAMPLE 14-2: Ring beam-supported shell—The shell shown in Fig. 14-6 is to be analyzed for the dead load stresses. The dimensions of the shell are as shown in the figure. The thickness of 3 in. is used as a reasonable minimum. The dome is constrained by a ring beam which is in turn supported on closely spaced slender columns. The solution is achieved first from the membrane state then corrected for the boundary conditions. The load is assumed to be 50 psf to include roofing, etc. in the dead weight.

SOLUTION:
Membrane forces and displacements:

$$N_\phi = -\frac{p_E r}{1+\cos\phi} = -2168 \text{ lbs/ft (edge)}$$
$$= -1915 \text{ lbs/ft (crown)}$$

$$N_\theta = p_E r\left(\frac{1}{1+\cos\phi} - \cos\phi\right) = -766 \text{ lbs/ft (edge)}$$
$$= -1915 \text{ lbs/ft (crown)}$$

$$\Delta_H = (r\sin\phi_c)\,\epsilon_\theta = \frac{rN_\theta}{Et}\sin\phi_c$$

TABLE 14-4 Table of Functions F_1 to F_4

$\beta\phi$	F_1	F_2	F_3	F_4
0	1.0000	1.0000	1.0000	0
0.1	0.9907	0.8100	0.9003	0.0903
0.2	0.9651	0.6398	0.8024	0.1627
0.3	0.9267	0.4888	0.7077	0.2189
0.4	0.8784	0.3564	0.6174	0.2610
0.5	0.8231	0.2415	0.5323	0.2908
0.6	0.7628	0.1431	0.4530	0.3099
0.7	0.6997	0.0599	0.3798	0.3199
0.8	0.6354	-0.0093	0.3131	0.3223
0.9	0.5712	-0.0657	0.2527	0.3185
1.0	0.5083	-0.1108	0.1988	0.3096
1.1	0.4476	-0.1457	0.1510	0.2967
1.2	0.3899	-0.1716	0.1091	0.2807
1.3	0.3355	-0.1897	0.0729	0.2626
1.4	0.2849	-0.2011	0.0419	0.2430
1.5	0.2384	-0.2068	0.0158	0.2226
1.6	0.1959	-0.2077	-0.0059	0.2018
1.7	0.1576	-0.2047	-0.0235	0.1812
1.8	0.1234	-0.1985	-0.0376	0.1610
1.9	0.0932	-0.1899	-0.0484	0.1415
2.0	0.0667	-0.1794	-0.0563	0.1230
2.1	0.0439	-0.1675	-0.0618	0.1057
2.2	0.0244	-0.1548	-0.0652	0.0895
2.3	0.0080	-0.1416	-0.0668	0.0748
2.4	-0.0056	-0.1282	-0.0669	0.0613
2.5	-0.0166	-0.1149	-0.0658	0.0492
2.6	-0.0254	-0.1019	-0.0636	0.0383
2.7	-0.0320	-0.0895	-0.0608	0.0287
2.8	-0.0369	-0.0777	-0.0573	0.0204
2.9	-0.0403	-0.0666	-0.0534	0.0132
3.0	-0.0423	-0.0563	-0.0493	0.0071
3.1	-0.0431	-0.0469	-0.0450	0.0019
3.2	-0.0431	-0.0383	-0.0407	-0.0024
3.3	-0.0422	-0.0306	-0.0364	-0.0058
3.4	-0.0408	-0.0237	-0.0323	-0.0085
3.5	-0.0389	-0.0177	-0.0283	-0.0106
3.6	-0.0366	-0.0124	-0.0245	-0.0121
3.7	-0.0341	-0.0079	-0.0210	-0.0131
3.8	-0.0314	-0.0040	-0.0177	-0.0137
3.9	-0.0286	-0.0008	-0.0147	-0.0140

$F_1 = e^{-\beta\varphi}(\cos\beta\varphi + \sin\beta\varphi)$
$F_2 = e^{-\beta\varphi}(\cos\beta\varphi - \sin\beta\varphi)$
$F_3 = e^{-\beta\varphi}(\cos\beta\varphi)$
$F_4 = e^{-\beta\varphi}(\sin\beta\varphi)$

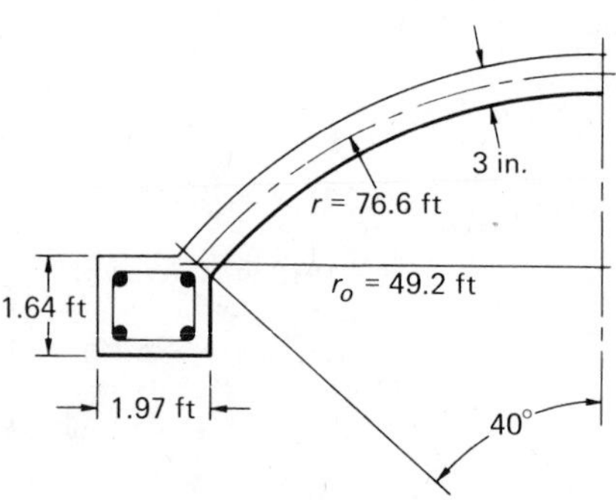

Fig. 14-6 Ring-beam supported dome.

$$E\Delta_H = \frac{(49.2)12}{3}(-766) = -150{,}749 \text{ lbs/ft (inward)}$$

$$\Delta_\phi = \frac{1}{R}\left(\frac{dw}{d\phi} + u\right) = -\frac{2p_E r}{Et}\sin\phi_c$$

$$E\Delta_\phi = -\frac{2(50)(76.6)(12)}{3}(0.643)$$

$$= -19{,}700 \text{ lbs/ft}^2 \text{ clockwise or flattening}$$

Horizontal thrust at edge:

The edge or ring beam is placed so that the thrust line for N_ϕ passes through the centroid of the beam.

$$\bar{H} = N_\phi \cos\phi_c = (2168)(\cos 40°) = 1661 \text{ lbs/ft}$$

Horizontal displacement of ring beam from $\bar{H}$ outward thrust:

$$E\Delta_H^B = \frac{\bar{H}r_c^2}{bd} = \frac{(1661)(49.2)^2}{(1.97)(1.64)} = 1{,}244{,}300 \text{ lbs/ft (outward)}$$

$$E\Delta_\phi^B = 0 \text{ Since load goes thru centroid of ring beam}$$

Compatibility:

The shell and the ring beam no longer have compatible displacements at junction. Additional edge moments and horizontal forces are applied to both the shell and the beam to enforce compatibility (Fig. 14-7)

$$E\Delta_H^S + E\Delta_H^{SH} + E\Delta_H^{SM} = E\Delta_H^B + E\Delta_H^{BH} + E\Delta_H^{BM}$$
$$E\Delta_\phi^S + E\Delta_\phi^{SH} + E\Delta_\phi^{SM} = E\Delta_\phi^B + E\Delta_\phi^{BH} + E\Delta_\phi^{BM}$$

The displacements and rotations of the ring beam as a result of applying horizontal force and moment along the top connection with the shell are

$$E\Delta_H^{BH} = \frac{4Hr^2}{bd} = \frac{4(49.2)^2}{(1.97)(1.64)}H = 2997H \text{ (inward)}$$

$$E\Delta_H^{BM} = -\frac{6Mr^2}{bd^2} = -\frac{6(49.2)^2}{(1.97)(1.64)^2}M = -2741M \text{ (outward)}$$

$$E\Delta_\phi^{BH} = -\frac{6Hr^2}{bd^2} = -2741H \text{ clockwise}$$

$$E\Delta_\phi^{BM} = \frac{12Mr^2}{bd^3} = 3342.8M \text{ counterclockwise or raises}$$

The displacements and rotations at the edge of the shell due to the horizontal force and moment are (from Table 14-3)

$$E\Delta_H^{SH} = -\frac{2\beta r}{t}H\sin^2\phi_k = -\frac{2(23.0)(76.6)}{0.25}(0.643)^2 H$$
$$= -5827H \text{ (outward)}$$

$$E\Delta_H^{SM} = -\frac{2\beta^2 M}{t}\sin\phi_k = -\frac{2(23)^2}{0.25}M(0.643)$$
$$= -2721.2M \text{ (outward)}$$

$$E\Delta^{SH} = -\frac{2\beta^2 H}{t}\sin\phi_k = -2721.2H \text{ clockwise or flattening}$$

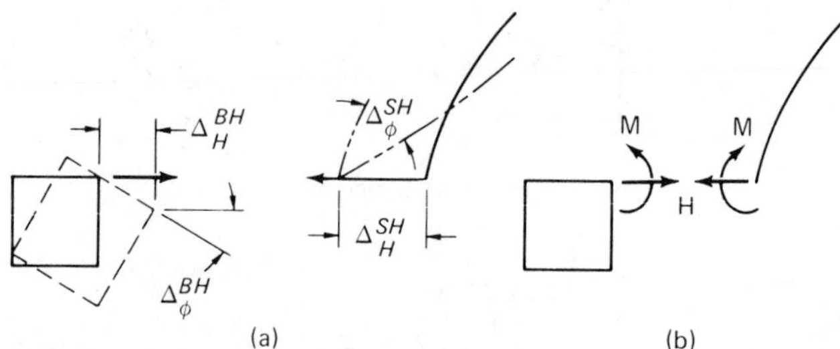

Fig. 14-7 Compatibility analysis at edge. (a) Displacements due to $H = 1$; and (b) forces to restore compatibility.

$$E\Delta_{\phi}^{SM} = -\frac{4\sqrt{3}\,\beta}{t^2} M = -\frac{4\sqrt{3}(23)}{(0.25)^2} M$$

$$= -2550M \text{ clockwise or flattening}$$

Writing the compatibility equations for the shell beam junction line

$$-150{,}749 + 5827H + 2721M = 1{,}244{,}300 - 2997H + 2741M$$

$$-19{,}700 - 2721H - 2550M = 0 - 2741H + 3343M$$

Thus

$$H = 158.5 \text{ lbs/ft}$$

$$M = -2.8 \text{ lbs-ft/ft}$$

The stress resultants anywhere in the shell can now be determined using Tables 14-3 and 14-4 plus the membrane solution.

Hoop force (N_θ) at the edge is

$$N_\theta = 2\beta H \sin \phi_k + \frac{2\sqrt{3}}{t} M - 766 \text{ lbs/ft}$$

$$= 4688 - 39 - 766 = +3883 \text{ lbs/ft}$$

This hoop tension decays to approximately zero when

$$2\beta F_3 H \sin \phi_k = 766 \quad \therefore \ F_3 = 766/4688 = 0.1633$$

$$\text{or } \beta\phi \cong 1.0 \quad \therefore \ \phi = 0.0435 \text{ or } 2^\circ 30'$$

This means that the tension stresses do not get beyond $3^1/_3$ ft from the edge.

This large tension force can be suppressed by prestressing the ring beam. The inclusion of a prestressed ring beam is readily accomplished by considering the prestressing force of F as a ring pressure of F/r.

14.3 CYLINDRICAL SHELLS

A special form of the shell of revolution is that of the cylindrical shell. It is used for a variety of structures ranging from pressure vessels and containment, or storage, units to roof structures. There are several reasons for singling out the cylindrical shell for special discussion. The prime reason is the extensive use made of this shell configuration. From a structural point of view the more important aspect is the fact that the shell is developable and, as stated in section 14.1.1, this can have a marked effect on the manner in which the shell responds to load. Bending stresses are much more important for this shell geometry.

In order for the bending effects of an edge disturbance to rapidly damp out, the structural behavior of strips parallel to the edge must act as a stiff, elastic restraint on the normal displacement of sections, or strips, running perpendicular to the edge. If the edge is a straight or asymptotic line of the surface, the elastic restraining or foundation effect of strips parallel to the edge is small, and bending penetrates deeply. If the cylinder is closed, the restraining sections are rings, and disturbances applied to the curved edges damp out rapidly as described in section 14.2.

Although the cross section of the cylinder can have any configuration, the most common, by far, is the circular cylinder.

The analysis of the cylindrical shell can be achieved by solution of the governing differential equations or by approximate methods based in part upon structural intuition. In order to be computationally feasible these procedures frequently require extensive systems of tabulated numerical data if the analysis is to be performed with a desk calculator. An excellent source of tabulated data for cylindrical shells is Ref. 14-31. However, with the wealth of computer programs available capable of solving the cylindrical shell

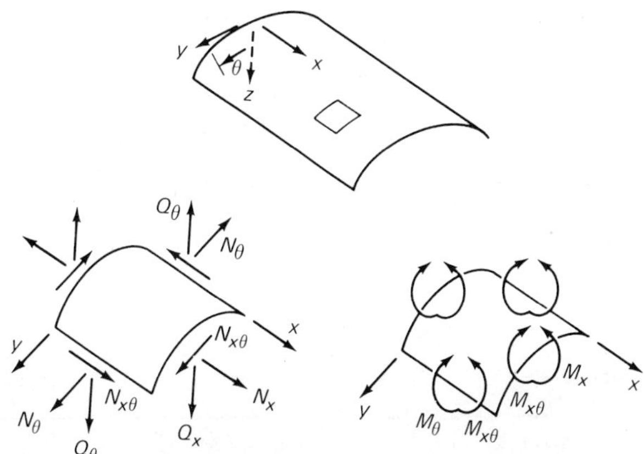

Fig. 14-8 Positive stress resultants.

problem, i.e., NASTRAN, EASE, MULEL among others, their use is recommended. This is particularly true when investigating problems involving the superposition of a large number of load components such as occurs with the wind load on a closed cylinder. See Ref. 14-39 for wind load description.

The positive definitions of stress resultants and displacements to be used in the analysis of the cylindrical shell are shown in Fig. 14-8. The nomenclature used is as shown in that figure.

14.3.1 Membrane Theory

The importance of membrane theory varies with the configuration. For closed or complete, circular cylindrical shells under symmetric or asymmetric load, membrane stresses predominate. For cylindrical panels stresses computed by membrane theory may represent only a small part of the total stress picture. The membrane stresses for a number of loading conditions are given in Table 14-5. For the liquid container, for example, only hoop stress in the amount

$$N_\theta = p_z r = \gamma r x \tag{14-10}$$

is developed. The resulting displacements are

$$w = p_z r^2/Eh = \gamma r^2/Eh \tag{14-11}$$

$$\Delta\phi = \frac{dw}{dx} = \gamma r^2/Eh$$

The structure is thus assumed to function as a system of rings sitting atop each other and carrying the fluid pressure by hoop tension. At the base some bending correction will be necessary in order to return the shell edge to its proper position.

For the antisymmetric load, such as shown as case 3 in Table 14-5b and considered as a quite gross approximation to the wind load, the membrane solution yields

$$N_x = \frac{p_w}{2r}(L - x)^2 \cos\theta$$

$$N_\theta = -p_w r \cos\theta \qquad (14\text{-}12)$$

$$N_{xy} = -p_w (L - x)\sin\theta \qquad (14\text{-}12)$$

The corresponding displacements are

$$u = \frac{p_w x}{6r}[3L^2 - 3Lx + x^2]\cos\theta$$

TABLE 14-5 Membrane Stresses in Developable Shells

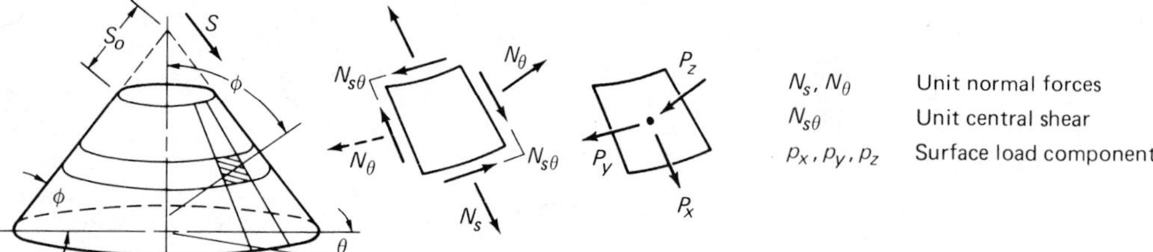

(a) Conical shell

System	Loading	N_s	N_θ	$N_{s\theta}$
	$p_x = p_E \sin\varphi$ $p_z = p_E \cos\varphi$	$-p_E \dfrac{s^2 - s_o^2}{2s} \dfrac{1}{\sin\varphi}$	$-p_E s \dfrac{\cos^2\varphi}{\sin\varphi}$	0
		For $s_o = 0$ (complete cone)		
		$-p_E \dfrac{s}{2} \dfrac{1}{\sin\varphi}$	$-p_E s \dfrac{\cos^2\varphi}{\sin\varphi}$	0
	$p_x = p_s \sin\varphi \cos\varphi$ $p_z = p_s \cos^2\varphi$	$-p_s \dfrac{s^2 - s_o^2}{2s} \cot\varphi$	$-p_s s \dfrac{\cos^3\varphi}{\sin\varphi}$	0
		For $s_o = 0$ (complete cone)		
		$-p_s \dfrac{s}{2} \cot\varphi$	$-p_s s \dfrac{\cos^3\varphi}{\sin\varphi}$	0

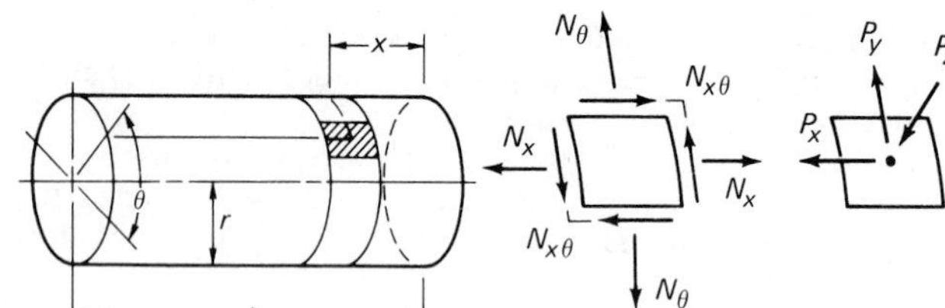

(b) Circular cylindrical shell

System	Loading	N_x	N_θ	$N_{x\theta}$
	$p_x = p_E$ $p_z = p$	$-p_E x$	$-pr$	0
	$p_z = -\gamma x$	0	$\gamma r x$	0
	$p_z = p_w \cos\theta$	$p_w \dfrac{x^2}{2r} \cos\theta$	$-p_w r \cos\theta$	$-p_w x \sin\theta$
	$p_y = -p_E \cos\theta$ $p_z = p_E \sin\theta$	$-p_E \dfrac{x}{r} (l - x) \sin\theta$	$-p_E r \sin\theta$	$-p_E (l - 2x) \cos\theta$
	$p_y = -p_E \cos\theta$ $p_z = p_E \sin\theta$	$p_E \left[\dfrac{l^2}{6r} - \nu r - \dfrac{x}{r}(l - x)\right] \sin\theta$ ν = Poisson's Ratio	$-p_E r \sin\theta$	$-p_E (l - 2x) \cos\theta$

TABLE 14-6 Edge Corrections for a Closed Cylindrical Shell

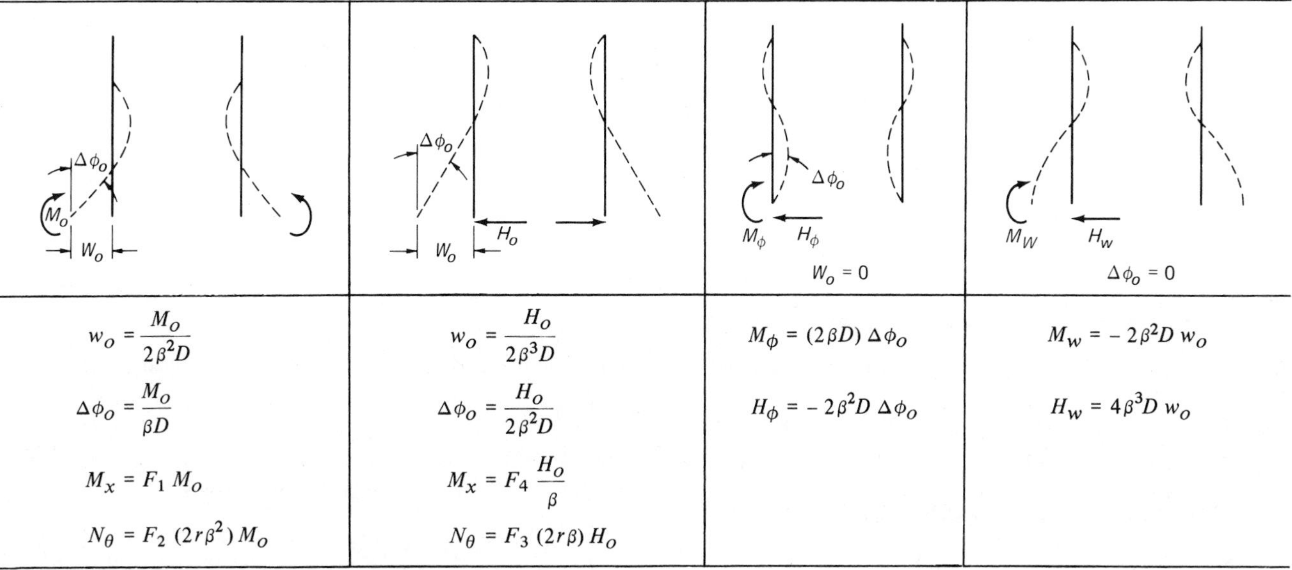

$w_o = \dfrac{M_o}{2\beta^2 D}$	$w_o = \dfrac{H_o}{2\beta^3 D}$	$M_\phi = (2\beta D)\,\Delta\phi_o$	$M_w = -2\beta^2 D\, w_o$
$\Delta\phi_o = \dfrac{M_o}{\beta D}$	$\Delta\phi_o = \dfrac{H_o}{2\beta^2 D}$	$H_\phi = -2\beta^2 D\,\Delta\phi_o$	$H_w = 4\beta^3 D\, w_o$
$M_x = F_1 M_o$	$M_x = F_4 \dfrac{H_o}{\beta}$		
$N_\theta = F_2\,(2r\beta^2)\,M_o$	$N_\theta = F_3\,(2r\beta)\,H_o$		

$\beta^4 = \dfrac{3(1-\nu^2)}{r^2 t^2}$ and F_1 to F_4 as defined in Table 14-4 with ϕ replaced by x.

$$v = \frac{p_w}{24r^2}\,[12r^2\,(4Lx - 2x^2) + (6L^2 - 4Lx + x^2)]\,\sin\theta \qquad (14\text{-}13)$$

$$w = -\frac{p_w}{24r^2}\,[12r^2\,(4Lx - 2x^2) + x^2\,(6L^2 - 4Lx + x^2)]\,\cos\theta$$

where L is the length of the shell. This solution is equivalent to a cantilever beam of circular cross section. N_x performs the function of the axial fiber stresses and N_{xy} plays the role of the beam shear.

14.3.2 Bending Theory

For the closed cylindrical shell, bending theory is used primarily to adjust the normal displacement, w, and the rotation Δ_θ at the edge of the segment to make those displacements compatible with the adjacent segments or with known displacement conditions. These bending solutions are of the beam on elastic foundation-type. They are listed in Table 14-6 for the case where a disturbance is applied to one end and this disturbance has a negligible effect at the far end. Although these equations were developed for the symmetric case, they can also serve as a first approximation to the antisymmetric case. For more waves of loading in the circumferential direction use of a computer program is recommended. Application of the bending analysis to symmetric cylindrical shells is demonstrated in later sections of this chapter.

The application of bending theory to cylindrical panels or roof structures, frequently referred to as barrel shells, is much more complex. A number of books[14-10,14-16,14-21] have been written on the development and solution of the equations. Even texts covering the broad class of shell structures[14-2,14-14] place a good bit of emphasis on the cylindrical roof structure. *ASCE Manual 31* was published to provide a tabulated form of solution of the complex differential equation system in order to make the analysis feasible within a design office.

With current computer capabilities the use of such elaborate tables is now considered unnecessary. Present analysis methods fall into two categories—beam methods and direct stiffness methods. Included in this latter grouping is the finite element method.

14.3.3 Barrel Shells

Barrel shells are known to carry the loads applied to them by two types of actions. Because of the stiffness of an arch, the first tendency of the barrel is to try to transfer the load from the crown toward the edges by transverse arch action. Ultimately the load must be transferred longitudinally and this is done by beam action. If the arch direction is short relative to the longitudinal span, the so-called long barrel, the transverse arch very quickly feels the absence of support along the longitudinal edge. For this case arching plays a minor role and ordinary, or shallow, beam theory is a good approximation (Fig. 14-9a). As the transverse arch increases relative to the longitudinal span, a short barrel results. Here transverse arching has equal importance to longitudinal beam action. The stress patterns that develop are similar to those of a deep beam (Fig. 14-9b).

The important stress resultants in the design of a barrel shell are: (1) the longitudinal stress, N_x, which is equivalent to the fiber stresses in a beam; (2) the membrane shear, $N_{x\theta}$, which is equivalent to the usual shear forces in a beam; (3) the hoop force, N_θ, significant primarily for short shells; and (4) the hoop moment, M_θ.

-1. Beam method—For the long barrel the assumption that plane sections remain plane is reasonable. Therefore, the linear variation of longitudinal stress, N_x, with vertical position in the cross section is assumed to exist. The application of ordinary beam theory for the determination of stresses in the cylinder was first presented by Lundgren. Chinn[14-7] prepared the general formulas while Parme and Connor[14-22] adapted correction factors and tables to make the design of long-barrel shells a relatively easy task. The beam method can be used for symmetric shells under uniform load for the following cases:

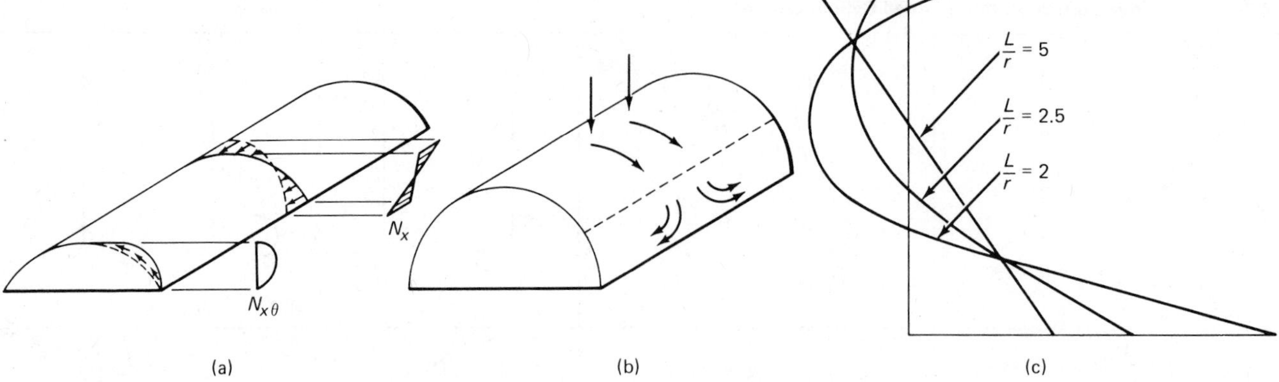

Fig. 14-9 Load-carrying characteristics of barrel shell. (a) Long barrel, ordinary beam theory; (b) short barrel, deep beam theory; and (c) longitudinal stress, N_x.

a. single shells without edge beams, if the span to radius ratio is greater than 5;
b. single shells with ordinary edge beams, if $L/r > 3$;
c. typical interior shell of a multiple shell system with butterfly edges, if $L/r > 2$; and for
d. a typical interior shell of a multiple shell system with edge beams if $L/r > 3$.

Computation of the longitudinal and shear stress is by standard beam equations

$$N_x = \frac{M_y t}{I} \tag{14-14}$$

$$N_{x\theta} = \frac{VQ}{Ib} t \tag{14-15}$$

where

I = cross section moment of inertia (Table 14-7)
M = beam bending moment, $\frac{1}{8} qL^2$
V = total shear on the cross section
b = thickness widths of shell at a horizontal section through the shell

To compute the transverse hoop force and moments, the equations presented by Chinn or a tabular form presented in the example can be used. These tend to be laborious so Parme and Connor developed Tables 14-8 for single and for typical interior shells for L/r ratios above.

As the shell gets longer there is a tendency for the shell to flatten under uniform load. Tables 14-8b, c allow this flattening to be corrected out of the solution.

For longitudinally continuous shells the above limits should be raised slightly. They are reasonable approximations for the intermediate regions between inflection points; however, the regions near the interior supports function like deep beams since the spacing between inflection points there is small relative to the clear span.

In order for the $N_{x\theta}$ forces to exist as predicted by beam theory, a stiff diaphragm or transverse arch must be placed between the column supports. If these transverse members are not provided and what is commonly called a ribless shell is used, then the shell must provide its own transverse arch. Analysis methods to handle such cases are presented in Refs. 14-26 and 14-38.

When reinforcing long-barrel shells the philosophy of the steel placement also follows that of the beam. Longitudinal reinforcement is provided to carry the total tensile, N_x, force and is placed as low near the longitudinal edge of the shell, or, if an edge beam is used, as deep in the beam as possible. Diagonal steel at approximate 45° is placed near the support to function as stirrups as in a beam. Bent-up bars can also be used to carry the shear. Hoop reinforcement is provided for the transverse shell moments, M_θ.

If the barrel shell is too shallow, that is, if the span to depth ratio is outside the region considered good beam proportions, edge beams are added to form T-beam action. These edge beams can be prestressed if architectural requirements force a still too shallow section. The prestressing also serves to control deflections and cracking. If architectural considerations demand no edge beams be used, so that feather or butterfly edges are employed, use of prestressing cables placed in the shell surface is highly desirable to control cracking and minimize the possibility of leakage.

TABLE 14-7 Moment of Inertia

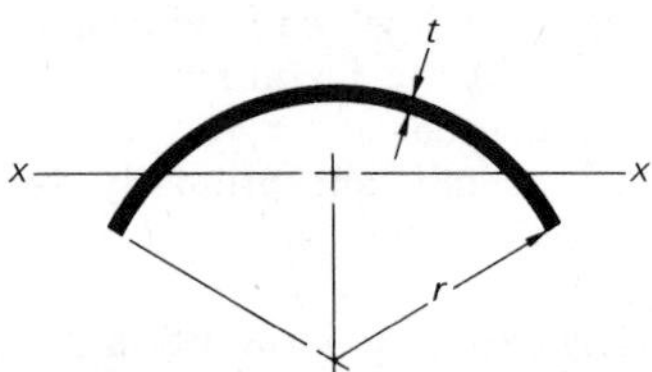

ϕ_k, deg	(11)	ϕ_k, deg	(11)
22.5	0.00041	37.5	0.00502
25.0	0.00068	40.0	0.00687
27.5	0.00110	45.0	0.01216
30.0	0.00168	50.0	0.02017
32.5	0.00249	55.0	0.03174
35.0	0.00358	60.0	0.04782

-2. *Direct stiffness method*—For the short and intermediate length barrel shells the analysis must utilize the solution of the complex differential equation system such as described in Ref. 14-2. Restricting consideration to only shells simply supported at the transverse ends, the theory of Ref. 14-16 was developed into a computer program described in Ref. 14-35. This program entitled MULEL[14-36] is available with instructions from ACI Headquarters.

EXAMPLE 14-3: A multiple barrel shell is to be used to span 50 ft. The loads for this shell, shown in Fig. 14-10, are:

Shell (P_d) (dead load plus roofing)	= 40 psi
Snow (P_u)	= 30 psi
Total	= 70 psi

TABLE 14-8a Stresses in Barrel Shells from Vertical Loads*

$$N_x = \frac{E}{r}\,[p_u \text{ col } (1) + p_d \text{ col } (5)]$$

$$N_\theta = r\,[p_u \text{ col } (2) + p_d \text{ col } (6)] + \frac{p_u t^2}{6r}\,[\text{col } (9) \times \Delta H/(p_u r^2/Et)]$$

$$N_{x\theta} = -L\,[p_u \text{ col } (3) + p_d \text{ col } (7)]$$

$$M_\theta = r^2\,[p_u \text{ col } (4) + p_d \text{ col } (8)] + \frac{p_u t^2}{6}\,[\text{col } (10) \times \Delta H/(p_u r^2/Et)]$$

Uniform transverse load — Dead weight load

θ deg.	θ	N_x (1)	N_θ (2)	$N_{x\theta}$ (3)	M_θ (4)	N_x (5)	N_θ (6)	$N_{x\theta}$ (7)	M_θ (8)
	θ_k	-6.010	-1.411	0.000	-0.00292	- 6.167	-1.433	0.000	-0.00309
	$0.75\theta_k$	-4.875	-1.189	2.211	-0.00112	- 5.003	-1.205	2.269	-0.00118
22.5	$0.50\theta_k$	-1.482	-0.614	3.533	0.00232	- 1.521	-0.615	3.626	0.00245
	$0.25\theta_k$	4.137	0.049	3.084	0.00235	4.245	0.065	3.165	0.00249
	0	11.927	0.361	0.000	-0.00662	12.239	0.384	0.000	-0.00702
	θ_k	-4.855	-1.402	0.000	-0.00353	- 5.012	-1.430	0.000	-0.00378
	$0.75\phi_k$	-3.937	-1.182	1.985	-0.00135	- 4.064	-1.202	2.049	-0.00145
25.0	$0.50\phi_k$	-1.193	-0.612	3.170	0.00280	- 1.232	-0.613	3.273	0.00300
	$0.25\phi_k$	3.342	0.044	2.765	0.00282	3.451	0.064	2.855	0.00304
	0	9.617	0.347	0.000	-0.00797	9.929	0.374	0.000	-0.00857
	θ_k	-4.000	-1.393	0.000	-0.00417	- 4.158	-1.426	0.000	-0.00453
	$0.75\phi_k$	-3.242	-1.175	1.799	-0.00159	- 3.370	-1.199	1.869	-0.00173
27.5	$0.50\phi_k$	-0.980	-0.609	2.871	0.00331	- 1.018	-0.610	2.985	0.00360
	$0.25\phi_k$	2.755	0.038	2.503	0.00332	2.863	0.063	2.602	0.00363
	0	7.908	0.331	0.000	-0.00938	8.220	0.363	0.000	-0.01025
	θ_k	-3.350	-1.383	0.000	-0.00482	- 3.508	-1.422	0.000	-0.00533
	$0.75\phi_k$	-2.714	-1.166	1.643	-0.00183	- 2.842	-1.195	1.720	-0.00203
30.0	$0.50\phi_k$	-0.817	-0.606	2.622	0.00384	- 0.856	-0.607	2.746	0.00424
	$0.25\phi_k$	2.308	0.032	2.284	0.00383	2.417	0.061	2.392	0.00426
	0	6.608	0.314	0.000	-0.01082	6.920	0.352	0.000	-0.01204
	ϕ_k	-2.844	-1.372	0.000	-0.00548	- 3.002	-1.418	0.000	-0.00618
	$0.75\phi_k$	-2.303	-1.158	1.511	-0.00207	- 2.431	-1.191	1.595	-0.00235
32.5	$0.50\phi_k$	-0.691	-0.603	2.410	0.00438	- 0.729	-0.603	2.544	0.00492
	$0.25\phi_k$	1.960	0.026	2.098	0.00434	2.069	0.660	2.215	0.00492
	0	5.596	0.297	0.000	-0.01227	5.908	0.339	0.000	-0.01393
	θ_k	-2.442	-1.361	0.000	-0.00615	- 2.601	-1.414	0.000	-0.00707
	$0.75\phi_k$	-1.977	-1.148	1.397	-0.00232	- 2.105	-1.186	1.488	-0.00268
35.0	$0.50\phi_k$	-0.591	-0.599	2.227	0.00491	- 0.629	-0.600	2.372	0.00565
	$0.25\phi_k$	1.684	0.019	1.938	0.00484	1.793	0.058	2.064	0.00561
	0	4.794	0.278	0.000	-0.01370	5.105	0.326	0.000	-0.01591
	θ_k	-2.118	-1.349	0.000	-0.00679	- 2.278	-1.409	0.000	-0.00800
	$0.75\phi_k$	-1.714	-1.138	1.298	-0.00255	- 1.842	-1.181	1.396	-0.00302
37.5	$0.50\phi_k$	-0.510	-0.596	2.069	0.00544	- 0.548	-0.596	2.224	0.00640
	$0.25\phi_k$	1.461	0.012	1.798	0.00532	1.571	0.057	1.933	0.00632
	0	4.146	0.260	0.000	-0.01509	4.458	0.312	0.000	-0.01796
	θ_k	-1.853	-1.335	0.000	-0.00742	- 2.013	-1.404	0.000	-0.00897
	$0.75\phi_k$	-1.498	-1.127	1.211	-0.00277	- 1.627	-1.176	1.315	-0.00337
40.0	$0.50\phi_k$	-0.443	-0.592	1.929	0.00595	- 0.482	-0.592	2.095	0.00719
	$0.25\phi_k$	1.279	0.005	1.675	0.00578	1.389	0.055	1.819	0.00705
	0	3.616	0.241	0.000	-0.01641	3.928	0.297	0.000	-0.02006
	ϕ_k	-1.449	-1.307	0.000	-0.00853	- 1.610	-1.393	0.000	-0.01096
	$0.75\phi_k$	-1.170	-1.104	1.065	-0.00316	- 1.299	-1.165	1.183	-0.00408

TABLE 14-8a (cont.)

θ deg.	θ	N_x (1)	N_θ (2)	$N_{x\theta}$ (3)	N_θ (4)	N_x (5)	N_θ (6)	$N_{x\theta}$ (7)	M_o (8)
45.0	$0.50\phi_k$	-0.343	-0.585	1.694	0.00688	-0.381	-0.583	1.882	0.00883
	$0.25\phi_k$	1.001	-0.011	1.468	0.00657	1.112	0.052	1.630	0.00854
	0	2.809	0.202	0.000	-0.01872	3.120	0.266	0.000	-0.02437
50.0	ϕ_k	-1.160	-1.276	0.000	-0.00939	-1.322	-1.380	0.000	-0.01301
	$0.75\phi_k$	-0.935	-1.079	0.947	-0.60344	-1.065	-1.152	1.079	-0.00480
	$0.50\phi_k$	-0.271	-0.578	1.504	0.00762	-0.308	-0.574	1.713	0.01054
	$0.25\phi_k$	0.802	-0.029	1.300	0.00714	0.914	0.049	1.481	0.01002
	0	2.232	0.164	0.000	-0.02042	2.543	0.234	0.000	-0.02871
55.0	ϕ_k	-0.946	-1.242	0.000	-0.00989	-1.109	-1.367	0.006	-0.01506
	$0.75\phi_k$	-0.761	-1.053	0.849	-0.00358	-0.892	-1.139	0.995	-0.00549
	$0.50\phi_k$	-0.217	-0.572	1.347	0.00807	-0.255	-0.563	1.578	0.01277
	$0.25\phi_k$	0.655	-0.048	1.161	0.00742	0.767	0.045	1.360	0.01144
	0	1.805	0.128	0.000	-0.02130	2.115	0.201	0.000	-0.03293
60.0	ϕ_k	-0.783	-1.205	0.000	-0.00992	-0.947	-1.352	0.000	-0.01705
	$0.75\phi_k$	-0.629	-1.025	0.766	-0.00355	-0.761	-1.124	0.927	0.00613
	$0.50\phi_k$	-0.177	-0.566	1.213	0.00815	-0.214	-0.552	1.467	0.01398
	$0.25\phi_k$	0.543	-0.068	1.043	0.00734	0.656	0.042	1.261	0.01275
	0	1.481	0.095	0.000	-0.02118	1.790	0.167	0.000	-0.03688

From Ref. (14-7).

TABLE 14-8b Effect of a Unit Horizontal Displacement

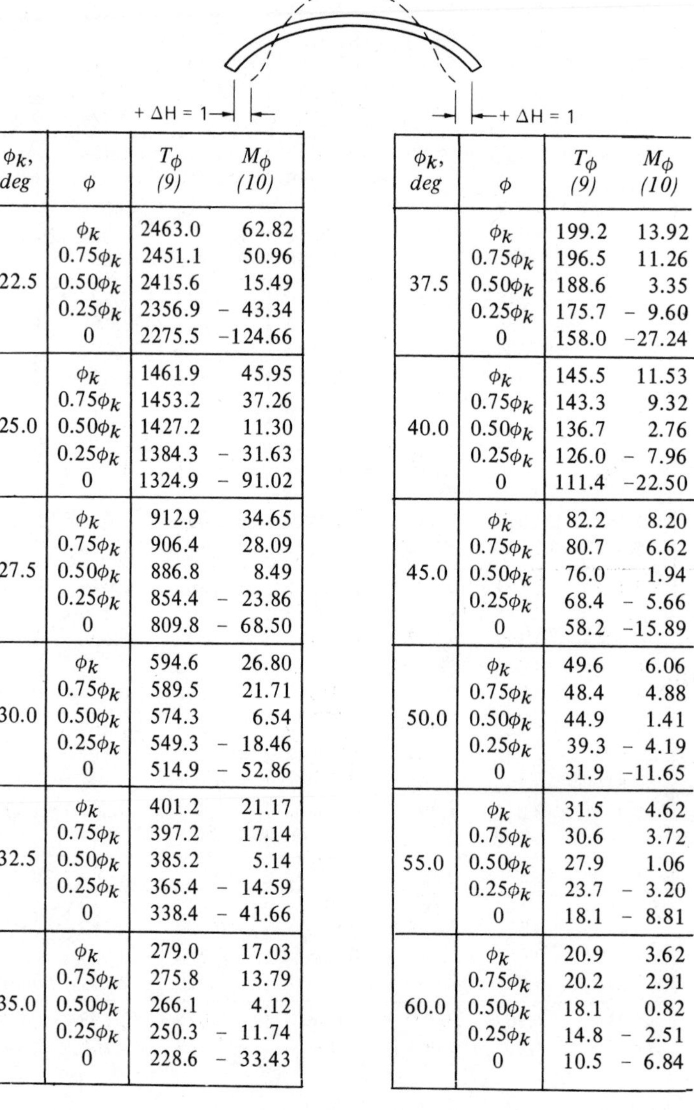

ϕ_k, deg	ϕ	T_ϕ (9)	M_ϕ (10)
22.5	ϕ_k	2463.0	62.82
	$0.75\phi_k$	2451.1	50.96
	$0.50\phi_k$	2415.6	15.49
	$0.25\phi_k$	2356.9	-43.34
	0	2275.5	-124.66
25.0	ϕ_k	1461.9	45.95
	$0.75\phi_k$	1453.2	37.26
	$0.50\phi_k$	1427.2	11.30
	$0.25\phi_k$	1384.3	-31.63
	0	1324.9	-91.02
27.5	ϕ_k	912.9	34.65
	$0.75\phi_k$	906.4	28.09
	$0.50\phi_k$	886.8	8.49
	$0.25\phi_k$	854.4	-23.86
	0	809.8	-68.50
30.0	ϕ_k	594.6	26.80
	$0.75\phi_k$	589.5	21.71
	$0.50\phi_k$	574.3	6.54
	$0.25\phi_k$	549.3	-18.46
	0	514.9	-52.86
32.5	ϕ_k	401.2	21.17
	$0.75\phi_k$	397.2	17.14
	$0.50\phi_k$	385.2	5.14
	$0.25\phi_k$	365.4	-14.59
	0	338.4	-41.66
35.0	ϕ_k	279.0	17.03
	$0.75\phi_k$	275.8	13.79
	$0.50\phi_k$	266.1	4.12
	$0.25\phi_k$	250.3	-11.74
	0	228.6	-33.43

ϕ_k, deg	ϕ	T_ϕ (9)	M_ϕ (10)
37.5	ϕ_k	199.2	13.92
	$0.75\phi_k$	196.5	11.26
	$0.50\phi_k$	188.6	3.35
	$0.25\phi_k$	175.7	-9.60
	0	158.0	-27.24
40.0	ϕ_k	145.5	11.53
	$0.75\phi_k$	143.3	9.32
	$0.50\phi_k$	136.7	2.76
	$0.25\phi_k$	126.0	-7.96
	0	111.4	-22.50
45.0	ϕ_k	82.2	8.20
	$0.75\phi_k$	80.7	6.62
	$0.50\phi_k$	76.0	1.94
	$0.25\phi_k$	68.4	-5.66
	0	58.2	-15.89
50.0	ϕ_k	49.6	6.06
	$0.75\phi_k$	48.4	4.88
	$0.50\phi_k$	44.9	1.41
	$0.25\phi_k$	39.3	-4.19
	0	31.9	-11.65
55.0	ϕ_k	31.5	4.62
	$0.75\phi_k$	30.6	3.72
	$0.50\phi_k$	27.9	1.06
	$0.25\phi_k$	23.7	-3.20
	0	18.1	-8.81
60.0	ϕ_k	20.9	3.62
	$0.75\phi_k$	20.2	2.91
	$0.50\phi_k$	18.1	0.82
	$0.25\phi_k$	14.8	-2.51
	0	10.5	-6.84

TABLE 14-8c Membrane Horizontal Displacement for Uniform Loads

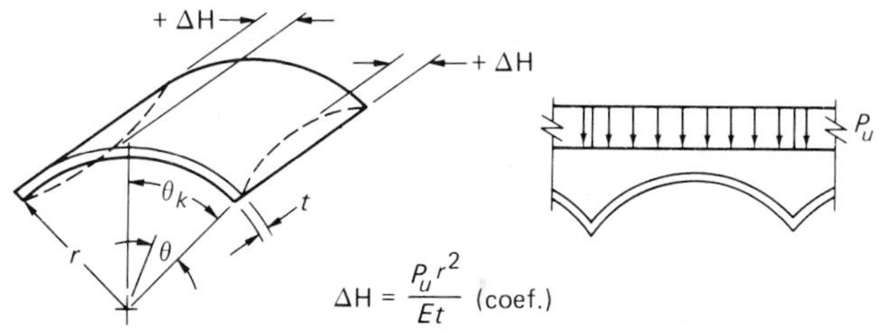

	r/L						
	0.100	0.125	0.150	0.175	0.200	0.225	0.250
θ_k deg	$\Delta H/(P_u r^2/Et)$						
22.5	- 91.83	- 38.37	- 18.88	- 10.37	- 6.16	- 3.87	- 2.52
25.0	- 123.80	- 51.79	- 25.54	- 14.09	- 8.42	- 5.33	- 3.52
27.5	- 161.58	- 67.67	- 33.43	- 18.49	-11.09	- 7.07	- 4.71
30.0	- 205.26	- 86.02	- 42.55	- 23.59	-14.20	- 9.08	- 6.09
32.5	- 254.82	-106.85	- 52.91	- 29.38	-17.72	-11.38	- 7.66
35.0	- 310.10	-130.69	- 64.47	- 35.84	-21.66	-13.94	- 9.42
37.5	- 370.84	-155.63	- 77.17	- 42.94	-25.99	-16.77	-11.37
40.0	- 436.65	-183.31	- 90.94	- 50.65	-30.70	-19.83	-13.47
45.0	- 581.48	-244.21	-121.26	- 67.61	-41.05	-26.59	-18.13
50.0	- 739.50	-310.68	-154.34	- 86.14	-52.37	-33.98	-23.22
55.0	- 904.36	-380.03	-188.87	-105.47	-64.18	-41.70	-28.54
60.0	-1068.79	-449.20	-223.31	-124.77	-75.97	-49.41	-33.85

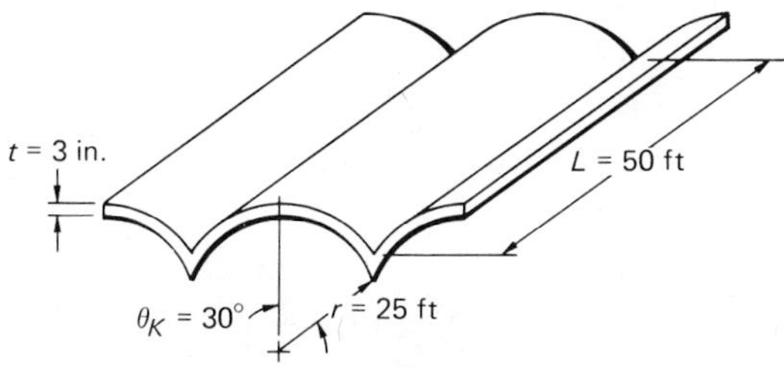

Fig. 14-10 Typical interior shell.

The dead load is distributed uniformly on the surface while the snow load is taken uniform on the horizontal projection. For this shell Tables 14-8 apply.

SOLUTION: The ΔH correction is negligible. Therefore

$$N_x = \frac{L}{r}\ (L)\ [p_u \text{ col }(1) + P_d \text{ col }(5)] = 100\ [30 \text{ col }(1) + 40 \text{ col }(5)]$$

$$N_{x\theta} = -L\ [p_u \text{ col }(3) + P_d \text{ col }(7)] = -50\ [30 \text{ col }(3) + 40 \text{ col }(7)]$$

$$M_\theta = r^2\ [p_u \text{ col }(4) + P_d \text{ col }(8)] = 625\ [30 \text{ col }(4) + 40 \text{ col }(8)]$$

TABLE 14-9 Stresses in Example 14-3 Shell

	@ $x = L/2$ N_x	@ $x = 0, L$ $N_{x\theta}$	@ $x = L/2$ M_θ
θ_k	-24,082	0	-223.6
$\frac{3}{4}\theta_k$	-19,510	-5904	-85.1
$\frac{1}{2}\theta_k$	-5875	-9425	+178.0
$\frac{1}{4}\theta_k$	16,592	-8210	+178.3
0	47,504	0	-503.9

The shear force, $N_{x\theta}$, is handled by steel placed on a 45° diagonal from the transverse supports down toward the valley of the shell. This means #5 @ 8 in the region of maximum shear.

The total tensile steel can be approximated by assuming linear variation from the valley to the neutral axis. The neutral axis is located at

$$\bar{y} = \frac{r \sin \theta_k}{\theta_k} - r \cos \theta_k = 2.2225 \text{ ft}$$

from the valley edge. This represents a tension zone of 5.56 ft of arc from the valley. A conservative estimate of total tensile force is therefore

$$T_s = \frac{1}{2}\ (5.56)\ (47.5) = 132 \text{ ft-kips}$$

The total area of steel necessary is 6.6 sq in. placed as low in the shell as cover and spacing limits allow. The reinforcing is shown in Fig. 14-11.

EXAMPLE 14-4: The typical single shell with edge beams to be analyzed is shown in Fig. 14-12. The shell is to be analyzed for a loading of live load plus dead load of 50 psf of shell surface and the beam weight of 600 plf.

SOLUTION: The analysis is performed using a beam method. The properties of the cross section are as listed in Table 14-10. The shell is divided into 12 segments, each of length 2.182 (Fig. 14-13). Properties of cross section determined for unit thickness then corrected. Values listed in the table are to the midpoint of each segment.

TABLE 14-10 Section Properties of Example 14-4 Shell

Seg.	θ	Sin θ	Cos θ	η	y	y^2
Beam	30	0.5000	0.8660	0	-1.9036	3.6237
1	27.5°	0.4618	0.8870	0.5245	0.6209	0.3855
2	22.5°	0.3827	0.9239	1.4462	1.5426	2.3796
3	17.5°	0.3007	0.9537	2.1922	2.2886	5.2377
4	12.5°	0.2164	0.9763	2.7568	2.8532	8.1408
5	7.5°	0.1305	0.9914	3.1353	3.2317	10.4439
6	2.5°	0.0436	0.9991	3.3255	3.4219	11.7094
Totals				13.3805		38.2969

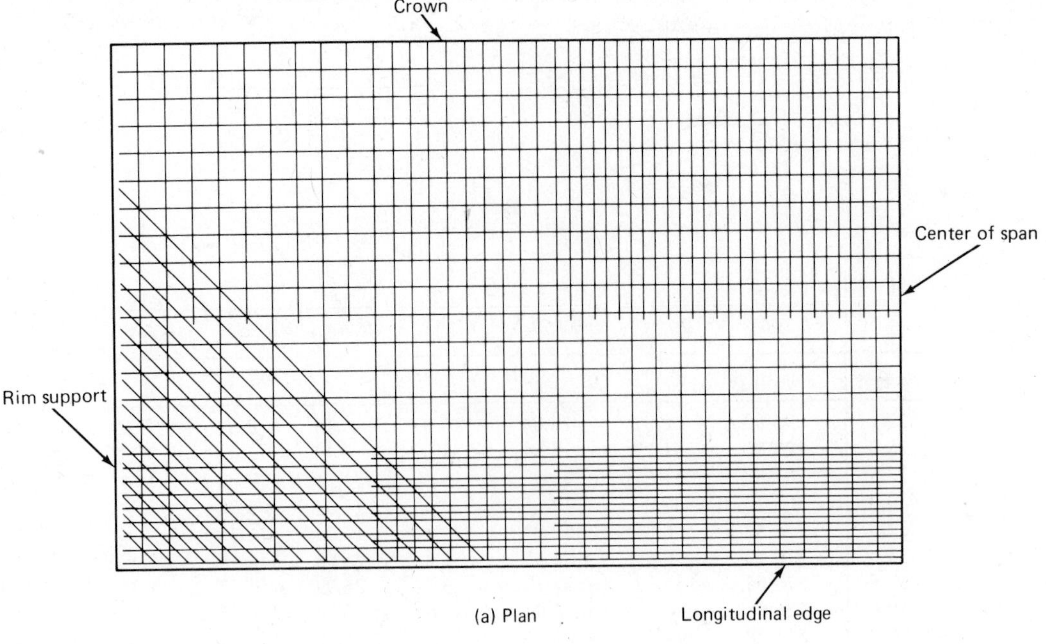

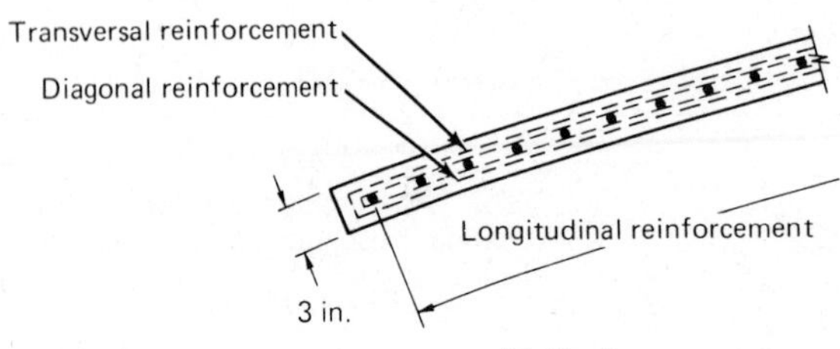

Fig. 14-11 Reinforcing.

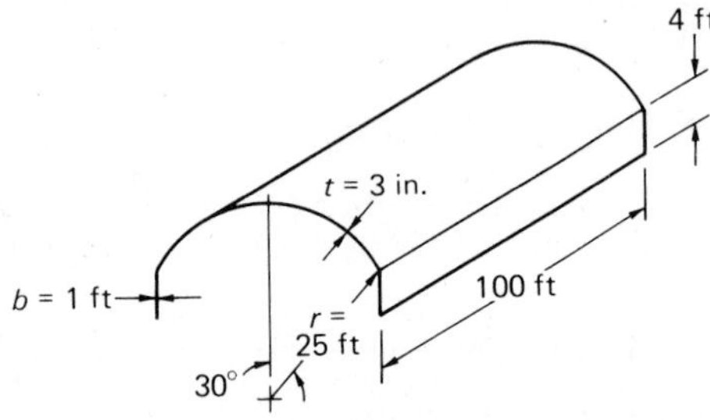

Fig. 14-12 Single shell.

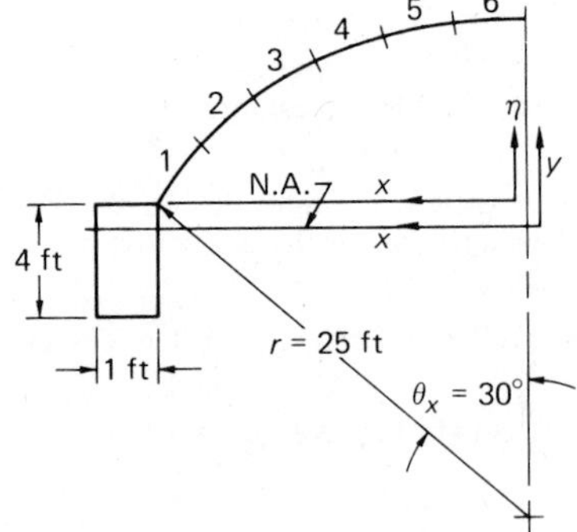

Fig. 14-13 Cross section.

δ Area = arc length (thickness) = (2.182) (0.25) = 0.5455 sq ft

Area cross section = 6 (0.5455) + 4 = 7.273 sq ft

Centroid, $\bar{\eta}$: Area $\times\ \bar{\eta} = (\delta \text{ Area} \times \eta - 8) = 0.5455\ (13.3805) - 8$

Area $\times\ \bar{\eta} = -0.701 \qquad \bar{\eta} = 0.0964$

It is desirable for the neutral axis to fall within the beam to minimize chances of flexural cracking causing leakage. The beam moment of inertia is

$$I = \Sigma\,(y^2)\,\delta\ \text{Area} + I_{0\,\text{Beam}} + A_{\text{Beam}}\,y^2 = 40.83\ \text{ft}^4$$

Load:

Shell + LL = 50 (6) (2.182) =	654.6 plf.
Beam =	600 plf.
Total =	1254.6 plf.

Moment:

$$\frac{1}{8}\,WL^2 = \frac{1}{8}\,(1.255)\,(100)^2 = 1568\ \text{in.-kip}$$

$$N_{x\,\text{Crown}} = \frac{My}{I}\,t = \frac{1568}{40.83}\,(0.25)\,[25\,(1 - 0.866) + 0.096] = 33.09\ \text{kip/in.}$$

$$N_{x\,\text{Beam}} = \frac{1568}{40.83}\,(1)\,(3.9036) = 149.91\ \text{kip/in.}$$

The shear is computed at the center of each segment by using the unbalanced horizontal force.

TABLE 14-11 Increment Forces

Seg.	$\frac{M}{I} y\ \delta$ Area	Shear, S	$\frac{dS}{dx}\ \delta$ Arc	$\Delta V'$	ΔV	Total V ΔV + Load	H
Beam	292.416	8.920 11.696		0.6317	0.6362	0.0362	
1	-13.007	11.436	0.4991	0.2305	0.2321	0.1231	0.4427
2	-32.316	10.529	0.4595	0.1758	0.1771	0.0681	0.4245
3	-47.944	8.924	0.3894	0.1171	0.1179	0.0089	0.3714
4	-59.772	6.770	0.2954	0.0639	0.0644	-0.0446	0.2884
5	-67.701	4.220	0.1842	0.0240	0.0242	-0.0848	0.1826
6	-71.685	1.433	0.0625	0.0027	0.0027	-0.1063	0.0624
Total	+0.009			1.2457	1.2546	~0	

Shear:

$$\frac{1}{2}\frac{L}{2} S = \Sigma \frac{M}{I} y\ \delta \text{ Area} \quad \therefore S = \frac{4}{L} \Sigma \frac{M}{I} y\ \delta \text{ Area}$$

$$S_{\text{seg 3}} = \frac{4}{L} [292.416 - 13.007 - 32.316 - \tfrac{1}{2}\, 47.944] = 8.924$$

$$\Delta V' = \frac{dS}{dx} (\delta \text{ arc}) \sin \theta$$

Since $\Delta V'$ does not sum up to total vertical load of 1.2546 klf, a new shear ΔV is calculated by

$$\Delta V = \frac{1.2546}{1.2467} \Delta V'$$

The magnitude of the loading within one segment is

$$(50)\ (2.182) = 0.109 \text{ kip}$$

This is added to ΔV to find V.

With the net horizontal and vertical forces per unit length established, it is possible to compute shell moment M_θ.

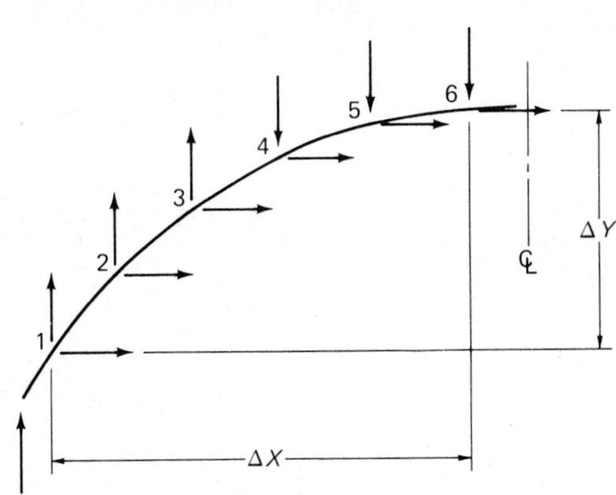

In order to compute the transverse moment M_θ, it is necessary to assume moment conditions at the junction of the shell and the beam. If the torsional moment of the beam is negligible and its resistance to horizontal forces is negligible, the moment starts at zero and the system is determinate. Multiplying the V- and the H-forces of Table 14-11 by the moment arms Δx and Δy of Tables 14-12, their sum yields

TABLE 14-12 Increment Distances

a) Horizontal Distance, Δx

From \ To	6	5	4	3	2	1	x
Edge	11.4095	9.2368	7.0890	4.9822	2.9330	0.9562	12.5
1	10.4533	8.2806	6.1328	4.0260	1.9768		11.5438
2	8.4765	6.3038	4.1560	2.0492			9.5670
3	6.4273	4.2546	2.1068				7.5178
4	4.3205	2.1478					5.4110
5	2.1727						3.2632
6							1.0905

b) Vertical Distance, Δy

From \ To	6	5	4	3	2	1	x
Edge	5.3255	5.1353	4.7568	4.1922	3.4462	2.5245	-1.9036
1	2.8010	2.6108	2.2323	1.6677	0.9217		0.6209
2	1.8793	1.6891	1.3106	0.7460			1.5426
3	1.1333	0.9431	0.5646				2.2886
4	0.5687	0.3785					2.8532
5	0.1902						3.2317
6							3.4219

M_θ at center of segment

Seg	1	2	3	4	5	6	Units
M_θ	0.0346	−0.0585	−0.2394	−0.4409	−0.6072	−0.7001	Ft. Kips/Ft.

If the beams along the edge were very stiff against rotation and/or horizontal movement, or, if the shell was a typical interior shell, the above method is still applicable. Using the above computed M_θ, compute M_θ/EI, the horizontal displacement and rotation of the edge with M_θ/EI as load. Apply corrective forces as in any typical flexibility approach. The moment at the junction of the shell and the beam is like the fixed-end moment in a beam. It is not necessary for statics but occurs from compatibility. The magnitude depends upon how much reinforcement is provided. Examples in literature range from the equivalent of full-fixity to pin-supported.

Three networks of reinforcing are necessary for this shell. (See Fig. 14-14.)

a. Longitudinal steel to carry the usual beam tension force. Total area of steel

$$\frac{1}{2}\,(149.91)\,(3.9036)/20 = 14.63 \text{ sq in.}$$

Use 19–#8 bars placed as low in beam as possible, i.e., (4, 4, 4, 4, 3)

b. Shear steel placed in shell running diagonally from the transverse support down to the beam.
Steel @ seg.

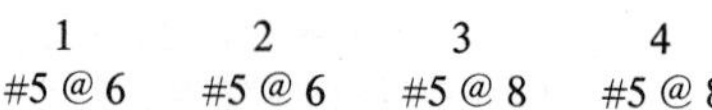

1	2	3	4
#5 @ 6	#5 @ 6	#5 @ 8	#5 @ 8

c. Steel placed in the hoop direction to reinforce against the hoop moment M_θ.
Use #4 @ 12
This steel should be placed as the top layer. It should be placed over full length of shell.

14.4 TRANSLATIONAL SHELLS

One form of shell used almost exclusively for roof systems is the translational shell. There have been cases where these shells have been used as foundation structures. This category of shell classification contains probably the largest variety of different shell surfaces. Principal shells of translational type are the elliptical parabola, hyperbolic parabola, and the conoid. One feature that explains the popularity of these shells is the range and variety of appearances that can be achieved with the same basic shell configurations.

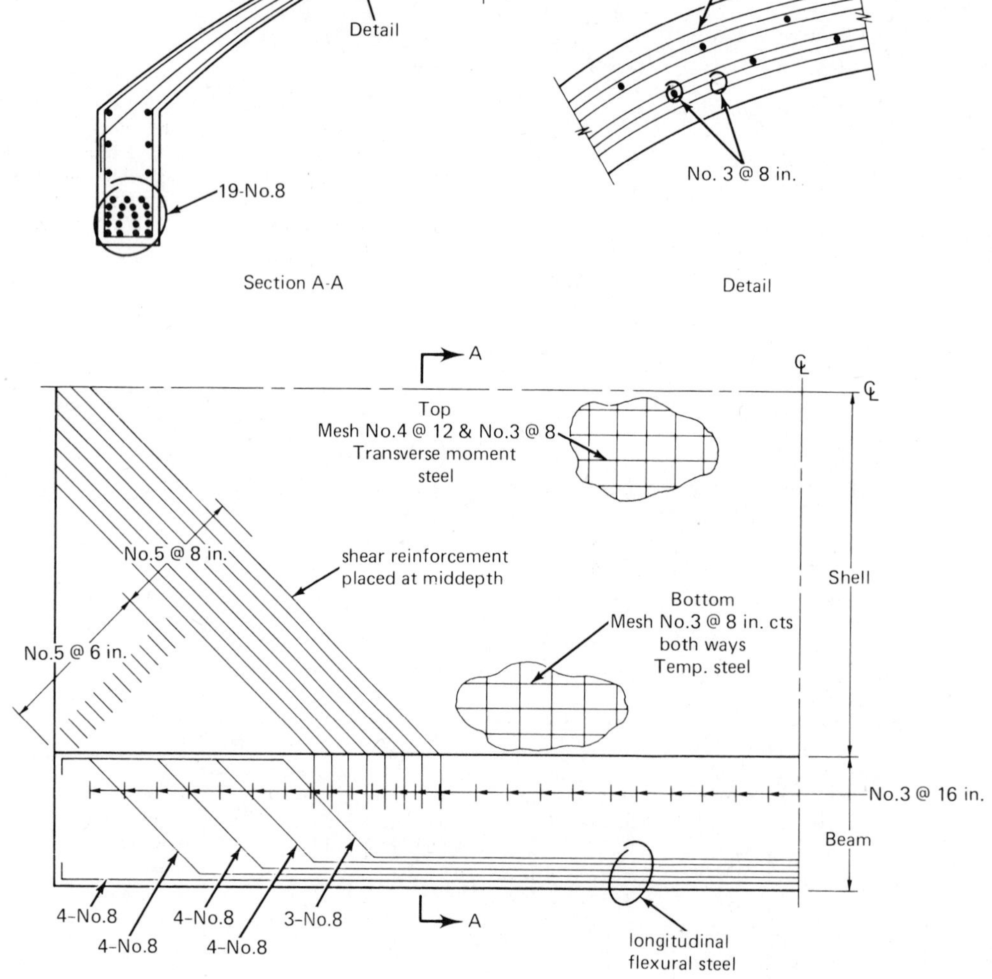

Fig. 14-14 Sketch of main reinforcing of cylindrical shell of ex. 14-4.

14.4.1 Introduction

The basic construction of the shells in this category is to take a curve fixed in space then translate a second curve along the first. This sweeps out the desired shell surface. Depending upon the selection of curves the surface can have positive, zero, or negative curvature. The surface thus can have zero, one, or two systems of real asymptotic lines.

14.4.2 Elliptical Parabola

-1. Surface equation—This shell surface is formed by taking two identical parabolic arches setting them in parallel planes a given distance apart, and placing a third parabolic arch to span the distance between the origin arches. This third parabola is then moved along the origin parabolas, sweeping out the continuous surface shown in Fig. 14-15. The equation of the surface is the sum of the equations of the parabolas:

$$z = h_x \left(\frac{x}{a}\right)^2 + h_y \left(\frac{y}{b}\right)^2 \tag{14-16}$$

where h_x and h_y are the rise of the arches in the x- and y-directions respectively and $2a$ and $2b$ are the spans of the parabolas in the x- and y-directions. The surface thus formed has a dome-like appearance.

-2. Membrane theory—Structurally the elliptical parabola is a very efficient shell. This shell carries load primarily by membrane stresses as long as there is some form of support along its edges. The shell functions as a system of interconnecting arches which transfer the load toward the edges of the shell. The membrane equation specialized to the elliptical parabola loaded by a vertical load is

$$\frac{2h_y}{b^2}\frac{\partial^2 F}{\partial x^2} + \frac{2h_x}{a^2}\frac{\partial^2 F}{\partial y^2} = p \tag{14-17}$$

One solution of this equation was achieved by Parme[14-25] in the form of a series of trigonometric and hyperbolic functions. The series solution for F is then integrated to determine the stress resultants which are expressible as

$$\begin{aligned} N_x &= -\frac{pa^2}{h_x} k \left[\text{function I of } \frac{h_x}{h_y}, \frac{x}{a}, \frac{y}{b}\right] \\ N_y &= -\frac{pb^2}{kh_y} \left[\text{function II of } \frac{h_x}{h_y}, \frac{x}{a}, \frac{y}{b}\right] \\ N_{xy} &= -\frac{pab}{\sqrt{h_x h_y}} \left[\text{function III of } \frac{h_x}{h_y}, \frac{x}{a}, \frac{y}{b}\right] \end{aligned} \tag{14-18}$$

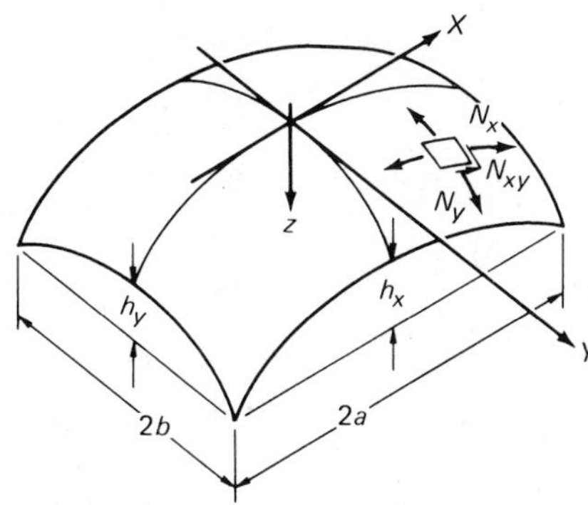

Fig. 14-15 Elliptical parabola.

where

$$k = \sqrt{\frac{1 + 4h_x^2\left(\dfrac{x}{a^2}\right)^2}{1 + 4h_y^2\left(\dfrac{y}{b^2}\right)^2}}$$

and

p = vertical load uniformly distributed over the horizontal projection.

Tables 14-13 and 14-14 give the numerical values for the functions in the brackets evaluated at several points on the shell for various h_x/h_y ratios.

The membrane theory presumes edge-supporting members in the plane of the edge while flexible normal to that plane. Such a supporting edge is obtainable, for example, by a truss. Frequently, for architectural reasons, the edge beams are tied arches, thus departing some from membrane boundaries.

The membrane solution results in just shear forces existing around the boundary of the shell surface. These shear forces accumulate as direct forces in the edge supporting members.

-3. Edge corrections—Because the edge of the membrane shell and its supporting members want to deflect in the vertical plane in different amounts, some flexure develops in the region of the edge. It has been shown that this flexure region closely resembles that of a beam on a stiff, elastic foundation. Thus the application of the Geckeler approximation, discussed earlier for shells of revolution and closed cylinders, is a reasonable means to assess the bending magnitude and penetration into the shell.

The application of edge shear and moment to the shell along the edge $x = \pm a$ results in a deflection and rotation of

$$\begin{aligned} w &= -2\beta^2 \frac{\Gamma_y^2}{Et} M_{xo} - 2\beta \frac{\Gamma_y^2}{Et} Q_{xo} + \frac{p\Gamma_y^2}{Et} \\ \frac{dw}{dx} &= 2\beta^2 \frac{\Gamma_y^2}{Et} Q_{xo} + 4\beta^3 \frac{\Gamma_y^2}{Et} M_{xo} \end{aligned} \tag{14-19}$$

where Γ_y is the radius of curvature of the shell along the edge and the remaining terms are as defined in section 14.2.4. Compatibility demands these displacements match the corresponding quantities for the supporting member. Since there is a significant difference in stiffness of the support members and the shell, conservative assumptions are to consider the edge clamped in the region near the column support while simply supported over the region of midspan of the edge. Near the column support therefore $w = dw/dx = 0$ and find

$$\begin{aligned} Q_{xo} &= \frac{p}{\beta} \approx pc \\ M_{xo} &= -\frac{p}{2\beta^2} \approx -\frac{1}{2} pC^2 \end{aligned} \tag{14-20}$$

where $C = \dfrac{1}{\beta} = 0.76\sqrt{tr_y}$. The penetration of the moment is $M_x = M_{xo} F_2$. While in the midspan region

$$\begin{aligned} Q_{xo} &= \frac{P}{2\beta} \approx \frac{1}{2} pC \\ M_{xo} &= 0 \end{aligned} \tag{14-21}$$

TABLE 14-13 Coefficients for Computing Force Components of Elliptical Paraboloid Shell*

x/a	Force component	(a) $h_x/h_y = 1.0$					(d) $h_x/h_y = 0.8$				
		Value of y/b									
		0	0.25	0.50	0.75	1.0	0	0.25	0.50	0.75	1.0
0.00	N_y	0.250	0.233	0.182	0.101	0	0.289	0.270	0.213	0.119	0
	N_x	0.250	0.267	0.318	0.399	0.500	0.211	0.230	0.287	0.381	0.500
	N_{xy}	0	0	0	0	0	0	0	0	0	0
0.25	N_y	0.267	0.250	0.199	0.111	0	0.304	0.285	0.228	0.130	0
	N_x	0.233	0.250	0.301	0.389	0.500	0.196	0.215	0.272	0.370	0.500
	N_{xy}	0	0.029	0.068	0.096	0.108	0	0.034	0.069	0.100	0.114
0.50	N_y	0.318	0.301	0.250	0.150	0	0.347	0.331	0.277	0.169	0
	N_x	0.182	0.199	0.250	0.350	0.500	0.153	0.169	0.223	0.331	0.500
	N_{xy}	0	0.068	0.140	0.210	0.244	0	0.065	0.139	0.215	0.255
0.75	N_y	0.399	0.389	0.350	0.250	0	0.416	0.406	0.369	0.270	0
	N_x	0.101	0.111	0.150	0.250	0.500	0.084	0.094	0.131	0.230	0.500
	N_{xy}	0	0.096	0.210	0.356	0.465	0	0.091	0.201	0.353	0.480
1.0	N_y	0.500	0.500	0.500	0.500	0	0.500	0.500	0.500	0.500	0
	N_x	0	0	0	0	0	0	0	0	0	0
	N_{xy}	0	0.108	0.243	0.465	∞		0.101	0.229	0.443	∞
		(b) $h_x/h_y = 0.6$					(c) = $h_x/h_y = 0.4$				
0.00	N_y	0.336	0.316	0.252	0.143	0	0.395	0.374	0.307	0.180	0
	N_x	0.164	0.184	0.248	0.357	0.500	0.105	0.126	0.103	0.320	0.500
	N_{xy}	0	0	0	0	0	0	0	0	0	0
0.25	N_y	0.348	0.329	0.267	0.555	0	0.403	0.383	0.319	0.192	0
	N_x	0.152	0.171	0.233	0.345	0.500	0.097	0.117	0.181	0.308	0.500
	N_{xy}	0	0.031	0.067	0.103	0.120	0	0.026	0.060	0.101	0.125
0.50	N_y	0.383	0.367	0.312	0.197	0	0.425	0.410	0.357	0.235	0
	N_x	0.117	0.133	0.188	0.304	0.500	0.075	0.090	0.143	0.265	0.500
	N_{xy}	0	0.060	0.132	0.216	0.265	0	0.049	0.115	0.208	0.274
0.75	N_y	0.436	0.426	0.392	0.296	0	0.459	0.451	0.419	0.331	0
	N_x	0.064	0.074	0.108	0.204	0.500	0.041	0.049	0.081	0.169	0.500
	N_{xy}	0	0.081	0.185	0.342	0.494	0	0.065	0.156	0.316	0.506
1.00	N_y	0.500	0.500	0.500	0.500	0	0.500	0.500	0.500	0.500	0
	N_x	0	0	0	0	0	0	0	0	0	0
	N_{xy}	0	0.089	0.208	0.413	∞	0	0.070	0.173	0.363	∞
		(c) $h_x/h_y = 0.2$									
0.00	N_y	0.462	0.446	0.388	0.248	0					
	N_x	0.038	0.054	0.112	0.252	0.500					
	N_{xy}	0	0	0	0	0					
0.25	N_y	0.465	0.451	0.396	0.261	0					
	N_x	0.035	0.049	0.104	0.239	0.500					
	N_{xy}	0	0.014	0.040	0.088	0.128					
0.50	N_y	0.473	0.462	0.414	0.303	0					
	N_x	0.027	0.038	0.086	0.197	0.500					
	N_{xy}	0	0.027	0.074	0.174	0.280					
0.75	N_y	0.485	0.480	0.456	0.383	0					
	N_x	0.015	0.020	0.044	0.117	0.500					
	N_{xy}	0	0.034	0.098	0.246	0.510					
1.00	N_y	0.500	0.500	0.500	0.500	0					
	N_x	0	0	0	0	0					
	N_{xy}	0	0.038	0.108	0.262	∞					

*From Ref. 14-25.

TABLE 14-14 Shear Along the Edges of Elliptical Paraboloid Shell*

y/b	h_x/h_y				
	1.0	*0.8*	*0.6*	*0.4*	*0.2*
	At $x = \pm a$				
0.0	0.0000	0.0000	0.0000	0.0000	0.0000
0.1	0.0419	0.0389	0.0342	0.0307	0.0137
0.2	0.0854	0.0793	0.0701	0.0550	0.0286
0.3	0.1319	0.1231	0.1096	0.0872	0.0481
0.4	0.1836	0.1721	0.1546	0.1254	0.0731
0.5	0.2432	0.2294	0.2081	0.1728	0.1075
0.6	0.3204	0.3066	0.2859	0.2493	0.1818
0.7	0.4071	0.3897	0.3627	0.3173	0.2296
0.8	0.5363	0.5178	0.4887	0.4400	0.3443
0.85	0.6279	0.6090	0.5791	0.5292	0.4306
0.9	0.7570	0.7378	0.7074	0.6667	0.5659
0.95	0.9777	0.9582	0.9276	0.8763	0.7741
1.0	∞	∞	∞	∞	∞
x/a	At $y = \pm b$				
0.0	0.0000	0.0000	0.0000	0.0000	0.0000
0.1	0.0419	0.0444	0.0468	0.0488	0.0500
0.2	0.0854	0.0903	0.0950	0.0990	0.1014
0.3	0.1319	0.1391	0.1460	0.1519	0.1553
0.4	0.1836	0.1930	0.2019	0.2095	0.2140
0.5	0.2432	0.2545	0.2652	0.2743	0.2798
0.6	0.3204	0.3317	0.3425	0.3516	0.3571
0.7	0.4071	0.4213	0.4348	0.4463	0.4532
0.8	0.5363	0.5515	0.5659	0.5782	0.5855
0.85	0.6279	0.6434	0.6582	0.6707	0.6782
0.9	0.7570	0.7728	0.7878	0.8005	0.8081
0.95	0.9777	0.9935	1.0087	1.0215	1.0290
1.0	∞	∞	∞	∞	∞

*From Ref. (14-25).

The penetration of the bending into the shell is obtained by using these values of Q_{xo} and M_{xo} in the equations of Table 14-6 with $Q_{xo} = H_o$. Thus the positive moment is

$$M_x = F_4 \frac{p}{2\beta^2} = \frac{1}{2} pC^2 F_4 \qquad (14\text{-}22)$$

where F_4 is found from Table 14-4 by replacing $\beta\phi$ by βy as was done for the cylinder.

It should be remembered these stress resultants are necessary for compatibility but not equilibrium.

-4. *Buckling considerations*—Since this shell form carries the load primarily through compressive arching forces, buckling of the shell surface must be considered (see section 14.2.5). A conservative estimate of the buckling pressure 14-32 is

$$p_{cr} = 0.06 \frac{Et^2}{\Gamma_x \Gamma_y} \qquad (14\text{-}23)$$

A generous factor of safety should be applied to this pressure to assure that imperfections, creep displacements and other such effects do not result in premature shell buckling.

If the boundary conditions represent a significant departure from the diaphragm supports of the membrane theory, a nonlinear bending solution using the finite element procedure is recommended.

-5. *Bending solutions*—Sharp departures from the membrane theory result if edge supports do not receive adequate horizontal restraint at the column heads. The shell acts as a very flexible cross-sectioned curved beam. Reliable stress evaluations can only be obtained from solutions to the equations of shallow shell theory. Numerical procedures such as finite element present the most direct solution procedure. This is most easily achieved using one of the general purpose programs such as NASTRAN, STRUDL, ASKA, EASE, or other such programs.

EXAMPLE 14-5: Given an elliptical parabola spanning a 100 × 80 ft rectangular plan. The rise is 20 ft with $h_x = h_y = 10$ ft. Load is 70 psf uniform on the horizontal projection. The maximum variation of k is from 0.89 to 1.08. Assume $k = 1$

SOLUTION:

$$N_x = -\frac{70(50)^2}{10} \text{(Table Coeff.)} = -17{,}500 \text{ (Coeff.)}$$

$$N_y = -\frac{70(40)^2}{10} \text{(Table Coeff.)} = -11{,}200 \text{ (Coeff.)} \qquad (14\text{-}24)$$

$$N_{xy} = \frac{70(50)(40)}{10} \text{(Table Coeff.)} = -14{,}000 \text{ (Coeff.)}$$

Computed maximum normal stress resultants are

$$M_x = -8750 \text{ lbs/ft @ } x = 0, \quad y = 40 \text{ ft}$$

$$N_y = -5600 \text{ lbs/ft @ } x = 50, \quad y = 0$$

The membrane shear for points along the edge $x = a$, and the principal tension stress

$$N = \frac{N_y}{2} + \sqrt{\left(\frac{N_y}{2}\right)^2 + (N_{xy})^2} \qquad (14\text{-}25)$$

at the same points given in Table 14-15.

The approximate area of steel to be placed on a diagonal line between adjacent edges is also shown in the table. A minimum reinforcement of at least 0.0018 bt should be provided (ACI 318, Sec. 7.13).

The edge bending effects are found from Eqs. 14-20 & 14-21. The radius of curvature is

$$R_x = \frac{\left[1 + \left(\frac{\partial z}{\partial x}\right)^2\right]^{3/2}}{\partial^2 z/\partial x^2} \qquad (14\text{-}26)$$

For the edge $y = b$, the curvature at the crown ($x = 0$) is 125 ft while at the support the curvature is 156 ft ($x = a$). Therefore the maximum bending moment near the crown is

$$M_x = \tfrac{1}{2} p\,(0.76)^2 t R_x F_2 = 0.16\,(70)\,(0.76)^2 (0.25)\,(125)$$

$$= 202 \text{ ft-lbs/ft}$$

TABLE 14-15 Membrane Shear and Principal Tension Stress Along Edge $x = a$

y/b	0	0.1	0.2	0.3	0.4	0.5	0.6	0.7	0.8	0.85	0.9	0.95
N_{xy}	0	−587	−1196	−1847	−2570	−3405	−4486	−5699	−7508	−8791	−10598	−13688
N	0			554	1001	1609	2488	3550	5210	6430	8510	11160
A_s								#4 @ 12	#4 @ 9	#4 @ 7	#4 @ 5	#4 @ 4

The maximum negative edge moment is

$$M_x = -\tfrac{1}{2}p\,(0.76)^2 tR_x = -0.5\,(70)\,(0.76)^2(0.25)\,(156)$$
$$= -788 \text{ ft-lbs/ft}$$

Steel to sustain these moments should be provided. The tie force that must be sustained at the corners can be found using Simpson's integration on the normal stress resultants at the crown (using the N_x values computed from Table 14-13, $x/a = 0$)

$$P_{tie} = \tfrac{1}{3}\,\tfrac{40}{4}\,[4375 + 8750 + 4(4672 + 6982) + 2(5565)]$$
$$= 236{,}200 \text{ lbs}$$

14.4.3 Hyperbolic Parabola

-1. Surface equation—The equation defining the hyperbolic paraboloid surface is

$$z = h_y\left(\frac{y}{b}\right)^2 - h_x\left(\frac{x}{a}\right)^2 \qquad (14\text{-}26)$$

when the coordinate lines are aligned with the principal curvatures. This surface is generated by translating a parabola spanning in the y-direction along a parabola spanning in the x-direction. The y-parabola is concave downward while the x-parabola is concave upward.

If the coordinate lines are selected to match the asymptotic or straight characteristic lines of the surface, the defining equation is

$$z = k\,xy \qquad (14\text{-}27)$$

where $k = f/ab$. This equation applies regardless of the angle of intersection ω_1 of the asymptotic lines. Figure 14-16 shows the shell formed in this manner.

To be effective the shell should have the product kt (warping times the thickness) not less than 0.003. Some shells have been built with this ratio as low as 0.002.

The hyperbolic parabola is one of the most versatile of the shell forms being adaptable into the shapes shown in Fig. 14-17. Further more complex arrangements are shown in Fig. 14-18.

-2. Membrane theory—With the surface defined by eq. (14-27), one membrane equilibrium equation relates the shear force directly as a function of the loading[14-6,14-37]

$$N_{xy} = \left(\frac{y}{2}p_x + \frac{x}{2}p_y - p_z\frac{ab}{2f}\right)\sin\omega \qquad (14\text{-}28)$$

where p_x, p_y, and p_z are the load intensities in the x-, y-, and z-directions. The shear force is determined by the load intensity at the point and by the warping geometry of the

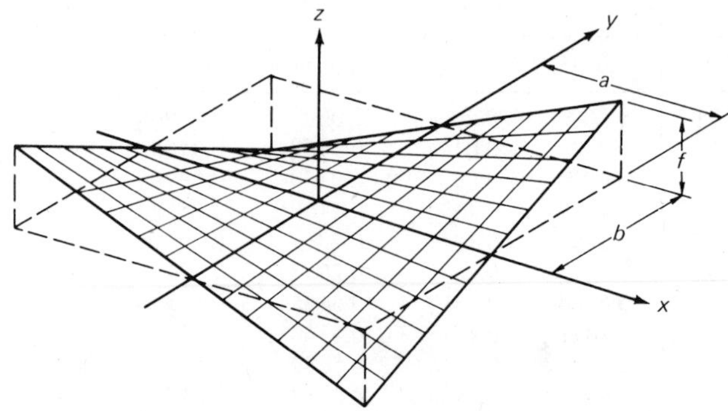

Fig. 14-16 Geometry of hyperbolic paraboloid, $z = kxy$, ($k = f/ab$).

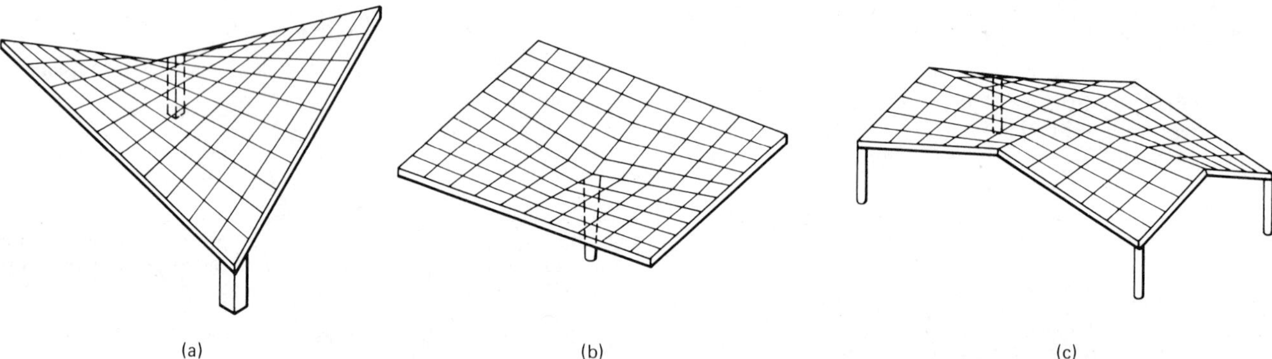

Fig. 14-17 Some fundamental configurations of hyperbolic paraboloids. (a) Saddle shell; (b) inverted umbrella; and (c) hipped roof.

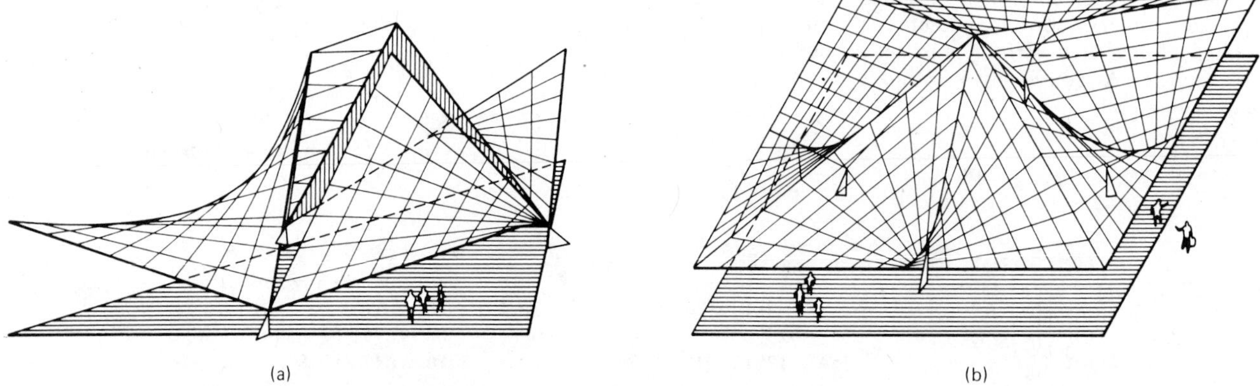

Fig. 14-18 More complex surfaces formed from simple panels. (a) Three-panel shell; and (b) eight-panel shell.

shell surface. The only way to change the magnitude of the shear force is to change the geometry of the shell.

The normal forces are determined from

$$N_x = -\int\left(\frac{\partial N_{xy}}{\partial y} + p_x \sin\omega\right) dx + f_1(y) \quad (14\text{-}29)$$

$$N_y = -\int\left(\frac{\partial N_{xy}}{\partial x} + p_y \sin\omega\right) dy + f_2(x) \quad (14\text{-}30)$$

These direct forces are the true normal forces only for an ω of 90°. If the axes are nonorthogonal these are forces parallel to the axes directions.

For a shallow shell under a constant vertical load, N_{xy} is essentially constant; $N_{xy} = p_z ab/2f$. The direct forces, N_x and N_y, depend upon the boundary restraint and are normally negligible. The principal stresses for this case are

$$N_1 = -N_2 = p_z \frac{ab}{2f} \quad (14\text{-}31)$$

if ω is 90°. In this case they represent a system of intersecting arches; a parabolic compression arch under uniform stress and a draping, or inverted, parabolic arch in uniform tension.

The membrane theory predicts the existence of shear forces around the boundary of the shell surface and assumes edge members capable of accepting these forces as a reaction, Fig. 14-19. These shear forces accumulate in the edge members as axial forces

$$N_A = \int N_{xy}\, dx + N_{\text{corner}} \quad (14\text{-}32)$$

where N_{corner} is the initial axial force existing in the edge member at its beginning. This initial axial force is normally negligible.

If the loads on the shell vary significantly over the shell surface, or if the shell has a very large k-value, then a bending analysis is necessary.

The membrane stresses for various load conditions are given in Table 14-16.

-3. *Unsymmetrical load*—An unsymmetrical load on a hyperbolic shell cannot be resisted solely by membrane forces since such a theory predicts a sharp change in forces in the supporting members. For example the inverted umbrella of Fig. 14-17b would have an axial force of magnitude N in the edge member of the load panel. Just beyond that point, in the unloaded panel the continuation of the same edge member has no load. To account for such unbalanced loads therefore Ref. 14-41 suggests ribs between the loaded panels be designed as cantilever beams for a load at least one half the unbalanced load.

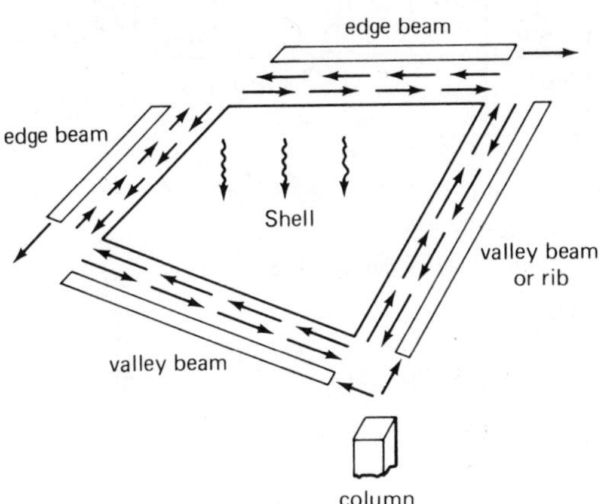

Fig. 14-19 Shell-beam interaction forces.

-4. *Edge corrections*—An edge correction similar to that discussed for elliptical parabolas can also be applied to the hyperbolic parabola (H.P.), however this correction must be restricted to the moderate and small shells. This procedure is particularly useful in determining the bending moments in the shell in the region near a rib separating loaded panels.

A bending analysis is particularly important to establish the stresses in the supporting beams. The supporting member should initially be sized to sustain the forces predicted by membrane theory; however, then a bending analysis must be performed (see 14.4.3.6). For most H.P. shells the axial force in the supporting beams is usually close to that predicted by membrane theory. This is not true of the ridge beam of a hipped roof H.P. An exception is in the hipped-roof configuration of Fig. 14-17c where the crown beams do not receive the axial force but only one-half to one-third that predicted by membrane theory.[14-33]

-5. *Buckling considerations*—The buckling load of a hyperbolic parabola has been investigated by Reissner[14-30] who predicted the critical load for a simply supported H.P. as

$$p_{cr} = 2E(kt)^2/\sqrt{3(1-\nu^2)} \quad (14\text{-}33)$$

This equation is adequate to predict the buckling load of H.P. shells provided account is taken of imperfections, and creep deformations on the geometry. The supporting members can be designed against buckling by ignoring any stabilizing influence of the shell. The membrane shear load causing buckling of the edge member is

$$N_{xy,\text{crit}} = 18.95 \frac{EI}{a^3} \quad (14\text{-}34)$$

For shells covering larger spans, even as little as 100 X 100 ft, more complete studies should be performed to properly establish the possibility of buckling.

-6. *Bending analysis*—Although membrane theory has been found to be adequate for small- to moderate-span H.P. shells, bending increases in importance as the spans of the shell increase. This is particularly true for the supporting or edge members. The shell and the edge members should no longer be considered as separate structural members. The weight of the supporting members can have a significant effect upon stresses.[14-34] The critical aspect in the design of the structure becomes the design of the support members. Design of the edge member to carry its own weight is an impossible task. The shell assists in carrying some of this weight. To evaluate the bending stresses in the supporting members requires a complex computer program utilizing finite element,[14-28], finite difference[14-23] or other similar numerical method. Reference[14-8] provides a more complete discussion of the available computer methods.

EXAMPLE 14-6: Hyperbolic paraboloid in the form of an inverted umbrella. Has plan of 40 X 50 ft with a rise of 5 ft. Shell thickness is 3.5 in. Average thickness including edge beams is approximately 4.5 in.; $f_c' = 3000$ psi; $f_s = 20{,}000$ psi; loading D.L. + L.L. + roofing is 80 psf.

SOLUTION: Stresses in the umbrella are

$$N_{xy} = \frac{pab}{2f} = \frac{(80)(20)(25)}{2(5)} = 4000 \text{ lbs/ft}$$

TABLE 14-16 Membrane Stresses for Hyperbolic Paraboloid

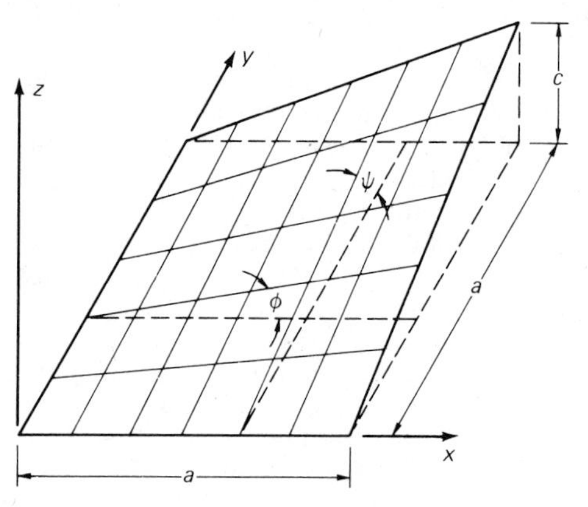

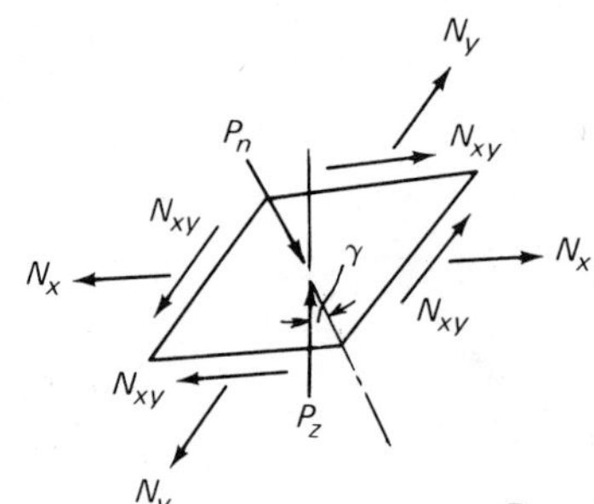

Equation of the surface

$$z = xy/n;\ n = a^2/c$$

N_x, N_y	Unit normal forces
N_{xy}	Unit control shear
$P_n P_z$	Surface load components along the normal and the z axis

Boundary Conditions: $x = 0, N_x = 0, y = 0, N_y = 0$

System	*Loading*	N_x	N_y	N_{xy}
	$p_z = -p_E$	$-p_E \frac{y}{2} \ln \frac{x + \sqrt{x^2 + y^2 + n^2}}{\sqrt{y^2 + n^2}} \cdot \frac{\cos\psi}{\cos\phi}$	$-p_E \frac{x}{2} \ln \frac{y + \sqrt{x^2 + y^2 + n^2}}{\sqrt{x^2 + n^2}} \cdot \frac{\cos\phi}{\cos\psi}$	$\frac{p_E}{2} \sqrt{x^2 + y^2 + n^2}$
	$p_z = p_s \cos\gamma$	0	0	$p_s \frac{a^2}{2c}$
	$p_n = \gamma \left(h - \frac{xy}{n}\right)$	$\frac{\gamma}{2n^2}\left[\frac{x^4}{4} + \frac{x^2}{2}(5y^2 + n^2) - 4nhxy\right] \frac{\cos\psi}{\cos\phi}$	$\frac{\gamma}{2n^2}\left[\frac{y^4}{4} + \frac{y^2}{5}(5x^2 + n^2) - 4nhxy\right] \frac{\cos\phi}{\cos\psi}$	$\frac{\gamma}{2n}\left(h - \frac{xy}{n}\right) \cdot (x^2 + y^2 + n^2)$
	$p_n = p$	$-p \frac{2xy}{n} \frac{\cos\psi}{\cos\phi}$	$-p \frac{2xy}{n} \frac{\cos\phi}{\cos\psi}$	$p \frac{x^2 + y^2 + n^2}{2n}$
	$p_n = p_w \frac{y^2}{x^2 + y^2 + n^2}$	$-p_w \frac{y}{n}\left(x + \frac{y^2}{y^2 + n^2} \cdot \arctan \frac{x}{\sqrt{y^2 + n^2}}\right) \frac{\cos\psi}{\cos\phi}$	$-p_w \frac{x}{y}\left(y - \sqrt{x^2 + n^2} \cdot \arctan \frac{y}{\sqrt{x^2 + n^2}}\right) \frac{\cos\phi}{\cos\psi}$	$p_w \frac{y^2}{2n}$

Required tensile reinforcement in the directions parallel to the edges

$$A_s = \frac{N_1}{f_s} = \frac{1}{2} \times \frac{4000}{20000} \times 1.414 = 0.14 \text{ in.}^2 \text{ both ways}$$

Shell reinforcement: use #3 @ 9 cts. both ways (0.146 in.^{2})

Outer edge beams

$$T = 4000 \times 100{,}000 \text{ lbs}$$

which requires

$$A_s = \frac{100000}{20000} = 5 \text{ in.}^2 \quad \text{(4-\#10 or 5-\#9)}$$

For a larger shell it might be desirable to prestress the edge beams to eliminate the tensile stresses.

Valley beam designed as tied column

Max comp.

$$c = 2(4000)\,(25) \frac{\sqrt{25^2 + 5^2}}{25} = 204{,}000 \text{ lbs}$$

assuming $p_g = 0.01$;

$$A_g = \frac{c}{0.8(.225 f'_c t\, p_g f_s} = \frac{204{,}000}{0.8(.225 \times 3000 + 0.01 \times 20{,}000}$$

$$= 291 \text{ in.}^2$$

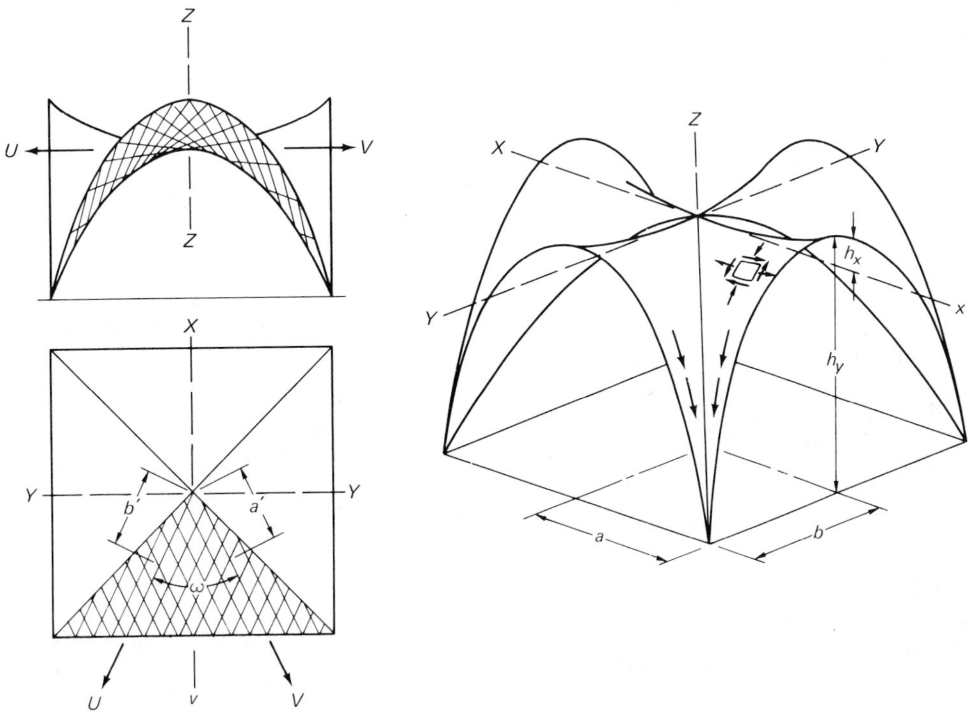

Fig. 14-20 Groined vault.

Required reinforcement

$$0.01(291) = 2.91 \text{ in.}^2 \text{ use 5 \#7}$$

with #3 ties @ 12 in cts. with 16 × 18 section at column head. For an unsymmetrical load on one-half the shell (one-half the design live load assumed).

$$\text{Moment} = p_L\,(2ab)\left(\frac{b}{2}\right) = \frac{1}{2}\,p_L ab^2$$

$$= \tfrac{1}{2}\,(10)\,(20)\,(25)^2 = 62{,}500 \text{ ft-lbs or } 750{,}000 \text{ in.-lbs}$$

Required

$$bd^2 = M/K = 750{,}000/236 = 3180 \text{ in.}^3$$

use d = 16 in., then

$$b = 3180/K^2 = 12.4 \;\therefore\; \text{o.k.}$$

$$A_s = M_s/f_s jd = 750{,}000/20{,}000 \times 0.866 \times 16 = 2.7 \text{ in.}^2$$

requires another 5-#7 or a total of 10-#7. Use 7-#8; 4 at the bottom; 3 at the top

Check secondary stresses per Ref. 14-11.

Check punching shear around column head.

-7. *Groined vault*—One particularly interesting form of the hyperbolic paraboloid is the groined vault configuration shown in Fig. 14-20. This arrangement allows the designer to have a stress-free edge. Therefore a thin trim edge can be exposed to the observer and the dramatic thinness of this form of construction can be displayed. The roof acts as a system of cross arches with the valleys acting as the primary structure to get the loads to the supports. Tables for stress evaluation are given in Ref. 14-11. Also, Ref. 14-6 presents an analysis procedure capable of handling the membrane stresses even for deep groins.

14.5 COMPUTER METHODS

Currently a number of computer programs are available for solving shell problems as elastic structure problems. An elastic solution is the commonly accepted basis for determining stresses, displacements and stability of thin reinforced concrete shells. The majority of the available programs utilize the finite-element method. A large variety of element choices exists. In general for shells, curved elements are preferable to flat elements. Also the higher the number of degrees of freedom per element, usually the more efficiently that element performs the stress analysis. To be useful to the shell designer the program must contain the ability to include beam members placed eccentric to shell. The analysis of the supporting beam members is normally the critical aspect of the structure design.

Because of the reluctance of some finite elements to deform into the deflection shapes that some shells bend into, the accuracy of any finite-element solution should be checked with static checks on gross equilibrium. The suitability of the elements to the type of shell problem should be verified. Many elements are based upon shallow-shell theory and should not be applied to problems well outside that class. Extreme care should be exercised in interpreting results in regions of rapid changes in curvature, or at abrupt changes in direction of the surfaces' tangent planes, such as groins, etc.

If two opposite edges of the shell can be considered simply supported the most efficient solution is by the direct stiffness method described in Refs. 14-35 and 14-36. MULEL[14-35] is such a program written specially for cylindrical shells.

14.6 PROPORTIONING AND REINFORCEMENT

Shell thickness is frequently not based on strength considerations but rather on construction, stability, code, or reinforcing steel cover requirements. Three inches is about the minimum. Four inches may be required if the enclosed environment has high humidity as, for example, over a swimming pool.

Stress concentrations resulting from abrupt changes in

thickness should be avoided. Gradual thickening should be used.

Guidelines on procedures for reinforcing shells were reported by the ACI committee in their *Practice and Commentary*.[14-9]. These guide lines are:

"403. Shell reinforcement

(a) The stress in the reinforcement may be assumed at the allowable value independently of the strain in the concrete.

(b) Where the tensile stresses vary greatly in magnitude over the shell, as in the case of cylindrical shells, the reinforcement capable of resisting the total tension may be concentrated in the region of maximum tensile stress. Where this is done, the percentage of crack control reinforcing in any 12 in. width of shell shall be not less than 0.35 percent throughout the tensile zone.

(c) The principal tensile stresses shall be resisted entirely by reinforcement.

(d) Reinforcement to resist the principal tensile stresses, assumed to act at the middle surface of the shell, may be placed either in the general direction of the lines of principal tensile stress (also referred to as parallel to the lines of principal tensile stress), or in two or three directions. In the regions of high tension it is advisable, based on experience, to place the reinforcing in the general direction of the principal stress.

(e) The reinforcement may be considered parallel to the lines of principal stress when its direction does not deviate from the direction of the principal stress more than 15 deg. Variations in the direction of the principal stress over the cross section of a shell due to moments need not be considered for the determination of the maximum deviation. In areas where the stress in the reinforcing is less than the allowable stress a deviation greater than 15 deg can still be considered parallel placing; a stress decrease of 5 percent shall be considered to compensate for each additional degree of deviation above 15 deg. Wherever possible, such reinforcing may run along lines considered most practical for construction, such as straight lines.

(f) Where placed in more than one direction, the reinforcement shall resist the components of the principal tension force in each direction.

When the reinforcement is placed in two directions at right angles to each other, the principal stress resultant N_p is

$$N_p = f_{sy}(A_{sy} \cos^2 \delta + A_{sx} \sin^2 \delta \tan \delta) \quad (14\text{-}35)$$

where δ is the angle measured in a counterclockwise direction from the face on which N_x acts (angle from the x-axis to the principal stress plane)

$$\tan 2\delta = \frac{2N_{xy}}{N_x - N_y} \quad (14\text{-}36)$$

A_{sy} is the area of steel in the y-direction, A_{sx} that in the x-direction.

(g) In those areas where the computed principal tensile stress in the concrete exceeds 300 psi, placement of at least one layer of the reinforcing shall be parallel to the principal tensile stress, unless it can be proven that a deviation of the reinforcing from the direction parallel to the lines of principal tensile stress is permissible because of the geometrical characteristics of a particular shell and because for reasons of geometry only insignificant and local cracking could develop.

(h) Where the computed principal tensile stress (psi) in the concrete exceeds the value $2\sqrt{f_c'}$ (where f_c' is also in psi), the spacing of reinforcement shall not be greater than three times the thickness of the thin shell. Otherwise the reinforcement shall be spaced at not more than five times the thickness of the thin shell, nor more than 18 in.

(i) Minimum reinforcement shall be provided as required in the Building Code (ACI 318) even where not required by analysis.

(j) The percentage of reinforcement in any 12 in. width of shell shall not exceed 30 f_c'/f_s. However, the maximum percentage shall not exceed 6 percent if f_s = 20,000 psi, 5 percent if f_s = 25,000 psi, or 4 percent if f_s = 30,000 psi when the latter values are acceptable. If the deviation of the reinforcing from the lines of principal stress is greater than 10 deg, the maximum percentage shall be one-half of the above values.

(k) Splices in principal tensile reinforcement shall be kept to a practical minimum. Where necessary they shall be staggered with not more than one-third of the bars spliced at any one cross section. Bars shall be lapped only within the same layer. The minimum lap for shell reinforcing bars, where draped, shall be 30 diameters with a 1 ft 6 in. minimum unless more is required by the Building Code (ACI 318), except that the minimum may be 12 in. for reinforcement not required by analysis. The minimum lap for welded wire fabric shall be 8 in. or one mesh, whichever is greater, except that Building Code requirements shall govern where the wire fabric at the splice must carry the full allowable stress.

(l) The computed stress of the shell reinforcing at the junction of shell and supporting member or edge member shall be developed by anchorage within or beyond the width of the member.

(m) Reinforcement to resist bending moments shall be proportioned and provided in the conventional manner with proper allowance for the direct forces.

(a) The concrete cover over reinforcement at surfaces protected from weather and not in contact with the ground shall be at least 1/2 in. for bars (3/8 in. when precast), 3/8 in. for welded wire fabric, and 1 in. for prestressed tendons. In no case shall the cover be less than the diameter of bar, prestressed tendon, or duct.

(b) If greater concrete cover is required for fire protection, such cover requirements shall apply only to the principal tensile and moment reinforcement whose yielding would cause failure."

14.79 CONSTRUCTION

Since thin shells are frequently used because of their high structural efficiency due care must be exercised in their construction so that conditions not anticipated in design do not occur. In order to accommodate the steel requirements it is rare that a shell less than 3 in. can be cast economically. The design engineer should investigate the effect of variations in shell surface as a result of possible variations in the formwork. Tolerances should be set to define acceptable variations.

Of particular importance is the sequence of decentering. This should be specified by the engineer in order to avoid the occurrence of an unanticipated support condition resulting in a highly concentrated force system. Generally decentering should begin at points of maximum deflections and proceed toward points of minimum deflection. The decentering of edge members should proceed along with that of the adjoining shell. If the edge members are to be reshored, the spacing and sequence should be specified by the design engineer. Support members should be decentered after decentering the shell.

If the shell surface has maximum slopes of less than 45°, use of formwork on both faces is unnecessary. In those regions where the slope exceeds 45° the use of a top form is desirable. In addition, slump of the concrete used, the method of placement, the amount of steel, and the steel pattern have a considerable influence on the behavior of concrete placed on slopes. For shells with very steep slopes, shotcreting or plastering can be used to eliminate the need for the top form.

Relative cost figures for shell construction can be found in Ref. 14-15.

If the shell can not be cast in one day, it should be sectioned into two or more regions. Concrete placed in one region should be allowed two or three days curing before the next region is cast. This procedure is necessary to reduce shrinkage stresses and cracking.

Construction joints should be shown on the design drawings and should be located if possible in regions of compressive stress.

The construction aspects suggested by the ACI committee are:

"501. Aggregate size

(a) The maximum size aggregate shall not exceed one-half the shell thickness, nor the clear distance between reinforcement bars, nor 1½ times the cover. Where formwork is required for two faces, the maximum size of aggregate shall not exceed one-quarter the minimum clear distance between the forms nor the cover over the reinforcement.

502. Forms

(a) Removal of thin shell concrete forms shall be considered a matter of design and the form removal sequence shall be specified or approved by the engineer.

(b) The minimum strength of concrete f'_c, based on field-cured cylinders, at the time of decentering and of reshoring, when required, shall be designated by the engineer.

(c) Where, in the opinion of the designing engineer, stability of short- or long-time deflections are important factors, the modulus of elasticity at the time of decentering, based on field cured beams, shall be specified by the engineer. The proportions and loading of these specimens shall insure action which is primarily flexural.

(d) The batter on vertical elements, or other elements, if desired or required for stripping, and the construction tolerances shall be designated by the engineer. When movable forms are used, a batter of $^1/_8$ in. per ft minimum is recommended for vertical surfaces to permit easy removal."

NOTATION

- a, b = one half the span of elliptical and hyperbolic paraboloids in the x- and y-directions, respectively
- E = Youngs' modulus
- F = Airy stress function used in the analysis of shallow shells
- F_1 to F_4 = functions expressing the decay of boundary disturbances with distance along a coordinate line of a shell of revolution
- H = horizontal component of edge force
- $h_x h_y$ = rise of elliptical parabola in the x- and y-directions respectively
- I = moment of inertia of cross section of beam or cylindrical shell analyzed as a beam
- k = conversion factor to transform projected force components back into shell surface for elliptical parabola also twist of surface for hyperbolic parabola
- L, l = span of shell
- M_φ = bending moment in the shell
- M_θ = bending moment in the shell
- N_φ = normal face in meridian direction
- N_θ = normal force in hoop or circumferential direction
- $N_{\theta\varphi}$ = in plane shear force
- R_1, R_2 = principal radii of curvature
- Q = transverse shear force
- r = radius of curvature
- s = coordinate along generator of conical shell
- t = shell thickness
- u, v = displacement components in tangent plane
- w = displacement component normal to shell surface
- x = coordinate along generator of cylindrical shell
- β = constant whose value is function of the radius to thickness ratio
- γ = specific weight
- Δ_H = horizontal displacement of edge of shell
- Δ_ϕ = rotation of normal at edge of shell
- θ = angular coordinate in circumferential direction for shell of revolution and in hoop direction for cylinder
- ν = Poisson ratio
- φ = angular coordinate measuring the inclination of the normal to the shell surface from the axis of revolution
- ω = angle between referenced axes for hyperbolic parabola

REFERENCES

14-1 Baker, et al., "Shell Analysis Manual," North American Aviation, *NASA Contractor Report CR-912* available through National Technical Information Service, Springfield, Virginia.

14-2 Billington, D. P., *Thin Shell Concrete Structures*, McGraw-Hill, New York, 1965.

14-3 Bleich, H. H., and Salvadori, M. G., "Bending Moments on Shell Boundaries," *Journal of the Structural Division, ASCE*, **85** (ST8), Oct. 1959.

14-4 Bouma, A. L., "On Approximate Methods of Shell Analysis, A General Survey," World Conference on Shell Structures, San Francisco 1962.

14-5 Budiansky, B., and Radkowski, P. P., "Numerical Analysis of Unsymmetrical Bending of Shells of Revolution," *AIAA Journal* **1**, (8), August 1963.

14-6 Candela, F., "General Formulas for Membrane Stresses in Hyperbolic Paraboloidical Shells," *Journal of Amer. Concrete Inst.*, **32** (4), Oct. 1960.

14-7 Chinn, J., "Cylindrical Shell Analysis Simplified by Beam Method," *Journal American Concrete Institute*, **30** (11), May 1959. Also discussion by Parme and Conner, **31** (6), Dec. 1959.

14-8 Clough, R. W. and Johnson, C. P., "Finite Element Analysis of Arbitrary Thin Shells," *American Concrete Institute Special Publication SP-28*, Symposium on Concrete Thin Shells.

14-9 "Concrete Shell Structures, Practice and Commentary," *Journal American Concrete Institute, Proceedings*, **61** (9), Sept. 1964.

14-10 "Design of Cylindrical Concrete Shell Roofs," *ASCE Manual of Engineering Practice No. 31*, ASCE, New York, 1952.

14-11 Elementary Analysis of Hyperbolic Paraboloid Shells, Portland Cement Association, Chicago, 1960.

14-12 Esquillan, N., "The Shell Vault of the Exposition

Palace Paris," *Journal of the Structural Division, ASCE*, **86** (ST1), January 1960.

14-13 Fischer, L., *Theory and Practice of Shell Structures*, Wilhelm Ernst & Sohn, Munich, 1968.

14-14 Flugge, W., *Stresses in Shells*, Springer Verlag, Berlin, 1960.

14-15 Gensert, R. M.; Kirsis, U.; and Peller, M., "Economic Proportioning of Cast-in-Place Concrete Thin Shells," *ACI Special Publication SP-28*, Symposium on Concrete Thin Shells.

14-16 Gibson, J. E., *The Design of Cylindrical Shell Roofs*, E. & F. N. Spon, Ltd., London, 1961.

14-17 Griggs, P. H., "Buckling of Reinforced Concrete Shells," *J. of EMD, ASCE*, **97** (EM3), June 1971.

14-18 Haas, A. M., and Van Koten, H., "The Stability of Doubly Curved Shells Having a Positive Curvature Index," *Herron*, **17** (4), 1970–71.

14-19 Hetenyi, M., "Spherical Shells Subjected to Axial Symmetrical Bending, *IABSE Publications*, **4**, 1937–38.

14-20 Hoff, N. J., "The Effect of Meridian Curvature on the Influence Coefficients of Thin Spherical Shells," *Problems of Continuum Mechanics*, Soc. Ind. and Applied Math., 1961.

14-21 Holland, I., *Design of Circular Cylindrical Shells*, Oslo University Press, Oslo, 1957.

14-22 Jones, R. E., and Strome, D. R., "Direct Stiffness Method Analysis of Shells of Revolution Utilizing Curved Elements," *AIAA Journal*, **4** (9), Sept. 1966.

14-23 Mohraz, B., "A Lumped Parameter Element for the Analysis of Hyperbolic Paraboloid Shells," *Int. J. for Numerical Methods in Engineering*, **4**, 1972.

14-24 NASA, "Collected Papers on Instability of Shell Structures-1962." *NASA Technical Note D-1510.*

14-25 Parme, A. L., "Shells of Double Curvature," *Transactions ASCE*, **123**, 1958.

14-26 Parme, A. L., and Conner, H. W., "Design Constants for Ribless Concrete Cylindrical Shells," *IASS Bulletin No. 18*, June 1964.

14-27 Parme, A. L., and Conner, H. W., "Design Constants for Interior Cylindrical Concrete Shells," *Journal American Concrete Institute*, July 1961.

14-28 Pecknold, D. A., and Schnobrich, W. C., "Finite Element Analysis of Skewed Shallow Shells," *Journal of the Structural Division of ASCE*, **95** (ST4), April 1969.

14-29 Ramaswany, G. S., *Design and Construction of Concrete Shell Roofs*, McGraw-Hill, New York, 1968.

14-30 Reissner, E., "On Some Aspects of the Theory of Thin Shells," *Journal of Boston Society of Civil Engineers*, **42** (2), April 1955.

14-31 Rudiger, D., and Urban, *J. Circular Cylindrical Shells*, B. G. Teubner, Verlag Leipzig, 1955.

14-32 Schmit, H., "Ergebnisse von Beulversuchen mit Doppelt Gekrümmten Schalenmodellen aus Aluminium," Proceedings of the Symposium on Shell Research, North Holland Publishing Co., Amsterdam, 1961.

14-33 Schnobrich, W. C., "Analysis of Hipped Roof Hyperbolic Paraboloid Structures," *Journal of the Structural Division of ASCE*, **97** (ST7), July 1972.

14-34 Schnobrich, W. C., "Analysis of Hyperbolic Paraboloid Shells," *American Concrete Institute Special Publication SP-28*, Symposium on Concrete Thin Shells.

14-35 Scordelis, A. C., and Lo, K. S., "Computer Analysis of Cylindrical Shells," *Journal American Concrete Institute of Proceedings*, **61** (5), May 1969.

14-36 Scordelis, A. C., "Analysis of Cylindrical Shells and Folded Plates," *Concrete Thin Shells*, American Concrete Institute Publication SP 28.

14-37 Scordelis, A. C.; Ramirez, H. D.; and Ngo, D., "Membrane Stresses in Hyperbolic Paraboloid Shells Having an Arbitrary Quadrilateral Shape in Plan," *Journal American Concrete Institute, Proceedings*, **67** (1), Jan. 1970.

14-38 Tedesko, A., "Multiple Ribless Shells," *Journal of Structural Division ASCE*, **87** (ST7), Oct. 1961.

14-39 Toebes, G. H., Disc. of "Wind Stresses in Domes," *Trans. ASCE*, Part I, pp. 854–861, 1962.

14-40 Tester, K. G., "Bertrag zur Berechnung der Hyperbolishen Paraboloidschale," *Ingenieur Archiv*, **16**, 1947.

14-41 Yu, C. W., and Kriz, L. B., "Tests of a Hyperbolic Paraboloid Reinforced Concrete Shell," Proceedings of World Conference on Shell Structures, San Francisco, 1962.

15

Reinforced Concrete Chimneys

WADI S. RUMMAN, Ph.D.*

15.1 INTRODUCTION

The construction of tall, reinforced concrete chimneys has been on the increase in the last two decades, due primarily to the increasing demand for air pollution control. Chimneys in the range of 1250 ft in height have been built (see Figs. 15-1 and 15-2); there is every reason to believe that this trend toward the construction of taller chimneys will continue.

This increasing demand for tall chimneys together with the use of slipform construction has emphasized the advantages of reinforced concrete construction. One primary advantage in a reinforced concrete chimney is that its resistance to wind vibrations is much greater than that of a corresponding steel chimney. In fact very few self-standing steel chimneys exceed 350 ft in height.

Because of the recent changes in the proportions of chimneys, many structural problems such as the response to earthquake and wind forces become more critical. However, the availability of the digital computer makes it possible to analyze chimneys for such forces and to compute stresses for many loading conditions in a quick and economical manner.

Some, but not all, of the advances in chimney design in the last ten years are now incorporated in the recent *ACI Chimney Specification 307-69* "Specification for the Design and Construction of Reinforced Concrete Chimneys".[15-1] This specification contains the basic information necessary to compute stresses due to dead weight, static wind, earthquake forces, and temperature. Reference to the *Specification* is made throughout the chapter. However, many problems, which are not now incorporated in this *Specification*, will be presented.

*Associate Professor of Civil Engineering, University of Michigan, Ann Arbor.

15.2 PROPORTIONING

The height of the chimney as well as the diameter at the top are normally chosen so that exit velocity and dispersion of gases are within the specified limits. The bottom diameter, however, is more frequently controlled by the structural requirements of both the concrete shell and the foundation. For example, a small base diameter can create design problems in the overhang part of the foundation and can also cause high stresses in the concrete stack. Experience has shown that, in general, a ratio of height-to-outside-base-diameter in the range of 12–13 will provide good proportions for the design of both the stack and the foundation. Once the top and bottom diameters are established, the outside diameters can vary linearly along the chimney to provide easier construction without significant loss in economy.

Fig. 15-1 The 1250-ft tall chimney of the International Nickel Co. of Canada, in Sudbury, Ontario. The chimney was designed and constructed by the M. W. Kellogg Co.

The thicknesses of the concrete are usually determined by satisfying the stresses due to static wind and by satisfying the minimum thickness as specified by the *ACI Chimney Specification* which states, "The minimum shell wall thickness for any chimney with an internal diameter of 20 ft or less shall be 7 in. When the internal diameter exceeds 20 ft, the minimum thickness shall be increased 1/4 in. for each 2 ft increase in internal diameter." Although considerable adjustments in the reinforcing steel may be required for earthquake and dynamic wind loading, yet the thicknesses of the concrete as controlled by the static wind loading and the minimum requirements, will remain practically unchanged.

Table 15-1 gives a listing of typical existing chimneys giving the height, the outside top diameter, the outside bottom diameter, and the total weight of the concrete.

15.3 LINERS

The liner, whose primary function is to protect the concrete from the corrosive elements of the flue gases and to provide an adequate exit velocity of the gas, is a very important part of a reinforced concrete chimney. Most of the original concrete chimneys were constructed as shown in Fig. 15-3a using the corbel-supported brick lining and fiber glass or fused silica insulation between the lining and the concrete shell. However, many chimneys of this type have shown signs of acid attack on the concrete shell which is the result of hot gases working their way through the brick lining and their condensation on the relatively cold concrete. The condensate is acidic due to the sulfuric content of the fuel. A good solution to the acid problem is the use of independent steel liners as shown in Fig. 15-3b. Many chimneys have been built with steel liners in the last ten years or so and their performance insofar as furnishing the protection against acid attack has been excellent.

Most of the steel liners have been supported at, or near, the bottom although a few are supported near the top. One of the advantages of top support is that the liner is in tension rather than compression; but a disadvantage is that the expansion due to temperature will have to be taken up by a joint near the flue opening. Bottom-supported liners, on the other hand, are susceptible to buckling above the flue opening not only due to the loading, but also due to temperature differentials between one side of the liner and another.

Other types of liners are the independent concrete liner and the independent brick liner. The independent brick liner is not commonly used in tall chimneys.

TABLE 15-1 Basic Data for Typical Chimneys

Chimney No.	*Height (ft)*	*Top Outside Diameter (ft)*	*Bottom Outside Diameter (ft)*	*Total Weight of Concrete Stack (kips)*	*Ratio of Height to Outside Base Diameter*
1	352	23.58	30.90	4530	11.4
2	450	16.33	35.79	6740	12.6
3	534	18.67	35.03	8370	15.2
4	622	23.33	47.26	12530	13.2
5	800	36.56	65.00	25030	12.3
6	825	25.00	63.96	22970	12.9
7	828	28.71	67.80	22850	12.2
8	900	34.25	72.00	33800	12.5
9	1000	33.67	83.00	–	12.0
10	1200	37.00	95.29	64550	13.1

15.4 WIND LOADING

The wind loading, both static and dynamic, is an important factor in chimney design. The *ACI Chimney Specification* provides for the design of chimneys for static loading but does not cover the dynamic effects of the wind.

15.4.1 Static Wind

The design static wind pressures for circular chimneys are specified in the *ACI Chimney Specification* and are given in Table 15-2.

The *Specification* specifies that "in special locations where records or experience indicate that these pressures are in-

TABLE 15-2 Design Static Wind Pressures for Chimneys with Circular Cross Section

	Wind Pressure–Map Areas, psf				
Height Zone (ft)	*30 or less*	*35*	*40*	*45*	*50*
0–100	23	23	26	29	32
100–500	31	33	36	42	45
500–1000	34	36	42	48	54

Fig. 15-2 The chimney in Fig. 15-1 during slipforming.

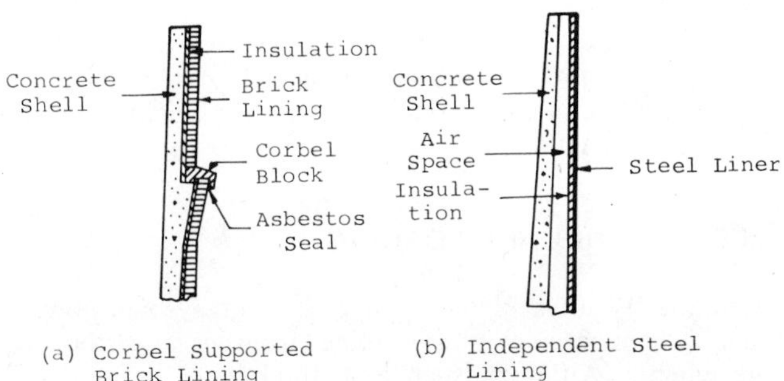

Fig. 15-3 Concrete chimneys with two types of lining.

adequate, higher pressures may be used at the discretion of the owner's engineers," and also adds that "the job specifications should indicate the wind pressure zone to be used for design of the chimney, to avoid misinterpretations of borderline map locations."

15.4.2 Radial Pressure

In addition to static wind pressures which are applied on the projected area of the chimney, the designer is also confronted with the effects of the radial pressure distribution around the circumference of the chimney. Experiments conducted on circular cylinders and stacks,[15-2,15-3] have shown that the distribution of the wind pressures around the circumference is as shown in Fig. 15-4.

The radial forces, which will have a resultant force (F) acting to the right, are resisted by shearing forces around the circumference as shown in the sketch. These shearing forces which are assumed to vary sinusoidally around the circumference should have a resultant equal to F and acting opposite to it.

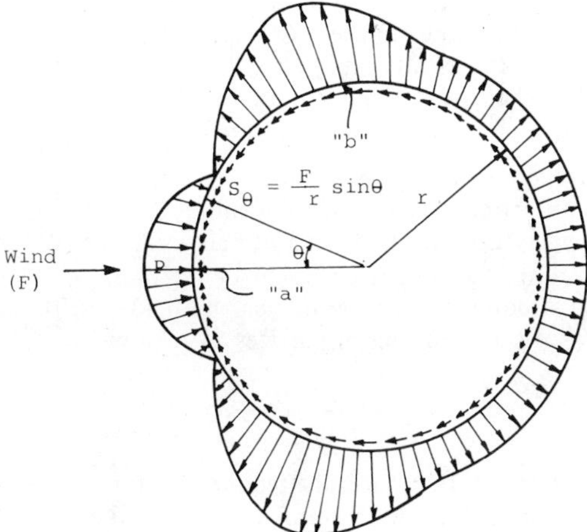

Fig. 15-4 Distribution of wind pressure on ring.

Setting

$$S_\theta = A \sin \theta \qquad (15\text{-}1)$$

we obtain from the equilibrium condition

$$F = 2 \int_0^\pi (A \sin \theta)\, r\, d\theta \qquad (15\text{-}2)$$

TABLE 15-3 Maximum Ring Moments

Source	Section a (Fig. 15-4) Tension on Inside	Section b (Fig. 15-4) Tension on Outside
Erdie & Ghosh	0.354 pr^2	0.311 pr^2
Diver	0.284 pr^2	0.256 pr^2
Rumman	0.314 pr^2	0.272 pr^2

from which

$$A = \frac{F}{\pi r} \tag{15-3}$$

The critical bending moments in the ring as obtained by different sources are tabulated in Table 15-3. Erdei and Ghosh[15-4] used a radial distribution of pressure based on the experiments of Roshko.[15-3] Diver[15-5] used the radial pressure distribution recommended by the French *Code*,[15-6] while the writer has used radial pressure distribution based on the experiments of Dryden and Hill.[15-2] The ASCE Task Committee on Wind Forces[15-7] recommends similar pressure distribution.

15.4.3 Lateral Swaying and Ovalling

The lateral swaying (vibration perpendicular to the direction of the wind) and ovalling of the circular cross section, due to periodic shedding of vortices have occurred in steel stacks especially in welded stacks without gunite. Although serious problems have occurred in steel stacks due to resonance conditions no such problems have been reported in concrete chimneys. The variable diameter, the larger mass, and the higher damping are factors contributing to the resistance of concrete chimneys to the lateral swaying and ovalling. Although a condition of resonance is not as likely to occur in concrete chimneys as it is in the more flexible structures, nevertheless, it is important that some computations be made to estimate possible resonance conditions. In other words it is advisable that concrete chimneys be designed for unusual wind conditions or site locations that could cause adverse lateral vibrations.

-1. Lateral swaying–The following procedures which are based on articles written on the subject[15-7 to 10] have been successfully used in the design of many chimneys.[15-11,15-12]

The chimney is considered to be in resonance when its natural frequency, $\omega/2\pi$ in cps, coincides with the frequency of the shedding of vortices SV/D_c, or

$$\frac{\omega}{2\pi} = \frac{SV}{D_c} \tag{15-4}$$

in which ω = the natural frequency of the chimney in radians/sec; S = Strouhal number (can be taken as 0.2); V = velocity of wind in ft/sec; and D_c = critical outside diameter of chimney.

A periodic forcing function, f, is assumed based on a critical velocity, V_c, computed from eq. (15-4). This force will take the form

$$f = C_L \frac{\rho V_c^2}{2} D \sin \omega t \tag{15-5}$$

in which f = transverse force per unit length; ρ = mass density of air; D = outside diameter; and C_L = lift coefficient.

It can be shown that if a chimney is subjected to the above forcing function, f, whose frequency coincides with that of the fundamental frequency of the chimney, then the maximum displacements, shears, and moments are obtained by multiplying their respective modal values by the multiplier, q, given by

$$q = \frac{\rho}{16\pi^2 S^2} \frac{C_L}{\beta} D_c^2 \frac{\int_0^H D\phi dx}{\int_0^H m\phi^2 dx} \tag{15-6}$$

in which β = fraction of critical damping; m = mass/unit length along the chimney; and ϕ = mode shape.

It should be mentioned that the approximate response as outlined above is a first-mode response based on a uniform forcing function. A research project which is being undertaken now at the University of Michigan under the direction of the author is treating the response as a combination of three modes and using a variable lateral forcing function. Preliminary results seem to indicate that correction for higher modes can be achieved by using a variable factor by which the bending moments should be multiplied to obtain the multimode response. This variable factor is expected to be higher than one in the upper portions of the chimney and lower than one in the lower portion. The value of the multiplier would depend primarily on the specified design wind velocity.

-2. Ovalling of circular cross sections–If the period of a circular ring coincides with half of the period of the shedding of vortices, then ovalling of the cross section could occur. It has been shown[15-12] that the velocity that could cause ovalling vibrations in the fundamental mode is given by the following equation:

$$V = 366 \left(\frac{t}{r}\right) \tag{15-7}$$

in which V = the wind velocity in miles/hr; r = the radius of the chimney cross section in ft; and t = the thickness of the cross section in in.

15.5 EARTHQUAKE LOADING

With the use of a digital computer, chimneys can now be analyzed for accelerograms obtained from actual earthquake movements. Although such a method or a method that utilizes response spectra for earthquakes is recommended for final design, yet the designer has now at his disposal certain simplified procedures which are useful for preliminary design. One such procedure is contained in *ACI Specification 307-69* and another similar procedure which is based on studies made on the response of many actual reinforced concrete chimneys to recorded accelerograms of actual earthquakes[15-13,15-14] will be presented in this chapter.

The main steps in this procedure are to determine the period of the fundamental mode of vibration and the base shear, and then to distribute the base shear as lateral forces along the chimney from which bending moments can be computed. This bending moment curve is then transformed by a variable multiplier into design earthquake moments.

15.5.1 Period

The period of the fundamental mode can be computed by hand using the Stodola process. If the chimney is assumed to be linearly tapered in both diameter and thickness, then the curves obtained by the writer[15-15] or those obtained by Housner and Keightley[15-16] can be used. The design example at the end of the chapter will utilize the Stodola process in computing the fundamental mode.

15.5.2 Base Shear

From studying the response of many chimneys to accelerograms of many earthquakes,[15-13,15-14] a relationship was established between the ratio of base shear-to-total weight and the period of the fundamental mode. This relationship can be expressed approximately by the following equation:

$$\frac{V_b}{W} = \frac{0.21}{\sqrt[3]{T^2}} \times R \tag{15-8}$$

in which V_b = base shear; W = total weight; T = first mode period in sec/cycle; and R = a factor which depends on the location of the chimney and on the risk that the owner is willing to take. A value for this factor ranging approximately from 0.6 to 1.6 is recommended.

15.5.3 Shear Distribution

The shear distribution along a chimney due to earthquakes is represented approximately by the curve shown in Fig. 15-5, which is given in terms of the base shear. From the analysis of many tapered, reinforced concrete chimneys, it was found that the most critical section for shearing stresses is about one tenth from the top. As shown in Fig. 15-5, the shearing force at this critical section would be about 25% of the base shear. In many cases this percentage is on the high side especially for short chimneys.

15.5.4 Earthquake Bending Moments

Using the results of the studies referred to earlier, the following steps for the determination of approximate bending moments are recommended.

Step 1 Distribute the base shear along the chimney so that the lateral force at any level is proportional to the weight at that level times the square of the height from the base. Referring to Fig. 15-6 the force at any distance, h, from the base is given by

$$F_h = V_b \frac{W_h h^2}{\sum W_h h^2} \tag{15-9}$$

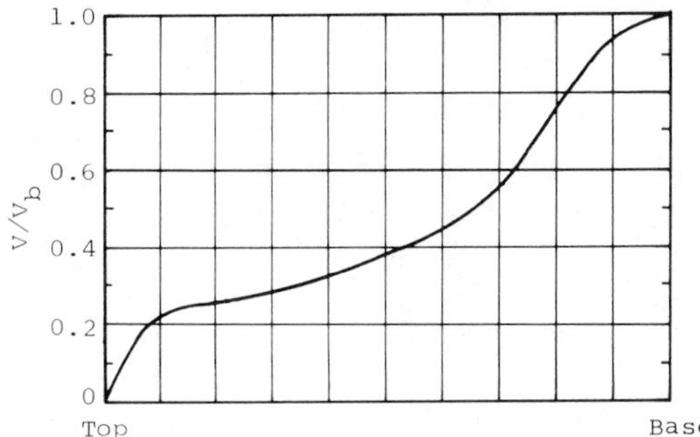

Fig. 15-5 Normalized approximate earthquake shear curves.

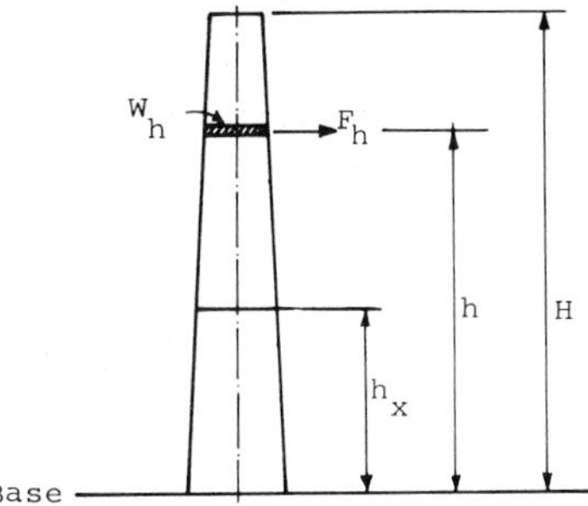

Fig. 15-6 Representation of symbols for earthquake analysis.

Step 2 The bending moment at any level, h_x, is then computed by the following formula:

$$M_x = J_x \sum_{h=h_x}^{H} F_h (h - h_x) \tag{15-10}$$

Step 3 The term J_x is represented approximately by the curves in Fig. 15-7. It is a variable multiplier whose magnitude is such that, for the same base shear, the moment curve of eq. (15-10) will coincide with the moment curve obtained from actual earthquake response. Although no two chimneys will have identical J_x curves, the results of modal response calculations on many chimneys[15-13] give J_x curves that follow the general trend and approximate values of Fig. 15-7.

15.6 STRESSES

The following formulas are incorporated in the *ACI Specification* for chimneys and are repeated here for the convenience of the reader. For a complete coverage the reader should refer to the *Specification.*

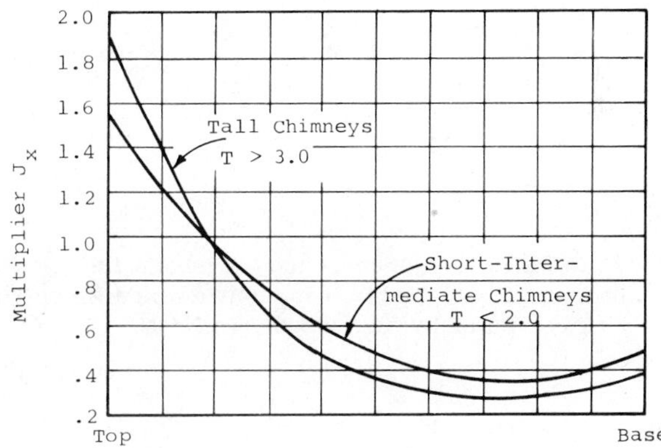

Fig. 15-7 Multiplier for earthquake moments.

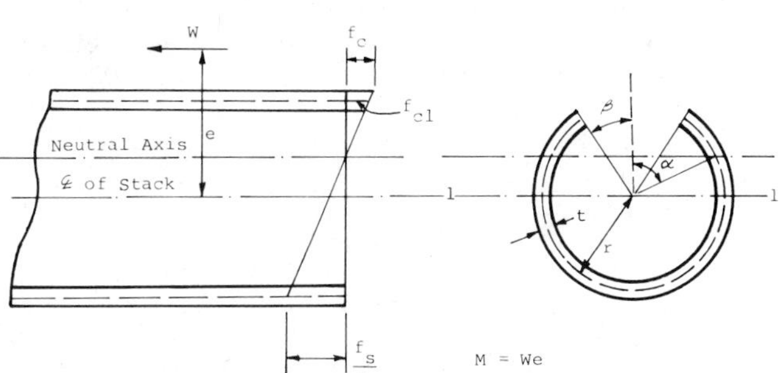

Fig. 15-8 Cross section subjected to bending and axial forces.

15.6.1 Stresses Due to Dead Weight and Moment

For a horizontal section subjected to a bending moment about axis 1-1 in Fig. 15-8 and also to a normal force (W), the maximum compressive stress in the concrete and the maximum tensile stress in the steel are computed by the following formulas:

$$f_{c1} = \frac{W(\cos\beta - \cos\alpha)}{2rt\,[(1-p)(\sin\alpha - \alpha\cos\alpha) - (1-p+np)(\sin\beta - \beta\cos\alpha) - np\pi\cos\alpha]} \qquad (15\text{-}11)$$

$$f_c = f_{c1}\left[1 + \frac{t}{2r\cos\beta(\cos\beta - \cos\alpha)}\right] \qquad (15\text{-}12)$$

$$f_s = nf_{c1}\left[\frac{1+\cos\alpha}{\cos\beta - \cos\alpha}\right] \qquad (15\text{-}13)$$

where

W = total normal load acting on the section under consideration
r = mean radius of chimney shell at section under consideration
t = thickness of chimney shell at section under consideration
p = ratio of total area of vertical reinforcement to total area of concrete of chimney shell at section under consideration
n = ratio of modulus of elasticity of the reinforcement to the modulus of elasticity of the concrete
α = one-half of the central angle subtended by the neutral axis as a chord on the circle of radius, r
β = one-half the central angle subtended by the opening as a chord on the circle of radius, r
$e = M/W$

The angle α is determined from the following equation:

$$\frac{e}{r} = \frac{(1-p)(\alpha - \sin\alpha\cos\alpha) - (1-p+np)(\beta + \sin\beta\cos\beta - 2\sin\beta\cos\alpha) + np\pi}{2[(1-p)(\sin\alpha - \alpha\cos\alpha) - (1-p+np)(\sin\beta - \beta\cos\alpha) - np\pi\cos\alpha]} \qquad (15\text{-}14)$$

The formulas listed above can also be used when a second opening diametrically opposite to the one in Fig. 15-8 exists. In case of no openings the value of β is equal to zero and the above equations are considerably simplified.

15.6.2 Stresses Due to Temperature

The vertical stress in the concrete (f_{CTV}) and the vertical stress in the steel (f_{STV}) due to a temperature differential, T_x, can be computed by the following formulas:

$$f_{CTV} = Lk\,T_xE_c \qquad (15\text{-}15)$$

and

$$f_{STV} = L(z-k)T_xE_s \qquad (15\text{-}16)$$

where

L = thermal coefficient of expansion of the concrete and of the reinforcing steel, to be taken as 0.0000065 per °F
E_c = modulus of elasticity of the concrete
E_s = modulus of elasticity of the steel
$k = -pn + \sqrt{pn\,(pn + 2z)}$
p = ratio of total area of vertical temperature reinforcement to total area of concrete of chimney shell at section under consideration
z = ratio of distance between inner surface of chimney shell and vertical temperature reinforcement to total shell thickness, t

T_x, which is the magnitude of the temperature differential across the concrete, depends on such factors as the type of liner, the insulation, the thickness of the concrete and the space between the liner and the concrete shell. The *ACI Chimney Specification* gives formulas for different conditions. For example, the temperature differential for the lined chimney with unventilated space between the lining and the shell, may be computed from the formula

$$T_x = \frac{tD_{bi}}{C_cD_c}\left[\frac{T_g - T_o}{\frac{1}{K_1} + \frac{t_bD_{bi}}{C_bD_b} + \frac{D_{bi}}{K_rD_s} + \frac{tD_{bi}}{C_cD_c} + \frac{D_{bi}}{K_2D_{co}}}\right] \qquad (15\text{-}17)$$

where

T_g = maximum temperature of gas inside chimney, °F
T_o = minimum temperature of outside air surrounding chimney, °F
D_{bi} = inside diameter of lining, ft
D_b = mean diameter of lining, ft
D_s = mean diameter of space between lining and shell, ft
D_c = mean diameter of concrete chimney, ft
D_{co} = outside diameter of concrete chimney, ft
C_c = coefficient of thermal conductivity of the concrete chimney shell, Btu per sq ft per in. of thickness per hr per °F difference in temperature (use 12 for concrete)
K_1 = coefficient of heat transmission from gas to inner surface of chimney lining, Btu per sq ft per hr per °F difference in temperature
C_b = coefficient of thermal conductivity of chimney lining, Btu per sq ft per in. of thickness per hr. per °F difference in temperature
t = thickness of concrete shell, in.
t_b = thickness of lining, in.
K_r = coefficient of heat transfer by radiation between outside surface of lining and inside surface of concrete chimney shell, Btu per sq ft per hr per °F difference in temperature ($K_r = T_g/120$)
K_2 = coefficient of heat transmission from outside surface of chimney shell to surrounding air, Btu per sq ft per hr per °F difference in temperature (K_2 = 12)

The horizontal stresses in the concrete and the steel due to temperature are computed by similar formulas as those used for vertical stress computations.

15.6.3 Stresses Due to Combined Temperature and Load

The maximum vertical stresses in the concrete due to the bending moment and the axial load combined with temperature are computed by the following formulas:

$$f_{c,comb} = \frac{f_{CTV}\,k_{comb}}{k};\quad k_{comb} \leqslant 1 \qquad (15\text{-}18)$$

$$f_{c,comb} = f_{c1} + \frac{f_{CTV}}{k}\left[\frac{2\,pnz + 1}{2\,(1+pn)}\right];\quad k_{comb} \geqslant 1 \qquad (15\text{-}19)$$

where

$$k_{comb} = -pn + \sqrt{pn(pn + 2z) + 2k(1 + pn)\frac{f_{c1}}{f_{CTV}}} \quad (15\text{-}20)$$

The maximum combined vertical stress in the steel is calculated from

$$f_{s,comb} = \frac{f_{STV}}{z - k}\left[z + pn - \sqrt{pn(pn + 2z) - 2pn(z - k)\frac{f_s}{f_{STV}}}\right] \quad (15\text{-}21)$$

except that when $f_s/f_{STV} \geqslant z/(z - k)$, the stress is the same as that given by eq. (15-13).

15.7 WORKING STRESS AND MAXIMUM STRESS DESIGNS

The moments and allowable stresses as specified in the *Chimney Specification*, as well as the moments discussed in this chapter, are those used in conjunction with a working stress design. The designation 'maximum stress design' is used in this chapter to designate a design that will keep the maximum tensile stress in the steel and the maximum compressive stress in the concrete within certain upper limits when the bending moments used in the working stress design are multiplied by a load factor.

The reason why a maximum stress design is desirable is due to the nonlinear variation of the stresses (especially in the steel) with respect to the change in the bending moment. This nonlinearity of the stresses is due to a change in the bending moment without any corresponding change in the normal force. Figure 15-9 illustrates how the stress in the steel can increase at a much faster rate than the increase in the bending moment. The stresses for a maximum stress design are computed by the same basic formulas that are used for the working stress design.

For a maximum stress design, it is recommended that the working load bending moments be multiplied by a factor of approximately 1.5 except for the resonant wind condition where a factor of 1.25 is recommended. The allowable stresses in the concrete and the steel due to the maximum loading condition can be increased considerably to values close to the yield stress. (About 0.85 f_c' for the concrete and about f_y for the steel.)

EXAMPLE 15-1: Properties of example chimney–To illustrate the use of the equations and curves presented in this chapter, the following example of chimney computations is given. The chimney to be analyzed is 900 ft high and has a constant taper with outside diameters varying from 34.25 ft at the top to 72.0 ft at the bottom. The ratio of height to outside bottom diameter is 12.7. The outside diameters, thicknesses and moments of inertia are tabulated at eleven equally spaced locations in Table 15-4.

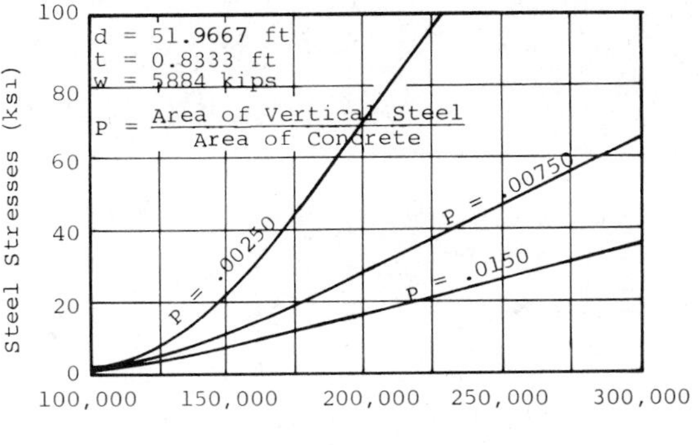

Fig. 15-9 Steel stresses vs moments for steel ratio, *P*.

TABLE 15-4 Properties of Example Chimney

Distance from Top (ft)	*Outside Diameter (ft)*	*Thickness (in)*	*Moment of Inertia (ft^4)*
0	34.25	9.00	11400
90	38.03	9.50	16600
180	41.80	9.88	23000
270	45.58	10.75	32400
360	49.35	12.13	46300
450	53.13	14.00	66400
540	56.90	16.75	96800
630	60.68	18.75	131000
720	64.45	22.00	182700
810	68.23	39.00	362300
900	72.00	32.00	360900

SOLUTION:

Step 1 –Static wind moments–This chimney will be designed for wind pressures corresponding to Zone 30 or less as given by Table 15-2. A complete set of the computations for wind moments, wind shears and wind deflections is given in Fig. 15-10. The computations are primarily based on Newmark's "Numerical Procedure"[15-17] which is very well adaptable to this type of problem. Because this procedure will be used again in the example in conjunction with the first mode computations, it will be helpful to cover the basic elements of the method.

If a beam, as for example, a cantilever beam, is subjected to a distributed loading as shown in Fig. 15-11a, then the shears and bending moments in the beam can be obtained by dealing with the equivalent concentrated loads as shown in Fig. 15-11b.

Although the shapes of the shear and bending moment diagrams are not the same in the two systems, it should be apparent that the shears and moments at locations 1, 2, 3, and 4 in the actual system would be the same as those in the equivalent system. Once the values of the equivalent concentrated loads are determined, the computations for the shears and bending moments become rather simple. Another simplification is used in which the equivalent concentrated load at a location is given in whole rather than in parts. In this case, although the shear values can not be determined at the locations, their values at sections midway between the locations would be approximately equal to the constant shear value in the equivalent system.

The equivalent concentrated loads for any load distribution can be obtained by assuming a parabolic curve for the distribution and by using equal elements. The formulas for these concentrated loads are given in Fig. 15-12.

In Fig. 15-10, the 900 ft chimney is divided into ten equal segments. Lines 1 and 2 give the wind pressure values and the outside diameters respectively. The equivalent concentrated loads are given in line 4. Shears and moments are then calculated in lines 5 and 6. The second phase of the computations is to calculate the deflections which is accomplished by using a conjugate beam loaded by the M/EI diagram and fixed at the right end. Lines 9, 10 and 11 give the equivalent concentrated loads, the shears and the bending moments in the conjugate beam which will be equivalent to angle changes, slopes and deflections in the real beam. By using the proper multiplier, the bending moments in ft-kips are given in line 12, the deflections in ft are given in line 13 and the shears in kips are given in line 14.

Step 2 First mode; shape, shears, and moments–The computations for the first mode of vibration are given in Fig. 15-13. The chimney is subjected to the weight of the concrete given in line 1 and also to additional concentrated loads (line 2) resulting from a steel liner.

The process involves assuming a mode shape (line 3) and computing inertia loads (lines 4 and 5). The inertia load is equal to the load

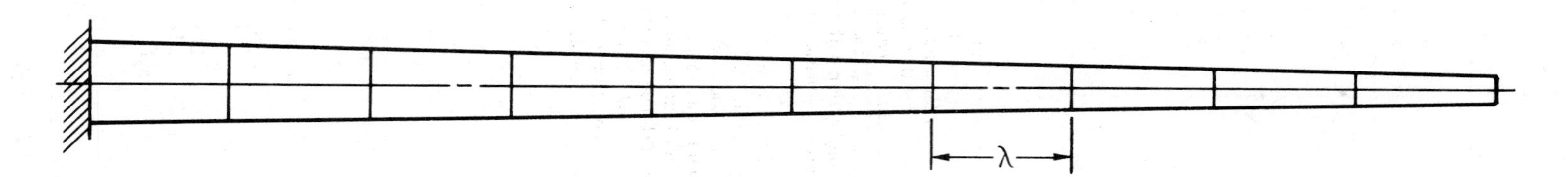

												Multiplier	
P	0.023	0.023	0.031	0.031	0.031	0.031	0.034	0.034	0.034	0.034	0.034	k/ft^2	1
D	72.00	68.23	64.45	60.68	56.90	53.13	49.35	45.58	41.80	38.03	34.25	ft	2
PD	1.656	1.569	1.998	1.881	1.764	1.647	1.678	1.550	1.421	1.293	1.165	k/ft	3
Equiv. Conc. Load		19.344	23.430	22.572	21.168	19.912	19.977	18.599	17.053	15.516	7.246	$\frac{\lambda}{12} \cdot \frac{k}{ft}$	4
V	184.817	165.473	142.043	119.471	98.303	78.391	58.414	39.815	22.762	7.246		$\frac{\lambda}{12} \cdot \frac{k}{ft}$	5
M	916.7	731.9	566.4	424.4	304.9	206.6	128.2	69.8	30.0	7.2	0	$\frac{\lambda^2}{12} \cdot \frac{k}{ft}$	6
I	36.09	36.23	18.27	13.10	9.68	6.64	4.63	3.24	2.30	1.66	1.14	I_0	7
$\frac{M}{EI}$	25.40	20.20	31.00	32.40	31.50	31.11	27.69	21.54	13.04	4.34	0	$\frac{\lambda^2}{12\, EI_0} \cdot \frac{k}{ft}$	8
Angle Change	134.00	258.40	362.60	386.50	378.51	370.29	329.55	256.13	156.28	56.44		$\frac{\lambda^3}{144\, EI_0} \cdot \frac{k}{ft}$	9
Chord Slope	134	392	755	1142	1520	1890	2220	2476	2632	2689		$\frac{\lambda^3}{144\, EI_0} \cdot \frac{k}{ft}$	10
Defl.	0	134	526	1281	2423	3943	5833	8053	10529	13161	15850	$\frac{\lambda^4}{144\, EI_0} \cdot \frac{k}{ft}$	11
B.M.	618773	494033	382320	286470	205808	139455	86535	47115	20250	4860	0	ft – k	12
Defl.	0	0.011	0.043	0.105	0.198	0.323	0.477	0.659	0.862	1.077	1.297	ft	13
Shear	1386	1241	1065	896	737	588	438	299	171	54		k	14

$\lambda = 90'$ $E = 3865200$ psi $I_0 = 10000$ ft^4

Fig. 15-10 Static wind computations.

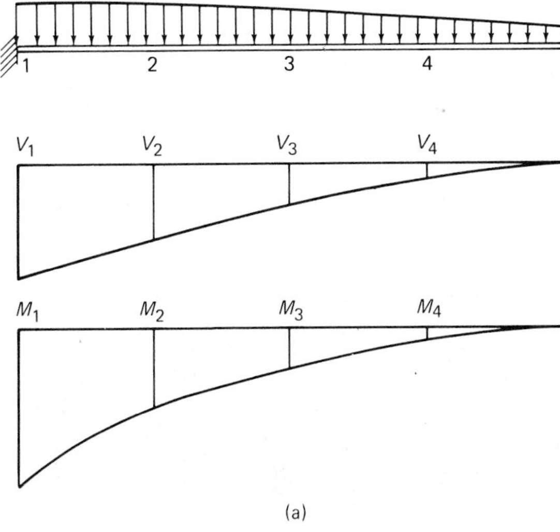

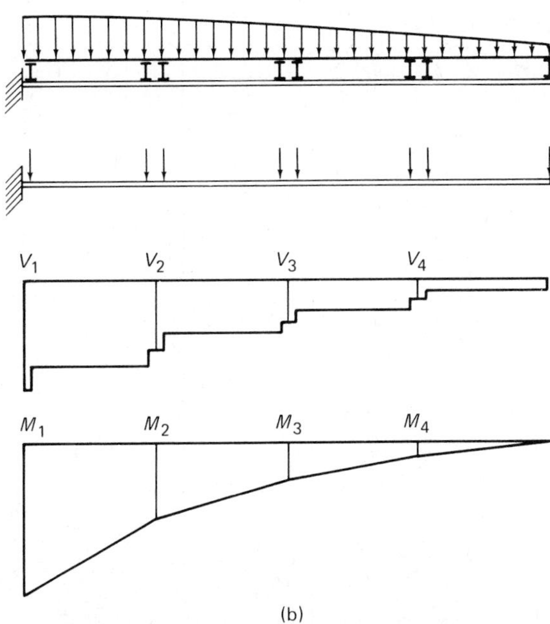

Fig. 15-11 Actual and equivalent loading systems.

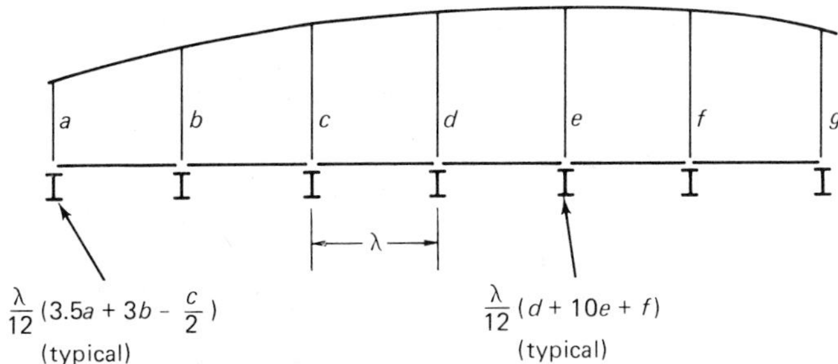

Fig. 15-12 Formulae for equivalent concentrated loads.

times the deflection times the square of the frequency ω (radians per second) and divided by the acceleration of gravity, g. The total equivalent concentrated load given in line 8 is used to calculate the deflections given in line 15. This deflection curve is then normalized (line 16) by making the top displacement equal to unity. This normalized shape is then compared to the assumed shape of line 3. If line 16 agrees reasonably well with line 3, then the process is terminated and the frequency, ω, is obtained by equating the values of line 3 to those of line 15. If the shape of line 16 does not agree with the shape of line 3, then the process is repeated starting with line 4 with an assumed shape equivalent to line 16.

In most cases one cycle is enough, especially if a reasonable shape is assumed and if ω is computed as the average of values obtained from different locations along the chimney.

Step 3 Lateral vibrations due to vortex shedding–A resonant wind velocity, based on the diameter of the chimney at about one third the distance from the top, is computed by eq. (15-4). Using the value of ω of 1.82 radians/second and a diameter of 46 ft, one obtains for a Strouhal number of 0.2 a resonant wind velocity of 66.5 ft/sec, or 45.5 mph.

The next step is to compute the value of the quotient $\int_o^H D\,\phi\,dx / \int_o^H m\,\phi^2\,dx$. The computations are made in a tabular form and are given in Table 15-5. The value of the integral $\int D\,\phi\,dx$ is the area under the function $D\phi$ and is obtained by summing up the equivalent concentrated loads computed using the formulas of Fig. 15-12. The integral $\int m\,\phi^2\,dx$ is obtained in a similar manner.

From the table

$$\int_o^H D\,\phi\,dx = \frac{\lambda}{12} \times 1689.22 = 12669 \text{ ft}^3$$

and

$$\int_o^H m\,\phi^2\,dx = \left[\frac{\lambda}{12} \times 375.20 + 367.24\right] \frac{1}{32.2} = 98800 \text{ lb-ft-sec}^2$$

TABLE 15-5 Calculations of $\int_o^H D \cdot \phi\,dx$ and $\int_o^H m \cdot \phi^2 dx$

D (ft)	*φ (ft)*	*Load Intensity W, (k/')*	*Conc. Load p (k)*	*Dφ*	*Equiv. Dφ / λ/12*	*Wφ²*	*pφ²*	*Equiv. Wφ² / λ/12*
34.25	1.000	11.84	50	34.25	199.83	11.84	50.00	66.10
38.03	.820	13.89	95	31.18	373.22	9.34	77.90	111.95
41.80	.650	15.89	116	27.17	324.76	6.71	75.40	80.79
45.58	.480	18.86	116	21.88	262.75	4.35	55.68	52.87
49.35	.340	23.02	120	16.78	201.90	2.66	40.80	32.46
53.13	.230	28.57	139	12.22	146.95	1.51	31.97	18.48
56.90	.140	36.51	139	7.97	96.17	.72	19.46	8.92
60.68	.070	43.53	147	4.25	52.40	.21	10.29	2.87
64.45	.030	54.10	166	1.93	24.03	.05	4.98	.71
68.23	.007	99.51	108	.48	6.73	0	.76	.05
72.00	0	87.13	0	0	.48	0	0	0
				Σ	1689.22		367.24	375.20

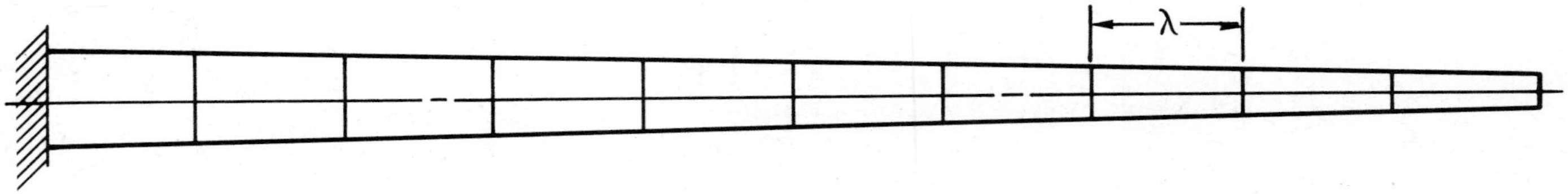

												Multiplier	
Load Intensity	87.13	99.51	54.10	43.53	36.51	28.57	23.02	18.86	15.89	13.89	11.84	k/ft	1
Conc. Load	0	108	166	147	139	139	120	116	116	95	50	k	2
Assumed Defl. Curve	0	0.008	0.03	0.07	0.15	0.23	0.35	0.50	0.65	0.82	1.00	y_t	3
Inertia Load Intensity	0	0.80	1.62	3.05	5.48	6.57	8.06	9.43	10.33	11.39	11.84	$\frac{\omega^2}{g} y_t \cdot \frac{k}{ft}$	4
Inertia Conc. Load	0	0.86	4.98	10.29	20.85	31.97	42.00	58.00	75.40	77.90	50.00	$\frac{\omega^2}{g} y_t \cdot k$	5
Inertia Conc. Load	0	0.11	0.66	1.37	2.78	4.26	5.60	7.73	10.05	10.39	6.67	$\frac{\lambda}{12} \frac{\omega^2}{g} y_t \cdot \frac{k}{ft}$	6
Equiv. Conc. Load of (4)	0	9.62	20.05	37.60	64.42	79.24	96.6	112.69	124.12	136.07	70.45	$\frac{\lambda}{12} \frac{\omega^2}{g} y_t \cdot \frac{k}{ft}$	7
(6 + 7)		9.73	20.71	38.97	67.20	83.50	102.20	120.42	134.17	146.46	77.12	$\frac{\lambda}{12} \frac{\omega^2}{g} y_t \cdot \frac{k}{ft}$	8
V	800.48	790.75	770.04	731.07	663.87	580.37	478.17	357.75	223.58	77.12		$\frac{\lambda}{12} \frac{\omega^2}{g} y_t \cdot \frac{k}{ft}$	9
M	5473	4673	3882	3112	2381	1717	1137	658	301	77	0	$\frac{\lambda^2}{12} \frac{\omega^2}{g} y_t \cdot \frac{k}{ft}$	10
I	36.09	36.23	18.27	13.10	9.68	6.64	4.63	3.24	2.30	1.66	1.14	I_0	11
$\frac{M}{EI}$	151.6	129.0	212.5	237.6	246.0	258.6	245.6	203.1	130.9	46.4	0	$\frac{\lambda^2}{12\,EI_0} \frac{\omega^2}{g} y_t \cdot \frac{k}{ft}$	12
Angle Change	811	1654	2492	2835	2956	3078	2918	2408	1559	595		$\frac{\lambda^3}{144\,EI_0} \frac{\omega^2}{g} y_t \cdot \frac{k}{ft}$	13
Chord Slope	811	2465	4957	7792	10748	13826	16744	19152	20711	21306		$\frac{\lambda^3}{144\,EI_0} \frac{\omega^2}{g} y_t \cdot \frac{k}{ft}$	14
Deflection	0	811	3276	8233	16025	26773	40599	57343	76495	97206	118512	$\frac{\lambda^4}{144\,EI_0} \frac{\omega^2}{g} y_t \cdot \frac{k}{ft}$	15
Defl. Shape	0	0.007	0.03	0.07	0.14	0.23	0.34	0.48	0.65	0.82	1.00	y_t	16
M	380899	325222	270172	216583	165708	119496	79131	45794	20948	5359	0	$y_t \cdot k$	17
V	619	611	595	565	513	449	370	277	173	60		$y_t \cdot \frac{k}{ft}$	18

$\lambda = 90'$ $\quad E = 3865200$ psi $\quad I_0 = 10000$ ft^4

$$118512 \frac{\lambda^4}{144\,EI_0} \frac{\omega^2}{g} y_t = y_t \qquad \omega^2 = 3.32\ (\text{radians/sec})^2 \qquad \omega = 1.82\ \text{radians/sec}$$

Fig. 15-13 First mode computations.

$$\frac{\int_o^H D\,\phi\,dx}{\int_o^H m\,\phi^2\,dx} = 0.128\,\frac{\text{ft}^2}{\text{lb-sec}^2}$$

The value of C_L/β to be used in eq. (15-6) to compute the multiplier, q, is perhaps the most uncertain of all the variables in the equation. Based on present day knowledge, it is suggested that a value of 16 be used. Using 0.00238 lb-sec^2/ft^4 for ρ, the mass density of the air, eq. (15-6) gives 1.64 for the multiplier q.

The deflections, shears, and bending moments caused by lateral vibrations due to vortex shedding are then computed by multiplying the mode shape, the modal shears and the modal moments of Fig. 15-13 by the multiplier, q.

Step 4 Earthquake moments–The procedure outlined in this chapter will be used to compute design earthquake moments in the chimney. The computations are summarized in Table 15-6. The total weight of the concrete shell is equal to 34826 kips and the weight of the additional concentrated loads due to the liner is equal to 1196 kips. The total weight therefore is equal to 36022 kips.

With the period $T = 3.45$ sec/cycle, and assuming $R = 1$, the value of the base shear from eq. (15-8) will be

$$\frac{V_b}{36022} = \frac{0.21}{\sqrt[3]{(3.45)^2}} \times 1; \quad \text{or} \quad V_b = 3320 \text{ kips}$$

The sum of the lateral loads should be equal to the base shear of 3320 kips. Therefore the loads in column 8 in Table 15-6 should be multiplied by the ratio 3320/6226 = 0.533, to obtain the design lateral loads tabulated in column 9. Using the values in column 9, the bending moments due to the lateral loads are computed and tabulated in column 11. These moments are then multiplied by the coefficients obtained from Fig. 15-5 and tabulated in column 12. The earthquake moments are listed in column 13.

The shears along the chimney due to earthquakes are approximated by multiplying the base shear by the shear coefficients of Fig. 15-3. The results are tabulated in column 14 of Table 15-6.

Step 5 Stresses–The compressive stresses in the concrete and the tensile stresses in the steel will be computed for both the working stress moments and the maximum stress moments. In the case of the maximum stress moments, a load factor of 1.5 is used for static wind and earthquake moments while a factor of 1.25 is used for resonant wind moments. The design moments the axial forces, W, and the temperature gradient, T_x, are tabulated in Table 15-7. The chimney

TABLE 15-6 Earthquake Moments Computations

Weight Intensity W (k/ft) (1)	*Conc. Loads p (kips)* (2)	*Equiv. Conc. Load for W (kips)* (3)	*Height* $h' = \frac{h}{900}$ (4)	$W \cdot h'^2$ (5)	*Equiv. Conc. Load for* Wh'^2 (6)	$p\,h'^2$ (7)	*Cols. 6 + 7* (8)	*Corrected Lateral Loads (k)* (9)	*Shear Due to Loads of Col. 9* (10)	*M (ft-k)* (11)	J_x (12)	M_e *(ft-k)* (13)	*V (kips)* (14)
11.84	50	564	1.0	11.84	526	50	576	307	307	0	1.9	0	0
13.89	95	1250	.9	11.25	1009	77	1086	579	886	27630	1.4	38682	714
15.89	116	1437	.8	10.17	916	74	990	528	1414	107370	.96	103075	847
18.86	116	1706	.7	9.24	831	57	888	473	1887	234630	.65	152510	946
23.02	120	2082	.6	8.29	745	43	788	420	2307	404460	.47	190096	1062
28.57	139	2589	.5	7.14	641	35	676	360	2667	612090	.37	226473	1245
36.51	139	3279	.4	5.84	521	22	543	289	2956	852120	.31	264157	1477
43.53	147	3944	.3	3.92	354	13	367	196	3152	1118160	.29	324266	1859
54.10	166	5130	.2	2.16	199	7	206	110	3262	1401840	.29	406534	2523
99.51	108	8522	.1	1.00	91	1	92	49	3311	1695420	.32	542534	3104
87.13	0	4323	0	0	14	0	14	7		1993410	.39	777430	3320
	1196^k	34826^k					6226^k						

TABLE 15-7 Design Moments, Weights and Temperature Gradient

Distance from Top (ft)	*Static Wind (ft-k)*	*1.5 x Static Wind (ft-k)*	*Resonant Wind (ft-k)*	*1.25 x Resonant Wind (ft-k)*	*Earthquake (ft-k)*	*1.5 x Earthquake (ft-k)*	T_x *(°F)*	*Weight Above Section (k)*
0	0	0	0	0	0	0	33	0
90	4860	7290	8789	10986	38682	58023	30	1160
180	20250	30375	34355	42944	103075	154613	29	2500
270	47115	7063	75102	93878	152510	228765	29	4050
360	86535	129803	129775	162219	190096	285144	30	5910
450	139455	209183	195973	244966	226473	339710	33	8230
540	205808	308712	271761	339701	264157	396236	37	11150
630	286470	429705	355196	443995	324266	486399	39	14800
720	382320	573480	443082	553853	406534	609801	43	19200
780	456800	685200	503300	629125	497246	745869	67	23000
810	494033	741050	533364	666705	542534	813801	67	25400
885	597941	896912	609425	761781	738202	1107303	0	33000
900	618773	928160	624674	780843	777430	1166145	0	34800

Flue Opening, 16′ wide × 30′ high–bottom of opening at 90′ from base
Door Opening, 12′ wide × 15′ high–bottom of opening at base

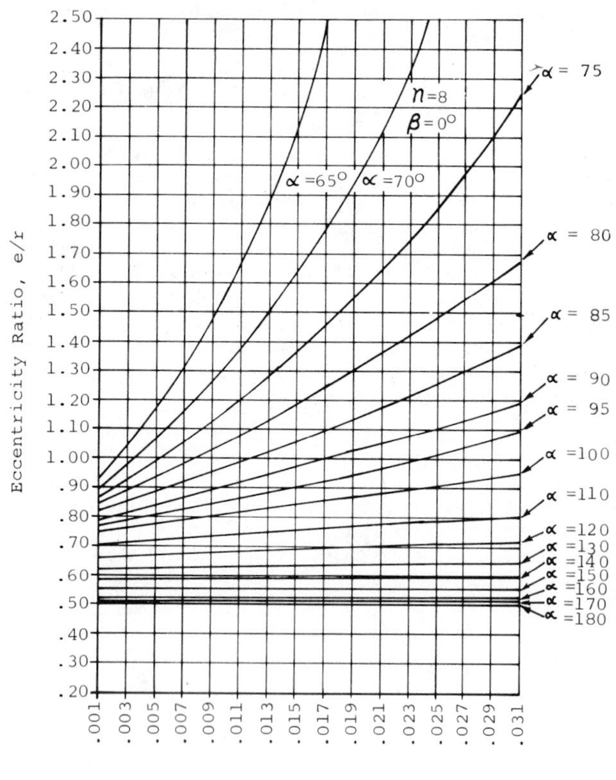

Fig. 15-14 Values of α for $n = 8$ and $\beta = 0°$.

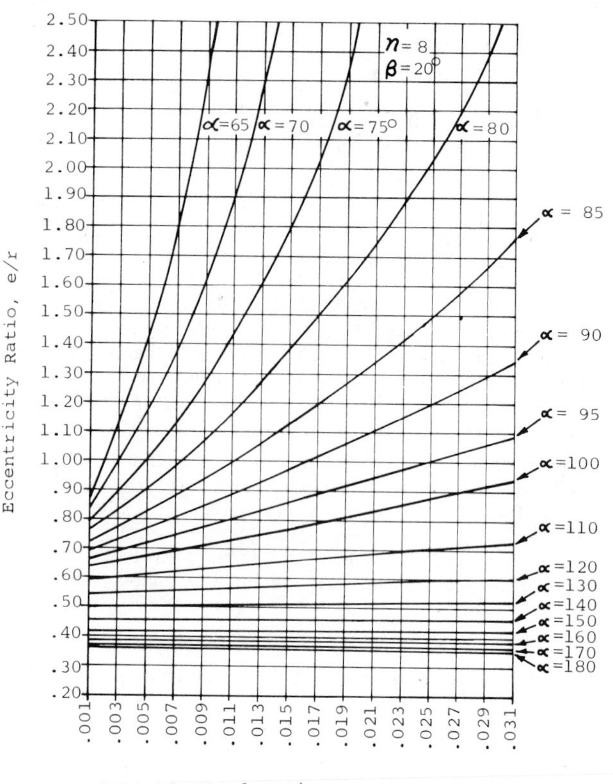

Fig. 15-15 Values of α for $n = 8$ and $\beta = 20°$.

TABLE 15-8 Concrete Stresses (psi)

Distance from Top (ft)	Vertical Steel Ratio			Static Wind		1.5 × Static Wind		Resonant Wind		1.25 × Resonant Wind		Earthquake		1.5 × Earthquake	
	Outside	Total	f_{CTV}	f_c	f_c-comb.	f_c	f_c-comb.	f_c	f_c-comb.	f_c	f_c-comb.	f_c	f_c-comb.	f_c	f_c-comb
0	.00250	.00375	119	0	120	0	120	0	120	0	120	0	120	0	120
90	.00435	.00655	140	122	441	141	472	153	490	170	513	683	1033	1046	1368
180	.01150	.01720	202	266	638	331	704	365	737	450	820	1086	1428	1612	1930
270	.01160	.01740	206	415	789	572	942	607	976	762	1125	1241	1583	1844	2159
360	.00800	.01200	189	568	942	847	1219	847	1219	1087	1448	1291	1642	1963	2277
450	.00425	.00635	163	705	1117	1152	1543	1055	1452	1417	1791	1281	1664	2089	2414
540	.00250	.00375	148	778	1236	1376	1797	1123	1563	1593	1995	1074	1517	1980	2345
630	.00250	.00375	158	854	1336	1497	1938	1110	1579	1576	2011	980	1456	1812	2225
720	.00250	.00375	177	868	1401	1498	1988	1018	1543	1415	1912	922	1452	1652	2128
780	.00250	.00375	284	574	1386	989	1782	632	1449	859	1669	623	1440	1138	1909
780*	.00250	.00375	284	797	1621	1435	2190	897	1714	1252	2031	883	1701	1636	2363
810*	.00250	.00375	284	820	1646	1443	2206	893	1714	1220	2009	911	1731	1673	2405
810	.00250	.00375	284	600	1417	1000	1799	642	1463	845	1662	653	1475	1168	1944
885	.00250	.00375	–	794	794	1183	1183	804	804	966	966	936	936	1653	1653
885*	.00250	.00375	–	964	964	1520	1520	979	879	1214	1214	1171	1171	2121	2121
900*	.00250	.00375	–	1014	1014	1559	1559	1020	1020	1246	1246	1240	1240	2232	2232
Allowable			1800	1125	3015	3600	3600	1690	3015	3600	3600	1690	3015	3600	3600

*Net Section

f_{CTV} = concrete stresses due to temperature f_c = concrete stresses due to load

f_c-comb. = combined concrete stress f'_c = 4500 psi

is assumed to have a flue opening 16 ft wide and 30 ft high, the bottom of which is at 90 ft from the base. Another door opening is assumed at the bottom of the stack whose width is 12 ft and depth is 15 ft.

The total vertical steel in the chimney is determined by a trial and error procedure. With the ratio e/r known and a steel ratio assumed, the value of α, which locates the neutral axis, is determined from eq. (15-14). The solution of eq. (15-14) for α is best accomplished by a computer program or the use of the curves given in the *ACI Chimney Specification.* Those curves are drawn for different values of openings (β) and for different values of the modulus ratio (n). Some of these curves are reproduced here in Figs. 15-14 and 15-15.

TABLE 15-9 Steel Stresses (psi)

Distance from Top (ft)	*Vertical Steel Ratio*			*Static Wind*		*1.5 × Static Wind*		*Resonant Wind*		*1.25 × Resonant Wind*		*Earthquake*		*1.5 × Earthquake*	
	Outside	*Total*	f_{STV}	f_s	f_s-*comb.*	f_s	f_s-*comb.*	f_s	f_s-*comb.*	f_s	f_s-*comb.*	f_s	f_s-*comb.*	f_s	f_s-*comb.*
0	.00250	.00375	3426	0	3426	0	3426	0	3426	0	3426	0	3426	0	3426
90	.00435	.00655	2971	0	2971	0	2971	0	2971	12	2973	17949	17949	34900	34900
180	.01150	.01720	2430	0	2430	331	2510	769	2622	2201	3050	17952	17952	32431	32431
270	.01160	.01740	2520	75	2537	2233	3139	2870	3375	6090	6090	17987	17987	34256	34256
360	.00800	.01200	2956	688	3094	5887	5887	5887	5887	12109	12109	18053	18053	39424	39424
450	.00425	.00635	3775	1445	3993	12567	12567	9534	9534	22002	22002	16944	16944	50231	50231
540	.00250	.00375	4718	1862	4933	19782	19782	10414	10414	29390	29390	8875	8875	48930	48930
630	.00250	.00375	5089	1859	5299	20510	20510	7375	7375	23767	23767	4222	5647	34333	34333
720	.00250	.00375	5768	1572	5940	18636	18636	4254	6303	15612	15612	2429	6044	24779	24779
780	.00250	.00375	9606	898	9697	11190	11414	1786	9791	7066	10469	1646	9776	16681	16681
780*	.00250	.00375	9606	2545	9875	17946	17946	4274	10080	12614	12614	4006	10047	24402	24402
810*	.00250	.00375	9606	2157	9832	16340	16340	3293	9962	10236	11099	3608	9999	23460	23460
810	.00250	.00375	9606	697	9676	9817	10993	1271	9736	5372	10223	1438	9754	15675	15675
885	.00250	.00375	–	206	206	6940	6940	298	298	2448	2448	1970	1970	21911	21911
885*	.00250	.00375	–	944	944	11055	11055	1102	1102	4536	4536	3788	3788	29057	29057
900*	.00250	.00375	–	782	782	10145	10145	850	850	3889	3889	3794	3794	29795	29795
	Allowable		24000	15000	32400	50000	50000	18000	32400	50000	50000	18000	32400	50000	50000

*Net Section

f_{STV} = steel stress due to temperature f_s = steel stress due to load

f_s-comb. = combined steel stress f_y = 60,000 psi

TABLE 15-10 Circumferential Steel Areas

(sq. in. per lin. ft.)

Distance from Top (ft)	*Inside Steel*	*Outside Steel*
0	0.291	0.252
90	0.335	0.290
180	0.386	0.334
270	0.412	0.357
360	0.414	0.359
450	0.368	0.319
540	0.340	0.294
630	0.339	0.294
720	0.318	0.275
810	0.468	0.468
900	0.386	0.386

Once the value of α is known, then the stresses in the concrete and the steel are computed from eqs. (15-12) and (15-13).

Table 15-8 gives the stresses in the concrete due to temperature gradient, and also due to the loading with and without temperature. Table 15-9 is the same except that the stresses are given for the vertical steel. The steel ratios as given in Tables 15-8 and 15-9 are such that the stresses in the concrete and the steel are kept within the allowable limits. A minimum value of 0.00250 is used for the outside vertical reinforcement, and a minimum of 0.00125 is used for the inside vertical reinforcement. It should be mentioned that the total steel ratio (sum of inner and outer vertical steel ratios) is used to compute stresses due to load, while the outside steel ratio only is used to compute temperature stresses.

Step 6 Circumferential reinforcement–The selection of the circumferential reinforcement is primarily controlled by the circumferential bending moments caused by the wind pressure distribution shown in Fig. 15-4. These bending moments are given in Table 15-3 where p is the design wind pressure (as given in Table 15-2) divided by the drag coefficient. Table 15-10 gives the required circumferential steel based on the following:

Maximum bending moment causing tension on the inside = 0.314 pr^2
Maximum bending moment causing tension on the outside = 0.272 pr^2
Drag coefficient = 0.7
Minimum steel ratio for outside or inside reinforcement = 0.001
Allowable stress in the steel, f_s = 30,000 psi

The above value for the allowable stress is used because of the rare occurrence of the maximum wind pressure. The steel required is computed by the approximate formula

$$A_s = \frac{M}{0.87\, d f_s}$$

where $d = t - 2.25$ in.

REFERENCES

15-1 *Specification for the Design and Construction of R/C Chimneys (ACI 307-69)*, American Concrete Institute, Detroit, Michigan, 1969.

15-2 Dryden, H. L., and Hill, G. C., "Wind Pressure on Circular Cylinders and Chimneys," *Research Paper No. 221*, Bureau of Standards Journal of Research, 5, Sept. 1930.

15-3 Roshko, A., "Experiments on the Flow Past a Circular Cylinder at Very High Reynolds Numbers," *Journal of Fluid Mechanics*, Cambridge University Press, May 1961.

15-4 Erdei, C., and Ghosh, J., "The Effects of Wind on Large-Diameter Chimneys and Shafts," *Concrete*, Sept. 1967.

15-5 Diver, M., "Étude des Cheminées en Béton Arme," *Annales de l'Institute du Bâtiment et des Travaux Publics*, Paris, May, 1966.

15-6 *Règles Définissant les Effects de la Neige et du Vent Sur les Constructions*, Société de Diffusion des Techniques du Bâtiment, Paris, 1965.

15-7 ASCE Task Committee on Wind Forces, "Wind Forces on Structures," *Transactions, ASCE*, **126**, Part II, 1124–1198, 1961.

15-8 Dockstader, E. A.; Swiger, W. F.; and Ireland, E., "Resonant Vibration of Steel Stacks," *Transactions, ASCE*, **121**, 1088–1111, 1956.

15-9 "Wind Effects on Buildings and Structures," Proceedings, Conference held at the National Physical Laboratory, Teddington, Middlesex, England, **II**, June 1963.

15-10 Walter, L. D., and Woodruff, G. B., "The Vibrations of Steel Stacks," *Transactions, ASCE*, **121**, 1054–1087, 1956.

15-11 Maugh, L. C.; and Rumman, W. S., "Dynamic Design of Reinforced Concrete Chimneys," *ACI Journal, Proceedings*, **64** (9), 558–567, Sept. 1967.

15-12 Rumman, Wadi S., "Basic Structural Design of Concrete Chimneys," *Journal of the Power Division, ASCE*, **96** (PO3), Proc. Paper 7334, 309–318, June 1970.

15-13 Rumman, Wadi S., "Earthquake Forces in Reinforced Concrete Chimneys," *Journal of the Structural Division, ASCE*, **93** (ST6), Proc. Paper 5650, 55–70, Dec. 1967.

15-14 Rumman, W. S., and Maugh, L. C., "Earthquake Forces Acting on Tall Concrete Chimneys," International Association for Bridge and Structural Engineering, Sept. 1968.

15-15 Rumman, W. S., "Vibrations of Steel-lined Concrete Chimneys," *Journal of the Structural Division, ASCE*, **89** (ST5), Proc. Paper 3661, 35–63, Oct. 1963.

15-16 Housner, G. W.; and Keightley, W. O., "Vibrations of Linearly Tapered Cantilever Beams," *Journal of the Engineering Mechanics Division, ASCE*, **88** (EM2), Proc. Paper 3101, 95–123, April, 1962.

15-17 Newmark, N. M., "Numerical Procedure for Computing Deflections, Moments, and Buckling Loads," *Transactions, ASCE*, **108**, 1943.

16

Silos and Bunkers

SARGIS S. SAFARIAN and ERNEST C. HARRIS*

16.1 INTRODUCTION

Bins for storing granular materials are of two main types—silos (also called deep bins), and bunkers (or shallow bins). The important difference between the two is in the behavior of the stored material. This behavior difference is influenced by both bin geometry and characteristics of the stored material. Material pressure against the walls and bottom are usually determined by one method for silos and by another for bunkers.

Silos and bunkers are made from many different structural materials. Of these, concrete is probably the most frequently used. Concrete can offer the necessary protection to the stored materials, requires little maintenance, is aesthetically pleasing, and is relatively free of certain structural hazards (such as buckling or denting[16-1]) which may be present in silos or bunkers of thinner materials.

Silos and bunkers may be of various plan shapes and may occur singly or connected in groups. (See Fig. 16-1.) Figure 16-2 shows typical group arrangements. All of these arrangements are used for silos, but only (a) and (e) are suited to bunkers.

Many spectacular silo collapses have occurred.[16-2] The causes of collapse have been many, but the most common have been: (1) occurrence of operational pressures much higher than assumed in the design; (2) construction errors; and (3) combinations of these and other causes. In silo groups, failures have also resulted from incorrect design or detail of interstice cells and pocket bins (or blister bins), or the neglect of severe stresses and deformations under various combinations of loaded and empty cells.

*Sargis S. Safarian, P.E., President SMH Engineering, Inc., Lakewood, Colorado. Ernest C. Harris, Ph.D., P.E., Professor, University of Colorado, Denver, Colorado.

16.2 DESIGN CONSIDERATIONS

The design process for silos and bunkers is of two types—functional and structural. Functional design must provide for adequate volume, proper protection of the stored materials, and satisfactory methods of filling and discharge. Structural considerations are stability, strength, and control (minimizing) of crack width and deflection. Loads to be considered include the following:

1. Dead load of the structure itself and items supported by the structure.
2. Live loads, as follows—
   (a) forces from stored material,
   (b) changes in above due to filling and emptying,
   (c) wind,
   (d) snow, and
   (e) seismic forces on structure and stored material.
3. Thermal stress due to stored material (especially important in long silo groups).

Fig. 16-1 Grain elevator—group of hexagonal and circular silos, Denver, Colorado. (Courtesy of Cargill, Inc.)

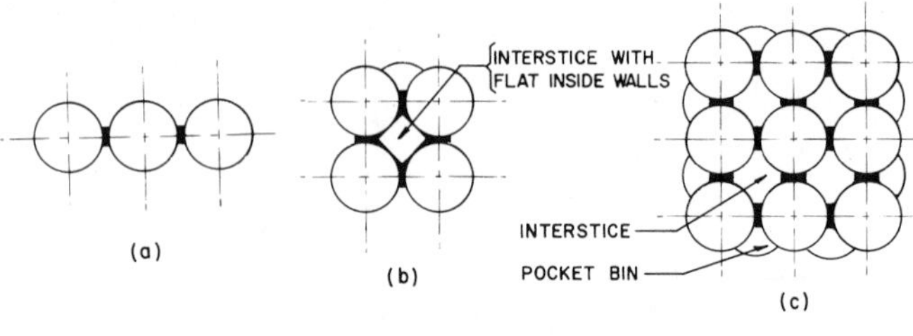

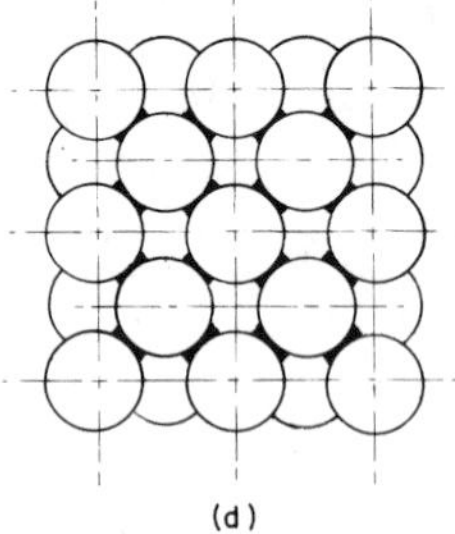

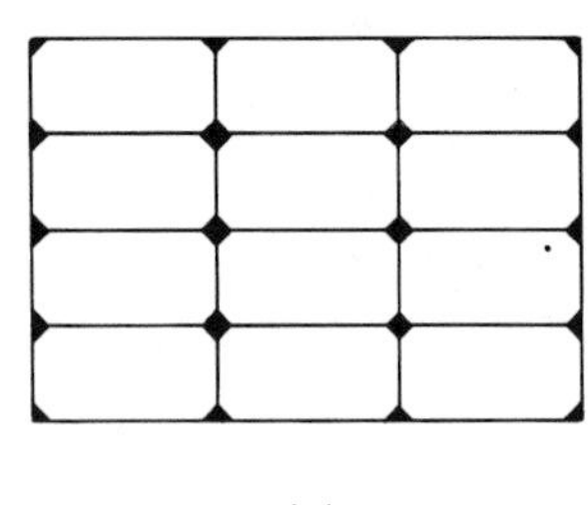

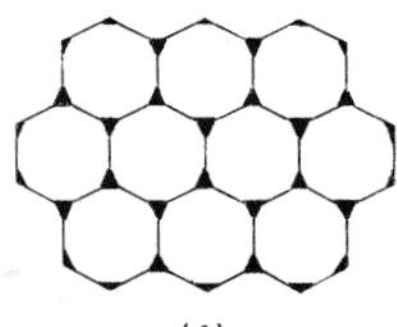

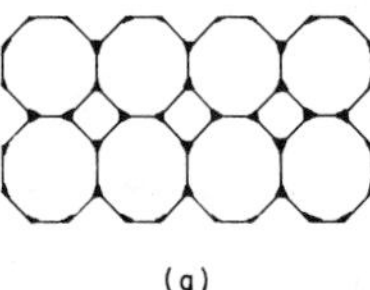

Fig. 16-2 Typical silo or bunker groups.

Fig. 23a-2 Typical silo or bunker groups

16.3 DESIGN METHOD—USD OR WSD

Either design method—ultimate strength (USD) or working stress (WSD)—may be used for silos and bunkers. In the materials which follow, most formulas for forces or moments give ultimate values of force or moment. However, by merely setting K_{ll} and K_g equal to 1.0, these same formulas may be used to solve for WSD forces and moments.

The *ACI Code* values of load factor, K, for dead and live loads should be used for silo or bunker design by USD. Capacity reduction factors, shown as ϕ in the *ACI Code*, are represented herein by ψ. (The symbol ϕ is reserved for angle of repose.) The *ACI Code* capacity reduction factors are suitable for bunker or silo components cast in stationary forms, but lower factors are suggested for slip-formed concrete. (See Table 16-1.)

TABLE 16-1 Capacity Reduction Factors, ψ

Stationary formed
concrete $\psi = 1.00 \times \phi$ from Section 9.2, ACI 318-71
Slip-formed concrete:
with continuous
inspection $\psi = 0.95 \times \phi$ from Section 9.2, ACI 318-71
(except use ϕ given by ACI 318-71 for vertical bearing and compression)

The authors personally prefer USD, and in all numerical examples herein, USD is used.

16.4 PROPERTIES OF STORED MATERIALS

The properties of the material to be stored affect the intensity of pressure loadings. In addition, they influence material flow and must be considered in selecting the outlet shape and size and the type of unloading system.[16-3,16-4,16-5]

Table 16-2 shows the properties of commonly stored materials. These values should be used only in the absence of test data for the actual material to be stored. Caution and good judgment must be used in selecting the properties to be used, as pressures are quite sensitive to variations of those properties. The designer must be alert to the possibil-

TABLE 16-2 Typical Design Properties of Granular Materials (Use only in the absence of the actual values.)

MATERIAL	WEIGHT lb/cu ft	ANGLE OF REPOSE ϕ	COEFFICIENT OF FRICTION μ' AGAINST CONCRETE	AGAINST STEEL
CEMENT, PORTLAND	100	25°	0.466	0.30
CEMENT, CLINKER	88	33°	0.6	0.3
PEAS	50	25°	0.296	0.263
WHEAT	50	25°	0.444	0.414
BEANS	46	31.5°	0.442	0.366
BARLEY	39	37°	0.452	0.376
CORN	44	27.5°	0.423	0.374
OATS	28	28°	0.466	0.412
SUGAR GRANULAR	62.5	35°	0.431	
SAND DRY	100	35°	0.70	0.50
SAND MOIST	112.5	40°	0.65	0.40
SAND SATURATED	125	25°	0.45	0.35
FLOUR	37.5	40°	0.30	0.30
LIME, BURNED (PEBBLES)	56.2	35°	0.50	0.30
LIME, POWDER	44	35°	0.50	0.30
COAL	50	35°	0.50	0.30
COAL, ANTRACITE	62.5	35°	0.50	0.30
COKE	37.5	40°	0.80	0.50
GRAVEL DRY	113	35°	0.45	
GRAVEL WET	125	25°	0.40	0.75
MANGANESE ORE	125	40°		
IRON ORE	165	40°	0.50	0.364
CLAY DRY	106	40°	0.5	0.7
CLAY DAMP	113	25°	0.3	0.4
CLAY WET	138	15°	0.2	0.3
LIME BURNED, FINE	57	35°	0.5	0.3
LIME BURNED, COARSE	75	35°	0.5	0.3
GYPSUM IN LUMPS, LIMESTONE	100	40°	0.5	0.3

ity of large variations from the tabulated values—for example, the possibility of a material ranging from dry, to damp, to saturated. If the expected condition is not certain, the designer should assume the worst condition for that material.

The designer should be alert to possible varying usage—the storage of different materials in the same silo or bunker. Which material should be used in design? Perhaps more than one, since one material may exert the greater lateral pressure while another may cause the more severe vertical pressure. For each type of pressure sought, the "worst" material should be considered.

16.5 SILO OR BUNKER?

Unless he elects to use *silo* static pressure equations for each, the designer must classify the structure as either a *silo* (deep bin) or *bunker* (shallow bin).

Empirical approximations are preferred by many engineers. Two such approximations are

a. by Dishinger,[16-6] $H > 1.5\sqrt{A}$
b. by the Soviet Code,[16-7]
   $H > 1.5D$ for circular silos
   $H > 1.5a$ for rectangular silos

If the bin in question satisfies either of the above, it is considered as a silo. If it satisfies neither rule, it is considered to be a bunker.

A second method is based on the position of the plane of rupture of the stored material. Figure 16-3 shows bins of two different depths. The plane of rupture is determined by the Coulomb theory. If friction against the wall is neglected for the case of a vertical wall and horizontal top surface, the Coulomb plane of rupture is midway between the angle of repose (ϕ) and the vertical wall. If the rupture plane intersects the top surface of the stored material, the bin is a bunker; otherwise it is a silo.

However, engineers do not agree on the location of the plane of rupture. Some would start the plane at the bottom of the hopper, point C of Fig. 16-3b, while others would pass it through point D. Thus, by one interpretation the bin would be a silo—by the other, a bunker. Fortunately, for such borderline cases, exact classification is not critical.

16.6 SILOS

16.6.1 Silo Loadings from Stored Material

Material stored in a silo applies lateral forces to the side walls, vertical force (through friction) to the side walls, vertical forces to horizontal bottoms and both normal and frictional forces to inclined surfaces. The static values of these forces, resulting from materials at rest, are all modified during withdrawal of the material. In general, all forces will increase, so that loads during withdrawal tend to control the design.

Forces applied by stored materials may also be affected by moisture changes, by compaction, and by settling which may accompany alternate expansion and contraction of the walls during daily or seasonal temperature change.

A rigorous approach to calculation of silo loads would involve the conditions of material flow during emptying. Until such approach is perfected, equations derived for static forces may be combined with experimental data to approximate the pressure increases occurring during material withdrawal. The procedure involves determining the static pressures (or forces) and then multiplying these by an "overpressure" factor, C_d to obtain design pressures (or forces).

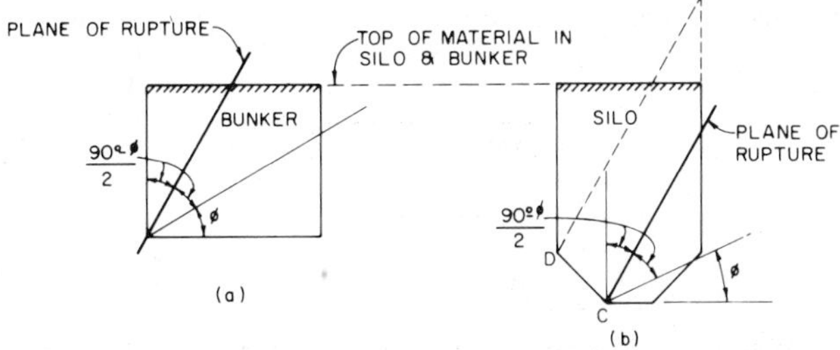

Fig. 16-3 Classification of bins using plane of rupture.

16.6.2 Computation of Static Pressures—Lateral and Vertical

Two methods for determining static pressures are Janssen's classic method[16-8] and Reimbert's method.[16-9] Figure 16-4 shows the silo dimensions used for each. Janssen's method is the more popular in the U.S. However, experiments[16-9] show Janssen's method to be unconservative in some cases, whereas Reimbert's method is reported to give pressures agreeing closely with test results.

Janssen's Method:
Vertical static unit pressure at depth Y below the surface is

$$q = \frac{\gamma R}{\mu' k}\left(1 - e^{-\mu' k Y/R}\right) \tag{16-1}$$

Lateral static unit pressure at depth Y is

$$P = \frac{\gamma R}{\mu'}\left(1 - e^{-\mu' k Y/R}\right) = qk \tag{16-2}$$

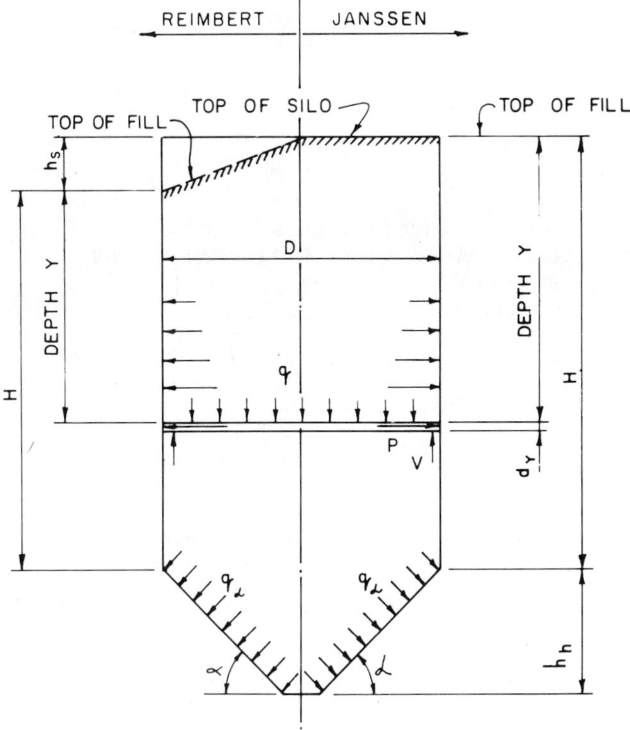

Fig. 16-4 Silo dimensions for use in Reimbert's and Janssen's equations.

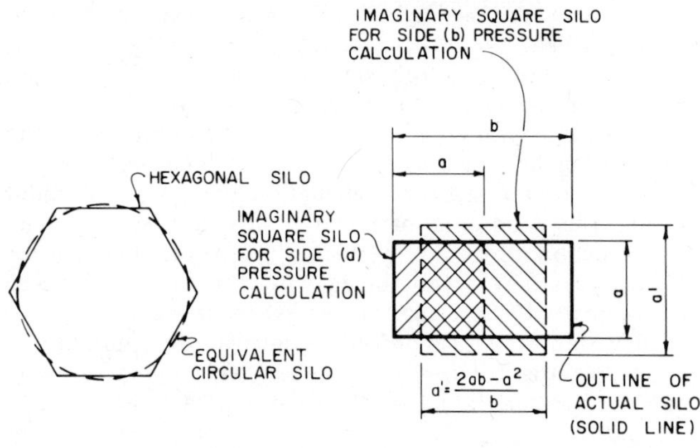

Fig. 16-5 Equivalent silo shapes.

where k is assumed to be

$$k = \frac{P}{q} = \frac{1 - \sin\phi}{1 + \sin\phi} \qquad (16\text{-}3)$$

In the above, R is the hydraulic radius (area/perimeter) of the horizontal cross section of the inside of the silo.

for circular silos, $R = D/4$

for polygonal silos, $R = D/4$ for a circular shape of equivalent area may be used (Fig. 16-5).

for square silos or the shorter wall, 'a', of rectangular silos, use $R = a/4$.

for the long wall 'b' of rectangular silos, use $R = a'/4$ where a', the length of side of an imaginary square silo,[16-9] is

$$a' = \frac{2ab - a^2}{b} \qquad (16\text{-}4)$$

Reimbert's Method:
Vertical static unit pressure at depth Y below the surface is

$$q = \gamma\left[Y\left(\frac{Y}{C} + 1\right)^{-1} + \frac{h_s}{3}\right] \qquad (16\text{-}5)$$

Lateral static unit pressure at depth Y is

$$P = P_{max}\left[1 - \left(\frac{Y}{C} + 1\right)^{-2}\right] = zP_{max} \qquad (16\text{-}6)$$

P_{max} (maximum lateral unit pressure) and C ("characteristic abscissa") for use in eqq. (16-5) and (16-6) vary by silo shape, as follows:

for circular silos

$$P_{max} = \frac{\gamma D}{4 \tan\phi'} \qquad (16\text{-}7)$$

$$C = \frac{D}{4\tan\phi' \tan^2\left(\frac{\pi}{4} - \frac{\phi}{2}\right)} - \frac{h_s}{3} \qquad (16\text{-}8)$$

for polygonal silos of more than four sides

$$P_{max} = \frac{\gamma R}{\tan\phi'} \qquad (16\text{-}9)$$

$$C = \frac{L}{\pi}\left[\frac{1}{4\tan\phi' \tan^2\left(\frac{\pi}{4} - \frac{\phi}{2}\right)}\right] - \frac{h_s}{3} \qquad (16\text{-}10)$$

(Use R as defined above for Janssen's method.)

for rectangular silos—on shorter wall, 'a'

$$(P_{max})_a = \frac{\gamma a}{4\tan\phi'} \qquad (16\text{-}11)$$

$$C_a = \frac{a}{\pi\tan\phi' \tan^2\left(\frac{\pi}{4} - \frac{\phi}{2}\right)} - \frac{h_s}{3} \qquad (16\text{-}12)$$

for rectangular silos—on longer wall, 'b'

$$(P_{max})_b = \frac{\gamma a'}{4\tan\phi'} \qquad (16\text{-}13)$$

$$C_b = \frac{a'}{\pi\tan\phi' \tan^2\left(\frac{\pi}{4} - \frac{\phi}{2}\right)} - \frac{h_s}{3}, \qquad (16\text{-}14)$$

where

$$a' = \frac{2ab - a^2}{b} \qquad (16\text{-}15)$$

For design purposes, the granular material is usually assumed level at the top of the silo ($h_s = 0$).

16.6.3 Static Pressure on Flat Bottoms

Pressure on flat bottoms is given by eqq. (16-1) and (16-5). For rectangular silos, however, these equations give different bottom pressures for areas next to the short and long sides. An approximation frequently used is to assume the pressure, q_a—computed using R (Janssen) or C (Reimbert) for the short side, 'a'—to act on the area A_a, as shown in Fig. 16-6. Similarly, pressure q_b, computed using R or C for side 'b', is assumed to act on area A_b.

16.6.4 Static Forces—Vertical Friction

For round, square, and regular polygonal silos, the total static frictional force per foot-wide vertical strip of wall above depth Y is approximately

By Reimbert's method

$$V = (\gamma Y - q)\,A/L \qquad (16\text{-}16)$$

By Janssen's method

$$V = (\gamma Y - 0.8q)\,A/L \qquad (16\text{-}17)$$

For rectangular silos, lateral pressures differ for the long and short walls; hence, side friction and vertical pressures also differ. The friction loads may be approximated by eqq. (16-16) and (16-17) when terms q, A, and L, respectively, are substituted by q_a, A_a, and a for side 'a', and by q_b, A_b, and b for side 'b'. (See Fig. 16-6.)

16.6.5 Static Pressures on Silo Hoppers

Static horizontal pressures, P, and vertical pressures, q, on inclined hopper walls are calculated by the Janssen or

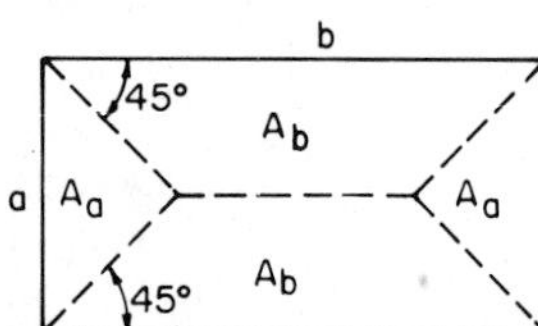

Fig. 16-6 Assumed distribution of vertical pressure on horizontal plane of rectangular silo.

Reimbert formulas. The hydraulic radius, R, may be reduced within the hopper depth, but usually is assumed constant and equal to that of the silo. The static unit pressure normal to the inclined surface at depth Y from the top of the fill is

$$q_\alpha = P \sin^2 \alpha + q \cos^2 \alpha \tag{16-18}$$

16.6.6 Overpressure and Overpressure Factors, C_d

Equations (16-1) through (16-18) above, are for static pressures only, due to stored material at rest, before withdrawal is begun. During withdrawal, these pressures may increase.[16-9,16-10,16-11,16-15,16-16] The increases are sometimes called dynamic effects, but the term 'overpressure' is preferred since the increases include both static and dynamic effects. Among these are arching of the material (causing higher wall pressure and vertical friction), and collapse of the arched material (causing increased vertical pressure due to impact).

Locally, the total pressure may be as much as three to four times[16-12,16-13] the static pressure computed by the Janssen or Reimbert equations. Silos have been designed ignoring overpressure. Many silos so designed have failed and many others not yet showing signs of distress may have dangerously low margins of safety.

Overpressure is not yet well enough understood to consider by rational methods. However, its effect can be approximated using overpressure factors, C_d, to convert from computed static pressures to design pressures. In general

$$\text{design pressure} = C_d \times \text{static pressure} \tag{16-19}$$

Table 16-3 shows tentative overpressure factors, C_d. The factors for use with Janssen's method are from the *Soviet Silo Code*[16-14] but with slight modification. Those for use with Reimbert's method are computed from those for Janssen's method.

To promote better flow of material, designers sometimes use a flow-guiding insert (the Bühler Nase) in the silo or bunker directly above the hopper. Tests show that this insert may cause large additional local overpressures, beyond the normal overpressures without the insert. When the insert is used, overpressure factors, C_d, for walls at the level of the insert should be 50% higher than shown by Table 16-3.

16.6.7 Design Pressures

For silos with centrally located discharge openings, design pressures due to stored material are:

$$q_{des} = C_d q \tag{16-20}$$

$$P_{des} = C_d P \tag{16-21}$$

$$q_{\alpha,des} = P_{des} \sin^2 \alpha + q_{des} \cos^2 \alpha \tag{16-22}$$

The maximum wall friction force per unit length of wall is simultaneous with the minimum, or static, q at depth Y Therefore, for concrete silos

$$V_{des} = V \tag{16-23}$$

16.6.8 Effect of Eccentric Discharge

Eccentric discharge may be considered by adding a correction, P_e, to the lateral design pressure, P_{des}, computed at depth Y by either the Janssen or Reimbert formula. P_e is assumed to vary from zero at the top of the silo to a maximum at depth $Y = H$ (Fig. 16-4) and to remain at that maximum for the full depth of the hopper. Within height

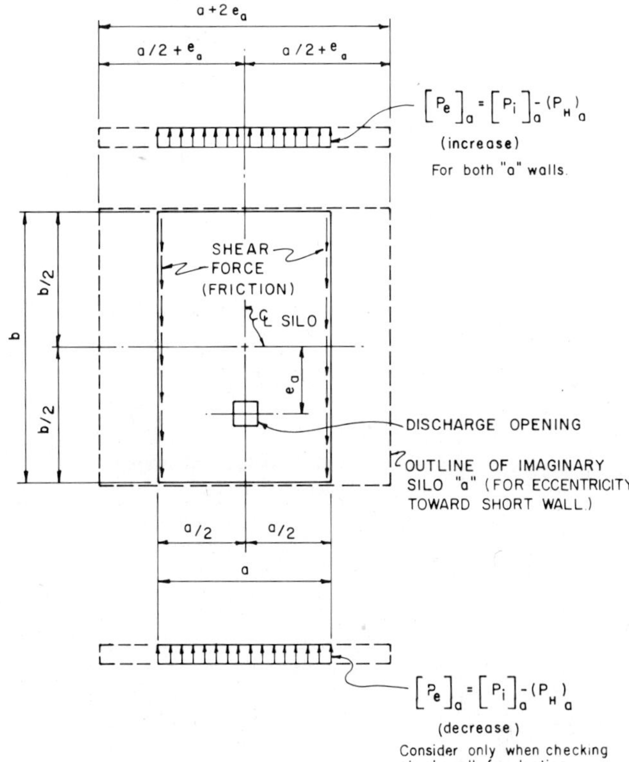

Fig. 16-7 Pressure change due to eccentric discharge in rectangular silo.

H, the lateral design pressure at depth Y is then

$$P_{des} = C_d P + \frac{Y}{H}(P_e) \tag{16-24}$$

The correction P_e at depth H is

$$P_e = P_i - P\,(\text{at } Y = H) \tag{16-25}$$

where P_i is the lateral static pressure at depth H in an imaginary silo, as shown by Fig. 16-7 or Fig. 16-8.

For rectangular silos, the imaginary silo is determined as shown by Fig. 16-7. When the opening is displaced toward side 'a', correction P_e for sides 'a' is computed using an imaginary silo measuring $(a + 2e_a) \times b$. (If e_a is larger than a, the imaginary silo should measure $3a \times b$.) Similarly, if the opening is eccentric toward side 'b', the imaginary silo will measure $(b + 2e_b) \times a$. If both eccentricities occur, each correction is computed separately, using the first described imaginary silo to determine P_e for sides 'a', and the second for sides 'b'.

The imaginary circular silo of Fig. 16-8 is centered on the discharge opening and has a radius equal to that of the actual silo plus the eccentricity.

Where multiple discharge openings occur, even though the group is centrally located, eccentric discharge is always possible and should be considered.

16.6.9 Lateral Pressure Design Curve

Table 16-3 shows different C_d values for different depth zones,[16-17] as follows:

Zone 1: The upper portion of depth $H_1 = D \tan \phi$. It is a zone of negligible 'arching' of the stored material, one for which some authorities[16-17,16-18] consider the Rankine for-

TABLE 16-3 Values of Overpressure Factor, C_d

Note: For funnel flow only. With mass flow, higher C_d factors than shown here are recommended.

Description of area of application	For concrete silo		For steel silo	
	Reimbert	*Janssen*	*Reimbert*	*Janssen*
I. Correction factor C_d for use in calculating horizontal reinforcing in silo walls (see lateral pressure design curve)				
1. Single circular and polygonal silos				
Upper H_1 portion of silo height	1.00	1.35	1.10	1.50
Lower ⅔ of silo height				
$H/D = 1.5$	1.50	1.75	1.65	1.95
See Note				
$H/D \geqslant 4.5$	1.75	1.75	1.95	1.95
2. Group of circular or polygonal silos in straight row and checker board arrangements				
(a) Exterior silos				
(Without exterior pocket bins)				
Upper H_1 portion of silo	1.00	1.35	1.10	1.50
Lower ⅔ of silo height				
$H/D = 1.5$	1.50	1.75	1.65	1.95
See Note				
$H/D \geqslant 4.5$	1.75	1.75	1.95	1.95
(b) Interior silos				
Upper H_1 portion of silo height	1.00	1.35	1.10	1.50
Lower ⅔ of silo height				
$H/D = 1.5$	1.25	1.50	1.40	1.65
See Note				
$H/D \geqslant 4.5$	1.50	1.50	1.65	1.65
3. Interstice bins				
Upper H_1 portion of silo	1.00	1.35	1.10	1.50
Lower ⅔ of silo height				
$H/D = 1.5$	1.25	1.50	1.40	1.65
See Note				
$H/D \geqslant 4.5$	1.50	1.50	1.65	1.65
4. Pocket bins in silo groups in any arrangement and exterior silos connected to pocket bins				
Upper H_1 portion of silo height	1.00	1.35	1.10	1.50
Lower ⅔ of silo height				
$H/D = 1.5$	1.75	2.00	1.95	2.20
See Note				
$H/D \geqslant 4.5$	2.00	2.00	2.20	2.20
5. Square and rectangular single silo and silo groups	1.00	1.35	1.10	1.50
Upper H_1 portion of silo				
Lower ⅔ of silo height				
$H/(a \text{ or } b) = 1.5$	1.25	1.50	1.40	1.65
See Note				
$H/(a \text{ or } b) \geqslant 4.5$	1.50	1.50	1.65	1.65
$H_1 = D \tan\phi \leqslant \frac{1}{3}H$				
$(H_1)_a = b \tan\phi \leqslant \frac{1}{3}H$				
$(H_1)_b = a \tan\phi \leqslant \frac{1}{3}H$				
II. Correction factor C_d for use in calculating bottom pressures in silos				
1. Flat slabs with or without hopper forming concrete fill on top of slab; concrete slab and beam system; concrete hoppers; concrete ring-beam and columns				
a) For flour and bran	1.50	1.25	1.50	1.25
b) For all types of grain	1.35	1.10	1.35	1.10
c) All types of granular material except (a) and (b)				
(1) Slabs with hopper forming concrete fill	1.35	1.10	1.35	1.10
(2) Slabs without hopper forming concrete fill, concrete hoppers, concrete ring-beams and columns supporting silo bottoms	1.75	1.50	1.75	1.50

TABLE 16-3 (Continued)

Description of area of application	For concrete silo Reimbert	For concrete silo Janssen	For steel silo Reimbert	For steel silo Janssen
2. Steel hoppers and ring-beams; steel beams in reinforced concrete and steel silos; steel columns				
a) For flour and bran	1.75	1.50	1.75	1.50
b) All types of grain	1.50	1.25	1.50	1.25
c) All types of granular material except (a) and (b)	2.00	1.75	2.00	1.75

NOTE: For Reimbert's equations C_d for H/D between 1.5 and 4.5 should be determined by interpolation.
For both methods, if $H/D > 5$, C_d values 15% higher than shown above are recommended.

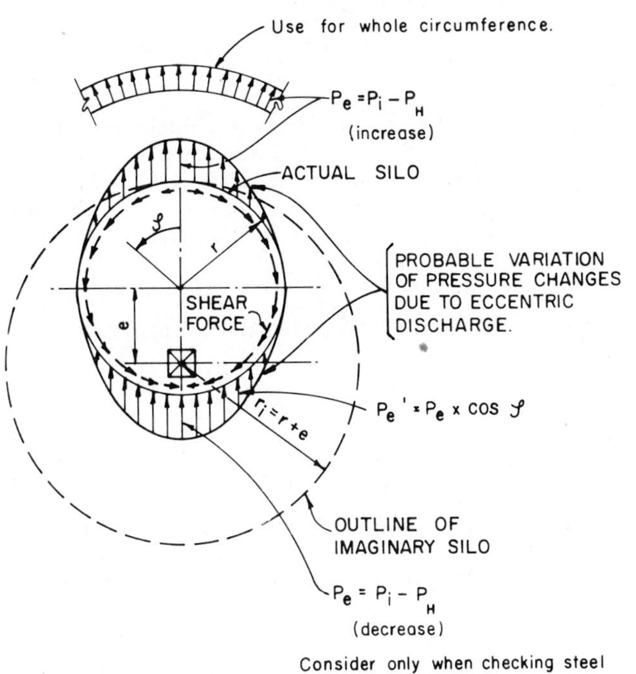

Fig. 16-8 Pressure change due to eccentric discharge in circular silo.

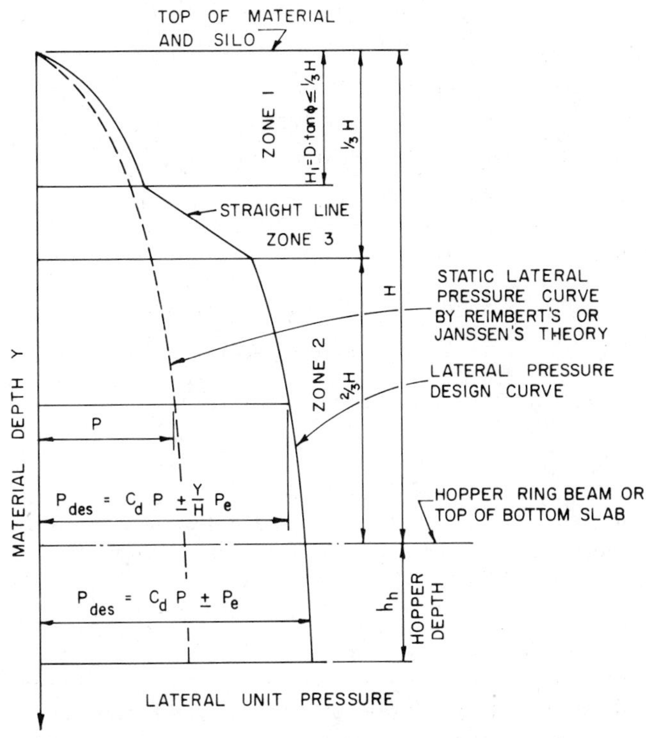

Fig. 16-9 Proposed lateral pressure design curve.

mula valid for computing lateral static pressure provided it is modified for silo shape.

Zone 2: This zone, the lower two-thirds, has significant arching of the stored material and, consequently, large lateral pressures. In this zone, C_d varies with the type of structure.

Zone 3: The pressure curve in this zone is assumed as a straight line. It joins the maximum pressure for Zone 1 and the minimum for Zone 2 and is horizontal when $H_1 = H/3$.

Figure 16-9 shows the resulting curve of lateral design pressures. Note that pressure increases, P_e, due to eccentric discharge (if any) are added to the product C_dP.

16.6.10 Wind and Earthquake Loads

Wind may affect the stability of empty silos, particularly tall, narrow silos or silo groups. Foundation pressures and column stresses, however, may be worse with wind acting on the full silo. Wind load reduction for cylindrical shape may be applied to single circular silos, but not to silo groups.

Earthquake loads may affect stability and strength. Columns and walls supporting silos may be particularly vulnerable and foundations may also be affected. In the absence of better codes covering the seismic design of silos, the authors suggest the earthquake requirements of the latest *Uniform Building Code*.[16-19]

There is test evidence[16-20] that only a fraction of the stored material weight need be considered when computing earthquake forces. The authors use not less than 80% of the weight of the stored material as an effective live load, W_{eff}, from which to determine seismic forces. Lower percentages have been suggested, but the authors believe more research is needed to substantiate the lower values.

When silo bottoms are on supports independent from the walls, the silo structure and independent bottom support will share the lateral force from seismic action on the effective weight of the stored materials. The authors compute the above distribution as follows: To the independent bottom structure, the *smaller* of the following–

1. A portion of the force due to W_{eff}, divided according to relative stiffness of the two structures.
2. Force due to the effective weight within the hopper (if any) plus the product of coefficient of internal friction times effective weight of the stored material above.

To the silo structure, the *larger* of the following–

1. The total force due to W_{eff}, minus that computed above for the independent bottom structure.
2. Fifty % of the total force due to W_{eff}.

The authors present the above only as approximations, to be improved upon or supplanted as soon as technical advances permit.

16.7 CIRCULAR SILOS

16.7.1 Wall Design Procedure

Ordinarily the following sequence is applicable:

(1) Determine hoop reinforcing steel required for design pressure from stored material;
(2) determine the wall thickness required;
(3) check crack width and modify the design as necessary;
(4) modify the hoop steel for temperature effect, if necessary; and
(5) determine the vertical steel.

16.7.2 Horizontal (Hoop) Reinforcing

In design, the radial pressure from stored material is considered uniform along the inside circumference at any given depth. The circular silo is then treated as though subject to tension only.
The ultimate hoop tensile force per unit height is

$$F_u = K_{II} P_{des}(D/2) \qquad (16\text{-}26)$$

The required steel area per unit height is

$$A_s = F_u/(\Psi f_y) \qquad (16\text{-}27)$$

Hoop steel can be in one layer or two. The required area will vary with height, being porportional to the design pressure. Hoop steel area, as required at the bottom of the pressure zone, but excluding temperature steel, should be continued at least 3 ft below that zone (i.e., below the top of the hopper, ring beam, or slab) or to the bottom of the slab or ring beam, whichever is farther. (Where hopper-forming fill is used, the full required hoop steel area should be continued to at least 3 ft below the top of the fill, and one-half of that steel area per unit width to the bottom of the fill.)

Circular silos having cross walls or joined to other silos, blister bins, interstices, etc., can have significant bending moments even under uniform radial load. These moments can be computed.

16.7.3 Wall Thickness

An isolated circular silo under uniform radial load gets all its strength from the horizontal steel wherever the concrete is cracked. Were the loading truly uniform, wall thickness would matter little, provided it was sufficient for vertical load and for lap splicing of the steel. However, since transient nonuniform load may occur, but the moments are not computable, some other approach is needed. One approach is the PCA formula,[16-21] eq. (16-28), below.

$$h_{min} = \left(\frac{mE_s + f_s - n f_{c,\,ten}}{f_s f_{c,\,ten}}\right) PD/2 \qquad (16\text{-}28)$$

In this equation, f_s is the allowable (WSD) steel stress. PCA suggests using $0.1\, f_c'$ for allowable stress $f_{c,\,ten}$. Consistent units must be used.

Unless steps are taken to reduce friction between the forms and concrete, the authors recommend that slip-formed silo walls be not less than 6 in. thick. Walls 9 in. or thicker should be doubly reinforced.

Vertical compressive stresses should also be checked. Suggested limits for circular silos are

$$f_{c,\,vert} = 0.225 f_c' \text{ for WSD methods, or} \qquad (16\text{-}29)$$

$$f_{c,\,vert} = 0.385 f_c' \text{ for USD methods} \qquad (16\text{-}30)$$

Having a tentative thickness, the next step is to check for crack width. A limit of 0.008 in. is suggested for grain or cement storage silos and other silos exposed to the weather.[16-2] One method of checking crack width due to axial tension is presented by Lipnitski.[16-22] In this method, the walls are assumed subject to pure tension and to have centrally located reinforcement. The total outside width w_{cr} of a vertical crack caused by simultaneous action of short- and long-term loadings is

$$w_{cr} = w_1 - w_2 + w_3 \qquad (16\text{-}31)$$

where

w_1 = crack width due to short-term application of total loading, F_{tot} (including overpressure)
w_2 = that portion of w_1 due to static loading F_{st} alone, during the time that F_{tot} acts. (w_2 is included in w_3, thus $w_1 - w_2$ is the width increase due to short-term overpressure.)
w_3 = crack width due to long-term static loading only

Values of w_1, w_2, and w_3 are each calculated using the formula

$$w_n = \Psi_1 s_{cr} f_s / E_s \qquad (16\text{-}32)$$

In the above, s_{cr} (crack spacing) is

$$s_{cr} = A\beta/\Sigma o \qquad (16\text{-}33)$$

where A is the area of a unit height of wall (sq in. per ft) and $\beta = 0.7$ and 1.0 for deformed and plain bars, respectively. Factor Ψ_1 for eq. (16-32) is determined as follows:

$$\Psi_1 \text{ (for short term)} = 1 - 0.7 \left[\frac{0.8\, A\, f_{tu}}{F \text{ or } F_{st}}\right] \qquad (16\text{-}34)$$

$$\Psi_1 \text{ (for long term)} = 1 - 0.35 \left[\frac{0.8\, A\, f_{tu}}{F_{st}}\right] \qquad (16\text{-}35)$$

where F includes overpressure and is computed using P_{des}; and F_{st} is the static tension only, computed using the static pressure P.

Note: When the term in brackets of eq. (16-34) or (16-35) exceeds 1.0, it should be set equal to 1.0. The authors use $f_{tu} = 4.5\sqrt{f_c'}$.

16.7.4 Temperature Reinforcement for Circular Silos

Temperature and shrinkage steel requirements of the *ACI Code* apply to silos. In addition, hot stored materials may cause thermal stresses too high to be ignored.

The approximate method illustrated below was developed specifically for cement storage silos. Its principles, however, should apply also to silos for storage of other hot granular materials. In this method

1. tensile strength of the concrete is neglected, and
2. wall temperatures are assumed to vary only radially.

In buildings, the usual practice is to ignore a certain amount of inside-outside temperature difference. For silos and bunkers, the authors usually neglect not more than the first 80°F of difference. The design inside temperature of the stored material is then

$$T_{i,\,des} = T_i - 80 \qquad (16\text{-}36)$$

The temperature of hot granular material in silos may drop appreciably near the inside face of the wall.[16-23] With hot cement, for example, the authors suggest that an 8-in. thickness of cement be considered as insulating material across which the temperature varies linearly. Figure 16-10 shows the temperature variation in the cement and the silo wall. The temperature difference ΔT between inside and outside of the wall is then

$$\Delta T = K_t (T_{i,\,\mathrm{des}} - T_0) \qquad (16\text{-}37)$$

K_t is determined by heat transfer principles; for stored cement and various wall thicknesses its value is given by Fig. 16-11.

The horizontal ultimate bending moment due to ΔT using plane-strain analysis[16-25] with Poisson's ratio = 0.3 is

$$M_{xt,u} = 1.43\, E_c h^2\, \alpha_t \Delta T\, K_g \text{ (in.-lb/ft)} \qquad (16\text{-}38)$$

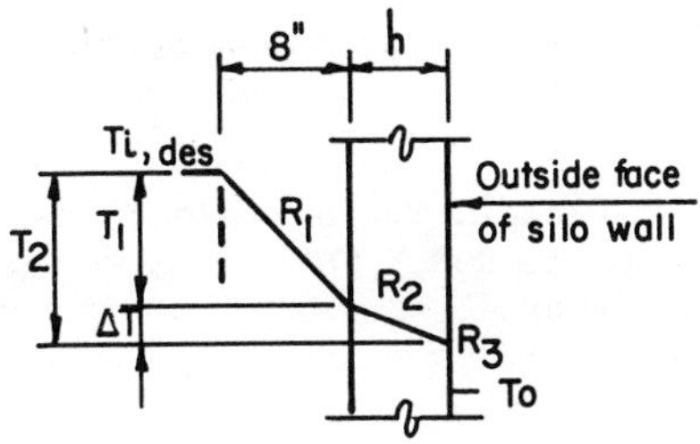

$\Delta T = \left[T_{i,\mathrm{des}} - T_o\right] \times K_t$

T_i = Temperature of stored cement (°F)

$T_{i\ \mathrm{des}}$ = (T_i−80°F) Design temperature. of stored cement.

T_0 = Design winter dry-bulb temperature (F)

h = Silo wall thicknes

Fig. 16-10 Computation of ΔT for wall of cement storage silo.

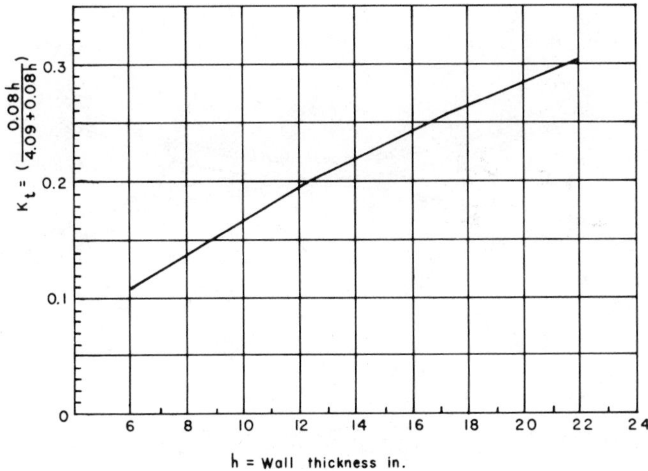

Above curve is based on the following assumptions[16-24]

1. Resistance of 8" (20.3 cm) cement (considered to act as insulating material) = 3.92.
2. Resistance of 1" (2.54 cm) thick concrete = 0.08.
3. Resistance of outer surface film = 0.17.

Fig. 16-11 Determination of K_t for use in computing ΔT for wall of cement storage silo.

for walls in which the radius is considerably larger than the thickness.[16-26] (The approach of Sect. 9.3.7 of the *1971 ACI Code*[16-27] is not recommended since, if hot material is stored, the worst combination of dead, pressure and temperature loadings is not improbable or short-term, but an *assured* combination.)

The bending moment due to temperature gradient is derived for an uncracked section. The added reinforcing, however, is determined for a cracked section. The required additional horizontal steel area, $(A_s)_t$, to resist moment $M_{xt,u}$ should be added to that required for material pressure alone.

The added steel $(A_s)_t$ should be located near the colder face. In single-reinforced walls, it should be added to the main hoop steel, ordinarily near the outer face. In doubly-reinforced walls, the entire amount $(A_s)_t$ should be added to the colder layer.

In the lower two-thirds of the silo, the entire computed $(A_s)_t$ should be provided. In the upper one-third, pressure varies rapidly with height. Steel spacings usually vary in a 'step' pattern, and when the step pattern is used the authors reduce the temperature steel in the upper third to one-half the computed $(A_s)_t$.

16.7.5 Vertical Reinforcement for Circular Silos

Both exterior and interior silo walls should have vertical reinforcement. Vertical bars serve several purposes: cantilever bending of the silo under wind or seismic load; distribution of lateral load concentrations to ensure participation of hoop steel above and below; supports for hoop steel and embedded items; and resistance to vertical bending due to radial temperature gradient.

The authors suggest that vertical bars in the storage zone be #4 or larger; that the vertical steel area be not less than the wall area (bh) times 0.0020 for exterior walls, or 0.0015 for interior walls; and that the spacing of vertical bars in each layer not exceed $4h$ nor 18 in. for exterior, or 24 in. for interior walls. Each permanently-encased jack rod may replace one vertical bar. The area of vertical steel in silo walls below the pressure zone should be not less than used above, nor less than 0.0020 bh.

16.8 GROUPED CIRCULAR SILOS

Circular silos in connected groups are shown by Figs. 16-1 and -2. The individual circular cells of such groups are usually analyzed as if isolated from each other, structural interaction being ignored. The circular cells are designed assuming the interstice cells to be empty. Those portions of wall comprising the interstice are then modified, considering each interstice wall to be fixed at the point where the cylinders meet. While it would be better to treat the entire cross section of the silo group as a rigid frame, complexity of analysis where all members are curved encourages the use of approximate methods.

16.8.1 Interstice Cells—(Fig. 16-2)

The portion of wall common to both the circular cell and the interstice must be designed for the following conditions:

1. Circular cell full and interstice empty. This causes ring tension if the wall is circular; tension and bending if it is flattened.
2. Circular cells empty and interstice full. If the four intersection points are treated as points of fixed position, each curved interstice wall will behave as a fixed-end

arch and will be subject to both axial compression and bending.

If the walls are flattened, they will have bending moment and axial tension. Their edges could be considered fixed at the junction to the circular walls or as members of a rectangular rigid frame.

Design pressures in interstice cells are determined by the same procedure as for silos. The interstice cross section is replaced by an imaginary circle or rectangle of area equal to that of the interstice.

16.8.2 Pocket Bins

Pocket bins (Fig. 16-2) complicate structural analysis of the horizontal frame and cause appreciable bending moment and axial load in the walls of the silos to which they attach. The main silo walls will almost certainly require more horizontal steel in some locations, and perhaps greater thickness than required if no pocket bin were present.

16.8.3 Circular Silos With Internal Crosswalls

Both the cylindrical walls and crosswalls of such silos are subject to combined horizontal tension and bending. In estimating static pressures due to stored material, each cell may be assumed replaced by a circle or rectangle of equal area. The static pressures are then multiplied by factors, C_d (as given for interstices), to obtain design pressures, q_{des} and p_{des}.

The walls of the cross section can be analyzed as rigid frames[16-28] to determine horizontal bending moments, $M_{x,u}$, and the horizontal axial tensions, F_u. Various load patterns (certain cells full, others empty) must be used. Crosswalls will have vertical dead load and friction load. If not supported at their bottoms, they must act as deep beams to transfer those loads to the main walls.

16.9 DETAILS–CIRCULAR SILOS AND SILO GROUPS

16.9.1 Horizontal Reinforcement

Horizontal reinforcement in circular silo walls should provide for the direct tension, bending moment (if any) and temperature gradient (if any). When only one layer of steel is used, it should be placed nearer the outside face of the wall. Walls 9 in. or thicker should have two layers. Hoop reinforcing should be continued below the pressure zone as noted in section 16.7.2.

16.9.2 Horizontal Reinforcement at Wall Intersections

Reinforcement at wall intersections must be properly anchored to prevent separation of the walls. Bar forces are usually maximum at wall intersections; thus, bars must extend *each way* from the intersection a sufficient distance for full anchorage by bond. It is generally a good practice to determine reinforcement requirements at the intersection by analyzing the frame formed by the group of cells. However, the complexity of analysis frequently dictates the use of approximations. In such cases, extra horizontal steel should be provided at areas such as wall intersections where bending moment is expected in combination with the computed tension.

16.9.3 Vertical Reinforcement

Vertical reinforcement should be provided as described in section 16.7.5. In addition, the thickened portion of walls at intersections below the hopper support or silo bottom slab should have vertical steel equal to not less than 0.0025 times the column cross section area. These 'intersection columns' usually have horizontal ties spaced at no more than 24-in. centers.

16.9.4 Reinforcing Bar Splices

Where fixed forms (nonmoving) are used, bar splices should conform to the *ACI Code*. In slip-formed concrete, however, the authors use a 20% increase of splice length for horizontal bars in the pressure zone.

Bar splices, both horizontal and vertical, should be staggered in circular silos. Hoop-reinforcing splices in the pressure zone should be offset by not less than one lap length nor 3 ft and should not coincide in position any more frequently than every third bar.

16.9.5 Dowels

Engineers do not agree as to the value of dowels joining silo walls to the foundation. The authors' preferences are as follows:

1. Always provide dowels for separate columns and for 'intersection columns'. The dowels should match the column steel and have sufficient embedment for complete development in both column and foundation.

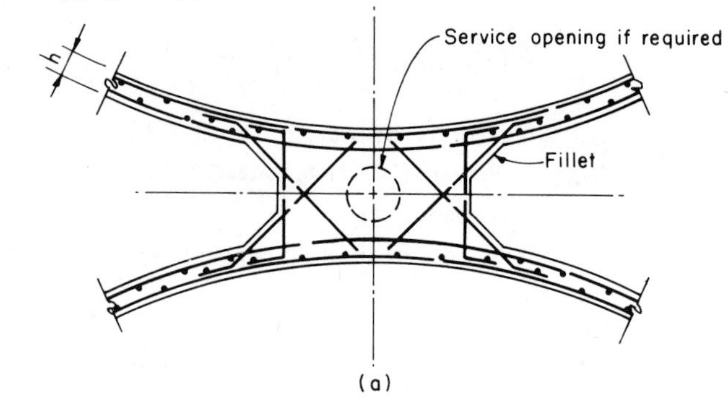

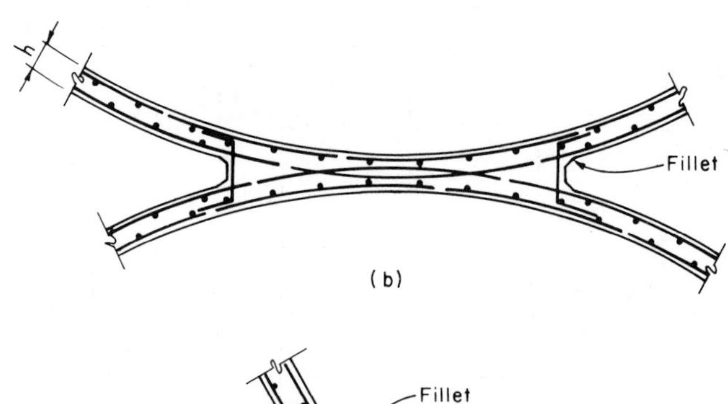

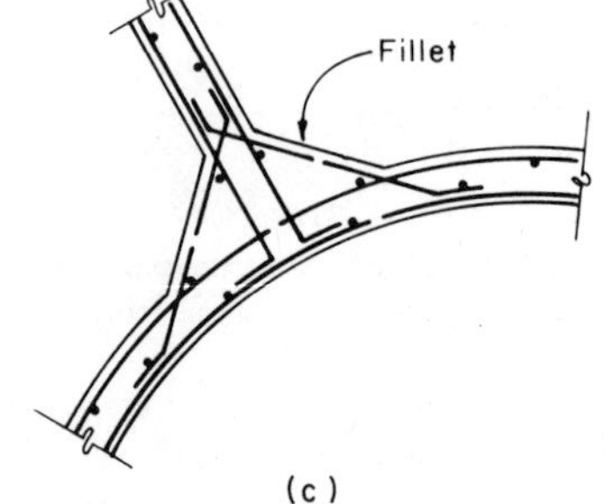

Fig. 16-12 Typical reinforcement pattern at intersecting walls of circular silos.

2. Include dowels at other points along the bottom of the wall as needed to prevent uplift of the wall from earthquake or wind loading.

16.9.6 Fillets

Fillets serve several purposes. They eliminate sharp corners, permitting smoother flow of the stored material and reducing the volume of dead storage. They make slipforming easier. They reduce stress concentrations and bending moments, helping to prevent cracks along wall intersection lines.

At intersecting walls, fillets should be provided, as shown by Fig. 16-12. Fillets 9 X 9 in. or larger, especially in a tensile zone, should be properly reinforced. Within the pressure zone, fillet steel should be of area per unit width at least matching that of the near face of the adjoining walls and should be extended into each wall at least the length required for anchorage.

16.9.7 Reinforcement at Wall Openings

Wall openings in the pressure zone should be kept small. Where such openings exist, wall reinforcement must be added, as follows:

1. Horizontal (hoop) reinforcing terminated by the opening must be replaced by adding at least 1.2 times the terminated hoop steel area, one-half above and one-half below the opening. The added steel should be as close as possible to the opening and extend both ways beyond the opening at least sufficiently to develop its full yield strength through bond. For openings whose height exceeds the anchorage length, the bars shall be extended a distance not less than one-half the opening height.
2. Horizontal reinforcement must aid the wall above to span the opening. For narrow openings, the normal hoop steel may suffice; for wider openings, reinforcement must be added to provide a lintel effect.
3. Added vertical reinforcement adjacent to each side of the opening should reinforce a narrow strip of wall designed as a column to carry its own vertical load plus one-half of the vertical load from the concrete wall above the opening.

Openings below the pressure zone range from very small ones to those large enough to pass trucks or railroad cars. Around openings below the pressure zone, walls should have extra reinforcement, as follows: (See Fig. 16-13.)

1. Horizontal steel should be added above the opening to enable the wall above, acting as a curved deep beam, to span the opening. For openings of width less than the silo inside radius, the amount of this added horizontal steel is sometimes arbitrarily chosen as equal to the normal horizontal steel area for a height of wall equal to one-half the opening width.
2. A narrow strip of wall, $3h$ in width, on each side of the opening should be designed as a column carrying its own share of vertical load plus that from one-half the wall width above the opening. Added steel for each column strip should be at least two bars of the size in the wall and should have area equal to at least half of that eliminated by the opening.

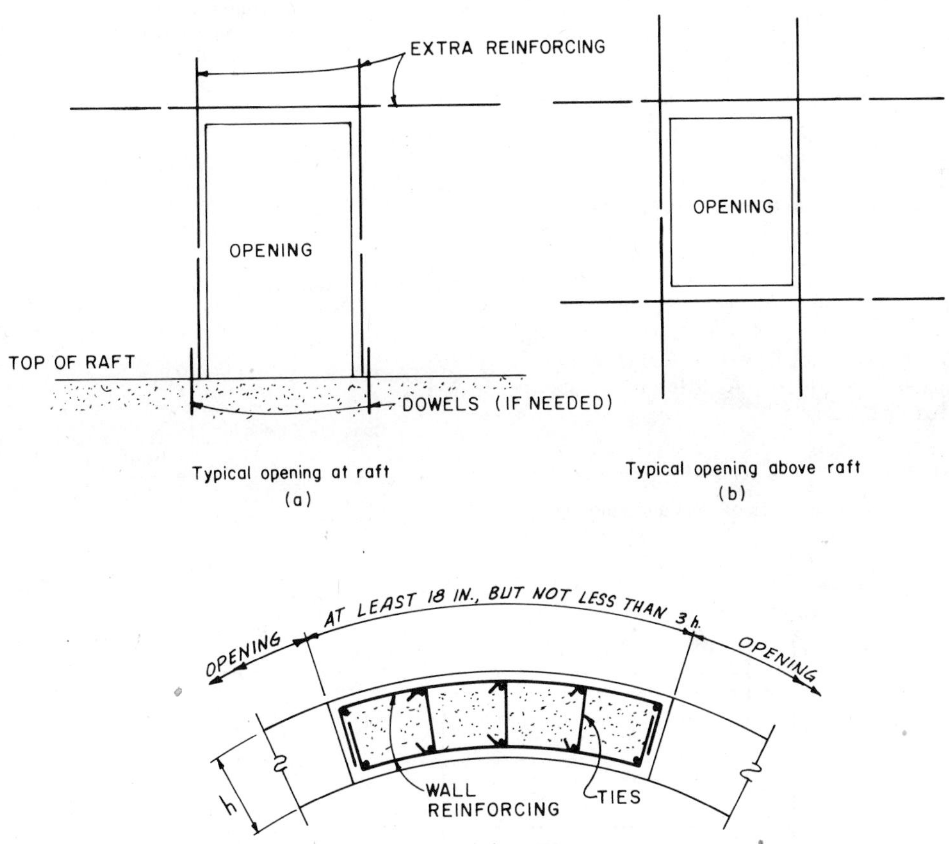

Fig. 16-13 Extra steel at openings.

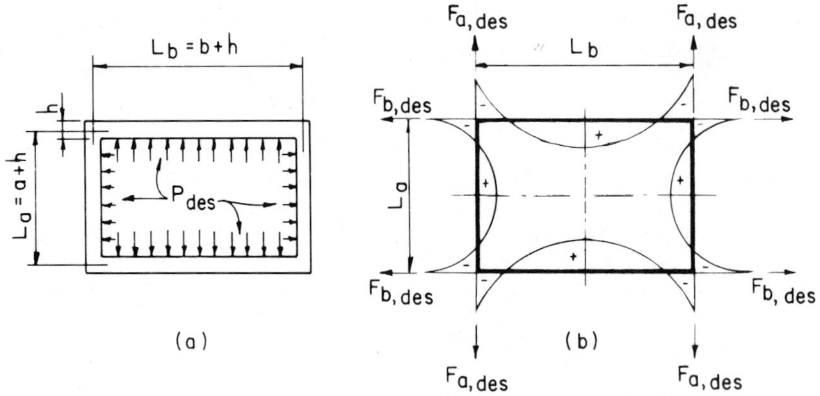

Fig. 16-14 Design loads, bending moment, and axial loads for rectangular silo.

3. Walls between closely spaced openings, as shown in Fig. 16-13c, should be designed as tied columns.

16.10 SQUARE AND RECTANGULAR SILOS

16.10.1 Horizontal Stresses Due to Stored Material

The cross section of a rectangular silo, Fig. 16-14a, is a rigid frame subject to outward pressure, P_{des}, varying with depth, as shown by Fig. 16-9. The resulting moment diagram will be as shown by Fig. 16-14b, each wall having axial tension, bending moment, and shear.

The horizontal ultimate tensile forces per unit height at depth Y are

$$F_{a,u} = K_{ll}P_{b,des}(b/2) \quad \text{(for wall 'a')} \tag{16-39}$$

$$F_{b,u} = K_{ll}P_{a,des}(a/2) \quad \text{(for wall 'b')} \tag{16-40}$$

The ultimate bending moments, $M_{x,u}$, may be reduced from the value computed at the frame centerline to the value at the inner face of the wall.

16.10.2 USD Approach for Combined Tension and Bending

The silo wall must be reinforced to resist combined tension and bending. The combination may be replaced by an eccentric tension, with $e = M_u/F_u$. The eccentric force, F_u, may lie in the space between the layers of steel (Fig. 16-15) or outside that space (Fig. 16-16). A different design approach is needed for each case.

CASE I: Small eccentricity; $e < \frac{h}{2} - d''$ (Fig. 16-15):

By taking moments about either layer of steel and ignoring the concrete strength

$$\text{Reqd } A_s = \frac{F_u e'}{\psi f_y (d - d')} \tag{16-41}$$

$$\text{Reqd } A_s' = \frac{F_u e''}{\psi f_y (d - d')} \tag{16-42}$$

CASE II: Large eccentricity, $e > \frac{h}{2} - d''$ (Fig. 16-16):

Determine *Code* limit for ratio y/d from Table 16-4. Compute y_L, limiting depth of compression block.

Assuming compressive steel to be needed, compute f_s' (effective compressive steel stress) in ksi.

$$f_s' \text{ effective} = 87\left(\frac{y_L - \beta d'}{y_L}\right) - 0.85 f_c' \tag{16-43}$$

but not over

$$f_y - 0.85 f_c'$$

If f_s' is negative, compression steel will be ineffective. If a singly-reinforced member would not be adequate, either depth d must be increased, or A_s' moved to a location where it will be effective.

$$\text{Required } A_s' = \frac{F_u(e''/\psi) - 0.85 f_c' b y_L \left(d - \frac{y_L}{2}\right)}{(f_s')_{\text{eff}}(d - d')} \tag{16-44}$$

If eq. (16-44) gives a positive A_s', compressive steel is required. In that case, the required tensile steel is

$$\text{Reqd } A_s = \frac{F_u/\psi + 0.85 f_c' b y_L + A_s'(f_s')_{\text{eff}}}{f_y} \tag{16-45}$$

TABLE 16-4 Code-Limit Values of y/d

Concrete Strength f_c' psi	Steel Yield Strength, psi		
	40,000	50,000	60,000
Up to 4,000	0.436	0.405	0.378
5,000	0.411	0.381	0.355
6,000	0.386	0.357	0.333

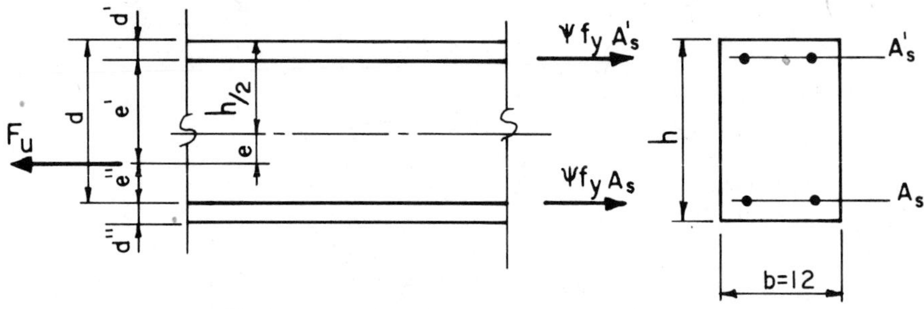

Fig. 16-15 Tension with small eccentricity (case I).

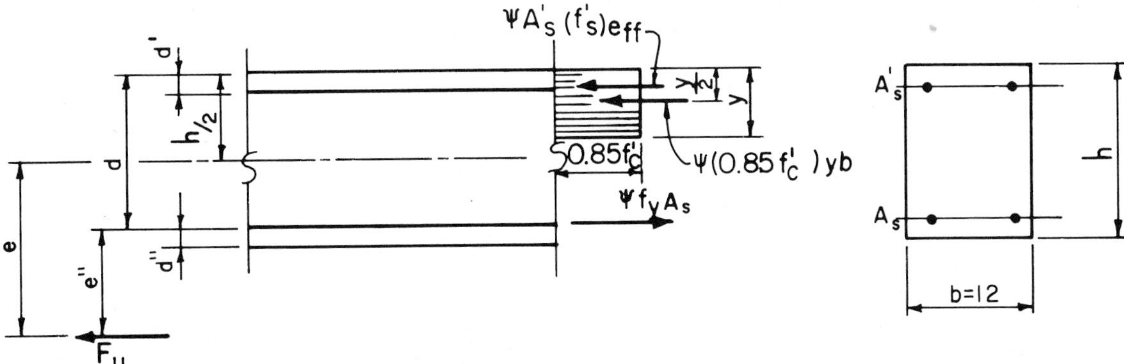

Fig. 16-16 Tension with large eccentricity (case II).

If eq. (16-44) gives a negative A_s', no compressive steel is needed. Then, regardless of whether or not negative steel is provided, the wall is designed as singly reinforced. In this case, the required tensile steel area is

$$A_s \text{ reqd} = \frac{F_u/\psi + 0.85 f_c' by}{f_y} \tag{16-46}$$

where y, actual depth of the compression block, is

$$y \simeq d - \sqrt{d^2 - \frac{2 F_u e''}{0.85 \psi f_c' b}} \tag{16-47}$$

16.10.3 Wall Thickness—Rectangular Silos

Walls of rectangular silos must have sufficient thickness to satisfy the following requirements:

1. For vertical compression, the USD requirements of Section 14.2 of the *ACI Code*.[16-27]
2. Flexural strength under combined tension and bending.
3. Shear. The ultimate shear stress at the face of the support for each wall should not exceed

$$v_c = 2(1 - 0.002 F_u/A_g) \sqrt{f_c'}, \text{ where} \tag{16-48}$$

$$v_u = \frac{V_u}{\psi bd} \tag{16-49}$$

4. In addition, the thickness should not be less than 6 in. for slipform construction unless special precautions are taken to reduce friction and avoid lifting the concrete.

16.10.4 Crack Width—Rectangular Silos

Crack width limits should be the same as for circular silos. Where the wall tensile force is between the two layers of steel (Case I, Fig. 16-15), the method shown for circular silos may be used to estimate the actual crack width.

With larger eccentricity (Case II, Fig. 16-16), the cracks are larger on one side but are closed where the concrete is in compression. Thus, the stored material should receive adequate protection at sections of large bending moment.

16.10.5 Temperature Effect—Rectangular Silos

If hot material is to be stored, extra steel may be needed. Computation of the thermal moments is complex. However, for the simplest case, a single square silo cross section, edge moments will keep the walls flat and the same moment will exist at all points between the edges. The horizontal ultimate moment, with Poisson's ratio = 0.3, will be

$$M_{xt,u} = M_{yt,u} = 1.43 K_g E_c h^2 \alpha_t \Delta T \text{ (in.-lb/ft)} \tag{16-50}$$

16.10.6 Other Vertical Forces and Moments

Vertical bending moments will occur near the bottom of walls which are continuous with a silo bottom slab or with hopper walls. The wall thickness and vertical steel must provide for such bending combined with the vertical forces from dead load, roof load and friction from the stored material. Bottom slabs or hoppers, suspended from the walls, will transfer vertical tensile forces to the lower portions of the silo walls. In such cases, the walls must also act as deep beams, transferring their vertical loads to corner columns by shear and in-plane bending.

16.10.7 Groups of Rectangular Silos

Figure 16-17 shows a typical group of rectangular silos with pyramidal hoppers. To compute the wall tensions, shears, and bending moments for such groups, various combinations of loaded and unloaded cells must be considered.

16.11 DETAILS—RECTANGULAR SILOS AND SILO GROUPS

All detail requirements given earlier for circular silos apply equally to rectangular silos.

16.11.1 Horizontal Steel

Ordinarily the horizontal steel (designed for combined tension and bending) will be in two layers. For all walls, the

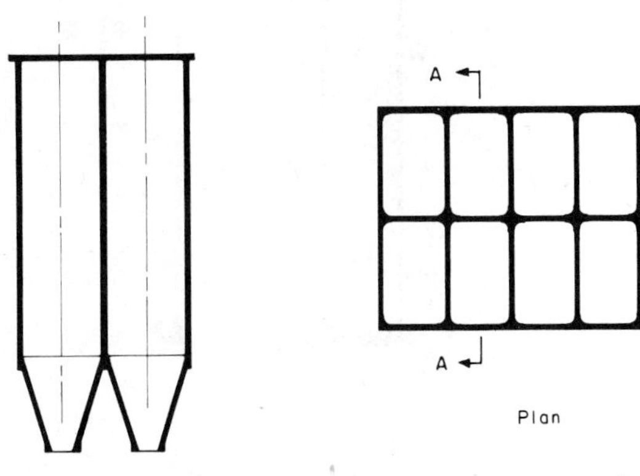

Fig. 16-17 Typical group of rectangular silos.

ACI requirements for continuing a portion of the positive steel over the full span and for bar extension and anchorage at supports and points of contraflexure must be observed. In slip-formed work, a 20% increase in splice lap length is recommended.

16.11.2 Corner Details

Fillets and corner reinforcing patterns are shown by Fig. 16-18. The fillet size should be at least equal to the wall thickness (but not less than 9 in.) and should have reinforcement equal to that computed for combined bending and axial load at the face of the wall—line x on Fig. 16-18a. The fillet steel should extend past its intersection with the inner wall steel a sufficient distance for anchorage.

16.12 REGULAR POLYGONAL SILOS

The principles of design and detail of single or grouped polygonal silos are similar to those for rectangular silos.

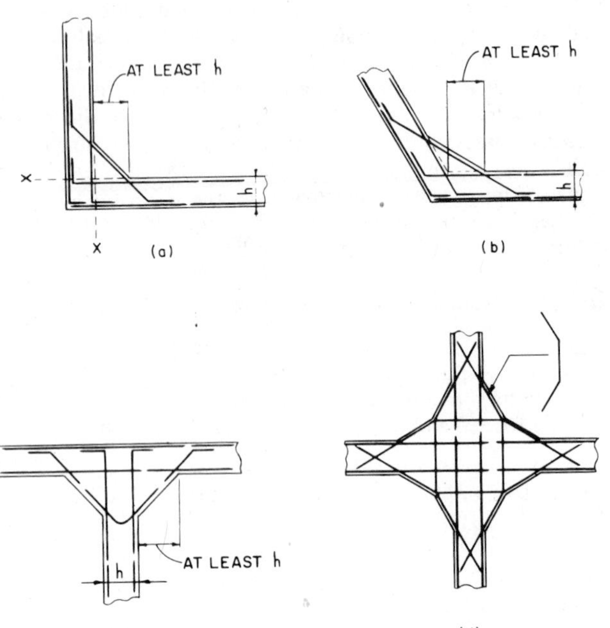

Fig. 16-18 Corner and intersection details for rectangular and polygonal silos.

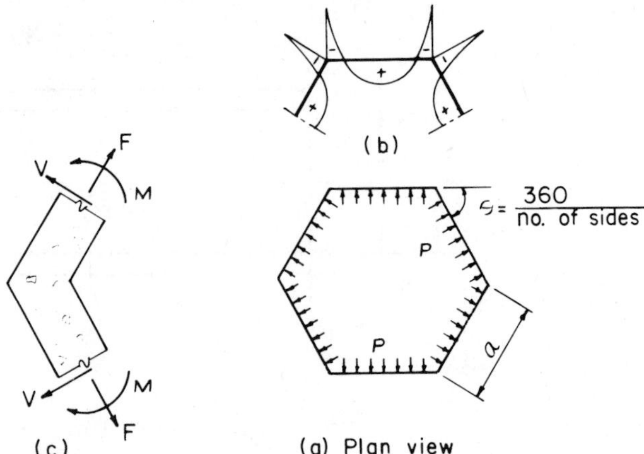

Fig. 16-19 Polygonal silo wall forces.

Their walls are subject to similar loading and to similar horizontal tension, shear and bending moment and vertical forces.

The horizontal ultimate tensile force in the wall of any single regular polygonal silo (Fig. 16-19) is

$$F_u = K_{ll} P_{des} \left[\frac{a}{2}\right]\left[\frac{\sin\theta}{1-\cos\theta}\right] \qquad (16\text{-}51)$$

This equation reduces to

$$F_u = 0.866\, K_{ll}\, a\, P_{des} \quad \text{(for hexagonal silos)} \qquad (16\text{-}52)$$

$$F_u = 1.207\, K_{ll}\, a\, P_{des} \quad \text{(for octagonal silos)} \qquad (16\text{-}53)$$

Bending moments $M_{x,u}$ and $M_{y,u}$ are determined as for rectangular silos. Reinforcement and details are as for rectangular silos, except that, for polygons with eight or more sides, wall intersection fillets may be omitted.

16.13 SILO BOTTOMS

Silo bottoms are supported by the silo walls extended down to the foundation or independently supported by columns resting either on the same foundations as the walls or independent foundations. See Fig. 16-20.

16.13.1 Loads

Silo bottom structures should be designed for load imposed by the stored material, their own dead load, the weight of

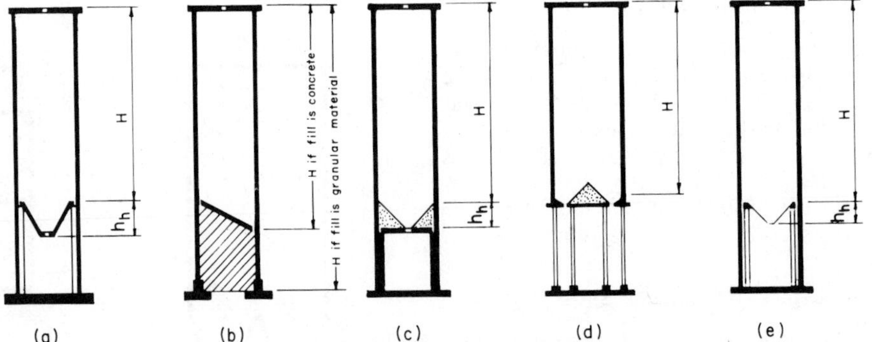

Fig. 16-20 Typical vertical cross sections of silos. (a) Silo on raft foundation, independent hopper resting on pilasters attached to wall; (b) silo with wall footings and independent bottom slab supported on fill; (c) silo with hopper-forming fill and bottom slab supported by thickened lower walls; (d) silo with multiple discharge openings and hopper-forming fill resting on bottom slab, all supported by columns—raft foundation has stiffening ribs on top surface; and (e) silo on raft foundation, with hopper independently supported by a ring-beam and column system.

any equipment, platforms, etc, hung from the silo bottom, and occasionally for earthquake and temperature effects.

16.13.2 Effect of Eccentric Discharge

The authors suggest that eccentric discharge be considered in the design of hoppers or of bottoms with hopper-forming fill since they are subjected to lateral pressures. It is recommended to assume the maximum increase for both the side near and the side opposite the eccentric discharge for square or rectangular silos and all around the perimeter for other shapes. The influence of eccentric discharge on vertical pressures is unknown, and current practice is to ignore eccentric discharge in the design of flat bottoms.

The maximum value of P_e (at depth H) is computed based on silo dimensions at the top of the hopper and is assumed constant for the depth of the hopper or hopper-forming fill.

Eccentric discharge should be considered in any case with multiple openings.

16.13.3 Earthquake Forces on Silo Bottoms

Where independent bottom supports are provided, lateral earthquake forces should be divided between the walls and the bottom support structure as suggested earlier. In addition, since wall friction may be destroyed by quake action, the entire vertical weight of the stored material should be considered applied to the bottom structure.

16.13.4 Flat Bottoms

The simplest flat bottom is a slab of uniform thickness. The flat bottom may also be a ribbed slab or a beam-and-slab system. To assure good flow of material, hopper-forming fill is usually constructed around the discharge opening. Concentric slab openings should not exceed one-third of the silo diameter.

For a slab without hopper-forming fill, the design loads are dead load and pressure, q_{des}, computed at the top of the slab. For USD methods, the vertical load per unit area would be

$$w_u = K_g g + K_{ll} q_{des} \qquad (16\text{-}54)$$

With earthquake, vertical friction at the wall is assumed to be zero, so that the ultimate vertical pressure on the bottom is

$$w_u = 0.75(K_g g + K_{ll} \gamma H) \qquad (16\text{-}55)$$

For a slab with hopper-forming fill, the weight of the fill itself and of the material stored in the hopper are added as dead load and q_{des}, for use in eq. (16-54), is computed at the top of the fill.

Slab shear stresses should be checked carefully.

16.13.5 Moments and Deflections for Circular Slab Bottoms

Radial and tangential bending moments for circular slabs with (or without) a central hole can be computed by the methods of Timoshenko[16-29] or from tables.[16-28] When a concentric opening is not over one-tenth of the silo diameter (or, for rectangular silos, the largest side of the opening does not exceed one-tenth of the short side of the silo), the opening is frequently ignored in computing slab bending moments. In this case, extra bottom reinforcement to replace that eliminated by the hole and added top reinforcement should be provided around the opening, or the methods of exs. 16-3 and 16-4 can be used.

16.13.6 Rectangular Flat Bottoms

Economy usually limits the wall length in rectangular silos to a maximum of about 20 ft. Thus, the bottom slab span is small and such slabs are usually supported along their boundaries only.

Moments and deflections of simple and fixed-edge rectangular slabs may be calculated using the methods presented by the *ACI Building Code* for analysis of two-way slabs. The effect of circular or rectangular discharge openings in the bottom slabs should be considered in the design. The extra reinforcing added to compensate for that intercepted by the opening should be continuous from support to support.

16.13.7 Additional Load at Openings

The edges of bottom slab openings usually receive added line loads comprising the weight of the hopper and equipment supported by it, the weight of stored material in the hopper and the design vertical pressures of material above the top of the hopper. Equations by Griffel[16-30] and tables by the authors[16-28] and others[16-22,16-31] may be used.

16.13.8 Conical Hoppers

The design pressure, $q_{\alpha,\,des}$, may be computed from eq. (16-22). In computing $q_{\alpha,\,des}$, engineers usually use the dimensions at the top of the hopper to obtain R or C for the Janssen or Reimbert formulas, ignoring the reductions of cross section within the hopper.

The conical hopper shell is subject to two tensile membrane forces. The meridional force, F_m, is parallel to the generator line of the cone. The tangential force, F_t, is in the plane of the shell and horizontal. The meridional force per unit width at depth Y is computed from equilibrium of the loads on the cone below that depth. These loads, shown on Fig. 16-21a, are the resultant of vertical pressures, q_{des} (at depth Y) and W, the combined weights of the hopper itself and material stored below depth Y plus any equipment supported by the hopper. (See Notation.)

For design by USD

$$F_{mu} = K_{ll}\left[\frac{q_{des} D}{4 \sin \alpha} + \frac{W_{ll}}{\pi D \sin \alpha}\right] + K_g \left[\frac{W_g}{\pi D \sin \alpha}\right] \qquad (16\text{-}56)$$

$$F_{tu} = K_{ll}\left[\frac{q_{\alpha,\,des} D}{2 \sin \alpha}\right] \qquad (16\text{-}57)$$

Both forces are maximum at the upper edge of the hopper, and approach zero at the lower edge.

The minimum acceptable thickness for the cone should be determined considering the acceptable crack width.[16-2,16-26] The authors prefer, however, that the shell thickness never be less than 5 in.

The required reinforcement area per unit width of shell is

$$A_s \text{ reqd} = F_{mu}/(\psi f_y) \quad \text{(meridional direction)} \qquad (16\text{-}58)$$

$$A_s \text{ reqd} = F_{tu}/(\psi f_y) \quad \text{(horizontal)} \qquad (16\text{-}59)$$

A conical hopper is usually supported at its upper end by a ring beam. This is merely a thickened portion of the conical shell with a cross section to satisfy the supporting condition and the loading. The ring beam depth should not be less than one-tenth of the hopper diameter.

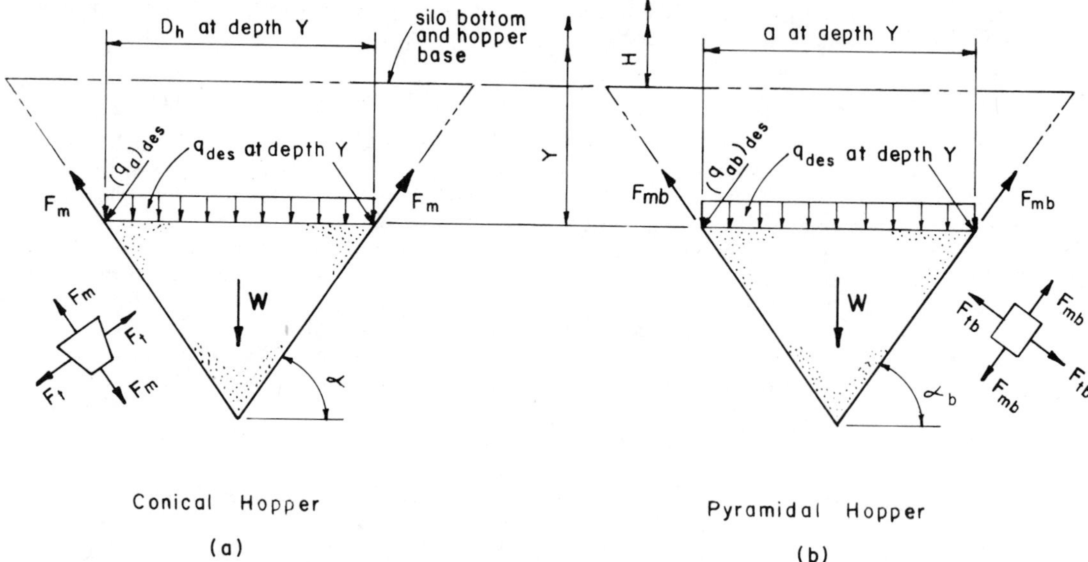

Fig. 16-21 Forces in conical and pyramidal hoppers.

If simply supported all around by the silo walls, ring beams are usually designed for the horizontal component of F_{mu} only. If the hopper wall is eccentric to the centroid of the ring beam, the beam will also receive uniform bending moment. The monolithically cast ring beam and conical shell is very stiff, however, and this moment is usually neglected in design of the ring beam with all-around support. The area of longitudinal steel in such ring beams is arbitrary, but should be not less than 0.5% of the cross-sectional area of the ring beam. (See Fig. 16-22.)

If the upper edge of the hopper is keyed all around to the silo walls or monolithic with the walls, adequate negative reinforcement should be provided in the hopper wall at the intersection of hopper and silo walls. The negative steel should be extended into the silo walls for complete anchorage by bond.

The ring beam and upper edge of a conical hopper supported at isolated points along its boundary by columns, pilasters or wall pockets may be designed in the manner shown later herein for a concrete ring beam supporting a steel hopper. If desired, the stiffness and strength added by the concrete hopper shell may be approximated by considering a width of hopper shell (the authors suggest four times the thickness) to act as a part of the ring beam.

16.13.9 Pyramidal Hoppers

Loads for pyramidal hoppers are the same as for conical hoppers. (See Fig. 16-21b.) Pyramidal hopper walls, however, are subject to bending as well as tensile membrane forces. The bending always includes two-way, plate-type bending and may include significant in-plane bending.

All of the following is for symmetrical (or nearly symmetrical) hoppers only. The angles of slope for walls 'a' and 'b' are α_a and α_b, respectively.

The membrane tensile forces in the inclined walls are assumed to consist of horizontal forces, F_t, and forces, F_m, in the plane of the wall and normal to F_t. Forces F_m per unit width vary in intensity along the width of the wall. For simplicity, however, they are usually assumed to be uniform along any one wall. Material pressures for each wall are assumed as those computed using the properties of the silo wall to which it is attached, plus a share of the dead load. In the hoppers of Fig. 16-23, for example, material pres-

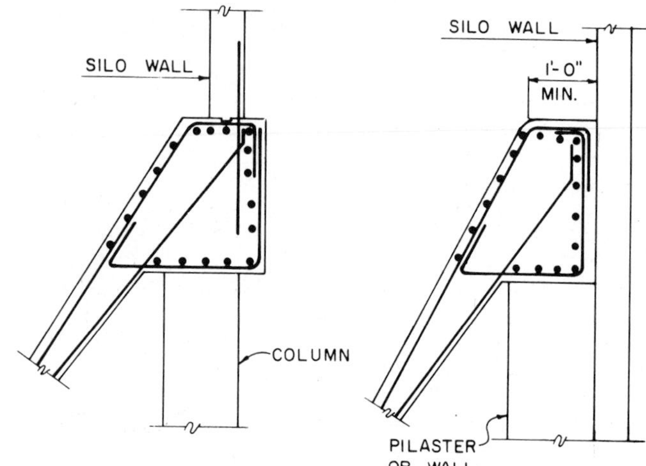

Fig. 16-22 Typical details of hopper-supporting beam.

sures applied to area A_a are assumed to affect the value of F_m for wall 'a', and those applied to area A_b to affect F_m for wall 'b'. If the area of the discharge opening is ignored and if some arbitrary factors c_a and c_b are selected to define the division of the load to area A_a and A_b (such that, for a symmetrical case, $2c_a + 2c_b = 1.0$), then the ultimate meridional forces per unit width of wall are

$$F_{mau} = \frac{K_{ll}(c_a W_{ll} + A_a q_{a,\mathrm{des}}) + K_g c_a W_g}{a \sin \alpha_a} \tag{16-60}$$

$$F_{mbu} = \frac{K_{ll}(c_b W_{ll} + A_b q_{b,\mathrm{des}}) + K_g c_b W_g}{b \sin \alpha_b} \tag{16-61}$$

(In the above, $q_{a,\mathrm{des}}$ is computed using the properties for the short wall and $q_{b,\mathrm{des}}$ using the imaginary square silo. For simplicity, some engineers use the average of $q_{a,\mathrm{des}}$ and $q_{b,\mathrm{des}}$ for all walls of the hopper.)

The ultimate horizontal forces, F_{tau} and F_{tbu}, per unit width of wall at depth Y may be approximated by

$$F_{tau} = K_{ll}\left(\frac{b}{2}\right) q_{ab,\mathrm{des}} \sin \alpha_a \tag{16-62}$$

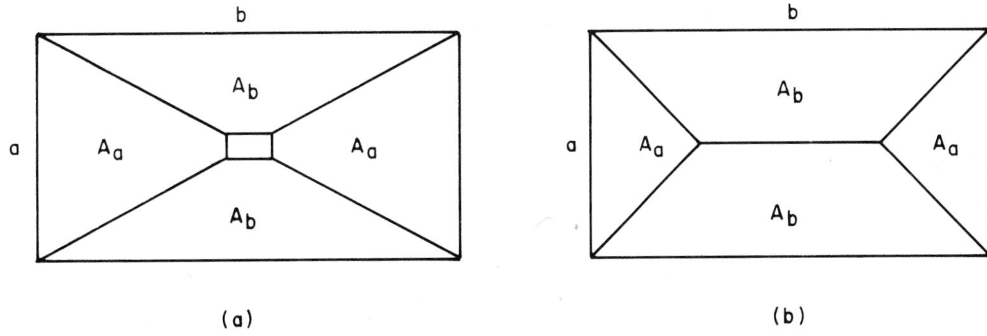

Fig. 16-23 Plan view of pyramidal hoppers.

TABLE 16-5 Isosceles Triangular Slab, Fixed Edges, Uniform Load

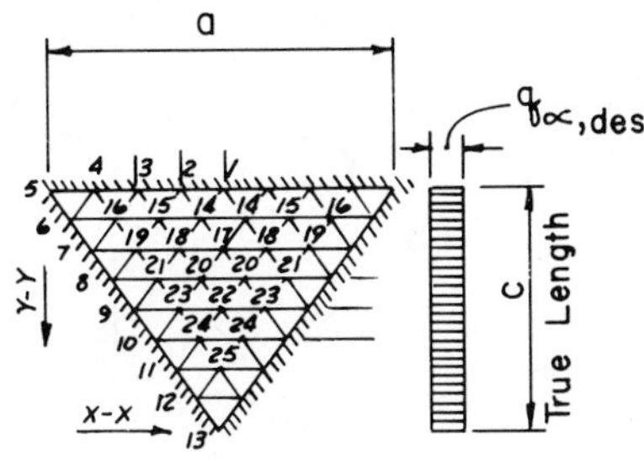

Bending Moments: $M_x = n_x \, q_{\alpha,des} \cdot a^2/64$ (per unit width)

$M_y = n_y \, q_{\alpha,des} \cdot a^2/64$ (per unit width)

Reaction per unit length of support: $R = \gamma \, q_{\alpha,des} \cdot a/8$

* Deflection: $\Delta = \dfrac{\lambda \, q_{\alpha,des} \, a^4 (1-\nu^2)}{4096 \, EI}$

* Consistent units must be used in above equations. To obtain Δ in inches when a = feet, q_α = lb/sq ft, E = psi and I = in.^{4}/ft, multiply the above expression by 1,728.

Ratio a/c = 0.75				
Point	λ	n_x	n_y	γ
1	0	−0,2330	−1,3977	2,444
2	0	−0,2008	−1,2044	2,184
3	0	−0,1185	−0,7111	1,428
4	0	−0,0343	−0,2055	0,442
5	0	0	0	0
6	0	−0,5758	−0,1710	1,148
7	0	−1,3155	−0,3559	2,175
8	0	−1,6064	−0,4476	2,531
9	0	−1,3521	−0,3939	2,240
10	0	−0,8259	−0,2561	1,633
11	0	−0,3147	−0,1180	1,016
12	0	−0,0600	−0,0330	0,493
13	0	0	0	0
14	1,24235	0,3990	0,3901	—
15	0,89879	0,2337	0,3524	—
16	0,37532	−0,1088	0,2113	—
17	1,99139	0,8284	0,6706	—
18	1,62283	0,5979	0,5655	—
19	0,73625	−0,1062	0,2396	—
20	1,58247	0,8117	0,3348	—
21	0,80196	0,0384	0,0702	—
22	0,98360	0,7271	0,0487	—
23	0,61384	0,2260	−0,0681	—
24	0,33246	0,3069	−0,0981	—
25	0,11205	0,2037	−0,0551	—

Ratio a/c = 1.00				
Point	λ	n_x	n_y	γ
1	0	−0,2091	−1,2547	2,219
2	0	−0,1787	−1,0721	1,968
3	0	−0,1039	−0,6232	1,294
4	0	−0,0297	−0,1784	0,481
5	0	0	0	0
6	0	−0,2241	−0,0958	0,577
7	0	−0,6748	−0,2598	1,515
8	0	−1,0164	−0,4016	2,118
9	0	−1,1018	−0,4276	2,204
10	0	−0,8148	−0,3305	1,796
11	0	−0,4169	−0,1754	1,116
12	0	−0,0984	−0,0502	0,481
13	0	0	0	0
14	0,62733	0,2066	0,1750	—
15	0,44479	0,1141	0,1954	—
16	0,17844	−0,0600	0,1530	—
17	1,17677	0,5347	0,5865	—
18	0,95205	0,3819	0,5125	—
19	0,42236	−0,0605	0,2531	—
20	1,10230	0,6009	0,4303	—
21	0,55797	0,0119	0,1719	—
22	0,81779	0,6222	0,1576	—
23	0,51131	0,2013	0,0131	—
24	0,32460	0,2989	−0,1003	—
25	0,11584	0,2065	−0,1122	—

Ratio a/c = 1.50				
Point	λ	n_x	n_y	γ
1	0	−0,1536	−0,9215	1,823
2	0	−0,1288	−0,7729	1,603
3	0	−0,0715	−0,4291	1,061
4	0	−0,0195	−0,1169	0,478
5	0	0	0	0
6	0	−0,0526	−0,0122	0,318
7	0	−0,1901	−0,1073	0,780
8	0	−0,3791	−0,2513	1,313
9	0	−0,5271	−0,3761	1,708
10	0	−0,5128	−0,3998	1,774
11	0	−0,3970	−0,2961	1,393
12	0	−0,1747	−0,1108	0,613
13	0	−0,0858	0,0375	0
14	0,20177	0,0622	−0,0120	—
15	0,13276	0,0284	0,0193	—
16	0,05195	−0,0201	0,0811	—
17	0,45050	0,2466	0,3756	—
18	0,35592	0,1670	0,3112	—
19	0,14802	−0,0242	0,2013	—
20	0,49613	0,3154	0,4033	—
21	0,24086	0,0210	0,2103	—
22	0,45203	0,3834	0,2662	—
23	0,27768	0,1209	0,1227	—
24	0,22829	0,2211	−0,0051	—
25	0,10763	0,1902	−0,1145	—

NOTE: For trapezoidal loading, use results from Table 16-5 ± results from Table 16-6. Table reconstructed after Lipnitski, Ref. 16-22.

$$F_{tbu} = K_{ll} \left(\frac{a}{2}\right) q_{\alpha a,\mathrm{des}} \sin \alpha_b \qquad (16\text{-}63)$$

Moments from two-way, plate-type bending (sometimes called local bending) may be computed using Tables 16-5 and 16-6 for inclined triangular walls with fixed sides and subject to either uniform or triangular loading, and (by superposition) for trapezoidal loading. Tables 16-5 and 16-6 may also be used for trapezoidal walls in which the length of the bottom edge is not over one-fourth of the top edge, by substituting for the trapezoid an imaginary triangle of slant height[16-22]

$$c_i = \frac{ca}{a - a_1} \qquad (16\text{-}64)$$

In the above, c is the actual slant height of the trapezoid and a_1 and a are the bottom and top edge lengths, respectively.

Approximate solutions for bending moments in trape-

TABLE 16-6 Isosceles Triangular Slab, Fixed Edges, Uniformly Varying Load

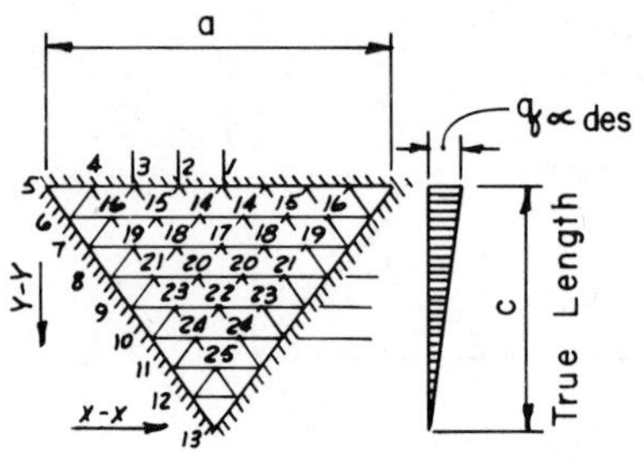

Bending Moments: $M_x = n_x \, q_{\alpha,des} \cdot a^2/64$ (per unit width)

$M_y = n_y \, q_{\alpha,des} \cdot a^2/64$ (per unit width)

Reaction per unit length of support: $R = \gamma \, q_{\alpha,des} \cdot a/8$

* Deflection: $\Delta = \dfrac{\lambda \, q_{\alpha,des} \cdot a^4}{4096 \, EI}$

Ratio a/c = 0.75				
Point	λ	n_x	n_y	γ
1	0	−0,2330	−1,3977	2,444
2	0	−0,2003	−1,2044	2,184
3	0	−0,1185	−0,7111	1,428
4	0	−0,0343	−0,2055	0,442
5	0	0	0	0
6	0	−0,5758	−0,1710	1,148
7	0	−1,3155	−0,3559	2,175
8	0	−1,6064	−0,4476	2,531
9	0	−1,3521	−0,3939	2,240
10	0	−0,8259	−0,2561	1,633
11	0	−0,3447	−0,1180	1,016
12	0	−0,0600	−0,0330	0,493
13	0	0	0	0
14	1,24235	0,3990	0,3901	—
15	0,89879	0,2337	0,3524	—
16	0,37532	−0,1088	0,2113	—
17	1,99139	0,8284	0,6706	—
18	1,62283	0,5979	0,5655	—
19	0,73625	−0,1062	0,2396	—
20	1,58247	0,8117	0,3348	—
21	0,80196	0,0384	0,0702	—
22	0,98360	0,7271	0,0487	—
23	0,61334	0,2260	−0,0681	—
24	0,33246	0,3069	−0,0981	—
25	0,11205	0,2037	−0,0551	—

Ratio a/c = 1.00				
Point	λ	n_x	n_y	γ
1	0	−0,2091	−1,2547	2,219
2	0	−0,1787	−1,0721	1,968
3	0	−0,1039	−0,6232	1,294
4	0	−0,0297	−0,1784	0,481
5	0	0	0	0
6	0	−0,2241	−0,0958	0,577
7	0	−0,6748	−0,2598	1,515
3	0	−1,0464	−0,4016	2,118
9	0	−1,1018	−0,4276	2,204
10	0	−0,8148	−0,3305	1,796
11	0	−0,4169	−0,1754	1,116
12	0	−0,0984	−0,0502	0,481
13	0	0	0	0
14	0,62733	0,2066	0,1750	—
15	0,44479	0,1141	0,1954	—
16	0,17844	−0,0600	0,1530	—
17	1,17677	0,5347	0,5865	—
18	0,95205	0,3819	0,5125	—
19	0,42236	−0,0605	0,2531	—
20	1,10230	0,6009	0,4303	—
21	0,53797	0,0119	0,1719	—
22	0,81779	0,6222	0,1576	—
23	0,51131	0,2013	0,0131	—
24	0,32460	0,2989	−0,1003	—
25	0,11534	0,2065	−0,1122	—

Ratio a/c = 1.50				
Point	λ	n_x	n_y	γ
1	0	−0,1536	−0,9215	1,823
2	0	−0,1288	−0,7729	1,603
3	0	−0,0715	−0,4291	1,061
4	0	−0,0195	−0,1169	0,478
5	0	0	0	0
6	0	−0,0526	−0,0122	0,318
7	0	−0,1901	−0,1073	0,780
8	0	−0,3794	−0,2513	1,313
9	0	−0,5271	−0,3761	1,708
10	0	−0,5128	−0,3998	1,774
11	0	−0,3970	−0,2961	1,393
12	0	−0,1747	−0,1108	0,613
13	0	−0,0858	0,0375	0
14	0,20177	0,0622	−0,0120	—
15	0,13276	0,0284	0,0493	—
16	0,05195	−0,0201	0,0811	—
17	0,45050	0,2466	0,3756	—
18	0,35592	0,1670	0,3412	—
19	0,14802	−0,0242	0,2013	—
20	0,49613	0,3154	0,4033	—
21	0,24086	0,0210	0,2103	—
22	0,45203	0,3834	0,2662	—
23	0,27768	0,1209	0,1227	—
24	0,22829	0,2211	−0,0051	—
25	0,10763	0,1902	−0,1145	—

NOTE: For trapezoidal loading, use results from Table 16-5 ± results from Table 16-6. Table reconstructed after Lipnitski, Ref. 16-22.
**See note on Table 16-5.*

zoidal walls with bottom edges exceeding $a/4$ (either symmetrical or slightly unsymmetrical) may be obtained by solving for equivalent rectangular plates of width, a_{eq}, and height, c_i, as follows:[16-32]

$$a_{eq} = \frac{2(a^2 + 2a + a_1)}{3(a + a_1)} \tag{16-65}$$

$$c_i = c - \frac{a(a - a_1)}{6(a + a_1)} \tag{16-66}$$

16.13.10 Thickness and Reinforcement of Pyramidal Hopper Walls

Wall thickness and reinforcement must be selected to keep crack width within acceptable limits. However, in pyramidal hoppers, combined bending and axial tension in both the meridional and horizontal directions must be considered. The authors prefer that the minimum wall thickness be not less than required for bending alone nor less than 6 in. Walls of pyramidal hoppers should be checked for shear.[16-27] See eqq. (16-48) and (16-49).

16.13.11 Pyramidal Hopper Details

Pyramidal hoppers should have fillets at the inside of hopper wall intersections. The fillet size usually matches that of the silo walls at the top of the hopper and tapers to zero at the discharge opening. See Fig. 16-24.

Hopper wall steel is placed either

1. at both faces of the wall, or
2. continuously at the outer face with inside-face steel near the corners and upper and lower edges only, where required for negative moment.

The first approach seems more practical in the U.S., although the second is common in Europe.

The steel areas (each way) should be not less than 0.0025 times the wall area. Concrete cover for reinforcing bars near the inside face should be at least 1 in., but may be reduced to 0.75 in. if a special inside lining is provided. Cover for other bars should conform to the *ACI Code*.

The discharge end should be properly reinforced and detailed. Concrete flanges are usually provided around the opening with embedded anchoring devices to fasten discharge gates, chutes, etc.

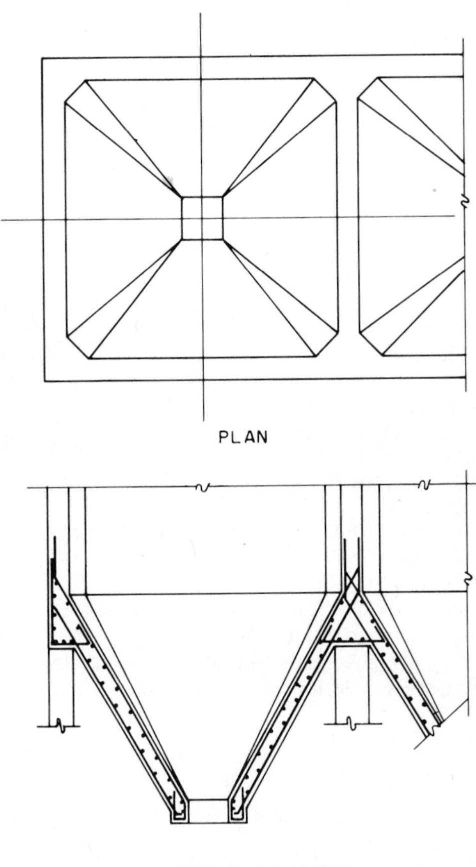

Fig. 16-24 Typical pyramidal hopper.

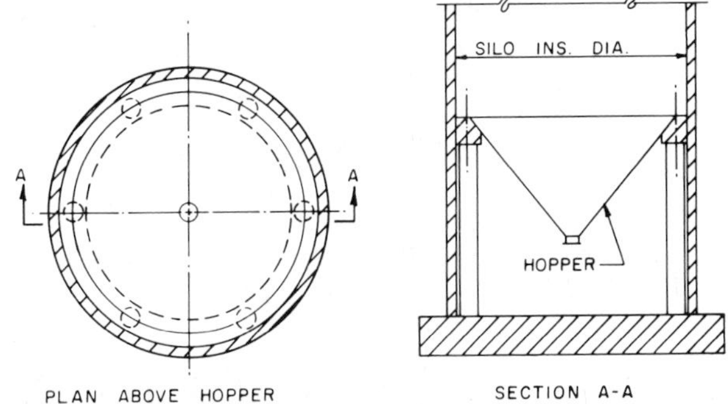

Fig. 16-25 Silo bottom—steel hopper supported on concrete ring-beam-and-column system.

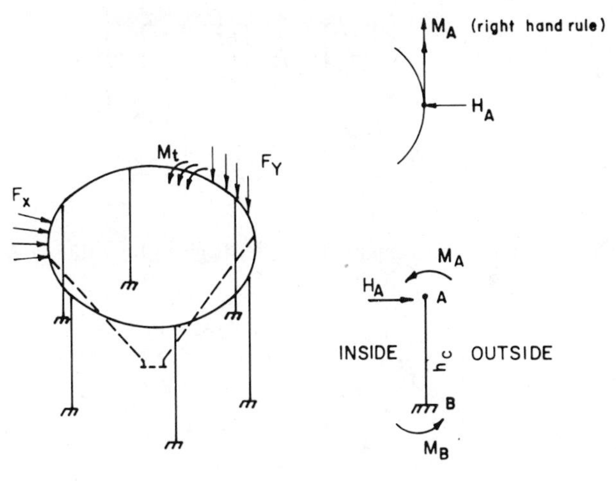

Fig. 16-26

16.13.12 Circular Concrete Ring-Beam-and-Column System Supporting a Steel Hopper

A steel hopper can do little to stiffen a concrete ring beam, therefore, all of the loads it applies to the beam—torsion included—must be considered. The tables and equations which follow cover the design of such ring beams and columns having an even number (4 to 12) of equally-spaced columns, fixed at their bases, and rigidly connected to the ring beam. (See Fig. 16-25.)

Equations 16-67 to 16-77 are for working design loads (WSD) rather than ultimate. In this type of structure live load is much larger than dead load. Consequently, it is simpler and sufficiently accurate to apply a common load factor, K_{II}, to the final answers in order to determine the USD values.

The external design loads acting on the ring beam (Figs. 16-26 and 16-27) are approximately

$$F_x = \frac{F_{mu} \cos \alpha}{1.7}, \quad \text{using } D = \text{silo inside diameter in eq. (16-56) to compute } F_{mu}. \tag{16-67}$$

$$F_y = g_r + \frac{F_{mu} \sin \alpha}{1.7} \tag{16-68}$$

where g_r is the weight of the ring beam per unit length. If the vertical pressure (by stored material) acting on the face of the ring beam is significant, it should be added to F_y (See ex. 16-2.)

The WSD uniform torsional moment is

$$M_t = F_m e \quad \text{(ft-lb/linear ft)} \tag{16-69}$$

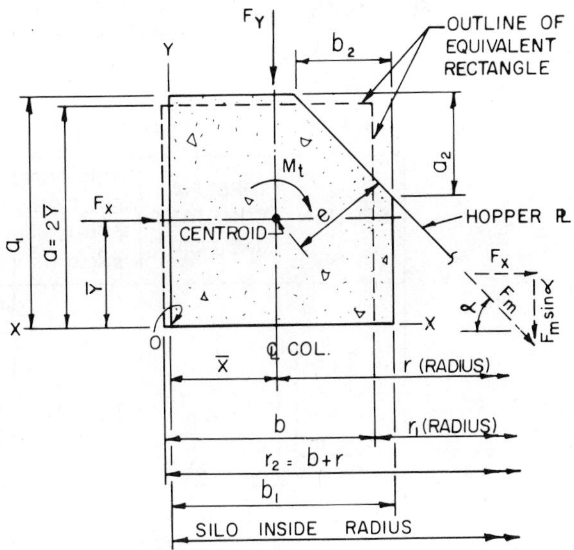

Fig. 16-27 Ring beam cross section.

The cross-sectional area of ring beam (Fig. 16-27) is

$$A_r = a_1 b_1 - \frac{b_2 a_2}{2} \tag{16-70}$$

Coordinates of the centroid measured from origin O are

$$\bar{x} = \frac{a_1 b_1^2/2 - (a_2 b_2/2)(b_1 - b_2/3)}{A_r} \qquad (16\text{-}71)$$

$$\bar{y} = \frac{a_1^2 b_1/2 - (a_2 b_2/2)(a_1 - a_2/3)}{A_r} \qquad (16\text{-}72)$$

An equivalent rectangle (dotted on Fig. 16-27) of height 'a' and width 'b' is substituted for the pentagon.

$$a = 2\bar{y} \qquad (16\text{-}73)$$

$$b = A_r/a \qquad (16\text{-}74)$$

The column shear, H_A, and upper end moment, M_A, are found by solving simultaneously eqq. (16-75) and (16-76).

$$\frac{F_x r^2}{A_r} = M_A \left[\frac{h_c^2}{2I_c}\right] - H_A \left[\frac{h_c^3}{3I_c} + \frac{\eta r^3}{2I_r}\right] \qquad (16\text{-}75)$$

$$\frac{12 M_t r}{a^3 \ln\left(\frac{r_2}{r_1}\right)} = M_A \left[\frac{h_c}{I_c} + \frac{\pi r z}{6.8 b^4 \lambda}\right] - H_A \left[\frac{h_c^2}{2I_c}\right] \qquad (16\text{-}76)$$

where η and z are numerical coeffients given by Table 16-7 and λ is a torsional property of the equivalent rectangular section, given by Table 16-8.

The WSD moment at the column base is then

$$M_B = H_A h_c - M_A \qquad (16\text{-}77)$$

The maximum WSD values of vertical shear, thrust, torque, and bending moment at the supports, midspans, and points of maximum torsion, for ring beams on 4, 6, 8, 10, or 12 supports are tabulated in Table 16-9.

For USD, the final forces and moments given by the preceeding equations or Table 16-9 should be multiplied by the load factor, K_{ll}. In all of the above equations, consistent units must be used. For example: F_x = lb/ft; M_t = ft-lb/ft; r, h_c, a, and b = ft; $I = \text{ft}^4$; E = lb/sq ft; A = sq ft; and M_T = ft-lb.

16.13.13 Column Design

Columns supporting silos or silo bottoms have large ratios of live to dead load. When unloaded after long periods of full loading, such columns may crack horizontally. To minimize such cracking, Lipnitski[16-22] suggests 1.5% for the upper limit of vertical reinforcing. The authors use a 2% limit unless a higher limit can be justified by analysis of the concrete behavior under long-term load.

TABLE 16-7 Numerical Coefficients η and z

NUMBER OF SUPPORTS	η	z
4	.012159	1.000
6	.003364	1.1153
8	.001387	1.3633
10	.000701	1.6199
12	.000404	1.8972

TABLE 16-8 Torsional Properties of Rectangular Cross Sections

SECTION	TORSIONAL CONSTANT in.^{4}, (cm)^4	TORSIONAL SECTION MODULUS in.^{3}, (cm)^3	POINTS OF MAX SHEAR STRESSES lb/in.^{2} (kg/cm^2)	VALUES OF COEFFICIENTS λ, β & γ			
				$m = \frac{a}{b}$	λ	β	γ
a > b (solid rectangle, height a, width b)	$J = \lambda b^4$	$Z_t = \beta b^3$	MIDDLE OF LONG SIDES $\tau_{max} = \frac{Torque}{Z_t}$; MIDDLE OF SHORT SIDES $\tau = \gamma \times \tau_{max}$; AT CORNER $\tau = 0$	1.0	0.140	0.208	1.00
				1.5	0.294	0.346	0.859
				2.0	0.457	0.493	0.795
				3.0	0.790	0.801	0.753
				4.0	1.123	1.150	0.745
a > b (narrow rectangle, height a, width b)	$J = \frac{(m-0.63)b^4}{3}$	$Z_t = \frac{(m-0.63)b^3}{}$	ALL POINTS ON LONG SIDE EXCEPT CORNERS $\tau_{max} = \frac{Torque}{Z_t}$; MIDDLE OF SHORT SIDES $\tau = 0.74 \times \tau_{max}$	6.0	1.789	1.789	0.743
				8.0	2.456	2.456	0.742
				10.0	3.123	3.123	0.742

TABLE 16-9 Summary of Shears, Thrusts, Torques, and Bending Moments in Ring Beam

NUMBER OF SUPPORTS	LOCATION OF FORCE	FORCE						
		VERTICAL SHEAR	RING COMPRESSIVE FORCE DUE TO F_X	BENDING MOMENT DUE TO M_t	BENDING MOMENT DUE TO F_Y	BENDING MOMENT DUE TO M_A	TORSIONAL MOMENT DUE TO F_Y	TORSIONAL MOMENT DUE TO M_A
4	SUPPORT	$.7854\ r\ F_Y$	$F_X\ r$	$M_t\ r$	$-.2146 r^2 F_Y$	$-.5000\ M_A$	0	$.5000\ M_A$
	MIDSPAN	0	$F_X\ r$	$M_t\ r$	$.1107\ r^2 F_Y$	$-.7071\ M_A$	0	0
	*19°-12' FROM SUPPORT	$.4503\ r\ F_Y$	$F_X\ r$	$M_t\ r$	0	$-.6366\ M_A$	$.0333\ r^2 F_Y$	$.3069\ M_A$
6	SUPPORT	$.5236\ r\ F_Y$	$F_X\ r$	$M_t\ r$	$-.0931 r^2 F_Y$	$-.8660\ M_A$	0	$.500\ M_A$
	MIDSPAN	0	$F_X\ r$	$M_t\ r$	$.0472\ r^2 F_Y$	$-1.000\ M_A$	0	0
	*12°-44' FROM SUPPORT	$.3014\ r\ F_Y$	$F_X\ r$	$M_t\ r$	0	$-.9549\ M_A$	$.0095\ r^2 F_Y$	$.2982\ M_A$
8	SUPPORT	$.3927\ r\ F_Y$	$F_X\ r$	$M_t\ r$	$-.0520 r^2 F_Y$	$-1.2071\ M_A$	0	$.500\ M_A$
	MIDSPAN	0	$F_X\ r$	$M_t\ r$	$.0261\ r^2 F_Y$	$-1.3071\ M_A$	0	0
	*9°-33' FROM SUPPORT	$.2260\ r\ F_Y$	$F_X\ r$	$M_t\ r$	0	$-1.2880\ M_A$	$.0040\ r^2 F_Y$	$.2928\ M_A$
10	SUPPORT	$.3142\ r\ F_Y$	$F_X\ r$	$M_t\ r$	$-.0331\ r^2 F_Y$	$-1.5390\ M_A$	0	$.500\ M_A$
	MIDSPAN	0	$F_X\ r$	$M_t\ r$	$.0167\ r^2 F_Y$	$-1.6170\ M_A$	0	0
	7°-37' FROM SUPPORT	$.1812\ r\ F_Y$	$F_X\ r$	$M_t\ r$	0	$-1.5915\ M_A$	$.0020\ r^2 F_Y$	$.2916\ M_A$
12	SUPPORT	$.2618\ r\ F_Y$	$F_X\ r$	$M_t\ r$	$-.0229\ r^2 F_Y$	$-1.8660\ M_A$	0	$.500\ M_A$
	MIDSPAN	0	$F_X\ r$	$M_t\ r$	$.0119\ r^2 F_Y$	$-1.9313\ M_A$	0	0
	*6°-21' FROM SUPPORT	$.1510\ r\ F_Y$	$F_X\ r$	$M_t\ r$	0	$-1.9049\ M_A$	$.0011\ r^2 F_Y$	$.2905\ M_A$

* POINT OF MAXIMUM TORSION DUE TO F_Y

16-14 DESIGN AND DETAILING OF REINFORCED CONCRETE BUNKERS

Bunkers may be square, rectangular, circular, or polygonal —single or in groups. (Fig. 16-2.) Bunkers may be symmetrical or nonsymmetrical about either, or both, principal axes, according to the location of the discharge opening. Circular bunkers usually have conical bottoms and single round discharge openings. Rectangular ones usually have pyramidal bottoms with one or more square, circular or elongated discharge openings. Flat bottoms are seldom used since the undesirable dead storage reduces the efficiency of material flow. This can be overcome, however, by hopper-forming concrete fill.

A bunker may consist of the pyramidal or conical hopper only, without vertical walls. In this case a continuous horizontal supporting beam is located along the upper edge and resists both the vertical and horizontal forces and torsion applied by the hopper shell.

Bunkers generally require a protective surface to guard against abrasion by falling sliding material. Usually, lining is necessary only on the sloping bottoms, whose surfaces receive the greatest abrasion and impact.

16.14.1 Loads and Forces

Bunkers are designed for loads imposed by stored material, their own weight and the weight of equipment and platforms carried by the structure. Impact, roof loads, wind, and earthquake may also need to be considered.

Material pressures on bunker walls and bottoms may be determined by the Janssen or Reimbert equations for silos, or by the Rankine method, which follows. This theory is not entirely accurate since it is based on assumptions not fully met and ignores boundary conditions. Friction forces on the walls are neglected and pressures are assumed to be normal to the surface against which they act.

If the top surface of the stored material is horizontal:

Vertical static unit pressure at depth Y below the surface is

$$q = \gamma Y \tag{16-78}$$

Horizontal static unit pressure at depth Y is

$$P = k\gamma Y \tag{16-79}$$

where $k = (1 - \sin\phi)/(1 + \sin\phi)$

Unit static pressure normal to an inclined surface at depth Y is

$$q_\alpha = \gamma Y(\cos^2\alpha + k\sin^2\alpha) \tag{16-80}$$

If the surface of the stored material slopes at the angle of repose, approximately equal to ϕ (see Fig. 16-28):

Vertical static unit pressure at depth Y is

$$q = \gamma(Y + a_0\tan\phi) \tag{16-81}$$

Horizontal static unit pressure at depth Y is

$$P = \gamma Y\cos^2\phi \tag{16-82}$$

Unit static pressure normal to the inclined surface of the hopper wall at depth Y is

$$q_\alpha = \gamma\cos^2\alpha(Y + a_0\tan\phi) + \gamma Y\cos^2\phi\sin^2\alpha \tag{16-83}$$

(Note: Many engineers prefer to compute P, q and q_α by the methods given for design values for silos.)

For small trapezoidal walls, it is sometimes sufficiently accurate to use average pressures. The average pressure on an inclined trapezoid of upper edge a and lower edge a_1 is

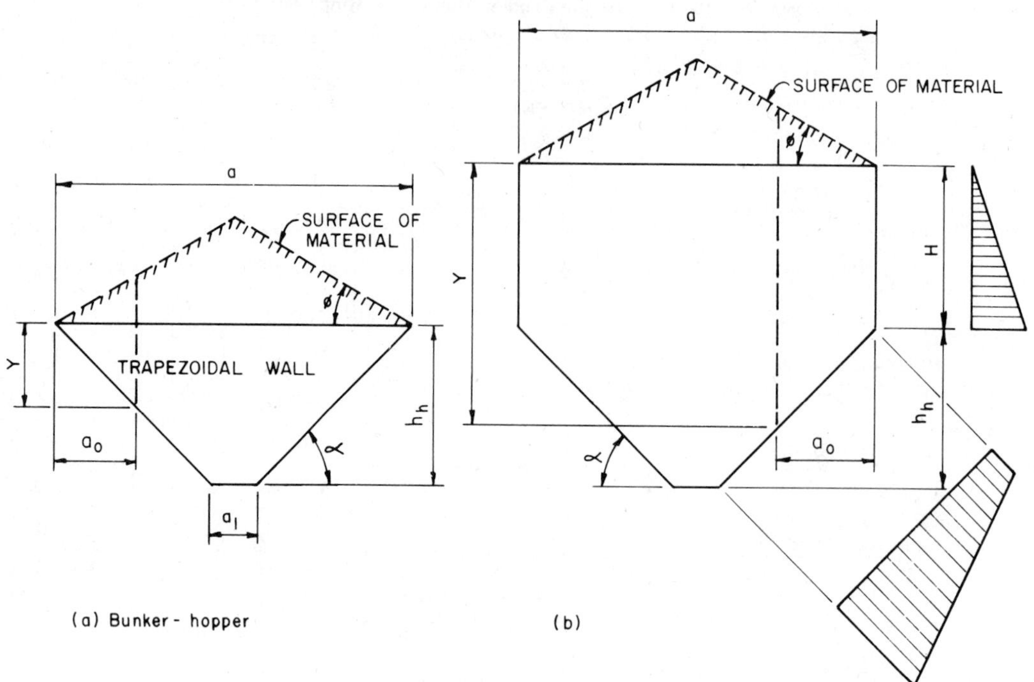

Fig. 16-28 Bunker and hopper dimensions.

$$q_{\alpha,\,avg} = \frac{a\,(5q_{\alpha,\,top} + 4q_{\alpha,\,bot}) + a_1\,(q_{\alpha,\,top} + 2q_{\alpha,\,bot})}{6\,(a + a_1)} \qquad (16\text{-}84)$$

P_{avg} for a vertical trapezoid is computed by the same equation, but with $q_{\alpha,top}$ and $q_{\alpha,bot}$ substituted by P-values computed from eq. (16-79) or (16-82).

16.14.2 Design Pressures

Design pressures are

$$P_{des} = C_d P \qquad (16\text{-}85)$$

$$q_{des} = C_d q \qquad (16\text{-}86)$$

$$q_{\alpha,\,des} = C_d q_\alpha \qquad (16\text{-}87)$$

C_d for bunkers is usually considered to be unity. However, significant impact may occur when the volume of material suddenly dumped into a bunker is large compared to the bunker capacity. In this case, C_d will be larger. Values[16-22] for use in bunker design are shown by Table 16-10.

TABLE 16-10 Values of Dynamic Coefficient C_d for Bunkers (by S. P. Abramovitsch, Ref. 16-22). NOTE: C_d in table is for concrete bunkers.

Ratio of volume dumped in one load to total bunker capacity	1:2	1:3	1:4	1:5	1:6 and less
Dynamic Coefficient, C_d	1.4	1.3	1.2	1.1	1.0

16.14.3 Wall Forces—Circular Bunkers with Symmetrical Conical Hoppers

Vertical walls of circular bunkers are subject to horizontal (hoop) tensile forces due to lateral pressure from the stored material, and frequently also to in-plane bending, vertical shear, and vertical membrane forces, depending upon the manner of support for the bunker.

The required ultimate hoop tensile force per unit width of vertical wall is

$$F_u = K_{ll} P_{des}\,(D/2) \qquad (16\text{-}88)$$

in which P_{des} is computed from eq. (16-85) and either eq. (16-79) or (16-82).

For the conical hopper, the required ultimate tensile forces per unit width are:

Meridional

$$F_{mu} = K_{ll}\left[\frac{q_{des}D}{4\sin\alpha} + \frac{W_{ll}}{\pi D \sin\alpha}\right] + K_g\left[\frac{W_g}{\pi D\sin\alpha}\right] \qquad (16\text{-}89)$$

Horizontal

$$F_{tu} = K_{ll}\left[\frac{q_{\alpha,\,des}D}{2\sin\alpha}\right] \qquad (16\text{-}90)$$

In the above, q_{des} and $q_{\alpha,des}$ are computed using eqq. (16-78) to (16-87) of this subsection, as applicable.

16.14.4 Wall Forces—Rectangular Bunkers and Symmetrical Pyramidal Hoppers

Vertical walls of rectangular bunkers are subject to horizontal membrane tensile forces and two-way bending, and frequently also to in-plane bending, vertical shear, and vertical membrane forces, depending on the manner of support used.

The ultimate horizontal tensile forces per unit width of

vertical wall are

$$F_{au} = K_{ll} P_{des} (b/2) \quad (16\text{-}91)$$

$$F_{bu} = K_{ll} P_{des} (a/2) \quad (16\text{-}92)$$

where P_{des} is from eq. (16-85) and either eq. (16-79) or (16-82).

WSD bending moments for rectangular panels may be computed using tables published by the Portland Cement Association,[16-21] Fischer,[16-31] or Lipnitski,[16-22] and correcting to USD by multiplying by the appropriate load factors, K.

When bunkers are not supported continuously, the vertical walls will have vertical tensile forces and also in-plane bending due to their action as horizontal beams in transferring vertical loads to the supports. Where the sloping walls are monolithic with the vertical walls, both work together in providing this beam action. In-plane bending should also be considered in the trapezoidal walls of bunkers having no vertical walls.

In addition to in-plane bending or the participation with the vertical walls in resisting vertical bending, inclined walls of pyramidal hoppers are subject to horizontal and inclined membrane tensile forces, F_t and F_m, and to two-way bending moment. For symmetrical hoppers, the ultimate forces are

Horizontal

$$F_{tau} = K_{ll} \left(\frac{b}{2}\right) [q_{\alpha,b}]_{des} \sin \alpha_a \quad (16\text{-}93)$$

$$F_{tbu} = K_{ll} \left(\frac{a}{2}\right) [q_{\alpha,a}]_{des} \sin \alpha_b \quad (16\text{-}94)$$

Meridional

$$F_{mau} = \frac{K_{ll} (c_a W_{ll} + A_a q_{des}) + K_g c_a W_g}{a \sin \alpha_a} \quad (16\text{-}95)$$

$$F_{mbu} = \frac{K_{ll} (c_b W_{ll} + A_b q_{des}) + K_g c_b W_g}{b \sin \alpha_b} \quad (16\text{-}96)$$

WSD values of bending moment may be computed using Tables 16-5 and 16-6, and eqq. (16-64) and (16-84). These moments may be converted to ultimate moments by multiplying by K_{ll}.

16.14.5 In-Plane Bending and Combined Action of Vertical and Inclined Walls

Engineers often disregard in-plane bending and combined action of vertical and inclined walls. The authors feel, however, that they should be disregarded *only* when their effects are shown to be negligible. Unfortunately, their analysis is complex and many simple approximations are substituted for rational methods.

In the case of a simple bunker-hopper, having no vertical walls, the trapezoidal wall may be assumed (conservatively) to act as a simple beam, spanning from support to support, and carrying the in-plane forces, F_{mu}, computed from eq. (16-95) or (16-96). The USD bending moment at the centerline, for wall 'a', would then be

$$M_u = F_{mau} a^2/8 \quad (16\text{-}97)$$

The tensile force (per unit width of wall) computed for the in-plane bending must be added to the membrane force F_{tau}, and the reinforcement must be chosen for the combined force acting together with bending moment M_{xu}.

The upper edge of the trapezoidal wall is subject to compression due to bending moment M, but this compressive force is probably more accurately estimated by considering the equilibrium of forces at the corner. Referring to Fig. 16-29, column force F_c is balanced by the resultant of a tensile force F_d along the intersection of walls 'a' and 'b' and compressive forces F_a and F_b at the upper edges of the walls.

$$F_d = F_c/\sin \alpha_i \quad (16\text{-}98)$$

where α_i is the angle between the horizontal plane and the wall intersection line.

$$F_a = F_c \cot \alpha_i \cos \Psi_a \quad (16\text{-}99)$$

$$F_b = F_c \cot \alpha_i \cos \Psi_b \quad (16\text{-}100)$$

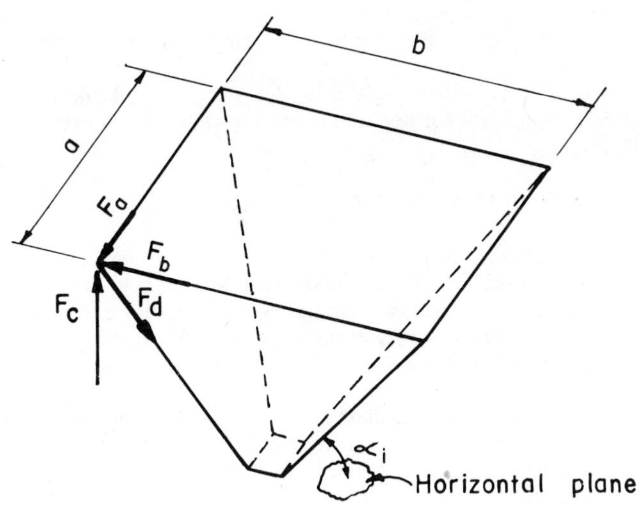

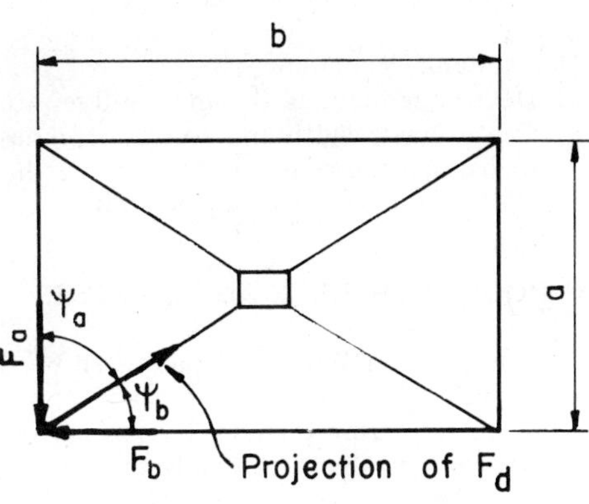

Fig. 16-29 Corner forces in pyramidal bunker-hopper.

where Ψ_a and Ψ_b are the angles between the horizontal projection of F_d and edges 'a' and 'b', respectively.

Multiplying by appropriate load factors, K, these forces may be converted to ultimate values, F_{au} and F_{bu}. These ultimate compressive forces may be assumed applied to a narrow strip (of width not exceeding one-tenth of the edge length) near the upper edge, or to an edge beam if one is provided (preferable). If applied to a narrow strip, the compressive force is combined with the tensile force F_{tu}. That upper edge strip must also carry the bending moments M_{xu} due to normal pressure loading. If a top edge beam is used, it will carry not only a compressive force (F_{au} or F_{bu} as shown by Fig. 16-29) but also a share of the tensile forces F_{tau} or F_{tbu} determined for the hopper shell. In addition, the edge beam will carry vertical shear and bending, horizontal shear and bending and (if the inclined wall is eccentric to the beam) torsion.

Extra steel to carry force F_{du} should be located parallel to the intersection of the inclined walls.

When the depth of the vertical wall exceeds one-half its span (from column to column), participation of the inclined wall in resisting vertical-plane bending is usually neglected. In this case, the vertical walls are assumed to transfer all vertical loads (dead, live, and impact) from the bunker and hopper to the columns.

16.14.6 Suggested Procedure for Design of Cast-in-Place Bunkers

The following procedure is recommended for designing cast-in-place bunkers:

1. Determine the properties of the material to be stored.
2. Establish basic bunker size and supporting system. (For preliminary purposes the wall thickness may be assumed equal to 1/25 of the span but not less than 6 in. for hoppers for vertical walls.)
3. Determine necessary geometric data—angle of inclined walls, wall sizes, center of gravity of material, etc.
4. Determine design pressures of granular material on walls and bottom, considering dynamic effect if any. Sketch pressure diagrams.
5. Determine tensile forces in all vertical and inclined walls.
6. Determine bending moments in the walls due to local bending.
7. Determine moments and shears due to in-plane bending.
8. Design bunker components—
   a) Determine required thickness and reinforcement.
   b) Check crack width and modify thickness, bar size, or bar spacing as required to ensure that the crack width not exceed a tolerable limit.

16.15 ROOFS FOR SILOS AND BUNKERS

The silo roof system should be supported by the walls in a manner permitting free expansion and contraction and slight movement during earthquakes. Means must be provided, however, to limit the movement to a tolerable amount. Possible means include fixing the roof at one interior location (such as an elevator tower), shear pins between roof and walls, an inverted curb with clearance to the outer wall surface, or a thickened portion of slab with clearance to selected inside surfaces. The latter two methods are illustrated by Fig. 16-30.

Steel beams, if used, are usually incorporated in the slipform platform serving also as the bottom form for the roof.

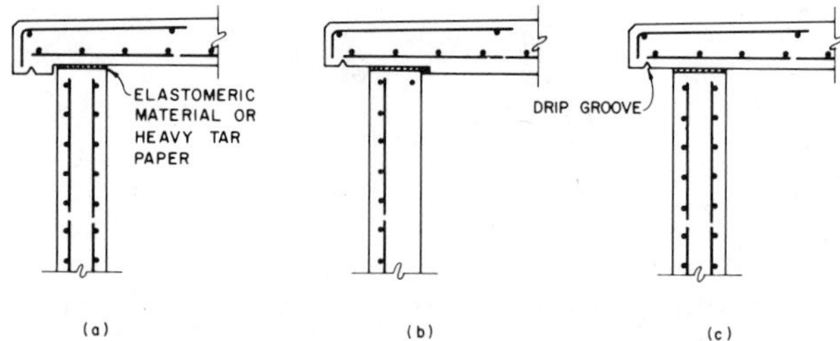

Fig. 16-30 Typical silo roof-to-wall details.

When a bunker is cast with stationary forms and has a concrete slab roof, the slab is usually keyed or dowelled to the wall, so that it provides lateral support to the top of the wall. In this case, the slab is subjected to combined bending and tension. If the tension is significant, it must be considered in design of the slab.

16.16 FOUNDATIONS FOR BUNKERS AND SILOS

Chapter 5 of this handbook, "Foundations and Retaining Walls," covers the selection of foundation type and design methods. This section gives only those requirements peculiar to bunker and silo foundations.

Loads to be considered are.

1. dead load—weight of structure, equipment, and supported backfill (if any)
2. live load—full weight of stored material and snow. Overpressure does not effect silo foundations.
3. wind
4. earthquake

Load combinations to be considered in design of the foundation for strength, stability, and reactive pressure include

1. dead + live (all cells full)
2. dead + live (all cells full) + earthquake
3. dead + wind, all cells empty
4. Unsymmetrical loading of silo or bunker groups. Assume half of the cells full and the other half empty.
   a. Dead + live
   b. Dead + live + wind or earthquake (whichever controls)
5. Dead + live in all cells but one; one cell empty. (May control for bending for the foundation slab.)

6-17 SLIPFORM CONSTRUCTION OF SILOS

Slipforming is well adapted to construction of silos. (See Fig. 16-31.) However, it is essential that specifications for slipform construction recognize the special problems of silo slipforming. Specifications for conventional reinforced concrete construction are usually inadequate for slipform work.

Within the pressure zone and for approximately 3 ft below, the horizontal bars should be tied to the vertical reinforcing or jack rods during construction. Laying hoop bars directly on freshly poured concrete results in inaccuracy of bar spacing and bar lapping, especially when vibrators are used. It also makes bar placing completely dependent on the crew depositing concrete into forms.

Continuous, around-the-clock inspection is recommended during slipforming. The inspection should be by the Engineer or a competent person appointed by the Engineer. Continuous inspection is desirable since the work to be in-

Fig. 16-31 Around-the-clock slipforming of cement storage silos, Portland, Colorado. (Courtesy of Ideal Cement Co.)

spected will be covered very quickly, making errors hard to correct. Inspection to assure proper lap length and vertical location of horizontal reinforcing is particularly important.

EXAMPLES

(Equation and table numbers in parentheses at right refer to equations and tables in text.)

EXAMPLE 16-1: Figure 16-32 shows the basic dimensions of a single circular silo for coal storage. Its conical hopper is supported by eight pilasters, monolithic with the lower walls. The total load from the roof (dead, live and equipment) is 2,000 lb/ft. Design the walls and hopper, using f'_c = 4,000 psi and f_y = 60,000 psi (40,000 psi for ties and stirrups) and assuming that the silo walls will be slip-formed.

SOLUTION: For coal

$$\gamma = 50 \text{ pcf}; \quad \phi = 35^\circ; \quad \mu' = 0.50 \qquad \text{(Table 16-2)}$$

For use in Janssen's equation

$$k = (1 - \sin 35)/(1 + \sin 35) = 0.271 \qquad (16\text{-}3)$$

$$R = D/4 = 40/4 = 10.0 \text{ ft}$$

For lateral design pressure curve

$$H_1 = D \tan \phi = 28.0 \text{ ft}; \quad H/3 = 120/3 = 40 \text{ ft}$$

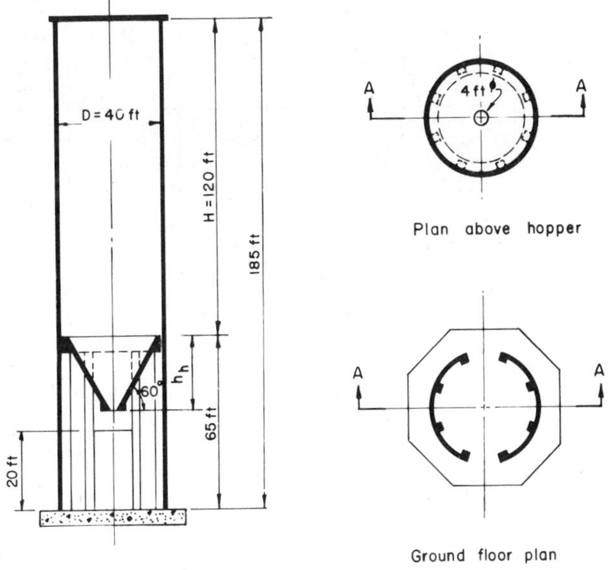

Fig. 16-32 Silo for example 16-1.

Overpressure factors for lateral pressure (Table 16-3)

upper H_1, $C_d = 1.35$

lower 2/3 of H, $C_d = 1.75$

hopper, $C_d = 1.50$

At bottom of silo, $Y = 120$ ft; $x = \mu' kY/R = 1.63$

$$(1 - e^{-x}) = 0.803$$

Basic vertical pressure $= q = 2{,}960$ psf (16-1)

Basic lateral pressure $= qk = 803$ psf (16-2)

$P_{des} = C_d P = 1.75 \times 803 = 1{,}406$ psf (16-21)

In the hopper, at depth $Y = 130$ ft

$x = 1.76$; $(1 - e^{-x}) = 0.828$

$q = 50 \times 10 \times 0.828/(0.50 \times 0.271) = 3{,}050$ psf (16-1)

$P = qk = 828$ psf (16-2)

$q_{des} = C_d q = 4{,}575$ psf; $P_{des} = C_d P = 1{,}242$ (16-20, 16-21)

$q_{\alpha,des} = 1{,}242 \sin^2 60 + 4{,}575 \cos^2 60 = 2{,}076$ psf (16-22)

Similarly, design pressures are calculated at various depths and are tabulated on Table 16-11 (silo) and Table 16-12 (hopper).

Silo Wall Design:
Wall forces and hoop steel areas at $Y = 120$ ft

$F_u = 1.7 \times 1{,}406 \times 20 = 47{,}800$ lb/ft (16-26)

Reqd $A_s = 47{,}800/(0.95 \times 0.90 \times 60{,}000) = 0.932$ sq in./ft (16-27)

Try #6 bars @ 5 in. c/c

Wall thickness

$$h_{min} = \left[\frac{0.0003 \times 29 \times 10^6 + 24{,}000 - 8 \times 400}{12 \times 24{,}000 \times 400}\right] \frac{803 \times 40}{2}$$

$= 4.12$ in. (16-28)

Assume $h = 8$ in. Check vertical compressive stress at foundation.

TABLE 16-11 Silo Wall Lateral Pressures and Hoop Reinforcing (Example 16-1)

DEPTH Y ft	BASIC LATERAL PRESS. P lb/sq ft	CORRECTION COEFFICIENT C_d FROM TABLE 3	DESIGN LATERAL PRESS. P_{des} lb/sq ft	T_u lb/ft	REQUIRED HOOP STEEL A_s in²/ft
10	127	1.35	171	5,810	0.114
20	237	1.35	320	10,900	0.214
$H_1 = 28$	316	1.35	426	14,500	0.284
$H/3 = 40$	418	1.75	732	24,900	0.488
50	492	1.75	861	29,300	0.574
60	556	1.75	974	33,000	0.649
70	613	1.75	1,072	36,400	0.715
80	662	1.75	1,158	39,300	0.770
90	705	1.75	1,233	42,000	0.824
100	742	1.75	1,299	44,100	0.865
110	775	1.75	1,356	46,100	0.905
120	803	1.75	1,406	47,800	0.940

Vertical loads

wt of wall $= 185 \times 150 \times 8/12 = 18{,}500$ lb/ft

friction, V_{des} (at $Y = 120$) $= (50 \times 120 - 0.8 \times 2{,}960)(400\pi/40\pi)$

$= 36{,}300$ lb/ft (16-17, 16-23)

Roof (given) $= 2{,}000$ lb/ft (treat as live load)

$F_{vert} = 1.4(18{,}500 + 1.7(38{,}300) = 90{,}900$ lb/ft

TABLE 16-12 Conical Hopper Pressures, Forces, and Reinforcing (Example 16-1)

DEPTH Y ft	OVERPRESS. FACTOR C_d	BASIC LATERAL PRESS. P lb/sq ft	DESIGN LATERAL PRESS. P_{des} lb/sq ft	DESIGN VERTICAL PRESS. q_{des} lb/sq ft	DESIGN PRESS. ON INCLINED FACE q_α lb/sq ft	ULT. TANG. FORCE F_{tu} lb/ft	ULT. MERID. FORCE F_{mu} lb/ft	REQ'D TANG. STEEL in.²/ft	REQ'D MERID. STEEL in.²/ft
120	1.50	803	1,205	4,446	2,015	46,542	60,000	FOR RING-BEAM REINF. SEE RESPECTIVE CALCULATIONS	
125	1.50	816	1,224	4,518	2,048	40,460	50,700	1.28	1.60
130	1.50	828	1,242	4,575	2,076	55,200	70,300	1.02	1.30
135	1.50	839	1,259	4,647	2,106	44,400	55,500	0.83	1.03
140	1.50	850	1,275	4,705	2,132	32,800	41,400	0.61	0.76
145	1.50	860	1,290	4,759	2,157	20,700	26,300	0.39	0.49
150	1.50	870	1,303	4,810	2,180	8,600	11,700	0.16	0.22

$$\psi = 0.95(0.70) = 0.665$$

$$f_{c,\text{vert}} = 90{,}900/(12 \times 8)\psi = 1{,}440 \text{ psi}$$

$$\text{permissible ult } f_{c,\text{vert}} = 0.385 \times 4{,}000 = 1{,}540 \text{ psi O.K.} \qquad (16\text{-}30)$$

(The above assumes that all hopper vertical load is carried by the pilasters.)

Check crack width at $Y = 120$ ft.

$$F_{tot} = P_{des}D/2 = 28{,}120 \text{ lb/ft}; \quad F_{st} = PD/2 = 16{,}060 \text{ lb/ft}$$

$$s_{cr} = (8 \times 12)(0.7)/5.65 = 11.9 \text{ in.} \qquad (16\text{-}33)$$

For short-term total load

$$\psi_1 = 1 - 0.7\left[\frac{0.8(8 \times 12)(285)}{28{,}120}\right] = 0.455 \qquad (16\text{-}34)$$

$$f_s = F_{tot}/A_s = 28{,}120/1.05 = 26{,}800 \text{ psi}$$

$$w_1 = (0.455 \times 11.9 \times 26{,}800)/(29 \times 10^6) = 0.005 \text{ in.} \qquad (16\text{-}33)$$

Similarly, for short-term static load

$$\psi_1 = 0.045; \quad f_s = 15{,}300 \text{ psi}; \quad w_2 = 0.00028 \text{ in.} \qquad (16\text{-}34, 16\text{-}33)$$

and for long-term static load

$$\psi_1 = 0.524; \quad w_3 = 0.0033 \text{ in.} \qquad (16\text{-}35, 16\text{-}33)$$

Total crack width

$$w_{cr} = w_1 - w_2 + w_3 = 0.0080 \text{ in.} \quad \text{O.K.} \qquad (16\text{-}31)$$

Use 8-in. wall thickness.

Table 16-11 may now be completed to determine the required hoop steel areas for all levels of the silo wall above the hopper ring beam. The data is also plotted on Fig. 16-33.

Vertical reinforcement in pressure zone

$$\text{Reqd } A_s = 0.002 \times 8 \times 12 = 0.192 \text{ sq in./ft}$$

Use #5 bars @ 18 in. c/c

Conical Hopper Design:

Wall forces at depth $Y = 130$ ft (20 ft above outlet, 23.5 ft above inside vertex of cone). Assume an 8-in. average wall thickness.

Inside $D = 27.1$ ft; $D_{avg} = 27.8$ ft

$$W(\text{approx}) = \frac{\pi}{3}\left(\frac{27.1}{2}\right)^2 (23.5)(50) + \frac{8\pi}{12}(150)\left(\frac{27.8}{2}\right)^2 /\cos 60$$

$$= 226{,}000 + 121{,}500$$

$$= 347{,}500 \text{ lb}$$

Meridional force and required reinforcing

$$F_{mu} = \left[\frac{(4{,}575)(27.8)}{4(0.866)} + \frac{347{,}500}{\pi(27.8)(0.866)}\right] \times 1.7 = 70{,}300 \text{ lb/ft} \qquad (16\text{-}56)$$

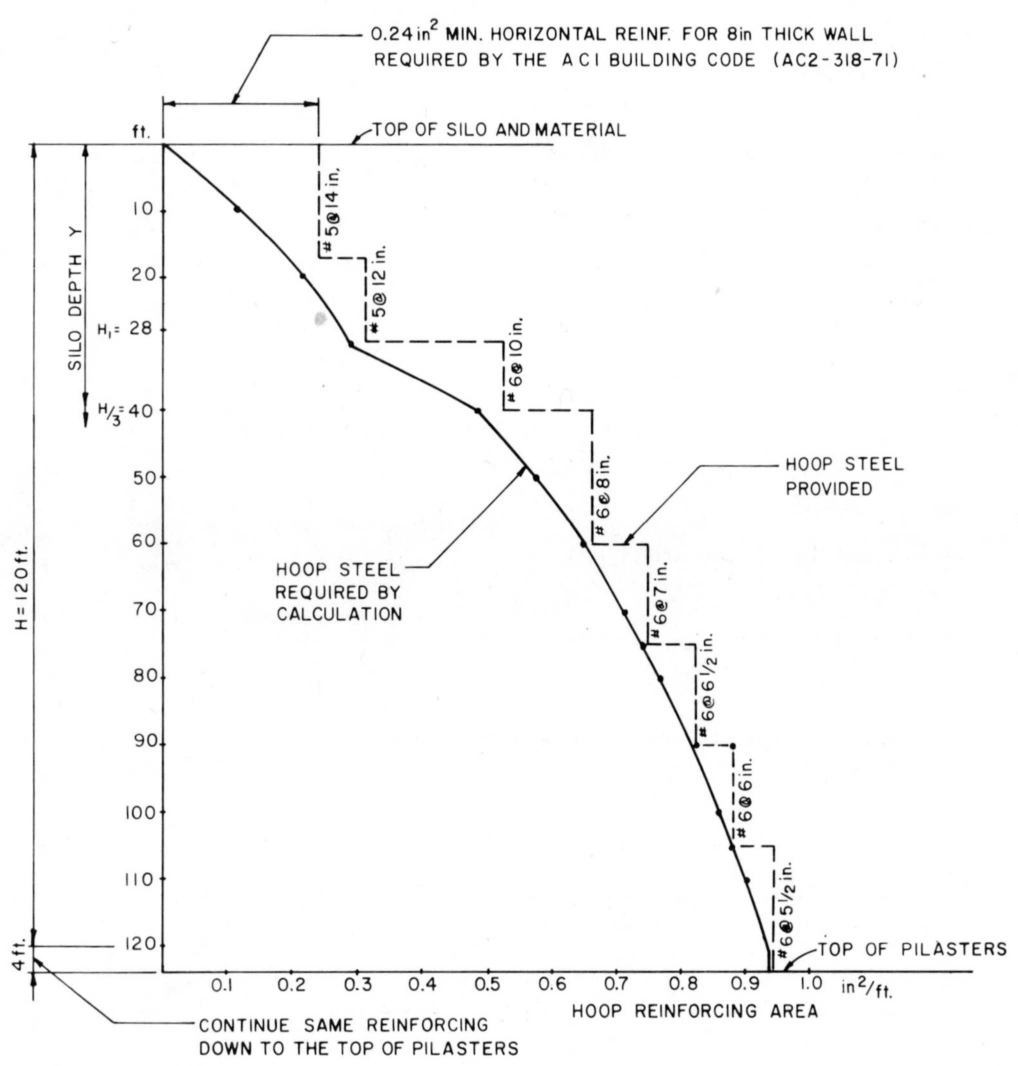

Fig. 16-33 Required and furnished hoop steel for silo walls (example 16-1).

$$\text{Reqd } A_s = 70{,}300/(0.9 \times 60{,}000) = 1.30 \text{ sq in./ft} \quad (16\text{-}58)$$

Tangential force and required tangential reinforcing

$$F_{tu} = 1.7\,(2{,}076)(27.1)/2 \sin 60 = 55{,}200 \text{ lb/ft} \quad (16\text{-}57)$$

NOTE: Above uses inside diameter.

$$\text{Reqd } A_s = 55{,}200/(0.9 \times 60{,}000) = 1.02 \text{ sq in./ft} \quad (16\text{-}59)$$

The results of similar calculations at several elevations are tabulated on Table 16-12.

Check hopper wall for crack width at $Y = 130$ ft. Both directions must be considered.

Meridional force (horizontal cracks)

$$A_s(64\text{–}\#8 \text{ and } 64\text{–}\#9) = 114.6/27.8\pi = 1.31 \text{ sq in./ft}$$

$$\Sigma o = 4.84 \text{ in./ft}$$

$$F_{m,tot} = 41{,}300 \text{ lb/ft}$$

$$F_{m,st} = 36{,}700/C_d + 4{,}590 = 29{,}090 \text{ lb/ft}$$

$$s_{cr} = (8 \times 12)(0.7)/4.84 = 13.9 \text{ in.} \quad (16\text{-}33)$$

For short-term total load, $\psi_1 = 0.63; f_s = 21{,}500$ psi;

$$w_1 = 0.0095 \text{ in.} \quad (16\text{-}34, 16\text{-}33)$$

For short-term static, $\psi_1 = 0.47; f_s = 22{,}200$ psi;

$$w_3 = 0.0050 \text{ in.} \quad (16\text{-}34, 16\text{-}33)$$

For long-term static, $\psi_1 = 0.74; f_s = 22{,}200$ psi;

$$w_3 = 0.0079 \text{ in.} \quad (16\text{-}35, 16\text{-}33)$$

Total crack width, $w_{cr} = w_1 - w_2 + w_3 = 0.0124$ in. (16-31)

The width exceeds 0.008 in. and seems excessive. If the remaining outside bars are extended by about another 2 ft, the steel at this height will be 128 – #8 and 64 – #9. Recomputation of crack width then gives $w_{cr} = 0.0058$ in. O.K.

Tangential force (vertical cracks)

$$F_{t,tot} = 32{,}500; \quad F_{t,st} = 32{,}500/C_d = 21{,}600 \text{ lb/ft}$$

$$A_s = 1.23 \text{ sq in./ft}; \quad \Sigma_0 = 6.6 \text{ in./ft}$$

$$s_{cr} = 96(0.7)/6.6 = 10.2 \text{ in.} \quad (16\text{-}33)$$

Short-term total, $\psi_1 = 0.53$; $f_s = 26{,}400$ psi;

$$w_1 = 0.0049 \text{ in.} \quad (16\text{-}34, 16\text{-}33)$$

Short-term static, $\psi_1 = 0.30$; $f_s = 17{,}600$ psi;

$$w_2 = 0.0019 \text{ in.} \quad (16\text{-}34, 16\text{-}33)$$

Long-term static, $\psi_1 = 0.65$; $f_s = 17{,}600$ psi;

$$w_3 = 0.0040 \text{ in.} \quad (16\text{-}35, 16\text{-}33)$$

Total crack width, $w_{cr} = w_1 - w_2 + w_3 = 0.0070$ in. < 0.008 in. O.K. (16-31)

Similar crack width checks should be made at several locations within the hopper, especially at the bottom of the ring beam and points where meridional bars are terminated. See Fig. 16-34 for hopper detail. The ring beam is designed by the procedure in section 16.13.12. The ring beam forces are summarized in Table 16.13.

EXAMPLE 16-2: The cement storage silo of Fig. 16-35 has a conical steel hopper, supported by a concrete ring beam on six independent columns. The walls extend down to the raft foundation and have two opposite openings. Cement is stored at 200°F while the outdoor design temperature is –5°F.

Design the silo and hopper supports using: $f'_c = 5{,}000$ psi for the slipformed walls and hopper supports; $f_y = 60{,}000$ psi (40,000 psi for stirrups and ties). Assume the total service load from the roof to be 1,000 lb per ft of wall.

SOLUTION:

For cement

$$\gamma = 100 \text{ pcf}; \quad \phi = 25°; \quad \mu' = 0.466 \quad \text{(Table 16-2)}$$

Use Reimbert's method. Assume material surface flat ($h = 0$).

$$P_{max} = 100(36)/(4 \tan 25°) = 1{,}930 \text{ lb/sq ft} \quad (16\text{-}7)$$

$$C = \frac{36}{4 \tan 25° \tan^2/(45° - 12.5°)} = 47.5 \text{ ft} \quad (16\text{-}8)$$

At depth $Y = 108$ ft, $Y/C = 108/47.5 = 2.27; z = 0.906$

$$q = 100(108)/3.27 = 3{,}300 \text{ psf} \quad (16\text{-}5)$$

$$P = zP_{max} = 1{,}750 \text{ psf} \quad (16\text{-}6)$$

Similarly, P and q are computed at various depths Y. Within the hopper, $C = 47.5$ ft is used throughout, even though the diameter changes with Y. (See Tables 16-14 and 16-15.)

Depth $H_1 = 36 \tan 25° = 16.8$ ft (for lateral pressure design curve). Overpressure factors are as follows: (Table 16-3)

For horizontal pressures

Upper height H_1, $C_d = 1.10$

Lower 2/3 of H, $C_d = 1.63$ (interpolated, $H/D = 3.0$)

For ring beam and columns

$$C_d = 1.75$$

For steel parts of hopper

$$C_d = 2.00$$

Design pressures, computed by eqq. (16-20, 16-21 and 16-22), are tabulated in Tables 16-14 and 16-15.

TABLE 16-13 Summary of Forces in Ring Beam (Example 16-1)

LOCATION	VERTICAL SHEAR (KIPS)	COMPRESSIVE FORCE DUE TO F_x (KIPS)	BENDING MOMENT (ft-k) DUE TO M_t	BENDING MOMENT (ft-k) DUE TO F_y	TORSIONAL MOMENT DUE TO F_y (ft-k)
SUPPORT	394	565	677	– 983	0
MIDSPAN	0	565	677	483	0
9°33' FROM SUPPORT	226	565	677	0	75

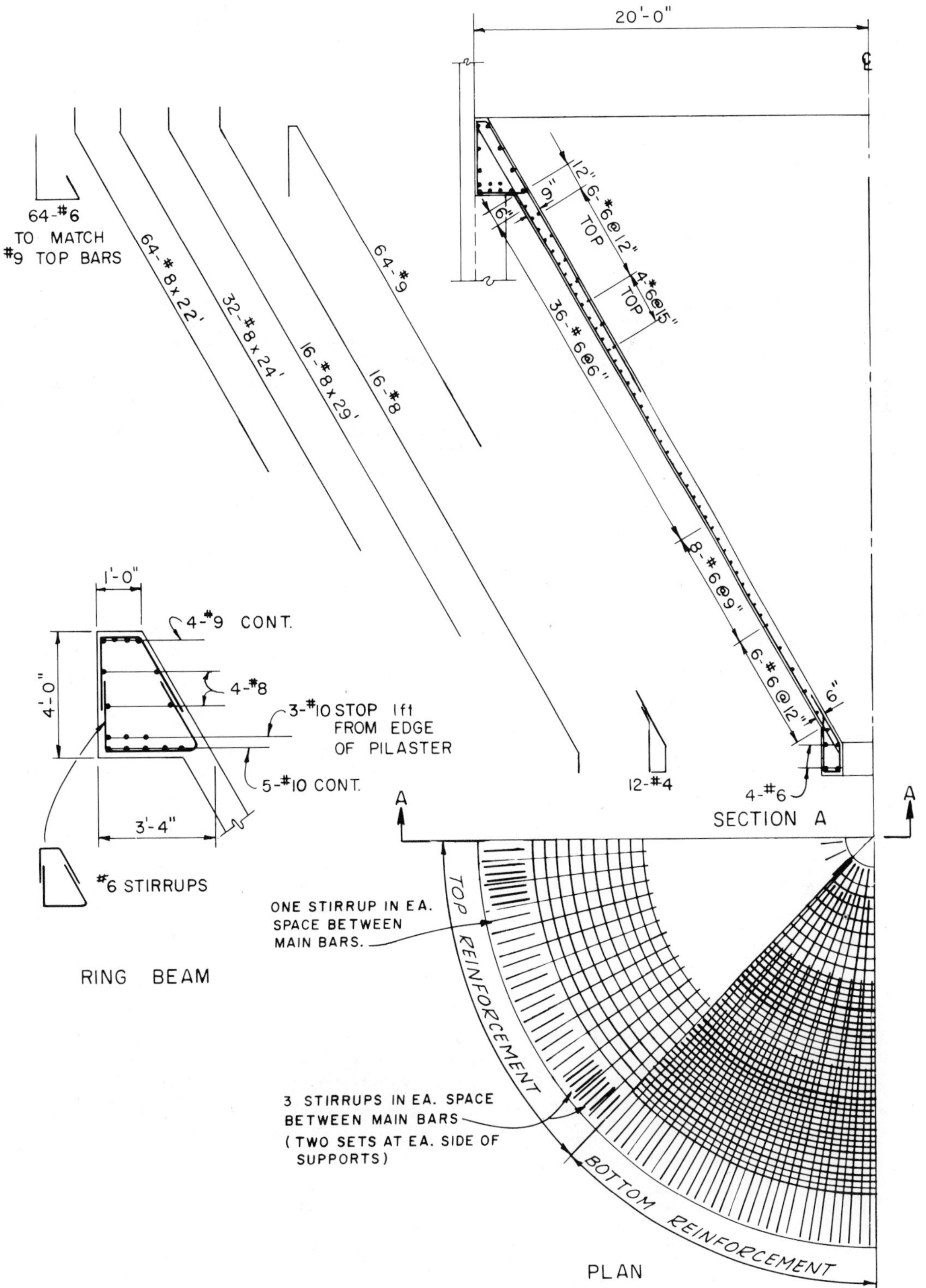

Fig. 16-34 Hopper detail (example 16-1).

Silo Wall Design:

Wall forces and tentative hoop steel areas (computed as in ex. 16-1) are shown by Table 16-14. Two layers of hoop steel are needed to permit practical spacing. Therefore, assume a 9-in. wall thickness.

Check vertical compressive stress at bottom ($Y = 138$ ft).

Wall wt = 138 × 150 × 9/12 = 15,500 lb/ft

Friction load,

$$V_{des} \text{ (at } Y = 108 \text{ ft)} = 67{,}500 \text{ lb/ft} \tag{16-16}$$

From roof (treat as live)	1,000 lb/ft
Total live load	= 68,500 lb/ft

Ultimate load = 1.7 (68,500) + 1.4 (15,500) = 138,200 lb/ft

Away from openings, neglecting steel

$$\psi = 0.95\ (0.70)$$

$$f_{c,\text{vert}} = 138{,}200/(\psi \times 12 \times 9) = 1{,}830 \text{ psi}$$

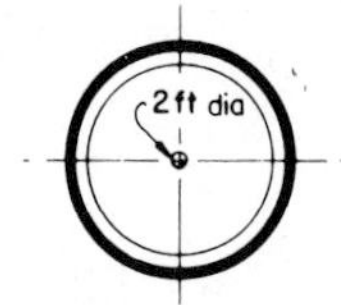

Plan above hopper

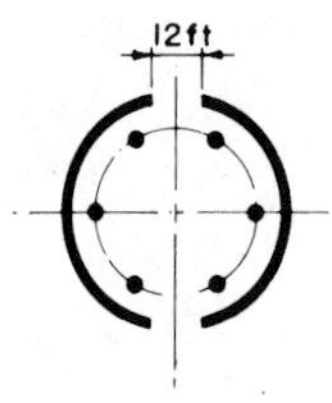

Ground floor plan

Section A

Fig. 16-35 Dimensions of silo for example 16-2.

Permissible ult $f_{c,\text{vert}} = 0.385 \times 5{,}000 = 1{,}930$ psi O.K. (16-30)

Check crack width

At $Y = 108$ ft, using the procedure shown in ex. 16-1

$$w_{cr} = 0.0042 \text{ in.} < 0.008 \qquad \text{O.K.}$$

Design temperature steel

$$T_{i,\text{des}} = 200 - 80 = 120°\text{F} \qquad (16\text{-}36)$$

$$K_t = 0.15; \quad \Delta T = [120 - (-5)]\ (0.15) = 18.8°\text{F}$$

[Figs. 16-10 and 16-11 and eq. (16-37)]

$$M_{xt,u} = 1.43 \left[\frac{29 \times 10^6}{8}\right] (9^2)(6 \times 10^{-6})(18.8)(1.4)$$

$$= 66{,}300 \text{ in.-lb/ft} \qquad (16\text{-}38)$$

$$\text{Reqd } (A_s)_t = 0.185 \text{ sq in./ft}$$

This amount is added to the hoop steel in each face in the bottom two-thirds of the silo height. One-half as much is added to each face for the upper one-third. The required total hoop steel areas are plotted on Fig. 16-36.

For vertical reinforcing, the procedure shown by ex. 16-1 is used.

Design of Hopper Supports (Ring beam and columns):

Assume ring beam and column sizes as in Fig. 16-37. Section properties are

$$A_r = 13.62 \text{ sq ft}; I_r = 15.86 \text{ ft}^4; I_c = 3.16 \text{ ft}^4 \text{ (for 36-in. dia)}$$

$$\eta = 0.003364; \quad z = 1.1153 \qquad \text{(Table 16-7)}$$

Torsional constant

$$\lambda = 0.159 \qquad \text{(Table 16-8)}$$

Ring beam loads, neglecting wt of hopper (all WSD) (to compute WSD value of F_m, use eq. 16-56, but without K_{ll} and K_g).

TABLE 16-14 Silo Wall Lateral Pressures and Hoop Reinforcing (Example 16-2)

DEPTH Y ft/m	BASIC LATERAL PRESS. P lb/sq ft	CORRECTION COEFFICIENT C_d FROM TABLE 3	DESIGN LATERAL PRESS. P_{des} lb/sq ft	T_u lb/ft	REQUIRED HOOP STEEL A_s in²/ft
5	350	1.0	350	10,700	0.22
10	612	1.0	612	18,700	0.37
H_1 = 16.8	875	1.0	875	26,800	0.53
$H/3$ = 36	1,310	1.63	2,130	65,100	1.28
40	1,360	1.63	2,220	68,000	1.34
45	1,420	1.63	2,320	71,000	1.40
50	1,470	1.63	2,400	73,500	1.45
55	1,515	1.63	2,470	75,600	1.49
60	1,550	1.63	2,530	77,500	1.52
70	1,615	1.63	2,630	80,500	1.56
80	1,660	1.63	2,700	82,700	1.62
90	1,700	1.63	2,770	84,800	1.65
100	1,730	1.63	2,820	86,300	1.69
108	1,750	1.63	2,850	87,200	1.70

TABLE 16-15 Ring-Beam and Hopper Pressures and Forces (Example 16-2)

DEPTH Y ft	CORRECTION COEFFICIENT C_d	BASIC LATERAL PRESS, P lb/sq ft	DESIGN LATERAL PRESS. P_{des} lb/sq ft	BASIC VERTICAL PRESS. q lb/sq ft	DESIGN VERTICAL PRESS. q_{des} lb/sq ft	DESIGN PRESS. ON INCLINED FACE $q_{\alpha des}$ lb/sq ft	TANG. STRESS F_t lb/ft	MERID. STRESS F_m lb/ft	
108	1.75	1,750	3,060	3,300	5,780	4,180	87,500	67,000	FOR RINGBEAM ONLY
111	2.25	1,755	3,940	3,330	7,500	5,410	93,600	69,600	FOR STEEL HOPPER DESIGN ONLY
116	2.25	1,760	3,960	3,360	7,560	5,460	64,200	46,500	
121	2.25	1,775	4,000	3,420	7,700	5,540	34,300	25,550	
126	2.25	1,785	4,020	3,450	7,760	5,570	7,260	5,050	

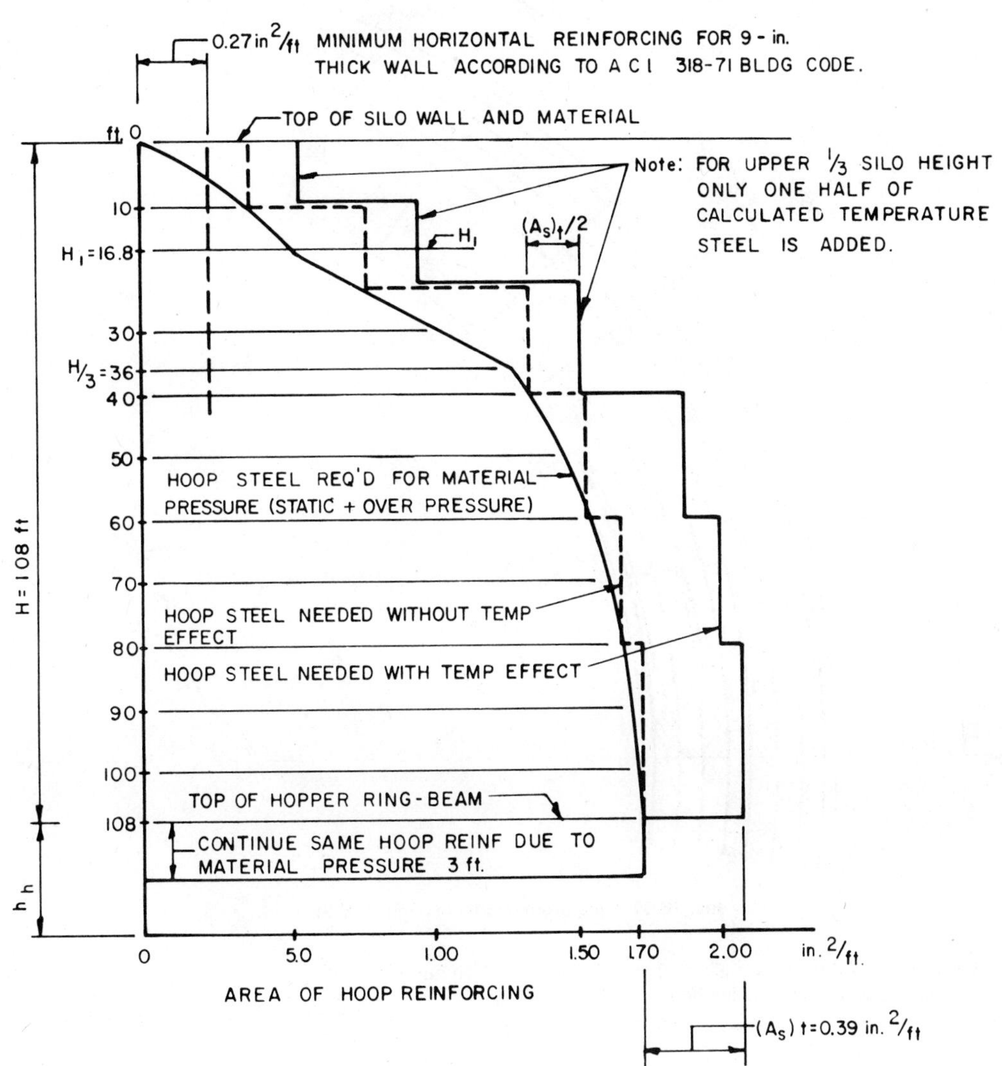

Fig. 16-36 Hoop reinforcement (example 16-2).

$$F_m = \frac{5{,}780(2 \times 16.23)}{4 \sin 50^\circ} + \frac{(100)(16^2)(19\pi)/3}{\pi(2 \times 16.23) \sin 50^\circ} = 67{,}800 \text{ lb/ft}$$

$$F_x = 67{,}800 \cos 50^\circ = 43{,}600 \text{ lb/ft} \qquad (16\text{-}67)$$

$$F_y = (150 \times 13.62) + 67{,}800 \sin 50^\circ + 5{,}780(18^2 - 16.23^2)/(2 \times 16.23) = 64{,}730 \text{ lb/ft} \qquad (16\text{-}68)$$

$$M_t = 67{,}800 \times 1.6 = 108{,}480 \text{ ft-lb/ft} \qquad (16\text{-}69)$$

Substitution in eqq. (16-75) and (16-76), using h_c = 27.79 ft, gives

$$H_A = 42.1 \text{ kips, and } M_A = 787 \text{ ft-kips}$$

These values are used with Table 16-9 to compute the shears, bending moments, ring compressive forces, and torsion values shown

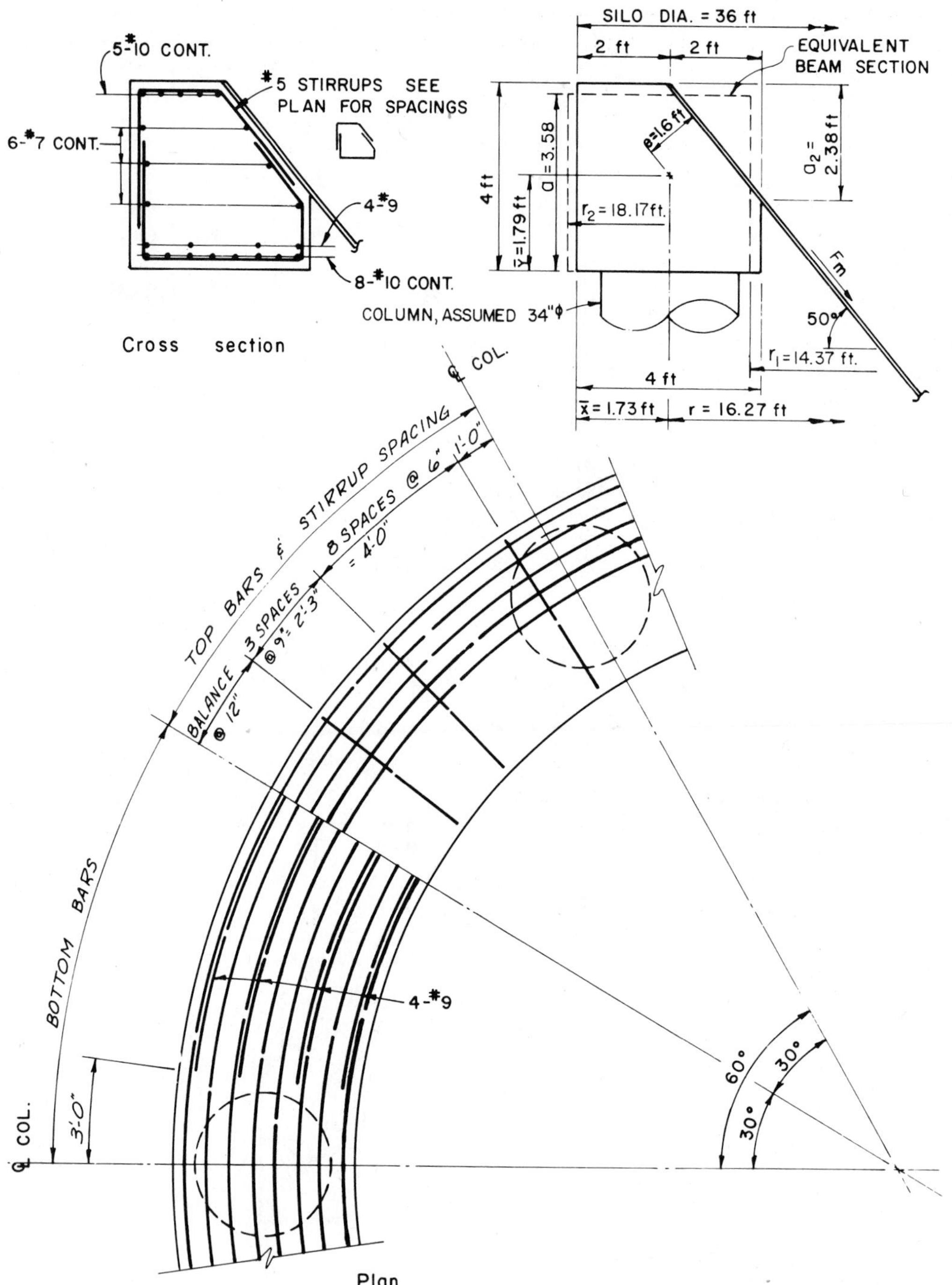

Fig. 16-37 Ring beam details (example 16-2).

in Table 16-16. All values shown are WSD values; they must be multiplied by K_{II} to obtain values for the USD method.

Column design:
Moment at top

$$M_A = 787 \text{ ft-kip}$$

Bottom moment

$$M_B = H_A(h_c) - M_A = 382 \text{ ft-kip}$$

M_A and M_B are both positive, meaning that they are in the direction shown by Fig. 16-26. The column is bent in double curvature.

Column axial load,

$$F = 2 \times \text{shear at support of ring beam} = 1{,}092 \text{ kip}$$

Column reinforcing

$$M_u = K_{II} M_A = 1{,}340 \text{ ft-kip}; \quad F_u = K_{II} F = 1{,}858 \text{ kip}$$

Reinforcing required for 34 in. column = 27.2 sq in.
For this area, $p = 0.03 > 0.02$; ∴ Try larger column.

For 36 in. diameter column, p reqd = $0.017 < 0.0020$ O.K.

$$A_s \text{ reqd} = 17.3 \text{ in.}$$

Use 36 in. dia column with 14–#10 bars (spirally reinforced).

EXAMPLE 16-3: Figure 16-38 shows dimensions of a circular silo for storing wheat. The silo bottom is a circular slab with central opening, simply-supported on the lower walls and carrying hopper-forming concrete fill. Roof loads, total, are 1,000 lb per ft of wall.

TABLE 16-16 Summary of Forces on Ring Beams (Example 16-2)

LOCATION OF SHEAR	FORCES (FOR DL & LL)						
	VERTICAL SHEAR KIP	COMPRESSIVE FORCE DUE TO F_x KIP	BENDING MOMENT DUE TO M_t ft-K	BENDING MOMENT DUE TO F_y ft-K	BENDING MOMENT DUE TO M_A ft-K	TORSIONAL MOMENT DUE TO F_y ft-K	TORSIONAL MOMENT DUE TO M_A ft-K
SUPPORT	546	709	1,769	- 1,587	- 682	0	394
MID-SPAN	0	709	1,769	805	- 787	0	0
12° 44 FROM ℄ OF	315	709	1,769	0	- 752	162	235

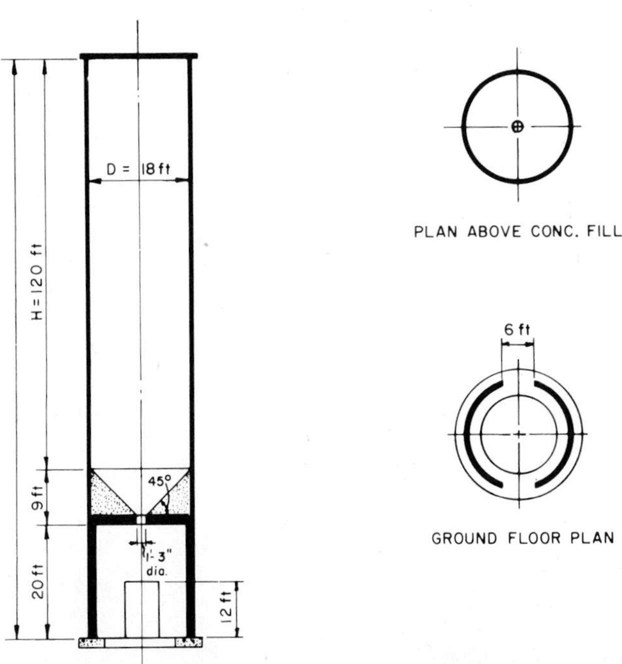

Fig. 16-38 Dimensions for silo of example 16-3.

TABLE 16-17 Silo Wall Lateral Pressures and Hoop Reinforcing (Example 16-3)

DEPTH Y ft	BASIC LATERAL PRESSURE (P) lb/sq ft	CORRECTION COEFFICIENT C_d FROM TABLE 3	DESIGN LATERAL PRESSURE (P_{des}) lb/sq ft	F_u lb/ft	REQUIRED HOOP STEEL (A_s) in²/ft
4	74	1.35	100	1,530	0.045
H_1 = 8	139	1.35	187	2,850	0.084
$H/3$ = 40	406	1.75	710	10,850	0.32
48	435	1.75	760	11,600	0.34
56	455	1.75	795	12,20	0.36
64	470	1.75	821	12,60	0.37
72	481	1.75	840	12,80	0.38
80	489	1.75	855	13,10	0.38
88	495	1.75	865	13,200	0.39
100	501	1.75	875	13,40	0.39
120	505	1.75	885	13,500	0.40

Design the slipform walls using $f_c' = 3{,}000$ psi and $f_y = 40{,}000$ psi, and the slab using $f'_c = 4{,}000$ psi and $f_y = 60{,}000$ psi.

SOLUTION:

For wheat: $\gamma = 50$ pcf; $\phi = 25°$; $\mu' = 0.444$ (Table 16-2)

Silo Wall Design:

Table 16-17 shows design pressures, computed by Janssen's formulas. (Note that C_d factors are increased by 15 percent since H/D exceeds 5.) A 6-in. wall thickness was assumed and steel selected as shown. A check showed crack width to be 0.005 in., which is acceptable. The procedures for wall design are all as illustrated by ex. 16-1. Final hoop steel areas show on Fig. 16-39. The minimum vertical reinforcement area is

$$A_s = 0.002\,(6)\,(12) = 0.144 \text{ sq in./ft}$$

Bottom Slab Design:

Vertical pressure at top of fill (Y = 120 ft)

$$k = (1 - \sin\phi)/(1 + \sin\phi) = 0.41 \tag{16-3}$$

$$q = P_{120}/k = 505/0.41 = 1{,}230 \text{ psf} \tag{16-1}$$

$$q_{des} = C_d q = 1.26 \times 1{,}230 = 1{,}550 \text{ psf} \quad \text{(Table 16-3, eq. 16-20)}$$

Load from wheat in hopper (assumed uniform)

$$= \frac{\pi(9^2)\,(9)\,(50)}{3\pi(9^2)} = 150 \text{ lb/sq ft}$$

Total design live load = 1,550 + 150 = 1,700 lb/sq ft

Load from weight of hopper-forming fill (assumed uniform)

$$= \frac{2\pi(9^2)\,(9)\,(150)}{3\pi(9^2)} = 900 \text{ lb/sq ft}$$

Slab weight–Assume 18-in. slab, wt. = 225 lb/sq ft

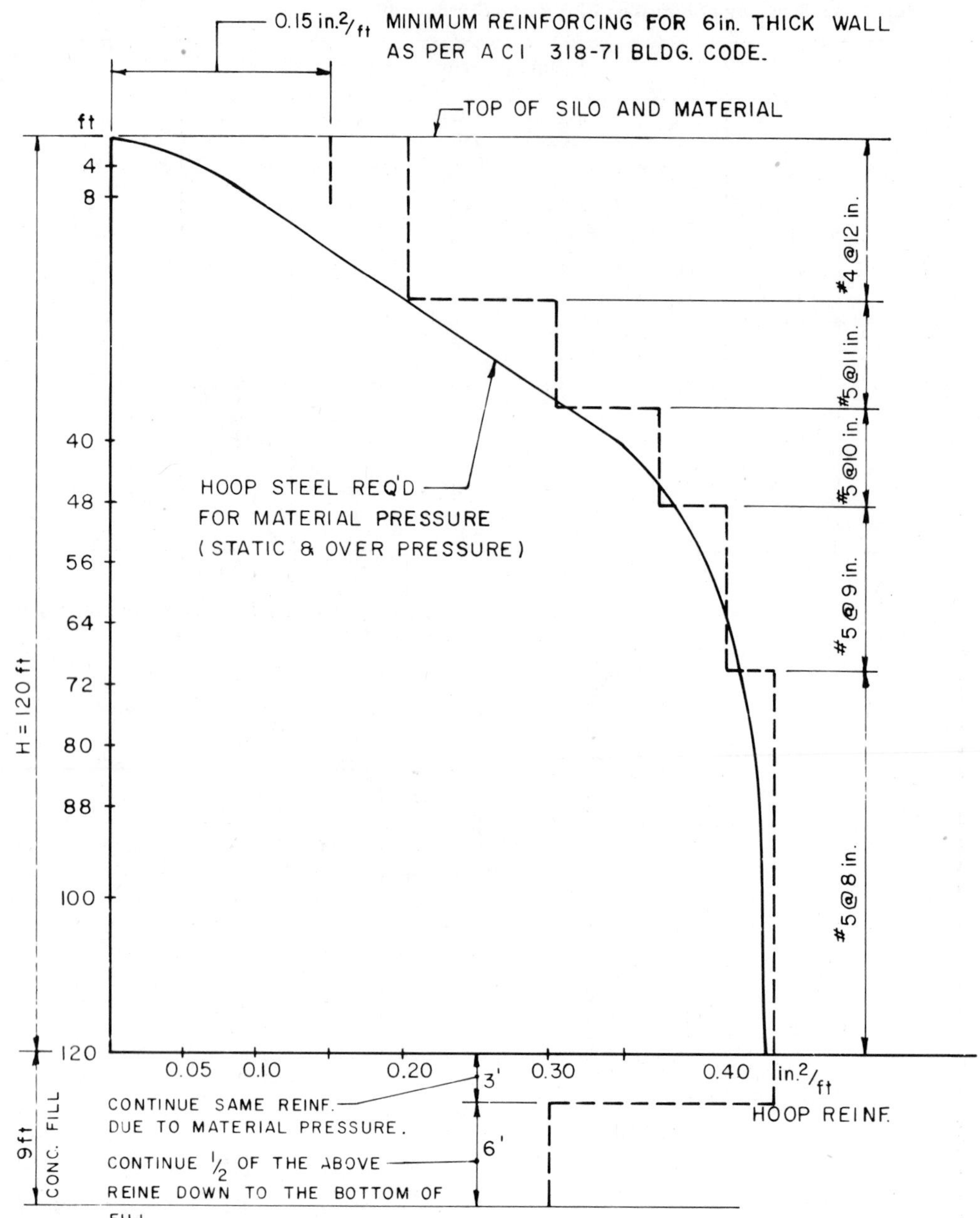

Fig. 16-39 Hoop reinforcement (example 16-3).

Total dead load = 900 + 225 = 1,125 lb/sq ft

Ultimate load = w_u = 1.7 (1,700) + 1.4 (1,125) = 4,475 lb/sq ft

Slabs with holes may be designed in two ways:

(1) By computing bending moments for a slab with no hole, and reinforcing with a steel member of adequate strength and of stiffness equal to that of the removed slab; or

(2) By considering the hole, and reinforcing for bending moments obtained using tables of radial and tangential bending moments[16-28] or Timoshenko's equations.[16-29]

When the hole is small, the first method may be the better. That method is shown here.

Ultimate bending moments per unit width, neglecting the hole, are tabulated in Table 16-18.

The slab stiffness and strength at the hole are replaced by the steel plate ring shown by Fig. 16-40. To provide plate stiffness equalling that of the missing concrete, the plate thickness is chosen using the transformed-area principle.

$$n = E_s/E_c = 8$$

$$\text{total plate thickness} = \text{opening width}/n$$

$$= 15/8 \text{ in. or } 15/16 \text{ in. per side.}$$

Use 15/16 plate.

The ring cross section (two rectangular sections) must have strength to resist the total radial moment on the 15-in. opening width.

$$M \simeq 1.25 \times 74.8 = 93.6 \text{ ft-kip}$$

$$S = 2 \times \tfrac{1}{6} \times \tfrac{15}{16} \times 18^2 = 101 \text{ in.}^3$$

$$f_b = 12\,M/S = 1.12 \text{ ksi} << F_b \qquad \text{(Plate thickness O.K.)}$$

Radial reinforcing shown by Table 16-18 must transmit its inner end load to the center ring plate. This transfer is through lap splicing to radial dowels which are welded to the ring plate. (See Fig. 16-40) The total A_s required at the connection to the ring is approximately 1.17 × 15 π/12, or 4.58 sq. in. For #7 dowels, the number needed is then 4.58/0.60, or 8. Use eight dowels (or more) as shown.

TABLE 16-18 Silo Bottom Slab Moment and Reinforcing Steel (Example 16-3)

r/r_o	$(M_r)_u$ k - ft	REQ'D STEEL in^2/ft	$(M_t)_u$ k - ft	REQ'D STEEL (A_s) in^2/ft
0.0	68.9	1.08	68.9	1.08
0.1	68.1	1.07	68.5	1.07
0.2	66.1	1.03	67.1	1.05
0.3	62.7	0.98	65.5	1.02
0.4	57.9	0.90	62.5	0.98
0.5	51.8	0.80	59.1	0.92
0.6	44.1	0.67	54.8	0.85
0.7	35.1	0.54	49.4	0.76
0.8	24.8	0.38	43.4	0.67
0.9	13.1	0.20	36.7	0.56
1.0	0	0	29.5	0.45

Elastic deflection of the center of the plate is

$$\Delta = \frac{0.058\ (4{,}475)\ (9^4)\ (1{,}728)}{(29/8)\ (10^6)\ (15^3)\ (12/12)} = 0.24 \text{ in.}$$

The long-term deflection is approximately 2.0 × 0.24, or 0.48 in.
Shear per unit width for critical section at distance, d, from the edge is

$$V_u = \frac{4.475\ (9 - 1.25)^2 \pi}{2\pi\ (9 - 1.25)} = 17.4 \text{ k/ft}$$

$$v_u = 17{,}400/(0.85 \times 12 \times 15) = 114 \text{ psi} < 2\ \sqrt{f_c'} = 127 \text{ psi} \quad \text{O.K.}$$

Lower Wall Design:
Assume a 12-in. thickness. Bearing width of slab on wall = 6 in. Compression stress at bearing of slab on wall =

$$f_{c,\text{ vert}} = \frac{4{,}475\ (9^2)\pi}{(6 \times 12)\ (18\pi)\ (0.70)} = 400 \text{ psi} < 0.385\ f_c' \quad \text{O.K.}$$

Total reaction at bottom of wall–

	Weight	K	Ultimate
from roof	56		
material above hopper	1,525	1.7	2,750 kips
material in hopper	38		
hopper-forming fill	229		
bottom slab	57	1.4	1,397
upper wall	563		
lower wall, 0.15 (20)(18π) – 0.15 (6 × 12 × 2)	149		
		Total =	4,147 kips

Net wall width = 18π – 12 = 44.5 ft

Compressive stress =

$$f_{c,\text{vert}} = 4{,}147{,}000/(12 \times 12 \times 44.5 \times 0.70) = 922 \text{ psi}$$

$$\text{Permissible} = 0.385\ (f_c') = 1{,}155 \text{ psi} \quad \text{O.K.} \qquad (16\text{-}30)$$

Vertical steel–lower wall

$$A_s = 12 \times 12 \times 0.002 = 0.29 \text{ sq in./ft}$$

Use #4 @ 16 in. c/c.

Horizontal steel–lower wall

$$A_s = 12 \times 12 \times 0.0025 = 0.36 \text{ sq in./ft}$$

Use #4 @ 12 in. c/c.

At lower wall openings
Vertical–Add (to each side) $A_s = 3 \times 0.29 = 0.87$ sq in.
Add 2-#7 each side of opening.
Horizontal–Add 2-#7 above opening.

EXAMPLE 16-4: Design a single rectangular concrete silo for storing peas. Dimensions are shown on Fig. 16-41. The bottom is a symmetrical pyramidal hopper. The silo walls rest on the hopper base, which is supported by four columns. The roof applies a total load of 3,000 lb/ft to the top of the walls. The walls are to be slipformed. Use $f_c' = 4{,}000$ psi and $f_y = 60{,}000$ psi.

SOLUTION:

Silo Wall Design:
For peas

$$\gamma = 50 \text{ lb/cu ft}; \quad \phi = 25°; \quad \text{and } \mu' = 0.296 \qquad \text{(Table 16-2)}$$

Using Janssen's equations

$$k = (1 - \sin 25°)/(1 + \sin 25°) = 0.406 \qquad (16\text{-}3)$$

For the shorter wall

$$R_a = a/4 = 3.0 \text{ ft}$$

For the longer wall

$$a' = (2 \times 12 \times 20 - 12 \times 12)/20 = 16.8 \text{ ft} \qquad (16\text{-}4)$$

$$R_b = a'/4 = 4.2 \text{ ft}$$

Using these R-values and the above material properties, static lateral pressures are computed at various depths, Y, using eq. (16-2). The pressure values are tabulated in Table 16-19. Next, overpressure factors, C_d, are selected from Table 16-3, and design pressures computed by eq. (16-19) are added to Table 16-19.

Vertical wall loads due to friction–

$$A_a = 12 \times \tfrac{6}{2} = 36; \quad A_b = \tfrac{1}{2}\ (240 - 2A_a) = 84$$

At the bottom, $Y = 80$ ft, so that for wall a

$$V_{a,\text{ des}} = (50 \times 80 - 0.8 \times 485/0.406)\ 36/12 = 9{,}130 \text{ lb/ft} \qquad (16\text{-}17, 16\text{-}23)$$

and for wall b

$$V_{b,\text{ des}} = (50 \times 80 - 0.8 \times 636/0.406)\ 84/20 = 11{,}550 \text{ lb/ft} \qquad (16\text{-}17, 16\text{-}23)$$

Wall tension and bending moment–

For the short wall at depth $Y = 80$ ft

$$F_{a,u} = 1.7 \times 1{,}100 \times 20/2 = 18{,}700 \text{ lb/ft} \qquad (16\text{-}39)$$

Similarly,

$$F_{b,u} = 8{,}550 \text{ lb/ft} \qquad (16\text{-}40)$$

Assume wall thickness $h = 10$ in. with $d = 8.5$ in. Bending moments are computed by moment distribution for an idealized rectangular frame measuring 20′–10″ by 12′–10″. Outward pressures on the frame are the design pressures of Table 16-19. Corner moments so solved are $M_u = 50.0$ ft-kip/ft. Reduced to the edge of the fillet, the moments are

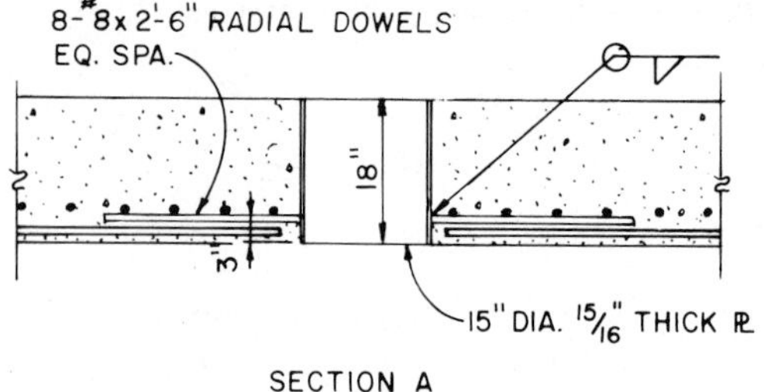

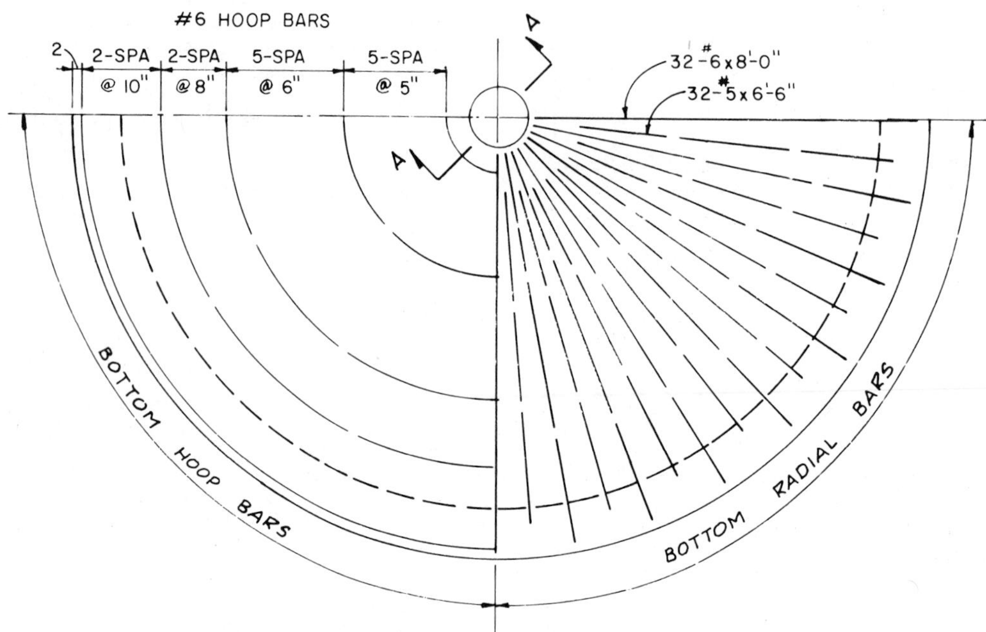

Fig. 16-40 Reinforcing of silo bottom slab (example 16-3).

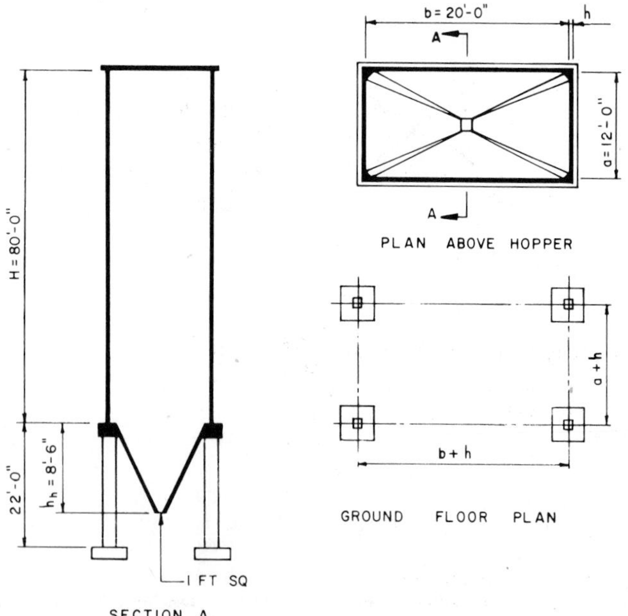

Fig. 16-41 Single rectangular silo with pyramidal hopper, supported on four columns (example 16-4).

Wall a,

$$M_{au} = -37.8 \text{ ft-kip/ft}$$

Wall b,

$$M_{bu} = -22.9 \text{ ft-kip/ft}$$

At midspan

$$\text{wall } a, M_{au} = -22.0 \text{ ft-kip/ft}; \quad \text{wall } b, M_{bu} = +52.6 \text{ ft-kip/ft}$$

These values, and similar values computed for other depths, are tabulated in Table 16-20.

Horizontal reinforcement for combined tension and bending–

For wall 'a', at the toe of the fillet, at depth Y = 80 ft

$$e = M_{au}/F_{au} = 37.8 \times 12/18.7 = 2.43 \text{ in.}$$

$$e > h/2 - d'', \therefore \text{Case II} \qquad \text{(Fig. 16-16)}$$

$$\text{Limiting } y/d \text{ value} = 0.378 \qquad \text{(Table 16-4)}$$

$$y_L = 0.378\,(8.5) = 3.22 \text{ in.}$$

$$\psi = 0.95 \times 0.90 = 0.85 \qquad \text{(Table 16-1)}$$

TABLE 16-19 Pressures on Silo Walls (Example 16-4)

DEPTH Y ft	COEFF. C_d FROM TABLE 2	WALL "a" STATIC PRESSURE lb/sq ft	WALL "a" DESIGN PRESSURE lb/sq ft	WALL "b" STATIC PRESSURE lb/sq ft	WALL "b" DESIGN PRESSURE lb/sq ft
3	1.35	57	77	59	78
5.6	1.35	112	152	105	142
9.3	1.35	157	212	167	226
27	1.5	336	505	380	570
35	1.5	382	574	448	679
40	1.5	405	609	485	727
50	1.5	438	657	540	810
60	1.5	460	690	582	873
70	1.5	476	715	614	920
80	1.5	485	728	636	953

$$f_s' \text{ effective} = 87{,}000 \left[\frac{3.22 - 0.85\,(1.5)}{3.22}\right] - 0.85\,(4{,}000) = 49{,}100 \text{ psi} \qquad (16\text{-}43)$$

$$< (f_y - 0.85\,f_c')$$

Compressive steel will be effective.

$$e'' = 24.3 - 3.5 = 20.8 \text{ in.}; \quad b = 12 \text{ in.}$$

$$\text{Reqd } A_s' = \frac{18.7\,(20.8/0.85) - 0.85\,(4)\,(12)\,(3.22)\,(8.5 - 3.22/2)}{49.1\,(8.5 - 1.5)}$$

$$= \text{negative} \qquad (16\text{-}44)$$

Even though negative steel would be effective (in tension), none is required, as shown by the negative answer.

Tensile reinforcement (inside face)

$$\text{Actual } y = 8.5 - \sqrt{(8.5)^2 - \frac{2 \times 18.7 \times 20.8}{0.85 \times 0.85 \times 4 \times 12}} = 0.45 \text{ in.} \qquad (16\text{-}47)$$

$$\text{Reqd } A_s = \frac{18.7/0.85 + 0.85\,(4)\,(12)\,(0.45)}{60} = 1.34 \text{ sq in./ft} \qquad (16\text{-}46)$$

For midspan of wall '*a*', F_{au} = 18.7 kip/ft and M_{au} = −20.0 ft-kip/ft.

By computations like those above, compressive steel is not needed and the required tensile steel area is A_s = 0.78 sq in./ft. (There is no point of contraflexure in wall '*a*'; thus, this steel, too, is near the inside face.)

Similar computations for wall '*b*' show no compressive steel to be required either at the fillet or at midspan. The required tensile steel areas are

$$A_s \text{ (at corner, inside)} = 0.76 \text{ sq in./ft}$$

$$A_s \text{ (midspan, outside)} = 1.77 \text{ sq in./ft}$$

In this case there *is* a point of contraflexure, so the latter steel area must be provided near the outside face.

The results of similar computations at various depths *Y* are tabulated in Table 16-20. The minimum permissible horizontal reinforcing is 0.025 × 10 × 12 = 0.30 sq in./ft. Bar sizes and spacings are shown by Fig. 16-42.

TABLE 16-20 Ultimate Forces and Moments and Horizontal Steel for Vertical Walls (Example 16-4)

DEPTH Y ft	WALL "a" TENSION F_{au} k/ft	WALL "a" AT SUPPORT M_u ft-k/ft	WALL "a" AT SUPPORT A_s sq in/ft	WALL "a" AT MID-SPAN M_u ft-k/ft	WALL "a" AT MID-SPAN A_s sq in/ft	WALL "b" TENSION F_{au} k/ft	WALL "b" AT SUPPORT M_u ft-k/ft	WALL "b" AT SUPPORT A_s sq in/ft	WALL "b" AT MID-SPAN M_u ft-k/ft	WALL "b" AT MID-SPAN A_s sq in/ft
3	1.33	-3.5	0 / 0.12	-1.9	0 / 0.08	0.78	-1.6	0 / 0.06	+4.2	0.13 / 0
5.6	2.42	-6.9	0 / 0.22	-3.7	0 / 0.15	1.55	-2.9	0 / 0.10	+7.0	0.24 / 0
9.3	3.84	-9.7	0 / 0.33	-5.2	0 / 0.20					
27	9.70	-22.8	0 / 0.77	-12.2	0 / 0.46	5.15	-11.7	0 / 0.41	+31.2	1.01 / 0
35	11.5	-25.8	0 / 0.89	-13.8	0 / 0.52	5.85	-13.8	0 / 0.46	+33.2	1.13 / 0
40	12.4	-27.6	0 / 0.96	-14.7	0 / 0.56	6.20	-14.8	0 / 0.50	+35.4	1.15 / 0
50	13.8	-29.8	0 / 1.03	-16.0	0 / 0.62	6.70	-16.5	0 / 0.54	+39.5	1.30 / 0
60	14.9	-31.2	0 / 1.09	-16.7	0 / 0.66	7.05	-17.7	0 / 0.61	+42.5	1.39 / 0
70	15.7	-32.2	0 / 1.12	-17.2	0 / 0.68	7.30	-18.7	0 / 0.64	+44.8	1.48 / 0
80	16.2	-32.8	0 / 1.16	-17.5	0 / 0.69	7.42	-19.9	0 / 0.67	+45.7	1.52 / 0

POSITIVE MOMENT INDICATES TENSION ON OUTSIDE SURFACE; NEGATIVE – TENSION INSIDE.
UPPER HALF OF BLOCK FOR STEEL AREA SHOWS REQUIRED STEEL AREA NEAR OUTER FACE; LOWER HALF SHOWS AREA FOR INNER FACE.

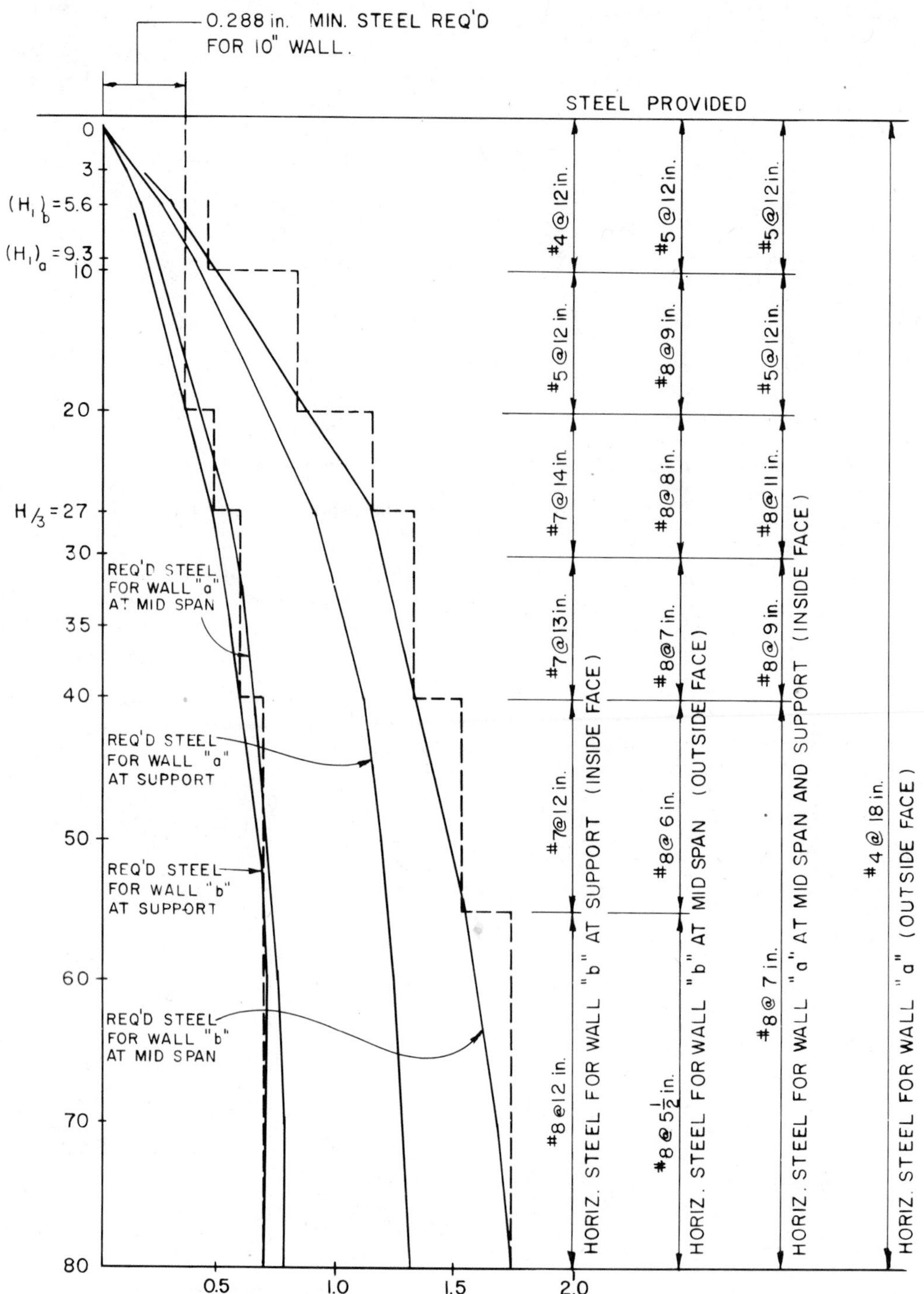

Fig. 16-42 Horizontal wall steel for rectangular silo (example 16-4).

Vertical reinforcement–

$$A_s \text{ reqd} = 0.02 \times 12 \times 10 = 0.24 \text{ sq in./ft}$$

Use #5 @ 16 in. c/c.

Hopper Design

For all depths, use R equal to the hydraulic radius at the top of the hopper. In this case, the pressure changes very little with depth, so use the pressures for Y = 80 ft throughout.

Vertical design pressures at Y = 80 ft are

$$q_{a,\text{des}} = 840/0.406 = 2{,}060 \text{ lb/sq ft} \qquad (16\text{-}1, 16\text{-}3, 16\text{-}20)$$

$$q_{b,\text{des}} = 1{,}100/0.406 = 2{,}710 \text{ lb/sq ft} \qquad (16\text{-}1, 16\text{-}3, 16\text{-}20)$$

Angles of slope are

$$\alpha_a = 45° \text{ and } \alpha_b = 60°$$

Design pressures normal to the inclined walls are

$$q_{\alpha a,\text{des}} = (840 + 2{,}060)(0.707)^2 = 1{,}450 \text{ lb/sq ft} \qquad (16\text{-}18)$$

$$q_{\alpha b,\text{des}} = (1{,}100)(0.866)^2 + 2{,}710\,(0.5)^2 = 1{,}503 \text{ lb/sq ft} \qquad (16\text{-}18)$$

Horizontal ultimate membrane forces (tensile) are

$$F_{tau} = 1.7\,(20/2)\,(1{,}503) \sin 45° = 18{,}100 \text{ lb/ft} \qquad (16\text{-}62)$$

$$F_{tbu} = 1.7\,(12/2)\,(1{,}450) \sin 60° = 12{,}800 \text{ lb/ft} \qquad (16\text{-}63)$$

TABLE 16-21 Membrane Forces in Wall of Pyramidal Hopper (Example 16-4)

DEPTH Y ft	WALL "a"					WALL "b"				
	DESIGN PRESSURES psf			ULT. FORCES k/ft		DESIGN PRESSURES psf			ULT. FORCES k/ft	
	LATERAL $P_{a,des}$	VERTICAL $Q_{a,des}$	NORMAL $Q_{\alpha a,des}$	HORIZONTAL F_{tau}	MERIDIANAL F_{mau}	LATERAL $P_{b,des}$	VERTICAL $q_{b,des}$	NORMAL $q_{\alpha b,des}$	HORIZONTAL F_{tbu}	MERIDIANAL F_{mbu}
80	728	1,790	1,260	15.7	25.4	953	2,350	1,300	11.1	15.7
82				11.8	19.6				9.2	12.5
84				8.5	14.0				7.0	9.5
86				5.1	8.3				4.6	6.2
88				1.4	3.0				2.3	1.9

The weight of the hopper and its contents are estimated to be 41,200 lb and 34,000 lb, respectively. A_a and A_b are each one-fourth of the hopper area, or (20 X 12)/4 = 60 sq ft.

Meridional ultimate forces (tensile) are computed assuming c_a and c_b to each be 1/4.

$$F_{mau} = \frac{1.7\,(0.25 \times 34{,}000 + 60 \times 2{,}060) + 1.4\,(0.25 \times 41{,}200)}{12 \sin 45^\circ}$$

$$= 28{,}200 \text{ lb/ft} \quad (16\text{-}60)$$

$$F_{mbu} = \frac{1.7\,(0.25 \times 34{,}000 + 60 \times 2{,}710) + 1.4\,(0.25 \times 41{,}200)}{20 \sin 60^\circ}$$

$$= 17{,}600 \text{ lb/ft} \quad (16\text{-}61)$$

Table 16-21 shows the membrane forces at various depths Y.

Hopper wall bending moments are computed using Table 16-5. For hopper wall 'a', a = 13 ft (measured to center of walls), and c = 13 ft. Use a/c = 1.0.

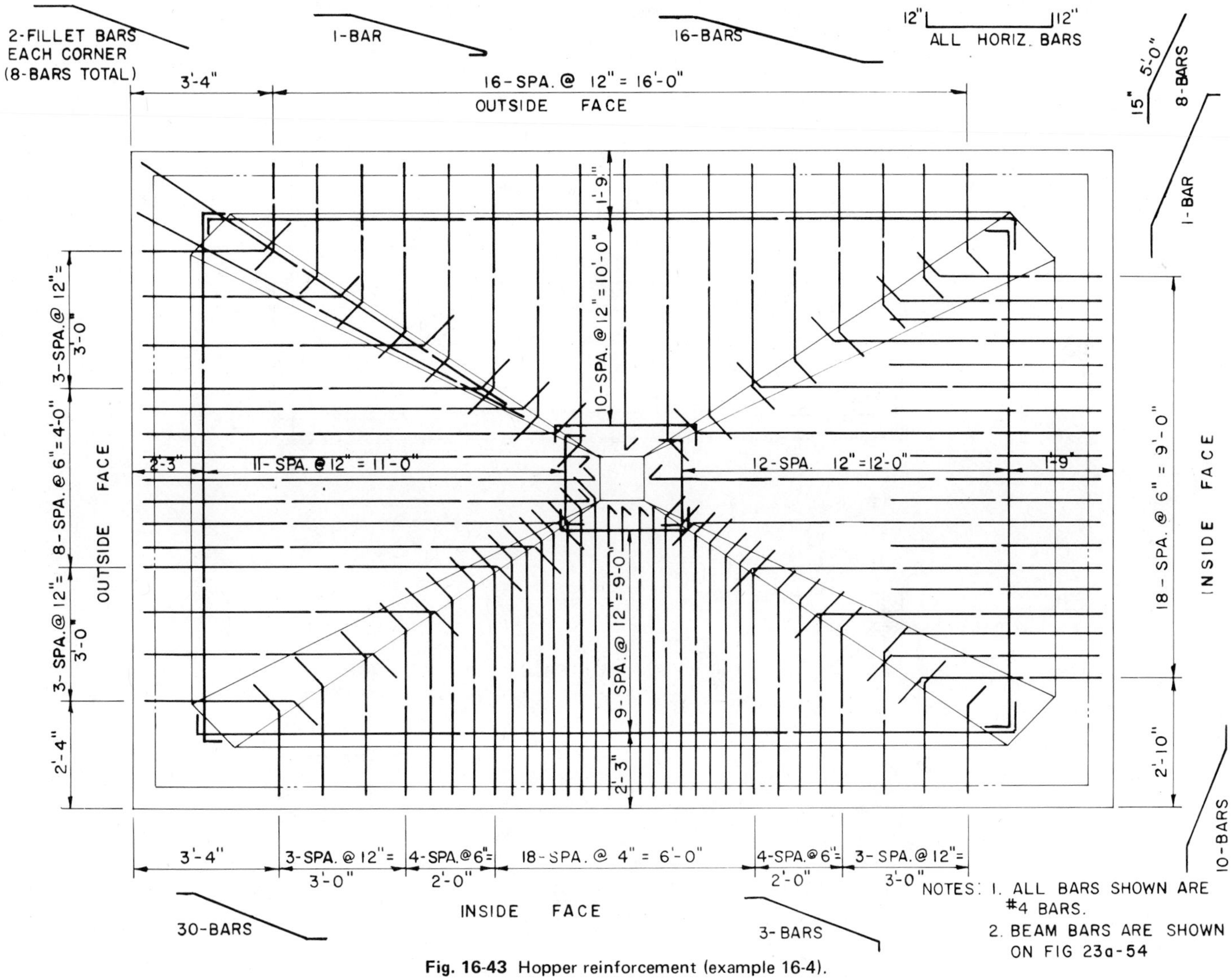

NOTES: 1. ALL BARS SHOWN ARE #4 BARS.
2. BEAM BARS ARE SHOWN ON FIG 23a-54

Fig. 16-43 Hopper reinforcement (example 16-4).

At the center of the upper edge (point 1)

$$n_x = -0.209; \quad n_y = -1.255 \qquad \text{(Table 16-5)}$$

$$M_{xau} = 1.7\ [0.209\ (1{,}450)\ (13)^2/64] = 1{,}360 \text{ ft-lb/ft}$$

$$M_{yau} = 1.7\ [1.255\ (1{,}450)\ (13)^2/64] = 8{,}150 \text{ ft-lb/ft}$$

Similar moment computations were made for several points of the triangular wall. From these and the corresponding membrane forces (Table 16-21) the reinforcement was computed following the procedure shown for combined tension and bending. The final pattern of reinforcing steel selected is shown by Fig. 16-43.

Hopper Edge Beam and Bottom Reinforcement of Vertical Walls: Dowels are provided to transfer the vertical forces from the hopper beam into the vertical walls. For wall '*a*', for example, the vertical load to be transferred is

$$P_u = F_{mau} \sin \alpha_a = 19.9 \text{ kip/ft}$$

$$A_s \text{ reqd} = 19.9/(0.95 \times 0.9 \times 60) = 0.39 \text{ sq in./ft}$$

By scaling from a detail sketch of the walls and edge beam, it was determined that force F_{mau} is about 6 in. from the centroid of the edge beam. The resulting twisting moment can be resisted by the vertical wall provided extra dowels are used. For this purpose, dowels having $A_s = 0.34$ sq in./ft are needed.

The total dowel steel required near the inner face is

$$A_s = 0.39/2 + 0.34 = 0.53 \text{ sq in./ft}$$

and near the outer face

$$A_s = 0.39/2 = 0.19 \text{ sq in./ft.}$$

The dowel requirement is met by using #7 "hairpin bars" @ 14 in. c/c as shown in Fig. 16-44.

Because of dowels joining it to the vertical wall, the edge beam has negligible vertical shear and torsion. The upper wall shear and horizontal components of the hopper wall meridional force and shear are assumed to be in equilibrium; thus, no horizontal transverse load is carried by the edge beam. Its only purpose is to simplify construction. Minimum longitudinal steel and stirrups are provided (Fig. 16-44).

The vertical walls are analyzed as deep girders to carry vertical load as follows:

For wall '*a*'–

from dowels to hopper,	19.9 kip/ft
friction, 1.7 (9.13) =	15.5
wall weight, 1.4 (0.83) (80) (0.15) =	13.9
Total =	49.3 kip/ft

Similarly, for wall '*b*', the ultimate vertical load is 47.1 k/ft.

In this example, the vertical walls were analyzed as deep beams. That analysis showed bottom steel to be needed, as follows:

Bottom of wall '*a*', 2–#9

Bottom of wall '*b*', 3–#9

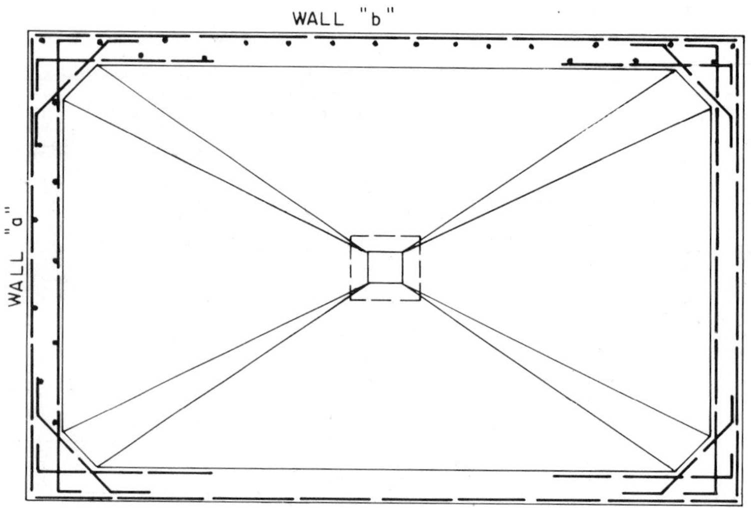

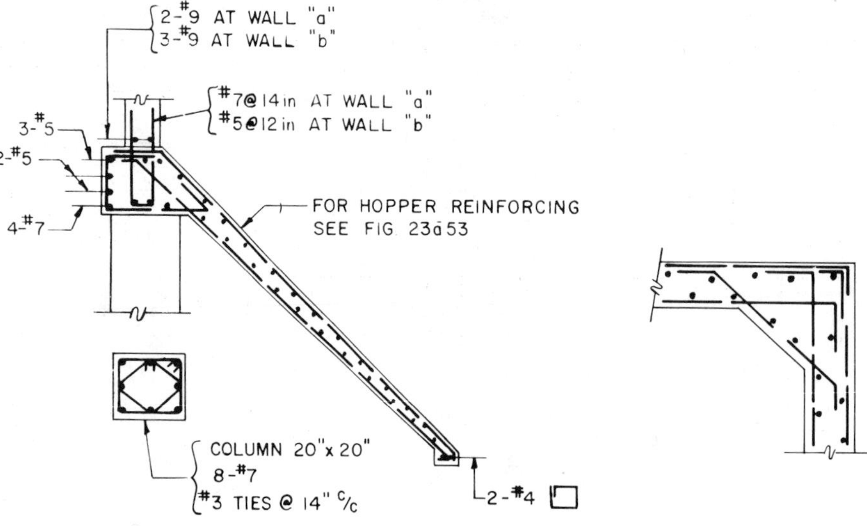

Fig. 16-44 Hopper reinforcement details (example 16-4).

Column Design
Load per column–

	Dead	Live
material stored in hopper		8,500 lb
material stored above hopper		240,000
walls	168,750 lb	
roof (principally dead load)	17,000	
hopper	10,300	
Totals	196,050 lb	248,500 lb

Ultimate load = P_u = 1.4 (196.1) = 1.7 (248.5) = 698 kips

A 20 in. X 20 in. tied column with 8-#7 bars is satisfactory. (Fig. 16-44)

Earthquake zone 0 was assumed, therefore seismic forces were not considered. If seismic forces were to be considered, further analysis would be necessary. A stronger column, preferably spirally reinforced would be required.

EXAMPLE 16-5: Figure 16-45 shows the basic scheme and dimensions of a hexagonal silo for storing dry sand. The bottom is a horizontal slab with one eccentric discharge opening. The structure is supported on six columns. Roof loads total 1,200 lb per foot of wall. Slipforming is to be used. Design the walls, bottom and columns using f'_c = 4,000 psi and f_y = 60,000 psi.

SOLUTION:
For dry sand

$$\gamma = 100\ pcf;\quad \phi = 35°;\quad \mu' = \tan\phi' = 0.70 \quad \text{(Table 16-2)}$$

Using Reimbert's method and assuming a flat top surface

$$R = \frac{6\,(\tfrac{1}{2})\,(12)\,(12)\sin 60°}{6\,(12)} = 5.2\text{ ft}$$

$$P_{max} = \frac{100 \times 5.2}{\tan 35°} = 743\text{ lb/sq ft} \qquad (16\text{-}9)$$

The characteristic abscissa is

$$C = \frac{6\,(12)}{\pi}\left[\frac{1}{4\,(0.7)\tan^2(45° - 17.5°)}\right] - 0 = 30.3\text{ ft} \qquad (16\text{-}10)$$

$$H_1 = 2 \times 10.4 \tan 35° = 14.6\text{ ft}$$

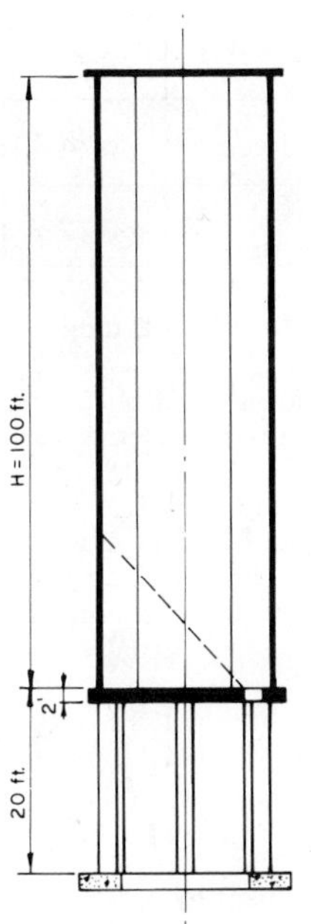

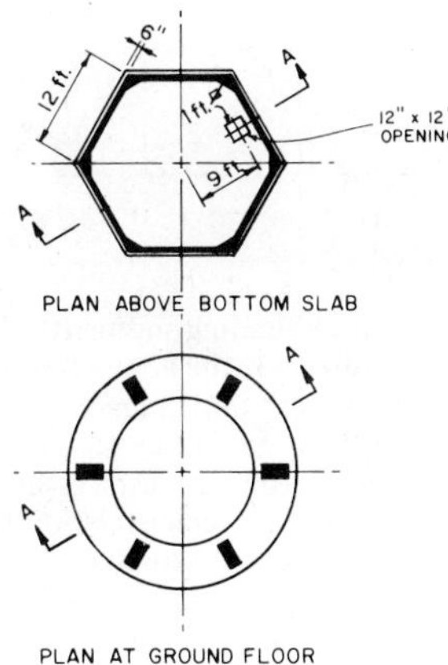

Fig. 16-45 Hexagonal silo dimensions (example 16-5).

Overpressure factors with H/D = 100/20.8 = 4.8 (Table 16-3)

upper 1/3 of depth	C_d = 1.21
lower 2/3 of depth	C_d = 1.93
for bottom slab	C_d = 1.48

Table 16-22 shows overpressure factors and lateral pressures at various depths. The computations are illustrated below for depth Y = 100 ft.

Y/C = 100/30.3 = 3.3; z = 0.946

$$P = 0.946\,(743) = 702\text{ lb/sq ft} \qquad (16\text{-}6)$$

$$P_{des} = 1.93\,(702) = 1{,}350\text{ lb/sq ft (lateral)} \qquad (16\text{-}21)$$

$$q_{des} = 1.48\,(100)\,[100/(3.3 + 1)] = 3{,}440\text{ lb/sq ft (vertical)} \qquad (16\text{-}20, 16\text{-}5)$$

$$V_{des}\text{ (frictional load)} = (100 \times 100 - 3{,}440/1.48)\,(5.2) = 45{,}100\text{ lb/ft} \qquad (16\text{-}16)$$

All of the above are for a silo with central discharge.

The lateral design pressure must now be corrected for eccentricity of discharge. At Y = 100 ft

$$Y = H;\quad \therefore \text{Use full correction, } P_e$$

An imaginary silo will be selected by substituting a circle for the hexagon and proceeding as shown by Fig. 16-8. A circle of area equal to that of the hexagon has a radius of 10.9 ft. The radius of the imaginary silo is then

$$r_i = 10.9 + 9 = 19.9\text{ ft} \qquad \text{(Fig. 16-8)}$$

TABLE 16-22 Lateral Pressures on Silo Walls (Example 16-5)

DEPTH Y (ft)	C_d	BASIC LATERAL PRESS. P lb/sq ft	LATERAL PRESSURES DUE TO ECCENTRICITY lb/sq ft	DESIGN LATERAL PRESS. P_{des} lb/sq ft
5	1.0	195	24	219
10	1.0	323	48	371
H_1=14.5	1.0	404	70	474
$H/3$=33	1.75	574	159	1,163
40	1.75	604	193	1,249
50	1.75	636	242	1,355
60	1.75	660	290	1,445
70	1.75	674	338	1,518
80	1.75	686	386	1,586
90	1.75	696	435	1,655
100	1.75	702	483	1,713

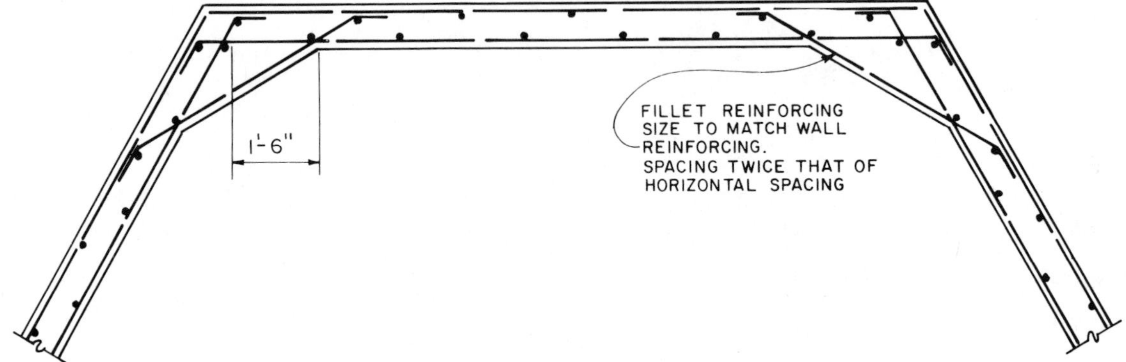

Fig. 16-46 Hexagonal silo wall reinforcing (example 16-5).

The corresponding hexagonal imaginary silo has all linear dimensions equal to (19.9/10.9) times those of the actual hexagon. Thus

$$R_i = 5.2\,(19.9/10.9) = 9.5 \text{ ft}$$

$$C = 30.3\,(19.9/10.9) = 55.3 \text{ ft}$$

$$Y/C = 100/55.3 = 1.81; \quad z = 0.873$$

$$P_{max} = 100\,(9.5)/\tan 35° = 1{,}360 \text{ lb/sq ft} \qquad (16\text{-}9)$$

$$P_i = 0.873\,(1{,}360) = 1{,}185 \text{ lb/sq ft} \qquad (16\text{-}6)$$

$$P_e = 1{,}185 - 702 = 483 \text{ lb/sq ft} \qquad (16\text{-}25)$$

The corrected P_{des} at $Y = 100$ ft is then

$$P_{des} = 1{,}350 + \frac{100}{100}(483) = 1{,}833 \text{ lb/sq ft} \qquad (16\text{-}24)$$

At depths Y of less than 100 ft, the correction to P_{des} at that depth is

$$P_{e,y} = 483\,(Y/100)$$

Corrections and corrected design pressures for various depths are tabulated on Table 16-22.

Wall force and bending moment at $Y = 100$ ft

$$F_u = 0.866\,(1.7)\,(12)\,(1{,}833) = 32{,}400 \text{ lb/ft} \qquad (16\text{-}52)$$

Assume a 9-in. wall thickness. The span, intersection-to-intersection, is then 12.43 ft. The edge of the fillet (Fig. 16-46) is 1.71 ft from the intersection of walls. Fillet toes are 12.43 - 3.42, or 9.01 ft apart. Assuming a parabolic moment diagram, the bending moment at any depth, reduced to the edge of the fillet, is

$$M_u = \frac{1.7\,P_{des}}{8}(9.01^2 - 12.43^2/3) = 6.31\,P_{des}$$

or at $Y = 100$ ft

$$M_u = 6.31\,(1{,}833) = 11{,}600 \text{ ft-lb/ft}$$

At midspan

$$M_u = (12.43)^2\,P_{des}/24$$

Ultimate wall forces and bending moments for various depths, as well as the required areas of positive and negative steel (selected for combined tension and bending) are shown by Table 16-23. The reinforcing chosen shows on Fig. 16-47.

Bottom Slab Design

By hanging the bottom slab from the vertical walls and taking advantage of their deep-beam action to transfer vertical loads to the six columns, the bottom slab can be designed as though supported all along the wall line. Since the discharge opening is small, tables for circular slabs may be used to obtain approximate radial and tangential bending moments. To do this, a circle of 11.5 ft radius is substituted for the actual hexagonal shape.

Bottom slab design loads, assuming a 24-in. slab, are q_{des} + 300, or 3,740 lb/sq ft. Table 16-24 shows the bending moments and steel areas required (computed assuming simple support). Fig. 16-48 shows the slab reinforcement. Note that a circular/radial pattern is actually used. Around the discharge opening, added reinforcement must be provided to (at least) substitute for that which is intercepted by the hole. Dowels to suspend the slab from the wall must carry an ultimate load of

$$F_u = 1.7\,(3.74)\,(374)/72 = 33.0 \text{ kips per foot of wall}$$

The required dowel area is

$$A_s = 33.0/(60 \times 0.85) = 0.65 \text{ sq in. per ft of wall}$$

The dowels must have anchorage for full development by bond, both in the slab and in the wall.

Finally, the walls are checked for bending as deep beams and the columns are designed.

TABLE 16-23 Horizontal Forces, Bending Moments, and Reinforcing Steel Areas (Example 16-5)

DEPTH Y ft	ULTIMATE TENSILE FORCE F_u K/ft.	ULTIMATE MOMENT AT EDGE OF FILLET ft-K/ft.	AREA OF REQ'D STEEL in.²/ft INSIDE FACE	ULTIMATE MOMENT AT MID-SPAN ft-K	AREA OF REQ'D STEEL in²/ft OUTSIDE FACE
5	3.87	-1.38	0.10	1.41	0.09
10	6.55	-2.34	0.16	2.39	0.16
H_1=14.5	8.38	-2.97	0.22	3.05	0.23
$H/3$=33	20.6	-7.35	0.51	7.50	0.52
40	22.0	-7.87	0.54	8.04	0.55
50	23.9	-8.55	0.59	8.73	0.61
60	25.5	-9.12	0.64	9.30	0.64
70	26.8	-9.57	0.66	9.76	0.69
80	28.0	-9.95	0.69	10.2	0.71
90	29.2	-10.45	0.74	10.65	0.74
100	30.3	-10.81	0.75	11.0	0.77

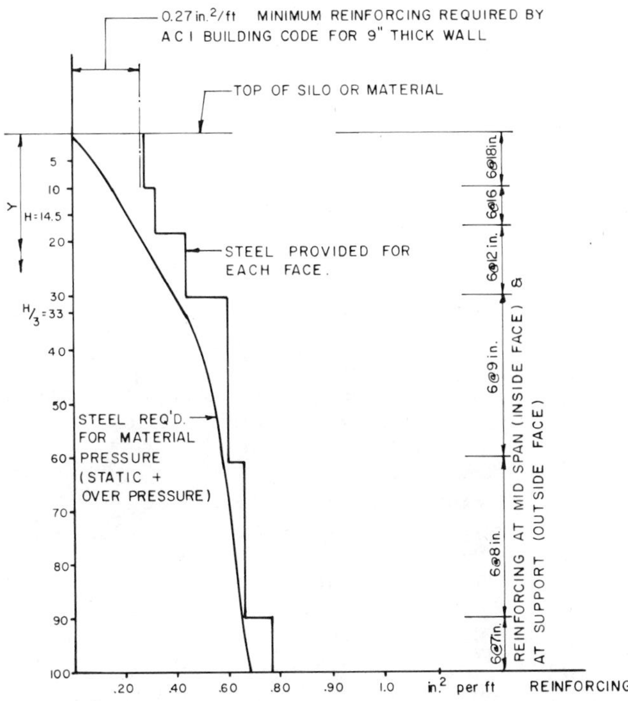

Fig. 16-47 Horizontal reinforcing for silo wall (example 16-5).

TABLE 16-24 Ultimate Bending Moments and Required Steel Areas for Bottom Slab (Example 16-5)

$\frac{r}{r_o}$	$(M_r)_u$ ft.-K	REQ'D STEEL in/ft	$(M_t)_u$ ft.-K	REQ'D STEEL in/ft
0.0	121.5	1.35	121.5	1.35
0.1	120.5	1.34	121.0	1.34
0.2	116.5	1.29	118.5	1.32
0.3	111.5	1.23	115.5	1.29
0.4	102.0	1.12	110.5	1.22
0.5	91.5	1.00	104.5	1.15
0.6	78.0	0.85	97.0	1.07
0.7	62.0	.68	87.3	0.96
0.8	43.6	.47	76.6	0.84
0.9	23.0	.44	65.0	0.71
1.0	0.0	.44	52.0	0.57

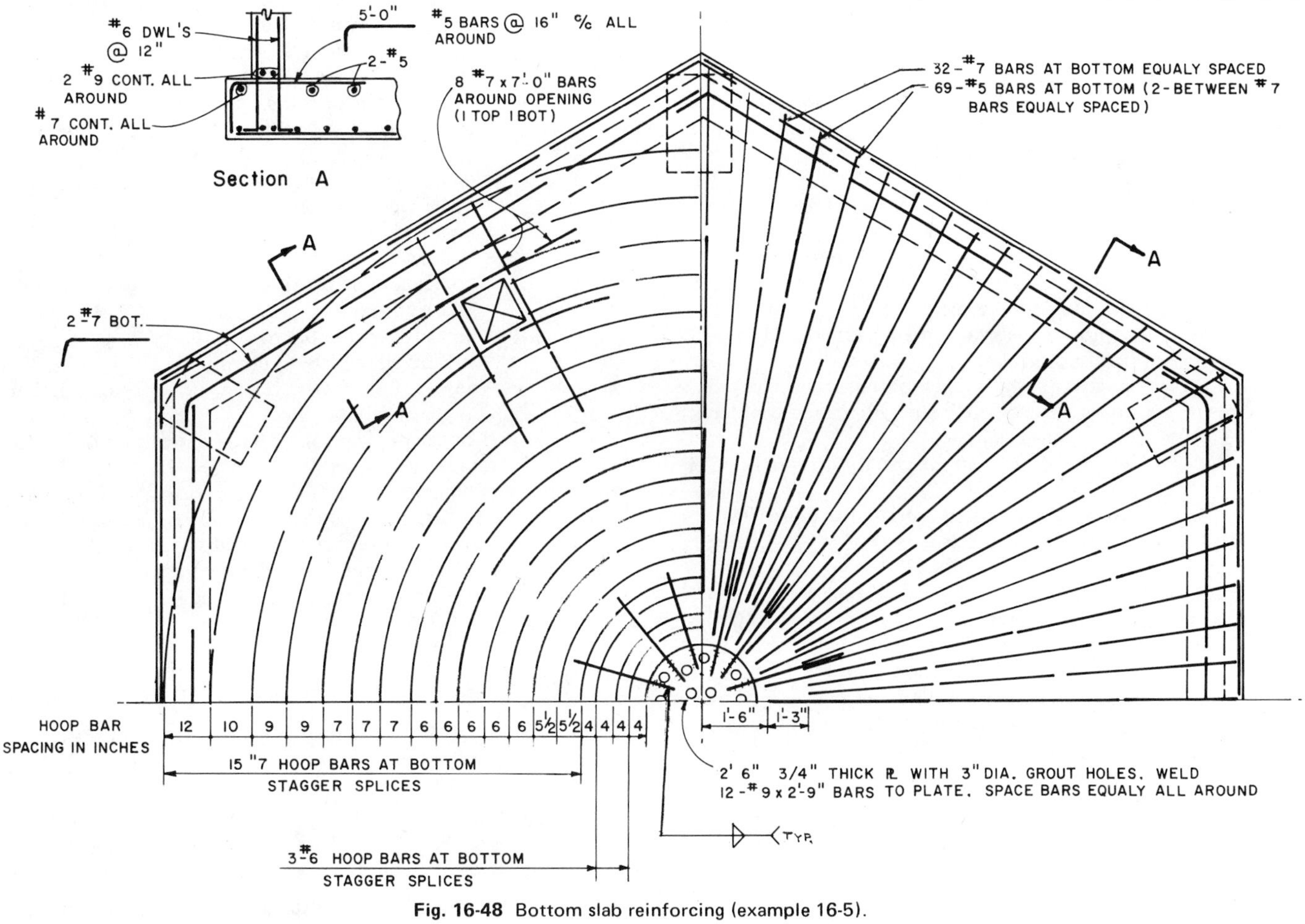

Fig. 16-48 Bottom slab reinforcing (example 16-5).

NOTATION

A = area
A_r = area of cross section of ring beam
A_s = reinforcing steel area (usually per unit width of wall)
C = characteristic abscissa for Reimbert's formula
C_d = overpressure factor, for converting static pressures to design pressures
D = diameter (inside, unless noted)
E_s, E_c = modulus of elasticity for steel and concrete, respectively
F = force, tensile or compressive
F_b = allowable bending stress for structural steel
F_m = meridional membrane force per unit width of hopper wall
F_t = horizontal membrane force per unit width of hopper wall
H = height of silo (see Fig. 16-4)
H_a = shear in column supporting ring beam
I = moment of inertia
K = load factor (*ACI Code*)
K_t = factor for computing temperature difference (Fig. 16-11)
L = perimeter of horizontal (inside) cross section
M = bending moment, usually per unit width of wall
P = horizontal static pressure due to stored material
P_e = a correction to P_{des} for effect of eccentricity of discharge
P_i = value of P in imaginary silo
R = hydraulic radius of horizontal cross section of storage space
T_i = temperature inside mass of stored material
T_0 = temperature of air outside of silo wall
T_u = torque, ultimate
V = in eqq. (16-16) and (16-17), total vertical frictional force on a unit width of wall above the section in question
V_u = ultimate shear force
W = total weight of stored material, or weight of a hopper and its contents
W_g = that portion of W due to hopper weight
W_{ll} = that portion of W due to weight of material stored in the hopper
Y = depth from surface of stored material to point in equation
a = length of short side of rectangle; length of upper edge of inclined wall; inside wall length of polygonal silo
a' = length of side of imaginary square silo, eq. (16-4)
a_{eq} = width of equivalent rectangular plate
a_1 = length of bottom (shorter) edge of trapezoidal wall
b = length of long side of rectangle; width of a flexural section
c = slant height of hopper wall
c_a, c_b = factors defining division of total hopper load between inclined walls a and b
c_i = slant height of imaginary trapezoidal area
d = effective depth of concrete flexural member
des = a subscript indicating 'design' force or pressure
e = eccentricity
f_b = actual bending stress in structural steel
$f_{c,\text{vert}}$ = vertical compressive stress
f'_c = ultimate compressive strength of concrete
f_s = steel stress, tension
f'_s = effective compressive steel stress
f_{tu} = ultimate tensile strength of concrete
f_y = yield strength of steel
g = dead load per unit area (or a subscript indicating gravity load)
g_r = weight of ring beam, per unit length
h = wall thickness
h_c = height of column
h_h = height of hopper
h_s = height of sloping top surface of stored material
k = ratio of P to q
ll = a subscript meaning 'live load'
m = shrinkage coefficient, generally 0.0003
n = modular ratio, E_s/E_c
q = static vertical pressure due to stored material
q_α = unit static pressure normal to surface inclined at angle α to horizontal
s_{cr} = crack spacing
st = a subscript meaning 'static'
t = a subscript indicating 'temperature' or 'thermal'
u = a subscript meaning 'ultimate'
v_c = permissible ultimate shear stress
v_u = shear stress due to computed ultimate load
w_1, w_2, w_3 = crack width increments, defined in section 16.7.3
w_{cr} = crack width, total
w_u = ultimate load per unit area, normal to the wall surface in question
y = depth of compression block in USD
z = coefficient
α = angle of inclination of hopper wall
α_i = angle between hopper wall intersection line and a horizontal plane
α_t = thermal coefficient of expansion of concrete (in./in./°F)
β = ratio, depth of rectangular compression block to depth from compression face to neutral axis
γ = weight per unit volume for stored material
η = coefficient, Table 16-7
θ = angle of rotation, radians
λ = torsional property, Table 16-8
μ' = coefficient of friction between stored material and wall = tan ϕ'
τ_c = normal permissible torsion stress
τ_u = torsion stress due to computed ultimate load
ϕ = angle of internal friction, or (approximately) angle of repose
ϕ' = angle of friction between walls and stored material
ψ = capacity reduction factor (used here to substitute for ϕ of the *ACI Code*)
ψ_1 = factor for use in crack width computation
ψ_a, ψ_b = angle between horizontal projections of hopper wall intersection line and edges 'a' and 'b' respectively

REFERENCES

16-1 Jenike, A. W., "Denting of Circular Bins with Eccentric Drawpoints," *Proceedings, ASCE*, **93** (ST1), 27–35, Feb. 1967.

16-2 Theimer, O. F., "Failures of Reinforced Concrete Grain Silos," *Publication 68-MH-36*, American Society of Mechanical Engineers, New York, 1968.

16-3 Jenike, A. W., Flow of Bulk Solids," *Bulletin 64*,

Utah Engineering Experiment Station, University of Utah, March 1954.

16-4 Theimer, O. F., "Ablauf fördernde Trichterkonstructionen von Silozellen," *Aufberechmungs–Technik No. 10*, 547–556, Oct. 1969.

16-5 Kvapil, R., and Taubmann, H. J., "Flow and Extraction of Solids from Bins," *Publication 68-MH-32*, ASME, New York, 1968.

16-6 Fischer, W., "*Silos und Bunkers in Stahlbeton*," Veb Verlag fur Bauwesen, Berlin, 1966.

16-7 *Soviet Concrete and Reinforced Concrete Code*, (Stroitelnie Normi i Pravila Tshast II Razdel B Glava 1 Betonnie i Zelezobetonnie Konstruktsi SNKP) II-B.1-62. Publication of the Committee of Soviet Ministers USSR of Construction Works, Moscow, 1962.

16-8 Janssen, H. A., "Versuche Über Getreidedruck in Silozellen," *VDI Zeitschrift* (Düsseldorf), **39**, 1045–1049, Aug. 31, 1895.

16-9 Reimbert, M. and Reimbert, A., *Silos–Traité Théoretique et Pratique*, Editions Eyrolles, Paris, 1961, 255 pp.

16-10 Petrov, B. A., "Experimental Determination of Cement Pressure on Reinforced Concrete Silo Walls (Experimentalnoe Opredelenie Davlenia Cementa na Stenki Zhelezobetonnich)," *Cement*, No. 2, 21–25, 1958.

16-11 Pieper, K., and Wenzel, F., *Druckverhältnisse in Silozellen*, Verlag van Wilhelm Ernst and Sohn, Berlin, Munich, 1968.

16-12 Deutsch, G. P., and Schmidt, L. C., "Pressures on Silo Walls," *Publication 68-MH-24*, ASME, New York, 1968.

16-13 Takahashi, K., "Silo Design," *Concrete Journal*, **5** (9), Aug. 1969, Japanese National Council on Concrete (in Japanese).

16-14 "Instructions for Design of Silos for Granular Materials (Ukazania Po Proectirovaniu Silosov Dlia Siputschich Materialov," *Soviet Code CH-302-65*, Gosstroy, USSR, Moscow, 1965.

16-15 Pieper, K., "Investigation of Silo Loads in Measuring Models," *Publication 68-MH-30*, ASME, New York, 1968.

16-16 Pieper, K., and Wagner, K., "Der Einfluss Verschiedsher Auslaufarten aur die Saitendrucke in Silozellen," *Aufbereitungs–Technik No. 10*, 542–546, Oct. 1969.

16-17 Platonov, P. N., and Kovtun, A. P., "The Pressure of Grain on Silo Walls (Davlenie Zerna na Stenki Silosov Elevatorov)," *Mukomolno Elevatornaia Promyschlennost* (Moscow, **25** (12), 22–24, Dec. 1959.

16-18 Leonhardt, Dr. F., "Zur Frage der Sicheren Bemessung von Zement-Silos," *Beton-und Stahlbetonbau* (Berlin), **55** (3), 48–58, March 1960.

16-19 *Uniform Building Code*, Vol. I, International Conference of Building Officials, Pasadena, California, 1970.

16-20 Chandrasekaran, A. R., and Jain, P. C., "Effective Live Load of Storage Materials Under Dynamic Conditions," *Indian Concrete Journal*, 369–385, Sept. 1968.

16-21 "Circular Concrete Tanks without Prestressing," *Publication ST-57*, Structural Bureau, Portland Cement Association, Chicago.

16-22 Lipnitski, M. E., and Abramovitsch, S. P., *Reinforced Concrete Bunkers and Silos*, (*Zhelezobetonnie Bunkera i Silosi*), Izdatelstvo Literaturi Po Stroitelstvu, Leningrad, 1967.

16-23 Bohm, F., "The Calculation of Circular Silos for the Storage of Cement," *Beton-und Stahlbetonbau* (Berlin), **51** (2), 29–36, Feb. 1956.

16-24 *ASHRAE Guide and Data Book*, American Society of Heating, Refrigerating and Air Conditioning Engineers, New York, 1963.

16-25 Boley, B. A., and Weiner, J. H., *Theory of Thermal Stresses*, John Wiley and Sons, Inc., New York, 1960.

16-26 Safarian, S. S., and Harris, E. C., "Determination of Minimum Wall Thickness and Temperature Steel in Conventionally Reinforced Circular Concrete Silos," *ACI Journal, Proceedings*, **67** (7), 539–547, July, 1970.

16-27 ACI Committee 318, *Building Code Requirements for Reinforced Concrete (ACI-318-71)*, American Concrete Institute, Detroit, 1971.

16-28 Safarian, S. S., and Harris, E. C., Book on silo design, now in preparation.

16-29 Timoshenko, S., and Winowsky-Krieger, S., *Theory of Plates and Shells*, Second Ed., McGraw-Hill, New York, 1959.

16-30 Griffel, W., *Plate Formulas*, Frederick Ungar Publishing Company, New York, 1968.

16-31 Fischer, W., *Silos und Bunker in Stahlbeton*, Veb Verlag fur Bauwesen, Berlin, 1966.

16-32 Ciesielski, R., et al., *Behalter, Bunker, Silos, Schornsteine, Fernsehturme und Freileitungsmaste*, Verlag von Wilhelm Ernst and Sohn, Berlin, 1970.

17

Concrete Masonry Construction

WALTER L. DICKEY*

17.1 INTRODUCTION

17.1.1 General

This section is to assist the engineer in the general design of modern masonry with emphasis on reinforced concrete masonry. Although masonry is historically one of man's oldest, permanent construction materials, the introduction of newer engineering methods has been relatively slow, overshadowed somewhat by the arbitrary and customary uses, or the traditional design.

17.1.2 Reinforced vs Unreinforced

The historically poor seismic performance of traditionally constructed masonry, as well as the growing sophistication in earthquake- or tornado-resistant design, has introduced reinforcing, which forms a new and distinct type of masonry insofar as performance and resistance to loads; some of these methods are clarified in the following. The historically poor performance of traditional materials has served to give masonry a negative image where earthquake resistance is concerned. Reinforced masonry, however, is a material that will perform well in earthquakes and is not subject to the same failures as previous unbraced masonry, or any unbraced material for that matter. It has performed so well that there are indications that it has considerable great damping and ductility that should be evaluated.

*Consulting Structural Engineer, Masonry Institute of America, Los Angeles, California.

17.1.3 Masonry Materials

Among the materials being used today we find the following:

a. Stone masonry and clay brick masonry have been the materials used the longest in construction and demonstrate great permanence.
b. Concrete brick is used in manners similar to the older uses of (clay) brick, with certain improvements in shape that are possible in casting or fabrication.
c. Concrete block is a newer development and is especially well adapted for economy and efficiency of use, both in reinforced and in unreinforced masonry.
d. Hollow (clay) brick is an even newer development and is similar.
e. Tile is another traditional masonry for which there are many reinforced techniques possible that may make this a very different and effective material. GSU (Glazed Structural Units) provide not only beautiful, durable finish, but excellent structural capacity.

However, masonry has been modified by the addition of a strong-bonding portland cement type of mortar for the "new masonry", as well as a bonding portland cement grout. The earlier mortars were not intended to bond the masonry together but merely to keep the units apart the right distance and take up the spaces due to inequalities of size and dimension.

17.1.4 Design Considerations

Proper masonry design must include consideration of the following:

a. Detailing, which is a very important aspect in the proper use of masonry elements. One of the advantages of masonry unit construction is the flexibility of construction shape. However, due to the fact that hand work is involved, thought must be given to proper placement and modular dimension. Proper detailing requires that the structural considerations and requirements be interpreted into proper placement of elements for the field construction, that is, a combination of theory and practice.
b. Waterproofing is an important aspect of masonry and details are included as an important facet of proper design.
c. Fire resistance and endurance are two very desirable features of masonry construction.
d. Seismic design considerations in this section are based on the *Uniform Building Code* provisions. They are included because structural engineers are increasingly aware of seismic factors. Also, it is recognized that all areas in the world are liable to some seismic activity. In the eastern part of the United States there has generally been a feeling that seismic factors were not important in structural design. However, records show that one of the largest earthquakes in the North American continent was in the central part of the United States. Therefore, seismic resistance should be considered in any area, even though traditionally and economically it has not been considered feasible to include it in the so-called nonseismic areas. When one considers the factors for Zone 1 and Zone 2, proper consideration of adequate seismic resistance will add relatively little to the cost of the structure, since about the only difference will be in the provisions of connections and in details. Seismic-resistant design is becoming more common; therefore, this factor is expanded in the design example of masonry buildings.

17.2 MASONRY ASSEMBLAGES

17.2.1 General

Masonry construction has been defined as a type of construction in which masonry units are placed by hand with mortar between them. The classification of types used in the *Uniform Building Code** is as follows: unburned clay masonry, gypsum masonry, glass masonry, stone masonry, cavity wall masonry, hollow unit masonry, solid masonry, grouted masonry, reinforced grouted masonry, and reinforced hollow unit masonry. These have developed in different ways and are used in various manners due to their varying and specific characteristics.

These items are clarified more in detail as follows for compliance with the *Uniform Building Code.*

*Although this chapter refers to the *Uniform Building Code*, the reader is referred to other building codes which may be applicable in the location of his project.

17.2.2 Unburned Clay Masonry

This is the descendant of the early masonry such as used in the early California Missions. It is also an older type of masonry that has been used in many parts of the world. However, there have been some modifications, such as binders, particularly for durability in seismic areas—also, for weather resistance. This type masonry has been used effectively for one-story structures and for homes. The unfired clay units, or so-called adobe, are laid in a running bond and, for compliance with *UBC*, with Type M or S mortar, that is, a good cement mortar which will add strength. Its limits and dimensions are such that the stresses will be relatively low under the normally anticipated loadings.

There have been examples of excellent use in homes where these were used as filler walls, engineered with reinforcement in the horizontal bed joints spanning to the posts for lateral support. There have also been examples in which reinforced concrete elements were incorporated within the thickness of the wall by grout poured into spaces provided within the clay masonry, as studs or as bond beams.

17.2.3 Gypsum Masonry

Gypsum masonry is made with gypsum block or tile and gypsum mortar. Gypsum masonry is not to be used in bearing walls, nor where exposed directly to the weather, nor where subject to wetting because the gypsum tends to disintegrate under those conditions.

The gypsum mortar is composed of one part gypsum and not more than three parts sand by weight.

There have been certain examples of gypsum partitions reinforced by the application of mesh reinforcing bonded to the surface by plaster.

Their most effective use is in partitions for separation of rooms. The stresses are low in engineered masonry to provide for a safe use of this material.

17.2.4 Glass Masonry

Glass block masonry consists of glass blocks manufactured for this purpose, which are used in nonload-bearing exterior or interior walls and in openings which otherwise might be filled with windows. It might be used either isolated or in continuous spans, provided, however, that the glass block panels have a minimum thickness, in the interest of safety, of 3.5 in. at the mortar joint and that the mortared surfaces of the blocks are treated properly for mortar bonding.

Glass block panels for exterior walls should not exceed 144 sq ft of unsupported wall surface and should not extend more than 15 ft in any dimension. For interior walls, glass block panels shall not exceed 250 sq ft of unsupported area nor 25 ft in any dimension.

Exterior glass block panels should be provided with expansion joints at the sides and top of not less than $1/2$ in. These expansion joints should be free of mortar and should be filled with resilient material to provide for the expansion and contraction of the panel without restraining stress.

Glass block should be laid in Type M or S mortar with vertical and horizontal mortar joints from $1/4$ to not more than $3/8$ in. thick. Wire reinforcement is sometimes placed in the joints for structural design.

17.2.5 Stone Masonry

Stone masonry is that form of construction made with natural or cast stones laid and set in mortar.

In ashlar masonry of unreinforced masonry, bond stones shall be provided uniformly distributed to provide not less than 10% of the area of exposed facets.

Rubble stone masonry 24 in. or less in thickness should have bond stones with a maximum spacing of 3 ft vertically and 3 ft horizontally. If it is of greater thickness than 24 in. it shall have at least one bond stone for each 6 sq ft of wall surface on both sides.

Stone masonry walls shall not be less than 16 in. in thickness and with stresses not exceeding those outlined in the code tables for stress.

Stone masonry is sometimes used as solid masonry. If the units are laid in two wythes with grout between, they will function in a manner similar to solid grouted masonry of brick. In general, these would be designed as full-thickness masonry walls with the allowable stresses of masonry. However, in some instances wythes have been specified as separate wythes between which a grout was poured and the wall was designed on the basis of the grout serving as a reinforced concrete wall, with the outer wythes merely as nonstructural forming and veneer.

17.2.6 Cavity Walls

Cavity wall masonry is that type of construction made with brick, structural clay tile, or concrete masonry units, or any combination of such units in which facing and backing are completely separated, except for the ties, which serve as bonding.

In these walls the facing and backing walls are generally from 3 in. to 4 in. thick, but may be thicker. There are arbitrary requirements and limits of spacing of ties, type of ties, thicknesses, and heights as limited by the *UBC*, which should be referred to for detail.

These are generally designed as unreinforced elements, however, they can be designed as reinforced or partially-reinforced masonry. They provide excellent weather-resistant walls.

17.2.7 Hollow Unit

Hollow unit masonry is that type of construction made with hollow masonry units in which units are laid in mortar. Generally, this is for unreinforced masonry construction. However, reinforcement may be embeded in the mortar joints. See "Reinforced Hollow Unit Masonry."

17.2.8 Solid Masonry

Solid masonry consists of clay or concrete brick or solid load-bearing concrete masonry units laid with full mortar joints. There are arbitrary requirements outlined in the *UBC* for the thicknesses, bonds, headers, ties, and so forth. This also is a conventional type of construction that has performed satisfactorily and is subject to traditional limitations, generally as unreinforced masonry, or partially-reinforced.

17.2.9 Grouted Masonry

Grouted masonry is constructed of clay brick or concrete brick units in which interior joints or areas are filled by pouring grout between the outer and inner wythes.

In the low-lift grouting method, the grout fill is placed as the wythes are installed.

In the high-lift grouting, the outer wythes are built to their full height and grout placed later. The end result is very nearly the same, the difference is merely that of a method of installation. Grouted construction is an effective method of incorporating reinforcing into masonry, either high-lift, low-lift, solid, hollow, or what not, and is especially applicable to earthquake areas or tornado areas.

-1. The low-lift grouting method—Developed initially in California after the Long Beach earthquake, this consists of laying up the two outer tiers as separate wythes. One tier may be carried up 18 in. before grouting, but the other tier shall be laid and grouted in lifts not to exceed six times the width of the grouting space or a maximum height of 8 in. The rate at which the wythe should be raised is governed by the size of the units and the absorption of the units.

When work is stopped for over an hour, the horizontal joints are formed by stopping the tiers all at the same elevation and keeping the grout approximately 1 in. below the top, so that there is a keyed joint.

-2. High-lift grouting—One of the newer types of masonry construction, this procedure is particularly applicable to reinforced masonry. It is outlined in more specific detail for two wythe-grouted walls as follows:

a. Tiers are laid up full height.
b. The two tiers are tied with ties of not less than #9 wire in the form of a rectangle 4 in. wide and 2 in. less than the overall thickness of wall. One wythe or tier shall be built up not more than 18 in. higher than the other. Ties shall be spaced not more than 24 in. on center horizontally and 18 in. on center vertically for running bond, and not more than 12 in. on center vertically for stack bond.
c. Cleanouts shall be provided by leaving out every other unit in the bottom tier of the section being poured, or by holes in the foundation. The inside shall be cleaned by a high pressure stream of air or water.
d. The longitudinal vertical grout space shall be not less than 3 in. in width if it contains steel, not less than 2 in. if it contains no horizontal steel, and not less than the thickness required by proper clearances of reinforcing within the grout space.
e. Vertical grout barriers or dams shall be built of solid masonry across the grout space for the entire height of the wall to control the flow of grout horizontally at not more than 30 ft apart.
f. Grout shall be a plastic mix suitable for placing without segregation.
g. Grouting shall be done in a continuous pour in lifts not exceeding 4 ft. It shall be consolidated at time of placing and reconsolidated after excess moisture has been absorbed but before plasticity is lost.

17.2.10 Reinforced Grouted Masonry

This is grouted masonry with reinforcement added. The space between masonry units and reinforcement shall be not less than 1/4 in. except that 1/4 in. bars may be laid in horizontal mortar joints at least 1/2 in. thick. Steel wire reinforcement may be laid in horizontal mortar joints at least twice the thickness of the wire diameter.

17.2.11 Reinforced Hollow Unit Masonry

Made of hollow masonry units in which certain cells are continuously filled with grout and in which reinforcement is embeded. Only Type S and Type M mortar shall be used. This is generally assumed to be of concrete masonry units but may be of clay units. The hollow shapes used most, for a considerable period of time, have been the hollow two cell units which provide for easy placement of grout and reinforcement.

Hollow unit masonry is built to provide unobstructed vertical continuity of the cells that are to be filled with grout. The head joints shall be solidly filled with mortar for a distance in from the face of the wall or unit not less than the thickness of the longitudinal face shells. This is done to provide greater water resistance by breaking capillary action, and also, because the stress in the face shells is transmitted better this way. Bond shall be provided by lapping units in successive courses or by providing equivalent mechanical anchorage.

Cells to be filled shall be not less than 2 X 3 in.

Vertical reinforcement shall be held in position at top and bottom and at intervals not exceeding 192 diameters of the reinforcement.

All cells containing reinforcement shall be solidly filled with grout. Units shall be laid up 4 ft maximum in height and filled with grout, except for high-lift grouting. Here the walls are built full height, cleanouts are provided and then the grout is poured in successive 4 ft lifts which are puddled or vibrated upon pouring and are then reconsolidated later.

17.2.12 Veneer

Veneer is generally classified as a nonstructural material because it must not carry load other than itself; however, it is a frequently used masonry material and the structural engineer may be required to consider design, particularly since the *UBC* requires that veneers be designed to certain criteria. In the revised *Uniform Building Code* "Veneer" chapter, there are two types of veneer defined, 'adhered veneer' and 'anchored veneer.' The requirements for the design are that the adhered veneer develop 50 psi in shear and that the anchored veneer attachments develop twice the weight of the veneer being attached.

The *UBC* also provides that "in lieu of the design required by the section . . . veneer may be applied by one of the methods specified in the *UBC Standard*." In that *Standard* there are listed many of the types of veneer application which have proven satisfactory for many years and hence would not require design. They are spelled out in adequate detail in the *Standard*. The anchored types are (see Figs. 17-1, 17-2).

a. masonry or stone veneer (5 in. max.) on concrete or masonry
b. masonry or stone veneer (5 in. max.) on stud construction
c. slab type (2 in. max.) on concrete or masonry
d. slab type (2 in. max.) on stud construction
e. terra cotta on concrete or masonry

(The chapter on Veneer was chapter 29 prior to 1970 when it was changed to Chapter 30, with no major change of content.)

17.3 MASONRY MATERIALS

The list of masonry units available to the engineer for masonry construction is repeated here, with certain comments pertinent to the design, to call attention to certain items that might otherwise be overlooked. Also, the appropriate ASTM number is mentioned, and certain of the physical properties of each type.

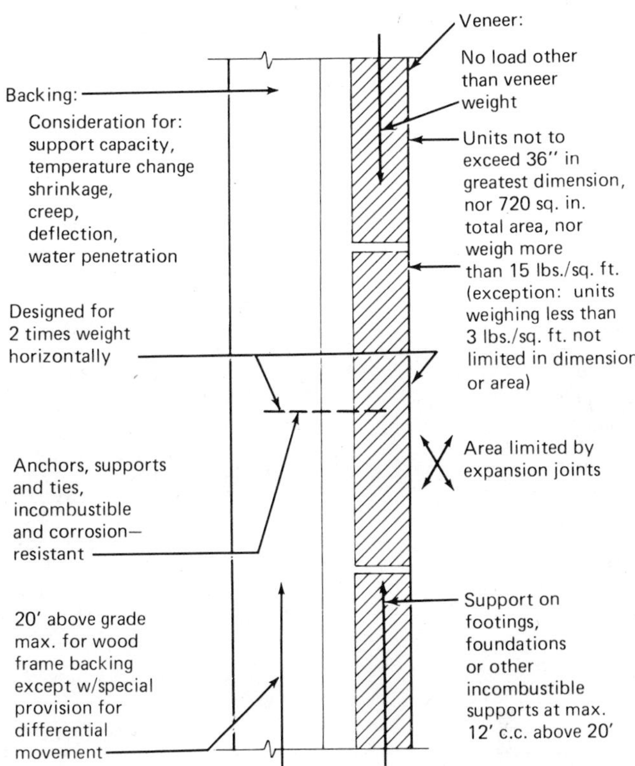

Fig. 17-1 Veneer units anchored to backing—general requirements. Reference: Uniform Building Code, Chapter 30—Veneer.

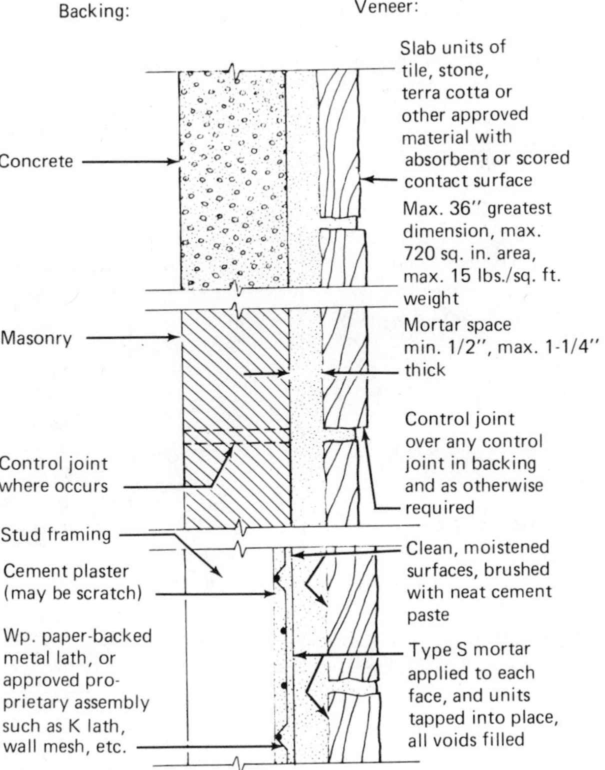

Fig. 17-2 Masonry veneer units anchored to concrete, masonry or plaster backing. Reference: Uniform Building Code, Standard no. 29-1-67, sec. 29.103, 1.

TABLE 17-1 Requirements for Concrete Masonry Units

Type of Concrete Masonry Unit	ASTM Designation	Grade of Unit	*Minimum Compressive Strength in psi on Average Gross Area*		*Maximum Water Absorption in lb/ft³ for Units of Different Weight Classifications Based upon Oven Dry Unit Weight*			
					Lightweight		*Medium Weight*	*Normal Weight*
			Average of Three Units	*Individual Unit*	*Less than 85 lb/ft³*	*Less than 105 lb/ft³*	*105 to 125 lb/ft³*	*125 lb/ft³ or more*
Concrete Brick	C 55	N	3500*	3000*		15	13	10
		S	2500*	2000*		18	15	13
Solid Load-bearing units	C 145	N	1800	1500	20	18	15	13
		S	1200	1000				
Hollow Load-bearing units	C 90	N	1000	800	20	18	15	13
		S	700	600				
Hollow Non-load-bearing units	C 129		350	300	NO ABSORPTION LIMIT			

**Concrete Brick Tested Flatwise*

Physical Requirements for Building Brick**

Grade	*Compressive Strength, flat, min, psi*		*Water Absorption, 5-hr Boil, Max, %*		*Saturation* Coefficient, Max, %*	
	Avg of 5	*Individual*	*Avg of 5*	*Individual*	*Avg of 5*	*Individual*
SW, severe weathering	3,000	2,500	17.0	20.0	0.78	0.80
MW, moderate weathering	2,500	2,200	22.0	25.0	0.88	0.90
NW, no exposure	1,500	1,250	No limit	No limit	No limit	No limit

**Ratio of 24-hr cold absorption to 5-hr boil absorption.*
***Facing Brick, Grades SW and MW similar.*

Physical Requirements for Structural Clay Tile

Type and Grade	*Absorption, % (1 hr Boiling)*		*Compressive Strength, psi (Based on Gross Area)*			
			End-Construction Tile		*Side-Construction Tile*	
	Avg of 5 Tests	*Individual*	*Min Avg of 5 Tests*	*Individual*	*Min Avg of 5 Tests*	*Individual*
Load-bearing (ASTM C34):						
LBX	16	19	1,400	1,000	700	500
LB	25	28	1,000	700	700	500
Non-load-bearing (ASTM C56),						
NB	–	28				
Floor tile (ASTM C57):						
FT1	–	25	3,200	2,250	1,600	1,100
FT2	–	25	2,000	1,400	1,200	850
Facing tile (ASTM C212):						
FTX	9	11				
FTS	16	19				
Standard	–	–	1,400	1,000	700	500
Special-duty	–	–	2,500	2,000	1,200	1,000
Glazed units (ASTM C126)	–	–	3,000	2,500	2,000	1,500

17.3.1 Building Brick from Clay or Shale

These are fired from clay or shale and are to be in compliance with ASTM C62. A summary of physical properties is listed in Table 17-1. There are three grades, SW, MW, NW, i.e., for Severe-Weathering, Moderate-Weathering, and Negligible-Weathering conditions. The physical properties and requirement classifications are based on these basic areas of use. It is to be noted that certain of the requirements of durability are waived for SW brick in areas where severe weathering may not occur, that is, where the weathering index is less than 100. In particular, the CB ratio is waived. The ratio of the cold water absorption to the boiling water absorption is a measure of the durability, and, if this is not required for freezing and thawing resistance, it is waived.

It is also noted that these units are considered solid even when they have core holes up to a volume of 25% of the total. When the volume is more than 25%, up to 40% coring, the units are known as Hollow-Brick and are to comply with ASTM C652, or with standards developed by the Western States Clay Products Association. If the void is greater than 40% of the units they may be designated by one of the tile specifications, i.e., C34, C126, C212.

17.3.2 Concrete Block

These are the conventional hollow units or concrete block. These are as specified in ASTM "C90–Hollow Load-Bearing Concrete Masonry Units." For conditions of lesser stress requirements they are under "C129–Hollow Non-Load-Bearing Concrete Masonry Units." It is to be noted that the compressive stress required in C90 is 1,000 or 700 psi on the gross area. However, the designer's concern is with the net, or actual strength of masonry. This testing and approval at 1,000 psi on the gross area permits a manufacturer to supply a shape of minimum area at high strength or a maximum area of lower strength material. This appears to add some confusion, but in practice it does not cause too much difficulty.

The following is an example to clarify the varying strength relationships. A conventional 8 in. concrete block develops 1,000 psi on the gross area. This is 2,200 psi on net area since they are less than one half solid. These then develop 1,350 psi in masonry assemblages, even though the mortar and grout are also stronger than 2,000 psi. The engineer must carefully distinguish between the strength of block units, the strength of grout, and the strength of masonry required for his design.

17.3.3 Concrete Brick

These are solid units of portland cement and appropriate aggregate and would be governed by ASTM "C55–Concrete Building Brick" or "C145–Solid Load-Bearing Concrete Masonry Units." These units are classified for use as Type 1 (moisture-controlled units), and Type II (non-moisture-controlled units). They come in grade N, for use as architectural veneer and facing units in exterior walls, for use where high strength and resistance to moisture penetration and severe frost are desired; and grade S, for general use where moderate strength and resistance to frost action and moisture penetration are required.

17.3.4 Hollow Clay Units

These are sometimes called clay block and generally would be specified under ASTM C652 "Hollow Brick." This is similar to face brick but has 40% void permitted rather than 25% coring.

There are also hollow units of Grade LB, similar to ASTM C34 except that face shells are 1¼ in. minimum thickness. These are included in *UBC* and considered in design the same as hollow concrete block units.

17.3.5 Partition Tile

Although they are generally nonbearing they must be considered by the structural designer as to safety, stability, and anchorage.

17.3.6 Mortar

Mortar consists of portland cement and/or masonry cement, lime and sand in proper proportions and generally specified under ASTM C270 or C476. However, the specification strength of the mortar is not as important to the designer as the strength of the masonry assemblage which will be constructed with that mortar afterward. Mortar is covered further under "Testing and Control," section 17.4.

17.3.7 Grout

This consists of portland cement and aggregate to which is added enough water to make a fluid mixture. Care must be exercised to avoid excess water which might cause a segregation of the particles during pouring. This is specified according to ASTM C476 and is covered further under "Testing and Control."

17.3.8 Reinforcing Bars

The reinforcing bars are as specified for reinforced concrete, i.e., ASTM A615, etc.

17.3.9 Wire Reinforcing

This is used for joint reinforcing and is specified according to ASTM A82, i.e., for the higher design stress of 30,000 psi.

17.3.10 Precast Concrete

Mo-sai and Schokbeton are examples of panels frequently classed and installed as masonry. These are discussed in detail elsewhere.

17.4. TESTING AND CONTROL

17.4.1 General

After the design, the next most important step is the control of the construction in the field to see that it accomplishes the goals of the design. The control is especially important in reinforced masonry because, in general, the designer has utilized the materials more efficiently than in some of the unreinforced uses which have been on the basis of traditional limits of size and use. If proper testing and control is not effective, the efforts of the design may be wasted.

17.4.2 Mortar

This is one of the items that causes considerable question and confusion in enforcement and interpretation.

ASTM C270 and the *UBC* have certain requirements for

TABLE 17-2 Mortar Proportions

(Parts by Volume)

Mortar Type	Minimum Compressive Strength at 28 Days (psi)	Portland Cement	Hydrated Lime or Lime Putty[1]		Masonry Cement	Damp, Loose Aggregate
			Min.	Max.		
M	2500	1	–	¼	–	Not less than 2¼ and not more than 3 times the sum of the volumes of the cement and lime used
		1	–	–	1	
S	1800	1	¼	½	–	
		½	–	–	1	
N	750	1	½	1¼	–	
		–	–	–	1	
O	350	1	1¼	2½	–	

[1] *When plastic or waterproof cement is used as specified in Section 2403 (p), hydrated lime or putty may be added but not in excess of one-tenth the volume of cement.*

mortar. These are (1) proportions required, as shown by Table 17-2; or (2) development of compressive strength as shown by the second column of the table. Either proportion or performance may govern.

The performance, or strength, as shown in the table, is not applicable to the field control. The psi noted in the table is for a cube (not a cylinder) of the material at a consistency which could not be used in the field. Consequently, there developed a compromise method, one recognizing that mortar should be wet to provide a good, intimate bond with the masonry units. While the mortar is in place, the masonry units will absorb the excess water and reduce the water/cement ratio; the end result will be much stronger.

It is emphasized that the values in the table are not the strengths of mortar that one might attain from field tests; there is no field testing provision in ASTM for mortar or for grout. Those strengths in the table are the strengths that would be developed by a laboratory mix of the proposed material, mixed to a specified flow which, incidentally, would not be used in the field. It would be too stiff for proper field use and for proper bond. Also, the values in Table 17-2 are those which will be developed by 2 in. cubes of the laboratory mix, which would be quite different from 2 in. × 4 in. cylinders that are taken as mortar samples in the field.

This change in water/cement ratio, influenced by various types of masonry units, and its influence on the strength of a wall, caused the California State Division of Architecture (now the Office of Architecture and Construction) to develop a field sampling method that is now incorporated in the *UBC Standards.* The ASTM did not have such a method thus the architects, engineers and building officials, concerned with quality control of field construction, had to develop a method that would give them more than the specified strength for a certain mix. They wanted to know if the actual wall assembly would be of proper strength for the desired function.

17.4.3 Grout

It is to be noted also that the ASTM Standards for grout do not contain any requirements for determining the field strength of that grout. The premise is simply that the proposed material mixed in those proportions will provide adequately strong grout.

The then State Division of Architecture also developed a field grout sampling method whereby they could be assured

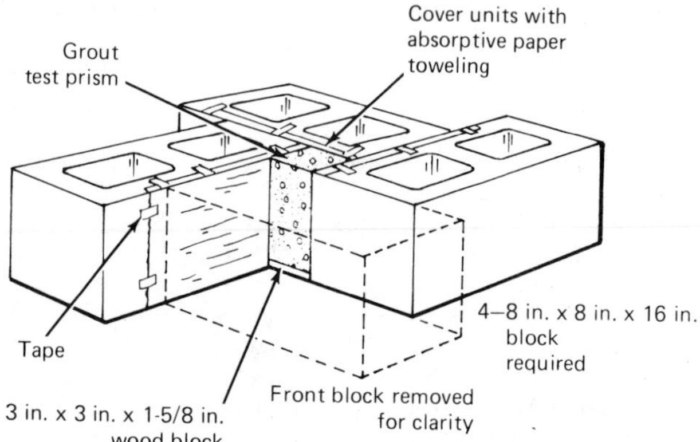

Fig. 17-3 Principal details of absorptive mold for making test prisms of grout.

that the grout going into a project would be satisfactory rather than merely that it had the right proportions. This is also included as a *UBC Standard* and is essentially a field sampling method for obtaining grout samples that will approach a simulation of the grout as placed in a wall. The samples are cast in molds of the same material as the masonry in the wall (Fig. 17-3). The grout so sampled and tested is required by OAC to meet a compressive strength of 2,000 psi.

17.4.4 Prism Testing

The f'_m, or ultimate compressive strengths of masonry at 28 days, upon which the design stresses are based as certain fractions thereof, are normally achieved without difficulty by the conventional use of materials for f'_m as shown in Tables 17-3 and 17-4. For masonry of 2600 psi and higher strength it is necessary that prism tests be made. A prism test is the making and load-testing of an actual assemblage of mortar, grout, brick, block and workmen that will be intended for the walls of the structure. This method has been specified in the *Uniform Building Code* for some time in Section 2404. Although the detail method of making, storing and testing the samples is well described, the specifier or designer must make sure that the specimens are handled and transported in a proper manner, otherwise the results of that procedure may give erroneous results.[17-1]

TABLE 17-3 Allowable Working Stresses in Reinforced Masonry

Table No. 24 H, 1970 UBC, lists the limits on stress in Section 2404 (c) 3 and these have been extended in the following table

MAXIMUM WORKING STRESSES (*pounds per square inch*)
Reinforced Solid & Hollow[1] *Unit Masonry*

Type of Unit / Type of Stress		*Hollow Clay Units*[1] *Grade LB or Hollow Concrete Units Grade A*		*Grouted Solid, Hollow Units: Concrete, Grade A Clay, Grade LB or Solid Units 2500 psi on Gross*		*Solid Units 3000 psi on Gross*		*Special Testing*[5] f'_m *Established by Prism Tests*								
Ultimate compressive strength	f'_m	675	1350	750	1500	900	1800	2000	2700	3000	3500	4000	4500	5000	5300	6000
Special inspection required		No	Yes	No	Yes	No	Yes	Yes	Yes	Yes	Yes	Yes	Yes	Yes	Yes	Yes
Compression–Axial, walls	$0.2\ f'_m$	135	270	150	300	180	360	400	540	600	700	800	900	1000	1060	1200
Compression–Axial, columns	$.18\ f'_m$	122	244	135	270	162	324	360	486	540	630	720	810	900	954	1080
Compression–Flexural	$.33 f'_m$	225	450	250	500	300	600	667	900*	–	–	–	–	–	–	–
Shear																
No shear reinforcement[2]	$.02\ f'_m$	15	27	15	30	15	36	40	50*	–	–	–	–	–	–	–
Reinforcement taking entire shear;																
Flexural members	$.05\ f'_m$	50	67	50	75	50	90	100	120	–	–	–	–	–	–	–
Shear walls	$.04\ f'_m$	30	54	30	60	30	72	75	–	–	–	–	–	–	–	–
Shear as revised for 1973 edition of UBC[6]																
No shear reinforcement, Flexural[2]	$1.1\sqrt{f'_m}$	25	40	25	43	25	47	49	50*	–	–	–	–	–	–	–
Shear walls $M/Vd \geqslant 1$	$.9\sqrt{f'_m}$	17	33	17	34	17	34*	–	–	–	–	–	–	–	–	–
$M/Vd = 0$	$2.0\sqrt{f'_m}$	25	50	25	50	25	50*	–	–	–	–	–	–	–	–	–
Reinforcing taking all shear,																
Flexural	$3.0\sqrt{f'_m}$	75	110	75	116	75	127	134	150*	–	–	–	–	–	–	–
Shear walls $M/Vd \geqslant 1$	$1.5\sqrt{f'_m}$	35	55	35	58	35	64	67	75*	–	–	–	–	–	–	–
$M/Vd = 0$	$2.0\sqrt{f'_m}$	52	73	55	77	60	85	89	104	110	118	120	–	–	–	–
Modulus of elasticity[3]	$1{,}000\ f'_m$	$.675 \times 10^6$	1.35×10^6	$.75 \times 10^6$	1.5×10^6	$.9 \times 10^6$	1.8×10^6	2.0×10^6	2.7×10^6	3.0×10^6	–	–	–	–	–	–
Modular ratio–$n = Es/Em$	$30{,}000/f'_m$	44	22	40	20	33	17	15	11	10*	–	–	–	–	–	–
Modulus of rigidity[3]	$400\ f'_m$	$.27 \times 10^6$	$.54 \times 10^6$	$.3 \times 10^6$	6×10^6	$.36 \times 10^6$	$.72 \times 10^6$	$.8 \times 10^6$	1.08×10^6	1.2×10^6	–	–	–	–	–	–
Bearing on full area[4]	$.25\ f'_m$	170	340	187	375	225	450	500	675	750	875	1000	1125	1250	1325	1500
Bearing on ⅓ or less of area[4]	$.30\ f'_m$	200	400	225	450	270	540	600	810	900	1050	1200	1350	1500	1590	1800
Bond–Plain bars		30	60	30	60	30	60	60	–	–	–	–	–	–	–	–
Bond–Deformed		100	140	100	140	100	140	140	–	–	–	–	–	–	–	–

[1] *Stresses for hollow unit masonry are based on net section.*
[2] *Web reinforcement shall be provided to carry the entire shear in excess of 20 psi whenever there is required negative reinforcement and for a distance of one-sixteenth the clear span beyond the point of inflection.*
[3] *Where determinations involve rigidity considerations in combination with other materials or where deflections are involved, the modulii of elasticity and rigidity under columns entitled "yes" for special inspection shall be used.*
[4] *This increase shall be permitted only when the least distance between the edges of the loaded and unloaded areas is a minimum of one-fourth of the parallel side dimension of the loaded area. The allowable bearing stress on a reasonably concentric area greater than one-third, but less than the full area, shall be interpolated between the values given.*
[5] *Special testing shall include preliminary tests conducted as specified in Section 2404 (c) to establish "f'_m" and at least one field test during construction of walls per each 5000 square feet of wall but not less than three such field tests for any building.*
[6] *For seismic stresses in shear walls, these stresses shall be halved, then increased ⅓ i.e., multiplied by ⅔.*
**Maximum value permitted by UBC.*

TABLE 17-4 Allowable Working Stresses in Unreinforced Unit Masonry

	Type M	*Type S*	*Type M or Type S Mortar*				*Type N*		
Material	*Compression*[1]	*Compression*[1]	*Shear or Tension in Flexure*[2,3]		*Tension in Flexure*[4]		*Compression*[1]	*Shear or Tension in Flexure*[2,3]	
Special inspection required	No	No	Yes	No	Yes	No	No	Yes	No
Solid brick masonry									
4500 plus psi	250	225	20	10	40	20	200	15	7.5
2500–4500 psi	175	160	20	10	40	20	140	15	7.5
1500–2500 psi	125	115	20	10	40	20	100	15	7.5
Solid concrete unit masonry									
Grade A (N)	175	160	12	6	24	12	140	12	6
Grade B (S)	125	115	12	6	24	12	100	12	6
Grouted masonry									
4500 plus psi	350	275	25	12.5	50	25			
2500–4500 psi	275	215	25	12.5	50	25			
1500–2500 psi	225	175	25	12.5	50	25			
Hollow unit masonry[5]	170	150	12	6	24	12	140	10	5
Cavity wall masonry solid units[5]									
Grade A or 2500 psi plus	140	130	12	6	30	15	110	10	5
Grade B or 1500–2500 psi	100	90	12	6	30	15	80	10	5
Hollow units[5]	70	60	12	6	30	15	50	10	5
Stone masonry									
Cast stone	400	360	8	4	–	–	320	8	4
Natural stone	140	120	8	4	–	–	100	8	4
Gypsum masonry	20	20	–	–	–	–	20		
Unburned clay masonry	30	30	8	4	–	–			

[1] *Allowable axial or flexural compressive stresses in pounds per square inch gross cross-sectional area (except as noted). The allowable working stresses in bearing directly on concentrated loads may be 50 per cent greater than these values.*

[2] *This value of tension is based on tension across a bed joint, i.e., vertically in the normal masonry work.*

[3] *No tension allowed in stack bond across head joints.*

[4] *The values shown here are for tension in masonry in the direction of running bond, i.e., horizontally between supports.*

[5] *Net area in contact with mortar or net cross-sectional area.*

The above is Table No. 24-B of the 1973 edition of the UBC.

The sample of wall to be tested was originally specified as 16 in. long by 16 in. high, however, it was recognized that in some areas there may not be test facilities available to test a prism that large and it now permits the length of the prism to be as short as 4 in. It is recommended, however, that the prism not be made shorter than its width or thickness because that would change the orientation of the shorter dimension of the prism and might change the type of failure. This would give an erroneous, probably lower (or more conservative) value than the true one.

It must be emphasized that f'_m is the masonry assemblage strength. Certain fractions of this are used for various allowable stresses. This is not the grout strength, the brick strength, the mortar strength, nor the block strength; it is generally less than the strength of those component portions. An example of this has been shown in the older versions of *UBC* Table No. 24-H. For example, brick of 3,000 psi, mortar of 2,000 psi, and grout of 2,000 psi was assumed to have, and generally did have, an f'_m of slightly over 1,800 psi. Hollow concrete block masonry strength, or f'_m, was to be 1,350 psi. This was provided by blocks of from 2,000 to 2,200 psi on the net area, (the gross area strength was 1,000 psi minimum) and the mortar and grout were assumed at 2,000 psi. The prisms would develop two-thirds to three-fourths of the unit strength, i.e., 1,350 psi.

17.5 DETAILING

If the detailing of masonry is, perhaps, more important than that of other materials, it is partly because the units are hand placed. Not only must the structural and aesthetic needs be fulfilled, but the practicality of placement must be proper.

One of the early developments was the modular size, e.g., the original 2, 4, 8 in. size of the common brick. Modern units, as concrete block, are generally units for 4 in. modular dimensions, i.e., a unit plus a mortar joint is a multiple of 4 in.

Examples of workable details are shown to aid detailing. (See Figs. 17-4 to 17-9.)

Another important part of detailing is the assurance of weather resistance, flashing or waterproofing, and the

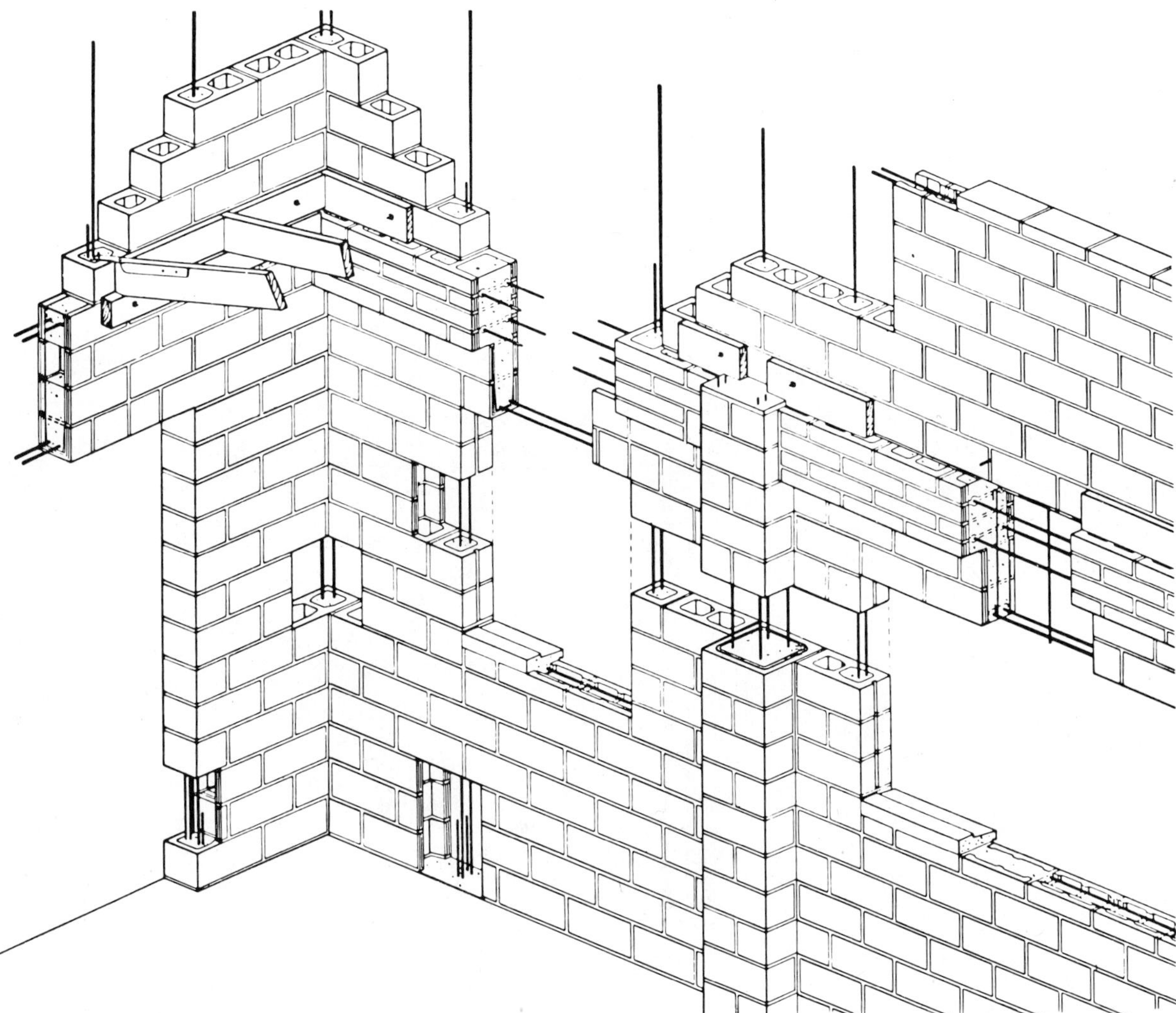

Fig. 17-4 Typical commercial construction.

following recommendations are made to help complete the masonry design.

17.5.1 Recommended Waterproofing

-1. Parapets and masonry projections above the roof

a. Provide a metal cap at the top of the wall or a dense capping unit sloped inward.
b. Where cast-in-place concrete caps are specified they shall be a minimum of 2 in. thick and shall be reinforced.
c. The inside face of the parapet and all masonry above the roof shall be waterproofed by hot mopping with roofing asphalt or by covering with roofing paper or other approved waterproofing.
d. No through-wall flashing shall be allowed (in seismic areas).
e. Flashing and counterflashing shall be set one-half inch into the mortar joint and the portion of wall above the flashing shall be hot-mopped with roofing asphalt or other approved waterproofing.
f. The exterior face of masonry above the roof shall be waterproofed in the same manner as the exterior wall.
g. All joints shall be checked for tightness and where cracks are visible mortar shall be chipped out, tuck-pointed and tooled.
h. All flashing, heads, jambs, sills, inserts and similar points shall be thoroughly caulked.
i. Waterproofing shall be compatible with the masonry units and shall be guaranteed by the applicator and waterproofing manufacturer with full knowledge of the units used and the condition of the wall at the time of the application of the waterproofing.
j. Waterproofing shall consist of a minimum of two coats with at least 24 hrs between applications.
k. Walls shall be waterproofed as frequently as recommended in the guarantee by the waterproofing manufacturer.
l. The first coat of waterproofing shall be a filler coat unless otherwise specified.

17.5.2 Walls Below Grade

a. All walls below grade shall be waterproofed on the exterior surface extending from the foundation pad to above grade.
b. Waterproofing shall be either asphalt conforming to ASTM D449, Type A, or coal-tar pitch conforming to ASTM D450, Type B, or as otherwise approved.

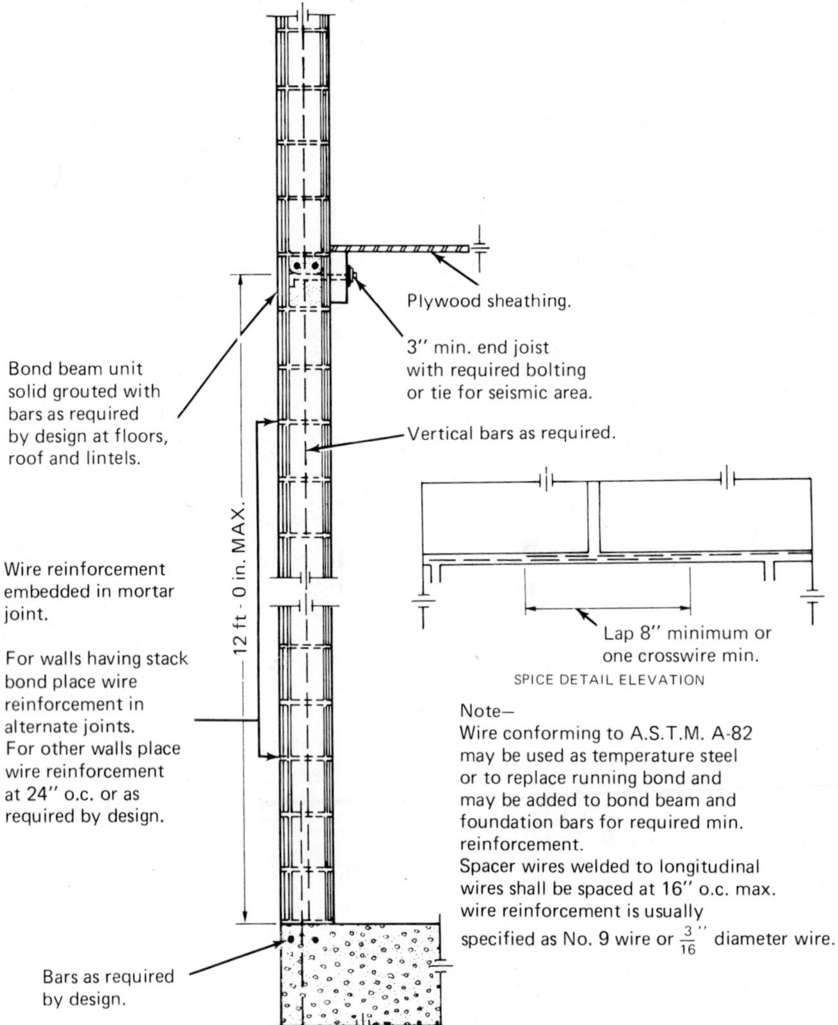

Fig. 17-5 Typical wall with wire reinforcement in mortar joints.

c. Surfaces to be waterproofed shall be clean, and dry, and shall be given either a priming coat of creosote oil conforming to ASTM D43 and two mop coats of hot coal-tar pitch, or a priming coat of asphalt primer conforming to ASTM D41 and two mop coats of hot asphalt. Mop coats shall be applied uniformly using not less than 20 lb of tar or asphalt per 100 sq ft per coat and shall provide a continuous impervious coating free from pinholes or other voids.
d. Where known water is present, membrane waterproofing shall be used as specifically noted on the plans.

17.5.3 Exposed Walls Above Grade

a. All exterior masonry walls shall be waterproofed, except as approved otherwise.
b. At the time of waterproofing the wall shall have been completed at least one month and shall be in a dry condition.

17.6 STRUCTURAL DESIGN

There are two rather different areas of masonry design, reinforced and unreinforced. Reinforced is based on principles similar to the elastic method of reinforced concrete, masonry functioning in compression and the steel in tension. Reinforced is used in areas subject to earthquakes because the steel provides for the excellent ductility and utilization of damping and energy absorption. The reinforcing changes the type of failure from a brittle type to a ductile type, desirable for earthquake resistance. There are certain arbitrary requirements for steel percentage in order to assure such performance.

Partially reinforced masonry is similar, but requires only that the steel be put in the places and amounts required for performance.

These two types are also recommended for areas subject to tornadoes.

Unreinforced masonry may be considered in two manners. One is as traditional masonry, following certain arbitrary rules and limits; another is as the so-called engineered masonry in which loads and stresses are calculated based on fundamental principles, modified somewhat by correlation with test data.

These areas are discussed in the following, with primary emphasis on reinforced masonry as it gives promise of strength assurance with construction economies. A number of design aids are available.[17-2 & 17-12]

The most common or simple design procedure is to estimate, analyze to verify the adequacy, revise if necessary, and analyze again. This is frequently simple with reinforced

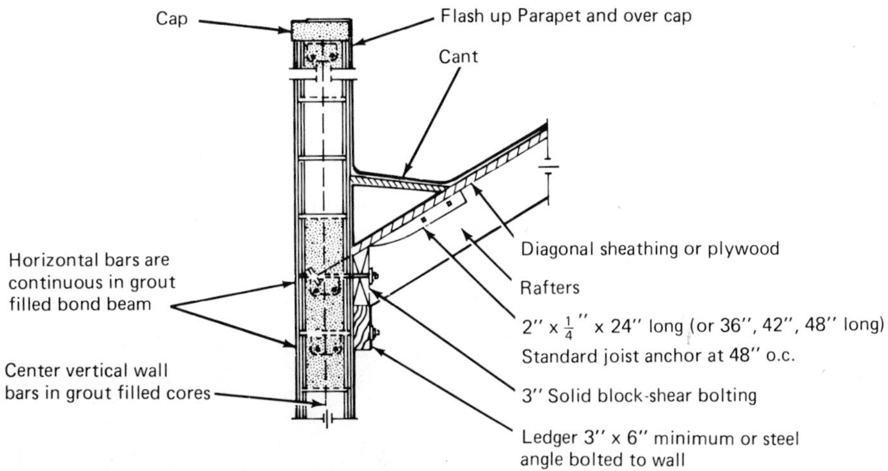

(1) Wood ledger—Note: above detail is typical and shall apply to sections 2, 3, 4 and 5 unless otherwise indicated

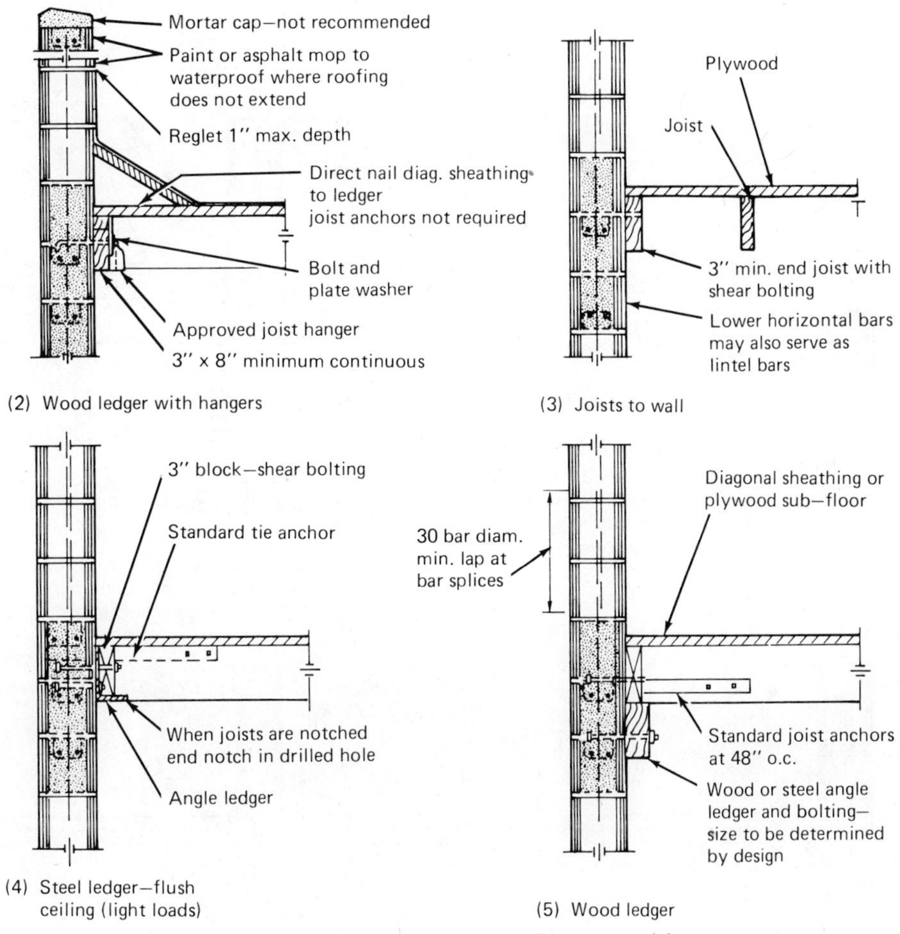

(2) Wood ledger with hangers

(3) Joists to wall

(4) Steel ledger—flush ceiling (light loads)

(5) Wood ledger

Fig. 17-6 Roof and floor details—commercial.

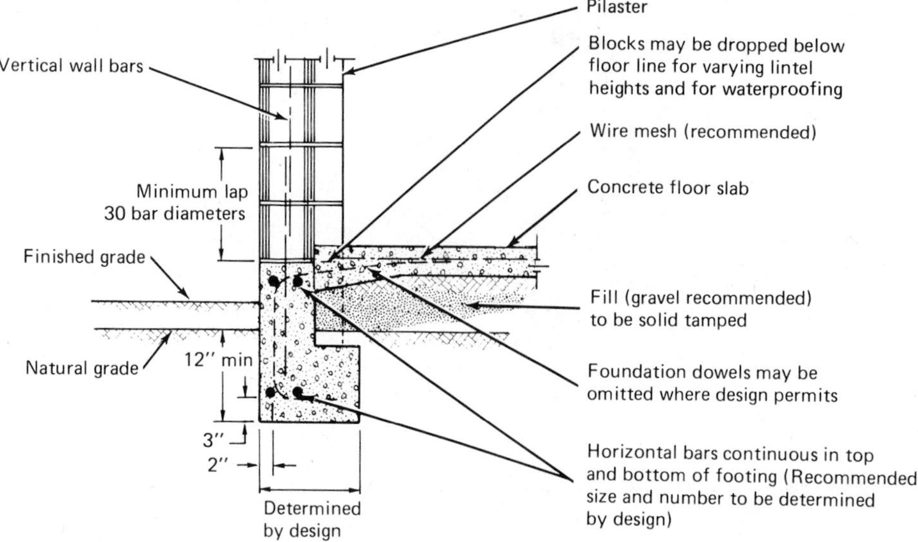

(1) Detail shown above is typical and shall apply for details 2, 3, 4, and 5 unless shown otherwise—above commercial type footing shows property line condition—center footing under wall when not on property line—

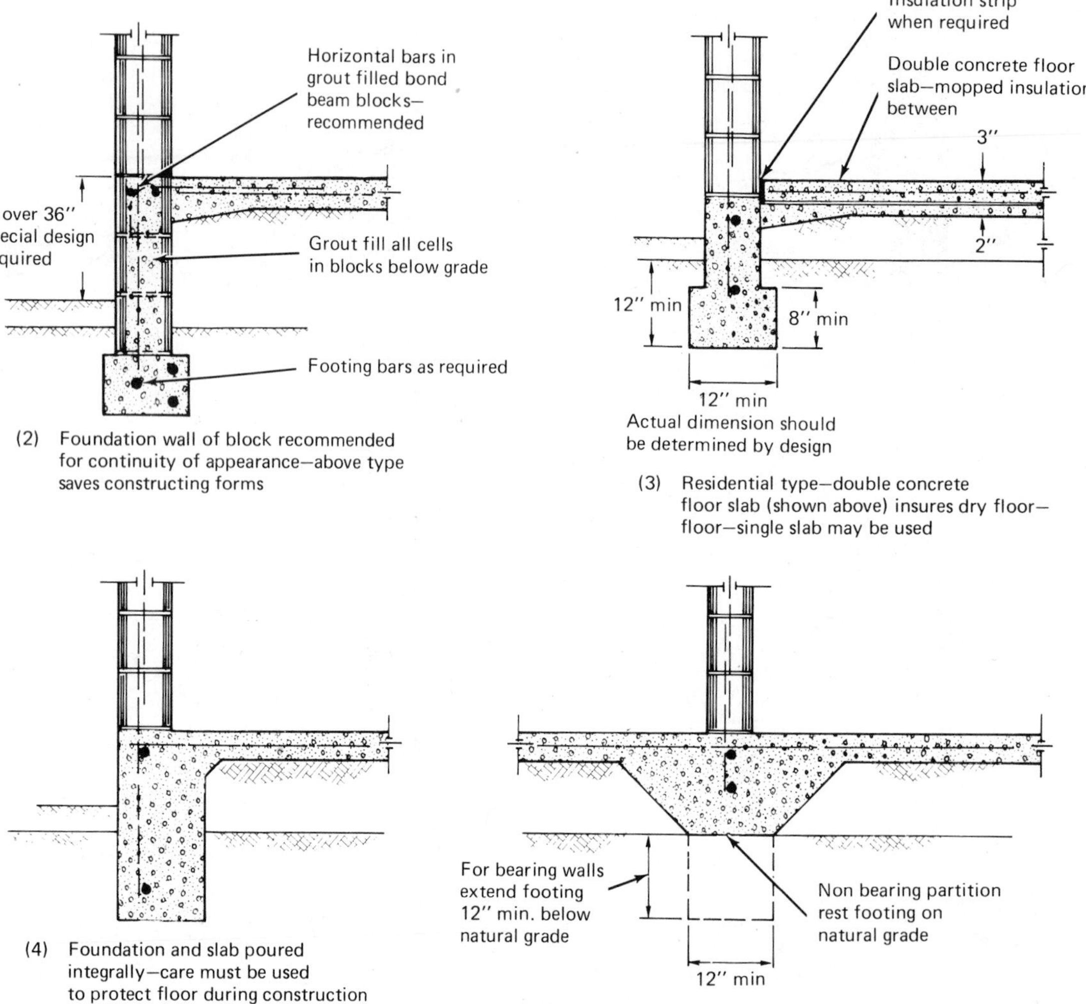

(2) Foundation wall of block recommended for continuity of appearance—above type saves constructing forms

(3) Residential type—double concrete floor slab (shown above) insures dry floor—floor—single slab may be used

Actual dimension should be determined by design

(4) Foundation and slab poured integrally—care must be used to protect floor during construction

(5) Interior wall footing

Fig. 17-7 Foundation details—concrete floor.

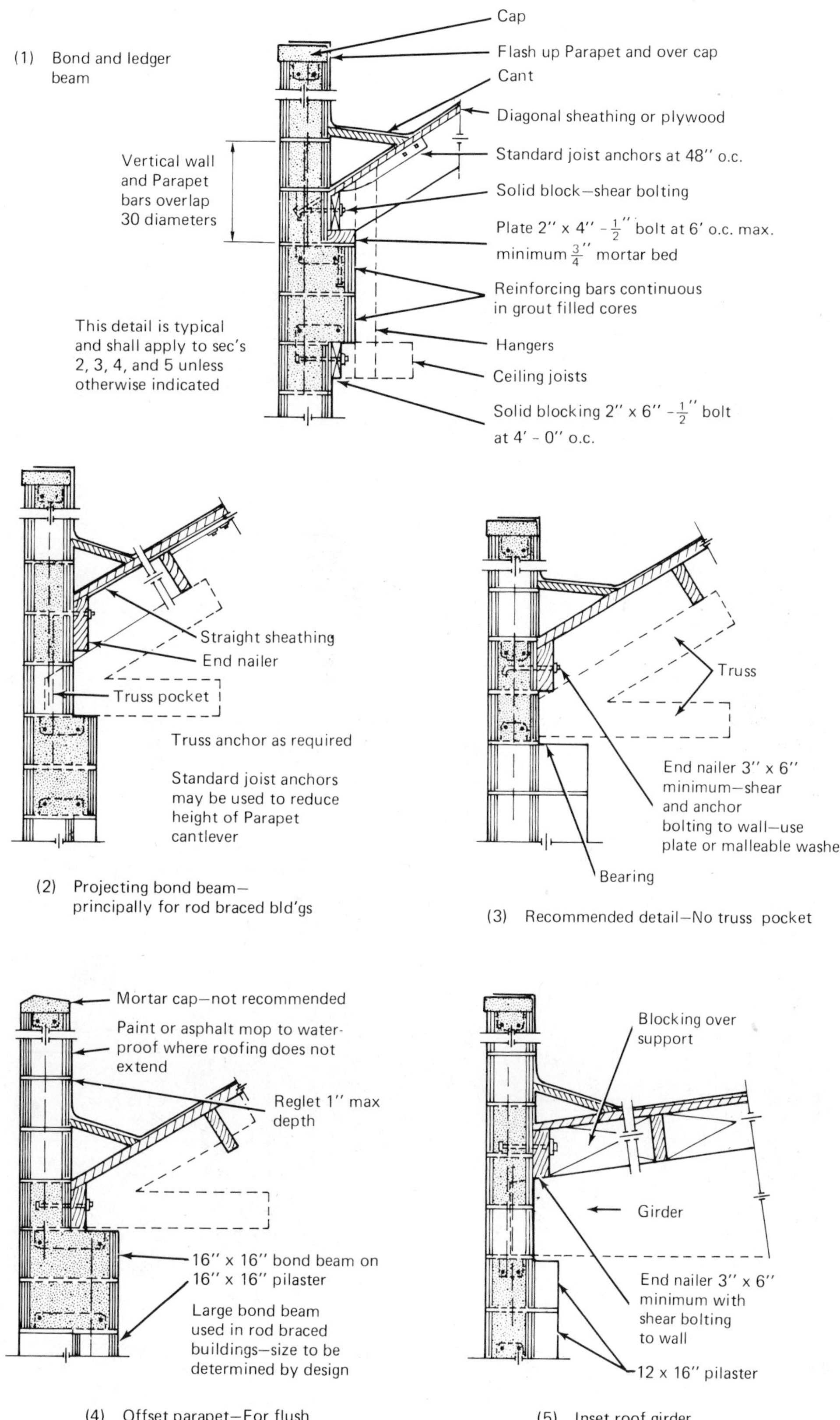

(1) Bond and ledger beam
Cap
Flash up Parapet and over cap
Cant
Diagonal sheathing or plywood
Standard joist anchors at 48" o.c.
Vertical wall and Parapet bars overlap 30 diameters
Solid block—shear bolting
Plate 2" x 4" - 1/2" bolt at 6' o.c. max. minimum 3/4" mortar bed
Reinforcing bars continuous in grout filled cores
This detail is typical and shall apply to sec's 2, 3, 4, and 5 unless otherwise indicated
Hangers
Ceiling joists
Solid blocking 2" x 6" - 1/2" bolt at 4' - 0" o.c.
Straight sheathing
End nailer
Truss pocket
Truss anchor as required
Standard joist anchors may be used to reduce height of Parapet cantlever
(2) Projecting bond beam—principally for rod braced bld'gs
Truss
End nailer 3" x 6" minimum—shear and anchor bolting to wall—use plate or malleable washer
Bearing
(3) Recommended detail—No truss pocket
Mortar cap—not recommended
Paint or asphalt mop to water-proof where roofing does not extend
Reglet 1" max depth
16" x 16" bond beam on 16" x 16" pilaster
Large bond beam used in rod braced buildings—size to be determined by design
(4) Offset parapet—For flush interior walls
Blocking over support
Girder
End nailer 3" x 6" minimum with shear bolting to wall
12 x 16" pilaster
(5) Inset roof girder

Fig. 17-8 Roof details—commercial.

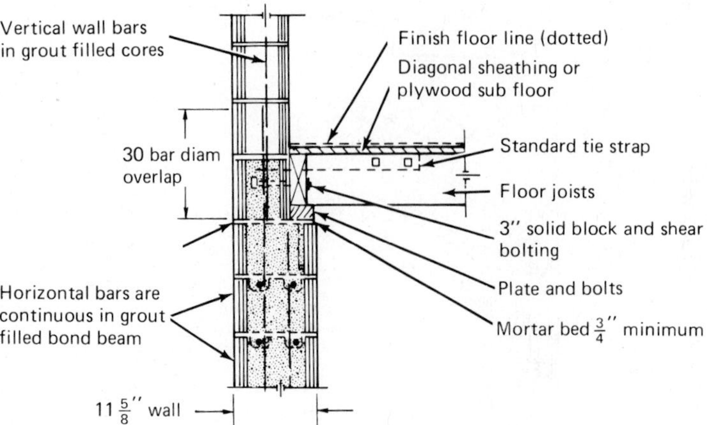

(1) Floor on 12" wide wall—Note: above detail is typical and shall apply to sections 2, 3, 4 and 5 unless otherwise indicated

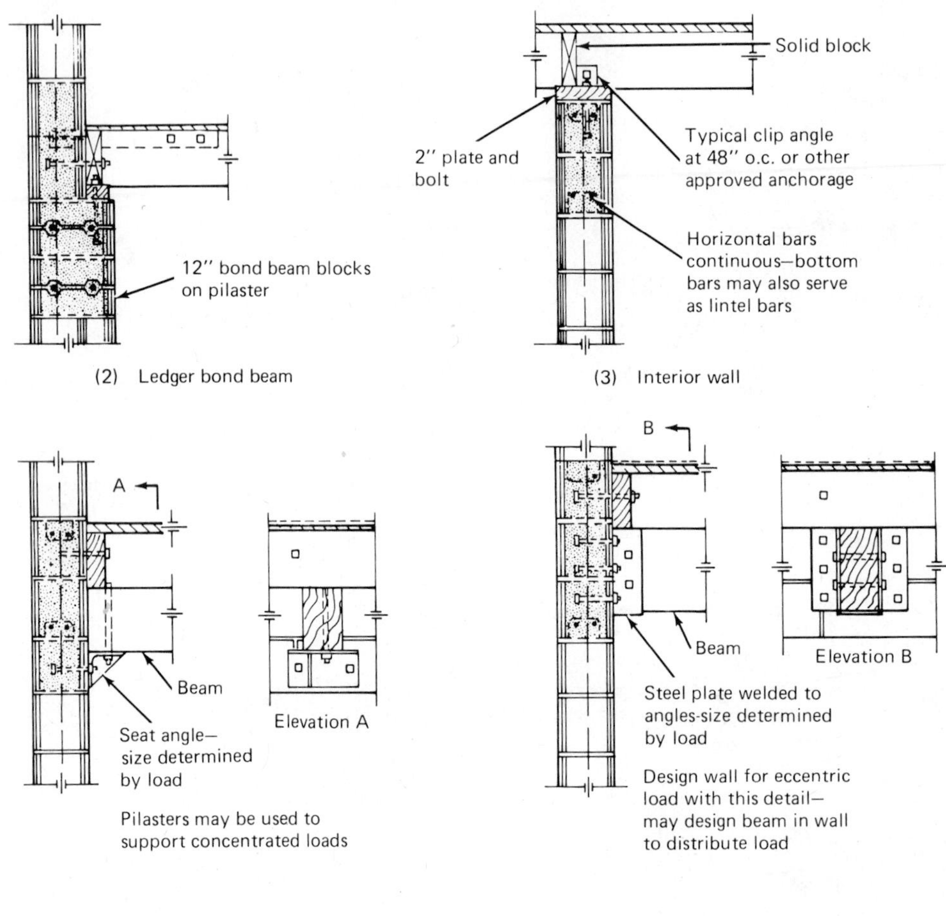

(2) Ledger bond beam

(3) Interior wall

(4) Beam connection (light loads)

(5) Beam connection (heavy loads)

Fig. 17-9 Floor and beam details.

masonry because one must have a certain amount of steel due to arbitrary requirements and often, this is adequate.

17.6.1 Reinforced Masonry Principles

The design principles have been based on the same assumptions as the earlier development of reinforced concrete design, on the basis of elasticity assumptions.[17-17]

There has been a reluctance to accept ultimate design principles in masonry because of the lack of adequate data for proof and refinement of the methods and also because of the scatter of results that frequently occurs in masonry, partly due to the individual workmanship on items and partly due to variations in the ingredient materials.

The assumptions of elastic design used herein are:

1. Plane sections before bending remain plane and stress is directly proportioned to strain.
2. The modulus of elasticity of the masonry, mortar and grout are constant within the member in the range of working stresses.
3. Stress in reinforcing is uniform over its area.
4. The member is straight and of uniform cross section.
5. External forces are in equilibrium.
6. In reinforced masonry the masonry carries no tensile stress.
7. For bending, the span of the member is large compared to the depth.

This elastic assumption is adequately conservative in view of the low working stress limits imposed on masonry. Based on the above, the following equations are developed from straight-line theory for reinforced flexural members:

$$n = E_s/E_m \doteq 30{,}000{,}000/1000\, f'_m = 30{,}000/f'_m$$

$$v = \frac{V}{bjd}; \quad u = \frac{V}{\Sigma ojd}$$

$$A_v = \frac{V_s}{f_v jd}; \quad A_v = \frac{V}{f_v \sin \alpha}$$

$$k = \frac{1}{1 + \dfrac{f_s}{nf_m}}; \quad j = 1 - \frac{k}{3}; \quad A_s = \frac{M}{f_s jd}$$

$$M = K\, bd^2; \quad K = \tfrac{1}{2} f_m jk$$

$$p = \frac{A_s}{bd}; \quad k = \sqrt{2np + (np)^2} - np; \quad f_m = \frac{M}{bd^2}\frac{2}{jk}$$

17.6.2 Reinforced Bearing Walls

Bearing walls are considered for combined stresses due to axial load and bending. Eccentricity of load application as well as stress due to lateral forces of wind and earthquake must be considered.

The code limitation on h/t in the so-called buckling formula is obviously an empirical type of equation. For one thing, it applies to horizontal spacing of lateral support as well as vertical spacing of lateral support. Also it makes no clear consideration of end conditions, although it is recognized in strength of materials that the end conditions influence very greatly the capacity of any wall element to resist buckling forces and lateral bending loads.

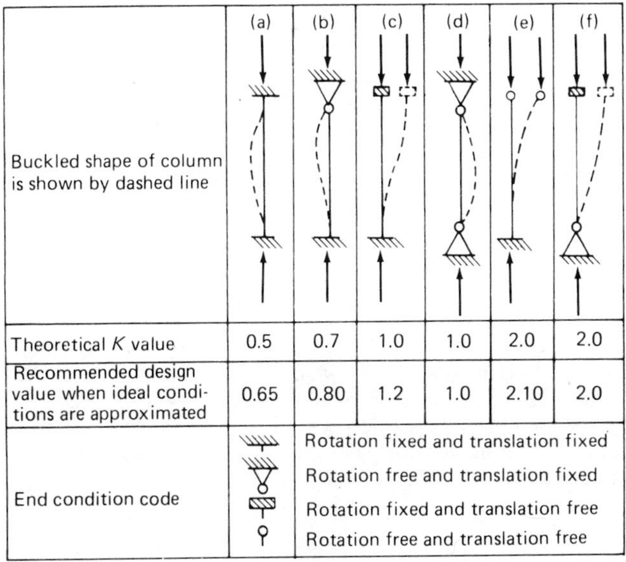

	(a)	(b)	(c)	(d)	(e)	(f)
Buckled shape of column is shown by dashed line						
Theoretical K value	0.5	0.7	1.0	1.0	2.0	2.0
Recommended design value when ideal conditions are approximated	0.65	0.80	1.2	1.0	2.10	2.0
End condition code	Rotation fixed and translation fixed Rotation free and translation fixed Rotation fixed and translation free Rotation free and translation free					

Fig. 17-10

Recognizing this, the *Uniform Building Code* contains an exception to those empirical limits which states:

> *EXCEPTION:* The height or length to thickness ratio may be increased and the minimum thickness may be decreased when data is submitted which justifies a reduction in the requirements specified in this Section.

The most common use of this exception has been the proper consideration of end restraint of walls as shown in theory of strength of materials and consideration of Euler's principle, just as has been done in steel and in concrete design. This consideration of end conditions requires assurance of such end restraint or calculation of actual end rotation be made. Also, other items may govern rather than buckling and must also be calculated, e.g., wind, quakes, eccentricity, etc., may cause excessive stresses.

Wall vertical capacity is based on the masonry material allowable capacity reduced by a reduction factor, i.e.

$$f_m = 0.20\, f'_m \left[1 - \left(\frac{h}{40t}\right)^3\right]$$

Both columns and walls must be designed with consideration of the lateral or bending loads which may occur simultaneously. The philosophy might be considered as providing that the fraction of strength the member has consumed in vertical load plus the proportion it has consumed in bending must not be more than 1, or fully consumed i.e.

$$\frac{f_a}{F_a} + \frac{f_b}{F_b} \text{ shall not exceed } 1$$

where

f_a = computed axial unit stress due to total axial load on effective area

F_a = permitted axial unit stress, for axial load only, including stress reduction, or increase due to combination with seismic or wind load

f_b = computed flexural unit stress

F_b = bending unit stress permitted if member were carrying bending only, including increase due to combined stresses of wind or seismic

Stresses may be increased one-third for combination of vertical load with seismic or wind stresses, except that the capacity so indicated must not exceed the capacity when not so combined.

The design is, in practice, simplified by this factor. In general, if the wall is adequate for bending alone, the increase of vertical capacity due the one-third increase will keep the member satisfactory. Or, if the vertical capacity is adequate for the static loads, it will still be adequate for addition of the lateral, generally.

Another saving factor for combined stresses is that when the compressive load on the wall is great enough so that the compressive unit stress is greater than the tension produced by the bending the section will be all in compression, i.e., there will be no tension. The full actual section of the wall will act and

$$\frac{P}{A} + \frac{Mc}{I} = f$$

The steel at the center will not act and hence is not a design problem.

For more precise design one is referred to charts or tables to simplify the work.[17-4, 17-6, 17-13] These investigate the actual stress condition and provide a more correct, and generally greater, allowable capacity.

17.6.3 Reinforced Concrete Masonry Columns

The following provisions from UBC guide the design of reinforced columns:

Every portion of a bearing wall whose length is less than three times its thickness must be designed as a column.

The least dimension of every masonry column shall be not less than twelve inches, unless designed for one-half the allowable stresses, in which case the minimum least dimensions shall be eight inches. No masonry column shall have an unsupported effective length greater than 20 times its least dimension.

The reinforcing ratio (p_g) shall be not less than 0.5% nor more than 4%. The number of bars shall be not less than four, nor the diameter less than three-eighths inch (3/8").

Where lapped splices are used, the amount of lap shall be sufficient to transfer the working stress by bond, but in no case shall the length of splice be less than 30 bar diameters, and welded splices shall be full butt welded.

Lateral ties shall be at least one-fourth inch (1/4") in diameter and shall be spaced apart not over 16 bar diameters, 48 tie diameters, or the least dimension of the column. Lateral ties shall be placed not less than one and one-half inches (1 1/2") and not more than five inches (5") from the surface of the column, and may be against the vertical bars, or placed in the horizontal bed joints. For seismic considerations and anchorages, additional ties must be provided.

The allowable axial load on columns shall not exceed:

$$P = A_g\,(.18 f_m' + 0.65\, p_g f_s)\left[1 - \left(\frac{h}{40t}\right)^3\right]$$

where

P = maximum concentric column axial load
A_g = the gross area of the column
f_m' = ultimate compressive masonry strength; the value of f_m' shall not exceed 6000 psi
p_g = ratio of the effective cross-sectional area of vertical reinforcement to A_g
f_s = allowable stress in reinforcement
t = least thickness of column, in.
h = clear height, in. (or effective height if calculated)

TABLE 17-5 Concrete Masonry Column Loads: P, in kips = $0.18 f_m' + 0.65 f_s A_s$; for columns 1-ft high

Column Size			Load on Steel Min = $0.005\, f_s A_g$ Max = $0.04\, f_s A_g$ f_s = 16 ksi		f_s = 24 ksi		Load on Masonry = $0.18\, f_m' A_g$												
nominal		A_g	Min	Max	Min	Max	675	750	900	1350	1500	1800	2000	2700	3000	35000	4000	5000	6000
	8	58	3	24	56	36	7	7.8	9.4	14.1	15.7	18.8	20.9	28.2	31.4	36.6	41.8	52.2	62.6
8	16	119	6	49	9	74	14	16	19	29	32	39	43	58	64	75	86	107	128
	24	180	9	75	14	112	22	24	29	44	48	58	65	87	97	113	130	162	194
	12	135	7	56	11	84	16	18	22	33	36	44	48	65	74	85	97	121	145
12	16	182	9	76	14	113	22	24	29	44	49	59	65	88	98	115	131	164	196
	24	275	14	114	22	171	33	37	44	67	74	89	99	134	148	173	198	247	297
	32	368	19	153	29	230	44	49	59	89	99	119	132	179	198	231	265	330	397
	16	244	12	101	19	152	29	33	39	59	66	79	88	118	131	153	175	220	263
16	24	368	19	153	29	230	44	49	59	89	99	119	132	179	198	231	265	330	397
	32	494	25	206	39	310	60	67	80	120	133	160	178	241	267	312	356	446	533
	40	620	32	264	49	388	75	83	100	151	167	201	223	301	335	390	447	558	670
	24	558	29	232	44	349	67	75	90	135	151	181	201	271	302	353	402	503	604
24	32	747	39	311	59	467	90	101	121	181	202	242	269	363	404	471	538	673	807
	40	936	48	390	74	585	113	126	151	227	253	303	337	442	505	590	673	841	1010
	48	1136	59	472	90	710	137	153	184	276	307	368	409	553	613	715	818	1022	1226

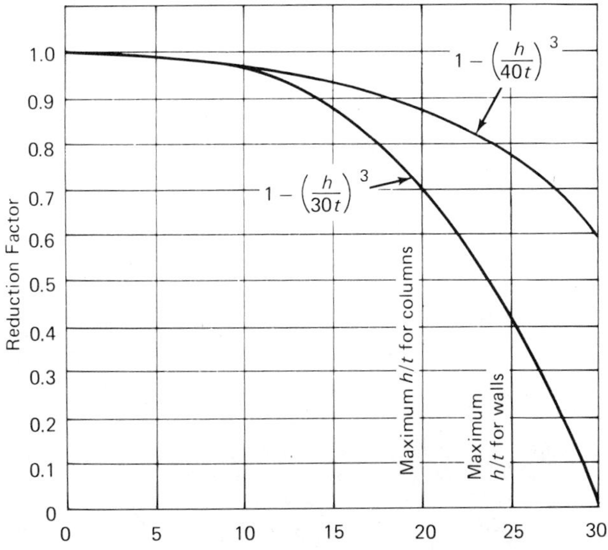

Fig. 17-11 Axial-load reduction factor curve.

The above equation is easily solved by adding the basic capacity of the masonry to the capacity of the steel, i.e., $(0.18 f'_m A_g) + (0.65 A_s f_s)$, and applying the appropriate reduction factor depending on height, end restraint and thickness. The thickness for use in h/t consideration is the nominal thickness, whereas the thickness for area or for stress determination is the actual or net thickness.

For solution of the above equation including exponential reduction factors, the rather simple design procedures shown in Table 17-5 may be used as follows:

a. Adjust the concentric load for the column to a basic or equivalent load for slenderness ratio 0 by dividing the load by the reduction factor, $[1 - (h/40t)^3]$, obtained from Fig. 17-11.
b. Enter Table 17-5 for the size of column and design prism strength, and select the load carried by the concrete masonry.
c. Subtract the load carried by the concrete masonry from the total load to determine the amount that must be carried by the steel reinforcement.
d. On the left side of Table 17-5 check to see that the amount of load to be carried by the steel falls between the minimum and maximum values and select appropriate bars.

See design example following.

EXAMPLE 17-1 Reinforced Masonry Column
Given:

load = 150,000 lbs
column = 16 × 16 in. (nominal)
height = 20 ft (= 240 in.)
material, or prism, strength, f'_m = 1,500 psi
reinforcing, f'_s = 24,000 psi

SOLUTION:

h/t = 240/16 = 15
reduction factor = 0.94 (from curve 17-11)
equivalent or basic load = 150/(0.94) = 160 kips
masonry capacity = 66 kip (from table)

Steel must carry 160 – 66 = 94 kips. This is between the limits of maximum and minimum of 152 and 19 (from table).

Select 8–#8 bars @ 99 kip (from table) to carry the 94 kips.

17.6.4 Combined Loads

Combined bending and direct stresses in reinforced masonry columns is considered most easily and conservatively by the simple unity equation as in walls, i.e.,

$$\frac{f_a}{F_a} + \frac{f_b}{F_b} \text{ shall not exceed } 1$$

For occasional design the procedure is to assume the size of column and reinforcing and check compliance, revising as indicated by the analysis. For greater ease of such tedious calculations of walls and columns there are available large books of tables, such as Ref. 17-6 and 17-13.

17.6.5 Fire Endurance

Fire endurance is one of the excellent properties of masonry and will have an influence on the design, e.g., certain fire ratings for various occupancies require specific minimum thicknesses of material; these will then be considered in the stress analysis. Also, one must check whether the structure as designed for stress provides the thickness required by the functional design for fire resistance, sound resistance, thermal insulation, etc.

TABLE No. 17-6 Rated Fire-Resistive Periods for Various Walls and Partitions*

		Minimum Finished Thickness Face-to-Face (in.)			
Material	*Construction*[1]	*4 hr*	*3 hr*	*2 hr*	*1 hr*
Concrete masonry units[2]	expanded slag or pumice	4.7	4.0	3.2	2.1
	expanded clay or shale	5.7	4.8	3.8	2.6
	limestone, cinders or air cooled slag	5.9	5.0	4.0	2.7
	calcareous or siliceous gravel	6.2	5.3	4.2	2.8

*From UBC.
[1] *Thicknesses shown for concrete masonry units are "equivalent thicknesses" as defined in U.B.C. Standard No. 24-4. Thickness includes plaster, lath and gypsum wallboard where mentioned and grout when all cells are solidly grouted.*
[2] *See also Footnote No. 1. The equivalent thickness may include the thickness of portland cement plaster or 1.5 times the thickness of gypsum plaster applied in accordance with the requirements of Chapter 47 of the Code.*

TABLE 17-7. Minimum Thickness of Masonry Walls

Type of Masonry	*Maximum Ratio–unsupported height or length to thickness*	*Nominal Minimum Thickness (in.)*
Bearing Walls:		
unburned clay masonry	10	16
stone masonry	14	16
cavity wall masonry	18	8
hollow unit masonry	18	8
solid masonry	20	8
grouted masonry	20	6
reinforced grouted masonry	25	6
reinforced hollow unit masonry	25	6
Nonbearing Walls:		
exterior unreinforced walls	20	2
exterior reinforced walls	30	2
interior partitions unreinforced	36	2
interior partitions reinforced	48	2

The ratings for hollow block of various aggregate types for various thicknesses is shown in the Table 17-6. The thickness is the "equivalent thickness" of hollow units, as listed in Table 43B of *UBC*.

The *UBC* provides the valid factor of including cement plaster, lath, and wall board and 1.5 times the thickness of gypsum as part of the required thickness. This latter factor is because the heat penetration resistance of gypsum is about twice as good as masonry of the same thickness. Grout is considered as contributing to equivalent thickness only when all cells are filled solid, not when filled intermittently. This is generally a conservative assumption.

17.6.6 Unreinforced Masonry Design

Unreinforced masonry is constructed either according to traditional empirical limits as shown in Table 17-7 (see also Ref. 17-14), or as engineered to take greater advantage of certain of the desirable characteristics. To assure that it is constructed as designed, it should be carefully inspected during construction.

The permitted design stresses may be determined either by prism tests or by unit tests and use of Table 17-8.

TABLE 17-8 Assumed Compressive Strength of Unreinforced Brick Masonry[1]

Compressive Strength of Units, PSI	*Assumed Compressive Strength of Unreinforced Brick Masonry f'_m, PSI*					
	Type M Mortar[2]		*Type S Mortar*[2]		*Type N Mortar*[2]	
Special Inspection Required	*Yes*	*No*	*Yes*	*No*	*Yes*	*No*
14,000 plus	4600	2300	3900	1950	3200	1600
10,000	3400	1700	2900	1450	2400	1200
6,000	2200	1100	1900	950	1600	800
2,000	1000	500	900	450	800	400

[1] *See Section 2419 (c) 1 of UBC.*
[2] *See Section 2419 (c) 2 of UBC.*

TABLE 17-9 Stress Reduction Factors, *c* for Unreinforced Masonry[1]

Slenderness Ratio[2]	*Virtual Eccentricity Normal to Plane of Member as a Proportion of the Thickness of the Member*[3]		
	0 to 1/20	*1/6*	*1/3*
5	1.00	0.66	0.32
10	0.92	0.63	0.27
15	0.79	0.56	0.22
20	0.64	0.42	0.16
25	0.49	0.36	0.12
30	0.38	0.27	0.08

[1] *Linear interpolation between values for the stress factors is permissible.*
[2] *Except as provided in Section 2419 (c) 5 A, the slenderness ratios of walls and columns shall be limited to 25 and 20, respectively.*
[3] *See Section 2419 (c) 5 E. of UBC.*

However, for f'_m values of 2600 and above it is desirable that prism tests be used because of the different influence of absorption upon the grout and mortars in masonry assemblages.

The bearing capacity of walls is based on the capacity of the masonry times a reduction factor, which is influenced by the slenderness ratio and the amount of eccentricity of load application. Thus, it takes into effect the bending due to wind or eccentric loads.

$$P = c\,(0.20\, f'_m)\, A_g$$

where:

c = stress reduction factor given in Table 17-9 corresponding to the slenderness ratio and
A_g = actual gross cross-sectional area of the wall

Design shall be such that no calculated tension shall occur in shear walls for loading in the direction of the plane of the wall.

Axial load on unreinforced masonry columns is similar, the equation being:

$$P = c\,(0.18\, f'_m)\, A_g$$

where:

c = stress reduction factor given in Table 17-9 corresponding to the slenderness ratio
A_g = actual gross cross-sectional area of the column

The coefficient, c, was developed from plotting of results of tests, with consideration of the theory of strength of materials.

17.6.7 Chimneys

Although residential chimneys and fireplaces are built generally by traditional methods, and codes consider them as nonstructural items, they are an important field of masonry use. The arbitrary reinforcing, anchorages, limits of dimension, etc., are summarized in the Table 17-10 and Fig. 17-12.

In addition to the arbitrary traditional construction of chimneys and fireplaces, there are custom or specially engineered designs. The elements may be structural if so designed. Free-standing chimneys beyond the code limits must be specifically designed structurally, see Ref. 17-15.

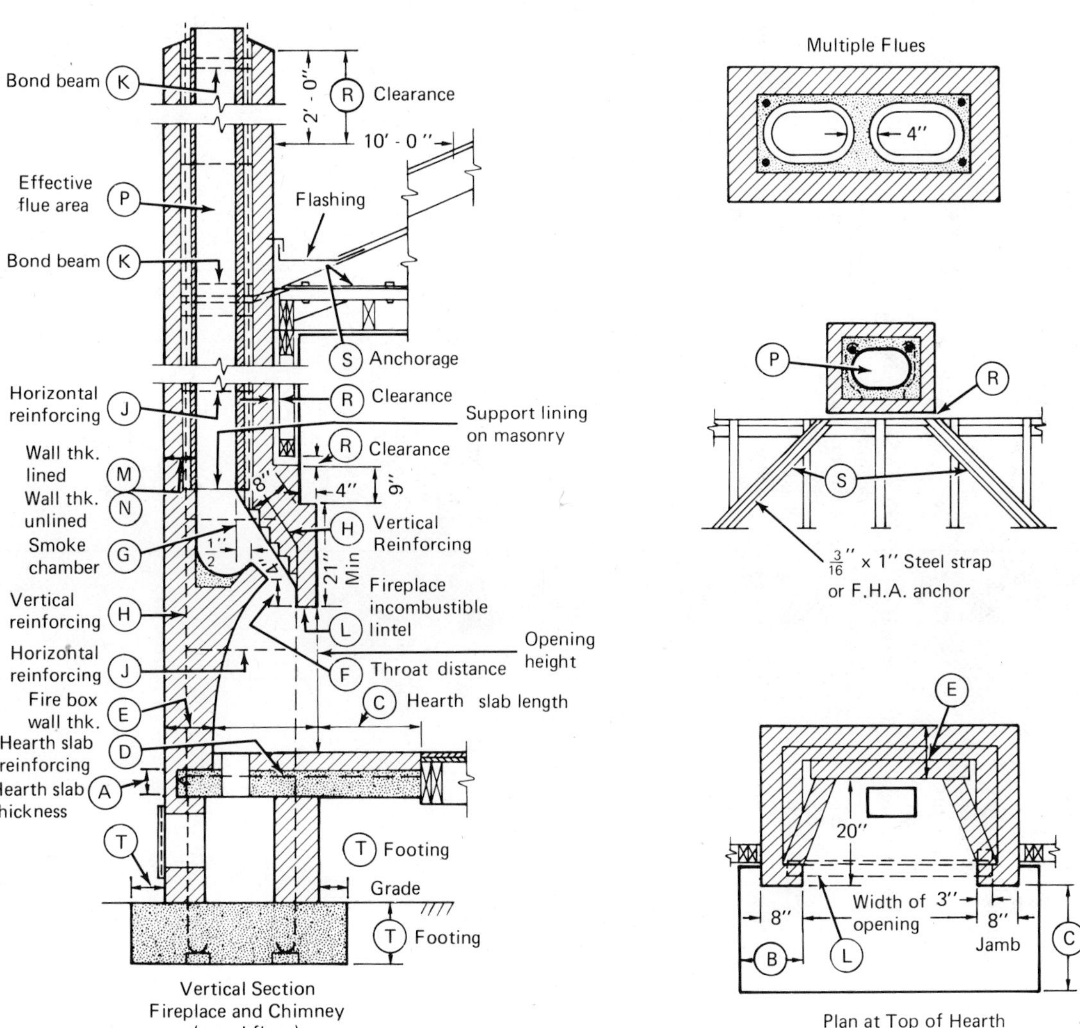

Fig. 17-12 Chimney and fireplace details.

TABLE 17-10 General Code Requirements

Item	*Letter*	*Los Angeles City*	*Los Angeles County & Unif. Bldg. Code*	*FHA & VA-L. A. Area*	*City of Long Beach**
Hearth slab thickness	A	3″ at outer edge 5″ at chimney edge	4″	4″	4″
Hearth slab width (Each side of opening)	B	8″	12″	8″	12″
Hearth slab length (front of opening)	C	20″	18″	16″	20″
Hearth slab reinforcing	D	#3 at 6″ perpendicular to chimney #3 at 12″ parallel to chimney	Reinforced to carry its own weight and all imposed loads	Required if cantilevered in connection with raised wood floor construction	Reinforced to carry its own weight and all imposed loads
Thickness of walls of firebox	E	8″ common brick may include 4″ firebrick lining	10″ common brick or 8″ where a firebrick lining is used	8″ including minimum 2″ firebrick lining–12″ when no lining is provided	8″ including a min. 4″ firebrick lining
Distance from top of opening to throat	F	6″	6″	6″ min.; 8″ recommended	
Smoke chamber edge of shelf	G	$\frac{1}{2}''$ offset		$\frac{1}{2}''$ offset	
Rear wall-Thickness			6″	6″ plus paraging; may be omitted if wall thickness is 8″ or more of solid masonry. Form damper is required	8″
Chimney Vertical Reinforcing	H	Four #4 full length bars for chimney up to 40″ wide. Add Two #4 bars for each additional 40″ width. No splice in bars. Maximum slope of bent bars 1:2	Four #4 full length bars for chimney up to 40″ wide. Add Two #4 bars for each additional 40″ of width.	Four #4 bars full length, no splice unless welded	Four #4 bars
Horizontal Reinforcing	J	$\frac{1}{4}''$ bars at 24″ and 2 at bend in vertical steel	$\frac{1}{4}''$ bars at 24″	$\frac{1}{4}''$ bars at 24″	$\frac{1}{4}''$ bars at 18″
Bond beams	K	Two $\frac{1}{4}''$ bars at top bond beam 4″ high Two $\frac{1}{4}''$ bars at anchorage	No specified requirements. L. A. City requirements are good practice	Two $\frac{1}{4}''$ bars at top bond beam 4″ high Two $\frac{1}{4}''$ bars at anchorage bond beam 5″ high	
Fireplace lintel	L	$2\frac{1}{2}'' \times 3'' \times \frac{3}{16}''$ angle with 3″ end bearing	Incombustible material	$2\frac{1}{2}'' \times 3'' \times \frac{3}{16}''$ angle with 3″ end bearing	Incombustible material
Walls with flue lining	M	1″ grout around lining. 4″ min. from flue lining to outside face of chimney	Brick with grout around lining. 4″ min. from flue lining to outside face of chimney	Brick with grout around lining. 4″ min. from outside flue lining to outside face of chimney	Brick with grout around lining. 4″ min. from flue lining to outside face of chimney

Walls with unlined flue	N	8" Solid masonry	8" Solid masonry	8" Solid masonry	8" Solid masonry
Distance between adjacent flues	O	4" including flue liner thickness	4" including flue liner	4" wythe for brick	4" including flue liner thickness
Effective flue area (Based on area of fireplace opening)	P	$\frac{1}{10}$	Round lining - $\frac{1}{12}$ 50 sq. in. min. Rectangular lining $\frac{1}{10}$ or 64 sq. in. min. Unlined or lined with firebrick - $\frac{1}{8}$ 100 sq. in. min.	$\frac{1}{10}$ for chimneys over 15' high and over. $\frac{1}{8}$ for chimneys less than 15' high	Round lining - $\frac{1}{12}$ 50 sq. in min. Rectangular lining $\frac{1}{10}$ or 64 sq. in. min.
Clearances	R				
Wood frame		1" clear	1" when outside of wall or $\frac{1}{2}$" gypsum board 2" when entirely within structure	$\frac{3}{4}$" from subfloor or floor or roof sheathing 2" from framing members	2" clear
Combustible material		6" to fireplace opening	6" min. to fireplace opening 12" from opening when material projects more than $\frac{1}{8}$" for each 1" from opening	$3\frac{1}{2}$" to edge of fireplace 12" from opening when projecting more than $1\frac{1}{2}$"	6" min. to fireplace opening 12" from opening when material projects more than $\frac{1}{8}$" for each 1" from opening
Above roof		2' at 10'	2' at 10'	2' at 10'	2' at 10'
Anchorage	S				
Strap		$\frac{3}{16}$" × 1"	$\frac{3}{16}$" × $\frac{1}{2}$"	$\frac{1}{4}$" × 1"	Shall be anchored at each floor and ceiling line more than 6' above grade
Number		2		2	
Embedment into chimney		Min. 18"		18" hooked around outer bar	
Fasten to		3 joists		3 joists	
Bolts		Two $\frac{1}{2}$" dia.		Two $\frac{1}{2}$" dia.	
Logs		Two $\frac{3}{8}$" × 3"			
Nails		Six–16d	Six–16d	Six–16d	
Footing	T				
Thickness		12" min.	12" min.	8" min. for 1 story chimney 12" min. for 2 story chimney	As per Chapter 23 and 28
Width		6" each side of fireplace wall	6" each side of fireplace wall	6" each side of fire place wall	

**By the end of 1970 the City of Long Beach may adopt by reference the 1970 Edition of the Uniform Building Code.*

EXAMPLE 17-2 Given the following:

8 in block wall, nonbearing, 18 ft high, exterior,
reinforcement #5 @ 4 ft in direction of vertical span
wind = 15 psf, seismic = $0.2w$ (Zone 3)
stresses for hollow concrete masonry construction, not continuously inspected

with h/t limits of:

25, for bearing walls
30, for exterior non bearing
48, for interior non bearing

Check if wall is satisfactory.

SOLUTION:

Calculation	Notes to designer
$h/t = \dfrac{18 \times 12}{8 \text{ in. (nominal)}} = 27$, OK	less than 30
$w = 46$ psf	assume grout @ 4 ft o.c. in reinforced cells
$0.2w = 9.2$ psf	seismic factor for Zone 3
∴ wind @ 15 psf governs	
$f'_m = 675$	from Table 17-3, for no continuous inspection,
$0.2\ f'_m = 135$	bearing on wall
$0.33 f'_m = 225 \times \frac{4}{3} = 300$	compression in flexure, increased one-third for wind stress
$n = 44$	$n = E_s/E_m = 30{,}000{,}000/1000\ f'_m$
$f_s = 24{,}000 \times \frac{4}{3} = 32{,}000$	for steel with $f_y = 60{,}000$, increase of $\frac{1}{3}$ for wind stress
A_s = #5 @ 4 ft = 0.077 sq. in/ft.	A_s reqd = 12 × 7.65 in. × 0.007 = 0.064, OK
$M = \dfrac{15 \times 18 \times 18 \times 12}{8} = 4000$ in. lb	vertical moment for wind of 15 psf, since wind governs
$p = 0.077/12 \times 3.8 = 0.00168$	
$np = 44 \times 0.0017 = 0.074$	
$j = 894$	referring to np Table 17-11
$2/kj = 7.04$	
$f_s = \dfrac{4000}{0.077 \times 0.894 \times 3.8} = 15{,}700$ psi	$f_s = \dfrac{M}{A_s jd}$
$f_m = \dfrac{4000 \times 7.04}{12 \times 3.8^2} = 165$ psi	$f_m = \dfrac{M\ (2/kj)}{bd^2}$

Therefore, the wall is O.K. for wind load imposed, since stresses are less than the maximum permitted.

EXAMPLE 17-3 Assume the above wall has vertical load of five kips per foot imposed from framing above. It is now a load bearing wall, thus

Calculation	Notes to designer
$h/t = \dfrac{18 \times 12}{8} = 27$	This is greater than 25 permitted by code for simply-supported bearing walls. Therefore wall thickness must be increased or the end conditions revised for more support, i.e., assume fixed at foundations or effective h = 0.7 clear height.
$\dfrac{h'}{t} = \dfrac{0.7 \times 18 \times 12}{8} = 18.9$	
$\therefore F_a = 0.89 \times 135 = 120$	from Fig. 17-11.
$f_a = \dfrac{P}{A} = \dfrac{5000}{12 \times 7.6} = 54.5$	
$\dfrac{54.5}{120} + \dfrac{165}{300} = 0.45 + 55 = 1$	$\dfrac{f_a}{F_b} + \dfrac{f_b}{F_b} = 1$ Wall is O.K. especially since some code wording on combined stresses indicates that all allowable stress is increased one-third when combined with wind.
$\dfrac{54.5}{135} + \dfrac{165}{300} = 0.40 + 55 = 0.95$	It is also to be noted that fixing of the base will reduce the M at the mid-height for combination with reduced F_a value. The maximum M occurring as it does with P/A at the bottom will be with an F_a not reduced by h/t factor.

The above stress-check for bending could also have been determined by the method of Fig. 17-13, i.e.,

1. enter chart @ d = 3.8 in.
2. go horizontally to between $pn = 0.06$ and $pn = .08$
3. go down to M ft-lb (= 4065 ÷ 12) = 340
4. go left to f_m = 165 psi

17.7 MULTISTORY LOAD-BEARING WALLS

17.7.1 High Rise

The high-rise load-bearing concept has become a widely used method. Masonry bearing wall structures have been used since early mankind. They are his earliest permanent construction material with many tried and true benefits and advantages, and this new scheme is merely the adaptation of the newer techniques of design to this old method of construction.

The design principles utilized here might impose certain design disciplines on the design team; however, by recognizing these limitations and taking advantage of them, a more efficient functional solution can be provided for the client. The scheme uses the high compressive bearing values of masonry for high-rise construction to aid the architect in his goal of providing the client with the most value and construction facility for the least cost. This scheme takes advantage of the sound insulation of masonry and its fireproofing qualities. Thermal inertia utilizes the heat absorption of the masonry due to its mass as an important factor in leveling off the load requirements, hence, requiring less mechanical equipment and cost.

17.7.2 Structural Design

The vertical load requirements are simply those of ordinary statics as in so-called nonseismic areas.

There are two basic facets of earthquake-resistant design. One is that of the so-called general considerations, that is, providing for symmetry of location of resisting elements, consideration of relative deflections, discontinuities, and similar factors. The other facet is in the calculation of stresses with the proper equations to establish the magnitude of the assumed design forces to be imposed on the structure.

The general aspects of the scheme are to provide rigidities in both directions and to avoid those arrangements that may cause trouble during an earthquake, that is, to visualize what might occur if the structure is shaken by ground motion in various directions. For example, we might consider the influence of the intersection of wings of buildings of varying periods of vibration or the introduction of

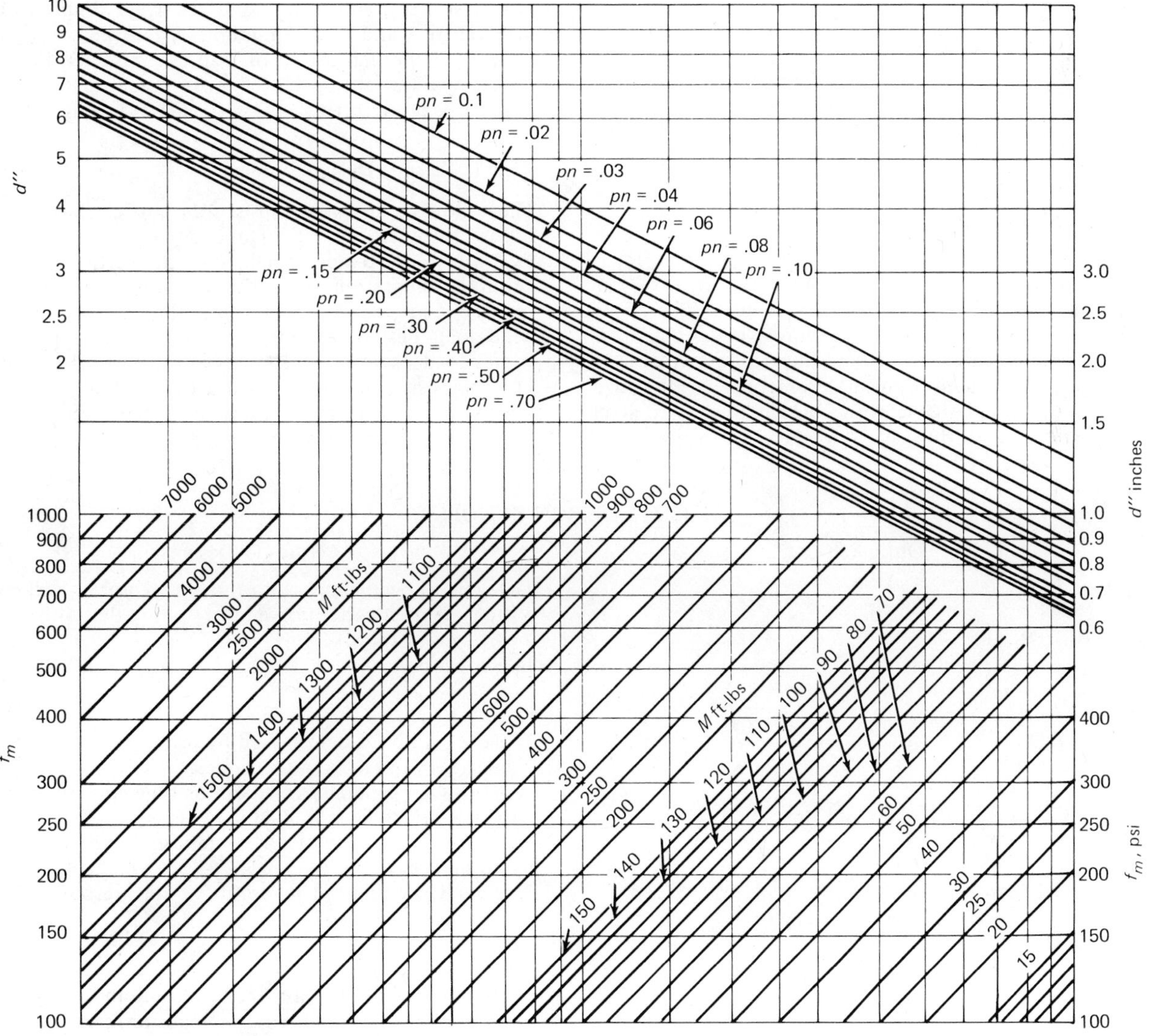

Fig. 17-13 Beam chart.

Starting from d'', to horizontal to pn line, go vertical down to $M'^{\#}$, go horizontal to f_m#/sq in.

For d'' larger than 10'' use $d''/10$ and $M'^{\#}/100$, then read f_m#/sq in. direct.

Check shear, axial, bond, and steel stresses separate from this chart.

discontinuities and variation in pier sizes or shear walls, the stiffening effect of towers, stairs and so forth. If potential elements of distress can be eliminated before design, the solution is much simpler.

The calculations herein are concerned primarily with the manipulation of numbers and equations in accordance with the *Uniform Building Code* requirements. There are several types of structural systems which respond differently to earthquake motion, and accordingly, there are different code coefficients. The load-bearing type of building is a "box" system, that is, bearing and shear wall systems for the vertical loads, with diaphragms to carry the horizontal loads to those vertical shear walls. The assumptions of design are that loads are applied laterally to the structure at their centers of gravity and carried down through the system to the foundation.

The factors determining the magnitude of those forces are largely set on the basis of probability.

We may accept the summaries in the governing codes of earthquake design as the best judgement to follow at present. These, however, are to be considered as economic minimums, and the engineer is advised to use his judgement in designing to those numerical minimums. In some cases, he may find it advisable to increase them. In other instances, he may wish to revise a placement or location of elements to minimize the possibility of damage.

Most of the local code provisions, such as in major cities like Los Angeles, and the *Uniform Building Code*, are based on the Lateral Force Recommendations of the Structural Engineers Association of California, based on many years of study, exercise of judgement and compromise.

17.7.3 Wall Design

Walls and parapets are designed to resist the lateral loads which would be due to their own inertia (seismic loads at right angles to the plane of the wall), while subjected to simultaneous vertical loads. This involves the problem of combining bending and direct stress. Superficially, this may seem to be a complicated problem; in practice, we recognize that the walls are designed with considerably more steel than minimum, and they are subjected to considerable compressive stress which will minimize possibility of tension. Also, they are relatively short in vertical spans for bending, say, a story height of 8 to 10 ft; therefore, the bending stress that might cause tension on one face and the assumption of a cracked masonry section which would complicate the design, will not generally occur due to this type of loading, especially in the lower stories.

A check should be made of the upper story walls, say, those carrying roof load only, or for an exceptionally high story, to confirm that the tension effect caused by wind or seismic load normal to the face would not overcome the compressive value contributed by the P/A, or direct stress.

Essentially, the designer will not be concerned with the forces perpendicular to the wall. His major concern will be those walls which serve as cantilever beams acting in the direction of the wall plane. This involves combined stresses of bending and compression. The *Code* stipulates that we may use the equation

$$\frac{f_a}{F_a} + \frac{f_b}{F_b} \text{ shall not exceed } 1$$

Although it is recognized that this equation may not be correct, particularly when considering cracked sections, it is adequately on the safe side, within the range of permitted design stresses, and is safe and relatively simple to use. A more correct and precise method is, generally, not justified in view of the inaccuracy of load assumptions. For special conditions warranting closer more precise calculation the designer is referred to books with charts such as Ref. 17-13.

17.7.4 Seismic Design Synopsis

The following design synopsis is an outline that one might follow through the earthquake portions of the *Uniform Building Code*. It does not include the vertical load design detail consideration since this is similar to conventional vertical load consideration. Also, this does not cover the wind load design procedure since that is simply a static lateral load assumption similar to the vertical load design. This synopsis is in the form of a CSI outline specification, that is, with the steps to follow on the left hand side with the "Notes to Designer" on the right hand side.

SEISMIC DESIGN SYNOPSIS

Outline	Notes to Designer
Occupancy Area Type of construction	The type of occupancy, with the area required, will govern the type of construction and fire rating that must be used throughout the structure.
Required fire rating Walls Floors	These will be based on the above as defined in UBC
Floor load—live load dead load	These would be as determined by probable loads or those listed in load tables of the Code.
Partition load	Partitions are generally fixed in this type construction so the specific load as applied may be used in lieu of an average floor load which is sometimes imposed.
Equipment load	Based on actual loads
Roof load—live load	This would be reduced according to the appropriate formula.
dead load	Actual loads.
Vertical wall load	
Wind load	As determined by the site, height of building, shape and so forth, either based on required loads or upon loads calculated by velocity determination.

Shape factor

This will generally be appropriate only in specific instances for special structures.

Z or Zone factor

The Zone factors in *UBC* and *SEAOC* are Zone 3 = 1, Zone 2 = 0.5, Zone 1 = .25 California is generally Zone 3, so that Z would be 1.

K (Table 23-H in *UBC Code*)
shearwall $K = 1.33$
25% frame $K = .80$
100% frame $K = .67$

This factor K is determined according to the scheme of framing. In the box system, considered most appropriate for masonry, the K-factor is 1.33, on the assumption that a stiff building will be subject to higher seismic forces than some other structural systems.

$$T = \frac{0.05h_n}{\sqrt{D}}$$

This is the empirical determination of T, or period, based on dimension. Other type of calculation to determine the building period could be used in lieu of this simplification, and may be more correct, particularly in special structures.

$$C = \frac{0.05}{\sqrt[3]{T}}$$

T–	4	3	2.5	2.0	1.5	1.0	.8	.6	.4	.2
C–	.0316	.0348	.037	.040	.0435	.050	.054	.060	.068	.086

Having determined T, coefficient C is determined, influencing the magnitude of force. Table below is provided to simplify this calculation. The C-factor for exterior bearing and non bearing walls themselves is 0.20, or 20%g.

$$V = ZKCW$$

The above factors can now be substituted in this basic equation for the total base shear, V. This is to be distributed up the building according to the forces determined in the equations below.

$$\left(\frac{h_n}{D_s}\right)$$

This represents the ratio of height and plan dimension in the direction of loading considered.

$$F_t = .004V\left(\frac{h_n}{D_s}\right)^2$$

This represents the assumed portion of the shear which is to be concentrated at the top, because of the so-called whiplash effect.

$$F_x = \frac{(V - F_t)\,w_x h_x}{\sum_{i=1}^{n} w_i h_i}$$

This represents the assumed force to be applied to any level x and may be simply calculated by tabular form or by summation.

$$M = F_t h_n + \sum_{i=1}^{n} F_i h_i$$

This represents the overturning moment, at the base.

$$M_x = \left[F_t(h_n - h_x) + \sum_{i=x}^{n} F_i(h_i - h_x)\right]$$

Given a load distribution and magnitude, the design is simply that of structural design of the elements such as the cantilever piers carrying the lateral load.

It is to be noted that the above semidynamic technique of seismic design is changing rapidly with improvements in appropriate dynamic analysis.

One method of simplifying the problem and probably obtaining a more stable structure is to stagger the openings in one floor above the other so that the long wall would become a perforated plate rather than a series of coupled cantilevers, and in this way, the stresses will generally be much less throughout the wall.

17.8 TABLES AND CHARTS

Some examples of design aids are shown here to be of help in design, but also to serve as suggestions for charts the designer may make for his own special needs. The table for column design is in Example 17-1, in Structural Design section 17.6.

Two techniques of flexural aids are shown. One is the '*np*'-table and one is the K-chart. This recognizes that some engineers prefer tables and some prefer curves.

17.8.1 Reinforcing steel area (Table 17-13)

Table 17-13 is to aid in determining the areas of steel required for masonry for various jurisdictions. The *Uniform Building Code* requires the area of steel in masonry to be not less than 0.002 times the wall area, with *not less than one-third* of the steel in either direction.

Title 21 of the *California Administrative Code* requires 0.003 times the wall area for total steel.

The values in the upper table are for the conditions of 1/3, 2/3, and *total area* as required above.

Having determined the desirable area of steel for one direction by use of lower spacing table, this amount can be subtracted from the total area required, giving the amount of steel required in the other direction. The most desirable spacing and size can then be selected.

The double wires listed are for 'ladder bar' type of reinforcement, and the area given is the area of both wires, i.e., the contribution of the double wire to the total steel in that direction.

The wall thickness shown is actual, not nominal, so it is on the safe side. This may be refined further by using the actual thickness.

EXAMPLE 17-4. Given a nominal 8 in. block wall. Select ladder bar reinforcing and vertical bar reinforcing to be suitable for walls under the *Uniform Building Code.*

SOLUTION: Enter the table of "Area of Steel Required" under 8 in. thickness and on the line for factor of 1/3 × 0.002 find the value of 0.064 (or @ 7 in. = 0.056 + 1/2 in. = 0.004 which equals 0.060).

Provide 2 3/16 in. round wires at a spacing of 16 in. which, from the "Area of Steel per Foot Chart" is 0.041. This leaves 0.064 – 0.041 or 0.023 sq in. of steel required per foot. Provide also #5 bars horizontally at 8 ft for A_s = 0.038 in./ft, which is more than the 0.023 required, therefore it is okay. (0.038 + 0.041 = 0.079)

Enter the table of "Area of Steel Required" for total; namely 1.0 × 0.002. For the 8 in. wall find 0.192 as total steel required (or @ 7 in. = 0.168 + 1/2 in. = 0.012 = 0.180). Subtracting the 0.079 provided horizontally from the total of 0.192 gives 0.113 required for the vertical steel. Entering the area table for a desirable 4 ft spacing we find that #6 bars at 4 ft will give 0.110, which is hence satisfactory.

This could have been on the basis of steel required for 7 in. plus 1/2 in. with a more economical end result as shown in the parenthesis.

TABLE 17-11 Table of Flexure Coefficients

$$p = \frac{A_s}{bd} \quad n = \frac{E_s}{E_m} \quad k = \sqrt{2np + (np)^2} - np$$

$$j = 1 - \frac{k}{3} \quad f_m = \frac{M}{bd^2}\left(\frac{2}{kj}\right)$$

np	*k*	*j*	2/*kj*	*np*	*k*	*j*	2/*kj*
0.050	0.2702	0.910	8.14	0.180	0.4464	0.851	5.26
0.060	0.2914	0.903	7.60	0.185	0.4507	0.850	5.22
0.070	0.3106	0.897	7.19	0.190	0.4550	0.848	5.18
0.080	0.3279	0.891	6.85	0.195	0.4592	0.847	5.14
0.090	0.3437	0.885	6.58	0.200	0.4633	0.846	5.11
0.100	0.3583	0.881	6.34	0.220	0.4789	0.480	4.96
0.105	0.3651	0.878	6.24	0.240	0.4932	0.836	4.85
0.110	0.3718	0.876	6.14	0.260	0.5066	0.831	4.75
0.115	0.3782	0.874	6.04	0.280	0.5190	0.827	4.66
0.120	0.3844	0.872	5.97	0.300	0.5307	0.823	4.58
0.125	0.3904	0.870	5.89	0.350	0.5569	0.814	4.41
0.130	0.3962	0.868	5.82	0.400	0.5799	0.807	4.27
0.135	0.4019	0.866	5.75	0.450	0.6000	0.800	4.17
0.140	0.4074	0.864	5.68	0.500	0.6181	0.794	4.07
0.145	0.4127	0.862	5.62	0.550	0.6342	0.789	4.00
0.150	0.4179	0.860	5.56	0.600	0.6490	0.784	3.94
0.155	0.4229	0.859	5.51	0.650	0.6624	0.779	3.88
0.160	0.4278	0.857	5.46	0.700	0.6748	0.775	3.82
0.165	0.4326	0.856	5.41	0.750	0.687	0.770	3.78
0.170	0.4374	0.854	5.36	0.800	0.700	0.767	3.74
0.175	0.4419	0.853	5.31				

TABLE 17-12 Average Weight of Completed Wall

		Lightweight Aggregate			*Sand-gravel Aggregate*		
Wall Thickness		6"	8"	12"	6"	8"	12"
Solid grouted wall		56	77	118	68	92	140
Vertical cores grouted at	16" o.c.	46	60	90	58	75	111
	24" o.c.	42	53	79	53	68	99
	32" o.c.	40	50	73	51	65	93
	40" o.c.	38	47	70	50	62	89
	48" o.c.	37	46	68	49	61	87
No grout in wall		31	35	50	43	50	69

Note: The above table gives the average weights of completed walls of various thickness in pounds per square foot of wall face area. An average amount has been added into these values to include the weight of bond beams and reinforcing steel. Grout and mortar are assumed to use sand-gravel aggregates.

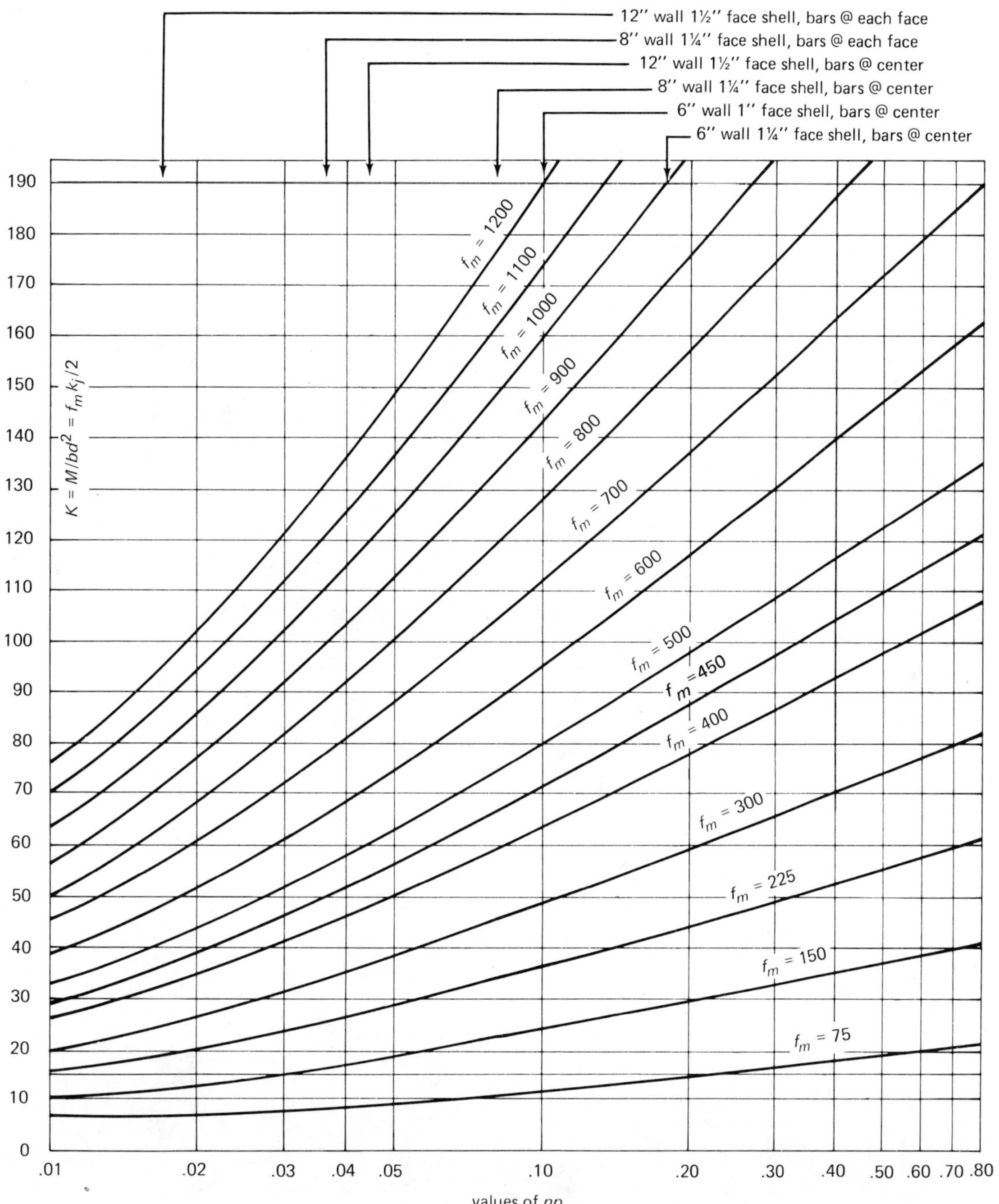

Fig. 17-14 Flexural coefficient 'K'. This is developed for rectangular beam sections. The limits of hollow or partially-grouted walls, where the face shells would function as T-beams, are shown at the top. The area to the right of the arrow requires consideration of T-beam action. For closer design in these cases, T-beam design should be used, though the difference will not be great and is not frequently justified.

TABLE 17-13 Steel Areas

Factor UBC @ .002 TITLE 21 @ .003	Area of Steel Required (by UBC or Title 21) — Wall Thickness												
	1/2"	1"	2"	3"	4"	5"	6"	7"	8"	9"	10"	11"	12"
1/3 × .002	.004	.008	.016	.024	.032	.040	.048	.056	.064	.072	.080	.088	.096
1/3 × .003	.006	.012	.024	.036	.048	.060	.072	.084	.096	.108	.120	.132	.144
2/3 × .002	.008	.016	.032	.048	.064	.080	.096	.112	.128	.144	.160	.176	.192
1.0 × .002													
2/3 × .003	.012	.024	.048	.072	.096	.120	.144	.168	.192	.216	.240	.264	.288
1.0 × .003	.018	.036	.072	.108	.144	.180	.216	.252	.288	.324	.360	.396	.432

Bar			Area of Steel per Foot — Bar Spacing																
			8"	12"	16"	20"	24"	28"	32"	36"	40"	44"	48"	56"	64"	72"	80"	88"	96"
Size	Diam.	Area	0'-8"	1'-0"	1'-4"	1'-8"	2'-0"	2'-4"	2'-8"	3'-0"	3'-4"	3'-8"	4'-0"	4'-8"	5'-4"	6'-0"	6'-8"	7'-4"	8'-0"
2 - #9	.148	.0345	.052	.034	.026	.021	.017	.015	.013	.012	.010	.009	.0086	–	–	–	–	–	–
2 - #8	.162	.0412	.062	.041	.031	.025	.021	.018	.015	.014	.012	.011	.010	–	–	–	–	–	–
2 3/16	.1875	.0552	.083	.055	.041	.033	.028	.024	.021	.018	.017	.015	.014	–	–	–	–	–	–
2 1/4	.250	.098	.147	.098	.073	.059	.049	.042	.037	.033	.029	.027	.024	–	–	–	–	–	–
2 5/16	.312	.152	.229	.15	.114	.092	.076	.065	.057	.051	.046	.042	.038	–	–	–	–	–	–
#2	1/4	.049	.073	.05	.036	.029	.024	.021	.018	.016	.015	.013	.012	–	–	–	–	–	–
#3	3/8	.110	.165	.11	.083	.066	.055	.047	.041	.037	.033	.030	.027	.024	.021	.018	.016	.015	.014
#4	1/2	.196	.293	.20	.147	.118	.098	.084	.073	.065	.059	.054	.049	.042	.037	.033	.029	.027	.024
#5	5/8	.307	.460	.31	.230	.184	.154	.132	.115	.102	.092	.084	.077	.066	.057	.051	.046	.042	.038
#6	3/4	.442	.663	.44	.332	.265	.221	.189	.166	.147	.133	.120	.110	.095	.083	.074	.066	.060	.055
#7	7/8	.601	.900	.60	.450	.361	.300	.258	.226	.200	.180	.164	.150	.129	.112	.100	.090	.082	.075
#8	1.0	.786	1.180	.79	.590	.471	.392	.337	.295	.261	.236	.214	.196	.168	.147	.131	.118	.107	.098
#9	1.128	1.000	1.50	1.00	.750	.600	.500	.428	.375	.333	.300	.273	.250	.214	.187	.167	.150	.136	.125

17.9 SPECIFICATIONS

The following are guide specifications to aid and guide the specifier in development of a spec for a specific project. It is written in the form of a CSI Guide Spec, i.e., with the spec statement on the left hand side and "Notes" on the right. Care in providing the proper spec is especially important in masonry where there are certain deviations from the past or traditional methods. The following is for reinforced grouted masonry, unreinforced masonry specs would be implied by omissions of items (see Ref. 17-11 and 17-16).

Division 4—Guide Specification, Reinforced Grouted Masonry

Guide Specifications	Notes to Specifier
Division 4—Masonry The General Requirements of the General Conditions, Supplementary Conditions and Division I of the specifications apply to this section.	Division I, General Requirements is that portion of the CSI format for building specifications which contains all provisions pertaining to the job as a whole, including the Special Conditions pertinent to that particular job. In the CSI format, General Conditions and Supplementary General Conditions are not placed within the divisions of the specifications.
1. *Work Included* as part of the reinforced masonry contract.	1. *Work Included* and *Work not Included* are listed here to define these facets, though they are sometimes not made part of the Division Specifications.
a. The work of this section shall include all labor, materials, equipment, and appliances required to complete the masonry work, including the setting and incorporating of the steel reinforcing thereof, as indicated on the drawings and as herein specified or necessary, so that other materials and work may be installed and/or performed and the whole work completed in accordance with the contract document.	If these paragraphs are included, the designer and specifier must check that there are not portions overlapping, or not covered elsewhere and whether local jurisdiction or custom dictates specific items, e.g., furnishing of reinforcing bars.
b. The setting and incorporating into the masonry of all bolts, anchors, metal attachments, nailing blocks, inserts, etc. as indicated on the drawings, as furnished by, and as located by, the General Contractor or others.	
c. The building in of all door and window frames, vents, conduits, pipes, and so forth, as furnished and set by the General Contractor or others.	
d. (Specify here if any plain or unreinforced masonry is included, such as paving, and if furnishing and/or setting of stone, veneer, terra cotta, or similar materials is included.)	d. These items are masonry work but are not incorporated into all projects.
e. The removal of and the repair of such sections of the masonry for inspection as specified.	
f. The cleaning of the masonry as necessary.	
g. The removal of all surplus masonry materials and waste upon completion of the masonry work.	
2. *Work not Included* as part of the reinforced masonry contract.	
a. Carpenter work or concrete work of any description, except setting of precast concrete elements.	
b. The furnishing or placing of any anchorage set in concrete.	b. The masonry contractor may choose to furnish veneer tie inserts so that he may be assured the ties and inserts are compatible.
c. The furnishing or placing of reinforcing or dowel steel in concrete or in other material.	
d. The furnishing or locating of any bolts or anchors for door frames, window frames, partitions, trusses, beams, girders or other attachments.	
e. The furnishing or placing of any shoring, templates or similar carpentry work.	
f. The furnishing or placing of any structural steel.	
g. Water and electricity used during construction.	
h. Continuous inspection or testing.	
3. Materials	
a. Cement—The cement shall conform to the specification for portland cement ASTM Designation C150 Type 1 or Type II, low alkali; or shall conform to the specification for masonry cement ASTM Designation C91.	a. The low-alkali requirement is included in order to reduce tendency for efflorescence. It has less soluble content to dissolve and contribute to the material that will be transported to the surface for efflorescence deposit.

Guide Specifications	Notes to Specifier
	Air entraining portland cement complying with ASTM C175 may be used satisfactorily but is not listed here because it is not customarily used.
b. White Cement—This shall be white portland cement.	
c. Aggregate (1) Sand for mortar The aggregate used in mortar shall conform to the "Standard Specifications for Aggregate for Masonry Mortar, ASTM Designation: C144."	This is generally modified to use the maximum limit of the fines to give a more workable mortar.
(2) Sand for GSU The aggregate for pointing mortar of Glazed Structural Units shall be 80 mesh ground silica sand.	
(3) Sand and Gravel for Masonry Grout The aggregate for masonry grout shall conform to the "Standard Specifications for aggregate for masonry grout ASTM C404."	
d. Water—Water used in mortar and grout shall be taken from a supply distributed for domestic purposes, and at the time of mixing shall be clean and free of acids, alkalies, or organic materials.	
e. Quick Lime—The quick lime shall conform to the "Standard Specifications for Quick lime for Structural Purposes." ASTM Designation: C5.	
f. Hydrated Lime—Hydrated lime shall conform to the "Standard Specifications for Hydrated Lime for Masonry Purposes, ASTM C207," Type S.	
g. Mortar Coloring—The color shall be a standard commercial brand of chemically inert coloring material accurately measured in a definite manner for each batch of mortar to produce a consistently even (specify color) ________ color of mortar.	g. The order of placing into the batch may affect the shade of some colors, as will climatic conditions and water content. The best are mineral oxides, either natural or synthetic.
h. Admixture—No admixture shall be used in mortar or grout except by specific consent of the architect or engineer and the local building department.	
i. Building Brick—Building brick shall comply with ASTM Designation C62.	
(1) Grade: Brick shall be Grade (NW, MW, or SW).	(1) The specifier shall check which of the three is to be used.
(2) Type: Brick shall be type (FBS, FBX or FBA). (Specify type required.)	(2) The specifier should check the local market to determine the availability and proper name and size, actual or nominal, so he may specify correctly.
(3) Color: Shall be approved in job samples which include the range or variation in color.	(3) The color will be normally terra cotta red unless otherwise specified and approved.
j. Face Brick—Face brick shall comply with ASTM Designation C216.	
(1) Grade: Brick shall be grade (MW or SW). (Specify only one grade).	
(2) Type: Brick shall be type (FBS, FBX or FBA). (Specify type required.)	(2) FBS, or Face Brick Standard, has a wider color range and greater variation of size, and may be more economical. FBX, Face Brick Excellent, has a narrow color range, minimum variation of size, and high degree of mechanical perfection for high quality face brick work. FBA, Face Brick Architectural, has a non-conformity of size, color and texture for architectural blends, clinkers and other specialties. When a type is not specified, FBS will be supplied.
(3) Size: Face brick shall be ________ (specify either nominal or actual size as supplied by the manufacturer) and as indicated on the drawings.	(3) The specifier must check the local market to determine sizes commercially available and specify accordingly.
(4) Colors and Textures: (Must be specified as by one of the options below.) Same as approved wall sample constructed at job site. Equal to that made by a certain manufacturer ________ Similar to that used on ________ buildings as made by ________	(4) The construction of a wall sample is a recommended procedure. It can serve as a sample for joints, bonding, workmanship, tolerance of units. The specific brick desired should be named. The reference to a building or part of a building is similar to the use of the sample panel.
k. Glazed Structural Units—Where the term "Glazed Structural Units" is used in these specifications or indicated	k. The subject of Glazed Structural Units was made subject of a Limited Scope Study by CSI.

Guide Specifications

on the drawings, it shall mean a GSU made of burned fire-clay body with a ceramic glazed surface. Quality and physical requirements shall be in accordance with ASTM Designation C126.

(1) Grade: GSU shall be Grade S.

(2) Type: GSU shall be Type I.

(3) Size: GSU shall be nominal ____ X ____ X ____ inches.

(4) Glazed: GSU shall be ceramic glazed color as selected from Standard Palette.

l. Load bearing Tile–Tile shall be in accordance with ASTM C34. Type ______ and Grade ______.

m. Hollow brick shall have face shells of fired clay or shale in accordance with (MW, SW) grade brick as described in C652.

Color and texture shall be according to approved sample in the architect's office (or other designated location).

n. Hollow Concrete Units–shall comply with (ASTM C90, if load bearing) Grade (N-I) (N-II) (S-I) (S-II)

(ASTM C129, if non load bearing) (Type I if moisture controlled) (Type II, if non moisture controlled).

o. Concrete Brick–shall comply with (ASTM C145) (ASTM C55) (Grade N-I) (N-II) (S-I) (S-II)

p. All concrete masonry units shall have a maximum linear shrinkage of not less than ( ) when tested in accordance with ( ).

q. Reinforcing Steel

(1) All reinforcing steel bars conform to ASTM A615.

(2) Wire shall conform to ASTM A82.

(3) Ladder Bar prefabricated reinforcing shall be manufactured of wire conforming to ASTM A82.

(4) Detailing, fabrication and furnishing shall be in accordance with "Recommended Practice for Supplying Masonry Steel" of the Western Reinforcing Steel Institute.

r. Storage of Materials

(1) All mortar materials shall be stored under cover in a dry place.

(2) Units shall be piled off the ground in a dry location. During freezing weather all masonry units shall be protected with tarpaulins or other suitable materials.

(3) No damaged or contaminated material shall be used in this work.

4. Mortar

a. Proportions–Mortar shall be type M or S.

The ingredients shall be measured accurately.

When partial batches are mixed care shall be used in proportioning all the ingredients to maintain the same ratios.

b. Mixing–Mortar shall be mixed in a mechanically operated mortar mixer for at least three minutes after all ingredients are in the drum and at least long enough to make a thorough, complete intimate mix of the materials.

Notes to Specifier

(1) Only in rare instances will Grade G, or ground joints, be necessary since the manufacture of these units is such that Grade S is generally satisfactory for even the highest quality work.

l. These are not used customarily to a great extent in seismic areas.

m. Samples shall consist of five specimens showing the range of color, size and texture.

n. The specifier shall select the appropriate Grade desired, i.e., N for Unprotected exposure, S for Protected, and whether I, Moisture controlled, or II, Non-Moisture Controlled.

o. C55 is generally used because a stronger unit is required for compliance. The specifier shall check which is appropriate.

p. This should be selected according to the custom of the area, e.g., 0.06 of 1% from saturated to oven-dry condition when tested in accordance with the Quality Control Standards of the California Concrete Masonry Association.

(4) The industry has outlined methods and details in a "Recommended Practice" which should be included for reference.

If the project is for public works (or some other agency requiring it) the measuring box must be specified. For most work, however, that extra time (and cost) is to be avoided by use of shovel count. The Inspector may calibrate a shovel count initially by using a box.

b. Some specifications limit the time that mortar may be used after mixing. This is not valid because it makes no consideration of the temperature nor dryness of the day. On a cold day mortar can be used for a long time with actually an increase in strength on remixing, whereas, on an extremely hot dry day the time limit set by some specifications may not be short enough. Actually, tests have shown that mortar and grout will increase in strength with continued mixing.

Guide Specifications

c. Tempering–The consistency of mortar shall be adjusted to the satisfaction of the mason, and water shall be added as is necessary or convenient in using the mortar. This should be done by forming a basin in the mortar, adding water and mixing it in–not by splashing water over the surface.

Mortar in which a final set has begun so that it has become harsh shall not be used.

5. Grout (low-lift)

(a) Proportions

(1) Grout for wall spaces 2 in. or less shall be a mixture of one part cement, three parts sand, with water added to produce a consistency for pouring without separation of the materials (not more than one tenth part of lime may be added).

(2) Grout for wall spaces exceeding two inches may be a mixture of one part cement and three parts masonry sand and two parts 3/8 in. maximum pea gravel (not more than 5% of the pea gravel shall pass a number 8 sieve). As much water as possible shall be added to produce a consistency for pouring without segregation of ingredients.

(3) Grout for a wall space exceeding 4 in. shall be a mixture of one part cement and four to four and a half parts combined aggregate conforming to ASTM C404.

(b) Mixing–All ingredients shall be measured according to the specified proportions for the batch and mixed in a mechanically operated batch mixer. The grout shall be mixed for a period of at least three minutes after all ingredients for the batch are in the drum. The drum must be completely emptied before the succeeding batch of materials is placed therein.

c. Tempering–The consistency of grout shall be adjusted so it will flow immediately into place without segregation of ingredients. Water may be added to compensate for loss, but grout that has begun final set and become harsh shall not be used.

6. Grout (high lift)

a. Proportions–The mix shall be one part portland cement, three parts sand and not more than two parts pea gravel, with approved admix; or may be one part portland cement, two and one half parts sand with not more than one and one half parts pea gravel, measured by volume, damp loose.

b. Mixing–Grout shall be mixed for at least three minutes after all ingredients for the batch are in the drum. In any event, it shall be mixed long enough for complete intimate mixing. The drum must be completely empty before the succeeding batch of material is placed therein.

Transit-mixed grout shall be used whenever possible. Transit-mix shall be continually rotated from the time the water is added until the grout is placed.

c. Admix–When the leaner mix is used, and when the brick are high absorption, an admix must be used to reduce water requirement and reduce volume loss.

7. Workmanship

a. Foundations–The foundation on which the wall is to be built shall have a clean, level surface prior to start of work by the masonry contractor. Sandblasting may be necessary if the surface contains laitance or other foreign material lodged in the pores or on the surface.

The foundation elevation shall be such that the bed joint is not to vary more than from 1/4″ to 3/4″. The foundation edge shall be true to line so that the masonry does not project over more than 1/4″.

Notes to Specifier

c. This factor is influenced greatly by temperature, humidity, etc., as mentioned above.

This factor is the governing one for length of time.

(1) Lime is not recommended, but may be permitted.

(2) This will generally result in a slump of from 8 to 10 in.

c. The time of set is dependent upon temperature and humidity. Also, if grout is kept continually agitated, as in the mixer, it may be used for relatively long periods, even increasing in ultimate strength.

a. An approved admix, such as Suconem Grout Aid, must be used for brick with a high initial rate of absorption in order to overcome the factors of excessive grout loss. Such admix will generally improve the quality of the grouted spaces in hollow unit masonry also even of lower absorption, but must be the proper type.

c. Suconem Grout Aid is one admix approved for this use as it reduces the excess water and provides for expansion to offset the volume loss.

a. The foundation may be brought to proper elevation by chipping, by pouring additional concrete, or, possibly the unit may be cut to fit.

Guide Specifications

Projecting dowels shall be clean of loose scale, dirt, concrete or other material that will inhibit bond. Dowels shall be in proper location prior to start of work by masonry contractor.

b. Wetting of Brick—All clay brick shall be wetted until they have an initial absorption rate not exceeding 0.25 ounces per square inch per minute determined in accordance with ASTM C67. When being laid, the brick shall have suction sufficient to hold the mortar and to delete the excess water from the mortar and grout. The brick shall be sufficiently damp so that the mortar will remain plastic enough to permit the brick to be levelled and plumbed immediately after being laid, without destroying bond.

c. Concrete brick and block shall not be wet, except in extremely hot dry weather, and then only by light spray.

d. Laying

(1) All masonry shall be plumb, level and true to line and all corners and angles shall be square unless otherwise indicated on the drawings.

(2) Line blocks shall be used whenever possible. When it is absolutely necessary to use a line pin, the hole in the joint shall be filled with mortar immediately when the pin is withdrawn.

(3) All units shall be clean and free of dust, dirt or other foreign materials before laying.

(4) All pattern work, bonds or special details indicated on the drawings shall be accurately executed. Bonding headers shall not be used in grouted masonry.

(5) Mortar for all bed joints shall be bevelled. Bevelled bed joints shall be sloped toward the center of the wall in such a manner that the bed joint will be filled when the unit is finally brought to line. Furrowing of bed joints will not be permitted. Fins of bed joint mortar that protrude into the grout space are to be avoided. If they occur, they shall be left in place, if not projecting more than the bed joint thickness. In no case shall they be cut off and dropped onto the grout below.

(6) All head joints of brick work, regardless of thickness, shall be completely filled with mortar or grout.

Head joints of hollow units shall be full for the thickness of the face shells.

(7) All units in stretcher courses shall be shoved into place.

(8) Both outer wythes of a brick wall shall be laid to a line. Units that are moved or shifted shall be relaid in fresh mortar.

(9) Except at the finishing course, all grout shall be stopped approximately 1 in. below the top of the last course. At the finished course, the last grout pour shall be brought flush with the top of the brick.

(10) Whenever possible, grouting shall be done from the inside face of the masonry. Extreme care shall be used to prevent any grout or mortar from staining the face of masonry to be left exposed or unpainted. If any grout or mortar does contact the face of such masonry, it shall be removed immediately. All sills, ledges, offsets, etc. shall be protected from droppings of mortar. Door jambs and corners shall be protected from damage during the masonry work.

8. Low-Lift Grouting

a. One tier of two wythe walls may be carried up 18 in. before grouting, but the other exterior tier shall be laid up and grouted in lifts not to exceed 6 times the width of the grout space, with a maximum of 8 in.

Notes to Specifier

Dowels may be sloped not more than 1:6 where bending is necessary. Dowels improperly located shall be removed, and replaced with approved inserts.

b. This rate may be checked approximately, by drawing a circle with a wax pencil around a quarter, dropping 20 drops of water therein and noting the time for the water to disappear. This should take approximately 90 sec. If the time is much less the brick should be wetted to reduce its initial rate of absorption.

c. Wetting will cause expansion with subsequent shrinkage.

(5) This is to prevent void spaces in the bed. The bevelled bed may leave some space at the back unfilled with mortar, but the grout will fill that. This is better than having excess fins protruding into the grout space.

(6) The grouting can fill the back portion of the head joint with grout, as well as or better than the mortaring.

(7) This is done to improve the bond of the head and bed joints.

a. Minimum grout space $1 5\frac{3}{4}''$ or the thickness of steel plus clearances.

Guide Specifications	Notes to Specifier
b. Hollow units shall be laid up not higher than 4 ft before grouting.	
9. High-Lift Grouting	
a. Masonry shall be laid full story prior to grouting.	
b. Cleanout holes shall be provided at the bottom of all cores containing vertical reinforcement in hollow unit masonry and in two wythe masonry shall be provided by omitting alternate units on the first course of one wythe.	b. The cleanout holes may be provided in the foundation and they may be closed by a form board or by masonry units. Sand or polyethelene film has been used as bond prevention to aid the cleaning of the bottom.
c. Mortar projections and mortar droppings shall be washed out of the grout space and off the reinforcing steel with a jet stream of water as required to clean the space, or by a stick and air.	
d. All grout shall be consolidated at time of pouring by puddling or vibrating and then reconsolidated by later puddling before the plasticity is lost.	d. If the timing is proper the consolidation of one course can serve as the reconsolidation of the pour placed earlier.
e. The minimum dimension of the grout space shall be 2 in.	
f. Two wythe masonry shall cure at least three days and hollow unit masonry shall cure at least 24 hours before grouting.	f. This is to assure that the wall wythes will not be too weak to resist the pressure since blowouts are to be avoided.
g. Grout shall be poured to not more than four foot depths, then wait approximately one hour and pour another four foot depth. The full height in each section of the wall shall be poured in one day.	g. The actual rate of pouring depends upon the absorption of the masonry and on the climate of the day.
h. Vertical grout barriers or dams shall be built across the grout space of two wythe masonry the entire height of the wall to control the flow of the grout horizontally. These barriers shall be less than 30 ft on center.	
i. All reinforcing steel shall be inspected in place before grouting and there shall be continuous inspection during the grouting operation.	
j. In two wythe masonry, wire ties, consisting of #9 wire rectangles, shall connect the wythes and shall be spaced not more than 12 in. o.c. vertically for stacked bond, not more than 24 in. o.c. vertically for running bond, and not more than 24 in. o.c. horizontally.	
k. Grout spaces shall not be wetted down prior to pouring the grout.	k. Wetting will reduce the suction of the brick and reduce the strength of the grout.
l. Unless otherwise indicated on the working drawings, the grout shall have 8 to 10 in. of slump.	l. This depends upon the fluidity of the grout and its flow into place.
m. Approved admix shall be used for high lift grouting in highly absorbant masonry units.	m. This may be determined in the sample panel. The admix should reduce water, retain water and provide for expansion to reduce volume loss, as Suconom Grout Aid.
10. Joinery	
a. General—At the time of laying, all masonry which is not to be plastered shall have the mortar joints finished as specified. The mason shall cut out and repoint defective joints. Those having holes such as made by line pins shall be pointed and tooled properly. Regardless of other jointing herein specified, all jointing in all masonry surfaces which are exposed to the weather shall be tooled, making solid, smooth, watertight compacted joints. Tooling shall be done when the mortar is partially set but still sufficiently plastic to bond. All tooling shall be done with a tool which compacts the mortar, pressing the excess mortar out of the joint rather than dragging it out.	a. The specifier shall select the appropriate paragraph below for a specific project.
b. Flush Joints—All joints (or specify which ones) ______ shall be "Flush Joints" made by cutting off the mortar flush with the face of the work with a trowel.	
c. Tooled Joints—All joints (or specify which ones) __________ shall be "Tooled Joints" made by striking them with a metal jointing tool to produce (specify "Concave" or "Flush" or "V" or other) ____________ finished joints.	c. Tooled joints provide best strength and weather resistance.
d. Raked Joints—All joints (or specify which ones) __________ shall be recessed by raking out the mortar a	d. The raked joint is the one most likely to cause trouble from leakage. If the deep shadow of the raked joint is de-

Guide Specifications	Notes to Specifier
distance of $^3/_8$" from the face and then tooling with a square jointer or the blade of a tuck pointer.	sired, a deep V, 70° instead of 90° apex, may be used as an alternate to give the depth while compacting the surface and pressing it tightly against the brick.
e. Struck Joints f. (Specify Other Types of Joints) ______	f. This is to provide for special or unusual joints that may be desired in design, such as beaded or weeping.

11. Pointing and Cleaning

a. At the completion of the work, all holes or defective mortar joints in exposed masonry shall be pointed. Where necessary, defective joints shall be cut out and repointed.

b. Exposed masonry shall be protected against staining from wall grouting or other sources and excess mortar shall be cleaned off the surfaces as the work progresses.

c. At the completion of the work, all exposed masonry shall be clean.

d. Sandblasting—In the event ordinary cleaning is not adequate, special methods such as sandblasting, chipping and so forth must be used to clean the surface.

12. Cleanup

All waste and surplus masonry materials shall be removed from the job, and all stains or dirt from this operation affecting other materials adjacent shall be cleaned satisfactorily.

REFERENCES

17-1. "Specification for the Design and Construction of Load-Bearing Concrete Masonry," National Concrete Masonry Association.

17-2. "Concrete Masonry Structures—Design and Construction," reported by American Concrete Institute Committee 531, *Journal of the ACI, Proc.*, **67**, 1970.

17-3. "Design Manual—The Application of Non-Reinforced Concrete Masonry Load Bearing Walls in Multi-Storied Structures," National Concrete Masonry Association, June 1969.

17-4. "Non-Reinforced Concrete Masonry Design Tables," NCMA, 1971.

17-5. "Design Manual—The Application of Reinforced Concrete Masonry Load-Bearing Walls in Multi-Storied Structures," NCMA, Mar. 1968.

17-6. Reinforced Concrete Masonry Design Tables, NCMA, 1971.

17-7. "Design Manual—Plain Concrete Block Masonry Bearing Walls in High Rise Buildings," National Concrete Producers Association of Canada, Jan. 1969.

17-8. "Reinforced Load Bearing Concrete Block Walls for Multi-story Construction," Concrete Masonry Association of California, 1969.

17-9. "Concrete Masonry Design," by Grant, A., and Clark, John H., Unit Masonry Association of Washington, 1966.

17-10. Masonry Structural Design for Buildings, Department of the Army and the Air Force, TM 5-809-3, Mar. 1970.

17-11. "A Guide for the Design and Construction of Unit Masonry," Canadian Standards Association, CSA Standard A224-1970.

17-12. Masonry Design Manual, Masonry Institute of America, 1969.

17-13. "Reinforced Masonry Engineering Handbook," by Amrhein, James E., Masonry Institute of America, 1972.

17-14. American Standard Building Code Requirements for Masonry, A.41.1-1953, National Bureau of Standards Miscellaneous Publication 211.

17-15. "Residential Fireplace and Chimney Handbook—Design, Specifications Construction, Code Requirements," Masonry Institute of America, April 1970.

17-16. "Guide Specification for Concrete Masonry," NCMA, 1971.

17-17. Building Code Requirements for Reinforced Masonry A41.2-1960, National Bureau of Standards Handbook 74.

17-18. "Reinforced Masonry Design and Practice." *Journal ASCE*, 1961.

18

Sanitary Structures-Tanks and Reservoirs

JOHN F. SEIDENSTICKER*

18.1 GENERAL DESIGN CONSIDERATIONS

Since the purpose of these structures is to contain water, the watertightness of the basins is of paramount consideration. For water treatment plants, contamination of the water supply by ground water must be prevented. For waste water treatment plants, contamination of the ground water is of concern, in addition to environmental effects, freezing and thawing, etc. on the exposed containment structure. Thus, watertight containments of maximum serviceability are required, leading to the choice of high-quality concrete as structural material.

The impermeability of the containment structure is also affected by construction sequence and joint details. Shrinkage is normal behavior for fresh concrete, and since it cannot usually be eliminated, the construction sequence should be devised to minimize this effect. Joints should be detailed to accommodate and control shrinkage, and thus to control any resultant cracking. However, the most effective way to control shrinkage is to design quality concrete mixes to minimize this effect, and to establish quality control procedures in the field to assure that quality is maintained as the concrete is placed.

Water and waste water treatment structures are designed to perform similar unit processes, i.e., sedimentation, filtration, storage, etc., and are somewhat similar hydraulically. Thus, the structural design is based on loading conditions which are also similar. In general, the raw water or waste water flows through the plant under the force of gravity, with a resultant hydraulic grade line. Pumping is used where necessary to raise the hydraulic grade line to suitable elevation. For most treatment processes, where velocity is low, the hydraulic grade line represents the top of the water level in the plant, and drops slowly due to energy losses as water moves through the various stages (except at filters where there is a greater loss of head under normal operating conditions). See Fig. 18-1 for schematic representation of hydraulic grade lines in a water treatment plant. The major design parameters are based on water containment structures where the water is under atmospheric pressure and dynamic conditions of flow are secondary. Certain elements, particularly where water is transferred between processes, such as in piping and flumes, will involve pressures above atmospheric, dynamic loadings and energy-velocity relationships. In addition, supporting functions, such as pumping and chemical treatment, will have their own type of loadings.

In order to establish the various loading conditions, a general description of each type of plant is given. From consideration of the different functions, the loadings will result as logical consequences. Careful consideration need be given to the inclusion of all loading elements, and with

*John F. Seidensticker, Principal Structural Engineer, Greeley and Hansen, Engineers, Chicago, Illinois.

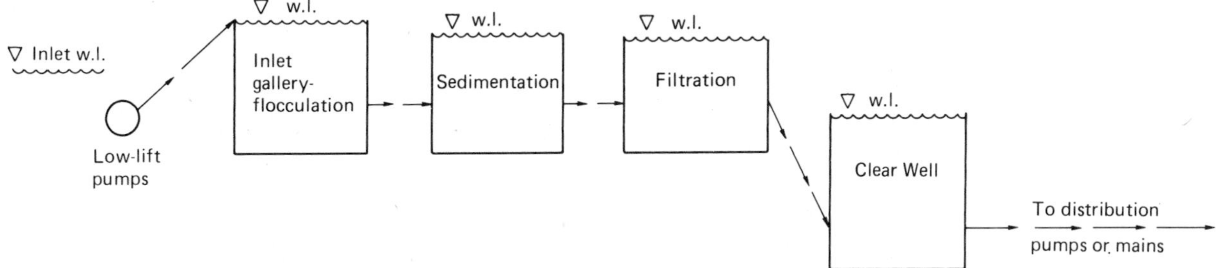

Fig. 18-1 Flow diagram—water treatment plant.

careful attention to principles of statics, errors can be avoided.

18.1.1 Water Treatment Plant–General Description[18-1, 18-2]

The function of a water treatment plant is to provide a potable and aesthetically pleasing water supply by removal of turbidity, hardness or other undesirable minerals and the elimination of pathogenic micro-organisms. This is accomplished by sedimentation, softening, filtration, and disinfection. Due to present day requirements for efficiency and water quality, virtually all current designs are based on rapid sand filtration, and water flow through the plant is by gravity. After the raw water is raised by pumps to a suitable elevation for flow through the plant, it passes through the processes of coagulation and flocculation, sedimentation, filtration, disinfection, and is stored in a clear well. Some chemical pretreatment is generally given to aid in flocculation and softening. A commonly used flow diagram is illustrated in Fig. 18-1.

A typical cross section through a water treatment plant is shown in Fig. 18-2. After flocculation, the sedimentation is handled in one or more passes, and is often an over-under arrangement for two passes as shown in Fig. 18-2. The tank size and arrangement should be determined by economic considerations and the available land. From the hydraulic grade line, the water pressure on the containing walls of the structure will be as shown in Fig. 18-3. Note that although normal operation of filters involves head loss, the structural design condition is based on a full hydrostatic head which occurs during backwashing of the filters.

Water treatment plants are generally enclosed to protect the quality of the water, particularly in the clear water well. The filters will be enclosed with a one-story structure to provide for inspection and control, whereas the sedimentation tank roof will be just over the inside water level, and commonly covered with a foot or so of dirt to protect the structure and to provide for landscaping. A waterproof membrane can be used to prevent ground water from entering the tanks, and this is protected with two or three inches of concrete with light mesh reinforcement before backfilling.

Other elements of the plant include water conduits, flumes, etc., subject to hydraulic loads which may include dynamic forces, and pipe galleries, laboratories, chemical storage, etc. The loads to be considered in the design of these elements are included in the respective sections.

18.1.2 Waste Water Treatment Plant–General Description[18-3, 18-4]

The function of a waste water treatment plant is to improve the sanitary quality of waste water by the reduction of suspended solids, biochemical oxygen demand (BOD), pathogenic micro-organisms, and other matter such that the effluent quality is improved sufficiently for discharge into the waterway. The efficiency of the plant is commonly referred to in terms of percent removal of suspended solids and reduction of BOD. This is accomplished by treatment processes which can be described in terms of primary, secondary, and advanced treatment. The processes involved in each category have certain similarities, although the holding tanks required may vary. The sludge generated by treatment is passed through digesters or thickeners to reduce volume and aid in sludge disposal processes.

Primary treatment involves a single pass settling basin (see Fig. 18-4). The settling basin includes a sludge collection trough at one end of rectangular basins or at the center or periphery of circular basins. Settled sludge is removed by a

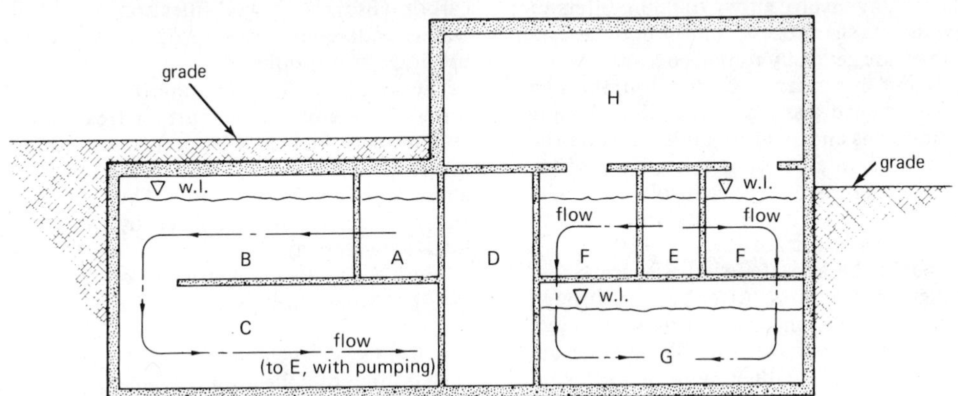

Fig. 18-2 Cross section—water treatment plant. (A) Intake gallery and flocculation; (B) upper story of sedimentation basin; (C) lower story of sedimentation basin; (D) pipe gallery; (E) filter inlet gallery and wash-water trough; (F) filters; (G) clear water well; and (H) filter gallery.

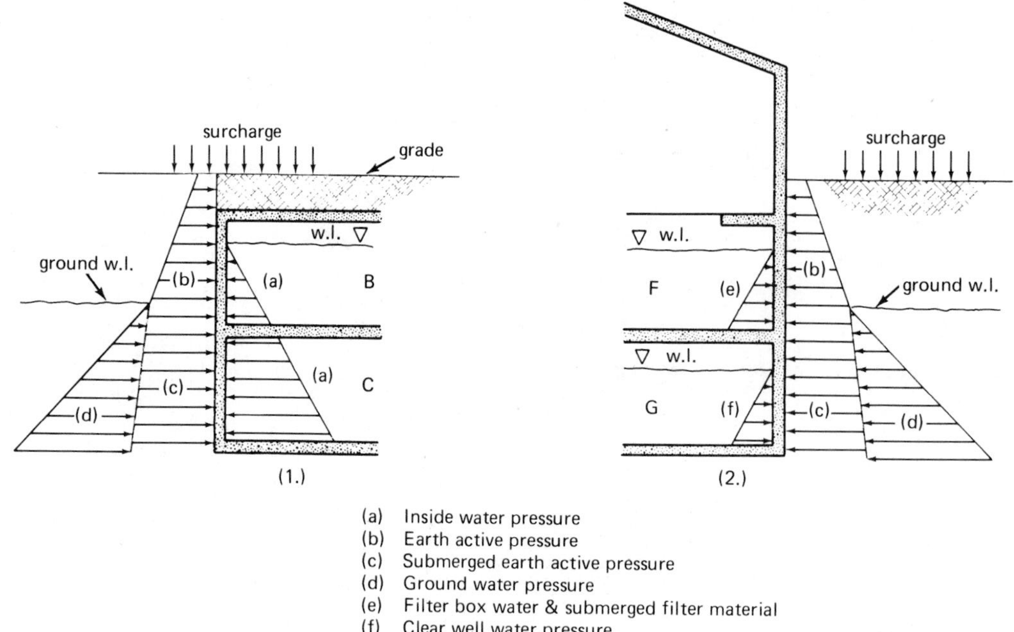

(a) Inside water pressure
(b) Earth active pressure
(c) Submerged earth active pressure
(d) Ground water pressure
(e) Filter box water & submerged filter material
(f) Clear well water pressure

Fig. 18-3 Pressures on walls—water treatment plant.

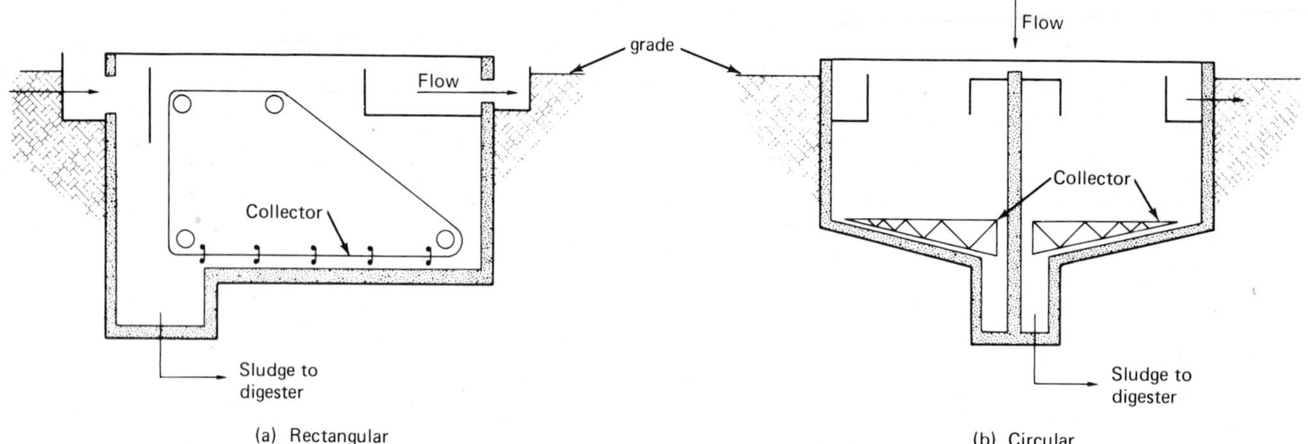

Fig. 18-4 Settling or sedimentation tanks.

collector or scraping device which moves it into the collection trough. This basin is usually open-top.

Secondary treatment may involve either trickling filters or some form of activated sludge process, and in both cases is aerobic. Trickling filters are generally round, and waste water from primary treatment is spread over the top through rotating arms. Hydraulic head losses are greater in trickling filters than in settling basins and aeration tanks. Despite the loss of head during normal operation, they are designed for full hydrostatic head, as standard practice involves periodic flooding for control of flies and other insects which tend to collect and breed on the water surface.

Aeration tanks used in the activated sludge process may be circular or rectangular structures. After passing through trickling filters or aeration tanks, the effluent passes through sedimentation tanks for secondary settling. These tanks are similar to primary settling tanks, either round or rectangular, with collection troughs to gather sludge.

Advanced waste water treatment involves either biological, chemical, or physical processes to achieve treatment beyond that available by conventional secondary processes. These processes may include nitrification-denitrification, carbon absorption, sand-filtration, etc. Tanks to hold waste water undergoing each process are similar in nature to previous descriptions.

The sludge collected from the treatment processes described above undergoes further treatment for reduction of volume and increased concentration of solids. Anaerobic digesters are covered for process control, and may have either fixed or floating roofs. The effluent from the various treatments is usually sterilized by chlorine before discharge into the waterway.

Some general characteristics of waste treatment plants may be listed as follows:

a. Tanks will generally have sloping or hopper bottoms leading to collection troughs to facilitate the collection of sludge. Settling or sedimentation tanks will have power-driven equipment to move settled sludge to such troughs.

b. Where aerobic processes are involved, tanks will be open-top. Anaerobic digestion tanks will be closed-top for control of the process.
c. Waste water treatment plants are usually at relatively low elevations in respect to water table. They will therefore be subject to water pressures outside of tank walls as well as inside, and consequently, flotation of the entire plant may be a problem. Where dead load of tank is not sufficient to overcome buoyancy, relief valves must be provided in basin floor or walls.

Liquid wastes from industrial processes vary in characteristics and strength according to the type of industry. Some common waste characteristics requiring treatment are acidity or alkalinity, high temperature, toxicity, grease and oil, and bacteria. In general, the hydraulic problems for industrial process waste water treatment are similar to those for municipal sewage plants.

TABLE 18-1 Summary of Waste-Water Treatment Processes

Preliminary Treatment
 Screens, comminutors, and grit removal
Primary Treatment
 Settling basins–round or rectangular tanks with hopper bottoms.
Secondary Treatment
 a. Trickling filters–circular tanks.
 b. Aeration tanks–circular tanks or banks of rectangular tanks, with flat bottom.
 c. Secondary sedimentation tanks.
Advanced Treatment
 Biological or physical processes to achieve maximum improvement in sanitary water quality.
 Tanks similar to above.
Chlorine Contact Tanks
 Rectangular tanks
Anaerobic sludge digestion
 Circular tanks with fixed or floating roof.

18.2 WATERTIGHTNESS

Good quality concrete is practically impervious to liquids. Leaks in basins will occur either at cracks or at joints in the concrete. Cracks are generated by stresses due to drying shrinkage and/or temperature variations. A close spacing of contraction joints, such as 25 ft apart would presumably eliminate cracking.[18-5] However, since every joint is itself a potential source of leakage, their number should be minimized, and the solution is thus a compromise. The elements which the designer can control to minimize cracking are as follows:

a. concrete quality and curing
b. joint type and spacing
c. reinforcing percentage

The total drying shrinkage of unreinforced concrete has been shown to vary from 4×10^{-4} to 8×10^{-4} in./in. when exposed to air at 50% relative humidity.[18-6] Approximately 34% of the total shrinkage is realized in the first month, and 90% in the first year. Given reasonable limits for construction time, it is apparent that most of the potential drying shrinkage will take place before the basin is filled. After filling, at least one face of basin walls will be continuously wet and swelling will occur, although not usually to exceed the extent of the previous shrinkage. It has been reported that the swelling that occurs during continuous wetting over a period of years is about one-third of the shrinkage of air-dried concrete for the same period.

The most important factor affecting drying shrinkage is the amount of water per unit volume of concrete. Water can be reduced by use of both air-entraining and plasticizing admixtures, and by using the minimum amount of cement consistent with the desired structural quality. The concrete mix should have the largest practical coarse aggregate, as this will increase the total aggregate content, and reduce the required amount of cement. Harsh graded aggregates should be avoided.

Volume changes from temperature variations will also induce shrinkage cracking. Concrete placement during excessively warm weather or with excessively high fresh concrete temperatures increases shrinkage. Too rapid filling of a basin with water cooler than the basin concrete will induce shrinkage cracks.

18.2.1 Concrete Quality and Curing[18-7]

It has been recommended that concrete for tanks should have a water-cement ratio no higher than 0.53 (equivalent to 6 gal/bag); for thin sections this is reduced to 0.44 (equivalent to 5 gal/bag).[18-8] The use of water-reducing and air-entraining admixtures will enable the water-cement ratio to be minimized such that a ratio of 0.45 can be established as maximum.[18-9] Although it is desirable from a shrinkage standpoint to minimize the cement content, requirements for durability, workability and density will establish a minimum cement content of 5.5 to 6 bags (518 to 564 lbs) per cubic yard.

Concrete mixes proportioned according to the above requirements should result in the following properties (based on 0.75–1.0 in. coarse aggregate):

compressive strength:	4000 psi minimum
air content:	$6 \pm 1\%$
slump:	4 in. maximum

Variations from the recommended concrete mix may be considered, based on local conditions or requirements, but should produce concrete of equivalent density and durability.

During the curing period, strength is gained and porosity is reduced very rapidly during the first few days. Continuous wet-curing will minimize shrinkage during the time of strength build-up, and should be considered preferable to the use of curing compounds during this initial period. Wall forms may be loosened and left in place so that a continuous flow of water may pass over the fresh concrete surface. For horizontal surfaces, burlap or other water-retaining coverings may be used, continuously moistened. After the initial curing period (seven days for normal cements), the use of membrane curing may be used for the subsequent curing.

18.2.2 Joint Details and Placement

Given good quality concrete with a minimum of drying shrinkage, some shrinkage stresses will still exist. To control cracking from volume change, the designer can use steel reinforcing and joints. These two elements must be considered together, as an increase in spacing between control joints will require an increase in steel percentage. Three types of joints are defined as follows: (See Fig. 18-5.)[18-10]

1. Construction joint. Placed to define the end of a pour, this joint is basically a cold joint in the concrete. Reinforcing is continuous, and either a water stop is used or the new concrete is bonded to the old.

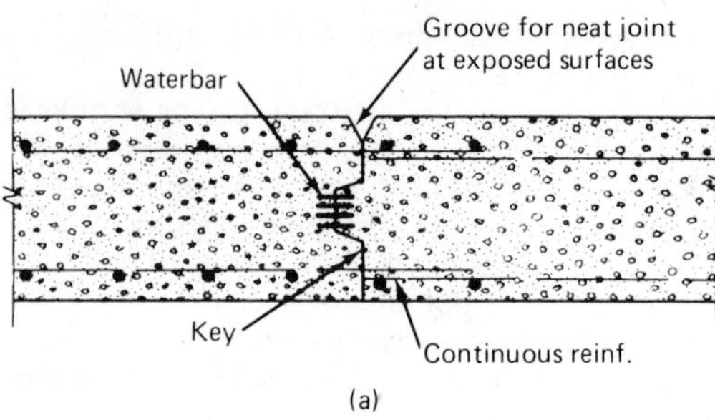

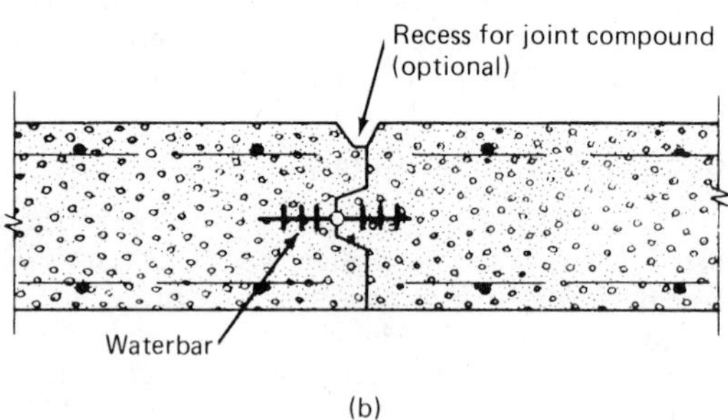

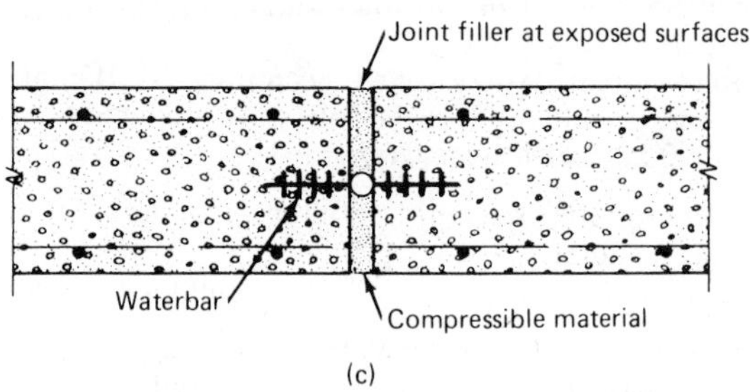

Fig. 18-5 Joint details. (a) Construction joint; (b) contraction or control joint; and (c) expansion joint.

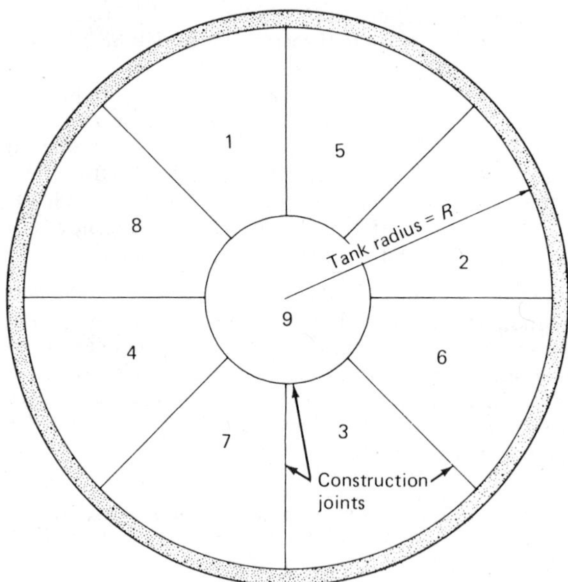

Fig. 18-6 Construction joint layout—circular basin. All areas are approximately equal. Sequence of concrete placement indicated by numbers.

Properly bonded horizontal construction joints can be made watertight. Bonding of vertical construction joints is difficult, and waterstops are used instead. Although reinforcing is continuous across a vertical construction joint, the absence of bonding presumably creates a weak link, and defines the location of shrinkage cracks rather than random cracking. This theory has met with limited success. A recess can be placed at the joint to create a neat line.

2. Contraction joint. Reinforcing is omitted and curing compound or other bond breaker is used to prevent restraining forces. Water bars are used across the joint. Unbonded dowels or keys may be used to transfer shear forces. Properly spaced contraction joints will presumably interrupt restraining forces such that other random cracking is eliminated. The choice of dumbell or labyrinth waterbar depends on the amount of movement anticipated at the joint. The labyrinth is easier to install, and less easily displaced by concrete placing operations. However, large joint movements can cause failure in concrete surrounding the labyrinth waterbar.
3. Expansion joint. This joint eliminates both tension and compression forces. Its use in walls of tanks is limited, since swelling would not normally exceed the drying shrinkage. Unless compressive stresses become excessive, good quality concrete should be able to withstand them. This joint should be used to separate structures or portions of structures of different mass, such as the connection of an inlet flume to a tank, or a lightweight slab at a heavy wall.

These joints are listed in order of increasing complexity. The potential failure of a joint is in the order of its complexity. The author has witnessed numerous failures in expansion joints, and recommends they be used sparingly and with great attention to detail and construction.

Large circular basins are usually detailed with construction joints only. The center of the base slab is one pour, and the remainder of the tank base slab and walls is divided into a number of equal segments. (See Fig. 18-6.) Basins up to 150 ft in diameter have been built in this manner, and have proved to be watertight. The practice in rectangular basins is not as uniform. Three separate design approaches can be described as follows:

1. Expansion joints placed 100 to 150 feet apart, in both directions. Between these joints, construction joints are placed at about 25 ft intervals.
2. Expansion or contraction joints placed 50 to 60 ft apart, with no other joints.
3. 'Continuous designs' with construction joints placed about 30 ft apart. Adjacent elements concreted in a skip pattern, with reinforcing lapped in the unconcreted section. Expansion joints used sparingly, such as at changes in mass or where basin length exceeds 300 ft.

The use of method (3) recognizes that random shrinkage cracking will occur. This cracking is controlled by using higher steel reinforcing percentages than in the other two methods. The job specific conditions of basin size, environment and function will determine the design method used. Where practical, the author prefers the use of method (3) for the following reasons:

a. It is the most consistent with the concept of concrete as a structural material.
b. It balances leakage through random cracking vs. failed

contraction or expansion joints. Controlled random cracks will seal themselves by the process known as autogenous healing, whereas leaky joints are difficult to repair, if the source of leaking can be found.

c. Lateral pressures due to fluid in a rectangular basin can be carried by tension reinforcing between end walls. Severing the basin with contraction joints means lateral pressures must be carried by the soil in base friction or lateral wall pressure. Spectacular failures have occurred when this support was removed during plant expansion.

18.2.3 Reinforcing for Shrinkage

The amount of reinforcing steel necessary to completely control cracking when contraction joints are eliminated has been estimated at 0.65%.[18-11] The use of 0.25% as specified by the *ACI 318 Code*[18-12] is not adequate in most cases. Observation by the author on structures and conduits built with 0.25% or less reinforcement indicates shrinkage cracks will form at about 25-ft intervals with a width of about 0.06 in. This crack width is excessive; it will not permit autogenous healing and if exposed to freezing may severly damage the area around the crack. Experience with reinforcing ratios of 0.45% minimum indicate this is a good compromise on cost vs. serviceability. For a 12-in. wall, #5 @ 12 in both faces yields a ratio of 0.43%.

The designer should also consider cost in his decision on construction method. For cases where leaks may be very damaging, the additional reinforcement required to ensure a watertight system may be more expensive than installation of additional contraction joints. For very large basins, the elimination of expansion joints is not practical.

18.3 DESIGN PARAMETERS

18.3.1 Loads

Live loads for use in structural design of the various elements can usually be determined by rational consideration of the type of material. For instance, water and raw sewage will weigh 62.4 lbs per cu ft. Sediment in a water treatment basin (area B and C, Fig. 18-2) may vary from 40 to 70 lbs per cu ft. Sludge may vary from 65 to 75 lbs per cu ft. A detailed list of weights of chemicals is given in the report of ACI Committee 350, "Concrete Sanitary Engineering Structures."

Live loads in offices and laboratories will often be governed by local building codes, but these should be recognized as minimum. The designer should make a load analysis based on use and proposed equipment to determine if heavier loads are justified. Some commonly used minimums are as follows (from the *American Standard Building Code*):[18-13]

offices	80 psf
public spaces	100 psf
railings	50 plf (applied horizontally at top of railing)

Heavy equipment loads should be determined from the manufacturer. Where such equipment rests on the floor, consideration should be given to the potential two-way structural action of the slab system,[18-14] and to the necessary space provided around each machine. For instance, a machine with a 4 X 5-ft base weighing 20,000 lbs will load the floor at 1000 psf over its base area only. However, if no other machine is within 6 ft, and the slab can be shown to be capable of spreading the load, then the floor area can be taken as (4 + 6) by (5 + 6) or 110 sq ft, and the design load is more nearly 180 psf. It can also be shown that an HS 20-44 highway loading on a 20-ft span slab is equivalent to about 250 psf over the entire span, thus again illustrating the ability of properly designed concrete slabs to spread concentrated loads.

In general, when equivalent uniform live loads exceed 100 psf, a careful analysis of slab behavior and actual loads is in order. Arbitrarily excessive live loads will result in a structure of excessive cost.

Dead loads can be determined from known densities of structural materials. For other materials, such as block walls, tile floors, roofing, etc., reference can be made to manufacturers catalogs, or to other lists of weights of building materials.

Knowledge of ground water elevations and of floods of adjacent bodies of water is essential for the design of exterior walls and for analysis of uplift on the entire basin. Where the basins rest on compressible soils, lateral earth pressures are commonly taken as the active soil pressure plus maximum water level. Where basins are founded on rock or where the walls are otherwise restrained, earth pressures will be greater than the active pressure.

Buoyancy of the basin may be resisted by the dead weight of the structure and soil alone or combined with tie downs such as rockbolts, tension piles, or other soil anchors. The effective soil mass may be increased by extending the foundation mat beyond the exterior walls. Where this method is not practical, provisions for flooding of the basin should be made by relief valves in the basin floor or side walls. When buoyancy is controlled by flooding of the entire basin, a back-up system involving planned failure of a portion of the base slab should be provided to prevent loss of the entire basin.

When buoyancy of the structure is resisted by dead weight of structure and soil alone or combined with soil anchors, a safety factor of 1.5 should be provided against flotation based on high ground water level or recurring floods. When checking the design for floods exceeding a 50-yr period, this safety factor may be reduced to 1.2. Similarly, when designing walls for exterior lateral pressures based on greater than 50-yr floods, the allowable stresses may be increased by one-third or the load factor reduced by one-fourth, depending on design method used.

18.3.2 Foundations

In the selection of foundation type, it is important to consider the minimizing of differential settlement, as this could lead to cracking in basin walls and loss of watertightness. Uniform settlement, if not excessive, may not be harmful. Under certain conditions, if the structure rests on a stiff-or hard-clay stratum of sufficient uniform thickness underlain by a soft strata, some small uniform settlement may be anticipated and may be acceptable. However, connecting piping and conduits must be detailed to accommodate such settlement.

In general, it would appear unwise to combine foundation types, except under extreme subsoil conditions and carefully controlled investigation. Combining rigid foundations, such as piles, caissons, or rock bearing with foundations on sandy or clayey soils can be expected to yield some differential settlement.

A good design approach is to include in the design team a professional who specializes in soils investigation to determine foundation requirements and to influence the choice of foundation type. Borings should always be taken, under the guidance of this professional as to number, depth and spacing.

Basin foundations on compressible soils require special attention to the resultant pressure distribution under the base slab and to the variations in soil loadings caused by alternate or partial use of adjoining basins.[18-15, 18-16]

18.3.3 Structural Design Constants

Since the impermeability, or watertightness, of the concrete walls and slabs of sanitary structures is of prime importance, it is necessary to provide reinforcement ratios in the direction of moment stress such that flexural cracking is controlled. This can be accomplished by using conservative design stresses and methods and by using larger numbers of smaller bars for main reinforcement. The flexural crack width is a function of bond stress as well as axial stress in the reinforcing. It is recommended by ACI Committee 350 that bar diameter be less than 6% of member thickness, and that maximum bar spacing should be on the order of member thickness rather than as permitted by ACI Building Code.

The choice of design method is at the present a source of controversy. The general design provisions of the *ACI Building Code 318-71* are referred to as strength design. Many engineers prefer to design water-containing basins according to the alternate method outlined in Section 8.10 of this Code, which is similar to the working stress design method of previous Codes. Whichever method is used, the 1971 Code edition is the governing document, and the general serviceability requirements of this Code, such as the requirements for deflection control and crack control, must be met.

In the report prepared by ACI Committee 350, it is recommended that "ultimate strength design methods . . . not be applied to sanitary engineering structures." Recognizing this report as representing the best opinion of competent engineers of considerable experience, the design examples presented herein will use working stress design procedures. However, this report recommends stress levels, particularly in reinforcing, lower than those commonly used in other structural designs. It is the opinion of the author, based on review of discussion of the above report, and knowledge of the standards of practice of some governmental bodies, that these stresses are conservative. Serviceable designs have been built with reinforcing stress levels as stipulated in the 1963 edition of ACI Building Code. Accordingly, concrete and steel stresses used herein will be as listed in *ACI Building Code 318-63*, Chapter 10.

To maintain control of concrete production quality, the design is often based on a concrete strength of 3000 psi, rather than the actual strength of at least 4000 psi resulting from the design mix requirements mentioned previously. However, since the basins consist predominately of walls and slabs, and within the height/thickness or span/thickness ratios listed below, little significant economy would result from designing to the higher concrete strength.

Since crack width depends primarily on steel stress and distribution (spacing), it follows that steel stress will be limited to levels which can be handled by steels conforming to ASTM A615, grade 40. The use of higher strength steels under these conditions is of no advantage. As an additional check for control of flexural cracking, some engineers recommend that tension in the adjacent concrete be limited to 350 psi, including the effects of shrinkage and temperature where applicable.

It is certain that within a short period of time, the state of the art will reach a point where strength design procedures in conjunction with serviceability requirements will supplant working stress for the design of tanks and reservoirs. At

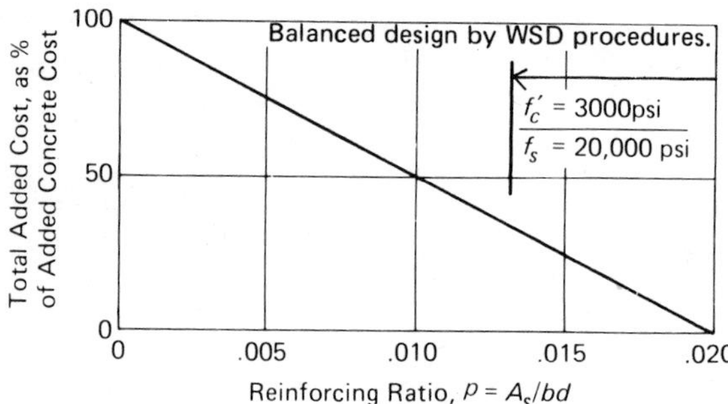

Fig. 18-7. First cost vs. reinforcing ratios.

the present, recommended design parameters for use in strength design are being discussed in the literature,[18-17] and it is of interest to compare the results of the two methods.

Present discussion indicates strength design can conservatively be based on the following parameters:

concrete: $f_c' = 3000$ psi
reinforcing steel: $f_y = 40$ ksi
load factor, walls containing water = 1.8
maximum reinforcing ratio: $q = pf_y/f_c' = 0.18$
thus: $p = 0.18 \times 3/40 = 0.0135$

Compare design of 12-in. wall for moment of 15.0 ft-kips. Clearance to reinforcing steel is 2 in., thus $d = 9.5$ in.

By working stress design procedures

Concrete capacity

$$RMC = 0.226 \times 9.5^2 = 20.4 \text{ ft-kips}$$

$$\textit{Ratio applied to capacity} = 15.0/20.4 = 0.808$$

Reinforcing required

$$A_s = 15.0/1.44 \times 9.5 = 1.10 \text{ sq in.}$$

By strength design procedures

Concrete capacity (at $q = 0.18$)

$$M_u = 0.434 \times 9.5^2 = 39.2 \text{ ft-kips}$$

$$\textit{Ratio applied to capacity} = 1.8 \times 15.0/39.2 = 0.690$$

Reinforcing required ($a \approx 0.690 \times 2.01 = 1.38$)

$$A_s = 12 \times 1.8 \times 15.0/0.90 \times 40(9.5 - 0.69) = 1.03 \text{ sq. in.}$$

The difference between the two methods for the chosen parameters is within a few percent. As more information, experience and testing with strength design applied to basins is gathered, the engineer will be able to use strength design coupled with serviceability checks with increasing confidence.

The choice of wall thickness is also more conservative than is customary in building design. Thickness to span ratios may vary between one-twelfth for cantilever elements and one-sixteenth for propped cantilever elements, and minimum thickness should be 10 in. In this regard, it is interesting to study the economics of using the larger or smaller of the above thickness-span ratios. Assuming the change in concrete cost at $35 per cubic yard in place, and reinforcing at $0.15 per pound, then from Fig. 18-7 it is seen that increasing wall thickness increases first cost for reinforcing ratios less than 0.020, which is a higher ratio than for WSD 'balanced' design. The use of thicker walls thus will increase first cost, but is done for increased serviceability.

Concrete protection for reinforcement should also be conservatively specified. Section 808 of ACI 318-71 specifies clearances under various conditions of service. For tanks with open tops, both faces of walls and walkway slabs should be considered as exposed to the weather. For ease in design and construction, the required protection is often set at 2 in. for all reinforcing, rather than specifying 1.5 in. for #5 and smaller bars. The reinforcing is placed within tolerances, and the consequences of a bar being placed too close to a water-containing surface may result in corrosion of the bar, and spalling of the surface. In addition, greater protection is of value for these surfaces where flexural cracking may occur, providing a path for water to reach and corrode reinforcing.

18.3.4 Vibrations

The mechanical equipment associated with water and wastewater treatment plants does not generally involve impact loads under normal operating conditions. The impact loads involved in hoists and cranes for removal or maintenance of mechanical equipment will be discussed later (section 18.6.3 on pump houses). Much mechanical equipment, such as scrapers and clarifiers, operates at slow speeds such that dynamic loadings to the structure are negligible. Other equipment, such as centrifugal pumps, centrifuges, and compressors, operate at high rates of speed and the dynamic loading to the structure is important.[18-18, 18-19]

The determination of the dynamic load-structure interaction is a very complicated process due primarily to the difficulty in accurately predicting the response of the structure. The major thrust of the problem is to avoid conditions where the machinery frequency approaches that of the supporting structure or foundation. Design methods have been developed to estimate the natural frequency of the structure, but these should be used with caution because of the necessary simplifications and assumptions in the methods.

Because of these difficulties and the potential severity of problems, some rules have received wide acceptance. Where possible, the foundation is insulated or otherwise separated from the structure. Where the machine rests on structure rather than independent foundation, attempts should be made to determine natural frequencies of the structure in order to avoid resonance. In addition, the machine base may be supported on suitable vibration isolation equipment where practical.

For existing structures where problems exist, or for testing of newly built structures, it is possible to determine a structure transfer function with the use of excitation devices. Together with the knowledge of the machine dynamic input, a computer simulation of the interaction of machine and structure may be made. Variations in machine base and vibration isolation equipment may then be set into the computer, leading to a problem solution.

18.4 DESIGN OF RECTANGULAR TANKS

The shape of the tank is determined by its function. Circular tanks are structurally more efficient, both in terms of wall area per unit volume and economy of materials. However, for tanks within buildings or for batteries of tanks, the rectangular shape is preferred. Sidewalls of rectangular tanks and reservoirs can be designed as cantilever, propped cantilever, or two-way slabs, depending on the tank configuration. A typical project will often involve combinations of these structural systems.

18.4.1 Loadings and Analysis of Walls

Basin walls are designed for hydrostatic pressures on the inside, and earth pressures on the outside, plus hydrostatic pressure where the water table is high (Fig. 18-2, 18-3). The inside water pressure has a triangular distribution, but the designer is cautioned to watch for possible overpressure, which may arise from vapor pressure in a closed tank, or from surges in a conduit (such as intake piping) producing a liquid level above the top of the tank.

For rectangular tanks, design is usually based on full continuity between the various elements. For small tanks, walls may be supported on three or four sides, depending on the existence of a top slab. Procedure for the design of such tanks including coefficients and design examples is given in "Rectangular Concrete Tanks,"[18-20] by the Portland Cement Association. The walls are designed as two-way elements with varying edge supports, depending on construction details and the designer's choice. Coefficients in this publication are tabulated for wall length to height ratios up to 3, but it is recommended that their use be limited to ratios ⩽2. For ratios greater than 2, design should be based on one-way action. For single-story open basins, the wall will cantilever from the base except for the vicinity of corners, or may be a propped cantilever with lateral support at the top provided by a walkway slab-beam. For single-story closed basins, the wall will act as a vertical

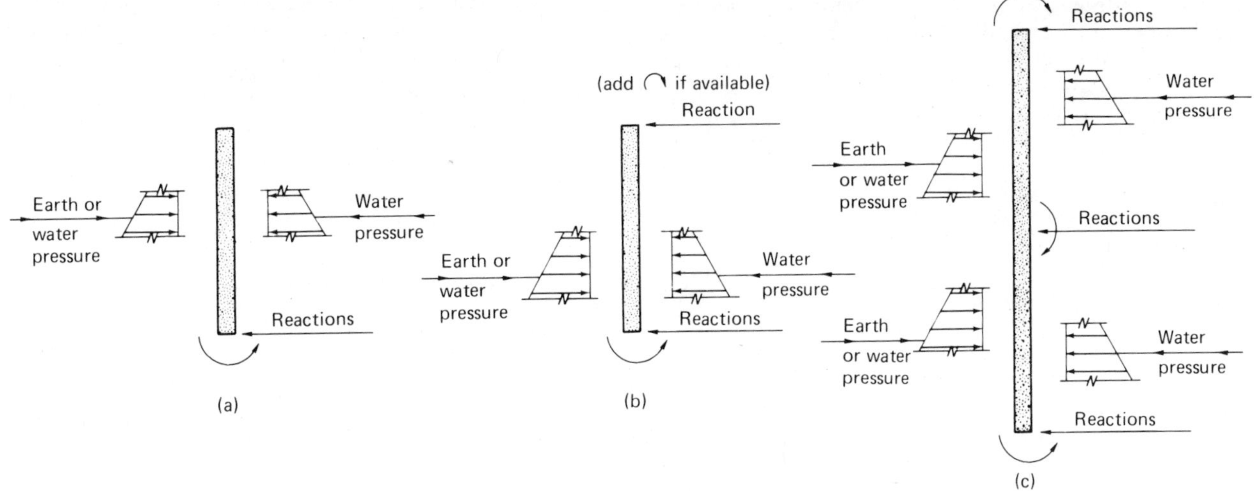

Fig. 18-8 Loading on basin walls. (a) Cantilever; (b) propped cantilever; and (c) two-span slab.

slab with continuity generally at both top and bottom. For two-story basins, the wall will become a two-span vertical slab (see Fig. 18-8). Moment distribution can be readily used to obtain effects of continuity with base and top slabs. Reinforcing details at corners should be considered to handle the local effects of continuity around the corner, where the walls will revert to two-way action.[18-21]

Interior walls of a basin may or may not be designed for water pressure. Depending on the design or operating criteria: if the interior walls are to be baffles only, it is not necessary for the wall to support full hydrostatic head on either side alone. In waste water treatment plants, for flexibility in operation and the requirement of servicing equipment in the basin, interior walls should be designed for such pressures. In water treatment plants, some part of the plant must always be in use, and this consideration will determine design criteria for interior walls.

The exterior walls are subjected to earth and ground water pressures on the outside, and water pressure inside (see Fig. 18-8). These walls should be designed for either loading condition, as it is reasonable to anticipate that the basin may be empty when backfilled, or that backfill may not be in place (e.g., expansion of facility, etc.). Often testing of the structure is desirable to check for watertightness, and such testing should be done before backfilling so that leakage could be observed. Construction specifications for the project should indicate that backfilling operations take place after top slabs are in place, so that walls are not required to act as cantilevers, unless, of course, they are designed for this condition.

It is important for the designer to recognize the fixity that actually exists at the base of the wall. For a cantilever wall, which is statically determinate, full fixity at the base is required to provide stability. However, this does not mean that the joint itself does not rotate. Any rotation of the joint causes added deflection of the cantilever. Thus, for the propped cantilever, which is once redundant, continuity at the base will induce a moment generally less than the full fixed-end moment. The amount of the rotation at the joint, due to deformation in the bottom slab, can be determined from slope-deflection equations directly, or, using the results of a moment-distribution analysis, but is of little practical interest. However, this structural behavior should be considered in the design of basin end walls, in determining the moment between wall and slab.

18.4.2 Loadings and Analysis of Floors and Roof

From the standpoint of functional structure, the most efficient form for the design of these elements is usually some form of flat plate-flat slab. It is advantageous to design intermediate and roof slabs as thin as practical in order to limit excavation, and to avoid creation of air pockets which may become stagnant. For both these reasons, the flat slab is often the best choice for structure. The roof is designed to support itself, membrane, topping and fill where applicable, and a live load which reflects probable maximum loading condition. This live load should be no less than 100 lbs per sq ft. for plants and people, unless it can be positively established that such loads will never exist. Consideration should be given to the possibility that trucks or other heavy equipment may cross the area, requiring additional live load capacity.

The intermediate slab of a two-story setting basin will be subjected to a live load of settled material, plus incidental people load. The amount of sediment is determined by operational design of the basin and when it must be emptied for cleaning. This slab is not subject to a water load, except as a tie between exterior walls, as the water pressure above and below the slab is roughly the same. Where the intermediate slab forms the base of filter boxes, the loading is vastly different, and is discussed later in the section on Filter Boxes.

18.4.3 Foundations

Where the foundation is piles, caissons, or other discrete elements, the base slab will be designed as a structural element carrying all basin loads, i.e., concrete, water, superstructure, etc., to the foundation. The magnitude of these loads may require incorporating large beams in the slab.

Where the base slab rests directly on sand or clay, it can be designed as a mat foundation carrying all loads except for the water, as the water load will be carried directly by the soil through the slab. Where possible, the bottom of this slab should be reasonably level for ease in excavation, and the top surface must drain, leading to the flat slab design. Drop panels can be incorporated in the top surface by sloping up or raising the slab at columns. The actual distribution of soil pressure will depend on relative soil-slab stiffness, and the conservative approximation is to assume uniform soil pressure. A closer approximation would be to concentrate a certain percentage of the column load in uniform pressure within the drop panel only, with remaining load spread uniformly over the entire base slab.

The design of foundations for open basins is handled in similar manner as for closed basins. If the foundation is piles or caissons, all loads must be supported on these elements, and the base slab becomes the structure to deliver the load to the piles. Where the soil has adequate bearing capacity, the only vertical loads to be supported are from the walls and any supported mechanical equipment on top walkways, and associated live loads. These loads can be distributed into soil pressure similarly to foundation loads for closed basins. The effect of relative soil stiffness to foundation stiffness may again be considered by assigning higher soil pressures near the column or wall load, and a sloping haunch may be provided in top or bottom of the slab. Effects of continuity between walls and base slab will require careful attention to corner details. On such structures as aeration tanks, where interior walls are designed for loading on one side only, alternate loading conditions are considered in the determination of maximum base slab design moments.

EXAMPLE 18-1: The example given in this section is a multicell rectangular tank, such as may be part of a battery of aeration tanks. The system consists of four tanks 25 ft wide and 140 ft long with a design water depth of 16.5 ft. The tank is designed for a fluid weighing 62.4 lbs per cu ft. General tank layout and cross section is given in Figs. 18-9 and 18-10. For this design example, inlet and outlet structures, wiers and gates are omitted in order to simplify the problem. The steps in the analysis are as follows:

1. Determine thickness of a vertical wall.
2. Determine thickness of the base slab.
3. Analyze tank–determine design moments for walls and base.
4. Determine reinforcing for walls and base.
5. Analyze and design walkways on long walls.
6. Analyze and design lateral walkways and tie beams.

Thicknesses of the walls and slabs in this example have been determined to be 14 in. and 18 in., respectively. The resulting dead load of the tank is found to be 6950 kips. Including a small average live load on the walkways and 16.5 ft of water, the average soil pressure under the base slab is 1400 psf. In order to use a mat foundation, the soil must then have a safe minimum bearing capacity of 1400 psf.

The dead load available to resist uplift is equivalent to 6950/ $(105.83 \times 142.33 \times 0.0624 \times 1.5) = 4.9$ ft of water displaced by

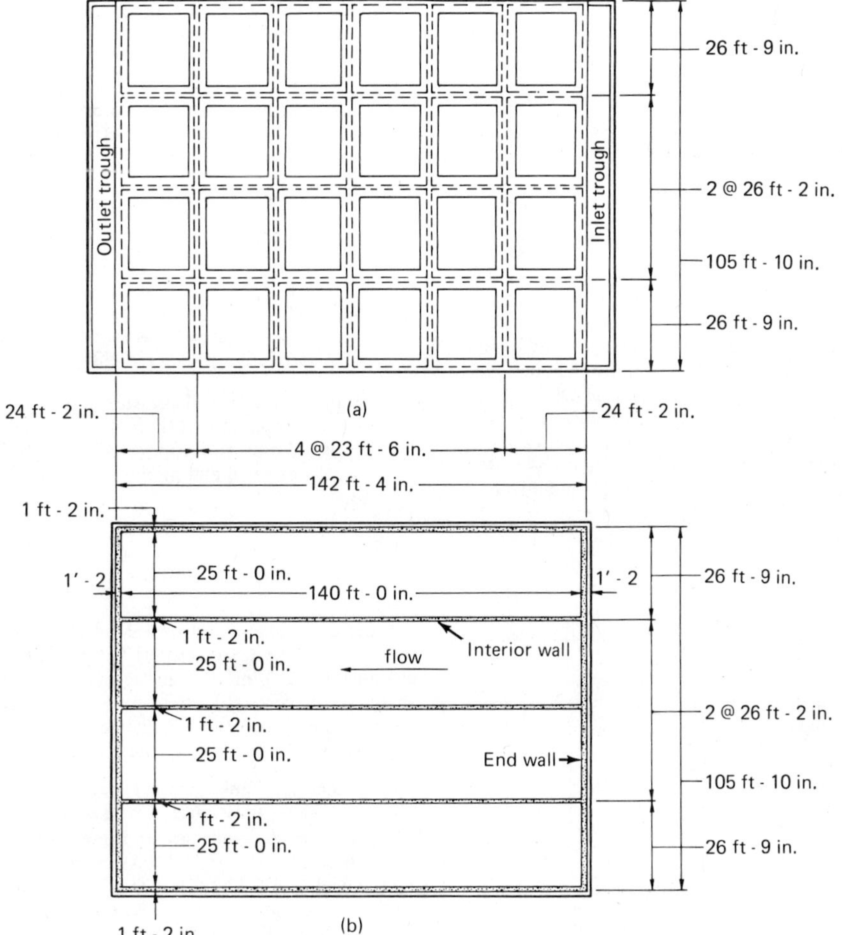

Fig. 18-9 Tank plans. (a) Top plan and (b) foundation plan.

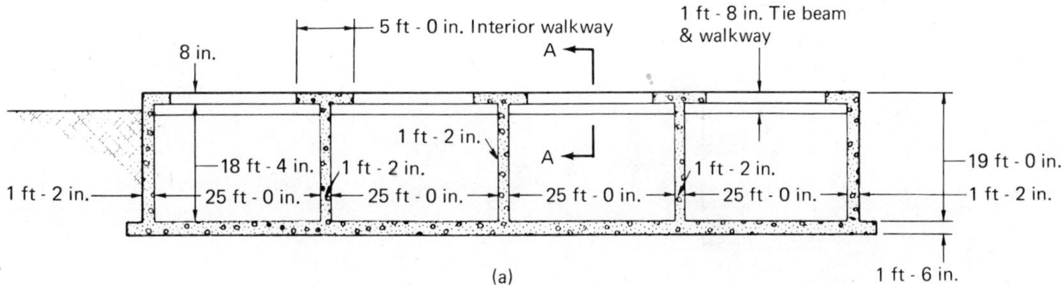

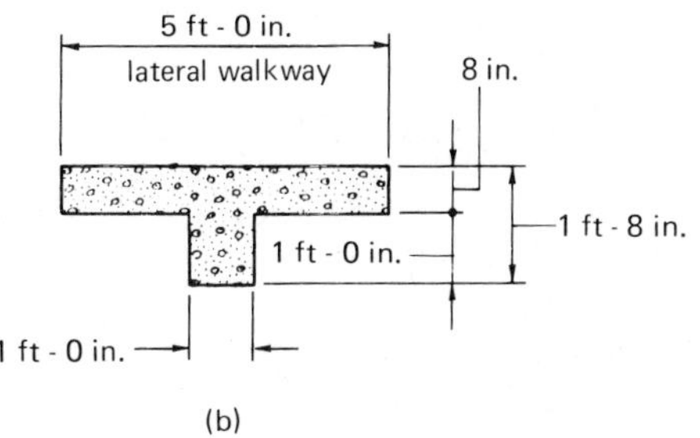

Fig. 18-10 Tank details. (a) Tank cross section and (b) section A-A.

the tank, where 1.5 is a load factor. If ground water level is higher than this, additional hold-down capacity may be obtained by extending the base slab to engage surrounding soil. For instance, extending the base slab to 6 ft from outside walls increases the tank dead load to 7650 kips, and 18 ft of soil above the slab yields an additional 6700 kips. Allowable ground water level is now 14350/ (117.83 × 154.33 × 0.0624 × 1.5) = 8.4 ft above bottom of base slab.

For this example problem the ground water level is assumed to be below 4.9 ft above bottom of base slab. To accommodate a possible future expansion program, the exterior wall should be investigated for assumed loading outside same as inside. A detailed analysis for interior and end walls only is given. The moment due to exterior soil loading is shown to be less than that due to water.

Design parameters for this example will be concrete: $f_c' = 3000$ psi, and steel: $f_y = 40$ ksi. Working stress design procedures will be used based on $f_c = 1350$ psi and $f_s = 20$ ksi. The long wall panels have an aspect ratio of 140/18.33 = 7.65, and for the end wall panels, the ratio is 25/18.33 = 1.36. Thus the long walls will be analyzed as plane frames, spanning across the tank (Fig. 18-10a). The end walls will be analyzed as two-way panels, using the PCA publication "Rectangular Concrete Tanks." Clearance for all wall reinforcing is set at 2 in. All moments and shears are in terms of a one-ft width of wall or base slab.

SOLUTION:

Step 1 Determine vertical wall thickness. The wall loadings are as follows:

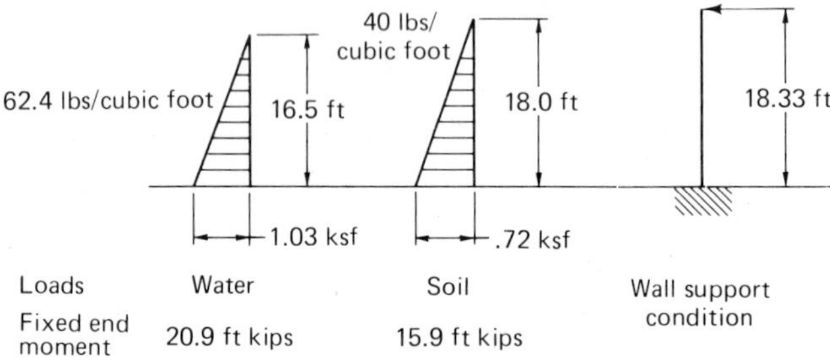

Formulas for fixed end moment for propped cantilever for partial uniform and triangular loadings are given in Fig. 18-11. As mentioned above, the exterior wall loading will be assumed same as water load. The wall span is 18.33 ft, and for a thickness to span ratio of 1/16, the thickness is 18.33/16 = 1.14 ft. For a 14-in. wall, with d of 11.5 inches, the concrete capacity is $0.226 \times 11.5^2 =$ 29.9 ft-kips. Since the final end moment for the walls will be less than 20.9 ft-kips (the maximum fixed-end moment), the 14-in. wall thickness will provide a more than adequate concrete section.

Step 2 Determine base slab thickness. The total load causing moments in the mat foundation equals tank dead load (*not* including the base slab), and the small live load on walkways, or 3560 + 120 = 3680 kips. The assumed uniform base pressure is 3680/(106.83 × 143.33) = 0.240 ksf. An approximate value for the maximum base moment may be determined by adding together:

a. Moment due to soil pressure using an assumed maximum moment coefficient of 1/11.
b. Moment due to support of interior walls, assuming three-fourths of fixed end moment to remain after distribution.

Thus

$$M_{max} \approx \frac{1}{11} \times 0.240 \times 25^2 + \frac{3}{4} \times 20.9 = 29.3 \text{ ft-kips}$$

A depth to span ratio of 1/16 yields 18.8-in. slab. For an 18-in. slab, the concrete capacity is $0.226 \times 14.5^2 = 47.5$ ft-kips, or approximately 60% more than required. It is determined that the tank analysis will be based on 14-in. walls and 18-in. base slab, but the designer should be ready to change these thicknesses if the following analysis suggests this to improve the design.

Step 3 Analyze tank. Normal tank operation will be with all cells filled (loading condition a, Fig. 18-12). To provide for the possibility of partial tank usage, three partial loadings are analyzed (loading conditions b1, b2, b3). Note that only two analyses need be made, since b3 = b1 − b2. The foundation cases are handled separately; the soil pressure becomes case c1 and lateral pressure on walls case c2 (assumed same as inside water pressure). Note that case c2 is the negative of case a for this analysis. The results of analysis of

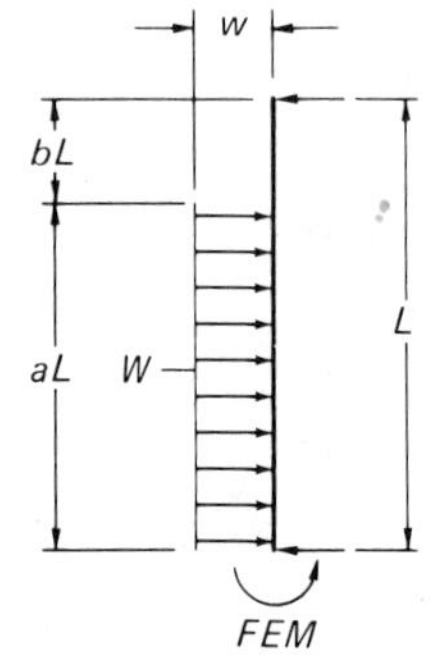

$$FEM = \frac{WaL}{60}(b+1)^2$$

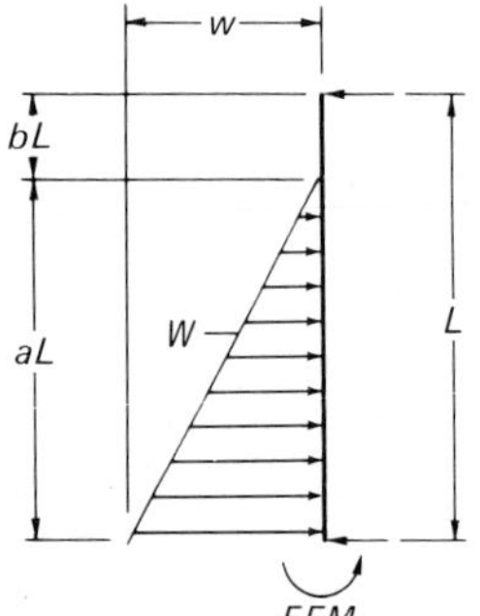

$$FEM = \frac{WaL}{60}[8 + 3b(b+3)]$$

Fig. 18-11 Fixed-end moment formulas.

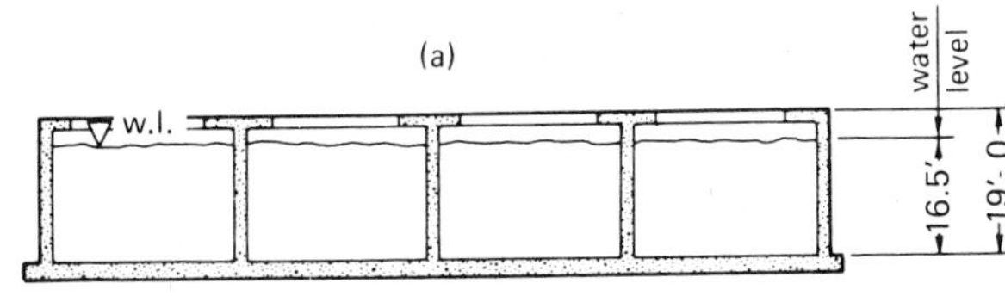

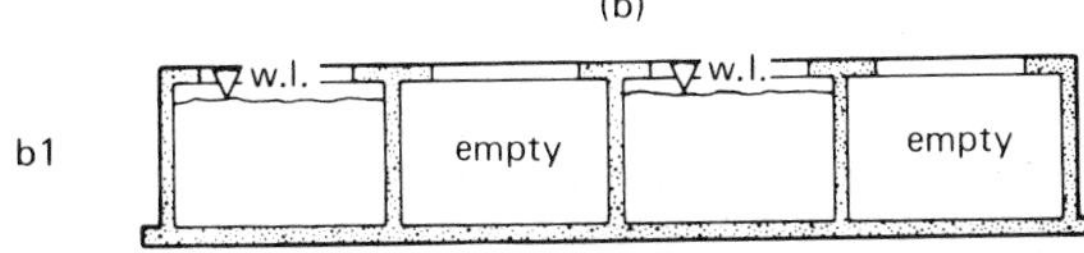

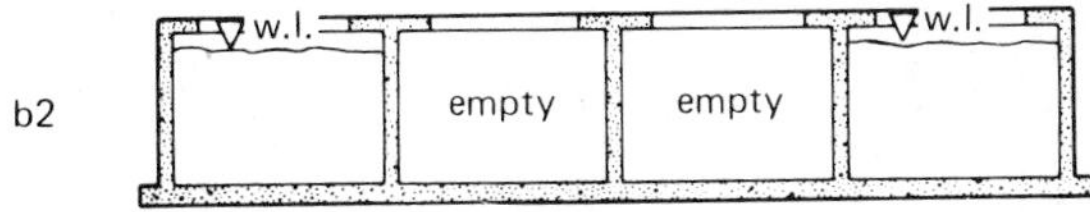

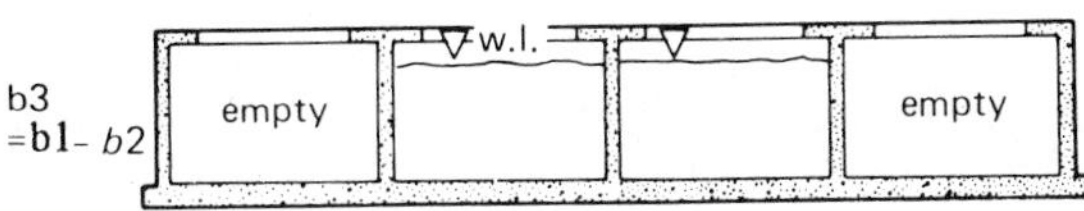

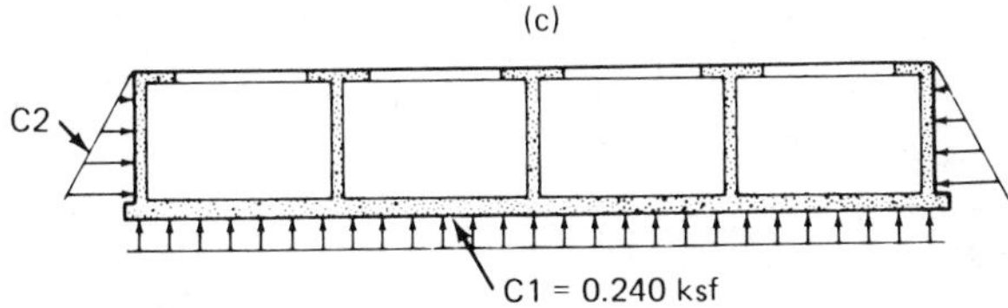

Fig. 18-12 Loading conditions. (a) Full load; (b) partial load; and (c) foundation load.

these loading conditions are given in Fig. 18-13, in terms of ft-kips per ft of width of long walls and slab.

Long Wall Design–From Fig. 18-13, the moments at the base of the long walls are as follows:

maximum, exterior walls	13.6 ft-kips	(case a or c2)
minimum, exterior walls	6.9 ft-kips	(case b1 + c1)
maximum, intermediate walls	17.6 ft-kips	(case b3 + c1)
minimum, intermediate walls	12.7 ft-kips	(case b1 + c1)
maximum, center wall	14.5 ft-kips	(case b1)

The interior walls are engineered for one design to simplify construction. Maximum positive moment in these three walls is established for an end moment of 12.7 ft-kips. The moment curves are plotted for comparison in Fig. 18-14. To illustrate the potential range of moment in these walls, curves for hinged and fixed condition at the bottom are included. Also shown is a curve for the hinged bottom condition based on coefficients from Table VI of "Rectangular Concrete Tanks." These coefficients were taken for the largest aspect ratio tabulated (b/a = 3), and yield moments approximately 10% less than those determined by plane frame action.

The designer can see that a wall design based on a fixed bottom may be unsafe. The reinforcing at the bottom would be more than adequate, but the midspan region would be underdesigned, and in particular, the lateral reaction at the top would be much smaller than the correct value, leading to a serious underdesign of the long walkways. These errors may be avoided by designing for the maximum of both simple and fixed end conditions, but resulting in an overdesign and unnecessary extra cost. The correct approach is to analyze the separate loading conditions as plane frames, as illustrated in this example, to determine accurately the maximum design moments and reactions.

Step 4 Determine reinforcing for walls and base. Reinforcing for the interior wall is determined as follows:

Maximum bottom moment = 17.6 ft-kips

$A_s$reqd = 17.6/(1.44 × 11.5) = 1.06 sq in. per ft.; use #7 @ 6½ dowels

Maximum span moment = 12.9 ft-kips

$A_s$reqd = 12.9/(1.44 × 11.50) = 0.78 sq in. per ft.; use #6 @ 6½

Dowel splice length: at 30 bar diameters, 30 × $\frac{7}{8}$ = 26.3 in. Dowels extended 30 in. above construction joint.

Longitudinal reinforcement in these walls can be determined for two conditions:

a. The set of tanks is considered a continuous design (see section 18.2.2), with recommended minimum shrinkage reinforcement ratio of 0.0045. To this should be added additional reinforcement area to resist tension from water load on end walls. Assuming deflection of a vertical section of end wall to be similar to that of the long walls, the maximum tension load will occur near midheight or at about 8-ft water depth (see Fig. 18-4). The added load in lower portion of end wall will be delivered to the base slab from continuity with base slab. Using coefficients from Table VII of "Rectangular Concrete Tanks," the assumed maximum value of this tension load is 0.295 × 16.5^2 × 62.4 = 5000 lbs per ft for one side, or 10000 lbs per ft total. Required reinforcing, A_s = 10000/20000 = 0.50 sq in. per ft. To this is added shrinkage reinforcing, as discussed in section 18.2.2, of 0.45 × 12 × 14 = 0.75 sq in. per ft. Total reinforcing for both faces is 1.25 sq in. per ft.
b. Moment at the intersection with end walls, where two-way action will occur. The coefficient in Table VI of "Rectangular Concrete Tanks" for walls with long aspect ratio (b/a = 3) is 0.015, and can be considered reasonably accurate as the wall approaches the corner. The value of a must be adjusted somewhat to enter the table due to freeboard of wall above water level: a reasonable approximation is to use 16.5 + ⅓(18.33 − 16.5) = 17.11. Then c/a = 25.0/17.11 = 1.46. The longitudinal moment is 0.028 × 62.4 × 17.11^3 = 15.6 ft-kips, and A_s reqd = 0.95 sq in. per ft. The solution will be to provide #6 @ 8½ horizontally in both faces. At the corner, add #5 @ 8½ dowels to carry the locally high bending moment from two-way structural action. The placement of reinforcing for interior walls is shown in Fig. 18-14.

The design of the exterior walls will follow a similar procedure. The maximum bottom moment is 13.6 ft-kips and the maximum span moment is 16.0 ft-kips. The value of the horizontal reaction at the top varies for interior and exterior walls. For the interior walls it is 2.55 − 12.7/18.33 = 1.86 kips, and for exterior walls it is 2.17 kips (based on interior water load). These reactions will be used in the design of the walkways.

Reinforcing for the base is determined as follows:

From Fig. 18-13, the maximum moment conditions are:

Tension top of slab

20.2 ft-kips (case b1 + c1)

A_s reqd = 20.2/(1.44 × 15.5) = 0.91 sq in. per ft.

Tension bottom of slab

27.4 ft-kips (case b2 + c1)

A_s reqd = 27.4/(1.44 × 14.5) = 1.31 sq in. per ft.

Exterior corner

13.6 ft-kips (case a or c2)

A_s reqd = 13.6/(1.44 × 14.5) = 0.65 sq in. per ft.

To these reinforcement areas must be added area for tension steel to carry bottom thrust, equal to maximum shear at bottom of wall:

$$V_{max} = 5.95 + 17.6/18.33 = 6.91 \text{ kips}$$

$$A_s \text{ reqd} = 6.91/(2 \times 20) = 0.17 \text{ sq in. per ft.}$$

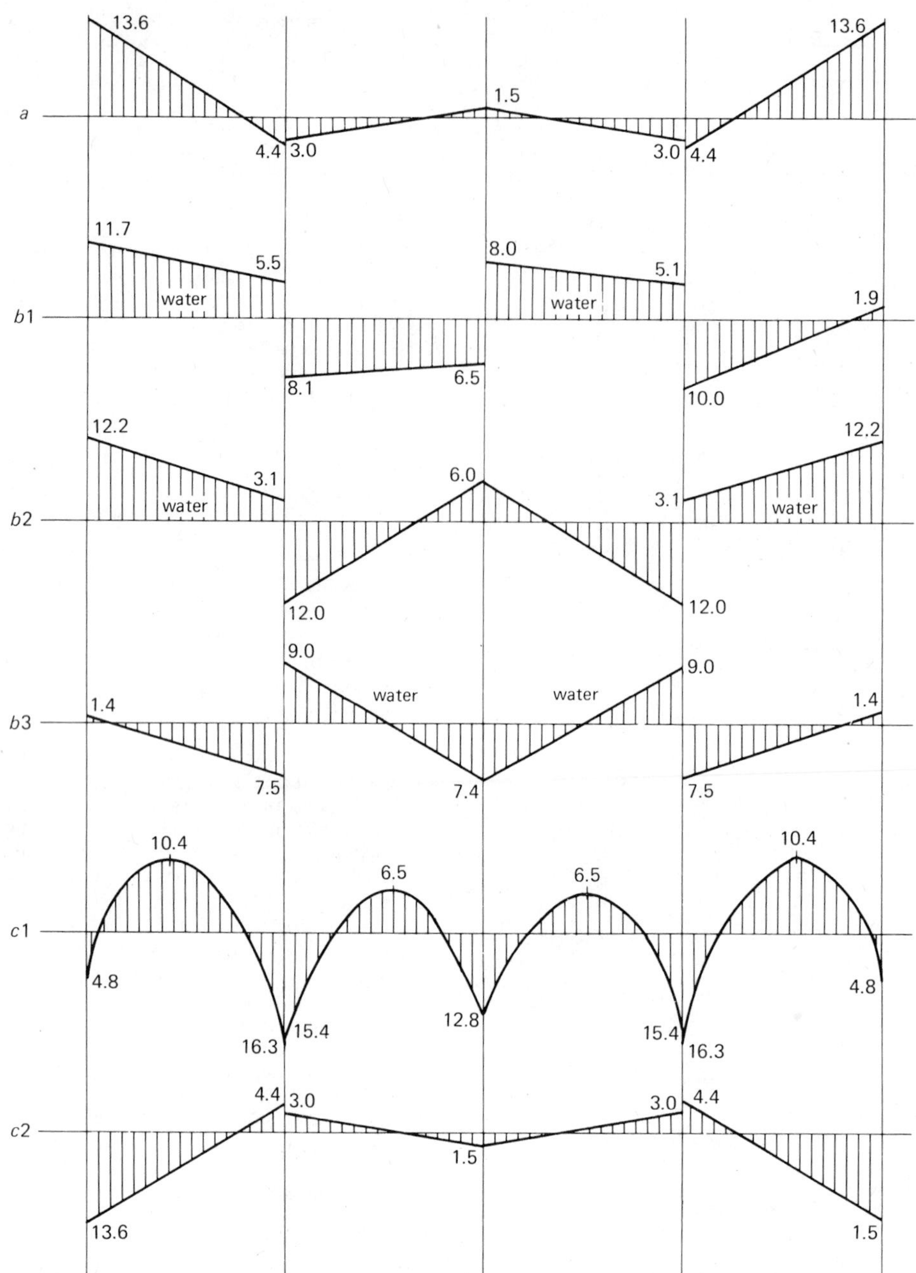

Fig. 18-13 Summation of loading conditions. Moments in base slab plotted on tension side.

in top and bottom layers. The required reinforcement areas are then:

Top

$$A_s \text{reqd} = 0.91 + 0.17 = 1.08 \text{ sq in. per ft.; use \#8 @ } 8\frac{1}{2}$$

Bottom

$$A_s \text{reqd} = 1.31 + 0.17 = 1.48 \text{ sq in. per ft.; use \#8 @ } 6$$

Reinforcement arranged as shown in Fig. 18-15 will provide #8 @ 12 in bottom of slab in center of span and at exterior edge, and these areas must be checked. For the center of span, the moment capacity is (0.79 − 0.17) × 1.44 × 14.5 = 12.9 ft-kips and should be checked against case b1 + c1 and case b2 + c1 to determine bar length. For the exterior corner, the A_s required is 0.65 + 0.17 = 0.82 sq in. However, the reinforcement in top of slab is not fully stressed at this point, and can pick up the 0.03 sq in. shortage from bottom reinforcement.

Consistent with wall design (see section 18.2.2), bottom slab longitudinal reinforcement should be 0.0045-ratio shrinkage reinforcement plus tension for end walls.

Shrinkage

$$0.0045 \times 12 \times 18 = 0.86 \text{ sq in.}$$

Tension

$$6.91/20 = 0.34 \text{ sq in.}$$

Total required reinforcement = 1.20 sq in. per ft.; use #7 @ 12 top and bottom. See Fig. 18-15 for details of base slab reinforcing.

End wall design: Reinforcing for the end walls and wall intersections can be determined with the aid of coefficients in Tables I, VI and VII in "Rectangular Concrete Tanks." The final solution will be somewhat overdesigned, however, since the available coefficients are for hinged bottom walls, and the effect of continuity must be

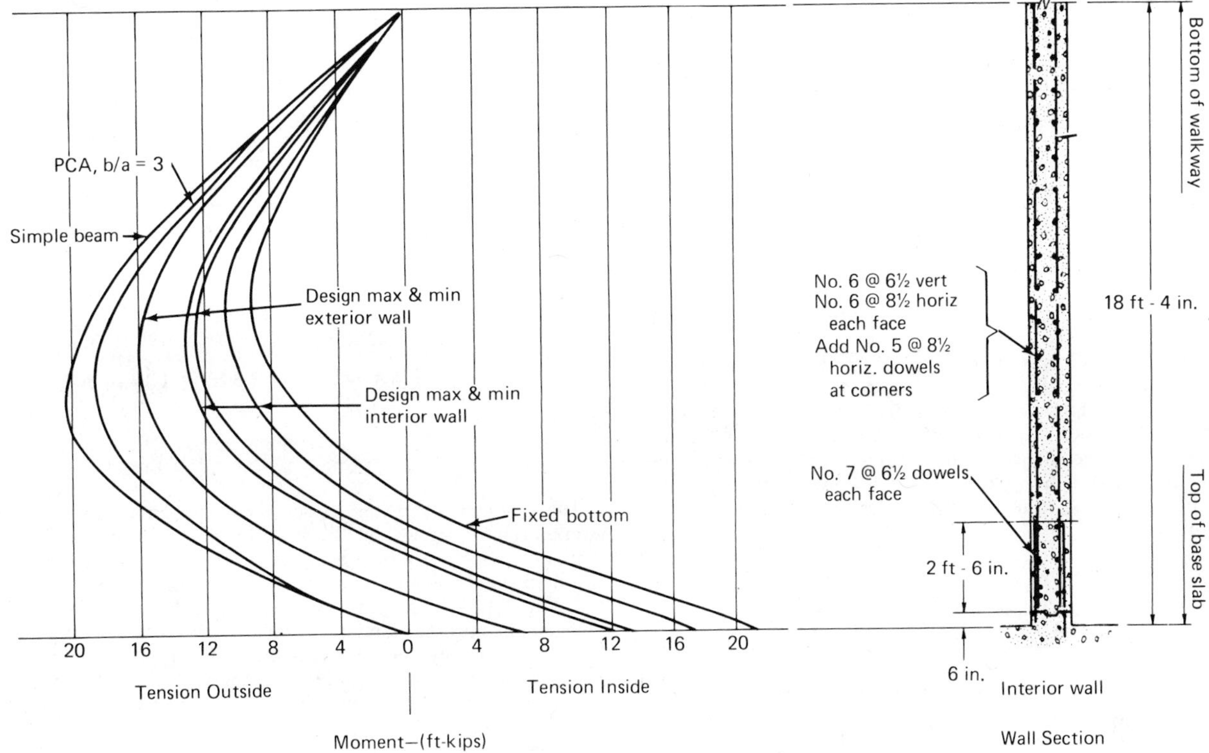

Fig. 18-14 Long wall (interior shown).

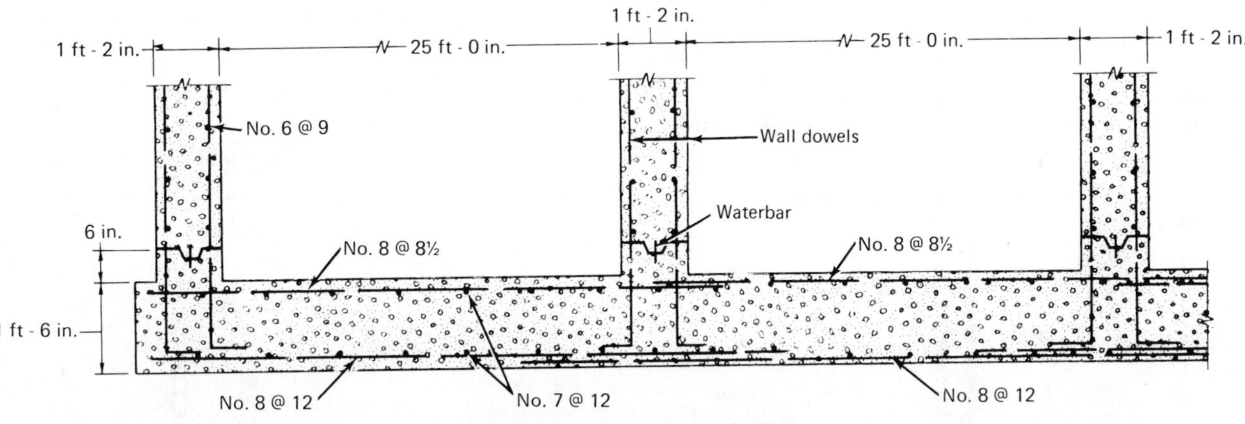

Fig. 18-15 Slab reinforcing details.

ignored. The actual reinforcing detail can develop continuity by repeating the pattern used in the long walls.

As developed earlier, the effective depth used in the tables is 17.11 ft, then b/a for Table 1 is 25.0/17.11 = 1.46, and the coefficients are as follows:

	Midspan ($y = 0$)		*Interior Intersection* ($y = b/2$)		*Exterior Corner*	
x/a	M_y	M_x	M_y	M_x	M_y	M_x
¼	+.015	+.013	−.006	−.032	−.006	−.027
½	+.028	+.020	−.010	−.052	−.010	−.045
¾	+.030	+.017	−.010	−.048	−.010	−.041

M = coeff. × wa^3

M_x = horizontal span moments

M_y = vertical span moments

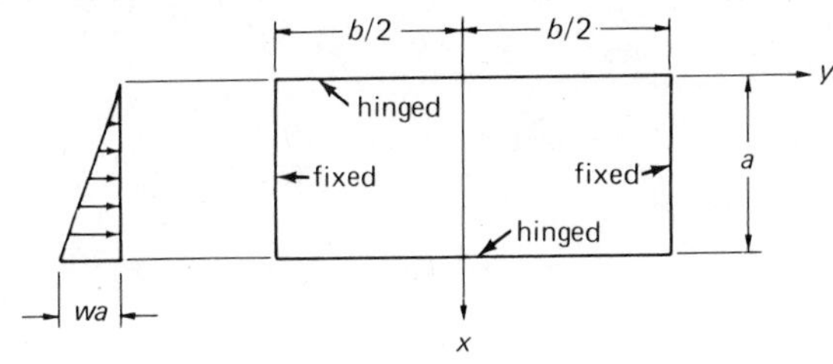

The coefficients for the exterior corner are taken from Table VI. The wall moments and reinforcing are as follows:

vertical span

$$M_{max} = 0.030 \times 0.0624 \times 17.11^3 = 9.35 \text{ ft-kips}$$

$$A_s \text{ reqd} = 9.35/(1.44 \times 11.5) = 0.57 \text{ sq in. per ft.; use \#5 @ 6½}$$

horizontal span, outside corner

$$M_{max} = 0.045 \times 0.0624 \times 17.11^3 = 13.8 \text{ ft-kips}$$

$$A_s \text{ reqd} = 13.8/(1.44 \times 10.75) = 0.88 \text{ sq in. per ft.}$$

horizontal span, inside corner

$$M_{max} = 15.9 \text{ ft-kips}; \ A_s \text{ reqd} = 1.03 \text{ sq in. per ft.}$$

midspan

$$M_{max} = 6.24 \text{ ft-kips}; \ A_s \text{ reqd} = 0.40 \text{ sq in. per ft.}$$

For the vertical span, #5 @ 6½ are used for positive moment and one size larger, or #6 @ 6½, are used as bottom dowels, repeating the pattern used on the long walls.

For the horizontal span, it is necessary to add some tension area to account for lateral reaction from long walls. From Table VII, the value for this reaction at $x/a = ½$ is $0.3912 \times 0.0624 \times 17.11^2 = 7.15$ kips and added area $= 7.15/(20 \times 2) = 0.18$ sq in. each wall face. Total area required for the continuous reinforcing is then $0.40 + 0.18 = 0.58$ sq in. and #6 @ 8½ are used. At the intersections, bars are added to make up the difference between 0.40 sq in. and required area:

inside corner

0.63 sq in. per ft.; use #7 @ 9

outside corner

0.48 sq in. per ft.; use #6 @ 9

The details of this reinforcing are shown in Fig. 18-16, for inside pressure only. It is assumed in this layout that outside design pressure was the same, and that horizontal reinforcing in both faces is then the same. Note that horizontal reinforcing is moved to the outer layer in the end walls since these moments are larger than the vertical moments.

Design of longitudinal walkways: The design of interior walkways is presented; exterior walkways follow a similar pattern. These walkways provide lateral support for the top of the long walls, and span between the lateral walkways and tie beams. The maximum horizontal reaction for the interior walkways is 1.86 kips per foot of length. This is a uniform constant load, thus *ACI Code* coefficients are adequate for analysis:

$$\text{Maximum moment} = 1/10 \times 1.86 \times 22.5^2 = 94.2 \text{ ft-kips}$$

$$\text{Maximum shear} = 1.15 \times 1.86 \times 22.5/2 = 24.0 \text{ kips}$$

The minimum practical thickness for these elements is 8 in. to provide room for two bars vertically in each edge in addition to ties. The reinforcing clearance is set at 1.5 in. to ties. Then, assuming an 8 × 60 section, the capacities are:

Moment

$$RMC = 0.226 \times 8 \times 57.5^2/12 = 499 \text{ ft-kips}$$

$$A_s \text{ reqd} = 94.2/(1.44 \times 57.5) = 1.14 \text{ sq in.}$$

Shear

$$V_c = 60 \times 8 \times 57.5 = 27.6 \text{ kips}$$

The moment and shear capacities are more than adequate. Minimum reinforcing is $0.0033 \times 8 \times 57.5 = 1.52$ sq in. Since this minimum exceeds the 1.14 sq in. from analysis an alternate minimum requirement of one-third more than analysis may be supplied, or $1.33 \times 1.14 = 1.52$ sq in. Note that partial loads may reverse bending moments, thus same reinforcing is provided in both edges, and laps should fully develop the reinforcing and preferably be made at quarter-point of span. Additional longitudinal reinforcing is added in the 60 in. width to provide a total reinforcing ratio of 0.0045, consistent with minimum used for basin walls and slab.

The overhanging slab is designed for a uniform load of 100 lbs per sq ft or a concentrated load at the edge of the slab, and it is determined to use #4 @ 12 ties. A reinforcing detail is shown in Fig. 18-17a.

Design of lateral walkways and tie beams: These elements provide the support for longitudinal walkways, in addition to their own weight and a reasonable live load. Since they are continuous across the tank cells, the maximum condition will occur in the exterior

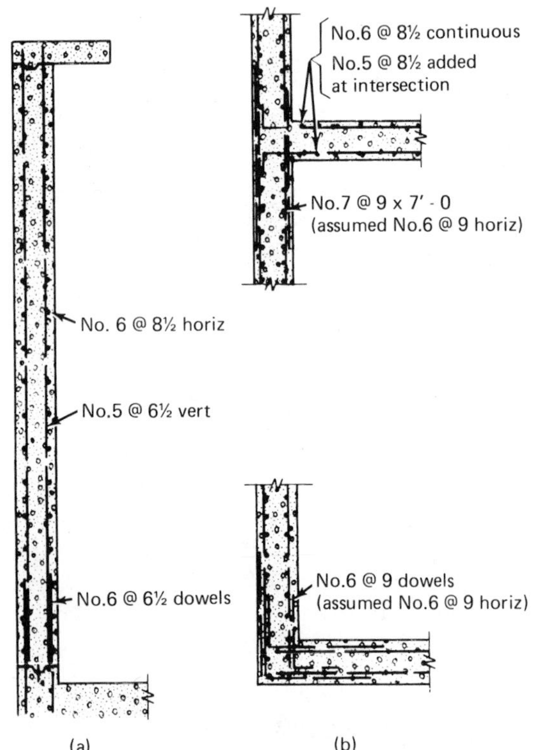

Fig. 18-16 End wall reinforcing details. Reinforcing noted for inside pressure only. (a) Vertical section and (b) plans at corners.

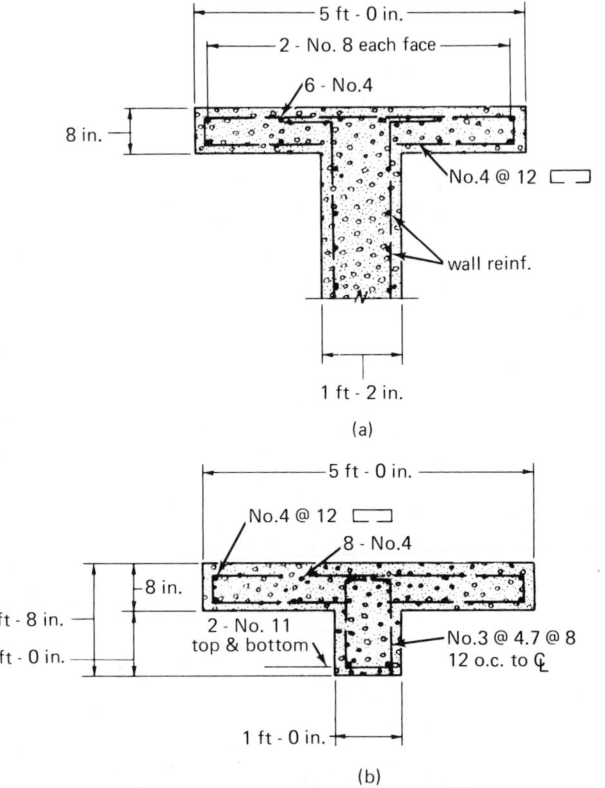

Fig. 18-17 Walkway reinforcing details. (a) Longitudinal walkway and (b) lateral walkway.

bays, and the lateral reaction from the exterior wall will govern. Further, the critical condition may occur when the exterior cell is empty, and the beam is required to act as a column.

In previous discussion, it was suggested that the exterior wall may be designed for outside water load same as interior, and that loading is used here to determine the required thrust. The wall reaction at the top is 2.17 kips/ft, and the thrust is then 54.6 kips.

The significant live load on this beam will be caused by temporary loading from equipment removal or installation, etc. A minimum of 100 lbs per sq ft should be considered acceptable, but potential concentrated loads should be checked. Allowing for a 12 × 20 beam section, the maximum support moment is 7.20 ft-kips, and shear = 16.5 kips.

The combination of thrust and moment yields an eccentricity of 72.0/54.6 = 1.32 ft. It has been shown that working stress design procedures based on the NE-KF approach may be significantly in error on the unsafe side when the eccentricity of the beam column is less than member depth. The designer should use strength design procedures in accordance with the *ACI Code 318-71*. It is suggested that a load factor of 1.8 be used, in accordance with the discussion in section 18.3.3, in order to achieve a design consistent with the results of working stress design used in the rest of the example. The proper size of concrete section and required reinforcing may be determined with the aid of available charts and graphs.

Ties to satisfy column requirements would be #3 @ 12. Beam shear requirements will require #3 bars (deformed), at a spacing not to exceed $d/2$, if required. For the 12 × 20 section, V_c = 12.6 kips. The shear at d out is 16.5 − 1.7 = 14.8 kips, and V' = 2.2 kips. For maximum spacing, #3 @ 8 yields 9.6 kips, and the tie or stirrup spacing is #3 @ 4, 7 @ 8, 12 o.c. to center line. The reinforcing detail for this element is shown in Fig. 18-17b.

EXAMPLE 18-2: This example concerns the design of a two-story settling basin, part of a water treatment plant (areas B and C, Fig. 18-2). The major point of this example is to illustrate the loads on the structure and the general approach to the wall and slab design. The detailed analysis of the slab design is omitted, but follows *ACI Code (318-71)* procedures, and differs little from an analysis of any other flat slab structure according to Chapter 13 of this *Code*.

The layout of the settling basin is shown in Fig. 18-18. The slabs should be designed for the following loads:

Top slab

12-in. concrete slab	150 psf
water proof membrane	10 psf
3-in. concrete protection	38 psf
2-ft dirt cover	240 psf
Total DL	438 psf
LL	100 psf (minimum)

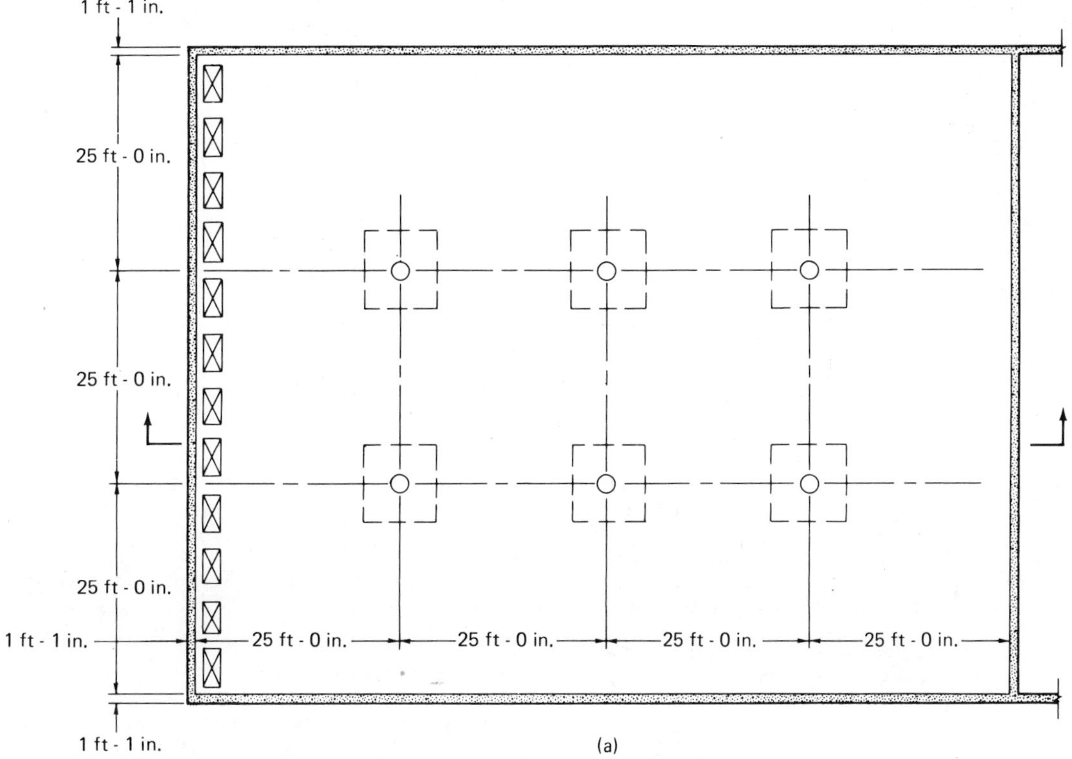

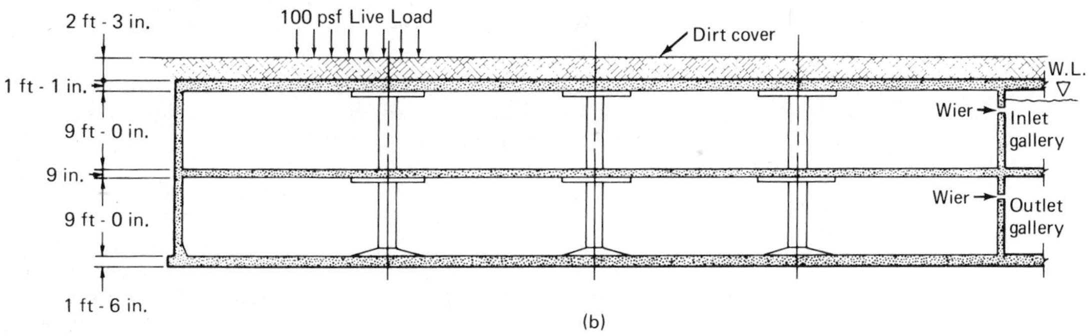

Fig. 18-18 Two-story settling basin. (a) Plan—intermediate level and (b) section.

Intermediate slab

DL–9-in. concrete slab	113 psf
LL–sludge, misc.	75 psf

Base slab (design pressure)

DL of supported slabs	551 psf
add for walls, columns	140 psf
Total DL	691 psf
LL	175 psf

SOLUTION: To determine total base soil pressure, add:

18-in. base slab	225 psf
17.75 ft water	1109 psf
Total pressure	2200 psf

For an underground basin such as this one, where the final ground level is about the same as original ground surface, the weight of the removed soil = 22.5 × 120 = 2700 psf, and exceeds the final total soil pressure. Thus, settlement problems would not usually occur, but a soil investigation should nevertheless be carefully performed to verify uniform and adequate soil support.

The basin is analyzed as a continuous structure. Plane frames are considered in both directions, with the width of each frame 25 ft, to conform with the flat slab analysis procedures. It is recommended that the basin be filled before backfilling, and an analysis should be performed for this loading. During normal operation, the basin will be periodically emptied for cleaning, establishing a second major loading condition. Careful consideration should be given to the proper lateral soil pressure on the walls, to act together with the soil load on the top slab. If this lateral pressure may be removed at a future date, another loading condition is established. Possible flotation of the empty basin should be checked as described in ex. 18-1. The dead load of this basin alone is equivalent to a ground water level 9 ft above the base slab, or 13 ft below grade.

The base slab may be established at uniform thickness, or drop panels may be incorporated in the top of the slab to provide additional shear capacity at the columns. In this example, drop panels are used, sloping up from the edges to the column in order to provide for easier cleaning of the basin. A collection trough leading to sump pit should be provided to facilitate the removal of settled material.

The wall loadings and design moments are shown in Fig. 18-19. Lateral soil pressure is assumed to be greater than active due to the increased wall stiffness from support at top and intermediate levels. Wall reinforcing for a section in the middle of the basin is given in Fig. 18-20. Tension resulting from inside water pressure is carried in the slabs. In the vicinity of the corners, two-way action will deliver tension reaction to the wall; but it can be shown that the wall acting as a vertical plate element will easily deliver this load to the slabs. Thus lateral reinforcement in the wall is for shrinkage crack control only, at the ratio of 0.0045 established previously for continuous structures (see section 18.2.3).

The effect of two-way action at the wall corners requires special consideration, as described in ex. 18-1. Additional horizontal wall reinforcing is required to handle the corner moments, similar to that shown in Fig. 18-16b.

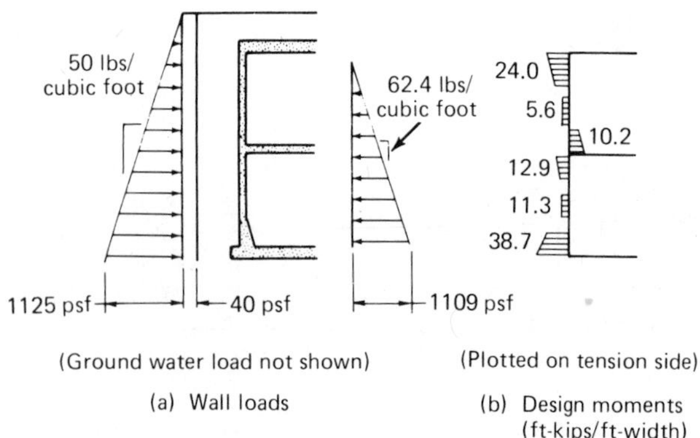

Fig. 18-19 (a) Wall loads and (b) design moments (ft-kips/ft of width).

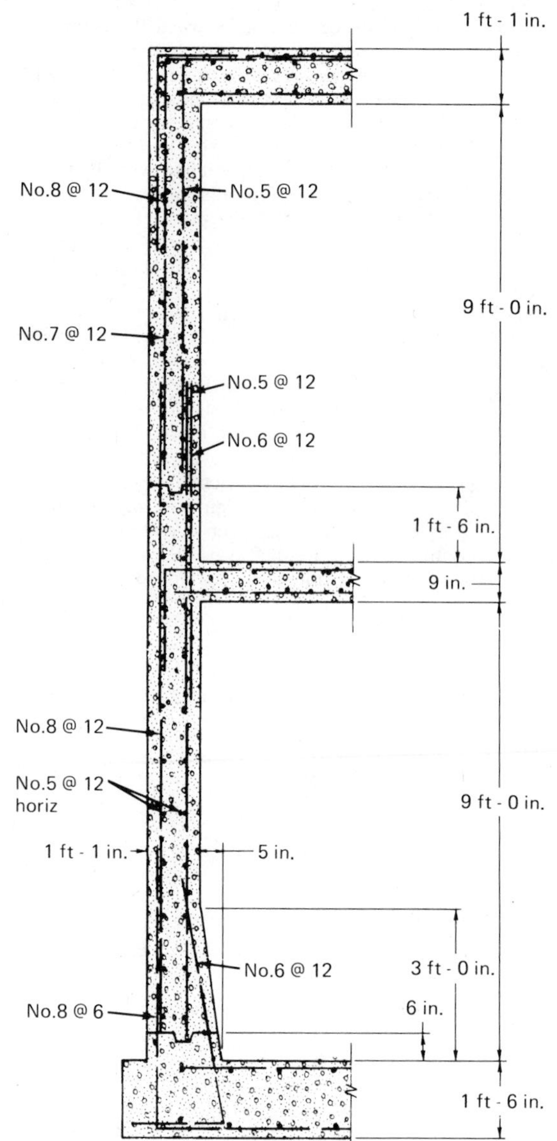

Fig. 18-20 Wall reinforcing detail.

18.4.4 Filter Boxes

Under normal operating conditions, water or waste water passing through filters will be subject to loss of hydraulic head (see Fig. 18-3). However, for backwashing or flushing cycles, the walls will be subjected to normal hydraulic pressure plus the submerged lateral pressure from filter material. Thus walls will be designed in similar manner to other basins except for somewhat higher lateral pressures. For instance, with sand as filter media with unit dry weight of 120 lbs per sq ft and ϕ of 30°, the maximum unit lateral pressure will be 0.25(120-62.4) + 62.4, or about 80 lbs per sq ft per ft.

Filter boxes in water treatment plants are placed in banks of square or rectangular tanks. The sediment collected on the filter media is removed by backwashing into the supply gullets, and then removed from the plant. These boxes are often supported above clear wells, into which the filtered water drains by gravity. Thus weight of filter media and water is of primary concern. A common rapid sand filter will be 10 to 11 ft deep, with 4 ft of filter media (sand and gravel) underlain by cast iron underdrains. The total weight of media and water will then be 1035 lbs per sq ft, as follows:

6 ft water @ 62.4	=	375 psf
4 ft sand @ 120	=	480 (70% solid)
4 ft water @ 0.30 X 62.4	=	80
cast iron underdrain system and gravel; approx.	=	100
		1035 psf

Some newer media and proprietary media will have different specific gravities, which may increase the vertical load. For instance, in a mixed media filter, the following materials may be used:

	spec. grav.
garnet	4.0
sand and gravel	2.65
anthrafilt	1.60

These boxes require walkways for inspection and gullets for moving raw water in and filtered particles out. Walkways are cantilevered off of walls and designed for loads as described previously in ex. 18-1.

Waste water trickling filters are generally round tanks with sloping bottoms. In order to accomplish even dispersal of the waste water over the filter surface, rotating distributors are used. These tanks should be designed as outlined in the section on circular basins, but the designer is cautioned to be aware that they will be flooded periodically for fly control, at which time maximum lateral pressures will be achieved.

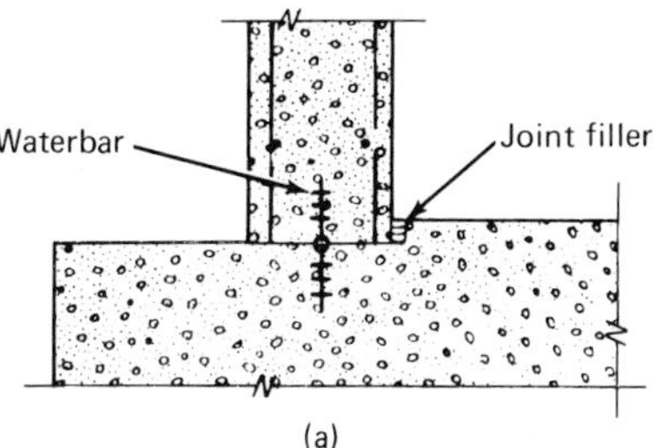

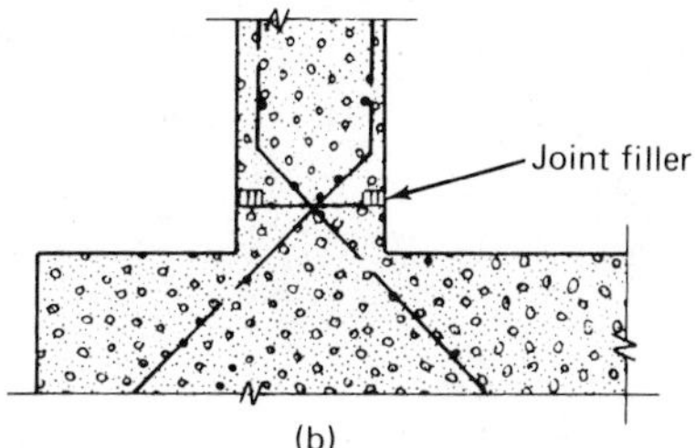

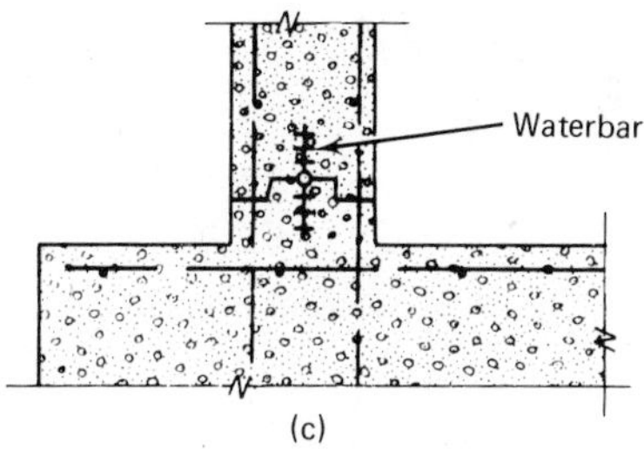

Fig. 18-21 Base joint details. (a) Sliding base; (b) hinged base; and (c) continuous base.

18.5 DESIGN OF CIRCULAR TANKS

Circular tanks are, in general, more economical than rectangular tanks. The circle has least perimeter in terms of area enclosed, and a major portion of the structural action is in hoop tension. However, whereas the circular tank may be preferred for economy, the tank shape used will largely be based on function and plant layout.

18.5.1 Edge Conditions

The detail and performance of the joint between side wall and base slab has a significant effect on the structural behavior of the wall. Three types of joints are defined, as illustrated in Fig. 18-21. The sliding joint enables all lateral pressure to be resisted by hoop tension, whereas the hinged or continuous joint creates two-way action in the wall. The governing equation is of the same form as that for beams on elastic foundation. Tables of coefficients are available[18-22] to aid in determination of hoop tension and wall moments.

When the design utilizes a hinged or continuous base joint, the ability of the foundation or base slab to deliver the required restraint must be carefully checked. Since axial tension is a relatively stiff structural system, a small radial displacement in the foundation or base slab will relieve the base shear, and in effect the base joint becomes more nearly a sliding joint. As an example, the radial displacement in ex. 18-3 to follow may be studied (see Fig. 18-22). The maximum hoop tension for a sliding joint would be 62.4 X 33 X 55 = 113,200 lbs per ft, and the reinforcement area is 113,200/20,000 = 6.70 sq in. per ft. The transformed section is 12 X 22 + 9.2 X 6.70 = 325.6 sq in. and the unit stress is 113,200/325.6 = 406 psi. The corresponding unit strain equals $f_c/E_c = 406/3{,}160{,}000 = 0.00013$, and the radius has received an elongation of 55 X 12 X 0.00013 = 0.086 in., or a little over 1/16 inch. Thus, if the wall is designed for a sliding base joint, the joint must accommodate this movement. Conversely, if designed as hinged or continuous base joint, the foundation or base slab must be capable of restraining this movement.

18.5.2 Side Walls—Shrinkage and Tension

The design of circular tank walls is handled in two steps. First, the horizontal steel is provided for all the ring tension at an allowable stress, f_s, as though designing for a cracked section. Then the concrete stress is determined for the transformed section due to combined shrinkage and tension.

The formula for combined shrinkage and tension stress in a reinforced concrete ring can be developed as follows.[18-11] At a point near the center of an uncracked section of the wall, the shortening per unit length of steel and concrete must be equal, and strain capatibility yields

$$\frac{f_{ss}}{E_s} = C - \frac{f_{cs}}{E_c}$$

where f_{ss} and f_{cs} are steel and concrete stresses due to shrinkage, and C is the coefficient of shrinkage of concrete.

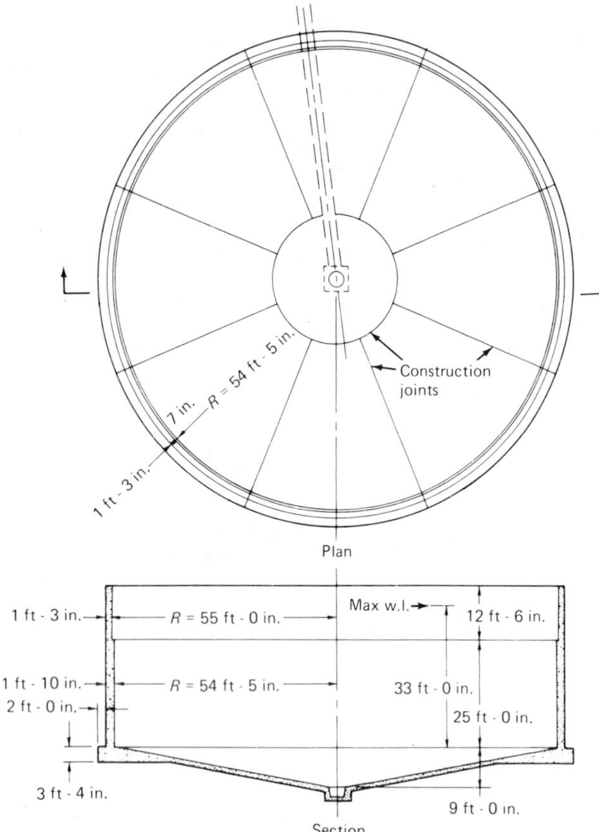

Fig. 18-22 Tank in example 18-3.

From stress consistency

$$A_c f_{cs} = A_s f_{ss}$$

Combining these two expressions yields the following concrete tension due to shrinkage:

$$f_{cs} = C E_s \frac{A_c}{A_c + nA_s}$$

The concrete stress in the transformed section due to tension, T, is very nearly equal to $T/(A_c + n\,A_s)$, and the combined concrete stress becomes

$$f_c = \frac{C E_s A_s + T}{A_c + n\,A_s}$$

The limitation on concrete stress in the transformed section is established at 350 psi, in consideration of the quality concrete mix recommended in section 18.2.1. The coefficient of shrinkage of concrete is generally assumed to to be 0.0003. The allowable steel stress in ring tension has been subject to very conservative limitations in the past. Unless bond and crack width are carefully examined, simply reducing steel stresses may not prevent leakage. For this example, the absence of special problems such as corrosive atmosphere, etc., is assumed, and a stress of 20 ksi is used. It can be shown that using lower steel stress, resulting in more reinforcing, will increase the concrete stress in the transformed section, causing the concrete to crack sooner. However, since the lower steel stress gives smaller cracks, it is the choice of many designers. In the opinion of the author, this advantage is subordinate to that of obtaining maximum bond.

EXAMPLE 18-3: This example concerns the design of a medium-sized digester tank. Tank plan and cross section are shown in Fig. 18-22. The lower wall thickness has been determined to be 22 in. Corbels are provided for support of clarifier equipment, and above this point the wall thickness is reduced to 15 in. The wall projects 4.5 ft above maximum water level, and it is assumed that this projection corrects for the loss of circumferential stiffness in the thinner top section. Thus, for design purposes, the top of the tank is assumed to be at the maximum water level, with a uniform wall thickness of 22 in. Design parameters for this example will be, concrete: $f_c' =$ 3000 psi, concrete tension on transformed section: 350 psi maximum, reinforcing: $f_s = 20$ ksi, and clearance to reinforcing = 2 in. except at bottom of bottom slab, 3 in.

The design of tank wall and slab is accomplished with the aid of coefficients from the Portland Cement Association publication, "Circular Concrete Tanks Without Prestressing." When coefficients are used, the applicable Table is listed. The steps in the analysis are as follows:

1. Establish preliminary wall thickness.
2. Determine ring reinforcing, check concrete stress and verify wall thickness.
3. Determine vertical wall reinforcing.
4. Determine base slab thickness and reinforcing.

The design is given for interior water pressure only. The design for loading from exterior soil and ground water pressure is not shown. However, it can be handled easily, as the tank ring will be in compression, and vertical wall moments only need be determined. For a tank of this size, design to resist flotation with tank dead load is not practical. Relief valves should be provided together with a back-up system as described in section 18.3.1.

Step 1 Establish wall thickness. The formula for concrete stress due to shrinkage and tension may be used to establish preliminary wall thickness. Substituting $A_s = T/f_s$ and $A_c = 12\,t$ yields

$$t = \frac{C E_s + f_s - n f_c}{12 f_c f_s} T$$

It is determined that this tank design shall utilize a continuous joint at wall and base slab. Then the maximum hoop tension will occur at about 0.65 × water depth, and is 0.65 × 62.4 × 33 × 55 = 74,700 lbs per ft. The estimated wall thickness is:

$$t = \frac{0.0003 \times 29{,}000 + 20 - 9.2 \times .350}{12 \times .350 \times 20} \times 74.7$$

$$= 22.3 \text{ in.; use 22 in.}$$

To enter the Tables in Ref. 18-22, compute the quantity $H^2/D\,t =$ $33^2/(110 \times 1.83) = 5.4$.

Step 2 Determine ring reinforcing. With coefficients from Tables I and II, the hoop tension is determined for both fixed and hinged base, and is shown together with the tension for a sliding base in Fig. 18-23a. Due to the difficulty in maintaining perfectly fixed conditions at the base, as discussed previously, the conservative design approach is to design for the maximum condition. Thus, $T_{max} =$ 0.626 × 62.4 × 33 × 55 = 71,000 lbs per ft. Reinforcing at this level is $A_s = 71.0/20 = 3.55$ sq in. per ft.; use #7 @ 4 in two curtains ($A_s = 3.60$ sq in.) The concrete tensile stress based on the transformed section is

$$f_c = \frac{0.0003 \times 29 \times 10^6 \times 3.60 + 71{,}000}{22 \times 12 + 9.2 \times 3.60} = 345 \text{ psi}$$

Since 350 psi is considered allowable, the 22-in. wall thickness is acceptable.

Hoop reinforcing is reduced above and below the point of maximum tension in accordance with requirements in Fig. 18-23a. The details of this reinforcing are shown in Fig. 18-24.

Step 3 Determine vertical wall reinforcing. With coefficients from Tables VII and VIII, vertical wall moments are determined for both fixed and hinged base, and are plotted in Fig. 18-23b. The designer is again required to choose proper design values. Due to the continuous joint design, water pressure on the base can be expected to provide rotational restraint at the joint. However, some rotation

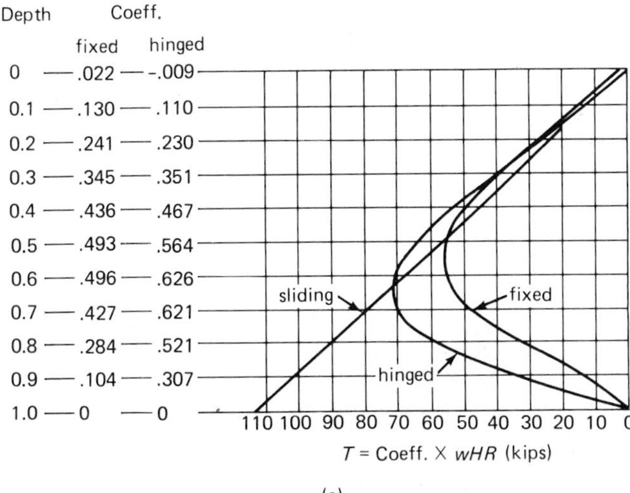

(a)

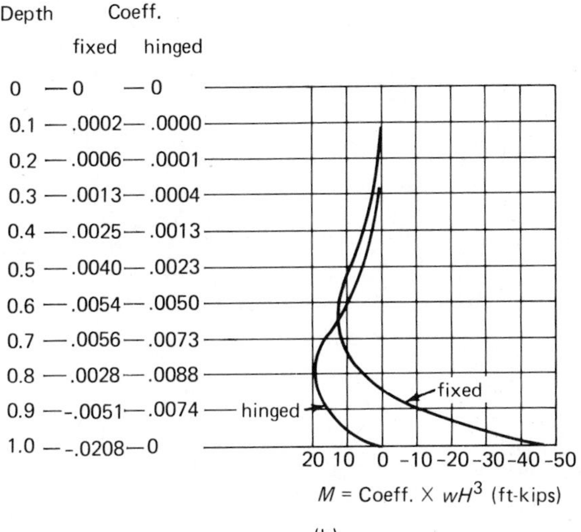

(b)

Fig. 18-23 (a) Hoop tension and (b) moment vs. coefficients for determining ring reinforcing in example 18-3.

must occur, together with some strain in the base slab from the horizontal reaction at the wall base. The conservative design approach is again to provide reinforcing for maximum conditions. At the inside of the wall at the base, the moment per foot is $M = 0.0208 \times 62.4 \times 33^3 = 46.6$ ft-kips. Reinforcing, $A_s = 46.6/(1.44 \times 19.5) = 1.66$ sq in. use #9 @ 7. For the outside of the wall, the moment per foot is $M = 0.0088 \times 62.4 \times 33^3 = 19.7$ ft-kips. Reinforcing, $A_s = 19.7/(1.44 \times 19.5) = 0.70$ sq in.; use #6 @ 7. This exterior vertical reinforcing is extended to the base, and should be checked for moment due to exterior soil and ground water load. The reinforcing is reduced to #6 @ 14 where applicable. The details of this reinforcing are shown in Fig. 18-24.

Step 4 Determine base slab thickness and reinforcing. The thickness of the base slab at the edge must satisfy the following requirements:

a. provide anchorage for #9 vertical dowels
b. provide stiffness and moment capacity for continuity with wall
c. act as footing for wall

It is determined that a 3 ft-4 in. thickness will be adequate. The wall shear delivered to the base, from Table XVI for fixed base, is $0.207 \times 62.4 \times 33^2 = 14.1$ kips/ft. Reinforcing is determined as for hoop tension, $A_s = 14.1/20 = 0.71$ sq in. per ft; use #6 @ 14 in two layers. The minimum base thickness, near the center of the tank, is

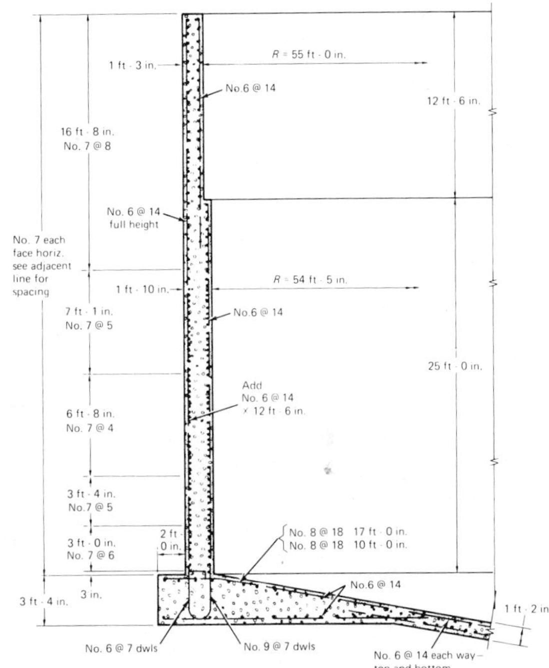

Fig. 18-24 Reinforcing details.

set at 14 in., consistent with tank size and minimum reinforcement ratio.

The top reinforcing in the thickened edge of base slab is determined for wall base moment = 46.6 ft-kips: $A_s = 46.6/(1.44 \times 37.5) = 0.97$ sq in. per ft.; use #8 @ 9. The length of this reinforcement is set by cantilever slab span required using base water pressure = 33 × 62.4 = 2060 psf, with safety factor of 2: $L = 2 \times \sqrt{2 \times 46.6/2.06} = 13.5$ ft. Alternate #8 @ 18 bars are extended to $R = 40$ ft. and lapped with #6 @ 14 bars from center portion. Bottom reinforcement in the thickened edge is provided by the wall dowels.

18.6 OTHER STRUCTURES

18.6.1 Superstructures

Superstructures for offices, laboratories, pump houses, etc., are designed for normal live loads, in addition to any special loads such as crane bridges, in similar manner to other buildings and structures. Roof design for covering of tanks, filter beds, etc., should take into consideration the exposure to high humidity and possible corrosive atmosphere. Concrete members are usually preferred to steel for reasons of permanence and maintenance in such environments. If steel framing is used, purlins are preferable to bar joists for ease in painting and to limit inaccessible pockets. Similarly, precast planks are preferred over formed metal decks.

When precast, prestressed members are used, adequate covering should be provided for prestressing strands, and particular caution should be used in areas exposed to chlorine vapor. Steel connection and bearing plates should be adequately protected against corrosion. When cored slabs are used, the ends should be sealed to prevent the accumulation of moisture.

18.6.2 Control Structures

Sluice gates are used to control flow in large channels or to shut off or divert flow through various parts of the treatment plant. The operation of these gates is generally slow; thus, the final static head in closed position is the maximum lateral design load. As the gate is lowered into position, the final movement will be to wedge the gate tight within the guides at sides and bottom. The gate stand is supported on a bracket or similar device and this bracket is designed to support the full vertical load of the gate and controls. An additional important design criteria, which is sometimes forgotten, is the possibility of upward force on the bracket as the gate is wedged into position. This force can be computed from the mechanical advantage of the screw, the diameter of the control wheel, and the driving force. Motor driven gates should be equipped with automatic shut-off device to prevent this upward force from becoming excessive.

Bar racks and screens are designed for lateral loads equivalent to a completely clogged condition, although whether mechanically or hand cleaned the operating head loss should not exceed a couple of feet. Since the bar rack precedes the screen, and its purpose is to stop heavy items, the screen can be designed for full hydrostatic head. Consideration should be given, however, to the design of the bar rack to absorb energy required to stop heavy items. This requirement can be partially avoided in water treatment plants by careful planning of the intake. In rivers the bar rack and intake can be located perpendicular to the flow, and thus any blows to the rack will not be direct, that is, the rack will not be required to completely stop any item. Then, from knowledge of stream velocity and assuming a large log (at least equal to a railroad tie,) it is possible to compute energy required to redirect such a log through a reasonable angle, such as 45°.

Since waste treatment plants must accept the entire flow of the incoming sewer, bar screen location cannot be employed to reduce impact loadings. Velocities of flow in sewers generally range from 2 to 6 ft per sec. depending on sewer diameter, with the larger velocities designed for larger sewers. Again, assumptions must be made regarding size and weight of impact item.

Wiers as control devices will not have any particular loads as major design parameter; rather the connection detail for surface weirs or the quality control for all such elements subject to higher water velocities will be the major consideration. For submerged wiers, used to create uniform flow in a basin with a minimum of head loss, consideration must be given to construction details so that a dense uniform surface will be obtained throughout the length of the wier.

18.6.3 Pumping Stations

The structural design of the wet well is governed, under operating conditions, by the same sort of hydraulic loads described previously. However, given a considerable length of intake piping, there will be momentum of the intake water such that at shut-off of intake pumps a surge will develop. The height of this surge is one of the major hydraulic design elements of pumping stations. It can be moderated by overflow provisions, but will still result in considerable increased head applied to the wet well. For instance, a plant designed with 3600 ft intake in 60-in. diameter pipe, with water velocity set at 5 ft per sec, the surge without overflow reached 19 ft above static water level. A 36-in. overflow was provided with invert 5 ft above static water level, and the surge was reduced to 10 ft.

Although not strictly a structural design criteria, the layout and arrangement of baffles around the pump intakes is of prime importance for long and efficient plant life. Cavitation, eddy formation, and excessive local velocities can erode the highest quality concrete. In large plants, such as Chicago's Central District Filtration Plant (capacity 150 mgd), model studies have been employed to indicate areas of redesign or the effect of improved contours.

Ventilation of the wet well is a necessary element, primarily because of the potential surge in the intake line. In water treatment plants, this can be accomplished through connection to the dry well and following air vents, and the volume of air to be moved must be considered as part of the surge calculation. On plants with direct lake intakes, surge due to wind and wave action has been known to cause sufficient air pressure in the dry well to break windows. Specifications for waste treatment plants[18-4] prescribe separation of wet well and dry well. In this case, any surge in the intake line and resultant air pressure must be dissipated through air vents in the wet well. However, normal design of waste treatment plants is such that pumping is a result of high flows and surges are not of the same order of magnitude as in long line water treatment plant intakes.

Dry wells and pump houses contain pumps, controls and associated elements, and the design for structural support of these elements follows standard practice of AISC[18-23] and similar applicable codes. Other than providing structural support for the entire pump housing, it is necessary to provide superstructure with clearance and capability for removal of pumps, etc., for maintenance. Crane rail forces will depend on weight of equipment to be moved, and the designer should use his judgement in assessing impact forces depending on the type and capacity of power driven elements. In general, hand operated hoists and travelers need little or no impact, whereas motor-driven equipment should include impact percentages in accordance with applicable codes. Standard practice in design of crane bridges is to use double flanged wheels to provide better alignment of the bridge: in this case half of the lateral impact force is carried on each rail.

18.7 CORROSION PROTECTION

18.7.1 Chlorination

In virtually all plants, the final treatment is sterilization by chlorination. In water supplies, where the desired residual is low, chlorination is accomplished by diffusers in the clear well, and vapor problems in the well are not anticipated. For waste treatment, where the desired residual is 20 to 40 ppm, chlorine contact tanks are used, with detention times ranging from 15 to 30 minutes. It is important that these tanks be either open to the atmosphere, or if closed the hydraulic line should be above the top of the tank so that the top concrete surface contacts water. The possibility of accumulation of chlorine vapor must be given careful consideration in waste treatment plants. Thus if the contact tanks are open, they should not be covered with superstructure.

Chlorine can be obtained in three size containers: 150-lb cylinders, ton containers or tank cars. Rooms for storage of these containers must have ventilation capabilities to preclude accidents from accumulation of chlorine gas. Interlocks should be provided with the doors and lighting such that positive ventilation is provided whenever the room is occupied. Emergency alarms are also recommended. Other than these safety precautions, no special consideration need be given to the structural design of the room, except for ease in handling and weighing the containers.

The design of rooms for handling of other chemicals is handled in similar manner. Consideration must be given to ease of handling, and to safety where required. The conduit or piping for the chemicals is specially designed for that purpose, such as PVC, rubber, or stainless steel, and if properly installed should serve to separate the chemicals from the structure.

18.7.2 Coatings

Concrete in water treatment and domestic sewage plants seldom requires special protection, except for the handling of chemicals such as chlorine and liquid alum. A detailed list of chemicals used in treatment plants, including their effect on concrete, is given in the report of ACI Committee 350,[18-9] and in Portland Cement Association publications.[18-24]

Industrial waste treatment often involves high temperature and low pH, such that special protection is required. Coatings commonly used are special brick or tile, epoxies, and plastic or rubber sheets. Care should be taken to create a clean, sound surface for bonding of the protective coating. Detailed recommendations of the choice and application of proper coating may be obtained from the report of ACI Committee 515,[18-25] or from the Manufacturer.

REFERENCES

18-1 "Water Treatment Plant Design," *Manual of Engineering Practice*, No. 19, American Soc. of Civil Eng., New York, 1940.

18-2 *Recommended Standards for Water Works*, Great Lakes–Upper Mississippi River Board of State Sanitary Engineers, 1968 edition.

18-3 "Sewage Treatment Plant Design," *Manual of Engineering Practice*, No. 36, American Soc. of Civil Eng., New York, 1959.

18-4 *Recommended Standards for Sewage Works*, Great Lakes–Upper Mississippi River Board of State Sanitary Engineers, 1971 edition.

18-5 "Causes, Mechanism, and Control of Cracking in Concrete," *ACI Special Publication SP-20*, American Concrete Inst., Detroit, 1968.

18-6 "Volume Changes of Concrete," *Information Sheet IS 018.02T*, Portland Cement Assn., Skokie, 1967.

18-7 *Manual of Concrete Practice*, Part I, 1970; Parts II and III, 1968, American Concrete Inst., Detroit.
NOTE: This *Manual* contains many papers and reports on concrete mix design, admixtures, aggregates, curing, placement, etc., not listed here separately.

18-8 "Watertight Concrete," *Information Sheet IS002. 02 T*, Portland Cement Assn., Skokie, 1969.

18-9 ACI Committee 350, "Concrete Sanitary Engineering Structures," *ACI Journal, Proc.* **68** (8), 560, Aug. 1971, Also Discussion, *Proc.* **69** (2), 125, Feb. 1972.

18-10 Wallace, G. B., "Joints and Cracks in Concrete Water-holding Structures," Proc. ACI Specialty Conf. on Concrete Construction in Aqueous Environment, Washington, 1962, p. 21.

18-11 Vetter, C. P., "Stresses in Reinforced Concrete Due to Volume Changes," *Transactions, ASCE*, **98**, 1039, 1933.

18-12 *Building Code Requirements for Reinforced Concrete*," ACI 318-71, American Concrete Inst., Detroit, 1971.

18-13 *American Standard Building Code Requirements for Minimum Design Loads in Buildings and Other Structures*, ANSI A 58.1-1955.

18-14 Morris, C. T., and Carpenter, S. T., *Structural Frameworks*, John Wiley and Sons, New York, 1943.

18-15 Davies, J. D., "Influence of Support Conditions on the Behavoir of Long Rectangular Tanks," *ACI Journal, Proc.* **59** (4), 601, April 1962.

18-16 Davies, J. D., "Analysis of Long Rectangular Tanks Resting on Flat Rigid Supports," *ACI Journal, Proc.* **60** (4), 487, April 1963.

18-17 Gogate, A. B., "Structural Design Considerations for Settling Tanks and Similar Structures," *ACI Journal, Proc.* **65** (12), 1017, Dec. 1968.

18-18 Whitman, R. V., and Richart, F. E., Jr., "Design Procedures for Dynamically Loaded Foundations," *Journal, Soil Mechanics Div.* ASCE, **93** (SM 6), 169, Nov. 1967.

18-19 Rothbart, H. A., editor, *Mechanical Design and Systems Handbook*, McGraw-Hill, New York, 1964.

18-20 "Rectangular Concrete Tanks," *Information Sheet IS 003.02D*, Portland Cement Assn., Skokie, 1969.

18-21 Davies, J. D., and Cheung, Y. K., "Bending Moments in Long Walled Tanks," *ACI Journal, Proc.* **64** (10), 685, Oct. 1967.

18-22 "Circular Concrete Tanks Without Prestressing," *Information Sheet IS 072.01D*, Portland Cement Assn., Skokie, 1942.

18-23 "Specification for the Design, Fabrication and Erection of Structural Steel for Buildings," American Institute of Steel Construction, New York, 1969.

18-24 "Effect of Various Substances on Concrete and Protective Treatments, Where Required," *Information Sheet IS 001.03T*, Portland Cement Assn., Skokie, 1968.

18-25 ACI Committee 515, "Guide for the Protection of Concrete Against Chemical Attack by Means of Coatings and Other Corrosion-Resistant Materials," *ACI Journal, Proc.* **63** (12), 1305, Dec. 1966.

18-26 "Underground Concrete Tanks," *Information Sheet IS 071.02D*, Portland Cement Assn., Skokie, 1942.

18-27 "Concrete for Hydraulic Structures," *Information Sheet IS 012.03*, Portland Cement Assn., Skokie, 1969.

18-28 "Concrete for Waste-Water Treatment Works," *Information Sheet PA 063.01W*, Portland Cement Assn., Skokie, 1962.

18-29 "Concrete for Water Treatment Works," *Information Sheet PA 069.01W*, Portland Cement Assn., Skokie, 1963.

18-30 ACI Committee 311, "Manual of Concrete Inspection," *ACI Special Publications SP-2*, 5th Edition, American Concrete Inst., Detroit, 1967.

18-31 "Shotcreting," *ACI Special Publication SP-14*, American Concrete Inst., Detroit, 1968.

18-32 Hurd, M. K., "Formwork for Concrete," *ACI Special Publications SP-4*, 2nd edition, American Concrete Inst., Detroit, 1969.

18-33 ACI Committee 344, "Design and Construction of Circular Prestressed Structures," *ACI Journal, Proc.* **67** (9), 657, Sept. 1970.

18-34 ACI Committee 504, "Guide to Joint Sealants for Concrete Structures," *ACI Journal, Proc.* **67** (7), 489, July 1970.

18-35 "Design and Construction of Sanitary and Storm

Sewers," *Manual of Engineering Practice*, No. 37, American Soc. of Civil Eng., New York, 1969.

18-36 "Concrete Manual," U.S. Bureau of Reclamation, Denver, 1963.

18-37 Allen, E. A., and Higginson, E. C. "Waterstops in Articulated Concrete Construction," *ACI Journal, Proc.* **52** (1), 83, Sept. 1955. Also, Discussion, p. 1149.

18-38 Betz, J. M., "Repair of Corroded Concrete in a Waste-Water Treatment Plant," *Journal, Water Poll. Cont. Fed.*, 332, March 1964.

18-39 Biczok, I., *Concrete Corrosion and Concrete Protection*, Chemical Publishing Co., New York, 1967.

18-40 Davies, J. D., and Long, J. E., "Behavior of Square Tanks on Elastic Foundations," *Journal, Engineering Mechanics Div.*, ASCE **94** (EM 3), 753, June 1968.

18-41 Hetenyi, M., *Beams on Elastic Foundation*, Univ. of Michigan Press, Ann Arbor, 1946.

18-42 Norris, C. H., et al, *Structural Design for Dynamic Loads*, McGraw-Hill, New York, 1959.

18-43 Richart, F. E., Jr.; Woods, R. D.; and Hall, J. R., Jr., *Vibrations of Soils and Foundations*, Prentice-Hall, Englewood Cliffs, N.J., 1970.

18-44 Timoshenko, S., and Woinowski-Krieger, S., *Plates and Shells*, McGraw-Hill, New York, 1959.

19

Concrete Pipe

JOHN G. HENDRICKSON, JR.*

19.1 INTRODUCTION

19.1.1

Present concrete pipe is made in sizes ranging from 4 in. inside diameter to over 16 ft. Pipes are used to carry fluids in gravity flow such as in highway culverts, storm drains, and sanitary sewers. They are used for pressure lines such as sewer force mains, irrigation lines, and water supply lines.

Gravity flow pipe of smaller sizes are either plain or reinforced, but all larger sizes (greater than 24 in. i.d.) are reinforced against crushing. Some pipe are made with telescoping ends designed for mortar or mastic as a jointing material. However, to satisfy demands for leakproof but flexible joints, close tolerance pipe ends designed for use with rubber gaskets are commonly produced.

Nonreinforced pressure pipe is made largely for carrying irrigation water at low pressures. Concrete pressure pipe for water lines is either reinforced or prestressed. Some designs incorporate a light-gauge steel cylinder in the wall of the pipe as a waterproof diaphragm. For high internal pressure, pipe ends are formed with steel end rings. When these are telescoped together, a round rubber gasket is compressed to form a flexible, watertight seal.

*Principal Specifications Engineer, Greeley and Hansen Engineers, Chicago, Illinois.

19.1.2

Most concrete pipe are made to *ASTM Specifications* except pressure water pipe which is usually made in accordance with specifications of the American Water Works Association. Many agencies such as state highway departments and others have their own specifications which are closely patterned after *ASTM* or *AWWA Specifications*.

Commonly used sizes of concrete pipe are made on ingenious machines which mechanically compact concrete of a low water-cement ratio. Large sizes and more elaborate designs are made by a casting process utilizing external vibration. Curing with moist steam is more commonly used but water curing and curing compounds are used, sometimes in conjunction with steam curing.

19.1.3

The commonly used acceptance tests for concrete pipe designed to withstand backfill loading are the absorption test and the three-edge bearing test, as set forth in *ASTM Specification C497* "Standard Methods of Test for Determining Physical Properties of Concrete Pipe or Tile." The absorption test is intended to check the density and imperviousness of the concrete while the three-edge-bearing test verifies the structural strength of the pipe.

In the three-edge-bearing test the test pipe is placed horizontally on two closely spaced wooden strips. The test load is applied to the top of the pipe through a rigid longitudinal bearing beam.

Test load requirements and absorption limits are listed in the particular pipe specification. For nonreinforced pipe the minimum strength is specified which the pipe must withstand before failure. For reinforced pipe the load is specified at which a longitudinal crack 0.01 in. wide and continuous over a one foot length will occur; also ultimate loads are specified. Some specifications require only tests to the 0.01-in. crack.

Pipe which carries both internal pressure and the weight of the backfill may be required to pass a three-edge-bearing test and a hydrostatic test.

19.2 NONREINFORCED CONCRETE PIPE

19.2.1

Concrete pipe for use in culverts, sanitary and storm sewers is made in conformance with *ASTM Specification C14* "Standard Specifications for Concrete Sewer, Storm Drain and Culvert Pipe." Standard sizes of the inside diameter range from 4 in. thru 24 in. Lengths are usually 5 or 6 ft. However, the largest sizes may be supplied in longer lengths by some manufacturers.

Three strength classes are available under the current specification. The pipe may be supplied for use with mortar or mastic jointing material. If pipe is used with rubber gaskets the pipe ends must also conform to the tolerances of *ASTM Specification C443.*

19.2.2 Irrigation and Drainage Pipe

Concrete pipe for irrigation lines where mortar joints are to be used is made in conformance with *ASTM Specification C118* "Standard Specifications for Concrete Pipe for Irrigation and Drainage." The same pipe installed with open joints is used for farm or highway drainage.

Standard concrete irrigation pipe under the specifications is produced in sizes from 4 through 24 in. internal diameter and is used to carry irrigation water under pressure. Minimum wall thicknesses, three-edge bearing test strengths and hydrostatic test pressures for each size are set forth in the specification.

Standard and Heavy Duty Drainage pipe produced under the specifications are also made in sizes from 4 thru 24 in. Wall thickness and three-edge-bearing test strengths are set forth in the specification for each class. This pipe is laid with open joints to provide underground drainage. The heavy duty pipe is stronger and can be laid at greater depths.

19.2.3 Nonreinforced Concrete Pipe

Nonreinforced concrete pipe to be used with rubber gasket joints to carry irrigation water at low pressures is made in conformance with *ASTM C505* "Standard Specification for Nonreinforced Concrete Irrigation Pipe with Rubber Gasket Joints." Inside diameters are from 6 through 24 in. The specified working pressure for this pipe is 30 feet of head, but the required hydrostatic test pressure is 40 psi. Three-edge bearing test loads are also specified.

19.2.4 Concrete Pipe for Subsurface Drainage

Three types of concrete pipe are made for substrate drainage: concrete drain tile, perforated concrete pipe, and porous concrete pipe.

-1. Concrete drain tile is made in conformance with *ASTM Specification C412*, "Standard Specifications for Concrete Drain Tile." This pipe is widely used for agricultural drainage and also for foundation drains and highway underdrains. The pipe is available in three classes; Standard Quality, Extra Quality, and Special Quality. Sizes range from 4 through 24 in. Butt end joints are commonly used.

-2. Perforated concrete pipe is made in conformance with *ASTM Specification C444*, "Standard Specification for Perforated Concrete Pipe." Sizes are from 4 thru 27 inches or larger in inside diameter. Type I has circular perforations while Type II has slotted perforations. The size and spacing of the perforations is set forth in the specification for each size of pipe

-3. Porous concrete pipe is made in conformance with *ASTM Specification C654* "Standard Specification for Porous Concrete Pipe." The pipe is usually made with tongue-and-groove ends. The barrel section is made with a special concrete mix with the aggregate so graded as to produce a porous or honeycombed texture through which ground water can pass into the pipe interior. The user is warned that the pipe may be subject to detrimental leaching where soft or acid water occurs.

19.3 REINFORCED CONCRETE PIPE—NONPRESSURE

19.3.1

Reinforced concrete pipe for the construction of culverts, storm drains, and sanitary sewers is made in conformance with *ASTM Specification C76* "Standard Specification for Reinforced Concrete Culvert, Storm Drain, and Sewer Pipe."

The size range covered by the specification is 12 in. through 108 in. However, special designs for larger sizes are accepted under this specification and pipe 204 in. in diameter have been made.

Five strength classes, Class I through V, are provided for. The 0.01-in. and ultimate strength requirements are set forth by D-Load specified for each class. D-Load is the load per foot of pipe diameter per linear foot for a pipe tested in three-edge bearing. The specified D-Loads are:

	0.01-inch	USD
Class I	800 D	1200 D
II	1000 D	1500 D
III	1350 D	2000 D
IV	2000 D	3000 D
V	3000 D	3750 D

Three wall thicknesses are available under the specification; Wall A is the thin wall design; Wall B, the most widely used; and Wall C, the heavy wall design. Standard designs are not available in all sizes, strengths, and wall thicknesses. Wall A designs for Class V strengths are not shown in the tables of standard designs since a special design is required for thin wall pipe in the high strength range.

Small diameter pipe up to about 30 to 33 in. are reinforced with a single circular cage or with a cage made slightly elliptical so as to better reinforce for the tensile stresses developed under load. Larger sizes are reinforced with an inner and outer circular cage or a single elliptical cage. Pipe 102 and 108 in. in diameter, if reinforced with an elliptical cage, also have an inside circular cage. The circular cage facilitates handling pipe of this size and type.

Reinforced concrete pipe is also supplied in accordance with *ASTM C655*, "Standard Specification for Reinforced Concrete D-Load Culvert, Storm Drain, and Sewer Pipe." Under this specification the pipe manufacturer supplies the design of the pipe for the acceptance of the purchaser. Acceptance is based on tests or empirical evaluations.

19.3.2

Reinforced concrete arch pipe is made for use in culverts, sanitary sewers, and storm sewers where vertical head room is limited. This pipe is made in conformance with *ASTM Specification C506*.

Standard sizes range from the hydraulic equivalent of a 15-in. circular pipe to that of a 108-in. circular pipe. The required dimensions and proportions of the arch pipe sections are set forth in the specification.

Three strength classes are specified. As for C76 pipe, these are specified by D-load. For arch pipe the D-load is the load per foot of nominal inside span per linear foot in three-edge bearing. The three standard strength classes for this type of pipe are:

	0.01 inch	USD
Class II	1000 D	1500 D
III	1350 D	2000 D
IV	2000 D	3000 D

19.3.3

Reinforced concrete elliptical pipe is also made for use in sewers and culverts. When placed with the long axis horizontal, it is called horizontal elliptical pipe or HE-pipe. Like arch pipe it is used where vertical headroom is limited.

The pipe may also be placed with the long axis vertical and is correspondingly called vertical elliptical or VE-pipe. Its use is in locations with limited lateral space or where very high backfill loading occurs.

-1. Horizontal elliptical pipe—The pipe is made in conformance with *ASTM Specification C507*. Standard sizes range from the hydraulic equivalent of an 18-in. circular pipe through 108-in. equivalent circular.

Minimum wall thicknesses and steel areas are specified for five strength classes as follows:

	0.01 inch	USD
Class HE-A	600 D	900 D
HE-I	800 D	1200 D
HE-II	1000 D	1500 D
HE-III	1350 D	2000 D
HE-IV	2000 D	3000 D

As with arch pipe, the D-load applies to the nominal inside horizontal diameter in feet.

-2. Vertical elliptical pipe—The pipe is also made in conformance with *ASTM Specification C507*. The wall thickness and shape of vertical elliptical pipe are the same as those for horizontal elliptical pipe. However, since load requirements are quite different, the steel areas and concrete strengths are set forth in different tables.

Standard sizes for this pipe are from 36-in. equivalent round to 108-in. equivalent round.

Strength requirements are given for five classes as follows:

	0.01 inch	USD
Class VE-II	1000	1500
VE-III	1350	2000
VE-IV	2000	3000
VE-V	3000	3750
VE-VI	4000	5000

19.4 REINFORCED CONCRETE PIPE FOR LOW INTERNAL PRESSURES

19.4.1

This type of reinforced concrete pipe is intended for use with low internal hydrostatic pressure and is made in conformance with *ASTM Specification C361* "Reinforced Concrete Low-Head Pressure Pipe." The maximum hydrostatic pressure usually does not exceed 125 ft of head. The use of this type of pipe is largely for irrigation and sewer force mains.

The specification lists standard designs for hydrostatic heads of 25, 50, 75, 100, and 125 ft. in combination with earth fill loading over the top of the pipe for 5, 10, 15, and 20 ft. Specific installation conditions for which these designs apply are given in the appendix of the specification.

Sizes covered in this specification are 12 through 72 in. in internal diameter.

Joints for this type of pipe utilize a solid rubber gasket to seal the joint against leakage and still provide some flexibility. The ends of the pipe may be all concrete. A characteristic of the pressure type of joint is that the gasket is confined in a groove in the spigot end of the pipe when the joint is closed (Fig. 19-1). The gasket may also be confined between shoulders cast in the bell and spigot (Fig. 19-2). Some joints utilize a concrete collar with each end of the pipe made as spigot.

19.4.2

Reinforced concrete water pipe for use with low internal pressures is made in conformance with AWWA Specification C302 "American Water Works Association Standard for Reinforced Concrete Water Pipe—Noncylinder Type, Not Prestressed." This type of pipe is largely used for low-head transmission lines, reservoir connections, and lines not

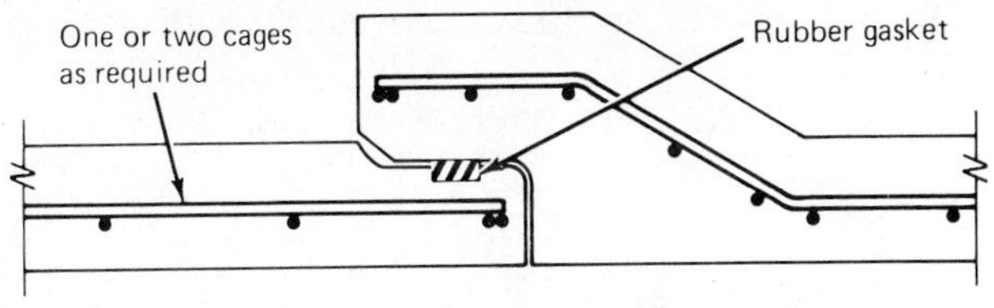

Fig. 19-1

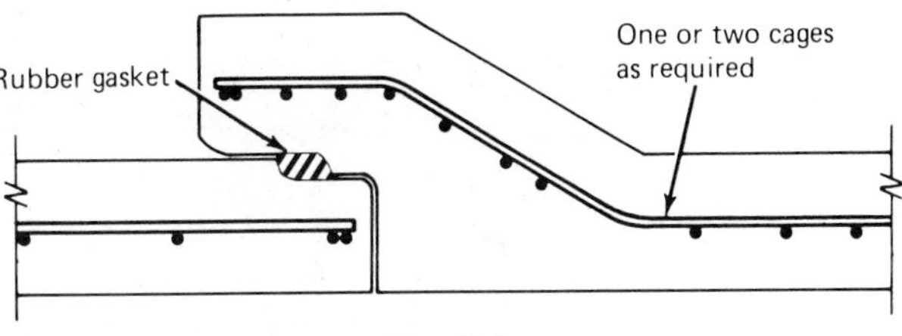

Fig. 19-2

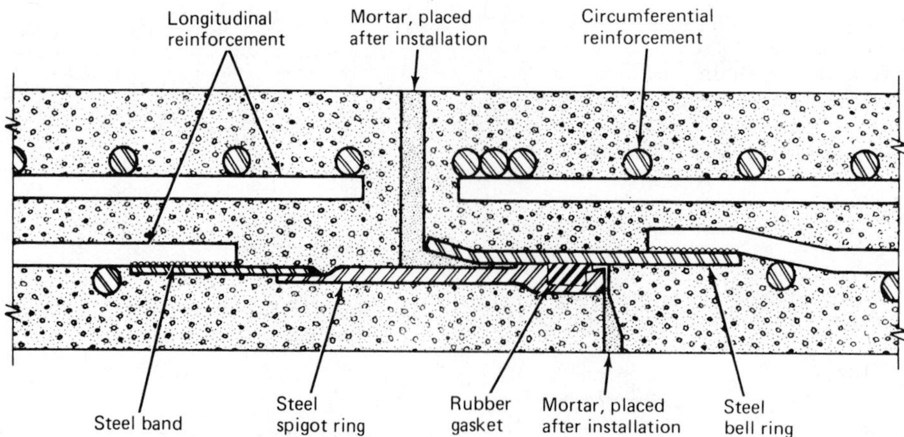

Fig. 19-3 Reinforced concrete pressure pipe with rubber and steel joint.

subject to possible higher pressures or shock. Maximum design pressure for this type of pipe is 45 psi.

The sizes covered by this specification are 12 through 96 in. in internal diameter. Minimum lengths are usually 8 ft. Maximum lengths are as follows:

Size, in.	Max Length, ft
12-15 (inclusive)	$8^1/_2$
16-21 (inclusive)	12
24 and larger	16

Joints for this type of pipe commonly employ a steel bell ring and a steel spigot ring with a groove in which a rubber gasket is confined as shown in Fig. 19-3. However, all-concrete joints are also used.

Pipe designs are based on allowable stresses set forth in the specification. However, minimum designs are specified.

19.5 REINFORCED CONCRETE PIPE FOR MODERATE AND HIGH INTERNAL PRESSURES

19.5.1

Concrete water pipe for design pressures varying from a minimum of 40 psi to 260 psi is manufactured according to *AWWA Specification C300*, "Reinforced Concrete Water Pipe–Steel Cylinder Type, not Prestressed." Sizes covered by the standard are 20 through 90 in. inclusive. Larger sizes have been made using special designs.

This pipe consists of a thin steel cylinder with steel joint rings welded to its ends. The cylinder is surrounded with one or more cages of reinforcing made from bars, wire or welded wire fabric. This assembly is encased in a wall of dense concrete covering the cylinder both inside and out. The steel joint rings form a self-centering joint sealed with a round rubber gasket as shown in Fig. 19-4.

Manufacturers of this type of pipe supply design data on the allowable backfill loading combined with internal pressure. Such information has been developed largely by tests.

19.5.2 Prestressed Concrete Water Pipe

Prestressed concrete water pipe is manufactured according to *AWWA Specification C301* "Prestressed Concrete Pressure Pipe, Steel Cylinder Type for Water and Other Liquids." Standard sizes covered by this *Specification* are from 16 through 96 in., but prestressed pipe as large as 201 in. in internal diameter has been built. Design pressures depend on the design of the pipe but are in the range of 200 to 275 psi. Standard lengths are 16 or 20 ft.

Essential features of this pipe include a light gauge steel cylinder with steel joint rings welded to each end. This unit is either lined with concrete or embedded in concrete to form a concrete core. High-tensile strength wire is wound around the outside of the core at a predetermined stress. A dense coating of cement mortar or concrete is placed on the exterior of the core to protect the prestress wires.

Cores formed by placing the concrete entirely within the steel cylinder are used in sizes up to and including 20 in. and may be used in sizes up to and including 48 in. Cores

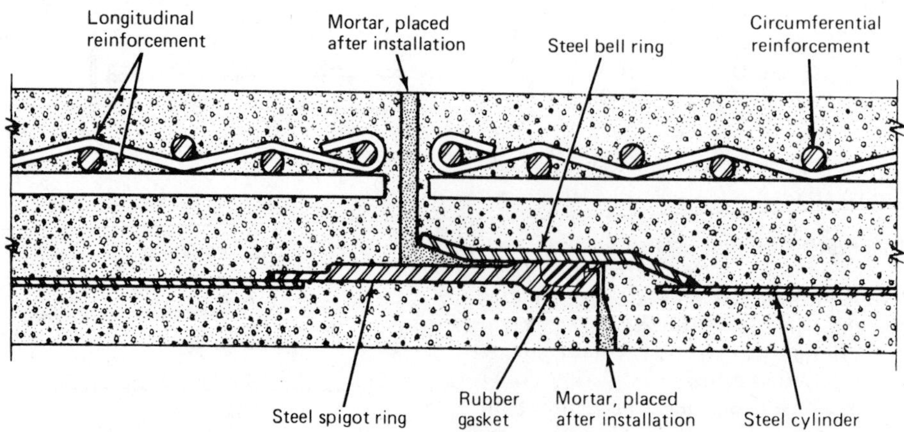

Fig. 19-4 Reinforced concrete cylinder pipe with rubber and steel joint.

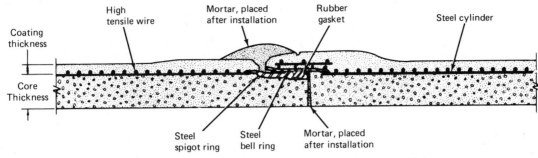

Fig. 19-5 Prestressed concrete cylinder pipe with rubber and steel joint.

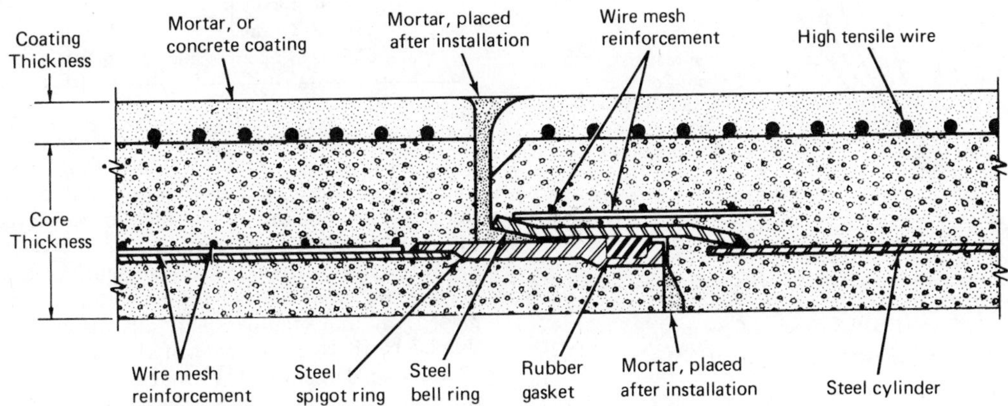

Fig. 19-6 Prestressed concrete embedded cylinder pipe with rubber and steel joint.

with the steel cylinder embedded in concrete may be used in pipe sizes 24 through 48 in. and must be used in larger sizes of pipe.

The steel joint rings form a self-centering joint sealed with a rubber gasket as shown in Figs. 19-5 and 19-6. This is the general type of joint used with all concrete pressure pipe.

19.5.3 Pretensioned Concrete Cylinder Pipe

This type of pipe is made according to *AWWA Specification C303* "Reinforced Concrete Water Pipe—Steel Cylinder Type, Pretensioned." Standard sizes are from 10 to 42 in., inclusive. Larger sizes can be furnished by agreement with the purchaser. Maximum design pressures are 400 psi. Lengths range from 24 to 40 ft. However, 21-in. and smaller pipe cannot exceed 32 ft in length.

The pipe is made with a welded thin-steel cylinder with joint rings welded to each end. This assembly is lined with a layer of cement mortar centrifugally spun into place. Reinforcement consisting of steel rod is wound on the outside of the cylinder at a predetermined stress of around 11,000 psi. The rod is welded to the steel joint ring at each end of the cylinder. A dense mortar coating is placed over the cylinder and rods.

The joint is formed by steel joint rings similar to that for other types of concrete pressure pipe as shown in Fig. 19-5.

19.6 LOADS ON PIPE IN TRENCHES

Most sewers, waterlines and irrigation lines are constructed by excavating a trench in undisturbed soil. The pipe is placed on the bottom of the trench and covered with earth backfill to the original ground surface. See Fig. 19-7.

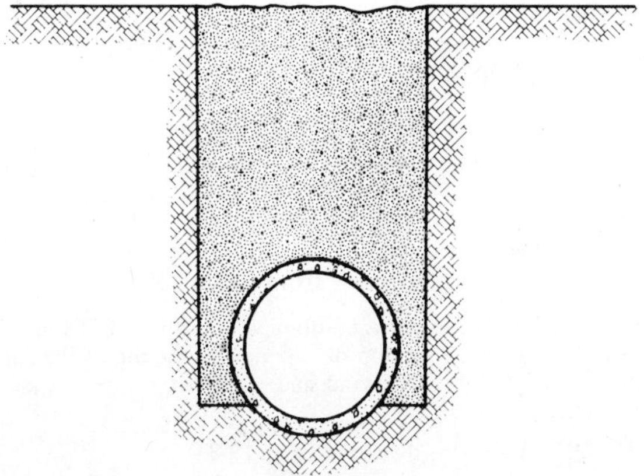

Fig. 19-7

19.6.1

In calculating backfill loads on rigid pipe it is assumed that the load develops as the backfill settles since it is not compacted to the density of the original ground.

The load on the pipe equals the weight of the backfill material above the top of the pipe minus the shearing or friction forces developed along the sides of the trench as the backfill settles. These shearing forces are calculated in accordance with Rankine's Theory. Cohesion between the backfill and the sides of the trench is assumed negligible.

The backfill along the sides of the pipe is assumed to carry none of the vertical load.

The load is calculated according to the Marston Formula

$$W_d = C_d \, w \, B_d^2$$

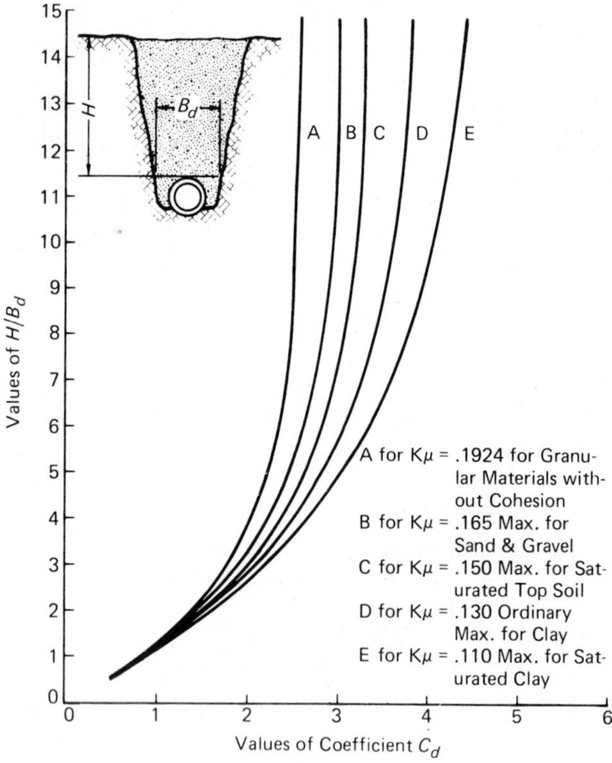

Fig. 19-8 Trench installations.

where

W_d = backfill load in lbs per linear ft of conduit
C_d = load coefficient
w = unit weight of the backfill in lbs per cu ft
B_d = width of the trench in feet at the elevation of the top of the pipe.

The coefficient C_d depends on the coefficient of internal friction of the fill material, μ; the coefficient of sliding friction along the sides of the trench, μ'; and the ratio of active lateral pressure to vertical pressure, K; as calculated from Rankine's formula. Values of C_d for various values of $K\mu$ or $K\mu'$ can be determined from Fig. 19-8.

EXAMPLE 19-1: Consider a 60-in. pipe in a trench 8 ft wide, with a depth of 30 ft from the top of the pipe to the top of the natural ground. Backfill material is sand and gravel with a density of 120 lbs per cu ft.

SOLUTION: Using the diagram of Fig. 19-8

$$\frac{H}{B_d} = \frac{30}{8} = 3.75$$

$$C_d = 2.15$$

$$W_d = C_d\, W\, B_d^2$$

$$= 2.15 \times 120 \times 8^2$$

$$= 16{,}000 \text{ lbs per ft}$$

19.6.2

The load carrying capacity of a rigid pipe depends on the distribution of the bedding on the underside of the pipe and the development of active lateral pressure at the sides of the pipe. For rigid pipe in trenches no active lateral pressure is assumed.

The effectiveness of the bedding is given by a load factor. This is applied to the three edge bearing strength of the pipe to calculate its field load supporting strength. For reinforced pipe, the three-edge-bearing strength of the pipe at the 0.01-in. crack is used to calculate its supporting strength. For nonreinforced pipe the ultimate three-edge-bearing strength with a factor of safety of from 1.25 to 1.5 is used.

The four commonly used types of bedding are shown in Table 19-1 with the load factors determined experimentally at Iowa State College.*

TABLE 19-1: Bedding Classes and Experimentally Determined Load Factors

Class of Bedding	*Load Factor*
Class A–Concrete Cradle	2.25-3.4
Class B–First Class	1.9
Class C–Ordinary	1.5
Class D–Impermissible	1.1

Bedding methods for pipes in trenches as determined at Iowa State College are illustrated in Fig. 19-9.

Class A–Concrete Cradle Bedding–requires embedment of the lower part of the pipe in plain or reinforced concrete of suitable thickness and extending up the sides of the pipe a distance not less than 25% of the depth of the pipe.

Class B–First Class Bedding–requires the pipe to be placed on fine granular materials on an earth foundation shaped to fit the underside of the pipe for at least 60% of its outside diameter. The trench is then filled to 1 ft over the top of the pipe with granular materials placed in 6-in. layers and tamped to fill all the space around the pipe.

Class C–Ordinary Bedding–requires placing the pipe on an earth bedding shaped to fit the exterior of the pipe for 50% of its external diameter. The trench is then filled to 6 in. over the top of the pipe with granular material placed and shovel tamped to fill all spaced around the pipe.

Class D–Impermissible Bedding–allows the pipe to be placed on the bottom of the trench with no effort to shape the trench to fit the pipe. Fill is placed around the pipe.

Alternate bedding procedures are illustrated in Fig. 19-10. In addition, excellent bedding has been obtained by filling under and around the pipe with soil cement grout. On other installations lean concrete has been used to insure uniform and solid support.

EXAMPLE 19-2: From the previous example the load on a 60-in. pipe was 16,000 lbs per ft. Installing the pipe with Class B bedding, the required 0.01-in. strength would be:

$$\frac{16000}{1.9} = 8422 \text{ lbs per ft}$$

$$\text{D-load} = \frac{8422}{5} = 1684 \text{ lbs}$$

A Class IV pipe is required.

19.6.3

In accordance with the preceding analysis the backfill load on a pipe will increase as the trench width increases. However, there is a limiting width beyond which there is no further increase in load. This width depends on the pipe diameter, the depth of the trench, and possible settlement of the pipe into the bedding at the bottom of the trench as discussed for embankment conduits in section 19.7.1.

*"The Supporting Strength of Rigid Pipe Culverts" by M. G. Spangler, Iowa State College *Engineering Experiment Station Bulletin 112*, 1933.

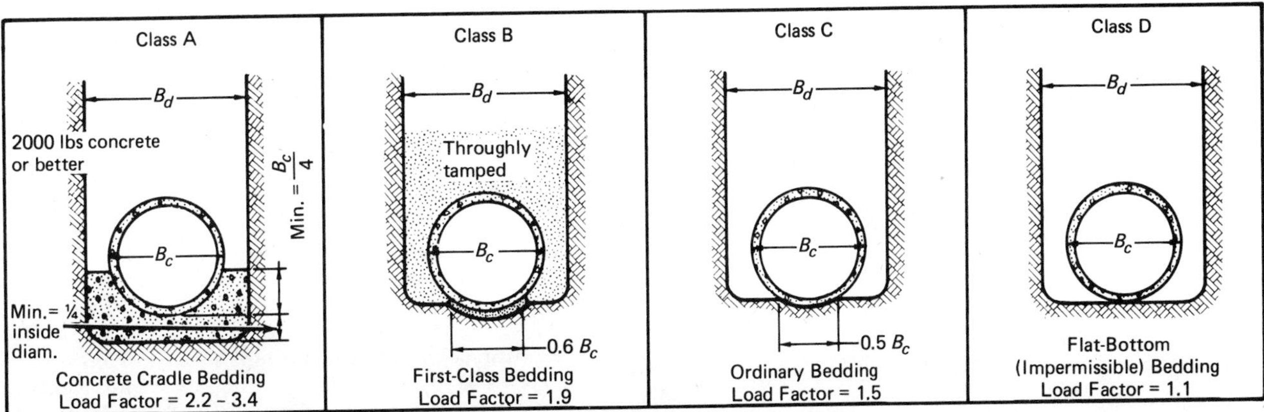

Fig. 19-9 Bedding method for trench conduits.

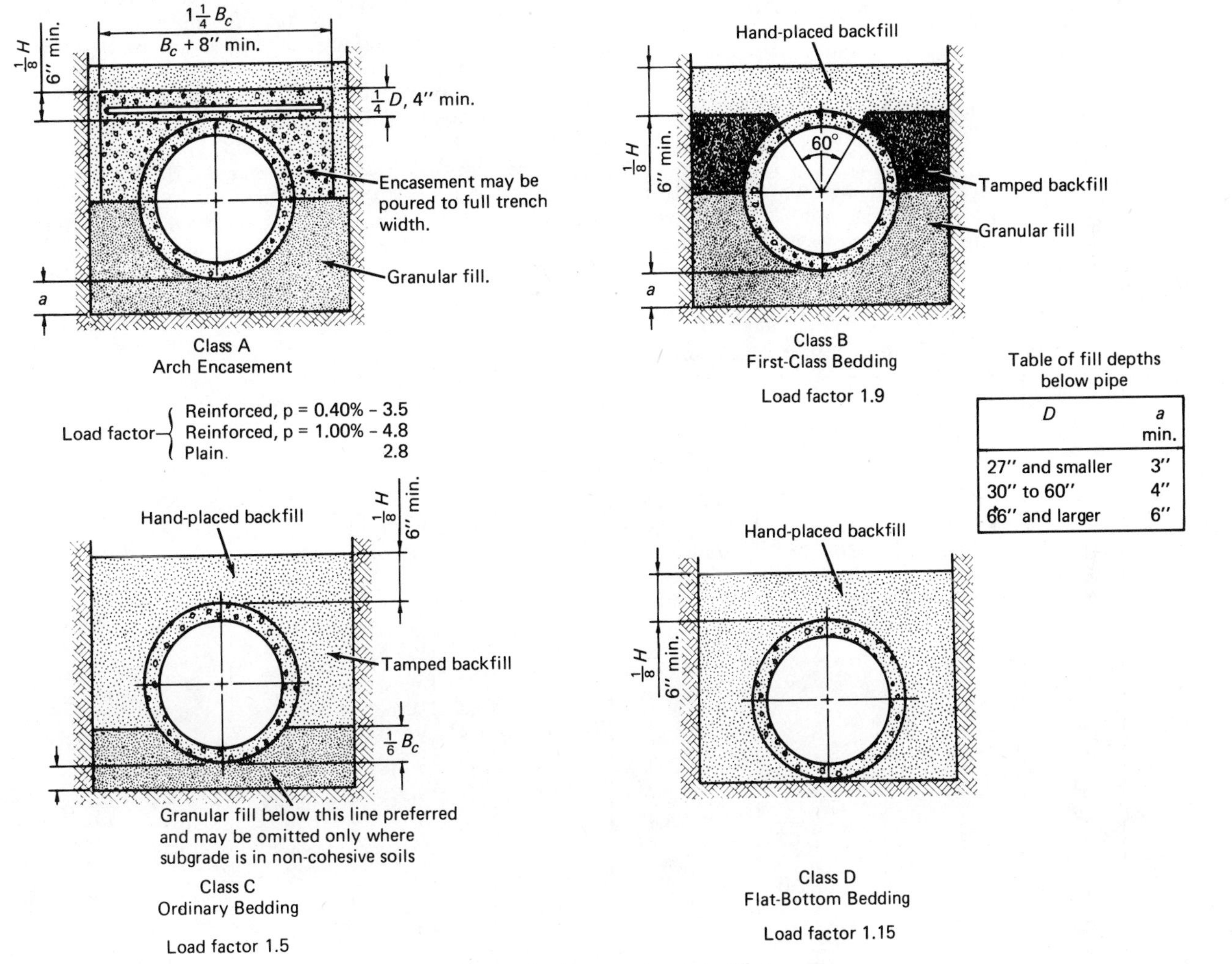

Table of fill depths below pipe

D	a min.
27" and smaller	3"
30" to 60"	4"
66" and larger	6"

Fig. 19-10 Alternate trench bedding.

Figure 19-11 is used to determine the limiting trench width.

EXAMPLE 19-3: In the preceeding example, H/B_c for the 60-in. pipe was 3.75. Assuming Class B bedding in granular material, a reasonable value of $r_{sd}p$ is 0.50. From Fig. 19-11, these values give a B_d/B_c of about 2.0. Since the ratio in the example of B_d/B_c is 8.0/6.0 = 1.33 and less than 2.0, the transition-width ratio; the installation is correctly considered a trench type installation.

The ratio of B_d/B_c greater than the transition-width ratio would indicate that the load calculation should have the followed the procedure for embankment installations.

19.6.5

Concrete pressure pipe is designed to carry simultaneously the external backfill load and the internal pressure. Since pipelines carrying liquid under pressure do not have to maintain a uniform grade line, it is customary to install

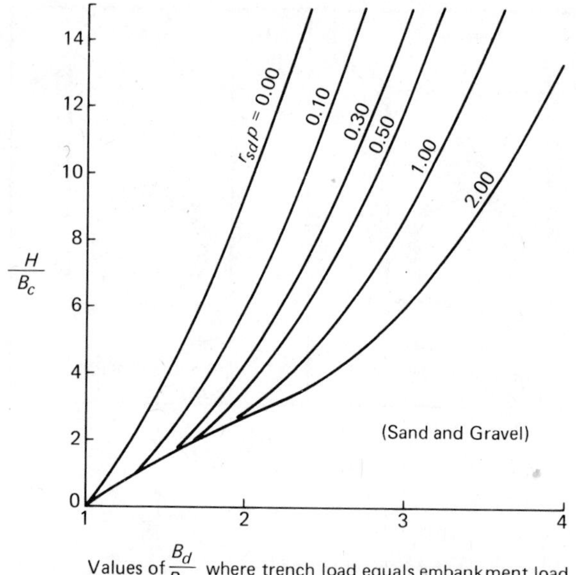

Fig. 19-11 Transition-width ratio curves.

them in relatively shallow trenches where the backfill is in the range of 5 to 6 ft. In such cases the internal pressure will be the major factor in the design. However, some installations may carry 15 to 20 ft of cover and must be designed accordingly.

-1. Reinforced concrete pressure pipe made in conformance with ASTM Specification C361 is designed for relatively low pressure (about 50 psi or less) and backfill cover varying from 5 to 20 ft.

Tables in the *Specification* give the required steel areas for pipe sizes varying from 12 through 108 in., with backfill in 5-ft intervals from 5 to 20 ft, and for pressure heads varying in 25-ft intervals from 25 ft to 125 ft. These standard designs are provided for both circular and elliptical reinforcing cages and assume about a Class B bedding.

Two bedding procedures which are satisfactory for these designs are described in the Appendix of ASTM C361.

-2. Reinforced concrete water pipe made in conformance with AWWA C302 must satisfy minimum requirements set forth in the following table taken from the specification.

The minimum longitudinal steel for each pipe shall be six bars and shall provide an area equal to six 0.25-in. round bars for pipe 8 ft long, six 0.375-in. round bars for 12 ft long, and six 0.5-in. bars for pipe 16 ft long. The minimum longitudinal area shall also not be less than the equivalent of 0.5-in. round bars spaced 42 in. on centers. In pipe with two reinforcing cages, the longitudinals are to be divided equally between the two cages.

The area of circumferential steel in the reinforcement cage or cages is based on a stress of 12,500 psi when the pipe is subjected to the design pressure specified by the purchaser with no allowance for tension in the concrete.

When external load is the determining factor in the design of the pipe the design is based on the combined internal pressure and external load designated by the purchaser.

TABLE 19-2 Requirements for Pipe of Various Sizes*

	6,000-psi Centrifugal Concrete		*4,500-psi Poured Concrete*		*Circumferential Reinforcement Spacing*		
Pipe ID (in.)	*Min. Pipe Wall Thickness* (in.)	*Min. No. of Cages*	*Nominal Pipe Wall Thickness* (in.)	*Min. No. of Cages*	*Min.* (in.)	*Max.* (in.)	*Min. Total Steel Area per Lineal Foot* (sq in.)
12	2	1			1¼	4	0.08
15	2	1			1¼	4	0.11
16	2⅛	1			1¼	4	0.12
18	2¼	1			1¼	4	0.14
20	2⅜	1			1¼	4	0.16
21	2⅜	1			1¼	4	0.17
24	2½	1	3	1	1¼	4	0.20
27	2⅝	1	3¼	1	1¼	4	0.23
30	2¾	1	3½	1	1¼	4	0.25
33	2⅞	1	3¾	2	1¼	4	0.28
36	3	1	4	2	1¼	4	0.30
42	3½	2	4½	2	1¾	5	0.35
48	4	2	5	2	1¾	5	0.40
54	4½	2	5½	2	1¾	5	0.45
60	5	2	6	2	1¾	5	0.50
66	5½	2	6½	2	2¼	6	0.61
72	6	2	7	2	2¼	6	0.71
78	6½	2	7½	2	2¼	6	0.81
84	7	2	8	2	2¼	6	0.90
90	7½	2	8	2	2¼	6	1.00
96	8	2	8½	2	2¼	6	1.09

**For pipe larger than 96 in. in diameter, dimensions and details of design shall be subject to approval by the purchaser.*

TABLE 19-3 Requirements for Pipe of Various Sizes*

Pipe ID (in.)	*Min. Thickness* Pipe Wall (in.)	Concrete Lining (in.)	*Circumferential Reinforcement Spacing* Min. (in.)	Max (in.)
20	3¼	1	1¼	4
24	3½	1	1¼	4
30	3½	1	1¼	4
36	4	1	1¼	4
42	4½	1	1¾	5
48	5	1¼	1¾	5
54	5½	1¼	1¾	5
60	6	1¼	1¾	5
66	6½	1½	2	6
72	7	1½	2	6
78	7½	1½	2	6
84	8	1½	2	6
90	8	1¾	2¼	6
96	8½	1¾	2¼	6

****For pipe larger than 96 in. in diameter, dimensions and details of design shall be subject to approval by the purchaser.***

Tensile stress in the tension steel is not to be greater than 22,000 psi and compressive stresses in the concrete are not to be greater than 45% of the 28-day compressive strength. The minimum circumferential steel cannot be less than given in Table 19-2 taken from AWWA C302.

-3. Reinforced concrete pressure pipe made in conformance with *AWWA Specification C300* combines a light-gage steel cylinder with one or more reinforcing cages. Certain minimum requirements are given in the Table 19-3.

For pipe designs based on internal pressure, the steel stresses are not to exceed 12,500 psi for the design pressure specified by the purchaser. The area of reinforcing bars is not to be less than 40% of the combined area of the cylinder and reinforcing cages.

When the design of the pipe must be based on combined internal pressure and external load, the design is developed by the manufacturer subject to approval by the purchaser.

-4. Prestressed concrete cylinder pipe can be designed for internal pressures as high as 200 psi, or higher, in combination with backfill loads. Standard design procedures provide for water hammer equal to at least 40% of the design pressure and for live load, including impact, equal to AASHTO H-20 loading.

According to AWWA C301, the wrapping stress in the high tensile wire is not to exceed 75% of the ultimate strength of the wire. The initial compression in the concrete at the time of wrapping is not to exceed 55% of the compressive strength of the concrete at the time of wrapping.

The design procedure for prestressed concrete cylinder pipe is described in detail in the paper, "Standardization of Design Procedures for Prestressed Concrete Pipe," by R. E. Bald, *AWWA Journal*, November 1960.

Tests have shown that for prestressed concrete cylinder pipe a cubic parabola best shows the relationship between internal pressure combined with external load. The design curve is defined by the equation

$$w = \frac{W_o}{\sqrt[3]{P_o}} \left[\sqrt[3]{P_o - p} \right]$$

where

w = the three-edge bearing load equivalent of the design backfill loading

p = maximum design pressure occurring in combination with w but not greater than $0.8\ P_o$.

W_o = nine-tenths of the three-edge bearing load producing incipient cracking in the core with no internal pressure

P_o = the internal pressure required to overcome all compression in the core concrete with no external load

The wire area, tensile stress, wire spacing, and concrete stress in the core can be varied to produce a desired P_o. W_o must be determined from the manufacturers test data. A transient load capacity beyond the design curve is available to carry water hammer and live load impact. These details are discussed in Appendix A of AWWA C301.

Appendix B of AWWA C301 also provides a Stress Analysis Method for combined load design.

-5. The design procedure for pretensioned, concrete, cylinder pipe is discussed in detail in Appendix A of AWWA C303. For combined loads the design of the pipe must consider its structural characteristics under combined loading together with the bedding and compaction of material around the pipe to prevent injurious deflection. The maximum deflection under ground after installation should not exceed $D^2/4000$ in. Based on the loads on the pipe and estimates of the effectiveness of the bedding and backfilling the required wall stiffness can be calculated. This can be compared with the calculated wall stiffness for the wall section chosen.

19.7 LOADS ON PIPE UNDER EMBANKMENTS

19.7.1

The typical case of a load on a pipe under an embankment is the highway culvert where the pipe is placed on the natural

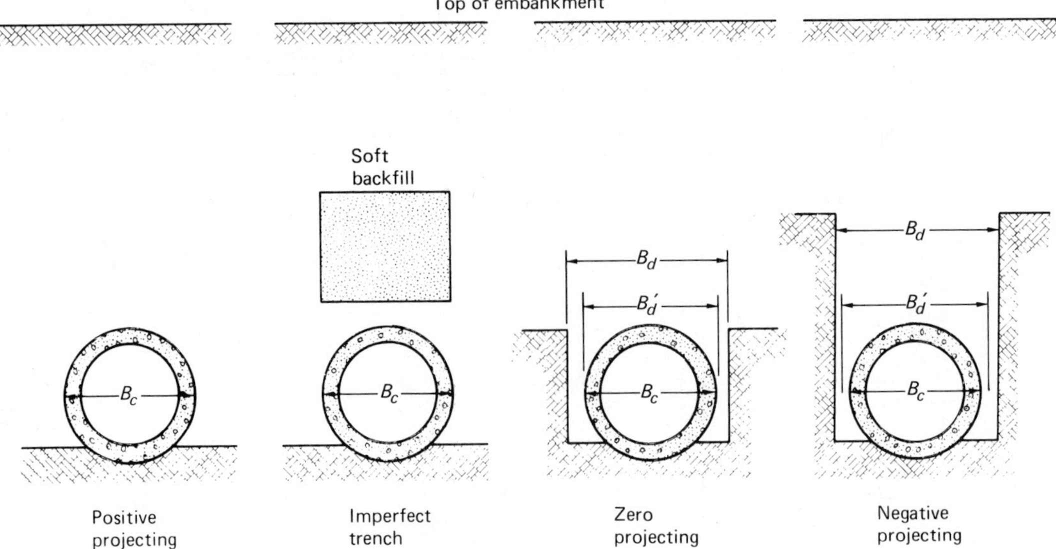

Fig. 19-12 Pipe culvert installation methods.

ground and the fill built up over it. Pipe culverts may be constructed by any of four different installation methods (see Fig. 19-12). The load will vary with each method. Therefore, the method of installation must be selected and the pipe designed accordingly.

-1. Positive projecting conduits—By this method of installation the pipe culvert is installed on the surface of the natural ground with the top of the pipe above the adjacent surfaces of the ground. The embankment is then built up over the culvert to the predetermined elevation.

-2. Imperfect trench conduit—In this type of installation the pipe is installed and the fill built up over the pipe as discussed for positive projecting conduits. However, the fill is only built to an elevation of one-half, one, or two outside pipe diameters plus 1 ft above the top of the pipe.

A trench equal in width to the outside pipe diameter is then dug with vertical sides to within 1 ft of the top of the pipe. Immediately after excavation the trench is refilled with loose compressible material. The remainder of the embankment is constructed in the normal method.

This method of installation reduces the backfill loading on the pipe by inducing arching of the embankment over the pipe. It is economically feasible only with high fills which would require special pipe designs with most other methods of installation.

-3. Zero projecting conduit—As the name implies, the pipe is installed in a narrow trench in the natural ground of such depth that the top of the pipe is approximately level with the adjacent natural ground surface. After the trench is backfilled the embankment is constructed in the normal manner.

-4. Negative projecting conduit—The pipe is installed in a narrow trench in the natural ground of such depth that the top of the pipe is below the adjacent natural ground surface. The trench is backfilled with material which is thoroughly compacted to the elevation of the top of the pipe. Between the top of the pipe and the surface of the natural ground loose highly compressible fill is used. The balance of the embankment is built up in the normal manner.

It will be noted that this method is somewhat similar to the imperfect trench method and similarly the load on the pipe is reduced. However, the reduction is not as great as for the imperfect trench method.

19.7.2

The formulas used to calculate the load on embankment conduits have been developed at Iowa State College by research and field tests. For each type of installation the formulas are as follows:

positive-projecting

$$W_c = C_c\, w\, B_c^2$$

imperfect-trench

$$W_c = C_n\, w\, B_c^2$$

zero-projecting

$$W_c = C_c\, w\, B_d\, B_d'$$

where

W_c = load per unit length of pipe
w = unit weight of fill material
B_d = width of trench at top of pipe
B_c = outside diameter of pipe
B_d' = average of B_c and B_d
C_c and C_n = load coefficient

19.7.3

The values of C_c and C_n can be determined from the curves of Figs. 19-13, 19-14, 19-15 and 19-16 for the various installation methods. The values depend on the ratio of height of fill over the conduit to the horizontal width of conduit or trench, projection ratio, settlement ratio and the internal friction coefficient of the soil.

19.7.4

The projection ratio, p, for positive projecting conduits is illustrated in Fig. 19-17(a). It may vary from 1 to 0 but for most installations with Class B bedding it is assumed to be 0.7. For imperfect trench conduits and negative projecting conduits the projection ratio, p', can have any convenient

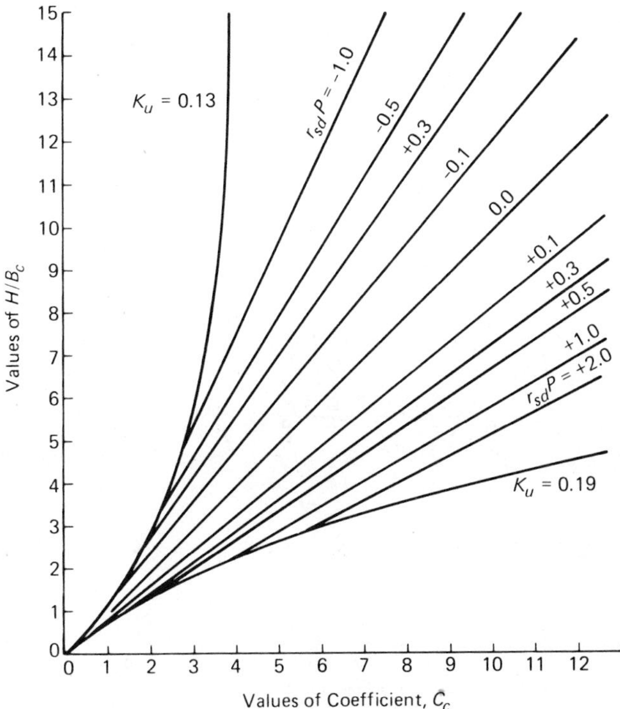

Fig. 19-13 Positive and zero projecting installations.

value as shown in Fig. 19-17b and c. The curves of Figs. 19-14, 19-15 and 19-16 are for values of p' of 0.5, 1.0, and 2.0.

19.7.5 Settlement Ratio

The settlement ratio, r_{sd}, is an abstract quantity depending on the settlement and deflection of the pipe itself and the compression of the soil adjacent to the pipe. The ratio determines whether friction forces acting on the column of soil over the pipe will add to or reduce the load on the pipe. Based on limited test data the following values are suggested for concrete pipe culverts:

Pipe on rock or unyielding soil	$r_{sd} = +1.0$
Pipe on ordinary soil foundation	$r_{sd} = +0.5$ to $+0.8$
Pipe on yielding foundations as compared to the adjacent natural ground	$r_{sd} = 0$ to $+0.5$
For negative or imperfect trench conduits	$r_{sd} = 0$ to -1.0

19.7.6 Supporting Strength

The supporting strength of concrete pipe culverts is dependent on the bedding under and around the pipe and on the active lateral pressure on the sides of the pipe. Four types of bedding are commonly used and are described as follows (see Fig. 19-18):

Class A bedding: The pipe is embedded in a continuous concrete cradle constructed of 2000-lb concrete or better; with a thickness under the pipe of one-fourth the inside diameter and extending up alongside the pipe a minimum of one-fourth the outside diameter.

Class B bedding: The pipe is carefully bedded on fine granular materials over an earth foundation carefully shaped to fit the lower part of the pipe exterior for at least 15% of its overall height. Compactible soil material is tamped in layers not more than 6 in. thick for the remainder of the lower 30% of its height.

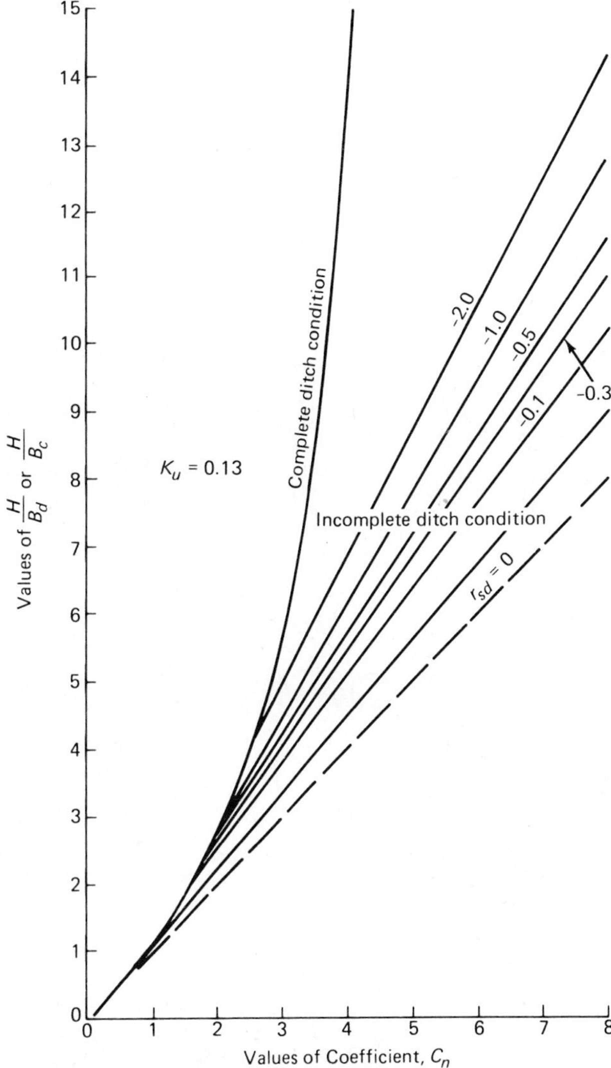

Fig. 19-14 Diagram for coefficient C_n for negative projecting conduits and imperfect ditch conduits when $p' = 0.5$.

Class C bedding: The pipe is bedded with 'ordinary' care in a soil foundation shaped to fit the lower part of the pipe with reasonable closeness for at least 10% of its overall height. The remainder of the pipe is surrounded by material placed by hand to completely fill all spaces under and adjacent to the pipe.

Class D bedding: The pipe is placed without any effort to shape the bedding to fit the pipe. Material is placed around the pipe by hand to completely fill all spaces under and adjacent to the pipe.

19.7.7 Load Factor

In culvert construction practice, unlike trench construction, it is necessary to calculate the load factor for each installation. The active lateral pressure acting on the sides of the pipe is taken into account in the load factor formula. The formula is

$$L_f = \frac{1.431}{N - xq}$$

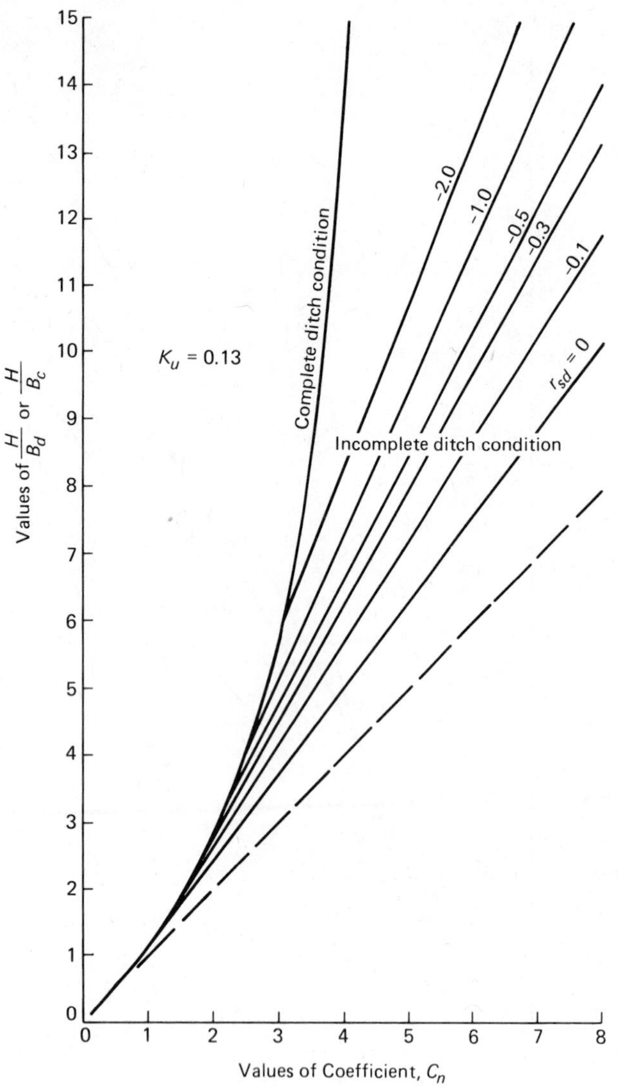

Fig. 19-15 Diagram for coefficient C_n for negative projecting conduits and imperfect ditch conduits when $p' = 1.0$.

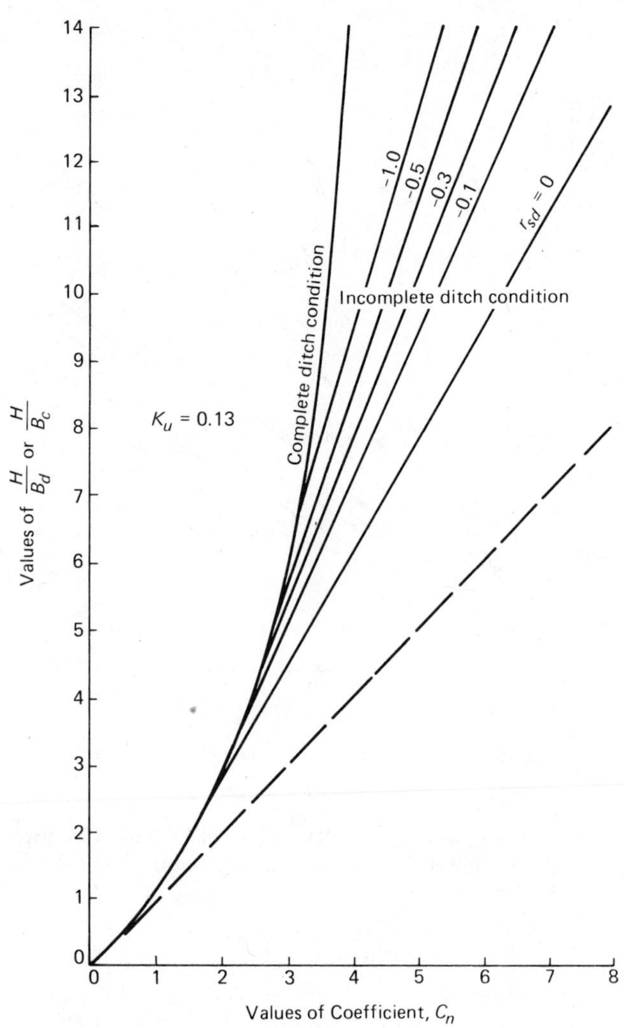

Fig. 19-16 Diagram for coefficient C_n for negative projecting conduits and imperfect ditch conduits when $p' = 2.0$.

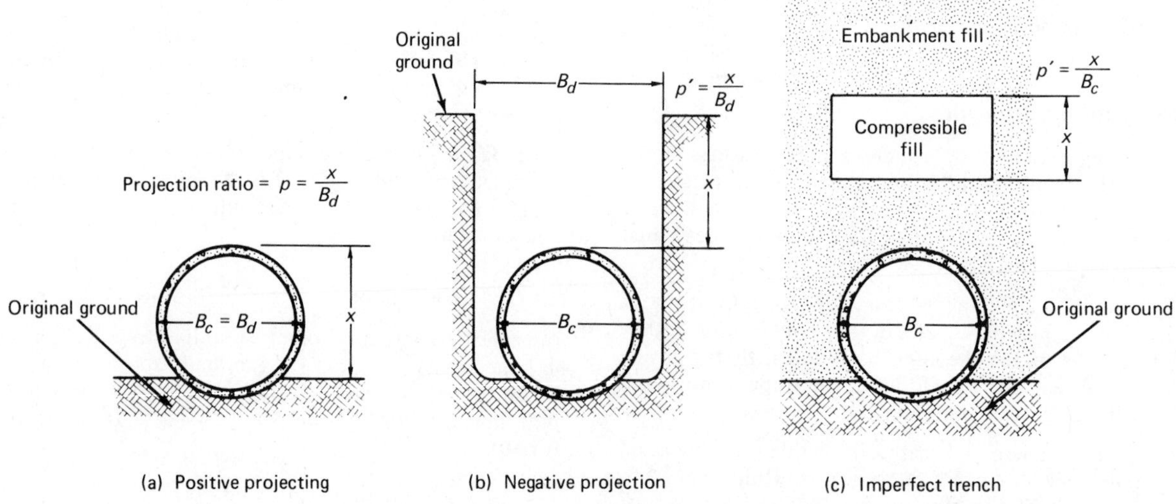

Fig. 19-17 Projection ratio for embankment conduits.

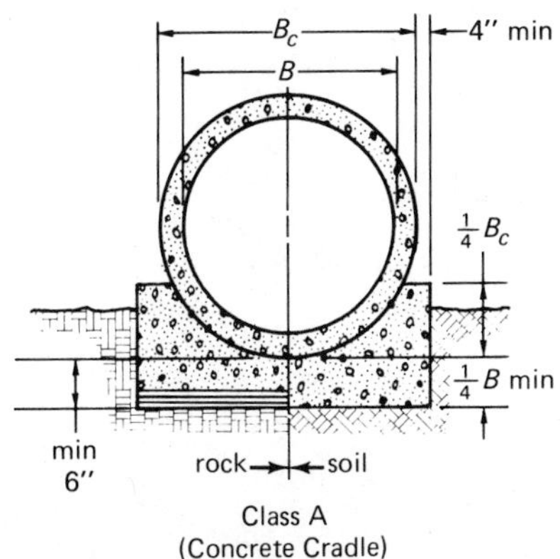

Class A
(Concrete Cradle)

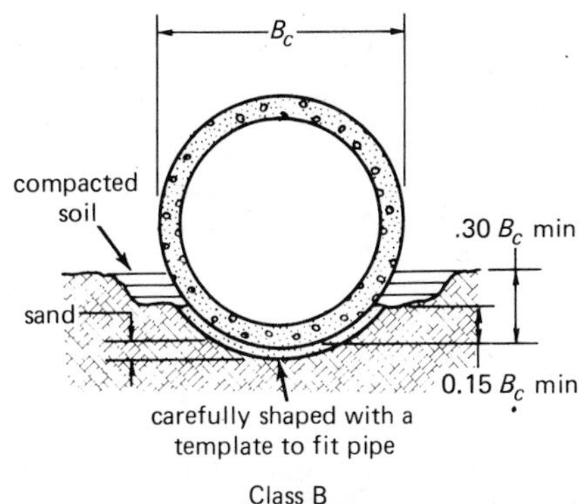

Class B

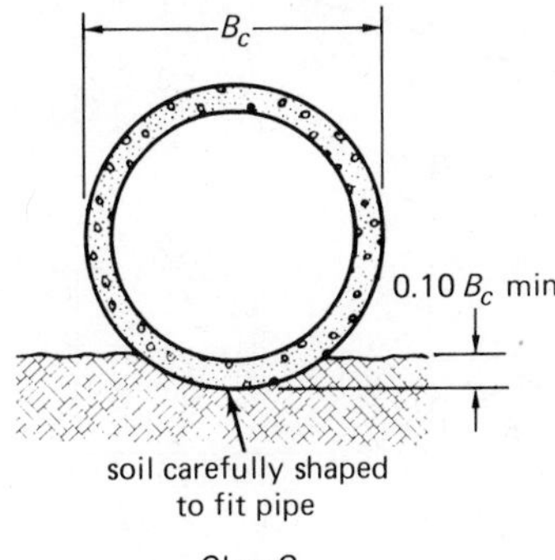

Class C

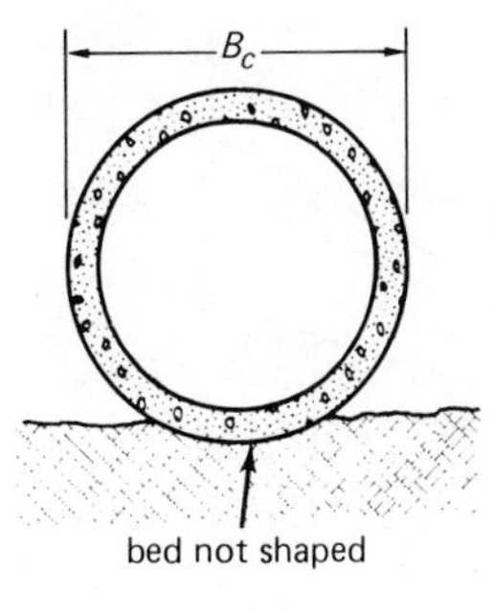

Class D

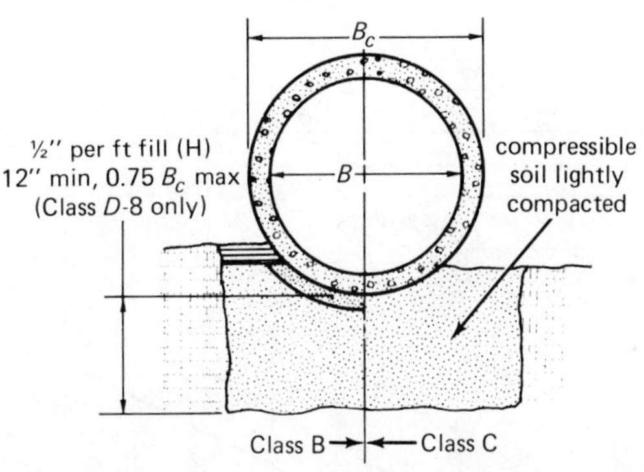

(On rock or unyielding material)

Fig. 19-18 Classes of bedding for embankment conduits.

where

N = a function of the distribution of vertical load and vertical reaction

x = a function of the area of the vertical projection of the pipe on which the active lateral pressure of the fill material acts

q = the ratio of the total lateral pressure to the vertical load

$$q = \frac{MK}{C_c}\left(\frac{H}{B_c} + \frac{M}{2}\right)$$

Where M is the fractional part of B_c over which the lateral pressure acts. It may or may not be equal to the projection ratio, p.

Values of N for Various Classes of Bedding

Bedding Class	N
Class A	0.505
Class B	0.707
Class C	0.840
Class D	1.310

Values of X and X' for Various M-Ratios

M	X	X'
0.0	0.000	0.150
0.3	0.217	0.743
0.5	0.423	0.856
0.7	0.594	0.811
0.9	0.655	0.678
1.0	0.638	0.638

When Class A (Concrete Cradle) Bedding is used, X' rather than X is used in the load factor formula.

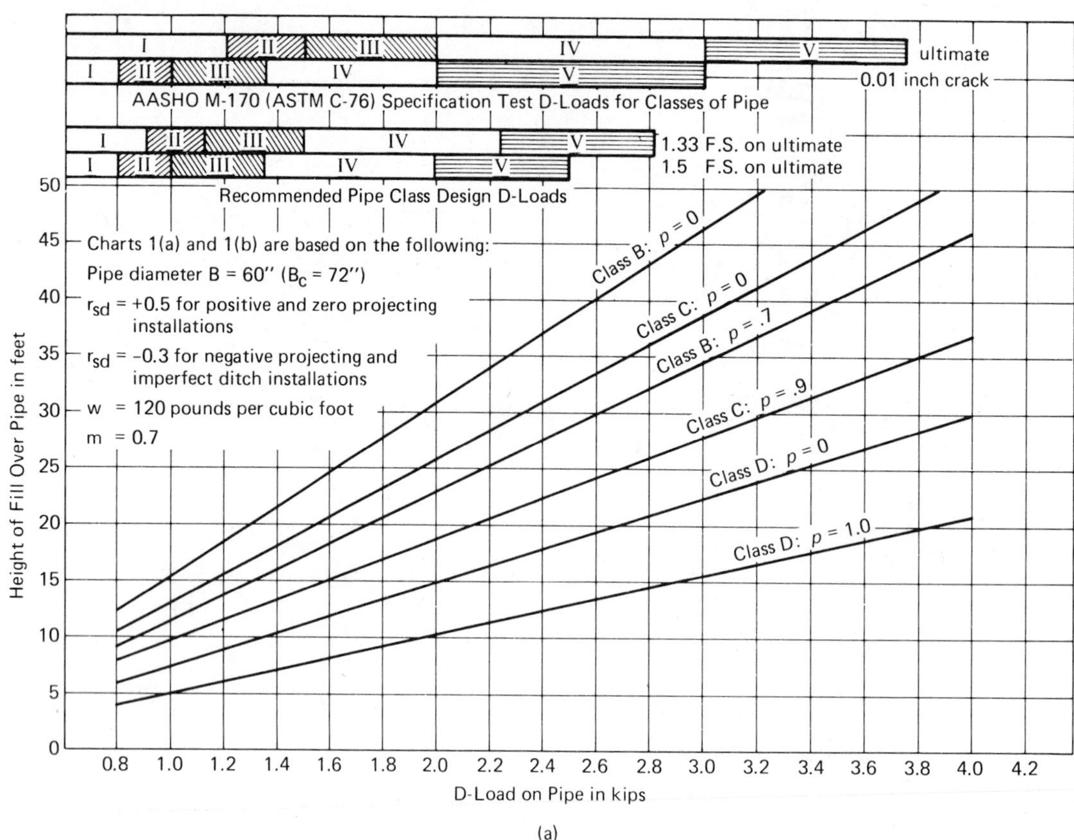

(a)

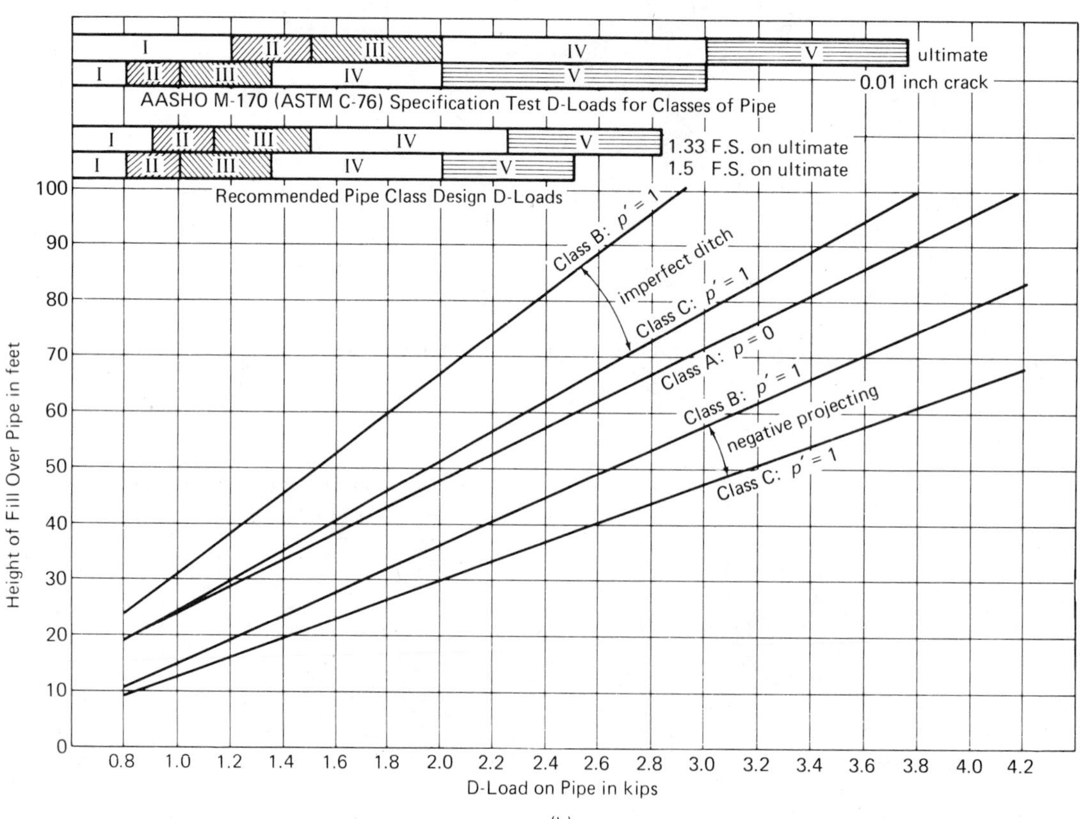

(b)

Fig. 19-19 (a) Positive and zero projecting installations and (b) Class A, imperfect ditch, and negative projecting installations.

EXAMPLE 19-4: Using data of previous examples, calculate loads and supporting strengths.

SOLUTION: Load calculation for 30 ft of fill over a projecting 60-in. culvert with $p = 0.7$

$$r_{sd} = 0.7; \quad w = 120 \text{ lbs per cu ft}$$

$$H = 30; \quad B_c = 60 + 2(6) = 72 \text{ in.} = 6 \text{ ft}$$

$$\frac{H}{B_c} = \frac{30}{6} = 5.0$$

$$pr_{sd} = 0.7 \times 0.7 = 0.49$$

From Fig. 19-13, $C_c = 7.5$.

$$W_c = C_c W B_c^2$$

$$W_c = 7.5 \times 120 \times 6^2$$

$$= 32400 \text{ lbs per linear foot of pipe}$$

Supporting strength calculation:

$$L_f = \frac{1.431}{N - xq}, \text{ where } q = \frac{MK}{C_c}\left(\frac{H}{B_c} + \frac{M}{2}\right)$$

From preceding problem with $K = 0.33$

$$q = \frac{0.7 \times 0.33}{7.5}\left(5.0 + \frac{0.7}{2}\right)$$

$$= 0.165$$

For Class B bedding, N = 0.707, x = 0.594

$$L_f = \frac{1.431}{0.707 - 0.549 \times 0.165}$$

$$= 2.35$$

Assuming a factor of safety of 1.5, the required three-edge bearing strength for the installation is

$$\text{three-edge} = \frac{W_c}{L_f x \text{ FS}} = \frac{32400}{2.35 \times 1.5} = 9200 \text{ lbs per ft}$$

$$\text{three-edge D-load} = \frac{9200}{5.0} = 1840 - D$$

19.7.8 Charts of Allowable Fill Heights

Figures 19-19a and b are taken from Charts I(a) and (b) of "Reinforced Concrete Pipe Culverts—Criteria for Structural Design and Installation" by the Bridge Division, Office of Engineering and Operations, U.S. Bureau of Public Roads, Aug. 1963.

These charts can be used to rapidly determine the required D-Load strength for concrete pipe culverts under various fill heights and for various installation procedures.

19.8 LIVE LOADS ON BURIED PIPE

In addition to the weight of the backfill material, surface wheel loads from trucks, aircraft, railroad locomotives, etc. produce pressure on the top of buried conduits. This must be added to the backfill load to determine the total load on the pipe. The magnitude of the live load decreases rapidly as the cover over the pipe increases. For fills greater than 8 ft over highway culverts, the live load effect can be neglected. Rigid pavements also greatly reduce the effect of the live load.

Live load pressures on buried pipes can be calculated using the tables from the Portland Cement Association publication, "Vertical Pressures on Culverts under Wheel Loads on Concrete Pavement Slabs." No distinction is made between trench conduits and embankment conduits in calculating live load pressures.

Supporting strengths of pipe for live loads can be conservatively calculated using the load factors for trench-type installations.

19.9 HYDRAULIC CAPACITY OF CONCRETE PIPE SEWERS

Flow by gravity through a sewer depends on the slope of the sewer, the cross section of the sewer, depth of flow, character of the interior surface, whether flow is steady or intermittent, and if there are any obstructions, bends, or connections.

Storm sewers are normally designed to flow just full but not under pressure for periods of maximum flow. Sanitary sewers are normally designed to flow half full. It is generally assumed that the hydraulic grade line is parallel to the pipe invert in calculating flow through sewers. It is also assumed that uniform flow will occur.

When the velocity of flow is known, the capacity of the sewer can be calculated from the formula

$$Q = VA$$

where

Q = quantity of flow in cubic feet per second
V = mean velocity of flow in feet per second
A = cross sectional area of the conduit in square feet

The velocity of flow in gravity-flow sewers is usually calculated by the Manning formula

$$V = \frac{1.486}{n} R^{2/3} S^{1/2}$$

where

R = hydraulic radius, ft
S = slope of the hydraulic grade line (usually the slope of the sewer)
n = Manning roughness coefficient

The value of n commonly used for concrete pipe storm sewers is 0.012. For sanitary sewers a value of 0.013 is used.

The Manning formula is used for sewers for any cross section flowing either full or partly full. A nomograph shown in Fig. 19-20 is used to solve the Manning formula for circular sewers flowing full. Figure 19-21 can be used to determine the hydraulic properties of circular sections for any depth of flow. Using Figs. 19-20 and 19-21, the velocity in circular sewers for any depth of flow can be determined.

19.10 HYDRAULIC CAPACITY OF CONCRETE PRESSURE PIPE

Flow in irrigation lines operating under pressure is calculated by the Scobey formula

$$V = C_5 H_f^{0.5} d_i^{0.625}$$

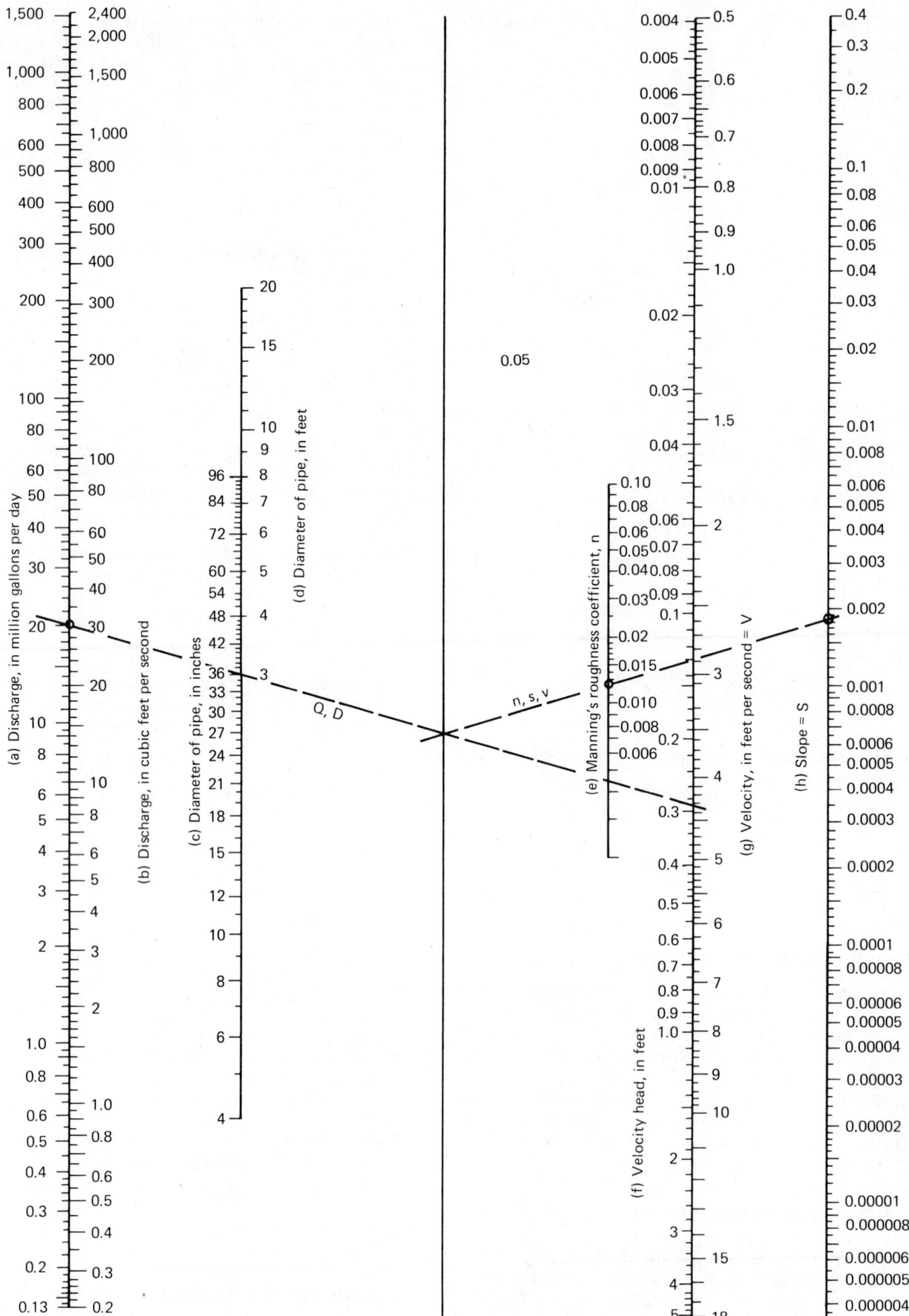

Fig. 19-20 Nomograph for flow in round pipe—Manning's formula.

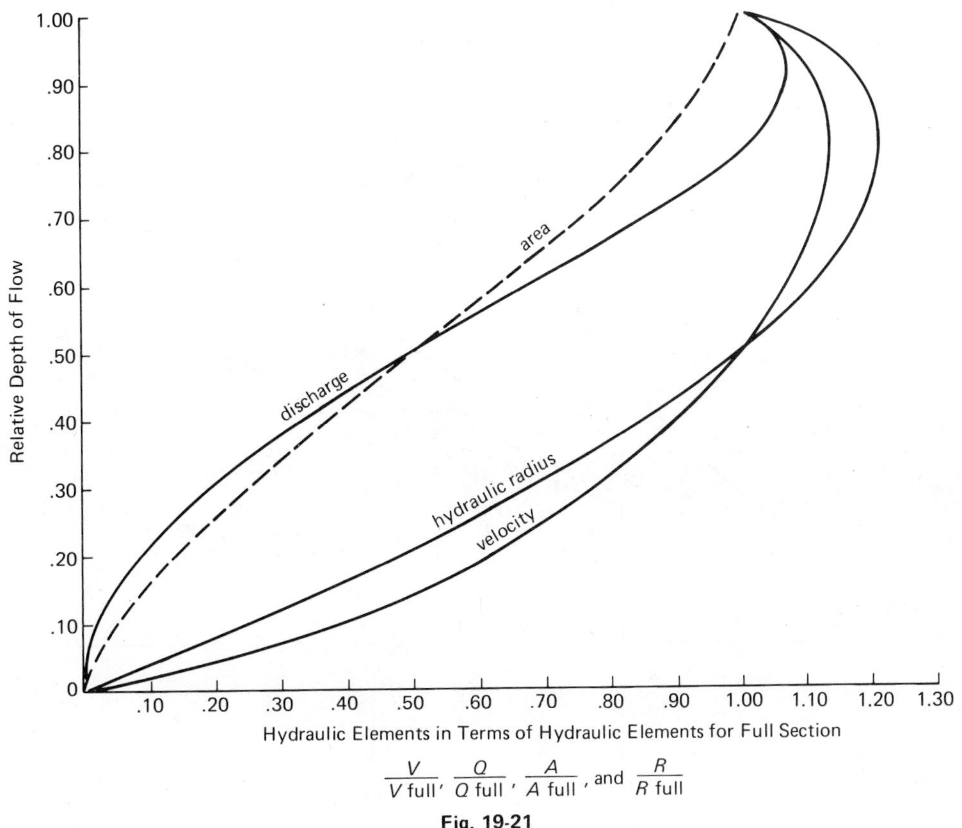

Fig. 19-21

where

C_s = roughness coefficient
d_i = inside diameter of the pipe, in.
H_f = loss of head in feet per 1000 ft of pipe

Values of C_s for concrete irrigation pipe are 0.345 for average pipe and 0.370 for smooth pipe.

The Hazen-Williams formula is widely used to compute the velocity of flow in concrete pressure water pipe.

$$V = 1.318\ C_i\ R^{0.63}\ S^{0.54}$$

where

C_i = the Hazen-Williams roughness coefficient
R = mean hydraulic radius
S = slope of the hydraulic grade line

For concrete pressure pipes the value of C_i commonly used is 140.

19.11 HYDRAULIC CAPACITY OF CULVERTS

The hydraulic analysis of a culvert is complicated in that different types of flow can occur depending on such factors as

type of inlet,
length and slope of culvert,
size of culvert opening,
barrel roughness,
depth of headwater, and
depth of tailwater.

19.11.1 Inlet Control

If a concrete pipe culvert is on a sufficiently steep slope (0.5% ± depending on size and length), it operates with 'inlet control'. With inlet control due to the slope of the culvert, the water is carried away from the inlet faster than it enters the culvert. Thus the inlet acts as an orifice. Flow thru the culvert is governed by hydraulic characteristics of the inlet, culvert size, and the headwater depth. The barrel of the culvert always flows partly full under Inlet Control even with the inlet submerged.

19.11.2 Outlet Control

For flatter slopes the flow thru the culvert barrel will more nearly equal the rate of flow thru the inlet. If the slope is sufficiently flat the flow thru the barrel will be less than that which can theoretically enter the culvert. When this occurs the pipe begins to flow full if the inlet is submerged. The discharge thru the culvert then depends on all the variables such as type of inlet, culvert size, headwater depth, length, slope, and barrel roughness. The culvert is then said to operate with 'Outlet Control'.

19.11.3 Unsteady Flow

Between 'inlet control' and 'outlet control', an unsteady flow condition exists where the culvert rapidly vascillates from one to the other. The average discharge lies between the two and can be obtained from charts by interpolation.

19.11.4 Hydraulic Analysis Charts

To simplify the hydraulic analysis of culverts, charts and nomographs have been developed and presented in various publications.

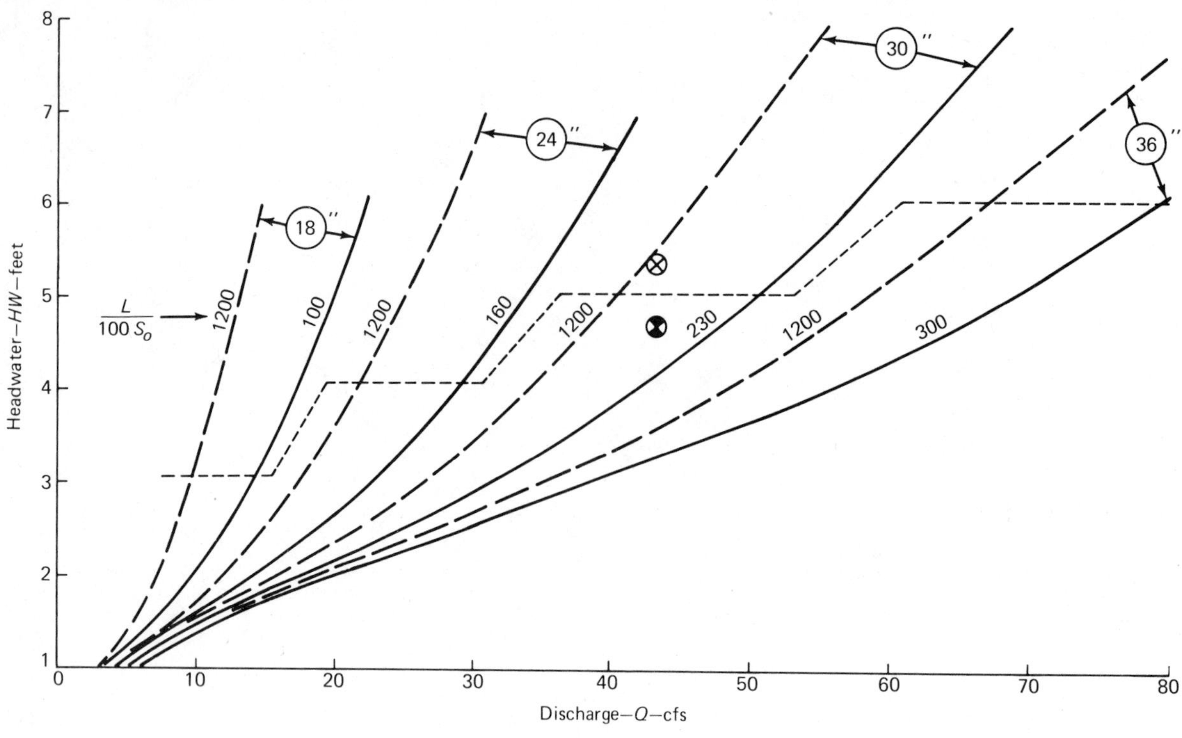

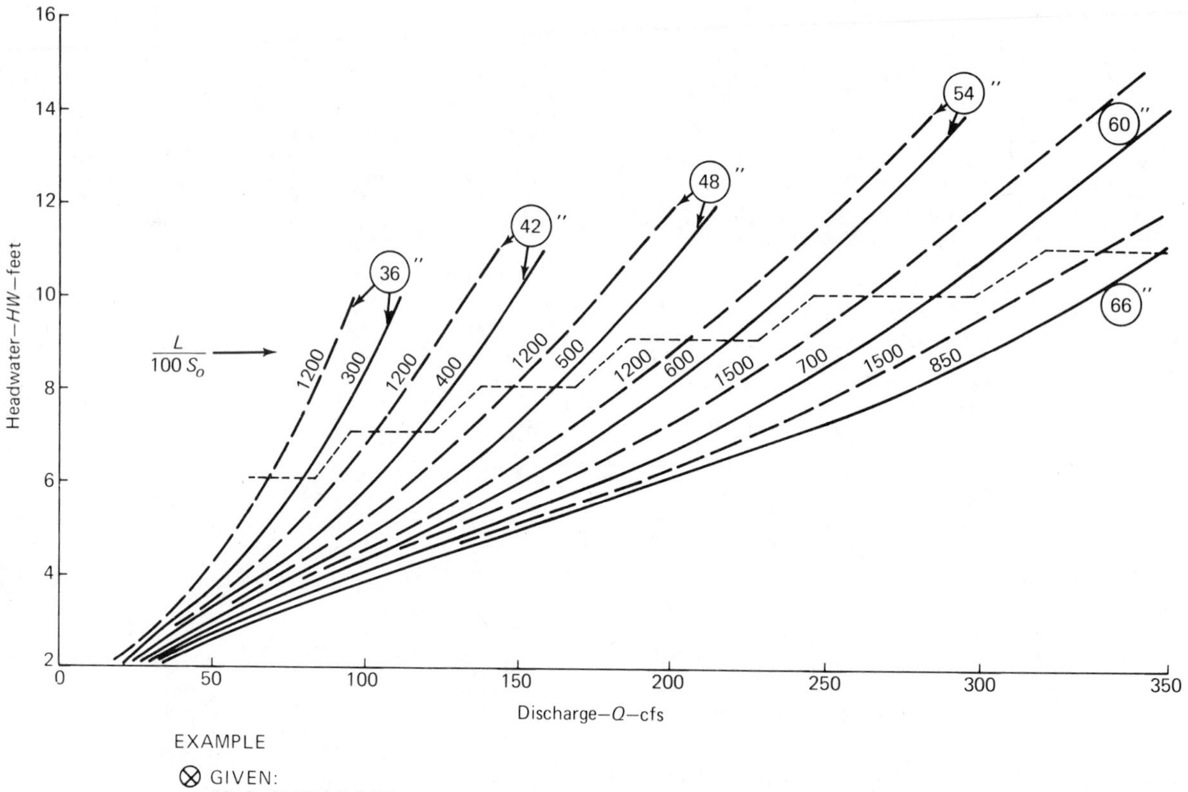

EXAMPLE

⊗ GIVEN:
43 cfs; AHW = 5.4 FT
L = 120 FT; S_o = 0.002

SELECT 30"
HW = 4.7 FT

Fig. 19-22 Culvert capacity for circular concrete pipe, groove-edged entrance, 18 to 60 in.

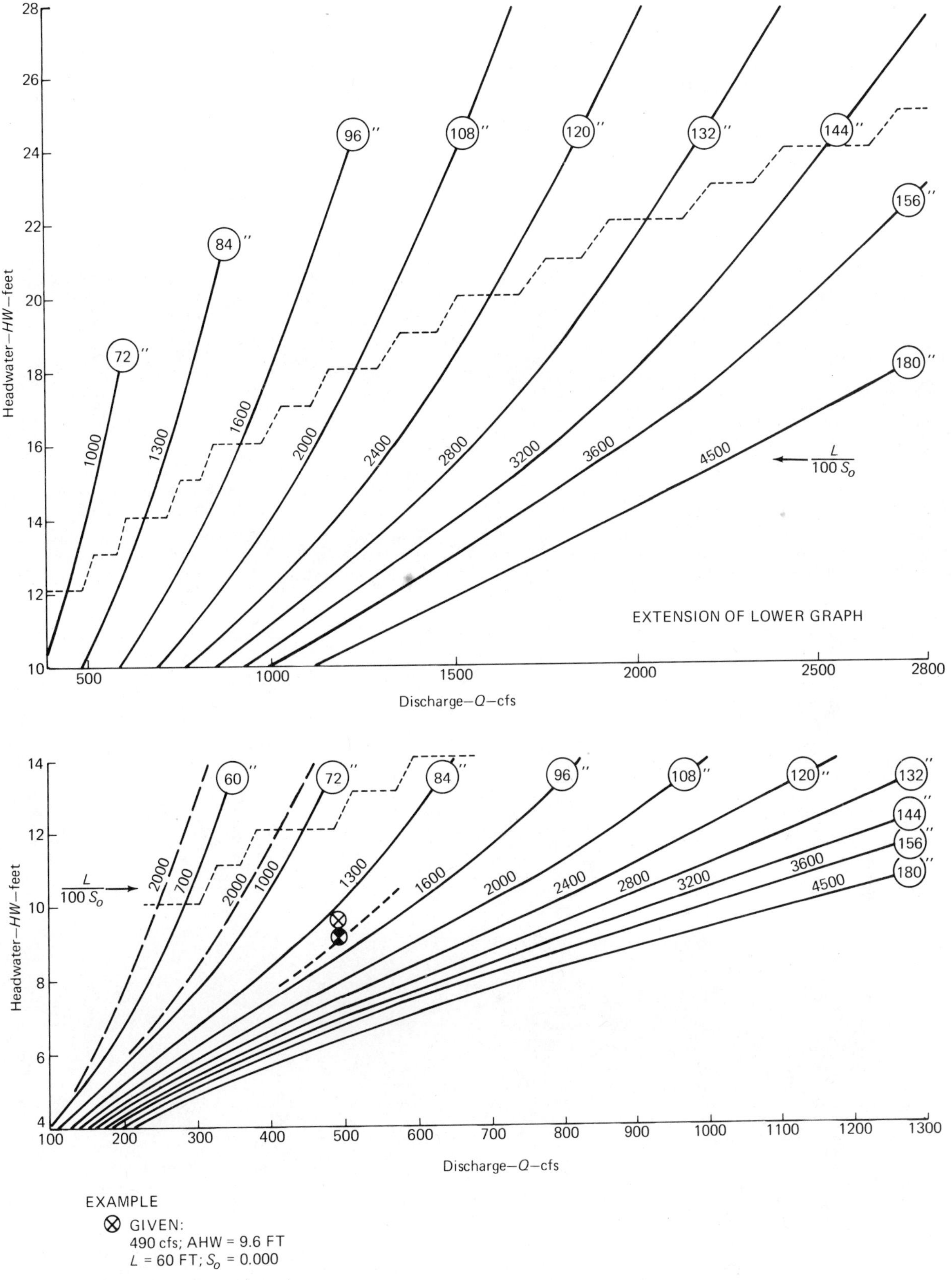

EXAMPLE

⊗ GIVEN:
490 cfs; AHW = 9.6 FT
L = 60 FT; S_o = 0.000

SELECT 90" $\left(\frac{L}{D} = 8\right)$
HW = 9.2 FT

Fig. 19-23 Culvert capacity for circular concrete pipe, groove-edged entrance, 60 to 180 in.

Figures 19-22 and 19-23 are Charts 19-2 and 19-3, taken from "Capacity Charts for the Hydraulic Design of Highway Culverts" Bureau of Public Roads, *Hydraulic Engineering Circular No. 10*, U.S. Government Printing Office, Washington D.C., 1965, and can be used for most concrete culvert pipe installations.

Each chart has a vertical scale giving the depth of headwater above the bottom of the pipe in feet. The horizontal scale gives the discharge in cubic feet per second. On Fig. 19-22 each pipe size is represented by two curves, a solid-line curve and a dashed-line curve. The solid-line curve gives the headwater-discharge relationship for that size culvert operating under inlet control as previously described. The dashed-line curve represents the headwater-discharge relationship for that size culvert operating under outlet control.

An index number is given on each curve. The index number equals the length (L) divided by the slope in percent ($100S_o$). If the index number calculated for the installation is equal to or less than that on the solid-line curve, the culvert operates with inlet control and the headwater-discharge relationship is given by that curve. If the index number is between that on the solid line curve and that on the dashed line curve, the headwater-discharge relationship can be determined by interpolation.

If the index number is greater than that on the dashed line curve, the culvert operates with outlet control. The headwater-discharge relationship can be approximately determined by an extrapolation, but a more rigorous analysis is desirable, particularly if the difference in index numbers is large. Procedure is given in *Hydraulic Engineering Circular No. 10* previously referred to in section 19.11.4.

On Fig. 19-23, each pipe size is represented by a single solid-line curve. For the large diameters shown on this chart and for practical headwater depth and culvert lengths, the curves representing inlet control and outlet control are very nearly alike. Consequently, a single curve is given as most culverts in these sizes will operate with Inlet Control.

For most culverts, the length, slope, type of pipe, and allowable headwater depth are determined by the topography at the installation site. The problem is to determine the required size. For example, for a given location we might have $L = 400$ ft, $Q = 180$ cfs, and $S = 0.004$. Then, allowable $HW = 7$ ft.

Using Fig. 19-22, these quantities indicate a 54-in. pipe as a trial size. The index number is

$$\frac{400}{100 \times .004} = 1000$$

This value falls between the solid-line curve and dashed-line curve. Interpolation however shows that an allowable headwater of 7.0 ft will have a corresponding discharge of only 165 cfs. A 60-in. pipe will carry the required discharge with a headwater depth of about 6.4 ft and therefore is a satisfactory solution.

20

Marine Structures

BEN C. GERWICK, JR.*

20.1 GENERAL CONSIDERATIONS

20.1.1 Adaptability of Concrete for Marine Structures

Concrete is an almost ideal material for marine structures of all types: harbor, coastal, and ocean, and for all marine environments; submerged, partially and completely exposed, temperate, tropical, and arctic. It is economical, durable, strong. As will be shown in later sections, the design and placement techniques for marine application assume great importance for optimum results.

The basic present limitations on the use of concrete for marine structures have been essentially related to those subjected to high bending forces, such as ocean storm waves. Even in the deep ocean, however, concrete structures are now gaining increased acceptance through utilization of different structural concepts, e.g., caissons, in lieu of the more typical jacket frames. Concrete appears to have a major role to play in the development and exploitation of the oceans, temperate, tropical, and arctic.

Concrete inherently has rigidity and weight. The interior surface of submerged concrete structures is generally free from weeping and condensation. Reinforced and particularly prestressed concrete has excellent fatigue properties under cyclic loading, such as waves. Under overload, due to impact, grounding, or explosion, it does not tear and rip, but cracks and crushes locally. At extreme low temperatures, it maintains a nonbrittle, ductile behavior.

Repairs to damaged sections are easily effected: in many cases repairs can be accomplished underwater without the need for dewatering or drydocking. Concrete is readily molded into any desired form or shape, including double-curved surfaces, and it has high resistance to abrasion from moving gravel, sand, and ice.

Finally, concrete is inherently economical. It can be locally produced in most parts of the world, using predominantly indigenous materials and labor. Thus, it has been inevitable that concrete would attain a dominant role in marine structures. Like all materials, it requires the employment of proper design and construction techniques if the full advantages of this material are to be realized in service.

20.1.2 Environmental Considerations

Concrete for marine structures is subjected to structural criteria and performance requirements that are common to all concrete structures. However, the environment in which it must perform differs significantly—the marine environment exerts extreme demands on the concrete from both durability and loading aspects.

-1. Submergence and pressure—Concrete structures that are submerged in seawater must consider the effects of the

*Consulting Engineer, and Professor of Civil Engineering, University of California at Berkeley.

continuous presence of water under hydrostatic pressure. This not only exerts a structural loading–it also means that the concrete face, at least, will be continuously wet, and that water will penetrate the pores of the concrete as well as any cracks, moving towards the inner face. If the inner face is at atmospheric pressure, which occurs in many structures, then there is a continuous slow movement inward.

The concrete will thus absorb water and increase its density. This absorption of water can have a drastic effect on structures which are designed for flotation, with the dead weight being approximately offset by the buoyancy. In such a case, an increase in density of 2% to 3% due to absorption can have a 100% or greater impact on the net (buoyant) weight.

The uniform temperature of seawater in a particular locality practically eliminates the short-term expansion and contraction of concrete.

The continuously saturated environment of submerged concrete eliminates drying shrinkage, and permits long term chemical crystallization processes to be completed, thus generally leading to higher strengths. Autogenous healing of microcracks and even hairline cracks may also take place due to secondary crystallization. Concrete placed under pressure, i.e., poured in a submerged state, benefits greatly in strength and density. The concrete is given additional compaction by the hydrostatic loading.

-2. *Seawater chemistry*–The chloride ions in seawater are of particular concern for concrete structures because they reduce the *pH* of the concrete and thus offset the passivation effect of the cement. As a result, corrosion of embedded steel becomes more likely to occur.

When seawater moves through permeable concrete and evaporates from the other face, salt cells are formed in the concrete pores. These set up a battery action with the reinforcement, leading to electrolytic corrosion.

A piece of steel, suspended bare in the sea-air splash zone, will corrode less than the same steel, encased in highly permeable concrete, hung in the same environment. This emphasizes the importance of making the concrete as impermeable as possible.

Seawater can also attack the concrete itself. Magnesium ions from the seawater can replace calcium ions present in the cement as C_3A. This leads to a general softening, expansion, and loss of strength. Sulfate ions can similarly cause disruption of the concrete. These effects can be minimized by using low alkali cement. The aggregates must be selected for durability and soundness under sulphate attack. The dissolved gases in seawater, particularly oxygen, contribute to corrosion. Since the oxygen content decreases with depth, it follows that corrosion of the steel is usually also reduced with depth.

In some seas, H_2S is present, dissolved in the seawater. Its attack on the concrete can be lessened by making the concrete dense and impermeable and by using a rich cement factor.

Where acid, sewage, or other aggressive chemicals are discharged directly through or onto concrete structures, additional coatings such as epoxies or bitumastic may be required.

-3. *Movement of seawater*–This gives rise to many of the design loadings on the concrete structures, affecting their overall stability.

Currents in the ocean generally run from 0.5 knot to a maximum of perhaps 4 knots; although higher velocities occur in restricted channels and inlets. Contrary to popular belief, it is frequently found in specific locations that the bottom current may be almost as high as the surface current. In estuaries where fresh and salt water meet, it is not unusual to find the salt water flooding in at the bottom while the surface is still ebbing.

At some locations, the direction of the currents rotate during the tidal cycle. Prolonged winds in one direction may produce substantial currents even in the open sea.

The total current load may be computed by the formula

$$P = \frac{Av^2wK}{2g}$$

where

P = total current load, lbs
A = area exposed normal to the current, sq ft
v = velocity of current, fps
g = 32.2 ft/sec^2
w = unit weight of water, 62.4 or 64 lb/cu ft
K = shape constant (1.33 for rectangular, 0.75 for circular)

The current forces can alternatively be computed by a naval architecture approach in which the components of surface friction (P_f) and eddy action (P_e) are added

$$P = P_f + P_e = v^2 \sin\theta\,(sc + 3.2\,A)$$

in which

θ = angle that direction of flow makes with the face
c = a constant, taken as 0.5
s = immersed surface, sq ft
v = velocity of current in knots

and the remainder of symbols are as stated above.

Current exerts pressure on the structure. It also may cause an eddy action by flow over the top, tending to lift a structure that is on the bottom. Conversely, when a structure is suspended just above the bottom, the current tends to increase its effective weight, pulling it down. The increased velocity of current underneath and around the structure may also cause erosive scour of the bottom sediments.

Scour may also take place at the corners of a structure set on the bottom. Erosion may be more serious at the downstream corners than at the upstream nose. Waves exert a dynamic and cyclic loading on a submerged structure. At the surface, the waves may actually break against the surface with enormous force. Below the surface, large pressure differentials may occur. For structures completely submerged, dynamic uplift forces occur during the wave cycle which may or may not be in phase with the maximum lateral forces.

The pressure differentials on the two sides of a bottom-supported structure may lead to tunneling underneath and consequent serious erosion of the bottom sediments.

If the pressure differential can penetrate into a composite structure, e.g., a breakwater, the back pressure may lift out the individual components with almost explosive force. This same phenomenon is being used intentionally to absorb and dissipate energy. Holes have been provided in the vertical walls of caissons for quay walls, and breakwaters, and offshore storage vessels in the North Sea in order to absorb a portion of the wave energy.

Nonbreaking waves reflecting from a structure may form standing waves (clapotis) directly in front of the structures, leading to erosion of the soils below the crest of the standing wave. Splashing waves deposit salt above the waterline and lead to alternate wetting and drying (see section 20.1.2.7). The waves also throw salt into the atmosphere where it is carried onto the exposed superstructure by spray and fog.

Surf produces cavitation in the surface of concrete structures which can only be minimized by extreme care to achieve a hard, dense surface, free from fins and projections. The use of special treatments such as epoxies or polymer-impregnation may be useful in extreme cases.

Moving water also moves sand and gravel, causing abrasion. Even silt laden waters moving at high speeds can cause abrasion of the concrete surface. Moving water, whether due to waves or currents, seriously affects the underwater placement of fresh concrete. It tends to pull the cement paste out of the mix. Small gaps in caissons or sheet pile cofferdams frequently have led to serious leaching of cement.

-4. Marine organisms—A great deal of publicity has been given to the boring clam, the sea urchin, and the rock-boring mollusc. However, these organisms are normally able to attack only weak, porous concrete. The barnacle exerts high local pressure on the concrete and may destroy bitumastic coatings and roughen the concrete surface.

Fouling (growth of marine bio-organisms) of the hulls of concrete barges and ships is apparently less serious than with steel hulls but, nevertheless, does occur. It is reportedly easier to remove such fouling from concrete surfaces.

Fouling of concrete intake and discharge structures has been given considerable attention. Periodic heating of the discharge may kill young mussels. Chemical impregnation of the concrete is under study but has the obvious problem posed by progressive leaching out of the chemicals.

Barnacles and mussels may increase the effective diameter of concrete piles and do produce an adverse shape factor, thus increasing the resistance to the waves and the consequent load on the piles.

-5. Ice and debris—Moving ice and debris may exert direct impact loads on concrete structures. Such material may pile up, increasing the effective area and thus the load. If the ice exists in solid sheet form, then it must fail by crushing, unless suitable ice breakers are installed to cause it to ride up and fail in tension.

Ice cakes and debris frequently become wedged between adjoining piles. The resulting torsion may break the piles. In one case in Alaska, a large ice mass that had formed on the piles slid down at low tide during the Spring thaw and broke the adjoining batter piles. Ice cakes may also adhere to concrete breakwater units, wedging them apart and lifting them at high tide.

While the loading of ice and debris is much the same on concrete structures as on those of steel and timber, the resultant effects differ substantially, depending on the structural and physical characteristics in the particular case. Concrete piles may be more vulnerable to impact and torsion whereas concrete caissons or sea walls may be more resistant due to their mass (longer natural frequency), local rigidity, and ductile behavior at the low temperatures involved.

When water freezes inside a concrete structure it exerts a typical bursting pressure. The design of a hollow-core cylinder pile or caisson must incorporate means of preventing freezing, such as a wood or styrofoam log floated inside and/or vents beneath the surface connecting to the unfrozen water. Otherwise, serious bursting may take place, as actually happened in a bridge project in South Dakota.

-6. Seismic effects, tsunamis, and explosions—A marine structure, whether of concrete or other material, is subjected to seismic accelerations like any other structure, plus the acceleration of the mass of water displaced. Because concrete is frequently chosen for very large structures (offshore oil storage, caissons, bridge piers, etc.) this dynamic effect must be given special consideration.

Tsunamis, or tidal waves, produce great uplift forces. Concrete is an ideal material for the decks of structures in areas subject to tsunamis because of its dead weight.

Explosive forces cause serious overpressures in the water which tend to collapse a submerged structure. Concrete, by virtue of its inherent rigidity, is an ideal material to resist such overpressure. However, the design for such structures should take care to eliminate, insofar as possible, reentrant angles and holes, or cavities, which might concentrate the blast effect.

Recent studies, still in a very preliminary phase, indicate that earthquakes may produce overpressure effects up to 100% of the hydrostatic loads, with the maximum effect occurring at about 600 ft of water depth.

All submerged structures are subjected to an additional virtual mass effect under dynamic loading due to the acceleration of the surrounding water by the structure. While each case must be studied individually this virtual mass generally varies from a factor of 1 (100% additional) for small structures (e.g., 5-ft diameter cylinder) down to about 0.2 (20%) for very large and rigid concrete structures (e.g., 400-ft diameter structures)

-7. Sea-air interface: the splash zone—This zone is subjected to both the salt-laden waters and the oxygen and CO_2 of the atmosphere. With alternate wetting and drying, salt cells are formed in permeable concrete, leading to severe corrosion of the reinforcement. The reinforcement then expands due to the corrosion products, spalling off more concrete and allowing corrosion to proceed further. Obviously, concrete designed for use in this zone must be made as impermeable as possible.

This zone is also subject to severe freeze-thaw attack. The combination of salt water splash and freeze-thaw is particularly disastrous. Air-entrainment is essential for this exposure. Prestressing is also beneficial in eliminating hairline and microcracking due to shrinkage.

-8. Ship impact—Until recently, little attention was given in concrete structure design to damage from ship impact, other than to provide fendering systems as required. The growth in size of vessels and the importance of critical marine structures has now led to greater consideration of this problem and to the ways to absorb energy and minimize damage. Heavy mass helps, as does the provision of diaphragm action, prestressing, and well-distributed reinforcement.

20.2 DURABILITY AND CORROSION PROTECTION

20.2.1 Corrosion of Reinforcement

This is the most common source of distress in concrete marine structures. Concrete acts to protect reinforcement by passivating it with a cement paste, giving a *pH* of 13 or 14 adjacent to the steel. The *pH* of seawater is about 8. Corrosion of reinforcement occurs below a *pH* of 11.

The *pH* is lowered by the presence of chloride ions in the concrete. These may have been introduced through salt in the mixing water or the curing water. Chlorides may have been present on the surface of the aggregates. Beach sands and desert sands subjected to salt fogs (as in some parts of Arabia) are particularly bad in this regard. Corrosion is accelerated by the presence of oxygen and by the existence of voids along the surface of the steel.

Salt-cell electrolytic corrosion is caused by salt water migrating into and through permeable concrete, then evaporating from the surface, leaving tiny deposits of salt which set up a potential with the reinforcement.

The corrosion process is much more rapid in hot climates and warm water than in cold. The proper protection against corrosion of the reinforcement by factors described above can be provided by the following means:

a. Dense impermeable concrete. This is achieved by a low water/cement ratio (below 0.45), a dense but workable mix design, good consolidation and curing, and a smooth dense surface finish.
b. Adequate cover. For marine exposures, a cover of 2.5 to 3 in. is recommended by most authorities. However, a cover as low as 1.5 in. has been used successfully on many prestressed concrete piles, etc. and has apparently been satisfactory due to the dense concrete and rich mix employed. Even smaller covers, down to 0.375 in., have been used on concrete ships and boats, where especial care was taken to achieve a dense cement paste cover.
c. Rich cement factor. Six to eight sacks per cubic yard is recommended.
d. Salt-free aggregates. Beach and desert sands may require washing with fresh water.
e. Minimum chloride content in mixing and curing water and in admixtures. The use of calcium chloride in marine structures is not recommended, and is prohibited for prestressed constructions.
f. Selecting the proper type of cement. While low C_3A content may enhance the durability of the concrete, it causes a reduced resistance to chloride permeability and thus may not give as good a corrosion protection for the reinforcing steel. Thus, a moderate C_3A content is the best solution for marine exposure of reinforced or prestressed concrete: ASTM Type II cement is recommended.
g. Consolidation of cement around the steel. Vibration techniques should be adopted which eliminate voids near the steel while producing as impermeable a surface as possible. When grouting posttensioned ducts of marine structures, particular pains must be taken to prevent the formation of voids (see section 20.3.4).
h. Adequate curing
i. In special cases, it may be desirable to seal the surface, as with bitumastic paint, epoxy paint, or other protective treatment.[20-5]

Embedded metals other than steel, such as copper or aluminum, may set up electrolytic corrosion, especially where an electrolyte such as seawater is present. Galvanized and cadmium-coated inserts have been successfully employed on a great many marine structures with satisfactory results.

Cracks in the concrete are undesirable in marine structures. In the atmospheric and splash zones, they permit the penetration of CO_2 (which lowers the *pH*), chlorides, and oxygen. However, small hairline cracks at the surface (0.005 in., 0.12 mm width) apparently seal themselves with secondary crystallization in many cases and have not proven nearly as serious as general permeability. Similarly, cracks located well beneath the waterline (20 ft or more) are generally less serious than would appear; however, it is highly desirable to seal them with an underwater setting cement or epoxy.

When corrosion of reinforcement has been a serious problem in marine structures, it has usually shown up within a very short time. One of the most notorious of such cases was the original San Mateo-Hayward Bridge across San Francisco Bay. The concrete was very porous, leading to continuous salt-cell electrolytic corrosion. Serious problems developed within one year, even before the bridge was completed. During the subsequent years, extensive repairs were carried out, amounting in cost to more than the original value of the bridge. However, the salt-cells were spread throughout the concrete and repairs merely served to change cathodic zones to anodic and vice versa.

However, another project in San Francisco Bay, a wharf, was successfully repaired by chipping off the undersurface, guniting it, and coating it with heavy coal tar paint. The upper surface had always been sealed by asphalt. In this case, apparently, the moisture electrolyte was effectively sealed off and passivated.

20.2.2 Chemical Attack on Concrete

Protection of concrete against chemical attack in a marine environment may utilize both internal and external means.

Internally, protection consists of selecting components which will remain inert and resist the seawater attack. These steps include:

a. Use of low alkali cement (less than 0.65% Na_2O and K_2O), such as ASTM Type II.
b. Use of nonreactive aggregates (ASTM tests C289 and C277).
c. Use of cement containing moderate C_3A content, 6% to 8%, such as ASTM Type II.
d. Aggregates to be sound—not subject to sulfate attack (see ASTM C88).
e. Adequate curing. Optimum results appear to be obtained with normal moist curing followed by a period of drying before immersion in salt water.
f. Good consolidation of concrete, and good surface finish.
g. Rich cement factor, low water-cement factor.
h. Use of air-entrainment, 4% to 6%, appears to be beneficial although it is not always felt to be essential.

External means of protection include:

a. Bitumastic coatings. A hot-applied coal tar has been found especially effective. However, for practical reasons, most bitumastic coatings are sprayed or painted on cold. Some of these compounds are very toxic, requiring adequate protection for personnel engaged in applying the compound.
b. Epoxy coatings. There are numerous formulations available, including some which may be applied underwater.
c. Other coatings and protective treatments are described in Ref. 20-5.

Acids, even highly diluted, as from discharges of refineries and smelters, will attack concrete surfaces. Under such exposure, an external coating giving complete encasement is the only suitable answer.

20.2.3 Freeze-Thaw Salt-Water Attack

This is an extremely severe environmental condition requiring special consideration. It occurs in the tidal and splash zones.

Air entrainment 6% to 8%, is essential for durability. Water-cement ratios should be low (0.45 or less) and cement factors high (eight sacks per cu yd). Also, particular care should be taken to achieve complete consolidation and a smooth dense surface free from fins, rock pockets, etc. Any pockets should be filled with epoxy mortar.

In specific localities, some aggregates are known to be especially durable while others have a less satisfactory history. The relevant factors are not completely known; therefore, it is best to use aggregates which have given satisfactory service.

Jackets of treated timber, and wrought iron, have been extensively used in the past, as has granite facing. When properly done, these give excellent protection, but their cost is high. It is believed that careful attention to the internal steps listed above will give entirely satisfactory results.

With posttensioned anchorages, these should be recessed, so that the resultant patch ends up flush. The patch should be of epoxy mortar. The exposed ends of pretensioning tendons should be cut back and patched by epoxy.

20.3 CONSTRUCTION METHODS AND PROCEDURES

20.3.1 Concrete Mix Design

-1. Normal weight concrete—Normal weight concrete for marine use must be proportioned so as to achieve maximum durability for both the concrete and its embedded reinforcement.

Aggregates should be selected in accordance with the normal test requirements, i.e., ASTM C33, and should be nonreactive, sound, and abrasion resistant. They should be free from salt and dust deposits. The total amount of chlorides in sand and aggregates should not exceed 0.02%.

Satisfactory concrete has been produced from beach sands by washing with fresh water. Properly selected coral aggregates can be utilized but generally produce concrete of reduced strength. Proportioning should be such as to insure a dense compacted mix of minimum permeability.

Water must be free of excessive chlorides (not more than 500 ppm) and sulphates (not more than 1000 ppm).

Seawater has been used successfully for unreinforced concrete in a marine exposure and for reinforced concrete that is fully and permanently submerged. The cement content must be especially rich. However, seawater has generally proven unsatisfactory for reinforced concrete construction, the chlorides having lowered the *pH* near the steel below the corrosion point. Prior coating of reinforcement with a cement slurry made with fresh water has proven helpful. While these steps help in prolonging the life of reinforced concrete exposed to the action of both air and sea (or salt fog), seawater is definitely a choice of last resort for all reinforced marine concrete.

Cement should have a moderate C_3A content, low alkali content (less than 0.65% Na_2O and K_2O). ASTM Type II is generally to be preferred. Cement content should be high; six to eight sacks per cu yd is recommended.

The water/cement ratio must be kept as low as possible to minimize permeability. (w/c ratio of 0.45 or less by weight.) (See section 20.3.2.3 for special requirements for concrete placed underwater.)

Air entrainment of 4% to 8% is beneficial, and is essential for concrete that will be exposed to freezing and thawing.

Admixtures may often be beneficial in reducing the water/cement ratio, in preventing segregation, and in reducing the heat of hydration in mass pours. Admixtures should not contain more than a trace of chlorides.

Reinforcing steel should be well-distributed.

Adequate cover of 2 to 3 in. should be provided over the reinforcement. In special cases, lesser covers have been satisfactorily employed, by paying particular attention to a high cement factor, small aggregates, and a dense surface finish. For concrete cast in the dry or precast, thorough curing is essential. Steam curing, if used, should be followed by water curing for three to seven days. Water curing alone should be 7 to 14 days. A period of drying out afterward is desirable.

-2. Heavyweight concrete—Heavyweight concrete may be employed in certain marine structures to give maximum net negative buoyancy. Typical examples are pipeline encasement, to keep the cover from getting excessively thick and heavy in air; and bottom-founded structures where the heavy weight is used to offset uplift and tension due to wave forces. Thus, it is possible to produce concrete having a submerged effective weight of 160 lb/cu ft as compared with the 90 lb/cu ft obtained with normal concrete. See ASTM C637.

High density can also be effective for radioactive shielding. It should be of particular advantage in breakwater units (tetrapods, etc.) where the high density will greatly improve resistance to displacement by waves.

The heavyweight mix can be obtained by using magnetite ore as aggregate. Other materials are ilmenite, steel shot, steel punchings, barite, and limonite. The mix must be carefully proportioned, and checked by trial mixes to insure that excessive segregation will not occur. Good grading, rich cement content, low water/cement ratio, and air-entrainment are useful in preventing segregation.

-3. Lightweight concrete—Lightweight concrete is of especial interest for floating structures. The net submerged weight may be one-half or less, as compared with normal concrete. Since minimum dimensions frequently control design, e.g., wall thickness for impermeability, rigidity, ability to place, space and cover for reinforcement, etc., lightweight concrete may show substantial benefits.

Lightweight concrete made with expanded shales, slates, or clays, has a successful history of durability and performance at sea. Its impermeability and durability may in fact be made equal to that of normal concrete with good engineering control.

Certain specific steps are required with lightweight structural aggregate concrete. Air-entrainment, 4% to 6%, is definitely desirable to prevent segregation. The maximum size of coarse aggregate particles should be one-half the cover over the reinforcing steel. The use of approximately 30% to 100% natural sand will improve consistency and performance in diagonal tension in the dry and semi-dry state. Fully submerged, the diagonal tension is approximately equal to that of normal concrete. For typical marine structural elements, the cement factor should be at least seven sacks per cu yd and the water/cement ratio kept below 0.45.

-4. Cyclopean concrete—Cyclopean concrete employs a combination of large natural rock with underwater concrete binder, in order to form seawalls, quay walls, mat foundations and caisson fill.

Clean rocks, all greater than 12 in. diameter, are placed in layers approximately 3 ft thick. The voids, usually running 30%, are then filled with tremie concrete placed by tremie pipe or bottom-dump bucket. Such methods produce considerable laitance, for which removal is usually accomplished by overflowing or jetting of the surface with a diver immediately after initial set.

-5. Sacked concrete—Underwater walls and slope protection have been successfully built using bagged or sacked concrete. In the author's opinion, the best results are obtained with burlap sacks, two-thirds filled with concrete of

a low water-cement ratio. Once placed, these tend to knit together.

Dry mix in burlap sacks has also been employed; however, hydration is not always complete due to the impermeability of the dry mix. Sacks and bags may be placed in stretcher and header courses so as to interlock. Fabric forms of nylon have been developed which are spread over the surface to be protected and then pumped full of grout. The fabric itself may be porous to act as a combined filter and protective layer.

20.3.2 Underwater Concrete

-1. General—High-quality concrete can be constructed underwater, provided the required engineering and construction techniques are exactly followed. With care, the resultant concrete structure can have strengths and durabilities equal to those obtained in the best practice in the dry. The problem is, carelessness and inexpertise can lead to much greater disaster underwater than in the dry. If the mix and placement techniques are violated, then results underwater will be very bad. The cost of repair or replacement will be excessive.

All techniques for underwater concrete placement are designed so as to prevent the water from mixing with the concrete and washing out the cement paste. Since the concrete placed underwater normally must be moved downward through a considerable height, steps must be taken to prevent segregation.

The first requirement may be met by conveying the concrete in a closed bucket or tube and depositing it immediately on top or *under* the already poured concrete, so that the new concrete has minimum contact with the water.

The second requirement is additionally aided by selection of a mix that flows easily within itself yet has great cohesiveness and richness.

Concrete containing normal portions of coarse aggregate, has been successfully placed underwater to depths approaching 200 ft. Grout-intruded concrete and grout itself has been successfully placed to even greater depths, 500 ft and more. Major structural constructions such as bridge piers and drydocks, with embedded reinforcement and structural steel have been accomplished on a regular and consistent basis. Precast concrete elements have been joined underwater in full composite action by means of underwater-placed concrete.

-2. Forms for underwater concrete—Side forms for underwater structures may usually best be formed from precast concrete. These are rigid enough to withstand excessive deflection from the liquid head of the concrete as it is poured. When the joining face between precast and underwater concrete is steeply inclined or vertical, the two concretes will obtain a high degree of bond and may even chemically knit.

Steel side forms have also been extensively used, especially on bridge piers. They are most efficient for cylindrical configurations where the steel can be used in tension. Steel sheet piles are commonly and extensively employed as side forms for bridge piers, intake structures, etc.

Stone-filled or steel-weighted timber cribs have been much used in the past. Even dikes of well-graded rock can be used to restrain a mass pour. Sand bags or concrete-filled burlap sacks or bags of cement may be placed to form dikes.

Bottom forms must be adequate to support the weight of the fresh concrete, prevent leakage out, or water erosion in. The most common bottom form is a blanket of graded gravel. Precast concrete bottom forms are frequently employed where the natural bottom is soft mud or exposed to the water.

Top forms are seldom required. When needed, e.g., to form a special seat for a gate, they should be of precast concrete or a combination of structural steel and precast concrete. The under surface must be convex downward, so as to promote a flow of the concrete up and around. Otherwise, laitance and weak material may be trapped underneath.

Frequently, all that is necessary to achieve a suitable top surface is to over-flow the top and then screed it off, working to rigid screeds. In most bridge piers and cofferdams the top surface will be later exposed in the dry. Then it can be chipped to proper grade and a topping lift placed in the dry.

Underwater forms must be tight, especially if the water is moving. Currents and waves will leach cement out of even small cracks, producing a weak zone which may extend to a considerable depth. Sealing between side forms and the bottom is often accomplished by diver-placed sand bags, or by weighted canvas. Canvas and rubber hose rings have been used to seal horizontal joints. Pressures at such joints are usually very small: the full fluid head of the fresh concrete generally is dissipated by the friction loss through the joint.

The minimum cover provided over reinforcement for underwater placed concrete should be of the order of 6 in., taking into account the nature of any permanent forms, the size of aggregate, and the accuracy of construction.

-3. Underwater concrete mixes

-3.1 Tremie concrete—The recommended mix is as follows:

Coarse aggregate: For very large pours—1.5 in. maximum; for normal pours—0.75 in. maximum; for very restricted pours—0.375 in. maximum.

Always use rounded aggregates (gravels of spheroidal shape). Avoid elongated or sharp-edged aggregates. Maximum size of coarse aggregate should also be related to space between reinforcement so as to permit ready flow without vibration.

Fine aggregate: Use 42 to 45% fine sand. *Cement*—A rich mix is required, seven to eight sacks per cu yd, ASTM Type II cement. *Water*—To produce 6 to 8 in. slump. w/c 0.45 max.

Admixtures: Air-entrainment of 4% is frequently desirable. Water-reducing and plasticizing admixtures may be employed with excellent results; however, not all those commercially available and marketed for this purpose are satisfactory. Further, specific admixtures may be incompatible with certain brands of cement. Therefore admixtures with a satisfactory history of performance with the cements in question are to be preferred. Where inadequate prior experience is available, then a field test should be made to verify performance.

-3.2 Grout-intruded aggregate—Coarse aggregate is selected to ensure approximately 20% to 30% voids. Fines below the minimum size, e.g., 0.625 in., *must* be positively eliminated. This may require rescreening just prior to placement. On one major project where the aggregate was transported by barge, fines accumulated from load to load and worked their way to the deck of the barge. Every so often, an overzealous operator cleaned off the deck with his clamshell bucket, placing it in the structure and thereby forming a layer of relatively impervious fines which later did not get properly cemented.

The grout may be a sand-cement or neat-cement grout. Sand-cement grouts usually are richer than a one-to-one mix, sometimes two parts cement to one part sand. Various admixtures are employed to prevent segregation. One process uses a colloidal mixing technique to prevent segregation and improve flowability.

Various wetting agents may be employed to promote penetration of the grout. These admixtures and agents are proprietary and it is essential the manufacturer's recommendations be followed in detail. Techniques and admixtures should either have had extensive satisfactory performance on past structures or be thoroughly proven in field tests prior to use on the structure itself.

-3.3 A typical tremie grout may have the following mix:

Aggregate: Well-graded sand with a moisture content of 7% or less, from 500 to 1500 lbs per cu yd. Sufficient fines should be present to "carry" the coarse particles and prevent segregation. *Cement*–Type II, from 10 to 16 sacks per cu yd. *Water*–To give the consistency of thick soup.

Admixtures: Air-entraining, water reducing, and platicizing.

Neat-cement grout slurries (15 or 16 lbs cement per gallon of water) have been successfully placed at great depths underwater.

-4. *Underwater placement*[20-9] –

-4.1 Tremie process–The concrete is placed through a tube whose lower end is always kept immersed in the fresh concrete. The tube may run from 2.5-in. diameter for tremie grout, up to 10 or 12-in. diameter for large tremie concrete pours. Pipes are generally of steel, with flanged joints. The joints must be well-gasketed and sealed, otherwise the water may leach cement from the concrete. Occasionally, rubber hoses are employed, especially with tremie grout.

The bottom end of the tube must be kept below the surface a minimum practical distance, say 2 ft, yet must not become fixed in set concrete. Thus, it must be carefully raised. The raising must be done smoothly and gradually.

On one major bridge pier, due to errors of sounding, the pipe was raised several feet above the surface of the previously-poured concrete. This resulted in a layer of sand and gravel, the cement being completely washed out. Then when the gravel once more mounded around the end of the tube, good underwater concrete was obtained, thus covering the loose sand and gravel with a structural shield. Upon dewatering the cofferdam, the hydrostatic head caused a blow in the bottom. The eventual removal and replacement cost many hundreds of thousands of dollars.

On another project, the tremie tube was left too low too long, and the fresh concrete was injected below an already set plug. This acted as a hydraulic ram, raising the heavy cage of reinforcing steel.

The above examples indicate the importance of controlling the tubes' height.

For initial starting of the pour, a seal (e.g., gasketed plywood) is affixed to the bottom, the tube lowered to the bottom, filled with concrete (in the dry), then the tube is raised a few inches, the weight of concrete breaks the bottom seal, and the concrete flows out. If the concrete surface drops too far in the tube, the tube is lowered back. Flow into and through the tube should be as continuous and regular as possible. Thus, the feed to the hopper on the tremie pipe should be by conveyor, or by a controlled bucket (hydraulic or air controlled), or by pump–not by sudden opening of a bucket.

When the water is deep, the tremie tube may float while empty. The tube may be weighted. Alternatively, it may be placed open-ended, and a ball–such as a backetball or volleyball–placed in the top. The added concrete pushes the ball down, expelling the water. The ball pops out the bottom and usually comes to the surface. On very deep pours, the ball gets progressively smaller and may even collapse due to hydrostatic pressure. A pipeline pig, (a steel cylinder with neoprene "skirts" which loosely seal the pipe as the pig moves through it under unbalanced pressure), is thus more suitable.

The flow of water ahead of the ball or pig may scour the bottom or erode fresh concrete that has been previously placed. Thus, the filling and pushing down of the ball must be done slowly.

A newly-developed means is to place the main tremie pipe and seat it firmly on the bottom, still full of water; then a smaller tremie pipe, with plugged end, is inserted in the main tremie pipe, and the small pipe is filled with concrete. It is then slowly withdrawn, filling the larger pipe, which is then subsequently used in the regular manner.[20-2]

A variation of this has been developed for extremely deep (400-ft) pours planned for the North Sea. The main tremie pipe (12 in.) is placed with a pig at mid-height (-200 ft). It is open below the pig, sealed above. The concrete is fed into the upper compartment by a small (6 in.) tremie pipe. This forces the pig down and eventually out the lower end. Fresh concrete is fed continuously in through the 6 in. pipe and the column of concrete in the lower half of the main tremie balances the external water head, thus achieving a self-regulating flow.

In placing tremie concrete, the seal may be lost if the pipe is raised too high, or if the control of the pipe and the feed of the concrete is erratic and jerky. Then it is necessary to go through the starting process once again, taking particular pains not to erode or disturb the fresh concrete previously placed by a rush of water or agitation of the pipe.

A recent development in the Netherlands is a rubber tremie tube. The tube is collapsed flat and held this way by the water pressure. As a slug of concrete is fed into the top, it opens the tube ahead of itself as it slides down. Above the slug, the tube is closed again by the water pressure. For this case, the concrete may have a somewhat lower slump, say 4 in. Progressive slugs are fed in at the top.

Valves have often been proposed and tried on the bottom of tremie pipes. They are not successful due to the fact that the concrete is a mix rather than fluid. They should not be used.

-4.2 Tremie grout–This should be well mixed in a machine mixer and placed through a pipe at relatively low pressure, sufficient only to accomplish uniform flow. When placing at great depths, the gravity head may cause the concrete to run out too fast, creating a partial vacuum and sucking air or water through the joints. Therefore, all joints should be sealed tight, and the pipe should be properly sized to pass the grout while developing a high enough friction head to keep the pipe full.

-4.3 Underwater buckets–Special bottom-dump buckets are available which have covers over the top so as to prevent contact with sea water. They are lowered to the bottom, then opened so that the concrete flows out over the surface. This method produces much more laitance than does concrete that is properly placed through tremie tubes: thus, its use is primarily adapted to mass pours, cyclopean walls, etc. The bucket method is generally not suitable where there are piles extending up from the bottom, as these will force the concrete to fall through the water. The bucket method is particularly useful for placing concrete in moving water or the surf zone, etc., as for filling a pipe trench. It can be placed with much lower slump and larger aggregate, then covered with canvas, etc. to prevent erosive scour.

-4.4 Grout-intruded aggregate—Grout pipes and recording pipes are preplaced in the forms or tied to the reinforcing steel, then the coarse aggregate, having no fines, is placed. Grout is then pumped in. The recording tube signals (by an electrical resistance gauge) that the grout has reached the prescribed level.

For best results, a blanket of aggregate should be kept 2 to 3 ft above the maximum grout level for that stage. This is particularly important where the surface is exposed to moving water. For deep pours, pipes, aggregate, and grout are placed in several stages.

It is essential to protect the preplaced aggregate from contamination by silt or sewage, etc., between the time of placement and grouting. In any event, the time interval should be kept as short as possible.

Grout-intruded aggregate is particularly well-adapted to concreting structural elements containing many inserts and embedded items.

-4.5 Underwater-setting cement—A number of proprietary cements have been developed which set rapidly underwater. These are mixed in small batches, then lowered in a covered can to the diver who uses them for patches and repairs. They may gain their set in a matter of minutes, thus requiring no external protection against the water.

20.3.3 Precast Concrete

This technique is particularly well adapted for marine structures of all types: piling, sheet piles, breakwaters, overwater, and submerged structures, etc.

The conceptual design should consider the available size of equipment for transport and erection or installation. Precast elements must be designed for adequate structural strength during these phases as well as for service in the permanent structure.

Connections of precast elements require extremely careful detailing. Not only must the connections perform adequately as in conventionally-reinforced concrete, but they must accommodate the greater tolerances generally imposed by the marine environment (waves, current, survey difficulties, etc.) and must be durable in themselves against corrosion.

a. Cast-in-place connections, with overlapping or welded reinforcing steel have been widely employed in the past. The surfaces should be sand-blasted or wire brushed to remove laitance prior to setting. After setting, these surfaces and the reinforcing steel should be washed with fresh water to remove any salt. The concrete should have a very low water-cement ratio to minimize shrinkage, and should be well vibrated. Particular care should be taken with curing in order to prevent shrinkage cracking.
b. Thin concrete joints are much used in English practice. These are generally 3 in. thick, poured with a fine concrete with 0.375 in. maximum coarse aggregate, well-vibrated.
c. Matching joints (dry joints) in which the faces are cast against each other in the casting yard, produces excellent results. The surfaces may be coated with a thin coat of epoxy just before joining in the field.
d. Grout pads—cement mortar spread thinly over a joint, are useful only in horizontal joints where the weight aids in achieving uniform compaction. Even here, however, some degree of non-uniform bearing may result.

20.3.4 Prestressed Concrete

Prestressed concrete assures the highest quality of performance in the marine environment. The resultant structure will be essentially crackfree, and have high resistance to cyclic and impact loadings.

With pre-tensioned concrete, thorough vibration will cause the cement paste to fill the interstices between the wires of strands. Exposed ends of pre-tensioning strands should be daubed with epoxy, although corrosion will normally only proceed about 0.125 to 0.5 in. deep.

With posttensioned concrete, the ducts should be washed clean with fresh water, then blown out with air before placing the tendons. The tendons should be clean, dry, and free from salt. The author recommends the dusting on of Shell Vapor Phase Inhibitor powder (VPI) as the tendon is pulled into the duct, if the tendon is to be left ungrouted more than a few hours. The ends should then be sealed until the time of stressing.

As soon as possible after stressing, the ducts shoud be grouted. Strands are preferable to wire tendons because the grouting is usually more thorough.

The end anchorages should be flush-type (recessed) anchorages. These recesses should be filled flush with an epoxy mortar. Particular care should be taken to seal the joint.

Posttensioning may be very usefully employed to achieve continuity across joints and to make the structure act as a whole. The ducts must be made continuous across the joints by means of sleeves. These should be taped with a waterproof tape.

20.3.5 Ferro-Cement

Ferro-cement is extensively utilized in concrete boat building and more recently is being applied to ocean-going barges.[20-6] Essentially, it consists of many layers of wire mesh, each coated with a thin layer of cement mortar placed by hand or by shotcrete. The total amount of steel is proportionally large, 800 lbs per cu yd or so.

Ferro-cement is essentially crack-free (the tiny hair cracks are arrested by the mesh). It is durable, although minor rust staining may occur. It can readily be formed to complex double-curved surfaces. It behaves extremely well under impact, giving local crushing only. Thin ferro-cement panels are very flexible.

A recent related development is the product known as Wirand, developed by Battelle Development Corporation, in which finely divided wires are mixed with the concrete.[20-7] Nylon fibers have been similarly mixed; these are not as effective in resisting cracks but do give excellent toughness under impact.

20.3.6 Polymer-Impregnated Concrete

Polymer-impregnated concrete holds forth excellent promise for marine structures, particularly vessels, deep-sea habitats, and arctic structures subjected to ice abrasion. Precast panels are impregnated with a monomer, preferably under vacuum, and then subjected to either irradiation or thermal treatment to change the monomer to a polymer.

The resultant material is extremely strong (15,000 psi and higher), impermeable, very durable against freeze-thaw and salt-water attack, and highly abrasion-resistant.

The technology is just beginning to develop into practical application, but holds great promise.

20.4 SPECIAL DESIGN CONSIDERATIONS FOR MARINE STRUCTURES

Structural design should be carried out in accordance with the standard provisions of ACI-318, with specific consideration given the matters discussed below.

Piling must be designed for column action, with consideration of the degree of fixity at the head and in the soil; accidental eccentricity; lateral loading, as from waves or soils; and dead weight bending in the case of batter piles. In certain framing schemes, the piles may be subjected to combined moment and axial compression or tension.

Because of the corrosive nature of the marine environment and because of the frequent need to prevent leakage, cracking must be strictly limited. This may dictate prestressing or a lower than normal stress level in reinforcing steel. The frequency of cracking loads must also be considered in evaluating stress levels for design.

The amount of cover and special requirements for materials, mix, placement, and protective coatings must be specified, as outlined in section 20-2. All connections and fittings must be properly protected from corrosion.

Hollow structures, partially or wholly submerged, must be designed for hydrostatic loading, taking into account not only the basic shape of the structure but also the effect of tolerances in out-of-roundness, etc. due to form tolerances and plastic flow.

Deep-submerged vessels must consider implosion and in-plane lamination (cracking) as failure mechanisms.[20-17]

20.5 SPECIFIC APPLICATIONS

The application of concrete to marine structures has been so widespread that it is impracticable to attempt to discuss all past applications in detail. However, a brief description of main categories may be useful in indicating the prime considerations, precautions, and results in specific areas, as well as furnishing a background for the future applications to the entire field of harbor, coastal, and ocean structures.

More detailed and extensive information for specific areas is contained in the references listed at the end of this chapter.

20.5.1 Wharves, Jetties, Moles, and Trestles

These structures represent one of the largest uses of marine concrete.

-1. Piling—Prestressed precast concrete piling, ranging in size from 12 to 36 in. square and from 24 to 72 in. round have been extensively utilized for bearing piles, batter piles, and moment-resisting, vertical structural columns. The larger sizes (generally those above 20 in.) are usually hollow-cored with wall thicknesses of 4.5 to 6 in.

Prestressing is usually designed to give a minimum effective prestress of at least 750 psi, in order to permit handling and driving without damage. For very long piles, or batter piles, or moment-resisting piles, higher prestress values are desirable, usually 1000 to 1200 psi.

Hollow-core piles must have sufficient spiral (0.4% steel area or more) to prevent vertical cracking, which may arise from a wide variety of causes. Vents should be provided, below the water line to prevent freezing, near the top to prevent hydraulic ram and hydrostatic effects.[20-15] Prestressed, concrete cylinder piles, 54 in. in diameter, up to 160 ft in length were employed at Baton Rouge. See Fig. 20-1.

Fig. 20-1 Prestressed cylinder piles, Port of Baton Rouge, La.

-2. Deck structures—The concrete caps must accommodate driving tolerance in the piling, yet assure proper structural behavior, e.g., fixed-end or tension connections to the piles, in accordance with the design. When precast concrete caps are used, it is good practice to make them over-width and over-length so as to enable them to be used under the most adverse combination of tolerances in pile location.

Concrete deck slabs are usually designed for continuity. They often incorporate the pile cap within the deck, thus minimizing formwork. One of the most efficient schemes of wharf construction, for example, utilizes an absolutely flat soffit and constant thickness deck, the cap being obtained by reinforcement within the confines of the slab.

Precast deck slabs are extensively employed. These may be full-depth slabs, or simple-span construction, or half-depth slabs, designed to work in composite action with a poured-in-place top slab, thus permitting the attainment of continuity over the cap.

-3. Concrete edge beams, curbs, pipe beams, fire walls, and catwalks fill out most of the remaining components of a modern wharf structure.

-4. Fender piles are a relatively new but well proven application of prestressed concrete. They are installed on major wharves in Kuwait, Singapore, and Los Angeles (Fig. 20-2). They should be designed for energy absorption and deflection.[20-15]

The over-water extensions of La Guardia Airport in New York represented one of the largest and most successful applications of precast, pre-tensioned, and posttensioned

Fig. 20-2 Prestressed concrete fender piles.

concrete, designed and constructed to meet extremely severe design criteria in an adverse marine environment.

20.5.2 Offshore Platforms and Terminals

Caissons appear to have a renewed role as breasting dolphins for the new large offshore terminals, as well as for oil production and storage. It is contemplated that they will be initially cast to partial height in a shallow basin, then launched at high tide, moored to an outfitting dock, completed afloat, towed to the site and sunk onto a prepared foundation.

The lower portion is frequently posttensioned. The upper walls may be slip-formed or cast with successive panel form methods.

For exploitation of the huge Ekofisk oil field in the North Sea, a very large caisson has recently been completed in Norway. It has double walls, the outer wall being perforated

Fig. 20-3 Ekofisk Oil Storage Caisson under construction afloat in Stavanger Harbor, Norway; 86,000 cu m of prestressed, precast, and cast-in-place concrete destined for service in the North Sea.

Fig. 20-4 Concrete caisson for lighthouse, Irish Sea.

to absorb and dissipate wave energy as can be seen in Fig. 20-3. Much of the structure is prestressed.

Concrete caissons with tapered superstructures, both fixed and telescoping have been installed as lighthouses in Scandinavia, the Irish Sea (Fig. 20-4), and Eastern Canada. These were built in successive basins, in order to facilitate launching of the partially completed structures.

For the Arctic Ocean concrete caissons with conical tapered superstructures (Fig. 20-5) appear to have major advantages over other types of structures, (Ref. 20-11). The ice will ride up the cone, thus failing in radial and circumferential tension. The concrete caisson has a very low natural frequency of vibration due to its mass, well below the 1- to 2-sec frequency of crushing ice. The concrete is rigid and is not adversely affected by the low temperatures.

20.5.3 Floating Structures

The first recorded use of reinforced concrete was the concrete boat which Lambot built about 1858 in France.

Reinforced concrete ships were constructed in substantial numbers in World Wars I and II, and a considerable number of concrete floating drydocks and moored floating docks are in service throughout the world. Concrete caissons for bridge piers have been launched, towed to position, to be later sunk into the bottom. Some, such as the Martinez-Benecia Bridge in California and the 1st Avenue Bridge in Seattle, Washington, utilize permanent buoyancy to carry much of the dead load.

The more recent advent of prestressing makes possible more efficient structural designs and has given new impetus to this development since prestressing offers superior performance along with substantial economy.

With prestressing, the entire wall thickness is able to work

Fig. 20-5 Concrete caissons for Arctic Ocean platforms.

to resist local bending stresses. Cracking, with its implications for durability, fatigue, and water-tightness is essentially prevented. The ability to resist impact from collision or other accident is generally improved. Concrete placement and consolidation is facilitated. Most modern floating structures are therefore prestressed to a greater or less degree.

-1. Barges and ships—Almost 20 large pretensioned concrete barges, designed by Honolulu-based engineer Al Yee, have been constructed in the Philippines and have seen practically continuous ocean service since 1964. First costs were reported competitive with steel, actual maintenance has been substantially less, experience in collision, etc. has been definitely superior. These barges are generally of 2000-ton capacity and carry both dry cargo and petroleum products.

Recently a number of barges and dredge hulls have been built in New Zealand of prestressed concrete for service in the South Pacific.

Serious design studies and proposals have been developed for prestressed concrete barges for the transport of cryogenic materials. The favorable behavior of concrete and especially prestressed concrete at very low temperatures promises added security for this type of usage.

Reinforced and partially-prestressed concrete compressor and production barges have been built in substantial numbers in Louisiana, towed as far as Mexico and Nigeria, and sunk in shallow water as permanent stations (Fig. 20-6).

Economic studies show that the greatest percentage savings can be achieved in large hulls, where the ratio of skin surface to volume offsets to a considerable degree the adverse effect of concrete's higher dead load. Thus, a number of designers are currently investigating very large barges and even oil tankers and ore carriers.

Structural lightweight concrete was utilized with excellent results and durability in some of the ships from World Wars I and II; it appears that prestressed lightweight concrete may be an ideal material for concrete vessels.

In designing hulls for marine use, it is considered desirable to add the hydrostatic and local loadings to the design moment (from hog and sag), and then to compare these combined stresses to the tensile (cracking) strength in the concrete. Then the ultimate condition must also be checked to ensure an adequate factor of safety against failure. Such vessels must of course also be designed to survive collision just as must steel vessels.

Fig. 20-6 Concrete barge for service as production and processing platform.

-2. Offshore Airports—The LaGuardia airport extension was constructed over water on a piled platform, using prestressed concrete for its 40 acres of runways, (Fig. 20-7). Many proposals have since been prepared along with feasibility studies, for floating airports. These generally propose the use of concrete pontoons, prestressed together so as to

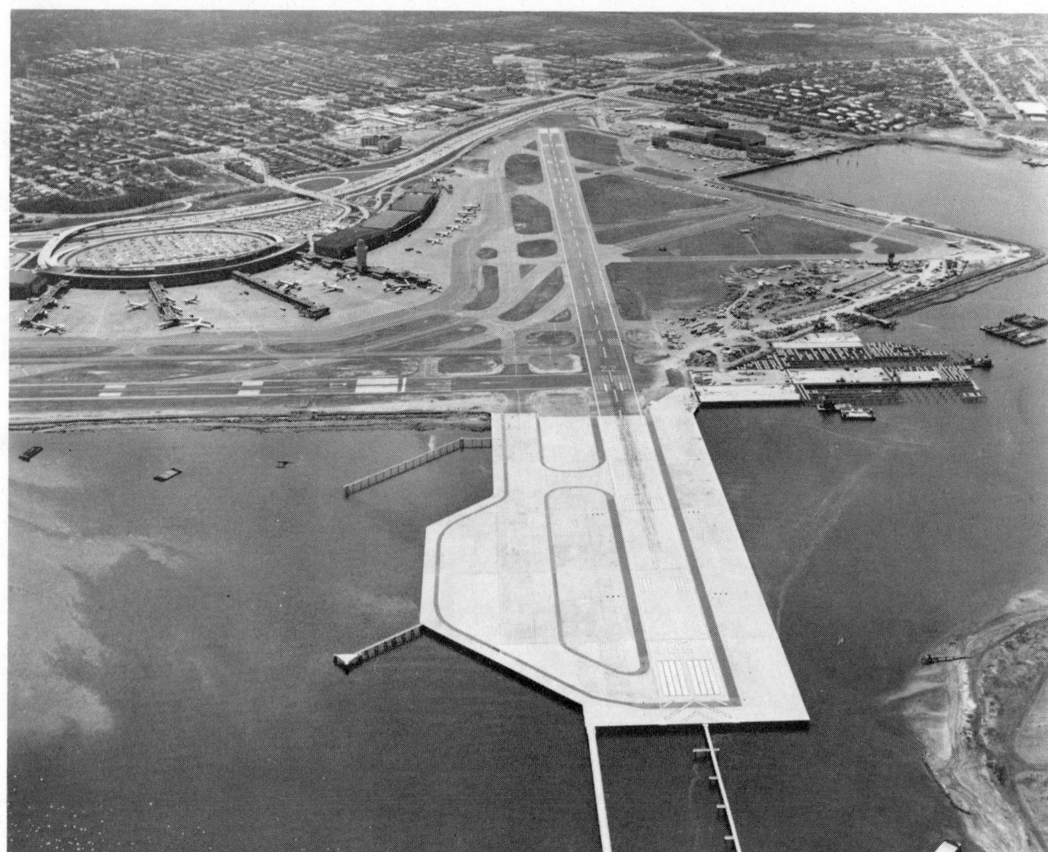

Fig. 20-7 Overwater extension of runways at LaGuardia Airport, New York.

Fig. 20-8 Proposed floating airport of prestressed concrete, England.

Fig. 20-9 Prestressed concrete floating bridge, Hood Canal, Washington

act as a huge flexible mat, thus distributing concentrated loads from aircraft over a large area, (Fig. 20-8).

The pontoons may well be prestressed within themselves to care for local stresses and hydrostatic loading. They are then posttensioned together using tendons in ducts in the concrete or in separate tubes inside the pontoons. In the latter case, the tubes of course must be protected from corrosion.

-3. Floating bridges—Floating bridges have been built across Lake Washington and Hood Canal, both near Seattle. The most recent of these are prestressed longitudinally. The Hood Canal Bridge initially gave trouble due to excessive movement between units during high waves and wind. It was subsequently strengthened by making the individual units up into groups of four, with epoxy concrete joints and heavy longitudinal posttensioning, (Fig. 20-9).

A somewhat similar bridge in Hobart, Tasmania was less successful, suffering continual minor damage during storms. After some 10 years, it was replaced by a high level bridge.

-4. Stable ocean platforms—For many years, ocean structures have been proposed which achieved their stability from being supported on long vertical cylinders. These cylinders, with or without enlarged bases, move the mass below the waterline and reduce the effect of the waves on the structure. Roll and pitch are substantially reduced, and the heave period can be lengthened beyond the area where the structure can respond resonantly with the sea. The problem has been largely one of economics, since the cylindrical legs are subjected to high external hydrostatic forces as well as structural loading.

Recently, this concept has been revived, with the cylindrical legs to be constructed of high strength concrete. Such legs would be manufactured horizontally, longitudinally posttensioned. They would then be launched, pairs or groups joined structurally, and then up-ended by water filling. During the up-ending, high bending stresses occur. This procedure must be verified by model tests as well as calculations.

20.5.4 Marine Bridge Piers

Piers for major overwater bridges have long been constructed of concrete. A few of the types most widely employed in the present day are discussed in the following sub-sections:

-1. Cofferdams—The conventional construction method is to drive sheet piles or install cribs, then seal the bottom with a course of underwater concrete placed either by the tremie or grout-intruded method. Then the cofferdam is dewatered, the seal cleaned, and the structural footing block poured in the dry.

A newer method which makes substantial economies possible, combined the seal course and footing block. After the sheet piles are installed and the interior excavated to grade, reinforcing steel cages, often combined with structural steel, are lowered into position and high quality underwater concrete placed. Such a procedure reduces the required penetration on the sheet piles, reduces the excavation, reduces the bracing, eliminates the seal, and may reduce the area to be dewatered to just that of the pier shafts.

-2. Caissons for bridge piers—These are generally constructed in stages. The lower lift or cutting edge is constructed on a ways or in a shallow basin, then launched and floated out to an outfitting dock. Here the walls are raised using panel forms, slip forms, or precast panels. Then it is towed to the site where it is progressively sunk by raising the top walls and excavating from within. When it reaches its design depth, the bottom is usually plugged with a course of underwater concrete.

Large concrete caissons have been used for such major bridges as the Lillaebelt Bridge in Denmark, the Carquinez Bridge in San Francisco, and are proposed for the Intercontinental Peace Bridge across Bering Straits (Fig. 20-10).

-3. Cylinder piles—Very large cylinder piles of prestressed concrete are increasingly being employed to support major bridges. These are then capped above water to form the bridge pier. The prestressed cylinder piles employed for the

Fig. 20-10 Concrete caissons for the proposed Intercontinental Peace Bridge across the Bering Straits.

San Diego-Coronado Bridge were 54 in. in diameter. Fourteen-foot diameter prestressed concrete cylinder piles weighing 660 tons were utilized for the Oosterschelde Bridge in The Netherlands. These were sunk by internal excavation and external weighting, then plugged with a course of grout-intruded aggregate.

Recently 2000-ton caisson-piles were placed by a huge shearlegs for the piers for the bridge at Hiroshima, Japan (Fig. 20-11).

-4. *Bell piers*—This is a form of box-caisson construction utilizing precast concrete or steel shells to form the outer portion of the pier. These are lowered into place, fixed, then filled with structural tremie concrete.

The pier itself is never dewatered.[20-4] This scheme has been used on a number of very large projects: the Narragansett Bay Bridge at Newport, Rhode Island; the Richmond-San Rafael and San Mateo-Hayward Bridges in California; and two bridges across the Columbia River (Fig. 20-12).

Extensive cores from the San Mateo-Hayward Bridge proved that with proper materials, mixes, and placement techniques and control, high quality structural concrete can be consistently obtained underwater.

Fig. 20-11 2000-ton concrete caissons for the Hiroshima Bridge, Japan.

Fig. 20-12 Concrete bell-pier units for the Columbia River Bridge, Astoria, Oregon.

20.5.5 Subaqueous Tunnels

These include some of the largest and most challenging structural achievements of modern engineering: the John F. Kennedy tunnel at Antwerp (Fig. 20-13), the Chesapeake Bay tunnels, the Oakland-Alameda and BART tubes in San Francisco Bay, the LaFontaine tunnel at Montreal, and the East River, New York, and Mobile, Alabama tunnels presently under construction in the USA.

The general pattern has been to construct these in segments, either in a dewatered basin or afloat, then tow them to position and accurately lower them to grade by controlled flooding. Individual segments may consist of 30,000 tons of concrete or more. Once in place on the bottom, they are joined, usually by a procedure which includes a tremie concrete seal. The huge size of these units and the great effect that minor deviations in wall thickness, water absorption, etc. may have on the net buoyancy just before sinking, makes it desirable to provide means for making positive adjustment, e.g., weight boxes to be filled with concrete to a computed depth, prior to final positioning and sinking.

20.5.6 Other Subaqueous Structures

Similar in principle to the above are the many underwater structures currently being actively engineered and proposed: underwater oil storage vessels, sub-sea oil well completion chambers, habitats, mining chambers, etc.

Because of the hydrostatic pressures involved, great care must be exercised to achieve tolerance and quality control, high strength, and impermeability. External or internal liners of steel or epoxy mastic may be required. Manufacturing methods must be chosen that will minimize the effects of creep and shrinkage during construction, when the support conditions and effective weight are basically different from those in the final location. Such structures may fail by implosion and lamination rather than by conventional compression, so this phenomena requires careful consideration.

Submarines are now being proposed for commercial use such as subsea oil transport. The design criteria for such structures is very severe because of the consideration of overpressure conditions which may occur during accident. Concrete is an ideal material for this purpose, but must be of the highest quality achievable under present technology. Special reinforcement may be required for transitions and for local stress conditions.

20.5.7 Coastal Structures

Concrete has been extensively utilized for seawalls and to a lesser extent for breakwaters. It is abrasion-resistant, heavy in unit weight, and durable.

-1. Seawalls—These frequently employ concrete sheet piles. In modern construction, these are usually prestressed, to provide greater durability and crack-resistance, as well as inherent economy. Prestressing may be somewhat eccentric or may be supplemented by mild steel to take care of the moment conditions, particularly negative moment at the wale or tie-back level.

Tie-backs to the anchor are usually steel rods but more recently prestressed concrete has been employed because of its reduced elongation under load and its durability. These concrete ties act as combination deadmen as well, gaining their pull-out resistance from embedment in the backfill, without special deadmen. Such a principle was employed on the anchorage for the Lillaebelt suspension bridge in Denmark.

-2. Breakwaters—Where extreme exposure conditions exist or where natural quarried rock is unavailable, precast con-

Fig. 20-13 Prestressed concrete segments for the John F. Kennedy Subaqueous Tunnel, Antwerp, Belgium.

crete breakwater units are employed. Special shapes have been developed to give proper keying action, wave energy dissipation, resistance to erosion, and prevention of back pressure. These include the tetrapod, tribar, stabit, hollow-block, dolose, etc.[20-1,20-11]

Consideration in the future should be given to the use of heavy-weight aggregate for such units in order to increase their stability.

-3. Seawall caissons—Concrete caissons have been used as cribs, breakwater and seawall units. At Baie Comeau in Quebec, the walls of the caisson were perforated to absorb wave energy, the same concept being employed on the Ekofisk oil storage caisson (section 20.5.2).

20.5.8 Underwater Pipelines

-1. Concrete pipe is extensively employed for outfalls for disposal of waste and for intakes for salt water for the cooling of power plants or industrial uses. Such pipes are generally precast, the largest sizes being cast vertically, then set underwater, snugged tight, and joined.

To reduce the weight of these units for ease in handling, lightweight concrete has been employed. The bell and spigot ends were made of standard hard rock concrete, blended into the lightweight, in order to minimize spalling.

Often, pipe segments have been prejoined to facilitate setting. For the Hyperion Outfall in Los Angeles, the units were prejoined to 96-ft lengths.

Structural box sections, prestressed longitudinally, were used for an underwater interceptor near Seattle. The long units were set on underwater pile-supported piers.

In New Zealand, concrete pipe sections have been assembled into continuous pipelines by posttensioning, then floated or pulled into position just like welded steel pipe.

-2. Concrete encased steel pipe is extensively employed for submarine oil, water, and gas pipelines. The concrete coating serves as weight coating and as protection to the inner corrosion coating of the steel.

Both conventional concrete and heavy weight concrete coating are used. The latter is useful in keeping down the total outside diameter of pipelines, and thus facilitating shipment, installation, and burial.

20.5.9

The strong surge of interest in the development of the resources of the continental shelf have in turn intensified interest in the utilization of concrete for major sea structures. On the one hand, the facilities desired and the environmental criteria are demanding ever larger and more sophisticated structures. Fortunately new technological developments are making available ever better materials and techniques. It is clear that concrete is destined to play a major role in the sea.

REFERENCES

20-1 Myers, J. J.; Holm, C. H.; and McAllister, R. F., *Handbook of Ocean and Underwater Engineering,* McGraw-Hill, New York, 1969.

20-2 Gjørv, Odd E., "Durability of Reinforced Concrete Wharves in Norwegian Harbours," Ingeniørforlaget A/S Oslo, 1968.

20-3 Gerwick, Ben C. Jr., "Placement of Tremie Concrete," Symposium, Concrete Construction in Aqueous Environments, *American Concrete Institute Publication SP-8,* 1964.

20-4 Havers, J. A., and Stubbs, F. W., Jr., Section 28, "Cofferdams and Caissons," Handbook of Heavy Construction, Second Edition, McGraw-Hill, New York, 1971.

20-5 "Effect of Various Substances on Concrete and Protective Treatments, Where Required," *Concrete Information Bulletin 1 S 001.03T,* Portland Cement Assoc. Skokie, Ill., 1968.

20-6 "Ferro-Cement Boats," Concrete Report (CR010.-01G), Portland Cement Assoc., 1969.

20-7 Monfore, G. E., "A Review of Fiber Reinforcement of Portland Cement Paste, Mortar, and Concrete," *Research Department Bulletin 226,* Portland Cement Assoc., 1968.

20-8 Gerwick, B. C. Jr., "Controlled Sinking of Large Concrete Ocean Structures," *American Society of Mechanical Engineers Publication 71-UnT-6,* 1971.

20-9 Gerwick, B. C. Jr., "Underwater Concrete Construction," *American Society of Mechanical Engineers Publication 71-WA/UnT-8,* 1971.

20-10 Gerwick, B. C. Jr., "Construction of Large Concrete Shells for Ocean Structures, Techniques and Methods," Proceedings, Symposium on Hydromechanically-Loaded Shells, International Assoc. of Shell Structures, 1971.

20-11 Proceedings, "Conference on Port and Ocean Engineering," Norwegian Technical University, Trondheim, Norway, 1971.

20-12 Gerwick, B. C. Jr., "Ocean Structures of Prestressed Concrete," *Journal of the Prestressed Concrete Institution,* March–April, 1971.

20-13 Singer, R. H., "Prestressed Concrete Used for Offshore Barges, Storage," *Offshore,* November 1970.

20-14 "Study of Construction Methods for Large Undersea Concrete Structures," CR72.002, Naval Civil Engineering Laboratory, Port Hueneme, California, 1971.

20-15 Gerwick, B. C. Jr., *Construction of Prestressed Concrete Structures,* Wiley-Interscience, New York 1971.

20-16 "Concrete Polymer Material," *Fifth Topical Report, BNL 50275(T-602),* Brookhaven Natural Laboratory and Bureau of Reclamation.

20-17 Proceedings of the Symposium on "Concrete Sea Structures," Federation Internationale de la Precontrainte, London, April 1973. Available through Cement and Concrete Association, Wexham Springs, Slough SL3 6PL, England.

20-18 Jackson, G. W., and Sutherland, W. M., *Concrete Boatbuilding,* John de Graff, Inc., New York, 1969.

20-19 Gerwick, B. C. Jr., "Design and Construction of Prestressed Concrete Vessels," *Preprints, 1973* Offshore Technology Conference, Houston, May 1973.

21

Concrete Dams

MERLIN D. COPEN*

21.1 INTRODUCTION

Dams are usually constructed to store or control the flow of water. A few have been built to retain silt, mine tailings, or debris. The water stored by a dam may be used for municipal and industrial purposes, power production, flood control, irrigation, navigation, or a combination of these.

Dams are classified according to the material used in their construction, i.e., concrete or earth. Concrete dams have grace and beauty, are watertight, and are adaptable to openings in or through them. They require smaller, but better foundations, and smaller quantities of construction materials than earth dams. On the other hand, earth dams can be built on poor rock or earth foundations of materials which are available at or near the site by large, highly mechanized equipment. The type of dam selected for a site must be based on safety, economy and performance.

21.2 TYPES OF CONCRETE DAMS

Concrete dams may be divided into three principal types: gravity, arch, and buttress.

*Consulting Engineer, Aurora, Colorado, formerly Head, Concrete Dams Section, Bureau of Reclamation, Denver, Colorado.

The term gravity dam applies to a solid concrete dam which depends primarily on its weight to resist applied loads (see Fig. 21-1). In cross section gravity dams are roughly triangular in shape. The base width-to-height ratio may vary from 0.6 to 0.9 depending on site conditions, height of dam, and applied loads. Gravity dams are generally straight in plan although some deviate from a straight line in

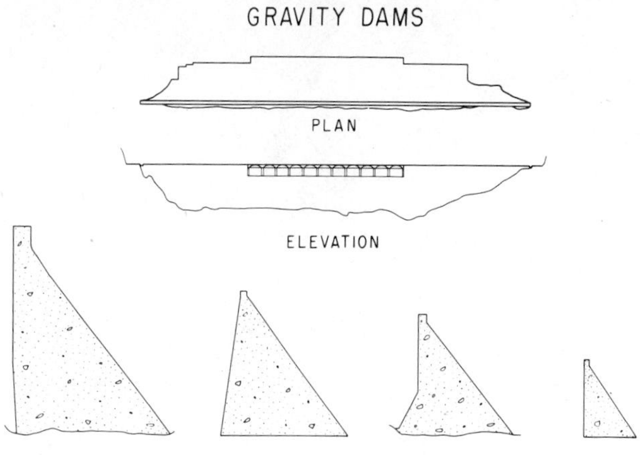

Fig. 21-1 Gravity dams—plan, elevation, and sections.

Fig. 21-2 Grand Coulee Dam, Washington.

Fig. 21-3 Grand Coulee Dam, Washington.

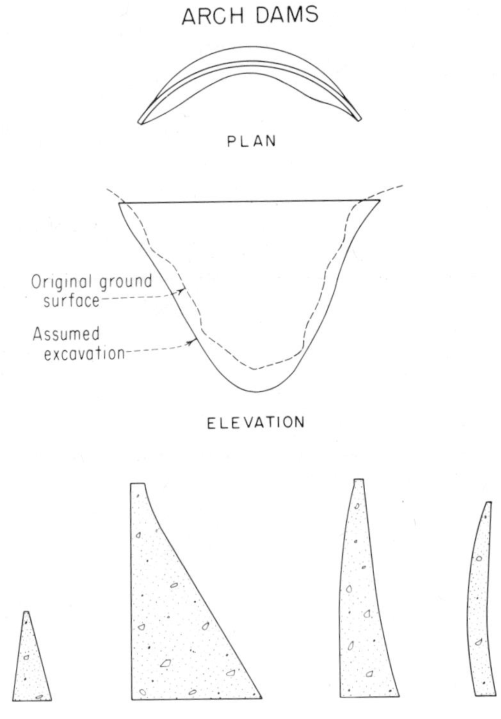

Fig. 21-4 Arch dams—plan, elevation, and sections.

order to take advantage of the topography of a site. Figures 21-2 and 21-3 are photographs of Grand Coulee Dam, an example of a gravity dam.

An arch dam is a solid concrete dam, curved upstream in plan, which transmits a major part of the imposed load to the canyon walls by thrust (see Fig. 21-4). The shape of most arch dams in plan is circular, but some have been designed and constructed with elliptic, parabolic, multi-centered and other complex shapes. In cross section arch dams may have vertical, inclined or curved upstream and/or downstream faces. The base width to height ratio may

Fig. 21-5 Swift Dam on Birch Creek, Montana—view of upstream face.

Fig. 21-6 Yellowtail Dam, Montana.

Fig. 21-7 Glen Canyon Dam, Arizona.

vary from less than 0.1 to about 0.5. Examples of arch dams are illustrated in Figs. 21-5, 21-6 and 21-7.

Buttress dams are comprised of two principal structural elements: the sloping water-supporting deck and the buttresses which support the deck. Buttress dams are classified according to type of water-supporting deck. Slab-and-buttress dams are composed of flat slabs supported on transition sections at the upstream edge of the buttress (see Fig. 21-8). Watertight joints are constructed between the slabs and their supports which permit the dam to adapt itself to variable foundation movements without serious cracking. Stoney Gorge Dam is an example of a slab-and-buttress dam, Fig. 21-9. The water-supporting deck for a multiple-arch dam consists of a series of arch barrels. Each of these is supported by buttresses (see Fig. 21-10). The multiple arch dam has the advantage of greater load carrying ability, for a given span between buttresses, than is economically feasible for flat slab construction. Bartlett Dam, Figs. 21-11 and 21-12, is a multiple-arch dam. The massive-head buttress dam is formed by flaring the upstream edges of the buttresses to span the distance between buttress walls (see Fig. 21-13). The terms round-head and diamond-head, which refer to the shape of the enlargement at the upstream face, more fully describe this type. The concrete portion of Pueblo Dam (Fig. 21-14) is a massive (diamond) head buttress dam.

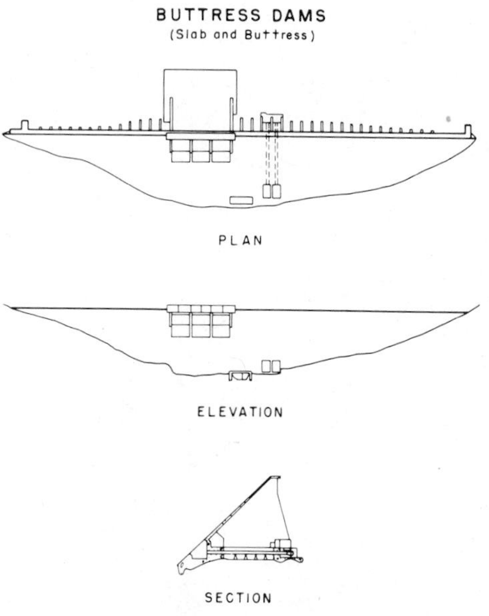

Fig. 21-8 Buttress dams (slab-and-buttress)—plan, elevation, and section.

Fig. 21-11 Bartlett Dam, Arizona.

Fig. 21-9 Stoney Gorge Dam, California.

Fig. 21-12 Bartlett Dam, Arizona.

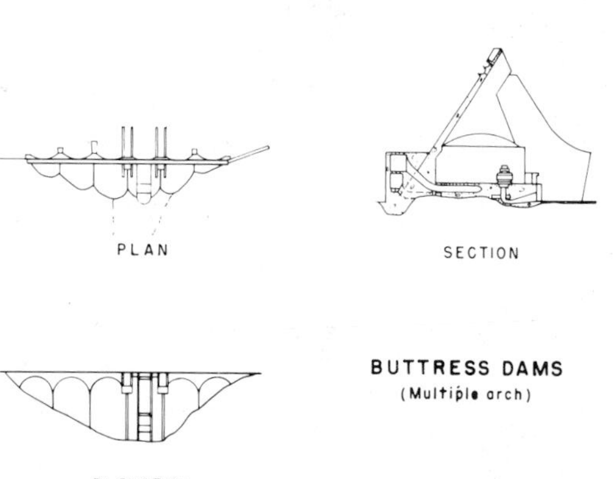

Fig. 21-10 Buttress dams (multiple-arch)—plan, elevation, and section.

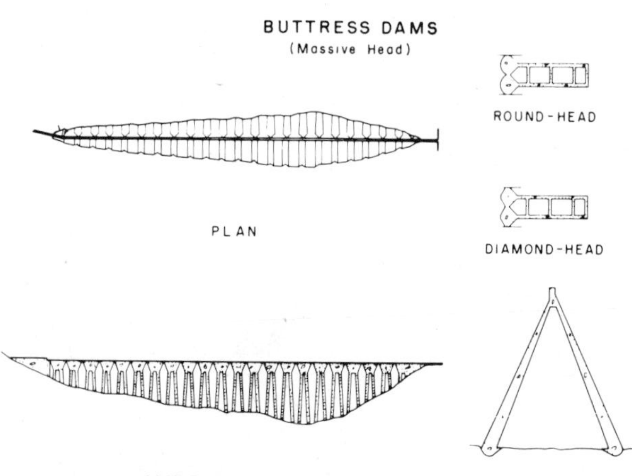

Fig. 21-13 Buttress dams (massive-head)—plan, elevation, and sections.

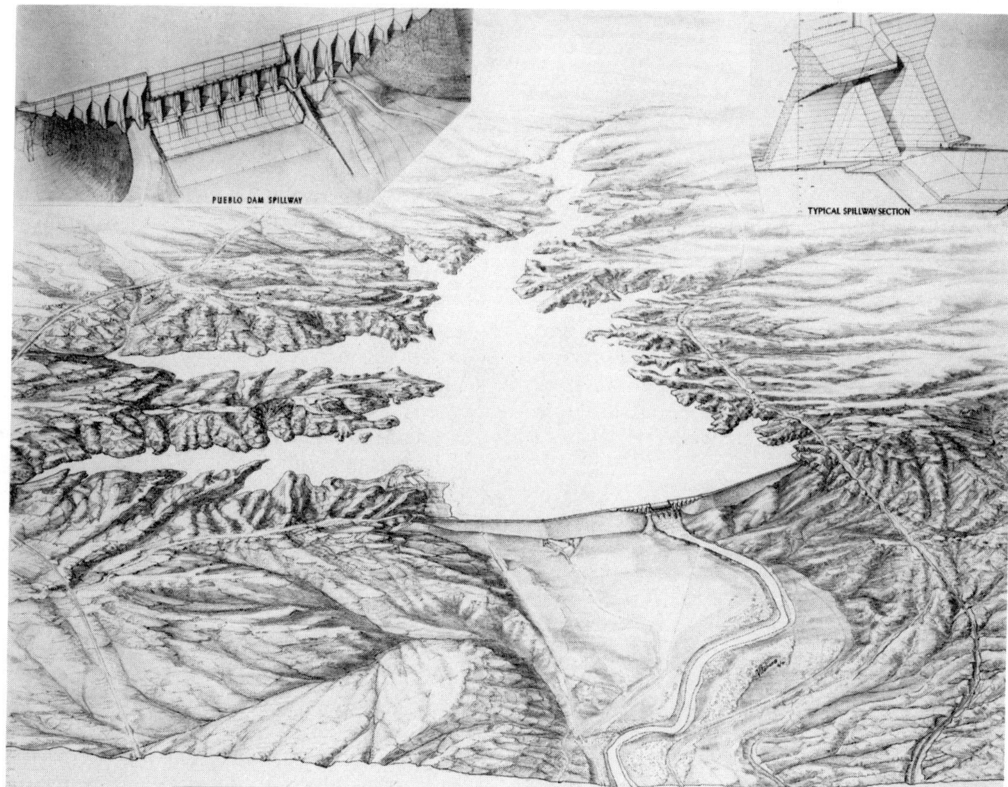

Fig. 21-14 Pueblo Dam, Colorado (artist's concept).

21.3 DETERMINATION OF TYPE

The determination of the type of dam to be used for a particular site involves the consideration of many factors. Those factors which will generally have important bearing include: site conditions, hydraulic considerations, climatic effects, and traffic conditions.

Site conditions which must be considered in the choice of type of dam are topography, geology, availability of construction materials, and accessibility to the damsite. Narrow valleys and good rock foundations are generally preferable for concrete dams. Very narrow canyons are especially adaptable to arch dams since arch action is most effective for crest-to-height ratios of less than 5:1. Rock foundations are necessary for all but the lowest gravity dams; the quality of rock depending on the size of dam to be constructed. A rock quality sufficient to resist the arch thrusts at their abutments is essential for arch dams. Buttress dams, because of their comparatively light weight, can be built on foundations which do not have sufficient strength to support gravity dams. A greater degree of variability in foundation properties can be tolerated by massive-head or slab and buttress dams than by other types of concrete dams.

Selection of the type of dam to be constructed depends, to some extent, on the availability of aggregate and other materials required for its construction. If concrete aggregate sources are located at considerable distance from the damsite, a thin arch or buttress type may be indicated because of their smaller concrete volume requirement. Accessibility to the damsite is closely related to the availability of materials. Cost of materials delivered to the site will be increased if access is difficult and expensive.

Hydraulic factors which affect the choice of type of dam for a particular site are spillway requirements, river diversion, outlet works, and penstocks. In cases where very large spillway capacity is required, economy may be achieved by choosing a gravity (Figs. 21-2 and 21-3) or buttress dam (Fig. 21-14) and designing an overflow spillway in the river channel. With arch dams, overflow spillways (Fig. 21-5) are usually limited to small capacities and low heads. Side channel and tunnel spillways are generally adaptable to all types of dams. The method used for diverting a river during construction depends on the anticipated maximum river flows during the construction period, type of dam and spillway, and space limitations at the damsite. If there is sufficient space in the river bottom, diversion can be most economically accomplished over low concrete blocks in an arch or gravity dam, regardless of the quantity of water to be diverted. Likewise, the economical method of diversion for buttress dams is between buttresses, through temporary openings in the deck.

Diversion by means of tunnels around the damsite is used when space limitations require it or when a tunnel-type spillway is planned which would utilize a portion of the diversion tunnel. Costs of diversion schemes will usually not be enough in themselves to be a major factor in type of dam determination unless river flows are of very great magnitude. Outlet works and penstocks seldom influence the type of dam except where especially thin structures are contemplated. Thin-arch dams are not well adapted to large openings or to great numbers of openings, especially in regions of high stress. If a large outlet or penstock capacity is required for a thin-arch dam, it may be preferable to provide a tunnel through the foundation rock. The expense of such tunneling might favor the construction of another type of dam with less expensive outlet design.

Climatic factors sometimes influence the selection of type of dam. Under freezing and thawing conditions concrete tends to spall at the surface. Gravity or thick-arch dams are little affected in their strength and stability by such action if they are repaired periodically. However, spalling may affect the safety of very thin arch dams, slab-and-buttress, or multiple-arch dams. In these cases even small reductions in thickness may have considerable effect on the magnitude and distribution of stresses. Also, if reinforcing steel is exposed by spalling it may soon rust out. Large temperature variations, extremely cold or extremely hot temperatures, influence greatly the behavior of thin-arch dams and may influence their consideration for sites where such conditions exist.

The necessity for carrying traffic over a dam sometimes effects the choice of type. A gravity or thick-arch dam may be easily adapted to a highway over the top of the dam. A highway over a thin-arch or buttress dam usually involves expensive construction of cantilevers, bracing, or bridging. This additional cost may effect the choice of type to be used.

21.4 LOADING CONDITIONS

Good concrete dam design requires the consideration of all possible loads or combinations of loads which may occur at a site. Loading conditions to be evaluated will usually include combinations of part or all of the following: concrete weight, water pressure, temperature change, earthquake, silt, and ice load.[21-1, 21-2, 21-5]

The unit weight of concrete is generally determined by laboratory tests involving the use of aggregate selected for use in each particular dam. The range of unit weights is usually from 135 to 160 pounds per cubic foot. The effects of the concrete weight on the stresses in the dam and foundation must be carefully evaluated. Care must be taken to assure that proper consideration is given to construction sequence. The relationship of block placements to contraction joint grouting and water storage may have appreciable significance.

Water pressure is applied to the upstream face of a dam by the impounded reservoirs and to the downstream face by tailwater. The pressure is proportional to the depth of water. Uplift pressures are produced in the interior of a dam or foundation by water in pores, cracks or porous formations in the rock. The intensity of uplift pressure depends on the reservoir depth, the distance from the faces, drains provided in the dam and foundation, and other measures used to reduce the rate or extend the path of percolation.

Temperature change in concrete may result from both internal and external factors. Internally the effects of heat of hydration must be included in all types of concrete dams. The need for reinforcement, joint spacing, and depth of individual placements of concrete are particularly affected. The ambient air temperature, solar radiation, and reservoir water temperatures produce changes in concrete temperatures externally. These changes are of special importance to arch dams and massive-head buttress dams. The effects of both maximum and minimum concrete temperatures combined with appropriate reservoir levels should be included for the most efficient design of all arch dams. Generally minimum concrete temperatures will produce the most critical stress condition in massive-head buttress dams.

Investigations indicate that essentially all concrete dams respond dynamically to earthquake shocks. The result of such response is to accentuate the movement of the dam and correspondingly increase the stresses in the structure. To compute the magnitude of stress resulting from this vibration, it is necessary to determine the natural frequency of the dam in its various modes, relate this to an appropriate earthquake, then compute the response of the dam to this earthquake. The response is usually computed in terms of accelerations from which movements and stresses may be determined.

Most streams carry some silt, especially during flood flows. If this silt is permitted to accumulate at the face of any concrete dam, its effect should be included in the analyses for stresses and stability. A unit horizontal pressure of 85 lbs per cu ft and a vertical weight of 120 lbs per cu ft are satisfactory assumptions if more definite information is not available. Silt accumulations may be reduced or eliminated by the use of properly located sluiceways or outlets through the dam.

Ice loads may cause considerable difficulty if permitted against gates or other thin membered structures. So far as the dam is concerned, ice is a local concentrated load and of importance only at or near the point of application. The pressure produced by ice is dependent on temperature change, ice sheet thickness, wind conditions, canyon shape and slope of the upstream face of the dam. Depending on these conditions, pressures ranging between 10,000 and 20,000 pounds per linear foot are considered satisfactory.

21.5 STRESS AND STABILITY ANALYSES

To evaluate the safety of a dam, it is necessary to determine the stresses and stability in both the dam and the foundation upon which it is placed. A safety factor of four or more, based on the ratio of concrete strength to computed stress, is usually required for all concrete dams. This factor should be maintained for all load combinations which might be expected to occur frequently during the life of the structure. Such load combinations include normal full reservoir, weight of dam, uplift, and silt and ice, where applicable. In addition, with arch dams and multiple-arch or massive-head buttress dams, minimum concrete temperatures anticipated with full reservoir should be included. Arch dams should also be analyzed with the minimum reservoir expected and maximum temperatures which might occur at that time.

Maximum computed compressive stresses are usually limited to 1000 to 1200 psi. Higher stresses may be permitted if concrete and foundation material are of sufficient quality to support them. Tensile stresses should be avoided in gravity and massive-head buttress dams under normal loading conditions. In multiple-arch and slab-and-buttress dams, tensile stresses may be permitted if adequate reinforcement is provided. An effort should be made to avoid tensile stresses in arch dams. Where this is not possible, tension should be confined to local areas with magnitudes well below the strength of the concrete, Larger tensile stresses may be permitted on the downstream face of an arch dam than can be tolerated where the concrete is exposed to water pressure.

All concrete dams should also be examined for unusual loading conditions such as earthquake and flooding. Under such conditions a reduced factor of safety could be permitted.

Dams and their foundations should also be analyzed for resistance to sliding. A safety factor similar to that required for stresses is also necessary to assure safety against sliding. This can easily be accomplished in the dam proper but is sometimes more difficult to obtain in the foundation. Usually if the design for a dam meets the required stress criteria, sliding is not a problem. If sufficient safety is not found, a modification in section will correct it. Foundation blocks, defined by structural features of the rock mass, should be loaded with forces and moments from the dam, hydrostatic load from the reservoir, and earthquake where applicable. Joints, faults, or shear zones which define the boundaries of a rock mass should be assigned shear strengths based on tests performed on these structural features. The stability of the rock mass may then be determined by appropriate computations. If sufficient safety is not indicated, steps should be taken to improve the stability of the rock mass.

Drainage to reduce seepage and uplift in the foundation is usually good practice in all concrete dams. Deep curtain grouting combined with drainage will extend the water percolation path, reduce uplift and thus contribute to the safety of the foundation. Special joint treatment, rock bolting, or prestressing techniques may be required to stabilize local areas of the foundation. If local areas of weak or undesirable foundation material are found, it is usually beneficial to excavate the material and replace with concrete.

The analysis of stresses in a gravity dam is closely related to the treatment of transverse contraction joints. If these joints are neither keyed nor grouted, stresses may be computed by assuming all loads are carried vertically to the foundation. This is a simple procedure which is also used in determining the preliminary section for any gravity dam. If transverse joints are keyed then load may be transmitted horizontally by shear and torsion. The distribution of load is determined by first dividing the dam into representative independent vertical and horizontal elements. The load is divided between these elements so that their downstream deflections and slopes are in agreement at points of intersection. If transverse joints are both keyed and grouted, then in addition to horizontal transmission of forces by shear and torsion, flexure also occurs in the horizontal elements. The load distribution in this case is also determined by using a system of representative horizontal and vertical elements and obtaining slope and deflection agreement at points common to both elements. Vertical, horizontal and principal stresses are computed from the loads obtained by the above procedures.

The accurate determination of stresses in arch dams is a difficult and complex problem. A variety of methods have been used by designers with varying degrees of success. Several comprehensive methods for the analysis of arch dams are available. These include structural models, trial-load (computerized version referred to as "Arch Dam Stress Analysis System," ADSAS), shell theory, three-dimensional finite element, and dynamic-relaxation methods.

Large-scale structural models of micro-concrete or smaller models of plaster-celite (or other suitable material) are loaded to simulate water loads, concrete weights, temperature change or dynamic loads or combinations of these. The model displays, by means of gages and meters, the effect of applied loads. Models of rubber or similar material are sometimes used as a check against other models and analytical methods.

The trial-load method (ADSAS) assumes that the dam is composed of a system of horizontal arch elements and a system of vertical cantilever elements, each system occupying the entire volume of the dam. Representative elements of the two systems are loaded such that movements at conjugate points are identical. These loads were formerly determined by trial, hence the name of the method. Computers may now be used to solve the simultaneous equations involved. From the load distribution required to achieve deflection agreement, stresses are determined.

The finite element technique divides the whole volume of the dam into small but finite elements for which a stiffness matrix can be derived. A computer is then used to solve the simultaneous equations thus formed. Shell theory may be applied by reducing the partial differential equations arising from thin-shell theory to a set of ordinary differential equations using approximation techniques. The resulting equations are solved by the matrix progression technique.

Dynamic relaxation applies a damping technique to solve the three dimensional equations of elasticity for an arch structure by a procedure involving iteration of equations of compatibility and equilibrium. Any of these methods can be used by responsible engineers if they thoroughly understand the technique, the assumptions included in it, and how to interpret the results.

Buttress dams are composed of two principal elements, a sloping upstream and supporting piers, or buttresses. For the slab-and-buttress type of structure, the deck consists of a simply-supported flat slab spanning between the buttresses, each being structurally independent. Both the slab and buttress can be easily designed and analyzed by methods available in any structures text. Multiple-arch dams are continuous structures with the deck made up of a series of arches, usually having a 180° central angle and circular in shape. The buttresses are designed to carry the loads imposed upon them by the arch sections. The arches may be analyzed by trial-load or one of the other methods discussed for arch dams. The movements of the buttresses caused by arch loadings must be included in the analysis of the arch sections.

Massive-head buttress dams are formed by flaring the upstream edges of the buttresses to span the distance between the buttress walls. Each buttress and head form a variable-shaped cantilever, independent of adjacent cantilevers. Estimates of stresses in these structures may be made by taking vertical, horizontal and inclined unit sections and applying appropriate loads to them. Three-dimensional finite element or photoelastic analyses should be conducted to determine stresses accurately.

21.6 FOUNDATION REQUIREMENTS

The essential requirements for the foundation of a concrete dam are stable support for the structure, under all conditions of loading, and adequate resistance to loss of stored water. Joints, faults, shears and other anomolies in foundation rock are important considerations in the design of concrete dams. If such rock structures are so oriented that loads from the dam merely close, compress, or consolidate them, their effects can easily be included in the design of the dam. If, however, joints or faults are oriented in such a way that blocks of rock might be displaced by loads from the dam, these conditions must be recognized, analyzed, and treated as the circumstance requires.

Suitable support for concrete dams requires removal of all material above competent rock followed by treatment to make the foundation sound and water-tight. During the removal of overburden, loose, weathered and fractured rock, blasting must be carefully controlled to prevent damage to the underlying rock which will provide the foundation for the dam. The abutments for arch dams are generally excavated along radial or semiradial lines so that arch thrusts will act approximately normal to the rock surface. If local areas of weak or undesirable foundation materials are encountered, it may be beneficial to excavate the areas and fill them with concrete, or confine and cover them with concrete.

To reduce seepage and improve the homogeneity of rock foundations, it is generally necessary to treat them by injecting grout under pressure into openings in the foundation. This grouting is generally performed in two stages. The preliminary low pressure grouting is performed before any concrete is placed, to provide consolidation of surface rock and seal major surface seams and joints. Grout holes are usually drilled on 20-ft centers to depths of from 20 to 50 ft. After sufficient concrete has been placed in the dam to permit higher pressures, deep holes on a line extending across the channel near the upstream face of the dam are drilled and grouted at high pressure to form the principal barrier against seepage under the structure.

The spacing and depth of these holes depend on the characteristics of the foundation and the height of the dam. Generally for a major structure they are spaced on 5-ft centers to a depth of about 40% of the hydrostatic head. The drilling and grouting is generally done from a gallery provided in the dam near the upstream face. However, in some cases the grouting may be done from the heel of the dam before water is stored in the reservoir.

The final treatment of the foundation consists of drilling a line of drainage holes downstream from the grout curtain to intercept any water that may percolate through the grout curtain. The drainage holes are drilled after all grouting in an area is complete. The spacing and depth of holes is governed by the height of the dam and extent of foundation grouting. Drainage holes are generally 3 in. in diameter, spaced on 5- to 10-ft centers to a depth of 20 to 40% of the hydrostatic head.

21.7 MATERIAL REQUIREMENTS

The essential requirements of concrete for dams are strength, durability, and watertightness. These properties should be achieved with the greatest possible economy, not only in the first cost but also in terms of ultimate service.

The concrete compressive strength at one year's age should be at least four times the maximum compressive stress in the dam. This usually means that one year strengths of 4,000 psi or less are adequate. Frequently, when requirements for durability are met, more than sufficient strength is available for design considerations. Concrete strengths greater than required for stress and durability should be avoided. Higher strength concretes generally are more brittle and have higher elastic moduli than do moderately strong concretes. These properties may lead to cracking as the structure adapts itself to its environment, whereas a more plastic material would be less likely to crack. If the minimum compressive strengths indicated above are obtained, sufficient tensile strength is usually available to satisfy the needs of the dam. This assumes the structure has been properly designed and constructed.

Concrete for dams should be sufficiently durable to withstand the conditions to which it will be subjected, such as weather, chemicals and erosion. Disintegration of concrete by weather is caused primarily by freezing and thawing and expansion and contraction resulting from temperature variations. Concrete mixes should be carefully designed to satisfy these conditions for the location where the dam is to be constructed. Chemical deterioration may result from reactions between alkalies in cement and minerals in concrete aggregates or from contact with various chemical agents in the water or foundation. Alkali-reactive aggregates should be avoided or low-alkali cements used where such aggregate cannot be completely eliminated. Erosion of concrete may be caused by abrasive material in flowing water, cavitation, traffic, and floating ice. Where any of these conditions are expected, care should be taken in both the design of the dam and the concrete to eliminate or reduce their effects to a minimum.

From a practical standpoint, well designed, homogeneous concrete can be considered watertight. Voids exist in all concretes which will eventually be filled with water. This will rarely, however, actually cause seepage through the dam. Cracking produced by volume change under restraint is usually the source of leakage into or through a dam. This can be avoided by proper spacing of contraction joints, control of concrete temperatures and well designed, homogeneous concrete.

Concrete aggregate usually consists of natural sand and gravel, crushed rock, or mixtures of these materials. For economic reasons the choice of aggregate is usually limited to local deposits. The aggregate must be sound, well-graded and free of contaminating substances. Chemically-reactive and lightweight aggregate should be avoided. Volume change in concrete caused by temperature change or wetting and drying can be damaging to concrete. It is important that these properties are known in order to properly design and construct the dam.

The selection of cement type and concrete admixtures is dependent upon the requirements of the structure, local conditions, aggregate and other variables. Pozzolans, set-controlling admixtures, air-entraining agents, water-reducing admixtures, or combinations of these may be used as the circumstances indicate. Early strength requirements, control of heat generation, durability needs, and chemically reactive aggregates are important considerations in this connection.

21.8 DAM CONSTRUCTION

All concrete dams require good, uniform, well-consolidated concrete. Therefore after designs have been completed and materials selected, the methods of batching, mixing, handling, placing, and curing must be controlled. Materials, particularly aggregate, having uniform properties must be supplied, batch after batch, to the mixer. Accurate control of the ingredients entering the mixer is essential. Adequate and uniform mixing will then produce uniform concrete. Proper and careful handling and placing will put the concrete in final position for consolidation. Thorough consolidation is absolutely necessary if maximum strength, durability, and uniformity are to be attained. Vibration equipment is available, if properly applied, to adequately accomplish this very essential requirement.

The condition of forms is important not only to the appearance of the structure but also its quality. Good form materials and proper form construction and maintenance are important to the satisfactory construction of the dam. The forms must be clean, tight, and strong enough to support the plastic concrete placed against them without deforming.

Proper hydration of cement to form hard, durable concrete requires that concrete be maintained in a moist condition for a suitable period of time. Within practical limits, most of the desirable properties of concrete are improved as the effectiveness and duration of curing treatment are increased. Special precautions are necessary to insure adequate curing when the weather reaches extremes of either heat or cold.

Rapid placement of mass concrete prevents the escape of heat of hydration of the cement. The resulting temperature rise and subsequent drop to stable conditions would produce tensile stresses and cracking unless steps are taken to reduce or eliminate such temperature changes. Much of the temperature rise may be prevented by use of low-heat cements, reduction of cement content, limitation of rate of placement, placement during cool weather, precooling concrete ingredients, or artificially cooling the concrete by circulating cool water through tubing placed in the concrete. Since it is generally not practical to completely prevent volume change, contraction joints are formed in the mass of the dam to relieve the tensile stresses resulting from restrained contraction of the concrete. These joints are formed vertically, normal to the axis of the dam, and vary from 50 to 100 ft apart depending on site and construction conditions. In thick dams, especially of the gravity type, it may also be necessary to provide longitudinal joints to avoid cracking.

Concrete is placed in blocks formed by the contraction joints in 5- to 10-ft lifts. After these lifts harden, the surface is thoroughly cleaned of any debris or laitance, roughened and a new lift of concrete bonded to it. Careful preparation of lift lines is essential to the proper construction and behavior of the dam.

Galleries, shafts and chambers are formed in concrete dams to provide drainage for water percolating through the upstream face or foundation; provide space to drill drainage holes and grout the foundation and contraction joints; provide access into the interior of the dam for inspection; and provide access to and room for necessary mechanical and electrical equipment. Most major structures require one or more elevators for ready access from the top to the bottom of the dam. Many times outlet works penstocks, and spillways also require openings through the dam. All such openings formed in the dam must be reinforced if excessive tensions are anticipated around them.

21.9 INSTRUMENTATION

Measuring devices for monitoring their behavior should be installed on all concrete dams. Some of these instruments give information which is used in controlling construction operations, and others provide for the continued observation of the structure after construction is completed. By this continued observation, not only can the safety of the structure be ascertained at all times, but also much actual behavior data can be compiled which will be useful in the design of future dams.

Perhaps the most dependable, simple, and inexpensive measurements of movements of dams are obtained by plumbline and collimation measurements. Deflections of the structure may also be obtained by precise survey methods using targets on the face of the dam; precise level surveys to determine settlements at various locations; and tiltmeters to measure rotations and deflections in the dam.

Many electrical measuring instruments to determine the behavior of dams are embedded in the concrete. These include resistance thermometers for measuring the temperature of the concrete at various locations; strainmeters for measuring strains in the mass of the dam, or in reinforcement steel around openings in the dam, thereby making computation of stresses possible in the concrete and steel; joint meters for measuring contraction joint openings; stress meters for determining stresses in the concrete; and hydrostatic pressure gages to measure uplift pressure in and under the dam.

Also of importance are measurements of the behavior of the foundations of dams. These may be obtained by precise survey methods on the surface of the foundation adjacent to the dam or in tunnels extending from the dam into the foundation. Measurements of movements in tunnels in the foundation can also be made by use of invar tapes installed in them. Deformation meters installed in shafts in the foundation around the perimeter of the dam will also provide valuable data regarding the behavior of the foundation.

21.10 ENVIRONMENTAL CONSIDERATIONS

Generally, construction of a dam and the subsequent reservoir will enhance the environment of an area. However, unless such construction includes care and planning before, during and after the time the dam is built, the environment may be seriously damaged.

Access roads to the damsite should be carefully planned and constructed to cause minimum change or defacement of the area. Aggregate sources should be carefully selected, in the reservoir area if possible, so as not to disturb excessively the natural appearance of the site. Where such excavations will not be covered by the reservoir they should be restored, as nearly as possible, to their natural condition. During construction care should be taken to control excessive turbidity in the stream, and to limit air and noise pollution, particularly if the site is near an inhabited area. All excavations for the dam should be performed with care to protect or if possible enhance the appearance of the area. Habitat for fish and wildlife should be preserved in every way possible. Frequently these conditions may actually be improved with the filling of a reservoir if sufficient planning has preceded the construction of the dam.

After construction is completed, the area around the damsite may be restored to near its original condition by filling or grading excavated surfaces and reseeding where vegetation has been removed. Parks, marinas, and other visitor and recreational facilities are often constructed to enhance the beauty and usefulness of the area surrounding dams and reservoirs.

REFERENCES

21-1 "Arch Dams," Chapter 1, *Design Standards No. 2, Concrete Dams*, United States Department of the Interior, Bureau of Reclamation, Wash., D.C., March 24, 1965.

21-2 "Gravity Dams," Chapter 2, *Design Standards No. 2, Concrete Dams*, United States Department of the Interior, Bureau of Reclamation, Wash., D.C., August 19, 1966.

21-3 "Symposium on Arch Dams," given at Colorado State University, Fort Collins, Colorado, 1957. Also reprinted from the *Proc., ASCE, Journal of the Power Division*, **82**, 1956 and **83**, 1957.

21-4 Rydzewski, J. R., editor, *Theory of Arch Dams*, Pergamon Press, Elmsford, New York, 1965.

21-5 "Joint ASCE-USCOLD Committee on Current U.S. Practice in the Design and Construction of Arch Dams, Embankment Dams, and Concrete Gravity Dams," ASCE-USCOLD, 1967.

21-6 "A Review of British Research and Development," Symposium on Arch Dams at the Institution of Civil Engineers, London, 20-21 March, 1968.

22

Concrete Pavement Design

ROBERT G. PACKARD*

22.1 CONCRETE PAVEMENT DESIGN

Since several aspects of concrete pavement design are interrelated, it is essential to review their functions in minimizing the causes of uncontrolled cracking of slabs, loss of uniform support in the foundation, and loss of surface integrity. Control of these factors involves four aspects of design

1. Concrete quality–selection of suitable materials and proportions of materials for a concrete pavement that will have adequate durability and strength.
2. Subgrade-subbase design–Proper subgrade preparation techniques are required to ensure reasonably uniform support for the slab and control subgrade volume changes caused by expansive soils or frost action. Where traffic volume is heavy, the use of a subbase layer is required to prevent mud-pumping; this layer must be a nonconsolidating granular or stabilized material to prevent joint faulting.
3. Thickness design–Slab thickness is determined so that flexural stresses due to traffic loads are kept within safe limits.
4. Joint design–The jointing arrangement determines where cracks will form due to restrained shrinkage and temperature stresses. These stresses are minimized by the use of short joint spacings in plain pavements; for reinforced pavements with longer joint spacings, intermediate cracks will form between the joints but these are not detrimental since they are held tightly together by the reinforcing steel.

Details of these aspects of design for highway, street and airport pavements are presented in the following sections.

*Principal Paving Engineer, Paving and Transportation Department, Portland Cement Association, Skokie, Illinois.

22.2 CONCRETE QUALITY

Fundamentals of mix design for quality concrete are given in Chapter 6, "Properties of Materials for Reinforced Concrete," in addition to those fundamentals, the following recommendations are made specifically for paving mixtures to obtain (1) the durability needed to resist the effects of climate and traffic, and (2) the flexural strength required to carry the expected weights and numbers of traffic loads.

In frost-affected areas, concrete pavements are subjected to many cycles of freezing and thawing and to the application of deicing salts. To protect concrete pavements against the action caused by these agents it is essential to have a mix with a low water-cement ratio, an adequate cement factor, and sufficient quantity of entrained air. The amounts of entrained air needed to produce weather-resistant concrete vary with the maximum size aggregate.* Recommended percentages of entrained air are

Maximum size aggregate, in.	*Entrained air, percent*
$1^1/_2$, 2, $2^1/_2$	5 ± 1
$^3/_4$, 1	6 ± 1
$^3/_8$, $^1/_2$	$7^1/_2$ + 1

The amount of mixing water also has a critical influence on the durability and weather resistance of hardened concrete. The least amount of mixing water with a given cement content that will produce a plastic, workable mix will result in the greatest durability in the hardened concrete. Laboratory and field experience with air-entrained concrete shows that for satisfactory pavement durability the water-cement ratio should not exceed 0.54. The cement factor should be not less than 517 lbs per cu yd. In areas where severe frost and deicing agents are common, the water-cement ratio should not exceed 0.49 gal of water with a minimum cement factor of 564 lb per cu yd.

22.3 SUBGRADES AND SUBBASES

For satisfactory performance a concrete pavement must be provided with a uniform foundation. The basic objective is to construct a foundation that is, and will remain, reasonably uniform under the climatic and traffic conditions that will prevail. This objective is achieved economically by proper soil grading operations and subgrade treatments during construction and, where needed, by the use of a subbase layer of high-quality material placed directly beneath the pavement. Building up strong support in the foundation is not essential and is usually not economically justified.

The principle of uniform support, rather than strong support, is explained by consideration of the properties of the concrete slab. The rigidity of concrete enables it to distribute loads over large areas of the subgrade; deflections are small and pressures on the subgrade are low.[22-1, 22-2, 22-3, 22-4]

To design a subgrade and subbase that provide reasonably uniform support for the slab, the three major causes of nonuniform support that need to be controlled are

1. expansive soils
2. frost action
3. mud-pumping

Effective control of high-volume-change soils and frost action is most economically achieved through appropriate subgrade preparation techniques, while prevention of mud-pumping requires a thin subbase layer. Although a subbase also provides some control of high-volume-change soils and frost action, the use of thick subbase layers for substantial control of these factors is no more effective than subgrade work and usually costs more.

22.3.1 Subgrades

Where the subgrade conditions are not reasonably uniform, correction is most economically achieved by proper subgrade preparation techniques such as selective grading, crosshauling, mixing at abrupt transitions, and moisture-density control of subgrade compaction. Particular attention is needed for the control of expansive soils and differential frost heave.

-1. Expansive soils—Identification of the types of soils that are expansive and the mechanism of soil volume change has been gained through research and experience. A summary of the technology of expansive soils, including tests for identification, is given in Ref. 22-5. The simpler tests provide indices, such as plasticity index, shrinkage limit, and bar shrinkage, that serve as useful guides to identify approximate volume change potential of soils. For example, the following table shows approximate expansion plasticity relationships:

Degree of Expansion	*Percentage of Swell (ASTM D1883)*	*Approximate Plasticity Indices (ASTM D424)*
nonexpansive	2 or less	0 to 10
moderately expansive	2 to 4	10 to 20
highly expansive	more than 4	more than 20

Most soils sufficiently expansive to cause distortion of concrete pavements are in the AASHO A-6 or A-7 groups. By the Unified Soil Classification system, soils classified as CH, MH, or OH are considered highly expansive.

Expansive soils have been found to be controlled effectively and economically by the following subgrade treatments:

a. Subgrade grading operations—Excessive swell can be controlled by placing the more expansive soils in the lower parts of the embankments and crosshauling less expansive soils to form the upper part of the subgrade in both embankments and excavations. Selective grading also makes it possible to have reasonably uniform soil conditions in the upper part of subgrade, with gradual transitions between soils with varying volume change properties.

In deep cuts into highly expansive soils, considerable expansion may occur due to the removal of the natural surcharge load and the consequent absorption of additional moisture. Since this expansion takes place slowly, it is essential to excavate these deep cuts well in advance of other grading work.

b. Compaction and moisture control—Volume changes are further reduced by adequate moisture and density controls during subgrade compaction. It is critically important to compact highly expansive soils at 1 to 3% above optimum moisture, *AASHO T99*. Where embankments are of considerable height, compaction moisture contents can be increased from slightly below optimum in the lower part of the embankment to above optimum in the top 1 to 3 ft. Expansive soils compacted slightly wet of optimum expand less, and have higher strengths after wetting and absorb less water.

After pavements are placed in service, most subgrades reach a moisture content approaching their plastic limit. That is, the natural moisture content reached is close to and slightly above the standard optimum (*AASHO T99*). When this moisture content is obtained in construction, the subsequent changes in moisture will be much less and the subgrade will retain the reasonably uniform stability needed for good pavement performance.

c. Nonexpansive cover—In areas with prolonged periods of dry weather, highly expansive subgrades may require a cover layer of low-volume-change soil placed full width over the subgrade. This will minimize changes in the moisture

*Generally, maximum size aggregate should not exceed one-fourth the pavement thickness.

content of the underlying expansive soil and will also have some surcharge effect. A low-volume-change layer with a low to moderate permeability, is not only more effective but usually less costly than a permeable, granular soil. Highly permeable open-graded subbase materials are not recommended as cover for expansive soils since they permit greater changes in subgrade moisture content.

Local experience with expansive soils is the best guide for adequate depth of cover.

-2. *Frost action*—Uniquely, field experience with concrete pavements has shown that frost action damage due to inadequate design is a result of frost heave—in the form of abrupt, differential heave. Subgrade softening on thaw is not a design consideration for concrete pavements since a strong subgrade support is not required. Design for concrete is concerned with reducing the nonuniformity of subgrade soil and moisture conditions that lead to objectionable differential heaving—especially where subgrade soils vary abruptly from nonfrost susceptible sands to the highly frost susceptible silts at cut-fill transitions, where the groundwater is close to the surface, or where water-bearing strata are encountered.

-2.1 Frost heave—For frost heave to occur, all of three conditions must be present: (1) a frost-susceptible soil; (2) freezing temperatures penetrating the subgrade; and (3) a supply of water.

Heaving is caused by the growth of ice lenses in the soil. When freezing temperatures penetrate a subgrade, water from the unfrozen portion of the subgrade is attracted to the frozen zone. If the soil is susceptible to high capillary action, the water moves to ice crystals initially formed, freezes on contact and expands. If a supply of water is available, the ice crystals will continue to grow, forming ice lenses of appreciable thickness that lift or heave the overlying pavement.

-2.2 Frost-susceptible soils—Criteria and soil-classifications used for identifying frost-susceptible soils usually reflect susceptibility to softening on thaw as well as to heaving. The major concern is to reduce heaving, especially differential heaving. Control of spring softening is not a consideration for concrete pavements. Thus, specific criteria or classifications identifying frost-susceptible soils should be reviewed with differentiation in classification between soils susceptible to heave and those susceptible to thaw softening.

The worst heaving usually occurs in fine-grained soils subject to capillary action. Low-plasticity soils with a high percentage of silt-size particles (0.05 mm to 0.005 mm) are particularly susceptible to frost heave. These soils have pore sizes small enough to develop capillary potential but large enough for rapid travel of water to the frozen zone. Coarser soils have higher rates of flow, but do not have the potential to lift enough moisture for heaving. More cohesive soils, although developing high capillarity, have low permeability, and moisture moves too slowly for growth of thick ice lenses.

-2.3 Spring subgrade softening—A frozen subgrade thaws both from the surface downward and from the bottom upward. As a result, thawing is usually more rapid than freezing. When thawing starts, the moisture content of the subgrade may be high due to the previous moisture increase during freezing and due to surface water infiltration. This water in the upper thawed layer cannot drain downward because of the frozen zone below. In addition, ice lensing or simple expansion may have caused a loss in density. Under these conditions, there is a sharp reduction in subgrade support during the thaw period.

The periods of reduced subgrade support that accompany thawing have very little effect on concrete pavements. This is because concrete reduces pressures on soft subgrades to safe limits by distributing loads over large areas and because concrete pavements are designed for fatigue stresses due to load repetitions.

Concrete pavements designed on the basis of normal weather subgrade strengths have ample reserve capacity for the periods of reduced support during spring thaws. Because of their reserve load-carrying capacity, concrete pavements are exempted from load restrictions during periods of spring thaw.

Further evidence that concrete pavements designed with uniform support are not influenced by spring thaw is shown by the results of the AASHO Road Test.[22-6] The pavement performance and the equations written to relate the design variables to traffic loads show that concrete pavements, with or without a subbase, were not affected by the spring thaw periods.

-2.4 Control of frost heave—As in the case of expansive soils, a large degree of control of frost heaving is accomplished most economically by appropriate grading operations and by controlling subgrade compaction and moisture. These include

a. *Grade elevation.* Grade lines are set high enough and side ditches are constructed deep enough so that highly frost-susceptible soils are above the capillary range of groundwater tables. Where groundwater is near the surface, the grade is kept 4 or 5 ft above ditch bottom in cuts and natural ground in fills.

b. *Selective grading.* Highly frost-susceptible soils are placed in the lower portions of embankments and less susceptible soils are crosshauled to form the upper portion of the subgrade. Crosshauling and mixing are also used at cut-fill transitions to correct abrupt changes in soil type.

c. *Mixing.* Where soils vary widely or frequently in texture and where nonuniform conditions are not clearly defined, mixing of the soils is effective in preventing differential frost heave. With modern construction equipment, the mixing of nonuniform soils to form a uniform subgrade is often more economical than importing select materials from borrow pits.

d. *Removal of silt pockets.* Where highly frost-susceptible soils are pocketed in less susceptible soils, they are excavated and backfilled with soils like those surrounding the pocket. Moisture and density conditions for the replacement soil should be as similar as possible to those of the adjacent soils. At the edges of the pocket, the replacement soil should be mixed with the surrounding soil to form a tapered transition zone, just as in cut-fill transitions.

e. *Compaction and moisture control.* After reasonable uniformity has been achieved through the grading operations, additional uniformity is obtained by proper subgrade compaction at controlled moisture contents. The permeability of most fine-grained soils is substantially reduced when they are compacted slightly wet of AASHO T99 optimum moisture. Reducing soil permeability retards the rate of moisture flow to the frozen zone and consequently reduces frost heaving. Research[22-7, 22-8] confirms that less frost heave occurs at the wetter condition.

f. *Drainage.* Where high grades are impractical, subgrade drains are used to lower ground water tables. Where wet spots are encountered in the grade due to seepage through a permeable strata underlain by an impervious material, intercepting drains are used. Pipe backfill should meet filter criteria[22-9] so that neither soil infiltration nor clogging of pipe openings will occur.

g. *Nonfrost susceptible cover.* Layers of clean gravel or sand will reduce frost heave but are not required solely for this purpose when the less costly grading operations are properly employed. The benefit of the use of thick subbase layers is somewhat diminished since coarse soils permit slightly deeper frost penetration than do fine-grained soils at their higher moisture content.[22-10, 22-11] When a subbase layer is required to prevent mud-pumping, it also provides some protection against frost action. However, these layers are more effective in preventing loss in subgrade support on thaw which is not a design consideration for concrete pavements.

22.3.2 Subbases

A subbase is not required for the sole purpose of obtaining strong or uniform support for the pavement. Also, a subbase is not necessarily required for purposes of control or expansive subgrades or for control of frost action. For these purposes, although a subbase layer is beneficial, proper subgrade preparation techniques should also be utilized as the most economical means of obtaining good pavement performance. The essential function of a subbase is to prevent mud-pumping of fine-grained soils.

-1. Prevention of mud-pumping—Mud-pumping is the forceful displacement of a mixture of soil and water that occurs under slab joints, cracks and pavement edges. It is caused by the frequent deflection of slab edges by heavy wheel loads when fine-grained plastic subgrade soils are saturated. Continued uncontrolled mud-pumping eventually leads to the displacement of enough soil so that uniformity of support is destroyed and slab edges are left unsupported.

Subbase studies show that three factors are necessary for mud-pumping to occur:

1. a subgrade soil that will go into suspension
2. free water between the pavement and subgrade, or subgrade saturation
3. frequent passage of heavy loads

These studies also showed that mud-pumping did not occur on natural subgrades or granular subbases with less than about 45% passing the No. 200 sieve, and a plasticity index of 6, or less.

The performance of test sections with no subbase at the AASHO Road Test shows that adequately designed pavements without subbases are suitable for many city streets, county roads, light-traffic highways, and light-duty airports. Subbase surveys conducted by highway departments show that

1. pavements designed to carry not more than 100 to 200 heavily loaded trucks per day do not require subbases to prevent pumping damage;
2. soils with less than 45% passing a No. 200 sieve and with a *PI* of 6, or less, are suitable for moderate volumes of heavy truck traffic; and
3. subbases meeting *AASHO Specification M155* effectively prevent mud-pumping in pavements carrying the greatest volumes of traffic.

AASHO Specification M155 to prevent pumping states:

"Granular material for use as subbase under concrete pavement may be composed of sand, sand-gravel, crushed stone, crushed or granulated slag, or combinations of these materials. The material shall meet the following requirements:

Maximum size:	not more than 1/3 the thickness of the subbase
Passing No. 200 sieve:	15% maximum
Plasticity index:	6 maximum
Liquid limit:	25 maximum

NOTE: Materials with a higher percentage passing No. 200 sieve or with a higher plasticity index than 6 or a higher liquid limit than 25 may be used, provided that a stabilization method found to be locally suitable is used. The material shall be graded suitably to permit compaction to such a density that a minimum increase in densification will occur after the pavement is in service."

-2. Subbase thickness—Since the primary purpose of a subbase is to prevent pumping, it is neither necessary nor economical to use thick subbases. Experimental projects have shown that a 3-in. depth of subbase will prevent mud-pumping under very heavy traffic. When a subbase is required, a depth of 4 to 6 in. is commonly specified for highway pavements and a depth of 6 to 8 in. for heavy-duty airport pavements.

-3. Subbase gradation control—While a wide variety of locally available materials have performed well as subbases for concrete pavements, the subbase for an individual project should have a reasonably constant gradation. This makes it possible for compaction equipment to produce the uniform support that is essential for good pavement performance. *AASHO Specification M155* is suitable for establishing minimum subbase requirements, but it does not afford acceptable gradation control for an individual project.

One method for overcoming this problem is to adopt a specification such as *AASHO M147*. This specification is shown in Table 22-1. The subbase for a particular project is kept within the limits of a single gradation.

-4. Subbase compaction—Granular materials are subject to consolidation from the action of heavy truck traffic after pavements are placed in service. To prevent a detrimental amount of consolidation, subbases must be compacted to very high densities.

Performance experience and research furnish convincing support for the following recommendations:

1. Subbases should have a minimum of 100% standard density (*AASHO T99*). On projects that will carry large volumes of traffic, the specified density should be not less than 105% of standard or 98 to 100% of modified density (*AASHO T180*).
2. When subbase depths are increased beyond the minimum needed to prevent pumping, there is an increasing risk of subbase consolidation from heavy traffic.

TABLE 22-1 Grading Requirements for Subbase Materials

Sieve designation	*Percentage by weight passing square mesh sieves*					
	Grading A	*Grading B*	*Grading C*	*Grading D*	*Grading E*	*Grading F*
2-inch	100	100	–	–	–	–
1-inch	–	75–95	100	100	100	100
3/8-inch	30–65	40–75	50–85	60–100	–	–
No. 4	25–55	30–60	35–65	50–85	55–100	70–100
No. 10	15–40	20–45	25–50	40–70	40–100	55–100
No. 40	8–20	15–30	15–30	25–45	20–50	30–70
No. 200	2–8	5–20	5–15	5–20	6–20	8–25

-5. *Cement-treated subbases*—The use of cement-treated subbases* for concrete pavement has become a standard practice in many parts of the United States and Canada. Several factors are responsible for their increasing use. An important one is the growing scarcity of good granular subbase materials at a time when tremendous quantities of these materials are needed for building new pavements.

The principal benefits to be derived from use of cement-treated subbases are:

1. A uniform and strong, non-consolidating support is provided for the pavement.
2. Firm support for the slipform paver or side forms contributes to the construction of smoother pavements.
3. A stable working base expedites construction operations.

22.4 THICKNESS DESIGN FOR HIGHWAY AND STREET PAVEMENTS

The thickness design procedure for the Portland Cement Association[22-16] is presented as appropriate for a wide range of conditions with proper assessment of the design variables.

22.4.1 Basis for Design

The thickness design method is based on knowledge from the following sources:

1. Theoretical studies of pavement slab behavior by H. M. Westergaard,[22-17–22-20] G. Pickett and G. K. Ray,[22-21, 22-22] and others.
2. Model and full-scale tests such as the Arlington Test conducted by the Bureau of Public Roads[22-23] and tests made at the PCA laboratories.[22-1–22-4, 22-24]
3. Experimental pavements subjected to controlled test traffic, such as the Bates Test Road,[22-25] the Pittsburg, Calif. Test Highway,[22-26] the Maryland Road Test,[22-27] and the AASHO Road Test.[22-6, 22-28]
4. The performance of normally constructed pavements subject to normal mixed traffic.

All these sources of knowledge are useful. However, the knowledge gained from performance of normally constructed pavements is the most important.

22.4.2 Subgrade-Subbase Strength

The degree of subgrade or subgrade-subbase support is in terms of k, Westergaard's modulus of subgrade reaction. The k-value is determined by in-situ plate bearing tests with a 30-in. diameter plate. Procedures are described in *ASTM Method D1196* and the Department of the Army (U.S.), *Technical Manual TM-5-824-3* and are discussed in Ref. 22-29.

If time and equipment are not available for plate-load testing, k can be estimated by correlation to soil classification or laboratory strength test as shown in Fig. 22-1. Tables 22-2 and 22-3 show the effect of subbase thickness on the design k-value.

*Design and construction practices for cement-treated subbases are given in Refs. 22-12 and 22-13. Data on the properties of cement-treated subbases are given in Refs. 22-2, 22-14, 22-15, and 22-24.

TABLE 22-2 Effect of Untreated Subbase on k Values

Subgrade k value	Subbase k value, pci			
	4 in.	6 in.	9 in.	12 in.
50	65	75	85	110
100	130	140	160	190
200	220	230	270	320
300	320	330	370	430

TABLE 22-3 Design k Values for Cement-Treated Subbases

Subgrade k value	Subbase k value, pci			
	4 in.	6 in.	8 in.	10 in.
50	170	230	310	390
100	280	400	520	640
200	470	640	830	–

22.4.3 Load Stresses

Bending of a concrete pavement under wheel loads produces both compressive and tensile (or flexural) stresses. Although compressive stresses are not critical, the ratio of flexural stresses to flexural strength is high, often exceeding values of 0.5. As a result, slab thickness design is controlled by keeping flexural stresses due to traffic loads within safe limits.

For street and highway pavements with the usual 12-ft pavement lanes the critical stress location is considered to be the transverse joint edge rather than at the outside pavement corner or edge. This conclusion is based on studies of truck placements[22-30] across pavement lanes and an analysis of the effect of load position.[22-31, 22-32] Accordingly, the charts shown in Figs. 22-2 and 22-3 were developed from influence charts[22-22] to be used for determining stresses at this load position.

While load stresses are critical at the transverse joint edge, shrinkage and temperature stresses are low at this location and usually are not additive to load stresses. For this reason, load stresses only are considered. Conservatively, no beneficial allowance is made for load transfer across the joint.

22.4.4 Flexural Strength of Concrete

Flexural strength is determined by modulus of rupture (M-R) tests, usually made on 6 x 6 x 30-in. beams. The modulus of rupture is determined by third-point loading tests (ASTM C78). The 28-day M-R values are currently used for thickness design of streets and highways, However, this practice is conservative since concrete continues to gain strength for many years (see Fig. 22-14).

The following approximate relationship between flexural and compressive strength is sometimes useful in preliminary design stages; however, the final design should be based on modulus of rupture test data:

$$MR = K\sqrt{f_c'}$$

where

MR = flexural strength (modulus of rupture), psi
K = a constant between 8 and 10
f_c' = compressive strength, psi

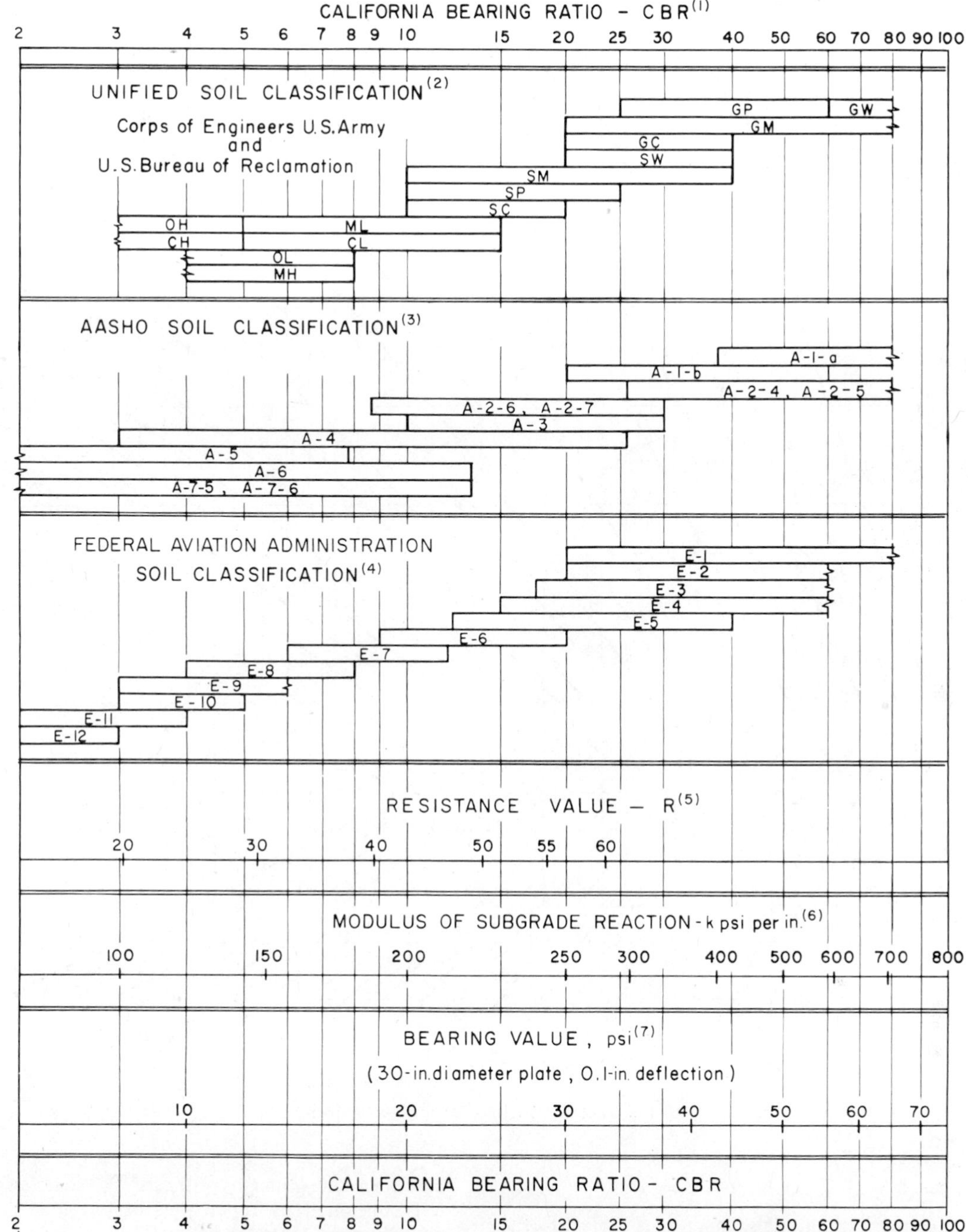

Fig. 22-1 Soil classification, resistance value, *k*-value, and bearing value vs. California bearing ratio.

(1) For the basic idea, see Porter, O. J., "Foundations for Flexible Pavements," Highway Research Board, *Proceedings of the Twenty-second Annual Meeting*, **22,** 100–136, 1942.

(2) "Characteristics of Soil Groups Pertaining to Roads and Airfields," Appendix B, *The United Soil Classification System*, U.S. Army Corps of Engineers, Technical Memorandum 3-357, 1953.

(3) "Classification of Highway Subgrade Materials," Highway Research Board, *Proceedings of the Twenty-fifth Annual Meeting*, **25,** 376–392, 1945.

(4) *Airport Paving*, U.S. Department of Commerce, Federal Aviation Agency, pp. 11–16, May, 1948. Estimated using values given in FAA *Design Manual for Airport Pavements.*

(5) Hveem, F. N., "A New Approach for Pavement Design," *Engineering News-Record*, **141** (2), 134–139, July 8, 1948. *R* is factor used in California Stabilometer-Method of Design.

(6) See Middlebrooks, T. A., and Bertram, G. E., "Soil Tests for Design of Runway Pavements," Highway Research Board, *Proceedings of the Twenty-second Annual Meeting*, **22,** 152, 1942. *k* is factor used in Westergaard's analysis for design of concrete pavement.

(7) See (6), page 184.

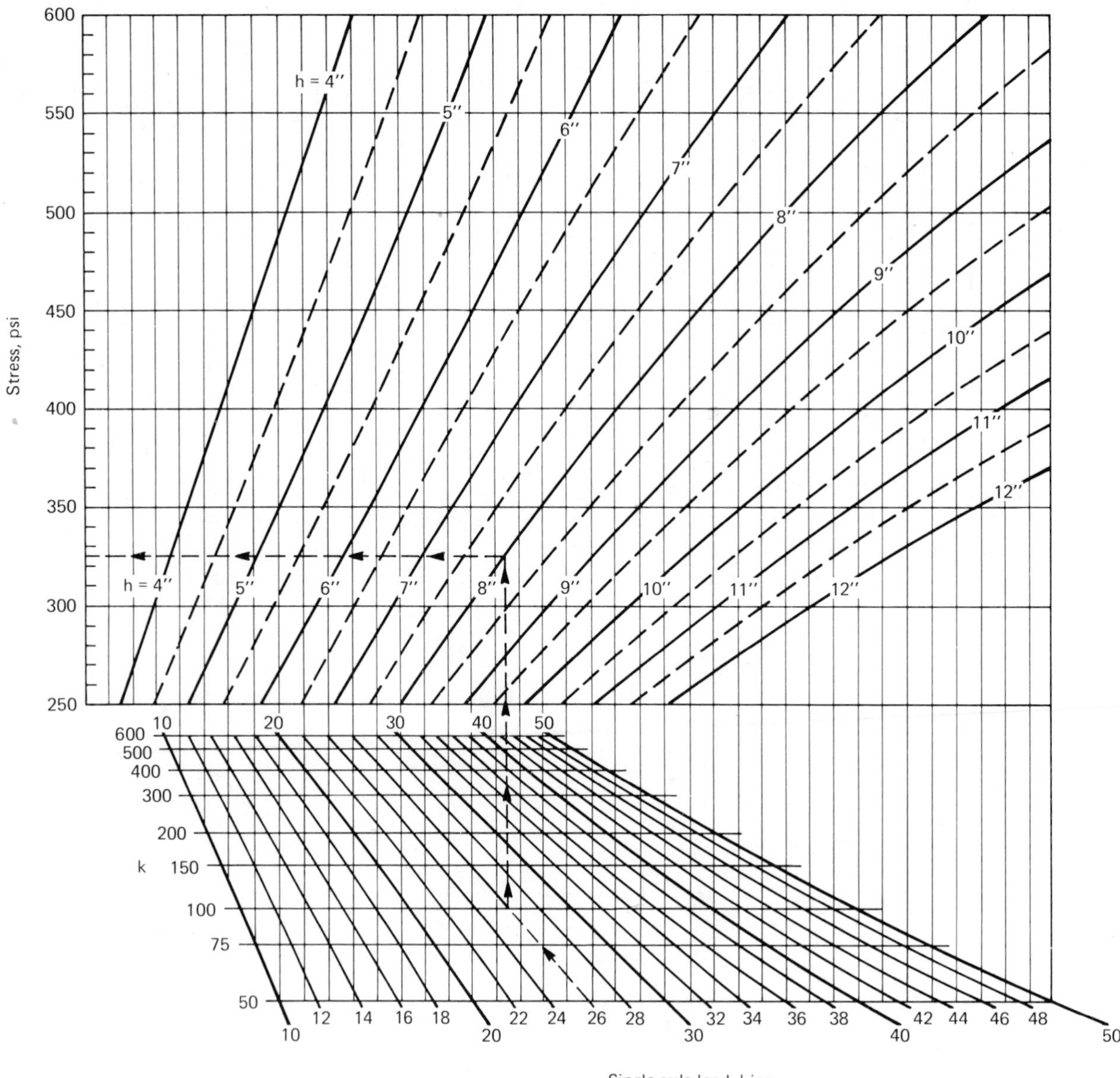

Fig. 22-2 Stress chart for single-axle loads.

22.4.5 Fatigue

Like other construction materials, concrete is subject to the effects of fatigue. Since the critical stresses in concrete are flexural, fatigue due to flexural stress is used for thickness design. Stress ratio is defined as the ratio of flexural stress to modulus of rupture. For example, if an axle load causes a flexural stress of 500 psi and the modulus of rupture is 700 psi, then

$$\text{stress ratio} = \frac{500}{700} = 0.71$$

Flexural fatigue research on concrete has shown that, as ratios decrease, the number of stress repetitions to failure increases. Allowable load repetitions for stress ratios between 0.50 and 0.85 are shown in Table 22-4. The values are conservatively based on fatigue research on concrete.[22-33, 22-34]

The design example in this section shows how Table 22-4 fatigue data are used for thickness design. This usage is based on the Miner hypothesis[22-35] that fatigue resistance not consumed by repetitions of one load is available for repetitions of other loads.

Theoretically, the total fatigue used should not exceed 100%. For designs based on the 28-day modulus of rupture, fatigue consumption can be increased to about 125%. This increase takes account for the strength gain after 28 days.

22.4.6 Load Safety Factors

Research[22-6,22-27] shows that moving loads cause less stress than static loads. Therefore, the formerly used concept of assigning an impact factor to increase the effect of loads is more accurately classified as a load safety factor. In the design procedure axle loads are multiplied by load safety factors of 1.0, 1.1, or 1.2 depending on pavement class as follows:

1.0 = for highways, residential streets, and other streets that will carry small volumes of truck traffic

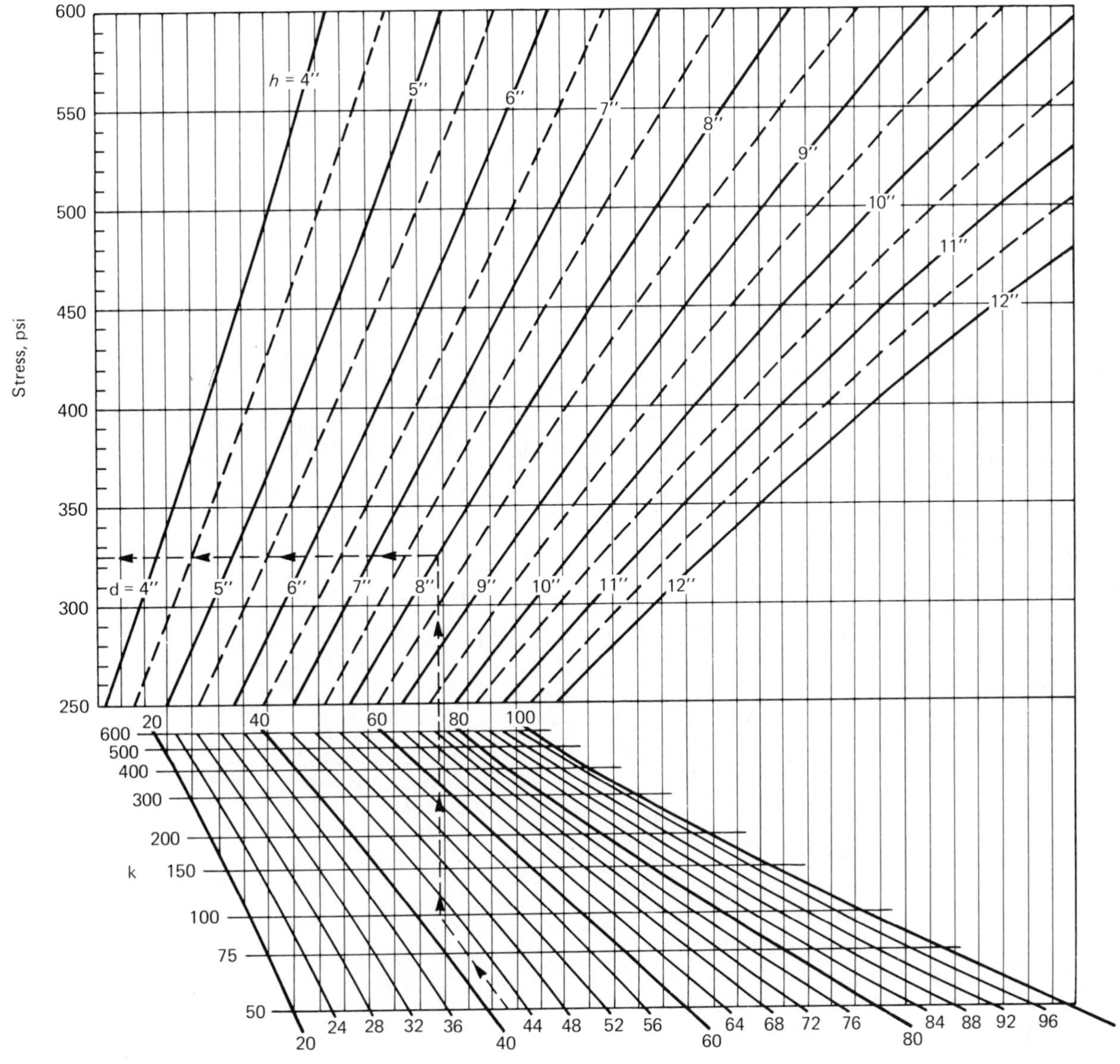

Fig. 22-3 Stress chart for tandem-axle loads.

1.1 = for highways and arterial streets where there will be moderate volumes of truck traffic

1.2 = for Interstate and other multilane projects where there will be uninterrupted traffic flow and high volumes of truck traffic.

22.4.7 Traffic Analysis Period

The term 'pavement life' is not subject to precise definition. Some engineers and highway agencies consider the life of a concrete pavement ended when the first overlay is placed. Based on this concept, the life of concrete pavements may vary from less than 20 years on some projects with design, material, or construction defects, to more than 50 years on other projects where these defects are absent.

A more reasonable basis for engineering analysis and pavement design is to recognize that the useful life of a concrete pavement does not end when the first–or second–overlays are placed. Instead, the concrete continues to serve as the primary load-carrying element in the pavement structure. Accordingly, a period of 40 years is assumed as the traffic analysis period for design purposes.

22.4.8 Traffic Analysis

The numbers and weights of heavy axle loads are major factors in the design procedure. The total number of trucks anticipated during the traffic analysis period is estimated. These are distributed into axle weight groupings by the use of special traffic loadometer studies made for the project, or by using annual loadometer surveys made by the highway department. Methods for detailed traffic analysis are given in Ref. 22-16. The end result of the traffic analysis is to estimate the numbers of single and tandem axle loads in separate weight categories that will use the pavement during the design period.

22.4.9 Design Procedure

An example of the design procedure and calculations is shown in Fig. 22-4 for an urban, Interstate pavement. The

TABLE 22-4 Stress Ratios and Allowable Load Repetitions

Stress ratio*	*Allowable repetition*	*Stress ratio*	*Allowable repetition*
0.51**	400,000	0.69	2,500
0.52	300,000	0.70	2,000
0.53	240,000	0.71	1,500
0.54	180,000	0.72	1,100
0.55	130,000	0.73	850
0.56	100,000	0.74	650
0.57	75,000	0.75	490
0.58	57,000	0.76	360
0.59	42,000	0.77	270
0.60	32,000	0.78	210
0.61	24,000	0.79	160
0.62	18,000	0.80	120
0.63	14,000	0.81	90
0.64	11,000	0.82	70
0.65	8,000	0.83	50
0.66	6,000	0.84	40
0.67	4,500	0.85	30
0.68	3,500		

**Load stress divided by modulus of rupture*
***Unlimited repetition for stress ratios of 0.50 or less.*

axle loads and expected number of load repetitions (columns 1 and 6) were first estimated for a period of 40 years.

From the axle load data, stress and fatigue calculations are made for the trial thickness. Load stresses are determined from Figs. 22-2 and 22-3 and allowable load repetitions are listed in Table 22-4. The design thickness is selected to prevent flexural fatigue failure of the slab. In this case, at a trial thickness of 8.5 in., the total fatigue resistance used is 83%. This would be an adequate design. Normally, a slab thickness is selected so that fatigue consumption does not exceed 125% for the 28-day modulus of rupture.

22.5 JOINT DESIGN FOR HIGHWAY AND STREET PAVEMENTS

Joints are placed in concrete pavements to control cracking due to (1) tensile stresses resulting from restrained shrinkage; (2) compressive stresses that may occur when the slab expands; and (3) the combined effects of loads and restrained warping. Restrained warping may occur when the slab attempts to warp upward or downward due to temperature and moisture differences between the upper and lower part of the slab. Stress occurs when the slab is restrained from warping by its weight or by frictional resistance of the subgrade or subbase.

Joints are also used to divide the pavement into suitable increments for construction purposes.

22.5.1 Basis for Design

Joint design is based on knowledge from the following three major sources.

1. theoretical studies of concrete behavior[22-18, 22-23, 22-36–22-43]
2. experimental pavements and model and full-scale laboratory tests[22-6, 22-23, 22-25, 22-27]
3. the performance of highway and street pavements subject to normal mixed traffic.

Slab dimensions and jointing details for individual projects should reflect careful study of performance of pavements similar to the project being designed.

A detailed discussion of joint design is given in Ref. 22-44.

22.5.2 Transverse Joints

In the range of slab thicknesses used for highway and street pavements, transverse joint spacing for plain pavements (unreinforced) should be about 15 to 20 ft to control cracking due to restrained contraction and restrained warping. Surveys of pavements in service show that the correct spacing will vary in different localities, climate, and with different types of aggregate. Because of these variations, local experience is the best guide as to whether spacings greater than 15 to 20 ft can be used to effectively control cracking. Where experience is not available, the spacings shown in Table 22-5 are used. These spacings are for average soil and climatic conditions; the maximum values are used only where environmental conditions are favorable.

TABLE 22-5 Spacing of Contraction Joints for Unreinforced Pavements

Type of coarse aggregate	
Crushed granite	25–30 ft
Crushed limestone	20–30 ft
Crushed flinty limestone	20–25 ft
Calcareous gravel	20–25 ft
Siliceous gravel	15–20 ft
Gravel less than 3/4-in. size	15–20 ft
Slag	15–20 ft

For reinforced pavements (mesh-dowel pavements) longer joint spacings are used and cracking is controlled by placing distributed steel between the joints. The purpose of the distributed steel is not to prevent cracking but to hold slab edges firmly together after cracks have formed. Based both on the economics and performance of mesh-dowel pavements, it is desirable to limit joint spacing to about 30 or 40 ft. Cost studies[22-45] for highway pavements indicate an optimum spacing in this range. In addition, the performance of all pavements depends strongly on the effectiveness of the joint seals. At joint spacings greater than 40 ft, the joint seals are usually not effective because of the greater fluctuations in joint width.

There are three general types of transverse joints: contraction joints, construction joints, and expansion joints. In addition to their special functions, expansion joints and untied construction joints also function as contraction joints.

-1. Contraction joints—The purpose of a contraction joint is to control cracking caused by shrinkage and the combined effect of loads and warping. Since contraction joints must be free to open, distributed steel reinforcement, if any, is interrupted at these joints. For the same reason, contraction joints are not tied by deformed tiebars.

Contraction joints also provide some relief from expansive forces since the initial shrinkage of the concrete opens the joint slightly and thereby provides for subsequent expansion.

Recommended details for contraction joints are given in the following paragraphs and are illustrated in Fig. 22-5.

In these joints a plane of weakness is created by means of a groove formed while the concrete is plastic, or by inser-

Project Design Two-B

Type Urban Interstate-Level Terrain No. of Lanes 4

Subgrade k 100 pci., Subbase 4-in. Granular Untreated

Combined k 130 pci., Load Safety Factor 1.2 (L.S.F.)

Procedure

1. Fill in Col. 1, 2, and 6, listing axle loads in decreasing order.
2. Assume 1st trial depth. Use ½-in. increments.
3. Analyze 1st trial depth by completing columns 3, 4, 5, and 7.
4. Analyze other trial depths, varying M.R., slab depth and subbase type.

1	2	3	4	5	6	7
Axle Loads kips	Axle Loads × 1.2 L.S.F. kips	Stress psi	Stress Ratios	Allowable Repetitions No.	Expected Repetitions No.	Fatigue Resistance Used** percent

Trial depth 8.5 in. M.R.* 700 psi k 130 pci

Single Axles

1	2	3	4	5	6	7
30	36.0	367	.52	300,000	3,700	1
28	33.6	353	.51	400,000	3,700	1
26	31.2	328	<.50	Unlimited	7,400	0
24	28.8		"	"	195,000	0
22	26.4		"	"	764,000	0
		[From Fig. 22-2]		[From Table 22-4]		[Col. 6 ÷Col. 5 × 100]

Tandem Axles

1	2	3	4	5	6	7
54	64.8	413	.59	42,000	3,700	9
52	62.4	398	.57	75,000	3,700	5
50	60.0	387	.55	130,000	36,270	28
48	57.6	375	.54	180,000	36,270	20
46	55.2	361	.52	300,000	57,530	19
44	52.8	346	<.50	Unlimited	179,790	0
42	50.4		"	"	"	0
40	48.0		"	"	"	0
		[From Fig. 22-3]		[From Table 22-4]		[Col. 6 ÷Col. 5 × 100]

Total = 83

*M.R. = Modulus of Rupture for 3rd pt. loading.
**Total fatigue resistance used should not exceed about 125%.

Fig. 22-4 Design procedure and calculations.

tion of a strip in the plastic concrete, or by sawing a groove after the concrete has hardened. The concrete subsequently cracks and the irregular slab faces below the groove provide load transfer by aggregate interlock.

The depth of the groove should not be less than one-fourth of the slab thickness. Also, the depth of sawed grooves should not be less than the diameter of the largest size coarse aggregate. If the groove is formed, the corners should be edged to a radius of 0.25 in.

The width of the groove should be such that horizontal slab movements will not increase or decrease the width excessively. (See section 22.5.5.) This is necessary to facilitate sealing of the joint and to ensure that the joint sealant is not unduly strained. In some mild climate areas, the sealing of closely-spaced joints may not be required.

Mechanical load-transfer devices are usually recommended where the joint spacing exceeds 20 ft, and even at a lesser spacing where service conditions are severe. The requirements for mechanical load transfer devices are given in section 22.8.2. Dowels are most commonly used as the load transfer device. The sliding end of the dowel is coated or lubricated to permit free movement. Dowels are installed at mid-depth of the slab.

Experience has shown that dowels, or other load-transfer devices, are not needed under certain favorable conditions. The need for dowels depends on climate, subgrade conditions and the amount of heavy truck traffic. The use of undoweled joints with plain pavements having short joint spacings is generally limited to (1) pavements in areas where there is little or no frost action, and (2) secondary roads

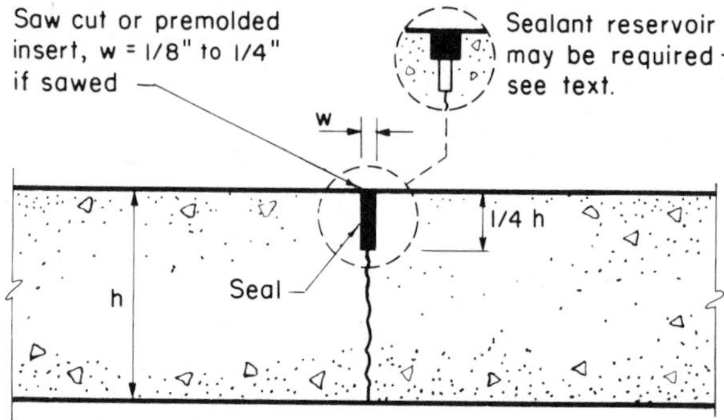

Undoweled Joint, Plain Pavement

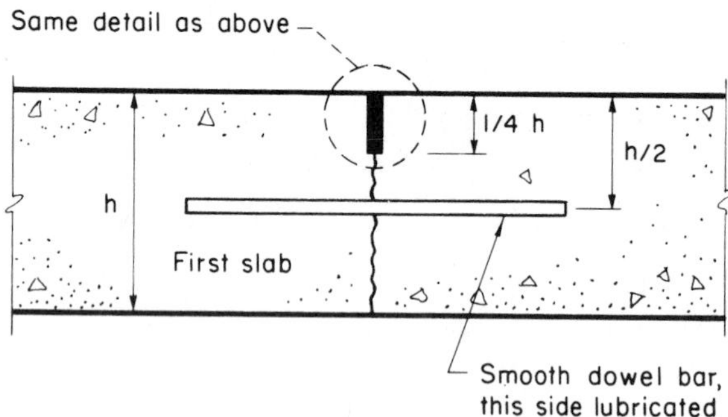

Doweled Joint, Plain Pavement

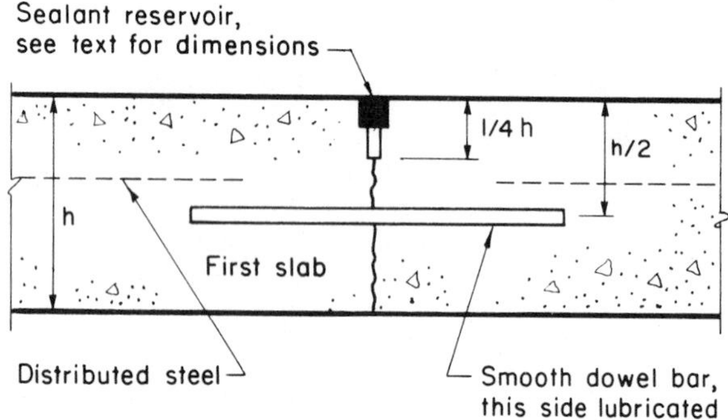

Doweled Joint, Reinforced Pavement

Fig. 22-5 Contraction joints.

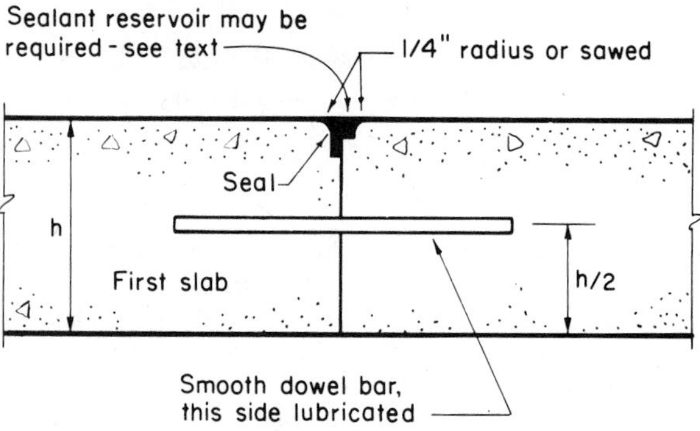

Butt Joint

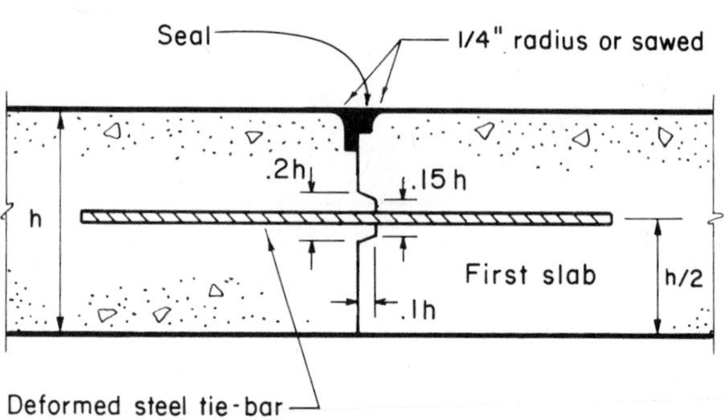

Keyed & Tied Joint

(Use only in middle third of normal joint interval)

Fig. 22-6 Construction joints.

and residential city streets that carry only small volumes of heavy trucks.

A cement-treated subbase provides joint support[22-15] because its rigidity offers considerable resistance to deflection, bending and shear. This adds to the aggregate-interlock load-transfer capacity.

-2. Transverse construction joints—Transverse construction joints are designed for planned interruptions to paving operations such as at the end of each day's paving or for emergency interruptions caused by inclement weather or equipment breakdown.

Planned construction joints are installed at a normal joint interval and consist of a butt-type joint as shown at the top of Fig. 22-6. In an emergency, such as equipment breakdown or non-delivery of materials, it may be necessary to install a joint that is not located at the regular joint spacing. There may be two types of these emergency joints:

a. for lane-at-a-time construction, a keyed (tongue and groove) joint is required as shown at the bottom of Fig. 22-6. This joint is tied with deformed steel tie-bars to prevent joint movement that would otherwise cause sympathetic cracking in the adjacent lane.
b. for full-width paving, either a doweled butt joint or a tied keyed joint may be used.

Whenever construction joints are tied, it is essential to edge or saw the joint so that spalling will not develop.

-3. Expansion joints—Design details for expansion joints are shown in Fig. 22-7. Studies[22-46, 22-47, 22-48] of pavements in service have shown that, except at structures and unsymmetrical intersections, expansion joints are not needed if:

a. the pavement is built with materials that have normal expansion characteristics,
b. the pavement is constructed during periods when temperatures are well above freezing.
c. contraction joints are closely spaced so that intermediate cracks do not form, or
d. contraction joints are properly maintained to prevent infiltration of incompressible materials.

22.5.3 Longitudinal Joints

For control of longitudinal cracking, longitudinal joints are usually spaced at 12 ft to coincide with traffic lane mark-

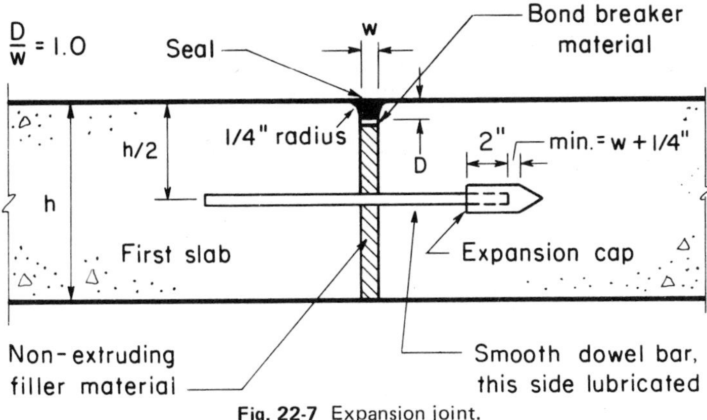

Fig. 22-7 Expansion joint.

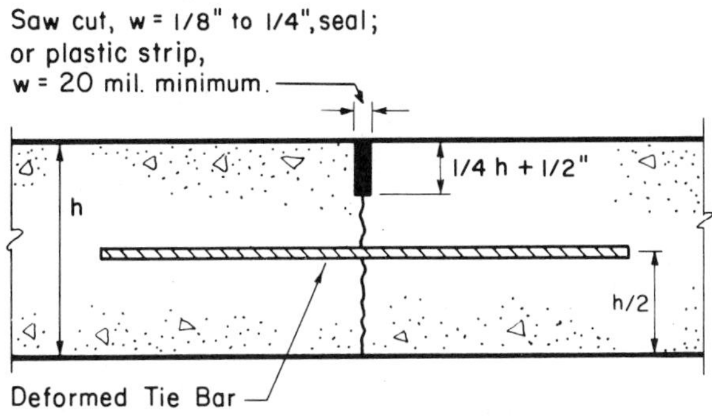

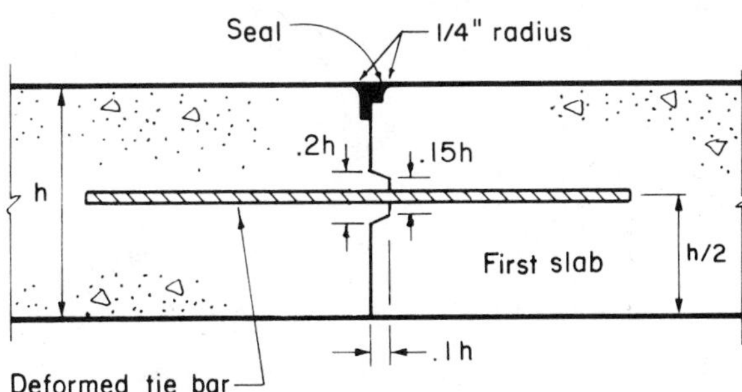

Fig. 22-8 Longitudinal joints.

ings. Recommended details for longitudinal joints are illustrated in Fig. 22-8.

Where two or more pavement lanes are constructed at one time, a weakened plane joint as shown at the top of Fig. 22-8 is specified. This is normally used for full-width construction and is called a longitudinal center joint. In these joints the groove is made by inserting a strip of plastic, metal or other material in the fresh concrete, or by sawing a groove after the concrete has hardened. To ensure that the pavement cracks at the joint it is essential that the depth of the groove be at least one-fourth of the slab thickness plus 0.5 in.

A keyed joint is specified for lane-at-a-time paving and where the total pavement width is not placed in one operation. This is called a longitudinal construction joint and is shown at the bottom of Fig. 22-8.

Both of these are hinged joints tied together with deformed tiebars that hold abutting slabs together, thus assuring load transfer at the joint by aggregate interlock, in the first case, and by the keyway, in the latter case. Recommended tiebar sizes and spacings for these joints are given in section 22.8.3.

Tiebars may be omitted from longitudinal construction joints where there is enough transverse confinement, such as curbs and gutters, or paved shoulders at pavement edges, to prevent slab separation. Tiebars are also omitted from the interior joints of parking lots and other wide pavement areas since these are confined by surrounding slabs. The usual practice is to tie only the longitudinal joints within 37.5 ft of the outside pavement edge.

22.5.4 Jointing Layout for Pavement Intersections

A typical jointing arrangement for pavement intersections is shown in Fig. 22-9. This figure illustrates the location of various joint types and the application of the principles presented previously in this section.

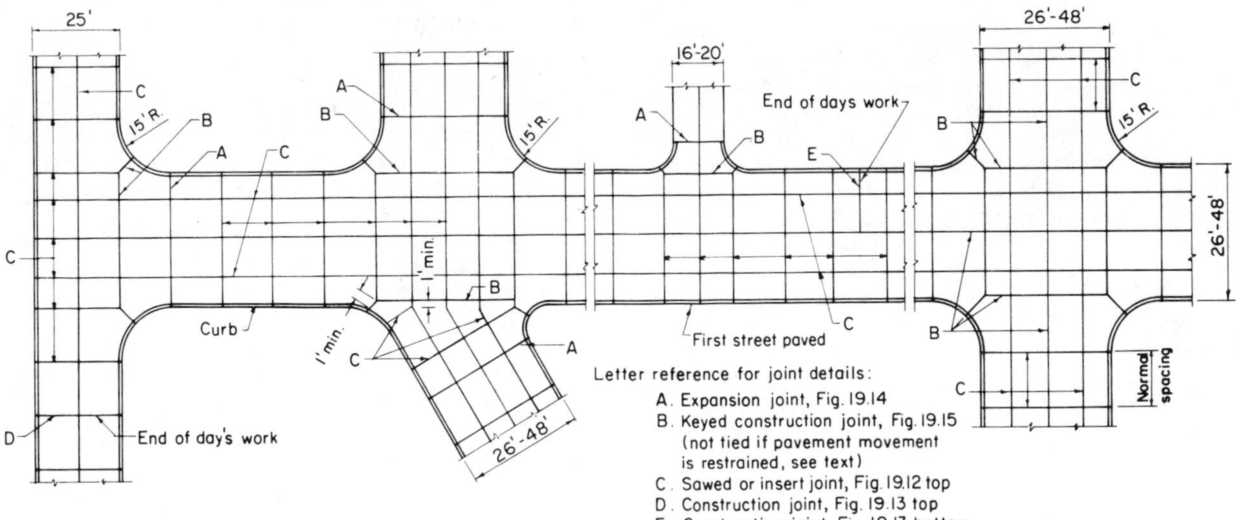

Fig. 22-9 Typical jointing arrangement for pavement intersections.

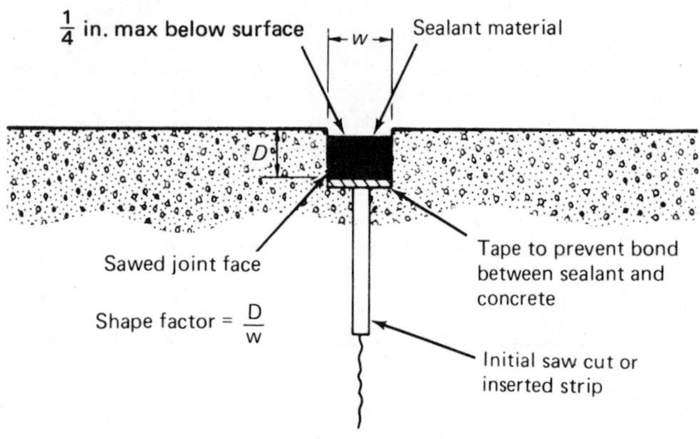

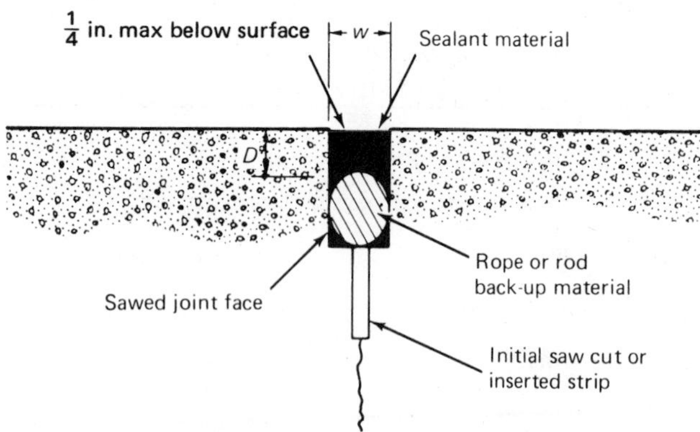

See text for dimensions of sealant reservoir

Fig. 22-10 Joint sealant reservoir and shape factor.

22.5.5 Joint Shapes Required for Effective Joint Sealing

As a pavement expands and contracts due to temperature and moisture changes, the pavement joints open and close. As a result, sealants placed in the joints must be capable of withstanding this repeated extension and compression[22-49 through 22-52] while maintaining their function of preventing the intrusion of incompressible solids. In order to maintain an effective seal, the joint width must be large enough so that later changes in joint width will not put undue strain on the sealant. To obtain the required joint width, the top of the joint is sawed wider forming a reservoir, usually rectangular, for the sealant material as shown in Fig. 22-10.

For poured joint sealants, the shape factor (depth-to-width ratio) of the sealant reservoir has a critical effect on the sealant's capacity to withstand extension and compression. Within certain practical limitations, the lower the depth to width ratio, or shape factor value, the lower the strain on the sealant under a given joint movement. The required shape factor will depend on the properties of the sealant and the amount of joint movement; the latter in turn is related to the panel length or joint spacing and to the maximum expected seasonal temperature change in the slab.* Table 22-6 lists recommended depths and widths of the sealant reservoir for poured sealants frequently used such as those meeting current Federal and ASTM specifications. A stiff, self-adhering plastic strip is applied to the

TABLE 22-6 Joint Width and Depth for Poured Sealants

Joint Spacing ft	*Sealant Reservoir Shape*	
	Width in.	*Depth in.*
15 or less	1/4	1/2 min.
20	3/8	1/2 "
30	1/2	1/2 "
40	5/8	5/8
50	3/4	3/4

TABLE 22-7 Joint and Seal Width for Preformed Seals

Joint Spacing, ft	*Joint Width, in.*	*Seal Width, in.*
25 or less	1/4	7/16
30	3/8	5/8
50	1/2	7/8
70	3/4	1 1/4

bottom of the reservoir to break the bond between the sealant and the bottom concrete surface.

Frequently, a butyl or polypropylene rope is placed in the bottom of the sealant space to break bond and prevent loss of sealant into the crack below the joint filler. In this case, it is necessary to saw the reservoir deeper an extra amount equal to the rope diameter so that the shape factors of the sealant listed in Table 22-6 are maintained.

For preformed, compression-type seals, recommended joint width and width of seals are listed in Table 22-7. The depth of saw cut is such that the compression seal is installed about 0.125 in. below the pavement surface. With special equipment, the seals are installed, without stretching, to a compressed width of about half of their uncompressed width.

22.6 THICKNESS DESIGN FOR AIRPORT PAVEMENTS

In determining the thickness of concrete pavements for airports, several major factors are considered: the types and weights of aircraft, and their effect on flexural stresses in the slab; the flexural strength of the concrete; the supporting strength of the subgrade (and subbase if one is used); the type of facility being designed (taxiway, runway, apron, hangar floor, etc.); and the frequency of aircraft operations expected on the pavement.

Complete descriptions of the current design procedures of several agencies are given in Refs. 22-53 through 22-55 and 22-61; the methods and results are quite similar as pointed out in Ref. 22-56. The procedure of the Portland Cement Association is described in this section.

22.6.1 Load Stresses

Flexural stresses caused by aircraft loads are computed on the assumption that the load is placed near the middle of a large expanse of pavement. This can be assumed when adequate load transfer is provided at all joints. Aircraft wheel loads do not travel near the free outside edges of

*Actual joint movements are reported[22-50, 22-51] as approximately the same as theoretical values when a value of 5×10^{-6} in./in./°F is used for the thermal coefficient of concrete.

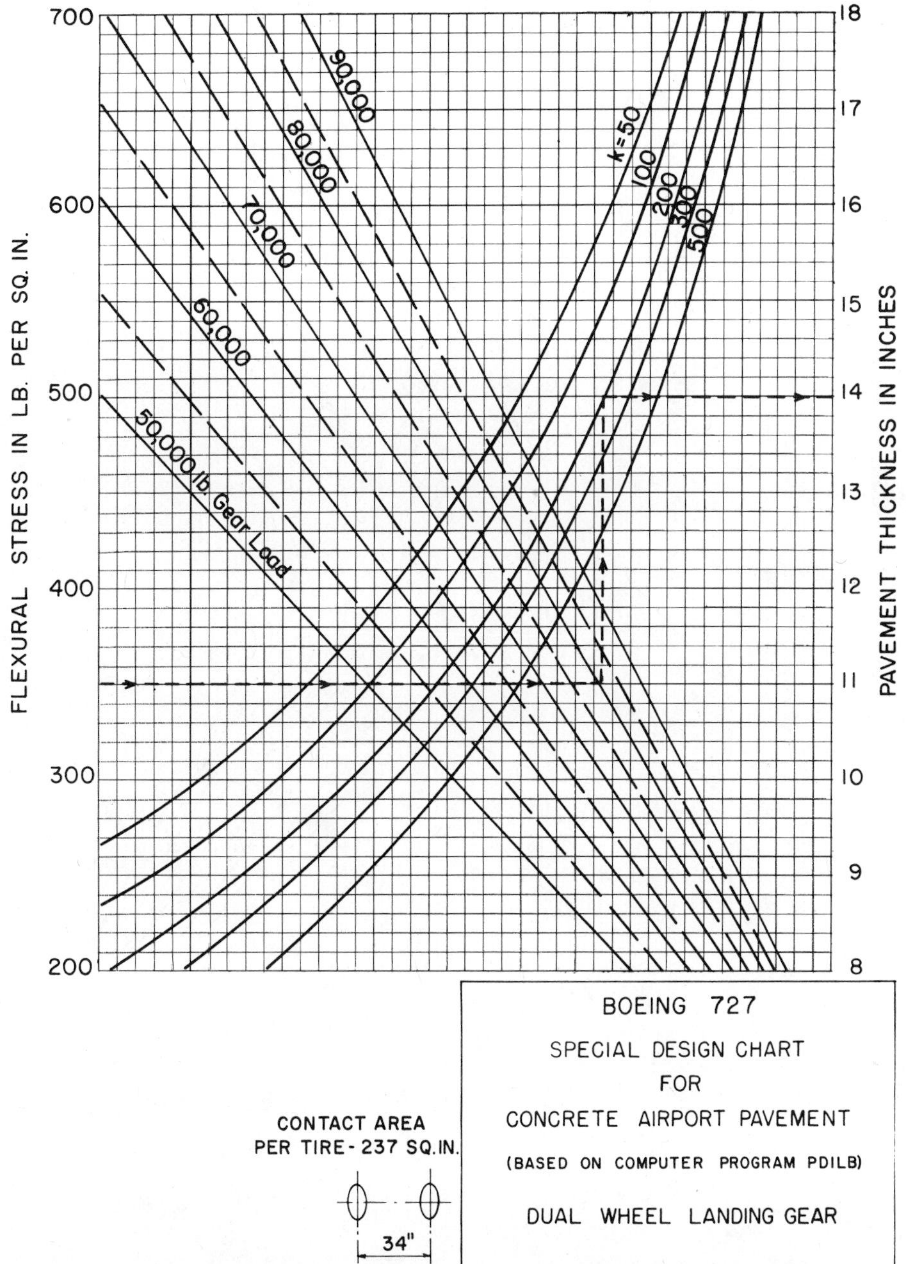

Fig. 22-11 Concrete airport pavement design chart for Boeing 727.

pavements; if free edge butt joints are used in the interior of a pavement expanse, the edge thickness is increased to compensate for the lack of load transfer.

The flexural stress caused by a given aircraft load is determined by the use of influence charts[22-22] or by computer program.[22-57] Aircraft manufacturers publish the required data on loads, wheel spacings and tire contact areas that are needed in the analysis. Design charts are then prepared for the specific aircraft that will use the pavement facility. Charts for several aircraft are shown in Figs. 22-11 through 22-13.

Design charts for most civil and military aircraft are published by the Portland Cement Association.

22.6.2 Flexural Strength

Concrete strength is determined by third-point flexural strength tests (ASTM C78) performed at 90 days for airport pavement design, rather than the 28-day strength test used for highway design. Even in 90 days, an airport pavement is subjected to much fewer operations with less channelization of traffic compared to highway pavements for a 28-day period.

Because of the continued strength increase during the early years of pavement life, the modulus of rupture soon exceeds the 90-day value. Figure 22-14 shows a conservative relationship between flexural strength and age of concrete.

22.6.3 Subgrade-Subbase Strength

The supporting strength of the subgrade and subbase, Westergaard's modulus of reaction (k) is determined by plate loading tests or by correlation to soil type or laboratory strength tests as described in section 22.4.2. Where subbase layers* will be used, the k-value on top of the sub-

*For pavements serving heavy, multiple-wheel gear aircraft, the effect of stabilized subbases is discussed in Appendix B of reference 22-53.

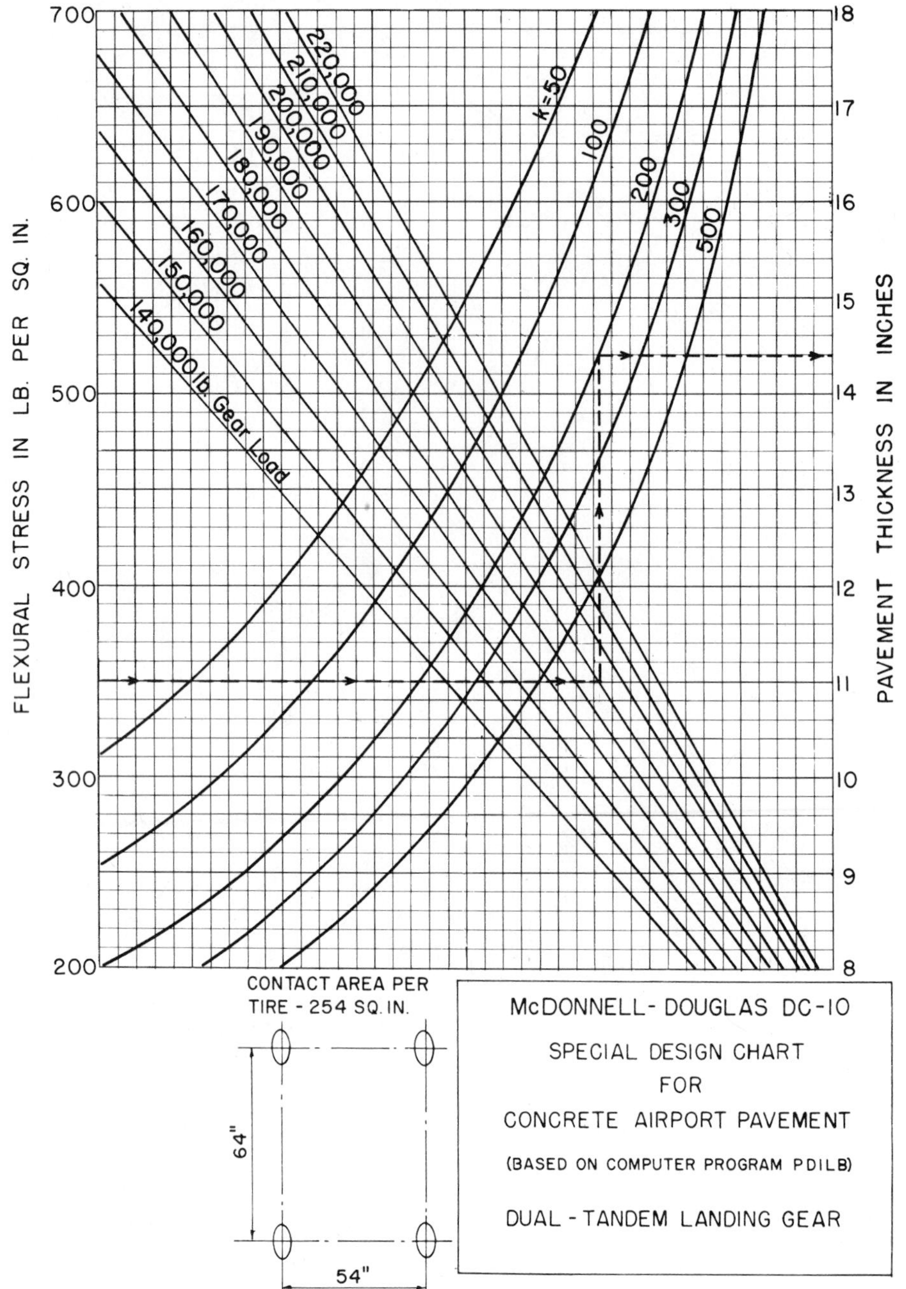

Fig. 22-12 Concrete airport pavement design chart for DC-10.

base layer may be estimated from Tables 22-2 and 22-3. On large projects, these values should be verified by plate-loading tests.

22.6.4 Safety Factor

The safety factors (ratios of working stress to design modulus of rupture) used for airport pavement design represent an assessment of the expected frequency of traffic operations and the channelization of these on runways, taxiways, and aprons. Undoubtedly, one of the most important factors in airport pavement design is the estimation of the loads of future, heavier aircraft and increased numbers of operations that the pavements will serve. Good design practice requires the gathering of data on expected future operating and load conditions from several sources including commercial airline forecasts, information from airport operating officials, and loading data for future aircraft projected by aircraft manufacturers.*

Based on this information, a conservative safety factor is selected and used to determine the allowable working stress in the design charts. The following ranges of safety factors are recommended:

Installation	Safety Factor
Aprons, taxiways, hard standings, runway ends (for distance of 1000 ft), hangar floors	1.7–2.0
Runways (central portion)	1.4–1.7

*When a specific forecast is made of the mixed aircraft that will operate during the design life, the fatigue methods discussed in Appendix A of Reference 22-53 may be used for the design or evaluation of pavements.

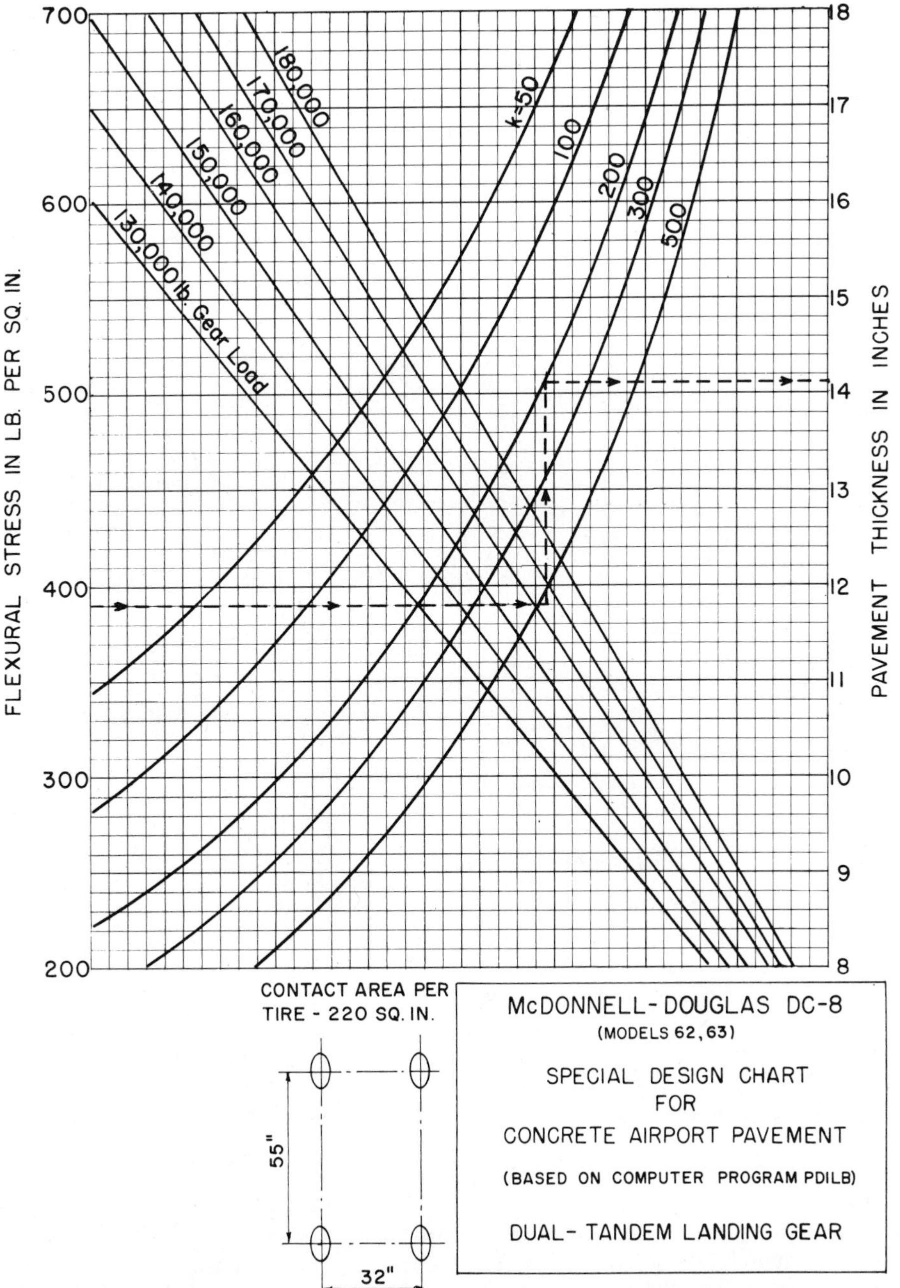

Fig. 22-13 Concrete airport pavement design chart for DC-8.

The lower safety factors for runways are permissible because most runway traffic consists of fast-moving loads which are partly airborne. In addition, the aircraft wheel loads are distributed transversely over a wide pavement area in such a manner that the number of stress repetitions in any one spot is quite small. Also, the number of stress repetitions at one location on a runway will be much lower than on taxiways, even on a one-runway airport.

Where taxiways intersect runways, the runway for a short distance each way should be of the same thickness as the taxiway. Any portions of runways which will serve as taxiways must also be designed as taxiways. Because of traffic channelization and spot parking on aprons, the frequency of stress repetitions in one spot may approach that in the taxiways.

At airports with a large number of operations by planes with critical loads, safety factors near the top of the suggested range should be used. On fields with only occasional operations by planes with critical loads, safety factors near the bottom of the range should be used. Those fields with a few daily operations by these loads should use some intermediate value. Even though there may be a large number of operations by lighter aircraft, the fatigue limit of the concrete will not be used up. The safety factor of 2.0 results in pavement adequate for an unlimited number of operations.

For heavy-duty runways serving large volumes of traffic, designers sometimes select a 'keel-section' design where the center section of the pavement is thicker than the outside pavement edges.

22.6.5 Design Procedure

Determination of slab thickness is done in the following steps:

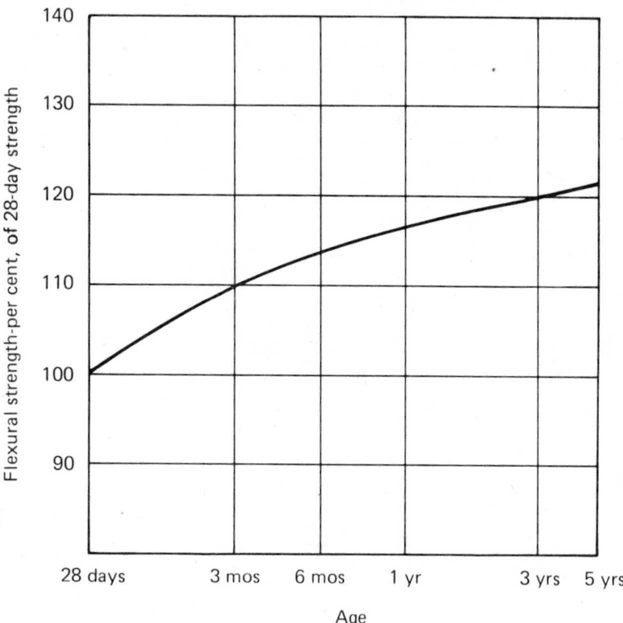

Fig. 22-14 Flexural strength-age relationship.

Step 1 The k-value is determined by plate loading tests or by correlation to subgrade soil test data (see section 22.4.2).

Step 2 A careful estimate of future, as well as present, operating and load conditions is made and an appropriate, conservative safety factor is selected as discussed in section 22.6.4.

Step 3 Working stress for a specific aircraft is determined by dividing the modulus of rupture of the concrete by the safety factor chosen.

Step 4 From the design chart for the specific aircraft, determine the pavement thickness. Enter the chart with the allowable working stress and proceed to the right to intersect the gear load line; then proceed up or down to the k-value; from this point project to the right to read slab thickness.

Step 5 Repeat the process for other aircraft of critical loads, again selecting new, appropriate safety factors for the level of operations expected for these aircraft, and select a design thickness for the most critical condition.

The following example illustrates the use of the design procedure.

EXAMPLE 22-1: Assume that a new runway and taxiway are to be designed to serve frequent operations of aircraft of which the B-727 and DC-10 produce the most critical loading conditions. In addition, the runway is expected to carry occasional operations of DC-8-63's.

SOLUTION:

Step 1 The existing subgrade is a pumpable sandy clay soil on which several plate loading tests have indicated a k-value of 170 pci. As discussed in section 22.3.2, a subbase will be required to prevent the traffic from pumping the subgrade soil. Since it is not necessary or economical to use a thick subbase layer, 6 in. is selected as an effective and practical subbase thickness. Based on these data, a preliminary design k-value of approximately 200 pci is estimated from Table 22-2 (This is later verified by plate tests on subbase test sections constructed during the design stages of the project.)

Steps 2-4 Data on concrete strengths made with local aggregates indicate that it is reasonable to specify a 90-day design modulus of rupture of 700 psi. Based on the previous discussion appropriate safety factors are selected and shown in columns 4 and 7 of Table 22-8. Working stresses are computed (columns 5 and 8) and the required slab thicknesses (columns 6 and 9) are determined from published stress charts (Figs. 22-11, 12, and 13 in this example).

Based on these data, a slab thickness of 14.5 in. is required for the taxiway and runway ends, while 13.0 in. is required for the central portion of the runway (thicknesses increased to next half-inch).

Step 5 Pavement stresses induced by other, less critical aircraft in the traffic forecast are next determined by the use of published design charts for these specific aircraft. (For this purpose, use of the charts is in the reverse order to that indicated previously; start with pavement thicknesses on the right-hand scale, and find slab stress on the left-hand scale.) If stresses for these other aircraft are less than 350 psi, these aircraft will not fatigue the pavements (safety factor of 2.0 or greater) and the slab thicknesses established above are adequate.

22.6.6 Pavement Evaluation

In addition to designing new pavements, it is sometimes desired to evaluate existing airport pavement installations to determine what gear loads can be carried without overstressing the slab.

The steps required in making such an evaluation are as follows:

Step 1 Determine the k-value, or values, of the subgrade beneath the present pavement. This may be done by loading the existing pavement as explained in Refs. 22-58 through 22-60, removing slab panels and making plate-bearing tests directly on the subgrade or subbase course (see discussion of subgrade-subbase strength, section 22.4.2). The results should be based on several tests, the number depending on the engineer's judgment concerning the requirements of the site.

TABLE 22-8 Example Calculations for Thickness Design

Design k-value - 200 pci | Design MR = 700 psi

			Pavement Facility					
			Taxiway & Runway Ends			*Runway, central portion*		
Aircraft	*Gear Load, Lbs.*	*Operations*	*Safety Factor*	*Working stress, psi (MR ÷ Col. 4)*	*Slab Thickness, in.*	*Safety Factor*	*Working stress, psi (MR ÷ Col. 7)*	*Slab Thickness, in.*
(1)	(2)	(3)	(4)	(5)	(6)	(7)	(8)	(9)
B-727	80,000	Frequent	2.0	350	14.0	1.7	412	12.5
DC-10	190,000	Frequent	2.0	350	14.4	1.7	412	12.7
DC-8-63	165,000	Occasional	1.8	389	14.1	1.5	467	12.4

Step 2 Determine the modulus of rupture of the concrete by cutting beams from the pavement and making flexural tests. If this is impractical, strengths may be estimated by making suitable age-strength increase on the 28-day strengths taken from construction records. Figure 22-14 may be used to estimate the increase in flexural strength with age. If there is any evidence of deterioration of the concrete, specimens cut from the pavement should be tested.

Step 3 Find the thicknesses of the different installations (runways, taxiways, etc.) from construction plans and records or from concrete cores taken with a core drill.

In determining the strengths of the concrete and subbase-subgrade, use of correlative materials tests is increasing. These include compressive strength tests and tensile splitting tests on concrete specimens, dynamic, nondestructive tests on the surface of the pavement and field and laboratory soil strength tests on the subgrade and subbase materials. These data supplement and reduce the number of conventional tests required.

With the thickness and flexural strength of the concrete and the k-value for the subgrade known, the load-carrying capacity of the pavement may be determined from the design charts. Either an allowable load for a given safety factor, or the safety factor which will be in effect under a given load may be determined.

22.6.7 Future Pavement Design

Designers of airport pavements are faced with the need for evaluating existing design procedures and experience to apply to designs for future aircraft loadings and operational conditions.

Current design procedures are based on the prevention of flexural failure of the concrete slab. The safe limits or working stresses have been established by considerable performance experience on both civil and military airports for a wide range of slab thicknesses and aircraft loadings—extending to slab thickness of from 18 to 26 in. on military airfields designed for the more than 500,000-lb gross weight B-52 aircraft.

This experience indicates that heavy loads, per se, and large numbers of load repetitions are conditions that have been successfully handled with present design methods. This experience suggests that the flexural stress is a valid and critical design criteria—that when these stresses are kept within defined limits that other possible modes of distress such as slab deflections, subgrade pressures and shearing stresses at joints are also kept within safe, but undefined, limits.

However, this experience is for single, dual, and dual-tandem gear aircraft—where an effective load area represented by four wheels, or fewer, affects the pavement and subgrade. For future aircraft and airport pavements, there are three factors which greatly increase the effective size of loaded area:

1. Increase in number of wheels so that as many as 8 or 12 wheels may be acting together to affect stresses and deflections.
2. Increase in pavement thickness; stiffer pavement systems will spread out the zone of interaction to encompass additional wheel loads.
3. Increase in size of tire contact areas.

As the gross weights of aircraft have increased, increases in the number of wheels and size of tires have been made to satisfy the allowable flexural stress in the concrete slab. With further increases in the number of wheels interacting the question arises whether deflections and subgrade strains should be considered in addition to flexural stresses. It thus seems appropriate, for heavy, multiwheel aircraft, to also consider deflections and subgrade pressures and, through experience and research, establish safe working values.

As an aid to designers, deflection charts are published by PCA for heavy, multiwheel aircraft. The magnitudes of subgrade pressures and comparative load transfer requirements for different aircraft can be closely related to relative deflection values. For this reason, deflections may be selected as the supplemental design factor.

22.7 JOINT DESIGN FOR AIRPORT PAVEMENTS

A discussion of the function of joints and the general basis for joint design is given in sections 22.1 and 22.5 and cited references. Jointing practices for airport pavements are given in the design manuals of the Portland Cement Association,[22-53] the U.S. Army Corps of Engineers,[22-54] the Federal Aviation Administration[22-54] and the U.S. Navy.[22-61]

The purpose of joints in concrete pavements is to control transverse and longitudinal cracking due to restrained shrinkage, and the combined effects of restrained warping and aircraft loads. Joints are also used to divide the pavement into suitable increments for construction, and to accommodate slab movements at intersections with other pavements and at structures.

Adequate transfer of loads across the joint must be provided to satisfy basic thickness design principles. Depending on the type of joint, load transfer is obtained by dowels, keyways or by the aggregate interlock of slabs with short joint spacings. Substantial joint support is also provided by the use of a stiff, cement-treated subbase under the slab. Where no load transfer mechanism is provided, the joint edges are thickened to compensate so that stresses and deflections at these free edges are controlled within the design limits.

Joint widths must be made large enough so that the joints may be effectively sealed, as discussed in section 22.5.5.

22.7.1 Transverse Joints

In airport pavements, especially aprons, intersections, and turnarounds, it is sometimes difficult to distinguish between longitudinal and transverse joints. Transverse joints are considered to be those joints perpendicular to the lanes of construction; their direction within the pavement will depend upon the plan of construction.

-1. Contraction joints, unreinforced pavements—The spacing of transverse contraction joints, as well as longitudinal joints depends on the shrinkage properties of the concrete, subgrade soil conditions, climatic conditions, and slab thickness.

Table 22-5, for unreinforced (plain) pavements, may be used as a guide in the selection of joint spacings to reflect the shrinkage properties of concretes made with different aggregate types. These spacings are for average soil and climatic conditions and for slab thicknesses of 12 in. and less.

Pavement performance shows that the allowable joint spacing for crack control increases directly with slab thickness. Unless environmental and shrinkage conditions are especially severe, a joint spacing (in ft) not greatly exceeding twice the slab thickness (in in.) will control cracking. For example: a 15-ft spacing would probably be used for an 8-in. slab under average conditions, and 25 ft for a 12-in. slab.

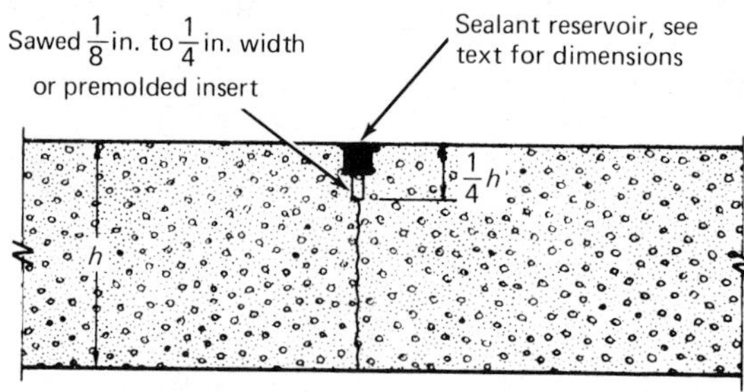

Sawed or Premolded Insert

For reinforced pavements, smooth dowel bars installed at depth $h/2$

See text for use of dowel bars at certain locations in unreinforced pavements

CONTRACTION JOINT

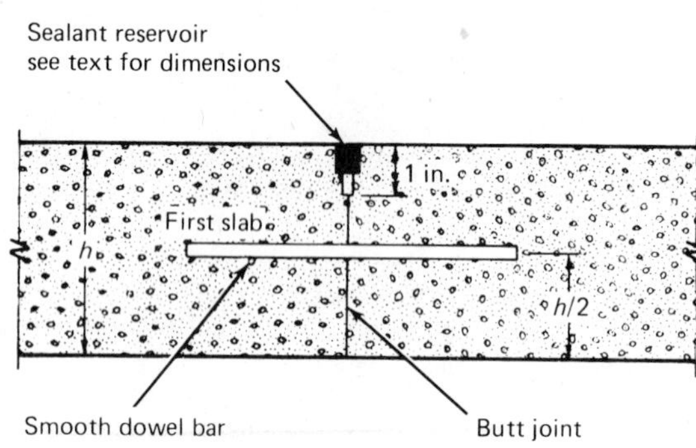

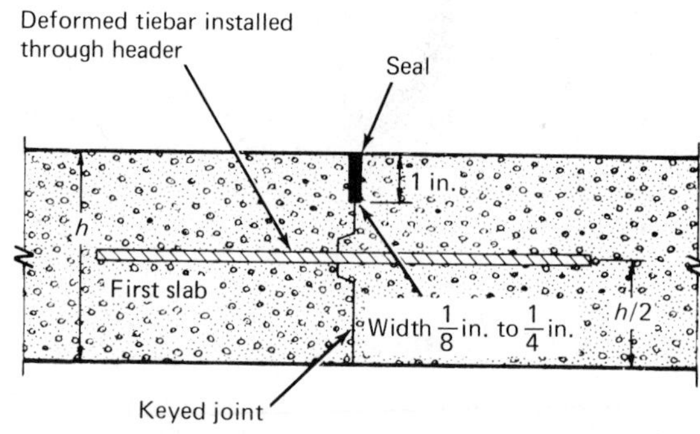

(This joint used only in middle third of normal joint interval)

Joints formed with header shaped to cross-section

CONSTRUCTION JOINTS

Fig. 22-15 Transverse joints, airport pavements.

Panels 25 by 25 ft have given very satisfactory service for thick pavements under extremely heavy traffic conditions. Generally, it is desirable to have panels with approximately equal transverse and longitudinal joint spacings. Experience[22-54] indicates that adequately designed slabs are not likely to develop an intermediate crack if the length-to-width ratio does not exceed 1.25.

Transverse contraction joint details for unreinforced pavements are shown at the top of Fig. 22-15. Load transfer at contraction joints will be provided by the aggregate interlock between the fractured faces below the groove. This will be true only if the joints are kept quite tight by omission of expansion joints within the pavement and by using a contraction-joint spacing not exceeding those recommended above.

Airport performance studies have shown that dowels are not required at transverse joints with short joint spacing. However, dowels must be placed across contraction joints near the free ends of the pavement and near expansion joints if any are used. Observations have shown that for a distance of about 100 ft back from each free end and 60 ft back from each expansion joint, the joints will gradually open to a point where the aggregate interlock may not be effective. Dowels are required in these few joints to ensure load transfer. The size and spacings of dowel bars are given in section 22.8.2.

-2. Contraction joints, reinforced pavements—Reinforced, jointed pavements (mesh-dowel pavements) have been used with joint spacings ranging from 30 to 70 ft. Design of these joints is the same as contraction joints for unreinforced pavements, except that dowels are required because joints open wider, making load transfer by aggregate interlock less effective. (Design for the amount of reinforcement and for dowels is given in section 22.8.)

Based on the performance of mesh-dowel pavements, it is desirable to limit joint spacing to about 30 or 40 ft for airport pavements less than 12 in. thick and to about 50 ft for thicker pavements. Importantly, the performance of all pavements depends strongly on the effectiveness of the joint seals. For longer joint spacings joint seals are usually not effective because of the greater fluctuations in joint width.

-3. Construction joints—Transverse construction joints are necessary at the end of each day's run or where paving operations are suspended for 30 minutes or more. If the construction joint occurs at or near the location of a transverse contraction joint, the butt-type joint with dowels is recommended. If the joint occurs in the middle third of the normal joint interval, a keyed joint with tie bars should be used. These construction joints should be constructed in accordance with one of the details shown at the bottom of Fig. 22-15.

22.7.2 Longitudinal Joints

The spacing of longitudinal joints depends on the construction equipment used, the overall width of the pavement and pavement thickness.

In the past, equipment has been best suited to paving widths of 20 to 25 ft and these joint spacing have been commonly used for pavement thicknesses of 12 in. or more. Intermediate longitudinal joints are necessary in thinner pavements to prevent the formation or irregular longitudinal cracks. Pavement performance experience indicates the following as a guide to longitudinal spacings:

a. for all pavements thinner than 12 in., and for pavements 12 to 15 in. thick carrying channelized traffic, longitudinal joints should not be more than 12.5 ft. apart
b. for pavements thicker than 15 in., and for pavements 12 to 15 in. thick not carrying channelized traffic, intermediate joints at 12.5 ft. are not required. In this case, convenient joint spacings are selected not to exceed contraction joint spacings for unreinforced pavements.

Recent developments in equipment permit paving widths up to 50 ft. This provides the opportunity to select joint spacings to satisfy specific design situations. For example, 37.5 or 50-ft wide construction lanes may be used with intermediate joints at 12.5, 18.75, or 25 ft, depending on pavement thickness. These intermediate joints are often called center joints and may be the surface-groove type such as a saw kerf or premolded insert.

The longitudinal joints within 37.5 ft of pavement edges have generally been hinged (held together by deformed steel tiebars). It is necessary to tie these outside joints to prevent progressive joint opening. Figure 22-16 illustrates the common types of intermediate longitudinal joints.

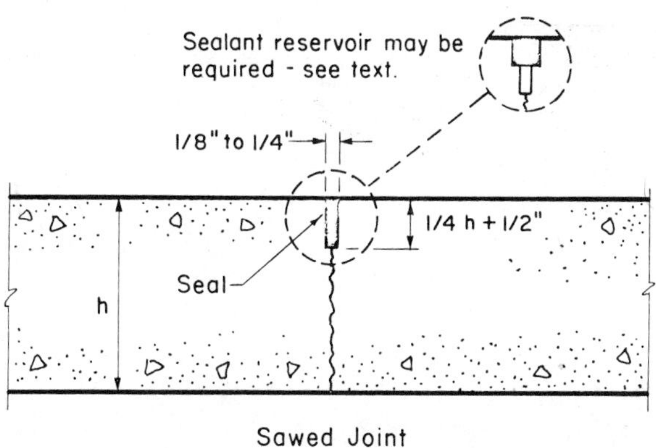

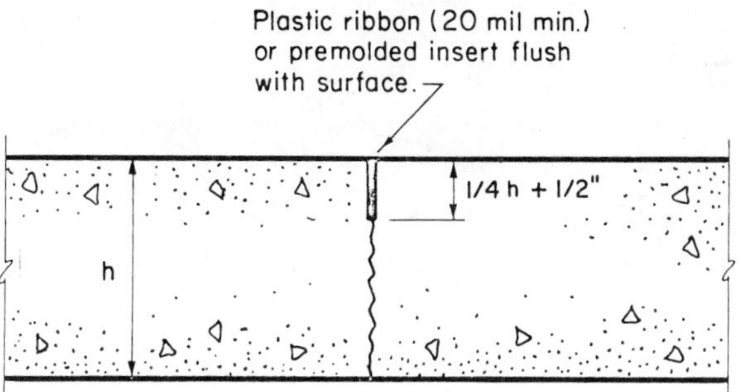

Note: Deformed tiebars at depth h/2 should be used across these joints where called for in text.

Fig. 22-16 Intermediate longitudinal joints, airport pavements.

Longitudinal construction joints at the edge of each construction lane may be of the keyed (tongue-and-groove) type to provide load transfer at this location. Details of these joints are shown in Fig. 22-17. The keyed joint may be extruded with a slipform paver or formed with a shaped metal strip attached to the forms to produce a groove along the edge of the slab. When the adjacent slabs are placed, the new concrete will form the key portion of the joint. Performance experience and research[22-62] has shown that it is important to use the keyway dimensions shown in Fig. 22-17, upper right. Larger keys reduce the strength of the joint and may result in keyway failures. The key must also be located at the mid-depth of the slab.

A keyway should be constructed along the outside edge of all pavement areas to provide for load transfer for future pavement expansions. A thickened outside edge may be used for the same purpose. One-half of a tie bolt may be installed to permit tieing the added lanes at a later date.

In narrow taxiways (75 ft or less) all longitudinal joints should be tied (provided with deformed tiebars or tiebolts) to prevent excessive opening and loss of load transfer. In wider pavement areas it is necessary to tie only longitudinal joints with 37.5 ft of the free edge. Recommended sizes and spacings of tiebars are given in section 22.8.3.

22.7.3 Expansion Joints

When contraction joints are spaced as outlined previously, expansion joints are not required transversely or longitudinally within airport pavements except under special conditions. The practice of omitting expansion joints tends to hold the interior of the pavement areas in restraint, prevents cracks and joints from opening and adds to the strength of the joint.[22-46]

Expansion joints (0.75 to 1.5 in.) must be provided between concrete pavements and all buildings or other airfield structures. Expansion joints may sometimes be required at intersections of runways, taxiways, and aprons (see section 22.7.4). Expansion joints used within the pavement may be doweled or provided with thickened edges. Figure 22-18 shows types of expansion joints to be used in the pavement and between pavements and structures.

Requirements of filler materials for expansion joints are given in current specifications of ASTM, AASHO and federal agencies.

22.7.4 Jointing Arrangements for Runways and Intersections

Figure 22-19 shows typical jointing arrangements for concrete airport pavement. This figure illustrates the location of various joint types and shows where tiebars and dowels are required.

The layout of joints at pavement intersections in airports presents many unusual problems. Because of the large, irregularly shaped areas of pavement involved and the possibility of any angle of intersection between two or more facilities, it is impossible to establish any universal joint pattern. Figure 22-20 illustrates some typical intersections and shows suggested jointing layouts for each.

The main body of pavement (runway or apron) will not contain expansion joints but will be free to move at each end. Intersecting pavements (taxiways, runways, etc.) must be provided with expansion joints to allow for longitudinal expansion without damage to either pavement. Since expansion joints are omitted in the main body of pavement

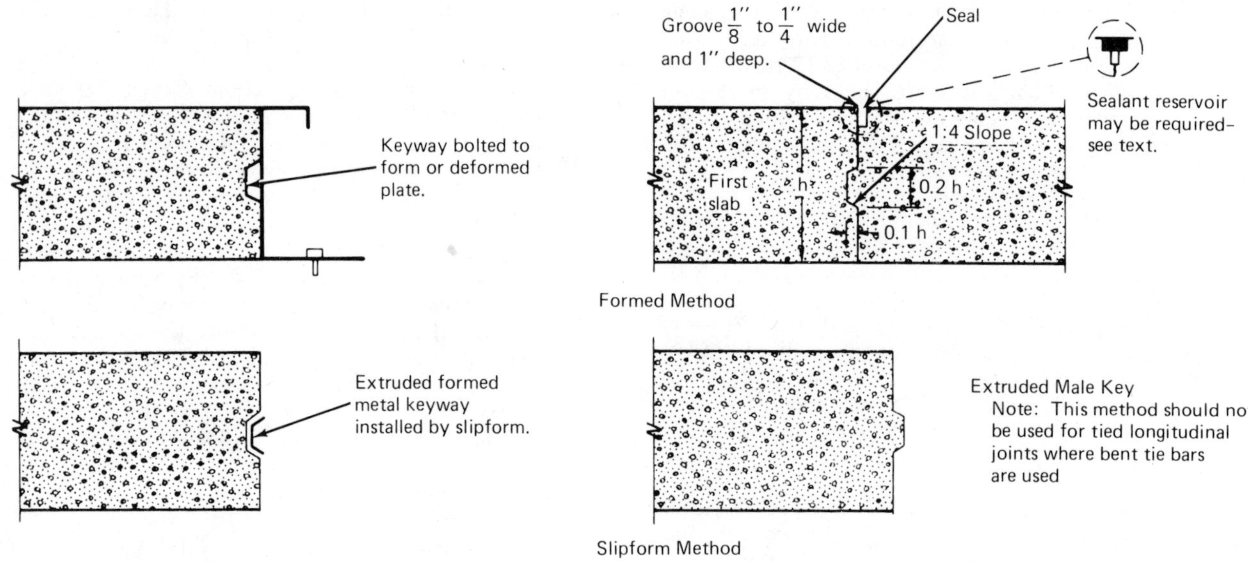

UNITED KEYED JOINTS
(All dimensions, seal and sealant reservoir as shown upper right)

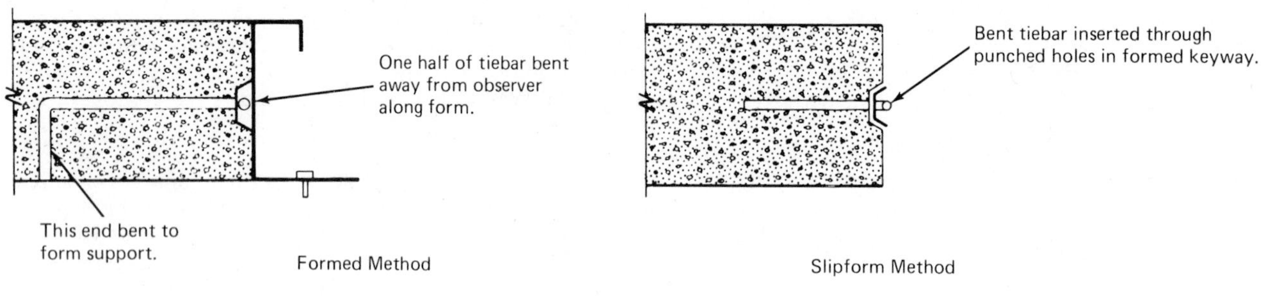

KEYED JOINTS WITH TIEBARS
(All dimensions, seal and sealant reservoir as shown upper right)

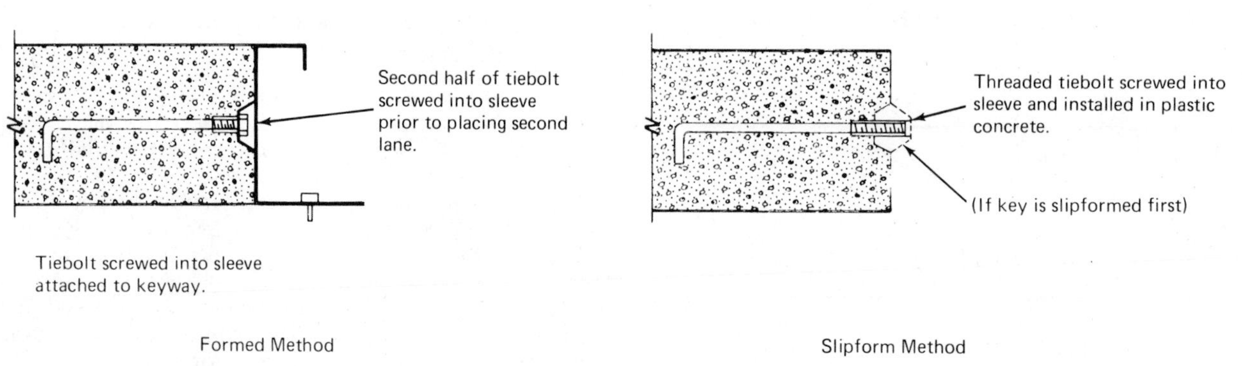

KEYED JOINTS WITH TIEBOLTS
(All dimensions, seal and sealant reservoir as shown upper right)

Notes: Tiebars or tiebolts are used only at certain locations - see text
Keyway and tiebars at depth h/2

Fig. 22-17 Longitudinal construction joints, airport pavements.

Seal with plastic material
1/4 in. radius
h/2
2 in.
min = w + 1/4 in.
h
Non-extruding filler
Expansion cap
6 to 10h
w
Extra edge thickness h/5 if dowels not used
Smooth dowel bar painted & lubricated to prevent bond with concrete

Seal with plastic material
1/4 in. radius if trafficked
Non-extruding filler
h
Building, hangar floor or drainage structure.
h/5
6 to 10h
Edge thickening used only at joints abutting hangar floors, at doorways, old pavement or flush structures (wherever wheel loads cross edge).

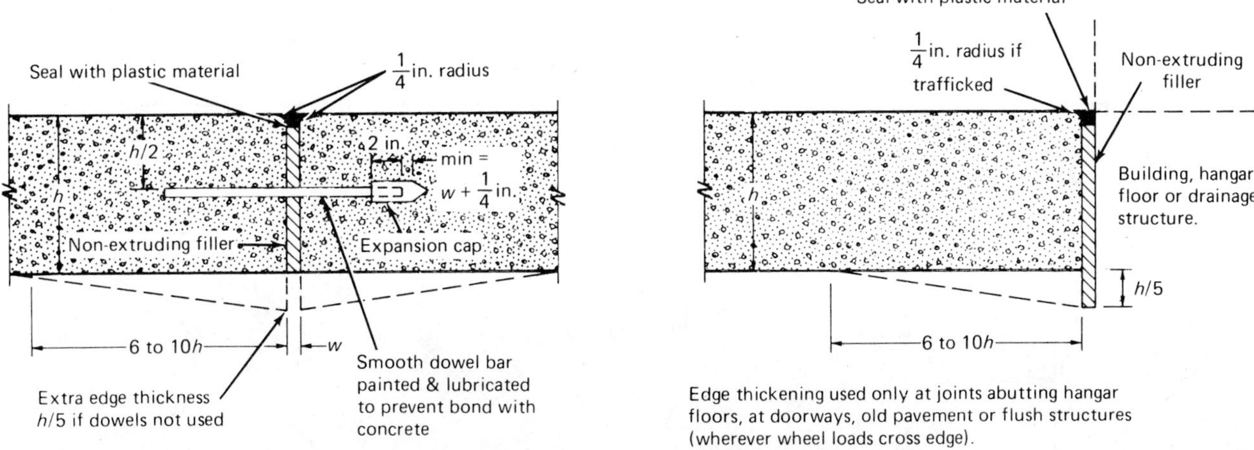

Fig. 22-18 Expansion joints, airport pavements.

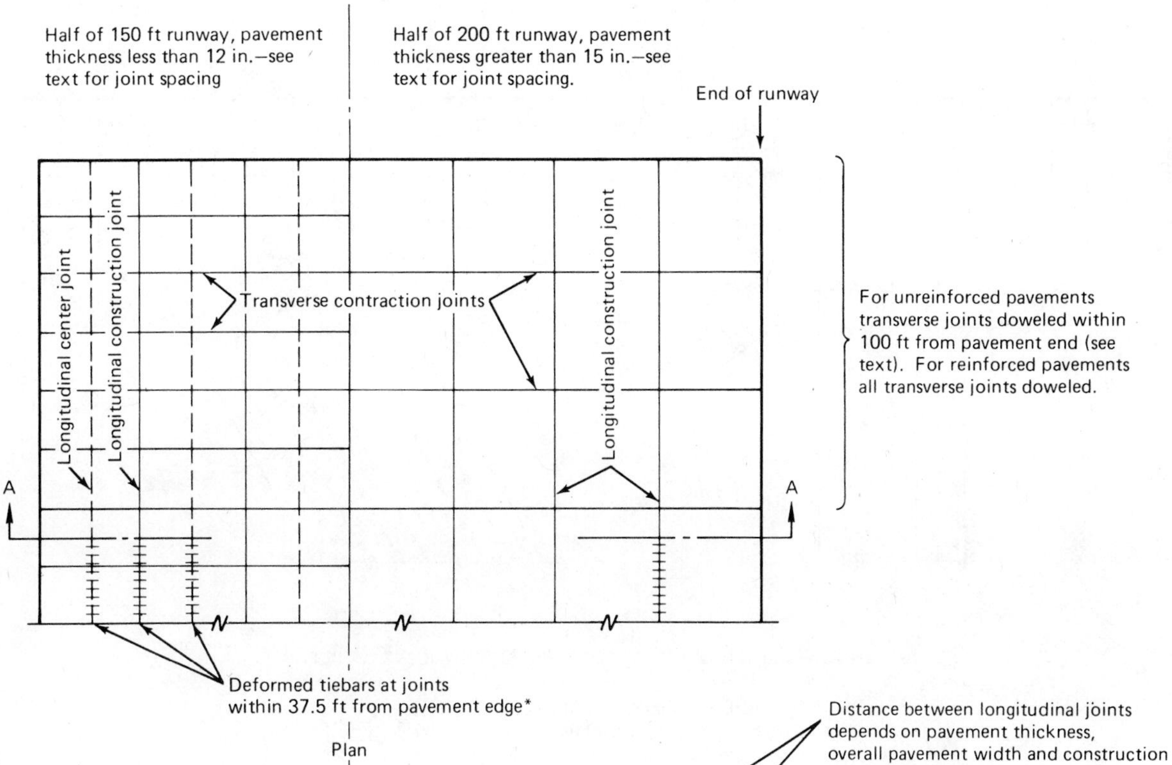

Half of 150 ft runway, pavement thickness less than 12 in.—see text for joint spacing
Half of 200 ft runway, pavement thickness greater than 15 in.—see text for joint spacing.
End of runway
Longitudinal center joint
Longitudinal construction joint
Transverse contraction joints
Longitudinal construction joint
For unreinforced pavements transverse joints doweled within 100 ft from pavement end (see text). For reinforced pavements all transverse joints doweled.
A
A
Deformed tiebars at joints within 37.5 ft from pavement edge*
Plan

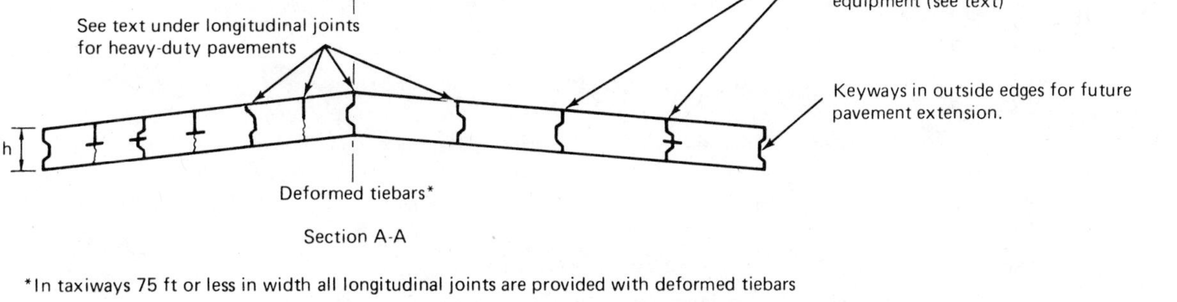

Distance between longitudinal joints depends on pavement thickness, overall pavement width and construction equipment (see text)
See text under longitudinal joints for heavy-duty pavements
Keyways in outside edges for future pavement extension.
h
Deformed tiebars*
Section A-A

*In taxiways 75 ft or less in width all longitudinal joints are provided with deformed tiebars

Fig. 22-19 Jointing plan for airport pavements.

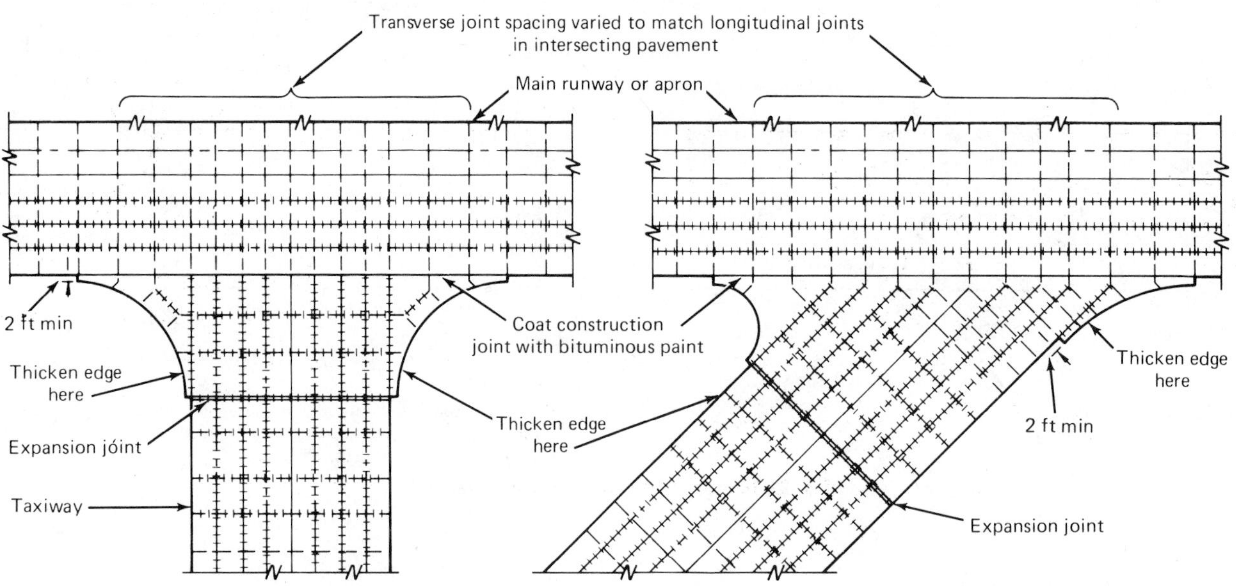

LAYOUTS EMPLOYING UNTIED KEYED CONSTRUCTION JOINT AT INTERSECTION

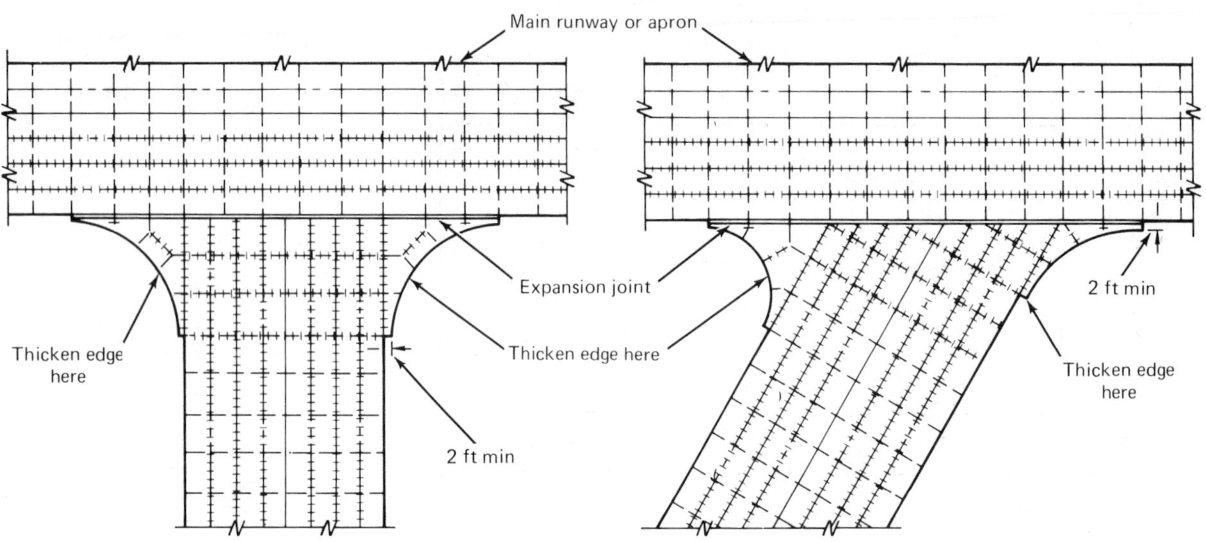

LAYOUTS EMPLOYING UNDOWELED THICKENED–EDGE EXPANSION JOINT AT INTERSECTION

Longitudinal joints tied within 37.5 ft of free pavement edges
Unreinforced pavements-transverse joints doweled on each side of expansion joint (reinforced pavements-all transverse joints doweled).

LEGEND

Symbol	Meaning	Symbol	Meaning
———	Keyed longit.construction joint	– – – –	Transverse contraction joint
++++++++++	Keyed longit construction joint with tiebars	+++ ++++	Transverse contraction joint with dowels
—— - ——	Lontitudinal center joint	++++++++++	Transverse expansion joint with dowels
+++++ ++++	Longitudinal center joint with tiebars	═══	Thickened-edge expansion joint at intersection

Note: For conditions requiring dowels, tiebars, expansion joints and thickened edges–see text.

Fig. 22-20 Typical intersections and suggested jointing layouts.

(runways or aprons), longitudinal movement may occur along both edges. Both the expansion joint requirements and the free longitudinal movement may be provided by the expansion joint arrangements shown in Fig. 22-20.

22.8 USE OF STEEL IN JOINTED PAVEMENTS

Steel in jointed concrete pavements may be used in the form of distributed steel, i.e., welded wire fabric or bar mats distributed throughout the concrete; and in the form of deformed tiebars and smooth dowels across certain joints.* The basis for design of distributed steel, tiebars and dowels is given in Ref. 22-38, 22-42, 22-63 thru 22-66. The use of steel in continously reinforced pavements is discussed in section 22.9.

Where the pavement is jointed to form short panel lengths which will control intermediate cracking, distributed steel is not necessary; however, where joints are placed to form longer panels in which some intermediate cracking may be expected, distributed steel is used. In this case, dowels are used at all transverse joints to ensure adequate load transfer since larger joint openings will result.

22.8.1 Distributed Steel

The function of distributed steel in jointed pavements is to hold together the fractured faces of slabs if cracks should form. The quantity of steel used may vary from 0.06 to 0.20% of cross-sectional area of the pavement depending on joint spacing, slab thickness and other factors.

Structural capacity across the cracks is achieved by the interlocking action of the rough faces of the slabs, and the infiltration of foreign material into the cracks is minimized. Distributed steel does not significantly increase flexural strength when it is used in quantities that are considered to be within the range of practical economy. When joint spacings are in excess of those that will control cracking, distributed steel is used to control the opening of intermediate cracks.

Since the steel is intended to keep cracks tightly closed, it must have sufficient strength to hold the two slabs together during contraction of the concrete. The maximum tension in the steel members across a crack is computed as equal to the force required to overcome friction between pavement and subgrade from the crack to the nearest free joint or edge. This force will be the greatest when the crack occurs at the middle of a slab. For practical reasons, the steel is usually made the same weight throughout the length of the slabs.

The factors that must be considered in the design of distributed steel include the weight of the concrete that must be moved, the coefficient of subgrade resistance and the tensile strength of the steel to be used. The amount of steel required per 1-ft width of slab is given in Fig. 22-21 as computed by the following formula:

$$A = \frac{L C_f w h}{24 f_s} \tag{22-1}$$

where

A = area of steel required per foot of width of slab, sq.in.
L = distance between free (untied) joints, ft
C_f = coefficient of subgrade (or subbase) resistance to slab movement

*Joint locations requiring tiebars or dowels are discussed in sections 22.5 and 22.7.

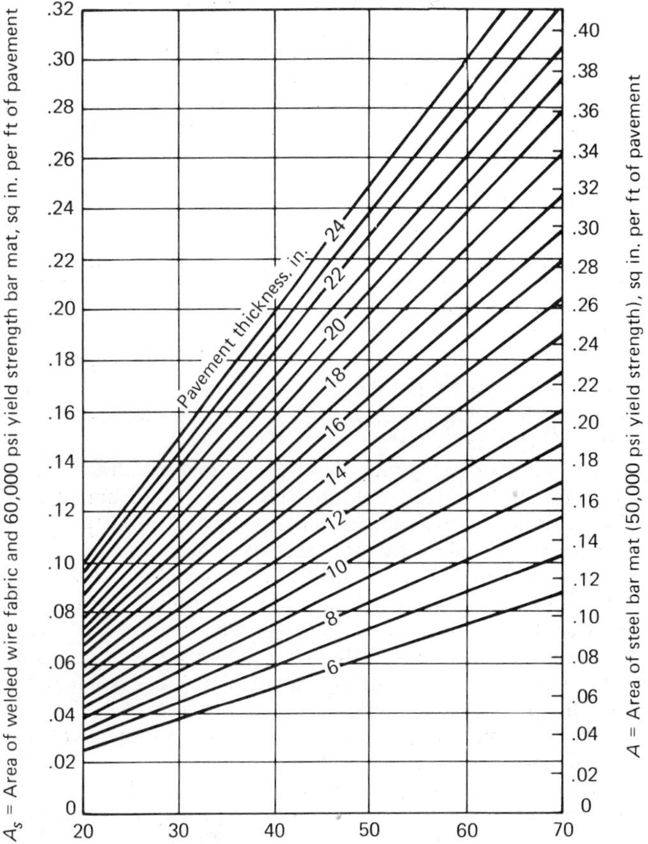

Fig. 22-21 Design chart for distributed steel.

w = weight of concrete, lb. per cubic foot (150 lb per cu ft for normal weight concrete)
h = slab thickness, in.
f_s = allowable working stress in the steel, psi

When this formula is used to calculate longitudinal steel, L will be the distance between free transverse joints. A value of C_f of 1.5 is most commonly used for design.

The allowable working stress in the steel, f_s, will depend on the type of steel used and should provide a small factor of safety. However, safety factors need not be as high as those used for building and bridge design. Figure 22-21 shows steel areas for working stresses of 45,000 (welded wire fabric and 60,000 psi yield strength bar mats) and 35,000 psi (50,000 psi yield strength bar mats).

Because longitudinal joint spacings are sufficiently close to control intermediate cracking (see section 22.5.3 and 22.7.2) transverse steel does not have to be as heavy as that required by eq. 22-1. Transverse steel needs to be only a sufficient amount to serve as spacers for the longitudinal steel.

Since distributed steel is not intended to act in flexure, its position within the slab is not critical, except that it should be adequately protected from corrosion, with a minimum cover of 2 in. of concrete. The steel may be placed at mid-depth of the slap (the neutral axis) or higher–up to one-third the slab thickness below the top surface.

22.8.2 Design of Dowels

Dowels are installed across joints in a concrete pavement to act as load-transfer devices which permit the joint to

TABLE 22-9 Recommended Size and Spacing of Dowel Bars

Slab Depth, in.	*Dowel Diameter, in.*	*Total Dowel Length*, in.*	*Dowel Spacing, in. c. to c.*
5–6	¾	16	12
7–8	1	18	12
9–11	1¼	18	12
12–16	1½	20	15
17–20	1¾	22	18
21–25	2	24	18

**Allowance made for joint openings and minor errors in positioning of dowels.*

open and close. Their function is to distribute part of the load to the adjacent slab, thus reducing the deflection and stress in the loaded slab.

The locations at which dowels are used are discussed in sections 22.5 and 22.7 on jointing. Correct alignment and lubrication of dowels are essential for proper joint function.

Several methods of theoretical analysis have been proposed for the design of dowels. Most of these design methods will result in dowel sizes and spacings which give satisfactory service. Condition surveys of existing pavements and extensive tests on full-scale slabs have shown no clear cases of dowel failures where the pavement slab itself is adequate for the loads carried.

For an economical system of load transfer the dowel size should be in correct proportion to the load for which the pavement is designed. Since the pavement thickness is also in proportion to the load, dowel design may be related to pavement thickness.

Table 22.9 lists suggested dowel sizes and spacings. It is based on studies of highway pavements and airport pavement experience.

Dowels are installed at the mid-depth of the slab.

22.8.3 Design of Tiebars

Tiebars or tiebolts are deformed steel bars. They are used across the joints of concrete pavement where it is necessary to hold the faces of the slab in firm contact. Tiebars by themselves are not designed to act as load-transfer devices. The load transfer across a joint in which they are used is provided by aggregate interlock or a keyway.

Tiebars are designed to overcome the resistance of the subgrade (or subbase) to horizontal slab movement when the pavement is contracting. The required cross-sectional area of tiebar per foot length of joint is given by the following formula:

$$A = \frac{bC_f wh}{12 f_s} \tag{22-2}$$

where

A = cross-sectional area of steel required per foot length of joint, sq in
b = distance between joint in question and the nearest untied joint or free edge, ft
C_f = coefficient of resistance between pavement and subgrade, or subbase (usually taken at 1.5)
w = weight of concrete, lb per cu ft (150 lb per cu ft for normal weight concrete)
h = slab thickness, in.

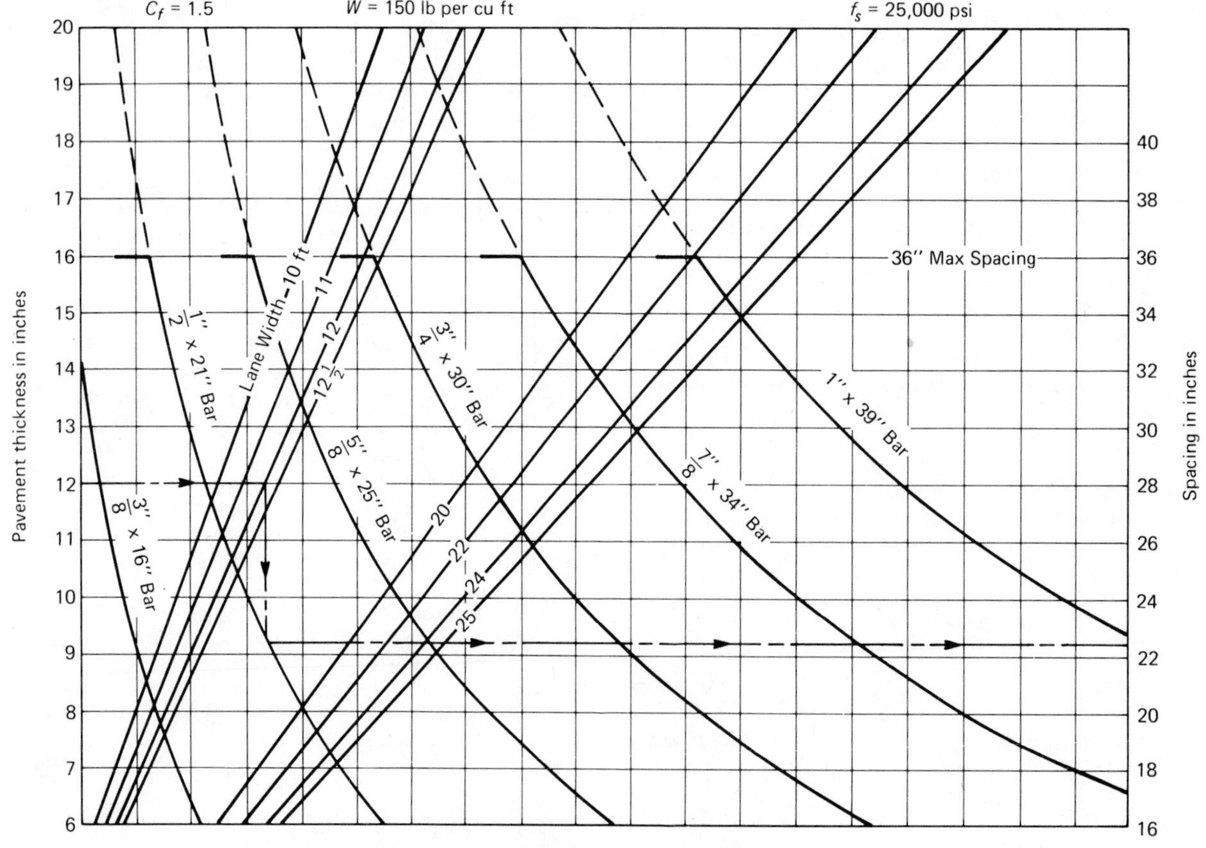

Fig. 22-22 Recommended tiebar dimensions and spacings.

f_s = allowable working stress in the steel, psi (usually taken at 25,000 psi)

Tiebars should be long enough so that anchorage on each side of the joint will develop the allowable working strength of the tiebar. In addition an allowance of about 3 in. should be made for inaccurate centering of the tiebar. Expressed as a formula, this becomes

$$L_t = \frac{1}{2} \frac{f_s \times d_b}{350^*} + 3 \qquad (22\text{-}3)$$

where

L_t = length of tiebar, in.
f_s = allowable working stress in steel, psi (same as that used in eq. 22-2)
d_b = diameter of tiebar, in.

Recommended tiebar dimensions and spacings are given in Fig. 22-22. Since a major use for tiebars is across longitudinal joints it is worthwhile to standardize the length and spacing of tiebars in order to simplify construction procedures and reduce overall pavement costs.

The tiebar dimensions given in Fig. 22-22 satisfy eqs. 22-2 and 22-3 when the following factors are used:

$C_f = 1.5$; w = 150 lb per cu ft; and f_s = 25,000 psi**

22.9 CONTINUOUSLY-REINFORCED PAVEMENTS

A continuously-reinforced concrete pavement contains a relatively large amount of continuous steel, usually 0.6% or more, in the longitudinal direction. It is built without transverse joints throughout its length except at the locations where it intersects or abuts existing pavements or structures. Thus, there are no transverse joints to seal and maintain. In this type of pavement, transverse cracks develop at close intervals (averaging between 3 and 7 ft) and are held tightly together by the continuous, longitudinal steel.

The fundamental principles of design for this type of pavement are: (1) to provide adequate pavement thickness for traffic loads; and (2) to provide enough longitudinal reinforcing steel so that the transverse cracks are kept tightly closed. Sources of information on the design of continuously-reinforced pavements are given in Refs. 22-67 through 22-71 and 22-37.

22.9.1 Pavement Thickness

Thickness design procedures for continuously-reinforced pavements are the same as those for jointed pavements described in sections 22.4 and 22.6.

22.9.2 Longitudinal Steel

-1. Amount—The amount of reinforcing steel required to control volume changes is dependent primarily on the thickness of the concrete, tensile strength of the concrete and yield strength of the reinforcement. Other factors which influence the amount of steel are contraction due to temperature drop, shrinkage due to drying, and modulus of elasticity of concrete and steel.

The controlling factor is the crack width. If insufficient steel is used, the crack widths will be too wide permitting intrusion of solids and water. Crack width criteria have not been firmly established but good performance has resulted when crack spacings average between 3 and 7 ft. Since crack spacing is directly related to crack width and is more readily observed, the design has indirectly become a matter of determining the amount of steel required to obtain desirable crack spacings. Formulas relating the amount of longitudinal steel and theoretical crack spacing have been proposed by several authors.[22-69 through 22-71]

In general, however, the amount of steel is based on data obtained from experimental pavements and pavements in service. It is the usual practice to specify steel having a minimum yield strength of 60,000 psi, and to require 0.6% of the gross cross-sectional area of the pavement. In severe climates, where freezing and thawing occur or where unusually heavy traffic prevails, it may be desirable to give consideration to the use of a somewhat higher percentage.

The following formula may be used to check the minimum amount of steel, 0.6% under normal conditions, or to compute a minimum amount based on special concretes or steels that might be selected.

$$\rho = \left(\frac{f_t'}{f_s - n f_t'} \right) 100 \qquad (22\text{-}4)$$

where

ρ = steel area ratio, percent (total cross-sectional area of steel divided by gross cross-sectional area of concrete times 100)
f_t' = tensile strength of concrete, in psi (assumed to equal 0.4 times the modulus of rupture)
f_s = allowable working stress of steel, in psi (0.75 times the yield strength)
$n = E_s/E_c$ (ratio of elastic moduli, steel-to-concrete)

Having established the required percentage of longitudinal steel, the steel area may be computed by the formula.

$$A_s = \frac{bh\rho}{100} \qquad (22\text{-}5)$$

where

A_s = total cross-sectional area of longitudinal steel, sq in.
b = width of slab, in.
h = slab thickness, in.
ρ = steel area ratio, percent

-2. Size and spacing—Size and spacing of longitudinal steel members are interrelated and dependent on a number of factors. The minimum size should be such that the spacing between the bars will be large enough to permit easy placement of concrete. The minimum size should be that which will provide a clear space between bars of at least twice the size of the aggregate being used, but in no case less than 4 in.

The maximum size is governed by the percentage of steel, the maximum spacing permitted, bond strength, and load transfer considerations. For good load transfer and bond strength, it is believed that the spacing should not exceed 9 in.

*The maximum working stress for the bond in deformed bars is generally taken as 0.10 of the compressive strength of the concrete, up to a maximum of 350 psi. It is permissible to use this maximum value in the design of tiebars because paving concrete should have a compressive strength in excess of 3,500 psi.

**A working stress of 25,000 psi is used for steels with yield strength of 40,000 psi, which has been normally specified if tiebars are to be bent and later straightened.

Size and spacing are related by

$$S_w = \frac{A_b}{h} \cdot \frac{100}{\rho} \qquad (22\text{-}6)$$

where

S_w = spacing, center to center, in.
A_b = cross-sectional area of one steel bar or wire, sq in.
h = slab thickness, in.
ρ = steel area ratio, percent

At the locations where longitudinal steel is spliced it is important that the length of splice be adequate to resist the tensile forces caused by shrinkage of the concrete at early ages. Recommendations for the length of lap splices are given in Refs. 22-67, 22-68, and 22-69.

-3. Position of longitudinal steel—Since the primary function of reinforcement in continuously reinforced pavements is to hold transverse cracks tightly closed, its design position vertically in the slab is not extremely critical. Practice has varied somewhat in this respect; pavements have been built with the center of the longitudinal steel ranging from 2.5 in. below the surface to mid-depth of the slab. Placement at mid-depth results in less steel stress at cracks due to wheel loads and temperature drops than for other placement positions. Another approach used is that the steel should be placed above mid-depth to reduce the surface width of cracks. The steel, of course, must have sufficient cover to preclude the development of cracks over the longitudinal reinforcement and prevent corrosion.

Although the mid-depth of slab has generally been chosen as the maximum depth of the longitudinal reinforcement, this has not been firmly established by experience; the suggested minimum design depth is one-third of slab depth and should provide a 2.5 in. cover over the longitudinal steel.

22.9.3 Transverse Steel

A relatively small amount of transverse steel is commonly, but not always,* used in continuously reinforced pavements to maintain the spacing of longitudinal bars to which it is tied or welded. In the case of pre-set reinforcement the transverse steel aids in supporting the longitudinal steel above the subbase. Transverse steel may also serve the purpose of tie bars across longitudinal joints.

The subgrade drag theory, eq. (22-1), is used to compute the required amount of transverse steel; it is based on providing a sufficient amount of steel to hold chance longitudinal cracks tightly closed.

22.9.4 Transverse Joints for Continuously-Reinforced Pavements

Some aspects of the design of transverse joints for continuously-reinforced pavements are discussed in this section, since they apply in particular to this pavement type.

The two types of transverse joints in continuously-reinforced pavements are construction joints which are placed at the end of a day's work or when paving operations are temporarily stopped, and expansion joints which are located at intersections with other pavements and at fixed objects. Jointing practices and design for longitudinal joints are the same as for jointed pavements and are discussed in sections 22.5 and 22.7.

Construction joints, because of their smooth faces, do not have as high load transfer capacity as natural cracks in which aggregate interlock supplements the shear strength of the longitudinal steel. Therefore, it is necessary to strengthen them. This is done by installing additional deformed bars of the same size as the longitudinal reinforcement. It is recommended that these extra bars be installed at a reasonably uniform spacing across the pavement, and in sufficient number to increase the area of steel across the joint at least one-third. These bars should be at least 3 ft long.

Expansion joints for continuously-reinforced pavements are installed only at bridges, structures, and where the pavement adjoins a section of jointed pavement. These joints must accommodate large seasonal movements. Two types of expansion joints that have been successfully used are the wide-flange beam and the 'bridge-finger' type. The joints should be designed to accommodate seasonal movements of 2 to 3 in. depending on climatic conditions. If the pavement ends are anchored by lugs or piles built into the subgrade, experience shows that about 1 in. of end movement needs to be accommodated by conventional expansion joints.

A less expensive terminal provision is to install several conventional expansion joints at pavement ends. However, this treatment should be used only where experience has indicated that sufficient movement is accommodated to prevent pavement growth that will damage adjoining structures or pavement.

Additional details on the design of terminal joints and end anchorages are given in Refs. 22-67, 22-68, and 22-69.

22.10 CONCRETE OVERLAYS

A pavement overlay is required when the condition of the existing pavement is no longer serviceable, or when a pavement is being strengthened to carry greater loads than it was designed for. In such cases it is both practical and economical to strengthen by the use of a concrete overlay.

This section discusses the design of the three basic types of overlays (partially bonded, unbonded, and bonded) on existing concrete pavements. Concrete overlays for flexible pavement are also discussed. Basic information on the design of the overlays is given in Refs. 22-75 through 22-81.

22.10.1 Flexural Strength

As explained in sections 22.4.4 and 22.6.2, the design for a new pavement is based on a 28-day flexural strength for highway pavement and a 90-day strength for airport pavement. For overlay design, strengths at these ages are also used for determining the single-slab thickness, h, in eqs. (22-7, 22-8, and 22-9).

This practice is conservative since the design of a concrete overlay should involve consideration of the relative flexural strength of the overlay and base slabs. Since the base slab may be several years old, the strength gain during this period should result in flexural strengths well above the 28- or 90-day design value (see Fig. 22-14). If desired, the strength values of the two layers may be considered in design by a method proposed by Mellinger.[22-81]

22.10.2 Condition of Existing Pavement

The condition of the existing pavement is an important factor in the selection of the type of concrete overlay. In

*Transverse steel may not be required where the longitudinal reinforcement is placed in fresh concrete by tube-feed or a method which will insure its proper spacing and depth.

subsequent overlay design charts and formulas, a coefficient C is used to express the structural condition* of the pavement as follows:

$C = 1.0$, when the extising pavement is in good overall structural condition
$C = 0.75$, when the existing pavement has initial joint and corner cracks, but no progressive structural distress or recent cracking
$C = 0.35$, when the existing pavement is generally badly cracked or structurally shattered

Careful consideration must be given to the assignment of the C-value in terms of which cracks or other structural defects will influence the performance of the restructured pavement. Cracking may or may not represent a failed condition. For example, cracking due to warping stresses is not progressive and not structurally detrimental since load transfer is provided by aggregate interlock.

Cracking due to nonuniform foundation support (subgrade pumping, subbase consolidation, subgrade settlement) may not be detrimental if the condition has reached equilibrium and, through cracking, the slabs have settled so that uniform support is again provided.

Progressive structural defects, cracks, or joints where load transfer has been lost, rocking slabs, or progressive foundation settlement are conditions which will seriously affect performance of the overlay and should be carefully evaluated.

*C-values apply to structural condition only, and should not be influenced by surface defects.

Thus, the selection of type and design of overlay, and preliminary repair work, should be based on a thorough knowledge of the pavement condition and the causes of structural defects.

22.10.3 Partially Bonded Overlays

Experience both with actual pavements in service and full-scale test pavements has shown that the use of a separation course** between the existing slab and the overlay slab leads to greater deflections and more breaking in the overlay slab than where such separation courses are not used. As a result, pavements with an overlay slab placed directly over the existing slab are stronger when no separation course is used. This direct overlay is termed a 'partially-bonded' overlay.

Based on studies of overlay pavements, the U.S. Army Corps of Engineers[22-76] uses this formula for design of overlays placed directly on the existing pavement:

$$h_r = \sqrt[1.4]{h^{1.4} - Ch_e^{1.4}} \tag{22-7}$$

where

h_r = thickness of overlay, in.
h = thickness of required single slab, in. (as determined in sections 22.4 and 22.6).
h_e = thickness of existing slab, in.

**'Separation course,' as used here, refers to any material between the two slabs which will break the bond (bituminous coating, plastic sheet, granular layer, or asphaltic concrete layer).

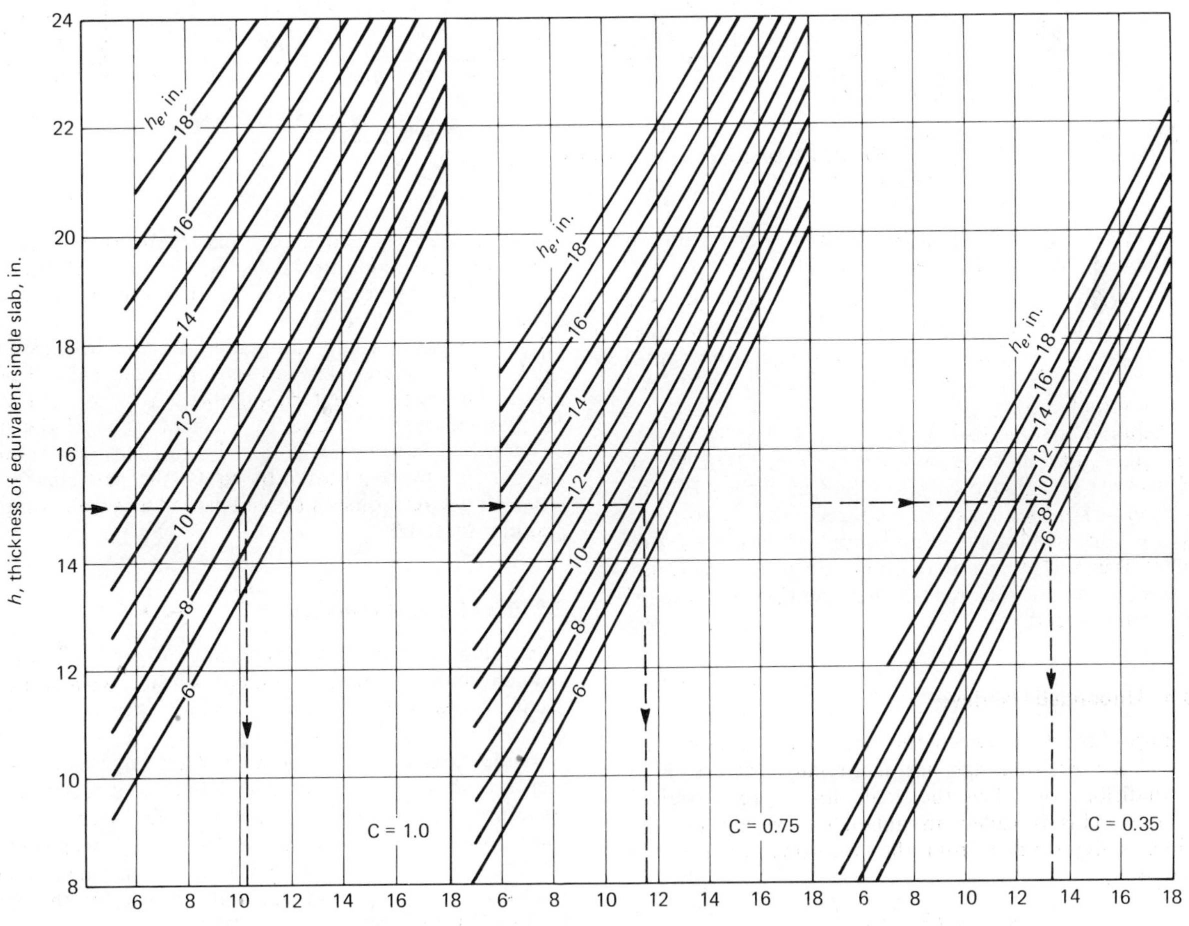

Fig. 22-23 Design chart for partially bonded overlays, $h_r = \sqrt[1.4]{h^{1.4} - Ch_e^{1.4}}$.

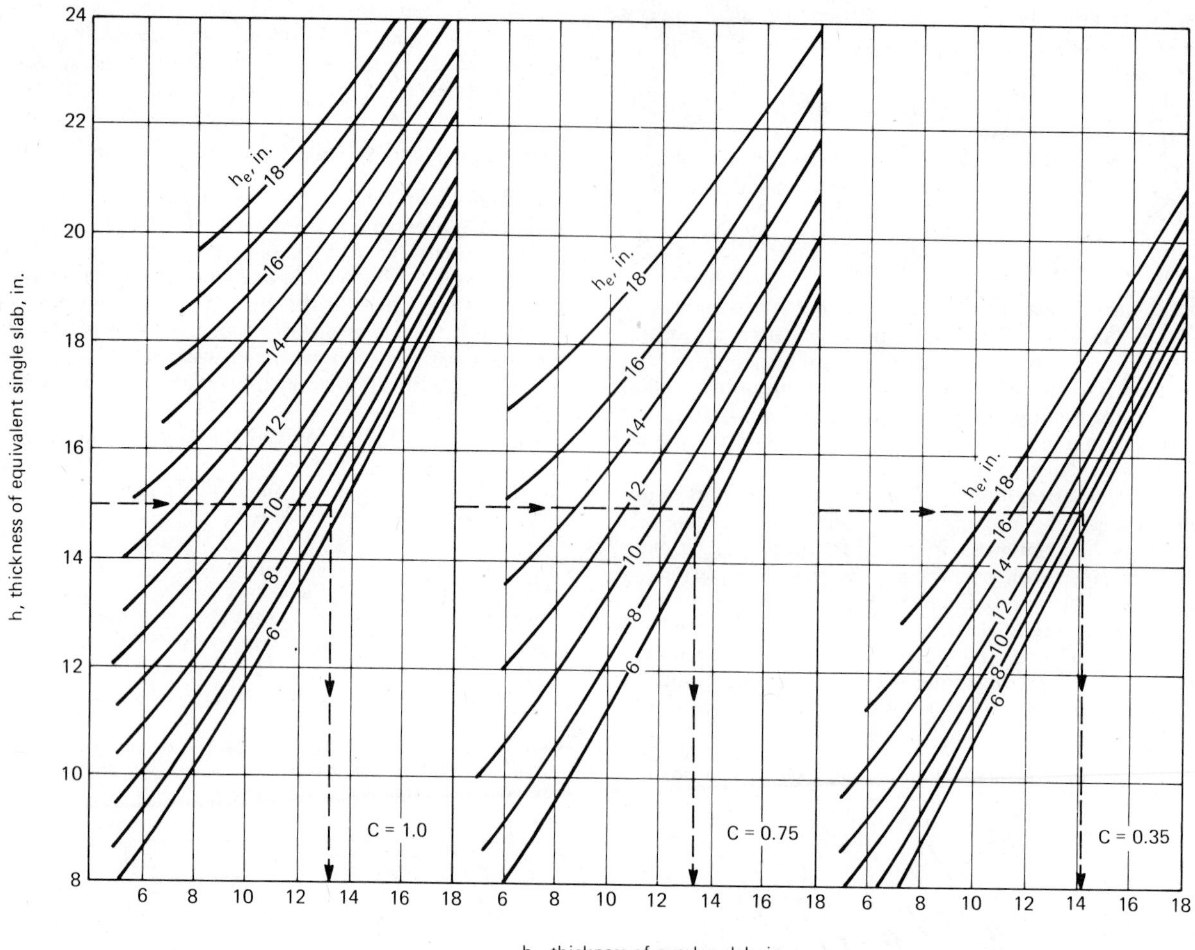

Fig. 22-24 Design chart for unbonded overlays, $h_r = \sqrt{h^2 - Ch_e^2}$.

C = a coefficient indicating the structural condition of the existing pavement (section 22.10.2)

The equation recognizes that friction between the two slabs or the development of some degree of bond provides somewhat greater capacity than when a separation course is used. As a result, thinner overlay sections are obtained than for unbonded overlays.

The required thicknesses for concrete overlay pavements may be taken directly from the curves of Fig. 22-23. Where thicknesses are required for other values of C they may be found by interpolation between the three sets of curves.

Partially bonded overlays are not used when the existing pavement is in extremely poor condition unless the structural defects can be repaired so that the C-value is significantly better than 0.35.

22.10.4 Unbonded Overlays

A separation course between the slabs may sometimes be needed if the existing slab has an irregular surface, is in poor condition, or when the grade line is to be raised appreciably. This is called an unbonded overlay and the thickness is determined from the following equation (or Fig. 22-24):

$$h_r = \sqrt{h^2 - Ch_e^2} \tag{22-8}$$

where the symbols have the same meaning as in eq. (22-7).

Equation (22-8) recognizes that there is not as much interaction between the slabs in the form of friction or bond and gives greater thicknesses than for partially bonded overlays.

In some cases where the grade line is to be raised and a thick layer of material is necessary between slabs, it may be more economical to determine a new k on top of the layer by plate-bearing tests and to design a new full-depth slab as described in sections 22.4 and 22.6. This may result in a thinner slab than indicated by eq. (22-8), particularly if the separating layer consists of well-graded and well-compacted granular material.

22.10.5 Bonded Overlays

A bonded overlay is a resurfacing placed on a carefully cleaned and prepared pavement surface with a bonding agent consisting of a sand-cement grout or epoxy mixture.[22-77, 22-78]

Bonded overlays have been used on large areas of pavements both to correct surface defects and to increase the structural capacity of pavements. When used on a structurally sound slab to increase its load-carrying capacity, the overlay and base slab should have a combined thickness equal to a single slab of adequate design for the planned loading. In this case

$$h_r = h - h_e \tag{22-9}$$

Of the different overlay types, the thinnest section will be obtained for bonded overlays since the resurfacing and the existing pavement act together as a monolithic slab. The economy of the thinner section, however, is somewhat offset by the extra cost of surface preparation and grouting required for the bonded overlay.

For bonded overlays, design thickness may be based on flexural strength of the existing pavement because of the monolithic slab action. Bonded overlays are recommended for use only where the existing slabs are in good structural condition, or structural defects have been repaired. Reports on the performance of bonded overlays are given in Refs. 22-79 and 22-80.

22.10.6 Slab Replacement or Repair

For a pavement with a few localized areas of structural defects, the *C*-value may sometimes be raised appreciably by a limited program of repair, patching or slab replacement. The increased *C*-value results in a thinner overlay for the entire pavement which may more than offset the costs of localized repairs.

22.10.7 Joint Location

Joints and random cracks in the base pavement will be reflected in partially-bonded and bonded overlays unless some preventative measures are taken. The practice used to prevent joint reflection is to match the joint locations in the overlay with those in the existing slab.

Many old concrete pavements were built with expansion joints at relatively short intervals. Expansion joints can usually be omitted in the overlay with contraction joints placed over the expansion joint location. This applies only to partially bonded overlays. When a bonded or monolithic overlay is used, it is imperative that joints in the overlay match those in the base pavement precisely in location, and that the same type and width of joint be used (expansion or contraction).

When a separated or unbonded overlay is used in which the separation course is of a substantial thickness, or when a continuously reinforced overlay is used, it is not necessary to match the transverse joints in the overlay concrete to those in the existing concrete either in the location or type. This is one of the advantages of the separated overlay and this type is required frequently when the joint pattern of the existing pavement is irregular and it is not desirable to repeat it in the overlay pavement.

In unbonded overlays with thin separation courses and partially-bonded overlays that are not continuously-reinforced, contraction joints can be placed directly over existing expansion joints, contraction joints, or construction joints. If this does not result in slab lengths short enough to control cracking, additional intermediate contraction joints should be placed to form equal slab lengths which are short enough to control cracking, as described in sections 22.5 and 22.7. Load transfer at transverse joints is provided by aggregate interlock, except near the ends of pavements where dowels should be used. A plain (unreinforced) overlay with short joint spacings may be placed on a reinforced slab with long joint spacings provided that the intermediate cracks in the existing pavement are tightly closed and in good condition. In this case, dowels are used in the overlay pavement only at locations matching the existing doweled joints.

For all overlay types, longitudinal joints in the resurfacing should also match the joints in the base. The longitudinal construction joints should be provided with a key* for load transfer to the adjacent slab. Tiebar requirements at longitudinal joints in concrete resurfacing are the same as explained in sections 22.5 and 22.7 for full-depth pavement.

22.10.8 Tiebars, Dowels, and Distributed Steel

The need and use of tiebars, dowels and reinforcing steel are the same as discussed in sections 22.5 and 22.7. Where used, tiebars, dowels and reinforcing steel are designed as outlined in section 22.8. If they exist in the base slab at the proper location and are functioning adequately, tiebars, dowels and distributed steel in the overlay are designed based on the overlay slab thickness. Otherwise, tiebar and distributed steel design, in bonded and partially-bonded overlays, is based on the total thickness of old and new slab; dowel design is based on the thickness of an equivalent slab, h_e.

Reinforcement serves the same purpose in concrete overlays as it does in original construction. It is not required when short joint spacings are used but is needed for pavements with longer joint spacing to restrain cracks that occur so they do not open sufficiently to present a maintenance problem.

If the old pavement is extensively cracked, the use of distributed steel or continuous reinforcement may be the most dependable method of minimizing uncontrolled cracking in unbonded or partially-bonded overlays.

22.10.9 Continuously-Reinforced Overlays

Because they are less susceptible to reflective cracking, continuously reinforced overlays appear to offer an advantage over other overlay types. For these overlays, a separation course is normally used over the existing pavement and the overlay thickness is determined as described in section 22.10.4. The amount of reinforcing steel is based on the overlay pavement thickness requirements; other design details are described in section 22.9.

A few partially-bonded (no separation course), continuously-reinforced overlays have been constructed. Partially-bonded overlays should be used only if the existing pavement is in fairly good condition, and joint spacings are short so that joint movements will not greatly affect the overlay. In this case, a thinner overlay could be used on the basis of the partially bonded overlay. However, additional steel may be required to prevent excessive crack opening in the overlay.

22.10.10 Separation Courses

Some success has been experienced in preventing reflective cracking by the use of separation courses between the base slab and overlay. There are not sufficient data available, however, to indicate the minimum thickness of separation course that will be completely effective. There are indications that any type of bond breaker will reduce the amount of reflective cracking. As discussed previously, the use of a bond-breaker or separation course creates the requirement for a thicker overlay.

22.10.11 Concrete Overlays for Flexible Pavement

Concrete overlays on flexible pavements have been used for several years. These overlays have performed well and demonstrated the feasibility of this type of construction.[22-82]

*For thin, bonded overlays, keyways or other load-transfer devices, are usually omitted.

Concrete overlays for flexible pavement should be designed in the same manner as a concrete pavement on grade. The modulus of subgrade reaction, k, should be determined by plate bearing tests made on the surface of the flexible pavement. Several agencies apply the limitation that no k-value greater than 500 lb per cu in. will be used in designing rigid overlays for flexible pavements. This appears, however, to be a rather arbitrary limitation and more development work should be done on this point to realize fully the advantages of this composite design.

REFERENCES

22-1 Childs, L. D., and Kapernick, J. W., "Tests of Concrete Pavements on Gravel Subbases," Paper 1800, *Proceedings of the American Society of Civil Engineers*, **84** (HW3), Oct. 1958.

22-2 Childs, L. D., and Nussbaum, P. J., "Pressures at Foundation Soil Interfaces under Loaded Concrete and Soil-Cement Highway Slabs," *Proc., ASTM*, **62**, 1243-1263, 1962.

22-3 Childs, L. D., and Kapernick, J. W., "Tests of Concrete Pavements on Crushed Stone Subbases," *Proc., ASCE*, Paper 3497, **89** (HW1), 57-80, April 1963.

22-4 Childs, L. D.; Colley, B. E.; and Kapernick, J. W., "Tests to Evaluate Concrete Pavement Subbases," *Research and Development Laboratory Bulletin D11*, Portland Cement Association. Reprinted from Paper 1297, *Proc., ASCE*, **83** (HW3), July 1957.

22-5 "A Review Paper on Expansive Clay Soils," Woodward-Clyde & Associates, Los Angeles, Calif., 1967.

22-6 "The AASHO Road Test, Report 5, Pavement Research," *Highway Research Board Special Report 61E*, Wash., D.C., 1962.

22-7 Springenschmid, R., "The Influence of Water Content on the Frost Behavior of Consolidated and Stabilized Soils," (in German) *Rock Mechanics and Engineering Geology*, **3** (3/4), 113-121, 1965. Abstract in *Highway Research Abstracts*, **36** (9), 5, Sept. 1965.

22-8 Beskow, G., "Soil Freezing and Frost Heaving with Special Attention to Roads and Railroads," The Swedish Geological Society, Series C, NO. 375, 26th Yearbook, No. 3, 1935. Translated by J. O. Osterberg, Northwestern University, Published by Technological Institute, Northwestern University, 1947.

22-9 "Investigation of Filter Requirements for Underdrains," U.S. Army Corps of Engineers, *Technical Memorandum 183-1*, U.S. Waterways Experiment Station, Vicksburg, Mississippi, 1941.

22-10 "Frost Action in Roads and Airfields," *A Review of Literature 1765-1951, Highway Research Board Special Report 1*, 1952.

22-11 Miyakawa, I., and Koyama, M., "On the Residual Frost Subgrade Underneath the Select Fill in the Alleviation Practice for Frost Damage," Civil Engineering Institute, Hokkaido Development Bureau, Hiragishi, Sapporo, Japan. Reprinted from *Soil and Foundation*, **3** (1), Sept. 1962.

22-12 "Subgrades and Subbases for Concrete Pavements," Portland Cement Association, Skokie, 1971.

22-13 Kawala, E. L., "Cement-Treated Subbase Practice in U.S. and Canada," ASCE Proceedings Paper 4947, *Journal of the Highway Division*, **92** (HW2), 75-98, Oct. 1966.

22-14 Colley, B. E., and Nowlen, W. J., "Performance of Concrete Pavements Under Repetitive Loading," *Research and Development Laboratory Bulletin D13*, Portland Cement Association, Skokie, 1958.

22-15 Colley, B. E., and Humphrey, H. A., "Aggregate Interlock at Joints in Concrete Pavements," *Highway Research Record, No. 189*, 1-18, 1967.

22-16 "Thickness Design for Concrete Pavements," Portland Cement Association, Skokie, 1966.

22-17 Westergaard, H. M., "Computation of Stresses in Concrete Roads," Highway Research Board Proceedings, Fifth Annual Meeting, Part 1, 90-112, Wash., D.C., 1925.

22-18 Westergaard, H. M., "Stresses in Concrete Pavements Computed by Theoretical Analysis," *Public Roads*, 7 (2), 25-35, April 1926.

22-19 Westergaard, H. M., "Theory of Concrete Pavement Design," Highway Research Board Proceedings, Seventh Annual Meeting, Part 1, 175-181, Wash., D.C., 1927.

22-20 Westergaard, H. M., "Analytical Tools for Judging Results of Structural Tests of Concrete Pavements," *Public Roads*, **14** (10), 185-188, Dec. 1933.

22-21 Pickett, G.; Raville, M. E.; Janes, W. C.; and McCormick, F. J., "Deflections, Moments, and Reactive Pressures for Concrete Pavements," *Kansas State College Bulletin No. 65*, Kansas State College, Manhattan, Kansas, Oct. 1951.

22-22 Pickett, G., and Ray, G. K., "Influence Charts for Concrete Pavements," Paper No. 2425, *Transactions, ASCE*, **116**, 49-73, 1951.

22-23 Teller, L. W., and Sutherland, E. C., "The Structural Design of Concrete Pavements," *Public Roads*, **16** (8-10), 1935; **17** (7, 8), 1936; and **23** (8), 1943.

22-24 Childs, L. D., "Tests of Concrete Pavement Slabs on Cement-Treated Subbases," *Highway Research Record, No. 60*, 39-58, 1964.

22-25 Older, C., "Highway Research in Illinois," *Proc., ASCE*, 175-217, Feb. 1924.

22-26 Aldrich, L., and Leonard, I. B., "Report of Highway Research at Pittsburg, California, 1921-1922," California State Printing Office, Sacramento, Calif., 1923.

22-27 "Road Test One–MD," *Highway Research Board Special Report 4*, Highway Research Board, 1952.

22-28 Fordyce, P., and Teske, W. E., "Some Relationships of the AAHSO Road Test to Concrete Pavement Design," *Highway Research Record No. 44*, 35-70, 1963.

22-29 Fordyce, P., and Yrjanson, W. A., "Modern Design of Concrete Pavements," Proceedings Paper 6726, *Transportation Engineering Journal, ASCE*, **95** (TE3), 407-438, Aug. 1969.

22-30 Taragin, A., "Lateral Placement of Trucks on Two-Lane and Four-Lane Divided Highways," *Public Roads*, **30** (3), 71-75, Aug. 1958.

22-31 Fordyce, P., and Packard, R. G., "Concrete Pavement Design," paper presented at the Forty-ninth Annual Meeting, American Association of State Highway Officials Committee on Design, Portland, Ore., Oct. 1963. (May be obtained from PCA, Skokie, Illinois).

22-32 "Load Stresses at Pavement Edge," Portland Cement Association, Skokie, 1969.

22-33 Kesler, C. F., *Fatigue and Fracture of Concrete*,

Stanton Walker Lecture Series of the Materials Sciences, National Sand and Gravel Association and National Ready-Mixed Concrete Association, Wash., D.C., 1970.

22-34 Ballinger, C. A., "The Cumulative Fatigue Damage Characteristics of Plain Concrete," Highway Research Record 370, 48–60 Highway Research Board, Washington, D.C., 1971.

22-35 Miner, M. A., "Cumulative Damage in Fatigue," *Transcation, ASME*, **67**, A159–A164, 1945.

22-36 Westergaard, H. M., "Analysis of Stresses in Concrete Roads Caused by Variations in Temperature," *Public Roads*, 8 (3), 201–215, May 1927.

22-37 Friberg, B. F., "Frictional Resistance Under Concrete Pavements and Restraint Stresses in Long, Reinforced Slabs," *Proc., Highway Research Board*, Wash., D.C., 1954.

22-38 Kelley, E. F., "Applications of Results of Research to the Structural Design of Concrete Pavements," *Public Roads*, **20** (5), July, 1939.

22-39 Harr, M. E., and Leonards, G. A., "Warping Stresses and Deflections in Concrete Pavements," *Proceedings of the Thirty-eighth Annual Meeting of the Highway Research Board*, Wash., D.C., 1959.

22-40 Wiseman, J. F.; Harr, M. E.; and Leonards, G. A., "Warping Stresses and Deflections in Concrete Pavements, Part 2," *Proceedings of the Thirty-ninth Annual Meeting of the Highway Research Board*, Wash., D.C., 1960.

22-41 ACI Committee 325, Subcommittee II, "Considerations in the Selections of Slab Dimensions," *ACI Journal*, **28** (5), Nov. 1956.

22-42 Bradbury, R. H., *Reinforced Concrete Pavements*, The Wire Reinforcement Institute, Washington, D.C., 1938.

22-43 Moore, J. H., with discussions by E. C. Sutherland and W. Harwood, "Thickness of Concrete Pavements," Paper 2834, *Transactions, ASCE*, **121**, 1956.

22-44 "Joint Design for Concrete Highway and Street Pavements," Portland Cement Association, Skokie, 1975.

22-45 Halm, H. J., "An Analysis of Factors Influencing Concrete Pavement Cost," *Highway Research Bulletin 340*, Highway Research Board, 1962.

22-46 Anderson, A. A., "Expansion Joint Practice in Highway Construction," Paper 2384, *Transactions, ASCE*, **114**, 1949.

22-47 Finney, E. A., and Oehler, L. T., "Final Report on Design Project, Michigan Test Road," Proceedings of the Thirty-eighth Annual Meeting of the Highway Research Board, 241–285, 1959.

22-48 "Joint Spacing in Concrete Pavements," Highway Research Board, *Research Report 17-B*, 1956.

22-49 Schutz, R. J., "Shape Factors in Joint Design," *Civil Engineering*, **32** (10) Oct., 1962.

22-50 Tons, E., "Factors in Joint Seal Design," *Highway Research Record No. 80*, Highway Research Board, 1965.

22-51 Tons, E., "Materials and Geometry in Joint Seals," *Adhesive Ages*, 8 (9), 22–28, Sept. 1965.

22-52 Cook, J. P., and Lewis, R. M., "Evaluation of Pavement Joint and Crack Sealing Materials and Practices," *National Cooperative Highway Research Program Report No. 38*, 1967.

22-53 Packard, R. G., *Design of Concrete Airport Pavement*, Portland Cement Association, Skokie, 1973.

22-54 "Rigid Pavements for Airfields other than Army," *TM 5-824-3, AFM 88-6* Chap. 3, Departments of the Army and Air Force, Jan. 1970.

22-55 "Airport Paving," AC 150/5320-6A, CHMU 4 Department of Transportation, Federal Aviation Administration, Incorporate changes up through Change 3 issued April 1, 1970.

22-56 Ray, G. K.; Cawley, M. L.; and Packard, R. G., "Concrete Airport Pavement Design—Where Are We?" presented at ASCE/AOCI Airports Specialty Conference, Atlanta, Ga., April 16, 1971. Published by ASCE in *Airports, Key to the Air Transportation System*, ASCE, New York, 1972.

22-57 Packard, R. G., "Computer Program for Concrete Airport Pavement Design," Portland Cement Association, Skokie, 1967.

22-58 Palmer, L. A., "Field Loading Tests for the Evaluaation of the Wheel-Load Capacities of Airport Pavements," *Symposium on Load Tests of Bearing Capacity of Soils*, ASTM Special Technical Publication No. 79, 5-40, June 1947.

22-59 "The AASHO Road Test, Report 6, Special Studies," *Highway Research Board Special Report 61F, Highway Research Board*, 1962.

22-60 "Results of Modulus of Subgrade Reaction Determination at The Road Test Site by Means of Pavement Volumetric Displacement Test," U.S. Army Engineer Division, Ohio River, Corps of Engineers, Ohio River Division Laboratories, Cincinnati, Ohio, April, 1962.

22-61 "Design Manual-Airfield Pavements," NAVFAC DM-21, Department of the Navy, Naval Facilities Engineering Command, Alexandria, Va., June 1973.

22-62 "Model Tests to Determine Optimum Key Dimensions for Keyed Construction Joints," U.S. Army Corps of Engineers, Ohio River Division Laboratory, Cincinnati, Ohio, 1954.

22-63 ACI Committee 325, Subcommittee III, "Structural Design Considerations for Pavement Joints," *ACI Journal*, **53** (1), July 1956.

22-64 Friberg, B. F., "Design of Dowels in Transverse Joints of Concrete Pavements," *Transactions, ASCE*, **66** (5), May 1940.

22-65 Bradbury, R. D., "Design of Joints in Concrete Pavements," *Proc., Highway Research Board*, 1932.

22-66 Timoshenko, S., and Lessels, J. M., *Applied Elasticity*, Westinghouse Technical Night School Press, Pittsburgh, Pa., 1925.

22-67 "Design and Construction, Continuously-Reinforced Concrete Pavement," Continuously-Reinforced Pavement Group, sponsored by Concrete-Reinforcing Steel Institute and Committee of Steel Bar Producers of the American Iron and Steel Institute, 1968.

22-68 ACI Committee 325, Subcommittee VII, "A Design Procedure for Continuously-Reinforced Concrete Pavements for Highways," Title 69-32, *ACI Journal*, **69** (6), 309–319 June 1972.

22-69 McCullough, B. F., *Design Manual for Continuously-Reinforced Concrete Pavement*, United States Steel Corporation, Pittsburgh, Pa., 1970.

22-70 Vetter, C. P., "Stresses in Reinforced Concrete Due to Volume Changes," *Transactions, ASCE*, **98**, 1933.

22-71 Zuk, W., "Analysis of Special Problems in Continuously-Reinforced Concrete Pavements," *Highway Research Board Bulletin 214*, Highway Research Board, 1959.

22-72 AASHO Committee on Design, "AASHO Interim Guide for the Design of Rigid Pavement Structures," American Association of State Highway Officials, Wash., D.C., April 1962.

22-73 McCullough, B. F., and Ledbetter, W. B., "LTS Design of Continuously-Reinforced Concrete," Paper 3357, *Transactions, ASCE*, 127, Part IV, 1962.

22-74 "Test Investigations of Lap Splices of Reinforcing Steel in Continuously-Reinforced Concrete Pavement," CRSI Committee on Continuously-Reinforced Pavement, *Bulletin No. 3*, Chicago, Ill., May 1963.

22-75 ACI Committee 325, Subcommittee VIII, "Design of Concrete Overlays for Pavements," Title No. 64-40, *ACI Journal*, **64**, 470-474, Aug. 1967.

22-76 "Rigid Airfield Pavements," *Engineering and Design Manual*, EM1110-45-303, U.S. Department of the Army, Office of the Chief of Engineers, Feb. 1958.

22-77 "Bonded Concrete Resurfacing," Portland Cement Association, Skokie, 1960.

22-78 ACI Committee 503, "Guide for the Use of Epoxy Compounds with Concrete," *ACI Journal, Proceedings*, **59** (9), 1121–1142, Sept. 1962.

22-79 Gillette, R. W., "A Ten-Year Report on the Performance of Bonded Concrete Resurfacings," *Highway Research Record No. 94*, Highway Research Board, 61–76, 1965.

22-80 Gillette, R. W., "Performance of Bonded Concrete Overlays," *ACI Journal, Proceedings*, **60** (1), 39–50, Jan. 1963.

22-81 Mellinger, F. M., "Structural Design of Concrete Overlays," *ACI Journal, Proceedings*, **60** (2), 225–236, Feb. 1963.

22-82 Westall, W. G., "Concrete Overlays on Asphalt Pavements," *Highway Research News* No. 22, 52–57, Highway Research Board, Feb. 1966.

23

Preparation of Structural Drawings as Related to Detailing of Reinforced Concrete*

PAUL F. RICE** and W. C. BLACK***

23-1 GENERAL CONSIDERATIONS

All contract documents include a complete set of plans and specifications. Contract plans or drawings consist of architectural, structural, heating/ventilating, and electrical drawings. These drawings, along with the specifications, plus bid proposal, form the documents on which the General Contractor makes his bid.

For all major building requiring reinforced concrete, separate structural drawings are required to show the reinforced concrete design. Generally accepted practice in building construction requires the structural drawings to show number or spacing, size, and position of all reinforcement, with sufficient dimensions so that length and bend points can be computed for each bar. In bridge design, usual practice is for the designer also to list for each type of bar required, the mark numbers, bending dimensions and lengths, and number required. In all reinforced concrete design, structural drawings should also indicate location of all splices, and construction, expansion and contraction joints. Many of these requirements can be transmitted for typical conditions by means of general notes on the drawing combined with specification provisions. Grade of bars is usually designated in specifications—perhaps only in specifications where only one grade is to be used, although best practice is to repeat this requirement in general notes on the structural drawings. Where more than one grade of bars is employed, structural drawings must also indicate location of each.

This chapter will concern itself with the structural drawings and specifications on the drawings. The civil and/or structural engineer is generally responsible for the structural drawings and the part of the specifications that are pertinent. Basically, structural drawings are prepared for the direction of construction by the general contractor. The first use will be for the preparation of a cost estimate for bidding. The precision of cost estimates depends upon the precision of drawings. Maximum benefits from competitive bidding can be achieved only by providing complete information required for the estimates. The information necessary for an estimate includes dimensions of the concrete for computation of form areas and concrete quantities. Strength and type of concrete required must be prescribed, and if more than one, elements utilizing each must be identified so that quantities may be computed.

All concrete outlines must be clearly defined and coordinated with the architectural drawings. Errors or omissions in drawings will encourage faulty estimates and can

*Illustrations in this chapter are courtesy of Concrete Reinforcing Steel Institute.

**Paul F. Rice, Technical Director, Concrete Reinforcing Steel Institute, Chicago, Illinois.

***William C. Black, Chief Engineer, Reinforcing Bars, Piling and Construction Specialties, Bethlehem Steel Corporation, Bethlehem, Pennsylvania.

result in costly delays of subsequent construction. One common error in drawings is contradictory or incompatible structural and architectural details. Columns must be properly oriented—schedules must show relative location. Wall and column footing elevations must be shown; steps in wall footings must be completely defined. Brick ledges properly need to be shown in wall elevations, to define clearly the exact elevations and steps at changes in elevation.

All openings, whether in walls or floor systems, must be completely defined, that is, sizes of opening and relative location must be dimensioned. Usually, this information is provided on the architectural drawings, but if not, should be included on the structural drawings.

The designer should indicate locations at which construction joints will be acceptable to him, over-all limits on length cast at one time, if any, and any other limitations imposed by the design on extent of work that may be scheduled for one casting operation. Each general concontractor's bid will be predicated on a schedule best suited to his operation, where the choice is open.

23.2 ALTERNATE DESIGNS

Where the contract documents call for a base bid plus alternates, it is most important to delineate the extent of the base bid and then define each alternate, whether it is an addition or deduction from the base bid. Frequently, confusion arises because the alternate not only adds or deducts, but also modifies some element already covered under the base bid. The estimator must then, in effect, deduct the original element from the base bid and add the modified element effect before adding or deducting the other elements that merely add or deduct. An illustration of this is shown in Fig. 23-1.

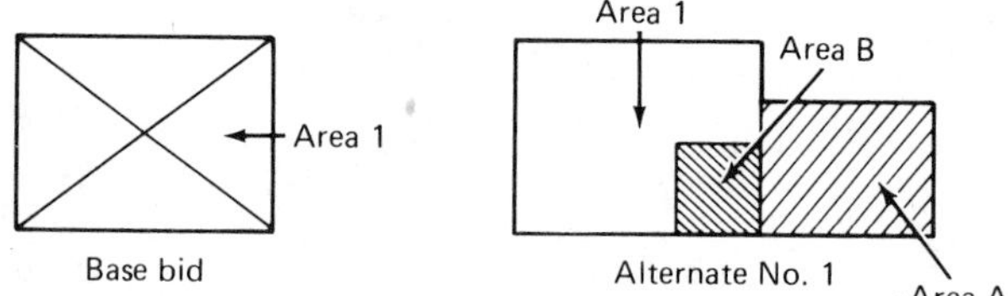

Fig. 23-1 Alternate bid areas.

The area covered under the Base Bid is shown as Area 1. Alternate #1 added Area A plus modified Area B from the original. The estimator would have to first estimate Area 1 for the Base Bid. Then he would estimate Area A for addition under Alternate #1. Then he would separate costs/quantities for Area B as originally designed and deduct this amount under Alternate #1. Finally, he would estimate Area B as modified under Alternate #1, for addition to same. The bid for Alternate #1 would then be computed as follows:

Alternate #1 = Area A + modified Area B − original Area B

Reinforcing bars and welded wire fabric need to be completely defined for the reinforced concrete estimate. This requirement can be presented in plan, in section, in separate schedules, or in any combination of the three—the important point is that complete data is mandatory.

23.3 REINFORCING BARS, SPIRALS, AND WELDED-WIRE FABRIC

All reinforcing bars and related material should be shown on the structural drawings, including the grade of steel. Even though the architectural drawings and specifications are part of the contract documents, complete reinforcing information should be shown on the structural drawings, as the reinforcing steel estimator is working primarily from structural drawings, and, under pressure of bid date deadlines, could easily overlook critical information located elsewhere.

Reinforcing materials are almost always furnished, and sometimes placed, under subcontracts. Different types of reinforcing steel, such as prestressing steel and fittings, conventional reinforcing bars, welded wire fabric, structural steel shearheads, etc., are often supplied and/or placed under separate subcontracts. Each subcontract supplier must estimate his specialty and the chance of error is heightened by the sheer diversity. Omissions in estimates can be minimized if structural drawings are complete and clear, even if repetition of some specification provisions is necessary. Erroneous estimates can be costly to all parties involved. Ambiguous or incomplete drawings result in varying 'allowances' by the estimators, and usually result in a wide scatter of subcontract bid prices.

23.4 ESTIMATING

The single item of reinforcing material is usually the largest subcontracted. Separate estimates to furnish reinforcement are ordinarily prepared by supplier subcontractors. The information required for a complete reinforcing steel estimate is needed quickly. It must be shown completely and without ambiguity if close estimates are to be secured. Time usually is not available to secure any information lacking on the drawings or clarifications of details on drawings. For the estimator, the drawings must clearly show:

a. Grade and size of all bars. Where more than one grade of bars is specified, the extent of each must be clearly shown.
b. Quantity as in columns and beams, or spacing as in slabs and walls. It is important to indicate clearly limits of reinforcing where spacing is given, especially where the reinforcing may change size and/or spacing (extent of each welded wire fabric style must also be shown).

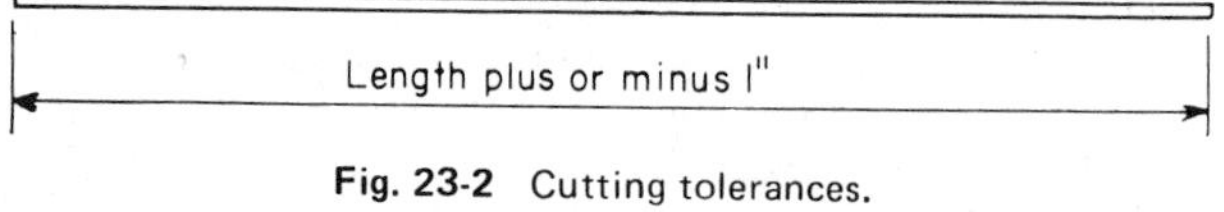

Fig. 23-2 Cutting tolerances.

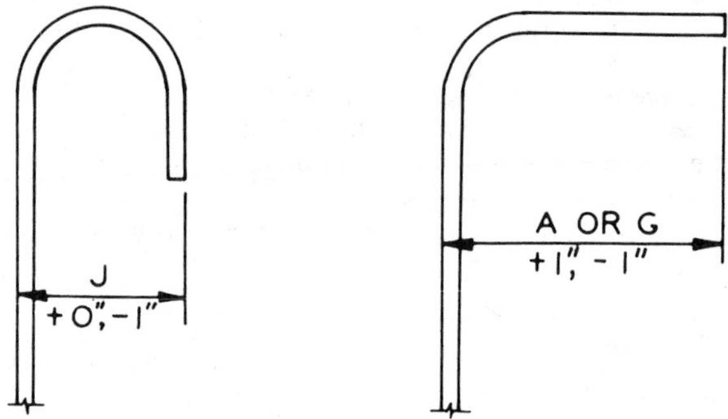

Fig. 23-3 Tolerances for standard hooks.

STANDARD FABRICATING TOLERANCES

For bar sizes #3 through #11

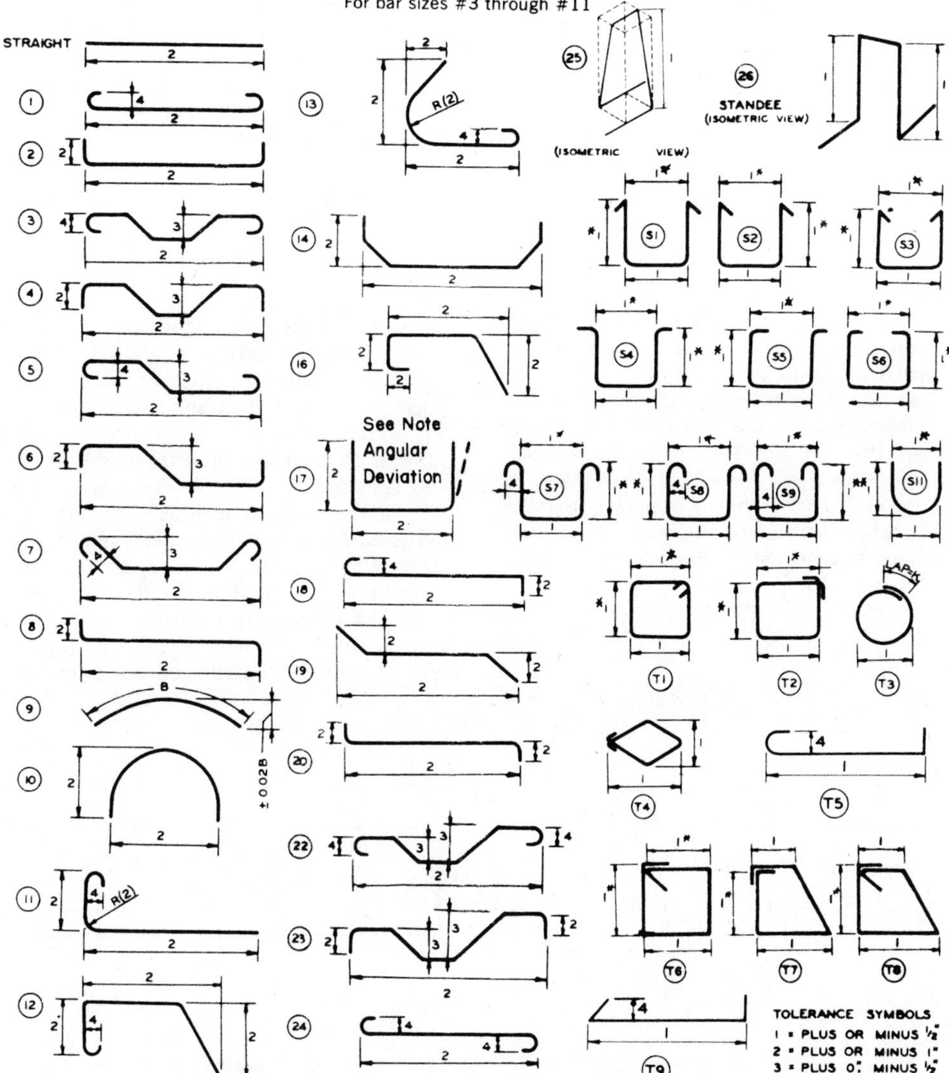

Note: Entire shearing and bending tolerances are customarily absorbed in the extension past the last bend in a bent bar.

*Dimensions on this line are to be within tolerance shown but are not to differ from the opposite parallel dimension more than ½".
Angular Deviation — maximum ±2½° or ±½"/ft., but not less than ½", on all 90° hooks and bends.

STANDARD FABRICATING TOLERANCES

For bar sizes #14 and #18

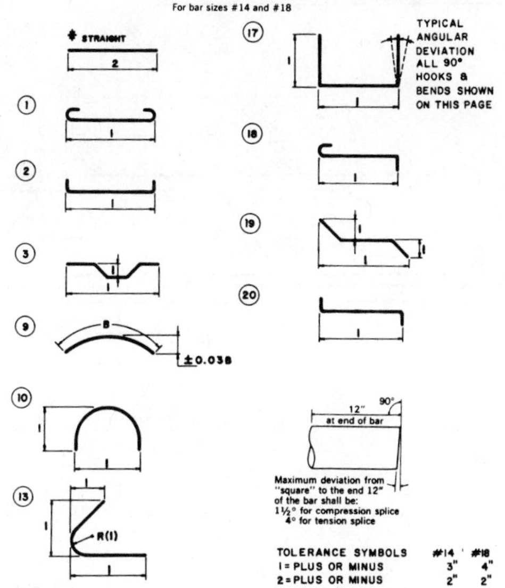

Angular Deviation — maximum ±2½° or ±½"/ft., but not less than ½", on all 90° hooks and bends.
*Saw-cut both ends — overall length ±½".

Fig. 23-4

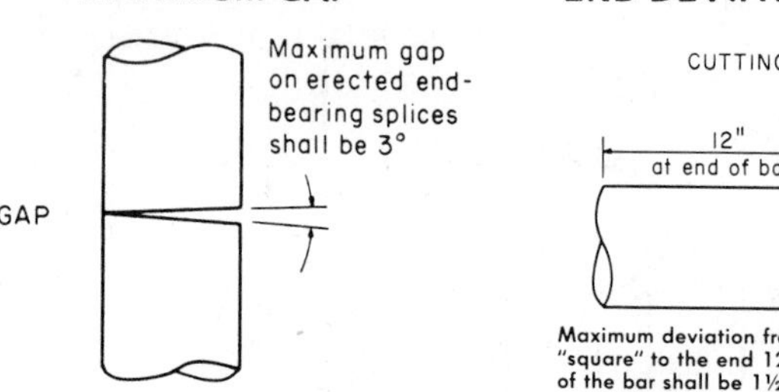

Fig. 23-5 Nominally square saw-cut ends. Maximum gap tolerance for spliced bars which transmit compressive stresses through direct end-bearing: for adequate structural performance, the total angular deviation of the gap shall not exceed 3°, as shown.

In order to achieve a proper fit in the field, the ends of the bars must be saw-cut, or otherwise cut in such a manner as to provide a reasonably flat surface. It is recommended that the end deviation of an individual bar from 'square' not exceed 1.5° when measured from a right angle to the end 12 in. of the bar as shown. Relative rotation or other field adjustment of the bars may be necessary during erection to secure a fit which falls within the recommended gap limits.

For tension butt splices with saw-cut ends, out-of-square tolerance is 4° on each bar end.

c. Length of bars that do not run continuously, such as negative moment reinforcing in a beam.
d. Any unusual tolerance requirements in fabrication or placing. See Figs. 23-2 through 23-6 for standard tolerances and certain special tolerances for bar fabrication and placing.
e. Required bending. The estimator must separately compute quantities of rebars, straight in each size, light and heavy bending, and special fabrication for bending to nonstandard radii. See Fig. 23-7 for standard end hooks and bend radii.

Clear and specific details are mandatory on splicing of reinforcing bars and welded wire fabric. Under ACI 318-71, it is necessary for the designer to show first, the type of splice required, i.e., butt or lapped; in tension or compression; and, if butt spliced, whether mechanical and/or arc welded splices are permitted. In the case of lapped splices, he must also indicate length of lap for all conditions and specifically identify where each applies. For long runs of continuous bars, such as temperature reinforcement, the designer must indicate where splices are permitted, if restricted; arrangements for stagger, if required; or conversely, state that splices may occur at contractor's option.

Welded (plain wire) fabric splices must be defined as to full tension or, when permitted, half tension. Locations where they will be allowed must be clearly stated. Splices

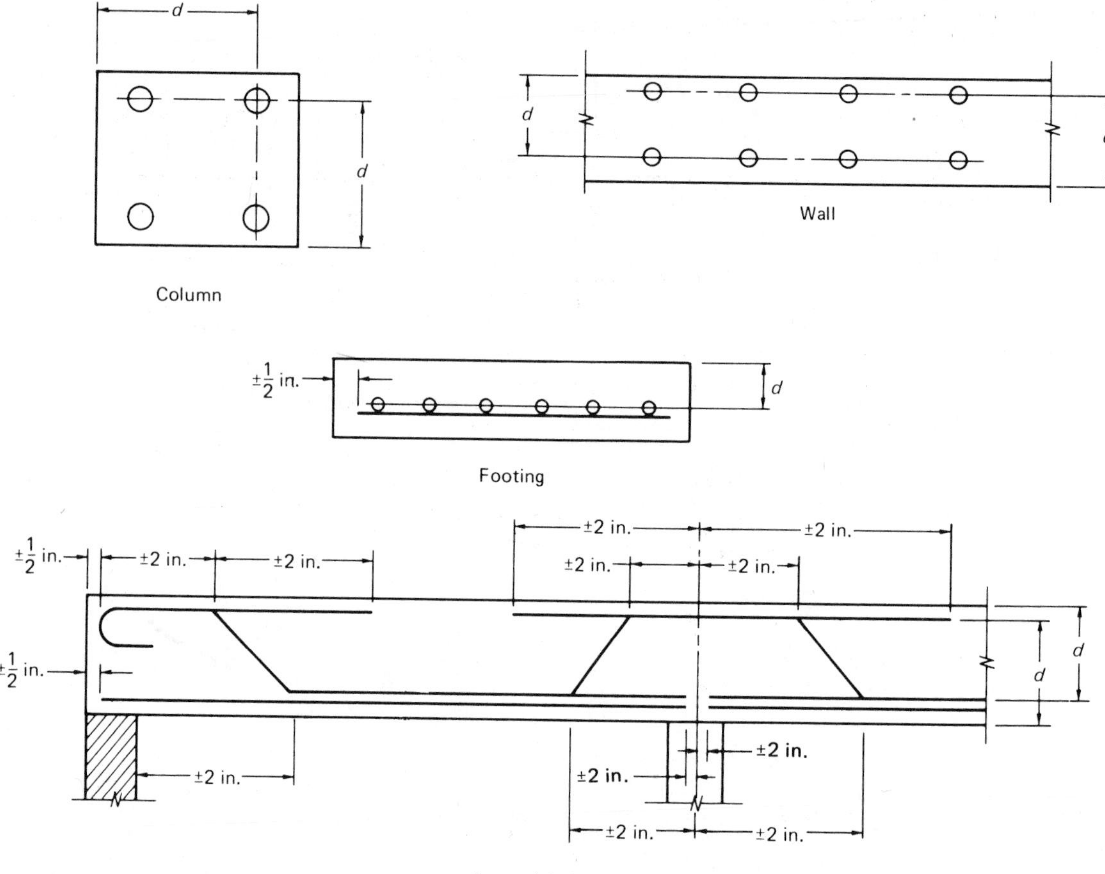

Beam, slab, joist, etc.

Tolerance on dimension "d" and clear cover where:

$d \leq 8$ in. $\pm\frac{1}{4}$ in.

8 in. $< d <$ 24 in. $\pm\frac{3}{8}$ in.

$d >$ 24 in. $\pm\frac{1}{2}$ in.

Fig. 23-6 Standard placing tolerances. (ACI 318-71)

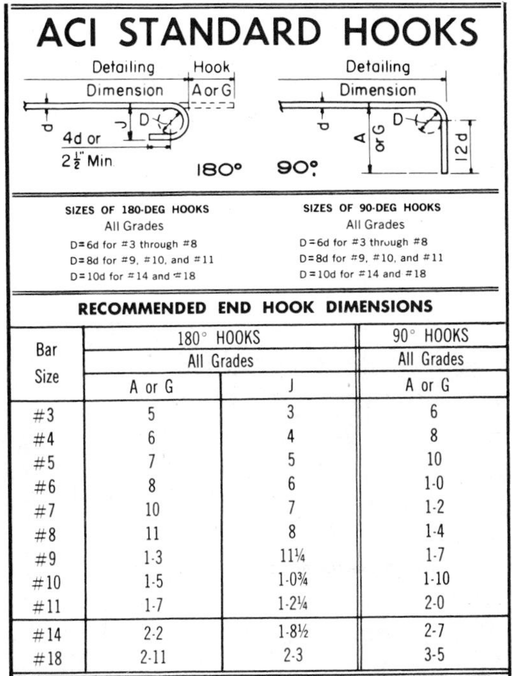

SIZES OF 180-DEG HOOKS
All Grades
D = 6d for #3 through #8
D = 8d for #9, #10, and #11
D = 10d for #14 and #18

SIZES OF 90-DEG HOOKS
All Grades
D = 6d for #3 through #8
D = 8d for #9, #10, and #11
D = 10d for #14 and #18

RECOMMENDED END HOOK DIMENSIONS

Bar Size	180° HOOKS All Grades		90° HOOKS All Grades
	A or G	J	A or G
#3	5	3	6
#4	6	4	8
#5	7	5	10
#6	8	6	1-0
#7	10	7	1-2
#8	11	8	1-4
#9	1-3	11¼	1-7
#10	1-5	1-0¾	1-10
#11	1-7	1-2¼	2-0
#14	2-2	1-8½	2-7
#18	2-11	2-3	3-5

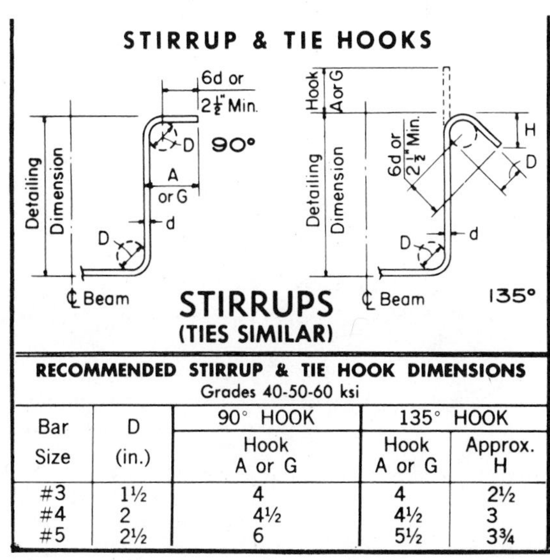

RECOMMENDED STIRRUP & TIE HOOK DIMENSIONS
Grades 40-50-60 ksi

Bar Size	D (in.)	90° HOOK Hook A or G	135° HOOK Hook A or G	135° HOOK Approx. H
#3	1½	4	4	2½
#4	2	4½	4½	3
#5	2½	6	5½	3¾

Fig. 23-7 Standard end hooks.

for welded (deformed wire) fabric and end anchorages of all fabric should be shown in details.

When prestressing is involved, the structural drawings must indicate quantity, type, any special fittings, conduits, or sheaths needed, as well as general requirements such as initial force, assumed losses, final force, grouting, etc., which might restrict use of various proprietary systems.

23.5 REINFORCING STEEL PLACING DRAWINGS

Once an agreement has been reached between the General Contractor and the reinforcing Steel Supplier, the contract documents are turned over to the Supplier's Engineering Department for processing. The exact sequence of building the structure must now be determined to establish the sequence in which the placing drawings are prepared for the rebar or other reinforcing materials and the sequence for fabrication and shipment.

The reinforcing steel detailer needs complete detail information to prepare the placing drawings and bills of material. It is his task to define each and every reinforcing bar, spiral, piece of welded wire fabric, or prestressing unit required, as the case may be, for the entire contract. This precision is only possible when the contract drawings are clear, complete and concise. Typical details should really be 'typical'! Details which are similar to the typical detail drawn should all have dissimilarities noted as exceptions.

Many difficulties requiring redesign, due to inadequate or improper design, become apparent at this stage as the detailer is actually calculating the individual bar dimensions and lengths, and must check for shipping limitations (see Fig. 23-8), clearances, and interference with crossing bars. The structural drawings should always be checked for these points before release to bidders. Typical situations which the structural engineer preparing the drawings or responsible for same should check are: (a) width clearance for girder bars passing through columns; (b) depth clearance and cover of crossing slab and beam, joist and beam, or beam and girder bars; (c) cover and depth for end hooks on truss bars in shallow slabs or joists; (d) interference with conduit by other trades; and (e) bar detail for holes in wall or slab.

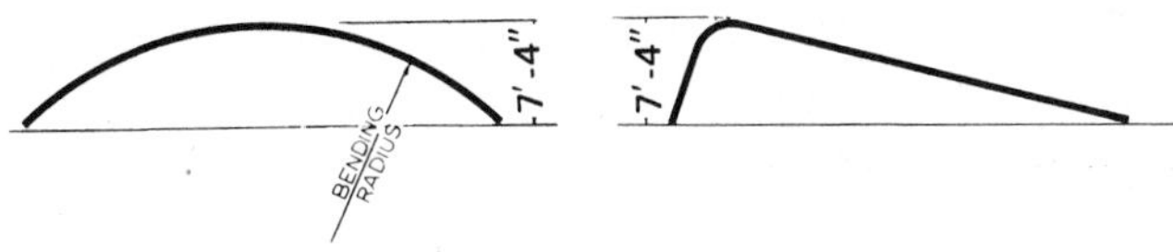

Fig. 23-8 Maximum dimension of bent bars for truck delivery.

23.6 APPROVALS

Usual practice by structural engineers is to insist upon submission of placing drawings for approval prior to fabrication.

Placing drawings are submitted by the reinforcing steel fabricator (Supplier) to the General Contractor for approval by the Architect/Engineer. At this time, the designer reviews the interpretation of his design by the detailer and either approves or notes any corrections, changes or additions, Two important considerations at this time are: (a) changes necessary to revise, complete, or refine the design; and (b) delays in approval.

If changes in the design require more reinforcing material or additional time for reestimating or redetailing, the question of extra payment is involved. Undue delay in return of approved placing drawings may result in extended construction schedules which also increase costs. Changes during construction are always costly. If necessary, due to changes in the Owner's requirements or necessary changes in design, the cost is borne by the Owner. Cost of changes to correct errors in material, field practice, ambiguous details, etc., are usually negotiated. The obvious best policy is to minimize problems by preparing complete, clear, concise contract drawings.

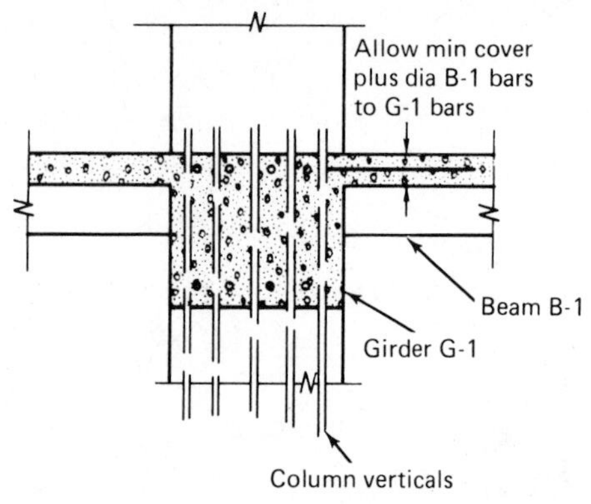

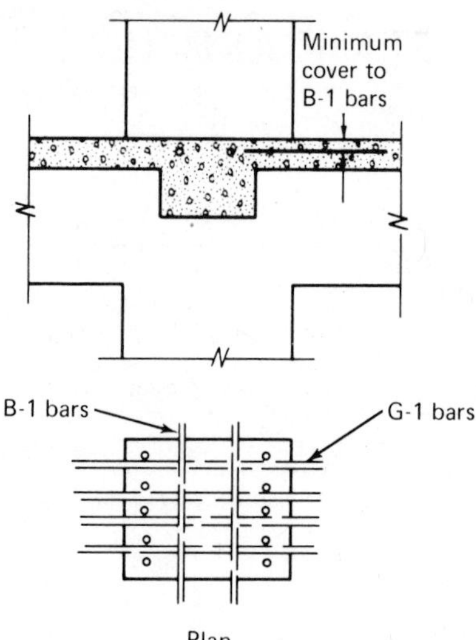

Fig. 23-9 Clearances for intersecting beam-girder-column bars.
NOTE: 1. If odd number of column verticals, use even number (spaces) for girder bars.
2. Allow for depth of beam bars in cover specified on girder bars.

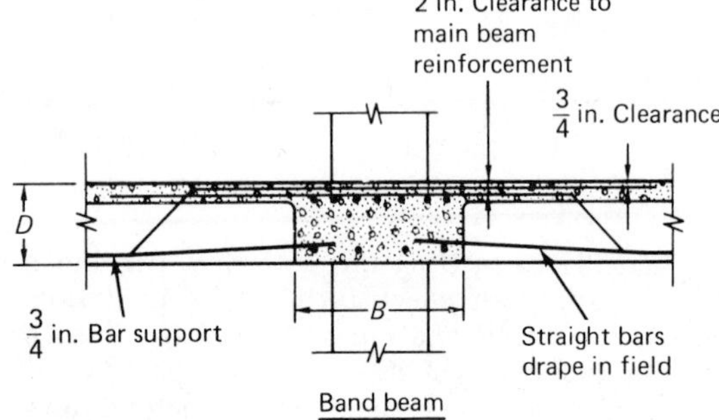

Fig. 23-10 Clearances for joist-band beam connections.

Fig. 23-12 Conduit in place, ready for reinforcement.

23.7 PRACTICAL CONSIDERATIONS FOR MOST EFFICIENT PRESENTATION OF DESIGNS

At this point various reinforced concrete elements will be discussed in some detail:

23.7.1 Foundations

Wall and column footing elevations must be shown and clearly marked as to whether they are top or bottom of footing. If intended to be cast separately, construction joints should be shown, reinforcement designed accordingly, and casting sequence indicated if essential to design. The designer must show whether continuous bars in wall foot-

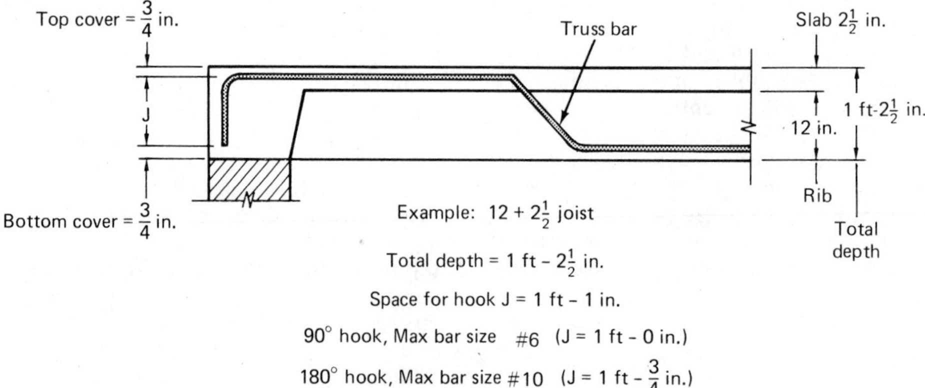

Fig. 23-11 Truss bar size limited by depth of end hook.

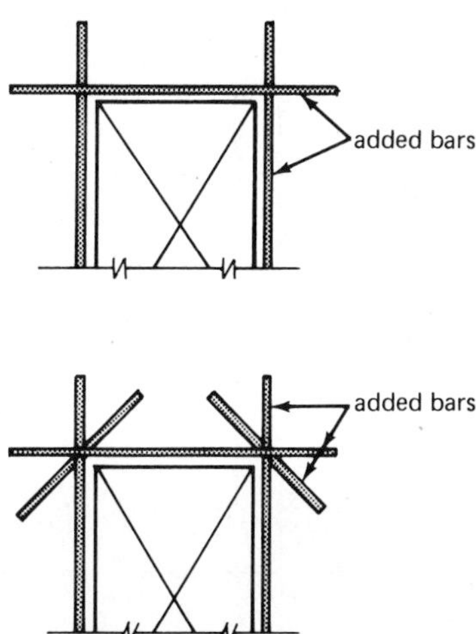

Fig. 23-13 Two common details of special reinforcement at openings in walls or slabs.

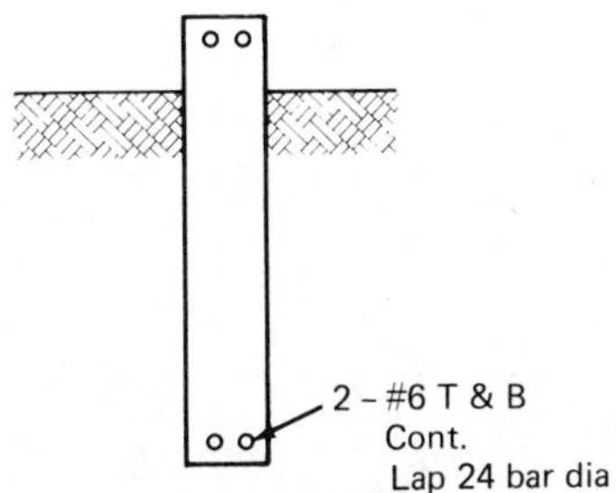

Fig. 23-14 Grade beam. NOTE: Whenever word 'continuous' appears in bar description, additional explanation is required in General Notes, or in description of bar.

ings run (a) through the column footing, (b) extend into the column footing some specific distance, or (c) whether separate dowels are to be provided out of the column footing. If corner bars are required in continuous footings or walls, they should be shown. Continuous large size bars are frequently shown at top and bottom in foundation walls. Where pockets occur at top of wall, the designer should indicate how the continuous reinforcement is to be detailed at these points. Brick ledges in foundation walls affect the reinforcing and, therefore, the detailer must know the elevation and extent of each. Columns and piers in foundation walls with brick ledges must be carefully lo-

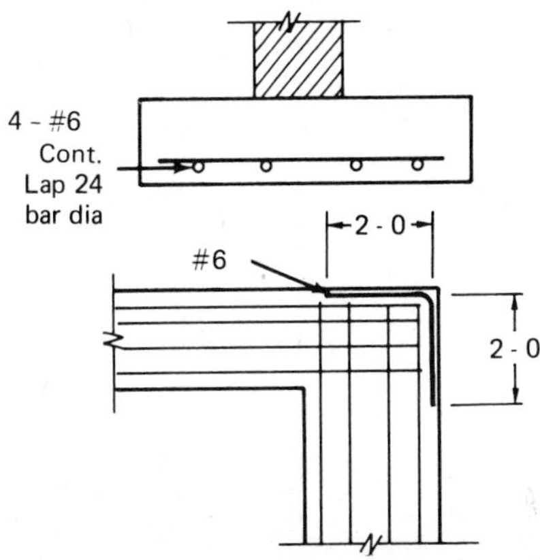

Fig. 23-15 Wall footing; corners. NOTE: Where continuous bar is to be continuous; around a corner in grade beam, wall footings, wall, etc., show corner detail desired.

cated by the designer. The contract drawings must clearly show whether the column size includes the brick ledge width or not. Notes on typical sections through walls and section arrows on plan or elevation should be used to show the exact extent of each typical section. Cantilever walls must clearly indicate where splices are permitted and arrangement of any alternating vertical bars. (See Figs. 23-14–23-18, inclusive.)

23.7.2 Columns

Columns are usually shown on contract drawings in schedule format with cross sections, typical or otherwise. It is important to identify clearly column faces, especially

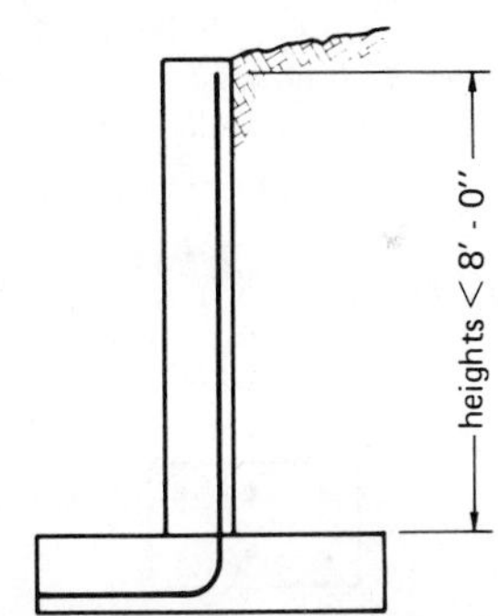

Fig. 23-17 Low retaining wall. Avoid splicing main steel.

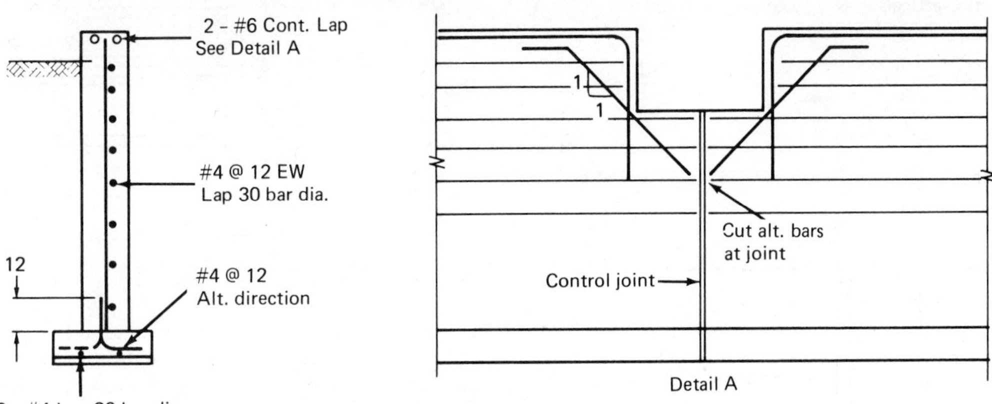

Fig. 23-16 Continuous bars require explanatory detail at interruptions.

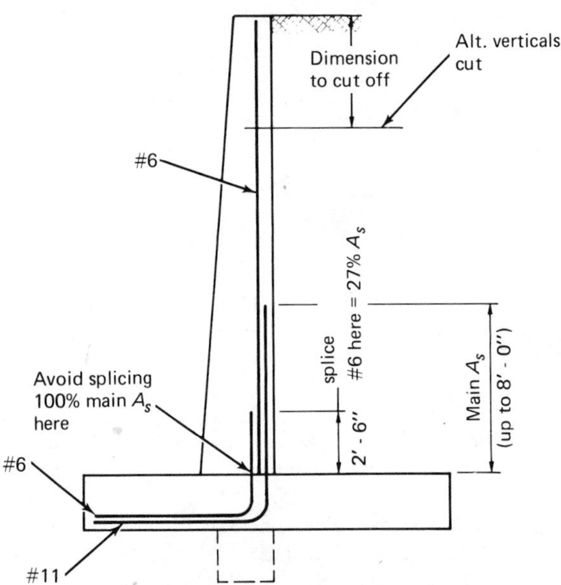

Fig. 23-18 Retaining wall requiring splices in main steel. Simple pattern to stagger splices.

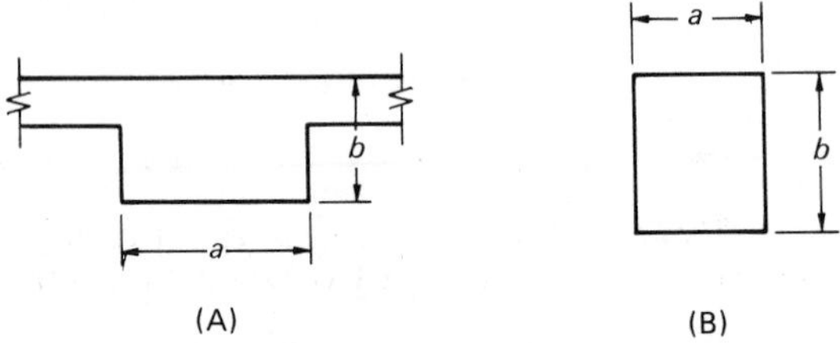

Fig. 23-19 Typical orientation of (scheduled) column dimensions. (A) Wall column orientation and (b) interior column. The dimensions are a = first dimension scheduled, and b = second dimension scheduled.

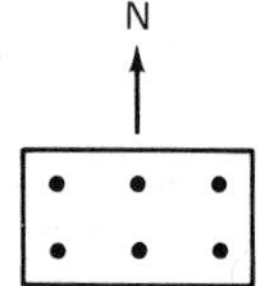

Fig. 23-20 Rebar locations in columns. NOTE: If nontypical, rebar plan sketch must be shown for each column at each change in rebars in schedule. If typical, General Note may be used, such as, "All column verticals in N-S faces of columns."

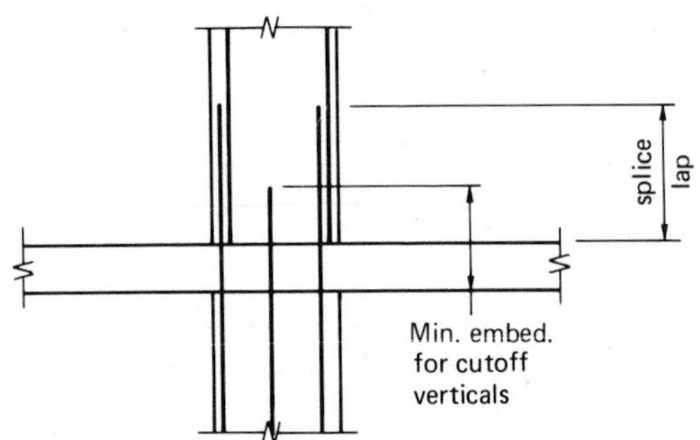

Fig. 23-21 Show minimum embedment required where standard cutoff 3 in. below finished floor will not satisfy design.

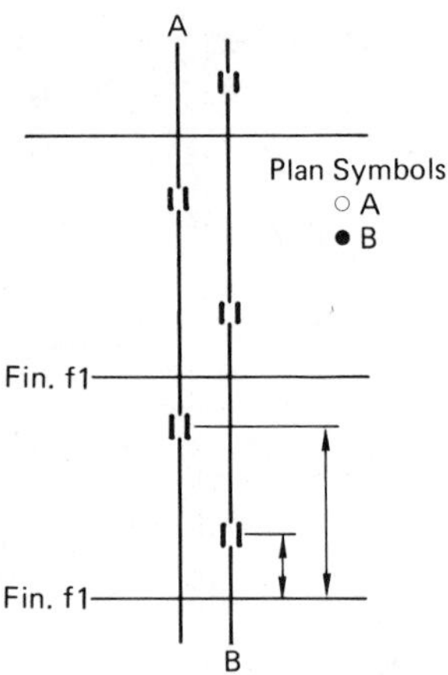

Fig. 23-22 Column elevation to show staggered splices. (Show dimensions to locate from finished floor or top of footing in schedule.)

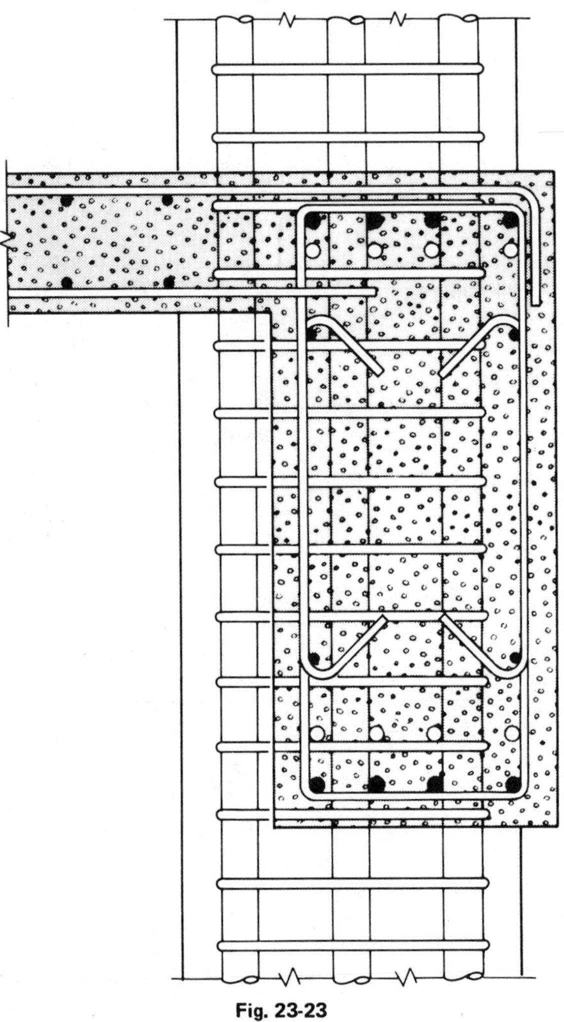

Fig. 23-23

for rectangular columns. One scheme that facilitates designation of direction for long side-short side in a separate schedule is shown in Fig. 23-19. The small key drawing like Fig. 23-19 should be placed on each sheet containing the schedule or portions of a schedule to which it refers. It is also important for the designer to indicate number of

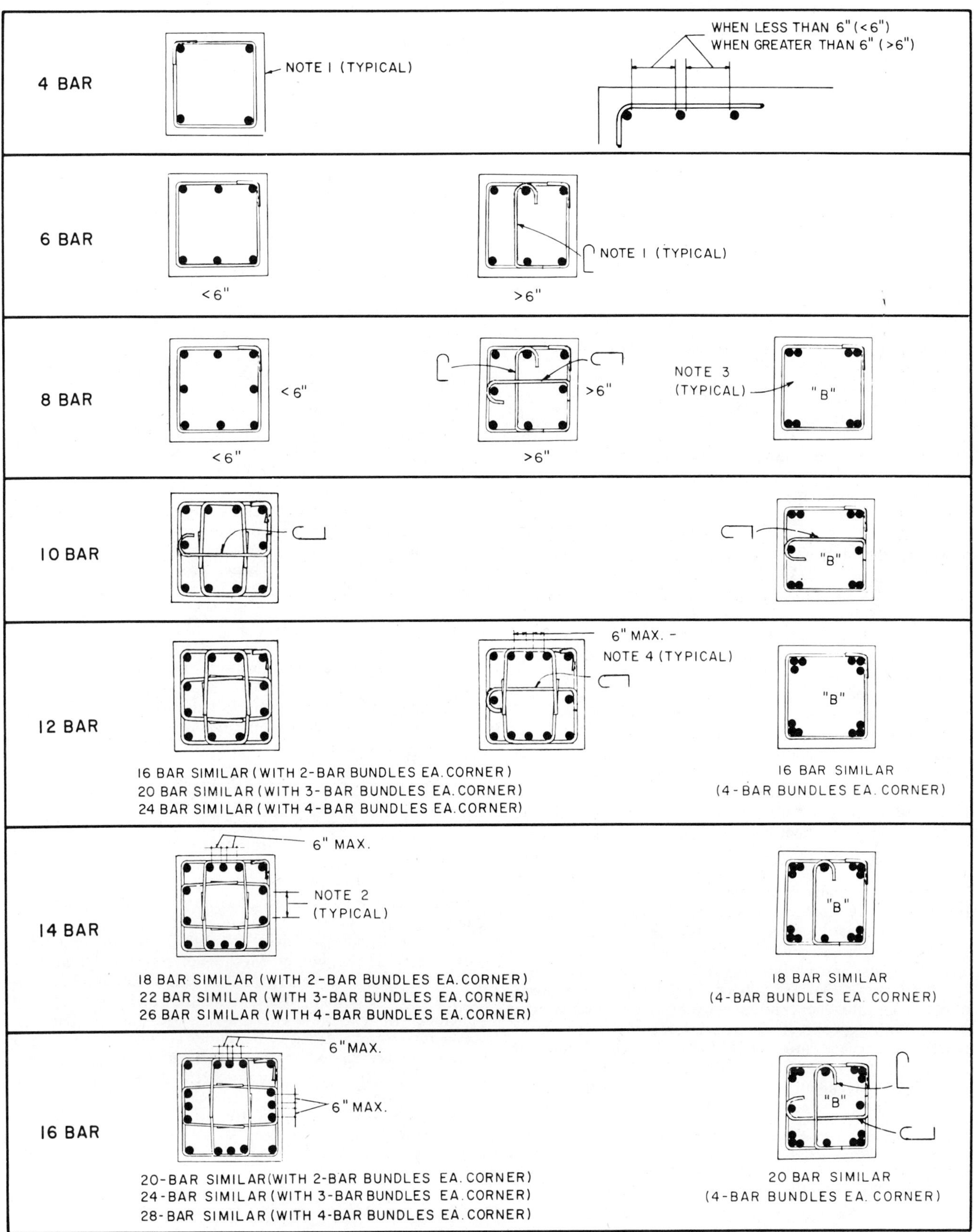

*Applicable for either preassembled cages or erection in place on free-standing butt spliced vertical bars.

Fig. 23-24 Universal standard column ties.*

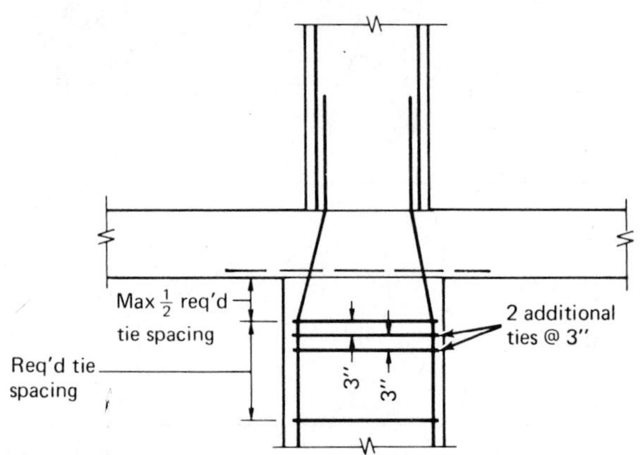

Fig. 23-25 Additional ties at offset bend. (Show number, size, and spacing, if required.)

vertical bars in each face. Where more than one bar size is used, location in a plan detail is essential. Such plan details should usually be in the schedule at each story where bar arrangement changes. (See Fig. 23-20.)

Where more than one size of rebar is required in a column, an elevation detail is necessary to locate the bars and splices. Usually a typical joint splice detail will suffice; however, the designer should provide for nonstandard conditions (Fig. 23-21) and show exactly where dowels are to be cut off for embedment or continuation by splices. He must specify splice and anchorage lengths for all bar sizes involved, and indicate when and how far to anchor terminated bars from a column below into one above.

Butt-spliced columns frequently require a schematic elevation of each column bar (Fig. 23-22) to show the relationship of staggered splices. As the column schedule is filled in, the designer should check spaces and number of spaces at the beam-column intersection to insure that the bars will physically pass by each other. Figure 23-22 shows a careful scale drawing to illustrate such a check at a difficult intersection. Where such a study is required by the designer simply to select bar sizes, cover, etc., it should be added to his structural drawing for information to detailer and placer.

Column tie arrangements must be specifically shown by the designer, as there are several permissible (ACI) arrangements. It is usually preferable to use the ACI universal tie arrangement for ease of erection. (See Fig. 23-24.) Ninety-degree hooks are preferred for closed ties. Normally, column verticals that are not terminated at a floor level are offset bent into the column above. If the offset occurs below confining concrete on all sides, recommended design practice is to place three sets of ties at 3 in. (see Figure 23-25) right at the offset bend. If desired, the designer should show this arrangement.

23.7.3 Beams, Girders, and One-Way Joists

Beams, girders, and one-way joists are usually shown on schedules, but sometimes are drawn in elevation and/or section. Whichever method the designer uses, it is incumbent for him to show the quantity. Extent of reinforcing as a percentage of the span is satisfactory as long as the detail really applies to *all* beams. The designer must check to determine if multiple layers of principal bars are required, due to spacing limitations, and so indicate on his drawings. When stirrups are required, the designer must show quantity, size and spacing, and ACI-bend types (see Fig. 23-26).

It is acceptable and often convenient, particularly in design schedules to designate bent bars by ACI Bend Type. When closed stirrups are required, the designer should indicate whether he will accept the ACI recommended two-piece units, or preferably, show same on his drawing (Fig. 23-27). If support bars at the top of the stirrups are desired, (in addition to bars required by design) they should be indicated by the designer (Fig. 23-28).

23.7.4 Slabs

Slab reinforcing may be shown right on the plan, in section, or in schedule format for one-way slabs. See Fig. 23-29 for design of slab in schedule and in section, and Fig. 23-30 for design of same slab on plan. It is important to clearly indicate the alternating bars in one-way slabs. Figure 23-31 shows the wrong ways to show alternating truss bars. The bottom straight bars in the plan view should not be indicated as continuous with truss bars as the truss bars should be staggered in adjacent bays to provide uniformly spaced negative moment reinforcing. The schedule does not make it clear that the slab design requires alternate straight and truss bars at 8 in. on center. Temperature reinforcement should be clearly located–if in the bottom full width of slab–it should be so shown. If the designer wants the temperature bars to follow the tension face of the concrete, he must specifically show it thus (Fig. 23-32).

The designer's requirements where a slab changes direction should be clearly shown (see Fig. 23-33).

The detailer must know whether to terminate the top bar and hook into the support or extend a specific anchorage distance into the adjacent span.

Two-way flat slabs have only minor differences from flat plates. Flat slabs with drop panels generally follow the ACI/CRSI standard rebar arrangement (see Fig. 23-34). It is usually satisfactory for the designer to show typical column and middle strips and use the foregoing arrangement.

Flat plates may or may not follow the standard rebar arrangement. Frequently, these involve erratic column layouts and the designer then must show clearly the exact extent of the column and middle strips and the reinforcing included therein. He may show this right on the plan or in schedule format. Straight bars only are customary for flat plates. Figure 23-35 shows the *ACI Building Code* requirements for a straight bar design. All the dimensions shown in tabular form (as Kl_c) are to the clear span. When column sizes do not vary significantly, it is practical to convert these dimensions to column center-line dimensions as follows:

$$K'l_1 = Kl_c + 1/2c$$

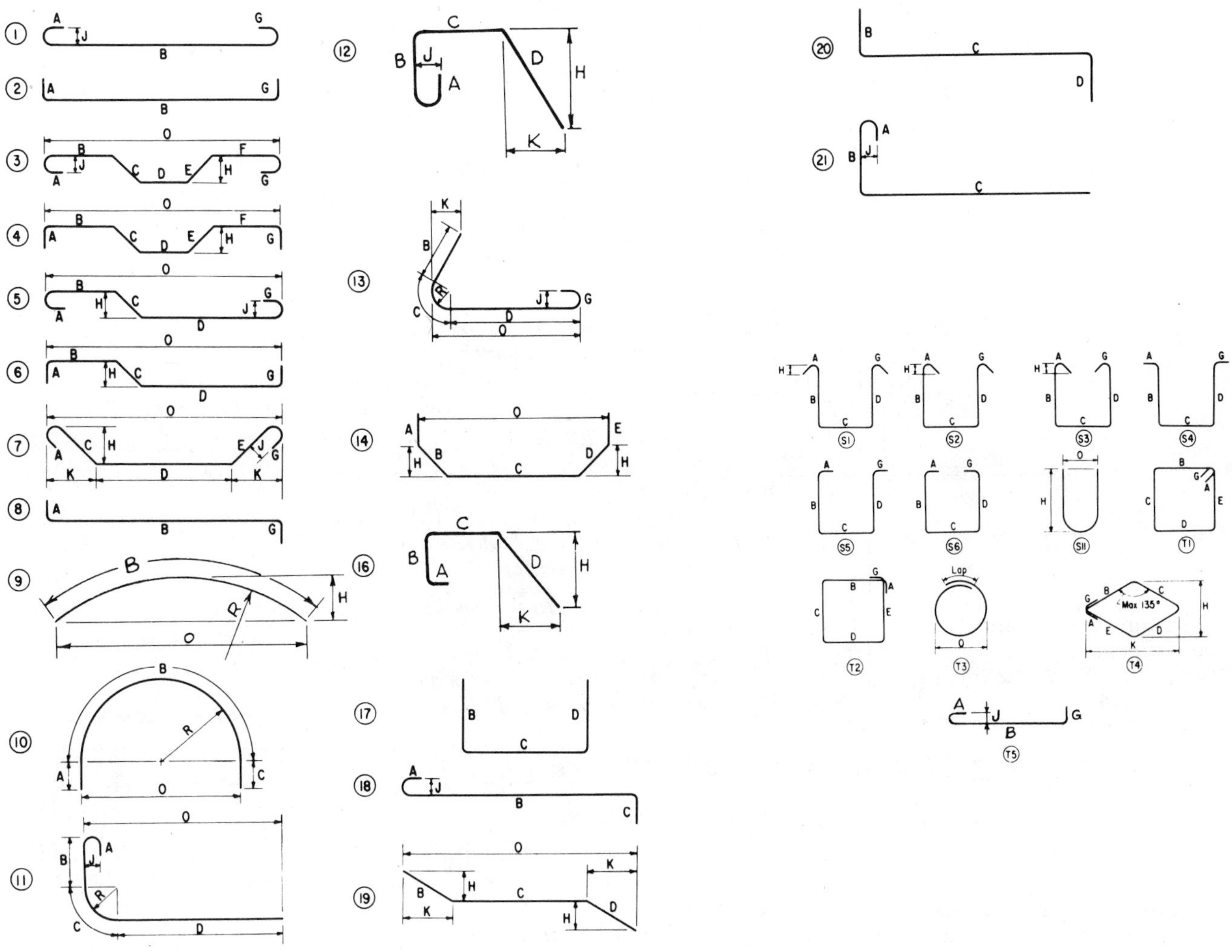

Unless otherwise noted diameter D is the same for all bends and hooks on a bar.

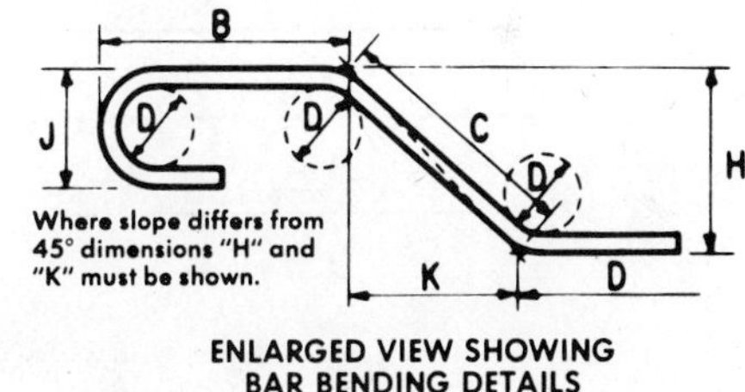

ENLARGED VIEW SHOWING BAR BENDING DETAILS

NOTES:

1. All dimensions are out to out of bar except "A" and "G" on standard 180° and 135° hooks.
2. "J" dimension on 180° hooks to be shown only where necessary to restrict hook size, otherwise standard hooks are to be used.
3. Where "J" is not shown, "J" will be kept equal to or less than "H" Where "J" can exceed "H", it should be shown. 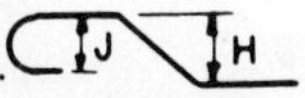
4. "H" dimension stirrups to be shown where necessary to fit within concrete.
5. Where bars are to be bent more accurately than standard bending tolerances, bending dimensions which require closer working should have limits indicated.
6. Figures in circles show types.
7. For recommended diameter "D", of bends, hooks, etc. see tables.

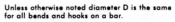

Fig. 23-26 Typical bar bends.

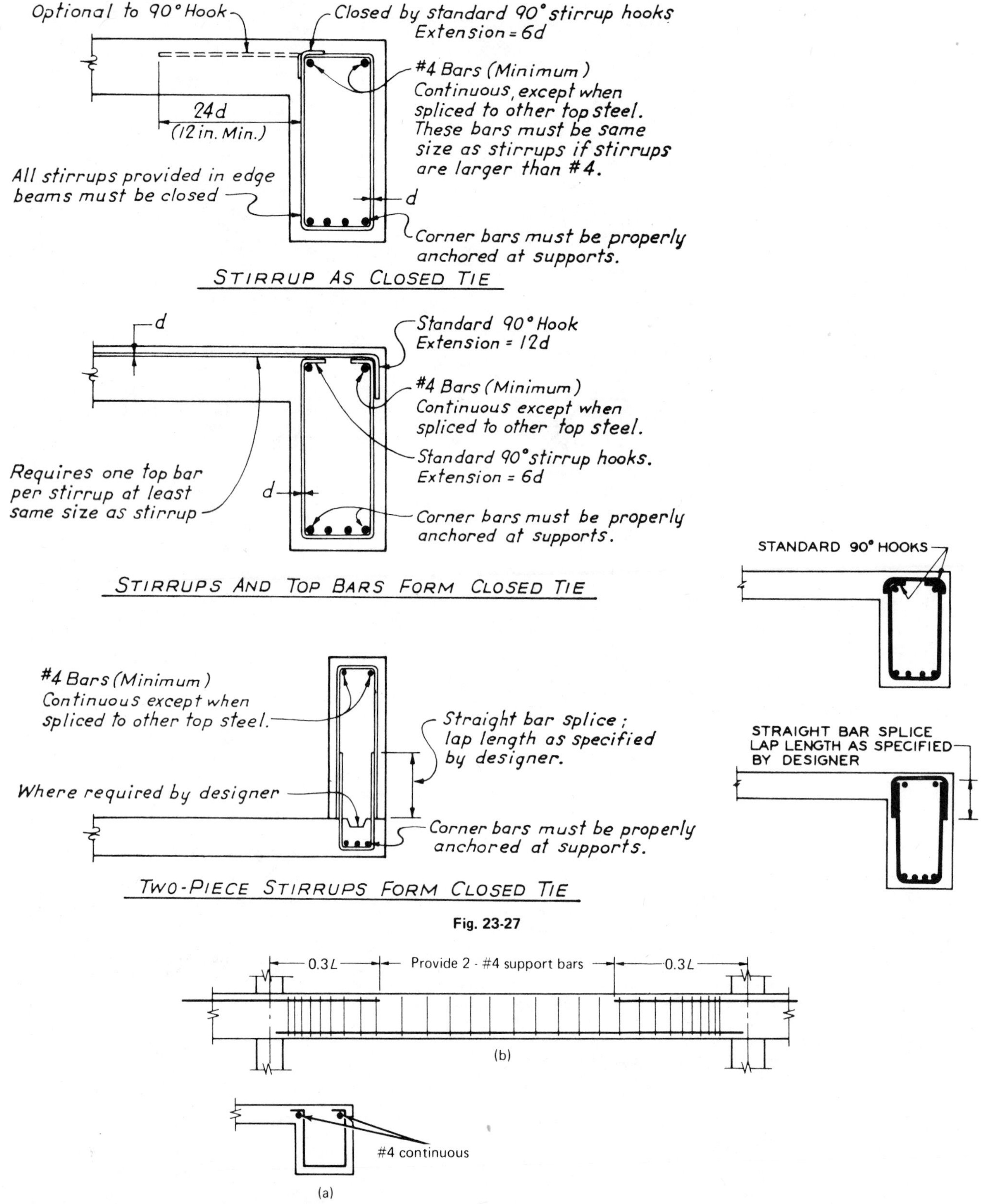

Fig. 23-27

(b)

(a)

Fig. 23-28 If designer wishes to provide positive support for stirrups between cutoff points for main top bars, two alternative provisions are shown: (a) Smaller corner bars, continuous as part of main steel; and (b) note for top 'support bars' which will not be lap spliced for continuity.

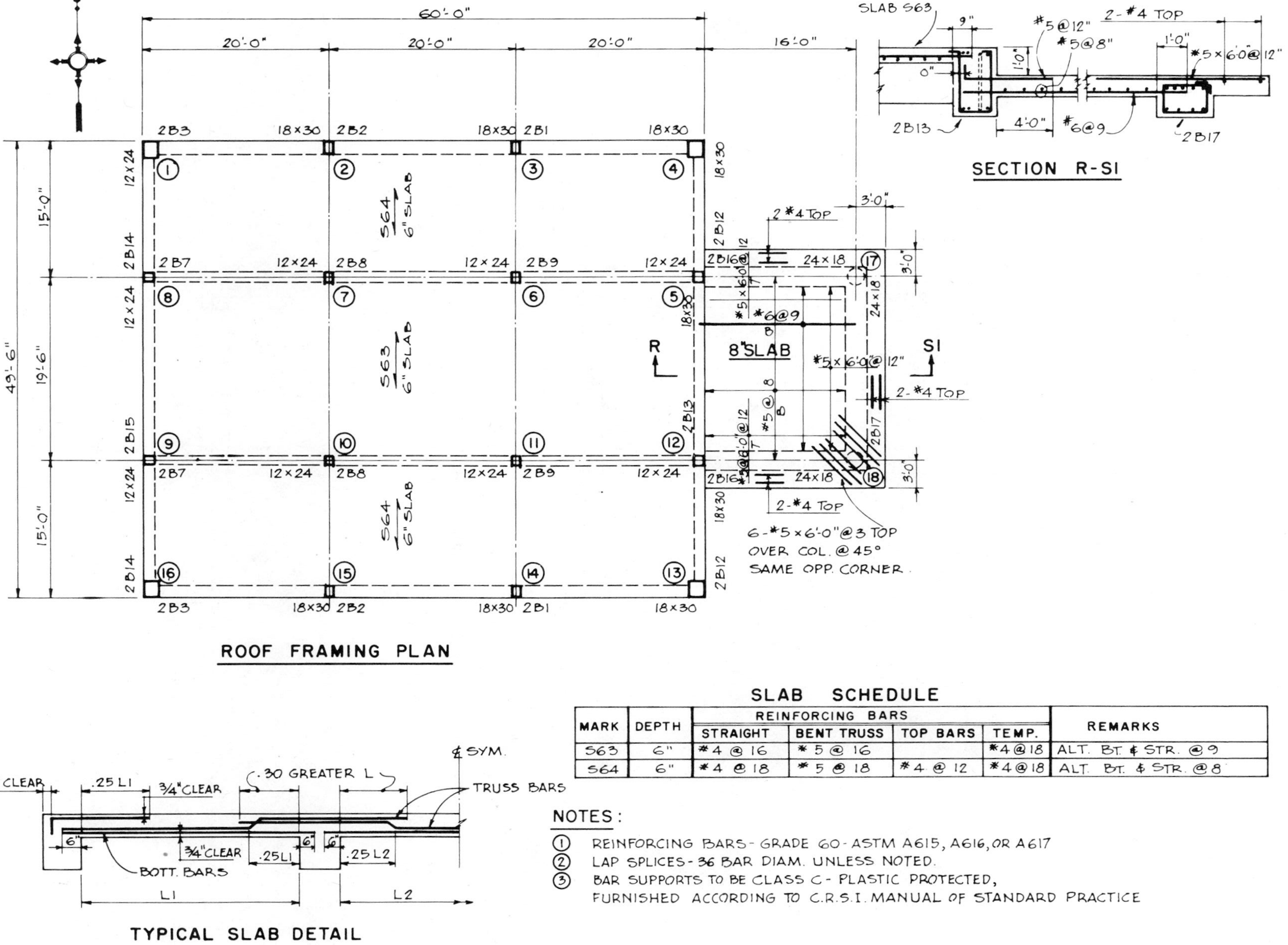

SLAB SCHEDULE

MARK	DEPTH	REINFORCING BARS				REMARKS
		STRAIGHT	BENT TRUSS	TOP BARS	TEMP.	
S63	6"	#4 @ 16	#5 @ 16		#4 @ 18	ALT. BT. & STR. @ 9
S64	6"	#4 @ 18	#5 @ 18	#4 @ 12	#4 @ 18	ALT. BT. & STR. @ 8

NOTES:

1. REINFORCING BARS - GRADE 60 - ASTM A615, A616, OR A617
2. LAP SPLICES - 36 BAR DIAM. UNLESS NOTED.
3. BAR SUPPORTS TO BE CLASS C - PLASTIC PROTECTED, FURNISHED ACCORDING TO C.R.S.I. MANUAL OF STANDARD PRACTICE

Fig. 23-29 Main slab reinforcement in schedule and special slab (RS-1) on detail section.

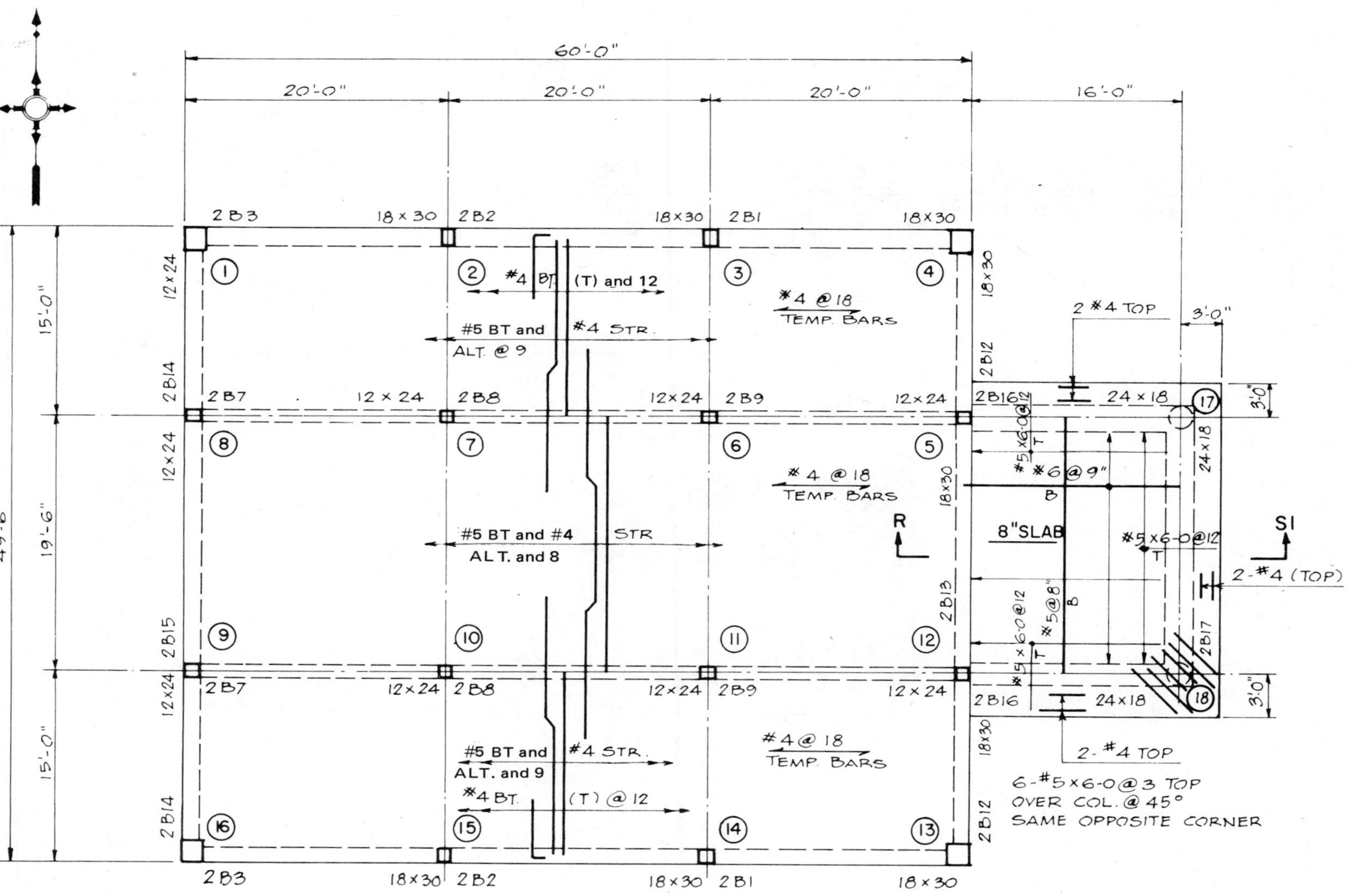

ROOF FRAMING PLAN DESIGN

Fig. 23-30 Main slab reinforcement shown on plan.

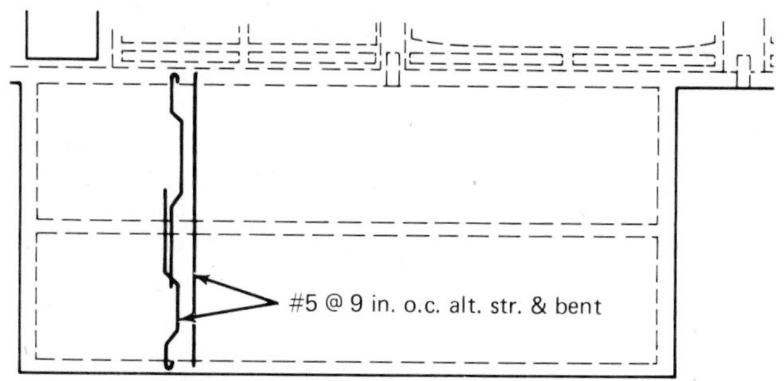

Floor Plan

Slab Schedule			
Mark	Main	Temp	Remarks
S7	#5 @ 8 in.	#3 @ 11 in.	truss bars

Fig. 23-31 Wrong way to show alternate bars.

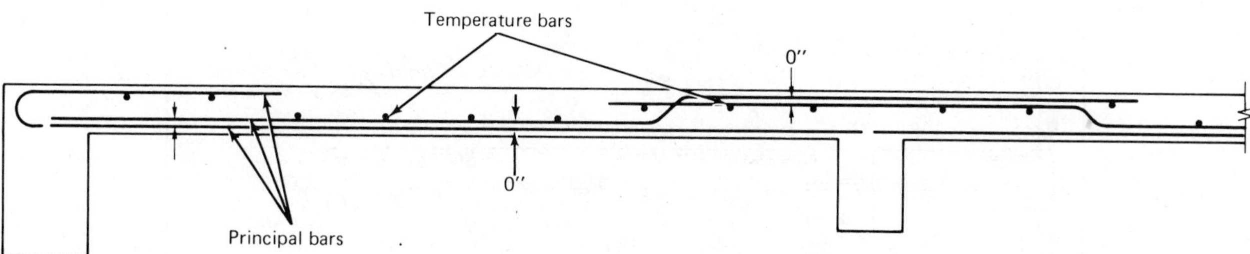

Fig. 23-32 Typical one-way solid slab.

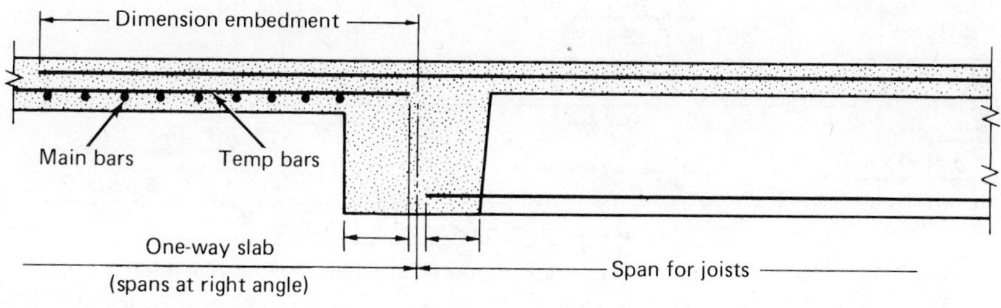

Fig. 23-33 Changing direction of span and/or floor system used. NOTE: Indicate all extensions, embedments, etc. required for design.

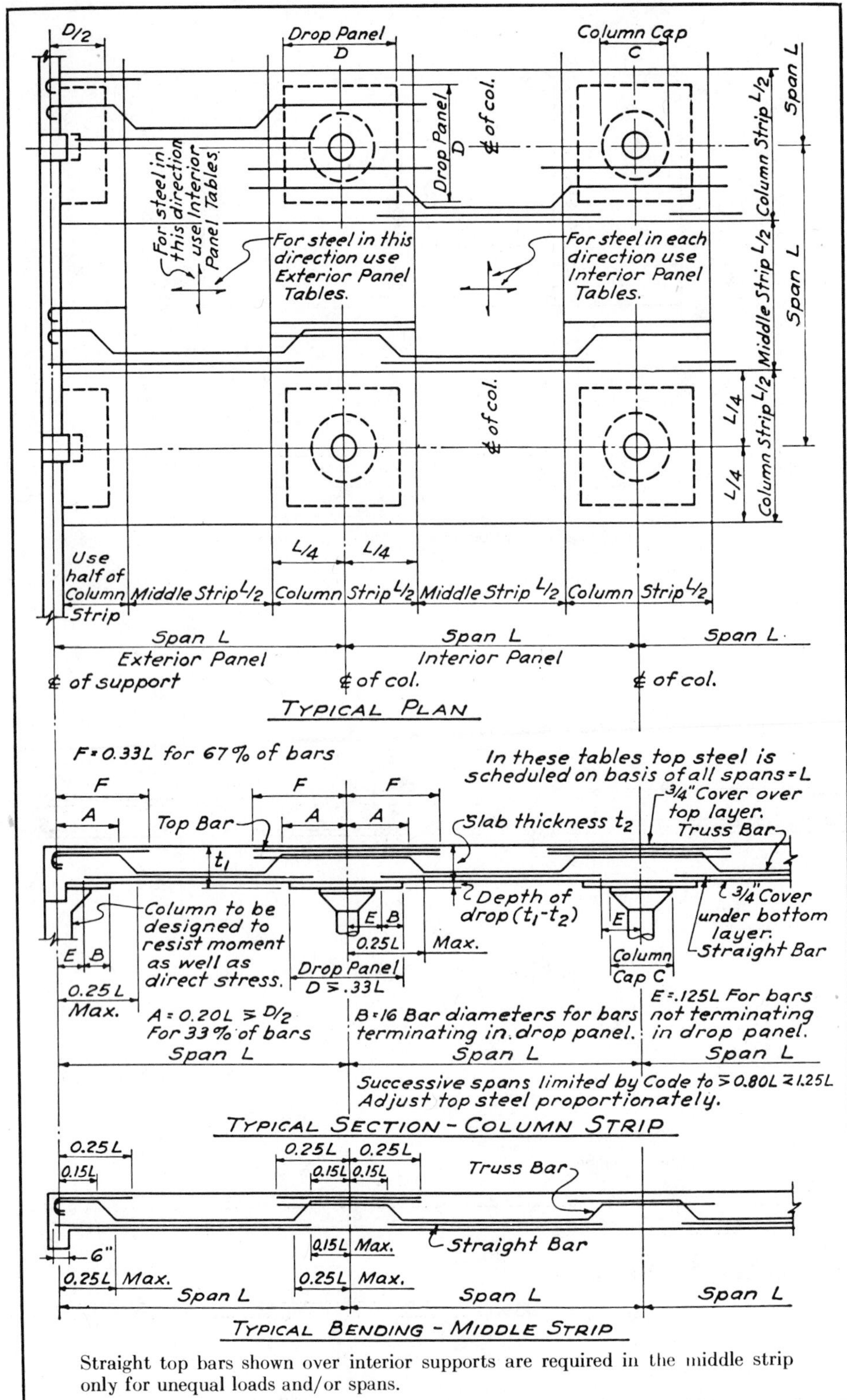

Straight top bars shown over interior supports are required in the middle strip only for unequal loads and/or spans.

Fig. 23-34 Solid flat slabs with drop panels.

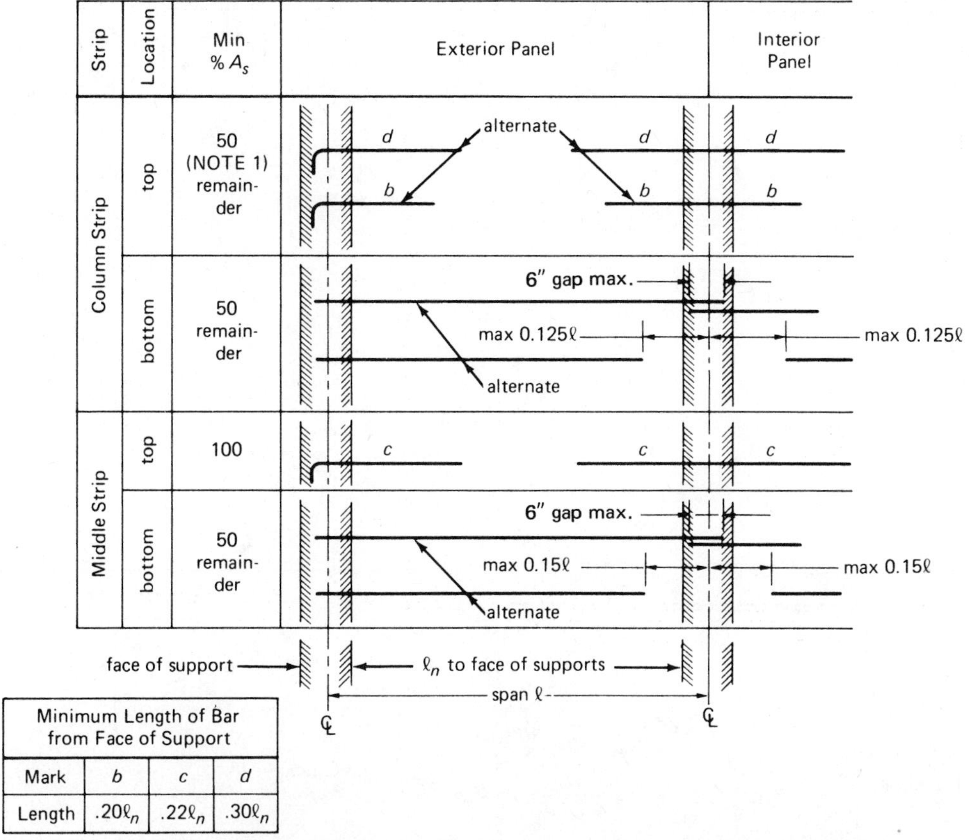

Minimum Length of Bar from Face of Support			
Mark	*b*	*c*	*d*
Length	$.20\ell_n$	$.22\ell_n$	$.30\ell_n$

Fig. 23-35 Code requirements for length of straight bars (without drop panels).

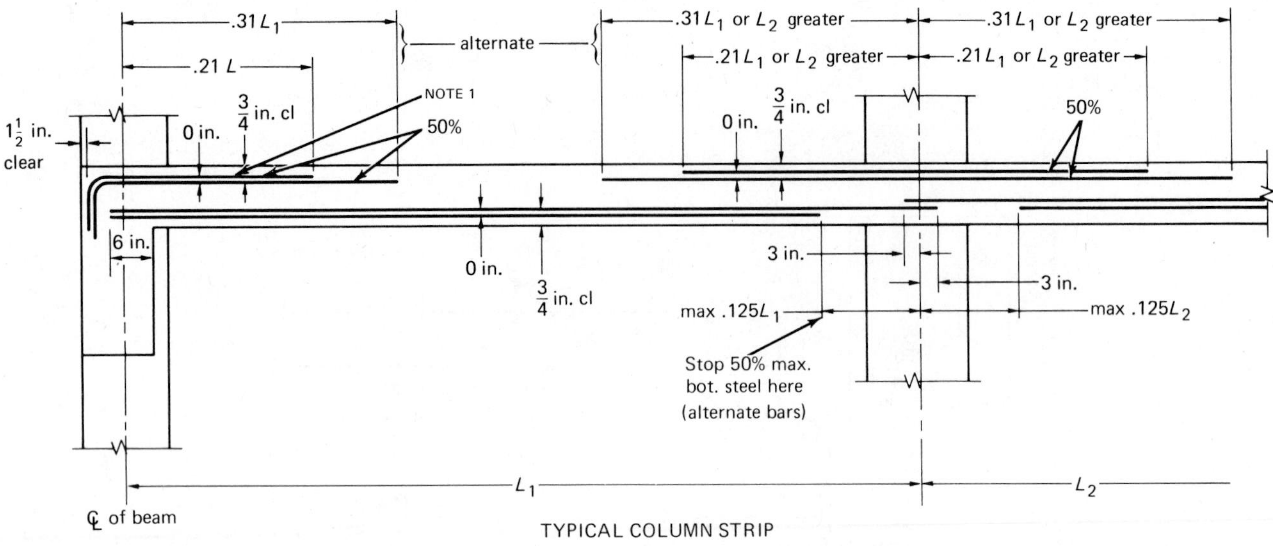

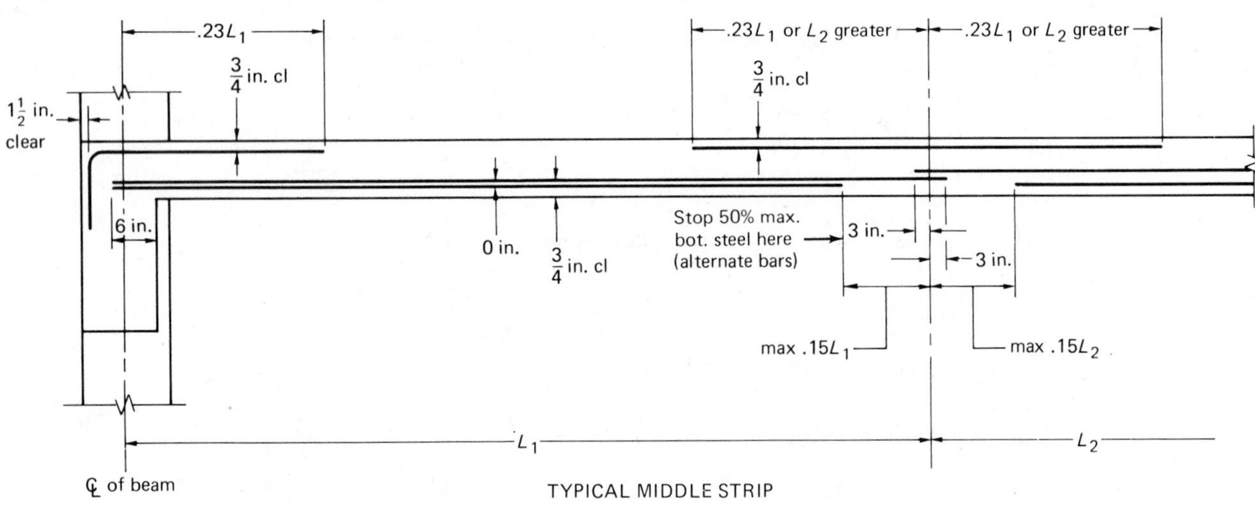

NOTE 1: At exterior columns place half of the column strip top bars within middle third of the column strip.

Fig. 23-36 Bar length details for flat plates.

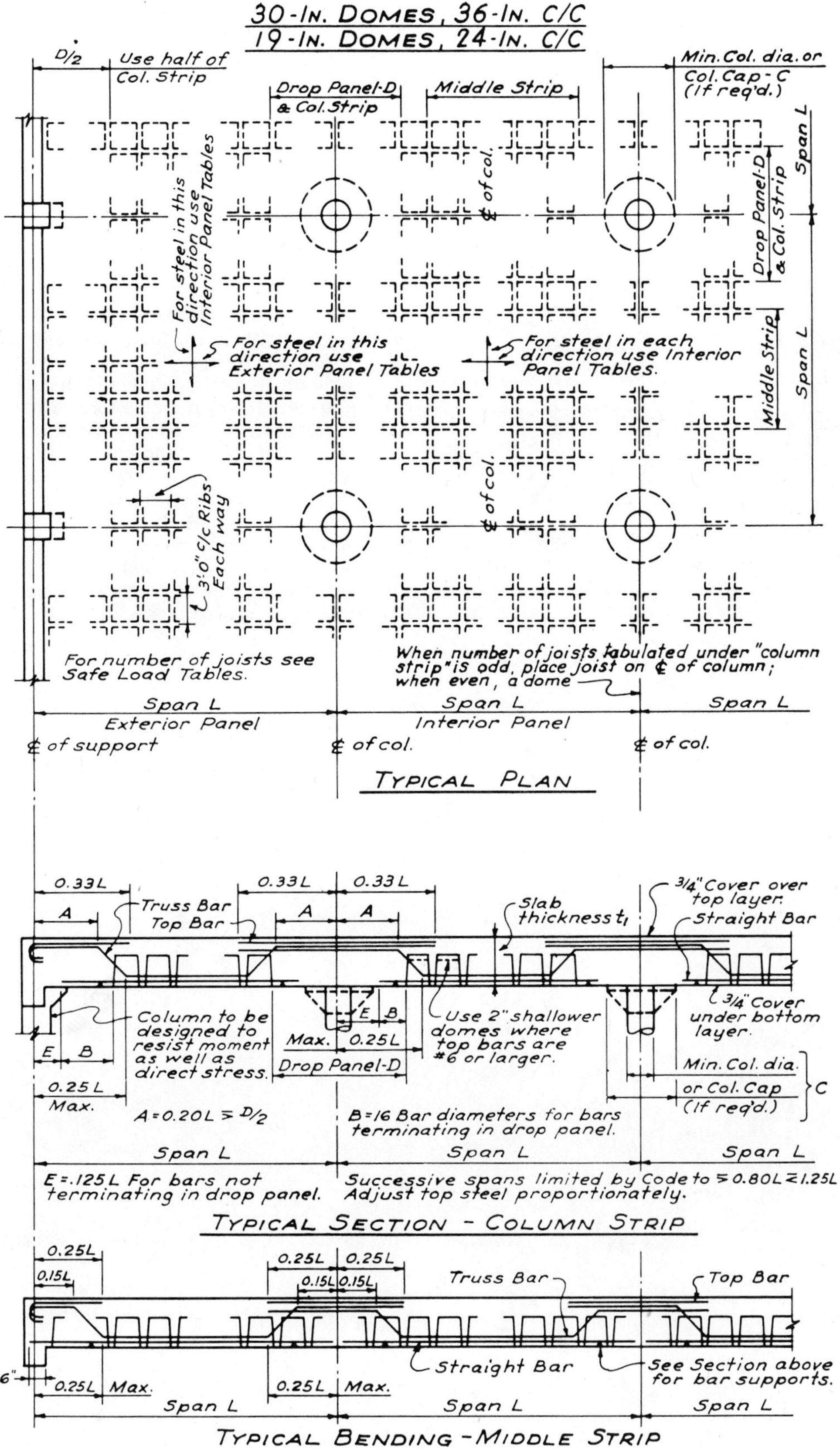

Fig. 23-37 Two-way waffle flat-slab.

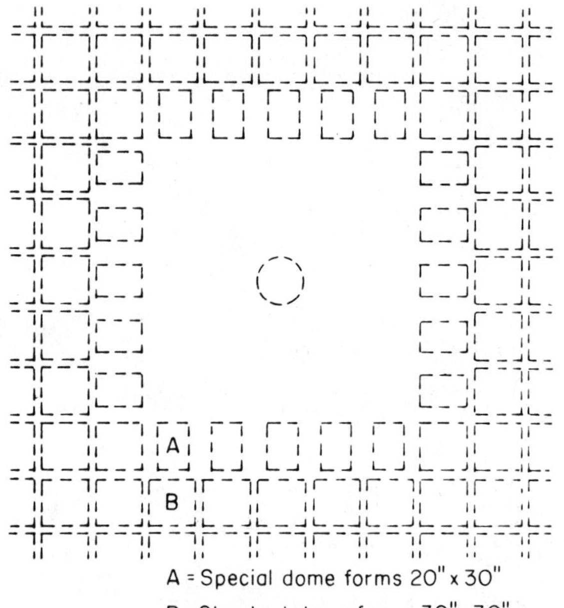

Fig. 23-38 Use of special forms around solid area. (Narrower to widen joist for shear, or shallower to thicken top slab for clearance and cover on top steel.)

where K' is the modified constant, l_1 the column centerline span dimension, K is the tabulated constant (Fig. 23-35), l_c is the clear span, and c is the average column dimension. Assuming a 20-ft typical span and average 12 in. square columns, the typical detail on the structural drawing would appear as in Fig..23-36.

In some cases it may be more convenient not to introduce column and middle strips formally on the plans. In such cases a uniform reinforcing mat over the entire slab is used for the bottom reinforcement and loose bars added on column lines only where necessary; the negative reinforcement is handled in the customary way with column and middle strips laid out on the plan view.

Waffle slabs, although designed as flat slabs, are really made up of joists in two directions and are usually scheduled. (See Fig. 23-37.)

In addition, typical column and middle strip section are shown to indicate extent of reinforcing and location of bend point (truss bars). Truss bar designs generally require additional loose bars in the column strip at the top over the column head. When the top bars (loose or truss) in the column strip extend beyond the solid head, the designer must check to see that the bars have proper clearance at the top of the domes in both directions and call for shallower domes around the solid head if necessary, or when shear at edge of solid head area is critical, narrower domes may be used to increase joist width (Fig. 23-38).

24

Construction Methods and Equipment

J. F. CAMELLERIE*

24.1 CONSTRUCTION ORIENTATION OF DESIGN

The responsibility of the engineer, as well as the architect, is to provide a building or installation to meet the client's need in the required time and without undue cost. To design beams, walls, and footings sufficient to take the anticipated loads is only the beginning and most elementary part of the engineer's responsibility; unintelligent computers can perform much of this function. The science of engineering also includes combining the structural elements with each other, with the environment, and with the attitudes, capabilities and tools available to the constructors. Structures must be built by men who cannot be controlled like robots, and they must be built in a natural environment that is not yet subservient to man.

Historically, man learned to build before he learned to draw or to compute. In our highly specialized society we have developed a breed of engineers who can compute and draw, but have only hazy notions of how to build. Structures designed by such engineers are 'successfully' completed only after much anguish and modification in the field and at unnecessary cost in time and money to the owner and society.

The engineer must orient his design to the men and machines who will build the structure and to the materials, times, and environment in which it will be built. He must not complete the design on the assumption that the contractor will have the construction knowledge which he himself lacks, or that changes such as going from precast to cast-in-place concrete, or vice-versa, are minor field modifications. Nor is the engineer's responsibility finished once the building department or other regulatory agency has approved the drawings for construction. The construction of the structure is, in fact, the implementation of his design. He must participate in this implementation, not merely as an inspector assuring compliance to the letter of the contract, but as a cooperative, guiding intelligence taking advantage of, or overcoming, unexpected developments as the work progresses.

Some of the specific aspects involved in the construction orientation of design will be discussed briefly in the pages that follow, but the concept and philosophy of the engineer and architect as 'master builders' underlies all of the specific facets that follow.

24.1.1 Simplicity and Repetition

Simplicity and repetition are much to be desired in design. There is a natural tendency to make each beam and column the smallest size that meets stress requirements. This procedure will result in maximum economy of material but will

*Principal Engineer, Camellerie/Bialkowski Association, Huntington, New York.

certainly not result in minimum cost of the finished product. A great number of sizes increases the time and cost for shop drawing production, purchasing, bar fabrication, form construction, erection and stripping, rebar placing, and job inspection. Reuse of materials and subassemblies are prohibited; errors and antagonism are encouraged in the field.

Considerable expenditure of materials is economically justifiable in obtaining simplicity and repetition of design; this standardization will allow the use of assembly-line methods, prefabrication, subassemblies, specialized equipment, and reuse of more expensive materials, resulting in faster construction and lower cost. Figure 24-1 shows an actual forming study used to guide design.

24.1.2 Quality Control

Quality control, as related to field conditions, must also be integrated into the design. There is a great tendency to design walls to the least possible thickness and to obtain the required strength by using high-strength concrete and very high percentages of steel. In the first place, placing concrete in a 12-ft lift of 6-in. wall with 4% steel is like passing the concrete through a sieve. The mix will be completely separated by the time it reaches the bottom. To help matters along, the steel will physically prevent the proper use of vibrators. Structural draftsmen should be required to show reinforcing steel diameters to scale and with proper

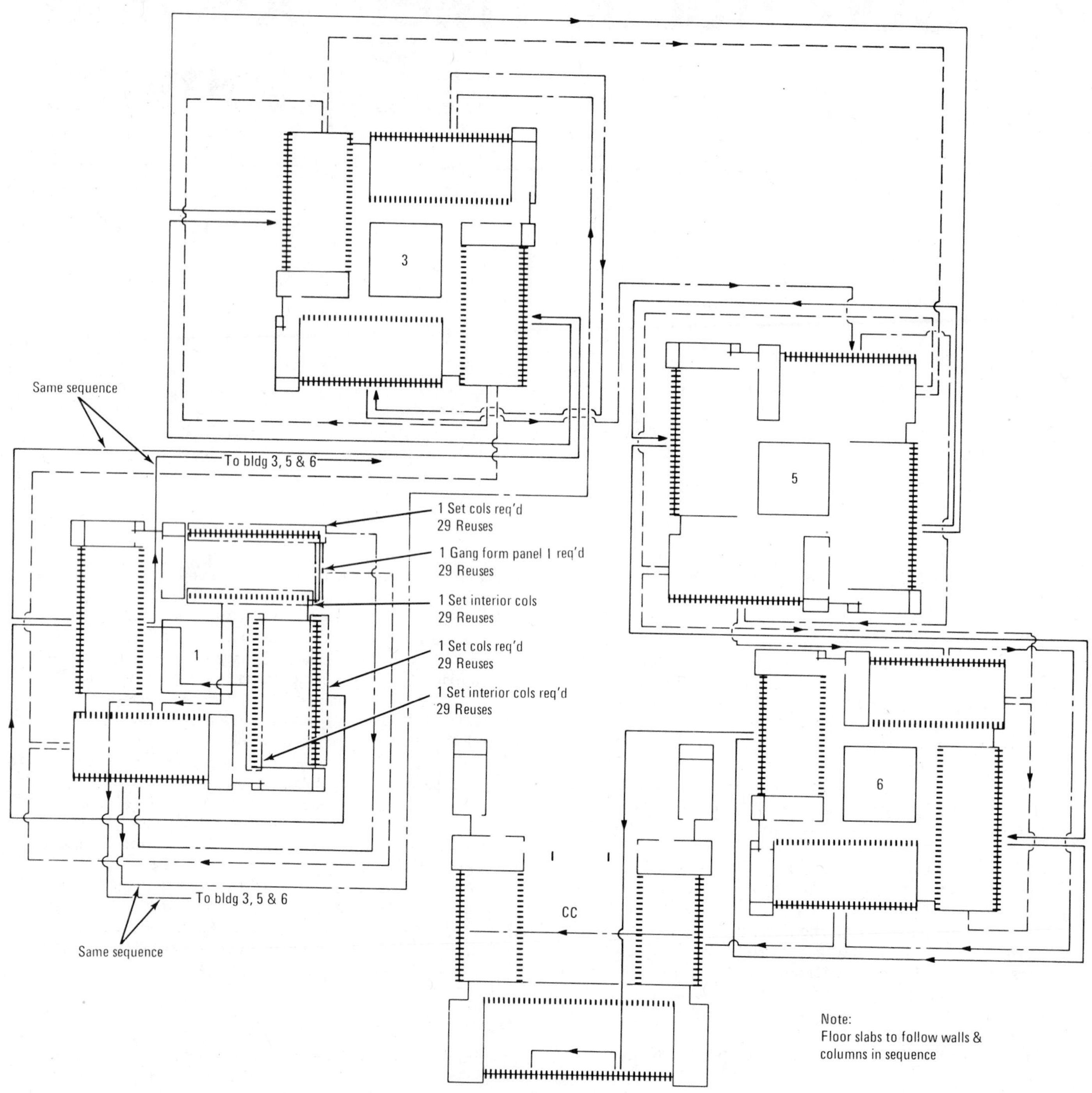

Fig. 24-1 Forming sequence and reuse study. (*Courtesy of C.I.P.*)

bend radii rather than the sharp pencil lines and right angle bends so neatly appearing on the drawings.

In addition, high-strength concrete is faster setting; on a hot day it is very difficult to place this concrete before it takes initial set. It is not an ideal concrete to run through the steel 'sieve'. The physical problems associated with strength of concrete, percentage of steel, height of lift, and ambient temperatures must all be considered in the structural design or the strength requirements will be satisfied only in theory but not in actuality.

The characteristics of concrete specified must be considered in relation to the type of structure being cast, climatic condition, placing techniques, method of mixing, and distance of transport from mixer to form. An increase of cement content will increase the strength of concrete in laboratory-cured cylinders. The effect may be opposite for the same concrete in the field, caught in downtown traffic, retempered with water, dumped into hoppers, raised up on material tower, dropped into a buggy, wheeled a few hundred feet, and finally 'placed' in a form. The designer's mind must be diverted from that 28-day moist-cured cylinder in the lab to the concrete that will be physically in his structure. A preoccupation with low slumps can also be very dangerous if not considered in the light of field placing.

24.1.3 Tolerances

Tolerances are a critical segment of design. It is absolutely essential that the design be such as to allow for size variation of members as fabricated. Many engineers have never taken the trouble to add up the various tolerances allowed by ASTM for steel beam fabrication. They include camber, sweep, rolling, out-of-square, web off-center, and length tolerances. A 40-ft beam set perfectly to plan and elevation at its ends can be 0.75 in. out of plan and elevation at the center, be 0.625 in. too long, and still be within acceptable ASTM tolerances. It is also interesting to note AISC tolerance for plumb in a steel building. This tolerance is 500 to 1 or 1 in. in 42 ft, with a maximum of 3 in. away from and 2 in. toward the building line for buildings 36 stories high and over. This tolerance is an erection tolerance and is additive to the fabrication tolerance.

The *ACI Standard Recommended Practice for Concrete Formwork* (ACT 347-68) recommends a tolerance in plan of 0.5 in. in any 20 ft and 1 in. for structures 40 ft and over. These tolerances are quite optimistic for tall buildings when one compares them to AISC tolerances and to tolerances actually obtained on concrete buildings. The same *ACI Standard* suggests a tolerance of 1 in. in 50 ft without height limitation for slip-formed concrete. In the opinion of the author, this standard is more practical except that total tolerances may be limited to 2 or 3 in. for very tall buildings.

To illustrate the effect of tolerance on design, we may imagine a 40-story building with steel-column framing on the outside and a concrete core at the center. Structural steel beams span from columns to concrete core walls, a span of 40 ft. The tolerances at the top of the building may add up as follows:

℄ steel columns	3 in. toward building ℄
Fabrication tolerances for columns	0.75 in. toward building ℄
Beam length tolerances	0.625 in. too long
Concrete core tolerance	2 in. away from building ℄
Span may be actually	6.375 in. shorter than theoretical

The span can also be 5.375 in. longer than theoretical and still be within acceptable tolerances.

Actually this is a very pessimistic and somewhat exaggerated analysis. The author does not suggest designing for a tolerance of 6 in. It is obvious, however, that provision for span differentials of at least 2 in. should be considered.

Structural steel tolerances, plate glass tolerances, and all other fabrication and erection tolerances must be considered in concrete design as the concrete elements must be coordinated with and must accommodate the other structural and functional elements comprising the total structure. In addition to this, let the engineer not forget the dramatic movements that occur in the structure under construction due to sun or wind on one side, or seasonal thermal expansions and contractions.

It is essential that all concrete elements and joints be designed to provide the minimum amount of tolerance and flexibility required to meet the expected field variations. Whenever possible, additional tolerance should be provided to meet unexpected problems and to allow for human error. An engineer must first determine that the tolerances he specifies are actually attainable in the field. No amount of construction care and no type of legal coercion can result in a tolerance beyond the capabilities of Man and his tools working in a 'cruel' world. Beyond this, fabrication and erection costs are directly related to the tolerance refinement required. It is therefore economically desirable and practically safer to design with maximum flexibility and to keep tolerance requirements as liberal as possible.

24.1.4 Design to Suit Available Equipment

An important construction aspect that must be considered in design is the increased capabilities offered and the limitations imposed by available construction equipment.

Availability of equipment should mean that the equipment is not only available in time for use on the job, but also that site conditions are such as to make the equipment usable at all required locations; that there are no government ordinances prohibiting or limiting the use of the equipment; that there is acceptance by local unions; and finally, that maintenance and spare parts are available. The arsenal of construction equipment is increasing daily in variety, capacity, and reliability. There is also great capability for designing and fabricating equipment as required for special operations. This potential should certainly be considered in the design concept since it is now possible to produce structures which were economically impossible to produce a few years ago.

Great as this equipment arsenal is, the limitations are very real and must be carefully investigated as many an engineer discovered when his structure went out for bids. Confined urban construction sites on main traffic arteries require the most careful methods and equipment studies.

24.1.5 Labor Considerations

In a democratic country with well developed trade unions, the strength, attitude, political influence, and business affiliations of the unions in any particular area must be carefully considered in the design of a project. In countries of strong government control, appraisals of official assistance and/or hindrance must be made.

In all areas, the labor market must be studied as to availability of men in the various trades, skill level of the labor, prevailing wages, fringe benefits, shifting and overtime practices, and competition for labor with other projects. The possibilities and economics of importing skilled and unskilled labor may also be considered. Such studies of the

TABLE 24-1 Contractor Reaction Summary by Region Progress City, A.S.

Contractor	*Interest by Region*									*Remarks*
	I	*II*	*III*	*IV*	*V*	*VI*	*VII*	*VIII*	*IX*	
Contractor "A" New York, N.Y.	X	X	X	X	X	X	X	X	X	
Contractor "B" Pittsburgh, Pa.	X	X	X	X	X	X	X	X	X	
Contractor "C" Kansas City, Mo.	X	X	X	X	X	X	X	X	X	
Contractor "D" Buffalo, N.Y.		X								Both slip form and precast
Contractor "E" Chicago, Illinois	X	X	X	X	X	X	X	X	X	
Contractor "F" Minneapolis, Minn.	X	X	X	X	X	X	X	X	X	
Contractor "G" Philadelphia, Pa.		X								
Contractor "H" Hutchinson, Kansas	X	X	X	X	X	X	X	X	X	
Contractor "J" Dallas, Texas				X		X	X			
Contractor "K" Philadelphia, Pa.	X	X	X	X	X	X	X			
Contractor "L" Spokane, Washington								X	X	

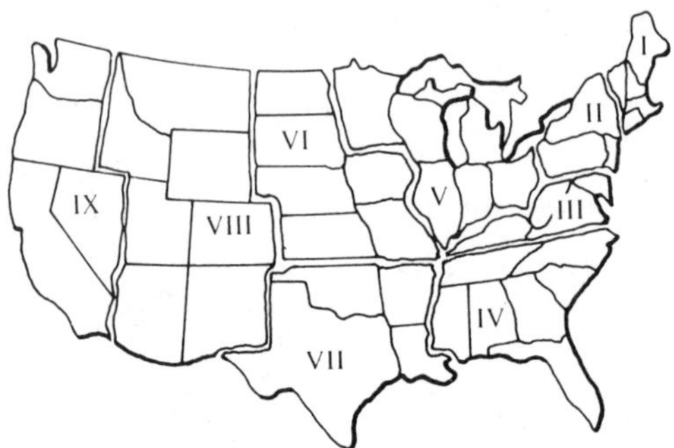

labor market may very well effect design decisions as to materials, type of structure, construction techniques, sequence of construction and starting and completion dates.

The contractor is normally the person or organization that brings the equipment and labor together to actually perform the work. A check is advisable to determine the attitude of local contractors to the work in general and to the techniques in particular. A willingness and capability of specialist contractors to work in the specific area must also be looked into. There are many ways of doing this and none of them are guaranteed as to accuracy. Nevertheless, this type of study is very valuable. Table 24-1 shows a typical study made to determine contractor reactions to a specific project and to various techniques. Table 24-2 shows a typical study to determine availability to specialist contractors by region.

24.1.6 Construction Schedules

Construction schedules are often academic, reflecting the wishful thinking of the designers and completely ignored by the contractors. Schedules should be carefully arrived at considering construction methods, operation sequences, and climatic handicaps. There is no point in setting a schedule if the schedule is not implemented by meaningful contractual penalties and rewards. Completion behind schedule may actually cost the owner considerable sums in rental, force him to miss a seasonal market opportunity or an educational semester or get him into serious problems in the financing of the project. On the other hand, early completion may increase the cost of the project because of limiting reuse of forms and equipment and possibly involving payment of premium wages to labor. The cost value

TABLE 24-2 Contractor Reaction Summary by Speciality

	Interest					*Preference*				*Increase in Contingency*			
Contractor	*General*	*Conventional Forming*	*Precast*	*Slip Form*	*Prestress*	*Conv. Form*	*Precast*	*Slip Form*	*Prestress*	*Pre-Cast %*	*Slip Form %*	*Pre-stress %*	*Remarks*
Contractor "A"	Great	Yes	Yes	Yes	Yes	–	–	√	–	–	–	–	
Contractor "B"	Some	Yes	Yes	Yes	Yes	–	–	–	–	5	10	5	
Contractor "C"	Great	Yes	Yes	Yes	Yes	√	–	–	–	2	5	5	
Contractor "D"	Some	Yes	Yes	No	No	√	√	–	–	3	5	5	
Contractor "E"	Some	Yes	Yes	Yes	Yes	√	–	–	–	2	6	3	
Contractor "F"	None	–	–	–	–	–	–	–	–	–	–	–	
Contractor "G"	Some	Yes	Yes	Yes	Yes	–	–	–	–	–	–	–	
Contractor "H"	Some	Yes	Yes	No	Yes	√	–	–	–	5	10	5	
Contractor "I"	Some	Yes	Yes	Yes	Yes	√	–	–	–	10	10	10	
Contractor "J"	Great	Yes	Yes	Yes	Yes	√	–	–	–	5	5	8	
Contractor "K"	None	–	–	–	–	–	–	–	–	–	–	–	not active

of early and late completion should certainly be considered in the early design stages.

24.2 SCHEDULING AND CONTROL OF CONSTRUCTION

24.2.1 Subdivision of Construction into Phases

Small projects automatically fall into phases by virtue of the necessary sequences of operation and the trades involved. Large, complicated projects have more problems, more possible choices of phasing, and more engineering decisions to make. The major considerations effecting the phasing of the job area follow:

a. availability of funds
b. methods of financing
c. capability of contractors
d. impact on labor market
e. fabrication time
f. critical prerequisite sequences
g. interference between concurrent phases
h. use of equipment
i. weather considerations
j. economic and use alternates to determine completion date
k. design time

On very large projects to be constructed for government agencies or large private institutions, fiscal policies often impose very rigid limitations on how much of the total commitment will be made available in each fiscal year and usually require that the money be spent during the year that it is budgeted. Failure to spend the money in time could mean loss of the money.

Jobs financed by private money, such as bank or insurance company investments, often dovetail the construction investment with other investments. In addition, payments of funds for construction are often tied into a series of completions. Obviously the construction schedule must be tied into the financial aspects of funding.

Some projects are of such a size that few, if any, contractors have the financial and personnel capacity to undertake them. Those contractors physically capable of doing the job may feel it is too much of a risk to take on one project alone. They would rather divide that amount of risk among three or four different projects. Too large a project may also drain the labor market and cause 'labor piracy', inefficiency, and premium cost. In order to obtain enthusiastic bids in sufficient numbers, large construction jobs are often divided into several smaller sections and separate bids solicited for each phase.

In solving the above problems, construction management problems are created which must be skillfully solved. Those operations which are prerequisite to each other must be completed before the next operation is started. Parallel and supporting operations must be phased in so as to be completed without holding up the critical work or adding time to the intermediate completion dates. Certain items that require long fabrication and/or delivery dates may have to be ordered very early and contract documents must allow and insure such early purchases.

The most popular tool for controlling and dovetailling these operations is the critical-path chart. By showing all planning, procurement, fabrication, construction, and testing operations as vectors to a time dimension, these operations are so arranged that all operations follow immediately after prerequisite operations and as concurrently as possible. When properly done, one can tell the minimum construction time required, critical schedules that must be held, slack time in operations not on the critical path, effect of speeding certain operations, etc. This chart is no better than the input it receives, the contract documents governing the work, or the construction managers implementing it.

Design is prerequisite to all operations. Usually the time required for completion of the entire design will set back the construction schedule excessively. It is therefore necessary to carry out the design concurrent with construction with only a limited lead time. This necessitates dovetailling the design into the critical-path chart and following design sequences catering to construction schedules rather than those sequences more natural to design methods.

24.2.2 Dovetailing of Operations

Once a job is divided among several contractors, each having his own interests to promote, it becomes necessary to schedule and to cover the work by document and field management in such a way that interferences between operations are eliminated or minimized. There are, also, the interests of the client, such as an airport authority, which must maintain operation during construction, transportation and municipal agencies responsible for the streets and highways which the contractors must use and, of course, the neighbors.

If the interaction of these problems is left to be solved "in the field," the result will be litigations, delays, and excessive extra costs. The architect or engineer must plan the construction procedures in considerable detail, with the help of construction consultants, if necessary. He must indicate in the contract documents

a. what site areas will be available, and when;
b. what access roads are available, and when;
c. what limitations exist on hours of work;
d. what limitations exist on noise and other by-products;
e. what equipment may be used, where, and when;
h. what equipment may not be used;
i. what cooperation is required to others;
j. what access is required to others;
k. firm completion dates with penalty;
l. safety requirements for public and work force;
m. protection for existing installation and installation by others: and
n. predictable delays or discontinuation of the work.

24.2.3 Common Use of Site and Equipment by Contractors

It is almost always necessary, especially in vertical construction, that contractors use the same site area on the job. Hopefully this can be managed in a chronological order; often it is concurrent or at least with a time overlap. This division by time and area must be clearly defined, taking into consideration the realistic needs of the contractors involved. Very often equipment such as cranes, hoists, personnel elevators, utility lines, and permanent installations temporarily used for construction are required by more than one contractor in sequence, or concurrently.

If each contractor can have his own equipment fully under his own control, this is the ideal condition. Often space requirements prohibit this. There is obviously an unnecessary cost which always accrues to the owner if each contractor erects and dismantles the same tower crane on the same spot as he performs his particular phase of the work. Economy and site efficiency can best be served by specifying equipment capable of performing all phases and operations of the work, scheduling the hours of use to the contractors involved, and arranging for an equitable sharing of cost through a rental or allowance arrangement.

24.2.4 Good Administration Procedures and Payment Policies

In Section 24.2.1 above, we touched on the financial capabilities of contractors. Obviously, the financing required to do work for a client who pays promptly as the work is completed is not as great as the financing required to do work for a client who withholds 10% of all payments until total completion, delays paying large vouchers on the basis of small disagreements, and generally has administration procedures that take months to process the simplest unquestioned voucher. Such clients eliminate many competent construction firms from bidding, encourage brokers whose staff is composed of lawyers but few, if any, construction men, and definitely incurs the cost of financing which the contractor passes on. Very often, such organizations also fail to make decisions on problems holding up work in the field. Architects and engineers should make every effort to promote contract documents that insure reasonably quick payment for work performed, rapid approval of shop drawings, and helpful inspection procedures. Good administrative procedures attract better contractors, improve their cooperation, lower costs, and help meet schedules.

24.2.5 Inspection

Inspection is more difficult and more critical than ever. The industry is using new materials, new techniques, higher allowable stresses, and lower safety factors (ultimate strength design). There is more room for error and less allowance for it. All too often, the only qualification for an inspector is that he is willing to work for a very low salary. This is a very dangerous procedure for modern sophisticated construction. Inspectors must have some training in construction technology, must be given special instruction by the architect and engineer in charge, and must have open lines of communication to their superiors and to the contractor.

The design architect and engineer must visit the job often to check on and to instruct his resident staff. In some instances, when specialized knowledge is not available in-house, it is necessary to obtain the field assistance of construction consultants, manufacturers' technicians, or engineering specialists.

The serious inspector will spend any free time he gets on the job studying the plans and specifications and reading relevant technical books and papers. A little homework would not hurt either. But remember, if this type of intelligent conscientious person is required on the job, the remuneration must be commensurate. Very often inspection degenerates into enforcing the letter of the law without regard to the way in which the instructions of the contract documents are effecting the work. The inspector is, and must be, a policeman. To be really effective however, he must know the limits within which he may make his own decisions and must know when to request help from the design engineer. In his turn, the design engineer must clearly instruct his inspector in the field as to what is inflexible, what is flexible, and to what extent, and when to call for help.

One item that receives very little attention (except in cold weather) is concrete temperature. In the author's opinion, concrete thermometers should be commonplace on the job and records of temperatures and slump at the truck discharge and in the forms should be kept all year round. Also keep in mind that slump control at the truck discharge does not necessarily result in proper slump at the placement area.

A few representative hints for new inspectors:

a. Keep a daily diary.
b. Check grade marking on rebars, bolts, wood, plywood, and accessories.
c. Check concrete delivery tickets carefully to insure truck is at right job.
d. Prevent unauthorized persons from requesting and getting water added at truck mixers.
e. Check that concrete is not kept in truck mixers too long (check delivery ticket).
f. Check that concrete is not moved laterally by vibrators or dropped more than 5 ft vertically without drop chutes or elephant trunks.
g. If vibrators are allowed to sink into concrete under their own weight without additional force, the vibrator penetration will be about right.
h. Check curing procedures to insure no misses and no discontinuity.
i. Check that no debris (wood shavings, oil, coffee containers, bottles, etc.) falls into concrete or is left in forms before filling.
j. Check forms for loose ties, improperly wedged shores and/or undue movement during placing.
k. Spot check layouts and elevations.
l. Check that screeds, rebars, and prestress conduit are properly secured against movement during placing.
m. Be especially careful about concrete cover on reinforement.

n. Check bent rebars, especially smaller sizes with high yield points, for cracks.
o. Check rebar laps and connections.
p. Check for broken conduit in posttension work.
q. Beware of contractors buying coffee at critical times.
r. In fact, be careful of diversionary tactics always.

24.2.6 Climatic Considerations

Certain operations, such as steel erection, can be carried out in any season without undue cost or delay. Other operations, such as concreting, are vulnerable to both extreme heat and cold. Excavation and foundation work is vulnerable to heavy rain. As obvious as this seems, many construction schedules ignore the seasonal effects and start 'as soon as possible' and then allow no additional time for the resulting handicaps. In some instances, starting the job sooner, delaying the start, or even closing down the job for a period may result in considerable savings in cost—possibly without changing the completion date. Different systems, materials and/or construction techniques should be considered in order to 'beat the season', 'close in the job', or otherwise adapt to the adverse conditions. Often, the cost of overtime is justified in order to avoid the cost of inefficiency due to inclement weather.

24.3 CONCRETE CONSTRUCTION TECHNIQUES

As long as men have been on earth, they have been devising ways of moving things, raising them, connecting them together, and generally building things where they wanted them to be. In less organized societies each community devised and passed on from man to man 'tricks of the trade.' In our times we have great capabilities for the exchange of knowledge, equipment, and materials, bringing these together from near and far, and erecting structures in the most difficult locations. Construction technology is the science of construction involving the use of the most advantageous materials and equipment available, including the necessary planning, preparation and execution. Some of the more sophisticated techniques commonly used in this country are discussed in the following paragraphs.

24.3.1 Precast Concrete

Concrete may be cast in place in its final shape and location, or it may be precast, i.e., cast somewhere other than where it will be used and ultimately assembled in the final location (see Fig. 24-2). There are several advantages in using precast sections, the most significant being the application of factory methods to an otherwise custom-type industry.

Precasting concrete members and panels at central plants permits the use and reuse of highly productive automated equipment, weather protection during casting, stabilized experienced labor, concurrent operations removed from the site and from each other, excellent quality control, simplification, and standardization. The last item, standardization, is the key to ultimate success in the use of precasting. The industry is in the process of developing off-the-shelf items such as single-tees, double-tees, highway girder sections, slabs, and panels. As the various manufacturers coordinate their operations more and more closely together, a catalog similar to the AISC catalog of available shapes will develop.

Precasting can be accomplished at plants as described above and shown in Fig. 24-3, or at the job site. On large jobs, when economic considerations so indicate, a temporary plant can be set up at the job site. This casting yard may be equipped with very sophisticated equipment or may utilize very simple forming and casting procedures.

Because precasting often enables vertical surfaces to be

Fig. 24-2 Placing precast concrete.

cast in a horizontal position, wall panels of very excellent aesthetic qualities can be produced in unlimited variations. These panels may have any type of troweled or scarified finish, molded designs, exposed aggregate finish, or open-space filligree effects. They will be free from tie holes and air bubbles. It is more difficult to obtain the same quality with vertical in-situ casting. In planning precast construction, the following aspects must be carefully studied:

proximity of precast suppliers
transportation costs
access to site
equipment required to erect sections
possibility of on-site precasting
use of standard available sections
simplicity of sections for manufacture

Fig. 24-3 A precasting yard.

largest pieces manageable
simple but effective jointry
lifting devices and strength during placing
adequate tolerances
weatherproofing of building

The four most common pitfalls are

1. difficulty or impossibility of placing due to size, weight, site obstructions or height.
2. difficulty of placing due to insufficient tolerance and/or inflexible jointry.
3. intricate jointry requiring extensive welding or difficult grouting.
4. failure to specify lift procedures and to design for loads that occur during lifting.

Summarizing the advantages of precast concrete, they may include lower cost, quicker construction, early structural strength, less congested construction sites, high quality, and relatively easy rejection of imperfect pieces. The disadvantages may include, difficult and expensive jointry of the component parts, heavy transportation costs, weight limitations, and labor jurisdictional problems.

24.3.2 Prestressed Concrete

Prestressed concrete is widely used in this country in both cast-in-place construction (posttensioning) and in precast construction (pre-tensioning). It allows longer spans without excessive depths, more crack-free structures, lighter members, and better joining of concrete subsections of buildings, whether cast-in-place or precast. In some cases prestressing will allow the construction of structures that otherwise could not have been built. In all cases, it takes advantage of the higher strength steels and concrete available thereby effecting reductions in concrete and weight and the costs related to handling the plastic or precast concrete.

Prestressing is a technique that is well mated to precasting and helps to solve many of the weight and jointry problems that militate against precast concrete. By the same token, the casting yard operation used for precasting brings out the best economics in the use of prestressed concrete. The precasting yard lends itself to the economical use of sophisticated tensioning equipment and anchors, high strength concrete, and accelerated curing.

One difficulty that is associated with the use of prestressed concrete is deflection control during stressing and during the later life of a member. It is difficult to accurately estimate the amount of deflection to be expected at various ages of a prestressed member, since the deflection depends on a varying modulus of elasticity, shrinkage and creep of the concrete, relaxing of the steel tendons, etc. This may result in connection difficulties and in unsightly malclosures or arch effects. The use of shear connections is often advisable.

One major problem in post-tensioning work is the construction time lost when the tendons are tensioned in place. Posttensioning cannot be effected until the concrete has reached sufficient strength. Unless the construction sequences embody enough time for this concrete strength gain, the successive delays will lead to serious lengthening of the construction time and high project costs.

Another problem, particularly on high rise building, is the effect of posttensioning on other components of the structure. For instance, the posttensioning of floor systems may lead to serious flextural stress in the columns. This possibility must always be carefully studied.

24.3.3 Slip-form Construction

Vertical elements such as walls, columns and piers lend themselves to construction by the slip form method. In this method a form 3.5 to 6 ft high is erected and coupled to a jacking device capable of lifting the form together with working decks and scaffolds. The advantage of this system lies in the fully-automated continuous reuse of the same form without stripping or re-erecting. There is also a considerable economy in the reduction of scaffolding, automatic templating of rebars, and frequent elimination of hoist towers.

Probably the most important advantage is the speed of construction which is a floor-a-day on daily slides, and up to 24 ft per day for round-the-clock slides. The low cost per square foot contact area of the forms is offset by the premium time associated with slip-forming. Even on daily slips, a skeleton crew on overtime is required to clear the forms.

Like all the other techniques discussed here, slip forming requires that the design be oriented to the construction method. Most often slip-form operations are performed on structures not specifically designed for this method with the contingent loss in speed and economy. Those items which contribute most significantly to the economy of this method are

a. maximum concrete yardage per foot of height (5 to 20 yds/ft and more is in competitive range)
b. maximum height of structure
c. minimum number of inserts and openings
d. maintenance of uniform wall dimensions throughout height
e. willingness of trades to accept shifting for round the clock operation

24.3.4 Tilt-up Construction

Wall elements of limited heights may be economically and quickly constructed by the tilt-up method. This method of construction is actually a site precasting operation. A form is erected on the ground or floor of the structure close to the final position of the element in the wall. The panel is then cast in a horizontal position, and once the concrete has acquired sufficient strength, it is tilted up into the vertical position usually by rotating it upward about its bottom edge. These wall panels are then connected to each other or to columns to form a continuous structure. The panels may be load-bearing or non-load-bearing.

The obvious advantage of tilt-up construction is the simplicity of the formwork, the reuse factor, the absence of form ties, the ease of placing concrete, openings, inserts and utilities, and finally, the relatively simple equipment required to raise it into position. The columns may be cast-in-place concrete, precast concrete, or steel with many possible arrangements. Some of the economic advantage of this system is lost in the teaspoon-type concrete placing required for casting the columns. Precast columns are also somewhat expensive because of the fussy jointry. However, the trend in this type construction is towards columnless walls in which the entire panel acts as a column, or towards incorportaion of the column into the panel during horizontal casting.

Tilt-up construction has been used mainly in the construction of one-story buildings having extensive wall areas; in this field the economy is quite good. Clear working space is of course required all along the wall. Tilt-up construction can also be used in multistory structures by casting and tilting-up the wall panels on the floor slabs. A very interesting modification of this method was used in the Westyard Distribution Center in New York City (Fig. 24-4). A very large number of panels were required. With three sides on heavily trafficked streets and one side over the entrance to the Lincoln Tunnel, the parking of trucks and the stationing of cranes around the building was a serious problem. A movable tilt-up casting table was designed which ran on rollers, cast

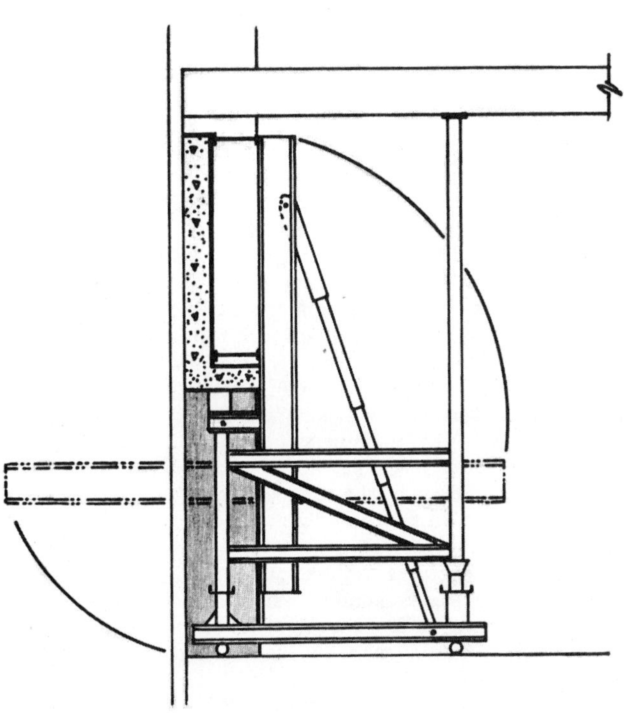

Westyard Distribution Center

Casting Operations

1. Adjustable tilt-up form is rolled into position.
2. The form is tilted to the vertical position for fit to the concrete frame.
3. It is then rotated down to the horizontal position for re-bar and concrete placing.
4. The concrete is steam cured overnight and the form with the panel is tilted up into position and the panel bolted to the frame.
5. The form is then moved to the next bay.

Placing Concrete in Forms

Fig. 24-4 Precast, tilt-up construction. (*Courtesy of C.I.P.*)

the wall panels at hip level and then raised them into position using a built in hydraulic system.

24.3.5 Lift-slab Construction

This technique is used for casting several stories of floor slabs at ground level and then jacking them into position at the proper floor height (Fig. 24-5). This system normally utilizes steel columns which are temporarily braced in place. The lift slabs are cast one upon another on top of the basement or first floor slab. A bond breaker is used between slabs; only side forms are required. Structural steel shear heads are cast into the slabs around the columns to provide a lift point for raising into position and to act as the final device for stress transfer from the slabs into the columns. When the concrete in the slabs have reached sufficient strength, the slabs are jacked into position at the proper floor levels following a carefully preplanned sequence.

This technique offers considerable advantages cost wise and time wise when the floor system lends itself to sectioning off into independent wings that can be raised straight up into position. An elevator and stair core is usually used to furnish lateral stability; cast-in-place floor sections being used for connection. This system is somewhat limited as to height; it lends itself very well to medium-rise structures to 10 stories, although buildings up to 20 stories have been built. Lift-slabs also lend themselves very well to posttensioning techniques. All the slabs may be stressed in one operation while they are still at ground level which results in a very economical post-tensioning operation.

24.3.6 Drop-slab Construction

This method of casting slabs is similar and opposite in technique to the lift-slab method. A core is constructed the full height of the building and some form of transfer girders are placed at the top, cantilevering out over the core. Rods are hung from the girders and the slab forms are attached to the rods. Once the slab is placed and has acquired sufficient strength, the form is dropped downward into place for the next floor. This method can be used for high-rise structures, as there is no height limitation. Transfer girders may be provided at several levels in very high buildings. One such system is shown in Fig. 24-6.

The form reuse without the necessity of disassembly and reassembly offers considerable economic advantage. In conventional forming the forms must be disassembled into small units, carried out past the building line, brought up over the cast slab and into the building for reassembly. This is a slow expensive operation. There are however some expenses involved in the drop-slab system which militate against the time- and cost-saving features of this system. These disadvantages include

a. necessity for floor closure strip all around

Fig. 24-5 A lift-slab operation.

b. necessity for waiting for strength build-up
c. difficulties in breaking bond
d. possibility of a tricky re-shore system
e. necessity to bring concrete in from side
f. difficulty in raising girders into position
g. possibility of an expensive core

24.3.7 Composite Construction

This system consists of combining a concrete slab with structural steel beams, bar joists or steel forms. The concrete takes the compression stresses while the steel takes the tension stresses. This system offers good opportunities for economy by using each material in the most efficient manner. Care must be taken during construction, remembering that the steel elements must take deadload plus construction loads without the help of the concrete until such time as the concrete has gained sufficient strength. In addition, the steel sections used tend to be somewhat unstable laterally without the concrete. As a result, additional shoring and bracing is required and adds to the cost along with other special requirements of the system such as placing shear studs. These tend to reduce the savings which accrue from the decrease in concrete and steel quantities. Engineers must be careful to show complete instructions for shoring and bracing on the structural drawings.

24.3.8 Combination of Techniques

Practically all structures are built using a combination of techniques. In the planning and design of a structure, it is not a particular technique that emerges as the most desirable but rather the proper combination of techniques to obtain optimum time and cost economy. In comparing one technique against other techniques usable in the construction of a particular structure, the parameters should not be cost per yard of concrete in place. Rather, one must consider the cost of the entire structure based on combined techniques and including the values of early occupancy, tax and insurance advantages, overhead costs, financing costs, etc.

There is no value in slip-forming walls very quickly if the slab-casting system is not compatible and loses the time and economy gained. On the other hand, it certainly pays to spend a few more dollars per yard for slip-formed walls if the overall floor cycle is reduced, or if significant economies are effected in the floor system. There will be no value in reducing the size of concrete members by posttensioning if the strand-tensioning cycles will result in serious extension of the construction time or inefficient discontinuous use of the labor force. On the other hand, considerable additional cost will be warranted in using prestressing techniques to reduce weight of the superstructure in areas of serious foundation problems. The entire structure must be considered as a whole and the designer must be careful that in solving his problems in one design area, he does not create more costly ones in other areas.

Figure 24-7 shows a system study made for a high rise building. This study included materials, forming system, equipment, hoisting and placing methods and was in sufficient detail to yield reasonably accurate labor, schedule and cost prediction.

Figure 24-8 shows a combination of high structural steel bents, low cast-in-place concrete bents, and precast prestressed single-tees that will be used to support a two-mile long cross-country conveyor enclosed in a light guage metal housing.

Figure 24-9 shows a combination of concrete construction techniques used to erect a concrete building in New York City. Included are a slip-formed core, architectural precast concrete load-bearing column-spandrel units, and a cast-in-place slab-and-girder floor system.

24.4 UTILIZATION OF FORMWORK

24.4.1 Planning

Formwork costs represent the major and most variable portion of the cost of concrete in place, normally 33 to 55%. As there is very little room for savings in the purchase of rigidly-specified concrete and reinforcing steel, the differences in material prices are usually very small between bidding contractors and the engineer's estimate. The placing costs can vary depending on good scheduling, proper crew planning and sequencing, efficient working spaces, and adequate materials handling. All of these items are related to and dependent on forming system selected and on how the this system is implemented. The impact of the forming system is therefore greater than indicated by the 33 to 55% figure noted above.

On concrete structures, the successful bidder is that contractor who has come up with the most economical forming system. Furthermore, the probability of the bids coming in within the engineer's estimate is predicated on a design which allows the contractor to come in with a sufficiently economical forming system. All of this emphasizes the importance of adequate study and analysis of the forming systems available so as to select the most economical system or systems for the particular project, and to do all that is possible to optimize the economy of these systems. This planning is necessary during the design, bidding, and construction phases. Table 24-3 and 24-4 show a typical comparative study.

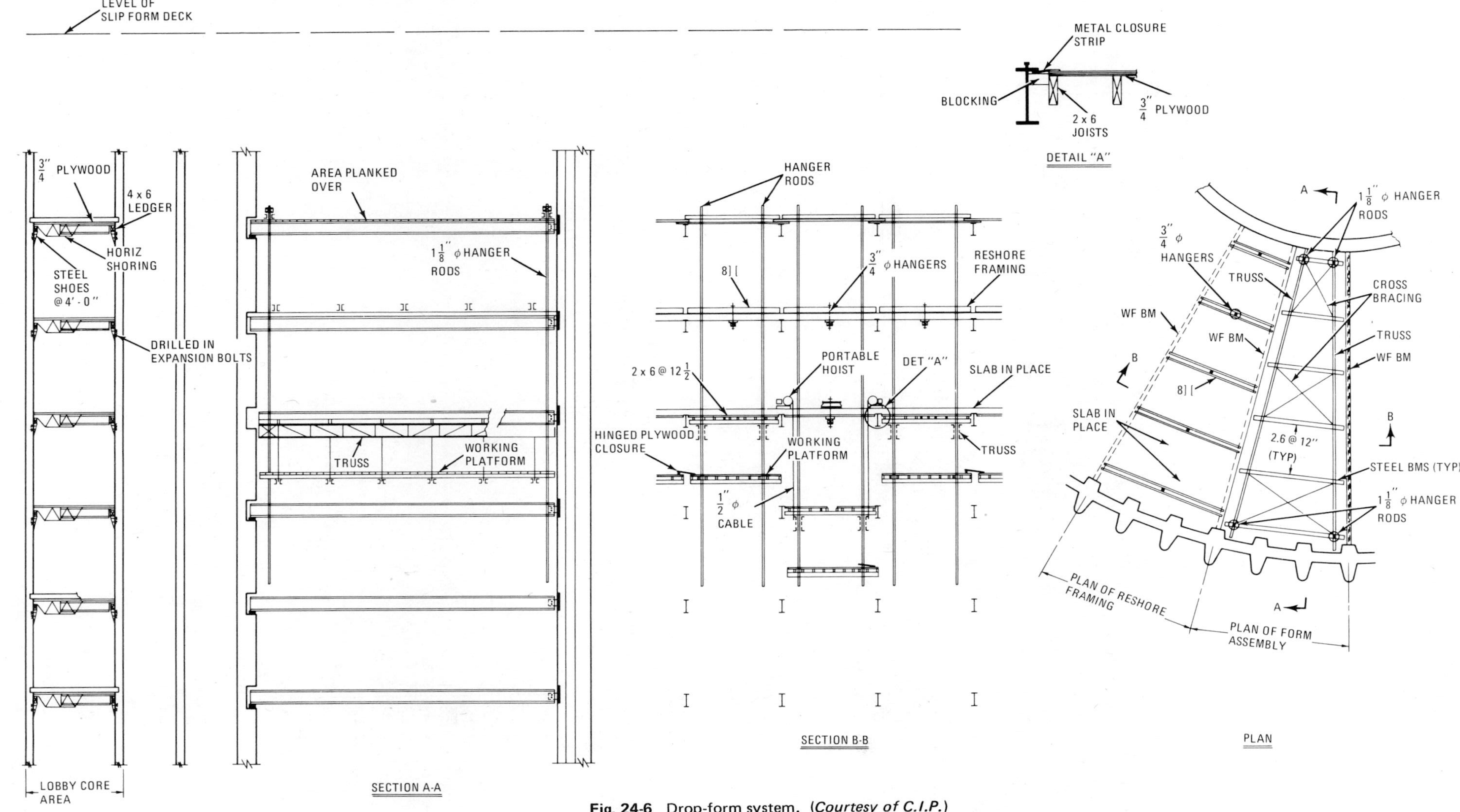

Fig. 24-6 Drop-form system. (*Courtesy of C.I.P.*)

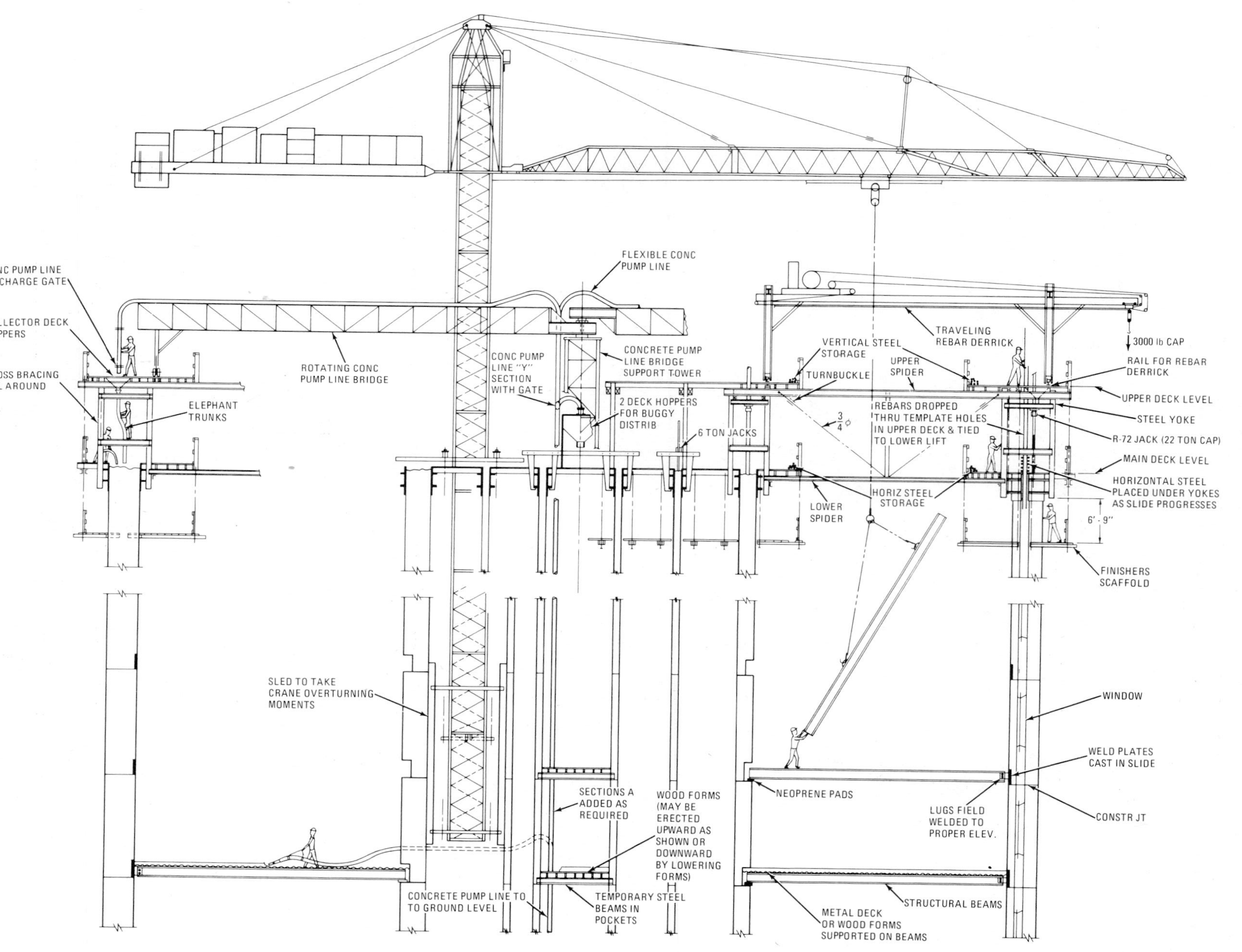

Fig. 24-7 Systems study for high-rise building. (*Courtesy of C.I.P.*)

Fig. 24-8 Combined materials in cross-country housing conveyor.

Fig. 24-9 American Bible HQ building. (*Courtesy of Brennan & Sloan.*)

24.4.2 Factors Affecting Economy of Formwork

There is no doubt that the one item having maximum effect on the economy of concrete formwork is the number of uses that will be made of each form section. Since this reuse item is intrinsically tied up with the stripping ability of the forms, it is imperative to design and detail the forms in such a way as to make stripping quick and nondestructive. Economical reuse may also be increased by designing and sequencing structural components so that forms may be reused partially-filled, bulkheaded or slightly modified. The cheapest form obtainable will usually be more expensive per square foot of contact area when used only once, as compared to very expensive forms having many reuses.

The cost of concrete formwork in place is composed of two items. The first is the make-up item which is the cost of the form ready for erection at the job site. This is a one-time cost; hence, the total form cost per square foot of contact area is inversely proportional to the number of uses. The second item consists of stripping, cleaning, repairing, erection, and expendable hardware. This item tends to remain constant for each reuse up to a ceratin point at which the cost of repairing and cleaning starts rising rapidly. This is the point of diminishing return at which further reuse of that form section becomes uneconomical. This point of diminishing return is of course dependent on the strength and durability of the particular form. It is also dependent on the care with which the forms are handled, cleaned, and stored. The most expensive form can be destroyed in one use by improper handling. See Tables 24-5 and 24-6 showing comparative costs for formwork.

As stated above, the forming system and reuse factors do affect the efficiency of the placing crews. Thus, the final forming system and material cost study must include full and efficient use of placing crews and the ability to meet the required construction schedule. The overall optimum reuse factor will probably be less than the diminishing-return factor discussed above. If quality requirements do not require otherwise, the forming material selected will be that which is most economical at the reuse factor selected.

Ganging reduces the stripping and erection cost much like the reuse factor reduces the make-up cost of formwork. By combining small forming sections into large assemblies, the forms can be stripped and re-erected at substantially lower cost when conditions allow (Fig. 24-10). The size of the ganged form should be the maximum size that can be lifted by the equipment available, considering accessibility to the areas involved. The ganging size will also depend on the building configuration and the reuse cycle. Ganging is possible in both vertical and horizontal casting. In order to use ganged forms, the surfaces being cast must be free of projecting dowels, brackets, and complications which will prevent easy release and lifting of the forms.

In considering the reuse factor relative to the required production, a determination will be made as to the actual number of sets of forms required. Obviously the number of sets must be kept down to a minimum consistent with minimum time required before stripping. Frequently the length of time that the forms must remain in place is rather arbitrarily arrived at without the realization that overconservatism will adversely effect the cost of the structure. The use of minimum concrete strength rather than an arbitrary time requirement is recommended as this will allow the contractor to take economic advantage of early strength and high strength concrete. It will also furnish safety against too little strength in cold weather. When stripping time is based on concrete strength as indicated by field-cured cylinders, the use of high strength concrete may be economically advantageous by reducing stripping time and therefore the number of form sets. It may also allow better utilization of the placing crews.

Another consideration that will promote economy in formwork is complete and detailed design of the formwork. Contractors are often overly anxious about the cost of detailed design not realizing that this cost will be saved many times over in the field through less waste, less reworking, better production, and timely completion. In

TABLE 24-3 Cost Comparison by Casting Systems Type of Building and Labor Cost

Forming System	*Average Comparative Cost Factors*	*High-Rise Building*			*Medium-Rise Building*		
		Average Labor Cost	*High Labor Cost*	*Low Labor Cost*	*Average Labor Cost*	*High Labor Cost*	*Low Labor Cost*
Cast-in-Place Gang Forming	1.00	.96	1.14	.75	1.04	1.26	.82
Conventional Form + Precast Panels	1.02	.98	1.16	.79	1.06	1.26	.86
Combination of Slip and Gang Form	.98	.94	1.12	.75	1.02	1.21	.82
Slipform	.96	.86	1.02	.69	1.06	1.20	.83
Precast	.96	.95	1.04	.78	.97	1.09	.80

TABLE 24-4 System Comparison for Particular Building

Placing System	*Comparative Cost Factors*	*Finish**	*Concrete Joints*	*Construction Time Factor*	*Safety**	*Construction Tolerance*		*Special Features*
						Plumb-ness	*Wall th'k*	
Cast-in-Place Gang Forming	1.00	A–	horizontal and vertical	1.00	B	1″	-1_4″, $+1_2$″	No Special Skills Required
Conventional Form + Precast Panels	1.02	A+	horizontal and vertical	1.00	B+	1″	-1_4″, $+1_2$″	Precasting at Yard
Combination of Slip and Gang Form	.98	A–	horizontal and vertical	.75	B+	3″	-1_2″, $+1_2$″	Forming Operation + Conventional Slip Forming
Slipform	.97	A+	none	.40	A–	2″	-1_4″, $+1_2$″	Special Skill and Equipment Required
Precast	.94	A+	horizontal	.50	A	1_2″	-1_4″, $+1_2$″	1. Winter Protection at Grade. 2. Inspection at Grade.

**Finish rated as follows: Uniform finish without tie holes— A+*
Uniform finish with tie holes— A–
***Safety rated as follows: Work mostly on ground or solid platform—A*
Work mostly on scaffolding— B

TABLE 24-5 Effect of Reuse on Concrete Formwork Based on One Use Equal to 1.00.

Number of Uses	*Cost Per Sq Ft CA*	
	Wood Form	*FRP*
One	1.00	1.00
Two	0.62	0.54
Three	0.50	0.38
Four	0.44	0.31
Five	0.40	0.26
Six	0.37	0.23
Seven	0.36	0.21
Eight	0.35	0.19
Nine	0.33	0.18
Ten	0.32	0.17

making the design, sizing of form components must be based on actual placing rates, expected temperatures and standardized stocks on hand. If external vibration is to be used, the forms must have the necessary mass. Lifting devices and scaffold brackets should also be integrated. Every effort should be made to keep form members to a minimum consistent with strength requirements, not so much to reduce the make-up cost, but because weight is so important in the handling of the forms.

24.4.3 Form Options

Forms, shores, and accessories may be rented from companies specializing in this operation. Column forms, floor domes, modular panels, shoring, etc. are available for rent at competitive rates. If the design allows, use of standard, rented form components can be the most economical solu-

TABLE 24-6 Comparative Cost of Forms per sq ft Contact Area for Different Materials and Reuses. (Based on field built plywood forms as base equal to 1.00)

Form Type	*Make-Up Cost*	*Assumed Reuses*	*Weight (lb/sq ft)*	*Cost (one use)*	*Cost (assumed uses)*
Plywood Forms, field built	1.00	4	10	1.50	.75
Overlaid Plywood, shop built	1.50	10	10	2.00	.65
Woodfaced, Steel Frame	2.50	10	12	3.00	.75
Epoxied Steel	3.75	40	15	4.15	.50
Fiberglass Reinforced Plastic	6.00	30	8	6.40	.60

Fig. 24-10 Ganged form being placed.

tion. Conversely, whenever possible, designs should allow for use of standard form equipment. Forms may be manufactured in shops specializing in woodwork, steel, or fiberglass. The quality thus obtained is usually very high, resulting in forms capable of holding the tolerances and yielding excellent reuse. Quality required, availability and distance of shops, local labor conditions, and site congestion are factors in decision making.

On large projects, a shop may be erected on the site. If sufficient area, skill, equipment, and shelter are provided, the product will equal that of permanent shops, eliminating shipping problems and obtaining better control. The shop can be used for repair and other functions once the make-up is completed.

Lastly, the forms can be built in place. Some fabrication and storage facilities of a minimum standard must be provided. The quality may not be as high but this may be the best option available in the case of highly complicated nonreusable forming or the opposite extreme, simple, undemanding formwork. Of course, combinations of rented, shop, and in-place fabrication are used as economics dictate.

24.4.4 Vertical Forming

There are basically four types of forming systems available for vertical work such as walls, columns, piers, and civil structures. They are conventional, jump, slip, and cantilever forming.

In conventional forming, an entire story or lift is constructed including the flat work. The vertical work is placed first, and after a delay of two hours or more, the floors are placed. After the concrete has attained the required strength, the forms are disassembled and passed up to the next floor from outside the building or through shafts. This entails a maximum of manual work and a minimum of equipment requirements. It is a good idea to keep form sizes down to 2 ft in width and 80 lbs in weight for best efficiency. Since the floor cycles tend to be slow, a lateral subdivision is required for full continuous use of labor crews.

In jump-forming, the walls and columns usually proceed ahead of the flat work so that the wall and column forms are out of the way of the floor forming and both systems are almost independent of each other. These forms are released and lifted straight up to the next position for reuse. They are often ganged and almost always handled by cranes (Fig. 24-11). They can be one-floor high, supported on inserts set in the lift below; or, two sets, each one-floor high, which leapfrog past each other, the lower one supporting the upper. Pockets, box-outs and keys are formed for receiving the floors. Where sufficient reuse is involved with a minimum of change in wall, column, and floor dimensions, this system will have considerable merit. It should be noted here that concrete is cheaper than voids, that is to say, unnecessary openings and reductions in dimensions are to be avoided.

Slip-forming is basically the ultimate extension of the jump-form system. The entire vertical forming system is combined into one ganged form which is raised as a unit using hydraulic jacks (Fig. 24-9) The requirements for both systems are similar but more stringent for slip-forming. One or the other will be more economical based on building configuration, height, simplicity, and labor conditions. Slip-forming is always faster.

Cantilevered forming lends itself to forming structures that have mild variation in sizes and slope. The forms for each lift are cantilevered by use of strongbacks from the lift below. The systems provides for relatively easy adjustment of the forms against the strong back. This system has good built-in control for holding tolerance. It is used extensively in the construction of dams and gravity cooling towers as illustrated in Figs. 24-12 and 13. Figure 24-14 shows a carefully designed system for forming exterior

Fig. 24-11 Jump forming.

columns and core walls for a 40-story building. Note the mobile lifting device (Fig. 24-15) developed for raising the column forms. The work proceeds in a clockwise sequence of operations divided into four quadrants for a four-day cycle.

24.4.5 Flat-Forming

There are basically six systems of floors and/or floor forming, namely, flat-slab, beam-and-slab, flying, metal-pan, leave-in-place, and cellular-steel forms.

It is basic to floor forming that the underside be perfectly level, to simplify fabrication, erection, and stripping. In flat-slab work, drop panels should be eliminated at the cost of thicker slabs and/or shear-transfer devices. In beam-and-slab work the bottom of all beams and, if possible, girders should also be kept level.

For light loadings and short to medium spans, it is hard to beat flat-slab forming for economy. Arrangement of the columns to facilitate stripping and erection or to allow the use of flying forms must be considered.

Beam-and-slab floors are used to suit structural requirements. For the sake of easy stripping, all beams should be of same depth in any one floor and of the same width in vertical projection floor to floor. Congestion at beam and girder intersections must be kept down to a minimum.

Flying forms are really a system of ganging floor forms together with their supporting shores. (See Fig. 24-16). This system can be used with either flat-slab or beam-and-slab. The form can have drop panels which fold down for handling. In order to use flying forms efficiently, one side of the building must be completely open to allow removal of the flying-form sections. Adequate crane facilities are required and column arrangements must be regularly lined up in the direction of form extraction. The economy of this system lies in designing flying sections of the maximum possible size that the cranes can handle and keeping the closure panels to a minimum. The use of flying forms have much merit and should be carefully considered.

Pan and dome floors find their area of maximum economy in forming long spans. With this system the weight of the concrete is materially decreased with little loss in strength. The forms can be steel, fiberglass reinforced plastic, hardboard, fiberboards, or even cardboard as conditions allow. To obtain maximum economy, the pan and dome sizes must all be standard and if possible the same size. As in any other form system maximum reuse is desirable.

Leave-in-place forms derive their economy from saving the cost of stripping and cleaning. Whenever this cost exceeds the cost of the leave-in-place forms, this system becomes economical. It pays to use a slightly heavier gage than is required structurally if a line of shoring can be eliminated. These forms can be steel, concrete, molded fiberboard, gypsum, or insulation board, depending on architectural requirements. Leave-in-place steel forms are especially economical with a steel beam system. Cellular-steel forms are a composite type of steel form which have the advantage of greater span without shoring. They are capable of longer spans between beams and may provide raceways for telephone and electricity.

24.4.6 Shoring

Since the load imposed by wet concrete is usually greater than the live load that will be finally placed on a hardened floor, carrying this weight to ground is an important con-

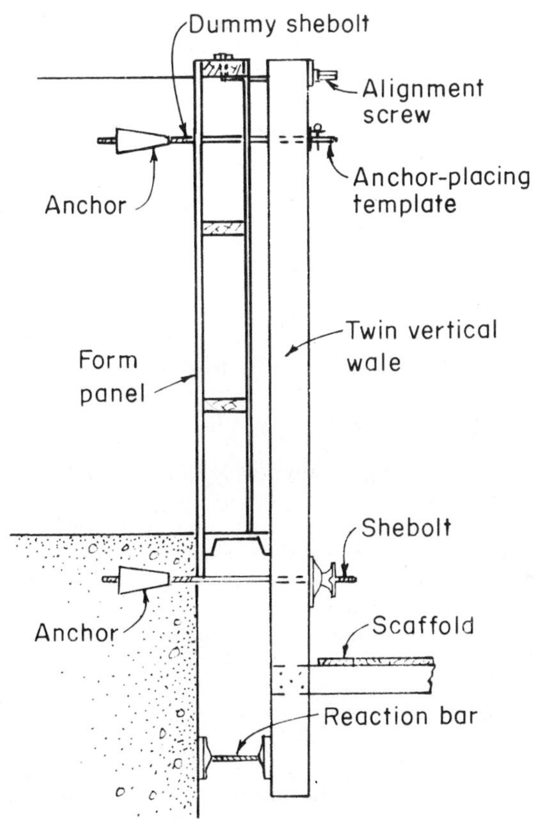

Fig. 24-12 Typical details of cantilevered form construction on a dam.

Fig. 24-14 C.B.S. HQ building. (*Courtesy of C.B.S.*)

sideration in formwork systems The two basic shoring systems are vertical and horizontal. The vertical shores, wood, steel, etc. are in common use as shores and reshores. They have the advantage of directness, simplicity and low

Fig. 24-13 Cantilever forms on a cooling tower.

Fig. 24-15 C.B.S. HQ Building. (*Courtesy of C.B.S.*)

Fig. 24-16 Flying forms in action.

cost. Their disadvantage lies in the necessity to wait until the floor below can support them, the clutter they create interfering with other operations, and the tricky reshoring and relieving operations involved.

Horizontal shores, most of them adjustable, are more expensive and have lighter capacities. They do, however, make each floor-forming system independent and leave the decks below free for other construction operations. The economy of the two systems can only be compared on the basis of the total construction operation.

24.5 CONCRETE PLACING EQUIPMENT

Next to formwork, concrete placing methods have the most significant effect on concrete economy and quality. It is important that the optimum system be selected which guarantees ample capacity in yards per hour, minimum labor requirements, proper handling of concrete to prevent segregation, and maximum protection from the elements. The equipment should be as nearly trouble-free as possible and provisions should be made for spare parts and equipment and possibly for alternate handling in case of breakdown.

The method of placing must be planned in conjunction with concrete mix, admixture use, site conditions, climatic conditions, forming system, and handling of other materials. The following placing methods are in common use. None stands out under all conditions as superior to other methods; each has applications depending on specific conditions. Table 24-7 shows some characteristics of placing methods.

24.5.1 Belt Conveyors

Belt conveyors are generally divided into three types: portable, or self-contained conveyors; feeder, or series conveyors; and side discharge, or spreader conveyors. Portable conveyors are excellent for short reach and/or lift situations up to 35 ft. These conveyors require almost no set-up cost, require only one man to operate them, and are often economical in very small placements that cannot be placed directly out of a ready-mix truck but are within the 35-ft reach.

Placements beyond the reach of a portable conveyor require a number of feeder conveyors in series. These conveyors come in standard sections of 24, 32, 48 and 56 ft in length and can be used for lifting concrete at inclines up to 20–35°, depending on concrete mix, belt type, and capacity required (Fig. 24-17). These conveyor sections may be manhandled or self-propelled and may be brought to the job under their own power, loaded on trucks, or towed. Steel sections vary in weight from 800 to 3000 lbs; aluminum sections weigh 15 to 30% less than steel sections of the same size.

Concrete placements often require considerable spreading

TABLE 24-7 Relative Placing Costs Using Cranes, Belt Conveyors, and Pumps

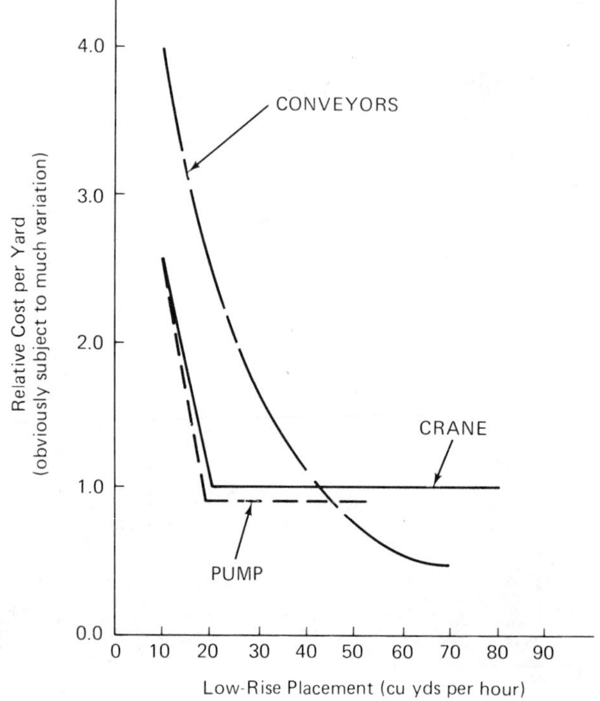

Fig. 24-17 A set of belt conveyors in action.

making it advantageous to use side discharge conveyors or cantilevered radial spreaders for the final depositing of concrete. Feeder and spreader conveyors are at their best when used in large placements with fast rates of placing. Although they may be used efficiently in placements as low as 50 yds, their best efficiencies are reached in placements of 200 yds, or more, with placing rates of over 25 yds per hr. Actually, the same crews are required whether placing 30 or 100 yds. Although some conveyors are rated to 150 yds per hr, such capacities are very difficult to attain in practice. With 16-in. belts (size in general use) and belt speeds of 300 to 650 fpm, rates of 40 to 100 yds per hr are more realistic. Belts 24-in. wide make rates increased up to 67% possible.

Special concrete mixes are not required for belt handling. A slump of 2.5 to 3 in. is usually used as this results in maximum capacity without overloading the equipment. Higher slumps are acceptable but will result in lower rates of handling. Aggregate sizes are usually limited to 2 or 3 in. for 16-in. belts (4 to 6 in. for 24-in. belts). Concrete weight will run 10 to 17 lbs per ft on a 16-in. belt and 20 to 36 lbs per ft on a 24-in. belt. With proper operation, the quality of the concrete will not be adversely effected by belt handling as no significant segregation should result if concrete is discharged against or onto concrete that is already in place.

Conveyor belts are quite dependable and are not overly subject to breakdown if care and good maintenance includes

1. flushing belts of equipment as necessary,
2. adjusting and replacing belt scrapers as required,
3. proper lubrication,
4. protection of motor and other equipment from concrete and dust, and
5. no continuous overloads.

Rental of belt conveyors runs 40 to 50 dollars per day per 32-ft section. This will result in an average rental cost of from one to two dollars per yard. It will take a five to eight man crew for erection and four to six men for operation of feeder conveyors. The spreading and consolidating crew will be in addition to this.

24.5.2 Pumping Concrete

Concrete can be economically pumped at rates up to 100 yds per hr, for horizontal distances up to 2000 ft and vertical distances to 500 ft without staging. At present, no equipment is capable of accomplishing all this concurrently and only the largest pumps and equipment can accomplish the maximum rate or distance at all. A large variety of equipment is available with capacities ranging from 15 to 100 yds per hr and rated distances varying from 300 to 2000 ft horizontally and 100 to 400 ft vertically. (Actually concrete has been pumped vertically for over 500 ft without use of booster pumps.) Although good economy is often obtained at shorter runs, the most effective pumping distances vary from about 300 to 1000 ft horizontally and from 100 to 300 ft vertically. The pumps come skid-, trailer-, or truck-mounted, weigh 1500 to 25,000 lbs, and are generally very mobile.

Hydraulic booms, as illustrated in Fig. 24-18 are available for low-rise work with reaches approaching 100 ft vertically. The pipe used runs in sizes from 3 to 8 in. in diameter (4 in. is most common) and includes rigid and flexible sections. Pipe that is used on booms or that must be manhandled is limited to 4 or 5 in. because of the weight. Lines over 5 in. in diameter must be a stationary installation.

The pumping of concrete is especially advantageous in placements that are physically difficult to approach with ready-mix trucks and show best economy in the range of 20 to 80 yds per hr. Equipment may be purchased for $400 to $700 per yard, rated capacity. Rental, including operator, runs $1.50 to $4.00 per yard with a minimum yardage requirement.

Some of the problems associated with pump concrete are pump breakdowns, line clogging and ruptures, and abrasion of equipment. The equipment must be rugged and wear resistant. The flow of concrete must be as continuous as possible with movement at least every ten minutes and clearing of the lines required for stops over 30 minutes to 1 hour depending on temperatures, concrete admixtures, etc. Water is required for cleaning the system and some provision is

Fig. 24-18 Combined vertical and horizontal pumping.

needed to collect the resultant discharge so as not to contaminate the freshly placed work. Placements should always start at the furthest point of deposit and work towards the pump location as this makes it possible to remove pipe sections with minimum interruption. Adding sections of pipe, on the other hand, is difficult and causes restart problems.

Concrete pumping is normally conducted at optimum line pressures of 200 psi with 300 psi as an intermittent maximum. Pressures in pump will run about three times line pressure. Flexible hose tends to increase pressure and its use is therefore kept down to a minimum. In high-rise work the pressures tend to get very high as the weight of the column of concrete is added to the pressure required to overcome friction. For very high lifts, extra heavy lines and connections will be required. In addition, substantial anchorages will be required to tie the pipeline to the structure to keep the pump from 'walking away'.

Normally two ready-mix trucks are unloaded simultaneously into the pump hopper. A minimum of five men is required, two at the truck and three at the nozzle. The spreading and consolidating crew will be in addition to this.

In order to overcome friction in the lines, a rich film of cement, sand, and water is required as a lubricant between the pipe walls and the moving slug of concrete. In order to establish this film, a grout is first pumped through the lines. Good aggregate gradation with enough cement paste and fine aggregate particles is absolutely required. Poor gradation, especially in the sand, will result in mechanical interlocking and insufficient matrix, and will not support the grout film. Excessive water can wash away this film; angular sharp aggregate can scrape it away; leaks in the line can cause loss of the film. Loss of film will cause a plug. The following minimum requirements must therefore be met in a concrete to be pumped:

1. Cement content–450 lb/yd, minimum.
2. Well-graded, fine aggregate with 15 to 30% of sand passing No. 50 sieve and 3 to 6% passing No. 100; fineness modulus of 2.40 to 3.00.
3. Maximum volume of coarse aggregate may be reduced up to 10% from ACI 211.1-70, depending on equipment and mix.
4. Slump 2 to 6 in., depending on mix, length of run, and equipment.
5. No bleeding mixtures.

Although some pump manufacturers specify minimum slumps less than 2 in., the pumping rate is slower at these low slumps and additional slump may be required for losses during handling. It may be difficult to maintain proper moisture content with lightweight concrete. This difficulty can be solved by presoaking the aggregate; by using a vacuum- or thermal-saturation process to treat the aggregate; or by using coated pellets, adding one to two dollars per yard to the cost of the concrete. Note that the pressure in the lines tends to increase the temperature and accelerate the setting time of the concrete, which may be an advantage or disadvantage depending on ambient temperatures.

The use of admixtures (especially those having a hydroxylated polymer base, 3 to 5%), air-entrainment, and additional paste will often prove economical by increasing production and reducing or eliminating plugs. Some difficulty has been experienced with pumping through aluminum pipe. The lime in the cement reacts with the aluminum abraded from the pipe, releasing hydrogen gas, thus resulting in lower strengths; an effect not unlike an excessive amount of entrained air.

24.5.3 Cranes

The use of cranes and buckets in placing concrete is general and widespread. In low-rise work, cranes find greatest usage where placements have to be made at scattered locations in the work, especially relatively small placements which do not justify installation of conveyors or pump lines (Fig. 24-19). They are especially useful in placements above ground such as columns and piers. The relatively low cost of bringing mobile cranes on the job, setting up, and demobilizing is often an important advantage.

In high-rise work, cranes are often very effective because of their capability to deliver concrete at various locations and

Fig. 24-19 Crew placing concrete from a bucket.

at various levels without movement of the crane. One of the major economic advantages of crane usage is of course its versatility in handling other materials such as rebars, forms, inserts, and structural steel. The disadvantages of using many types of cranes lies in their inability to move to all parts of the work and in limitation in height of lift, reach, and capacity. In congested areas and on city streets, the working room they need may be a serious problem.

There are many types of cranes of widely varying characteristics and capacities. The most common type is the conventional crawler or truck-mounted crane. These cranes are either tread-mounted (must be transported over roads on trailers) or truck-mounted and self-propelled. Some smaller ones are skid-mounted for installation on the structure. These cranes rely on a boom which may be anywhere in length from a short hydraulic telescoping arm to a 450-ft stick. The cranes revolve 360° horizontally and boom to within 15° and 20° of level and plumb. They are available in capacities from 2.5 to 175 tons, and more. This designation of capacity is rather a poor system of classification since it is the capacity when the boom is nearly vertical and quite short; a moment-capacity designation, such as used in tower cranes, would be more useful. Although cranes may reach to 400 ft in height and reach, at this reach the payload is greatly reduced. For instance, a 110-ton crane with a 250-ft boom has a capacity of 1 ton at 220-ft radius. Outriggers are required to realize the capacities of many cranes.

One problem of cranes operating on the ground is that they must stay away from the building in order to accommodate the boom angle. Thus they have difficulty reaching across the face of the building to interior locations. This problem is solved by another type of crane, the tower crane. (See Figs. 24-11 and 24-20.) This crane has a vertical mast with a rotating horizontal jib at the top; the mast is stationary. The load is hoisted on a trolley which travels radially on the jib; the jib itself rotates 360°. This means that as long as the jib is high enough to clear the work and surrounding obstructions, loads can be picked up and deposited anywhere within the radius capacity of the particular crane.

These cranes take a minimum of space on the ground and can climb up with, and on, the building. They can free stand as much as 200 ft without bracing, and can be extended to several hundred more feet when braced against the building or guyed. They are sometimes mounted on railroad carriages. In smaller capacities they are truck-mounted and have in several instances been suspended from slip-form decks. They are readily available in capacities of from 230 to 1800 ft tons with capacities up to 30 tons at short radii and 3 tons at 200 ft radius. They are also available with luffing booms.

Tower cranes are versatile, efficient tools and are superior to conventional cranes in situations where loads must be carried up and over the construction or other obstruction, and at heights above the capacity of the conventional cranes. They are not as efficient when all points cannot be reached from one position. The main disadvantage is the cost of installing, climbing, and dismantling the cranes. Sometimes there is a clearance problem as to the jib clearing adjacent buildings. Generally speaking, cranes compete very well in economical placing of concrete when the job conditions allow free access and movement, when crane capacities are not exceeded, and when the placing area is open for vertical placement. Rentals run from $1000 to $15,000 per month depending on capacity and reach required. An operator and oiler is required in addition to rental.

Helicopters with capacities up to nine tons have been used successfully to place concrete, to handle formwork, and to erect and dismantle cranes. Careful planning is required as the hover time is usually limited to less than one hour. In inaccessible spots they can prove very economical in terms of time and money. Figure 24-21 shows a helicopter placing the jib sections on a tower crane atop a 535-foot concrete core.

24.5.4 Materials Hoist Towers

In studying placing methods for medium- and high-rise work, the hoist tower must always be considered. It has no capability for horizontal movement of concrete either at the base or at the discharge point, but it can be used in conjunction with conveyors, pump lines, chutes, and buggies to do the job. It takes a minimum of space to operate, has ample capacity and relatively low rental and erection cost. It is not mobile. For heavy concrete placing requirements, a hoist tower should be considered including the additional cost of required additional placing equipment and labor. Rental will run 500 to 3500 dollars per month, depending on capacity and height including hoist, but not operator.

24.5.5 Buggies

Hand and power buggies are used for horizontal distribution of concrete. Sometimes hand buggies are lifted on materials hoist platforms to the level of discharge and then wheeled across. Hand buggies are rubber tired and have capacities of 6 cu ft. and higher. Some are made very narrow (30 in.) for tight situations. In some cases wheelbarrows may be used for the same purpose but the capacity will be considerably less. Depending on site conditions, labor and run involved, a hand buggy can place up to 5 yds per hr.

Motorized power buggies with up to 1 cu yd capacity are available, although most units have capacities of 9 to 12 cu ft (Fig. 24-22). They are faster, very maneuverable and are capable of climbing slopes up to 35%. Power buggies will place 15 to 20 cu yds per hr. When power buggies are used, forming and shoring must be designed for at least 50% more live load than normally (75 psf) and lateral bracing must be carefully designed. Runways are required over reinforcing steel when any type buggy is used. Rental for buggies runs approximately $25 per month for plain carts to $250 for power carts.

24.5.6 Pneumatic Placing of Concrete

This method which is illustrated in Fig. 24-23 is also known as shotcrete and gunite. Concrete is shot at high velocity against a form or other surface under pressure. Because this is a no-slump material and because of the force of the jet, the mix adheres and is firmly compacted in place. Only one-sided forms are required and these forms will be subject to gun pressure only, which is less than the normal hydrostatic force of concrete in conventional placing. Shotcreting is advantageously used in relatively thin sections such as shell roofs, walls, tanks, chimney and hydraulic linings and cover and repair applications for all types of structures. It is especially useful for casting curved and three-dimensional sections.

The pneumatic mortar or concrete is applied in layers 1 to 1.5 in. thick, the total thickness obtained by successive placements usually being up to 4 in. Placements considerably thicker than this are obtainable but the cost can be high. Two to six yards of shotcrete per day per gun can be placed, depending on equipment used. For a 3-in. thick placement the cost may run $2.00 to $4.50 per sq ft subcontracted. Normal mixtures used are one part portland cement to three parts sand with a maximum aggregate size of 0.25 to 0.375 in., although mixes of one to four and

TABLE "B"

MINIMUM REQUIREMENTS FOR SLABS SUPPORTING CRANES

Δ = AREA OF STEEL IN SQ. IN. PER LIN. FT. (TYP.)

* = SHEAR GOVERNS

STANDARD WEIGHT CONC. f'c = 3000 p.s.i.						STANDARD WEIGHT CONC. f'c = 1500 p.s.i.						LIGHT WEIGHT CONC. f'c = 3000 p.s.i.						LIGHT WEIGHT CONC. f'c = 1500 p.s.i.					
SPAN DIM. X	FREE END SLAB		FIXED END SLAB			SPAN DIM. X	FREE END SLAB		FIXED END SLAB			SPAN DIM. X	FREE END SLAB		FIXED END SLAB			SPAN DIM. X	FREE END SLAB		FIXED END SLAB		
	SLAB THICK.	BOTTOM STEEL	SLAB THICK.	BOTTOM STEEL	TOP STEEL		SLAB THICK	BOTTOM STEEL	SLAB THICK.	BOTTOM STEEL	TOP STEEL		SLAB THICK	BOTTOM STEEL	SLAB THICK.	BOTTOM STEEL	TOP STEEL		SLAB THICK.	BOTTOM STEEL	SLAB THICK.	BOTTOM STEEL	TOP STEEL
1'-0"	*19½"	0.64	*19½"	0.113	0.55	1'-0"	*26"	0.47	*26"	0.09	0.40	1'-0"	*28½"	0.40	*28½"	0.08	0.34	1'-0"	*35½"	0.32	*35½"	0.07	0.27
2'-0"	*19½"	1.21	*19½"	0.26	0.98	2'-0"	*26"	0.87	*26"	0.20	0.64	2'-0"	*28½"	0.75	*28½"	0.17	0.60	2'-0"	*35½"	0.59	*35½"	0.14	0.47
3'-0"	*19½"	1.78	*19½"	0.44	1.38	3'-0"	*26"	1.29	*26"	0.34	0.99	3'-0"	*28½"	1.11	*28½"	0.28	0.86	3'-0"	*35½"	0.86	*35½"	0.23	0.66
4'-0"	*19½"	2.34	*19½"	0.65	1.74	4'-0"	*26"	1.71	*26"	0.50	1.26	4'-0"	*28½"	1.46	*28½"	0.42	1.08	4'-0"	*35½"	1.13	*35½"	0.34	0.84
5'-0"	*19½"	2.95	*19½"	0.89	2.13	5'-0"	*26"	2.13	*26"	0.67	1.54	5'-0"	*28½"	1.83	*28½"	0.57	1.32	5'-0"	*35½"	1.42	*35½"	0.46	1.03
7'-0"	22½"	3.55	*19½"	1.42	2.84	7'-0"	31"	2.43	*26"	1.03	2.03	7'-0"	*28½"	2.58	*28½"	0.90	1.76	7'-0"	*35½"	2.10	*35½"	0.71	1.37
10'-0"	27½"	4.40	22½"	1.95	3.37	10'-0"	38"	2.95	31"	1.41	2.37	10'-0"	*28½"	3.77	*28½"	1.44	2.46	10'-0"	36½"	2.88	*35½"	1.16	1.92

WHEN TWO OR MORE SLABS ARE USED TO SUPPORT THE CRANE THE MINIMUM COMBINED SLAB THICKNESS MUST BE AS FOLLOWS;

WHEN SHEAR GOVERNS:

$$t_{COMB} = t_{REQUIRED} + n - 1$$

FOR ALL OTHER CASES:

$$t_{COMB} = \sqrt{(t_1-1)^2 + (t_2-1)^2 + (t_3-1)^2} + 1 \geq t_{REQUIRED}.$$

THE COMBINED STRENGTH OF THE CONCRETE SLABS SHALL BE ASSUMED AS FOLLOWS;

$$f_{COMB} = \frac{t_1 f_1 + t_2 f_2 + t_3 f_3}{t_1 + t_2 + t_3}$$

THE COMBINED REINFORCING STEEL MUST BE EQUAL TO OR GREATER THAN THE REQUIRED STEEL AND IS COMPUTED AS FOLLOWS;

$$As_{COMB} = \frac{As_1 (t_1 - 1) + As_2 (t_2 - 1) + As_3 (t_3 - 1)}{t_{COMB}}$$

n = NUMBER OF SLABS USED IN SUPPORT.
t_1, t_2, t_3, ETC. ARE THE RESPECTIVE THICKNESS OF THESE SLABS.
f_1, f_2, f_3, ETC. ARE THE RESPECTIVE STRENGTH OF THESE SLABS.
f COMB = IS AVERAGE STRENGTH OF SLABS.
t COMB = SUM OF THICKNESSES OF SLABS USED.
t REQUIRED = THICKNESS OF ONE SLAB AS SHOWN ON TABLES.
As COMB = SUM OF STEEL AREAS OF SLABS USED.
As REQUIRED = STEEL AREAS SHOWN ON TABLES.
As_1, As_2, As_3, = STEEL AREAS OF RESPECTIVE SLABS USED.
t, n MEASURED IN INCHES.
f MEASURED IN POUNDS PER SQ. IN.
As MEASURED IN SQ. IN.

MAXIMUM LOADING (INCLUDING PAYLOAD and COUNTERWEIGHT)			SYMBOL	UNIT	H = 82'-0"				H = 65'-8"				H = 49'-3"			
					JIB "K"	JIB "N"	JIB "L"	JIB "LL"	JIB "K"	JIB "N"	JIB "L"	JIB "LL"	JIB "K"	JIB "N"	JIB "L"	JIB "LL"
TOTAL LOAD		VERTICAL LOAD	V	KIPS	185	190	185	191	185	190	185	191	185	190	185	191
TOTAL LOAD		LONGITUDINAL OVERTURNING MOMENT	M_L	FT. KIPS	1075	1065	999	1039	1075	1065	999	1039	1075	1065	999	1039
TOTAL LOAD		TRANSVERSE OVERTURNING MOMENT	M_T	FT. KIPS	471	501	550	580	369	393	434	458	269	287	318	336
TOTAL LOAD		SLEWING MOMENT (TOP SUPPORT ONLY)	M_S	FT. KIPS	174	225	272	335	174	225	272	335	174	225	272	335
TOTAL LOAD		TRANSVERSE HORIZONTAL SHEAR (TOP SUPPORT ONLY)	S_T	KIPS	6.6	7.0	7.6	7.9	6.2	6.6	7.2	7.5	5.8	6.1	6.7	7.1
VERTICAL LOAD PER LEG			V/4	KIPS	46.2	47.5	46.2	47.8	46.2	47.5	46.2	47.8	46.2	47.5	46.2	47.8
SHEAR PER BOLT GROUP IN ANY DIRECTION	TOP SUPPORT	D = 21'-4"	St	KIPS	15.6	17.4	19.3	21.8	14.8	16.7	18.0	21.0	14.0	15.8	17.3	19.9
		23'	"	"	14.9	16.9	18.3	21.3	13.9	15.9	17.6	20.3	13.1	15.4	16.7	19.2
		25'	"	"	14.2	16.3	17.8	20.1	13.5	15.3	17.1	19.7	12.6	14.6	16.3	18.5
		27'	"	"	13.5	15.8	17.5	19.9	12.9	14.8	16.5	19.3	12.2	14.2	15.9	18.2
		29'	"	"	12.9	15.4	16.9	19.7	12.5	14.2	16.1	18.8	11.7	13.7	15.4	18.0
		31'	"	"	12.6	15.2	16.5	19.1	12.2	13.8	15.7	18.5	11.3	13.1	14.8	17.7
		33'	"	"	12.2	14.7	16.3	18.8	12.0	13.5	15.5	18.0	11.0	12.8	14.7	17.5
		35'	"	"	12.0	14.3	15.9	18.3	11.4	13.3	15.4	17.9	10.8	12.6	14.6	17.3
	BOTTOM SUPPORT	D = 21'-4"	Sb	KIPS	12.6	12.5	11.7	12.2	12.6	12.5	11.7	12.2	12.6	12.5	11.7	12.2
		23'	"	"	11.7	11.6	10.9	11.3	11.7	11.6	10.9	11.3	11.7	11.6	10.9	11.3
		25'	"	"	10.8	10.7	10.0	10.4	10.8	10.7	10.0	10.4	10.8	10.7	10.0	10.4
		27'	"	"	10.0	9.9	9.3	9.6	10.0	9.9	9.3	9.6	10.0	9.9	9.3	9.6
		29'	"	"	9.3	9.2	8.6	8.9	9.3	9.2	8.6	8.9	9.3	9.2	8.6	8.9
		31'	"	"	8.7	8.6	8.1	8.4	8.7	8.6	8.1	8.4	8.7	8.6	8.1	8.4
		33'	"	"	8.2	8.1	7.6	7.9	8.2	8.1	7.6	7.9	8.2	8.1	7.6	7.9
		35'	"	"	7.7	7.6	7.2	7.4	7.7	7.6	7.2	7.4	7.7	7.6	7.2	7.4

$$Sb = \frac{\sqrt{M_L^2 + M_T^2}}{5.33D}$$

BUT NOT LESS THAN $\frac{M_L}{4D}$

$$St = \frac{\sqrt{\left(\frac{M_L}{D}\right)^2 + \left(\frac{M_T}{D} + S_T + \frac{1.33 M_S}{6.3}\right)^2}}{5.33}$$

BUT NOT LESS THAN

$$\frac{1}{4}\sqrt{\left(\frac{M_L}{D}\right)^2 + \left(\frac{M\ SLEWING\ W/o\ WIND}{6.3}\right)^2}$$

M SLEWING W/o WIND = 237[K-FT](LL), 207[K-FT](L), 168[K-FT](N), 145[K-FT](K).

NOTE: M_S, S_t, S_b — LOADS HAVE BEEN REDUCED BY 25% WHERE APPLICABLE TO ALLOW FOR 33% INCREASE IN ALLOWABLE STRESSES.

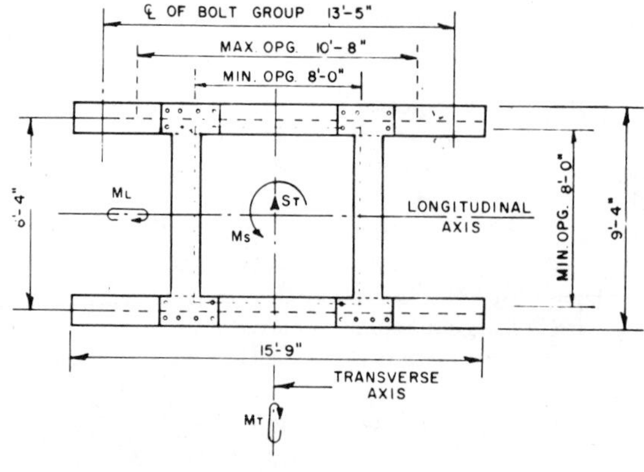

PLAN OF TOP SUPPORT
BOTTOM SUPPORT SIMILAR

Fig. 24-20 Design criteria for a climbing crane. (*Courtesy of American Pecco*)

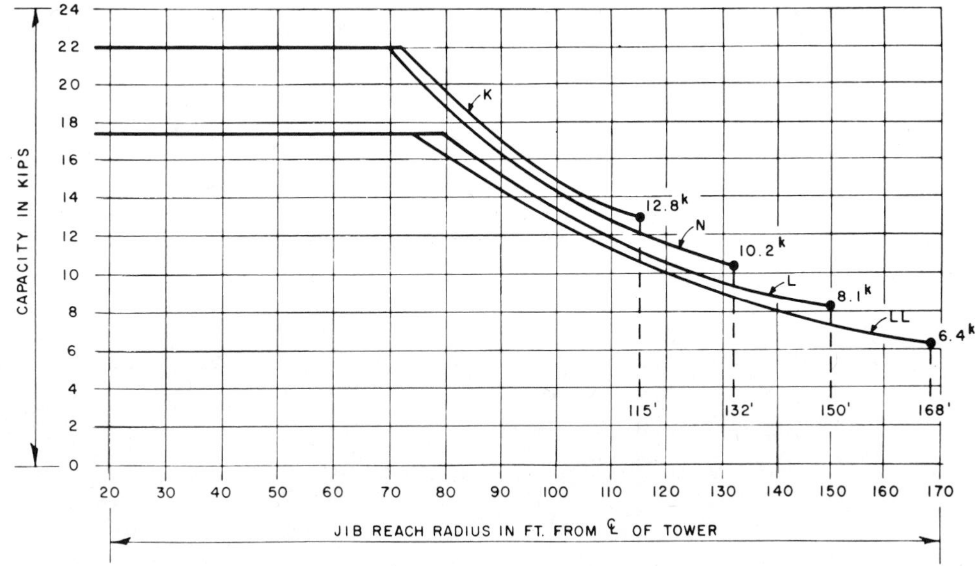

LOADING DIAGRAM

DESIGN CRITERIA

1. DESIGN BASED ON OPERATIONAL LOADS PLUS WIND LOADING FROM A 40 M.P.H. WIND.
2. WINDS OVER 40 M.P.H. ARE CONSIDERED STORM CONDITION AND NO OPERATION WILL BE PERMITTED DURING STORM CONDITIONS.
3. DURING STORM CONDITION (OVER 40 M.P.H.) THE JIB MUST BE FREE TO WEATHERVANE.
4. WHEN NOT IN USE AND DURING STORM CONDITION, THE HOIST TROLLEY MUST BE AT THE LONGEST REACH RADIUS POSITION.
5. DURING STORM CONDITIONS WITH WINDS 65 M.P.H. AND OVER A BALANCING WEIGHT (NOT INCLUDING TROLLEY) IS REQUIRED AT THE END OF THE JIB OF THE WEIGHT INDICATED IN TABLE "A". THIS LOAD SHOULD HAVE A MINIMUM PROJECTED AREA.
6. THE BALANCE LOAD SHOWN IN TABLE "D" MIGHT BE SLIGHTLY LESS DUE TO THE WIND CONDITIONS.
7. SLEWING TORQUE IS BASED ON A MAXIMUM SLEWING SPEED OF 1.0 REVOLUTIONS PER MINUTE AND A MINIMUM STOPPING TIME OF 5 SECONDS AT MAXIMUM SPEED.
8. MINIMUM SLAB THICKNESS AS GIVEN IN TABLE "B" IS BASED ON CONCRETE HAVING A MINIMUM COMPRESSIVE STRENGTH AS SHOWN BASED ON FIELD CURED CYLINDERS.
9. SLABS NOT MEETING REQUIREMENTS SHOWN MUST BE SHORED TO SLABS OR OTHER STRUCTURAL MEMBER HAVING PROPER STRENGTH.
10. IF SUPPORT CONDITIONS ARE DIFFERENT THAN THOSE SHOWN ON THIS DRAWING AN EXPERIENCED PROFESSIONAL ENGINEER SHALL BE CONSULTED.

Fig. 24-20 *(Continued)*

aggregates to 0.75 in. are often used. Water included is usually three to six gallons per sack of cement. The shotcrete nozzle can be operated at distances up to 500 ft horizontally from the mix equipment and about half this distance vertically.

Care must be taken to keep 'rebound' to a minimum (20% is good), blow off excess rebound material, keep form and steel surfaces clean, and prevent voids behind reinforcing bars. Definite means of checking the thickness and moist-curing are essential.

24.5.7 Other Placing Methods

Other methods of placing concrete requiring special equipment are in limited use. One such method is preplaced aggregate concrete in which the coarse aggregate is placed and compacted and then a grout of sand, cement, and water is intruded into the aggregate mass. Another placing method is dry-casting in which stone, sand and cement is placed dry and compacted and water is intruded at a later time.

24.5.8 Selection of Equipment

In order to determine the placing equipment to use, the various types and combinations of equipment are studied and those which cannot be used or are obviously uneconomical are eliminated. A detailed cost study of each method is then made including variation in concrete mix, formwork, set-up, rental, dismantling, operators, fuel, placing labor, and any other related costs.

24.6 SAFETY PAYS

The moral obligation to safeguard human life is so obvious that it will not be belabored in this chapter. As is usually the case, the moral approach is, in the long run, the best pragmatic approach. Accidents cost money; insurance against accidents costs money; court litigations cost money; job delays cost money. Workmen's Compensation Insurance, which amounts to 2 to 12% of payroll, is based on the manual rate established for that particular classification of work and upon the individual contractors relative safety record within the classification. A low accident history will, for instance, tend to improve the competitive position of concrete construction versus steel erection as a whole and the competitive position of a particular contractor within his own classification. Liability insurance will likewise be based on accident records.

Anyone who has been involved in an accident investigation and/or court litigation cannot help but wonder at

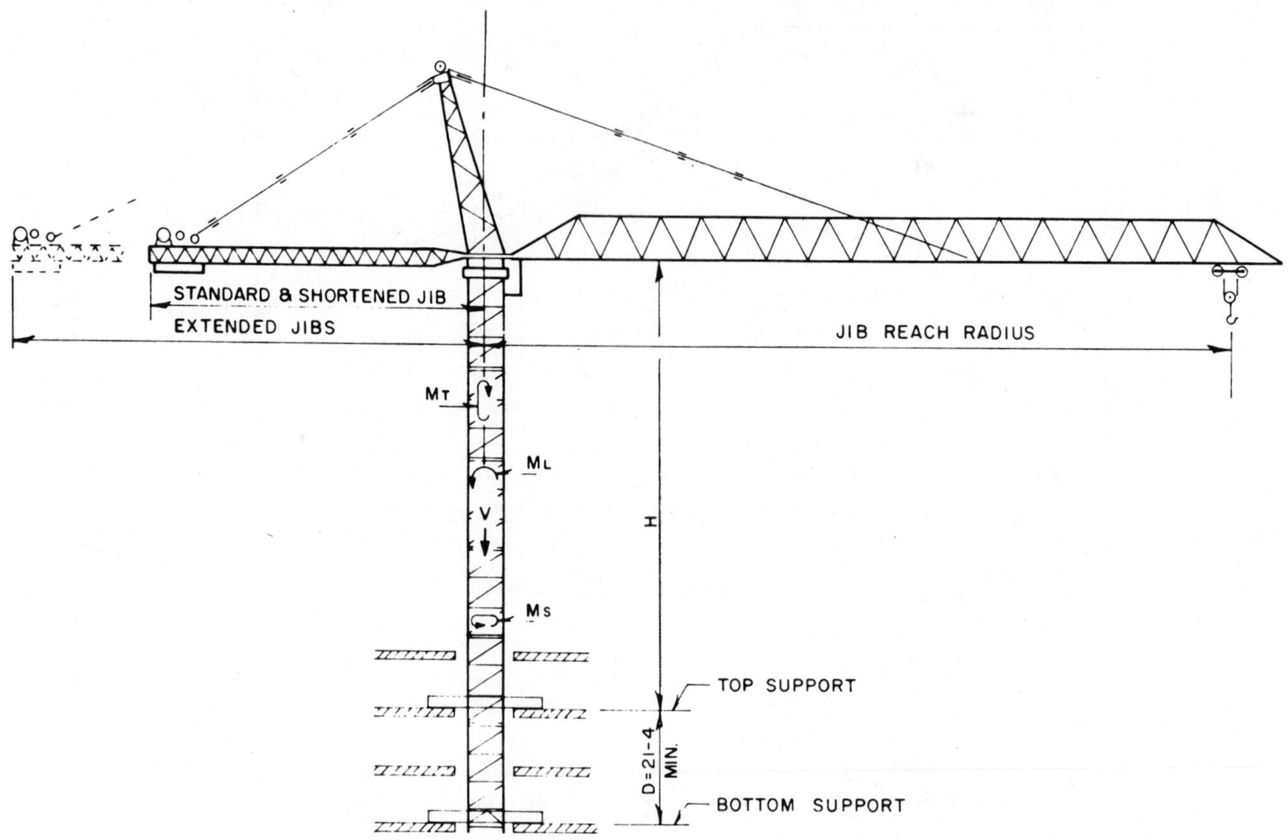

TABLE "D"			
BALANCE LOAD DURING CLIMBING			
JIB	MK	LENGTH	BALANCE LOAD
SHORT	K	115'	12,800 lbs.
NORMAL	N	132'	10,200 lbs.
EXTENDED	L	150'	8,100 lbs.
DB'LE. EXT.	LL	168'	6,400 lbs.

Fig. 24-20 *(Continued)*

the extent and costs to which tests, computations, searches, and expert opinions are carried. One cannot help but wish that some of this thought and expense had been used to prevent the accident in the first place.

This chapter will not attempt to go into the organization and administration of good safety programs, nor into safety codes and practices. Information on these matters is easily available from many private sources such as publishers and the Association of General Contractors and is mandated in a statutory manner by municipalities, states, government agencies at all levels and very recently by the federal government. We will, however, mention general attitude and some specific hazards which seem worthy of special note.

In the first place, safety is more a state of mind than a set of rules. Top management of the owners, designers, and builders must indicate in a way that cannot be misunderstood that they consider safety of paramount importance. They must commend those people who promote safety and discourage or remove completely from their organization those people who are careless. This attitude must filtrate through the whole hierarchy from president to night watchman. Above all, placing incompetent persons in safety positions just to make a show of safety-mindedness is criminal.

24.6.1 Good Housekeeping

One of the simplest and most basic aspects of safe construction is good housekeeping. Materials, debris, lumber scraps, empty cans, grease spots, forms and hardware lying about waiting to be tripped over, slipped on, walked into or kicked or blown off onto someone's head below are not only a serious hazard but an indication of indifferent supervision and an encouragement to workmen to be careless and to add to the lethal litter. Any construction site which shows poor housekeeping also indicates a lax supervision and a disregard for human welfare. Good housekeeping not only indicates good safe operation, it also pays dividends in better material utilization and labor efficiency.

Fig. 24-21 Erecting crane by helicopter.

Fig. 24-22 Motorized buggies in action.

Fig. 24-23 Shotcreting in action.

24.6.2 Competent Personnel

Another important consideration in safety-mindedness is the use of personnel who are, in fact, competent and skilled to perform the functions assigned to them. Don't ask a carpenter who doesn't know how to place a cable clamp correctly to guy a hoist tower. Don't ask a carpenter foreman to put up shoring or bracing for a derrick or other piece of heavy equipment if the man has no knowledge of the loads and stresses involved. Don't assign a form-watcher who hasn't the slightest idea of what he is watching for. This may seem rather rudimentary, but it is very surprising how often people are assigned important tasks for which they are not qualified and for which they have not received proper instructions. If a special skill or just normal skill is required, supervisors must insure that such skill is actually used. If skill or knowledge is missing, it is false economy to make do rather than pay the price to get proper help.

24.6.3 Scaffolding and Runways

These items tend to be very well covered by codes. Normally the sizing of the members are properly designed and furnished; it is not the member that usually fails but the connections. This is especially true of wood. The designers will very often fail to design or specify the type, number, and location of nails or bolts required in connections. Minimum and maximum laps are ignored and cleating, bracing, and hold-downs are left to the field. Drilling holes for hanger rods will often result in very unsafe situations. Designers, supervisors, and inspectors must give their utmost attention to connections.

Another important facet often overlooked is the capacity of the scaffold or runway. This capacity, in pounds per square foot, must be carefully selected on the basis of actual field loadings expected, included storage of materials, heavy equipment handled, and moving loads such as power buggies. Once selected and designed for, this capacity must be prominently stated on the drawings and every effort taken to insure that supervisors using the scaffolds and runways fully appreciate and understand the limitations.

24.6.4 Shoring and Formwork

Formwork and, more importantly, shoring are probably the major killers on construction sites. The importance of proper design, especially design for lateral loading and moving equipment cannot be overemphasized. Deck formwork almost always fails laterally due to insufficient lateral bracing of the whole system or of the shores to each other. The lateral bracing and lacing must be completely designed including connections, and again, capacities and limitations must be prominently shown on the drawings; on wall and column pours, show placing rates as related to temperatures. During placing of concrete, forms and shoring must be carefully watched for telltale danger signals such as excessive deflection.

One serious hazard involved in placing decks is premature or accidental removal of shoring or improper reshoring procedures. A workman accidentally hitting a shore and knocking it out of position can trigger a widespread collapse. Shores therefore must be as rigidly connected as possible and if possible protected form other operations. Education of the workers in this matter is an important safety requirement. An item that is systematically abused is the wedging at the base of shores. This wedging is required to give full bearing as the full load of the shore must pass through the wedges. In so many instances this bearing is not provided; I have seen wedging with barely an inch of overlap and not nailed into position. *Inspectors please note.*

24.6.5 Cranes and Derricks

Cranes and derricks are being mounted more and more on partially-completed buildings, and even on formwork. As the buildings get taller in height or more extensive in area, convential boom cranes working from outside the building line can no longer reach the work. The 400-ft booms have been pushed to the limit and must be used with care.

Climbing cranes and derricks exert, in addition to the vertical loads, severe lateral loadings, that is to say overturning moments. Very few failures of climbing cranes have been failures of the equipment itself; almost always the accident was caused by improper support to the building. When one considers that these cranes exert overturning moments up to 2000 ft-kips (and will get larger yet), erecting such cranes without proper engineering design is certainly entertaining disaster. Yet, this has been done and serious accidents have been experienced.

Some substantial improvements are in order in the field of crane safety, both climbing and ground-supported types. There is a remarkable lack of codes in this area, and a national code similar to the European codes is urgently required. Some crane manufacturers provide users and engineers with excellent and meaningful loadings and support data. Many more supply little or no data, or supply data that is incomplete or subject to interpretation. The manufacturers representatives may be nontechnical people who do not understand the loadings themselves and who often lead undiscriminating clients into serious problems. Crane users, engineers, and approving agencies should give preference and encouragement to those manufacturers who do furnish adequate data. Figure 24-9 shows a good data sheet.

Lastly, the buildings must be checked by a competent structural engineer to insure that the building members are not overloaded. This would preferably be the engineer of record or an engineer specializing in this field and approved by the engineer of record. The manufacturer is responsible for furnishing data but not for structural analysis of the support on the building. The cost for this engineering is properly borne by the contractor who benefits by the use of the crane.

Failures in derricks are often caused by improper guying; by damaged, rusted, or otherwise impaired parts; or by overloading. The comments made above in regard to cranes also apply here. On derricks as well as on cranes, capacities must be clearly stated and posted for the operator to see; loading must be clearly shown.

Guying failures usually result from improper use of or failure to use hardware such as clamps and thimbles, or attachment of guys to objects incapable of taking the loads. It is not uncommon to see guys subject to potentially high loads casually attached to 0.5-in. reinforcing dowels, pieces of formwork, small footings, and at ridiculously ineffective angles, and slack, at that. When this happens, three things are indicated. First, an engineer did not locate the anchors and specify the connections. Secondly, an experienced, knowledgeable rigger was not used to make the installation. Thirdly, the inspector was incompetent. Engineering design, workmanlike installation, and intelligent, alert inspection are all required.

All the design in the world will not help equipment that is not in proper working order. Rusted connections, loose bolts, cracked welds, embrittled notches, bent compression members, and damaged cables do not show on manufacturers catalogs or on engineering drawings. You can find them with startling frequency in the field. Rigid, continuous, and periodic inspection is required by the equipment owners, users, operators, and inspection agencies. This inspection must be made by persons who know what to guard against.

One last point on cranes, derricks, and all hoisting equipment; this equipment is designed primarily to operate at normal allowable operating stresses. Capacity charts take full advantage of the stress capabilities of the equipment without considering the dynamic nature of the loading and prolonged use of the equipment; the metal will become fatigued. At the present time there is little consideration given to this problem in the way of lower allowable stresses, accurate use records, or testing programs. Such consideration seems to be in order.

24.6.6 Fire, Smoke, and Toxic Gases

These items contribute heavily to construction injuries, property damage, and insurance costs. The codes covering these hazards are normally explicit and comprehensive. Accidents occur from failure to comply with the codes. The main requirement in this safety area seems to lie in education and frequent refreshing of the minds of workers and supervisors at all levels as to the dangers of welding sparks, salamanders, unventilated areas, improper combustion, leaky lines, damaged electrical lines or connections, and sprayed-on chemicals and materials. Exhaust from some equipment may also be dangerous.

25

Structural Analysis

KURT H. GERSTLE, Ph.D.*

25.1 INTRODUCTION

25.1.1 Purpose of Structural Analysis

Structural analysis is the process of determining the forces and deformations in structures due to specified loads so that the structure can be designed rationally, and so that the state of safety of existing structures can be checked.

In the design of structures, it is necessary to start with a concept leading to a configuration which can then be analyzed. This is done so members can be sized and the needed reinforcing determined, in order to: a) carry the design loads without distress or excessive deformations (serviceability or working condition); and b) to prevent collapse before a specified overload has been placed on the structure (safety or ultimate condition).

Since normally elastic conditions will prevail under working loads, a structural theory based on the assumptions of elastic behavior is appropriate for determining serviceability conditions. Collapse of a structure will usually occur only long after the elastic range of the materials has been exceeded at critical points, so that an ultimate strength theory based on the inelastic behavior of the materials is necessary for a rational determination of the safety of a structure against collapse. Nevertheless, an elastic theory can be used to determine a safe approximation to the strength of ductile structures (the lower bound approach of plasticity), and this approach is customarily followed in reinforced concrete practice. For this reason only the elastic theory of structures is pursued in this Chapter.

*Professor of Civil Engineering, University of Colorado, Boulder, Colorado.

25.1.2 Modeling of Structures

Looked at critically, all structures are assemblies of three-dimensional elements, the exact analysis of which is a forbidding task even under ideal conditions and impossible to contemplate under conditions of professional practice. For this reason, an important part of the analyst's work is the simplification of the actual structure and loading conditions to a model which is susceptible to rational analysis.

Thus, a structural framing system is decomposed into a slab and floor beams which in turn frame into girders carried by columns which transmit the loads to the foundations. Since traditional structural analysis has been unable to cope with the action of the slab, this has often been idealized into a system of strips acting as beams. Also, long-hand methods have been unable to cope with three-dimensional framing systems, so that the entire structure has been modeled by a system of planar subassemblies, to be analyzed one at a time. The modern matrix-computer methods have revolutionized structural analysis by making it possible to analyze entire systems, thus leading to more reliable predictions about the behavior of structures under loads.

Actual loading conditions are also both difficult to determine and to express realistically, and must be simplified for purposes of analysis. Thus, traffic loads on a bridge structure, which are essentially both of dynamic and random nature, are usually idealized into statically moving standard trucks, or distributed loads, intended to simulate the most severe loading conditions occurring in practice.

Similarly, continuous beams are sometimes reduced to simple beams, rigid joints to pin-joints, filler-walls are neglected, shear walls are considered as beams; in deciding how to model a structure so as to make it reasonably realistic but at the same time reasonably simple, the analyst must remember that each such idealization will make the solution more suspect. The more realistic the analysis, the greater will be the confidence which it inspires, and the smaller may be the safety factor (or factor of ignorance). Thus, unless code provisions control, the engineer must evaluate the extra expense of a thorough analysis as compared to possible savings in the structure.

25.1.3 Relation of Analysis and Design

The most important use of structural analysis is as a tool in structural design. As such, it will usually be a part of a trial-and-error procedure, in which an assumed configuration with assumed dead loads is analyzed, and the members designed in accordance with the results of the analysis. This phase is called the preliminary design; since this design is still subject to change, usually a crude, fast analysis method is adequate. At this stage, the cost of the structure is estimated, loads and member properties are revised, and the design is checked for possible improvements. The changes are now incorporated in the structure, a more refined analysis is performed, and the member design is revised. This project is carried to convergence, the rapidity of which will depend on the capability of the designer. It is clear that a variety of analysis methods, ranging from 'quick and dirty' to 'exact', is needed for design purposes.

An efficient analyst must thus be in command of the rigorous methods of analysis, must be able to reduce these to shortcut methods by appropriate assumptions, and must be aware of available design and analysis aids, as well as simplifications permitted by applicable building codes. An up-to-date analyst must likewise be versed in the bases of matrix structural analysis and its use in digital computers as well as in the use of available analysis programs or software.

25.1.4 Approaches in Structural Analysis

Structural analysis is based on the following building blocks:

1. Statics–Any portion of a structure must be in static equilibrium; in the more general case of dynamic analysis, it must satisfy Newton's Laws of Motion.
2. Geometry–The deformations of adjacent portions of a structure must be compatible with each other, so as to eliminate any jags, kinks, or other discontinuities which are inconsistent with the nature of the structure and its supports.
3. Force-deformation relations–In the methods presented here, elastic behavior is assumed; the classical relations of the theory of elastic bars therefore apply. In a more extended analysis which goes beyond the elastic range, inelastic force-deformation relations must be considered.

These ideas form the bases of the methods presented in this section. They are discussed further in any text on mechanics of materials, or mechanics of deformable bodies. In the presentation which follows, section 25.2 is concerned with the statics of structures; in section 25.3, deformations of statically determinate structures are analyzed. In section 25.4, the force method is introduced for the analysis of statically indeterminate structures, and the displacement method is covered in section 25.5. In section 25.6, the analysis for moving loads is presented. Section 25.7, finally, introduces the powerful finite-element method which is an extension of the displacement method applicable to a wider range of structures than can be handled by classical means.

25.1.5 References and Design Aids

References applicable to the study of structural analysis can conveniently be split into four distinct groups.

1. Those concerned with the principles underlying the topic, in particular, books on statics, mechanics of materials, and the theories of elasticity and inelasticity:

Shames, I. H., *Engineering Mechanics*, Vol. 1, Prentice-Hall, Englewood Cliffs, 1966.

Merriam, J. L., *Statics*, John Wiley and Sons, New York, 1966.

Crandall, S. H., and Dahl, N. C., *Mechanics of Solids*, McGraw-Hill, New York, 1959.

Popov, E. P., *Introduction to Mechanics of Solids*, Prentice-Hall, Englewood Cliffs, 1968.

2. The classical texts on the analysis of civil engineering structures, stressing mainly longhand computational methods:

Sutherland, H., and Bowman, H. L., *Structural Theory*, Wiley, New York, 1950.

Norris, C. H., and Wilbur, J. B., *Elementary Structural Analysis*, McGraw-Hill, New York, 1960.

Timoshenko, S. P., and Young, D. H., *Theory of Structures*, McGraw-Hill, New York, 1965.

3. The modern texts on the theory of structures, stressing matrix and computer methods, but still restricting their coverage to classical civil engineering structures such as beams, trusses, arches, and frames:

Martin, H. C., *Introduction to Matrix Methods of Structural Analysis*, McGraw-Hill, New York, 1966.

Laursen, H. I., *Structural Analysis*, McGraw-Hill, New York, 1969.

Gere, J. M., and Weaver, W. W., *Analysis of Framed Structures*, Van Nostrand Reinhold, New York, 1965.

4. Books dealing with the extension of matrix-computer methods to more general structures such as two- and three-dimensional elements, plates, and shells by means of the finite-element method:

Zienkiewicz, O. C., *The Finite Element Method in Engineering Science*, McGraw-Hill, New York, 1971.

Przemieniecki, J. S., *Theory of Matrix Structural Analysis*, McGraw-Hill, New York, 1968.

Design aids for structural analysis are often too limited to do justice to the wide variety of structural configurations faced by the designer, may, however, be useful for conceptual purposes, or for parameter studies. Several attempts to encompass results for a wide variety of framed structures have been made:

Kleinlogel, A., *Rigid Frame Formulas*, 11th Edition, Ungar, New York, 1952.

Leontovich, V., *Frames and Arches*, McGraw-Hill, New York, 1959.

Such compilations have been rendered somewhat obsolete by advances in computational methods and particularly by the availability of efficient computer programs. Tables of beam moments, influence ordinates, flexibilities, and stiffnesses continue to be useful and will be cited at appropriate places.

Since the presentation of this chapter aims at a general approach valid for a wide variety of structures, it does not provide scope for specialized treatment of specific structures. Thus, basic considerations and analysis methods which may be applicable to certain classes of structures such as arches, or high-rise buildings, have not been covered, but may be studied in specialized treatises such as:

McCullough, C. B., and Thayer, E. S., *Elastic Arch Bridges*, Wiley, New York, 1931.
Amirikian, A., *Analysis of Rigid Frames*, U.S. Government Printing Office, 1942.

The Portland Cement Association has published a series of valuable pamphlets dealing with specific methods and structures:

"Continuous Concrete Bridges,"
"Analysis of Rigid Frame Concrete Bridges,"
"Continuity in Concrete Building Frames," and
"Analysis of Small Reinforced Concrete Buildings for Earthquake Forces."

25.2 INTERNAL FORCES IN STRUCTURES

25.2.1 Introduction

The satisfaction of the equations of equilibrium is the most important aspect of structural analysis, as well as the simplest. If a structure cannot be in equilibrium with the applied loads, it cannot stand; conversely, if a sufficiently ductile structure is in static equilibrium, it will stand even though geometrical conditions may be violated. Furthermore, all of the subsequent material requires thorough familiarity with statics of structures.

25.2.2 Internal Forces

Analysis of framed structures is concerned with the determination of the internal member forces necessary for design. In the general case of three-dimensional structures, any member section may be subjected to three force components (two shear forces and one axial force), and to three moment components (two bending moments and one torsional moment), as shown in Fig. 25-1a. All of these quantities will be designated as forces, or stress resultants.

Special cases arise in planar structures, in which all out-of-plane forces are disregarded, leaving only one axial force, one shear force, and one bending moment, as shown in Fig. 25-1b. Other combinations are possible such as the case of combined torsion, shear, and bending moment in the bow girder of Fig. 25-1c, or the simple case of bending moment and shear in the beam of Fig. 25-1d. Differential relations between loads and internal forces are useful in the determination of these stress resultants.

Statically determinate structures are those in which all member forces can be found by statics alone. This requires that the number of available equilibrium equations matches the number of unknown forces. For such structures, it is convenient to determine the reactions first by application of appropriate equilibrium equations to a free body of the entire structure, or, in the case of articulated structures such as three-hinged arches, to free bodies consisting of portions of the structure. Once the reactions are determined, the internal forces are found by writing the equilibrium equations of free bodies containing these desired quantities as external forces; this requires cutting the structure at the section containing these forces.

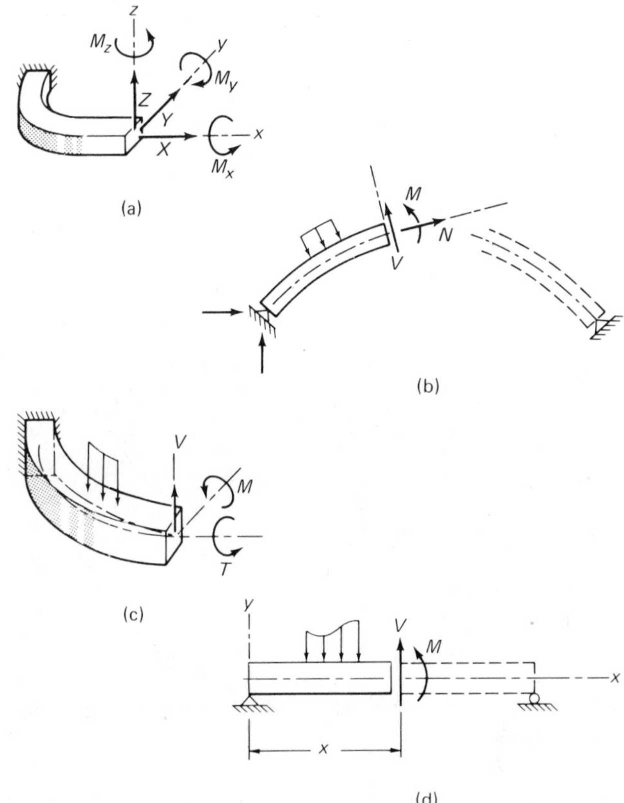

Fig. 25-1 Internal forces.

Once found, these forces are conveniently plotted along the structure to guide the designer in the sizing of members and layout of reinforcing. The most commonly used plots of this type are shear and moment diagrams, but the variation of other forces such as axial thrusts and torsional moments can be plotted similarly.

In the following examples, two different types of structures are analyzed.

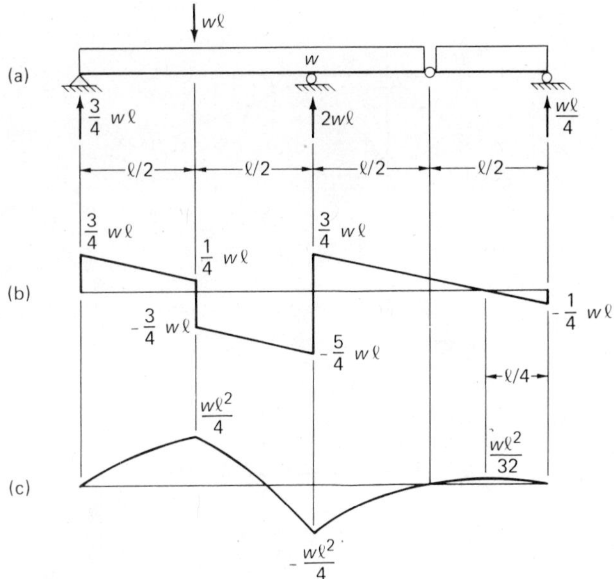

Fig. 25-2 (a) Beam and reactions; (b) shear diagram; and (c) moment diagram for example 25-1.

EXAMPLE 25-1: Draw shear and moment diagrams for the articulated beam of Fig. 25-2a.

SOLUTION: The reactions are found first and indicated in Fig. 25-2a; internal shear and moment along the beam are then determined by use of appropriate free bodies and drawn as in Fig. 25-2b and c. Note that at all points, the slope of the moment diagram is equal to the shear, and the extreme moment values occur at points of zero shear.

EXAMPLE 25-2: Calculate the internal forces in the horizontal cantilever bow girder of Fig. 25-3a due to its dead load, of intensity w lbs/ft.

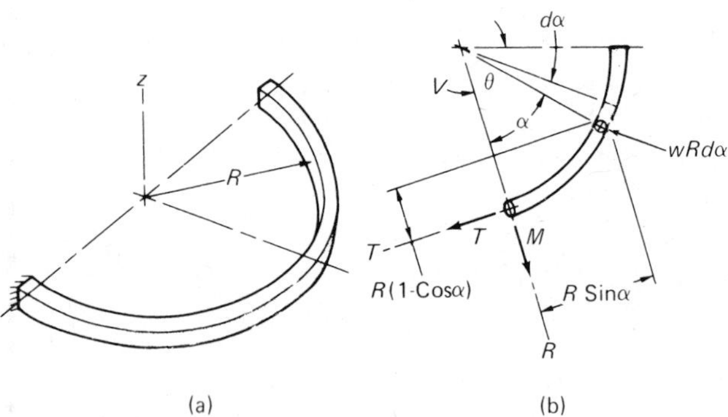

Fig. 25-3 (a) Cantilever bow girder and (b) free body diagram for example 25-2.

SOLUTION: We refer to the plan view of the free body of Fig. 25-3b, and write the equilibrium conditions

$$\Sigma F_Z = 0: -\int_{\alpha=o}^{\theta} wRd\alpha + V = 0; \quad V = wR\theta$$

$$\Sigma M_R = 0: -\int_{\alpha=o}^{\theta} (wRd\alpha)\,(R \sin \alpha) + M = 0;$$

$$M = wR^2\,(1 - \cos\theta)$$

$$\Sigma M_T = 0: \int_{\alpha=o}^{\theta} (wRd\alpha)\, R\,(1 - \cos\alpha) + T = 0;$$

$$T = -wR^2\,(\theta - \sin\theta)$$

25.2.3 Design Approximations, Aids, and References

-1. Some aids for calculation—The principle of superposition is a powerful tool in simplifying structural computations. In complicated loading systems, it often pays to consider only a part of the load at one time, to compute its effects, and add to find the total effect due to all parts. In this process, each part should be so simple that its contribution can be found with minimum effort.

EXAMPLE 25-3: Compute the reactions due to the distributed load shown in Fig. 25-4a.

SOLUTION: It is convenient here to break the load up into the two triangular distributions shown in Fig. 25-4b, indicate their resultants, calculate their individual effects on the reactions at A B, and add, leading to

$$R_A = 0; \quad R_B = -\tfrac{1}{2}\, w_o L.$$

Any loading applied to a symmetric structure can be decomposed into a symmetric and an antisymmetric part, and symmetry considerations applied to each portion, as shown in the next example.

EXAMPLE 25-4: Compute the reactions due to the distributed load shown in Fig. 25-5a.

SOLUTION: The load is broken up into the symmetric part of Fig. 25-5b, and the antisymmetric part of Fig. 25-5c, the reactions (or any other required functions) are calculated, and added to yield

$$R_A = \left(\tfrac{3}{8} + \tfrac{5}{24}\right) w_o L = \tfrac{7}{12}\, w_o L; \quad R_B = \left(\tfrac{3}{8} - \tfrac{5}{24}\right) w_o L = \tfrac{1}{6}\, w_o L$$

Many types of distributed loads can be expressed in the form of polynomials. With appropriate choice of origin, such distributions can be expressed in the form

$$p(x) = p_o \left(\frac{x}{l}\right)^n$$

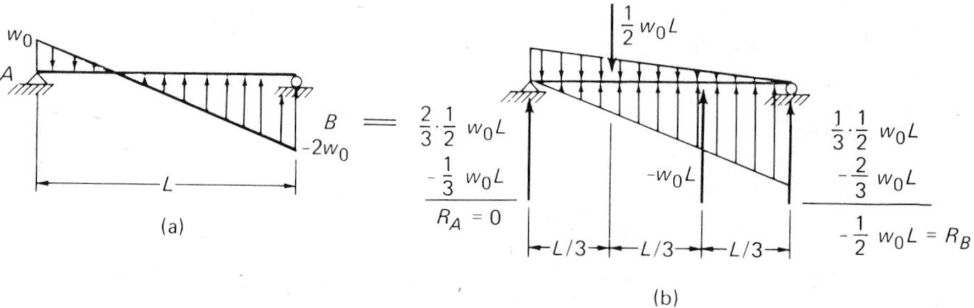

Fig. 25-4 Example 25-3.

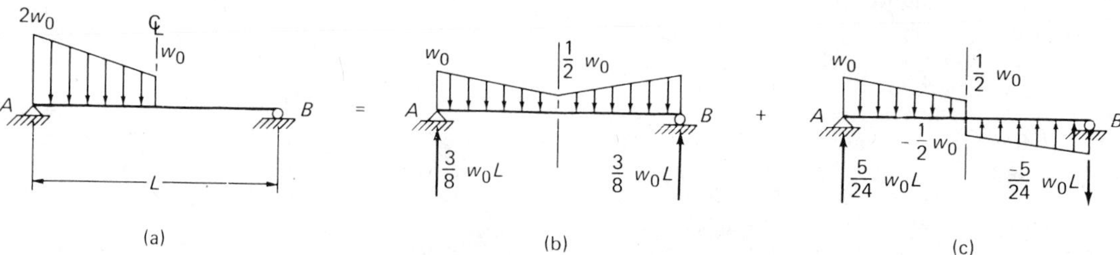

Fig. 25-5 Example 25-4.

TABLE 25-1 Beam Forces for Polynomial Loads

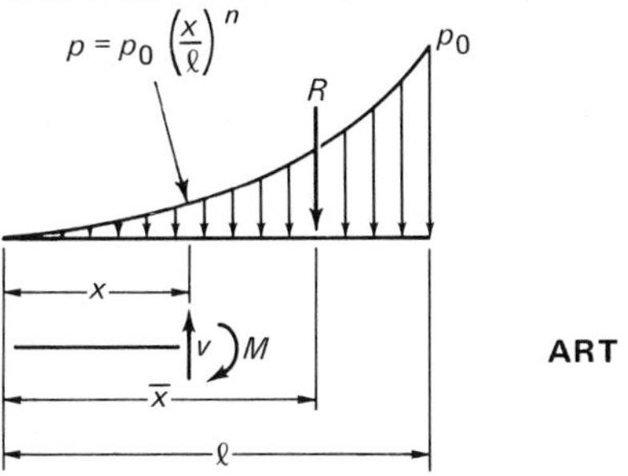

ART

Formula \ Multiplier for	$n = 0$ Const. Ld.	$n = 1$ Linear Ld.	$n = 2$ Parab. Ld.	$n = 3$	Common Factor
$R = \frac{1}{n+1} p_o l$	1	$\frac{1}{2}$	$\frac{1}{3}$	$\frac{1}{4}$	$\times p_o l$
$\bar{x} = \frac{n+1}{n+2} l$	$\frac{1}{2}$	$\frac{2}{3}$	$\frac{3}{4}$	$\frac{4}{5}$	$\times l$
$p = \left(\frac{x}{l}\right)^n p_o$	1	(x/l)	$(x/l)^2$	$(x/l)^3$	$\times p_o$
$V = \frac{1}{n+1}\left(\frac{x}{l}\right)^{n+1} p_o l$	(x/l)	$\frac{1}{2}(x/l)^2$	$\frac{1}{3}(x/l)^3$	$\frac{1}{4}(x/l)^4$	$\times p_o l$
$M = \frac{1}{n+2}\left(\frac{x}{l}\right)^{n+2} p_o l^2$	$\frac{1}{2}(x/l)^2$	$\frac{1}{3}(x/l)^3$	$\frac{1}{4}(x/l)^4$	$\frac{1}{5}(x/l)^5$	$\times p_o l^2$

where p_o represents a reference load intensity, l represents a loaded length, and n represents the order of the polynomial. In this case, the magnitude and location of the resultant, as well as the resultant internal forces, can be calculated by the formulas of Table 25-1, and used to compute reactions and forces. A more extensive tabulation of coefficients of the type shown in Table 25-1 can be found in *Structural Analysis*, by J. J. Tuma, Schaum's Outline Series, 1969, pp. 26–33.

Another useful device for handling difficult loadings is their replacement by appropriately placed concentrated load resultants, computing their effects individually, and summing. As the number of concentrated replacement loads increases, the approximate converges toward the exact solution. Such discretizing procedures form an important part of the field of numerical analysis. The approach is indicated by Fig. 25-6, which shows a uniform load on a cantilever beam replaced by its statically equivalent quarter-point resultants, and the individual moment diagrams superposed to form the final, approximate curve.

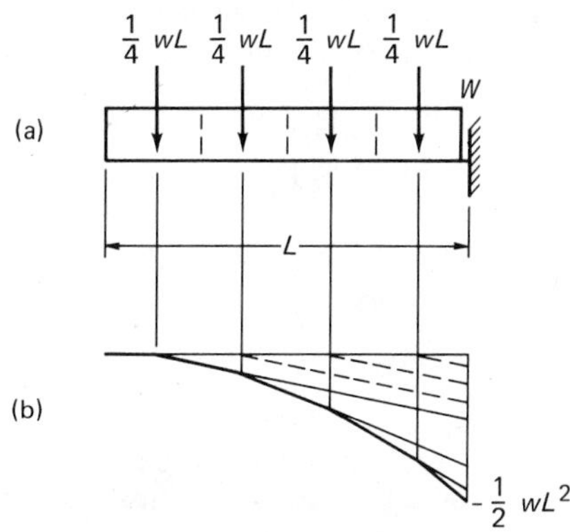

Fig. 25-6 (a) Discretized load, and (b) linearized moment diagram.

-2. References—For the principles and basic applications of the laws of statics, any one of the mechanics texts listed in section 25-1.5 will do.

The construction of shear and moment diagrams for beams, as well as the differential relations between loads, shear, and moments, is covered in texts of Mechanics of Materials such as the ones listed in section 25.1.5. The structures books listed in that section contain more advanced applications and calculations for member forces.

Design aids for the reactions and critical internal forces of statically determinate structures of standard configurations are available, for instance, see:

"Manual of Steel Construction," American Institute of Steel Construction, New York.

"Design Handbook," Concrete Reinforcing Steel Institute, Chicago.

25.3 DEFORMATIONS OF STATICALLY DETERMINATE STRUCTURES

25.3.1 Introduction

Knowledge of the deformations of structures is necessary for two purposes: a) to determine stiffness and deformation characteristics, and b) to supplement the equilibrium equations with geometrical conditions for the analysis of indeterminate structures.

Two main methods are available for the calculation of deformations:

1. The geometric methods—these include the double-integration method, and the methods of integration of curvature, such as the curvature-area method and the conjugate-frame methods. Only the integration of curvature method will be pursued in this section.
2. The energy methods—these include the methods of least work, strain-energy, complementary and potential

energy methods, and the method of virtual work. Only the latter will be outlined here, and extended to a matrix formulation suitable for computer work.

As presented here, all methods will be based on the assumption of small deformations which is adequate for most purposes.

25.3.2 The Geometric Approach for Flexural Deformations

-1. Angle change and tangential deviation—We define the following terms:

a. Curvature, Φ: The change of slope per unit length along the member. Fig. 25-7 shows that Φ can also be interpreted as the relative inclination of two transverse planes unity apart.
b. Angle change, θ_{BA}: The change of slope between two points A and B of a member.
c. Tangential deviation, δ_{BA}: The normal distance of point B on the deformed member from the tangent drawn to the deformed member at point A.

With these definitions, we consider the initially straight member AB of Fig. 25-8, subjected to a curvature Φ over an infinitesimal length ds. The geometry of this figure shows that due to this kink, of magnitude Φds

$$d(\theta_{BA}) = \Phi\, ds.$$

$$d(\delta_{BA}) = (\Phi\, ds)\, s.$$

Note that the origin of s is at point B.

Due to the curvature of all points between A and B

$$\theta_{BA} = \int_A^B \Phi\, ds \qquad (25\text{-}1)$$

$$\delta_{BA} = \int_A^B \Phi\, s\, ds \qquad (25\text{-}2)$$

With these relations, all deformations of flexural structures can be determined once the curvature Φ is known; this requires visualization of the geometry of the deformed shape. Note also that only geometrical relations are involved, so we conclude that these deformations can be due to any effect, such as elastic or inelastic bending, temperature or shrinkage; it is only necessary that the curvature Φ can be calculated.

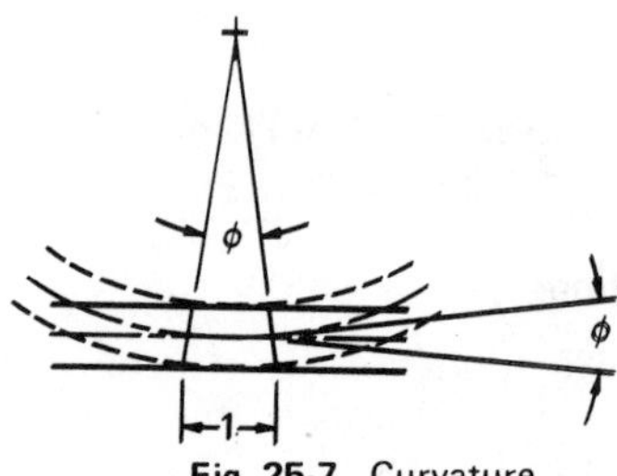

Fig. 25-7 Curvature.

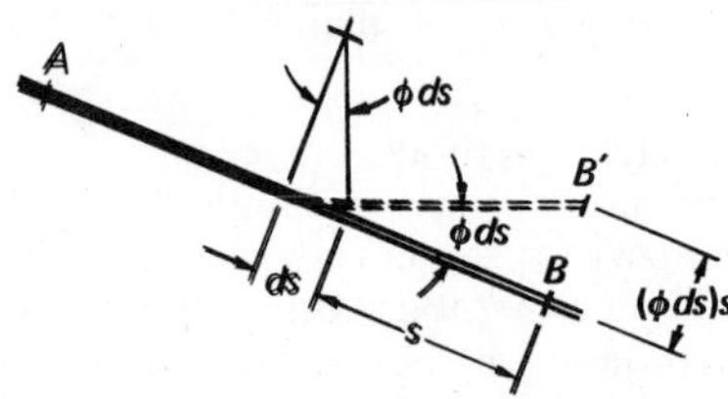

Fig. 25-8 Infinitesimal angle change and tangential deviation.

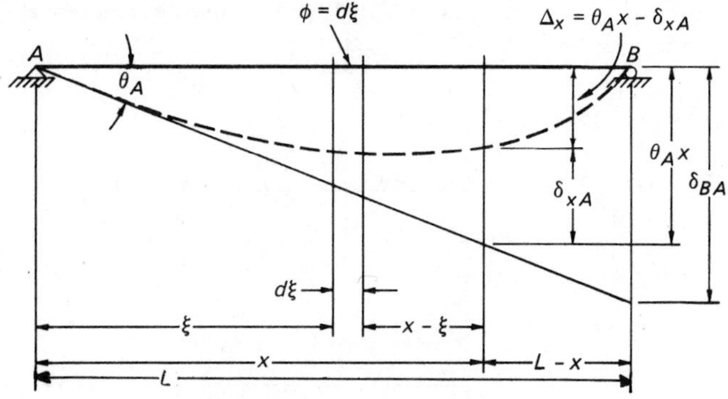

Fig. 25-9 Example 25-5.

EXAMPLE 25-5: The beam shown in Fig. 25-9 is subjected to a curvature $\Phi = ax$. Determine the end slope θ_A, and the equation of the elastic curve.

SOLUTION: From the geometry of the deformed shape of the beam of Fig. 25-9:

$$\theta_A = \frac{\delta_{BA}}{L} = \frac{1}{L}\int_{x=0}^{L} \Phi\,(L-x)\,dx = \frac{a}{L}\int_{x=0}^{L}(Lx - x^2)\,dx = \frac{aL^2}{6}$$

$$\Delta_x = \theta_A x - \delta_{xA}$$

$$= \frac{aL^2}{6}x - \int_{\xi=0}^{x}(a\xi)\,(x-\xi)\,d\xi = \frac{aL^3}{6}\left[\left(\frac{x}{L}\right) - \left(\frac{x}{L}\right)^3\right]$$

-2. The curvature-area theorems—We can interpret eqs. (25-1) and (25-2) by considering the plot of the curvature Φ along the member AB, as shown in Fig. 25-10. We observe

$$\theta_{BA} = \int_A^B \Phi\, ds = \text{Area under curvature diagram between } A \text{ and } B = \theta_{BA}.$$

$$\delta_{BA} = \int_A^B \Phi\, s\, ds = \text{Static moment of area under curvature diagram between } A \text{ and } B \text{ about point } B = \delta_{BA}.$$

These relations are called the first and second curvature-area theorems. The integration can be performed either analytically or by subdividing the area under the curvature diagram into suitable portions and evaluating the integrals numerically or graphically.

If the curvature is due to bending of an elastic beam of stiffness EI, then it is linearly related to the moment M:

$$\Phi = \frac{M}{EI} \qquad (25\text{-}3)$$

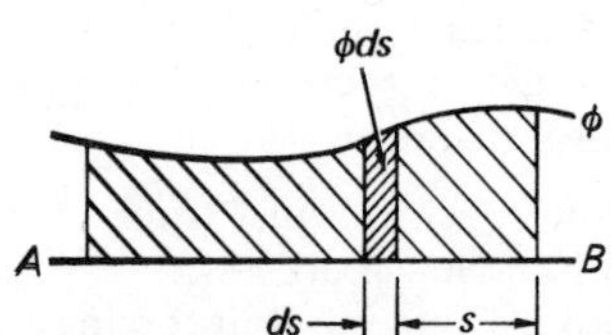

Fig. 25-10 Curvature-area concept.

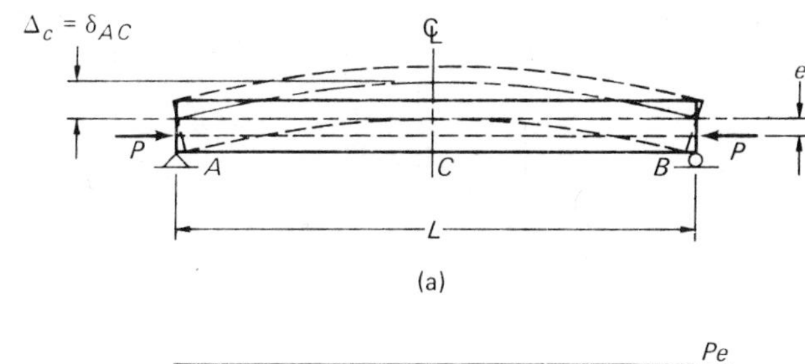

(a)

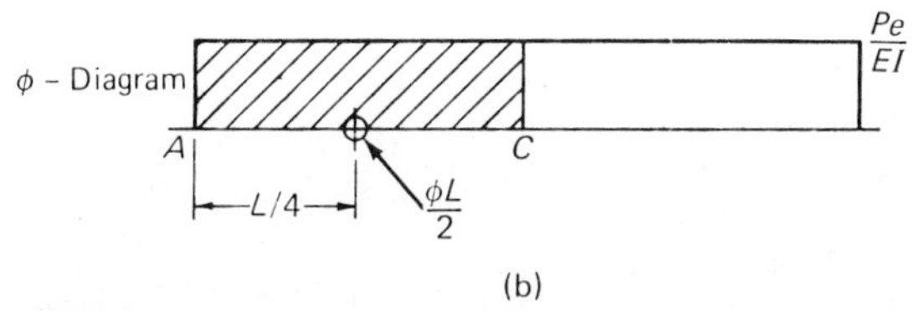

(b)

Fig. 25-11 Example 25-6. (a) Deformed shape, and (b) curvature diagram (ϕ-diagram).

EXAMPLE 25-6: The simple beam of flexural stiffness EI shown in Fig. 25-11a is prestressed by a tendon at constant eccentricity, leading to a constant curvature Φ (= Pe/EI). Determine the maximum deflection.

SOLUTION: From the geometry of the deformed shape of Fig. 25-11b, and the curvature diagram of Fig. 25-11b:

$$\Delta_C = \delta_{AC} = \left(\Phi \cdot \frac{L}{2}\right)\left(\frac{L}{4}\right) = \frac{\Phi L^2}{8} = \frac{PeL^2}{8EI}$$

EXAMPLE 25-7: The simple beam shown in Fig. 25-12a is subjected to a cranked-in end moment M at point B. Determine the end slope θ_A, and the maximum deflection.

SOLUTION: We draw the deflected shape of the beam as in Fig. 25-12a, and recognize the tangential deviation $\delta_{BA} = \theta_A \cdot L$, which can be computed by the second curvature-area theorem as the static moment of the area under the curvature diagram of Fig. 25-12b about point B; therefore

$$\theta_A = \frac{\delta_{BA}}{L} = \frac{1}{L}\left(\frac{1}{2}\frac{M}{EI}\cdot L\right)\left(\frac{L}{3}\right) = \frac{ML}{6EI}$$

(Compare this method to that used in ex. 25-5.)

The maximum deflection Δ_C occurs at point C, defined by $\bar{x}$, where the slope is zero. We use the first curvature-area theorem to find this point.

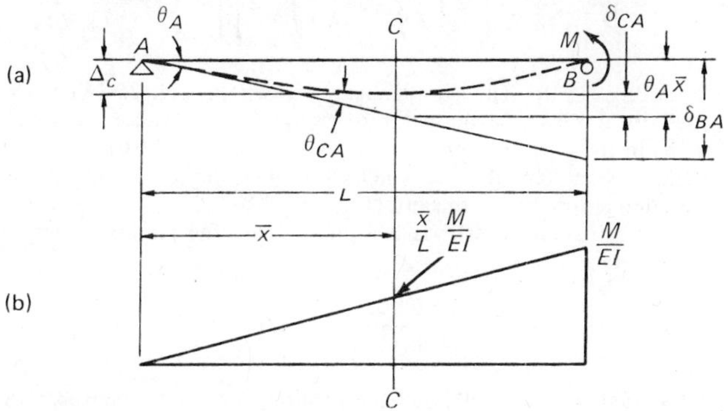

Fig. 25-12 Example 25-7. (a) Beam with EI = constant, and (b) curvature diagram (ϕ-diagram).

$$\theta_C = 0 = \theta_A - \theta_{CA} = \frac{ML}{6EI} - \frac{1}{2}\left(\frac{\bar{x}}{L}\frac{M}{EI}\right)\bar{x} = 0;\quad \bar{x} = \frac{L}{\sqrt{3}} = .578L$$

Having found the location $\bar{x}$ of the point of maximum deflection, we use the geometrical relations of Fig. 25-12a to find Δ_C.

$$\Delta_C = \theta_A \bar{x} - \delta_{CA}$$

Using again the second curvature-area theorem applied to the properties of Fig. 25-12b, and the previously determined values of θ_A and $\bar{x}$, we set

$$\Delta_C = \frac{ML}{6EI}\bar{x} - \left(\frac{1}{2}\frac{\bar{x}}{L}\frac{M}{EI}\cdot\bar{x}\right)\left(\frac{\bar{x}}{3}\right) = 0.0643\frac{ML^2}{EI}$$

25.3.3 Deformations by Virtual Work

-1. The theorem of virtual work—This theorem provides a tool for a unified formulation of both statical and geometrical equations. Before stating this theorem, we define the following:

a. Statically compatible force system: A set of external forces P^* and corresponding internal stresses σ^* which are in equilibrium with each other.
b. Geometrically compatible displacement system: A set of deformations Δ^{**} and corresponding internal strains ϵ^{**} which are related by the geometry of the deformed structure.

The theorem of virtual work states that for any structure in equilibrium

$$\Sigma P^* \cdot \Delta^{**} = \int (\sigma^* dA)\cdot(\epsilon^{**} ds) = \int_{\text{vol.}} \sigma^* \epsilon^{**}\, dV \tag{25-4a}$$

(ΣP^* and $\sigma^* dA$: Stat. Comp. Force System; Δ^{**} and $\epsilon^{**} ds$: Geom. Comp. Displace. System)

We define $\Sigma P^* \Delta^{**}$ as the external virtual work,

$$\int_{\text{vol.}} \sigma^* \epsilon^{**}\, dV$$

as the internal virtual work, so that the theorem can also be expressed as

External Virtual Work = Internal Virtual Work (25-4b)

There need be no relationship between the force system P^*, σ^*, and the displacement system Δ^{**}, ϵ^{**}, and the theorem is therefore not restricted to elastic structures.

-2. The theorem of virtual forces—For the purpose of finding deformations, the displacement system Δ^{**}, ϵ^{**} is real, but the force system P^*, σ^* can be assumed for convenience, that is to say, it is *virtual*; in this form, eq. (25-4) is called the Theorem of Virtual Forces. If the load P^* is assumed to be a unit load in the sense and location of the desired deformation, then this technique is called the Unit Load Method.

The internal virtual work, defined by $\int_{\text{vol.}} \sigma^* \epsilon^{**}\, dV$, will now be calculated for a linear member under different loading conditions; in all cases the single-starred forces are due to the unit virtual load, the double-starred deformations are due to whatever is the real cause of the distortions.

(1) axial load, N:

$$\text{Int. V. W.} = \int \left(\frac{N^*}{A} dA\right)(\epsilon^{**} ds) = \int_{\text{length}} N^* \epsilon^{**} ds$$

(2) bending moment, M:

$$\text{Int. V. W.} = \int_{\text{length}} M^* \Phi^{**} ds$$

(3) shear force, V:

[25-5(a–d)]

$$\text{Int. V. W.} = \int_{\text{length}} V^* \gamma^{**} ds$$

(γ^{**} is real shear strain)

(4) torsion moment, T:

$$\text{Int. V. W.} = \int_{\text{length}} T^* \theta^{**} ds$$

(θ^{**} is real unit angle of twist)

The internal virtual work is now inserted into eq. (25-4), with the virtual external force $P^* = 1$, so that, in the absence of specified support displacements

$$\Delta^{**} = \text{Internal Virtual Work} \qquad (25\text{-}6)$$

The internal virtual work due to the different types of distortion is additive. We also note that no assumption of elasticity has been made so that at this stage the method is valid for deformations due to any cause, such as elastic or inelastic bending, shrinkage, temperature gradients, or any other effect. It is only necessary to know the value of the double-starred distortions.

-3. The theorem of virtual forces for elastic structures— The distortions due to loads applied to elastic structures are related to the real internal forces:

$$\epsilon^{**} = \frac{N^{**}}{EA}; \quad \Phi^{**} = \frac{M^{**}}{EI};$$
$$\gamma^{**} = \frac{V^{**}}{G(kA)}; \quad \theta^{**} = \frac{T^{**}}{GC} \qquad [25\text{-}7(a\text{–}d)]$$

where E and G represent the material stiffnesses, and A, I, kA, and C represent, respectively, the axial, flexural, shear, and torsional stiffness factors of the member section.

In much of the literature, the real double-starred forces in eq. (25-7) are denoted by the upper-case symbols N, M, V, and T, and the virtual single-starred forces of eq. (25-5) are denoted by the lower-case symbols n, m, v, and t. Using this notation, inserting eq. [25-7(a–d)] into eq. [25-5(a–d)] and applying eq. 25-6, we find

$$\Delta = \int_L \frac{Nn\,ds}{EA} + \int_L \frac{Mm\,ds}{EI} + \int_L \frac{Vv\,ds}{G \cdot kA} + \int_L \frac{Tt\,ds}{GC} \qquad (25\text{-}8)$$

for an elastic structure under load causing internal axial and shear forces, and bending and twisting moments. For planar structures, no twisting is involved. In many cases, axial and shear deformations are sufficiently small so they can be neglected.

EXAMPLE 25-8: Calculate midspan deflection and end slope θ_B due to the uniform load on the restrained beam of Fig. 25-13a with the indicated end moments. Beam of constant stiffness *EI*.

SOLUTION: The real curvature diagram is drawn as in Fig. 25-13b, and the real curvature is written conveniently by parts

$$\Phi^{**} = \frac{M}{EI} = \frac{wL^2}{48\,EI}\left[23\left(\frac{x}{L}\right) - 24\left(\frac{x}{L}\right)^2 - 2\right]$$

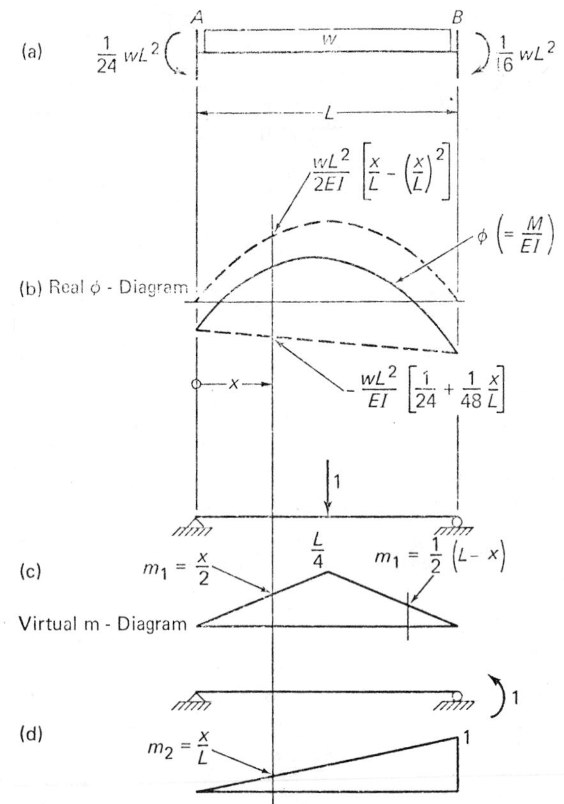

Fig. 25-13 Example 25-8. (a) Beam; (b) real curvature; (c) virtual load and moment for $\Delta_{℄}$; and (d) virtual load and moment for θ_B.

For the midspan deflection, we use the virtual unit force of Fig. 25-13c, due to which the virtual moment $m_1 = x/2$ for $0 \leqslant x \leqslant L/2$, and $m_1 = 1/2\ (1 - x)$ for $L/2 \leqslant x \leqslant L$. Use of eq. (25-8) leads to

$$\Delta = \frac{wL^2}{48\,EI}\left\{\int_{x=0}^{L/2}\left[23\left(\frac{x}{L}\right) - 24\left(\frac{x}{L}\right)^2 - 2\right]\left[\frac{x}{2}\right]dx + \int_{x=L/2}^{L}\left[23\left(\frac{x}{L}\right) - 24\left(\frac{x}{L}\right)^2 - 2\right]\cdot\frac{1}{2}\left[1 - x\right]dx\right\}$$

$$= \frac{5}{768}\frac{wL^4}{EI}$$

For the end slope θ_B, we use the virtual unit moment and moment diagram of Fig. 25-13d, so that equating external and internal work leads to

$$\theta_B = \int_{x=0}^{L}\frac{wL^2}{48\,EI}\left[23\left(\frac{x}{L}\right) - 24\left(\frac{x}{L}\right) - 2\right]\left[\frac{x}{L}\right]dx = \frac{1}{72}\frac{wL^3}{EI}$$

EXAMPLE 25-9: The simple beam shown in Fig. 25-14a is exposed to a transverse temperature gradient which causes increased strains in the extreme concrete surface of value + 0.001 in./in., and in the tension steel of value – 0.0005 in./in. Compute the maximum beam deflection due to this effect.

The real curvature Φ^{**} is computed from the strains shown in Fig. 25-14a as

$$\Phi^{**} = -\frac{0.0010 + 0.0005}{24\text{ in.}} = -62.5 \times 10^{-6}\text{ rad/in.}$$

The virtual unit load P^*, and the resulting virtual moment M^* are shown in Fig. 25-8b, from which

$$M^* = 0.5\,x\ (0 \leqslant x \leqslant 20\text{ ft})$$

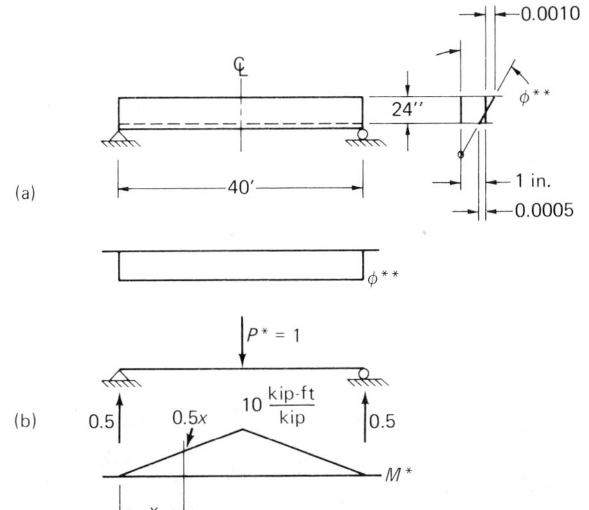

Fig. 25-14 Example 25-9. (a) Beam, strains, and curvature; and (b) virtual load and moments.

From Eq. 25-4, we equate external and internal virtual work; using symmetry:

$$1 \cdot \Delta = 2 \int_{x=0}^{20 \times 12} (-62.5 \times 10^{-6})\,(0.5\,x)\,dx = -1.8 \text{ in. (upward)}$$

25.3.4 Flexibilities

-1. The member flexibility—The flexibility of all or a portion of a structure is defined as the displacement due to a unit value of applied load.

The member flexibility, f, of a straight prismatic member under axial load, for instance, is its elongation due to a unit load, or $f = L/AE$.

For a beam segment, the rotational end flexibilities are of importance. We consider the simple prismatic beam shown in Fig. 25-15. We define the flexibility f_{ij} as the rotation at end i due to a unit moment at end j. There is a total of four flexibilities, the rotations at ends i = 1, 2, due to the unit moments at ends j = 1, 2. These four quantities can be found by the outlined methods of deformation analysis as

$$f_{11} = f_{22} = \frac{L}{3EI};\quad f_{12} = f_{21} = \frac{L}{6EI}$$

It is convenient to express the relationship between the end rotations θ and the end moments M in matrix form.

$$\begin{Bmatrix} \theta_1 \\ \theta_2 \end{Bmatrix} = \frac{L}{6EI}\begin{bmatrix} 2 & 1 \\ 1 & 2 \end{bmatrix}\begin{Bmatrix} M_1 \\ M_2 \end{Bmatrix} = [f]\,\{M\} \qquad (25\text{-}9)$$

in which the rotational member flexibility matrix $[f]$ is seen to contain the previously computed flexibilities.

-2. The structure flexibility—The structure flexibility, δ_{ij} represents the displacement i of a structure due to an applied unit load j. To define the quantities clearly, we introduce the concept of nodes. In Fig. 25-16, for instance, node 1 represents the applied moment as well as the rotation of the left end, and node 2 stands for an applied upward load as well as a vertical upward displacement at midspan. These loads can be related to the corresponding displacements by the structure flexibility matrix $[\delta]$:

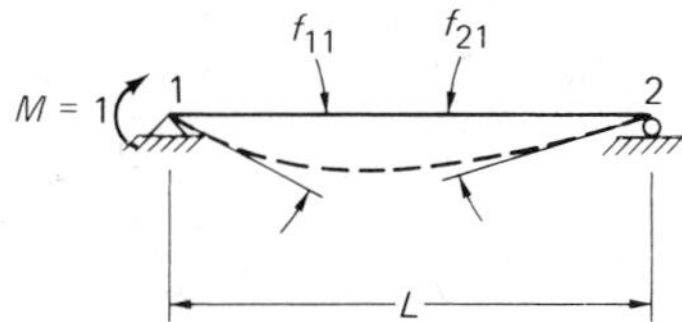

Fig. 25-15 Rotational end flexibilities.

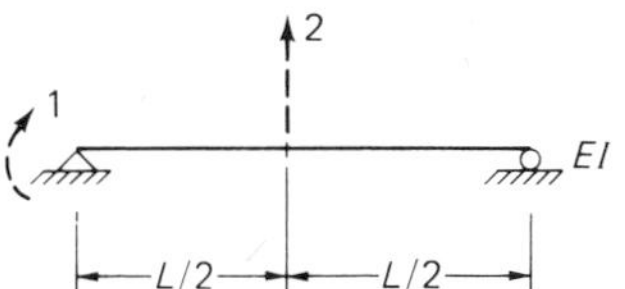

Fig. 25-16 Node designation.

$$\begin{Bmatrix} \theta_1 \\ \Delta_2 \end{Bmatrix} = \frac{L}{48EI}\begin{bmatrix} 16 & -3L \\ -3L & L^2 \end{bmatrix}\begin{Bmatrix} M_1 \\ P_2 \end{Bmatrix}$$

or

$$\{\Delta\} = [\delta]\ \{P\} \qquad (25\text{-}10)$$

where the elements of the structure flexibility method $[\delta]$ have been computed by any method. Once this matrix is known, the displacements $\{\Delta\}$ due to any combination of loads $\{P\}$ can be computed by matrix multiplication. If the displacements due to several loading conditions must be computed, the appropriate number of columns is added to the $[\Delta]$ and $[P]$ matrices.

We note that the flexibility matrices $[f]$ and $[\delta]$ are symmetric and have positive diagonal terms. The former fact is due to Maxwell's Law of reciprocal deformations, the latter follows from the fact that for $i = j$, applied load and the resulting displacement at the same node must be of the same sense.

25.3.5 Beam Deformations by Matrix Method

The matrix calculation of structure deformations is suitable for computer use, and forms one of the building blocks for computer analysis of indeterminate structures. We will develop the method by use of the theorem of virtual forces, discussed earlier, in matrix form. A convenient virtual applied load will be a unit load at the point and in the sense of the desired deformation. The internal moments due to this virtual load will be called m_i, where i defines the location of the moment m. The matrix of the moments m_{ij} due to applied unit loads at j is a force transformation matrix to be determined by statics and will be called $[b] \equiv [m_{ij}]$.

The real moments, M_{ij}, are the moments at i due to the real loads $[P]$ applied at j, and can also be represented by the force transformation matrix

$$[M_{ij}] = [b]\ [P]$$

EXAMPLE 25-10: For the beam of Fig. 25-17, write the $[b]$ matrix relating the moments at nodes 2 and 3 to vertical unit loads applied at these nodes. Then, evaluate the moments at these points due to P_2 = 2 kips, P_3 = 3 kips.

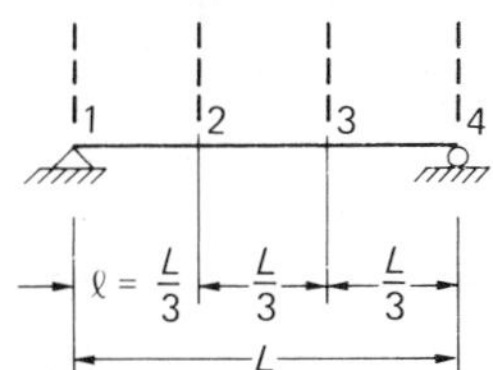

Fig. 25-17 Examples 25-10 and 25-11.

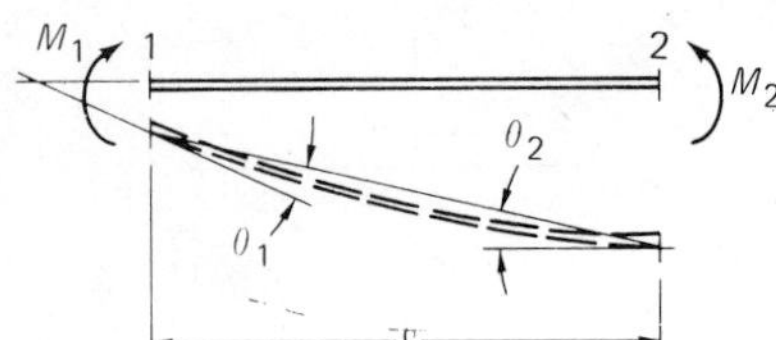

Fig. 25-18 Virtual work of beam element.

SOLUTION:

$$[b] = \begin{bmatrix} m_{22} & m_{23} \\ & \\ m_{32} & m_{33} \end{bmatrix} = \begin{bmatrix} \frac{2}{9} & \frac{1}{9} \\ & \\ \frac{1}{9} & \frac{2}{9} \end{bmatrix} L$$

$$\begin{Bmatrix} M_2 \\ \\ M_3 \end{Bmatrix} = [b] \begin{Bmatrix} P_2 \\ \\ P_3 \end{Bmatrix} = [b] \begin{Bmatrix} 2 \\ \\ 3 \end{Bmatrix} = \begin{Bmatrix} \frac{7}{9} \\ \\ \frac{8}{9} \end{Bmatrix} L \text{ (kips)}$$

We now consider the internal virtual work done in one beam element, subjected only to end moments, as shown in Fig. 25-18. This virtual work is given by the real end rotations θ_1 and θ_2 (measured with respect to the chord) multiplied by the corresponding virtual end moments m_1, m_2; this can be done in matrix form

$$\text{Internal Virtual Work} = [m_1 \quad m_2] \begin{Bmatrix} \theta_1 \\ \theta_2 \end{Bmatrix}$$

We now find the real end rotations θ_1 and θ_2 from the real moments M_1 and M_2 by recalling the solution for a simply supported prismatic beam of length l, subject to end moments M_1 and M_2.

$$\begin{Bmatrix} \theta_1 \\ \theta_2 \end{Bmatrix} = \frac{l}{6EI} \begin{bmatrix} 2 & 1 \\ 1 & 2 \end{bmatrix} \begin{Bmatrix} M_1 \\ M_2 \end{Bmatrix} \equiv [f] \; \{M\}$$

where the element flexibility matrix $[f]$ is given by eq. (25-9).

The internal virtual work in one element is then, denoting the element by i:

$$\text{Internal V.W.} = [m_1 \quad m_2] \; [f] \begin{Bmatrix} M_1 \\ M_2 \end{Bmatrix}$$

$$= [b^i]^T \; [f^i] \; [b^i] \; \{P\}$$

For the virtual work in the entire beam, we add the contributions from all elements.

$$1 \cdot \Delta = \sum_{\text{Elm'ts}} [b^i]^T \; [f^i] \; [b^i] \; \{P\}$$

$$= [[m_1^1 \quad m_2^1][m_1^2 \quad m_2^2] \cdots] \cdot \frac{l}{6EI} \begin{bmatrix} \begin{bmatrix} 2 & 1 \\ 2 & 2 \end{bmatrix} & & \\ & \begin{bmatrix} 2 & 1 \\ 1 & 2 \end{bmatrix} & \\ & & \ddots \end{bmatrix} \begin{Bmatrix} \begin{Bmatrix} m_1^1 \\ m_2^1 \end{Bmatrix} \\ \begin{Bmatrix} m_1^2 \\ m_2^2 \end{Bmatrix} \\ \vdots \end{Bmatrix} \{P\}$$

or

$$\{\Delta\} = [b]^T \; [f] \; [b] \; \{P\}. \qquad (25\text{-}11)$$

If we consider a beam with n elements, and desire the displacements at d points due to any combination of loads of these points, then these displacements will be contained in a $[\Delta]$ matrix of order $(d \times d)$, the $[b]$ matrix will be $(2n \times d)$, the $[f]$ matrix will be $(2n \times 2n)$, and the $[P]$ matrix will be $(d \times d)$.

If we consider displacements due to unit values of applied loads, we set the $[P]$ matrix $= \begin{bmatrix} 1 & & & \\ & 1 & & \\ & & \cdot & \\ & & & \cdot \end{bmatrix} = [I]$, and recognize the resulting matrix

$$[\delta] = [b]^T \; [f] \; [b] \qquad (25\text{-}12)$$

as the structure flexibility matrix already discussed earlier.

EXAMPLE 25-11: For the beam of Fig. 25-17, calculate the structure flexibility matrix for points 2 and 3. Then calculate the displacements at nodes 2 and 3 due to $P_2 = 2$ kips, $P_3 = 3$ kips.

SOLUTION:

$$[b] = \frac{L}{9} \begin{bmatrix} \begin{bmatrix} 0 & 0 \\ 2 & 1 \end{bmatrix} \\ \begin{bmatrix} 2 & 1 \\ 1 & 2 \end{bmatrix} \\ \begin{bmatrix} 1 & 2 \\ 0 & 0 \end{bmatrix} \end{bmatrix}; \quad [b]^T = \frac{L}{9} \left[\begin{bmatrix} 0 & 2 \\ 0 & 1 \end{bmatrix} \begin{bmatrix} 2 & 1 \\ 1 & 2 \end{bmatrix} \begin{bmatrix} 1 & 0 \\ 2 & 0 \end{bmatrix} \right]$$

$$[f] = \frac{\left(\frac{L}{3}\right)}{6EI} \begin{bmatrix} \begin{bmatrix} 2 & 1 \\ 1 & 2 \end{bmatrix} & & \\ & \begin{bmatrix} 2 & 1 \\ 1 & 2 \end{bmatrix} & \\ & & \begin{bmatrix} 2 & 1 \\ 1 & 2 \end{bmatrix} \end{bmatrix};$$

$$[\delta] = [b]^T \; [f] \; [b] = \frac{L^3}{486EI} \begin{bmatrix} 8 & 7 \\ 7 & 8 \end{bmatrix}$$

$$\begin{Bmatrix} \Delta_2 \\ \Delta_1 \end{Bmatrix} = [\delta] \begin{Bmatrix} P_2 \\ P_3 \end{Bmatrix} = \frac{L^3}{486EI} \begin{bmatrix} 8 & 7 \\ 7 & 8 \end{bmatrix} \begin{Bmatrix} 2 \\ 3 \end{Bmatrix} = \frac{L^3}{486EI} \begin{Bmatrix} 37 \\ 38 \end{Bmatrix}$$

Equation (25-12) for the determination of the structure flexibility matrix is perfectly general. For instance, in the case of a member subject to bending about one axis, torsion and axial load and the corresponding displacements at each end, the $[b]$ matrix will consist of four forces for each element, and the $[f]$ matrix will consist of one (4×4) matrix for each element.

The structure flexibility matrix is an important building block for the force method of indeterminate analysis which will be discussed in the next section.

25.3.6 Design Considerations and References

-1. Stiffness of concrete sections—In planar reinforced concrete structures, the question of appropriate choice of the flexural stiffness EI arises. While theoretically both variation of reinforcing and the effect of cracking should be considered, the *ACI Code* allows use of the gross concrete section for deformation calculations. The stiffening effect of slabs in floor systems has been discussed by:

Khan, F. R., and Sbarounis, J. A., "Interaction of Shear

Walls and Frames," *ASCE Proceedings, Journal of the Structural Division*, **90** (ST3), June 1964.

When torsional deformations must be considered, the torsional stiffness factor C is needed. For rectangular cross sections, the reference below lists this factor assuming elastic behavior of the gross section. For open cross sections composed of rectangular components, such as T- or L-shaped members, the total stiffness can be obtained by adding the stiffnesses of the individual rectangular components. Further information on this topic is in

Gerstle, K. H., *Basic Structural Design*, McGraw-Hill, New York, 1967.

-2. Design aids—For complicated load systems, the principle of superposition is useful, specially when tabulated or graphical information for the deflections due to simple component loads is available. Numerous sources of such information are available, for instance:

"Manual of Steel Construction," American Institute of Steel Construction, N. Y.

"Design Handbook," Concrete Reinforcing Steel Institute, Chicago, 1973.

"Tables of Deflection Coefficients for Simple Beams," *Engineering Experiment Station Bulletin No. 87*, Oklahoma State University, Stillwater, Okla., 1953.

Most of this information is in the form $\Delta = C \cdot PL^3/EI$ or $C \cdot wL^4/EI$, where C is called the deflection coefficient. Most of these data apply only to prismatic beams under standard load systems and support conditions. Other quick design information is available in form of nomographs, such as

Rose, F. O., "More on Deflections of Steel Beams," *Civil Engineering*, **39** (12), p. 49, Dec. 1969.

For other standard types of structures, some information on deformations is also available; for instance, the end deformations of circular bow girders of various types under loading normal to their plane are represented graphically in

Hogan, M. B., "Circular Beams Loaded Normal to the Plane of Curvature," *Journal of Applied Mechanics*, 5 (2), June 1938.

Beam deflections can be controlled by specifying an allowable length-to-depth ratio, L/d, according to the following reasoning: For a balanced beam design, the maximum curvature occurs at the point of maximum moment, and is, according to standard reinforced concrete theory,

$$\Phi_{max} = \frac{1}{E_c d}\left(f_{c\,All} + \frac{f_{s\,All}}{n}\right)$$

The maximum beam deflection, Δ, can, by geometric methods, be linearly related to the maximum curvature:

$$\Delta = K\,\Phi_{max}\,L^2$$

or

$$\left(\frac{L}{d}\right)_{All} = \frac{1}{K}\,\frac{E_c}{f_{c\,All} + \dfrac{f_{s\,All}}{n}} \cdot \frac{\Delta}{L} \qquad (25\text{-}13)$$

Values for the factor K, which depends on the loading and support conditions, are given in Table 25-2, and $E_c/[f_{c\,All} + (f_{s\,All}/n)]$ can be considered as a balanced design constant. With the allowable deflection ratio Δ/L

TABLE 25-2 Factors for eq. 25-13

Beam and Loading	*K*
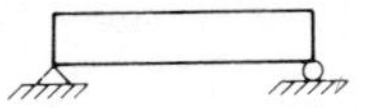 Simply supported, uniform load	5/48
 Fixed both ends, uniform load	1/32
 Fixed one end, simply supported other, uniform load	8/185
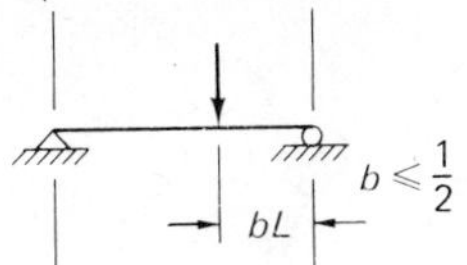 Simply supported, concentrated load at bL, $b \leqslant \frac{1}{2}$	$\frac{1}{27}(1+b)\sqrt{3(1-b^2)}$
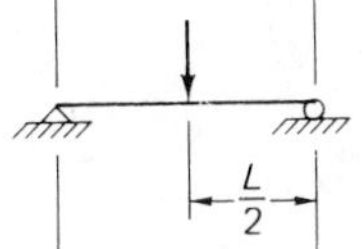 Simply supported, concentrated load at $\frac{L}{2}$	1/12
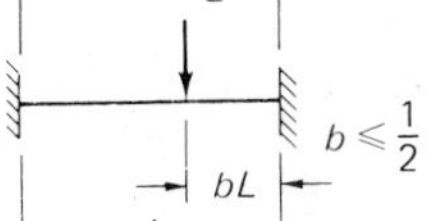 Fixed both ends, concentrated load at bL, $b \leqslant \frac{1}{2}$	$\frac{2}{3}\frac{(1-b)b}{(3-2b)^2}$
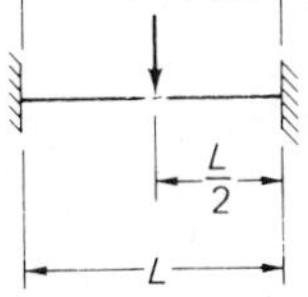 Fixed both ends, concentrated load at $\frac{L}{2}$	1/24

specified, the allowable span-depth ratio of the beam can be found from eq. 25-13.

25.4 THE FORCE METHOD

25.4.1 Static Indeterminacy of Structures

Most reinforced concrete structures are statically indeterminate, that is, the number of unknown forces exceeds the number of available equilibrium equations, so that additional equations of geometry are needed for a rigorous analysis. The excess of unknown forces over the number of independent equilibrium equations is called the degree of static indeterminacy or redundancy. For an analysis by the force method, the degree of static indeterminacy must first be found.

An externally statically indeterminate structure is one which has a larger number of external reactions than equilibrium equations; for such a structure, the indeterminate analysis deals with the determination of the redundant reactions; when these are found, the remaining forces can be found by statics. Continuous beams, single-story rigid frames, single and multiple arches fall in this category, as shown by the examples of Fig. 25-19. Under each of the structures shown in this figure, e denotes the number of equilibrium equations, f denotes the number of reactions, and $r = f - e$ denotes the degree of static in terminacy of the structure.

Internally statically indeterminate structures have a larger number of internal forces than available independent equilibrium equations. Among this type are ring- or circuit-type structures, such as shown in Fig. 25-20. The structures of Fig. 25-19 can also be treated as internally indeterminate structures if internal forces are considered as redundants, as discussed in the next section. For the determination of the degree of indeterminacy of such structures, it is often convenient to introduce sufficient *cuts* or *releases* to reduce the structure to static determinacy, and to count up the number of redundant forces removed. In the three-story, three-bay frame of Fig. 25-21, three internal forces (shear, moment, and axial force) are removed with each of the cuts shown, so that the structure is revealed as statically indeterminate to the 27th degree.

25.4.2 The Method of Consistent Displacements

In the force method, the equations of statics are supplemented by a number of equations of geometry equal to the degree of indeterminacy, r, of the structure. The r un-

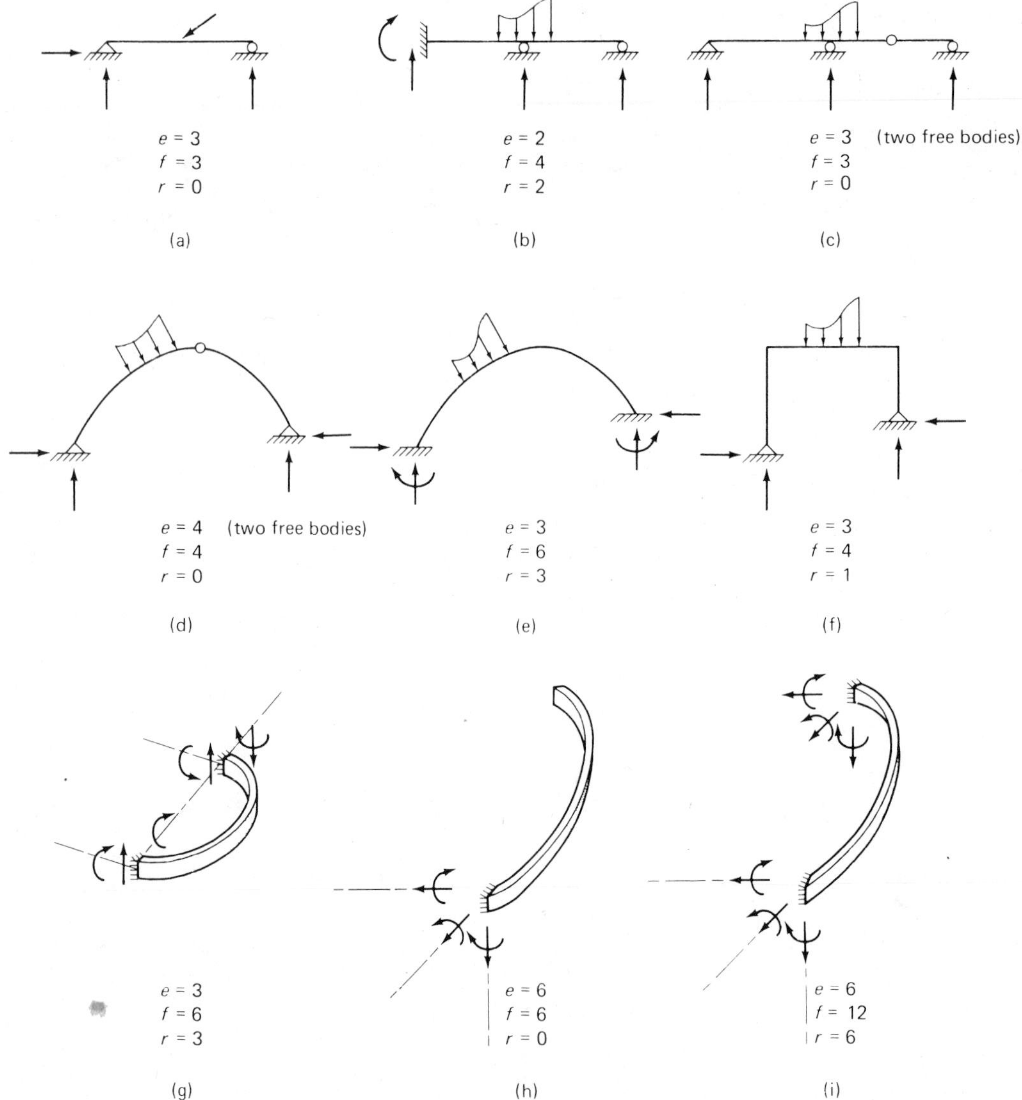

Fig. 25-19 External indeterminancy.

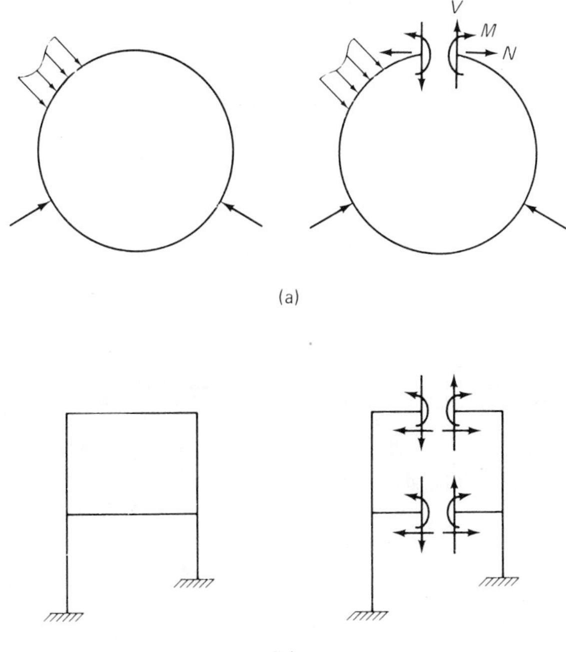

Fig. 25-20 Internal indeterminacy. (a) *left*—Ring structure; (a) right—cut structure; (b) *left*—two-story, single-bay frame; and (b) right—cut structure.

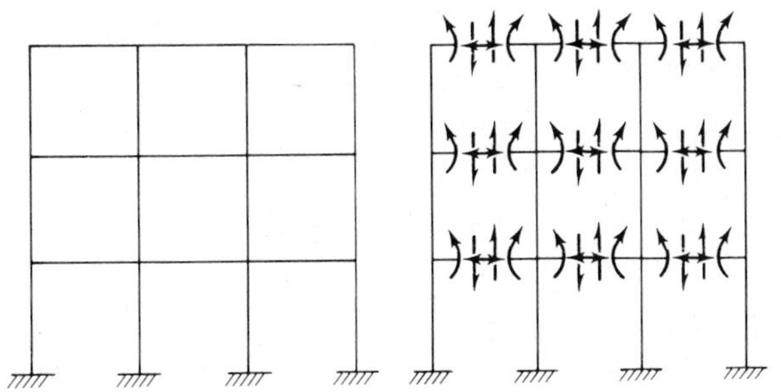

Fig. 25-21 Three-story, three-bay frame.

known, or redundant, forces are the unknowns in these equations which express the consistency of the displacements with the specified support and continuity conditions. Hence the method in its classical formulation is called the method of consistent displacements.

The method will be outlined step by step in terms of the two times statically indeterminate continuous beam of Fig. 25-22a.

Step 1 Reduce the structure to static determinacy by removing r forces (the redundants). The r points at which forces have been removed are called the cuts. The resulting structure is called the primary structure, shown in Fig. 25-22b.

Note that the conditions of geometry of the actual structure are violated at the r cuts. The aim of the subsequent calculations is to calculate those values of the r redundants necessary to restore the continuity at the cuts.

Step 2 Calculate the displacements Δ_i^0 at cuts i of the primary structure due to the applied loads. In general, r displacements Δ^0 will have to be computed for each loading condition.

Step 3 Calculate the displacements at the cuts i of the primary structure due to unit values of the redundants at j,

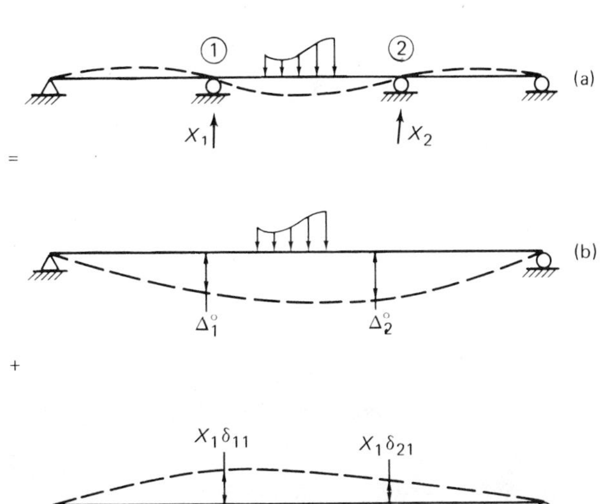

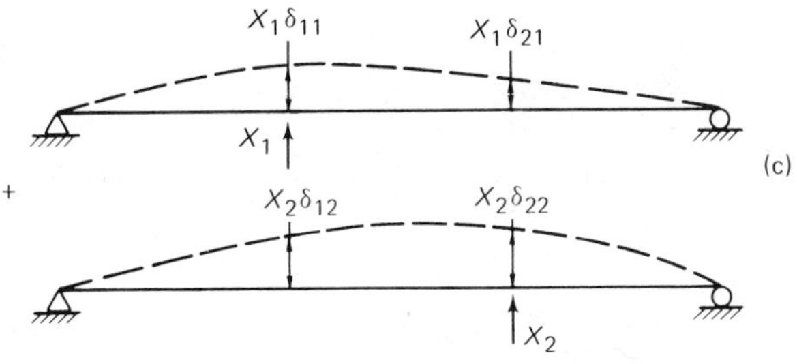

Fig. 25-22 Force method. (a) Real beam, two times statically indeterminate; (b) primary system; and (c) effects of redundants.

shown in Fig. 25-22c. These are the flexibilities δ_{ij}, of total number $r \times r$.

Note that the flexibilities are properties of the structure, not of the loading, and will therefore be the same irrespective of the loading condition. By the principle of superposition, the displacement at point i due to the actual redundant at point j, X_j will be $X_j\delta_{ij}$.

Step 4 Restore continuity at the r cuts by adding the displacements for each cut i due to the applied loads, Δ_i^0, and due to all redundants, $X_j\delta_{ij}$, and equating to the specified displacements, which, for the problem illustrated, are zero.

For $r = 2$:

$$\begin{aligned} \sum \Delta_1 &= 0: \Delta_1^0 + X_1\delta_{11} + X_2\delta_{12} = 0 \\ \sum \Delta_2 &= 0: \Delta_2^0 + X_1\delta_{21} + X_2\delta_{22} = 0 \end{aligned} \tag{25-14}$$

These r continuity equations have to be solved for the r unknown redundants, X_j. Once the redundants are found, all other forces can be determined by statics.

In the case of specified support displacements Δ_1 and Δ_2 at the points of redundancy, these values must be substituted for the zeros on the right-hand side of the equations. We note that in the case of statically indeterminate structures (as contrasted with determinate structures), support displacements can cause internal forces which must be considered in design.

An important consideration in solving statically indeterminate structures is the choice of the primary system. In general, the closer the structural action of the primary system resembles that of the real structure, the better conditioned will be the resulting set of simultaneous continuity equations, and the less the roundoff errors of the numerical solution. Another consideration is the ready availability of flexibilities, so that decomposition of the structure into standard components with known flexibilities (such as the rotational end flexibilities of prismatic members) should be considered. In this light, the force method of analysis of a

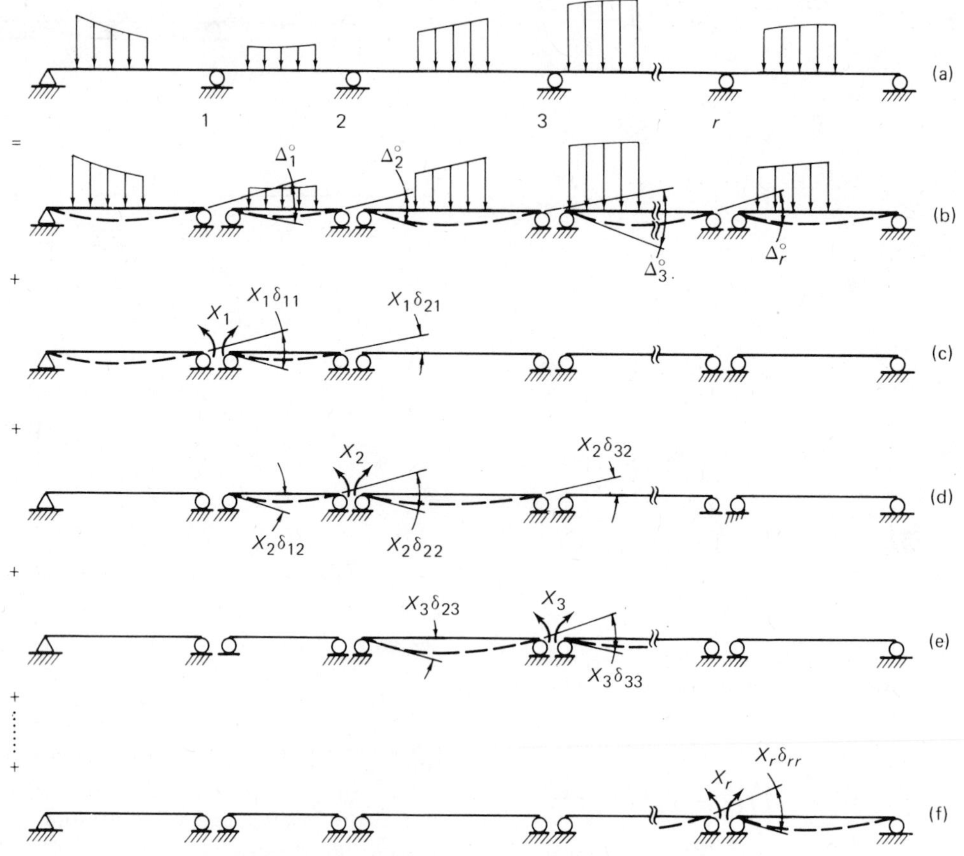

Fig. 25-23 Force method analysis of continuous beam.

continuous beam of $(r + 1)$ spans might be undertaken as shown in Fig. 25-23; the beam is cut over each interior support, releasing the r support moments which become the redundant forces; the requirement of slope continuity of the real beam must now be satisfied by adding the slope changes (or 'kinks') over the supports due to all effects on the primary system, and equating their sum to zero.

Figs. 25-23b through 25-23f show the primary system under the applied load, and the redundant moments, applied one at a time. With the redundants and displacements as defined in these figures, the r continuity conditions become:

The Δ_s^0 are the differences of the end rotations of the simple spans due to the specified loads.

It is also noted that, in contrast to the action of the redundants chosen in Fig. 25-22, in this case the effect of each redundant is felt only in the adjacent beam spans, so that each continuity equation contains only three redundants, all other terms being zero. For this reason this version of the force method is called the 'three-moment equation'. When expressed in matrix form, it leads to a 'tri-diagonal matrix' which can be solved very efficiently.

We observe that the labor of analysis can be vastly reduced by appropriate choice of the primary system.

$$
\begin{array}{llllllll}
\sum\theta_1 = 0: & X_1\delta_{11} & + X_2\delta_{12} & + 0 & + 0 & + 0 + \cdots + 0 & = -\Delta_1^0 \\
\sum\theta_2 = 0: & X_1\delta_{21} & + X_2\delta_{22} & + X_3\delta_{23} & + 0 & + 0 + \cdots + 0 & = -\Delta_2^0 \\
\sum\theta_3 = 0: & 0 & + X_2\delta_{32} & + X_3\delta_{33} & + X_4\delta_{34} & + 0 + \cdots + 0 & = -\Delta_3^0 \\
\vdots & & & & & & \\
\sum\theta_r = 0: & 0 & + 0 & + 0 & + 0 & + 0 + \cdots + X_r\delta_{rr} & = -\Delta_r^0.
\end{array}
$$

In this case, the flexibilities are the differences of the appropriate end rotations due to cranked-in moments on prismatic members, that is

$$\delta_{ij} = \left(\frac{L}{3EI}\right)_i + \left(\frac{L}{3EI}\right)_{i+1} \quad \text{for } i = j, \quad \text{and} \quad \delta_{ij} = \frac{L}{6EI} \text{ for } i \neq j$$

EXAMPLE 25-12: Analyze the beam of Fig. 25-24a due to the following effects: (a) a uniform load w over the entire beam, and (b) a support settlement Δ of Point B.

SOLUTION: The beam is statically indeterminate to the second degree; it is decomposed into the primary system by removing the moments M_1 and M_2 which become the redundants. Figs. 25-24b,

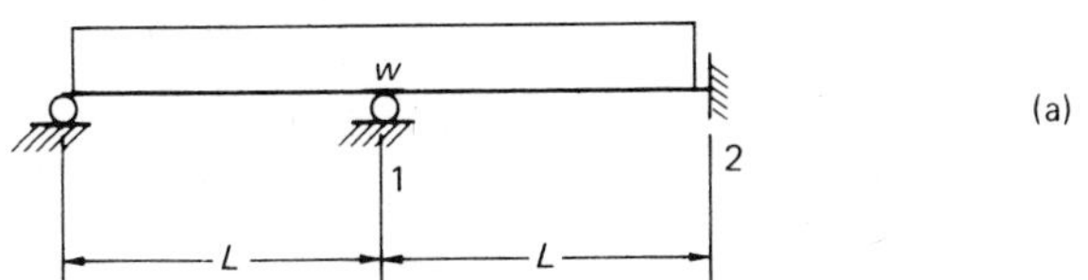

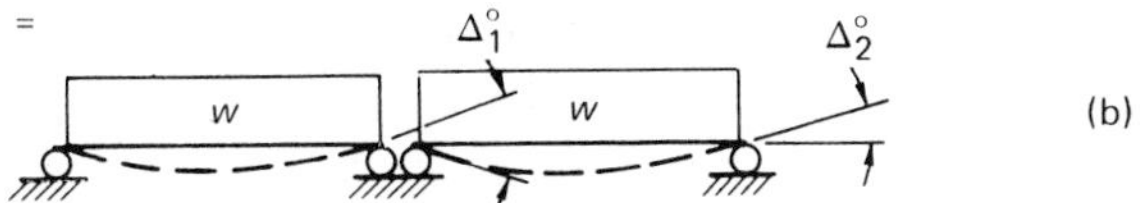

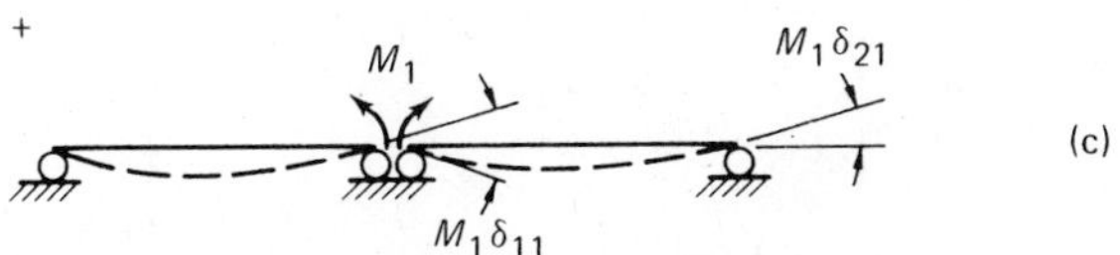

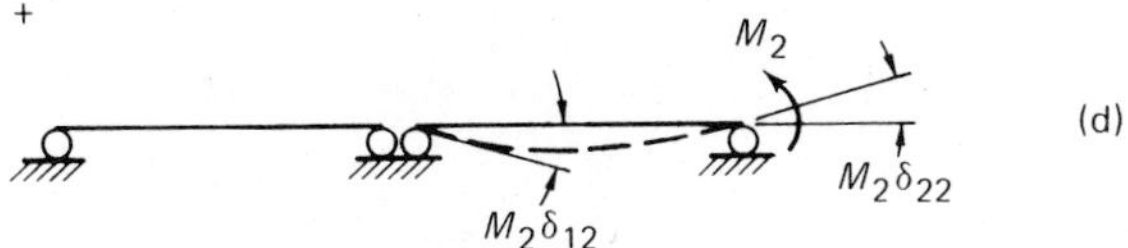

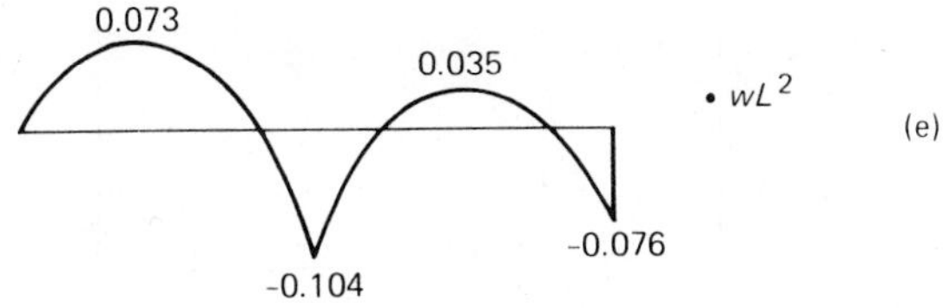

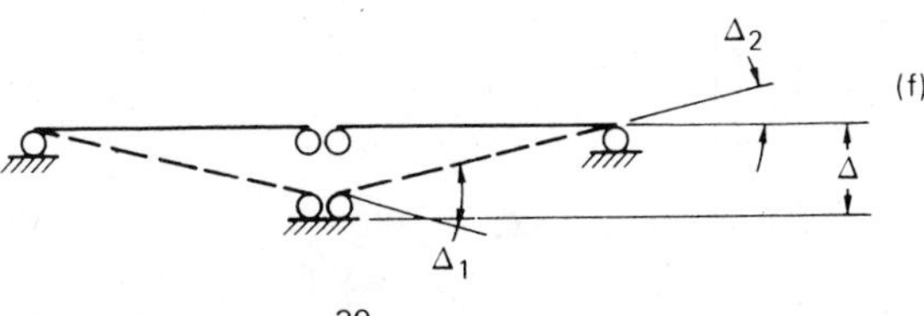

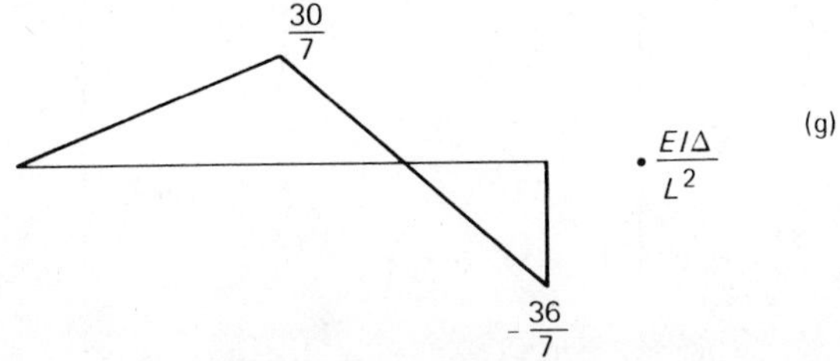

Fig. 25-24 Example 25-12.

c, and d show the relative slope changes which are set equal to zero:

$$\sum_{\theta 1} = 0: \Delta_1^0 + M_1\delta_{11} + M_2\delta_{12} = 0$$

$$\sum_{\theta 2} = 0: \Delta_2^0 + M_1\delta_{21} + M_2\delta_{22} = 0$$

where the displacements are calculated for part (a) by any method as

$$\Delta_1^0 = 2 \cdot \frac{wL^3}{24EI} = \frac{1}{12}\frac{wL^3}{EI}; \quad \Delta_2^0 = \frac{1}{24}\frac{wL^3}{EI}$$

$$\delta_{11} = 2 \cdot \frac{L}{3EI}; \quad \delta_{22} = \frac{L}{3EI}; \quad \delta_{12} = \delta_{21} = \frac{L}{6EI}.$$

so that

$$\frac{2}{3}M_1 + \frac{1}{6}M_2 = -\frac{1}{12}wL^2$$

$$\frac{1}{6}M_1 + \frac{1}{3}M_2 = -\frac{1}{24}wL^2$$

from which $M_1 = -0.104\ wL^2$, and $M_2 = -0.076\ wL^2$.

The moment diagram is shown in Fig. 25-24e.

Fig. 25-24f shows the deformed primary system due to the support settlement of part (b), from which the kinks at supports 1 and 2 are calculated as

$$\Delta_1 = -2\frac{\Delta}{L}, \quad \Delta_2 = +\frac{\Delta}{L}$$

The kinks due to the redundants are as before, so that the conditions of geometry become

$$\frac{2}{3}M_1 + \frac{1}{6}M_2 = 2\frac{EI}{L^2}\Delta$$

$$\frac{1}{6}M_1 + \frac{1}{3}M_2 = -\frac{EI}{L^2}\Delta.$$

from which $M_1 = \frac{30}{7}\frac{EI}{L^2}\Delta$; $M_2 = -\frac{36}{7}\frac{EI}{L^2}\Delta$.

The moment diagram due to the specified support settlement Δ is shown in Fig. 25-23g.

25.4.3 Fixed-End Moments

Fixed-end moments (FEMs) are the end-moments of a structure when all ends are fixed against rotation and translation. They are calculated by the force method. Fixed-end moments are one of the building blocks used in the displacement method of analysis.

EXAMPLE 25-13: Calculate FEMs at ends A and B of the prismatic beam of Fig. 25-25 due to a load P at a from one end, b from the other.

SOLUTION: The structure is two times indeterminate; we consider the desired end moments as redundants. Calculate deformations of the primary structure by any method.

$$\Delta_A^0 = \frac{1}{6}\frac{Pab}{EIL}(a + 2b); \quad \Delta_B^0 = \frac{1}{6}\frac{Pab}{EIL}(2a + b)$$

$$\delta_{AA} = \frac{1}{3}\frac{L}{EI} = \delta_{BB}; \quad \delta_{BA} = \frac{1}{6}\frac{L}{EI} = \delta_{AB}$$

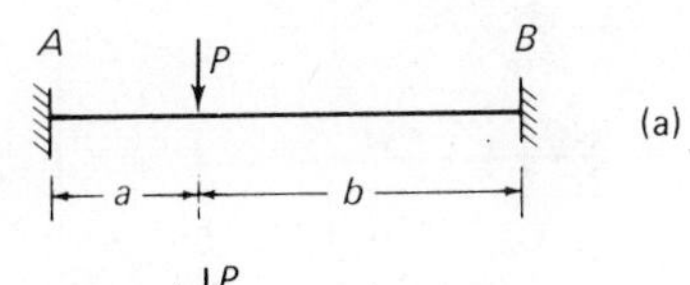

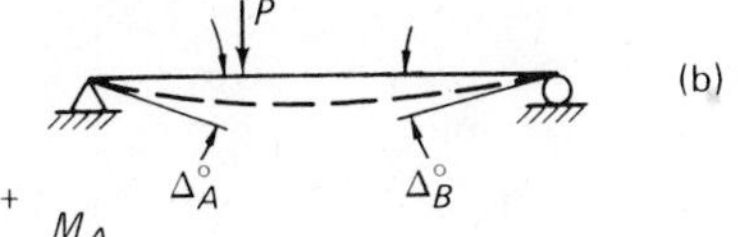

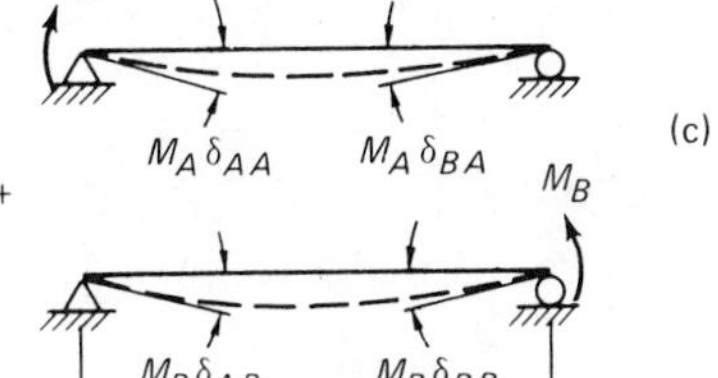

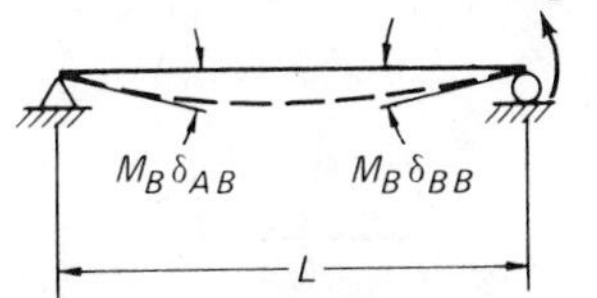

Fig. 25-25 Example 25-13. Fixed-end moments. (a) Real beam; (b) primary system; and (c) flexibilities.

Set up equations of geometry:

$$\sum \theta_A = 0: M_A \frac{1}{3}\frac{L}{EI} + M_B \frac{1}{6}\frac{L}{EI} = -\frac{1}{6}\frac{Pab}{L}(a + 2b)$$

$$\sum \theta_B = 0: M_A \frac{1}{6}\frac{L}{EI} + M_B \frac{1}{3}\frac{L}{EI} = -\frac{1}{6}\frac{Pab}{L}(2a + b)$$

Solving simultaneously

$$M_A = -\frac{Pab^2}{L^2} = \text{FEM at } A$$

$$M_B = -\frac{Pab^2}{L^2} = \text{FEM at } B. \tag{25-15}$$

TABLE 25-3 Fixed-End Moments for Prismatic Beams

Load Type	M_A^F	M_B^F
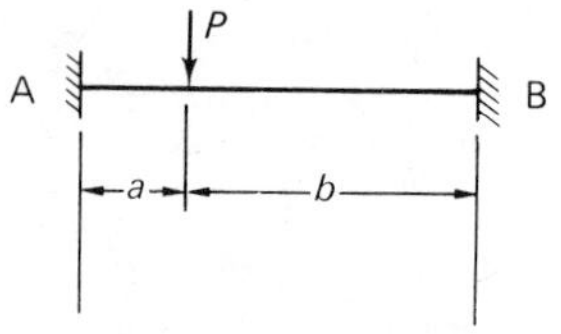 Concentrated load P at distances a from A and b from B	$+\frac{Pab^2}{L^2}$	$-\frac{Pa^2b}{L^2}$
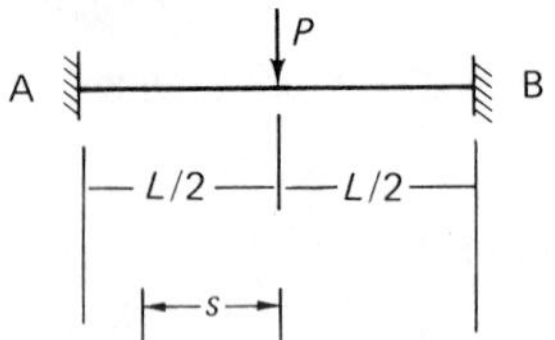 Concentrated load P at midspan ($L/2$, $L/2$)	$+\frac{PL}{8}$	$-\frac{PL}{8}$
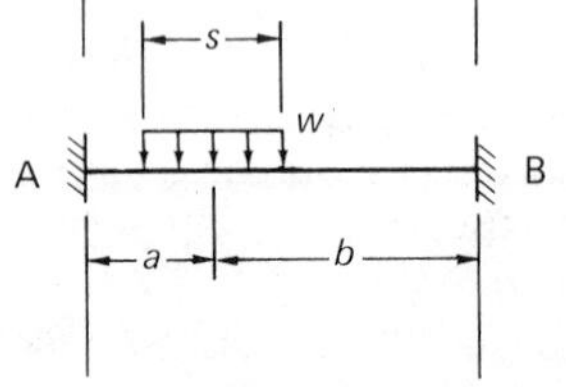 Uniform load w over length s centered at distance a from A, b from B	$+\frac{ws}{12L^2}[12ab^2 + s^2(L - 3b)]$	$-\frac{ws}{12L^2}[12a^2b + s^2(L - 3a)]$
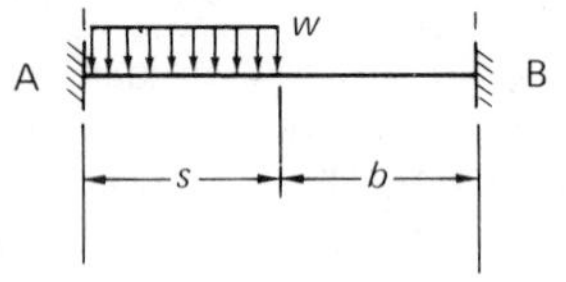 Uniform load w over length s from A; b to B	$+\frac{ws^2}{12L^2}[2L(3L - s) + 3s^2]$	$-\frac{ws^3}{12L^2}(4L - 3s)$
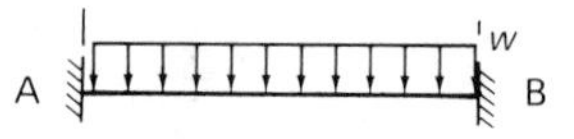 Uniform load w over full span	$+\frac{wL^2}{12}$	$-\frac{wL^2}{12}$
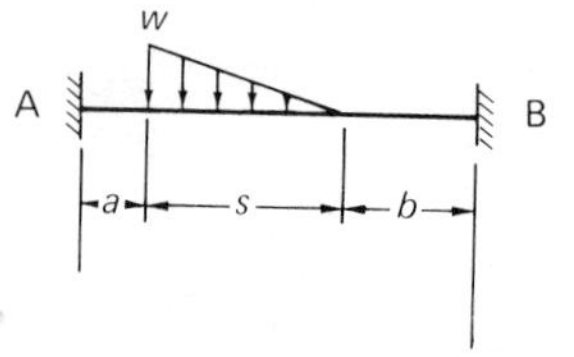 Triangular load w over length s, starting at a from A; b to B	$+\frac{ws}{60L^2}[10b^2(3a + s) + s^2(15a + 10b + 3s) + 40abs]$	$-\frac{ws}{60L^2}[10a^2(3b + 2s) + s^2(10a + 5b + 2s) + 20abs]$
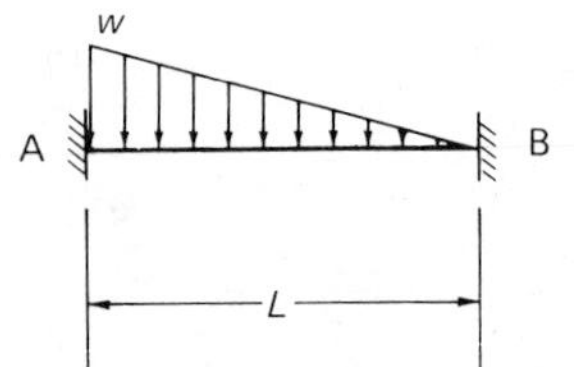 Triangular load w at A decreasing to zero at B, span L	$+\frac{wL^2}{20}$	$-\frac{wL^2}{30}$
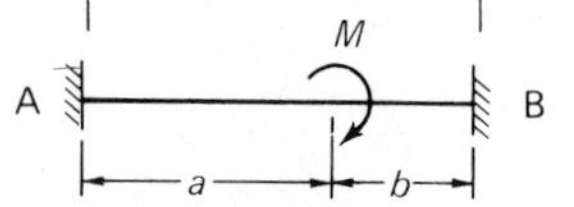 Applied moment M at a from A, b from B	$-M\frac{b}{L}\left(2 - \frac{3b}{L}\right)$	$-M\frac{a}{L}\left(2 - \frac{3a}{L}\right)$

Identical methods can be used for other structures and loads. Results of such calculations lead to the values for fixed-end moments for prismatic beams under various loading conditions shown in Table 25-3. Section 25.4.6.2 also lists references for tabulated information about fixed-end moments and stiffness factors for nonprismatic beams of interest to concrete engineers.

25.4.4 Stiffness Factors and Stiffness Matrix

Stiffness factor is the force needed to achieve a unit displacement at one node when all other nodes are fixed against displacement. Stiffness factors are one of the building blocks of the stiffness, or displacement, method for analyzing indeterminate structures.

The stiffness factor k_{ij} is the force i needed to achieve a unit displacement j, all other displacements equal to zero. Stiffness factors can be calculated by the force method.

EXAMPLE 25-14: Given a prismatic straight beam, length L, stiffness EI, shown in Fig. 25-26a, calculate the end stiffness factors.

SOLUTION: Four displacements are considered, two end rotations θ_1 and θ_2, and two displacements, Δ_3 and Δ_4. Corresponding to each displacement, there are support forces M_1 and M_2, and V_3 and V_4. For instance, for $\theta_1 = 1$, $\theta_2 = \Delta_3 = \Delta_4 = 0$: $M_1 \equiv k_{11}$, $M_2 \equiv k_{21}$, $V_3 \equiv k_{31}$, $V_4 \equiv k_{41}$.

We will calculate these stiffnesses by the force method. The primary structure is the simple beam of Fig. 25-26B; k_{11} and k_{21} are the redundant end moments due to $\theta_1 = 1$.

$$\Sigma\theta_A = 1:\ k_{11}\frac{L}{3EI} - k_{21}\frac{L}{6EI} = 1$$

$$\Sigma\theta_B = 0: -k_{11}\frac{L}{6EI} + k_{21}\frac{L}{3EI} = 0$$

Solving simultaneously

$$k_{11} = \frac{4EI}{L};\quad k_{21} = \frac{2EI}{L}$$

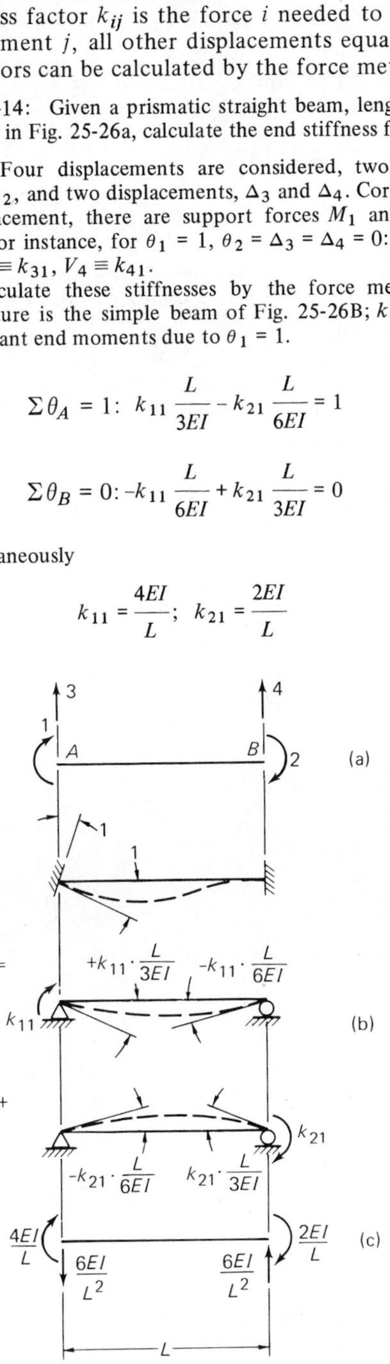

Fig. 25-26 Example 25-14. Stiffnesses. (a) Real beam; (b) primary structure; and (c) final forces.

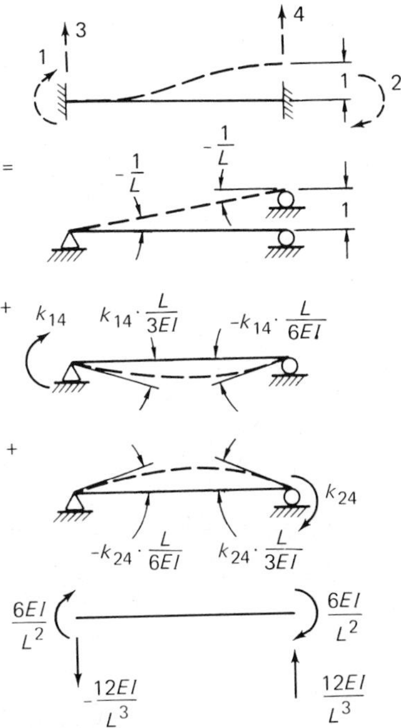

Fig. 25-27 Example 25-14 (cont'd.). Stiffnesses (cont'd.).

The reactions, k_{31} and k_{41}, are found by statics.

$$k_{31} = -\frac{6EI}{L^2};\quad k_{41} = +\frac{6EI}{L^2}$$

The stiffness factors k_{i3} and k_{i4} are the nodal forces corresponding to unit values of the end translations Δ_3 and Δ_4. To calculate these stiffnesses, we use the same primary structure. For k_{i4}, for instance, we refer to Fig. 25-27

$$\Sigma\theta_A = 0:\ k_{14}\frac{L}{3EI} - k_{24}\frac{L}{6EI} = \frac{1}{L}$$

$$\Sigma\theta_B = 0: -k_{14}\frac{L}{6EI} + k_{24}\frac{L}{3EI} = -\frac{1}{L}.$$

$$k_{14} = k_{24} = \frac{6EI}{L^2}.$$

By statics:

$$k_{34} = -k_{44} = -\frac{12EI}{L^3}.$$

It will be recognized that the stiffness factors k_{ij} can be easily assembled into a member, or element, stiffness matrix $[k]$, relating the nodal forces to the nodal displacements; for ex. 25-14, for instance,

$$\begin{Bmatrix} M_1 \\ M_2 \\ Y_3 \\ Y_4 \end{Bmatrix} = \frac{EI}{L^3}\begin{bmatrix} 4L^2 & 2L^2 & -6L & 6L \\ 2L^2 & 4L^2 & -6L & 6L \\ -6L & -6L & 12 & -12 \\ 6L & 6L & -12 & 12 \end{bmatrix}\begin{Bmatrix} \theta_1 \\ \theta_2 \\ y_3 \\ y_4 \end{Bmatrix} \tag{25-16}$$

or

$$\{X\} = [k]\ \{\Delta\} \tag{25-17}$$

Comparing eqs. (25-10) and (25-17), it can be deduced that if for a given structure the stiffness and flexibility matrices $[k]$ and $[\delta]$ are expressed in terms of the same nodes, then

$$[k] = [\delta]^{-1} \tag{25-18}$$

that is, the stiffness matrix is the inverse of the flexibility matrix; it is therefore a symmetric matrix, as demonstrated by eq. 25-16.

Stiffness factors are a property of the structure, not of the load. For haunched members of interest in concrete design, values of stiffness factors are given in the references listed in section 25.4.6.2.

25.4.5 Matrix Formulation of the Force Method

We recall that the force method of analysis of an r-times statically indeterminate structure demands satisfaction of the r conditions of geometry:

$$\begin{aligned} X_1\delta_{11} + \cdots + X_r\delta_{1r} + \Delta_1^0 &= \Delta_1 \\ &\vdots \\ X_1\delta_{r1} + \cdots + X_r\delta_{rr} + \Delta_r^0 &= \Delta_r \end{aligned} \tag{25-14}$$

where δ_{ij} are the influence, or flexibility, coefficients at point i of the primary structure due to a unit redundant force at j; Δ_i^0 are the displacements of point i of the primary structure due to the applied loads; and Δ_i are the specified displacements at points of redundancy.

If we label the p points of known applied concentrated loads consecutively with the r points of redundancy, that is, $(r+1), \cdots (r+p)$, then the displacements Δ_i^0 due to applied loads X_j can be written as

$$\Delta_i^0 = X_{(r+1)}\delta_{i(r+1)} + \cdots + X_{(r+p)}\delta_{i(r+p)}$$

Similarly, the unknown displacements Δ_i [$i = (r+1)$, etc.] at the p load points are obtained by superposition of the effects of all redundant and applied forces:

$$\begin{aligned} \Delta_{r+1} &= X_1\delta_{(r+1)1} + \cdots + X_r\delta_{(r+1)r} + X_{(r+1)}\delta_{(r+1)(r+1)} + \cdots + X_{(r+p)}\delta_{(r+1)(r+p)}. \\ &\vdots \\ \Delta_{r+p} &= X\ \delta_{(r+p)1} + \cdots \qquad \cdots + X_{(r+p)}\delta_{(r+p)(r+p)} \end{aligned}$$

We add these p equations to the r equations of geometry and get eq. 25-19.

$$\begin{array}{l} \\ r \text{ eqs. of geometry for unknown reactions, } X \left\{ \begin{array}{l} \\ \\ \\ \end{array} \right. \\ p \text{ eqs. for unknown displacements, } \Delta \left\{ \begin{array}{l} \\ \\ \\ \end{array} \right. \end{array}
\begin{array}{l}
\overbrace{X_1\delta_{11} + \cdots + X_r\delta_{1r}}^{r\text{ unknown forces (redundants)}} \overbrace{+ X_{(r+1)}\delta_{1(r+1)} + \cdots + X_{(r+p)}\delta_{1(r+p)}}^{p\text{ known forces (loads)}} = \Delta_1 \\
\vdots \\
X_1\delta_{r1} + \cdots + X_r\delta_{rr} + X_{(r+1)}\delta_{r(r+1)} + \cdots + X_{(r+p)}\delta_{r(r+p)} = \Delta_r \\
X_1\delta_{(r+1)1} + \cdots + X_r\delta_{(r+1)r} + X_{(r+1)}\delta_{(r+1)(r+1)} + \cdots + X_{(r+p)}\delta_{(r+1)(r+p)} = \Delta_{(r+1)} \\
\vdots \\
X_1\delta_{(r+p)1} + \cdots + X_r\delta_{(r+p)r} + X_{(r+1)}\delta_{(r+p)(r+1)} + \cdots + X_{(r+p)}\delta_{(r+p)(r+p)} = \Delta_{(r+p)}
\end{array}
\begin{array}{l} \left.\right\} \text{known displacements of redundants} \\ \left.\right\} \text{unknown displacements of loads} \end{array} \tag{25-19}$$

In matrix form, this set of equations can be written as

$$\begin{bmatrix} \delta_{\alpha\alpha} & \delta_{\alpha\beta} \\ \delta_{\beta\alpha} & \delta_{\beta\beta} \end{bmatrix} \begin{bmatrix} X_\alpha \\ X_\beta \end{bmatrix} = \begin{bmatrix} \Delta_\alpha \\ \Delta_\beta \end{bmatrix}; \quad \begin{array}{l} (X_\alpha \text{ unknown}; \Delta_\alpha \text{ known}) \\ (X_\beta \text{ known}; \quad \Delta_\beta \text{ unknown}) \end{array} \tag{25-20}$$

The square flexibility matrix $[\delta]$, of order $(r+p)$, can be determined by any of the methods for finding deformations of statically determinate structures, for instance, by the matrix formulation

$$[\delta] = [b]^T [f]\ [b] \tag{25-12}$$

The flexibility matrix has been partitioned into $\delta_{\alpha\alpha}$ (order $r \times r$), $\delta_{\alpha\beta}$ (order $r \times p$), $\delta_{\beta\alpha}$ (order $p \times r$), $\delta_{\beta\beta}$ (order $p \times p$) in anticipation of future operations.

The matrix of forces $\{X\}$, of order $(r+p) \times 1$ for a single loading condition, or of order $(r+p) \times l$ for l different loading conditions, contains r unknown redundants, denoted by the subscript α, followed by p known loads, subscripted β.

The matrix $\{\Delta\}$, of order $(r+p) \times 1$ for a single, of order $(r+p) \times l$ for l loading conditions, contains r specified redundant displacements Δ_α, followed by p unknown load displacements Δ_β.

The solution of this set of equations is accomplished in two steps. First, solve r equations for the redundants X_α.

$$[\delta_{\alpha\alpha}]\ [X_\alpha] + [\delta_{\alpha\beta}]\ [X_\beta] = [\Delta_\alpha]$$

or

$$[X_\alpha] = [\delta_{\alpha\alpha}]^{-1}[[\Delta_\alpha] - [\delta_{\alpha\beta}]\ [X_\beta]] \tag{25-21}$$

This corresponds to the earlier solution for the r redundants by classical methods. If all specified support displacements are zero, the Δ_α matrix drops out.

After the redundants $[X_\alpha]$ are known, the remaining p equations can be solved for the displacements at the load points.

$$[\Delta_\beta] = [\delta_{\beta\alpha}]\ [X_\alpha] + [\delta_{\beta\beta}]\ [X_\beta] \tag{25-22}$$

The internal forces, S, can be found by using the force transformation matrix $[b]$:

$$[S] = [b]\ [X]. \tag{25-23}$$

EXAMPLE 25-15: Find the redundant X_1, the displacements Δ_2 and Δ_3, and the moments at points 1 to 4 of the beam of Fig. 25-28a due to two conditions:

(a) loading condition 1: $X_2 = 3$ kips; $X_3 = 2$ kips and
(b) loading condition 2: A support settlement $\Delta_1 = 1$.

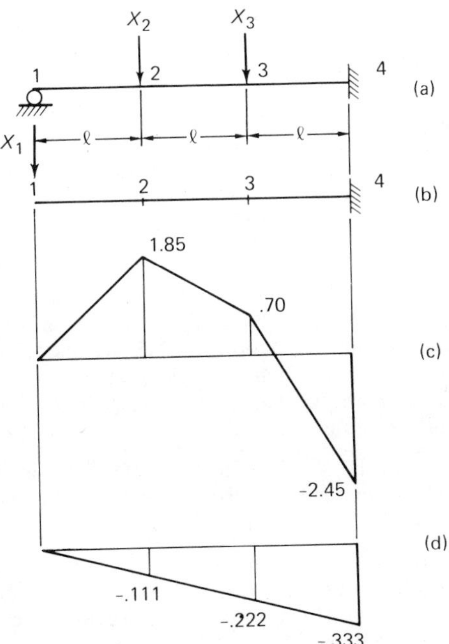

Fig. 25-28 Example 25-15. (a) Structure, constant EI; (b) primary structure; (c) moment diagram, loading condition 1 ($\times\ l$); and (d) moment diagram, loading condition 2 ($\times\ EI/l^2$).

SOLUTION: The primary structure is as shown in Fig. 25-28b.

Step 1 Determine flexibility matrix $[\delta] = [b]^T[f]\ [b]$.

$$[f] = \frac{l}{6EI}\begin{bmatrix} 2 & 1 & & & & \\ 1 & 2 & & & & \\ & & 2 & 1 & & \\ & & 1 & 2 & & \\ & & & & 2 & 1 \\ & & & & 1 & 2 \end{bmatrix}; \quad [b] = l\begin{bmatrix} 0 & 0 & 0 \\ -1 & 0 & 0 \\ -1 & 0 & 0 \\ -2 & -1 & 0 \\ -2 & -1 & 0 \\ -3 & -2 & -1 \end{bmatrix}$$

$$\underset{(3\times 3)}{[\delta]} = \underset{(3\times 6)}{[b]^T}\ \underset{(6\times 6)}{[f]}\ \underset{(6\times 3)}{[b]}$$

$$= \frac{l^3}{6EI}\begin{bmatrix} 54 & 28 & 8 \\ 28 & 16 & 5 \\ 8 & 5 & 2 \end{bmatrix} \quad \text{(note symmetry)}$$

Step 2 Set up force-displacement equations for two loading conditions.

$$\begin{bmatrix} \delta_{\alpha\alpha} & \delta_{\alpha\beta} \\ \delta_{\beta\alpha} & \delta_{\beta\beta} \end{bmatrix}\begin{bmatrix} X_\alpha \\ X_\beta \end{bmatrix} = \begin{bmatrix} \Delta_\alpha \\ \Delta_\beta \end{bmatrix}$$

(The first column of the $[X]$ and $[\Delta]$ matrices denotes the first, the second, the second loading condition.)

$$\frac{l^3}{6EI}\begin{bmatrix} 54 & 28 & 8 \\ 28 & 16 & 5 \\ 8 & 5 & 2 \end{bmatrix}\begin{bmatrix} X_{11} & X_{12} \\ X_{21} = 3 & X_{22} = 0 \\ X_{31} = 2 & X_{32} = 0 \end{bmatrix} = \begin{bmatrix} \Delta_{11} = 0 & \Delta_{12} = 1 \\ \Delta_{21} & \Delta_{22} \\ \Delta_{31} & \Delta_{32} \end{bmatrix}$$

Step 3 Solve for redundant X_1 due to both loading conditions (α-eqs.).

$$\frac{l^3}{6EI}\left\{(54)\,[X_{11} \quad X_{12}] + [28 \quad 8]\begin{bmatrix} 3 & 0 \\ 2 & 0 \end{bmatrix}\right\} = [0 \quad 1]$$

Now, solve for $[X_{11} \quad X_{12}]$.

$$[X_{11} \quad X_{12}] = \frac{1}{54}\left\{\frac{6EI}{l^3}\,[0 \quad 1] - [100 \quad 0]\right\} = \left[-1.85 \quad 0.111\frac{EI}{l^3}\right]$$

Note that the redundant due to the applied loads (condition 1) is 1.85 kips upward, that due to the support settlement (condition 2) depends on the beam stiffness.

Step 4 Solve for displacements Δ_2 and Δ_3 due to both loading conditions (β-eqs.).

$$\begin{bmatrix} \Delta_{21} & \Delta_{22} \\ \Delta_{31} & \Delta_{32} \end{bmatrix} = \frac{l^3}{6EI}\begin{bmatrix} 28 \\ 8 \end{bmatrix}\left[-1.85 \quad 0.111\frac{EI}{l^3}\right] + \begin{bmatrix} 16 & 5 \\ 5 & 2 \end{bmatrix}\begin{bmatrix} 3 & 0 \\ 2 & 0 \end{bmatrix}$$

$$= \begin{bmatrix} 1.016\dfrac{l^3}{EI} & 0.519 \\ 0.700\dfrac{l^3}{EI} & 0.148 \end{bmatrix}$$

Note that the displacements due to the applied loads depend on the beam stiffness, those due to the support settlement ($\Delta_1 = 1$) do not.

Step 5 Solve for the moments at points 1 to 4 due to both loading conditions.

$$[S] = [b]\ [X]$$

(We contract the $[b]$ matrix to list only *one* moment at each point.)

$$\begin{bmatrix} M_{11} & M_{12} \\ M_{21} & M_{22} \\ M_{31} & M_{32} \\ M_{41} & M_{42} \end{bmatrix} = l\begin{bmatrix} 0 & 0 & 0 \\ -1 & 0 & 0 \\ -2 & -1 & 0 \\ -3 & -2 & -1 \end{bmatrix}\begin{bmatrix} -1.85 & 0.111\dfrac{EI}{l^3} \\ 3.0 & 0 \\ 2.0 & 0 \end{bmatrix}$$

$$= \begin{bmatrix} 0 & 0 \\ 1.85l & -0.111\dfrac{EI}{l^2} \\ 0.70l & -0.222\dfrac{EI}{l^2} \\ -2.45l & -0.333\dfrac{EI}{l^2} \end{bmatrix}$$

The moment diagrams for the two loading conditions can now be plotted, as shown in Figs. 25-28c and d.

25.4.6 Approximate Analysis and Design Aids

-1. Preliminary analysis—The nature of structural design as a trial-and-error process demands some facility on part of the engineer to perform quick, approximate analyses in order to be able to estimate preliminary costs, and to determine approximate member dead loads and stiffnesses to be used in more exact analysis.

Such approximate analysis methods rely often on the ability to visualize the deformation of the structure under load; a physical feel for structural behavior is necessary for this. After the members have been sized preliminarily, an exact analysis should always follow before the design is finalized.

One approach to approximate analysis of statically indeterminate structures consists of the following steps:

Step 1 Introduce a number of assumptions regarding values of forces equal to the degree of static indeterminacy, r, of the structure. Among such assumptions might be the location of inflection points (that is, location where $M = 0$), relative distribution of shears, magnitudes of reactions as bounded by some known limiting values, etc.

Step 2 Having introduced values for r forces, the structure is reduced to static determinacy. Thus, all other forces can be computed by statics.

The procedure is illustrated in the following example:

EXAMPLE 25-16: Perform a preliminary analysis of the two-story frame under gravity and lateral loads shown in Fig. 25-29a.

SOLUTION: The vertical loads are treated first, as shown in Fig. 25-29b. The extreme moment diagrams, one for the simply-supported case, the other for the fixed-ended case, are shown dashed for the uniform load on each of the beams. The actual end condition lies somewhere in between, leading to the beam moment diagrams shown in solid line. The beam end moments are transferred to the adjacent columns such that moment equilibrium of the joints, and shear equilibrium of the frame is maintained, leading to the moments shown in solid line in Fig. 25-29b; these moments satisfy equilibrium.

In a similar fashion, the moments of Fig. 25-29 are in equilibrium with the lateral loads; in drawing them, the following guidelines were followed.

1. Points of inflection were assumed at the midpoints of the second-story columns, and the column, and top level beam moment diagrams drawn accordingly.
2. Of the total shear of 8 kips in the bottom level, the shorter, stiffer column was assumed to transmit the greater amount of 5 kips, the longer, more flexible column the rest. The bottom, fixed end of each column was assumed to carry more moment than the upper, elastically restrained ends, leading to the lower story column moment diagrams shown. With the column end moments drawn, the lower beam moment diagram can be drawn.

The complete moment diagram of Fig. 25-29d represents an equilibrium set of forces. Whether or not these forces also satisfy geometrical compatibility can only be ascertained after member stiffnesses has been assumed. A slight lack of such compatibility can be accommodated if the members and joints possess sufficient ductility; it is therefore important to design the structure for both strength and ductility to avoid cracking.

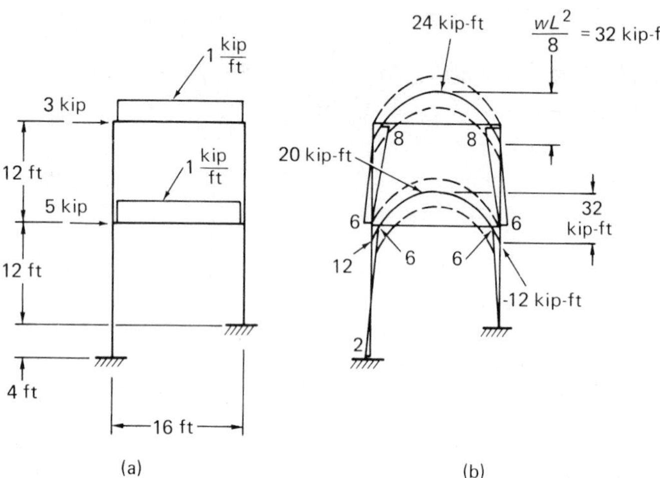

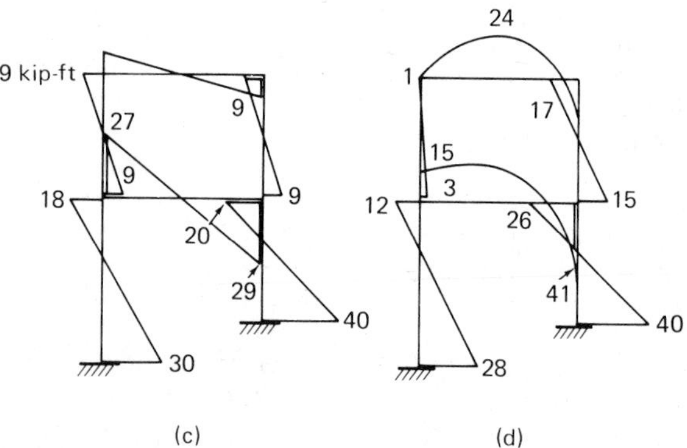

Fig. 25-29 Example 25-16. (a) Structure and loads; (b) approx. analysis for vertical loads; (c) approx. analysis for lateral loads; and (d) approx. moment diagram for preliminary design.

-2. *Design aids*—The solutions of indeterminate structures of standard form collected in the works by Kleinlogel and Leontovich cited in section 25.1.5 are useful for some purposes. For the design of rigid frames of rectangular, gabled, and parabolic shape, the following contains graphs, tables, and examples:

Griffiths, J. D., "Single Span Rigid Frames in Steel," American Institute of Steel Construction, New York, 1948.

Tables of fixed-end moments are available in many standard references; Table 25-3 lists these for some common loading conditions on prismatic beams. For fixed-end moments as well as stiffness factors for haunched members of the type often found in concrete structures, the first of the following provides tabular, the second graphical aids:

Portland Cement Association, "Handbook of Frame Constants," Chicago, 1958.
Portland Cement Association, "Concrete Members with Variable Moment of Inertia," *Concrete Information ST 103*, Chicago, 1964.

By inverting the rotational end stiffnesses given in this reference in accordance with eq. 25-18, the member end flexibilities can be found for use in a force method analysis as outlined in sections 25.4.2 or 25.4.5.

A large amount of design information for arches, including influence lines for circular and parabolic arches, are contained in

Michalos, J., *Theory of Structural Analysis and Design*, The Ronald Press, New York, 1958.

For circular bow girders, the following references contain graphical results:

Hogan, M. B., "Circular Beams Loaded Normal to the Plane of Curvature," *ASME Transactions (Journal of Applied Mechanics)*, **5** (2), June 1938, and **11** (1), March 1944.

25.5 THE DISPLACEMENT METHOD

25.5.1 Basic Formulation

In the force method for analysis of indeterminate structures, the redundant forces are unknowns. A set of equations of geometry, equal to the number of unknown forces (degree of static indeterminacy r), has to be solved to determine these forces.

In the displacement method, (also called the stiffness method) the nodal displacements are unknowns. A set of equations of equilibrium, equal to the number of unknown displacements (degree of kinematic indeterminacy k), has to be solved to determine these displacements.

If the number of redundant forces, r, is less than the number of unknown displacements, k, the force method is advantageous. If, on the other hand, $r > k$, then the displacement method is preferred. Furthermore, the systematic formulation of the displacement method is often easier, so that for computer solutions the displacement method is commonly used irrespective of the degrees of static or kinematic indeterminacy.

Basically the displacement method is carried out in the following steps:

Step 1 Force-displacement relations (stiffness)

Step 2 Geometrical relations

Step 3 Equilibrium relations to determine unknown displacements.

Step 4 Determination of forces by substituting displacements into stiffnesses.

These steps are illustrated by a simple example:

EXAMPLE 25-17: Calculate the moments in the structure shown in Fig. 25-30 due to the cranked-in moment M.

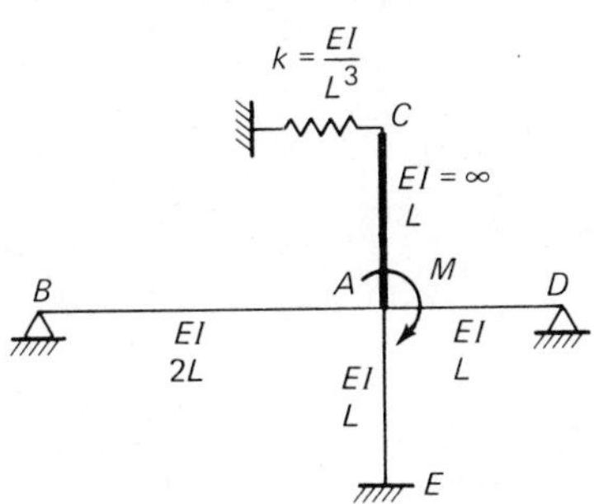

Fig. 25-30 Example 25-17. Displacement method.

This structure is statically indeterminate to the fifth degree, but its kinematic degree of indeterminacy is only one, since all forces can be determined in terms of the unknown rotation θ_A. We proceed along the outlined steps:

Step 1 The end moments associated with a unit end rotation (the rotational stiffness factors) are found for all members as outlined in section 25.4.

$$k_{AB} = \frac{M_{AB}}{\theta_{AB}} = \frac{3}{2}\frac{EI}{L}; \quad k_{AC} = \frac{M_{AC}}{\theta_{AC}} = \frac{EI}{L}; \quad k_{AD} = \frac{M_{AD}}{\theta_{AD}} = 3\frac{EI}{L};$$

$$k_{AE} = \frac{M_{AE}}{\theta_{AE}} = 4\frac{EI}{L}$$

Step 2 Since joint A is rigid, all member ends turn through the same angle θ_A:

$$\theta_{AB} = \theta_{AC} = \theta_{AD} = \theta_{AE} = \theta_A$$

Step 3 The single equilibrium equation needed is that of rotational equilibrium of joint A.

$$\Sigma M_A = 0: \; M_{AB} + M_{AC} + M_{AD} + M_{AE} = M$$

Expressing the moments in terms of θ_A to obtain the equilibrium equation in terms of the displacements:

$$\frac{EI}{L}\left[\frac{3}{2} + 1 + 3 + 4\right]\theta_A = M$$

from which the unknown displacement is

$$\theta_A = \frac{2}{19}\frac{L}{EI}M$$

Note that the total stiffness against rotation at point A of the entire structure

$$\frac{M}{\theta_A} = \frac{19EI}{2L}$$

is obtained by adding the member stiffnesses at point A.

Step 4 To obtain the member end moments at A, we substitute the value of the unknown displacement θ_A found in Step 3 into the stiffness relations of Step 1.

$$M_{AB} = \frac{3}{19}M; \quad M_{AC} = \frac{2}{19}M; \quad M_{AD} = \frac{6}{19}M; \quad M_{AE} = \frac{8}{19}M$$

Note that the cranked-in moment M is distributed to the adjacent members in direct proportion to their stiffness. The fraction of the total moment resisted by member i is called the distribution factor and is equal to the ratio of the rotational stiffness of member i to the sum of the stiffnesses of all members entering the joint:

$$D.F._i = \frac{k_i}{\sum_i k} \tag{25-24}$$

25.5.2 The Element Stiffness Matrix

As already discussed in section 25.4.4, the element stiffness matrix $[k]$ contains the stiffness k_{ij}, defined as the forces which must be applied to the structural element at node i in order to insure a unit displacement at node j, all others zero.

The calculation of the stiffnesses for beam elements by the force method was outlined in section 25.4.4, and eq. (25-16) gives the flexural stiffness matrix for a planar, prismatic, straight, elastic beam element. If axial deformations are to be considered as well, it is necessary to add nodes at both member ends in the axial direction, as shown in Fig. 25-31; the corresponding stiffness matrix is then given by eq. (25-25).

$$[k] = \begin{array}{c} \\ 1 \\ 2 \\ 3 \\ 4 \\ 5 \\ 6 \end{array} \begin{array}{c} \begin{array}{cccccc} 1 & 2 & 3 & 4 & 5 & 6 \end{array} \\ \begin{bmatrix} \frac{AE}{L} & & & & & \\ 0 & \frac{12EI}{L^3} & & & \text{Sym.} & \\ 0 & \frac{6EI}{L^2} & \frac{4EI}{L} & & & \\ -\frac{AE}{L} & 0 & 0 & \frac{AE}{L} & & \\ 0 & -\frac{12EI}{L^3} & -\frac{6EI}{L^2} & 0 & \frac{12EI}{L^3} & \\ 0 & \frac{6EI}{L^2} & \frac{2EI}{L} & 0 & -\frac{6EI}{L^2} & \frac{4EI}{L} \end{bmatrix} \end{array} \tag{25-25}$$

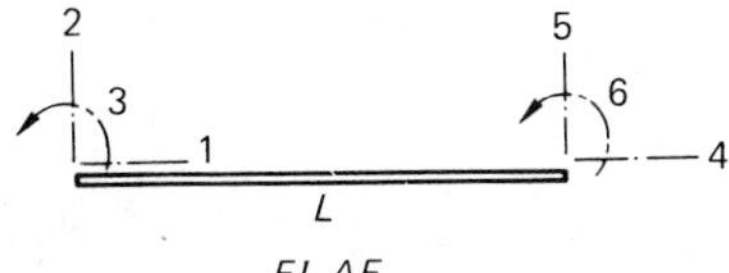

Fig. 25-31 Plane beam element.

$$[k] = \begin{array}{c} \\ 1 \\ 2 \\ 3 \\ 4 \\ 5 \\ 6 \\ 7 \\ 8 \\ 9 \\ 10 \\ 11 \\ 12 \end{array}
\begin{array}{c} \begin{array}{cccccccccccc} 1 & 2 & 3 & 4 & 5 & 6 & 7 & 8 & 9 & 10 & 11 & 12 \end{array} \\
\left[\begin{array}{cccccccccccc}
\frac{EA}{L} & & & & & & & & & & & \\
 & \frac{12EI_z}{L^3} & & & & & & & & & \text{Sym.} & \\
 & & \frac{12EI_y}{L^3} & & & & & & & & & \\
 & & & \frac{GC}{L} & & & & & & & & \\
 & & -\frac{6EI_y}{L^2} & & \frac{4EI_y}{L} & & & & & & & \\
 & \frac{6EI_z}{L^2} & & & & \frac{4EI_z}{L} & & & & & & \\
-\frac{EA}{L} & & & & & & \frac{EA}{L} & & & & & \\
 & -\frac{12EI_z}{L^3} & & & & -\frac{6EI_z}{L^2} & & \frac{12EI_z}{L^3} & & & & \\
 & & -\frac{12EI_y}{L^3} & & \frac{6EI_y}{L^2} & & & & \frac{12EI_y}{L^3} & & & \\
 & & & -\frac{GC}{L} & & & & & & \frac{GC}{L} & & \\
 & & -\frac{6EI_y}{L^2} & & \frac{2EI_y}{L} & & & & \frac{6EI_y}{L^2} & & \frac{4EI_y}{L} & \\
 & \frac{6EI_z}{L^2} & & & & \frac{2EI_z}{L} & & -\frac{6EI_z}{L^2} & & & & \frac{4EI_z}{L}
\end{array}\right] \end{array} \tag{25-26}$$

For the more general case of a three-dimensional prismatic beam element, we must consider the twelve nodes shown in Fig. 25-32, that is, axial and transverse translatory nodes along three axes at each member end and rotatory nodes about each of these axes. With these nodes, axial forces and deformations, transverse shears and corresponding translations, twisting, and bending about two axes can be accounted for. With the numbering of Fig. 25-32, elementary beam and torsion theory leads to the stiffness matrix of eq. (25-26).

Another standard element stiffness matrix is the one for the inclined planar truss member with the nodal numbering of Fig. 25-33. Simple calculations lead to the stiffness matrix

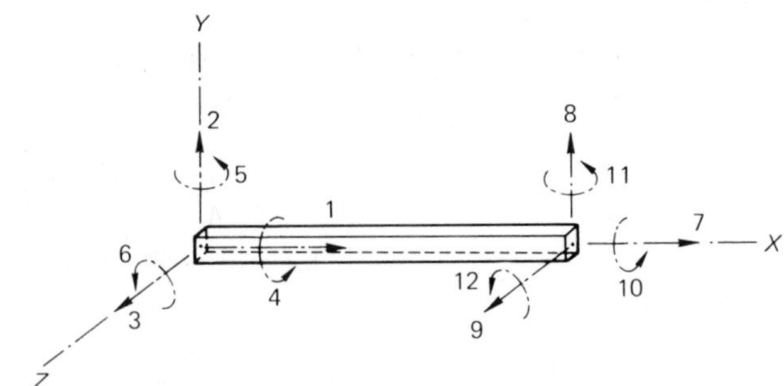

Fig. 25-32 Three-dimensional beam element.

$$[k] = \frac{AE}{L} \begin{array}{c} \\ 1 \\ 2 \\ 3 \\ 4 \end{array} \begin{array}{c} \begin{array}{cccc} 1 & 2 & 3 & 4 \end{array} \\ \left[\begin{array}{cccc}
\cos^2\alpha & & \text{Sym.} & \\
-\sin\alpha\cos\alpha & \sin^2\alpha & & \\
-\cos^2\alpha & +\sin\alpha\cos\alpha & \cos^2\alpha & \\
+\sin\alpha\cos\alpha & -\sin^2\alpha & -\sin\alpha\cos\alpha & \sin^2\alpha
\end{array}\right] \end{array} \tag{25-27}$$

A catalogue of stiffness matrices of the type shown in eqs. (25-25) to (25-27) is necessary to perform efficient computer analysis by the displacement method. While computer programs available for production runs (see section 25.5.7) have matrices such as eq. 25-26 built in, more specialized work may require stiffnesses of other members,

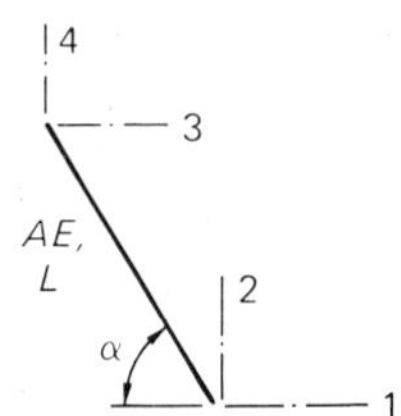

Fig. 25-33 Truss member.

such as tapering or curved members, or more general structural elements such as in the finite-element method. Otherwise, it will be necessary to model these elements as an assembly of straight prismatic members to which eq. 25-26 can be applied.

25.5.3 The Structure Stiffness Matrix

The structure stiffness matrix $[K]$ relates the forces and displacements of a structure composed of elements:

$$\{X\} = [K]\,\{\Delta\} \tag{25-28}$$

Each element K_{ij} of this stiffness matrix is defined as the force which must be applied to the complete structure at node i in order to insure a unit displacement at node j, all others zero.

We recall from section 25.5.1 that the total, or structure stiffness at a node is obtained as the sum of the stiffnesses of all members attached to this node. This summing of element stiffnesses results from expressing the equilibrium equations in terms of the compatible displacements.

For instance, it may be recalled from the discussion of section 25.5.1 that the rotational structure stiffness at node 1 of the structure of Fig. 25-34 is obtained by summing the individual member stiffnesses.

$$K_{11} = k_{11}^1 + k_{11}^2 + k_{11}^3 = \left(\frac{4EI}{L}\right)^{①} + \left(\frac{4EI}{L}\right)^{②} + \left(\frac{4EI}{L}\right)^{③}$$

if the members are all prismatic.

Similarly, considering the translational structure stiffness along node 2, and considering axial and flexural deformations of the prismatic members

$$k_{22} = k_{22}^1 + k_{22}^2 + k_{22}^3 = \left(\frac{AE}{L}\right)^{①} + \left(\frac{AE}{L}\right)^{②} + \left(\frac{12EI}{L^3}\right)^{③}$$

In general it is essential that each node of the structure be carefully labeled, and that the nodal numbering of each element correspond to that of the structure. The element stiffness matrices for all members are then written, and superimposed, or assembled, to form the structure stiffness matrix, as shown in the following example.

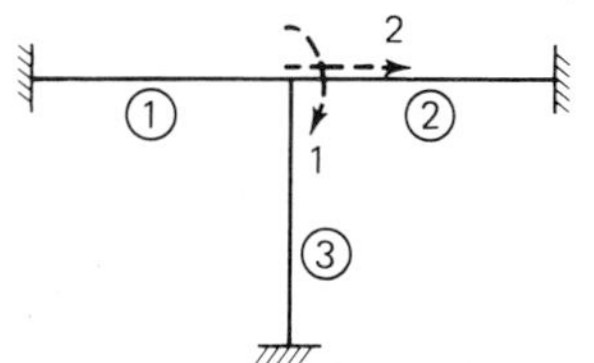

Fig. 25-34 Structure stiffness nodes.

EXAMPLE 25-18: Assemble the stiffness matrix for the structure and nodes shown in Fig. 25-35.

SOLUTION: We first write out the element stiffness matrices, member by member, taking care to adhere to the established numbering. These matrices follow eq. 25-26 for the two-force members 1 and 3, and eq. 25-24 for the planar beam element 2; accordingly, using the structure properties given in Fig. 25-35, we obtain

$$[k]^{①} = \begin{array}{c} \\ 1 \\ 2 \\ 4 \\ 5 \end{array}\begin{array}{c} \begin{array}{cccc} 1 & 2 & 4 & 5 \end{array} \\ \left[\begin{array}{cccc} 89 & & & \\ -89 & 89 & & \text{Sym.} \\ -89 & 89 & 89 & \\ +89 & -89 & -89 & 89 \end{array}\right] \end{array}$$

$$[k]^{②} = \begin{array}{c} \\ 6 \\ 7 \\ 8 \\ 1 \\ 2 \\ 3 \end{array}\begin{array}{c} \begin{array}{cccccc} 6 & 7 & 8 & 1 & 2 & 3 \end{array} \\ \left[\begin{array}{cccccc} 267 & & & & & \\ 0 & 0.298 & & & \text{Sym.} & \\ 0 & 35.8 & 5{,}700 & & & \\ -267 & 0 & 0 & 267 & & \\ 0 & -0.298 & -35.8 & 0 & 0.298 & \\ 0 & 35.8 & 2{,}850 & 0 & -35.8 & 5{,}700 \end{array}\right] \end{array}$$

$$[k]^{③} = \begin{array}{c} \\ 1 \\ 2 \\ 9 \\ 10 \end{array}\begin{array}{c} \begin{array}{cccc} 1 & 2 & 9 & 10 \end{array} \\ \left[\begin{array}{cccc} 81 & & & \\ 47 & 27 & & \text{Sym.} \\ -81 & -47 & 81 & \\ -47 & -27 & 47 & 27 \end{array}\right] \end{array}$$

Assembling these element stiffnesses into the locations of the 10×10 structure stiffness matrix as indicated by the numbering of the rows and columns, we obtain

$$[K] = \begin{array}{c} \\ 1 \\ 2 \\ 3 \\ 4 \\ 5 \\ 6 \\ 7 \\ 8 \\ 9 \\ 10 \end{array}\begin{array}{c} \begin{array}{cccccccccc} 1 & 2 & 3 & 4 & 5 & 6 & 7 & 8 & 9 & 10 \end{array} \\ \left[\begin{array}{cccccccccc} 437 & & & & & & & & & \\ -42 & 116.3 & & & & & & & & \\ 0 & -35.8 & 5{,}700 & & & & & & & \\ -89 & 89 & 0 & 89 & & & \text{Symmetric} & & & \\ +89 & -89 & 0 & -89 & 89 & & \text{Matrix} & & & \\ -267 & 0 & 0 & 0 & 0 & 267 & & & & \\ 0 & -0.3 & 35.8 & 0 & 0 & 0 & 3 & & & \\ 0 & -35.8 & 2{,}850 & 0 & 0 & 0 & 35.8 & 5{,}700 & & \\ -81 & -47 & 0 & 0 & 0 & 0 & 0 & 0 & 81 & \\ -47 & -27 & 0 & 0 & 0 & 0 & 0 & 0 & 47 & 27 \end{array}\right] \end{array}$$

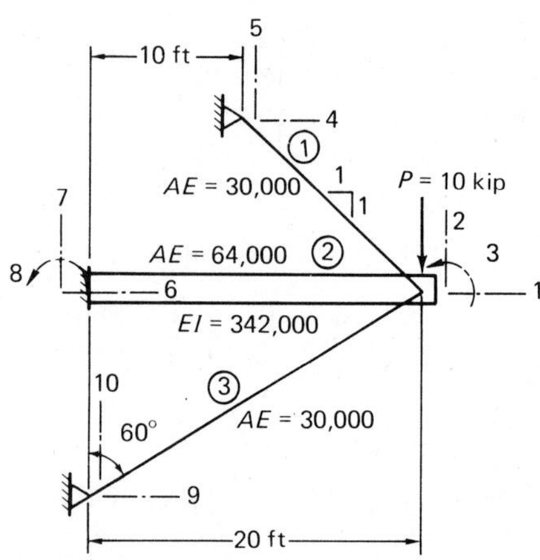

Fig. 25-35 Example 25-18 and 25-19.

The physical meaning of these numbers should be clearly understood; element K_{11}, for instance, is about four times the value of K_{22}, indicating that the horizontal force required to stretch the structure a specified amount horizontally is four times as much as a vertical force at the same point causing the same amount of vertical displacement. The contribution of the beam member, k_{22}^2, is only a small fraction of the total value of the stiffness K_{22}, indicating that this stiffness is mainly due to the inclined truss members. Considerations of this type can be of great value to the designer in evaluating the effectiveness of a structural configuration.

25.5.4 Matrix Formulation of the Displacement Method

The matrix displacement method, if used in conjunction with appropriate computing equipment, is the most powerful of the various methods of analysis. In its generalization as the finite-element method it is capable of analyzing any solid body; in this section, however, its application will be restricted to the analysis of framed structures.

We recall that the force method related the displacements $\{\Delta\}$ to the forces $\{X\}$ by means of the structure flexibility matrix $[\delta]$:

$$[\Delta] = [\delta]\,[X]. \tag{25-27}$$

Analogously, the displacement method relates the forces to the displacements by means of the structure stiffness matrix $[K]$:

$$[X] = [K]\,[\Delta]. \tag{25-28}$$

As in the force method, each node will have a force and a corresponding displacement, one of which is known, the other unknown. For a structure with k degrees of kinematic indeterminacy (or degrees of freedom), we collect the k nodes with unknown displacements and known forces (loads) at the top, the remaining f nodes corresponding to the known support displacements and unknown forces (reactions) at the bottom, call the former nodal values α, the latter β, and partition the matrix equation:

$$\begin{matrix}\text{known forces (loads)} \\ \\ \text{unknown forces (reactions)}\end{matrix}\ \left\{\begin{bmatrix} X_\alpha \\ \hline X_\beta \end{bmatrix}\right\} = \left[\begin{array}{c|c} K_{\alpha\alpha} & K_{\alpha\beta} \\ \hline K_{\beta\alpha} & K_{\beta\beta} \end{array}\right] \left\{\begin{bmatrix} \Delta_\alpha \\ \hline \Delta_\beta \end{bmatrix}\right\}\ \begin{matrix}\text{unknown displacements at load pts.} \\ \\ \text{known (specified) displacements}\end{matrix} \tag{25-29}$$

The solution is obtained in two steps: first, the first k equations are solved for the unknown displacements:

$$[X_\alpha] = [K_{\alpha\alpha}]\,[\Delta_\alpha] + [K_{\alpha\beta}]\,[\Delta_\beta]$$

from which

$$[\Delta_\alpha] = [K_{\alpha\alpha}]^{-1}\,[[X_\alpha] - [K_{\alpha\beta}]\,[\Delta_\beta]]. \tag{25-30}$$

Note that a $k \times k$ square matrix has to be inverted, corresponding to a solution of k simultaneous equations.

The displacements $[\Delta_\alpha]$ thus found are then substituted in the f remaining equations to solve for the unknown forces:

$$[X_\beta] = [K_{\beta\alpha}]\,[\Delta_\alpha] + [K_{\beta\beta}]\,[\Delta_\beta] \tag{25-31}$$

With these basic concepts in mind, we outline the steps necessary for the formulation of the matrix displacement method:

Step 1 Identify the separate elements of the structure. The interconnections between these elements are called joints.

Step 2 At each joint, identify and number the nodes for which forces and corresponding displacements exist. In the most general three-dimensional case, there may be six nodes (three forces and corresponding translations, and three moments and corresponding rotations) at each joint. In other cases, there may be less. Number the nodes with unknown displacements first.

Step 3 Calculate and write the stiffness matrix for each element, adhering to the numbering established in Step 2.

Step 4 Assemble the structure stiffness matrix by superposition of the element stiffness matrices.

Step 5 Write the matrix equation $[X] = [K][\Delta]$, substitute known values of forces and displacements, partition into eqq. 25-30 and 25-31 and solve for the unknown displacements Δ_α and the unknown reactions, X_β.

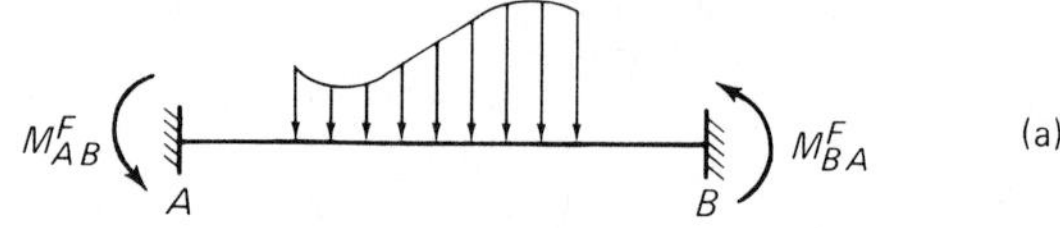

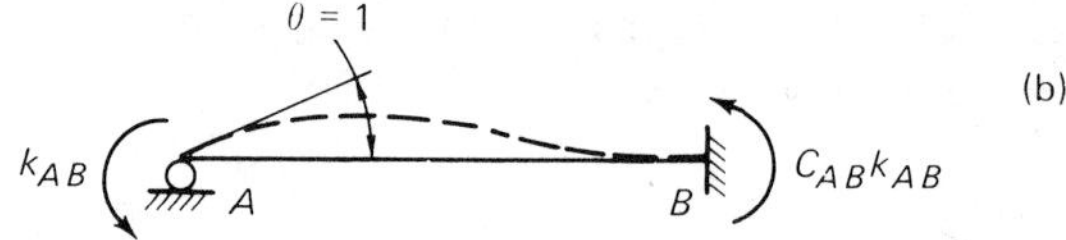

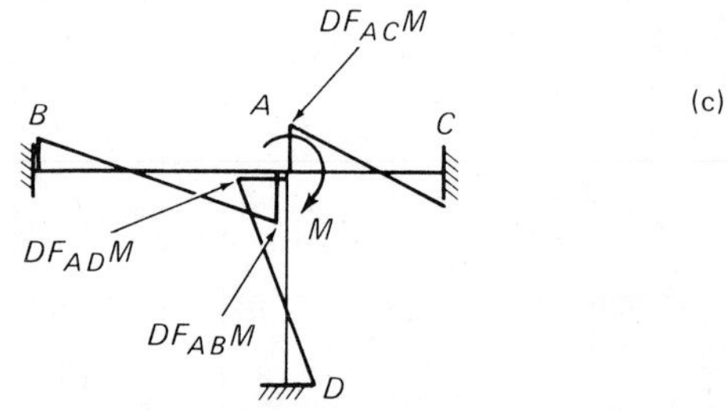

Fig. 25-36 Basic concepts for moment distribution. (a) Fixed-end moment; (b) rotational stiffness and carry-over factor; and (c) distribution factor.

Step 6 To find the element (internal) forces, use the force-displacement relations for each element i:

$$[X^i] = [k^i]\,[\Delta^i] \qquad (25\text{-}32)$$

This concludes the analysis by the displacement method.

EXAMPLE 25-19: Analyze the structure shown in Fig. 25-35 by the matrix displacement method. Members 1 and 3 are two-force, while member 2 is subject to flexural and axial deformations.

SOLUTION: The nodes are identified and numbered as shown in Fig. 25-35, remembering to number the nodes for which the displacements are unknown first. The element stiffnesses were already written and assembled into the appropriate structure stiffness matrix in Ex. 25-18. The given values of the applied loads, X_1 to X_3, and the specified support displacements, Δ_4 to Δ_{10} (all zero), are now inserted into the force-displacement eq. 25-29 to yield

$$\begin{Bmatrix} 0 \\ -10 \\ 0 \\ \hline X_4 \\ X_5 \\ X_6 \\ X_7 \\ X_8 \\ X_9 \\ X_{10} \end{Bmatrix} = \left[\begin{array}{ccc|ccccccc} 437 & & & & & & & & & \\ -42 & 116.3 & K_{\alpha\alpha} & & & K_{\alpha\beta} & \text{Sym.} & & & \\ 0 & -35.8 & 5{,}700 & & & & & & & \\ \hline -89 & 89 & & 89 & & & & & & \\ 89 & -89 & & -89 & 89 & & & & & \\ -267 & 0 & 0 & & & 267 & & & & \\ 0 & -.3 & 35.8 & & & 0 & .3 & & & \\ 0 & -35.8 & 2{,}850 & & & 0 & 35.8 & 5{,}700 & & \\ -81 & -47 & K_{\beta\alpha} & & & & K_{\beta\beta} & & 81 & \\ -47 & -21 & & & & & & & 47 & 27 \end{array}\right] \begin{Bmatrix} \Delta_1 \\ \Delta_2 \\ \Delta_3 \\ \hline 0 \\ 0 \\ 0 \\ 0 \\ 0 \\ 0 \\ 0 \end{Bmatrix}$$

Equations 25-30 and 25-31 can now be solved to yield the matrix of unknown deformations and the matrix of unknown reactions:

$$\begin{Bmatrix} \Delta_1 \\ \Delta_2 \\ \Delta_3 \end{Bmatrix} = \begin{Bmatrix} -0.0086 \text{ in.} \\ -0.0893 \text{ in.} \\ -0.0006 \text{ rad.} \end{Bmatrix}$$

$$\begin{Bmatrix} X_4 \\ X_5 \\ X_6 \\ X_7 \\ X_8 \\ X_9 \\ X_{10} \end{Bmatrix} = \begin{Bmatrix} -7.1802 \text{ kips} \\ 7.1802 \text{ kips} \\ 2.2904 \text{ kips} \\ 0.0067 \text{ kips} \\ 1.5977 \text{ kip-inches} \\ 4.8898 \text{ kips} \\ 2.8131 \text{ kips} \end{Bmatrix}$$

Note that these reactions satisfy the equilibrium conditions. The interior element forces, that is, the end forces and moment of members 1, 2, and 3 along nodes 1, 2, and 3, can now be found by applying eq. 25-32 to each of the members. This concludes the example.

25.5.5 Moment Distribution

The method of moment distribution is a numerical application of the displacement method in which the desired quantities are determined by a method of successive approximation. For this reason it is extremely useful for longhand analysis. Because the procedure lends itself to simple physical interpretation, it can be used for quick approximate solutions, and the thought process involved can guide the designer's judgement.

Before outlining the procedure, we introduce (or review) some preliminary concepts:

a. Fixed-end moment—The member end moments due to loads when the member ends are prevented from rotating, shown in Fig. 25-36a. The calculation of fixed-end moments was discussed in section 25.4.3.
b. Member end stiffness (Fig. 25-36b)—The member end moment necessary to cause unit rotation of the end when all other supports are prevented from movement. Their calculation was discussed in section 25.4.4, and it was shown there that, in particular, the the rotational stiffness of a prismatic member is proportional to the member characteristic EI/L.
c. Distribution factor (Fig. 25-36c): Defines the member end moment at a joint as a fraction of the total unbalanced moment applied to the joint. In section 25.5.1, this factor was shown to be equal to the ratio of the member end stiffness to the sum of stiffnesses of all members at the joint; that is, for member i,

$$\text{D.F.}_i = \frac{k_i}{\sum_i k} \qquad (25\text{-}24)$$

the sum of the distribution factors for any joint equals 1.
d. Carry-over factor C_{AB} (Fig. 25-36b)—The ratio of the moment at the far, fixed end B of a member to the cranked-in moment at the near end A of the member, that is, the ratio of the rotational end stiffnesses: $C_{AB} = k_{BA}/k_{AB}$; for a prismatic beam, this value is 1/2; for haunched sections, the stiffnesses can be calculated by the force method, or the references listed in section 25.4.6.2 can be consulted for values.
e. Sign convention—Moments are positive when acting counter-clockwise on the member or clockwise on the joint.

The moment distribution procedure begins with the moments due to loads on the geometrically determinate structure, that is, all joints prevented from movement. These are the fixed-end moments. The structure next is gradually eased into its final deformed shape by allowing one joint at a time to rotate. Each time a joint is released in this fashion, the unbalanced moment on the joint is distributed to the adjacent members (whose far ends are fixed at this stage) in accordance with their distribution factor; a fraction of the moment thus distributed to the near member end, equal to the carry-over factor, is carried over to the far, fixed member end. This carry-over moment will in turn become an unbalanced moment to be distributed to the adjacent members upon release of that joint. As the joints are successively released, the residual unbalanced moments become smaller and smaller, allowing the total moments, obtained by addition of all incremental moments, to converge to the correct solution. In general, the procedure converges rapidly; furthermore, premature termination of the scheme will result in approximate results which may be useful for various design purposes. The successful application of the calculations depends on an efficient tabular scheme, as shown in the following example; in this scheme, all joints are released simul-

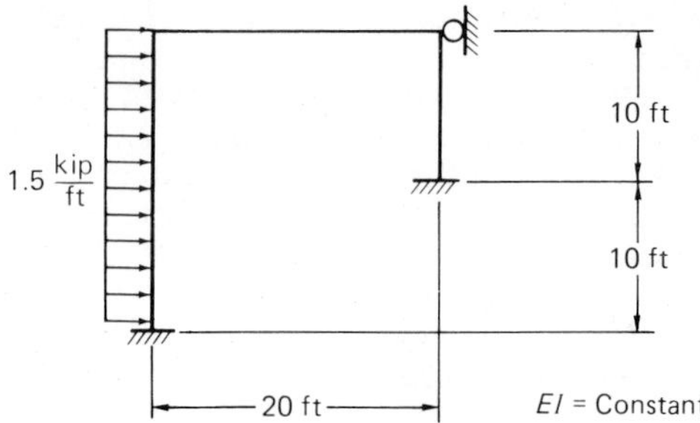

Fig. 25-37 Example 25-20.

taneously in the 'balancing cycle,' following which all residual moments are carried over to the far member ends in the 'carry-over cycle.' This process is repeated to the required degree of convergence.

EXAMPLE 25-20: Analyze the rigid-jointed structure of Fig. 25-37 by moment distribution.

SOLUTION: We note that the joints of this frame can only rotate, but are constrained against translation. The solution is carried out in the following table (counter-clockwise moments on member ends are positive):

Fixed-End Moments: $M^F_{AB} = -M^F_{BA} = \dfrac{wL^2}{12} = 50$ k-ft

Joint	A	B		C		D	
Member	AB	BA	BC	CB	CD	DC	
k	1	1	1	1	2	4	$\times \dfrac{EI}{5}$
D.F.	0	0.5	0.5	0.33	0.67	0	
FEM.	+50.0	−50.0					k-ft
Bal.	0	+25.0	+25.0				
C.O.	+12.5	0	0	+12.5			
Bal.	0	0	0	−6.3	−6.2		
C.O.			−3.1	0	0	−3.1	
⋮							
Final Mom.	+63	−24	+24	+8	−8	−4	k-ft

All reactions and internal forces can be obtained by statics from these end moments.

When translation of joints is possible, the moment distribution can be carried out in the following steps:

Step 1 Restrain all joints against translation, and carry out the moment distribution as before; this leads to a set of moments M_L.

Step 2 Corresponding to each translatory degree of freedom of the structure, introduce a unit displacement, calculate the resulting fixed-end moments, and distribute them in accordance with the method outlined earlier. This involves a number of separate moment distributions equal to the number of translatory degrees of freedom, each one resulting in moments M_i due to a unit displacement. The moments due to an actual (but as yet unknown) displacement Δ_i are $\Delta_i M_i$.

Step 3 Corresponding to each translatory degree of freedom, introduce a force equilibrium equation, called the shear equation. These equations are written in terms of the member end moments, and will therefore contain the unknown Δ_is. The equations are solved for the Δ_is.

Step 4 The moments M_i of Step 2 are multiplied by the appropriate translations Δ_i, and added to the moments M_L of Step 1 in order to obtain the final moments. The remaining forces can then be found by statics.

EXAMPLE 25-21: Analyze the single-story, single-bay rigid frame of Fig. 25-38a.

SOLUTION: This structure has one degree of translatory freedom (neglecting axial deformations), the sway of the top joints. We first restrain the top against this sway by introducing an imaginary restraint, and perform a moment-distribution. This is identical to the preceding example, Ex. 25-20 and we will therefore only list here the resulting moments M_L:

Joint	A	B		C		D	
Member	AB	BA	BC	CB	CD	DC	
D.F.	0	0.50	0.50	0.33	0.78	0	
M_L	+63	−24	+24	+8	−8	−4	k-ft

We next introduce a unit sway to the right, as shown in Fig. 25-38b, compute the fixed-end moments of the columns, which we can identify as the translatory stiffness k_{13} and k_{23} of eq. 25-16, $(= 6EI/L^2)$, and perform the moment distribution; the actual moments due to the sway Δ, will be Δ times the resulting moments:

FEM	+60	+60			+240	+240	$\times \dfrac{EI}{4000}\Delta$
Bal.	0	−30	−30	−80	−160	0	
C.O.	−15	0	−40	−15	0	−80	
Bal.	0	+20	+20	+5	+10	0	
⋮							
M_{Sway}	+55	+50	−50	−80	+80	+160	$\times \dfrac{EI}{4000}\Delta \equiv \Delta'$

The actual value of the sway is determined by the shear equation; referring to Fig. 25-38b.

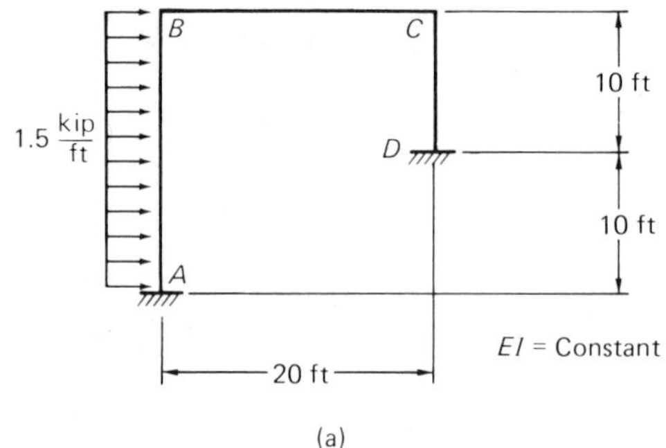

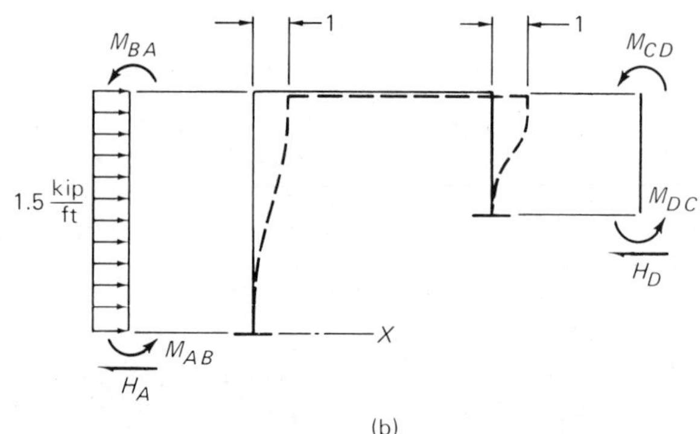

Fig. 25-38 (a) Example 25-21; and (b) effects of joint translation.

$$\overset{+}{\Sigma \rightarrow} F_x = 0: (1.5 \text{ kip/ft})(20 \text{ ft}) - H_A - H_D = 0$$

where

$$H_A = \frac{1.5 \times 20}{2} + \frac{1}{20}(M_{AB} + M_{BA})$$

$$= 15.0 + \frac{1}{20}[(63 + 55\,\Delta') + (-24 + 50\,\Delta')]$$

and

$$H_D = \frac{1}{10}(M_{DC} + M_{CD}) = \frac{1}{10}[(-4 + 160\,\Delta') + (-8 + 80\,\Delta')]$$

so that

$$\Sigma F_x = 0: \quad 30.0 - (16.95 + 5.25\,\Delta') - (-1.20 + 25.00\,\Delta') = 0;$$

$$\Delta' = 0.487$$

We continue the tabular calculation by prorating the sway moments, and summing to determine the final moments.

M_L	+63	−24	+24	+8	−8	−4	
M_{Sway}	+27	+24	−24	−39	+39	+78	
Final Mom.	+90	0	0	−31	+31	+74	k-ft

The remaining forces can be determined by statics.

25.5.7 Design Aids and References

The presentation of the displacement method in its matrix form in section 25.5.4 can serve only as a short introduction to the approach. For a more thorough treatment, the reader is referred to the book by Gere and Weaver mentioned in section 25.1.5, or the following:

Hall, A. S., and Woodhead, R. W., *Frame Analysis*, Wiley, New York, 1961.

Rubinstein, M. F., *Matrix Computer Analysis of Structures*, Prentice-Hall, Englewood Cliffs, N.J., 1966.

All of these books contain some of the standard stiffness matrices for structural members, as well as discussion of important topics not covered here, such as treatment of loads applied between nodal points, and transformation of matrices from member to structure coordinates, points essential to the effective application of the method.

To incorporate the effect of member haunches in the analysis, the stiffness method permits the idealization of the nonprismatic member by a series of prismatic elements of appropriate properties in order to utilize the standard prismatic member stiffness matrices (eqq. 25-25 and 25-26), but this will lead to a vastly larger number of unknowns with consequent increase of the required computer storage and time. To avoid this, the values listed in the "Handbook of Frame Constants" referred to in section 25.4.6.2 can be used in the required stiffness matrices.

For the application of any of the matrix methods, appropriate computing equipment, including both hardware (the machinery) and software (the programs) are essential. For routine analysis of framed elastic structures, a number of canned programs are available which contain all the calculations. The engineer supplies structure, member, and load characteristics, and the program yields all desired forces and deformations. For details regarding such a program, see

Fenves, S. J., et al., "STRESS, User's Manual," M.I.T. Press, Cambridge, 1964.

Of all longhand methods, the moment-distribution procedure is probably of most value for the analysis of reinforced concrete frames. A number of schemes are available to speed up the convergence, including modification of the stiffness factors to account for actual support conditions, symmetry and antisymmetry considerations, block, and overrelaxation techniques, and the like. Many of these methods are covered in

Gere, J. M., *Moment Distribution*, Van Nostrand Reinhold, New York, 1963.

This book also contains an excellent bibliography and plotted design aids. The information contained in Portland Cement Association's "Handbook of Frame Constants" referred to in section 25.4.6.2 is intended particularly for use in moment distribution. If it becomes necessary to evaluate fixed-end moments, stiffnesses, and carry-over factors longhand, the method of column analogy covered in textbooks such as

Kinney, J. S., *Indeterminate Structural Analysis*, Addison-Wesley, Reading, 1957

becomes very convenient. This method can be derived by orthogonalization of the equations of the force method discussed in section 25.4.

Of particular importance is the application of the moment-distribution method to the analysis of lateral force effects on tall building frames with many degrees of side-sway. In this case, the shear forces associated with these side-sway effects can be solved by a method of successive corrections of unbalanced shears, analogously to the operations performed on the unbalanced moments indicated in section 25.5.5. This very useful device for longhand lateral force analysis is described in

Morris, C. T., Discussion of H. Cross, "Analysis of Continuous Frames by Distributing Fixed-End Moments," Transactions ASCE **96** (1), 1932.

The thought process underlying the displacement method can be useful to the designer. The magnitude of relative stiffnesses of different elements can help in visualizing the flow of forces through the structure, with the stiffest elements in general the most effective.

The limiting case of the geometrically determinate (or fixed-ended) structure forms a useful counterpart to the other limiting case of the statically determinate (or pin-ended) structure and can help bracket the correct design forces. The forces on the geometrically determinate structure (the fixed-end forces) can then be relaxed, or distributed, an arbitrary number of times to arrive at improved design values. A two-cycle moment distribution based on this concept is suggested for preliminary design of building frames in

Portland Cement Association, *Continuity in Concrete Building Frames*, 4th Edition, Chicago, 1959.

A useful concept apparent from the moment distribution process is the attenuation, or damping out, of the effects of local loadings. Thus, the *ACI Code* permits analysis of building frames for gravity loads by fixing the far ends of the columns of the story under consideration, and analyzing one story at a time.

25.6 ANALYSIS FOR MOVING LOADS

25.6.1 Introduction

In many structures a number of different loading conditions is possible. For instance, a traffic load moving across a bridge will cause different forces depending on its location; similarly, a warehouse structure designed to carry heavy live loading may have this load distributed over different bays or different stories at any time, and it is not always obvious which of these distributions will cause the maximum stresses in a member. In such cases, the analyst must determine that load position or combination which will be critical in its effect on the forces; influence lines are a useful tool to accomplish this; several methods of determining these influence lines are covered in section 25.6.2. In section 25.6.3 these influence lines are used to determine critical member forces at given sections of the structure. Finally, section 25.6.4 presents the concept of envelope of forces by which the designer can determine the maximum design forces at all sections of the structure due to critical positioning of the live loads.

25.6.2 Influence Lines

-1. Definitions—An 'influence function' (also called 'Green's function' in mathematics) denotes the effect at one specified point as function of the position of the cause, of unit value. For instance, the value of a simple beam reaction due to a unit load as a function of the position of this load would be an influence function.

An 'influence line' denotes the plot of the influence function. Thus, the influence line of a specified structural effect (such as internal force, reaction, or deflection) is the value of this effect plotted as function of the position of the unit load which causes it. For forces in statically determinate structures, influence lines for forces can be found by simple statics. In the following, several ways of performing these statical calculations will be shown.

-2. Influence lines by free bodies—In this basic approach, the influence line for the desired force is obtained by considering the equilibrium of appropriate free bodies to which the unit load is applied at different discrete points. More conveniently, the unit load is applied at a point defined by the variable x, and the desired force is computed as a function of x, and plotted along the structure.

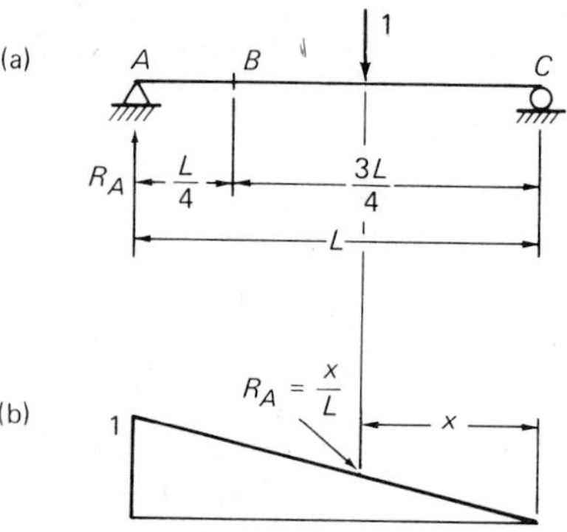

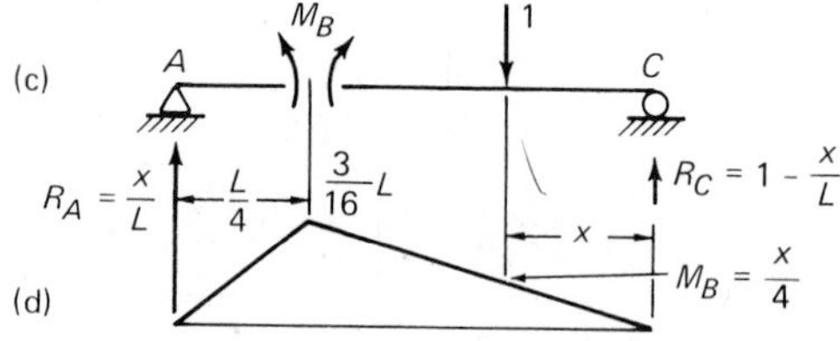

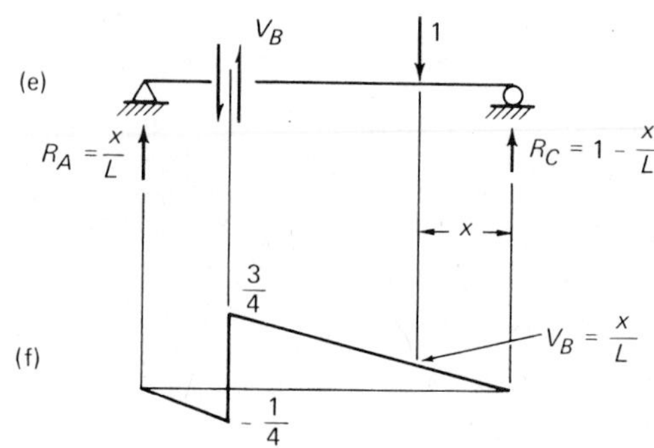

Fig. 25-39 Example 25-22. (a) Beam; (b) influence line for R_A; (c) free bodies for M_B; (d) influence line for M_B; (e) free bodies for V_B; and (f) influence line for V_B.

EXAMPLE 25-22: Determine the influence lines for: (a) the reaction at A; (b) the moment at B; and (c) the shear at B, of the simple beam shown in Fig. 25-39a.

a. We consider the entire beam as free body and sum moments about B.

$$\Sigma M_B = 0: \; R_A \cdot L - 1 \cdot x = 0; \; R_A = \frac{x}{L}$$

The plot of this function is shown in Fig. 25-39b.

b. We consider two free bodies AB and BC, shown in Fig. 25-39c, such that the desired moment is an external force. The reactions are determined first. It is convenient to take the equilibrium of the unloaded free body; thus, for a unit load between A and B

ΣM_B (of free body BC) = 0:

$$M_B = \frac{3}{4}L\left(1 - \frac{x}{L}\right); \; \left(\frac{3}{4} \leqslant x \leqslant L\right)$$

and for the load between B and C

$$\Sigma M_B \text{ (of free body } AB) = 0: \; M_B = \frac{x}{4}; \; \left(0 \leqslant x \leqslant \frac{3}{4}L\right)$$

The influence line is shown in Fig. 25-39d. Note that the mathematical discontinuity at B requires two separate calculations.

c. We consider again two free bodies AB and BC, as shown in Fig. 25-39c, and sum vertical forces of the unloaded body.

For $\frac{3}{4}L \leqslant x \leqslant L$:

$$\Sigma F_Y \text{ (of } BC) = 0: \; V_B = \left(\frac{x}{L} - 1\right)$$

For $0 \leqslant x \leqslant \frac{3}{4}L$:

$$\Sigma F_Y \text{ (of } AB) = 0: \; V_B = \frac{x}{L}$$

The influence line is plotted in Fig. 25-39f; note that it has a unit jump in shear as the unit load crosses the critical section.

-3. Alternate methods for determination of influence lines—The theorem of virtual displacements is particularly convenient for establishing influence lines. From it, the principle of Mueller-Breslau is easily established:

> The influence line for any force is generated by the displaced shape of the structure resulting from a virtual unit displacement at the point and in the sense of the desired force.

The influence lines shown in Figs. 25-39 and 25-40 should

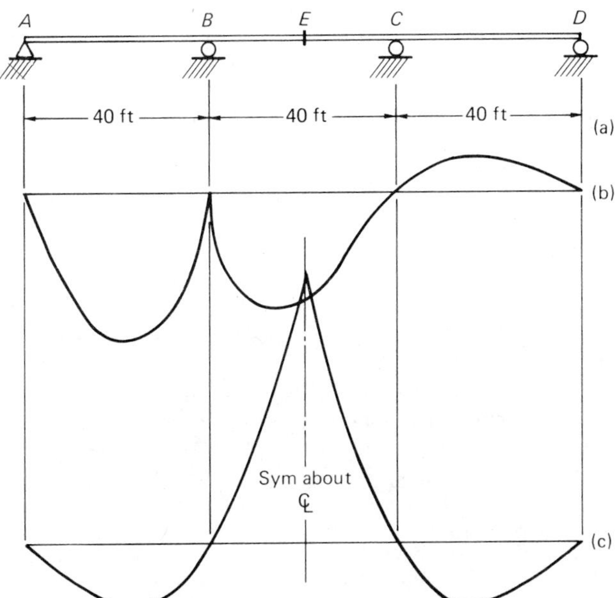

Fig. 25-40 Statically indeterminate influence lines. (a) Beam; (b) influence line for M_B; and (c) influence line for M_E.

be envisioned as such displaced shapes for an understanding of this concept.

The force transformation matrix $[b]$ can also serve to determine influence lines; the elements of the ith row (b_{i1}, b_{i2}, . . . , b_{ij}, . . .) represent the influence ordinates for the force i due to a unit load at points 1, 2, . . . , j, ...

-4. *Influence lines for statically indeterminate structures*—All of the methods which have been outlined are valid for statically indeterminate structures, but the reader should be aware that a much larger amount of work is required to determine such influence lines. Here, two methods will be indicated, one suitable for longhand, the other for computer calculations.

The moment-distribution method, together with the principle of superposition, can be used for the longhand determination of influence lines. The procedure follows these steps:

Step 1 Compute the member end moments due to unit fixed-end moments, taken one at a time. This involves a number of moment-distributions equal to twice the number of loaded members.

Step 2 Compute the fixed-end moments of each member due to a moving unit load. For a straight, prismatic member, for instance, these fixed-end moments are given by the result of ex. 25-13.

Step 3 Compute the actual member end moments by multiplying the end moments determined in Step 1 by the appropriate actual fixed-end moment values determined in Step 2.

Step 4 Once the member end moments have been determined in Step 3, any desired force quantity can be found by statics alone.

A well-organized tabular procedure is necessary to carry out this scheme.

The force or displacement methods in their matrix form, with appropriate computer program, are very suitable for the determination of statically indeterminate influence lines. The following features are required for this use:

a. A nodal numbering system which permits application of the unit load at all desired locations.
b. A program which permits the solution for enough loading conditions to accommodate all load positions considered.

The output of such a computer routine should include all force quantities for which influence lines are desired.

Fig. 25-40 shows two typical influence lines for a three-span continuous beam. The reader should visualize these influence lines as the deformed beam shape due to a unit kink corresponding to the moment considered, in the sense of Mueller-Breslau's theorem discussed earlier. The use of this theorem enables the analyst to visualize the shape of even very complicated influence lines without calculations, thus enabling him to anticipate critical loading positions.

25.6.3 Design Forces Due to Moving Loads

-1. *Use of influence lines to compute forces due to arbitrary loads*—With the influence line for a certain force, F, drawn, we can find the value of this force due to any given loading condition, such as the one shown in Fig. 25-41a. The influence line for the force whose value is desired has been drawn as in Fig. 25-41b. The influence ordinate at any point x is denoted by y.

The effect of the concentrated load P_1 on the desired force F equals P_1 times the effect of a unit load, y_1

$$F = P_1\, y_1$$

The resultant of the distributed load $w(x)$ acting over the element of length dx is $w \cdot dx$. The contribution of this elementary load to the force F is $(w \cdot dx) \cdot (y)$, and that of the entire distributed load between two points A and B is

$$F = \int_A^B w\, y\, dx$$

This integral can always be evaluated, but another, and sometimes very time-saving method is to replace the variable $w(x)$ by its average value w_{ave}, and take it outside the integral:

$$F = w_{\text{ave}} \int_A^B y\, dx.$$

The integral $\int_A^B y\, dx$ can be interpreted geometrically as the area under the influence line between A and B, so that the value of the function due to this distributed load equals

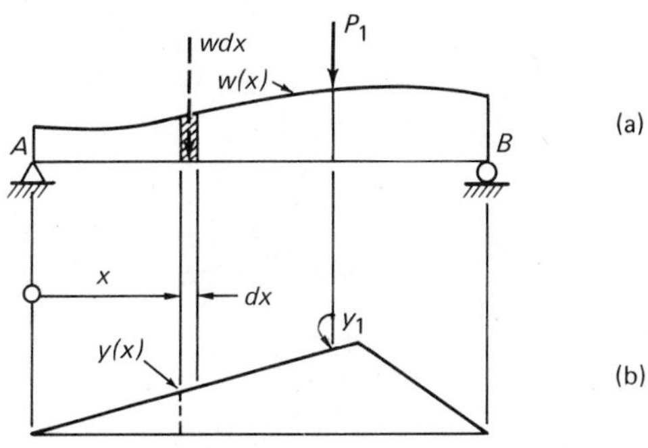

Fig. 25-41 Use of influence lines for determination of force. (a) Loads, and (b) influence line.

the average load intensity multiplied by the area under the portion of the influence line over which it acts. For uniform loads, this is very convenient. For linearly-varying loads and influence lines, it can be shown that the average load w_{ave} is the value of the load intensity at the centroid of the area under the influence line.

The total value of the desired force due to the entire load is then

$$F = \int w\,y\,dx + \Sigma P \cdot y \qquad (25\text{-}33)$$

Note that depending on the sign of the influence ordinate y, the contributions of the various forces can be either additive or canceling.

EXAMPLE 25-23: The influence line for shear at the quarter point of the simple beam of Fig. 25-42a has been drawn in Fig. 25-42b; it is required to determine this shear force due to the uniform beam dead load w kip/ft and the concentrated midspan load of value wL.

SOLUTION: The net area under the influence line = $L/32 - 9L/32 = -L/4$. Using eq. 25-33, the shear force is

$$V_B = -w \cdot \frac{L}{4} - wL \cdot \frac{1}{2} = -\frac{3}{4}wL$$

Note that a loading between points B and C only would lead to a larger value of shear, which indicates that partial loading can sometimes be more critical than full loading.

-2. *Use of influence lines to compute critical loading conditions*—With moving or variable loads on a structure, it becomes important to determine that loading condition which causes the critical value of the forces to be used in design. This is done by use of influence lines. A number of theoretical criteria are available for the calculation of the critical load position.

Here, however, we will only indicate some simple cases in which this determination can be done by common sense or a simple trial-and-error procedure. As a general rule, the heaviest concentrated load should be positioned at the point of maximum influence ordinate, and the remaining loads clustered around this point. Any variable distributed loads should be applied over the portion of the structure with influence ordinate of appropriate sign. In cases of sign reversal of the influence diagram, both positive and negative critical values must be determined.

The process is illustrated in ex. 25-24 of the following section.

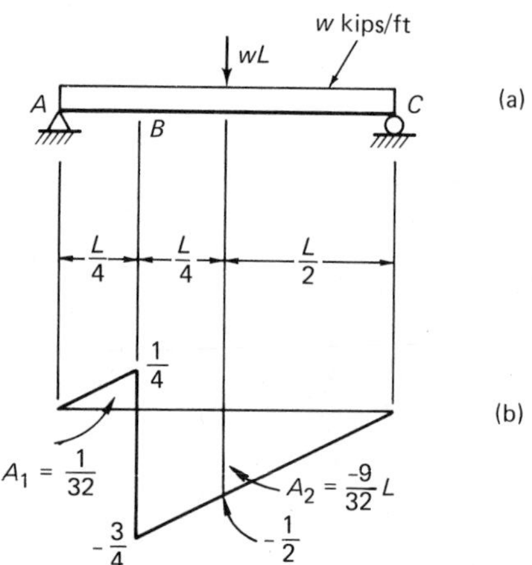

Fig. 25-42 Example 25-23. (a) Beam and load; and (b) influence line for V_B.

-3. *Envelopes of forces*—The outlined method enables the determination of the extreme values of one force, say, the moment at one section. This process now has to be repeated for every section of the structure (in practice, a finite number of sections is considered, such as $1/10$th points along each span of a bridge girder, so that all sections can be designed for the critical positive and negative moment values which can possibly arise under any possible location of the design load. The curve which connects the extreme moment values at all sections is called the moment envelope. For a complete design of bridge structure, both moment and shear envelopes are necessary. Repetitive work of this type is best assigned to the computer, but a simple example will be presented here.

EXAMPLE 25-24: Compute the envelope of moments for the simple span bridge shown in Fig. 25-43a. The design live load consists of the truck shown in Fig. 25-43a and the dead load is 2 kip/ft.

SOLUTION: Because of the symmetry of the bridge, only half of the span needs to be designed. The left half is split into four 10-ft segments as shown in Fig. 25-43b, and the influence lines for their end points are calculated by any method and plotted as in Fig. 25-43c. Next, the critical truck positions are determined for each

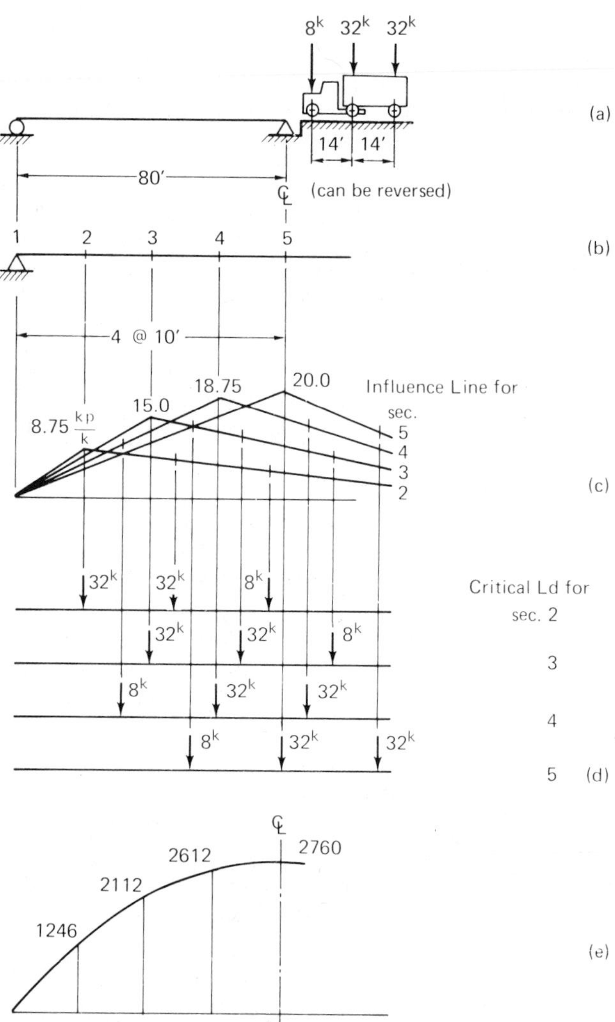

Fig. 25-43 Example 25-24. (a) Structure and load; (b) design sections; (c) influence lines; (d) critical load positions; and (e) moment envelope.

TABLE 25-4

Axle	Section / Axle Ld. P	2 y	2 P · y	3 y	3 P · y	4 y	4 P · y	5 y	5 P · y
1	8 kip	5.25	42.0	8.00	64.0	10.00	80.0	13.00	104.0
2	32 kip	7.00	224.0	11.50	368.0	18.75	600.0	20.00	640.0
3	32 kip	8.75	280.0	15.00	480.0	13.50	432.0	13.00	416.0
L. L. Total			546.0		912.0		1112.0		1160.0
D. L. Intensity	*w*	*A*	*A* · *w*	*A*	*A* · *w*	*A*	*A* · *w*	*A*	*A* · *w*
	2 k/ft	350	700.0	600	1200.0	750	1500.0	800	1600.0
D. L. + L. L. Total (in kip-ft)			1246.0		2112.0		2612.0		2760.0

section, as shown in Fig. 25-43d. The critical live load moments are found by the calculations shown in Table 25-4, and the dead load moments are calculated and superposed, leading to the moment envelope ordinates at the bottom of this table. Lastly, the moment envelope is plotted as shown in Fig. 25-43e; the span can now be designed to resist the critical moments at all sections.

25.6.4 Design Aids and References

Influence Coefficients are available in tabulated or plotted form for many types of structures. For prismatic and haunched continuous beams, some of the available references are:

"Steel Construction Manual," American Institute of Steel Construction, 7th Edition, New York, 1970.

Griot, G., *Influence Line Tables*, Ungar, New York, 1954.

Influence lines for some types of arches are contained in the book by Michalos cited in section 25.4.6.2.

The load systems to be considered for the design of bridges are usually contained in the appropriate building code, such as

American Association of State Highway Officials, "Standard Specifications for Highway Bridges," Washington, D.C.

This set of specifications calls for consideration of two different load types, an oversize single truck at its critical location, and an equivalent uniform lane load to simulate a continuous lane of traffic. With such loading specified, it is possible to compute moment and shear envelope values for standard bridges of different spans; some of this information is also available in the above mentioned AASHO Specifications.

For the design of building frames, the live load is usually applied spanwise. The critical design forces in continuous beams and frames are not due to loads over the entire structure, rather, the theorem of Mueller-Breslau cited in section 25.6.2.3 can be used to determine the critical load conditions. Figure 25-44 shows the influence line for a typical support moment sketched quantitatively as the deformed shape of the structure resulting from an appropriate unit kink. This influence line shows that the critical loading condition, shown in Fig. 25-44b, consists of loads on the spans with negative influence ordinates adjacent to the support in question, and alternate spans. Similarly, Figs. 25-44c and d show that the span in question, and alternate spans should be loaded for critical moments in the span.

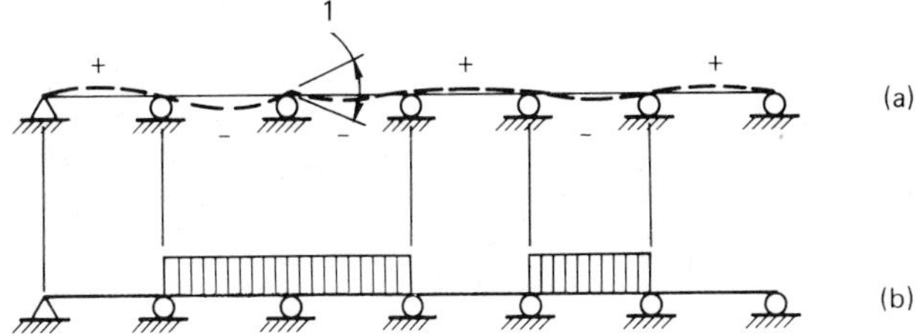

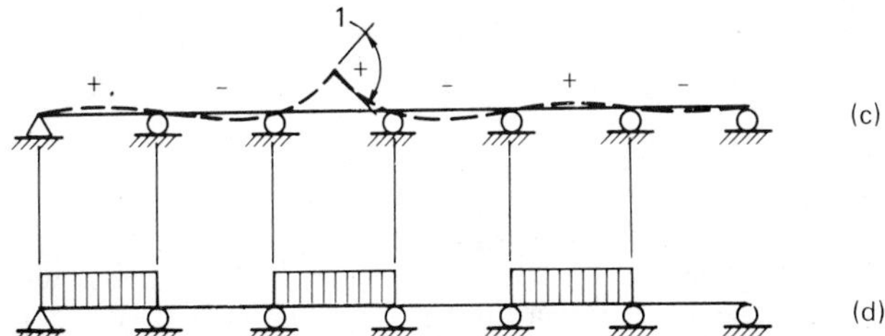

Fig. 25-44 Critical loading conditions. (a) Influence line, and (b) critical loading, for support moment; (c) influence line, and (d) critical loading, for midspan moment.

The design moments at different points of a continuous beam are thus due to different loading conditions. The moment coefficients specified in the *ACI Code* for the design of concrete beams (subject to certain restrictions) have been computed from a series of such analyses. They should thus always be shown as points on moment envelopes, with

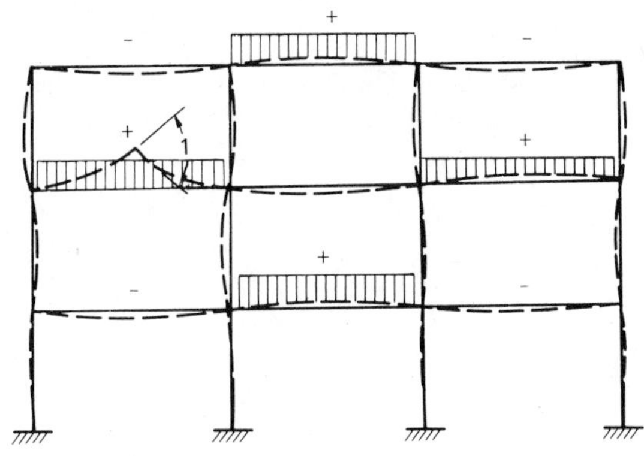

Fig. 25-45 Checkerboard loading on frame.

other points of these envelopes determined from the statics of the appropriate loading condition.

The same reasoning is also commonly applied to building frames. Figure 25-45 shows the influence line for a positive midspan moment resulting from an appropriate unit kink, and the corresponding "checkerboard" loading of the spans with positive influence ordinates. Similar procedures show that the critical column moments are also due to such checkerboard loadings.

25.7 THE FINITE ELEMENT METHOD

25.7.1 Introduction

The finite element method, as presented here and as commonly used, is an extension of the matrix displacement (or stiffness) method as covered in section 25.5.4, applicable to the analysis of any solid; it has been used for the analysis of all kinds of bodies in their elastic and inelastic range, including ultimate strength and time-dependent behavior. Because of its versatility and ease of application when appropriate computer programs are available, it promises to become the foremost tool for analysis of slabs, walls, shells, and other structural elements which are intractable by conventional methods. Furthermore, it can handle combinations of different kinds of such elements in a unified fashion.

This chapter furnishes a short introduction to the underlying concepts, as well as some references for further study and use of the method.

25.7.2 Outline of the Finite-Element Approach

-1. Basic approach—The first step in a finite-element analysis is to model the body to be analyzed, as shown in Fig. 25-46a, as an assembly of finite elements, interconnected at specified nodal points, as shown in Fig. 25-46b. In the case of the plane-stress body shown here, a decomposition into triangular elements is convenient because of the ease with which such elements can simulate irregular boundaries. They also permit variation of the element size, with the smallest elements in regions of stress concentration, and larger elements in areas of more regular stress variation.

As the specified loads are applied to the model of Fig. 25-46b, the boundaries of adjacent elements between the nodal points would tend to open up or to overlap. It is therefore to be expected that the analysis of such a finite-element model would lead to a less stiff solution, with larger deformations, than the exact solution of the prototype body. In order to avoid such discontinuities, it is necessary to model the element behavior in such a fashion that the common boundaries of adjacent elements will deform together; such elements are called compatible elements.

In the stiffness method of analysis, the displacements of the nodal points (or joints) are the primary unknowns. In order to label these unknowns clearly, it is necessary to introduce a node for each displacement component, as shown in Fig. 25-46c. According to the discussion of section 25.5.4, there will also be a force quantity called nodal force, associated with each of these nodes; at each node, either the displacement, or the force, will be specified, the other quantity being unknown. In assigning a numbering system to the nodes, those with unknown displacement should be numbered first.

Let us now assume that it is possible to determine structural stiffness values, K_{ij}, defined as in section 25.5.3, that is, K_{ij} is the force along node i associated with a unit displacement of node j, all other nodal displacements equal to zero. The determination of the structure stiffness matrix $[K]$ consisting of elements K_{ij} will be mentioned in section 25.7.2.2.

The structure stiffness matrix $[K]$ relates the nodal forces and displacements according to eq. (25-29), in which the nodal numbering, as mentioned above, has been so arranged that the unknown displacements, Δ_α, are on top, and the unknown (reactive) nodal forces, X_β at the bottom:

$$\begin{bmatrix} K_{\alpha\alpha} & K_{\alpha\beta} \\ K_{\beta\alpha} & K_{\beta\beta} \end{bmatrix} \begin{bmatrix} \Delta_\alpha \\ \Delta_\beta \end{bmatrix} = \begin{bmatrix} X_\alpha \\ X_\beta \end{bmatrix} \qquad (25\text{-}29)$$

The solution for the unknown nodal displacements, and the unknown nodal forces takes place according to the two-step scheme outlined in section 25.5.4, leading to eqq. (25-30) and (25-31):

$$[\Delta_\alpha] = [K_{\alpha\alpha}]^{-1} \left[[X_\alpha] - [K_{\alpha\beta}]\ [\Delta_\beta]\right] \qquad (25\text{-}30)$$

$$[X_\beta] = [K_{\beta\alpha}]\ [\Delta_\alpha] + [K_{\beta\beta}]\ [\Delta_\beta] \qquad (25\text{-}31)$$

The solution of eq. (25-30) involves the inversion of a matrix of order equal to the number of unknown displacements. For realistic problems, the number of unknown displacements can range from the order of tens to the order of thousands. It is apparent that problems of this type can be tackled only with the help of adequate computing equipment.

It should be clear from the above outline that the only difference between the matrix displacement method of framed structures as outlined in section 25.5.4 and the finite element method is the choice of the stiffness matrix; this enables different types of elements to be incorporated into the analysis in a unified fashion, and contributes greatly to the usefulness of the approach.

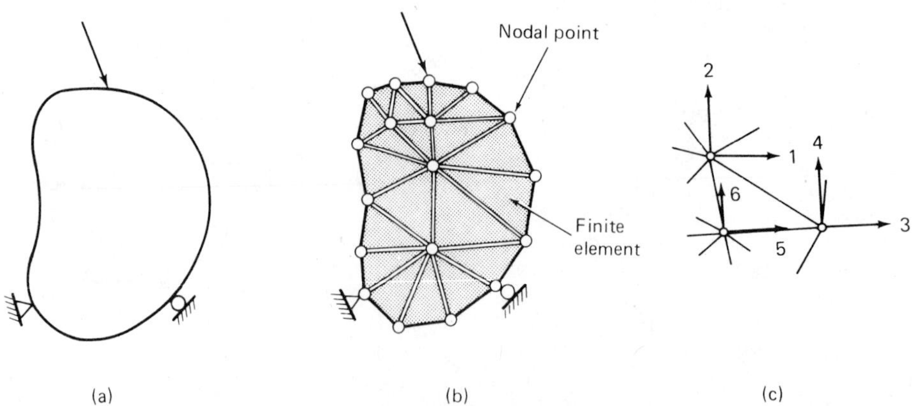

Fig. 25-46 Finite element model. (a) Prototype body; (b) finite element model; and (c) nodal numbering.

-2. *Further details*—For finite-element analysis of elastic structures, stiffness matrices should be determined in accordance with the laws of the theory of elasticity. Since an exact analysis of elastic elements, such as the triangular plane-stress elements of Fig. 25-46b is a forbidding task, an approximate method is used which is completely general in its basic formulation and whose results will converge to the exact solution with decreasing element size; accordingly, the larger the number of elements, the more exact will be the solution.

This method can be based on the theorem of virtual displacements, eq. (25-4), in which the virtual strains as well as the real stresses are computed by means of an assumed displacement function with unknown coefficients. These coefficients are expressed in terms of the nodal displacements; the strains are obtained by appropriate differentiation of the displacement function, and the stresses are computed from the strains by Hooke's law. The element stiffness matrix can then be calculated by insertion of these values into eq. 25-4a in its matrix form.

Space does not permit any further details here; suffice it to say that the procedure can be expressed in standard matrix form, and is suitable for computer programming.

With the element stiffness matrices $[k]$ computed, they can be assembled into the structure stiffness matrix $[K]$ according to the procedure of section 25.5.3. This matrix $[K]$ can now be inserted into eq. 25-29, which in turn is solved for the unknown displacements $[\Delta]$. The desired end result of a finite-element analysis is usually the distribution of stresses throughout the body; they can be obtained from the nodal displacements by manipulation of matrices already available from earlier computations. For further details, the reader is referred to one of the references cited in section 25.7.4.

25.7.3 Examples of Finite-Element Analysis

We demonstrate the effectiveness of the finite-element method by means of the deep cantilever beam shown in Fig. 25-47, loaded by a linearly varying end load. It is well known that conventional beam theory, based on the plane section assumption, is unreliable for such members.

To show the convergence of the solution with decreasing element size, we consider three solutions for different mesh sizes shown in Figs. 25-47a to c. The element used in this case was a rectangular plane-stress element which admits linearly varying normal stresses but constant shear stress. The nodal numbering used is shown in the case of the 2 × 3 mesh, with the free nodes listed first (actually, the computer program used did this ordering automatically).

Figure 25-48 shows the mid-depth deflection at the free beam end as function of the mesh size. We note that the curve becomes asymptotic with increasing number of elements, but even for the very coarse 2 × 3 mesh the answer is sufficiently good for design purposes, though it underestimates the exact deflection (that this is always so in approximate solutions by the stiffness method can be proved by the theorem of minimum potential energy).

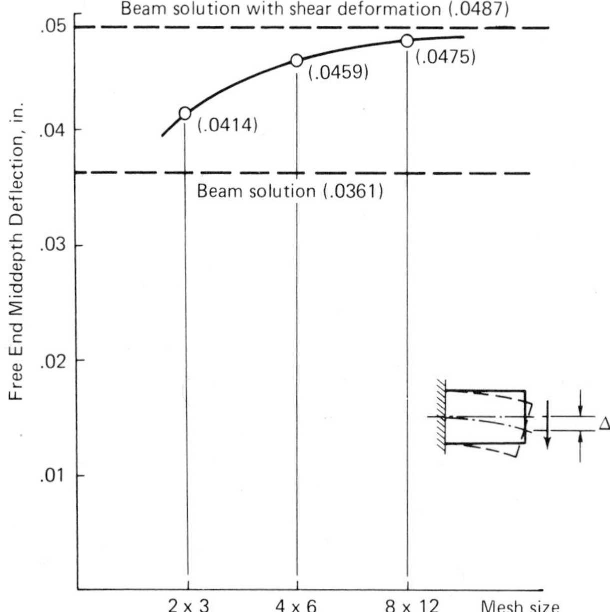

Fig. 25-48 Convergence of finite element solution.

Figure 25-49 shows the stress distribution near the fixed end of the beam. It is clear that the strength of materials solution based on the plane section assumption would give a completely incorrect picture of the shear stress variation, and would underestimate the peak flexure stress. With stresses of this type known, the principal stress trajectories can be plotted by the usual means.

To indicate the power of the finite element method, we show in Fig. 25-50 the results of an inelastic finite-element solution which includes the effects of tensile cracking and compressive crushing of the concrete, and yielding of the steel under increasing load. The comparison of actual and calculated results shows the detail with which the complete structural response can be predicted by suitable analysis.

25.7.4 Additional Aspects of the Finite-Element Method

-1. *Stiffness matrices for different structures*—The use of eq. (25-29) is entirely general; it is only necessary to assemble

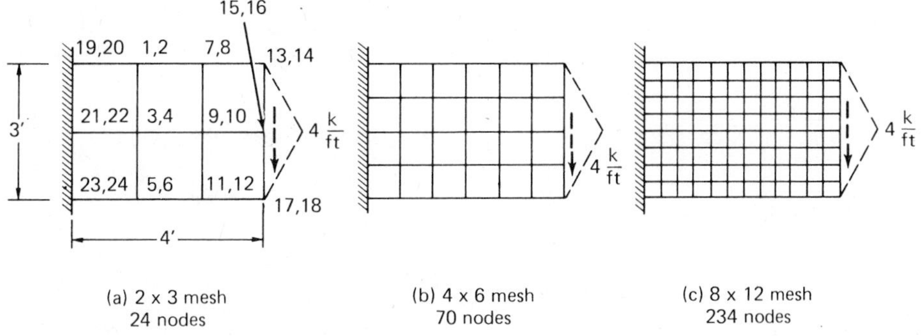

Fig. 25-47 Finite element models of deep cantilever beam. E = 3000 ksi, μ = 0.3, thickness = 12 in.

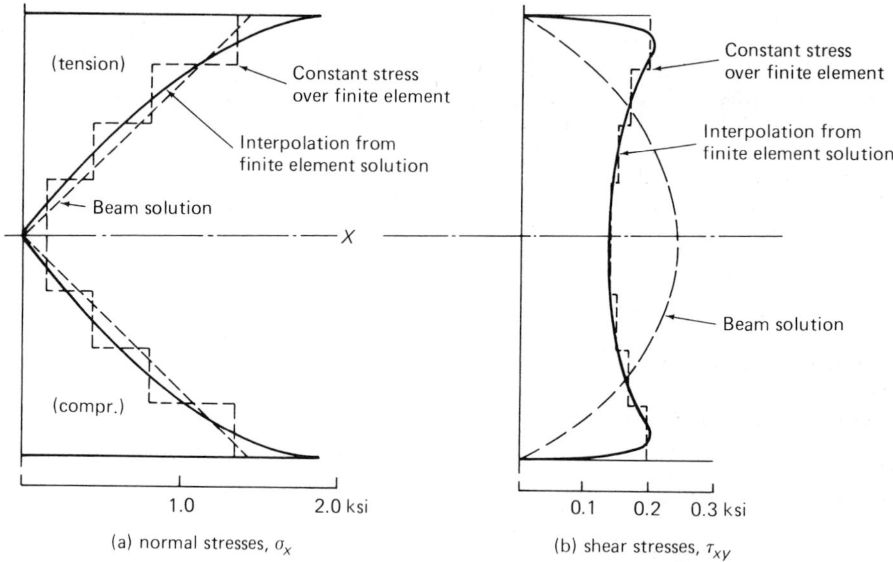

Fig. 25-49 Stress distribution near fixed end of cantilever beam.

the structure stiffness matrix $[K]$ from element stiffness matrices $[k]$ which are capable of representing the response of the particular structure under investigation.

The accuracy of stiffness matrices can be improved by the introduction of additional nodes, for instance, nodal points could be added at the midpoints of the three sides of the triangular elements of Fig. 25-46, leading to six additional nodes in each element. This in turn would permit the use of more accurate displacement functions which would be better able to reproduce the actual stress variation than, for instance, the constant shear stress element used in the example of section 25.7.3. Whether the improved accuracy

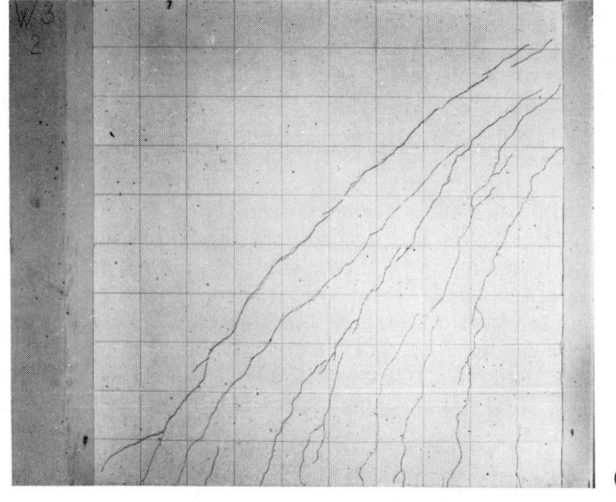

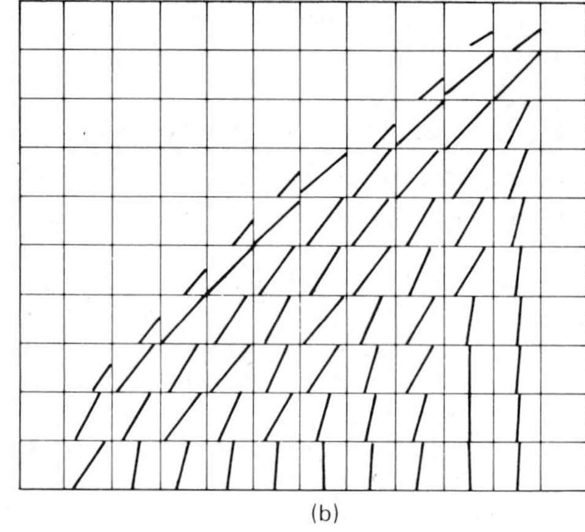

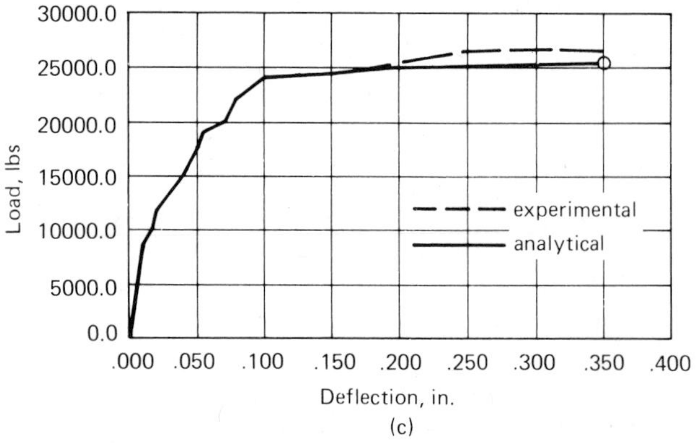

Fig. 25-50 Response of reinforced concrete panel at ultimate stage. (a) Experimental cracking; (b) analysis; and (c) load-deflection curve.

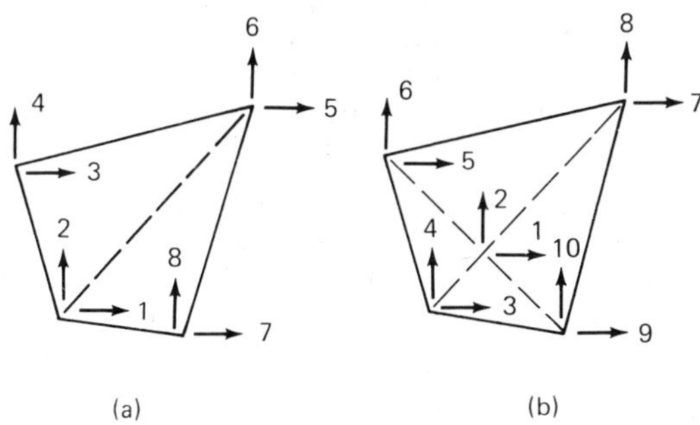

Fig. 25-51 Quadrilateral elements, (a) by superposition of two triangles; and (b) by superposition of four triangles.

merits the additional computations is a matter to be investigated from case to case.

In the following, element stiffness matrices for different structures are discussed briefly.

a. Beam element—Figure 25-32 of section 25.5.2 shows an appropriate nodal system for the general case of beams. By superimposing the stiffness matrix of eq. (25-26) upon other stiffness matrices, structures composed of combinations of beam- and other elements can be modeled.
b. Quadrilateral plane-stress element—General quadrilateral elements are very versatile and widely used. As shown in Fig. 25-51, the element stiffness matrix can be obtained by superposition of two triangular stiffness matrices, or by considering the combination of four triangles as a structure, and calculating the stiffness matrix for the exterior corners. More exact results are obtained by a direct solution.
c. Plane-strain element—The only difference between a plane-stress solution and a plane-strain solution as required, for instance, in tunnel design, lies in the use of the appropriate form of Hooke's law as obtained from books on the theory of elasticity.
d. Flat-plate bending element—In plate bending, the bending moments about two axes, and the corresponding rotations, and transverse forces and corresponding deflections, are taken as nodal quantities. The internal quantities are internal bending and twisting moments, and the corresponding curvatures and twist. Figure 25-52 shows a rectangular plate bending element with 12 nodes, permitting a transverse deflection displacement function containing 12 terms.
e. Finite-element analysis of shells—Conceptually, it is simple to think of shell structures as being composed of flat elements oriented at appropriate angles to each other. This will require a transformation to rotate the nodal quantities from the local to a common structure coordinate system.

   If only membrane action is to be considered, it is sufficient to include only the in-plane nodal quantities as shown in Fig. 25-46c. In the more general case of shell bending, the in-plane nodes and the plate-bending nodes are required, and the total element stiffness matrix is obtained by superposition of the in-plane and bending stiffnesses.

   A more accurate representation of the shell behavior is furnished by curved shell elements. In this case, the in-plane and the bending action cannot be decoupled as in the case of flat elements, and the formulation of the stiffness matrix must follow the rules of classical shell theory. In the general case of a quadrilateral doubly-curved shell element as shown in Fig. 25-53, there will be five nodal quantities at each joint, for a total of 20 element nodes. A number of specialized shell elements for cylindrical and axisymmetric shells are available.

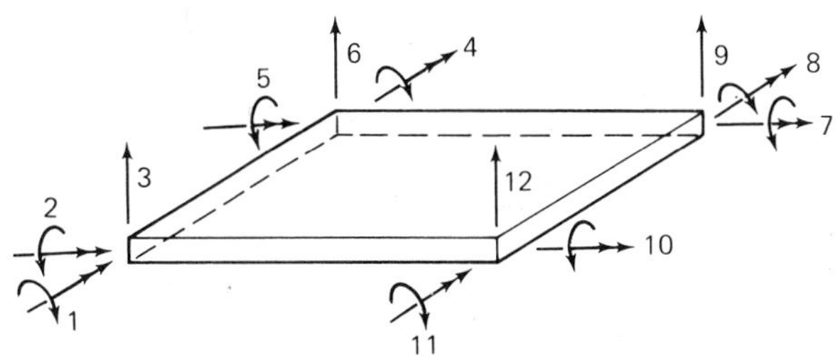

Fig. 25-52 Plate bending element.

f. Three-dimensional solids—The three-dimensional analogue to the planar triangular element of Fig. 25-46 is the tetrahedron shown in Fig. 25-54. This element has 12 nodes, and can accommodate a four-term linear displacement function for each of the three internal displacement components. Appropriate differentiation of these displacements will lead to constant strain and stress within the element. The simulation of realistic three-dimensional situations usually requires a larger number of nodal quantities than can be handled by available computing equipment. On the other hand, if axial symmetry is present, the formulation can be reduced to a quasi-planar case by appropriate procedures.

-2. References—To supplement the presentation of this section, the reader is in the first place referred to the texts of Zienkiewicz and Przemieniecki listed in section 25.1.5.

The former of these presents an excellent overview of the state of the art, including discussions of anisotropic media, non-linear problems, buckling considerations, and computer programming. The latter tends to go more into detail, and contains a number of useful elementary stiffness matrices. It also has a bibliography of over 400 items.

The literature of the finite-element method has in recent

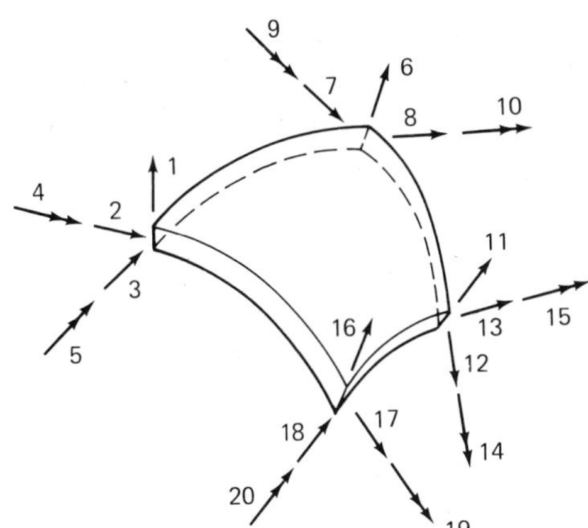

Fig. 25-53 Curvilinear shell finite element.

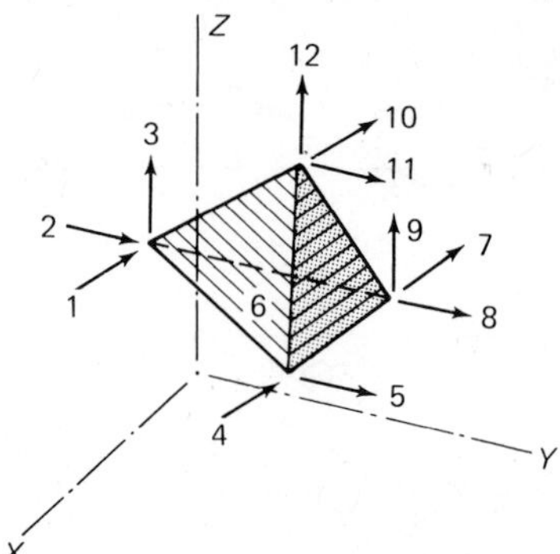

Fig. 25-54 Tetrahedral element.

years become so voluminous that any systematic listing is almost impossible. However, two real treasure chests of information may be mentioned:

"Matrix Methods in Structural Mechanics," (Proc., Conference held at Wright-Patterson A.F.B., Ohio, 26–28 Oct. 1965), U.S. Dept. of Commerce, Nov. 1966. Available through Clearing House for Federal Scientific and Technical Information.

Among the many useful papers in this collection is one by Clough and Tocher on finite elements for plate bending, and several which extend the finite-element method to nonlinear problems involving plasticity and creep.

A more recent compilation of papers is

"Proc. of the Symposium on Application of Finite Element Methods in Civil Engineering," Nov. 13–14, 1969, Vanderbilt Univ., Nashville, Tenn., A.S.C.E., New York, 1969.

Among the notable contributions in this work may be listed survey papers on three-dimensional finite elements by Clough and on plate and shell elements by Gallagher (with a 119-item bibliography), several papers on analysis of shear walls, and on various aspects of concrete behavior.

A number of computer systems with capability for finite-element analysis is commercially available. As example of these, Program ELAS, developed at the Jet Propulsion Laboratory, Pasadena, Calif., and available from Duke University, Durham, N.C., may be cited. This program can analyze framed, plate, shell, or solid structures in various combinations. It appears that at present such 'canned' programs provide the easiest means of access to the power of the finite-element analysis for structural engineers in professional practice.

26

Computer Applications

STEVEN J. FENVES, Ph.D.*

26.1 BASIC CONCEPTS

Computer applications are in daily use in essentially every branch of concrete engineering. These applications cover not only the principal design processes of analysis, proportioning, and detailing, but all auxilliary activities such as preparation of design documents (specification text, bar schedules, drawings, etc.), quantity takeoff, and estimating. Finally, a large portion of analytical research in concrete behavior and concrete structures involves extensive use of computers.

The range of computer applications in concrete engineering is continuously expanding. New programs are being developed for problems which were inconceivable in the past, either because of sheer magnitude of the numerical calculations involved (e.g., the exact analysis of large, complex structures), or because of logical complexity of the possibilities involved (e.g., the direct production of design drawings). In all present and future applications, it has to be kept in mind that the computer is no more than a tool in the hands of its user, much like the slide rule, design tables, etc., used in the past to simplify or speed up certain activities. However, there is a fundamental difference between the computer and other design tools which profoundly affects the manner and extent of its use. To understand this difference, we must first describe the tool more precisely.

*Professor , Department of Civil Engineering, Carnegie - Mellon University, Pittsburgh, Pennsylvania.

26.1.1 Description of the Tool

Computers used in engineering practice can be defined as general-purpose, stored-program, electronic digital computers. Digital means that, within the computer, numbers are represented by discrete digits, as in a desk calculator, in contrast to analog computers where numbers are represented by continuously varying physical quantities; digital computers can also represent and manipulate symbols other than numbers, such as alphabetic characters or geometric entities. Electronic means that the internal operations are performed by electronic circuits, rather than mechanical devices such as counting wheels. Stored-program means that, for each application, the computer is provided with a sequence of steps or instructions, called the program, which defines the process of solution. General-purpose means that the computer is not built specifically for one type of application, so that by using different programs it is capable of solving a wide variety of problems.

A computer's primary function is information-processing, of which computation is only one part. Its internal storage device, or memory, contains two types of information: (1) information which directs the processing, consisting of the instructions comprising the program; and (2) information that is processed, namely, the data pertaining to the particular problem being processed. Even though there is no internal difference between the two types of information, both being stored in identical coded forms, this distinction affects to a great degree the manner in which use of the computer differs from use of other computational or processing tools.

The manner in which the computer operates differs from the desk calculator in the following ways:

1. Increase in speed—Since the calculations are performed electronically rather than mechanically, their speed is greatly increased; while this feature does not change the manner of computation, it enables the computer to process problems which are impractical by manual methods.
2. Internal storage of data—The original data and the partial and final results are stored in the computer's storage device; and therefore, the process does not have to be stopped for the operator to enter new data, or copy results.
3. Internal storage of instructions—In addition to the data, the instructions specifying the process of computation are stored within the storage device and are executed automatically in the prescribed sequence.
4. Ability to make decisions—Besides arithmetic operations, the computer can perform logical operations such as testing a number for negative sign, or nonzero digits; thus, it makes elementary decisions by selecting a path from among several alternatives.

26.1.2 The Concept of Programming

From the previous description, it becomes clear that the user's primary interaction with the computer is through a program, which consists of the entire sequence of computational and decision-making steps comprising the solution process. In fact, it is convenient to think about the computer in terms of a composite machine, encompassing both the actual *hardware* of the computer machinery, and the *software*, or programs, which make the hardware perform a specific function. However, the composite machine does not come into existence until the appropriate program has been developed and loaded into the hardware storage device. Thus, computer-aided data processing requires a complete separation of the problem-solving process into two stages.

-*1. Program development*—This stage is concerned with describing the procedure to be followed in its entirety and providing the computer with the information to control the processing. Each operation and each possible alternative is rigorously defined in terms of the algebraic quantities and logical possibilities involved. The description of the steps in the solution, translated into instructions the computer is capable of executing, becomes the program for the processing of the problem. The development process is described in greater detail in the next section.

-*2. Production*—The second stage involves the use of a previously developed program to obtain results based on particular input data at hand. Generally, the program user will be a different person or organization than the one which developed the program. Therefore, the user is one step removed from the original problem, and is required to operate within the boundaries of the options or alternatives provided in the available program.

As will be discussed in section 26.2.1, the distinction between development and production is considerably reduced when problem-oriented languages are available, where the user can specify the execution of complex sequences of problem solution steps with convenient commands. But even in this case, the alternatives available to the user are only those which have been provided for in the design and implementation of the language.

26.1.3 The Program Development Process

The process of going from the inception of a computer program to actual use consists of a sequence of steps. The degree to which the process outlined below is followed will depend largely on the importance of the program and its intended distribution.

Step 1 Problem statement—The process starts with a complete statement of the problem to be programmed. The variables are defined, the scope of the program outlined, and the required results listed. This step includes the review of the procedure currently employed; the evaluation of the methods, assumptions, and limitations used in the procedure; the exploration of possible alternate procedures; and the extrapolation of potential future applications. Since a successful program is likely to change significantly the organization's method of handling future problems of the type being considered, the problem statement must be worked out as a joint effort between the programmers, management and supervisory personnel, and the design personnel who will eventually use the program.

Step 2 Problem analysis—The definition of the program proceeds by a careful designation of the input and output data, involving the determination of the range of the variables, the precision required, as well as the layout of the format in which these quantities are to be presented to the computer or generated by the program. This step has a definite bearing on the usefulness of the finished program: a program requiring hours of preparing extraneous data for input and producing pages of unorganized numerical results will be unpopular with users, and therefore its value to the organization will be greatly reduced.

The method of solution must be selected with the greatest care, and may involve radical departures from manual techniques. Particularly in design applications, the alternatives that may present themselves must all be foreseen and the appropriate procedure for each case defined. Provisions for programmed checks on the consistency of the data, as well as for possible expansions, alterations, and changes in specifications must also be considered at this stage.

Finally, whenever the mathematical solution selected involves processes which cannot be performed directly by algebraic operations, methods of numerical analysis must be employed to reduce the computations to suitable numerical techniques. Most common operations, such as taking the square root or obtaining the sine of an angle are available in the form of groups of instructions, or subroutines, provided as part of the computer software. In many cases, it is worthwhile to develop alternate formulations which make greater use of existing subroutines.

Step 3 Programming—This step involves essentially the charting of the process of computation and can be most readily accomplished through the use of flow diagrams (see section 26.2.2). A flow diagram is a graphical representation of the logic (dictionary definition: chain of reasoning) of the program and shows each operation and decision performed, as well as the interrelationship between the steps. The flow diagram is a basic means of communication and should be used as such in reviewing the scope of the program by all parties involved in the eventual use of the program. Changes in the flow diagram can be easily made, while reprogramming of complete programs may be quite expensive and time-consuming.

Step 4 Coding—This involves the translation of the program steps into the actual instructions the computer can execute. In the past, this step has been tedious and prone to a large number of blunders, but procedure-oriented programming languages (see section 26.2.1) have considerably reduced the time-consuming aspects of this step.

Step 5 Checking—The computer program must be subjected to extensive testing on the computer. Testing is needed to remove coding blunders (which are almost unavoidable in programs of any size), programming or logical errors resulting from improper programming, and numerical errors produced by incorrect numerical techniques. Most software systems provide various aids to make program checking (debugging) more efficient. Test solutions obtained by the program must be compared against manual solutions, to verify the correctness and accuracy of the results. In selecting test cases to be run, it is important that all possible alternates provided for in the program be thoroughly investigated.

Step 6 Program documentation—If a program is to be used by persons other than the programmer himself, it is essential that it be properly documented before it is put into production. This document, or writeup, should consist essentially of four parts: (1) a brief application description giving the method of solution, approximations and limitations involved, etc., so that a potential user can understand the process and decide whether or not it is applicable to his particular problem; (2) a user's manual, describing exactly how the data are to be prepared and options specified, and interpreting the results produced; (3) an operator's manual specifying the procedures to be used in running the program on the computer; and (4) a programmer's manual describing in detail the methods used and containing the flowcharts and the program listings.

26.2 PROGRAMMING TOOLS

26.2.1 Programming Languages

A computer program exists in two forms: the external form as coded by the programmer, and the internal form as used by the hardware. The internal form is rigidly fixed by the design of the hardware, as the hardware can only execute a limited number of instructions built into its repertoire, or order code. To execute any program, the external form of the instructions comprising the solution sequence must be converted into the internal, or machine language, form and entered into the storage device. However, the software system can provide considerable assistance in this conversion process, and direct machine language coding is practically never used today.

-1. Symbolic languages—At the lowest level of coding, the programmer still specifies each instruction separately, but uses convenient symbols (mnemonics or abbreviations) for the operations (e.g., *ADD*, *MPY*, etc.) and arbitrary symbols for the operands. A relatively simple program, called an assembler, converts the operation names to their internal representations, assigns storage locations to the operands and then substitutes addresses for the names, thus producing an executable program. Symbolic coding is still extremely time-consuming and tedious, and should be used only as a last resort where extremely high efficiency is needed or highly hardware-dependent functions are involved.

-2. Procedural languages—At the next level, a major increase in programmer efficiency and productivity is achieved by allowing the programmer to specify units of processing in the form of statements, without concern for the number of instructions necessary to execute the statement. Statements may be algebraic (i.e., $SUM = A + B + C$), logical (*IF(A.EQ.0) GO TO 100*), or input/output (i.e., *READ A, B, C*). The translator which converts the procedural source program into the executable object program is called a compiler. On most systems, the compiled object program may be saved, so that on subsequent production runs the compilation can be bypassed. Procedural languages such as FORTRAN, PL/I, ALGOL and BASIC account for the majority of application programming in present use. Furthermore, for the more popular languages compilers exist for a number of different computers, so that a given source program can be compiled and executed with little or no reprogramming.

-3. Problem-oriented languages—At the topmost level, problem-oriented languages permit the user, who is not a programmer, to describe a problem directly in the terminology of his discipline, without concerning himself with the details of the actual procedures needed to execute the problem. Thus, for example, a statement of the form *MEMBER PRISMATIC 15 AREA 10. INERTIA 200.* is sufficient to describe a member and its properties to an analysis program, including the fact that a procedure applicable to prismatic members is to be used. Similarly, a statement of the form *CHECK BOND* may be used to invoke the application procedure. It is to be noted that a problem-oriented language presupposes the existence of a translator, which converts the user's source statements into the appropriate processing steps, and that the language can be used to specify only those data and procedures for which processing steps have been implemented.

The availability of problem-oriented languages further improves the productivity of the users, and in fact can eliminate much of the separation of the problem-solving described in section 26.1, as the user can go directly from his problem to its specification to the computer by means of the available problem-oriented statements. However, the programming of the procedures incorporated into the software becomes even more crucial, as these procedures will tend to be used more frequently, and under a greater variety of circumstances, than procedures contained in conventional programs.

26.2.2 Flow Diagramming

As mentioned in section 26.1.4, a flow diagram, or flow chart, is a graphic representation of the flow or logic of the problem solution procedure, consisting of the operations and decisions involved and of the order in which these must be executed. The common flow-diagramming symbols are shown in Fig. 26-1.

Flow diagrams are valuable tools both during the development of a program and after it has been turned over for production use. During development, they are the primary communication tool between the individuals working on a common problem. For all but the simplest problems, it is common practice to prepare a one-page block-diagram, showing only the major steps. Each of the boxes on this diagram can then be expanded into a more detailed diagram, and so on down several layers, depending on the complexity of the program.

Flow diagrams are also fundamental communication tools for testing, explaining, reviewing and modifying programs.

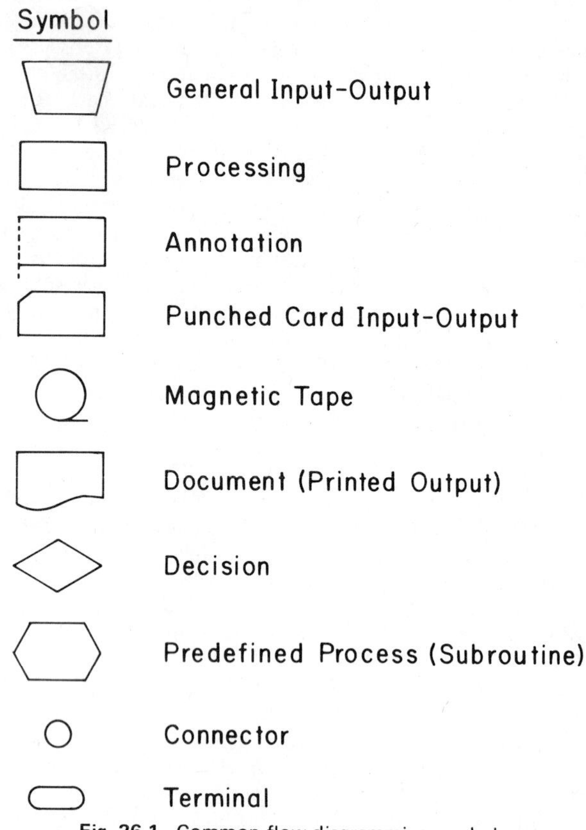

Fig. 26-1 Common flow-diagramming symbols.

Their usefulness is greatly enhanced by liberal use of annotation boxes to provide additional information, and by using problem-oriented rather than procedure-oriented statements in the boxes (e.g., use 'last beam?' rather than '$i = n$?' to describe a test).

Several flow diagrams are shown in the succeeding examples.

26.2.3 Decision Tables

As the complexity of the program logic increases, flow diagrams become less well suited to represent all possible paths and to check that all logically possible cases are properly accounted for. The technique of tabular decision logic, or decision tables for short, can be advantageously applied to problems of this type.

A decision table is a concise tabulation of the logical rules of the type described applying to a given problem. The decision table is laid out consisting of four areas as shown below:

Condition Stub	Condition Entry
Action Stub	Action Entry

The condition stub is a list of the logical variables involved in the problem. The action stub is a list of the actions involved. The condition entry lists the pertinent combinations of the logical variables in columns, each column specifying a rule. The action entry gives the actions to be taken corresponding to the specified rule.

A decision table is called limited-entry if all logical variables have only two values, so that the elements of the condition entry are limited to yes (Y), no (N), and immaterial (I), and the elements of the action entry are Y or blank, specifying whether a given action is to be taken or not. By contrast, in an extended-entry decision table, the logical variables may be multivalued, the elements of the condition entry may bear any kind of logical relationship (e.g., $=$, $\neq$, $<$, $>$, etc.) to the elements of the condition stub, and any number and kind of action may be specified. Extended-entry tables can always be transformed into limited-entry ones by defining additional rules.

A decision table is complete if there are rules for all possible combinations of the logical variables. The presence of an I for immaterial automatically makes the decision table incomplete. Incomplete decision tables often contain one additional rule, called the *ELSE* rule. The action associated with the *ELSE* rule is to transfer to an error procedure to handle a combination of parameters for which no specific rule has been provided.

The salient advantages of decision tables are: (1) the decision table displays at a glance all the logical possibilities for which provisions have been made; thus, incomplete and inconsistent cases can be located before the sequential conversion to a flow diagram is undertaken; (2) the formalized, tabular display permits the application of formal methods for checking the consistency of the logic; (3) the simple format of the decision table can be understood, evaluated, and generated by laymen not familiar with computer programming more readily than involved flow diagrams; experience has shown that decision tables are also useful in checking, documenting, and updating programs; and (4) decision tables are valuable programming aids for procedural languages such as FORTRAN, as well as being convenient programming languages by themselves.

An application of decision tables is shown in section 26.2.3.

26.2.4 Tools for Organizing Programs

All but the smallest programs consist of several discrete and separate processing steps. The separate steps may be: (1) sequentially related (i.e., input, processing, output); (2) in parallel (e.g., processes for prismatic, tapered, curved, etc., beams); or (3) laid out in the fashion of a tree if many logical possibilities are built into a program.

-1. Subroutines—The major tool for organizing the separate steps into the overall program is the use of subroutines. A subroutine is a self-contained set of instructions or statements designed to perform a specific task. The subroutine may be *called* from another program; when this occurs, control is transferred from the calling program to the subroutine. When the subroutine is completed, a *return* is made to the calling program and processing resumed in the calling program at the statement following the subroutine call. Thus, as far as the programmer is concerned, the composite machine behaves as if the function of the subroutine were available in the actual hardware.

The primary reasons for using subroutines are: (1) to combine procedures used repeatedly within a program into a single entity; (2) segmenting large program steps into more manageable parts; (3) incorporating previously developed processes into new programs; and (4) organizing the overall process.

-2. Organization—The organization of large programs into an efficient logical structure is accomplished through modularization and hierarchical arrangement. By modularization is meant that each subroutine is designed to perform a single, clear-cut logical function. The statement of a

single, clear-cut logical function should not be taken literally to mean that separate subroutines should be written for procedures that are the least bit dissimilar. Upon some inspection it often turns out that procedures which initially appear dissimilar are, in fact, very closely related and may be combined into a single subroutine.

The second, and more important, organizational tool is that of hierarchical arrangement. A well-organized program should be laid out much like a pyramid, with the lowest-level subroutines on the bottom and succeedingly higher-level subroutines on top, culminating in a single main or control program. The latter need not (and should not) consist of more than a series of calling statements to the subroutines under it. The hierarchical arrangement must be maintained throughout, so that no program calls on an equal-level or higher-level program as a subroutine. Also, the logical interconnection between subroutines should be kept to a minimum.

In a system of subroutines adhering to the two principles above, changes and modifications can be easily implemented by simply replacing the subroutine or subroutines affected. There is an additional bonus accruing from the modular and hierarchical organization of programs, namely that the program can be made much more versatile than originally intended with essentially no increase in programming effort. This versatility is achieved by rearranging, under control of input data, the sequence of calling the various subroutines. If a program is properly segmented in the manner described, it is only a small step to convert it into a problem-oriented system, by combining it with a control program which reads the user's commands, decodes them, and transfer control to to the subroutine(s) corresponding to that command.

26.2.4 Tools for Organizing Data

The proper organization of data is as important a consideration in the design of a major program as the organization of procedures. Indeed, the most serious limitation of many otherwise useful programs is the inflexibility of their data structure, in terms of the size and variety of problems that may be handled. Also, in many cases possible program modifications and extensions are limited by the fact that the data are poorly organized or the data organization is improperly documented.

The tools for data organization and management available on most operating systems fall into two categories: (1) those dealing with the management of data in the primary of high-speed storage device; and (2) those dealing with secondary storage devices (disks, drums, etc.). On many newer computing systems, the distinction between these two largely disappears, as dynamic memory allocation or paging techniques allow the programmer to deal with extremely large virtual memories.

Primary storage can be managed efficiently by placing data needed by several segments into a common area available to all segments. Further economies may be achieved by sharing an area among several mutually exclusive uses.

When the program and its data exceed the available primary storage space, use must be made of secondary storage. Secondary storage must also be used when data must be saved on a long-term basis, such as in a project file containing all information about a project, which is filled, updated, and modified as the design progresses. Management of secondary storage is based on the recognition that only a small part of the complete file must be brought into primary storage for reference or processing, and can then be written back out into secondary storage. Most operating systems provide subroutines for the convenient management of secondary storage.

26.3 COMPUTATIONAL METHODS

The computational method to be used for a particular application depends on many factors. The selection will, of course, depend on the programmer's engineering and mathematical background, and also on his personal preference, when several alternate methods are available. The frequency of expected use of the program and the available time for programming also enter in the selection. For a one-shot problem where answers are needed quickly, the simplest formulation will be sufficient, while for a program intended to be used daily, with large volumes of data, it is worthwhile to spend considerable time in developing efficient methods. It is not possible to give absolute rules concerning selection from among alternate methods. For example, moment-distribution and slope-deflection, as well as many other techniques, have proven to be useful in frame-analysis applications. However, almost invariably, all these methods require certain modifications from the standard manual techniques in order to use the computer efficiently.

26.3.1 General Methods

Computational methods generally fall into a number of distinct categories discussed below. In general, most programs of any complexity will contain more than one of these categories.

-1. Direct methods—In the simplest case, a closed (explicit) formula can be obtained for the desired answer (output) in terms of the known quantities (input), and a program can be written to evaluate the formula for various values of input. The flow diagram is linear and consists of a number of operations in a direct sequence, as shown in Fig. 26-2.

-2. Iterative methods—A much more effective use of the computer, and a more commonly used method, involves a number of instructions used repetitively to process different sets of data. The method is characterized by a flow diagram comprising a closed loop, consisting of the four basic parts shown in Fig. 26-3. Iterative procedures commonly involve operations on tabular data stored in memory and yield results which are stored back in new tables.

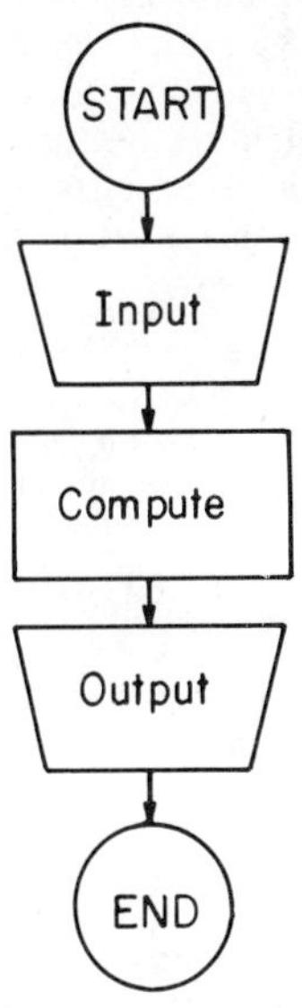

Fig. 26-2 Flow diagram for direct solution.

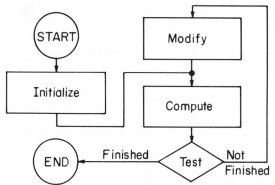

Fig. 26-3 Flow diagram for iterative solution.

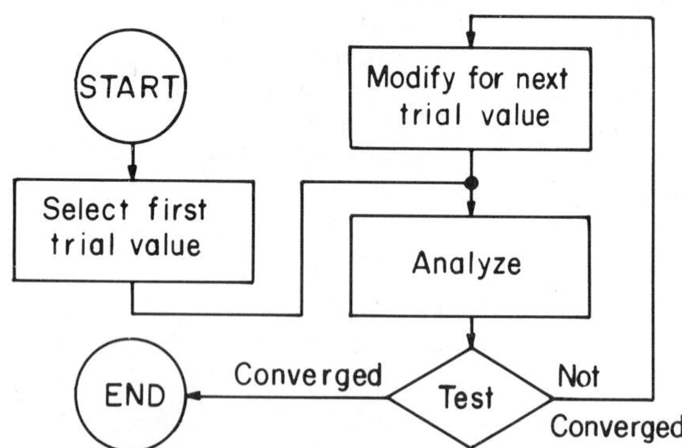

Fig. 26-4 Flow diagram for successive approximations.

-3. *Successive approximations*—A special case of the iterative method occurs when the instructions in the loop operate on successive approximations of a single final answer. Engineers are familiar with many procedures of successive approximations; for example, the moment-distribution procedure is a solution of the slope-deflection equations by successive approximations. In computer applications, successive approximations are used even more extensively because of the need of replacing exact methods involving integrals, trigonometric functions, etc., by numerical methods.

A typical flow diagram for a program involving successive approximations is shown in Fig. 26-4. The steps involved in initializing, modifying and testing are far from trivial, and require that the programmer specify: (1) an initial value that is simple to compute, yet is a reasonable estimate of the final answer; (2) a testing procedure that will positively detect when the final answer is reached; and (3) a modification scheme that assures the best rate of convergence toward the final answer.

EXAMPLES 26-1 to 4: The types of numerical methods mentioned above may be more clearly illustrated by means of an example problem. We wish to construct a moment-curvature (M-ϕ) curve for a reinforced concrete beam. For many practical purposes, the curve may be approximated by straight lines connecting computed M-ϕ coordinates for cracking, yield, and ultimate loads. The determination of these three M-ϕ values requires no knowledge of the concrete stress-strain (f_c-ϵ_c) curve, instead depending solely upon the solution of an algebraic equation relating moments and curvatures to the f_c and ϵ_c assumed to be acting at each level of loading. At ultimate loading, for example:

$$M_u = A_s f_y d\,(1 - 0.42k_u)$$

where A_s is the area of tension steel; d, the effective depth of the steel; f_y, the yield stress of the steel; and $k_u = A_s f_y / f_{ave}$, where f_{ave} is the average compressive stress of the concrete.

For the purpose of this example, however, we will also construct the real M-ϕ curve, generating several actual (as opposed to approximate) coordinate values of M-ϕ. Each point will be the result of considering equilibrium between tension and compression forces and the geometry of strain distributions at even intervals of loading. This latter, real analysis requires the actual f_c, ϵ_c values rather than the assumed ones of the former approximate analysis.

The development of the procedure for each M-ϕ construction, along with more detailed explanation of each analysis, follows.

EXAMPLE 26-1: Approximation analysis by direction solution—Because of the straightforward computations involving only one equation in one unknown for each moment or curvature, the generation of an approximate M-ϕ curve is a typical application of the direct solution method.

For conciseness, only the M-ϕ values at ultimate will be derived. All symbols used in this and succeeding examples are defined in the notation list and are illustrated in Fig. 26-5.

SOLUTION: Assumptions:

1. $\epsilon_{cu} = 0.003$ rad/in.
2. linear strain distribution
3. $\epsilon_s > \epsilon_y$ at ultimate
4. flat-top steel
5. at ultimate, concrete strain is not proportional to stress
6. beyond yield, the tensile strength of the concrete is negligible ($T_c \simeq 0$)

Equating compression and tension, (see Fig. 26-5a)

$$C = T_s$$

or

$$0.7\,f'_c k_u d b = A_s f_y \tag{26-1}$$

thus

$$k_u d = \frac{A_s f_y}{0.7 f'_c b} \tag{26-2}$$

and

$$k_u = \frac{A_s f_y}{0.7 f'_c b d} \tag{26-3}$$

Summing moments about C

$$M_u = A_s f_y d\,(1 - 0.42k_u) \tag{26-4}$$

finally

$$\phi_u = \frac{\epsilon_{cu}}{k_u d} = \frac{0.003}{k_u d} \tag{26-5}$$

Equations (26-4) and (26-5) represent expressions for the two output variables (results) sought, given the input variables (data) A_s, f_y, f_c, b, and d; and the intermediate result k_u defined by eq. (26-3). The flow diagram for the computational process is shown in Fig. 26-3. Because input and output of data has no bearing upon the subject of this example problem, their boxes in the flow diagram contain only very general expressions. Note the similarity of the flow diagram with the generalized one for a direct solution shown in Fig. 26-2.

Similar, but distinct, expressions for M_y, ϕ_y and M_{cr}, ϕ_{cr} may be incorporated into the same flow diagram, changing only the computation box.

EXAMPLE 26-2: Real analysis by iteration—Generate real values at M and ϕ between yield and ultimate failure ($\epsilon_y \leqslant \epsilon_s \leqslant \epsilon_{su}$).

SOLUTION: This is quite similar to the preceding approximate analysis in that it, too, requires the direct solution of algebraic expressions in a single unknown (ϵ_c in this case). This is the result of examining equilibrium and geometry of the strain distribution, and remembering that $f_s = f_y$ is assumed. However, in contrast to the unique expressions for M and ϕ at cracking, yield, and ultimate in the approximate analysis, the real analysis will result in common expressions (as functions of ϵ_c) for M and ϕ and all intermediate results, valid for generating any point on the curve between yield and

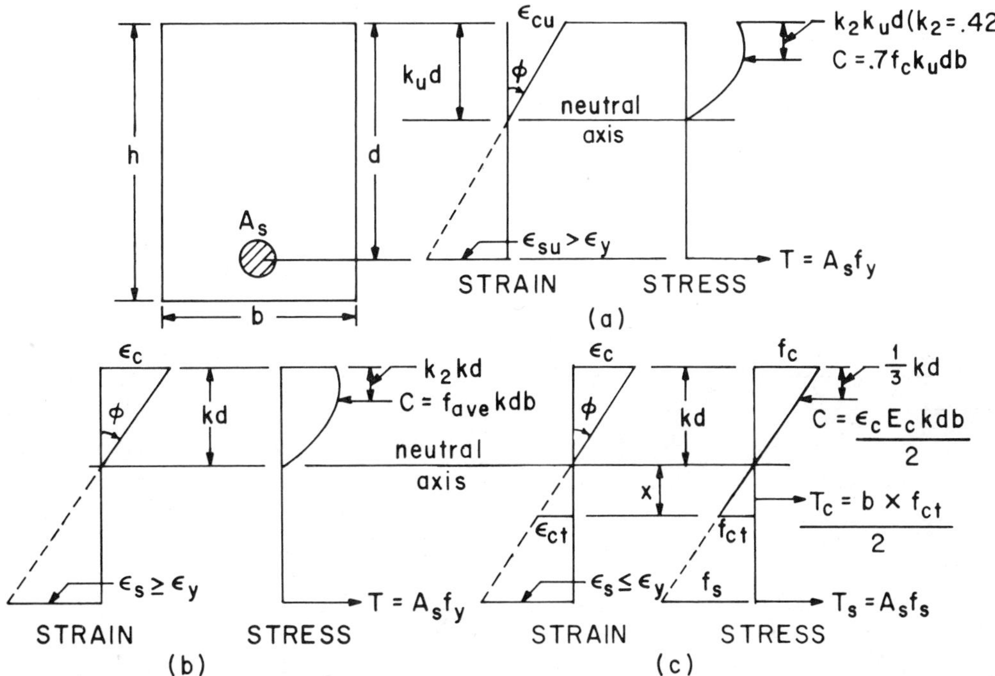

Fig. 26-5 Definition of variables used in examples.

ultimate. Thus, we have only to repeat or iterate a series of computations, each time modifying the value of ϵ_c. The procedure described is the method of iteration.

From the geometry of the strain distribution (see Fig. 26-5b)

$$\frac{\epsilon_c}{kd} = \frac{\epsilon_s}{d - kd} \tag{26-6}$$

therefore

$$k = \frac{\epsilon_c}{\epsilon_c + \epsilon_s} \tag{26-7}$$

From equilibrium

$$C = T_s \tag{26-8}$$

or

$$f_{\text{ave}} kbd = A_s f_y \tag{26-9}$$

Solving eq. (26-9) for k

$$k = \frac{pf_y}{f_{\text{ave}}} \tag{26-10}$$

where

$$p = \frac{A_s}{bd} \tag{26-11}$$

Equating the results of eqq. (26-7) and (26-10) gives

$$\frac{pf_y}{f_{\text{ave}}} = \frac{\epsilon_c}{\epsilon_c + \epsilon_s} \tag{26-12}$$

so that

$$\epsilon_s = \frac{f_{\text{ave}} \epsilon_c}{pf_y} - \epsilon_c \tag{26-13}$$

Then, for a given value of ϵ_c and knowing f_{ave}, p, and $f_y = E_s \epsilon_y$ we can solve for ϵ_s. Once both ϵ_s and ϵ_c are known, we can substitute into (26-7) to obtain kd. Finally, summing moments about C

$$M = A_s f_y (d - k_2 kd) \tag{26-14}$$

$$\phi = \frac{\epsilon_c}{kd} \tag{26-15}$$

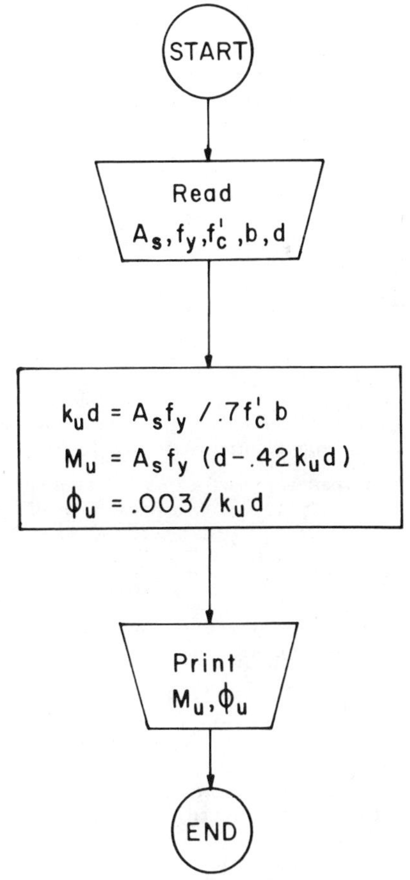

Fig. 26-6 Flow diagram for approximate method.

Finding f_{ave} and $k_2 kd$ requires knowing the area under the $f_c - \epsilon_c$ curve. One solution might be to read in as data the coordinates of several points along the curve and then perform numerical integration to find the total area. In our procedure we will not waste vital

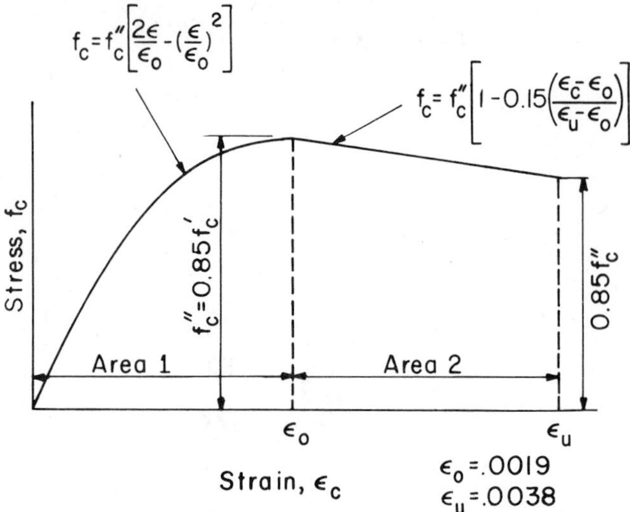

Fig. 26-7 Assumed stress-strain relationship.

storage area on coordinates and, instead will approximate the $f_c - \epsilon_c$ curve by Hognestad's equation (see Fig. 26-7).* In addition to saving storage space, these approximations make the program more flexible, as it is applicable for any strength (f_c') concrete. The curve is shown in Fig. 26-6.

For any value of ϵ_c, we can integrate over the curve and dividing the resulting area (total force) by ϵ_c, find the average compressive stress, f_{ave}, on the cross section. Furthermore, taking moments of the areas and dividing this result by the total area, we can find the depth, $k_2 kd$ where the resultant force acts. A flow diagram for the iterative process is shown in Fig. 26-8. The omission of input and output suggests that this diagram represents a segment of a larger main program. We assume that as we enter the segment all material properties and dimensions are available in storage. Furthermore, so as to avoid time-consuming transfer back to the main program after each iteration, we will store the results in arrays M and ϕ. Hence the use of subscript, i. Note that as an alternate to the direct evaluation used, we could integrate under the $f_c - \epsilon_c$ curve numerically in a sub-loop within the loop on ϵ_c.

EXAMPLE 26-3: Real analysis by successive approximation–Prior to yielding of the steel ($\epsilon_s < \epsilon_y$), we do not know the depth, kd, to the neutral axis, nor ϵ_c or f_c. We are given only ϵ_s so that ϵ_c and f_c cannot be found by direct algebraic substitution as before. Because this is a real analysis we cannot assume values for them. We do however, know that $C = T$ from equilibrium, and the relationship between ϵ_s, ϵ_c, and kd from the strain distribution. (See Fig. 26-5c.)

SOLUTION: The method of solution we will use is one of trial and error, where we successively approximate the value of kd, and find the resulting tensile and compressive forces. The test for convergence is whether $C = T$ or differs by some small tolerance, δ. δ need be no smaller than 10^{-2} kips, or 10 lbs, depending upon which units are being used.

We organize our attempt at approximating kd by noting that we know both limiting values of kd, namely

$$0 \leqslant kd \leqslant d$$

Thus we can use the interval halving technique to minimize the number of trials necessary.

By this procedure, kd is set equal to $(LL + UL)/2$, where LL is the lower limit and UL, the upper. Initially $LL = 0$ and $UL = d$. After the first trial if $C > T$, we know that we have overestimated the area in compression. This implies that for the next trial $i + 1$ $LL < kd_{i+1} \leqslant kd_i$. Therefore we set $UL = kd$ and go through the procedure again. If, on the other hand, $T > C$, we have underestimated kd, so that $kd_i \leqslant kd_{i+1} < UL$, and we set $LL = kd$.

*Hognestad, E., "A study of Combined Bending and Axial Load in Reinforced Concrete Members," *Bulletin 399*, University of Illinois, Engineering Experimental Station, November, 1951.

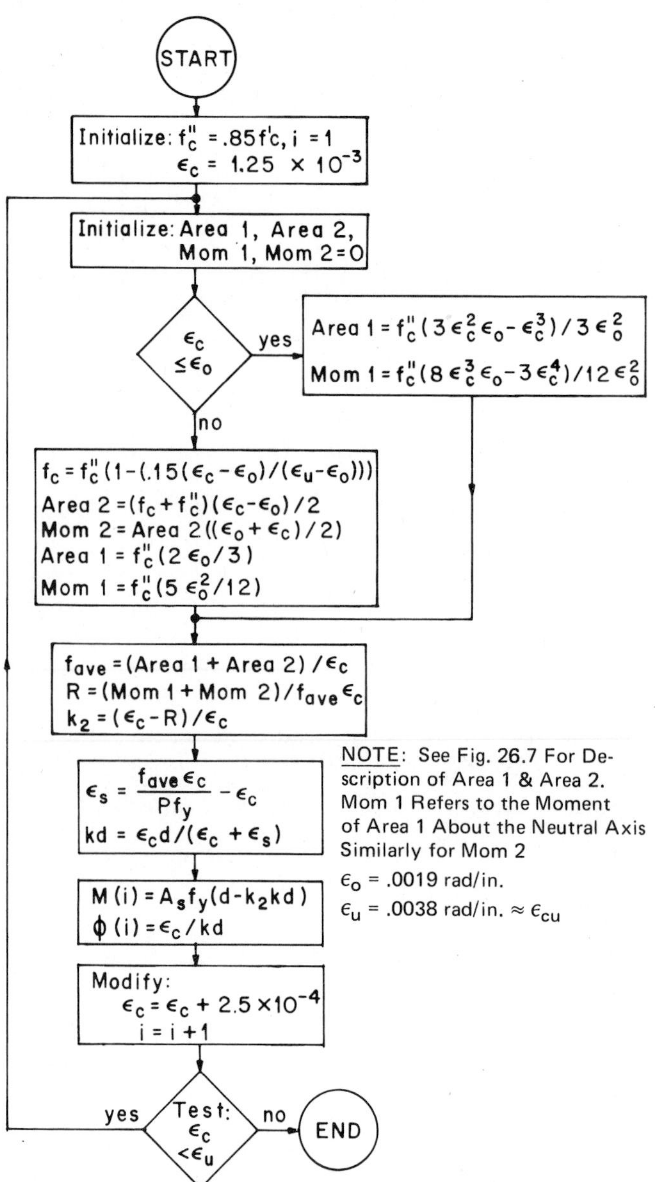

Fig. 26-8 Flow diagram for iterative solution.

Note: See fig. 26-7 for description of Area 1 and Area 2. Mom 1 refers to the moment of Area 1 about the neutral axis; similarly for Mom 2. ϵ_0 = .0019 rad/in., ϵ_u = .0038 rad/in ≈ ϵ_{cu}.

The flow diagram for the process is shown in Fig. 26-9. Once again the diagram represents a segment or subprogram. The main program generates a value of ϵ_s, followed by a call to this segment where the value of kd will be successively approximated until $C = T$ to the required tolerance, δ.

EXAMPLE 26-4: Decision tables–As a simple example of decision tables, assume that subroutines based on the flow diagrams of Figs. 26-6, 26-8, and 26-9 are all available. A control program, which is to call the appropriate subroutines, is to be developed. The logic of such a control program may be represented by the following decision table:

Approximate method desired:	*Y*	*N*	*N*
$f_s < f_y$	*I*	*Y*	*N*
Use Fig. 26-6	*Y*		
Use Fig. 26-8			*Y*
Use Fig. 26-9		*Y*	

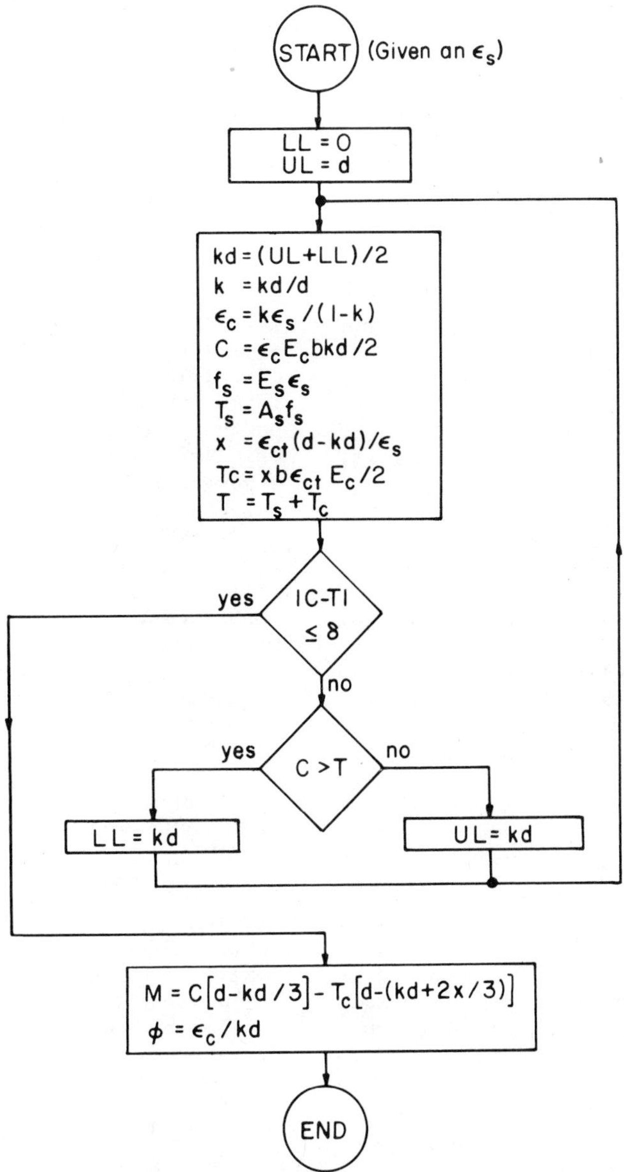

Fig. 26-9 Flow diagram for trial-and-error method.

The last column (rule) is to be read: "*If* approximate method *not* desired *and* f_s *not* less than f_y, *then* use the procedure shown on Fig. 26-8."

A more meaningful example of decision tables, which also illustrates the type of ambiguities or omissions that are discovered when decision logic is applied, is shown in the following example. Paragraph 9.3.2 of the 1970 *ACI Code* prescribes the strength factor, which we shall call U_W, to be used when wind loading, W, occurs in combination with dead load, D, and live load, L. The pertinent text is: "9.3.2. In the design of a structure or member, if resistance to the structural effects of a specified wind load W must be included in the design, the following combinations of D, L, and W shall be investigated in determining the greatest required strength U:

$$U = 0.75\ (1.4D + 1.7L + 1.7W) \qquad (9.2)$$

Where the cases of L having its full value or being completely absent shall both be checked to determine the most severe condition, and

$$U = 0.9D + 1.3W \qquad (9.3)$$

but in any case the strength of the member or structure shall not be less than required by Eq. (9.1)."

The resulting decision table is as follows:

	(1)	(2)	(3)	(4)
W must be included?	Y	Y	Y	N
D, W have same sign?	Y	Y	N	I
D, L have same sign?	Y	N	I	I
$U_W = 0.75(1.4D + 1.7L + 1.7W)$	Y			
$U_W = 0.75(1.4D + 1.7W)$		Y		
$U_W = 0.9 + 1.3W$			Y	
$U_W = 0$				Y

Rules (1) and (2) are a straightforward representation of the qualification attached to formula (9.2) in the text, and rule (4) is included for completeness. However, the immaterial in rule (3) may not completely represent the intent of the code: in most cases where formula (9.3) governs (e.g., column tension due to uplift), D and L can be expected to have the same sign, and thus it is conservative to omit L; however, there may be unusual cases where L and W both act in a direction opposite to D, but the *Code* makes no provisions for such a case.

26.4 APPLICATIONS IN CONCRETE ENGINEERING

The range of computer applications in concrete engineering is enormous, covering procedures from preliminary conceptual design to construction control on the site. It is safe to say that every conceivable procedure used in concrete engineering has been programmed by some person or organization to some degree of completeness and generality. Even though this large range exists, most computer applications intended for producting use can be classified into the major categories discussed below. This classification purposely excludes the large number of "one-shot" and research programs intended to solve a specific problem and not designed for routine production use.

26.4.1 Analysis

In section 25.1.1 of this handbook, structural analysis is defined as "the process of determining the forces and and deformations in structures due to specified loads." The present discussion will be more specific, and stipulate that analysis is performed on a fully prescribed mathematical model of the structure, i.e., one for which all relevant member properties (dimensions, stiffnesses, etc.) are assumed to be known.

Historically, analysis has been the first class of applications to be programmed, and even today it comprises the bulk of production use of computers. The reasons for this preponderance are: (1) analysis is a clearly identifiable, time-consuming task in the design process; (2) the increased size and complexity of structures requires modeling and analytical techniques impossible to perform by manual methods; and (3) analysis of a model is based on rational principles of mathematics and behavior, and thus not subject to individual interpretation.

-1. Methods—For the reasons discussed in section 25.1.1, analysis applications in concrete engineering are based, almost without exception, on the elastic theory of structures. Furthermore, the majority of applications involve structures idealized as frames, with slabs replaced by equivalent beams and shear walls represented by equivalent columns. However, with the development of programs based on the Finite Element Method (see section 25.7) such gross idealization is no longer necessary, and two and three-dimensional elements can be accurately modeled and readily incorporated into the overall model to be analyzed.

Analysis programs, again almost without exception, use the deformation or stiffness method (see section 25.5), rather than the force or flexibility method (see section 25.4) commonly used for hand calculations. The reasons for this are: (1) in the deformation method the assembly of the governing simultaneous equations [eg. (25-29) in section 25.5] requires a minimum of input data, follows a simple, logical sequence independent of the choice of redundants, and the resulting equations can be solved quite accurately; whereas (2) in the flexibility method, the governing equations [eq. (25-20) in section 25.4] depend on the analyst's choice of redundants, their assembly is less prone to automation, require considerably more input or precomputation, and are prone to large roundoff errors if an improper primary structure is selected.

In this connection, it is worthwhile to point out that whereas matrix algebra is a convenient and concise tool to describe the analysis problem, it is not to be used directly to program the procedure. To cite just two examples: (1) the triple-product matrix $[\delta] = [b]^T[f]\ [b]$ defined by eq. (25-12) of section 25.3 should not be evaluated by matrix transposition and multiplication subroutines; since $[f]$ is diagonal, and many submatrices of $[b]$ are zero, a direct generation of the nonzero coefficients of $[\delta]$ is much more efficient; and (2) it is much more efficient to solve a set of simultaneous equations, even with multiple right-hand sides, than to invert the coefficient matrix and post-multiply by the right-hand side vector, as implied by eqq. (25-21) and (25-30) of Chapter 25.

-2. Assumptions—Many of the early frame analysis programs were based on the slope deflections and moment-distribution methods, with their built-in assumption of no axial distortions. More recent programs tend to incorporate a full deformation method formulation, including axial and shearing distortions, for two reasons: (1) in tall, slender buildings, column elongations must be taken into account for realistic modeling; and (2) the efficiency and simplicity of programming the general deformation method far outweighs the penalty of carrying along the additional degrees of freedom associated with the axial deformations. Most general-purpose analysis programs permit the input of constraints where required by the model (e.g., constraining beam elongations to be zero while still allowing column elongations).

Between plane frame programs based on the slope-deflection method and general-purpose displacement method programs applicable to two- and three-dimensional frames (and, with finite element capabilities, to structures comprising plate, shear and solid elements as well), there is a host of analysis programs for special structural types, such as shear walls, shells of various types, footings, foundation mats, etc.

26.4.2 Proportioning

Proportioning is the process which, given the analysis results (forces and deformations) acting on a section or structural element, selects the size of the element and the required reinforcement, including prestressing, so as to insure that the member behaves adequately both under service and ultimate conditions. Frequently, the combination of independent loading patterns to produce the critical design condition on the element being designed is considered part of proportioning, rather than analysis. On the other hand, the selection of actual bar sizes and their placement is frequently considered to be part of detailing, and will be discussed in the following section.

-1. The iterative process—If the element to be proportioned is statically determinate (e.g., a simple-span bridge girder), analysis and proportioning can be combined into a direct design procedure. If, however, the element is a component of an indeterminate structure, analysis and proportioning form an iterative, or trial-and-error, loop, where initial sizes or stiffnesses are assumed, an analysis performed, the elements reproportioned, and the entire cycle repeated until satisfactory results are obtained. The degree to which this cycle is automated, i.e., performed by a program without a designer's intervention, depends both on hardware or cost limitations and the design office's preferences and mode of operation. Only a limited number of fully-automatic design-proportioning software systems have been reported so far, due to: (1) the difficulty of specifying and programming the criteria for satisfactory performance; (2) the very large number of variables which may be adjusted in order to achieve such performance; and (3) the difficulty of automatically generating an initial solution. These three problems are precisely those discussed in connection with successive approximation methods in section 26.2.3.

-2. *Optimization*—Since the number of possible designs satisfying the applicable criteria is theoretically infinite, the iterative design-proportion process may be controlled so as to minimize some desired feature, such as material volume or cost. This is the motivation for the techniques of structural optimization and for programs based on such techniques. Structural optimization is still in its infancy as far as realistic reinforced-concrete structures are concerned, but holds a great deal of promise. The present difficulties are those mentioned above for any successive approximation process, coupled with the difficulty of defining realistic objective functions.

-3. *Design assumptions*—Proportioning programs differ in several major respects from the analysis programs discussed previously. First, the determination of strength and serviceability of elements, and thus their proportioning, is governed to a great extent by the *Code* provisions. Unfortunately, these provisions are not presented to the profession in a format suitable for direct conversion to computer programs. The major problem is not that the formulas and methods in the *Code* are not optimal for computer formulation, but that the logic of the *Code*, such as interactions between various provisions scattered through the text and the specification of the limitations or range of applicability of provisions, is not directly discernible. As a consequence, 'programming according to the *Code*' requires a great deal of individual interpretation, and therefore limits the applicability of the program to those users who agree with the interpretations embodied in it. The recent interest in representing design specifications in the form of decision tables (See section 26.2.3) promises to provide a basis for a more rigorous representation of the logic of codes, and thus for easier identification of assumptions embodied in specific programs.

It is universally accepted that the *Code* by itself is not a complete design guide, and that additional design logic, including assumptions, limitations, shortcuts, search strategies, etc., must be incorporated into any proportioning program. To cite just one example, one seldom, if ever, designs every beam of a frame separately for the most critical combination of load effects acting on it; rather, one chooses a typical or critical beam to be proportioned, and then replicates its design for all similar beams. The definition of what is typical, critical, or similar will generally vary widely among designers or design organizations. It is for reasons such as these that proportioning programs lack the generality of analysis programs, and tend to incorporate, often in a slavish way, the assumptions and practices of the originating organization. As a consequence, programs acquired from other organizations (see section 26.5.2) must be carefully reviewed to ascertain whether their assumptions and limitations agree with those in use by the acquiring organizations.

-4. *Data-processing requirements*—From a data-processing standpoint, proportioning programs differ from those for analysis in that they generally involve fewer and simpler calculations, but vastly larger volumes of data. This is particularly true if the programs are integrated into a software system, and access all or portions of their data in a project file, containing the description of all the important aspects of an entire project.

In closing this section, mention should be made of programs developed exclusively for checking designs, that is, programs which accept the complete description of the structure and its elements, including size of members and amount and position of reinforcement, and which then verify whether the structure satisfies the applicable code provisions. This approach is used by some building officials in Europe, and will undoubtedly be more widely used elsewhere as more computer-generated designs are submitted for approval.

26.4.3 Detailing

In concrete engineering applications, detailing refers primarily to the determination of the number, size, layout, and location of reinforcement, given the element dimension and the areas of steel required. Frequently, the above calculations are combined with printing of schedules and bar bending diagrams (see next section), fabrication, inventory, shipping and placement control, and the printing of bundling information and shipping tags. Some programs of this type, developed by reinforcing-steel fabricators, produce directly control tapes for driving numerically controlled cutting and bending machines.

The comments made on proportioning programs in the previous section apply equally to detailing programs. Thus, while certain details, such as lap and development lengths, hook requirements, etc., are covered by the *Code*, the logic of many other operations has to be developed individually on the basis of the practices of the particular design firm, fabricator, or client.

The data-processing aspects of detailing programs tend to be even more involved and complex than those of proportioning programs. This is especially true when the actual detailing is combined with inventory and shipping control, so that the data file contains information on an entire structure or project, and reinforcing for specific components is automatically ordered when required by the construction schedule.

As discussed above, detailing programs, especially those used by reinforcing-bar suppliers, incorporate many aspects of production control besides the actual calculation of bar sizes, dimensions, and placing. The input to such programs is precisely the output generated by proportioning programs, namely concrete dimensions and required bar areas. Unfortunately, the contractual arrangement between the design firms responsible for proportioning and the contractors or subcontracts responsible for detailing is often such that the design firm must first transcribe the computer-generated data onto design drawings, and the contractor must then "take-off" the same data to provide the input to his detailing program. It is to be expected that in the near future, the same information will also be transmitted in computer-readable form so that a major source of cost, delay and errors due to this double transcription can be eliminated.

26.4.4 Preparation of Final Documents

As discussed in section 26.4.1, analysis problems became an early target of computer applications because of their clearly identifiable, time-consuming nature. At the other extreme of the spectrum, the most time-consuming and expensive operation in a design office is the preparation of specifications, schedules, and drawings comprising the final documents leaving the office. Thus, these operations also were recognized as fruitful computer applications, and, today, many aspects of final document preparation are routinely performed by computer programs.

In contrast to analysis, the preparation of final documents involves practically no calculations, but is essentially a task in information processing in the broad sense. Not surprisingly, therefore, the techniques applicable to these tasks are based primarily on data-processing and logical methods, rather than classical numerical analysis. At the risk of repetition, it is to be emphasized again the applications in this area contain, by necessity, assumptions, and procedures which may not be applicable to every organization. However, in many instances such differences can be handled by differences in the data only, leaving the procedures essentially unchanged.

The majority of applications in the area considered may be classified into one of the three categories discussed below.

-1. Specifications—The preparation of written job specifications has always been a time-consuming task. Most offices maintain a set of standard or master specifications, and then produce specifications for specific jobs by incorporating additions, modifications, and deletions. There are several computer programs in use which perform exactly the same function: the master specification is stored in computer-readable form, the specification writer produces a set of exceptions keyed to lines or paragraphs of the master, and the program merges the exceptions with the master and produces the text directly on multilith or other medium for direct production. Indexing, cross-referencing, etc., of the text may also be performed.

While this application has been extremely successful, its conceptual basis is highly questionable. From an information theory standpoint, the master, or boilerplate, is noise, i.e., it carries no information specific to the project in question, the only information being contained in the exceptions. Yet the program disguises this small amount of information within the large body of text copied from the master. A sounder approach would be to supply to the contractor separately the master and the exceptions, both in computer-readable form, and let him reconstruct the portions of interest or consequence to him.

-2. Drawings—With the steadily decreasing cost and increasing accessibility of plotter devices attached to the computer, the direct production of drawings is becoming more popular. There are several problems, however, which have so far prevented the wide-scale use of computer graphics. Technical problems, such as the quality and reproducibility of the plots being produced, are being solved with improved hardware and materials. The programming problems, on the other hand, are quite severe, and successful applications tend to cover only: 1) schematic plots, such as erection diagrams, where little detail is to be presented; or 2) highly repetitive structural types, such as footings, beams, bridge-piers, etc., where the high cost of programming all possible cases can be economically justified. Furthermore, there is a great deal of disparity between the available plotting devices and their related systems software (plotting and control routines), so that programs designed for one type of plotter are difficult to transfer to another type of device. It is to be expected that further development and standardization will significantly reduce the cost of computer plotting. However, computer-generated drafting will become truly justifiable economically only when it is realized that the data generated for the purpose of graphic display have other uses as well, namely, as discussed in section 26.4.3, as direct input to programs dealing with the manufacturing and erection aspects of the structure.

-3. Schedules—Between the two extremes of pure text and to-scale drawings is the large class of tables, schedules, and schematic diagrams often used in contract documents. In concrete engineering, much of the design information, such as beam-and-column schedules, bar-bending schedules, etc., has traditionally been presented in this form.

The presentation of information in such formats is a natural for computer-generated data, and much of the output of proportioning and detailing programs is presented in tabular formats ready for direct inclusion into the design documents. It is common practice in many firms to key such tables to not-to-scale generalized sketches and diagrams.

While many of the computer-produced tables and schedules are painstakingly designed to duplicate to the utmost detail the layout and contents of tables previously produced by hand, there is a definite trend, wherever contractual arrangements permit, to rearrange the contents and format to best fit the total production process. Just to cite one example, some design firms display on their beam, slab, and column schedules, in addition to the traditional data, sufficient placement information so that the bars can be placed in the field without any detailed placement drawings. It is to be expected that this trend will continue in the future, and that much information traditionally displayed on drawings will only be tabulated on appropriate forms for direct inclusion in contract documents.

26.5 ASPECTS OF COMPUTER USE

Computer hardware, access mechanisms, programming tools, applications, and related services have grown so rapidly that the average designer is justifiably bewildered by the onslaught of claims, advertisements, and offers. The following list, while by no means inclusive, is an attempt to classify programs and services so as to provide a basis for further search and evaluation.

26.5.1 Types of Programs

As described in section 26.1.3, the development of a new program is a tedious and time-consuming process. A prudent engineer would therefore do well to investigate whether an available program may be directly used or modified to suit. In performing such a search, it must be kept in mind that programs must be evaluated on several dimensions or categories, one obviously being the technical suitability of the program. The principal categories are briefly described in the following paragraphs.

-1. Scope—In terms of the scope of available capabilities, programs can be categorized as special-purpose, general-purpose, or software systems, although the distinctions at the interfaces are by no means clearcut. A special-purpose program addresses itself to a clearly delineated, specific situation, both as far as the type of problem handled and the method used. A typical program in this category would be the analysis and proportioning of a two-legged brigde pier. Programs in this class are quite economical to use if the problem at hand fits exactly the scope. By contrast, a general-purpose program deals with broader classes of both problems and methods, as, for example, an analysis program for arbitrary plane frames, or a proportioning program for beams and beam-columns allowing the optional use of either the working stress or the ultimate method. While such programs require more input preparation than special-purpose ones, since the problem type, options to be used, etc., must be specified, they tend to be more economical in the long run because of their broader scope and increased flexibility. Finally, application software systems provide flexible mechanisms to interconnect separate processing

steps (each of which may be a substantial program on its own) into automatic sequences.

-2. Hardware-dependent aspects—The principal hardware-dependent aspects to consider are the programming language used, the program size, and its data requirements. As far as language is concerned, programs written in a symbolic language can be run only on the specific machine or class of machines for which they were written, whereas programs written in a procedural language can be used on other machines as well, possibly with some reprogramming. The size determines the minimal hardware configuration on which the program can run. Although it is possible to segment large programs and data into smaller units, this process is extremely tedious if the programs have not been designed with such segmentation in mind. The data requirements, that is, the volume, format, and organization of the data on which the program operates, are among the prime factors in determining the difficulty of transferring a program from one machine type to another, inasmuch as the data-management operations tend to be the least standardized between different hardware systems. The data requirements are also generally the limiting factors on the size of problem a given program can handle.

-3. Access mode—A major factor in evaluating programs is the mode of access for which they have been designed. This mode may be either: (1) batch processing, in which the entire problem is submitted at once, typically on punched cards, and run as a single job, typically producing its output on a page printer; and (2) time-sharing, in which the user is connected directly to the computer typically through a typewriter-like terminal and inputs data or examines results at his own pace. An intermediate mode is remote-batch processing, in which the data to a batch-processing program are transmitted either through a remote input/output device or by means of files built up and examined through a time-sharing terminal.

While batch processing is unquestionably more economical in computer charges in a production environment, it has some major disadvantages, due to the fact that the *turnaround time* between submission of data and receipt of results (or error messages) is of the order of hours or days. As a consequence, input formats must be very rigid, to minimize the chance of errors and consequent lost runs, and output tends to be voluminous, to assure that all conceivably needed results are made available. In contrast, input to time-sharing programs may be much less rigid, as the user is immediately informed of errors or omissions and may make the appropriate corrections, and output is much less structured, as the user is able to request additional results while still connected to the program. Thus, the increase in total productivity may be considerably higher in a time-shared environment.

-4. Type of interaction—One of the crucial aspects of a program, independent of scope and access mode, is the type or level of interaction between the user and the program. The two extreme levels are: (1) programs with absolutely rigid formats, where every column of every input card must be filled according to specific rules; and (2) programs with completely free input, where commands, data labels, and data values may be entered in free format, and (within reason) in any order.

It is generally accepted that in a high-volume, repetitive production environment, rigid formats lend themselves to greater production rates and easier checking, and that well-documented input forms are useful in assisting the occasional user in preparing his data. On the other hand, rigid formats tend to stifle creativity and experimentation, and have the serious disadvantage that the data are not self-descriptive, in the sense that without the input forms and instructions, it is not possible to read the input data for a previous problem.

A compromise solution adopted in many programs is to allow control statements, options, indicators, etc., to be entered in a free-form, problem-oriented language, but require that voluminous sets of detailed data be entered in fixed formats.

26.5.2 Sources of Programs

Computer programs, with various degrees of completeness and reliability, are available from a variety of sources. The most important sources are described below.

-1. Public domain—Many newer textbooks and technical articles contain source listings of programs for various applications in concrete engineering. It is to be kept in mind, however, that many such programs are essentially appendices to technical reports, and are therefore to be classified as special-purpose programs, likely to require major modifications before being suitable for production use. Columns specifically devoted to general-purpose algorithms were recently initiated in the *ACI Journal* and in *Computers and Structures* (Pergamon Press).

-2. Program libraries—Most computer manufacturers, some government agencies, and many users' groups maintain program libraries, from which copies of programs and their documentation may be obtained at nominal charges. Most libraries (except for some users' groups discussed below) maintain essentially no control over the scope, reliability, and quantity and quality of documentation of programs accepted into the libraries. Thus, testing and verification of such programs may be quite time-consuming and expensive.

-3. Cooperative users' groups—Some users' groups, notably CEPA (Civil Engineering Program Applications) and APEC (Automated Procedures for Engineering Consultants) maintain libraries of programs developed or commissioned by the group's members, with stringent controls on the quality of programs and documentation. Many users belonging to these organizations have found programs in these libraries directly applicable to their needs, or at least as suitable bases for local modifications. Other groups in this category are HEEP (Highway Engineering Exchange Program), consisting of most state highway departments, and the ICES User's Group.

-4. Trade and service organizations—Many trade and service organizations, and even individual manufacturers, provide programs under a variety of access or use mechanisms as essentially an extension of their traditional policy of providing handbooks, design charts, etc., to facilitate the use of the products or services represented by the organizations. In concrete engineering, the most important organization in this category is PCA, which is developing a series of programs for various aspects of analysis and design of concrete structures.

-5. Proprietary sources—There is a growing number of firms, sometimes related to traditional engineering service organizations or to computer service firms, which undertake,

on a contract basis, the development of new programs to client's specifications. Most contracts of this type specify a maintenance period, typically one year, during which the supplier is responsible for providing corrections for errors, or bugs, discovered in program use. It is to be expected that this type of professional software service will increase considerably in the future, often in conjunction with other types of services discussed in the next section.

26.5.3 Other Services

Access to programs is only one type of service needed by an organization contemplating to use or currently using computers for its production work. Some of the other services available are briefly described below.

-1. Assistance for in-house installations—The relative merits and disadvantages of an in-house computer installation vs. use of outside machines cannot be discussed here, due to the host of considerations involved. For organizations owning or contemplating an in-house computer, there are consulting firms providing installation and workload analyses to determine the optimal use of existing equipment or to recommend equipment to be installed. Recently, many manufacturers and software organizations have begun to offer software products, that is, standard proprietary programs installed on the user's machine on either a lump-sum or royalty basis. An even more recent service is that of site management, where an outside organization undertakes the complete management of an in-house computer at guaranteed cost and performance.

-2. Machine access—Access to computers, on a charge basis for time actually used, is available from a multiplicity of sources, ranging from hardware manufacturers' service bureaus through independent service organizations to other users with excess machine capacity. Time-shared and remote-batch access to computers is available on an even broader basis, and, with further reductions in terminal costs and communication charges, will undoubtedly develop into the primary mode of computer use for most engineering organizations, except perhaps a few of the largest ones.

-3. Program access—Computer access on a charge basis usually covers use of compilers and other systems software, as well as access to 'free' programs available through the supplier. In addition, many service bureaus and software firms provide access to their own proprietary programs, usually on the basis of a software surcharge multiplier applied to the basic machine rate. Such arrangements are usually accompanied by some form of performance guarantee, frequently limited to reruns at no charge if software malfunctioning can be established. Many engineering firms, even those with small in-house computing facilities, find this type of program access highly advantageous for infrequent use of large-scale or highly specialized programs.

-4. Problem solution services—Finally, organizations with no in-house capability for computer use may avail themselves of complete problem solution services offered by many organizations. Under a typical arrangement, the user supplies a diagram of the structure with all the member properties and loads identified, and receives back a complete listing of analysis results. Similar services are also offered in the other application areas described in section 26.3. While many engineering firms have found this approach suitable for initial familiarization or for highly unusual and infrequent problems, they eventually tend to adopt other approaches because of the lack of professional control implicit in such arrangements.

ACKNOWLEDGEMENT

The illustrative examples (exx 26-1 to 4) were prepared by Mr. William L. Nelson, Research Assistant in Civil Engineering, University of Illinois, Urbana, Illinois.

NOTATION

A_s = area of tension reinforcement
b = cross-sectional width of member
d = distance from top of beam to centroid of steel
h = cross-sectional depth of member
kd = depth from top of section to neutral axis
ϵ_c = strain in the concrete at the top face
ϵ_s = steel strain
x = depth from neutral axis to depth at which $f_c = f_{ct}$
f_s = steel stress
f_c = concrete stress
f_{ct} = failure stress of concrete in tension
f_y = yield strength of steel
$k_2 kd$ = depth at which C acts, measured from top face of beam
C = concrete compression force
T_s = steel tensile force
T_c = concrete tensile force
E_c = modulus of elasticity of concrete
E_s = modulus of elasticity of steel

NOTE: the subscript u on terms such as k and ϵ_c, refers to that quantity at ultimate.

REFERENCES

26-1 Bashkow, T. R., ed., *Engineering Applications of Digital Computers*, Academic Press, New York, 1968.

26-2 Beaufait, F. W., et al., *Computer Methods of Structural Analysis*, Prentice-Hall, Englewood Cliffs, N.J., 1970.

26-3 Coleman, C. W., "Computer Graphics for Architects and Civil Engineers," *Graphic Science*, **13** (5), May 1971.

26-4 "Computer Applications in Concrete Design and Technology," *ACI Special Publication SP-16*, American Concrete Institute, Detroit, 1967.

26-5 Conference Papers, "ASCE Conferences on Electronic Computation," (First Conference, 1957; Second Conference, 1960, Third Conference, *Journal of the Structural Division, ASCE*, **89** (ST4), Aug. 1963; Fourth Conference, **93** (ST6), Dec. 1966; Fifth Conference, **97** (ST1), Jan. 1971).

26-6 Fenves, S. J., *Computer Methods in Civil Engineering*, Prentice-Hall, Englewood Cliffs, N.J., 1967.

26-7 Haberman, C. M., *Use of Digital Computers for Engineering Applications*, C. E. Merrill, Columbus, Ohio, 1966.

26-8 Harper, G. N., ed., *Computer Applications in Architecture and Engineering*, McGraw-Hill, New York, 1968.

26-9 Prager, W., *Introduction to Basic FORTRAN Programming and Numerical Methods*, Blairsdell Publishing Co., New York, 1965.

26-10 Roos, D., ed., "*ICES* System: General Description," *Report R67-49*, Department of Civil Engineering, MIT, Cambridge, Mass., 1967.

26-11 Spindell, P., *Computer Applications in Civil Engineering*, Van Nostrand Reinhold, New York, 1971.

26-12 Ural, O., *Matrix Operations and Use of Computers in Structural Engineering*, International Textbook Co., Scranton, Pa., 1971.

26-13 Weaver, W., Jr., *Computer Programs for Structural Analysis*, Van Nostrand Reinhold, New York, 1967.

26-14 "Impact of Computers on the Practice of Structural Engineering in Concrete," ACI Special Publication, SP-33, American Concrete Institute, Detroit, 1972.

27

Computer Software for Concrete Structures

A. M. LOUNT*

27.1 GENERAL

A computer program may be defined as a set of instructions enabling an electronic computer to perform specific operations using data supplied to it and producing results which will then be interpreted by the user. This data must be provided by some other process and the results utilized outside the program itself.

The computer program, then, must be viewed as part of the overall design process and even, in the larger picture, as part of the overall construction activity. It must serve a useful role in that context and the selection of a particular program for a particular problem solution must include consideration of all steps in the process.

The previous chapter has considered the computer as a tool and has outlined various programming considerations of this tool in the areas of analysis, proportioning, and detailing. In this chapter we will consider the computer in relationship to its *environment* and in particular two aspects of this relationship.

1. the data processing chain in concrete engineering, and
2. selection of computer programs for this chain.

Selection of equipment is, of course, a major aspect of the overall problem, but is only mentioned briefly in this particular chapter which is considered with assisting the engineer in establishing his computational needs.

Experience has shown that any application of computers to concrete analysis, proportioning, or detailing functions requires as much planning of the manual work procedures pertinent to the program as in this development or selection of the program itself.

*Consulting Engineer and Managing Director, Taskmaster Computing Systems, Edmonton, Alberta, Canada.

27.2 THE DATA PROCESSING CHAIN

27.2.1 The Nature of Data

Data Processing implies a broader concept then merely computation, which covers only one aspect of the entire design process. This process is outlined in the flow chart in Fig. 27-1.

Note that this process is concerned with the gathering and interpretation of information (or data) over a period of time. For instance, in the design of a warehouse, the needs are first established so that the size of the warehouse can be established. We say, therefore, that the data, i.e., the 'amounts to be stored,' leads to the determination of another item of data, i.e., the 'area.' From the area is derived the overall dimensions, usually expressed in terms of the coordinates of work points (or nodes). From this *nodal geometry*, members may be defined. The existence (or inci-

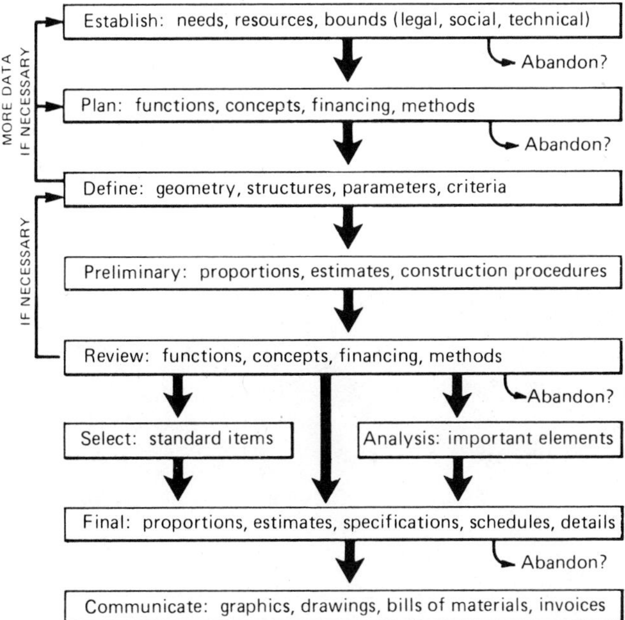

Fig. 27-1 Flow chart for design process.

dence) of a member may thus be established by declaring the work points, or *nodes*, at the end of the member.

From the above, it may be seen that data must consist not only of values, e.g., the x-, y-, and z-coordinates of a node, but also of identification. A node must be identified, a member must be identified, etc. This can be done by referring to its location in a list (e.g., we point to a position in a list). We may point to this position by means of a *pointer*. We can also identify an item in a data list by means of a name or label, e.g., node A703, or member BG33.

This process, of course, is inherent in any design procedure, manual or automated. The difference lies in that the elements of the process tend to be taken for granted in the manual process, whereas they must be clearly defined in developing an automated process.

In the manual process, a work point exists in the mind of the designer. He must visualize the position of the beginning and end of a member in order to determine the length of the member; so must a technician, another engineer, the draftsman, the architect, the detailer, the contractor and the workman building the structure. It is, therefore, identified by a mental process, e.g., "The intersection of the centre line of member B44 with girder G4 at the third floor." This statement, mental or not, is in fact the label for the point. The designer may, or may not, refer to it by name and it may, or may not, be defined graphically on a drawing. Hopefully, all members of the design team will be able to understand the reference and define the point location in the same place.

In an automated process, a specific label and values must be assigned to each point, so that the work point will exist in the memory of the computer system.

The human brain has an amazing capacity to retain information and we all use stored, or filed, information or data without being consciously aware of it. However, there is a limit to an individual's ability to retain data accurately and the design process must provide some means to record, transmit, and store relevant information along the line. Moreover, it is not acceptable for such data to remain locked in the mind of the designer. It must be recorded in some manner for future reference. Therefore, the manual process provides sketches, design notes, drawings, tables, etc. in which are recorded the main data for the project. In the automated process this may be core storage, disk, datacells, magnetic tape, or punched cards. No matter what process is used, records must be kept of important data generated during the course of the process.

It follows, then, that any combination of automated and manual process must ensure compatibility of and completeness of the various manual and computerized records. It also follows that the efficiency of such a combined process will depend on the efficiency of the data transfer process as well as on that of the computations themselves.

Whether we use a programmable desk calculator, or a large scale fully integrated system, the pertinent records must be kept in some manner by either man or machine and the information must be passed from one to the other. These four elements: computation, record keeping, data transfer, and display all cost resources regardless of the system used and must not be overlooked in overall planning for the use of a given computer program.

27.2.2 The Functions Within the Design Process

The functions have been described in the flow chart (Fig. 27-1). Their relative importance may be estimated for a typical project in accordance with the chart in Fig. 27-2. The amount of actual effort spent on analysis and proportioning is actually quite small, but tends to be emphasized because of the prestige associated with it in the past.

On the same chart has been plotted an estimate of the long range potential for automation in the function outline,

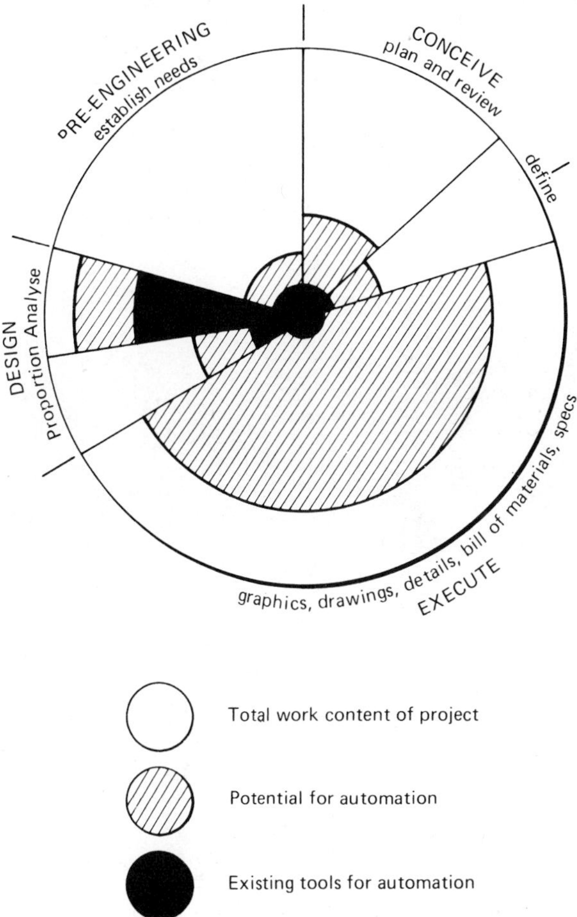

Fig. 27-2 Approximate relationship of work on a typical project.

as well as the extent to which tools have been developed to fill the potential.

Within each function there are a number of operations which, as stated previously, may be classified as:

a. data transfer
b. computation
c. filing or storage
d. display

The proportion of the work contents of these operations will vary. Some typical ratios are shown in Fig. 27-3. We may use such a chart form to assess the extent to which a program can assist in a given function.

There is no fundamental difference between the manner in which a programmable desk calculator and a large scale central processor carry out a multiplication or logical comparison. However, due to size limitation in the case of the calculator, the operation encompassed by it is of necessity limited and most, if not all, of the transfer and storage of data remains a manual process.

Figure 27-4 attempts to show the effect, on the overall cost of a project, of introducing automation into the design process. It can be seen that, while the reduction in computation time is important, the whole cost is actually more sensitive to the efficiency of the data transfer links. It is also obvious that the effective use of any machine must, therefore, involve consideration of the data transfer and storage links.

In fact, it may well be better to have an efficient manual transfer between computations on a desk calculator than an inefficient integrated system on a computer requiring considerable communication cost to access.

On the other hand, to base a choice of system solely on computational and accessibility consideration without thoroughly investigating the question of data transfer and filing efficiency is equally wrong.

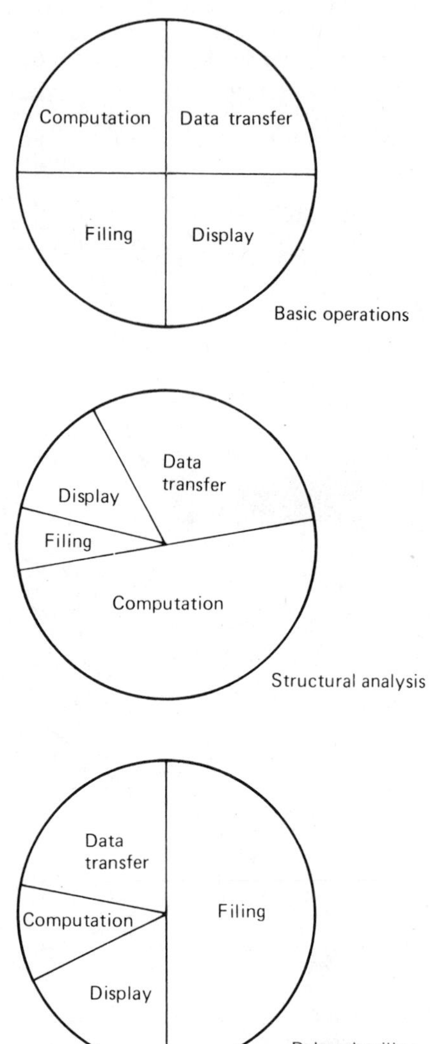

Fig. 27-3 Operational ratios of various processes.

Nodal Geometry	x-, y-, z-coordinates labels and identification type–defines member, slab, load workpoint, other references?
Member Incidence	labels and identification start, end, and intermediate nodes. member segment identification segmentation support declarations member type end and side offsets connectivity
Member Properties	moments of inertia, areas, depths, widths materials, modulus of elasticity torsional and shear moduli attitude centroid offsets at segment's
Central Loads	labels and identification load-case declarations probabilistic factors, return periods intensities distributions locations in terms of nodal geometry type frequency in repetitive loads, period
Member Loads	details of above as applicable to given members
Member Reactions and Forces	force and reaction matrices at specific points, releases, restraints

Fig. 27-4 Typical data processed in the design of a structure.

27.3 TYPICAL DATA IN THE DESIGN PROCESS

Typical of the data flowing through the process of designing a concrete structure would be that shown in Fig. 27-3. In addition, data such as unit prices, weight of material, costs, labor content, etc. are all part of the mass of information considered by the designer or design team. He, or they, must also keep track of code requirements, regulations, previous projects, references, design tables and charts, as well as any office design policies, etc.

It is clear that one of the main problems facing the designer is the task of organizing, storing, retrieving and revising data specific to this project. This will be done in design briefs, or it may also be done in *files*. This concept of files should be thoroughly understood by the user of computer programs if he is to make effective use of the system at his disposal.

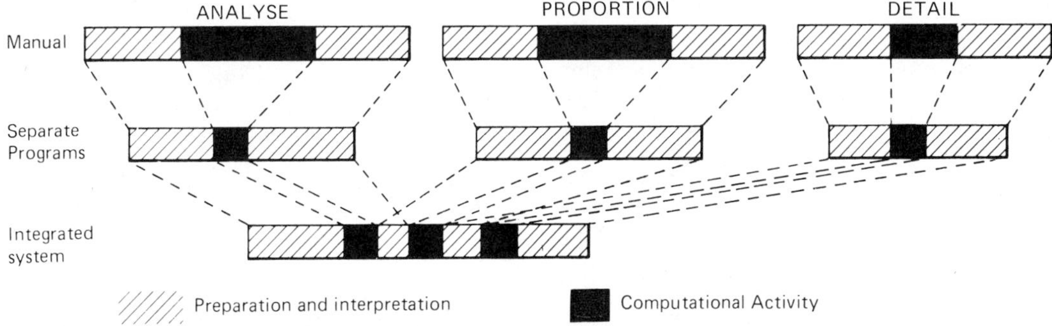

Fig. 27-5 Effect of automation on cost of design.

27.4 THE CONCEPT OF DATA FILES

27.4.1 What is a Data File

A data file is an ordered set of information pertaining to a specific subject. Examples of engineering data files are calculation sheets, loading summaries, bills of materials, reinforcing schedules, progress payment certificates, etc. The traditional form of such files is that of a sheet of note paper held in a folder which in turn, hopefully, can be located when wanted in a filing cabinet, or in a microfilm library.

A common form of file when we apply the full sense of our definition is the engineering drawing. This document has two purposes:

a. The transmission of information relating to a specific project from one party to another, and
b. the recording of such information for future reference and possible future elaboration of the design.

The data file, be it a manually prepared record or some form of digital record, bridges the time period between the completion of one operation (or execution of one program) and the beginning of another. Its existence in time extends beyond the duration of the operations (or programs) involved.

27.4.2 The Data File and its Relation to the Design Process

The overall design process is thus concerned with data files and their manipulation, in whatever form they may exist. Since each operation must reference a data file it follows that the form, or layout, of files must be clearly organized. Just as carefully prepared design notes, sketches, and drawings are part of good design office procedure, well planned files are essential to efficient integration of any data processing operation. Files that can be understood by only one or two programs are no better than drawings that can be understood only after extensive deciphering. Data files which are specific to a particular program system tend to restrict the interaction of disparate programs, retard the evolution of a system to handle broader objectives, and prevent a collective approach by the profession to a network of interlocking programs.

Data files may be discussed in terms of their physical and logical aspects.

27.5 PHYSICAL ASPECTS OF FILES

When we discuss the physical aspects of files, we are concerned with their actual physical form, the devices on which they are held, and the physical attributes by which we recognize them and their contents.

The *physical form* may be a deck of cards, paper tape, magnetic tape, cassette, datacell, disk, or drum.

Magnetic tapes may exist in a variety of track formats, densities of information, and internal arrangement of records. Magnetic tapes must be mounted on drive units each time they are used and they are, of course, sequential in nature since they must be read past a single read-head.

Disks are units rotating at high speed past a series of multiple read-heads. Each band of disks serviced by a read head is referred to as a track. Thus a disk is random in nature, that is, any record stored on a disk may be accessed without having to read more than the track on which it is located.

Disks and tapes are at opposite ends of the accessibility-cost relationship. Large amounts of data may be stored for great lengths of time at low storage cost on tape; whereas, the cost and time delay in retrieving information is high. Conversely, disks may be accessed at a minimal cost, but present the highest cost for long term storage. In between in cost are datacells, drums, and laser storage. We also have variations such as cassettes, minitapes, disk packs, replaceable surface drums, etc.

The design of a data processing system must include a cost-benefit evaluation of the storage medium to be used. It must be remembered that filing cabinets in manual systems cost money, occupy space and are, in fact, quite expensive in employee time to operate.

27.6 THE LOGICAL ASPECTS OF DATA FILES

When we discuss the logical nature of files, we are primarily concerned with the arrangement and ordering of the contents of the file regardless of the medium. Thus we may be concerned with labelling, organization, and internal structure.

27.6.1 Labels

Just as a file folder must be labelled and quite often indexed for retrieval purposes, a data file must be labelled and indexed. This usually takes the form of a set of up to 4, 6, 8, or 16 characters depending on the system. The name of a file may be linked to a project and/or the owner (or user), e.g.

JONES: MILLS: NODES

Many systems or conventions may be devised to label files and most are system dependent.

27.6.2 Kinds of Files

Files may be specific, general, or standardized. Files particular to a given program or system are referred to as being specific. Such files, when applied to large integrated systems, tend to occupy much space and usually require experienced programming staff for modification.

General files have the advantage that only relevant information need be stored, but this is accomplished at the cost of overhead computations required to define the meaning of each item stored.

Standardized files follow certain conventions that are generally understood and permit reduction of storage space without a large amount of computation. They do, however, require the acceptance of standards.

27.7 EXAMPLE OF THE INTERNAL STRUCTURE OF A FILE

An example of a standardized file to define a member might be as follows:

$$\langle \text{File} \rangle ::= \{\langle \text{page} \rangle\}_0^n$$

where

$$\langle \text{Page} \rangle ::= \{\langle \text{record} \rangle\}_0^n$$

where

$$\langle \text{record} \rangle ::= \{\langle \text{Level or set of words} \rangle\}_0^6$$

where

$$\langle \text{level or set of words} \rangle ::= \{\langle \text{Level 1} \rangle | \langle \text{Level 2} \rangle | \langle \text{Level 3} \rangle | \langle \text{Level 4} \rangle | \langle \text{Level 5} \rangle | \langle \text{Level 6} \rangle | \langle \text{Empty} \rangle\}$$

The above syntactical notation has become generally accepted for describing the logic of such information structures. The term Automata is the most general term applied to the whole set of grammar, languages, and information sets related to automated procedures. The meaning of the above can be most precisely and concisely set down in this notation, first suggested by Backus (1959). The above notation may be fully interpreted as follows:

a ⟨file⟩ may be defined as 'zero' to '*n*' sets of data each consisting of a ⟨page⟩.

where ⟨page⟩ may be defined as 'zero' to '*n*' sets of data each consisting of a ⟨record⟩.

where a ⟨record⟩ may be defined as 'zero' to '6' sets of data each consisting of a ⟨level or set of words⟩.

where a ⟨level or set of words⟩ may be defined as a set of data consisting of one or more sets of data referred to as ⟨level 1⟩, *or* a set ⟨level 2⟩, *or* a set ⟨level 3⟩, etc.

In the same notation we may complete our description:

⟨Level 1⟩ ::= [⟨pointer to start node⟩ [⟨pointer to end node⟩]]
⟨Level 2⟩ ::= [⟨pointer to member index⟩]
⟨Level 3⟩ ::= [⟨member name⟩ [⟨start node name⟩ [⟨end node name⟩]]]
⟨Level 4⟩ ::= [⟨member offsets start end⟩]
⟨Level 5⟩ ::= [⟨member offsets end end⟩]
⟨Level 6⟩ ::= [⟨user defined information⟩]

where

⟨member offsets start end⟩ ::=

= ⟨*x*-offset⟩ [⟨*y*-offset⟩ [⟨*z*-offset⟩]]

The square brackets [ ] denote optional entries and the 'nesting' specifies permissible truncation from the right provide that the number of words in each level has been specified in a general file specification.

A particular user might define his files as being 5 pages of 100 records. Each record consisting of 0, 0, 3, 0, 0, and 0 words in each of levels 1 thru 6 respectively. Anyone working in this system would recognize the words in a record in a file so defined as being

record *n* ::= {⟨word 1⟩⟨word 2⟩⟨word 3⟩}

where ⟨word 1⟩ ::= ⟨member name⟩
⟨word 2⟩ ::= ⟨start node name⟩
⟨word 3⟩ ::= ⟨end node name⟩

Other definitions to complete ::=

⟨pointer to start node⟩ ::= {location or address in 'node' file of the record containing information pertinent to the start node of the member defined.}

⟨pointer to end node⟩ ::= {similarly for 'end' node}

⟨pointer to member index⟩ ::= {location or address in a 'member index file' of the record containing information relating to cross referencing to all files involving this member properties, member loads, member reactions etc.}

Note that this is an example of a typical standard file system. Many forms and arrangements may be devised, but all can be described in a precise manner with this notation.

27.8 FILE HIERARCHY

In developing a system operating on standard files each page of such a system would probably coincide with a particular version of the common block within the program. Thus, a page may be turned whenever required, and we can equate this concept of files to a library.

A large *data bank* for a project may be a *library*. The library is made up of *sections*; sections are made up of *works* (or opus); works are made up of *volumes*; volumes of *chapters*; chapters of *pages*; pages of *paragraphs*; paragraphs of *records*; records of *words*.

Jackson has mentioned that a high-rise building could produce up to 100,000,000 words.

Such a volume of information must be organized carefully so that it can be largely stored on tape and only the most active information held on high cost disk storage. The cost would become prohibitive of it were kept on disk under commonly prevailing storage cost rates.

27.9 INPUT AND OUTPUT DATA

In using or developing programs, it is important to consider the form of input. There are, in general, three main forms.

27.9.1 Fixed Format

Each line of data consists of characters in fixed positions. column

1	2	3	4	5	6	7	8	9	10	11	12	13	14	15
	1	0	.	5	B	E	A	M	S			4	0	2

Each character or digit must be in the correct column. Fixed format input has the advantage that each field of information is self defining. The disadvantage is that input forms tend to become complex, hard to adapt to varying conditions and are sensitive to misalignment due to mispunching. Fixed format is particularly suitable to volume data processing with keypunching and verification equipment. The internal computing overheads of interpreting data is at a minimum and the cost of data preparation is minimal. However for engineering data processing purposes, the proliferation of input formats, restricts the expansion and development of new program applications. Fixed format is quite unsuitable to low- and medium-volume terminal operation, and requires specially trained keypunch personnel.

27.9.2 Free Format

Data may be located at any point along a line and the meaning of a particular word must be defined by its context, e.g.

ITEM 402, 'BEAMS', DEPTH = 10.5

This format has the advantages of great flexibility. It permits expansion of program applications and is generally easy to follow and understand for beginners. It can be, however, wordy and expensive to process for volume data processing. In long data lists, there is a great deal of unnecessary repetition. This can be reduced by eliminating keywords,

0, 3.3, 7, 9.8, 10.3, 108.4, 0.0, 0.0, 507.3

The above is much more efficient but is very difficult to follow and check. Commas can be accidently omitted or, even if the system permits omission of commas, redundant zeroes become a hazard and involve extra words.

27.9.3 Floating Format or Tabular

Input consists of free format within specified column bounds.

*TABULAR

LOAD	NODE	NODE	NODE	NODE	TYPE	INTENSITY	UNITS
L1	A4	A5	A6	A7	UDL	40	RSF
L2	A5				CONC	100	KIPS
L3	A7	A8	A9	A10	UDL	40	PSF
L4	A9	A10	A11		=	=	
L5	=	=	=		=	70	

As long as the data is reasonably aligned beneath the headings it is identified with the heading.

This format has the advantage of most closely resembling typed tabular lists, not having unnecessary zeroes obscuring rapid visual checking of the tables, and permitting great flexibility in tabular arrangements of input data since the headings may be selected to suit the specific circumstances. Columns which would otherwise produce zeroes are omitted.

This type of input is particularly suited to preparation for slow- and medium-speed data terminals for medium-size design offices, or design teams in large offices which operate around a terminal. It is also very suitable for optical scanners. In that context, any competent secretary can prepare such data as it requires little or no special training. Misalignment by up to two or three columns is not critical. This can be vital in operations involving remote offices where qualified data preparations personnel and keypunchers are expensive and scarce.

27.9.4 Input standards

Great progress could be accomplished if general agreement could be reached on methods of data preparations. The British Institution of Structural Engineers has developed a pioneer standard in this area. While probably not suitable to North American use as is, it does illustrate what can be done. Acceptance of such a standard would greatly assist the development of technician level personnel essential to the success of large-scale design office use of computers.

27.10 THE TIME ELEMENT IN THE DESIGN PROCESS

The relative time element may represent the elapsed time required to perform certain functions, but it also reflects the period over which a project develops, the points in time at which crucial decisions are made, the relationship of the sequence of events, and reference points reflecting the status of the design at given stages of revision or evolution.

Since the elapsed time required to perform a specific computation or series of computations, becomes relatively insignificant, the timing and sequencing of the design operations and design decisions assume more importance and may actually govern the design schedule. Many of the decisions points are governed by external factors and therefore, the much dreamed of design of a structure in a day or two becomes a practical unreality. It is necessary for the computer process to be a start–stop–start–stop process, to match the external decision making processes.

This factor coupled with a probable increase in the number of alternates investigated makes it most important to ensure that data at all points reflects accurately the status of the design at that stage, and that the design process be capable of readily accommodating revisions.

The time factor must also be considered in the use of integrated systems. Systems that require an entire file library to be passed through the process each time there is a revision, will obviously have problems in expanding to large projects.

A project starts at day one, but it may only be finished at day n, where n may be 2, 10, 100, or 1000. And the records must be kept for years thereafter in some form.

During the design process, many alternatives will have been chosen and, if the project follows normal patterns, it will be subject to revisions, and changes over its course. It will need to be subdivided for estimating and scheduling and possibly summarized in many ways for many purposes. Introducing the concepts of a common data base varying with time will help make the records simple and easy to use.

27.11 SELECTION AND DEVELOPMENT OF COMPUTER PROGRAMS

27.11.1 Basic Considerations in Selection and Development Program

Computer programs will replace, or make practical, many of the operations in this flow of information. Indeed, many studies and evaluations of alternatives, or methods of analysis which would be bypassed in a manual process become practical in an automated process. In many cases, the feasibility of the project itself may hinge on the ability to carry out within economic limits certain computations. An example of this type of computation might be, for instance, the calculation of a complex space frame structure.

Note, however, that it may be more important to evaluate whether the space frame structure is the most desirable solution rather than to accept the challenge of undertaking a previously unthinkable calculation. It is thus very important to keep a clear view of objectives in designing a data processing operation. Is it to produce solutions to isolated complex problems, or is it to aid in the routine processing of data arising out of routine projects? Programs which may be excellent for one requirement, may not be at all suitable for the other.

Programs designed to solve complex problems may be classed as scientific in nature; whereas, many problems associated with routine work, e.g., reinforcing-steel detailing are really commercial or business-oriented in nature. In the first case, reasonable care in techniques and office procedures will generally suffice; whereas, in the second case great care must be taken. In many ways it is the routine usage that presents the greatest challenge to the user's organizing ability, because, to be successful, consideration of the people involved is often more important than the machines or the programs. A program which may be an excellent teaching tool may be economically unviable in a design office environment. Conversely a program which may appear to be highly efficient for constant commercial usage by skilled technicians may present serious difficulties in staffing and staff training.

Revisions are an inescapable attribute of the design process. Programs which require total reruns each time an element is changed, are obviously more costly than programs that permit reworking only the affected elements.

27.11.2 Classes of Programs

As mentioned before there are two general classes of programs.

1. Those designed to solve specific complex problems. e.g.,
   hyperbolic paraboloid shells
   shear wall-frame interaction
   space frames
2. Those designed to process routine work. e.g.,
   reinforcing steel detailing
   beam proportioning
   retaining wall design

In addition programs may handle isolated parts of the design process, e.g., the design of a column; or may be meant to cover many aspects of the design process, e.g.,*AMECO*, *ICES*, *GENESYS*, and *TASKMASTER*. At one end of the scale we have short programs designed for programmable desk calculators, at the other we have the large integrated systems requiring the power of a major computer installation. They all have one element in common—they are all meant to be part of the design process. While, in one case the computer performs only certain computations, with the designer manually keeping track of and transcribing data, in the other case, the computer performs much of this executive work.

In the integrated programs, there are also differences. In some systems such as *AMECO*, the entire system is self-contained. A program revision entails change at the original program level. In *ICES* and *GENESYS*, provisions are made for some revision by users and the development of additional components by means of facilities such as *ICETRAN* and *GENTRAN*.

In other systems, particularly some now under development, the integrating system is merely executive, linking external programs together through common interface files and providing means for standardization of input and output data.

In all programs it is essential to ask oneself who is to be responsible for updating and maintenance? All too often this gets overlooked. It is an essential part of any agreement to use a program.

27.12 EXAMPLES OF AVAILABLE PROGRAMS

27.12.1 Programmable Desk Calculators

There are of course, hundreds of programs in design offices developed for programmable desk calculators. To give an idea of the type and kinds of programs available we may cite the work carried out by the Portland Cement Association, Old Orchard Road, Skokie, Illinois, 60076. Programs developed by PCA include

1. Flat Plate floors—analysis and design.
2. Flat Slats and Waffle slabs—analysis and design.
3. Eight-span continuous beam—analysis and design.
4. Tied rectangular columns—investigation and design.
5. Spiral and tied round columns—Investigation and design.
6. Eight-span, continuous beam—influence lines.
7. Moment distribution for 32-joint frame without sidesway.

These programs are available from PCA subject to a licensing agreement. These programs are, in effect, somewhat limited versions of the corresponding IBM 1130-programs.

27.12.2 Typical in-House Computer Production Programs

Many engineering offices have produced production packages of varying size. These are usually highly related to the type and volume of practice of given firms and to the specific priorities assigned by the developer to various aspects of the design process. Typical examples of such programs are:

a. Crain & Crouse, Miami	program for high-rise owner-built apartments
b. Russell Fling, Columbus	programs for design and scheduling of beams and girders
c. Bureau of Reclamations	suite of programs for powerhouse structures
d. U.S. Steel, Bethlehem	programs for reinforcing steel
e. Associated Engineering, Edmonton	program for analyzing structures in contact with flexible foundations
f. Erdman & Anthony	automated bridge pier analysis

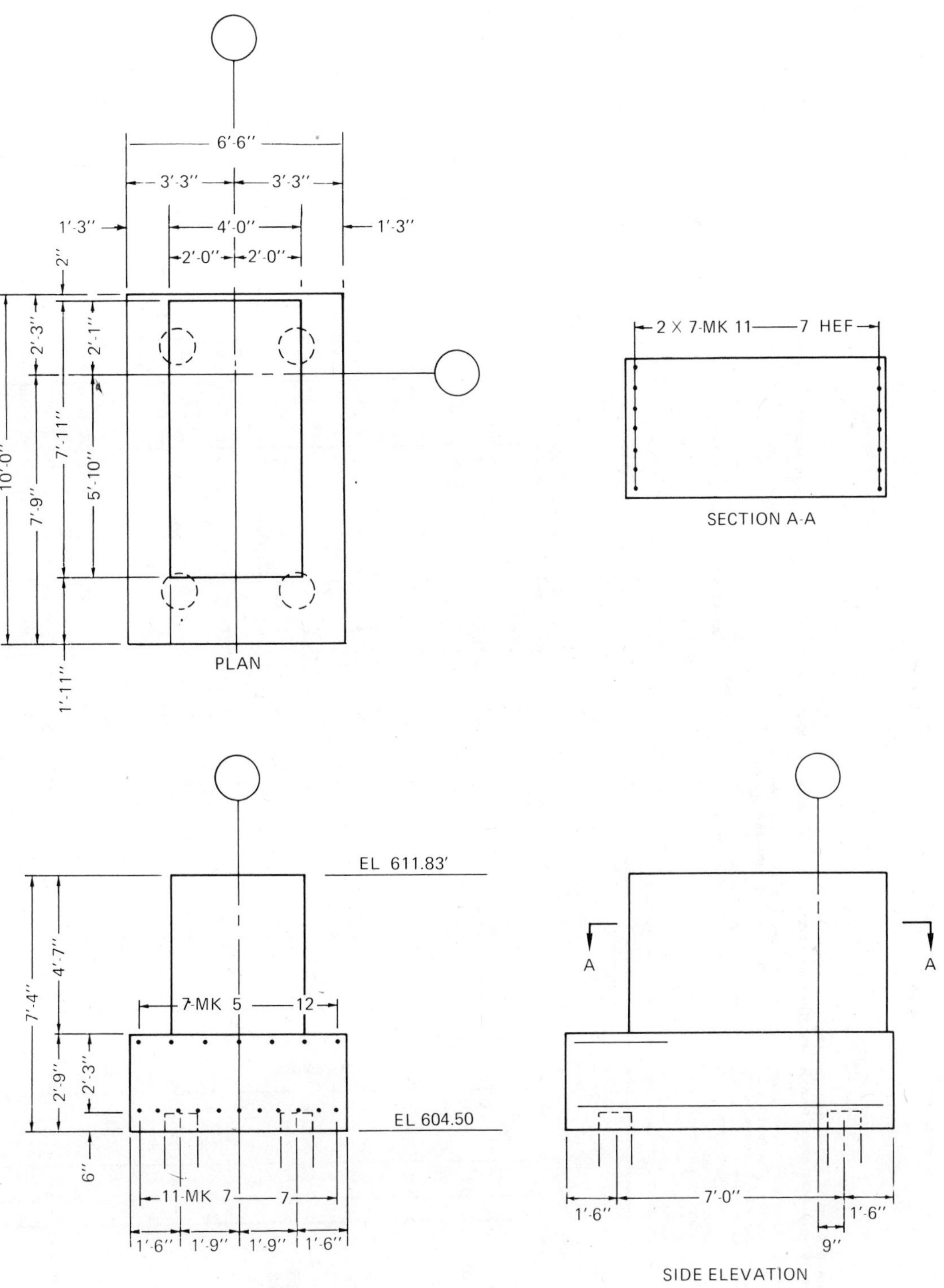

Fig. 27-6 (From: Trygve Hoff and Associates, *ACI Special Publication SP-33*, see Ref. 27-2.)

BEAM SCHEDULE

R.S. FLING & PARTNERS INC DEC 70

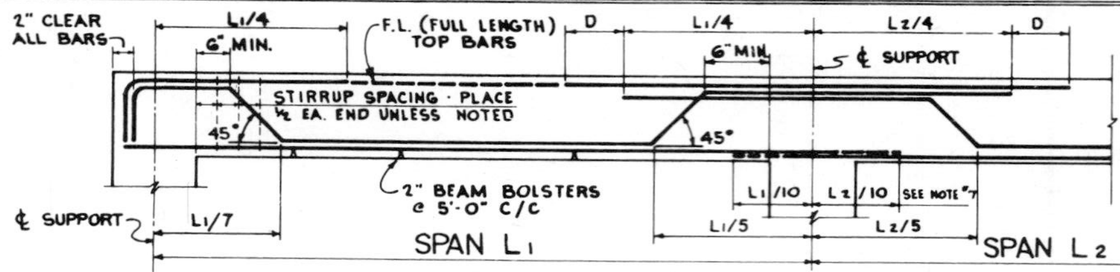

BENDING DIAGRAM

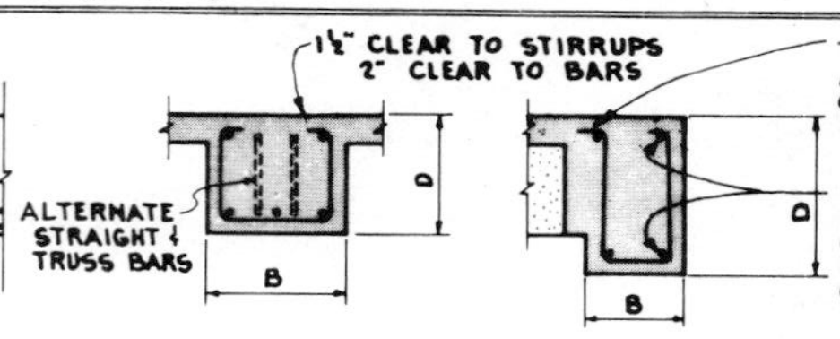

TYPICAL BEAM CROSS SECTIONS

NOTES

1. FURNISH BARS PER BENDING DIAGRAM WHERE BAR DIMENSIONS ARE NOT SCHEDULED.
2. VERIFY THAT ALL SCHEDULED BAR DIMENSIONS PROVIDE SPECIFIED CONC. COVERAGES AT ALL CONCRETE SURFACES AND ADEQUATE CLEARANCES AT MEMBER INTERSECTIONS.

3. DETAIL AND PLACE BARS FOR INTERIOR SPANS SYMMETRICALLY. USE LONGEST ADJACENT SPAN TO FIGURE LENGTH.
4. PROVIDE STANDARD HOOKS ON TOP AND TRUSS BARS AT ALL DISCONTINUOUS MEMBER ENDS. MAX. J = D - MINUS 4"
5. SCHEDULED BAR LENGTHS DO NOT INCLUDE HOOKS.
6. EXTEND TOP BARS 12 BAR DIAMETERS PAST MIDSPAN IF NOTED "CONT." IN SCHEDULE.
7. LAP BOTTOM BARS 1'-0" MIN. AT SUPPORTS. EXTEND L/10 INTO ADJACENT SPANS IF BARS NOTED WITH AN ASTERISK (*) IN SCHEDULE.
8. PLACE STRAIGHT TOP AND BOTTOM BARS IN STIRRUP CORNERS.
9. STIRRUP QUANTITY IS THE TOTAL NUMBER OF PIECES (NOT SETS) FOR THE BEAM. "SGL" MEANS ⊔. "DBL" MEANS ⊔⊔. "TPL" MEANS ⊔⊔⊔.
10. STIRRUP SPACINGS ARE LISTED FROM FACE OF SUPPORT TOWARD CENTER OF SPAN.
11. WHERE REQ'D. STIRRUP SPACINGS DIFFER AT BEAM ENDS, SPACINGS ARE LISTED FROM INDICATED FACE OF SUPPORT TOWARD CENTER OF SPAN.
12. F.L. DENOTES TOP BARS FULL LENGTH OF SPAN AND TO L/4 + D OF ADJACENT SPANS.

MARK	B	D	BOTTOM BARS QUAN. & SIZE		BOTTOM BARS LENGTH	TRUSS BARS QUAN. & SIZE		B	H + K (IN.)	D	$H_1 + K_1$ (IN.)	F	O/O	TOP BARS QUAN. & SIZE		TOP BARS LENGTH	TOP BARS LOCATION	STIRRUPS QUAN. & SIZE		STIRRUPS TYPE	STIRRUPS SPACINGS (IN.)				REMARK NO.
B 1														4	4	4- 6	EXT SPT								
B 1	30	16	4	5	13- 3													12	3	DBL	3@13				
B 2	30	16	4	5	18- 7	1	10	6-10	12 3/4	10- 6	12 3/4	6-10	26- 4	4	4	28- 8	F.L.	28	3	DBL	7@ 7				
B 3	30	16	5	5	18- 7	1	10	6-10	12 3/4	10- 6	12 3/4	3- 3	22- 9	4	4	6- 4	LINE G	28	3	DBL	7@ 7				
B 4														4	4	4- 6	EXT SPT								
B 4	30	16	6	5	13- 3													24	3	DBL	6@ 7				
B 5	30	16	4	5	18- 7	2	10	6-10	12 3/4	10- 6	12 3/4	6-10	26- 4	5	4	28- 8	F.L.	52	3	DBL	4@ 5	9@ 7			
B 6	30	16	4	7	13- 3									5	5	18- 3	F.L.	32	3	DBL	8@ 7				
B 7	30	16	4	5	18- 7	2	10	5- 6	12 3/4	10- 6	12 3/4	5- 6	23- 8	4	4	11- 0	LINE F	52	3	DBL	4@ 5	9@ 7			
B 8	30	16	4	7	18- 7	3	7	6-10	12 3/4	10- 6	12 3/4	3- 3	22- 9	4	5	6- 4	LINE G	40	3	DBL	10@ 7				
B 13														4	4	7- 6	EXT SPT								
B 13	30	16	4	7	24- 2	1	11	4- 2	12 3/4	13- 9	12 3/4	9- 7	29- 8	4	7	15- 8	LINE E	48	3	DBL	12@ 7				
B 14	30	16	6	5	25- 6	2	11	9- 7	12 3/4	14- 8	12 3/4	4- 6	30-11	4	4	7-11	LINE G	60	3	DBL	4@ 6	11@ 7			
B 15	24	36	4	5	13- 3									7	8	24- 3	LINE D	18	4	DBL	9@16				
B 16	24	36	4	10	36- 3									4	4	12- 2	LINE G	36	4	DBL	9@16				
B 17	24	36	4	5	13- 3									5	6	24- 3	LINE D	18	4	DBL	9@16				
B 18	24	36	4	6	36- 3	2	11	17- 1	32	23- 1	32	3- 2	48- 8	4	4	12- 4	LINE G	36	4	DBL	9@16				
B 19	24	36	4	5	13- 3									6	10	25- 6	LINE D	38	4	DBL	9@ 6	6@ 7	4@13		
B 20	24	36	5	11	38- 9	1	9	17- 1	32	23- 1	32	3- 2	48- 8	4	4	12- 4	LINE G	48	4	DBL	4@11	8@16			
B 21	12	24	2	10	26- 0	1	11	8-11	20	14- 8	20	3-10	30- 9	2	4	8- 6	LINE G	58	3	SGL	13@ 3	8@ 4	8@ 9		
B 22														4	4	7- 6	EXT SPT								
B 22	30	16	4	7	24- 2	1	9	2-10	12 3/4	15- 1	12 3/4	9- 7	29- 6					44	3	DBL	11@ 7				
B 23	30	16	5	7	25- 6	1	10	9- 7	12 3/4	16- 1	12 3/4	3- 1	30-11	4	4	32- 2	F.L.	52	3	DBL	13@ 7				
B 24	24	36	4	5	13- 3									4	11	24- 3	LINE D	18	4	DBL	9@16				
B 25	VARI	ES	5	11	36- 3	1	10	17- 1	32	23- 1	32	3- 2	48- 8	4	5	12- 4	LINE G								
B 25	24	36															LINE D	34	4	DBL	7@ 7	4@10	6@16		
B 25	24	36															LINE C	20	4	DBL	3@10	7@15			
B 26	24	36	4	5	13- 3									8	9	24- 3	LINE D	28	4	DBL	4@ 8	4@10	6@12		
B 27	24	36	8	11	38- 9	1	10	17- 1	32	23- 1	32	3- 2	48- 8	4	4	12- 4	LINE G	80	4	DBL	7@ 7	5@ 9	8@16		
B 28	24	36	4	5	13- 3									6	11	24- 3	LINE D	32	4	DBL	6@ 6	4@ 8	6@14		
B 29	24	36	8	11	38- 9	2	9	17- 1	32	23- 1	32	3- 2	48- 8	4	4	12- 4	LINE G	100	4	DBL	9@ 5	7@ 6	9@15		
B 30	24	36	4	5	13- 3									7	7	22- 0	LINE D	18	4	DBL	9@16				
B 31	24	36	5	7	26- 2									4	4	9- 1	LINE FA	24	4	DEL	6@16				

Fig. 27-7 (From: Sadler, John, *ACI Special Publication SP-33*, see Ref. 27-2.)

Groups such as CEPA and APEC provide clearing house for information on such programs and for the actual interchange of programs.

Much of the most advanced and promising work has been going on within the design groups of large design offices such as Holmes & Narver, Sargent & Lundy, Giffels and Associates, Skidmore Owings & Merrill, Parsons Brunkerhoff, Quade and Douglas, and many others.

Nor are Europeans in any way behind North Americans in this regard. In fact, British firms of W.V. Zinn & Associates, and W. S. Atkins have been driving forces behind the British GENESYS system providing large comprehensive design subsystems for the overall system. French, German, Swiss, Swedish, and Spanish groups have all been very active.

27.12.3 Institution-Provided Programs—the PCA Packages

Reference must, of course, be made to the PCA programs developments. In addition to the desk calculator programs, PCA has released some 11 programs for the IBM 1130, as well as supporting the ICES-STRUDL package at MIT. These programs include the following:

1. Concrete airport pavement design.
2. Ultimate strength design of reinforced-concrete columns.
3. Analysis and design of flat plates and continuous concrete frames. A new edition (1971) conforms to the provisions of the *1971 ACI-318 Building Code*, for the analysis and design of flat plates, flat slabs, and waffle slabs. Wind moments are also considered in the 1971 edition.
4. Analysis and design of simple-span precast-prestressed highway or railway bridges.
5. Analysis of floor systems supported by central core and exterior columns.
6. Analysis and design of staggered wall-beam frames.
7. Load analysis and design of concrete-column stacks.
8. Preliminary designs, itemized quantities, and cost estimates for reinforced-concrete building frames. (This is one of the few programs available for development of preliminary estimates.)
9. Analysis of plane multistory frame-shearwall structures under lateral and gravity loads.
10. Analysis and design of foundation mats and combined footings.
11. Preliminary analysis of multicenter arch dams with variable thickness.

Of course, the British Genesys center, and similar so-called software centers now being established, will also become sources of programs.

27.12.4 Major Commercial Programs

There are many large scale analysis and design programs available such as *REBAR 70*, *STAN*, *STRIP*, *GASP*, *STARDYNE*, *ICES–STRUDL*, *AMECO*. These programs come from such diverse countries as the US, Canada, Japan, and Switzerland. Major commercial packages will do much, but sometimes they may be expensive to operate and run, because of their diversity and flexibility in situations not requiring same.

There seems to be no problem in acquiring programs to do specific chores. Such disparate programs large or small, must be fitted into the design process in one way or another. By far the most costly aspect of the use of such programs can be the data preparation, processing, and interpretation associated with them.

The question of availability, lease cost, royalty, turnaround time, etc. must all be considered in selecting a program. Large programs tend to produce large costs unless well integrated into the overall routine.

27.12.5 Integrated Packages

Currently systems such as *ICES*, *AMECO*, *STRUCTIV* (Omnidata Corp, New York City), and *GENESYS* are advertised as being available and as having integrated system features. Others under development include *TASKMASTER*. Integration is really a meaningless word unless it is carefully understood in the context that the designer of a particular system meant it. For instance, it may be possible to detail a building as soon as it is designed, but is it possible to detail a portion of it a month or two later with revisions thrown in also? What has been the cost of making these revisions in terms of dollars and manpower? Could it be done by technicians or was the most experienced man in the office tied up doing 'donkey work' to make it work? What was the true cost? What work already designed was unnecessarily redesigned just because the system was not designed in a modular fashion?

Such questions and many more should be asked when considering an integrated system. The right approach should be the first consideration. Individual subsystems come later.

27.12.6 Whither Computer Graphics?

Reference is made to the report of ACI Committee 118 on the state of the art of computer graphics. This comprehensive report reviews the progress in this highly significant area. Examples of the possible output are shown in Figs. 27-6 and 27-7.

REFERENCES

27-1 "Computer Applications in Concrete Design and Technology," *ACI Special Publication SP-16*, American Concrete Institute, Detroit, 1967.

27-2 "Impact of Computers on the Practice of Structural Engineering in Concrete," *ACI Special Publication SP-33*, American Concrete Institute, Detroit, 1972.

27-3 *Standardization of Input Information for Computer Programs Structural Engineering*, The Institution of Structural Engineers, London, 1967.

27-4 Fenves, S. J., *Computer Methods in Civil Engineering*, Prentice-Hall, Englewood Cliffs, N. J., 1967.

27-5 *ICES STRUDL II–Engineering Users Manual*, Report 68-91, "Frame Analysis," MIT Press, Cambridge, Mass., Nov., 1968.

27-6 Klotz, L. H., "On the Application of ICES and Universal Software Systems," *Civil Engineering*, ASCE, **39** (2), Feb. 1969.

27-7 Struble, G., *Assembler Language Programming, The IBM System/360*, Addison-Wesley, Reading, Mass., 1971.

27-8 McCracken, D. D., *A Guide to FORTRAN Programming*, John Wiley and Sons, New York, 1965.

27-9 Knuth, D. E., The Art of Computer Programming, vols. 1–3, Addison-Wesley, Reading, Mass., 1973.

27-10 Biggs, J. M.; Logcher, R. D.; and Wenke, H. N., "Use of *STRUDL* in Reinforced Concrete Building Design," *ACI Special Publication SP-33*, American Concrete Institute, Detroit, 1972.

27-11 Allwood, R. J., "Problem Oriented Language for the Design of Reinforced Concrete Structures," *ACI Special Publication SP-33*, American Concrete Institute, Detroit, 1972.

27-12 "*Genesys* Reference Manual," The Genesys Center, Institute of Technology, Loughborough, Leicestershire, U.K., March, 1970.

27-13 *Report of Ministry of Public Buildings and Works, U.K.*, Proceedings of the Conference on Genesys, Sponsored by the Subcommittee in Computers in Structural Engineering, Institution of Electrical Engineers, Savoy Place, London, WC 2, Nov., 1968.

27-14 Palejs, A. A., "Command Language Interface for Automatic Design of Structures in Three Dimensions," *ACI Special Publication SP-33*, ACI, Detroit, 1972.

27-15 Lount, A. M., "Integrated Systems—Some Fundamental Considerations in Engineering Computer Usage," *ACI Special Publication SP-33*, ACI, Detroit, 1972.

27-16 Lount, A. M., and Simmonds, S. H., *The Role of Files in Automating the Design Process*, Paper No. 73-355, 14th Structures, Structural Dynamics and Materials Conference, Williamsburg, Virginia, March, 1973.

27-17 Fintel, M., "Computer Program Developments by the Portland Cement Association," Portland Cement Association, Skokie, 1971.

Index